HEAT TRANSFER REVIEWS
1976-1986

HEAT TRANSFER REVIEWS 1976-1986

Edited by

E. R. G. ECKERT
R. J. GOLDSTEIN
T. F. IRVINE, Jr.
J. P. HARTNETT

A WILEY-INTERSCIENCE PUBLICATION
John Wiley & Sons, Inc.
NEW YORK / CHICHESTER / BRISBANE / TORONTO / SINGAPORE

Library of Congress Cataloging in Publication Data:

Heat transfer reviews, 1976-1986 / edited by E.R.G. Eckert . . . [et al.].
p. cm.
"A Wiley-Interscience publication."
Reviews originally published in the International journal of heat and mass transfer
Includes bibliographical references.
1. Heat – Transmission. I. Eckert, E.R.G. (Ernst Rudolf Georg), 1904- . II. International journal of heat and mass transfer.
TJ260.H3965 1990 89-21469
621.402'2 – dc20 CIP
ISBN 0-471-52458-1

Printed in the United States of America

10 9 8 7 6 5 4 3 2 1

CONTENTS

	Review Year	Page
PREFACE		*i*
Heat Transfer Review from *International Journal of Heat and Mass Transfer* Vol. 20, Page 1097 (1977) **E.R.G. Eckert, E.M. Sparrow, R.J. Goldstein, C.J. Scott, E. Pfender, S.V. Patankar and J.W. Ramsey.**	1976	1
Heat Transfer Review from *International Journal of Heat and Mass Transfer* Vol. 21, Page 1269 (1978) **E.R.G. Eckert, E.M. Sparrow, R.J. Goldstein, C.J. Scott, E. Pfender, S.V. Patankar and J.W. Ramsey.**	1977	37
Heat Transfer Review from *International Journal of Heat and Mass Transfer* Vol. 22, Page 1469 (1979) **E.R.G. Eckert, E.M. Sparrow, R.J. Goldstein, C.J. Scott, E. Pfender, S.V. Patankar and J.W. Ramsey.**	1978	73
Heat Transfer Review from *International Journal of Heat and Mass Transfer* Vol. 24, Page 1 (1981) **E.R.G. Eckert, E.M. Sparrow, R.J. Goldstein, E. Pfender, S.V. Patankar, J.W. Ramsey and K.Y. Teichman.**	1979	111
Heat Transfer Review from *International Journal of Heat and Mass Transfer* Vol. 24, Page 1863 (1981) **E.R.G. Eckert, R.J. Goldstein, S.V. Patankar, E. Pfender, J.W. Ramsey, T.W. Simon, E.M. Sparrow and K.Y. Teichman.**	1980	151
Heat Transfer Review from *International Journal of Heat and Mass Transfer* Vol. 25, Page 1783 (1982) **E.R.G. Eckert, R.J. Goldstein, S.V. Patankar, E. Pfender, J.W. Ramsey, T.W. Simon, and E.M. Sparrow.**	1981	199
Heat Transfer Review from *International Journal of Heat and Mass Transfer* Vol. 26, Page 1733 (1983) **E.R.G. Eckert, R.J. Goldstein, E. Pfender, W.E. Ibele, S.V. Patankar, J.W. Ramsey, T.W. Simon and N.A. Decker.**	1982	235

	Review Year	*Page*
Heat Transfer Review from *International Journal of Heat and Mass Transfer* Vol. 27, Page 2179 (1984) **E.R.G. Eckert, R.J. Goldstein, S.V. Patankar, E. Pfender, J.W. Ramsey, T.W. Simon, N. Decker and T.H. Kuehn.**	1983	281
Heat Transfer Review from *International Journal of Heat and Mass Transfer* Vol. 28, Page 2181 (1985) **E.R.G. Eckert, R.J. Goldstein, S.V. Patankar, E. Pfender, J.W. Ramsey, T.W. Simon, N. Decker, T.H. Kuehn and H. Lee.**	1984	323
Heat Transfer Review from *International Journal of Heat and Mass Transfer* Vol. 29, Page 1767 (1986) **E.R.G. Eckert, R.J. Goldstein, E. Pfender, W.E. Ibele, J.W. Ramsey, T.W. Simon, N.A. Decker, T.H. Kuehn, H.O. Lee and S.L. Girshick.**	1985	379
Heat Transfer Review from *International Journal of Heat and Mass Transfer* Vol. 30, Page 2449 (1987) **E.R.G. Eckert, R.J. Goldstein, E. Pfender, W.E. Ibele, S.V. Patankar, J.W. Ramsey, T.W. Simon, N.A. Decker, T.H. Kuehn, H.O. Lee and S.L. Girshick.**	1986	469
Author Index		551
Subject Index		653

PREFACE

Since 1954, the faculty connected with the Heat Transfer Laboratory of the University of Minnesota has prepared annual reviews of the previous year's literature in heat transfer. For many years now these HEAT TRANSFER REVIEWS have appeared in the International Journal of Heat and Mass Transfer published by Pergamon Press. Comments are continuously received from practicing engineers, research workers, and students, testifying to the usefulness of these reviews in locating existing literature on a particular topic or in keeping up with recent developments in special research areas.

The Reviews for the years 1953 through 1975 have been collected and published in two volumes of the series Progress in Heat and Mass Transfer by Pergamon Press. Volume 3 covers the years from 1953 through 1969 and Volume 8 from 1970 to 1975. The wide acceptance of these volumes has caused the editors to prepare again a volume which included the reviews for the years 1976 through 1986 together with a comprehensive author and subject index.

The editors would like to express their appreciation to the authors of the yearly reviews since without their original contributions this book would not have been possible. We are also grateful to Pergamon Press for granting permission to publish these reviews. Last but not least, our thanks is extended to those who worked on the project, especially David Balderas and John B. Lau of the University of Illinois at Chicago and to S.C. Kim and B. Seo of the State University of New York at Stony Brook.

It is our hope that this volume, like the previous ones will prove useful to the heat transfer community in its task to provide the tools for continuous improvements in the efficiency of engineering devices. The recent awareness of the necessity to conserve energy and to find new energy sources, for instance, has opened up a large new field to heat transfer research.

E. R. G. Eckert
R. J. Goldstein
T. F. Irvine, Jr.
J. P. Hartnett

HEAT TRANSFER REVIEWS
1976-1986

HEAT TRANSFER REVIEW - 1976

E. R. G. ECKERT, E. M. SPARROW, R. J. GOLDSTEIN, C. J. SCOTT, E. PFENDER, S. V. PATANKAR and J. W. RAMSEY

Heat Transfer Laboratory, Department of Mechanical Engineering, University of Minnesota, Minneapolis, Minnesota, U.S.A.

INTRODUCTION

THIS review surveys results that have been published in various fields of heat transfer during 1976. As in the past, the number of papers published during that period was such that only a selection can be included in this review. A more complete listing of papers is available in the heat-transfer bibliographies published periodically in this journal.

The Sixteenth National Heat Transfer Conference was held 8–11 August 1976 in St. Louis, Missouri. The recipient of the Max Jakob Memorial Award, R. G. Deissler, presented an invited lecture, "Tornadoes and other atmospheric vortices". A lecture entitled "An inquiry of selected topics on heat-exchange design" was given by A. C. Mueller, who received the 1975 Donald Q. Kern Award. A third invited lecture on "Water reactor safety research programs" was presented by L. S. Tong. Twenty-eight sessions treated, among other subjects, interfacial phenomena, internal heat generation, process heat transfer, thermonuclear power, and heat transfer in foodstuffs. The papers presented at the conference are available as preprints and many of these will be published in the *Journal of Heat Transfer* or in the publication series of the American Institute of Chemical Engineers.

The 25th Heat Transfer and Fluid Mechanics Institute was held 21–23 June 1976 at the University of California, Davis, California. An invited lecture by P. Bradshaw discussed progress and problems in the development of turbulence models; another lecture by R. Davis treated numerical methods for interacting boundary layers. Five sessions were concerned with thermal convection, two-phase flow and boiling, compressible and turbulent flow, combustion, and aerospace heat transfer problems. Proceedings are available through Stanford University Press.

The Scientific Affairs Division of the North Atlantic Treaty Organization assembled a group of experts to review research needs on thermal energy storage and to make recommendations for future research. One of the panels discussed heat transfer and thermal energy transport. A report of the conference is available through the Scientific Affairs Division, North Atlantic Treaty Organization, Brussels, Belgium.

An International Solar Energy Conference was held 15–20 August 1976 at Winnipeg, Canada under the title "sharing the Sun 76". Many of the papers presented included heat-transfer problems.

An advanced course and the 1976 International Seminar were organized by the International Centre for Heat and Mass Transfer 23 August through 4 September 1976 at Dubrovnik, Yugoslavia. The summer course was entitled "Heat Disposal from Power Generation", and dealt primarily with cooling towers and climatic modifications by energy production. The seminar, "Turbulent buoyant convection", discussed plumes, stratified fluids, air and smoke movements in buildings, and combustion phenomena. The proceedings of these conferences will be published by Scripta Book Company Washington, D.C.

The American Institute of Chemical Engineers focussed attention on plasma chemical processes in four sessions of its Eighty-Second National Meeting held 29 August through 1 September 1976 at Atlantic City, New Jersey, with heat-transfer information included in some of the papers. Other sessions dealt with transport processes in the oceans and with interfacial phenomena.

The 97th ASME Winter Annual Meeting, held 5–10 December 1976 at New York City, included in its program eleven sessions organized by the Heat Transfer Division. Topics ranged from application of computer technology through electric effects, liquid metal fast breeder reactors, thermal energy storage, pipe line heat transfer, heat exchangers, to solar collectors. Reprints of the papers are available from ASME Headquarters and many of them will also be published in the *Journal of Heat Transfer*. The dinner speaker, P. E. Glaser, talked on "The future of power from space". Heat Transfer Memorial Awards were presented to W. H. Giedt and R. Viskanta.

A number of books dealing with heat transfer or including heat-transfer topics have appeared on the market. They are listed in the bibliographic portion of this review. Volume 12 of the series, *Advances in Heat Transfer*, published by Academic Press, New York, became available in 1976. It contains contributions on dry cooling towers, heat transfer in flows with drag reduction, molecular gas band radiation, and a perceptive on electrochemical transport phenomena.

Developments in heat-transfer research during 1976 can be characterized by the following highlights: solid–liquid phase change, fins, and contact resistance found special attention in the area of heat conduction. A number of numerical solutions have also been reported. Flow and heat transfer in complex passages and solidification of liquids flowing through ducts were investigated. The entrance region of ducts was treated analytically for laminar flow, and various turbulence models were applied to the analysis of turbulent duct flow.

Impulsively started boundary layers were considered and detailed measurements and analyses in free and impinging jets were reported. Studies continue to be made on flow and heat transfer over cylinders with special attention to the near critical Reynolds number region. The number of papers on heat transfer in fluidized beds increased sharply. Measurements of the fluctuating velocities and temperatures, new models of turbulence and the application of kinetic and statistical theory attempted to improve our understanding of the turbulent transfer mechanism.

Many papers are concerned with natural convection, expecially in inclined fluid layers at low and moderate Rayleigh numbers, with non-Newtonian or micropolar fluids and with porous media. Flow and heat transfer in enclosures with partially rotating surfaces were investigated analytically and experimentally. Emphasis in the field of combined heat and mass transfer was on applications and on detailed local measurements. Surfaces which enhance boiling and condensation were developed for various engineering uses.

Attention was given to radiative heat transfer in nongray media in combination with convection and conduction. Emissivities are reported for various surfaces and cavity shapes. Detailed measurements of the bubble growth clarified boiling heat transfer in liquid metals. Interest in plasma heat transfer shifted from aerospace to chemical and material processing applications. Electric arcs were used as a heat source. The laser-Doppler method and hot wire instruments were frequently used, and radiative transport and thermophysical properties were measured.

Heat exchangers found considerable attention with studies concentrating on local measurements by the use of the heat and mass transfer analogy and on more sophisticated computer analyses. Pollution in lakes and in the urban atmosphere, heat transfer in rock drilling, prevention of freeze-up of pipe lines, polymer moldings, reactor accidents, measurement of insolation parameters, flat plate and concentrating solar collectors, and energy storage found attention.

To facilitate the use of this review, a listing of the subject headings is made below in the order in which they appear in the text. The letter which appears adjacent to each subject heading is also attached to the references that are cited in the category.

Conduction, A
Channel flow, B
Boundary-layer and external flows, C
Flow with separated regions, D
Transfer mechanisms, E
Natural convection, F
Convection from rotating surfaces, G
Combined heat and mass transfer, H
Change of phase, J
Radiation
  Radiation in participating media, K
  Surface radiation, L
Liquid metals, M
Measurement techniques, P
Heat-transfer applications
  Heat exchangers and heat pipes, Q
  Aircraft and space vehicles, R
  General, S
  Solar energy, T
Plasma heat-transfer, U

CONDUCTION

Problems involving phase change evoked considerable interest among recent papers on heat conduction. A substantial interest in fins, in thermal contact resistance, and in numerical solution methods is also reflected in the current literature.

For phase change problems, available analytical methods of solution, both exact and approximate, have been reviewed and a corresponding bibliography assembled [42A]. The enthalpy model, which had previously been applied only for cases with no density change, has been extended to accommodate substances which undergo a change of density upon phase change

[53A]. Biot's variational method was successfully applied to yield a lengthy closed-form solution for a phase change problem with radiative heating [46A]. A Laplace transform solution of the Neuman phase change problem agreed well with the classical solution [25A]. A two-region moving boundary problem with conduction in one region and coupled conduction–mass diffusion in the other can be transformed into a Stefan-type problem with error-function solutions [41A]. The heat balance integral has been generalized to accommodate phase change problems with variable thermophysical properties [13A]. Phase change with variable properties was solved numerically by a two time-level implicit method in conjunction with Taylor's forward projection method [40A].

The analysis of solid–liquid phase change in a binary alloy requires that mass diffusion be considered along with heat conduction [8A]. In the analysis of transient ablation of a teflon layer, account is taken of crystalline-amorphous phase transition, thermal expansion, depolymerization, and formation of higher-molecular products [26A]. The most efficient power level for melt-through of a metal target due to laser heating minimizes the radial heat flow away from the melted region and does not vaporize the melt [61A]. In an investigation of salt redistribution phenomena during the freezing of a concentrated salt solution, it was found that the solute concentration increases with time [24A]. Simultaneous heat and mass transfer in the presence of a moving interface is also encountered in the freezing of foodstuffs [59A].

A Stefan-like problem in the *in vivo* freezing of a biological tissue took account of blood profusion, metabolic heat, and tissue heat capacity [49A], Thermal responses of biological systems undergoing freezing, as in cryosurgery, were determined through the application of the finite-element method [14A]. Other applications of the finite element method included frost heave of an embankment, ground freezing, and the stress distribution in a solidifying bar [35A]. A one-dimensional model of freezing and thawing ground with a seasonally varying surface cover was used to study the factors that produce differences between the yearly means of the air and ground temperatures [23A]. Transient heat transfer in a hot granular material being poured into a cylindrical container is a moving boundary problem which can be solved quasi-statically by letting the coordinate system move with the filling level [33A]. The final stages of freezing of an inward solidifying sphere have been described by an asymptotic theory for the case in which the ratio of the latent heat to the sensible heat is large [57A].

Fins continue to be an active applications area. In an optimization study of circumferential and longitudinal conducting–radiating fins based on maximizing heat transfer per unit weight, fins of triangular profile were found to be superior to those of rectangular profile [51A]. Use of an individually optimized fin profile does not necessarily result in an optimum multi-finned surface [18A]. It has been proposed that the fin of infinite length serve as the basis of an alternative definition of the fin efficiency [6A]. The transient response of a fin has been evaluated for temperature variations in a fluid flowing adjacent to the back side of the wall to which the fins are attached [58A].

Variational techniques were used to find the fin shape which maximizes the heat loss for a given fin volume, with the added constraint that the fin thickness is bounded between upper and lower limits [3A]. A variational technique yielded fin efficiency results for simultaneous convective–radiative heat loss [9A]. When applied to a fin with a linear, temperature-dependent thermal conductivity, the Galerkin method provided fin efficiency results that agreed well with those from a numerical solution [43A]. An extension of the heat balance integral is made which encompasses steady two-dimensional heat-conduction problems where the conduction is almost one-dimensional, but for which the one-dimensional theory becomes unreliable [52A]. Upper and lower bounds on the temperature solutions for unsteady heat transfer in a fin-like rod were obtained in a mathematically oriented paper [11A].

Various aspects of thermal contact have been studied. The constriction resistance at an annular contact area supplying heat to a coaxial circular tube has been derived from heat conduction solutions [67A]. Insertion of interstitial plates between contacting surfaces is an effective means of decreasing thermal contact resistance [63A]. Analysis of the thermal contact between wavy surfaces shows a directional effect which is stronger at low applied pressures and high heat fluxes [20A]. The thermal contact conductances of lead ferrite and boron nitride exhibit a significant dependence on temperature and pressure, and increase as temperature and pressure increase [22A]. From an investigation of thermal bonds between miniaturized electronic devices and their heat sinks, diamond has been found to be highly effective [28A].

Numerical solution methods have already been mentioned in prior paragraphs of this section, and other numerically orientated papers will now be cited. General orthogonal curvilinear coordinate systems have been incorporated into finite element heat-conduction analysis [50A]. As examples of the viability of the finite element method for solving problems of simultaneous heat and mass transfer, results have been obtained for the drying of brick and for moisture transfer through a basement foundation [15A]. Special finite-difference equations have been derived to improve the accuracy of solutions near sharp corners [4A]. The error in a finite difference solution associated with a discontinuous boundary condition is smaller for a nine-point difference approximation than for a five-point approximation [48A]. A highly efficient finite-difference computer program for solving Poisson's equation suppresses all possible operations involving zero and one and uses minimum storage [34A]. An improved Monte Carlo method was proposed in which results are obtained only on the boundaries of standard figures inscribed within a solid; temperatures at interior points of the inscribed figures are then calculated by analytical techniques [69A].

Some attention has been given to other types of solution methodologies. Perturbation methods were shown to be highly effective for steady conduction problems with temperature-dependent thermal conductivity and specific heat [1A]. A hybrid method, involving the use of a digital computer and an RC network, has been proposed for solving transient problems with variable properties [10A]. It has been demonstrated that Gauss' principle of least constraint merits inclusion among the available variational methods for solving linear and nonlinear transient heat-conduction problems [64A]. To achieve a more accurate solution, the standard heat balance integral is supplemented by an additional integral condition derived by multiplying the energy equation by the temperature and then integrating over the solution domain [68A]. To assist in the evaluation of the transient response of a slab with nonuniform initial temperature, several relevant series have been evaluated, and the values are presented graphically [62A].

Certain aspects of particulate and composite materials were investigated. The thermal conductivity of hollow glass microspheres was found to be about one and a half times that of stagnant air. The experiments also showed the microspheres to be an effective opacifier of thermal radiation when added to powder insulations [55A]. The cooling rate of a solid particle with irregularly shaped surfaces is much higher than that of a corresponding particle having the same volume but a smooth surface [56A]. A microcontinuum model of heat conduction in materials with inner structure successfully predicted thermal conductivities for such diverse substances as human blood and aluminium spheres in silicon rubber [47A]. Starting with a variational equation governing the micro-conduction, an equation has been derived for the macro-conduction in composite materials which contain microstructure effects [5A]. By the use of a method developed for studying wave propagation problems, a continuum theory was developed for diffusion-type processes in a laminated composite with periodic microstructure [38A].

Several papers dealt with specific problems of steady and transient heat conduction. The steady conduction of heat as it flows from an infinite medium into a slender highly conducting disk and then into a line sink at the disk center is solved by perturbation analysis [65A]. For a thermally conducting solid containing a crack, the perturbed temperature field in the vicinity of the crack has been determined [2A]. The eigenfunctions corresponding to the zero eigenvalue in a separation of variables solution in rectangular coordinates can be employed to yield a meaningful temperature solution [21A]. The concept of the critical radius of insulation has been generalized to account for simultaneous convection and radiation, with account being taken of the possible variations of the convection coefficient with radius and temperature difference [54A].

A study of transient heat transfer in insulation-coated wires took account of a variety of thermal boundary conditions at the external surface of the insulation [29A]. A solution for the temperature rise produced by a heat-generating layer in contact with a large body was used to study the temperature distribution in thin-film nichrome resistors under transient conditions [19A]. The available analytical solution for the time-varying temperature at the heated bore of a large cylindrical solid has been extended to a larger range of times and generalized via a superposition integral to accommodate prescribed variations in bore heating [16A]. Heat pump applications motivated solutions for transient heat conduction between an array of buried pipes and the earth [45A].

For solids subjected to successive periods of surface heating and cooling with different surface heat-transfer coefficients for the respective parts of the cycle, a proper cycle-average heat-transfer coefficient has been formulated [31A, 32A]. In experiments in which heat was transferred through the interface between two solids that were cyclically brought into contact and separated, it was found that a significant reduction in thermal resistance could be achieved by increasing the impact with which the surfaces made contact [27A]. The thermal analysis of a slab moving from a chamber at one temperature to a second chamber at a different temperature is similar to the Graetz problem for slug flow in a duct [66A]. Automatic welding was modeled as a moving point source of heat, with conduction being the dominant mode of transfer [39A].

Theoretical work has continued on the hyperbolic heat equation and on materials with memory. According to the theory of thermoelasticity, a temperature wave can propagate at a finite wavespeed in a rigid heat conductor [36A]. A finite heat propagation speed induces two stress waves in an elastic medium rather than the single wave predicted for conventional infinite-speed thermal propagation [30A]. The deficiencies of wave solutions of the conventional transient (i.e. parabolic) heat-conduction equation are explored and, by this, the derivation of the hyperbolic heat-conduction equation is motivated [7A, 37A]. Sufficient conditions are given for the hyperbolicity of the equation of linear heat conduction for materials with memory [17A]. The method of energy integrals was used to establish the stability characteristics of temperature fields in heat conductors with memory [44A].

In a highly mathematical treatment, the stability characteristics of the solutions for nonlinear internal heat generation have been investigated [12A]. The conventional heat-conduction laws were deduced by expressing the temperature at a point as a space and time average of temperatures at adjacent points and at previous times [60A].

CHANNEL FLOW

Papers dealing with heat transfer in ducts covered a broad range of topics. Particular attention was given to complex flow passages (encompassing augmentation devices), to solidification of flowing liquids in ducts, to laminar entrance region analyses, and to turbulent analyses involving various turbulence models.

Studies of complex flow passages were motivated by heat exchanger applications. Detailed measurements of the transfer coefficients on the fins of a fin and tube heat exchanger revealed the presence of vortex systems which enhanced the values of the coefficients [36B]. In a corrugated-wall fin and tube heat exchanger, the heat-transfer coefficients on the windward corrugations are appreciably higher than those on the leeward corrugations [16B]. Mass-transfer experiments have demonstrated the presence of secondary-flow vortices in corrugated wall channels [17B]. Heat-transfer coefficients for laminar flow of water and glycol in a tube with twisted-tape inserts exceeded those for an unaugmented tube by as much as a factor of nine [21B]. The use of a helically twisted inner body in an annular turbulent airflow resulted in heat-transfer augmentation whose extent was very sensitive to the pitch of the helix [33B]. A finite element solution for fully developed laminar heat transfer in an internally finned tube revealed that there is an optimum number of fins which maximizes the heat transfer [28B].

The turbulent heat-transfer coefficients in a heated tube downstream of a tee are substantially larger than those in a conventional turbulent pipe flow [45B]. The Nusselt number at the point of reattachment of the separated flow downstream of a partial blockage in a duct is significantly affected by the extent of the blockage [39B]. A study of the turbulent velocity field for longitudinal flow in a rod bundle indicated the presence of secondary flow [5B]. It was demonstrated that the method of asymptotic expansions is applicable to the calculation of flow distributions in rod bundles or in coupled, parallel channels [26B].

A number of papers dealt with the freezing of flowing liquids. Water, Wood's metal, naphthalene, paraffin wax, and olive oil were employed in experiments to determine how far into a cooled tube a liquid could penetrate before freezing and plugging occurred [6B]. A solidification process involving a warm liquid flowing over a cold wall was characterized by a frozen layer thickness which grew at first, but later diminished when the convection from the liquid exceeded the conduction into the wall [11B]. Liquid phase subcooling appears to be unavoidable in internal solidification studies involving filtered tap water and does not appear to be influenced by the presence of additional particulates in the liquid, tube surface roughness, or vibration [31B]. With the assumption that the axial variations in the liquid–solid interface are gradual, the equations for liquid solidification in the combined hydrodynamic and thermal entrance region of a tube reduce to those for single phase flow with a modified pressure gradient [22B]. The blockage due to freezing which occurs as a hot fluid is poured into a cold-walled cavity or mold has been analyzed [27B]. The solidification characteristics of a flowing liquid are an important consideration in the analysis of a hypothetical disruptive accident in the core of a liquid-metal-cooled fast breeder reactor [12B].

The laminar entrance region continues to attract interest. Both the entrance region and the fully developed Nusselt numbers for laminar flow in a semi-circular tube are higher than those in a circular tube [20B]. A finite difference solution for laminar flow and heat transfer between concentric spheres indicated the presence of flow separation when the Reynolds number attains a sufficiently large value [2B]. Entrance region and fully developed laminar Nusselt numbers, obtained via the Kantorowich method of variational calculus, are tabulated as a function of cross sectional aspect ratio for rectangular and elliptical ducts [23B]. A finite element method for laminar flow and convective heat-transfer problems was applied to a diverging channel with plane walls [25B]. Linearization of the inertia and convection terms yields satisfactory entrance region results for laminar flow and heat transfer in a porous-walled tube [42B].

The thermal entrance region for combined Couette/Poiseuille flow is another variant of the classical Graetz problem [9B]. The laminar entrance region in a parallel plate channel was re-solved using an integral (Karman-Pohlhausen) method with a higher order polynomial than that used in previous integral analyses [32B]. A study of simultaneously developing laminar velocity and temperature distributions in a tube underscored the importance of radial convection [24B]. The coupled laminar heat and mass transfer in the entrance region of a duct was solved as an eigenvalue problem [29B], as was the laminar mass transfer in the presence of a heterogeneous chemical reaction [3B]. In a study carried out to evaluate concentration and temperature profiles in a tubular flow reactor, it was found that the results were markedly affected by the heat of reaction [15B].

Turbulent flow analyses and experiments have also been reported. The $k-\varepsilon$ turbulence model, when applied to developing flow and heat transfer in tubes and ducts, predicted a minimum in the streamwise variation of the Stanton number [40B]. Solutions for turbulent channel flow based on a modified mixing length model showed many of the same features as the results from the $k-\varepsilon$ model [10B]. Turbulent heat transfer in an annulus with time-varying wall heat flux was solved by superposition; the general solution was applied to the case where the heat flux varies exponentially with time [13B]. Various parameters in a mixing length formulation for heat transfer in rough pipes were deduced by comparison with experiment [19B]. Turbulent heat transfer in a parallel-plate channel was analyzed as an eigenvalue problem (Graetz-type solution) using an algebraic model for the eddy diffusivity [37B].

Results of numerical solutions for turbulent pipe flows near the thermodynamic critical point revealed that the thermal entrance region is longer than that for constant property fluids [38B]. In an experimental study of turbulent heat transfer to supercritical and subcritical CO_2 in a horizontal pipe, the effects of buoyancy caused enhanced heat transfer at the bottom of the pipe and reduced heat transfer at the top [1B]. Entrance region experiments for airflow in a uniformly heated rectangular duct with simultaneously developing velocity and temperature profiles revealed the presence of laminar and transition regions near the duct inlet whose lengths diminished wth increasing Reynolds number [41B].

The Petukhov correlation equation for turbulent pipe-flow Nusselt numbers has been generalized to accommodate low Reynolds numbers, entrance region effects, and property variations [14B]. To obtain an accurate predictive equation for the heating or cooling of superheated steam, a variable-property factor should be appended to the conventional Nusselt–Reynolds–Prandtl correlation (i.e. Dittus–Boelter) [30B].

Non-Newtonian and particle-laden flows have been investigated. Laminar heat-transfer coefficients for Newtonian and viscoelastic liquids flowing in helical coils were found by experiment to be higher than those in straight tubes under comparable flow conditions [34B]. Experimental studies of laminar heat transfer have been performed for several non-linear viscoplastic greases in tubes having several diameters and lengths [8B]. The non-Newtonian behavior of laminar blood flow was modeled by the Casson stress–strain relation in a study of heat transfer in the entrance region of a tube [43B]. Enhanced laminar heat-transfer coefficients in the entrance region of a tube containing a flowing suspension are primarily due to increased tube wall shear rates [44B]. The introduction of a spray of cold liquid droplets into a duct is an effective method for the rapid cooling of a hot gas discharge [18B].

Some theoretical work involved stability and control theory. Linear stability theory, when applied to water flow in a concentric annulus, indicated an earlier onset of instability when the inner cylinder was heated and account is taken of the temperature dependence of the viscosity [7B]. The analysis of thermal instability of a horizontal fluid layer confined between two plates subjected to a vertical magnetic field was extended to the case in which there is a throughflow (i.e. Hartmann flow) [46B]. Optimal control theory was employed to solve the problem of attaining a specified temperature distribution at the exit of a tube for laminar flow conditions [4B].

A partially parabolic flow is one in which influences are transmitted from downstream regions to upstream regions only by way of pressure. A numerical scheme for handling partially parabolic duct flows has been devised by Pratap and Spalding [35B].

BOUNDARY LAYER AND EXTERNAL FLOWS

The reported work in the area of boundary layer and external flows includes laminar and turbulent boundary layers, free jets, impinging jets, and flow over a cylinder.

An experimental study of a turbulent boundary layer with alternating adverse and favorable pressure gradients is reported [52C]. The drag and heat transfer on rough plates have been measured for various Prandtl numbers [16C]. Measurements have been made to determine the local anisotropy in heated and cooled turbulent boundary layers [38C]. A study of spanwise nonuniformity of a nominally two-dimensional turbulent boundary layer is reported [18C]. A high Mach number turbulent boundary layer on the wall of a conical nozzle has been experimentally investigated [4C]. Measurements are made for the free-surface temperature distribution of a falling water film [7C].

Boundary layers on curved surfaces have formed the focus of a number of investigations. A method for calculating curved shear layers has been described [50C]. A study of heat transfer in a laminar wall jet over a curved surface has been reported [46C]. The flow field of a laminar cylindrical wall jet has been studied [45C]. An analysis is presented of the second-order effects associated with curvature and other factors in self-similar laminar boundary layers [1C]. The steady laminar flow through circular pipes having varying curvature has been analyzed [39C]. Measurements are reported for the growth of a turbulent boundary layer in a bend; the behaviour on the convex side is found to be quite different from that on the concave side [37C].

Numerous papers are aimed at predicting the details of laminar and turbulent boundary layers. A comprehensive correlation has been presented for forced convection on flat plates [13C]. An expression is developed for the mean temperature profile in the near-wall region of a turbulent boundary layer for various Prandtl numbers [28C]. Skin-friction coefficients for turbulent boundary layers at low Reynolds numbers are predicted by using a form of the mixing-length hypothesis [43C]. An integral method has been used to calculate turbulent boundary-layer development for certain external-velocity distributions; it is found that different values of the initial momentum thickness produce a range of equilibrium boundary layers [21C]. The effect of variable physical properties on drag and heat transfer on a flat plate has been studied through numerical solutions [53C]. An approximate method has been described for the solution of non-similar laminar boundary layers [41C]. The two-dimensional laminar boundary layer has been calculated by a least-squares finite-element method [35C]. An inverse method (i.e. prescribed wall shear stress rather than the external velocity) is presented for compressible laminar and turbulent boundary layers [9C]. Calculations for heat transfer through a turbulent boundary layer with an unheated starting length are performed on the basis of the surface renewal model [51C]. A two-equation turbulence model is used to predict the velocity field of two-dimensional turbulent boundary layers; situations with a variety of pressure gradients, namely those featured in the 1968 Stanford Conference, have been predicted with good agreement with experimental data [42C]. Semi-similar solutions have been reported for three-dimensional laminar boundary layer [56C]. A solution procedure for three-dimensional boundary layers is described; it uses orthogonal curvilinear coordinates [10C]. Solutions for *unsteady* boundary layers have been obtained [6C, 12C, 55C]. The compressible laminar boundary layer on a yawed cone is solved by a finite-difference method [47C]. A two-equation turbulence model is used to predict the development of a compressible turbulent boundary layer [36C]. The interaction of a shock wave with a boundary layer is analyzed [24C, 49C]. Numerical solutions are presented for the flat-plate boundary layer of a micro polar fluid [20C]. Heat transfer across turbulent liquid films is predicted by combining the standard heat-transfer concepts with the results of film hydrodynamics [8C]. An analysis is made for two parallel adjacent streams with interfacial mass transfer [34C].

Among the free jet investigations, an experimental study reports the velocity and concentration profiles for a carbon dioxide jet injected into air [11C]. Experimental data are presented for the velocity and temperature profiles resulting from rectangular jets of different aspect ratios; the jets are found to behave initially like two-dimensional jets, but become similar to axisymmetric jets in the far downstream region [48C]. Studies of the mean and conditional temperature and velocity profiles are reported for a hot turbulent jet [26C, 27C]. A turbulence model has been applied to horizontal free-shear flows under the influence of buoyancy [19C]. Heat transfer to laminar falling liquid jets has been analyzed by an integral method [40C].

The impinging jets have been studied experimentally and analytically. Impinging jet experiments involving air and carbon dioxide have been reported [17C]. The shear stress distribution resulting from an impinging jet is measured; the results are related to the heat transfer on the impingement surface [5C]. Local mass-transfer coefficients resulting from the impingement of a row of circular jets are measured by means of the naphthalene sublimation technique [32C]. An impinging laminar jet with suction at the impingement surface is analyzed by a finite-difference method [54C]. A correlation has been developed for the stagnation region heat transfer in particle laden hypersonic flows [22C].

The flow over a circular cylinder is the subject of several investigations. Measurements are reported for the mean velocity and turbulence quantities in an axisymmetric turbulent boundary layer along a circular cylinder [3C]. A study has been presented for the heat transfer from a cylinder in crossflow [23C]. The local heat transfer from a circular cylinder is experimentally investigated for Reynolds numbers around the critical value [57C]. The effect of free-stream turbulence on the heat transfer from a circular cylinder has been studied [31C, 30C]. The axisymmetric turbulent boundary layer along a circular cylinder is described by the method of matched asymptotic expansions; the existence of a log-law region and a defect-law region is shown [2C]. An analysis is made of the heat transfer from an isothermal hot wire in a low Reynolds number flow having a small sinusoidally fluctuating velocity superimposed on the mean velocity [15C]. Unsteady laminar boundary layer on an impulsively started circular cylinder has been investigated [25C, 29C]. Heat-transfer coefficients from spheres were measured in the naturally turbulent outdoor environment; the value of the Nusselt number obtained outdoors was up to 2.2 times the value found in low turbulence wind tunnels [33C].

Among diffusion-related studies, the diffusion from a line source in a turbulent boundary layer has been experimentally investigated [14C]. A linear stability criterion is developed for the case of a nonlinear concentration-dependent diffusion process [44C].

FLOW WITH SEPARATED REGIONS

Single bodies

A direct method of calculating through regions of separated flow is given [32D]. This problem is complicated by the existence of a separation point singularity which occurs whenever the pressure gradient is prescribed in the neighbourhood of separation. When a circular cylinder, distorted into a wavy form, moves transversely in a fluid at low Reynolds number, the wake of the cylinder is remarkably different [16D] from a straight one, but the periodic character of wake shedding seems to remain unchanged. An analysis is given [21D] for the velocities and temperatures in the turbulent wake of a heated cylinder with intermittency taken into account. Numerical solutions of the Navier–Stokes equations were obtained [22D] for separated flows around a circular cylinder at $Re = 40, 80$ and 200 using three finite different techniques. In [37D] the behaviour of the heat flux rate in the separation region near a double step at the entrance to an enlarged flat duct is correlated with measurements of heat transported by turbulence $\overline{v'T'}$. Both experimental and numerical results on laminar boundary layers on a sphere show that increasing the sphere temperature above that of the surrounding water shifts the laminar separation point slightly to the rear [40D].

Amr and Kassoy [1D] analyze reattachment of a blownoff supersonic boundary layer subsequent to injection cutoff. Fluid injection through a cavity has significant effects on the flow pattern and laminate heat-transfer rates. The greatest reduction in heat transfer is evidenced at the reattachment point on the downstream cavity wall [18D]. An experimental investigation of the turbulent near-wake regions of several two-dimensional and axially-symmetric bluff bodies helps explain the similarity and the mechanisms of formation of vortices [41D]. Reference [27D] describes rear stagnation point location in the subsonic near wake.

Packed and fluidized beds

Transport phenomena in non-homogeneous porous media are described [29D]. Recent research in the field of capillary-porous material transport processes at the Institute of Heat and Mass Transfer of the BSSR Academy of Sciences is summarized [23D]. Accurate bounds for the effective diffusion constant and for Darcy's constant in porous media are obtained by a new variational formula [10D]. A theoretical analysis of an internally-energized, porous reactor is presented [28D] for both constant and temperature-dependent rate of generation. The concept of generating temperature dependent rate of heating within a prous element is aimed at gaining stability in operation and protecting the solid from burnout. A boundary-layer integral balance method is applied to determine the transfer potentials of heat and mass in porous bodies of cylindrical geometry [34D]. Pismen [31D] studies convective currents induced by chemical reactions in partially-filled porous media. Reference [39D] concludes that for up to 48% solid phase volume fraction, it is impossible, from a heat transfer point of view, to construct a thermal conductivity model that accurately represents heterogeneous porous materials.

Reference [6D] investigates the transient temperature distribution within packed beds after being subjected to single blow heating. The boundary conditions do not require that fluid or solid temperatures be specified at the initial and final bounding surfaces. The Knudsen equation for flow through circular pores may be used for most porous materials without fear that a noncircular pore cross section will alter the conclusions significantly [11D]. The evaporative heat-transfer process in porous media is not clearly understood. Using a flow visualization rig, [9D] shows that both vapor and liquid regions coexist at the heating surface over a wide range of heat fluxes. General solutions for small Peclet number heat or mass transfer in concentrated two-phase particulate systems submerged in arbitrary Stokesian flows with arbitrary temperature or concentration boundary conditions are given [45D] that were obtained using regular perturbation expansions of the dependent variables in powers of the Peclet number. The use of a contact resistance as a technique to match predictions with practical results is very common in granular media heat transfer. In [5D] the use of contact resistance is shown to be a source of error due to wide variability with geometry.

Methods for determination of heat and mass transfer in packed beds are surveyed and compared [14D]. The influence of thermal conductivity in the reactor wall on total axial conductivity and the reduction in overall heat transfer due to axial heat conductivity is correlated [13D] in the form of an effectiveness factor. Reference [3D] describes the effective radial thermal conductivity of various forms of particulate feeds through which gas is flowing. Lou and Wang [24D] give temperature distributions in cylindrical packed beds. The agreement of heat- and mass-transfer data in packed beds confirms that particle-to-particle transfer is negligible in the turbulent region [17D]. A model to simulate heat transfer accompanying melting in a layer of granular particles is proposed [30D]. Reference [2D] describes application of the limiting current method to mass transfer in packed beds at very low Reynolds numbers ($Re \ll 1$).

A generalized analysis to describe the overall process of gas-particle heat transfer in shallow fluidized-bed heat exchangers is developed [25D] and a laboratory scale cross-flow fluidized bed heat exchanger was built and operated to investigate new heat-transfer correlations [26D]. Saxena and Vogel [36D] describe the properties of a dolomite fluidized bed reactor normally used as a combustor for a range of particle sizes and shapes at minimum fluidization. General theories on fluidized bed heat transfer appear in [46D] and [42D]. Heat transfer in the thermal stabilization zone of a liquid-fluidized bed involves the distance from the grating to the point where the heat-transfer coefficient is constant [19D].

Pulsation of fluidizing air leads to an 80% increase in the heat transfer over normal fluidization for large particles at air flow rates a little above incipient fluidization [4D]. The enhanced heat transfer for a vertical heater plate immersed in a vibrated powder bed is attributed [15D] to the scouring of the gas sublayer on the surface of the heater by the particles. Other studies of heat transfer from immersed surfaces in liquid fluidized beds are presented [35D]. Entrainment of solid particles by gas jets discharged downward through slotted nozzles into bubble-free beds is considered [8D]. The electrical resistance of fluidized beds is determined [43D] along with the electrode-temperature in a unit with cylindrical electrode geometry.

Time-averaged temperature profiles are measured downstream in liquid-fluidized beds by injecting hot water. Eddy thermal diffusivities are computed using Taylor's theory of eddy diffusion [33D]. Siegel discusses the effect of distributor plate-to-bed resistance ratio on the onset of fluidized bed channeling [38D]. A model of bubbles rising in an unbounded fluidized bed is presented [7D]. The influences of bubble-induced heat transfer in gas fluidized beds is given in [20D]. For gas fluidized beds of small particles operating below the radiative temperature level, transient conduction into the emulsion phase is responsible for at least 90% of the heat transfer and the remainder is contributed by the superimposed gas convection. The first statistically significant measurements of propagation properties of voidage waves in fluidized beds [12D] have shown that the breakup of these waves is followed by a mode of motion suggestive of bubble formation. Finally, Werther [44D] explains the significance of bubble coalescence for the design of gas/solids fluidized beds.

TRANSFER MECHANISMS

The transfer mechanisms in turbulent flows form the focus of most of the work in this category.

Detailed measurements of the fluctuating velocity

and temperature fields have been reported in a variety of situations. A trace particle method for describing three-dimensional turbulent velocity field has been presented [11E]. Bursts in a turbulent boundary layer in zero pressure gradient and in an axisymmetric jet have been experimentally studied [2E]. Measurements have been reported for a turbulent boundary layer subjected to random fluctuations in the external stream [5E]. A study has been made of high-frequency wall-pressure fluctuations in turbulent boundary layers [4E]. A mixing layer with unequal free-stream turbulence levels in the two streams has been investigated [16E]. Turbulent shear stresses in hypersonic boundary layers have been measured [12E]. A study is presented of the decay of turbulence in a closed vessel [25E]. An experimental investigation of special turbulent flows is reported; these are turbulent shear flows with quadratic mean-velocity profiles generated by using a honeycomb and parallel-rod grids [20E]. Turbulent Prandtl numbers have been measured in the near-wall region in small tubes [14E]. An examination of the wall structure of the turbulent boundary layer has been made by the use of hot-wire rakes and conditional sampling techniques [3E]. Studies of dispersion from a heated wire in a turbulent boundary layer [21E] and from a line source in grid turbulence [22E] have been reported.

Some work has been published on the phenomenological models of turbulence. A modified law of the wake for turbulent shear flows is presented [9E]. The surface renewal model of turbulence is discussed by reference to interfacial conditions [23E]. An algebraic stress model has been proposed for stratified turbulent shear flows [15E]. A Reynolds stress model for turbulent corner flows has been presented [7E], and its results have been compared with experimental data [8E]. Buleev's model for turbulence has been examined; the model has been found to be successful in only simple configurations [18E]. A turbulence model involving differential equations for the Reynolds stresses is developed for low-Reynolds-number turbulence [10E]. A second-order closure for turbulence is used to compute variable-density flows [13E]. The variation of experimentally determined turbulent Prandtl and Schmidt numbers is examined for free turbulent flows; for jets and wakes, these numbers are found to vary by a factor of about 2 across the layer [19E].

A method is presented for calculating the temperature profile in a turbulent flow from the amplitude distribution of measured temperature fluctuations at a fixed position in the flow [17E]. A theory of turbulence decay for an axisymmetric turbulence temperature wake is described [6E]. A kinetic theory of turbulence correlations has been developed for turbulent compressible flows [26E]. Thc temperature fluctuations in a turbulent shear flow are modeled by a ramp-like signal with superimposed Gaussian fluctuations (1E). Burger's model of turbulence is used to calculate the energy decay with various initial energy spectrums [27E]. A statistical theory of one-dimensional turbulence in a compressible flow has been presented [24E]. A statistical formulation of unsteady homogeneous turbulence has been described [28E].

NATURAL CONVECTION

The outpouring of papers on various aspects of natural convection continues unabated. As usual, there is considerable interest in flows within enclosures, in boundary-layer flows, in thermal plumes, and in aspects of combined convection wherein buoyancy effects are superimposed on a flow field producing alterations in the velocity distribution and in the heat transfer. Other studies consider flows in porous media, including energy recovery from geothermal reservoirs, flows in which the forces inducing the flow result not only from buoyancy effects due to temperature but also from buoyancy effects due to concentration difference, and surface tension induced flows as well. Thermal diffusion effects and flow of non-Newtonian and micropolar fluids are also examined.

For the case of a horizontal layer heated from below in what is called Bénard convection or Rayleigh–Benard convection, studies covering stability, laminar flow and turbulent flow have been reported. The critical Rayleigh number is obtained from consideration of the dissipative process in a fluid [83F]. The prediction of the critical Rayleigh number for the non-linear density distribution obtained in a horizontal layer of water cooled from below to the maximum density point agrees well with experimental measurement [59F]. A numerical simulation [54F] finds three-dimensional (twice periodic) disturbances starting at a Rayleigh number of about 5600; the predicted Nusselt numbers in this study agree well experiments up to a Rayleigh number of 25 000. Another three-dimensional numerical analysis of laminar natural convection considers heating from below in fluids in different shaped enclosures [66F]. Temperature oscillations were measured in a mercury bath heated from below [67F].

The range of validity of the Boussinesq approximation has been examined as applied specifically to natural convection in air and water [34F]. An analysis of the heat transfer at high Rayleigh number fits earlier experimental data very well [55F]. Measurements of the heat flux and velocity and temperature fluctuations have been carried out in a layer of air up to a Rayleigh number of about 7×10^9 [30F].

Linear stability theory is used [88F] to obtain the onset of flow in a melting layer heated from below in which the fluid's density decreases with temperature. Linear stability criteria are also applied to predict the onset of convection in a porous medium with internal heat generation [32F] and with a periodic temperature variation on the lower boundary of the porous surface [17F]. A temperature dependent viscosity can significantly affect the critical Rayleigh number for the onset of convection in a porous medium [46F].

Measurements of the heat transfer in a porous layer slightly above the critical Rayleigh number agree with earlier estimates of the upper bounds for the heat transfer [14F]. A computer simulation of the natural convection in a porous medium as might apply in geothermal energy recovery includes effects of confinement, variable permeability, and thermal boundary conditions [73F].

Thermal diffusion in a binary gas mixture in a porous medium results in a critical Rayleigh number at the onset of flow, which can be either greater than or less than that for the case of a pure fluid [52F]. The influence of the Soret effect on the thermal instability of a two component fluid layer has been analyzed [90F]. A non-linear analysis for the case of a positive Soret coefficient gives the relationship of the Nusselt number to the Rayleigh number [68F]. The influence of a negative Soret coefficient on stability of the flow of a real salt solution has been examined [15F].

The measured critical Rayleigh number for the onset of thermohaline convection compares favorably to predictions by both linear and energy methods [94F]. Molecular diffusion is the dominating transport process in the thermohaline region between two convecting horizontal layers at different concentrations [56F].

Instability induced by combined buoyancy and surface tension has been studied in a region where the fluid has a maximum density as a function of temperature [96F]. A micropolar fluid is one in which the length of the molecules is not negligible compared to the characteristic geometric length of the flow. The stability of such fluids in a horizontal layer when heated from below has been examined in two studies [2F, 28F].

The heat transfer from a single heated horizontal surface facing upwards is often comparable to the heat transfer across horizontal fluid layers. The convection from horizontal plates of different shapes has been measured in both the laminar and turbulent regimes [5F]. Laminar flow analysis of the mass-transfer convection equivalent to a heated plate facing either upward or downard has been described [8F]. The instability following transient heating of a plate facing upwards has been obtained [74F]. The importance of free convection phenomena in controlling the burning from a horizontal fuel surface has been alanyzed [24F]. A numerical study of buoyancy driven flows as would occur in the atmosphere has been reported [97F].

Studies on heat transfer to enclosures include a number of different geometries. The turbulent natural convection of liquified gases has been studied in containers of different shapes with a resulting simple correlation of Nusselt number as a function of Rayleigh number [26F]. Numerical predictions and experimental measurements were carried out for natural convection of a heat generating fluid within different enclosures [47F]. Correlations for the heat transfer to fluids in horizontal cylinders were obtained using simple parameters [36F].

An experimental and numerical study of natural convection in the annulus between horizontal concentric cylinders gives excellent agreement of measurement and analysis [50F]. Correlating equations have been obtained over a range of conditions for natural convection heat transfer between concentric horizontal cylinders [51F]. An analysis of free convective heat transfer between two horizontal concentric cylinders shows the effect of a density inversion [79F]. Three-dimensional cellular convection in a cylinder with a free boundary has been analyzed [43F]. The streamlines in the transient cooling of a horizontal cylinder of water through the maximum density point have been determined [20F].

A study of the influence of natural convection in flat plate solar collectors shows the importance of spacing between the collector's surface and the glass cover [13F]. Heat-transfer across inclined fluid layers has been experimentally investigated over a range of Rayleigh numbers and angles of inclination [6F, 38F]. Laminar natural convection resulting from localized heating on the wall of a rectangular enclosure has been analyzed [23F].

A model has been analyzed to indicate the flow instabilities in a vertical layer of fluid enclosed by two plates at different temperatures [61F]. Wave-like temperature variations were found above a critical Rayleigh number for fluids enclosed between two vertical plates

[91F]. The flow in an open vertical cylinder whose surface is at a temperature different from that of the surrounding fluid has been calculated including thermal entrance effects [37F]. The effect of a magnetic field on the convection of a ferromagnetic fluid in a vertical layer has been examined [10F]. Combined radiation-convection heat transfer in an electrically conducting fluid in a vertical layer under the presence of a transverse magnetic field has been described [27F].

A number of studies have considered aspects of the natural convection boundary layer along a vertical plate. A simple correlation for the effect of Prandtl number on Nusselt number has been obtained from earlier theoretical analyses [31F]. Local non-similar solutions have been obtained for the laminar flow adjacent to such a surface [45F]. The effect of discontinuities in surface conditions on the heat transfer from a vertical plate has been reported [58F]. Universal velocity profiles have been used in an integral analysis to predict the turbulent flow heat transfer from such a surface [53S]. A local non-similarity analysis has been applied to convection when a vertical surface is placed in a stratified fluid [18F]. The relationship between artificially induced and naturally occurring transition of a natural convection boundary layer has been described [42F].

Numerical calculations of the growth and non-linear interaction of secondary motions superimposed on the boundary layer on a vertical plate have been made [7F]. Combined heat and mass transfer produce buoyancy forces which affect flow transitions when the Lewis number is not unity [12F]. Numerical integration of the flow equations for an electrically conducting fluid adjacent to a hot vertical plate in a strong magnetic field has been reported [92F].

An analysis of the convection about a thin vertical cylinder immersed in a fluid of low Prandtl number delineates the flow in the outer and inner regions around the cylinder [25F]. Experimental results for the natural convection heat transfer from a horizontal tube to a fluid near the critical point can be correlated using standard dimensionless heat-transfer parameters; such measurements were performed using a number of different fluids [11F, 84F].

Experimental measurements of the heat transfer from a melting ice layer to surrounding water indicate a minimum rate of heat transfer occurs not when the surrounding water is at the lowest temperature studied, but when it is approximately 5.6°C [9F]. Transient natural convection mass transfer from a plate yields an overshoot with a step change in boundary conditions as had been observed earlier [16F]. Both steady-state and non-steady-state natural convection heat transfer to a power law fluid at high Prandtl number have been considered [82F]. Non-linear analysis of flow in a porous medium adjacent to a horizontal surface has application to heat transfer in geothermal reservoirs [19F].

Experiments and an approximate analysis for heat transfer from prolate and oblate spheroids agree well with each other [71F]. Extensions of that study in which the prolate spheroid is taken to the limit of a vertical needle have been presented [70F]. A prediction of laminar and turbulent free convection from elliptic cylinders agrees well with other investigators' results at the extremes of eccentricity, i.e. from vertical plates and horizontal cylinders [69F]. Laminar natural convection from downward-facing heated blunt bodies to liquid metals has been analyzed for application to a core catcher in a fast nuclear reactor [35F].

Convection in some three-dimensional flows including that about inclined plates and circular cylinders has been examined numerically [64F] and in the case of inclined cylinders experimentally [63F]. Local similarity and non-similarity methods are used to determine the convection about a vertical cylinder embedded in a porous medium [60F]. Temperature and velocity profiles in polymer solutions around heated spheres have been obtained [4F]. Experiments and an integral analysis yield the heat transfer by convection to non-Newtonian dilatant fluids surrounding a horizontal cylinder with different boundary conditions [48F].

Several authors consider buoyancy-induced convection in rotating systems. Small disturbance theory is used to predict the stability of convection in a narrow rotating annulus with a heated inner wall [49F]. The effect of rotation on a fluid layer heated from below has been analyzed [41F]. Experiments on the natural convection in a rotating differentially heated cylindrical annulus have been conducted with fluids of different Prandtl number [29F].

The rise of a buoyant laminar plume emanating from a discrete heated source has been studied using a shadowgraph system [81F]. An approximate solution has been obtained in closed form for the laminar-convection velocity and density fields due to a buoyant source [89F]. The interaction of two laminar plumes originating from line sources of unequal strength are observed using an interferometer [33F]. The swaying motion of a plume above a horizontal heated plate has been postulated to be self-excited [3F].

Experiments on turbulent flow in buoyant jets agree with predictions from an integral analysis using an entrainment coefficient approximated from isothermal jet studies [78F]. A model for the flow of a heated liquid plume including surface effects is compared to earlier models and experiments [80F].

Studies of combined convection in which the flow field is the result of both forced flow and buoyancy effects have been conducted in a number of different situations. These include boundary-layer flows, duct flows, and flows within porous media. The effect of starting length on non-similar solutions for mixed convection on a vertical plate has been calculated [1F]. The use of asymptotic results at large Prandtl number are found to apply even at Prandtl number of one for heat transfer by mixed convection about a semi-infinite isothermal plate [93F]. Aiding flow of a Bingham plastic on a vertical flat plate has been analyzed [57F]. The effect of freestream oscillations adjacent to a vertical porous plate with suction has been analyzed with the plate either cooled or heated [86F].

A power series solution yields the velocity and temperature distributions for combined convection in a laminar wall jet along a vertical plate [72F]. Increased heat transfer is found for combined convection in aiding flow along a vertical cylinder with uniform heat flux [62F]. Boundary-layer analysis has been applied to aiding flow about a horizontal cylinder [87F].

Combined convection in water near its maximum of density flowing over a horizontal heated plate has been studied to determine the critical Rayleigh number [21F]. Longitudinal vortices above a critical Rayleigh number are found with a Blasius flow over a heated horizontal plate [95F].

The flow of air in a horizontal channel is found to be more sensitive to heating from below in the thermal entrance region than in the fully developed region [44F]. Another study of stability in the entrance region for horizontal flow in a channel includes the effect of axial heat conduction [22F]. With combined convection in a horizontal tube, the variation of temperature with viscosity can give a significant increase in the rate of heat transfer [40F]. A numerical analysis gives the heat flow for combined laminar convection in inclined rectangular channels [65F]. Experiments on combined convection in inclined circular tubes agree with predictions at low Rayleigh numbers and low Reynolds numbers [75F]. Heat-transfer experiments have been performed on water flowing in horizontal and inclined tubes at supercritical pressure conditions [85F].

A study of combined forced and natural convection in wet geothermal formations includes the effect of water reinjected into the reservoir [77F]. The stability of throughflow in a porous media when heated from below has been determined [39F]. Combined convection in a vertical annulus containing a micro-polar fluid has been described [76F].

CONVECTION FROM ROTATING SURFACES

The distributions of the velocity and temperature, or Reynolds stresses, and of heat flux and temperature fluctuations were determined [4G] in turbulent incompressible flow through a tube rotating around its axis. An analysis [7G] was performed to obtain friction factors and heat-transfer coefficients on a rotating circular cylinder.

The recirculating, swirling, turbulent flow was predicted [2G] in the cavity between a rotating disc and a stationary shrouded disk. Good agreement with experimental results was obtained for laminar flows; a two-equation turbulence model was used for turbulent flow. Local measurements of the temperature field and of heat transfer for the same type of enclosure with coolant throughflow entering through the rotating shaft and leaving through an annular space at the rim of the rotating disk were reported [8G]. A computer analysis [1G] determined the velocity and temperature field and local Nusselt numbers in an enclosure of the same type with turbulent throughflow.

Heat transfer through a laminar compressible boundary layer on a sphere rotating in a quiescent fluid was analyzed [3G]. An asymptotic solution was obtained [6G] describing heat transfer for closed streamline flow past small rotating particles at large Peclet numbers and small Reynolds numbers. Flow and heat transfer were calculated in a hydromagnetic Ekman layer on a rotating porous plate [5G] under the influence of a transverse magnetic field.

COMBINED HEAT AND MASS TRANSFER

Interest in combined heat and mass transfer centers on applications of mass addition, either local or distributed over a surface, to protect a surface from a hot mainstream. A number of experimental and analytical studies have been reported in this area.

A review of two-dimensional film cooling using slots shows the ranges of validity of previously reported correlating equations for the local film cooling effectiveness [12H]. An examination of the relative sizes of the thermal and velocity boundary layers in two-dimensional film cooling is useful in predicting the effectiveness [11H]. A recent study confirms previous findings that, over a moderate range, an adverse pressure gradient has little influence on two-dimensional film cooling [7H]. The effects of multiple rows of staggered film cooling jets were predicted from the results of a single row of staggered holes using the principle of superposition [13H]. In this study, the lateral average effectiveness was determined. A finite difference procedure [2H] predicts film cooling performance from a single row of inclined jets with laminar flow. The details of the flow field in the vicinity of a jet entering a turbulent boundary layer was measured to enhance the understanding of film cooling performance close to an injection hole [1H]. Photos of helium-filled soap bubbles put in with the secondary jets showed the trajectory of the injected fluid in three-dimensional film cooling. From these streak lines, qualitative estimates of the effects of changes in the hole spacing and inclination were made [4H].

Film cooling applications include high speed flows where compressible aerodynamic effects can be quite important. The effects of the slot height and lip thickness on two-dimensional film cooling in a Mach 6 airstream were measured [8H]. A recent analysis shows that the angle of injection into a hypersonic laminar boundary layer significantly influences the heat transfer [9H].

The flow following uniformly-distributed injection through a surface with the injected fluid having a velocity component parallel to the mainstream has been examined [3H]. An integral solution to the laminar boundary-layer equations with transpiring flow gives simple, yet apparently reliable, predictions of the wall heat flux [17H].

The effect of blowing on heat transfer to a turbine blade has been approximated [14H]. Optimization of the film coolant flow in a gas turbine cascade has been studied including the effects on pressure loss and flow separation [10H].

A model has been established for the flow and heat transfer in a multi-component boundary layer to include the influence of multi-component diffusion [6H]. Another study [15H] shows the effects of thermodynamic coupling on flow in a compressible multi-component laminar boundary layer.

The combined effect of heat and mass transfer from a freely falling drop has been studied [16H]. The transport of heat and salt in a layer across which temperature and concentration gradients have been established has been studied including the influence of turbulence produced by vertically oscillating grids [5H].

CHANGE OF PHASE

Boiling

Studies of evaporation in vertical grooves and capillary tubes [39J] support the hypothesis that fluid flow in and to an evaporating menescus results from a curvature gradient. A theoretical model to determine the required superheat for stable boiling from a cavity has been developed [47J] considering the liquid penetration, the effects of transient heat flux, and the inertial and viscous effects of liquid moving inside the cavity. The initial-to-final vapor density ratio, when significantly greater than unity, strongly influences vapor bubble growth rates in the inertial and the inertial-heat-transfer regimes [51J]. A heat-transfer controlled bubble model has yielded three semi-empirical correlations to predict bubble growth rate, maximum bubble diameter, and maximum bubble growth time for subcooled nucleate flow boiling of water up to 17.7 MN/m^2 [54J]. The problem of bubble growth in the presence of a magnetic field is examined [28J] in the heat-transfer region. The work reveals the existence of a new nondimensional number which physically represents the ratio of pondermotive forces over pressure forces. Heat transfer in nucleate boiling is reduced in the presence of magnetic fields—substantially more for potassium than for mercury. A microlayer mechanism supported by extensive experimental evidence is used to explain the high heat fluxes observed during nucleate boiling [32J]. The data in [23J] comprise a comprehensive set of measurements which substantiates a model for predicting boiling heat flux incorporating microlayer evaporation, natural convection, and nucleate boiling mechanisms.

The hydrodynamic stability approach is used to explain the effect of electrostatic fields on the maximum heat flux [4J]. Heat-transfer crisis in annuli is studied [35J] in which the rate of heat transfer varies longitudinally either sinusoidally or decreases exponentially. Rao and Andrews [40J] describe the effect of heater diameter on the critical heat flux from horizontal cylinders in pool boiling. A large inlet break loss-of-coolant accident in a pressurized-water reactor would cause the flow through the core to reverse within milliseconds. Reference [48J] gives an analysis of critical heat flux with flow reversal transients. Canon and Park [7J] develop a stability criterion for transition boiling and demonstrate that, when it is satisfied for boiling equipment, then stable boiling can be obtained over the entire transition region. The dynamics of the transition region are studied [38J] for flow boiling using an apparatus that includes a feedback control system that permits operation at any point on the boiling curve. Impaction of volatile liquid droplets onto a hot liquid surface is studied [57J]. The transplosion phenomenon (inverse Leidenfrost phenomenon) is examined experimentally for a molten copper–water system [63J].

A three-dimensional variational approach to stability of the vapor film during film boiling from horizontal cylinders leads to results [5J] which agree with data. Marschall analyzes binary, laminar film boiling [31J]. Basu [2J] also deals with film boiling of organic binaries on a horizontal wire at atmospheric pressure. Experiments of film boiling of carbon dioxide generate curves of heat-transfer coefficient vs pressure that take on minimum values at the triple and critical points [21J]. Localized contact between liquid and heating surface was found to exist during stable film boiling that resulted in surface scaling [15J].

A new empirical correlation equation is developed [16J] which permits one to predict the performance of a cylindrical heater in the presence of cross flow and boiling, provided only that its performance in the absence of crossflow (pool boiling) is known. Burnout is investigated with flow of a steam–water mixture in a round tube [26J]. The effects of certain organic additives on boiling are discussed [11J]. Noncircular duct hot patch data, extensively referred to in the literature, have been found to be in error. When these data are corrected and compared with predictions of hot patch dryout using the most recent form of the Tong F factor [30J], unintelligible results are obtained. In many heat exchangers operating at supercritical pressures, heat transfer in turbulent pipe flow occurs with very large temperature gradients such that the flow core temperature is much below critical and a process known as "pseudo-boiling" occurs [24J].

Condensation

Wilson has shown [61J] that, within the framework of thin film approximations, the solution of unsteady and three dimensional condensation problems can be reduced to the solution of a system of ordinary differential equations, thereby adding to the class of problems that can be solved in closed form. Reference [10J] extends the pioneering work of Sparrow and Gregg on the phenomena of condensation on a rotating disk to include constant axial suction at the permeable condensing surface. Sage and Estrin [43J] apply the analytical methods for treating multi-component systems to film condensation of a condensable component from its mixture with two noncondensables. Epstein and Cho [14J] analyze laminar film condensation on a vertical melting surface—obtaining an integral solution of the boundary-layer equations for one and two component systems. Laminar film condensation equations with combined body force and forced convection of a binary vapor over a flat plate are solved by finite difference and two integral methods [27J]. Generally speaking, the whole region of condensation may be analyzed with good accuracy by approximate integral techniques.

The synthesis of single-droplet constriction resistance solutions with the known droplet distribution leads to a simple correlation for the effect of the condenser material thermal properties on the dropwise condensation heat-transfer coefficient [17J]. In a companion paper [18J], the magnitude of the surface thermal conductivity effect on dropwise condensation heat transfer is determined experimentally. An expression is suggested [56J] for the total condensation length in pipes if the heat-transfer coefficients over a short tube segment is known. The theoretical dependance of heat-transfer coefficients on maximum drop size is incorporated into a new condensation theory giving the dependence of sweeping frequency and heat-transfer coefficient on distance along the condensing surface [41J]. The stability of drops as a function of their size and the slope of the surface upon which they reside has shown that when the angle of inclination of the surface to the horizontal is $\theta = \arctan \pi$, the drops are least stable [1J]. At present there is no commercially acceptable dropwise condensing surface although the motivation for producing such a surface is significant. There is general agreement that dropwise condensation is a heterogeneous process. Reference [12J] demonstrates the direct effect of nucleation site dimensions on dropwise condensation heat-transfer rate. Several attempts have been made to use surface tension forces to reduce the thickness of the condensate films on vertical surfaces and, thereby, reduce the heat-transfer rate. Reference [8J] studies condensation on static and rotating pinned tubes since the addition of pins produces only a small area increase.

Reference [13J] gives a two-dimensional analysis of the effect of precursory cooling on conduction controlled rewetting of a vertical surface. Precursory cooling refers to the cooling caused by the droplet–vapor mixture in the region ahead of a wet front in surface rewetting. Experiments are summarized [42J] that investigate transient radial heat transport effects created by oscillatory turbulent subcooled flow in uniformly heated tubes. Shock [46J] analyzes evaporation of binary mixtures in upward annular flow. Entrance effects increase heat transfer by 20% in a vertical tube with vaporization of water [50J]. A simple procedure is given [59J] to obtain heat-transfer coefficients in the interline region (junction of vapor, absorbed evaporating thin film, and adsorbed non-evaporating wetting thin film). In [19J] the transport reaction mechanism of mist formation in a convective field is discussed on the basis of a "critical supersaturation model". Analyses are performed for mist formation conditions, mist region, and vaporization rate, and are verified using a naphthalene sublimation rate. Kosky [25J] studied the flow of a steam–air mixture containing a fraction (0.5–20 wt %) of mist being swept across a roughened, heated tube.

Two-phase flow

It is often necessary to use simplified (one-dimensional) forms of the conservation equations for many multiphase, multi-component flows. In [62J] the effects of radial distributions of velocities, enthalpies, and void fractions are explained. In [52J] the theoretical study of time varying, two-phase-flow problems in several space dimensions introduced such a complex set of non-linear partial differential equations that numerical solution procedures using high speed computers are required. The ability of one-dimensional balances to predict two-phase pressure drop across abrupt area changes in oscillatory flow has been examined [60J]. New experiments data on the onset of so-called "density wave oscillations" and their frequence are available [44J]. Significant time lag between the system pressure drop and the inlet flow was observed during the unstable operation.

In comparing a theoretical laminar analysis of a horizontal condenser–evaporator tube with data, a 50% discrepancy is reduced to about 15% when the effects of ripples are taken into account [33J]. The flow pattern in a horizontal agitated thin-film evaporator are studied [58J]. Mogro [34J] studies the separation of gas-vapor mixtures by condensation in a vertical tube at Reynolds numbers as high as 450 000. Condensation heat transfer of steam on multiple free falling water jets enveloped by air streams are investigated experimentally [3J]. An integral method is used [49J] to obtain a closed-form expression for the hydrodynamic entrance length for falling liquid films emerging from a flat slit and flowing in laminar flow along an inclined flat plate. Liquid spindle oil falling films are much more laminar-like than turbulent for Reynolds numbers below 6000 [53J]. Reference [20J] predicts the interfacial drag and the height of the wall layer in annular flows. Infrared radiometry provides a convenient technique for measurement of surface temperatures at the free liquid interface of falling liquid film [6J]. Large interfacial disturbances have been observed in falling film evaporators [55J] that are caused by surfactant in sea water that enhance heat transfer. Experiments [45J] are used to show that a turbulence model used by Limbeig successfully predicted evaporation coefficients with turbulent falling liquid films using a modification of the eddy diffusivity in the outer region. Reference [22J] also incorporated new gas absorption data into an improved eddy diffusivity model to demonstrate an engineering method for calculating heat transfer across a freely falling turbulent film and a turbulent film with interfacial shear due to cocurrent vapor flow. Heat transfer to an air suspension of solids flowing in a horizontal pipe is proportional to the pipe length [37J]. This fact must be kept in mind when comparing results of different investigators. Surface tension driven flow in cylindrical melt suspended between two solid rods—commercially called "floating zone melting"—is analyzed numerically [9J]. In [36J] some similarity solutions are presented for velocity and temperature profiles within the thin film of fluid that forms between a moving hot interface and a melting solid. The laser-Dopler anemometer has been successfully utilized in velocity measurements of the individual phases in two-phase, gas liquid flow in the bubbly flow regime. The results show that the slip velocity can be measured without disturbing the flow [63J].

RADIATION

Radiation in participating media

Radiative transport in emitting, absorbing and scattering non-gray media continues to attract interest.

An analytical method is proposed for solving problems of radiative transport in an absorbing and emitting medium which does not require time-consuming numerical integrations [6K]. A study of the mean beam lengths for spheres and cylinders indicates that large values of L have a smaller effect for the cylinder and only at somewhat greater optical depths. For the sphere, where the largest L is the diameter, the deviations of the ratio of mean beam length to geometric beam length from unity is smallest [23K].

A kinetic model for the interaction of diatomic molecules with non-equilibrium radiation allows rotational and vibrational relaxation at different internal temperatures [17K]. An asymptotically exact expression for radiation heat transfer through planar media of molecular gases is derived for large optical depth using an exponential band model [15K]. By introducing an emissivity correlation parameter for the water vapor emissivity, it is shown that only a single exponential is required for describing this emissivity [4K]. Calculated band absorption applied to CO and nitric oxide shows very good agreement with experimental data [8K].

The effect of axial radiation on the cartesian Graetz problem depends strongly on the properties of the boundaries. For mirror boundaries axial radiation is negligible, even at low Peclet numbers if radiation is appreciable. In the case of black boundaries, axial radiation is negligible only when radiation effects are small and the Peclet number is large [5K]. A study of simultaneous convective and radiative heat transfer in a cylindrical geometry with uniform wall temperature considering non-gray gases such as H_2O, CH_4, CO_2 and CO shows that the optically thin approximation may be used for the calculation of heat transfer with reasonable accuracy if the dimensionless path length is less than 0.4 [9K]. Calculations of the temperature profile and of the radiative and convective wall-heat fluxes for a hydrodynamically established flow of a non-gray gas in a cylinder considering laminar as well as turbulent flow are in qualitative agreement with the constant-volume heat-source case [24K]. A linearized problem of a steady one-dimensional slow flow of a gray inviscid non-heat conducting gas past a gray semi-transparent heat source is analyzed by a method of matched asymptotic expansions assuming that the magnitude of heat transfer by radiation is dominating [10K].

If the effects of coupled thermal and non-gray radiation can be separated, the number of absorption–emission bands should be irrelevant and super-position is expected to give reasonable estimates for combined conduction–radiation through multiband, planar media [14K]. In connection with the development of a barothermal gas cell for measuring small absorption coefficients it was found that sufficient heat transfer by thermal conduction from the weakly absorbing windows into the gas produces a measurable pressure rise in the gas [2K]. In a follow-up paper a photoacoustic theory is developed for cylindrical configurations assuming that bulk absorption in the solid is the mechanism by which energy is transferred from, for example, a laser beam [1K].

Geometrically thin shock layers are, in general, advantageous in reducing the radiation heat transfer to a high-speed body [7K]. Radiative heat transfer slows down the rate of decrease of the velocity of propagation of a pressure shock [19K]. The absorption coefficients for air at high temperatures and pressures ($T \geqslant 1.6 \times 10^6$ K, $p \geqslant 30$ atm) measured behind 8–12 km/s shock waves reveal values lower than those calculated using the Biberman–Norman theory [21K].

Studies of the radiation of the fuel particles in a p.f. flame show that the index of refraction and the absorption coefficients are sensitive to a change in the content of carbon in the combustible mass of the fuel [3K]. It is shown that the two-flux spherical harmonic modelling of two-dimensional radiative transfer in furnaces or combustors provides accuracy, computational simplicity and economy compared with the zone method of analysis [18K].

A simple atmospheric transmittance model for an absorbing gas is based on temperature differences and holds for monochromatic as well as polychromatic radiation. Transmittances derived from this model agree with line-by-line calculations to within 0.0031 for the cases studied [12K].

Quantitative measurements of the ratio of the IR spectral reflectance of standard sea water at near-normal incidence to the corresponding spectral reflectance of pure water at 27°C are in general agreement with earlier results obtained at large angles of incidence [16K].

A loss formula for radiation losses of curved optical fibres with constant radius of curvature is derived as an extension of the curvature-loss formula for slab wave guides [11K]. An expression derived for the fraction of incident radiation transmitted through a right angle bend in a rectangular light pipe demonstrates that light attentuation is due only to reflection losses at the walls if the launching angle is 45° [22K].

A theoretical description of the optical properties of aggregated gold films attaining modal sizes up to 1500 Å for the visible and IR range of the spectrum is in satisfactory agreement with experiments [20K]. Single-layer, gradient refractive index anti-reflection films have been developed with a reduction of the reflectance from 8% to less than 0.5% in the wavelength regime from 0.35 to 2.5 μm [13K].

Surface radiation

Monochromatic normal emissivities have been reported [10L] for gold at wavelengths between 1 and 30 μm at temperatures of 300, 79 and 6K and have been compared with predictions. Total normal emissivities for copper and silver are also included. Spectral directional emissivities of rough metal surfaces were calculated [2L] assuming a Gaussian distribution of the slope and height of the surface roughnesses and including shadowing by adjacent peaks. Results are obtained for gold and chrome.

Shape factors for radiative exchange between parallel ring sectors sharing a common centerline have been calculated [8L]. Optical projection offers a means to determine shape factors [6L].

Local and integrated hemispherical emissivities are reported [1L] for cavities in the shape of a cone and a truncated cone touching at their bases. The emissivities increase considerably as a ring shaped lid is attached to the opening of the truncated cone. A closed form solution has been presented [5L] for the radiosity at the edge of a rectangular cavity.

A previously published theory describing radiative heat transfer in packed beds of spheres with high conductivity has been verified by measurements [7L] at temperatures between 350 and 930 K. Variational techniques have been applied [3L] to radiative and convective straight fins with steady and unsteady conditions. A computer analysis treats heat transfer in a channel formed by two vertical flat plates with asymmetric heating by combined radiation and laminar free convection [4L]. Flat and wave-shaped radiant energy receptors in straight tubes are of advantage for high temperature air preheaters because they increase heat transfer by a factor between 1.48 and 2 with an increase in pressure drop by a factor between 1.6 and 3 [9L].

LIQUID METALS

Local heat-transfer coefficients and the temperature field were measured [6M] in natural convection from a horizontal cylinder to mercury under the influence of a magnetic field. Nusselt numbers were found to decrease with increasing Hartmann number. Velocity and temperature measurements by sheathed thermocouples and by an anemometer were made [9M] for free convection flow in mercury adjacent to a vertical cylinder. An expanded Rayleigh equation was used [7M] to study the behaviour of inert argon bubbles in liquid sodium under forced convection.

An analysis [4M] of the evaporation of a hemispherical bubble in nucleate boiling of liquid metals shows that a dry patch grows near the center by evaporation of the microlayer but that its effect can usually be neglected. The analysis was extended [2M] to describe the growth of the bubble and its separation. A simple theory [3M] predicts the growth rates of hemispherical bubbles in nucleate boiling of liquid metals on heated surfaces. Experiments [5M] dealing with boiling of potassium in niobium pipes at high pressure agree with previous results and broaden the range of parameters.

Precision measurements [8M] of film condensation of mercury on a nickel-plated copper surface resulted in a temperature drop from the vapor to the film surface and in condensation coefficients which vary between 1 and 0.6 depending on condensation rate and vapor pressure. Measurements [1M] of heat transfer in condensation of potassium–vapor streams indicate that at velocities close to the molecular velocity the heat transfer decreases by an order of magnitude.

MEASUREMENT TECHNIQUES

Topics of major interest in the area of measurement techniques include laser-Doppler anemometry, hot wire anemometry, flow and velocity metering, temperature measurement, thermophysical property measurement and radiation property measurement.

Atmospheric wind speeds have been remotely measured at distances up to 60 m using a dual beam laser-Doppler anemometer (LDA) and it is projected that the techniques can be used for remote wind measurements at ranges of several hundreds of meters [8P]. The effects of finite-size probe volume on LDA measurements has been investigated and corrections for the effects are proposed which brings the LDA data into good agreement with previous hot-film data [27P]. The use of transparencies based on a few simple drawings facilitates the production of Moiré fringe patterns that visually simulate the origin of many possible LDA signals and these simulations help to clarify the physical basis for these signals [16P].

The benefits that are derived from processing the photon-resolved signal of a laser velocimeter include an increased data rate, the improvement in the particle lag problem and the capability of extracting data from low signal-to-noise ratio environments [26P].

Advances in other optical measurement techniques such as holography and interferometry have been reported. Tests have been conducted on a complete photographic system for performing real-time holographic interferometry and which has the feature that the expose-process-view cycle is fully automated [43P]. A proposed method of measuring shifts of spectral lines employing a Fabry–Perot spectrometer was analytically shown to be significantly superior to existing methods using grating instruments and the method has been employed to measure the flow velocity of an argon plasma [14P]. The production rate and diameter of water drops formed at the tips of capillaries can be measured along with their free-fall velocities with an optical detector [15P]. The feasibility of an acoustical shadowgraph technique for determining turbulent properties was demonstrated successfully by using a coherent collimated acoustic beam propagating perpendicular to a two-dimensional turbulent jet [23P].

A number of papers dealt with hot wire and hot film anemometry. An artificial fluctuating velocity field has been produced with an electrodynamic shaker to test the ability of hot wire anemometers to accurately measure higher-order moments of velocity and evidence was found to suggest that a constant-temperature anemometer may be more suitable than a constant-current anemometer for measuring odd-order moments of velocity and that both types of anomemeters are in good agreement for even-order moments [21P]. Experimental studies demonstrate that major errors are encountered when trying to measure turbulent shear stress near a wall using hot wire techniques [40P]. Wall shear stresses near a stagnation point were measured for stationary and moving surfaces using a flush mounted, hot film probe. However, the response for the dynamic case proved inadequate and an alternative technique was developed in which the probe was displaced from the wall [7P]. An experimental comparison of two generally accepted methods of Reynolds stress measurement by hot wires/films and the use of fixed vs rotating probes, lead to the result that the two methods give good agreement and the choice of which to use becomes a matter of personal preference [37P].

Preliminary results indicate that the use of a water-cooled film-type sensor is promising for measuring turbulence in high-temperature gas flows [51P]. Heat-transfer measurements for a constant-temperature hot wire anemometer showed that the Nusselt number of the wire is proportional to the Reynolds number aised to an exponent whose value varies with the gas temperature [30P]. The measurement of all three mean velocity components can be made conveniently using a single hot film anemometer [17P].

A modest number of papers described measurement techniques which involve the use of Pitot probes, flow meters, and flow visualization. An improved fast-response bell type micromanometer with a Pitot tube has been developed for measuring air speeds in the range of approximately 1 m/s [4P]. A study of the behaviour of Pitot tubes in the turbulent pipe flow of a dilute polymer solution found that errors that occurred in total pressure measurements were negative and in some cases were as high as 40% [18P]. A pair of papers in the Russian literature deal with increasing the precision of flowmeters. Formulas are recommended for engineering calculations to determine the stream coefficients, position of greatest stream contraction, local resistance coefficients of orifices, and recommended lengths of straight pipe upstream of the meter [46P]. Generalized graphs are provided which permit the selection of flowmeter parameters that are optimal for given conditions and that insure the maximum measurement precision in measuring gas flow with orifice or nozzle type meters [32P]. It has been shown that the effective hydrodynamic length of flow through a jet viscometer orifice may be used to calculate the largest admissible Reynolds number for 95% approach to fully developed flow [41P]. A probe that mixes liquid nitrogen and superheated steam produces smoke for visualization studies in low speed wind tunnels [36P].

A variety of temperature measurement techniques have been reported for a wide range of temperatures. A general discussion of the measurement of temperatures below 1.5 K is presented in [5P]. The true temperatures of gray bodies, non-gray bodies with known radiation properties, and a limited set of non-gray bodies with unknown radiation properties can be determined by measuring the radiation from the bodies in four wavelength bands [24P]. The bulk temperature of a flowing polymer stream can be measured to within approximately 1°C by using a pair of sensors, one with a highly conducting and one with a poorly conducting face, placed in a plane metal wall bounding the flow [34P]. The measurement of temperature difference by means of automatic bridges with semiconductor thermistors is discussed in [50P].

A thermometer system which uses a platinum coated quartz wire as the sensor is capable of measuring temperature fluctuations of less than 0.1 K with a frequency of 500 Hz in air flows [35P]. It has been analytically and experimentally shown that a device with a 5.3 μm dia platinum wire can be used to record temperatures which vary as rapidly as 0.05 s [45P].

The error in temperature measurement caused by the step-like movement of the mercury in a conventional mercury-in-glass thermometer can be reduced to less than 1 mK by measuring the expansion of the mercury capacitively [39P]. An analysis was conducted to estimate the error in the measurement of the temperature of a solid body or a porous material due to conduction along the sensor leads and included the effects of an arbitrary temperature distribution along the sensor length and the conduction to the external surroundings [42P].

A CO_2 laser can be used to simultaneously heat an absorbing gas and measure its temperature [28P]. Simultaneous determinations of the temperature and species concentrations of a combustion flow were achieved using laser absorption techniques [48P]. The temperature and enthalpy of high-temperature flames issuing from burners have been determined with an accuracy of 1.5% using a simple two-thermocouple probe and the predicted temperature distribution for the probe [33P].

The determination of very high heat fluxes (on the order of 10^6 W/cm^2) to surfaces can be achieved by a comparison of calculated and measured temperature–time histories of the surface [44P].

A Pirani manometer with linear response for gas pressures between 3×10^{-4} and 0.5 torr, a time constant of 20 ms, and automatic temperature compensation for room temperature variations of ± 10°C can be built from inexpensive standard electronic components [22P].

A number of devices and techniques have been described for the determination of the specific heat, thermal conductivity and viscosity of materials. Preliminary test results indicate that a newly designed closed loop calorimeter is suitable for measuring the specific heat of liquids up to pressures of 100 MPa and temperatures of 400°C [10P]. A method has been developed for simultaneously determining the thermal conductivity and specific heat of electrically nonconducting solids with an accuracy of 1% and a precision of 0.1% on samples as small as 5×7 mm [9P]. The measurement of the thermal conductivity of solids under pressure using a transient hot wire technique is aided by the use of an electronic circuit which maintains constant power to the wire [2P]. The thermal conductivity of granular substances can be measured with high precision by using a spherical probe [38P]. The minimization of measurement errors involved in a probe method of determining soil thermal conductivity requires a minimum acceptable initial measurement time [20P].

An apparatus, which is a simple modification of a capillary extrusion rheometer, provides measurements of viscosity at much lower shear rates than can be obtained with former viscometers [11P]. Measurements of the viscosity of a slurry or suspension can be achieved by using a specially designed spindle in a conventional Brookfield viscometer [47P].

An isothermal flow calorimeter is suitable for measuring heats of reaction and heats of mixing over a temperature ange of 0–70°C and a pressure range of 1–400 atm [12P].

A method which employs total heat balances on two geometrically similar objects, one with known emittance and one with unknown emittance, has been proposed as a technique for determining the emittance of materials [3P]. A system for measuring the emittance of refractory materials in vacuum for temperatures up to 3000°C in the wavelength range of 0.4–5.0 μm has been shown to be accurate to ± 0.02 for emittances above 0.50 [25P]. An IR hemi-ellipsidal mirror reflectometer has been shown to function with very tolerable errors (less than 2%) in measuring the hemispherical-directional reflectance of room temperature samples in the 2–34 μm wavelength range [49P].

A specially designed shutter-aperture section which is placed between the test specimen and the detector in an evacuated emissometer permits the elimination of errors in the emittance measurements which are introduced by the window through which the radiation exits the evacuated system [13P]. A reference black body operating at approximately 2700 K with an emittance in the range of 0.997 to 0.999 has been built using vitreous carbon for the cavity wall [31P].

Measurements of transmittance at normal incidence have been used to simultaneously determine the refractive index and thickness of a transparent film [29P]. A determination of the temperature dependence of the refractive index of transparent liquids and solids can be achieved by an interferometric examination of the material as its temperature is changed linearly with time [1P]. A method is described for the dynamic measurement of the state of polarization of a light beam [6P].

An IR field optics system has been designed that achieves the maximum flux concentration allowed by the Abbe sine inequality and provides efficient coupling to bolometer-type detectors [19P].

HEAT-TRANSFER APPLICATIONS

Heat exchangers and heat pipes

The use of heat- and mass-transfer coefficients in the design of heat- and mass-transfer equipment simplifies, in general, the analysis and has become well established [11Q]. Some examples are shown, however, in which it is better to base calculations on heat flux. A finite difference method was used to calculate [16Q] the temperature distribution and the heat flux for tubes with rectangular fins. The results are presented in diagrams in the way that the overall heat-transfer coefficient of the tube with fins is related to the heat-transfer coefficient of a smooth tube. The mass-transfer analogy was used [10Q] to measure local transfer coefficients near the third tube in a three-row fin and tube heat exchanger. Heat transfer around the second tube in a row of three tubes was measured [1Q] for spacing to diameter ratios between 1.3 and 5 at $R = 40\,000$. The electrochemical method was used [5Q] to obtain shell side heat-transfer coefficients in a baffled shell and tube exchanger. A relationship of the form

$$j = A\,Re^{-1} + B\,Re^{-0.5} + C$$

for Colburn's exchange coefficient gave best fit with experimental results at Reynolds numbers between 1 and 10^4. Heat transfer in wave-shaped air preheater tubes was found [9Q] to be two times that in straight tubes with a pressure loss increase of 80%.

An analysis of a large dry cooling tower [8Q] was based on spine-fin exchanger geometry. A surface heat exchanger built into the skin of an aircraft with and without longitudinal fins removes heat from a hot air stream [7Q]. An analysis [2Q] considering heat transfer between a stream of water and a surface moving normally to the mean water velocity was based on penetration theory. The results agree well with experimental ones for high water and surface velocities but differ at low velocities. Heat transfer was measured [13Q] for pseudoplastic fluids (dilute aqueous polymer solutions) in a turbine agitated vessel for the jacket and the heat-transfer coil. Heat and mass transfer in an evaporative plate-fin heat exchanger were found [4Q] considerably larger than in equivalent single phase flow. A method of calculating heat transfer with condensation from the flow of a steam–air mixture on heat exchanger tube bundles is presented [3Q]. A design example for a typical condenser and evaporator shows (12Q) the effect of fouling factors.

The energy equations for the fluid and the solid in a regenerator can be transformed into a single one by application of an equivalence principle [14Q]. Mathematical models are described for the analysis [6Q, 15Q] of the performance of thermal regenerators including the effect of fluid hold-up. This effect increases the regenerator effectiveness at small values of the dimensionless bed length.

A design parameter for assessing wicking capabilities of heat pipes was established [17Q] as a function of a wick parameter by a large number of measurements.

Aircraft and space vehicles

Heat transfer associated with space shuttle and ablation is still of great concern although there is a decline in publication activity in this field.

By studying the effects of surface temperature and Reynolds number on leeward shuttle heating it was found that the average separated-flow Stanton number for the fuselage leeward surface depends moderately on the windward-wall to free-stream total temperature ratio [1R]. Leeward heat-transfer experiments of the shuttle orbiter fuselage indicate a steep decrease of the heat-transfer rate from an attached flow value at $\phi = 90°$ to a nearly constant level within the separated region [9R]. Heating rates for highly swept wing leading edges (shuttle orbiter) may be described by a Stanton number which depends on the local Reynolds number but is independent of the free-stream flow conditions and of the surface temperature [2R]. A successful spacecraft thermal design depends on an accurate determination of the net energy transport across the vehicle control surfaces which requires a calculation of the orbit averaged albedo as a function of orbit inclination, beta angle, solar declination, surface geometry and orbit altitude [14R].

Results of numerical calculations of the heat and mass transfer within ablating teflon layers are in good agreement with ablation measurements in arc jet facilities [6R]. Tungsten and molybdenum ablation modeling for re-entry compares favorably with high-and-low pressure arc jet tests, with ballistic range firings, and with sound rocket flights [12R]. Attempts to correlate the analysis with experimental data of charring ablators show that the effective thermal conductivity of char is strongly dependent on the wall heat flux [11R].

The Hypersonic Arbitrary Body Program (HABP) in conjunction with accepted, easily used heat-transfer prediction techniques gives reasonable overall agreement with experimental results and predicts the type of heating information needed to generate nominal thermal protection system weights [10R]. Extensive investigations of the heating and pressure distribution on corrugation roughened surfaces in thick supersonic and hypersonic turbulent boundary layers show that the peak heat flux is directly proportional to the Reynolds number, whereas the thin boundary-layer data are inversely proportional to the Reynold's number [3R].

Quantitative peak heating data such as those caused by interfering flow fields can be obtained using the phase change coating technique [7R]. By introducing an effective thermophysical property defined as $(k\rho c)^{\frac{1}{2}}$, the accuracy of the phase-change coating technique for estimating local heat-transfer coefficients can be substantially improved [4R].

Both the magnitude and the distribution of calculated radiative heating rates to outer planet entry probes (Jupiter, Saturn, Uranus) is strongly dependent on the atmospheric model and the free-stream conditions [17R]. The thermal design of the Viking/Mars/Lander is based on maximum use of passive techniques and is integrated into the overall vehicle design operation [13R].

A study of cryogenic heat pipe systems for spacecrafts indicates that for special applications a solid cryogen/heat pipe/radiant cooler system provides a weight savings when compared to a two stage solid cryogenic cooler [16R]. In the Satellite Solar Power Station (SSPS) concept it is proposed to reject waste heat associated with solar energy conversion and microwave generation direct into space [5R].

The effect of preheating of solid propellants on the burning rate seems to be associated with the thermal decomposition of the propellant [18R]. Studies of the ignition dynamics for a packed bed of solid propellant grains indicate that vigorous ignition, i.e. dumping of hot exploded gases into a propellant bed within 100 μs results in extreme pressures in the bed interior [8R].

Local heat flux calculations in rocket-engine nozzle walls assuming a linear temperature distribution perpendicular to the wall surface may be substantially in error [15R]. An analysis of the process of filling vehicle-borne, high-pressure gas tanks from gas reservoirs permits a prediction of the temperature overshoot resulting from loading a fixed weight of gas to a predetermined pressure [19R].

General

Several papers were concerned with pollution in lakes, in rivers, and in the atmosphere. A computer study [15S] of the velocity and temperature field considered a rectangular reservoir with inflow at one end and outflow at the other; windshear was taken into account. One-dimensional temperature predictions [3S] in unsteady flow were applied to the Conowingo reservoir; a one-dimensional two layer model describes seasonal variation of the temperature distribution in stratified lakes [12S]. Turbulence created by wind, by buoyancy, by temperature gradients, and by solar radiation absorption was considered. The analysis was applied to a calculation of the thermocline and the temperature distribution in Cayuga Lake, New York. The thermal regime of rivers was numerically predicted [16S] for the Mississippi River. Temperature and pollutant concentrations were calculated [22S] in urban atmospheres with prescribed emissions along the city. Numerical calculations were made for a typical Midwestern town. Data collected in Calgary show multiple-cell heat islands at street level and the effect of wind speed on the urban boundary layer [14S].

Thermal instabilities in discharging gas from a cylindrical vessel were studied [10S] with schlieren cinematography. A thermal analysis [8S] of the melting of rocks by drills was compared with experiments. Pipelines are heated by steam carrying tubes arranged between the pipeline and its insulation. Performance of this scheme was analyzed [19S]. Tables are available [1S] for the thermal conductance of air caused by radiation, conduction, and convection. The analysis [18S, 23S] of heat transfer in brakes should be helpful for the improvement of the shoes. The temperature distribution within tools was measured [11S] using powders with known melting point by establishing the boundary line between melted and unmelted powder. A sudden drop in heat transfer between the ingot and the mold has been observed and analyzed [21S] occurring when the surfaces of both separate because of shrinkage. A theoretical model for heat transfer in polymer molding [4S] results in temperature and velocity profiles. The velocity and temperature field for flow through an axisymmetric furnace with coaxial burner having a swirl number 0.5 was calculated [9S] and the results were experimentally verified. Predicted exit temperature profiles from gas turbine combustors agree well with measured ones [5S].

Results in the literature on post accident heat removal in nuclear reactors were cast into a form convenient for use [2S] and were studied experimentally [7S] by measuring heat-transfer between glycerin heated by microwave radiation to simulate the molten fuel and paraffin as a melting barrier. Heat transfer in rod bundles under emergency cooling conditions was calculated [20S] considering combined radiation and convection from the fuel rod to the steam and droplet mixture. Subcooled decompression analysis [13S] was presented for a loss of coolant accident.

Three papers [6S, 17S, 24S] considered biological heat transfer in sweating, burn injury, and in the human temporal bones.

Solar energy

The number of heat transfer related solar energy papers was approximately the same in 1976 as in the previous year. Topics of major interest among these papers include insolation, materials development, flat plate collectors, concentrating collectors, energy storage, systems studies for heating and cooling, and systems studies for central electric power generation.

The present status of knowledge of the total and spectral insolation is reviewed and an overview of the solar monitoring program of NASA is presented [25T]. A conclusion that emerges from a detailed survey of solar radiation measurement, its techniques and instrumentation, is that no single method can supply the type of information needed for the large variety of solar energy applications [26T]. It is shown that it is possible to determine an average attenuation of direct-beam radiation for a single cloud and a net transformation of direct to diffuse radiation per unit of cloud area and that mean values of these quantities make it possible to reconstruct total, direct and diffuse insolation values using standard surface observations [30T].

Two papers were published dealing with the optical properties of special materials for solar applications. Transparent heat-mirror films, which transmit solar radiation but reflect IR radiation, have been made which, when used on a solar collector, produce an effect equivalent to a selective solar absorber coating with a solar absorptance of 0.9 and an IR emittance of 0.08 [9T]. Tests and calculations indicate an increase rather than a decrease in solar absorptance of second surface mirrors at angles of incidence greater than 80° [24T].

Flat plate collector study areas that have been reported include thermal tests, performance predictions, and performance improvement efforts. A standard test procedure was proposed for determining the thermal efficiency of solar collectors [10T]. The accuracy of the tests when using the proposed method would be primarily limited by the accuracy with which the insolation could be measured. The measured thermal efficiency and evaluation of 23 flat plate collectors which differ according to absorber material, absorber coating, type of glazing material, the use of honeycomb material, and the use of vacuum are presented in reference [21T].

An analytical expression was obtained which makes it possible to calculate the water outlet temperature from a flat plate collector as a function of the insolation, temperature of surroundings, wind velocity, and water flow rate [3T]. The response times of flat plate collectors increase significantly with increases in the number of covers, the thickness of the covers, the absorber thickness, and the absorber temperature, while only small changes in response time result from large variations of the emissivity of the absorber plate [31T]. For a flat plate solar collector of the sheet-and-tube design and with a serpentine tube configuration, the heat removal factor can be expressed in terms of three dimensionless groups [1T].

A comparison of collector efficiency characteristics indicated that a collector with a noselective black absorber and a glass honeycomb core between the absorber and cover plate was markedly superior to single-glazed selective-black and double-glazed non-selective-black collectors, especially at higher collector temperatures [5T]. Experimental studies indicate that a well-designed honeycomb core will give good performance in a solar collector even with clearance gaps of 1.5 mm between the core and its coverglass and between the core and the absorber plate [8T].

Concentrating collectors have also been studied. A series of papers dealing with the optics of linear concentrators resulted from work being performed on the Compound Parabolic Concentrator (CPC). The optical performance of a CPC was described in detail [16T]. A differential equation has been derived that describes the reflector of an ideal two-dimensional radiation concentrator with an absorber of arbitrary convex shape and the equation has been solved in closed form for the special case of an absorber of circular cross section [17T]. Design procedures have been described for ideal concentrators to be used for conditions of finite sources and restricted exit angles [18T].

The operational performance of cylindrical–parabolic solar collectors was the topic of three papers. It has been predicted that the performance can be completely specified by a dimensionless temperature (ratio of the absorber operating temperature to the absorber temperature in the absence of heat removal [22T]). The performance predicted for a somewhat indealized collector agrees to within approximately 20% with measured results [7T]. The maximum concentration for a cylindrical parabolic solar collector is more than 40% higher if the shape of the absorber tube is an elliptic cylinder rather than a circular cylinder [14T].

A proposed receiver, designed to operate with one-axis tracking or linear collectors, is predicted to be able to deliver 80 and 53% of the energy incident on the receiver to a working fluid for operating temperatures of 130 and 370°C respectively [4T].

A comparison of the long-term behavior of a packed bed storage system, as predicted by complex and simplified models, indicates the simple model is adequate for analyzing reasonable gravel bed heat storage systems [11T]. Na_2SO_4 is predicted to be superior to magnetise brick, alumina brick, NaCl, KCl, $CaCl_2$ and some eutectic mixtures of inorganic salts, for use as a sensible heat storage material [28T]. The transient response of a specific heat storage configuration composed of a number of rectangular cross-sectional channels for the flowing fluid connected in parallel and separated by solid storage material is described in [20T]. The decomposition of a number of classes of salts (ammonium salts, Group IA and IIA hydroxides, carbonates, sulfates, and oxides) have been evaluated in regard to their use for storage of solar energy and a number of specific salts have been selected for more detailed consideration [29T]. An analysis of the use of adsorbent beds for energy storage indicates that they can be much smaller than non-adsorbing beds, are relatively insensitive to thermal losses, and are more suited to energy storage for drying purposes than for simple heating purposes [6T].

Solar heating and cooling systems have been investigated analytically and experimentally. The application of a simulation of solar space and water heating systems for residences resulted in a simple graphical method that architects and heating engineers can use to design economical solar heating systems [13T]. Reference [15T] describes a general method for determining the most economically desirable combination of solar gain, structural energy loss, thermal energy storage, and auxiliary energy use. A solar air conditioning plant that is based on open evaporation from a lithium chloride solution flowing on a roof has been tested [12T].

Solar-thermal electric power generation studies have been reported. A solar-thermal power system using a central receiver is predicted to be more efficient than distributed collector systems because high temperatures of the working fluid can be achieved while maintaining lower radiative and convective energy losses [27T]. To achieve this efficiency requires compact boiler and superheater designs to effectively concentrate the solar heat input and to improve system power cycle efficiency [23T]. A model of the performance of large-area solar concentrators for central receiver power plants, which was formulated using a continuum field representation of ideal heliostat arrays, established theoretical limits of performance against which realistic solar power systems can be compared [19T]. The optimum efficiency predicted for the use of flat plate collectors to generate electrical power was approximately 3.5% [2T].

PLASMA HEAT TRANSFER

Heat-transfer studies in ionized gases and in MHD channels reported in 1976 refer to fundamental investigations as well as to applications, in particular to those using electric arcs as a heat source.

A survey paper on Miscellaneous Arc Devices reviews a number of more recent publications dealing with the interaction of arcs with flow and/or magnetic fields [23U]. Time-resolved laser-Doppler measurements in high-current arc plasmas yield not only information about the time evolution of the plasma flow velocity over the arc cross section, but provide also an estimate of the spatial distribution of globular electrode material in the arc as a function of time [21U]. A method is proposed for estimating axial variations in high-current free burning, quasi-steady arcs assuming that the arc energy balance is governed by axial convection and optically thin radiation [19U]. The behaviour of laminar arcs in which radiation is negligible can be predicted by a set of equations which comprise the overall integral equations derived earlier [M. D. Cowley, *J. Phys. D*: *Appl. Phys.* **7**, 2218 (1974)], the core energy equation, and empirical curves relating shape factors of the arc to a non-dimensional heat flux parameter [3U]. An analysis of the instability of electric arcs burning axially in accelerated flows demonstrates a strong similarity with the instabilities in fluid jets. It seems that the source of this instability is the shear layer between the cold flow and the hot, more highly accelerated arc column [7U].

Temperature measurements in magnetically balanced cross-flow rail electrode arcs indicate that forced convection and magnetic field interaction may produce different arc shapes. Depending on the parameter setting (flow velocity and balancing magnetic field strength) straight and S-shaped arcs have been observed [1U]. Thermal and electrical conductivities of water vapor plasmas are determined for temperature intervals from 2000 to 3500 K and from 7700 to 8000 K which are of particular interest in connection with transport of reaction energy in the plasma [22U]. A reduction of anode cooling in a nitrogen plasma torch using a hollow tungsten anode increases the size of the anode root and increases its travel velocity thereby reducing the anode erosion rate. At the same time the temperature distribution within the jet becomes more uniform and the power loss to the anode increases [13U],

The dynamic behaviour of an AC arc in a laminar accelerating flow depends primarily on the product of frequency and time scale of the flow which controls the relative importance of the energy transport by thermal conduction and that by convection in the conducting core of the arc around current zero [4U]. A new method for solving the energy equation for moving-boundary AC arcs should be applicable to various types of arcs including the three-dimensional case with both radial and axial convection and the case where radiation cooling dominates [10U].

Heat transfer from a hot, dense, magnetized plasma brought into sudden contact with a cold wall is in approximate agreement with classical theory. Measurement for $T = 1.5 \times 10^6$ K, $n = 1.5 \times 10^{16}$ cm^{-3}, and $B = 0.9\,T$ reveal initial heat fluxes of approximately 3×10^5 W/cm^2 [6U]. A new method for measuring extremely high local heat fluxes to surfaces is based on a comparison of calculated and measured temperature–time histories on the rear of the surfaces exposed to these heat fluxes [16U]. The application of this method for measuring peak anode heat fluxes in 5 kA pulsed arcs using nitrogen and argon at 1 atm and various anode materials reveals peak heat fluxes in the order of 100 kW/cm^2 [17U]. Heat transfer to a sphere in an argon plasma flow may be described by the same type of dimensionless relations pertaining to low temperature flows ($T_f < 2000$ K) provided that corresponding allowances for the flow properties are introduced [9U].

Performance data are reported of a nominal 250 kW Huels-type arc heater using hydrogen in a pressure range from 10 to 50 atm and helium in a range from 25 to 100 atm with currents ranging from 240 to 400 A. Comparison of these data with data derived from arc operation in air and computer-coded correlations permit scaling of the arc heater [15U]. Spectrometric measurements of the temperature distribution in a three-phase arc gas heater reveal substantial temperature variations over the arc chamber cross section. The bulk temperature of the heated gas, however, is in satisfactory agreement with the spectrometrically determined average temperatures [14U].

A model proposed for the calculation of the two-dimensional flow and temperature fields in an inductively coupled plasma demonstrates the existence of a magnetic pumping effect which leads to the formation of two recirculating eddies, one at each end of the induction coil. At higher gas flow rates, the downstream eddy is swept away and simultaneously there is a substantial reduction of the heat flux to the plasma confining tube [2U].

A plasma produced by a 15 A arc in argon at atmospheric pressure is not in thermal equilibrium, even on the arc axis. The electron temperature is approximately 15% higher than T_{Saha} and 30% higher than the excitation temperature of the first argon ion. This will have a significant effect on heat-transfer calculations in such plasmas [5U]. Calculations of the radiation emission coefficients of SF_6 arc plasmas show that at 1 atm line radiation can be an order of magnitude higher than continuous radiation and radiation for wavelengths > 2000 Å is less than 10% of the total radiation for temperatures exceeding 15 000 K [11U]. Approximate relations describing the radiation escape factor and the line source function of a nothermal plasma slab are in satisfactory agreement with numerical results of other authors [20U]. A method of computing radiative cooling of a hydrogen plasma in a shock tube assuming quasi-isothermal radiative transfer, provides results which are in good agreement with non-isothermal calculations for a plane parallel slab if the temperature gradients are ⩽ 300 K/cm [18U].

The pressure work and the wall conductance can have a significant influence on the limiting, fully developed heat-transfer conditions in an MHD channel [8U]. An analysis of combined free and forced convective hydromagnetic flow through a channel shows that Hall currents exert a profound influence on the flow and heat-transfer characteristics [12U].

REFERENCES

Books

1. S. W. Chi, *Heat Pipe Theory and Practice*. Hemisphere, Washington, D.C. (1976).
2. E. R. G. Eckert and R. J. Goldstein (editors), *Measurement in Heat Transfer*, 2nd Edn. Hemisphere, Washington, D.C. (1976).
3. J. P. Holman, *Heat Transfer*, 4th Edn. McGraw-Hill, New York (1976).
4. Y. Y. Hsu and R. W. Graham, *Transport Processes in Boiling and Two-Phase Systems Including Near-Critical Fluids*. Hemisphere, Washington, D.C. (1976).
5. T. F. Irvine, Jr. and J. P. Hartnett (editors), *Advances in Heat Transfer*, Vol. 12. Academic Press, New York (1976).
6. S. Whitaker, *Elementary Heat Transfer Analysis*. Pergamon Press, New York (1976).

Conduction

1A. A. Aziz and J. Y. Benzies, *Int. J. Heat Mass Transfer* **19**, 271 (1976).
2A. J. R. Barber, *Int. J. Heat Mass Transfer* **19**, 956 (1976).
3A. E. R. Barnes, *Q. Appl. Math.* **34**, 1 (1976).
4A. G. E. Bell and J. Crank, *Int. J. Num. Meth. Engng* **10**, 827 (1976).
5A. M. Ben-Amoz, *Z. Angew. Math. Phys.* **27**, 335 (1976).
6A. L. Bolle, *Int. J. Heat Mass Transfer* **19**, 443 (1976).
7A. V. A. Bubnov, *Int. J. Heat Mass Transfer* **19**, 175 (1976).
8A. R. Bung and H. M. Tensi, *Wärme- und Stoffübertragung* **9**, 203 (1976).

9A. A. Campo, *Wärme- und Stoffübertragung* **9,** 139 (1976).
10A. B.-H. Caussade, *Int. J. Heat Mass Transfer* **19,** 303 (1976)
11A. C. Y. Chan, *Q. Appl. Math.* **34,** 173 (1976).
12A. P. J. Chen, R. R. Nachlinger and J. W. Nunziato, *Q. Appl. Math.* **34,** 311 (1976).
13A. B. T. F. Chung and L. T. Yeh, *AIAA Jl* **14,** 388 (1976).
14A. G. Comini and S. Del. Giudice, *J. Heat Transfer* **98,** 543 (1976).
15A. G. Comini and R. W. Lewis, *Int. J. Heat Mass Transfer* **19,** 1387 (1976).
16A. L. Y. Cooper, *Int. J. Heat Mass Transfer* **19,** 575 (1976).
17A. P. L. Davis, *SIAM J. Appl. Math.* **30,** 75 (1976).
18A. P. L. Dhar and C. P. Arora, *J. Franklin Inst.* **301,** 379 (1976).
19A. H. Domingos and D. Voelker, *J. Heat Transfer* **98,** 329 (1976).
20A. J. Dundurs and C. Panek, *Int. J. Heat Mass Transfer* **19,** 731 (1976).
21A. A. Feingold, *J. Heat Transfer* **98,** 527 (1976).
22A. L. S. Fletcher and W. R. Ott, *J. Heat Transfer* **98,** 331 (1976).
23A. R. R. Gilpin and B. K. Wong, *J. Heat Transfer* **98,** 537 (1976).
24A. B. W. Grange, R. Viskanta and W. H. Stevenson, *Int. J. Heat Mass Transfer* **19,** 373 (1976)
25A. O. Hoffer, *Wärme- und Stoffübertragung* **9,** 79 (1976).
26A. B. Holzknecht and K. Stephan, *Wärme- und Stoffübertragung* **9,** 73 (1976).
27A. J. R. Howard, *Int. J. Heat Mass Transfer* **19,** 367 (1976).
28A. P. R. W. Hudson, *J. Phys. D: Appl. Phys.* **9,** 225 (1976).
29A. Y. Jaluria, *J. Heat Transfer* **98,** 127 (1976).
30A. T. T. Kao, *AIAA Jl* **14,** 818 (1976).
31A. J. Kern, *Int. J. Heat Mass Transfer* **19,** 869 (1976).
32A. J. Kern, *Int. J. Heat Mass Transfer* **19,** 879 (1976).
33A. J. Kern and J. O. Hansen, *Int. J. Heat Mass Transfer* **19,** 707 (1976).
34A. H. H. King, *Int. J. Num. Meth. Engng* **10,** 799 (1976).
35A. R. W. Lewis and B. R. Bass, *J. Heat Transfer* **98,** 478 (1976).
36A. K. A. Lindsay and B. Straughan, *Z. Angew. Math. Phys.* **27,** 653 (1976).
37A. A. V. Luikov, V. A. Bubnov and I. A. Soloviev, *Int. J. Heat Mass Transfer* **19,** 245 (1976).
38A. A. Maewal, T. C. Bache and G. A. Hegemier, *J. Heat Transfer* **98,** 133 (1976).
39A. N. D. Malmuth, *Int. J. Heat Mass Transfer* **19,** 349 (1976).
40A. K. Mastanaiah, *Int. J. Num. Meth. Engng* **10,** 833 (1976).
41A. M. D. Mikhailov, *Int. J. Heat Mass Transfer* **19,** 651 (1976).
42A. A. Mori and K. Araki, *Int. Chem. Engng* **16,** 734 (1976).
43A. A. Muzzio, *J. Heat Transfer* **98,** 680 (1976).
44A. R. R. Nachlinger and J. W. Nunziato, *Int. J. Engng Sci.* **14,** 693 (1976).
45A. J. Neiss and E. R. F. Winter, *Wärme- und Stoffübertragung* **9,** 39 (1976).
46A. A. Prasad and S. N. Sinha, *AIAA Jl* **14,** 1494 (1976).
47A. P. Řiha, *Int. J. Engng Sci.* **14,** 529 (1976).
48A. J. B. Rosser, *Z. Angew. Math. Phys.* **27,** 249 (1976).
49A. B. Rubinsky and A. Shitzer, *J. Heat Transfer* **98,** 514 (1976).
50A. G. E. Schneider, *J. Heat Transfer* **98,** 525 (1976).
51A. N. M. Schnurr, A. P. Shapiro and M. A. Townsend, *J. Heat Transfer* **98,** 643 (1976).
52A. A. A. Sfeir, *J. Heat Transfer* **98,** 466 (1976).
53A. N. Shamsundar and E. M. Sparrow, *J. Heat Transfer* **98,** 550 (1976).
54A. L. D. Simmons, *J. Heat Transfer* **98,** 150 (1976).
55A. E. M. Sparrow and N. Cur, *J. Heat Transfer* **98,** 232 (1976).
56A. K. Stephan and W. Bohnstedt, *Wärme- und Stoffübertragung* **9,** 145 (1976).
57A. K. Stewartson and R. T. Waechter, *Proc. R. Soc.* **348,** 415 (1976).
58A. N. V. Suryanarayana, *J. Heat Transfer* **98,** 324 (1976).
59A. W. Tarnawski, *Int. J. Heat Mass Transfer* **19,** 15 (1976).
60A. A. G. Tempkin, *Int. J. Heat Mass Transfer* **19,** 785 (1976).
61A. P. J. Torvik, *AIAA Jl* **14,** 417 (1976).
62A. V. Vacek, *Wärme- und Stoffübertragung* **9,** 49 (1976).
63A. T. N. Veziroğlu, H. Yüncü and S. Kakaç, *Int. J. Heat Mass Transfer* **19,** 959 (1976).
64A. B. Vujanovic and B. Bačlić, *Int. J. Heat Mass Transfer* **19,** 721 (1976).
65A. A. S. Williamson, *Int. J. Engng Sci.* **14,** 869 (1976).
66A. L. S. Yao, C. L. Tien and S. A. Berger, *J. Heat Transfer* **98,** 326 (1976).
67A. M. M. Yovanovich, *AIAA Jl* **14,** 882 (1976).
68A. T. F. Zien, *AIAA Jl* **14,** 404 (1976).
69A. G. E. Zinsmeister and S. S. Pan, *Int. J. Num. Meth. Engng* **10,** 1057 (1976).

Channel flow

1B. G. A. Adebiyi and W. B. Hall, *Int. J. Heat Mass Transfer* **19,** 715 (1976).
2B. K. N. Astill, *J. Heat Transfer* **98,** 601 (1976).
3B. H. F. Bauer, *Int. J. Heat Mass Transfer* **19,** 479 (1976).
4B. D. E. Bourne, E. Huntley, I. J. Hendrie and B. P. Hunter, *Int. J. Heat Mass Transfer* **19,** 1045 (1976).
5B. P. Carajilescov and N. E. Todreas, *J. Heat Transfer* **98,** 262 (1976).
6B. F. B. Cheung and L. Baker, Jr., *Nucl. Sci. Engng* **60,** 1 (1976).
7B. Y. K. Choo and M. C. Potter, *Physics Fluids* **19,** 1676 (1976).
8B. S. Yu. Danilevich and G. B. Froishteter, *Int. Chem. Engng* **17,** 376 (1976).
9B. A. S. El-Ariny and A. Aziz, *J. Heat Transfer* **98,** 427 (1976).
10B. A. F. Emery and F. B. Gessner, *J. Heat Transfer* **98,** 594 (1976).
11B. M. Epstein, *Int. J. Heat Mass Transfer* **19,** 1281 (1976).
12B. M. Epstein, M. A. Grolmes, R. E. Henry and H. K. Fauske, *Nucl. Sci. Engng* **61,** 310 (1976).
13B. D. Gärtner, *Wärme- und Stoffübertrangung* **9,** 179 (1976).
14B. V. Gnielinski, *Int. Chem. Engng* **16,** 359 (1976).

15B. J. A. Golding and R. Dussault, *Int. J. Heat Mass Transfer* **19,** 493 (1976).
16B. L. Goldstein, Jr. and E. M. Sparrow, *J. Heat Transfer* **98,** 26 (1976).
17B. L. Goldstein, Jr. and E. M. Sparrow, *Int. J. Heat Mass Transfer* **19,** 1337 (1976).
18B. I. S. Habib, *J. Heat Transfer* **98,** 421 (1976).
19B. A. P. Hatton and P. Walklate, *Int. J. Heat Mass Transfer* **19,** 1425 (1976).
20B. S. W. Hong and A. E. Bergles, *Int. J. Heat Mass Transfer* **19,** 123 (1976).
21B. S. W. Hong and A. E. Bergles, *J. Heat Transfer* **98,** 251 (1976).
22B. G. J. Hwang and J.-P. Sheu, *Can. J. Chem. Engng* **54,** 66 (1976).
23B. V. Javeri, *Wärme- und Stoffübertragung* **9,** 85 (1976).
24B. V. Javeri, *Int. J. Heat Mass Transfer* **19,** 943 (1976).
25B. T. E. Laskaris, *Physics Fluids* **18,** 1639 (1976).
26B. K. O. Lund, *J. Heat Transfer* **98,** 19 (1976).
27B. J. Madejski, *Int. J. Heat Mass Transfer* **19,** 1351 (1976).
28B. J. H. Masliyah and K. Nandakumar, *J. Heat Transfer* **98,** 257 (1976).
29B. M. D. Mikhailov and B. K. Shishedjiev, *Int. J. Heat Mass Transfer* **19,** 553 (1976).
30B. Z. L. Miropol'skii, *Thermal Engng* **22** (3), 91 (1975).
31B. J. C. Mulligan and D. D. Jones, *Int. J. Heat Mass Transfer* **19,** 213 (1976).
32B. E. Naito, *Int. Chem. Engng* **16,** 162 (1976).
33B. M. A. Nyamira and Yu. V. Vityamas, *Int. Chem. Engng* **16,** 412 (1976).
34B. D. R. Oliver and S. M. Asghar, *Trans Inst. Chem. Engrs* **54,** 218 (1976).
35B. V. S. Pratap and D. B. Spalding, *Int. J. Heat Mass Transfer* **19,** 1183 (1976).
36B. F. E. M. Saboya and E. M. Sparrow, *Int. J. Heat Mass Transfer* **19,** 41 (1976).
37B. M. Sakakibara and K. Endo, *Int. Chem. Engng* **16,** 728 (1976).
38B. N. M. Schnurr, V. S. Sastry and A. B. Shapiro, *J. Heat Transfer* **98,** 609 (1976).
39B. N. Seki, S. Fukusako and T. Hirata, *Int. J. Heat Mass Transfer* **19,** 700 (1976).
40B. P. L. Stephenson, *Int. J. Heat Mass Transfer* **19,** 413 (1976).
41B. A. S. Sukomel, V. I. Velichko, Yu. G. Abrosimov and D. F. Gutsev, *Thermal Engng* **22** (3), 99 (1975).
42B. R. C. H. Tsou and Y. P. Chang, *Int. J. Heat Mass Transfer* **19,** 445 (1976).
43B. S. A. Victor and V. L. Shah, *Int. J. Heat Mass Transfer* **19,** 777 (1976).
44B. R. W. Watkins, C. R. Robertson and A. Acrivos, *Int. J. Heat Mass Transfer* **19,** 693 (1976).
45B. D. A. Wesley and E. M. Sparrow, *Int. J. Heat Mass Transfer* **19,** 1205 (1976).
46B. R. S. Wu and K. C. Cheng, *Int. J. Heat Mass Transfer* **19,** 1343 (1976).

Boundary layer and external flows

1C. N. Afzal, *Int. J. Engng Sci.* **14,** 415 (1976).
2C. N. Afzal and R. Narasimha, *J. Fluid Mech.* **74,** 113 (1976).
3C. N. Afzal and A. P. Singh, *Aeronaut. Q.* **27,** 217 (1976).
4C. E. Backx and B. E. Richards, *AIAA Jl* **14,** 1159 (1976).
5C. W. D. Baines and J. F. Keffer, *Int. J. Heat Mass Transfer* **19,** 21 (1976).
6C. A. Bianchini, L. de Soçio and A. Pozzi, *J. Appl. Mech.* **43,** 396 (1976).
7C. A. S. Borodin and J. J. C. Picot, *Can. J. Chem. Engng* **54,** 59 (1976).
8C. L. K. Brumfield and T. G. Theofanous, *J. Heat Transfer* **98,** 496 (1976).
9C. T. Cebeci, *J. Aircraft* **13,** 709 (1976).
10C. T. Cebeci, K. Kaups and A. Moser, *AIAA Jl* **14,** 1090 (1976).
11C. P. Chassaing and A. Claria, *Int. J. Heat Mass Transfer* **19,** 249 (1976).
12C. F. J. Chung and J. N. Chiou, *Computers Fluids* **4,** 1 (1976).
13C. S. W. Churchill, *A.I.Ch.E. Jl* **22,** 264 (1976).
14C. D. J. Collins and C. W. Gorton, *A.I.Ch.E. Jl* **22,** 610 (1976).
15C. H. G. Davies, *J. Fluid Mech.* **73,** 49 (1976).
16C. M.-R. M. Drizhyus, S. L. Bartkus, A. S. Shlanchyauskas and A. A. Zhukauskas, *Int. Chem. Engng* **16,** 552 (1976).
17C. Y. Era and A. Saima, *Bull. J.S.M.E.* **19,** 800 (1976).
18C. Y. Furuya, I. Nakamura, H. Osaka and T. Shimizu, *Bull. J.S.M.E.* **19,** 1922 (1976).
19C. M. M. Gibson and B. E. Launder, *J. Heat Transfer* **98,** 81 (1976).
20C. A Goodarz, *Int. J. Engng Sci.* **14,** 639 (1976).
21C. M. R. Head, *J. Fluid Mech.* **73,** 1 (1976).
22C. D. T. Hove and E. Taylor, *AIAA Jl* **14,** 1486 (1976).
23C. H. M. Hua and P. S. Lykoudis, *Int. J. Heat Mass Transfer* **19,** 698 (1976).
24C. G. R. Inger and W. H. Mason, *AIAA Jl* **14,** 1266 (1976).
25C. P. C. Jain and B. S. Goel, *J. Heat Transfer* **98,** 303 (1976).
26C. P. E. Jenkins, *J. Engng Pwr* **98,** 501 (1976).
27C. P. E. Jenkins and V. W. Goldschmidt, *Physics Fluids* **19,** 613 (1976).
28C. F. I. Kalbaliev, *Int. Chem. Engng* **16,** 460 (1976).
29C. M. Katagiri, *J. Phys. Soc. Japan* **40,** 1171 (1976).
30C. V. I. Katinas, S. A. Shvegzhda, I. I. Zhyugzhada and A. A. Zhukauskas, *Int. Chem. Engng* **16,** 469 (1976).
31C. V. I. Katinas, I. I. Zhyugzhada, A. A. Zukauskas and S. A. Shvegzhda, *Int. Chem. Engng* **16,** 283 (1976).
32C. R. N. Koopman and E. M. Sparrow, *Int. J. Heat Mass Transfer* **19,** 673 (1976).
33C. C. J. Kowalski and J. W. Mitchell, *J. Heat Transfer* **98,** 649 (1976).
34C. C. K. Law, *AIAA Jl* **14,** 1497 (1976).
35C. P. P. Lynn and K. Alani, *Int. J. Num. Meth. Engng* **10,** 809 (1976).
36C. C. Mari, D. Jeandel and J. Mathieu, *Int. J. Heat Mass Transfer* **19,** 893 (1976).
37C. R. N. Meroney and P. Bradshaw, *AIAA Jl* **13,** 1448 (1975).
38C. P. G. Mestayer, C. H. Gibson, M. F. Coantic and A. S. Patel, *Physics Fluids* **19,** 1279 (1976).
39C. S. Murata, Y. Miyake and T. Inaba, *J. Fluid Mech.* **73,** 735 (1976).

40C. N. S. Murty and V. M. K. Sastri, *Int. J. Heat Mass Transfer* **19,** 115 (1976).
41C. G. Nath, *J. Fluids Engng* **98,** 292 (1976).
42C. K. H. Ng and D. B. Spalding, *Int. J. Heat Mass Transfer* **19,** 1161 (1976).
43C. R. H. Pletcher, *AIAA Jl* **14,** 696 (1976).
44C. A. B. Ponter, S. Vijayan and W. Peier, *Int. J. Heat Mass Transfer* **19,** 954 (1976).
45C. R. S. Reddy Gorla, *J. Fluids Engng* **98,** 116 (1976).
46C. R. S. Reddy Gorla, *Appl. Scient. Res.* **31,** 437 (1976).
47C. B. Roux and B. Forestier, *AIAA Jl* **14,** 1069 (1976).
48C. A. A. Sfier, *Int. J. Heat Mass Transfer* **19,** 1289 (1976).
49C. J. S. Shang, W. L. Hankey, Jr. and C. H. Law, *AIAA Jl* **14,** 1451 (1976).
50C. R. M. C. So, *J. Fluids Engng* **98,** 328 (1976).
51C. L. C. Thomas, *J. Heat Transfer* **98,** 334 (1976).
52C. Y. Tsuji and Y. Morkawa, *Aeronautic Q.* **27,** 15 (1976).
53C. P. P. Vaitekunas, A. A. Shlanchyauskas and A. A. Zhukauskas, *Int. Chem. Engng* **16,** 403 (1976).
54C. A. R. P. van Heiningen, A. S. Mujumdar and W. J. M. Douglas, *J. Heat Transfer* **98,** 654 (1976).
55C. C. B. Watkins, Jr., *Int. J. Heat Mass Transfer* **19,** 395 (1976).
56C. J. C. Williams III, *Appl. Scient. Res.* **31,** 161 (1975).
57C. G. B. Zdanavichyus, B. A. Chesna, I. I. Zhygzhda and A. A. Zhukauskas, *Int. Chem. Engng* **16,** 121 (1976).

Flow with separated regions

1D. A. Amr and D. R. Kassoy, *Physics Fluids* **19,** 1312 (1976).
2D. P. W. Appel and J. Newman, *A.I.Ch.E. Jl* **22,** 979 (1976).
3D. R. Bauer and E. U. Schlünder, *Chem.-Ingr.-Tech.* **48,** 227 (1976).
4D. S. C. Bhattacharya and D. Harrison, *Trans Instn Chem. Engrs* **54,** 281 (1976).
5D. J. Broughton and J. Kubie, *Int. J. Heat Mass Transfer* **19,** 232 (1976).
6D. D. M. Burch, R. W. Allen and B. A. Peavy, *J. Heat Transfer* **98,** 221 (1976).
7D. Yu. A. Buyevich, *Int. J. Multiphase Flow* **2,** 337 (1976).
8D. R. Clift, M. Filla and L. Massimilla, *Int. J. Multiphase Flow* **2,** 549 (1976).
9D. K. Cornwell, B. G. Nair and T. D. Pattan, *Int. J. Heat Mass Transfer* **19,** 236 (1976).
10D. M. Doi, *J. Phys. Soc. Japan* **40,** 567 (1976).
11D. B. D. Eldridge and L. F. Brown, *A.I.Ch.E. Jl* **22,** 942 (1976).
12D. M. M. El-Kaissy and G. M. Monsy, *Int. J. Multiphase Flow* **2,** 379 (1976).
13D. D. Gelbin, K.-H. Radeke, B. Rosahl and W. Stein, *Int. J. Heat Mass Transfer* **19,** 987 (1976).
14D. D. Gelbin, K.-H. Radeke, W. Stein and H.-J. Wolff, *Int. J. Heat Mass Transfer* **19,** 9 (1976).
15D. R. G. Gutman, *Trans Inst. Chem. Engrs* **54,** 251 (1976).
16D. H. Honji and K. Ishi-I, *J. Phys. Soc. Japan* **41,** 1089 (1976).
17D. G. A. Hughmark, *A.I.Ch.E. Jl* **22,** 198 (1976).
18D. R. W. Johnson and A. M. Dhanak, *J. Heat Transfer* **98,** 226 (1976).
19D. A. I. Karpenko, N. I. Syromyatnikov and L. K. Vasanova, *Heat Transfer, Soviet Res.* **7** (4), 107 (1975).
20D. J. Kubie, *Int. J. Heat Mass Transfer* **19,** 1441 (1976).
21D. P. A. Libby, *Physics Fluids* **19,** 494 (1976).
22D. C. L. Lin, D. W. Pepper and S. C. Lee, *AIAA Jl* **14,** 900 (1976).
23D. A. V. Luikov, *Int. Chem Engng* **18,** 54 (1976).
24D. Y. S. Lou and C. Wang, *Int. J. Engng Scq.* **14,** 467 (1976).
25D. D. R. McGaw, *Int. J. Heat Mass Transfer* **19,** 657 (1976).
26D. D. R. McGaw, *Int. J. Heat Mass Transfer* **19,** 665 (1976).
27D. R. A. Merz, R. H. Page and C. E. G. Przirembel, *J. Spacecraft Rockets* **13,** 310 (1976).
28D. D. Moalem, *Int. J. Heat Mass Transfer* **19,** 529 (1976).
29D. G. Neale, H. Anaka and D. Hampel, *Can. J. Chem. Engng* **53,** 691 (1975).
30D. M. Nishimura, Y. Hirabayashi and S. Sugiyama, *Int. Chem. Engng* **16,** 169 (1976).
31D. L. M. Pismen, *Chem. Engng Sci.* **31,** 693 (1976).
32D. R. P. Pletcher and C. L. Dancey, *J. Fluids Engng* **98,** 568 (1976).
33D. K. Ragunathan and K. Subba Raju, *I/EC Process Des. Dev.* **15,** 82 (1976).
34D. K. N. Rai and R. N. Pandey, *Int. J. Engng Sci.* **14,** 975 (1976).
35D. J. F. Richardson, M. N. Roman and K. J. Shakiri, *Chem. Engng Sci.* **31,** 619 (1976).
36D. S. C. Saxena and G. J. Vogel, *Can. J. Chem. Engng* **54,** 453 (1976).
37D. N. Seki, S. Fukusako and T. Hirata, *J. Heat Transfer* **98,** 588 (1976).
38D. R. Siegel, *A.I.Ch.E. Jl* **22,** 590 (1976).
39D. I. I. Sherif, A. S. A. Ammar and S. A. El-Messih, *Int. J. Heat Mass Transfer* **19,** 227 (1976).
40D. M. C. Smith, M. C. Potter and S. Kapur, *Int. J. Heat Mass Transfer* **19,** 33 (1976).
41D. R. K. Sullery, A. K. Gupta and C. S. Moorthy, *AIAA Jl* **13,** 1425 (1975).
42D. G. I. Tardos, G. Gutfinger and N. Abauf, *A.I.Ch.E. Jl* **22,** 1147 (1976).
43D. G. G. Tyukhay, Ye. V. Kravchenko and G. P. Mankevich, *Heat Transfer, Soviet Res.* **7** (4), 101 (1975).
44D. J. Werther, *Chem. Ingr.-Tech.* **48,** 339 (1976).
45D. I. Yaron, *Int. J. Heat Mass Transfer* **19,** 61 (1976).
46D. S. S. Zabrodsky, N. V. Antonishin and A. L. Parnas, *Can. J. Chem. Engng* **54,** 52 (1976).

Transfer mechanisms

1E. R. A. Antonia and J. D. Atkinson, *Physics Fluids* **19,** 1273 (1976).
2E. R. A. Antonia, H. O. Danh and A. Prabhu, *Physics Fluids* **19,** 1680 (1976).
3E. R. F. Blackwelder and R. E. Kaplan, *J. Fluid Mech.* **76,** 89 (1976).
4E. M. K. Bull and A. S. W. Thomas, *Physics Fluids* **19,** 597 (1976).
5E. G. Charnay, J. Mathieu and G. Comte-Bellot, *Physics Fluids* **19,** 1261 (1976).
6E. P. Freymuth, *Physics Fluids* **19,** 1475 (1976).
7E. F. B. Gessner and A. F. Emery, *J. Fluids Engng* **98,** 261 (1976).

8E. F. B. Gessner and J. K. Po, *J. Fluids Engng* **98,** 269 (1976).
9E. P. S. Granville, *J. Fluid Engng* **98,** 578 (1976).
10E. K. Hanjalic and B. E. Launder, *J. Fluid Mech.* **74,** 593 (1976).
11E. R. R. Johnson, R. E. Elins III, E. R. Lindgren and J. K. Yoo, *Physics Fluids* **19,** 1422 (1976).
12E. A. J. Laderman, *AIAA Jl* **14,** 1286 (1976).
13E. W. S. Lewellen, M. E. Tske, C. duP. Donaldson, *AIAA Jl* **14,** 382 (1976).
14E. D. M. McEligot, P. Pickett and M. F. Taylor, *Int. J. Heat Mass Transfer* **19,** 799 (1976).
15E. R. N. Meroney, *Computers Fluids* **4,** 93 (1976).
16E. S. Mohammadian, M. Saiy and S. J. Peerless, *J. Fluids Engng* **98,** 229 (1976).
17E. E. Ohlmer and D. Schwaim, *Int. J. Heat Mass Transfer* **19,** 765 (1976).
18E. A. Quarmby, *Int. J. Heat Mass Transfer* **19,** 1319 (1976).
19E. A. J. Reynolds, *Int. J. Heat Mass Transfer* **19,** 757 (1976).
20E. H. K. Richards and J. B. Morton, *J. Fluid Mech.* **73,** 165 (1976).
21E. D. J. Shlien and S. Corrsin, *Int. J. Heat Mass Transfer* **19,** 285 (1976).
22E. P. J. Sullivan, *Physics Fluids* **19,** 159 (1976).
23E. L. C. Thomas, *Chem Engng Sci.* **31,** 787 (1976).
24E. H. Tokunaga, *J. Phys. Soc. Japan* **41,** 328 (1976).
25E. M. Tsuge, H. Kido, K. Kato and S. Nishimura, *Bull. J.S.M.E.* **19,** 676 (1976).
26E. S. Tuge and K. Sagara, *Physics Fluids* **19,** 1478 (1976).
27E. K. Yamamoto and I. Hosokawa, *Physics Fluids* **19,** 1423 (1976).
28E. A. Yoshizawa, *J. Phys. Soc. Japan* **40,** 274 (1976).

Natural convection

1F R M Abdel-Wahed, E M Sparrow and S V. Patankar, *J. Heat Transfer* **98,** 576 (1976).
2F. G. Ahmadi, *Int. J. Engng Sci.* **14,** 81 (1976).
3F. S. Aiba and N. Seki, *Int. J. Heat Mass Transfer* **19,** 1075 (1976).
4F. W. S. Amato and C. Tien, *Int. J. Heat Mass Transfer* **19,** 1257 (1976).
5F. M. Al-Arabi and M. K. El-Reid, *Int. J. Heat Mass Transfer* **19,** 1399 (1976).
6F. J. N. Arnold, I. Catton and D. K. Edwards, *J. Heat Transfer* **98,** 67 (1976).
7F. T. Audunson and B. Gebhart, *Int. J. Heat Mass Transfer* **19,** 737 (1976).
8F. J. Bandrowski and W. Rybski, *Int. J. Heat Mass Transfer* **19,** 827 (1976).
9F. M. S. Bendell and B. Gebhart, *Int. J. Heat Mass Transfer* **19,** 1081 (1976).
10F. B. M. Berkovsky, V. E. Fertman, V. K. Polevikov and S. V. Isaev, *Int. J. Heat Mass Transfer* **19,** 981 (1976).
11F. K. Bier, H. R. Engelhorn, D. Gorenflo and A. P. Solodov, *Wärme- und Stoffübertragung* **9,** 193 (1976).
12F. A. Boura and B. Gebhart, *A.I.Ch.E. Jl* **22,** 94 (1976).
13F. H. Buchberg, I Catton and D. K. Edwards, *J. Heat Transfer* **98,** 182 (1976).
14F. R. J. Buretta and A. S. Berman, *J. Appl. Mech.* **43,** 249 (1976).
15F. D. R. Caldwell, *J. Fluid Mech.* **74,** 129 (1976).
16F. G. D. Callahan and W. J. Marner, *Int. J. Heat Mass Transfer* **19,** 165 (1976).
17F. J. P. Caltagiorone, *Int. J. Heat Mass Transfer* **19,** 815 (1976).
18F. C. C. Chen and R. Eichhorn, *J. Heat Transfer* **98,** 446 (1976).
19F. P. Cheng and I. D. Chang, *Int. J. Heat Mass Transfer* **19,** 1267 (1976).
20F. K. C. Cheng and M. Takeuchi, *J. Heat Transfer* **98,** 581 (1976).
21F. K. C. Cheng and R.-S. Wu, *Appl. Scient. Res.* **31,** 465 (1976).
22F. K. C. Cheng and R.-S. Wu, *J. Heat Transfer* **98,** 564 (1976).
23F. H. H.-S. Chu and S. W. Churchill, *J. Heat Transfer* **98,** 194 (1976).
24F. J. F. Clarke and N. Riley, *J. Fluid Mech.* **74,** 415 (1976).
25F. L. J. Crane, *Z. Angew. Math. Phys.* **27,** 61 (1976).
26F. D. E. Daney, *Int. J. Heat Mass Transfer* **19,** 431 (1976).
27F. N. Datta and R. N. Jana, *Int. J. Heat Mass Transfer* **19,** 1015 (1976).
28F. A. B. Datta and V. U. K. Sastry, *Int. J. Engng Sci.* **14,** 631 (1976).
29F. J. S. Fein and R. L. Pfeffer, *J. Fluid Mech.* **75,** 81 (1976).
30F. D. E. Fitzjarrald, *J. Fluid Mech.* **73,** 693 (1976.
31F. T. Fujii and M. Fujii, *Int. J. Heat Mass Transfer* **19,** 121 (1976).
32F. R. D. Gasser and M. S. Kazimi, *J. Heat Transfer* **98,** 49 (1976).
33F. B. Gebhart, H. Shaukatullah and L. Pera, *Int. J. Heat Mass Transfer* **19,** 751 (1976).
34F. D. D. Gray and A. Giorgini, *Int. J. Heat Mass Transfer* **19,** 545 (1976).
35F. G. M. Harpole and I. Catton, *J. Heat Transfer* **98,** 208 (1976).
36F. W. Hauf and U. Grigull, *Wärme- und Stoffübertragung* **9,** 21 (1976).
37F. H. J. Hetherington and T. D. Patten, *Int. J. Heat Mass Transfer* **19,** 1121 (1976).
38F. K. G. T. Hollands, T. E. Unny, G. D. Raithby and L. Konicek, *J. Heat Transfer* **98,** 189 (1976).
39F. G. M. Homsy and A. E. Sherwood, *A.I.Ch.E. Jl* **22,** 168 (1976).
40F. S. W. Hong and A. E. Bergles, *J. Heat Transfer* **98,** 459 (1976).
41F. C. Hunter and N. Ciahi, *J. Fluid Mech.* **72,** 433 (1976).
42F. Y. Jaluria, *Int. J. Heat Mass Transfer* **19,** 1057 (1976).
43F. C. A. Jones, D. R. Moore and N. O. Weiss, *J. Fluid Mech.* **73,** 353 (1976).
44F. Y. Kamotani and S. Ostrach, *J. Heat Transfer* **98,** 62 (1976).
45F. T.-T. Kao, *J. Heat Transfer* **98,** 321 (1976).
46F. D. R. Kassoy and Z. Zebib, *Physics Fluids* **18,** 1649 (1975).
47F. R. J. Kee, C. S. Landram and J. C. Miles, *J. Heat Transfer* **98,** 55 (1976).
48F. C. B. Kim and D. E. Wollersheim, *J. Heat Transfer* **98,** 144 (1976).

49F. S. A. Korpela and V. S. Arpaci, *J. Fluid Mech.* **75,** 385 (1976).
50F. T. H. Kuehn and R. J. Goldstein, *J. Fluid Mech.* **74,** 695 (1976).
51F. T. H. Kuehn and R. J. Goldstein, *Int. J. Heat Mass Transfer* **19,** 1127 (1976).
52F. M. L. Lawson, W.-J. Yang and S. Bunditkul, *J. Heat Transfer* **98,** 35 (1976).
53F. E. J. LeFevere, *Int. J. Heat Mass Transfer* **19,** 1215 (1976).
54F. F. B. Lipps, *J. Fluid Mech.* **75,** 113 (1976).
55F. R. R. Long, *J. Fluid Mech.* **73,** 445 (1976).
56F. T. R. Mancini, R. I. Loehrke and R. D. Haberstroh, *Int. J. Heat Mass Transfer* **19,** 839 (1976).
57F. W. J. Marner and J. Kleppe, *J. Fluids Engng* **98,** 119 (1976).
58F. A. F. Messiter and A. Linan, *Z. Angew Math. Phys.* **27,** 633 (1976).
59F. G. P. Merker, P. Waas, J. Straub and U. Grigull, *Wärme- und Stoffübertragung* **9,** 99 (1976).
60F. W. J. Minkowycz and P. Cheng, *Int. J. Heat Mass Transfer* **19,** 805 (1976).
61F. J. Mizushima and K. Gotoh, *J. Fluid Mech.* **73,** 65 (1976).
62F. A. Mucoglu and T. S. Chen, *J. Heat Transfer* **98,** 523 (1976).
63F. P. H. Oosthuizen, *J. Heat Transfer* **98,** 672 (1976).
64F. P. H. Oosthuizen, *J. Heat Transfer* **98,** 570 (1976).
65F. J.-W. Ou, K. C. Cheng and R.-C. Lin, *Int. J. Heat Mass Transfer* **19,** 277 (1976).
66F. H. Ozoe, K. Yamamoto, S. W. Churchill and H. Sayama, *J. Heat Transfer* **98,** 202 (1976).
67F. B. R. Pamplin and G. H. Holt, *J. Phys. E: Appl. Phys.* **9,** 145 (1976).
68F. J. K. Platten and G. Chavepeyer, *Int. J. Heat Mass Transfer* **19,** 27 (1976).
69F. G. D. Raithby and E. G. T. Hollands, *J. Heat Transfer* **98,** 72 (1976).
70F. G. D. Raithby and K. G. T. Hollands, *J. Heat Transfer* **98,** 522 (1976).
71F. G. D. Raithby, A. Pollard, K. G. T. Hollands and M. M. Yovanovich, *J. Heat Transfer* **98,** 3, 452 (1976).
72F. R. S. Reddy Gorla, *Appl. Scient. Res.* **31,** 455 (1976).
73F. R. J. Ribando and K. E. Torrance, *J. Heat Transfer* **98,** 42 (1976).
74F. J. L. Robinson, *Physics Fluids* **19,** 778 (1976).
75F. S. A. Sabbagh, A. Aziz, A. S. El-Ariny and G. Hamad, *J. Heat Transfer* **98,** 322 (1976).
76F. V. U. K. Sastry and G. Maiti, *Int. J. Heat Mass Transfer* **19,** 207 (1976).
77F. V. E. Schrock and A. D. K. Laird, *J. Heat Transfer* **98,** 213 (1976).
78F. R. A. Seban and M. M. Behnia, *Int. J. Heat Mass Transfer* **19,** 1197 (1976).
79F. N. Seki, S. Fukusako and M. Nakaoka, *J. Heat Transfer* **98,** 670 (1976).
80F. M. A. Shirazi and L. R. Davis, *J. Heat Transfer* **98,** 367 (1976).
81F. D. J. Shlien, *Physics Fluids* **19,** 1089 (1976).
82F. Z. P. Shulman, V. I. Baikov and E. A. Zalsgendler, *Int. J. Heat Mass Transfer* **19,** 1003 (1976).
83F. P. Singh, *Int. J. Heat Mass Transfer* **19,** 581 (1976).
84F. A. P. Solodov and D. Gorenflo, *Wärme- und Stoffübertragung* **9,** 151 (1976).
85F. V. M. Solomov and V. Lokshin, *Thermal Engng* **22,** 74 (1975).
86F. Y. M. Soundalgekar, *Wärme- und Stoffübertragung* **9,** 111 (1976).
87F. E. M. Sparrow and L. Lee, *Int. J. Heat Mass Transfer* **19,** 229 (1976).
88F. E. M. Sparrow, L. Lee and N. Shamsundar, *J. Heat Transfer* **98,** 88 (1976).
89F. P. J. Sullivan and P. J. Sutherland, *Z. Angew Math. Phys.* **27,** 671 (1976).
90F. M. Takashima, *J. Phys. Soc. Japan* **41,** 1394 (1976).
91F. Y. Tauchiva, *J. Phys. Soc. Japan* **40,** 280 (1976).
92F. G. Wilks, *Z. Angew. Math. Phys.* **27,** 621 (1976).
93F. G. Wilks, *Int. J. Heat Mass Transfer* **19,** 951 (1976).
94F. J. H. Wright and R. I. Loehrke, *J. Heat Transfer* **98,** 558 (1976).
95F. R.-S. Wu and K. C. Cheng, *Int. J. Heat Mass Transfer* **19,** 907 (1976).
96F. R.-S. Wu and K. C. Cheng, *Int. J. Heat Mass Transfer* **19,** 559 (1976).
97F. O. Zeman and J. L. Lumley, *J. Atmos. Sci.* **33,** 1974 (1976).

Convection from rotating surfaces

1G. L. M. de Socio, E. M. Sparrow and E. R. G. Eckert, *Int. J. Heat Mass Transfer* **19,** 345 (1976).
2G. A. D. Gosman, F. C. Lockwood and J. N. Loughhead, *J. Mech. Engng Sci.* **18,** 142 (1976).
3G. M. Y. Hussaini and M. S. Sastry, *J. Heat Transfer* **98,** 533 (1976).
4G. O. G. Martynenko, I. A. Vatutin and I. V. Skutova, *Int. J. Heat Mass Transfer* **19,** 1143 (1976).
5G. B. S. Mazumder, A. S. Gupta and N. Datta, *Int. J. Heat Mass Transfer* **19,** 523 (1976).
6G. G. G. Poe and A. Acrivos, *Int. J. Multiphase Flow* **2,** 365 (1976).
7G. A. N. Sherstyuk, *Thermal Engng* **22,** 41 (1975).
8G. E. M. Sparrow and L. Goldstein, Jr., *J. Heat Transfer* **98,** 387 (1976).

Combined heat and mass transfer

1H. G. Bergeles, A. D. Gosman and B. E. Launder, *J. Heat Transfer* **98,** 373 (1976).
2H. G. Bergeles, A. D. Gosman, and B. E. Launder, *J. Heat Transfer* **98,** 379 (1976).
3H. T. S. Chen and E. M. Sparrow, *J. Heat Transfer* **98,** 674 (1976).
4H. R. S. Colladay and L. M. Russell, *J. Heat Transfer* **98,** 245 (1976).
5H. P. F. Crapper, *Int. J. Heat Mass Transfer* **19,** 1371 (1976).
6H. S. Endrenyi and B. Palancz, *Int. J. Heat Mass Transfer* **19,** 1117 (1976).
7H. G. W. Haering, *Int. J. Heat Mass Transfer* **19,** 117 (1976).
8H. J. N. Hefner, *AIAA Jl* **14,** 817 (1976).
9H. G. R. Inger and T. F. Swean, *AIAA Jl* **14,** 589 (1976).
10H. T. Maeda and K. Shimura, *Bull. J.S.M.E.* **19,** 1149 (1976).

11H. R. E. Mayle and F. C. Kopper, *J. Heat Transfer* **98,** 240 (1976).
12H. D. K. Mukherjee, *J. Engng Pwr* **98,** 556 (1976).
13H. J. F. Muska, R. W. Fish and M. Suo, *J. Engng Pwr* **98,** 457 (1976).
14H. I. T. Shvets and V. M. Repukhov, *Heat Transfer, Soviet Res.* **7** (4), 40 (1975).
15H. Y. Tambour and B. Gal-Or, *Physics Fluids* **19,** 219 (1976).
16H. H.-C. Yao and V. E. Schrock, *J. Heat Transfer* **98,** 120 (1976).
17H. T.-F. Zien, *Int. J. Heat Mass Transfer* **19,** 513 (1976).

Change of phase

1J. R. Bairamov, M. Mamedov and L. E. Rybakova, *Int. Chem. Engng* **16,** 325 (1976).
2J. D. K. Basu, *Iseael J. Technol.* **13,** 386 (1976).
3J. S. Bedenek, *Int. J. Heat Mass Transfer* **19,** 448 (1976).
4J. J. Berghams, *Int. J. Heat Mass Transfer* **19,** 791 (1976).
5J. J. Berghmans, *Int. J. Multiphase Flow* **2,** 319 (1976).
6J. A. S. Borodin and J. J. C. Picot, *Can. J. Chem. Engng* **54,** 59 (1976).
7J. R. M. Canon and E. L. Park, Jr., *Int. J. Heat Mass Transfer* **19,** 696 (1976).
8J. R. Chandran and F. A. Watson, *Trans Instn Chem. Engrs* **54,** 65 (1976).
9J. C. E. Chang and W. R. Wilcox, *Int. J. Heat Mass Transfer* **19,** 355 (1976).
10J. S. P. Chary and P. K. Sarma, *J. Heat Transfer* **98,** 682 (1976).
11J. P. I. Chashchin, L. F. Shigina, N. S. Shvab and A. D. Sobol', *Thermal Engng* **22** (8), 105 (1975).
12J. C. M. Detz and R. T. Vermesh, *A.I.Ch.E. Jl* **22,** 87 (1976).
13J. S. S. Dua and C. L. Tien, *J. Heat Transfer* **98,** 407 (1976).
14J. M. Epstein and D. H. Cho, *J. Heat Transfer* **98,** 108 (1976).
15J. L. C. Farrar and E. Marschall, *J. Heat Transfer* **98,** 173 (1976).
16J. R. M. Fand, K. K. Keswani, M. M. Jotwani, and R. C. C. Ho, *J. Heat Transfer* **98,** 395 (1976).
17J. R. J. Hannemann and B. B. Mikic, *Int. J. Heat Mass Transfer* **19,** 1299 (1976).
18J. R. J. Hannemann and B. B. Mikic, *Int. J. Heat Mass Transfer* **19,** 1309 (1976).
19J. Y. Hayashi, A. Takimoto and M. Kanbe, *J. Heat Transfer* **98,** 114 (1976).
20J. W. H. Henstock and T. J. Hanratty, *A.I.Ch.E. Jl* **22,** 990 (1976).
21J. G. Hesse, E. M. Sparrow and R. J. Goldstein, *J. Heat Transfer* **98,** 166 (1976).
22J. G. L. Hubbard, A. F. Mills and D. K. Chung, *J. Heat Transfer* **98,** 319 (1976).
23J. R. L. Judd and K. S. Hwang, *J. Heat Transfer* **98,** 623 (1976).
24J. N. L. Kafengaus, *Heat Transfer, Soviet Res.* **7** (4), 94 (1975).
25J. P. G. Kosky, *Int. J. Heat Mass Transfer* **19,** 539 (1976).
26J. L. L. Levitan and F. P. Lantsman, *Thermal Engng* **22** (1), 102 (1975).
27J. K. Lucas, *Int. J. Heat Mass Transfer* **19,** 1273 (1976).
28J. P. J. Lykoudis, *Int. J. Heat Mass Transfer* **19,** 1357 (1976).
29J. R. Mahaligam, R. S. Limaye and J. A. Brink, Jr., *A.I.Ch.E. Jl* **22,** 1152 (1976).
30J. C. C. Maneri and R. E. Schneider, *J. Heat Transfer* **31.98,** 332 (1976).
31J. E. Marschall, *Wärme- und Stoffübertragung* **9,** 167 (1976).
32J. R. Mesler, *A.I.Ch.E. Jl* **22,** 246 (1976).
33J. D. Moalem and S. Xideman, *Int. J. Heat Transfer* **19,** 259 (1976).
34J. J. Schütt Mogro, *Wärme- und Stoffübertragung* **9,** 129 (1976).
35J. A. P. Ornatskiy, V. A. Chernobay and A. F. Vasil'yev, *Heat Transfer, Soviet Res.* **7** (4), 27 (1975).
36J. J. R. A. Pearson, *Int. J. Heat Mass Transfer* **19,** 405 (1976).
37J. Yu. Ya. Pechenegov and V. G. Kashirskiy, *Heat Transfer, Soviet Res.* **7** (4), 113 (1975).
38J. W. C. Peterson, M. G. Zaslouk and S. I. Güceri, *Nucl. Sci. Engng* **61,** 250 (1976).
39J. G. Preiss and P. C. Wayner, Jr., *J. Heat Transfer* **98,** 178 (1976).
40J. P. K. M. Rao and D. G. Andrews, *Can. J. Chem. Engng* **54,** 403 (1976).
41J. J. W. Rose, *Int. J. Heat Mass Transfer* **19,** 1363 (1976).
42J. R. P. Roy and G. Yadigaroglu, *J. Heat Transfer* **98,** 630 (1976).
43J. F. E. Sage and J. Estrin, *Int. J. Heat Mass Transfer* **19,** 323 (1976).
44J. P. Saha, M. Ishii and N. Zuber, *J. Heat Transfer* **98,** 616 (1976).
45J. R. A. Seban and A. Faghri, *J. Heat Transfer* **98,** 315 (1976).
46J. R. A. W. Shock, *Int. J. Multiphase Flow* **2,** 411 (1976).
47J. A. Singh, B. B. Mikic and W. M. Rohsenow, *J. Heat Transfer* **98,** 401 (1976).
48J. R. A. Smith, F. A. Price and P. Griffith, *J. Heat Transfer* **98,** 153 (1976).
49J. A. Stücheli and M. N. Özişik, *Chem. Engng Sci.* **31,** 369 (1976).
50J. S. R. Tailby, E. K. Clutterbuck and A. S. Bakshi, *Trans. Instn Chem. Engrs* **54,** 84 (1976).
51J. T. G. Theofanous and P. D. Patel, *Int. J. Heat Mass Transfer* **19,** 425 (1976).
52J. J. R. Travis, F. H. Harlow and A. A. Amaden, *Nucl. Sci. Engng* **61,** 1 (1976).
53J. T. Ueda and H. Tanaka, *Int. J. Multiphase Flow* **2,** 261 (1976).
54J. H. C. Unal, *Int. J. Heat Mass Transfer* **19,** 643 (1976).
55J. V. G. Van der Mast and L. A. Bromley, *A.I.Ch.E. Jl* **22,** 533 (1976).
56J. K. D. Voskresenskiy and Ye. S. Turilina, *Heat Transfer, Soviet Res.* **7** (4), 79 (1975).
57J. K. L. Waldram, H. K. Fauske and S. G. Bankoff, *Can. J. Chem. Engng* **54,** 456 (1976).

58J. T. Watanabe, M. Toyama and K. Nakamura, *Bull. J.S.M.E.* **19,** 1047 (1976).
59J. P. C. Wayner, Jr., Y. K. Kao and I. V. LaCroix, *Int. J. Heat Mass Transfer* **19,** 487 (1976).
60J. J. Weisman, T. Ake and R. Knott, *Nucl. Sci. Engng* **61,** 297 (1976).
61J. S. D. R. Wilson, *J. Heat Transfer* **98,** 313 (1976).
62J. G. Yadigaroglu and R. T. Lahey, Jr., *Int. J. Multiphase Flow* **2,** 477 (1976).
63J. W. Zyszkowski, *Int. J. Heat Mass Transfer* **19,** 625 (1976).

Radiation in participating media
1K. H. S. Bennett and R. A. Forman, *Appl. Optics* **15,** 1313 (1976).
2K. H. S. Bennett and R. A. Forman, *Appl. Optics* **15,** 347 (1976).
3K. A. G. Blokh and L. D. Burak, *Thermal Engng* **22,** 65 (1975).
4K. R. D. Cess and M. S. Lian, *J. Heat Transfer* **98,** 676 (1976).
5K. G. C. Davis, *Int. J. Heat Mass Transfer* **19,** 157 (1976).
6K. L. M. de Socio and Gaffuri, *Z. Angew. Math. Phys.* **27,** 591 (1976).
7K. J. B. Elgin and J. R. Baron, *AIAA Jl* **14,** 1492 (1976).
8K. A. Hashemi, T. C. Hsieh and R. Greif, *J. Heat Transfer* **98,** 432 (1976).
9K. D. R. Jeng, E. J. Lee and K. J. DeWitt, *Int. J. Heat Mass Transfer* **19,** 589 (1976).
10K. S. Kubo, *J. Phys. Soc. Japan* **40,** 1186 (1976).
11K. D. Marcuse, *J. Opt. Soc. Am.* **66,** 216 (1976).
12K. L. M. McMillin and H. E. Fleming, *Appl. Optics* **15,** 358 (1976).
13K. M. J. Minot, *J. Opt. Soc. Am.* **66,** 515 (1976).
14K. D. A. Nelson, *J. Heat Transfer* **98,** 638 (1976).
15K. D. A. Nelson, *J. Heat Transfer* **98,** 310 (1976).
16K. L. W. Pinkley and D. Williams, *J. Opt. Soc. Am.* **66,** 554 (1976).
17K. J. R. Saraf, *Physics Fluids* **19,** 1492 (1976).
18K. N. Selçuk and R. G. Siddall, *Int. J. Heat Mass Transfer* **19,** 313 (1976).
19K. S. Srinivasan and R. Ram, *Int. J. Heat Mass Transfer* **19,** 297 (1976).
20K. V. V. Truong and G. D. Scott, *J. Opt. Soc. Am.* **66,** 124 (1976).
21K. V. A. Volkov, V. V. Zhurin, A. B. Karasev and B. K. Tkachenko, *AIAA Jl* **14,** 371 (1976).
22K. M. D. Waugh and B. V. Rao, *Appl. Optics* **15,** 1331 (1976).
23K. A. T. Wassel and D. K. Edwards, *J. Heat Transfer* **98,** 308 (1976).
24K. A. T. Wassel and D. K. Edwards, *J. Heat Transfer* **98,** 101 (1976).

Surface radiation
1L. R. E. Bedford and C. K. Ma, *J. Opt. Soc. Am.* **66,** 724 (1976).
2L. R. C. Birkebak and A. Abdulkadir, *Int. J. Heat Mass Transfer* **19,** 1039 (1976).
3L. A. Campo, *Wärme- und Stoffübertragung* **9,** 139 (1976).
4L. J. R. Carpenter, D. G. Briggs and V. Sernas, *J. Heat Transfer* **98,** 95 (1976).
5L. A. L. Crosbie, *J. Heat Transfer* **98,** 148 (1976).
6L. R. Farrell, *J. Heat Transfer* **98,** 311 (1976).
7L. G. Kaspatek and D. Vortmeyer, *Wärme- und Stoffübertragung* **9,** 117 (1976).
8L. C. P. Minning, *AIAA Jl* **14,** 813 (1976).
9L. S. V. Shalvarova and N. T. Spodyryak, *Heat Transfer, Soviet Res.* **7** (4), 54 (1975).
10L. W. M. Toscano and E. G. Cravalho, *J. Heat Transfer* **98,** 438 (1976).

Liquid metals
1M. I. T. Alad'yev and N. S. Kondrat'yev (dec.), *Heat Transfer, Soviet Res.* **7** (4) 74 (1975).
2M. S. G. Bankoff and H. K. Choi, *Int. J. Heat Mass Transfer* **19,** 87 (1976).
3M. O. E. Dwyer, *Chem Engng Sci.* **31,** 187 (1976).
4M. O. E. Dwyer and C. J. Hsu, *Int. J. Heat Mass Transfer* **19,** 185 (1976).
5M. I. G. Gorlov, A. I. Rzayev and V. F. Khudyakov, *Heat Transfer, Soviet Res.* **7** (4), 60 (1975).
6M. I. Michiyoshi, O. Takahashi and A. Serizawa, *Int. J. Heat Mass Transfer* **19,** 1021 (1976).
7M. W. J. Minkowycz, D. M. France and R. M. Singer, *J. Heat Transfer* **98,** 5 (1976).
8M. S. Necmi and J. W. Rose, *Int. J. Heat Mass Transfer* **19,** 1245 (1976).
9M. J. R. Welty and D. R. Peinecke, *J. Heat Transfer* **98,** 146 (1976).

Measurement techniques
1P. G. Abbate, A. Attanasio, U. Bernini, E. Ragozzino and F. Somma, *J. Phys. D.: Appl. Phys.* **9,** 1945 (1976).
2P. P. Andersson and G. Bäckström, *Rev. Scient. Instrum.* **47,** 205 (1976).
3P. A. A. Anufriev, *Measmt Tech., Pittsb.* **19,** 308 (1976).
4P. N. Araki, K. Nakagawa and K. Hakomori, *Rev. Scient. Instrum.* **47,** 745 (1976).
5P. D. N. Astronov, L. B. Belyanskii, V. A. Pavlov and **V. T. Shkraba,** *Measmt Tech., Pittsb.* **19,** 701 (1976).
6P. R. M. A. Azzam, *Rev. Scient. Instrum.* **47,** 624 (1976).
7P. W. D. Baines and J. F. Keffer, *Rev. Scient. Instrum.* **47,** 440 (1976).
8P. K. G. Bartlett and C. Y. She, *Appl. Optics* **15,** 1980 (1976).
9P. R. H. Bruce and D. S. Cannell, *Rev. Scient. Instrum.* **47,** 1323 (1976).
10P. E. A. Bruges, *J. Mech. Engng Sci.* **18,** 121 (1976).
11P. K. K. Chee and A. Rudin, *Can. J. Chem. Engng* **54,** 129 (1976).
12P. J. J. Christensen, L. D. Hansen, D. J. Eatough, R. M. Izatt and R. M. Hart, *Rev. Scient. Instrum.* **47,** 730 (1976).
13P. D. K. Deardorf, *Rev. Scient. Instrum.* **47,** 1279 (1976).
14P. H. F. Lobele and J. H. Massig, *Appl. Optics* **15,** 69 (1976).
15P. W. E. Dunn, B. T. Chao and A. M. Clausing, *Rev. Scient. Instrum.* **47,** 321 (1976).
16P. F. Durst and W. H. Stevenson, *Appl. Optics* **15,** 137 (1976).
17P. D. Favier and N. Pellerin, *Rev. Scient. Instrum.* **47,** 368 (1976).
18P. N. A. Halliwell and A. K. Lewkowicz, *Physics Fluids* **18,** 1617 (1975).

19P. D. A. Harper, R. H. Hildebrand, R. Stiening and R. Winston, *Appl. Optics* **15,** 53 (1976).
20P. J. G. Harley and W. Z. Black, *J. Heat Transfer* **98,** 530 (1976).
21P. K. N. Helland and C. W. Van Atta, *Physics Fluids* **19,** 1109 (1976).
22P. H. R. Hidberg and G. Süss, *Rev. Scient. Instrum.* **47,** 912 (1976).
23P. C.-M. Ho and L. S. G. Kovasznay, *Physics Fluids* **19,** 1118 (1976).
24P. M. H. Horman, *Appl. Optics* **15,** 2099 (1976).
25P. J. O. Hylton and R. L. Reid, *AIAA Jl* **14,** 1303 (1976).
26P. H. T. Kalb and V. A. Cline, *Rev. Scient. Instrum.* **47,** 708 (1976).
27P. M. E. Karpul and W. G. Tiederman, Jr., *AIAA Jl* **14,** 1099 (1976).
28P. Y. J. Kaufman and A. Gersten, *Appl. Optics* **15,** 353 (1976).
29P. E. E. Khawaja, *J. Phys. D.: Appl. Phys.* **9,** 1939 (1976).
30P. A. M. Koppius and G. R. M. Trines, *Int. J. Heat Mass Transfer* **19,** 967 (1976).
31P. A. F. Kotyuk, I. N. Samoilov, L. S. Lovinskii and V. I. Sapritskii, *Measmt. Tech., Pittsb.* **19,** 269 (1976).
32P. I. M. Kovela, Yu. I. Sitnitskii and Z. N. Teplyukh, *Measmt. Tech., Pittsb.* **19,** 677 (1976).
33P. A. A. Kozhevnikov, *Heat Transfer, Soviet Res.* **7** (4), 149 (1975).
34P. R. L. Laurence and J. R. A. Pearson, *Int. J. Heat Mass Transfer* **19,** 901 (1976).
35P. C. Nieuwvelt, J. M. Bessem and G. R. M. Trines, *Int. J. Heat Mass Transfer* **19,** 975 (1976).
36P. A. G. Parker and J. C. Brusse, *J. Aircraft* **13,** 57 (1976).
37P. F. J. Pierce and C. I. Ezekwe, *AIAA Jl* **14,** 412 (1976).
38P. H. Rohatschek, *Int. J. Heat Mass Transfer* **19,** 433 (1976).
39P. R. Saarimaa and P. Wallin, *Rev. Scient. Instrum.* **47,** 195 (1976)
40P. V. A. Sandborn, *AIAA Jl* **14,** 400 (1976).
41P. R. Schnurmann and P. A. Davies, *Proc. R. Soc.* **347,** (1648) 47 (1975).
42P. B. S. Singh and A Dybbs, *J. Heat Transfer* **98,** 491 (1976).
43P. H. M. Smith, M. H. Sewall and J. R. King, *Appl. Optics* **15,** 729 (1976).
44P. J. L. Smith and E Pfender, *Rev. Scient. Instrum.* **47,** 1056 (1976).
45P. G. P. Titov, A. A. Tarin, N. V. Bashkatov and P. E. Suetin, *Measmt Tech., Pittsb.* **19,** 311 (1976).
46P. A. A. Tupichenkow, P. A. Garshin and R. E. Smirnov, *Measmt Tech., Pitsb.* **19,** 669 (1976).
47P. V. M. Underwood, *Rev. Scient. Instrum.* **47,** 1079 (1976).
48P. J. Y. Wang, *Appl. Optics* **15,** 768 (1976).
49P. B. E. Wood, J. G. Pipes, A. M. Smith and J. A. Roux, *Appl. Optics* **15,** 940 (1976).
50P. L. V. Zanichkovskaya and V. A. Kochan, *Measmt Tech., Pittsb.* **19,** 301 (1976).
51P. V. N. Zmeykov, B. P. Ustimenko and A. A. Shishkin, *Heat Transfer, Soviet Res.* **7** (4), 134 (1975).

Heat exchangers and heat pipes
1Q. S. Aiba and Y. Yamazaki, *J. Heat Transfer* **98,** 503 (1976).
2Q. T. R. Bott and D. V. Law, *J. Heat Transfer* **98,** 659 (1976).
3Q. V. T. Buglaev and M. M. Andreev, *Thermal Engng* **22** (8) 99 (1975).
4Q. V. D. Dobud'ko, V. S. Kortikov and L. S. Aksel'rod, *Thermal Engng* **22** (8), 126 (1975).
5Q. B. Gay, N. V. Mackley and J. D. Jenkins, *Int. J. Heat Mass Transfer* **19,** 995 (1976).
6Q. P. J. Heggs and K. J. Carpenter, *Trans. Instn Chem. Engrs* **54,** 232 (1976).
7Q. R. LeClaire, *Aero. Q.* **27,** 40 (1976).
8Q. F. K. Moore and C. Ncubizu, *J. Heat Transfer* **98,** 345 (1976).
9Q. I. M. Rudenko and N. Spodyryak, *Heat Transfer, Soviet Res.* **7** (4), 48 (1975).
10Q. F. E. M. Saboya and E. M. Sparrow, *J. Heat Transfer* **98,** 520 (1976).
11Q. E. U. Schlünder, *Chem.Ingr.-Tech.* **48,** 212 (1976).
12Q. K. E. Starner, *ASHRAE Jl* **18** (5), 39 (1976).
13Q. S. Suryanarayanan, B. A. Mujawar and M. R. Rao, *I/EC Process Des. Dev.* **15,** 564 (1976).
14Q. D. Vortmeyer and Sl Le Mong, *Wärme- und Stoffübertragung* **9,** 29 (1976).
15Q. A. J. Wilmott and C. Hinchcliffe, *Int. J. Heat Mass Transfer* **19,** 821 (1976).
16Q. K. Wörrlein, *Forsh Geb. Ing Wes.* **42,** 37 (1976).
17Q. F. C. Yip, *J. Spacecraft Rockets* **13,** 237 (1976).

Aircraft and space vehicles
1R. J. J. Bertin and W. D. Goodrich, *J. Spacecraft Rockets* **13,** 473 (1976).
2R. J. J. Bertin, S. J. Mosso, D. W. Barnette, and W. D. Goodrich, *J. Spacecraft Rockets* **13,** 540 (1976).
3R. H. J. Brandon, R. V. Masek, and J. G. Dunavant, *AIAA Jl* **13,** 1460 (1975).
4R. J. P. Drumond, R. A. Jones, and R. L. Ash, *AIAA Jl* **10,** 1476 (1976).
5R. P. E. Glaser, *J. Spacecraft Rockets* **13,** 573 (1976).
6R. B. Holzknecht and K. Stephan, *Wärme- und Stoffübertragung* **9,** 73 (1976).
7R. J. W. Keyes, *J. Spacecraft Rockets* **13,** 61 (1976).
8R. H. Krier and S. S. Gokhale, *Int. J. Heat Mass Transfer* **19,** 915 (1976).
9R. J. P. Lamb and G. K. Mruk, *J. Spacecraft Rockets* **13,** 701 (1976).
10R. P. L. Lawing and J. L. Hunt, *J. Aircraft* **13,** 1018 (1976).
11R. K. Mastanaiah, *J. Heat Transfer* **98,** 139 (1976).
12R. H. L. Moody, D. H. Smith, R. L. Haddock, and S. S. Dunn, *J. Spacecraft Rockets* **13,** 746 (1976).
13R. T. F. Morey and D. N. Gorman, *J. Spacecraft Rockets* **13,** 229 (1976).
14R. F. C. Prenger, Jr. and W. C. Patterson, *J. Spacecraft Rockets* **13,** 244 (1976).
15R. J. Reinkenhof and R. Schmucker, *J. Spacecraft Rockets* **13,** 63 (1976).
16R. A. Sherman and P. Brennan, *J. Spacecraft Rockets* **13,** 288 (1976).
17R. K. Sutton, *J. Spacecraft Rockets* **13,** 294 (1976).
18R. V. R. P. Verneker, K. Kishore, and G. Prasad, *AIAA Jl* **14,** 1330 (1976).
19R. L. D. Wing, *J. Spacecraft Rockets* **13,** 727 (1976).

General applications
1S. L. Adams, *ASHRAE Jl* **18,** (3), 37 (1976).
2S. L. Baker, Jr., *Nucl. Sci. Engng* **61,** 222 (1976).
3S. D. N. Brocard and D. R. F. Harleman, *J. Hydraul. Div. Am. Soc. Civ. Engrs* **102,** No. HY 3, 227 (1976).
4S. E. Boyer and C. W. Macosko, *A.I.Ch.E. Jl* **22,** 268 (1976).
5S. G. B. Cox, Jr., *J. Aircraft* **13,** 630 (1976).
6S. A. F. Emery, R. E. Short, A. W. Guy, K. K. Kraning and J. C. Lin, *J. Heat Transfer* **98,** 284 (1976).
7S. R. E. Faw and L. Baker, Jr., *Nucl. Sci. Engng* **61,** 231 (1976).
8S. H. N. Fisher, J. Heat Transfer 98, 485 (1976).
9S. P. Hutchinson, E. E. Khalil, J. H. Whitelaw and G. Wigley, *J. Heat Transfer* **98,** 276 (1976).
10S. S. C. Johnston and H. A. Dwyer, *J. Heat Transfer* **98,** 360 (1976).
11S. S. Kato, K. Yamaguchi, Y. Watanabe and Y. Hiraiwa, *J. Engng Ind.* **98,** 607 (1976).
12S. A. M. Mitky and M. N. Ozişik, *Int. J. Heat Mass Transfer* **19,** 201 (1976).
13S. K. Namatame and K. Kobayashi, *J. Heat Transfer* **98,** 12 (1976).
14S. L. C. Nkemdirim, *J. Appl. Meteorol.* **15,** 818 (1976).
15S. W. L. Oberkampf and L. I. Crow, *J. Heat Transfer* **98,** 353 (1976).
16S. P. P. Paily and E. O. Macagno, *J. Hydraul. Div. Am. Soc. Civ. Engrs* **102,** No. HY 3, 255 (1976).
17S. D. C. Ross and K. R. Diller, *J. Heat Transfer* **98,** 292 (1976).
18S. D. A. Secrist and R. W. Hornbeck, *J. Engng Ind.* **98,** 385 (1976).
19S. A. Steiner, *J. Heat Transfer* **98,** 509 (1976).
20S. K. H. Sun, J. M. Gonzalez-Santalo and C. L. Tien, *J. Heat Transfer* **98,** 414 (1976).
21S. K. Urakawa and I. Morioka, *Bull. J.S.M.E.* **19,** 1054 (1976).
22S. R. Viskanta, R. O. Johnson and R. W. Bergstrom, *J. Heat Transfer* **98,** 662 (1976).
23S. H. R. Wetenkamp and R. M. Kipp, *J. Engng Ind.* **98,** 453 (1976).
24S. J. H. Young, H. Merte, Jr. and G. A. Gates, *Int. J. Heat Mass Transfer* **19,** 503 (1976).

Solar energy
1T. S. I. Abdel-Khalik, *Solar Energy* **18,** 59 (1976).
2T. R. E. Athey, *Solar Energy* **18,** 143 (1976).
3T. R. Bairamov, K. Toiliev and M. Khodzhiev, *Int. Chem. Engng* **16,** 127 (1976).
4T. D. A. Boyd, R. Gajewski and R. Swift, *Solar Energy* **18,** 395 (1976).
5T. H. Buchberg and D. K. Edwards, *Solar Energy* **18,** 193 (1976).
6T. D. J. Close and T. L. Pryor, *Solar Energy* **18,** 287 (1976).
7T. M. W. Edenburn, *Solar Energy* **18,** 437 (1976).
8T. D. K. Edwards, J. N. Arnold and I. Catton, *Solar Energy* **18,** 253 (1976).
9T. J. C. C. Fan and F. J. Bachner, *Appl. Optics* **15,** 1012 (1976).
10T. J. E. Hill and E. R. Streed, *Solar Energy* **18,** 421 (1976).
11T. P. J. Hughes, S. A. Klein and D. J. Close, *J. Heat Transfer* **98,** 336 (1976).
12T. A. Kababaev, A. Khandurdyev, O. Klyschaeva and N. Kirbanov, *Int. Chem. Engng* **16,** 60 (1976).
13T. S. A. Klein, W. A. Beckman and J. A. Duffie, *Solar Energy* **18,** 113 (1976).
14T. E. Lumsdaine and J. C. Cherng, *Solar Energy* **18,** 157 (1976).
15T. R. T. Nash and J. W. Williamson, *Solar Energy* **18,** 11 (1976).
16T. A. Rabl, *Solar Energy* **18,** 93 (1976).
17T. A. Rabl, *Appl. Optics* **15,** 1871 (1976).
18T. A. Rabl and R. Winston, *Appl. Optics* **15,** 2880 (1976).
19T. M. R. Riaz, *J. Engng Pwr* **98,** 375 (1976).
20T. F. W. Schmidt and J. Szego, *J. Heat Transfer* **98,** 471 (1976).
21T. F. F. Simon, *Solar Energy* **18,** 451 (1976).
22T. P. Singh and L. S. Cheema, *Solar Energy* **18,** 135 (1976).
23T. A. Sobin, W. Wagner and C. R. Easton, *Solar Energy* **18,** 21 (1976).
24T. J. W. Stultz, *J. Spacecraft Rockets* **13,** 57 (1976).
25T. M. P. Thekaekara, *Appl. Optics* **15,** 915 (1976).
26T. M. P. Thakaekara, *Solar Energy* **18,** 309 (1976).
27T. L. L. Van-Hull and A. F. Hildebrandt, *Solar Energy* **18,** 31 (1976).
28T. A. Verma, K. E. Johnson and E. O. Sherman, *Can. J. Chem. Engng* **54,** 285 (1976).
29T. W. E. Wentworth and E. Chen, *Solar Energy* **18,** 205 (1976).
30T. M. L. Wesely and R. C. Lipschutz, *Solar Energy* **18,** 467 (1976).
31T. N. E. Wijeysundera, *Solar Energy* **18,** 65 (1976).

Plasma heat transfer
1U. D. M. Benenson and J. J. Nowobilski, *AIAA Jl* **14,** 408 (1976).
2U. M. I. Boulos, *IEEE Trans. Plasma Sci.* **PS-4,** 28 (1976).
3U. S. K. Chan, M. D. Cowley and M. T. C. Fang, *J. Phys. D: Appl. Phys.* **9,** 1085 (1976).
4U. S. K. Chan, M. T. C. Fang and M. D. Cowley, *Proc. Instn Elect. Engrs* **122,** 585 (1975).
5U. P. W. Chan, M. Niinura, R. J. Churchill and B. C. Schanberg, *IEEE Trans. Plasma Sci.* **PS-3,** 174 (1975).
6U. B. Feinberg, *Plasma Physics* **18,** 265 (1976).
7U. A. M. Howatson and D. R. Topham, *J. Phys D: Appl. Phys.* **9,** 1101 (1976).
8U. V. Javeri, *Wärme- und Stoffübertragung* **8,** 261 (1975).
9U. I. V. Kalganova and V. S. Klubnikin, *Teplofizika* **14,** 408 (1976).
10U. H. E. Lee, H. K. Messerle and A. D. Stokes, *Proc. Instn Elect. Engrs* **122,** 227 (1975).
11U. R. W. Liebermann and J. J. Lowke, *Jl Quantre Spectros. & Radiat. Transf.* **16,** 253 (1976).
12U. B. S. Mazumder, A. S. Gupta and N. Datta, *Int. J. Engng Sci.* **14,** 285 (1976).
13U. W. Neumann, *Beitr. Plasma Phys.* **16,** 97 1976).
14U. D. A. Nusupbekova, Sh. Ibrayev and K. K. Shalina, *Heat Transfer, Soviet Res.* **7** (4), 143 (1975).
15U. J. H. Painter and J. F. Shaeffer, *AIAA Jl* **14,** 1161 (1976).

16U. J. L. Smith and E. Pfender, *Rev. Scient. Instrum.* **47,** 1056 (1976).
17U. J. L. Smith and E. Pfender, *IEEE Trans. Power Apparatus Systems* **PAS-95,** 704 (1976).
18U. G. H. Stickford, Jr., *AIAA Jl* **14,** 421 (1976).
19U. D. C. Strachan and M. R. Barrault, *J. Phys. D: Appl. Phys.* **9,** 429 (1976).
20U. S. Suckewer and A. Kuszell, *Jl Quatre Spectros. & Radiat Sci.* **16,** 53 (1976).
21U. P. S. Todorovic, G. R. Jones and M. R. Barrault, *J. Phys. D: Appl. Phys.* **9,** 423 (1976).
22U. M. Todorovic, M. Rekalic, N. Ikonomov, B. Pavlovic and V. Vukanovic, *Beitr. Plasma Phys.* **15,** 333 (1975).
23U. J. Uhlenbusch, *Physica, 's Grav.* **82C,** 61 (1976).

HEAT TRANSFER REVIEW - 1977

E. R. G. ECKERT, E. M. SPARROW, R. J. GOLDSTEIN, C. J. SCOTT, E. PFENDER, S. V. PATANKAR and J. W. RAMSEY

Heat Transfer Laboratory, Department of Mechanical Engineering, University of Minnesota, Minneapolis, Minnesota, U.S.A.

INTRODUCTION

THIS review surveys results that have been published in various fields of heat transfer during 1977. As in the past, the number of papers published during that period was such that only a selection can be included in this review. A more complete listing of papers is available in the heat-transfer bibliographies published periodically in this journal. The reviews published during the years 1970–1975 have also been reprinted as Volume 8 of the series *Progress in Heat and Mass Transfer* and include a comprehensive author and subject index.

A symposium on turbulent shear flows held on 18–21 April, 1977 at Pennsylvania State University included papers on the effect of turbulence in jets, boundary layers, wakes, plumes, and channels, including turbulent energy transfer. Proceedings of the conference are available. The Third International Symposium on Plasma Chemistry was held on 13–19 July, 1977 in Limoges, France. Plasma heat transfer was discussed in connection with processing of ceramics, chemical synthesis, and coal gasification. Diagnostics of thermal plasmas using spectroscopy and laser methods found attention. Proceedings can be obtained from the University of Limoges, Allée André-Maurois, Limoges, France.

The 17th National Heat Transfer Conference brought together researchers and engineers interested in various fields of heat transfer. The majority of the sessions were organized around engineering applications. Jerry Taborek, who received the Donald Q. Kern Award, presented an invited lecture on the evolution of heat exchanger design techniques, and E. M. Sparrow, the recipient of the Max Jakob Memorial Award, discussed heat transfer in complex duct flows in a second invited lecture. The papers presented at the Conference are available as reprints and many of these will be published in the *Journal of Heat Transfer* or in the volume *17th National Heat Transfer Conference, AIChE* papers.

The 1977 International Seminar organized by the International Center for Heat and Mass Transfer in August, 1977 in Dubrovnik, Yugoslavia was devoted to heat transfer in buildings and covered thermal and moisture transfer properties of structural materials, solar radiation, district heating techniques, air movement inside and outside of buildings, and heat storage. The proceedings of the conference will be published by Scripta Book Company, Washington, D.C.

The 98th Winter Annual Meeting of the American Society of Mechanical Engineers included 14 sessions on heat transfer, mainly organized around various engineering developments. Only two of the sessions were devoted to basic transfer processes, one dealing with heat transfer in solids and the other on turbulent flows. A panel discussion dealt with mechanical design problems associated with heat-transfer equipment. Reprints of the papers are available from ASME Headquarters in New York and many of them will be published in the *Journal of Heat Transfer*.

The 1977 International Solar Energy Congress was somewhat delayed and was held on 16–21 January, 1978 in Vigyan Bhawan, New Delhi, India. The papers presented at the conference were organized around the subjects of international and national programs, solar radiation and energy storage, flat plate collectors including solar ponds and concentrating systems, space heating and cooling, solar thermal power, wind power, and agricultural and industrial applications.

A number of books dealing with heat transfer or including heat transfer topics has appeared on the market. They are listed in the bibliographic part of this review. Volume 13 of *Advances in Heat Transfer*, published by Academic Press, New York in 1977, contains survey articles on heat and mass transfer between impinging jets and solid surfaces, in rivers, bays, lakes and estuaries, in porous media, and on viscous dissipation in shear flows of molten polymers.

Trends and developments in heat-transfer research during 1977 are characterized by the following highlights: Heat conduction was studied analytically and numerically, especially in situations including phase change. Complex flow configurations in channel flow caused by corrugations, spherical and cylindrical blockages, and partially constricted inlets, as well as fluids containing additives and particulate materials, found special attention. Variable property effects and freezing of flowing liquids in ducts were investigated.

Unsteady flow and heat transfer in laminar boundary layers has been studied by various analytical techniques. Heat and mass transfer from cylinders in crossflow continues to be a topic of interest. Experimental and theoretical studies were concerned with the effect of surface roughness and free stream turbulence. Packed and fluidized bed heat transfer found worldwide attention in several papers. Detailed measurements of the fluctuating velocities and temperatures in turbulent flow have been made and sophisticated models of turbulence were developed and applied to various flow situations. A very large number of papers considered natural convection, especially in porous media, with the development of insulations and geothermal energy extraction as a focus. A large number of papers, especially from the Soviet Union, considered heat transfer in boiling or condensation occurring in various applications.

Radiative transport in emitting, absorbing, and scattering media including particle suspensions found attention. Research activities increased substantially in heat transfer associated with thin films. Radiation properties of various solid substances were reported and the effect of surface roughness was studied. Measurement techniques were developed for flow and temperature measurements, especially in two-phase flow.

The literature on heat exchangers is still growing with attention focused on the development of surfaces leading to high performance devices, on improved calculation methods, and on the analysis of heat exchanger networks. Research activities, on the other hand, declined on ablation and thermal protection systems. Models to analyze flat plate and concentrating solar collectors and the solar flux arriving on the earth's surface were developed. Plasma heat transfer found particular interest in connection with chemical and material processing, as well as in arc circuit-breaker research.

To facilitate the use of this review, a listing of the subject headings is made below in the order in which they appear in the text. The letter which appears adjacent to each subject heading is also attached to the references that are cited in this category.

Conduction, A
Channel flow, B
Boundary-layer and external flows, C
Flow with separated regions, D
Transfer mechanisms, E
Natural convection, F
Convection from rotating surfaces, G
Combined heat and mass transfer, H
Change of phase, J
Radiation
- Radiation in participating media, K
- Surface radiation, L

Liquid metals, M
Measurement techniques, P
Heat-transfer applications
- Heat exchangers and heat pipes,Q
- Aircraft and space vehicles, R
- General, S
- Solar energy, T

Plasma heat transfer, U

CONDUCTION

The current literature on heat conduction contains a strong focus on problems of phase change. A substantial interest in solution methodology, both analytical and numerical, is reflected in recent publications.

Heat conduction has traditionally been assumed to be the sole transport mechanism in phase change problems. However, in an analysis of melting about a vertical cylinder, it was demonstrated that natural convection induced by temperature variations in the melt region is the dominant mode of transport [61A]. An extensive survey of analytical methods for solving heat conduction problems with moving boundaries was focused on the Soviet literature and contains over 250 references [35A]. The introduction of high conductivity particles into a solidifying low conductivity material aids the solidification rate but diminishes the heat of fusion per unit volume [58A]. If a hot liquid is suddenly poured over a relatively cold solid, simultaneous freezing of the liquid and melting of the solid may occur [25A]. A sphere of molten heat-generating material, initially at its fusion temperature and thereafter subjected to surface cooling, may either completely solidify, temporarily solidify and then remelt, or have a solid outer crust with an inner molten core [21A]. A sudden temperature change at the boundary of a moist porous material activates a moving interface which separates wet and dry portions of the material [54A]. In the analysis of teflon ablation, account was taken of crystalline–amorphos phase transition, thermal expansion, and formation of higher-molecular products [29A].

Solutions for transient conduction without phase change can be employed to predict the surface heat flux and interface position in phase change problems provided that the conduction solution is evaluated using suitably defined effective properties [17A]. Biot's variational principle was employed in solving a one-dimensional phase change problem with surface convection and radiation [72A]. The method of matched asymptotic expansions successfully solved the problem of solidification of a slab of finite thickness [66A]. With the Stefan problem reformulated as a set of coupled nonlinear integral equations, a large-time expansion for the position of the phase boundary was developed [36A]. For the solution of moving boundary problems, a stationary grid of finite elements was used in conjunction with special elements involving finite element models defined by discontinuous interpolation [67A]. The treatment of thermal boundary conditions in the finite difference solution of freezing problems has been examined via experiments on model and actual biological systems [18A]. Conventional integral transform techniques have to be generalized to accommodate the unknown interface position in phase change problems [48A]. The success of the isotherm migration method in tracking the movement of a two-dimensional quench front underscores its potential applicability to phase change problems [24A]. An implicit finite difference scheme was also shown to be effective in solving the two-dimensional heat equation during quenching [5A].

The Heat Balance Integral solution for solidification of a continuous casting shows that the neglect of axial conduction can lead to serious errors at low casting speeds [56A]. By application of simple transformations to available Heat Balance Integral solutions for the freezing of circular cylinders, spheres, and uniform prisms, results can be obtained for ellipses, rectangular prisms, and ellipsoids [23A]. Two Heat Balance Integrals, respectively obtained by integrating the heat equation and its first moment, were applied to the freezing of a cubit [51A]. An upper bound on the time required to freeze a layer of given thickness has been deduced from the Heat Balance Integral [37A]. A Heat Balance Integral formulation suitable for modeling polymer melting processes in a screw extruder took account of sensible heat, temperature dependent viscosity, and viscous heat generation [28A]. As an innovation to the Heat Balance Integral as applied to finite regions without phase change, the integral condition is used to find an exponent in the approximating polynomial [22A].

Various transient problems were solved in direct response to applications. In order to diminish the heat leak through current-carrying cables which energize superconducting magnets, the cable may be cooled axially by a stream of helium gas which runs counter to the conduction heat flow. The temperature in the cable can be determined from a one-dimensional transient conduction model [8A]. Metallurgical changes during heat treatment of alloys can result in significant changes of thermal conductivity [26A]. As an alternative to experiment, finite differences have been used to calculate the temperature history of fire-exposed protected steel columns [38A]. The time-dependent temperature distribution in multilayer metal oxide—semiconductor structures with internal heat sources was determined by an implicit numerical formulation [59A]. A Laplace-transform solution for the transient thermal behaviour of a sphere situated in a large surrounding medium was evaluated for liquid sodium entrained in UO_2—the motivation being a hypothetical accident in a fast breeder reactor [19A]. For a UO_2 fuel rod with zirconium cladding, the conjugate heat-conduction problem for the rod and cladding was solved by Laplace transforms, with account being taken of non-ideal contact [57A].

Other studies of transients were concerned with methodology as well as with potential applications. A finite-difference technique using data from a single embedded thermocouple was added to the arsenal of methods available for solving the inverse problem of transient heat conduction [69A]. Another recently published method is based on iterating the analytical solution which relates the time varying temperature at the internal measurement point with the Biot number corresponding to the unknown surface heat-transfer coefficient [42A]. The Kirchhoff transform followed by a series expansion enabled the unsteady heat equation to be solved for the case of variable thermal conductivity [63A]. Variable thermal conductivity effects in transient conduction can also be dealt with via an iterative scheme which is based on the (separation of variables) series solution for the constant property case [41A]. The variational embedding method enabled an approximate analytical solution for transient conduction in a semi-infinite solid with variable thermophysical properties [10A]. Upper bounds have been found for transient conduction in a rod, one end of which exchanges heat by radiation (i.e. nonlinearly) with the surroundings [49A]. The temperature response of a cylindrical surface to periodic time variations of the surface heat flux was determined via Duhamel's integral [20A]. Periodic steady state solutions can be obtained directly, without having to determine the decaying initial transient which precedes the onset of the periodic steady mode [45A]. Transient heat-conduction problems with special classes of source terms can be subdivided into a source-free transient plus a steady state part [44A]. Optimal control theory was employed to find the heat input to the face of a semi-infinite solid in order to keep the face temperature close to a prescribed value [47A].

Several papers dealing with steady conduction have been published. The heat conducted through a stack of parallelograms separated by thin partition walls increases as the parallelograms become more inclined [16A]. For fluid flow through a tube or over a wire with a heat-transfer coefficient which depends linearly on the wall temperature, axial conduction in the tube wall can combine with the heat-transfer coefficient variation to produce nonuniform steady-state axial temperature distributions [65A]. The thermal signature on an exposed surface due to a buried spherical heat source was determined with the aid of a transformation to a bispherical coordinate system [60A]. Exact and approximate solutions were obtained for a buried point or line source for the case of convective exchange at the exposed surface [43A]. Results for the steady-state temperature distribution in a solid cylinder moving in the direction of its axis through a crossflow of hot gas are applicable to the moving wire technique used to measure the heat-transfer characteristics of flames [32A]. The conformal transformation of an eccentric annulus into a concentric annulus facilitated steady-state heat-conduction solutions in the former [33A].

Fins continue to evoke some interest. The assumption of a uniform fin base temperature equal to that of the unfinned surface can lead to significant errors in the predicted fin heat-transfer rates [62A]. Optimum fin dimensions corresponding to internal heat generation have been determined for the rectangular fin [6A] and for the circular fin [50A]. To handle the time-varying temperature distribution in a radiating and convecting fin, a numerical procedure was employed whereby the spatial derivatives are represented by finite differences whereas the time derivatives are retained as such [11A]. Use of the theory of dual extremum principles led to approximate solutions for fins with variable heat-transfer coefficients and with radiative exchange and variable thermal conductivity [3A].

The complexity of conduction in anisotropic materials has called forth powerful mathematical methods as well as numerical solution procedures. Green's functions were used to good effect to solve for the transient and steady heat conduction in an anisotropic medium of a monocline system which is homogeneous in a circular cylindrical coordinates [14A, 15A]. Furthermore, Green's functions were also found to be effective for anisotropic media which are homogeneous in rectangular coordinates [13A]. A method which satisfies the boundary conditions in the least-squares sense was proposed for analyzing three-dimensional conduction in anisotropic solids of arbitrary shape [46A]. Finite difference solutions simulating the microwave freeze-drying of food underscored the importance of the anisotropic character of the material [4A].

Various aspects of thermal contact have been studied. When stacks of thin layers of tin-coated steel were compressed, the thermal resistance was inversely proportional to the compression per layer for freshly assembled stacks throughout the loading range corresponding to macroscopic conformity [2A]. The thermal constriction resistance of nonsymmetric contact areas is less than the corresponding resistance for circular contact [73A]. A correlation of all presently available data enables the prediction of thermal contact conductance for stacks of thin metallic layers in high vacuums [1A]. With a view to determining the constriction resistance, solutions have been obtained for the transient response of two semi-infinite bodies at initially different temperatures which communicate thermally through a small circular contact area [55A]. Measured contact conductances were compared with predicted values for cones pressing against plane surfaces [40A].

Numerical solution methods have already been mentioned in prior paragraphs of this section, and other numerically oriented papers will now be cited. Many algorithms used in transient heat conduction which are unconditionally stable for linear problems lose this property when applied to nonlinear problems [31A]. In predicting temperatures with an implicit finite difference method, there is a tendency for the surface temperature to oscillate if a non-linear radiation boundary condition is imposed; the oscillations were eliminated by an iterative procedure [70A]. The Crank–Nicolson noise encountered in the stiff equations resulting from the finite element discretization of the heat equation can be reduced by proper choice of the forcing function [71A]. New coarse-mesh methods for the numerical solution of steady conduction problems were developed by use of local Green's functions which convert the original differential equation into local integral equations defined in subregions of the original physical system [30A]. A square-grid finite difference procedure for solving Poisson's equation in a rectangle having a non-rational side ratio uses multi-point difference equations [52A]. A hybrid method based on the linking of a resistance–capacitance network and a digital computer has been proposed for solving problems governed by the nonlinear diffusion equation [12A]. The method of weighted residuals, which is used for obtaining approximate solutions for steady and transient heat conduction problems, has been augmented by the incorporation of Lagrange multipliers for satisfaction of both initial and boundary conditions [64A].

Theoretical work has continued on the hyperbolic heat equation and on materials with memory. Whereas the hyperbolic heat equation effectively explains the finite propagation velocity of heat waves, another explanation has been suggested on the basis of the finite time required for the molecular distribution function to approach equilibrium [9A]. Use of the method of characteristics for solving the hyperbolic heat equation enables greater flexibility with regard to boundary conditions and material properties than do available analytical solutions [68A]. The hyperbolic heat equation has been modified to accommodate distributed heat sources and change of phase [27A]. The conduction boundary-layer concept was used in the investigation of non-Fourier effects in solids having surface curvature [34A]. Non-Fourier effects in the melting of a semi-infinite solid were found to be confined to very small times and to a very thin layer near the surface [53A]. The mathematical attributes of linear and non-linear integro-differential equations for heat flow in materials with memory have been explored [7A, 39A].

CHANNEL FLOW

Recent studies of heat transfer in ducts reflect a strong interest in complex flow configurations, in fluids containing additives and particulate material, in variable properties, and in the freezing of flowing liquids. The nature of the thermally developed regime was re-examined, and new correlations have been proposed.

There are four types of thermal boundary conditions that are compatible with the existence of a thermally developed regime in constant area ducts. They are: (a) uniform wall temperature, (b) uniform wall heat flux, (c) convective heat transfer from the external surface of the duct and (d) axial exponential variation of the wall heat flux. The Nusselt numbers for uniform wall temperature and uniform wall heat flux are bounds for the Nusselt numbers for external convection. Furthermore, cases (a)–(c) are special forms of (d) [52B]. Generalized concepts of fully developed flow and heat transfer have been formulated for ducts of periodically varying cross section. The new concepts were illustrated by application to laminar flow and heat transfer in a heat-exchanger configuration consisting of successive ranks of plate segments placed transverse to the flow direction [38B]. A single correlation for the pipe flow Nusselt number, encompassing all Reynolds numbers (i.e. laminar, transitional, and turbulent) and all Prandtl numbers has been proposed by Churchill and compared with available experimental data [8B]. All the available turbulent heat-transfer results for tubes and channels having roughness elements in the form of regularly repeated ridge-like protrusions perpendicular to the stream direction have been brought together and correlated satisfactorily [24B].

A wide variety of complex flow configurations have been dealt with. Experiments showed that in the low-Reynolds-number turbulent regime, corrugation of a channel wall caused large increases in the heat transfer coefficient compared to a parallel plate channel, but only moderate increases occurred in the laminar regime [17B]. The wake behind spherical blockages in a turbulent pipe flow has a major effect on the downstream heat transfer at the pipe wall [28B]. Measured heat-transfer coefficients for a crossflow cylinder situated in a turbulent pipe flow are substantially larger than those for a crossflow cylinder in an unbounded low-turbulence flow [39B]. Mass-transfer experiments in laminar channel flow downstream of a partially constricted inlet indicated the presence of separation zones of different length on the respective walls. In addition, downstream of the shorter separation zone, a previously undiscovered region of high mass transfer was encountered [51B]. For radial outflow in a passage formed by two parallel disks, the heat transfer at the two disks differs because of flow separation adjacent to the inlet aperture on one of the disks [18B].

Solutions for laminar flow and heat transfer in channels whose walls are interrupted periodically along the stream direction demonstrated that such passages can provide superior heat-transfer performance when compared to a conventional parallel plate channel [50B]. Finite difference solutions have added laminar Nusselt numbers for I-shaped ducts to the available store of fully developed results [37B]. The Schwarz–Neumann technique enables fully developed laminar flow and heat-transfer results for complex duct shapes to be obtained from available series solutions for simpler shapes [59B]. Numerical solutions for laminar flow and heat transfer in internally finned helical coils indicate that $f\,Re$ is a function of both the Dean number and the radius of curvature, whereas the Nusselt number is a function of Dean number alone [32B]. The gain in heat transfer and the penalty in pressure drop have been measured in airflow for a longitudinal fin with various augumentation treatments which subdivides a rectangular channel into two parts [56B]. With the gas-cooled reactor as the motivating application, a transformation has been developed to generalize heat-transfer and friction data for a single roughened rod contained in a concentric smooth tube to predict results for a cluster of roughened rods [9B]. Flow acceleration caused by the sloping wall of a duct brought about the laminarization of an initially turbulent flow [55B].

Investigations were performed of liquid flows with additives, non-Newtonian and fluids with microstructure, and particle-laden gas flows. Predictions of heat transfer in drag reducing viscoelastic fluids in turbulent pipe flow were based on generalizations of Reichardt's model [25B] and of Cess' model [14B]; in both cases the predictions agreed well with experimental data. Electrochemical mass-transfer techniques were used to study the effects of drag reducing additives in a turbulent pipe flow. At a given volumetric flow rate, the percent change in the fully developed mass transfer was greater than the percent change in the pressure drop [33B]. From numerical solutions for laminar flow of a non-Newtonian power-law fluid in a square duct, it was found that when the non-Newtonian flow behaviour index is less than one, the heat-transfer coefficients are higher than those for a corresponding Newtonian flow [6B]. The laminar Nusselt number for fluids with microstructure depends on the ratio of the rotational viscosity coefficient to the shear viscosity coefficient [27B]. Droplet evaporation augments the laminar heat-transfer coefficients in the thermal entrance region of a parallel plate channel [3B]. The turbulent heat-transfer characteristics of particle-laden gas flows in ducts were explored in a series of papers that appeared in *Heat Transfer, Soviet Res.* [13B, 36B, 46B, 53B, 54B].

Variable properties and supercritical fluids have been studied. Numerical solutions for laminar flow of inert gas mixtures in a uniformly heated tube indicated that for design predictions, the property ratio method is better than the film temperature method for heat transfer while the latter method is preferable for apparent wall friction [34B]. The temperature dependence of the rheological properties can have a significant effect on heat-transfer rates and pressure drop when viscous fluids are heated or cooled in tubes [40B]. The implications of variable conductivity and specific heat on the correlation of turbulent pipe-flow heat-transfer results have been examined, but the assumptions of constant viscosity and Prandtl number limit the practical utility of the findings [21B]. Numerical solutions for turbulent pipe flow of supercritical helium gave higher heat-transfer coefficients at higher heat fluxes and when the inlet bulk enthalpy was above the pseudocritical value [44B]. On the other hand, measured heat-transfer coefficients for supercritical helium in both vertical upflow and downflow agreed with predictions of the Dittus–Boelter equation to within 20% [5B].

A number of papers with the freezing of flowing liquids. When a superheated liquid in laminar flow in a parallel plate channel is subjected to convective cooling at the walls, the thermal entrance region consists of a solidification-free zone followed by a zone where a solid grows inward toward the center of the channel [7B]. An analysis of the steady state freezing of a turbulent liquid metal flow adjacent to the cooled walls of a parallel plate channel led to a coupled convection–conduction problem for the liquid core and the frozen layers [47B]. To solve the related problem of transient freezing in a tube for a liquid initially at its fusion temperature, it was assumed that the turbulent velocity profile is adequately represented by the slug flow profile, the Blasius friction factor formula is applicable, and the shape of the frozen layer can be represented by a simple functional form [12B]. The latter assumption was subsequently lifted by solving an integro-differential equation [11B]. A variant of the tube or channel solidification problem is the freezing which occurs when a cold continuous sheet (or wall) is caused to move through a liquid environment [29B]. Dendritic ice forms in a pipe when there is no mainflow through the pipe during the freezing process. This ice form occurs because the quiescent water supercools considerably below 0°C before ice nucleation begins [15B]. Studies were made of the effect of this ice form on the pressure gradient required to start flow in a partially frozen pipe. It was found that under some conditions, the growth of dendritic ice in a pipe can result in a flow blockage when only a small fraction of the water in the pipe is frozen [16B].

Some interest still remains in axial conduction. An analysis of laminar flow in a parallel plate channel with uniform but unequal temperatures at the bounding walls shows that axial conduction at low Peclet numbers can cause a considerable increase in the thermal development length [57B]. Axial conduction in the channel walls bounding a turbulent flow was found to have heightened effects at low Prandtl numbers [43B]. Axial diffusion was also investigated for mass transfer in laminar pipe flow with chemical reaction, but no account was taken of possible diffusion upstream of the inlet cross-section [10B].

The effects of flow pulsations were investigated. The time-averaged fully developed Nusselt number for laminar pulsatile flow in curved pipes is larger than that for pulsatile flow in straight pipes [49B]. For pulsatile laminar airflow in a tube, low frequency pulsations caused a decrease in the heat-transfer coefficient whereas high frequency pulsations brought about an increase [31B]. Pulsatile laminar flow and heat transfer in a parallel plate channel with uniform injection and suction at the respective walls was studied with reference to the dialysis of blood in artificial kidneys [41B].

Several papers covered a miscellany of pipe flow studies. When the heating rate at the wall of a turbulent pipe flow is rapidly increased from an initial value to another value, the time variation of the fully developed heat-transfer coefficient takes on a maximum value before attaining steady state [26B]. The turbulent Graetz problem for a parallel plate channel was solved for Prandtl numbers between zero and 10^4. For Prandtl numbers greater than 0.1, the fully developed Nusselt numbers based on the hydraulic diameter agreed well with those for the circular tube, with lesser agreement at lower Prandtl numbers [48B]. The presence of constant heat sources and frictional energy dissipation in turbulent pipe flow reduces the rate of heat transfer [1B]. Laminar flow experiments in a heated tube with stepped streamwise variations of heat flux yielded Nusselt numbers that deviated by about 10% from analytical predictions [42B]. Local turbulent heat-transfer coefficients for the thermal entrance region of a pipe were measured for air flow under constant wall temperature conditions [30B].

General topics in laminar flow continue to be analyzed. An integral momentum–energy treatment of the simultaneously developing laminar velocity and temperature fields in a parallel plate channel employed a stronger integral constraint for the velocity, but the Nusselt numbers are very little affected [4B]. Extensive tabulations of local Nusselt numbers have been made available for laminar channel flow with finite external resistance, with [22B] and without [23B] magnetohydrodynamic effects. A method has been evolved for determining the wall temperature distribution required for the attainment of a uniform bulk temperature in a laminar pipe flow having temperature-dependent heat generation [35B].

Mass transfer in pipes, with special emphasis on high Schmidt number fluids, has been studied both experimentally and analytically. Experiments for turbulent mass transfer in pipes at high Schmidt number (1000–6000) indicate that neither the Colburn equation nor the Dittus–Boelter equation are good representations of the data in that range of Schmidt (or Prandtl) number [2B]. Other experiments encompassing Schmidt numbers up to 37 000 have established the dependence of the mass-transfer coefficient on the Schmidt number [45B], and this dependence has been affirmed by an algebraic turbulence model [19B]. The spreading of solute that has been injected into fully developed laminar pipe flow is governed by both diffusion and convection [20B, 58B].

BOUNDARY LAYER AND EXTERNAL FLOWS

Theoretical and experimental investigations of laminar and turbulent boundary layers form the bulk of the studies in this category. Attention is also given to free-stream turbulence, impinging jets and flow over a cylinder.

The conjugate heat-transfer problem involving the laminar boundary layer on a plate of finite thickness is studied by the use of Lighthill's method of boundary-layer analysis [24C]. The laminar flow and heat transfer on a flat plate situated in the laminar wake of an upstream plate is analyzed by a numerical method [27C].

Among the theoretical analyses of turbulent boundary layers, an approximate local analysis is performed to solve the integral momentum equation for two-dimensional compressible boundary layers [17C]. Numerical solutions have been obtained for a turbulent boundary layer with variable physical properties of the fluid [31C]. The predictions are made by using a mixing length model and are compared with experimental data. Analytical expressions have been obtained for the temperature distribution in terms of the local velocity for turbulent boundary layers with non-unity Prandtl numbers [35C]. A theoretical analysis has been reported for heat transfer in turbine blades with longitudinal cooling slots [4C]. A formula is developed for the friction factor in turbulent flow of liquid films over vertical surfaces; then, by the Reynolds analogy, an expression for heat transfer for the same situation is obtained [5C]. A numerical solution of the flow and concentration field has been presented for a point source at the ground level of a turbulent shear layer. The effective viscosity is governed by one differential equation. The computed results are shown to agree well with available experimental data [25C]. Mass transfer between a turbulent flow and a wall with first-order chemical reaction is analyzed for large Schmidt numbers [14C].

An experimental study of the velocity and temperature fields is reported for the thermal layer that grows inside a turbulent boundary layer which is subjected to a step change in the wall heat flux [3C]. An interferometric investigation has been made of the temperature field created by a cold air jet issued inclined to an adiabatic flat plate in hotter ambient air [11C]. Detailed measurements of the turbulent flow field have been reported for a subsonic compressible turbulent boundary layer. These show that the ratio of the Reynolds stress to the turbulence kinetic energy depends on the Mach number [1C]. Reference [34C] presents an experimental study of a thick hypersonic turbulent boundary layer on a cooled wall. Experimental data on high-temperature free jets produced by burning the methane gas have been used to confirm a proposed extension of Prandtl's eddy diffusion model [15C]. Heat transfer from developing water film over cylinders has been experimentally investigated [23C].

Many papers have dealt with unsteady boundary layers. An anlysis has been made of the unsteady heat transfer from a plate in an incompressible flow [8C]. In another unsteady-heat-transfer study, the finite-difference solution of an integro-differential equation is obtained for a laminar boundary layer over a plate of finite thermal capacity and thermal resistance [32C]. The laminar thermal boundary layer in response to harmonic oscillations in velocity imposed on the free-stream velocity is analyzed. It is shown that the heat transfer is reduced for Prandtl numbers less than unity, and increased for Prandtl numbers greater than unity, due to the presence of the oscillatory free stream [30C]. Unsteady heat transfer in a thick thermal boundary layer over a spherical surface in random motion is studied in terms of the inner and outer regions of a boundary layer by a method of successive approximations [36C]. Approximate solutions using integral methods and the method of characteristics have been obtained for laminar boundary layers with a steady velocity field and a transient temperature field caused by a sudden change in the wall temperature [29C]. An analysis has been made for the transient heat and mass transfer at the surface of a dielectric drop in another dielectric liquid under the action of an electric field, for the low Reynolds number, high Peclet number situation [22C]. Heated laminar boundary layers have been experimentally studied for their stability characteristics by subjecting them to linear sinusoidal disturbances created by a vibrating ribbon [28C].

Among the investigations of the effect of free-stream turbulence, a laminar boundary layer is experimentally studied to find the influence of the free-stream turbulence on the velocity profile, the drag coefficient and other quantities [9C]. An analysis is performed for the effect of free-stream turbulence on heat and momentum transfer to axisymmetric bodies in supersonic and hypersonic turbulent streams. The turbulence is characterized by the turbulence kinetic energy (obtained from a differential equation) and a length scale (prescribed algebraically) [33C]. An experimental study is reported of the local heat-transfer coefficients influenced by the free-stream turbulence in a flow over a cylinder around the critical Reynolds number [37C].

Several investigations are directed towards the impinging-jet situation. A similarity solution has been obtained for steady laminar compressible three-dimensional stagnation-point flow with variable transport properties [20C]. An analysis is reported for the heat transfer in axisymmetric stagnation flow on an infinite circular cylinder for the constant wall temperature and constant wall heat flux boundary conditions [10C]. Experimental data have been obtained for the heat transfer from a continuously moving belt to an air jet impinging normally. The maximum heat-transfer coefficients are about 1.5–2.0 times higher than those predicted for the stationary surface [26C]. Mass transfer from a flat surface due to an impinging free jet from a rectangular duct has been measured by laser holographic interferometry [19C]. An experimental study has been made of the heat transfer from impinging circular jets to investigate the effect of the intensity of turbulence [13C].

The flow over a circular cylinder continues to be a subject of considerable interest. A single comprehensive equation is developed for the rate of heat transfer from a cylinder in cross flow covering the entire range of the Prandtl number and the Reynolds number [6C]. A correlation is presented for the recovery factor for the compressible flow of air over an infinite cylinder placed normal to the flow [12C]. From experimental data for drag on spheres and cylinders, the effective Reynolds number based on the Eckert reference temperature is found to correlate the drag data well [18C]. The heat- and mass-transfer coefficients to disks and cylinders in cross flow have been found from new experiments and compared with earlier results and available correlations [16C]. A model for the eddy diffusivity induced by free-stream turbulence is constructed and applied to the stagnation point of a cylinder in cross flow. The predicted Nusselt number and friction factor agree well with available experimental results [21C]. An experimental investigation of the heat transfer from a circular cylinder in cross flow has been reported, in which pyramidal roughness elements were used on the cylinder surface. The transition to turbulence has been determined as a function of the Reynolds number and a roughness parameter [2C]. An analysis has been presented for the mass transfer from a circular cylinder performing harmonic oscillations in a fluid at rest [7C].

FLOW WITH SEPARATED REGIONS

Single bodies

A systematic, unified, comparative analysis is made [9D] of all possible similitude rules pertaining to incipient separation-viscous interaction for both laminar and turbulent flows. A new regime of heat-transfer augmentation [24D] occurs at the walls of a channel with a partially constricted inlet; it occurs on the side washed by the short stall and is downstream of both the short and long stall reattachments. The laminar boundary layer and separation has been studied [28D] for a steady outer inviscid flow over an upstream moving wall. Consideration is given [23D] to the nature of the separation and subsequent reversed flow occurring when a jet-like boundary layer on a wall encounters a concave corner of finite angle. Two papers [18D, 19D] give analyses of the effects on the thermal boundary layer of the separation–reattachment zone that is induced in a supersonic flow by a slight hollow in the surface of a flat plate. Three-dimensional heat- and mass-transfer effects across high-speed reattaching flows are analyzed on the basis of a compressible, small disturbance flow model of the local vortex instability mechanisms involved [8D]. An experimental study of the turbulent near wake of a flat plate is given [1D].

Packed and fluidized beds

Much of the data for heat transfer from packed beds includes a length effect resulting from higher transfer coefficients near the inlet [13D]. The effective thermal conductivity of sheared suspensions is discussed [16D]. New dryout heat flux data for inductively heated particulate beds have now been obtained and semi-theoretical correlations suggested [4D]. Heat-transfer data in flowing, packed beds [3D] closely fit the predictions of the Mickley model which includes the role of the circulating particles. Půst [20D] analyzes sintered heat exchangers for throughflow helium cryostats. Concentration profiles are used to obtain actual driving forces in packed beds for $(0.1 < Re < 100)$ in order to evaluate axial and radial dispersion contributions to mass-transfer factors [7D]. Two papers [29D, 30D] describe and give a critique on the determination of packed bed particle-to-fluid heat-transfer coefficients from steady, radial measurements. Another paper [12D] distinguishes between two types of temperature excursions in adiabatic packed bed reactors; the hot spot and the unstable temperature runaway. A numerical analysis scheme is developed [10D] for the heat transfer in packed bed recuperators.

Much potentially useful energy is wasted because it is carried away by effluent gases containing contaminants which cause rapid fouling. A solution to this problem [21D] might lie in the use of a falling cloud of dense particles to extract heat in combination with a fluidized bed to remove the particle heat. Shallow fluidized beds have good potential as heat exchangers since the pressure drop is low enough not to require the use of large fans [2D]. McGaw [14D] describes the development of a mechanism for gas-particle heat transfer in shallow fluidized beds on the basis of experiments. A two-resistance model for heat transfer in aggregative and particulate liquid-fluidized beds [17D] agrees well with data. Mass-transfer data in liquid fluidized beds are in disagreement with some recent theories that predict an important decrease in mass-transfer rates at low Reynolds number [27D]. Experimental results are given [11D] for three phase fluidization in which solid particles are fluidized by the cocurrent flow of liquids and gases. The magnetic field was found [22D] to be an important parameter governing the fluidized bed structure when the solid particles are ferromagnetic. The application of an alternating electric field to cause movement of particles charged naturally in a fluidized bed has been demonstrated to increase heat transfer up to 140% for frequencies up to 100 Hz [5D].

The world-wide interest in various fluidized bed combustor systems is surveyed [6D, 25D]. In [15D] fluidization and gas combustion are studied in a rotating fluidized bed with centripetal acceleration of ten times gravity. Fluidized bed combustion is a possible method for providing the heat source for a Stirling engine because it is nearly isothermal at low temperatures and high heat flux rates [26D].

TRANSFER MECHANISMS

Various aspects of turbulent flows have been dealt with in papers on transfer mechanisms. These include detailed measurements, turbulence models and other theoretical analyses.

An experimental study reports measurements of the structure of isotropic turbulence generated by a water fall of low head; the measurements are made by the use of the hot-film technique [16E]. Hot-wire anemometer data are reported on some aspects of the fine-scale structure of turbulence in grid flows, in zero-pressure-gradient boundary layers, and in relaminarizing boundary layers [5E]. The spectral energy transfer of turbulent velocity fields has been examined over a wide range of Reynolds numbers by experimental and empirical methods [12E]. Measurements have been made of the turbulent flow in a rectangular duct of 12:1 aspect ratio using conditional-sampling techniques [7E]. Another rectangular-duct flow has been investigated by the use of a laser-Doppler anemometer; the axial mean velocity and turbulent intensity were measured in the developing flow, and all three mean velocity components and five of the six Reynolds stresses were obtained in the nearly fully developed flow [18E]. An experimental study of turbulence field in the near-wall region of a circular pipe reports the fluctuations of temperature and velocity, the length scales, and other correlations. The flow is found to be intermittent even within the viscous sublayer [9E]. Two-point turbulence data from hot and cold wire anemometry are presented which describe the temperature and velocity fields downstream of a heated grid in a low-speed wind tunnel [24E]. Statistics of temperature fluctuations are measured in a slightly heated boundary layer well downstream of a step change in wall heat flux [2E]. Reference [26E] reports the three components of the dissipation of the fluctuating temperature measured simultaneously in the inner region of a fully developed turbulent boundary layer. In a related study [3E], the probability density of the dissipation function is found to be log-normal even at relatively small Reynolds numbers. The eddy diffusivity near the free surface in an open channel flow has been determined by experiments around a Prandtl number of 3, and a model has been proposed for the calculation of the variation of the eddy diffusivity [27E]. Measurements have been presented for the influence of the Schmidt number on the frequency of the mass-transfer fluctuations at a wall in a turbulent flow [25E]. The effect of a drag-reducing additive on the structure of wall turbulence in a pipe flow has been investigated by hologram interferometry [1E]. Turbulence measurements in a rough-walled pipe flow have been reported and analyzed in a particular manner to show that the turbulence structure in a rough pipe follows the same scaling laws as for a smooth pipe [21E]. Hot-film anemometer measurements in fully developed pipe flow have been described; a procedure has been developed for identifying "bursts" in the wall layer [10E, 11E]. Artificially initiated turbulent spots in a Blasius boundary layer were investigated using hot-wire anemometers; the spots were generated by electric discharges [30E].

A number of papers are concerned with various types of turbulence models. For three-dimensional turbulent boundary layers, some simple algebraic expressions have been proposed for the eddy viscosity in the outer region [19E]. The predicted mixing lengths in the outer region of a turbulent boundary layer have been compared with those derived from velocity profiles [29E]. Various similarity theories for the atmospheric boundary layer have been critically examined in the light of numerical results and recent observations [4E]. A method for calculating turbulent diffusion in the planetary boundary layer is developed on the basis of Donaldson's second-order closure. The predictions of the model show good agreement with the laboratory data for diffusion in a free convection mixing layer [14E]. Detailed examination and comparison of four two-equation turbulence models have been made. The models have been applied to equilibrium boundary layers under various pressure gradients and to flow over a convex wall [6E]. Similarity solutions have been obtained for various free turbulent flows and duct flows by the use of a two-equation turbulence model; a set of universal constants has been obtained which gives satisfactory agreement with experimental data [28E]. For the general case of turbulent plane flow, local equilibrium solutions for a Reynolds-stress-transport turbulence model have been presented. The local equilibrium implies that the production of the stress balances its dissipation [20E]. Reynolds-stress models of turbulence are examined with reference to realizability conditions. It is shown that certain existing models do not guarantee realizable solutions. Suggestions have been made for changing the existing models into realizable ones [23E].

Among more fundamental studies of turbulence, a kinetic theory of turbulent flows is further developed [15E]. A discussion of the mechanism of entrainment in turbulent free shear flows concludes that entrainment is the folding of the turbulent and nonturbulent fluids by the rotation of the large eddies [8E]. The return to isotropy of homogeneous turbulence is considered in a theoretical investigation. The main assumption is that the changes are slow relative to turbulence time scales. Constraints on the governing functions are found by theoretical arguments. A convenient form is proposed for the functions, and it is shown that the predictions based on these are in excellent agreement with all available data [17E]. An estimate of the coefficient of self-diffusion is derived from kinematical consideration [22E]. A non-Newtonian viscosity is obtained for macromolecular liquids from nonlinear constitutive laws. The viscosity is found to decrease with increasing magnitude of the velocity gradient [13E].

NATURAL CONVECTION

Considerable research activity in natural convection is evident with a large number of analytical, numerical and experimental works considering various problems in pure natural convection and combined natural and forced convection both in internal flows and in boundary layer flows. As has been the case for some years now, there is much interest in heat transport in flow in plane layers heated from below. There seems to be more work being done on buoyant plumes. Considerable research is in evidence on natural convection in porous media.

An analysis presents lower bounds for the onset of instability in buoyancy-driven convection following an impulsive change in surface heat flux [106F]. Convection in a low Prandtl number fluid, in the limit as Prandtl number approaches zero, has been analyzed in a circular geometry [86F]. The effects of surface tension on the onset of flow in a fluid layer over an ice surface has been examined [91F]. A numerical analysis of Benard convection in a two-component fluid includes the influence of the Soret effect [83F]. The effect of sidewalls in Bénard convection with roataion has been examined [29F]. A numerical solution of highly truncated modal equations yields the flow and temperature distributions over a range of Rayleigh number and Prandtl number for a horizontal layer heated from below [102F].

Natural convection in a volumetrically heated horizontal fluid layer with an adiabatic lower boundary has been examined at high Rayleigh number [20F]. In experiments using a similar geometry, the transient as well as the steady state heat transport are determined [55F]. The heat transfer and stability in volumetrically heated layers are analyzed with heating from below [23F].

The stability of longitudinal rolls in inclined plane layers has been analyzed [24F]. The reduction of natural convection heat transfer with the introduction of square and hexagonal structured honeycomb panels in inclined layers has been reported [8F]. A related study finds the ability of rectangular honeycomb to inhibit natural convection improves with tilt from the horizontal [1F].

The heat transfer and temperature distribution in an open tank of water heated from below have been measured [52F]. The heat transfer for moderate Rayleigh number convection of liquid metals in enclosures of different geometries has been calculated [28F]. The stability of a fluid in an inclined slot with simultaneous diffusion of heat and mass has been studied [13F]. Numerical predictions [80F] agree with flow visualization [81F] of the natural convection patterns in an inclined rectangular box heated from below.

A transverse temperature gradient influences the flow field in a low Prandtl number fluid heated from below [99F]. The temperature fields and heat transfer have been obtained for convection in an enclosure with heating on a side wall and/or on the bottom wall [44F]. A numerical study of natural convection in a box with differential side heating predicts three-dimensional motion [60F]. Finite element analysis has been applied to free convection in enclosures [100F].

An analysis of heat transfer by natural convection across vertical layers has been given over a range of conditions from conduction to turbulent flow [88F]. The effect of nonlinear variation of density with temperature in natural convection flows between two parallel vertical walls has been examined [104F]. A correlation has been obtained for the Nusselt number following the initiation of transient convection in vertical enclosures [4F]. An earlier analysis of convection between differentially heated side walls has been reconsidered [87F]. The finite element method has been used to study the natural convection flow between two vertical parallel plates [78F]. The presence of a solute can influence the heat transfer throughout a fluid in an enclosure with differentially heated side walls [108F]. A linearized solution of the natural convection in a vertical open tube has been presented [62F]. Unsteady natural convection in a vertical cylindrical cavity has been examined [68F].

Natural circulation within a horizontal pipe from one end has been described [45F]. Convection of low Prandtl number fluids within a horizontal cylindrical annulus gives a flow different from that found at higher Prandtl number; the center of rotation moves downward at higher Rayleigh numbers in the low Prandtl number fluid [27F]. Different flow regimes are observed over a range of Grashof and Prandtl numbers in the region between a sphere and its cubical enclosure [85F]. Different modes of motion are also described for convection in a hemispherical cavity tipped at various inclinations to the horizontal [75F]. Convection in helium I has been studied to simulate the heat transfer in rotating super-conducting windings [10F].

Interest in various stages of convection within porous media have expanded considerably this year. These include studies related to stability, developed convection, and applications. A linear stability analysis indicates the onset of convection in a fluid layer overlying a layer of porous media filled with the same fluid [77F]. Convective instability in porous media has been compared to that in a fluid with a concentrated suspension that is heated from below [105F]. Temperature dependent viscosity and thermal coefficient of expansion have been included in determining the critical Rayleigh number in an elastic porous matrix [67F]. The importance of aspect ratio has been demonstrated in determining the critical Rayleigh number in a water-saturated porous medium [111F]. The onset of flow and its characteristics in an anisotropic porous layer have been examined [34F]. A finite amplitude analysis treats thermal convection in a horizontal porous layer with internal heat sources [103F]. Another study finds anisotropy has little effect on natural convection in a porous layer other than in a nonhomogeneous stratified layer [11F]. Solutions for uniform heat flux to a fluid in a porous layer with natural convection are reported [15F]. A pH indicator method is used to visualize the flow in a Hele–Shaw cell to model thermal convection in porous media [43F].

Measurements of convection in a vertical slot filled with porous insulation indicate the heat transfer in common insulated building walls can be dominated by wall leakage [6F]. The effect of lateral mass flux in vertical free convection in a porous medium has been examined [17F]. A perturbation analysis is used to study natural convection flows in geothermal reservoirs [56F].

Numerical procedures are used to calculate natural convection from a confined heated horizontal plate facing upwards [107F]. Analysis has been applied to natural convection heat transfer from a hot plate facing upwards [2F] and downwards [3F]. The dependence on Grashof number of transitions and different plume modes have been established for natural convection from heated strips on a horizontal surface [5F].

The rough surface at a salt-brine interface yields a higher natural convection mass transfer rate than a smooth surface does [12F]. Thermohaline convection in the form of "salt fingers" is observed in the freezing of a saline solution [41F].

A perturbation technique is used to describe low Grashof number flow around a sphere in a thermally stratified fluid [94F]. Heat transfer from horizontal cylinders to different fluids is represented by somewhat different correlations depending on the choice of the reference temperature for evaluating the fluid properties [37F]. Laminar convection from horizontal cylinders of elliptical cross-section has been analyzed [63F]. The natural convection around a horizontal cylinder of ice with a density inversion in the surrounding fluid has been studied both theoretically and experimentally [89F]. The natural convection around a suddenly heated cylinder has been described [42F].

A number of studies relate to flow and temperature distribution in buoyant plumes both in the atmosphere and in the laboratory. Temperature and velocity profiles were measured in an axisymmetric buoyant plume [39F]. Velocity measurements in a plume above a horizontal line source were 20–25% smaller than that predicted by analysis [75F]. Prediction of turbulent plumes issuing into stratified or uniform ambients includes the effect of buoyancy on the turbulent transport model [59F]. The plume above a falling point source has been analyzed [26F].

The structure of free convection in the atmosphere near the ground surface is modeled as a field of plumes rising regularly from the surface [61F]. The initial Froude number is important in determining the entrainment characteristics of buoyant plumes in atmospheric inversions [35F]. A companion study considers turbulence properties to get entrainment into a plume [36F]. The release of thermal plumes appears to dominate the flow in the atmospheric boundary near the ground surface [101F].

A simple turbulence model is used in a numerical analysis [58F] to describe the curved trajectory of a buoyant jet. Another model is used to predict the trajectory and dilution of a buoyant jet in a cross flow [109F]. A model of a surface buoyant jet that enters normal to the stream is used to study the flow from the effluent of a power plant [74F]. A detailed study of the flow of a vertical plane turbulent buoyant jet includes measurements of the mean properties [54F] and the turbulent structure [53F] of the jet.

A technique is developed to predict the natural convection on a vertical plate with arbitrary wall temperature or heat flux [51F]. The effects of a very long vertical plate on laminar convection have been considered [73F]. The temperature and flow fields above a local heat source on a vertical adiabatic wall have been measured [9F]. An approximate solution considering variable gas properties has been obtained for laminar convection with uniform heat flux from a vertical plate [69F]. The effect of stratification in the surrounding fluid on the heat transfer from a vertical wall has been examined [49F]. An analysis of the flow arising from a line thermal source at the leading edge of a vertical adiabatic plate has been described [47F].

Two studies consider the transient natural convection from a vertical plate. In one [65F], the influence of variable properties is examined. The other considers time-dependent surface temperature or heat flux [66F].

A turbulence model has been applied to predict heat transfer from a vertical isothermal surface [84F]. The frequencies for growth of disturbances that influence transition on a vertical plate are found to be determined by frequency filtering [48F].

The natural convection mass transfer following an injection of pure water through a porous plate into an aqueous solution of sodium chloride has been measured [57F]. Unsteady natural convection flow past an infinite vertical plate has been analyzed with constant suction [97F] and with variable suction [98F] at the plate surface. The mass transfer from an upward facing, downward facing and an inclined flat plate has been studied using an electrochemical system [82F]. Experiments yield the temperature and velocity fields with natural convection over a vertical plate with a forward-facing step [46F].

Solutions for natural convection heat transfer from a vertical plate to a power law non-Newtonian fluid include the effects of variable gravity forces [40F]. Natural convection in curly-viscous fluids have been studied for constant wall temperature [92F] and for variable wall heat flux [93F].

Experimental data has been used to develop expressions for optimizing the geometry of fins heating the surroundings by natural convection [33F].

A number of studies consider combined forced and natural convection around a horizontal cylinder. A finite difference scheme was employed to solve equations for such a flow [71F]. For vertical forced flow across a cylinder, an analysis covers the range from pure natural to pure forced convection [70F]. With a vertical free-stream, heating the cylinder is found to delay separation while cooling the cylinder brings the separation point nearer to the stagnation region [64F]. The influence of free stream turbulence on mixed convection in a horizontal forced flow has been examined [90F]. With downward forced flow around a heated cylinder, periodic variations of heat transfer are found [76F]. Equations have been presented [22F] to correlate mixed convection from flat plates, spheres, and horizontal cylinders in the case of aiding flow.

The effect of natural convection Taylor–Görtler vortices along concave walls has been studied [50F]. For a forced flow parallel to an open heated cylinder, the effect of buoyancy on the convection on both the outer and inner surfaces of the cylinder has been analyzed [110F].

Local similarity and non-similarity methods have been used to predict the combined natural and forced convection boundary layer flow over a horizontal plate [14F]. The critical Rayleigh number for convection in a liquid film flowing over a horizontal plate has been determined [31F]. The effects of natural convection on the flow around an impulsively started plate has been analyzed [96F].

Similarity solutions are used to predict the mixed convection from horizontal impermeable surfaces in a saturated medium with both parallel and stagnation flows [16F]. A related study examines solutions in aiding and opposed flows [18F].

Studies on combined natural and forced convection in internal flows include a number of works with flow inside horizontal tubes. In the entrance region the influence of secondary flow [79F] and temperature-dependent viscosity [21F] on heat transfer have been studied. A finite-difference procedure has been applied to fully developed turbulent flow through horizontal tubes [95F]. Viscous dissipation is found to have a significant stabilizing influence on the flow of a high Prandtl number fluid in a horizontal channel heated from below [19F]. The enhancement of heat transfer with combined convection in internally finned square ducts has been demonstrated [38F].

For Coutte flow heated from below, instabilities can occur either as stationary waves or as waves which propagate along the convection rolls [25F]. The modal structure for combined convection with Couette flow has been described [30F].

An experimental study of heat transfer in a vertical tube with forced flow either upwards or downwards has been reported [72F]. A simple correlation for combined convection inside vertical tubes with either aiding or opposing flow has been proposed [32F].

CONVECTION FROM ROTATING SURFACES

Heat transfer in the entrance length of a horizontal rotating tube with laminar throughflow and cooling of the wall was found to be strongly reduced at rotational speeds above 40 per min [9G]. The effect of variable density on the stability of flow between two rotating cylinders was studied [6G]. It was found that instability starts with an oscillatory secondary flow and that Prandtl number has an effect on its onset. The electrochemical technique was used [3G] to study mass transfer in Taylor vortex flow. Axial throughflow was found to reduce the Sherwood number on the outer cylinder by 30–50%. Turbulent heat transfer was measured [5G] for flow through an annulus with inner cylinder rotation. A correlation between rotational and non-rotational Nusselt numbers based on a parameter characterizing the flow helix gave results in excellent agreement with the experiments. Laminar flow and heat transfer in a shrouded cylindrical cavity with a rotating and a stationary disk and a stationary cylindrical wall was analyzed [12G] with the result that heat-transfer coefficients increase with increasing coolant flow, decreasing rotational speed, and decreasing length to diameter ratio. A prediction [14G] of local heat transfer on a rotating disk using a two-equation model of turbulence agrees with experiments within 3%. Heat-transfer rates between a rotating disk and an eccentrically impinging jet were found [10G] essentially independent of jet flow rate in a rotationally dominated regime but to increase strongly with impingement rate in an impingement dominated regime. Fully developed heat transfer for steady, laminar, incompressible flow through a pipe rotating around an axis normal to the pipe axis was analyzed in [15G].

Two papers [4G, 11G] studied local heat transfer from rotating wedge shaped radial blades by an integral analysis and by mass-transfer experiments using naphthalene ablation. The cooling problem of alternators instigated experiments on pool boiling of liquid helium and heat transfer in supercritical helium in acceleration fields of 180–1720 g [13G]. The results could be described by the equation $Nu = 0.29\ Ra^{0.29}$. The desire to understand MHD flow in the earth's interior resulted in two papers [7G, 1G], on the thermal stability in a rotating spherical fluid shell with different distributions of heat sources. An exact solution [2G] of magnetohydrodynamic Couette flow and heat transfer with rotating surfaces indicates that on the moving surface the shear increases but heat transfer decreases with increasing magnetic or rotation parameters. Separation of gas mixtures can be obtained in the Lindstrom–Lang vortex tube. It was found [8G] that maximum separation is connected with a critical Reynolds number.

COMBINED HEAT AND MASS TRANSFER

Research in combined heat and mass transfer continues with considerable emphasis on cooling problems related to high temperature systems such as gas turbine plants. In this respect, considerable emphasis is focused on film cooling with somewhat reduced interest in transpiration cooling. Other efforts in the area of combined heat and mass transfer include the effects of thermodynamic coupling in boundary layers, combined heat and mass transfer in evaporation problems, and the influence of suction on wall heat transfer.

Studies of two-dimensional film cooling in which a coolant is ejected through a spanwise continuous slot concentrated on the effects of high temperature, streamline curvature, and high speed flow. When compared to the film cooling effectiveness on a flat plate, a convex surface yielded better results and a concave surface relatively poorer film cooling performance [11H]. The presence of blades was found to have little effect on the film cooling effectiveness over a cylindrical surface [17H]. For slot injection used to cool plates exposed to a high temperature gas stream, even very simple previously-developed semi-empirical relations could correlate the data well [15H]. Film cooling was found to be much more effective in a rapidly converging nozzle than in a gradually converging one [20H]. In a Mach 6 flow, a series of slots at different positions in the mainstream direction gave a better cooling effect than did a single slot for the same total mass flow rate [6H]. Numerical analysis and experiments on the heat transfer to a wall jet included the effects of injection through a slot as would occur in two-dimensional film cooling [9H]. Detailed measurements of concentration and mass flux in the turbulent boundary layer following slot injection of helium give insight into the processes that would affect two-dimensional film cooling [8H].

In many practical applications, the film coolant would be introduced through a series of discrete holes across the span of the surface to be protected. The characteristics of the flow and cooling effectiveness following injection through a round hole inclined at 30° to the surface, shed light on the film cooling performance very close to injection [2H]. A mass-transfer analogy is used to obtain experimental results on film cooling effectiveness over a large range of density differences between the mainstream and the injected fluid [16H]. The effect of using spanwise-angled holes at the leading edge of a turbine blade to introduce a film coolant has been investigated [5H].

Local non-similar solutions were obtained for the flow and heat transfer in a plane wall jet with mass addition along the surface [4H]. Experiments [3H] and an approximate analysis [19H] indicate the temperature profile and flow in the turbulent boundary layer over a porous wall with suction.

Comparison of the diffusion of mass and heat using the laminar and turbulent boundary-layer equations indicates the feasibility of using heat transfer to simulate mass transfer and vice versa [10H].

The heat transfer and skin friction with mass transfer in the region of a three-dimensional stagnation point [13H] were determined by the numerical solution of the steady laminar compressible flow boundary-layer equations. The effect of geometry on the heat transfer to impinging jets as occurs at the inlet edge of a turbine blade has been measured [1H].

Thermodynamic coupling in boundary layers has been studied with applications including injection of low molecular weight gases into boundary layers, ablation, and chemically reacting flows [18H]. Mass transfer in the vapor phase rather than heat transfer is found to be the controlling mechanism in distillation [7H]. An analytical study of heat and mass transfer includes processes involved in the evaporation of a liquid drop [12H]. Local non-similarity solutions have been obtained for heat and mass transfer from a subliming particle [14H].

CHANGE OF PHASE

Boiling

A step input of reactivity in a nuclear reactor will result in a power excursion in which reactor power rises exponentially in time. Two papers [42J, 43J] study incipient boiling superheat, boiling heat transfer, and burnout for such exponentially increasing heat inputs. A generalized equation for critical heat flux in free-convection boiling appears in [50J]. Overshoot temperature differences far in excess of normal boiling temperature differences were required to initiate boiling of refrigerant R-113 and methane in [20J]. The large superheats appear to be a characteristic of organic liquids of low surface tension and relatively low conductivity compared with water. The nucleation criterion hitherto used in predicting incipient boiling superheat seems wrong if gases exist [32J]. A possible mean number of nucleation sites is discussed [11J]. Naturally formed boiling site cavities are described [4J]. In [26J] the research work of the past ten years on the subject of incipient superheats in alkali metals is examined in view of the large discrepancies. A theoretical model has been developed [48J] to predict the frequency of bubble departure in nucleate boiling considering the effects of surface superheat and cavity size. Experiments [54J] verify the bubble growth model of van Stralen, *et al.* that combines the initial Rayleigh solution with a diffusion-type solution accounting for both a relaxation and an evaporating microlayer. A lengthy report [44J, 45J] is given on the formation of superheated steam bubbles, their size, shape and frequency of formation at high system P and T. The second part describes heat transfer during bubble formation. Experimental data on nucleate boiling in binary mixtures [47J] suggest that diffusion resistance plays a significant role in the reduction of heat transfer in aqueous systems and is the controlling factor in non-aqueous systems. Zÿl [58J] employs numerical methods to potential and unsteady viscous flow fields around a bubble. Subcooled nucleate boiling experiments [39J] demonstrate that heat transfer is controlled by bubble formation on the metal surface and not influenced by bubble implosions away from the surface. The effects of phase interface and surface tension on entropy, free energy, internal energy, enthalpy, free enthalpy and their differentials are reduced for droplets and vapor bubbles. These results are used to give the thermodynamic interpretation of vapor bubble collapse [37J]. A theoretical analysis of one-component vapor bubble collapse with translatory motion in a uniformly subcooled liquid has been carried out [6J]. Photographs demonstrate [31J] that there is nucleation associated with bubble venting when boiling occurs in thin liquid films. This has previously gone unrecognized. A correlation of nucleate boiling heat transfer is given [35J] based on bubble population density. Recent Russian correlations of burnout in annuli are given in [36J]. In [3J] a technique is given to construct a boiling curve from quenching data. A transient boiling heat-transfer crisis is described in [51J]. Measurement of transient surface temperature during nucleate boiling of water at high heat fluxes supports the view of many investigators that a liquid film, called the macrolayer, exists beneath the agglomeration of vapor bubbles and is paramount in transferring heat [57J]. A comprehensive experimental and theoretical study of dispersed flow heat transfer (post dryout) is given in [12J]. The mechanisms include transition and film boiling. Transition boiling with flow boiling inside a horizontal tube may always be operated under stable conditions and burnout does not occur [49J].

The pressure excursions which follow the sudden exposure of a hot body to a liquid pool have been the subject of several recent studies with the safety of nuclear reactors an important example. In [24J] the heat conduction arising from the saturation temperature's dependence upon pressure is shown to be important. In binary film boiling it has been shown [28J] that the liquid–vapor interface temperature depends on liquid mixture diffusion and omission of these diffusion processes might lead to erroneous heat-transfer results. Reference [7J] reports the rates of vaporization of liquid nitrogen and methane spilled on a water surface. Dua and Tien [8J] extend the works of Tien and Yao and develop a generalized two-parameter relation for conduction-controlled rewetting on a hot vertical surface. Ünal [53J] uses high speed photography to obtain good correlation of void fraction and incipient points of boiling during subcooled nucleate flow boiling of water. Effective cooling of a fast rotating drum can be obtained by evaporating a liquid flow as a thin film on a conical internal surface [40J]. The role of Taylor instability on the sublimation of a horizontal slab of dry ice placed beneath a pool of warm water or benzene has been observed [5J] and classified as pseudo-film boiling. In [9J] an analysis is given for stable, laminar, natural convection film boiling heat transfer from an isothermal flat circular plate facing downward and submerged in a saturated liquid. In [1J], vaporization effects of mercury droplets in Leidenfrost film boiling are measured—diffusion from the upper surface of the drop is a dominant mode of mass transfer.

Current literature describes forced convection boiling heat transfer as primarily convective heat transfer. The original evidence for this concept, reported by Dengler and Addoms is re-examined [30J] and found to be misinterpreted. Development strategy is given [14J] for model development of benchmark solutions to specially designed flow-boiling equations that closely approximate the true mathematical solutions. The heat transfer–controlled collapse of large, cylindrically-shaped water vapor bubbles rising due to buoyancy effects inside a vertical tube with constant wall temperature is dependent on tube diameter and fluid subcooling [38J]. Gradon and Selecki [13J] analyze evaporation of a liquid drop immersed in another immiscible liquid when the dispersed phase surface tension is greater than the continuous phase surface tension. When the drop is critically overheated, a thermal explosion takes place.

Condensation

In all practical situations, the average heat-transfer coefficients for laminar film condensation on semi-infinite helically-wound circular cylinders lie within 10% of values for inclined circular cylinders [22J]. Selected experiments on film condensation, including a noncondensable gas in an enclosed chamber [33J], have been carried out to clarify the effects of molecular weights of vapor and noncondensable gas and the convective motion induced by the vapor flow. An analysis is given [34J] for the condensation of saturated vapor on its laminar falling liquid sheet in which velocity and temperature variations in the transverse direction are included. A numerical method is used to calculate heat transfer into thermally-governed transient helium flow surges in two-phase condensing flow [56J]. Heat transfer during annular flow condensation of steam inside tubes is studied [52J], and with steam condensing on a cylindrical, laminar jet of water [18J].

The quasi-steady Nusselt number for a drop of condensate is calculated for an arbitrary contact angle in the range $(0, \pi/2)$ using a spherical segment geometry [41J]. One of the promising means of attaining dropwise condensation is to periodically introduce the required quantity of water-repellent into the stream [17J]. A new model is proposed (and the corresponding algorithm developed) to describe the temperature field of a suddenly exposed droplet experiencing moisture condensation from an air-stream mixture [27J, 59J]. Other experiments on direct contact condensation of steam bubbles in water at high pressure [2J] show that the successive shapes of the bubbles during their collapse histories proceeded from a sphere to a hemisphere to an ellipsoid to a sphere to collapse; short-lived bubbles collapse as ellipsoids. McCoy and Hanratty [29J] critically examine measurements of the rate of disposition of droplets in annular two-phase flow.

Two-phase flow

The frost formation process was studied by photographic observations [15J] and frost formation types and thermal properties were classified into several groups according to their structures in the temperature range of 0 to −25°C. Data are reported [46J] for flashing expansion flowrate and pressure profiles in convergent–divergent nozzles fed with high pressure subcooled water. Thirty micron glass particles increased turbulent thermal entry lengths and fully developed Nusselt numbers whereas 62 and 200 μm particles did not [55J]. Heat and mass transfer in rising two-phase flows in rectangular channels [23J] and tubes [10J] are described. Transient boiling flow instabilities in a multi-channel upflow system received attention [21J]. Heat transfer during cooling of the inner surface of a brass cap with boiling liquid nitrogen [25J] is described. Eighteen two-phase friction pressure drop models and correlations were tested against 2220 measured values under adiabatic conditions and 1230 diabatic conditions [9J]. The Baroczy correlation, the Thom correlation, and the homogeneous model multipliers gave "best" performances. A critical review is given [19J] of transient and statistical measurement techniques for two-phase flows.

RADIATION

Radiation in participating media

Radiative transport in emitting, absorbing and scattering gray and non-gray media, including in particle suspensions is still of considerable interest. There is a substantial increase of the research activities associated with thin films.

A mathematical relation is presented for the gas-to-surface radiative exchange area for a rectangular parallelpiped gas volume and an adjacent face [10K]. The generalized zone method is used for numerical calculation of the radiative heat exchange in a system consisting of a long square duct filled with an absorbing medium [39K]. The governing equations are derived for flows of a radiating gas at velocities close to the isentropic and isothermal speeds of sound in channels with slowly changing temperature and curved walls. This nonlinear system of equations defines the flow of gas in channels whose transverse optical thickness is of the order of unity [3K].

A method based on integral Fourier transform technique is used for obtaining both the total and the angular radiation intensity in a non-conservative, isotropically scattering, finite slab with an internal source and with general transparent boundary conditions [12K]. In preliminary studies the expansion of the simple one-dimensional, single scattering model is reported into a two-dimensional, multiple scattering model [29K]. The exponential kernal approximation is used to develop a closed-form expression which describes the apparent directional and hemispherical emittances as a function of the optical thickness, substrate reflectance and emittance, and scattering albedo of an isothermal, isotropically scattering medium [6K]. The influence of substrate reflectance on the radiative intensity and flux distribution and on the apparent reflectance and transmittance of an isothermal, isotropically scattering medium is examined by using an exponential kernal approximation to obtain a simple algebraic expression from which the bidirectional, directional hemispherical, hemispherical directional, and hemispherical reflectances can be evaluated [7K]. An "effective absorptivity in the presence of scattering" is introduced for the solution of the radiative heat-transfer equations through absorbing, emitting and scattering media. Comparisons of this method with results calculated with conventional methods shows that it is accurate and faster than present techniques [21K].

The principal advantages and disadvantages are discussed of modeling radiation heat transfer in participating media by means of a visible-light model [1K]. In a radiative fluid, the Doppler effect due to differential motions of neighbouring fluid elements causes photon mean-free paths to deviate from the static values. The radiative flux becomes in this situation a tensorial property [16K].

Studies of the interaction of radiative transfer with convective transfer for slug flow of an absorbing–emitting gas in a circular tube with an isothermal black wall demonstrate that inclusion of radiative transfer influences the gas temperature distribution for the conditions treated [31K]. Studies of the non-steady thermal response of a one-dimensional transpiration cooled system in a radiative and convective environment indicate that the surface solid and fluid temperature and the back-face heat transfer quickly respond to the incident pulse [25K]. Transient combined laminar free convection and radiation is considered in a rectangular enclosure containing a fire or other high temperature heat sources. Solutions of the conservation equations indicate that radiation dominates the heat transfer in the enclosure and alters the convective flow pattern significantly [26K].

Investigations of the applicability of the Milne–Eddington approximation for radiation heat transfer in combustion systems reveal that this approximation applies well to hydrocarbon soot particles. For the calculation of total band radiation from gases, however, this approximation appears questionable [18K]. Studies of the radiative-convective interaction in a laminar methane-air diffusion flame revealed excellent agreement between results based on the non-gray wide-band model and experimental data [32K]. Predictions of the homogeneous gray model are in reasonable agreement with spectral measurements of the radiance and transmittance of radiating flames with polystyrene, Delrin and Plexiglas as fuels [14K]. Studies of the effect of molecular gas radiation on a planar, two-dimensional, turbulent-jet-diffusion flame indicate that realistic non-gray radiation transfer analysis can be coupled to an implicit numerical method for solution of the coupled, highly nonlinear conservation equations without undue expenditure of computation time [17K].

Local nonsimilarity techniques are employed to analyze radiation–convection interaction in a liquid in natural convection flow adjacent to a vertical surface with constant wall heat flux. Solutions are in reasonable agreement with available data [8K]. Although studies of the IR spectrum of water as a function of temperature provide significant information regarding the water structure, no sudden spectral changes have been found at 30°C or at any other temperature in the range from 1 to 50°C [35K]. Using adiabatic laser calorimetric techniques the absorption coefficient of pure water is found to be 1.7×10^{-4} cm^{-1} at 488 nm and 2.9×10^{-4} cm^{-1} at 541.5 nm. The values are in good agreement with the generally accepted long-path transmission spectra of Clarke and James [22K]. Measurements of the optical constants of liquid H_2O and D_2O are now possible over a greatly extended frequency range from 6 to 450 cm^{-1} [2K].

Investigations of the spectral absorption characteristics of water vapor and carbon dioxide mixtures in the 2.7 μm band show that spectral distributions, predicted from published data for spectral H_2O and CO_2 absorption coefficients and the line half-width to spacing ratios, are in good agreement with measured distributions [36K]. A generalized direct exchange factor is derived for the analysis of radiative heat transfer in isothermal, molecular gases. These exchange factors are valid for any molecular-gas band absorption model [33K]. Measurements of the IR emission of the 4.5 μm combination band of nitrous oxide are reported for temperatures of 1200, 1700 and 1900 K, a temperature range for which no previous data are available [34K].

Light scattering and absorption measurements by opaque particles on a per unit mass basis indicate that the most efficient light scattering and absorption occur in the size range where the particle changes from a Rayleigh to a Mie scatterer [17K]. A mathematical description is given of combined radiative and convective heat transfer to and from a turbulent gas-suspension flow. Experiments indicate that with increasing wall temperature at constant solid loading ratio, Nu increases due to the rising contribution of radiative transfer [38K]. Measurements of the visible and near IR imaginary refractive index of size fractioned atmospheric dust samples show that this index can vary with particle size nearly two orders of magnitude at some fixed wavelengths [27K]. Imaginary refractive indices are measured in atmospheric dust in the 2.0–3.0 μm spectral range using a modified Cary 171 spectrophotometer [19K].

A method is described which allows the determination of the optical constants of thin films from measurements of normal incidence, reflectance, and transmittance even in regions near the critical points where successive approximations become very inaccurate [13K]. In connection with the optical characterization of thin films, a computer inversion method is used to show that a thin inhomogenous film can usually be approximated by one homogenous film [4K]. The minimum thickness of a metallic film for infrared reflectors is calculated and the reason why metallic films cannot be used for making Fabry–Perot filters in the IR is shown [15K]. The optical properties of single-layer gradient refractive-index films offer considerable advantage over conventional interference films in many applications [30K]. A high degree of spectral selectivity can be achieved with absorptive thin films of titanium or zirconium compounds on a reflective silver film. Solar absorptance in the range from 0.8 to 0.88 can be obtained while achieving low thermal emittances from 0.04 to 0.084 at 327°C [11K].

A simple method for calculating the absorptance, reflectance (R), and phase disortion of multilayer-dielectric reflectors is based on the zeroth-order ($R = 1$) electric field distribution in the coating [37K]. Linear programming is used for the design of antireflection coatings [35K]. There is a considerable increase in surface reflectance (approximately 9 times) and a drop in transmittance (46%) if there is dropwise condensation of the H_2O on the glass surface of a solar collector system [23K].

Laboratory measurements of radiative heating in ice are generally in agreement with the prediction of an analytical model which includes the effects of anisotropic scattering as well as the spectral dependence of the absorption coefficient of ice [20K]. A radiation probe is used for studying radiant heat transfer at the exit surface of a furnace which is part of a 300 MW steam generator [5K].

The spectral radiant intensity of burning jet fuel (JP-4) is measured in the spectral region from 200 nm to 18 μm and simulated altitudes ranging from sea level to 10.7 km. An extremely intense CO_2 emission band at 4.4 μm is of primary interest for possible use in connection with fire detection [28K].

SURFACE RADIATION

Three papers [2L, 7L, 8L] report emissivities of copper, aluminium, and silver at temperatures between 150 and 1000 K measured by a transient calorimetric method. Some values at high temperature deviate from theoretical predictions and previous measurements. The optical constants n and k were measured for gold film evaporated onto a base metal by transmission interferometry [5L]. The parameter n was found to decrease with λ in the range 4000–5000 Å from 1.50 to 0.2, whereas k increases from 2 to 4 cm^{-1} in good agreement with previous work. The same constants were measured [11L] for copper and nickel at 5461 and 6328 Å after cleaning the surface by ion bombardment. Ellipsometry (the study of surfaces by analysis of polarization of reflected light), was used [3L] to study the optical constants of selective surfaces.

The effect of large pyramidal surface roughness on spectral directional emittance was calculated [1L] for a ratio of roughness height to wave length much larger than one (applying geometric optics).

Local temperatures in semi-gray, non-diffuse cones and V-grooves, calculated [9L] with radiosity integral equations, were found sensitive to higher order interreflections near the apex. Equivalent emissivities describing radiative heat transfer in three and four row systems of gray cylinders were calculated [4L]. Diagrams present [6L] shape factors for radiative energy transfer between rings and inverted cones sharing a common axis. A synoptics presents an analysis [10L] of thermal radiation effects on a normal shock wave incident on a plane wall.

LIQUID METALS

A simple model of the movement of an inert gas bubble in a liquid metal describes very well the actual behaviour [3M]. The analysis should be useful for design studies of the liquid-metal fast-breeder reactor. Experiments [1M] showed that a bubble moving in mercury changes shape from spherical to elliptical to spherical cup as the radius changes from 0.1 to 3 mm. A magnetic field generally decreases the rise velocity.

The results of heat-transfer measurements [2M] during the dropwise condensation of mercury agree well with previous studies.

MEASUREMENT TECHNIQUES

The development of instrumentation related to heat-transfer studies continues to draw the attention of a number of investigators. Areas of primary interest include temperature and heat flux measurement, radiation property evaluation, hot wire anemometry, and laser-Doppler anemometry. In addition, special instrumentation has been described for measuring quantities such as thermal diffusivity, wall shear and void fraction.

Differences in the thermal history of low-temperature glass-ceramic capacitance thermometers result in irreproducibilities in their equilibrium calibrations [43P]. An apparatus has been developed for thermal cycling germanium thermometers and evaluating their reproducibility at approximately 20K [37P]. A method of the remote measurement of temperature fields of bodies, within approximately 0.2°C in the −30–200°C range, is based on the detection of their infrared radiation with subsequent transformation of that radiation into electric signals which are displayed on a CR tube [14P]. An inexpensive recording thermometer has been developed which is accurate to ±0.4°C in the 0–30°C temperature range [27P].

The temperature distortion caused by a cavity drilled into a disk to accommodate a thermocouple can be eliminated by a properly chosen combination of the ratio of thermocouple diameter to cavity diameter and the thermocouple material [10P]. Voltage spikes generated across a 12-cm length of alumel wire in a 60 Hz sinusoidal magnetic field with an intensity of 200 A/m have been observed which would correspond to an instantaneous temperature error of 35°C if the wire were one leg of a chromel–alumel thermocouple [34P]. This effect was further demonstrated in tests which showed that chromel–alumel thermocouples used in a magnetic field indicated temperatures in error by about ±150% at 100°C [29P]. A mechanically pulsed suction thermocouple has been developed and has proved to be a reliable and fast tool for the measurement of local gas temperatures within the combustor of an aircraft gas turbine [30P].

A number of unique temperature measuring devices have been described. Two types of surface thermometers, produced using micro-electronic thin film technology, have been successfully applied to the difficult experimental task of measuring surface temperatures during high conductance processes such as condensation and boiling [20P]. Thin films of collodial graphite applied as an aerosol spray, which is commercially available, were tested for use as thermometers and exhibited good stability at low temperatures but were affected by humidity at room temperature [13P]. Analytical and experimental investigations predict that fluidic capillary temperature sensors could be developed which would have resolution capabilities of 0.001°C [15P]. A method for determining temperatures of solid particles from changes in their magnetic properties can be employed for tracing the motion of a single particle in a fluidized bed [2P]. A new technique, based on rotational Raman scattering, has been described for simultaneously measuring temperature and concentration profiles in diatomic gases during transient flows [40P]. A sensible heat flux detector has been evaluated [12P].

A relatively simple experimental procedure can be used to determine the temperature dependence of the response of instrumentation used to measure radiation, thus minimizing the errors incurred when such measurements are performed in the field or other uncontrolled temperature environments [31P]. The absorption of infrared radiation in thin film coatings can be determined by the separation of surface and bulk absorption using laser calorimetric measurements [21P]. Transmission, emittance, and calorimetric techniques of measuring the absorption of materials are described, with special emphasis on how calorimetry can be used to simultaneously determine bulk and surface absorption [24P]. A system is described which makes it possible to measure the spectral reflectance of materials in the 0.3–2 μm wavelength range for a hydrostatic pressure range of 0–70 kbar [46P]. In an attempt to improve the accuracy of reflectance determinations of high vapor pressure liquids, a double ionization chamber has been built and tested which measures the ionization from a photon beam in the vapor (or an added gas) above the liquid before and after reflection [6P].

The evaluation, calibration and application of hot-wire anemometers was the topic of several papers. The calibration and use of hot-wire probes in low density flows was discussed [18P]. Experimental studies are reported on hot-wire anemometers with wires of short relative lengths (L/D varying from 300 to 540) for use in low velocity streams and a simple iterative procedure is proposed for determining true velocities from the probe readings [41P]. Frequency response measurements show, contrary to presently used theoretical models, that for wires with substantial end conduction losses, a single time constant response is not obtained [8P]. The heat transfer from cylindrical anemometer probes in CO_2 air mixtures has been studied [9P]. Modified sets of response equations have been derived for an inclined hot-wire probe which permit the determination of the components of the three dimensional velocity vectors, the turbulence intensity, and the Reynolds stresses at any location of a swirling flow [4P]. For the application of hot-wire anemometry in transonic flow, it has been established that (for a sensor Reynolds number greater than 20 and high sensor overheat ratios) the velocity sensitivity remains independent of Mach number and is equal to the density sensitivity [25P]. The application of a dual hot-wire temperature sensing technique to transient, supersaturated gas flows facilitates previously impossible quantitative observations of deviations from ideality of supersaturated gases [17P].

The theory and the application of the laser-Doppler anemometers (LDA) continue to be areas of interest. The signal visibility characteristics of a dual beam LDA operated in a backscatter mode depends on the Mie scattering size parameter, the refractive index, the beam polarization and the size, shape and location of the light collecting aperture [1P]. Measured correlations of scattering powers and signal-to-noise ratio as function of angle for an LDA are generally in excellent agreement with Mie theory [3P]. Errors which occur when using LDA within turbulent phase boundaries (e.g. turbulent flames) have been analyzed and experimental techniques to minimize such errors have been proposed [23P]. Coherent differential Doppler measurements of a transverse velocity at a remote point have been investigated [39P].

A dual-plate holographic interferometer is purported to have several advantages over the traditional single-plate dual-exposure apparatus including insensitivity to mechanical vibration, availability of Schlieren and shadowgraph images from the same holograms, and the ability to arbitrarily position the fringes during reconstruction [19P]. Holographic measurements offer a means to investigate the fundamental structure of boundary-layer turbulence and transition for hypersonic, high Reynolds number flow around various shaped bodies at both zero and small angles of attack [22P].

Special instrumentation has been applied to boiling studies. A rotating-field conductance gauge provides a useful and versatile, yet relatively simple, technique for measuring void fractions [35P]. The application of nuclear magnetic resonance [NMR] to the measurement of void fractions of two-phase flows has also been studied. Experiments on both static and flow systems have established the linearity of the NMR method for the full range of void fractions and have demonstrated the feasibility of directly using the NMR method for measuring void-fractions [33P]. An investigation of two methods of detecting boiling crisis on a non-isothermal surface, resistance bridge and thermocouple, indicates that the bridge approach is more reliable since it responds faster and responds to crisis onset at any point [28P].

A porous plate floating element was used to accurately measure the shear stress in a transpired zero pressure gradient boundary layer in the range from laminar asymptotic suction to blow-off [11P]. Wall shear stress measurements can be made in a variable temperature flow field with a hot-film probe which has been calibrated to account for flow temperature changes [38P].

A limited number of papers. describe the measurements of transport properties. A thermal diffusivity apparatus can perform measurements in transient and periodic temperature modes on electrically conducting solids which have medium range thermal diffusivities [36P]. A comprehensive method exists for measuring the specific heat and thermal diffusivity of liquid and free-flowing materials [16P]. A modification of the transient hot-wire method of measuring thermal conductivities provides a simplified means for obtaining engineering data of moderate precision and accuracy [45P].

English-language translations of Russian papers dealing with the following topics appeared in Measurement Techniques: (a) ultrasonic resonance methods of temperature measurement [42P]; (b) apparatus for studying and calibrating resistance thermometers, thermodiodes and thermocouples [5P, 32P, 44P]; (c) the effects of thermoelectric nonuniformity of electrodes on differential thermocouples [7P]; and (d) the determination of the convective heat-transfer coefficient from a gas to a thermocouple [26P].

HEAT TRANSFER APPLICATIONS

Heat exchangers and heat pipes

The utility of the use of heat- and mass-transfer coefficients is investigated in [19Q]. It is concluded that in general the use of such coefficients offers advantages in design calculations. There are, however, exceptions where the analyses become simpler if based on heat flux.

Experiments demonstrated [8Q] that oblique flow and an intermediate tube arrangement cause a decrease in the material required for a heat exchanger by approximately 10%. Local heat-transfer coefficients and pressures around the circumference of a tube at various locations in a tube bundle were measured [24Q] at Reynolds numbers from 6×10^4 to 1.7×10^6 and at Prandtl numbers between 3 and 6. The large variations (56%) occurring in the first row of the bundle decrease to 27% in the fourth row. A method [13Q] to calculate heat transfer in tube bundles with external longitudinal fins affected by heat conduction in the fins is presented. The effects of non-uniform passages on the performance of plate fin heat exchangers were statistically modeled [16Q]. Results agree well with measurements. The deposition of paraffin wax from kerosene in cooled heat exchanger tubes generates a heat-transfer resistance which fluctuates in time after a short initial period [3Q].

The performance of cross-flow heat exchangers with variable physical properties was calculated [23Q] based on the assumption that the heat-transfer coefficient varies according to a power law with distance or temperature. The results indicate that conventional methods are not accurate for large exponents in the power law. A short note [7Q] presents a relation for the mean temperature difference in odd-tube-pass heat exchangers. A new rotary plate evaporator was described [18Q] where the evaporation occurs from a thin film carried by the rotation along the heat exchangers surface. A heat exchanger design method [1Q] makes it possible to design for prescribed or minimum entropy production.

Two papers [11Q, 22Q] present analyses of regenerators in which condensation occurs, based on the assumption of local thermodynamic equilibrium. The results of an investigation [2Q] of heat transfer to spiral coils during the mixing of a non-Newtonian liquid by anchor mixers are presented as a function of Re, Pr, μ/μ_0 in a Reynolds number range from 21 to 28 200, for Prandtl numbers between 19.2 and 8.264, and an exponent in the shear deformation relation between 0.65 and 0.87.

Mathematical models [6Q, 9Q] are offered for the calculation of spray cooling systems. The Ranz–

Marshall correlation serving the same purpose was found [5Q] to agree within ±5% with experiments on droplet temperatures.

Three papers [4Q, 12Q, 17Q] offer methods for the synthesis of heat exchanger networks. Design considerations are discussed [15Q] for heat exchangers in a gas turbine power plant. The temperature distribution in a cooled radial in-flow turbine rotor is reported in [10Q].

An efficient numerical method for the analysis of nuclear reactor cores was developed [14Q] and integrated into the COBRA-IIIC Code. Measured values for natural circulation in an experimental sodium cooled breeder reactor [20Q] agreed well with predictions of an analytical model. Fluctuations were detected [21Q] in the tubes of a large capacity, straight-tube, once-through sodium heated steam generator with a frequency depending on transit time, Froude number, and the ratio of the length of the boiling region to the total heated length.

AIRCRAFT AND SPACE VEHICLES

Ablation and thermal protection systems, in particular for space shuttle applications are still of considerable interest.

There have been a number of technological uncertainties concerning the applicability of radiative, metallic thermal protection systems (TPS) to the multimission environment of the space shuttle [3R]. A broad-based technology program at NASA Langley demonstrates that metallic TPS are competitive with those for shuttle reusable surface insulation. Numerical computations of flow fields and laminar heating rates around the Space Shuttle orbiter windward surface, including the root of the wing leading edge, illustrate the sensitivity of these calculations to several flow field modeling assumptions [4R]. In a similar study exact inviscid flow field codes are used, together with a quasi-three dimensional boundary-layer analysis to estimate windward surface heating and streamline patterns of the shuttle orbiter vehicle under laminar flow conditions [10R]. Studies of re-entry vehicle stagnation region heat transfer in particle environments indicate that heating augmentation is dominated by surface roughness effects. This roughness is due to surface erosion caused by impinging particles [7R]. The effects of sealed heatshield tile misalignment producing "roughness elements" have been studied with respect to orbiter boundary-layer transition [5R]. Wind-tunnel results indicate that the effects of roughness on the location of boundary-layer transition for the orbiter is strongly affected by the ratio of wall to total temperature.

No definite conclusions can be drawn from comparisons of wind-tunnel and flight test heat-transfer measurements on pylon-mounted stores because of the scatter of the flight test data [9R]. Studies of ablation effects under laminar flow conditions in a wind-tunnel simulation facility at $M = 10$ on a sharp 5° half angle cone indicate that care must be taken in fabricating porous mass-injection models because the mass-injection distribution can have an effect on the measured stability or drag [6R]. Transient ablation studies of Teflon hemispheres show that the effect of thermal expansion is significant and that the internal temperature distribution changes abruptly in a very thin layer near the body surface [1R]. Reliable predictions of high heating rate pyrolysis phenomena cannot be made from data obtained in conventional low heating rate apparatus based upon our present understanding of the pyrolysis process [2R].

Coating systems consisting of plasma-sprayed layers of zirconia stabilized with either yttria, magnesia, or calcia over a thin alloy bond coat have been investigated as thermal barrier coatings for cooled turbines [8R]. On the basis of durability and processing cost, the yttria-stabilized zirconia is considered to be superior. Electroless metallic plating techniques have been used for making advanced models for aeroheating tests. Mach 7 test results using several simple models demonstrate that this plated slab approach provides valid aeroheating data [11R].

In order to provide simple engineering approaches for predicting turbulent boundary-layer heating, data obtained in both ground and flight tests are compared with results from several prediction methods in the respective incompressible planes [12R].

GENERAL

A number of papers with numerous contributions from the Soviet Union deal with heat and mass transfer in the ground. Analyses on the effect of permafrost are presented [4S, 5S, 22S] and compared with experiments. Experimental data are available on the effect of snow cover [3S]. A computer program [8S] considers unsteady nonlinear heat conduction in pavement structures. The results are compared with experimental findings in Norway. Polymeric films are considered [16S] for artificial soil thawing. An analysis [18S] of transient heat and mass transfer in soils close to heated porous pipes accounts for liquid and vapor transport. Seasonal soil temperatures for three climates are presented.

Analyses [1S, 2S, 21S] study geothermal energy extraction by fracturing of rocks. An emprical correlation of experimental data [12S] describes the relation of grassland surface temperature and air temperature.

A number of papers deals with heat transfer in homes. It is found [17S] that the pitch of the roof should be around 15° to achieve optimum insulation effect. The results of an analysis [9S] describing the temperature change between air inlet and outlet of a room agree well with experiments on a two-dimensional model. A mathematical model [20S] predicts multilayer chimney-liner temperatures for unsteady state. A compressive mechanical load reduces the heat-transfer resistance of clothing fabrics considerably [15S]. Experiments [11S] established heat absorption rates of a sprinkler spray used to cool room fires.

Thermal resistance of metal/diamond interfaces used at microwave diodes was measured [6S] at 400K. Prestressing decreases the thermal contact resistance of laminated transformer stacks [14S]. Surface temperatures in sliding contacts were studied analytically [7S] by a Fourier transform method and experimentally [13S] by an infrared technique. Correlations [19S] predict quenching rates and sputtering temperatures for rewetted nuclear fuel elements. An analysis [10S] of submerged multiple port discharges is based on a gradual transition of the jets from axisymmetric to merging to pseudo-slot two-dimensional profiles and the results are compared with experimental findings.

SOLAR ENERGY

Solar energy continues to be an active applications area. Topics of major interest among the heat-transfer related solar energy publications include solar flux, flat plate collectors, concentrating collectors, solar absorber coatings, and thermal energy storage.

Methods of estimating the average daily solar radiation incident on a surface have been reviewed and the methods extended to allow predictions to be made for a wider range of surface orientation [11T]. Experimental results have been used to modify and improve a model for predicting the flux of solar radiation incident on a sloping surface [29T]. Instrumentation has been developed which has the accuracy and precision needed to determine the spectral distribution of daylight at any geographical location such that comparisons between various locations can be made with uncertainties of less than 2% [12T].

In many applications, a portion of the total solar flux incident on a collector is energy which has been reflected from the ground. The average solar reflectivity of winter landscapes varies from as low as 0.16 in areas containing open water to as high as 0.73 in typical rural areas [9T].

The performance analysis of flat plate collectors was the subject of several publications. A steady-state, two-dimensional, nodal, heat-transfer analysis was used to demonstrate that operating a collector with constant outlet temperature can be as efficient as operating with a constant mass flow rate [24T]. The same model was used to establish that long term flat plate collector performance calculations based on averaged meteorological data will only correlate with calculations based on hourly data if the weather conditions are not highly variable [23T]. However, in the case where weather conditions are not varying rapidly, the use of an integrated collector equation makes the calculation of monthly and annual solar collector performance a simple task if the average collector temperature is known [7T, 13T]. Another two-dimensional steady-state model of flat plate solar collectors was used to evaluate the sensitivity of collector performance to changes in design parameters [19T]. A flat plate collector analysis, suitable for a programmable desk calculator, was used to predict that the reduction of radiation losses which can be achieved by the application of low emittance coatings for the absorber plate results in an increase in the convection losses [30T]. The mass flow rate and temperature distribution in a natural-circulation solar water heater were evaluated using both linear and non-linear temperature distribution models with the result that the flow rates predicted by the two models differ by less than 10% [31T].

Performance measurements demonstrated that a single-glazed flat-plate solar collector with a glass honeycomb mounted between the absorber plate and the cover glass is capable of operating at efficiencies as high as 55% with the working fluid 65°C above the ambient temperature [14T].

Two of the fundamental heat-transfer processes which occur in flat-plate solar collectors have been examined in deeper detail than in the past. Experimental studies of forced convection heat transfer at an inclined and yawed square plate demonstrated that the standard computational equation substantially over estimates the wind related heat-transfer coefficient on the outer surface of a flat-plate solar collector [27T]. An analysis of the heat-transfer processes in a solar collector tube revealed that for realistic dimensions and thermal properties of the plate and tube, circumferential variations of the ouside temperature and bore heat flux can be neglected, provided the tube flow is laminar [26T].

Analytical and experimental studies dealing with concentrating solar collectors have been reported. The optimum concentration ratio for a solar collector has been shown to be a function of a dimensionless temperature (receiver temperature/ambient temperature) and a dimensionless receiver number which is proportional to the ratio of the receiver heat loss to the absorbed fraction of the solar insolation [2T]. The results of a closed form analysis of the solar flux distributions produced on the surface of a central tower receiver are presented in the form of universal curves which can be used readily to study the effects of changes in system parameters [21T]. A study of different designs of central tower, air heating receivers has shown that high thermal conversion efficiencies (80–90%) may be achieved with cavity receivers while substantially lower efficiences (54–70%) were predicted for exposed surface receiver designs [10T]. A cylindrical parabolic solar collector was shown experimentally to be capable of achieving thermal efficiencies of 50 and 30% respectively for selectively coated and black painted absorber tubes while operating at a temperature of 300°C [18T].

The use of selective coatings which have high solar absorptance and low infrared (IR) emittance is an effective means of improving the thermal efficiency of solar collectors. A survey of selective absorber coatings for solar energy technology describes both the past accomplishments and promising new materials [15T]. A coating with a solar absorptance as high as 92% and an IR emittance as low as 3% has been produced by electroplating black-chrome onto a copper substrate [6T]. A model which treats the black-chrome of this type of coating as a collection of small particles, either as metal in a dielectric or dielectric particles within the metal, is capable of predicting the range of observed radiation properties, especially the variations in the IR emittance [8T]. The solar transmittance of a gold black film has been shown to depend on the gold particle size and packing fraction in the film [17T]. A five parameter empirical Fermi function can be used to describe the spectral radiation properties of selective solar absorber coatings and the model predicts that the IR emittance of typical coatings will increase with temperature at the rate of approximately $2–5 \times 10^{-4}$ emittance units/°C [25T]. The optical constants of silicon single crystals and sputtered amorphous silicon films at temperatures up to 800°C have been determined and, based on these values, the solar-thermal efficiencies of silicon–metal selective absorbers have been predicted to be as high as 79% at 400°C [1T].

There was a significant increase in the number of energy storage publications. The majority of these described sensible energy storage systems. The development of closed-form solutions for the transient response of a packed bed provides a simple and convenient means for estimating the long-term dynamic performance of rock-bed thermal storage systems [20T]. A technique, suitable for the optimization of the design of a solid sensible heat storage unit, can be used to establish possible operating ranges for the fraction of the maximum heat stored and the fraction of the available energy stored at a given set of conditions and system constraints [22T]. An analysis of unsteady heat conduction in a composite hollow cylinder with the heat source at the inner surface of the cylinder revealed that the thermophysical properties of each layer of the composite affect the temperature distribution, rate of temperature increase, and the amount of heat stored in the material [28T]. The temperature profiles and streamlines in a liquid filled thermal energy storage tank can be predicted with a two-dimensional model and the temperature profile along the vertical axis of the tank can be predicted to within approximately 15% using a simplified one-dimensional analysis [4T].

A preliminary guide, including several monograms, has been developed to aid the designer in the assessment of sensible and latent heat storage systems for both short and long term applications [16T]. Initial tests of a latent heat storage concept indicate that the major problem preventing the use of sodium sulfate decahydrate (i.e. the degradation of capacity with cycling) can be avoided by using a mixture of 68.2 wt% $NaSO_4$-$10H_2O$ and 31.8 wr% H_2O [3T].

A proposed system which uses adsorbent beds for energy storage has a much higher stored energy density than non-adsorbent systems, requires no thermal insulation and has an indefinite storage period [5T].

PLASMA HEAT TRANSFER

Heat-transfer studies in ionized gases reported in 1977 refer to fundamental investigations as well as to applications, in particular to those using electric arcs or rf-discharges as a heat source.

High speed photographs of the luminous core of an

AC arc in an orifice air flow demonstrate that the luminous core area can be correlated by a parameter which is derived from the enthalpy flow, the kinetic energy transport, and an empirical radiation term [20U]. Temperature measurements in AC cross-flow arcs of 61.1 A RMS shows that the temperature reaches a minimum after current zero [21U]. In magnetic-blast circuit breakers the current distribution over the cross section of a narrow insulating slot may cause thermal overload of the arc chamber. Results of pertinent studies show that the current density is higher on the arc "front" and lower on the "tail" [18U]. Results of a numerical simulation of a decaying air plasma column show a non-exponential decay of the temperature at the axis in the beginning caused by pressure oscillations. The overall decay is exponential and agrees well with experimental results [10U].

A free-burning high intensity, pulsed arc ($I < 23$ kA, $t = 10$ ms) in air on steel cathodes changes from a convection ($t < 3$ ms) to a radiation dominated regime ($3\text{ ms} < t < 6\text{ ms}$). For $t > 6$ ms, the arc becomes unstable which is probably associated with diminishing cathode jet convection [17U]. Time-dependent plasma temperature measurements in a 60 Hz high-pressure sodium arc correlate well with the steady-state and time dependent plasma characteristics calculated from arc dimensions, power density, thermal conductivity, radiative heat transfer, and other arc parameters [16U].

Threshold current densities which are associated with the corresponding threshold heat fluxes leading to the formation of anode spots are studied for Al and Cu anodes in triggered vacuum gaps. They are found to depend upon the duration of arcing time, on the electrode material, surface conditions, electrode size and separation [8U]. An energy balance at the arc root of a Cu-cathode covered with a thin film indicates that Joule heating is more important than positive ion bombardment which confirms a filamentary switching model [7U].

Measurements of the continuum emission coefficient in a wall-stabilized argon arc in the range from 2 600 to 7000 Å at pressures from 1 to 30 atm and temperatures from 11 200 to 13 200 K show discrepancies with theoretical predictions. These discrepancies at high pressures cannot be entirely explained by a shift of recombination thresholds [3U]. Spectrometric studies of the boundary layer in front of a plane anode perpendicular to the axis of a wall-stabilized, high-intensity arc in argon atmosphere indicate substantial deviations from LTE [6U]. Calculated transport coefficients of a two-temperature argon plasma are presented for electron temperatures ranging from 5 000 to 20 000 K using electron temperature to heavy particle temperature ratios from 1 to 3 and pressures from 10^{-3} to 10^3 atm [9U]. Plasmas heated uniformly by inverse bremsstrahlung approach a constant electron–ion temperature difference which is only a function of the laser wavelength and intensity [15U].

Studies of the radiative energy loss from an argon plasma in a shock tube show that for the primary shock, emissivity effects outweigh reabsorption, whereas for the reflected shock region, the relative influence of reabsorption increased sufficiently to reduce radiation cooling of the plasma [11U]. Atomic and ionic temperatures in argon and helium plasma jets are derived from laser-Doppler measurements of the flow velocity coupled with dynamic and static pressure measurements [4U].

With specialized instrumentation, the problems experienced during actual welding operations, such as arc fluctuations induced by the moving weld pool, can be overcome and useful determinations made of the heat flow geometry within the arc [14U]. In electro-discharge machining removal of material is due to melting and evaporation of the metal. For a given heat input and a maximum surface temperature, the center depth and the radius of the heat source are correlated by a universal function [12U].

A survey on plasma technology and its application to extractive metallurgy stresses the potential of plasma furnaces for metallurgical processing, in particular the use of plasma heating for the production of ferro-alloys [5U]. When mixtures of WO_3 and TiO_2, ZrO_2, or HfO_2 are heated with graphite in an argon plasma arc furnace, cubic solid solutions in the W–Ti–C, W–Zr–C or W–Hf–C systems are formed which turn out to be superconductors. Their transition temperature varies with composition [13U]. Numerical predictions of the behaviour of iron particles with radii from 20 to 80 μm injected into an rf-plasma indicate that particles with radii less than 70 μm will melt under the chosen plasma conditions. Previous models seem to overestimate this potential of rf-plasmas for powder processing [22U]. Graphite materials exposed to arc-heated jets at stagnation pressures from 2.2 to 10.3 MPa and bulk enthalpies from 4700 to 5800 kJ/kg reveal non-uniform surface ablation which can be correlated with porosity, fiber density, and the fiber-matrix interfacial spacing [1U].

Studies of the design of a closed-cycle MHD generator employing nonequilibrium ionization and of the overall performance of a power plant which uses a MHD topper for a steam bottoming plant demonstrate the feasibility and adaptability of the analysis for optimizing the thermal efficiency [19U]. Investigations of the Hall effect on the combined free and forced convection, full-developed flow in a parallel plate channel with perfectly conducting walls reveal that the heat transfer at both plates decreases with increasing Grashof number and increasing Hall parameter [2U].

REFERENCES

Books

1. A. A. Bishop, F. C. Kulacki, S. G. Bankoff and O. C. Jones, Jr. (editors), *Nuclear Reactor Safety Heat Transfer*. ASME, New York (1977).
2. J. P. Hartnett and T. F. Irvine, Jr. (editors), *Advances in Heat Transfer*, Vol. 13. Academic Press, New York (1977).
3. J. R. Howell and T. C. Min (editors), *Heat Transfer in Solar Energy Systems*. ASME, New York (1977).
4. O. C. Jones, Jr. and S. G. Bankoff (editors), *Symposium on Thermal and Hydraulic Aspects of Nuclear Reactor Safety*, Vol. 1, *Light Water Reactors*; Vol. 2, *Liquid Metals*. ASME, New York (1977).
5. B. V. Karlekar and R. M. Desmond, *Engineering Heat Transfer*. West, St. Paul, Minnesota (1977).
6. M. Necati Özişik, *Basic Heat Transfer*. McGraw-Hill, New York (1977).
7. D. R. Pitts and L. E. Sissom, *Heat Transfer*. Schaum's Outline Series in Engineering, McGraw-Hill, New York (1977).
8. R. C. Reid, J. M. Prausnitz and T. K. Sherwood, *The Properties of Gases and Liquids*, 3rd Edn. McGraw-Hill, New York (1977).
9. D. B. Spalding, *GENMIX: A General Computer Program for Two-Dimensional Parabolic Phenomena*. Pergamon Press, London (1977).
10. D. B. Spalding and N. Afgan (editors), *Heat Transfer and Turbulent Buoyant Convection*. Hemisphere, Washington, D.C. (1977).
11. T. N. Veziroglu (editor), *Two-Phase Flow and Heat Transfer Symposium-Workshop*. Pergamon Press, Oxford (1977).
12. H. Y. Wong, *Heat Transfer for Engineers*. Longman, London (1977).

Conduction

1A. F. R. Al-Astrabadi, P. W. O'Callaghan, S. D. Probert and A. M. Jones, *J. Heat Transfer* **99**, 139 (1977).
2A. F. R. Al-Astrabadi, S. D. Probert, P. W. O'Callaghan and A. M. Jones, *J. Mech. Engng Sci.* **19**, 167 (1977).
3A. N. Anderson and A. M. Arthurs, *Q. Appl. Math.* **35**, 337 (1977).
4A. T. K. Ang, J. D. Ford and D. C. T. Pei, *Int. J. Heat Mass Transfer* **20**, 517 (1977).
5A. P. Archambault and J. C. Chevrier, *Int. J. Heat Mass Transfer* **20**, 1 (1977).
6A. A. Aziz, *Int. J. Heat Mass Transfer* **20**, 1253 (1977).
7A. C. E. Beevers, *Q. Appl. Math.* **35**, 111 (1977).
8A. A. Bejan, *J. Heat Transfer* **99**, 689 (1977).
9A. B. M. Berkovsky and V. G. Bashtovoi, *Int. J. Heat Mass Transfer* **20**, 621 (1977).
10A. V. P. Bhatkar and B. Rao, *Z. Angew. Math. Phys.* **28**, 301 (1977).
11A. A. Campo, *Wärme- und Stoffübertragung* **10**, 203 (1977).
12A. B. H. Caussade and G. Renard, *Int. J. Heat Mass Transfer* **20**, 555 (1977).
13A. Y. P. Chang, *Int. J. Heat Mass Transfer* **20**, 1019 (1977).
14A. Y. P. Chang and R. C. H. Tsou, *J. Heat Transfer* **99**, 41 (1977).
15A. Y. P. Chang and R. C. H. Tsou, *J. Heat Transfer* **99**, 132 (1977).
16A. K. C. Chung and K. N. Astill, *J. Heat Transfer* **99**, 127 (1977).
17A. S. W. Churchill and J. P. Gupta, *Int. J. Heat Mass Transfer* **20**, 1251 (1977).
18A. A. C. Cleland and R. L. Earle, *Int. J. Heat Mass Transfer* **20**, 1029 (1977).
19A. F. Cooper, *Int. J. Heat Mass Transfer* **20**, 991 (1977).
20A. L. Y. Cooper, *Int. J. Heat Mass Transfer* **20**, 527 (1977).
21A. V. K. Dhir, K. Wong and W. E. Kastenberg, *Nucl. Sci. Engng* **63**, 350 (1977).
22A. A. A. Dolinskiy and G. K. Ivanitskiy, *Heat Transfer, Soviet Res.* **8**(1), 23 (1976).
23A. P. W. Duck and D. S. Riley, *Int. J. Heat Mass Transfer* **20**, 297 (1977).
24A. D. Durack and B. Wendraff, *Nucl. Sci. Engng* **64**, 187 (1977).
25A. M. Epstein, *J. Heat Transfer* **99**, 527 (1977).
26A. K. Farnia and J. V. Beck, *J. Heat Transfer* **99**, 471 (1977).
27A. A. G. Goloveyko and A. D. Khudokormov, *Heat Transfer, Soviet Res.* **8**(4), 138 (1976).
28A. O. M. Griffin, *Int. J. Heat Mass Transfer* **20**, 675 (1977).
29A. B. Holzknecht, *Int. J. Heat Mass Transfer* **20**, 661 (1977).
30A. W. C. Horak and J. J. Dorning, *Nucl. Sci. Engng* **64**, 192 (1977).
31A. T. J. R. Hughes, *Computer Meth. Appl. Mech. Engng* **10**, 135 (1977).
32A. L. W. Hunter and S. Favin, *J. Heat Transfer* **99**, 668 (1977).
33A. A. A. Ivlev, *Heat Transfer, Soviet Res.* **8**(3), 164 (1976).
34A. T. T. Kao, *J. Heat Transfer* **99**, 343 (1977).
35A. E. M. Kartashov and B. Ya. Lyubov, *Heat Transfer, Soviet Res.* **8**(2), 1 (1976).
36A. H. Katz, *SIAM J. Appl. Math.* **32**, 1 (1977).
37A. J. Kern, *Int. J. Heat Mass Transfer* **20**, 467 (1977).
38A. T. T. Lie, *J. Heat Transfer* **99**, 113 (1977).
39A. R. C. Maccamy, *Q. Appl. Math.* **35**, 1 (1977).
40A. S. J. Major and A. Williams, *AIAA Jl* **15**, 419 (1977).
41A. R. C. Mehta, *J. Heat Transfer* **99**, 137 (1977).
42A. R. C. Mehta, *AIAA Jl* **15**, 1355 (1977).
43A. G. P. Merker, R. Schallig and U. Grigull, *Wärme- und Stoffübertragung* **10**, 145 (1977).
44A. M. D. Mikhailov, *Letters Heat Mass Transfer* **4**, 163 (1977).

45A. M. D. Mikhailov, G. Comini, S. del Giudice and G. P. Runchi, *Int. J. Heat Mass Transfer* **20**, 195 (1977).
46A. G. P. Mulholland and B. P. Gupta, *J. Heat Transfer* **99**, 135 (1977).
47A. W. E. Olmstead and W. E. Schmitendorf, *Z. Angew. Math. Phys.* **28**, 697 (1977).
48A. M. N. Özişik and S. I. Guceri, *Can. J. Chem. Engng* **55**, 145 (1977).
49A. C. A. Pao, *Q. Appl. Math.* **34**, 429 (1977).
50A. A. Razani and G. Ahmadi, *J. Franklin Inst.* **303**, 211 (1977).
51A. D. S. Riley and P. W. Duck, *Int. J. Heat Mass Transfer* **20**, 294 (1977).
52A. J. B. Rosser, *Z. Angew. Math. Phys.* **28**, 185 (1977).
53A. M. H. Sadd and J. E. Didlake, *J. Heat Transfer* **99**, 25 (1977).
54A. H. Saito and N. Seki, *J. Heat Transfer* **99**, 105 (1977).
55A. G. E. Schneider, A. B. Strong and M. M. Yovanovich, *Int. J. Heat Mass Transfer* **20**, 301 (1977).
56A. A. A. Sfeir and J. A. Clumpner, *J. Heat Transfer* **99**, 29 (1977).
57A. K. M. Shukla and R. N. Pandey, *Int. J. Heat Mass Transfer* **20**, 1351 (1977).
58A. R. Siegel, *Int. J. Heat Mass Transfer* **20**, 1087 (1977).
59A. A. Silard, M. Bodea, M. Luca, M. Lita and N. Iosif, *Letters Heat Mass Transfer* **4**, 149 (1977).
60A. R. D. Small and D. Weihs, *J. Heat Transfer* **99**, 47 (1977).
61A. E. M. Sparrow, S. V. Patankar and S. Ramadhyani, *J. Heat Transfer* **99**, 520 (1977).
62A. N. V. Suryanarayana, *J. Heat Transfer* **99**, 129 (1977).
63A. M. Suzuki, S. Matsumoto and S. Maeda, *Int. J. Heat Mass Transfer* **20**, 883 (1977).
64A. T. R. Tauchert, *J. Heat Transfer* **99**, 495 (1977).
65A. A. Watson, *Int. J. Heat Mass Transfer* **20**, 65 (1977).
66A. S. Weinbaum and L. M. Jiji, *J. Appl. Mech.* **44**, 25 (1977).
67A. L. C. Wellford, Jr. and R. M. Ayer, *Int. J. Num. Meth. Engng* **11**, 933 (1977).
68A. D. C. Wiggert, *J. Heat Transfer* **99**, 35 (1977).
69A. S. D. Williams and D. M. Curry, *J. Spacecraft Rockets* **14**, 632 (1977).
70A. S. D. Williams and D. M. Curry, *Int. J. Num. Meth. Engng* **11**, 1605 (1977).
71A. W. L. Wood, *Int. J. Num. Meth. Engng* **11**, 1059 (1977).
72A. L. T. Yeh and B. T. F. Chung, *J. Spacecraft Rockets* **14**, 178 (1977).
73A. M. M. Yovanovich and S. S. Burde, *AIAA Jl* **15**, 1523 (1977).

Channel flow

1B. I. T. Alad'yev, K. D. Voskresenskiy, Ye. S. Turilina and N. N. Trifonova, *Heat Transfer, Soviet Res.* **8**(2), 55 (1976).
2B. F. P. Berger and K. F. F. L. Hau, *Int. J. Heat Mass Transfer* **20**, 1185 (1977).
3B. M. S. Bhatti, *J. Heat Transfer* **99**, 574 (1977).
4B. M. S. Bhatti and C. W. Savery, *J. Heat Transfer* **99**, 142 (1977).
5B. D. J. Brassington and D. N. H. Cairns, *Int. J. Heat Mass Transfer* **20**, 207 (1977).
6B. A. R. Chandrupatla and Y. M. K. Sastri, *Int. J. Heat Mass Transfer* **20**, 1315 (1977).
7B. K. C. Cheng and S. -L. Wong, *Can. J. Chem. Engng* **55**, 149 (1977).
8B. S. W. Churchill, *I/EC Fundamentals* **16**, 109 (1977).
9B. M. Dalle Donne and L. Meyer, *Int. J. Heat Mass Transfer* **20**, 583 (1977).
10B. Vi-D. Dang and M. Steinberg, *Chem. Engng Sci.* **32**, 326 (1977).
11B. M. Epstein and G. M. Hauser, *J. Heat Transfer* **99**, 687 (1977).
12B. M. Epstein, A. Yim and F. B. Cheung, *J. Heat Transfer* **99**, 233 (1977).
13B. N. I. Gel'perin, V. G. Aynshteyn, L. I. Krupnik, Z. N. Memedlyayev and S. V. Yefimova, *Heat Transfer, Soviet Res.* **8**(5), 77 (1976).
14B. A. J. Ghajar and W. G. Tiederman, *A.I.Ch.E. Jl* **23**, 128 (1977).
15B. R. R. Gilpin, *J. Heat Transfer* **99**, 419 (1977).
16B. R. R. Gilpin, *Int. J. Heat Mass Transfer* **30**, 693 (1977).
17B. L. Goldstein, Jr. and E. M. Sparrow, *J. Heat Transfer* **99**, 187 (1977).
18B. V. S. Grigor'yev and A. F. Polyakov, *Heat Transfer, Soviet Res.* **8**(3), 151 (1976).
19B. G. A. Hughmark, *A.I.Ch.E. Jl.* **23**, 601 (1977).
20B. B. Hunt, *Int. J. Heat Mass Transfer* **20**, 393 (1977).
21B. M. B. Ibrahim and V. Walker, *Int. J. Heat Mass Transfer* **20**, 291 (1977).
22B. V. Javeri, *Int. J. Heat Mass Transfer* **20**, 543 (1977).
23B. V. Javeri, *Wärme- und Stoffübertragung* **10**, 137 (1977).
24B. B. A. Kader and A. M. Yaglom, *Int. J. Heat Mass Transfer* **20**, 345 (1977).
25B. D. D. Kale, *Int. J. Heat Mass Transfer* **20**, 1077 (1977).
26B. H. Kawamura, *Int. J. Heat Mass Transfer* **20**, 443 (1977).
27B. M. S. Khader and R. I. Vachon, *J. Heat Transfer* **99**, 684 (1977).
28B. S. K. Končar-Djurdjević and A. P. Duduković, *A.I.Ch.E. Jl* **23**, 125 (1977).
29B. H. K. Kuiken, *Int. J. Heat Mass Transfer* **20**, 309 (1977).
30B. V. M. Legkii and V. D. Burley, *Thermal Engng* **23**(9), 73 (1976).
31B. V. V. Mamayev, V. S. Nosov and N. I. Syromyatnikov, *Heat Transfer, Soviet Res.* **8**(3), 111 (1976).
32B. J. H. Masliyah and K. Nandakumar, *Can. J. Chem. Engng* **55**, 27 (1977).
33B. G. A. McConaghy and T. J. Hanratty, *A.I.Ch.E. Jl* **23**, 493 (1977).
34B. D. M. McEligot, M. F. Taylor and F. Durst, *Int. J. Heat Mass Transfer* **20**, 475 (1977).
35B. S. Mori, T. Inoue and A. Tanimoto, *Can. J. Chem. Engng* **55**, 138 (1977).
36B. A. T. Nikitin, V. V. Sapozhnikov and V. V. Bogachev, *Heat Transfer, Soviet Res.* **8**(5), 87 (1976).
37B. V. O'Brien and L. W. Ehrlich, *J. Heat Transfer* **99**, 485 (1977).
38B. S. V. Patankar, C. H. Liu and E. M. Sparrow, *J. Heat Transfer* **99**, 180 (1977).
39B. R. J. Pederson and E. M. Sparrow, *J. Heat Transfer*

99, 425 (1977).
40B. F. Popovska and W. L. Wilkinson, *Chem. Engng Sci.* **32**, 1155 (1977).
41B. G. Radhakrishnamacharaya and M. K. Maiti, *Int. J. Heat Mass Transfer* **20**, 171 (1977).
42B. O. I. Reisman and R. O. Parker, *Nucl. Sci. Engng* **63**, 188 (1977).
43B. M. Sakakibara and K. Endoh, *Int. J. Heat Mass Transfer* **20**, 507 (1977).
44B. N. M. Schnurr, *J. Heat Transfer* **99**, 580 (1977).
45B. D. A. Shaw and T. J. Hanratty, *A.I.Ch.E. Jl* **23**, 28 (1977).
46B. V. K. Shchukin, V. A. Filin and A. I. Mironov, *Heat Transfer, Soviet Res.* **8**(5), 117 (1976).
47B. A. A. Shibani and M. N. Özişik, *J. Heat Transfer* **99**, 20 (1977).
48B. A. A. Shibani and M. N. Özişik, *Int. J. Heat Mass Transfer* **20**, 565 (1977).
49B. H. A. Simon, M. H. Chang and J. C. F. Chow, *J. Heat Transfer* **99**, 590 (1977).
50B. E. M. Sparrow, B. R. Baliga and S. V. Patankar, *J. Heat Transfer* **99**, 4 (1977).
51B. E. M. Sparrow and J. P. Kalejs, *Int. J. Heat Mass Transfer* **20**, 1241 (1977).
52B. E. M. Sparrow and S. V. Patankar, *J. Heat Transfer* **99**, 483 (1977).
53B. F. Ye. Spokoynyy, Z. R. Gorbis, N. V. Svyatetskiy, and R. V. Zagaynova, *Heat Transfer, Soviet Res.* **8**(5), 94 (1976).
54B. A. S. Sukomel, V. A. Rizhko, R. V. Kerimov and F. F. Tsvetkov, *Heat Transfer, Soviet Res.* **8**(5), 103 (1976).
55B. H. Tanaka and J. I. Shimizu, *J. Heat Transfer* **99**, 682 (1977).
56B. Z. V. Tishchenko and N. D. Butskii, *Int. Chem. Engng* **17**, 498 (1977).
57B. R. -S. Wu, K. C. Cheng and J. W. Ou, *Can. J. Chem Engng* **54**, 526 (1976).
58B. J. S. Yu, *J. Appl. Mech.* **43**, 537 (1976).
59B. J. P. Zarling, *J. Heat Transfer* **99**, 332 (1977).

Boundary layer and external flows

1C. M. Acharya, *AIAA Jl* **15**, 303 (1977).
2C. E. Achenbach, *Int. J. Heat Mass Transfer* **20**, 359 (1977).
3C. R. A. Antonia, H. Q. Danh and A. Prabhu, *J. Fluid Mech.* **80**, 153 (1977).
4C. Ya. I. Blyumin, *Heat Transfer, Soviet Res.* **8**(1), 155 (1976).
5C. M. N. Chepurnoy, V. E. Shnayder and A. D. Berkuta, *Heat Transfer, Soviet Res.* **8**(1), 62 (1976).
6C. S. W. Churchill and M. Bernstein, *J. Heat Transfer,* **99**, 300 (1977).
7C. P. Demargue and D. Kalache, *Int. J. Heat Mass Transfer* **20**, 773 (1977).
8C. A. Sh. Dorfman, *Int. Chem. Engng* **17**, 505 (1977).
9C. E. P. Dyban, E. Ya. Epik and T. T. Surpun, *Int. Chem. Engng* **17**, 501 (1977).
10C. R. S. Reddy Gorla, *Appl. Scient. Res.* **32**, 541 (1976).
11C. G. P. Hammond, W. S. Fung, P. W. O'Callaghan and S. D. Probert, *J. Mech. Engng Sci.* **19**, 47 (1977).
12C. A. Hirschberg and J. C. van Muiswinkel, *Int. J. Heat Mass Transfer* **20**, 668 (1977).
13C. C. J. Hoogendoorn, *Int. J. Heat Mass Transfer* **20**, 1333 (1977).
14C. B. A. Kader and A. A. Gukhman, *Int. J. Heat Mass Transfer* **20**, 1339 (1977).
15C. K. Kataoka and T. Takami, *A.I.Ch.E. Jl* **20**, 889 (1977).
16C. V. Kottke, H. Blenke and K. G. Schmidt, *Wärme- und Stoffübertragung* **10**, 89 (1977).
17C. L. Lakshminarayana, *J. Spacecraft Rockets* **14**, 219 (1977).
18C. J. F. Marchman, III, *AIAA Jl* **15**, 907 (1977).
19C. J. H. Masliyah and T. T. Nguyen, *Can. J. Chem. Engng* **55**, 156 (1977).
20C. B. K. Meena and G. Nath, *AIAA Jl* **15**, 1041 (1977).
21C. H. Miyazaki and E. M. Sparrow, *J. Heat Transfer* **99**, 614 (1977).
22C. F. A. Morrison, Jr., *J. Heat Transfer* **99**, 269 (1977).
23C. P. H. Oosthuizen and T. Cheung, *J. Heat Transfer* **99**, 152 (1977).
24C. P. Payvar, *Int. J. Heat Mass Transfer* **20**, 431 (1977).
25C. K. S. Rao, V. W. Nee and K. T. Yang, *J. Heat Transfer* **99**, 433 (1977).
26C. K. S. Raju and E. U. Schlünder, *Wärme- und Stoffübertragung* **10**, 131 (1977).
27C. E. M. Sparrow, R. M. Abdel-Wahed and S. V. Patankar, *Int. J. Heat Mass Transfer* **20**, 975 (1977).
28C. A. J. Strazisar, E. Reshotko and J. M. Prahl, *J. Fluid Mech.***83**, 225 (1977).
29C. J. Sucec, *J. Engng Pwr* **99**, 567 (1977).
30C. H. S. Takhar and V. M. Soundalgekar, *Appl. Scient. Res.* **33**, 101 (1977).
31C. M. M. Tamonis, L. I. Dagis and A. A. Zhukauskas, *Int. Chem. Engng* **17**, 740 (1977).
32C. R. C. C. Wang, B. T. F. Chung and L. C. Thomas, *J. Heat Transfer* **99**, 513 (1977).
33C. A. T. Wassel and V. E. Denny, *J. Spacecraft Rockets* **14**, 212 (1977).
34C. R. D. Watson, *AIAA Jl* **15**, 1455 (1977).
35C. D. L. Witfield and M. D. High, *AIAA Jl* **15**, 431 (1977).
36C. W. J. Yang, *Letters Heat Mass Transfer* **4**, 75 (1977).
37C. G. B. Zdanavichyus, Y. V. Survilla and A. A. Zhukauskas, *Int. Chem. Engng* **17**, 115 (1977).

Flow with separated regions

1D. M. L. Agrawal, P. K. Pande and R. Prakash, *AIAA Jl* **15**, 740 (1977).
2D. B. M. Al Ali and J. Broughton, *Appl. Energy* **3**, 101 (1977).
3D. A. O. O. Denloye and J. S. M. Botterill, *Chem. Engng Sci.* **32**, 461 (1977).
4D. V. Dhir and I. Catton, *J. Heat Transfer* **99**, 250 (1977).
5D. R. Elsdon and C. J. Shearer, *Chem. Engng Sci.* **32**, 1147 (1977).
6D. J. Gibson, *Appl. Energy* **3**, 87 (1977).
7D. T. H. Hsiung and G. Thodos, *Int. J. Heat Mass Transfer* **20**, 331 (1977).
8D. G. R. Inger, *AIAA Jl* **15**, 383 (1977).
9D. G. R. Inger, *AIAA Jl* **15**, 619 (1977).
10D. S. H. Jury and M. C. Berbano, *J. Franklin Inst.* **303**, 285 (1977).
11D. S. D. Kim, C. G. J. Baker and M. A. Bergougnou,

Chem. Engng Sci. **32**, 1299 (1977).
12D. S. Kimura and O. Levenspiel, *I/EC Process Des. Dev.* **16**, 145 (1977).
13D. C. -H. Li and B. A. Finlayson, *Chem. Engng Sci.* **32**, 1055 (1977).
14D. D. R. McGaw, *Chem. Engng Sci.* **32**, 11 (1977).
15D. C. I. Metcalfe and J. R. Howard, *Appl. Energy* **3**, 65 (1977).
16D. A. Nir and A. Acrivos, *J. Fluid Mech.* **78**, 33 (1976).
17D. R. D. Patel and J. M. Simpson, *Chem. Engng Sci.* **32**, 67 (1977).
18D. B. Prunet-Foch, F. Legay-Desesquelles and G. B. Diep, *AIAA Jl* **15**, 1295 (1977).
19D. B. Prunet-Foch, F. Legay-Desesquelles and G. B. Diep, *Int. J. Heat Mass Transfer* **20**, 1211 (1977).
20D. L. Půst, *Int. J. Heat Mass Transfer* **20**, 1255 (1977).
21D. P. R. Sanderson and J. R. Howard, *Appl. Energy* **3**, 115 (1977).
22D. S. Kh. Shumkov and D. G. Ivanov, *Int. Chem. Engng* **17**, 480 (1977).
23D. F. T. Smith and P. W. Duck, *Q. J. Mech. Appl. Math.* **30**, 143 (1977).
24D. E. M. Sparrow and J. P. Kalejs, *Int. J. Heat Mass Transfer* **20**, 1241 (1977).
25D. N. I. Sul'zhik, A. I. Il'chenko, A. V. Stepanov and G. P. Kuchin, *Int. Chem. Engng* **17**, 122 (1977).
26D. R. H. Thring, *Int. J. Heat Mass Transfer* **20**, 911 (1977).
27D. P. Tournie, C. Laguerie and J. P. Couderc, *Chem. Engng Sci.* **32**, 1259 (1977).
28D. D. T. Tsahalis, *AIAA Jl* **15**, 561 (1977).
29D. N. Wakao, S. Kaguei and H. Nagai, *Chem. Engng Sci.* **32**, 1261 (1977).
30D. N. Wakao, S. Kaguei and B. Shiozawa, *Chem. Engng Sci.* **32**, 451 (1977).

Transfer mechanisms

1E. B. U. Achia and D. E. Thompson, *J. Fluid Mech.* **81**, 439 (1977).
2E. R. A. Antonia and H. Q. Danh, *Physics Fluids* **20**, 1050 (1977).
3E. R. A. Antonia and K. R. Sreenivasan, *Physics Fluids* **20**, 1800 (1977).
4E. S. P. S. Arya and A. Sundarajan, *Boundary-Layer Meteorology* **10**, (1976).
5E. M. A. Badri Narayanan, S. Rajagopalan and R. Narasimha, *J. Fluid Mech.* **80**, 237 (1977).
6E. T. L. Chambers and D. C. Wilcox, *AIAA Jl* **15**, 821 (1977).
7E. R. B. Dean and P. Bradshaw, *J. Fluid Mech.* **78**, 641 (1976).
8E. P. M. Bevilaqua and P. S. Lykoudis, *AIAA Jl* **15**, 1194 (1977).
9E. M. Elena, *Int. J. Heat Mass Transfer* **20**, 935 (1977).
10E. T. R. Heidrick, S. Banerjee and R. S. Azad, *J. Fluid Mech.* **82**, 705 (1977).
11E. T. R. Heidrick, S. Banerjee and R. S. Azad, *J. Fluid Mech.* **81**, 137 (1977).
12E. K. N. Helland, C. W. Van Atta and G. R. Stegen, *J. Fluid Mech.* **79**, 337 (1977).
13E. S. Hess, *Physica* **86A**, 383 (1977).
14E. W. S. Lewellen and M. E. Taske, *Boundary-Layer Meterology* **10**, 69 (1976).
15E. M. B. Lewis, *Physics Fluids* **20**, 1058 (1977).
16E. S. C. Ling and A. Saad, *Physics Fluids* **20**, 1796 (1977).
17E. J. L. Lumley and G. R. Newman, *J. Fluid Mech.* **82**, 161 (1977).
18E. A. Melling and J. H. Whitelaw, *J. Fluid Mech.* **78**, 289 (1976).
19E. G. L. Mellor and H. J. Herring, *AIAA Jl* **15**, 886 (1977).
20E. D. Naot, *Physics Fluids* **20**, 546 (1977).
21E. A. E. Perry and C. J. Abell, *J. Fluid Mech.* **79**, 785 (1977).
22E. Y. Salu and D. Montgomery, *Physics Fluids* **20**, 1 (1977).
23E. U. Schumann, *Physics Fluids* **20**, 721 (1977).
24E. P. Sepri, *Physics Fluids* **19**, 1876 (1976).
25E. D. A. Shaw and T. J. Hanratty, *Chem. Engng Jl* **23**, 160 (1977).
26E. K. R. Sreenivasan, R. A. Antonia and H. Q. Danh, *Physics Fluids* **20**, 1238 (1977).
27E. H. Ueda, R. Möller, S. Komori and T. Mizushina, *Int. J. Heat Mass Transfer* **20**, 1127 (1977).
28E. H. Vollmers and J. C. Rotta, *AIAA Jl* **15**, 714 (1977).
29E. R. D. Watson, *AIAA Jl* **15**, 591 (1977).
30E. I. Wygnanski, M. Sokolov and D. Friedman, *J. Fluid Mech.* **78**, 785 (1976).

Natural convection

1F. J. N. Arnold, D. K. Edwards and I. Catton, *J. Heat Transfer* **99**, 120 (1977).
2F. J. Bandrowski, *Int. Chem. Engng* **17**, 164 (1977).
3F. J. Bandrowski and J. Rybski, *Int. Chem. Engng* **17**, 179 (1977).
4F. A. Bar-Cohen and A. Herman, *Letters Heat Mass Transfer* **4**, 111 (1977)
5F. R. F. Boehm and D. Kamyab, *J. Heat Transfer* **99**, 294 (1977).
6F. P. J. Burns, L. C. Chow and C. L. Tien, *Int. J. Heat Mass Transfer* **20**, 919 (1977).
7F. A. Cabelli, *Appl. Scient. Res.* **33**, 45 (1977).
8F. R. L. D. Cane, K. G. T. Hollands, G. D. Raithby and T. E. Unny, *J. Heat Transfer* **99**, 86 (1977).
9F. V. P. Carey and J. C. Mollendorf, *Int. J. Heat Mass Transfer* **20**, 1059 (1977).
10F. S. Caspi, H. Y. Lee and T. H. K. Frederking, *J. Engng Ind.* **99**, 723 (1977).
11F. G. Castinel and M. Combarnous, *Int. Chem. Engng* **17**, 605 (1977).
12F. C. Chang, G. C. Vliet and A. Saberian, *J. Heat Transfer* **99**, 603 (1977).
13F. C. F. Chen and R. D. Sandford, *J. Fluid Mech.* **83**, 83 (1977).
14F. T. S. Chen, E. M. Sparrow and A. Mucoglu, *J. Heat Transfer* **99**, 66 (1977).
15F. P. Cheng, *Letters Heat Mass Transfer* **4**, 119 (1977).
16F. P. Cheng, *Int. J. Heat Mass Transfer* **20**, 893 (1977).
17F. P. Cheng, *Int. J. Heat Mass Transfer* **20**, 201 (1977).
18F. P. Cheng, *Int. J. Heat Mass Transfer* **20**, 807 (1977).
19F. K. C. Cheng and R. -S. Wu, *Appl. Scient. Res.* **32**, 327 (1976).

20F. F. B. Cheung, *Int. J. Heat Mass Transfer* **20**, 499 (1977).
21F. E. B. Christiansen and T. Shinohara, *Can. J. Chem. Engng* **54**, 626 (1976).
22F. S. W. Churchill, *A.I.Ch.E. Jl* **23**, 10 (1977).
23F. R. M. Clever, *Z. Angew Math. Phys.* **28**, 585 (1977).
24F. R. M. Clever and F. H. Busse, *J. Fluid Mech.* **81**, 107 (1977).
25F. R. M. Clever, F. H. Busse and R. E. Kelley, *Z. Angew Math. Phys.* **28**, 771 (1977).
26F. L. J. Crane, *Z. Angew Math. Phys.* **28**, 599 (1977).
27F. J. R. Custer and E. J. Shaughnessy, *J. Heat Transfer* **99**, 596 (1977).
28F. J. R. Custer and E. J. Shaughnessy, *J. Heat Transfer* **99**, 675 (1977).
29F. P. G. Daniels, *Z. Angew. Math. Phys.* **28**, 577 (1977).
30F. A. Davey and W. H. Reid, *J. Fluid Mech.* **80**, 527 (1977).
31F. E. J. Davis and C. K. Choi, *J. Fluid Mech.* **81**, 565 (1977).
32F. R. Dodd, Trans. *Instn. Chem. Engrs* **55**, 223 (1977).
33F. M. K. El-ridi, I. G. Chumak and L. G. Kalinin, *Heat Transfer, Soviet Res.* **8**(1), 33 (1976).
34F. J. -F. Epherre, *Int. Chem. Engng* **17**, 615 (1977).
35F. M. I. O. Ero, *J. Heat Transfer* **99**, 335 (1977).
36F. M. I. O. Ero, *Int. J. Heat Mass Transfer* **20**, 403 (1977).
37F. R. M. Fand, E. W. Morris and M. Lum, *Int. J. Heat Mass Transfer* **20**, 1173 (1977).
38F. M. K. Gangal and B. D. Aggarwalla, *Z. Angew Math. Phys.* **28**, 85 (1977).
39F. W. K. George, Jr., R. L. Alpert and F. Tamanini, *Int. J. Heat Mass Transfer* **20**, 1145 (1977).
40F. R. S. R. Gorla, *Letters Heat Mass Transfer* **4**, 155 (1977).
41F. B. W. Grange, R. Viskanta and W. H. Stevenson, *Letters Heat Mass Transfer* **4**, 85 (1977).
42F. A. S. Gupta and I. Pop, *Physics Fluids* **20**, 162 (1977).
43F. B. K. Hartline and C. R. B. Lister, *J. Fluid Mech.* **79**, 379 (1977).
44F. J. Hiddink, J. Schenk and S. Bruin, *Appl. Scient. Res.* **32**, 217 (1976).
45F. S. W. Hing, *Int. J. Heat Mass Transfer* **20**, 685 (1977).
46F. C. K. Hsieh and R. W. Coldeway, *J. Heat Transfer* **99**, 439 (1977).
47F. Y. Jaluria and B. Gebhart, *Int. J. Heat Mass Transfer* **20**, 153 (1977).
48F. Y. Jaluria and B. Gebhart, *Int. J. Heat Mass Transfer* **20**, 434 (1977).
49F. M. C. Jischke, *J. Heat Transfer* **99**, 676 (1977).
50F. R. Kahawita and R. Meroney, *J. Appl. Mech.* **99**, 11 (1977).
51F. T. -T. Kao, G. A. Domoto and H. G. Elrod, Jr., *J. Heat Transfer* **99**, 72 (1977).
52F. K. B. Katsaros, W. T. L.., J. A. Businger and J. E. Tillman, *J. Fluid Mech.* **83**, 311 (1977).
53F. N. E. Kosovinos, *J. Fluid Mech.* **81**, 45 (1977).
54F. N. E. Kosovinos and E. J. List, *J. Fluid Mech.* **81**, 25 (1977).
55F. F. A. Kulacki and A. A. Emara, *J. Fluid Mech.* **83**, 375 (1977).
56F. K. H. Lau and P. Cheng, *Int. J. Heat Mass Transfer* **20**, 1205 (1977).
57F. G. B. Lewis, J. L. Novotny and K. T. Yang, *J. Heat Transfer* **99**, 446 (1977).
58F. K. K. Madni and R. H. Pletcher, *J. Heat Transfer* **99**, 641 (1977).
59F. I. K. Madni and R. H. Pletcher, *J. Heat Transfer* **99**, 99 (1977).
60F. G. D. Mallinson and G. DeVahl Davis, *J. Fluid Mech.* **83**, 1 (1977).
61F. M. J. Manton, *Boundary-Layer Meteorology* **10**, 347 (1976).
62F. R. A. Meric, *Int. J. Heat Mass Transfer* **20**, 429 (1977).
63F. J. H. Merkin, *J. Heat Transfer* **99**, 453 (1977).
64F. J. H. Merkin, *Int. J. Heat Mass Transfer* **20**, 73 (1977).
65F. M. Miyamoto, *Int. J. Heat Mass Transfer* **20**, 1258 (1977).
66F. K. Mizukami, *Int. J. Heat Mass Transfer* **20**, 981 (1977).
67F. L. W. Morland, A. Zebib and D. R. Kassoy, *Physics Fluids* **20**, 1255 (1977).
68F. H. Mouton and H. De Roeck, *Int. J. Heat Mass Transfer* **20**, 627 (1977).
69F. H. Mouton and M. Roiland, *Letters Heat Mass Transfer* **4**, 93 (1977).
70F. A. Mucoglu and T. S. Chen, *J. Heat Transfer* **99**, 679 (1977).
71F. A. Mucoglu and T. S. Chen, *Can. J. Chem. Engng* **55**, 265 (1977).
72F. T. E. Mullin and E. R. Gerhard, *J. Heat Transfer* **99**, 586 (1977).
73F. L. G. Napolitano, G. M. Carlomagno and P. Vigo, *Int. J. Heat Mass Transfer* **20**, 215 (1977).
74F. A. Natalizio and W. D. Baines, *J. Hydraul. Div. Am. Soc. Civ. Engrs* **103**, No. HY2, 123 (1977).
75F. H. J. Nawoj and R. S. Hickman, *J. Heat Transfer* **99**, 609 (1977).
76F. S. K. Nayak and V. A. Sandborn, *Int. J. Heat Mass Transfer* **20**, 189 (1977).
77F. D. A. Nield, *J. Fluid Mech.* **81**, 513 (1977).
78F. O. Ofi and H. J. Hetherington, *Int. J. Heat Mass Transfer* **20**, 953 (1977).
79F. J. W. Ou and K. C. Cheng, *Int. J. Heat Mass Transfer* **20**, 953 (1977).
80F. H. Ozoe, H. Sayama and S. W. Churchill, *Int. J. Heat Mass Transfer* **20**, 123 (1977).
81F. H. Ozoe, K. Yamamoto, H. Sayama and S. W. Churchill, *Int. J. Heat Mass Transfer* **20**, 131 (1977).
82F. M. A. Patrick, A. A. Wragg and D. M. Pargeter, *Can. J. Chem. Engng* **55**, 432 (1977).
83F. J. K. Platten and G. Chavepeyer, *Int. J. Heat Mass Transfer* **20**, 113 (1977).
84F. O. A. Plumb and L. A. Kennedy, *J. Heat Transfer* **99**, 79 (1977).
85F. R. E. Powe, J. A. Scanlan and T. A. Larson, *Int. J. Heat Mass Transfer* **20**, 159 (1977).
86F. M. R. E. Proctor, *J. Fluid Mech.* **82**, 97 (1977).
87F. C. Quon, *J. Heat Transfer* **99**, 340 (1977).
88F. G. D. Raithby, K. G. T. Hollands and T. E. Unny, *J. Heat Transfer* **99**, 287 (1977).
89F. T. Saitoh, *Appl. Scient. Res.* **32**, 429 (1976).
90F. T. S. Sarma and S. P. Sukhatme, *Int. J. Heat Mass Transfer* **20**, 51 (1977).
91F. N. Seki, S. Fukusako and M. Sugawara, *J. Heat*

Transfer **99**, 92 (1977).
92F. Z. P. Shul'man, V. I. Baykov and E. A. Zal'tsgendler, *Heat Transfer, Soviet Res.* **8**(2), 69 (1976).
93F. Z. P. Shul'man, E. A. Zal'tsgendler and V. I. Baykov, *Heat Transfer, Soviet Res.* **8**(2), 59 (1976).
94F. S. N. Singh, *Int. J. Heat Mass Transfer* **20**, 1155 (1977).
95F. D. Skiadaressis and D. B. Spalding, *Letters Heat Mass Transfer,* **35** (1977).
96F. V. M. Soundalgekar, *J. Heat Transfer* **99**, 499 (1977).
97F. V. M. Soundalgekar and P. D. Wavre, *Int. J. Heat Mass Transfer* **20**, 1375 (1977).
98F. V. M. Soundalgekar and P. D. Wavre, *Int. J. Heat Mass Transfer* **20**, 1363 (1977).
99F. D. Sweet, E. Jakeman and D. T. J. Hurle, *Physics Fluids* **20**, 1412 (1977).
100F. B. Tabarrok and R. C. Lin, *Int. J. Heat Mass Transfer* **20**, 945 (1977).
101F. C. L. Ting and D. R. Hay, *Boundary-Layer Meteorology* **11**, 243 (1977).
102F. J. Toomre, D. O. Gough and E. A. Spiegel, *J. Fluid Mech.* **79**, 1 (1977).
103F. M. Tveitreid, *Int. J. Heat Mass Transfer* **20**, 1045 (1977).
104F. K. Vajravelu and K. S. Sastri, *Int. J. Heat Mass Transfer* **20**, 655 (1977).
105F. K. Walker and G. M. Homsy, *J. Heat Transfer* **99**, 338 (1977).
106F. P. C. Wankat and G. M. Homsy, *Physics Fluids* **20**, 1200 (1977).
107F. H. P. Weiss, *Z. Angew. Math. Phys.* **28**, 409 (1977).
108F. R. A. Wirtz, *Int. J. Heat Mass Transfer* **20**, 841 (1977).
109F. S. J. Wright, *J. Hydraul. Div. Am. Soc. Civ. Engrs* **103**, No. HY5, 499 (1977).
110F. L. S. Yao and I. Catton, *J. Heat Transfer* **99**, 122 (1977).
111F. Z. Zebib and D. R. Kassoy, *Physics Fluids* **20**, 4 (1977).

Convection from rotating surfaces

1G. I. A. Eltayer and S. Kumar, *Proc. R. Soc.* **A352**, 145 (1977).
2G. R. N. Jana, N. Datta and B. S. Mazumder, *J. Phys. Soc. Japan* **42**, 1034 (1977).
3G. K. Ktoka, H. Dor and T. Komai, *Int. J. Heat Mass Transfer* **20**, 57 (1977).
4G. H. Koyama, A. Nakayama, K. Sato and T. Shimizu, *J. Heat Transfer* **99**, 634 (1977).
5G. T. M. Kuzay and C. J. Scott, *J. Heat Transfer* **99**, 12 (1977).
6G. C. H. Li, *J. Lubric. Tech.* **99**, 318 (1977).
7G. A. K. Macpherson, *Proc. R. Soc.* **A353**, 349 (1977).
8G. J. Marshall, *Int. J. Heat Mass Transfer* **20**, 227 (1977).
9G. J. E. McElhiney and G. W. Preckshot, *Int. J. Heat Mass Transfer* **20**, 847 (1977).
10G. D. E. Metzger and L. D. Grochowsky, *J. Heat Transfer* **99**, 663 (1977).
11G. A. K. Mohanty, T. S. Raghavachar and R. S. Nanda, *Int. J. Heat Mass Transfer* 20, 1417 (1977).
12G. J. P. Narain, *Computers Fluids* **5**, 61 (1977).
13G. R. G. Scurlock and G. K. Thornton, *Int. J. Heat Mass Transfer* **20**, 31 (1977).
14G. B. I. Sharma, *J. Heat Transfer* **99**, 151 (1977).
15G. V. Vidyanidhi, V. V. S. Suryanarayana and V. C. Chenchu Raju, *J. Heat Transfer* **99**, 148 (1977).

Combined heat and mass transfer

1H. L. V. Arsen'yev and I. B. Mitrayayev, *Heat Transfer, Soviet Res.* **8**(3), 122 (1976).
2H. G. Bergeles, A. D. Gosman and B. E. Launder, *AIAA Jl* **15**, 499 (1977).
3H. L. Fulachier, E. Verollet and I. Dekeyser, *Int. J. Heat Mass Transfer* **20**, 731 (1977).
4H. R. S. Gorla, *AIAA Jl* **15**, 1514 (1977).
5H. G. J. Hanus and M. R. L'Ecuyer, *J. Energy* **1**, 44 (1977).
6H. F. C. Howard and A. J. Srokowski, *AIAA Jl* **15**, 1366 (1977).
7H. F. Kayihan, O. C. Sandall and D. A. Mellichamp, *Chem. Engng Sci.* **32**, 747 (1977).
8H. J. C. LaRue and P. A. Libby, *Physics Fluids* **20**, 192 (1977).
9H. B. Leduc and A. Jaumotte, *Int. Chem. Engng* **17**, 592 (1977).
10H. M. Ledoux and P. Valentin, *Int. J. Heat Mass Transfer* **20**, 961 (1977).
11H. R. E. Mayle, F. C. Kopper, M. F. Blair and D. A. Bailey, *J. Engng Pwr* **99**, 77 (1977).
12H. D. M. Moalem and T. Miloh, *Appl. Scient. Res.* **32**, 395 (1976).
13H. G. Nath and B. K. Meena, *AIAA Jl* **15**, 747 (1977).
14H. N. Nishikawa and H. Oguchi, *J. Heat Transfer* **99**, 342 (1977).
15H. M. A. Paradis, *J. Engng Pwr* **99**, 11 (1977).
16H. D. R. Pedersen, E. R. G. Eckert and R. J. Goldstein, *J. Heat Transfer* **99**, 620 (1977).
17H. V. M. Repukhov and K. A. Bogachuk-Kozachuk, *Heat Transfer, Soviet Res.* **8**(1), 56 (1976).
18H. Y. Tambour and B. Gal-Or, *Physics Fluids* **20**, 880 (1977).
19H. E. Verollet, L. Fulacher and I. Dekeyser, *Int. J. Heat Mass Transfer* **20**, 107 (1977).
20H. J. J. Williams and W. H. Giedt, *J. Heat Transfer* **99**, 386 (1977).

Change of phase

1J. K. J. Baumeister, G. J. Schoessow and C. E. Chmielewski, *Can. J. Chem. Engng* **55**, 521 (1977).
2J. G. G. Brucker and E. M. Sparrow, *Int. J. Heat Mass Transfer* **20**, 371 (1977).
3J. S. C. Cheng, K. T. Heng and W. Ng, *Int. J. Multiphase Flow* **3**, 495 (1977).
4J. K. Cornwall, *Letters Heat Mass Transfer* **4**, 63 (1977).
5J. V. K. Dhir, J. N. Castle and I. Catton, *J. Heat Transfer* **99**, 411 (1977).
6J. Dimić, *Int. J. Heat Mass Transfer* **20**, 1325 (1977).
7J. A. K. Dinger, E. M. Drake and R. C. Reid, *Int. J. Heat Mass Transfer* **20**, 176 (1977).
8J. S. S. Dua and C. L. Tien, *Int. J. Heat Mass Transfer* **20**, 174 (1977).
9J. M. M. Farahat and E. E. Madbouly, *Int. J. Heat Mass Transfer* **20**, 269 (1977).

10J. L. F. Fedorov and E. N. Voropayeva, *Heat Transfer, Soviet Res.* **8**(3), 91 (1976).
11J. I. M. Fedotkin and N. A. Pidsukha, *Heat Transfer, Soviet Res.* **8**(3), 41 (1976).
12J. E. N. Ganic and W. M. Rohsenow, *Int. J. Heat Mass Transfer* **20**, 855 (1977).
13J. L. Gradon and A. Selecki, *Int. J. Heat Mass Transfer* **20**, 459 (1977).
14J. W. T. Hancox and S. Banerjee, *Nucl. Sci. Engng* **64**, 106 (1977).
15J. Y. Hayashi, A. Aoki, S. Adachi and K. Hori, *J. Heat Transfer* **99**, 239 (1977).
16J. W. Idsinga, N. Todreas and R. Bowring, *Int. J. Multiphase Flow* **3**, 401 (1977).
17J. V. P. Isachenko, A. P. Solodov and A. P. Mal'tsev, *Thermal Engng* 23(12), 65 (1976).
18J. V. P. Isachenko, S. A. Soskov and E. V. Yakusheva, *Thermal Engng* **23**(8), 60 (1976).
19J. O. C. Jones, Jr. and J. M. Delhaye, *Int. J. Multiphase Flow* **3**, 89 (1976).
20J. K. A. Joudi and D. D. James, *J. Heat Transfer* **99**, 398 (1977).
21J. S. Kakac, T. N. Veziroglu, N. Özboya and S. S. Lee, *Wärme- und Stoffübertragung* **10**, 175 (1977).
22J. A. Karmi, *Int. J. Heat Mass Transfer* **20**, 1137 (1977).
23J. A. N. Khoze, S. V. Dynayev and V. A. Sparin, *Heat Transfer, Soviet Res.* **8**(3), 87 (1976).
24J. S. E. Kibbee, J. A. Orozco and L. C. Burmeister, *Int. J. Heat Mass Transfer* **20**, 1069 (1977).
25J. O. F. Kinevskiy and Yu. N. Ostrovskiy, *Heat Transfer, Soviet Res.* **8**(3), 99 (1976).
26J. H. M. Kottowski and C. Svatteri, *Int. J. Heat Mass Transfer* **20**, 1281 (1977).
27J. E. Kulic and E. Rhodes, *Can. J. Chem. Engng* **55**, 131 (1977).
28J. E. Marschall and L. L. Moresco, *Int. J. Heat Mass Transfer* **20**, 1013 (1977).
29J. D. D. McCoy and T. J. Hanratty, *Int. J. Multiphase Flow* **3**, 319 (1977).
30J. R. B. Mesler, *A.I.Ch.E. Jl* **23**, 448 (1977).
31J. R. Mesler and G. Mailen, *A.I.Ch.E. Jl* **20**, 954 (1977).
32J. K. Mizukami, *Letters Heat Mass Transfer* **4**, 17 (1977).
33J. Y. Mori, K. Hijikata and K. Utsunomiya, *J. Heat Transfer* **99**, 257 (1977).
34J. N. S. Murty and V. M. K. Sastri, *Can. J. Chem. Engng* **54**, 633 (1976).
35J. K. Nishkawa and Y. Fujita, *Int. J. Heat Mass Transfer* **20**, 233 (1977).
36J. A. P. Ornatskiy, V. A. Chernobay, S. V. Perkov and A. F. Vasil'yev, *Heat Transfer, Soviet Res.* **8**(3), 37 (1976).
37J. E. Pattantyus-H, *Int. J. Heat Mass Transfer* **20**, 1307 (1977).
38J. D. R. Pitts, H. C. Hewitt and B. R. McCullough, *J. Heat Transfer* **99**, 392 (1977).
39J. A. B. Ponter, W. Peier and R. Klimkiezwicz, *Int. J. Heat Mass Transfer* **20**, 1261 (1977).
40J. W. Roetzel, *Int. J. Heat Mass Transfer* **20**, 549 (1977).
41J. S. S. Sadhal and W. W. Martin, *Int. J. Heat Mass Transfer* **20**, 1401 (1977).
42J. A. Sakurai and M. Shiotsu, *J. Heat Transfer* **99**, 554 (1977).
43J. A. Sakurai and M. Shiotsu, *J. Heat Transfer* **99**, 547 (1977).
44J. H. Schmidt, *Int. J. Heat Mass Transfer* **20**, 647 (1977).
45J. H. Schmidt, *Int. J. Heat Mass Transfer* **20**, 635 (1977).
46J. V. E. Schrock, E. S. Starkman and R. A. Brown, *J. Heat Transfer* **99**, 263 (1977).
47J. R. A. W. Shock, *Int. J. Heat Mass Transfer* **20**, 701 (1977).
48J. A. Singh, B. B. Mikic and W. M. Rohsenow, *J. Heat Transfer* **99**, 246 (1977).
49J. L. Stephan and E. G. Hoffman, *Int. J. Heat Mass Transfer* **20**, 1381 (1977).
50J. V. I. Tolubinskiy, A. M. Kichigin and S. G. Povsten, *Heat Transfer, Soviet Res.* **8**(3), 23 (1976).
51J. V. N. Tolubinskiy, Yu. N. Ostrovskiy and V. Ye. Pisarev, *Heat Transfer, Soviet Res.* **8**(3), 17 (1976).
52J. T. Ueda and M. Inoue, *Bull. J.S.M.E.* **20**, 1174 (1977).
53J. H. C. Ünal, *Int. J. Heat Mass Transfer* **20**, 409 (1977).
54J. S. J. D. van Stralen, W. Ziji and D. A. De Vries, *Chem. Engng Sci.* **32**, 1189 (1977).
55J. M. K. Wahi, *J. Heat Transfer* **99**, 145 (1977).
56J. G. L. Wedekind and B. L. Bhatt, *J. Heat Transfer* **99**, 581 (1977).
57J. C. -L. Yu and R. B. Mesler, *Int. J. Heat Mass Transfer* **20**, 827 (1977).
58J. W. Zÿl, *Int. J. Heat Mass Transfer* **20**, 487 (1977).
59J. L. M. Zysina-Molozhen, I. N. Soskova and V. B. Mitenkov, *Heat Transfer, Soviet Res.* **8**(3), 135 (1976).

Radiation in participating media

1K. V. N. Adrianov, *Heat Transfer, Soviet Res.* **8**(1), 94 (1976).
2K. M. N. Afsar and J. B. Hasted, *J. Opt. Soc. Am.* **67**, 902 (1977).
3K. V. V. Aleksandrov and A. A. Frolova, *J. Appl. Math. Mech.* **40**, 785 (1976).
4K. W. J. Anderson and W. N. Hansen, *J. Opt. Soc. Am.* **67**, 1051 (1977).
5K. V. I. Antonovskii and O. V. Kiselev, *Thermal Engng* **23**(2), 22 (1976).
6K. B. F. Armaly and H. S. El-Baz, *J. Heat Transfer* **99**, 208 (1977).
7K. B. F. Armaly and H. S. El-Baz, *AIAA Jl* **8**, 1180 (1977).
8K. J. D. Bankston, J. R. Lloyd and J. L. Novotny, *J. Heat Transfer* **99**, 125 (1977).
9K. P. Baumeister, R. Moore and K. Walsh, *J. Opt. Soc. Am.* **67**, 1039 (1977).
10K. H. B. Becker, *J. Heat Transfer* **99**, 203 (1977).
11K. R. Blickensderfer, D. K. Deardorff and R. L. Lincoln, *Solar Energy* **19**, 429 (1977).
12K. V. C. Boffi, F. Santarelli and C. Stramigioli, *Int. J. Engng Sci.* **15**, 607 (1977).
13K. R. D. Bringans, *J. Physics D: Appl. Phys.* **10**, 1855 (1977).
14K. R. O. Buckins and C. L. Tien, *Int. J. Heat Mass Transfer* **20**, 93 (1977).
15K. B. Carli, Reflectivity of metallic films in the infrared, *J. Opt. Soc. Am.* **67**, 908 (1977).
16K. K. L. Chan, *Physics Fluids* **20**, 1202 (1977).

17K. F. R. Faxvog, *J. Opt. Soc. Am.* **67**, 1376 (1977).
18K. J. D. Felske and C. L. Tien, *J. Heat Transfer* **99**, 458 (1977).
19K. J. B. Gillespie and J. D. Lindberg, *J. Opt. Soc. Am.* **67**, 1363 (1977).
20K. R. R. Gilpin, R. B. Robertson and B. Singh, *J. Heat Transfer* **99**, 227 (1977).
21K. D. Y. Goswami and R. I. Vacon, *Int. J. Heat Mass Transfer* **20**, 1233 (1977).
22K. M. Hass and J. W. Davisson, *J. Opt. Soc. Am.* **67**, 622 (1977).
23K. C. K. Hsieh and A. K. Rajvanshi, *Solar Energy* **19**, 389 (1977).
24K. R. K. James and D. K. Edwards, *J. Heat Transfer* **99**, 221 (1977).
25K. H. Kubota, *J. Heat Transfer* **99**, 628 (1977).
26K. D. W. Larson and R. Viskanta, *J. Fluid Mech.* **78**, 65 (1976).
27K. J. D. Lindberg and J. B. Gillespie, *J. Opt. Soc. Am.* **67**, 1364 (1977).
28K. R. M. F. Linford, C. F. Dillow and T. M. Trumble, *J. Aircraft* **14**, 481 (1977).
29K. D. C. Look and J. T. Tripses, *Letters Heat Mass Transfer* **4**, 129 (1977).
30K. W. J. Minot, *J. Opt. Soc. Am.* **67**, 1046 (1977).
31K. N. K. Nakra and T. F. Smith, *J. Heat Transfer* **99**, 60 (1977).
32K. D. E. Negrelli, J. R. Lloyd and J. L. Novotny, *J. Heat Transfer* **99**, 212 (1977).
33K. D. A. Nelson, *Int. J. Heat Mass Transfer* **20**, 437 (1977).
34K. M. Nikanjam and R. Greif, *Int. J. Heat Mass Transfer* **10**, 1262 (1977).
35K. L. W. Pinkley, P. P. Sethna and D. Williams, *J. Opt. Soc. Am.* **67**, 494 (1977).
36K. K. Saido and W. H. Giedt, *J. Heat Transfer* **99**, 53 (1977).
37K. M. Sparks, *J. Opt. Soc. Am.* **67**, 1590 (1977).
38K. F. Ye. Spokoynyy, Z. R. Gorbis, N. V. Svyatetskiy and R. V. Zagaynova, *Heat Transfer, Soviet Res.* **8**(5), 94 (1976).

Surface radiation

1L. A. Abdulkadir and R. Birkebak, *Wärme- und Stoffubertragung* **10**, 23 (1977).
2L. E. A. Estalote and K. G. Ramanathan, *J. Opt. Soc. Amer.* **67**, 39 (1977).
3L. E. M. Lushiku and K. R. O'Shea, *Solar Energy* **19**, 271 (1977).
4L. H. Masuda and K. Kitazawa, *Bull. J.S.M.E.* **20**, 834 (1977).
5L. T. A. McMath, R. A. D. Hewko, O. Singh, A. E. Curzon and J. C. Irwin, *J. Opt. Soc. Amer.* **67**, 639 (1977).
6L. C. P. Minning, *J. Heat Transfer* **99**, 492 (1977).
7L. K. G. Ramanathan and S. H. Yen, *J. Opt. Soc. Amer.* **67**, 32 (1977).
8L. K. G. Ramanathan, S. H. Yen and E. A. Estalote, *Appl. Opt.* **16**, 2810 (1977).
9L. S. Shimoji, *AIAA Jl* **15**, 289 (1977).
10L. M. L. Smith, *AIAA Jl* **15**, 1057 (1977).
11L. T. Smith, *J. Opt. Soc. Amer.* **67**, 48 (1977).

Liquid metals

1M. Y. Mori, K. Hijikata and I. Kurlyama, *J. Heat Transfer* **99**, 404 (1977).
2M. S. Necmi and J. W. Rose, *Int. J. Heat Mass Transfer* **20**, 877 (1977).
3M. R. M. Singer, D. M. France and W. J. Minkowycz, *Int. J. Heat Mass Transfer* **20**, 87 (1977).

Measurement techniques

1P. R. J. Adrian and K. L. Orloff, *Appl. Optics* **16**, 677 (1977).
2P. V. Ye. Babenko, M. B. Grinbaum and G. B. Panter, *Heat Transfer, Soviet Res.* **8**(5), 56 (1976).
3P. E. A. Balik and J. H. C. Chen, *Appl. Optics* **16**, 674 (1977).
4P. N. Bank and W. H. Gauvin, *Can. J. Chem. Engng* **55**, 516 (1977).
5P. L. B. Belyanskii and D. I. Sharevskaya, *Measmt Tech., Pittsb.* **20**, 527 (1977).
6P. R. D. Birkhoff, L. R. Painter and J. M. Heller, Jr., *Appl. Optics* **16**, 2576 (1977).
7P. B. K. Bragin and B. P. Pavlov, *Measmt Tech., Pittsb.* **19**, 890 (1976).
8P. K. Bremhorst, L. Krebs and D. B. Gilmore, *Int. J. Heat Mass Transfer* **20**, 315 (1977).
9P. P. Chassaing, *Physics Fluids* **20**, 1260 (1977).
10P. C. J. Chen and P. Li, *AIAA Jl* **15**, 869 (1977).
11P. K. Depooter, E. Brundrett and A. B. Strong, *J. Fluids Engng* **99**, 580 (1977).
12P. R. L. Desjardins, *Boundary-Layer Meteorology* **11**, 147 (1977).
13P. B. Dodson, T. Low and J. Michel, *Rev. Scient. Instrum.* **48**, 290 (1977).
14P. O. A. Drozdov, E. A. Kurmazenko and A. V. Revyakin, *Heat Transfer, Soviet Res.* **8**(1), 129 (1976).
15P. T. M. Drzewiecki and R. M. Phillippi, *J. Engng Pwr* **99**, 406 (1977).
16P. B. N. Egorov, V. S. Kilesso and A. G. Komarov, *Thermal Engng* **23**(7), 70 (1976).
17P. T. I. Eklund and R. A. Dobbins, *Int. J. Heat Mass Transfer* **20**, 1051 (1977).
18P. S. L. Gai, *Wärme- und Stoffübertragung* **10**, 107 (1977).
19P. B. W. Hannah and W. I. King, Jr., *AIAA Jl* **15**, 725 (1977).
20P. R. J. Hannemann, *J. Engng Pwr* **99**, 385 (1977).
21P. J. A. Harrington, M. Braunstein and J. E. Rudisill, *Appl. Optics* **16**, 2843 (1977).
22P. A. G. Havener, *AIAA Jl* **15**, 592 (1977).
23P. N. S. Hong, A. R. Jones and F. J. Weinbert, *Proc. R. Soc.* **A353**, 77 (1977).
24P. A. Hordvik, *Appl. Optics* **16**, 2827 (1977).
25P. C. C. Horstman and W. C. Rose, *AIAA Jl* **15**, 395 (1977).
26P. Z. G. Idiatulin, O. G. Tsibinogin and O. M. Petashvili, *Measmt Tech. Pittsb.* **20**, 260 (1977).
27P. A. P. Jackman, R. D. Noble and H. J. Bonouvrie, *Rev. Scient. Instrum.* **48**, 865 (1977).
28P. A. M. Kichigin, *Heat Transfer, Soviet Res.* **8**(3), 32 (1976).
29P. T. G. Kollie, R. L. Anderson, J. L. Horton and M. J. Roberts, *Rev. Scient. Instrum.* **48**, 501 (1977).

30P. D. Kretschmer, J. Odgers and A. F. Schlader, *J. Engng Pwr* **99**, 1 (1977).
31P. R. J. Landry and R. W. Peterson, *Appl. Optics* **16**, 2968 (1977).
32P. S. P. Logvinenko, V. I. Eremenko, V. D. Sukhikh and G. F. Mikhina, *Measmt Tech. Pittsb.* **20**, 414 (1977).
33P. G. F. Lynch and S. L. Segel, *Int. J. Heat Mass Transfer* **20**, 7 (1977).
34P. D. W. McDonald, *Rev. Scient. Instrum.* **48**, 1106 (1977).
35P. M. Merilo, R. L. Dechene and W. M. Cichowias, *J. Heat Transfer* **99**, 330 (1977).
36P. V. V. Mirkovich, *Rev. Scient. Instrum.* **48**, 560 (1977).
37P. H. H. Plumb, L. M. Besley and W. R. G. Kemp, *Rev. Scient. Instrum.* **48**, 419 (1977).
38P. J. K. Reichert and R. S. Azad, *Rev. Scient. Instrum.* **48**, 341 (1977).
39P. R. L. Schwiesow, R. E. Cupp, M. J. Post and R. F. Calfee, *Appl. Optics* **16**, 1145 (1977).
40P. J. R. Smith and W. H. Giedt, *Int. J. Heat Mass Transfer* **20**, 899 (1977).
41P. D. -K. Son, M. Satge and J. Coutanceau, *Int. Chem. Engng* **17**, 617 (1977).
42P. B. I. Stadnyk, Ya. T. Lutsik and V. I. Lakh, *Measmt Tech., Pittsb.* **20**, 257 (1977).
43P. C. A. Swenson, *Rev. Scient. Instrum.* **48**, 489 (1977).
44P. V. I. Telin, *Measmt Tech. Pittsb.* **19**, 896 (1976).
45P. W. N. Trump, H. W. Luebke, L. Fowler and E. M. Emery, *Rev. Scient. Instrum.* **48**, 47 (1977).
46P. B. Welber, *Rev. Scient. Instrum.* **48**, 395 (1977).

Heat exchangers and heat pipes

1Q. A. Bejan, *J. Heat Transfer* **99**, 374 (1977).
2Q. H. Blasinski, A. Heim and C. Kuncewicz, *Int. Chem. Engng* **17**, 548 (1977).
3Q. T. R. Bott and J. S. Gudmundsson, *Can. J. Chem. Engng* **55**, 381 (1977).
4Q. V. Cena, C. Mustacchi and F. Natali, *Chem. Engng Sci.* **32**, 1227 (1977).
5Q. K. H. Chen and G. J. Trezek, *J. Heat Transfer* **99**, 381 (1977).
6Q. K. H. Chen and G. J. Trezek, *J. Heat Transfer* **99**, 274 (1977).
7Q. R. Crozier, Jr. and M. Samuels, *J. Heat Transfer* **99**, 487 (1977).
8Q. V. I. Evenko and A. K. Anisin, *Thermal Engng* **23**, (7), 21 (1976).
9Q. H. A. Frediani, Jr. and N. Smith, *J. Engng Pwr* **99**, 279 (1977).
10Q. A. Hamed, E. Baskharone and W. Tabakoff, *J. Aircraft* **14**, 173 (1977).
11Q. R. B. Holmberg, *J. Heat Transfer* **99**, 196 (1977).
12Q. R. C. Kelahan and J. L. Gaddy, *A.I.Ch.E. Jl* **23**, 816 (1977).
13Q. R. Leenaerts, *Int. Chem. Engng* **17**, 140 (1977).
14Q. R. E. Masterson and L. Wolf, *Nucl. Sci. Engng* **64**, 222 (1977).
15Q. C. F. McDonald, T. Van Hagan and K. Vepa, *J. Engng Pwr* **99**, 237 (1977).
16Q. J. R. Mondt, *J. Engng Pwr* **99**, 657 (1977).
17Q. N. Nishida, Y. A. Liu and L. Lapidus, *A.I.Ch.E. Jl* **23**, 77 (1977).
18Q. W. Roetzel, *Wärme- und Stoffübertragung* **10**, 61 (1977).
19Q. E. U. Schluender, *Int. Chem. Engng* **17**, 245 (1977).
20Q. R. M. Singer and co-workers, *Nucl. Sci. Engng* **63**, 75 (1977).
21Q. H. C. Unal, M. L. G. Van Gasselt and P. W. P. H. Ludwig, *Int. J. Heat Mass Transfer* **20**, 1389 (1977).
22Q. J. G. van Leersum and P. J. Banks, *Int. J. Heat Mass Transfer* **20**, 927 (1977).
23Q. H. Yamashita, R. Izumi and S. Yamaguchi, *Bull. J.S.M.E.* **20**, 1008 (1977).
24Q. A. A. Zhukauskas, R. V. Ulinskas and K. F. Martsinauskas, *Int. Chem. Engng* **17**, 744 (1977).

Aircraft and space vehicles

1R. N. Arai, K. Karshima and K. Sato, *AIAA Jl* **15**, 1655 (1977).
2R. A. D. Baer, J. H. Hedges, J. D. Seader, K. M. Jayakar and L. H. Wojcik, *AIAA Jl* **15**, 1398 (1977).
3R. H. L. Bohon, J. L. Shideler and D. R. Rummler, *J. Spacecraft Rockets* **14**, 626 (1977).
4R. W. D. Goodrich, C. P. Li, C. K. Houston, P. B. Chiu and L. Olmedo, *J. Spacecraft Rockets* **14**, 257 (1977).
5R. W. D. Goodrich and C. J. Staimach, Jr., *J. Spacecraft Rockets* **14**, 638 (1977).
6R. B. J. Griffith, E. E. Edenfield and W. T. Strike, *J. Spacecraft Rockets* **14**, 593 (1977).
7R. D. T. Hove and W. C. L. Shih, *AIAA Jl* **15**, 1002 (1977).
8R. C. H. Lierert and F. S. Stepka, *J. Aircraft* **14**, 487 (1977).
9R. R. K. Matthews and J. C. Key, Jr., *J. Aircraft* **14**, 565 (1977).
10R. J. V. Rakich and M. J. Lanfranco, *J. Spacecraft Rockets* **14**, 265 (1977).
11R. C. J. Staimach, Jr. and W. D. Goodrich, *J. Spacecraft Rockets* **14**, 600 (1977).
12R. E. V. Zoby and R. A. Graves, Jr., *AIAA Jl* **15**, 901 (1977).

General

1S. I. T. Alad'yev, A. D. Perederiy, V. M. Sugrobov and V. P. Trusov, *Heat Transfer, Soviet Res.* **8**(3), 158 (1976).
2S. I. T. Alad'yev, Ye. V. Saperov, V. P. Trusov and V. K. Fardzinov, *Heat Transfer, Soviet Res.* **8**(2), 86 (1976).
3S. A. L. Are, *Heat Transfer, Soviet Res.* **8**(2), 157 (1976).
4S. A. L. Are and R. Ya. Demchenko, *Heat Transfer, Soviet Res.* **8**(2), 117 (1976).
5S. V. T. Balobayev, *Heat Transfer, Soviet Res.* **8**(2), 134 (1976).
6S. E. A. Burgemeister, *J. Phys. D: Appl. Phys.* **10**, 1923 (1977).
7S. A. Floquet, D. Play and M. Godel, *J. Lubric. Tech.* **99**, 277 (1977).
8S. P. E. Frivik, E. Thorbergsen, S. Del Giudice and G. Comini, *J. Heat Transfer* **99**, 533 (1977).
9S. S. Fujii, N. Kanda, T. Kikuta and N. Takayama, *Bull. J.S.M.E.* **20**, 40, (1977).

10S. L. D. Kannberg and L. R. Davis, *J. Heat Transfer* **99**, 648 (1977).
11S. H. -C. Kung, *J. Heat Transfer* **99**, 353 (1977).
12S. A. A. Mandarov, *Heat Transfer, Soviet Res.* **8**(2) 127 (1976).
13S. H. S. Nagaraj, D. M. Sanborn and W. O. Winer, *J. Lubric. Tech.* **99**, 254 (1977).
14S. P. W. O'Callaghan, A. M. Jones and S. D. Probert, *Appl. Energy* **3**, 13 (1977).
15S. P. W. O'Callaghan and S. D. Probert, *Appl. Energy* **1**, 3 (1977).
16S. B. A. Olovin, *Heat Transfer, Soviet Res.* **8**(2), 176 (1976).
17S. S. D. Probert and T. J. Thirst, *Appl. Energy* **3**, 41 (1977).
18S. D. L. Slegel and L. R. Davis, *J. Heat Transfer* **99**, 541 (1977).
19S. S. K. Yu, P. R. Farmer and M. W. E. Coney, *Int. J. Multiphase Flow* **3**, 415 (1977).
20S. R. G. Siddall and J. K. Pearson, *Int. J. Heat Mass Transfer* **20**, 815 (1977).
21S. A. S. Tsyrul'nikov, I. A. Ryzhenko, E. I. Merzlyakov and V. T. Beloborod'ko, *Heat Transfer, Soviet Res.* **8**(1), 13 (1976).
22S. I. A. Zolotar', *Heat Transfer, Soviet Res.* **8**(2), 92 (1976).

Solar energy

1T. D. E. Ackley and J. Tauc, *Appl. Optics* **16**, 2806 (1977).
2T. J. W. Baughn and J. B. Bergquam, *J. Engng Pwr* **99**, 490 (1977).
3T. D. R. Biswas, *Solar Energy* **19**, 99 (1977).
4T. A. Cabelli, *Solar Energy* **19**, 45 (1977).
5T. D. J. Close and R. V. Dunkle, *Solar Energy* **19**, 233 (1977).
6T. P. M. Driver, R. W. Jones, C. L. Riddiford and R. J. Simpson, *Solar Energy* **19**, 301 (1977).
7T. R. Gicquel, *Int. Chem. Engng* **17**, 575 (1977).
8T. S. W. Hogg and B. G. Smith, *J. Phys. D. Appl. Phys.* **10**, 1863 (1977).
9T. B. D. Hunn and D. O. Calafell, II, *Solar Energy* **19**, 87 (1977).
10T. P. O. Jarvinen, *Solar Energy* **19**, 139 (1977).
11T. S. A. Klein, *Solar Energy* **19**, 325 (1977).
12T. W. H. Klein, B. Goldberg and W. Shropshire, Jr., *Solar Energy* **19**, 115 (1977).
13T. P. J. Lunde, *ASHRAE Jl* **19**, 55 (1977).
14T. J. C. McMurrin, N. A. Djordjevic and H. Buchberg, *J. Heat Transfer* **99**, 169 (1977).
15T. L. Melamed and G. M. Kaplan, *J. Energy* **1**, 100 (1977).
16T. P. W. O'Callaghan and S. D. Probert, *Appl. Energy* **3**, 51 (1977).
17P. P. O'Neill, C. Doland and A. Ignatiev, *Appl. Optics* **16**, 2822 (1977).
18T. J. W. Ramsey, B. P. Gupta and G. R. Knowles, *J. Heat Transfer* **99**, 163 (1977).
19T. P. P. Rao, J. E. Francis and T. J. Love, Jr., *J. Energy* **1**, 324 (1977).
20T. M. Riaz, *J. Heat Transfer* **99**, 489 (1977).
21T. M. Riaz and T. Gurr, *Solar Energy* **19**, 185 (1977).
22T. F. W. Schmidt, R. R. Somers, II, J. Szego and D. H. Laananen, *J. Heat Transfer* **99**, 174 (1977).
23T. D. L. Siebers and R. Viskanta, *Solar Energy* **19**, 163 (1977).
24T. D. L. Siebers and R. Viskanta, *Solar Energy* **19**, 411 (1977).
25T. D. E. Soule and D. W. Smith, *Appl. Optics* **16**, 2818 (1977).
26T. E. M. Sparrow and R. J. Krowech, *J. Heat Transfer* **99**, 360 (1977).
27T. E. M. Sparrow and K. K. Tien, *J. Heat Transfer* **99**, 507 (1977).
28T. T. Tanaka, T. Tani, S. Sawata, K. Sakuta and T. Horigome, *Solar Energy* **19**, 415 (1977).
29T. R. C. Temps and K. L. Coulson, *Solar Energy* **19**, 179 (1977).
30T. T. P. Woodman, *Solar Energy* **19**, 263 (1977).
31T. Y. Zvirin, A. Shitzer and G. Grossman, *Int. J. Heat Mass Transfer* **20**, 997 (1977).

Plasma heat transfer

1U. I. Auerbach, M. L. Lieberman, K. E. Lawson and H. O. Pierson, *J. Spacecraft Rockets* **14**, 19 (1977).
2U. N. Datta and R. N. Jana, *Int. J. Engng Sci.* **15**, 561 (1977).
3U. C. Goldbach, G. Nollez and P. Plomdeur, *J. Phys. B: Atom. Molec. Phys.* **10**, 1181 (1977).
4U. G. Gouesbet and M. Trinité, *Letters Heat Mass Transfer* **4**, 141 (1977).
5U. S. M. Hamblyn, *Mineral Sci. Engng* **9**, 151 (1977).
6U. J. V. Heberlein and E. Pfender, *IEEE Trans. Plasma Sci.* **PS-5**(3), 171 (1977).
7U. A. H. Hitchcock and A. E. Guile, *Electronics Letters* **13**(7), (1977).
8U. S. Kamakshaiah and R. S. Rau, *IEEE Trans. Plasma Sci.* **PS-5**(1) 1 (1977).
9U. D. Kannapan and T. K. Bose, Physics Fluids 22, 1668 (1977).
10U. W. Kollmann, H. Sommer and C. G. Stojanoff, *J. Non-Equil. Thermodyn.* **2**(1), 17 (1977).
11U. P. F. Logan, R. J. Stalker and M. K. McIntosh, *J. Phys. D: Appl. Phys.* **10**, 323 (1977).
12U. C. C. Marty, *J. Engng Ind.* **99**, 682 (1977).
13U. O. Matsumoto, H. Hidetoshi, Y. Kawara and T. Ohue, *High Temp. Sci.* **9**, 27 (1977).
14U. G. S. Mills, *Welding J. (Welding Res. Suppl.)* **56**(3), 93-S (1977).
15U. Z. A. Pietrzyk, D. W. Scudder and G. Nelson, *Physics Fluids* **20**, 1204 (1977).
16U. T. H. Rautenberg, Jr. and P. D. Johnson, *J. Appl. Phys.* **48**, 2270 (1977).
17U. D. C. Strachan, *J. Phys. D: Appl. Phys.* **10**, 361 (1977).
18U. A. Tslaf, *IEEE Trans. Plasma Sci.* **PS-5**(2), 53 (1977).
19U. R. E. Voshall, R. J. Wright and R. W. Liebermann, *IEEE Trans. Plasma Sci.* **PS-5**(2) 110 (1977).
20U. H. L. Walmsley, G. R. Jones, F. Haji and D. C. Strachan, *J. Phys. D: Appl. Phys.* **10**, 383 (1977).
21U. J. L. Wu and D. M. Benenson, *IEEE Trans. Plasma Sci.* **PS-5**(4), 286 (1977).
22U. T. Yoshida and K. Akashi, *J. Appl. Phys.* **48**, 2252 (1977).

HEAT TRANSFER REVIEW - 1978

E. R. G. ECKERT, E. M. SPARROW, R. J. GOLDSTEIN, C. J. SCOTT, E. PFENDER, S. V. PATANKAR and J. W. RAMSEY

Heat Transfer Laboratory, Department of Mechanical Engineering, University of Minnesota, Minneapolis, Minnesota, U.S.A.

INTRODUCTION

THIS review surveys results that have been published in various fields of heat transfer during 1978. As in the past, the number of papers published during that period was such that only a selection can be included in this Review. A more complete listing of papers is available in the Heat Transfer Bibliographies published periodically in this journal.

The twenty-third Annual International Gas Turbine Conference was held 9–13 April 1978 at the Wembley Conference Centre, London, England. It included three sessions on heat transfer in which fifteen papers were presented on film and liquid cooling of turbine blades, cooling of disks, and on heat exchangers. The second AIAA/ASME Thermophysics and Heat Transfer Conference was in its greater part devoted to heat transfer. Papers may be obtained from the AIAA or from the ASME order department. Volumes on environmental effects of atmospheric heat/moisture release covering two sessions of the meeting can also be ordered from the ASME.

The twenty-sixth meeting of the Heat Transfer and Fluid Mechanics Institute was held at Washington State University 26–28 June 1978. About half of the papers presented were concerned with heat transfer problems. Three invitation lectures dealt with "Recent Trends in Experimental Heat Transfer and Fluid Mechanics" (A. E. Bergles), "Structure and Transport Properties in Turbulent Shear Flows" (L. S. G. Kovasznay), and "An Approach to Numerical Heat Transfer" (S. V. Patankar). The conference proceedings are available from Stanford University Press.

The Sixth Internation Heat Transfer Conference was held 7–11 August 1978 in Toronto, Canada. The Canadian Executive Committee chose a novel way for the organization of the conference by scheduling a large portion of the available time for discussion. The thirty-six keynote papers were presented by authors from many countries. They surveyed the present state of knowledge in the various fields of heat transfer, whereas the authors of the general papers were available to interested attendants for discussion of their papers during seventeen poster sessions, each covering ten to forty papers. Ten panel sessions were held on selected subjects. An exhibition of heat transfer equipment was open to attendants during the conference. The conference proceedings can be obtained from Hemisphere Publishing Corporation, Washington, DC in eight hardbound volumes.

The 1978 International Seminar, organized by the International Center for Heat and Mass Transfer, 1–5 September 1978, was devoted to two-phase momentum, heat and mass transfer in chemical processes and energy engineering systems. The conference proceedings are being published by the Hemisphere Publishing Corporation, Washington, DC.

The 1978 Annual Meeting of the American Underground Space Association was also held in Toronto, Canada, 21–23 October 1978. The presentations and discussions dealt primarily with energy demands of earth sheltered buildings. Information on the conference can be obtained through the American Underground Space Association, Department of Civil and Mineral Engineering, University of Minnesota. A Conference on Physicochemical Hydrodynamics, 6–8 November 1978 in Washington, DC, honored B. G. Levich. Twenty-five lectures discussed topics such as two-phase flow, change of phase, heat and mass transfer in chemical reactions, thermal diffusion effects and surface tension.

The Annual Meeting of the ASME, 10–15 December 1978, in San Francisco, included seventeen sessions devoted to heat transfer. Sessions on solar energy and gas turbines also included heat transfer effects. Chang-Lin Tien discussed heat transfer in the Peoples' Republic of China at the Heat Transfer Luncheon. Heat transfer memorial awards were presented to R. J. Goldstein and J. A. Clark. Dr. Benjamin Gebhart received the Freeman Scholar Award for his review of buoyancy introduced fluid motion characteristic of applications in technology. Reprints of the papers presented at the annual meeting are available at ASME Headquarters in New York and many of them will be published in the *Journal of Heat Transfer.*

A number of new journals deal with heat transfer subjects. The *Regional Journal of Energy, Heat and Mass Transfer* can be subscribed to at the Scientific Secretary, Regional Centre for Energy, Heat and Mass Transfer for Asia and the Pacific, Department of Mechanical Engineering, Indian Institute of Technology, Madras, India. The journal, *Numerical Heat Transfer*, started its publication in January 1978 and is available from Hemisphere Publishing Corporation and McGraw-Hill International Book Company. The journal, *Energy Developments in Japan*, started in July 1978 and can be subscribed to at Rumford Publishing Company, Chicago, Illinois. A number of books dealing with heat transfer or including heat transfer topics has appeared on the market. They are listed in the bibliographic part of this review.

Trends and developments in heat transfer research during 1978 are characterized by the following highlights: Studies in heat conduction included phase change and moving boundaries. Strong interest appeared in solution methodology. Channel flow was studied in complex configurations, and techniques to enhance heat transfer found attention. Also included were various types of fluid property effects, surface mass transfer and freezing.

Published work on boundary layers consists of experimental and numerical investigations. There is a greater emphasis on three-dimensional boundary layers, rough surfaces, stagnation point and conjugate heat transfer. Interest in heat transfer in separated regions on single bodies appears to be diminishing, whereas the literature on packed and fluidized bed heat transfer experienced an explosive growth.

The work on transfer mechanisms is almost exclusively concerned with turbulence. Detailed measurements of fluctuating quantities in shear flows have been reported. Statistical and phenomenological models for turbulent flows have been proposed. Natural convection studies concentrated on transient effects, non-Newtonian fluids and the use of finite element methods. Heat transfer from rotating single and multiple disks, in axially and radially rotating pipes and annuli was studied analytically and experimentally. Research on phase change appears to be reverting to fundamental analytic work.

Studies on radiative energy transport in non-gray emitting, absorbing, and scattering media are of particular interest. Results of measurements of radiation properties for various media have been published. Analysis is concerned with radiative exchange between surfaces including absorption, reflection, and polarization. The effect of scattering was studied in fibrous media. Papers concerned with measurement techniques concentrate on signal processing techniques to extract temperature and velocity information from hot wire measurements. Publications on the laser-Doppler method have decreased in number.

Interest in heat exchangers is still growing. Experiments are directed toward a detailed study of the performance of heat exchange surfaces using, for instance, the naphthalene sublimation technique. Analysis optimizes the performance of single heat exchangers and of arrays of them. Five papers investigate processes in heat pipes. Plasma heat transfer studies were primarily concerned with problems of arc circuit interruption and with applications in plasma chemistry.

To facilitate the use of this review, a listing of the subject headings is made below in the order in which they appear in the text. The letter which appears adjacent to each subject heading is also attached to the references that are cited in that category:

Conduction, A
Channel flow, B
Boundary-layer and external flows, C
Flow with separated regions, D
Transfer mechanisms, E
Natural convection, F
Convection from rotating surfaces, G
Combined heat and mass transfer, H
Change of phase, J
Radiation
  Radiation in participating media, K
  Surface radiation, L
Measurement techniques, P
Heat-transfer applications
  Heat exchangers and heat pipes, Q
  Aircraft and space vehicles, R
  General, S
  Solar energy, T
Plasma heat transfer, U.

CONDUCTION

The recent heat conduction literature is dominated by phase change and related moving boundary problems. There is, in addition, a strong interest in solution methodologies.

The basic assumption that Stefan-type melting problems are purely heat conduction problems was severely shaken by recent experiments. Those experiments, which involved a horizontal cylindrical heater embedded in a phase change medium, demonstrated that after a brief initial period when conduction prevails, the melting process is dominated by natural convection in the liquid melt [42A]. A similar finding was encountered in experiments in which the melting occurred about a vertical cylindrical heater [35A]. Experiments on the freezing of water passing in crossflow over a cooled tube showed that the thickness of the frozen layer is a minimum at the forward stagnation point and increases around the circumference, attaining a constant value in the rear portion of the cylinder [30A]. An analysis of solidification in a continuous casting process showed that the thickness of the solidified material increases approximately as the square root of the distance along the mold from the location where solidification begins [39A]. Increasing the ingot withdrawal rate or the heat addition at the solidification surface causes the interface to move downward within the mold and have a nonplanar shape [40A]. A metal sphere, when immersed in its own liquid, will at first be enveloped in a freezing annulus which will subsequently melt, as will the sphere itself [10A]. A numerical solution for simultaneous melting and evaporation was motivated by the need to determine the propagation of the melting crater and the material erosion in electrical discharge machining [50A]. The inward freezing in an annulus was found to be markedly influenced by the thermal boundary condition at the inner bounding cylinder [16A].

A novel numerical approach, which focuses on the melting (freezing) of particles rather than of the bulk solid, has been formulated and applied [13A]. Another innovative numerical scheme for one-dimensional phase change problems continuously follows the position of the moving interface [12A]. To neutralize the difficulties of dealing with a moving boundary in phase change problems, a change of variable has been introduced which fixes the melting (freezing) front [38A]. To facilitate an analysis of the melting of a subcooled solid surrounding a heated cylinder, coordinate transformations were made which immobilize both the moving interface between the melt region and the solid and the moving temperature wave that diffuses into the solid. Depending on the degree of subcooling, the surface heat transfer can be several times greater than that for no subcooling [41A]. The application of the enthalpy method to multidimensional phase change problems is once again described, thereby repeating earlier work published in the *Journal of Heat Transfer* **97**, 333 (1975) [9A]. In a finite-element method, the difficulty of handling sharp peaks in the heat capacity–temperature curve has been mitigated by use of the enthalpy as the prime variable, with the specific heat being evaluated as a simple backward difference involving enthalpies and temperatures [27A]. For substances which solidify over a range of temperatures, an analytical method which uses an apparent heat capacity which includes the latent heat was proposed and evaluated experimentally [14A]. A refinement of the Heat Balance Integral for phase change problems is based on subdividing the surface-to-melting (freezing) temperature range into equal intervals and applying the HBI to each individual interval [4A].

The liberation (or absorption) of heat during solidification (or melting) can be treated as a moving heat source (or sink) located at the liquid–solid interface [31A]. The Landau melting problem, which is characterized by an applied heat flux at the moving interface, was numerically solved by working with an integrodifferential equation for the position of the interface [36A]. Solidification of a binary mixture of finite thickness does not admit an exact solution, and a series method has been used [5A]. A generalized Neumann problem involving arbitrary initial conditions and arbitrary boundary conditions does not yield a similarity solution as do the basic Stefan and Neumann problems, but the similarity variable $x/\sqrt{t}$ can be used in constructing the solution for the generalized problem[46A]. The zero- and finite-order moments of the heat equation, used in conjunction with physically motivated inequalities relating to the temperature distribution, were employed to determine bounds on the movement of the interface in Stefan-type problems [11A]. A moving boundary problem with combustion at the moving interface was studied from the existence-uniqueness viewpoint, but useful solutions are not presented [44A]. The governing equations and solution algorithm for one-dimensional transient heat conduction in an expanding rod with one free end are similar to those for Stefan-type phase change problems [24A].

Several papers dealing with topics in steady heat conduction have been published. The critical radius of insulation has been generalized to accommodate experimental correlations of the natural convection heat transfer coefficient which account for variations of temperature difference and body dimensions [3A]. A further generalization included the effects of uniform surface heat generation due to a phase change at the external convection boundary [19A]. A new algorithm has been formulated for evaluating the performance of arrays of extended surfaces: the performance of the array involves cascading the parameters for the individual fins [17A]. Exact (series) solutions were derived for the steady state temperature field in an infinitely long cylinder in which there is a non-axisymmetric heat source [29A]. A series method was used for determining the steady temperature field in an eccentric annulus with uniform inner thermal conditions and circumferentially nonuniform outer conditions [2A]. Solutions were obtained via Green's functions for the steady temperature field in a nonmoving liquid metal contained in a long duct in which there is a heated centerbody of finite length [25A]. The Picard iterative technique is shown to be effective in solving steady one-dimensional heat conduction problems with internal heat sources [26A]. In the conduction limit (zero Reynolds number) for the Nusselt number for flow external to bodies, the effect of variable thermal conductivity in the fluid is independent of the shape of the body [7A].

Transient problems were treated in a number of papers. The orthogonal collocation method, as applied to transient heat conduction with variable conductivity, involves the approximation of the spatial derivative terms in the energy equation by an orthogonal polynomial, thereby transforming the equation into a set of ordinary differential equations that can be integrated numerically [20A]. Temperature-dependent thermal conductivity in transient conduction problems can also be handled by a series solution in which the expansion parameter is related to the conductivity variation [45A]. Correction formulae have been developed to extend the applicability of the Heisler transient response charts to solids with temperature-dependent thermal conductivity [43A]. The use of a vectored separation of variables approach enables solution of the coupled partial differential equations which express energy conservation for the fluid and solid components of a packed-bed thermal storage tank [22A]. The temperature field created by a moving three-dimensional heat source was determined by integration of a solution for a one-dimensional source [49A]. Upper and lower bounds on the transient temperature distribution in the region surrounding a buried cylindrical heat source were determined by application of the theory of differential inequalities [23A].

Interest persists in composite and anisotropic solids. A mixture theory for diffusion in a fiber-reinforced composite converts what is essentially a three-dimensional problem into a problem involving a single spatial variable [21A]. In another paper, it was shown that a number of anisotropic heat conduction problems can be transformed into isotropic ones [33A]. The idea of using steady equivalent thermal properties as an alternative to unsteady heat conduction analysis was experimentally examined for layered composites and found to be satisfactory only when the thermal conductivities of the two constituents do not differ widely [48A]. An analysis of the transient temperature field in an aerodynamically heated laminated composite conical shell utilized Galerkin's method in conjunction with the Laplace transform technique and showed that the resulting temperature gradients were large enough to warrant concern about thermal stresses [34A]. The heat propagation in bilaminates depends critically on the behavior at the interfaces [15A]. A steady state solution for an infinite anisotropic sheet surrounding an insulated elliptical hole indicates the presence of rapid temperature changes adjacent to the hole [37A].

Numerical methods have already been cited in earlier paragraphs of this section, and other numerically oriented papers will now be reported. Computationally useful methods have been devised for determining the critical time step to prevent oscillatory results in a finite-element marching solution for two-dimensional transient heat conduction [28A]. A finite-element procedure for elements with several temperature-dependent thermal parameters was formulated by resolving element matrices into component matrices, one component for each thermal parameter [47A]. Through examples, it was demonstrated that a finite-element method can be effectively used to treat the inverse problem of transient heat conduction [18A]. In a novel approach to the solution of transient heat flow problems in two dimensions, the movements of isotherms along orthogonal lines are treated in successive small intervals of time [8A]. The use of a Galilean transformation to immobilize a moving thermal interface leads to a finite element method where the element stiffness is no longer symmetric [32A].

There are a few rather mathematical papers which deal with topics of engineering relevance. Mathematical theorems are employed to devise a method for determining the thermal conductivity from overspecified initial and boundary data [6A]. Some progress has been made toward formulating the problem of heat conduction in the presence of random thermal conductivity, but neither numerical solutions nor results have been obtained [1A].

CHANNEL FLOW

The current literature on heat transfer in ducts reflects a strong interest in flows in complex configurations and in techniques to enhance heat transfer. Various types of fluid property effects, including non-Newtonian behavior, were investigated. A modest interest in turbulence models persists, and a number of papers concerned with mass transfer and with freezing have been published. The venerable area of laminar developing and fully developed heat transfer continues to draw some attention. On the basis of a new compilation of data, the exponent of the Reynolds number in a Nusselt–Reynolds–Prandtl correction for turbulent pipe flow has been shown to be a function of both Re and Pr, for Prandtl numbers below 100 [15B].

Research involving complex flows and enhanced heat transfer surfaces will now be described. Augmentation is accomplished in a static mixer by alternately subdividing the flow into separate parallel streams and then recombining the streams [60B]. A finite difference solution for turbulent heat transfer in a wavy-wall tube shows that positive results of the augmentation can be attained at moderate and small Pr, but not at large Pr [25B]. The use of sonic oscillations to increase laminar and turbulent heat transfer in a tube may have a side benefit of preventing salt deposits on heat transfer surfaces in commercial heat exchangers [51B]. Turbulent airflow experiments in a parallel plate channel with rib-roughened surfaces showed that ribs at a 45° angle of attack have superior heat transfer performance at a given friction power than ribs at a 90° angle of attack or than sand-grain roughness [26B]. For a two-plate colinear array aligned parallel to the flow, airflow experiments showed that under most practical considerations the effect of increasing plate thickness is to increase the Nusselt number [16B].

Measurements of heat transfer in an array of wall-attached cylinders situated in a crossflow of air in a flat rectangular duct showed that fully developed conditions prevail for the fourth and all subsequent rows [56B]. Experimentally determined heat transfer coefficients on the upstream and downstream faces of a transversely positioned blockage in a square duct were found to be quite insensitive to the extent of the blockage [57B]. Heat transfer experiments in helical coiled tubes showed that the overall heat transfer coefficients are relatively insensitive to whether the boundary condition is uniform wall temperature or uniform heat transfer per unit length [31B]. Heat transfer results for fully developed laminar flow in a pipe containing a longitudinal rod assembly were found to be sensitive to the size of the gap between the outermost ring of rods and the pipe wall [4B]. A new form of Reynolds analogy for a rough surface, derived on the assumption that the eddy Prandtl number is constant, is purported to supersede existing relationships [38B].

When a tube or duct flow is subjected to a streamwise periodic variation of wall temperature or wall heat flux, a periodic thermally developed regime is established at sufficient distances from the inlet [44B]. An analysis of laminar forced convection heat transfer from a shrouded fin array with tip clearance shows that the minimum fin heat loss occurs adjacent to the fin base and the maximum heat loss occurs at the tip [54B]. In turbulent duct flows in which the walls are nonuniformly heated and in which wall conduction plays a role, the Reynolds number exponent in the Nusselt–Reynolds relation is modified by the interaction of the wall conduction with the fluid convection [5B]. For pipe flow experiments involving circumferentially varying thermal boundary conditions, criteria are formulated which facilitate the determination of boundary conditions on the inner wall of the tube from measurements made on the outer wall [3B].

A strong interest has been evidenced in various types of fluid property effects. An analysis of the effects of temperature-dependent viscosity on fully developed turbulent pipe flow for liquids expresses the heat transfer coefficient ratio for variable to constant viscosity in terms of the friction factor ratio for variable to constant viscosity as well as of the wall and bulk temperatures and of a fluid viscosity–temperature parameter [27B]. An experimental investigation of heat transfer to laminar flow of oil in a horizontal pipe gave results which agree with the Sieder–Tate correlation at small temperature differences, but deviate at larger temperature differences [42B]. The accounting of variable liquid viscosity in laminar pipe flow shows that the Nusselt number is increased by heating and decreased by cooling [35B]. A linear stability analysis studying the onset of secondary flows in a parallel plate channel dealt with water in a temperature range which encompassed the maximum density condition [13B]. The transient response of a turbulent supercritical water flow in a tube is different for increases of wall heat flux than for decreases in wall heat flux [53B]. The correlation of heat transfer data for supercritical helium is facilitated by a harmonic mean Prandtl number and the use of a Reynolds number which is the lesser of the Reynolds numbers corresponding to wall and bulk conditions [8B]. The turbulent friction and heat transfer coefficients for dilute suspensions of asbestos fibers in water may be much lower than those for water flow alone [41B]. Analysis and experiments for electrorheological laminar flow in an annulus showed that the heat transfer is enhanced with increasing electric intensity [52B]. When the Joule–Thompson effect is important, the pressure work and viscous dissipation have to be included in the energy equation, as in an analysis of the turbulent Graetz problem [11B].

The interest in property effects is further substantiated by research on non-Newtonian flows. From numerical solutions for simultaneously developing flow and heat transfer in a square duct, it is found that for the same Graetz number, Prandtl number, and thermal boundary condition, a non-Newtonian fluid with flow behavior index <1 gives a higher heat transfer coefficient than a Newtonian fluid [9B]. Correction factors have been formulated to relate temperature-dependent dissipative non-Newtonian flows to simpler flows where these effects are absent [62B]. A study of the thermal entrance region for combined Couette–Poisseuille flow of a power-law non-Newtonian fluid showed that dilatant fluids attain developed temperature profiles in shorter lengths than do pseudoplastic fluids [39B]. Numerical solutions enabled correlation of variable property effects for laminar non-Newtonian flow in a tube with uniform wall temperature [24B]; the uniform heat flux case is treated in a companion paper [23B].

A few papers deal with turbulent transport characteristics. The fully developed heat transfer for turbulent flow in concentric annuli, for which prior solutions have been obtained by eddy diffusivity models, was resolved by employing a surface rejuvenation model [14B]. Best agreement of analysis and experiment for radial and circumferential diffusion in turbulent flow in an annulus occurs for a ratio of circumferential to radial eddy diffusivity equal to two [28B]. Experiments on turbulent airflow in the entrance region of an isothermal-walled tube gave a value of 0.97 for the turbulent Prandtl number in the turbulent part of the wall region [29B]. A penetration model for the fluid boundary layer adjacent to a pipe wall enabled calculation of the effect of chemical reaction on the heat transfer coefficient for turbulent pipe flow [43B].

Mass transfer effects drew some attention. Injection of fluid at the inner wall of an annular duct leads to an increase in the turbulent heat transfer coefficient corresponding to heating at the outer wall [19B]. Laminar flow in a parallel plate channel with injection of a reactant was studied to aid in interpreting experiments on diffusion flames in counterflow configurations above liquid and solid fuels [50B]. The axial heat conduction problem for laminar pipe flow has been re-examined in the context of mass transfer [17B]. Fully developed heat transfer in a tube in the presence of an external thermal resistance was revisited in the analogous problem of mass transfer with a first-order chemical reaction at a catalytic wall [30B]. Mass transfer measurements for turbulent airflow in an annulus with non-axisymmetric injection of nitrous oxide gas at the outer wall demonstrated that an isotropic eddy diffusivity model underestimates the magnitude of the circumferential diffusion [48B]. The Leveque solution, which is valid in the immediate neighborhood of a duct inlet, has been extended to larger downstream distances for laminar flow in a parallel plate channel in which the wall mass (heat) flux is proportional to the wall concentration (temperature) [10B].

The freezing of liquids in pipes was investigated. Experiments on the freezing of water flowing turbulently in a cooled isothermal tube agreed well with analytical predictions based on a simple turbulence model [59B]. In the flow of hot Freon 112A through a melting ice pipe, two distinct modes of Freon freezing were observed [63B]. Previous solutions for solidification in a tube having a given wall temperature were extended to the case where there is a convective heat transfer to a fluid environment outside the tube [20B]. The classical Graetz-type analysis was extended to the case of laminar liquid flow in a parallel plate channel in which there is sufficient cooling at one of the walls to cause freezing at that wall [12B].

Papers on laminar duct flows continue to appear. In a pair of companion papers, the Galerkin–Kantorowich variational method was used to solve the laminar thermal entrance region problem for a rectangular duct and a circular tube, the outside surface of which exchanges heat with a surrounding fluid medium; the tube study also took MHD effects into account [32B, 33B]. Finite difference and Galerkin solutions for the laminar thermal entrance region of an annulus agree very well with each other [40B]. A series solution starting at the pipe inlet and proceeding in the downstream direction was used for solving the laminar entrance region problem for a variety of thermal boundary conditions [47B]. A refined integral momentum and energy analysis for simultaneously developing laminar velocity and temperature in a parallel-plate channel gave Nusselt numbers that are in good agreement with literature values [6B]. Heat transfer measurements in the entrance region of a rectangular channel with laminar or transitional flow of air resulted in a correlation [36B]. Circumferentially nonuniform heating in the entrance region of a laminar pipe flow cannot be handled by the methods used in the Graetz problem because the eigenfunctions are not orthogonal [21B]. In the thermally developed regime for laminar pipe flow where there is heat exchange with the surroundings via a circumferentially varying external heat transfer coefficient, the average heat transfer coefficient is quite insensitive to the imposed circumferential variations [55B].

A variety of topics are treated in papers published in *Heat Transfer, Soviet Research.* These include: transient heat transfer in tube flows [34B, 49B], variable property effects [1B, 61B], effect of turbulence level in the fluid entering a tube or duct [18B, 58B], axial wall temperature variations [45B], high frequency pressure fluctuations [22B], crossflow heat transfer to a flattened tube situated in a duct inlet [2B], high Prandtl number experiments using electrochemical techniques [7B], deviations from Reynolds analogy due to surface roughness [46B], and longitudinal turbulent flow in a rod bundle [37B].

BOUNDARY LAYER AND EXTERNAL FLOWS

Publications on boundary layers deal with laminar and turbulent flows on flat and curved surfaces, flows over cylinders and spheres, free jets, impinging jets, and falling liquid films. Some new experimental data have been presented, and a considerable amount of activity has been directed towards the development and application of analytical and numerical techniques.

A calculation method has been reported for the analysis of two-dimensional boundary layers [5C]. For the same general situation, a Galerkin finite-element method using Hermitian polynomials has been applied, and the results have been compared with those from other methods [6C]. A new numerical solution scheme is described for the multi-equation systems encountered in local nonsimilarity boundary layer analysis. It employs integrated forms of the governing equations, which already satisfy the boundary conditions. The method is shown to be very efficient compared with the usual forward integration/shooting method [40C]. A numerical scheme is presented for boundary layers with massive injection through a porous surface. It is based on the exponential difference operator and on the general approach of the Keller box scheme. Applications were made to self-similar boundary layers [14C]. A calculation method which is capable of handling regions of sharp pressure gradients in two-dimensional laminar boundary layers is described [11C]. Calculations are reported for turbulent compressible boundary layers on a permeable plate [3C]. An integral method has been used to predict friction and heat transfer from compressible turbulent boundary layers in a convergent–divergent nozzle. The predicted results are shown to agree well with experimental data [35C]. An integral analysis of the thermal turbulent boundary layer is described; it uses the inner region parameters based on the friction velocity [58C].

Three-dimensional turbulent boundary layers have been computed by a finite-difference method, the turbulence being characterized by a two-equation model. The situations chosen for the computation are: the boundary layer on a flat plate approaching a circular cylinder mounted on the plate, the boundary layer beneath the leading-edge vortex on a delta wing, and the flow near the upper surface of a curved duct. The predicted results show reasonable agreement with available experimental data [45C]. A three-dimensional parabolic procedure employing non-orthogonal coordinates has been developed for the prediction of turbulent boundary layers on bodies of arbitrary shape [34C].

The second-order laminar boundary layer over a parabola has been analyzed [21C]. The Reynolds analogy is discussed with reference to boundary layers with pressure gradients [39C]. Squire's temperature profile for a turbulent boundary layer on an adiabatic flat plate has been extended to the case of finite heat transfer. The Reynolds analogy factor has thereby been deduced, and the results have been compared with experimental data [60C]. A transport equation for the turbulent heat flux in a boundary layer has been derived on the basis of local similarity [2C]. The diffusion in a turbulent boundary layer downstream of a line source is discussed. Two different eddy diffusivity distributions are considered, and their implications are compared with experimental data

[37C].

An analysis has been made of turbulent boundary layers subjected to adverse pressure gradients. A general family of velocity profiles was derived from the assumption that the boundary layer is in local equilibrium. The family agrees well with almost all available experimental data [23C]. Solutions have been presented for the boundary layer equations with prescribed variations of the free-stream velocity. The method of inner and outer expansions is used [38C]. An approximate analysis has been reported for the thermal response of a laminar boundary layer over a smooth object of arbitrary shape. The results show good agreement with available experimental data for Prandtl numbers of unity or larger [31C]. Dimensionless equations are given for laminar flow heat transfer with variable physical properties [33C].

Some work has dealt with boundary layers on rough walls. An algebraic eddy–viscosity model modified for wall roughness has been used to predict turbulent boundary layers in adverse, zero, and favorable pressure gradients. The results show satisfactory agreement with experimental data [8C]. A method is described for the calculation of heat transfer from rough surfaces [54C]. An integral method has been developed for the prediction of heat transfer from a fully rough turbulent boundary layer. The predictions show good agreement with experimental data over a range of conditions including variations of free-stream velocity, surface temperature, and blowing [10C].

Some new analytical results are presented for conjugate heat transfer in a flat plate boundary layer [25C]. For the same problem, an approximate method for solving an integral equation is described [26C].

The results of numerical calculations are reported for turbulent boundary layers with significant variations of physical properties [56C]. A companion experimental investigation presents results for the heat transfer rates in a channel through which combustion products flow [12C].

The problem of heat/mass transfer to a finite strip at low Peclet numbers was solved by the method of matched asymptotic expansions. Experimental data obtained by the limiting-current method show excellent agreement with the theoretical predictions [1C].

Among other experimental investigations, heat transfer measurements were performed in a two-dimensional convergent–divergent nozzle. In the cooled boundary layer with increasing favorable pressure gradient, the heat transfer was found to decrease when the dissipation was small, and to increase when the dissipation was large [61C]. An investigation is reported of the mean and fluctuating characteristics of the thermal turbulent boundary layers in divergent flows [30C]. Measurements have been made of the mean-flow profiles in a turbulent boundary layer at a Mach number of three for various wall-to-total temperature ratios [29C]. The effect of an upstream adiabatic zone on the heat transfer in the downstream boundary layer has been investigated [41C]. The local and average rates of mass transfer have been determined for thick plates placed in parallel flow [27C]. An experimental investigation is reported for the laminar flow in a rectangular streamwise corner [13C]. Experiments have been performed on a free turbulent jet, in which the mean fluxes of mass, momentum and total enthalpy were measured. An extension of the Reichardt inductive theory is used to adequately predict the measured quantities [50C].

Measurements in high-Reynolds-number boundary layers have shown that the effect of free-stream turbulence is to increase heat transfer by about 5% for each 1% rms increase of the longitudinal intensity [52C]. The effect of free-stream turbulence on the heat transfer to curved surfaces has been investigated [7C].

Other studies related to curved surfaces include the use of second-order boundary layer theory to determine the influence of the Prandtl number on the heat transfer at the stagnation point [17C]. Also, the governing equations for a supersonic turbulent boundary layer over a wavy surface have been solved numerically. The eddy–viscosity model of Cebeci and Smith has been used, and the predictions are compared with experimental data [44C].

Flow over cylinders and spheres is the topic of a number of papers. An integral analysis of the thermal boundary layer on a continuous cylinder travelling through stagnant ambient fluid has been reported [24C]. Laminar forced convection from non-isothermal thin needles was analytically investigated, in which similarity temperature fields have been obtained [9C]. An analysis has been made of the effect of suction and slip velocity of a non-Newtonian fluid flowing over a circular cylinder [47C]. A series method has been used to analyze the thermal boundary layer on an isothermal cylinder in cross flow of a micropolar fluid [36C]. Boundary layer solutions are presented for the effect of viscous dissipation on the steady-state heat transfer in an axisymmetric stagnation flow on an infinite cylinder. The boundary conditions of isothermal and adiabatic wall have been treated for a wide range of the Prandtl number [18C]. Experimental data are reported for local heat transfer rates from a cylinder undergoing transverse oscillations while being subjected to a cross flow. Local heat transfer coefficients were found to increase up to 60% [46C]. The local heat transfer from a sphere in a turbulent flow has been experimentally investigated for Reynolds numbers ranging from 2600 to 6100 [20C]. An analysis has been made of the heat transfer from a porous sphere in a low Reynolds number flow [55C]. Also, the creeping flow around a sphere has been analyzed [48C].

Numerical studies of impinging jets are reported. An analysis has been made of the heat transfer from a laminar water jet impinging on a surface. The potential flow region was solved by a finite-element method, while the boundary layer problem was solved by an integral method. The solutions include the effect of melting at the impingement surface [32C]. The heat and mass transfer to a surface from a laminar impinging jet has been calculated by a finite-difference method. The effect of initial jet velocity profile has been investigated [15C].

Some work has been associated with unsteady boundary layers. Finite-difference solutions have been obtained for the unsteady laminar boundary layer flow at a three-dimensional stagnation point. The velocity of the incident stream is taken to vary arbitrarily with time [28C, 43C]. A method is described for solving laminar boundary layer equations with periodic fluctuations of the outer-flow velocity and temperature. The method combines numerical analysis and perturbation techniques and leads to a significant reduction of computer storage [57C]. Two examples of nonseparating unsteady boundary layers with local flow reversal have been constructed from semi-similar solutions. The associated thermal boundary layers imply no drastic change of heat transfer at the locations of flow reversal [59C]. A simple analysis has been presented for unsteady heat transfer in impulsive Falkner–Skan flows. The results are compared with those from an earlier numerical solution [19C]. An analytical investigation deals with the flow and heat transfer along a porous plate subjected to a transverse sinusoidal suction velocity distribution fluctuating with time [53C]. Experimental data are reported for the augmentation of convective heat transfer by mechanical scraping of a surface boundary layer [22C].

Heat transfer from liquid films continues to be a topic of research interest. The local transfer rates between a liquid film and a horizontal tube have been measured by an electrochemical technique. The transfer rate was found to depend on the mode of liquid feeding [51C]. An exact solution of the convective diffusion equation for the case of dissolution of a solid wall into a falling laminar liquid film has been obtained. The results show satisfactory agreement with experimental data for the dissolution of gypsum into water [4C]. Existing data for the transport to falling liquid films are reviewed to show the effect of augmented transport by surface waves [49C]. An analysis has been made of the heat transfer to a non-Newtonian laminar falling liquid film on an inclined wall [42C]. The influence of heat transfer and interfacial shear on the hydrodynamic stability of thin liquid films has been analyzed. The linear stability of the flow is determined by a successive perturbation solution of the governing Orr–Sommerfeld equation [62C]. An experimental study is reported for the heat transfer coefficient and film breakdown heat flux in subcooled water films flowing down on the outside of a heated vertical tube [16C].

FLOW WITH SEPARATED REGIONS

Single bodies

Transfer coefficients on both surfaces of a blockage plate situated in a duct flow are insensitive to the extent of the blockage [20D]. The k–ε turbulence model is used to predict successfully the recirculating flow in a cylindrical enclosure [18D]. Williams [21D] discusses the nature of unsteady three-dimensional laminar separation. Two papers by Simpson and Collins [6D, 19D] deal with flowfield predictions of separating turbulent shear layers. A numerical scheme is given [16D] for the simultaneous solution of a boundary layer and an outer inviscid stream with separated flow. LaRue and Libby [14D] deal with detailed similarity in the turbulent wake of a heated cylinder.

Packed and fluidized beds

Interesting details on visualization of blob mechanics in flow through transparent porous media are given in [17D]. Natural convection experiments in rectangular cavities packed with porous media revealed [9D] that Prandtl number and aspect ratio have significant effects on the heat transfer. In a theoretical model of heat transfer to a packed or quiescent fluidized bed [5D], the bed is divided into two regions of constant bed voidage and gas velocity. The low Peclet number behavior of the Sherwood number in a packed bed is dependent upon its defining equation [8D]. A recently developed variational principle of virtual dissipation was applied [4D] to coupled heat and mass transfer in a porous solid containing a fluid.

The basis for a method for estimating the effective radial thermal conductivity of packings in gas flow [2D, 3D] is the sum of a convective fraction and a packing fraction through which the gas does not flow but includes conduction, radiation, and the Smoluchowski effect. Another analysis [11D] deals with the effective shear viscosity of a regular array of suspended spheres. Experimental measurements of particles in fixed beds show [10D] either a constant Nusselt number as the Reynolds number is reduced or, if axial dispersion is neglected, the Nusselt number decreases to zero. Literature correlations for predicting axial mass dispersion coefficients [13D] may lead to adequate accuracy in technical equipment, but analyses of packed bed processes in a laboratory or bench scale study should include separate investigations of dispersion behavior. An extensive review of heat transfer in porous media is given in [15D]. Transient heat transfer in liquid flow through permeable subsurface strata is described [12D]. The surface tension of the liquid was shown to be an important variable in mechanically agitated gas-liquid dispersions in a helical coil [7D]. A design equation is given for gas-to-particle heat transfer in vertical pneumatic conveying of granular materials [1D].

TRANSFER MECHANISMS

Detailed measurements of turbulent flows and formulation of mathematical models for turbulent transport have been the main research activities in transfer mechanisms.

An experimental study has been made of the decay of temperature fluctuations in grid-generated turbulence. The decay rate was found to be a function of the initial temperature fluctuation intensity [27E]. Another experimental investigation showed that the magnitude of the skewness of the temperature derivative in a turbulent flow downstream of a heated grid is small and can be reduced to zero by perturbing the flow with a second ambient temperature grid [3E]. Measurements are reported for the turbulence produced by a grid which simultaneously provides a mean temperature profile varying linearly with height. The downstream evolution of the temperature fluctuations was found to disagree with an earlier theory; the discrepancy is attributed to the neglect of molecular diffusivity [26E].

Detailed turbulence measurements are reported in heated turbulent jets and mixing layers [21E, 22E]. Structure functions of turbulent temperature and velocity fluctuations have been measured in the atmosphere and in the laboratory. The laboratory measurements were performed for the inner region of a thermal boundary layer and on the axis of a heated jet [5E]. Both conventional and conditional measurements for the temperature and velocity fields in a round heated turbulent jet are reported. The measurements show that the large-scale turbulent motions are responsible for the bulk of momentum and heat transfer and that small scales are more efficient in transporting heat than momentum [9E]. Correlation measurements were made in both broad and narrow frequency bands of the longitudinal velocity fluctuations in fully developed pipe flow. The data show that the low-frequency, large-scale fluctuations extend over most of the radial region and are highly correlated [8E]. The mean velocity field and the six components of the Reynolds stress tensor were measured in a three-dimensional turbulent boundary layer with moderate skewing relative to the free-stream direction. The stresses were measured by a rotating probe technique [11E].

Calculations are reported for the turbulent heat transfer in a stabilized pipe flow [18E]. A model is described for the prediction of turbulent heat transfer in gas-liquid flows [20E]. From the statistical characteristics of turbulence, calculations have been performed for the convective heat transfer in turbulent flow [10E]. A method is described for the prediction of turbulent shear flows in which non-equilibrium chemical reactions take place [13E]. A numerical method has been formulated for the solution of turbulent flow problems [30E]. The algebraic stress modeling technique has been applied to the prediction of a round turbulent buoyant jet. Transport equations were solved for the turbulence kinetic energy, its dissipation, and the mean square temperature fluctuations [23E]. Numerical solutions have been presented for compressible boundary layers. An integral form of the equation for the turbulence kinetic energy is employed. The results are obtained for relaminarization, various pressure gradients, and surface roughness, and are compared with experimental data [1E].

A theoretical discussion deals with the role of the Prandtl number in turbulent flow. A simple model based on spectral considerations has been used to obtain an approximate solution [19E]. A periodic intermittent model has been developed for the wall region of a turbulent boundary layer. The model accounts for the intermittent bursts and the longer viscous flow that prevails between the bursts [25E]. A phenomenological model of intermittency has been presented. It employs dynamical quantities such as velocity amplitudes, eddy turnover times and energy transfer [12E]. The properties of constitutive heat transfer equations for materials with memory have been derived from thermodynamic restrictions. A number of inequalities have been obtained for restricting the behavior of the constitutive equations [15E].

A theoretical procedure is described for the interline heat transfer at the junction of liquid–solid–vapor. The analysis assumes that the interline transport is controlled by the London–van der Waals dispersion force between the condensed phases (solid and liquid) [29E]. The heat sink capability of the interline region has been determined by an approximate procedure which includes the effect of the solid resistance to heat transfer. The limiting case in which the solid conductivity is the controlling factor is given primary attention [28E].

The influence of buoyancy on turbulent transport is discussed with reference to a transport model derived from the application of the eddy-damped quasi-Gaussian approximation to the equations for the third moments. The model requires that a flux of one quantity depends on the gradients of all relevant quantities [17E]. Some correlations among turbulent shear stress, kinetic energy of turbulence and axial turbulence intensity have been developed by reference to experimental data for pipe and channel flows [2E]. The surface renewal model has been employed to formulate expressions for the eddy viscosity and the eddy diffusivity [24E].

A stochastic model has been proposed for the calculation of turbulent diffusion [16E]. A theoretical discussion deals with the accuracy of the structure function of temperature in turbulent shear flows [4E]. A number of experimental studies of wall turbulence have been re-examined to evaluate the ratio of the time scale of a scalar-fluctuation field to that of the velocity-fluctuation field. The ratio was found to be nearly uniform and equal to about 0.5 [6E]. A quantitative model is described for the momentum transfer by coherent motion in turbulent boundary layers. The large-scale eddies are supposed to synchronize with the bursts near the wall. The results show reasonable agreement with experimental data [7E]. A system of linear equations are presented for the determination of the fields of velocity and of a general scalar variable in a turbulent flow. The system characterizes the large-scale coherent structures found experimentally in turbulent flow [14E].

NATURAL CONVECTION

Publications on convective heat transfer in which buoyancy plays a major role in determining the fluid motion appear to be unbounded. As has been true for some time, there is active interest in boundary layer flows, flow in enclosures, and combined natural and forced convection phenomena. Interest also continues in transient effects and in convection with non-Newtonian fluids. Interest seems to be growing in convection in porous media as well as in the possibility of using a number of modifications of numerical techniques, such as finite element methods, to study natural convection.

Almost as numerous as ever are studies related to natural convection heat transfer from a vertical flat plate. One study [61F] compares different analytical approximations to predict laminar convection along a plate whose wall surface temperature has a power law distribution. Tables of Nusselt numbers have been prepared for natural convection in fluids of different Prandtl number along a non-isothermal plate [69F]. The influence on heat transfer of induced flow near the leading edge of a plate has been studied [38F]. Higher order approximations to the flow over a surface with uniform heat flux indicate corrections may be necessary in the normal boundary layer solutions at moderate and low Raleigh numbers [59F]. Use of an integral technique [14F] indicates the relative importance of viscous dissipation on natural convection heat transfer. An approximate solution for the heat transfer to a stably-stratified fluid agrees well with earlier results [77F]. Measurements of the thermal boundary layer indicate that at high Prandtl number, an enclosure can significantly influence the temperature profile around a vertically-heated plate [16F]. Natural convection effects in the laminar region above a flame along a vertical surface have been investigated [3F]. Variable suction velocity along a porous vertical plate introduces three-dimensional natural convection effects [91F].

Several investigations considered transient effects when the surface temperature of a vertical surface is changed with time. Measurements in the flow near such a wall which is suddenly heated, indicate the time when the influence of the leading edge is first felt at any given height [58F]. A finite difference solution indicates several extremes can occur in the heat transfer coefficient before a steady state is reached [49F]. The influence of the leading edge of a plate within a uniform transverse magnetic field was analyzed to determine the time at which pure conduction no longer dominates the heat transfer [64F]. Other studies considered the influence of plate motion on heat transfer to a rarefied gas with the plate oscillating [92F] or impulsively started [93F]. The time to go through a transient process of natural convection heat transfer has been considered [72F].

Several studies considered turbulent natural convection along a vertical plate. Use of one turbulence model indicates that the Nusselt number would vary as the Rayleigh number is raised to the 0.367 power [57F]. Space-time correlation measurements of the temperature and velocity fluctuations have been made [17F] on a constant temperature plate up to Rayleigh numbers of 10^{10}. An analysis [78F] on the turbulent heat transfer from a rough vertical plate predicts the Nusselt number is proportional to the Grashof number raised to the one-half power. An analysis indicates the turbulent convective heat transfer from a vertical plate to a power law fluid [89F]. The heat transfer and structure of the turbulent thermal plume along a vertical isothermal wall were measured to approximate the flow that would occur above a fire [55F].

Considerable interest is still directed toward problems of natural convection in horizontal layers, in which the lower boundary is maintained at a higher temperature than the upper boundary. An examination of the convection rolls in low Prandtl number fluids indicates an increase in the wavelength of these rolls with an increase in Rayleigh number [22F]. At low Rayleigh numbers, the velocity field has been mapped using a laser-Doppler anemometer [27F]. A post-stability analysis of the amplitude of the velocity fields in convective rolls agrees with measurements obtained with a laser-Doppler anemometer [28F]. The effects of spatially-periodic thermal boundary conditions are greatly amplified near the classical critical Rayleigh number [52F]. The influence of variable viscosity on the convective heat transport can be obtained by considering the ratio of the actual critical Rayleigh number to that for a constant viscosity fluid [12F]. The effect of cylindrical sidewalls on convection near the critical Rayleigh number has been predicted [15F]. Surface tension was found to enhance the heat transfer in a horizontal layer of small thickness [41F]. Application of a magnetic field amplifies low temperature thermal convection in oxygen gas [42F]. The influence of small but finite heat transfer at the side walls in such a layer tends to influence the transitions in the flow [24F]. The effect of a horizontal temperature gradient on the stability of a fluid heated from below has been investigated [101F]. An experiment demonstrated the importance of density inversion on the heat transfer and onset of the flow in a layer of water over a surface of ice [86F]. A linear stability analysis found the density anomaly in water at 4°C has a stabilizing effect on flow transition [62F]. Numerical solutions provide a picture of the steady flow of non-Newtonian fluid in thermal convection [73F].

Natural convection in a cavity with parabolic sidewalls has been studied with reference to applications in solar collectors [1F]. Models have been developed to explain heat transfer across a horizontal cavity with differentially-heated endwalls [9F]. In a related study, an integral analysis was used to predict the convection through a horizontal duct connecting two fluid reservoirs at different temperatures [8F].

Some studies consider the heat transfer in horizontal layers of fluids in which volumetric energy sources are present. Situations such as this might occur in fluid layers undergoing chemical or nuclear reactions and have been of some interest to those examining nuclear reactor safety problems. In one study [21F], a simple model has been postulated to obtain correlations for Nusselt number with uniform energy sources. Stable hexagons are found in such layers at Rayleigh numbers up to somewhat more than three-and-one-half times the critical value [96F]. An analysis of the hydrodynamic stability confirms the applicability of an approximate empirical relation for determining the instability limit [71F]. In a layer undergoing a phase change at the upper surface, increased melting was found to destabilize the layer [107F].

A number of papers report on studies of convection in vertical layers. Different types of laminar cellular flow and the transition to turbulence have been examined visually in a vertical layer with differentially-heated sidewalls [84F]. A numerical analysis of buoyancy effects in a vertically-heated tube agrees well with experimental results [37F]. With a restricted entry in an open vertical duct, it is possible to have no throughflow at moderate Rayleigh numbers [29F]. The instability of a vertical layer, with differentially-heated sidewalls, has been studied with stably-stratified fluid [11F] and unstably-stratified fluid

[35F].

The temperature profile has been measured in a vertical layer of air with one heated and one cooled sidewall [109F]. Laser-Doppler measurements show the three-dimensional flow in a vertical cavity [65F]. Flow visualization was used to obtain a model for the flow of fluids of different Prandtl number in a differentially-heated vertical cavity with different aspect ratios [85F]. A finite difference solution predicts natural convection in a vertical square cavity [80F]. Finite element methods have also been used to study natural convection in a square enclosure with differentially-heated sidewalls [60F]. A slightly conducting sidewall has significant influence on transitions in a two-dimensional box [39F]. The importance of a density inversion in a vertical cavity has been demonstrated by heat transfer measurements and flow visualization [86F]. A finite-difference solution for the convection in a two-dimensional enclosure with heat sources and heat sinks over the region of the upper surface has been demonstrated [23F]. The oscillatory flow in the plume above a line source in a rectangular enclosure has been measured [47F].

Numerical analysis was used to predict the natural convection across a vertical layer confined between one flat wall and a parallel wavy wall [97F]. For inclined layers between a flat wall and a parallel corrugated wall, a larger heat flux is obtained than if both walls are flat [31F]. The natural convection flow in an insulated horizontal pipe has been studied when the ends of the pipe are at different temperatures [10F].

Measurements have been made on free convection from a downward-facing heated inclined flat plate to water over a range of conditions from laminar to turbulent flow [32F]. A horizontal corrugated plate was found to lose more heat by natural convection than a horizontal finned plate [4F]. An analysis describes the natural convection around a horizontal oscillating heated plate [99F]. A holographic system was used to observe periodic fluctuations in free convection mass transfer from a horizontal surface [5F].

A three-dimensional similarity solution simulates the effect of buoyancy and variable viscosity on the flow over a heated cone in water [102F]. Non-linear solutions were found for the buoyancy-induced flow adjacent to an inclined heated wall immersed in a stably-stratified fluid [50F]. The time-dependent convective flow in a stably-stratified fluid around an inclined wall has been predicted [18F].

Finite element methods are being used to predict convective heat transfer. In one study [33F], the free convection flow around a hexagonal cylinder has been calculated. In another work [98F], the importance of the proper choice of finite element has been demonstrated.

Many studies consider natural convection in a porous material. Similarity solutions for steady and unsteady natural convection from a flat plate adjacent to a fluid in a porous media are summarized [51F]. The onset of longitudinal vortices in a porous media due to natural convection effects when the medium is over a horizontal heated surface of non-uniform temperature has been studied [45F].

The frequency of oscillation in a porous medium heated from below was found to vary as the Rayleigh number raised to the three-halves power [44F]. The onset of natural convection in a cylinder of water-saturated porous medium when heated from below has been analyzed [111F]. Integral methods were used to predict the heat transfer in porous layers adjacent to vertical and horizontal plates [19F].

Other studies considered flow in porous media in enclosures. A simple model has been used [112F] to derive flow and temperature fields for three-dimensional natural convective motion in a confined porous medium. Instability can be strongly influenced by large variations in viscosity and thermal expansion coefficients due to large temperature differences in a porous medium [43F]. Convection in a long, shallow cavity containing a fluid in a porous medium has been studied when heating occurs from one side [100F]. Conditions for the presence of three-dimensional flow with natural convection in a porous medium have been examined [95F]. Natural convection in a horizontal porous medium subjected to different sidewall temperatures has been studied, including the effects of permeability of endwalls, Rayleigh number and aspect ratio [7F]. A perturbation analysis predicts the natural convection in a porous medium containing a concentrated heat source [6F]. The effect of a layer of liquid above a porous bed of inductively-heated particles on the stability and heat transfer in the porous layer has been examined [79F].

Partial spectral expansions were used to predict natural convection flow in the regions between concentric cylinders and between concentric spheres [88F]. An experiment showed eccentricity has little effect on the average heat transfer across an annulus between two cylinders [54F]. Transient natural convection in a horizontal pipe containing water includes the effect of maximum density [20F].

Analyses using a series solution [56F] and a non-similarity technique [68F] appear to agree well with the results from finite difference solutions for the heat transfer from horizontal cylinders. Experiments have been performed on heat transfer from horizontal cylinders of different cross sections [70F]. Experimental measurements on a suddenly-heated horizontal wire indicate the different stages of transient flow at relatively low Rayleigh number [74F]. An approximate integral method predicts the convection of a visco-elastic fluid on a curved surface [90F]. Three-dimensional equations are reduced to a two-dimensional form for heat transfer from an isothermal infinite inclined circular cylinder [76F].

The conditions for inertial damping and resonance of cellular convection in a rotating fluid annulus have been demonstrated [25F]. Experiments on centrifugally-driven thermal convection in a rotating cylinder have been performed with moderate and high Prandtl number fluids [46F].

Several studies have been performed on free convection plumes and buoyant jets. The development of the plume above a vertical heated plate has been studied under laminar flow conditions [94F]. The heat transfer from a ridge in the form of an inverted "v" has been studied experimentally, including the separation and flow in the resulting plume [64F]. The dilution of buoyant jets entering a stagnant fluid has been measured [26F]. The influence of a transverse magnetic field on a laminar plume above a line heat source has been reported [36F]. Interferometric techniques for studying temperature fields in axisymmetric buoyant plumes have been described [13F]. The effect of buoyancy on laminar axisymmetric jets has been studied experimentally and analytically [53F]. Integral equations have been used to study the motion of a buoyant jet in a crossflow [82F].

Mixed convection, in which both a forced flow and buoyancy-induced flow are present, is the subject of several investigations. The relative strength of these forced and buoyancy-induced flows can vary, and this is usually an important parameter in determining the form of the heat transport variation.

Papers that consider mixed convection in boundary layer flows include one using a series truncation to obtain the solutions for both aiding and opposing flows on a semi-infinite vertical flat plate [2F]. The instability of a mixed convection along a vertical flat plate has been analyzed [66F]. Experiments on flow over a heated horizontal plate indicated that longitudinal vortices that are formed are the first stage in laminar-to-turbulent boundary layer transition [34F]. The ratio of the Grashof number divided by the Reynolds number raised to the three-halves power was found to correlate heat transfer for laminar forced flow over a heated horizontal plate [48F]. With forced flow along a horizontal cylinder, free convection effects become dominant some distance downstream from the leading edge [108F]. For such a flow, a variable viscosity can enhance the heat transfer and stabilize the boundary layer [106F]. The local Nusselt number and friction factor was found for aiding and opposing flow over a range of conditions from pure natural to pure forced convection for a flow over a sphere with uniform surface heat flux [67F]. A semi-empirical equation for mixed convection was obtained using values of Nusselt number for pure forced and pure free convection [81F]. Experiments with gas flowing down a heated vertical pipe at low turbulent Reynolds number indicated a reduction of friction factor and an increase in heat transfer due to buoyancy as compared to constant property conditions [30F]. Buoyancy effects on the velocity development of a magnetohydrodynamic flow in a vertical channel have been studied [110F].

Combined natural and forced convection in the entry region of a heated straight tube have been studied [103F]. In a related study, an analytical approximation was used to show different regimes of flow in the entry region [104F]. Simplified expressions have been obtained for the Nusselt number for laminar mixed convection in horizontal tubes [40F]. Mixed convection has also been studied in a heated curved pipe [105F].

Experiments with a horizontal tube heated only over 180 degrees of the circumference showed that buoyancy has a significant effect on Nusselt number at low Reynolds number if heated from the bottom, but little effect if the heated position is on top [83F]. Numerical techniques are used to analyze the laminar flow in a tube over a portion of its circumference [75F].

CONVECTION FROM ROTATING SURFACES

The equation

$$Sh\,Sc^{-1}(0.5657 + Sc^{2/3}) = 0.620\,Re^{1/2}$$

was found [7G] by experiments to describe mass transfer from spinning disks up to a critical Reynolds number $Re_{crit} = 2.6 \times 10^5$. Experiments [6G] on the transfer mechanism from a rotating disk to non-

Newtonian fluids established that the mass and heat transfer boundary layer thicknesses are always less than the hydrodynamic boundary layer thicknesses.

Equations were derived from an experimental study [4G] of heat transfer from an inward flow of air to coaxial disks rotating in the same direction and with the same angular velocity. The equations hold for turbulent and transitional flow. Asymptotic solutions were obtained [11G] describing convective heat transfer and frictional heating of a cone resting with its apex on a flat plate and rotating around an axis normal to the plate. The solution is for small Reynolds number and small cone angles. It includes variable properties. Spatial oscillations and convection celles appear in the flow of a fluid enclosed in an annulus with end walls when the upper wall is insulated, whereas the lower wall displays a temperature gradient in radial direction [2G]. This occurs when critical values of the Ekman number are exceeded.

Pulsating temperatures and heat fluxes were calculated [12G] for turbulent non-isothermal flows in a rotating cylindrical pipe. Buoyancy induced laminar flow in a rotating spherical annulus was investigated [10G] for the condition that the inner surface or the outer surface rotates or where both move in opposite directions. Experiments [8G] determined heat transfer from the outer cylinder to a liquid flowing through an annulus for the condition that the inner cylinder rotates.

A numerical analysis [13G] established the temperature distribution in a rotating cylindrical body with a line heat source located at the surface with its axis parallel to the cylinder axis and with cooling provided along a line diametrically opposed. Exact solutions are presented [3G] for the equation of the thermal boundary layer for axisymmetric rotating bodies with arbitrary surface temperature distribution and self-similar velocity distribution. Momentum and heat transfer for forced flow moving with a laminar boundary layer over a rotating body can be expressed as a set of coupled ordinary differential equations by appropriate coordinate transformations [5G].

Very high heat transfer coefficients were found [1G] in experiments on the vaporization of water films in rotating radial pipes at centrifugal accelerations 10^3–10^4 times the gravitational acceleration. Wetting and burnout were also studied. A solution [9G] of the laminar compressible boundary layer equations describing swirling flow through a nozzle established a strong effect of rotation on axial skin friction, whereas the effect on heat transfer is small.

COMBINED HEAT AND MASS TRANSFER

A review [4H] of heat and mass transfer for laminar and turbulent flow over plates, cylinders and spheres was published. Extensive experiments [15H] studied heat transfer in a turbulent boundary layer on a rough surface with air injection through a porous plate. Sand grain roughness was simulated by spheres with 1.25 mm diameter. A correlating equation for smooth walls could be adapted to describe heat transfer on the rough wall for transitional and fully rough condition. Experimental results and a semi-empirical analysis [21H] describe heat and mass transfer on a permeable surface in the zone of transition from laminar to turbulent flow. Optimization procedures [8H] make it possible to minimize the coolant flow requirements by optimizing the distribution of injection holes over the film cooled surface. Superposition methods were used to describe the heat transfer effectiveness of the holes.

It was demonstrated by experiments [2H] that the film cooling effectiveness decreased with an increase of the turbulence intensity of the cooling air ejected through a slot of the film cooled wall into a hot gas stream. The effect of swirl of the main air flow over a rotating cylinder on the effectiveness of film cooling with slot injection was measured [16H] for swirl angles between 0 and 75° and for blowing rates from 0.45 to 1.4.

The ejection of a turbulent jet into a crossflow was treated by a three-dimensional numerical analysis [1H]. Anisotropic transport coefficients had to be introduced to obtain agreement for a blowing rate smaller than 0.1 at 90° injection and for a blowing rate smaller than 0.5 at 30° injection. No agreement was obtained beyond these values. Film cooling with ejection through two rows of holes is shown [10H] to be considerably more effective than ejection through one row. The heat transfer coefficients were found to be very close to those without ejection for blowing rates smaller than 1 when defined with the adiabatic wall temperature. Effects of positive and negative pressure gradients on the efficiency of film cooling were studied [3H] on a stationary gas turbine blade with ejection close to its nose.

Local and average heat transfer from arrays of impinging jets to a surface were measured [9H] for a square array of holes 10–15 diameters apart.

A simple analytical approximation [6H] describing laminar binary boundary-layer flow along a vaporizing layer agrees with numerical calculations and experiments on benzene evaporation. A Reynolds number based on the number of roughness elements per unit area is shown [12H] to dominate the boundary layer behavior for laminar boundary layer flow over a rough subliming surface. Carbon vaporization into a nonequilibrium stagnation-point boundary layer was analyzed [19H] for hydrogen and helium in the mainstream assuming frozen flow of the carbon species. The analysis simulates conditions for a Jupiter entry probe. A computer analysis of heat, mass, and momentum transfer for sheer flow over a wavy boundary layer can be applied to the flow of air over water [14H]. It includes the possibility of circulating eddies. Integral methods [20H] were used to analyze heat transfer on a spray-cooled cylinder by a water mist approaching normal to the cylinder axis and creating a liquid film on the cylinder surface. Numerical and asymptotic solutions [5H] describe laminar condensation heat, and mass transfer in the vicinity of the forward stagnation point of a spherical droplet. Transfer between steam, air, and absorbing fission products was considered. The Stefan problem describing freezing, melting, or drying on a plate or cylinder was extended [18H]. An analytic solution [7H] was obtained for a certain class of multicomponent mass transport situations with heterogeneous surface reactions using Lévêques approximation.

The non-linear problem of heat and mass transfer in capillary porous bodies was analyzed [17H] postulating that the phase transition does not occur in a sharp plane but in a zone of finite thickness. A method is presented [13H] for the calculation of transient heat and mass transfer in colloidal capillary porous bodies with internal heat sources. Tapered cylinders were used [11H] as models of a pore space to investigate heat and mass transfer in capillary systems.

CHANGE OF PHASE

Boiling

The nature of the emulsifying agent strongly affects pool boiling heat transfer in oil-in-water emulsions [49J]. Internal microconvection is the important mechanism for the enhanced heat transfer observed in highly subcooled liquids [56J]. Ünal [83J] determined the void fraction, incipient point of boiling, and the initial point of net vapor generation in sodium-heated, helically-coiled steam generator tubes. In [66J] the Alad'ev correlation for wall superheat is modified using a combination of surface and fluid properties. The unsteady-quenching method for determining boiling curves is evaluated [55J]. Sharma and Varshney [67J] discuss the Mikic–Rohsenow pool boiling correlation for pressures less than atmospheric. The correlation of Happel and Stephan correctly predicts nucleate boiling of liquified gases and their binary mixtures [85J]. The orientation of the heater surface is of major importance in nucleate boiling [12J]. There are contradictory views as to the effects of the thickness of the heater wall on boiling curves. Chuck and Meyers [17J] explain that previously observed effects were at opposite ends of the ΔT scale. The concept of bubble flux density, which is a function of both the active site density and the frequency of bubble departure, is used to correlate surface effects on nucleate boiling [48J, 68J]. Microlayer evaporation phenomenon is a significant heat transfer mechanism, especially at low pressures, since up to 40 per cent of the total heat transport is accounted for by microlayer evaporation [25J]. The size distribution of naturally occurring nucleation sites on a heater surface is determined [20J] by gas bubble nucleation and vapor bubble nucleation. More sites were active in gas diffusion than in boiling. In [76J] solutions for bubble growth rates including the effects of inertia, heat transfer, and microlayer evaporation were obtained in generalized coordinates. Formation and stability of a liquid microlayer is described by a balance of surface tension and deceleration force—in contrast to the usual boundary layer treatment [97J]. A model [4J], based on microlayer evaporation and drying-out phenomena, is proposed for interpretation of the deposition rate of suspended iron oxide on walls. The growth, departure, and implosion of small and large bubbles were analyzed [47J] utilizing semi-differential forms for the heat and mass equations. Bubbles grow more slowly in power law liquids the smaller the characteristic exponent [89J]. The size and shape of a bubble formed slowly on a sharp or round-edged orifice were derived with the help of a new analytical solution for the bubble profile [15J]. A closed-form, unsteady heat conduction solution, obtained by a double transformation, was used [73J] to estimate the evaporation rate at the perimeter of a growing bubble. The growth of vapor bubbles was studied in superheated liquids [58J], in combined gravitational and nonuniform temperature fields [34J], and for effects of internal circulation velocities and noncondensable gases [54J].

When a binary surface tension positive system boils, a maximum nucleate boiling heat flux is exhibited at a low concentration of the more volatile component [57J]. A DNB rivulet model was constructed [16J] to describe DNB temperature oscillation phenomena. Grigor'ev [28J] gives a theory of critical heat flux. In [37J] the available works on prediction of pool boiling stability that deal with infinitesimal temperature excursions on isothermal surfaces are extended to deal with nonisothermal surfaces and finite temperature deviations. Reference [21J] introduces a method by which the length of the annular flow regime in a straight vertical tube steam generator can be predicted and, therefore, the dry-out position located. The critical heat flux obtained with sudden increases in heater power depends on the initial condition in the wall layer produced by the initial power level [78J, 79J]. Heat transfer enhancement devices such as vanes, twisted tapes, or annular projections increase the critical heat flux up to 60% [62J]. Boiling crisis was studied [77J] in concentric and eccentric annuli with turbulence promoters on the unheated surface. Two papers [51J, 52J] report on boiling crisis in annuli with different axial heat flux distributions. Comparisons of calculated and experimental results are given for dryout in rod bundles [91J] and in vertical rod assemblies [2J]. A number of papers deal with the deterioration of heat transfer in the post dry-out region [41J, 75J, 86J]. An experimental study of surface rewetting of a copper tube by a falling film indicated the variations of surface heat flux behind the wet front to be similar to those observed in nucleate and film pool boiling [18J]. The Leidenfrost points for freon C51-12, 13, carbon tetrachloride, and chloroform were determined [22J] for pressures up to the critical pressure. A variable property analysis is available [42J] for laminar film boiling on vertical plates. Film boiling from the upper or lower sides of a horizontal plate was studied experimentally [64J].

A high thermal capacity test section was used to obtain flow boiling curves [13J]. Most existing correlations overpredict the heat transfer rates. The flow direction, i.e. the buoyancy forces, proved to have no effect on the critical heat flux [44J]. In [48J] surface tension is shown to play an important role for the onset of burnout phenomena, not only in ordinary pool boiling, but also in boiling systems with forced flow. Accurate analyses of the thermal behavior of a hot surface during rewetting (e.g. a nuclear fuel bundle during emergency core cooling) require reliable transition boiling data [14J]. A high thermal inertia hot patch was used for measuring film boiling at low qualities [29J]. An analysis is given for pressure excursions in transient film boiling [11J].

A number of papers examine the mechanics of evaporation of liquid droplets [95J] with heated surface whose temperature is below the maximum boiling rate point [93J]. In [43J] the entire boiling curve for a single droplet on a heated plate was measured and evaluated. In [65J] experiments also probed the transient temperature profile of a hot wall struck by an impinging liquid drop. Knowledge of the temperature field inside a vaporizing droplet is applied to a spray drying problem [94J]. Surfactants are a powerful tool to control evaporation processes [33J]. Features of boiling heat transfer in thermosyphons are determined that are different from pool boiling [6J, 7J]. Other papers deal with boiling with capillary porous structures [3J, 69J, 70J].

Two of the most important unknowns for the designers of boiling heat transfer equipment; the effect of multicomponent mixtures and multitube bundle geometry, were studied experimentally [38J, 87J].

Condensation

Two fundamental papers on kinetic theory of evaporation and condensation describe linear and nonlinear problems [71J] and the hydrodynamic equation and slip boundary conditions [72J]. Results of linear stability analysis [84J] of film condensation indicate that surface tension, viscosity, and the overall effect of condensation mass transfer tend to stabilize while gravity tends to destabilize the flow. Reference [90J] summarizes five negative effects that a noncondensable gas exerts on condensation heat transfer [9J]. Heat transfer coefficients were determined for turbulent film condensation in a vertical channel [36J] using Deissler's expression for the eddy diffusivity [59J]. Finally, the effects on film condensation of surface force enhancement [60J], where the dynamic effect of the condensable on the condensate film is significant [5J], and of a chemically reacting gas ($N_2O_4 \rightleftarrows 2NO_2 \rightleftarrows 2NO + O_2$) [45J] are available.

Experimental data [50J] for dropwise condensation of steam on well prepared and characterized surfaces show that the heat transfer strongly depends on the hysteresis of the contact angle. Rose [61J] proposes that the extremely high rate of coalescence during dropwise condensation leads to space-and-timewise temperature fluctuations of such high frequency that the surface temperature is essentially steady—so that the material of the condensing surface plays a negligible role in determining vapor-side heat transfer. Hannemann [30J] gives an approximate analysis for the effect of condensing surface thickness on the constriction resistance and examines various dropwise thermal resistances as to their importance in design. Best rates of condensation heat transfer are obtained with polar fluids [8J]. Vibration affects the flow and heat transfer in film condensation [10J]. Condensation of wet steam on a widely spaced, staggered tube bundle placed in a foam bed [35J] is described. A surface renewal model has been developed for condensation inside a circular tube with in-line static mixers [23J]. Additional condensation papers from the Soviet Union are found in references [46J, 80J, 81J, 98J].

Two-phase flow

The only liquid metal paper of the year [92J] concerns magnetic field effects on bubble growth in boiling liquid metals. Two-phase flows in capillary-porous bodies is the subject of five papers [27J, 39J, 40J, 82J, 96J]. The interactions that occur between a water downflow and a gas upflow on the rate of wetting of a hot vertical surface are described [19J]. Reference [53J] suggests using spectral noise analysis for determining the heat transfer mode in high rate boiling and for diagnosing the thermal state of the equipment. Intermittency functions have been defined [1J] which describe liquid, vapor, and interface regions at a fixed point in order to investigate statistical properties of temperature fluctuations in a boiling boundary layer. Two papers are given [24J, 74J] on melting thermodynamics. Two papers [31J, 32J] on bubble nucleation and growth in variable pressure fields deal with pressure oscillations. A new model [88J] for flashing of choked flow is more consistent in its assumptions than earlier models and predicts observed critical flow rates competitively. From measurements on a chilled tube [63J] the frost thickness turns out to be independent of the variables commonly significant in mass transfer such as Reynolds number and vapor pressure difference. A water pipe with no main flow in it may undergo 4–7°C of supercooling before ice nucleation occurs [26J].

RADIATION

Radiation in participating media

Radiative transport in emitting, absorbing and scattering non-gray media seems to attract increasing attention. The radiative flux and its divergence was formulated in terms of the total band absorptance for multidimensional problems involving non-gray radiating gases. Thus the frequency integration is eliminated, and the expressions are more compact [37K]. Approximate solutions obtained for radiative transfer in scattering absorbing plane-parallel media are compared with exact numerical solutions for both the isotropic and anisotropic cases. They agree within 10–25% which is adequate for most engineering applications [4K]. Approximate closed-form solutions are presented for the effects of optical thickness, anisotropic scattering, incident angle, polar angle and azimuthal angle for collimated radiation in a planar, conservative anisotropic scattering medium. Comparisons with exact solutions indicate that these dependencies are accurately predicted [2K]. A new method for calculating the spectrally integrated emission characteristics provides high accuracy without requiring excessive computer time [29K].

A two-dimensional model is used to predict the behavior of a semi-infinite isotropically scattering cylindrical medium exposed to collimated radiation. Comparisons with experimental results using a He–Ne laser beam incident normal to the surface of distilled water which contains latex paint as scattering centers show good agreement with the predictions of the model [18K]. A 2-D numerical algorithm is formulated for calculating radiative transfer in two dimensions through fog. The computer code generated from this algorithm is used to examine multiple scattering effects at wavelengths of 1.06, 3 and 10.6 μm [8K]. A solution is given to the radiative transfer of the solar irradiation through a turbid atmosphere bounded by a Lambert surface based on the single-scattering approximation [24K]. A comparison of predictions with measurements for radiative transfer in an algal suspension (scattering–absorbing medium) shows good agreement for the radiative fluxes if predictions are employed based on a six-flux model and the method of discrete ordinates [6K].

Spectral water vapor transmission data of the bands between 0.7 and 1 μm differ appreciably from values published by Moskalenko, which are the only spectral water vapor data below 1 μm applicable to real atmospheres [13K]. Measurements of water vapor continuum absorption in the 3.5–4.0 μm region show agreement with previous measurements by others for 65°C but the 23°C results indicate a continuum absorption which is larger than expected by about a factor of 2 [40K].

Solutions to the system of combined conduction and radiation for both non-gray and gray media having either conductive or convective boundary conditions at the top interface show that effects of top interface reflectance and internal scattering dominates the heat transfer behavior [27K]. By an analysis of combined conduction–radiation heat transfer through an irradiated semi-transparent plate the validity was checked of neglecting the interaction of conduction with radiation in the solid [38K]. Analytical and experimental studies of combined transient conduction–radiation heat transfer in selectively absorbing media are in satisfactory agreement [3K].

Combined radiative–convective–conductive heat transfer in the boundary layer of a combustion gas was analyzed by considering two infinite plates one of them exposed to a flow of hot gases. The results show that heat transfer to the wall will be increased if the latter plate exceeds the ignition temperature of the combustible gas and the combustion in the flow remains incomplete [31K]. The method of effective heat fluxes was used for investigating combined convection–radiation heat transfer in a closed system of gray bodies with a two-layer model of the gas medium [32K]. In a study of combined radiative–convective–conductive heat transfer in a dust-laden gas flowing in a flat duct, the effects are evaluated of particle size and density and of the scattering function on heat transfer [14K]. Measured temperature profiles for laminar convection in a CO_2 forced upflow are in good agreement with a simplified analysis which includes the effects of buoyancy and thermal radiation [9K].

The spectral and total emissivities of diatomic gases were calculated in the rigid rotor approximations using the Elsasser model of absorption bands. Results for CO, NO and OH are in satisfactory agreement with experimental data [25K]. For a treatment of vibrational nonequilibrium radiation of diatomic gases the authors derive in the first paper [15K] the macroscopic equations which describe the radiation field and the internal state of the gas. In the second paper [16K] they introduce approximations to the master equation for vibrational energy distributions of a diatomic gas. The calculated diffusion coefficient which is related to the mean absorption coefficient differs substantially from Planck and Rosseland values. Calculated infrared band absorption for nonrigid rotation (applied to CO, NO, N_2O and CO_2) compares favorably with experimental data for small and moderate optical depths, but substantial differences exist for very large optical depths [5K]. Direct line-by-line integration and quasi-random band model techniques are used to calculate the spectral transmittance and total band absorptance of the 4.7 μm CO, 4.3 μm CO_2, 15 μm CO_2 and 5.35 μm NO bands. In almost all cases, the line-by-line results are in excellent agreement with experimental values [36K].

There is a close agreement between the various calculated mean absorption coefficients for luminous flames and for smoke which demonstrates the usefulness of the mean absorption coefficient concept for applications [11K]. Comparisons of calculated and measured radiative heat fluxes to the floor in a full-scale room fire show reasonable agreement [23K]. Measurements of the radiation in a methanol furnace showed that the radiative emission character of such a furnace is not significantly different from a furnace using other clean burning fuels, such as coke-oven gas and city gas [10K]. Results of absorbtivity measurements of enclosed industrial flames of gasoline, kerosene, diesel fuel, and oil in the visible and near infrared show that they are wavelength independent. The total absorptivities of these flames depend only on the excess air ratio [33K]. Investigations of radiative heat transfer in a layer of a gray gas impinging onto a heated surface may be of interest for the application of flat-flame burners [20K].

In the temperature range from -3 to $+80$°C the refractive index of water shows an exponential behavior as a function of the temperature [1K]. Studies of the effective infrared optical depths of the clear ocean indicate that this depth is better approximated by the reciprocal of the absorption coefficient as the wavelength increases [12K]. The use of mean (wavelength independent) radiation absorption properties may result in serious errors when predicting the absorption of solar radiation in water [39K]. Refractive indices of some common solvents are reported over their energy range of transparency from 2.1 eV (5,900 Å) in the visible up to 6.2 eV (2,000 Å) in the region of the UV absorption edge with an accuracy of $\pm 10^{-3}$ at room temperature [19K].

In a hypersonic flow of a radiating gas around the nose of a body which is subject to intensive evaporation, the flow region consists of the shock layer and a vapor layer in which radiation is partially absorbed. A similarity law is described for the radiative heat transfer coefficient at the stagnation point [22K]. On the basis of the mean properties of air at temperatures from 10^4 to 2×10^4 K and pressures from 0.1 to 100 atm the radiative cooling of the air behind strong shock waves was calculated [41K]. Investigations of the shock wave structure of dusty gases which contain radiating particles show that ahead of the shock front, the particles are heated by radiative energy transfer from behind the shock front [21K]. Studies of how radiative heat transfer under optically thin conditions affects the process of sharpening or flattening of acceleration waves with planar, cylindrical and spherical symmetry demonstrate that radiative heat transfer delays the formation of a shock wave and has a stabilizing effect so that not all compressive acceleration waves will grow into shock waves [26K].

Thermal emissivity (ε) measurements of cholesteric liquid crystal films as a function of the film thickness (d) showed that ε increases from 0.2 to 0.4 as d increases from 13 to 33 μm and finally reaches a constant value as high as 0.87 for $d \approx 1$ mm [35K]. Considering photosensitized reactions in an absorbing–scattering medium within a plane slab it may be concluded that scattering will reduce conversion in all those situations where a good radial mixing occurs. Therefore conversion is more sensitive to the power absorbed within the slab per unit interspacial area than to the local rate of energy absorption [34K]. Studies of the effects of substrate reflectance and medium depth in a two-dimensional multiple scattering experiment reveal that the substance reflectance character and medium depth effects may be quite severe [17K].

A 300 W halogen lamp with a color temperature of 3200 K was utilized for studying radiative melting of a horizontal clear ice layer. Backmelting was clearly observed and predicted results of melting rates are in good agreement with measurements [28K]. Studies of the spectral reflectance of particulate materials using a Monte Carlo model including asperity scattering reveal that the theoretical values of reflectance, using the experimentally determined optical complex indices of refraction, compare well with the experimental reflectance observations [7K].

Modelling of heat and mass transfer in monolithic honeycomb catalysts, assuming an exothermic catalytic reaction occurring on a surface of a cylindrical passage, shows that radiation exchange can cause significant changes of the temperature distribution in the passage [30K].

Surface radiation

The total hemispherical emissivity of tungsten with an electropolished surface was measured [8L] for the first time at temperatures from 180 to 1000 K. The results depart significantly from theoretical predictions. Monochromatic bidirectional reflectance in the visible and near infrared range of water cryofilms on specular and diffuse surfaces of copper and epoxy were measured [5L]. Film thickness and incidence angles were varied and it was found that the reflected radiation retains the specular peak up to a film thickness of 500 μm. Generalized equations are presented [1L] for the calculation of absorptance, reflectance, and transmittance of any number of parallel surfaces. Radiation configuration factors are discussed for arrangements of disks [3L].

An analysis [4L] of solar and infrared radiation properties of parallel plate transparent honeycombs includes scattering and polarization. Calculations were performed for glass and mylar. A formula [9L] describes geometric factors for radiative exchange between plane specular v-shaped reflectors and collectors.

Scattering was analyzed [7L] for thin dielectric fibers with a diameter much smaller than the wave length. Planck's radiation law was generalized [2L] to anisotropic dispersive media.

The temperature of an injection-cooled and radiatively-heated perforated wall was analyzed [10L] considering air, argon, and helium as coolant. A numerical analysis [6L] describes radiative and convective heat transfer from a specularly reflecting surface with fins using the Monte Carlo method.

MEASUREMENT TECHNIQUES

The development of measurement techniques and instrumentation related to heat-transfer studies continues to draw the interest of a number of investigators. Primary areas of activity included temperature and heat flux measurements, hot wire anemometry and transport property measurements. In addition, specialized techniques and equipment have been described for evaluating parameters such as heat and mass transfer coefficients and void fraction.

Calibration models have proven useful for describing and correcting measurements of the enthalpy of high temperature gas streams using gas-aspirated liquid-cooled calorimetric probes [29P]. Temperature measurements have been made inside a high-pressure vessel (pressures to 3000 bar) by enclosing a platinum resistance thermometer in a soft glass tube with sealed platinum contacts [45P]. The transfer function for a platinum wire used as a temperature sensor has been calculated in order to establish that the heat transfer between the sensor and its supports can appreciably effect the measurements [35P]. It was demonstrated that when a thermocouple is suitably interfaced with a microcomputer, a range of new possibilities arises for increasing the measurement accuracy of the thermocouple, particularly for such applications as high-frequency temperature measurements in turbulent combustion [48P].

A review of the phenomena and relationships encountered in surface temperature measurements is given which outlines methods of radiothermometry and proposes a number of methods of avoiding problems caused by unknown emission factors [17P]. The use of a high-sensitivity heat flux detector provides a contactless method of measuring the temperature of rotating metal surfaces at 313–353 K with an error not exceeding 1 K [22P]. The fabrication and principal characteristics of an electrically calibrated absolute radiometer are described and an instrument of this type is demonstrated to have an overall precision of at least 0.5% for the power level of 50 μW [8P]. Methods of obtaining quantitative thermometric data with a thermographic camera are presented, including a description of calibration techniques and an assessment of obtainable accuracies [11P].

The number of papers describing the evaluation, calibration and application of hot-wire anemometers continues to be large. A two-wire probe technique is proposed and analyzed for the simultaneous and continuous measurement of velocity and temperature in nonisothermal flows using a method that provides automatic compensation, even for large velocity and temperature fluctuations, over a frequency range of *d–c* to 6 kHz [26P]. Another technique for simultaneous measurement of velocity and temperature involves the conversion of analogue hot-wire anemometer voltage signals into digital forms which are processed on a digital computer in accordance with the anemometer response equations [30P]. When a hot-wire anemometer is calibrated at low velocities with flow in the horizontal direction, the heat transfer from the probe may be less than when the imposed velocity is zero [42P]. A relatively simple technique has been developed to evaluate the circulation associated with various spatial contours utilizing an X-geometry hot-wire probe [18P].

Errors occurring in hot-wire measurements in the vicinity of a wall and which are associated with the proximity of a heat-conducting surface in high velocity and temperature gradients were analyzed [40P]. The conduction between the sensor and the prongs can also lead to significant errors to measurements made in a turbulent flow near a wall [28P]. Application of a novel cross-correlation dynamic calibration technique has shown the significance of hot-wire end conduction losses for the measurement of stream temperature fluctuations under nonisothermal flow conditions and leads to recommendations of wire selection to minimize the effects of the losses [9P]. The X-geometry hot-wire probe with prongs perpendicular to the mean flow was found to be unacceptable for turbulence measurements [47P].

The number of reported investigations dealing with the theory and development of laser-Doppler anemometers has decreased significantly as the associated technology matures and the instrumentation becomes more prevalent in the laboratory. Recent observations demonstrate that CW laser-Doppler measurements can be made on smoke-stack plumes (and similar sources of pollution) at ranges approaching a kilometer. The good agreement between the laser measurements of both the exit speeds of the plume and the vortex shedding frequency of the stack and more conventional measurements of the quantities, provides confidence in the laser-Doppler data [10P]. At the other end of the spectrum, laser-Doppler measurements have been made in spatially restricted flow [14P]. A disadvantage of the laser-Doppler velocimeters which are currently on the market is their expense. One paper describes the construction of a comparatively inexpensive and simple velocimeter and presents experimental results to verify its effectiveness [41P].

The need to investigate the flow structure and turbulence beneath surface water waves led to the development and testing of a miniature drag sphere velocity probe [16P]. A pressure probe has been studied which can be rotated up to $\pm 30°$ from the velocity vector with little error in the static pressure reading [27P]. Analytical and experimental studies of ASME flow nozzles were performed which resulted in new correlations for static pressure tap errors which, in turn, were applied to the theoretical nozzle discharge coefficients [6P].

Development continues in the area of transport property measurements. The steady-state method for determining the thermal conductivity of materials by measuring the one-dimensional heat transfer through cylindrical specimens has been improved by accounting for the thermal expansion of the specimen [34P]. A laser flash method which was previously developed for the measurement of thermal conductivity of solids has been successfully applied to liquids of low thermal conductivity (e.g. the thermal conductivities of water and toluene were measured by this method with a mean deviation of 2.6% [43P]).

Two papers address the measurement of thermal conductivity via the transient hot-wire technique. The first provides a contribution to the theory of the technique [31P] and the other describes its application to measuring the thermal conductivity of liquids [44P].

A transient method for measuring thermal properties of soils is superior to steady-state approaches in that it is faster, reduces the problem of moisture migration and simultaneously measures thermal conductivity and thermal diffusivity [2P]. An apparatus is described in which temperature and moisture content are simultaneously measured in a granular material through which one-dimensional heat transfer occurs, thus allowing the thermal conductivity to be established as a function of moisture content [25P]. A simulator for remote sensing and its application to solid moisture measurements are described [19P, 20P].

An instrument has been developed for the measurement of the product of density, specific heat and thermal conductivity of solid materials, a parameter which frequently occurs in analyses of unsteady heat conduction processes [1P]. The accuracy of a micro-transient thermal diffusivity measuring technique for liquids has been improved [23P]. The energy problem has been formulated and solved for biconical and concentric spherical viscometers [7P].

Instrumentation for the evaluation of radiation properties has been studied. A method is described for making *in situ* reflectance measurements of diffuse surfaces by using a GaAs laser and off-the-shelf optical components and the method does not involve the usual radiation integration over a hemisphere [12P]. A confocal mirror assembly was used to measure reflectances in excess of 0.99 with high precision [4P]. The measurement of radiation properties of selective surfaces using a solar calorimeter is described [46P]. The accuracy of portable instruments manufactured by Gier–Dunkle Instruments, Inc. and Willey Corporation for the measurement of solar absorptance and infrared emittance was evaluated in order to establish their applicability for testing solar selective surfaces [38P].

Two papers report measurements in two-phase flows. An optical probe, which has been miniaturized to reduce disturbances, can be used to evaluate local void fractions and interface velocities in liquid–vapor two-phase flows [3P]. A method of measuring the diameter of a transparent droplet in the absence of backlighting is described and its accuracy is experimentally investigated [13P].

Special instrumentation has been applied to measuring heat and mass transfer coefficients and shear stresses. The use of a differential interferometer with half-wavelength compensation has proven to be a good method for comparative investigations of convective heat transfer with such diverse fluids as nitrogen gas and silicone oil [36P]. The principal methods of measuring the convective transfer of momentum, heat and mass are surveyed on the basis of the three-fold analogy that exists between their transfer processes, and it is concluded that the simplest and most sensitive method is the mass transfer technique based on the principle of absorption with chemical reaction and coupled color reaction [32P]. Experiments of mass transfer in the presence of free and forced convection demonstrated that the camphene sublimation technique in connection with holographic interferometry is well suited for measuring local transport coefficients [24P]. As the result of new calibration techniques, the application of Preston tubes for the measurement of shear stress is extended to transpired turbulent boundary layers from near relaminarization to near blow-off conditions [15P].

English-language translations of Russian papers dealing with the following topics appeared in *Measurement Techniques*: (a) devices for the alignment [39P] and signal drift compensation [37P] of radiation pyrometers; (b) apparatus for measuring radiant flux density [21P, 33P]; and (c) resistance-thermometer errors caused by operating in still air [5P].

HEAT TRANSFER APPLICATIONS

Heat exchangers and heat pipes

The study of the performance of heat transfer surfaces for heat exchangers has found wide interest in recent years. The mass transfer analogy using naphthalene sublimation was used [8Q] to study heat transfer and pressure drop for a pair of colinear interrupted plates aligned with the flow. It was found that heat transfer increases with the thickness of the plates in the Reynolds number range which is usually used but that the pressure drop increases even more. A computer analysis [26Q] studied the details of laminar forced convection heat transfer from a shrouded longitudinal fin array with and without tip clearance. The presentation of heat exchanger performance by Kays and London was modified [25Q] to permit a ready comparison of plate fin surfaces. Experimental results [19Q] reveal the influence of various parameters on heat transfer in longitudinally finned tubes.

Local heat transfer rates at a tube in a bundle with crossflow were measured [20Q] at Reynolds numbers between 10^5 and 10^6. Drag coefficients for the first to fifth row are also reported. A numerical prediction

[16Q] of the details of laminar flow and heat transfer in tube bundles used a cylindrical network close to the tubes and a Cartesian system for the rest. Agreement with experimental results by Bergelin was found satisfactory at Reynolds numbers between 1 and 1000. Heat transfer performance of tubes in longitudinal and crossflow was compared [12Q] based on the ratio of heat transferred to power required to drive the fluid. Naphthalene sublimation was used [27Q] to measure local heat transfer and pressure drop for a staggered wall attached array of cylinders with tip clearance to air flow in a rectangular channel. Developed conditions were attained after the fourth row. A method for calculating heat transfer [18Q] to staggered tube bundles in cross flow is described. Results of experiments [17Q] on the effect of thermal conductivity of fins arranged transversely on a tube in air flow are presented as a relation $Nu = C\,Re^m$ with C and m depending on the thermal conductivity of the fin metal. Heat transfer information is available for tubes with slotted transverse fins [13Q] and for shaped fins [1Q].

Several papers deal with the analysis of heat exchanger performance and with their dynamic characteristics. The deterioration in the thermal performance of cross flow heat exchangers due to the flow nonuniformity [4Q] and due to unequal temperature distribution [30Q] was analyzed. Longitudinal heat conduction in cross flow heat exchangers was found [5Q] to deteriorate the heat exchanger effectiveness. Design calculations of heat exchangers were developed [14Q] for heat exchangers with parallel-mixed flow including multi-shell heat exchangers. A number-of-entropy-production-units is proposed [2Q] as a general criterion for the rating of heat exchangers. Analyses consider the dynamic response of heat exchangers [28Q, 29Q] to step changes of velocity and temperature. Experimental results are compared [15Q] with standard predictions of the transient behavior of heat exchangers.

A few papers deal with regenerative heat exchangers. Charts are provided [11Q] from which the conditions can be determined for which heat conduction is important or can be neglected relative to convection. The transient behavior of regenerators can be studied in two papers [3Q, 22Q]. Improvements and new designs of Harrison plate fin regenerators for industrial gas turbines are discussed [7Q].

A one-dimensional model for the analysis of heat pipes [23Q] gives results in satisfactory agreement with experimental data, especially when the effects of suction and injection are included [24Q]. Experimental and analytical studies [10Q] on heat pipe performance were extended to supersonic velocities. Information is also presented [6Q] for heat pipes with a central tube artery. A small quantity of a noncondensable gas reduces heat transfer in a rotating heat pipe strongly [9Q] depending on the molecular weight of the noncondensing and working fluids. The response of a heat pipe to a step change in heat input was analyzed [21Q] using Fourier transformation.

Aircraft and space vehicles

Over the past year there has been a substantial decline in publication activity in this field although space shuttle oriented work is still evident.

Analytical and experimental studies of roughness-induced boundary layer transition for the shuttle orbiter demonstrated that heat shield tile misalignment has only a slight effect on heat transfer and transition location for $T_w = 0.42T_t$. Cooling of the boundary layer causes the tile-induced disturbances to increase significantly, promoting premature transition [2R]. Flow visualization studies complemented by extensive heat transfer measurements at the leeside of an 80-degree swept delta wing at $M_\infty = 10$ revealed that the dominant contributor to the heating load are vortices generated by boundary layer separation, either at the leading edge or inboard [3R].

Taking the effect of temperature-dependent heat capacity on aerodynamic ablation of melting bodies into account, the surface temperature history and melting distance history have been calculated [5R]. An experimental study of transient ablation of Teflon showed that the heat flow inside the heat shield material parallel to the surface cannot be neglected which invalidates a one-dimensional analysis [1R].

Radiative heat transfer studies within a solid-propellant rocket motor indicate that most solid rocket exhaust flows are optically thick in the subsonic portion of the flow for all but the smallest motor, lowest chamber pressures, or lowest aluminum content propellants [4R].

General

Experiments and numerical calculations considered processes in furnaces and steam boilers. A three-dimensional mathematical model [24S] used the Patankar–Spalding method to predict furnace heat transfer. The results describe the velocity and temperature pattern as well as the heat flux distribution in good agreement with experimental results obtained at the test facility of the International Flame Research

Foundation in Holland. Heat transfer at various gas flow patterns in furnaces was simulated [12S] in channels. Experiments studied [28S] convective heat transfer in a vertical cyclone chamber. A model for transient heat flow to a liquid fuel droplet in combustion gas predicted [6S] the limit for micro-explosions. Local stability was examined [25S] by a nonlinear model of a fluidized bed steam generator. Measurements on 38 power plants [20S] revealed the effect of the thermal resistance of a slag layer on the efficiency of furnace water walls. Experimental results on temperature conditions in compact once-through steam generators are reported [26S].

A method for rapid numerical simulation of transient heat transfer to fuel pins of a nuclear reactor was developed [14S]. Laser holographic interferometry and analysis revealed the natural convection pattern in nuclear spent-fuel shipping casks [17S]. An analytic model describes [1S] surface heating by photon and ion irradiation in laser fusion reactors.

A new surface [7S] for cooling towers has water flowing down in grooves, whereas the rest of the surface is dry and acts like a fin. Experiments and analyses determined the performance of this heat transfer surface. Several papers [5S, 29S, 30S] deal with the performance of heat storage units.

A finite difference calculation [18S] and semi-empirical equations [3S] provide information on heat transfer in superconducting coils of electric machines. Flow and heat transfer in convectively cooled underground electric cables was determined experimentally [2S, 10S] and the results are presented by correlations of the form $Nu = f(Re\,Pr^{0.4})$ with the parameters based on the hydraulic diameter and including the effect of surface roughness. Laplace equations were used to analyze [21S] the effect of an elevated pipeline on moss covered ground on temperature. The temperature distribution in deep and shallow lakes was treated [9S] as a heat conduction problem with corrections for convection in some regions. Heat transfer for cased wells with perforations in shallow geothermal systems is described [15S] by a thermosyphon model. The correlation $St\,Pr^{2/3} = 0.0627 Syl^{-0.02} Re^{-0.498}$ with Syl indicating the ratio of sensible to latent heat, describes the heat transfer characteristics of cooling and dehumidifying coils [22S] according to experimental results obtained in an air conditioning tunnel. New full scale tests [19S] provided data for heat transfer in insulated air duct systems.

Correlations for the Nusselt number in turbine cascades [32S] describe the effect of turbulence, rotation, and pressure gradient. Short duration measurements on gas turbine surfaces [11S] performed in a light-piston tunnel determined heat transfer to flat plates and curved surfaces simulating gas turbine surfaces. The effect of a concave curvature is found to be well described by Bradshaw's prediction. Transient heat transfer was measured [33S] on nozzle blades of an operating turbine. An analytic study [13S] considered heat transfer between a thin water film and the surface of a radial cooling channel of a gas turbine blade under the influence of centrifugal and Coriolis forces.

Several papers deal with heat transfer in the process industry. Calculation methods [16S] and optimization analysis [27S] considered heat transfer by convection and radiation in drying equipment and fermentation systems [8S]. The heat removal by cooling water in a mill used to reduce particle size in a slurry was studied [4S] with the aid of a mathematical model. Improved equations describing the relation between velocity and strain-rate fields in forging of rings were obtained [23S] by a numerical method. Wax or polyurethane liners of gun barrels reduce heat transfer caused by friction [31S]. In single firings the heat transfer is reduced by 10%. The reduction is even larger with multiple firings explaining a 25-times increase in wear life.

Solar energy

The number of heat transfer related solar energy papers that appeared in the archival literature decreased. Two topics, collector performance and thermal energy storage, dominated the field.

The papers dealing with collector performance are divided nearly equally between the studies of flat plate and concentrating collectors. A definition of efficiency based on "convertable energy" or "potential work", which include the restrictions imposed by both the first and second laws of thermodynamics, has been proposed for solar collectors [12T]. The theory and the experimental procedure are described for simultaneously obtaining a number of points on the efficiency–temperature performance curve for a flat plate collector and a standard day is suggested to provide a basis of comparison between collectors [18T]. It was found that ambient temperature, flow rate and wind are the variables which most influence the performance of a water-trickle collector and the use of improved glazing material and double glazing are the design changes which significantly improve the collector's performance [2T]. Based on a model for the heat transfer in a flat plate collector with a rectangular flow passage, it was established that for a collector with circulation driven by natural convection the flow rate increases but the efficiency decreases when the flow passage is enlarged [5T]. Operational data is presented for a system using evacuated-tube solar collectors [10T].

An analytical study of a flat-sided rectilinear trough as a solar concentrator predicts that the practical concentration factor for such a device ranges from 1.5 to 4 [3T]. A similar study, by the same authors, investigated the concentrating properties of specularly reflecting pyramids, hexagons and circular cones [4T]. The radiative performance of a trapezoidal groove solar collector has been determined as a function of various directional distributions of insolation and the results are presented in the form of nomographs [1T]. An analysis of catenary shaped solar collectors indicates that moderate concentrations can be achieved over a range of incidence angles from 0 to 50° [17T]. The use of total internal reflection prismatic reflectors appears promising for a number of applications [15T]. An exact analytical method for determining the flux density of reflected sunlight from heliostats having polygonal boundaries has been developed [9T].

The performance of solar collectors can be enhanced by the use of selective solar absorber coatings to reduce the radiation heat losses, and by the use of transparent honeycomb materials to suppress natural convection. Solar selective surfaces consisting of a substrate of copper, silver, or nickel underlying a single thin layer of material were compared using the criterion of maximum net radiation transfer at a given surface temperature and copper was found to be the best substrate for a 200°C surface temperature [13T]. The residual absorption observed in sputtered amorphous silicon films makes them inefficient as selective absorbers for solar collectors [20T]. Globular all-metal films can have poor broadband selective properties; however, the addition of very thin dielectric layers between the globules and the substrate permits surface-wave resonances, giving broadband absorption and a higher solar selectivity [7T].

An approximate equation for predicting the solar transmittance of transparent honeycombs was found to be in reasonable agreement with measured results and should be very useful in the evaluation of systems, the interpretation of experimental results and the screening of candidate honeycomb materials [6T].

Both sensible and phase change approaches for storing thermal energy were investigated. A one-dimensional single-phase conductivity model of packed beds in which air and rock are at the same temperature was used to develop closed-form solutions for the transient responses produced by time-varying air inlet temperatures and these solutions were found to be in close agreement with long-term predictions based on more complex models [16T]. Heat losses from a packed bed thermal storage tank, due to imperfect insulation, dominate over internal thermal conduction effects in causing the degradation of the temperature stratification within the tank [11T]. An experimental study showed that, even at very large flow rates, thermal stratification can be maintained in cylindrical water tanks [8T]. A pressurized-liquid concept for solar-thermal energy storage is presented for continuous operation of power systems and it is predicted that the useful stored energy extraction efficiency would exceed 88% while operating in the neighborhood of 300°C [19T]. Storage systems using phase change materials require roughly one-half to one-fourth the storage volume of a rock bed unit or one-half to slightly more than the storage volume of a liquid-based unit [14T].

The use of passive solar heating and cooling systems also received attention. A general discussion of various types of passive solar heating and cooling systems is presented along with procedures for estimating the net daily energy gains for the systems [21T].

PLASMA HEAT TRANSFER

Heat transfer studies in ionized gases continue to be of interest. Many papers are concerned with arc applications for circuit interruption. A numerical analysis of the Joule heating effect on plasma heat transfer to a positively biased body exposed to a plasma flow indicates that a certain fraction of the heat dissipated by the current flow in the vicinity of the body is transferred to the body [8U]. An analysis of the energy transfer to a negatively biased probe immersed into a plasma shows that at sufficiently high potentials the surrounding gas is heated and heat transfer to the probe surface is enhanced. Approximately half of the electrical power input is transferred to the probe surface [1U].

The behavior of high current density arcs ($>10^3$ A/cm^2) burning in narrow channels of solid materials is governed by (1) arc induced evaporation of wall material, (2) axial convection and (3) radiation heat transfer [12U]. Studies of the arc behavior in narrow insulating channels showed that, in contrast to an open arc with convective cooling, the mean electrical conductivity of an arc burning in a narrow channel depends strongly on the currents [16U].

The reduction in radiation found in vertical mercury arcs at high pressures (up to 13 atm) is caused by self-absorption and not by increased radial heat losses because of turbulence or shear-induced eddies [19U]. Calculations of the radiative transport in wall-stabilized nitrogen arcs ($p = 1$ and 10 atm) showed that radiative losses account for a significant fraction of the power input. Losses for wavelengths > 2000 Å are most important for a wide range of conditions and the dominant contribution to these losses are due to line radiation [15U]. Measured values of electrical conductivity, emission coefficient for optically thin radiation, and thermal conductivity as a function of the temperature using a wall-stabilized arc in air at $p = 1$ atm compare favorably with theoretical predictions [5U].

Two forms of arcing damages are observed on carbon cathodes in atmospheric air: that due to the vapor (cold cathode) arc with current densities around 6×10^7 A/cm^2 and relatively low erosion rates, and that due to the thermionic arc (thermal arc mode) with current densities around 2×10^3 A/cm^2 and higher erosion rates caused by increased heating of the surface [9U]. For establishing a theory for the cathode mechanism in low-current vacuum arcs, the energy-balance equation, space charge equation, electron-emission equation, current equation, MHD equation of motion, and the minimum principle of Joule heating were taken into account. The calculated relationship between arc current and evaporation rate is in qualitative agreement with experimental data [10U]. A new non-stationary cathode spot model for metal vapor arcs differs from other known models by its emphasis on ion mass generation by Joule heating and the separation of energy flows associated with electron and ion currents in the cathode spot [4U]. Studies of the recovery of dielectric strength after current interruption in vacuum suggest that the effect of gross anode heating is important to the neutral decay because the highest temperatures reduce the tendency for vapor condensation at high currents where both anode and cathode surfaces become hot [6U].

Experimental results of gas blast arc studies near current zero show an enhanced electrical conductance decay at the orifice inlet. It is not yet clear whether this decay is due to radial heat conduction losses or whether additional processes need to be invoked [3U]. There is no need to incorporate radial turbulent cooling in the model description of an air blast arc circuit breaker in order to explain experimental results. If radial turbulent cooling is present, however, the interruption capability of the arc is substantially enhanced [17U]. Studies of power frequency scaling and of electrode vapor effects in gas blast arc circuit breakers showed that a small degree of vapor contamination (1%) by a low ionization potential material (Cu) can increase the electrical conductivity of the arc enough to halve its thermal recovery speed [7U].

Systematic measurements of the energy balance of thermal induction plasmas (argon) under different operating conditions using a 12 kW 500 Hz power supply clearly indicated that improvements of the coupling could increase the overall efficiency to more than 60%. Thus, the process-efficiency for plasma processing would be close to that of d.c. arcs [18U]. Calculations of the trajectories and temperature histories of alumina particles (10–250 μm) injected into the fire ball of an inductively coupled plasma indicate that internal plasma recirculation in the coil region is responsible for the observed rejection of the particles by the fire ball [2U].

Anode heat flux measurements in a quasisteady MPD plasma accelerator showed that for arc currents from 5.5 to 44 kA and argon mass flow rates from 1 to 48 g/s, the fraction of the total arc power transferred to the anode decreases from 50% at 200 kW to 10% at 20 MW [14U]. An analysis of electrothermal instability in the seeded combustion gas (MHD) boundary layer near cold electrodes demonstrates that the ratio of Joule heating to heat conduction by the gas governs the transition from a diffuse discharge mode to an arc mode triggered by small perturbation of gas temperature, electron temperature, current density and electron density [13U].

Investigations of the radiative transfer in pulsed laser plasma-target interactions indicate that the radiative contribution to the observed enhanced thermal coupling between plasma and target may be appreciable [11U].

REFERENCES

Books

1. B. Berkovsky (editor), *Thermomechanics of Magnetic Fluids*, Hemisphere, Washington, DC (1979).
2. R. W. Bryers (editor), *Ash Deposits and Corrosion due to Impurities in Combustion Gases*, Hemisphere, Washington, DC (1978).
3. E. R. G. Eckert and T. F. Irvine, Jr. (editors), *Heat Transfer Reviews* 1970–1975, Pergamon, Oxford (1977).
4. J. J. Ginoux (editor), *Two-Phase Flows and Heat*

Transfer with Application to Nuclear Reactor Design Problems, Hemisphere, Washington, DC (1978).
5. M. Hirsh and H. Oskam (editors), *Gaseous Electronics*, Vol. 1, Chapter 5: Electric Arcs and Arc Gas Heaters, Academic Press, New York (1978).
6. C. J. Hoogendoorn and N. H. Afgan (editors), *Energy Conservation in Heating, Cooling and Ventilating Buildings*, Hemisphere, Washington, DC (1978).
7. F. Kreith and J. F. Kreider, *Principles of Solar Engineering*, Hemisphere, Washington, DC (1978).
8. S. S. Lee and S. Sengupta (editors), *Waste Heat Management and Utilization*, Hemisphere, Washington, DC (1979).
9. E. F. Nogotov (editor), *Applications of Numerical Heat Transfer*, Hemisphere, Washington, DC (1978).
10. E. M. Sparrow and R. D. Cess, *Radiation Heat Transfer* (augmented edition), Hemisphere/McGraw-Hill, Washington/New York (1978).
11. S. van Stralen and R. Cole, *Boiling Phenomena*, Hemisphere, Washington, DC (1979).
12. T. N. Veziroğlu and S. Kakac (editors), *Two-Phase Transport and Reactor Safety*, Hemisphere, Washington, DC (1978).
13. H. J. Wirz and J. J. Smolderen (editors), *Numerical Methods in Fluid Dynamics*, McGraw-Hill, New York (1978).
14. Z. P. Zarić (editor), *Thermal Effluent Disposal from Power Generation*, Hemisphere, Washington, DC (1978).

Conduction

1A. G. Ahmadi, *Letters Heat Mass Transfer* **5,** 167 (1978).
2A. R. W. Alperi, *J. Heat Transfer* **100,** 548 (1978).
3A. R. T. Balmer, *AIChE Jl* **24,** 547 (1978).
4A. G. E. Bell, *Int. J. Heat Mass Transfer* **21,** 1357 (1978).
5A. B. A. Boley, *Int. J. Heat Mass Transfer* **21,** 821 (1978).
6A. J. R. Cannon and P. C. DuChateau, *J. Heat Transfer* **100,** 503 (1978).
7A. J.-S. Chang and J. G. LaFramboise, *Int. J. Heat Mass Transfer* **21**, 360 (1978).
8A. J. Crank and A. B. Crowley, *Int. J. Heat Mass Transfer* **21**, 393 (1978).
9A. A. B. Crowley, *Int. J. Heat Mass Transfer* **21**, 215 (1978).
10A O. Ehrlich, Y.-K. Chuang and K. Schwerdtfeger, *Int. J. Heat Mass Transfer* **21,** 341 (1978).
11A. D. Glasser and J. Kern *AIChE Jl* **24,** 161 (1978).
12A. L. E. Goodrich, *Int. J. Heat Mass Transfer* **21,** 615 (1978).
13A. D. Greenspan, *Computer Meth. Appl. Mech. Engng* **13,** 95 (1978).
14A. M. Hattori, *Bull. JSME* **21,** 1507 (1978).
15A. G. Horvay, B. Gold and E. S. Kaczenski, *J. Heat Transfer* **100**, 281 (1978).
16A. L. M. Jiji and S. Weinbaum, *Int. J. Heat Mass Transfer* **21**, 581 (1978).
17A. A. D. Kraus, A. D. Snider and L. F. Doty, *J. Heat Transfer* **100,** 288 (1978).
18A. G. W. Krutz, R. J. Schoenhals and P. S. Hore, *Numerical Heat Transfer* **1,** 489 (1978).
19A. T. H. Kuehn, *J. Heat Transfer* **100,** 374 (1978).
20A. S. H. Lin, *Letters Heat Mass Transfer* **5,** 29 (1978).
21A. A. Maewal, G. A. Gurtman and G. A. Hegemier, *J. Heat Transfer* **100,** 128 (1978).
22A. S. B. Margolis, *Q. Appl. Math.* **36,** 97 (1978).
23A. W. W. Martin and S. S. Sadhal, *Int. J. Heat Mass Transfer* **21,** 783 (1978).
24A. G. H. Meyer, *Int. J. Heat Mass Transfer* **21,** 824 (1978).
25A. T. Miloh, *J. Heat Transfer* **100,** 555 (1978).
26A. D. Moalem-Maron and Y. Meinhardt, *Letters Heat Mass Transfer* **5,** 269 (1978).
27A. K. Morgan, R. W. Lewis and O. C. Zienkiewicz, *Int. J. Num. Meth. Engng* **12,** 1191 (1978).
28A. G. E. Myers, *J. Heat Transfer* **100,** 120 (1978).
29A. A. K. Naghdi, *J. Heat Transfer* **100,** 558 (1978).
30A. M. Okada, *Bull. JSME* **21,** 1514 (1978).
31A. M. N. Özişik, *J. Heat Transfer* **100,** 370 (1978).
32A. J. Padovan, *AIAA Jl* **15,** 1811 (1977).
33A. K. C. Poon and Y. P. Chang, *Letters Heat Mass Transfer* **5,** 215 (1978).
34A. B. B. Raju, R. Chandra and M. S. Rao, *AIAA Jl* **16,** 547 (1978).
35A. J. W. Ramsey and E. M. Sparrow, *J. Heat Transfer* **100,** 368 (1978).
36A. J. D. Randall, *Int. J. Heat Mass Transfer* **21,** 1447 (1978).
37A. M. H. Sadd and I. Miskleglu, *J. Heat Transfer* **100,** 553 (1978).
38A. T. Saitoh, *J. Heat Transfer* **100,** 294 (1978).
39A. R. Siegel, *Int. J. Heat Mass Transfer* **21,** 1421 (1978).
40A. R. Siegel, *J. Heat Transfer* **100,** 3 (1978).
41A. E. M. Sparrow, S. Ramadhyani and S. V. Patankar, *J. Heat Transfer* **100,** 395 (1978).
42A. E. M. Sparrow, R. R. Schmidt and J. W. Ramsey, *J. Heat Transfer* **100,** 11 (1978).
43A. J. Sucec and S. Hedge, *J. Heat Transfer* **100,** 172 (1978).
44A. A. I. Suslov, *Appl. Math. Mech.* **41,** 87 (1977).
45A. M. Suzuki and S. Maeda, *Int. J. Heat Mass Transfer* **21,** 653 (1978).
46A. L. N. Tao, *Q. Appl. Math.* **36,** 223 (1978).
47A. E. A. Thornton and A. R. Wieting, *J. Heat Transfer* **100,** 551 (1978).
48A. H. V. Truong and G. E. Zinsmeister, *Int. J. Heat Mass Transfer* **21,** 905 (1978).
49A. R. Weichert and K. Schönert, *Q. J. Mech. Appl. Math.* **31,** 363 (1978).
50A. Y. Zvirin and M. Toren, *Wärme- und Stoffübertragung* **11,** 29 (1978).

Channel flow

1B. A. B. Ambrazevicius, P. J. Valatkevicius and P. M. Kezelis, *Heat Transfer, Soviet Res.* **9** (4), 156 (1977).
2B. A. M. Baklastov, A. L. Yefimov and V. A. Gorbenko, *Heat Transfer, Soviet Res.* **9** (3), 101 (1977).
3B. J. W. Baughn, *J. Heat Transfer* **100,** 537 (1978).
4B. R. W. Benodekar and A. W. Date, *Int. J. Heat Mass Transfer* **21,** 935 (1978).
5B. R. A. Berezhinskii and V. A. Orlov, *Thermal Engng* **25** (2), 25 (1978).
6B. M. S. Bhatti and C. W. Savery, *J. Heat Transfer* **100,** 539 (1978).
7B. R. D. Borisova, A. A. Gukhman, V. V. Dil'man and B.

A. Kader, *Heat Transfer, Soviet Res.* **9** (4), 133 (1977).
8B. D. J. Brassington, *Int. J. Heat Mass Transfer* **21**, 76 (1978).
9B. A. R. Chandrupatla and V. M. K. Sastri, *Num. Heat Transfer* **1**, 243 (1978).
10B. T. W. Chapman, W. W. Collins and S. D. Troyer, *AIChE Jl* **24**, 338 (1978).
11B. K. C. Cheng and J. W. Ou, *Can. J. Chem. Engng* **56**, 31 (1978).
12B. K. C. Cheng and S. L. Wong, *Appl. Sci. Res.* **33**, 309 (1977).
13B. K. C. Cheng and R. S. Wu, *Appl. Sci. Res.* **33**, 405 (1977).
14B. B. T. F. Chung, L. C. Thomas and Y. Pang, *J. Heat Transfer* **100**, 92 (1978).
15B. S. W. Churchill and J. P. Gupta, *I/EC Proc. Des. Dev.* **17**, 351 (1978).
16B. N. Cur and E. M. Sparrow, *Int. J. Heat Mass Transfer* **21**, 1069 (1978).
17B. V. D. Dang, *Chem. Engng Sci.* **33**, 1179 (1978).
18B. Ye. P. Dyban, E. Ya. Epik and V. Ye. Filipchuk, *Heat Transfer, Soviet Res.* **9** (4), 123 (1977).
19B. A. M. El-Nashar, *I/EC Fundamentals* **17**, 213 (1978).
20B. M. Epstein and G. M. Hauser, *Letters Heat Mass Transfer* **5**, 19 (1978).
21B. M. Faghri and J. R. Welty, *Int. J. Heat Mass Transfer* **21**, 317 (1978).
22B. B. M. Galitseyskiy and Yu. A. Ryzhov, *Heat Transfer, Soviet Res.* **9** (4), 178 (1977).
23B. F. Gori, *J. Heat Transfer* **100**, 220 (1978).
24B. F. Gori, *Int. J. Heat Mass Transfer* **21**, 247 (1978).
25B. J. Gosse and R. Schiestel, *Int. Chem. Engng* **18**, 1 (1978).
26B. J. C. Han, L. R. Glicksman and W. M. Rohsenow, *Int. J. Heat Mass Transfer* **21**, 1143 (1978).
27B. O. T. Hanna and O. C. Sandall, *J. Heat Transfer* **100**, 224 (1978).
28B. M. R. F. Heikal and A. P. Hatton, *Int. J. Heat Mass Transfer* **21**, 841 (1978).
29B. M. Hishida, Y. Nagano and Y. Nakamura, *Bull. JSME* **21**, 1175 (1978).
30B. J. L. Houzelot and J. Villermaux, *Chem. Engng Sci.* **32**, 1465 (1977).
31B. L. A. M. Janssen and C. J. Hoogendoorn, *Int. J. Heat Mass Transfer* **21**, 1197 (1978).
32B. V. Javeri, *Int. J. Heat Mass Transfer* **21**, 1035 (1978).
33B. V. Javeri, *Int. J. Heat Mass Transfer* **21**, 1029 (1978).
34B. E. E. Kalinin and G. A. Dreytser, *Heat Transfer, Soviet Res.* **9** (4), 168 (1977).
35B. K. N. Krishnan and V. M. K. Sastri, *Wärme- und Stoffübertragung* **11**, 73 (1978).
36B. V. M. Legkii, *Int. Chem. Engng* **18**, 74 (1978).
37B. V. I. Lel'chuk, K. F. Shuyskaya, A. G. Sorokin and O. N. Bragina, *Heat Transfer, Soviet Res.* **9** (4), 100 (1977).
38B. D. C. Leslie, *Letters Heat Mass Transfer* **5**, 99 (1978).
39B. E. M. Mitwally, *AIChE Jl* **24**, 1108 (1978).
40B. A. Mojtabi and M. P. Caltagirone, *Int. J. Heat Mass Transfer* **21**, 261 (1978).
41B. A. L. Moyls and R. H. Sabersky, *Int. J. Heat Mass Transfer* **21**, 7 (1978).
42B. D. R. Oliver and S. S. Rao, *Trans. Inst. Chem. Engrs* **56**, 62 (1978).
43B. G. Ooms, G. Groen, D. P. de Graag and J. F. Ballintijn, *Chem. Engng Sci.* **33**, 357 (1978).
44B. S. V. Patankar, C. H. Liu and E. M. Sparrow, *Int. J. Heat Mass Transfer* **21**, 557 (1978).
45B. B. S. Petukhov, V. S. Grigor'yev, A. F. Polyakov and S. V. Rosnovskiy, *Heat Transfer, Soviet Res.* **9** (4), 114 (1977).
46B. V. N. Pilipenko, *Heat Transfer, Soviet Res.* **9** (4), 78 (1977).
47B. A. S. Popel and J. F. Gross, *Int. J. Heat Mass Transfer* **21**, 1133 (1978).
48B. D. P. Robinson and V. Walker, *Int. J. Heat Mass Transfer* **21**, 1299 (1978).
49B. V. L. Rvachev, A. P. Slesarenko and V. I. Popishvili, *Heat Transfer, Soviet Res.* **9** (4), 105 (1977).
50B. K. Seshadri and F. A. Williams, *Int. J. Heat Mass Transfer* **21**, 251 (1978).
51B. V. M. Sheptun, *Thermal Engng* **24** (9), 47 (1977).
52B. Z. P. Shulman and E. V. Korobko, *Int. J. Heat Mass Transfer* **21**, 543 (1978).
53B. O. K. Smirnov and S. N. Krasnov, *Thermal Engng* **25** (4), 70 (1978).
54B. E. M. Sparrow, B. R. Baliga and S. V. Patankar, *J. Heat Transfer* **100**, 572 (1978).
55B. E. M. Sparrow, S. V. Patankar and H. Shahrestani, *Num, Heat Transfer* **1**, 117 (1978).
56B. E. M. Sparrow and J. W. Ramsey, *Int. J. Heat Mass Transfer* **21**, 1369 (1978).
57B. E. M. Sparrow and K. P. Wachtler, *Int. J. Heat Mass Transfer* **21**, 761 (1978).
58B. A. S. Sukamel, D. F. Guntsev and V. I. Velichko, *Heat Transfer, Soviet Res.* **9** (4), 128 (1977).
59B. S. B. Thomason, J. C. Mulligan and J. Everhart, *J. Heat Transfer* **100**, 387 (1978).
60B. Th. H. Van der Meer and C. J. Hoogendoorn, *Chem. Engng Sci.* **33**, 1277 (1978).
61B. J. W. Vilemas and M. A. Nemira, *Heat Transfer, Soviet Res.* **9** (4), 151 (1977).
62B. K. Wichterle, *Int. Chem. Engng* **18**, 305 (1978).
63B. A. Yim, M. Epstein, S. G. Bankoff, G. A. Lambert and G. M. Hauser, *Int. J. Heat Mass Transfer* **21**, 1185 (1978).

Boundary layer and external flows
1C. R. C. Ackerberg, R. D. Patel and S. K. Gupta, *J. Fluid Mech.* **86**, 49 (1978).
2C. R. A. Antonia and H. Q. Danh, *Int. J. Heat Mass Transfer* **21**, 1002 (1978).
3C. G. G. Astashenkova, V. P. Motulevicy and E. D. Sergievskiy, *Heat Transfer, Soviet Res.* **9** (3), 42 (1977).
4C. V. Beschkov, C. Boyadjiev and G. Peev, *Chem. Engng Sci.* **33**, 65 (1978).
5C. S. Biringen and J. Levi, *AIAA Jl* **16**, 1016 (1978).
6C. M. N. Bismarck-Nasr, *AIAA Jl* **15**, 1813 (1977).
7C. A. Brown and R. C. Burton, *J. Engng Pwr* **100**, 159 (1978).
8C. T. Cebeci and K. C. Chang, *AIAA Jl* **16**, 730 (1978).
9C. J. L. S. Chen and T. N. Smith, *J. Heat Transfer* **100**, 358 (1978).
10C. H. W. Coleman, M. M. Pimenta and R. J. Moffat, *AIAA Jl* **16**, 78 (1978).

11C. N. Curle, *Aero Q.* **28,** 149 (1977).
12C. L. I. Dagis, M. M. Tamonis and Z. Z. Zhukauskas, *Int. Chem. Engng* **18,** 350 (1978).
13C. H. A. El-Gamal and W. H. Barclay, *Aero Q.* **24,** 75 (1978).
14C. T. M. El-Mistikawy and M. J. Werele, *AIAA Jl* **16,** 749 (1978).
15C. M. I. O. Ero, *AIAA Jl* **16,** 611 (1978).
16C. T. Fujita and T. Ueda, *Int. J. Heat Mass Transfer* **31,** 97 (1978).
17C. K. Gersten, H.-D. Papenfuss and J. F. Gross, *Int. J. Heat Mass Transfer* **21,** 275 (1978).
18C. R. S. R. Gorla, *Letters Heat Mass Transfer* **5,** 121 (1978).
19C. J. C. Gottifredi and O. D. Quiroga, *Int. J. Heat Mass Transfer* **21,** 662 (1978).
20C. G. L. Hayward and D. C. T. Pei, *Int. J. Heat Mass Transfer* **21,** 35 (1978).
21C. K. Inque and A. Tate, *J. Physical Soc. Japan* **44,** 1995 (1978).
22C. G. H. Junkhan, *J. Heat Transfer* **100,** 25 (1978).
23C. B. A. Kader and A. M. Yaglom, *J. Fluid Mech.* **89,** 305 (1978).
24C. J. Karnis and V. Pechoc, *Int. J. Heat Mass Transfer* **21,** 43 (1978).
25C. R. Karvinen, *Int. J. Heat Mass Transfer* **21,** 1261 (1978).
26C. R. Karvinen, *Letters Heat Mass Transfer* **5,** 197 (1978).
27C. V. Kottke, H. Blenke and K. G. Schmidt, *Wärme- und Stoffübertragung* **10,** 217 (1977).
28C. M. Kumari and G. Nath, *J. Fluid Mech.* **87,** 705 (1978).
29C. A. J. Laderman, *AIAA Jl* **16,** 723 (1978).
30C. A. I. Leont'yev, Ye. V. Shishov, V. M. Belov and V. N. Afans'yev, *Heat Transfer, Soviet Res.* **9** (4), 48 (1977).
31C. F. N. Lin, *Int. J. Heat Mass Transfer* **21,** 683 (1978).
32C. A. W. Lipsett and R. R. Gilpin, *Int. J. Heat Mass Transfer* **21,** 25 (1978).
33C. V. I. Makarevicius, *Heat Transfer, Soviet Res.* **9** (4), 145 (1977).
34C. N. C. G. Markatos, D. B. Spalding, D. G. Tatchell and N. Vlachos, *Computer Meth. Appl. Mech. Engng* **15,** 161 (1978).
35C. K. Mastanaiah, *Int. J. Heat Mass Transfer* **21,** 1403 (1978).
36C. M. N. Mathur, S. K. Ojha and P. S. Ramachandran, *Int. J. Heat Mass Transfer* **21,** 923 (1978).
37C. R. E. Mayle, *Int. J. Heat Mass Transfer* **21,** 364 (1978).
38C. J. H. Merkin, *J. Fluid Mech.* **88,** 309 (1978).
39C. V. K. Migay, *Heat Transfer, Soviet Res.* **9** (4), 96 (1977).
40C. W. J. Minkowycz and E. M. Sparrow, *Numerical Heat Transfer* **1,** 69 (1978).
41C. B. P. Mironov, V. N. Vasechikin and N. I. Yarygina, *Heat Transfer, Soviet Res.* **9** (4), 57 (1977).
42C. M. V. Narayan and P. K. Sarma, *Int. J. Multiphase Flow* **4,** 413 (1978).
43C. G. Nath and M. Kumari, *AIAA Jl* **16,** 1007 (1978).
44C. A. Polak and M. J. Werle, *J. Heat Transfer* **100,** 678 (1978).
45C. A. K. Rastogi and W. Rodi, *AIAA Jl* **16,** 151 (1978).
46C. U. C. Saxena and A. D. K. Laird, *J. Heat Transfer* **100,** 684 (1978).
47C. M. Schmal and A. M. Figueiredo, *Int. J. Heat Mass Transfer* **21,** 175 (1978).
48C. R. Schmitz and B. U. Felderhof, *Physica* **92A,** 423 (1978).
49C. R. A. Seban and A. Faghri, *J. Heat Transfer* **100,** 143 (1978).
50C. P. M. Sforza and R. F. Mons, *Int. J. Heat Mass Transfer* **21,** 371 (1978).
51C. S. Sideman, H. Horn and D. Moalem, *Int. J. Heat Mass Transfer* **21,** 285 (1978).
52C. J. C. Simonich and P. Bradshaw, *J. Heat Transfer* **100,** 671 (1978).
53C. P. Singh, V. P. Sharma and U. N. Misra, *Int. J. Heat Mass Transfer* **21,** 1117 (1978).
54C. A. A. Slianciauskas and M. R. M. Drizhus, *Heat. Transfer, Soviet Res.* **9** (4), 40 (1977).
55C. S. Swarup, *Z. angew Math. Phys* **29,** 147 (1978).
56C. M. M. Tamonis, L. I. Dagis and A. A. Zhukauskas, *Int. Chem. Engng* **18,** 343 (1978).
57C. D. P. Telionis and M. S. Romaniuk, *AIAA Jl* **16,** 488 (1978).
58C. L. Thomas, *J. Heat Transfer* **100,** 744 (1978).
59C. J. C. T. Wang and S. F. Shen, *AIAA Jl* **16,** 1025 (1978).
60C. R. E. Wilson, *Int. J. Heat Mass Transfer* **21,** 1167 (1978).
61C. W. Winkler and U. Grigull, *Wärme- und Stoffübertragung* **10,** 281 (1977).
62C. S.-M. Yih and R. D. Seagrave, *AIChE Jl* **24,** 803 (1978).

Flow with separated regions

1D. J. Bandrowski and G. Kaczmarzyk, *Chem. Engng Sci.* **33,** 1303 (1978).
2D. R. Bauer and E. U. Schlünder, *Int. Chem. Engng* **18,** 181 (1978).
3D. R. Bauer and E. U. Schlünder, *Int. Chem. Engng* **18,** 189 (1978).
4D. M. A. Biot, *Q. Appl. Math.* **36,** 19 (1978).
5D. J. S. M. Botterill and A. O. O. Denloye, *Chem. Engng Sci.* **33,** 509 (1978).
6D. M. A. Collins and R. L. Simpson, *AIAA Jl* **16,** 291 (1978).
7D. E. De Maerteleire, *Chem. Engng Sci.* **33,** 1107 (1978).
8D. P. Fedkiew and J. Newman, *Chem. Engng Sci.* **33,** 1043 (1978).
9D. S. Fukusako and H. Inaba, *Int. J. Heat Mass Transfer* **21,** 985 (1978).
10D. D. J. Gunn, *Int. J. Heat Mass Transfer* **21,** 467 (1978).
11D. R. Kapral and D. Bedeaux, *Physica* **91A,** 590 (1978).
12D. O. A. Kremnev, A. V. Shurchkov, N. A. Aronova and Ye. M. Kozlov, *Heat Transfer, Soviet Res.* **9,** (1), 51 (1977).
13D. G. Langer, A. Roethe, K. P. Roethe and D. Gelbin, *Int. J. Heat Mass Transfer* **21,** 751 (1978).
14D. J. C. LaRue and P. A. Libby, *Physics Fluids* **21,** 891 (1978).
15D. V. A. Mairovo, *Thermal Engng* **25,** 59 (1978).
16D. H. L. Moses, R. R. Jones III, W. F. O'Brien, Jr. and R. S. Peterson, *AIAA Jl* **16,** 61 (1978).
17D. K. M. Ng, H. T. Davis and L. E. Scriven, *Chem. Engng Sci.* **33,** 1009 (1978).

18D. S. V. Patankar, E. M. Sparrow and M. Ivanovic, *Int. J. Heat Mass Transfer* **21,** 269 (1978).
19D. R. L. Simpson and M. A. Collins, *AIAA Jl* **16,** 289 (1978).
20D. E. M. Sparrow and K. P. Wachtler, *Int. J. Heat Mass Transfer* **21,** 761 (1978).
21D. J. C. Williams, *J. Fluid Mech.* **88,** 241 (1978).

Transfer mechanisms

1E. J. C. Adams, Jr. and B. K. Hodge, *AIAA Jl* **16,** 643 (1978).
2E. K. M. M. Alshamani, *AIAA Jl* **16,** 859 (1978).
3E. R. A. Antonia, A. J. Chambers, C. W. Van Atta, C. A. Friehe, and K. N. Helland, *Physics Fluids* **21,** 509 (1978).
4E. R. A. Antonia and C. W. Van Atta, *Physics Fluids* **21,** 1096 (1978).
5E. R. A. Antonia and C. W. Van Atta, *J. Fluid Mech.* **84,** 561 (1978).
6E. C. Béguier, I. Dekeyser and B. E. Launder, *Physics Fluids* **21,** 307 (1978).
7E. A. C. M. Beljaars, *Letters Heat Mass Transfer* **5,** 231 (1978).
8E. K. J. Bullock, R. E. Cooper and F. H. Abernathy, *J. Fluid Mech.* **88,** 585 (1978).
9E. R. Chevray and N. K. Tutu, *J. Fluid Mech.* **88,** 133 (1978).
10E. Ye. P. Dyban and E. Ya Epik, *Heat Transfer, Soviet Res.* **9** (4), 11 (1977).
11E. C. L. Ezekwe, F. J. Pierce and J. E. McAllister, *AIAA Jl* **16,** 645 (1978).
12E. U. Frisch, P. L. Sulem and M. Nelkin, *J. Fluid Mech.* **87,** 719 (1978).
13E. J. Janicka and W. Kollmann, *Wärme- und Stoffübertragung* **11,** 157 (1978).
14E. B. A. Kolovandin, V. A. Sosinovich and S. V. Kravar, *Letters Heat Mass Transfer* **5,** 253 (1978).
15E. V. L. Kolpashchekov and A. I. Schnipp, *Int. J. Heat Mass Transfer* **21,** 155 (1978).
16E. N. Lee and A. E. Dukler, *Chem. Engng Sci.* **33,** 1169 (1978).
17E. J. L. Lumley, O. Zeman and J. Siess, *J. Fluid Mech.* **84,** 581 (1978).
18E. P. L. Maksin, B. S. Petukhov and A. F. Polyakov, *Heat Transfer, Soviet Res.* **9** (4), 1 (1977).
19E. C. Rey and J. Mathieu, *Int. J. Heat Mass Transfer* **21,** 1009 (1978).
20E. A. A. Shrayber, *Heat Transfer, Soviet Res.* **9** (3) 35 (1977).
21E. K. R. Sreenivasan and R. A. Antonia, *AIAA Jl* **16,** 867 (1978).
22E. K. R. Sreenivasan, R. A. Antonia and S. E. Stephenson, *AIAA Jl* **16,** 869 (1978).
23E. F. Tamanini, *J. Heat Transfer* **100,** 659 (1978).
24E. L. C. Thomas, *AIChE Jl* **24,** 101 (1978).
25E. F. G. van Dongen, A. C. M. Beljaars and D. A. DeVries, *Int. J. Heat Mass Transfer* **21,** 1099 (1978).
26E. K. S. Venkataramani and R. Chevray, *J. Fluid Mech.* **86,** 513 (1978).
27E. Z. Warhaft and J. L. Lumley, *J. Fluid Mech.* **88,** 659 (1978).
28E. P. C. Wayner, Jr., *Int. J. Heat Mass Transfer* **21,** 362 (1978).
29E. P. C. Wayner, Jr., *J. Heat Transfer* **100,** 155 (1978).
30E. J. C. Wu and A. Sugavanam, *AIAA Jl* **16,** 948 (1978).

Natural convection

1F. S. I. Abdel-Khalik and K. R. Randall, *J. Heat Transfer* **100,** 199 (1978).
2F. N. Afzal and N. K. Banthiya, *Z. angew. Math. Phys* **28,** 993 (1977).
3F. T. Ahmad and G. M. Faeth, *J. Heat Transfer* **100,** 112 (1978).
4F. M. Al-Arabi and M. M. El-Refaee, *Int. J. Heat Mass Transfer* **21,** 357 (1978).
5F. G. Antonini, G. Guiffant and D. Geiger, *Letters Heat Mass Transfer* **5,** 187 (1978).
6F. A. Bejan, *J. Fluid Mech.* **89,** 97 (1978).
7F. A. Bejan and C. L. Tien, *J. Heat Transfer* **100,** 191 (1978).
8F. A. Bejan and C. L. Tien, *J. Heat Transfer* **100,** 725 (1978).
9F. A. Bejan and C. L. Tien, *J. Heat Transfer* **100,** 641 (1978).
10F. A. Bejan and C. L. Tien, *Int. J. Heat Mass Transfer* **21,** 701 (1978).
11F. R. F. Bergholz, *J. Fluid Mech.* **84,** 743 (1978).
12F. J. R. Booker and K. C. Stengel, *J. Fluid Mech.* **86,** 289 (1978).
13F. R. L. Boxman and D. J. Shlien, *Appl. Optics* **17,** 2788 (1978).
14F. A. Brown, *Trans., Inst. Chem. Engrs* **56,** 77 (1978).
15F. S. N. Brown and K. Stewartson, *Proc. R. Soc.* **360A,** 455 (1978).
16F. V. P. Carey and J. C. Mollendorf, *Int. J. Heat Mass Transfer* **20,** 481 (1978).
17F. R. Cheesewright and K. S. Doan, *Int. J. Heat Mass Transfer* **21,** 911 (1978).
18F. C. F. Chen, *J. Heat Transfer* **100,** 653 (1978).
19F. P. Cheng, *Letters Heat Mass Transfer* **5,** 243 (1978).
20F. K. C. Cheng, M. Takeuchi and R. R. Gilpin, *Numerical Heat Transfer* **1,** 101 (1978).
21F. F. B. Cheung, *J. Heat Transfer* **100,** 416 (1978).
22F. R. M. Clever and F. H. Busse, *Z. angew Math. Phys.* **29,** 711 (1978).
23F. L. A. Clomburg, Jr., *J. Heat Transfer* **100,** 205 (1978).
24F. P. G. Daniels, *Proc. R. Soc.* **358A,** 173 (1978).
25F. P. G. Daniels, *J. Fluid Mech.* **85,** 193 (1978).
26F. L. R. Davis, M. A. Shirazi and D. L. Siegel, *J. Heat Transfer* **100,** 442 (1978).
27F. M. Dubois and P. Bergé, *J. Fluid Mech.* **85,** 641 (1978).
28F. M. Dubois, C. Normand and P. Berge, *Int. J. Heat Mass Transfer* **21,** 999 (1978).
29F. J. R. Dyer, *Int. J. Heat Mass Transfer* **21,** 1341 (1978).
30F. J. P. Easby, *Int. J. Heat Mass Transfer* **21,** 791 (1978).
31F. S. M. ElSherbiny, K. G. T. Hollands and G. D. Raithby, *J. Heat Transfer* **100,** 410 (1978).
32F. D. E. Fussey and I. P. Warneford, *Int. J. Heat Mass Transfer* **21,** 119 (1978).
33F. D. K. Gartling, Computer Meth. *Appl. Mech. Engng* **12,** 365 (1977).
34F. R. R. Gilpin, H. Imura and K. C. Cheng, *J. Heat Transfer* **100,** 71 (1978).
35F. K. Gotoh, S. Yanase and J. Mizushima, *J. Physical*

Soc. Japan **43,** 1773 (1977).
36F. D. D. Gray, *Appl. Sci. Res.* **33,** 437 (1978).
37F. R. Greif, *J. Heat Transfer* **100,** 86 (1978).
38F. J. Gryzagoridis, *Letters Heat Mass Transfer* **5,** 203 (1978).
39F. P. Hall and I. C. Walton, *Proc. R. Soc.* **358A,** 199 (1978).
40F. N. Hattori and S. Kotake, Bull. *JSME* **21,** 861 (1978).
41F. T. E. Hinkebein and J. C. Berg, *Int. J. Heat Mass Transfer* **21,** 1241 (1978).
42F. W. K. Honeywell and J. E. Vevai, *AIChE Jl* **24,** 1035 (1978).
43F. R. N. Horne and M. J. O'Sullivan, *J. Heat Transfer* **100,** 448 (1978).
44F. R. N. Horne and M. J. O'Sullivan, *Physics Fluids* **21,** 1260 (1978).
45F. C. T. Hsu, P. Cheng and G. M. Homsy, *Int. J. Heat Mass Transfer* **21,** 1221 (1978).
46F. J. L. Hudson, D. Tang and S. Abell, *J. Fluid Mech.* **86,** 147 (1978).
47F. T. Igarashi, Bull. *JSME* **21,** 1022 (1978).
48F. H. Imura, R. R. Gilpin and K. C. Cheng, *J. Heat Transfer* **100,** 429 (1978).
49F. D. B. Ingham, *Int. J. Heat Mass Transfer* **21,** 67 (1978).
50F. P. A. Iyer and R. E. Kelly, *J. Heat Transfer* **100,** 648 (1978).
51F. C. H. Johnson and P. Cheng, *Int. J. Heat Mass Transfer* **21,** 709 (1978).
52F. R. E. Kelly and D. Pal, *J. Fluid Mech.* **86,** 433 (1978).
53F. K. Küblbeck, J. Straub, S. Bloss and U. Grigull, *Wärme- und Stoffübertragung* **11,** 131 (1978).
54F. T. H. Kuehn and R. J. Goldstein, *J. Heat Transfer* **100,** 635 (1978).
55F. J. A. Lidburdy and G. M. Faeth, *J. Heat Transfer* **100,** 177 (1978).
56F. F. N. Lin and B. T. Chao, *J. Heat Transfer* **100,** 160 (1978).
57F. S.-J. Lin and S. W. Churchill, *Numerical Heat Transfer* **1,** 129 (1978).
58F. R. L. Mahajan and B. Gebhart, *J. Heat Transfer* **100,** 731 (1978).
59F. R. L. Mahajan and B. Gebhart, *Int. J. Heat Mass Transfer* **21,** 549 (1978).
60F. R. S. Marshall, J. C. Heinrich and O. C. Zienkiewicz, *Numerical Heat Transfer* **1,** 315 (1978).
61F. B. K. Meena and G. Nath, *J. Heat Transfer* **100,** 163 (1978).
62F. G. P. Merker, *Wärme- und Stoffübertragung* **10,** 255 (1977).
63F. R. M. Miller and B. Gebhart, *Int. J. Heat Mass Transfer* **21,** 1229 (1978).
64F. S. P. Mishra and D. G. Sahoo, *Appl. Sci. Res.* **24,** 1 (1978).
65F. G. L. Morrison and V. Q. Tran, *Int. J. Heat Mass Transfer* **21,** 203 (1978).
66F. A. Mucoglu and T. S. Chen, *Numerical Heat Transfer* **1,** 267 (1978).
67F. A. Mucoglu and T. S. Chen, *J. Heat Transfer* **100,** 542 (1978).
68F. M. A. Muntasser and J. C. Mulligan, *J. Heat Transfer* **100,** 165 (1978).
69F. T. Y. Na, *Appl. Sci. Res.* **33,** 519 (1978).
70F. H. Nakamura and Y. Asako, *Bull. JSME* **21,** 471 (1978).
71F. K. S. Nin, R. E. Faw and T. W. Lester, *J. Heat Transfer* **100,** 729 (1978).
72F. Yu. N. Ostrovskiy and V. Ye. Pisarev, *Heat Transfer, Soviet Res.* **9** (3), 84 (1977).
73F. E. M. Parmentier, *J. Fluid Mech.* **84,** 1 (1978).
74F. J. R. Parsons, Jr. and J. C. Mulligan, *J. Heat Transfer* **100,** 423 (1978).
75F. S. V. Patankar, S. Ramadhyani and E. M. Sparrow, *J. Heat Transfer* **100,** 63 (1978).
76F. J. L. Peube and D. Blay, *Int. J. Heat Mass Transfer* **21,** 1125 (1978).
77F. G. D. Raithby and K. G. T. Hollands, *J. Heat Transfer* **100,** 378 (1978).
78F. K. Ramakrishna, K. N. Seetharamu and P. K. Sarma, *J. Heat Transfer* **100,** 727 (1978).
79F. S. J. Rhee, V. K. Dhir and I. Catton, *J. Heat Transfer* **100,** 78 (1978).
80F. B. Roux, J. C. Grondin, P. Bontoux and B. Gilly, *Numerical Heat Transfer* **1,** 331 (1978).
81F. E. Ruckenstein, *AIChE Jl* **24,** 940 (1978).
82F. M. Schatzmann, *Z. angew. Math. Phys.* **29,** 608 (1978).
83F. R. R. Schmidt and E. M. Sparrow, *J. Heat Transfer* **100,** 403 (1978).
84F. N. Seki, S. Fukusako and H. Inaba, *J. Fluid Mech.* **84,** 695 (1978).
85F. N. Seki, S. Fukusako and H. Inaba, *Bull. JSME* **21,** 246 (1978).
86F. N. Seki, S. Fukusako and H. Inaba, *Wärme- und Stoffübertragung* **11,** 145 (1978).
87F. N. Seki, S. Fukusako and M. Sugawara, *Wärme- und Stoffübertragung* **10,** 269 (1977).
88F. E. J. Shaughnessy, J. Custer and R. W. Douglass, *J. Heat Transfer* **100,** 435 (1978).
89F. A. V. Shenoy and R. A. Mashelkar, *AIChE Jl* **24,** 344 (1978).
90F. A. V. Shenoy and R. A. Mashelkar, *Chem. Engng Sci.* **33,** 769 (1978).
91F. P. Singh, V. P. Sharma and U. N. Misra, *Appl. Sci. Res.* **34,** 105 (1978).
92F. V. M. Soundalgekar and R. N. Aranake, *Appl. Sci. Res.* **34,** 49 (1978).
93F. V. M. Soundalkgekar, S. G. Pohancrkar and M. R. Patil, *J. Appl. Mech.* **43,** 697 (1978).
94F. E. M. Sparrow, S. V. Patankar and R. M. Abdel-Wahed, *J. Heat Transfer* **100,** 184 (1978).
95F. J. M. Straus and G. Schubert, *J. Fluid Mech.* **87,** 385 (1978).
96F. M. Tvietereid, *Int. J. Heat Mass Transfer* **21,** 335 (1978).
97F. K. Vajravelu and K. S. Sastri, *J. Fluid Mech.* **86,** 365 (1978).
98F. J. G. van Steeg and P. Wesseling, *Computers Fluids* **6,** 93 (1978).
99F. R. L. Verma and P. Singh, *Australian J. Phys.* **30,** 335 (1977).
100F. K. L. Walker and G. M. Homsy, *J. Fluid Mech.* **87,** 449 (1978).
101F. J. E. Weber, *J. Fluid Mech.* **87,** 65 (1978).
102F. L.-S. Yao, *J. Appl. Mech.* **45,** 481 (1978).
103F. L.-S. Yao, *J. Fluid Mech,* **88,** 465 (1978).

104F. L.-S. Yao, *J. Heat Transfer* **100,** 212 (1978).
105F. L.-S. Yao and S. A. Berger, *J. Fluid Mech.* **88,** 339 (1978).
106F. L.-S. Yao and I. Catton, *Int. J. Heat Mass Transfer* **21,** 407 (1978).
107F. L.-S. Yao and I. Catton, *J. Heat Transfer* **100,** 376 (1978).
108F. L.-S. Yao, I. Catton and J. M. McDonough, *Numerical Heat Transfer* **1,** 255 (1978).
109F. S. H. Yin, T. Y. Wung and K. Chen, *Int. J. Heat Mass Transfer* **21,** 307 (1978).
110F. C. P. Yu and C. E. Hendrix, *Appl. Sci. Res.* **33,** 369 (1977).
111F. A. Zebib, *Physics Fluids* **21,** 699 (1978).
112F. A. Zebib and D. R. Kassoy, *Physics Fluids* **21,** 1 (1978).

Convection from rotating surfaces

1G. J. T. Dakin, *Int. J. Heat Mass Transfer* **21,** 1325 (1978).
2G. P. G. Daniels and K. Stewartson, *Q. J. Mech. Appl. Math.* **31,** 113 (1978).
3G. A. Sh. Dorfman and G. F. Selyavin, *Heat Transfer, Soviet Res.* **9** (2), 105 (1977).
4G. V. M. Kapinos, V. N. Pustovalov, A. P. Rud'ko and L. A. Gura, *Heat Transfer, Soviet Res.* **9** (4), 73 (1977).
5G. M.-H. Lee, D. R. Jeng and K. J. DeWitt, *J. Heat Transfer* **100,** 496 (1978).
6G. P. Mishra and P. C. Singh, *Chem. Engng Sci.* **33,** 1463 (1978).
7G. P. Mishra and P. C. Singh, *Chem. Engng Sci.* **33,** 1449 (1978).
8G. A. A. Mosyak, B. G. Rykova, P. D. Kostov and G. I. Gruzintsev, *Heat Transfer, Soviet Res.* **9** (4), 110 (1977).
9G. G. Nath and M. Muthanna, *Int. J. Heat Mass Transfer* **21,** 1213 (1978).
10G. E. J. Shaughnessy and R. W. Douglass, *Int. J. Heat Mass Transfer* **21,** 1251 (1978).
11G. R. M. Turian and W. Aung, *Int. J. Heat Mass Transfer* **21,** 1087 (1978).
12G. I. A. Vatutin, O. G. Martynenko and I. V. Skutova, *Heat Transfer, Soviet Res.* **9** (4), 35 (1977).
13G. T. Yamada and K. Kawashimo, *Bull. JSME* **21,** 266 (1978).

Combined heat and mass transfer

1H. G. Bergeles, A. D. Gosman and B. E. Launder, *Numerical Heat Transfer* **1,** 217 (1978).
2H. R. Best and H. Beer, *Wärme- und Stoffübertragung* **11,** 175 (1978).
3H. E. N. Bogomolov and S. M. Piotukh, *Thermal Engng* **25** (1), 6 (1978).
4H. H. Brauer and D. Sucker, *Int. Chem. Engng* **18,** 375 (1978).
5H. J. N. Chung and P. S. Ayyaswamy, *Int. J. Heat Mass Transfer* **21,** 1309 (1978).
6H. H. Eickhoff, *Wärme- und Stoffübertragung* **11,** 103 (1978).
7H. R. Ghez, *Int. J. Heat Mass Transfer* **21,** 745 (1978).
8H. J. K. Hedrick, D. E. Metzger and D. I. Takeuchi, *J. Energy* **2,** 210 (1978).
9H. B. R. Hollworth and R. D. Berry, *J. Heat Transfer* **100,** 352 (1978).
10H. M. Y. Jabbari and R. J. Goldstein, *J. Engng Pwr* **100,** 303 (1978).
11H. O. A. Kiseleva, Ya. I. Rabinovich, V. D. Sobolev, V. M. Starov and N. V. Churayev, *Heat Transfer, Soviet Res.* **9** (1), 23 (1977).
12H. P. A. Libby, *AIAA Jl* **16,** 130 (1978).
13H. A. V. Lykova, A. M. Medvedev and V. D. Kosoy, *Heat Transfer, Soviet Res.* **9** (1), 60 (1977).
14H. N. C. G. Makatos, *Computer Meth. Appl. Mech. Engng* **14,** 323 (1978).
15H. R. J. Moffat, J. M. Healzer and W. M. Kays, *J. Heat Transfer* **100,** 134 (1978).
16H. V. M. Repukhov and K. A. Bogachuk-Kozachuk, *Heat Transfer, Soviet Res,* **9** (2), 100 (1977).
17H. S. G. Romanovskiy, K. D. Lukin and L. I. Margolin, *Heat Transfer, Soviet Res.* **9** (1), 94 (1977).
18H. G. S. Shubin, *Heat Transfer, Soviet Res.* **9** (1), 65 (1977).
19H. T. Suzuki, *AIAA Jl* **16,** 754 (1978).
20H. S. D. Wilson and A. F. Jones, *I/EC Fundamentals* **17,** 183 (1978).
21H. V. M. Yeroshenko, A. A. Klumov and Yu. N. Terent'yev, *Heat Transfer, Soviet Res.* **9** (4), 66 (1977).

Change of phase

1J. N. H. Afgan and L. A. Jović, *Int. J. Heat Mass Transfer* **21,** 427 (1978).
2J. P. A. Andreev, *Thermal Engng* **25** (4), 29 (1978).
3J. N. V. Antonishin and V. S. Nikitin, *Heat Transfer, Soviet Res.* **9** (1), 11 (1977).
4J. Y. Asakura, M. Kikuchi, S. Uchida and H. Yusa, *Nucl. Sci. Engng* **67,** 1 (1978).
5J. A. M. Baklastov, L. I. Arkhipov and A. A. Ivanov, *Heat Transfer, Soviet Res.* **9** (3), 96 (1977).
6J. M. K. Bezrodnyi and D. V. Alekseenko, *Thermal Engng* **24** (7), 71 (1977).
7J. M. K. Bezrodnyy and D. V. Elekseyenko, *Heat Transfer, Soviet Res.* **9** (5), 14 (1977).
8J. M. K. Bologa and A. B. Didkovskiy, *Heat Transfer, Soviet Res.* **9** (1), 147 (1977).
9J. V. M. Borishanskiy, *Heat Transfer, Soviet Res.* **9** (2), 35 (1977).
10J. Yu. M. Brodov, R. Z. Savel'yev, V. A. Permyakov, V. K. Kuptsov, and A. G. Gal'perin, *Heat Transfer, Soviet Res.* **9** (1), 152 (1977).
11J. L. C. Burmeister, *Int. J. Heat Mass Transfer* **21,** 1411 (1978).
12J. L.-T. Chen, *Letters Heat Mass Transfer* **5,** 111 (1978).
13J. S. C. Cheng, W. W. L. Ng and K. T. Heng, *Int. J. Heat Mass Transfer* **21,** 1385 (1978).
14J. S. C. Cheng, W. W. Ng, K. T. Heng and D. C. Groenveld, *J. Heat Transfer* **100,** 382 (1978).
15J. A. K. Chesters, *Int. J. Multiphase Flow* **4,** 279 (1978).
16J. C. L. Chu, J. M. Roberts and A. W. Dalcher, *J. Engng Pwr* **100,** 424 (1978).
17J. T. L. Chuck and J. E. Myers, *Int. J. Heat Mass Transfer* **21,** 187 (1978).
18J. S. S. Dua and C. L. Tien, *Int. J. Heat Mass Transfer* **21,** 955 (1978).
19J. R. B. Duffey, M. C. Ackerman, B. D. G. Piggott and S. A. Fairbairn, *Int. J. Multiphase Flow* **4,** 117 (1978).
20J. R. I. Eddington, D. B. R. Kenning and A. I. Kor-

neichev, *Int. J. Heat Mass Transfer* **21,** 855 (1978).
21J. M. El-Shanawany, A. A. El-Shirbini and W. Murgatroyd, *Int. J. Heat Mass Transfer* **21,** 529 (1978).
22J. G. S. Emmerson and C. W. Snoek, *Int. J. Heat Mass Transfer* **21,** 1081 (1978).
23J. L. T. Fan, S. T. Lin and N. Z. Azer, *Int. J. Heat Mass Transfer* **21,** 849 (1978).
24J. R. Farhadieh and L. Baker, *J. Heat Transfer* **100,** 305 (1978).
25J. H. S. Fath and R. L. Judd, *J. Heat Transfer* **100,** 48 (1978).
26J. R. R. Gilpin, *Can. J. Chem. Engng* **56,** 466 (1978).
27J. A. S. Ginzburg, *Heat Transfer, Soviet Res.* **9** (1), 1 (1977).
28J. V. A. Grigor'ev, *Thermal Engng* **25** (2), 4 (1978).
29J. D. C. Groeneveld and S. R. M. Gardiner, *Int. J. Heat Mass Transfer* **21,** 664 (1978).
30J. R. J. Hannemann, *Int. J. Heat Mass Transfer* **21,** 65 (1978).
31J. K. Hijikata, Y. Mori and T. Nagatani, *J. Heat Transfer* **100,** 460 (1978).
32J. O. C. Jones, Jr. and N. Zuber, *J. Heat Transfer* **100,** 453 (1978).
33J. F. C. H. Jongenelen, F. Groeneweg and J. H. Gouda, *Chem. Engng Sci.* **33,** 777 (1978).
34J. J. G. H. Joosten, W. Zul and S. J. D. van Stralen, *Int. J. Heat Mass Transfer* **21,** 15 (1978).
35J. A. N. Khoze, Yu. V. D'yachenko and A. S. Zakharov, *Heat Transfer, Soviet Res.* **9** (2), 32 (1977).
36J. S. Kotake, *Int. J. Heat Mass Transfer* **21,** 875 (1978).
37J. S. A. Kovalev and G. B. Rybchinskaya, *Int. J. Heat Mass Transfer* **21,** 691 (1978).
38J. V. A. Kravchenko and L. F. Tolubinskaya, *Heat Transfer, Soviet Res.* **9** (3), 9 (1977).
39J. P. D. Lebedev (deceased), D. P. Lebedev and V. V. Uvarov, *Heat Transfer, Soviet Res.* **9** (1), 7 (1977).
40J. V. G. Leitsina, N. V. Pavlyukevich and G. I. Rudin, *Int. J. Heat Mass Transfer* **21,** 399 (1978).
41J. M. I. Marinov and I. P. Kabanov, *Thermal Engng* **24** (7), 68 (1977).
42J. E. Marschall and L. L. Moresco, *Num. Heat Transfer* **1,** 285 (1978).
43J. I. Michiyoshi and K. Makino, *Int. J. Heat Mass Transfer* **21,** 605 (1978).
44J. L. S. Midler and M. Ye. Shitsman, *Heat Transfer, Soviet Res.* **9** (2), 1 (1977).
45J. A. A. Mikhalevich and V. B. Nesterenko, *Int. J. Heat Mass Transfer* **21,** 385 (1978).
46J. Z. L. Miropol'skiy and L. R. Khasanov, *Heat Transfer, Soviet Res.* **9** (1), 157 (1977).
47J. D. Moalem-Maron and W. Zijl, *Chem. Engng Sci.* **33,** 1339 (1978).
48J. M. Monde and Y. Katto, *Int. J. Heat Mass Transfer* **21,** 295 (1978).
49J. Y. H. Mori, E. Inui and K. Komotori, *J. Heat Transfer* **100,** 613 (1978).
50J. A. W. Neumann, A. H. Abdelmessih and A. Hameed, *Int. J. Heat Mass Transfer* **21,** 947 (1978).
51J. A. P. Ornatskiy, V. A. Chernobay, A. F. Vasil'yev and S. V. Perkov, *Heat Transfer, Soviet Res.* **9** (3), 1 (1977).
52J. A. P. Ornatskiy, V. A. Chernobay, A. F. Vasil'yev and G. V. Struts, *Heat Transfer, Soviet Res.* **9** (1), 105 (1977).
53J. A. P. Ornatskiy and I. G. Sharayevskiy, *Heat Transfer, Soviet Res.* **9** (3), 28 (1977).
54J. M. N. Özişik and T. S. Kress, *Nucl. Sci. Engng* **66,** 397 (1978).
55J. W. Peyayopanakul and J. W. Westwater, *Int. J. Heat Mass Transfer* **21,** 1437 (1978).
56J. M. S. Plesset and A. Prosperetti, *Int. J. Heat Mass Transfer* **21,** 725 (1978).
57J. A. B. Ponter and W. Peier, *Int. J. Heat Mass Transfer* **21,** 1025 (1978).
58J. A. Prosperetti and M. S. Plesset, *J. Fluid Mech.* **85,** 345 (1978).
59J. M. D. Razavi and A. S. Damle, *Trans. Inst. Chem. Engrs* **56,** 81 (1978).
60J. V. G. Rifert, P. A. Barabash, A. B. Golubev, G. G. Leont'yev and S. I. Chaplinskiy, *Heat Transfer, Soviet Res.* **9** (2), 23 (1977).
61J. J. W. Rose, *Int. J. Heat Mass Transfer* **21,** 835 (1978).
62J. A. N. Ryabov, *Heat Transfer, Soviet Res.* **9,** 112 (1977).
63J. H. W. Schneider, *Int. J. Heat Mass Transfer* **21,** 1019 (1978).
64J. N. Seki, S. Fukusako and K. Torikoshi, *J. Heat Transfer* **100,** 624 (1978).
65J. M. Seki, H. Kawamura and K. Sanokawa, *J. Heat Transfer* **100,** 167 (1978).
66J. P. R. Sharma, S. C. Gupta and B. S. Varshney, *Letters Heat Mass Transfer* **5,** 223 (1978).
67J. P. R. Sharma and B. S. Varshney, *J. Heat Transfer* **100,** 733 (1978).
68J. M. Shoukri and R. L. Judd, *J. Heat Transfer* **100,** 618 (1978).
69J. M. S. Sifaoui and A. Perrier, *Int. J. Heat Mass Transfer* **21,** 629 (1978).
70J. G. F. Smirnov, *Thermal Engng* **24** (9), 55 (1977).
71J. Y. Sone, *J. Physical Soc. Japan* **45,** 315 (1978).
72J. Y. Sone and Y. Onishi, *J. Physical Soc. Japan* **44,** 1981 (1978).
73J. D. A. Spence and D. B. R. Kenning, *Int. J. Heat Mass Transfer* **21,** 719 (1978).
74J. F. D. Stacey and R. D. Irvine, *Australian J. Phys.* **30,** 631 (1977).
75J. M. A. Styrikovich, A. I. Leont'yev, V. S. Polonskiy and I. I. Malashkin, *Heat Transfer, Soviet Res.* **9** (1), 123 (1977).
76J. T. G. Theofanous, T. H. Bohrer, M. C. Chang and P. D. Patel, *J. Heat Transfer* **100,** 41 (1978).
77J. V. I. Tolubinskiy, Ye. D. Domashev, A. K. Litoshenko and A. S. Matorin, *Heat Transfer, Soviet Res.* **9** (1), 132 (1977).
78J. V. I. Tolubinskiy, Yu. N. Ostrovskiy and V. Ye. Pisarev, *Heat Transfer, Soviet Res.* **9** (2), 73 (1977).
79J. V. I. Tolubinskiy, Yu. N. Ostrovskiy and V. Ye. Pisarev, *Heat Transfer, Soviet Res.* **9** (2), 83 (1977).
80J. L. I. Tovazhyanskiy, V. I. Atroshchenko and M. S. Kedrov, *Heat Transfer, Soviet Res.* **9** (2), 28 (1977).
81J. P. A. Tselishchey, T. M. Bogacheva and G. G. Abayev, *Heat Transfer, Soviet Res.* **9** (2), 66 (1977).
82J. L. B. Tsimermanis and F. Kh. Tsimermanis, *Heat Transfer, Soviet Res.* **9** (1), 15 (1977).
83J. H. C. Ünal, *J. Heat Transfer* **100,** 268 (1978).
84J. M. Unsal and W. C. Thomas, *J. Heat Transfer* **100,**

629 (1978).
85J. T. von Hoffmann, *Wärme- und Stoffübettragung* **11,** 189 (1978).
86J. V. A. Vorob'ev, O. V. Remizov and V. V. Sergeev, *Thermal Engng* **25** (2), 23 (1978).
87J. K. W. Wall and E. L. Park, Jr., *Int. J. Heat Mass Transfer* **21,** 73 (1978).
88J. G. B. Wallis and H. J. Richter, *J. Heat Transfer* **100,** 595 (1978).
89J. D. F. Warner, E. L. Park, Jr. and K. G. Mayhan, *Int. J. Heat Mass Transfer* **21,** 137 (1978).
90J. H. Weiss and C. Gutfinger, *Israel J. Tech.* **15,** 255 (1977).
91J. P. B. Whalley, *Int. J. Multiphase Flow* **4,** 427 (1978).
92J. C. P. C. Wong, G. C. Vliet and P. S. Schmidt, *J. Heat Transfer* **100,** 466 (1978).
93J. W. J. Yang, *Letters Heat Mass Transfer* **5,** 151 (1978).
94J. I. S. Yefremova and M. S. Smirnov, *Heat Transfer, Soviet Res.* **9** (1), 90 (1977).
95J. M. C. Yuen and L. W. Chen, *Int. J. Heat Mass Transfer* **21,** 537 (1978).
96J. A. S. Zelepuga and V. S. Karpenko, *Heat Transfer, Soviet Res.* **9** (1), 19 (1977).
97J. W. Ziji and D. Moalem-Maron, *Chem. Engng Sci.* **33,** 1331 (1978).
98J. N. V. Zozulya, V. A. Karkhu and V. P. Borovkov, *Heat Transfer, Soviet Res.* **9** (2), 18 (1977).

Radiation in participating media
1K. G. Abbate, U. Bernini, E. Ragozzino and F. Somma, *J. Phys. D: Appl. Phys.* **11,** 1167 (1978).
2K. R. O. Buckius and D. C. Hwang, *J. Heat Transfer* **100,** 665 (1978).
3K. A. L. Burka, *Heat Transfer, Soviet Res.* **9** (5), 42 (1977).
4K. Y. S. Chou, Appl. Optics 17, 364 (1978).
5K. K. H. Chu and R. Greif, *J. Heat Transfer* **100,** 230 (1978).
6K. K. J. Daniels, N. M. Laurendeau and F. P. Incropera, *Int. J. Heat Mass Transfer* **21,** 1379 (1978).
7K. W. G. Egan and T. Hilgeman, *Appl. Optics* **17,** 245 (1978).
8K. B. W. Fowler and C. S. Sung, *Appl. Optics* **17,** 1797 (1978).
9K. R. Greif, *Int. J. Heat Mass Transfer* **21,** 477 (1978).
10K, W. L. Grosshandler and R. F. Sawyer, *J. Heat Transfer* **100,** 247 (1978).
11K. G. L. Hubbard and C. L. Tien, *J. Heat Transfer* **100,** 235 (1978).
12K. C. S. Kelley, *Appl. Optics* **17,** 3054 (1978).
13K. P. Koepke and H. Quenzel, *Appl. Optics* **17,** 2114 (1978).
14K. L. A. Konyukh, *Heat Transfer, Soviet Res.* **9** (5), 96 (1977).
15K. V. I. Kruglov and Yu. V. Khodyko, *Int. J. Heat Mass Transfer* **21,** 163 (1978).
16K. V. I. Kruglov and Yu. V. Khodyko, *Int. J. Heat Mass Transfer* **21,** 169 (1978).
17K. D. C. Look, *Letters Heat Mass Transfer* **5,** 175 (1978).
18K. D. C. Look, H. F. Nelson, A. L. Crosbie and R. L. Dougherty, *J. Heat Transfer* **100,** 480 (1978).
19K. R. A. MacRae, E. T. Arakawa and M. W. Williams, *J. Chem. Engng Data* **23,** 189 (1978).
20K. M. M. Mel'man and A. S. Nevskiy, *Heat Transfer, Soviet Res.* **9** (5), 104 (1977).
21K. T. Minota, *J. Phys. Soc. Japan* **45,** 1025 (1978).
22K. V. N. Mirky and V. P. Stulov, *Heat Transfer, Soviet Res.* **9** (5), 55 (1977).
23K. A. T. Modak and M. K. Mathews, *J. Heat Transfer* **100,** 544 (1978).
24K. J. Otterman, *Appl. Optics* **17,** 3431 (1978).
25K. Yu. A. Popov, *Heat Transfer, Soviet Res.* **9** (5), 112 (1977).
26K. R. Ram, *Appl. Sci. Res.* **34,** 93 (1978).
27K. J. A. Roux and A. M. Smith, *J. Heat Transfer* **100,** 98 (1978).
28K. N. Seki, M. Sugawara and S. Fukusako, *Wärme- und Stoffübertragung* **11,** 207 (1978).
29K. V. G. Sevast'yanenko, *Heat Transfer, Soviet Res.* **9** (5), 36 (1977).
30K. J. Sinkule and V. Hlaváček, *Chem. Engng Sci.* **33,** 839 (1978).
31K. B. S. Soroka, *Heat Transfer, Soviet Res.* **9** (5) 72 (1977).
32K. B. S. Soroka and L. I. Val', *Heat Transfer, Soviet Res.* **9** (2) 134 (1977).
33K. M. V. Stradomskiy, Ye. P. Vasil'yev, V. I. Kozlenko and Ye. A. Yefremova, *Heat Transfer, Soviet Res.* **9** (2) 87 (1977).
34K. C. Stramigioli, G. Spandoni and F. Santarelli, *Int. J. Heat Mass Transfer* **21,** 660 (1978).
35K. N. Tamura and S. Takata, *Appl. Optics* **17,** 3051 (1978).
36K. S. N. Tiwari and S. K. Gupta, *J. Heat Transfer* **100,** 240 (1978).
37K. S. S. Tsai and S. H. Chan, *J. Heat Transfer* **100,** 486 (1978).
38K. R. Viskanta and E. D. Hirleman, *J. Heat Transfer* **100,** 169 (1978).
39K. R. Viskanta and J. S. Toor, *Solar Energy* **21,** 17 (1978).
40K. K. O. White, *Appl. Optics* **17,** 2711 (1978).
41K. V. P. Zaurayev, *Heat Transfer, Soviet Res.* **9** (5), 87 (1977).

Surface radiation
1L. T. W. Cadman and D. Sadowski, *Appl' Optics* **17,** 531 (1978).
2L. K. D. Cole, *Australian J. Phys.* **30,** 671 (1977).
3L. A. Feingold, *J. Heat Transfer* **100,** 742 (1978).
4L. J. R. Felland and D. K. Edwards, *J. Energy* **2,** 309 (1978).
5L. A. M. Smith, B. E. Wood and L. S. Fletcher, *AIAA Jl* **16,** 510 (1978).
6L. S. Tanaka and T. Kunitomo. *Bull. JSME* **21,** 258 (1978).
7L. N. K. Uzunoglu, N. G. Alexopoulos and J. G. Fikioris, *J. Opt. Soc. Am.* **68,** 194 (1978).
8L. D. P. Verret and K. G. Ramanathan, *J. Opt. Soc. Am.* **68.** 1167 (1978).
9L. N. E. Wijeysundera, *Solar Energy* **20,** 81 (1978).
10L. B. A. Zhestkov and V. P. Likash, *Heat Transfer, Sov. Res.* **9** (5), 50 (1977).

Measurement techniques
1P. R. M. Abdel-Wahed, T. P. Bligh and E. R. G. Eckert, *Int. J. Heat Mass Transfer* **21,** 967 (1978).

2P. R. M. Abdel-Wahed, E. Pfender and E. R. G. Eckert, *Wärme- und Stoffübertragung* **11,** 1 (1978).
3P. N. Abuaf, O. C. Jones, Jr. and G. A. Zimmer, *Rev. Sci. Instrum.* **49,** 1090 (1978).
4P. O. Arnon and P. Baumeister, *Appl. Optics* **17,** 2913 (1978).
5P. V. M. I. Bardylo, N. G. Koval'chuk and V. I. Lakh, *Measmt Tech.* **21,** 102 (1978).
6P. R. P. Benedict and J. S. Wyler, *J. Fluids Engng* **100,** 265 (1978).
7P. A. R. Bestman, *J. Heat Transfer* **100,** 750 (1978).
8P. L. P. Boivin and T. C. Smith, *Appl. Optics* **17,** 3067 (1978).
9P. K. Bremhorst and D. B. Gilmore, *Int. J. Heat Mass Transfer* **21,** 145 (1978).
10P. A. Brown, E. L. Thomas, R. Foord and J. M. Vaughan, *J. Phys. D: Appl. Phys.* **11,** 137 (1978).
11P. T. C. Cetas, *Rev. Sci. Instrum.* **49,** 245 (1978).
12P. W. S. Chan and S. U. Khan, *Appl. Optics* **17,** 2335 (1978).
13P. J. I. Chen, J. H. Lienhard and R. Eichhorn, *Int. J. Multiphase Flow* **4,** 233 (1978).
14P. T. Cochrane and J. C. Earnshaw, *J. Phys. D: Appl. Phys.* **11,** 1509 (1978).
15P. K. Depooter, E. Brundrett and A. B. Strong, *J. Fluids Engng* **100,** 10 (1978).
16P. M. A. Donelan and J. Motycka, *Rev. Sci. Instrum.* **49,** 298 (1978).
17P. F. Esvignes, *Int. Chem. Engng* **18,** 7 (1978).
18P. M. S. Francis, D. A. Kennedy and G. A. Butler, *Rev. Sci. Instrum.* **49,** 617 (1978).
19P. H. Genda and H. Okayama, *Appl. Optics* **17,** 807 (1978).
20P. H. Genda and H. Okayama, *Appl. Optics* **17,** 3439 (1978).
21P. O. A. Gerashchenko, *Measmt Tech.* **21,** 249 (1978).
22P. O. A. Gerashchenko, T. G. Grishchenko, O. V. Luchay and V. I. Nikandrov, *Heat Transfer, Soviet Res.* **9** (3), 143 (1977).
23P. F. J. Goldner, *Int. J. Heat Mass Transfer* **21,** 69 (1978).
24P. H. Gross-Wilde and J. Uhlenbusch, *Int. J. Heat Mass Transfer* **21,** 677 (1978).
25P. K. G. Gupta, M. J. Laubitz and A. Feingold, *Letters Heat Mass Transfer* **5,** 89 (1978).
26P. M. Hishida and Y. Nagano, *J. Heat Transfer* **100,** 340 (1978).
27P. L. J. Huey, *J. Fluids Engng* **100,** 229 (1978).
28P. F. Jullien, *Letters Heat Mass Transfer* **5,** 259 (1978).
29P. A. Kannel, *Rev. Sci. Instrum.* **49,** 955 (1978).
30P. J. F. Keffer, R. S. Budny and J. G. Kawall, *Rev. Sci. Instrum.* **49,** 1343 (1978).
31P. J. Kestin and W. A. Wakeham, *Physica* **92A,** 102 (1978).
32P. V. Kottke and H. Blenke, *Chem.-Ing.-Tech.* **50,** 81 (1978).
33P. O. A. Kraev and V. M. Shul'ga, *Measmt Tech.* **21,** 401 (1978).
34P. F. F. Lezhenin, P. I. Shurgay, A. G. Kostornov and M. S. Shevchuk, *Heat Transfer, Soviet Res.* **9** (2), 144 (1977).
35P. F. Milton, P. Paranthoen and M. Trinite, *Int. J. Heat Mass Transfer* **21,** 1 (1978).
36P. H. Oertel, Jr. and K. Buhler, *Int. J. Heat Mass Transfer* **21,** 1111 (1978).
37P. G. A. Padalko and D. A. Rapoport, *Measmt Tech.* **21,** 246 (1978).
38P. R. B. Pettit, *J. Engng Pwr* **100,** 489 (1978).
39P. V. S. Pikahsov, A. E. Erinov and V. A. Klevchishkin, *Measmt Tech.* **21,** 395 (1978).
40P. A. F. Polyakov and S. A. Shindin, *Letters Heat Mass Transfer* **5,** 53 (1978).
41P. Y. Sato, K. Yamamoto and T. Mizushina, *Int. Chem. Engng* **18,** 26 (1978).
42P. H. Shaukatullah and B. Gebhart, *J. Heat Transfer* **100,** 381 (1978).
43P. Y. Tada, M. Harda, M. Tanigaki and W. Eguchi, *Rev. Sci. Instrum.* **49,** 1305 (1978).
44P. S. Takizawa, H. Murata and A. Nagashima, *Bull. JSME* **21,** 273 (1978).
45P. N. J. Trappeniers, S. N. Biswas and P. van't Klooster, *Rev. Sci. Instrum.* **49,** 1007 (1978).
46P. H. Willrath and R. B. Gammon, *Solar Energy* **21,** 193 (1978).
47P. I. Wygnanski and C. M. Ho, *Rev. Sci. Instrum.* **49,** 865 (1978).
48P. A. J. Yule, D. S. Taylor and N. A. Chigier, *J. Energy* **2,** 223 (1978).

Heat exchangers and heat pipes

1Q. A. M. Baklastov, A. L. Yefimov and V. A. Gorbenko, *Heat Transfer, Soviet Res.* **9** (3), 105 (1977).
2Q. A. Bejan, *Int. J. Heat Mass Transfer* **21,** 655 (1978).
3Q. A. Burns and A. J. Willmott, *Int. J. Heat Mass Transfer* **21,** 623 (1978).
4Q. J. P. Chiou, *J. Heat Transfer* **100,** 580 (1978).
5Q. J. P. Chiou, *J. Heat Transfer* **100,** 346 (1978).
6Q. D. Chisholm, *Int. J. Heat Mass Transfer* **21,** 1207 (1978).
7Q. K. W. Cuffe, P. K. Beatenbough, M. J. Daskavitz and R. J. Fowler, *J. Engng Pwr* **100,** 576 (1978).
8Q. N. Cur and E. M. Sparrow, *Int. J. Heat Mass Transfer* **21,** 1069 (1978).
9Q. T. C. Daniels and R. J. Williams, *Int. J. Heat Mass Transfer* **21,** 193 (1978).
10Q. I. G. Goryachko and G. V. Zhizhin, *Heat Transfer, Soviet Res.* **9** (2), 54 (1977).
11Q. P. J. Heggs and K. J. Carpenter, *Trans. Inst. Chem. Engrs* **56,** 86 (1978).
12Q. D. D. Kalafati and V. V. Popalov, *Thermal Engng* **24** (9), 44 (1977).
13Q. B. I. Kokorev, *Thermal Engng* **25** (2), 33 (1978).
14Q. L. F. Krasnoshchekov, *Thermal Engng* **24** (8), 54 (1977).
15Q. Yu. N. Kuznetsov, V. N. Oyvin and V. I. Pevzner, *Heat Transfer, Soviet Res.* **9** (4), 160 (1977).
16Q. B. E. Launder and T. H. Massey, *J. Heat Transfer* **100,** 565 (1978).
17Q. V. M. Legkiy, *Heat Transfer, Soviet Res.* **9** (2), 120 (1977).
18Q. V. K. Migai, *Thermal Engng* **25** (2), 29 (1978).
19Q. V. G. Morozov, *Thermal Engng* **25** (3), 14 (1978).
20Q. P. S. Poshkas, V. Yu, Survila and A. A. Zhukauskas,

Int. Chem. Engng **18,** 237 (1978).
21Q. A. Rajakumar and P. R. Krishnaswamy, *Int. J. Heat Mass Transfer* **21,** 1333 (1978).
22Q. P. Razelos and M. K. Benjamin, *Int. J. Heat Mass Transfer* **21,** 735 (1978).
23Q. V. Ya. Sasin and V. N. Fedorov, *Heat Transfer, Soviet Res.* **9** (3), 65 (1977).
24Q. V. Ya. Sasin and V. N. Fedorov, *Heat Transfer, Soviet Res.* **9** (3), 70 (1977).
25Q. J. G. Soland, W. M. Mack, Jr. and W. M. Rohsenow, *J. Heat Transfer* **100,** 514 (1978).
26Q. E. M. Sparrow, B. R. Baliga and S. V. Patankar, *J. Heat Transfer* **100,** 572 (1978).
27Q. E. M. Sparrow and J. W. Ramsey, *Int. J. Heat Mass Transfer* **21,** 1369 (1978).
28Q. K. S. Tan and I. H. Spinner, *I/EC Fundamentals* **17,** 353 (1978).
29Q. I. Todo, *Bull. JSME* **21,** 644 (1978).
30Q. H. Yamashita, R. Izumi and S. Yamaguchi, *Bull. JSME* **21,** 1168 (1978).

Aircraft and space vehicles
1R. N. Arai, K. Karashima and K. Sato, *AIAA Jl* **15,** 1655 (1977).
2R. J. J. Bertin, E. S. Idar, III and W. D. Goodrich, *J. Spacecraft Rockets* **15,** 113 (1978).
3R. K. Y. Narayan, *AIAA Jl* **16,** 160 (1978).
4R. B. E. Pearce, *J. Spacecraft Rockets* **15,** 125 (1978).
5R. A. Prasad, *AIAA Jl* **16,** 1004 (1978).

General
1S. S. I. Abdel-Khalik and T. O. Hunter, *J. Heat Transfer* **100,** 311 (1978).
2S. R. S. Abdulhadi and J. C. Chato, *J. Heat Transfer* **100,** 36 (1978).
3S. I. T. Alad'yev, K. D. Voskresenskiy, Ye. S. Turilina and A. A. Ivlev, *Heat Transfer, Soviet Res.* **9** (4), 138 (1977).
4S. D. I. Baskir, J. B. Hunter and C. B. Schlesinger, *I/EC Proc. Des. Dev.* **17,** 318 (1978).
5S. A. Bejan, *J. Heat Transfer* **100,** 708 (1978).
6S. H. S. Bennett and R. Kayser, Jr., *I/EC Fundamentals* **17,** 8 (1978).
7S. J. M. Bentley, T. K. Snyder, L. R. Glicksman and W. M. Rohsenow, *J. Heat Transfer* **100,** 520 (1978).
8S. N. Blakebrough, W. J. McManamey and K. R. Tart, T rans. Inst. Chem. Engrs **56,** 127 (1978).
9S. S. Bloss and U. Grigull, *Wärme- und Stoffübertragung* **11,** 119 (1978).
10S. J. Chato and R. S. Abdulhadi, *J. Heat Transfer* **100,** 30 (1978).
11S. H. Consigny, B. E. Richards and J. P. Ville, *J. Engng Pwr* **100,** 439 (1978).
12S. M. A. Denisov, *Heat Transfer, Soviet Res.* **9** (5), 134 (1977).
13S. M. A. El-Masri and J. F. Louis, *J. Engng Pwr* **100,** 586 (1978).
14S. M. D. Green and J. Kornfilt, *Nucl. Sci. Engng* **65,** 385 (1978).
15S. D. B. Kreitlow, G. M. Reistad, C. R. Miles and G. G. Culver, *J. Heat Transfer* **100,** 713 (1978).
16S. P. S. Kuts, *Int. J. Heat Mass Transfer* **21,** 567 (1978).
17S. D. W. Larson, D. K. Gartling and W. P. Schmiddel, Jr., *J. Energy* **2,** 147 (1978).
18S. T. E. Laskaris, *J. Heat Transfer* **100,** 702 (1978).
19S. T. L. Lauvray, *ASHRAE Jl* **20** (6), 69 (1978).
20S. P. L. Magidey, *Heat Transfer, Soviet Res.* **9** (5), 142 (1977).
21S. R. L. Mussulman, *J. Heat Transfer* **100,** 363 (1978).
22S. S. Nagaraja and M. V. Krishna Murthy, *Int. J. Heat Mass Transfer* **21,** 87 (1978).
23S. V. Nagpal, G. D. Lahoti and T. Altan, *J. Engng Indust.* **100,** 413 (1978).
24S. B. R. Pai, S. Michelfelder and D. B. Spalding, *Int. J. Heat Mass Transfer* **21,** 571 (1978).
25S. A. Ray, D. A. Berkowitz and V. H. Sumaria, *J. Energy* **2,** 269 (1978).
26S. O. V. Remizov, *Thermal Engng* **25** (2), 11 (1978).
27S. L. N. Ryzhkov and V. I. Rychkov, *Heat Transfer, Soviet Res.* **9** (5) 117 (1977).
28S. E. N. Saburov and S. V. Karpov, *Heat Transfer, Soviet Res.* **9** (5), 21 (1977).
29S. F. W. Schmidt and J. Szego, *J. Heat Transfer* **100,** 740 (1978).
30S. J. Szego and F. W. Schmidt, *J. Heat Transfer* **100,** 148 (1978).
31S. J. R. Ward and T. L. Brosseau, *J. Heat Transfer* **100,** 697 (1978).
32S. L. M. Zysina-Molozhen, A. A. Dergach, M. A. Medvedeva and E. G. Roost, *Heat Transfer, Soviet Res.* **9** (4), 88 (1977).
33S. L. M. Zysina-Molozhen, M. M. Ivashchenko, A. A. Dergach and Ya. M. Fel-dshteyn, *Heat Transfer, Soviet Res.* **9** (4), 184 (1977).

Solar energy
1T. R. B. Bannerot and J. R. Howell, *Solar Energy* **19,** 549 (1977).
2T. J. T. Beard, F. A. Iachetta, L. U. Lilleieht, F. L. Hickstep and W. B. May, Jr., *J. Engng Pwr* **100,** 497 (1978).
3T. D. G. Burkhard, G. L. Strobel and D. R. Burkhard, *Appl. Optics* **17,** 1870 (1978).
4T. D. G. Burkhard, G. L. Strobel and D. L. Shealy, *Appl. Optics* **17,** 2431 (1978).
5T. G. Grossman, A. Shitzer and Y. Zvirin, *Solar Energy* **19,** 493 (1977).
6T. K. G. T. Hollands, K. N. Marshall and R. K. Wedel, *Solar Energy* **21,** 231 (1978).
7T. C. M. Horwitz, *J. Opt. Soc. Am.* **68,** 1032 (1978).
8T. Z. Lavan and J. Thompson, *Solar Energy* **19,** 519 (1977).
9T. F. W. Lipps and M. D. Walzel, *Solar Energy* **21,** 113 (1978).
10T. W. C. Louis and D. C. Miller, *ASHRAE Jl* **20** (5), 39 (1978).
11T. S. B. Margolis, *J. Heat Transfer* **100,** 371 (1978).
12T. E. Marschall and G. Adams, *Solar Energy* **20,** 413 (1978).
13T. D. R. McKenzie, *Appl. Optics* **17,** 1884 (1978).
14T. D. J. Morrison and S. I. Abdel-Khalik, *Solar Energy* **20,** 57 (1978).
15T. A. Rabl, *Solar Energy* **19,** 555 (1977).
16T. M. Riaz, *Solar Energy* **21,** 123 (1978).

17T. G. A. Rottigni, *Appl. Optics* **17,** 969 (1978).
18T. H. Tabor, *Solar Energy* **20,** 293 (1978).
19T. M. E. Talaat, *J. Energy* **2,** 136 (1978).
20T. J. O. White, T. R. Kirst and J. Tauc, *Appl. Optics* **17,** 2427 (1978).
21T. J. L. Yellott, *ASHRAE Jl* **20** (1), 60 (1978).

Plasma heat transfer
1U. M. Ali, D. Bradley and M. L. Gupta, *J. Phys. D: Appl. Phys.* **11,** 1638 (1978).
2U. M. I. Boulos, *IEEE Trans. Plasma Sci.* **PS-6** (2), 93 (1978).
3U. A. Chapman and G. R. Jones, *IEEE Trans. Plasma Sci.* **PS-6** (3), 300 (1978).
4U. J. E. Daalder, *J. Phys. D: Appl. Phys.* **11,** 1667 (1978).
5U. R. S. Devoto, N. H. Bauder, J. Cailleteau and E. Shires, *Physics Fluids* **21,** 552 (1978).
6U. G. A. Farrall, *IEEE Trans. Plasma Sci.* **PS-6** (4), 360 (1978).
7U. G. Frind, R. E. Kinsinger and R. D. Miller, *IEEE Trans. Plasma Sci.* **PS-6** (1), 43 (1978).
8U. A. Kanzawa and E. Pfender, *IEEE Trans. Plasma Sci.* **PS-6** (1) 33 (1978).
9U. E. W. Gray, *IEEE Trans. Plasma Sci.* **PS-6** (4), 384 (1978).
10U. T. Kubono, *J. Appl. Phys.* **49,** 3863 (1978).
11U. R. W. Mitchell, R. W. Conrad, E. L. Roy, D. Keeper and C. W. Mathews, *J. Quant. Spectrosc. Radiat. Transfer* **20,** 519 (1978).
12U. L. Niemeyer, *IEEE Trans. Power Appl. Syst.* **PAS-97** (3), 950 (1978).
13U. K. Okazaki, Y. Mori, K. Hijikata and K. Ohtake, *AIAA Jl* **16,** 334 (1978).
14U. A. J. Saber and R. G. Jahn, *AIAA Jl* **16,** 328 (1978).
15U. P. J. Shayler and M. T. C. Fang, *J. Phys. D: Appl. Phys.* **11,** 1743 (1978).
16U. A. Tslaf, *IEEE Trans. Plasma Sci.* **PS-6** (4), 442 (1978).
17U. D. T. Tuma and E. Fong, *IEEE Trans. Plasma Sci.* **PS-6** (4), 527 (1978).
18U. R. Treptow, D. Gold and A. El-Shamy, *IEEE Trans. Plasma Sci.* **PS-6** (2), 121 (1978).
19U. R. J. Zollveg, *J. Appl. Phys.* **49** 1077 (1978).

HEAT TRANSFER REVIEW - 1979

E. R. G. ECKERT, E. M. SPARROW, R. J. GOLDSTEIN, E. PFENDER, S. V. PATANKAR, J. W. RAMSEY and K. Y. TEICHMAN

Heat Transfer Laboratory, Department of Mechanical Engineering, University of Minnesota, Minneapolis, Minnesota, U.S.A.

INTRODUCTION

THIS review surveys results that have been published in various fields of heat transfer during 1979. As in the past, the number of papers published during that period was such that only a selection can be included in this review. A more complete listing of papers is available in the Heat Transfer Bibliographies published periodically in this journal.

The Fourteenth AIAA Thermophysics Conference was held 4–6 June 1979 at Orlando, Florida. Twelve sessions dealt with subjects like radiation, thermal protection systems, thermophysics, heat pipes, contact resistance, thermal control, entry environment and technology, nonintrusive diagnostics, computing technology and thermophysical analysis. Papers presented may be obtained from the American Institute of Aeronautics and Astronautics.

The Eighteenth National Heat Transfer Conference organized jointly by ASME and AIChE was held 6–8 August 1979 at San Diego, California. An invited paper by Kenneth J. Bell treated design techniques for shell and tube exchangers. The Max Jakob Memorial Award was given to Niichi Nishiwaki, the Donald Q. Kern Award to Kenneth J. Bell and a Conference Award Certificate to Joseph S. W. Chi for the best paper presented at the Seventeenth National Heat Transfer Conference. Thirty-six sessions were devoted to process heat transfer, solidification and melting, heat transfer in heat generating fluids and turbulent flows, radiation and process heat transfer, rotating flows, heat-transfer enhancement, interfacial transport, boiling, contact heat transfer, natural convection, waste heat recovery, solar energy, heat exchangers for fluidized beds, heat transfer in catalytic combustion, in porous media and packed beds, and in thermal systems. An open forum at which short presentations can be given has become a standing feature of this conference. Newcomers to the field or those who want to refresh their knowledge of heat transfer can participate in a Short Course Program of the ASME with courses in augmentation of heat transfer, heat transfer in fires, and numerical solution of heat transfer and fluid flow. The AIChE organized its Today Series Courses on practical heat transfer, in shell and tube heat exchangers, solar energy applications, fluidized bed heat transfer and applications in combustion. Papers are available through ASME and AIChE in bound volumes.

A Short Course on Fluid Mechanics Measurements, held 10–13 September 1979 at the University of Minnesota, brought together experts to present, after an introduction to physical laws and measurement, lectures on flow visualization, hot wire and hot film anemometry, laser velocimetry, measurements in two-phase flow, in non-Newtonian fluids, volume flow measurements, and computers in flow measurements. The lectures will be published in book form by Hemisphere Publishing Corporation, Washington, D.C.

The 1979 Seminar of the International Centre for Heat and Mass Transfer was devoted to heat and mass transfer in metallurgical systems and was held 3–7 September 1979 at Dubrovnik, Yugoslavia. Ten sessions covered heat transfer in blast furnaces, in the processing of iron and steel, during crystallization, in welding and cutting, in turbines and combustors, diffusion treatment, and corrosion. The proceedings will be published by Hemisphere Publishing Corporation.

The First National Conference on Numerical Methods in Heat Transfer convened 24–26 September at College Park, Maryland. Twenty-two lectures discussed finite difference and element methods.

The Winter Annual Meeting of the American Society of Mechanical Engineers was held 2–7 December 1979 at New York and included eighteen sessions on heat transfer. S. Levy discussed, as luncheon speaker, the importance of fundamentals and real time in heat transfer analysis. The Heat Transfer Memorial Award was presented to Arthur E. Bergles. Reprints of the papers presented at the meeting are available at ASME Headquarters in New York and many of them will be published in the *Journal of Heat Transfer*.

The Fourth International Symposium of Plasma Chemistry was held in Zurich, Switzerland from 27 August to 1 September 1979 and attracted more than 250 participants from 20 countries. Discussions of heat transfer in thermal plasmas generated by means of arcs or rf discharges were augmented by specific applications in the area of extractive metallurgy, the production and purification of solar grade silicon, the gasification and desulphurization of coal, and plasma spheroidization. One session was devoted to diagnostic techniques including LDA and thermogravimetric methods.

Trends and developments in heat-transfer research during 1979 are characterized by the following highlights: As in recent years, the conduction literature continues to be dominated by phase change and related moving boundary problems. Other active areas of heat conduction include solution methodologies (both analytical and numerical), the inverse problem, composite and anisotropic materials, variable properties, thermal contact, and fins. In general, this was an active year for heat conduction. The number of conduction papers in this year's review is the largest ever.

Recent studies of heat transfer in channel flows reflected a strong interest in complex flow configurations, fluids of various properties, and oscillating and pulsating flows. Significant attention was also given to the nature of the thermal entry region and specific techniques for augmenting heat transfer.

Boundary layer literature covers turbulent jets, flow over a cylinder, and laminar and turbulent boundary layers on walls. Effects of curvature and of Coriolis forces have been given attention. Numerical solutions as well as experimental work have been reported.

Literature on transfer mechanisms is largely concerned with mathematical models for turbulence. New models are proposed, existing ones are discussed and applied to new situations. Some new experimental data on the structure of turbulence have been reported.

A considerable number of papers considered natural convection in an enclosure with two vertical walls of different temperatures. This geometry was selected as a test case for an evaluation of various numerical schemes. Natural convection in porous media and in plumes also found interest.

Experimental and analytical results were published for heat transfer in rotating flows considering geometries like disks, rotating freely or in enclosures, tubes, and nozzles. Combined heat and mass transfer papers included chemical reaction, thermal diffusion, change of phase, and presence of particulates. Film and transpiration cooling found attention in a number of papers.

Boiling is a process which still needs clarification for a complete understanding. Accordingly, a large number of papers treat fundamental and practical aspects of boiling with strong participation by authors from the U.S.S.R. Film evaporators have found special attention.

Literature devoted to the study of heat transfer during condensation included investigation into the problem of film condensation, the effect of material properties on dropwise condensation, and condensation in two-phase and multi-component flows. A significant number of papers were concerned with the problem of heat transfer during melting and freezing. Melting topics included: the melting of ice and paraffins, the influence of natural convection on the melting process, and the determination of melting fronts and local heat-transfer coefficients during the melting process. The problems of freezing in a low gravity environment, salt rejection during the freezing process of saline solutions, and the effect of natural convection on freezing were topics included in the literature of heat transfer during freezing.

Some experimental results on radiation properties have been reported. The analysis of radiative energy transfer considered various geometries and surface properties and included phase change.

Laser Doppler anemometer and hot-wire hot-film techniques continue to be developed for a large variety of heat transfer subject areas including the study of the flow distributions in two-phase processes, arrays of heat exchanger tubes, subsonic and supersonic free jets, mixing layers, and engines. The development of specialized techniques and equipment for the experimental investigation of heat transfer in two-phase flows and plasmas was reported.

Heat transfer characteristics of a number of geometries of heat-transfer surfaces have been reported, and improved calculation methods for heat exchangers account for leakage, bypass, and irregular flow pattern. Heat pipes are treated in a number of papers, including characteristics of wicks and degradation.

The largest number of papers on solar energy continues to deal with the experimental and analytical evaluation of the performance of active solar collectors. The studies are distributed approximately equally between flat-plate and concentrating collectors. There was an increase in the number of papers describing the thermal performance of passive solar heating systems and components.

In 1979, there has been a substantial increase in research activities in plasma heat transfer associated with new developments in arc circuit interruption and in plasma processing.

To facilitate the use of this review, a listing of the subject headings is made below in the order in which they appear in the text. The letter which appears adjacent to each subject heading is also attached to the references that are cited in that category.

Conduction, A
Channel flow, B
Boundary layer and external flows, C
Flow with separated regions, D
Transfer mechanisms, E

Natural convection, F
Convection from rotating surfaces, G
Combined heat and mass transfer H,
Change of phase, J
Radiation
Radiation in participating media, K
Surface radiation, L
MHD, M
Measurement techniques, P
Heat-transfer applications
Heat exchangers and heat pipes, Q
Aircraft and space vehicles, R
General, S
Solar energy, T
Plasma heat transfer, U

CONDUCTION

As in recent years, the conduction literature continues to be dominated by phase change and related moving boundary problems. Other active areas of heat conduction include solution methodologies (both analytical and numerical), the inverse problem, composite and anisotropic materials, variable properties, thermal contact, and fins.

Experimental studies of solid–liquid phase change indicate that the commonly accepted dominance of heat conduction may not be valid. Experiments have demonstrated that freezing in the presence of a superheated liquid can be drastically slowed and ultimately terminated by natural convection in the liquid [72A]. Photographs of melting about a horizontal heated cylinder affirmed the dominant effects of natural convection [1A]. Heat-transfer measurements for melting about horizontal heated cylinders showed that the convection-dominated heat-transfer coefficients for a multi-tube array are similar in magnitude to those for a single tube [57A]. Another experimental study of melting about horizontal cylinders affirms the importance of natural convection [10A].

The classical Stefan problem, when converted from a constant surface temperature boundary condition to a constant heat flux boundary condition, is characterized by an interface position which is described by a power series in $t^{1/2}$ (t = time) [75A]. Another form of solution for the constant flux boundary condition leads to a first-order ordinary differential equation relating the interface position with the time [19A]. Exact solutions of freezing or melting of a polymorphous material in a semi-infinite region are expressed in polynomials and functions in the error integral family [76A]. A similar form of solution was employed for the Neumann problem subjected to convection at the heating (or cooling) surface [77A]. For phase change, the effects of density change are usually omitted in classical solutions. By an appropriate transformation of the space and time variables and the material constants, it is possible to transform a variable density solidification problem into one with equal densities in both phases [78A]. The periodic variation of the surface temperature bounding a phase change medium gives rise to multiple phase boundaries [14A].

Solidification due to the presence of a line sink of heat has been solved exactly, subject to a model characterizing the distribution of the solid fraction within the two-phase zone which results from an extended freezing range between the solidus and liquidus temperatures [48A]. Short time solutions for cylindrical freezing were obtained via Boley's embedding technique, with the governing partial differential equations being converted to integro-differential equations and then solved by series expansion [27A]. A perturbation method has been employed to analyze the problem of freezing of a biological tissue when a predetermined constant cooling rate is imposed on its outer surface [61A]. A problem of growth and decay of the frozen layer that forms in forced flow on a finite non-melting wall served as the vehicle for the presentation of a refined version of the heat balance integral method [20A, 21A]. Another refinement of the heat balance integral method was used to solve for solidification about a cooled cylinder [12A].

To provide design data for phase-change thermal storage systems, information on total melt time and average surface heat flux was determined for a phase-change material subject to a constant temperature at its boundary [69A]. In a related paper, a simple relation is presented between the wall temperature of a semi-infinite phase-change material and the time when this temperature is attained during a melting process with a convective boundary condition [70A]. Convective heating prior to melting and subsequent melting of a semi-infinite solid were both analyzed by employing Biot's variational principle [54A]. Teflon ablation has been modeled as a two-layer problem, one layer being a gel and the other a solid [6A]. Phase change driven by radiation and aerodynamic heating can be reduced to an initial value problem by use of Biot's variational method [84A].

The isotherm migration method has been rephrased in an implicit format and applied to a two-dimensional Stefan problem [18A]. A numerical scheme for conduction phase change treats each phase alternately [60A]. An application of the method of fractional steps and of the method of lines to the numerical solution of two-dimensional Stefan-type problems is described [44A]. To reduce the computational effort needed to solve multi-dimensional phase change problems, a similarity rule has been evolved which applies to problems in which sensible heat contributions are much smaller than latent heat contributions and where the heat flux distribution on the surface of the phase change substance is more or less uniform [65A]. Monte Carlo simulation was employed as a tool to investigate the sensitivity of a Stefan moving boundary problem to the uncertainty associated with the input information [45A].

Various topics in steady conduction, encompassing both new solutions and methodologies, have been treated. For steady heat-conduction problems with oblique-derivative boundary conditions, the use of Green's functions reduces the boundary value problem to equivalent integral equations [55A]. Another approach is to decompose the second-order partial differential equations into first-order systems [56A]. Steady conduction in the circle-in-square geometry was solved by a conformal mapping method [36A]. Conformal mapping also proved useful in solving a steady conduction problem in a slab with segments of convection and insulation on one face and uniform temperature on the other [49A]. An isoperimetric lower bound was obtained for the steady heat flow across a strip-like region when the temperature is constant on one boundary and satisfies a convection condition on the other [2A]. Numerical solutions gave results for the temperature distribution along the exposed surface of a convectively cooled wall whose rear side is heated by a periodic array of groove-embedded heating elements [71A]. To limit the heat transfer into a volume from the boundary surface, a control array of heat sources or sinks may be positioned near the surface [52A]. A thermocouple model was employed to derive an estimate of temperature measurement errors associated with the presence of the thermocouple [62A]. Study of a steady state heat-conduction problem involving a thin disk affixed to a convectively cooled solid plate was motivated by the measurement of heat fluxes from convectively cooled surfaces by thermopile heat flux gages [83A].

Several papers dealt with thermal contact. Calculations based on existing models have provided values of thermal contact conductance as a function of applied load, both for elastic and plastic modes of surface deformation [16A]. The macroscopic constriction experienced by a heat flux in passing between stainless steel and copper surfaces can be alleviated by insertion of a copper gauze [4A]. Theory, corroborated by experiments, shows that the contact resistance is reduced when the slipping speed between two solids in friction is increased [81A]. Predicted conductances of stacks of laminations corresponding to a wide range of contact pressures and with various interstitial fluids agreed well with measurements involving disks of stainless steel, brass, and transformer core steel [80A]. In the theory of contact conductances, a single contact on the surface of a semi-infinite body is frequently used as a building block for the case of surfaces which have many contacting points. The validity of this approach has been recently re-examined [11A].

Fins continue to draw some attention. Fin-to-fin, fin-to-base, and fin-to-environment radiant interactions were accounted in determining the efficiency of a convecting–radiating fin array subject to a timewise-periodic base surface temperature [24A]. Computer-aided extension of the regular perturbation series for a radiating fin yielded highly accurate results [8A]. Solutions based on the heat balance integral have shown that the efficiency of a triangular fin departs from that of the traditional one-dimensional model as the apex angle increases [15A]. A generalized approach to the one-dimensional modeling of fins enabled identification of the proper dimensionless parameters which govern the problem [58A].

Interest persists in composite and anisotropic materials. The governing differential equation for three-dimensional steady conduction in solid and hollow cylinders of general anisotropic media can be reduced to Kummer's equation by use of the Fourier transform and a change of variables [17A]. It was also shown that certain heat-conduction problems for anisotropic media can be transformed into corresponding problems for isotropic media [51A]. Some solutions for transient head conduction in cylinders with three-dimensional anisotropy have been presented [50A]. For steady state heat transfer in a heat generating composite solid composed of two materials, Green's functions were used to convert the boundary value problem into a Fredholm integral equation of the second kind [25A]. The model for heat conduction in which a two-phase composite material is regarded as spheres of the dispersed phase situated in a lattice of the continuous phase has been solved for the case where the spheres are infinitely conducting and nearly touching [74A]. In order to explore longitudinal heat propagation in matrix-filler bilaminates, it is necessary to augment it into a trilaminate [28A].

Transient conduction papers have dealt with a variety of topics. The linear one-dimensional diffusion equation may be solved by the method of variation of parameters, which yields a solution in the form of an infinite series of products of eigenfunctions and time-varying coefficients [13A]. A collection of Green's functions has been assembled for transient conduction problems whose boundaries vary linearly with time [29A]. A variational method for the solution of the transient heat-transfer problem in nuclear reactor elements was used to analyze the consequences of a shutdown of the reactor [37A]. A solution for transient heat conduction in the ground due to buried radioactive nuclear waste takes account of orthotropic thermal conductivity and of the time-varying nature of the heat generation rate due to radioactive decay [82A].

Analytical solutions are given for transient heat transfer between a semi-infinite solid and a perfect conductor with internal heat generation, taking into account a thermal contact resistance at the interface [46A].

There was special emphasis on transients in the presence of variable thermal conductivity. The orthogonal collocation method, which involves the approximation of the spatial derivative term by an orthogonal polynomial of the Legendre type, was shown to be a highly effective tool for solving transient problems having variable conductivity [39A]. Existing variable property solutions for transient conduction in a semi-infinite solid have been supplemented by a new set of solutions obtained via perturbation methods [64A]. A perturbation method was also used to solve transient heat-conduction problems with spatially random thermal conductivity [3A]. For a semi-infinite solid with periodically time-varying surface temperature, variable thermal conductivity causes the propagation of a given harmonic to be affected by the amplitude and phase of the other harmonics [68A]. For a thermal conductivity which varies as a non-negative power of the temperature, a perturbation method yielded a similarity solution of the transient heat-conduction equation [9A].

The inverse problem of transient conduction was treated in several papers. Timewise temperature information at an interior point of a solid cylinder can be used to determine temperatures at other interior points for given boundary conditions [31A]. A method for determining the transient surface temperature of a solid when the temperature variation at an interior point is known employs approximate iterative techniques suitable for the solution of one-dimensional transient thermal conduction problems in homogeneous or composite solids [7A]. It was also found necessary to use an iterative technique to solve the inverse problem of heat conduction when variations of thermal conductivity are considered [41A]. Through examples it was demonstrated that a finite-element method can be effectively used to treat the inverse heat-conduction problem [35A]. A method was presented that enables the quantitative evaluation of the resolving power and accuracy of surface temperature predictions generated from inaccurate and discrete interior measurements [30A]. The key ingredient in an approximate method for solving both the direct and inverse variable property heat-conduction problems is to replace the original non-linear partial differential equation with a succession of equivalent linear ones in some optimum fashion [32A].

In addition to numerical work already cited in earlier paragraphs, other numerically oriented papers will now be reported. Use of the finite element method for transient thermal problems can lead to the anomalous result that, even in the absence of local sources of heat, temperatures near suddenly cooled boundaries show an initial increase [23A]. A numerical method for transient heat conduction with variable conductivity uses a finite element discretization in space and the Crank–Nicolson discretization in time [47A]. To extend the applicability of finite difference techniques for the solution of transient heat-conduction problems, a finite difference approximation was formulated in generalized, nonorthogonal coordinates [59A]. Two new methods of introducing nonlinear derivative boundary conditions for ADI methods have been developed [73A]. An explicit numerical method for solving transient combined heat-conduction and convection problems uses the DuFort–Frankel method for the conduction equations and the fully implicit scheme for the convection equations [26A]. A five-parameter family of two-level 10-point difference approximations to the two-dimensional heat operator is given and its characteristics explored [33A]. The orthogonal collocation method proved to be highly effective for solving transient conduction in a composite slab with variable thermal conductivity [38A].

There are interesting control and statistical aspects of heat conduction. There are many industrial processes in which it is necessary to control the temperature distribution in a given material. The resulting optimal control problem was reformulated as a mathematical programming problem by discretizing the space coordinate by finite elements while the Runge–Kutta method was utilized for time integrations [43A]. The heating of a slab by the ambient temperature is formulated as an optimal control problem with the view of attaining a certain temperature level in the slab while keeping the ambient temperature as low as possible [42A]. In certain industrial processes concerned with the manufacture of steel, it is necessary to determine control temperatures to be applied at a number of time steps to the surface of the steel ingot to induce a specific temperature variation within the ingot. The resulting dynamic control problem was converted to one of static optimization and then solved using a partial quadratic interpolation technique [22A]. Parameters in the heat-conduction equation should be estimated by Aitken's method when the errors are correlated [63A].

In a highly mathematical treatment, the hyperbolic form of the heat equation for pure heat conduction is generalized to combined heat- and mass-transfer processes [66A]. Nonlinear thermoelasticity theory allows heat to travel with finite wavespeed [40A].

Some heat-conduction-related papers that were published in *Heat Transfer, Soviet Research* include: moving boundary problems [34A, 53A], inverse problem of transient heat conduction [5A, 79A], and analytical techniques in nonlinear transient heat conduction [67A].

CHANNEL FLOW

Recent studies of heat transfer in channel flows reflect a strong interest in complex flow configurations, fluids of various properties, and oscillating and pulsating flows. The nature of the thermal entry region concerned many authors, and others addressed specific techniques for augmenting heat transfer.

The problem of convective heat transfer in tubes once again was an area of active research. A numerical analysis for laminar heat transfer in a pipe with convective and radiant wall fluxes indicated that axial conduction is strongly altered by the parameters responsible for the convection and radiation, and increases as the heat flux at the wall is decreased [4B]. Numerical calculations performed for the problem of hydrodynamic steady flow in a round tube for varying Reynolds numbers and for different laws governing wall temperature demonstrated the influence of longitudinal diffusion on *Nu* [14B]. Assuming hydrodynamically stabilized flow and constant fluid properties, solutions were obtained for convective heat transfer in tubes for conditions when heating varies arbitrarily over the perimeter and length at the same time [13B]. Local and average heat-transfer characteristics were studied for turbulent air flow in an asymmetrically heated tube. The results show a strong interaction between circumferential wall conduction and fluid convection under non-uniform heating and a significant effect of wall conduction at low *Re* [20B]. A finite difference procedure was used to predict turbulent flow and heat transfer in horizontal, inclined, and vertical pipes when influenced by buoyancy. Results were presented for the velocity and temperature fields, and the associated flow resistance and heat-transfer coefficients agreed well with experimental results [1B]. Using an eddy conductivity model, an analytical solution to the problem of turbulent heat transfer in pipes with internal heat generation and insulated walls provided results which agreed closely with the available data over a wide range of *Pr* [25B]. An experimental investigation of the turbulent structure of velocity and temperature fields made in fully developed pipe flow of air demonstrated that in the turbulent part of the wall region, the axial heat flux, as well as the intensity of the velocity and temperature fluctuations, reach their maximum. In the turbulent core, the intensity of the fluctuations decreased, indicating the velocity and temperature fields are approaching isotropy [17B]. An experimental investigation was performed on the basis of a vortex statistical model of turbulence to determine losses of energy in channels. Losses were shown to include the disruption of the small scale part of turbulent structure which is brought about by viscous forces in tiny vortices, and a semi-empirical relationship was presented for these losses [46B].

Research was also actively pursued for channel flows of different geometries. An analytical study of momentum and heat transfer for turbulent forced convection in rectangular channels demonstrated that only a single secondary current occurs in the trapezoidal symmetry element of the rectangular duct. This study accounted for secondary anisotropic turbulent transport properties and turbulence-induced secondary flow, and included the effects of peripheral wall conduction as well as radiation [26B]. An experimental study of the heat-transfer rate from a plate to accelerating streams of air, water, and transformer oil in rectangular channels determined that the heat-transfer rate deviated from that corresponding to a turbulent mechanism. This deviation was shown to occur gradually for favorable pressure gradients, and the magnitude of the deviation was shown to be a function of velocity and geometry [35B]. Double beam interferometry was used to determine the development of local mass-transfer boundary layers for laminar flow in a rectangular channel. The application of an overall mass balance permitted the distinction between different concentration contours associated with optically indistinguishable interference fringes. A Pohlhausen function was found best to describe the concentration profiles in the boundary layers under steady laminar forced convection [23B]. An experimental study showed that tubes of eliptical cross-section give higher laminar flow heat-transfer coefficients than those of circular cross-section but with an accompanying penalty in pressure drop. These experiments were performed for constant wall temperature conditions and over various aspect ratios [31B]. A numerical analysis was developed to predict friction factors and heat-transfer coefficients in a pipe of irregular cross-section. Finite element solutions substantially agreed with experimental results and showed that *Nu* varies with *Ra* and the cable configuration [6B]. Using a simplified differential energy transport equation for turbulent flow, a new method was presented for the transformation of turbulent Stanton numbers measured in an annular channel geometry. The thermal performances of rough surfaces were evaluated for a single rod and when extended to a rod bundle, agreed well with experimental results [18B]. An experimental study was performed of local heat transfer from the inner and outer tubes of an annulus under various heating conditions. For the case of two sided heating and moderate heating rates, the effect of the variability of physical properties on heat transfer for varying *Re*, diameter ratio, length to diameter ratio, and heat flux ratio, is virtually independent of the boundary conditions at the other tube, and reduces to the identical effect described for one sided heating [28B, 29B].

The effects of curvature on fluid flow and heat transfer in channel flows with different boundary conditions was also addressed. A numerical study of fully developed viscous flow and heat transfer in curved semicircular sectors was studied for thermal boundary conditions of (i) axially uniform heat flux with uniform peripheral temperatures and (ii) axially uniform heat flux with mixed conditions at the periphery. The fractional increase in heat- or mass-transfer coefficients to the fractional increase in friction factor for curved and straight semi-circular ducts was greater than unity for all cases studied and increased with both Dean and *Pr* numbers [22B]. The effect of curvature on fluid flow and heat transfer in a tube with axial uniform heat flux and peripherally uniform wall temperature was analyzed numerically, and demonstrated that secondary flow circulations become more complex as the Dean number increases. The Nusselt number was shown to increase with increasing Prandtl and Dean numbers, and a method was provided to find the optimum convergence parameter which substantially reduced computational time [36B]. An experimental investigation was also performed for the hydrodynamics and heat transfer in a curved channel with a rectangular cross-section which accounted for not only curvature but also the shape factor of the cross-section [11B].

Many authors addressed the problem of characterizing momentum and heat transfer in the entrance region of channel flows. An experimental study of the velocity and temperature profiles in a high temperature (up to 3000 K) air stream in the initial section of a tube with turbulent flow showed that the velocity and temperature of the air stream at the tube axis remained practically constant for ratios of length to diameter which were less than or equal to 10 [19B]. A step-by-step analysis of developing laminar flow in straight ducts of arbitrary but constant shape was performed by modifying a finite element code for non-linear heat conduction [10B]. In a related paper, a finite element technique was used to solve velocity and heat-transfer characteristics in the thermal entrance region of hydrodynamically developed laminar duct flows of arbitrary but constant cross-section [41B]. A numerical simulation of asymmetrically heated pipe flows suggested that flows require 150 diameters from the start of heating to reach full development. In this study, particular attention was given to the anisotropy of the turbulent thermal diffusivities [7B]. By considering the droplet heat-transfer contribution as a distributed heat sink, convective heat transfer of laminar droplet flow was calculated numerically for the thermal entry region of circular tubes with constant wall temperature [44B]. From an analytical investigation conducted for the random entrance region heat transfer of a fluid in a parallel plate channel with boundary conditions of randomly varying temperature oscillations, it was concluded that as convective heat-transfer increases, the variance increases and the thermal field must be described by both the mean and the variance [5B].

Several papers were published dealing with channel flows of fluids with variable properties. The results obtained from an implicit finite difference scheme used to determine a solution for heat transfer in laminar non-Newtonian Couette flow with a pressure gradient were in good agreement with a previously obtained semi-analytical solution [21B]. Experimental results were performed for mass transfer in a pipe for an aqueous 4% NaCl solution under Newtonian and non-Newtonian conditions. For the non-Newtonian flows, there was a significant reduction in mass transfer [42B]. An adaptation of the Lévêque solution for the case of mass transfer from a flowing fluid to the core of an annulus in the fully developed streamline flow of a non-Newtonian fluid was compared to experimentally-determined mass-transfer coefficients as measured using an electrochemical technique. The results showed that mass transfer in non-Newtonian fluids following the power law model can be well represented by a modified version of the simplified Lévêque solution [37B]. The average rate of heat transfered to the channel wall when a hot incompressible third order fluid in rectilinear motion flows (i) between two parallel plates and (ii) in a circular tube was investigated. Heat-transfer coefficients were evaluated for several values of a non-Newtonian parameter, and were enhanced for those fluids which exhibited large strain rate gradients in the region of the channel wall [12B]. A numerical and experimental investigation of friction and heat-transfer parameters for turbulent flow of helium–argon mixtures (Pr = 0.4–0.49) was performed in smooth electrically heated vertical circular tubes and demonstrated that popular existing correlations developed using gases with Prandtl numbers on the order of 0.7 overpredict observed Nusselt numbers. By comparing numerical calculations and measured constant property Nusselt numbers, Prandtl numbers were determined in the wall region which were valid for conditions where properties varied greatly [33B]. Asymptotic methods were used to analyze channel flow with temperature dependent viscosity and internal viscous dissipation. Flow was shown to evolve from a Poiseuille flow with a uniform temperature distribution to a plug flow with hot boundary layers. An asymptotic solution was obtained for the transition flow region, and an explicit expression was derived for the pressure gradient in terms of the local downstream coordinate [30B]. The Lévêque solution for the high Graetz number (thermal entry) flow of a Newtonian fluid in a circular pipe or rectangular channel was extended to derive higher order terms in the power series expansion of temperature for the flow of a power law fluid [38B]. Laser anemometry and thermocouple measurements were used to compare turbulent heat transfer in dilute solutions of polymers in a rectangular channel with theoretical results. The results demonstrated that velocity and temperature distributions are steeper for the dilute solutions than they would have been for the universal curve for water [34B]. Turbulent heat transfer in dilute polymer solutions was also studied to evaluate the effect and extent of drag reduction and deterioration of heat transfer [45B]. A numerical calculation was performed for turbulent flow in rough pipes using a mixing length model in the turbulent core and a roughness drag element coefficient and sublayer Stanton number to characterize heat transfer to the wall. Variable property effects were more pronounced for rough pipes as compared with flow in smooth pipes, and terms on the order of the roughness height divided by pipe radius were shown to be significant [43B].

Momentum and heat transfer in complex channel geometries was also an area of strong interest. Experiments performed to study how fluid withdrawal at a branch point in a tube affects the turbulent heat-transfer characteristics of the main line flow downstream of the branch showed an augmentation of the circumferential Nusselt number of the main line flow. The augmentation increased at any fixed Reynolds number when the ratio of withdrawn flow to main line flow was increased [40B]. Experimentally determined heat-transfer coefficients for turbulent air flow in a circular tube were studied for flow conditions situated downstream of a mixing tree. Results demonstrated that mixing and turning of the flow gives rise to significant augmentation of heat-transfer coefficients as compared with those of a conventional thermal entrance region, and that the thermal entrance length is also substantially elongated [39B]. A finite difference numerical analysis was used to investigate laminar transport phenomena in constricted parallel ducts with fully developed flow and temperature profiles. Thermal conditions of uniform temperature and uniform heat flux were used to determine thermal and hydrodynamic effects at both the entrance and exit of the constricted flow geometry [3B]. Experimental measurements of developing and fully developed heat-transfer coefficients along a periodically interrupted surface affirmed the preclusion of the obtainment of hydrodynamic and thermal development. The presence of the interruptions served to augment the heat-transfer coefficient, and in the case of fully developed flow, the coefficients are on the order of twice those of a conventional duct flow [8B]. The use of circumferential grooves was shown to be of advantage in an electrochemical study which compared the enhancement of mass-transfer coefficients for horizontal conduits with various grooved surfaces [24B]. Heat transfer and turbulent flow in internally finned tubes and annuli were analyzed using a mixing length model, and in general, the fins were found to be as effective a heat-transfer surface as the wall (per unit area). The local heat-transfer coefficient exhibited a substantial variation along the fin height and lesser, more gradual, variation on the tube wall or annulus [32B]. The electrochemical analog technique was used to measure mass and heat-transfer distributions in pipes roughened with small square ribs. The mass-transfer distributions were less uniform at high Reynolds numbers and were virtually independent of the Schmidt number over a wide range. An analytical and numerical study of augmented heat transfer and thermal homogeneity was performed by comparing motionless mixing by both flow division and inversion. Nearly optimal results were achieved with a practical no pressure drop two channel partial flow invertor with an increase of 20–30% in Nusselt number [27B].

Oscillating and pulsating flows in channels were also studied. Using the electrolytic method, the transition from laminar to turbulent flow was measured for oscillating and pulsating flow in a pipe. The critical Reynolds number was shown to be dependent on the amplitude and frequency of the oscillation, and for pulsating flows, there was an additional dependence on the superimposed stationary flow [16B]. The density wave stability boundary and the growth of temperature perturbations were experimentally observed for supercritical helium in a long heated channel of high aspect ratio. During the density wave oscillation, the channel exit temperature and inlet mass flow were observed to be in phase and the oscillation period was close to twice the fluid transit time [9B]. Mass transfer in a hydraulic fully developed flow within a tube with steady pulsating and oscillating flows was studied using the electrolytic method. The mass transfer was increased by up to five times the value of the steady flow condition, and experimental records showed details of turbulence, flow separation, and backflow [15B].

BOUNDARY LAYER AND EXTERNAL FLOWS

Boundary-layer research continues to focus on experimental and theoretical studies of laminar and turbulent flows on rough and smooth, flat and curved surfaces. Jet impingement on surfaces and flow around cylinders and spheres are also topics of considerable interest.

The Prandtl number dependence of the Nusselt number for the laminar flat-plate boundary layer has been expressed by a correlation that is very accurate over a very large range of the Prandtl number [35C]. In a similar study, approximate formulas of very high accuracy have been obtained for heat transfer from a flat-plate boundary layer with blowing or suction [36C]; the formulas cover almost the entire range of the Prandtl number and the blowing/suction parameter. It is shown that the average Nusselt numbers for external flows for uniform wall temperature and uniform heat flux turn out to be almost equal if they are appropriately defined [32C]. Analytical solutions have been presented for the two-dimensional boundary layer flow of a power-law non-Newtonian fluid [28C].

A number of calculation methods have been presented for laminar and turbulent boundary layers. A simple analytical method is proposed for heat transfer in turbulent boundary layer with a hydrodynamic starting length [21C]. Turbulent boundary layer with separation has been calculated by an integral method [1C]. Calculations have been presented for the thermal layer developing within a velocity boundary layer with a step change in wall temperature [9C]. A modified integral method is described for boundary layers with prescribed surface heat flux [43C]. Approximate methods are developed for calculating the properties of laminar boundary layer of heated water [22C].

In other boundary layer studies, the growth of the two-dimensional mixing layer from a turbulent and nonturbulent boundary layer is investigated [8C]. A study is made of the turbulent boundary layer with large freestream to wall temperature ratio [6C]. Influence of the external flow on heat transfer has been determined for the boundary layer of a continuously moving plate [20C]. The hydromagnetic flow and heat transfer over a stretching sheet is the subject of one investigation [13C].

An analysis has been presented for the growth of an ice layer on an isothermal plate in a forced-convection parallel flow. Results for laminar flow are compared with experimental measurements [23C]. Also, experiments for transition and turbulent flow are reported for the ice layer [24C]. The stabilizing and destabilizing effects of Coriolis force on two-dimensional laminar and turbulent boundary layers have been investigated [27C]. Heat transfer from a rough plate has been measured in the flow of air and transformer oil [17C]; further data on the drag and heat transfer on a rough wall have been presented in [16C].

A number of studies have been conducted to determine the effect of surface curvature on boundary layer heat transfer. Heat-transfer measurements for turbulent boundary layers on convex and concave isothermal surfaces have been presented [30C]; the concave surface is found to transfer more heat than a flat surface, while a convex surface transfers less. Detailed measurements have been reported for the turbulent boundary layer on a flat surface downstream of short convex or concave regions [41C]. The heat transfer in the boundary layer of a micropolar fluid flowing past a curved surface has been studied [34C]. A discussion deals with the calculation of laminar and turbulent boundary layers on longitudinally curved surfaces [11C]. In [3C], the transonic laminar boundary layer flow near convex corners has been analyzed.

Flow and heat transfer over cylinders and spheres is the subject of a number of papers. Heat transfer from a slender cylinder in axial air flow has been measured [5C]. An investigation deals with skin friction and heat transfer on a cylinder moving through a stationary fluid [2C]. A numerical method has been used to determine the heat transfer from a circular cylinder impulsively started from rest [10C]. An experimental study deals with the effect of surface turbulizers on the local heat-transfer rate on a cylinder [14C]. The finite-element method has been applied to the flow around a circular cylinder [47C]. A two-equation turbulence model is used to investigate the effect of free-stream turbulence on skin friction and heat transfer around a circular cylinder in cross flow [44C]; good agreement with experiment is achieved.

A correlation is presented for the heat transfer from subcooled water films on horizontal tubes [39C]. A numerical method is used to analyze the variable property heat transfer to a single sphere in high temperature surroundings [37C]. Another study deals with the laminar boundary layer on an impulsively started rotating sphere [15C]. Heat transfer through the interface of spherical particles has been investigated for an unsteady situation [7C].

Impinging jets and stagnation-point heat transfer have been extensively studied. Experimental data and a simple correlation are presented for the heat transfer from an inclined rectangular plate of finite width [42C]. Measurements are also reported for the local and average heat-transfer coefficients for a square plate placed at various orientations to the flow direction [46C]. The stagnation-point heat transfer is studied with the inflow of a turbulized jet at a barrier [19C]. A discussion deals with the response of a stagnation boundary layer to a change in the external velocity [12C]. The mass transfer near the front stagnation point has been investigated in a fluctuating flow [4C]. A study describes heat and mass transfer from an axisymmetric jet impinging on a baffle [33C]. In a related work, the heat transfer in the vicinity of the stagnation point is measured for the situation in which a turbulized jet impinges onto a baffle [18C]. Local mass-transfer coefficients resulting from the impingement of a two-dimensional laminar jet have been measured with a holographic technique, and a correlation of the data has been obtained [29C]. A study has been made of the two-dimensional stagnation-point flow impinging obliquely on a plane wall [45C].

Another work deals with the heat exchange between a blunt body and a high-temperature gas jet [38C].

In a detailed experimental study of turbulent annular jets, the mean and fluctuating properties of the inner region have been measured [26C]. Measurements of temperature and velocity profiles are presented for heated three-dimensional turbulent jets from rectangular nozzles [40C]. Analysis is made of the stability and mixing of a vertical buoyant jet in a channel of confined depth [23C]. Measurements are reported for the heat-transfer characteristics of the impinging flow from a two-dimensional array of jets [31C].

FLOW WITH SEPARATED REGIONS

In a number of applications, the heat transfer in regions where the flow separates from a surface can be quite important. Several studies have been conducted which examine the heat transfer in a cavity placed in a wall over which there is a boundary layer flow. Others consider separation and reattachment and the resulting heat transfer following an abrupt change in the contour of a wall over which a fluid is flowing. Still other works consider the wake region behind a cylinder or other blunt body.

Different flow regimes have been visualized in a rectangular cavity of variable depth; heat-transfer measurements show significant variation of heat transfer with depth [12D]. An electro-chemical technique has been used to determine the mass transfer in a semicircular cavity in a wall [1D].

The heat transfer to a cylinder on which tripping wires are placed is closely related to the width of the wake region behind the cylinder [2D]. The statistical properties of the wake behind a heated cylinder have been examined [5D]. Experiments on heat transfer from a smooth cylinder [14D] and a rough cylinder [13D] demonstrate significant changes in heat transfer near the critical Reynolds number. Local measurements around a cylinder in crossflow show the influence of a spray of liquid on the cylinder [4D].

The nose shape on the front of a flat plate of finite thickness has a significant effect on the heat transfer in the separated and reattaching regions [7D]. The heat transfer downstream of a disc mounted co-axially on a smaller diameter cylinder arranged with axes parallel to the mainstream flow is a maximum at the reattachment point downstream of the disc [9D]. The heat transfer has been measured downstream of a rearward-facing step when the flow is generated by pulsating jets [3D]. Use of the electrochemical technique indicates that the maximum heat transfer downstream of a sudden expansion is well within the recirculation zone [10D].

Experiments related to wake regions include a measurement of the temperature characteristics in the turbulent wake behind an asymmetrically-heated plate [6D]. The effect of roughness on the separated region behind a flat plate has been studied [11D]. An analysis of the wake region behind a blunted cone in a hypersonic flow has been presented [8D].

TRANSFER MECHANISMS

Papers on transfer mechanisms deal with turbulence models, detailed measurements of turbulence, and theoretical analyses of turbulence growth and decay.

An algebraic model has been developed for stresses and heat fluxes in a turbulent shear flow with streamline curvature [12E]. A discussion deals with the prediction of turbulent Prandtl and Schmidt numbers from modeled transport equations [14E]. Reacting Couette flow is calculated by using an approximate theory developed to describe the combined effects of heat and mass transfer (30E). A version of a one-equation turbulence model has been presented for the calculation of free shear flows [11E]. A statistical model of turbulence is developed for two-dimensional mixing layers on the assumption that the turbulence is in a state of quasi-equilibrium [29E]. The use of subgrid-scale models for predicting turbulent flow is described in [6E]; the outcome of these models is compared to those used for large-eddy simulation, and it is suggested that the method could be used to predict the constants for the conventional models. Numerical modeling concepts are given for the turbulent mixing of large flows [21E]. Entropy generation due to finite temperature gradients and due to viscous effects is analyzed for pipe flow, flat-plate boundary layer, cylinder in cross flow, and entrance region of a duct [2E]. Development of a second-moment approximation to turbulent convection is discussed in [18E]; the turbulent heat fluxes are used as dependent variables of transport equations, and thus no assumption needs to be made for the turbulent Prandtl number. A discussion deals with the invariance of turbulence closure models [27E].

A number of papers report extensive turbulence measurements. Experimental data for turbulence are presented for a boundary layer on a rough wall with blowing and heat transfer [23E]. Experiments and analysis given in [28E] deal with the local isotropy and large structures in a heated turbulent jet. Hot-wire anemometry is used to measure the distortion of grid-generated turbulence by a circular cylinder [5E]. Measurements of turbulence in a mixed layer and through a density interface are made in the absence of a mean flow [20E]. Experiments are conducted to measure the fluctuations of turbulent shear in the shear layer of a circular jet [7E]. Detailed turbulence measurements have been made by introducing strong external disturbances into a mixing layer to observe the formation of coherent eddies [32E]. Experimental data are presented for various terms in the turbulent heat-transfer equation for a retarded boundary layer [26E]. Turbulent diffusion coefficients have been measured in mixing vessels [22E]. Experimental measurements have been performed for turbulence produced by a shearing instability in a two-fluid system [17E].

A theoretical study is made of thermal diffusion in an inclined column [1E]. Thermoconvective waves have been studied in a compressible fluid [16E] and in a conducting and radiating fluid [15E]. Theoretical discussions and analyses of turbulence fields have appeared in many publications [8–10E, 13E, 19E]. A statistical approach to turbulence has been developed in [31E], while the boundary layer transition is dealt with in [3E]. Properties of large structure in a turbulent duct flow are discussed [24E].

Also reported are the experimental data for turbulence in the continuous phase of fluid–particle systems [4E]. Another investigation deals with heat transfer in a ferromagnetic suspension under the action of a magnetic field [25E]; heat transfer is found to be enhanced in the direction of the field and decreased in the perpendicular direction.

NATURAL CONVECTION

Interest remains high in heat transfer by natural convection. As usual, there is a large number of papers related to convection in enclosures, particularly those heated from below. There is a good deal of activity in the numerical calculation of convection in vertical enclosures heated from one side and cooled on the opposing side. There also appears to be growing interest in natural convection in porous media.

A conference report on convection in fluid layers with buoyancy-driven flows has been published [56F]. An analytical and experimental study describes the instability of convection rolls in a layer heated from below [14F]. Hexagonal and roll cells are predicted for post-critical buoyancy and surface-driven flows [72F]. Introduction of a small amount of a third component can alter the stability criterion in doubly-diffusive convection [49F]. With a large variation in the viscosity of a fluid, stable rolls occur similar to those in a constant viscosity fluid [111F]. A numerical calculation indicates that the mode of circulation in a layer heated from below depends on the thermal boundary conditions and the aspect ratio [15F]. Measurements of the velocity distribution at low Rayleigh number, using scattered laser radiation, show the development of rolls [124F]. The instability in a layer of nematic liquid crystals has been observed [51F].

In a layer of fluid heated from below, the Nusselt number was found to increase somewhat at low aspect ratio [89F]. In a cylinder, the onset of flow at low Rayleigh number has been found to be very different from that in a large aspect-ratio horizontal layer [97F]; in the same system; use of a binary gas mixture yields a lower critical Rayleigh number [98F].

Reconsideration of turbulent convection data for a fluid layer heated from below suggests that more concern should be directed toward aspect ratio effects even at large Rayleigh number [29F]. A model to generalize results for both heating from below and horizontal layers with internal energy sources has been described [11F].

The unsteady temperature distribution has been measured in an evaporating water layer to analyze mixing at a water–vapor interface [6F]. Heat and mass transfer across an interface has been studied with application to film boiling [57F]. Heating a stably-stratified layer from below can result in a series of convecting layers [59F]. The heat transfer to a stably-stratified salt solution subjected to a steady lateral temperature gradient has been measured [140F].

Experiments on convection with internal energy sources and with different top and bottom layer temperatures yielded a correlation for the heat transfer [13F]. The onset of convection with non-uniform energy sources, as would occur in a fluid absorbing ionizing radiation, was predicted [145F].

Unsteady convection in horizontal channels with arbitrary temperatures has been calculated through to steady state flow [91F]. The reasponse to fluctuating thermal boundary conditions in horizontal layers has been measured [37F].

The cumulative effects of surface tension at an upper liquid–gas interface and heating of a horizontal layer from below have been examined [76F]. The surface tension-driven flow pattern has been calculated for a differentially-heated vertical layer open at the top [30F]. The influence of combined surface and buoyancy-driven forces on the stability of a liquid film flowing down a heated inclined wall has been analyzed [131F].

A number of studies consider convection in an inclined layer. Spatial perturbations are found to trigger instability in such a layer heated from below [45F]. The effect of boundary conditions on the stability of an inclined layer has been calculated [100F], as has the flow in such a layer with a square cross-section [101F]. Consideration of radiation and convective boundary conditions leads to critical parameters for the onset of the flow [54F]. Use of the Galerkin method to predict the Nusselt number at small angles of inclination yields results which are in good agreement with experimental data [35F]. Correlations have been obtained for convection in an inclined array of cells [36F]. Little effect of aspect ratio on heat transfer is found at different inclinations down to moderate aspect ratio [110F].

Natural convection in vertical layers with two opposing walls maintained at different temperatures and the upper and lower bounding planes adiabatic has been examined using a number of numerical schemes. Emphasis is on two-dimensional flow where no variation is assumed in the horizontal dimension parallel to the heated and cooled walls. A comparison of different numerical solutions for a square cavity has been presented [31F]. The maximum Nusselt number is found to occur at an aspect ratio near unity with the layer being slightly higher than it is wide [114F]. Specific calculations for the flow in square enclosures have been done using a finite element method [85F] and a high-order finite difference calculation [112F]. Other computational studies examine convection in a high aspect ratio slot simulating the vertical layer between window panes [52F] and compare flows in slots over a range of aspect ratios [64F].

A semi-empirical analysis provides Nusselt numbers for heat transfer across vertical enclosures [7F]. Experiments with a vertical layer indicate that an additional stable, vertical gradient retards the flow in a differentially-heated channel [99F]. The effect of finite conduction in the side walls of a vertical channel on the heat transfer has been analyzed [41F]. Additional numerical analyses include the development of algorithms for three-dimensional natural convection flows [103F] and a solution of three-dimensional transients in rectangular enclosures [16F].

The critical Grashof number for the onset of transverse rolls in a vertical layer is found to be almost independent of Prandtl number [113F]. Velocity and temperature fields have been measured for turbulent flow in a differentially-heated vertical layer [75F].

Several works consider the convection in a horizontal annulus formed by a solid circular cylinder inside a circular cylindrical cavity with the two surfaces at different temperatures. Multicellular flow has been found in such an annulus, particularly with a low Prandtl number fluid [18F]. A critical Rayleigh number has been found for the stability of a flow through an annular gap [92F]. Flow under transient conditions has been measured in an annulus [136F]. Convection with a non-uniform temperature on the outer cylinder in the form of a vertical gradient has been analyzed [125F]. The natural convection along an annulus when the temperatures at the two ends of the annulus are different has been studied to predict the horizontal flow in such geometry [10F]. The flow in both a cylindrical and spherical annulus with a porous media present has been examined.

The temperature distribution in a triangular enclosure has been measured using an interferometer [38F]. The effect of gas compressibility on thermal convection in a circular annulus has been predicted [71F]. Measurements of heat transfer have been made in partially-filled toroidal vessels [69F] and spherical cavities [70F].

A model of an open-loop thermosyphon has been used to simulate ground-water flow in aquifers [134F]. Flow in a toroidal thermosyphon has been studied experimentally and numerically with different regions heated and cooled [25F]. The effect of dissipation in a free natural convection loop has been analyzed [148F]. Transient behavior in a natural circulation loop has been studied with heated bottom and cooled top regions [48F] and with point heat sources and sinks [149F].

Work on natural convection boundary layers includes studies of the flow along a vertical heated plate. The use of a film temperature satisfactorily takes into account the effects of variable properties on heat transfer in a laminar boundary layer of water [119F]. The Grashof number by itself is found to be insufficient to define transition to turbulence in a boundary layer on a vertical surface [83F]. In a related study, different criteria for determining transition of a natural convection boundary layer on a vertical plate are compared [107].

Models for the outer and inner portions of a turbulent boundary layer are used to predict the Nusselt number along a vertical plate [44F]. Hot-wire measurements have been made in the transition region along a flat plate with a constant heat flux boundary condition [12F]. Analysis of the heat transfer from a vertical plate to a fluid near its thermodynamic critical point includes the effect of variable properties [117F]. Analyses for the transient boundary layer on a vertical plate include the effect of step changes in heat input [115F] and of surface temperature oscillation [94F].

Local non-similarity solutions show that the velocity field on an inclined surface is greatly affected by inclination while the temperature field and heat transfer are not very different from those with a vertical plate [53F]. Temperature and velocity distributions have been measured in the longitudinal vortex flow along an inclined heated surface [120F].

Studies related to the natural convection heat transfer from a heated horizontal circular cylinder include the development of simplified equations to predict the average Nusselt number over a range of conditions [40F]. Higher heat transfer from a cylinder can occur when it is immersed in a shallow pool of liquid as contrasted to a deeper pool [60F]. Different heat-transfer relations are required for cooling as compared to heating a horizontal cylinder in supercritical CO_2 [104F]. A boundary-layer analysis predicts the flow around a suddenly-heated cylinder [61F]; a numerical study of a similarly heated cylinder has been carried out to steady-state conditions [67F].

A boundary-layer analysis of the convection above a

circular horizontal disc indicates the total heat transfer is chiefly due to the high rate of heat transfer at or near the outer edge of the disc [146F]. Heat transfer from a conical surface has been analyzed when the cone has a horizontal axis [142F] and when the cone has a vertical axis [2F]. A numerical study indicates the transverse curvature of a vertical frustrum of a cone significantly increases the rate of heat transfer [95F]. Approximate solutions have been obtained for the heat transfer from heated spheres and cylinders immersed in a thermally-stratified fluid [19F].

There is some interest in the natural convection in a fluid (usually water) in which the range of temperatures encountered includes the temperature at which the density is maximum. An analysis of buoyancy-induced flows in water includes the effects of both thermal and saline gradients [43F]; a related study considers flows adjacent to horizontal surfaces [42F]. Other studies in pure water which include the density maximum relate to heat transfer in a horizontal layer heated from below [86F] and convection in a porous medium [141F].

The influence of surface mass transfer on the heat transfer from a vertical cylinder increases with the Prandtl number of the fluid [90F]. Combined heat and mass transfer has been analyzed on an inclined surface [23F] and, in another study, on vertical, horizontal, and inclined surfaces [126F]. The effect of suction and injection on heat transfer from a heated cylinder has been examined [74F]. A good correspondence of the heat- and mass-transfer analogy is found when using the sublimation of naphthelene to study heat transfer even in the presence of humid air [129F].

A turbulence model is used to predict the flow and temperature distribution in the transition of a heated vertical jet into a buoyant plume [20F]; a related study examines the flow near the origin of the buoyant jet [21F]. An interferometric measurement agrees with similarity predictions of the flow in a plume above a point source, if one assumes an apparent source position [122F]. Streak photographs show the velocity field of a rising plume from a heat source [123F]. Related experiments on a rising plume fit a model for the initiation of the plume and motion of the plume cap [121F]. A similarity analysis of a transient plume predicts that the velocity of the leading edge is less than the velocity of fluid following it [28F]. Swaying motion is observed before transition in the plume above a horizontal line source [144F].

A local heat source on a horizontal surface below a stably-stratified fluid can lead to either plume flow or layers of flow cells, depending upon the specific boundary and initial conditions [133F]. Motion of a suddenly-released thermal in the form of a vortex ring has been studied analytically [17F]. The motion at a liquid–liquid interface upon which a submerged buoyant jet impinges has been examined [81F]. An analysis predicts the flow in the turbulent plume above a line source in a horizontal wall [84F]. The turbulence properties have been measured in a thermal plume rising adjacent to an isothermal wall [80F].

A number of studies have examined convection in horizontal layers of saturated porous media. Conducting side boundaries have a significant influence on the onset of convection in such a layer [82F]. The presence of either two- or three-dimensional flow at low Rayleigh number in such porous media depends on the initial conditions [132F]. The effects of isotropy on supercritical motion and heat transfer has been examined [78F]. Experiments have been performed with distributed energy sources within a porous horizontal layer [73F]. The transient behavior of a geothermal reservoir has been examined by considering the heat transfer through three horizontal layers of different porosity [109F]. The stability of a Hele–Shaw cell heated from below can be compared to non-linear convection in a porous medium [79F].

Flow regimes are predicted within an inclined porous layer, heated either from above or below [139F]. The Nusselt number has been calculated for a differentially-heated vertical slot containing a porous medium [8F].

High Rayleigh number boundary-layer predictions for convection in a saturated porous media are also useful at moderate Rayleigh numbers [24F]. The effect of surface mass flux on heat transfer from a vertical wall adjacent to a saturated porous medium has been analyzed [87F]. An analysis predicts the onset of longitudinal vortices in a porous media adjacent to a flat inclined surface [58F]. A relationship has been derived for the heat flow from two-dimensional and axisymmetric bodies of arbitrary shape immersed in a saturated porous medium [88F]. Under some conditions, jet-type flow can occur in a heated vertical wedge of saturated porous medium [46F]. Numerical solutions predict the natural convection in porous media around plates, spheres, and cylinders [4F]. A boundary-layer model predicts the flow of a vapor adjacent to a wall immersed in a liquid-saturated porous medium [102F].

A shadowgraph technique has been used to measure local heat-transfer coefficients on a cylinder immersed in a melting solid [5F]. Experiments have been performed on melting around a horizontal array of heating cylinders [108F]. The effect of natural convection on freezing about a vertical cooled cylinder has been studied experimentally [130F].

Several studies examine natural convection in rotating systems. Analysis on the natural convection in the annulus between two rotating spheres shows the possibility of strong secondary flows [32F, 34F]. The effect of rotation on thermal convection in water near its point of maximum density has been studied numerically [106F]. The heat transfer from an isolated rotating plate has been examined with a finite difference technique [138F]. Rotation is generally found to have a stabilizing effect on the potential motion of a two-component fluid layer heated from below [3F]. The temperature field has been measured in a heated spherical cavity which is partially filled with liquids and rotated [50F].

Electric and magnetic fields can play a role in buoyancy-driven convection. The influence of an alternating electric field on natural convection in an organic fluid is strongly dependent on the dielectric constant of the fluid [147F]. A magnetic field appears to decrease the Nusselt number for convection in a rectangular container [118F]. A transverse magnetic field has only a modest effect on the flow field above a line source on a vertical adiabatic wall [47F]. The influence of a magnetic field on convection in a layer heated from below has been analyzed [105F]. Combined natural convection with rotation in magnetic fields has been studied in an electrically-conducting fluid [128F] and in a ferromagnetic fluid [26F].

When heat transfer takes place with flow which is generated partly by some mechanism external to the region of interest, such as a pump or fan, and partly by buoyancy forces, the term mixed convection—or alternately, combined free and forced convection—is used. Such flows are common in many heat-transfer situations. Generally, any forced convection flow in a gravitational field will have some buoyancy forces due to temperature differences required to transfer heat. However, in cases where these buoyancy forces and their effects on the flow are small compared to the external forces, they can often be neglected. In other forced convection situations, the flow due to buoyancy is of the same order of magnitude as the forced flow and therefore cannot be neglected. When the buoyancy forces tend to generate flow in more-or-less the same direction as the forced flow, we say the flows are aiding; conversely, when the buoyancy-driven induced flows tend to be in the opposite direction from the forced flow, we say we have opposing flow.

Mixed convection with opposing flow on a vertical isothermal plate has been examined [135F]. The influence of dissipation on aiding flow along a semi-infinite vertical plate has been analyzed numerically [127F]. A similarity solution has been obtained for mixed flow over a horizontal plate with a non-uniform wall temperature [116F]. The critical Reynolds number for instability of mixed convection over a horizontal plate has been obtained [22F]. On an inclined flat surface, buoyancy increases the Nusselt number in aiding flow and decreases it in opposing flow [93F]. Unsteady mixed convection has been studied to simulate phenomena on a hot-wire anemometer exposed to a low speed flow [62F]. A numerical solution for unsteady flow on an isothermal circular cylinder has been obtained for aiding flow [68F].

A number of mixed flow studies have been concerned with heat transfer in enclosed regions, such as ducts. In the thermal entrance region of a duct, the heat transfer can be augmented by a factor greater than four due to natural convection instabilities [66F]. Combined convection has been studied in vertical ducts of arbitrary cross-section [27F]. The effect of combined convection on sublimation from a wall of a rectangular duct has been examined [143F]. The influence of through-flow on heat transfer in a long, slender, horizontal cavity differentially-heated at the ends has been examined [9F]. Combined convection has been examined for heat transfer in a partially-filled vertical cylinder [65F] and in a rotating spherical annulus [33F].

Natural convection is often of importance in problems related to the heating and cooling of buildings. The influence of buoyancy on the temperature distribution and flow in an enclosure for which there is through-flow have been studied with application to a ventilated room [96F]. An open rectangular cavity with different temperatures at the side wall and the bottom wall has been examined to simulate the flow in a passively-heated solar-thermal energy-storage system within a building [1F].

Several important techniques have been used for flow visualization or quantitative determination of the flow field in natural convection flows, which are usually at low speed. The thymol blue technique has been used in a glycerol–water solution to study motion over a range of Prandtl number from 7–1000 [137F]. The difficulty of measuring velocity in a natural convection field with a hot-wire anemometer has been described [63F]. Laser–Doppler techniques are particularly suited for measurements in low-speed flows, and they have been used to determine the velocity field in natural convection along a vertical flat wall [39F] and around a vertical cylinder [55F]. The deflection of a laser beam through a liquid heated from below has been used to indicate transitions in the flow [77F].

CONVECTION FROM ROTATING SURFACES

Heat transfer from axisymmetric line sources at the surface of a rotating disk into a laminar constant property flow has been calculated [8G]. A similar process of jet cooling at the rim of a rotating disk depends primarily on the ratio of jet flow rate to disk pumping flow [7G]. Experiments [12G] determined local heat transfer from a disk rotating in an enclosure. A linear theory [14G] considers how the boundary layers on the surfaces of two coaxial disks rotating with a slightly different angular velocity are modified when blowing occurs on one disk and suction on the second disk. Results of flow and heat transfer studies in the inlet zone of a porous tube with injection into the swirling flow were published [11G]. An empirical equation describing the decay of circulation of swirling flow in a pipe has been obtained from experimental results [9G]. Rotation was found [5G] to increase heat transfer in the inlet section of turbine blade cooling channels by a factor of 1.2 to 1.4. Experiments were performed [10G] to study heat transfer from a cylinder placed in a coaxial rotating flow. A Reynolds analogy solution [13G] describes heat transfer for combined Taylor vortex and axial flow through the annular space between an inner rotating and an outer stationary cylinder.

An experimental and numerical study [1G] considers the spin-up of a thermally stratified rotating fluid in a cylinder, the average velocity of which increases linearly in time. The spin-up adjustment was found slower than previously reported. Velocity profiles obtained by an analytical model [2G] for concentric flow between coaxial rotating cylinders with a stationary lower boundary agree well with experiments. A numerical solution [6G] of the steady axisymmetric incompressible laminar boundary-layer equations for swirling flow with heat and mass transfer in conical nozzles and diffusers agrees with solutions by integral methods except near the exit. The compressible boundary layer equations describing swirling flow in a nozzle or diffuser were transformed using Lee's transformation and solved numerically [4G]. It was found that the variation of the product density times velocity and mass transfer have a strong effect on heat transfer. The analytical solution describing the temperature distribution in laminar boundary layers on rotating bodies with non-uniform surface temperature are presented in the form of universal functions [3G].

COMBINED HEAT AND MASS TRANSFER

The concepts of eddy viscosity and mixing length were used to predict numerically [4H] the effects of tangential slot injection on turbulent boundary layer flow. The turbulence model in the shear layer near the slot had to be improved to obtain agreement with experiments. Measurements of film cooling effectiveness with injection from a single hole or a row of holes with 8, 5.33 and 2.67 pitch to diameter ratio and an inclination of 30° indicated the effects of stream turbulence and velocity gradients [3H]. The boundary-layer thickness, Reynolds number, stream turbulence intensity (0.3–20.6%) and turbulence scale (0.06 to 0.33 of the jet diameter) can affect film cooling performance significantly [9–11H]. Experimental studies considered film cooling in supersonic axisymmetric nozzles [26H] and in hypersonic flow [18H] with injection through a tangential slot to an isothermal wall. Hydrogen was found to be the most efficient coolant for constant mass injection rate because the mixing layer stays discrete and laminar.

A new approach [20H] correlates boundary-layer mass-transfer rates with thermal diffusion and variable properties for laminar flow and a wide range of Lewis numbers. Equations are presented [6H] describing heat-transfer properties for mass flow through perforated plates. Solutions [25H] for laminar flow at larger Schmidt or Prandtl numbers and for turbulent flow describe heat and mass transfer between a flowing fluid and a solid wavy surface for large and small wave numbers; experiments clarified the wave induced variation of the turbulent transport. Heat transfer in turbulent boundary layers was studied on a porous plate with a permeable zone of varying length and with injection of a cooling gas [2H], as well as with a porous wall with alternating injection and suction zones at a Mach number of 2 [12H]. Correction factors describe [19H] the influence of the Stefan flow on the Lewis factor for one-dimensional heat and mass transfer. Mass transfer studies [16H] in boundary layers of a non-Newtonian power law fluid included the effect of mass injection. Numerical solutions of mass transfer with chemical surface reaction are reported [22H] for flat plates. A computer technique for the study [13H] of steady and unsteady, turbulent, chemically reacting flows in axisymmetric situations uses the Patankar–Spalding method. Analytic methods are compared [5H] with experimental results describing gas turbine combustor cooling augmented by backside convection enhanced with transverse-rib turbulence promoters.

Binary laminar boundary layer flow along a vaporizing liquid layer is studied analytically [17H] for a co- or counter current liquid film. Turbulent heat and mass transfer by convection, radiation, and evaporation was studied [21H] on a rough air–water interface; a simple theory is used and compared with experimental results. Nusselt's theory is confirmed by an experimental investigation [14H] of local mass transfer in a laminar and turbulent falling liquid film at $Re = 3.86$ to 2496. A laminar stagnation flow boundary layer model [24H] predicts experimental results for surface ablation by melting in the impingement region of a liquid jet. A physical–mathematical model was used [15H] to calculate the time evolution of spherical droplets in a vapor environment. Low Peclet number heat and mass transfer from a dielectric drop suspended in an electric field is influenced by circular motion within the drop [7H]; an analysis using perturbations gives results valid to $Pe = 60$. Heat transfer from a wedge to air flow was increased from 2 to 14 times [1H] when water mist was mixed with the cooling air. Particle trajectories near an airfoil with a film cooled leading edge were studied [23H] analytically and experimentally by flow visualization in a cascade. They depend on two parameters. Most important of them is the Stokes number

$$St = \frac{\rho_p U_1 D_p^2}{18\mu_1 b_x},$$

in which the subscript p refers to particles and the subscript 1 to cascade inlet conditions; b_x denotes the chord length. It is found that for $St < 0.1$ the particles follow essentially the fluid path, whereas for $St > 10$, most particles will impact. The deflection of the particles by leading edge film cooling is also shown. An analysis [8H] of heat and moisture transfer in concrete slabs considers the effect of material characteristics and the hardening of concrete.

CHANGE OF PHASE

Boiling

Similarity numbers were introduced to describe heat transfer in pool boiling [43J].

The heterogeneous nucleation theory can describe not only the onset of nucleate boiling but also the boiling crisis [22J]. The mechanical energy stability criterion is extended [47J] to describe pool boiling burnout. Previously published correlations for the critical heat flux in forced convection boiling in vertical tubes are compared with new experiments [34J] and are found useful. An extension of the previous analysis to annuli results in generalized correlations [32J]. Previously published equations for the critical heat flux density in free convection in boilers have been verified [84J]. Burnout in four-rod bundles with flow of water and Freon 12 was studied experimentally [60J]. Flow turbulization increases the critical heat flux density and the pressure drop in annuli [56J]. The critical heat flux is increased by a factor 1.2 to 1.5 with turbulence promoters in the form of transverse fins [55J]. The pressure drop also increases by a factor of 1.2 to 1.4. A comparison of experimental and analytical results for critical heat loads connected with boiling helium in tubes indicates that this process is still poorly understood [4J]. The hydraulic resistance and burnout with helium boiling in tubes was studied [15J]. Thermal fluctuations in the tube wall caused by transition boiling in sodium-heated steam-generator tubes were measured [19J]. An empirical formula was derived describing the limiting vapor quantity in the second boiling crisis [53J]. Analysis [25J] of nucleation processes in large scale vapor explosions is based on the assumption that spontaneous nucleation cannot occur until the thermal boundary layer is sufficiently thick to support a critical size cavity. An analysis of dryout and rewetting was published [73J] for a model reactor core.

A previous correlation for the critical heat flux of forced convection boiling in vertical tubes with inlet subcooling was used for an analysis of the process [33J]. The upper limit for the critical heat flux in saturated forced convection boiling on a heated disk with a small impingement jet was studied [35J]. A graphical method [68J] to predict the critical heat flux in vertical tubes agrees with a large number of published data. A post dryout analysis [31J] for upward flow in a vertical tube considers heat transfer from the wall to the vapor, from the vapor to droplets, and from the wall to dropets (a small contribution). The Leidenfrost phenomenon was studied [7J] for water, n-octane and carbon tetrachloride drops. Interfacial phenomena were observed [88J] in the interline region of a rewetting hot spot of a falling thin film. Large interline temperature gradients and superheats were found possible with a stationary or moving interline.

An analysis of boundary layer flow in forced convection film boiling [87J] demonstrates that the pressure gradient in the liquid dominates the dynamics of the flow of the vapor boundary layer. Transition and film boiling may coexist [26J] in boiling of nitrogen on a vertical fin. Useful considerations are discussed [46J] for film boiling of helium below the λ point.

The quasisteady assumption was useful in the analysis [91J] of transient film boiling on a sphere with forced convection. Transient heat transfer with boiling was also studied [82J, 83J] for a stepwise increase in heating. Transient heat transfer from liquid-surrounded electrically heated wires in natural convection was found [81J] quite different from steady heat transfer.

A blowing parameter was used to correlate experimental results [29J] for evaporation and breakdown of thin water films driven by shear stresses. Boundaries of the region of existence of nucleate boiling should be useful for the design of thin film evaporators [80J]. A critical thickness for breakup of films of boiling water was found [79J] to be a function only of the heat flux density. Fluid evaporation and boiling heat transfer were studied [86J] in the grooves of thin film evaporators.

Experimental results for the pressure drop in horizontal finned tube evaporators for boiling refrigerants were compared [10J] with calculations. Evaporative heat transfer was enhanced by 30–760% by internal fins in tubes when refrigerant 22 was boiled [41J]. The pressure drop increased by 10–290%. The effect of internal fins on heat transfer and pressure drop was also studied [16J] for Freon 22. The effect of surface roughness was found significant [37J] for boiling heat transfer to light hydrocarbons and nitrogen. Experiments [42J] established contributions under which heat transfer is improved in boiling liquid nitrogen or helium in capillary-porous bodies. Experimental data have been correlated [3J] for heat transfer with boiling of dissociating nitrogen tetroxide in vertical tubes.

The two dimensional and axisymmetric growth and breakoff of bubbles with small contact angle was analyzed [71J]. A model for the effect of droplet interactions on vaporization [77J] considers evaporation of a droplet inside a bubble. An interferometer was used to study a stationary evaporating ethanol meniscus [63J]. An analysis [64J] results in very high local heat fluxes near the interline. An analysis [38J] of vaporization of a disperse system of atomized liquid droplets flowing in line with an air stream indicates that serious errors may result when the system is simplified to one with an averaged droplet size. The effect of contact angle on bubble nucleation was studied [17J]. Integral forms of the heat transport equation are used [94J] to obtain numerical solutions for the growth and departure of a vapor bubble at a horizontal superheated wall. An analysis [51J] of single evaporating droplets in an immiscible liquid is compared with published experimental data. The classical theory of homogeneous vapor nucleation is extended [61J] to multi-component liquids. Random explosive destabilization of vapor film boiling on a surface at 850°C was studied in subcooled water [18J]. Experiments formed the basis of a mathematical model [8J] of dropwise evaporation with very high heat fluxes.

Condensation

Research devoted to the investigation of heat transfer during condensation included studies of film condensation, dropwise condensation, and condensation in two phase and multi-component flows.

Many authors addressed the problem of film condensation inside pipes and different geometries. A dimensionless correlation for predicting heat-transfer coefficients during film condensation in pipes was presented which was applicable for different fluids and over ranges of reduced pressure, saturation temperature, vapor velocity, vapor quality, mass flux, heat flux, Reynolds number, and Prandtl number [69J]. An experimental study of heat transfer with condensation of a moving vapor (Freon-21) on a bundle of plain horizontal tubes within a wide range of vapor velocities and heat fluxes demonstrated the influences of the condensate and vapor velocities must be taken into account [45J]. The approximate integral method was used to study laminar film condensation of pure vapors in tubes. The results demonstrated that the assumed velocity profile at the beginning of condensation is not very significant for the calculated condensation performance in the tube; however, increasing the external heat transfer and putting the tube in a vertical position improved the performance of the tube [49J]. An accumulation and analysis of experimental data recommended a correlation for the problem of condensation of flowing vapor on horizontal tubes in downward vapor flow [6J]. The results of an experimental study of condensation heat transfer in tubes suggested a new correlation for calculating heat transfer in complete condensation of steam in vertical tubes [9J]. A study of the influence of film condensation on the flow of the entrance region of a tube showed that the inlet region becomes shorter with increasing condensation rate. This was attributed to a dramatic reduction of the vapor speed which tends to support the growth of the vapor boundary layer in spite of the opposing effect of suction [48J]. An experimental investigation of steam condensation in an inclined bundle of tubes demonstrated the inclination of a multirow bundle intensifies heat transfer on the condensing steam side by eliminating the flooding of tubes by condensate running off from top to bottom tubes [70J]. An experimental investigation of steam condensation inside profiled horizontal tubes demonstrated improved heat transfer in a horizontal tube film evaporator [65J]. A similarity solution was formulated and numerically evaluated for the problem of film condensation on a vertical plate or cylinder. The calculated fin heat transfer was found to be markedly less than that which would have been predicted by an isothermal fin model [59J]. Based on experimental evidence, a curve was suggested for film condensation of vapor in the absence of an appreciable external friction on the gas–liquid interface. The curve is quantitatively very similar for different geometries and reveals the regions of laminar wave motion, quasi-selfsimilar heat transfer, and fully developed turbulent heat transfer [44J].

The influence of material properties on dropwise condensation was also an area of strong interest. An experimental study performed to determine the least thickness required for an electroplated gold surface to promote dropwise condensation of steam at atmospheric pressure demonstrated that gold deposits of 500 layers or more resulted in dropwise condensation. Deposits of a hundred layers or less resulted in filmwise condensation and a transition region occurred between these two limits [93J]. For droplets condensing on or evaporating from a solid surface, the effects of the solid material properties and the droplet contact angle were analyzed by solving the steady heat conduction equation for the geometry of a spherical segment droplet on a semi-infinite solid. By assuming heat is transferred to or from the solid, only through the droplet, and that there is perfect thermal contact at the solid–liquid interface, solutions were found for the overall heat flow through the droplet [66J]. An experimental investigation of dropwise condensation of different steam/air mixtures on copper, aluminum, and nickel substrates, showed that the heat flux, the heat-transfer coefficient, and the surface temperature were observed to be highest on copper and lowest on nickel substrates for the same identical test conditions. Experimental results also demonstrated that the presence of non-condensable gases is an inhibiting influence on the heat-transfer performance in the condensation of steam. From a mathematical model of the problem, it was concluded that the minute fraction of the substrate surface covered by liquid droplets was most effective for the flow of heat and that the contribution to the heat transfer by the bare area was negligible [1J, 2J]. The high rate of heat transfer of dropwise condensation, as well as its limits, were explained on the basis of the behaviors of submicroscopic active drops. Numerical analysis and consideration of the non-dimensionalized forms of the basic equations resulted in an expression for the Nusselt number for dropwise condensation in terms of a few characteristic parameters [76J].

Investigators also studied the problem of heat and mass transfer during condensation of multicomponent and two phase flows. The problem of binary vapor condensation in the presence of an inert gas was studied to analyze the effect of the nature of the inert gas and its composition on the condensation process. The results of the computation show that diffusional interaction effects are significant for small concentrations of inert gas [40J]. A numerical study of laminar flow transport of binary vapor–gas mixtures in channels with porous walls and of removal of the condensing component by suction through these walls showed that the transport processes depended significantly on variations in physical properties, variability of the composition of the flowing gas mixture along the channel, and on transverse flow [72J]. A simplified analysis was presented for calculating condensation fluxes from multi-component vapor mixtures, taking into account the diffusional interactions in the vapor phase [39J]. The hydrodynamics of condensation of the stream in a moving two phase film were studied and working equations were obtained for determining the average fraction voids, hydraulic drag, and the volumetric heat-transfer coefficient [36J]. A significant reduction in computing time and memory requirements was obtained for the numerical computation of non-equilibrium condensation in one and two dimensional supersonic two phase flows [50J]. A method presented for solving the two phase boundary-layer equations for the condensation of flowing vapor on a horizontal cylinder provided numerical solutions for the distribution of the local values of the Nusselt number on the periphery of the cylinder as a function of different governing parameters [20J].

A miscellany of papers defied simple categorization. The results from a theoretical and experimental investigation of the hydrodynamics and heat transfer with steam condensation on vibrating horizontal tubes demonstrated that tube vibrations can lead to a 12% decrease in heat exchanger performance [67J]. Data presented for the boundary conditions of constant heat flux and constant wall temperature for a rotating heat pipe condenser demonstrated that an efficient spray cooling mechanism can be used to increase heat-transfer coefficients. However, there appears to be an upper cooling water flow rate above which dry saturated conditions cease to exist over the entire condenser length [13J]. A theoretical solution was presented for the problem of steam jet condensation in a parallel flow of liquid. An accompanying experimental investigation of steam condensing in a cocurrent and in a countercurrent flow of turbulent flowing water demonstrated general agreement with theoretical predictions of the level of the intensity of heat transfer [30J].

Two-phase flow

An analytical study was performed to evaluate the surface heat flux and the boundary-layer structure over an isothermally heated circular cylinder subjected to evaporating particle–vapor flow in forced convection. The results showed that enhancement of heat transfer is comparatively small in cases where there is large temperature difference, but distinctly large in cases involving large particles [54J]. The effects of the flow rate and the concentration of solid particles in the slurry on the pressure drop were investigated in solid–liquid two-phase flow in a horizontal pipe with low flow rates [78J]. The deposition motion of liquid drops in dispersed flow with heat addition was analyzed and an expression was suggested for the reaction force on the drop due to non-uniform drop evaporation inside the laminar sublayer. The effects of the drop diameter, deposition velocity, slip velocity, and wall temperature on the drop trajectory were examined [21J]. A method was presented for predicting local Nusselt numbers for heat transfer to a stratified gas–liquid flow for turbulent liquid/turbulent gas conditions. Liquid Nusselt numbers predicted by an analogy between momentum transfer and heat transfer were in agreement with experimental results for wavy interfacial conditions, and the circumferential average gas phase Nusselt numbers were in rough agreement with the Dittus–Boelter equation for single-phase flow based on flow through a conduit of irregular shape [14J]. A technique was developed for ascertaining thermocapillary movement of gas bubbles in various organic melts contained in a heated horizontal tube. In a related paper, the vapor pressure of the liquid and the composition of the gases in the bubble were shown to have a profound effect on thermal transport in the bubble, and hence, on the temperature gradient on the surface of the bubble [58J, 90J]. A theoretical analysis based on the premise that fluid flow results from the London–van der Waals dispersion force was used to evaluate the changing profile of an evaporating meniscus as a function of heat flux. The useful capillary pressure of the meniscus was significantly reduced as the interline (junction of solid–liquid–vapor) heat flux was increased [89J]. Heat- and mass-transfer mechanisms were theoretically studied for several capillary groove configurations and the effects of the solid–liquid–vapor interline heat transfer were isolated and evaluated. A method was provided for predicting surface heat-transfer coefficients for surfaces wetted to capillary grooves and basic information about heat transfer in the vicinity of the meniscus attachment region [27J]. Experimental data were obtained using propane to quantitatively determine the deterioration of temperature separation when the condition at the inlet fluid to a Ranque–Hilsch tube becomes a saturated mixture. Observations indicated that temperature separation diminished rapidly for inlet qualities of less than 80%, largely due to the increase of the "hot side" temperature [12J]. Experimental measurements were used to propose an empirical formula for predicting the pressure drops due to pipe bends in single-phase and air–solids two-phase flows in circular and eliptical bends [52J]. Numerical calculations over a wide range of temperatures characterized the effects of heat transfer on the behavior of the bubble and the impulse pressure in a viscous compressible liquid [85J].

Melting and freezing

A significant number of papers were concerned with the problem of heat transfer during melting. An experimental study performed on a horizontal flow of equally spaced heating cylinders melting a mixture of sodium nitrate and sodium hydroxide presented results for the timewise distribution of the melting coefficient for both single and multiple cylinders [62J]. The shape of the melting fronts and local heat-transfer coefficients were determined for melting from an array of electrically heated cylinders embedded in a paraffin (*n*-octadecane). Results indicated that natural convection effects are an important influence on the melt shape, surface temperature, and heat transfer coefficients [5J]. Visual observations from an experimental study of the melting of a horizontal slab of frozen olive oil placed underneath a pool of warm water showed that the melt removal is governed by Taylor instability and that melt releasing nodes lie at about a Taylor wavelength apart. Predictions of the growth of the interface based on equilibrium between surface tension and buoyant forces compare well with experimental data [75J]. An experimental study was made of the velocity profiles which result from the free convective melting of a vertical ice sheet into fresh water at temperatures in the range of 2–7°C. Results suggest that: (i) below 4.7°C, upward flow exists; (ii) above 7°C, entirely downward flowing boundary layers are suggested, and (iii) for intermediate temperatures, an oscillatory dual flow regime is indicated [92J]. An experimental investigation of the melting of a paraffin (*n*-octadecane) from a heated vertical wall of a rectangular test cell provided evidence of the importance of natural convection on heat transfer in the melt region. These conclusions were based on the resulting melt shapes, the temperatures in the paraffin, and the heat-transfer coefficients at the solid–liquid interface, and in their timewise variation [23J]. The problem of turbulent flow past freezing or melting ice was formulated for the case of random boundary waves. The methodology presented provides a means for determining the phase shifts between the streamwise distribution of turbulent transfer between wavy boundaries and the flows past them [28J].

Experiments were performed for freezing under conditions where the liquid phase was either above or at the fusion temperature. Freezing occured on a vertical tube housed in a constant temperature cylindrical vessel and the phase change medium was *n*-eicosane. In the presence of liquid superheating, the freezing process was dramatically slowed and ultimately terminated by natural convection in the liquid [74J]. A study of the freezing of a liquid in turbulent flow inside a circular tube whose wall is kept at a uniform temperature lower than the freezing temperature of the liquid was solved. The radius of the solid–liquid interface and the local wall heat flux were determined as a function of time and position along the tube for different values of the Prandtl number and the freezing parameter [11J]. The phenomena of salt rejection during the freezing process of saline solutions in cells were investigated both analytically and experimentally. When freezing proceeds, in the solid–liquid region, freezing fronts were found to occur stepwise within several adjacent cells, due to the depression of the freezing point caused by salt rejection in the cells [24J]. An experiment performed to investigate the behavior of bubbles at a solidification interface and in the melt ahead of the interface in a low gravity environment observed that a larger number of trapped bubbles were present in the low gravity specimens, indicative of easier bubble nucleation, and that the morphology of the grown-in voids was dependent upon the applied temperature gradient [57J].

RADIATION

Radiation in participating media

Exact radiative transfer solutions are obtained for an isothermal medium confined within an isothermal enclosure using band radiation and diffuse reflections [20K]. An asymptotically exact approximate method of analysis for band radiation makes direct use of band absorption formulations for absorbing–emitting media. The method has acceptable errors even for low order approximations [21K]. The advantage of the P_3 approximation for solving higher order differential equations of radiative transfer is discussed and this approximation is used for solving the radiative exchange for the case of a participating medium between plane parallel plates, concentric cylinders, and concentric spheres [4K]. Based on an iterative solution for anisotropic radiative transfer in a slab, results have been obtained for the hemispherical reflectivity and transmissivity of the slab over a wide range of parameters [25K]. A new diffusion model for radiative transfer in particulate media extrapolates the conventional Eddington approximation to problems involving large absorption [18K].

A new six-flux model for radiative transfer should prove to be a useful component for incorporation into existing procedures for complete mathematical modeling of rectangular furnaces and heaters [23K]. A collocation method using B-splines as approximation functions appears to be an attractive, non-iterative candidate for the solution of combined radiation–conduction problems pertaining, for example, to LMFBR core disruptive melt down accident analyses [7K].

The band absorption of radiating gases can be incorporated into the three-dimensional differential approximation without increasing its complexity [19K]. Considering the effect of molecular gas radiation upon the thermal development downstream from a step change in wall temperature, one finds that even in the entrance region, self absorption by wall layer gas blocks significantly the radiation exchange between the gas core and the wall [3K].

A study of the interaction of radiative and convective transfer for slug flow of an absorbing–emitting gas in a black wall circular tube with specified wall heat flux shows that the outlet surface temperature has a strong influence on the temperature distribution particularly those for the wall and is even felt by the wall zones near the duct inlet [8K]. Calculations of the emissive power and of the radiative fluxes at the boundaries of a two-dimensional, finite, planar, gray, absorbing–emitting–scattering medium bounded by non-isothermal black walls demonstrate that any step variation in the wall temperature can be expressed in terms of universal functions for the semi-infinite step variation in wall temperature [12K]. Considering an absorbing–scattering slab bounded by emitting and reflecting parallel surfaces, the authors of this paper show how the emissivity and reflectivity of the bounding surfaces and the scattering properties of the participating medium affect radiative transfer within this medium [5K]. Studies of the radiative properties of a finite scattering medium bounded by a diffuse substrate indicate that the scattering layer acts as a source if the substrate emittance is less than the emittance of a semi-infinite scattering medium. Conversely when the substrate emittance is greater, the scattering layer acts as a shield and decreases the apparent emittance [10K]. Experimental studies of two-dimensional scattering in a medium of finite thickness show fair agreement with theoretical predictions [16K].

For predicting radiative transfer in aqueous suspensions a three-flux model is proposed consisting of refraction at the air–water interface, the highly anisotropic scattering in the suspension, and the diffuse reflection at the bottom. Excellent agreement is obtained between predictions of the net radiative flux in the suspension and measurements obtained for a suspension of unicellular algae [14K]. Solutions of the radiative transfer equation in shallow ponds show that scattering is not important when the albedo is small ($\omega \leq 0.5$) and the phase angle function is highly forward peaked. Therefore, even the simple Beer's law expression provides accurate predictions [13K]. By using a tunable far IR optically pumped laser the reflection and transmission of water at 25°C have been determined from which the complex index of refraction can be derived [24K]. Predictions based on two-dimensional multiple scattering with absorption are in fair agreement with experiments using a medium consisting of water and various concentrations of white latex paint [17K].

A proposed method for calculating radiative heat transfer in rod bundles relies on combining similar surfaces for the radiative heat-transfer calculations and identifying droplets and vapor as groups of absorbing media. The results show that the cooling of the hot rods is strongly influenced by radiation absorption of droplets when the droplet size is small [27K]. An iterative method for solving the radiation transport equation is very effective in the case of a two-phase medium of cylindrical geometry [1K].

Measurements with a newly developed system (differential absorption lidar) for remotely determining profiles of H_2O vapor concentrations in the atmosphere compare favorably with results obtained with rawinsondes [6K]. Studies of the aerosol extinction in various weather situations show that the air humidity affects the aerosol extinction mainly through modification of the particle size distribution. The humidity influence on the refractive index is less important [22K]. Recent advances in electro-optical instrumentation and in computer technology resulted in the elucidation of the effects of ozone absorption, the identification of a cloud bow on Venus, and the interpretation of Rayleigh scattering on Jupiter [9K].

In connection with radiative transport in liquids and solids a new formulation of the directional emittance of a semi-infinite scattering medium is suggested and numerical results are presented for a wide range of albedos and refractive indices [11K]. Calculations of the temperature distribution in a semi-transparent slab (optical window) reveal the importance of radiation. Results obtained for opaque boundaries are in good agreement with exact formulations found in the literature [2K].

It is shown that radiative interaction in laminar compressible boundary layers for moderate Mach numbers leads to a considerable increase of the thickness of the thermal boundary layer and tends to flatten temperature profiles [26K]. Transport of radiative energy at cryogenic temperatures is treated by an approximate method. Calculations of the time required for thermal stabilization of a system of metal screens is in agreement with experimental findings [15K].

Surface radiation

Total hemispherical emissivity values have been reported [9L] for stainless steel AlSl 304 with a polished and electroplated surface in a temperature range from 340 to 1100 K. The measured values are very close to those published by Davisson and Weeks. Closed form expressions are presented [2L] for spectral hemispherical radiative properties. Modelling of the atmospheric effects in space measurements of ground reflectances has been discussed [12L].

Shape factors have been calculated [5L] for radiative exchange between coaxial annular disks of different radius ratio separated by a solid cylinder using the contour integral method. The same method was used [6L] to obtain shape factors in semi-closed form for radiative exchange between an end plane and the outer wall of concentric tubular enclosures. The results of a computer program [7L] using the Monte Carlo method for the analysis of three-dimensional radiative exchange factors for non-gray, nondiffuse surfaces compare well with the results of some space shuttle experiments. An efficient technique [8L] considers heat conduction in a wall with radiation across a gap. Experiments on convective heat transfer between cylindrical objects located in a channel and the channel wall have to account for radiative exchange. This problem is treated in [1L]. An analysis [3L] considers radiative exchange between diffusely emitting and specularly reflecting axisymmetric surfaces of enclosures, passages, and cavities. It is used to calculate radiative exchange between parabolic reflectors and a heat source located at their focus. Experiments [4L] considered transient heat exchange between a radiating plate and a high temperature gas flow.

Melting of an ice layer adhering to a vertical surface produced a rough surface when heated by short wavelength radiation, but a smooth surface when heated with long wavelength radiation [11L]. The rate of melting was well predicted by an analysis using a band model extinction coefficient. Back melting on a substrate covered by a horizontal cloudy ice layer was observed [10L] when it was exposed to radiative heating. Perturbation solutions [13L] to one-dimensional phase change on a fixed boundary subject to convection and radiation agree well with earlier work. The radiative term was approximated by a Taylor series.

MHD

Increasing emphasis on MHD generators seems to revive research activities in this field.

Radiative heat transfer in a MHD generator increases the heat flux to the walls and alters the gas temperature distribution in the channel. Real gas results differ substantially from those for a gray gas [9M]. In closed-cycle MHD generators a small amount of an alkaline metal is added to the working fluid (inert gas). Fog formation greatly reduces the deposition of the metal from the gas onto cold surfaces of the condensing heat exchanger [1M]. Experimental studies of residual disturbances in turbulent MHD flows after laminarization show that an almost undisturbed laminarized flow is obtained by eliminating entry effects [4M].

Studies of MHD Couette flow including Hall and ion-slip effects show that the Nusselt number depends in a rather involved way on the Hartmann number, the Hall parameter, and the ion-slip parameter [10M]. The flow of a hot, ionized fluid in a rectangular MHD channel is considered with significant heat transfer by thermal radiation. The temperature profiles are considerably distorted from those of uniform wall temperature when the walls are differentially and non-uniformly heated [6M]. Calculations for incompressible MHD flow between inclined walls indicate that in the approximation of constant fluid properties, the onset of flow separation is independent of heat transfer [5M].

Considering the effect of massive blowing rates on the steady laminar hypersonic boundary layer flow of an electrically conducting fluid in the stagnation region of an axisymmetric body in the presence of a magnetic field, one finds that massive blowing rates remove the viscous layer from the boundary, whereas the effect of the magnetic field is just the opposite [7M]. There are substantial magnetic field effects on the response of a cylindrical Langmuir probe when the probe's sheath is comparable to or greater than the Larmor radius [11M].

Ferromagnetic fluids can be used for controlled cooling of current-carrying conductors [2M]. Convective heat transfer in ferromagnetic fluids can be controlled by the application of homogeneous magnetic fields [3M]. Heat transfer to a cylinder in cross flow of a magnetic fluid can be substantially increased by applying an inhomogeneous magnetic field [8M].

MEASUREMENT TECHNIQUES

The development of instrumentation and measurement techniques related to heat-transfer studies are described by a number of researchers. The continued development of hot-wire anemometry and laser–Doppler anemometry for a wide variety of convection heat-transfer studies has been reported. There has been an increase in the number of papers which describe specialized techniques and instrumentation for performing measurements in two-phase flow and plasma processes.

A data reduction technique has been presented which permits constant-temperature hot-film and hot-wire probes to be used to determine flow angles and velocities for a wide range of supersonic and hypersonic flow conditions [24P]. An improved technique is proposed for making velocity measurements in unsteady flow studies where accuracy is the dominant consideration rather than frequency response [9P]. The attenuation of turbulence and mean velocity signals due to the line averaging imposed by hot-wires when used in the wake of a cylinder can be minimized by the proper selection of wire length [44P]. The flow patterns in highly-turbulent regions between cylinders have been determined using a process whereby a single hot-wire probe is rotated to three discrete positions and the three equations thus obtained are solved iteratively [48P]. The dynamic calibrations of hot-film and modified hot-wire probes were studied in order to assess their suitability for use in the measurement of turbulent fluctuations in compressible boundary-layer flows [2P]. If short probe cables are used on a simple constant-temperature hot-wire anemometer, the frequency response is limited by the finite open-loop gain of the amplifier used in the anemometer system [33P]. The calibration of probes at Reynolds numbers less than two at various Knudsen numbers show that a previously proposed relationship between the Nusselt and Reynolds numbers is valid, even for short wires [47P]. Temperature-sensitive constant-current hot wires operating at very low resistance ratios exhibit a significant increase in their temperature fluctuation response at a frequency of approximately 1/6 Hz [38P]. The theory of the transient hot-wire cell for measuring the thermal conductivity of gaseous mixtures has been reworked to include the coupling between the transport of energy and mass that exists in a mixture [28P].

Results of a calculation of the axisymmetric thermal gradient and viscous shear layers over a heated spanwise strip on an insulated cylinder in a periodically-reversing flow have been applied to the calibration of a hot-film probe used for *in vivo* blood flow velocity measurements [50P]. The frequency response on conical and wedge-shaped hot-film probes is strongly dependent on the geometry of the backing material [11P]. The calibration of cylindrical hot-film velocity sensors with aspect ratios near 20 obey a somewhat different heat-transfer relation than do large aspect ratio hot wires [5P].

A number of papers discuss the theory and application of laser–Doppler anemometry (LDA). An experimental method is proposed for estimating and correcting the effect of interaction of laser beams with turbulent temperature fluctuations along the beam paths in order to apply LDA to turbulence measurement in non-isothermal flows [35P]. Correction equations have been derived for the sensing volume and the individual realization biasing, and the equations have been applied to two-component flow fields measured by a one-dimensional LDA system [45P]. An analysis of Gaussian beam effects indicated that both axial and lateral frequency gradients can exist in the probe volume of improperly-aligned LDA systems [17P]. A technique is reported that facilitates measurement of the flow field in the cylinder of an internal combustion engine [16P]. The temperature, concentration of species (neutral as well as ionized) and velocity in jets, flames, and shock tubes can be obtained by the combined use of Raman scattering and laser–Doppler velocimetry [31P, 32P].

Three other velocity measuring approaches were described. The theory of the measurement of velocity by electrodiffusive sensors with conical and wedge-shaped working surfaces have been analyzed, and theoretical equations for determining amplitude and frequency characteristics have been suggested [39P]. A compact three-tube offset-geometry pressure probe, which is a combination of a total head tube and a Conrad probe, can be successfully used to determine the velocity vector in a subsonic flow [40P]. The use of a white light fringe image velocimeter (WFIV) has been shown to deserve serious consideration as a practical technique for laminar flow measurements [13P].

Instruments and techniques for performing radiation measurements have been described. A continuous-temperature IR calibrator provides reference video signals for temperature measurements which are made using an infrared scanner [26P]. A broad-range radiometer (0.25–20 μm) has been developed which incorporates a split-disk, windowed-design bolometer to facilitate measurements in the solar spectral range and is easily convertible to IR usage by the removal of the window [41P]. Energy photometry and spectral measurements of radiation detectors have been discussed [42P]. Three new surface treatments have been described that are satisfactory diffusers of infrared radiation for use in an integrating sphere pyrradiometer [10P]. Two sensitive optical techniques are useful for the study of small reflectance or absorptance differences between two interfaces [27P]. The absorption coefficient of low-loss materials (e.g. KCl) can be measured by calorimetry using pulsed CO_2 lasers [36P].

Development continues in the area of transport property measurements. A comparative method of measuring thermal conductivity is described which does not require calibrated thermometers and which involves temperature rises of less than 1 K while maintaining an uncertainty of $\pm$ 5% [30P]. The use of an apparatus for measuring the thermal conductivity, thermal diffusivity, and specific heat of reactor materials at elevated temperatures has been demonstrated [6P]. A transient hot-strip method has been developed for simultaneously measuring the thermal conductivity and the thermal diffusivity of solids and fluids with low electrical conductivity [22P]. A new contactless method has been used to measure the thermal conductivity of thin layers of solids with results deviating less than 15% from tabulated values [14P]. The thermal conductivity of a number of liquids has been measured using a newly-proposed spherically symmetrical apparatus [49P]. A thermal relaxation method can be used to measure heat capacities of samples which are as small as 10 mg [23P]. The steady shear viscosity of a viscoelastic fluid at low shear rates can be determined using a falling ball viscometer if appropriate analytical and experimental procedures are adopted [15P].

A variety of measurement techniques have been applied in experimental investigations of plasma processes. The plasma diagnostic applications of a pulsed, watt-level, low-pressure far-IR laser have been discussed [21P]. An improved circuit for the measurement of the electron energy distribution in a plasma makes it possible to simultaneously eliminate the fluctuation of plasma space potential, the deviation of dc probe voltage, and the attenuation of the second derivative signal during measurements [34P]. Using microwave techniques, a fast-scanning heterodyne receiver has been developed that scans 60–90 GHz every 10 ms, and the receiver was interfaced to a computer for completely automated calibrated temperature measurements of a high-temperature plasma [19P]. A four-framing Mach–Zehnder interferometer was used to observe the behavior of a plasma in a plasma-focus discharge [25P]. A modified foil-transmission technique facilitates the measurement of the temperature and the fractions of thermal and superthermal electrons in laser-produced plasmas [20P]. A feedback stabilization technique has been described for a fractional fringe interferometer for measuring plasma electron densities [43P].

There was an increase in the number of papers that described procedures and instruments for the study of two-phase flows. A radio frequency probe can be used to measure local void fractions and interface velocities in a gas–liquid two-phase flow [1P]. The principles of ultrasonic thermometry were used to detect and locate critical heat flux in a tube bundle with non-uniform axial heat generation [8P]. A technique based on scattering and transmission of fast neutron beams provides good sensitivity for the measurement of void fractions and phase distributions in transient flow boiling studies [7P]. A modified Gardon-type heat flux meter can be used effectively to provide reproducible measurements of the entire boiling curve during quenching experiments [46P]. The use of electrodiffusion measurement techniques was proven effective in the study of nucleate boiling heat transfer [12P]. Optical methods for the measurement of temperature in high-temperature flows of gas–solid suspensions were found to be sensitive to the concentration of solids but not sensitive to particle size [3P]. Direct measurements of bubble sizes in a fluidized bed can be obtained with an optical probe which uses light-emitting diodes as light sources and photocells as detectors [18P].

The use of universal extension leads for high-temperature thermocouples has been described [29P]. The influence of conduction between the sensor and its supports on the response of a resistance thermometer used to measure temperature fluctuations has been evaluated [37P]. An automated hybrid digital–analog system for processing the data of transient thermal experiments was examined [4P].

HEAT TRANSFER APPLICATIONS

Heat exchangers and heat pipes

Heat transfer and pressure drop was measured for inline and staggered tube bundles in cross flow [29Q] for Reynolds numbers between 5×10^4 and 2×10^6 and for Prandtl numbers between 3 and 7. Average heat transfer and pressure drop data for the same tube and flow arrangement were obtained [28Q] for high Prandtl numbers (between 100 and 14 000) and small Reynolds numbers (from 1 to 2×10^4). The velocity and temperature field as well as the turbulence were also measured [27Q] for $Pr = 2540$ and $Re = 500$. Equations for a single cylinder were adopted [4Q] to describe heat transfer in single rows and in bundles of tubes for transverse flow by addition of proper correction factors so that agreement with published data is obtained. An approximate calculation procedure [9Q] for axial coolant flow through rod bundles considers the space as interconnected cells. Convective heat transfer from an inline tube bundle to flow inclined towards the tube axis has been reported [13Q]. Equations describing the resistances of finned tube bundles were developed [11Q] using extensive available data. A generalized procedure [23Q] for the optimization of wavy condensing surfaces extends the work of Zener and Lavi.

Limitations on the heat exchanger performance imposed by the second law of thermodynamics are discussed [24Q] using the effectiveness-NTU method. A method to determine the relation of heat flux per transfer area or volume of a heat exchanger to the required power consumption to move the fluid is presented in [26Q]. An approximate equation [17Q] describes the mean temperature difference for eight air-cooled cross-flow heat exchanger arrangements using the effectiveness of the process stream and NTU of the air side. Heat transfer, pressure drop, and performance relations have been obtained for inline, staggered, and continuous plate heat exchangers [20Q] by numerical solutions of the flow and heat-transfer equations. The results are found to depend on one dimensionless parameter. Under most conditions, but not always, staggered arrays were found to have better performance. An improved procedure [5Q] is described which corrects the equations for mean transfer coefficients on the shell side of shell and tube heat exchangers with segmented baffles with transverse flow through the tube bank for leakage, bypass, and irregular flow pattern. A computer calculation [12Q] optimizes the fin side design of compact tube-fin heat exchangers with rippled fins for Reynolds numbers lower than 1200. Equations are developed [14Q] for the calculation of high performance finned tube heat exchangers for flue gases. Work on two-phase flow in shell and tube heat exchangers (condensers, boilers) at NEL, UK is reviewed [6Q] and correlations are obtained for the pressure drop.

The model of a thick-walled recuperator in continuous operation is used [25Q] as an analog for the transient performance of thermal regenerators. An analytic solution to the electric analog simulation of a regenerative heat exchanger with time varying fluid inlet temperature [16Q] considers discrete space steps but continuous time. Good agreement with Hausen's simplified method is obtained. A computer analysis [18Q] of the regenerator with periodic heat storage is presented. A finite difference calculation [8Q] considers combined heat and mass transfer in regenerators with hygroscopic material. Sulphur attack was found to be detrimental to rotary ceramic heat exchangers [2Q].

An analysis [15Q] considers turbulent flow and pressure losses behind V-shaped oblique high resistance heat exchangers.

An analysis [22Q] of steady, laminar, incompressible flow in a flat plate heat pipe with adiabatic top plate agrees well with experiments on a porous plate model. Experiments [1Q] with gravity-assisted copper–water heat pipe showed that the heat-transfer capability is limited by the onset of film boiling in the evaporator zone. The heat-transfer characteristics of 35 wicks were investigated by experiments in the heating zone [21Q]. For low temperature heat pipes, no boiling occurred inside the wick at high heat flux densities. The Marangoni effect is proposed [3Q] to explain capacity degradation in axially grooved heat pipes when used with a control gas. Analytic predictions for a liquid-trap heat pipe using ammonia [7Q] agree well with experiments. Formation of noncondensible gas was observed in commercially available helium and stainless steel heat pipes filled with methanol or ammonia after 11 000 h [10Q]. Variable conductance heat pipe radiators for spacecraft maintained temperatures between 30 and 40°C at a temperature difference within 5°C [19Q].

Aircraft and space vehicles

An analysis of thermal louvers used for active temperature control in spacecrafts provides a description of the thermal features of non-isothermal louvered panels and yields convenient procedures for evaluating thermal gradients and heat-transfer rates [3R]. An analysis aimed towards optimization of movable louvers yields closed-form expressions for calculating effective emittance, solar rejection capability, and base surface temperature required for active thermal control. Comparisons with experimental results obtained for thermal control louvers used on the ATS-F & G spacecraft show good agreement [1R].

Studies of temperature sensitivity and erosive burning in solar rocket motors show that for typical composite propellants erosive burning increases both temperature sensitivity and pressure exponent [2R]. A highly simplified mathematical model is proposed for a numerical solution of the non-steady temperature distribution in solid propellant rocket motors [4R].

Laser-heated thrusters require protective cooling. It appears that the thruster walls can be protected by a combination of mass injection into the boundary layer and forced convection water cooling [5R].

General

Experiments [26S] on a *turbine* nozzle cascade established that the turbulence intensity of the approach flow (0.52–12%) may double the Nusselt number. Thin film gages were used [7S] to measure heat transfer to a gas turbine in a helium driven shock tube. An investigation was performed [25S] of heat transfer to turbine blades under transient operational conditions.

The heat-transfer resistance of slag and ash deposits on the panel walls of a *steam boiler* was measured [11S] with a water-cooled heat flow meter. Spectral absorptance and emissivity of ash and slag deposits are reported [13S, 14S]. The radiation heat transfer was measured [1S] on the tubes in the furnace exit of steam boilers including its circumferential distribution. A method of predicting the temperatures in the superheater of a drum boiler was discussed [22S] for the condition of accelerated start.

A simple model is proposed [16S] which predicts the effects of flow maldistribution on radial flow fixed bed *reactors*. Models describing the fuel-to-cladding heat-transfer coefficients in a reactor fuel element are reviewed [12S] and a new model is proposed.

A technique is presented [20S] which measures heat transfer in the cylinder of large *diesel engines*. Heat-transfer measurements were also performed during a single stroke piston compression [9S] and results of such measurements in a low RPM marine diesel engine have been reported [19S].

The following equation has been developed [18S] to describe natural convection heat transfer in horizontal annular cavities containing air

$$Nu_\delta = (0.181\,(r_o/r_i) - 0.215)\,Gr_{di}^{0.25}.$$

The ratio between the outer radius r_o and the inner radius r_i varied between 1.3 and 7.5. The Grashof number Gr varied between 3×10^3 and 10^8. The subscript δ stands for the difference in the two radii and the subscript di on the Grashof number indicates that this characteristic length is used. The heat transfer was minimized when the inner cylinder was moved somewhat upward from the concentric position. An internal layer of small permeability is proposed to improve the stability of the flow in evaporative cooling of a porous heat producing element [24S]. A new method [2S] is useful for the optimization of thermal insulation systems by minimizing thermodynamic irreversibility. Design data were established [23S] from experiments studying heat transfer from argon to helium II. A maximum heat flux of 1.7 W/cm^2 was observed at 1.91 K. The data should be useful for the design of superconducting magnetic energy storage. Temperature fields were studied [5S] in hybrid integrated circuits.

Computer calculations [3S] describe soil freezing by seasonally operating refrigerators. An idealized model is used [10S] to arrive at conclusions consistent with current practice on the response of building components to heating by a fire. Published data are correlated [15S] which describe heat transfer in agitated vessels containing single-phase Newtonian or non-Newtonian fluids.

Analysis and numerical solutions [8S] agree with plant observation on heat transfer in a ladle of molten steel during pouring. Thermocouples were used [6S] to measure dye temperatures during production drop forging. Equations are developed [21S] which describe convective heat transfer in rotary kilns with limestone and Ottawa sand. Air blowing out of the grinding wheel was found [17S] to effect the local heat-transfer coefficient and the temperature of the workpiece during the grinding process. A numerical solution of an alloy solidification problem was presented [4S].

Solar energy

Solar energy continued to be an active applications area. Topics of major interest among the heat-transfer related solar energy publications include: solar radiation; thermal properties of materials used in solar equipment; active collectors; passive collection techniques; and system performance.

Analytical expressions are presented which can be used to accurately calculate the direct and diffuse solar flux components using only the long-term daily-average total solar radiation on a horizontal surface [8T]. An atmospheric model has also been presented for computing direct and diffuse solar radiation [3T]. In a universal model, the direct solar transmittance of the sky is expressed in terms of fundamental measurable quantities [19T].

Studies of the radiation properties of materials used in solar collectors were reported in a number of papers. Calculations were used to demonstrate that in the 0.32–206 μm wavelength range, commonly used values of the reflectance of glass can be in error as much as 50% if the incident radiation is diffuse [16T]. Results are presented graphically for the transmittance of isotropic and non-isotropic diffuse radiation through one-, two-, and three-glass plates as a function of the position of the sun relative to the plate surface [23T]. Detailed calculations reaffirmed that the transmission of diffuse solar flux through a multi-plate glass cover configuration is accurately estimated using the transmittance of the cover system at an effective angle of incidence of 60° [21T]. A method of selected ordinates facilitates rapid determination of solar transmittance, absorptance, or reflectance with improved accuracy [38T]. An electrochemical conversion technique has been developed to deposit selective black nickel coatings (solar absorptance 0.94 and IR emittance 0.08) on galvanized iron and zincated or zinc-electroplated aluminum surfaces [15T]. An empirical relation was developed for the solar transmittance of a partially-absorbing mixture of water and Direct Black EX dye as a function of dye concentration and mixture depth [17T]. Suspensions of solid particles, of the appropriate diameter, in gases have been shown to have selective absorption properties [1T].

The performance of active solar collectors continues to be the dominant topic in the heat-transfer related solar energy literature. Based on the utilizability concept of Hottel, Whillier, Liu and Jordan, an analytical model has been developed to predict the long-term average energy delivered by almost any solar collector [9T, 10T]. Analysis has shown that steady-state salt-gradient solar ponds can be more efficient than common flat-plate collectors when the ratio of the difference between inlet and ambient temperatures to the incident solar flux is large [20T].

Flat-plate collectors were investigated both analytically and experimentally. The efficiency of a flat-plate collector whose absorber plate was fabricated from corrugated steel sheets was found to be somewhat lower than the efficiency of collectors fabricated with the more standard tube-in-sheet absorber plate [30T]. An examination of the radiant exchange for a collector with a tube-in-sheet absorber plate revealed that maximum collector efficiencies occur when the ratio of tube spacing to tube diameter is between 8 and 20 [32T]. The heat loss from a flat plate placed inside a partially-evacuated glass tube is approximately equal to that of a conventional flat-plate collector [25T]. Analysis shows that neither the use of honeycomb structures placed in the air gap between the absorber and cover plate to suppress free convection nor the evacuation of the air gap to eliminate conduction and free convection is as effective a method of improving collector performance as the use of selective surfaces [31T]. Neglecting axial conduction in the analysis of flat-plate collectors can result, for certain conditions, in over-predicting the heat transfer by up to 30% [22T]. The efficiency of a collector with the heat-transfer fluid undergoing phase change has been shown to increase linearly with the liquid level in the collector [33T]. Experimental results show that the wind-related heat-transfer coefficient on a flat-plate collector is greatest adjacent to the edges of the plate [34T].

A variety of concentrating solar collectors have been investigated. Vee-Trough/Vacuum Tube solar collectors were found to have efficiencies of 35–40% at an operating temperature of about 175°C [27T]. A permanently mounted (no-tilt adjustment) grooved collector appears to provide little benefit to the performance of a conventional flat-plate collector placed at the base of a groove [2T]. A high-temperature collector with a non-focusing Fresnel lens and a secondary concentrator has an efficiency of 48% for a solar flux level of 850 W/m^2 when the average fluid temperature is 200°C above the ambient temperature [7T]. Moderately-focusing heat-pipe solar collectors were found to have efficiencies comparable to typical flat-plate collectors [26T]. The effects of thermal optical design parameters on the performance of representative point-focusing solar power plants were assessed [36T]. Total heat losses from annular receiver geometries can be reduced up to 50% by varying the receiver dimensions and the gas pressure or composition in the annular space [24T].

Five papers reported studies on passive solar energy topics. A computer program for analyzing the performance of direct gain passive solar-heated enclosures accurately predicts the conditions of a test cell [35T]. Less than 10% difference was found between estimates of the solar fraction provided by a passive water–wall solar system when analyzed using the *f*-chart and solar load ratio models [11T]. The performance of a passive solar heating system using a partially-transparent thermal storage wall was found to be approximately equal to direct gain or Trombe wall systems [13T]. A theoretical study of laminar free convection in one-dimensional solar-induced flows, similar to that which exists in many passive solar collectors, has been conducted [5T]. Experimental results were reported for a typical Israeli water heating system with thermosyphonic flow [29T].

A number of models for predicting the performance of active solar collection systems were proposed. A climatic design method has been presented which uses the long-term outdoor temperature distribution as well as the utilizability of solar radiation to predict the solar fraction for building heating [6T]. An algorithm was derived for choosing the insulation levels and the solar collection areas which will minimize the overall cost of constructing and heating a building [4T]. A Markov model approach to the generalized solar energy space heating performance analysis problem requires approximately 5% of the computer time required by other dynamic simulation approaches of comparable accuracy [18T]. An estimation of the fraction of a building heating load supplied by solar energy is made by approximating the heat storage, solar flux, temperature, and hot water demands by sinusoids and then solving the heat-transfer differential equations in closed form [12T].

An exact solutions approach for the analysis of a packed-bed thermal storage system appears to be more effective than the use of finite difference techniques [37T]. The use of thermal stratification in the sensible heat storage unit of a residential solar heating or cooling system can improve the overall system performance by up to 15% [28T]. The effect of buoyancy is greater on the local heat transfer than on the local mass transfer in a solar generator [14T].

PLASMA HEAT TRANSFER

Increased activities in the area of gas blast or vacuum arcs for circuit interruption and plasma processing are reflected by a relatively large number of papers in this section.

Measurements in an orifice-type air-blast model circuit breaker indicate that cooling depends much less on turbulence during the current zero process than predicted by turbulent theory [12U]. Studies of the flow situation in arc gas blast nozzles show that arc heating moves the sonic point a considerable distance downstream from the nozzle throat which may restrict the entire flow outside of the arc core to subsonic values [28U]. Diagnostic results for the thermal region surrounding the electrically conducting core of a gas-blast arc in a circuit-breaker model configuration should enable empirical laws to be derived as the basis for a boundary layer analysis of gas-blast arcs [31U].

A simplified arc model based on the integral method provides an adequate description of flow dominated arcs in supersonic nozzles [9U]. A boundary layer integral method applied to a DC arc burning in a convergent–divergent nozzle of a gas-blast circuit breaker shows that for a given nozzle geometry the arc model predicts the current density at the nozzle throat where the mass flow approaches zero [6U]. Measured radial temperature profiles and electric fields in high-current (1–17kA), supersonic ($M = 1.5$) SF_6 interrupter arcs are used to evaluate axial and radial components of the energy balance. Axial losses may be as high as 75% of the input. A significant fraction of the input at 17 kA is lost by optically thin impurity (Cu) line radiation [2U]. A short magnetically driven arc in air produces an abrupt change in refractive index in the gas preceding the luminous region of the subsonic arc. The refractive index change is due to a front of high-pressure air sustained primarily by molecular heat conduction from the arc core [21U].

Although the cathode plays an essential role in the development of breakdown, the performance of a vacuum gap in terms of switching speed is much more affected by the material of the anode than that of the cathode[5U]. An improved model for anode spot formation in vacuum arcs shows that both magnetic constriction and anode evaporation must be taken into account [25U]. Measurements of the thermal balance of vacuum arc cathodes are in agreement with a previous theory for currents exceeding 100A [35U]. Erosion structures on cathodes arced in vacuum with currents up to a few hundred amps depend upon local conditions of energy input. The basic erosion pattern consists of craters ranging in diameter from less than 1 μm up to several μm [7U]. The application of a magnetic field of 0.35 T to a vacuum arc at 500A using Cu electrodes leads to an increased energy dissipation in the arc. The energy input to both electrodes increases with magnetic field strength, and the input to the anode is by a factor of 1.7 to 1.9 greater than the energy collected by the cathode [11U]. The formation of "hot spots" in a vacuum spark may be caused by electron beam heating of the plasma. It appears that the plasma pressure is $\geq 10^7$ atm and that the "hot spots" require heating rates $\geq 10^{15}$ W/cm^3 [23U]. Measurements in wall-stablilized arcs with superimposed axial flow demonstrate that the electric field strength increases with mass flow rate as turbulent effects become more pronounced [15U]. Employing a simple theory for the free-burning arc, predictions for arcs in air for currents ranging from 1 to 20 000A are in fair agreement with experiments for a position of 1 cm from the lower electrode [20U]. A simple model of metal–halide arcs has been used to assess the effects of adding different elements and mixtures of elements on the temperature profile and other properties of metal halide lamps [8U]. The improved color of the pulsed high-pressure sodium arc is due to the development of above steady-state plasma temperatures causing enhanced excitation of higher energy levels of sodium [18U]. The charged particle density in a cylindrical plasma boundary layer changes rather abruptly from "frozen" chemistry to thermochemical equilibrium in the neighborhood of the flow separation point [29U].

Studies of the anode region of high intensity arcs show that the electron enthalpy transport is the dominating heat transfer mechanism. The corresponding enthalpy flux may reach extremely high values if arc contraction occurs in front of the anode [19U]. Electrode erosion is one of the major reasons for the failure of high current sparkgaps used for switching. A new reliable sparkgap [130 kA, 0.7C] has been developed with electrodes consisting of a special type of graphite [1U]. Initial anode heat fluxes in pulsed high-current arcs may be an order of magnitude higher than the following quasi-steady heat fluxes [17U]. An arc operated in a flash tube at currents from 1 to 3 kA and with pulse lengths of approximately 1 ms interacts strongly with the enclosing tube. Oscillations of the luminous boundary and dark contractions of the arc channel have been observed [27U].

Calculations of the contribution of lines to the total emission coefficient of a thermal xenon plasma at temperatures between 10^4 and 1.4×10^4 K and pressures between 200 and 4000 kPa show the best agreement with measurements when a previously derived line factor is used [13U]. Radiation emitted from the axis of a wall-stablilized, low-current cylindrical arc in pure oxygen at temperatures around 9000 K and wavelengths from 2000 to 9000 Å consists primarily of contributions due to the affinity of the negative oxygen ions, the Kramers–Unsöld continuum and O_2^+ – association radiation [16U].

The gasdynamic and thermal parameters of non-isothermal, conventional plasma jets, both free and interacting with a baffle are studied experimentally. A semi-empirical relation derived for the stagnation-point heat flux on the baffle compares favorably with experimental results [4U]. An axisymmetric, supersonic ($M_2 = 2.32$), super-expanded, high temperature ($T \approx 3000$ K) gas jet impinges on permeable or impermeable surfaces arranged normal or inclined to the axis of the jet. If the heat-transfer intensifying effect of compression shocks is not taken into account, the error in calculating heat transfer may be as high as a factor of three [32U].

A numerical technique is described for determining the temperature distribution of a conical cathode in a MPD arc configuration [22U]. The glow-to-arc transition in a positive point-to-plane discharge can be explained by the Coulomb interaction heating process which arises whenever the collision frequency between electrons becomes greater than that between electrons and neutrals [3U]. A corona discharge gives rise to a large increase in heat transfer at low flow velocities and this gain in heat transfer decreases to zero at high stream velocities [30U].

Measured heat-transfer coefficients to stationary spheres of molybdenum (2–5.6 mm in diameter) in a confined argon plasma jet are in reasonable agreement with previous theoretical predictions [24U]. Numerical results for the heat transfer and phase change of spherical alumina particles heated in an argon–hydrogen plasma jet are in good agreement with measurements of surface temperature, particle velocity, and diameter [10U]. An experimental study of heat, mass, and charge transfer with chemical reactions on metal surfaces during transient heating with an argon plasma jet containing 4% oxygen, demonstrate that chemical reactions affect the heating of the metal, the convective heat transfer, and the transport of energy by radiation [33U].

Deviations from kinetic equilibrium (Te > Tg) in an argon plasma jet are diminished by mixing with nitrogen. The difference between electron and atom temperatures decreases with vaporization of metal particles injected into the plasma jet [34U]. Metallic particles with mean diameters as small as 0.005 mm may be produced by plasma jet heating of a porous graphite cylinder plated with the material from which the particles are generated [26U]. A multiple arc system capable of producing a large volume discharge may be useful for chemical and material processing in plasmas [14U].

REFERENCES

Conduction

1A. R. M. Abdel-Wahed, J. W. Ramsey and E. M. Sparrow, *Int. J. Heat Mass Transfer* **22,** 171 (1979).
2A. A Acker, *Z. Angew Math. Phys.* **29,** 840 (1978).
3A. G. Ahamdi, *J. Appl. Mech.* **34,** 933 (1978).
4A. F. R. Al-Astrabadi, S. D. Probert, P. W. O'Callaghan and A. M. Jones, *Appl. Energy* **5,** 23 (1979).
5A. O. M. Alifanov, Ye. A. Artyukhin and B. M. Pankratov, *Heat Transfer, Soviet Res.* **10**(3), 63 (1978).
6A. N. Arai and K. Karashima, *AIAA Jl* **17,** 191 (1979).
7A. R. G. Arledge and A. Haji-Sheikh, *Numerical Heat Transfer* **1,** 365 (1978).
8A. A. Aziz, *Letters Heat Mass Transfer* **6,** 199 (1979).
9A. D. K. Babu and M. Th. van Genuchten, *Q. Appl. Math.* **37,** 11 (1979).
10A. A. G. Bathelt, R. Viskanta and W. Leidenfrost, *J. Heat Transfer* **101,** 453 (1979).
11A. J. V. Beck, *J. Heat Transfer* **101,** 132 (1979).
12A. G. E. Bell, *Int. J. Heat Mass Transfer* **22,** 1681 (1979).
13A. J. E. Bernard and T. M. Post, *Numerical Heat Transfer* **1,** 517 (1978).
14A. J. Bransier, *Int. J. Heat Mass Transfer* **22,** 875 (1979).
15A. L. C. Burmeister, *J. Heat Transfer* **101,** 562 (1979).
16A. A. W. Bush and R. D. Gibson, *Appl. Energy* **5,** 11 (1979).
17A. Y. P. Chang and K. C. Poon, *J. Heat Transfer* **101,** 548 (1979).
18A. J. Crank and A. B. Crowley, *Int. J. Heat Mass Transfer* **22,** 1331 (1979).
19A. M. S. El-Genk and A. W. Cronenberg, *Letters Heat Mass Transfer* **6,** 321 (1979).
20A. M. S. El-Genk and A. W. Cronenberg, *Int. J. Heat Mass Transfer* **22,** 1719 (1979).
21A. M. S. El-Genk and A. W. Cronenberg, *Int. J. Heat Mass Transfer* **22,** 167 (1979).
22A. T. M. El-Gindy and P. Townsend, *Int. J. Num. Meth. Engng* **14,** 227 (1979).
23A. A. F. Emery, K. Suginara and A. T. Jones, *Numerical Heat Transfer* **2,** 97 (1979).
24A. R. G. Eslinger and B. T. F. Chung, *AIAA Jl* **17,** 1134 (1979).
25A. L. Feijoo, H. T. Davis and D. Ramkrishna, *J. Heat Transfer* **101,** 137 (1979).
26A. C. R. Gane and P. L. Stephenson, *Int. J. Num. Meth. Engng* **14,** 1141 (1979).
27A. S. C. Gupta and A. K. Lahiri, *Int. J. Engng Sci.* **17,** 401 (1979).
28A. B. Gold and G. Horvay, *J. Appl. Mech.* **46,** 557 (1979).
29A. K. Hawlitschek, *Z. Angew Math. Phys.* **29,** 777 (1978).
30A. R. G. Hills and G. P. Mulholland, *Int. J. Heat Mass Transfer* **22,** 1221 (1979).
31A. M. Imber, *AIAA Jl* **17,** 91 (1979).
32A. M. Imber, *AIAA Jl* **17,** 204 (1979).
33A. H. M. Khalil, *Numerical Heat Transfer* **1,** 377 (1978).
34A. P. M. Kolesnikov, L. V. Grishanova, V. N. Abrashin,

L. N. Degtereva, N. S. Kolesnikova and V. A. Tsurko, *Heat Transfer, Soviet Res.* **10**(2), 33 (1978).
35A. G. W. Krutz, R. J. Schoenhals and P. S. Hore, *Numerical Heat Transfer* **1**, 489 (1978).
36A. P. A. A. Laura and G. Sanchez Sarmiento, *Int. J. Heat Mass Transfer* **22**, 341 (1979).
37A. G. Lebon and Ph. Mathieu, *Int. J. Heat Mass Transfer* **22**, 1187 (1979).
38A. S. H. Lin, *Int. J. Num. Meth. Engng* **14**, 1726 (1979).
39A. S. H. Lin, *Int. J. Engng Sci.* **17**, 373 (1979).
40A. K. A. Lindsay and B. Straughan, *Z. Angew. Math. Phys.* **30**, 477 (1979).
41A. R. C. Mehta, *Int. J. Heat Mass Transfer* **22**, 1149 (1979).
42A. R. A. Meric, *Int. J. Heat Mass Transfer* **22**, 1347 (1979).
43A. R. A. Meric, *Int. J. Num. Meth. Engng* **14**, 1851 (1979).
44A. G. H. Meyer, *Numerical Heat Transfer* **1**, 351 (1978).
45A. J. O. Mingle, *Numerical Heat Transfer* **2**, 387 (1979).
46A. E. Muchowski, *Wärme- und Stoffübertragung* **12**, 161 (1979).
47A. S. Orivuori, *Int. J. Num. Meth. Engng* **14**, 1461 (1979).
48A. M. N. Özişik and J. C. Uzzell, Jr., *J. Heat Transfer* **101**, 331 (1979).
49A. P. Payvar, *Q. Jl Mech. Appl. Math.* **32**, 253 (1979).
50A. K. C. Poon and Y. P. Chang, *Letters Heat Mass Transfer* **6**, 293 (1979).
51A. K. C. Poon, R. C. H. Tsou and Y. P. Chang, *J. Heat Transfer* **101**, 340 (1979).
52A. A. S. Popel, *J. Heat Transfer* **101**, 560 (1979).
53A. A. A. Poznyak, *Heat Transfer, Soviet Res.* **10**(2), 117 (1978).
54A. A. Prasad, *J. Spacecraft Rockets* **16**, 445 (1979).
55A. D. Ramkrishna and N. R. Amundson, *Chem. Engng Sci.* **34**, 301 (1979).
56A. D. Ramkrishna and N. R. Amundson, *Chem. Engng Sci.* **34**, 309 (1979).
57A. J. W. Ramsey, E. M. Sparrow and L. M. C. Varejao, *J. Heat Transfer* **101**, 732 (1979).
58A. P. Razelos, *Wärme- und Stoffübertragung* **12**, 113 (1979).
59A. S. R. Robertson, *Numerical Heat Transfer* **2**, 61 (1979).
60A. J. C. W. Rogers, A. E. Berger and M. Ciment, *Siam J. Num. Analysis* **16**, 563 (1979).
61A. B. Rubinsky and E. G. Cravalho, *J. Heat Transfer* **101**, 326 (1979).
62A. P. Satyamurthy, R. K. Marwah, N. Venkatramani and V. K. Rohatgi, *Int. J. Heat Mass Transfer* **22**, 1151 (1979).
63A. I. P. Schisler and J. V. Beck, *Letters Heat Mass Transfer* **6**, 181 (1979).
64A. V. Seeniraj and M. Arunachalam, *Int. J. Heat Mass Transfer* **22**, 1455 (1979).
65A. N. Shamsundar and R. Srinivasan, *J. Heat Transfer* **101**, 585 (1979).
66A. S. Sieniutycz, *Int. J. Heat Mass Transfer* **22**, 585 (1979).
67A. A. P. Slesarenko, *Heat Transfer, Soviet Res.* **10**(3), 55 (1978).
68A. M. S. Sodha, I. C. Goyal, S. C. Kaushik, G. N. Tiwari, A. K. Seth and M. A. S. Malik, *Int. J. Heat Mass Transfer* **22**, 777 (1979).
69A. A. D. Solomon, *Solar Energy* **22**, 251 (1979).
70A. A. D. Solomon, *Letters Heat Mass Transfer* **6**, 189 (1979).
71A. E. M. Sparrow and C. A. C. Altemani, *Numerical Heat Transfer* **2**, 129 (1979).
72A. E. M. Sparrow, J. W. Ramsey and R. G. Kemink, *J. Heat Transfer* **101**, 578 (1979).
73A. B. Ström, *Int. J. Heat Mass Transfer* **22**, 1375 (1979).
74A. W. M. Suen, S. P. Wong and K. Young, *J. Phys. D: Appl. Phys.* **12**, 1325 (1979).
75A. L. N. Tao, *Z. Angew. Math. Phys.* **30**, 416 (1979).
76A. L. N. Tao, *J. Appl. Mech.* **46**, 789 (1979).
77A. L. N. Tao, *Q. Appl. Math.* **37**, 1 (1979).
78A. L. N. Tao, *Q. Jl Mech. Appl. Math.* **32**, 175 (1979).
79A. A. G. Temkin, *Heat Transfer, Soviet Res.* **10**(2), 20 (1978).
80A. T. N. Veziroglu, A. Williams, S. Kakac and P. Nayak, *Int. J. Heat Mass Transfer* **22**, 447 (1979).
81A. J. J. Vullierme, J. J. Lagarde and H. Cordier, *Int. J. Heat Mass Transfer* **22**, 1209 (1979).
82A. P. B. Wells and E. A. Bathke, *J. Energy* **3**, 227 (1979).
83A. D. A. Welsey, *J. Heat Transfer* **101**, 346 (1979).
84A. L. T. Yeh and B. T. F. Chung, *J. Heat Transfer* **101**, 592 (1979).

Channel flow

1B. A. M. Abdelmeguid and D. B. Spalding, *J. Fluid Mech.* **94**, 383 (1979).
2B. F. P. Berger and K. F., F. L. Hau, *Int. J. Heat Mass Transfer* **22**, 1645 (1979).
3B. S. Bundikul and W. -J. Yang, *J. Heat Transfer* **101**, 217 (1979).
4B. A. Campo and J.-C. Auguste, *Int. J. Heat Mass Transfer* **21**, 1597 (1978).
5B. A. Campo and T. Yoshimura, *Int. J. Heat Mass Transfer* **22**, 5 (1979).
6B. S. Y. Chern and J. C. Chato, *Numerical Heat Transfer* **1**, 453 (1978).
7B. C. C. Chieng and B. E. Launder, *Numerical Heat Transfer* **2**, 359 (1979).
8B. N. Cur and E. M. Sparrow, *J. Heat Transfer* **101**, 211 (1979).
9B. D. E. Daney, P. R. Ludtke and M. C. Jones, *J. Heat Transfer* **101**, 9 (1979).
10B. S. Del Giudice, *Numerical Heat Transfer* **2**, 291 (1979).
11B. K. V. Dement'eva and I. I. Telegina, *Thermal Engng* **26**, 38 (1979).
12B. N. T. Dunwoody and T. A. Hamill, *Z. Angew. Math. Phys.* **30**, 587 (1979).
13B. N. M. Galin, *Thermal Engng.* **26**, 247 (1979).
14B. N. M. Galin and A. I. Groshev, *Thermal Engng* **26**, 263 (1979).
15B. P. P. Grassmann and M. Tuma, *Int. J. Heat Mass Transfer* **22**, 799 (1979).
16B. P. Grassmann and M. Tuma, *Wärme- und Stoffübertragung* **12**, 203 (1979).
17B. M. Hishida and Y. Nagano, *J. Heat Transfer* **101**, 15 (1979).
18B. M. Hudina, *Int. J. Heat Mass Transfer* **22**, 1381

(1979).
19B. R. M. Kezhyalis, P. Yu. Valatkyavichyus and A. B. Ambrazyavichyus, *Int. Chem. Engng* **19,** 281 (1979).
20B. G. R. Knowles and E. M. Sparrow, *J. Heat Transfer* **101,** 635 (1979).
21B. S. H. Lin, *Int. J. Heat Mass Transfer* **22,** 1117 (1979).
22B. J. H. Masliyah and K. Nandakumar, *A.I.Ch.E. Jl* **25,** 478 (1979).
23B. F. R. McLarnon, R. H. Muller and C. W. Tobias, *I/EC Fundamentals* **18,** 97 (1979).
24B. D. Moalem-Maron, S. Sideman and H. Horn, *Chem. Engng Sci.* **34,** 420 (1979).
25B. T. Y. Na and J. P. Chiou, *Wärme- und Stoffübertragung* **12,** 55 (1979).
26B. H. Nagel, *Wärme- und Stoffübertragung* **12,** 89 (1979).
27B. E. B. Nauman, *A.I.Ch.E. Jl* **25,** 246 (1979).
28B. M. A. Nemira, J. V. Vilemas and M. Simonis, *Heat Transfer, Soviet Res.* **11**(1), 101 (1979).
29B. M. A. Nemira, J. V. Vilemas and V. M. Simonis, *Heat Transfer, Soviet Res.* **11**(1), 112 (1979).
30B. H. Ockendon, *J. Fluid Mech.* **93,** 737 (1979).
31B. D. R. Oliver and S. S. Rao, *Trans. Inst. Chem. Engrs.* **57,** 104 (1979).
32B. S. V. Patankar, M. Ivanovic and E. M. Sparrow, *J. Heat Transfer* **101,** 29 (1979).
33B. P. E. Pickett, M. F. Taylor and D. M. McEligot, *Int. J. Heat Mass Transfer* **22,** 705 (1979).
34B. I. L. Povkh, A. B. Stupin, S. N. Maksyutenko, P. V. Aslanov, A. P. Simonenko and A. T. Musienko, *Heat Transfer, Soviet Res.* **10,** (1), 98 (1978).
35B. A. A. Pyadishyus, I. R. Kazlauskas and A. A. Zhukauskas, *Int. Chem. Engng.* **19,** 714 (1979).
36B. N. J. Rabadi, J. C. F. Chow and H. A. Simon, *Numerical Heat Transfer* **2,** 279 (1979).
37B. M. R. Remorino, R. D. Tonini and U. Bohm, *A.I.Ch.E. Jl* **25,** 368 (1979).
38B. S. M. Richardson, *Int. J. Heat Mass Transfer* **22,** 1417 (1979).
39B. E. M. Sparrow and R. G. Kemink, *Int. J. Heat Mass Transfer* **22,** 909 (1979).
40B. E. M. Sparrow and R. G. Kemink, *J. Heat Transfer* **101,** 23 (1979).
41B. M. Strada, S. Del Giudice and G. Comini, *Numerical Heat Transfer* **1,** 471 (1978).
42B. J. T. Teng, R. Greif, I. Cornet and R. N. Smith, *Int. J. Heat Mass Transfer* **22,** 493 (1979).
43B. A. T. Wassel and A. F. Mills, *J. Heat Transfer* **101,** 469 (1979).
44B. S. -C. Yao, *J. Heat Transfer* **101,** 480 (1979).
45B. V. A. Yoselevich and V. N. Pilipenko, *Heat Transfer, Soviet Res.* **10,**(1), 45 (1978).
46B. A. E. Zaryankin, E. E. Likherzak and B. V. Baranovskii, *Thermal Engng* **25,** 34 (1978).

Boundary layer and external flows

1C. G. M. Assassa and K. D. Papailiou, *J. Fluids Engng* **101,** 110 (1979).
2C. E. Beese and K. Gersten, *Z. Angew Math. Phys.* **30,** 117 (1979).
3C. R. J. Bodonyl, *Q. Jl Mech. Appl. Math.* **32,** 63 (1979).
4C. Yu. Ye. Bogolyubov and L. P. Smirnova, *Heat Transfer, Soviet Res.* **11**(1), 59 (1979).
5C. V. R. Borovskiy, V. A. Shelimanov and N. A. Sharkova, *Heat Transfer, Soviet Res.* **11**(1), 80 (1979).
6C. T. K. Bose, *Wärme- und Stoffübertragung* **12,** 211 (1979).
7C. H. Brauer, *Wärme- und Stoffübertragung* **12,** 145 (1979).
8C. F. K. Browand and B. O. Latigo, *Physics Fluids* **22,** 1011 (1979).
9C. L. W. B. Browne and R. A. Antonia, *J. Heat Transfer* **101,** 144 (1979).
10C. T. Cebeci, *Numerical Heat Transfer* **1,** 557 (1978).
11C. T. Cebeci, R. S. Hirsh and J. H. Whitelaw, *AIAA Jl* **17,** 434 (1979).
12C. T. Cebeci, K. Stewartson and P. G. Williams, *SIAM J. Appl. Math.* **36,** 190 (1979).
13C. A. Chakrabarti and A. S. Gupta, *Q. Appl. Math.* **37,** 73 (1979).
14C. B. A. Chesna, V. Yu. Survilla and E. I. Adomaitis, *Int. Chem. Engng* **19,** 276 (1979).
15C. S. C. R. Dennis and D. B. Ingham, *Physics Fluids* **22,** 1 (1979).
16C. M. -R. Drižius, S. I. Bartkus and A. A. Slančiauskas, *Heat Transfer, Soviet Res.* **10**(3), 21 (1978).
17C. M. R. Drižius, S. I. Bartkus and A. A. Slančiauskas, *Heat Transfer, Soviet Res.* **11**(1), 126 (1979).
18C. Ye. P. Dyban and A. I. Mazur, *Heat Transfer, Soviet Res.* **11**(1), 52 (1979).
19C. E. P. Dyban and A. I. Mazur, *Int. Chem. Engng* **19,** 170 (1979).
20C. O. A. Grechannyi and A. Sh. Dorfman, *Int. Chem. Engng* **19,** 708 (1979).
21C. C. Haberland and W. Nitsche, *Wärme- und Stoffübertragung* **12,** 45 (1979).
22C. G. M. Harpole, S. A. Berger and J. Aroesty, *J. Appl. Mech.* **46,** 9 (1979).
23C. T. Hirata, R. R. Gilpin and K. C. Cheng, *Int. J. Heat Mass Transfer* **22,** 1425 (1979).
24C. T. Hirata, R. R. Gilpin and K. C. Cheng, *Int. J. Heat Mass Transfer* **22,** 1435 (1979).
25C. G. H. Jirka and D. R. F. Harleman, *J. Fluid Mech.* **94,** 275 (1979).
26C. N. W. M. Ko and W. T. Chan, *J. Fluid Mech.* **93,** 549 (1979).
27C. H. Koyama, S. Masuda, I. Ariga and I. Watanabe, *J. Engng Pwr* **101,** 23 (1979).
28C. F. N. Lin and S. Y. Chern, *Int. J. Heat Mass Transfer* **22,** 1323 (1979).
29C. J. H. Masliyah and T. T. Nguyen, *Int. J. Heat Mass Transfer* **22,** 237 (1979).
30C. R. E. Mayle, M. F. Blair and F. C. Kopper, *J. Heat Transfer* **101,** 521 (1979).
31C. D. W. Metzger, L. W. Florschuetz, D. I. Takeuchi, R. D. Behee and R. A. Berry, *J. Heat Transfer* **101,** 526 (1979).
32C. A. F. Mills, *J. Heat Transfer* **101,** 734 (1979).
33C. V. Ye Nakoryakov, B. G. Pokusayev and Ye. N. Troyan, *Heat Transfer, Soviet Res.* **10**(5), 102 (1978).
34C. P. Subhadra Ramachandran, M. N. Mathur and S. K. Ojha, *Int. J. Engng Sci.* **17,** 625 (1979).
35C. J. W. Rose, *Int. J. Heat Mass Transfer* **22,** 969 (1979).

36C. J. W. Rose, *Int. J. Heat Mass Transfer* **22,** 1243 (1979).
37C. N. N. Sayegh and W. H. Gauvin, *A.I.Ch.E. Jl* **25,** 522 (1979).
38C. V. L. Sergeyev and V. P. Veselov, *Heat Transfer, Soviet Res.* **10**(1), 126 (1978).
39C. V. Sernas, *J. Heat Transfer* **101,** 176 (1979).
40C. P. M. Sforza and W. Stasi, *J. Heat Transfer* **101,** 353 (1979).
41C. A. J. Smits, S. T. B. Young and P. Bradshaw, *J. Fluid Mech.* **94,** 209 (1979).
42C. E. M. Sparrow, J. W. Ramsey and E. A. Mass, *J. Heat Transfer* **101,** 199 (1979).
43C. J. Sucec, *Int. J. Heat Mass Transfer* **22,** 771 (1979).
44C. B. Sundén, *Int. J. Heat Mass Transfer* **22,** 1125 (1979).
45C. K. Tamada, *J. Phys. Soc. Japan* **46,** 310 (1979).
46C. K. K. Tien and V. M. Sparrow, *Int. J. Heat Mass Transfer* **22,** 349 (1979).
47C. S.-Y. Tuann and M. D. Olson, *Computers & Fluids* **6,** 219 (1978).

Flow with separated regions

1D. J. K. Aggarwal and L. Talbot, *Int. J. Heat Mass Transfer* **22,** 61 (1979).
2D. S. Aiba, T. Ota and H. Tsuchida, *Wärme- und Stoffübertragung* **12,** 221 (1979).
3D. J. C. Blaise and M. Martin, *Letters Heat Mass Transfer* **6,** 449 (1979).
4D. G. I. Gimbutis, V. J. Săpola and V. J. Šimkiavičius, *Heat Transfer, Soviet Res.* **10**(5), 25 (1978).
5D. J. G. Kawall and J. F. Keffer, *Physics Fluids* **22,** 31 (1979).
6D. R. Morel, C. Rey , M. Awad, J. Mathieu and J. P. Schon, *Physics Fluids* **22,** 623 (1979).
7D. T. Ota and N. Kon, *Int. J. Heat Mass Transfer* **22,** 197 (1979).
8D. L. I. Skurin and A. V. Yurkin, *Heat Transfer, Soviet Res.* **10**(5), 95 (1978).
9D. R. Smyth, *Letters Heat Mass Transfer* **6,** 405 (1979).
10D. D. J. Tagg, M. A. Patrick and A. A. Wragg, *Trans. Inst. Chem. Engnrs* **57,** 176 (1979).
11D. A. A. Vasil'yev and P. G. Itin, *Heat Transfer, Soviet Res.* **10**(5), 33 (1978).
12D. H. Yamamoto, N. Seki and S. Fukusako, *J. Heat Transfer* **101,** 475 (1979).
13D. A. A. Žukauskas, V. S. Simanavičius and P. M. Daujotas, *Heat Transfer, Soviet Res.* **10**(6), 79 (1978).
14D. A. A. Žukauskas, I. I. Zyugzda and Y. Yu. Survila, *Heat Transfer, Soviet Res.* **10**(5), 1 (1978).

Transfer mechanisms

1E. F. Akoum and C. Klapisz, *Int. J. Heat Mass Transfer* **22,** 1147 (1979).
2E. A. Bejan, *J. Heat Transfer* **101,** 718 (1979).
3E. R. F. Blackwelder, *Physics Fluids* **22,** 583 (1979).
4E. H. Brauer, *Chem. -Ing -Tech.* **51,** 934 (1979).
5E. R. E. Britter, J. C. R. Hunt and J. C. Mumford, *J. Fluid Mech.* **92,** 269 (1979).
6E. R. A. Clark, J. H. Ferziger and W. C. Reynolds, *J. Fluid Mech.* **91,** 1 (1979).
7E. M. R. Davis and P. O. A. L. Davis, *J. Fluid Mech.* **93,** 281 (1979).
8E. T. D. Dickey and G. L. Mellor, *Physics Fluids* **22,** 1029 (1979).
9E. M. Doi and T. Imamura, *J. Phys. Soc. Japan* **46,** 1358 (1979).
10E. C. Dopazo, *Physics Fluids* **22,** 20 (1979).
11E. M. A. El-Hawary, *AIAA Jl* **17,** 303 (1979).
12E. M. M. Gibson, *Int. J. Heat Mass Transfer* **21,** 1609 (1978).
13E. K. N. Helland and M. Rosenblatt, *Physics Fluids* **22,** 819 (1979).
14E. M. Jischa and H. B. Rieke, *Int. J. Heat Mass Transfer* **22,** 1547 (1979).
15E. H. Keck, *Int. J. Heat Mass Transfer* **22,** 1513 (1979).
16E. H. Keck and W. Schneider, *Int. J. Heat Mass Transfer* **22,** 1501 (1979).
17E. C. G. Koop and F. W. Browand, *J. Fluid Mech.* **93,** 135 (1979).
18E. B. E. Launder and S. S. A. Samaraweera, *Int. J. Heat Mass Transfer* **22,** 1631 (1979).
19E. J. Lee, *Physics Fluids* **22,** 40 (1979).
20E. T. J. McDougall, *J. Fluid Mech.* **94,** 409 (1979).
21E. A. N. Nahavandi, G. D. Holder and M. A. Borhani, *Can. J. Chem. Engng* **57,** 425 (1979).
22E. Y. Okamoto, M. Nishikawa and K. Matsuda, *Int. Chem. Engng* **19,** 639 (1979).
23E. M. M. Pimenta, R. J. Moffat and W. M. Kays, *J. Heat Transfer* **101,** 193 (1979).
24E. S. Rajagopalan and R. A. Antonia, *Physics Fluids* **22,** 614 (1979).
25E. Z. P. Shulman, V. I. Kordonsky and S. A. Demchuk, *Int. J. Heat Mass Transfer* **22,** 389 (1979).
26E. E. V. Shishov, A. I. Leontiev and P. S. Roganov, *Int. J. Heat Mass Transfer* **22,** 1627 (1979).
27E. C. G. Speziale, *Physics Fluids* **22,** 1033 (1979).
28E. K. R. Sreenivasan, R. A. Antonia and D. Britz, *J. Fluid Mech.* **94,** 745 (1979).
29E. C. K. W. Tam and K. C. Chen, *J. Fluid Mech.* **92,** 303 (1979).
30E. R. Tal Thàu and B. Gal-Or, *Int. J. Heat Mass Transfer* **22,** 857 (1979).
31E. A. Yoshizawa, *J. Phys. Soc. Japan* **46,** 669 (1979).
32E. I. Wygnanski, D. Oster and H. Fiedler, *J. Fluid Mech.* **93,** 325 (1979).

Natural convection

1F. H. Akbari and T. R. Borgers, *Solar Energy* **22,** 165 (1979).
2F. M. Alamgir, *J. Heat Transfer* **101,** 174 (1979).
3F. J. C. Antoranz and M. G. Velarde, *Physics Fluids* **22,** 1038 (1979).
4F. V. S. Avduyevskiy, V. N. Kalashnik and R. M. Kopyarkevich, *Heat Transfer, Soviet Res.* **10**(5), 136 (1978).
5F. A. G. Bathelt, R. Viskanta and W. Leidenfrost, *J. Fluid Mech.* **90,** 227 (1979).
6F. M. Behnia and R. Viskanta, *Int. J. Heat Mass Transfer* **22,** 611 (1979).
7F. A. Bejan, *J. Fluid Mech.* **90,** 561 (1979).

8F. A. Bejan, *Letters Heat Mass Transfer* **6,** 93 (1979).
9F. A. Bejan and J. Imberger, *J. Heat Transfer* **101,** 417 (1979).
10F. A. Bejan and C. -L. Tien, *Int. J. Heat Mass Transfer* **22,** 919 (1979).
11F. R. F. Bergholz, M. M. Chen and F. B. Cheung, *Int. J. Heat Mass Transfer* **22,** 763 (1979).
12F. R. G. Bill, Jr. and B. Gebhart, *Int. J. Heat Mass Transfer* **22,** 267 (1979).
13F. P. Boon-Long, T. W. Lester and R. E. Faw, *Int. J. Heat Mass Transfer* **22,** 437 (1979).
14F. F. H. Busse and R. M. Clever, *J. Fluid Mech.* **91,** 319 (1979).
15F. A. M. C. Chan and S. Banerjee, *J. Heat Transfer* **101,** 233 (1979).
16F. A. M. C. Chan and S. Banerjee, *J. Heat Transfer* **101,** 114 (1979).
17F. S. S. -H. Chang, *J. Appl. Mech.* **45,** 711 (1978).
18F. M. C. Charrier-Mojtabi, A. Mojtabi and J. P. Caltagirone, *J. Heat Transfer* **101,** 171 (1979).
19F. C. C. Chen and R. Eichhorn, *J. Heat Transfer* **101,** 566 (1979).
20F. C. J. Chen and C. H. Chen, *J. Heat Transfer* **101,** 532 (1979).
21F. C. J. Chen and C. P. Nikitopoulos, *Int. J. Heat Mass Transfer* **22,** 245 (1979).
22F. T. S. Chen and A. Mucoglu, *Int. J. Heat Mass Transfer* **22,** 185 (1979).
23F. T. S. Chen and C. F. Yuh, *Numerical Heat Transfer* **2,** 233 (1979).
24F. P. Cheng and I.-D. Chang, *Letters Heat Mass Transfer* **6,** 253 (1979).
25F. P. S. Damerell and R. J. Schoenhals, *J. Heat Transfer* **101,** 672 (1979).
26F. M. Das Gupta and A. S. Gupta, *Int. J. Engng Sci.* **17,** 271 (1979).
27F. S. Del Giudice, G. Comini and M. D. Mikhailov, *Int. J. Heat Mass Transfer* **21,** 1619 (1978).
28F. M. A. Delichatsios, *J. Fluid Mech.* **93,** 241 (1979).
29F. R. A. Denton and I. R. Wood, *Int. J. Heat Mass Transfer* **22,** 1339 (1979).
30F. L. M. de Socio, *Letters Heat Mass Transfer* **6,** 375 (1979).
31F. G. de Vahl Davis, I. P. Jones and P. J. Roache, *Computers & Fluids* **7,** 315 (1979).
32F. R. W. Douglass, E. J. Shaughnessy and B. R. Munson, *J. Heat Transfer* **101,** 427 (1979).
33F. R. W. Douglass, B. R. Munson and E. J. Shaughnessy, *Int. J. Heat Mass Transfer* **21,** 1543 (1978).
34F. R. W. Douglass, B. R. Munson and E. J. Shaughnessy, *Int. J. Heat Mass Transfer* **21,** 1555 (1978).
35F. B. Dulieu, M. L. Bonniaud and J. P. Walch, *Int. J. Heat Mass Transfer* **22,** 739 (1979).
36F. D. K. Edwards, J. N. Arnold and P. S. Wu, *J. Heat Transfer* **101,** 741 (1979).
37F. V. Ye. Fertman, Uu. I. Barkov and A. R. Bayev, *Heat Transfer, Soviet Res.* **10**(6), 50 (1978).
38F. R. D. Flack, T. T. Konopnicki and J. H. Rooke, *J. Heat Transfer* **101,** 648 (1979).
39F. R. D. Flack and C. L. Witt, *J. Heat Transfer* **101,** 256 (1979).
40F. T. Fujii, M. Fujii and T. Matsunaga, *Numerical Heat Transfer* **2,** 329 (1979).
41F. L. B. Gdalevich, E. F. Nogotov and V. E. Fertman, *Int. J. Heat Mass Transfer* **22,** 1601 (1979).
42F. B. Gebhart, M. S. Bendell and H. Shaukatullah, *Int. J. Heat Mass Transfer* **22,** 137 (1979).
43F. B. Gebhart and J. C. Mollendorf, *J. Fluid Mech.* **89,** 673 (1979).
44F. W. K. George, Jr. and S. P. Capp, *Int. J. Heat Mass Transfer* **22,** 813 (1979).
45F. G. Z. Gershuni, E. M. Zhukhovitskii and S. M. Iorshina, *Appl. Math. Mech.* **42,** 310 (1979).
46F. K. P. Goyal and D. R. Kassoy, *Int. J. Heat Mass Transfer* **22,** 1577 (1979).
47F. D. D. Gray, *Int. J. Heat Mass Transfer* **22,** 1155 (1979).
48F. R. Greif, Y. Zvirin and A. Mertol, *J. Heat Transfer* **101,** 684 (1979).
49F. R. W. Griffiths, *J. Fluid Mech.* **92,** 659 (1979).
50F. A. A. Gukhman, A. A. Zaytsev and G. M. Solov'yev, *Heat Transfer, Soviet Res.* **10**(6), 23 (1978).
51F. E. Guyon, P. Pieranski and J. Salan, *J. Fluid Mech.* **93,** 65 (1979).
52F. J. T. Han, *Numerical Heat Transfer* **2,** 165 (1979).
53F. M. M. Hasan and R. Eichhorn, *J. Heat Transfer* **101,** 642 (1979).
54F. M. A. Hassab and M. N. Özişik, *Int. J. Heat Mass Transfer* **22,** 1095 (1979).
55F. C. F. Hess and C. W. Miller, *Int. J. Heat Mass Transfer* **22,** 421 (1979).
56F. E. J. Hopfinger, P. Atten and F. H. Busse, *J. Fluid Mech.* **92,** 217 (1979).
57F. D. Y. Hsieh, *Physics Fluids* **22,** 1435 (1979).
58F. C. T. Hsu and P. Cheng, *J. Heat Transfer* **101,** 660 (1979).
59F. H. E. Huppert and P. F. Linden, *J. Fluid Mech.* **95,** 431 (1979).
60F. F. P. Incropera and M. A. Yaghoubi, *J. Heat Transfer* **101,** 743 (1979).
61F. D. B. Ingham, *Z. Angew. Math. Phys.* **29,** 871 (1978).
62F. P. C. Jain and B. L. Lohar, *J. Heat Transfer* **101,** 126 (1979).
63F. Y. Jaluria, *J. Appl. Mech.* **46,** 231 (1979).
64F. I. P. Jones, *Numerical Heat Transfer* **2,** 193 (1979).
65F. E. K. Kalinin, G. A. Dreytser and A. S. Neverson, *Heat Transfer, Soviet Res.* **10**(6), 40 (1978).
66F. Y. Kamotani, S. Ostrach and H. Miao, *J. Heat Transfer* **101,** 222 (1979).
67F. M. Katagiri and I. Pop, *Wärme- und Stoffübertragung* **12,** 73 (1979).
68F. M. Katagiri and I. Pop, *Z. Angew. Math. Mech.* **59,** 51 (1979).
69F. Yu. A. Kirichenko and Zh. A. Suprunova, *Heat Transfer, Soviet Res.* **10**(6), 55 (1978).
70F. Yu. A. Kirichenko, V. N. Shchelkunov and V. N. Timon'kin, *Heat Transfer, Soviet Res.* **10**(6), 45 (1978).
71F. Ye. P. Kostorogov, E. A. Shtessel' and A. G. Merzhanov, *Heat Transfer, Soviet Res.* **10**(6), 36 (1978).
72F. J. R. Kraska and R. L. Sani, *Int. J. Heat Mass Transfer* **22,** 535 (1979).
73F. F. A. Kulacki and R. G. Freeman, *J. Heat Transfer* **101,** 169 (1979).

74F. V. S. Kupsova and V. G. Malinin, *Heat Transfer, Soviet Res.* **10**(5), 145 (1978).
75F. S. S. Kutateladze, V. P. Ivakin and A. G. Kirdyashkin, *Heat Transfer, Soviet Res.* **10**(5), 118 (1978).
76F. S. S. Kutateladze, A. G. Kirdyashkin and V. S. Berdnikov, *Heat Transfer, Soviet Res.* **10**(5), 126 (1978).
77F. S. S. Kutateladze, N. A. Rubtsov and V. A. Bazanov, *Heat Transfer, Soviet Res.* **10**(5), 154 (1978).
78F. O. Kvernovold and P. A. Tyvand, *J. Fluid Mech.* **90**, 609 (1979).
79F. O. Kvernvold, *Int. J. Heat Mass Transfer* **22**, 395 (1979).
80F. J. A. Liburdy, E. G. Groff and G. M. Faeth, *J. Heat Transfer* **101**, 249 (1979).
81F. J. J. Lorenz and P. A. Howard, *J. Heat Transfer* **101**, 538 (1979).
82F. R. P. Lowell and C.-T. Shyu, *Letters Heat Mass Transfer* **5**, 371 (1978).
83F. R. L. Mahajan and B. Gebhart, *J. Fluid Mech.* **91**, 131 (1978).
84F. P. C. Manins, *J. Fluid Mech.* **91**, 765 (1979).
85F. R. S. Marshall, J. C. Heinrich and O. C. Zienkiewicz, *Numerical Heat Transfer* **1**, 315 (1978).
86F. G. P. Merker, P. Waas and U. Grigull, *Int. J. Heat Mass Transfer* **22**, 505 (1979).
87F. J. H. Merkin, *Int. J. Heat Mass Transfer* **21**, 1499 (1978).
88F. J. H. Merkin, *Int. J. Heat Mass Transfer* **22**, 1461 (1979).
89F. B. A. Meyer, J. W. Mitchell and M. M. El-Wakil, *J. Heat Transfer* **101**, 655 (1979).
90F. W. J. Minkowycz and E. M. Sparrow, *Int. J. Heat Mass Transfer* **22**, 1445 (1979).
91F. H. K. Mohanty, *Int. J. Heat Mass Transfer* **22**, 383 (1979).
92F. A. Mojtabi and J. P. Caltagirone, *Physics Fluids* **22**, 1208 (1979).
93F. A. Mucoglu and T. S. Chen, *J. Heat Transfer* **101**, 422 (1979).
94F. P. K. Muhuri and A. S. Gupta, *Z. Angew. Math. Mech.* **59**, 117 (1979).
95F. T. Y. Na and J. P. Chiou, *Wärme- und Stoffübertragung* **12**, 83 (1979).
96F. P. V. Nielsen, A. Restivo and J. H. Whitelaw, *Numerical Heat Transfer* **2**, 115 (1979).
97F. J. M. Olson and F. Rosenberger, *J. Fluid Mech.* **92**, 609 (1979).
98F. J. M. Olson and F. Rosenberger, *J. Fluid Mech.* **92**, 631 (1979).
99F. S. Ostrach and C. Raghaven, *J. Heat Transfer* **101**, 238 (1979).
100F. M. N. Özişik and M. A. Hassab, *Numerical Heat Transfer* **2**, 251 (1979).
101F. H. Ozoe, K. Yamamoto and S. W. Churchill, *A.I.Ch.E. Jl* **25**, 709 (1979).
102F. E. M. Parmentier, *Int. J. Heat Mass Transfer* **22**, 849 (1979).
103F. D. W. Pepper and R. E. Cooper, *J. Heat Transfer* **101**, 739 (1979).
104F. V. P. Petrov, S. P. Beschastnov and V. I. Belozerov, *Heat Transfer, Soviet Res.* **10**(5), 150 (1978).
105F. M. R. E. Proctor and D. J. Galloway, *J. Fluid Mech.* **90**, 273 (1979).
106F. C. Quon, *J. Heat Transfer* **101**, 261 (1979).
107F. Z. H. Qureshi and B. Gebhart, *Int. J. Heat Mass Transfer* **21**, 1467 (1978).
108F. J. W. Ramsey, E. M. Sparrow and L. M. C. Varejao, *J. Heat Transfer* **101**, 732 (1979).
109F. R. Rana, R. N. Horne and P. Cheng, *J. Heat Transfer* **101**, 411 (1979).
110F. K. R. Randall, J. W. Mitchell and M. M. El-Wakil, *J. Heat Transfer* **101**, 120 (1979).
111F. F. M. Richter, *J. Fluid Mech.* **89**, 553 (1979).
112F. B. Roux, J. C. Grondin, P. Bontoux and B. Gilly, *Numerical Heat Transfer* **1**, 331 (1978).
113F. D. W. Ruth, *Int. J. Heat Mass Transfer* **22**, 1199 (1979).
114F. M. N. A. Said and A. C. Trupp, *J. Heat Transfer* **101**, 569 (1979).
115F. B. Sammakia and B. Gebhart, *Numerical Heat Transfer* **1**, 529 (1978).
116F. W. Schneider, *Int. J. Heat Mass Transfer* **22**, 1401 (1979).
117F. T. R. Seetharam and G. K. Sharma, *Int. J. Heat Mass Transfer* **22**, 13 (1979).
118F. M. Seki, H. Kawamura and K. Sanokawa, *J. Heat Transfer* **101**, 227 (1979).
119F. H. Shaukatullah and B. Gebhart, *Numerical Heat Transfer* **2**, 215 (1979).
120F. H. Shaukatullah and B. Gebhart, *Int. J. Heat Mass Transfer* **21**, 1481 (1978).
121F. D. J. Shlien, *Physics Fluids* **22**, 2277 (1979).
122F. D. J. Shlien and R. L. Boxman, *Physics Fluids* **22**, 631 (1979).
123F. D. J. Shlien and A. Brosh, *Physics Fluids* **22**, 1044 (1979).
124F. P. G. Simpkins and T. D. Dudderar, *J. Fluid Mech.* **89**, 665 (1979).
125F. S. N. Singh and J. M. Elliot, *Int. J. Heat Mass Transfer* **22**, 639 (1979).
126F. F. T. Smith and D. S. Riley, *Int. J. Heat Transfer* **22**, 309 (1979).
127F. V. M. Soundalgekar and H. S. Takhar, *Israel J. Technol.* **15**, 368 (1977).
128F. A. M. Soward, *J. Fluid Mech.* **90**, 669 (1979).
129F. E. M. Sparrow and J. E. Niethammer, *J. Heat Transfer* **101**, 404 (1979).
130F. E. M. Sparrow, J. W. Ramsey and R. G. Kemink, *J. Heat Transfer* **101**, 578 (1979).
131F. S. Sreenivasan and S. P. Lin, *Int. J. Heat Mass Transfer* **21**, 1517 (1978).
132F. J. M. Straus and G. Schubert, *J. Fluid Mech.* **91**, 155 (1978).
133F. K. E. Torrance, *J. Fluid Mech.* **95**, 477 (1979).
134F. K. E. Torrance, *J. Heat Transfer* **101**, 677 (1979).
135F. S. Tsuruno and I. Iguchi, *J. Heat Transfer* **101**, 573 (1979).
136F. E. van de Sande and B. J. G. Hamer, *Int. J. Heat Mass Transfer* **22**, 361 (1979).
137F. M. Vedhanayaygam, J. H. Lienhard and R. Eichhorn, *J. Heat Transfer* **101**, 571 (1979).
138F. B. J. Venkatachala and G. Nath, *J. Heat Transfer*

101, 745 (1979).
139F. J.-P. Walch and B. Dulieu, *Int. J. Heat Mass Transfer* **22,** 1607 (1979).
140F. R. A. Wirtz and C. S. Reddy, *J. Fluid Mech.* **91,** 451 (1979).
141F. R.-S. Wu, K. C. Cheng and A. Craggs, *Numerical Heat Transfer* **2,** 303 (1979).
142F. L. -S. Yao, I. Catton and J. M. McDonough, *J. Appl. Mech.* **45,** 952 (1978).
143F. T. Yonemoto, T. Chida, S. Takashashi and T. Tadaki, *Int. Chem. Engng* **19,** 159 (1979).
144F. H. Yosinobu, Y. Onishi and S. Amano, *J. Physical Soc. Japan* **47,** 312 (1979).
145F. A. Yucel and Y. Bayazitoğlu, *J. Heat Transfer* **101,** 666 (1979).
146F. Md. Zakerullah and J. A. D. Ackroyd, *Z. Angew. Math. Phys.* **30,** 427 (1979).
147F. V. A. Zheltukhin, Yu. K. Solomatnikov and D. M. Mikhaylov, *Heat Transfer, Soviet Res.* **10**(6), 10 (1978).
148F. Y. Zvirin, *Int. J. Heat Mass Transfer* **22,** 1539 (1979).
149F. Y. Zvirin and R. Greif, *Int. J. Heat Mass Transfer* **22,** 499 (1979).

Convection from rotating surfaces

1G. R. C. Beardsley, K. D. Saunders and A. C. Warn-Varnas, *J. Fluid Mech.* **93,** 161 (1979).
2G. T. P. Elson, *Chem. Engng Sci.* **34,** 373 (1979).
3G. D. R. Jeng, K. J. DeWitt and M. H. Lee, *Int. J. Heat Mass Transfer* **22,** 89 (1979).
4G. M. Kumari and G. Nath, *J. Appl. Mech.* **46,** 275 (1979).
5G. V. I. Lokai and E. I. Gunchenko, *Thermal Engng* **26,** 93 (1979).
6G. B. K. Meena and G. Nath, *J. Heat Transfer* **101,** 151 (1979).
7G. D. E. Metzger, W. J. Mathis and L. D. Grochowsky, *J. Engng Pwr* **101,** 68 (1979).
8G. D. L. Oehlbeck and F. F. Erian, *Int. J. Heat Mass Transfer* **22,** 601 (1979).
9G. S. Ito, K. Ogawa and C. Kuroda, *Int. Chem. Engng* **19,** 600 (1979).
10G. E. N. Saburov, S. V. Karpov and Y. L. Leukhin, *Heat Transfer, Soviet Res.* **11**(1), 67 (1979).
11G. V. K. Shchukin, A. A. Khalatov and A. V. Kozhevnikov, *Heat Transfer, Soviet Res.* **10**(5), 63 (1978).
12G. V. K. Shchukin and V. V. Olimpiev, *Heat Transfer, Soviet Res.* **10**(3), 5 (1978).
13G. D. A. Simmers and J. E. R. Coneu, *Int. J. Heat Mass Transfer* **22,** 679 (1979).
14G. S. H. Smith, *Q. Jl Mech. Appl. Math.* **32,** 135 (1979).

Combined heat and mass transfer

1H. T. Aihara, M. Taga and T. Haraguchi, *Int. J. Heat Mass Transfer* **22,** 51 (1979).
2H. Yu. V. Bàryshev, A. I. Leont'yev and N. K. Peyker, *Heat Transfer, Soviet Res.* **10**(5), 39 (1978).
3H. A. Brown and C. L. Saluja, *Int. J. Heat Mass Transfer* **22,** 525 (1979).
4H. A. M. Cary, Jr., D. M. Bushnell and J. N. Hefner, *J. Heat Transfer* **101,** 699 (1979).
5H. D. M. Evans and M. L. Noble, *J. Engng Pwr* **101,** 109 (1979).
6H. V. V. Glazkov, M. D. Gusev and B. A. Zhestkov, *Heat Transfer, Soviet Res.* **10**(5), 57 (1978).
7H. S. K. Griffiths and F. S. Morrison, Jr., *J. Heat Transfer* **101,** 484 (1979).
8H. C. L. D. Huang, H. H. Siang and C. H. Best, *Int. J. Heat Mass Transfer* **22,** 257 (1979).
9H. K. Kadotani and R. J. Goldstein, *J. Engng Pwr* **101,** 298 (1979).
10H. K. Kadotani and R. J. Goldstein, *J. Engng Pwr* **101,** 459 (1979).
11H. K. Kadotani and R. J. Goldstein, *J. Engng Pwr* **101,** 466 (1979).
12H. A. I. Leont'yev, V. I. Rozhdestvenskiy and Yu. A. Vinogradov, *Heat Transfer, Soviet Res.* **10**(5), 84 (1978).
13H. N. C. G. Markatos and A. Moult, *Trans. Inst. Chem. Engrs* **57,** 156 (1979).
14H. K. Meyer, *Wärme- und Stoffübertragung* **12,** 121 (1979).
15H. R. Monaco, *Int. J. Heat Mass Transfer* **22,** 805 (1979).
16H. A. Nachman and S. Taliaferro, *Proc. R. Soc.* **A365,** 313 (1979).
17H. K. Pientka, *Wärme- und Stoffübertragung* **12,** 165 (1979).
18H. B. E. Richards and J. L. Stollery, *J. Aircraft* **16,** 177 (1979).
19H. L. Speitkamp and H. Hartmann, *Wärme- und Stoffübertragung* **12,** 269 (1979).
20H. S. Srivastava and D. E. Rosner, *Int. J. Heat Mass Transfer* **22,** 1281 (1979).
21H. R. L. Street, *Int. J. Heat Mass Transfer* **22,** 885 (1979).
22H. D. Sucker and H. Brauer, *Wärme- und Stoffübertragung* **12,** 35 (1979).
23H. M. Suo, W. P. Patrick and B. V. Johnson, *J. Energy* **3,** 161 (1979).
24H. M. J. Swedish, M. Epstein, J. H. Linehan, G. A. Lambert, G. M. Hauser and L. J. Stachyra, *A.I.Ch.E. Jl* **25,** 630 (1979).
25H. C. B. Thorsness and T. J. Hanratty, *A.I.Ch.E. Jl* **25,** 686 (1979).
26H. E. P. Volchkov, V. K. Koz'menko and V. P. Lebedev, *Heat Transfer, Soviet Res.* **10**(5), 47 (1978).

Change of phase

1J. M. Issa Abdul-Hadi, *Can. J. Chem. Engng* **57,** 459 (1979).
2J. M. Issa Abdul-Hadi, *Can. J. Chem. Engng* **57,** 451 (1979).
3J. A. A. Andrizhiyevskiy, G. G. Dolzhenkova and V. A. Nemtsev, *Heat Transfer, Soviet Res.* **10**(6), 58 (1978).
4J. V. V. Arkhipov, S. V. Kvasnyuk, V. I. Deev and V. K. Andreev, *Thermal Engng* **26,** 286 (1979).
5J. A. G. Bathelt, R. Viskanta and W. Leidenfrost, *J. Heat Transfer* **101,** 453 (1979).
6J. L. D. Berman, *Thermal Engng* **26,** 274 (1979).
7J. V. Betta, P. Mazzei, V. Naso and R. Vanoli, *J. Heat Transfer* **101,** 612 (1979).
8J. C. Bonacina and S. Del Guidice, *J. Heat Transfer* **101,** 441 (1979).

9J. V. M. Borshanskiy, D. I. Volkov and N. I. Ivashchenko, *Heat Transfer, Soviet Res.* **11**(1), 35 (1979).
10J. F. M. Chistyakov, N. I. Frolova and S. G. Kuyshinov, *Heat Transfer, Soviet Res.* **10**(2), 1 (1978).
11J. C. Cho and N. M. Özişik, *J. Heat Transfer* **101,** 465 (1979).
12J. R. L. Collins and R. B. Lovelace, *J. Heat Transfer* **101,** 300 (1979).
13J. T. C. Daniels and R. J. Williams, *Int. J. Heat Mass Transfer* **22,** 1237 (1979).
14J. E. J. Davis, N. P. Chermisinoff and C. J. Guzy, *A.I.Ch.E. Jl* **25,** 958 (1979).
15J. V. I. Deev, V. I. Petrovichev and A. I. Pridantsev, *Thermal Engng* **26,** 45 (1979).
16J. F. N. D'yachkov, *Heat Transfer, Soviet Res.* **10**(2), 10 (1978).
17J. R. I. Eddington and D. B. R. Kenning, *Int. J. Heat Mass Transfer* **22,** 1231 (1979).
18J. G. Flament, F. Moreaux and G. Beck, *Int. J. Heat Mass Transfer* **22,** 1059 (1979).
19J. D. M. France, R. D. Carlson, T. Chiang and R. Priemer, *J. Heat Transfer* **101,** 270 (1979).
20J. E. S. Gaddis, *Int. J. Heat Mass Transfer* **22,** 371 (1979).
21J. E. N. Ganic and W. M. Rohsenow, *J. Heat Transfer* **101,** 288 (1979).
22J. E. Gerum, J. Straub and U. Grigull, *Int. J. Heat Mass Transfer* **22,** 517 (1979).
23J. N. W. Hale, Jr. and R. Viskanta, *Letters Heat Mass Transfer* **5,** 329 (1978).
24J. Y. Hayashi and T. Komori, *J. Heat Transfer* **101,** 459 (1979).
25J. R. E. Henry and H. K. Fauske, *J. Heat Transfer* **101,** 280 (1979).
26J. O. Herrera and Ye. V. Ametistov, *Heat Transfer, Soviet Res.* **11**(1), 156 (1979).
27J. F. W. Holm and S. P. Goplen, *J. Heat Transfer* **101,** 543 (1979).
28J. K.-S. Hsu, F. A. Locher and J. F. Kennedy, *J. Heat Transfer* **101,** 598 (1979).
29J. E. Ihnatowicz, S. Gumkowski and J. Mikieiewicz, *J. Heat Transfer* **101,** 712 (1979).
30J. V. P. Isachenko, A. P. Solodov and V. V. Sennikov, *Thermal Engng* **26,** 279 (1979).
31J. Y. Koizumi, T. Ueda and H. Tanaka, *Int. J. Heat Mass Transfer* **22,** 669 (1979).
32J. Y. Katto, *Int. J. Heat Mass Transfer* **22,** 575 (1979).
33J. Y. Katto, *Int. J. Heat Mass Transfer* **22,** 1567 (1979).
34J. Y. Katto, *Int. J. Heat Mass Transfer* **22,** 783 (1979).
35J. Y. Katto and M. Shimizu, *J. Heat Transfer* **101,** 265 (1979).
36J. A. N. Khoze and A. S. Zakharov, *Heat Transfer, Soviet Res.* **11**(1), 47 (1979).
37J. V. A. Kravchenko and Yu. N. Ostrovskiy, *Heat Transfer, Soviet Res.* **11**(1), 133 (1979).
38J. L. I. Kravetskiy, M. I. Pavlishchev and P. A. Khinkis, *Heat Transfer, Soviet Res.* **11**(1), 85 (1979).
39J. R. Krishna, *Letters Heat Mass Transfer* **6,** 439 (1979).
40J. R. Krishna, *Letters Heat Mass Transfer* **6,** 137 (1979).
41J. G. R. Kubanek and D. L. Miletti, *J. Heat Transfer* **101,** 447 (1979).
42J. Yu. I. Krokhin and A. S. Kulikov, *Heat Transfer, Soviet Res.* **10**(6), 141 (1978).
43J. S. S. Kutateladze, *Int. J. Heat Mass Transfer* **22,** 289 (1979).
44J. S. S. Kutateladze and I. I. Gogonin, *Int. J. Heat Mass Transfer* **22,** 1593 (1979).
45J. S. S. Kutateladze, I. I. Gogonin, A. R. Dorokhov and V. I. Sosunov, *Thermal Engng* **26,** 270 (1979).
46J. D. A. Labuntsov and E. V. Ametistov, *Thermal Engng* **26,** 283 (1979).
47J. J. H. Lienhard and Md. Mojibul Hasan, *J. Heat Transfer* **101,** 276 (1979).
48J. K. Lucas, *Wärme- und Stoffübertragung* **12,** 131 (1979).
49J. K. Lucas and B. Moser, *Int. J. Heat Mass Transfer* **22,** 431 (1979).
50J. P. Mathieu, *Numerical Heat Transfer* **2,** 319 (1979).
51J. M. R. Mokhtarzadeh and A. A. El-Shirbini, *Int. J. Heat Mass Transfer* **22,** 27 (1979).
52J. Y. Morikawa, Y. Tsuji, K. Matsui and Y. Jittani, *Int. J. Multiphase Flow* **4,** 575 (1978).
53J. V. G. Morozov, *Heat Transfer, Soviet Res.* **10**(2), 58 (1978).
54J. N. Nishikawa and H. Takase, *J. Heat Transfer* **101,** 705 (1979).
54J. N. Nishikawa and H. Takase, *J. Heat Transfer* **101,** 705 (1979).
55J. A. P. Ornatskiy, V. A. Chernobay and V. S. Furayev, *Heat Transfer, Soviet Res.* **11**(1), 119 (1979).
56J. A. P. Ornatskiy, V. S. Furayev and V. A. Chernobay, *Heat Transfer, Soviet Res.* **11**(1), 24 (1979).
57J. J. M. Papazian, R. Gutowski and W. R. Wilcox, *AIAA Jl* **17,** 1111 (1979).
58J. J. M. Papazian and R. L. Kosson, *AIAA Jl* **17,** 1279 (1979).
59J. S. V. Patankar and E. M. Sparrow, *J. Heat Transfer* **101,** 434 (1979).
60J. N. I. Perepelitsa, R. S. Pomet'ko and A. P. Sapankevich, *Thermal Engng* **26,** 42 (1979).
61J. E. L. Pinnes and W. K. Mueller, *J. Heat Transfer* **101,** 617 (1979).
62J. J. W. Ramsey, E. M. Sparrow and L. M. C. Varejao, *J. Heat Transfer* **101,** 732 (1979).
63J. F. J. Renk and P. C. Wayner, *J. Heat Transfer* **101,** 55 (1979).
64J. F. J. Renk and P. C. Wayner, *J. Heat Transfer* **101,** 59 (1979).
65J. V. G. Rifert and V. Yu Zadiraka, *Thermal Engng* **25**(8), 54 (1978).
66J. S. S. Sadhal and M. S. Plesset, *J. Heat Transfer* **101,** 48 (1979).
67J. R. Z. Savel'ev and Yu. M. Brodov, *Thermal Engng* **25**(9), 15 (1978).
68J. M. M. Shah, *Int. J. Heat Mass Transfer* **22,** 557 (1979).
69J. M. M. Shah, *Int. J. Heat Mass Transfer* **22,** 547 (1979).
70J. G. G. Shklover and A. V. Buevich, *Thermal Engng* **25**(6), 49 (1978).
71J. L. A. Slobozhanin and N. S. Shcherbakova, *Heat Transfer, Soviet Res.* **11**(1), 144 (1979).
72J. B. M. Smol'skiy and V. S. Bogachev, *Heat Transfer, Soviet Res.* **10**(5), 74 (1978).

73J. D. M. Snider, *J. Heat Transfer* **101,** 43 (1979).
74J. E. M. Sparrow, J. W. Ramsey and R. G. Kemink, *J. Heat Transfer* **101,** 578 (1979).
75J. K. Taghavi-Tafreshi, V. K. Dhir and I. Catton, *J. Heat Transfer* **101,** 318 (1979).
76J. H. Tanaka, *J. Heat Transfer* **101,** 603 (1979).
77J. J. M. Tishkoff, *Int. J. Heat Mass Transfer* **22,** 1407 (1979).
78J. M. Toda, J. Yonehara and S. Maeda, *Int. Chem. Engng* **19,** 646 (1979).
79J. V. I. Tolubinskiy, V. A. Antonenko and Yu. N. Ostrovskiy, *Heat Transfer, Soviet Res.* **10**(3), 1 (1978).
80J. V. I. Tolubinskiy, V. A. Antonenko and Yu. N. Ostrovskiy, *Heat Transfer, Soviet Res.* **11**(1), 30 (1979).
81J. V. I. Tolubinskiy, Yu. N. Ostrovskiy and V. Ye. Pisarev, *Heat Transfer, Soviet Res.* **10**(2), 74 (1978).
82J. V. I. Tolubinskiy, Yu. N. Ostrovskiy and V. Ye. Pisarev, *Heat Transfer, Soviet Res.* **11**(1), 1 (1979).
83J. V. I. Tolubinskiy, Yu. N. Ostrovskiy and V. Ye. Pisarev, *Heat Transfer, Soviet Res.* **11**(1), 18 (1979).
84J. V. I. Tolubinskiy, A. M. Kichigin and S. G. Povsten', *Heat Transfer, Soviet Res.* **11**(1), 6 (1979).
85J. Y. Tomita and A. Shima, *Z. Angew. Math. Mech.* **59,** 297 (1979).
86J. L. L. Vasiliev, A. N. Abramenko and L. E. Kanonchik, *AIAA Jl* **17,** 1395 (1979).
87J. S. K. Walsh and S. D. R. Wilson, *Int. J. Heat Mass Transfer* **22,** 569 (1979).
88J. P. C. Wayner, Jr., *Int. J. Heat Mass Transfer* **22,** 1033 (1979).
89J. P. C. Wayner, Jr., *AIAA Jl* **17,** 772 (1979).
90J. W. R. Wilcox, R. S. Subramanian and J. M. Papazian, *AIAA Jl* **17,** 1022 (1979).
91J. S. D. R. Wilson, *Int. J. Heat Mass Transfer* **22,** 207 (1979).
92J. N. W. Wilson and B. D. Vyas, *J. Heat Transfer* **101,** 313 (1979).
93J. D. W. Woodruff and J. W. Westwater, *Int. J. Heat Mass Transfer* **22,** 629 (1979).
94J. W. Zijl, F. J. M. Ramakers and S. J. D. van Stralen, *Int. J. Heat Mass Transfer* **22,** 401 (1979).

Radiation in participating media

1K. K. S. Adzerikho, V. I. Antsulevich, Ya. K. Lapko and V. P. Nekrasov, *Int. J. Heat Mass Transfer* **22,** 131 (1979).
2K. D. W. Amlin and S. A. Korpela, *J. Heat Transfer* **101,** 76 (1979).
3K. A. Balakrishna and D. K. Edwards, *J. Heat Transfer* **101,** 489 (1979).
4K. Y. Bayazitoglu and J. Higenyi, *AIAA Jl* **17,** 424 (1979).
5K. V. C. Boffi, F. Santarelli, G. Spiga and C. Stramigioli, *Int. J. Heat Mass Transfer* **22,** 1705 (1979).
6K. E. V. Browell, T. D. Wilkerson and T. J. McIlrath, *Appl. Optics* **18,** 3474 (1979).
7K. T. C. Chawla and S. H. Chan, *Int. J. Heat Mass Transfer* **22,** 1657 (1979).
8K. C. W. Clausen and T. F. Smith, *J. Heat Transfer* **101,** 376 (1979).
9K. D. L. Coffeen, *J. Opt. Soc. Am.* **69,** 1051 (1979).
10K. A. L. Crosbie, *J. Heat Transfer* **101,** 68 (1979).
11K. A. L. Crosbie, *AIAA Jl* **17,** 117 (1979).
12K. A. L. Crosbie and J. W. Koewing, *AIAA Jl* **17,** 196 (1979).
13K. K. J. Daniel, N. M. Laurendeau and F. P. Incropera, *J. Heat Transfer* **101,** 63 (1979).
14K. F. P. Incropera and W. G. Houf, *J. Heat Transfer* **101,** 496 (1979).
15K. S. S. Kutateladze, N. A. Rubtsov, Ya. A. Bal'tsevich and G. P. Yeremenko, *Heat Transfer, Soviet Res.* **10,** 106 (1978).
16K. D. C. Look, *J. Heat Transfer* **101,** 556 (1979).
17K. D. C. Look and M. F. Lackner, *Letters Heat Mass Transfer* **6,** 385 (1979).
18K. W. E. Meador and W. R. Weaver, *Appl. Optics* **18,** 1204 (1979).
19K. M. F. Modest, *J. Heat Transfer* **101,** 735 (1979).
20K. D. A. Nelson, *J. Heat Transfer* **101,** 81 (1979).
21K. D. A. Nelson, *J. Heat Transfer* **101,** 85 (1979).
22K. B. Nilsson, *Appl. Optics* **18,** 3457 (1979).
23K. R. G. Siddall and N. Selcuk, *Trans. Inst. Chem. Engrs* **57,** 163 (1979).
24K. O. A. Simpson, B. L. Bean and S. Perkowitz, *J. Opt. Soc. Am.* **69,** 1723 (1979).
25K. W. H. Sutton and M. N. Özişik, *J. Heat Transfer* **101,** 695 (1979).
26K. S. P. Venkateshan and K. K. Prasad, *J. Fluid Mech.* **90,** 33 (1979).
27K. S-C. Yao, L. E. Hochreiter and C. E. Dodge, *J. Heat Transfer* **101,** 736 (1979).

Surface radiation

1L. N. M. Galin, V. M. Yesin and N. A. Minyaylenko, *Heat Transfer, Soviet Res.* **11**(1), 138 (1979).
2L. D. C. Look, *AIAA Jl* **17,** 443 (1979).
3L. J. R. Mahan, J. B. Kingsolver and D. T. Mears, *J. Heat Transfer* **101,** 689 (1979).
4L. V. K. Mel'nikov and Ye. P. Sukhovich, *Heat Transfer, Soviet Res.* **10**(3), 11 (1978).
5L. C. P. Minning, *AIAA Jl* **17,** 318 (1979).
6L. C. P. Minning, *AIAA Jl* **17,** 1406 (1979).
7L. M. F. Modest, *Numerical Heat Transfer* **1,** 403 (1978).
8L. K. Rädle, *Wärme- und Stoffübertragung* **12,** 157 (1979).
9L. C. R. Roger, S. H. Yen and K. G. Ramanathan, *J. Opt. Soc. Am.* **69,** 1384 (1979).
10L. N. Seki, S. Sugawara and S. Fukusako, *J. Heat Transfer* **101,** 90 (1979).
11L. N. Seki, M. Sugawara and S. Fukusako, *Wärme- und Stoffübertragung* **12,** 137 (1979).
12L. D. Tanre, M. Herman and P. Y. Eschamps, *Appl. Optics* **18,** 3587 (1979).
13L. M. M. Yan and P. N. S. Huang, *J. Heat Transfer* **101,** 96 (1979).

MHD

1M. K. M. Aref'yev, V. M. Borishanskiy, N. I. Ivashchenko and N. M. Fishman, *Heat Transfer, Soviet*

Res. **10**(6), 132 (1978).
2M. V. G. Bashtovoy, S. V. Isayev and M. P. Pavlinov, *Heat Transfer, Soviet Res.* **10**(6), 5 (1978).
3M. B. M. Berkovskiy and A. N. Vislovich, *Heat Transfer, Soviet Res.* **10**(5), 160 (1978).
4M. H. Branover and P. Gershon, *J. Fluid Mech.* **94,** 629 (1979).
5M. S. H. Choi and H. E. Wilhelm, *Physics Fluids* **22,** 1073 (1979).
6M. J. B. Helliwell and M. F. Mosa, *Int. J. Heat Mass Transfer* **2,** 657 (1979).
7M. R. Krishnaswamy and G. Nath, *Physics Fluids* **22,** 1631 (1979).
8M. G. Ye Kron'kalns, E. Ya. Blums and M. M. Mayorov, *Heat Transfer, Soviet Res.* **10**(6), 1 (1978).
9M. T. F. Smith and P. H. Paul, *J. Heat Transfer* **101,** 502 (1979).
10M. V. M. Soundalgekar, N. V. Vighnesam and H. S. Takhar, *IEEE Trans. Plasma Sci* **PS-7**(3), 178 (1979).
11M. E. P. Szuszczewicz and P. Z. Takacs, *Physics Fluids* **22,** 2424 (1979).

Measurement techniques

1P. N. Abuaf, T. P. Feierabend and G. A. Zimmer, *Rev. Scient. Instrum.* **50,** 1260 (1979).
2P. M. Acharya, *Rev. Scient. Instrum.* **50,** 952 (1979).
3P. V. Ye. Alemasov, V. K. Maksimov and V. I. Sagadeyev, *Heat Transfer, Soviet Res.* **10**(3), 79 (1978).
4P. O. M. Alifanov, V. S. Kuznetsov and B. M. Pakratov, *Heat Transfer, Soviet Res.* **10**(3), 92 (1978).
5P. E. L. Andreas, *J. Appl. Mech.* **46,** 15 (1979).
6P. S. A. Balankin and D. M. Skorov, *Soviet Atomic Energy* **46,** 304 (1979).
7P. S. Banerjee, P. Yuen and M. A. Vanderbroek, *J. Heat Transfer* **101,** 295 (1979).
8P. A. R. Barber, K. E. Kneidel, C. S. Fitzgerald and L. C. Lynnworth, *J. Heat Transfer* **101,** 622 (1979)
9P. J. C. Bennett, *AIAA Jl* **17,** 215 (1979).
10P. R. Birkebak and M. Alamgir, *J. Heat Transfer* **101,** 379 (1979).
11P. J. F. Brison, G. Charnay and G. Comte-Bellot, *Int. J. Heat Mass Transfer* **22,** 111 (1979).
12P. A. P. Burdukov, V. Ye. Nakoryakov and G. G. Kuvshinov, *Heat Transfer, Soviet Res.* **10**(3), 116 (1978).
13P. J. H. Chan and E. A. Ballik, *J. Appl. Mech.* **46,** 218 (1979).
14P. Yu. A. Chistyakov, *Heat Transfer, Soviet Res.* **10**(3), 72 (1978).
15P. Y. I. Cho and J. P. Hartnett, *Letters Heat Mass Transfer* **6,** 355 (1979).
16P. J. B. Cole and M. D. Swords, *Appl. Optics* **18,** 1539 (1979).
17P. F. Durst and W. H. Stevenson, *Appl. Optics* **18,** 516 (1979).
18P. S. Dutta and C. Y. Wen, *Can. J. Chem. Engng* **57,** 115 (1979).
19P. P. C. Efthimion, V. Arunasalam and R. Bitzer, *Rev. Scient. Instrum.* **50,** 949 (1979).
20P. S. K. Goel, P. D. Gupta and D. D. Bhawalkar, *Rev. Scient. Instrum.* **50,** 1156 (1979).
21P. J. A. Grossman, W. A. Peebles and N. C. Luhmann, Jr., *Rev. Scient. Instrum.* **50,** 1341 (1979).
22P. S. E. Gustafsson, E. Karawacki and M. N. Khan, *J. Phys. D.: Appl. Phys.* **12,** 1411 (1979).
23P. I. Hatta, *Rev. Scient. Instrum.* **50,** 292 (1979).
24P. V. T. Helms III, *J. Spacecraft Rockets* **16,** 20 (1979).
25P. K. Hirano, K. Shimoda and S. Emori, *Rev. Scient. Instrum.* **50,** 1236 (1979).
26P. C. K. Hsieh and K. C. Su, *Rev. Scient. Instrum.* **50,** 888 (1979).
27P. E. D. Huber and S. O. Sari, *Rev. Scient. Instrum.* **50,** 438 (1979).
28P. H. E. Khalifa, J. Kestin and W. A. Wakeham, *Physica's Grav.* **A97,** 273 (1979).
29P. O. A. Knyazev, *Measmt Techniques* **22,** 279 (1979).
30P. W. N. Lawless and C. F. Clark, *Rev. Scient. Instrum.* **50,** 787 (1979).
31P. S. Lederman, A. Celentano and J. Glaser, *Physics Fluids* **22,** 1065 (1979).
32P. S. Lederman, A. Celentano and J. Glaser, *AIAA Jl* **17,** 1106 (1979).
33P. S. S. Lu, *Rev. Scient. Instrum.* **50,** 772 (1979).
34P. S. Matsumura and S. -L. Chen, *Rev. Scient. Instrum.* **50,** 1425 (1979).
35P. T. Mizushina, F. Ogino, H. Ueda and S. Komori, *Proc. R. Soc.* **A366**(1724), 63 (1979).
36P. L. C. Nistor, S. V. Nistor and V. Teodorescu, *Appl. Optics* **18,** 3517 (1979).
37P. P. Paranthoen and C. Petit, *Letters Heat Mass Transfer* **6,** 311 (1979).
38P. A. E. Perry, A. J. Smits and M. S. Chong, *J. Fluid Mech.* **90,** 415 (1979)
39P. N. A. Pokryvaylo, A. K. Nesterov and A. S. Sobolevskiy, *Heat Transfer, Soviet Res.* **10**(3), 106 (1978).
40P. A. K. Raghava, K. L. Kumar, R. C. Malhotra and D. P. Agrawal, *J. Fluids Engng* **101,** 143 (1979).
41P. R. S. Redman and C. M. Wolff, *J. Spacecraft Rockets* **16,** 350 (1979).
42P. L. N. Samoilov and M. I. Epshtein, *Measmt Techniques* **21,** 950 (1978).
43P. J. Schneider and S. Robertson, *Rev. Scient. Instrum.* **50,** 856 (1979).
44P. C. M. Sheih, J. J. Finnigan, E. F. Bradley and P. J. Mulhearn, *Rev. Scient. Instrum.* **50,** 528 (1979).
45P. L. Shemmer and S. Einav, *Rev. Scient. Instrum.* **50,** 879 (1979).
46P. J. Thibault and T. W. Hoffman, *Int. J. Heat Mass Transfer* **22,** 177 (1979).
47P. N. Van Thinh, *Letters Heat Mass Transfer* **6,** 103 (1979).
48P. J. R. Turner and T. D. Eastop, *Trans. Inst. Chem. Engrs* **57,** 139 (1979).
49P. G. Villoutreix and J. Martinet, *Letters Heat Mass Transfer* **6,** 301 (1979).
50P. C. B. Watkins and W. Aung, *Numerical Heat Transfer* **1,** 543 (1978).

Heat exchangers and heat pipes

1Q. H. Nguyen Chi and A. Abhat, *AIAA Jl* **17,** 1003 (1979).
2Q. J. P. Day, *J. Engng Pwr* **101,** 270 (1979).
3Q. J. E. Eninger and B. D. Marcus, *AIAA Jl* **17,** 797 (1979).

4Q. V. Gnielinski, *Int. Chem. Engng* **19,** 380 (1979).
5Q. V. Gnielinski and E. S. Gaddis, *Int. Chem. Engng* **19,** 391 (1979).
6Q. I. D. R. Grant and D. Chisholm, *J. Heat Transfer* **101,** 38 (1979).
7Q. M. Groll, W. D. Münzel and W. Supper, *J. Spacecraft Rockets* **16,** 195 (1979).
8Q. R. B. Holmberg, *J. Heat Transfer* **101,** 205 (1979).
9Q. R. I. Kalinin, V. I. Ivanov and M. M. Nazarchuk, *Heat Transfer, Soviet Res.* **11**(1), 75 (1979).
10Q. W. B. Kaufman and L. K. Tower, *J. Spacecraft Rockets* **16,** 98 (1979).
11Q. V. A. Lokshin and V. N. Fomina, *Thermal Engng* **25**(6), 31 (1978).
12Q. S. W. Mandel, M. A. Townsend and T. F. Parrish, Jr., *J. Heat Transfer* **101,** 514 (1979).
13Q. I. R. Mikk, A. Y. Veski and R. A. Kruus, *Heat Transfer, Soviet Res.* **10**(5), 20 (1978).
14Q. J. M. Mineur and G. F. Dunstan, *Appl. Energy* **5,** 89 (1979).
15Q. F. K. Moore and J. R. Ristorcelli, Jr., *Int. J. Heat Mass Transfer* **22,** 1175 (1979).
16Q. P. Razelos, *Wärme- und Stoffübertragung* **12,** 59 (1979).
17Q. W. Roetzel and J. Neubert, *J. Heat Transfer* **101,** 511 (1979).
18Q. F. E. Romie, *J. Heat Transfer* **101,** 726 (1979).
19Q. C. J. Savage, B. G. M. Aalders and H. Kreeb, *J. Spacecraft Rockets* **16,** 176 (1979).
20Q. E. M. Sparrow and C. H. Liu, *Int. J. Heat Mass Transfer* **22,** 1613 (1979).
21Q. V. I. Tolubinsky, V. A. Antonenko and Yu. N. Ostrovsky, *AIAA Jl* **17,** 1390 (1979).
22Q. H. van Ooijen and C. J. Hoogendoorn, *AIAA Jl* **17,** 1251 (1979).
23Q. R. L. Webb, *J. Heat Transfer* **101,** 335 (1979).
24Q. M. E. Weber, *Appl. Energy* **5,** 159 (1979).
25Q. A. J. Willmott and A. Burns, *Int. J. Heat Mass Transfer* **22,** 1107 (1979).
26Q. A. A. Zhukauskas, R. V. Ulinskas and A. A. Shvegzhda, *Int. Chem. Engng* **19,** 711 (1979).
27Q. Z. A. Żukauskas, R. V. Ulinskas, E. S. Bubelis and C. Sipavicius, *Heat Transfer, Soviet Res.* **10**(5), 29 (1978).
28Q. A. A. Žukauskas, R. V. Ulinskas and C.-S. J. Sipavicius, *Heat Transfer, Soviet Res.* **10**(6), 90 (1978).
29Q. A. A. Żukauskas and R. V. Ulinskas, *Heat Transfer, Soviet Res.* **10**(5), 9 (1978).

Aircraft and space vehicles

1R. M. Furukawa, *J. Spacecraft Rockets* **16,** 412 (1979).
2R. R. L. Glick, *J. Spacecraft Rockets* **16,** 58 (1979).
3R. R. D. Karam, *J. Spacecraft Rockets* **16,** 92 (1979).
4R. P. A. A. Laura, G. Sanchez Sarmiento and M. E. Olivetto, *Int. J. Heat Mass Transfer* **22,** 625 (1979).
5R. P. K. S. Wu, *J. Spacecraft Rockets* **16,** 56 (1979).

General

1S. V. K. Antonovskii and O. V. Keselev, *Thermal Engng* **26,** 20 (1979).
2S. A. Behan, *Int. J. Heat Mass Transfer* **22,** 219 (1979).
3S. N. A. Buchko, *Heat Transfer, Soviet Res.* **11**(1), 149 (1979).
4S. A. B. Crowley and J. R. Ockendon, *Int. J. Heat Mass Transfer* **22,** 941 (1979).
5S. O. I. Bukovskaya and L. A. Kozdoba, *Heat Transfer, Soviet Res.* **10**(3), 123 (1978).
6S. T. A. Dean and T. M. Silva, *J. Engng Ind.* **101,** 385 (1979).
7S. M. G. Dunn and F. J. Stoddard, *J. Engng Pwr* **101,** 275 (1979).
8S. P. Egerton, J. A. Howarth, G. Poots and S. Taylor-Reed, *Int. J. Heat Mass Transfer* **22,** 1525 (1979).
9S. R. Greif, T. Namba and M. Nikanjam, *Int. J. Heat Mass Transfer* **22,** 901 (1979).
10S. A. M. Kanury and D. J. Holve, *J. Heat Transfer* **101,** 365 (1979).
11S. E. S. Karasina, A. A. Abryutin and A. N. Efimenko, *Thermal Engng* **26,** 27 (1979).
12S. K. Lassmann, *Wärme- und Stoffübertragung* **12,** 185 (1979).
13S. I. R. Mikk and T. B. Tiykma, *Heat Transfer, Soviet Res.* **10**(3), 39 (1978).
14S. V. V. Mitor and I. N. Konopel'ko, *Heat Transfer, Soviet Res.* **10**(3), 45 (1978).
15S. R. Poggemann, A. Steiff and P.-M. Weinspach, *Chem.-Ingr-Tech.* **51,** 948 (1979).
16S. P. R. Ponzi and L. A. Kaye, *A.I.Ch.E. Jl* **25,** 100 (1979).
17S. Y. Saito, N. Nishiwaki and Y. Ito, *J. Engng Ind.* **101,** 97 (1979).
18S. M. J. Shilston and S. D. Probert, *Appl. Energy* **5**(1), 61 (1979).
19S. M. V. Stradomskiy, O. N. Alekseyev and Ye. A. Maksimov, *Heat Transfer, Soviet Res.* **10**(3), 129 (1978).
20S. M. V. Stradomskiy, Ye. A. Maksimov and V. A. Asmalovskiy, *Heat Transfer, Soviet Res.* **10**(3), 51 (1978).
21S. S. H. Tscheng and A. P. Watkinson, *Can. J. Chem. Engng* **57,** 433 (1979).
22S. A. I. Tugov and G. I. Moseev, *Thermal Engng* **26,** 208 (1979).
23S. S. W. Van Sciver and R. W. Boom, *J. Heat Transfer* **101,** 371 (1979).
24S. L. L. Vasiliev and V. A. Maiorov, *Int. J. Heat Mass Transfer* **22,** 301 (1979).
25S. L. M. Zysina-Molozhen, M. M. Ivashchenko, M. P. Polyak, A. A. Degach, and Ya. M. Fel'dshtein, *Thermal Engng* **25**(11), 37 (1978).
26S. L. M. Zysina-Molozhen and E. G. Roost, *Thermal Engng* **26,** 219 (1979).

Solar energy

1T. M. Abdierahman, P. Fumeaux and P. Suter, *Solar Energy* **22,** 45 (1979).
2T. R. B. Bannerot and J. R. Howell, *Solar Energy* **22,** 229 (1979).
3T. S. Barbaro, S. Coppolino, C. Leone and E. Sinagra, *Solar Energy* **22,** 225 (1979).

4T. C. D. Barley, *Solar Energy* **23,** 149 (1979).
5T. O. A. Barra and E. P. Carratelli, *Solar Energy* **23,** 211 (1979).
6T. R. H. Bushnell, *Solar Energy* **23,** 321 (1979).
7T. M. Collares-Pereira, *Solar Energy* **23,** 409 (1979).
8T. M. Collares-Pereira and A. Rabl, *Solar Energy* **22,** 155 (1979).
9T. M. Collares-Pereira and A. Rabl, *Solar Energy* **23,** 223 (1979).
10T. M. Collares-Pereira and A. Rabl, *Solar Energy* **23,** 235 (1979).
11T. M. S. Drew and R. B. G. Selvage, *Solar Energy* **23,** 327 (1979).
12T. M. S. Drew and R. B. G. Selvage, *Solar Energy* **23,** 435 (1979).
13T. R. Fuchs and J. F. McClelland, *Solar Energy* **23,** 123 (1979).
14T. P. Gandhidasan, V. Sriramulu and M. C. Gupta, *Solar Energy* **22,** 9 (1979).
15T. P. K. Gogna and K. L. Chopra, *Solar Energy* **23,** 405 (1979).
16T. C. K. Hsieh and K. C. Su, *Solar Energy* **22,** 37 (1979).
17T. B. J. Huang and S. Nieh, *Letters Heat Mass Transfer* **6,** 57 (1979).
18T. B. J. Huang, T. Y. Wung and S. Nieh, *Solar Energy* **22,** 221 (1979).
19T. R. King and R. O. Buckius, *Solar Energy* **22,** 297 (1979).
20T. C. F. Kooi, *Solar Energy* **23,** 37 (1979).
21T. R. Mastrullo and P. Mazzei, *Letters Heat Mass Transfer* **6,** 281 (1979).
22T. W. F. Phillips, *Solar Energy* **23,** 187 (1979).
23T. V. M. Puri, *Solar Energy* **22,** 183 (1979).
24T. A. C. Ratzel, C. E. Hickox and D. K. Gartling, *J. Heat Transfer* **101,** 108 (1979).
25T. G. T. Roberts, *Solar Energy* **22,** 137 (1979).
26T. C. C. Roberts, Jr., *J. Energy* **3,** 122 (1979).
27T. M. K. Selcuk, *Solar Energy* **22,** 413 (1979).
28T. M. K. Sharp and R. I. Loehrke, *J. Energy* **3,** 106 (1979).
29T. A. Shitzer, D. Kalmanoviz, Y. Zvirin and G. Grossman, *Solar Energy* **22,** 27 (1979).
30T. W. Shing-An, *Solar Energy* **23,** 333 (1979).
31T. D. L. Siebers and R. Viskanta, *J. Energy* **3,** 8 (1979).
32T. T. F. Smith and H. Y. Lee, *J. Heat Transfer* **101,** 185 (1979).
33T. R. S. Soin, K. Sangameswar Rao, D. P. Rao and K. S. Rao, *Solar Energy* **23,** 69 (1979).
34T. K. K. Tien and E. M. Sparrow, *Int. J. Heat Mass Transfer* **22,** 349 (1979).
35T. W. O. Waray and J. D. Balcomb, *Solar Energy* **23,** 421 (1979).
36T. L. Wen, *J. Energy* **3,** 82 (1979).
37T. H. C. White and S. A. Korpela, *Solar Energy* **23,** 141 (1979).
38T. J. A. Wiebelt and J. B. Henderson, *J. Heat Transfer* **101,** 101 (1979).

Plasma heat transfer

1U. D. Affimito, E. Bar-Avraham and A. Fisher, *IEEE Trans. Plasma Sci.* **PS-7**(3), 162 (1979).
2U. D. R. Airey, *J. Phys. D: Appl. Phys.* **12,** 113 (1979).
3U. F. Bastien and E. Marode, *J. Phys. D: Appl. Phys.* **12,** 249 (1979).
4U. I. A. Belov, I. P. Ginzburg and C. F. Gorshkov, *Heat Transfer, Soviet Res.* **10**(5), 87 (1978).
5U. I. D. Chalmers and B. D. Phukan, *J. Phys. D: Appl. Phys.* **12,** 1285 (1979).
6U. S. K. Chan and M. T. C. Fang, *J. Phys. D: Appl. Phys.* **12,** 1853 (1979).
7U. J. E. Daalder, *J. Phys. D: Appl. Phys.* **12,** 1769 (1979).
8U. G. Eardley, B. F. Jones and D. A. J. Mottram, *J. Phys. D: Appl. Phys.* **12,** 1101 (1979).
9U. M. T. C. Fang and D. Brannen, *IEEE Trans. Plasma Sci.* **PS-7**(4), 217 (1979).
10U. J. K. Fiszdon, *Int. J. Heat Mass Transfer* **22,** 749 (1979).
11U. J. A. Foosnaes and W. G. J. Rondeel, *J. Phys. D: Appl. Phys.* **12,** 1867 (1979).
12U. G. Frind, L. E. Prescott and J. H. Van Noy, *J. Phys. D: Appl. Phys.* **12,** 133 (1979).
13U. K. Günther, *J. Phys. D: Appl. Phys.* **12,** 1093 (1979).
14U. J. E. Harry and L. Hobson, *IEEE Trans. Plasma Sci.* **PS-7**(3), 157 (1979).
15U. R. J. Hill and G. R. Jones, *J. Phys. D: Appl. Phys.* **12,** 1707 (1979).
16U. H. Hoffmann, *J. A. S. R. T.* **21,** 163 (1979).
17U. D. Johnson and E. Pfender, *IEEE Trans. Plasma Sci.* **PS-7**(1), 44 (1979).
18U. P. D. Johnson and T. H. Rautenberg, Jr., *J. Appl. Phys.* **50,** 3207 (1979).
19U. C. H. Liu and E. Pfender, *Studies in Heat Transfer – A Festschrift for E. R. G. Eckert*, McGraw-Hill, New York 127 (1979).
20U. J. J. Lowke, *J. Phys. D: Appl. Phys.* **12,** 1873 (1979).
21U. H. H. Mai, K. Dimoff and B. Jean, *J. Appl. Phys.* **50,** 3944 (1979).
22U. R. C. Mehta, *AIAA Jl* **17,** 1272 (1979).
23U. C. R. Negus and N. J. Peacock, *J. Phys. D: Appl. Phys.* **12,** 91 (1979).
24U. N. N. Sayegh and W. H. Gauvin, *A.I.Ch.E. Jl* **25,** 1057 (1979).
25U. D. Schuöcker, *IEEE Trans. Plasma Sci.* **PS-7**(4), 209 (1979).
26U. P. Shipp and E. Pfender, *A.I.Ch.E. Symp. Ser. No. 186*, **75,** 1 (1979).
27U. Y. Skowronek, K. Benisty and M. M. Popovic, *J. Phys. D: Appl. Phys.* **12,** 1125 (1979).
28U. A. D. Stokes, *J. Phys. D: Appl. Phys.* **12,** 561 (1979).
29U. M. Suzuki and A. Kanzawa, *AIAA Jl* **17,** 1320 (1979).
30U. H. R. Velkoff and R. Godfrey, *J. Heat Transfer* **101,** 157 (1979).
31U. H. L. Walmsley, G. R. Jones and M. R. Barrault, *J. Phys. D: Appl. Phys.* **12,** 887 (1979).
32U. I. K. Yermalayev, V. G. Puzach and V. A. Fadeyev, *Heat Transfer, Soviet Res.* **10**(5), 78 (1978).
33U. M. V. Zake, V. N. Kovalev, V. E. Liyepinya and V. K. Mel'nikov, *Heat Transfer, Soviet Res.* **10**(2), 138 (1978).
34U. M. V. Zake and V. E. Liyepinya, *Heat Transfer, Soviet Res.* **10**(2), 128 (1978).
35U. M. P. Zekster and G. A. Lyubimov, *J. Phys. D: Appl. Phys.* **12**(5), 761 (1979).

HEAT TRANSFER REVIEW - 1980

E. R. G. ECKERT, R. J. GOLDSTEIN, S. V. PATANKAR, E. PFENDER, J. W. RAMSEY, T. W. SIMON, E. M. SPARROW, and K. Y. TEICHMAN
Heat Transfer Laboratory, Department of Mechanical Engineering, University of Minnesota, Minneapolis, Minnesota, U.S.A.

INTRODUCTION

THIS REVIEW surveys results that have been published in the open literature covering various fields of heat transfer during 1980. As in the past, the number of papers published during that period was such that a selection only can be included.

The 19th National Heat Transfer Conference, sponsored by the American Institute of Chemical Engineers and the American Society of Mechanical Engineers, was held at Orlando, Florida, 27–30 July 1980. Thirty-five sessions were devoted to research in basic modes of heat transfer and to heat transfer applications in coal conversion, solar energy, laser systems, radioactive waste handling, fusion reactor components, gas turbines, LNG shipping and storage, and passive solar systems. Heat exchangers for two-phase applications, for high temperatures, and air cooling were also discussed. Courses in the AIChE-Today Series and in the ASME Professional Development Series provided the opportunity to become acquainted with new developments in various fields of heat transfer. Invited lectures were given by E. R. G. Eckert on "Pioneering Research," by E. H. Young on "History and Development of Application of Extended Surface Tubes," and by S. W. Churchill on "Adventures in Heat Transfer." D. K. Slayton of NASA discussed during the Awards Luncheon the development of the insulation protecting the space shuttle vehicles during reentry. The Max Jakob Memorial Award was presented to S. W. Churchill and the D. Q. Kern Award to E. H. Young. J. M. Robertson was the recipient of the Conference Award Certificate for the best paper on "boiling heat transfer with liquid nitrogen in brazed aluminum plate fin heat exchangers." The papers presented at the conference are available as preprints or in the publication series of the American Institute of Chemical Engineers. Many will also be published in the Journal of Heat Transfer.

The 101st Winter Annual Meeting of the ASME provided lectures and discussions on heat transfer in 19 sessions or symposia. Application areas like frost and ice formation, arctic heat transfer, OTEC systems, MHD and fusion reactors, and nuclear waste disposal were covered. S. P. Kezios, the luncheon speaker, discussed "Perspectives in Heat Transfer—Highlights in the Development of ASME Heat Transfer Division." J. H. Lienhard received the Heat Transfer Memorial Award. The papers presented at the conference are available as preprints or in book form at ASME Headquarters. Many of them will also be published in the *Journal of Heat Transfer*.

The 27th Heat Transfer and Fluid Mechanics Institute was held during 23–25 June 1980 at the University of California. Proceedings of the conference, which presented several papers on heat transfer, are available through the University of Southern California.

The Annual Winter Meeting of the American Society for Heating, Refrigeration, and Air Conditioning Engineers (ASHRAE) was held on 3–7 February 1980 in Los Angeles and the ASHRAE Annual Meeting on 22–26 June 1980 at Denver, Colorado. Many of the papers presented dealt with heat transfer problems. They are available as preprints or in the conference proceedings through ASHRAE Headquarters.

The Sixth National Heat and Mass Transfer Conference in the USSR was held at Minsk, 11–13 September 1980. Information on the proceedings of the conference can be obtained at the Luikov Heat and Mass Transfer Institute, Minsk, USSR.

A joint U.S./Japan Heat Transfer Seminar was held 29 September–2 October 1980 in Tokyo, Japan sponsored by the Japan Society for the Promotion of Science and the National Science Foundation. Thirteen U.S. delegates attended and about twice that number of Japanese delegates. T. Mizushina and W. Aung discussed the present status of heat transfer research in Japan and in the United States, respectively. The subsequent papers dealt with basic heat transfer phenomena, and with applications like heat storage, heat recovery, and non-conventional energy.

Three sessions on heat transfer were included in the program of the 25th Annual International Gas Turbine Conference of the Gas Turbine Division of the ASME. The keynote paper was presented by R. W. Graham on the subject "The Impact of New Instrumentation on Advanced Turbine Research." Papers are again available either as preprints or in book form at ASME Headquarters.

A number of books dealing with heat transfer or including heat transfer topics have appeared on the market. They are listed in the bibliographic portion of this review.

The following highlights illuminate developments in heat transfer research during 1980.

The heat conduction literature was dominated by problems involving change of phase. There was considerable activity in the development of numerical solution methods, and analytical solution methods were also formulated.

Once again this year, heat transfer in channel flows continued to be an area of active experimental and theoretical research. Flows in channels of different geometries and/or flows of non-Newtonian fluids were an area of great interest. Many authors addressed ways to enhance heat transfer by internal finning, roughened walls, and various turbulence promoters. Finally, the problems of the coupled effects of the various heat transfer modes—natural and forced convection or convection and radiation were addressed.

The literature contains numerical and experimental investigations of wall boundary layers. Attention is also given to cylinders in cross flow, stagnation point, free jets, and impinging jets.

Heat transfer in flow with separated regions on single or multiple objects, in cavities or ducts were studied. Impinging jets found attention as a means to increase heat transfer. Considerable efforts were directed to studies of heat transfer in porous media in two or three component systems without or with phase change and with the solid as packed bed or in the fluidized state.

Papers on transfer processes dealt with measurements of turbulent fluctuations, diffusion and thermal diffusion mechanisms, transfer processes in fluidized beds, and moisture transport in soils.

Natural convection research continues to be of expanding interest among heat transfer researchers. A large portion of the papers published this year addressed the problems of combined heat and mass transfer adjacent to inclined plates or cylinders. Various flow conditions ranging from laminar steady flow to turbulent transient convection flows were studied for fluids of constant or variable properties. The following other subjects were also examined with respect to natural convection: thermal instability, periodic vortex formation and staggered and continuous plate arrays.

Analysis and mass transfer experiments increased our knowledge of heat transfer in various rotating systems.

Heat transfer with change of phase remains an active area of research and continues to be one where there is much yet to be learned. Boiling heat transfer literature is divided between papers dealing with fundamentals and papers on practical applications for specific geometries, fluids, or surface materials. Many of the "applications" papers deal with reactor heat transfer. Boiling of cryogenic fluids, boiling on structural surfaces, and dry patch formation and rewetting were discussed frequently. Evaporation of liquid films and condensation on liquid films covering horizontal tubes appear to be areas of increasing interest. Condensation of flow forced through porous media, with applications to secondary oil recovery, was an area of interest which promises to increase in popularity.

Radiation transport in emitting, absorbing, and scattering media is of increasing interest.

Laser–Doppler anemometry techniques continue to be developed for a large variety of heat transfer situations. A substantial decrease occurred in the number of papers describing new hot-wire/hot-film anemometry methods. The number of papers dealing with temperature measurement techniques increased.

Optimization of heat exchangers by new heat transfer surfaces and proper dimensioning of the units in groups of heat exchange equipment was studied.

The number of papers on heat pipes decreased this year.

Solar energy utilization continues to be an active area. The largest number of papers dealt with the experimental and analytical evaluation of the performance of solar collectors. The number of papers dealing with thermal energy storage increased while those describing studies of the radiation properties of materials used in solar collectors diminished.

Plasma heat transfer studies are attracting increasing attention, especially those related to applications of electric arcs.

To facilitate the use of this review, a listing of the subject headings is made below in the order in which they appear in the text. The letter which appears adjacent to each subject heading is also attached to the references that are cited in that category:

Conduction, A
Channel flow, B
Boundary layer and external flows, C
Flow with separated regions, D
Transfer mechanisms, E
Natural convection—internal flows, F
Natural convection—external flows, FF
Convection from rotating surfaces, G
Combined heat and mass transfer, H
Change of phase, J
Radiation
  Radiation in participating media, K
  Surface radiation, L
MHD, M
Measurement techniques, P
Heat transfer applications
  Heat exchangers and heat pipes, Q
  General, S
  Solar energy, T
Plasma heat transfer, U.

CONDUCTION

The conduction literature continues to be dominated by problems of phase change. Solution methodologies, both numerical and analytical, have evoked some interest, as have fins and composite materials.

Experiments on the melting of various paraffins in a vertical annulus affirmed the important role played by natural convection in the melt region [5A]. Interferometric observations added further insights into the role played by liquid-phase natural convection in the freezing of a liquid whose temperature is above the fusion temperature [49A]. In the convectively unstable situation of freezing from above, the convection may decrease the rate of solidification if the Rayleigh number is sufficiently large. For melting from below, natural convection develops rapidly and greatly influences the motion of the phase-change boundary [15A]. Heat transfer during melting at horizontal cylinders was studied experimentally for uniform surface temperature and uniform wall heat flux, for both *n*-heptadecane and *n*-octadecane as phase-change media [6A]. For the onset of surface tension-driven convection, the critical Marangoni number is always smaller for solidifying fluid layers than for those not undergoing a phase change [1A].

During experiments on the effects of turbulent flow on freezing in a tube, an unexpected oscillatory phenomenon was encountered [46A]. Solidification of molten metals flowing in pipes was studied both numerically and experimentally in response to the need for information about the castability of metals and alloys [23A]. The problem of freezing at the wall in a laminar pipe flow has been extended to include axial heat conduction, with solutions being obtained for values of Peclet number as low as unity [26A].

As it cools, the solidified part of an ingot contracts and creates a small gap between the ingot and its mold. Numerical solutions showed that the presence of the gap increases the time of complete solidification [7A]. An analysis of the solidification of metals in a mold takes account of a heat transfer coefficient at the metal—mold interface [10A]. A computer simulation has been performed of the heat transfer, mass transfer, and fluid flow during the growth of a single crystal [22A].

A modification of the heat balance integral enabled a solution to be obtained for inward freezing in cylindrical coordinates [48A]. Biot's variational method has been extended to a problem of melting due to radiative heating; it was assumed that the melt is removed immediately upon its formation [36A]. The fictitious-domain technique, used earlier for homogeneous, steady conduction, has been extended to the two-dimensional Stefan problem [20A]. The total melting time for simple bodies subject to convective boundary condition can be estimated by evaluating an algebraic equation, as can the melting time for a rectangular body with an imposed surface temperature [42A, 43A].

A model describing the heat transfer, solute redistribution, and interface stability during the planar solidification of an aqueous binary solution has been developed [32A]. A generalized solution for unidirectional planar freezing at constant temperature of a supercooled aqueous solution is valid for both dilute and non-dilute solutions and also at both short and long times [27A]. If, in the process of solidification, the liquid phase contains a gas phase in solution, the gas will be rejected at the phase interface due to the difference in solubility of the gas in the liquid and solid. Experiments to clarify the effects of the presence of the gas were performed for freezing of liquid water containing nitrogen gas [35A]. The solidification of a binary alloy has been solved via a series [45A].

Interest remains strong in numerical solution methods. A new algorithm is developed for the numerical solution of the multi-dimensional transient diffusion equation with an implicit formulation. This algorithm can be considered as a generalization of the tridiagonal algorithm commonly used for the one-dimensional transient diffusion problem [51A]. Specialized moving finite elements have been developed that can be employed to generate a nonlinear heat conduction model for situations involving traveling boundary and heat generation fields superposed on an initial state. To facilitate the solution of the resulting nonlinear finite-element formulation, a multilevel heuristic iterative solution strategy is developed [33A]. In a study of the use of finite elements for transient conduction—forced convection problems, the advantages and disadvantages of conventional versus upwind convective finite elements were evaluated [47A]. A grid refinement method was proposed for the finite element solution of transient temperature fields which result from a sudden change in the temperature of a fluid in contact with the boundary of a solid [34A]. For the solution of the Poisson equation by the finite element method, the use of a weighting function corresponding to the control volume method gave more accurate results than did a weighting corresponding to the Galerkin method [37A]. Long-time solutions to heat conduction transients with time-dependent inputs have been obtained by a method which begins with either a finite-difference or finite-element model of the heat transfer problem and produces response coefficients [8A]. A transport equation approximation to the heat conduction equation served as the basis of a new Monte Carlo approach for solving heat conduction problems [12A].

Fins continue to draw some attention. New parameters, the thermal transmission matrices or ratios, have been proposed to replace the efficiency as a performance measure of fins [24A]. The effects of temperature-dependent thermal conductivity and position-dependent heat transfer coefficient were taken into account in solutions aimed at obtaining optimal dimensions of circular fins of trapezoidal profile [38A]. The method of transformed groups was used to avoid the iterative procedure that is commonly used to integrate the nonlinear fin equation [24A]. In an analysis of steady periodic heat transfer in convecting fins of arbitrary profile, both the cases of periodic variation of the base temperature and periodic variation of the environment temperature are considered [3A]. Two-dimensional numerical solutions of a rectangular fin and its related base surface wall were compared with one-dimensional solutions, and the conditions were identified when the one-dimensional results either exceeded or fell below the two-dimensional results [16A].

Composites and layered material are treated in several papers. Green's functions were used to solve the transient heat-conduction problem for a two-layer composite consisting of a core of one material and an envelope annular ring of a second material [39A]. They were also used to obtain analytical solutions for heat conduction in composites of infinite, semi-infinite, and finite laminates in the unsteady, periodic, and steady modes [18A]. In models that were formulated to determine the influence of bonding materials on the heat diffusion in otherwise bi-laminated composites, the geometric arrangement of the composite with the bond was treated as a special type of tri-laminated composite in which each of its major constituents is sandwiched between two bonding layers [31A]. It was shown that two bodies that are pressed together and exchange heat by conduction may not necessarily remain in full contact and can separate locally if heat flows into the material with larger distortivity [11A].

Miscellaneous heat conduction solutions continue to appear. A problem of conjugate heat transfer was solved in which transient conduction in a wall interacted with boundary layer convective heat transfer from one face of the wall [9A]. The thermal stratification of a stagnant lake or tank, subject to surface heating, was solved as a problem of transient heat conduction, with convection and eddy viscosity being neglected [14A]. Charts have been prepared which give the surface temperature oscillations of electrical resistance heaters supplied with alternating current [19A]. In a problem of steady conduction in a rectangle, one of the surfaces was convectively coupled to an isothermal environment while the opposite surface was subjected to a non-zero flux distribution over part of its length. The problem was solved by standard separation of variables [40A].

For the inverse problem of transient heat conduction, an analytical solution was obtained for the case in which the given temperature information is periodic in time [13A]. In a variant of the inverse problem of transient heat conduction, a method is presented by which scanned surface temperature data from a body are used to predict the cavity lying beneath the surface [17A].

Analytical solution methods continue to be evolved. Approximate closed-form solutions to three-dimensional steady conduction problems can be obtained by a method borrowed from elasticity problems [4A]. A boundary-fitted coordinate system which transforms a region onto a fixed rectangular domain was employed as the basis for solving two-dimensional, steady, anistropic heat conduction problems [28A]. The use of a helical coordinate system was shown to be advantageous in solving for the steady temperature distribution in a heat-generating helical coil [50A]. A general steady state solution, obtained by the finite integral transform technique, contains one less infinite series than is encountered when all partial derivatives with respect to the space variables are removed from the differential equation by integral transformation [29A].

Theoretical and mathematical issues were addressed by several authors. In a mathematically oriented study of the nonlinear Fourier equation, it was shown how positive majorants and positive minorants can be used to furnish upper and lower bounds which govern the intensity of the heat generation in a certain exothermic process [30A]. Mathematical issues have been dealt with for a rigid heat conductor with memory [41A]. Second law considerations led to Lagrangian and Hamiltonian formulations of heat conduction [21A]. A thermodynamic approach to rigid heat conductors introduces the heat flux vector as independent variable while its temporal evolution is governed by a first order differential equation. The form of the second law is that wherein the entropy flux and the entropy source are not given *a priori* but are determined through constitutive equations [25A]. For a particular form of the hyperbolic heat equation, it was found that the velocity of a temperature wave in a simple medium depends inversely upon the temperature to the three-quarters power [44A]

CHANNEL FLOW

The study of heat transfer in channel flow configurations was an area of very active research. Experimental results were presented for turbulent heat transfer in large aspect channels. The results were compared to both theoretical and empirical results [25B]. Velocity and temperature profiles were computed for developing turbulent flow in a square duct with constant wall temperature, constant wall heat flux or asymmetric heating. The computations utilized an explicit numerical differencing scheme and an algebraic closure model based upon a three-dimensional mixing length, and the results were in good agreement with measured data [15B]. A numerical study of heat transfer in a 90°, constant cross-section curved duct, steady, laminar flow was presented which was aimed primarily at characterizing the effects on heat transfer of duct geometry and entrance conditions during the developing period of the flow. Calculations were based on fully elliptic forms of the transport equations and a comparison of results based on parabolic equations showed how the latter approach can lead to erroneous results for strongly curved flows [83B]. Experiments were performed to determine entrance-region and fully developed heat transfer characteristics for turbulent air flow in an unsymmetrically heated equilateral triangular duct; friction factors were also measured. The fully developed *Nu* numbers based upon hydraulic diameter did not compare favorably with circular tube correlations. However, excellent *Nu* number predictions were obtained by employing the Petukhov–Popov correlation in conjunction with the measured friction factor for the triangular duct [2B]. Naphthalene sublimation was used to measure the local and average heat transfer coefficients on the downstream face of an enlargement step in a pipe. The highest values of the local transfer coefficient were found to occur on the portion of the enlargement face adjacent to the aperture through which the flow enters the enlarged space. The lowest coefficients occur in the corner where the enlargement face meets the wall of the enlarged pipe [74B]. Naphthalene sublimation was used to study the heat transfer and pressure drop characteristics of an array of staggered plates aligned parallel to the direction of a forced convection air flow. For a given operating condition, the per-plate heat transfer coefficients were found to be the same for the second and all subsequent rows. The fully developed heat transfer coefficients increase with Reynolds number for all of the plate thicknesses investigated and, in general, thicker plates give rise to higher heat transfer coefficients [72B]. An analytical study was made of the laminar flow and heat transfer in ducts whose cross-section is bounded by a wall with periodic corrugations distributed across the span; the other bounding wall is parallel to the corrugated wall and is plane. The results determined numerically demonstrated that if the temperature of the duct wall is to be minimized, a corrugated duct can be highly effective but at the price of additional surface area and greater duct height [67B]. A finite difference formulation was used to analyze laminar fully developed flow and heat transfer for a duct in which there are cross-sectional non-uniformities, e.g. a spanwise periodic array of rectangular protuberances from one wall. The protuberances affected the convective heat transfer by both altering the velocity field and by serving as fins [69B]. Both mixing-related hydrodynamic processes and the effects of the thermal imbalance of the component air streams were studied in a tube downstream of a tee in which air streams of different temperature are mixed [70B]. Complementary heat transfer and fluid flow experiments were performed to determine transfer coefficients and pressure drops associated with the pressure of a slat-like blockage in a tube. The flow was turbulent in all cases, and the heat transfer coefficients in the region just downstream of the blockage were found to be several times as large as those for a corresponding conventional turbulent pipe flow [73B].

Analytical and experimental results were presented for a fully developed turbulent flow in a curved channel with a rectangular cross-section under the condition of a constant wall heat flux. A secondary flow appears in the channel due to centrifugal force, and the flow and temperature fields are strongly influenced. The coefficient of resistance and the Nusselt number were obtained analytically and experimentally for air, and were in good agreement with previous results [38B]. Regular perturbation techniques were employed to investigate the nature of the known equations governing fully developed forced and free convective flow between heated parallel walls. Particular attention was given to the solution near the critical Rayleigh numbers, at which infinite flow rates are predicted by linear theory; and some tentative suggestions were presented on the transitional mechanisms between the various flow regimes [6B]. Measurements of pipes and channels with non-circular cross-section were performed for laminar flow operating conditions [62B]. The problem of turbulent heat transfer in a revolving square-sectioned tube was studied [49B]. The onset of significant departure from isothermalities caused by viscous energy dissipation in flow through a slit was determined for isothermal and adiabatic walls. A series solution of the energy equation enabled calculation of dimensionless profiles for any power law fluid, and such solution provides useful standards for judging the performances of numerical schemes for solving complex non-isothermal flows [82B].

Experimental information was presented for single phase forced convection in a circular tube containing a two-dimensional rib roughness which examines the effect of the rib helix angle. Heat transfer and friction characteristics for air flow were reported for three helix angles (30, 49 and 70°). It was concluded that the preferred helix angle is 45° [19B]. Experimental investigations were performed to improve the concepts of the influence of mass velocity, enthalpy of the flow, and height and and pitch of the fins on the conditions leading to burnout in tubes with internal helical finning [39B]. A numerical analysis was presented for fully developed laminar convective heat transfer in tubes with internal longitudinal fins and uniform outside wall temperature. The distributions of fin temperature, fluid temperature, and local heat flux were shown to be strongly dependent upon the finned tube geometry and in some cases on the fin conductance parameter as well [66B]. A parametric analysis of the performance of internally finned tubes in turbulent forced convection was presented for application to heat exchangers. The best axial internal fins offer less than 10% material savings for equal pumping power; however, material savings are increased to 49 percent using internal fins having a 30° helix angle [79B]. The experimental data on heat transfer in tubes with twisted tape turbulence promotors was correlated in a manner which accounted for the actual pattern of the fluid flow in the spiral channel. It was concluded that sufficient accuracy can be obtained by using the well known equations for laminar and turbulent axial flow in a cylindrical channel [52B]. Experiments to assess the effectiveness of turbulence promoters in tubes with non-uniformly heated perimeters compared the temperature conditions of a smooth tube and tubes with twisted tapes at the same test parameters. The results showed that under conditions of "normal" turbulent heat transfer, the effectiveness of the twisted tape is comparatively low. In contrast, under conditions of impaired heat transfer, there is a high effectiveness attributed to the laminarization of the flow in the smooth tube [31B]. A numerical analysis carried out for laminar heat transfer in a circular tube subjected to an axially periodic variation of the external heat transfer coefficient found that external finning yields substantial heat transfer enhancement compared with unfinned tubes [68B]. Using a parallel plate duct in which one surface was heated and the other surface was insulated, experiments were carried out to examine the effects of a single roughness element on the insulated wall on the heat transfer of the opposite wall. For a variety of roughness elements and over a range of Reynolds numbers, local heat transfer coefficients, the velocity distribution, and the turbulence intensity after the roughness and drag coefficients by the single roughness element were measured. The local heat transfer coefficient distribution along the flow direction after the roughness element shows two or three peaking values, and these peaks seem to be explained by examining the velocity distribution and the turbulence intensity along the flow direction [23B]. Experimental data on hydraulic resistance and heat transfer in a stream of water passing through an electrically heated circular tube fitted on the inside with spiral vortexing bands of various pitches were presented. The results correlated with previously published dimensionless relationships within accuracy of 3–5% [13B]. Experiments on fully developed turbulent flow in a rectangular channel with various aspect ratio were performed to determine the parameters of the velocity profiles over two-dimensional rectangular roughnesses. Results showed that the slopes of the non-dimensional velocity profiles in the smooth and rough zones decrease with increasing relative roughness, height, and drag of the rough wall [46B].

A study was made of the average rate of heat transferred to the wall of a rectangular channel when a hot incompressible non-Newtonian fluid flowed through it. The results showed an increase in the heat transfer coefficient for those fluids which exhibit large strain-rate gradients in the wall region of an arbitrarily rectangular channel [14B]. An experimental study of laminar heat transfer to in-tube flow of non-Newtonian fluids provided results for the Nusselt number which were compared with numerical predictions and experimental data. Two correlations were recommended according to the temperature dependence of the rheological characteristics [30B]. Velocity profiles exhibiting maxima away from the axis of symmetry were calculated for supercritical carbon dioxide flowing vertically in a heated circular duct. The analysis was carried out at the two-dimensional level and the turbulent fluxes were represented by a two-equation model of turbulence [24B]. A calculation procedure was presented for flow and heat transfer in ducts of annular cross-section which was applied to both laminar and turbulent flows. The procedure was used to study heat transfer in the laminar flow of water and ethylene glycol in vertical ducts which predicted the onset of the flow reversal caused by the effect of buoyancy when only one wall is heated [45B]. A study of the heat transfer characteristics to a non-Newtonian flow between parallel plates was performed which accounted for variable fluid properties [42B]. The problem of laminar heat transfer in circular tubes to power law fluids with temperature dependent rheological properties was solved numerically including the effects of viscous dissipation but neglecting internal heat generation. A finite difference scheme applied for the boundary condition of constant wall temperature provided results in good agreement with those previously published [21B]. Numerical computations of the pressure drops in non-isothermal laminar flow of power law fluids in circular tubes with various wall conditions—both constant and variable wall heat flux and temperature cases—resulted in an extended friction factor formula similar to one previously reported. In addition, an approximate friction factor formula was proposed which enables the calculation of the local friction factor when only one value of the so-called film temperature has been previously measured [55B]. The energy equation for mechanically fully developed turbulent flow between parallel plates was numerically solved taking into account the term of streamwise heat conduction. In general, the influence of the streamwise molecular conduction results in a significant decrease of the Nusselt number in the thermal inlet [17B].

The problem of heat transfer in laminar and turbulent flows of non-Newtonian fluids was examined. It was shown that in slightly non-isothermal laminar flow and in the absence of internal heat sources, the dependence of viscosity on the shear rate has little effect on the heat transfer coefficient. In addition, it was concluded that the heat transfer coefficient for turbulent flows of non-Newtonian fluids not exhibiting elastic properties can be calculated from standard equations, employing Reynolds and Prandtl numbers based on viscosity corresponding to the shear stress at the channel wall [34B]. The principal equation of conjugate statements of problems of steady state heat transfer in non-Newtonian liquids flowing along thermally stabilized zones of flat channels were examined [64B]. A monotonic differencing method was used to solve the problem of heat transfer in the flow of non-Newtonian fluids between two infinite parallel plates, where the upper plate is moving in its plane at constant velocity and the lower plate is immobile. The development of the temperature field as well as the variation in the Nusselt number and in the bulk temperature with distance from the start of the thermal inlet were addressed [78B]. Stabilized heat transfer in the flow of rheologically complex fluids in a flat channel with allowance for energy dissipation and high temperature dependence of rheological properties in the fluid was investigated. The heat transfer taking place at the outer surfaces of the channel walls to the ambient medium was assumed to obey Newton's law, and it was shown that the variation in the ambient temperature (under the condition of a constant pressure gradient) may result in a sudden change in the flow rate in the channel [35B]. An analytical solution was obtained for low Peclet number heat transfer with heat generation between parallel plates for a power law fluid. The heat transfer rate increases when fluid behavior changes from dilatent to pseudo-plastic. Positive heat generation due to viscous dissipation and chemical or nuclear reaction increases the heat transfer rate [11B]. Variance and co-variance correlations and a one-dimensional power spectrum with respect to velocity and temperature fluctuations have been measured in an anisotropic turbulent flow of air with a uniform velocity and with uniform transverse temperature gradient. Under the condition that the temperature gradient is parallel to the principal access of the turbulence stress, the vector of the turbulent heat flux and that of the pressure temperature gradient co-variance are parallel to each other; but the ratio of the two vectors is dependent upon the degree of anisotropy in the velocity field and the direction of the two vectors. The ratio of the dissipation time scale for the temperature variance to that for the turbulent energy was found to be the same as the ratio in a completely self-preserved isotropic turbulence [37B]. Heat transfer and friction in a heated turbulent gas were analyzed using a two-equation model of turbulence. Local heat transfer coefficients obtained experimentally were compared with three existing models and a modified version of one of the models predicted the decrease in the heat transfer coefficient occuring in the turbulent gas and a laminarization which occurs at local Reynolds numbers in excess of the critical value for an adiabatic flow [33B]. An experimental study of unsteady mass transfer in a pipe carrying a laminar flow of a clay suspension that exhibits non-Newtonian properties was described. The results, reduced in the form of $Sh/Sh_o = f(Re, So)$ show that the structure properties of the fluids have a significant effect on the time to steady heat and mass transfer [85B]. Experiments made on a liquid metal turbulent channel flow under a transverse magnetic field focus on the magnetic entrance effect on the pressure drop along the channel, mean velocity profile, and turbulence fluctuation intensity. Under a dominant entrance effect, an M-shaped velocity profile is observed at high Hartmann number and the development of a flow under the uniform magnetic field is retarded by the collapse of the M-shaped velocity profile. On the other hand, when the entrance effect is weak, the flow develops slowly within a short distance from the inlet and the turbulent intensity becomes nearly zero [26B]. Correlations were presented for accurately calculating the local temperature of the walls of annuli based upon data of local heat transfer from the inner and outer tubes of annuli (heated from one or both sides) to a stable air flow. These correlations were shown to hold for high heat fluxes as well as arbitrary ratios of tube diameters and of heat fluxes at the wall, for the case of variable physical properties of the gas, assuming constant heat flux densities along the annulus [54B]. A numerical study of heat transfer to generalized Couette flow of a non-Newtonian fluid in annuli with a moving inner cylinder was applicable to the process of coating a wire with a layer of polymer [41B].

The temperature profile in a circular tube of infinite extent through which a fluid is moving under conditions of small Peclet numbers ε is determined by means of an asymptotic analysis in ε. The results show that the heated region extends an $O(\varepsilon^{-1})$ distance relative to the radius of the tube upstream of the point $\chi = 0$, and that convective effects remain important even when $\varepsilon \rightarrow 0$ [1B]. An entirely analytical solution is presented to the Graetz problem for the Dirichlet boundary condition based on a self-adjoint formalism resulting from a decomposition of the convective diffusion equation into a pair of first-order partial differential equations [56B]. The temperature at the top part of the wall of a horizontal pipe conveying a supercritical mixture of steam and water was computed from a dimensionless equation for heat transfer which corrects for the effect of buoyancy [81B]. An iterative numerical scheme was used to analyze the interactive heat transfer problem involving laminar forced convection flow in a vertical pipe and laminar natural convection boundary layer flow external to the pipe. The analysis demonstrated that the pipe Nusselt number is bounded between the values for uniform wall temperature and uniform heat flux, but that the external natural convection Nusselt number is highly sensitive and departs substantially from the uniform wall temperature results [71B]. An analysis was made for simultaneously developing laminar velocity and temperature fields in a parallel plate channel in which both convective and radiative heat transfer were taking place. One wall of the channel was heated and the other insulated as in an air-operated flat plate solar collector. The results showed that radiant interchange causes the task of convective heating of the fluid to be shared between the two walls with as much as 40% of the convective transfer taking place at the externally adiabatic wall [43B]. An analysis of laminar forced convection heat transfer in a horizontal pipe was performed for the case in which the flowing fluid loses heat to the external environment by natural convection and radiation. It was found that whereas the pipe Nusselt number is generally insensitive to the variation of the external convection coefficient, a constant Nusselt number thermally developed regime does not exist [18B]. Recent developments in convective heat transfer calculations using optimal control least-squares penalty finite elements have shown that convergence and accuracy can be better controlled by the least squares penalty formulation than by other methods [32B]. Mass transfer was studied in a two-dimensional channel with porous walls through which two species in solution were separately and uniformly injected. Concentration distributions for each solute were obtained by numerical integration of species conservation equations for $Pe \leq 300$ [58B]. The problem of mass transport for plug flow in a pipe of which the cross-section is given by a parallelogram of various side-ratios and oblique angles was solved. Tables and graphs of the eigenvalues are presented for various oblique angles and side-ratios of the parallelogram pipe [5B].

An analytical solution was obtained to the extended Graetz problem which describes wall flux, based on a self-adjoint formalism resulting from a decomposition of the convective diffusion equation, into a pair of first-order partial differential equations. The solution obtained was in striking contrast with incomplete numerical efforts in the past, and demonstrated the effect of the finite heating section condition. Although the Peclet number is the sole deciding factor for whether or not axial conduction or diffusion deserves inclusion, no global Peclet number criteria may be set for all problems [57B]. By plotting the advancing front theory in dimensionless form, it was shown that the theory can be extended to the non-reactive situation and gives a reasonable approximation for the classical Graetz and Leveque problems [75B]. A computational method was presented to calculate momentum and energy transport in two-dimensional viscous compressible duct flow. The flow in the ducts is partitioned into finite streams, and the difference equations are then obtained by applying momentum and energy conservation principles directly to each individual stream. The method was shown to be applicable to both laminar and turbulent flows [20B]. A more reliable solution was obtained to the problem of solving the transport equation for temperature including convection and diffusion in two dimensions for the thermal entry problem. This solution was obtained by using a calculation domain longer than the entry length and a flux boundary condition [8B]. A complete two-dimensional partial differential equation for developing laminar flow in a circular tube was treated by a finite-difference analysis which allowed for property variations with temperature. The continuity and momentum equation and then the energy equation are solved by direct elimination at each axial step and a marching procedure was used in the axial direction. The step-wise energy balance is rigidly satisfied throughout by using it as a constituent equation. The analysis predicts the complete developing hydrodynamic and thermal fields together with friction factors and heat transfer coefficients, and has been tested for a range of fluid velocity and thermal boundary conditions for various fluids [9B]. A generally applicable approach to developing laminar flow problems in straight ducts of arbitrary shape was presented. A three-level central algorithm which does not require iteration is used to march ahead with the solution in the axial direction. The calculation procedure can be implemented by utilizing with minor modifications any standard finite element code for linear heat conduction and several illustrated examples were presented to demonstrate the accuracy and versatility of the proposed technique [10B]. The assured spectral representation characteristics of differential operators and the method of separation of variables was used in certain cases to solve boundary value problems in transport with mixed and oblique derivative boundary conditions: these cases include (i) a cylinder cooled by a fluid flowing circumferentially around it and (ii) an infinite slab or infinite cylinder cooled by a fluid flowing along the axial direction. Solutions were obtained for both steady- and unsteady-state problems [60B]. A model was presented for heat transfer from an immersed cylinder within a packed bed of particles to gas flowing through the beds parallel to the cylinder's axis. Simplifications in the model permitted a closed form solution which agreed closely with the results of the previous numerical solution when similar correlations to the physical mechanism were employed. The results were shown to be sensitive to the values of the heat transfer coefficient at the cylinder wall [12B]. An analysis was presented to treat the heat exchanger problem between a heat-generating packed bed and an external turbulent flow using a two-dimensional model with one parameter. The analytical results agree well with previously obtained experimental data [48B]. A model was proposed for heat transfer in a turbine agitated vessel from theoretical principles based on the established flow pattern which included the contribution to the heat transfer coefficient from boundary layer mixing [4B].

Experimental studies were presented on heat transfer coefficients in two-phase cocurrent flow in a horizontal tube where liquid jet momentum was used for gas–liquid dispersion and heat transfer. Empirical equations were developed to predict the augmentation ratio of heat transfer as a function of the physical and dynamic variables of the system [50B]. The hydrodynamic and heat transfer characteristics of thin liquid films flowing in a horizontal heated channel were investigated experimentally and analytically. A physical criterion for film breakdown was formulated and incorporated into a model which successfully predicts the critical heat flux for water and organic liquids for $5 < Re < 100$ [63B]. Transient solidification of a liquid in axial flow into and through a cold rod bundle was investigated. Simple relations presented for the liquid mass displaced into or through the rod bundle were shown to be approximately consistent with experimental data [16B].

Experimental data of the critical heat flux of boiling nitrogen in horizontal tubes reported that the critical heat flux is strongly affected by the system pressure and the vapor quality whereas the effect of the mass velocity and the length to diameter ratio of the test section can be neglected. A wettability function was introduced for horizontal tubes that allows a reasonably accurate correlation with previously published results for the critical heat flux for pool boiling [7B]. An analytical study was made of the critical axial heat flux in heated closed and vertical tubes in the case of small length to diameter ratios. An expression for the central heat flux was obtained which showed the dependency of the dimensionless diameter and compared favorably with similar studies [53B]. An experimental and analytical study of heat transfer coefficients on pipe flows of hydrocarbons at transcritical pressures was performed. The experimental data yielded heat transfer coefficients to and from propane over ranges of pressure, heat flux, and Reynolds number and was compared with analytical results [36B].

Solutions to the steady advection-diffusion equation in a branching channel were obtained for both uniform and spatially varying flow fields and for two channel geometries. An interesting feature of the solutions is that anisotropy of the dispersion coefficients in the directions of the streamlines may be accounted for [22B]. The problem of the delayed arrival of a hot or cold front (relative to the flow speed) in a fluid flowing through a pipe was studied using perturbation solutions. The results provided substantiation of preliminary work which predicted that a temperature front propagates at a slower speed than the flowing fluid for small values of the Fourier number [44B]. The problem of intensifying heat exchange in the case of a nuclear reactor with longitudinal flow around the fuel rod elements was studied by applying regular microroughness to the surface of the fuel elements. Results were presented for the intensification of heat transfer and hydraulic resistance as a function of Reynolds number and the height of the protrusions [40B]. An approximate model has been recommended for calculating the non-steady temperature in a nuclear reactor channel which accounts for the height of the fuel element [77B]. Convective heat transfer of laminar mist flow was analyzed for both constant wall temperature and constant heat flux conditions. The diminishing of droplet size, the increasing of vapor velocity, and the dilution of droplet density along the tube were considered [80B]. Typical results were presented for the problem of two-dimensional steady-state heat conduction in coolant channels of different geometries subjected to the third kind of thermal boundary conditions as usually encountered in a regeneratively or dump-cooled liquid rocket engine. The solution method was based on the boundary collocation technique of linear least-squares matching using Householder reflections [76B].

The problem of drying granular solids in a fluidized bed was addressed and a description of the process was presented on the basis of mass transfer coefficients [27B]. The dependence of heat transfer upon the cross-sectional distribution of void fraction in a adiabatic two-phase bubble flow was explored experimentally. It was found that the void fraction distribution had a significant effect on the heat transfer coefficient and increasing the void fraction near the tube wall caused an increase in heat transfer coefficient [51B]. The problem of cocurrent stratified flow of liquid metal and water was investigated in a horizontal rectangular channel. A turbulence model was proposed to explain the experimental results and the analytical results showed good agreement with the experimental ones. The main results included (i) the position of the interface mainly depends upon the flow ratio of both liquids; (ii) the pressure drop is approximately proportional to the square of the average velocity of water; (iii) the heat transfer rate at the direct contact interface is several times higher than that of cocurrent flow separated by a thick solid plate [3B]. Experimental results were reported on the heat transfer and fluid friction of heated helium gas flow in a straight and helically coiled circular duct over a range of Reynolds numbers and the non-dimensional heat flux parameters. At high heating rates, a degradation of heat transfer characteristics was found to occur at local bulk Reynolds numbers well in excess of the minimum number for a fully turbulent adiabatic flow and the resulting heat transfer coefficients were much lower than those associated with the fully turbulent adiabatic flow at the same Reynolds number [47B]. The results of the study of unsteady heat transfer in the cooling of channels by water at supercritical pressure found that the reduced coefficients of unsteady heat transfer decreased with local deterioration in heat transfer. The heat transfer coefficients were determined under transient conditions produced by varying the internal heat release within the channel walls, and the temperatures of the inner surfaces of the channel were calculated by solving the inverse problem of heat conduction via the regularization technique [65B]. Both numerical and experimental studies of heat transfer in parallel nozzles with two-dimensional critical flows were extended to the case of convergent nozzles. According to the results, the critical flow rate increases with the decrease of the convergent ratio and the variation of local Nusselt numbers along the flow direction becomes small by the effect of accelerated boundary layers [28B]. The problem of the accelerated critical airflow and heat transfer characteristics in divergent nozzles in which a sonic point exists midway between the inlet and the exit of the nozzle was studied both theoretically and experimentally taking into consideration the viscous effect. The compressible laminar boundary layer equations were solved numerically using a finite difference method and a chart of the sonic line and the critical flow rate in the divergent nozzle were shown [29B]. A method for calculating visco-inertial flows of supercritical helium in circular pipes with correction for dependence of the turbulent transfer coefficient on variability of properties density fluctuations and thermal acceleration of the flow was developed. The results of the numerical calculations of heat transfer coefficients were in satisfactory agreement with experimental data and confirmed the reduction in the heat transfer coeffficient due to the non-isothermicity of the flow at different conditions [84B]. The results of an analytical calculation of turbulent flow of supercritical helium in a circular pipe under conditions of significant variability of critical properties and free convection (in uptake and downtake flows) were presented. The appearance of a wall temperature peak in the uptake flow, observed experimentally, was attributed to the transition to conditions with thermally turbulent flow [59B].

The thermal and inertial effects in turbulent flow thrust bearings was studied [61B].

BOUNDARY LAYER AND EXTERNAL FLOWS

Literature on boundary layers deals with theoretical and experimental studies of wall boundary layers. Attention is also given to flow over a cylinder, stagnation flows, free jets and impinging jets.

Expressions have been obtained for average heat transfer coefficients for forced convection on a flat plate with an adiabatic starting length [29C]. The Blasius solution has been extended to the case of non-isothermal wall [49C]. The non-similar solutions for compressible boundary layers have been discussed [8C]. Numerical solutions are presented for the heat transfer from a non-isothermal wedge [50C]. Finite difference solutions have been obtained for the boundary layer equations [13C] and for the developing laminar flow in mixed convection [11C].

Measurements of fluctuations of velocity and temperature are reported for the thermal boundary layer downstream of a line heat source [5C]. Holographic technique is used to visualize the flow in a transition situation [7C]; the laminar separation is avoided by surface heating, while the cooling is found to destabilize the boundary layer. Shock-tube experiments are performed for heat transfer to catalytic surfaces [34C] and for end-wall boundary layer characteristics [23C]. The free-stream turbulence is found to have only a weak influence on the heat transfer in an accelerated compressible boundary layer [9C]. An experimental investigation reports the heat transfer rates on end wall and blades in a turbine cascade [19C]. Turbulent spots are measured in a heated laminar boundary layer [51C]. Flat plate heat transfer is experimentally studied in a dissociated laminar flow [14C]. Results are reported for the turbulent boundary layer with slot injection [28C].

Among other studies, an analysis that includes chemical reaction has shown that heat transfer and interfacial shear have a profound influence on the mass transfer rate from a laminar falling liquid film [54C]. A mathematical model has been developed for the propagation of an optical beam across supersonic layers [15C]; the predictions show good agreement with experimental data. Heat transfer in a flow on a continuous moving plate has been considered in [41C] and [3C]. Film cooling is investigated in [17C], while [40C] deals with non-Newtonian boundary layers. The overall resistance concept is applied to conjugate heat transfer [33C].

Several papers deal with unsteady boundary layers and stability. Numerical solutions for the transient boundary layers behind a blast are presented in [30C]. The influence of wall capacitance and resistance on unsteady heat transfer is considered [27C]. The unsteady transport processes on a flat plate are discussed in [31C] and [21C]. An exact analytical solution has been obtained for the transient heat transfer from a plate with periodic variation of temperature [46C]. Influence of oscillatory flow on wall friction and heat transfer has been considered [25C]. An analysis is made of the effect of a time-dependent entry disturbance on a free interface film [32C]. Instability of flows with heat transfer is investigated in [35C] and [24C], while the stability of a liquid layer is discussed in [22C].

A number of investigations focus on jet flows. A finite-difference model has been used for the prediction of surface discharge jets [38C]; the calculations show good agreement with experimental data. The heat transfer in a confined jet has been studied experimentally [18C]. Surface cooling by an air–water jet has been investigated in [20C]. The structure of turbulent jets has been discussed [39C]. Turbulence characteristics of plane and circular jets have been measured [6C]. Heat transfer from in-line and staggered arrays of circular jets has been measured for an extensive range of the geometrical parameters [16C].

Another topic of interest is the impingement or stagnation flow. Heat transfer in a viscoelastic boundary layer at a stagnation point has been considered [43C]. Also, the stagnation point behavior of a micropolar fluid has been studied [45C]. An investigation of heat transfer to a non-Newtonian fluid at an unsteady stagnation point appears in [44C]. Cooling of a surface by a water jet has been considered [26C]. A finite-element analysis deals with the impingement of a free jet [36C]. Mass transfer coefficients have been measured by the naphthalene sublimation technique for an obliquely impinging circular jet [42C]; the point of maximum mass transfer is found to be displaced from the point of geometrical impingement.

Heat transfer from spheres, cylinders, and other bodies has stimulated some investigations. Particle-to-particle heat transfer in a fluidized bed is considered in [52C]. Heat transfer from a particle-laden gas flow has been studied [10C]. An investigation considers the heat transfer from a bed of heated spheres to a slowly moving fluid [2C]. Conjugate heat transfer from a droplet in a creeping flow has been studied [1C]. An experimental investigation has led to an empirical correlation for the effect of the finite length on the heat transfer from cylinders [37C]; for a length-to-diameter ratio of 4 or more, the finite length is found to have little effect. Conjugate heat transfer from a cylinder in low Reynolds number flow has been analyzed [47C]. A simplified model has been presented for heat transfer from a cylinder in a packed bed [12C]. Heat transfer from a tube bank is considered in [4C]. A study deals with the asymmetric boundary layer on a non-isothermally heated cone [53C]. An experimental investigation is reported on the heat transfer from a rectangular body to an external flow at various angles of attack [48C].

FLOW WITH SEPARATED REGIONS AND THROUGH POROUS MEDIA

Separated regions

Turbulent shear stress and heat flux measurements [26D] in a separated and reattached flow over a cylinder with plane leading surface demonstrate that downstream of the reattachment line heat transfer develops much faster than momentum transfer. The same observation was made on a cylinder in axial flow with a conical nose [18D] and with a blunt leading configuration [13D]. A computer calculation [4D] of turbulent heat transfer downstream of an abrupt pipe expansion using the h/ε model for turbulence and a transition function for the near wall region resulted in generally encouraging agreement with experiments. Correlations have been developed [20D] for convective heat transfer at back-steps, open cavities, and cylinders in plane supersonic separated flow. Measurements on a row of four in-line cylinders [1D] revealed that the heat transfer for forced flow changes drastically at a critical Reynolds number $Re_d = 1.14 \times 10^5 (c/d)^{-5.84}$ with c denoting the distance of the cylinders and d the diameter. Correlations were presented [16D] for continuous heat and mass transfer to spherical solids and bubbles in a flow field with high turbulence intensities. Relations are presented based on shear velocity.

Measurements [28D] of convective heat transfer on a plate from an impinging round hot gas jet found that at $x/D > 8$ the jet turbulence augments heat transfer. Two secondary peaks were found [27D] on the heat transfer distributions at a flat plate from a circular impinging jet at a small distance of the nozzle. The outer peak was caused by transition to a turbulent boundary layer and the inner one by turbulence in the jet. The scaling laws [25D] for the center line temperature and velocity of a heated jet discharged vertically into an ambient of uniform temperature were derived from dimensional analysis. Experimental data [21D] were reported for heat transfer from a heated and ventilated plane jet tangent to a plane wall. A computer analysis [22D] describes heat transfer from a gas jet to a porous baffle. Local heat transfer was measured [6D] for an array of round impinging jets with one-sided exhaust of the spent air. The paper reports also a relation of the optimum between the open surface of the perforations and the spacing with regard to heat transfer and pumping power. Experiments [17D] indicate that mass transfer from a gas flow to an array of liquid jets is about 30% higher than mass transfer for flow normal to a solid cylinder at $Re \sim 10$ and that the difference becomes smaller at higher Reynolds numbers.

A new finite-element formulation for convection–diffusion problems [2D] with three-node triangular elements and control volumes avoids difficulties of other methods.

Porous media

An analysis [9D] of heat and mass transfer in fixed beds at low Reynolds numbers agreed well with measured data when based on a realistic bed geometry model. A paper [10D] summarizes heat and mass transfer in packed beds. The electrochemical method was used (5D) to study overall liquid–solid mass transfer in packed beds with upward cocurrent gas and liquid flow. The results are compared with data obtained by other techniques. Experiments and analysis [32D] considered steady non-isothermal liquid flow and heat transfer in sintered fibrous metallic wicks. A closed form solution [24D] was obtained for the unsteady convective heat transfer in a porous medium after a step change of the inlet temperature of the fluid. An analysis [29D] considered buoyancy induced transport in a porous medium saturated with pure or saline water at temperature levels at which the density goes through a maximum. Two-phase, two-component heat and mass transfer in a porous medium subject to a fire was analyzed [30D] and the results are compared with experiments on an alumina powder system. A computer analysis considered moisture migration caused by temperature differences in an unsaturated porous medium with no mass flow through its surfaces for a step change of the temperature of one surface [7D] or for a periodic temperature fluctuation of one surface [8D]. Multi-phase moisture transfer in a drying porous medium considers the dependence on material characteristics [15D].

Fluidized beds

Measurements [31D] of the temperature distribution at the bed inlet and outlet are used to obtain heat transfer coefficients between the particles and the fluid in a fluidized bed. The coefficients are found to vary with bed height, particle diameter, and fluid velocity. A change of flow from bubbling to turbulent was observed [33D] to occur in fluidized beds. Heat transfer to immersed tube banks was also studied. An analysis [14D] considers heat and mass transfer from solids to a rising bubble in a fluidized bed used for drying granular solids. A model proposed by Gabor was used [19D] to predict radiative contribution in a high temperature fluidized bed.

Experiments [12D] were carried out in an air-solid fluidized bed to determine heat transfer to a horizontal tube. Available correlations were found inadequate to predict the measured data. Surface roughness can increase heat transfer from horizontal immersed tubes to a fluidized bed by up to 40% [11D]. Local heat transfer coefficients were found [3D] to change strongly around the periphery of horizontal tubes in air fluidized beds. The mean heat transfer coefficient increases with decreasing particle size and increasing system pressure. Previously reported correlations were found unsatisfactory. The solidity of heat exchangers for coal-fired fluidized beds can sometimes be significantly increased without degradation of the fluidization according to [23D].

TRANSFER MECHANISMS

Much of the published work related to transfer mechanisms is concerned with turbulent flow. A surface rejuvenation model is used to obtain the universal laws of the velocity and temperature distribution near a wall [23E]; the predicted results show good agreement with experimental data for different values of the Prandtl number. A theory of thermally induced surface fluctuations on simple fluids has been developed in [11E]. The homogenization of a transport process is discussed in [18E], while the influence of initial conditions on the properties of Burgers' turbulence is studied in [14E]. The temperature fluctuations resulting from grid-generated turbulence are investigated in [21E] and [24E]. The skewness of the temperature derivative is discussed in [20E]. A new variational principle is proposed for unsteady heat and mass transfer processes [19E]. The velocity and temperature fluctuations have been measured by a hot film probe placed in a flow of mercury [5E].

Among other studies related to transfer mechanisms, the flow processes and heat and mass transfer in fluidized beds have been investigated [3E, 12E]. Ignition and flame quenching of solid fuel dusts are considered in [2E], while results on the turbulent exchange in gas diffusion flames are presented in [17E].

Many papers deal with the transport processes on the molecular or atomic levels. Numerical treatment of free-molecular pure diffusion is outlined [13E]. A theoretical study is made of stagnation heat transfer by accounting for molecular vibrational relaxation [15E]. The transfer mechanisms in molecular Knudsen gas are considered [9E]. A discussion has been presented on a Ginzbourg–Landau constitutive equation for fluctuating heat flux [7E]. A statistical theory is developed for the thermal conductivity of composite materials [16E]. The thermodynamic implications of nonlinear constitutive relations in gasdynamics are discussed in [25E]. An analytical study deals with the molar and molecular processes of heat and mass transfer at high pressures and temperatures [6E].

Other investigations are concerned with a variety of transport mechanisms. An analysis has been made of moisture migration caused by temperature gradients in a porous material [4E]. New solutions are presented for the thermal diffusion effects in a boundary layer on a catalytic surface [22E]. Laminar flow dissolution and diffusion in a non-Newtonian fluid is considered in [10E]. Heat transfer coefficients for gas–solid transport in a pneumatic system have been presented [8E]. Attention is given to the rate of heat and mass transfer from a small particle that is freely suspended in a linear shear field [1E].

NATURAL CONVECTION—INTERNAL FLOWS

Natural convection in open and closed cavities continues to draw substantial attention. As has been true for at least a decade, there are many papers on various aspects of convection in horizontal layers heated from below. There is also interest in related problems of natural convection in inclined plane layers and in plane and other layers with volume heat sources. Papers on natural convection in cylindrical cavities with either horizontal or vertical axes are also in plentiful supply. Related phenomena which have been studied during the past year include combined heat and mass transfer, convective flows involved in melting and solidification, flows in porous media, capillary convection, and mixed natural and forced convection.

For a horizontal layer heated from below, experiments indicate the influence of very large variation in fluid viscosity across the layer; use of the properties at the mid-plane permits application of constant property solutions for both the critical Rayleigh number and the initial variation of Nusselt number above this critical value [1F]. Speckle photographs show constant velocity contours in a layer at low Rayleigh number [37F]. Very high Rayleigh number experiments [18F] show that the Nusselt number varies closely as $Ra^{1/3}$. Velocity and temperature profiles taken under turbulent thermal convection conditions seem to indicate agreement with previous studies [62F]. The onset of Rayleigh–Benard convection in horizontal layers was studied by considering random fluctuations within the fluid layer [27F].

Numerical analysis of convection in a shallow water layer with heating from an array of horizontal cylinders indicates a uni-cellular flow [66F]. Thermal instability in a micropolar fluid layer subject to transverse magnetic field has been analyzed [48F]. Galerkin technique [72F] is used to study the convection in the earth's mantle, where $Pr \simeq 10^{23}$.

Several studies consider the influence of thermal boundary conditions on convection in a horizontal layer primarily for conditions of nearly insulating boundaries—that is a Biot number based on layer thickness and fluid conductivity that is very small [51F]. In the limit of very low Prandtl number, a square pattern of convection cells is indicated [4F]. Large-scale motion for a low Prandtl number fluid is predicted [7F]. Another study considers all boundaries of the layer adiabatic except the bottom surface which has a variable temperature [29F].

Measurements of the velocity distribution in a horizontal layer show that flow regimes leading to the onset of turbulence are dependent on both aspect ratio of the layer and Prandtl number [19F]. The influence of aspect ratio on stability for both internal heating and heating from below has been demonstrated [17F]. An analysis of transient natural convection in a cavity containing water shows the importance of the maximum density point at 4°C [64F].

The influence of variable vertical spacing with a flat upper wall on thermal convection has been examined [12F]. A Wollanstan prime interferometer is used to study the heat transfer in triangular enclosures either heated or cooled from below [16F]. An analysis of convection in a trapezoidal enclosure [24F] agrees well with experimental measurements [25F]. Natural convection as it affects flow in rooms with small ventilation openings has been studied experimentally and numerically [20F].

Flow visualization in inclined air layers heated from below correlates well with heat transfer measurements [55F] and analysis [54F]. The influence of aspect ratio in a rectangular-shaped honeycomb on the critical Rayleigh number has been studied at different angles of inclination [59F].

With volume heat sources in a fluid layer, the layer may tend to be partially stable and partially unstable depending on the boundary conditions at the top and bottom surfaces. For an insulated lower boundary, a higher Nusselt number is observed for rigid–free velocity boundary conditions as compared to rigid–rigid boundary conditions [15F]. An analytical model predicts the form of relationship with Nusselt number as a function of an internal Rayleigh number and Prandtl number [8F]. For non-uniform heat generation, a linear stability analysis has been performed on a layer with a free surface [71F]. An eddy diffusivity model shows asymmetries with respect to the midplane when thermal radiation is significant in a layer heated from below [9F].

Characteristics of convection in a heat-generating fluid contained in a rectangular cavity with cooled side walls indicate the primary flow is bi-cellular circulation [2F]. With different thermal boundary conditions on the side walls, the heat flux in heat generating fluids has been examined using an interferometer [63F]. The temperature distribution following a step change in volumetric heating has been determined [35F].

Several studies examine convection in the annulus between two horizontal cylinders maintained at different temperatures. The development of cellular motion has been predicted in an annulus of fairly small thickness [65F]. Boundary layer approximation to the flow in the annulus appears to agree well with earlier numerical and experimental results [28F]. Numerical calculations over a range of Prandtl number and diameter ratios provide local and average heat transfer results [34F]. A related study [33F] of convection around a single cylinder indicates the problems in using simple boundary layer models. For slightly eccentric annuli, the heat transfer can be predicted from solutions for concentric annuli by a simple radial coordinate transformation [68F]. A correlation [52F] for boundary layer convection dependence upon Prandtl number may be useful within a large annulus.

Both single cells and double cells are found in the convective flow of a layer held between two spheres maintained at different temperatures [6F]. The numerical solution [58F] for such a convective flow agrees well with earlier experimental results. A visualization system has been used to show the flow in the cavity formed between an external spherical enclosure and different shaped bodies within the enclosure [46F].

Temperature and velocity distributions have been measured in a horizontal cylindrical cavity with the ends maintained at different temperature [30F]. Unsteady natural convection has been examined in a rectangular cavity with one side wall heated and the other side wall cooled [45F].

Special numerical techniques have been used to show the stability of convection up to a very high Rayleigh number in a differentially heated cavity [32F]. An analysis [39F] and an experiment [38F] consider the convection in adjacent vertical rectangular channels as would apply in certain solar energy systems.

Several papers have appeared on natural convection in vertical cavities. One considers a transient laminar flow field in a converging vertical channel [26F]. The effects of radiation on convection in a vertical channel have been examined [61F]. The instability of convection in the annulus between coaxial vertical cylinders has been examined [67F]. Another study [10F] predicts the breakdown of the conduction regime in a vertical annulus into multi-cellular flow.

Natural convection generally plays a significant role in determining the melting volume and local heat transfer around a heater immersed in a potentially melting solid. The importance of such natural convection around a horizontal heater has been demonstrated [69F]. The effect of the maximum density point of water on the convection of a melting ice layer has been measured. In freezing on a vertical surface, natural convection can also be very important [70F].

Rotation as well as gravity can supply the body force for natural convection. Several flows in rotating heated cavities have been studied analytically [47F]. A study of combined convection and radiation heat transfer in a rotating fluid considers the effect of radiation on stability [41F]. Rotation in a magnetic field has been studied with low amplitude thermal convection [60F].

Natural convection in saturated porous layers has been reported on with greater frequency in the past few years. Studies include analytical and experimental work on stability, convective flow patterns, and heat transfer in such layers. Lower bounds of the critical time for the onset of convection following a sudden rise in bottom surface temperature of a porous layer have been shown to be best determined by the energy method [5F]. Another study on the stability of flow in a porous medium following sudden heating includes the influence of distributed heat sources [53F]. The evolution of unsteady flow has been followed in a porous layer [23F].

An analysis indicates the onset of convective heat transfer in a vertical porous cavity heated from below [3F]. The onset of convection in a layered porous medium is quite similar to that in a homogeneous layer [35F].

The study of convection in a porous layer having an exothermic decomposition reaction at the bottom portion of the layer shows a thermal front passing through the layer [56F]. Convection due to a point source in a fluid saturated medium has been studied numerically [22F]. In a two-component fluid-saturated porous medium heated from below, the Soret effect appears to stabilize the layer [44F].

In a double-diffusive convection, two gradients—thermal and concentration—can drive the flow. Applications of such a flow in stellar bodies where the Prandtl number is very small have been examined [50F]. A transformation is used to show in a simple form the influence of the Soret and Dufour effects on convection [31F]. Instability in double-diffusive convection in an inclined layer has been examined analytically [43F] and with a flow-visualization system [42F]. The merging of layers in the early stages of double-diffusive convection has been demonstrated [49F].

Surface tension or capillary effects can influence convection in layers with free surfaces. Energy theory can predict the stability of such layers [11F]. Criteria for the instability of a spherical shell indicates the importance of surface tension forces [21F]. Experiments on capillary driven convection show the streamline flow about a growing crystal [57F].

Combined free and forced convection—also called mixed convection—can occur in cavities or channels. Turbulent effects have been measured [40F] in the channel formed between two vertical parallel plates. On one plate there is opposed flow while on the other side there is aiding flow. A mixed flow in the entrance region of a vertical annulus shows potential flow reversal with either aiding or opposing flow [14F]. Measurements show the different flow regime of mixed convection in a heated horizontal tube [13F].

NATURAL CONVECTION—EXTERNAL FLOWS

The method of using the local similarity solution of a free convection boundary layer was used to study steady two-dimensional conjugate heat transfer problems of free convection from a vertical flat plate. The effects of axial conduction in the flat plate on the interfacial temperature were significant in the case of constant heat flux at the outer surface of the flat plate [28FF]. An analytical study of laminar natural convection on both sides of a vertical conducting wall of finite height separating two semi-infinite fluid reservoirs of different temperatures reported representative streamlines and temperature and heat flux distributions [3FF]. A study of natural convection flow arising from a steady line thermal source embedded at the leading edge of a vertical surface was performed for moderately large values of Grashof number by the method of matched asymptotic expansions. Results presented for prescribed wall temperature and for an adiabatic convective wall plume showed that the structure of the wall plume depends strongly on the location of the horizontal plane [1FF]. Natural convection heat transfer from the face-to-face surface of parallel, square vertical plates was investigated experimentally. The experiments encompassed three types of hydrodynamic conditions along the lateral edges—fully open, blockage of one edge gap, blockage of both edge gaps—and the results obtained differed from those accepted in the literature [44FF]. The effect of a buoyant boundary layer spawned by a heated vertical plate on the natural convection heat transfer from an upper colinear vertical plate was determined analytically. Heat transfer at the upper plate was found to be affected both by the preheating and by the finite velocity imparted to the fluid by the first plate, respectively tending to degrade and to enhance heat transfer [46FF]. Local similarity and local non-similarity methods were applied to analyze the effects of buoyancy-induced streamwise pressure gradient on the momentum and heat transfer characteristics in the laminar boundary layer adjacent to a continuous, moving horizontal flat plate [8FF]. The combined buoyancy effects of thermal and mass diffusion on the heat and mass transfer characteristics of a laminar boundary layer adjacent to a continuous plate moving in a vertical or inclined direction through a quiescent ambient fluid was studied analytically. It was found that for the thermal boundary assisting condition, the wall shear stress and the surface heat/mass transfer rates are further increased when the buoyancy force from mass diffusion assists the thermal buoyancy force. Numerical results were presented for the local Nusselt number and the local Sherwood number for diffusion of common species into air and water [47FF, 10FF]. The buoyancy force effects on the heat transfer characteristics of the laminar boundary layer adjacent to an inclined, continuous flat sheet that moves in a fluid at rest were examined analytically. The results showed that a positive buoyancy force contributes to an increase in the local friction factor and the local Nusselt number [29FF]. Basic heat transfer results for natural convection in an array of vertical in-line plate segments were obtained by numerical finite difference solutions and compared to previously computed results for staggered and continuous plate arrays. Results included enhanced heat transfer rates by as much as 90% for fixed values of the wall to ambient temperature difference and heat transfer in the discrete plate array configurations [36FF].

A numerical analysis based on the integral equations of momentum and energy was performed to determine the distribution of velocity and temperature and a heat transfer coefficient for the problem of natural convective heat transfer from an arbitrarily curved plate. The specific case addressed was natural convective heat transfer from the inner surface of an isothermal plate whose leading edge is located vertically and whose end edge is horizontal and is subjected to a variation in the gravitational component in air [26FF]. Data obtained from a heat transfer experiment of a vertical falling liquid film provided new correlations for the heat transfer coefficient for moderate and low Reynolds numbers cases. The results reconfirmed the classification of flow regimes of falling liquid film previously proposed. In addition, it was concluded that the turbulent boundary layer theory was applicable only to high Reynolds number regions (greater than 10^3) and that for moderate and low Reynolds number regions the flow cannot be represented by the concept of eddy diffusivity [24FF].

A numerical solution to the Navier–Stokes equations for laminar natural convection about a horizontal isothermal circular cylinder was provided for $10^0 \leq Ra \leq 10^7$. The flow was shown to approach natural convection from a line heat source as $Ra \rightarrow 0$ and laminar boundary layer flow as $Ra \rightarrow \infty$. It was concluded that boundary layer solutions do not adequately describe the flow and heat transfer at low and moderate values of Ra because of the neglect of curvature effects and the breakdown of the boundary layer assumptions in the region of the plume [27FF]. Laminar natural convection heat transfer from the outside surface of a uniformly heated cylinder (constant heat flux condition) was investigated experimentally at different angles of inclination of the cylinder. General equations for the effect of inclination were determined for both the local and average heat transfer coefficients which increased with the angle of inclination of the cylinder [2FF]. The transverse curvature effect on the heat transfer in the turbulent natural convection flow from the outer surface of a slender vertical circular cylinder was studied by an improved integral method for various values of Prandtl numbers and for various values of transverse curvature parameter [30FF]. The onset of transient natural convection from a suddenly heated horizontal cylinder of finite diameter was studied both numerically and experimentally. The termination of the initial conductive and "locally" convective heat transfer regime which precedes the onset of global natural convection was treated as a thermal stability phenomenon and confirmed experimentally [32FF]. The naphthalene sublimation technique was used to perform experiments to determine the natural convection heat flux distribution on the faces of isothermal circumferential fins affixed to a horizontal heated (or cooled) tube. The findings suggested that the highest heat transfer coefficients occur adjacent to the periphery of the fin while the lowest is at the inner portion of the fin adjacent to the tube. These results are in contrast to the conventional model which assumes a uniform heat transfer coefficient across the face of the fin [45FF].

The problem of free convective flow from a heated sphere in the Boussinesq approximation at high Grashof number was studied numerically. The characteristics of the boundary layer close to the surface of the sphere were evaluated and the eruption of the fluid from the boundary layer into the plume above the sphere demonstrated that only far away from the surface of the sphere may the heated sphere be treated as a point source of heat [35FF]. The development of the boundary layer formed from a buoyancy-driven asymmetric water boundary layer along a heated cylinder was studied under two heating conditions—constant wall heat flux and constant wall temperature. In general, it was found that close to the leading edge, the magnitude of the secondary flow is small and the boundary layer flow is forced convection dominant. The secondary flow grows downstream, the interaction of the free and forced convection becomes important, and the flow becomes free convection dominant further downstream. The temperature dependent viscosity of water has the effect of thinning the heated boundary layer and thus the buoyancy effect and the variable viscosity effect enhance each other over the lower part of the cylinder and compete with each other over the upper part [51FF]. From experimental results, vertical distributions of temperature heat flux densities and heat transfer coefficients were determined for free convection heat transfer from a vertical thin wire to air. The results are in good agreement with those predicted by the approximate integral method taking into consideration the variation of electrical resistance with temperature [15FF]. A numerical analysis was presented of the boundary layer equation with curvature effect terms for free convection about a sphere in air and a simple expression for the average heat transfer coefficient was proposed. The results confirmed the role of curvature effect terms in the continuity equation and in the energy equation is dominant as far as heat transfer is concerned [14FF]. The axisymmetric plume that arises from the point heat source affected by the existence of an externally induced flow aligned with the plume flow was studied for the two Prandtl number values of 7 and 0.7. Of particular interest was the effect of the velocity field on the centerline temperature and the boundary layer thicknesses [4FF].

The series of several variables was applied to the solution of the boundary layer equations of free convection in laminar three-dimensional systems. Temperature profiles obtained from numerical computation were compared with experimental data for the case of free convection over an inclined circular cylinder [48FF]. An analytical expression for the heat transfer coefficient for a high Prandtl number fluid in turbulent natural convection adjacent to a heated vertical surface was presented which compared favorably with previous numerical studies [39FF]. A study of the boundary layer behavior in transient turbulent thermal convection flow concluded that the behavior of the thermal boundary layer is quite different from the corresponding steady-state behavior. This deviation may be an indication of the fact that the effect of flow restructuring in the turbulent core region is not negligible upon the thermal boundary layer [11FF]. A comparative study was performed of the stream function–vorticity formulation and penalty function formulation for natural convection in an enclosure. Numerical results were presented for several geometries, boundary conditions, and the effects of Rayleigh number and Prandtl number on the flow and heat transfer [37FF].

Using a time dependent numerical model, the anelastic–liquid equations were solved in two dimensions and a systematic investigation of compressible convection was presented using an approximate set of equations derived for a compressible liquid of infinite Prandtl number. Both marginal stability and finite amplitude convection were studied and it was concluded that a single parameter defined as the ratio of the depth of the convecting layer d to the temperature scale height of the liquid H_P governs the importance of the non-Boussinesq effects of compressibility, viscous dissipation, variable adiabatic temperature gradients, and non-hydrostatic pressure gradients. When $d/H_P \ll 1$, the Boussinesq equations result; when $d/H_p = O(1)$, the non-Boussinesq terms become important [25FF].

An analytical study was performed of a combined free and forced convection boundary layer problem with the two flows perpendicular to each other. The problem was formulated for flow over a wedge and results were presented for different Prandtl numbers over a range from forced convection dominant to free convection dominant flow [13FF]. The problem of laminar combined free and forced convective heat transfer with blowing along an isothermal vertical wall was studied. Special attention was given to clarify the limits between the combined convection and the effectively pure (either forced or free) convection [49FF]. The periodic vortex formation in combined free and forced convection was visualized experimentally using smoke. The shroud of heated air formed around an approximate point heat source was modeled by an expanding spherical vortex [41FF]. The boundary layer equations of mixed convection were examined in the vicinity of a point of zero skin friction.

The correlation between the uniform wall temperature case and that of the compressible boundary layer flow was outlined, and it was concluded that the Goldstein–Stewartson theory is appropriate for this case. In addition, the case of a uniform heat flux wall was examined theoretically. Significantly, it was concluded that the original Goldstein–Stewartson theory was also sufficient to describe the structure of the singularity at the separation in this case. Indeterminancies associated with the theory were determined via a reconciliation between an analytical and a numerical representation of skin friction and heat transfer coefficients near separation [21FF].

The combined effects of buoyancy forces from thermal and mass diffusion in a laminar boundary layer adjacent to a continuous, horizontal flat plate moving through an otherwise quiescent fluid was studied analytically by the local non-similarity method of solution. In general, it was found that wall shear stress and the surface heat and mass transfer rates increase with increasing thermal boundary force [7FF]. An analytical study was performed to examine the heat and mass transfer characteristics of natural convection flow along a vertical cylinder under the combined buoyancy force effects of thermal and species diffusion. Parameters varied during the study included cylinder curvature, surface boundary condition, Prandtl and Schmidt numbers, thermal and concentration Grashof numbers, and the relative buoyancy force effect between the species and thermal diffusion [9FF]. The conditions marking the onset of vortex instability in mixed convective flow over an inclined surface in a saturated porous medium were investigated by means of a linear stability analysis. The problem was solved numerically for the cases of (i) and inclined surface at constant wall temperature with free stream velocity at zero angle of incidence to the inclined surface and (ii) an inclined surface with constant heat flux with free stream velocity at 45° with respect to the inclined surface. Results showed that the effect of external flow is to suppress the growth of vortex disturbances in both aiding and opposing flows. For the same value of the mixed convection parameter, the opposing flow was found to be more unstable than the aiding flow [20FF].

A regular perturbation analysis was presented for three laminar natural convection flows in liquids with temperature dependent viscosity: (i) a freely-rising plane plume, (ii) the flow above a horizontal line source on an adiabatic surface, and (iii) a flow adjacent to a vertical uniform flux surface [6FF]. Numerical results were presented for the transient and steady-state velocity field, temperature field, and heat transfer characteristics for natural convection at a heated vertical plate in air accounting for internal heat generation and variable properties. One result indicated was the 30% enhancement of the wall heat transfer coefficient in the variable fluid properties case as opposed to constant properties case [50FF].

An equation was provided for predicting the combined convection heat transfer rates to power law fluids for the simple situation of flow past an isothermal vertical flat plate [42FF]. The correlating equations for combined laminar forced and free convective heat transfer to Newtonian fluids was extended to the case of visco-elastic fluids. The results of the analysis found that the effect of visco-elasticity is to increase the Nusselt numbers, especially at higher Prandtl numbers, and that in the stagnation region of a heated horizontal cylinder, natural convection constructively acts to increase the Nusselt number and thus ameliorates the heat transfer [43FF].

The problem of three dimensional thermal instability over a horizontal ice cylinder which occurs in a minimum heat transfer region was solved numerically. The appearance of a convexo-concave melting front which was predicted from experiment was explained by the convection pattern along the cylinder. The transient process of onset of stable vortices around a cylinder was clarified by streamlines and isotherms [40FF]. Experimental data were presented for the onset of manifest convection in water around a heated cylinder. Thin nickel coatings were applied to three Plexiglas cylinders of different diameter, and the nickel surface was used as a heater. Observations made by a holographic interferometer showed that initially conduction prevailed but after a period of time the manifest convection was detected around the cylinder [17FF]. Flow visualization and temperature measurements were used to study conditions in shallow water layers heated from below by cylindrical sources and cooled from above by an air interface. For Rayleigh numbers between 10^5 and 10^6, experiments were performed for a single horizontal cylinder and for an array of five cylinders which demonstrated that the free convection boundary layer which develops on the cylinder is without separation [23FF]. The results on an analytical/experimental study of submerged buoyant fresh water and salt water jets injected horizontally into a quiescent unstratified reservoir showed that fresh water jets of a fixed exit Froude number penetrated to a greater horizontal distance before surfacing as the temperature of the reservoir was lowered. Salt water jets of a fixed exit Froude number penetrated shorter horizontal distances into the reservoir as the same concentration was increased. The dependence of the flow characteristics upon temperature and salt concentration was shown to be a significant factor contributing to the wide discrepancy in previously reported data [38FF]. A linear small perturbation method was applied to the problem of cellular convection in minute droplets on a heated solid surface. It was concluded that (i) larger values of the Biot number lead to greater stability, (ii) a hemispherical droplet is more stable under surface tension variations than a flat liquid layer, and (iii) a parabolic steady thermal profile is more susceptible than a linear one to thermal disturbances due to surface tension force [18FF]. Experiments of free convection heat transfer from electrically heated platinum wires and a platinum strip to supercritical carbon dioxide demonstrated that the heat transfer can be predicted by a conventional Nusselt type correlation if the dimensionless numbers are based on integrated thermophysical properties [31FF]. An experimental study of free convection heat transfer from a copper sphere to supercritical helium presented empirical correlations accurate to $\pm 20\%$ between Nu, Pr, Gr, $\rho_b \Delta\rho$, T_b and ΔT. A boiling-like phenomenon was observed experimentally near the critical point in that the heat transfer coefficient h increases with temperature difference until h reaches a maximum after which h decreases for large ΔT [19FF]. Experiments conducted on the heat transfer from internally heated $ZnSO_4$–H_2O pools in curved surfaces extended the existing data for non-boiling pools to higher Rayleigh numbers. The data for convective downward heat transfer from non-boiling pools to a curved surface were reasonably close to the Mayinger correlation extrapolated to higher Rayleigh numbers and lower ratios of pool depth to the radius of curvature. Sideward heat transfer to a surface could be described by $Nu = 0.7Ra^{0.2}$ [16FF].

Experimental results quantified the fact that when a vertical ice surface melts into a stable salinity gradient, the melt water spreads out into the interior in a series of nearly horizontal layers. A puzzling feature of the results is the relatively weak dependence of the layer scale on local salinity, though the vigor of convection and the rate of melting are greater when the salinity is high [22FF]. Calculated numerical results were presented from laminar buoyancy induced flows driven by thermal transport to or from a vertical isothermal surface in cold pure pure and saline water wherein a density extremum arises. Buoyancy force reversals were shown to arise for a vertical ice surface at 0°C melting and fresh water between 4 and 8°C at atmospheric pressure. Temperature conditions for which buoyancy force reversals occur are of special interest because of the resulting anomalous flow behavior and low surface heat transfer rates. The transition from conditions with no buoyancy force reversal to those resulting in a large buoyancy force reversal is accompanied by as much as 50% decrease in surface heat transfer [5FF]. The intradiurnal heating and cooling cycle of the mixed layer of the tropical ocean was investigated through the use of a pseudo-two-dimensional numerical model. Particular emphasis was given to two-component diffusion resulting from dynamic instabilities in the water column. The resulting Rayleigh numbers indicate the possibility of double diffusive convection whereby the vertical transfer to salt and heat may proceed at rates far greater than can be accounted for by molecular diffusion alone [12FF]. The concept that the usual Navier–Stokes equations are not suitable for studying the flow of water at 4°C was extended to include the problem of unsteady free convection flow of water at 4°C past an infinite vertical plate following a depth change in temperature of the plate. It was found that the skin friction increases due to the increase of the temperature of the water when the plate is being cooled, and decreases when the plate is being heated due to free convection currents. The rate of heat transfer increases with the increase of the temperature of the water in both the cooling and heating cases [34FF]. The finite element scheme proposed by Galerkin was used to explore the effects of thermally induced property changes during the extrusion of a flowing, molten polymer. A new phenomenon, thermal extrudate swell, was discovered and it was found that extrudate expansions up to 70% of the die diameter may occur in a Newtonian fluid with thermal properties similar to those of low density polyethylene [33FF].

CONVECTION FROM ROTATING SURFACES

The temperature distribution was calculated [8G] for flow around a rotating disk including the effect of viscous heating. Buoyancy effects were analyzed [6G] for laminar flow and heat transfer on rotating hemispherical bodies in the range $0 < Gr/Re^2 < \infty$ and $0.72 < Pr < 100$. Natural convection mass transfer to Newtonian and non-Newtonian fluids from rotating vertical cylinders could be described [4G] by the following equation: $Sh\,Sc^{-1/3} = 0.62 f Re$ with f indicating the friction factor for a Schmidt number varying between 2.6 and 32,000 and a Reynolds number between 10^2 and 10^5. Mass transfer to non-Newtonian fluids was found to be lower than what would be predicted by the Newtonian relation. An analysis [3G] considered natural convection in a rotating spherical annulus with a uniform gravitational field parallel to the axis of rotation for Reynolds and Prandtl numbers larger than those studied before. A numerical solution [9G] describes laminar developed heat transfer in a rotating cylindrical tube for a wide range of Prandtl numbers. Heat transfer between eccentric rotating cylinders was also analyzed [5G]. Mass transfer experiments and analyses were published for rods and plane strips rotating around an axis normal to their length dimension [2G]. Particle motion in the flow around pitched blade turbines was observed [7G] stereoscopically. The spiraling flow of a viscous compressible gas surrounding a rotating porous cylinder with fluid ejection was calculated [1G].

COMBINED HEAT AND MASS TRANSFER

Much of the activity in combined heat and mass transfer continues to relate to film cooling and transpiration cooling. Interest in such flows comes, to a great extent, from the development of high temperature gas turbines. Today, most high performance engines use film cooling as well as other techniques for maintaining reasonable temperatures in the high temperature stages, certainly with stationary blades if not with the rotating blades as well. This film cooling usually occurs with a row or a series of rows across the span of the turbine blades, often in the stagnation region and sometimes along the pressure and suction sides of the blade as well. In addition to film and transpiration cooling, heat transfer to impinging jets, and combined heat and mass transfer in porous media has been studied.

Experiments on the influence of injection hole spacing and orientation with a single row of injection holes appear to agree with earlier studies [5H]. With two rows, one downstream of the other, a staggered pair of rows is better than an in-line pair [1H]. The use of ammonia injected from a turbine rotor blade surface with ozalid paper along it shows the jet flow attachment on a rotating system downstream of the film cooling holes [4H].

Injection of a film coolant can produce a loss in the aerodynamic efficiency of a gas turbine stage. An analysis of such loss considering the mixing of jets with the freestream agrees well with measurements of the head loss in the flow through a cascade [10H].

With full-coverage film cooling, there is an array of injection jets along the whole surface of the wall to be cooled. In the limit with increasing hole density per unit area (and of course decreasing hole size) full-coverage film cooling would approach transpiration cooling. The velocity fluctuations and turbulent shear stress have been measured [19H] in the region of full-coverage film cooling while a model [20H] predicts the flow recovery downstream of the injection region. Measurements of the average heat transfer in this system [2H] were correlated by a numerical procedure [3H]. Another study of full-coverage film cooling on a turbine blade indicates the cooling effectiveness as the system approaches transpiration cooling [7H].

Film cooling is used elsewhere in gas turbine systems, for example, in the combustor section. There, where the flow can approximate a two-dimensional film cooling more closely than on a turbine blade, the design of the slot geometry is still very important. The turbulent mixing that can occur following injection tends to decrease effectiveness [17H].

An analysis [12H], using a series expansion, finds the maximum heat transfer was not at the stagnation point for an axisymmetric flow with mass transfer. Analytical approximations have been developed to predict the heat transfer with transpiration cooling into a laminar boundary layer using different injection gases [11H].

Ablation cooling is used for protecting vehicles entering a planetary atmosphere. One study for the Galileo probe, which will be used in exploring the Jovian atmosphere, shows how turbulence can permit significantly increased surface radiation [13H].

An array of jets impinging on a surface containing vents to remove the spent airflow, gives higher heat transfer than a similar flow with the spent air leaving at the sides of the plate [9H].

Combined heat and mass transfer in porous media is important to such fields as drying, geothermal energy utilization, and heat transfer from underground buildings. A finite-element analysis [16H] has been found useful in studying applied heat and mass transfer in porous media including the influence of variable properties and composite porous media materials. Another study of coupled heat and mass flow in a composite medium shows the effect of the coupled flow on internal stresses [15H]. The influence of the Soret and Dufour effects on the diffusion of heat and moisture in solids has been analyzed [6H].

Simple models for combined heat and mass transfer have been applied to a laminar liquid film flowing down a vertical surface [14H]. Experiments show that the heat transfer over a flat plate may be augmented by a significant factor when mist flow is present due to evaporation of the water droplets [8H].

A "film penetration model" has been used to study convective heat and mass transfer between two phases [18H]

CHANGE OF PHASE

Boiling

Nucleate boiling. The influence of nucleation of one bubble on another during nucleate boiling was analyzed [57J] and photographs of the process were taken. Correlations for the onset of nucleate boiling were reviewed [41J], several of which were recommended. Boiling on a thin heated wire of pulsed current was investigated [104J] to study boiling with intermittent superheating.

A study of growth of bubbles on heated surfaces [134J] lead to an expression for the bubble breakoff diameter and departure frequency. The effect of hydrodynamics on the shape of vapor bubbles [108J] was analyzed. A model for the equilibrium radii of vapor bubbles and liquid droplets was devised [84J] using the Thomson–Helmholtz equation for vapor pressure on curved surfaces. It was concluded that, in many instances, the difference of pressure across the interface is significant. The conservation equations were solved [18J] to determine the rate of bubble growth on heated surfaces within a superheated liquid boundary layer. A correction to account for difference in refractive index of light in photographic studies of bubble growth was developed [143J]. With the correction, it appears that a bubble remains spherical from nucleation to breakoff.

Several studies dealt with boiling of non-standard fluids. The mass transfer of the liquid–vapor interface was evaluated analytically for two-phase multi-component mixtures [112J]. The mechanisms of helium boiling were experimentally investigated [102J]. Helium flow boiling in tubes [2J], low pressure boiling of cryogenic liquids [98J], and film boiling of the super fluid Helium II [76J] were studied. Evaporation (thermal decomposition) of foams was studied [43J] experimentally using a 0.6% surfactant solution in tap water.

Enhanced heat transfer using a surface with pores and interconnecting tunnels was studied experimentally [91J] and analytically [92J]. It was found that with this surface/subsurface geometry, wall superheat could be reduced 80–90%. A review of structured-surface enhancement literature was made [132J]. A study on surface effects of prolonged boiling [52J] showed no support of the ageing theory.

Enhancement by vibration was studied on a horizontal wire [93J]. The wire was mechanically vibrated; an increase of heat transfer was observed at certain frequencies. Various enhancement techniques were tested on boiling of cryogenic fluids [99J]. The heat transfer coefficient was increased 8–10 times for nucleate boiling by covering the tube with wick material, and was increased 5 times for film boiling by introduction of internal fins.

Forced convection boiling. A mathematical model of dispersed annular forced-convection flow was developed [60J] by applying conservation equations for each region of the flow. A model for the variation of flow regime along a vapor-generating channel [73J] was presented. An analysis of laminar dispersed (mist) flow under constant heat flux and constant wall temperature boundary conditions [137J] was used for discussion of the fundamentals for this flow regime. Spectral analysis of the acoustic noise of boiling has been employed. Forced convection boiling experiments in an annulus [131J] revealed information about boiling modes, both in the spectra and in total intensity. The onset and development of harmonic, high-intensity pressure fluctuations in steam-generating channels was studied [100J] by spectral composition. Flow stability in commercial boiler tubes was experimentally studied using spectral analysis [113J]. Stochastic fluctuations of void fraction in a flow-boiling channel were studied using spectral analysis [55J]. The results lead to a model for propagation of vapor fraction perturbations. An analysis of subcooled low quality boiling at high pressures resulted in a model for prediction of void fraction [31J]. An experimental study was made with boiling at low pressures [58J]. Flow instability measurements in multi-channel boilers were taken [59J]. Flow pattern instabilities were observed. The effect of tangential conduction and non-uniform heat flux along the perimeter of a horizontal tube with saturated flow boiling was studied [14J].

Annular dispersed flow boiling in tubes with flowing sodium [142J] showed that phase change was by evaporation from the liquid film surface without vapor generation at the wall. Injection of inert gas or the presence of artificial cavities could be used to fix the point where the vapor phase appears. The effects of centrifugal acceleration and subcooling on nucleate boiling of cryogenic fluids (nitrogen and hydrogen) were evaluated [69J]. Acceleration was a high as 2000 g; the effects on the onset of nucleate boiling and on CHF were observed. Boiling Freon-113 (both forced convection and pool boiling) was experimentally studied in an upflow vertical tube to observe nucleate, transition and film boiling [140J]. CHF increased with the root of velocity and transition boiling was very sensitive to velocity. Experiment with refrigerants and refrigerant – oil mixtures showed the effect of oil carry-over on evaporator performance [22J].

Turbulence promoters were evaluated for boiling flow in an annulus [21J] with uniform and distributed heat flux. Internal spiral ribbon mixers were found to be more effective for single-phase heat transfer than for subcooled nucleate boiling [3J]. The evaluation used the parameter pumping power/heat transfer as a descriptor of performance.

A correlation of data for commercial heater tubes and common heat transfer fluids was presented [119J]. A survey of forced convection boiling correlations was presented [44J].

Droplet dispersed flow boiling between rod bundles during reactor emergency cooling conditions was analytically studied [136J]. There was an extra dependence of Nusselt number upon mean vapor temperature; it was concluded that pipe flow is a poor approximation to this complex geometry. A two fluid (liquid and vapor) model of two-phase flow as proposed for nuclear reactor studies [20J]. The conservation equations were written for the two fluids separately; they were matched at the phase boundaries.

An effect of electric field on forced convection boiling was found [12J] for field strengths greater than 1000 kV/m.

Forced convection critical heat flux in a vertical tube was studied under various conditions. General features of CHF in flows in vertical heated tubes with zero inlet subcooling were presented [62J]. A test of previous correlations against new data was made [63J] and appropriate dimensionless groups were recommended. CHF with reference to dryout, inlet subcooling, and boiling length was evaluated analytically [64J]. A high-pressure correlation ($0.1 < \rho_v/\rho_l < 0.3$) was recommended [65J]. Burnout under conditions of subcooled boiling and forced convection was discussed [87]. A generalized CHF correlation based upon 800 data points for flow in an internally heated annulus [118J] was given. Annular CHF data with (i) the rod heated and shroud unheated with (ii) with the rod unheated and shroud heated was correlated [66J]. CHF data in parallel flow between heated plates was taken and correlated [67J]. The effects of channel length and dissolved gases on CHF were evaluated; [68J] and [32J], respectively. A state-of-the-art review of the boiling crisis with respect to reactor safety was given [130J].

A study of CHF on a heated plate with an impinging water jet [87J] revealed different boiling regimes depending upon volumetric flow rate.

Evaporative heat transfer to dispersed flow was analyzed [71J] as a three-path problem; (i) wall-to-vapor, (ii) vapor-to-liquid and (iii) wall-to-liquid. It was concluded that the wall-to-vapor heat transfer was predominant. The minimum heat flux and lower-bound for droplet size in post-dryout dispersed flow was analytically evaluated [25J]. The predicted minimum drop sizes were smaller than observed in tests. A non-equilibrium model for dispersed flow based on heat transfer effectiveness between phases [115J] was developed to describe heat transfer and droplet diameter. In an experimental study of the same flow type, vapor superheat was measured [95J]. Details of the superheat probe design were presented. The effect of roughness in dispersed flow and annular dispersed flow was analytically evaluated [125J]. A perturbation technique applied to the hydrodynamic boundary layer in subcooled film boiling showed that shear stress at the phase interface influences the potential flow only slightly when the subcooling is large [29J]. When the core flow is near the saturation temperature, the other extreme, the inviscid solution does not describe the liquid motion; but, for the heat transfer problem, the liquid boundary layer and the vapor are decoupled. Visual observations in an annulus with and without obstacles on the heated wall showed that dry patches form upstream of the obstacle for slug and froth flows and downstream for annular flows [30J]. In an analysis of rewetting [116J], it was determined that the process of removal of a non-condensible absorbed gas layer from the surface was controlling.

Natural convection boiling. Basic studies in the mechanisms of natural convection (pool) boiling included one on the dynamics of growth and breakoff; behavior of the microlayer, and boiling in slotted channels [6J]. Another investigated the bubble departure radius under low gravity (1 g and below) conditions [23J]. It was found that the critical radius increases considerably under "microgravity" conditions in boiling CBr_4. A study of the mechanisms of boiling and evaporative cooling showed fluctuations of surface temperature in the vicinity of a single nucleation site of 6–10°C [120J]. The mechanisms of boiling on a horizontal plate or horizontal wire in helium were experimentally studied with an emphasis on rates of bubble growth, break-off diameter and departure frequencies [103J]. Motion pictures were taken. Holographic interferometry was employed in the investigation of the initial stage of nucleate boiling in a superheated boundary layer [51J]. An analysis of thermal stability of liquid droplets on a heated surface was made [46J] using small-perturbation methods. Droplet stability increased with bi- and hemispherical droplets were more stable than flat layers. Taylor instabilities in boiling, melting, condensation or evaporation were analyzed and investigated experimentally [127J]. The effects of viscosity, fluid layer height and surface tension on Taylor wavelengths were of concern. Contributions of various boiling mechanisms at subatmospheric pressures were assessed for a flat horizontal surface [167J] and for a horizontal tube [28J]. Heat transfer in a narrow vertical channel could be categorized into three modes [121J]: (i) natural convection without boiling, (ii) intermittent vapor generation, and (iii) vapor lock.

Experiments in pool boiling of mercury on a horizontal plate with an imposed magnetic field parallel to the direction of gravity [128J], showed that increased magnetic flux decreases both the incipient boiling and burn-out heat fluxes.

The effect of surface boiling on the process of material vaporization using high intensity irradiation was studied [17J]. An analytic study of boiling heat transfer on slotted surfaces [72J] lead to some recommended correlations. Boiling heat transfer and CHF were experimentally studied on flat plates with slotted, capillary and capillary-porous (wick) structures [1J]. A thin film if silicone was added to an upward-facing copper surface making it an ill-wetted surface for water [138J]. The behaviour of bubble growth was similar to that of film boiling on an ordinary (wettable) surface which is quite different than that of normal nucleate boiling. Sublimation of a horizontal slab of dry ice was presented as an analog to pool boiling on a flat plate [26J]. This study was presented as a model for the problem of reactor core (UO_2) melt. The effect of wall curvature on boiling heat transfer from a hot pool to a cold wall was experimentally evaluated [39J].

Augmentation of pool boiling by electrolytic generation of hydrogen gas was found to be effective for higher rates of heat transfer [90J]. A regression analysis applied to 5000 natural convection boiling data points [129J] resulted in several recommended correlations. The domain of fluid types was separated into water, cryogenic fluids and refrigerants.

Several effects on boiling heat transfer with sea water were studied (106J). Surfactants can increase the heat transfer coefficient by making bubbles smaller and departure frequency greater; a correlation for the effect of scale formation with sea water was presented. A fluidized bed condition was found to have enhanced boiling heat transfer from a horizontal cylinder [126J].

Critical heat flux data were found experimentally in a natural convection flow chaneled between two parallel, horizontal disks, where the lower surface was heated [61J].

An analytical model for the prediction of the minimum film boiling temperature [78J] was developed. The minimum film boiling heat flux from flat plates and spheres in saturated and subcooled liquids was found [45J] to be variable over a wide range, depending upon properties of the heater surface and boiling fluid.

Condensation

Several studies of the mechanisms of condensation were made. Approximations of the isotherms of superheated vapor were used to calculate the pressure in an embryonic droplet [139J]. The theory of nonequilibrium phase change and the advisability of using simple methods to calculate mass transfer were discussed [81J]. The latent heat release due to condensation, vapor phase removal and change of vapor structure were taken into account [109J]; at higher pressure, the critical radius of droplets is comparable to the mean free path. Transient surges in tube-type condensers were analyzed [10J, 11J]. Condensation of a vapor into fog at high temperature is radiatively induced [19J] and greatly enhanced heat transfer communication from the gas to a boundary. Heat transfer by film condensation in a binary mixture was analyzed [74J]. There is species diffusion of the more volatile away from the wall.

Film condensation. Heat transfer on a vertical surface with film condensation of a stationary vapor was evaluated [75J]. Laminar film condensation in the presence of waves was analyzed [123J]. Perturbation methods were applied to study nonlinear stability of a vertical condensation film [133J] leading to an expression for the wave amplification rate. An experiment was made [40J] to study the film breakdown and dry patch formation process. Dropwise condensation of steam in a vertical tube was studied [54J]. The effect of surface thermal conductivity on dropwise condensation was evaluated experimentally [124J]. Condensation curves of Q vs ΔT were essentially the same for copper, bronze and ptfe-coated surfaces. An analysis of the effect of droplet trajectory on heat transfer from cylinders having a liquid film boundary layer at the surface and a vapor outer layer was made [79J]. A set of equations was recommended for laminar film condensation of a binary miscible vapor [37J].

Forced convection condensation. Forced convection condensation on a horizontal tube was treated theoretically [33J] and experimentally [35J]. Forced convection condensation on horizontal flat plates and in horizontal tubes in the presence of a non-condensing gas was investigated [114J]. Film behavior at breakdown on the flat plate has similarities with open channel critical flow. Forced convection laminar filmwise condensation on a vertical tube was found to be characterized by five dimensionless parameters [27J]. Centrifugal acceleration was found to enhance film condensation heat transfer considerably [86J].

Condensing flows in porous media were studied. An analysis of flows similar to those in steam-injection secondary oil recovery was analyzed [97J]. A condensing wave, which has a saturation shock discontinuity, propagates into the dry, porous matrix. For small ΔT there is no local temperature change but there is a discontinuity in quality. A similar flow was studied experimentally using Freon, [96J]. The axial distribution of condensate was wavelike and had a saturation-jump at the leading edge.

Natural convection condensation. A study of flow patterns for condensation on the outside of horizontal tubes [15J] showed that a key parameter is the ratio of shear to gravity forces. Laminar filmwise condensation on a horizontal tube with circumferential grooves was modeled [34J]; the results were compared with experimental data. A similar analysis was made for finned tubes with convection on both sides of the tube wall [111J]. An analytical and experimental program on finned tubes found regions of high heat transfer coefficient in the trough of vertical finned tubes [48J]. The film in the trough was thin due to suction of the liquid flowing within the trough. It was suggested that fins and troughs be made thinner. An analysis of a vertical fluted surface was made [101J] by first analyzing the hydrodynamics of the isothermal flow, then including condensation of vapor to find an expression for the average Nusselt number. A study of condensate film flow in horizontal tube bundles [122J] showed that a reduction in spacing reduces weight of liquid films and a reduction of thermal resistance. Various enhancement techniques used in condensing steams were reviewed [82J]. Condensation of steam in contact with cold water (pertinent to LWR safety) was reviewed [5J]. Instabilities leading to large and rapid variations of heat transfer coefficient are poorly understood.

Free condensation. Equations describing the dynamics of vapor bubble collapse were developed and an expression for the heat transfer between phases [94J] was formulated. This vapor bubble model could be used to predict shock waves in liquids containing collapsing vapor bubbles. A similar analysis [38J] was made to determine the pressure wave produced by the collapse of a bubble. Direct contact condensation of vapor on a spray of subcooled liquid droplets was studied analytically and experimentally [77J]. The most important parameter was the initial drop size. A "universal" flow map was developed [13J] for direct contact condensation problems. Measurements were made in the condensation region of a steam jet submerged in water [4J].

Heat transfer studies in two-phase closed-type thermosyphons [9J, 53J] lead to correlations of heat transfer coefficient and optimum quantity of working fluid. Rates of thermocapillary convection were measured and compared with analysis [7J].

Vaporization

Vaporization of films. Liquid vapor interaction and entrainment as found in falling-film evaporators were analytically studied [141J]. The tubes were horizontal and liquid entrainment was by stripping action of the crossflow. Stability of the falling film was experimentally evaluated [42J] leading to curves of minimum irrigation density versus heat flux. A mechanistic model for a liquid film draining down a heated wall countercurrent with its rising vapor [117J] showed that the liquid film downflow is limited by vapor upflow. The geometry simulated a PWR downcomer. A study of heat transfer to thin liquid films with low and high evaporation rates [135J] lead to some key parameters to the problem.

An analytical model for predicting moisture and temperature distribution in a moist, porous material during drying was developed [80J] which accounted for departure from thermodynamic equilibrium. A numerical study was made of evaporation of moisture on a plate maintained at a constant wet bulb temperature [89J]. A round, laminar hot-air jet impinging upon the surface effected the drying. Drying of a dielectric surface using radio frequency and microwave electromagnetic fields was studied [105J]. The process characteristics were determined by the power absorbed by the wet surface, pore structure of the surface and effective specific heats.

Free vaporization. Vaporization in free flashing liquid flows was studied. Flashing in flowing fluids is due to (i) static decompression and (ii) turbulent fluctuations [56J]. Spray flashing of superheated water was found to be faster than in a liquid jet or a suddenly depressurized pool [36J]. Experiments with isoenthalpic flashing of flowing water in pipes showed that the liquid temperature exceeded the vapor temperature [83J]. Comparisons were made between equilibrium and measured qualities.

Evaporation of free droplets in laminar tube flow at low pressures was studied [110J] under constant heat flux and constant wall temperature boundary conditions. A theoretical study of free droplet evaporation indicated that the process is unsteady and the droplet liquid is non-isothermal and is circulating [107J]. The effect of an ion wind on free surface evaporation into a laminar air stream was experimentally evaluated [47J]. Under laminar boundary layer conditions, the evaporation rate increased as the root of the current density. With high current density the boundary layer became turbulent and the evaporation rate became less sensitive to current density. Evaporation of a liquid drop on a hot liquid surface was experimentally studied [49J]. The evaporation time was comparable to that of the case of a droplet on a hot solid surface. The experiment allowed evaluation of liquid–liquid direct contact heat transfer. A similar experiment [50J] was used to identify twelve types of heat transfer modes. A numerical analysis of a droplet evaporating on a liquid surface [70J] showed that evaporation was limited by resistance to heat transfer within the droplet. A closed periodic condensation—evaporation cycle of a droplet (i) in a thermal gradient or (ii) between two stratified fluids, was analyzed [85J]. The droplet falls into a hotter, more dense liquid, vaporizes, then rises into the colder, less dense liquid and condenses, completing the cycle. Cinemicrophotography was used to study the boiling of water-in-oil emulsions [88J]. Two processes of bubble formation were found: (i) bubble formation on the solid surface and (ii) casual bubble formation away from the surface. The effects of sodium entrainment and heat transfer with two-phase UO_2 were studied [24J]. Taylor instabilities can cause entrainment of sodium coolant. The dominant heat transfer mechanism is radiation. The effect of heat transfer is to reduce fuel vapor pressure and reduce expansion work.

RADIATION

Radiation in participating media

Radiative transport in emitting, absorbing, and scattering media is of continuing interest. Exact solutions have been obtained for radiation heat transfer between parallel plates separated by an absorbing, emitting and Mie-anisotropically scattering medium [31K]. Using a differential approximation for analyzing an axially symmetric radiation field in a gray medium within a finite, cylindrical enclosure results in solutions which demonstrate that the accuracy of the differential approximation is of the same order for axially symmetric and one-dimensional problems [13K]. The radiative heat transfer problem for an isotropically scattering slab with specularly reflecting boundaries may be reduced to the solution of a set of algebraic equations by expanding the source function in Legendre polynominals in the space variable in the integral form of the equation of radiative transfer [19K]. A rigorous solution of radiative transfer in an absorbing and anisotropically scattering slab based on projectional methods results in computational formulae which can be numerically processed [23K]. There is a possibility for a unified presentation of experimental results obtained along isochores and isotherms and along the coexistence curves of CO_2 and Xe in the critical region, considering multiple scattering contributions to the depolarization of radiation [25K].

Neither narrow-band nor wide-band models are necessary to model non-homogeneous effects on radiative heat transfer in high temperature combustion gases. A simplified method based on total transmittance data reduces computational times over two orders of magnitude with a sacrifice in accuracy of less than 10% [11K]. A refined method for calculating the emissivities of individual components (H_2O, CO_2, CH_4) in the product gases of hydrocarbon combustion and of mixtures of these components indicates that this method can be employed for more complex calculations of the radiative energy transport [24K]. The efficient extraction of a high-temperature working fluid from a coal fired fluidized bed combustor depends on the design of the immersed heat exchanger, in particular, on the solidity of the cooling tubes immersed in the bed [16K]. There are certain similarities between the free vertical development of the vortex flame in a high power gas and oil flame in a cyclone furnace and the flame from bottom burners. The data obtained in this study can be used for the analysis of systems of boiler furnaces with bottom arrangement of gas and oil burners [17K]. Monte Carlo techniques for the stimulation of radiation in industrial furnaces can be conveniently used to compute the view factor matrix entering into Hottel's zone method [26K]. It is demonstrated that the introduction of an effective temperature is a useful tool for calculating radiative heat transfer in modern power plants [3K]. Studies of the heat transfer mechanisms in horizontal flame propagation indicate that in the early stages of a fire, heat conduction through the solid is dominant; radiation from the flame becomes of increasing importance as the size of the fire increases [21K]. Treating a flare as a surface radiator results in data which are close to the required accuracy for safety considerations [1K].

A collation method employing piecewise cubic splines as approximating functions and Gaussian points as the collation points provides a very convenient technique for the solution of thermally developing combined radiation–convection problems with and without scattering [7K]. Another collation method for the solution of linear and non-linear integral equations employs piecewise Hermite splines as approximating functions and the Gaussian quadrature points as the collation points. Since it is a high-order method, it requires only a small number of equations to produce the desired accuracy at low computational cost [9K]. Using the method of collation, results for thermally developing Poiseuille flow with scattering show that scattering tends to decrease the radiation component without affecting convection at low optical thicknesses [6K].

Studies of the coupled problem of radiative transport and turbulent natural convection in a volumetrically located, horizontal, gray fluid layer show that under severe radiation conditions, no boundary layer flow regime exists even at very high Rayleigh numbers [10K]. Studies of the effect of radiation in combination with turbulent natural convection on the rates of heat transfer in volumetrically heated fluid layers at relatively high temperatures demonstrate that even at high Rayleigh numbers the radiation mode is as effective as the turbulent natural convection mode in removing heat from the upper surface of molten pools with adiabatic lower boundary [8K]. An analysis of radiative energy transport in porous media reveals a convective-like radiative energy transport mechanism and it is suggested that this term is proportional to the void fraction gradient [30K]. A fluidized bed is utilized for checking the validity of the hypothesis that the various components in combined conduction, convection and radiation heat transfer are additive [20K]. Heat transfer by simultaneous conduction and radiation in an absorbing, emitting, and anisotropically scattering medium is analyzed by solving the integral–differential equations by a successive approximation technique. The results show that for a one-dimensional system, the common approach of treating the total heat transfer as a simple addition of separate independent contributions of conduction and radiation is in many cases very inaccurate [32K].

Expressions are developed for predicting the effect of volume and interface absorption and of scattering in high-reflectance or anti-reflectance multi-layer coatings. These expressions are used together with experimental data to predict the performance of quarter wave high reflectance and anti-reflectance multi-layer stacks at the design wavelength [2K]. By using a prism of high index of refraction as a substrate for depositing nearly transparent, thin layers, the index of refraction of these layers can be measured with high accuracy (to the third decimal place) [4K]. If an optically thick layer containing spherical pigments (TiO_2, Fe_2O_3) is painted on a specularly reflecting substrate, then the hemispherical reflectance will be affected by the absorption of the substrate (wavelength range from 0.35–20 μm) [14K]. An analysis of the heat transfer through irradiated layers of slag at high temperatures demonstrates that the treatment of slag as an opaque substance can result in large under-predictions of the total heat transfer in comparison to the situation where the material of the layer is semi-transparent to radiation [27K].

Studies of the thermal emission of a polydisperse aerosol medium indicate that ignoring the fog thermal emission can lead to large errors in the assessment of the performance of optical detection systems [29K]. The results of studies of the radiation properties for polydispersions of absorbing spherical particles may be applied to coal with wavelength dependent optical properties [5K]. Analytical and experimental results of studies of the visible radiation transfer in a black ink suspension are in good agreement for certain laboratory conditions [28K].

The conditions required for self-focusing of stationary Gaussian and circular high-power laser beams are derived for the case when radiation-induced heat generation contributes to temperature non-uniformities in the gas [22K]. Deflagration waves supported by thermal radiation may be classified as supercritical and subcritical deflagration. Subcritical deflagration leads to larger compression [18K]. Tables have been developed for radiative volume heating of spherical water droplets in black body surroundings up to 1450 K. From these tables one can find the time required to evaporate from one size droplet to another [12K]. Measurements of net heat losses to a clear sky as a function of surface temperature, ambient temperature, and humidity from both gray and spectrally selective surfaces are in good agreement with an analysis of sky-radiator-ambient heat exchange [15K].

Surface radiation

Results of a new calculation method of the local hemispherical effective emissivities along an isothermal, diffuse, cylindrical-inner-cone cavity demonstrate that this method provides high accuracy, especially for cavities with high surface emissivity, long cylinder length, and small aperture radius [14L]. The total hemispherical emissivity of a clean (100) surface of a single crystal tungsten over a temperature range from 340 to 1260 K is found to be only slightly sensitive to adsorption of mono-layers of oxygen on the surface [12L]. Measurements of the total normal emittance of AISI stainless steel samples oxidized at 1050 K show a remarkable influence on the surface finishing [6L].

An analysis of the angle factor between the inner halves of a cylinder which are assumed to be at different temperatures shows that the angle factor approaches $2/\pi$ as the aspect ratio (R/L) of the cylinder assumes values <0.01 [5L]. The shape factor between a differential plane area element and a three-dimensional object with straight edges can be expressed in terms of known shape factors using simple algebraic manipulation [13L]. Temperature calculations for shell enclosures under the influence of thermal radiation are rather complex due to non-linearity effects and the complicated view factors. Both difficulties can be circumvented through the use of upper and lower bound solutions which provide sufficient accuracy for determining thermal stresses in shell components [2L].

Measurements of the optical reflectivity spectrum of a $CuFeS_2$ single crystal at room temperature for a range between 0.025 to 6 eV allow the establishment of a schematic energy level diagram for $CuFeS_2$ [10L]. The results of an analysis of the reflectance of randomly rough surfaces show that the equations corresponding to the Fresnel approximation give the same results as those derived by means of the Fraunhofer approximation if certain geometrical conditions are valid which are determined in this study [11L].

By taking polarization effects into account, detailed numerical calculations of the radiative transfer for well known materials reveal up to 25% deviations from classical calculations [3L].

An adaptation of Kirchoff's and Draper's law for polarized radiation requires only two characteristic properties if radiative exchange is confined to smooth surfaces (no diffuse radiation) [1L]. A generalized temperature response chart has been developed for a body with negligible heat conduction resistance which is simultaneously under the influence of radiative/convective heat transfer in a medium with gray walls and a generally non-isothermal gas [7L]. An analysis of anisotropic conduction and surface radiation around a hollow cylinder is applied to a graphite-reinforced boom exposed to solar heating in space [4L].

Black fins attached to the base of the cavities of a V-groove cavity array promote both radiative heat transfer and collimation of energy in the direction normal to the base surface [9L]. Studies of the radiation properties of black chrome selective surfaces indicate their superior performance for solar collectors, especially when moderate operating temperatures are required [8L].

MHD

Renewed interest in MHD and its applications produced a relatively large number of papers during the past year.

It is shown that a set of two-dimensional turbulent boundary layer equations that describe the three-dimensional fluid flow in a MHD channel, fully satisfy the three-dimensional conservation laws [2M]. The effects of the Hall and the ion slip currents on forced convective heat transfer in the thermal entrance region of a MHD channel manifest themselves by a reduction of the heat transfer rates [3M]. A study of the effects of a transverse magnetic field on forced and free convection of an electrically conducting fluid flowing over a vertical semi-infinite plate indicates that a reverse flow of air exists near the plate when $Gr < 0$. In this study dissipative heat and stress work is taken into account [12M]. The study of thermally driven hydromagnetic convection in a rapidly rotating sphere may have possible applications to the Earth's fluid core where the work done by the buoyancy force may provide the necessary energy to drive the geodynamo [6M].

Viscous and Joule dissipation effects on MHD free convection flows past a semi-infinite vertical plate become more dominant with increasing values of the magnetic field parameter and the Prandtl number [14M]. A study of the combined buoyancy effects of thermal and mass diffusion on MHD natural convection flows reveals that for fixed Gr and Pr the value of X_t (dimensionless length parameter) decreases as the strength of the magnetic field increases [1M].

In order to reduce effects detrimental to the operation of a MHD power plant, a high ratio of volume of the MHD duct to its surface is required. The authors of this paper claim that efficiencies up to 60% are feasible in a conventional steam power plant equipped with a MHD topper [10M]. Studies of the operating region of MHD generators indicate that the optimal operating range for maximum electrical power output lies on the boundary between the transonic and supersonic flow regime at relatively low inlet stagnation pressures [15M]. In the development of a high-power prototype portable MHD generator system, a number of significant advances have been made, such as combustor heat release density, combustion efficiency, channel power density, and energy extraction rates [13M]. There is a possibility to produce short burst of power with electrolytes as the MHD medium because many electrolytes are soluble and ionized in water vapor above the critical point [9M].

A simplified version of the line using reversal technique is used for measuring the gas temperature and the potassium atom number density of MHD combustion systems [8M]. Experimental values of the electrostatic sheath thickness which develops over the cathode surface in a MHD combustion plasma seeded with K_2CO_3 are in qualitative agreement with analytical predictions for different seeding ratios and applied potentials between the electrodes [4M]. Studies of single-phase (Na) and two-phase (Na/N_2) MHD pipe flows show that the wall-voltage profiles of single-phase flows are symmetric with respect to the center of the applied magnetic field region; two-phase wall-voltage profiles are asymmetric because of the expansion of nitrogen along the length of the test section [5M].

By applying radial electric fields to a flow of a dielectric organic fluid passing through an annulus between a Cu tube and a concentric wire, it was found that for laminar flow the heat transfer rates to the outer wall increased in the thermally well developed region and the thermal development region was advanced [7M]. Stability conditions are derived for different thermal boundary conditions of an electrically conducting fluid layer subjected to volumetric heating and bounded between two rigid surfaces in the presence of a magnetic field [16M]. A liquid metal flowing within metal walls and exposed to a magnetic field is stirred thermoelectrically if the interfacial temperature is non-uniform [11M].

MEASUREMENT TECHNIQUES

Developments of measurement techniques and instrumentation related to heat transfer studies are described by a number of researchers. Primary areas of activity include temperature and heat flux measurements, laser–Doppler velocimetry, hot-wire anemometry and transport property measurements.

A comprehensive review of thermometry, which cites the contributions of eighty publications, is presented [17P]. The effect of heat leaks in platinum resistance thermometry is analyzed and an experimental method is proposed for estimating the magnitude of this effect [15P]. An effective thermal/vapor barrier for precision temperature baths has been made by floating on the liquid surface a pile of common packing material such as styrofoam pellets fused together by a solvent spray [16P]. A method which uses several thermocouples and the wire bridge technique of true temperature indication provides a means for measuring the temperature distribution in high-temperature furnaces that is independent of thermocouple calibration or inhomogeneity errors [24P]. A very simple, inexpensive procedure for installing thermocouples in i.r. spectroscopy or reaction cells is to insert fine thermocouple wires (0.13 mm dia.) through small holes (0.25 mm dia.) drilled in i.r. windows (NaCl, KCl, CsBr) 3.2 mm thick and seal the remaining opening with "Vacseal," a silicone resin [10P]. The recovery factor of a cylindrical thermocouple with plane frontal area in normal flow has been investigated [5P]. The frequency response of wires used as sensors for temperature fluctuation measurements has been experimentally investigated [22P]. In a subsequent paper, three different methods of measuring the frequency response are discussed and an explanation is proposed why there are great discrepancies in the time constants as determined by these methods [34P]. A procedure for measuring translational flame temperature utilizing the laser induced fluorescence spectrum of OH is demonstrated [6P]. A simple differential technique using a floating Langmuir probe and a hot probe has been used to obtain direct oscillographic display of electron temperature oscillation in a plasma [41P].

An optical thermometer has been constructed which is based on the fact that the time constant for fluorescent decay of the red R lines of Ruby is a monotonic function of temperature over a range of more than 200°C [42P]. Reflectance errors in i.r. thermography have been analyzed and corrections are provided for converting the apparent radiance temperature to actual surface temperature [37P]. A simple technique is presented for characterizing the non-linear responsivity of infrared detectors such as HgCdTe [45P].

Two companion papers [27P, 28P], describe the problems encountered in the measurements of radiant heat flux in large boiler furnaces that result from ash deposition and describe instruments that can accurately measure the fluxes. An instrument for the measurement of heat flux from a surface with uniform temperature was found to have an uncertainty of $\pm 2\%$ in the range of 420–1670 W/m^2 [19P]. The responses of heat flow transducers, used in the evaluation of convective body heat losses, have been investigated at hyperbaric environments [29P].

Several researchers report studies dealing with the theory and application of laser-Doppler velocimeters (LDV) [39P]. A new LDV has been proposed where the waveform of the Doppler signal changes according to the direction of the velocity vector, thus making it possible to measure selectively a specified velocity vector. Based on this new LDV, simultaneous measurement of the falling velocity and size of large droplets has been achieved by employing LDV to selectively measure velocity and employing two laser beams intersecting at right angles in the center of the LDV sampling volume to measure size [38P]. The azimuthal and axial velocity of the source-sink flow in a rapidly rotating cylinder were measured by using the back scattered mode of an LDV [43P]. Simple optical methods that can substantially increase signal detectability in LDV involve rotating the laser polarization to amplitude-modulate exclusively the Doppler signal [21P]. The fundamental limitations imposed by photo-detection noise on the accuracy with which flow velocity can be determined in laser anemometry has been investigated [31P].

An apparatus has been built which combines laser–Doppler velocimetry and optical plethysmography for a rapid, multi-parameter characterization of blood flow in cutaneous tissues [47P]. A laser crossed-beam anemometer has been applied to the study of capillary blood flow [20P]. In a new method of remote wind sensing (dual-frequency Doppler–lidar) two optical beams of unlike frequency are superimposed in the sensed volume and velocity information is obtained from the difference in the Doppler shifts of light scattered from the two beams by aerosols moving with the air [11P]. A two-color, two-spot laser velocimeter system has been shown to be an effective instrument for measuring atmospheric wind speeds at ranges up to 150 m [3P]. A laser beam manifold, a device which transforms a single laser beam into several uniform parallel beams, and particle photography system has been developed for use in fluid velocity measurements [33P].

A comprehensive bibliography of thermal anemometry papers published in 1977 and 1978 is presented [13P]. Both qualitative and quantitative features of unsteady and turbulent flow fields can be examined using combined simultaneous flow visualization and hot-wire anemometry [12P]. The error in mean and r.m.s. temperature measurements in the outer part of a heated jet discharged into still air using two standard hot wire–cold wire arrangements were found to be about 16% and 46% respectively at the half-temperature radius [1P]. A method is presented for the correction of measured higher-order moments of hot wire velocity and cold wire temperature signals for the respective parasitic sensitivities to temperature and velocity fluctuations [35P].

An improved feedback-stabilized Michelson interferometer is proposed in which the effects of mechanical vibrations are reduced by using a two-laser light source [48P]. Shearing interferograms of refractive indices around a flame and of spouting gas are presented to demonstrate the use of a lateral shear interferometer which uses twin three-beam holograms, the first of which plays the role of a beam splitter and the second of which eliminates the effect of tilt

[25P].

Two papers addressed the problem of measuring mass flow rate. The quasi-steady solution technique which has been widely used for transient turbine-meter analysis has been shown to be not generally valid for two-phase transients of practical concern and a suitable analytical model has been presented [18P]. Using the pulsation properties of gas bubbles, it is possible to measure acoustically the volumes of the gas bubbles and, hence, possible to measure very low gas flow rates (from 4 to 100 mm^3/s) [32P].

A new high sensitivity micromanometer is described where the downwards or upwards displacement of menisci under the action of a pressure differential in the U-tube arrangement is magnified by a factor of 10^4 or more [23P]. A glass membrane manometer, usable up to 820°C, has been devised which makes it possible to measure pressure contactlessly with a sensitivity of 0.01 mmHg [30P].

Development continues in the areas of transport property and transport coefficient measurements. The measurement of thermal conductivity of gases with a precision of the order of a few parts in a thousand using a transient hot wire technique requires corrections for the change in resistance per unit length of the wires produced by thermal expansion and the effect of adding a filter network to eliminate stray 60 Hz electric pulses [8P]. A steady state apparatus has been developed for measuring the thermal conductivity of cast materials by employing a novel thermal symmetry arrangement [46P]. The accuracy implications of the geometry of a guard heater used on a thermal conductivity measurement device has been investigated [4P]. An apparatus that is capable of measuring thermal diffusivity between 80 and 500 K using a modified Angström method is described [44P]. A diffusion cell has been used to measure the molecular diffusion coefficient of carbon monoxide in a carbon monoxide–nitrogen mixture through cigarette paper [9P]. A new method of correcting for intermolecular collisions in the measurements of Knudsen accommodation coefficients for energy, using a hot wire instrument, reduces the uncertainty in the measurements to a few percent or less [26P]. Calculations have been performed to estimate the feasibility of conducting plume transport and diffusion studies by creating a plume of fluorescent particles that are then interrogated by a lidar in both the fluorescent and elastically scattered modes [40P].

A specific measurement program was devised and taken to the Arctic, to determine some aspects of ice-pack transport [14P].

English translations of Russian papers dealing with the following topics appeared in "Measurement Techniques": An infrared scanning device for temperature measurement [2P], a device for calibrating thermometers in the range 78–350 K [36P], and a method of determining the angular dependence of emittance at high temperatures [7P].

HEAT TRANSFER APPLICATIONS

Heat exchangers and heat pipes

The mass transfer analogy with naphthalene sublimation was used [21Q] to obtain coefficients on in-line pin fin arrays. Conditions were developed after the fourth row and in-line arrays transferred more heat than staggered ones at equal pumping power and heat transfer area. Heat transfer rates on the shell side of shell-and-tube heat exchangers increase [4Q] with use of cross baffles. Means for the avoidance of leakage flow in bundled-tube heat exchangers were discussed [24Q]. The mass of convective fins with minimum mass differs little from the triangular fin [13Q]. A study on generalization of the methods of heat exchanger analysis from the year 1942 by London and Seban was now published [12Q] as a pioneer paper. The evaluation of compact heat exchangers regarding heat exchange rate and pumping power was discussed [11Q]. A calculation [7Q] for fully developed flow in a heat exchanger with superimposed velocity pulsations indicates that the interaction of velocity and temperature pulsations introduces a new term in the energy equation and produces thus higher heat transfer rates. A fast and accurate graphical method was published [20Q] for the design of air conditioning cooling coils in which sensible and latent heat is transferred. An approximate explicit equation with 10 sets of empirical coefficients [18Q] allows a rapid calculation of the mean temperature difference in air-cooled crossflow heat exchangers. An analysis [16Q] considers radiation effects on the performance of high temperature heat exchangers transferring energy between water vapor and carbon dioxide on one side and helium on the other side. An approximate analysis [10Q] of crossflow heat exchangers by the method of weighted residuals compares favorably with the finite-difference method. The optimal area allocation between two stages of a heat exchanger system was analyzed [1Q] with respect to interstage temperature and area. A thermodynamic approach [6Q] to the design of heat exchanger networks provides a listing of solutions with a minimum number of heat exchangers, heaters, and coolers. Experiments [5Q] were used to evaluate blocks of four cross flow elements to approximate cocurrent and countercurrent heat exchangers. Practical problems are discussed for process heat exchangers including fouling [22Q].

The trade between accuracy and time requirement was investigated [8Q] for finite difference methods to calculate the temperature changes in thermal regenerators. Variable time steps provided improvement. Refined closed methods were presented [23Q] for counterflow thermal regenerators.

NTU values predicted with the Ranz–Marshall correlations [17Q] agreed with experiments for atmospheric spray cooling systems. Local mean and bulk temperatures in the dispersed and continuous phases were measured [15Q] in a spray column by a computer-aided experiment. Local heat transfer coefficients could be obtained from the results. An analysis of heat transfer surface geometries for dry cooling towers [14Q] permits selection of the optimum geometry and in this way directs research towards better surfaces. Direct contact heat transfer to drops exposed to an intermittent electric field normal to the trajectory was studied experimentally [9Q] for water drops 3.9–5.9 mm in diameter rising in a silicone oil. The drops were found to elongate periodically in the direction of the field.

The characteristics of sintered powder wicks were measured in a study of the operating limits of a heat pipe [19Q]. A model developed to predict the maximum heat transfer rate agrees well with experiments. An analysis [3Q] was used to study axial and azimuthal dryout in gravity assisted heat pipes and discusses the methods to prevent both. The heat transfer and hydrodynamics of electrohydrodynamic heat pipes have been studied [2Q].

General

The heat release from reactor fuel elements for fast transients were calculated [23S] by series expansion of the equations. The stability of two-phase flow in vapor generators with parallel channels was analyzed [17S] by computer analysis.

The analysis [16S] of flow in an agitated horizontal thin film evaporator was compared with experimental results. A summary paper [21S] compared available relations for heat transfer in agitated vessels for multiphase systems. A model was developed [9S] for momentum and heat transfer in mixers with close-clearance agitators to obtain relations between the power number and Reynolds number. Experiments [8S] verified the analysis. Maximum heat transfer coefficients in an apparatus containing dispersed two-phase systems can be expressed [15S] as a function of Grashof number times Prandtl number. Equations were derived [4S] which correctly described the influence of the parameters on heat transfer in bubble column reactors. Pulsations caused by interrupting valves with 50–1000 cycles/min increase heat transfer to water flowing at $Re = 6600$ to 28,000 [10S].

Heat treatment using internal friction caused by ultrasonic waves produces materials with good mechanical properties [12S]. An analysis [13S] of polymer reaction molding was tested by experiments. A model for non-linear shear stress and thermal effects in elasto-hydrodynamic line contacts was derived [11S] in qualitative agreement with experimental data. Experiments were performed [18S] to correlate bulk body temperatures on point and line contacts. An analysis [7S] indicates that thermal effects in the film are dominant in describing lubrication of journal bearings in turbulent flow. Thermally induced whirl was found experimentally [14S] to exist whenever the lubricant inlet temperature is higher than that of the bearing surface in hydraulic journal bearings. The thermal behavior of the step thrust bearings was analyzed [22S].

A computer analysis [1S] modelled heat transfer to slab-shaped food products by passing cold air over the product which is wetted continuously by a spray of chilled water. Spray quenching in a ventilated-duct fire is modelled [6S] by considering heat transfer between a cold liquid spray, a hot gas, and the ignited duct wall.

A review [5S] considers methods to measure the thermal insulation properties of fabrics in the presence of circulating air. A quasi-steady approach is used [19S] for the thermal analysis of insulated structures in the ground with a melting or freezing zone. Buried pipes, infinite strips and circular disks are considered.

Good results have been obtained [3S] in measuring heat transfer in steam turbines by relating the time dependent temperatures measured just below the wall surface to the fluid temperature fluctuations. The increase of heat transfer in a spark ignition engine could be correlated [2S] with the arrival of the flame front at four positions on the cylinder head.

LOX-cooled thrust chamber technology is discussed in [20S]. The response of graphite epoxy sandwich panels to measure temperature transients was investigated [24S].

Solar energy

Topics of major interest among the heat-transfer related solar energy publications include: solar radiation, flat-plate and concentrating solar collectors, heat storage, and system performance. A review of collector and energy storage technology indicated that collector technology is commercially available to achieve delivery temperatures up to 175°F at averaged yearly efficiencies better than 30% in good solar climates [34T].

The amount of solar energy that is intercepted by surfaces of any orientation is estimated from a new model of the clear sky spacial distribution of solar radiation [31T]. A statistical procedure has been employed to develop correlations between the hourly global horizontal radiation and its diffuse component [10T]. A Lambert conformal projection of the sun's path and of shading obstacles, combined with an overlay of concentric circles, facilitates an expedient analysis of various solar collector sitings [14T].

A linearized model of the dynamic performance of a flat-plate solar collector is shown to be satisfactory in many respects [6T]. A combined experimental and analytical study of the transient response of thermosyphon solar collectors indicated that there are long time delays in the development of the flow in the collector but that these do not affect the energy collection [16T]. Thermal performance and sensitivity analyses were conducted for a variety of high temperature flat-plate collectors [28T]. Collectors can be analyzed in terms of the variables generally used in analyzing other types of heat exchangers [24T]. A flat-plate solar collector with an evacuated tubular cover is predicted to be less efficient than one with two flat covers which are not evacuated [3T]. Incorrect evaluation of the convective heat transfer coefficient on the outer surface of flat-plate collectors can lead to large errors in the predictions of useful energy collected [25T]. Based on measurements of wind speeds near flat-plate solar collectors, it was concluded that it is not valid to take a single wind speed measurement and expect this value to give an accurate indication of the average wind speed across the collector surface [19T]. The thermal contact conductance of aluminum/copper interfaces was measured in vacuo with and without high conductivity silicone grease between the metals [26T].

An analysis of the solar collection capabilities of a flat-plate solar collector with a flat mirror allows a determination of the optimal collector and mirror orientations for a variety of applications [13T]. The concentration ratios of a number of flat-plate/flat-mirror solar collector configurations with fixed and movable mirror panels were calculated [12T]. A cusp mirror—heat pipe evacuated tubular solar collector was found to be more efficient than conventional flat-plate solar collectors when operated at elevated temperatures [20T].

A generalized technique for the optical design of solar concentrators has been developed which considers limb darkening effects, collector placement error, concentrator optical errors and concentrator pointing errors [8T]. Information theory has been applied to the problem of solar radiation collection and it was found that the optimum solar concentrator corresponds to a perfect imaging system, i.e. one that images the entire sky on the absorber with no aberrations [22T]. A realistic class of collectors with maximal concentration, which can serve as a basis for design and optimality studies, has been defined [7T]. Statistical methods were used to predict the effects of tracking errors on the performance of point focusing solar collectors [9T]. It is shown that paraboloidal mirrors of short focal ratio and similar systems can have their flux concentration enhanced to near the thermodynamic limit by the addition of non-imaging compound elipitical concentrators [33T]. A new stepped spherical concentrator of short focal ratio is suitable for high temperature applications, and the cost/performance trade-off relationships of the design parameters can be approached through the optical simulation method presented in [2T].

There was an increase in the number of studies reported in which the problem of thermal storage was investigated. Good agreement was obtained between experimental and analytical studies of the performance of a rock-bed heat storage unit [23T]. A packed-bed thermal storage unit has been dynamically modelled using the procedure of calculating the output temperature as the sum of earlier input temperatures, each multiplied by a predetermined response factor [27T]. Four methods of reducing the mixing of the fluid in a storage tank have been examined critically with a view toward updating the effectiveness of each [29T]. A two-tank heat storage system, which provides a method of positive antiblending using a flexible diaphragm to separate supply and return water in each tank has been designed [30T]. A method for the determination of the optimum economic distribution of thermal insulation on inground annual heat storage tanks is described [32T]. Heat loss and storage functions are derived for a thermal well [18T]. In an investigation of nitrate salts for latent heat storage, a method of estimating the thermal conductivity of the mixture is described and the experimentally determined temperature history of the storage material is found to be in reasonably good agreement with numerical predictions [11T].

A variety of systems were investigated. A number of ASHRAE calculation procedures have been combined to develop a method for predicting the performance of solar energized systems for the heating or cooling of buildings [17T]. The National Bureau of Standards fabricated, instrumented and tested six typical solar hot water heating systems and compared the experimental results to predictions performed using three different computer programs [15T].

A mathematical correlation between collector area and auxiliary energy used in a solar hot water system was obtained using a numerical transient system analysis program [5T]. Solar assisted heat pumps should have capacity modulation in order to achieve optimum performance [4T]. The heat and mass transfer analysis of grain drying for a prototype system is described and the results of a survey showing the effects of various system parameters on drying time are presented [21T]. It was predicted that the operation of solar ponds in the southern part of Iran is not affected greatly by the normal daily and weekly changes of climatic conditions [1T].

PLASMA HEAT TRANSFER

There seems to be an increasing interest in basic studies as well as in applications of ionized gases and in the associated heat transfer aspect.

The results of a simplified analysis of the arc behavior in SF_6 Puffer Breakers are in good agreement with measurements and permit not only the design for optimum performance, but also an estimate of the thermal behavior of the arc as current zero is approached [21U]. An experimental investigation of the arc phenomena in SF_6 Puffer Breakers shows that the arc column remains stable on the center axis during the high-current period and becomes turbulent near current zero [19U]. Modelling results of circuit interruption arcs in SF_6 taking axial inhomogeneities into account are in agreement with the results of full-scale interruption tests [4U]. Results of a new optical method for measuring temperature and pressure profiles in a transient arc generated by separating Cu/W electrodes are in good agreement with electrical measurements [25U]. The good agreement between analysis and experiment allows the application of sealing laws to gas-blast circuit breaker arcs during the high current phase [11U]. Possibilities are delineated of using the additional energy absorption capacity of magnetic-blast chutes for arc circuit interruption by increasing the permissible temperature of their walls up to the boiling point [33U].

Studies of the free recovery of wall-stabilized arcs with superimposed laminar or turbulent flows (axial) indicate that the conductance decay rate increases with mass flow rate. This and other findings are explained in terms of a boundary layer arc model which includes turbulent diffusion effects [15U]. Analytical investigations of the free recovery of the gas blast arc column show that the temperature decay of the arc column in N_2 is much faster than in SF_6 [27U]. The thermal interruption capability of an arc in a double flow configuration is drastically reduced by the vortex superimposed to the axial flow [31U].

The results of interferometric and shadowgraph investigations of arc–flow interactions in a double-nozzle flow system allow for the first time a sharp distinction between different forms of arc–gas flow interactions [32U]. Analytical results of the response of a wall-stabilized air arc to step and sinusoidal current changes are in good agreement with experimental findings [23U]. An experimental study of non-equilibrium effects in wall-stabilized argon arcs indicates that the electrons are not in equilibrium with the heavy particles at low currents and low pressures [30U].

Temperatures in SF_6 arcs covering a range from 9000 to 20,000 K may be derived from a combination of measured SI, SII, FI, and FII absolute line intensities and computed equilibrium plasma composition data using an iterative procedure [29U]. Results of temperature measurements in turbulent arcs indicate relatively large fluctuations in temperature ($\approx 5\%$) across the entire arc column [8U]. A new spectroscopic method useful for temperature measurements in thermal air plasma is based on measurements of the integrated emission coefficients of 19 rotational lines concentrated around the band head at 391.5 nm (N_2^+ molecule) [28U]. Based on calculations of the index of refraction in a high pressure hydrogen plasma it should be possible by interferometric measurements to determine the electron density in a high pressure hydrogen arc with an accuracy of the order of 1% [26U].

A pulsed probe method may be useful for plasma flow boundary layer studies when physical conditions eliminate the possibility of using optical methods, and in some cases it may serve as a routine method for measuring free stream flow velocities of a plasma [12U]. A study of the effects of probe temperature and probe material on electrostatic probe measurements in a partially ionized, high density plasma shows that the ion current increases as the probe temperature increases and the floating potential shifts to more negative potentials [24U].

The electron density and the collision frequency in an over dense plasma cylinder produced inside capillary tubes may be determined by a two-frequency method measuring the reflection coefficient in a rectangular wave guide crossed by the plasma [35U]. A new method for atom temperature measurements in plasmas makes use of the thermal diffusion effect. For this purpose a light tracer gas is added to the plasma [13U].

The effects of cold gas injection on a confined arc column manifest themselves by an arc constriction at the location of gas injection raising the arc core temperature and resulting in additional resistance to the penetration of the cold flow into the arc [6U]. Considering the effect of convection on the stability of a two-dimensional arc, the results conform with those given by the classical Kaufmann criterion [9U]. The results of an empirically based model for a long, vertical d.c. arc in atmospheric air indicate a quasi-steady state between 20 and 5000 A and two distinct core temperature ranges around 7000 K and 12,000 K [22U].

A model for short arc thermal reignition indicates that this reignition takes place in a thermal layer near the anode [20U]. Modelling of the anode contraction region of an atmospheric pressure nitrogen arc at 250 A indicates that heat transfer close to the anode is dominated by electron enthalpy transport [7U]. The anode assumes different operating modes in a MPD arc configuration at low pressures (≈ 1 Torr) depending upon the parameter $I^2/\dot{m}$ (I = current, $\dot{m}$ = mass flow rate) [16U]. Considerable advances have been made in recent years in the understanding of the basic processes involved in the erosion of metal cathodes in electric arcs [14U].

Small additions of Cu-vapor to an atmospheric pressure nitrogen plasma produce a significant increase of the electrical conductivity for temperatures between 5000 and 7000 K [1U]. Both the thermal conductivities and the coefficients of thermal diffusion for a high temperature plasma in a uniform magnetic field have been calculated from an expansion in terms of Sonine's polynominals using 50 terms. The resulting accuracy extends to at least six decimal places for various magnetic field strengths [18U].

The percentage of ionization in a duoplasmatron in He reaches values up to 95% at 0.2 to 0.3 Torr and decreases to 15% when the pressure exceeds 0.5 Torr [10U]. A new approach for the ignition of pulverized coal makes use of a vortex-stabilized electric arc heater which saves substantial amounts of premium fuel. The arc igniter appears to be economically viable and competitive with conventional practice [2U]. Natural convection has a negligible effect on the flow and temperature fields of a confined, inductively coupled, thermal r.f. plasma, but a significant effect on a free r.f. discharge [5U].

For studying the heat and momentum transfer for alumina particles injected into a plasma jet, results are used derived from emission spectroscopy, laser–Doppler anemonetry, and in-flight radiation pyrometry [34U]. Mixing of an argon plasma jet with the surrounding air is studied by using conventional techniques and a novel form of laser Schlieren velocimetry. Results indicate that there is a gradual degradation of non-translational arc energy in the innermost zone while the flow in the outer zone appears to be independent of events within the nozzle [17U]. The efficiency of a conventional plasma torch operated with argon at atmospheric pressure is typically in the range of 40–50%. By adding a small concentration of a diatomic gas, the efficiency increases to approximately 65% and, at the same time, the power dissipation in the torch increases [3U].

REFERENCES

Books

John A. Duffie and William A. Beckman, *Solar Engineering of Thermal Processes.* John Wiley, New York (1980).

U. Grigull and H. Sandner, *Wärmeleitung.* Springer, Berlin (1980).

Frank Kreith and William Z. Black, *Basic Heat Transfer.* Harper & Row, New York (1980).

Peter J. Lunde, *Solar Thermal Engineering.* John Wiley, New York (1980).

S. V. Patankar, *Numerical Heat Transfer and Fluid Flow.* Hemisphere, Washington, and McGraw-Hill, New York (1980).

W. Schumann and M. Dubas, *Holographic Interferometry.* Springer, Berlin (1979).

Robert Siegel and John R. Howell, *Thermal Radiation Heat Transfer* (2nd edn.). Hemisphere, Washington, D.C. (1980).

C. M. Vest, *Holographic Interferometry.* Wiley–Interscience, New York (1979).

J. P. Hartnett, T. F. Irvine, Jr., E. Pfender and E. M. Sparrow

(editors), *Studies in Heat Transfer: A Festchrift for E. R. G. Eckert*. Hemisphere, Washington, D.C. (1979).
Sadik Kakac (editor), *Turbulent Forced Convection in Channels and Bundles. Theory and Applications to Heat Exchangers and Nuclear Reactors*. Hemisphere, Washington, D.C. (1979).
E. F. Somerscales (editor), *Fouling of Heat Transfer Equipment*. Hemisphere, Washington, D.C. (1980).
A. S. Mujumdar (editor), *Drying '80*. McGill University, Canada (1980).
Gary R. Hough (editor), *Viscous Flow Drag Reduction*. American Institute of Aeronautics and Astronautics, New York.

New Journals

Heat Transfer Engineering, Hemisphere, Washington, D.C.
Plasma Chemistry and Plasma Processing, Plenum Press, New York.
Latin American Journal of Heat and Mass Transfer, Editorial Office, No. 867, La Plata, Argentina.

Conduction

1A. B. N. Antar, F. G. Collins and G. H. Fichtl, *Int. J. Heat Mass Transfer* **23,** 191 (1980).
2A. A. Aziz and T. Y. Na, *Lett. Heat Mass Transfer* **7,** 15 (1980).
3A. A. Aziz and T. Y. Na, *Num. Heat Transfer* **3,** 331 (1980).
4A. M. N. Bapu Rao, *Int. J. Heat Mass Transfer* **23,** 443 (1980).
5A. M. Bareiss and H. Beer, *Letters Heat Mass Transfer* **7,** 329 (1980).
6A. A. G. Bathelt and R. Viskanta, *Int. J. Heat Mass Transfer* **23,** 1493 (1980).
7A. P. M. Beckett and N. Hobson, *Int. J. Heat Mass Transfer* **23,** 433 (1980).
8A. H. T. Ceylan and G. E. Myers, *J. Heat Transfer* **102,** 115 (1980).
9A. B. T. F. Chung and S. A. Kassemi, *J. Heat Transfer* **102,** 177 (1980).
10A. T. W. Clyne and A. Garcia, *Int. J. Heat Mass Transfer* **23,** 773 (1980).
11A. M. Comninou and J. Dunders, *J. Heat Transfer* **102,** 319 (1980).
12A. S. K. Fraley, T. F. Hoffman and P. N. Stevens, *J. Heat Transfer* **102,** 121 (1980).
13A. D. M. France and T. Chiang, *J. Heat Transfer* **102,** 579 (1980).
14A. S. S. Girgis and A. C. Smith, *Int. J. Engng Sci.* **18,** 69 (1980).
15A. N. W. Hale, Jr. and R. Viskanta, *Int. J. Heat Mass Transfer* **23,** 283 (1980).
16A. P. J. Heggs and P. R. Stones, *J. Heat Transfer* **102,** 180 (1980).
17A. C. K. Hsieh and K. C. Su, *J. Heat Transfer* **102,** 324 (1980).
18A. S. C. Huang and Y. P. Chang, *J. Heat Transfer* **102,** 742 (1980).
19A. F. A. Jeglic, K. A. Switzer and J. H. Lienhard, *J. Heat Transfer* **102,** 392 (1980).
20A. O. B. Kavaljov, N. A. Larkin, W. M. Fomin and N. N. Yanenko, *Computer Meth. Appl. Mech. Engng* **22,** 259 (1980).
21A. K. Kawashimo, *Bull. JSME* **23,** 231 (1980).
22A. N. Kobayashi, *Computer Meth. Appl. Mech. Engng* **23,** 21 (1980).
23A. M. Kölling and U. Grigull, Wärme- und Stoffübertragung **14,** 231 (1980).
24A. A. D. Kraus and A. D. Snider, *J. Heat Transfer* **102,** 415 (1980).
25A. G. Lebon, Int. *J. engng Sci.* **18,** 727 (1980).
26A. S. L. Lee and G. J. Hwang, *Can. J. chem. Engng* **58,** 177 (1980).
27A. R. L. Levin, *Int. J. Heat Mass Transfer* **23,** 951 (1980).
28A. J. C. McWhorter, III and M. H. Sadd, *J. Heat Transfer* **102,** 308 (1980).
29A. M. D. Mikhailov and M. N. Özişik, *Int. J. Heat Mass Transfer* **23,** 609 (1980).
30A. D. Meinköhn, *Int. J. Heat Mass Transfer* **23,** 833 (1980).
31A. A. H. Nayfeh, *J. Heat Transfer* **102,** 312 (1980).
32A. M. G. O'Callaghan, E. G. Cravalho and C. E. Huggins, *J. Heat Transfer* **102,** 673 (1980).
33A. J. Padovan, *Num. Heat Transfer* **3,** 259 (1980).
34A. Z. Pammer, *Int. J. Num. Meth. Engng* **15,** 495 (1980).
35A. D. J. Petrie, J. H. Linehan, M. Epstein, G. A. Lambert and L. J. Stachyra, *J. Heat Transfer* **102,** 784 (1980).
36A. A. Prasad, *J. Spacecraft Rockets* **17,** 474 (1980).
37A. S. Ramadhyani and S. V. Patankar, *Int. J. Num. Meth. Engng* **15,** 1395 (1980).
38A. P. Razelos and K. Imre, *J. Heat Transfer* **102,** 420 (1980).
39A. E. Schachinger and B. Schnizer, *Wärme- und Stoffübertragung* **14,** 7 (1980).
40A. G. E. Schneider, M. M. Yovanovich and R. L. D. Cane, *J. Spacecraft Rockets* **17,** 372 (1980).
41A. G. Seifert, *Q. Appl. Math.* **38,** 246 (1980).
42A. A. D. Solomon, *Letters Heat Mass Transfer* **7,** 183 (1980).
43A. A. D. Solomon, *Letters Heat Mass Transfer* **7,** 379 (1980).
44A. R. J. Swenson, *Letters Heat Mass Transfer* **7,** 465 (1980).
45A. L. N. Tao, *Q. J. Mech. Appl. Math.* **33,** 211 (1980).
46A. S. B. Thomason and J. C. Mulligan, *J. Heat Transfer* **102,** 782 (1980).
47A. E. A. Thornton and A. R. Weiting, *Num. Heat Transfer* **3,** 281 (1980).
48A. R. H. Tien, *J. Heat Transfer* **102,** 378 (1980).
49A. P. D. Van Buren and R. Viskanta, *J. Heat Transfer* **102,** 375 (1980).
50A. C.-Y. Wang, *J. Appl. Mech.* **47,** 951 (1980).
51A. W. W. Yuen and L. W. Wong, *Num. Heat Transfer* **3,** 373 (1980).

Channel flow

1B. A. Acrivos, *Appl. Sci. Res.* **36,** 35 (1980).
2B. C. A. C. Altemani and E. M. Sparrow, *J. Heat Transfer* **102,** 590 (1980).
3B. S. Aoki, M. Aritomi, A. Inoue and M. Takahaghi, *Bull. JSME* **23,** 376 (1980).
4B. M. Balakrishna and M. S. Murthy, *Chem. Engng Sci.* **35,** 1486 (1980).
5B. H. F. Bauer, *Wärme- und Stoffübertragung* **14,** 49 (1980).
6B. P. M. Beckett, *SIAM J. appl. Math.* **39,** 372 (1980).

7B. W. Bonn, R. Krebs and D. Steiner, *Wärme- und Stoffübertragung* **14,** 31 (1980).
8B. M. W. Chang and B. A. Finlayson, *Num. Meth. Engng* **15,** 935 (1980).
9B. M. W. Collins, *Num. Meth. Engng* **15,** 381 (1980).
10B. G. Comini, S. Del Guidice and M. Strada, *Num. Meth. Engng* **15,** 507 (1980).
11B. V. D. Dang, *Can. J. chem. Engng* **58,** 401 (1980).
12B. N. Decker and L. Glicksman, *Chem. Engng Sci.* **35,** 831 (1980).
13B. M. R. M. Drizhyus, R. K. Shkema and A. A. Shlanchyauskas, *Int. chem. Engng* **20,** 486 (1980).
14B. N. T. Dunwoody and T. A. Hamil, *Int. J. Heat Mass Transfer* **23,** 943 (1980).
15B. A. F. Emery, P. K. Neighbors and F. B. Gessner, *J. Heat Transfer* **102,** 51 (1980).
16B. M. Epstein, L. J. Strachyra and G. A. Lambert, *J. Heat Transfer* **102,** 330 (1980).
17B. S. Faggiani and F. Gori, *J. Heat Transfer* **102,** 292 (1980).
18B. M. Faghri and E. M. Sparrow, *Int. J. Heat Mass Transfer* **23,** 861 (1980).
19B. D. L. Gee and R. L. Webb, *Int. J. Heat Mass Transfer* **23,** 1127 (1980).
20B. M. S. Greywall, *Computer Meth. appl. Mech. Engng* **21,** 231 (1980).
21B. P. Gryglaszewski, Z. Nowak and J. Stacharska-Targosz, *Wärme- und Stoffübertragung* **14,** 81 (1980).
22B. P. F. Hamblin, *J. Fluid Mech.* **99,** 101 (1980).
23B. S. Hasegawa and K. Ichimiya, *Bull. JSME* **23,** 1940 (1980).
24B. E. G. Hauptmann and A. Malhotra, *J. Heat Transfer* **102,** 71 (1980).
25B. F. D. Haynes and G. D. Ashton, *J. Heat Transfer* **102,** 384 (1980).
26B. M. Hirata and K. Katamura, *Bull. JSME* **23,** 1563 (1980).
27B. J. H. B. J. Hoebink and K. Rietema, *Chem. Engng Sci.* **35,** 2135 (1980).
28B. N. Isshiki, *Bull. JSME* **23,** 60 (1980).
29B. N. Isshiki, *Bull. JSME* **23,** 902 (1980).
30B. S. D. Joshi and A. E. Bergles, *J. Heat Transfer* **102,** 397 (1980).
31B. B. Ya. Kamenetskii, *Thermal Engng* **27,** 222 (1980).
32B. G. R. Karr and T. J. Chung, *Num. Heat Transfer* **3,** 35 (1980).
33B. H. Kawamura, *Buil. JSME* **23,** 489 (1980).
34B. Ye. M. Khabakhpasheva, *Heat Transfer, Soviet Res.* **12**(1), 1 (1980).
35B. B. M. Khusid and S. L. Benderskaya, *Heat Transfer, Soviet Res.* **12,**(1), 80 (1980).
36B. A. N. Koblyakov, *Heat Transfer, Soviet Res.* **12**(3), 105 (1980).
37B. M. Kobyashi, *Bull. JSME* **23,** 150 (1980).
38B. H. Koizumi, *Bull. JSME* **23,** 1262 (1980).
39B. V. I. Lankevich, N. I. Perepelitsa and A. P. Sapankevich, *Thermal Engng* **27,** 203 (1980).
40B. V. L. Lel'chuk, Yu. M. Nikitin, E. I. Puchkov and V. P. Smirnov, *Thermal Engng* **27,** 98 (1980).
41B. S. H. Lin and D. M. Hsieh, *J. Heat Transfer* **102,** 786 (1980).
42B. S. H. Lin and W. K. Hsu, *J. Heat Transfer* **102,** 382 (1980).
43B. C. H. Liu and E. M. Sparrow, *Int. J. Heat Mass Transfer* **23,** 1137 (1980).
44B. P. C. Lu, *J. Heat Transfer* **102,** 570 (1980).
45B. M. R. Malik and R. H. Pletcher, *Num. Heat Transfer* **3,** 241 (1980).
46B. L. Meyer, *Int. J. Heat Mass Transfer* **23,** 591 (1980).
47B. Y. Mori, *Bull. JSME* **23,** 788 (1980).
48B. S. Mori and A. Tanimoto, *Can. J. chem. Engng* **58,** 279 (1980).
49B. W. D. Morris and F. M. Dias, *J. mech. Engng Sci.* **22,** 95 (1980).
50B. D. Mukherjee, M. P. Sinha, N. K. Purohit and A. K. Mitra, *Int. J. Heat Mass Transfer* **23,** 1351 (1980).
51B. M. Nakazatomi, *Bull. JSME* **23,** 1625 (1980).
52B. Yu. G. Nazmeev and N. A. Nikolaev, *Thermal Engng* **27,** 151 (1980).
53B. Z. Nejat and R. Mottaghian, *Wärme- und Stoffübertragung* **14,** 43 (1980).
54B. M. A. Nemira, J. V. Vilemas and V. M. Simonis, *Heat Transfer, Soviet Res.* **12**(1), 104 (1980).
55B. Z. Nowak, P. Gryglaszewski and J. Stacharska-Targosz, *Wärme- und Stoffübertragung* **14,** 281 (1980).
56B. E. Papoutsakis, D. Ramkrishna and H. C. Lim, *Appl. Sci. Res.* **36,** 13 (1980).
57B. E. Papoutsakis, D. Ramkrishna and H. C. Lim, *AIChE Jl* **26,** 779 (1980).
58B. J. L. Peube, J. L. Bousgaribies and M. J. Pierre-Eugene, *Int. J. Heat Mass Transfer* **23,** 813 (1980).
59B. V. N. Popov, V. M. Belyayev and Ye. P. Valuyeva, *Heat Transfer, Soviet Res.* **12**(2), 123 (1980).
60B. D. Ramkrishna and N. R. Amundson, *Chem. Engng Sci.* **35,** 577 (1980).
61B. Z. S. Safar, *J. mech. Engng Sci.* **22,** 45 (1980).
62B. G. Schenkel, *Chem.-Ing.-Tech.* **52,** 346 (1980).
63B. A. Sharon and A. Orell, *Int. J. Heat Mass Transfer* **23,** 547 (1980).
64B. Z. P. Shul'man and B. M. Khusid, *Heat Transfer, Soviet Res.* **12**(1), 31 (1980).
65B. O. K. Smirnov and S. N. Krasnov, *Heat Transfer, Soviet Res.* **12**(2), 135 (1980).
66B. H. M. Soliman, T. S. Chau and A. C. Trupp, *J. Heat Transfer* **102,** 598 (1980).
67B. E. M. Sparrow and M. Charmchi, *Int. J. Heat Mass Transfer* **23,** 471 (1980).
68B. E. M. Sparrow and M. Charmchi, *J. Heat Transfer* **102,** 605 (1980).
69B. E. M. Sparrow and A. Chukaev, *Num. Heat Transfer* **3,** 149 (1980).
70B. E. M. Sparrow, N. Cur and R. G. Kemink, *J. Heat Transfer* **102,** 568 (1980).
71B. E. M. Sparrow and M. Faghri, *J. Heat Transfer* **102,** 402 (1980).
72B. E. M. Sparrow and A. Hajiloo, *J. Heat Transfer* **102,** 426 (1980).
73B. E. M. Sparrow, K. K. Koram and M. Charmchi, *J. Heat Transfer* **102,** 64 (1980).
74B. E. M. Sparrow and J. E. O'Brien, *J. Heat Transfer* **102,** 408 (1980).
75B. P. Stroeve and R. Srinivasan, *AIChE Jl* **26,** 136 (1980).
76B. B. K. Sultanian and V. M. K. Sastri, *Wärme- und*

Stoffübertragung **14,** 245 (1980).
77B. A. S. Trofimov and A. V. Sobolev, *Soviet Atomic Energy* **48,** 129 (1980).
78B. V. F. Volchenok and I. M. Galkin, *Heat Transfer, Soviet Res.* **12**(1), 68 (1980).
79B. R. L. Webb and M. J. Scott, *J. Heat Transfer* **102,** 38 (1980).
80B. S.-C. Yao and A. Rane, *J. Heat Transfer* **102,** 678 (1980).
81B. L. A. Yaskin, *Heat Transfer, Soviet Res.* **12**(3), 43 (1980).
82B. R. M. Ybarra and R. E. Eckert, *AIChE Jl* **26,** 751 (1980).
83B. G. Yee, R. Chilukuri and J. A. Humphrey, *J. Heat Transfer* **102,** 285 (1980).
84B. V. M. Yeroshenko, A. D. Starostin and L. A. Yaskin, *Heat Transfer, Soviet Res.* **12**(2), 129 (1980).
85B. Yu. Ye. Zverkhovskiy, *Heat Transfer, Soviet Res.* **12**(1), 98 (1980).

Boundary layer and external flows

1C. B. Abramzon and I. Borde, *AIChE Jl* **26,** 536 (1980).
2C. A. Acrivos, E. J. Hinch and D. J. Jeffrey, *J. Fluid Mech.* **101,** 403, 1980.
3C. N. Afzal and I. S. Varshney, *Wärme- und Stoffübertragung* **14,** 289 (1980).
4C. S. Aiba, T. Ota and H. Tsuchida, *Bull. ASME* **23,** 1163 (1980).
5C. J. Andreopoulos and P. Bradshaw, *J. Heat Transfer* **102,** 755 (1980).
6C. R. A. Antonia, B. R. Satyaprakash and A. K. M. F. Hussain, *Phys. Fluids* **23,** 695 (1980).
7C. V. H. Arakeri, *Int. J. Heat Mass Transfer* **23,** 413 (1980).
8C. J. S. Bansal, *Appl. Sci. Res.* **36,** 117 (1980).
9C. K. Bauer, J. Straub and U. Grigull, *Int. J. Heat Mass Transfer* **23,** 1635 (1980).
10C. T. J. Bose, *Wärme- und Stoffübertragung* **14,** 165 (1980).
11C. M. W. Collins, *Num. Meth. Engng* **15,** 437 (1980).
12C. N. Decker and L. Glicksman, *Chem. Engng Sci.* **35,** 831 (1980).
13C. S. K. Dey, *Num. Heat Transfer* **3,** 505 (1980).
14C. R. A. East, R. J. Stalker and J. P. Baird, *J. Fluid Mech.* **97,** 673 (1980).
15C. S. E. Elgobashi and A. T. Wassel, *Int. J. Heat Mass Transfer* **23,** 1229 (1980).
16C. L. W. Florschuetz, R. A. Berry and D. E. Metzger, *J. Heat Transfer* **102,** 132 (1980).
17C. M. Fujii and S. Kikkawa, *Bull. JSME* **23,** 618 (1980).
18C. Y. Fujiwara and T. Miwa, *Bull. JSME* **23,** 1718 (1980).
19C. R. A. Graziani, M. F. Blair, J. R. Taylor and R. E. Mayle, *J. Engng Pwr* **102,** 257 (1980).
20C. T. Hachisu and T. Sasaki, *Bull. JSME* **23,** 788 (1980).
21C. J. Hanawa, *Bull. JSME* **23,** 1632, 1980.
22C. E. Hasegawa and Y. Iwasaka, *Bull. JSME* **23,** 224 (1980).
23C. A. Hirschberg, W. H. H. van Heugten, J. F. H. Willems and M. E. H. van Dongen, *Int. J. Heat Mass Transfer* **23,** 799 (1980).
24C. C. T. Hsu and P. Cheng, *Int. J. Heat Mass Transfer* **23,** 789 (1980).
25C. M. Ishida and T. Yamada, *Bull. JSME* **23,** 1467 (1980).
26C. S. Ishigai, *Bull. JSME* **23,** 2155 (1980).
27C. R. Karvinen, *Lett. Heat Mass Transfer* **7,** 385 (1980).
28C. J. C. LaRue and P. A. Libby, *Phys. Fluids* **23,** 1111 (1980).
29C. S. C. Lau and E. M. Sparrow, *J. Heat Transfer* **102,** 364 (1980).
30C. S. W. Liu and H. Mirels, *Phys. Fluids* **23,** 681 (1980).

Flow with separated regions and through porous media

1D. S. Aiba, T. Ota and H. Tsuchida, *Int. J. Heat Mass Transfer* **23,** 311 (1980).
2D. B. R. Baliga and S. V. Pantankar, *Num. Heat Transfer* **3,** 393 (1980).
3D. R. Chandran, J. C. Chen and F. W. Staub, *J. Heat Transfer* **102,** 152 (1980).
4D. C. C. Chieng and B. E. Launder, *Num. Heat Transfer* **3,** 189 (1980).
5D. G. Delaunay, A. Storck, A. Laurent and J.-C. Charpentier, *I/EC Process Des. Dev.* **19,** 514 (1980).
6D. E. P. Dyban, A. I. Mazur and V. P. Golovanov, *Int. J. Heat Mass Transfer* **23,** 667 (1980).
7D. E. R. G. Eckert and M. Faghri, *Int. J. Heat Mass Transfer* **23,** 1613 (1980).
8D. M. Faghri and E. R. G. Eckert, *Wärme- und Stoffübertragung* **14,** 217 (1980).
9D. L. R. Glicksman and F. M. Joos, *J. Heat Transfer* **102,** 736 (1980).
10D. V. Gnielinski, *Chem.-Ing.-Tech.* **52,** 228 (1980).
11D. N. S. Grewal and S. C. Saxena, *J. Heat Transfer* **101,** 397 (1979).
12D. N. S. Grewal and S. C. Saxena, *Int. J. Heat Mass Transfer* **23,** 1505 (1980).
13D. S. Hatakeyama, *Bull. JSME* **23,** 1639 (1980).
14D. J. H. B. J. Hoebink and K. Rietema, *Chem. Engng Sci.* **35,** 2135 (1980).
15D. C. L. D. Huang, *Int. J. Heat Mass Transfer* **22,** 1295 (1979).
16D. G. A. Hughmark, *I/EC Fundamentals* **19,** 198 (1980).
17D. H. T. Karlsson, S. Aberg and I. Bjerle, *Int. J. Heat Mass Transfer* **23,** 355 (1980).
18D. S. Kikuchi, *Bull. JSME* **23,** 402 (1980).
19D. A. D. Koler, N. S. Grewal and S. C. Saxena, *Int. J. Heat Mass Transfer* **22,** 1695 (1979).
20D. J. P. Lamb, *J. Heat Transfer* **102,** 351 (1980).
21D. G. F. Marsters, B. Howkins and E. Kortschak, *Int. J. Heat Mass Transfer* **23,** 301 (1980).
22D. S. K. Matveyev and G. V. Kocheryzhenkov, *Heat Transfer, Soviet Research* **10**(5), 107 (1978).
23D. G. Miller and V. Zakkay, *J. Heat Transfer* **102,** 731 (1980).
24D. A. Montakhab, *J. Heat Transfer* **101,** 507 (1979).
25D. F. Ogino, H. Takeuchi, I. Kudo and T. Mizushina, *Int. J. Heat Mass Transfer* **23,** 1581 (1980).
26D. T. Ota and N. Kon, *J. Heat Transfer* **102,** 749 (1980).
27D. B. N. Pamadi and I. A. Belov, *Int. J. Heat Mass Transfer* **23,** 783 (1980).
28D. Cz. O. Popiel, Th. H. Van Der Meer and C. J. Hoogendoorn, *Int. J. Heat Mass Transfer* **23,** 1055 (1980).

29D. J. M. Ramilson and B. Gebhart, *Int. J. Heat Mass Transfer* **23,** 1521 (1980).
30D. M. S. Sahota and P. J. Pagni, *Int. J. Heat Mass Transfer* **22,** 1069 (1979).
31D. N. Seki, S. Fukusako and K. Torikoshi, *Wärme- und Stoffübertragung* **14,** 173 (1980).
32D. B. S. Singh and A. Dybbs *Int. J. Heat Mass Transfer* **22,** 1049 (1979).
33D. F. W. Staub, *J. Heat Transfer* **101,** 391 (1979).

Transfer mechanisms

1E. A. Acrivos, *J. Fluid Mech.* **98,** 299 (1980).
2E. D. R. Ballal, *Proc. R. Soc., London* **369A,** 479 (1980).
3E. H. J. Bock and O. Molerus, *Chem.-Ing.-Tech.* **52,** 260 (1980).
4E. H. A. Dinulescu and E. R. G. Eckert, *Int. J. Heat Mass Transfer* **23,** 1069 (1980).
5E. L. O. Eyler and A. Sesonske, *Int. J. Heat Mass Transfer* **23,** 1561 (1980).
6E. Yu. T. Glazunov, *Int. J. Heat Mass Transfer* **23,** 759 (1980).
7E. D. Jou and C. Perez-Garcia, *Physica* **103A,** 320 (1980).
8E. G. Kaczmarzyk and J. Bandrowski, *Int. chem. Engng* **20,** 98 (1980).
9E. H. F. P. Knaap and I. Kuscer, *Physica* **103A,** 95 (1980).
10E. S. Kumar and S. N. Upadhyay, *I/EC Fundamentals* **19,** 75 (1980).
11E. R. Loudon, *Proc. R. Soc., London* **372A,** 257 (1980).
12E. H. Martin, *Chem.-Ing.-Tech.* **52,** 199 (1980).
13E. J. O. Mingle, *Num. Heat Transfer* **3,** 499 (1980).
14E. J. Mizushima and A. Segami, *Phys. Fluids* **23,** 2559 (1980).
15E. Y. Mori, N. Himeno, K. Hujikata and T. Miyauchi, *Int. J. Heat Mass Transfer* **23,** 1625 (1980).
16E. T. Nagatani, *Bull. JSME* **23,** 488 (1980).
17E. H. Seifert and R. Günther, *Chem.-Ing.-Tech.* **52,** 262 (1980).
18E. R. Sentis, *SIAM J. appl. Math.* **39,** 134 (1980).
19E. S. Sieniutycz, *Int. J. Heat Mass Transfer* **23,** 1183 (1980).
20E. K. R. Sreenivasan and S. Tavoularis, *J. Fluid Mech.* **101,** 783 (1980).
21E. K. R. Sreenivasan, S. Tavoularis, R. Henry and S. Corrsin, *J. Fluid Mech.* **100,** 597 (1980).
22E. Y. Tambour, *Int. J. Heat Mass Transfer* **23,** 321 (1980).
23E. L. C. Thomas, *Int. J. Heat Mass Transfer* **23,** 1099 (1980).
24E. Z. Warhaft, *J. Fluid Mech.* **99,** 545 (1980).
25E. L. C. Woods, *J. Fluid Mech.* **101,** 225 (1980).

Natural convection—internal flows

1F. G. Ahlers, *J. Fluid Mech.* **98,** 137 (1980).
2F. R. F. Bergholz, *J. Heat Transfer* **102,** 242 (1980).
3F. S. Bories and A. Deltour, *Int. J. Heat Mass Transfer* **23,** 765 (1980).
4F. F. H. Busse and N. Riahi, *J. Fluid Mech.* **96,** 243 (1980).
5F. J. P. Caltagiirone, *Q. J. Mech. appl. Math.* **33,** 47 (1980).
6F. J. P. Caltagirone, M. Combarnous and A. Mojtabi, *Num. Heat Transfer* **3,** 107 (1980).
7F. C. J. Chapman and M. R. E. Proctor, *J. Fluid Mech.* **101,** 759 (1980).
8F. F. B. Cheung, *J. Fluid Mech.* **97,** 743 (1980).
9F. F. B. Cheung and J. B. Novas, *Lett. Heat Mass Transfer* **7,** 171 (1980).
10F. I. G. Choi and S. A. Korpela, *J. Fluid Mech.* **99,** 725 (1980).
11F. S. H. Davis and G. M. Homsy, *J. Fluid Mech.* **98,** 527 (1980).
12F. P. M. Eagles, *Proc. R. Soc., London* **371A,** 359 (1980).
13F. M. A. El-Hawary, *J. Heat Transfer* **102,** 273 (1980).
14F. M. A. I. El-Shaarawi and A. Sarhan, *J. Heat Transfer* **102,** 617 (1980).
15F. A. A. Emara and F. A. Kulacki, *J. Heat Transfer* **102,** 531 (1980).
16F. R. D. Flack, *J. Heat Transfer* **102,** 770 (1980).
17F. H. Frick and R. M. Clever, *Z. angew Math. Phys* **31,** 502 (1980).
18F. R. J. Goldstein and S. Tokuda, *Int. J. Heat Mass Transfer* **23,** 738 (1980).
19F. J. P. Gollub and S. V. Benson, *J. Fluid Mech.* **100,** 449 (1980).
20F. A. D. Gosman, P. V. Nielsen, A. Restivo and J. H. Whitelaw, *J. Fluids Engng* **102,** 316 (1980).
21F. J. C. Han and W. J. Yang, *Lett. Heat Mass Transfer* **7,** 363 (1980).
22F. C. E. Hickox and H. A. Watts, *J. Heat Transfer* **102,** 248 (1980).
23F. R. N. Horne and J. P. Caltagiorone, *J. Fluid Mech.* **100,** 385 (1980).
24F. L. Iyican, Y. Bayazitoglu and L. C. Witte, *J. Heat Transfer* **102,** 640 (1980).
25F. L. Iyican, L. C. Witte and Y. Bayazitoglu, *J. Heat Transfer* **102,** 648 (1980).
26F. R. Izumi, *Bull. JSME* **23,** 150 (1980).
27F. B. Jhaveri and G. M. Homsy, *J. Fluid Mech.* **98,** 329 (1980).
28F. C. Jischke and M. Farschi, *J. Heat Transfer* **102,** 228 (1980).
29F. P. D. Killworth and P. C. Manins, *J. Fluid Mech.* **98,** 587 (1980).
30F. S. Kimura and A. Bejan, *Int. J. Heat Mass Transfer* **23,** 1117 (1980).
31F. E. Knobloch, *Phys. Fluids* **23,** 1918 (1980).
32F. K. Kübleck, G. P. Merker and J. Straub, *Int. J. Heat Mass Transfer* **23,** 203 (1980).
33F. T. H. Kuehn and R. J. Goldstein, *Int. J. Heat Mass Transfer* **23,** 971 (1980).
34F. T. H. Kuehn and R. J. Goldstein, *J. Heat Transfer* **102,** 768 (1980).
35F. R. L. Mahajan and B. Gebhart, *J. Heat Transfer* **102,** 368 (1980).
36F. R. McKibbin and M. J. O'Sullivan, *J. Fluid Mech.* **96,** 375 (1980).
37F. R. Meynart, *Appl. Optics* **19,** 1385 (1980).
38F. H. Mouton, J. C. Couanon and M. Roiland, *Lett. Heat Mass Transfer* **7,** 33 (1980).
39F. H. Mouton, J. C. Couanon and M. Roiland, *Lett.*

Heat Mass Transfer **7**, 447 (1980).
40F. M. Nakajima, K. Fukui, H. Ueda and T. Mizushina, *Int. J. Heat Mass Transfer* **23**, 1325 (1980).
41F. S. O. Onyegegbu, *J. Heat Transfer* **102**, 268 (1980).
42F. R. C. Paliwal and C. F. Chen, *J. Fluid Mech.* **98**, 755 (1980).
43F. R. C. Paliwal and C. F. Chen, *J. Fluid Mech.* **98**, 769 (1980).
44F. P. R. Patil and N. Rudriah, *Int. J. engng Sci.* **18**, 1055 (1980).
45F. J. Patterson and J. Imberger, *J. Fluid Mech.* **100**, 65 (1980).
46F. R. E. Powe, R. O. Warrington and J. A. Scanlan, *Int. J. Heat Mass Transfer* **23**, 1337 (1980).
47F. L. Rahm and G. Walin, *J. Fluid Mech.* **97**, 807 (1980).
48F. K. V. Rama Rao, *Int. J. engng Sci.* **18**, 741 (1980).
49F. C. S. Reddy, *J. Heat Transfer* **102**, 172 (1980)
49F. C. S. Reddy, *J. Heat Transfer* **102**, 172 (1980).
50F. N. Riahi, *Aust. J. Phys.* **33**, 59 (1980).
51F. N. Riahi, *Z. agnew. Math. Phys* **31**, 261 (1980).
52F. E. Ruckenstein, *AIChE Jl* **26**, 850 (1980).
53F. N. Rudraiah, B. Veerappa and S. B. Rao, *J. Heat Transfer* **102**, 254 (1980).
54F. D. W. Ruth, G. D. Raithby and K. G. T. Hollands, *J. Fluid Mech.* **96**, 481 (1980).
55F. D. W. Ruth, K. G. T. Hollands and G. D. Raithby, *J. Fluid Mech.* **96**, 461 (1980).
56F. E. Saaidjian and J. P. Caltagiorone, *J. Heat Transfer* **102**, 654 (1980).
57F. D. Schwabe and A. Scharmann, *Lett. Heat Mass Transfer* **7**, 283 (1980).
58F. S. N. Singh and J. Chen, *Num. Heat Transfer* **3**, 441 (1980).
59F. D. R. Smart, K. G. T. Hollands and G. D. Raithby, *J. Heat Transfer* **102**, 75 (1980).
60F. A. M. Soward, *J. Fluid Mech.* **98**, 449 (1980).
61F. E. M. Sparrow, S. Shah and C. Prakash, *Num. Heat Transfer* **3**, 297 (1980).
62F. H. Tanaka and H. Miyata, *Int. J. Heat Mass Transfer* **23**, 1273 (1980).
63F. B. L. Turner and R. D. Flack, *J. Heat Transfer* **102**, 236 (1980).
64F. P. Vasseur and L. Robillard, *Int. J. Heat Mass Transfer* **23**, 1195 (1980).
65F. I. C. Walton, *Q. J. Mech. appl. Math.* **33**, 125 (1980).
66F. M. A. Yaghoubi and F. P. Incropera, *Num. Heat Transfer* **3**, 315 (1980).
67F. S. Yanase, *J. Phys. Soc. Japan* **48**, 998 (1980).
68F. L. S. Yao, *J. Heat Transfer* **102**, 279 (1980).
69F. L. S. Yao and F. F. Chen, *J. Heat Transfer* **102**, 667 (1980)1.
70F. Y.-C. Yen, *J. Heat Transfer* **102**, 550 (1980).
71F. A. Yücel and Y. Bayazitoğlu, *Lett. Heat Mass Transfer* **7**, 391 (1980)1.
72F. Z. Zebib, G. Schubert and J. M. Straus, *J. Fluid Mech.* **97**, 259 (1980).

Natural convection—external flows

1FF. N. Afzal, *Int. J. Heat Mass Transfer* **23**, 505 (1980).
2FF. M. Al-Arabi and Y. K. Salman, *Int. J. Heat Mass Transfer* **23**, 45 (1980).
3FF. R. Anderson and A. Bejan, *J. Heat Transfer* **102**, 630 (1980).
4FF. A. V. Appalaswamy and Y. Jaluria, *J. appl. Mech.* **47**, 667 (1980).
5FF. V. P. Carey, B. Gebhart and J. C. Mollendorf, *J. Fluid Mech* **97**, 279 (1980).
6FF. V. P. Carey and J. C. Mollendorf, *Int. J. Heat Mass Transfer* **23**, 95 (1980).
7FF. T. S. Chen and F. A. Strobel, *Num. Heat Transfer* **3**, 115 (1980).
8FF. T. S. Chen and F. A. Strobel, *J. Heat Transfer* **102**, 170 (1980).
9FF. T. S. Chen and C. F. Yuh, *Int. J. Heat Mass Transfer* **23**, 451 (1980).
10FF. T. S. Chen, C. F. Yuh and A. Moutsoglou, *Int. J. Heat Mass Transfer* **23**, 527 (1980).
11FF. F. B. Cheung, *J. Heat Transfer* **102**, 373 (1980).
12FF. V. E. Delnore, *J. Fluid Mech.* **96**, 803 (1980).
13FF. R. Eichhorn and M. M. Hasan, *J. Heat Transfer* **102**, 775 (1980).
14FF. M. Fujii, T. Fujii and T. Honda, *Bull. JSME* **23**, 488 (1980).
15FF. M. Fujii, T. Fujii and S. Koyama, *Bull. JSME* **23**, 333 (1980).
16FF. J. D. Gabor, L. Baker, Jr., J. C. Cassulo, D. J. Erskine and J. G. Warner, *J. Heat Transfer* **102**, 519 (1980).
17FF. O. F. Genecli, *Wärme- und Stoffübertragung* **13**, 163 (1980).
18FF. J. C. Han and W-J Yang, *J. Heat Transfer* **102**, 581 (1980).
19FF. M. A. Hilal and R. W. Boom, *Int. J. Heat Mass Transfer* **23**, 697 (1980).
20FF. C. T. Hsu and P. Chen, *J. Heat Transfer* **102**, 544 (1980).
21FF. R. Hunt and G. Wilks, *J. Fluid Mech.* **101**, 377 (1980).
22FF. H. E. Huppert and J. S. Turner, *J. Fluid Mech.* **100**, 367 (1980).
23FF. F. P. Incropera and M. A. Yaghoubi, *Int. J. Heat Mass Transfer* **23**, 269 (1980).
24FF. S. Ishigai, *Bull. JSME* **23**, 787 (1980).
25FF. G. T. Jarvis and D. P. McKenzie, *J. Fluid Mech.* **96**, 515 (1980).
26FF. K. Kishinami, *Bull. JSME* **23**, 1261 (1980).
27FF. T. H. Kuehn and R. J. Goldstein, *Int. J. Heat Mass Transfer* **23**, 971 (1980).
28FF. M. Miyamoto, J. Sumikawa, T. Akiyoshi and T. Nakamura, *Int. J. Heat Mass Transfer* **23**, 1545 (1980).
29FF. A. Moutsoglou and T. S. Chen, *J. Heat Transfer* **102**, 371 (1980).
30FF. T. Y. Na and J. P. Chiou, *Wärme- und Stoffübertragung* **14**, 157 (1980).
31FF. R. J. Neumann and E. W. P. Hahne, *Int. J. Heat Mass Transfer* **23**, 1643 (1980).
32FF. J. R. Parsons, Jr. and J. C. Mulligan, *J. Heat Transfer* **102**, 636 (1980).
33FF. H. B. Phuoc and R. I. Tanner, *J. Fluid Mech.* **98**, 253 (1980).
34FF. I. Pop and V. M. Soundalgekar, *Chem. Engng Sci.* **35**, 750 (1980).
35FF. J. M. Potter and N. Riley, *J. Fluid Mech.* **100**, 769 (1980).
36FF. C. Prakash and E. M. Sparrow, *Num. Heat Transfer*

3, 89 (1980).
37FF. J. N. Reddy and A. Satake, *J. Heat Transfer* **102,** 659 (1980).
38FF. J. B. Riester, R. A. Bajura and S. H. Schwartz, *J. Heat Transfer* **102,** 557 (1980).
39FF. E. Ruckenstein and J. D. Felske, *J. Heat Transfer* **102,** 773 (1980).
40FF. T. Saitoh and K. Hirose, *J. Heat Transfer* **102,** 261 (1980).
41FF. P. A. Sandborn and V. A. Sandborn, *J. Heat Transfer* **102,** 174 (1980).
42FF. A. V. Shenoy, *AIChE Jl* **26,** 505 (1980).
43FF. A. V. Shenoy, *AIChE Jl* **26,** 683 (1980).
44FF. E. M. Sparrow and P. A. Bahrami, *J. Heat Transfer* **102,** 221 (1980).
45FF. E. M. Sparrow and P. A. Bahrami, *Int. J. Heat Mass Transfer* **23,** 1555 (1980).
46FF. E. M. Sparrow and M. Faghri, *J. Heat Transfer* **102,** 623 (1980).
47FF. F. A. Strobel and T. S. Chen, *Num. Heat Transfer* **3,** 461 (1980).
48FF. A. Suwono, *Int. J. Heat Mass Transfer* **23,** 53 (1980).
49FF. S. Tsuruno and I. Iguchi, *J. Heat Transfer* **102,** 168 (1980).
50FF. K. Vajravelu, *Num. Heat Transfer* **3,** 345 (1980).
51FF. L.-S. Yao, I. Catton and J. M. McDonough, *J. Fluid Mech.* **98,** 417 (1980).

Convection from rotating surfaces
1G. P. G. Bellamy-Knights, *Q. J. Mech. appl. Math.* **33,** 321 (1980).
2G. I. Cornet, R. Greif, J. T. Teng and F. Roehler, *Int. J. Heat Mass Transfer* **23,** 805 (1980).
3G. R. J. Dallman and R. W. Douglass, *Int. J. Heat Mass Transfer* **23,** 1303 (1980).
4G. P. C. Singh and P. Mishra, *Chem. Engng Sci.* **35,** 1657 (1980).
5G. M. Singh and S. C. Rajvanshi, *J. Heat Transfer* **102,** 347 (1980).
6G. A. Suwono, *Int. J. Heat Mass Transfer* **23,** 819 (1980).
7G. G. B. Tatterson, H.-H. Yuan and R. S. Brodkey, *Chem. Engng Sci.* **35,** 1369 (1980).
8G. N. R. Vira and D. N. Fan, *Num. Heat Transfer* **3,** 483 (1980).
9G. J. L. Woods and W. D. Morris, *J. Heat Transfer* **102,** 612 (1980).

Combined heat and mass transfer
1H. W. O. Afejuka, N. Hay and D. Lampard, *J. Engng Pwr* **102,** 601 (1980).
2H. M. E. Crawford, W. M. Kays and R. J. Moffat, *J. Engng Pwr* **102,** 1000 (1980).
3H. M. E. Crawford, W. M. Kays and R. J. Moffat, *J. Engng Pwr* **102,** 1006 (1980).
4H. R. P. Dring, M. F. Blair and H. D. Joslyn, *J. Engng Pwr* **102,** 81 (1980).
5H. N. W. Foster and D. Lampard, *J. Engng Pwr* **102,** 584 (1980).
6H. R. J. Hartranft and G. C. Sih, *Int. J. Engng Sci.* **18,** 1375 (1980).
7H. H. Hempel, R. R. Friedrich and S. Wittig, *J. Engng Pwr* **102,** 957 (1980).
8H. K. Hishida, M. Maeda and S. Ikai, *J. Heat Transfer* **102,** 513 (1980).
9H. B. R. Hollworth and L. Dagan, *J. Engng Pwr* **102,** 994 (1980).
10H. S. Ito, E. R. G. Eckert and R. J. Goldstein, *J. Engng Pwr* **102,** 964 (1980).
11H. W. Kast, *Wärme- und Stoffübertragung* **13,** 217 (1980).
12H. M. Kumari and G. Nath, *Int. J. Engng Sci.* **18,** 1285 (1980).
13H. J. N. Moss, A. L. Simmonds and E. C. Anderson, *J. Spacecraft Rockets* **17,** 177 (1980).
14H. V. Ye. Nakoryakov and N. I. Grigor'yeva, *Heat Transfer, Soviet Res.* **12**(3), 111 (1980).
15H. G. C. Sih, M. T. Shih and S. C. Chou, *Int. J. Engng Sci.* **18,** 19 (1980).
16H. M. Strada and R. W. Lewis, *Num. Heat Transfer* **3,** 429 (1980).
17H. G. J. Sturgess, *J. Engng Pwr* **102,** 524 (1980).
18H. R. K. M. Thambynayagam, P. Winter and S. J. Branch, *Trans. Inst. chem. Engrs* **58,** 277 (1980).
19H. S. Yavuzkurt, R. J. Moffat and W. M. Kays, *J. Fluid Mech.* **101,** 129 (1980).
20H. S. Yavuzkurt, R. J. Moffat and W. M. Kays, *J. Fluid Mech.* **101,** 159 (1980).

Change of phase
1J. B. A. Afanas'yev, V. V. Zrodnikov, A. L. Koba, T. F. Smirnov and A. K. Pimenoy, *Heat Transfer, Soviet Res.* **12**(2), 70 (1980).
2J. V. V. Arkhipov, V. V. Parygin, V. I. Deev and A. I. Pridantsev, *Thermal Engng* **27,** 200 (1980).
3J. N. Z. Azer, S. T. Lin and L. T. Fan, *I/EC Proc. Des. Dev.* **19,** 246 (1980).
4J. Yu. I. Balakleyevskiy, *Heat Transfer, Soviet Res.* **12**(3), 124 (1980).
5J. S. G. Bankoff, *Int. J. Multiphase Flow* **6,** 51 (1980).
6J. V. I. Baranenko, *Heat Transfer, Soviet Res.* **12**(2), 115 (1980).
7J. V. I. Baranenko and L. A. Chichkan, *Heat Transfer, Soviet Res.* **12**(2), 40 (1980).
8J. B. R. Bergel'son, *Thermal Engng* **27,** 48 (1980).
9J. M. K. Bezrodnyy, S. N. Faynzil'berg and Ye. A. Kondrusik, *Heat Transfer, Soviet Res.* **12**(1), 118 (1980).
10J. B. L. Bhatt and G. L. Wedekind, *J. Heat Transfer* **102,** 694 (1980).
11J. B. L. Bhatt and G. L. Wedekind, *J. Heat Transfer* **102,** 495 (1980).
12J. R. A. Blachowikz, B. W. Brooks and K. B. Tan, *Chem. Engng Sci.* **35,** 761 (1980).
13J. J. A. Block, *Int. J. Multiphase Flow* **6,** 113 (1980).
14J. W. Bonn, J. Iwicki, R. Krebs, D. Steiner and E. U. Schlünder, *Wärme- und Stoffübertragung* **13,** 265 (1980).
15J. G. Breber, J. W. Palen and J. Taborek, *J. Heat Transfer* **102,** 471 (1980).
16J. A. P. Burdukov, G. G. Kuvshinov and V. Ye. Nakoryakov, *Heat Transfer, Soviet Res.* **12**(2), 64 (1980).
17J. A. V. Burmistrov, *Heat Transfer, Soviet Res.* **12**(2), 22 (1980).
18J. P. Burow and H. Beer, *Wärme- und Stoffübertragung*

13, 253 (1980).
19J. S. H. Chan, D. H. Cho and D. W. Condiff, *Int. J. Heat Mass Transfer* **23,** 63 (1980).
20J. T. C. Chawla and M. Ishi, *Int. J. Heat Mass Transfer* **23,** 991 (1980).
21J. V. A. Chernobay, A. F. Vasil'yev, S. V. Perkov, V. A. Shpakovskiy and V. S. Furayev, *Heat Transfer, Soviet Res.* **12**(1), 113 (1980).
22J. S. Chongrungreong and H. J. Sauer, Jr., *J. Heat Transfer* **102,** 701 (1980).
23J. R. Cole, J. M. Papazian and W. R. Wilcox, *Int. J. Heat Mass Transfer* **23,** 219 (1980).
24J. M. L. Corradini, W. M. Rohsenow and N. E. Todreas, *Nucl. Sci. Engng* **73,** 242 (1980).
25J. C. F. Delale, *J. Heat Transfer* **102,** 501 (1980).
26J. V. K. Dhir, *J. Heat Transfer* **102,** 380 (1980).
27J. F. Dobran and R. S. Thorsen, *Int. J. Heat Mass Transfer* **23,** 161 (1980).
28J. V. G. Dobrzhanskiy, V. D. Chayka and Yu. V. Yakuborskiy, *Heat Transfer, Soviet Res.* **12**(2), 109 (1980).
29J. M. Epstein and G. M. Hauder, *Int. J. Heat Mass Transfer* **23,** 179 (1980).
30J. S. Esaki, *Bull. JSME* **23,** 152 (1980).
31J. S. Esaki, *Bull. JSME* **23,** 1475 (1980).
32J. V. V. Fisenko, Yu. D. Katkov, A. P. Lastochkin and V. I. Maksimov, *Soviet Atomic Energy* **48,** 335 (1980).
33J. T. Fujii and H. Honda, *Bull. JSME* **23,** 1563 (1980).
34J. T. Fujii and H. Honda, *Bull. JSME* **23,** 152 (1980).
35J. T. Fujii, H. Honda and K. Oda, *Bull. JSME* **23,** 1563 (1980).
36J. T. Fujii, Y. Ide and O. Miyatake, *Bull. JSME* **23,** 1415 (1980).
37J. T. Fujii and Y. Kato, *Bull. JSME* **23,** 1717 (1980).
38J. S. Fujikawa and T. Akamatsu, *J. Fluid Mech.* **97,** 481 (1980).
39J. J. D. Gabor, L. Baker, Jr., J. C. Cassulo, D. J. Erskine and J. G. Warner, *J. Heat Transfer* **102,** 519 (1980).
40J. E. N. Ganic and M. N. Roppo, *J. Heat Transfer* **102,** 342 (1980).
41J. A. I. Gavrilin, *Heat Transfer, Soviet Res.* **12**(3), 30 (1980).
42J. V. G. Genechev and A. Ye. Bokov, *Heat Transfer, Soviet Res.* **12**(3), 18 (1980).
43J. I. A. Gills and P. G. Udyma, *Heat Transfer, Soviet Res.* **12**(1), 133 (1980).
44J. G. Guglielmini, E. Nannei and C. Pisoni, *Wärme- und Stoffübertragung* **13,** 177 (1980).
45J. F. S. Gunnerson and A. W. Cronenberg, *J. Heat Transfer* **102,** 335 (1980).
46J. J. C. Han and Wen-Jei Yang, *J. Heat Transfer* **102,** 581 (1980).
47J. F. X. Hart and J. S. Massey, *Int. J. Heat Mass Transfer* **23,** 363 (1980).
48J. S. Hirasawa, K. Huikata, Y. Mori and W. Nakayama, *Int. J. Heat Mass Transfer* **23,** 1471 (1980).
49J. Y. Iida and T. Takashima, *Bull. JSME* **23,** 2156 (1980).
50J. Y. Iida and T. Takashima, *Int. J. Heat Mass Transfer* **23,** 1263 (1980).
51J. I. N. Il'in, V. P. Grivtsov, A. D. Amelin and S. R. Yaundalders, *Heat Transfer, Soviet Res.* **12**(2), 51 (1980).
52J. I. N. Il'in, D. P. Turlays and V. A. Grishin, *Heat Transfer, Soviet Res.* **12**(2), 101 (1980).
53J. H. Imura, *Bull. JSME* **23,** 151 (1980).
54J. V. P. Isachenko, F. Salomzoda and A. A. Shalakhov, *Thermal Engng* **27,** 193 (1980).
55J. P. K. Jain and R. P. Roy, *Num. Heat Transfer* **3,** 381 (1980).
56J. O. C. Jones, Jr., *J. Heat Transfer* **102,** 439 (1980).
57J. R. L. Judd and C. H. Lavdas, *J. Heat Transfer* **102,** 461 (1980).
58J. O. N. Kaban'kov and V. V. Yagov, *Thermal Engng* **27,** 297 (1980).
59J. T. Kan, *Bull. JSME* **23,** 788 (1980).
60J. V. M. Kashcheyev and Yu. V. Muranov, *Heat Transfer, Soviet Res.* **12**(3), 49 (1980).
61J. Y. Katto, *Bull. JSME* **23,** 1262 (1980).
62J. Y. Katto, *Int. J. Heat Mass Transfer* **23,** 493 (1980).
63J. Y. Katto, *Bull. JSME* **23,** 334 (1980).
64J. Y. Katto, *Bull. JSME* **23,** 1415 (1980).
65J. Y. Katto, *Int. J. Heat Mass Transfer* **23,** 1573 (1980).
66J. Y. Katto, *Bull. JSME* **23,** 787 (1980).
67J. Y. Katto and C. Kurata, *Int. J. Multiphase Flow* **6,** 575 (1980).
68J. M. G. Khor'kov, V. D. Glazko and N. M. Orlova, *Heat Transfer, Soviet Res.* **12**(3), 108 (1980).
69J. Yu. A. Kirichenko, N. M. Levchenko and K. V. Rusanova, *Heat Transfer, Soviet Res.* **12**(2), 57 (1980).
70J. C. A. Kodres, H. R. Jacobs and R. F. Boehm, *Num. Heat Transfer* **3,** 21 (1980).
71J. Y. Koizumi, *Bull. JSME* **23,** 489 (1980).
72J. A. D. Korneyev and S. D. Korneyev, *Heat Transfer, Soviet Res.* **12**(3), 47 (1980).
73J. B. P. Korol'kov and E. A. Tairov, *Heat Transfer, Soviet Res.* **12**(3), 36 (1980).
74J. S. Kotake and K. Oswatitsch, *Int. J. Heat Mass Transfer* **23,** 1405 (1980).
75J. S. S. Kutateladze, I. I. Gogonin, N. I. Grigor'eva and A. R. Dorokhov, *Thermal Engng* **27,** 184 (1980).
76J. D. A. Labuntsov and E. V. Ametistov, *Thermal Engng* **27,** 287 (1980).
77J. A. Lekic and J. D. Ford, *Int. J. Heat Mass Transfer* **23,** 1531 (1980).
78J. J. H. Lienhard and V. K. Dhir, *J. Heat Transfer* **102,** 457 (1980).
79J. C. C. Lu and J. W. Heyt, *AIChE Jl* **26,** 762 (1980).
80J. D. W. Lyons, C. T. Voliers and A. M. Elnasher, *J. Engng Ind.* **102,** 8 (1980).
81J. A. F. Mills, *Int. J. Multiphase Flow* **6,** 41 (1980).
82J. O. O. Mil'man and G. G. Shklover, *Heat Transfer, Soviet Res.* **12**(3), 131 (1980).
83J. Z. L. Miropol'sky, R. I. Shneyerova and V. Sh. Mekler, *Heat Transfer, Soviet Res.* **12**(3), 12 (1980).
84J. J. Mitrovic and K. Stephan, *Wärme- und Stoffübertragung* **13,** 171 (1980).
85J. D. Moalem-Maron, M. Sokolov and S. Sideman, *Int. J. Heat Mass Transfer* **23,** 1417 (1980).
86J. S. Mochizuki and T. Shiratori, *J. Heat Transfer* **102,** 158 (1980).
87J. M. Monde, *Bull. JSME* **23,** 333 (1980).
88J. Y. H. Mori, H. Sano and K. Komotori, *Int. J. Multiphase Flow* **6,** 255 (1980).

89J. A. S. Mujumdar, Y.-K. Li and W. J. M. Douglas, *Can. J. chem Engng* **58,** 448 (1980).
90J. A. Nakayama, *Bull. JSME* **23,** 1564 (1980).
91J. W. Nakayama, T. Kaikoku, H. Kuwahara and T. Nakajima, *J. Heat Transfer* **102,** 445 (1980).
92J. W. Nakayama, T. Kaikoku, H. Kuwahara and T. Nakajima, *J. Heat Transfer* **102,** 451 (1980).
93J. Ye. I. Nesis, V. I. Komorov, L. M. Kul'gina, A. A. Kul'gin, I. S. Sologab and S. Ye. Nesis, *Heat Transfer, Soviet Res.* **12**(2), 85 (1980).
94J. R. I. Nigmatulin, N. S. Khabeyev and F. B. Nagiyev, *Heat Transfer, Soviet Res.* **12**(2), 16 (1980).
95J. S. Nijhawan, J. C. Chen, R. K. Sundaram and E. J. London, *J. Heat Transfer* **102,** 465 (1980).
96J. R. H. Nilson and R. C. Montoya, *J. Heat Transfer* **102,** 489 (1980).
97J. R. H. Nilson and L. A. Romero, *Int. J. Heat Mass Transfer* **23,** 1461 (1980).
98J. V. K. Orlov and V. N. Savel'yev, *Thermal Engng* **27,** 229 (1980).
99J. V. K. Orlov, V. Ye. Poznak, V. N. Savel'yev and G. P. Kuz'menko, *Heat Transfer, Soviet Res.* **12**(2), 79 (1980).
100J. A. P. Ornatskiy and I. G. Sharayevskiy, *Heat Transfer, Soviet Res.* **12**(1), 137 (1980).
101J. C. B. Panchal and K. J. Bell, *Num. Heat Transfer* **3,** 357 (1980).
102J. Yu. M. Pavlov, S. A. Potekhin and G. M. Frolova, *Thermal Engng* **27,** 196 (1980).
103J. Yu. M. Pavlov, S. A. Potekhin and V. A. Shugayev, *Heat Transfer, Soviet Res.* **12**(2), 45 (1980).
104J. P. A. Pavlov, Ye. N. Sinitsyn, Ye. D. Nikitin and V. S. Uskov, *Heat Transfer, Soviet Res.* **12**(2), 34 (1980).
105J. R. M. Perkin, *Int. J. Heat Mass Transfer* **23,** 687 (1980).
106J. A. M. Podsushnyy, A. N. Minayev, V. N. Statsenko and Yu. V. Yakubovskiy, *Heat Transfer, Soviet Res.* **12**(2), 113 (1980).
107J. S. Prakash and W. A. Sirignano, *Int. J. Heat Mass Transfer* **23,** 253 (1980).
108J. Ye. M. Puzyrev, A. V. Kuz'min and V. V. Salomatov, *Heat Transfer, Soviet Res.* **12**(2), 11 (1980).
109J. R. Puzyrewski and W. Studzinski, *Int. J. Multiphase Flow* **6,** 425 (1980).
110J. A. Rane and S.-C. Yao, *Can. J. chem. Engng* **58,** 303 (1980).
111J. V. G. Rifert, *Heat Transfer, Soviet Res.* **12**(3), 142 (1980).
112J. H. J. Röhm and A. Vogelpohl, *Wärme- und Stoffübertragung* **13,** 231 (1980).
113J. T. M. Romberg and N. W. Rees, *Int. J. Multiphase Flow* **6,** 523 (1980).
114J. J. W. Rose, *Int. J. Heat Mass Transfer* **23,** 547 (1980).
115J. P. Saha, *Int. J. Heat Mass Transfer* **23,** 483 (1980).
116J. A. Segev and S. G. Bankoff, *Int. J. Heat Mass Transfer* **23,** 637 (1980).
117J. A. Segev and R. P. Collier, *J. Heat Transfer* **102,** 688 (1980).
118J. M. M. Shah, *Int. J. Heat Mass Transfer* **23,** 225 (1980).
119J. I. G. Shekriladze, *Heat Transfer, Soviet Res.* **12**(2), 120 (1980).
120J. I. G. Shekriladze, Sh. A. Mestvirishvili, D. G. Rusishvili, G. I. Zhorzholiani and V. G. Ratiani, *Heat Transfer, Soviet Res.* **12**(2), 91 (1980).
121J. A. G. Sheynkman and V. G. Zakharov, *Heat Transfer, Soviet Res.* **12**(3), 45 (1980).
122J. G. G. Shklover, V. P. Semenov and O. O. Mil'man, *Heat Transfer, Soviet Res.* **12**(3), 135 (1980).
123J. H. Sofrata, *Wärme- und Stoffübertragung* **14,** 201 (1980).
124J. S. A. Stylianou and J. W. Rose, *J. Heat Transfer* **102,** 477 (1980).
125J. M. A. Styrikovich, A. I. Leontiev, V. S. Polonsky and I. I. Malashkin, *Int. J. Heat Mass Transfer* **34,** 1045 (1980).
126J. N. I. Syromyatnikov, L. K. Vasanova and A. M. Tushin, *Heat Transfer, Soviet Res.* **12**(2), 111 (1980).
127J. K. Taghavi-Tafreshi and V. K. Dhir, *Int. J. Heat Mass Transfer* **23,** 1433 (1980).
128J. O. Takahashi and M. Nishida, *Int. J. Heat Mass Transfer* **23,** 27 (1980).
129J. K. Tephan and M. Abdelsalam, *Int. J. Heat Mass Transfer* **23,** 73 (1980).
130J. T. G. Theofanous, *Int. J. Multiphase Flow* **6,** 69 (1980).
131J. V. I. Tolubinskiy, A. M. Kichigin and S. G. Povsten, *Heat Transfer, Soviet Res.* **12**(2), 141 (1980).
132J. Ch. Trepp and T. Hoffman, *Wärme- und Stoffübertragung* **14,** 15 (1980).
133J. M. Ünsal and W. C. Thomas, *J. Heat Transfer* **102,** 483 (1980).
134J. A. A. Voloshko, *Heat Transfer, Soviet Res.* **12**(2), 1 (1980).
135J. L. Wassner, *Wärme- und Stoffübertragung* **14,** 23 (1980).
136J. S. Wong and L. E. Hochreiter, *J. Heat Transfer* **102,** 508 (1980).
137J. S.-C. Yao and A. Rane, *J. Heat Transfer* **102,** 678 (1980).
138J. Y. Yasukawa, *Bull. JSME* **23,** 1564 (1980).
139J. G. V. Yermakov and N. M. Semenova, *Heat Transfer, Soviet Res.* **12**(2), 28 (1980).
140J. S. Yilmaz and J. W. Westwater, *J. Heat Transfer* **102,** 26 (1980).
141J. D. Yung, J. J. Lorenz and E. N. Ganic, *J. Heat Transfer* **102,** 20 (1980).
142J. Yu. A. Zeigarnick and V. D. Litvinov, *Nucl. Sci. Engng* **73,** 19 (1980).
143J. L. V. Zyson, L. A. Fel'dberg, A. L. Dobkes and A. G. Sazhenin, *Heat Transfer, Soviet Res.* **12**(2), 6 (1980).

Radiation in participating media

1K. R. Becker, *Chem.-Ing.-Tech.* **52,** 162 (1980).
2K. H. E. Bennett and D. K. Burge, *J. Opt. Soc. Am.* **70,** 268 (1980).
3K. A. G. Blokh, K. S. Adzerikho, V. P. Trofimov and F. D. Lozhechnik, *Thermal Engng* **27,** 93 (1980).
4K. J. S. Bodenheimer, *Appl. Optics* **19,** 1031 (1980).
5K. R. O. Buckius and D. C. Hwang, *J. Heat Transfer* **102,** 99 (1980).
6K. T. C. Chawla and S. H. Chan, *J. Heat Transfer* **102,** 297 (1980).
7K. T. C. Chawla and S. H. Chan, *Num. Heat Transfer* **3,**

47 (1980).
8K. T. C. Chawla, S. H. Chan, F. B. Cheung and D. H. Cho, *J. Heat Transfer* **102,** 81 (1980).
9K. T. C. Chawla, W. J. Minkowycz and G. Leaf, *Num. Heat Transfer* **3,** 133 (1980).
10K. F. B. Cheung, S. H. Chan, T. C. Chawla and D. H. Cho, *Int. J. Heat Mass Transfer* **23,** 1313 (1980).
11K. W. L. Grosshandler, *Int. J. Heat Mass Transfer* **23,** 1447 (1980).
12K. G. M. Harpole, *Int. J. Heat Mass Transfer* **23,** 17 (1980).
13K. J. Higenyl and Y. Bayazitoglu, *J. Heat Transfer* **102,** 719 (1980).
14K. T. Kunitomo, *Bull. JSME* **23,** 1366 (1980).
15K. B. Landro and P. G. McCormick, *Int. J. Heat Mass Transfer* **23,** 613 (1980).
16K. C. N. Liu and T. M. Shih, *J. Heat Transfer* **102,** 724 (1980).
17K. I. R. Mikk, K. S. Grishutin and I. V. Zhukov, *Thermal Engng* **27,** 31 (1980).
18K. K. Nozaki and K. Nishihara, *J. Phys. Soc. Japan* **48,** 993 (1980).
19K. M. N. Özişik and W. H. Sutton, *J. Heat Transfer* **102,** 715 (1980).
20K. V. S. Pikashov, A. L. H'chenko, G. P. Kuchin and I. F. Zemskiy, *Heat Transfer, Soviet Res.* **12**(1), 124 (1980).
21K. S. R. Ray, A. C. Fernandex-Pello and I. Glassman, *J. Heat Transfer* **102,** 357 (1980).
22K. R. K. Soloukhin, O. G. Martynenko and N. E. Galich, *Int. J. Heat Mass Transfer* **23,** 1653 (1980).
23K. G. Spiga, F. Santarelli and C. Stramigioli, *Int. J. Heat Mass Transfer* **23,** 841 (1980).
24K. M. M. Tamonis, I. E. Sinkyavichyus and O. L. Tutlite, *Int. chem. Engng* **20,** 489 (1980).
25K. N. J. Trappeniers, A. C. Michels, H. M. J. Boots and R. H. Huijser, *Physica* **101A,** 431 (1980).
26K. H. A. J. Vercammen and G. F. Froment, *Int. J. Heat Mass Transfer* **23,** 329 (1980).
27K. R. Viskanta and D. M. Kim, *J. Heat Transfer* **102,** 388 (1980).
28K. T. R. Wagner, F. P. Incropera and W. G. Houf, *J. Heat Transfer* **102,** 709 (1980).
29K. P. L. Walker, *Appl. Optics* **19,** 2271 (1980).
30K. W. Whitaker, *I/EC Fundamentals* **19,** 210 (1980).
31K. W. W. Yuen and C. L. Tien, *J. Heat Transfer* **102,** 92 (1980).
32K. W. W. Yuen and L. W. Wong, *J. Heat Transfer* **102,** 303 (1980).

Surface radiation

1L. F. Collignon, *Letters Heat Mass Transfer* **7,** 107 (1980).
2L. I. G. Currie and W. W. Martin, *Computer Meth. Appl. Mech. Engng* **21,** 75–(1980).
3L. J. Y. Deschamps, *Lett. Heat Mass Transfer* **7,** 129 (1980).
4L. D. K. Edwards, *J. Heat Transfer* **102,** 706 (1980).
5L. E. Hahne and M. K. Bassiouni, *Lett. Heat Mass Transfer* **7,** 303 (1980).
6L. C. Isetti and E. Nannei, *Wärme- und Stoffübertragung* **14,** 211 (1980).
7L. J. Janata, *Lett. Heat Mass Transfer* **7,** 113 (1980).
8L. J. A. Manrique and R. Suarez, *Lett. Heat Mass Transfer* **7,** 25 (1980).
9L. H. Masuda, *J. Heat Transfer* **102,** 563 (1980).
10L. T. Oguchi, K. Sato and T. Teranishi, *J. Phys. Soc. Japan* **48,** 123 (1980).
11L. I. Ohlidal, *Appl. Optics* **19,** 1804 (1980).
12L. L. A. Wojcik, A. J. Sievers, G. W. Graham and T. N. Rhoom, *J. Opt. Soc. Am.* **70,** 443 (1980).
13L. W. W. Yuen, *J. Heat Transfer* **102,** 386 (1980).
14L. C. Zaixiang, C. Souren and C. Hongpan, *J. Opt. Soc. Am.* **70,** 1270 (1980).

MHD

1M. H. L. Agrawal, P. C. Ram and S. S. Singh, *Can. J. chem. Engng* **58,** 131 (1980).
2M. R. K. Ahluwalia and E. D. Doss, *Num. Heat Transfer* **3,** 67 (1980).
3M. A. N. Bhat and M. L. Mittal, *Int. J. Heat Mass Transfer* **23,** 919 (1980).
4M. A. Chandra, R. P. Dahiya, G. V. R. Raju and R. G. Gupta, *J. Phys D: Appl. Phys.* **13,** 1211 (1980).
5M. P. F. Dunn, *Int. J. Heat Mass Transfer* **23,** 373 (1980).
6M. D. R. Fearn, *Proc. R. Soc. London* **369A,** 227 (1979).
7M. T. Mizushina, F. Ogino, T. Matsumoto, M. Yokoyama and N. Kitano, *Int. J. Heat Mass Transfer* **23,** 1105 (1980).
8M. K. Onda, *Bull. JSME* **23,** 1263 (1980).
9M. A. E. Pozowolski, *J. Phys. D: Appl. Phys.* **13,** L81 (1980).
10M. A. E. Sheindlin, B. Ya. Shumyatskii, E. E. Shpil'rain, G. N. Morozov and G. M. Koryagina, *Thermal Engng* **27,** 119 (1980).
11M. J. A. Shercliff, *Int. J. Heat Mass Transfer* **23,** 1219 (1980).
12M. V. M. Soundalgekar and H. S. Takhar, *Wärme- und Stoffübertragung* **14,** 153 (1980).
13M. D. W. Swallom and O. K. Sonju, *J. Energy* **4**(3), 100 (1980).
14M. H. S. Takar and V. M. Soundalgekar, *Appl. Sci. Res.* **36,** 163 (1980).
15M. D. T. Trung and H. K. Messerle, *IEEE Trans. Plasma Sci.* **PS-8,** 268 (1980).
16M. C. P. Y. and Y. D. Shih, *Phys. Fluids* **23,** 411 (1980).

Measurement techniques

1P. R. A. Antonia, A. J. Chambers and A. K. M. F. Hussain, *Phys. Fluids* **23,** 871 (1980).
2P. V. V. Aristov and F. I. Konzhukov, *Meas. Techniques* **23,** 631 (1980).
3P. H. B. Barnes and W. M. Farmer, *Appl. Optics* **19,** 2930 (1980).
4P. K.-H. Bode, *Int. J. Heat Mass Transfer* **23,** 961 (1980).
5P. G. Breitkoff, W. Wittig and S. Kim, *Wärme- und Stoffübertragung* **13,** 287 (1980).
6P. C. Chan and J. W. Daily, *Appl. Optics* **19,** 1963 (1980).
7P. V. Ya. Chekhovskoi and L. N. Latyev, *Meas. Techniques* **23,** 743 (1980).
8P. A. A. Clifford, J. Kestin and W. A. Wakeham, *Physica* **100A,** 370 (1980).
9P. D. G. Drake, D. S. Riley, R. R. Baker and K. D. Kilburn, *Int. J. Heat Mass Transfer* **23,** 127 (1980).
10P. C. W. Draper and W. O. Pessel, *Rev. Sci. Instrum.* **51,**

258 (1980).
11P. W. L. Eberhard and R. M. Schotland, *Appl. Optics* **19,** 2967 (1980).
12P. R. E. Falco, *J. Fluids Engng* **102,** 174 (1980).
13P. P. Freymuth, *J. Fluids Engng* **102,** 152 (1980).
14P. B. Gebhart and T. Audunson, *Lett. Heat Mass Transfer* **7,** 293 (1980).
15P. E. Goldratt, Y. Yeshurun and A. J. Greenfield, *Rev. Sci. Instrum.* **51,** 361 (1980).
16P. J. S. Huang and M. W. Kim, *Rev. Sci. Instrum.* **51,** 852 (1980).
17P. R. P. Hudson, *Rev. Sci. Instrum.* **51,** 871 (1980).
18P. P. S. Kamath and R. T. Lahey, Jr., *J. Heat Transfer* **102,** 9 (1980).
19P. J. S. Kraebl, J. W. Baughn and A. A. McKillop, *J. Heat Transfer* **102,** 576 (1980).
20P. P. Le-Cong and R. H. Lovberg, *Rev. Sci. Instrum.* **51,** 565 (1980).
21P. P. Le-Cong and R. H. Lovberg, *Appl. Optics* **19,** 4222 (1980).
22P. J. C. Lecordier, C. Petit and P. Paranthoen, *Lett. Heat Mass Transfer* **7,** 311 (1980).
23P. W. Leidenfrost and J. Ku, *Rev. Sci. Instrum.* **51,** 1363 (1980).
24P. C. K. Ma, M. Ohtsuka and R. E. Bedford, *Rev. Sci. Instrum.* **51,** 52 (1980).
25P. K. Matsuda, *Appl. Optics* **19,** 2643 (1980).
26P. M. Miklavic and I. Kuscer, *Int. J. Heat Mass Transfer* **23,** 1297 (1980).
27P. S. B. H. C. Neal and E. W. Northover, *Int. J. Heat Mass Transfer* **23,** 1015 (1980).
28P. S. B. H. C. Neal and E. W. Northover, *Int. J. Heat Mass Transfer* **23,** 1023 (1980).
29P. M. L. Nuckois, *J. Engng Indust.* **102,** 247 (1980).
30P. T. Numata and J. Nishimura, *Rev. Sci. Instrum.* **51,** 991 (1980).
31P. C. J. Oliver, *P. Phys. D: appl. Phys.* **13,** 1145 (1980).
32P. G. G. Osterwegel and H. J. de Groot, *Rev. Sci. Instrum.* **51,** 201 (1980).
33P. R. B. Owen and C. W. Campbell, *Rev. Sci. Instrum.* **51,** 1504 (1980).
34P. P. Paranthoen, J. C. Lecordier and C. Petit, *Lett. Heat Mass Transfer* **7,** 437 (1980).
35P. J. T. Park and C. W. Van Atta, *Phys. Fluids* **23,** 701 (1980).
36P. Yu. L. Pilipchuk, D. P. Zaritskii, A. A. Tovma and V. M. Ignatenko, *Meas. Techniques* **23,** 732 (1980).
37P. J. C. Richmond, *Appl. Optics* **19,** 834 (1980).
38P. O. Sasaki, T. Shimizu and T. Abe, *Appl. Optics* **19,** 1151 (1980).
39P. O. Sasaki, T. Abe, T. Shimizu and M. Niwayama, *Appl. Optics* **19,** 289 (1980).
40P. B. G. Schuster and T. G. Kyle, *Appl. Optics* **19,** 2524 (1980).
41P. D. Sengupta, S. K. Saha, S. N. Sengupta and S. K. Mikherjee, *Rev. Sci. Instrum.* **51,** 1482 (1980).
42P. R. R. Sholes and J. G. Small, *Rev. Sci. Instrum.* **51,** 882 (1980).
43P. H. Tokoi, N. Ozaki and I. Harada, *Rev. Sci. Instrum.* **51,** 1318 (1980).
44P. J. W. Vandersande and R. O. Pohl, *Rev. Sci. Instrum.* **51,** 1694 (1980).
45P. D. L. Wall and A. Mantz, *Appl. Optics* **19,** 1569 (1980).
46P. C. A. Wilkins and R. L. Ash, *Rev. Sci. Instrum.* **51,** 998 (1980).
47P. R. W. Wunderlich, R. L. Folger, D. B. Giddon and B. R. Ware, *Rev. Sci. Instrum.* **51,** 1258 (1980).
48P. A. Yasuda, Y. Kanai, J. Kusonoki, K. Kawahata and S. Takeda, *Rev. Sci. Instrum.* **51,** 1652 (1980).

Heat transfer applications—heat exchangers and heat pipes

1Q. M. A. Ait-Ali and D. J. Wilde, *J. Heat Transfer* **102,** 199 (1980).
2Q. M. K. Bologa, I. A. Kozhukar', O. I. Mardarskiy and V. D. Shkilev, *Heat Transfer, Soviet Res.* **12**(3), 137 (1980).
3Q. C. A. Busse and J. E. Kemme, *Int. J. Heat Mass Transfer* **23,** 643 (1980).
4Q. S. A. Chaksh, *Int. chem. Engng* **20,** 498 (1980).
5Q. T. F. Degnan and J. Wei, *AIChE Jl* **26,** 60 (1980).
6Q. J. R. Flower and B. Linnhoff, *AIChE Jl* **26,** 1 (1980).
7Q. R. Hatami, *Wärme- und Stoffübertragung* **14,** 109 (1980).
8Q. H. Hausen, *Wärme- und Stoffübertragung* **14,** 1 (1980).
9Q. N. Kaji, Y. H. Mori, Y. Tochitani and K. Komotori, *J. Heat Transfer* **102,** 32 (1980).
10Q. H. Kanoh, *Bull. JSME* **23,** 432 (1980).
11Q. J. Kern, *Wärme- und Stoffubertragung* **13,** 205 (1980).
12Q. A. L. London and R. A. Seban, *Int. J. Heat Mass Transfer* **23,** 5 (1980).
13Q. I. Mikk, *Int. J. Heat Mass Transfer* **23,** 707 (1980).
14Q. A. Montakhab, *J. Engng Pwr* **102,** 807 (1980).
15Q. L. L. Moresco and E. Marschall, *J. Heat Transfer* **102,** 684 (1980).
16Q. Y. Mori, Y. Yamada and K. Hijikata, *Int. J. Heat Mass Transfer* **23,** 1079 (1980).
17Q. R. W. Porter, M. Jain and S. K. Chaturvedi, *J. Heat Transfer* **102,** 210 (1980).
18Q. W. Roetzel and J. Fürst, *Wärme- und Stoffübertragung* **13,** 131 (1980).
19Q. S. Shibayama and S. Morooka, *Int. J. Heat Mass Transfer* **23,** 1003 (1980).
20Q. H. M. Sofrata, *Wärme- und Stoffübertragung* **14,** 119 (1980).
21Q. E. M. Sparrow, J. W. Ramsey and C. A. C. Altemani, *J. Heat Transfer* **102,** 44 (1980).
22Q. A. P. Watkinson, *Can. J. chem. Engng* **58,** 553 (1980).
23Q. A. J. Willmott and R. C. Duggan, *Int. J. Heat Mass Transfer* **23,** 655 (1980).
24Q. D. Whittmann, *Chem.-Ing. Tech.* **52,** 249 (1980).

General heat transfer applications

1S. P. M. Abdul Majeed, *J. Heat Transfer* **102,** 761 (1980).
2S. A. C. Akidas, *J. Heat Transfer* **102,** 189 (1980).
3S. G. E. Breitkopf, *Wärme- und Stoffübertragung* **13,** 195 (1980).
4S. W. D. Deckwer, *Chem. Engng Sci.* **35,** 1341 (1980).
5S. E. Gardner, P. L. Grady and M. Mohamed, *J. Engng Ind.* **102,** 13 (1980).
6S. I. S. Habib, *J. Heat Transfer* **102,** 110 (1980).
7S. H. Hashimoto, *Bull. JSME* **23,** 773 (1980).
8S. A. Heim, *Int. Chem. Engng* **20,** 279 (1980).

9S. A. Heim, *Int. Chem. Engng* **20,** 271 (1980).
10S. R. C. Herndon, P. E. Hubble and J. L. Gainer, *I/EC Process Des. Dev.* **19,** 405 (1980).
11S. D. M. Heyes and C. J. Montrose, *J. Lubr. Tech.* **102,** 459 (1980).
12S. T. Kuno, *Bull. JSME* **23,** 1275 (1980).
13S. L. J. Lee and C. W. Macosko, *Int. J. Heat Mass Transfer* **23,** 1479 (1980).
14S. E. R. Maki and H. A. Ezzat, *J. Lubr. Tech.* **102,** 8 (1980).
15S. A. Mersman, H. Noth, D. Ringer and R. Wunder, *Chem.-Ing.-Tech* **52,** 189 (1980).
16S. K. Nakamura, *Bull. JSME* **23,** 1483 (1980).
17S. C. D. Rakopoulos, A. A. El-Shirbini and W. Murgatroyd, *Wärme- und Stoffübertragung* **13,** 275 (1980).
18S. H. A. Rubeye, *J. Lubr. Tech.* **102,** 107 (1980).
19S. R. Seshadri and A. V. G. Krishnayya, *Int. J. Heat Mass Transfer* **23,** 111 (1980).
20S. R. G. Spencer, D. C. Rousar and H. G. Price, *J. Spacecraft Rockets* **17,** 35 (1980).
21S. A. Steiff, R. Poggemann and P. M. Weinspach, *Chem.-Ing.-Tech.* **52,** 492 (1980).
22S. J. A. Tichy and R. N. Smith, *J. Lubr. Tech.* **102,** 34 (1980).
23S. G. West, *Wärme- und Stoffübertragung* **14,** 75 (1980).
24S. D. J. Zigrang and H. W. Bergmann, *J. Spacecraft Rockets* **17,** 219 (1980).

Solar energy

1T. A. Akbarzedeh and G. Ahmadi, *Solar Energy* **24,** 143 (1980).
2T. B. Authier and L. Hill, *Appl. Optics* **19,** 3554 (1980).
3T. Y. Bayazitoglu and S. Asgarpour, *Solar Energy* **24,** 105 (1980).
4T. W. F. Bessler and B. C. Hwang, *ASHRAE Jl* **22**(9), 59 (1980).
5T. K.-K. Chang and A. Minardi, *Solar Energy* **24,** 99 (1980).
6T. A. J. de Ron, *Solar Energy* **24,** 117 (1980).
7T. J. Guay, *Solar Energy* **24,** 265 (1980).
8T. T. E. Horton and J. H. McDermit, *J. Energy* **4**(2), 4 (1980).
9T. R. O. Hughes, *Solar Energy* **24,** 83 (1980).
10T. M. Iqbal, *Solar Energy* **24,** 491 (1980).
11T. M. Kamimoto, T. Tanaka, T. Tani and Y. Horigome, *Solar Energy* **24,** 581 (1980).
12T. D. C. Larson, *J. Energy* **4**(4), 170 (1980).
13T. D. C. Larson, *Solar Energy* **24,** 203 (1980).
14T. S. B. Leighton, *ASHRAE Jl* **22**(11), 16 (1980).
15T. S. T. Liu and A. H. Fanney, *ASHRAE Jl* **22**(5), 34 (1980).
16T. G. L. Morrison and D. B. J. Ranatunga, *Solar Energy* **24,** 55 (1980).
17T. A. B. Newton, *ASHRAE Jl* **22**(11), 58 (1980).
18T. R. L. Nicholis and T. N. Child, *Solar Energy* **24,** 593 (1980).
19T. M. V. Oliphant, *Solar Energy* **24,** 403 (1980).
20T. U. Ortabasi and F. P. Fehiner, *Solar Energy* **24,** 477 (1980).
21T. M. N. Özişik, B. K. Huang and M. Toksoy, *Solar Energy* **24,** 397 (1980).
22T. R. P. Patera and H. S. Robertson, *Appl. Optics* **19,** 2403 (1980).
23T. R. W. Persons, J. A. Duffie and J. W. Mitchell, *Solar Energy* **24,** 199 (1980).
24T. W. F. Phillips, *Solar Energy* **24,** 601 (1980).
25T. J. W. Ramsey and M. Charmchi, *J. Heat Transfer* **102,** 766 (1980).
26T. R. R. Somers II, J. W. Miller and R. H. McCafferty, *J. Energy* **4**(5), 233 (1980).
27T. E. F. Sowell and R. L. Curry, *Solar Energy* **24,** 441 (1980).
28T. J. G. Symons and R. Gani, *Solar Energy* **24,** 407 (1980).
29T. R. T. Tamblyn, *ASHRAE Jl* **22**(1), 65 (1980).
30T. R. T. Tamblyn, *ASHRAE Jl* **22**(11), 72 (1980).
31T. T. A. Weiss and G. O. G. Löf, *Solar Energy* **24,** 287 (1980).
32T. G. T. Williams, C. R. Attwater and F. C. Hooper, *Solar Energy* **24,** 471 (1980).
33T. R. Winston and W. T. Welford, *Appl. Optics* **19,** 347 (1980).
34T. C. Wyman, J. Castle and F. Kreith, *Solar Energy* **24,** 517 (1980).

Plasma heat transfer

1U. H. Agdelhakim, J. P. Dinguirard and S. Vacquie, *J. Phys. D: Appl. Phys.* **13,** 1427 (1980).
2U. P. R. Blackburn, *J. Energy* **4**(3), 98 (1980).
3U. A. Bokhari and M. Boules, *Can. J. chem. Engng* **58,** 171 (1980).
4U. E. V. Bonin, G. Brüggemann and H. G. Thiel, *IEEE Trans. Plasma Sci.* **PS-8,** 344 (1988).
5U. M. I. Boulos, R. Gagne and R. M. Barnes, *Can. J. chem. Engng* **58,** 367 (1980).
6U. D. M. Chen, K. C. Hsu, C. H. Liu and E. Pfender, *IEEE Trans. Plasma Sci. PS-8,* 425 (1980).
7U. D. M. Chen and E. Pfender, *IEEE Trans. Plasma Sci.* **PS-8,** 252 (1980).
8U. Y. K. Chien and D. M. Benenson, *IEEE Trans. Plasma Sci.* **PS-8,** 411 (1980).
9U. I. M. Cohen, P. S. Ayyaswamy and T. Sundararajan, *IEEE Trans. Plasma Sci.* **PS-8,** 390 (1980).
10U. F. F. Elakshar and K. Phillips, *J. Phys. D: Appl. Phys.* **13,** L57 (1980).
11U. M. T. C. Fang, S. Ramakrishnan and H. K. Messerle, *IEEE Trans. Plasma Sci.* **PS-8,** 357 (1980).
12U. C. R. Giles, R. M. Clements and P. R. Smy, *J. Phys. D: appl. Phys.* **13,** 33 (1980).
13U. G. Gouesbet and P. Valentin, *Phys. Fluids* **23,** 232 (1980).
14U. A. E. Guile and B. Jüttner, *IEEE Trans. Plasma Sci.* **PS-8,** 258 (1980).
15U. R. J. Hill and G. R. Jones, *J. Phys. D: Appl. Phys.* **13,** 593 (1980).
16U. H. Hügel, *IEEE Trans. Plasma Sci.* **PS-8,** 437 (1980).
17U. I. J. Jagoda F. J. Weinberg, *J. Phys. D: Appl. Phys.* **13,** 551 (1980).
18U. S. Kaneko and A. Yamao, *J. Phys. Soc, Japan* **48,** 2098 (1980).
19U. A. Kobayashi, S. Yanabu, S. Yamashita and Y. Ozaki, *IEEE Trans. Plasma Sci.* **PS-8,** 339 (1980).
20U. Z. Kolacinski, *IEEE Trans. Plasma Sci.* **PS-8,** 449 (1980).

21U. H. Kopplin, H. Motschmann, K. P. Rolff and K. Zückler, *IEEE Trans. Plasma Sci.* **PS-8,** 331 (1980).
22U. D. J. Latham, *Phys. Fluids* **23,** 1710 (1980).
23U. T. Matsumura, T. Sakakibara, Y. Kito and I. Miyachi, *IEEE Trans. Plasma Sci.* **PS-8,** 248 (1980).
24U. N. Negishi, *Bull. JSME* **23,** 1171 (1980).
25U. S. Okuda, Y. Ueda, Y. Murai, T. Miyamoto, Y. Doi and C. Uenosono, *IEEE Trans. Plasma Sci.* **PS-8,** 394 (1980).
26U. R. Radtke, K. Günther and R. Ulbricht, *J. Phys. D: Appl. Phys.* **13,** 1 (1980).
27U. E. Richley and D. T. Tuma, *IEEE Trans. Plasma Sci.* **PS-8,** 405 (1980).
28U. T. Sakuta, Y. Kito and I. Miyachi, *IEEE Trans. Plasma Sci.* **PS-8,** 29 (1980).
29U. E. Schulz-Gulde, *J. Phys. D: Appl. Phys.* **13,** 793 (1980).
30U. H. Shindo, T. Inaba and S. Imazu, *J. Phys. D: Appl. Phys.* **13,** 805 (1980).
31U. Ch. Sturzenegger, R. T. Reinhardt and H. J. Schötzhu, *IEEE Trans. Plasma Sci.* **PS-8,** 384 (1980).
32U. W. Tiemann, *IEEE Trans. Plasma Sci.* **PS-8,** 368 (1980).
33U. A. Tslaf, *IEEE Trans. Plasma Sci.* **PS-8,** 455 (1980).
34U. A. Vardelle, J. M. Baronnet, M. Vardelle and P. Fauchais, *IEEE Trans. Plasma Sci.* **PS-8,** 417 (1980).
35U. J. Vyskocil and J. Musil, *J. Phys. D: Appl. Phys.* **13,** L25 (1980).

HEAT TRANSFER REVIEW - 1981

E. R. G. ECKERT, R. J. GOLDSTEIN, S. V. PATANKAR, E. PFENDER, J. W. RAMSEY T. W. SIMON and E. M. SPARROW

Heat Transfer Laboratory, Department of Mechanical Engineering, University of Minnesota, Minneapolis, Minnesota, U.S.A.

INTRODUCTION

THIS review surveys results that have been published in the open literature covering various fields of heat transfer during 1981. As in the past, the number of papers published during that period was such that only a selection can be included.

An Advanced Study Institute on "Low Reynolds Forced Convection in Channels and Bundles" was held at Ankara, Turkey, 13–24 July 1981. Nine sessions dealt with forced convection and heat exchanger technology. Proceedings of the institute will be published by Hemisphere Publishing Corporation, Washington, DC.

The 20th National Heat Transfer Conference was held at Milwaukee, Wisconsin, 2–5 August 1981, and was sponsored by the American Society of Mechanical Engineers and the American Institute of Chemical Engineers. Thirty-eight sessions were devoted to various fields of fundamental and applied heat transfer, including sessions on porous media and mathematical modeling. Invited lectures were presented by G. F. Hewitt on "Two-Phase Flow and its Applications: Past, Present, and Future" and by R. A. Seban on "Some Aspects of the Heat Transfer in Reflooding of a Single Tube". The Donald Q. Kern Award was presented to G. F. Hewitt and the Max Jakob Memorial Award to R. A. Seban. The papers presented at the conference are available as preprints or in the published series of the American Institute of Chemical Engineers. Many will also be published in the *Journal of Heat Transfer*.

The Third Symposium on Turbulent Shear Flows sponsored by the U.S. Air Force Office of Scientific Research, Research Offices of the U.S. Army and Navy, and the National Science Foundation took place on 9–11 September 1981 at the University of California, Davis. One of the sessions was devoted to heat and mass transfer in boundary layers and various papers in the other sessions touched on heat transfer. A bound symposium volume is available.

The International Centre for Heat and Mass Transfer organized a Summer School on "Heat Exchangers" (31 August–5 September 1981) and an International Seminar on "Advancement in Heat Exchangers" at Dubrovnik, Yugoslavia, 7–12 September 1981. The Summer School was organized around the Heat Exchange Design Book which is published by Hemisphere Publishing Corporation in five parts. Proceedings of the International Seminar are also available through Hemisphere Publishing Corporation.

The Second National Symposium on "Numerical Methods in Heat Transfer" was sponsored by the National Science Foundation, the Office of Naval Research, and the University of Maryland and was held at College Park, Maryland, 28–30 September 1981. Twelve sessions dealt with finite difference and finite element methods used in modeling heat transfer processes. A short course for engineers and scientists on "Computation of Heat Transfer and Fluid Flow" was held at the University of Minnesota, 16–19 November 1981.

The 102nd Winter Annual Meeting of the American Society of Mechanical Engineers, held 15–20 November 1981 at Washington, D.C., contained in its program sixteen sessions on fundamental and applied aspects of heat transfer. At the Heat Transfer Luncheon, Frank Kreith gave a lecture on the topic "Is There a Solar Future?". Heat Transfer Memorial Awards were given to A. Cezairliyan, Kwang-Tzu Yang and Ivan Catton. The papers presented at the conference are available as preprints or in book form at ASME Headquarters. Many of them will also be published in the *Journal of Heat Transfer*.

A considerable number of books dealing with heat transfer or including heat transfer topics have appeared on the market. They are listed in the bibliographic portion of this review. The *Latin American Journal of Heat and Mass Transfer* is published in Argentina with the Editorial Office at Avenida 1, No. 867, La Plata, Argentina.

The following highlights illuminate developments in heat transfer research during 1981:

In heat conduction, problems of phase change appear to be the main focus of published work. Solution methodologies, both numerical and analytical, also continue to evoke interest.

Complex passages, as they occur in compact heat exchangers, have been primarily investigated as channel flow configurations. Laminar and turbulent boundary layers are of continuing interest. Free jets and flow across cylinders have also found attention.

Heat transfer in porous media and in fluidized beds found much attention, probably since these processes are not completely understood as yet. They are investigated experimentally and analytically with models simplifying the process. Studies on one-phase heat transfer, as well as change of phase, and combined heat and mass transfer are reported in the literature.

Porous media were also considered in natural convection studies. Double diffusion processes were under investigation. External natural convection studies concentrated on flat plates in steady and transient state. Some of the articles included the effects of variable properties. Mixed convection over plates, cylinders, and spheres was the topic of several papers.

Processes in rotating flows have been studied which led to a significant increase in heat transfer or to a favorable ratio of heat transfer to pressure drop.

Heat transfer with change of phase remains an active research topic. As in past years, the majority of the papers in this category are nuclear reactor heat transfer oriented. Present papers in the reactor heat transfer subcategory discuss: bubble nucleation in a superheated liquid medium upon depressurization; boiling outside of sodium-heated tubes; heat transfer, CHF, and bubble growth when liquid coolant comes in contact with molten nuclear fuel; heat transfer in falling liquid films with rewetting and quenching; and boiling in a porous medium. Topics of papers on basic phenomena of boiling and condensing include: the details of bubble formation, growth and departure and the microconvection associated with these processes; the effects of surface coatings, e.g. gold or plasma-coated polymers; and the effects of surface preparations, e.g. nucleation pits. Several papers were presented on droplet evaporation in superheated vapor or in a noncondensible gas.

Interest is still strong in studying the performance of heat exchangers, whereas heat pipes have found little attention the past year.

The largest number of papers on solar energy continue to deal with the experimental and analytical evaluation of the performance of flat plate solar collectors. A significant number of publications describe the thermal performance of passive solar heat systems.

There is renewed interest in plasma heat transfer associated with new developments in high temperature plasma chemistry and plasma processing.

To facilitate the use of this review, a listing of the subject headings is made below in the order in which they appear in the text. The letter which appears adjacent to each subject heading is also attached to the references that are cited in that category:

Conduction, A
Channel flow, B
Boundary layer and external flows, C
Flow with separated regions, D
Natural convection—internal flows, F
Natural convection—external flows, FF
Convection from rotating surfaces, G
Combined heat and mass transfer, H
Change of phase, J
Radiation
  Radiation in participating media, K
  Surface radiation, L
MHD, M
Numerical methods, N
Heat transfer applications
  Heat exchangers and heat pipes, Q
  General, S
  Solar energy, T
Plasma heat transfer, U.

CONDUCTION

Problems of phase change appear to be the main focus of the published work in heat conduction. Solution methodologies, both numerical and analytical, also continue to evoke interest.

In a method for immobilizing moving boundaries in phase-change problems, a convection-like transport appears in the governing equations due to the coordinate transformation [22A]. By immobilizing the moving boundary, the 2-dim. freezing on the outside of a coolant-carrying tube was solved by a finite-difference method [57A]. A closed-form solution has been obtained for the 2-dim. freezing on a wall that is convectively cooled from the rear [21A]. A simple

1-dim. model, which neglects the details of axial variations, was employed to solve for melting about a horizontal tube through which a hot fluid passes [48A]. When a heat-generating liquid layer, insulated at the bottom, is cooled from above by contact with a cold environment maintained at a fixed subfreezing temperature, a frozen crust may be formed at the top. If the top cooling is not strong enough, the crust may grow and decay periodically [8A]. For melting about a horizontal cylinder, short-time solutions (i.e. small natural convection) were obtained by both an integral method and a quasi-steady model [69A].

In an analysis of unidirectional freezing of aqueous solutions during cooling at subzero temperatures, under conditions where the solute is completely rejected by the advancing ice front, the conventional diffusion equation is invalid [32A]. The enthalpy method has been reinterpreted to increase its accuracy [63A]. In a companion paper, a modification of the enthalpy method has been extended to cylindrical problems containing a circular cross section and spatially uniform boundary conditions [64A]. A similarity rule which greatly simplifies the solution of solidification problems has been extended to take account of the volume change that accompanies the phase change [51A]. A conformal transformation method was applied for the determination of the shape of the interface between a solidified layer formed on the inside of a cooled pipe of rectangular cross section and a warmer flowing liquid which passes along the axis of the pipe [60A]. The freezing of a liquid passing through a pipe with highly cooled walls may result in blockage of the pipe [47A]. Numerical solutions for 2-dim. solidification in a rectangular region have reaffirmed the importance of accounting for natural convection in the melt [45A].

A series solution of the Stefan problem with a convectively heated surface has been formulated and is purported to be exact [34A]. Series solutions have also been constructed for the Stefan problem with prescribed surface heat flux. A similarity solution is possible if the heat flux varies as $t^{-1/2}$ [61A]. The accuracy of approximate solutions for the uniform heat flux boundary condition has been assessed [9A]. In the Stefan problem, the magnitude of the Stefan number serves as a measure of the ratio of the sensible heat to the latent heat [53A]. The analysis for the freeze-coating of a continuous moving sheet differs from the standard Stefan problem in that the frozen layer thickness varies with the space coordinate along the moving sheet rather than with time [50A].

For calculating transient heat conduction in an expanding solid, the governing energy equation can be transformed to that for a fixed boundary problem [24A]. Highly nonlinear steady and quasi-steady conduction problems in simple geometries, including the effect of material travel, can be solved effectively by an iterative adaptation of the SEPELI fast elliptic equation solver [43A]. A finite element algorithm was used for the prediction of the rate of freezing of fresh water for various boundary conditions [23A]. A Heat-Balance-Integral solution for freezing about a circular cylinder was applied to soil systems [35A].

In an applications-oriented study, a model was developed for predicting the characteristics of an array of phase-change cylinders arranged in crossflow with respect to a transfer fluid [54A]. A design-oriented computation procedure for phase-change storage yielded the melted fraction and the shape of the liquid–solid interface [19A].

A number of experimental papers have dealt with freezing and melting. For freezing on a finned vertical tube, the heat transfer enhancement is proportional to the fin area when there is strong natural convection in the liquid melt [58A]. Experiments demonstrated that fins provided greater enhancement for melting than for freezing [4A]. When a cooled surface is placed in a superheated liquid, the freezing is initially strongly affected by natural convection, but as the superheat wanes, conduction in the frozen solid becomes the dominant heat transfer mechanism [59A]. Local and average heat transfer rates at the ice–water interface of an ice layer grown on a circular cylinder were determined directly from a photograph showing the shape of the ice layer [7A]. In the melting of the vertical surface of a solid by a heated liquid pool, the melt and the external fluid did not intermix along their mutual vertical interface despite the fact that the two media were miscible [15A]. In experiments on melting about a heated vertical cylinder, the measured heat transfer coefficients were little affected by whether the upper surface of the melt was subjected to a slip or no-slip velocity boundary condition [27A].

Solutions for a variety of steady state conduction problems have been published. In response to problems encountered in the cold-rolling of flat metal products, steady state temperature distributions were determined for a rotating roll subject to constant heat input over one portion of the circumference and convective cooling over another portion of the circumference [42A]. In another steady state problem involving complex boundary conditions, a solution was obtained for the temperature distribution in the wall of a tube which is in contact with the wall of a larger tube over part of its circumference [1A]. Numerical values of the conduction shape factor have been determined for multi-hole prismatic bars [39A]. A simple rule is presented for conduction shape factors for bodies with a geometrical axis of symmetry and with boundaries that are maintained at one of two uniform temperatures or are insulated [33A]. Conduction from surfaces to the 3-dim. surrounding space is relevant as a limiting case for natural and forced convection, respectively, as the Rayleigh and Reynolds numbers approach zero [10A]. To analyze heat conduction in spheres packed in an infinite regular cubical array, a unit cube with a sphere at the center was selected as the typical module [13A]. An analytical model for the computation of steady conduction across rectangular-celled enclosures is based on the assumption of quasi-1-dim. conduction in the cell partitions [14A]. Classical series methods are employed to solve steady conduction problems in r, z cylindrical coordinates for boundary conditions appropriate to nuclear reactor fuel pins [67A]. A wire heated by ohmic dissipation can, under certain conditions, have multiple steady states. The sufficient conditions for a unique steady state have been identified [40A].

Fins continue to evoke interest. Numerical solutions of the momentum equation for the fluid and the energy equations for the fluid and the solid yielded local heat transfer coefficients along the principal faces and the tips of an array of rectangular fins. The averaged tip coefficients were not markedly different from those of a segment of the principal face adjacent to the tip [56A]. In the presence of condensation on a rectangular fin, the optimum fin length is smaller than for the case of no condensation [29A]. The optimum efficiency for a cylindrical pin fin is higher than that for the longitudinal rectangular fin [55A]. The effect of a timewise periodic variation of the base temperature of a radiating fin is to increase the mean heat transfer rate [2A]. In a companion paper, the analysis is extended to temperature-dependent thermal conductivity and to spatially varying heat transfer coefficient [3A]. The standard assumption of a uniform fin heat transfer coefficient was lifted in favor of a linearly varying coefficient from base to tip [18A]. A technique for analyzing an array of extended surfaces is based on principles of graph theory, employing a novel kind of incidence matrix [52A].

Some interest persists in anisotropic materials, composites, and variable properties. By the use of a numerical mapping technique, steady state heat conduction solutions for anisotropic composites of arbitrary shape have been obtained [44A]. The finite integral transform technique has been generalized to solve transient heat conduction in a 3-dim. anisotropic medium [37A]. In determining the thermal conductivities of fibrous composites, an analytical model is employed whereby the fibers are considered to be uniformly dispersed in a matrix of resin [17A]. To deal with temperature-dependent thermal conductivity in 1-dim. transient conduction problems, it is proposed to replace the nonlinear heat equation with a linear version which contains a pseudo heat generation term [25A].

Aspects of contact have been investigated. The problem of the transient response of two semi-infinite solids not making perfect contact is treated by modeling the interface as a series of equally spaced strips making perfect contact, with the remaining area assumed to be perfectly insulating [46A]. If an elastic half plane is indented by a perfectly conducting rigid flat punch which is at a different temperature from the half plane, the nature of the contact depends on the magnitude and direction of the heat flow [11A]. The basic nature of the planar Hertz contact problem differs depending on whether heat flows into the material with the larger distortivity or vice versa [12A]. The finite integral transform technique was used to obtain a quasi-steady solution for two finite periodically contacting regions, with imperfect thermal contact at the interface [62A].

Various specific problems of transient conduction were solved. The solution for the transient temperature field in a sphere with single-sided heating modeled a problem in magnetic thermonuclear reactor engineering [66A]. Transient heat flow from a disk into a half-space when the disk temperature undergoes a step change was investigated by a separation of variables technique using oblate spheroidal coordinates [36A]. A series solution is presented for the transient temperature response of a semi-infinite cylinder to a step change in heat flux at a disk centered in the end face of the cylinder [5A]. In a related paper, large time solutions are obtained for a semi-infinite body heated over a circular region on its exposed face [6A]. In a transient technique for measuring thermal conductivity, the ramp function offers advantages over the step and Dirac functions [38A]. A correction for heat conduction between a sensor wire and its supports should be made to attain high accuracy [30A].

In addition to the solution methodologies described in the foregoing papers, several other methodology-related papers have been published. The advantages of using body-fitted coordinates as the basis for the numerical solution of heat conduction problems was illustrated by a range of examples [16A]. The unsteady surface element method is an approach for determining the temperature and heat flow at the interface between contacting, conducting solids [26A]. A finite element procedure for unsteady conduction has been developed which utilizes conjugate base functions and a modified form of the secant method for solving the discretized equations [49A]. The main advantage of the boundary element method for solving transient heat conduction problems is the reduction by one of the dimensionality of the problem [68A]. A numerical method for steady heat conduction, the iterative boundary integral method, is purported to have no limitations with regard to geometric complexity, type of boundary condition, or thermal conductivity [28A]. Using a conformal transformation which maps the region of interest onto the upper half plane, the Heat Balance Integral is generalized to 2-dim. transient heat conduction [70A].

For situations with moving temperature fields (e.g. metal forming, welding, plastic processing), moving finite elements were found to be advantageous as the basis of a numerical solution [41A]. An electrical analogue for solving conduction problems with moving heat sources on the surface yielded good agreement with solutions in Carslow and Jaeger [31A].

The inverse problem of transient heat conduction was shown to be ill-posed, as the solution exhibits unstable dependence on the given data functions [65A]. As a variant of the inverse problem, a method is described for predicting the geometry of a cavity situated beneath a surface whose temperature is known (for example, by infrared scanning) [20A].

CHANNEL FLOW

Experimental and theoretical work has been reported for the fluid flow and heat transfer processes in channels of both simple and complex geometry.

Among the studies associated with circular pipes, the heat transfer to liquid metals in laminar and turbulent flow has been dealt with [6B]. The situation with drag-reducing non-Newtonian fluids has been considered [41B]. Ref. [20B] deals with heat transfer to pseudoplastic fluids, while [7B] describes mass transfer to viscoelastic fluids. The dispersion from a line source in a turbulent flow is handled in [13B]. An analysis has been done for the unsteady heat transfer to power-law fluids [25B]. The effect of cross flow at the entrance on the heat transfer in tubes has been investigated [31B]. Ref. [15B] deals with ice formation in a pipe containing flows in transition and turbulent regimes. Mechanical heating of non-Newtonian fluids has been discussed [12B]. The gun barrel wall heat transfer has been analyzed [1B].

Ref. [35B] deals with the heat transfer in a capillary flow. Fluidized bed heat transfer to a single vertical tube has been discussed [8B]. Entry lengths for heat and mass transfer to power-law fluids have been determined [47B]. Ref. [46B] presents a simple solution for heat transfer to a power-law fluid flowing in a pipe. Convection velocities have been determined for a turbulent pipe flow [30B]. Simultaneous diffusion and convection in laminar tube flow has been studied [51B]. Ref. [32B] deals with heat transfer in profiled tubes.

Axial conduction in the fluid and the solid walls of the channel plays an important role in some cases. Axial conduction is included in the analysis of laminar flow in a circular tube [4B, 33B]. The conjugate convective heat transfer is considered for viscoplastic fluids [40B]. The flow in a circular tube is analyzed with the effect of axial conduction [48B, 49B].

A numerical solution has been obtained for the heat transfer between concentric vibrating cylinders [17B]. The flow and heat transfer in rectangular cavities has been solved by a numerical scheme [24B]. Ref. [9B] deals with 3-dim. laminar flow in ducts. The entrance region of a flat plate duct has been analyzed [37B]. An analysis is presented for flow and heat transfer in rod bundles [50B]. Heat transfer in a converging channel is considered [10B]. Ref. [21B] deals with multicomponent mass transfer in turbulent flow, Nucleation and bubble growth has been discussed for immiscible liquid composites [18B]. Laminar flow heat transfer in triangular passages has been investigated [38B].

Critical heat flux in horizontal channels has been experimentally measured [23B]. The response of a turbulent boundary layer to a sudden decrease in wall heat flux has been studied [45B]. Experimental data are reported for heat transfer at an upstream facing surface of an aperture [43B]. A numerical solution has been obtained for interrupted-plate passages with finite plate thickness [36B]. Experimentally determined heat transfer coefficients are given for a tube downstream of a cylindrical plenum [22B]. Friction factors in internally finned channels are analytically predicted [39B]. Ref. [2B] considers heat transfer in a channel at supercritical pressure.

An analysis has been reported for flow in twisted

pipes [27B]; a related paper deals with heat transfer in the same situation [28B]. Turbulence models have been examined for ducts of annular cross section [26B]. Measurements are reported for the pressure drop in shrouded fin array with tip clearance [42B]. A study of flow and heat transfer in corrugated wall channels has been presented [19B]. Heat transfer is studied for a parallel-plate channel containing a cylinder [34B]. Ref. [3B] deals with heat transfer to pseudo-plastic fluids. The use of mass transfer method for heat transfer is discussed in [29B]. Experimental data for turbulized channels are correlated by the use of a turbulence model [14B]. The Rayleigh–Ritz method has been applied to forced convection [11B]. Laminar heat transfer in a duct has been considered with temperature-dependent properties [16B]. Ref. [5B] deals with the effect of mounting systems on heat transfer from inclined cylinders. The flow and heat transfer around a blockage in a duct have been considered [44B].

BOUNDARY LAYER AND EXTERNAL FLOWS

Most papers in the area of boundary layers are concerned with laminar and turbulent boundary layers on flat surfaces. Work has also been reported on turbulent jets, impingement flows, and flows over cylinders and spheres.

An integral method has been applied to the calculation of heat transfer in a turbulent incompressible boundary layer [32C]. The effect of a favorable pressure gradient on the heat transfer to a rough surface has been considered [6C]. A study is reported for effect of the Reynolds number on the turbulence structure of a slightly heated turbulent boundary layer [29C]. The combined free and forced convection on a vertical surface has been investigated [8C]. Boundary layers on turbine blades at different angles of attack have been studied [38C]. Ref. [23C] describes the shear stress and heat transfer characteristics at an infinite swept attachment line. The unsteady turbulent heat transfer from a flat plate has been investigated experimentally [12C] and numerically [13C]. The influence of adiabatic co-planar extension surfaces on the solar-collector heat transfer coefficients has been described [27C]. An experimental investigation of heat-stabilized laminar boundary layers in water has been reported [5C]. The influence of a density interface on a boundary layer has been discussed [21C]. A study has been reported on the forced convection near laminar separation [2C]. Laminar heat transfer has been investigated for the situation in which heat is transferred from a flat surface to a 2-dim. water jet [15C].

Among the studies related to free jets, the turbulent convective velocities in a plane jet have been measured [11C]. The structure of a slightly heated turbulent mixing layer has been studied [24C]. An experimental and theoretical investigation has been reported for a 2-dim. turbulent jet [14C]. An analysis of a laminar isothermal two-phase jet appears in [9C]. The interface heat transfer in a horizontal co-current stratified flow has been described [30C]. The complex flow and heat transfer phenomena arising from the impingement of a jet array with cross flow have been discussed [10C].

The flow and heat transfer at the axisymmetric stagnation region has been calculated [22C]. Experimentally determined heat transfer coefficients on a wall-attached cylinder are reported [28C]. Reference [1C] deals with the heat transfer in a tube bank. Laminar falling film characteristics have been determined for horizontal tubes [26C]. Heat transfer due to flow over rectangular bodies has been considered [31C]. Flow and heat transfer around a sphere has been discussed in a number of papers [18C, 35C, 34C, 7C].

Among studies concerned with time-dependent situations, a thermal boundary layer is considered [16C]. The temperature fluctuations in liquid-metal systems are analyzed [20C]. In a non-Newtonian flow, the unsteady thermal boundary layer is analyzed [25C]. Approximate convective heating equations have been proposed for hypersonic flows [37C].

Among the measurement and analysis of turbulence properties, the following topics have been investigated: decay of turbulence behind a grid [36C], turbulent scalar fields [17C], turbulent bursts leading to a prediction for turbulent Prandtl number [33C], origin of turbulence [4C], and transitional turbulent spot [3C]. The intrinsic scales in thermohydrogasdynamics have been identified [19C].

FLOW WITH SEPARATED REGIONS AND THROUGH POROUS MEDIA

Separated regions

Heat transfer was investigated from cylinders in unsteady flow [2D] and in high temperature surroundings [15D]. Experiments studied also heat transfer between a heated cylinder and an air stream with a water spray for steady and pulsating flow [3D]. An expression was derived [10D] for stagnation point heat transfer from a row of impinging jets to a concave cylinder surface. Similarity was verified for models scaled up by a factor of 10. A numerical 2-dim. analysis [6D] studied the flow and heat transfer in a rectangular cavity where the flow is starting from rest with some fluid entering at the upper left hand corner and leaving the cavity in the lower right hand corner.

Porous media

A numerical analysis [17D] studies the velocity and temperature field in a porous medium near an impermeable boundary. Flow and heat transfer are determined by three dimensionless parameters and the results reveal that errors occur when calculations are based on Darcy's law. Convection in a porous layer was also analyzed [5D] by integral relations at high Rayleigh number and for a fluid with temperature dependent viscosity. Mass transfer caused by natural convection in a porous medium consisting of packed spheres and screens was measured [7D] by the electrochemical method. The results were described by the following equation

$$Sh = 0.0228(Sc\,Gr)^{0.32}(R_h/d_p)^{0.22}$$

with d denoting the particle diameter and R_h the hydraulic radius of the flow cross section. Natural convection heat transfer was studied [4D] for a vertical impermeable partition between two porous media maintained at different temperatures. The results show that such a separation inserted in the middle of a porous slab reduces heat transfer drastically. Experiments on boiling heat transfer in porous wire mesh structures [14D] resulted in dimensionless expressions for the heat transfer coefficient.

Three papers deal with combined heat and moisture transfer in unsaturated porous media. The transfer processes in soil (loam and sand) surrounding a buried spherical heat source was studied [1D] analytically and experimentally. A rigorous solution describes the dynamic response of a porous medium or packed bed to an arbitrary varying inlet temperature using a two phase model [16D] and a steady state 1-dim. analysis [13D] considers condensation in a porous insulation occurring when the two surfaces are exposed to two different humid environments. Convective and diffusive transport as well as phase change are included.

Fluidized beds

Heat transfer to the particles of a fluidized bed is measured [11D] for the situation where small particles pass down through the bed of larger and heavier particles. Heat transfer is found to depend primarily on the particle residence time. An analysis of thermal dispersion and particle to fluid heat transfer [8D] considers small particles fluidized in the voids between larger stationary particles. Experiments on the maximum heat transfer coefficient between an electrically heated horizontal tube and a gas–solid fluidized bed with glass, dolomite, sand, silicon carbide, and alumina particles revealed that none of presently available correlations agree with the experimental results [9D]. New correlations are therefore proposed. Heat transfer between a fluidized bed of spherical and non-spherical particles with 2–3 mm size and immersed in-line and staggered bundles of horizontal tubes was measured at atmospheric temperature and pressure [19D]. Results were found to agree with Zabrodsky's theory for wide tube spacing. Heat transfer was also measured [18D] in fixed and fluidized beds through which a liquid is flowing. Mass and momentum transfer to Newtonian and non-Newtonian fluids (water and cellulose CMC solution) was measured in fixed and fluidized beds of uniform cylindrical pellets and spheres for *Re* from 0.038 to 6000 and *Sc* from 800 to 72 000 [12D].

NATURAL CONVECTION—INTERNAL FLOWS

Natural convection in enclosures continues to be of interest to many investigators. As has been true for some years, many papers relate either to convection in fluid layers heated from below or to flow in differentially-heated cavities that are generally heated on one vertical surface and cooled on another. This year there appears to be greater interest in heat transfer across shallow layers of fluid. The areas showing the largest increase in relative activity appear to be: (1) research concerned with thermohaline or double-diffusive convection, i.e. buoyancy forces due to differences in composition or concentration as well as variations in temperature; and (2) natural convection flow in porous media—many of these flows are of interest in geophysical phenomena as well as of importance to the fundamentals of fluid mechanics and heat transfer.

A number of studies considered special solutions for the flow and heat transfer in a horizontal layer heated from below. Boundary layer solutions are reported for non-linear thermal convection in a horizontal layer heated from below with stress free upper and lower boundaries [54F]. Steady numerical solutions in the form of 2-dim. rolls are found [15F] for convection in a low Prandtl number fluid; in a companion work [10F], an asymptotic model is described which agrees well with these numerical results. With poorly-conducting surfaces as the upper and lower boundaries, square cells are obtained as contrasted to the rolls found for conducting boundaries [50F]. Differences between buoyancy-driven and surface tension-driven stability in a horizontal layer has been studied in terms of the significance of the critical parameters [57F].

Other studies of buoyancy-driven flows in fluid layers heated from below include the influence of side-wall geometry. A numerical study of the post-stability flow pattern in a rectangular box has been presented [45F]. Growth rate calculations have been performed for flow in a cylindrical cavity [62F]. Measurements of the effect of temperature-dependent properties on low Rayleigh number thermal convection in a cylindrical cell show hystresis [73F]. Analysis on a fluid contained in a vertical cylinder with a free surface shows the relative importance of surface-tension-driven and buoyancy-driven phenomena [72F].

Experiments were reported on convection in a layer of a nematic liquid crystal which is heated from below [19F]. With this material there is a liquid–liquid phase change. A numerical study indicates the onset of thermal instability in a nematic liquid crystal heated from below [3F].

An analysis of convection in a Hele–Shaw cell shows that dissipation effects reduce the heat transfer [36F]. The effect of heat and mass transfer on Rayleigh–Taylor instability with a heavy fluid over a lighter one was analyzed [29F].

Many studies look at the natural convection from one surface completely enclosed within another with the fluid filling the space in between them. An example would be convection in the annulus between two cylinders maintained at different temperatures. A numerical analysis of the laminar convection between concentric and eccentric cylinders indicates an increase in Nusselt number when the center of the inner cylinder is below the center of the outer one [51F]. Electrochemical measurements are correlated by a power law variation of Sherwood number with Rayleigh number up to Rayleigh numbers greater than 10^{10} [60F]. Natural convection in the region between two concentric spheres was analyzed with a constant temperature or heat flux inner sphere boundary and with the outer sphere at variable temperature tending to give a stratified fluid layer [64F]. For an array of cylinders in a cubic enclosure, experiments indicate a higher heat transfer when the cylinders are horizontal as compared to when they are vertical [74F]. A procedure for fast numerical finite difference calculations has been developed and applied to predict the convection between spheres and cylinders [34F].

Several studies consider the natural convection in a vertical layer which is differentially heated by having one vertical wall at a different temperature from the opposite wall. Some numerical calculations of the heat transfer across vertical layers are said to be invalid outside a limited range of parameters due to points of instability [53F]. Other studies on differentially-heated vertical layers include constant heat flux boundary conditions [17F], non-uniform surface temperatures [20F] and the influence of Prandtl number over a large range of Rayleigh number [21F]. Extrapolation of finite-difference results for such flows can be made to zero grid size using only two different grid sizes [14F]. Experiments in vertical layers indicate single 2-dim. cells at low Rayleigh numbers and more complex flow at higher Rayleigh numbers [66F]. Calculations show the potential convection in the entrance region of a vertical concentric annulus [18F]. Transient convection in a vertical layer of water is analyzed to include the influence of maximum density point [56F].

Other studies of heat transfer in enclosures which are differentially heated on the sides, i.e. on the two vertical faces of a rectangular enclosure, include interferometric measurements in a shallow horizontal layer [61F], experiments up to high Rayleigh number [7F], and the effects of a vertical partition which partially obstructs the flow in a horizontal layer [43F]. A study on laminar high Rayleigh number convection in a shallow differentially heated layer found a power law variation of Nusselt number with Rayleigh number with the exponent decreasing with increasing Rayleigh number [63F]. Numerical calculations provide correlations for the heat transfer across a shallow rectangular cavity [33F]. Different solutions were obtained for the heat transfer across a shallow layer of low Prandtl number and high Prandtl number fluids [65F].

Using a thymol-blue technique, counterflow was observed in a horizontal cylinder connecting two fluid reservoirs maintained at different temperatures [9F]. An experimental and theoretical study considered the flow within a cavity extending from a reservoir [8F]. Penetration depth into an open cavity due to natural convection has been examined [2F]. An analytical and experimental study examined the flow in a shallow layer cooled from above with some fluid entering and leaving at one of the side boundaries [68F].

The natural convection in two separate chambers thermally connected by a wall of finite conductivity has been examined experimentally and analytically [71F]. Flow in an internal cavity connected thermally to a natural convection boundary layer on one of its outside walls has been examined [67F].

Natural convection has also been studied in inclined fluid layers. A power integral analysis [59F] indicates the parameters for instability and also shows the post-stability flow in both inclined and vertical layers. In an inclined plane air layer, the flow structure has been visualized over a range of inclinations and Rayleigh numbers [38F]. Analysis and experiments have been used in a study of the flow structure in an inclined annulus heated and cooled on its end plates [46F]. The influence of thermal resistance of wall boundaries on the onset of cellular convection in an inclined channel has been examined [23F].

Several studies consider double-diffusive convection in which the buoyancy forces that initiate and maintain the convection are due to both temperature variations and concentration variations in the fluid. A review [31F] of double-diffusive convection illustrates some important applications in astrophysics, engineering, and geology. An analysis of 2-dim. double-diffusive convection shows a transition between oscillatory and steady motion [16F]. An experimental study shows cell formation with a vertical concentration gradient when heated from the side [44F].

The criteria for stability of a thermohaline flow in horizontal layers of finite extent has been analyzed [48F]. Instability well below the Rayleigh number predicted by linear theories was found for thermo haline convection [52F]. The instability of salt-fingers in convection has been analyzed over a wide range of relative mass and thermal diffusivities [28F]. Under some conditions, three Rayleigh numbers are required to predict the stability of a rotating double-diffusive flow [47F].

As mentioned in the introduction, many papers report on natural convection in porous media. The 2-dim. convection pattern has been calculated in a region made up of two or three different layers heated from below [41F]. A general model for the stability of flow in a stratified porous layer has been proposed [55F]. The effect of geometry on critical Rayleigh number in convective modes for flow in an open fluid saturated box have been estimated [69F]. A finite element analysis predicts the flow in a porous layer heated differentially on the sides [25F]. A boundary layer analysis and experiments are compared for convection in a vertical porous layer differentially heated [40F]. The onset of flow as well as the preferred modes of flow in a permeable medium contained between vertical coaxial cylinders and heated from below have been calculated [4F].

Experiments on non-Darcy convection in a saturated porous medium show that earlier analyses overestimate the heat transfer rate [13F]. The importance of dissipation effects in such flows has been demonstrated analytically [12F].

Transient analyses have been performed for several geometries and boundary conditions with natural convection in a porous layer. An experimental and analytical study for a vertical layer heated differentially includes the effect of the difference in temperature between the liquid and solid [32F]. An analysis for individual localized heat sources has been generalized by superposition to get the effect of multiple heat sources [24F]. A simplified analysis of 3-dim. flow includes the influence of the heat transfer between the fluid and the solid matrix [11F]. The lateral motion of a flow into a saturated porous medium heated from one side has been analyzed [6F].

Analytical techniques have been developed to find the onset of thermohaline convection in a porous medium with properties varying significantly in the vertical direction [58F]. Other papers on double-diffusive convection in a porous medium include an analysis of the formation of the convecting layer [22F], and a study of the flow in a vertical layer heated from the side [35F].

Experiments on thermal convection in a differentially heated rotating annulus include the influence of a non-axisymmetric bottom surface [37F]. Depending on the input parameters, different flow regimes are observed in a rotating annulus with a variable bottom surface temperature [27F].

The transient and steady state flow in a toroidal thermosyphon heated from below and cooled from above has been examined including the influence of fluid addition [42F]. Experiments and analysis on the starting transient in an open loop heated from below and centrally has been examined [5F]. A study [76F] has considered the flow in a natural circulation system with parallel loops as might be used in the cooling of a light water nuclear reactor.

Several studies consider mixed convection (i.e. combined natural and convection) in ducts. The key parameter in correlating the bulk temperature rise with mixed convection in horizontal tubes is found to be the Rayleigh number raised to the 1/4th power divided by the Graetz number [26F]. An interferometer has been used to measure the mean temperature distribution along a horizontal isothermal tube with internal mixed convection [75F]. The effect of buoyancy on forced convection inside a vertical tube with radial internal fins has been analyzed numerically [49F]. Mixed convection in a square cavity in a channel wall with varying inclination has been examined [30F]. The flow pattern of mixed convection in a shrouded fin array depends on the relative temperatures of the fins and fluid surrounding them [1F].

The Galerkin method has been used to calculate natural convection in several arrays of vertical rod bundles [39F]. An orthogonal curvilinear coordinate system has been used for finite difference solutions of natural convection in non-rectangular enclosures [70F].

NATURAL CONVECTION—EXTERNAL FLOWS

Results of an experimental study of natural convection from a downward facing horizontal heated plate were reported and an explanation and correlation of the edge effects was made in terms of displacing the origin of a boundary layer solution [18FF]. The technique of live-fringe holographic interferometry has been used in an investigation of convective heat transfer beneath a heated horizontal plate in air [13FF]. Boundary layer flow of a thermomicroplar fluid past a non-isothermal vertical plate has been studied numerically [19FF]. For the class of natural convection flows on inclined surfaces in which inertial effects are unimportant (including flows at high Prandtl number) a local nonsimilarity analysis indicates that the effects of the surface-normal pressure gradient on the temperature profile can be characterized by a single local configuration parameter [25FF]. Two-dimensional buoyant clouds moving along inclined boundaries were investigated theoretically and experimentally and it was found that the "thermal theory" gives a good description of the flow in the slope angle range between approximately 5 and 90° [4FF]. A conjugate conduction–convection analysis has been made for a vertical plate fin which exchanges heat with its fluid environment by natural convection using a first-principles approach [29FF].

Transient natural convection adjacent to a vertical plate of finite thickness and heat capacity has been analyzed [26FF]. An investigation was conducted of the transient heat transfer from electrically heated conductors placed in cryogenic liquids [1FF].

Steady-state heat transfer from spheres and cylinders was the subject of several papers. A numerical analysis was carried out to describe the thick boundary layer around an isothermal sphere [14FF]. A study was made of steady state convection from a solid sphere to an incompressible Newtonian fluid with a Prandtl number of 0.72 for Grashhof numbers varying from 0.05 to 50 [15FF]. Numerical predictions of the flow patterns and heat transfer characteristics for laminar, steady-state, 2-dim. natural convection around a cylinder submerged in an unbounded Boussinesq fluid are in good agreement with previously obtained experimental data [12FF]. Experimental determinations were made of the heat transfer from circular cylinders as the cylinders were varied from horizontal to vertical [33FF]. A series solution method for laminar free convection boundary layer heat transfer over circular and elliptical cylinders has been extended to the treatment of nonisothermal objects [21FF]. In a study of the effects of vibrations on convection from horizontal cylinders, it was observed that for amplitude to diameter ratios exceeding 0.5, the relative vibrational heat-transfer coefficient increased almost linearly with the former irrespective of the frequency of vibration [10FF]. Experiments were performed to investigate the heat transfer characteristics of a short isothermal horizontal cylinder attached to an equi-temperature vertical plate and the Nusselt number was found to be rather insensitive to the cylinder length and the position of the cylinder along the plate [30FF]. Experiments were carried out on heat and mass transfer from air to horizontal tubes in the range 10^3–10^6 for $GrPr$ and $GrSc$ in order to investigate the analogy between heat and mass transfer and the effect of mass transfer on the mechanism of sensible heat transfer [11FF].

Experiments were performed to study the interactive natural convection from a pair of heated horizontal cylinders situated one above the other in a vertical plane [31FF]. Numerical solutions have been obtained for the upper-plate heat transfer response to a lateral offset of the plate from a position of precise alignment with a lower plate [32FF].

The temperature field of an axisymmetric laminar starting plume was measured for the first time using an interferometric measurement technique and Abel inversion data reduction [28FF]. The steady laminar plume above a horizontal laser beam, which is caused by absorption of thermal energy from the beam, was found to be 3-dim. [6FF]. An integral method was applied to the analysis of plume flow [17FF]. Round buoyant laminar and turbulent plumes in unstratified flow were analyzed [39FF]. The height to which dense vertical jets with two opposing buoyancy components rise was determined experimentally [34FF].

A study of the influences of property variations in natural convection from vertical surfaces showed that variable properties cause dramatic increases in heat transfer rates in the turbulent regime but have virtually no influence in the laminar regime [9FF]. A reference temperature method was developed by which heat transfer to fluids in the supercritical region under variable property conditions in laminar free convection on a vertical flat plate can be evaluated readily and accurately [16FF]. Plumes above line and point sources in pure and saline water have been analyzed, including the effects of nonlinear density variations [23FF]. A 2-dim. finite difference analysis has predicted the laminar flow and heat transfer characteristics for a vertical ice sheet at 0°C melting into fresh water by free convection under steady state conditions [37FF]. Time-exposure photographs are used to document the nature of the natural convection flow adjacent to a vertical ice surface melting in pure water at ambient water temperatures between 3.9 and 8.4°C [7FF].

The problem of mixed convection from a 2-dim. line heat source for both favorable and adverse buoyancy effects with respect to an oncoming vertical stream is analyzed in terms of two coordinate expansions, direct and inverse, valid for small and large streamwise distances from the heat source [2FF]. A buoyancy-extended version of the k–ε turbulence model is described which predicts well the main features of turbulent buoyant wall jets [22FF]. Experimental determinations were made of the local heat transfer rates of an electrically heated, vertical flat surface with combined free and forced convection of air in the same direction for the case of constant wall heat flux [3FF]. The cooling of a low-heat-resistance sheet that moves downwards at a velocity much smaller than the natural convection velocities that occur was analyzed [20FF]. The vortex instability of laminar, mixed-convection flow over an isothermal, horizontal flat-plate was investigated analytically using the linear stability theory [24FF]. The free-convection boundary layer along a partially heated infinitely long vertical cylinder disturbed by a steady horizontal flow was studied and the asymptotic solution indicates that the boundary layer is mainly induced by the buoyancy force near the thermal leading edge [38FF]. In a heat transfer and flow visualization investigation performed for a single cylinder and for single and double rows of cylinders submerged in an open channel flow, the regimes associated with free convection, combined convection and forced convection have been delineated [5FF]. In a pair of companion papers, nonlinear 2-dim. magneto-convection in a Boussinesq fluid has been studied in a series of numerical experiments both for the regime in which the development of nonlinear convection develops and for the dynamical regime that follows [35FF, 36FF]. Real-time holographic interferometry has been used to study the convective heat transport in a flat-plate solar collector [8FF]. Models of leaves and plants were used to study heat and mass transfer problems in agriculture, using an electrochemical technique [27FF].

CONVECTION FROM ROTATING SURFACES

Experiments and theory determined mass transfer to rotating disks and rings in laminar, transition, and developed turbulent flow [3G] on different portions of the surface. Measurements of heat transfer and friction loss in laminar radial flow through a stack of disks [4G] determined that centrifugal and Coriolis forces caused by rotation lead to a significant increase in heat transfer with simultaneous reduction in friction. The effect of impingement to increase heat transfer in condensation on a rotating disk was found [7G] to be larger experimentally than can be accounted for analytically. Naphthalene sublimation was used to measure mass transfer [2G] from a rotating inclined plate. The results revealed a tremendous increase in the transfer rate where the surface is exposed to wake flow. Computer results are presented [1G] for developing laminar free convection in an open ended vertical annulus with a rotating inner cylinder. The Prandtl number is 0.7, the radius ratio 0.5, one wall is isothermal and the other adiabatic. The Taylor stability problem was solved [5G] for a narrow gap between a hot rotating inner cylinder and a cool stationary one. The energy separation in vortex tubes can be increased [6G] when the tube is conical with a small opening angle. The length of the tube is also decreased.

COMBINED HEAT AND MASS TRANSFER

Combined heat and mass transfer is important in such diverse applications as cooling of high-temperature gas turbines, chemical processing, heat transfer in soils, and drying of paper. Many of the recent studies, motivated by a specific application, have even broader significance in terms of adding to our understanding of fluid mechanics and convective transport.

A numerical prediction of the film cooling

effectiveness for discrete hole injection compares favorably with experimental results at low injection rates [2H]. The effects of heat transfer in porous layers as might occur on transpiration-cooled turbine blades has been analyzed; this work includes the effect of surface curvature [8H].

The influence of the injection normal to a surface on external mass transfer has been analyzed [3H]. The Nusselt number distribution along a channel with surface injection or suction has been calculated [13H].

A number of analyses consider transient heat and mass transfer where there is either a sudden or gradual change in a boundary condition such as wall temperature. The influence of blowing or suction on the heat transfer in the forward stagnation region of a body following a step change in wall temperature or heat flux has been examined [7H]. A finite difference analysis considers buoyancy-driven convection due to concentration and temperature variation on an isothermal vertical flat plate [10H]. An analysis considers the drying near a cylindrical heat source embedded in a moist porous media [5H]. Another study on transient convection combines analysis and measurements of coupled heat and mass transfer in an unsaturated porous media [4H]. The thickness of a frost layer on a cold surface over which a humid stream flows increases with time raised to the one-half power [12H].

Combined heat and mass transfer has been examined using laminar boundary layer analysis [6H]. A film model was developed to analyze coupled heat and mass transfer in multi-component systems [11H]. The relative importance of thermal diffusion and diffusion thermo effects have been examined for different boundary conditions of wall temperature and free stream concentration [1H].

The effects of finite propogation of disturbances using coupled wave equations of heat and mass transfer have been reported [9H].

CHANGE OF PHASE

Boiling

Nucleate boiling. A discussion of the applicability of the bubble flux density concept was discussed [35J] then extended for use with methanol. A low nucleation site density model for surface quenching was extended to high densities [40J] where the areas of influence of nucleation sites overlap. The enhancement of eddy transport in shearing two-phase flows by microconvention was demonstrated [80J]. An analytical and experimental investigation of bubble waiting time showed that bubble nucleation in subcooled flows contributes significantly to convection [5J]. Vapor bubble growth on a heated wall in a stagnant liquid was analyzed and a simple relationship between time and the diameter of a bubble at departure was found [22J]. Stroboscopic photography was used to determine instantaneous size and velocity of evaporating bubbles in sprays [45J]. Bubble growth rate data taken in various mixtures of liquids documented the effect of liquid mixture makeup [86J]. Further support of the Forester–Zuber growth rate law was found and a universal correlation was presented for developed boiling heat transfer on surfaces of known microgeometry and on commercial heating surfaces [77J]. A photographic process was employed to create a regular array of pits in a copper surface [51J]; data from this surface showed enhancement of heat transfer and a significant effect on the boiling curve and burnout. The use of a gold film deposited on a glass substrate as a surface heater and resistance thermometer was demonstrated [59J]. Since the film was transparent, photographs could be taken from beneath. The evolution of enhanced surfaces was discussed [94J].

Forced convection boiling. A modification of an existing nucleate boiling correlation was proposed [81J]. The modification was the replacement of the bubble nucleation and growth term with one derived from natural convection nucleate boiling data. A previously formulated boiling heat transfer correlation for compact plate-fin heat exchangers was extended to include the effect of velocity [17J].

The results of a transition boiling study were summarized [63J]. Effects studied were: flow conditions; method of analysis; method of test operation, and equipment arrangement and construction.

A fully developed and adiabatic two-phase annular flow model with liquid entrainment was derived for flow in a pipe [47J]. A theory was presented for two-phase bubble flow in channels [68J]. An additional eddy diffusivity component was given for bubble agitation. The theoretical predictions of profiles of velocity and temperature as well as skin friction and heat transfer coefficient were compared to experimental data [69J].

Heat transfer measurements were made in the post-dryout region of an air–water dispersed flow [50J]. The effectiveness of wall-to-drop heat transfer depends mainly on the wall superheat for surface temperatures below the minimum film boiling temperature. The thermal entrance length exceeds that of single phase gas flow and decreases with increasing wall temperature. Turbulent droplet flow heat transfer was analyzed for the thermal entrance region of a tube [65J].

A model of the flow boiling crisis in high vapor quality annular flow was presented with some experimental results [73J]. Dryout measurements were made for the flow of boiling water in an annulus with various profiles of axial heat flux [11J]; for BWR geometries, total dryout power was only weakly dependent upon the profile shape. In [36J], previously published generalized CHF correlations were used to predict the well known relation between CHF and exit quality. A graphical method was used to present CHF data; three regimes were identified [37J]. These regimes were further discussed with respect to the flow patterns [38J]. A correlation of existing experimental data for CHF in uniformly heated rectangular channels was formulated [39J] and new experimental results with small heated length to diameter ratios were presented. CHF and liquid film flow rates were measured with upward flow in a uniformly heated tube [90J]. The exit film flow rate was found to be near zero at CHF conditions when the quality exceeded 50%. CHF in a helically-coiled tube was experimentally investigated [33J]; subcooled CHF decreases in the coil whereas high quality CHF is increased by coiling the tube. CHF experiments were performed in a LMFBR steam generator test facility [26J]; empirical correlations were presented. Dryout and pressure drop data were taken for helically coiled steam generator tubes [92J]. A criterion was developed to determine whether gravity influences burnout in upward flow or in downward flow over horizontal cylinders [29J]; low speed downflow runs indicated a buoyancy burnout mechanism that replaced the more common hydrodynamic burnout. A theory was given for the thermal-hydraulic phenomena, including dryout, during recovery of heated flow channels [84J]. The theory predicted the equivalent collapsed liquid level and the two-phase mixture level in a channel during certain nuclear reactor transients. A best-estimate prediction of transient CHF during reactor blowdown was presented [46J]. Several correlations based on local conditions were tested against recent blowdown heat transfer data.

A discussion of the spring model in superheated sodium boiling was given [27J] and density wave oscillations in once-through sodium-heated steam generator tubes were experimentally studied [91J].

Natural convection boiling. Transient boiling with various heater surfaces (wires and patches) and step power inputs was experimentally investigated [56J]; the onset of boiling was found to be time dependent. The variation of the local heat transfer coefficient around the circumference of a horizontal tube was found experimentally [12J]. Reynolds analogy was applied to film boiling on a horizontal plate and a heat transfer coefficient correlation was derived [42J].

Experimental and analytical studies of film boiling around small spheres [78J] were reported with a discussion of the relationship to vapor explosions. For subcooled liquid boiling, the ratio, not the difference, between the subcooled and saturated liquid film boiling Nusselt numbers is significant. Transient heat transfer from a horizontal cylinder initially in film boiling but responding to the passage of a shock wave was measured [31J] and analyzed [32J]. The transient heat transfer rate was as much as 20 times the steady state rate. The vapor mass is a key variable determining whether collapse is achieved. Implications for vapor explosions with liquid in contact with molten nuclear fuels were discussed.

An analysis was made of steady natural convection film boiling on a vertical plate in a porous medium [19J]; at a given Rayleigh number, the Nusselt number was found to be uniquely dependent upon the vapor film thickness. Contamination of nucleation sites by corrosion products was investigated [34J]. The effect of plasma-deposited polymers on nucleate boiling behavior of a copper surface was experimentally determined [30J]. A thin coating of TFE enhanced nucleate boiling while a thicker coating reduced it. A surface energy effect was postulated to explain this behavior.

Experimental results were presented for boiling of binary mixtures of varying compositions [23J]; bubble growth and shape change characteristics were as observed with pure liquids. The effects of heat flux, void ratio, and diameter ratio on boiling heat transfer in a concentric-tube open thermosyphon were examined [75J]; heat flux could be found as the summation of single phase free convection and boiling correlation values. Boiling of water drops superheated in a nonvolatile liquid [8J] is initiated at the water hydrocarbon interface by either growth of a single bubble or by streams of bubbles. The superposition of a magnetic field on boiling of mercury or mercury plus a wetting agent was shown to result in significant increase in the heat transfer coefficient [93J]. Increased field strength also lowered incipient boiling heat flux and encouraged transition to film boiling. Wave propagation of film or nucleate boiling over a heated surface was investigated analytically and experimentally [98J].

A new criterion of dryout was formulated for upward mist flow in tubes [49J]. An experimental investigation of CHF in horizontal channels with circumferentially varying heating [44J] showed that minimum CHF values with nonuniform heating were somewhat higher than those for uniformly heated channels. Measurements were made of CHF and rate of droplet entrainment when boiling a falling liquid film on the outside surface of a vertical tube [89J]; CHF was categorized according to film flow rate. CHF was measured in a countercurrent liquid vapor flow in a closed-end vertical tube [55J]. The length-to-diameter ratio of the test section was shown to influence the flooding CHF.

Condensation

Film condensation. A mixing length model was used to calculate velocity profiles and pressure drop in turbulent downward-directed flow with film evaporation or condensation [54J]. An experimental study of steam condensation on subcooled liquid film within an inclined duct [74J] showed that local heat transfer coefficients may be related to the turbulence intensity in the film. Verification was presented of multicomponent mass transfer models for condensation inside vertical tubes [95J]. Calculations for optimizing surface design were made of condensation rates on fluted surfaces [2J]. A numerical solution was found to wavy laminar film-wise condensation on vertical walls [79J]. Steady film condensation outside a wedge or cone embedded in a porous medium filled with dry saturated vapor was analyzed [18J]; a closed-form solution was found for Nusselt number as a function of Rayleigh number and film thickness. Methods were presented [71J] for determining suitable measures to prevent disturbances in heat exchangers due to periodic accumulation of condensate on tubes. Laminar film condensation of binary vapor mixtures was analyzed revealing some interface characteristics and showing the effect of linearly varying the wall temperature [72J].

Condensation rate and vapor and condenser surface temperature measurements were made during film condensation of mercury on vertical, nickel-plated, copper surfaces [58J]. Experimental heat transfer and hydrodynamic results in film condensation on vertical surfaces [25J] indicated the effects of the strength, frequency and uniformity of an electric field.

Gold surfaces were experimentally shown to promote dropwise condensation of steam [96J].

Free condensation. The validity of the basic assumptions of dropwise condensation was assessed [67J]. Results of an experimental study of surface characteristic and material property effects on dropwise condensation were presented [1J] and, in a technical note [85J], a discussion was presented on the effect of Knudson number on dropwise condensation. An analysis was made for the stagnation region of a spherical water drop moving in an environment composed of particulates and a saturated mixture of the following gases: condensable steam, noncondensable and nonabsorbable air, noncondensable but absorbable elemental iodine, and chemically reactive methyl iodide [20J]. Numerical analyses were made of (1) condensation of multicomponent vapor in the presence of inert components [9J] and (2) laminar condensation near the stagnation point of a droplet moving in a saturated mixture of steam, air and a noncondensable but absorbable gas [21J].

Vaporization

Vaporization of films. The system mean void fraction model was applied to problems involving incomplete vaporization [55J] and an analysis was made of the phenomenon of hydraulically-controlled dryout in a vertical channel [83J]. An analysis of laminar mist flow along a flat plate [87J] showed that the high heat flux resulted from a superposition of film evaporation and convective heat transfer. Data was taken on evaporating films on the inner wall of horizontal tubes [41J]; spiral grooves can double the evaporation heat transfer coefficient due to the capillary phenomenon. Local heat transfer coefficients were determined for evaporation of trickle films on vertical surfaces [70J]. A model of particle bed dryout based on the phenomenon of flooding [60J] predicted a dryout heat flux that depends on the square root of the particle diameter. Enhancement of evaporation initiated by the differential vapor recoil mechanism was experimentally studied [62J].

Experiments on refilling and rewetting hot horizontal tubes [14J] indicated that gravitational effects are important and lead to flow stratification; an analysis followed [15J]. A similar experiment [16J] showed a liquid tongue running along and rewetting the bottom surface; the phenomenon was thought to be hydrodynamically controlled. Quenching data were presented for reflooding of zircaloy and stainless steel clad rod bundles with cold water [64J] showing that zircaloy quenches faster under similar conditions.

Free vaporization. A numerical solution was presented [66J] for evaporation in a spray evaporator; a reflooding index is related exponentially to the droplet diameter and wetting ratio. Turbulent droplet flow heat transfer under post-dryout conditions was numerically analyzed [97J]; the droplets were distributed heat sinks and the eddy diffusivity function of Deissler was used for the vapor phase. An analytical study of both diffusion-controlled and radiation-controlled evaporation of a spray [6J] demonstrated that evaporation characteristics were correlated best by the droplet initial Sauter number. An analysis was presented for water evaporating in high temperature air [28J]. Experiments with liquid droplets evaporating into a flow of gas capable of condensing [43J] showed some surface condensation of the ambient gas and indicated that the process was limited by heat and mass transfer. An augmentation of evaporative cooling of liquid drops falling freely was attempted by coating the drops with an immiscible, volatile liquid [53J]; the dependence on initial drop temperature, volume ratio of volatile liquid to the drop liquid (water) and the air temperature and humidity was investigated. An analysis was made of unsteady evaporation and condensation heat and mass transfer in a single drop [64J] and an experiment was made of condensation or evaporation of a volatile liquid drop or condensible gas bubble in a liquid medium [82J]. An approximate 3-dim. numerical model for predicting characteristics of spray units was presented [61J] that predicted local distributions of dry-bulb temperature, absolute humidity and air streamlines within the flow.

An analysis of bubble growth during depressurization of liquid was made [13J] and a method for correlating the extent that the pressure undershoots the saturation pressure was given [3J]. The limit of homogeneous nucleation under these conditions was shown to lie close to the liquid spinodal line [48J]. The limits of superheat of some binary mixtures were measured at high pressure [7J] and a model was devised; the results apply to the burning of droplets of fuel blends which are mixtures of volatile and nonvolatile liquids. An analysis of thermal, mass and dynamic interaction between a vapor–gas bubble and a liquid during a rapid pressure change or within a sound field was presented [57J].

Other

Experiments were made for mass transfer to air bubbles adhering to the inside wall of a tube carrying supersaturated liquid [4J]. The effects of mass, momentum and thermal variations on the dynamics of inert gas bubbles moving in a channel flow of a liquid and dissolved gas solution were analyzed [52J]; differing axial distributions of pressure and temperature were presumed. A method was suggested [88J] for determining mass transfer coefficients at liquid–liquid interfaces. An analysis was made of the thermocapillary migration of a fluid droplet inside a drop assuming space laboratory environment [76J].

RADIATION

Radiation in participating media

Radiative transport in emitting, absorbing, and scattering media is of continuing interest. The basic theory for radiation transfer in an arbitrary medium has been derived using the general form of the Bethe–Salpeter equation [4K]. Time-dependent radiative transfer is studied in a semi-infinite medium with a reflecting boundary [8K]. The accuracy of an approximate method for calculating the intensity and fluxes of radiation from an inhomogeneous two-phase layer with emitting and reflecting boundaries has been ascertained for a wide range of initial parameters [1K]. Studies are reported of infrared (10 μm) induced evaporation of large water droplets [15K].

Considering gas enclosures the geometric mean transmittance and total absorptance between two finite areas can be reduced to single numerical integrations, thus eliminating much of the mathematical complexity [18K]. A new type of spectrometer has been developed for performing rocket measurements of the near infrared absorption spectrum of the upper atmosphere [11K]. It is shown that all gaseous admixtures in a radiation-absorbing non-equilibrium gas may be induced to drift along the light beam. This effect may be useful for separating sulphur isotopes [9K].

Results of an analysis of radiative–convective heat transfer of an axisymmetric blunt body in hypersonic air flow are presented for ablating graphite surfaces [13K]. An analysis of radiative transfer in a plane slab takes an azimuthal dependence of the radiation field into account, a situation which is of significant practical importance [16K].

Simple solutions for the effective thermal conductivity for combined radiation and free convection in an optically thick heated fluid layer compare favorably with numerical solutions of the governing energy equation incorporating both turbulent heat transport and thermal radiation [5K]. Considering turbulent flow of a multiphase medium with absorbing, emitting and anisotropically scattering particulates bounded by a heated or cooled constant-temperature wall, numerical results are presented for a wide range of the governing parameters [3K].

A study of the radiative heat flux in absorbing, emitting and linear anisotropically scattering cylindrical media shows that in engineering applications approximate methods may be used to accurately model the radiative contribution to overall heat transfer rates [2K]. Analytical studies of multidimensional, radiative transfer with anisotropic scattering are in satisfactory agreement with corresponding experiments [10K].

A new gray-gas approximation for carbon dioxide standard emissivity provides simplicity and reasonable accuracy (errors $<6\%$) for temperatures between 800 and 1800 K and for $1 \leqslant pL \leqslant 10^3$ cm atm [6K]. An experimental investigation of thermal radiation of certain gaseous hydrocarbons reveals that the emissivity is strongly dependent on the number of carbon atoms in a molecule [17K].

The $P-1$ approximation has been generalized for non-gray problems offering the advantage that all solutions can be expressed in terms of the spectral absorption coefficient. It is shown that this technique is superior for gases [19K]. The radial heat flux and its divergence are determined both exactly and approximately for homogeneous suspensions of small particles in an isothermal cylindrical medium. A closed-form approximate solution for the surface heat flux is in excellent agreement with exact results [7K].

Infrared measurements of the stratospheric composition indicate that measured NO and NO_2 concentrations are in disagreement with absolute values predicted by the Oxford 2-dim. diurnal model, but in close agreement with their day-time changes [14K]. By using a new non-steady measuring method, the different radiation fluxes, as well as the net radiation fluxes, can be established for rivers [12K].

Surface radiation

The calculation of shape factors for radiative transfer has been extended to rectangular, plane surfaces of arbitrary position and size. A computer facilitates a fast evaluation of the derived equations [2L]. Effective emissivities of conical cavities having diffuse surfaces are computed by an iterative procedure and effective reflectances are determined with a series technique which involves powers of the reflectance of the wall material [3L].

A perturbation analysis for periodic heat transfer from radiating fins shows that within the range of parameters of this study, the net effect is to decrease the mean temperature and increase the mean heat transfer rate [1L]. A technique for measuring reflectance factors under diurnal and intermittent cloud insolation conditions using hand-held radiometers has been developed. The results of such measurements indicate that this technique may be useful for achieving uniform reflectance factors for remote sensing studies [4L].

MHD

There is a continuing interest in MHD and its applications. In particular, MHD generators are considered.

Because of the high temperatures and the large sizes of large scale MHD generators, radiative heat transfer in the combustion gases becomes a significant energy transport mechanism [8M]. Studies of heat and mass transfer in MHD channels indicate that heat transfer by gas radiation almost equals that by convection for smooth walls, and amounts to 3/4 as much as the convective heat transfer for rough walls [3M]. A combined convection–gas radiation, two-zone flow model is formulated for studying heat transfer characteristics of MHD radiant boilers [4M]. Studies of combined conduction, convection, gas radiation and particle radiation in a reference MHD diffuser (1700 MW thermal) show that heat transfer by convection amounts to 25 MW and radiative heat transfer from 44 to 79 MW depending on the rate of ash carryover into the channel [2M].

The results of an analysis of a fully developed forced and free convective MHD flow between two electrically conducting vertical plates indicate that an increase of the thermal conductance ratio of the two plates leads to both an increase of the velocity and of the temperature [7M]. For establishing heat transfer scaling laws for MHD channels and diffusers, a quasi-3-dim. model has been developed for determining convective and radiative heat transfer characteristics [1M].

Open-cycle MHD generator channel development is discussed with emphasis on critical design criteria and limitations on channel operating parameters [6M]. The development of steam generator components for open-cycle MHD power plants requires, for example, the design of radiant furnaces [5M]. Results of performance studies are reported for a multiphase boiling MHD generator [9M].

NUMERICAL METHODS

Papers reporting the use of numerical methods for solving specific physical problems are listed in the appropriate category pertaining to the problem. Occasionally, some numerical innovations are embedded in such papers. The literature cited in this section is primarily aimed at the development of a numerical method rather than its application.

The inverse heat conduction problem has been treated [1N, 17N]. A calculation scheme for transient phase change around a cylinder is given [18N]. Variable time step methods for the Stefan problem have been examined [10N]. Ref. [4N] describes an efficient technique for calculating two-dimensional temperature distributions. Turbulent temperature fluctuations in liquid metals have been numerically simulated [8N]. A Monte Carlo technique has been adapted for the determination of radiation interception [16N]. An implicit Eulerian method has been used for steam generator analysis [5N]. Transient conjugate problems have been treated by a quasi-steady approach [15N]. Singular perturbation techniques using finite elements have been developed for 1-dim. [6N] and 2-dim. [7N] convection–diffusion problems. Ref. [12N] describes a generalized finite-difference method for heat transfer problems in irregular geometries. A general calculation procedure for 2-dim. fluid flow and heat transfer has been described [11N]. The finite analytic method has been used for the heat transfer in a cavity [3N].

A method has been developed for the solution of nonlinear boundary-layer equations [14N]. The concept of group transformations has been applied to the nonlinear heat diffusion equation [9N]. The technique of computer-extended series is developed in the context of natural convection [13N]. A variational approach has been employed in the prediction of laminar flow with dissipation [12N].

HEAT TRANSFER APPLICATIONS

Heat exchangers and heat pipes

A new method for the prediction of heat transfer coefficients is presented [1Q] in gas flow normal to banks of finned and smooth tubes with triangular pitch. A new design [14Q] of a shell and tube heat exchanger resulted in a slight increase in heat transfer coefficients with a significant reduction in pressure loss. Tests on pilot plants [6Q] revealed that measured heat transfer coefficients and pressure drops for air coolers do not satisfy available correlations for a small number of rows or forced draft. Experiments [2Q] studied the influence of fouling biofilms on heat transfer under controlled conditions. Results compare well with those of a simple mathematical model.

Several papers are concerned with the performance evaluation criteria of heat exchangers. Previous work by Bergles and Webb was extended to establish such criteria for single phase flow in tubes including the effect of shell side enhancement and fouling [19Q]. It was also investigated [13Q] under what conditions such criteria remain unchanged when the two fluid streams are switched. Some fundamental relations were developed [12Q] to describe the performance of a tubular heat exchanger based on NTU, heat capacity ratio, and a temperature efficiency parameter. A method of stochastic approximation [16Q] to identify parameters of parallel flow heat exchangers compares favorably with unsteady experimental results.

A study [9Q] concludes that tiny fins result in the best surface geometry for vertical condenser tubes with outside condensing surfaces. The design of multizone condensers for desuperheating, condensing, and subcooling is discussed [11Q]. A rotary dry cooling tower was developed [17Q] with disks rotating between a hot water bath and a forced air stream. A layer of oil on top of the water eliminates evaporation. An analysis was verified by experiments. The prediction of boiling heat transfer is discussed [5Q] for a compact plate-fin heat exchanger with frequent interruptions. In this extension of a previous study the assumption is made that local heat transfer coefficients are related to local metal-to-liquid temperature difference.

A parametric study [8Q] of a particulate direct contact heat exchanger shows that heat exchange improves as the particle diameter decreases, the heat capacity ratio, the flow rate ratio, and the approach temperature increase for laminar bulk flow and particle Reynolds numbers from 20 to 500. Experiments on heat transfer in tubeless evaporators [15Q] are compared with a theory assuming that uniformly distributed drops heat up and evaporate.

Three papers deal with the thermal performance of regenerators. Addition of 15% steam to combustion air doubled the heat transfer coefficients during the cooling period and increases them by 15% during the heating period [3Q]. It is demonstrated [7Q] how latitudinal conduction can be incorporated in lumped heat transfer coefficients for thermal regenerators. A theoretical model and experiments of heat and mass transfer in rotary regenerators with condensation, evaporation, and convection are described [18Q] and results obtained with an experimental facility are presented.

Two papers are concerned with heat pipes. From an experimental study of the transfer mechanism in the evaporator [10Q] it is concluded that the heat transfer mechanism in a wick of screens is identical to that in a sintered powder wick. Heat pipes filled with sodium reduced the wall temperature at the stagnation line of a space shuttle wing from 1500 to 900 K [4Q].

General

Experiments [13S] on water vapor contribution to the erosion of steel in high temperature flows are reported and interpreted by a surface reaction model. A theory is proposed [9S] to describe the effect of nozzle erosion on heat transfer in a ladle of molten steel during pouring. The nozzle temperature is calculated and compared with plant observation. Experiments established [14S] that the lowest temperature on a surface cooled by a liquid spray is a linear function of the spray mass flux. The vertical temperature distribution is calculated [2S] in a storage tank fed at its top with warm fluid in laminar flow.

Different designs of air cooled gas turbine blades have been discussed [10S]. Five computer programs used to calculate heat transfer rates to gas turbine blades have been compared with experiments [8S] at the University of Oxford free piston wind tunnel. The programs differ by their turbulence models. All calculated results show the right trend but quantitatively they differ considerably. An equation $Nu = cRe^n$ is developed [16S] from literature data describing average heat transfer coefficients to turbine blades. An analysis [6S] simulates the disturbance of a combustion chamber with film cooling by air jets with a model replacing the jets by cylinders normal to the chamber wall. The results are compared with experiments. Model tests determining heat transfer in the casings of turbines with a loop system of steam flow have been performed [22S] and the results are recommended for use in the design of similar casings.

An experimental and analytical study [7S] points to the conclusion that instantaneous heat transfer in an engine exhaust port is primarily caused by jet induced fluid motion.

The process in a liquid metal fast breeder reactor under natural circulation is simulated dynamically by a computer code [4S] using specifications for the Clinch River Breeder Reactor Plant. The results of experiments are reported [5S] for a circulating fluidized bed with coal combustion. A mathematical model of heat transfer from a large enclosed flame in a rotary kiln [15S] predicts the gas and refractory temperature profiles. It is based on the Hottel–Sarofim method and tested against experimental results.

Experiments on the cooling of high pressure rocket thrust chambers with liquid oxygen [20S] verified a supercritical heat transfer correlation developed from heated tube experiments. Design and off-design performance calculations of space radiators with liquid coolant circulation and radiating to the surroundings have been performed [12S] with and without heat pipes. An analysis shows [3S] that conventional methods do not describe adequately the unsteady convective heat transfer in a gun barrel.

A three-parameter equation is developed [18S] with the parameters Re, Pr, and $K = \varepsilon U/L$, using the improved equation

$$\tau = q\left[1+\varepsilon\left(\frac{du}{dt}\right)^n - 1\right]$$

for the shear. Velocity and temperature boundary layers are calculated with this equation. Frictionally generated heat in a polymer may lead to a heat explosion [19S] when this heat is not balanced by heat transfer to the surrounding. A model is proposed [11S] for the analysis of heat and mass transfer in hydroscopic capillary extruded products using equations for the liquid and vapor fluxes depending on temperature and equilibrium moisture content gradients.

An analysis [1S] of heat transfer during cooling of spherical food products establishes the parameter range for best effectiveness when the air passing over the product is continually wetted by a chilled water spray. The temperature separation in a Hartmann–Sprenger tube can be improved [17S] by coupling it to a resonator and by use of a nozzle with a thin rod along its axis. The cost of large scale heat transfer research and testing makes it necessary to carefully optimize the test program [21S].

Solar energy

Topics of major interest among the heat-transfer related solar energy publications include: solar radiation, flat-plate and concentrating collectors, thermal storage and system performance, both passive and active. The standard technique of analyzing solar photometer data to determine atmospheric optical depth and the spectral solar constant inadvertently weight the data unequally [10T]. Seasonal variations of atmospheric clearness numbers for use in clear day solar radiation modelling are proposed [9T].

Two analyses of a non-linear flat-plate collector are presented in which the overall loss coefficient is assumed to be a linear function of the temperature difference between the fluid in the collector and the environment [5T, 7T]. A brief analysis re-examines the original justification for linear and second-order efficiency characteristics [6T]. Simple stagnation-condition measurements for solar collectors can provide a direct measurement of the temperature dependence of the heat loss coefficient [8T]. Based on available correlations, relations were found for the local maxima's and minima's in heat transfer as the gap spacing is varied in flat-plate solar collectors [15T]. A simplified equation was proposed for evaluating the top loss coefficient for a flat-plate solar collector [2T]. New expressions for the heat transfer factor F_R based on inlet, outlet and mean fluid temperatures were derived for use with the Hottel–Whillier collector model by the use of restrictive heat transfer assumptions [14T]. The thermal boundary layer problems associated with flat-plate solar collectors have been analyzed using fourth degree polynomials [17T]. A fluorocarbon loaded solar collector system was found to have efficiencies as high as 83% for a collector-to-ambient temperature difference of approximately 10°C [23T]. Experimental thermal efficiencies for a distributed flow, subatmospheric pressure, flat-plate solar collector were reported for a wide range of environmental and operational conditions [24T]. New experimental results make it possible to draw more general conclusions about the effect of tilt and end clearance on the performance of honeycomb in suppressing convection in flat-plate collectors [30T]. The number of slabs and their thicknesses are both important parameters in the performance of thermal trap collectors [22T]. The energy absorbed by a fluid confined within a circular-cylindrical cover was calculated [26T].

The intermediate range of concentration ratios which can be achieved with compound parabolic concentrators (CPC) without diurnal tracking provides both economic and thermal advantages for solar collector design even when used with non-evacuated absorbers [21T]. Mathematical formulations were developed to study thermal processes in a CPC collector fitted with a concentric, evacuated double pipe to serve as a heat absorber [11T].

Stratification effects in a rock bed storage unit were analyzed in terms of a stratification coefficients which is shown to be a system constant that depends only on three dimensionless system parameters [20T]. Significant reductions in the collector area requirements for a particular system performance can be realized by employing seasonal sensible storage [3T]. The characteristic variation of the rate of heat transfer to and from a latent heat thermal energy storage capsule was investigated analytically and experimentally [12T]. The addition of phase change thermal storage for buffering will substantially improve the performance of parabolic dish solar thermal power plants [16T].

A design method for direct gain passive solar heating systems is given which is more general than the "Solar Load Ratio" method [18T]. The effect of air flow rate in collector-storage walls was analyzed [29T]. A heat transfer model of a parallelepiped tank full of water and covered with a translucent insulation of diffuse material to solar energy, was developed considering the multiple absorptions and reflections to evaluate the heat gain or loss by the water [27T]. A passive energy collection and storage system for greenhouses, based on the collection of energy from the greenhouse atmosphere and storage in the ground, is investigated [19T].

A simple periodic analysis of a basin-type solar still (both single as well as double) quite satisfactorily explains the thermal performance [25T]. Precipitation in a 'saturated' solar pond can increase the reflectance of the bottom of the pond and this can reduce the width of the nonconvective zone and, thus, seriously degrade the performance of the pond [13T]. A new technique was devised to make quantitative estimates of the value of the three major components of a solar-driven chemical heat pump [28T]. A second law efficiency analysis for solar water heaters was presented [1T]. Mounting heat-sink fins in a suitable funnel can contribute substantially to their effectiveness in cooling solar energy converters [4T].

PLASMA HEAT TRANSFER

There is a continuing, strong interest in plasma heat transfer. A 2-dim. analysis of free convection in horizontal, high pressure arcs consisting primarily of Hg vapor confined in quartz tubes, shows that the vertical location of the hot core is in good agreement with experimental findings [12U]. Similarity relations derived for electric arcs in forced axial flow indicate that a turbulent flow model is more consistent with experiments than a laminar flow model [16U]. Results of a two-temperature modeling of the anode contraction region of a high intensity argon arc indicate that the temperature discrepancy between electrons and heavy particles is very pronounced in the arc fringes and the region close to the anode [3U]. A similar study of an arc plasma reactor indicates that enhanced Joule heating in the constricted arc path raises the electron, as well as the heavy-particle temperatures resisting the penetration of the cold gas into the hot arc core [2U]. Calculations of the heating mechanism of cathode craters in vacuum arcs show that Joule heating is insufficient to explain the short time scale of their formation. Therefore, ion impact heating is considered to be the dominant process [9U]. A sectioned, rotating cylinder of 10 cm diameter is used for measuring local heat fluxes and current densities at the anode of a stabilized, rotationally symmetric atmospheric pressure argon arc for currents from 100 to 500 A. The results indicate maximum current densities of 340 A cm^{-2} and heat fluxes up to 7 kW cm^{-2} for the 500 A arc [1U].

The reduction and dephosphorization of molten iron oxide with hydrogen–argon plasmas shows that the efficiency of hydrogen utilization for the reduction is much higher than predicted by equilibrium values below 3000 K [14U]. Transition metal nitrides and alloys may be directly synthesized in a DC argon–nitrogen plasma from powders of the corresponding metals [17U]. Superconducting compounds such as cubic $\alpha-MoC_{1-x}$, etc. which are metastable at room temperature have been formed by heating and quenching of the corresponding equilibrium phases in a plasma jet [11U]. A reactive plasma zone melting process has been used for the purification of metallurgical grade silicon [13U]. Investigations of the heating of submicron particles (metals) in a thermal, optically thick plasma indicate that kinetic methods are required for calculating heating of the particles [18U]. A 100 kW three-phase plasma furnace has been used for spheroidization of aluminum silicate particles. A simple, 1-dim. model for particle heat transfer explains the maximum processing rates and the detrimental effect of an inhomogeneous particle size distribution [7U]. A new sample injection method, incorporated into the design of a r.f. plasma torch, is capable of complete evaporation of refractory materials at high feeding rates without interfering with the stability of the plasma [21U].

Transport properties of hydrogen, oxygen, and argon mixtures are calculated for a temperature range from 4000 up to 10000 K using the Chapman–Enskog approach [15U]. Studies of the total emission coefficient of an air plasma at temperatures from 17000 to 20000 K and pressures from 3 to 15 MPa indicate that calculated data in the literature are 2–4 times too high [6U].

Studies of the behavior of cold Langmuir probes immersed into a moving, compressible, high pressure plasma (MHD plasma) take the cooling-induced reduction of the local ion mobility and the distortion of the hydrodynamic flow pattern close to the probe into account [4U]. By measuring heat transfer rates from a rarefied argon plasma to a biased tungsten wire, accommodation coefficients of argon atoms were found to be 0.62 and for argon ions a value of 0.48 has been found [8U]. A new numerical method has been developed for determining local emission coefficients in asymmetric plasmas (extension of conventional Abel inversion) [20U]. Results of laser-induced fluorescence measurements in high-pressure mercury, mercury–metal halide, and sodium–mercury arcs indicate that this method offers interesting possibilities for the determination of local particle density ratios, local spectral line shifts (Stark shifts) and transition probabilities [19U].

The heat flux along a uniform magnetic field due to a temperature gradient is calculated using a Monte-Carlo solution to the Fokker–Planck equation. The calculated heat flux makes a smooth transition between the analytic expressions for the short and long mean-free-path limits [10U]. Small scale features (10–20 μm) in laser-produced plasmas may be influenced by an instability associated with the density dependence of radiative energy losses [5U].

REFERENCES

Books

D. K. Edwards, *Radiation Heat Transfer Notes.* Hemisphere, Washington, DC (1981).

E. F. C. Somerscales and J. G. Knudsen (editors), *Fouling of Heat Transfer Equipment.* Hemisphere, Washington, DC (1981).

S. Kakac, A. E. Bergles and F. Mayinger, *Heat Exchangers: Thermal-Hydraulic Fundamentals and Design.* Hemisphere, Washington, DC (1981).

Owen C. Jones, Jr. (editor), *Nuclear Reactor Safety Heat Transfer.* Hemisphere, Washington, DC (1981).

A. K. Agrawal and J. G. Guppy (editors), *Decay Heat Removal and Natural Convection in Fast Breeder Reactors.* Hemisphere, Washington, DC (1981).

G. Hetsroni (editor), *Handbook of Multiphase Systems.* Hemisphere, Washington, DC (1981).

G. F. Hewitt, J. M. Delhaye and N. Zuber (editors), *Multiphase Science and Technology: An International Series of Books,* Vol. 1. Hemisphere, Washington, DC (1981).

A. E. Bergles, J. G. Collier, J. M. Delhaye, G. F. Hewitt and F. Mayinger (editors) *Two-Phase Flow and Heat Transfer in*

the Power and Process Industries. Hemisphere, Washington, DC (1981).
P. J. Marto and R. H. Nunn (editors), *Power Condenser Heat Transfer Technology: Computer Modeling, Design, Fouling.* Hemisphere, Washington, DC (1981).
Frank W. Schmidt, *Thermal Energy Storage and Regeneration.* Hemisphere, Washington, DC (1981).

New journals

Engineering Thermophysics in China. Rumford, Chicago, Illinois.

Conduction

1A. R. W. Alperi, Tube wall temperatures of an eccentrically located horizontal tube within a narrow annulus with tube contact, *J. Heat Transfer* **103,** 186 (1981).
2A. A. Aziz and T. Y. Na, Perturbation analysis for periodic heat transfer in radiating fins, *Wärme- und Stoffübertragung* **15,** 245 (1981).
3A. A. Aziz and T. Y. Na, Periodic heat transfer in fins with variable thermal parameters, *Int. J. Heat Mass Transfer* **24,** 1397 (1981).
4A. A. G. Bathelt and R. Viskanta, Heat transfer and interface motion during melting and solidification around a finned heat source/sink, *J. Heat Transfer* **103,** 720 (1981).
5A. J. V. Beck, Transient temperatures in a semi-infinite cylinder heated by a disk heat source, *Int. J. Heat Mass Transfer* **24,** 1631 (1981).
6A. J. V. Beck, Large time solutions for temperatures in a semi-infinite body with a disk heat source, *Int. J. Heat Mass Transfer* **24,** 155 (1981).
7A. K. C. Cheng, H. Inaba and R. R. Gilpin, An experimental investigation of ice formation around an isothermally cooled cylinder in crossflow, *J. Heat Transfer* **103,** 733 (1981).
8A. F. B. Cheung, Periodic growth and decay of a frozen crust over a heat generating liquid layer, *J. Heat Transfer* **103,** 369 (1981).
9A. S. H. Cho and J. E. Sunderland, Approximate temperature distribution for phase change of a semi-infinite body, *J. Heat Transfer* **103,** 401 (1981).
10A. S. W. Churchill and S. M. Cheng, Approximate solutions for conduction from finite plane and solid surfaces to three-dimensional infinite surroundings, *Letters Heat Mass Transfer* **8,** 79 (1981).
11A. M. Comninou, J. R. Barber and J. Dundurs, Heat conduction through a flat punch, *J. Appl. Mech.* **48,** 871 (1981).
12A. M. Comninou, J. Dundurs and J. R. Barber, Planar Hertz contact with heat conduction, *J. Appl. Mech.* **48,** 549 (1981).
13A. M. E. Cunningham and K. L. Peddicord, Heat conduction in spheres packed in an infinite regular cubical array, *Int. J. Heat Mass Transfer* **24,** 1081 (1981).
14A. J. Effekhar, G. Darkazalli and A. Haji-Sheikh, Conduction of heat across rectangular cellular enclosures, *J. Heat Transfer* **103,** 591 (1981).
15A. R. Farhadieh, A study of the penetration of a hot liquid pool into a vertical solid surface, *Nucl. Sci. Engng* **77,** 84 (1981).
16A. A. Goldman and Y. C. Kao, Numerical solution to a two-dimensional conduction problem using rectangular and cylindrical body-fitted coordinate systems, *J. Heat Transfer* **103,** 753 (1981).
17A. L. S. Han and A. A. Cosner, Effective thermal conductivities of fibrous composites, *J. Heat Transfer* **103,** 387 (1981).
18A. P. J. Heggs, D. B. Ingham and M. Manzoor, The effects of non-uniform heat transfer from an annular fin of triangular profile, *J. Heat Transfer* **103,** 184 (1981).
19A. R. H. Henze and J. A. C. Humphrey, Enhanced heat conduction in phase-change thermal energy storage devices, *Int. J. Heat Mass Transfer* **24,** 459 (1981).
20A. C. K. Hseih and K. C. Su, A methodology of predicting cavity geometry based on the scanned surface temperature data—prescribed heat flux at the cavity side, *J. Heat Transfer* **103,** 42 (1981).
21A. C. F. Hsu and E. M. Sparrow, A closed-form analytical solution for freezing adjacent to a plane wall cooled by forced convection, *J. Heat Transfer* **103,** 596 (1981).
22A. C. F. Hsu, E. M. Sparrow and S. V. Patankar, Numerical solution of moving boundary problems by boundary immobilization and a control-volume-based finite-difference scheme, *Int. J. Heat Mass Transfer* **24,** 1335 (1981).
23A. T. R. Hsu and G. Pizey, On the prediction of fusion rate of ice by finite element analysis, *J. Heat Transfer* **103,** 727 (1981).
24A. L. W. Hunter, Transient thermal expansion of solids during inert heating, phase change, and surface gasification, *J. Heat Transfer* **103,** 601 (1981).
25A. M. Imber, Thermally symmetric nonlinear heat transfer in solids, *J. Heat Transfer* **103,** 745 (1981).
26A. N. R. Keltner and J. V. Beck, Unsteady surface element method, *J. Heat Transfer* **103,** 759 (1981).
27A. R. G. Kemink and E. M. Sparrow, Heat transfer coefficients for melting about a vertical cylinder with or without subcooling and for open or closed containment, *Int. J. Heat Mass Transfer* **24,** 1699 (1981).
28A. M. S. Khader and M. C. Hanna, An iterative boundary integral numerical solution for general steady heat conduction problems, *J. Heat Transfer* **103,** 26 (1981).
29A. A. Kilic and K. Onat, The optimum shape for convecting rectangular fins when condensation occurs, *Wärme- und Stoffübertragung* **15,** 125 (1981).
30A. J. C. Lecordier, C. Petit and P. Paranthoen, Influence de la conduction sur la fonction de transfert des froids dan les tres basses frequences; resultats experimentaux, *Letters Heat Mass Transfer* **8,** 103 (1981).
31A. D. Lenk and W. Zscherpel, Electrical analogue model for moving stripe sources of frictional heat on surfaces of solid bodies, *Int. J. Heat Mass Transfer* **24,** 439 (1981).
32A. R. L. Levin, The freezing of finite domain aqueous solutions: solute redistribution, *Int. J. Heat Mass Transfer* **24,** 1443 (1981).
33A. J. H. Lienhard, Heat conduction through "Yin-Yang" bodies, *J. Heat Transfer* **103,** 600 (1981).
34A. C. J. Lozano and R. Reemtsen, On a Stefan problem with an emerging free boundary, *Num. Heat Transfer* **4,** 239 (1981).
35A. V. J. Lundardini, Phase change around a circular cylinder, *J. Heat Transfer* **103,** 598 (1981).

36A. B. M. Marder and N. R. Keltner, Heat flow from a disk by separation of variables, *Num. Heat Transfer* **4,** 485 (1981).
37A. M. D. Mikhailov and M. N. Ozisik, On general solution of heat conduction in an anisotropic medium, *Letters Heat Mass Transfer* **8,** 329 (1981).
38A. S. S. Mohammadi, M. S. Graboski and E. D. Sloan, A mathematical model of a ramp forced hot-wire thermal conductivity instrument, *Int. J. Heat Mass Transfer* **24,** 671 (1981).
39A. A. K. Naghdi, Conduction shape factors for certain multi-hole prismatic bars, *J. Heat Transfer* **103,** 821 (1981).
40A. M. Nivoit, J. L. Profizi and D. Paulmier, Etude des equilibres thermiques d'un filament chauffe par effet Joule, *Int. J. Heat Mass Transfer* **24,** 707 (1981).
41A. J. Padovan, Transient finite-element analysis of conduction problems subject to variable-speed moving fields, *Num. Heat Transfer* **4,** 147 (1981).
42A. E. J. Patula, Steady-state temperature distribution in a rotating roll subject to surface heat fluxes and convective cooling, *J. Heat Transfer* **103,** 36 (1981).
43A. T. K. Pratt, F. Landis and A. F. Giamei, Iterative nonlinear thermal analysis of directional solidification, *Num. Heat Transfer* **4,** 199 (1981).
44A. U. Projahn, H. Rieger and H. Beer, Heat conduction in anisotropic composites of arbitrary shape (a numerical analysis), *Wärme- und Stoffübertragung* **15,** 223 (1981).
45A. N. Ramachandran, J. P. Gupta and Y. Jaluria, Two-dimensional solidification with natural convection in the melt and convective and radiative boundary conditions, *Num. Heat Transfer* **4,** 469 (1981).
46A. S. S. Sadhal, Unsteady heat flow between solids with partially contacting interface, *J. Heat Transfer* **103,** 32 (1981).
47A. P. Sampson and R. D. Gibson, A mathematical model of nozzle blockage by freezing, *Int. J. Heat Mass Transfer* **24,** 231 (1981).
48A. S. Saxena, S. Subrahmaniyam and M. K. Sarkar, On unsteady state heat transfer during melting on horizontal tubes with a convective boundary, *Letters Heat Mass Transfer* **8,** 155 (1981).
49A. H. L. Schreyer, Nonlinear finite-element heat conduction analysis with direct implicit time integration, *Num. Heat Transfer* **4,** 377 (1981).
50A. R. V. Seeniraj and T. K. Bose, Freeze-coating on continuous moving sheet and on an axially moving cylinder, *Wärme- und Stoffübertragung* **15,** 239 (1981).
51A. N. Shamsundar, Similarity rule for solidification heat transfer with change in volume, *J. Heat Transfer* **103,** 173 (1981).
52A. A. D. Snider and A. D. Kraus, A general extended surface analysis method, *J. Heat Transfer* **103,** 699 (1981).
53A. A. Solomon, A note on the Stefan number in slab melting and solidification, *Letters Heat Mass Transfer* **8,** 229 (1981).
54A. A. Solomon, Some approximation of use in perfecting the behavior of a PCM cylinder array, *Letters Heat Mass Transfer* **8,** 237 (1981).
55A. A. Sonn and A. Bar-Cohen, Optimum cylindrical pin fin, *J. Heat Transfer* **103,** 814 (1981).
56A. E. M. Sparrow and C. F. Hsu, Analytically determined fin-tip heat transfer coefficients, *J. Heat Transfer* **103,** 18 (1981).
57A. E. M. Sparrow and C. F. Hsu, Analysis of two-dimensional freezing on the outside of a coolant-carrying tube, *Int. J. Heat Mass Transfer* **24,** 1345 (1981).
58A. E. M. Sparrow, E. D. Larson and J. W. Ramsey, Freezing on a finned tube for either conduction-controlled or natural-convection-controlled heat transfer, *Int. J. Heat Mass Transfer* **24,** 273 (1981).
59A. E. M. Sparrow, J. W. Ramsey and J. S. Harris, The transition from natural-convection-controlled freezing to conduction-controlled freezing, *J. Heat Transfer* **103,** 7 (1981).
60A. J. L. Sproston, Two-dimensional solidification in pipes of rectangular section, *Int. J. Heat Mass Transfer* **24,** 1493 (1981).
61A. L. N. Tao, The exact solutions of some Stefan problems with prescribed heat flux, *J. Appl. Mech.* **48,** 732 (1981).
62A. B. Vick and M. N. Özişik, Quasi-steady-state temperature distribution in periodically contacting finite regions, *J. Heat Transfer* **103,** 739 (1981).
63A. V. Voller and M. Cross, Accurate solutions of moving boundary problems using the enthalpy method, *Int. J. Heat Mass Transfer* **24,** 545 (1981).
64A. V. R. Voller and M. Cross, Estimating the solidification/melting times of cylindrically symmetric regions, *Int. J. Heat Mass Transfer* **24,** 1457 (1981).
65A. C. F. Weber, Analysis and solution of the ill-posed inverse heat conduction problem, *Int. J. Heat Mass Transfer* **24,** 1783 (1981).
66A. K. A. Werley and J. G. Gilligan, The temperature distribution of a sphere placed in a directed uniform heat flux, *J. Heat Transfer* **103,** 399 (1981).
67A. G. West, Radial and axial heat transfer in reactor fuel pins (in German), *Wärme- und Stoffübertragung* **15,** 1 (1981).
68A. L. C. Wrobel and C. A. Brebbia, A formulation of the boundary element method for axisymmetric transient heat conduction, *Int. J. Heat Mass Transfer* **24,** 843 (1981).
69A. L. S. Yao and W. Cherney, Transient phase-change around a horizontal cylinder, *Int. J. Heat Mass Transfer* **24,** 1971 (1981).
70A. W. W. Yuen and R. A. Wessel, Application of the integral method of two-dimensional transient heat conduction problems, *J. Heat Transfer* **103,** 397 (1981).

Channel flow

1B. M. J. Adams and H. Krier, Unsteady internal boundary analysis applied to gun barrel wall heat transfer, *Int. J. Heat Mass Transfer* **24,** 1925 (1981).
2B. M. Adelt and J. Mikielewicz, Heat transfer in a channel at supercritical pressure, *Int. J. Heat Mass Transfer* **24,** 1667 (1981).
3B. H. Blasinski and C. Kuncewiecz, Heat transfer during the mixing of pseudo-plastic fluids with ribbon agitators, *Int. Chem. Engng* **21,** 679 (1981).
4B. H. Brauer, Heat transfer in tubes with laminar flow and axial heat conduction (in German), *Wärme- und Stoffübertragung* **15,** 137 (1981).

5B. K. Bremhorst, Effect of mounting systems on heat transfer from inclined cylinders in cross-flow, *Int. J. Heat Mass Transfer* **24,** 243 (1981).
6B. C-J. Chen and J. Shing Chiou, Laminar and turbulent heat transfer in the pipe entrance for liquid metals, *Int. J. Heat Mass Transfer* **24,** 1179 (1981).
7B. Y. I. Cho and J. P. Hartnett, Mass transfer in turbulent pipe flow of viscoelastic fluids, *Int. J. Heat Mass Transfer* **24,** 945 (1981).
8B. J. Cretu, L. Yang and S. C. Saxena, Heat transfer to a single smooth vertical tube immersed in an air fluidized bed, *Letters Heat Mass Transfer* **8,** 465 (1981).
9B. S. Del Giudice, M. Strada and G. Comini, Three-dimensional laminar flow in ducts, *Num. Heat Transfer* **4,** 215 (1981).
10B. J. Dey and G. Nath, Boundary-layer convective heat transfer in a convergent channel, *Letters Heat Mass Transfer* **8,** 371 (1981).
11B. L. A. Diz and N. V. Suryanarayana, Application of Rayleigh–Ritz method to forced convection turbulent heat transfer, *Wärme- und Stoffübertragung* **2,** 93 (1981).
12B. N. T. Dunwoody and T. Z. Hamill, Mechanical heating in lineal tube flow of non-Newtonian fluids, *Z. Angew. Math. Phys.* **32,** 97 (1981).
13B. P. A. Durbin and S. Frost, Dispersion from line sources in turbulent pipe flow, *Int. J. Heat Mass Transfer* **24,** 969 (1981).
14B. E. P. Dyban and E. Ya. Epik, Correlation of experimental data on heat transfer in turbulized channels on the basis of two-paramcter turbulence models, *Int. Chem. Engng* **21,** 342 (1981).
15B. R. R. Gilpin, Ice formation in a pipe containing flows in the transition and turbulent regimes, *J. Heat Transfer* **103,** 363 (1981).
16B. R. Gegorig, A contribution to the heat transfer in laminar duct flow with temperature-dependent properties (in German), *Wärme- und Stoffübertragung* **15,** 209 (1981).
17B. E. W. Haddon and N. Riley, The heat transfer between concentric vibrating circular cylinders, *Q. J. Mech. Appl. Math.* **34,** 345 (1981).
18B. J. C. Han and W. J. Yang, Transient nucleation and bubble growth in immiscible liquid composites induced by counter-diffusion of gases, *Letters Heat Mass Transfer* **8,** 11 (1981).
19B. R. Izumi, H. Yamashita and K. Oyakawa, Fluid flow and heat transfer in corrugated wall channels (1st Report, Analysis in the case where channels are bent perpendicularly two times), *Bull. JSME* **24,** 1425 (1981).
20B. S. D. Joshi and A. E. Bergles, Heat transfer to laminar in-tube flow of pseudo-plastic fluids, *A.I.Ch.E.Jl* **27,** 872 (1981).
21B. R. Krishna, Multicomponent mass transfer in turbulent flow, *Letters Heat Mass Transfer* **8,** 195 (1981).
22B. S. C. Lau, E. M. Sparrow and J. W. Ramsey, Effect of plenum length and diameter on turbulent heat transfer in a downstream tube and on plenum-related pressure losses, *J. Heat Transfer* **103,** 415 (1981).
23B. A. I. Leontiev, I. L. Mostinsky, V. S. Polonsky, M. A. Styrikovich and I. M. Chernika, Experimental investigation of the critical heat flux in horizontal channels with circumferentially variable heating, *Int. J. Heat Mass Transfer* **24,** 821 (1981).
24B. P. le Quere, J. A. C. Humphrey and F. S. Sherman, Numerical calculation of thermally driven two-dimensional unsteady laminar flow in cavities of rectangular cross section, *Num. Heat Transfer* **4,** 249 (1981).
25B. H.-T. Lin and Y.-P. Shih, Unsteady thermal entrance heat transfer of power-law fluids in pipes and plate slits, *Int. J. Heat Mass Transfer* **24,** 1531 (1981).
26B. M. R. Malik and R. H. Pletcher, A study of some turbulence models for flow and heat transfer in ducts of annular cross-section, *J. Heat Transfer* **103,** 146 (1981).
27B. J. H. Masilyah and K. Nandakumar, Steady laminar flow through twisted pipes: fluid flow in square tubes, *J. Heat Transfer* **103,** 785 (1981).
28B. J. H. Masilyah and K. Nandakumar, Steady laminar flow through twisted pipes: heat transfer in square tubes, *J. Heat Transfer* **103,** 791 (1981).
29B. A. I. Mazur, Investigation of heat transfer in a channel of complex shape by a mass transfer method, *Int. Chem. Engng* **21,** 688 (1981).
30B. P. J. McConachie, The distribution of convection velocities in turbulent pipe flow, *J. Fluid Mech.* **103,** 65 (1981).
31B. D. E. Metzger and J. V. Cordaro, The effect of crossflow at the entrance on heat transfer enhancement in tubes, *J. Heat Transfer* **103,** 178 (1981).
32B. V. K. Migai and P. G. Bystrov, Heat transfer in profiled tubes, *Thermal Engng* **28,** 178 (1981).
33B. H. Nagasue, Steady-state heat transfer with axial conduction in laminar flow in a circular tube with a specified temperature or heat flux wall, *Int. J. Heat Mass Transfer* **24,** 1823 (1981).
34B. K. Oyakawa and I. Mabuchi, Fluid flow and heat transfer in a parallel plate duct containing a cylinder, *Bull. JSME* **24,** 1795 (1981).
35B. E. Papoutsakis and D. Ramkrishna, Heat transfer in a capillary flow emerging from a reservoir, *J. Heat Transfer* **103,** 429 (1981).
36B. S. V. Patankar and C. Prakash, An analysis of the effect of plate thickness on laminar flow and heat transfer in interrupted-plate passages, *Int. J. Heat Mass Transfer* **24,** 1811 (1981).
37B. S. J. Rhee and D. K. Edwards, Laminar entrance flow in a flat plate duct with asymmetric suction and heating, *Num. Heat Transfer* **4,** 85 (1981).
38B. G. E. Schneider and B. L. LeDain, Fully developed laminar heat transfer in triangular passages, *J. Energy* **5,** 15 (1981).
39B. M. J. Scott and R. L. Webb, Analytic prediction of the friction factor for turbulent flow in internally finned channels, *J. Heat Transfer* **103,** 423 (1981).
40B. Z. P. Shul'man and B. M. Khusid, Conjugated convective heat transfer problems for viscoplastic fluids in plane-parallel channels, *Int. J. Heat Mass Transfer* **24,** 1045 (1981).
41B. R. Smith and M. F. Edwards, Heat transfer to non-Newtonian and drag-reducing fluids in turbulent pipe flow, *Int. J. Heat Mass Transfer* **24,** 1059 (1981).
42B. E. M. Sparrow and T. J. Becky, Pressure drop characteristics for a shrouded longitudinal-fin array with tip clearance, *J. Heat Transfer* **103,** 393 (1981).

43B. E. M. Sparrow and U. Gurdal, Heat transfer at an upstream-facing surface washed by fluid en route to an aperture in the surface, *Int. J. Heat Mass Transfer* **24,** 851 (1981).
44B. E. M. Sparrow, M. Moli and S. R. Chastain, Turbulent heat transfer coefficients and fluid flow patterns on the faces of a centrally positioned blockage in a duct, *Int. J. Heat Mass Transfer* **24,** 507 (1981).
45B. C. S. Subramanian and R. A. Antonia, Response of a turbulent boundary layer to a sudden decrease in wall heat flux, *Int. J. Heat Mass Transfer* **24,** 1641 (1981).
46B. R. D. Tonini and N. O. Lemcoff, A simple solution for the heat or mass transfer to a power law fluid in cylindrical pipes, *Letters Heat Mass Transfer* **8,** 45 (1981).
47B. R. D. Tonini and N. O. Lemcoff, Mass and heat transfer entry length for power law fluids in cylindrical geometry, *Letters Heat Mass Transfer* **8,** 425 (1981).
48B. B. Vick and M. N. Özişik, An exact analysis of low Peclet number heat transfer in laminar flow with axial conduction, *Letters Heat Mass Transfer* **8,** 1 (1981).
49B. B. Vick and M. N. Özişik, Effects of axial conduction and convective boundary conditions in slug flow inside a circular tube, *J. Heat Transfer* **103,** 436 (1981).
50B. S. Wong and K. Chen, On the two-dimensional analysis of fluid flow and heat transfer in rod bundles with arbitrary arrangement, *Nucl. Sci. Engng* **77,** 92 (1981).
51B. J. S. Yu, Dispersion in laminar flow through tubes by simultaneous diffusion and convection, *J. Appl. Mech.* **48,** 217 (1981).

Boundary layer and external flows

1C. S. Aiba, H. Tsuchida and T. Ota, Heat transfer around a tube in a bank, *Bull. JSME* **24,** 380 (1981).
2C. E. A. Akinrelere, Forced convection near laminar separation, *Aero. Q.* **33,** 212 (1981).
3C. R. A. Antonia, A. J. Chambers, M. Sokolov and C. W. Van Atta, Simultaneous temperature and velocity measurements in the plane of symmetry of a transitional turbulent spot, *J. Fluid Mech.* **108,** 317 (1981).
4C. A. Bejam, On the buckling property of inviscid jets and the origin of turbulence, *Letters Heat Mass Transfer* **8,** 187 (1981).
5C. S. J. Barker and D. Gile, Experiments on heat-stabilized laminar boundary layers in water, *J. Fluid Mech.* **104,** 139 (1981).
6C. H. W. Coleman, R. J. Moffat and W. M. Kays, Heat transfer in the accelerated fully rough turbulent boundary layer, *J. Heat Transfer* **103,** 153 (1981).
7C. H. Delmas and J. P. Couderc, Transfert de matiere autour d'une sphere, en phase liquide-influence du nombre de Schmidt, *Letters Heat Mass Transfer* **8,** 271 (1981).
8C. J. Dey and G. Nath, Mixed convection flow on vertical surface, *Wärme- und Stoffübertragung* **15,** 279 (1981).
9C. S. E. Elghobashi and I. E. A. Megahed, Mass and momentum transport in a laminar isothermal two-phase round jet, *Num. Heat Transfer* **4,** 317 (1981).
10C. L. W. Floreschuetz, C. R. Truman and D. E. Metzger, Streamwise flow and heat transfer distributions for jet array impingement with crossflow, *J. Heat Transfer* **103,** 337 (1981).
11C. V. W. Goldschmidt, M. F. Young and E. S. Ott, Turbulent convective velocities (broadband and wave number dependent) in a plane jet, *J. Fluid Mech.* **105,** 327 (1981).
12C. J. Hanawa, Transient convective heat transfer to turbulent flow from a flat plate with constant heat capacity (experimental study), *Bull. JSME* **24,** 1433 (1981).
13C. J. Hanawa, Transient convective heat transfer to turbulent flow from a flat plate with constant heat capacity (numerical analysis), *Bull. JSME* **24,** 124 (1981).
14C. J. Hoch and L. M. Jiji, Theoretical and experimental distribution in two-dimensional turbulent jet-boundary interaction, *J. Heat Transfer* **103,** 331 (1981).
15C. S. Inada, Y. Miyasaka and R. Izumi, A study on the laminar-flow heat transfer between a two-dimensional water jet and a flat surface with constant heat flux, *Bull. JSME* **24,** 1803 (1981).
16C. J. C. Keck, Thermal boundary layer in a gas subject to a time dependent pressure, *Letters Heat Mass Transfer* **8,** 313 (1981).
17C. B. A. Kolovandin and N. N. Luchko, Asymptotic modelling of turbulent passive scalar field dynamics, *Letters Heat Mass Transfer* **8,** 321 (1981).
18C. S. Krenk, Creeping viscous flow around a heat-generating solid sphere, *J. Appl. Mech.* **48,** 239 (1981).
19C. S. S. Kutateladze, Intrinsic scales in thermohydrogas-dynamics, *Int. J. Heat Mass Transfer* **24,** 803 (1981).
20C. C. C. Meek, M. J. Morris, R. C. Doerner and R. F. Hunt, Analysis of temperature fluctuations in liquid-metal systems, *Nucl. Sci. Engng* **79,** 202 (1981).
21C. J. F. Piat and E. J. Hopfinger, A boundary layer topped by a density interface, *J. Fluid Mech.* **113,** 411 (1981).
22C. P. Plostins and S. G. Rubin, The axisymmetric stagnation region full shock layer for large rates of injection, *Num. Heat Transfer* **4,** 359 (1981).
23C. D. I. A. Poll, Skin friction and heat transfer at an infinite swept attachment line, *Aero. Q.* **33,** 299 (1981).
24C. Rajagopalan and R. A. Antonia, Properties of the large structure in a slightly heated turbulent mixing layer of a plane jet, *J. Fluid Mech.* **105,** 261 (1982).
25C. Ramma Subba Reddy Gorla, Unsteady thermal boundary layer flow of a non-Newtonian fluid over a flat plate, *Int. J. Engng Sci.* **19,** 1391 (1981).
26C. J. T. Rogers, Laminar falling film flow and heat transfer characteristics on horizontal tubes, *Can. J. Chem. Engng* **59,** 213 (1981).
27C. E. M. Sparrow and S. C. Lau, Effect of adiabatic co-planar extension surfaces on wind-related solar-collector heat transfer coefficients, *J. Heat Transfer* **103,** 268 (1981).
28C. E. M. Sparrow and F. Samie, Measured heat transfer coefficients at and adjacent to the tip of a wall-attached cylinder in crossflow—application to fins, *J. Heat Transfer* **103,** 778 (1981).
29C. C. S. Subramanian and R. A. Antonia, Effect of Reynolds number on a slightly heated turbulent boundary layer, *Int. J. Heat Mass Transfer* **24,** 1833 (1981).
30C. M. Takahashi, A. Inoue, S. Oaki, M. Aritomi and H. Enro, Interface heat transfer of horizontal cocurrent liquid–liquid stratified flow (2nd report, Entrance region), *Bull. JSME* **24,** 1002 (1981).

31C. F. L. Test, R. C. Lessman and A. Johary, Heat transfer during wind flow over rectangular bodies in the natural environment, *J. Heat Transfer* **103,** 262 (1981).
32C. L. C. Thomas and M. M. Al-Sharif, An integral analysis for heat transfer in turbulent incompressible boundary layer flow, *J. Heat Transfer* **103,** 772 (1981).
33C. L. C. Thomas and M. B. Ibrahim, A model of the turbulent burst phenomenon: predictions for turbulent Prandtl number, *Letters Heat Mass Transfer* **8,** 357 (1981).
34C. K. Torii, M. Isobe, N. Miura and T. Horikoshi, Free-stream turbulence effects on heat and mass transfer from spheres (1st report, average transfer coefficients), *Bull. JSME* **24,** 131 (1981).
35C. K. Torii, M. Isobe, T. Horikoshi and N. Miura, Free-stream turbulence effects on heat and mass transfer from spheres (2nd report, local transfer coefficients), *Bull. JSME* **24,** 138 (1981).
36C. Z. Warhaft, The use of dual heat injection to infer scalar covariance decay in grid turbulence, *J. Fluid Mech.* **104,** 93 (1981).
37C. E. V. Zoby, J. N. Moss and K. Sutton, Approximate convective-heating equations for hypersonic flows, *J. Appl. Mech.* **18,** 64 (1981).
38C. L. M. Zysina-Molozhen and A. A. Dergach, Heat transfer in blade cascades at different angles of attach, *Thermal Engng* **28,** 346 (1981).

Flow with separated regions and through porous media

1D. J. Y. Baladi, D. L. Ayers and R. J. Schoenhals, Transient heat and mass transfer in soils, *Int. J. Heat Mass Transfer* **24,** 449 (1981).
2D. T. E. Base, J. M. Patel and G. C. Valaitis, Heat transfer from cylinders in unsteady flow, *Can. J. Chem. Engng* **59,** 247 (1981).
3D. C. Basilico, G. Jung and M. Martin, Etude du transfert convectif entre un cylindre chauffé et un ecoulement d'air chargé de gouttelettes d'eau, *Int. J. Heat Mass Transfer* **24,** 371 (1981).
4D. A. Bejan and R. Anderson, Heat transfer across a vertical impermeable partition imbedded between two porous media maintained at different temperatures, *Int. J. Heat Mass Transfer* **24,** 1237 (1981).
5D. P. A. Blythe and P. G. Simpkins, Convection in a porous layer for a temperature dependent viscosity, *Int. J. Heat Mass Transfer* **24,** 497 (1981).
6D. M. Farge and P. Duhamel, Simulation numérique du régime thermohydraulique transitoire d'un jet bidimensional injeté dans une cavité rectangulaire, *Int. J. Heat Mass Transfer* **24,** 1599 (1981).
7D. J. F. Gabitto and U. Böhm, Experimental study of free convective mass transfer in porous media, *Int. J. Heat Mass Transfer* **24,** 1675 (1981).
8D. D. J. Gunn and P. V. Narayanan, Axial thermal dispersion and particle to fluid transfer in packed-fluidized beds, *Int. J. Heat Mass Transfer* **24,** 1405 (1981).
9D. N. S. Grewal and S. C. Saxena, Maximum heat transfer coefficient between an electrically heated single horizontal tube and a gas-solid fluidized bed, *I/EC Fundamentals* **20,** 108 (1981).
10D. P. Hrycak, Heat transfer from a row of impinging jets to concave cylindrical surfaces, *Int. J. Heat Mass Transfer* **24,** 407 (1981).
11D. F. D. Hussein, P. P. Maitra and R. Jackson, Heat transfer to a fluidized bed using a particulate heat transfer medium, *I/EC Proc. Des. Dev.* **20,** 511 (1981).
12D. S. Kumar and S. N. Upadhyay, Mass and momentum transfer to Newtonian and non-Newtonian fluids in fixed and fluidized beds, *I/EC Fundamentals* **20,** 186 (1981).
13D. Y. Ogniewicz and C. L. Tien, Analysis of condensation in porous insulation, *Int. J. Heat Mass Transfer* **24,** 421 (1981).
14D. M. Rannenberg and H. Beer, Boiling heat transfer in capillary porous wire mesh structures, *Chemie–Ingr–Tech.* **4,** 290 (1981).
15D. N. N. Sayegh and W. H. Gauvin, Heat transfer to wires and cylinders in high temperature surroundings, *Can. J. Chem. Engng* **59,** 241 (1981).
16D. G. Spiga and M. Spiga, A rigorous solution to a heat transfer two-phase model in porous media and packed beds, *Int. J. Heat Mass Transfer* **24,** 355 (1981).
17D. K. Vafai and C. L. Tien, Boundary and inertia effects on flow and heat transfer in porous media, *Int. J. Heat Mass Transfer* **24,** 195 (1981).
18D. M. Wehrmann and A. Mersmann, Heat transfer in fixed and fluidized beds through which a liquid is flowing, *Chemie–Ingr–Tech.* **10,** 804 (1981).
19D. S. S. Zabrodsky, Y. G. Epanov, D. M. Galershtein, S. C. Saxena and A. K. Kolar, Heat transfer in a large-particle fluidized bed with immersed in-line and staggered bundles of horizontal smooth tubes, *Int. J. Heat Mass Transfer* **24,** 571 (1981).

Natural convection—internal flows

1F. S. Acharya and S. V. Patankar, Laminar mixed convection in a shrouded fin array, *J. Heat Transfer* **103,** 559 (1981).
2F. J. M. Auxilien, Evaluation de la profondeur de penetration des ecoulements de convection naturelle dans un espace semiconfine vertical, *Letters Heat Mass Transfer* **8,** 303 (1981).
3F. P. J. Barratt and D. M. Sloan, A numerical investigation into thermal instabilities in homeotropic nematics heated from below, *J. Fluid Mech.* **102,** 389 (1981).
4F. H. H. Bau and K. E. Torrance, Onset of convection in a permeable medium between vertical coaxial cylinders, *Physics Fluids* **24,** 382 (1981).
5F. H. H. Bau and K. E. Torrance, Transient and steady behavior of an open, symmetrically-heated, free convection loop, *Int. J. Heat Mass Transfer* **24,** 597 (1981).
6F. A. Bejan, Lateral intrusion of natural convection into a horizontal porous structure, *J. Heat Transfer* **103,** 237 (1981).
7F. A. Bejan, A. A. Alohmoud and J. Imberger, Experimental study of high-Rayleigh-number convection in a horizontal cavity with different end temperatures, *J. Fluid Mech.* **109,** 283 (1981).
8F. A. Bejan and S. Kimura, Penetration of free convection into a lateral cavity, *J. Fluid Mech.* **103,** 465 (1981).
9F. A. Bejan and A. N. Rossie, Natural convection in horizontal duct connecting two fluid reservoirs, *J. Heat*

Transfer **103,** 108 (1981).
10F. F. H. Busse and R. M. Clever, An asymptotic model of two-dimensional convection in the limit of low Prandtl number, *J. Fluid Mech.* **102,** 75 (1981).
11F. Y. T. Chan and S. Banerjee, Analysis of transient three-dimensional natural convection in porous media, *J. Heat Transfer* **103,** 242 (1981).
12F. P. Cheng, Thermal dispersion effects in non-Darcian convective flows in a saturated porous medium, *Letters Heat Mass Transfer* **8,** 287 (1981).
13F. P. Cheng, C. L. Ali and A. K. Verma, An experimental study of non-Darcian convection in a saturated porous medium, *Letters Heat Mass Transfer* **8,** 261 (1981).
14F. S. W. Churchill, P. Chao and H. Ozoe, Extrapolation of finite-difference calculations of laminar natural convection in enclosures of zero grid size, *Num. Heat Transfer* **4,** 39 (1981).
15F. R. M. Clever and F. H. Busse, Low-Prandtl-number convection in a layer heated from below, *J. Fluid Mech.* **102,** 61 (1981).
16F. L. N. DaCosta, E. Knobloch and N. O. Weiss, Oscillations in double-diffusive convection, *J. Fluid Mech.* **109,** 25 (1981).
17F. A. M. Dalbert, F. Penot and J. L. Peube, Convection naturelle laminaire dans un canal vertical chauffé à flux constant, *Int. J. Heat Mass Transfer* **24,** 1463 (1981).
18F. M. A. El-Shaarawl and A. A. Sarhan, Developing laminar free convection in a heated vertical open-ended concentric annulus, *I/EC Fundamentals* **20,** 388 (1981).
19F. D. E. Fitzjarrald, an experimental investiga.ion of convection in a fluid that exhibits phase change, *J. Fluid Mech.* **102,** 85 (1981).
20F. B. Gilly, P. Bontoux and B. Roux, Influence des conditions thermiques de paroi sur la convection, naturelle dans une cavité rectangulaire verticale, différentiellcmcnt chauffee, *Int. J. Heat Mass Transfer* **24,** 829 (1981).
21F. W. P. Grabebel, The influence of Prandtl number on free convection in a rectangular cavity, *Int. J. Heat Mass Transfer* **24,** 125 (1981).
22F. R. W. Griffiths, Layered double-diffusive convection in porous media, *J. Fluid Mech.* **102,** 221 (1981).
23F. M. A. Hassab and M. N. Özişik, Effects of thermal wall resistance on the stability of conduction regime in an inclined narrow slot, *Int. J. Heat Mass Transfer* **24,** 739 (1981).
24F. C. E. Hickox, Thermal convection at low Rayleigh number from concentrated sources in porous media, *J. Heat Transfer* **103,** 232 (1981).
25F. C. E. Hickox and K. K. Garling, A numerical study of natural convection in a horizontal porous layer subjected to an end-to-end temperature difference, *J. Heat Transfer* **!03,** 797 (1981).
26F. C. A. Hieber, Mixed convection in an isothermal horizontal tube: some recent theories, *Int. J. Heat Mass Transfer* **24,** 315 (1981).
27F. P. Hignett, A. Ibbetson and P. D. Killworth, On a rotating thermal convection driven by non-uniform heating from below, *J. Fluid Mech.* **109,** 161 (1981).
28F. J. Y. Holyer, On the collective instability of salt fingers, *J. Fluid Mech.* **110,** 195 (1981).
29F. D. Y. Hsieh and S. P. Ho, Rayleigh–Taylor stability with mass and heat transfer, *Physics Fluids* **24,** 202 (1981).
30F. J. A. C. Humphrey and E. W. Jacobs, Free-forced laminar flow convective heat transfer from a square cavity in a channel with variable inclination, *Int. J. Heat Mass Transfer* **24,** 1589 (1981).
31F. H. E. Hupper and J. S. Turner, Double-diffusive convection, *J. Fluid Mech.* **106,** 299 (1981).
32F. H. Inaba and N. Seki, Transient behaviors of natural convective heat transfer through vertical porous layer, *Appl. Sci. Res.* **3,** 257 (1981).
33F. H. Inaba, N. Seki, S. Fukusako and K. Kanayama, Natural convective heat transfer in a shallow rectangular cavity with different end temperatures, *Num. Heat Transfer* **4,** 459 (1981).
34F. D. B. Ingham, Heat transfer by natural convection between spheres and cylinders, *Num. Heat Transfer* **4,** 53 (1981).
35F. A. A. Khan and Z. Zebib, Double diffusive instability in a vertical layer of a porous medium, *J. Heat Transfer* **103,** 179 (1981).
36F. O. Kvernvold and P. A. Tyvand, Dissipation effects on thermal convection in a Hele–Shaw cell, *Int. J. Heat Mass Transfer* **24,** 887 (1981).
37F. H. Leach, Thermal convection in a rotating fluid: effects due to bottom topography, *J. Fluid Mech.* **109,** 75 (1981).
38F. S. J. M. Linthorst, V. M. M. Schinkel and C. J. Hoogendoorn, Flow structure with natural convection in inclined air-filled enclosures, *J. Heat Transfer* **103,** 535 (1981).
39F. D. Magallon, Méthode des elements finis appliquée au calcul due régime etabli laminaire de convection naturelle dans les faisceaux tubulaires verticaux de dimension finie, *Int. J. Heat Mass Transfer* **24,** 1191 (1981).
40F. T. Masuoka, Y. Yokote and T. Katsuhara, Heat transfer by natural convection in a vertical porous layer, *Bull. JSME* **24,** 995 (1981).
41F. R. McKibbin and M. J. O'Sullivan, Heat transfer in a layered porous medium heated from below, *J. Fluid Mech.* **111,** 141 (1981).
42F. A. Mertol, R. Greif and Y. Zvirin, The transient, steady state and stability behavior of a thermosyphon with throughflow, *Int. J. Heat Mass Transfer* **24,** 621 (1981).
43F. M. W. Nansteel and R. Greif, Natural convection in undivided and partially divided rectangular enclosures, *J. Heat Transfer* **103,** 623 (1981).
44F. U. Narusawa and Y. Suzukawa, Experimental study of double-diffusive cellular convection due to a uniform lateral heat flux, *J. Fluid Mech.* **113,** 387 (1981).
45F. C. Normand, Convective flow patterns in rectangular boxes of finite extent, *Z. Angew. Math. Phys.* **32,** 81 (1981).
46F. H. Ozoe, T. Shibata and S. W. Churchille, Natural convection in an inclined circular cylinderical annulus heated and cooled on its end plates, *Int. J. Heat Mass Transfer* **24,** 727 (1981).
47F. A. J. Pearlstein, Effect of rotation on the stability of a doubly diffusive fluid layer, *J. Fluid Mech.* **103,** 389 (1981).
48F. D. Pnueli, Lower bounds to thermohaline stability criteria for fluids confined inside arbitrary configur-

ations, *Israel J. Tech.* **18,** 117 (1981).
49F. C. Prakash and S. V. Patankar, Combined free and forced convection in vertical tubes with radial internal fins, *J. Heat Transfer* **103,** 566 (1981).
50F. M. R. E. Proctor, Planform selection by finite-amplitude thermal convection between poorly conducting slabs, *J. Fluid Mech.* **113,** 469 (1981).
51F. U. Projahn, H. Rieger and H. Beer, Numerical analysis of laminar natural convection between concentric and eccentric cylinders, *Num. Heat Transfer* **4,** 131 (1981).
52F. M. R. E Proctor, Steady subcritical thermohaline convection, *J. Fluid Mech.* **105** 507 (1981).
53F. G. D. Raithby and H. H. Wong, Heat transfer by natural convection across vertical air layers, *Num. Heat Transfer* **4,** 447 (1981).
54F. N. Riahi, Boundary-layer solutions of single-mode convection equations, *J. Fluid Mech.* **102,** 211 (1981).
55F. J. P. Richard and J. Gounot, Critère d'apparition de la convection naturelle dans des couches poreuses stratifées, *Int. J. Heat Mass Transfer* **24,** 1325 (1981).
56F. L. Robillard and P. Vasseur, Transient natural convection heat transfer of water with maximum density effect and supercooling, *J. Heat Transfer* **103,** 528 (1981).
57F. S. Rosenblat, G. M. Homsy and S. H. Davis, Eigenvalues of the Rayleigh–Benard and Marangoni problems, *Physics Fluids* **24,** 2115 (1981).
58F. H. Rubin, Onset of thermohaline convection in heterogeneous porous media, *Israel J. Tech.* **19,** 110 (1981).
59F. D. W. Ruth, Finite amplitude convection in inclined and vertical slots a power integral analysis, *Int. J. Heat Mass Transfer* **24,** 933 (1981).
60F. G. H. Sedahmed and L. W. Shemilt, Natural convection mass transfer in horizontal annuli, *Letters Heat Mass Transfer* **8,** 515 (1981).
61F. V. Sernas and E. I. Lee, Heat transfer in air enclosures of aspect ratio less than one, *J. Heat Transfer* **103,** 617 (1981).
62F. J. N. Shaumeyer, R. P. Behringer and R. Baierlein, Linear growth rates for the Rayleigh–Bénard instability in cylindrical geometry, *J. Fluid Mech.* **109,** 339 (1981).
63F. G. Shiralkar, A. Gadgil and C. L. Tien, High Rayleigh number convection in shallow enclosures with different end temperatures, *Int. J. Heat Mass Transfer* **24,** 1621 (1981).
64F. S. N. Singh and J. M. Elliott, Natural convection between concentric spheres in a slightly-thermally stratified medium, *Int. J. Heat Mass Transfer* **24,** 395 (1981).
65F. G. S. Shiralkar and C. L. Tien, A numerical study of laminar natural convection in shallow cavities, *J. Heat Transfer* **103,** 226 (1981).
66F. P. G. Simpkins and T. D. Dudderar, Convection in rectangular cavities with differentially heated end walls, *J. Fluid Mech.* **110,** 433 (1981).
67F. E. M. Sparrow and C. Prakash, Interaction between internal natural convection in an enclosure and an external natural convection boundary-layer flow, *Int. J. Heat Mass Transfer* **24,** 895 (1981).
68F. T. W. Sturm, Laminar gravitational convection of heat in dead-end channels, *J. Fluid Mech.* **110,** 97 (1981).
69F. P. K. Tewari and K. E. Torrance, Onset of convection in a box of fluid-saturated porous material with a permeable top, *Physics Fluids* **234,** 981 (1981).
70F. J. P. Van Doormaal, G. D. Raithby and A. B. Strong, Prediction of natural convection in nonrectangular enclosures using orthogonal curvilinear coordinates, *Num. Heat Transfer* **4,** 21 (1981).
71F. R. Viskanta and D. W. Lankford, Coupling of heat transfer between two natural convection systems separated by a vertical wall, *Int. J. Heat Mass Transfer* **24,** 1171 (1981).
72F. J. S. Vrentas, R. Narayanan and S. S. Agrawal, Free surface convection in a bounded cylindrical geometry, *Int. J. Heat Mass Transfer* **24,** 1513 (1981).
73F. R. W. Walden and G. Ahlers, Non-Boussinesq and penetrative convection in a cylindrical cell, *J. Fluid Mech.* **109,** 89 (1981).
74F. R. O. Warrington, Jr. and G. Crupper, Jr., Natural convection heat transfer between cylindrical tube bundles and a cubical enclosure, *J. Heat Transfer* **103,** (1981).
75F. W. W. Yousef and J. D. Tarasuk, An interferometric study of combined free and forced convection in a horizontal isothermal tube, *J. Heat Transfer* **103,** 249 (1981).
76F. Y. Zvirin, P. R. Jeuck III, C. W. Sullivan and R. B. Duffey, Experimental and analytical investigation of a natural circulation system with parallel loops, *J. Heat Transfer* **103,** 645 (1981).

Natural convection—external flows

1FF. G. I. Abramov, An investigation of unsteady heat transfer of conductors, heated by a current, with free connection of cryogenic liquids, *Thermal Engng* **28,** 547 (1981).
2FF. N. Afzal, Mixed convection in a two-dimensional buoyant plume, *J. Fluid Mech.* **105,** 347 (1981).
3FF. A. V. Babezba, G. I. Gimbutis and P. P. Shvenchyanas, Heat transfer at a vertical flat surface with the combined effect of forced and free convection in the same direction, *Int. Chem. Engng* **21,** 135 (1981).
4FF. P. Beghin, E. J. Hopfinger and R. E. Britter, Gravitational convection from instantaneous sources on inclined boundaries, *J. Fluid Mech.* **107,** 407 (1981).
5FF. W. D. Bennon and F. P. Incropera, Mixed convection heat transfer from horizontal cylinders in the crossflow of a finite water layer, *J. Heat Transfer* **103,** 540 (1981).
6FF. R. D. Boyd and C. M. Vest, Laminar natural convection above a horizontal laser beam, *Int. J. Heat Mass Transfer* **24,** 685 (1981).
7FF. V. P. Carey and B. Gebhart, Visualization of the flow adjacent to a vertical ice surface melting in cold pure water, *J. Fluid Mech.* **107,** 37 (1981).
8FF. A. Choudry, Digital holographic interferometry of convective heat transport, *Appl. Optics* **20,** 1240 (1981).
9FF. A. M. Clausing and S. N. Kempka, The influence of property variations on natural convection from vertical surfaces, *J. Heat Transfer* **103,** 609 (1981).
10FF. A. S. Dawood, B. L. Manocha and S. M. J. Ali, The effect of vertical vibrations on natural convection heat transfer from a horizontal cylinder, *Int. J. Heat Mass Transfer* **24,** 491 (1981).

11FF. M. K. El-Riedy, Analogy between heat and mass transfer by natural convection from air to horizontal tubes, *Int. J. Heat Mass Transfer* **24,** 365 (1981).
12FF. B. Farouk and S. I. Güceri, Natural convection from a horizontal cylinder—laminar regime, *J. Heat Transfer* **103,** 522 (1981).
13FF. R. E. Faw and T. A. Dullforce, Holographic interferometry measurement of convective heat transport beneath a heated horizontal plate in air, *Int. J. Heat Mass Transfer* **24,** 859 (1981).
14FF. T. Fujii, M. Fujii and T. Honda, A numerical analysis of laminar free convection around an isothermal sphere, *Num. Heat Transfer* **4,** 69 (1981).
15FF. F. Geoola and A. R. H. Cornish, Numerical solution of steady-state free convective heat transfer from a solid sphere, *Int. J. Heat Mass Transfer* **24,** 1369 (1981).
16FF. A. J. Ghajar and J. D. Parker, Reference temperatures for supercritical laminar free convection on a vertical flat plate, *J. Heat Transfer* **103,** 613 (1981).
17FF. R. Guillaume, X. Garonne and G. Salaun-Penquer, Application de la methode integrale a l'etude de la zone d'establissement d'in panache thermique humide, *Int. J. Heat Mass Transfer* **24,** 563 (1981).
18FF. D. W. Hatfield and D. K. Edwards, Edge and aspect ratio effects on natural convection from the horizontal heated plate facing downwards, *Int. J. Heat Mass Transfer* **24,** 1019 (1981).
19FF. S. K. Jena and M. N. Mathur, Similarity solutions for laminar free convection flow of a thermomicropolar fluid past a non-isothermal vertical flat plate, *Int. J. Engng Sci.* **19,** 1431 (1981).
20FF. H. K. Kuiken, A backward free-convective boundary layer, *Q. J. Mech. Appl. Math.* **34,** 397 (1981).
21FF. F. N. Lin and S. Y. Chern, Laminar free convection boundary-layer heat transfer over non-isothermal surfaces, *J. Heat Transfer* **103,** 819 (1981).
22FF. M. Ljuboja and W. Rodi, Prediction of horizontal and vertical turbulent buoyant wall jets, *J. Heat Transfer* **103,** 343 (1981).
23FF. J. C. Mollendorf, R. S. Johnson and B. Gebhart, Several plume flows in pure and saline water at its density maximum, *J. Fluid Mech.* **113,** 269 (1981).
24FF. A. Moutsoglou, T. S. Chen and K. C. Cheng, Vortex instability of mixed convection flow over a horizontal flat plate, *J. Heat Transfer* **103,** 257 (1981).
25FF. R. H. Nilson, Natural convective boundary-layer on two-dimensional and axisymmetric surfaces in high-*Pr* fluids or in fluid-saturated porous media, *J. Heat Transfer* **103,** 803 (1981).
26FF. B. Sammakia and B. Gebhart, Transient natural convection adjacent to a vertical flat surface: the thermal capacity effects, *Num. Heat Transfer* **4,** 331 (1981).
27FF. P. H. Schuepp, Electrochemical simulation of heat and mass transfer in agrometeorology, *Can. J. Chem. Engng* **59,** 164 (1981).
28FF. D. J. Shlien and R. L. Boxman, Laminar starting plume temperature field measurement, *Int. J. Heat Mass Transfer* **24,** 919 (1981).
29FF. E. M. Sparrow and S. Acharya, A natural convection fin with a solution-determined nonmonotonically varying heat transfer coefficient, *J. Heat Transfer* **103,** 218 (1981).
30FF. E. M. Sparrow and G. M. Chrysler, Natural convection heat transfer coefficients for a short horizontal cylinder attached to a vertical plate, *J. Heat Transfer* **103,** 630 (1981).
31FF. E. M. Sparrow and J. E. Niethammer, Effect of vertical separation distance and cylinder-to-cylinder temperature imbalance on natural convection for a pair of horizontal cylinders, *J. Heat Transfer* **103,** 638 (1981).
32FF. E. M. Sparrow and F. Samie, Natural convection analysis of two laterally misaligned vertical plates situated one above the other, *Num. Heat Transfer* **4,** 427 (1981).
33FF. W. E. Stewart, Jr., Experimental free convection from an inclined cylinder, *J. Heat Transfer* **103,** 817 (1981).
34FF. S. Takao and U. Narusawa, Dense, vertical jets with two opposing buoyancy components, *J. Heat Transfer* **103,** 135 (1981).
35FF. N. O. Weiss, Convection in an imposed magnetic field, Part I. The development of nonlinear convection, *J. Fluid Mech.* **108,** 247 (1981).
36FF. N. O. Weiss, Convection in an imposed magnetic field. Part II. The dynamical regime, *J. Fluid Mech.* **108,** 273 (1981).
37FF. N. W. Wilson and J. J. Lee, Melting of a vertical ice wall by free convection into fresh water, *J. Heat Transfer* **103,** 13 (1981).
38FF. L. S. Yao and F. F. Chen, A horizontal flow past a partially heated, infinite vertical cylinder, *J. Heat Transfer* **103,** 546 (1981).
39FF. C-S. Yih and F. Wu, Round buoyant laminar and turbulent plumes, *Physics Fluids* **24,** 794 (1981).

Convection from rotating surfaces

1G. M. A. I. El-Shaarawi and A. Sarhan, Developing laminar free convection in an open ended vertical annulus with a rotating inner cylinder, *J. Heat Transfer* **103,** 552 (1981).
2G. H. Koyama, S-I. Ohsawa and A. Nakayama, Mass transfer from a rotating inclined plate, *J. Heat Transfer* **103,** 204 (1981).
3G. C. G. Law, Jr., P. Pierini and J. Newman, Mass transfer to rotating disks and rotating rings in laminar, transition and fully-developed turbulent flow, *Int. J. Heat Mass Transfer* **24,** 909 (1981).
4G. S. Mochizuki and W. J. Yang, Heat transfer and friction loss in laminar radial flows through rotating annular disks, *J. Heat Transfer* **103,** 212 (1981).
5G. V. M. Soundalgekar, H. S. Takhar and T. J. Smith, Effects of radial temperature gradient on the stability of viscous flow in an annulus with a rotating inner cylinder, *Wärme- und Stoffübertragung* **15,** 233 (1981).
6G. H. Takahama and H. Yokosawa, Energy separation in vortex tubes with a divergent chamber, *J. Heat Transfer* **103,** 196 (1981).
7G. C. S. Wang and R. Greif, The effect of impingement on heat transfer in rotating condensation, *Int. J. Heat Mass Transfer* **24,** 1097 (1981).

Combined heat and mass transfer

1H. A. Atimtay and W. N. Gill, The effect of free stream concentration on heat and binary mass transfer with thermodynamic coupling in forced convection on a flat

plate, *Chem. Engng Sci.* **36,** 1001 (1981).
2H. G. Bergeles, A. D. Gosman and B. E. Launder, The prediction of three-dimensional discrete hole cooling processes. Part 2. Turbulent flow, *J. Heat Transfer* **103,** 141 (1981).
3H. J. C. Besse and P. Dumargue, Etude du transfert de matière dû a une injection pariétale normale localisée dans un écoulement laminaire à grand nombre de Schmidt, *Int. J. Heat Mass Transfer* **24,** 1119 (1981).
4H. P. Crausse, G. Bacon and S. Bories, Etude fondamentale des transferts couplés chaleurmasse en milieu poreux, *Int. J. Heat Mass Transfer* **24,** 991 (1981).
5H. J. G. Hartley and W. Z. Black, Transient simultaneous heat and mass transfer in moist, unsaturated soils, *J. Heat Transfer* **103,** 376 (1981).
6H. W. Kast, The change of heat transfer and mass transfer-coefficients by simultaneous heat and mass transfer (in German), *Wärme- und Stoffübertragung* **15,** 217 (1981).
7H. T. Sano, Unsteady stagnation point heat transfer with blowing or suction, *J. Heat Transfer* **103,** 448 (1981).
8H. R. Siegel and A. Snyder, Heat transfer in cooled porous region with curved boundary, *J. Heat Transfer* **103,** 765 (1981).
9H. S. Sieniutycz, The thermodynamic stability of coupled heat and mass transfer described by linear wave equations, *Chem. Engng Sci.* **36,** 621 (1981).
10H. V. M. Soudalgekar and P. Ganesan, Finite-difference analysis of transient free convection with mass transfer on an isothermal vertical flat plate, *Int. J. Engng Sci.* **19,** 757 (1981).
11H. R. Taylor, Coupled heat and mass transfer in multicomponent systems: solution of the Maxwell–Stefan equations, *Letters Heat Mass Transfer* **8,** 405 (1981).
12H. J. E. White and C. J. Cremers, Prediction of growth parameters of frost deposits in forced convection, *J. Heat Transfer* **103,** 3 (1981).
13H. V. M. Yeroshenko, L. I. Zaichik and B. Yu Bakhvalov, Heat transfer in laminar plane channel flow with uniform suction or injection, *Int. J. Heat Mass Transfer* **24,** 1649 (1981).

Change of phase

1J. A. H. Abdelmessih, Y. Rotenberg and A. W. Neumann, An experimental study of the effects of surface characteristics and material thermal properties on dropwise condensation, *Can. J. Chem. Engng* **59,** 138 (1981).
2J. T. Adamek, Calculation of condensation rates on fluted condensing surfaces for optimizing the surface design (in German), *Wärme- und Stoffübertragung* **15,** 255 (1981).
3J. Md. Alamgir and J. H. Lienhard, Correlation of pressure undershoot during hot-water depressurization, *J. Heat Transfer* **103,** 52 (1981).
4J. R. A. M. Al-Hayes and R. H. S. Winterton, Bubble growth in flowing liquids, *Int. J. Heat Mass Transfer* **24,** 213 (1981).
5J. A. Ali and R. L. Judd, An analytical and experimental investigation of bubble waiting time in nucleate boiling, *J. Heat Transfer* **103,** 673 (1981).
6J. A. C. Alkidas, The influence of size-distribution parameters on the evaporation of polydisperse dilute sprays, *Int. J. Heat Mass Transfer* **24,** 1913 (1981).
7J. C. T. Avedisian and I. Glassman, High pressure homogeneous nucleation of bubbles within superheated binary liquid mixture, *J. Heat Transfer* **103,** 272 (1981).
8J. C. T. Avedisian and I. Glassman, Superheating and boiling of water in hydrocarbons at high pressure, *Int. J. Heat Mass Transfer* **24,** 695 (1981).
9J. J. Bandrowski and A. Kubaczka, On the condensation of multicomponent vapours in the presence of inert gases, *Int. J. Heat Mass Transfer* **24,** 147 (1981).
10J. B. T. Beck and G. L. Wedekind, A generalization of the system mean void fraction model for transient two-phase evaporating flows, *J. Heat Transfer* **103,** 81 (1981).
11J. K. M. Becker, A. Letzter, O. Nylund and B. Schölin, An experimental study of the effect of the axial heat flux distribution on the dryout conditions in a 3650-mm long annulus, *Int. J. Multiphase Flow* **7,** 47 (1981).
12J. K. Bier, J. Goetz and D. Gorenflo, Angular distribution of the heat transfer coefficient for pool boiling at horizontal tubes (in German), *Wärme- und Stoffübertragung* **15,** 159 (1981).
13J. Y. S. Cha and R. E. Henry, Bubble growth during decompression of a liquid, *J. Heat Transfer* **103,** 56 (1981).
14J. A. M. C. Chan and S. Banerjee, Refilling and rewetting of a hot horizontal tube—Part I: Experiments, *J. Heat Transfer* **103,** 281 (1981).
15J. A. M. C. Chan and S. Banerjee, Refilling and rewetting of a hot horizontal tube—Part II: Structure of a two-fluid model, *J. Heat Transfer* **103,** 287 (1981).
16J. A. M. C. Chan and S. Banerjee, Refilling and rewetting of a hot horizontal tube, *J. Heat Transfer* **103,** 653 (1981).
17J. C. C. Chen, J. V. Loh and J. W. Westwater, Prediction of boiling heat transfer duty in a compact plate-fin heat exchanger using the improved local assumption, *Int. J. Heat Mass Transfer* **24,** 1907 (1981).
18J. P. Cheng, Film condensation along an inclined surface in a porous medium, *Int. J. Heat Mass Transfer* **24,** 983 (1981).
19J. P. Cheng and A. K. Verma, The effect of subcooled liquid on film boiling about a vertical heated surface in a porous medium, *Int. J. Heat Mass Transfer* **24,** 1151 (1981).
20J. J. N. Chung and P. S. Ayyaswamy, Material removal associated with condensation on a droplet in motion, *Int. J. Multiphase Flow* **7,** 329 (1981).
21J. J. N. Chung and P. S. Ayyaswamy, Laminar condensation heat and mass transfer to a moving drop, *A.I.Ch.E. Jl* **27,** 372 (1981).
22J. M. G. Cooper and T. T. Chandratilleke, Growth of diffusion-controlled vapour bubbles at a wall in a known temperature gradient, *Int. J. Heat Mass Transfer* **24,** 1475 (1981).
23J. M. G. Cooper and C. R. Stone, Boiling of binary mixtures—study of individual bubbles, *Int. J. Heat Mass Transfer* **24,** 1937 (1981).
24J. V. K. Dhir, R. B. Duffey and I. Catton, Quenching studies on a zircaloy rod bundle, *J. Heat Transfer* **103,** 293 (1981).
25J. A. B. Didkovsky and M. K. Bologa, Vapour film condensation heat transfer and hydrodynamics under the influence of an electric field, *Int. J. Heat Mass Transfer* **24,** 811 (1981).

26J. D. M. France, R. D. Carlson, T. Chiang and W. J. Minkowycz, Critical heat flux experiments and correlation in a long, sodium-heated tube, *J. Heat Transfer* **103**, 74 (1981).
27J. A. Galati, A spring model in superheated sodium boiling: General features and comparison with an experiment test, *Nucl. Sci. Engng* **79**, 1 (1981).
28J. G. M. Harpole, Droplet evaporation in high temperature environments, *J. Heat Transfer* **103**, 86 (1981).
29J. M. Z. Hasan, M. M. Hasan, R. Eichhorn and J. H. Lienhard, Boiling burnout during crossflow over cylinders, beyond the influence of gravity, *J. Heat Transfer* **103**, 478 (1981).
30J. T. Hinrichs, E. Hennecke and H. Yasuda, The effect of plasma-deposited polymers on the nucleate boiling behavior of copper heat transfer surfaces, *Int. J. Heat Mass Transfer* **24**, 1359 (1981).
31J. A. Inoue and S. G. Bankoff, Destabilization of film boiling due to arrival of a pressure shock—Part I: Experimental, *J. Heat Transfer* **103**, 459 (1981).
32J. A. Inoue, A. Ganguli and S. G. Bankoff, Destabilization of film boiling due to arrival of a pressure shock—Part II. Analytical, *J. Heat Transfer* **103**, 465 (1981).
33J. M. K. Jensen and A. E. Bergles, Critical heat flux in helically coiled tubes, *J. Heat Transfer* **103**, 660 (1981).
34J. K. A. Joudi and D. D. James, Surface contamination, rejuvenation and the reproducibility of results in nucleate pool boiling, *J. Heat Transfer* **103**, 453 (1981).
35J. R. L. Judd, Applicability of the bubble flux density concept, *J. Heat Transfer* **103**, 175 (1981).
36J. Y. Katto, On the heat-flux/exit-quality type correlation of CHF of forced convection boiling in uniformly heated vertical tubes, *Int. J. Heat Mass Transfer* **24**, 533 (1981).
37J. Y. Katto, General features on CHF of forced convection boiling in vertical concentric annuli with a uniformly heated rod and zero inlet subcooling, *Int. J. Heat Mass Transfer* **24**, 109 (1981).
38J. Y. Katto, On the relation between critical heat flux and outlet flow pattern of forced convection boiling in uniformly heated vertical tubes, *Int. J. Heat Mass Transfer* **24**, 541 (1981).
39J. Y. Katto, General features of CHF of forced convection boiling in uniformly heated rectangular channels, *Int. J. Heat Mass Transfer* **24**, 1413 (1981).
40J. D. B. R. Kenning and V. H. Del Valle M, Fully-developed nucleate boiling: Overlap of areas of influence and interference between bubble sites, *Int. J. Heat Mass Transfer* **24**, 1025 (1981).
41J. H. Kimura and M. Ito, Evaporating heat transfer in horizontal internal spiral-grooved tubes in the region of low flow rates, *Bull. JSME* **24**, 1602 (1981).
42J. V. V. Klimenko, Film boiling on a horizontal plate—new correlation, *Int. J. Heat Mass Transfer* **24**, 69 (1981).
43J. P. I. Lebedev, S. S. Rybanin and L. N. Stesik, Heat and mass transfer of a liquid drop evaporating into a flow of gas capable of condensing, *Int. J. Heat Mass Transfer* **24**, 611 (1981).
44J. A. I. Leontiev, I. L. Mostinsky, V. S. Polonsky, M. A. Styrikovich and I. M. Chernika, Experimental investigation of the critical heat flux in horizontal channels with circumferentially variable heating, *Int. J. Heat Mass Transfer* **24**, 821 (1981).
45J. Y. Lerner, A. Sonn and R. Letan, Studies in bubble evaporation by stroboscopic photography, *Int. J. Multiphase Flow* **7**, 703 (1981).
46J. J. C. M. Leung, K. A. Gallivan, R. E. Henry and S. G. Bankoff, Critical heat flux predictions during blowdown transients, *Int. J. Multiphase Flow* **7**, 677 (1981).
47J. S. Levy and J. M. Healzer, Application of mixing length theory to wavy turbulent liquid–gas interface, *J. Heat Transfer* **103**, 492 (1981).
48J. J. H. Lienhard and A. Karimi, Homogeneous nucleation and the spinodal line, *J. Heat Transfer* **103**, 61 (1981).
49J. J. Madejski, A new criterion of dry-out in two-phase flow, *Int. J. Heat Mass Transfer* **24**, 1657 (1981).
50J. K. Mastanaiah and E. N. Ganić, Heat transfer in a two-component dispersed flow, *J. Heat Transfer* **103**, 300 (1981).
51J. A. D. Messina and E. L. Park, Jr., Effects of precise arrays of pits on nucleate boiling, *Int. J. Heat Mass Transfer* **24**, 141 (1981).
52J. W. J. Minkowycz, N. J. Rabadi and D. M. France, Dynamics of inert gas bubbles in forced convective systems, *Int. J. Heat Mass Transfer* **24**, 1433 (1981).
53J. Y. H. Mori, K. Nagai, H. Funaba and K. Komotori, Cooling of freely falling liquid drops with a shell of an immiscible volatile liquid, *J. Heat Transfer* **103**, 508 (1981).
54J. Ch. Mostofizadeh and K. Stephan, Flow and heat transfer in surface evaporation and film condensation (in German), *Wärme- und Stoffübertragung* **15**, 93 (1981).
55J. Z. Najat, Effect of density ratio on critical heat flux in closed end vertical tubes, *Int. J. Multiphase Flow* **7**, 321 (1981).
56J. L. Nghiem, H. Merte, E. R. F. Winter and H. Beer, Prediction of transient inception of boiling in terms of a heterogeneous nucleation theory, *J. Heat Transfer* **103**, 69 (1981).
57J. R. I. Nigmatulin, N. S. Khabeev and F. B. Nagiev, Dynamics, heat and mass transfer of vapour-gas bubbles in liquid, *Int. J. Heat Mass Transfer* **24**, 1033 (1981).
58J. J. Niknejad and J. W. Rose, Interphase matter transfer: an experimental study of condensation of mercury, *Proc. R. Soc.* **378**, 305 (1981).
59J. E. Oker and H. Merte, Jr., Semi-transparent gold film as simultaneous surface heater and resistance thermometer for nucleate boiling surface, *J. Heat Transfer* **103**, 65 (1981).
60J. R. W. Ostensen and R. J. Lipinski, A particle bed dryout model based on flooding, *Nucl. Sci. Engng* **79**, 110 (1981).
61J. S. J. Palaszewski, L. M. Jiji and S. Weinbaum, A three-dimensional air-vapor-droplet local interaction model for spray units, *J. Heat Transfer* **103**, 514 (1981).
62J. H. J. Palmer and J. C. Maheshri, Enhanced interfacial heat transfer by differential vapor recoil instabilities, *Int. J. Heat Mass Transfer* **24**, 117 (1981).
63J. H. S. Ragheb, S. C. Cheng and D. C. Groeneveld, Observations in transition boiling of subcooled water under forced convective conditions, *Int. J. Heat Mass Transfer* **24**, 1127 (1981).
64J. I. Kh. Rakhmatulina, Nonsteady evaporation and growth of drops in gas(eous) medium, *Int. J. Engng Sci.* **19**, 1115 (1981).

65J. A. G. Rane and S.-C. Yao, Convective heat transfer to turbulent droplet flow in circular tubes, *J. Heat Transfer* **103,** 679 (1981).
66J. J. J. Rizza, A numerical solution to dropwise evaporation, *J. Heat Transfer* **103,** 501 (1981).
67J. J. W. Rose, Dropwise condensation theory, *Int. J. Heat Mass Transfer* **24,** 191 (1981).
68J. Y. Sato, M. Sadatomi and K. Sekoguchi, Momentum and heat transfer in two-phase bubble flow. I. Theory, *Int. J. Multiphase Flow* **7,** 167 (1981).
69J. Y. Sato, M. Sadatomi and K. Sekoguchi, Momentum and heat transfer in two-phase bubble flow. II. A comparison between experimental data and theoretical calculations, *Int. J. Multiphase Flow* **7,** 179 (1981).
70J. G. Schnabel, V. Gnielinski and E. U. Schlünder, Local heat transfer coefficients in the evaporation of trickle films on vertical profile surfaces, *Chemie—Ingr—Tech.* **3,** 187 (1981).
71J. H.-G. Schrey and J. Kern, Zum Rohrreiheneffekt bei gasbeaufschlagten Kondensatoren, *Int. J. Heat Mass Transfer* **24,** 335 (1981).
72J. J. Schröppel, Impact of wall temperature variation on the film condensation of binary gas–vapour mixtures, *Int. J. Heat Mass Transfer* **24,** 165 (1981).
73J. B. Sedler and J. Mikielewicz, A simplified model of the boiling crisis, *Int. J. Heat Mass Transfer* **24,** 431 (1981).
74J. A. Segev, L. J. Flanigan, R. E. Kurth and R. P. Collier, Experimental study of countercurrent steam condensation, *J. Heat Transfer* **103,** 307 (1981).
75H. N. Seki, S. Fukusako and K. Koguchi, An experimental investigation of boiling heat transfer of fluorocarbon R-11 refrigerant for concentric-tube thermosyphon, *J. Heat Transfer* **103,** 472 (1981).
76J. N. Shankar, R. Cole and R. S. Subramanian, Thermocapillary migration of a fluid droplet inside a drop in a space laboratory, *Int. J. Multiphase Flow* **7,** 581 (1981).
77J. I. G. Shekriladze, Developed boiling heat transfer, *Int. J. Heat Mass Transfer* **24,** 795 (1981).
78J. S. Chunkuan and M. M. El-Wakil, Film boiling and vapor explosions from small spheres, *Nucl. Sci. Engng* **77,** 470 (1981).
79J. H. Sofrata, Improved theoretical model for wavy filmwise condensation, *Wärme- und Stoffübertragung* **15,** 117 (1981).
80J. C. W. Sohn and M. M. Chen, Microconvective thermal conductivity in disperse two-phase mixtures as observed in a low velocity in disperse two-phase mixtures as observed in a low velocity Couette flow experiment, *J. Heat Transfer* **103,** 47 (1981).
81J. K. Stephan and H. Auracher, Correlations for nucleate boiling heat transfer in forced convection, *Int. J. Heat Mass Transfer* **24,** 99 (1981).
82J. B. Sudhoff, M. Phschke and P. M. Weinspach, The direct heat transfer in the condensation or evaporation of bubble drops, *Chemie–Ingr–Tech.* **6,** 471 (1981).
83J. K. H. Sun, R. B. Duffey and C. Lin, A nondimensional analysis of boiling dry a vertical channel with a uniform heat flux, *J. Heat Transfer* **103,** 667 (1981).
84J. K. H. Sun, R. B. Duffey and C. M. Peng, The prediction of two-phase mixture level and hydrodynamically-controlled dryout under low flow conditions, *Int. J. Multiphase Flow* **7,** 521 (1981).
85J. H. Tanaka, Effect of Knudson number on dropwise condensation, *J. Heat Transfer* **103,** 606 (1981).
86J. J. R. Thome and G. Davey, Bubble growth rates in liquid nitrogen, argon and their mixtures, *Int. J. Heat Mass Transfer* **24,** 89 (1981).
87J. M. Trela, An approximate calculation of heat transfer during flow of an air–water mist along a heated flat plate, *Int. J. Heat Mass Transfer* **24,** 749 (1981).
88J. A. Z. Trifonov, B. M. Nikolova, M. D. Mikhailov, B. K.Shishedjiev and R. B. Kuzmanova, Identification of the mass transfer coefficient at the liquid–liquid interface based on spectroscopic (ATR) data and mathematical modeling, *Wärme- und Stoffübertragung* **15,** 59 (1981).
89J. T. Ueda, M. Inoue and S. Nagatome, Critical heat flux and droplet entrainment rate in boiling of falling liquid films, *Int. J. Heat Mass Transfer* **24,** 1257 (1981).
90J. T. Ueda and Y. Isayama, Critical heat flux and exit film flow rate in a flow boiling system, *Int. J. Heat Mass Transfer* **24,** 1267 (1981).
91J. H. C. Ünal, Density-wave oscillations in sodium heated once-through steam generator tubes, *J. Heat Transfer* **103,** 485 (1981).
92J. H. C. Ünal, M. L. G. van Gasselt and P. M. van't Verlatt, Dryout and two-phase flow pressure drop in sodium heated helically coiled steam generator tubes at elevated pressures, *Int. J. Heat Mass Transfer* **24,** 285 (1981).
93J. L. Y. Wagner and P. S. Lykoudis, Mercury pool boiling under the influence of a horizontal magnetic field, *Int. J. Heat Mass Transfer* **24,** 635 (1981).
94J. R. L. Webb, The evolution of enhanced surface geometries for nucleate boiling, *Heat Transfer Engng* **2,** 46 (1981).
95J. D. R. Webb and R. G. Sardesai, Verification of multicomponent mass transfer models for condensation inside a vertical tube, *Int. J. Multiphase Flow* **7,** 507 (1981).
96J. D. W. Woodruff and J. W. Westwater, Steam condensation on various gold surfaces, *J. Heat Transfer* **103,** 685 (1981).
97J. S.-C. Yao and A. Rane, Numerical study of turbulent droplet flow heat transfer, *Int. J. Heat Mass Transfer* **24,** 785 (1981).
98J. S. A. Zhukov, V. V. Barelko and A. G. Merzhanov, Wave processes on heat generating surfaces in pool boiling, *Int. J. Heat Mass Transfer* **24,** 47 (1981).

Radiation in participating media

1K. K. S. Adzerikho and V. I. Antsulevich, Study of radiation properties of a plane inhomogeneous two-phase medium, *Int. J. Heat Mass Transfer* **24,** 57 (1981).
2K. F. H. Azad and M. F. Modest, Evaluation of the radiative heat flux in absorbing, emitting and linear-anisotropically scattering cylindrical media, *J. Heat Transfer* **103,** 350 (1981).
3K. F. H. Azad and M. F. Modest, Combined radiation and convection in absorbing, emitting and anisotropically scattering gas-particulate tube flow, *Int. J. Heat Mass Transfer* **24,** 1681 (1981).
4K. G. Diener, A derivation of the radiation transfer theory for random media, *Physica* **106A,** 598 (1981).
5K. M. Epstein, F. B. Cheung, T. C. Chawla and G. M.

Hauser, Effective thermal conductivity for combined radiation and free convection in an optically thick heated fluid layer, *J. Heat Transfer* **103,** 114 (1981).
6K. I. H. Farag and T. A. Allam, Gray-gas approximation of carbon dioxide standard emissivity, *J. Heat Transfer* **103,** 403 (1981).
7K. J. D. Felske and K. M. Lee, Non-gray particulate radiation in an isothermal cylindrical medium, *J. Heat Transfer* **103,** 121 (1981).
8K. B. D. Ganapol and G. Spiga, Time-dependent radiative transfer in a semi-infinite medium with a reflecting boundary, *Z. Angew. Math. Phys.* **32,** 555 (1981).
9K. G. A. Levin, The drift of foreign particles in a radiation-absorbing non-equilibrium gas, *J. Phys. D: Appl. Phys* **14,** L167 (1981).
10K. D. C. Look, H. F. Nelson and A. L. Crosbie, Anisotropic two-dimensional scattering: comparison of experiment with theory, *J. Heat Transfer* **103,** 127 (1981).
11K. A. Matsuzaki, Y. Nakamura and T. Itoh, Rocket-borne spectrometer to measure the near infrared absorption spectrum of the upper atmosphere, *Rev. Scient. Instrum.* **52,** 1685 (1981).
12K. E. Pruckner, The influence of the net radiation flux on the calculation of heat and mass transfer coefficient for rivers (in German), *Wärme- und Stoffübertragung* **15,** 181 (1981).
13K. M. N. Rolin, R. I. Soloukhin and F. B. Yurevich, Heat and mass transfer in an emitting compressed layer with reflection from the body surface and injection of ablation products, *Int. J. Heat Mass Transfer* **24,** 1771 (1981).
14K. H. K. Roscoe, J. R. Drummond and R. F. Jarnot, Infrared measurements of stratospheric composition. III. The daytime changes of NO and NO_2, *Proc. R. Soc.* **375A,** 507 (1981).
15K. K. Sassen, Infrared (10 μm) radiation induced evaporation of large water drops, *J. Opt. Soc. Am.* **71,** 887 (1981).
16K. C. Stramigioli, F. Santarelli, P. Vestrucci and G. Spiga, Hemispherical transmissivity and reflectivity for non-aximuthally symmetric radiative transfer in a plane slab, *Letters Heat Mass Transfer* **8,** 483 (1981).
17K. S. G. Tukhvatullin, L. M. Vasil'eva and N. Kh Akhunov, An experimental investigation of thermal radiation of certain gaseous hydrocarbons, *Thermal Engng* **28,** 242 (1981).
18K. W. W. Yuen, A simplified approach to the evaluation of the geometric-mean transmittance and absorptance for gas enclosures, *J. Heat Transfer* **103,** 808 (1981).
19K. W. W. Yuen and D. J. Rasky, Application of the $P-1$ approximation to radiative heat transfer in a nongray medium, *J. Heat Transfer* **103,** 182 (1981).

Surface radiation
1L. N. Azia and T. Y. Na, Perturbation analysis for periodic heat transfer in radiating fins, *Wärme- und Stoffübertragung* **15,** 245 (1981).
2L. U. Gross, K. Spindler and E. Hahne, Shape factor-equations for radiation heat transfer between plane rectangular surfaces of arbitrary position and size with parallel boundaries, *Letters Heat Mass Transfer* **8,** 219 (1981).
3L. Y. Ohwada, Numerical calculation of effective emissivities of diffuse cones with a series technique, *Appl. Optics* **20,** 3332 (1981).
4L. A. Richardson, Measurement of reflectance factors under daily and intermittent irradiance variations, *Appl. Optics* **20,** 3336 (1981).

MHD
1M. R. K. Ahluwalia and K. H. Im, Heat transfer scaling laws for MHD channels and diffusers, *J. Energy* **5,** 315 (1981).
2M. R. K. Ahluwalia and K. H. Im, Combined conduction, convection, gas radiation and particles radiation in MHD diffusers, *Int. J. Heat Mass Transfer* **24,** 1421 (1981).
3M. K. H. Im and R. K. Ahluwalia, Heat and mass transfer in MHD channels, *J. Energy* **5,** 22 (1981).
4M. K. H. Im and R. K. Ahluwalia, Convective and radiative heat transfer in MHD radiant boilers, *J. Energy* **5,** 308 (1981).
5M. T. R. Johnson, G. F. Berry and M. Petrick, Development of steam generator components for open-cycle MHD, *J. Energy* **5,** 208 (1981).
6M. R. Kessler, Open-cycle MHD generator channel development, *J. Energy* **5,** 178 (1981).
7M. M. Singh and V. M. Soundalgekar, Forced and free convective magnetohydrodynamic flow between vertical conducting walls, *Nucl. Sci. Engng* **79,** 374 (1981).
8M. T. F. Smith, F. Alipour-Haghighi and P. H. Paul, Turbulent MHD flow of a radiating gas, *Int. J. Heat Mass Transfer* **24,** 659 (1981).
9M. H. B. Urbach, J. G. Purnell and D. T. Knauss, Performance of a multiphase boiling MHD generator, *J. Engng Pwr* **103,** 582 (1981).

Numerical methods
1N. B. F. Blackwell, Efficient techniques for the numerical solutions of the one-dimensional inverse problem of heat conduction, *Num. Heat Transfer* **4,** 229 (1981).
2N. K. C. Chung, A generalized finite-difference method for heat transfer problems of irregular geometries, *Num. Heat Transfer* **4,** 345 (1981).
3N. C-J. Chen, H. Naeri-Neshat and K-S. Ho, Finite-analysis numerical solution of heat transfer in two-dimensional cavity flow, *Num. Heat Transfer* **4,** 179 (1981).
4N. A. Cichy, S. Lasocki and H. Marcak, Efficient numerical technique for determining two-dimensional temperature distribution for *N* arbitrarily located heat receivers with phase-change, *Int. J. Heat Mass Transfer* **24,** 1573 (1981).
5N. M. W. Crump and J. C. Lee, Nonlinear transient analysis of light water reactor steam generators using an implicit Eulerian method, *Nucl. Sci. Engng* **77,** 192 (1981).
6N. R. F. Diaz-Munio and L. C. Wellford, Jr., A finite-element singular-perturbation technique for convection–diffusion problems. Part 1. The one-dimensional case, *J. Appl. Mech.* **48,** 265 (1981).
7N. R. F. Diaz-Munio and L. C. Wellford, Jr., A finite-element singular-perturbation technique for convection–diffusion problems. Part 2. Two-dimensional problems, *J. Appl. Mech.* **48,** 272 (1981).

8N. G. Grötzbach, Numerical simulation of turbulent temperature fluctuations in liquid metals, *Int. J. Heat Mass Transfer* **24,** 475 (1981).
9N. A. Munier, J. R. Burgan, J. Gutierrez, E. Fijalkow and M. R. Feix, Group transformations and the nonlinear heat diffusion equation, *SIAM Appl. Math.* **40,** 191 (1981).
10N. R. S. Gupta and D. Kumar, Variable time step methods for one-dimensional Stefan problem with mixed boundary condition, *Int. J. Heat Mass Transfer* **24,** 251 (1981).
11N. S. V. Patankar, A calculation for two-dimensional elliptic situations, *Num. Heat Transfer* **4,** 409 (1981).
12N. M. I. Pavlovic and D. Z. Milojevic, A variational approach to the problem of stationary laminar convection with dissipation in the variable property fluid, *Int. J. Heat Mass Transfer* **24,** 343 (1981).
13N. T-H. Shih, Computer-extended series: natural convection in a long horizontal pipe with different end temperatures, *Int. J. Heat Mass Transfer* **24,** 1295 (1981).
14N. T. M. Shih and H. J. Huang, Numerical method for solving nonlinear ordinary and partial differential equations for boundary-layer flows, *Num. Heat Transfer* **4,** 159 (1981).
15N. J. Sucec, An improved quasi-steady approach for transient conjugated forced convection problems, *Int. J. Heat Mass Transfer* **24,** 1711 (1981).
16N. Modelling radiation interception by strongly clumped vegetation—a combined Monte Carlo and analytical model, *Can. J. Chem. Engng* **59,** 195 (1981).
17N. K. C. Woo and L. C. Chow, Inverse heat conduction by direct inverse Laplace transform, *Num. Heat Transfer* **4,** 499 (1981).
18N. L. S. Yao and W. Cherney, Transient phase-change around a horizontal cylinder, *Int. J. Heat Mass Transfer* **24,** 1971 (1981).

Heat transfer applications—heat exchangers and heat pipes

1Q. J. C. Biery, Prediction of heat transfer coefficients in gas flow normal to finned and smooth tube banks, *J. Heat Transfer* **103,** 705 (1981).
2Q. W. G. Characklis, M. J. Nimmons and B. F. Picologlou, Influence of fouling biofilms on heat transfer, *Heat Transfer Engng* **3,** 23 (1981).
3Q. O. P. Chawla and A. I. Khandwawala, Thermal performance of regenerators and waste heat recovery, *Int. J. Heat Mass Transfer* **24,** 1793 (1981).
4Q. C. J. Camarda and R. V. Masek, Design, analysis, and tests of a shuttle-type-heat-pipe-cooled leading edge, *J. Spacecraft Rockets* **18,** 71 (1981).
5Q. C. C. Chen, J. V. Loh and J. W. Westwater, Prediction of boiling heat transfer duty in a compact plate-fin heat exchanger using the improved local assumption, *Int. J. Heat Mass Transfer* **24,** 1907 (1981).
6Q. E. Gianolio and F. Cuti, Heat transfer coefficients and pressure drops for air coolers with different numbers of rows under induced and forced draft, *Heat Transfer Engng* **3,** 38 (1981).
7Q. C. Hinchcliffe and A. J. Willmott, Lumped heat-transfer coefficients for thermal regenerators, *Int. J. Heat Mass Transfer* **24,** 1229 (1981).
8Q. R. Letan, A parametric study of a particulate direct contact heat exchanger, *J. Heat Transfer* **103,** 586 (1981).
9Q. Y. Mori, K. Hijikata, S. Hirawawa and W. Nakayama, Optimized performance of condensers with outside condensing surfaces, *J. Heat Transfer* **103,** 96 (1981).
10Q. S. Morooka, T. Kuroki and T. Waki, Study on the heat pipe (3rd report, heat transfer mechanism in the evaporator of heat pipe with screen mesh wick), *Bull. JSME* **24,** 1811 (1981).
11Q. F. L. Rubin, Multizone condensers: Desuperheating, condensing, subcooling, *Heat Transfer Engng* **3,** 49 (1981).
12Q. K. P. Singh, Some fundamental relationships for tubular heat exchanger thermal performance, *J. Heat Transfer* **103,** 573 (1981).
13Q. K. P. Singh, On the necessary criteria for stream-symmetric tubular heat exchanger geometries, *Heat Transfer Engng* **3,** 19 (1981).
14Q. R. Smyth, A comparative assessment of rodbaffle shell-and-tube heat exchangers, *Heat Transfer Engng* **2,** 90 (1981).
15Q. K. Stephan and K. D. Stopka, Heat transfer in tubeless evaporators, *Chemie–Ingr–Tech.* **10,** 802 (1981).
16Q. H. Terasaka, H. Kanoh and M. Masubuchi, Parameter identification of heat exchangers by the method of stochastic approximation, *Bull. JSME* **24,** 1058 (1981).
17Q. J. A. Valenzuela and L. R. Glickman, Modeling of a rotary dry cooling tower, *J. Heat Transfer* **103,** 715 (1981).
18Q. L. G. van Leersum and C. W. Ambrose, Comparisons between experiments and a theoretical model of heat and mass transfer in rotary regenerators and nonabsorbing matrices, *J. Heat Transfer* **103,** 189 (1981).
19Q. R. L. Webb, Performance evaluation criteria for enhanced heat transfer surfaces in heat exchanger design, *Int. J. Heat Mass Transfer* **24,** 715 (1981.

General heat transfer applications

1S. P. M. Abdul Majeed, Analysis of heat transfer during hydrair cooling of spherical food products, *Int. J. Heat Mass Transfer* **24,** 323 (1981).
2S. K. Abu-Abdou and T. Hussien, Temperature distribution in a storage tank, *Wärme- und Stoffübertragung* **15,** 271 (1981).
3S. M. J. Adams and H. Krier, Unsteady internal boundary layer analysis applied to gun barrel wall heat transfer, *Int. J. Heat Mass Transfer* **24,** 1925 (1981).
4S. A. K. Agrawal, I. K. Madni, J. G. Guppy and W. L. Weaver III, Dynamic simulation of LMFBR plant under natural circulation, *J. Heat Transfer* **103,** 312 (1981).
5S. H. Beisswenger, G. Daradimos, K. Janssen and V. Peterson, Circulating fluidized bed as high-temperature reactor, *Chemie–Ingr–Tech.* **4,** 289 (1981).
6S. R. Best, Calculation of the disturbance of the combustion chamber film cooling through air jet injection by means of an analog consideration between the air jet and a cylinder (in German), *Wärme- und Stoffübertragung* **15,** 79 (1981).

7S. J. A. Caton and J. B. Heywood, An experimental and analytical study of heat transfer in an engine exhaust port, *Int. J. Heat Mass Transfer* **24,** 581 (1981).

8S. L. C. Daniels and W. B. Browne, Calculations of heat transfer rates to gas turbine blades, *Int. J. Heat Mass Transfer* **24,** 871 (1981).

9S. P. Egerton, J. A. Howarth and G. Poots, The effect of nozzle erosion on heat transfer in a ladle of molten steel during pouring, *Int. J. Heat Mass Transfer* **24,** 557 (1981).

10S. O. N. Favorskii and S. Z. Kopelev, Air cooling of gas turbine blades, *Thermal Engng* **28,** 435 (1981).

11S. M. Fortes and M. R. Okos, Heat and mass transfer in hygroscopic capillary extruded products, *A.I.Ch.E. Jl* **27,** 255 (1981).

12S. M. Furukawa, Design and off-design performance calculations of space radiators, *J. Spacecraft Rockets* **18,** 515 (1981).

13S. A. Gany, L. H. Cavany and J. W. Johnson, Water vapor contribution to the erosion of steel by high temperature flows, *J. Heat Transfer* **103,** 383 (1981).

14S. W. M. Grissom and F. A. Wierum, Liquid spray cooling of a heated surface, *Int. J. Heat Mass Transfer* **24,** 261 (1981).

15S. B. G. Jenkins and F. D. Moles, Modelling of heat transfer from a large enclosed flame in a rotary kiln, *Trans. Inst. Chem. Engrs* **59,** 17 (1981).

16S. V. M. Kapinos and A. F. Slitenko, Determination of average coefficients of heat transfer from turbine blades, *Thermal Engng* **28,** 458 (1981).

17S. M. Kawahashi and M. Suzuki, Temperature separation produced by a Hartmann–Sprenger tube couple a secondary resonator, *Int. J. Heat Mass Transfer* **24,** 1951 (1981).

18S. F. J. Kitte and I. Teipel, Transport phenomena for pseudo-plastic fluids (in German), *Wärme- und Stoffübertragung* **15,** 41 (1981).

19S. D. Meinkohn, Heat explosion theory and vibrational heating of polymers, *Int. J. Heat Mass Transfer* **24,** 645 (1981).

20S. H. G. Price, Cooling of high-pressure rocket thrust chambers with liquid oxygen, *J. Spacecraft Rockets* **18,** 338 (1981).

21S. A. L. Yarden, Large-scale heat transfer research and testing, *Heat Transfer Engng* **2,** 70 (1981).

22S. L. D. Zaberzhinskii, V. M. Pashnin and V. L. Pokhoriler, Investigation of heat-transfer conditions in casings of turbines with a loop system of steam flows, *Thermal Engng* **28** 461 (1981).

Solar energy

1T. C. G. Adler, J. W. Byrd and B. L. Coulter, The second law efficiency of solar heating, *Sol. Energy* **26,** 553 (1981).

2T. V. K. Agarwal and D. C. Larson, Calculation of the top loss coefficient of a flat-plate collector, *Sol. Energy* **27,** 69 (1981).

3T. J. E. Braun, S. A. Klein and J. W. Mitchell, Seasonal storage of energy in solar heating, *Sol. Energy* **26,** 403 (1981).

4T. J. W. Cochrane and H. J. Goldsmid, Enhanced air flow over cooling fins in solar energy convertors, *Sol. Energy* **27,** 89 (1981).

5T. P. I. Cooper and R. V. Dunkle, A non-linear flat-plate collector model, *Sol. Energy* **26,** 265 (1981).

6T. R. Gani, D. Proctor and J. G. Symons, Linear efficiency characteristics for high temperature flat-plate collectors, *Sol. Energy* **26,** 271 (1981).

7T. J. M. Gordon, On non-linear effects in flat-plate efficiency curves, *Sol. Energy* **26,** 265 (1981).

8T. J. M. Gordon, D. Govaer and Y. Zarmi, Temperature-dependent collector properties from stagnation measurements, *Sol. Energy* **25,** 465 (1980).

9T. D. Y. Goswami, E. E. Klett, E. K. Stafnakos and T. K. Goswami, Seasonal variations of atmospheric clearness numbers for use in solar radiation modelling, *J. Energy* **5,** 185 (1981).

10T. B. M. Herman, M. A. Box, J. A. Reagan and D. M. Evans, Alternate approach to the analysis of solar-photometer data, *Appl. Optics* **20,** 2925 (1981).

11T. C. K. Hsieh, Thermal analysis of CPC collectors, *Sol. Energy* **27,** 19 (1981).

12T. K. Katayama, A. Saito, Y. Utaka, A. Saito, H. Matsui, H. Maekawa and A. Z. A. Saifullah, Heat transfer characteristics of the latent heat thermal energy storage capsule, *Sol. Energy* **27,** 91 (1981).

13T. C. F. Kool, Salt gradient solar pond with reflective bottom: application to the "saturated" pond, *Sol. Energy* **26,** 113 (1981).

14T. P. J. Lunde, New heat transfer factors for flat plate solar collectors, *Sol. Energy* **27,** 109 (1981).

15T. A. Malhotra, H. P. Gard and U. Rani, Minimizing convective heat losses in flat plate solar collectors, *Sol. Energy* **25,** 521 (1980).

16T. R. Manvi, T. Fujita, B. C. Gajanana and C. J. Marcus, Thermal buffering of receivers for parabolic dish solar thermal power plants, *J. Energy* **5,** 381 (1981).

17T. G. M. Naidu and J. P. Agarwal, A theoretical study of heat transfer in a flat plate solar collector, *Sol. Energy* **26,** 313 (1981).

18T. W. A. Monsen, S. A. Klein and W. A. Beckman, Prediction of direct gain solar heating system performance, *Sol. Energy* **27,** 143 (1981).

19T. G. L. Morrison, Passive energy storage in greenhouses, *Sol. Energy* **25,** 365 (1980).

20T. W. F. Phillips, Effects of stratification on the performance of solar air heating systems, *Sol. Energy* **26,** 175 (1981).

21T. A. Rabl, J. O'Gallagher and R. Winston, Design and test of non-evacuated solar collectors with compound parabolic concentrators, *Sol. Energy* **25,** 335 (1980).

22T. T. D. M. A. Samuel and N. E. Wijeysundera, Optimum performance of thermal trap collectors, *Sol. Energy* **26,** 65 (1980).

23T. J. M. Schreyer, Residential application of refrigerant-charged solar collectors, *Sol. Energy* **26,** 307 (1981).

24T. T. F. Smith, P. A. Jensen and D. L. Spencer, Thermal performance of the distributed subatmospheric pressure, flat plate solar collector, *Sol Energy* **25,** 429 (1980).

25T. M. S. Sodha, J. K. Nayak, U. Singh and G. N. Tiwari, Thermal performance of a solar still, *J. Energy* **5,** 331 (1981).

26T. M. Sokolov and C. Saitiel, Numerical evaluation of radiation absorption by a fluid confined in a semi-transparent circular cylinder, *Sol. Energy* **27,** 31 (1981).

27T. I. S. Taha and S. M. Eldighidy, Analysis of outdoor solar storage tanks having translucent insulation, *Sol. Energy* **27**, 45 (1981).
28T. W. R. L. Thomas, Sensitivity analysis of a solar-driven chemical heat pump system, *J. Energy* **5**, 214 (1981).
29T. D. M. Utzinger, S. A. Klein and J. W. Mitchell, The effect of air flow rate in collector-storage walls, *Sol. Energy* **25**, 511 (1980).
30T. P. S. Wu and D. K. Edwards, Effect of combined tilt and end clearance upon natural convection in high *L/D* rectangular honeycomb, *Sol. Energy* **25**, 471 (1980).

Plasma heat transfer

1U. P. F. Bulanyi and S. P. Polyakov, Experimental investigation of heat-flux densities and electric current on anode of a high-current stabilized arc, *Teplofizika Vysokikh Temperatur* **19**, 497 (1981).
2U. D. M. Chen, K. C. Hsu and E. Pfender, Two-temperature modeling of an arc plasma reactor, *Plasma Chem. Plasma Proc.* **1**, 295 (1981).
3U. D. M. Chen and E. Pfender, Two-temperature modeling of the anode contraction region of high-intensity arcs, *IEEE Trans. Plasma Sci.* **PS-9**, 265 (1981).
4U. R. M. Clements and P. R. Smy, Transition from diffusion–convection to sheath-convection of a cold Langmuir probe in a moving compressible plasma, *J. Phys. D: Appl. Phys.* **14**, 1001 (1981).
5U. R. G. Evans, Radiation cooling instabilities in laser-heated plasma, *J. Phys. D: Appl. Phys.* **14**, L173 (1981).
6U. T. V. Gavrilova, Total emission coefficient of a dense air plasma, *Teplofizika Vysokikh Temperatur* **19**, 449 (1981).
7U. D. Gold, C. Bonet, G. Chauvin, A. C. Mathieu, G. Geirnaert and J. Millet, A 100 kW three-phase ac plasma furnace for spheroidization of aluminum silicate particles, *Plasma Chem. Plasma Proc.* **1**, 161 (1981).
8U. T. Honda, T. Hayashi and A. Kanzawa, Heat transfer from rarefied ionized argon gas to a biased tungsten fine wire, *Int. J. Heat Mass Transfer* **24**, 1247 (1981).
9U. B. Jüttner, Formation time and heating mechanism of arc cathode craters in vacuum, *J. Phys. D: Appl. Phys.* **14**, 1265 (1981).
10U. S. A. Khan and T. D. Rognlien, Thermal heat flux in a plasma for arbitrary collisionality, *Physics Fluids* **24**, 1442 (1981).
11U. O. Matsumoto, M. Konuma and Y. Kanzaki, Heating and quenching of metastable superconducting compounds with a plasma jet, *Plasma Chem. Plasma Proc.* **1**, 73 (1981).
12U. D. K. McLain and R. J. Zollweg, Convection in horizontal high-pressure mercury, mercury plus iodine, and metal halide additive arcs, *J. Appl. Phys.* **52**, 199 (1981).
13U. D. Morvan and J. Amouroux, Preparation of photovoltaic silicon by purification of metallurgical grade silicon with a reactive plasma process, *Plasma Chem. Plasma Proc.* **1**, 397 (1981).
14U. Y. Nakamura, M. Ito and H. Ishikawa, Reduction and dephosphorization of molten iron oxide with hydrogen–argon plasma, *Plasma Chem. Plasma Proc.* **1**, 149 (1981).
15U. A. E. Nasser, H. M. Sofrata and A. Sharaf, Transport phenomena of mixtures of hydrogen, oxygen and argon at high temperatures, *Wärme- und Stoffübertragung* **15**, 135 (1981).
16U. K. Ragaller and D. T. Tuma, Similarity relations for the electric arc in forced axial flow, *IEEE Trans. Plasma Sci.* **PS-9**, 75 (1981).
17U. P. Ronsheim, A. Mazza and A. N. Christensen, Thermal plasma synthesis of transition metal nitrides and alloys, *Plasma Chem. Plasma Proc.* **1**, 135 (1981).
18U. N. N. Rykalin, A. A. Uglov, Yu. N. Lokhov and A. G. Gnedovets, Properties of heating of submicron metal particles in a hot gas, *Teplofizika Vysokikh Temperatur* **19**, 557 (1981).
19U. W. J. van den Hoek and J. A. Visser, Diagnostics of high-pressure arc plasmas from laser-induced fluorescence, *J. Phys. D: Appl. Phys.* **14**, 1613 (1981).
20U. Y. Yasutomo, K. Miyata, S-I. Himeno, T. Enoto and Y. Ozawa, A new numerical method for asymmetrical Abel inversion, *IEEE Trans. Plasma Sci.* **PS-9**, 18 (1981).
21U. T. Yoshida, K. Nakagawa, T. Harada and K. Akashi, New design of a radio-frequency plasma torch, *Plasma Chem. Plasma Proc.* **1**, 113 (1981).

HEAT TRANSFER REVIEW - 1982

E. R. G. ECKERT, R. J. GOLDSTEIN, E. PFENDER, W. E. IBELE, S. V. PATANKAR, J. W. RAMSEY, T. W. SIMON, and N. A. DECKER

Heat Transfer Laboratory, Department of Mechanical Engineering, University of Minnesota, Minneapolis, Minnesota, U.S.A.

INTRODUCTION

THIS review surveys papers that have been published in the open literature covering various fields of heat transfer during 1982. As in the past, the number of papers published during that period was such that a selection only could be included.

A number of conferences dealing with heat transfer or including in their sessions some on heat transfer have been held during 1982.

The 24th Israel Annual Conference on Aviation and Astronautics held on 17 and 18 February 1982 at Tel-Aviv, included in its program papers on turbulence models to predict heat transfer and on film cooling of turbine blades. The papers will appear in the *Israel Journal of Technology*.

The 27th International Gas Turbine Conference and Exhibit held on 18–22 April 1982 in London, U.K., had sessions on external and internal blade heat transfer, on gas turbine waste heat recovery, on ceramic coatings for gas turbines, on film and on transpiration cooling. Copies of the ASME papers may be obtained by writing to the ASME Order Department.

The 3rd Joint AIAA/ASME Thermophysics, Fluids, Plasma, and Heat Transfer Conference, held on 7–11 June, in St. Louis, Missouri, devoted sessions to numerical heat transfer techniques, radiative heat transfer, thermal storage heat transfer, high temperature/high heat flux heat transfer, convective mass transfer, and to heat transfer in boundary layers. Papers are available at AIAA and ASME Headquarters.

The 17th Intersociety Energy Conversion Engineering Conference, held on 8–12 August, in Los Angeles, California, covered solar collector components, solar heating and cooling, and thermal energy storage in three sessions.

The big event of the year was the 7th International Heat Transfer Conference held during 6–10 September at the Technical University, Munich, Federal Republic of Germany. Welcoming addresses by representatives of the Assembly of International Heat Transfer Conferences, of the Bavarian government, and of the Technical University as well as an address by E. R. G. Eckert on "A Pioneering Era in Convective Heat Transfer Research", opened the conference. Technical sessions on all aspects of heat transfer were devoted to the presentation and discussion of more than 450 papers. Twenty-one keynote papers provided a detailed review of the present state of knowledge in flow boiling, heat transfer in turbulent flows, boiling of mixtures, numerical methods and visualization of heat transfer, radiative heat transfer, advances in compact heat exchanger technology, prediction of transport properties, boundary layer theory, forced convection heat transfer, particle heat transfer, heat transfer in combustion chambers, mass transfer in rheological systems, heat and mass transfer in rivers, cryogenic heat transfer, buoyancy effects in forced channel flow, natural convection in cavities and cells, heat transfer in the atmosphere, nuclear safety heat transfer, and heat transfer enhancement. The Max Jakob Award was presented to Chang-Lin Tien during the conference. The number of papers covered during the conference was such that they could not be included in the survey. They are, however, available as bound volumes in the *Proceedings—Heat Transfer 1982—München* from Hemisphere, Washington, D.C.

The 1982 Symposium of the International Centre for Heat and Mass Transfer, 30 August–3 September 1982 had as its topic heat and mass transfer in rotating machinery with ten invited lectures and 61 papers. The Proceedings of the Symposium are available from Hemisphere, Washington, DC.

The ASME 103rd Winter Annual Meeting, held on 14–19 November at Phoenix, Arizona, included in its program 19 sessions on various aspects of heat transfer including combustion-sprays, coal combustion, ice formation and arctic heat transfer, interaction of water chemistry and heat transfer in boilers and evaporators.

The Heat Transfer Memorial Award was presented to Yasuo Mori of the Tokyo Institute of Technology, Tokyo, Japan. Preprints of the papers are available at ASME Headquarters and many of the papers will also be published in the *Journal of Heat Transfer*.

A number of short courses were organized during 1982, among them a lecture series on Film Cooling and Turbine Blade Heat Transfer, held on 22–26 February at the Von Karman Institute for Fluid Dynamics, Belgium. Short courses on Computational Fluid Mechanics and Heat Transfer were held at the University of Minnesota and at the Pennsylvania State University, and a short course on Measurement of Building Fire Safety was held at the University of Wisconsin.

A considerable number of books dealing with heat transfer or including heat transfer topics have appeared on the market in 1982. They are listed in the bibliographic portion of this review. Of special interest to engineers will be the *Heat Exchanger Design Handbook* published at the beginning of 1983 by Hemisphere, Washington, DC. Volumes with the titles: heat exchanger theory, fluid mechanics and heat transfer, thermal and hydraulic design of heat exchangers, mechanical design of heat exchangers, and physical properties collect up-to-date information conveniently in five volumes. The *International Journal of Heat and Fluid Flow* resumed its quarterly publication in March, 1982.

The following highlights characterize the activity in heat transfer research in 1982.

Nearly half of the papers considering heat conduction were concerned with the presence of phase change. Some studies approached the problem analytically seeking verification of the models by comparison with suitable experimental data. Others report the results of experiments for specific systems. Many papers appear to have been prompted by a concern with actual heat transfer problems like freezing or thawing of soils, metal casting, or thermal oil recovery.

Channel flow papers dealt with uncommon property effects like non-Newtonian and low Prandtl number fluids, strongly available of properties and freezing. Turbulent flow studies included effects of flow separation.

Forced and natural flow and heat transfer were studied in saturated and unsaturated porous media and in fluidized beds. Attention in natural convection centered on heat transfer in horizontal layers heated from below or steady laminar or turbulent convection for constant and variable property fluids. Considerable interest is still shown in film cooling heat transfer.

The fundamental aspects of boiling heat transfer still need clarification. Studies in this direction covered nucleation at artificially produced sites, bubble growth, stability of film boiling and details of transition boiling. Papers dealing with applications centered on critical heat flux and post-CHF heat transfer or augmentation of evaporative heat transfer to strongly wetting refrigerants by use of finned tubes or tubes with porous coatings. Evaporation rates of liquid droplets in direct contact heat exchangers, in mist flows, and in electric fields as well as boiling in porous media have been examined.

Condensation has now become a sufficiently large topic to be treated separately with many of the papers dealing with the application to heat exchangers. Other papers correlated heat transfer behavior in liquid films with the state of the film. As a new topic, the collapse of a submerged vapor sheet in a sub-cooled liquid pool was investigated.

There is considerable interest in radiative transport in emitting, absorbing, and scattering media including surface effects.

Heat exchanger studies concentrated on augmentation by special surface shapes and flow arrangements as well as on optimization of complex heat exchanger systems. There were only a few papers on heat pipes. Heat transfer in new energy devices, in conservation, in computers, and in bioengineering is covered in the literature. Papers concerned with solar energy dealt with measurement and prediction of the availability of solar radiation, with the performance of flat plate solar collectors, and with heat transfer in thermal energy storage systems. Interest has been shown in plasma heat transfer connected with arc technology and with thermal plasma processing.

To facilitate the use of this review, a listing of the subject headings is made below in the order in which they appear in the text. The letter which appears adjacent to each heading is also attached to the references that are cited in that category:

Conduction, A
Channel flow, B
Boundary layer and external flows, C
Flow with separated regions, D
Natural convection—internal flows, F
Natural convection—external flows, FF
Convection from rotating surfaces, G
Combined heat and mass transfer, H

Boiling, J
Condensation, JJ
Radiation, K
MHD, M
Numerical methods, N
Heat transfer applications
  Heat exchangers and heat pipes, Q
  General, S
  Solar energy, T
Plasma heat transfer, U.

CONDUCTION

There continues to be considerable interest in specialized problems in heat conduction, particularly when the phenomena are accompanied by a change of phase.

Heat conduction from buried sources is analyzed for a fluid flowing in a pipe [4A] and for a sphere [3A]. For the pipe there are two cases: (1) involves mixed (convective) boundary condition with a uniform heat transfer coefficient at the pipe surface; and (2) laminar flow with linear temperature variation along the pipe, which couples the internal and external heat transfer processes, requiring simultaneous solutions of the energy equations for the two regions. The complicated geometry is handled neatly using the bicylindrical coordinate system, resulting in temperature distributions and shape factors in each case. The same coordinate system leads to an exact analytical solution for the steady-state temperature distribution in and around a heat generating sphere buried in a semi-infinite solid with an isothermal surface. The form factor and heat flux distribution along the medium surface are also reported and compared with an approximate source-sink solution.

At low pressures, consideration is given to the transition regime between hydrodynamic and Knudsen flow [18A]. Heat conduction through a diatomic, diamagnetic gas between parallel plates is represented by a simple model and explicit expansions found for the effect of a magnetic field. For the same parallel plate geometry, ref. [23A] considers the influence of accommodation coefficients and Maxwellian diffuse/specular reflection boundary conditions on the heat transferred through a rarefied gas, yielding results which agree within 3%.

Mathematical aspects of the conduction phenomena are the subject of a number of papers. Thus, ref. [35A] solves a non-self adjoint problem by an expansion in terms of the root vectors of the conduction operator and its adjoint using their orthogonality properties. Given a homogeneous isotropic elastic solid with memory, ref. [22A] treats the problem of time-dependent heat transfer by wave propagation due to the communication, at finite speed, that energy is being generated from a heat, chemical, or nuclear source. Effect of the wave is said to be noticeable in those practical situations characterized by short time and high heat flux. Noting the similarities between the vibrating string and certain heat conduction problems, ref. [34A] presents inequalities related to the norms of the solutions to the vibrating string and heat conduction partial differential equations; a maximum principle is shown to hold for both problems.

When two metallic surfaces are in contact, the question is raised about the heat transfer through the interstitial gas between the surfaces. The thermal conductance is measured in ref. [28A] by interpreting the phase data obtained by varying the laser source modulation frequency; experimental results confirm the hypothesis implicit in most theoretical formulations of the photoacoustic effect: that the surface resistance (thermal) is generally negligible.

A number of papers examine the stress in materials caused by thermal conditions. In solving the transient thermal stress problem for a hollow sphere under partial heating, ref. [56A] asserts that precise analysis requires that the corrected heat equation take into account the thermo-mechanical coupling term. The simultaneous solutions for stress and temperature distributions using a new numerical technique show a small lag in the case of certain materials. For an infinite, isotropic solid containing an annular crack, ref. [21A] seeks the axisymmetric thermal stress distribution; the problem reduces to triple integral equations, ultimately reduced to an infinite set of simultaneous equations. The relation between weak shocks and stress in heat conducting, thermoelastic materials is investigated using equations governing the growth and decay of wave amplitude obtained for shock waves of arbitrary form in a linear thermoelastic body [58A]. The stress 'jump' relation to shock strength is studied for second-order material.

For a spherical droplet moving into an unbounded volume of another immiscible liquid in creeping flow ($Re \ll 1$) the unsteady heat conduction is studied, with a result that an earlier solution to the problem is found to be unstable [16A]. Additional studies also find the techniques of analysis applied to problems emerging from technical practice. Thus ref. [26A] treats the problem of heat transfer from an elliptical cylinder moving through an infinite plate applied to electric beam welding. A general solution, in elliptical coordinates, is found for the two-dimensional steady-state temperature field from which the heat flux around the cylinder is determined, and the total heat flow rates found. The results are applied to evaluate the effectiveness of full penetration electron beam welding and to predict the relation between penetration depth and required power. In ref. [60A] heat transfer from wavy and straight wires of equal length are compared for surface temperatures maintained constant by forced convection. For non-uniformly heated wires, undulations in the wavy wire cause an increase in the total heat transfer, compared to the straight wire of the same length. In a nuclear reactor channel, ref. [20A] applies a finite-difference method to calculate heat transfer. The differential equations characterizing the separate system parts are connected by energy considerations involving fuel and cladding temperature. The method may be extended to other systems.

Several papers are concerned with heat conduction and chemical effects. Gaseous thermal explosion is treated in ref. [11A] where approximate solutions of the Arrhenius equation are evaluated. With general boundary conditions, critical ignition parameters are calculated in the case of finite Biot number. The rate of charring of a heated, one-dimensional wood-like slab is predicted in ref. [17A] by solving the transient conduction equation in the presence of a phase-change energy sink term. Rate and depth of charring are predicted for various boundary conditions and compared with experimental data. The sudden Joule heating of a system is used for studying chemical rate processes. Such an application at high pressure (2 kbar) is used to study fast reactions and verified by means of a well-studied reaction [8A]. In another instance, the same scheme is used to study chemical reaction kinetics in thin samples of electrically nonconducting systems [27A].

The resistance to heat conduction between surfaces in contact is the object of study for a number of papers. With roughness described by a stochastically varying local contact resistance, in a two-part study, general variational principles are derived, permitting the determination of lower and upper bounds of the macroscopic contact resistance [5A, 6A]. The method is illustrated by calculating an upper bound for a random arrangement of circular contact spots of equal size. The second work uses the method to determine upper bounds for arbitrary geometry and distribution of contact spots. Special attention is paid to a binary model which admits only two discrete values for the local contact resistance, inside and outside the direct contact areas. Again for two solids in contact, the thermal resistance is analyzed using as a model, a single heated channel of cylindrical contour and varying radius. The contact resistance of the interface is expressed as a function of the channel properties and extended to surfaces on the basis of materials used, number of contact areas, and a single geometrical property of the channel [47A]. The calculation of multilayer heat transfer is accomplished by a successive method which permits radiant heat exchange between interior surfaces to be considered if needed, giving results which compare favorably with field data [59A]. Turning to specific problems, ref. [51A] describes a procedure for predicting the recommended insert thickness which will result in minimizing the thermal resistance of a pressed contact between flat, non-wavy surfaces. The analysis is based on the assumption of ideal plastic deformations at interfaces between randomly high rough surfaces and insert material. Experimental measurements for several systems and theoretical predictions are in good agreement. For cladded fuels, the thermal gap conductance is studied by applying Kalman filter methods to an in-pile liquid metal fast breeder reactor simulation experiment [24A]. Interfacial resistance to mass transfer for dilute solutions is determined from the interfacial energy profile. Large resistance values appear to correspond to two energy maxima at the interface. The rate constant formulation used, in certain limiting cases, yields the same form as obtained by assuming local equilibrium directly across the interface [46A].

Heat conduction with finned surfaces features an investigation of two-dimensional heat conduction in heat exchanger fins since the difference between one- and two-dimensional analysis can be substantial. The complete assembly is considered, including supports, and the method is held to involve the solution of fewer equations than finite-difference or finite-element methods and yield accurate results for problems where these two approaches do not [13A]. The simultaneous solutions of heat conduction for the fin and convective heat transfer for flowing liquid obviates the need to use literature values of the convective heat transfer coefficient, customarily taken as uniform over the fin surface when the actual flow and boundary conditions are different from those in the literature [54A].

Heat transfer in the presence of a phase change is

studied from various aspects reflecting a wide-spread interest in understanding the phenomena and the variety of its occurrence. The melting of a horizontal solid surface by an overlying hot, liquid pool is studied experimentally for several pool–solid substrate pairs [9A]. The influence of density differences, melt-layer viscosity and latent heats are examined and the results compared with theory. For a spherical enclosure containing saturated material, the enclosure temperature is suddenly raised. The flow field resulting from density differences between solid and liquid phases is observed to concur with that predicted analytically. The energy storage characteristics of an enclosure are studied [29A]. In a vertical tube, the melting of a medium changing phase is studied experimentally. In contrast with the numerical solutions, which use a pure conduction model, this study shows that convective effects can lead to energy transfers 50% higher [53A]. The importance of convective effects during a phase change is also apparent in the system consisting of a heated horizontal circular cylinder embedded in a phase change material. Using a numerical mapping technique to handle the moving boundary, results indicate that natural convection is important in all cases [37A]. For a melting, or subliming, flat surface in direct contact with a hot fluid, free convection again occurs as an important consideration [2A]. Water film cooling of surfaces is an established practice. A new correlation for predicting quench front velocity in rewetting of hot, dry surfaces by falling water films is derived and proves accurate to within 2.5% compared to the accurate numerical solution [33A].

The phenomena of solidification attract wide interest. In a two-part study, the mathematical analysis of the steady, dendritic solidification of an aqueous solution is first developed to yield families of temperature, concentration, and dendrite shaped profiles in two zones, near the basal plane and near the dendrite tips [31A, 32A]. This allows heat and mass transfer processes to be analyzed near the basal plane, then near the tips of steadily advancing dendrites. Geometrical and thermal criteria are used to insure that the fields for the two zones match. Using pure water cooled by a surface, ref. [41A] studies experimentally the initiation and spread of ice formation. Observed thicknesses are in good agreement with the approximate quasi-steady-state solution. In a rectangular enclosure with isothermal sides and adiabatic top and bottom surfaces, the process of solidification is analyzed. Convective effects are considered as well as heat conduction through the solid and the idealized mold. Results give velocity profiles in the melt, temperature distributions in the melt, solid, and mold [36A]. The solidification process on a small scale is studied by considering the thermal evolution of liquid and solid deposited during the solidification of spherical droplets of liquid. The system geometry becomes a hollow sphere because of the density difference between the phases [14A].

For the laminar flow of a liquid in a circular tube whose external surface is convectively cooled, ref. [40A] examines analytically the factors which influence the length of the freeze-free zone and the freeze layer thickness as a function of time and tube length. One-dimensional, transient solidification with super-cooling is studied analytically and experimentally. Observations enable phase regions to be distinguished and classified; the numerical method proposed yields results in good agreement with experiment for the single substance studied [42A].

For a warm, flowing liquid planar solidification is examined under different boundary conditions. The formulation leads to a coupled set of two differential equations in temperature and solid thickness as functions of time, position and problem parameters. The results (temperature distributions in solid layer, solidification rate and time) agree with experimental results [43A].

The Stefan problem attracts the interest of several investigators. Considering the physical, mathematical, and numerical experience with problems of this kind, ref. [49A] advocates enthalpy (not temperature) as the desirable dependent variable. Discussion centers on the physical interpretation of the multi-valued function arising from the mathematical formulation. A one-dimensional problem, semi-infinite strip, with convective boundary conditions is considered; surface temperature, asymptotic behavior as the convective heat transfer coefficient grows large, uniqueness bounds on the phase change front and total system energy are discussed [52A]. In the Neumann solution of the two-phase Stefan problem, ref. [57A] examines the effect of replacing the coefficient in the free boundary condition by an inequality.

For solidification where conduction dominates and small temperature differences exist, ref. [48A] finds quasi-steady models an advantage. Shape factors are used to predict freezing rates and heat transfer, in various two-dimensional systems, from simpler equations. The method is not advised for determining local heat flux or details of interface shape.

Using a numerically based model, ref. [15A]

developed a single analog to represent the one-dimensional heat transfer equation, with and without phase change. Accurate solutions are obtained for test problems using a small (hand-held, programmable) calculator.

The shape of steady-state solidification interfaces formed when liquid above the freezing point circulates over a cold surface is given by ref. [45A]. The surface shape is related to the crystal structures formed during solidification, a matter of concern in casting, the development of naturally strengthened metallic composites, phase change and energy storage devices. A related paper analyzes the solidification interface shape resulting from applied, sinusoidal heating, particularly in the light of metal casting process requirements [44A].

Heat transfer in the liquid zone which precedes the advancing condensation in an oil reservoir undergoing a thermal recovery process is taken up in ref. [61A]. Various analytical expressions for the temperature distribution in the hot liquid zone are derived for high and low injection rates.

The freezing and thawing of soils in the presence of seepage flow is analyzed by means of a system of computer programs based on the finite-element method [10A]. The simultaneous solution for the coupled velocity and temperature fields, made possible by assumptions held to be reasonable, yields results which agree with the laboratory model of the system. For vertical cylinders in free convection, ref. [7A] measures the frost growth. Heat transfer correlation is found to be effective in predicting frost growth rates.

Other problems involving heat transfer with phase change range from a consideration of the thermal stresses developed during solidification due to the transient temperature distribution stresses which depend upon the Stefan number, to the investigation of the factors which influence the solidification process in systems of heterogeneous particles [39A]. The results find application in establishing the conditions for minimizing the frozen regions in moist coal loads during winter rail transport [38A].

Papers on thermal conductivity of materials and systems include measurements, calculation of effective values, and thermal conductivity of special systems. The transient hot-wire method for measuring the thermal conductivity of orthogonal, anisotropic materials is first theoretically investigated, then applied to samples of such materials. The results compare to within 5% of those obtained by the traditional one-dimensional steady-state method [55A]. A new, transient hot-strip method is examined theoretically and points to the need to include second-order terms for the hot-strip electrical resistivity dependence upon temperature [12A]. Considering a 'sphere-pac' nuclear fuel assembly, ref. [1A] reports a two-dimensional finite-difference method for calculating the effective thermal conductivity which gives results within $\pm 20\%$ of experimental values. For wet, fibrous sheets, hot-wire measurements at various moisture content, void fraction, and temperature values provide a basis for a porous structure model for wet fibers and the calculation of effective thermal conductivity for such a medium [19A]. Using five pulverized-coal samples, ref. [25A] applies a modified hot-wire method, and the principle or differentiation thermal analysis, to measure the effective thermal conduction of such systems from room temperature to 850°C. For three general types of soils, ref. [30A] calculates their thermal conductivity above and below the freezing point. Anisodiametric particle systems are studied for their effective thermal conductivity, particularly the case where the particle conductivity greatly exceeds that of the binding medium [50A]. Solutions with metallic powders of differing nature and dispersivity are also considered and the results compared with selected experimental data.

CHANNEL FLOW

Several studies of channel flow dominated by separation regions were given: heat transfer coefficients were found and a flow and surface temperature visualization study was made in the recirculation zone downstream of an orifice in a circular tube [37B]. It was found that the location of maximum heat transfer was not the reattachment point. Heat transfer and fluid flow analyses were made for laminar flow in cavities upstream of a forward-facing step and downstream of a backward-facing step [41B]. Experimental results were presented for heat transfer in rectangular ducts with sharp-edged inlets [36B] and in corrugated ducts [27B] where flow visualization showed large recirculation zones adjacent to the rearward-facing corrugation facets. Spirally-corrugated tubes of varying helix angle were experimentally studied where it was shown that the corrugated tubes gave a larger heat transfer to pumping power ratio than given by smooth tubes [11B]. Experiments were made for heat transfer in ducts containing uniformly-spaced arrays of circular pin fins [22B]. A criterion for evaluating heat transfer promoters which uses the heat transfer coefficients and the energy dissipation was proposed [32B]. Based on this criterion, promoters that increase turbulence in the boundary layers are more effective.

The following studies are for channel flows dominated by fluid property effects: the effects of thermal entry length, polymer degradation, and solvent chemistry were shown to be important in the evaluation of drag and heat transfer in turbulent pipe flow of viscoelastic fluids [17B]. The thermal entry region in a rectangular duct was theoretically and experimentally investigated for flow of a highly viscous Newtonian fluid [25B]. The governing equations for laminar fully-developed flow and heat transfer in a curved tube were numerically solved [14B] for non-Newtonian fluid behavior and approximate analytical solutions of channel flow of non-Newtonian fluids with viscous heating were presented [5B]. An experimental investigation was made showing pressure drop and mass transfer in aqueous solutions of polyox in turbulent pipe flow [35B]. It was found that centrifugal pumps cause rapid shear degradation which reduces the drag-reducing properties of the fluid. A multi-layered velocity profile applied to drag-reducing fluids flowing in a pipe was used to predict the effects of diameter on friction and heat transfer [20B]. Numerical predictions were made for heat transfer with constant wall heat flux to low Prandtl number fluids in the turbulent entrance region and in developed pipe flow [6B]. The influence of axial conduction in low Prandtl number turbulent flows was investigated [19B]. An analysis was made of the unsteady flow phenomenon between infinite cylinders where the inner cylinder rotates [2B]. The unsteadiness, called thermal runaway, is seen in injection molding where the fluid thermal conductivity is low and the viscosity decreases rapidly with temperature. The influence of tube dimensions and flow rates on supercooling-freezing flows was experimentally evaluated [24B], and data necessary for the design of waste-water treatment facilities was given. An experimental method was presented for measuring the flow pattern and heat transfer in liquid–liquid two-phase annular flows [28B]. A model was proposed for momentum and heat transfer in a two-phase upward bubble flow in an annulus [33B]. When void fraction profiles are given, the model can predict skin friction, wall heat transfer coefficients and profiles of temperature and velocity. Heat transfer and axial dispersion in batch bubble columns was investigated [4B]. Solidification of steady turbulent pipe flow was analyzed and a method was developed for predicting the onset of freeze-shut [8B]. Expressions for temperature and concentration profiles were analytically found for turbulent flow of a gas in which there is a single reversible reaction that does not go to equilibrium [23B]. Numerical solutions are presented for turbulent pipe flow of dissociating nitrogen tetroxide in equilibrium [1B]. Intraparticle forced convection was considered to explain experimentally-observed changes in effective diffusivity with flow rate in fixed-bed reactors [31B]. Such interactions can result in design errors of considerable magnitude. Combined radiative and convective heat transfer from laminar and turbulent flows was experimentally and analytically studied and a similitude parameter was presented [38B]. A system of equations for the unstable flow on an inclined plane channel with ion mass transfer was developed and solved [18B]. A system of three-dimensional vortices superposed on the flow was studied in the analysis.

Several channel flow studies which focused on turbulence effects were presented. In one, a method of calculating the characteristics of axisymmetric swirling and non-swirling turbulent recirculating flows in wide-angle diffusers was described and appraised [12B]. The differences between experiments and calculations were found to increase with swirl number. Fluid mechanics and mass transfer measurements were made for turbulent flow in triangular ducts [40B]. In the laminar or transitional flow near the apex, a remarkable change in turbulent intensity and space correlation was detected and the turbulent characteristics near the walls were observed to be different than those found in round tube flow. Temperature and velocity fields close to the free surface in an open channel flow were measured where it was found that in the fluid near the free surface vertical motions are damped while spanwise and streamwise motions are promoted [15B]. Experiments in an accelerating turbulent flow between converging parallel plates showed lower exit Nusselt number than at the inlet suggesting relaminarization due to acceleration [39B]. Prediction of these trends was successful with the K-KL model of turbulence.

Several of the channel flow studies dealt with buoyancy effects. One is a numerical study of turbulent vertical pipe flow of supercritical carbon dioxide [29B]. Numerical results were presented also for developing combined forced-free laminar boundary-layer flow in a vertical concentric annulus with a rotating inner cylinder [7B]. A semi-analytical correlation was developed for laminar mixed convection [13B] and a study of intermittent/slug flow heat transfer was made

[34B] in horizontal tube geometry.

Pulsating flow effects on heat transfer were seen in several papers. The influence of superimposed pulsation on heat transfer in a heat exchanger geometry was studied where pulsation was found to improve the overall Nusselt number appreciably [9B]. An analysis was presented for fully-developed laminar pulsating flow in curved tubes [30B]. There was considerable variation in peripherally-averaged Nusselt number with varying amplitude ratio, frequency, Reynolds number, curvature, and Dean and Prandtl numbers. The effect of induced fluctuations of the turbulent flow in a slotted channel was experimentally investigated [10B].

Other channel flow studies include one of mass transport in a two-dimensional channel due to shear-Couette flow including cases with a moving wall or chemical reaction [3B]. Laminar flow and heat transfer analyses were made in annular-type crevices of various inclinations and eccentricities [42B]. It was found that inclination can affect the flow and heat transfer rate significantly, especially when eccentricity is large. Experimental results were presented for laminar flow heat transfer downstream of a 180° bend [21B]. The effects of the bend were less important than natural convection effects in the flow downstream of the bend. A correlation of experimental data for annular channel turbulent flow of gases was presented [26B] and a presentation was made of several important problems where similarity methods had been successfully applied [16B].

BOUNDARY LAYER AND EXTERNAL FLOWS

The majority of papers in the area of boundary layers are concerned with laminar and turbulent boundary layers on flat and curved surfaces. In laminar boundary layers, the 'unsteady' situation is receiving increasing attention. Studies have also been reported on flows over cylinders and spheres.

The unsteady heat transfer from a flat plate to an oscillating airflow has been studied experimentally [20C]. Nonsimilarity numerical solutions have been obtained for heat transfer in unsteady laminar boundary layers [29C]. For a non-Newtonian fluid, the unsteady heat transfer in a boundary-layer flow over a wedge has been considered [23C]. Boundary layers on moving surfaces are often important; both laminar and turbulent boundary layers on nonisothermal moving surfaces have been studied [33C]. An investigation has been made for heat and mass transfer in falling liquid films [42C]. Conjugate heat transfer in a laminar boundary layer has also been considered [11C]. Reference [49C] deals with the influence of temperature-dependent properties on the heat transfer in a flat-plate laminar boundary layer. Heat transfer can influence transition to turbulence; this effect for a compressible boundary layer on a flat plate has been considered [48C]. A method of solution has been presented for an unsteady thermal boundary layer controlled by diffusion [37C]. Dispersion by laminar flow has been discussed [39C]. The interaction of a small drop with a solid plane has also been considered [30C]. Reference [7C] is concerned with heat transfer in a laminar flow with a phase-change boundary.

Literature dealing with turbulent boundary layers provides new experimental data or contributions to mathematical modeling of turbulence. The re-laminarization of a turbulent boundary layer under the action of a strong favorable pressure gradient has been studied experimentally to uncover the structure of the flow [31C]. On the other hand, the turbulent boundary layer under a strong adverse pressure gradient has also been considered [3C]. Reference [12C] discusses the influence of transition on heat transfer to the turbine blades. A turbulence model based on the concept of turbulent point heat transfer has been considered [22C]. Reference [27C] deals with the stagnation point region in the presence of blowing from the surface. Heat transfer from impinging jets has been studied in ref. [6C] while ref. [9C] is concerned with the stagnation point of a jet flowing over a baffle.

Some papers have considered the flow and heat transfer over cylinders and spheres. Reference [43C] describes a study of the unsteady heat transfer to a cylinder in cross flow. Heat transfer from fine wires and cylinders in a corona wind has been considered [26C]. A study is reported for heat transfer around a cylinder at high blockage ratios [24C]. The problem of variable properties is addressed in ref. [15C] with reference to the boundary layer on a moving continuous cylinder. Reference [25C] describes an investigation of flow and heat transfer around two circular cylinders. Since tube banks or arrays of cylinders are used in many applications, some studies have focused on their behavior. Heat transfer to tube banks is considered for in-line tubes in ref. [2C] and for staggered tubes in ref. [1C]. Low-velocity flow past arrays of cylinders has been studied [41C]. Heat transfer to horizontal tubes in a fluidized bed has been investigated [21C]. Reference [32C] deals with single rods with three-dimensional roughness. Heat transfer from single spheres in low

Reynolds number flow has been considered [13C]. Experimental measurements have been reported for heat transfer through a sheared suspension of rigid spheres [16C].

FLOW WITH SEPARATED REGIONS AND THROUGH POROUS MEDIA

Separated regions and jets

Fluid flow and heat transfer in the separated region of a circular cylinder at subcritical Reynolds numbers can be influenced by locating two small cylinders in the separated shear layers [13D]. The effect of a missing cylinder in the twelfth row of a bank of cylinders in cross flow was studied by measurements and visualization with oil lampblack at Reynolds numbers Re_d between 1270 and 8900 [34D]. It was found that the free space causes a change in the heat transfer coefficient on the neighboring cylinders up to 10% for staggered and up to 3% for an in-line element. Maldistribution of the flow created by partial blockage of the inlet cross section of a tube bank [35D] increased the heat transfer by 30–40% in the first few rows of tubes with an additional pressure drop of several velocity heads. An array of 6 × 6 pin fins on a flat surface exposed to an oncoming longitudinal flow which turns to a cross flow was studied experimentally in a Reynolds number range between 500 and 5000 with the result that the Nusselt number was found to be independent of the ratio of height or spacing to fin diameter when correlated with the longitudinal velocity [33D].

Heat transfer coefficients on the upstream side of a perforated plate positioned normal to an incoming flow can be described by the equation

$$Nu = 0.881\, Re^{0.476}\, Pr^{1/3},$$

in which the Nusselt number is based on the ratio of the modular area to the hole spacing as a characteristic length and the Reynolds number on the hole diameter [32D]. A numerical computation of wall jets flowing over the surface of a cone with axial upstream direction demonstrates that a two-equation model of turbulence is successful in predicting experimental results for various cone angles [29D]. A numerical study of wall heat transfer in the recirculating flow region of a confined jet demonstrates that the maximum heat transfer coefficient can occur ahead of the reattachment point [14D]. Liquid crystals were used to obtain the field of isotherms on a flat plate exposed to a normally impinging single jet or an array of impinging jets [11D]. Reference [20D] is a review of experimental and analytical data on heat and mass transport in turbulent liquid jets and shows that the turbulent transport has only a minor effect on mass transfer for the parameter range investigated. This includes new experiments for evaporation of turbulent jets.

Porous media

A linear stability analysis studies the influence of permeability upon the onset of convection in a porous medium [27D]. Exact solutions in ref. [23D] describe free convection in two-dimensional porous cavities of various cross sections at small Rayleigh number caused by a uniform temperature gradient normal to the gravitational field. Permeability of a porous medium contracts the cells and dampens the convection process according to a nonlinear analysis [24D]. A finite-difference technique was used to study convection in horizontally confined layers as they occur in fracture zones in the ground [18D]. Convection in a fluid-saturated porous layer is also analyzed [26D] as it is caused by sudden heating or cooling in the upper and lower boundaries or by distributed heat sources. The boundary shape is determined for a cooled porous region by conformal mapping, requiring that it result in a uniform surface temperature when the surface is subjected to nonuniform heating [31D]. Pulsatile flow in a heated horizontal or vertical porous channel is also analyzed [3D]. Inclusion of the thermal dissipation term in the energy equation describing flow in a porous medium demonstrates that its effect is negligible for natural convection [9D]. Finite-amplitude analysis is applied to thermohaline convection for a porous layer saturated with a two-component fluid [25D].

A number of papers dealt with natural convection in a porous material for conditions where the effect of capillary forces becomes important. A randomized three-dimensional network was used as a model of a porous medium to study two-phase flow through it [17D]. Permeability, relative permeability, and capillary pressure were calculated and compared with experimental results obtained on a granular bed of sand and glass spheres, and a sintered glass bed. A method was presented to calculate the hydraulic conductivity of porous media (soils) using a model considering several bundles of capillary tubes with statistically distributed diameters arranged in series [22D]. The results agreed best with experimental data when five bundles were used. The departure of the phase change of water in a porous medium from local thermodynamic equilibrium was investigated [2D]. The contribution of surface phenomena for porous bodies with unsteady heat and mass transfer was applied to the analysis of the wet bulb thermometer [8D]. A numerical analysis of heat and mass transfer within an intensely heated unsaturated porous medium indicates the existence of a dry and a wet zone and determines the temperature, pore pressure, and moisture distribution as a function of time as well as the vapor release from the surface [5D, 6D]. A continuum model and Darcy's equation with properly assumed vapor and liquid permeabilities are used. The stability of a moving condensation front in a porous layer was analyzed as a contribution to the understanding of the process of oil recovery by steam injection [37D]. It was found that lateral heat losses can have a significant influence. Sodium boiling was found to create non-rewetting hot spots and dryout as well as fluidization in a granular bed simulating debris in liquid metal reactors [7D]. An exact constant property solution of the desublimation processes in a constrained porous medium is based on heat conduction in the frozen and conduction and diffusion in the gaseous regime [16D].

A number of papers published in the Soviet Union dealt with heat and mass transfer in capillary-porous media including the phase composition of moisture [21D], vaporization in a porous membrane [15D], and thermal moisture conductivity of colloidal porous bodies [30D]. Generalized calculations of the kinetics of drying are also presented [28D].

Fluidized beds

Heat transfer between a wall and a two-dimensional fluidized bed was studied experimentally and the results were compared with several heat transfer models [12D]. Bubbling was recorded by a high-speed camera. The Adams–Welty model for heat transfer to a horizontal cylinder in a large particle fluidized bed was confirmed experimentally [4D]. It was found that the range of applicability of the model is wider than originally assumed. Coupled convection and conduction affecting heat transfer in a fluidized bed was studied based on a single particle model [1D]. The Adams–Welty model was used to describe convective heat transfer and was combined with transient conduction. Methods are presented which predict heat and mass transfer coefficients in a fluidized bed including a new model describing heat transfer from the solids in the bed to a wall using the kinetic theory of gases concept [19D]. A mechanistic theory for heat transfer between fluidized beds of large particles and immersed surfaces uses a model in which the solids are distributed around the heat transfer surface in unit orthorhombic cells [10D]. Heat transfer is composed of convection and unsteady-state conduction.

Heat transfer coefficients to a flowing granular material are found to increase with increasing velocity up to a maximum and then to decrease again [36D].

NATURAL CONVECTION—INTERNAL FLOWS

Studies on buoyancy-driven convection in enclosures continue unabated. There is apparently an increase in the number of papers related to heat transfer in horizontal layers heated from below in which a number of special effects can be considered, including nonlinear post-stability phenomena. The natural convection in layers in which there is a horizontal temperature gradient is an interesting proving ground for numerical studies, and both experiments and analysis on this geometry have been reported. Other important phenomena include the effect of internal energy sources, convection in porous regions, double-diffusive convection where temperature difference and concentration difference contribute to the buoyancy variations, and mixed convection where forced flows and buoyancy both influence the flow and heat transfer.

Reports of natural convection in horizontal layers of fluid heated from below include the effects of boundary conditions varying with time and position, the influence of maximum density points in fluids such as water, and the importance of sidewalls. A number of optical investigations of such convection have been described. Energy theory is used to predict stability in a suddenly-heated or cooled fluid layer [52F]. Following a step change in boundary temperature, the stability of a horizontal layer in a high Prandtl number fluid can be calculated using the depth of penetration of the thermal layer [34F]. An analysis shows variations in the critical Rayleigh number when the temperature difference varies across one of the horizontal surfaces [75F]. Slightly non-horizontal upper and lower surfaces lead to a material modulation of post-stability rolls [16F]. Large-scale circulation can occur with convection in a shallow sloping container [17F].

Thermal convection in a horizontal layer of water in which 4°C (the maximum density point) occurs within the layer can produce quite interesting phenomena. One study shows different regions of convection in such a layer heated from below [48F]. In another study in which all walls of a rectangular cavity are decreasing linearly with time, the influence of the maximum density point on flow patterns has been calculated [56F].

A perturbation analysis of Bénard flow with two free

surfaces finds the influence of the Rayleigh number on the number of cells present [26F]. In low Rayleigh number convection, the Prandtl number is found to set the size of the convection cell [45F]. The influence of variable viscosity on critical Rayleigh number has been analyzed over a large range of viscosity ratios [64F]. A numerical analysis of three-dimensional time-dependent equations shows the transition from periodic motion to a quasi-turbulent flow [47F]. Convection in a square box heated from below has been analyzed over a range of Rayleigh numbers considering different Prandtl number fluids [46F]. Finite-amplitude rolls are found in a long channel heated from below [24F].

Analysis of thermal convection in vertical cylinders heated from below includes a study of the transient phenomena in which a moving front of heated fluid moves up from the lower surface [43F]. In a low aspect ratio vertical cylinder, the convective motion is found to be non-axisymmetric [57F]. A detailed description is provided of precision measurements of thermal convection in a vertical cylinder in which a fluid is maintained at very low temperature [7F].

Experiments on Bénard convection include use of a laser-Doppler velocimeter to study the flow structure at low Rayleigh numbers [27F]. A holographic interferometer indicated the stream lines and rolls in such flow [71F]. Transitions in the flow pattern at various Rayleigh numbers were examined by optically studying the scale size of the flow pattern [60F]. An interferometer was used to study convection in a layer of visco-elastic fluid [25F].

Unsteady flow has been examined in a stratified water layer cooled from above [73F]. Natural convection in a heated open cavity becomes unsteady at low Rayleigh number when the cavity faces upward [53F]. An analysis considers the flow and heat loss from a free surface in an open channel extending from a large body of liquid [32F].

Numerous studies look at flow in a rectangular region which is heated on one side and cooled on the other. Some of the channels are completely enclosed while, occasionally, attention is directed toward an open channel. A non-boundary layer solution has been obtained for the flow in a vertical open channel [51F]. Optimum spacing to minimize heat flow through a double pane window is generally found to be that which gives transition to the multi-cellular flow regime [36F]. Low Rayleigh number flow in a differentially-heated open vertical well has been analyzed [18F].

Measurements on convection between vertical plates agree with earlier numerical predictions [78F]. A holographic interferometer has been used to study convection in a very narrow vertical layer heated from below [37F]. Convection across a vertical array of enclosures has been analyzed to examine the potential for designing a system that primarily permits heat flow in one direction [13F]. A vertical gas layer to be used as an insulating wall can be optimized by dividing it into a number of parallelogram-shaped enclosures [3F]. In a related work, the effect of radiation on experimental results in cavity flows has been analyzed [4F]. A numerical solution of convection in two adjacent vertical channels whose common wall acts as a thermal communication link has been presented [63F]. Partitions are found to have a significant influence on heat transfer across a differentially-heated enclosure [5F]. Linear theory has been applied to study the stability in a fluid found adjacent to a melting ice layer [28F].

The onset of convection in a vertical layer which is suddenly heated from the side has been analyzed [33F]. Experiments indicate a very slow approach to the steady state for transient differential heating of a shallow fluid layer [79F].

An approximate analysis for laminar flow in a differentially heated shallow layer compares favorably to numerical solutions [69F]. A numerical solution of flow in a tall narrow cavity shows a significant effect of aspect ratio on the rate of heat transfer across the cavity [31F]. The effect of variable temperature of the top and bottom walls on a differentially-heated rectangular enclosure has been examined [59F]. The Nusselt number along an open vertical square duct has been calculated for boundary conditions of constant temperature and constant heat flux [55F].

Heat transfer across differentially-heated inclined layers is important in a number of applications. Experiments with air in inclined and vertical layers lead to a correlation valid for Rayleigh numbers up to 10^7 and for a large range of aspect ratios [20F]. Radiation interaction can play a key role in measurements in layers of air [19F]. Numerical studies show the influence of conducting walls between neighboring inclined rectangular cells [50F]. Experiments using water in low aspect ratio cavities indicate different flow regimes at different angles of inclination [77F]. Finite-element analysis has been used to predict the natural convection in inclined rectangular cavities [65F].

Experiments on a horizontal layer with internal energy sources include the effect of external heating

[35F]. Bifurcations are found in a solution to internally-heated flow in a vertical channel [70F].

Flow set up in a rectangular cavity all of whose walls decrease with time has been analyzed [72F]. A study of stability in the flow of a fluid layer heated from below which is also rotating about a vertical axis includes a comparison between experimental and theoretical results [9F].

A steady-state solution is found for convection in a triangular cavity heated on one side and cooled on another [2F]. A combined numerical and experimental study shows that axial spacers can reduce the natural convection heat transfer across an annulus by as much as 70% [39F]. Bi-polar coordinates are used to analyze natural convection in the annulus between eccentric horizontal cylinders [12F]. Numerical calculations for natural convection in an annulus have been extended into the turbulent flow regime [23F]. Convection inside a heated horizontal cylinder has been studied to predict melting in a phase-change thermal energy storage device [58F]. Convection patterns in a spherical shell, as might occur in the earth's mantle, are explored for low Rayleigh number flow [10F]. Convection in a region bounded by two vertical concentric isothermal cylinders with the inner one shorter than the outer has been examined [11F].

A review of experiments and analysis led to a correlation for predicting the heat transfer from a buoyant plume to an unconfined ceiling [14F]. Plumes above a ring heater have been studied to simulate flow in the atmosphere, particularly as this relates to a specific fire [38F]. Computations of the natural convection from a two-dimensional source have been used to predict smoke flow from a fire [44F].

Studies of natural convection in porous media have application in geo-physical phenomena as well as to optimize insulation systems. Energy methods have been used to study the stability of natural convection flow in a porous substance [54F]. Measurement and analysis indicate the critical Rayleigh number and the Nusselt number at low Rayleigh numbers for a porous medium heated from below [6F]. Buoyancy effects in flow through porous media have been studied [41F]. An analysis shows the effect of natural convection on insulation in an attic-shaped enclosure [8F]. The onset of flow has been calculated for a saturated porous medium with fluid above it and with internal energy sources [61F].

Double-diffusive convection occurs when the buoyancy term has components due to two diffusive phenomena. These usually are thermal and mass diffusion and show up as variations in temperature and concentration. Mass diffusion effects are considered in analyzing the stability of a binary mixture heated from below [15F]. Double-diffusive convection has been studied at a sufficiently high Rayleigh number to simulate conditions in a solar pond [76F]. The development of flow in a salt-stratified solution heated from below with isothermal boundaries has been examined with a Mach–Zehnder interferometer [42F]. The Galerkin method is used to predict the flow in a differentially-heated inclined fluid layer with a stable concentration gradient [68F]. The critical parameters leading to the onset of cellular flow are found with a heated vertical wall adjacent to a stably-stratified fluid [67F]. In a related paper, convection cells are observed adjacent to the heated wall [66F].

Natural convection loops or thermo-syphons are useful in many systems including solar heaters; they also simulate natural convection cooling of nuclear reactors. A two-dimensional analysis of the flow and heat transfer in a toroidal natural convection loop agrees with earlier experiments [49F]. A one-dimensional analysis shows the transient thermal-hydraulic behavior in a toroidal thermo-syphon [74F]. An analysis has been conducted on a natural convection flow-driven system as might occur in an ocean thermal power device [1F]. Facilities have been designed to study natural convection in a liquid–metal fast-breeder reactor shut-down heat removal system [29F].

Mixed convection or combined natural and forced convection occurs when body forces affect the forced flow in a heat transfer system. A numerical solution of the laminar mixed convection in the entrance region of a horizontal pipe which is valid for arbitrary Prandtl numbers has been offered [30F]. Experiments with a laminar flow in such a geometry have been conducted over a range of Graetz, Reynolds, and Rayleigh numbers [80F]. Two-dimensional flow about a horizontal heated cylinder between two vertical walls includes the effect of mixed convection [22F]. Experiments on flow in an open vertical channel with one heated wall and a heavy transpiring gas passing through the wall have been conducted [40F]. Calculations describe the combined free and forced convection within a heated vertical cylinder with flow passing through small openings in the top and bottom walls of the cylinder [62F]. Combined convection around a vertical plate with a forced flow parallel to it shows different flow patterns in different regions along the plate [21F].

NATURAL CONVECTION—EXTERNAL FLOWS

An integral equation formulation for natural convection problems is developed and illustrated and theorems are proved which establish the properties of the eigenvalues and eigenfunctions of a linear integral operator that appears in the analysis [34FF]. By applying appropriate coordinate transformations and the Merk series, the governing momentum and energy boundary-layer equations for a vertical plate with nonuniform surface conditions are expressed as a set of coupled ordinary differential equations that depend on a wedge parameter and the Prandtl number [36FF]. The existence of multiple steady-state solutions in natural convection flow was demonstrated for the case of laminar flow adjacent to a vertical, heated or cooled, flat surface submerged in quiescent cold water [10FF]. The computed results for transient laminar natural convection flow adjacent to a vertical surface subject to a uniform heat flux boundary condition show good agreement with the experimental measurements [27FF]. Free convection heat transfer near the leading edge of a semi-infinite vertical plate with uniform heat generation has been analyzed numerically and the results were found to be in good agreement with experimental data [21FF]. In a combined theoretical and experimental study, techniques of fine wire thermocouple thermometry and real fringe laser-Doppler anemometry were used to study the boundary-layer flow adjacent to a vertical isothermal hot plate with high surface-to-ambient temperature difference [6FF]. The natural convection flow due to isolated heated elements located on a vertical adiabatic surface, a problem of particular relevance to electronic circuitary cooling, has been studied analytically [17FF]. The problem of laminar free convection adjacent to a vertical wall with an arbitrarily prescribed transpiration rate has been investigated using the method of strained coordinates [19FF].

The technique of live-fringe holographic interferometry has been used in an investigation of convective heat transfer in air beneath a heated horizontal circular plate [13FF]. Interferometric studies of free convection above a horizontal heated plate revealed periodical flow instabilities which resulted in large variations in local Nusselt numbers [37FF]. The vortex instability characteristics of laminar free convection flow over horizontal and inclined isothermal surfaces are studied analytically by linear theory [8FF]. It is shown that the principle of exchange of stability is not always accomplished in the Rayleigh–Bénard problem for fluids with internal structure [25FF]. A mathematical model has been developed for determining the dynamic response of a salt-stratified solution to heating from below, and predictions have been compared with experimental results [5FF]. Interfacial stability in liquid drops' evaporation of flat plates at room temperature was studied by both direct photography and laser shadowgraphy [38FF].

Numerical solutions for steady two-dimensional free convection from a horizontal circular cylinder are obtained for small Grashof numbers by the method of series truncation [30FF]. The evolution of the boundary layer formed by the sudden heating of a horizontal circular cylinder in air was analyzed assuming that the Grashof number is large [32FF]. A two-equation model has been adopted in obtaining numerical solutions of turbulent natural convection from an isothermal horizontal cylinder [12FF]. The method of extended perturbation series is applied to solve for laminar natural convection from an isothermal, thin vertical cylinder [3FF]. Experiments were carried out to determine the average and the local heat transfer by natural convection from the outside surface of isothermal cylinders of different diameters and lengths at different inclinations in both the laminar and the turbulent regions [1FF]. Unsteady low Grashof number natural convection about a sphere is studied when the surface temperature of the sphere is suddenly increased and it is shown that the solutions for the velocity and temperature are expressed in terms of three expansions reflecting the existence of three distinct regions in the plane of non-dimensional radius and time [28FF]. A numerical study of time-dependent free convection heat transfer from a solid sphere to an incompressible Newtonian fluid has been carried out for Grashof numbers between 0.05 and 12 500 and for Prandtl numbers of 0.72, 10 and 100 [14FF].

A study was made of the natural convection of cold water between two horizontal concentric cylinders with constant surface temperatures [24FF]. Nusselt numbers were measured for an in-line array of short horizontal cylinders that were affixed to a convectively participating vertical plate and, for the most part, the Nusselt numbers for the wall-attached cylinders fell below those for the classical horizontal cylinder of infinite length [33FF].

A theory is reported that explains the flickering motion of turbulent plumes as well as their large-scale sinuous structure [4FF]. An analytical study of the laminar flow in axisymmetric thermal plumes and buoyant jets in stably stratified media is reported

[16FF]. A detailed experimental investigation has been carried out to study the trajectory and other thermal characteristics of a buoyant jet discharged at an inclination to the vertical buoyant force [29FF]. Merging buoyant jets discharged in a cross flow were investigated experimentally using a unique visualization technique [15FF]. An experimental study of a vertical turbulent jet with a negative buoyancy showed that the width of the upflow spread linearly with respect to the distance from the nozzle exit, whereas the width of the whole jet was almost constant [23FF].

A matched asymptotic technique is used to construct inner and outer expansions including, for the first time, both mixed convection and higher-order boundary-layer effects in order to analyze mixed convection flow over a vertical semi-infinite surface with uniform heat flux [7FF]. Numerical calculations of the time-varying temperature and velocity profiles are presented for transient aiding mixed convection flow adjacent to a vertical flat surface [26FF]. The mixed convection flow that arises due to a finite heated element located on a vertical adiabatic surface in an external flow, aligned with the surface, is studied analytically [18FF]. A correlation was developed for the average Nusselt number for the case of laminar mixed convection from a vertical heated surface in a cross flow [11FF]. A heated flat plate was tested in a wind tunnel to study mixed convection in both upward and downward positions [35FF]. The buoyancy force plays an important role in defining the flow field in mixed convective flow over a semi-infinite horizontal plate with vectored mass transfer [9FF]. Measurements of the streamwise and vertical velocity fluctuations and temperature fluctuations in a thermally stratified flow in an open channel under unstable conditions indicate that the contribution to the total turbulent momentum flux from the sweep motion of large amplitude becomes greatest in the free surface region [22FF]. The combined effect of forced and free convection on the unsteady laminar incompressible boundary-layer flow with mass transfer at the stagnation point of a three-dimensional body with time-dependent wall temperature has been studied [20FF].

An analysis of laminar, free convection, diffusion flames burning on a vertical fuel surface is presented [31FF]. Flames that spread vertically down thermally thin fuels at the same Damkoehler number have the same dimensionless temperature fields although their physical size and hence temperature fields might be quite different [2FF].

CONVECTION FROM ROTATING SURFACES

Heat transfer by forced convection has been considered in a rotating concentric annulus [4G]. The situations involving eccentric cylinders or disks have been investigated [6G, 1G]. Reference [2G] presents a numerical solution obtained by the finite-element method for the rotating flow in a skewed shrouded rotor. Convection in a rotating layer has been considered [5G]. Reference [8G] deals with the laminar heat transfer from the surface of a rotating disk. Thermal cellular convection in rotating rectangular boxes has been considered [3G]. Heat transfer coefficients are reported for a cup-like cavity rotating about its own axis [7G].

COMBINED HEAT AND MASS TRANSFER

Systems in which both heat and mass are transferred are manifold. They include processes such as film cooling and transpiration cooling, where mass is fed into the boundary layer on a wall to protect the surface from a hot gas stream flowing over it, jet impingement cooling, and processes in which jets or equivalents are used for direct contact heat transfer. Other systems include applications in chemical reactors and in the drying of paper, film, and agricultural products.

A mixing length model is used to correlate the results for film cooling with near tangential injection through a discrete hole [11H]. With slot injection on a turbine blade, variations in the angle of attack of the blade affect the film cooling performance on both the pressure and suction surfaces [10H]. For film cooling injection through discrete holes, the heat transfer performance is strongly affected by the nature (laminar vs turbulent) of the injected flow [6H]. A high heat transfer coefficient in the immediate vicinity of film cooling injection holes has been demonstrated using a mass transfer system [5H]. High free stream turbulence significantly enhances heat transfer to a porous surface through which fluid is transpiring [12H]. A study of the influence of boundary-layer suction and injection on heat transfer in a supersonic boundary layer includes the effect of separation induced by a transpiring gas [3H].

In some direct contact heat transfer systems, a jet of one fluid enters another fluid in order to transfer heat between the two fluids. The hydrodynamics of a liquid jet as it would affect such heat transfer has been described in a companion pair of studies [8H, 9H]. A correlation has been obtained for heat transfer from an impinging jet to a solid surface on which it impinges with and without a cross flow [4H].

The heat and mass transfer across an air–water interface is strongly affected by the surface roughness produced by even moderate air speeds [7H]. The effect of swirl significantly increases the heat and mass transfer from an ablating surface [15H]. A model for combined heat and mass transfer shows the importance of high gas velocity in controlling the temperature of stored agricultural products [2H].

A correlation has been obtained for heat and mass transfer in tanks of liquid with mixing due to stirring and gas injection [1H]. The importance of simultaneous heat and mass transfer in gas absorption has been analyzed [14H]. A thermodynamic approach has been used to describe combined heat and mass transfer in a gas–solid system [13H].

BOILING

Nucleate boiling

The advantageous high heat transfer coefficients provided by nucleate boiling continue to promote interest in this phenomenon. Suggested equations governing the nucleate boiling curve have been compiled from a variety of sources, including only those relations dependent upon fluid properties alone so as to avoid bias due to the properties of the surface [40J]. Correlations for bubble departure diameters, bubble generation frequencies, and resultant heat transfer coefficients in nucleate boiling have been developed using the law of corresponding states and have been shown to fit well available data for several different fluids [10J]. Contact angle hysteresis within natural cavities, associated with micro-roughness and surface heterogeneity of cavity walls, has been argued to be the predominant barrier to boiling incipience [18J]. Measured deactivation superheat for an artificially prepared nucleation site in methanol was shown to agree well with the surface superheating required for the balance of pressure and surface tension forces [30J]. Previous investigations of this nature had been limited to water with its poorer wetting characteristics. Boiling characteristics of a natural surface were related to the real surface profile with attention paid to micro-roughness [38J]. In another study at the microscopic level, large changes of the local heat flux at a single isolated nucleation site were observed coincident with significant variations in the product of bubble departure diameter and bubble frequency, all with little change in the local wall superheat [69J]. A technique has been described to evaluate the stability of nucleate boiling on nonisothermal surfaces [39J]. Long-term temperature behavior of steel tube walls was reported, indicating the consequence of corrosion product accumulations [33J]. Certain sponge-like deposits were determined to be particularly undesirable.

Considerable interest has been shown in the development of commercially producible surfaces to enhance nucleate boiling. Carefully prepared, highly regular surfaces are still being studied for more control over the parameters which might be optimized. In one of these latter studies a fiber optic system provided bubble departure diameter and frequency information at low heat fluxes [56J]. Substantial improvement of nucleate boiling heat transfer coefficients were measured using porous metallic matrix surfaces [7J, 20J, 49J], although boiling curve hysteresis was cited as a potential problem in cases of very critical temperature control [7J].

The relative contributions of evaporative heat transfer and single phase convection promoted by bubble agitation remain a topic of interest. A simultaneous boiling and ion diffusion experiment was reported to demonstrate the dominance of the convective mechanism [11J]. Another experiment modeled the bubble-induced liquid circulation near the boiling surface by bubbling gas through a porous plate into a liquid. Tabulated data and some comments on the method are presented [47J]. Nucleate boiling of binary liquid nitrogen/argon mixtures were studied [68J]. Both evaporative and sensible heat transfer mechanisms were concluded to be impeded at maximum vapor–liquid mole fraction differences.

Most studies in the past have dealt with nucleate boiling at atmospheric or higher pressures. Subcooled nucleate boiling at subatmospheric pressures was observed to follow a substantially different relationship between wall superheat and liquid subcooling than at higher pressures [12J], while four distinctly different patterns of boiling were observed, dependent upon pressure and heat flux, for subatmospheric saturated boiling [25J].

Fundamental questions of bubble growth dynamics are still being addressed [17J]. A perturbation method was applied to the pre-asymptotic behavior of thermally controlled bubble growth [13J]. The method of characteristics was applied to gas bubble implosion [65J]. Results of the numerical analysis show the formation of jets and detachment of small bubbles. The effect of the liquid viscosity upon the displacement of liquid from the wall during bubbling was studied by means of a hydrodynamic model: an inert gas being bubbled through a porous plate into water/glycerol mixtures of varying viscosities [48J]. Liquid viscosity was concluded to be irrelevant to the departure from nucleate boiling.

Forced convection boiling

Forced convection boiling in tubes or channels is of great concern in the design of steam generating equipment and chemical process development as well as nuclear reactor design. Forced convection boiling data for flow in tubes has been correlated using a superposition of either of two single phase forced convection correlations (one for subcooled or low quality flows and the other for high quality annular flows) upon a pool boiling correlation. Agreement with available data for boiling of water is shown to be superior to the agreement of the widely used Chen correlation [9J]. Another forced convection correlation for boiling in tubes has been proposed which is independent of tube diameter and liquid viscosity except for their influence upon the transition from single phase convection to developed boiling [67J]. Existing literature on subcooled flow boiling heat transfer and hydrodynamics in pipes has been summarized [71J].

Widely diversified data for two phase pressure drop and heat transfer coefficients has been correlated for annular flows in vertical upward, horizontal, and vertical downward orientations [24J]. A technique for estimating the distribution of liquid flow between droplets and film in annular flow was presented and compared with data from high-speed motion picture photography [42J]. The extent of the non-evaporating region in downward flowing vertical liquid film evaporator tubes was determined experimentally for various flow rates and pressures [64J]. Heat transfer rates in the non-evaporating regions are estimated to be 4.1–4.5 times smaller than in the evaporating regions. A downward flowing vertical liquid annular film inside a tube with a counterflowing vapor core all in laminar flow was analyzed numerically and compared with experimental results [16J].

Good agreement has been demonstrated between a numerically resolved critical heat flux (CHF) estimation based on an annular flow model and the annular flow segment of a previously presented CHF correlation [35J]. Those same generalized correlation equations for CHF have been shown to be capable of predicting the 'critical steam quality' or 'boundary steam quality' phenomenon reported by many observers in which the critical heat flux falls precipitously beyond some given quality [36J]. The high pressure (HP) regime of that same generalized correlation was examined experimentally using Refrigerant 12 in tubes of 3 and 5 mm diameter [37J]. Data from the smaller diameter tube and from low mass flux experiments with the larger tube were consistent with the above-mentioned correlation and with another correlation in the literature, but at large mass fluxes in the larger tube an anomalous CHF event was detected with a smaller wall temperature excursion and at a location upstream of the end of the heated section of the apparatus. Vertical upward and downward evaporating flows of Refrigerant 113 in a small tube were examined for pressure drop, heat transfer, and CHF data [43J]. Heat transfer coefficients were consistent with those predicted by the Shah correlation originally developed around larger diameter tubes. Existing correlations for CHF, however, underpredicted the measured values. Critical heat flux variations were observed in helical coiled tubes with an imposed heat flux tilt toward the outside surface [31J]. Local or subcooled CHF could be improved or degraded by the flux tilt, or maldistribution, whereas CHF for flows with higher quality was generally reduced by the flux tilt.

A numerical model for post-dryout heat transfer considering three mechanisms: wall/vapor, vapor/droplets, and wall/droplets, was improved by the incorporation of a procedure for determining droplet size distribution [52J]. Wall temperature predictions agreed well with data, and wall-to-droplet heat transfer was estimated to be as much as 10% of the total near the location of dryout. The post-dryout region in horizontal tubes was observed to differ from that of vertical tubes [61J]: dryout appeared to be spread out axially along the tube as the upper surface of the tube dried out before the lower surface, the local wall temperature rise due to dryout was weakened, but wall temperature increased linearly with position downstream of the dryout as contrasted with the nearly uniform downstream wall temperature of the vertical tube. The rewet of dispersed flow film boiling was examined for vertical flow [26J]. Wall thermal properties were found to be insignificant below a threshold mass flux; above that mass flux the minimum temperature increased with mass flux for an Inconel section, but a copper test section was unaffected. A generalized 'heat transfer surface' technique was applied to the analysis of quenching in two phase forced convection water systems [57J]. Caution is recommended in the interpretation of data from quenching experiments. Transient wall temperatures in the post-dryout region were recorded experimentally in response to deliberately imposed system transients [22J]. The temperature–time histories were analyzed qualitatively using the boiling curve, and the dynamics were successfully modeled with a computer simulation. A stability analysis of film boiling has been reported using a steady-state small perturbation method. Below a critical vapor Reynolds number the film has a great tendency to collapse [62J]. Despite the use of a simple model for film thickness, the results reflected experimentally observed trends, and a better film thickness model is expected to improve the prediction of minimum wall temperature and the effects of subcooling.

The state of the art in design methods for multicomponent boiling and condensation in forced convection was summarized with the inclusion of some comparisons with data and some general recommendations for design procedures [60J]. An experimental evaluation of pressure drop in swirled flow boiling of Refrigerant 12 is reported [2J]. Satisfactory correlation was achieved by using an empirical function of the twist ratio of the twisted tape swirl inducer to modify the pressure drop predicted by the Martinelli–Nelson method.

Natural convection boiling

A standard experimental apparatus was proposed for pool boiling experiments so as to produce general, reproducible, and accurate heat transfer data [21J]. Transition boiling of water was the topic of three elements of a four article series on subcooled pool boiling: contamination on a platinum heat transfer surface was shown to improve the transition boiling heat transfer coefficients [27J], large subcooling (around 30 K) resulted in the formation of large voids which collapsed very rapidly, fragmenting into small voids and permitting large heat fluxes [28J], and a proposed model for subcooled transition boiling was supported by simultaneous measurement of heat flux and wall temperature with motion picture photography of bubble events [29J]. Evidence is presented for the existence of two distinct boiling curves in the transition region, transitional nucleate boiling and transitional film boiling [70J]. This hypothesis could explain the discrepancy between observed and hydrodynamically predicted minimum heat fluxes in film boiling since the selection of which transitional boiling curve will be obtained depends on physical parameters which have not been independently controlled in past experiments.

High-speed motion picture photography was used to observe the local behavior of bubbles on the surface of horizontal tubes in an evaporator tube bundle [19J]. A model, only accurate to within an order of magnitude, explained the higher heat fluxes observed at tubes in the upper center of the bundle. Higher heat fluxes were also observed in a study of a finned tube bundle in an evaporator [54J]. The greatest augmentation of heat transfer from these upper tubes was observed at heat fluxes corresponding to the transition from single phase natural convection to nucleate boiling.

In an effort to model boiling on finned surfaces, an experimental investigation of boiling in a narrow confined slot with nonisothermal walls was employed [58J]. Possibilities of improvement over boiling on plane surfaces were reported. Experiments with boiling in smaller gaps were reported using narrow annular spaces, some open at both top and bottom and others open only at the top [4J]. The nucleate boiling heat transfer coefficient was found to improve with decreasing gap width when the annulus was open at the top and bottom, and to be independent of annulus length. With a closed bottom, neither width nor length affected the heat transfer coefficient. For given dimensions, the closed bottom gave greater heat transfer coefficients, but went through a transition to unstable drying and rewetting at lower heat flux than the open annulus case. Critical heat flux was examined experimentally for natural convective boiling of several fluids in vertical rectangular channels heated on one side only [51J]. Little effect was visible for the channel spacing alone, but the aspect ratio, height-to-width, was found to be significant. Heat transfer to vertical and inclined boundaries of volumetrically heated boiling pools was modeled using a two zone scheme which correlated well with data [15J].

Other boiling and evaporation studies

Boiling in a bed of particles confined at the bottom and sides, flooded from above, and heated volumetrically or from the sides has been studied [1J]. A flow visualization experiment demonstrated a dry superheated layer formed at the bottom with a 'capillary' layer of counterflowing liquid and vapor above it. Large unsteady horizontal voids atop this region acted as plenums to feed vapor channels above them in turn. Simple physical models were presented for predicting the heights of the capillary and chimney zones and evaluated parametrically. In another study with a similarly confined bed heated only at the bottom, no dry region was observed, although it was suggested to be possible with sufficiently large heat transfer rates [6J]. Forced flow through a volumetrically heated porous layer was also examined [55J]. Vapor channeling was noted for particles smaller than 1.6 mm, and separate pressure drop models were suggested for particles larger and smaller than that diameter.

Direct contact evaporation was either the goal (evaluation of performance of direct contact evaporators) or an experimental method (determination of the limits of liquid superheat) in several studies. An analytical model for volumetric heat exchange coefficients was presented and compared favorably with data [66J]. A more detailed model examining only the heat exchange at a single droplet/bubble was reported with the principle resistance to heat transfer being a thermal boundary layer in the fluid external to the droplet [59J]. Liquid droplets heated by a surrounding immiscible liquid were filmed as they vaporized to determine qualitative aspects of boiling intensity and bubble growth data [5J]. Violent boiling at atmospheric pressure gave way to slower vaporization at elevated pressures. Results suggested that vapor explosions are unlikely in the case of thermally controlled bubble growth. Very rapid evaporation of butane droplets in heated ethylene glycol were observed at the superheat limit of 105°C [63J]. An interfacial instability was observed, possibly explaining how the rate of vaporization was two orders of magnitude greater than that expected for a stable interface. A correlation for spinodal pressure was developed based on an argument of corresponding states [45J]. This reduces the dependence of calculations upon difficult-to-obtain data on specific volume and surface tension.

Experiments were reported for heat transfer to air/water mist flows from horizontal circular cylinders with various surfaces [41J]. The surface conditions were intended to augment the heat transfer by spreading liquid over as much surface as possible to take best advantage of the evaporative cooling. The effect of incidence angle on mist flow heat transfer from an isothermal flat plate was determined [23J]. At low incidence angles, laser-Doppler anemometry demonstrated that droplet velocity distributions were self-similar and in keeping with the single phase velocity distributions. Heat transfer coefficients increased with incidence angle. Results of an experimental study were presented for mist cooling of very hot tubules [53J]. Three regions were identified; liquid film, dryout, and gas-phase forced convection. The mist flow was found to augment the heat transfer rate by as much as ten times. A technique for estimating the Leidenfrost temperature and the time required for vaporization of a droplet was presented [3J]. The influence of an electric field upon the vaporization of charged droplets was numerically determined [32J]. The augmentation is due to the acceleration of the droplet by the electric field. In another study the influence of the electric field upon the internal circulation of a droplet and the consequent enhancement of heat and mass transfer were addressed [14J]. It was determined that this effect is negligible except when electrical effects are much greater than gravitational effects, and that the apparent enhancement of heat and mass transfer in several preceding experiments must be attributable to effects other than internal circulation of the drops.

Turbulence and surface renewal associated with deep-water convection are important in mitigating the effects of surface active compounds on the evaporation rate of a body of water [34J]. Otherwise, there is an additional diffusive resistance to the evaporation of water through surfactant layers [50J]. Estimations were made of corrections to evaporation rates from a liquid to a flowing gas and vapor mixture [8J]. The transverse mass flux could have a significant effect on the heat and mass transfer, possibly resulting in a departure from the analogy between them.

A small-scale surface probe has been described using a miniature thermocouple in contact with a thin plated surface to detect liquid contact in transition boiling on high temperature surfaces [44J]. Experiments were performed to evaluate film breakdown and bundle depth effects on falling film evaporators proposed for evaporating ammonia in an Ocean Thermal Energy Conversion (OTEC) system [46J]. Film breakdown was observed for low film Reynolds numbers, but variations with depth were not observed.

CONDENSATION

Film condensation

A review of semi-empirical models for film condensation was given [16JJ] and predictions from these models were compared with available experimental data. A model based on absorption and capillary condensation was presented [28JJ] and then used to describe conditions at the contact line.

The case of upward vapor flow inside vertical tubes was analyzed where the results showed complete condensation on the laminar film in a short height of tube [22JJ]. The hydrodynamic and heat transfer processes of condensation or evaporation of a thin laminar wavy-film were analyzed and the results were compared with available data [9JJ, 10JJ]. Measurements were made of the heat transfer coefficients for dropwise condensation on a vertical wall where surface subcooling at the initiation of condensation of 0.03°C was found [13JJ]. Melting of a vertical wall as a result of condensation of a saturated vapor was analytically and experimentally studied [26JJ]. The presence of a melt or condensate layer was found to enhance heat transfer. A closed-form, approximate solution was presented for film condensation on a vertical rectangular fin [4JJ]. The steady flow and unsteady starting flow film condensation processes were investigated with surfaces of differing geometry [21JJ]. Measurements were made of the vapor-phase temperature profiles and rate of condensation for vertical wall condensation of binary vapor mixtures [1JJ]. The data were correlated with a modified Nusselt model. Experimental results were presented for film condensation on vertical surfaces [8JJ] and vertical tubes [3JJ].

A flow regime map was developed for predicting flow patterns [27JJ] and the annular-to-wavy flow pattern transition process was clarified [25JJ]. A theory was presented for condensation of wet vapor and uniformly distributed droplets on a horizontal tube [19JJ]. The heat transfer coefficient was dependent upon the droplet mass flux and velocity as well as the vapor velocity. The vapor condensation rate inside horizontal conduits was analyzed for co-current and counter-current flow of steam and condensate [18JJ]. A higher condensate production rate was found for co-current flow. Experimental results of condensation in finely-finned horizontal tubes were presented where the tube thermal conductivity was shown to have a significant effect [23JJ]. Results were also presented for condensation in a horizontal ten-tube bundle [17JJ] and in sloped single tubes [14JJ, 15JJ].

Condensation from a vapor–gas mixture was analyzed and experimentally investigated to study local characteristics of the process [11JJ]. Analytical and experimental results were presented for condensation in an inverted U-tube condenser where four modes of flow were identified and it was found that non-condensibles significantly aided flow stability [5JJ]. An analysis was made and a correlation was presented for condensation of sodium [20JJ].

Free condensation

Several studies of the problem of steam flowing through a pipe and out the submerged end into a pool of subcooled water were presented. High-speed movies showed that at high rates of steam flow an oscillating jet could be observed, and at low flow rates chugging could be seen [6JJ]. Three intervals associated with periodic flow characteristics of the condensing jet were identified and the dynamic behavior of each event was discussed [24JJ]. A theoretical and experimental investigation was made where it was found that conventional turbulence models established for single-phase flow are appropriate [7JJ]. Dimensionless temperature profiles in the steam jet were found to be independent of steam pressure or ambient liquid temperature [2JJ].

A numerical study was presented for the collapse of a bubble in a three-component, three-phase system [12JJ]. The model predicted collapse appropriately when the bubble is initially small and does not deform during its early life.

RADIATION

There is a substantial number of publications which are concerned with radiative transport in emitting, absorbing and scattering media.

A new approach to radiative heat transfer with scattering is presented and used to obtain the first general solution for radiative equilibrium in a non-gray, plane-parallel medium [46K]. The Galerkin method is applied to solve radiative heat transfer to an isotropically scattering, absorbing, and emitting plane-parallel medium with diffusely reflecting boundaries [41K]. Radiation heat transfer in a planar participating medium which scatters anisotropically is scaled to an isotropic scattering medium. The scaling that gives the best results is the one derived from the $P-1$ approximation [36K]. An approximate approach to radiative heat transfer in an isotropically scattering and non-gray absorbing medium indicates that the predictions for the radiative heat flux and intensity are accurate over large ranges in optical depth and scattering albedo [8K]. Radiative heat transfer in an isotropically scattering and non-gray absorbing planar medium is investigated using wide-band absorption quantities, including the effects of gray absorption and scattering by the scattering components and non-gray absorption by the gaseous components [7K].

The results of radiative heat transfer studies in an absorbing medium using complementary variational principles are in excellent agreement with those using more cumbersome solution procedures [19K].

Comparisons of three approximate numerical methods based on plane-symmetric, gray model (coal particle suspension) for evaluating radiant energy transfer in scattering and heat generating media indicate that the accuracy of flux methods decreases while the accuracy of the differential approximation increases with increasing optical thickness of the suspension [28K]. The coefficients for the weighted sum of the gray gas model are more accurate in terms of total properties over a wider range of temperatures and partial pressure–path length products than previously available coefficients [48K].

Theoretical investigations are reported of unsteady radiative heat transfer through an absorbing and emitting gray medium which occupies a semi-infinite space bounded by a semi-transparent gray plate [32K]. Time-dependent radiative transfer is studied in a semi-infinite medium with a reflecting boundary [18K].

A study of the role of various water clusters in i.r. absorption in the 8–14 μm window region indicates that homo-molecular water clusters larger than the dimer are not responsible for the continuum absorption [49K]. Infrared absorption measurements in the 6.3 μm band of H_2O vapor indicate that the dimerization effect on the monochromatic absorption coefficient and on the integrated absorption coefficient is negligible [34K]. Radiation intensity and emissivity of the 4.3 and 2.7 μm vibrational–rotational bands of CO_2 and of the 5.7 μm band of CO have been numerically calculated and measured over a wide range of parameters of vibrationally non-equilibrium mixtures expanding through the supersonic nozzle of a gas dynamic CO_2 laser [33K]. The K–ε model of turbulence is not completely satisfactory for predicting the temperature distribution and the dynamics of the mixed layer during simultaneous radiant heating and cooling or during cooling of thermally stratified water [54K].

Combined conductive and radiative heat transfer in an absorbing, emitting, and scattering cylindrical medium is studied using the Galerkin finite-element technique which is well suited to this problem because of the ease with which integrals with variable limits can be handled [16K]. The $P-1$ and $P-3$ methods yield results for combined conduction–radiation problems which are in close agreement with exact solutions for one-dimensional planar problems [44K]. Predictions of the effects of wall radiation and conduction on the stability of a fluid in a finite slot heated from below are in reasonable agreement with present and previous experimental data [24K]. Numerical solutions of the contact interface temperature of two semi-infinite media of different temperatures, considering internal thermal radiation and conduction effects, indicate that the technique is relatively simple and accurate [50K]. Results of a polynomial approximation method for the solution of heat transfer by conduction and radiation in an absorbing, emitting, and isotropically scattering medium show that the solution method is relatively simple and converges very quickly to the exact solution [12K].

An analysis of the effects of absorption of radiation by fluids during measurements of their thermal conductivity by the transient hot-wire technique shows that the contribution of radiative transport to the transient heating process means that the thermal conductivity derived from such measurements is systematically in error [38K]. Studies of combined radiation–convection in gray fluids enclosed in vertical cavities show that radiation decreases the intensity of the flow at low Rayleigh numbers and, in contrast, leads to an increased flow in convection regimes [35K].

From radiative flux measurements in the troposphere, heating rates for gases and for aerosols are calculated. Total heating rates of 0.175 and 0.377°C h^{-1} are determined for hazy and foggy atmospheres, respectively [51K]. In connection with scattering and absorption of solar radiation from aerosols in the atmosphere, the optical constants of aqueous solutions of sodium sulfate have been determined for a wavelength range from 2 to 20 μm [2K]. Experimental values of attenuation coefficients and transmissivities of radiant fluxes caused by aerosols consisting of fine metal powders differ significantly from those initially predicted [56K]. Studies of the absorption and scattering of thermal radiation within a dilute cloud of pulverized coal and char indicate that a simple hemispherical absorptivity of 0.89 may be used for heat transfer calculations if the particles can be assumed to act as Mie scatterers, and if the volume fraction of ash and soot particles is small [21K]. In order to truly represent scattering and radiative transfer by coal particles, a theory is required taking into account both the structure and the irregularities of coal particles [4K]. A method developed for determining the gray gas weighting coefficients for gas–soot mixtures from the coefficients which separately model the soot behavior and the behavior of the gray gases is accurate over wide ranges of temperature, path length and soot concentration [15K].

A Monte Carlo simulation is described for studying radiant transport through an adiabatic packed bed or porous solid [1K]. The results of radiative transfer studies in suspensions of 11 μm diameter latex particles using 0.6328 μm He–Ne laser light have direct application to the design of packed fluidized bed systems wherein thermal radiation is a significant heat transfer mode [6K]. Temperature profiles, total emissivity flux density distribution, and effective mean penetration distance have been determined from radiant heat transfer studies in high-temperature solar fluidized beds [17K].

A model designed and coded for application to radiative transfer in three-dimensional vegetative systems has been applied to a soybean row crop. The simulated directional reflectance data correspond well in gross trends with the measured data [29K]. A simple radiative transfer model has been used for establishing relationships between canopy biomass and reflectance which is of interest in agricultural remote sensing [42K]. In studies of i.r. thermography, calorimetric and radiation balance equations for an object in thermal equilibrium with its environment are combined to yield a graphic interpretation of exitance as a function of surface temperature, emittance, and background factors [47K].

A two-temperature i.r. pyrometer is suitable for monitoring the rapid heating of materials. The instrument covers a temperature range from 600 to 3000 K in approximately 100 ms with a spatial resolution of 1 mm and with an accuracy of a few percent [5K]. Infrared radiation has been used for sensitizing or desensitizing silver halide films for subsequent pulses of visible light [39K]. Physical properties of heat transfer sensors may be determined from vacuum heat loss observations [11K]. Infrared emission spectroscopy is a feasible technique over both a wide frequency and a large temperature range as illustrated by some experimental results [53K]. There is excellent agreement between CARS data and corrected thermocouple temperature measurements in diffusion flames in regions of relatively constant temperatures [14K].

The temperatures measured by the two-line fluorescence technique in plasmas are accurate to 3–5% and exhibit low random error [37K]. A semianalytic Monte Carlo radiative transfer model provides the framework for probing areas of uncertainty in laser fluorosensor performance [43K].

A simpler, semianalytical formulation for radiative view factors from a sphere to a class of axisymmetric bodies requires a simple numerical integration at most [10K]. A simple double-integral expression is derived for the diffuse radiation view factor between two parallel cylinders of finite length [26K]. A general formulation is developed for the radiation shape factors between a disk and a class of coaxial axisymmetric bodies such as a cylinder, a cone, an ellipsoid, and a paraboloid [40K]. It is shown that finite elements are quite versatile for calculating view factors for radiative heat transfer if extremely irregular geometries are involved [9K].

Simultaneous measurements of the emittance and electrical resistance of thin films (2000 Å) of Ni and stainless steel sputtered onto glass tubes have been made for a temperature range from 60 to 440°C [52K]. An interferometric method for the analysis of thin films has been extended to the i.r. region providing information on thickness, refractive index and absorption coefficient of a single layer [3K]. A twin-slope method for measuring the specific heat and coating absorptance of solid materials is based on monitoring of the heating and cooling rates of a sample irradiated with a standard lamp [25K]. Austenitic steels have a higher reflectance than the ferritic and martenistic steels, but their integrated solar reflectance of 68% is considered to be too low for solar reflectors [27K]. Equations are derived for analyzing the thermal effect of radiation on semiconductors and metals with sudden changes of the optical properties [20K].

Experimental and analytical studies are reported of the characteristics of heat transfer in a semi-transparent, laminar fluid flowing in an open channel which is heated uniformly by a radiant heat source from above [23K]. Experimental results on heat transfer augmentation by use of wall radiation for an annular channel where air flows between the inner tube with uniform heat input and an insulated outer tube are in good agreement with analytical predictions [55K].

The effect of a concentric radiating cylinder on liquid spray cooling of a hot gas discharge such as a product of combustion in a gas turbine is studied for a wide range of parameters [22K]. Studies of the effect of i.r. radiation on oriented polymer fibers and polymer fabrics show that the efficiency of radiative heating depends to a large degree on the direction of the incident radiation [31K].

Analytical studies of the effects of scattering of radiation within the medium, boundary surface reflectivity and initial optical thickness of the medium on the conversion occurring in a batch photochemical reaction in plane-slab geometry indicate that scattering reduces the conversion and tends to increase the uniformity of the distribution of the reactant within the medium [45K]. A comparison of experimental data with analytical results of radiant heat transfer in high-temperature furnaces shows satisfactory agreement [30K]. For freely radiating surfaces, the cooling power at ambient temperature lies between 58 and 113 W m^{-2} for different surfaces and model atmospheres [13K].

MHD

There is continuing interest in MHD related heat transfer, especially in connection with MHD generators.

The High Performance Demonstration Experiment (HPDE) in the field of MHD generator development produced 23 MW Faraday power at 9% enthalpy extraction efficiency using cold electrodes [3M]. Work is in progress to demonstrate that an MHD generator, simulating a commercial scale device, can convert 15% of the available thermal energy into electrical power [13M]. Experimental data derived from the HPDE facility for various electrical and gas dynamic variables are in good agreement with analytical predictions [4M]. Studies of the thermal behavior of capped vs solid frame electrodes in a coal-fired MHD generator show much lower heat fluxes on the capped electrodes as compared to the solid frame electrodes [5M]. Analytical studies of the same coal-fired MHD generator show that a quasi-three-dimensional code gives an excellent matching to experimental generator performance [6M]. The three-dimensional flow and heat transfer development in an MHD channel has been studied by solving the governing equations for the coupled electrical and flow fields [14M].

Studies of mass transfer and free convective MHD flow past an accelerated vertical porous plate show that the velocity increases when the modified Grashof number increases [9M]. Studies of the effect of Hall currents on MHD flow and heat transfer between two parallel porous plates indicate that in the absence of Hall currents, the change of the direction of the applied magnetic field does not affect the primary flow [1M].

Investigation of the effects of Hall current on Couette flow and heat transfer of a conducting fluid between two plates in a rotating system show that the role of each parameter on heat transfer is always opposite to its role on shear stress [7M]. The ADI method is most suitable for calculating the temperature distribution for steady MHD axial flow through a rectangular pipe with a discontinuity in the wall temperature [12M]. Experimental studies of local properties and heat transfer of Hg–Ar two-phase bubbly flow with small gas flow rate in a vertical annulus in the presence of a transverse magnetic field show that the two-phase Nusselt number decreases as well as the single phase Hg heat transfer with increasing Hartmann number at a fixed Péclet number [8M].

Nonlinear magnetic convection is studied for small Prandtl numbers using the modal equations for cellular convection [10M]. Studies of the effect of a magnetic field on heat flow in a gas in an intermediate pressure range indicate that the special features of the Senftleben–Beenakker effect observed with decreasing pressure arise not only due to spherically symmetric molecule–surface interaction, but also due to nonspherical scattering on walls [2M]. The effects of thermal radiation, the magnetic field strength, the finite electrical conductivity and the initial wave front curvature on the nonlinear breaking of modified magneto-gas dynamic waves are discussed [11M].

NUMERICAL METHODS

Since numerical methods are used for a variety of heat transfer problems, papers reporting the 'application' of numerical methods to specific physical problems are listed in the category pertaining to the problem. The literature cited in this section is primarily concerned with the development of new numerical techniques or with the evaluation of alternative methods.

The numerical methods for the solution of the heat conduction equation in curvilinear nonorthogonal coordinates are given a physical interpretation [15N]. An efficient sequential algorithm has been developed for the nonlinear inverse heat conduction problem [2N]. The use of the ADI methods for the polar coordinates is discussed [3N]. Solution of thermal problems by the boundary point least squares method is examined [11N]. An efficient algorithm is proposed for nonlinear heat transfer with phase change [14N]. Finite elements in time and space coordinates are used for heat conduction problems [21N]. Reference [18N] describes a splitting scheme for solving the heat conduction equation in cylindrical coordinates. The stability of mixed time integration schemes is examined for unsteady thermal problems [8N]. An artificial intelligence system called KNOWTRAN has been proposed for solving heat transfer problems [17N]. Optimal finite-analytic methods have been described [9N]. Reference [6N] establishes relationships between the truncation errors on uniform and nonuniform meshes. The use of a finite-element model for the Dirichlet boundary conditions in the Laplace equation is discussed [24N]. Self-adaptive solution algorithms are proposed for nonlinear heat conduction [12N]. The conjugate-gradient method is used to develop accurate integration formulas for unsteady heat conduction [23N].

New finite-difference schemes have been developed for parabolic differential equations [10N]. Numerical techniques are described for radiation heat transfer in participating media [16N]. A free convection flow is calculated by the finite-element method [4N]. In ref. [5N], the vorticity–velocity variables are used in obtaining numerical solutions for two-dimensional Navier–Stokes equations. A numerical method is proposed for the transonic cascade flow [22N]; the method is presented as thermodynamically consistent.

The representation of the combined convection and diffusion flux and the associated matter of false diffusion are given attention in several papers. An 'upwind' finite-element method is proposed via numerical integration [13N]. Reference [19N] presents a performance comparison of many methods for a standard convection–diffusion problem; the paper includes a commentary on the solutions submitted by a large number of researchers using different methods. An error analysis is performed for the finite-element method applied to convection–diffusion problems [1N]. Reference [7N] proposes boundary-layer refinements in convection–diffusion problems, while these problems are simulated in ref. [20N] by the influence schemes as well as standard finite-difference schemes.

HEAT TRANSFER APPLICATIONS

Heat exchangers and heat pipes

Heat transfer augmentation by proper roughness of the heat exchanger surface was studied, especially in the Soviet Union. Tubes with internal ridges generated by electrochemical etching is claimed to reduce the heat exchanger volume by a factor of 2.5 [9Q]; whereas periodic grooves on the tube surface effect a reduction of 30–50% [8Q]. The enhancement of heat transfer from a tube bundle in turbulent axial flow by spacer grids was studied experimentally [2Q]. The use of tubes with fins enhances heat transfer in film condensation as demonstrated by experiments with Freon 12 [3Q]. A microroughness parameter determines enhancement or suppression of nucleation in heat exchangers and steam generators [1Q]. The use of helical tubes in heat exchangers results in an appreciable increase in heat transfer and a corresponding reduction in heat exchanger dimensions [7Q]. Prediction equations describing heat transfer augmentation in rod bundles by grid spacers are compared with experimental results [17Q]. Swirling spacers are also considered. Evaporative cooling has been found to be the utmost heat transfer augmentation [12Q]. Experiments show that the performance of plate-fin and bare tube condensers can be predicted reliably. Experiments clarified contradictions found in data reported in the literature on the effect of pulsating flow in steam–water heat exchangers at Reynolds numbers between 500 and 2200 [10Q]. A multidimensional numerical model to predict the performance of shell and tube liquid-metal heat exchangers was developed and the results are compared with test data [14Q]. Steady laminar forced convection heat transfer in a spherical-annulus heat exchanger with water flowing on both sides was studied at Reynolds numbers between 41 and 465 by numerical analysis and by experiments [15Q].

Optimizations of heat exchangers with respect to construction cost, operating cost, or pay-back time are compared with the help of a new diagram [13Q]. A computer code developed for the optimization of air cooled heat exchangers should be a useful tool for heat exchanger design [6Q]. An enhanced tube heat exchanger for boiling of ammonia on the shell side and sea-water flowing on the tube side is considered for a case study of performance benefits and cost effectiveness [16Q]. A size reduction of 55–67% relative to a plane tube exchanger was made possible by the use of a porous surface on the boiling side.

An explicit procedure for the design of balanced regenerators has been developed [11Q].

The shut-down characteristics of an axial-groove heat pipe as thermal diode are determined by experiments and analysis [4Q, 5Q]. The shut-down times were found to depend on the shut-down power.

General

An analytical model simulating the cooling air flow in gas turbine components is discussed and compared with static bench tests [13S]. The heat flux and pressure at the end wall region and on the shroud in a turbine stage were measured with thin film heat transfer gages [7S]. The effects of the Mach number, Reynolds number, inflow angle, and turbulence level were measured on a gas turbine rotor blade in a piston tunnel using thin film gages [5S].

The time variation of the local heat flux through

various parts of the combustion chamber wall were measured on a pre-chamber diesel engine [19S]. The heat flux was mainly affected by the jet flow from the pre-combustion chamber. It was also primarily influenced by the engine speed and the volumetric efficiency [1S]. The highest heat transfer occurred near the stoichiometric mixture [2S]. An increase of the volumetric efficiency from 40 to 60% resulted in an increase of the peak heat flux of 30%.

Heat transfer from the combustion gases to a single row of closely spaced tubes of a Stirling engine heater was significantly enhanced by a swirling cross flow [35].

An analytical model predicts the temperature field in a sodium heated steam generator under oscillation induced at the critical heat flux [14S]. The model is based on the postulate that wet regions are swirling around the tube circumference.

Screening criteria were developed for the selection of fluids for long-distance heat transmission [18S]. Dowtherms, therminol, and water were considered among others. The transient heat transfer characteristics were studied by analysis and experiment for a vertical porous heat storage system imposing a constant heat flux through one wall whereas the other wall is insulated [12S]. It was found that natural convection strongly enhances the time for the approach to quasi-steady state. A similar study comes to the conclusion that a predetermined time for approach to steady state can be achieved by a proper combination of the properties of the porous medium and the dimension of the storage container [11S]. A two-phase heat transfer model simulates heat storage systems using a packed bed [17S].

Two papers are of interest for fire prevention. Experiments investigated the dynamics of smoke filling the upper hot air layer in a series of two or three rooms connected by open doorways [6S]. A numerical parametric study of the response of building components to a fire agree essentially with earlier dimensional and approximate analyses [10S].

Experiments determined the drag and heat transfer coefficients of climbing or horizontal thin films of an evaporating liquid as found in desalination plants for sea water [15S]. The real equilibrium temperature and the instantaneous cooling rate are determined for preheated river water [9S].

Heat transfer studies show that two temperature maxima can occur in large geologic repositories for spent fuel from nuclear reactors, one after 60 and one after 13 000 years [4S].

Regular arrays of heated areas arranged on a flat surface occur in electronic computers. Experiments clarified the heat transfer and pressure drop characteristics of such arrays of rectangular modules inserted into a flat wall forming one side of a rectangular duct with cooling air flowing through the duct [16S]. The effect of missing modules on the heat transfer of neighboring modules and heat transfer enhancement by protruding bars were especially investigated.

Experiments and analyses studied the thermal response of a human thigh exposed to diathermy treatment with cooling dependent on the blood flow rate [8S].

Solar energy

Calibrated terrestrial solar spectra data are presented in order to satisfy two objectives: to provide spectral data of high enough resolution and on a broad enough wavelength scale to make the data useful for studies where spectral position and band width are critical factors; and to provide in one location sets of spectra that illustrate several atmospheric effects on the solar spectrum [3T]. A series of total and spectral solar irradiance measurements were made on the ground (Table Mountain Facility, California, altitude 2.18 km) and the spectral data are presented for the 0.3–3.0 μm range for air mass 1.5 [27T]. The simultaneous measurement of solar radiation at several angular positions is accomplished with a newly designed multidirectional photodiode array [9T]. Hourly pyrheliometer and pyranometer data from four U.S. locations are used to establish a relationship between the hourly diffuse fraction and the hourly clearness index [10T]. The performance of models to estimate solar irradiance and its components is assessed using data for six Canadian stations for nine years [8T]. Four methods of estimating hourly diffuse irradiation from hourly global irradiation are compared, using global and diffuse irradiation data from five widely separated locations [38T]. An analytical solution is presented for the calculation of monthly average insolation on shaded surfaces at any tilt and azimuth [36T]. The annual average solar radiation incident on a variety of principal solar collector types, including shading effects, is calculated by approximating the annual variation of daily radiation by its value on one average day, equinox [13T]. A model for clear-day solar irradiance for a specific site is presented in a simple format utilizing parameters based on readily available long-term daily global irradiance data and the model results are found to be in excellent agreement with observations [21T]. A new empirical formula has been developed which permits the determination of equivalent nocturnal sky emissivity as a function of the site's altitude, ambient temperature, relative humidity and the degree of cloudiness, mist or haze present at the time [4T].

A fairly simple technique has been developed to calculate monthly utilizability or monthly collection efficiency for solar thermal collectors and the heart of the technique is an empirically determined performance map that provides quick evaluations of changes in collector design, geographic location, and collector inlet temperature [11T]. An explicit nonlinear model which predicts the performance of a solar collector, including effects of radiative losses, was obtained from the empirical correlation of simulation data from a total of 3969 separate computer simulations which included 50 different collector designs and a wide range of operating conditions [30T]. A survey is presented of the work done at the University of Melbourne with regard to the effectiveness of the insertion of a honeycomb structure into the air gap between the absorber plate and the cover of a flat-plate solar collector in order to suppress convection [15T]. The operation of a conventional solar air heater with two covers in a two-pass mode offers an inexpensive method of improving collector efficiency by about 10–15% [40T]. The energy absorbed by two types of tubular solar receiver elements were analyzed: (a) an inner tube with an absorbing coating surrounded by a semi-transparent cover tube, and (b) a semi-transparent inner tube filled with an absorbing fluid surrounded by a semi-transparent cover tube [34T]. Wind tunnel experiments were carried out to determine forced convection heat transfer coefficients for the leeward face of a roof-like structure [37T].

A comparison of the gain that can be achieved with a one- or two-stage concentrator, when the first stage is a Fresnel lens or parabolic mirror, shows that the achievable gain using a parabolic mirror is greater than that obtained using a flat or roof lens but is lower than that obtained using a curved lens [23T]. The conditions for achieving ideal and Lambertian symmetrical solar concentrators are analyzed [24T]. A general procedure for determining the optimum geometry of a reflector-augmented solar collector which produces a desired pattern of flux-augmentation is described [5T]. Natural convection heat transfer coefficients have been experimentally determined for trough-type collectors [28T].

The transmittance of unpolarized light through a solar collector cover made of cylindrical glass tubes in a coplanar parallel close-packed array is dependent on the orientation of the cylindrical axes of the tubes in the plane of the cover [16T]. A new apparatus that incorporates an integrating sphere is described which enables the solar transmittance of test samples to be measured as a function of both angle of incidence and azimuth angle [39T].

A mathematical model for simulating the dynamic temperature response of a packed column to an arbitrary time-dependent inlet air temperature is used to optimize heat storage in a rock bin system subject to a realistic transient inlet temperature [33T]. A general method for solving the differential equations describing the heat transfer process within a rock bed is presented and a numerical model accounting for secondary phenomena such as thermal losses and conduction effect is developed [7T]. A preliminary model for estimating possible thermal energy storage in a phase change shell and tube heat exchanger is presented [35T]. A comparison is made of three models for analyzing the thermal behavior of a paraffin wax thermal store powered by a liquid-type heat transfer fluid [26T]. An experimental study of the temperature decay in a thermally stratified water body with various initial temperature distributions indicate that the temperature field is largely one-dimensional [20T]. In a two-part paper, axisymmetric natural convection flows in a thermocline-type cylindrical enclosure are studied experimentally with a laser-Doppler velocimeter [17T] and a numerical model was developed using these experimental data [18T]. The effects of storage tank stratification on the instantaneous performance of a liquid-based solar heating system are presented in terms of a stratification coefficient which is shown to be a system constant that depends on only two dimensionless system parameters [31T].

After summarizing the methods for calculating the solar contribution for systems without thermal storage, this paper extends a previously proposed method which is based on using a frequency distribution of solar flux data [12T]. Using analytical methods it is shown that the effects of transition times of fluids in collector arrays and system piping on energy collection in solar systems with well-mixed storage tanks can be accounted for by a reduction factor multiplying the collector heat removal factor, F_R [41T]. The problem of nonuniform flow distribution on the thermal performance of solar collectors is treated and a method of determination of the deterioration of the collector efficiency due to the flow nonuniformity is developed and presented [6T].

A recently-developed analytic model for predicting passive solar heating performance is applied to the case of selective coatings on storage walls [14T]. Flow visualization studies on laboratory models of Trombe walls have given a deeper insight into the fundamental flow mechanisms, while air velocity and temperature measurements have been used to explore the natural convection heat transfer processes involved in the thermocirculation flow [1T]. The monthly-average auxiliary energy requirements of a building with a Trombe wall is estimated using upper and lower theoretical limits to system performance and an empirical correlation is presented for the fraction of the load met by the Trombe wall for systems with finite capacity [29T]. A paraffin Trombe wall with double glazing is studied experimentally with the help of a test-cell allowing several working conditions to be checked [2T]. Numerical studies were performed of the daily temperature fluctuations in a direct gain room [25T].

Events that induce or inhibit mixing layer development in a thermohaline system are simulated in laboratory experiments involving salt-stratified solutions heated from below, cooled from above and/or irradiated from above and a Mach–Zehnder interferometer was used to visualize mixing layer development [22T]. Calculations indicate that decreasing the reflectivity of the bottom of a solar pond from 0.20, typical of dirt, to 0.05 significantly increases pond thermal efficiency [19T]. A numerical calculation of the heat and mass transfer in solar stills is reported [32T].

PLASMA HEAT TRANSFER

Plasma heat transfer continues to attract interest, in particular in the areas of arc technology and plasma processing.

Experimental and analytical studies of the recovery of an axially blown SF_6 arc after zero current indicates that the temperature change is determined by flow turbulence and deviations from LTE have also some effect on the shape of the recovery curve [2U, 18U, 22U]. Experimental studies of a SF_6 arc plasma at atmospheric pressure indicate that there is collisional equilibrium for sulfur and demixing has a negligible effect on the plasma state for the chosen conditions [26U]. Investigations of the mechanisms for temperature decay in a freely recovering gas blast arc indicate that axial convection tends to heat the arc channel downstream of the stagnation region in both N_2 and SF_6. Radial convection is a dominant cooling mechanism for both N_2 and SF_6 [20U].

Measured isotherms in the vicinity of a thermionically emitting (thoriated W) cathode in a high intensity, atmospheric pressure argon arc reveal temperatures up to 26 000 K and temperature gradients in excess of 8.7×10^3°C mm^{-1} [1U]. A hot cathode arc discharge can be initiated with an auxiliary discharge which heats the cathode to the required temperatures [11U]. Using a Cu–W cathode in a pulsed free-burning arc (12.5 kA) at atmospheric pressure, the vapor supply from the cathode seems to eliminate the usually observed pumping action in the vicinity of the cathode and the associated enthalpy transport to the anode [28U].

Electrode heat transfer in vortex-stabilized arcs for both DC and AC operation is a linear function of the current and in the case of AC, electrode heat transfer is approximately half of the heat load measured at the anode of a DC arc under similar conditions [5U]. Anode spot formation in vacuum arcs is closely linked to anode melting and possibly to magnetic deformation of the molten surface. This lends support to the conjecture that localized heating of the anode can trigger spot formation [8U].

The application of the conventional multitemperature Saha equation to high pressure plasma at temperatures (heavy particles) below 8000 K may lead to substantial errors [19U]. Experimental and analytical studies of deviations from kinetic equilibrium in a decaying, atmospheric pressure argon arc plasma show that spectral line intensity measurements provide reliable data for the gas temperature [7U]. The temperature of a high-pressure gas discharge may be determined from the total intensity of a self-reversed spectral line if the plasma is in local thermodynamic equilibrium [9U].

A relatively simple method is presented by which the amount of self-absorption in laboratory-produced plasmas can be evaluated [17U]. A new electro-optical technique for measuring temperatures in thermal plasmas is based on the determination of the variations in the outputs of a two-wavelength pyrometer which views a small section of the plasma [15U]. A computer-controlled spectrometric system for measuring temperatures in thermal plasmas is particularly useful if fast scanning is necessary or desirable [6U].

Recent developments are reviewed associated with the extraction and refining of metals using thermal plasmas as the heat source [12U]. The two major heat transfer components in a nitrogen transferred arc, resembling an arc furnace, are enthalpy transport by the electrons to the anode and radiative heat transfer from the arc column [25U]. Theoretical and experimental investigations confirm that a hybrid plasma torch, in which a rf plasma is superimposed to an arc jet, is an effective reactor for the preparation of ultra fine refractory compounds [24U]. Comparisons of predicted and measured temperature and velocity fields in thermal plasma jets reveal good overall agreement. Radiation has a small effect on the overall heat transfer process [14U].

Evaporation severely reduces heat transfer to a particle exposed to a thermal plasma and, in general, this effect is more pronounced for materials with low latent heat of evaporation [3U]. Studies of the unsteady heating of small particles in thermal plasmas show that evaporation (or sublimation) is by far the slowest step among all the processes taking place in a plasma reactor [4U]. Experimental and analytical studies of the heat and momentum transfer to particles injected into a thermal plasma jet (plasma spraying) reveal good agreement between predicted and measured particle velocities. There is, however, a discrepancy between predicted and measured surface temperatures of the particles [27U]. An experimental study of the thermodynamic parameters, optical properties and heat transfer from a thermal plasma produced by reflection of a shock wave from a plane surface indicates that adding of solid particles to the air reduces the heat flux density at velocities in excess of 5 km s^{-1} [10U].

A reactive plasma process proved to be successful for the preparation of solar grade silicone. With this plasma zone melting technique, the iron content was, for example, reduced from 3500 to 0.5 ppm at displacement rates from 20 to 40 cm h^{-1} [16U]. When a mixture of Nb_2O_5 and B is heated in a thermal argon plasma to approximately 3000°C and subsequently annealed at approximately 2000 K, a superconducting NbB_2 is formed with a transition temperature $T_c > 4$ K [13U]. Hydrogen peroxide (H_2O_2) may be produced by dissociating water vapor in a rf discharge. Heating of the rf plasma and its spatial afterglow from 25 to 600°C has little effect on the percentage of H_2O dissociation and on the formation of H_2O_2 [21U]. A supersonic beam of atomic oxygen is produced by using a conventional plasma torch as the heat source. Dissociation efficiencies approaching unity are achieved with beam velocities in the range of 1.5–4 km s^{-1} [23U].

REFERENCES

Books

G. Bankoff (editor), *Heat Transfer in Nuclear Reactor Safety*, Hemisphere, Washington, DC (1982).

M. Cumo, *Termotechnica Sperimentale*, Comitato Nazionale per La Ricerca E Per Lo Sviluppo Dell'Energia Nucleare E Delle Energie Alternative, Luglio (1982).

P. Grassmann, *Physikalische Grundlagen der Verfahrenstechnik*, Otto Salle, Frankfurt am Main, Verlag Sauerländer AG, Aarau (1982).

G. F. Hewitt, J. M. Delhaye and N. Zuber (editors), *Multiphase Science and Technology*, Vol. 1. Hemisphere, Washington, DC (1982).

S. Kakav, R. K. Shan and A. E. Bergles (editors), *Low Reynolds Number Flow Heat Exchangers*. Hemisphere, Washington, DC (1982).

B. V. Karlekar and R. M. Desmond, *Heat Transfer* (2nd edn.). West, St. Paul, Minnesota (1982).

A. D. Kraus, *Analysis and Evaluation of Extended Surface*, Hemisphere, Washington, DC (1982).

J. F. Krieder and F. Kreith, *Solar Heating and Cooling: Active and Passive Design* (2nd edn.). Hemisphere, Washington, DC (1982).

W. Merzkirch (editor), *Flow Visualization II: Proceedings of the Second International Symposium*. Hemisphere, Washington, DC (1982).

T. Mizushina and Wen-Jei Yang (editors), *Heat Transfer in Energy Problems*, Hemisphere, Washington, DC (1982).

M. S. Plesset, N. Zuber and I. Catton, *Transient Two-phase Flow*. Hemisphere, Washington, DC (1982).

M. Van Dyke, *An Album of Fluid Motion*. Parabolic Press, Stanford, California (1982).

L. Van Wijngaarden (guest editor), *Applied Scientific Research* (an international journal on thermal, mechanical, and electromagnetic phenomena in continua), *Mechanics and Physics in Bubbles in Liquids*, Vol. 28. Martinus Niihoff, The Hague (1982).

Z. Zarić, *Structure of Turbulence in Heat and Mass Transfer*. Hemisphere, Washington, DC (1982).

Heat Transfer 1982: Proceedings of the 7th International Heat Transfer Conference, München, Federal Republic of Germany, September (1982) (edited by U. Grigull, E. Hahne, K. Stephan and J. Straub). Hemisphere, Washington, DC (1982).

New journals

The International Journal of Heat and Fluid Flow. IPC Science and Technology Press, Surrey, U.K.

Conduction

1A. M. J. Ades and K. L. Peddicord, A model for effective thermal conductivity of unrestructured sphere-pac fuel, *Nucl. Sci. Engng* **81**, 540 (1982).

2A. H. Alsmeyer, U. Müller and M. Reimann, Heat transfer from liquids to melting or subliming substances, *Chemie–Ingr–Tech.* **54**, 172 (1982).

3A. H. H. Bau, Temperature distribution in and around a buried heat generating sphere, *Int. J. Heat Mass Transfer* **25**, 1701 (1982).

4A. H. H. Bau and S. S. Sadhal, Heat losses from a fluid flowing in a buried pipe, *Int. J. Heat Mass Transfer* **25**,

1621 (1982).
5A. M. Bobeth and G. Diener, Variational bounds for the effective thermal contact resistance between bodies with rough surfaces, *Int. J. Heat Mass Transfer* **25,** 111 (1982).
6A. M. Bobeth and G. Diener, Upper bounds for the effective thermal contact resistance between bodies with rough surfaces, *Int. J. Heat Mass Transfer* **25,** 1231 (1982).
7A. C. J. Cremers and V. K. Mehra, Frost formation on vertical cylinders in free convection, *J. Heat Transfer* **104,** 3 (1982).
8A. H. Doss, R. van Eldik and H. Kelm, Construction and testing of a high-pressure Joule-heating temperature-jump apparatus, *Rev. Scient. Instrum.* **53,** 1592 (1982).
9A. R. Farhadieh, Downward penetration of a hot liquid pool into the horizontal surface of a solid, *J. Heat Transfer* **104,** 199 (1982).
10A. P. E. Frivik and G. Comini, Seepage and heat flow in soil freezing, *J. Heat Transfer* **104,** 323 (1982).
11A. K. E. Gustafson, Exact solutions and ignition parameters in the Arrhenius conduction theory of gaseous thermal explosion, *Z. Angew Math. Phys.* **33,** 392 (1982).
12A. S. E. Gustafsson, K. Ahmed, A. J. Hamdani and A. Maqsood, Transient hot-strip method for measuring thermal conductivity and specific heat of solids and fluids: second order theory and approximations for short times, *J. Appl. Phys.* **53,** 6064 (1982).
13A. P. J. Heggs, D. B. Ingham and M. Manzoor, The analysis of fin assembly heat transfer by a series truncation method, *J. Heat Transfer* **104,** 210 (1982).
14A. S. Heurtault, J. M. Badie, A. Rouanet and G. Arnaud, Solidification de sphères liquides surchauffées à caracteristiques physiques variables, *Int. J. Heat Mass Transfer* **25,** 1671 (1982).
15A. T. V. Hromadka, II and G. L. Guymon, A note on approximation of one-dimensional heat transfer with and without phase change, *Num. Heat Transfer* **5,** 223 (1982).
16A. S. R. K. Iyengar, Conjugate unsteady heat transfer from a droplet in creeping flow, *A.I.Ch.E. Jl* **28,** 700 (1982).
17A. A. M. Kanury and D. J. Holve, Transient conduction with pyrolysis (approximate solutions for charring wood slabs), *J. Heat Transfer* **104,** 338 (1982).
18A. H. F. P. Knaap and J. J. M. Beenakker, Heat conduction in a nearly collisionless molecular gas, *Physica* **112A,** 29 (1982).
19A. M. Kobari, Y. Shimizu, M. Endo and H. Inazumi, Effective thermal conductivity of fibrous sheets, *Heat Transfer—Jap. Res.* **10**(3), 48 (1981).
20A. J. Lach and W. Pieczka, Heat transfer at the cladding–cooling fluid boundary, *Int. J. Heat Mass Transfer* **25,** 1595 (1982).
21A. M. Lal, Some axially symmetric thermal stress distributions in an elastic solid containing an annular crack, *Int. J. Engng Sci.* **20,** 1261 (1982).
22A. E. Lorenzini and M. Spiga, Temperature in heat generating solids with memory, *Wärme- und Stoffübertragung* **16,** 113 (1982).
23A. S. K. Loyalka and J. R. Thomas, Jr., Heat transfer in a rarefied gas enclosed between parallel plates: role of boundary conditions, *Physics Fluids* **25,** 1162 (1982).
24A. C. C. Meek, R. C. Doerner and W. J. Adams, Determination of fuel-clad thermal gap conductance by use of Kalman filter methods, *Nucl. Sci. Engng* **81,** 560 (1982).
25A. T. Miura, H. Tajima, K. Miura and S. Ohtani, Measurement of effective thermal conductivity of packed coal during carbonization, *Heat Transfer—Jap. Res.* **10**(3), 81 (1981).
26A. T. Miyazaki and W. H. Giedt, Heat transfer from an elliptical cylinder moving through an infinite plate applied to electron beam welding, *Int. J. Heat Mass Transfer* **25,** 807 (1982).
27A. K. O. Möller and L. C. M. De Maeyer, Temperature-jump apparatus with infrared detection for relaxation studies in thin samples, *Rev. Scient. Instrum.* **53,** 1596 (1982).
28A. J. P. Monchalin, J. L. Parpal, L. Bertrand and J. M. Gagné, Investigation of gaseous heat transfer between surfaces by photoacoustic effect, *J. Appl. Phys.* **53,** 8525 (1982).
29A. F. E. Moore and Y. Bayazitoglu, Melting within a spherical enclosure, *J. Heat Transfer* **104,** 19 (1982).
30A. J. Neiss and E. R. F. Winter, Calculations of thermal conductivity of soils above and below the freezing point, *Chemie–Ingr–Tech* **2,** 176 (1982).
31A. M. G. O'Callaghan, E. G. Cravalho and C. E. Huggins, An analysis of the heat and solute transport during solidification of an aqueous binary solution—I. Basal plane region, *Int. J. Heat Mass Transfer* **25,** 553 (1982).
32A. M. G. O'Callaghan, E. G. Cravalho and C. E. Huggins, An analysis of the heat and solute transport during solidification of an aqueous binary solution—II. Dendrite tip region, *Int. J. Heat Mass Transfer* **25,** 563 (1982).
33A. E. Oliveri, F. Castiglia, S. Taibi and G. Vella, A new correlation for quench front velocity, *Int. J. Heat Mass Transfer* **25,** 1589 (1982).
34A. M. Pachter and R. I. Becker, Simple inequalities in the vibrating string and heat conduction problems, *J. Franklin Inst.* **313,** 165 (1982).
35A. D. Ramkrishna and N. R. Amundson, A non-self-adjoint problem in heat conduction, *J. Heat Transfer* **104,** 185 (1982).
36A. N. Ramachandran, J. P. Gupta and Y. Jaluria, Thermal and fluid flow effects during solidification in a rectangular enclosure, *Int. J. Heat Mass Transfer* **25,** 187 (1982).
37A. H. Rieger, U. Projahn and H. Beer, Analysis of the heat transport mechanisms during melting around a horizontal circular cylinder, *Int. J. Heat Mass Transfer* **25,** 137 (1982).
38A. B. Rubinsky, Solidification of a conglomerate of particles, *J. Heat Transfer* **104,** 193 (1982).
39A. B. Rubinsky, Thermal stress during solidification processes, *J. Heat Transfer* **104,** 196 (1982).
40A. M. S. Sadeghipour, M. N. Özişik and C. Mulligan, Transient freezing of a liquid in a convectively cooled tube, *J. Heat Transfer* **104,** 316 (1982).
41A. A. Saito, F. Kuriyama and R. Shimomura, Experimental research on the freezing process with supercooling, *Heat Transfer—Jap. Res.* **11**(2), 70 (1982).

42A. A. Saito, R. Shimomura, F. Kuriyama and K. Katayama, Heat conduction with supercooled solidification, *Bull. J.S.M.E.* **25,** 591 (1982).
43A. R. V. Seeniraj and T. K. Bose, Planar solidification of a warm flowing liquid under different boundary conditions, *Wärme- und Stoffübertragung* **16,** 105 (1982).
44A. R. Siegel, Analysis of solidification interface shape resulting from applied sinusoidal heating, *J. Heat Transfer* **104,** 13 (1982).
45A. R. Siegel and D. J. Sosoka, Cauchy integral method for two-dimensional solidification interface shapes, *Int. J. Heat Mass Transfer* **25,** 975 (1982).
46A. J. A. Shaelwitz and K. T. Raterman, Modeling and calculation of interfacial resistances with energy profiles, *Ind. Engng Chem. Fundam.* **21,** 154 (1982).
47A. I. Shai and M. Santo, Heat transfer with contact resistance, *Int. J. Heat Mass Transfer* **25,** 465 (1982).
48A. N. Shamsundar, Approximate calculation of multidimensional solidification by using conduction shape factors, *J. Heat Transfer* **104,** 8 (1982).
49A. R. E. Showalter, Mathematical formulation of the Stefan problem, *Int. J. Engng Sci.* **20,** 909 (1982).
50A. Z. P. Shulman, L. N. Novichyonok, E. P. Belskaya, B. M. Khusid and V. V. Melnichenko, Thermal conductivity of metal-filled systems, *Int. J. Heat Mass Transfer* **25,** 643 (1982).
51A. B. Snaith, P. W. O'Callaghan and S. D. Probert, Minimizing the thermal resistance of pressed contacts, *J. Mech. Engng Sci.* **24,** 183 (1982).
52A. A. D. Solomon, V. Alexiades and D. G. Wilson, The Stefan problem with a convective boundary condition, *Q. Appl. Math.* **40,** 203 (1982).
53A. E. M. Sparrow and J. A. Broadbent, Inward melting in a vertical tube which allows free expansion of the phase-change medium, *J. Heat Transfer* **104,** 309 (1982).
54A. E. M. Sparrow and M. K. Chyu, Conjugate forced convection–conduction analysis of heat transfer in a plate fin, *J. Heat Transfer* **104,** 204 (1982).
55A. E. Takegoshi, S. Imura, Y. Hirasawa and T. Takenaka, A method of measuring the thermal conductivity of orthogonal anisotropic materials by a transient hot wire method, *Heat Transfer—Jap. Res.* **11**(3), 74 (1982).
56A. Y. Tanigawa and Y. Takeuti, Coupled thermal stress problem in a hollow sphere under a partial heating, *Int. J. Engng Sci.* **20,** 41 (1982).
57A. D. A. Tarzia, An inequality for the coefficient σ of the free boundary $s(t) = 2\delta\sqrt{t}$ of the Neumann solution for the two-phase Stefan problem, *Q. Appl. Math.* **39,** 491 (1982).
58A. E. Ukeje, Weak shock waves in heat conducting thermoelastic materials, *Int. J. Engng Sci.* **20,** 1275 (1982).
59A. Y. Urano and T. Watanabe, An analysis of multi-layer wall heat transfer by state transition matrices. Part II. A successive calculation method and its accuracy, *Heat Transfer—Jap. Res.* **11**(2), 1 (1982).
60A. C. Y. Wang, Heat conduction in an undulating heating wire, *Appl. Scient. Res.* **39,** 181 (1982).
61A. Y. C. Yortsos and G. R. Gavalas, Heat transfer ahead of moving condensation fronts in thermal oil recovery processes, *Int. J. Heat Mass Transfer* **25,** 305 (1982).

Channel flow

1B. G. Ye. Aerov, A. N. Devoyno and L. I. Kolykhan, Experimental study of the mechanism of heat transfer in turbulent flow of a chemically reacting heat transfer fluid in a circular pipe, *Heat Transfer—Soviet Res.* **13**(5), 73 (1981).
2B. R. O. Ayeni, On the thermal runaway of variable viscosity flows between concentric cylinders, *Z. Angew. Math. Phys.* **33,** 408 (1982).
3B. H. F. Bauer, Mass transport for Couette flow in a flat conduit, *Wärme- und Stoffübertragung* **16,** 137 (1982).
4B. B. H. Chen and A. F. McMillan, Heat transfer and axial dispersion in batch bubble columns, *Can. J. Chem. Engng* **60,** 436 (1982).
5B. S. M. Dinh and R. C. Armstrong, Non-isothermal channel flow of non-Newtonian fluids with viscous heating, *A.I.Ch.E. Jl* **28,** 294 (1982).
6B. M. A. El Hadidy, F. Gori and D. B. Spalding, Further results on the heat transfer to low-Prandtl number fluids in pipes, *Num. Heat Transfer* **5,** 107 (1982).
7B. M. A. I. El-Shaarawi and A. Sarhan, Combined forced-free laminar convection in the entry region of a vertical annulus with a rotating inner cylinder, *Int. J. Heat Mass Transfer* **25,** 175 (1982).
8B. M. Epstein and F. B. Cheung, On the prediction of pipe freeze-shut in turbulent flow, *J. Heat Transfer* **104,** 381 (1982).
9B. M. Fallen, Heat transfer in a pipe with superimposed pulsating flow, *Wärme- und Stoffübertragung* **16,** 89 (1982).
10B. B. M. Galitseyskiy and A. A. Nozdrin, The effect of gas flow fluctuations on heat transfer coefficients in slotted channels, *Heat Transfer—Soviet Res.* **13**(4), 25 (1981).
11B. S. Ganeshan and M. Raja Rao, Studies on thermohydraulics of single and multi-start spirally corrugated tubes for water and time-independent power law fluids, *Int. J. Heat Mass Transfer* **25,** 1031 (1982).
12B. M. A. Habib and J. H. Whitelaw, The calculation of turbulent flow in wide-angle diffusers, *Num. Heat Transfer* **5,** 145 (1982).
13B. C. A. Hieber, Laminar convection in an isothermal horizontal tube: correlation of heat transfer data, *Int. J. Heat Mass Transfer* **25,** 1737 (1982).
14B. C.-F. Hsu and S. V. Patankar, Analysis of laminar non-Newtonian flow and heat transfer in curved tubes, *A.I.Ch.E. Jl* **28,** 610 (1982).
15B. S. Komori, H. Ueda, F. Ogino and T. Mizushina, Turbulence structure and transport mechanism at the free surface in an open channel flow, *Int. J. Heat Mass Transfer* **25,** 513 (1982).
16B. S. S. Kutateladze, Similarity analysis and asymptotic models, *Wärme- und Stoffübertragung* **16,** 3 (1982).
17B. E. Y. Kwack, J. P. Hartnett and Y. I. Cho, Turbulent heat transfer in circular tube flows of viscoelastic fluids, *Wärme- und Stoffübertragung* **16,** 35 (1982).
18B. M. E. Lago, Study of stability in ionic mass transfer in inclined plane channels, *Heat Transfer—Soviet Res.* **13**(6), 64 (1981).
19B. S.-L. Lee, Forced convection heat transfer in low Prandtl number turbulent flows: influence of axial conduction, *Can. J. Chem. Engng* **60,** 482 (1982).
20B. E. F. Matthys and R. H. Sabersky, A method of

predicting the 'diameter effect' for heat transfer and friction of drag-reducing fluids, *Int. J. Heat Mass Transfer* **25,** 1343 (1982).

21B. N. D. Mehta and K. J. Bell, Laminar flow heat transfer in a tube preceded by a 180° bend, *Heat Transfer—Soviet Res.* **13**(6), 71 (1981).

22B. D. E. Metzger, R. A. Berry and J. P. Bronson, Developing heat transfer in rectangular ducts with staggered arrays of short pin fins, *J. Heat Transfer* **104,** 700 (1982).

23B. L. V. Mishina, G. Z. Serebryannyy and N. N. Tushin, Correlations of heat and mass transfer parameters for turbulent pipe flow of gases not in chemical equilibrium (boundary conditions of the second kind), *Heat Transfer—Soviet Res.* **13**(5), 88 (1981).

24B. T. Mukushi and S. Takahashi, Supercooling of liquids in forced flow in pipes and the growth of ice crystals, *Heat Transfer—Jap. Res.* **10,** 58 (1981).

25B. H. Nakamura, A. Matsuura, J. Kiwaki, I. Yamada and M. Hasantani, Laminar heat transfer of high-viscosity Newtonian fluids in horizontal rectangular ducts, *Int. Chem. Engng* **22,** 479 (1982).

26B. M. A. Nyamira, Y. V. Vilyamas and V. M. Shimonis, Heat transfer in annular channels for turbulent flow of gases with variable physical properties. Correlations of results of experimental investigation, *Int. Chem. Engng* **22,** 165 (1982).

27B. J. E. O'Brien and E. M. Sparrow, Corrugated-duct heat transfer, pressure drop, and flow visualization, *J. Heat Transfer* **104,** 410 (1982).

28B. C. Pfender, J. Daubert and K. Stephan, A capacitance method to determine the position and structure of the interface in an annular flow, *Wärme- und Stoffübertragung* **16,** 45 (1982).

29B. V. N. Popov and Ye. P. Valuyeva, Heat transfer coefficient and pressure drop in buoyancy-dominated turbulent flow in a vertical tube, *Heat Transfer—Soviet Res.* **13**(4), 118 (1981).

30B. N. J. Rabadi, J. C. F. Chow and H. A. Simon, Heat transfer in curved tubes with pulsating flow, *Int. J. Heat Mass Transfer* **25,** 195 (1982).

31B. A. E. Rodrigues, B. J. Ahn and A. Zoulalian, Intraparticle-forced convection effect in catalyst diffusivity measurements and reactor design, *A.I.Ch.E. Jl* **28,** 541 (1982).

32B. Y. Sano and H. Usui, Evaluation of heat transfer promoters by the fluid dissipation energy, *Heat Transfer—Jap. Res.* **11**(1), 91 (1982).

33B. S. Saruwatari, Y. Sato and M. Sadatomi, Momentum and heat transfer in two-phase bubble flow in concentric annuli, *Bull. J.S.M.E.* **25,** 1746 (1982).

34B. O. Shoham, A. E. Dukler and Y. Taitel, Heat transfer during intermittent slug flow in horizontal tubes, *Ind. Engng Chem. Fundam.* **21,** 312 (1982).

35B. R. Smith, M. F. Edwards and H. Z. Wang, Pressure drop and mass transfer in dilute polymer solutions in turbulent drag-reducing pipe flow, *Int. J. Heat Mass Transfer* **25,** 1869 (1982).

36B. E. M. Sparrow and N. Cur, Turbulent heat transfer in a symmetrically or asymmetrically heated flat rectangular duct with flow separation at inlet, *J. Heat Transfer* **104,** 82 (1982).

37B. K. Suzuki, T. Sato, Y. Kang and T. Sugimoto, Circular tube turbulent heat transfer in the downstream of an orifice, *Heat Transfer—Jap. Res.* **11**(1), 70 (1982).

38B. M. M. Tamonis, Combined (radiative-convective) heat transfer from laminar and turbulent radiating flows in cooled ducts, *Heat Transfer—Soviet Res.* **13**(6), 1 (1981).

39B. H. Tanaka, A. Tateno, S. Hatamiya and H. Kawamura, Effect of laminarization and retransition on heat transfer for low Reynolds number flow through a converging to constant area duct, *J. Heat Transfer* **104,** 363 (1982).

40B. H. Usui, Y. Sano and H. Fukuma, Turbulence measurements and mass transfer in fully developed flow in a triangular duct with a narrow apex angle, *Int. J. Heat Mass Transfer* **25,** 615 (1982).

41B. H. Yamamoto, N. Seki and S. Fukusako, A numerical study of laminar heat transfer at bottom surface of a cavity submerged in separated flow region of duct, *Wärme- und Stoffübertragung* **16,** 219 (1982).

42B. S. Yao and Y. Hung, Heat convection in annular type crevices, *J. Heat Transfer* **104,** 403 (1982).

Boundary layer and external flows

1C. S. Aiba, H. Tsuchida and T. Ota, Heat transfer around tubes in staggered tube banks, *Bull. J.S.M.E.* **25,** 927 (1982).

2C. S. Aiba, H. Tsuchida and T. Ota, Heat transfer around tubes in in-line tube banks, *Bull. J.S.M.E.* **25,** 919 (1982).

3C. N. Afzal, Thermal turbulent boundary layer under strong adverse pressure gradient near separation, *J. Heat Transfer* **104,** 397 (1982).

4C. I. T. Alad'yev, A. A. Ivlev and Ye. S. Turilina, Effect of the direction of heat flux on forced convection in He-1 at 3 bar and 3 to 20 K, *Heat Transfer—Soviet Res.* **13**(4), 69 (1981).

5C. R. A. Altenkirch, M. Rezayat, R. Eichhorn and F. J. Rizzo, Boundary integral equation method calculations of surface regression effects in flame spreading, *J. Heat Transfer* **104,** 734 (1982).

6C. A. D. Aralov, S. I. Stepanov and B. N. Yudayev, Friction and heat transfer from plates to normally-incident axisymmetric turbulent jets, *Heat Transfer—Soviet Res.* **13**(4), 136 (1981).

7C. S. Asgarpour and Y. Bayazitoglu, Heat transfer in a laminar flow with a phase change boundary, *J. Heat Transfer* **104,** 678 (1982).

8C. T. Aytimur, B. H. Chen and A. F. McMillan, Time dependence in a three-dimensional model of thermal pollution in rivers, *Can. J. Chem. Engng* **60,** 699 (1982).

9C. I. A. Belov, G. F. Gorshkov and V. S. Terpigor'yev, Heat transfer and microstructure of the boundary layer in the vicinity of the stagnation point of a jet flowing over a baffle, *Heat Transfer—Soviet Res.* **13**(4), 142 (1981).

10C. J. J. Bertin, T. E. Hayden and W. D. Goodrich, Shuttle boundary-layer transition due to distributed roughness and surface cooling, *J. Spacecraft Rockets* **19,** 389 (1982).

11C. A. Brosh, D. Degani and S. Zalmanovich, Conjugated heat transfer in a laminar boundary layer with heat source at the wall, *J. Heat Transfer* **104,** 90 (1982).

12C. A. Brown and B. W. Martin, Flow transition phenomena and heat transfer over the pressure surfaces of gas turbine blades, *J. Engng Pwr* **104,** 360 (1982).

13C. P. O. Brunn, Heat or mass transfer from single spheres in a low Reynolds number flow, *Int. J. Engng Sci.* **20,** 817 (1982).
14C. R. K. Cheng and T. T. Ng, Some aspects of strongly heated turbulent boundary layer flow, *Physics Fluids* **25,** 1333 (1982).
15C. I. G. Choi, The effect of variable properties of air on the boundary layer for a moving continuous cylinder, *Int. J. Heat Mass Transfer* **25,** 597 (1982).
16C. Y. C. Chung and L. G. Leal, An experimental study of the effective thermal conductivity of a sheared suspension of rigid spheres, *Int. J. Multiphase Flow* **8,** 605 (1982).
17C. S. Crescitelli, G. Donsi, L. Massimilla and G. G. Volpicelli, Instantaneous heat transfer coefficient at the wall of fluidization columns, *Heat Transfer—Soviet Res.* **13**(6), 114 (1981).
18C. J. W. Deardorff and G. E. Willis, Investigation of the frozen-turbulent hypothesis for temperature spectra in a convectively mixed layer, *Physics Fluids* **25,** 21 (1982).
19C. Ye. P. Dyban and E. Ya. Epik, Effect of turbulent viscosity of the external flow on heat transfer in a turbulent boundary layer, *Heat Transfer—Soviet Res.* **13**(5), 1 (1981).
20C. N. Fujita and T. Tsubouchi, An experimental study of unsteady heat transfer from a flat plate to an oscillating air flow, *Heat Transfer—Jap. Res.* **11**(3), 31 (1982).
21C. S. E. George and J. R. Grace, Heat transfer to horizontal tubes in the freeboard region of a gas fluidized bed, *A.I.Ch.E. Jl* **28,** 759 (1982).
22C. R. S. R. Gorla, Stagnation point heat transfer augmentation due to free stream turbulence, *Appl. Scient. Res.* **39,** 143 (1982).
23C. R. S. R. Gorla, Unsteady heat transfer in laminar non-Newtonian boundary layer over a wedge, *A.I.Ch.E. Jl* **28,** 56 (1982).
24C. M. Hiwada and I. Mabuchi, Flow behavior and heat transfer around a circular cylinder at high blockage ratios, *Heat Transfer—Jap. Res.* **10,** 17 (1981).
25C. M. Hiwada, I. Mabuchi and H. Yanagihara, Fluid flow and heat transfer around two circular cylinders, *Bull. J.S.M.E.* **25,** 1737 (1982).
26C. H. Konno, T. Khoya, M. Toda, T. Saito, M. Kuriyama and E. Harada, Heat transfer from a horizontal fine wire and cylinders in a corona wind, *Heat Transfer—Jap. Res.* **10**(2), 20 (1981).
27C. R. Krishnaswamy and G. Nath, Compressible boundary-layer flow at a three-dimensional stagnation point with massive blowing, *Int. J. Heat Mass Transfer* **25,** 1639 (1982).
28C. C. K. Law and H. K. Law, A theoretical study on ignition in the laminar mixing layer, *J. Heat Transfer* **104,** 329 (1982).
29C. H.-T. Lin and Y.-P. Shih, Instant nonsimilarity solutions of unsteady laminar boundary-layer heat transfer, *Num. Heat Transfer* **5,** 299 (1982).
30C. C.-Y. Lin and J. C. Slattery, Thinning of a liquid film as a small drop or bubble approaches a solid plane, *A.I.Ch.E. Jl* **28,** 147 (1982).
31C. A. I. Leont'yev, Ye. V. Shipov, V. N. Afans'yev and V. P. Zabolotskiy, Experimental study of the structure of the turbulent thermal boundary layer in laminarized flows developed in strong favorable pressure gradients, *Heat Transfer—Soviet Res.* **13**(5), 43 (1981).
32C. L. Meyer, Thermohydraulic characteristics of single rods with three-dimensional roughness, *Int. J. Heat Mass Transfer* **25,** 1043 (1982).
33C. A. Moutsoglou and A. K. Bhattacharya, Laminar and turbulent boundary layers on moving, nonisothermal continuous flat surfaces, *J. Heat Transfer* **104,** 707 (1982).
34C. K. O'Neill, One-dimensional transport from a highly concentrated transfer type source, *Int. J. Heat Mass Transfer* **25,** 27 (1982).
35C. B. V. Perepelitsa and Ye. M. Khabakhpasheva, Correlations between temperature and velocity fluctuations in the wall region of turbulent flow, *Heat Transfer—Soviet Res.* **13**(4), 46 (1981).
36C. B. S. Petukhov, A. F. Polyakov and Yu. V. Tsypulev, Turbulent momentum and heat transfer in thermally-stratified flows, *Heat Transfer—Soviet Res.* **13**(4), 56 (1981).
37C. A. D. Polyanin, Method for solution of some non-linear boundary value problems of a non-stationary diffusion-controlled (thermal) boundary layer, *Int. J. Heat Mass Transfer* **25,** 471 (1982).
38C. D. Poulikakos and A. Bejan, Fin geometry for minimum entropy generation in forced convection, *J. Heat Transfer* **104,** 616 (1982).
39C. R. L. Powell and S. G. Mason, Dispersion by laminar flow, *A.I.Ch.E. Jl* **28,** 286 (1982).
40C. S. Rajagopalan and R. A. Antonia, Use of a quadrant analysis technique to identify coherent structures in a turbulent boundary layer, *Physics Fluids* **25,** 949 (1982).
41C. A. S. Sangani and A. Acrivos, Slow flow past periodic arrays of cylinders with application to heat transfer, *Int. J. Multiphase Flow* **8,** 193 (1982).
42C. G. I. Sergeyev and V. P. Kovalenko, Heat and mass transfer in falling films with low-frequency perturbations of the interface, *Heat Transfer—Soviet Res.* **13**(5), 96 (1981).
43C. B. M. Smol'skiy and L. A. Sergeyeva, Unsteady circumferential heat transfer to a cylinder in a transverse air flow, *Heat Transfer—Soviet Res.* **13**(4), 125 (1981).
44C. R. M. C. So, Heat transfer modeling for turbulent shear flows on curved surfaces, *Z. Angew. Math. Phys.* **32,** 514 (1981).
45C. C. S. Subramanian and R. A. Antonia, A model for the skewness of the temperature derivative in a turbulent boundary layer, *Physics Fluids* **25,** 957 (1982).
46C. L. C. Thomas, A turbulent burst model of wall turbulence for two-dimensional turbulent boundary layer flow, *Int. J. Heat Mass Transfer* **25,** 1127 (1982).
47C. I. M. Varfolomeyev, Yu. F. Gortyshev and V. K. Shchukin, Heat transfer and dynamics of supersonic air flow past cavities, *Heat Transfer—Soviet Res.* **13**(2), 86 (1981).
48C. A. R. Wazzan and H. Taghavi, The effect of heat transfer on three-dimensional spatial stability and transition of flat plate boundary layer at Mach 3, *Int. J. Heat Mass Transfer* **25,** 1321 (1982).
49C. F. Wehle and F. Brandt, Influence of temperature dependent physical properties on the heat transfer in the laminar boundary layer of a parallel affluxed flat plate, *Wärme- und Stoffübertragung* **16,** 129 (1982).

Flow with separated regions and through porous media

1D. R. L. Adams, Coupled gas convection and unsteady conduction effects in fluid bed heat transfer based on a single particle model, *Int. J. Heat Mass Transfer* **25,** 1819 (1982).

2D. J. C. Benet and P. Jouanna, Phenomenological relation of phase change of water in a porous medium: experimental verification and measurement of the phenomenological coefficient, *Int. J. Heat Mass Transfer* **25,** 1747 (1982).

3D. A. R. Bestman, Pulsatile flow in heated porous channel, *Int. J. Heat Mass Transfer* **25,** 675 (1982).

4D. N. M. Catipovic, T. J. Fitzgerald, A. H. George and J. R. Welty, Experimental validation of the Adams–Welty model for heat transfer in large-particle fluidized beds, *A.I.Ch.E. Jl* **28,** 714 (1982).

5D. A. Dayan, Self-similar temperature, pressure and moisture distributions within an intensely heated porous half space, *Int. J. Heat Mass Transfer* **25,** 1469 (1982).

6D. A. Dayan and E. L. Gluekler, Heat and mass transfer within an intensely heated concrete slab, *Int. J. Heat Mass Transfer* **25,** 1461 (1982).

7D. D. Droste and H. M. Kottowski, Sodium boiling pattern in porous blockages, *Nucl. Sci. Engng* **80,** 673 (1982).

8D. S. Endrenyi, Contribution to the theory of surface phenomena of porous bodies occurring due to heat and mass transfer processes, *Int. J. Heat Mass Transfer* **25,** 1457 (1982).

9D. H. I. Ene and Sanchez-Palencia, On thermal equation for flow in porous media, *Int. J. Engng Sci.* **20,** 623 (1982).

10D. V. L. Ganzha, S. N. Upadhyay and S. C. Saxena, A mechanistic theory for heat transfer between fluidized beds of large particles and immersed surfaces, *Int. J. Heat Mass Transfer* **25,** 1531 (1982).

11D. R. J. Goldstein and J. F. Timmers, Visualization of heat transfer from arrays of impinging jets, *Int. J. Heat Mass Transfer* **25,** 1857 (1982).

12D. M. Granier, J. P. Riba and J. P. Couderc, Transfert de chaleur entre une couche fluidisee bidimensionnele et une paroi, *Can. J. Chem. Engng* **60,** 593 (1982).

13D. T. Igarashi, Fluid flow and heat transfer in the separated region of a circular cylinder with wake control, *Heat Transfer—Jap. Res.* **11**(3), 1 (1982).

14D. Y. Kang and K. Suzuki, Numerical study of wall heat transfer in the recirculating flow region of a confined jet, *Heat Transfer—Jap. Res.* **11**(1), 44 (1982).

15D. V. I. Khvostov and S. K. Marinichenko, Investigation of vaporization in capillary-porous membranes to which liquid is supplied under capillary pressure, *Heat Transfer—Soviet Res.* **14**(1), 83 (1982).

16D. S. Lin, An exact solution of the desublimation problem in a porous medium, *Int. J. Heat Mass Transfer* **25,** 625 (1982).

17D. C.-Y. Lin and J. C. Slattery, Three-dimensional randomized, network model for two-phase flow through porous media, *A.I.Ch.E. Jl* **28,** 311 (1982).

18D. R. P. Lowell and H. Hernandez, Finite amplitude convection in a porous container with fault-like geometry: effect of initial and boundary conditions, *Int. J. Heat Mass Transfer* **25,** 631 (1982).

19D. H. Martin, Heat and mass transfer in fluidized beds, *Int. Chem. Engng* **22,** 30 (1982).

20D. A. F. Mills, S. Kin, T. Leininger, S. Ofer and A. Pesaran, Heat and mass transport in turbulent liquid jets, *Int. J. Heat Mass Transfer* **25,** 889 (1982).

21D. B. M. Mogilevskiy, F. Sh. Khodzhayev and A. F. Chudnovskiy, Determination of the phase composition of moisture transported in capillary-porous bodies by the two-phase conductivity method, *Heat Transfer—Soviet Res.* **14**(1), 87 (1982).

22D. J. Neiss and E. R. F. Winter, New methods for the calculation of hydraulic conductivity in porous media, *Wärme- und Stoffübertragung* **16,** 175 (1982).

23D. J. R. Philip, Free convection at small Rayleigh number in porous cavities of rectangular, elliptical, triangular and other cross-sections, *Int. J. Heat Mass Transfer* **25,** 1503 (1982).

24D. N. Rudraiah and S. Balachandra Rao, Nonlinear cellular convection and heat transport in a porous medium, *Appl. Scient. Res.* **39,** 21 (1982).

25D. N. Rudraiah, P. K. Srimani and R. Friedrich, Finite amplitude convection in a two-component fluid saturated porous layer, *Int. J. Heat Mass Transfer* **25,** 715 (1982).

26D. N. Rudraiah, B. Veerappa and S. Balachandra Rao, Convection in a fluid-saturated porous layer with non-uniform temperature gradient, *Int. J. Heat Mass Transfer* **25,** 1147 (1982).

27D. N. Rudraiah and D. Vortmeyer, The influence of permeability and of a third diffusing component upon the onset of convection in a porous medium, *Int. J. Heat Mass Transfer* **25,** 457 (1982).

28D. B. S. Sazhin, V. A. Reutskiy and S. M. Kovachev, Generalized calculation of the kinetics of drying of capillary-porous bodies, *Heat Transfer—Soviet Res.* **14**(1), 104 (1982).

29D. R. N. Sharma and S. V. Patankar, Numerical computation of wall-jet flows, *Int. J. Heat Mass Transfer* **25,** 1709 (1982).

30D. G. S. Shubin, Thermal moisture conductivity of colloidal capillary-porous bodies, *Heat Transfer—Soviet Res.* **14**(1), 65 (1982).

31D. A. Snyder and R. Siegel, Determination of boundary shape of cooled porous region, *J. Heat Transfer* **104,** 722 (1982).

32D. E. M. Sparrow and M. Carranco Ortiz, Heat transfer coefficients for the upstream face of a perforated plate positioned normal to an oncoming flow, *Int. J. Heat Mass Transfer* **25,** 127 (1982).

33D. E. M. Sparrow and E. D. Larson, Heat transfer from pin-fins situated in an oncoming longitudinal flow which turns to crossflow, *Int. J. Heat Mass Transfer* **25,** 603 (1982).

34D. E. M. Sparrow and M. Molki, Effect of a missing cylinder on heat transfer and fluid flow in an array of cylinders in cross-flow, *Int. J. Heat Mass Transfer* **25,** 449 (1982).

35D. E. M. Sparrow and R. Ruiz, Effect of blockage-induced flow maldistribution on the heat transfer and pressure drop in a tube bank, *J. Heat Transfer* **104,** 691 (1982).

36D. J. K. Spelt, C. E. Brennen and R. H. Sabersky, Heat

transfer to flowing granular material, *Int. J. Heat Mass Transfer* **25**, 791 (1982).
37D. Y. C. Yortsos, Effect of lateral heat losses on the stability of thermal displacement fronts in porous media, *A.I.Ch.E. Jl* **28**, 480 (1982).

Natural convection—internal flows

1F. K. Akagawa, T. Fujii, Y. Ito, S. Amano and T. Oka, A study of a power system by utilizing natural circulation effective utilization of heat sources (1st report, principles and theoretical performance analysis), *Bull. J.S.M.E.* **25**, 403 (1982).
2F. V. A. Akinsete and T. A. Coleman, Heat transfer by steady laminar free convection in triangular enclosures, *Int. J. Heat Mass Transfer* **25**, 991 (1982).
3F. Y. Asako and H. Nakamura, Heat transfer in a parallelogram shaped enclosure (2nd report, free convection infinitely stacked parallelogram shaped enclosure), *Bull. J.S.M.E.* **25**, 1412 (1982).
4F. Y. Asako and H. Nakamura, Heat transfer in a parallelogram shaped enclosure (3rd report, combined free convection and radiation heat transfer), *Bull. J.S.M.E.* **25**, 1419 (1982).
5F. S. M. Bajorek and J. R. Lloyd, Experimental investigation of natural convection in partitioned enclosures, *J. Heat Transfer* **104**, 527 (1982).
6F. H. H. Bau and K. E. Torrance, Low Rayleigh number thermal convection in a vertical cylinder filled with porous materials and heated from below, *J. Heat Transfer* **104**, 166 (1982).
7F. R. P. Behringer and G. Ahlers, Heat transport and temporal evolution of fluid flow near the Rayleigh–Bénard instability in cylindrical containers, *J. Fluid. Mech.* **125**, 219 (1982).
8F. A. Bejan and D. Poulikakos, Natural convection in an attic-shaped space filled with porous material, *J. Heat Transfer* **104**, 241 (1982).
9F. K. Bühler and H. Oertel, Thermal cellular convection in rotating rectangular boxes, *J. Fluid Mech.* **114**, 261 (1982).
10F. F. H. Busse and N. Riahi, Patterns of convection in spherical shells, part 2, *J. Fluid Mech.* **123**, 283 (1982).
11F. M. Charmchi and E. M. Sparrow, Analysis of natural convection in the space between concentric vertical cylinders of different height and diameter, *Num. Heat Transfer* **5**, 119 (1982).
12F. C. H. Cho, K. S. Chang and K. H. Park, Numerical simulation of natural convection in concentric and eccentric horizontal cylindrical annuli, *J. Heat Transfer* **104**, 624 (1982).
13F. K. C. Chung and L. M. Trefethan, Natural convection in a vertical stack of inclined parallelogrammic cavities, *Int. J. Heat Mass Transfer* **25**, 277 (1982).
14F. L. Y. Cooper, Heat transfer from a buoyant plume to an unconfined ceiling, *J. Heat Transfer* **104**, 446 (1982).
15F. E. Crespo and M. G. Velarde, Two-component Bénard convection in cylinders, *Int. J. Heat Mass Transfer* **25**, 1451 (1982).
16F. P. G. Daniels, Effects of geometrical imperfection at the onset of convection in a shallow two-dimensional cavity, *Int. J. Heat Mass Transfer* **25**, 337 (1982).
17F. P. G. Daniels, The onset of Bénard convection in a shallow sloping container, *Q. Jl Mech. Appl. Math.* **35**, 49 (1982).
18F. L. M. deSocio, G. Gaffuri and L. Musici, Stokes flow in a rectangle well: natural convection and boundary-layer function, *Q. Appl. Math.* **39**, 499 (1982).
19F. S. M. Elsherbiny, K. G. T. Hollands and G. D. Raithby, Effect of thermal boundary conditions on natural convection in vertical and inclined air layers, *J. Heat Transfer* **104**, 515 (1982).
20F. S. M. Elsherbiny, G. D. Raithby and K. G. T. Hollands, Heat transfer by natural convection across vertical and inclined air layers, *J. Heat Transfer* **104**, 96 (1982).
21F. G. H. Evans and O. A. Plumb, Numerical and approximate numerical solutions to a three-dimensional mixed convection boundary-layer, *Num. Heat Transfer* **5**, 287 (1982).
22F. B. Farouk and S. I. Güceri, Natural and mixed convection heat transfer around a horizontal cylinder within confining walls, *Num. Heat Transfer* **5**, 329 (1982).
23F. B. Farouk and S. I. Güceri, Laminar and turbulent natural convection in the annulus between horizontal concentric cylinders, *J. Heat Transfer* **104**, 631 (1982).
24F. H. Frick and R. M. Clever, The influence of side walls on finite-amplitude convection in a layer heated from below, *J. Fluid Mech.* **114**, 467 (1982).
25F. F. A. Garifulin, F. I. Zapparov, N. Z. Mingaleyev, P. A. Norden and F. Kh. Tazukov, Convection in a horizontal layer of a viscoelastic fluid, *Heat Transfer—Soviet Res.* **14**(1), 121 (1982).
26F. A. V. Getling, Types of evolution and extraction of preferred wave numbers of two-dimensional convective perturbations in a horizontal layer, *Heat Transfer—Soviet Res.* **13**(4), 106 (1981).
27F. J. P. Gollub, A. R. McCarriar and J. F. Steinman, Convective pattern evolution and secondary instabilities, *J. Fluid Mech.* **125**, 259 (1982).
28F. M. A. Hassab and M. M. Sorour, Onset of convection in a melted ice layer between vertical plates, *Int. J. Heat Mass Transfer* **25**, 909 (1982).
29F. M. P. Heisler, Development of scaling requirements for natural convection liquid-metal fast breeder reactor shutdown heat removal test facilities, *Nucl. Sci. Engng* **80**, 347 (1982).
30F. M. Hishida, Y. Nagano and M. S. Montesclaros, Combined forced and free convection in the entrance region of an isothermally heated horizontal pipe, *J. Heat Transfer* **104**, 153 (1982).
31F. H. Inaba, N. Seki and S. Fukusako, Heat transfer by natural convection in a shallow rectangular cavity with two opposing vertical walls at different temperatures, *Bull. J.S.M.E.* **25**, 939 (1982).
32F. S. C. Jain, Buoyancy-driven circulation in free-surface channels, *J. Fluid Mech.* **122**, 1 (1982).
33F. B. S. Jhaveri and G. M. Homsy, The onset of convection in fluid layers heated rapidly in a time-dependent manner, *J. Fluid Mech.* **114**, 251 (1982).
34F. K. D. Kihm, C. K. Choi and J. Y. Yoo, The onset of instability in a horizontal fluid layer due to a step change in temperature, *Int. J. Heat Mass Transfer* **25**, 1829 (1982).
35F. Y. Kikuchi, T. Kawasaki and T. Shioyama, Thermal convection in a horizontal fluid layer heated internally

and from below, *Int. J. Heat Mass Transfer* **25,** 363 (1982).

36F. S. A. Korpela, Y. Lee and J. E. Drummond, Heat transfer through a double pane window, *J. Heat Transfer* **104,** 539 (1982).

37F. J. N. Koster and U. Müller, Free convection in vertical gaps, *J. Fluid Mech.* **125,** 429 (1982).

38F. K. Kuwahara and Y. Oshima, Thermal convection caused by ring-type heat source, *Phys. Soc. Japan* **51,** 3711 (1982).

39F. S. S. Kwon, T. H. Kuehn and T. S. Lee, Natural convection in the annulus between horizontal circular cylinders with three axial spacers, *J. Heat Transfer* **104,** 118 (1982).

40F. T. S. Lee, P. G. Parikh, A. Acrivos and D. Bershader, Natural convection in a vertical channel with opposing buoyancy forces, *Int. J. Heat Mass Transfer* **25,** 499 (1982).

41F. B. LeFur, Solutions à hodographe linéaire des equations de la convection naturelle dans un milieu poreux, *Int. J. Heat Mass Transfer* **25,** 17 (1982).

42F. W. T. Lewis, F. P. Incropera and R. Viskanta, Interferometric study of stable salinity gradients heated from below or cooled from above, *J. Fluid Mech.* **116,** 411 (1982).

43F. E. Maeland, On non-linear heatup of a Boussinesq fluid in a circular cylinder, *Z. Angew. Math Phys.* **32,** 748 (1981).

44F. N. C. Markatos, M. R. Malin and G. Cox, Mathematical modelling of buoyancy-induced smoke flow in enclosures, *Int. J. Heat Mass Transfer* **25,** 63 (1982).

45F. J. M. McDonough and I. Catton, Accuracy of the mean field approximation and the physical effect of Prandtl number in Bénard convection, *Physics Fluids* **25,** 1502 (1982).

46F. J. M. McDonough and I. Catton, A mixed finite difference–Galerkin procedure for two-dimensional convection in a square box, *Int. J. Heat Mass Transfer* **25,** 1137 (1982).

47F. J. B. McLaughlin and S. A. Orszag, Transition from periodic to chaotic thermal convection, *J. Fluid Mech.* **122,** 123 (1982).

48F. G. P. Merker and J. Straub, Rayleigh–Bénard convection in water with maximum density effects, *Wärme- und Stoffübertragung* **16,** 63 (1982).

49F. A. Mertol, R. Greif and Y. Zvirin, Two-dimensional study of heat transfer and fluid flow in a natural convection loop, *J. Heat Transfer* **104,** 508 (1982).

50F. B. A. Meyer, J. W. Mitchell and M. M. El-Wakil, The effect of thermal wall properties on natural convection in inclined rectangular cells, *J. Heat Transfer* **104,** 111 (1982).

51F. H. Nakamura, Y. Asako and T. Naitou, Heat transfer by free convection between two parallel flat plates, *Num. Heat Transfer* **5,** 95 (1982).

52F. G. P. Neitzel, Onset of convection in impulsively heated or cooled fluid layers, *Physics Fluids* **25,** 210 (1982).

53F. F. Penot, Numerical calculations of two-dimensional natural convection in isothermal open cavities, *Num. Heat Transfer* **5,** 421 (1982).

54F. M. Quintard and L. Prouvost, Instabilitiés ze zones ze diffusion thermique instationnaires en milieu poreux, *Int. J. Heat Mass Transfer* **25,** 37 (1982).

55F. K. Ramakrishna, S. C. Rubin and P. K. Khosla, Laminar natural convection along vertical square ducts, *Num. Heat Transfer* **5,** 59 (1982).

56F. L. Robillard and P. Vasseur, Convective response of a mass of water near 4°C to a constant cooling rate applied on its boundaries, *J. Fluid Mech.* **118,** 123 (1982).

57F. S. Rosenblatt, Thermal convection in a vertical circular cylinder, *J. Fluid Mech.* **122,** 395 (1982).

58F. T. Saitoh and K. Hirose, High Rayleigh number solutions to problems of latent heat thermal energy storage in a horizontal cylinder capsule, *J. Heat Transfer* **104,** 545 (1982).

59F. G. S. Shiralkar and C. L. Tien, A numerical study of the effect of a vertical temperature difference imposed on a horizontal enclosure, *Num. Heat Transfer* **5,** 185 (1982).

60F. E. F. C. Somerscales and H. B. Parsapour, An application of the optical correlation computer to the detection of the Malkus transitions in free convection, *J. Heat Transfer* **104,** 255 (1982).

61F. C. W. Somerton and I. Catton, On the thermal instability of superimposed porous and fluid layers, *J. Heat Transfer* **104,** 160 (1982).

62F. E. M. Sparrow and F. Samie, Interaction between a stream which passes through an enclosure and natural convection within the enclosure, *Int. J. Heat Mass Transfer* **25,** 1489 (1982).

63F. E. M. Sparrow and W. Q. Tao, Buoyancy-driven fluid flow and heat transfer in a pair of interacting vertical parallel channels, *Num. Heat Transfer* **5,** 39 (1982).

64F. K. C. Stengel, D. S. Oliver and J. R. Booker, Onset of convection in a variable-viscosity fluid, *J. Fluid Mech.* **120,** 411 (1982).

65F. M. Strada and J. C. Heinrich, Heat transfer rates in natural convection at high Rayleigh numbers in rectangular enclosures: a numerical study, *Num. Heat Transfer* **5,** 81 (1982).

66F. Y. Suzukawa and U. Narusawa, Structure of growing double-diffusive convection cells, *J. Heat Transfer* **104,** 248 (1982).

67F. S. Takao, M. Tsuchiya and U. Narusawa, Numerical study of the onset of double-diffusive cellular convection due to a uniform lateral heat flux, *J. Heat Transfer* **104,** 649 (1982).

68F. S. Thangam, A. Zebib and C. F. Chen, Double-diffusive convection in an inclined fluid layer, *J. Fluid Mech.* **116,** 363 (1982).

69F. J. Tichy and A. Gadgil, High Rayleigh number laminar convection in low aspect ratio enclosures with adiabatic horizontal walls and differentially heated vertical walls, *J. Heat Transfer* **104,** 103 (1982).

70F. D. L. Turcotte, D. A. Spence and H. H. Bau, Multiple solutions for natural convective flows in an internally heated, vertical channel with viscous dissipation and pressure work, *Int. J. Heat Mass Transfer* **25,** 699 (1982).

71F. M. Ueda, K. Kegawa, K. Yamada, C. Yamaguchi and Y. Harada, Flow visualization of Bénard convection using holographic interferometry, *Appl. Optics* **21,** 3269 (1982).

72F. P. Vasseur and L. Robillard, Natural convection in a rectangular cavity with wall temperature decreasing at a uniform rate, *Wärme- und Stoffübertragung* **16,** 199

(1982).
73F. R. Viskanta and M. Behnia, Experimental and analytical study of heat transfer and mixing in thermally stratified buoyant flows, *Int. J. Heat Mass Transfer* **25,** 847 (1982).
74F. E. Wacholder, S. Kaizerman and E. Elias, Numerical analysis of the stability and transient behavior of natural convection loops, *Int. J. Engng Sci.* **20,** 1235 (1982).
75F. I. C. Walton, The effects of slow spatial variations on Bénard convection, *Q. Jl Mech. Appl. Math.* **35,** 33 (1982).
76F. I. C. Walton, Double-diffusive convection with large variable gradients, *J. Fluid Mech.* **125,** 123 (1982).
77F. R. A. Wirtz, J. Righi and F. Zirilli, Measurements of natural convection across tilted rectangular enclosures of aspect ratio 0.1 and 0.2, *J. Heat Transfer* **104,** 521 (1982).
78F. R. A. Wirtz and R. J. Stutzman, Experiments on free convection between vertical plates with symmetric heating, *J. Heat Transfer* **104,** 501 (1982).
79F. R. Yewell, D. Poulikakos and A. Bejan, Transient natural convection experiments in shallow enclosures, *J. Heat Transfer* **104,** 533 (1982).
80F. W. W. Yousef and J. D. Tarasuk, Free convection effects on laminar forced convective heat transfer in a horizontal isothermal tube, *J. Heat Transfer* **104,** 145 (1982).

Natural convection—external flows
1FF. M. Al-Arabi and M. Khamis, Natural convection heat transfer from inclined cylinders, *Int. J. Heat Mass Transfer* **25,** 3 (1982).
2FF. R. A. Altenkirch, D. C. Winchester and R. Eichhorn, Buoyancy effects on the temperature field in downward spreading flames, *J. Heat Transfer* **3,** 560 (1982).
3FF. A. Aziz and T. Y. Na, Improved perturbation solutions for laminar natural convection on a vertical cylinder, *Wärme- und Stoffübertragung* **16,** 83 (1982).
4FF. A. Bejan, Theory of instantaneous sinuous structure in turbulent buoyant plumes, *Wärme- und Stoffübertragung* **16,** 237 (1982).
5FF. T. L. Bergman, F. P. Incropera and R. Viskanta, A multi-layer model for mixing layer development in a double-diffusive thermohaline system heated from below, *Int. J. Heat Mass Transfer* **25,** 1411 (1982).
6FF. L. R. Cairnie and A. J. Harrison, Natural convection adjacent to a vertical isothermal hot plate with a high surface-to-ambient temperature difference, *Int. J. Heat Mass Transfer* **25,** 925 (1982).
7FF. V. P. Carey and B. Gebhart, Transport at large downstream distances in mixed convection flow adjacent to a vertical uniform-heat-flux surface, *Int. J. Heat Mass Transfer* **25,** 255 (1982).
8FF. T. S. Chen and K. L. Tzuoo, Vortex instability of free convection flow over horizontal and inclined surfaces, *J. Heat Transfer* **104,** 637 (1982).
9FF. J. Dey, Mixed convection flow over a semi-infinite horizontal plate with vectored mass transfer, *J. Heat Transfer* **104,** 558 (1982).
10FF. I. El-Henaway, B. Hassard, N. Kazarinoff, B. Gebhart and J. Mollendorf, Numerically computed multiple steady states of vertical buoyancy-induced flows in cold pure water, *J. Fluid Mech.* **122,** 235 (1982).
11FF. G. H. Evans and O. A. Plumb, Laminar mixed convection from a vertical heated surface in a crossflow, *J. Heat Transfer* **104,** 554 (1982).
12FF. B. Farouk and S. I. Guceri, Natural convection from a horizontal cylinder—turbulent regime, *J. Heat Transfer* **104,** 228 (1982).
13FF. R. E. Faw and T. A. Dullforce, Holographic interferometry measurement of convective heat transfer beneath a heated horizontal circular plate in air, *Int. J. Heat Mass Transfer* **25,** 1157 (1982).
14FF. F. Geoola and A. R. H. Cornish, Numerical simulation of free convective heat transfer from a sphere, *Int. J. Heat Mass Transfer* **25,** 1677 (1982).
15FF. M. Gregoric, L. R. Davis and D. J. Bushnell, An experimental investigation of merging buoyant jets in a crossflow, *J. Heat Transfer* **104,** 236 (1982).
16FF. K. Himasekhar and Y. Jaluria, Laminar buoyancy-induced axisymmetric free boundary flows in a thermally stratified medium, *Int. J. Heat Mass Transfer* **25,** 213 (1982).
17FF. Y. Jaluria, Buoyancy-induced flow due to isolated thermal sources on a vertical surface, *J. Heat Transfer* **104,** 223 (1982).
18FF. Y. Jaluria, Mixed convection in a wall plume, *Comp. Fluids* **10,** 95 (1982).
19FF. T. T. Kao, Laminar free convection with suction or blowing along a vertical surface, *A.I.Ch.E. Jl* **28,** 338 (1982).
20FF. H. Kumari and G. Nath, Unsteady mixed convection near the stagnation point in three-dimensional flow, *J. Heat Transfer* **104,** 132 (1982).
21FF. M. Miyamoto and T. Akiyoshi, Free convection heat transfer near leading edge of semi-infinite vertical flat plate (2nd report, uniform heat generation, $Pr = 0.72$), *Bull. J.S.M.E.* **25,** 583 (1982).
22FF. T. Mizushina, F. Ogino and N. Katada, Ordered motion of turbulence in a thermally stratified flow under unstable conditions, *Int. J. Heat Mass Transfer* **25,** 1419 (1982).
23FF. T. Mizushina, F. Ogino, H. Takeuchi and H. Ikawa, An experimental study of vertical turbulent jet with negative buoyancy, *Wärme- und Stoffübertragung* **16,** 15 (1982).
24FF. T. H. Nguyen, P. Vasseur and L. Robillard, Natural convection between horizontal concentric cylinders with density inversion of water for low Rayleigh numbers, *Int. J. Heat Mass Transfer* **25,** 1559 (1982).
25FF. C. Pérez-García and J. M. Rubi, On the possibility of overstable motions of micropolar fluids heated from below, *Int. J. Engng Sci.* **20,** 873 (1982).
26FF. B. Sammakia, B. Gebhart and V. P. Carey, Transient mixed convection adjacent to a vertical flat surface, *Int. J. Heat Mass Transfer* **25,** 835 (1982).
27FF. B. Sammakia, B. Gebhart and Z. H. Qureshi, Measurement and calculations of transient natural convection in water, *J. Heat Transfer* **104,** 644 (1982).
28FF. T. Sano, Unsteady natural convection about a sphere at small Grashof number, *Physics Fluids* **25,** 2204 (1982).
29FF. S. Satyanarayana and Y. Jaluria, A study of laminar buoyant jets discharged at an inclination to the vertical buoyancy force, *Int. J. Heat Mass Transfer* **25,** 1569

(1982).
30FF. Y. T. Shee and S. N. Singh, Natural convection from a horizontal cylinder at small Grashof numbers, *Num. Heat Transfer* **5,** 479 (1982).
31FF. M. Sibulkin and C. Y. Wang, Free convection diffusion flames with inert additives, *J. Heat Transfer* **104,** 723 (1982).
32FF. C. J. Simpson and K. Stewartson, A singularity in an unsteady free convection boundary layer, *Q. Jl Mech. Appl. Math.* **35,** 291 (1982).
33FF. E. M. Sparrow, D. S. Cook and G. M. Chrysler, Heat transfer by natural convection from an array of short, wall-attached horizontal cylinders, *J. Heat Transfer* **104,** 125 (1982).
34FF. J. S. Vrentas, C. M. Vrentas, R. Narayanan and S. S. Agrawal, Integral equation formulation for buoyancy-driven convection problems, *Appl. Scient. Res.* **39,** 277 (1982).
35FF. X. A. Wang, An experimental study of mixed, forced, and convection heat transfer from a horizontal flat plate to air, *J. Heat Transfer* **104,** 139 (1982).
36FF. J. Yang, D. R. Jeng and K. J. DeWitt, Laminar free convection from a vertical plate with non-uniform surface conditions, *Num. Heat Transfer* **5,** 165 (1982).
37FF. W. W. Yousef, J. D. Tarasuk and W. J. McKeen, Free convection heat transfer from upward-facing isothermal horizontal surfaces, *J. Heat Transfer* **104,** 493 (1982).
38FF. N. Zhang and W.-J. Yang, Natural convection in evaporating minute drops, *J. Heat Transfer* **104,** 656 (1982).

Convection from rotating surfaces

1G. B. Banerjee and A. K. Borkakati, Heat transfer in a hydrodynamic flow between eccentric disks rotating at different speeds, *Z. Angew. Math. Phys.* **33,** 414 (1982).
2G. P. Bar-Yoseph, J. J. Blech and A. Solan, Finite element analysis of rotating flow in a skewed shrouded rotor geometry, *Int. J. Num. Methods Engng* **18,** 351 (1982).
3G. K. Bühler and H. Oertel, Thermal cellular convection in rotating rectangular boxes, *J. Fluid Mech.* **114,** 261 (1982).
4G. A. T. Conlisk and J. D. A. Walker, Forced convection in a rapidly rotating annulus, *J. Fluid Mech.* **122,** 91 (1982).
5G. N. Riahi, Convection in a rotating layer with nearly insulating boundaries, *Z. Angew. Math. Phys.* **33,** 81 (1982).
6G. M. Singh and S. C. Rajvanshi, Heat transfer between rotating eccentric cylinders with different radii, *Int. J. Heat Mass Transfer* **25,** 1719 (1982).
7G. E. M. Sparrow and A. Chaboki, Heat transfer coefficients for a cup-like cavity rotating about its own axis, *Int. J. Heat Mass Transfer* **25,** 1333 (1982).
8G. S. E. Tadros and F. F. Erian, Generalized laminar heat transfer from the surface of a rotating disk, *Int. J. Heat Mass Transfer* **25,** 1651 (1982).

Combined heat and mass transfer

1H. J. C. Allan and R. Mann, Heat and mass transfer multiplicity in exothermic gas absorption, *Can. J. Chem. Engng* **60,** 566 (1982).
2H. K. J. Beukema, S. Bruin and J. Schenk, Heat and mass transfer during cooling and storage of agricultural products, *Chem. Engng Sci.* **37,** 291 (1982).
3H. J. Brandeis, On interactive computation of supersonic boundary layers with wall mass transfer, *J. Appl. Mech.* **49,** 651 (1982).
4H. R. J. Goldstein and A. I. Behbahani, Impingement of a circular air jet with and without crossflow, *Int. J. Heat Mass Transfer* **25,** 1377 (1982).
5H. R. J. Goldstein and J. R. Taylor, Mass transfer in the neighborhood of jets entering a crossflow, *J. Heat Transfer* **104,** 715 (1982).
6H. R. J. Goldstein and T. Yoshida, The influence of a laminar boundary layer and laminar injection on film cooling performance, *J. Heat Transfer* **104,** 355 (1982).
7H. B. M. Howe, A. J. Chambers, S. P. Klotz, T. K. Cheung and R. L. Street, Comparison of profiles and fluxes of heat and momentum above and below an air–water interface, *J. Heat Transfer* **104,** 34 (1982).
8H. J. Iciek, The hydrodynamics of a free, liquid jet and their influence on direct contact heat transfer. I. Hydrodynamics of a free cylindrical liquid jet, *Int. J. Multiphase Flow* **8,** 239 (1982).
9H. J. Iciek, The hydrodynamics of a free, liquid jet and their influence on direct contact with heat transfer. II. Conditions of change of liquid outflow type through sharp inlet edged orifice, *Int. J. Multiphase Flow* **8,** 251 (1982).
10H. S. Kikkawa and K. Kamino, Experimental investigation on film cooling of a gas turbine blade, *Heat Transfer—Jap. Res.* **10**(4), 57 (1981).
11H. S. Kikkawa and T. Onkura, Experimental and theoretical investigation on three-dimensional film cooling of a flat plate, *Heat Transfer—Jap. Res.* **10**(4), 40 (1981).
12H. B. P. Mironov, V. N. Vasechkin, V. M. Mamonov and N. I. Yarygina, Heat and mass transfer at high free-stream turbulence as a function of injection rate, *Heat Transfer—Soviet Res.* **13**(5), 54 (1981).
13H. S. Sieniutycz and J. Komorowska-Kulik, Thermodynamic approach to qualitative properties of trajectories of heat and mass transfer in gas–solid flow systems. II. Reacting systems, *Int. J. Heat Mass Transfer* **25,** 575 (1982).
14H. A. Steiff and P. M. Weinspach, Fluid dynamics and heat and mass transfer in stirred, aerated suspension reactors, *Chemie–Ingr–Tech.* **5,** 526 (1982).
15H. E. P. Volchkov, S. Yu. Spotar' and V. I. Terekhov, Turbulent heat and mass transfer in the inlet zone of a tube with swirled flow, *Heat Transfer—Soviet Res.* **13**(4), 92 (1981).

Boiling

1J. T. Abe, E. R. G. Eckert and R. J. Goldstein, A parametric study of boiling in a porous bed, *Wärme- und Stoffübertragung* **16,** 119 (1982).
2J. K. N. Agrawal, H. K. Varma and S. Lai, Pressure drop during forced convection boiling of R-12 under swirl flow, *J. Heat Transfer* **104,** 758 (1982).
3J. A. P. Andreyev and V. M. Borishansky, Calculation of the Leidenfrost temperature and of the time required for vaporization of spherical droplets at this temperature, *Heat Transfer—Soviet Res.* **13**(1), 23 (1981).
4J. S. Aoki, A. Inoue, M. Aritomi and Y. Sakamoto,

Experimental study on the boiling phenomena within a narrow gap, *Int. J. Heat Mass Transfer* **25,** 985 (1982).
5J. C. T. Avedisian, Effect of pressure on bubble growth within liquid droplets at the superheat limit, *J. Heat Transfer* **104,** 750 (1982).
6J. H. H. Bau and K. E. Torrance, Boiling in low-permeability porous materials, *Int. J. Heat Mass Transfer* **25,** 45 (1982).
7J. A. E. Bergles and M. C. Chyu, Characteristics of nucleate pool boiling from porous metallic coatings, *J. Heat Transfer* **104,** 279 (1982).
8J. L. D. Berman, Heat and mass transfer in the vapor-gas phase upon high-rate evaporation of liquids, *Heat Transfer—Soviet Res.* **13**(2), 37 (1981).
9J. R. W. Bjorge, G. R. Hall and W. M. Rohsenow, Correlation of forced convection boiling heat transfer data, *Int. J. Heat Mass Transfer* **25,** 753 (1982).
10J. V. M. Borshanskiy, G. N. Danilova, M. A. Gotovskiy, A. V. Borishanskaya, G. P. Danilova and A. V. Kupriyanova, Correlation of data on heat transfer in and elementary characteristics of the nucleate boiling mechanism, *Heat Transfer—Soviet Res.* **13**(1), 100 (1981).
11J. A. P. Burdukov and G. G. Kuvshinov, Investigation of the boiling mechanism by the electrodiffusion method, *Heat Transfer—Soviet Res.* **13**(3), 101 (1981).
12J. A. P. Burdukov, G. G. Kuvshinov and V. Ye. Nakoryakov, Nucleate boiling heat transfer with a pool of subcooled liquid at sub-atmospheric pressures, *Heat Transfer—Soviet Res.* **13**(3), 81 (1981).
13J. Yu. A. Buyevich, D. A. Kazenin and V. N. Mankevich, Perturbation methods in the problem of growth of a vapor bubble, *Heat Transfer—Soviet Res.* **14**(1), 53 (1982).
14J. L. S. Chang, T. E. Carleson and J. C. Berg, Heat and mass transfer to a translating drop in an electric field, *Int. J. Heat Mass Transfer* **25,** 1023 (1982).
15J. T. C. Chawla and S. H. Chan, Heat transfer from vertical/inclined boundaries of heat-generating boiling pools, *J. Heat Transfer* **104,** 465 (1982).
16J. J. R. Conder, D. J. Gunn and M. A. Shaikh, Heat and mass transfer in two-phase flow—a mathematical model for laminar film flow and its experimental validation, *Int. J. Heat Mass Transfer* **25,** 1113 (1982).
17J. M. G. Cooper, Growth and departure of individual bubbles at a wall, *Appl. Scient. Res.* **38,** 77 (1982).
18J. K. Cornwell, On boiling incipience due to contact angle hysteresis, *Int. J. Heat Mass Transfer* **25,** 205 (1982).
19J. K. Cornwell and R. B. Schüller, A study of boiling outside a tube bundle using high speed photography, *Int. J. Heat Mass Transfer* **25,** 683 (1982).
20J. V. A. Dyundin, G. N. Danilova and A. V. Borishanskaya, Heat transfer to refrigerants boiling on tubes covered with porous coatings, *Heat Transfer—Soviet Res.* **13**(1), 61 (1981).
21J. D. Gorenflo, J. Goetz and K. Bier, Proposal of a standard apparatus for the measurement of pool boiling heat transfer, *Wärme- und Stoffübertragung* **16,** 69 (1982).
22J. D. Hein, W. Kastner, W. Köhler and W. Krätzer, Heat transfer to post-dryout region during transients, *Wärme- und Stoffübertragung* **16,** 251 (1982).
23J. K. Hishida, M. Maeda and S. Ikai, Experimental study of heat transfer in binary mist flow (effect of incidence angles of a plate), *Heat Transfer—Jap. Res.* **10**(4), 71 (1981).
24J. G. A. Hughmark, Heat transfer with phase change in vertical upward, horizontal, and vertical downward tube flow, *Ind. Engng Chem. Fundam.* **21,** 339 (1982).
25J. Y. Iida, T. Tsuyuki and K. Kawabata, On the boiling patterns of nucleate pool boiling at reduced pressure, *Heat Transfer—Jap. Res.* **10**(2), 70 (1981).
26J. O. C. Iloeje, D. N. Plummer, W. M. Rohsenow and P. Griffith, Effects of mass flux, flow quality, thermal and surface properties of materials on rewet of dispersed flow film boiling, *J. Heat Transfer* **104,** 304 (1982).
27J. S. Inada, Y. Miyasaka, S. Sakumoto and R. Izumi, Study of boiling characteristic curves in subcooled pool boiling of water. Part II: Effect of surface contamination on boiling heat transfer and minutization of a vapor bubble, *Heat Transfer—Jap. Res.* **11**(2), 27 (1982).
28J. S. Inada, Y. Miyasaka and R. Izumi, Study of boiling characteristic curves in subcooled pool boiling of water Part III: Behavior of vapor bubbles and surface temperature variation in transitional pool boiling, *Heat Transfer—Jap. Res.* **11**(2), 42 (1982).
29J. S. Inada, Y. Miyasaka, R. Izumi and M. Kobayashi, Study of boiling characteristic curves in subcooled pool boiling of water (4th report, the mechanism of heat transfer in transition boiling), *Bull. J.S.M.E.* **25,** 1085 (1982).
30J. T. R. Jemison, R. J. Rivers and R. Cole, Incipient vapor nucleation of methanol from an artificial site-uniform superheat, *J. Heat Transfer* **104,** 567 (1982).
31J. M. K. Jensen and A. E. Bergles, Critical heat flux in helical coils with a circumferential heat flux tilt toward the outside surface, *Int. J. Heat Mass Transfer* **25,** 1383 (1982).
32J. G. Joffre, B. Prunet-Foch and M. Cloupeau, The evaporation of charged droplets in an electric field, *Heat Transfer—Soviet Res.* **13**(6), 81 (1981).
33J. V. I. Kashinskiy, Ye. I. Nevstruyeva and I. M. Romanovskiy, Effect of deposits of steel corrosion products on the temperature of tubes supporting nucleate boiling, *Heat Transfer—Soviet Res.* **13**(1), 43 (1981).
34J. K. B. Katsaros and W. D. Garrett, Effects of organic surface films on evaporation and thermal structure of water in free and forced convection, *Int. J. Heat Mass Transfer* **25,** 1661 (1982).
35J. Y. Katto, An analytical investigation on CHF of flow boiling in uniformly heated vertical tubes with special reference to governing dimensionless groups, *Int. J. Heat Mass Transfer* **25,** 1353 (1982).
36J. Y. Katto, A study on limiting exit quality of CHF of forced convection boiling in uniformly heated vertical channels, *J. Heat Transfer* **104,** 40 (1982).
37J. Y. Katto and S. Yokoya, CHF of forced convection boiling in uniformly heated vertical tubes: experimental study of HP regime by the use of Refrigerant 12, *Int. J. Multiphase Flow* **8,** 165 (1982).
38J. I. Z. Kopp, A new model of nucleation in the boiling of liquid, *Heat Transfer—Soviet Res.* **13**(1), 117 (1981).
39J. S. A. Kovalev and G. B. Rybchinskaya, Stability of heat transfer in nucleate boiling on nonisothermal surfaces, *Heat Transfer—Soviet Res.* **13**(1), 75 (1981).

40J. S. S. Kutateladze, Principle equations of thermohydrodynamics of nucleate boiling, *Heat Transfer—Soviet Res.* **13**(3), 1 (1981).
41J. H. Kuwahara, W. Nakayama and Y. Mori, Heat transfer from the heated cylinders with various surfaces in air/water mist flows, *Heat Transfer—Jap. Res.* **10**(2), 1 (1981).
42J. H. Langner and F. Mayinger, Droplet-spectrum and entrainment in heated vapour–liquid flow, *Wärme- und Stoffübertragung* **16**, 23 (1982).
43J. G. M. Lazarek and S. H. Black, Evaporative heat transfer, pressure drop and critical heat flux in a small vertical tube with R-113, *Int. J. Heat Mass Transfer* **25**, 945 (1982).
44J. L. Lee and J. C. Chen, Surface probe for measurement of liquid contact in film and transition boiling on high-temperature surfaces, *Rev. Scient. Instrum.* **53**, 1472 (1982).
45J. J. H. Lienhard, Corresponding states correlations of the spinodal and homogeneous nucleation limits, *J. Heat Transfer* **104**, 379 (1982).
46J. J. J. Lorenz and D. Yung, Film breakdown and bundle-depth effects in horizontal-tube, falling-film evaporators, *J. Heat Transfer* **104**, 569 (1982).
47J. I. G. Malenkov, Heat transfer in bubbling of gases through liquids. Experimental technique and results, *Heat Transfer—Soviet Res.* **13**(3), 21 (1981).
48J. I. G. Malenkov, Experimental data on the effect of viscosity on the displacement of liquid away from the wall during bubbling, *Heat Transfer—Soviet Res.* **13**(3), 15 (1981).
49J. P. J. Marto and Lt. V. J. Lepere, USN, Pool boiling heat transfer from enhanced surfaces to dielectric fluids, *J. Heat Transfer* **104**, 292 (1982).
50J. B. J. McCoy, Evaporation of water through surfactant layers, *A.I.Ch.E. Jl* **28**, 844 (1982).
51J. M. Monde, H. Kusuda and H. Uehara, Critical heat flux during natural convective boiling in vertical rectangular channels submerged in saturated liquid, *J. Heat Transfer* **104**, 300 (1982).
52J. R. A. Moose and E. N. Ganić, On the calculation of wall temperatures in the post dryout heat transfer region, *Int. J. Multiphase Flow* **8**, 525 (1982).
53J. Y. Mori, K. Hijikata and T. Yasunaga, Mist cooling of very hot tubules with reference to through-hole cooling of gas turbine blades, *Int. J. Heat Mass Transfer* **25**, 1271 (1982).
54J. J. Müller and E. Hahne, Boiling heat transfer in finned tube bundles, *Heat Transfer—Soviet Res.* **13**(6), 19 (1981).
55J. A. S. Naik and V. K. Dhir, Forced flow evaporative cooling of a volumetrically heated porous layer, *Int. J. Heat Mass Transfer* **25**, 541 (1982).
56J. W. Nakayama, T. Daikoku and T. Nakajima, Effects of pore diameters and system pressure on saturated pool nucleate boiling heat transfer from porous surfaces, *J. Heat Transfer* **104**, 286 (1982).
57J. R. A. Nelson, Forced-convection post-CHF heat transfer and quenching, *J. Heat Transfer* **104**, 48 (1982).
58J. B. S. Petukhov, S. A. Kovalev and V. D. Geshele, Heat transfer in boiling on finned surfaces, *Heat Transfer—Soviet Res.* **13**(1), 51 (1981).
59J. G. K. Raina and P. D. Grover, Direct contact heat transfer with change of phase: theoretical model, *A.I.Ch.E. Jl* **28**, 515 (1982).
60J. R. G. Sardesai, R. A. W. Shock and D. Butterworth, Heat and mass transfer in multicomponent condensation and boiling, *Heat Transfer Engng* **3**, 104 (1982).
61J. R. B. Schnittger and F. Mayinger, Studies on heat transfer in the post-dryout region in horizontal tubes, *Chemie–Ingr–Tech.* **54**, 915 (1982).
62J. A. Sharon, M. Bürger and W. Schwalbe, Stability analysis of film boiling, *Israel J. Technol.* **19**, 147 (1981).
63J. J. E. Shepherd and B. Sturtevant, Rapid evaporation at the superheat limit, *J. Fluid Mech.* **121**, 379 (1982).
64J. T. Shinkawa and H. Yoshida, Non-evaporating region in a downward flow vertical liquid film type evaporator, *Heat Transfer—Jap. Res.* **10**(2), 50 (1981).
65J. W. M. Sluyter, S. J. D. van Stralen and W. Zijl, The method of characteristics applied to numerical solutions of gas bubble implosion and fragmentation, *Int. J. Heat Mass Transfer* **25**, 1103 (1982).
66J. R. C. Smith, W. M. Rohsenow and M. S. Kazimi, Volumetric heat-transfer coefficients for direct-contact evaporation, *J. Heat Transfer* **104**, 264 (1982).
67J. N. G. Styushin, Forced-convection heat transfer to boiling liquids, *Heat Transfer—Soviet Res.* **13**(1), 84 (1981).
68J. J. R. Thome, Latent and sensible heat-transfer rates in the boiling of binary mixtures, *J. Heat Transfer* **104**, 474 (1982).
69J. V. I. Tolubinskiy, A. A. Kriveshko and Yu. N. Ostrovskiy, Investigation of microcharacteristics of boiling, *Heat Transfer—Soviet Res.* **13**(1), 93 (1981).
70J. L. C. Witte and J. H. Lienhard, On the existence of two transition boiling curves, *Int. J. Heat Mass Transfer* **25**, 771 (1982).
71J. A. F. Zaletnev and A. S. Simkin, Heat transfer and hydrodynamics in subcooled boiling in pipes, *Heat Transfer—Soviet Res.* **13**(2), 1 (1981).

Condensation

1JJ. K. Asano and S. Matsuda, Condensation of binary vapor mixtures on a vertical flat plate: an experimental study of the surface temperature of the condensate, *Heat Transfer—Jap. Res.* **11**(1), 1 (1982).
2JJ. Yu. I. Balakleyevskiy and V. Yu. Chekhovich, Condensation of a submerged steam jet, *Heat Transfer—Soviet Res.* **13**(3), 65 (1981).
3JJ. V. M. Borishanskiy, D. I. Volkov, N. I. Ivashchenko, O. P. Krektunov, L. A. Vorontsova and Yu. T. Illarionov, Heat transfer coefficients in steam condensation in a vertical tube, *Heat Transfer—Soviet Res.* **13**(2), 52 (1981).
4JJ. L. C. Burmeister, Vertical fin efficiency with film condensation, *J. Heat Transfer* **104**, 391 (1982).
5JJ. C. Calia and P. Griffith, Modes of circulation in an inverted U-tube array with condensation, *J. Heat Transfer* **104**, 769 (1982).
6JJ. C. K. Chan and C. K. B. Lee, A regime map for direct contact condensation, *Int. J. Multiphase Flow* **8**, 11 (1982).
7JJ. L. D. Chen and G. M. Faeth, Condensation of submerged vapor jets in subcooled liquids, *J. Heat*

Transfer **104,** 774 (1982).
8JJ. I. I. Gogonin, A. R. Dorakhov and V. I. Sosunov, Heat transfer coefficients in film condensation of stationary vapor on vertical surfaces, *Heat Transfer—Soviet Res.* **13**(3), 51 (1981).
9JJ. R. I. Hirshburg and L. W. Florschuetz, Laminar wavy-film flow: part 1, hydrodynamic analysis, *J. Heat Transfer* **104,** 452 (1982).
10JJ. R. I. Hirshburg and L. W. Florschuetz, Laminar wavy-flow: part 2, condensation and evaporation, *J. Heat Transfer* **104,** 459 (1982).
11JJ. O. P. Ivanov, An experimental and analytic study of local characteristics of condensation of vapor flows in channels, *Heat Transfer—Soviet Res.* **13**(2), 123 (1981).
12JJ. H. R. Jacobs and B. H. Major, The effect of noncondensible gases on bubble condensation in an immiscible liquid, *J. Heat Transfer* **104,** 487 (1982).
13JJ. K. Kaino, M. Izumi, S. Shimizu and T. Takeyama, Heat transfer near the initiation of dropwise condensation, *Heat Transfer—Jap. Res.* **11**(3), 44 (1982).
14JJ. V. B. Khabenskiy, V. S. Granovskiy and P. A. Morozov, Condensation in stratified two-phase flow inside horizontal or sloping tubes, *Heat Transfer—Soviet Res.* **13**(2), 85 (1981).
15JJ. V. B. Khabenskiy, V. S. Granovskiy and P. A. Morozov, Analysis of equations of condensation in horizontal or sloping tubes carrying stratified two-phase flows, *Heat Transfer—Soviet Res.* **13**(2), 98 (1981).
16JJ. S. S. Kutateladze, Semi-empirical theory of film condensation of pure vapours, *Int. J. Heat Mass Transfer* **25,** 653 (1982).
17JJ. S. S. Kutateladze, I. I. Gogonin, A. P. Dorokhov and V. I. Sosunov, Heat transfer in vapor condensation on a horizontal tube bundle, *Heat Transfer—Soviet Res.* **13**(3), 32 (1981).
18JJ. D. M. Maron and S. Sideman, Condensation inside near horizontal tubes in cocurrent and counter-current flow, *Int. J. Heat Mass Transfer* **25,** 1439 (1982).
19JJ. A. A. Nicol and M. J. Dempsey, Condensation of wet steam on a horizontal tube with additional droplet deposition, *Int. J. Multiphase Flow* **8,** 207 (1982).
20JJ. H. C. No and M. S. Kazimi, Wall heat transfer coefficient for condensation and boiling in forced convection of sodium, *Nucl. Sci. Engng* **81,** 319 (1982).
21JJ. U. Schaflinger, Steady and unsteady process of film condensation on a flat plate, a vertical coin, a horizontal pipe and a sphere, *Wärme- und Stoffübertragung* **16,** 149 (1982).
22JJ. R. A. Seban and J. A. Hodgson, Laminar film condensation in a tube with upward vapor flow, *Int. J. Heat Mass Transfer* **25,** 1291 (1982).
23JJ. G. G. Shklover, O. O. Mil'man, V. S. Baskov and G. A. Ankudinov, Heat transfer in condensation of steam on finely-finned horizontal tubes, *Heat Transfer—Soviet Res.* **13**(2), 108 (1981).
24JJ. M. E. Simpson and C. K. Chan, Hydrodynamics of a subsonic vapor jet in subcooled liquid, *J. Heat Transfer* **104,** 271 (1982).
25JJ. H. M. Soliman, On the annular-to-wavy flow pattern transition during condensation inside horizontal tubes, *Can. J. Chem. Engng* **60,** 475 (1982).
26JJ. K. Taghavi-Tafreshi and V. K. Dhir, Analytical and experimental investigation of simultaneous melting-condensation on a vertical wall, *J. Heat Transfer* **104,** 24 (1982).
27JJ. T. N. Tandon, H. K. Varma and C. P. Gupta, A new flow regimes map for condensation inside horizontal tubes, *J. Heat Transfer* **104,** 763 (1982).
28JJ. P. C. Wayner, Jr., Adsorption and capillary condensation at the contact line in change of phase heat transfer, *Int. J. Heat Mass Transfer* **25,** 707 (1982).

Radiation

1K. M. H. Abassi and J. W. Evans, Monte Carlo simulation of radiant transport through an adiabatic packed bed or porous solid, *A.I.Ch.E. Jl* **28,** 853 (1982).
2K. R. W. Adams, K. A. Fuller and H. D. Downing, Infrared optical constants of aqueous solutions of sodium sulfate, *J. Opt. Soc. Am.* **72,** 893 (1982).
3K. D. Apter and J. Shamir, Infrared interferometric analysis of thin films and optical surfaces, *Appl. Optics* **21,** 1512 (1982).
4K. R. K. Avery and A. R. Jones, Measurement of the complex refractive index of pulverised coal by light scattering: an attempt and some comments, *J. Phys.*, Series D, *Appl. Phys.* **15,** 1373 (1982).
5K. P. J. Beckwith and K. C. A. Crane, Two-wavelength infrared pyrometer for large and rapid temperature changes, *Rev. Scient. Instrum.* **53,** 871 (1982).
6K. M. Q. Brewster and C. L. Tien, Radiative transfer in packed fluidized beds: dependent versus independent scattering, *J. Heat Transfer* **104,** 573 (1982).
7K. R. O. Buckius, The effect of molecular gas absorption on radiative heat transfer with scattering, *J. Heat Transfer* **104,** 580 (1982).
8K. R. O. Buckius, Radiation heat transfer using mean photon path lengths, *Int. J. Heat Mass Transfer* **25,** 917 (1982).
9K. T. J. Chung and J. Y. Kim, Radiation view factors by finite elements, *J. Heat Transfer* **104,** 792 (1982).
10K. B. T. F. Chung and M. H. N. Naraghi, A simpler formulation for radiative view factors from sphere to a class of axisymmetric bodies, *J. Heat Transfer* **104,** 201 (1982).
11K. M. R. Davis, Determination of physical properties of heat transfer sensors from vacuum heat loss observations, *J. Heat Transfer* **104,** 213 (1982).
12K. I. E. Enoch, E. Ozil and R. C. Birkebak, Polynomial approximation solution of heat transfer by conduction and radiation in a one-dimensional absorbing, emitting, and scattering medium, *Num. Heat Transfer* **5,** 353 (1982).
13K. T. S. Eriksson and C. G. Granqvist, Radiative cooling computed for model atmospheres, *Appl. Optics* **21,** 4381 (1982).
14K. R. L. Farrow, P. L. Mattern and L. A. Rahn, Comparison between CARS and corrected thermocouple temperature measurements in a diffusion flame, *Appl. Optics* **21,** 3119 (1982).
15K. J. D. Felske and T. T. Charalampopoulos, Gray gas weighting coefficients for arbitrary gas–soot mixtures, *Int. J. Heat Mass Transfer* **25,** 1849 (1982).
16K. R. Fernandes and J. Francis, Combined conductive and

radiative heat transfer in absorbing, emitting, and scattering cylindrical medium, *J. Heat Transfer* **104,** 594 (1982).

17K. G. Flamant, Theoretical and experimental study of radiant heat transfer in a solar fluidized-bed receiver, *A.I.Ch.E. Jl* **28,** 529 (1982).

18K. B. D. Ganapol and G. Spiga, Time-dependent radiative transfer in a semi-infinite medium with a reflecting boundary, *Z. Angew. Math. Phys.* **32,** 555 (1981).

19K. G. Gaultieri, Complementary variational principles for radiation heat transfer in an absorbing medium, *Z. Angew. Math. Phys.* **33,** 826 (1982).

20K. G. Ye. Gorelik, O. G. Martynenko and N. V. Pavlyukevich, Effect of the temperature dependence of optical properties of materials on their radiant heating, *Heat Transfer—Soviet Res.* **14**(1), 49 (1982).

21K. W. L. Grosshandler and S. L. P. Monteiro, Attenuation of thermal radiation by pulverized coal and char, *J. Heat Transfer* **104,** 587 (1982).

22K. I. S. Habib, The effect of a concentric radiating cylinder on liquid spray cooling in a hot gas discharge, *J. Heat Transfer* **104,** 790 (1982).

23K. M. Hasatani, N. Arai, Y. Bando and H. Nakamura, Collection of thermal radiation by a semitransparent fluid layer flowing in an open channel, *Heat Transfer—Jap. Res.* **11**(3), 17 (1982).

24K. D. W. Hatfield and D. K. Edwards, Effects of wall radiation and conduction on the stability of a fluid in a finite slot heated from below, *Int. J. Heat Mass Transfer* **25,** 1363 (1982).

25K. C. K. Hsieh and X. A. Wang, Twin-slope method of measurement of specific heat and coating absorptance of solid materials, *Rev. Scient. Instrum.* **53,** 684 (1982).

26K. N. H. Juul, View factors in radiation between two parallel oriented cylinders of finite length, *J. Heat Transfer* **104,** 384 (1982).

27K. B. Karlsson and C. G. Ribbin, Optical constants and spectral selectivity of stainless steel and its oxides, *J. Appl. Phys.* **53,** 6340 (1982).

28K. H. Khalil, J. K. Shultis and T. W. Lester, Comparison of three numerical methods for evaluation of radiant energy transfer in scattering and heat generating, *Num. Heat Transfer* **5,** 235 (1982).

29K. D. S. Kimes and J. A. Korchner, Radiative transfer model for heterogeneous 3-D scenes, *Appl. Optics* **21,** 4119 (1982).

30K. V. Y. Klabukov, L. T. Grebenschchikov, V. A. Kuz'min, Ye. I. Maratkanova and V. I. Smirnov, Experimental and analytical study of radiant heat transfer in high-temperature furnaces, *Heat Transfer—Soviet Res.* **14**(1), 6 (1982).

31K. I. P. Kolchenogova and Yu. V. Zelenev, Effect of IR radiation on a system of oriented polymer fibers and materials woven from them, *Heat Transfer—Soviet Res.* **14**(1), 36 (1982).

32K. S. Kubo and E. Hayashiguchi, Theory of unsteady radiative heat transfer in a semi-infinite space, *Phys. Soc. Japan* **50,** 703 (1981).

33K. N. N. Kudryavtsev and S. S. Novikov, Theoretical and experimental investigations of i.r. radiation transfer in vibrationally nonequilibrated molecular gas containing CO_2 and CO, *Int. J. Heat Mass Transfer* **25,** 1541 (1982).

34K. T. Kunitomo, E. Ozaki and K. Kawahara, The effect of dimerization on infrared absorption in the 6.3 μm band and the total emissivity of water vapor, *Heat Transfer—Jap. Res.* **11**(3), 90 (1982).

35K. G. Lauriat, Combined radiation-convection in gray fluids enclosed in vertical cavities, *J. Heat Transfer* **104,** 609 (1982).

36K. H. Lee and R. O. Buckius, Scaling anisotropic scattering in radiation heat transfer for a planar medium, *J. Heat Transfer* **104,** 68 (1982).

37K. R. P. Lucht, N. M. Laurendeau and D. W. Sweeney, Temperature measurement by two-line laser-saturated OH fluorescence in flames, *Appl. Optics* **21,** 3729 (1982).

38K. J. Menashe and W. A. Wakeham, Effect of absorption of radiation on thermal conductivity measurements by the transient hot-wire techniques, *Int. J. Heat Mass Transfer* **25,** 661 (1982).

39K. G. R. Mitchel, Silver halide film as a linear detector for 10.6 μm radiation, *Rev. Scient. Instrum.* **53,** 111 (1982).

40K. M. H. N. Naraghi and B. T. F. Chung, Radiation configuration factors between disks and a class of axisymmetric bodies, *J. Heat Transfer* **104,** 426 (1982).

41K. M. N. Özişik and Y. Yener, The Galerkin method for solving radiation transfer in plane-parallel participating media, *J. Heat Transfer* **104,** 351 (1982).

42K. J. K. Park and D. W. Deering, Simple radiative transfer model for relationships between canopy biomass and reflectance, *Appl. Optics* **21,** 303 (1982).

43K. L. P. Poole, Radiative transfer model for airborne laser fluorosensors: inclusion of water Raman scattering, *Appl. Optics* **21,** 3063 (1982).

44K. A. C. Ratzel and J. R. Howell, Heat transfer by conduction and radiation in one-dimensional planar media using the differential approximation, *J. Heat Transfer* **104,** 388 (1982).

45K. F. Santarelli, C. Stramigioli, G. Spiga and M. N. Özişik, Effects of scattering and reflection of radiation on batch photochemical reaction in a slab geometry, *Int. J. Heat Mass Transfer* **25,** 57 (1982).

46K. A. Sharma and A. C. Cogley, Radiative heat transfer in a completely general plane-parallel environment, *Int. J. Heat Mass Transfer* **25,** 523 (1982).

47K. E. B. Singleton, Infrared thermography: a graphic interpretation of existence as a function of surface temperature, emittance, and background, *Appl. Optics* **21,** 782 (1982).

48K. T. F. Smith, Z. F. Shen and J. N. Friedman, Evaluation of coefficients for the weighted sum of gray gases model, *J. Heat Transfer* **104,** 602 (1982).

49K. S. H. Suck, A. E. Wetmore, T. S. Chen and J. L. Kassner, Jr., Role of various water clusters in IR absorption in the 8–14-μm window region, *Appl. Optics* **21,** 1610 (1982).

50K. C.-F. Tsai and S. H. Chan, Numerical solution of the contact interface temperature with internal thermal radiation and conduction effects, *Num. Heat Transfer* **5,** 359 (1982).

51K. F. P. J. Valero, W. J. Y. Gore and L. P. M. Giver, Radiative flux measurements in the troposphere, *Appl. Optics* **21,** 831 (1982).

52K. B. Window and G. L. Harding, Thermal emittance of films of nickel and stainless steel, *J. Opt. Soc. Am.* **72,**

1281 (1982).
53K. P. C. M. van Woerkom and R. L. deGroot, Infrared emission spectra from a heterogeneous catalyst system in reaction conditions, 2. Infrared spectroscopic aspects, *Appl. Optics* **21**, 3114 (1982).
54K. R. Viskanta and M. Behnia, Modeling of energy transfer in irradiated and cooled thermally stratified layers of water, *Heat Transfer—Soviet Res.* **13**(6), 51 (1981).
55K. Y. Yamada and Y. Mori, Effect of radiation heat transfer on the performance of high temperature heat exchangers, *Heat Transfer—Jap. Res.* **10**(4), 1 (1981).
56K. F. B. Yurevich, L. A. Konyukh, G. M. Yatskevich and V. A. Fogeleve, Attenuation of radiant fluxes by fine-particle aerosols, *Heat Transfer—Soviet Res.* **14**(1), 13 (1982).

MHD

1M. A. Bharali and A. K. Borkakati, The effect of hall currents on MHD flow and heat transfer between two parallel porous plates, *Appl. Scient. Res.* **39**, 155 (1982).
2M. V. D. Borman, S. Yu. Krylov, B. I. Nikolayev, V. I. Troyan and B. A. Frolov, Effect of magnetic field on heat flow in a polyatomic gas in intermediate pressure range, *Physica* **111A**, 491 (1982).
3M. E. D. Doss and Y. C. L. Wu, State-of-the-art of MHD generator development, *J. Energy* **6**, 161 (1982).
4M. Z. El Derini and E. D. Doss, Simulated and measured MHD generator performance, *J. Energy* **6**, 171 (1982).
5M. F. L. Galanga, J. T. Lineberry, Y. C. L. Wu, M. H. Scott, W. E. Clemons, Experimental results of the UTSI coal-fired MHD generator, *J. Energy* **6**, 179 (1982).
6M. J. T. Lineberry, F. L. Galanga, B. L. Liu, Y. C. L. Wu, M. H. Scott, Y. C. Pan and E. D. Doss, Comparison of experimental results from the UTSI coal-fired MHD generator to theoretical predictions, *J. Energy* **6**, 187 (1982).
7M. G. Mandal, K. K. Mandal and G. Choudhury, On combined effects of Coriolis force and hall current on steady MHD Couette flow and heat transfer, *Phys. Soc. Japan* **51**, 2010 (1982).
8M. I. Michiyoshi, M. Tanaka and O. Takahashi, Mercury–argon two-phase heat transfer in a vertical annulus under transverse magnetic field, *Int. J. Heat Mass Transfer* **25**, 1481 (1982).
9M. A. A. Raptis and G. J. Tzivanidis, Mass transfer and free convective MHD flow past an accelerated vertical porous plate, *Wärme- und Stoffübertragung* **16**, 145 (1982).
10M. N. Riahi, Nonlinear magnetic convection at small Prandtl numbers, *Aust. J. Phys.* **34**, 251 (1981).
11M. R. Shyam and D. V. Sharma, Nonlinear breaking of waves in an electrically conducting and radiating gas, *Z. Angew. Math. Phys.* **33**, 772 (1982).
12M. B. Singh and J. Lal, Heat transfer for MHD flow through a rectangular pipe with discontinuity in wall temperatures, *Int. J. Heat Mass Transfer* **25**, 1523 (1982).
13M. R. F. Starr, L. S. Christensen, G. W. Garrison and G. L. Whitehead, Preliminary Faraday performance of a large magnetohydrodynamic generator at high magnetic field, *J. Energy* **6**, 163 (1982).
14M. S. P. Vanka and R. K. Ahluwalia, Three-dimensional flow and thermal development in magnetohydrodynamic channels, *J. Energy* **6**, 218 (1982).

Numerical methods

1N. I. Babuska and W. G. Szymczak, An error analysis for the finite element method applied to convection diffusion problems, *Comp. Methods Appl. Mech. Engng* **31**, 19 (1982).
2N. J. V. Beck, B. Litkouhi and C. R. St. Clair, Jr., Efficient sequential solution of the nonlinear inverse heat conduction problem, *Num. Heat Transfer* **5**, 275 (1982).
3N. D. J. Evans and C. R. Gane, A.D.I. methods for the solution of diffusion problems in r–θ geometry, *Comp. Methods Appl. Mech. Engng* **31**, 281 (1982).
4N. M. K. Exeter, N. Hay and J. J. Webster, Finite element solutions for free convection flow around an LMFBR debris tray, *Int. J. Heat Mass Transfer* **25**, 999 (1982).
5N. T. B. Gatski, C. E. Grosch and M. E. Rose, A numerical study of the two-dimensional Navier–Stokes equations in vorticity-velocity variables, *J. Comp. Phys.* **48**, 1 (1982).
6N. J. D. Hoffman, Relationship between the truncation errors of centered finite-difference approximations on uniform and nonuniform meshes, *J. Comp. Phys.* **46**, 469 (1982).
7N. A. Kanarchos, Boundary layer refinements in convective diffusion problems, *Int. J. Num. Methods Engng* **18**, 167 (1982).
8N. W. K. Liu and J. I. Lin, Stability of mixed time integration schemes for transient thermal analysis, *Num. Heat Transfer* **5**, 211 (1982).
9N. R. Manohar and J. W. Stephenson, Optimal finite analytical methods, *J. Heat Transfer* **104**, 432 (1982).
10N. A. McD. Mercer, Some new finite-difference schemes for parabolic differential equations, *Num. Heat Transfer* **5**, 199 (1982).
11N. L. Ondris, Remarks concerning the solution of stationary plane thermal problems by the boundary point least squares method, *Comp. Methods Appl. Mech. Engng* **31**, 331 (1982).
12N. J. Padovan and S. Tovichakchaikul, Self-adaptive closed constrained solution algorithms for non-linear conduction, *Num. Heat Transfer* **5**, 253 (1982).
13N. G. Payre, M. de Broissia and J. Bazinet, An upwind finite element method via numerical integration, *Int. J. Num. Methods Engng* **18**, 381 (1982).
14N. W. D. Rolph III and K.-J. Bathe, An efficient algorithm for analysis of nonlinear heat transfer with phase changes, *Int. J. Num. Methods Engng* **18**, 119 (1982).
15N. G. E. Schneider, A physical approach to the finite-difference solution of the conduction equation in generalized coordinates, *Num. Heat Transfer* **5**, 1 (1982).
16N. A. Sharma and A. C. Cogley, Numerical techniques in radiative heat transfer for general, scattering, plane-parallel media, *Num. Heat Transfer* **5**, 21 (1982).
17N. A. Sharma and W. J. Minkowycz, KNOWTRAN: an artificial intelligence system for solving heat transfer problems, *Int. J. Heat Mass Transfer* **25**, 1279 (1982).
18N. M. M. Shorcri and J. P. Novak, A splitting scheme for the solution of the heat diffusion equation in cylindrical coordinates, *Int. J. Num. Methods Engng* **18**, 1249 (1982).

19N. R. M. Smith and A. G. Hutton, The numerical treatment of advection: a performance comparison of current methods, *Num. Heat Transfer* **5,** 439 (1982).
20N. G. D. Stubley, G. D. Raithby, A. B. Strong and K. A. Woolner, Simulation of convection and diffusion processes by standard finite difference schemes and by influence schemes, *Comp. Methods Appl. Mech. Engng* **35,** 153 (1982).
21N. L. G. Tham and Y. K. Cheung, Numerical solution of heat conduction problems by parabolic time–space element, *Int. J. Num. Methods Engng* **11,** 467 (1982).
22N. A. Thyagaraja, A thermodynamically consistent numerical method for transonic cascade flow simulation, *Int. J. Num. Methods Engng* **18,** 535 (1982).
23N. D. M. Trujillo and H. R. Busby, Investigation of highly accurate integration formulae for transient heat conduction analysis using the conjugate gradient method, *Int. J. Num. Methods Engng* **18,** 99 (1982).
24N. C. Yi-Heng, On a finite element model for solving Dirichlet's problem of Laplace's equation, *Int. J. Num. Methods Engng* **18,** 687 (1982).

Heat transfer applications—heat exchangers and heat pipes

1Q. V. M. Borishanskiy, G. N. Danilova, I. Z. Kopp, V. A. Dyundin and A. V. Borishanskaya, Division of roughness of heat exchangers and steam generators, *Heat Transfer—Soviet Res.* **13**(2), 26 (1981).
2Q. O. N. Bragina, V. L. Lel'chuk, A. G. Sorokin and K. F. Shuyskaya, Experimental study of enhancement of heat transfer from a tube bundle in turbulent axial flow, *Heat Transfer—Soviet Res.* **13**(4), 14 (1981).
3Q. I. I. Gogonin and A. R. Dorokhov, Enhancement of heat transfer in horizontal tubes condensers, *Heat Transfer—Soviet Res.* **13**(3), 119 (1981).
4Q. M. Groll and W. Supper, Transient shutdown of an axial-groove liquid trap heat pipe thermal diode, *Heat Transfer—Soviet Res.* **13**(6), 26 (1981).
5Q. M. Groll, W. Supper and C. J. Savage, Shutdown characteristics of an axial-groove liquid-trap heat-pipe thermal diode, *J. Spacecraft Rockets* **19,** 173 (1982).
6Q. C. P. Hedderich, M. D. Kelleher and G. N. Vanderplaats, Design and optimization of air-cooled heat exchangers, *J. Heat Transfer* **104,** 683 (1982).
7Q. V. M. Ievlev, B. V. Dzyubenko, G. A. Dreitser and Yu. V. Vilemas, In-line and cross flow helical tube exchangers, *Int. J. Heat Mass Transfer* **25,** 317 (1982).
8Q. E. K. Kalinin, G. A. Dreytser, S. G. Zakirov, A. A. Vakhabov, Sh. K. Akzamov and Ye. S. Levin, Improvement of heat transfer in tubular heat exchangers by the use of grooved tubes, *Heat Transfer—Soviet Res.* **13**(4), 30 (1981).
9Q. Yu. G. Kapatsyna, K. N. Semenov and M. K. Bologa, Augmentation of heat transfer by use of tubes with internal ridges generated by electrochemical etching of metal, *Heat Transfer—Soviet Res.* **13**(4), 41 (1981).
10Q. K. N. Krishnan and V. M. K. Sastri, Pulsating flows in heat exchangers—an experimental study, *Wärme- und Stoffübertragung* **16,** 169 (1982).
11Q. B. T. Kulakowski and F. W. Schmidt, Explicit design of balanced regenerators, *Heat Transfer Engng* **3,** 25 (1982).
12Q. W. Leidenfrost and B. Korenic, Evaporative cooling and heat transfer augmentation related to reduced condenser temperatures, *Heat Transfer Engng* **3,** 38 (1982).
13Q. P. Le Goff and M. Giuletti, Comparison of economic and energy optimizations for a heat exchanger, *Int. Chem. Engng* **22,** 252 (1982).
14Q. W. T. Sha, C. I. Yang, T. T. Kao and S. M. Cho, Multidimensional numerical modeling of heat exchangers, *J. Heat Transfer* **104,** 417 (1982).
15Q. D. B. Tufts and H. Brandt, Forced-convection heat transfer in a spherical annulus heat exchanger, *J. Heat Transfer* **4,** 670 (1982).
16Q. R. L. Webb, Performance cost effectiveness, and water-side fouling considerations of enhanced tube heat exchangers for boiling service with tube-side water flow, *Heat Transfer Engng* **3,** 84 (1982).
17Q. S. C. Yao, L. E. Hochreiter and W. J. Leech, Heat transfer augmentation in rod bundles near grid spacers, *J. Heat Transfer* **104,** 76 (1982).

General heat transfer applications

1S. A. C. Alkidas, Thermal loading of the cylinder head of a spark-ignition engine, *Heat Transfer Engng* **3,** 66 (1982).
2S. A. C. Alkidas and J. P. Meyers, Transient heat-flux measurements in the combustion chamber of a spark-ignition engine, *J. Heat Transfer* **104,** 62 (1982).
3S. C. P. Bankston and L. H. Back, Heat transfer from combustion gases to a single row of closely spaced tubes in a swirl crossflow stirling engine heater, *J. Heat Transfer* **104,** 55 (1982).
4S. S. W. Beyerlein and H. C. Claiborne, The possibility of multiple temperature maxima in geologic repositories for spent fuel from nuclear reactors, *J. Heat Transfer* **104,** 180 (1982).
5S. H. Consigny and B. E. Richards, Short duration measurements of heat transfer rate to a gas turbine rotor blade, *J. Engng Pwr* **104,** 542 (1982).
6S. L. Y. Cooper, M. Harkleroad, J. Quintiere and W. Rinkinen, An experimental study of upper hot layer stratification in full-scale multiroom fire scenarios, *J. Heat Transfer* **104,** 741 (1982).
7S. M. G. Dunn and A. Hause, Measurement of heat flux and pressure in a turbine stage, *J. Engng Pwr* **104,** 215 (1982).
8S. A. F. Emery and K. M. Sekins, The use of heat transfer principles in designing optimal diathermy and cancer treatment modalities, *Int. J. Heat Mass Transfer* **25,** 823 (1982).
9S. H. Glaser, The real equilibrium temperature and the instantaneous cooling velocity of river water, *Wärme- und Stoffübertragung* **16,** 9 (1982).
10S. D. J. Holve and A. M. Kanury, A numerical study of the response of building components to heating a fire, *J. Heat Transfer* **104,** 344 (1982).
11S. H. Inaba and N. Seki, A numerical study on transient heat transfer in a vertical porous layer, *Bull. J.S.M.E.* **25,** 934 (1982).
12S. H. Inaba and N. Seki, Heat transfer characteristics of a vertical porous heat storage saturated by a liquid and spherical solid particles, *Wärme- und Stoffübertragung* **16,** 209 (1982).
13S. H. F. Jen and J. B. Sobanik, Cooling air flow

characteristics in gas turbine components, *J. Engng Pwr* **104,** 275 (1982).

14S. T. T. Kao, S. M. Cho and D. H. Pai, Thermal modeling of steam generator tubing under CHF-induced temperature oscillations, *Int. J. Heat Mass Transfer* **25,** 781 (1982).

15S. V. N. Slesarenko, G. A. Gudakov, A. M. Belyy, V. M. Saverchenko and V. S. Nagornyy, Drag and heat transfer in thermal desalination of sea water, *Heat Transfer—Soviet Res.* **13**(1), 17 (1981).

16S. E. M. Sparrow, J. E. Niehammer and A. Chaboki, Heat transfer and pressure drop characteristics of arrays of rectangular modules encountered in electronic equipment, *Int. J. Heat Mass Transfer* **25,** 961 (1982).

17S. G. Spiga and M. Spiga, Analytical simulation in heat storage systems, *Wärme- und Stoffübertragung* **16,** 191 (1982).

18S. S. Szepe and J. M. Calm, Development of screening criteria for selecting long-distance heat transmission fluids, *J. Engng Pwr* **104,** 516 (1982).

19S. M. Yoshida, Y. Harigaya and K. Sato, Variation of heat flux through a combustion chamber wall of pre-chamber type diesel engine (heat flux through piston crown, cylinder head, suction valve, exhaust valve, precombustion chamber and exhaust port wall), *Bull. J.S.M.E.* **25,** 426 (1982).

Solar energy

1T. A. Akbarzadeh, W. W. S. Charters and D. A. Lesslie, Thermo-circulation characteristics of a Trombe wall passive test cell, *Solar Energy* **28,** 460 (1982).

2T. C. Benard, Y. Body, D. Gobin and B. Guerrier, Use of a variable parameter test-cell for the study of latent-heat solar walls, *Solar Energy* **29,** 101 (1982).

3T. R. E. Bird, R. L. Hulstrom, A. W. Kliman and H. G. Eldering, Solar spectral measurements in the terrestrial environment, *Appl. Optics* **21,** 1430 (1982).

4T. M. Centeno V, New formulae for the equivalent night sky emissivity, *Solar Energy* **28,** 489 (1982).

5T. M. F. Chiam, Stationary reflector-augmented flat-plate collector, *Solar Energy* **29,** 65 (1982).

6T. J. P. Chiou, The effect of nonuniform fluid flow distribution on the thermal performance of solar collector, *Solar Energy* **29,** 487 (1982).

7T. J. P. Coutier and E. A. Farber, Two applications of a numerical approach of heat transfer process within rock beds, *Solar Energy* **29,** 451 (1982).

8T. J. A. Davies and D. C. McKay, Estimating solar irradiance and components, *Solar Energy* **29,** 55 (1982).

9T. J. M. Davies, C. Vogel and S. K. Cox, Multidirectional photodiode array for the measurement of solar radiances, *Rev. Scient. Instrum.* **53,** 667 (1982).

10T. D. G. Erbs, S. A. Klein and J. A. Duffie, Estimation of the diffuse radiation fraction for hourly, daily and monthly average global radiation, *Solar Energy* **28,** 293 (1982).

11T. D. L. Evans, T. T. Rule and B. D. Wood, A new look at long term collector performance and utilizability, *Solar Energy* **28,** 13 (1982).

12T. J. P. Gerofi, E. Mannik and G. G. Fenton, Calculating the solar contribution to solar assisted systems, *Solar Energy* **28,** 377 (1982).

13T. I. M. Gordon and Y. Zarmi, A simple method for calculating the annual insolation on solar collectors, *Solar Energy* **28,** 483 (1982).

14T. J. M. Gordon, Selective coatings in passive solar heating, *Solar Energy* **29,** 13 (1982).

15T. K. I. Guthrie and W. W. S. Charters, An evaluation of a transverse slatted flat plate collector, *Solar Energy* **28,** 89 (1982).

16T. C. S. Herrick, Optical transmittance measurements on a solar collector cover of cylindrical glass tubes, *Solar Energy* **28,** 5 (1982).

17T. C. F. Hess and C. W. Miller, An experimental and numerical study on the effect of the wall in a thermocline-type cylindrical enclosure. I. Experiments, *Solar Energy* **28,** 145 (1982).

18T. C. F. Hess and C. W. Miller, An experimental and numerical study on the effect of the wall in a thermocline-type cylindrical enclosure. II. Numerical model, *Solar Energy* **28,** 153 (1982).

19T. J. R. Hull, Calculation of solar pond thermal efficiency with a diffusely reflection bottom, *Solar Energy* **29,** 385 (1982).

20T. Y. Jaluria and S. K. Gupta, Decay of thermal stratification in a water body for solar energy storage, *Solar Energy* **28,** 137 (1982).

21T. S. M. Jeter and C. N. Phan, Site-specific clear-day solar irradiance model from long-term irradiance data, *J. Energy* **6,** 115 (1982).

22T. W. T. Lewis, F. P. Incropera and R. Viskanta, Interferometric study of mixing layer development in a laboratory simulation of solar pond conditions, *Solar Energy* **28,** 389 (1982).

23T. E. Lorenzo and A. Luque, Comparison of Fresnel lenses and parabolic mirrors as solar concentrators, *Appl. Optics* **21,** 1815 (1982).

24T. A. Luque and E. Lorenzo, Conditions for achieving ideal and Lambertian symmetrical solar concentrators, *Appl. Optics* **21,** 3736 (1982).

25T. J. Maloney, T.-C. Wang, B. Chen and J. Thorp, Thermal network predictions of the daily temperature fluctuations in a direct gain room, *Solar Energy* **29,** 207 (1982).

26T. R. Marshall and C. Dietsche, Comparisons of paraffin wax storage subsystem models using liquid heat transfer media, *Solar Energy* **29,** 503 (1982).

27T. A. Mecherikunnel and C. H. Duncan, Total and spectral solar irradiance measured at ground surface, *Appl. Optics* **21,** 554 (1982).

28T. B. A. Meyer, J. W. Mitchell and M. M. El-Wakil, Convective heat transfer in vee-trough linear concentrators, *Solar Energy* **28,** 33 (1982).

29T. W. A. Monsen, S. A. Klein and W. A. Beckman, The unutilizability design method for collector-storage walls, *Solar Energy* **29,** 421 (1982).

30T. W. F. Phillips, A simplified nonlinear model for solar collectors, *Solar Energy* **29,** 77 (1982).

31T. W. F. Phillips and R. N. Dave, Effects of stratification on the performance of liquid-based solar heating systems, *Solar Energy* **29,** 111 (1982).

32T. J. Rheinländer, Numerical calculation of heat and mass transfer in solar stills, *Solar Energy* **28,** 173 (1982).

33T. A. E. Saez and B. J. McCoy, Dynamic response of a packed bed thermal storage system—a model for solar

air heating, *Solar Energy* **29,** 201 (1982).
34T. C. Saltiel and M. Sokolov, Optical analysis of solar energy tubular absorbers, *Appl. Optics* **21,** 4033 (1982).
35T. S. Saxena, S. Subrahmaniyam and M. K. Sarkar, A preliminary model for phase change thermal energy storage in a shell and tube heat exchanger, *Solar Energy* **29,** 257 (1982).
36T. K. Sharp, Calculation of monthly average insolation on a shaded surface at any tilt and azimuth, *Solar Energy* **28,** 531 (1982).
37T. E. M. Sparrow, J. S. Nelson and W. Q. Tao, Effect of leeward orientation, adiabatic framing surfaces, and eaves on solar-collector-related heat transfer coefficients, *Solar Energy* **29,** 33 (1982).
38T. J. W. Spencer, A comparison of methods for estimating hourly diffuse solar radiation from global solar radiation, *Solar Energy* **29,** 19 (1982).
39T. J. G. Symons, E. A. Christie and M. K. Peck, Integrating sphere for solar transmittance measurement of planar and nonplanar samples, *Appl. Optics* **21,** 2827 (1982).
40T. N. E. Wijeysundera, L. L. Ah and L. E. Tjioe, Thermal performance study of two-pass solar air heaters, *Solar Energy* **28,** 363 (1982).
41T. Y. Zarmi, Transition time effects in collector systems coupled to a well mixed storage, *Solar Energy* **29,** 3 (1982).

Plasma heat transfer

1U. P. Andanson and B. Cheminat, Temperature gradients within an electric arc close to the cathode, *J. Phys.*, Series D, *Appl. Phys.* **15,** L27 (1982).
2U. K. P. Brand, W. Egli, L. Niemeyer, K. Ragaller and E. Schade, Dielectric recovery of an axially blown SF_6-arc after current zero: Part III—Comparison of experiment and theory, *IEEE Trans. Plasma Sci.* **PS-10,** 162 (1982).
3U. Xi Chen and E. Pfender, Heat transfer to a single particle exposed to a thermal plasma, *Plasma Chem./Plasma Proc.* **2,** 185 (1982).
4U. Xi Chen and E. Pfender, Unsteady heating and radiation effects of small particles in a thermal plasma, *Plasma Chem./Plasma Proc.* **2,** 293 (1982).
5U. L. E. Gettel and F. L. Curzon, Energy transport in DC and AC vortex stabilised arcs, *J. Phys.*, Series D, *Appl. Phys.* **15,** 845 (1982).
6U. K. Etemadi and E. Pfender, Computer-controlled plasma emission spectroscopy, *Rev. Scient. Instrum.* **53,** 255 (1982).
7U. A. Gleizes, H. Kafrouni, H. Dang Duc and C. Maury, The difference between the electron temperature and the gas temperature in a stationary arc plasma at atmospheric pressure, *J. Phys.*, Series D, *Appl. Phys.* **15,** 1031 (1982).
8U. D. C. Jolly, Anode surface temperature and spot formation model for the vacuum arc, *J. Appl. Phys.* **53,** 6121 (1982).
9U. D. Karabourniotis, C. Karras, M. Drakakis and J. J. Damelincourt, Plasma temperature determination from the total intensity of a self-reversed spectral line, *J. Appl. Phys.* **53,** 7259 (1982).
10U. A. A. Kon'kov, G. A. Krymov, Ye. B. Eygenson and A. V. Vorontsov, Experimental investigation of radiative-conductive heat transfer between a thermal plasma and a solid, *Heat Transfer—Soviet Res.* **13**(5), 105 (1981).
11U. V. J. Kovarik, A. I. Herschcovitch and K. Prelec, Initiation of hot cathode arc discharges by electron confinement in Penning and magnetron configurations, *Rev. Scient. Instrum.* **53,** 819 (1982).
12U. K. U. Maske and J. J. Moore, The application of plasmas to high temperature reduction metallurgy, *High Temp. Technol.* **1,** 51 (1982).
13U. O. Matsumato, Y. Shioiri and Y. Kanzaki, Formation of niobium boride from NB_2O_5 and boron with plasma arc, *Plasma Chem./Plasma Proc.* **2,** 247 (1982).
14U. J. McKelliget, J. Szekely, M. Vardelle and P. Fauchais, Temperature and velocity fields in a gas stream exiting a plasma torch. A mathematical model and its experimental verification, *Plasma Chem./Plasma Proc.* **2,** 317 (1982).
15U. M. T. Mehmetoglu, F. Kitzinger and W. H. Gauvin, New temperature measurement technique for thermal plasmas, *Rev. Scient. Instrum.* **53,** 285 (1982).
16U. D. Morfan and J. Amouroux, Preparation of photovoltaic silicon by purification of metallurgical grad silicon with a reactive plasma process, *Plasma Chem./Plasma Proc.* **1,** 397 (1981).
17U. P. Pianarosa, J. M. Gagne, G. Larin and J. P. Saint-Dizier, Determination of self-absorption in emission lines for some optically thick plasmas, *J. Opt. Soc. Am.* **72,** 392 (1982).
18U. K. Ragaller, W. Egli and K. P. Brand, Dielectric recovery of an axially blown SF_6-arc after current zero: Part II—theoretical investigations, *IEEE Trans. Plasma Sci.* **PS-10,** 154 (1982).
19U. E. Richley, D. T. Tuma, On the determination of particle concentrations in multitemperature plasmas, *J. Appl. Phys.* **53,** 8537 (1982).
20U. E. Richley and D. T. Tuma, Mechanism for temperature decay in the freely recovering gas blast arc, *IEEE Trans. Plasma Sci.* **PS-10,** 2 (1982).
21U. S. Roychowdhury, U. K. Roychowdhury and M. Venugopalan, Effect of heating on dissociation of water vapor in high-frequency plasmas and formation of hydrogen peroxide in a cold trap downstream of the plasma, *Plasma Chem./Plasma Proc.* **2,** 157 (1982).
22U. E. Schade and K. Ragaller, Dielectric recovery of an axially blown SF_6-arc after current zero: Part 1—experimental investigations, *IEEE Trans. Plasma Sci.* **PS-10,** 141 (1982).
23U. J. A. Silver, A. Freedman and C. E. Kolb, Supersonic nozzle beam source of atomic oxygen produced by electric discharge heating, *Rev. Scient. Instrum.* **53,** 1714 (1982).
24U. T. Yoshida, T. Tani, H. Nishimura and K. Akashi, Characterization of a hybrid plasma and its application to a chemical synthesis, *J. Appl. Phys.* **51,** 640 (1982).
25U. P. Tsantrizos and W. H. Gauvin, Operating characteristics and energy distribution in a nitrogen transferred arc plasma, *Can. J. Chem. Engng* **60,** 822 (1982).
26U. S. Vacquie, N. Y. Amara, A. Gleizes and H. Kafrouni, Experimental study of a SF_6 arc plasma at atmospheric pressures, *J. Phys.*, Series D, *Appl. Phys.* **15,** 885 (1982).

27U. A. Vardelle, M. Vardelle and P. Fauchais, Influence of velocity and surface temperature of alumina particles on the properties of plasma sprayed coatings, *Plasma Chem./Plasma Proc.* **2,** 255 (1982).
28U. H. Yoshiyasu and S. Yamagata, Vapour flow of a free-burning arc, *J. Phys.*, Series D, *Appl. Phys.* **15,** L91 (1982).

HEAT TRANSFER REVIEW - 1983

E. R. G. ECKERT, R. J. GOLDSTEIN, S. V. PATANKAR, E. PFENDER, J. W. RAMSEY, T. W. SIMON, N. DECKER, and T. H. KUEHN

Heat Transfer Laboratory, Department of Mechanical Engineering, University of Minnesota, Minneapolis, Minnesota, U.S.A.

INTRODUCTION

THIS review surveys papers that have been published in the open literature covering various fields of heat transfer during 1983. As in the past, the number of papers published during that period was such that a selection only could be included.

A number of conferences dealing with heat transfer or including sessions on heat transfer have been held during 1983.

An International Symposium Workshop on Renewable Energy Sources held March 18–22 at Lahore, Pakistan was presented by the Clean Energy Research Institute of the University of Miami, Florida. Heat transfer had to be considered in many of the papers dealing with various forms of energy utilization.

A Thermal Engineering Conference in Honolulu, Hawaii was organized jointly by the American and the Japanese Societies of Mechanical Engineering and held March 20–24. Heat transfer topics dominated the sessions which dealt with all of the modes of heat transfer and with heat transfer applications, especially with heat exchangers. Keynote lectures were presented by P. J. Marto on "heat transfer and two-phase flow in shellside condensation", by K. E. Torrance on "boiling in porous media", by J. R. Lloyd on "separated forced convection", by K. Nishikawa on "boiling heat and its augmentation" and by G. M. Faeth on "recent advances in modeling particle transport properties and dispersion". A round table discussion was concerned with prospects for advanced open and closed Otec cycles. A number of sessions considered heat and mass transfer in underground media and in geothermal applications. Four short courses rounded out the program. The papers are available in four volumes through ASME and JSME.

The 28th International Gas Turbine Conference and Exhibit included in its program sessions on film cooling, turbine gas path heat transfer, and closed cycle gas turbine heat exchangers. Technical papers are available through ASME Headquarters.

The Third Multiphase Flow and Heat Transfer Symposium was held during April 10–20 at Miami Beach, Florida organized by the Clean Energy Research Institute of the University of Miami. It included sessions on heat and mass transfer, scale formation in boiling and various applications.

The 21st National Heat Transfer Conference sponsored by the American Institute of Chemical Engineers and the American Society of Mechanical Engineers was held at Seattle, Washington on July 24–27. The sessions were primarily organized according to the heat transfer modes but included, in addition, sessions on applications, for instance, analysis of steam generators, heat transfer in drying systems and in fusion energy systems. The topic "waste heat recovery" was considered in a round table discussion. Short presentations could be offered in an open forum. The papers presented at the conference are available as preprints or in the published series of the American Institute of Chemical Engineers. Many will also be published in the *Journal of Heat Transfer*. Invited lectures were given by J. G. Knudsen on "Fouling of heat exchangers—how we solve the problem" and by S. Ostrach on "Recent developments in transport phenomena". Both researchers also received the Donald Q. Kern Award and the Max Jakob Award respectively.

An International Conference on Energy Efficient Buildings with Earth Shelter Protection brought together an international group of experts in Sydney, Australia on August 1–10. Proceedings were published by Oklahoma State University.

Heat transfer in porous media was included in the sessions of the 20th Annual Meeting of the Society of Engineering Science at the University of Delaware, August 22–24.

The 15th Symposium of the International Center for Heat and Mass Transfer held at Dubrovnik, Yugoslavia, September 5–9, 1983 was devoted to heat and mass transfer measurement techniques.

The 1983 Tokyo International Gas Turbine Congress was organized by British, German, Japanese and American Engineering societies and held at Tokyo, Japan, October 23–29. Heat transfer was the topic of five sessions considering external, internal, and unsteady heat transfer, film cooling, and heat exchangers.

The Mechanical and Aerospace Engineering Department of the University of Tennessee and the National Bureau of Standards organized a course on Multi-component Condensation and Boiling at Gaithersberg, Maryland, November 8–10.

The Winter Annual Meeting of the American Society of Mechanical Engineers held at Boston, November 13–18, included 19 sessions dealing with various topics of heat transfer. Two symposia considered advances in the cooling of microelectronic devices and in electronic packages. A panel discussed advances in regenerator design and technology. A keynote address by C. L. Tien dealt with "recent advances in thermal insulation". J. Kestin discussed "false and creative fascination with accuracy" as the speaker at the heat transfer luncheon. E. R. G. Eckert selected the topic "experiments on energy separation in fluid streams" for the Thurston lecture. The Heat Transfer Memorial Award was presented to R. Eichhorn. Preprints of the papers are available at ASME Headquarters and many of the papers will also be published in the *Journal of Heat Transfer*.

The Third International Conference on Numerical Methods in Thermal Problems was held in Seattle, Washington, August 1–3, 1983.

Two journals on heat transfer were started in 1983. The *International Journal of Heat and Fluid Flow*, published quarterly by Butterworth Scientific—Journals Division, Guildford, Surrey, U.K., will include papers on engineering thermodynamics, heat transfer, and fluid dynamics together with measurements techniques and applications. The journal *Heat and Technology* was launched in August of 1983 and is published quarterly by Pitagora Editrice, Bologna, Italy. It will include basic and applied research in heat transfer and fluid mechanics and emphasize numerical, industrial and technical applications. A journal *Drying Technology*, published by Marcel Dekker, New York, should also be of interest to the heat transfer community.

The following highlights illuminate developments in heat transfer research during 1983.

Papers on channel flow considered complex thermal boundary conditions, conjugate heat transfer, non-Newtonian behavior, variation of fluid properties, and influence of buoyancy. Many of the configurations studied have relevance to heat transfer augmentation devices, compact heat exchangers, and cooling of electronic components. Experimental results were published on heat transfer to groups of cylinders and spheres, cavities, and tube banks. Thermal convection in saturated porous media filling containers of various shapes were treated analytically, whereas experiments were used to study heat and mass transfer in unsaturated media, including boiling, and in fluidized beds. Natural convection was studied for various flow conditions ranging from laminar steady to turbulent transient flows and for constant and variable properties.

Boiling heat transfer and the related conditions of two-phase flow continue to be areas of active research. The motivation for this research appears to be increasingly diffuse with studies directed towards cooling of nuclear reactors composing a diminishing but still major fraction of the literature. Investigations of critical heat flux [CHF] phenomena were reported and interest is evident in fundamentals of nucleate boiling including transient phenomena, boiling of mixtures, and fouling of heat transfer surfaces. The most frequent topic in condensation heat transfer in 1983 was film condensation, including studies evaluating various fin geometries and determining the effect of the interfacial waviness. A notable omission is the submerged vapor jet in a subcooled liquid pool, a frequent topic in last year's review.

A large number of papers presented calculation methods for radiative interchange in systems of gases, fluids, and solids including radiation–convection interaction. Methods to measure radiative flux and properties were described and applications included fibrous insulation and fluidized beds. Interest in MHD heat transfer has strongly decreased.

In the field of heat transfer applications, manufacturing processes have found increasing attention. The flights of the space shuttle provided the opportunity to check established procedures for the prediction of heat transfer in aerodynamic heating. Papers dealing with the solar energy application included predictions and measurements of the availability of solar radiation, the performance of flat plate solar collectors, and heat transfer in passive solar energy applications.

A growing number of papers on plasma heat transfer testify to the increasing importance of this field. They include particle heat transfer which is important in plasma chemistry and processing applications.

To facilitate the use of this review, a listing of the subject headings is made below in the order in which they appear in the text. The letter which appears adjacent to each subject heading is also attached to the references that are cited in that category:

Conduction, A
Channel flow, B
Boundary Layer and external flows, C
Flow with separated regions, D
Heat transfer in porous media, DP
Natural convection—internal flows, F
Natural convection—external flows, FF
Convection from rotating surfaces, G
Combined heat and mass transfer, H
Boiling, J
Condensation, JJ
Freezing and melting, JM
Radiation, K
MHD, M
Applications
  Heat exchangers and heat pipes, Q
  General, S
  Solar Energy, T
  Plasma heat transfer, U

CONDUCTION

The focus of the published work on conduction is contact resistance although other areas such as solution techniques and applications to various geometries are included.

Some work on fundamentals has been published including an application of irreversible thermodynamics to develop the theory of heat conduction [23A]. A property related to entropy that must decrease with time has been found that can be applied to solutions of the transient heat conduction equation [9A]. Classic heat diffusion theory was shown to underestimate the temperature and heat flux at a thermal front at short time intervals from the start of a thermal pulse [35A].

Various innovative solution techniques have been proposed. The use of the Chebyshev collocation operator for the heat equation was given with several mathematical proofs [14A]. Solution methods were also described for transient nonlinear heat conduction with absorption [21A]. Approximate solutions for parabolic nonlinear heat conduction problems were given [7A]. A fast approximate solution technique was developed for multi-dimensional transient problems with variable conductivity or specific heat capacity [5A].

A solution for a 2- or 3-layer composite slab with three boundaries insulated was given in a pair of papers [28, 29A]. Solutions to transient heat flow through a composite slab is useful in simulating the thermal performance of building walls. A fast root-finding technique for this problem was given [15A] that would help to determine the response factors used in practice. The smoothing of temperatures near the boundaries of a composite cylinder was used to remove short-circuiting phenomena that can appear in solutions [36A]. Treating a composite material as if it were homogeneous may simplify the solution procedure provided correct values for macroscopic conductivity and heat capacity are known [1A]. Numerical solutions were obtained for a slab penetrated by metal rods that behave as fins [17A]. A boundary integral method was used to compute the effective conductivity of an array of spheres embedded in a matrix in either a body-centered or face-centered cubic configuration [38A].

The steady heat flow through 1-D anisotropic, inhomogeneous solid was given using a boundary integral method [8A]. Two- and three-dimensional steady conduction through anisotropic materials was studied using a coordinate transformation [16A].

Contact resistance continues to pose challenging problems in a variety of applications. A crack in an elastic half-space produces a 2-D static thermal problem with radiative heat transfer through the crack [12A, 13A]. The contact resistance between two slabs changes with time if one of the slabs contracts as it cools [18A]. The conductance of materials used in reactor fuel rods is important to predict rod temperatures and to correlate experimental data [25A]. Fuel-cladding conductance has been analyzed using acoustic mismatch theory [10A] and simulated using a Kalman filter [34A]. The effect of a penney-shaped crack on the heat flow and local stress has been considered both within a solid [31A] and at the interface between two solids [26A, 2A]. A comprehensive mathematical treatment of steady heat flow through a thin, poorly conductive layer was presented [4A].

Measurement of the effective thermal conductivity of moist, capillary-porous bodies including phase change was described [37A]. The radiation errors encountered when measuring conductivities at cryogenic temperatures were analyzed [24A]. Corrections and improved measurement techniques were described. The development of an instrument to measure the thermal conductivity of an arbitrary shaped solid was given [20A]. A transient probe technique was described for measuring effective conductivity and specific heat in soils [19A].

Numerous applications have been considered, primarily for steady heat flow in particular geometries. The temperature distribution in an electrically heated thin foil shows that the foil does not provide a uniform heat flux as commonly assumed in experimental studies [32A]. The discrepancy was attributed to electromagnetic field variations. An optimum cylindrical pin fin was identified that includes a convective coefficient that is fin diameter dependent [22A]. A bicylindrical coordinate system was utilized to obtain a solution to steady conduction through an eccentric annulus with convective boundary conditions [11A]. A boundary integral method provided solutions to steady heat flow from a row of cylinders embedded in either a half-space or a homogeneous slab [6A]. Handbook solutions to this problem were shown to be in error by as much as 20%. A solution method to determine the optimum location of a fixed amount of insulation distributed on the surface of a 2-D conducting body was given [3A]. Numerical results were presented for a rectangular slab. A related study was performed in which exergy production in two parallel insulation systems is minimized [30A]. Transient internal heating due to torsional damping was analyzed to explain the premature failure of generator rotor shafts [27A]. Transient 1-D conduction with variable properties and internal heating was studied to simulate the thermal performance of salt gradient stabilized solar ponds [33A].

CHANNEL FLOW

Internal flows have been studied in simple configurations such as the circular tube or parallel plates and also in complex passages encountered in heat transfer augmentation devices. The flow conditions include laminar and turbulent regimes, single-phase and multi-phase systems, and Newtonian and non-Newtonian behavior. Both steady state and transient flows have been investigated.

The effect of viscous dissipation has been analyzed for the thermal entrance region in laminar pipe flow [30B]. Reference [18B] considers laminar heat transfer to Newtonian and non-Newtonian fluids. Also, high-viscosity non-Newtonian fluids are treated with different thermal boundary conditions in [16B]. A study of both heat transfer and friction in a laminar pipe flow of non-Newtonian fluids is reported in [14B]. In another investigation related to non-Newtonian flow, heat transfer to a power-law fluid is obtained at low Peclet numbers [15B].

Reference [29B] deals with liquid metal heat transfer with a uniform wall heat flux for a turbulent pipe flow. Enhanced heat transfer and the associated drag is considered for a turbulent pipe flow with a helical-wire-coil insert [41B]. The developing flow and heat transfer in a helical coil is studied in [24B], where transition from turbulent to laminar flow is included.

A number of papers focus on the unsteady nature of the duct flow. Reference [32B] considers the unsteady heat transfer in a laminar pipe flow subjected to a step change in the ambient temperature, while the response of the heat transfer to a step change in pumping pressure is dealt with in [31B]. The transient laminar heat transfer inside a circular tube is analyzed in [13B]. Experimental measurements for the flow field in an unsteady reacting nozzle exhaust flow are presented in [27B]. Reference [47B] deals with the large heat transport caused by the spontaneous gas oscillations in a tube.

Among more complex flows involving circular tubes are the studies of heat transfer downstream of an orifice [45B] and entrance heat transfer with deposit formation [46B]. A simple heat transfer model has been proposed for separated flows in tubes [10B]. Reference [26B] deals with heat transfer to gas–solid suspensions flowing downward in a circular tube, while heat transfer in a vertically upward gas–solid slug flow is considered in [5B]. Numerical solutions have been obtained for the wall temperature for near-critical para-hydrogen flow in vertical tubes [7B]. Reference [28B] provides empirical correlations for turbulent friction and heat transfer for visoelastic fluids. The Lockhart–Martinelli correlation has been discussed in [12B].

A study of convective heat transfer in a parallel plate channel with porous lining is presented in [11B]. Reference [2B] considers the effect of buoyancy in the entrance region of horizontal rectangular channels. For noncircular ducts the peripheral variation of the wall temperature is examined in [6B]. Correlations are provided in [8B] for convective heat transfer in vertical annular gas layers with constant heating from the inner wall. Reference [20B] deals with low Prandtl number convection between differentially heated end walls, while heat transfer coefficients are provided in [19B] for ducts with constant wall temperature. Thermal instability in plane Poiseuille flow heated from below has been investigated in [48B]. Reference [17B] considers the effect of the longitudinal vortex on the free and forced laminar convection between parallel plates. An experimental study of the low Reynolds number flow between interrupted flat plates is reported in [36B]. Another experimental investigation looks at the turbulent flow and heat transfer in a flat duct with maldistributed inlet flow [44B]. The heat transfer enhancement due to streamwise-periodic disturbances at one wall of a channel is considered in [42B].

The nonlinear convection in a sparsely packed porous medium has been studied by using spectral analysis [37B]. The heat transfer characteristics have been investigated for a 2-D T-shaped flow junction [25B]. Reference [34B] considers heat transfer in a parallel plate duct with circular cylinders inserted in a staggered arrangement. A study of flow and heat transfer in corrugated wall channels is reported in [22B] and [23B]. Heat transfer from concave and convex walls of a return bend of rectangular cross section is considered in [38B], [40B], and [39B].

Experimental measurements are reported for heat transfer and fluid flow in a tube fed by a plenum having nonaligned inlet and exit [43B]. In another investigation [33B], the tube heat transfer is measured for skewed inlet flow caused by competition among tubes fed by the same plenum. Reference [1B] deals with critical flashing flows in nozzles with subcooled inlet conditions. A solution for dispersed flow heat transfer is presented in [49B] under equilibrium fluid conditions. Heat transfer enhancement in an oscillatory cavity flow is studied in [4B]. Reference [21B] provides experimental measurements of heat transfer to a gaseous solid suspension. A numerical and experimental study of turbulent heat transfer downstream of a pipe expansion is presented in [3B]. Laminar forced convection heat transfer in spherical annuli is considered in [35B]. The influence of forced flow on the HeII–HeI transition has been investigated in the presence of heat transfer [9B].

BOUNDARY LAYER AND EXTERNAL FLOWS

Laminar flows

Heat transfer measurements were made on interrupted plates aligned with the flow. The effects of leading edge bluntness, plate spacing, and Reynolds number were evaluated. The heat transfer rate of a downstream plate was higher than predicted by theory, and may be higher than that for an upstream plate [35C]. A numerical heat transfer analysis was made on a low-heat-resistance stretching sheet moving through a fluid [1C]. The transient heat transfer problem for droplet heating at high Peclet number was computed. Results showed that the transient period of internal isothermal redistribution represents a significant part of the droplet total heating period; asymptotic heating results cannot be used with good accuracy [14C]. Heat transfer in micropolar boundary layer flow over a flat plate was studied [19C]. Numerical solutions to convective heat transfer problems in axisymmetric coordinates were presented. The method used a local analytic solution to minimize false diffusion [10C]. Gyarmati's principle for the evaluation of irreversible processes was applied to stagnation flows in two dimensions [5C]. An analysis for the unsteady conjugated heat transfer problem at the 2-D stagnation point on a high thermal conductivity plane wall was presented [13C]. Heat transfer from a constant-temperature hot-film on a flat wall was computed. Measurements were taken in pulsatile and linearly decelerating flows. The thermal wake was observed in high-frequency reversing flow and enhancement of the time-average heat transfer was found with high frequency pulsatile flow [22C]. An analysis of heat transfer from a thin needle in an accelerating flow of non-Newtonian fluid was presented. The more slender needles were shown to have higher wall heat flux, with a given Prandtl number [11C]. Heat transfer solutions were found for flow of a power-law fluid over 2-D and axisymmetrical bodies, including flows with a step in surface temperature. Geometries included wedge-flows and flows over circular cylinders [23C]. Nonisothermic flows of gas mixtures in a channel at intermediate Knudsen numbers were analyzed. The dependence of the kinetic coefficients of the Onsager matrix on the Knudsen number for one-component and binary mixtures was discussed [36C]. An analysis using Fourier transforms was made of thermal waves on a flat plate in a supersonic stream. The effect of variation in the wall temperature was studied [12C]. Analytic solutions were found for nonstationary boundary layers produced by planar, cylindrical and spherical detonation waves. The Prandtl number was shown to have a significant effect on the structure [26C]. Solutions were presented for a Fredholm integral equation of the first kind, which arises in forced convection. Existence and uniqueness of the solutions were discussed [4C].

Mixed [natural and forced] convection over a horizontal plate was analyzed. Fluids with high Prandtl numbers were shown to have a wider range of conditions for mixed convection than those of low

Prandtl numbers [30C].

A numerical model was presented for the conjugate problem of optimizing boundary heating of solids over a range of Biot numbers [27C]. Heat transfer from a surface coated with a thin film of high thermal resistance (e.g. paint) was analyzed. A model for an effective heat transfer coefficient for this conjugate problem was presented [9C].

Turbulent flows

Measurements of velocity and temperature profiles and momentum and heat fluxes were presented for air flow over a flat plate in the presence of an unheated starting length. Profiles of turbulent Prandtl number were deduced and support was found for a turbulent Prandtl number value of 0.9 in the turbulent core [32C]. Turbulent time scales for a thermal mixing layer were obtained via a transport equation for dissipation rates of temperature variance [16C]. Spectral relationships between velocity and temperature fluctuations were presented. The frequency dependence of a turbulent Prandtl number which is based on a shear stress and normal heat flux co-spectra was measured. Higher values of turbulent Prandtl number were associated with lower frequencies [17C]. The thermal structure of a diffusion flame was measured. Profiles of turbulence intensity of temperature fluctuations had two maxima and one minimum [34C]. A method for visualizing heat transfer was presented where "heatlines" and a "heat function", analogous to streamlines and the stream function, were used [24C].

Turbulent thermal and kinematic properties of a wake of a flat plate heated on one side were measured. Similarity coordinates were found and it was shown that profiles of Reynolds shear stress and turbulent heat transport were similar when the axes were properly oriented [28C]. Measurement of mean and fluctuating velocities and temperatures were taken in a boundary layer responding to a double step-change (up then down) in wall heat flux. Two boundary layers were identified; one for each step [3C]. Transient forced- and natural-convection average heat transfer coefficients were measured for helium flow in a channel with a step change in heat flux. For low heat fluxes, boiling heat transfer was superior to supercritical heat transfer [18C]. Heat transfer rates were measured in convex-curved boundary layers. A 50% reduction in Stanton number due to curvature was observed. Data was also presented for laminar and transitional boundary layers with streamwise convex curvature [31C]. Thermal and hydrodynamic measurements were made in an artificially thickened turbulent boundary layer on a rough surface. Mean velocity and turbulence profiles were 2-D and self-preserving. Heat transfer data showed the response to a large effective unheated starting length [25C]. Flow patterns were observed and local heat transfer rates were measured for flat-plate flow over a circular cylinder cavity. Maximum heat transfer was effected when the cavity had a depth-to-diameter ratio of 0.5 [20C]. Experimental results which show the effects of free-stream turbulence were presented. Skin friction and heat transfer increased 20% when the free-stream turbulence intensity was increased to 7% [7C]. The data was compared to existing correlations and new correlations were proposed. The effect of free-stream turbulence on the Reynolds analogy was discussed [8C].

Augmentation of heat transfer from a circular cylinder with increased turbulence created by swirl and impinging jets was measured. For a given pressure drop, impinging jet augmentation gave the greatest benefit [15C]. Heat transfer from a cylinder covered with close-fitting fabric was measured to learn the nature of wind penetration effects of clothing. Appropriate parameters were recommended [33C]. Heat transfer rates from an elliptic cylinder were measured at various angles of attack. The mean Nusselt number was shown to be higher than that of a circular cylinder [29C].

Measurements were presented for mean temperature in a nonbuoyant, heated jet issuing into a cold stream. The extra rate of strain due to streamwise curvature and longitudinal temperature gradients were shown to affect the rate of generation of temperature fluctuations. Measurements of turbulent Prandtl number indicated high near-wall values with a decline to 0.9 away from the wall [2C]. Measurements of heat transfer rates with an air jet impinging on a perforated target surface were presented. Spent air was withdrawn through the surface [21C]. Mass transfer into a turbulent liquid film was analyzed. Surface tension was shown to not affect the mass transfer coefficient [6C].

FLOW WITH SEPARATED REGIONS

Several papers considered the flow and heat transfer around groups of blunt bodies. Local heat transfer coefficients were measured [5D] on a circular cylinder exposed to flow normal to its axis with two small cylinders arranged downstream of the separation line for wake control. Results are reported at subcritical

Reynolds numbers. Local heat transfer rates on two spheres arranged sidewise side-by-side and exposed to a flow normal to their axis were influenced by the formation of a jet-like flow between the spheres as shown by experiments [10D] at a Reynolds number around 10^4 and a turbulence level of 1.1%. The ratio of the distance of the centers of the spheres to the sphere diameter was varied from 1.04 to 2. At small values of this ratio, the wake was deflected sidewise resulting in crosswise force on the spheres. Heat transfer and flow patterns were also studied [12D] for two contacting spheres at various angles of attack at Reynolds numbers between 4000 to 26,000, with both or only one sphere being heated. The interaction was found largest when the spheres were in line with a heat transfer coefficient which increased at higher Reynolds numbers. A perturbation method [15D] studied transient heat transfer from a wire with constant temperature in a fluctuating environment. Maximum heat transfer was found to occur at a certain frequency. Experimental results [6D] describe heat transfer from a cylinder in cross flow of a bubbly gas–liquid mixture. Heat transfer from fuel pins in gas cooled reactors could be improved by transverse, helical, or interrupted ribs [4D].

Two interferometric studies [2D, 1D] considered laminar heat transfer downstream of a step in the surface contour or in flow over a cavity. For the downstep, the heat transfer could be described by the equation

$$\overline{St} = 0.787\,(Re_s)^{0.55}(s/d)^{0.72}$$

where s – height of downstep, x = downstream distance from it.

In both geometries, heat transfer coefficients were less than on the plane surface. The turbulent Prandtl number profile across a plane jet was found [3D] to increase from a value 0.7 at the center plane towards 1 when approaching the border of the jet. A numerical study of wall heat transfer in the recirculating flow region of a confined jet [7D] was compared with the results of previous experiments.

Local mass transfer for cross-flow through tube banks was measured [9D] by the electro-chemical technique at intermediate Reynolds numbers. The per-tube heat transfer was also investigated [13D] in a tube bank with a constriction upstream of the tubes which created an abrupt enlargement of flow cross section. The results of experiments [D11] with an array of slabs exposed to flow normal to their plane could be described by the equation

$$\overline{Nu} = 1.19\,Re^{0.465}\,Pr^{0.375}\,(P/W)^{1/4}$$

in which P denotes the distance of the slabs and W their width. Oscillation of a grid of slabs which were parallel to the flow [8D] improved the sidewise spreading of a temperature profile in the approaching flow compared with the stationary grid. The time mean temperature profile with the shape of a probability curve approaching the grid was found to transform into a profile with two maxima downstream of the grid. The energy separation in a vortex tube was studied experimental [14D] for a range of inlet pressures. It is suggested that Goertler vortices, produced by tangential velocities in the vortex flow, are a major driving force for the energy separation.

HEAT TRANSFER IN POROUS MEDIA

Thermal convection in saturated porous media was studied almost exclusively by analysis. A number of papers considered a rectangular cavity with the vertical walls being at different temperatures or with a constant heat flux from one of the vertical walls to the other one. It was found [5DP] that the boundary layer description of the convection process is valid as long as the Darcy–Rayleigh number is large compared to 10^4L^2 where L is the height to depth ratio. An analysis [2DP] for constant heat flux at Rayleigh numbers between 100 and 5000 and a ratio of height to width between 1 and 10 established that the boundary layer thickness is constant along the height, the core is motionless, and the temperature gradient is also constant in the porous layer along the height in the central region. Numerical results agree with boundary layer analysis. Horizontal or vertical internal flow obstructions were found [1DP] to increase or decrease heat transfer. Several vertical layers of different permeability were studied [19DP] with regard to their influence on heat transfer. Analysis [5DP] of unsteady natural convection in the enclosure established the regions when vertical or horizontal boundary layers are formed. Three-dimensional natural convection in a cubic container with isothermal walls and internal heat generation was analyzed [4DP] for a porous medium modelling agricultural products. A numerical analysis [10DP] obtained temperature and velocity profiles as well as Nusselt numbers and established that boundary layer analysis sometimes underestimate Nusselt numbers at small Rayleigh numbers. An exact solution was obtained [23DP] for the problem to determine the shape of the coolant entrance boundary of a porous region which generates a desired distribution of the cooling along the exit boundary. A unit cell model predicts [14DP] the thermal conductivity of a granular medium containing an adhesive binder and demonstrates that the binder can increase the conductivity significantly. A transient analysis [22DP] of a packed bed thermal storage system includes longitudinal thermal dispersion and particulate conduction. The unsteady boundary layer developing on a horizontal cylinder imbedded in a porous medium is obtained [13DP] through solution of the boundary layer equations.

Experimental studies were concerned with heat and mass transfer in an unsaturated porous medium. A downward heat flux through a column of sand created three regions with vapor, liquid and vapor, and liquid filling the voids in each of the three respectively [24DP]. The temperature and pressure field in a porous plug with a pore size of 10 μm was measured [17DP] with vaporization occurring in the plug. A model [3DP] describing heat and mass transfer in unsaturated porous media was based on irreversible thermodynamics.

Experimental investigations contributed primarily to our knowledge in heat transfer in fluidized beds. A comparison of heat transfer data with flow visualization on a flat surface established [21DP] that heat transfer depends strongly on the inclination of the surface. Increasing pressure in a bed is beneficial [17DP] for the heat transfer to a immersed vertical wall. The following equation was found [15DP] to express heat transfer from a plane surface to a liquid–solid fluidized beds within $\pm 14\%$.

$$Nu = (0.0325\, Re + 1.19\, Re^{0.43})\, Pr^{0.37}(1-e)^{0.725}$$

Re, denoting the particle Reynolds number, is based on the terminal free fall velocity of the particle in the liquid, e is the bed voidage. A cavity is initially formed under a body immersed in an air-fluidized bed [9DP]. It rearranges the bed and increases heat transfer. Experimental results on heat transfer from single horizontal finned tubes in an air-fluidized bed of uniform particles [16DP] were compared with results on bare tubes. The fins improved heat transfer. Heat transfer on horizontal staggered smooth tube and fin-tube bundles was studied [18DP] in a fluidized bed of burning coal particles and in model beds. Heat transfer to vertical tube bundles in fluidized beds of particles with 0.8 to 3.1 mm dia. increased [8DP] with increasing pressure. Similar experiments [6DP] were concerned with heat transfer to the walls of the bed. Heat transfer in a three-phase fluidized bed was shown [11DP] to increase at the same rate as the decrease in the liquid holdup. A model was presented [12DP] to calculate heat transfer from immersed surfaces to large particle fluidized beds (1 mm or more). Conduction and convection at the interface is considered.

NATURAL CONVECTION—INTERNAL FLOWS

Interest in natural convection in enclosures remains high, as indicated by the large number of recent papers. As usual, there is considerable interest in convection in horizontal layers heated from below, and in differentially heated vertical and inclined layers. Subjects of growing interest include double diffusive convection where concentration gradients as well as temperature gradients provide buoyancy forces, flow in porous media with applications to geothermal heat recovery, convection in fires, and thermocapillary flows where surface tension plays an important role in driving the flow.

There is continued interest in convection in horizontal layers heated from below including analysis of the planform at low Rayleigh number and effects of side walls, variable properties, and potential internal energy sources. An analysis of convection in a high Prandtl number fluid shows conditions for bimodal and square patterns of convection and the impact of the flow on the Nusselt number [17F]. A one-equation model of turbulence has been applied to predict the Nusselt number at high Rayleigh number [26F]. The influence of variable fluid properties and boundary conditions on thermal convection has been demonstrated [10F]. Measurement of heat transfer across a horizontal layer of a fluid with highly temperature dependent viscosity can be predicted from the results for a constant viscosity fluid using a viscosity in the Rayleigh number determined at the average temperature of the boundaries [51F]. A convenient relation for the density variation of water has been used in calculating the onset of flow in a horizontal layer when the temperature is near the density extremum of water [35F]. The Galerkin method has been used to predict heat transfer in a vertical channel heated from below and having either perfectly conducting or perfectly adiabatic walls [16F]. The influence of side walls on stability was analyzed in another study [67F]. The onset of flow in a horizontal layer with variable temperature difference on the horizontal surfaces indicates either transverse or longitudinal roles, depending on the temperature difference [68F]. Observations of the convection pattern in nematic liquid crystals heated from below show the influence of stabilizing and destabilizing magnetic fields [54F]. Measurements of the time scale for transition in a layer of fluid with a step change in internal energy sources are used to check a boundary layer model for heat transport from the layer [27F]. Empirical relations for the heat transfer in a horizontal layer heated from below have been used to predict rollover of layers of different density in an LNG storage tank [23F].

A numerical solution for the multicellular convection in a vertical slot shows the influence of aspect ratio on heat transfer across such a differentially heated layer [30F]. Velocity measurements near the heated vertical wall in a rectangular enclosure indicate 2-D flow in the central region of the enclosure [40F]. The corner flows in a differentially heated cavity at high Rayleigh number and high Prandtl number exhibit a double structure with the outer layer dominated by convection [5F]. Secondary flows due to stratification in a differentially heated vertical layer have been observed [55F]. Analysis of the effect of nonlinear density stratification in a vertical layer of water shows the onset of a travelling wave [57F]. A partially divided differentially heated layer has been examined experimentally and numerically for low Rayleigh number flows [31F].

Measurements with inclined fluid layers indicate that the average heat transfer across a layer is closely represented by the heat transfer in the central section at sufficiently large aspect ratios and angles of inclination less than 60° from the horizontal [12F]. Finite difference calculations for the 3-D flow in an inclined layer have been performed for low Rayleigh number flow [39F]. Convection in an inclined rectangular box with partially heated opposing walls has been studied [9F]. Laminar flow in a tilted square cavity has been calculated for several angles of inclination [48F].

Flow visualization and heat transfer measurements have been obtained over a range of Rayleigh number and Prandtl number for convection across a channel with a cross section in the form of a parallelogram [56F]. The Nusselt number for convection across a triangular enclosure is found to vary as the Rayleigh number raised to the 0.59 power [44F]. A related analysis of convection in a triangular enclosure with a hot bottom wall shows that the flow is generally in the form of a single elongated cell [43F]. Heat transfer across a differentially heated cavity with a central vertical partition blocking part of the flow is very sensitive to the conductance of the partition [36F].

A number of studies considered the convection in the annular region between one body and the surrounding enclosure. A radial transformation was used to map the boundaries and aid in predicting the laminar convective flow in the annulus between two eccentric cylinders [47F]. A general relationship was developed to correlate early experimental results for laminar convection in horizontal annuli [6F]. Density inversion in water has a significant effect on the general flow pattern within a differentially heated horizontal annulus [66F]. In a vertical annulus, measurements of combined radiation and natural convection have been performed [28F] with either air or helium as the convecting fluid. Experiments using air to study convection in the annulus between vertical cylinders of different height indicate that the average Nusselt number is nearly independent of the eccentricity and the vertical position of the inner cylinder [61F]. Measurements of convection between a cube and a surrounding sphere are used to develop a general equation for heat transfer between a body of more or less arbitrary shape and a surrounding sphere [46F]. Finite element methods are used [8F] to determine the flow and heat transfer in the region between a square cylinder and a surrounding circular cylinder at low Rayleigh numbers. Calculations of the convection around a heated copper block immersed in a tank of water have been used to simulate the flow in a reactor core catcher [14F].

In double diffusive convection, buoyancy differences are due to two gradients, usually one being temperature, the other being concentration of one species in another. Experiments in double diffusive convection are compared to an analysis of flow and heat, momentum and mass transfer [32F]. Thermal diffusion effects on stability and formation of salt fingers in double diffusive convection have been analyzed [33F]. A linear stability analysis has been performed for a vertically stratified fluid in which mixed concentration and temperature gradients in the horizontal direction combine to provide a constant density in a given horizontal plane [24F]. Experiments in liquid mixtures of ^{3}He–^{4}He heated from below show effects akin to double diffusive convection [29F].

Buoyancy-driven convection in porous media occurs in insulation, in geothermal wells and in other systems. The analysis of convection in a porous medium bounded by two horizontal walls with linear temperature variations along the walls has been performed [42F]. A preferred mode of flow has been determined by stability analysis for finite amplitude convection in a porous layer with finite conducting boundaries [50F]. Both linear and nonlinear stability theory have been applied to determine the onset of flow in a differentially heated vertical porous layer [52F]. Numerical calculations for a differentially heated cavity of porous material with one of the sidewalls permeable, show reverse flow near this wall [22F].

Counterflow is found across the boundary between a vertical porous layer and the surrounding fluid [4F]. An analysis of transient convection in a fluid-saturated, porous medium contained between two concentric cylinders shows the time required for transient effects to die out [15F]. Experiments on convection in a porous medium filling the annulus between two vertical cylinders have been performed to simulate the heat loss from nuclear waste storage containers [49F]. Scaling arguments and numerical analyses are used to predict transient natural convection in a triangular shaped cavity filled with porous material [45F]. A developed code for predicting flow and heat transfer in porous media has been used to predict the heat loss from spent fuel elements in a storage pool [20F].

Buoyancy driven plumes are of interest because of their occurrence in the atmosphere, oceans, fires, and many engineering phenomena. The volume flow of a buoyant plume has been measured from the position of the interface formed with the entrained flow [1F]. An approximate solution for a laminar plume is valid for flow in a weak magnetic field [21F].

Experiments on the flow above a heated source located in a fluid layer stratified by a salinity gradient, show a hierarchy of plumes with a number of vortices [65F]. A special inlet manifold has been designed for introducing a buoyant jet into a liquid storage tank without changing the initial stratification [19F]. The importance of heat transfer and buoyancy processes in analyzing a fire in a large building have been demonstrated [13F]. The rate of growth of a developing fire in a compartment has been calculated [58F] using experimentally determined data for heat and smoke generation. An analysis shows the influence of buoyancy effects on turbulence characteristics [53F].

Thermosyphons are fluid loops in which the circulation is due to the difference in the buoyancy force in the upward and downward flowing legs. In some systems a forced convection is also imposed on the flow, and in others, boiling or the presence of vapor bubbles has a strong influence on the difference in pressure-head of the two legs of the loop. The influence of asymmetric heating on laminar mixed (forced and natural) convection in a thermosyphon has been studied both for aiding and opposing flows [41F]. Viscous dissipation and pressure work are shown to be of comparable magnitude in a free convection loop and should be considered together in an analysis of thermosyphons [3F]. Visualization of the flow in an inclined two-phase thermosyphon is used to identify key heat transfer mechanisms [37F]. The effects of geometry and working fluid on critical heat flux in a closed two-phase thermosyphon have been determined [25F]. The instability in a thermosyphon with boiling fluid is not greatly reduced by throttling the flow [64F].

Surface tension driven (thermocapillary) flows occur when a heated liquid layer has a free surface. The flow of a liquid in a one-quarter plane bounded on the upper surface by a passive gas and on one side by a semivertical wall shows the importance of capillary effects following a step-change of temperature on the vertical wall [11F]. Analyses of a layer with a free surface heated from below show the instabilities present with a horizontal temperature variation along the lower bounding surface [59F, 60F]. The influence of aspect ratio on the thermocapillary driven flow field in a differentially heated channel has been demonstrated [63F]. The stability of convection with a localized heat source on a free surface was examined [7F]. Linear stability analysis has been used to predict the flow with concentration and temperature-dependent surface tension for a horizontal layer [34F].

In mixed or combined convection, the flow is due to both buoyancy and externally driven or forced flow. An analytical solution for mixed convection in the entry region of a heated vertical channel indicates the presence of moving periodic cells when natural convection dominates [70F]. The heat transfer from a heated cylinder attached to a vertical plate is increased when the plate serves as a wall of a surrounding vertical duct [62F]. With mixed convection in an open horizontal channel the different flow regimes are delineated by values of the Grashof number divided by Reynolds number to the 3/2 power [69F]. Perturbation and finite difference methods indicate that both Prandtl number and Rayleigh number have a profound effect on the flow pattern for mixed convection in the region between horizontal concentric cylinders [38F]. The flow in thermal energy storage tanks is strongly dependent on the Froude number of the inlet flow [2F]. An analysis indicates the convective pattern around melting cladding after loss of sodium coolant in a fast reactor excursion [18F].

NATURAL CONVECTION—EXTERNAL FLOWS

The results of approx. 45 investigations of the natural convection heat transfer from flat plates in an infinite medium were reviewed and mean empirical relations were determined for the vertical- and horizontal-plate cases [26FF]. New correlations are presented for natural convective heat transfer from vertical isothermal surfaces in gases and the influences of variable properties are accounted for by using empirical equations that are of the form $Nu = g(Ra) \cdot f(Ra, Tw/T_\infty)$ [10FF]. For large Pr a simple relation for the propagation of the effect of the leading edge up a vertical plate generally agrees with results of past studies [6FF]. Space–time correlations of velocity and temperature fluctuations in the boundary layer of a vertical isothermal plate in air or N_2O at various pressurees provide estimates of turbulent scale [27FF]. The problem of how a free convection boundary layer on a vertical surface negotiates a small hump or indentation on the surface is discussed [29FF]. A transformation method is applied to study natural convection along irregular vertical surfaces and a sinusoidal surface is used as a specific example to demonstrate the advantages of the transformation method [44FF]. A numerical and experimental study of flow along an unheated plate above an isothermal plate demonstrates that the results are strongly influenced by the plate-fluid thermal conductivity ratio [24FF]. A new correlation is presented for natural convection heat transfer from a vertical isothermal surface in cold water and the correlation contains only one additional parameter beyond the Nusselt, Grashof and Prandtl numbers [5FF]. Calculations of the heat transfer from a flat plate in water include the effects of a melting ice surface and the density extremum [31FF]. The flow adjacent to a vertical isothermal surface immersed in saline water has been studied using a new density relation [19FF]. The linear stability of a steady plane parallel convective flow of a binary mixture in a plane vertical layer is considered with allowance for the thermal diffusion effect [15FF]. A numerical solution for the free-convection flow past a vertical semi-infinite flat plate embedded in a highly saturated porous medium has been developed by allowing the plate to have a nonuniform heat flux distribution [30FF].

A technical note describes an investigation of some aspects of the free convection in a corner whose walls consist of an isothermal vertical flat plate of finite height and an infinite adiabatic horizontal surface [35FF]. Unsteady free convective boundary layer flow over a semi-infinite horizontal plate is analyzed employing the method of matched asymptotic expansions for the case when the plate temperature oscillates with a nonzero mean and the free stream is at rest [42FF]. Laminar natural convection from a horizontal plate and the influence of plate-edge extensions were studied by a finite difference analysis and by experiments [16FF]. An analysis is performed to study the heat/mass transfer and vortex instability characteristics of buoyancy induced flows that result from simultaneous diffusion of heat and mass in laminar boundary layers adjacent to horizontal and inclined surfaces [9FF]. The breakup of a thin liquid layer heated from below has been studied [18FF].

Eight previously published correlation equations plus one new correlation for heat transfer by natural convection from horizontal isothermal cylinders are tested against a fairly extensive body of experimental data called from the literature for $10^{-8} < Ra < 10^{8}$ and $0.7 < Pr < 4 \times 10^{4}$ [13FF]. In a transient numerical analysis, the time histories of the isotherms and streamlines in the natural convection flow field around a horizontal cylinder were computed beginning with a step change in the cylinder temperature and ending when steady state was reached [36FF]. For natural convection of a coal oil mixture around a horizontal cylinder it was found that when the bulk temperature of the mixture was high the experimental heat transfer data was in good agreement with a similarity solution of the natural convection equations in a power-law fluid [2FF]. Two series of experiments were performed in order to measure the average and the local heat transfer coefficients in laminar natural convection around horizontal circular cylinders whose surface was in part isothermal and in part adiabatic [12FF]. An investigation was performed of the effects of vibration on the natural convection from horizontal cylinders [22FF]. Similarity velocity and temperature profiles, as well as heat transfer results, are presented for laminar natural convection from slender needles with axial power-law wall heat flux variation [7FF].

Empirical formulas are proposed from experiments on free convective heat transfer from a cylinder array arranged in a vertical line and subsequent experiments were used to examine the effect of using two parallel plates to enclose the array [40FF]. Experiments were carried out to determine the effects of transverse misalignment on the natural convection heat transfer characteristics of a pair of equi-temperature, parallel horizontal cylinders situated one above the other [38FF]. Both laminar and turbulent heat transfer from single and double rows of closely spaced isothermal heated cylinders were investigated numerically [14FF].

Theoretically, it is argued that the transition to turbulence in a buoyant plume occurs when the time of viscous penetration normal to the plume becomes comparable with the minimum time period with which the plume can fluctuate as an unstable inviscid stream [23FF]. A 2-D numerical model for predicting the steady surface thermal plume in the near field has been presented [11FF]. A steady state solution is obtained for the linear system of Navier–Stokes equations in the Boussinesq approximation for free axisymmetric convection flows from a point heat source in a stably stratified fluid [21FF]. An experimental study was conducted of plane plumes in water which is at or near the extremum temperature [20FF]. When a salinity gradient is heated at a single point, three kinds of layers are observed: the first kind is formed above the basic plume from a hierarchy of secondary plumes on top the basic one, the second develops around the upper part of the basic plume and the third forms around the lower part of the basic plume [41FF]. The thermal instability for 2-D stagnation flow for Prandtl numbers ranging from 0.7 to infinity was studied using an analysis that represents an exact solution since neither the boundary-layer approximation nor the parallel-flow assumption was invoked [8FF].

Both analytical and numerical techniques were employed to solve for the velocity and temperature fields in a 2-D mixed convection plume for the Prandtl number range from 0.72 to infinity [17FF]. The buoyancy effects on the thermal boundary layer induced by a horizontal forced flow along an infinite vertical cylinder with a step change of surface temperature is studied by a finite-difference method [43FF]. A theoretical study was performed of the laminar mixed convection from a horizontal cylinder in a cross stream [3FF]. Laboratory model data are presented for the 3-D time averaged thermal characteristics of a buoyant rectangular surface thermal plume in a crossflow [25FF]. A boundary-layer analysis of mixed convection about a point heat source is made in terms of the parameter Gr/Re^2 for the cases when the buoyancy effects are favorable or adverse with respect to the imposed vertical stream [34FF]. The mixed convection in an axisymmetric point heat source due to a vertical free stream is analyzed using the boundary layer equations for the cases of favorable and adverse buoyancy effects [1FF]. Two aspects of the influence of double diffusion on convection from point buoyancy sources were addressed; double-diffusive convection acting in conjunction with buoyancy from an isolated source in an unconfined environment and the problem of when fluxes of properties occur in the opposite direction to that of the initial buoyancy [28FF].

The authors of [37FF] state that their experimental evidence stands in opposition to the use of King's rule for evaluating the characteristic length for multidimensional natural convection problems. Unsteady 2-D free-convective flow through a porous medium bounded by an infinite vertical plate is considered when the temperature of the plate is oscillating with time about a constant nonzero mean value [33FF]. A numerical study was made of the free convection flow over isothermal and nonisothermal rotating flat plates subject to a linear variation of the gravity field; both the cases of hot and cold plates were considered [32FF]. The onset of convection in a cylindrical volume of fluid bounded above and below by rigid, perfectly conducting surfaces and laterally by a wall of arbitrary thermal conductivity was examined [4FF].

CONVECTION FROM ROTATING SURFACES

Heat transfer has been studied for the configuration in which there is a radially outward flow between two coaxial stationary or corotating disks [9G]. Laminar convection has been investigated for non-Newtonian power-law fluids in contact with rotating disks [11G]. Reference [5G] considers the effect of Prandtl number on heat transfer from an isothermal rotating disk with blowing at the wall. For a bounded cylindrical layer of fluid heated from below, the effect of rotation on stability has been determined [1G]. Experimental results have been presented for a rotating fluid heated from below [10G]. Reference [3G] deals with forced convection from a disk rotating in a vertical plane. Numerical solutions are given for the thermal instability of a rotating micropolar fluid layer [8G]. The thermal instability is also considered in [6G] for a horizontal fluid layer under rotation and internal heat generation. The stability of rotating cryogens is treated in [7G]. Reference [2G] reports experimental and theoretical results for convection in rotating spherical shells. A 3-D unsteady numerical simulation has been made for convection in rotating layers [4G].

COMBINED HEAT AND MASS TRANSFER

Studies involving combined transport of mass and heat are encountered in a number of applications. Some specific studies that we include in this category are film cooling and related processes in impingement heat transfer.

Measurements in a short duration wind tunnel indicate that the effect of the temperature difference between the gas and the surface is small when the wall temperature is less than the free-stream temperature [10H]. Numerical calculations along with experiments indicate the film cooling effect following injection through tangential multiple jets [6H]. The assumption of a Gaussian temperature profile and an entrainment function are used to study film cooling with 2-D tangential or normal injection [13H]. Mass transfer measurements indicate the film cooling effectiveness on suction and pressure sides of a turbine blade following injection through two rows of staggered film cooling holes [3H]. An analysis of a surface ablating with a time varying heat input shows strong influence of different approximations that are made to predict the ablation rate [1H].

An empirical relation has been developed for the heat transfer to a row of impinging circular jets with cross-flow [11H]. Heat transfer with an array of impinging jets within a duct flow has been determined experimentally [2H].

The heat and mass transfer occurring in the absorption of a gas or vapor into a laminar liquid film depend on the Peclet and Lewis numbers as well as on the equilibrium characteristics of the working media [4H]. Heat and mass transfer coefficients were measured and computed for steam absorption into a falling wavy film of liquid LiBr [12H]. Heat and mass transfer to a drop from a surrounding polyatomic gas has been analyzed [8H].

The overall balance for heat and mass transfer with parallel, counter and cross-flows in grain drying has been analyzed [9H]. A spouted bed heat and mass exchanger in which a jet of fluid enters a bed of tablets or other particular material has convenient drying characteristics [5H]. From another set of experiments on a spouted bed heat exchanger, optimization equations are developed for pressure drop and heat and mass transfer [7H].

BOILING

Nucleate boiling

The fundamentals of nucleate boiling are still under investigation. An analysis of the conditions of equilibrium for a vapor bubble [8J] showed support for earlier analytical conclusions and experimental evidence that a bubble can remain in a state of stable equilibrium in a finite system under certain circumstances. It was experimentally shown that a previously imposed pressure history had a strong deactivating effect on surface cavities serving as nucleation sites [19J]. In a high speed motion picture study, nucleation site density, bubble departure diameter, and frequency were determined for saturated atmospheric pressure water heated by a wire [48J]. Evaluation of the relative contributions to the boiling heat flux in those experiments suggested that the enhanced convection induced by bubble motion was never the dominant mode of heat transfer, but instead contributed less than natural convection at low heat fluxes and less than latent heat transport at high heat fluxes. In another study [38J], for which the goal was to predict interfacial transport terms to allow application of a two-fluid model, the nucleation site density was found to be of vital importance. A correlation for site density was developed and compared favorably with experimental data.

Shape and motion of bubbles are of concern. One investigation examined the effects of Marangoni convection [37J] on bubble motion while a few dealt with the bubble conditions at the time of separation from a surface. An equation was developed for the break-off radius of bubbles with attention to the vanishing of bubble and surface interactions at the instant of departure [44J]. Two unrelated studies considered the optical distortions imparted to our views and photographs of departing bubbles by the local temperature gradient at the heated surface [12J, 80J]. In an apparatus designed to minimize these distortions [13J], individually formed bubbles showed no evidence of retarding effects due to surface tension. Delay times of nucleation sites near a given site were shown [63J] to be related to the square of the separation distance for distances less than twice the bubble diameter. A model based on thermal diffusion through the heating solid fit the experimental data well. Laser interferometry was used [40J] to measure the thicknesses of microlayers during bubble growth in subcooled boiling of both water and ethanol. Analysis of the data implied that evaporation of the microlayer accounted for no more than one half of the rates of heat transfer measured.

The lower heat transfer rates associated with nucleate boiling of mixtures of substances were addressed in two different ways: a method for predicting heat transfer coefficients for boiling of binary mixtures was demonstrated using phase equilibrium data [65J], and another method was proposed [54J] using the liquid phase mass transfer coefficient.

Forced convection boiling

Experiments with both vertical and horizontal tube flows [5J] showed negligible differences in pressure drop between the regime of nucleate boiling and the regime of "forced convection boiling" with no vapor formation at the tube wall. The investigator concluded that the absence of a significant change in the wall shear stress implied an absence of effects of bubbles except as latent heat transport media. Theoretical models of heat transfer in these two regimes were constructed upon this conclusion. An experimental study [1J] determined that there was very little difference in heat transfer coefficients measured during nucleate boiling for upflow and for downflow in vertical systems despite obvious differences in void accumulation. Measured heat transfer coefficients are reported for flow boiling of refrigerant 114 in upflow and downflow in stainless steel tubes [17J]. Strong effects were observed of the influence of 180° bends as much as one meter upstream of the test section. A comparison of available data for subcooled boiling in annuli demonstrated the applicability of the Shah correlation (developed for flow in tubes) to boiling in cylindrical annuli heated by inner, outer, or both walls [56J]. An experimental investigation was carried out in an internally heated annular vertical flow of refrigerant 113 to study the statistical character of static pressure fluctuations and chordal-averaged vapor fractions [30J]. Results suggested that diagnosis of flow regimes should be possible based on measurements such as these.

Experiments were performed [49J] with analog feedback control to achieve a steady heater temperature during transition boiling. Results for nucleate boiling, transition, and film boiling were presented as Laplace transform transfer functions. Other investigators [78J], concluding that traditional experiments of transition boiling provide inadequate control of the heat generating element, performed three sets of experiments with one heating element by controlling either current, voltage, or temperature (resistance) to accumulate information on the true boiling curve. A simple homogeneous flow model was demonstrated to provide reasonable agreement with measured pressure drop oscillations and density wave oscillations in a single-channel boiling up-flow system [21J]. Experiments with five enhanced surface tubes and one bare tube explored the effects of these surfaces on two-phase flow instabilities in vertical boiling flows [43J]. Pressure drop vs flow rate characteristics were determined, from which the boundaries of instability to pressure drop and density wave instabilities were derived for each internal surface. An analytical model was presented [47J] for pressure drop in steady liquid/vapor countercurrent flow as a function of inlet vapor flow rate and applied heat flux. Experiments and analysis were reported [42J] concerning countercurrent stratified flow of condensing steam and water in a shallow rectangular channel. Models of slug formation from cocurrent horizontal flows predicted the onset of flooding of these systems for nearly horizontal flows. Measurements were made of film thickness and interfacial shear stress. It was determined that the condensation had little effect on the overall behavior. In a brief survey article of two-phase flow [26J] three current approaches to prediction of two-phase flows were described: empirical, analytical, and phenomenological, with the latter being suggested as the most promising avenue for further efforts.

Transient boiling heat transfer experiments were reported [35J] using an electrically heated platinum wire centered coaxially in an upward flow. Two patterns of change of wall superheat with heat flux were observed. In the more common pattern, wall superheat exceeded steady state values for intermediate heat fluxes, but coincided with steady values or the extrapolation of their curve at high heat fluxes. Maximum heat fluxes for these transients increased with decreasing transient period (increasing rate of change of flux) and with increasing pressure, velocity, and subcooling. In a temperature controlled transient boiling experiment with refrigerant 113 [74J], measured values of critical heat flux, CHF, were shown to be dependent upon the rate of change of the surface temperature of a thick heating element. Differences from steady conditions of critical heat flux were demonstrated to correlate well with the rate of change of the interface temperature. It was suggested that analysis of heat transfer in two coupled media can in general be decoupled using this rate of change. The CHF for quenching processes was shown to depend upon the direction of the motion of the quench front, the quenching being a complicated conjugate transient problem. In another investigation of transient conditions [55J] a model was formulated for the maximum heat flux based on the balance between liquid layer replenishment and evaporation.

A model was developed for predicting CHF in high velocity flow of low quality two-phase mixtures in heated tubes [70J]. In this model the breakdown of turbulent exchange between the bubbly layer and the core is considered to be the limiting phenomenon. Another model was presented as applicable to both pool boiling and forced convection CHF for saturated boiling at submerged bodies [24J]. In this latter model it was held that a stable liquid film of definite thickness during boiling is evaporated at CHF due to restrictions of liquid feed to the surface; predictions compared favorably with data for pool and flow boiling in several geometries. Dealing with the particular geometry of a horizontal tube with upward cross-flow [36J], CHF was modelled in two flow regimes: low velocity with bubble-like departure of vapor, and high velocity flow with sheet-like departure of vapor. Good agreement with limited existing data was reported. Steady film boiling in this same geometry was modelled using a boundary layer approach for the two-phase flow [58J]. The model tended to underestimate measured heat transfer coefficients. It was concluded [29J] on the basis of a consistent interpretation of previously reported experiments of liquid metal fast breeder reactor (LMFBR) natural convection cooling that a typical reactor subassembly can be safely cooled without dryout at decay heat power levels which are less than 8–10% of nominal average power.

An investigation of rewet heat transfer in vertical flow of refrigerant 113 was carried out using copper test sections of high thermal capacity [68J]. Maximum heat fluxes coincided well with measured critical heat fluxes, and wall superheats at the rewet condition were found to vary greatly with the flow quality. Experiments with rewetting of horizontal annular channels [50J] demonstrated that stratification effects are significant and stronger at low flow rates. Rewetting velocity was found to increase with increasing flow rate and with reduced initial surface temperature. Rewetting by a falling liquid film was investigated with refrigerant 113 flowing down the outside of a stainless steel tube [67J]. Spatial and temporal variations of temperature were measured with sufficient accuracy to derive the boiling curve from these experiments. The boiling curve and the rewet front velocity were little affected by the film flow rate; the boiling curve was likewise little influenced by initial wall temperature or wall thickness. Maximum heat fluxes observed were about twice the critical heat flux predicted for pool boiling. Experimentally determined post-dryout heat transfer coefficients inside a steam generator tube heated non-uniformly by the countercurrent flow of sodium were correlated with the more abundant data from experiments with uniform electric heating [69J].

In an investigation of two-phase blowdown of a vessel initially containing saturated liquid [27J, 28J] experiments showed that transient critical two-phase flow is similar to steady critical two-phase flow and that an initial depressurization was followed by a period of nearly constant pressure as the rate of vapor formation approximately balanced the volumetric discharge rate. Three models, to be used in series, were developed to describe the blowdown sequence: a single phase blowdown, then a period of pressure recovery with thermally governed vapor bubble growth, followed by attainment of thermodynamic equilibrium of liquid and vapor within the vessel. The model showed considerable sensitivity to the number of potential nucleation sites within the vessel and thus a sensitivity to the surface area of internals.

An iterative numerical technique was presented for determining distributions of two-phase flow amongst multiple paths between manifolds, accommodating homogeneous flow with and without heat transfer [15J]. Attention was paid to disturbance waves in thin liquid film annular flow in a vertical tube [64J]. A model was developed for heat transfer and pressure drop in these circumstances, capable of predicting heat transfer coefficients well but tending to overpredict film thickness. Wave amplitude and time rate of passage are input parameters in the model and were measured in accompanying experiments.

The effects of wire gauze-covered internal walls of horizontal steam generator tubes upon critical heat flux were explored [62J]. For large mass fluxes higher heat fluxes or higher flowing qualities at CHF were obtained. At lower mass fluxes, however, CHF occurred at lower heat fluxes or lower qualities. Two-phase pressure drop in the gauze-surfaced tube was correlated using a homogeneous flow model. The Martinelli–Nelson correlation was found to under-predict pressure drop data obtained for mixtures of refrigerants 12 and 13 in experiments in horizontal tubes [59J]. Fouling of argon-evaporating surfaces with carbon dioxide present in very small quantities in the argon was reported and explained by means of bubble growth mechanisms [45J].

Natural convection boiling

Effects of pool geometry and of spacing between heated tubes upon pool boiling characteristics were examined experimentally [22J] using identical finned heated tubes immersed horizontally in a small cylindrical and a large rectangular vessel containing refrigerant 11. Differences were substantial at moderate heat fluxes but vanished at larger heat fluxes. Experiments with various combinations of heated and unheated horizontal tubes in a bundle and a solitary heated finned tube were conducted in pool boiling of refrigerant 11 at 1 bar [23J]. It was tentatively shown that a simple vertical tube pair arrangement provides heat transfer coefficients representative of tubes in bundles of greater number. Influence of bundle spacing as well as effects of pressure were observed with bubble characteristics and heat transfer coefficients in a small boiling heat exchanger [60J].

A model was proposed for nucleate boiling at high heat flux, heat fluxes between the critical heat flux and 60% of the critical heat flux [4J]. The chief mechanism was hypothesized to be conduction through a macrolayer of liquid, as nucleate boiling was suggested to be suppressed under high heat flux conditions. A theoretical expression was proposed for the thickness of such a macrolayer and a mechanism for layer formation was described [3J].

Instabilities on heated wires were described [33J] and observations of the motion of the wave of transition from nucleate boiling to film boiling on inclined wires were reported [79J]. Temperature fluctuations and vapor formation at small heated surfaces flush with an unheated wall were experimentally observed [57J]. Those experiments showed no evidence of burnout despite heat fluxes of up to 12×10^6 W m^{-2}.

Experiments with boiling of water, acetone, and refrigerant 113 in narrow annular gaps displayed three boiling regimes [73J]. The Bond number, a ratio of gap size to bubble departure diameter, appeared to be a determinant for nucleation under slightly deformed bubbles, which occurred for Bond numbers slightly greater than unity. Experiments were performed in narrow eccentric annular gaps simulating the crevice between a heated tube and a support plate [31J, 32J]. The variations of eccentricities were found to have small effects relative to tube/support plate contact; separations of the order of 0.05 mm or less resulted in some transient dryout behavior. Large rates of vapor formation at a line of contact produced a heat transfer enhancing sweep of vapor at roughly 60° from the contact line. A correlation was developed for critical heat flux in narrow vertical annuli with closed bottoms [10J]. Based on countercurrent flooding conditions, the correlation is independent of liquid surface tension. Boiling of dielectrics in slots of 0.5 to 4 mm height displayed considerably different (enhanced) behavior from pool boiling on a plane surface [7J]. Boiling crisis in a thermosyphon was judged to be of hydrodynamic nature associated with the vapor phase achieving a critical velocity by which the liquid film is detached from the surface [2J].

Experiments with surfactants in water under saturated atmospheric boiling conditions showed improvements in both the heat transfer coefficient and the critical heat flux with increased concentration of the surfactants [72J]. Pool boiling experiments with three drag-reducing polymers in water produced enhanced boiling characteristics with two of the polymers but degraded heat transfer with the third [53J]. The investigators concluded that the changes had nothing to do with the drag-reducing effects of the additives. An empirical equation was developed for predicting heat transfer coefficients in a concentrator with natural convection flow through external boiling tubes [34J].

Other boiling and evaporation studies

Natural convection and surface tension effects were visualized in small evaporating droplets on an inclined surface with a temperature gradient through use of the motion of suspended solids and also by use of laser shadowgraphy [76J]. Bénard cells were observed in a ring of intermediate radius for droplets of some fluids evaporating on a mildly heated surface [77J]. Experimentally determined single droplet evaporation rates in high temperature air showed that higher temperatures, resulting in higher evaporation rates, resulted in lower Nusselt numbers [51J]. Numerical treatment of evaporation of single droplets in the Reynolds number range 10–100 agreed well with experimental data [52J]. Mass transfer reduced the friction drag but increased the pressure drag, with a net result of agreement with the standard drag curve if free stream density and a particular representative viscosity are used.

Local mass transfer coefficients (and by inference, heat transfer coefficients) were determined experimentally in the vicinity of a scraping blade in a thin film evaporator [71J]. An analytical model was shown to agree within a few percent with a finite difference approach for predicting film vaporization of hydrocarbons in a stream of hot air [16J]. Experiments conducted with a circular disk heating surface in natural convection with varying heights of saturated liquid above the surface [66J] suggested that at very small film thicknesses the heat transfer is solely by conduction. For greater thicknesses, the heat transfer coefficients appeared to be quite comparable with those for saturated pool boiling. A correlation was developed from experimental results of natural convection evaporation from a shallow unheated pan of water to atmospheric air [61J]. The Sherwood number varied with the Rayleigh number to the exponent 0.205, and the effects of pan sidewall height and of the extent of a horizontal frame at the sidewall height were measured, correlated, and found to induce no change in the Rayleigh number dependence. A numerical investigation of evaporation from a stationary body of water to a laminar stream of air, superheated steam, or a mixture of air and steam [11J] showed that at low temperature the evaporation rate decreased with increased humidity while above an inversion temperature of about 250°C the evaporation rate increased with humidity.

Consideration of the length and velocity scales of eddies formed in a steam vent led to an explanation of the onset of a roughened vapor/liquid interface during the collapse of the discharged steam and produced a theoretical estimate of the interface heat transfer coefficient [41J], the application of this work being modelling of BWR suppression pools. A 1-D experiment for the boiling and spreading of cryogenic liquids on water corroborated a model when reasonable, constant boiling rates were assumed in the model [9J]. Destabilization of film boiling by shock waves, investigated as a trigger for vapor explosions, was modelled in two ways, with and without assuming thermodynamic equilibrium between the phases [14J]. Both models displayed the same trends of observed data, and equilibrium appeared to be maintained if the shock rise time was longer than 100 μs.

A theoretical analysis of evaporation and condensation at the surface of a cylindrical body of liquid in its own vapor employed finite radial vapor velocities and demonstrated the need for a nonlinear analysis [46J]. Inadequacies of linearized theory were also demonstrated for half-space problems, as condensation and evaporation are not antisymmetric except for conditions at one particular value of a property parameter [75J]. A 1-D theoretical model was presented of evaporation and expansion of vapors from a metal surface with a high energy density source [25J]. The surface temperature was linked with the evaporation rate and hydrodynamic parameters.

The overall time-averaged heat transfer coefficient for a drop in an alternating electric field was shown to be governed by the Peclet number and a parameter defined as the thermal vibration number which related the frequency of the alternating field with the thermal relaxation time of the drop [20J]. At low Peclet number the alternating field enhances heat transfer, especially at low frequency, while at high Peclet number the field may enhance or detract from the time-averaged rate of heat transfer. The current state of understanding of the effects of electric fields on vaporization and condensation was assessed [6J], and areas are suggested for further exploration and future exploitation.

Centripetal acceleration and the corresponding hydrodynamic pressure gradient were charged with causing a decreased rate of desorption in boiling [39J]. Use of hydrophobic materials was demonstrated to counteract the depression of the desorption rate. A brief description was made of the variations in generation and transmission of acoustic emissions associated with different modes of boiling [18J].

CONDENSATION

Film condensation

Measured heat transfer coefficients and pressure drop data were presented for condensation of steam in tubes and annuli using several fin configurations. Fin efficiencies were presented [7JJ]. Numerical results for condensing on rectangular and triangular fins were compared and optimum fin dimensions were given in terms of the Biot number [13JJ]. Analytical comparisons of corruguated and tightly-ribbed tubes were presented. Slender and sharp surface profiles were shown to be superior to wide and round contours [1JJ]. A numerical method was presented for analysis of condenser tube heat transfer including condensation on porous, anisotropic surfaces. Complicated tube and baffle geometry was also incorporated [8JJ]. A comparison of the Chilton–Colburn and Von Karman analogies for condensation inside vertical tubes was presented. The analogies agreed with one another over the range of conditions analyzed [14JJ].

Reflux and natural circulation condensation in vertical inverted U-tube steam generators was experimentally investigated; various flow regimes were identified [2JJ]. Local heat transfer coefficients and condensation rates were measured in countercurrent, stratified flow of steam and subcooled water. Correlations were given for Nusselt numbers under smooth and wavy interface conditions. Basing the correlations on interfacial, rather than bulk, parameters was recommended [5JJ]. The effect of waves on vapor-side heat transfer was measured in the vertical tube film condensation process. The liquid Reynolds number was varied as a single vapor condensed from a noncondensable gas; no Reynolds number effect was observed over the Reynolds number range in the experiment [3JJ]. The effect of film waviness on condensation was experimentally investigated. Accurate simulation of large vertical condensers with laboratory setups was shown to be possible [9JJ].

An experimental study of condensation on cold walls during freeze-drying was made and the idea of regeneration of the surface by rapid heating was tested [4JJ].

Free condensation

The transition from dropwise to film condensation was experimentally investigated for ethanediol; transition mechanisms were proposed [11JJ]. The equilibrium shape and departure size of condensation drops on vertical surfaces were measured. The configuration of minimum total energy was used to develop a model which compared well with the measurements for acceleration fields up to 100 G [6JJ].

A theoretical analysis was presented for binary nucleation and subsequent heterogeneous droplet growth. Results were checked against water–ethanol data [10JJ]. Condensation rates were measured for dropwise condensation of steam. The heat transfer coefficient remained constant with an increase in surface subcooling to 10 K, then decreased. Maximum heat transfer rates of 10 MW m^{-2} were achieved [12JJ].

FREEZING AND MELTING

Theoretical and experimental studies of freezing or melting in cavities, on external surfaces or in porous media have been performed. Applications include continuous casting, reactor safety, latent thermal storage and thawing of frozen porous materials.

Analytical solutions to the Stefan problem continue to appear including a delineation of four possible cases depending upon the initial conditions [34JM]. A heat balance integral method was applied to the melting of a semi-infinite slab with prescribed surface heat flux and no subcooling [10JM]. A justification for the use of an enthalpy model in multidimensional phase change problems was given [37JM] in addition to a numerical solution technique. Interfacial wetting transition was examined in a 2-D melting problem using the three-state chiral clock model [17JM].

A pair of papers described the discrepancies between current models and experimental data for thermophysical properties of sodium near melting [1, 21JM]. An improved model was developed using a perturbation approach.

The 2-D continuous casting problem was transformed into a Cauchy boundary value problem which directly yields the solidification interface shape [27JM]. It was also shown that almost immediate solidification can occur at low velocities in continuous casting so that no free boundary exists [7JM]. An analysis of zone melting with multiple zones was presented [4JM] that indicated that the yield increases with the number of melting cycles.

Cavity problems can provide interesting results if the density changes and the volume is fixed. An analysis of water freezing inside a closed pipe indicated that as the water freezes, the pressure increases which lowers the freezing temperature [30JM]. Eventually the system can reach an equilibrium state without the liquid completely freezing in a constant temperature environment. The freezing of a liquid inside a sphere was analyzed without using boundary layer simplifications [13JM]. A numerical solution to the melting of a solid contained inside a horizontal tube was obtained [24JM] which included turbulent natural convection flow in the experimental phase. Experimental data was shown for comparison.

The shape of a vertical solid that melts when exposed to condensing vapor changes during the initial transient phase [32JM]. Eventually the shape becomes time invariant and is roughly a profile that shows the variation in local heat flux. Freezing around a cylinder was solved with an implicit numerical method that isolated the nonlinearity of the interface [22JM]. A computer model has been developed to predict ice growth on insulated cryogenic liquid storage tanks [16JM].

A 1-D exact solution for freezing of an infinite moist porous solid around a line heat sink was developed [3JM]. Approximate solutions were given for 2-D freezing of soil near a horizontal buried pipe [19JM]. Numerical studies were made on the thawing of moist spheres simulating coal particles [23] and cylinders simulating logs [29JM].

Exact solutions were developed for the Stefan

problem with contact resistance at the interface [33JM]. The solution is valid for arbitrary initial and boundary conditions. The effect of radiation on melting was considered which included a two-phase region [6JM]. Exact and approximate solutions were given for specific cases. The radiative melting of snow was analyzed with experimental data obtained for comparison [31JM]. An analysis of freezing a mushy-zone was made which can be applied to the rapid solidification of an alloy [14JM]. The drag on a hot sphere with internal heat generation melting its way through a solid was obtained using a simple analysis [8JM].

A number of experimental papers appeared that concern freezing or melting. Flow visualization of a double diffusion ice–salt water problem indicated that natural convection flow could occur in either or both vertical directions depending on the given parameters [26JM]. A pair of papers described flow visualization experiments during melting or freezing in a rectangular cavity [11, 12JM]. The freezing process of paraffin in a sealed vertical tube was documented [23JM] as was freezing and melting in vertical tubes of circular or square cross section [20JM]. Freezing rates were found to agree with a simple quasi-steady conduction model but natural convection dominated the paraffin melting process. Measurements and analysis were used to study freezing of three different liquids within a horizontal tube [36JM]. The freezing front was symmetric about the tube centerline in agreement with the results for the vertical tube configuration with paraffin. The freezing of a liquid-filled open rectangular cavity was also studied [15JM]. The cooling rate of water with supercooling determined the appearance of dendrites and the phase front movement [25JM] in contrast to the work on paraffin that did not show these effects. Two papers considered the problem of frost growth and the variation of its effective thermophysical properties with time [2JM, 35JM]. The melting of a solid under a hot liquid whose density is larger than the melt was studied by two independent groups [9JM, 5JM]. The melt rises through the liquid creating an unstable natural convective system. The solidification of solute from a binary aqueous salt solution was investigated [18JM]. The interface was smooth until dendrites formed during the final stages of freezing.

RADIATION

Radiative heat transfer continues to attract considerable attention as documented by the large number of publications during the past year.

A new integral function has been introduced for computing the spectral emissivity of an isothermal volume, containing either soot or gaseous species or both [11K]. The reflectivity and transmissivity of an isotropically scattering, plane-parallel slab with stepwise variation of the single scattering albedo within the medium are calculated for an isotropic radiation field [14K]. Radiant heat flux calculations in planar, absorbing, emitting, and isotropically scattering layers have been accurately reduced to calculations in nonscattering layers by scaling laws [24K]. Studies of the effect of a single scatter phase function distribution on radiative transfer in absorbing–scattering liquids are reported [6K]. A finite element solution of radiative heat transfer in a 2-D rectangular enclosure with gray participating media provides accurate temperature distributions in the medium and wall heat flux distributions, even for media with very small opacities where most of the approximation methods break down [34K]. Studies of 2-D radiation in absorbing–emitting media using the P–N approximation show that the P–3 approximation can be used to predict emission power distributions and heat transfer rates in 2-D media with opacities of unity or greater [33K].

Using a Monte Carlo method it is shown that the absorption, scatter and extinction efficiencies of a fly ash cloud have a primary dependence on the complex absorption index of the particles with a secondary dependence on temperature and particle size distribution [17K]. Experimental investigations of thermal radiation in soot-containing flames in a wavelength range from 0 to 23 μm indicate that internal absorption of thermal radiation in flames requires wavelength-dependent calculations if accurate modelling is desirable [15K]. A gas turbine combustor using heavy fuels is considered. Radiation, smoke and temperature measurements in an experimental combustor at various air pressure, inlet temperature, and air fuel ratios show a diminishing rate of increase of radiation with soot concentration and reduced sensitivity of smoke to fuel hydrogen content at higher combustor pressure [30K]. Results of infrared measurements in the 8–14 μm spectral region of two coal-fired power plant plumes and area haze in the Four Corners of New Mexico in Nov. 1980, show an increase with time of the haze bulk extinction coefficient during a persistent anticyclonic synoptic situation [44K]. Absorption and extinction measurements in smoke produced by atmospheric combustion of diesel fuel at 10 μm indicate a mass absorption coefficient of 0.84 ± 0.076 m^2 g^{-1}

and a total scattering coefficient of $0.15 \pm 0.014\ m^2\ g^{-1}$. Both values are normalized to the aerosol mass density [9K]. Radiative transfer through the atmosphere may be strongly affected by combustion-generated aerosols. A method for measuring the absorbing component is described [36K]. Temperature measurements in an automobile engine have been obtained by observing the CARS spectra of molecular nitrogen with an uncertainty of approximately 50–100°C [2K].

Using the modulation method of relative measurement of total emissivities, an instrument is described for determining the total emissivity of thin layers of various materials at temperatures between 223 and 423 K [13K]. An instrument for measuring spherical irradiance has a novel design feature consisting of a spherical, perforated receiving cavity [38K]. A portable instrument for measuring the emissivity of materials and coatings at room temperature is described in [35K]. The feasibility of absolute measurements of spatially inhomogeneous radiation is demonstrated and calculations show that the total errors in determining the local values of the density of radiant energy and of the incident radiant flux are 10.9 and 7%, respectively [3K]. A calorimetric method for measuring the emissivities and absorptivities of materials and fibers at temperatures between 170 and 370 K is described together with an error analysis [42K]. A generalized reflectance method for determining optical properties of absorbing materials overcomes the problem associated with nonideal surface conditions by applying transparent overcoats [29K]. The index of refraction in the near-millimeter wavelength range may be determined by using a mesh Fabry–Perot resonant cavity [40K]. A calculation method has been used to eliminate thermal radiation from effective thermal conductivities measured in a liquid within several layers of different thickness. The resulting "pure" thermal conductivities are independent of layer thickness which verifies the validity of the calculation method [8K].

Measured optical constants of fused silica and sapphire from 0.3 for 25 μm are in good agreement with data in the literature for the relatively small part of the spectral domain where they can be compared [23K]. Measurements of the temperature dependence of the refractive indices of fused silica and crystal quartz, in a wavelength range from 4500 for 16 000 Å and from room temperature up to 4000°C, indicate that the average thermal coefficients of the refractive indices of those materials show opposite signs [48K]. Infrared optical constants collected from the literature can be reasonable fitted to the Drude model for noble metals and Al, Pb, and W [32K]. Studies of the infrared reflectivity and crystal structure of InS in the wave number range from 50 to 900 cm^{-1} confirm that the measured dispersion parameters have more explicit values than those previously reported for twin crystals [45K]. Measured infrared specular reflectances of pressed crystal powders and mixtures agree with averaged crystal-optic reflectances if the low density of powders of harder substances is taken into account [49K]. Infrared optical properties of solid monomethyl hydrazine, N_2O_4, and N_2H_4 are reported for cryogenic temperatures [37K]. Studies of the refractive index of bismuth germanium oxide ($Bi_{12}GeO_{20}$) in the range 1.2–5.0 μm at room temperature show that a Herzberger formula completely describes data of $n(\lambda)$, up to the visible region [10K].

Heating of thin films of tin disulphide above 410 K converts the amorphous structure into crystallized films which have reduced transmission at wavelengths longer than the absorption edge and the absorption edge shifts to lower wavelengths [16K]. The development of an efficient and durable medium band antireflection coating on germanium for the 8.0–11.5 μm wavelength region is described [25K]. A new technique for determining the i.r. optical constants of thin films makes use of reflectance measurements [41K].

The results based on a new multidimensional model which makes it possible to calculate the spectrally-integrated total radiative flux for a molecular-gas is in excellent agreement with spectrally-integrated results from the P–1 approximation and with some exact results [28K]. The intensities, widths, and position of lines of three CO_2 bands near 3.6 μm have been measured and the results are in general agreement with other measured and estimated values [20K]. Investigations of the effect of dimerization on i.r. absorption in the 6.3 μm band and the total emissivity of H_2O vapor show that the effect is negligible [22K].

The results of a Monte Carlo finite difference method for coupled radiation–conduction heat transfer in semitransparent media are in excellent agreement with results obtained by other methods [1K]. Radiation–natural convection interactions in 2-D complex enclosures are considered taking CO_2 and NH_3 as participating gases into account [12K]. A modified Gardon heat flux probe is described for separate measurements of the components of radiative and convective heat fluxes [27K]. In the wavelength range 3–5 μm the attenuation factors due to CO_2 and H_2O in the atmospheric are 6 and 8%, respectively, at a 10 m range [39K]. Ground based vertical path absorption measurements were made up to a height of 1.5 km with a CO_2 linear transmitting alternatively on the $R(20)$ (10.247 μm) and $R(18)$ (10.260 μm) lines during daylight under conditions of both strong and weak temperature inversions [4K]. Measurements of aerosol and atmospheric i.r. emission spectra show that there is an auroral produced excitation for each band ($CO_2(\gamma_3)$, $NO(\Delta\gamma = 1)$, $O_3(\gamma_3)$ and $CO_2(\gamma_2)$) causing enhancements of $O_3(\gamma_3)$, 9.6 μm, and $CO_2(\gamma_2)$ 15 μm emissions [43K].

The thermal performance of radiative cooling panels has been calculated. Measurements show that, for most applications, white paint is a better radiator than aluminized polyvinyl fluoride films [5K]. Comparisons of equilibrium and nonequilibrium radiative heating studies are reported, associated with heat shield performance during outer planet atmospheric entry [31K].

Studies of radiant heat transfer in fibrous insulations indicate that the radiant heat flux can be minimized by making the mean radius of the fibers close to that which yields the maximum extinction coefficient [46K]. Corresponding experimental studies are compared with these predictions, indicating that the analytical models are useful in giving representative values for the radiative properties of typical light weight fibrous insulations [47K].

A model of radiative heat transfer in a dispersed medium is suggested which allows the calculation of the emissivity of an isothermal fluidized bed, the effective emissive ability of a nonisothermal bed and the temperature distribution near a heat transfer surface in a bed of a certain expansion when the radiative properties of the particles and the heat exchanger are prescribed [7K]. A Monte Carlo method is used for analyzing radiative heat transfer through a randomly packed bed of spheres. Qualitative agreement is shown in comparison with other work which used regular cubic packing, and with experimental data [50K].

The results of a study of the stability of horizontal fluid layers with radiative heat generation for two and three porous plates may be useful for solar applications [19K]. The radiation produced on impact of a high-velocity gas jet on an obstacle has been investigated [21K].

Experimental and analytical studies of the collection of thermal radiation by a semi-transparent fluid layer flowing in an open channel show that the results derived from the multiband model show better agreement with experimental data than those from the Planck mean model, independent of the optical thickness of the fluid [18K]. Numerical procedures based on Hottel's zone method are proposed for rapid solutions of view factors needed for radiative heat transfer [26K].

MHD

The number of publications in this field declined substantially during the past year.

Hydromagnetic flow and heat transfer between two horizontal parallel plates where the lower plate is a stretching sheet and the upper plate is porous have been studied in the presence of a transverse magnetic field [2M]. The response of flow and heat transfer to change of direction of the imposed magnetic field in steady MHD laminar free convection flow past an infinite vertical porous plate has been studied taking Hall effects into account [1M]. MHD flows have been studied, considering a circular cylinder in a channel [4M] and an elliptical cylinder for the case of an inviscid fluid [5M].

The effect of a uniform transverse magnetic field on the free convection flow of an electrically conducting fluid past an infinite vertical plate for both impulsive as well as uniformly accelerated motion of the plate is to increase the velocity field in both cases [3M].

APPLICATIONS

Heat exchangers and heat pipes

A comprehensive list of commercial enhanced tubes for shell and tube heat exchangers has been presented [18Q] and a standardized format of presenting geometric, flow, and heat transfer data for those is proposed. A new parameter is shown [32Q] to facilitate calculations for fins and an alternative to the fin efficiency in the analysis of fins has been proposed [17Q]. The conductivity ratio of the materials for a composite fin determines [9Q] the heat transfer rate and the time response. Experimental results were reported [31Q] of cross flow heat exchangers with tubes of lenticular shape. The performance of multiple parallel disk heat exchangers with radially outward flow is comparable [22Q] to high performance plate–fin surfaces. The conjugated heat transfer problem for the entrance region of laminar pipe flow with temperature variations in the surrounding was analyzed [14Q]. Turning off the flow at the inlet of a corruguated wall heat exchanger increases the heat transfer coefficient [33Q].

The exact solution of the Nusselt model of the cross-flow recuperator is presented [12Q] using the Mikusinski operator. Analytic expressions were developed for the mean temperature difference in multipass cross-flow heat exchangers [24Q]. Computer generated charts of the temperature difference in air coolers have been presented [25Q]. A simplified analysis [37Q] predicts the shell-side flow and temperature pattern in heat exchangers. A number of papers analyze the transient performance of heat exchangers for a step change in inlet temperature [8Q] and a periodic flow heat exchanger [5Q]. The method by Gaddis and Schlünder is used to model the time dependent heat exchanger performance [30Q]. Gas-to-gas cross-flow heat exchangers with neither gas mixed are studied [29Q] in their response to a unit step increase in the entering temperature. The influence of omitting some tubes in finned-tube bundles on their thermal and flow characteristics has been investigated [17Q]. Three papers [16Q, 35Q, 15Q] consider the economic aspects of the heat exchanger design.

Experiment [26Q] and analysis [13Q] are dealing with regenerators.

Laboratory experiments [28Q] clarified boiling heat transfer of Freon 11 in brazed aluminum heat exchangers at low mass and heat fluxes. The heat transfer performance of stainless steel metal fiber sintered surfaces to Freon 11 could be improved tenfold [10Q] over smooth surfaces. Experiment and analysis [2Q] determined the shell-side heat transfer coefficient for steam contaminated with noncondensible gases. The use of air–water mist flow improved the performance of a compact cooling unit for semiconductors from 1.8 to 20 times over the performance of a dry unit [26Q]. A mathematical model [11Q] describes the thermal performance of a wet surface plastic-plate heat exchanger with a change in temperature and humidity ratio of an air flow. A hybrid evaporative condenser built into a cooling tower saves heat transfer area or lowers the condensation temperature [7Q]. An exact analysis [34Q] is compared with the method by Merkel [1925] for a mechanical-draft air–water cooling tower and finds that the later method underestimates the performance from 5 to 15%. A computer code [VERA2D] was developed to describe the fields of air velocity, temperature, pressure, and moisture content in a wet cooling tower [19Q, 20Q].

Fouling processes are classified into five categories [36Q]. Particulate fouling on the gas side of finned tube heat exchangers was found [33Q] to increase the friction factor up to 1.4 to 2.5 times of the clean value, whereas the Stanton number decreased by 10 to 20%. Crystalline fouling studies [37Q] showed that the deposition rate is primarily a function of the supersaturation of the solution.

Experiments determined heat transfer and flow of Nak in a heat pipe with steel wool and metal fiber wicks [23Q]. The maximum heat flux through heat pipes with various cross sections was measured [21Q]. Experiments determined [4Q] turbulent heat and mass transfer in vacuum and gas loaded sodium heat pipes.

General

Advances in modelling turbulence and their influence on heat transfer is discussed [7S]. The effect of polymeric additives on the damping of the wall turbulence and on convective heat transfer was investigated experimentally [8S] for flow through a square channel using the electrodiffusion technique. A numerical analysis [12S] studied flow and heat transfer of polymeric fluids through a circular pipe and found that it depended primarily on the Weissenberg number.

Boiling in a two-phase thermosiphon occurs at low pressure in a pulsed mode [1S] with periods of quiet alternating with explosive evaporation. Periodic heat conduction into a semi-infinite medium with 3-D heat flow originating from a rectangular surface region is handled [2S] through Fourier transforms. Heat transfer to block-like aerodynamic components is enhanced by a factor 2 [13S] by fence-like barriers. The hydrodynamic and thermal situation in a steam-generating tube [14S] and rewetting of the tube by a falling film [8S] are studied experimentally to clarify the process in a loss of coolant accident [LOCA]. A survey of heat transfer problems in air cooled high temperature gas turbines is presented [3S]. Transpiration air-cooled turbine blades showed no evidence of erosion, corrosion, or deposition when exposed to combustion gases carrying solid particles with a temperature up to 1650°C [17S]. The cooling of casings, struts, and disks by steam was studied [10S]. An analysis was concerned with simultaneous transfer of heat and mass in a casting process [12S]. Another analysis [6S] provides a simplified prediction of the melting surface and of heat conduction in electro-discharge machining. The cooling of fibers by natural convection and radiation during the melt-spinning process is analyzed numerically [5S]. The flights of the shuttle orbiter provided the possibility to obtain heat transfer data during reentry [16S, 15S, 9S, 18S]. Generally, turbulent heat transfer and transition location compared well with currently established ways of prediction. Data during the second and third flight were generally lower than wind tunnel results. There were contradictions in the surface emissivity obtained from reflectance and emittance measurements.

Solar Energy

A statistical comparison was made of three solar flux estimating models [25T]. Data sets are presented for the direct normal and global solar spectra, both for Air Mass 1.5 [5T]. The Liu–Jordan model for solar flux has been modified to better account for the effects of altitude and snow cover (i.e. changes in the albedo of large portions of the terrain) [24T]. In developing a solar radiation model, a reasonable estimate of lower layer aerosol extinction can be determined using humidity, visibility, and mixing height [34T]. The spectral distributions of solar radiation at ground surface, computed from extraterrestrial solar spectral irradiance, are presented for several air mass values and for four levels of atmospheric pollution [28T]. The absorption of radiation by molecules such as O_3, H_2O and the uniformly-mixed gas is taken into account in an empirical model for the calculation of solar spectral diffuse and global irradiance under cloudless skys [6T]. A model of the spectral direct radiation of the sun at ground level compared with the spectral measurements performed by a monochromator has evidenced satisfactory agreement [13T]. A model has been developed to calculate the downwelling infrared radiation from a clear night atmosphere using LOWTRAN 5, an atmospheric transmittance/radiance model developed at the U.S. Air Force Geophysics Laboratory [37T]. Classical probability theory is used to derive expressions for the mean values of quantities such as the solar flux on inclined surfaces, collector output, and net gain through windows [17T].

The transient characteristics of a flat-plate solar collector were described [32T]. An efficient algorithm is presented for solving the set of nonlinear equations governing the total heat transfer across an arbitrary number of semitransparent flat-plate solar collector covers [18T]. An approximate calculation of the radiative–conductive flux of heat from the absorber to the cover plate of a flat-plate collector equipped with a convection-suppressing honeycomb is developed using an exponential kernal approximation to the passage transmittance function of the honeycomb cells [26T]. A mathematical model of a collector in which the working fluid passes through successive transparent layers indicates the collector performance is a function of the number of layers, the heat transfer coefficient across each layer and the absorption properties of the fluid [21T]. The single-pass as well as the two-pass mode of heat withdraw from a two-layer thermal trap collector with slabs of variable thermal and optical properties are modelled mathematically [33T]. A computer simulation of a regenerating-type solar collector is presented and optimum ranges of design parameters are indicated [20T]. An air cooled collector that uses a close packed array of evacuated tubes for a cover has been shown to produce high efficiencies at low solar flux levels [15T]. A system simulation study indicated that the collector with a cover made of close-packed evacuated tubes delivered three times more heat than a "typical flat plate" on a seasonal basis [16T]. Measurements are reported for assorted heat removal schemes in single-ended glass tubes with the nonuniform solar input around the tube simulated electrically [41T].

A review is presented of a wide variety of point-focusing solar concentrators that are under consideration for solar thermal energy use [19T]. The theory developed for the utilizability function has been applied to high concentration ratio focusing collectors [12T]. A comparison is made between fluidized and packed bed receivers for high temperature solar gas heating systems [11T]. The maximum efficiency of solar energy conversion devices is investigated [7T].

A finite element heat transport program has been modified to model the thermal behavior of a cylindrically symmetric salt gradient solar pond of the Rabl–Nielsen type [8T]. A computer model is described which may be used for predicting transient salinity and temperature profiles in salt gradient solar ponds which have large surface areas and where wind-mixing is expected to be important [4T]. If the temperature and salt concentration dependency of the thermal conductivity of the water in solar ponds is ignored, very little error is introduced into the thermal characteristics analysis [40T].

An extensive review of heat-of-fusion storage materials for low temperature heat storage (0–120°C) is presented [1T]. The heat and mass transfer characteristics of a thermochemical energy storage system based on solid–gas reactions have been studied through the example of the borax dehydration reaction in an apparatus simulating 1-D heat transfer [2T]. The losses of heat from a seasonal ground storage system due to conduction and natural convection in groundwater have been investigated [39T]. A simplified steady state model to evaluate the effect of floor coverings on the storage of solar energy indicates that even the thinnest of carpets interferes severely with storage in a floor slab [14T].

Based on a Fourier series solution of the heat conduction equation, a mathematical model has been developed to analyze the thermal performance of some typical passive heating concepts [30T]. A design method, based on monthly average temperatures, has been developed to approximate the need for shading and other strategies for passive cooling [31T]. The influences of solar radiation, environmental long-wave radiation and the radiation from internal walls were included in an analytic relation describing the thermal gain of a room [3T]. Thermal storage walls for use in passive solar collectors are analyzed with the purpose of establishing guidelines for the design of low-mass, high-efficiency walls [22T].

Tests conducted at the National Bureau of Standards Test Facility indicate that the performance of a typical solar domestic hot water system is relatively insensitive to various load profiles [10T]. The performance of six thermosyphon solar water heaters was measured while the systems were supplying typical domestic hot water loads [29T]. The thermal performance of an integral compact solar water heater is evaluated experimentally and numerically [36T]. A transient analysis is presented of the performance of a solar water heating system with natural circulation of hot water between the collector and an insulated storage tank which contains a heat exchanger to transfer heat to hot water which would be used for either domestic or industrial purposes [35T]. A transient simulation program is presented for analyzing a solar air heating system which has either rock bed or phase change energy storage [38T]. Simulations have shown that a solar driven absorption heat pump system can supply a significant fraction of a residential heating and cooling load with nonpurchased energy [37T]. A mathematical model for a flat-plate solar system integrated with a storage tank and an organic Rankine cycle loop was to used to determine the optimum flow rate through the collector [9T]. A 10% efficiency was obtained in a solar thermal engine system by augmenting a paraboloidal concentrator with a new type of nonimaging second stage concentrator which is shaped like the bell of a trumpet [23T].

Plasma heat transfer

Plasma heat transfer seems to be of particular interest in connection with arc technology and thermal plasma processing.

Studies of transient heat conduction of particles under plasma conditions reveal that, for example, the surface temperature of a 100 μm alumina particle can reach the boiling point (4000 K) while its center is still in the solid state [2U]. Heat transfer studies to particles immersed into a thermal plasma indicate that the Knudsen effect may play an important role, even at atmospheric pressure [5U]. Studies of the behavior of small particles injected into a thermal plasma flow taking convection into account demonstrate that the ratio of the heat fluxes with and without the Knudsen effect is almost identical to the ratio obtained for the case of pure heat conduction [4U]. Investigations of particle momentum and heat transfer in a thermal plasma show good agreement between measured and computed particle velocities. Measured particle temperatures, however, differ substantially from predicted values [25U].

A mathematical model has been developed for describing fluid flow and heat transfer in the cathode region of high intensity carbon arcs [20U]. An analysis of the cathode region of a free-burning high intensity argon arc reveals that a significant fraction of the total current (up to 18%) is carried by positive ions which is crucial for the heat balance at the cathode surface [14U]. A model for the thermionically emitting cathode of a 100 A argon arc has been proposed. At the very tip of the cathode, the current is predominantly carried by electrons and the surface temperature is determined by the balance of heating by plasma radiation and cooling by thermionic emission [7U].

Studies of the behavior of sustained high-current arcs on molten alloy electrodes during vacuum consumable arc remelting indicate that by introducing CO into the vertical alloy electrode system at larger electrode gaps, the cathode thermal power is reduced by approximately 50% [27U]. Studies of the fusion boundary energy transport during arc welding show that at current levels close to 300 A, significant weld pool convection occurs, especially at the (inner) stagnation point, causing a deeper penetration of the fusion boundary at this location [19U]. Investigations of the power dissipation in the column of a TIG welding arc at pressures up to 135 bar indicate that for a 100 A, 10 mm argon arc, convective dissipation accounts for about 70% of column power at all pressures. Radiation contributes 18% at 1 bar increasing to 26% at 20 bar [1U].

Measured and calculated temperature distributions in free-burning, high-intensity argon arcs show good agreement as long as the cathode tip remains intact

[13U]. Calculated temperature distributions of a free-burning, high-intensity arc are in better agreement with measurements if the calculations are based on a two-temperature model [12U]. Investigations of the operating characteristics and energy distribution in transferred plasma arc systems indicate that radiation from the plasma column to the chamber walls and transfer of energy to the anode by the arc current and convection arc the principal modes of transfer of the arc energy [6U]. Studies of the characteristics of a transferred-arc plasma show that the amount of energy release at the anode increases substantially with increasing current and power losses by radiation from the arc range between 41% and 53% of the total power input depending on the electrode separation [21U].

Studies of the effects of low-ionization-potential contaminants on thermal plasma show that for sufficiently high plasma temperatures, the behavior is dominated by the high-ionization potential component whereas low-ionization-potential properties determine the low-temperature plasma behavior [16U]. Experimental studies of the electrode vapor flow in a free-burning, pulsed (12.5 kA) arc in air with a Cu-anode and a Cu-W cathode show that the arc column is dominated by the electrode vapor during most of the arcing period [26U]. Investigations of d.c. nozzle arcs with mild wall ablation show that a theoretical model can adequately describe the nozzle arc behavior in the presence of wall ablation [10U].

Experimental studies of anode melting in a multicathode-spot vacuum arc show that peak currents for melting are in the range of 500–2 3000 A depending on the electrode materials [11U]. A new model for the anode spot formation in a vacuum arc is introduced, based on the assumption that anode spot formation is associated with a change of sign of the anode fall (from negative to positive) [9U].

Calculations of the specific heat flux to a cylinder in a thermal plasma cross flow are in general agreement with experimental data in spite of the use of the boundary layer approximation and the potential flow assumption for the outer region in these calculations [3U]. When W rods of 0.203 and 0.3175 cm dia. are exposed perpendicularly to an argon plasma jet up to the middle of the flame, special values of h and T_∞ have to be used in order to reproduce by calculation the measured temperature profiles [22U].

Rapid quenching of air plasma effluents (2350–2600 K) leads to higher nitric oxide concentrations than those predicted from stagnant equilibrium conditions [23U]. Studies of efficacy, arc tube temperature, and power dissipation in a high-pressure sodium lamp show that the proportion of the total radiation emitted in the D-lines decreases as the power input increases. But this is more than compensated by an increase in the total radiated power at the expense of thermal conduction [8U].

Studies of the efficiency and erosion characteristics of plasma jet igniters in I.C. engines indicate that the thermal energy delivered by the igniter to the adjacent gas corresponds to overall efficiencies of $\approx 10\%$ [24U]. Experimental and analytical studies of heat transfer from a thin horizontal wire and large cylinders in a corona wind suggest that the Nusselt number is a function of a new (complicated) dimensionless parameter [18U].

An infrared calorimeter for time-resolved plasma energy flux measurements has been developed with a response time of 12 μs and a peak-to-peak noise equivalent of 0.4 mJ/cm^2 [15U].

An extension of the Spitzer-Härm theory is presented for the case of thermal transport in steep temperature gradients [17U].

REFERENCES

Books

Richard J. Goldstein, *Fluid Mechanics Measurements*, Hemisphere Publ. Corp., Washington, D. C. (1983).

Tien-Mo Shih, *Numerical Heat Transfer*, Hemisphere Publ. Corp., Washington, D.C. (1983).

T. Netjat Veziroğlu (editor), *Alternative Energy Sources III*, Hemisphere Publ. Corp., Washington, D.C. (1983).

William Begell (editor), *Glossary of Terms in Heat Transfer, Fluid Flow, and Related Topics: English, Russian, German, French and Japanese*, Hemisphere Publ. Corp., Washington, D.C. (1983).

U. Grigull (editor), *Selected Publications of Wilhelm Nusselt and Ernst Schmidt*, Hemisphere Publ. Corp., Washington, D. C. (1983).

Allan D. Kraus, *Thermal Analyses and Control of Electronic Equipment*, Hemisphere Publ. Corp., Washington, D. C. (1983).

Tokuo Mizushina and Wen-Jei Yang (editors), *Heat Transfer in Energy Problems*, Hemisphere Publ. Corp., Washington, D.C. (1983).

Ernst U. Schlünder (editor-in-chief), K. J. Bell, D. Chisholm, F. G. Hewitt, F. W. Schmidt, D. B. Spalding, J. Taborek, Z. Žukauskas (editors) and V. Gnielinski (associate editor), *Heat Exchanger Design Handbook*, Hemisphere Publ. Corp., Washington, D.C. (1983).

J. Taborek, Geoffrey F. Hewitt and Naim H. Afgan (editors), *Heat Exchangers, Theory and Practice*, Hemisphere Publ. Corp., Washington, D. C. (1983).

T. Nejat Veziroğlu (editor), *Thermal Sciences 16: Proceedings of the 16th Southeastern Seminar*, Hemisphere Publ. Corp., Washington, D.C. (1983).

U. Grigull and H. Sandner, *Heat Conduction*, translated by Joseph Kestin, Hemisphere Publ. Corp., Washington, D.C.

(1983).

Arun S. Mujumdar (editor), *Drying '82*, Hemisphere Publ. Corp. (1983).

Zoram P. Zarić (editor), *Structure of Turbulence in Heat and Mass Transfer*, Hemisphere Publ. Corp., Washington, D.C. (1983).

J. P. Gupta, *Fundamentals of Heat Exchanger and Pressure Vessel Technology*, Hemisphere Publ. Corp., Washington, D.C. (1983).

J. Cook, *Award-winning Passive Solar Designs: Residential and Commercial*, McGraw-Hill, New York (1983).

Sadik Kadaç, R. K. Shah and Arthur E. Bergles (editors), *Low Reynolds Number Flow Heat Exchangers*, Hemisphere Publ. Corp., Washington, D.C. (1983).

D. E. Metzgar and N. H. Afgan (editors), *Heat and Mass Transfer in Rotating Machinery*. Springer Publ., Heidelberg, F. R. Germany (1983).

S. Patankar, A. Pollard, A. Singhal and S. Vanka (editors), *Numerical Prediction of Flow, Heat Transfer, Turbulence and Combustion–Selected Works of Professor D. Brian Spalding*, Maxwell House, Oxford, England (1983).

Heat Transfer, Publ. No. S-225, AIChE, Seattle (1983).

1983 Bound Conference Volumes in Heat Transfer Interfacial Transport Phenomena, HTD-Vol. 23, Bk. No. H00267, ASME (1983).

Fundamentals of Heat Transfer In-Fusion Energy Systems, HTD-Vol. 24, 1983 Bk. No. H00268, ASME.

Fire Dynamics and Heat Transfer, HTD-Vol. 25, Bk. No. H00269, ASME (1983).

Natural Convection in Enclosures, HTD-Vol. 26, Bk. No. H00270, ASME (1983).

Heat Exchangers for Two-Phase Applications, HTD-Vol. 27, Bk. No. H00271, ASME (1983).

Proceedings of the ASME/JSME Thernal Engineering Joint Conference 1983, ASME. *Vol. 1*: Two-Phase Flow and Boiling Phenomena, Heat Transfers in Enclosures, Augmentation Heat Transfer, High Temperature Heat Exchangers, Compact Heat Exchangers, Air-Cooled Heat Exchangers, Gas-side Fouling in Heat Exchangers, Bk. No. 10158A, ASME (1983).

Vol. 2: Heat Transfer in Fluidized/Packed Systems, Underground Conversion, Heat Transfer in Geothermal Utilization, Heat Transfer in Underground Media, OTEC Heat Transfer, Heat Exchangers for Alternate Energy Sources, Bk. No. 10158B, ASME (1983).

Vol 3: Free and Forced Convection, Advanced Thermal Power Generation Systems, Condensation, Heat Conduction, Heat Transfer in Nuclear Reactors, Bk. No. 10158C, ASME (1983).

Vol. 4: Thermal Radiation, Combustion for Fuel Economy, Furnace Combustion, Turbulent Combustion, Gas Turbine Components and Applications, Combustion in Diesel Engines. Alternative Fuel Engines, Heat Transfer in General, Combustion in General, Bk. No. 10158D, ASME (1983).

Conduction

1A. J. L. Auriault, Effective macroscopic description for heat conduction in periodic composites, *Int. J. Heat Mass Transfer* **26,** 861 (1983).

2A. J. R. Barber and M. Comninou, The penny-shaped interface crack with heat flow, Part 2: imperfection contact (83-WA/APM-13), *J. Appl. Mech.* **50,** 770 (1983).

3A. M. Bentwich, Optimal insulation distribution over a conducting body, *Q. J. Mech. Appl. Math.* **36,** 145 (1983).

4A. L. E. Bobisud, Heat conduction through a thin, poorly conductive layer, *Q. Appl. Math.* **41,** 85 (1983).

5A. A. Campo, Estimate of the transient conduction of heat in materials with linear thermal properties based on the solution for constant properties, *Wärme- und Stoffübertragung* **17,** 1 (1983).

6A. D. A. Caulk, Steady heat conduction from an infinite row of holes in a half-space or a uniform slab, *Int. J. Heat Mass Transfer* **26,** 1509 (1983).

7A. F. B. Cheung, Transient nonsimilarity nonlinear heat diffusion solutions, *J. Heat Transfer* **105**(2), 295 (1983).

8A. D. L. Clements and G. Rogers, A boundary integral equation for the solution of a class of problems in anisotropic inhomogeneous thermostatics and elastostatics, *Q. Appl. Math.* **41,** 99 (1983).

9A. W. A. Day, A decreasing property of solutions of parabolic equations with applications to thermoelasticity, *Q. Appl. Math.* **40,** 468 (1983).

10A. G. Dharmadurai, Acoustic mismatch theory for heat flow across fuel-cladding interfaces in fast reactors, *Nucl. Sci. Engng* **84,** 345 (1983).

11A. R. F. DiFelice Jr. and H. H. Bau, Conductive heat transfer between eccentric cylinders with boundary conditions of the third kind, *J. Heat Transfer* **105**(3), 678 (1983).

12A. G. M. L. Gladwell, J. R. Barber and Z. Olesiak, Thermal problems with radiation boundary conditions, *Q. J. Mech. Appl. Math.* **36,** 387 (1983).

13A. G. M. L. Gladwell and J. R. Barber, Thermoelastic contact problems with radiation boundary conditions, *Q. J. Mech. Appl. Math.* **36,** 403 (1983).

14A. D. Gottlieb and L. Lustman, The spectrum of the Chebyshev collocation operator for the heat equation, *SIAM J. Num. Analysis* **20,** 909 (1983).

15A. D. C. Hittle and R. Bishop, An improved root-finding procedure for use in calculating transient heat flow through multilayered slabs, *Int. J. Heat Mass Transfer* **26,** 1685 (1983).

16A. S. C. Huang and Y. P. Chang, Heat conduction in anisotropic wedge and elliptic cylinders, *J. Heat Transfer* **105**(3), 674 (1983).

17A. L. W. Hunter and S. Favin, The thermal resistance of an insulating slab penetrated by metal rods, *J. Heat Transfer* **105**(1), 208 (1983).

18A. L. W. Hunter and J. R. Kuttler, Cooling of a slab with thermal contraction and progressive loss of contact with a cold surface, *J. Heat Transfer* **105,**(4), 936 (1983).

19A. H. Inaba, Heat transfer behavior of frozen soils, *J. Heat Transfer* **105**(3), 680 (1983).

20A. A. N. Kalinin, Instrument with direct reading of the thermal conductivity of solids of arbitrary shape, *Heat Transfer—Soviet Res.* **14**(6), 15 (1982).

21A. R. Kersner, Nonlinear heat conduction with absorption: space localization and extinction in finite time, *SIAM J. Appl. Math.* **43,** 1274 (1983).

22A. C.-H. Li, Optimum cylindrical pin fin, *AIChE Jl* **29,** 1043 (1983).

23A. J. E. Llebot, D. Jou and J. Casas-Vásquez, A Thermodynamic approach to heat and electric

conduction in solids, *Physica A*, **121 A**(3), 552 (1983).
24A. M. Locatelli, J. Lopez and M. Núñez-Regueiro, Thermal radiation effects in thermal conductivity measurements : analysis and remedies, *J. Physics D: Applied Physics* **16**(4), 479 (1983).
25A. C. V. Madhusudana and L. S. Fletcher, Solid spot thermal conductance of zircaloy-2/uranium dioxide interfaces, *Nucl. Sci. Engng* **83,** 327 (1983).
26A. C. J. Martin-Moran, J. R. Barber and M. Comninou, The penny-shaped interface crack with heat flow, Part 1: perfect contact, *J. Appl. Mech.* **50,** 29 (1983).
27A. S. Panteliou and A. D. Dimarogonas, Heat propagation on a shaft due to transient torsional vibration, *Int. Comm. Heat Mass Transfer* **10**(2) 111 (1983).
28A. H. Salt, Transient conduction in a two-dimensional composite slab—I. Theoretical development of temperature modes, *Int. J. Heat Mass Transfer* **26,** 1611 (1983).
29A. H. Salt, Transient conduction in a two-dimensional composite slab—II. Physical interpretation of temperature modes, *Int. J. Heat Mass Transfer* **26,** 1617 (1983).
30A. W. Schultz and A. Bejan, Exergy conservation in parallel thermal insulation systems, *Int. J. Heat Mass Transfer* **26,** 335 (1983).
31A. K. Sridharan, The micropolar thermoelastic problem of uniform heat-flow interrupted by an insulated penny-shaped crack, *Int. J. Engng Sci.* **21,** 1459 (1983).
32A. J. D. Tarasuk and G. S. P. Castle, Temperature distribution in an electrically heated wide metallic foil, *J. Heat Transfer* **105**(1), 210 (1983).
33A. A. K. S. Thakur, Conductive heat transfer in salt gradient stabilized solar ponds, *Int. Comm. Heat Mass Transfer* **10**(4), 335 (1983).
34A. J. L. Tylee, Real-time nonlinear estimation of fuel-clad gap conductance, *Nucl. Sci. Engng* **84,** 58 (1983).
35A. B. Vick and M. N. Özisik, Growth and decay of a thermal pulse predicted by the hyperbolic heat conduction equation, *J. Heat Transfer* **105**(4), 902 (1983).
36A. H. P. Yagoda, Short-circuit phenomena: heat energy drains in composite cylindrical slabs, *Int. J. Heat Mass Transfer* **26,** 1847 (1983).
37A. Ye. D. Yershov, V. G. Cheverev, I. A. Komarov, Ye. N. Barkovskaya, T. G. Grishchencko and V. G. Rezhitskiy, Technique for measuring the thermal conductivity of wet soils and biological tissues, *Heat Transfer—Soviet Res.* **14**(6), 33 (1982).
38A. A. A. Zick, Heat conduction through periodic arrays of spheres, *Int. J. Heat Mass Transfer* **26,** 465 (1983).

Channel Flow

1B. N. Abuaf, O. C. Jones Jr. and B. J. C. Wu, Critical flashing flows in nozzles with subcooled inlet conditions, *J. Heat Transfer* **105**(2), 379 (1983).
2B. M. M. M. Abou-Ellail and S. M. Morcos, Bouyancy effects in the entrance region of horizontal rectangular channels, *J. Heat Transfer* **105**(4), 924 (1983).
3B. R. S. Amano, M. K. Jensen and P. Goel, A numerical and experimental investigation of turbulent heat transport downstream from an abrupt pipe expansion, *J. Heat Transfer* **105**(4), 862 (1983).
4B. G. Antonini, E. Brunier and G. Guiffant, Heat transfer enhancement in oscillatory cavity flow, *Int. Comm. Heat Mass Transfer* **10**(1), 13 (1983).
5B. D. Barnea and N. Yacoub, Heat transfer in vertical upward gas–liquid slug flow, *Int. J. Heat Mass Transfer* **26,** 1365 (1983).
6B. H. Barrow, A. K. A. Hassan and C. Avgerinos, Peripheral temperature variation in the walls of noncircular ducts, *Int. Comm. Heat Mass Transfer* **10**(3), 225 (1983).
7B. C. P. Bellmore and R. L. Reid, Numerical prediction of wall temperatures for near-critical para-hydrogen in turbulent upflow inside vertical tubes, *J. Heat Transfer* **105**(3), 536 (1983).
8B. R. Bhushan, M. Keyhani, R. N. Christsensen and F. A. Kulacki, Correlations for convective heat transfer in vertical annular gas layers with constant heat flux on the inner wall. *J. Heat Transfer* **105**(4), 910 (1983).
9B. S. Caspi and T. H. K. Frederking, Influence of forced flow on the HeII-HeI transition in the presence of heat flow, *J. Heat Transfer* **105**(4), 846 (1983).
10B. J. S. Caton, The use of a simple heat transfer model for separated flows in tubes, *J. Heat Transfer* **105**(4), 928 (1983).
11B. M. N. Channabasappa, K. G. Umapathy and I. V. Nayak, Convective heat transfer in a parallel plate channel with porous lining, *Wärme- und Stoffübertragung* **17,** 211 (1983).
12B. J. J. J. Chen, Comments on "A note on the Lockhart–Martinelli correlation in the turbulent-turbulent case, *Int. Comm. Heat Mass Transfer* **10**(5), 441 (1983).
13B. S. C. Chen, N. K. Anand and D. R. Tree, Analysis of transient laminar convective heat transfer inside a circular duct, *J. Heat Transfer* **105**(4), 922 (1983).
14B. O. Kh. Dakhin, V. A. Gerasimenko, N. Y. Tyabin and A. V. Baranov, Heat transfer and drag in laminar nonisothermal pipe flows of non-Newtonian fluids, *Heat Transfer—Soviet. Res.* **14**(2), 1 (1982).
15B. Vi-Duong Dang, Heat transfer of power low fluid at low Peclet number flow, *J. Heat Transfer* **105**(3), 542 (1983).
16B. G. B. Froyshteter, Heat transfer in pipe flow of high-viscosity non-Newtonian fluids under boundary conditions of the second and third kinds, *Heat Transfer—Sov. Res.* **14**(2), 10 (1982).
17B. K. Fukui, M. Nakajima and H. Ueda, The longitudinal vortex and its effects on the transport processes in combined free and forced laminar convection between horizontal and inclined parallel plates, *Int. J. Heat Mass Transfer* **26,** 109 (1983).
18B. J. C. Gottifredi, Q. D. Quiroga and A. F. Flores, Heat transfer to Newtonian and non-Newtonian fluids flowing in a laminar regime, *Int. J. Heat Mass Transfer* **26,** 1215 (1983).
19B. A. Haji-Sheikh, M. Mashena and M. J. Haji-Sheikh, Heat transfer coefficient in ducts with constant wall temperature, *J. Heat Transfer* **105**(4), 878 (1983).
20B. J. E. Heart, Low Prandtl number convection between differentially heated end walls, *Int. J. Heat Mass Transfer* **26,** 1069 (1983).
21B. S. Hasegaws, R. Echigo, K. Kanemaru, K. Ichimiya, Experimental study on forced convective heat transfer of flowing gaseous solid suspension at high temperature, *Int. J. Multiphase Flow* **9,** 131 (1983).
22B. R. Izumi, H. Yamashita, K. Oyakawa and N. Mori,

Fluid flow and heat transfer in corrugated wall channels (3rd report, Effects of bending angles in the case where channels are bent two times), *Bull. JSME* **26,** 1027 (1983).

23B. R. Izumi, H. Yamshita and K. Oyakawa, Fluid flow and heat transfer in corrugated wall channels (4th report, Analysis in the case where channels are bent many times), *Bull. JSME* **26,** 1146 (1983).

24B. C. E. Kalb and J. D. Seader, Entrance region heat transfer in a uniform wall-temperature helical coil with transition from turbulent to laminar flow, *Int. J. Heat Mass Transfer* **26,** 23 (1983).

25B. Y. Kawashima, M. Nakagawa and S. Iuchi, The heat transfer characteristics of a two-dimensional, right-angle, T-shaped flow junction, *Int. Chem. Engng* **23,** 510 (1983).

26B. J. M. Kim and J. D. Seader, Heat transfer to gas-solid suspensions flowing concurrently downward in a circular tube, *AIChE J.* **29,** 306 (1983).

27B. B. Klingenberg, H. Mach and G. Smeets, Flow field measurements of an unsteady reacting muzzle exhaust flow, *J. Heat Transfer* **105**(4), 884 (1983).

28B. W. Y. Kwack and J. P. Hartnett, Empirical correlations of turbulent friction factors and heat transfer coefficients for viscoelastic fluids, *Int. Comm. Heat Mass Transfer* **10**(5), 451 (1983).

29B. S.-L. Lee, Liquid metal heat transfer in turbulent pipe flow with uniform wall flux, *Int. J. Heat Mass Transfer* **26,** 349 (1983).

30B. T. F. Lin, K. H. Hawks and W. Leidenfrost, Analysis of viscous dissipation effect on thermal entrance heat transfer in laminar pipe flows with convective boundary conditions, *Wärme- und Stoffübertragung* **17,** 97 (1983).

31B. T. F. Lin, K. Hawks and W. Leidenfrost, Transient thermal entrance heat transfer in laminar pipe flows with step change in pumping pressure, *Wärme- und Stoffübertragung* **17,** 201 (1983).

32B. T. F. Lin, K. H. Hawks and W. Leidenfrost, Unsteady thermal entrance heat transfer in laminar pipe flows with step change in ambient temperature, *Wärme- und Stoffübertragung* **17,** 125 (1983).

33B. M. Molki and E. M. Sparrow, In-tube heat transfer for skewed inlet glow caused by competition among tubes fed by the same plenum. *J. Heat Transfer* **105**(4), 870 (1983).

34B. K. Oyakawa and I. Mabuchi, Heat transfer in parallel plate duct with circular cylinders inserted in staggered arrangement, *Bull. JSME* **26,** 537 (1983).

35B. S. Ramadhyani, M. M. Torbaty and K. N. Astill, Laminar forced convection heat transfer in spherical annuli, *J. Heat Transfer* **105**(2), 341 (1983).

36B. R. E. Roadman and R. I. Loehrke, Low Reynolds number flow between interrupted flat plates, *J. Heat Transfer* **105**(10), 166 (1983).

37B. N. Rudraian and S. B. Rao, Study of nonlinear convection in a sparsely packed porous medium using spectral analysis, *Appl. Sci. Res.* **40,** 223 (1983).

38B. N. Seki, S. Fukusako and M. Yoneta, Heat transfer from the heated concave wall of a return bend with rectangular cross section, *Wärme- und Stoffübertragung* **17,** 17 (1983).

39B. N. Seki, S. Fukusako and M. Yoneta, Heat transfer from the heated convex wall of a return bend with rectangular cross section, *J. Heat Transfer* **105**(1), 64 (1983).

40B. N. Seki, S. Fukusako and M. Yoneta, Turbulent heat-transfer characteristics along the heated convex wall of a rectangular cross-sectional return bend, *Wärme- und Stoffübertragung* **17,** 85 (1983).

41B. R. Sethumadhavan and M. Raja Rao, Turbulent flow heat transfer and fluid friction in helical-wire-coil inserted tubes, *Int. J. Heat Mass Transfer* **26,** 1833 (1983).

42B. E. M. Sparrow and W. Q. Tao, Enhanced heat transfer in a flat rectangular duct with streamwise-periodic disturbances at one principal wall, *J. Heat Transfer* **105**(4), 851 (1983).

43B. E. M. Sparrow and L. D. Bosmans, Heat transfer and fluid flow experiments with a tube fed by a plenum having nonaligned inlet and exit, *J. Heat Transfer* **105**(1), 56 (1983).

44B. E. M. Sparrow and N. Cur, Maldistributed inlet flow effects on turbulent heat transfer and pressure drop in a flat rectangular duct, *J. Heat Transfer* **105**(3), 527 (1983).

45B. K. Suzuki, Y. Kang, T. Sugimoto and T. Sato, Circular tube turbulent heat transfer in the downstream of an orifice, *Heat Transfer—Jap. Research* **11**(1), 70 (1982).

46B. E. J. Szetela and D. R. Sobel, Tube entrance heat transfer with deposit formation, *Heat Transfer Engng* **4**(2), 47 (1983).

47B. T. Yazaki, A. Tominaga and Y. Narahara, Large heat transport due to a spontanegous gas oscillation induced in a tube with steep temperature gradients, *J. Heat Transfer* **105**(4), 889 (1983).

48B. Y. K. Yeo and C. K. Choi, Thermal instability in plane Poiseuille flow heated from below, *Int. Chem. Engng.* **23,** 281 (1983).

49B. G. L. Yoder, Jr. and W. M. Rohsenow, A solution for dispersed flow heat transfer using equilibrium fluid conditions, *J. Heat Transfer* **105**(1), 10 (1983).

Boundary layer and external flows

1C. N. Afzal, The cooling of a low-heat resistance stretching sheet moving through a fluid: a rejoinder, *Wärme- und Stoffübertragung* **17,** 217 (1983).

2C. J. Andreopoulos, Heat transfer measurements in a heated jet-pipe flow issuing into a cold cross stream, *Physics Fluids* **26**(11), 3201 (1983).

3C. J. Andreopoulos, The response of a turbulent boundary layer to a double step-change in a wall heat flux, *J. Heat Transfer* **105**(4), 841 (1983).

4C. J. A. Belward, Further solutions of Fredholm integral equation of the first kind arising in elastic half-space and forced-convection problems, *Q. J. Mech. Appl. Math.* **36,** 237 (1983).

5C. D. K. Bhattacharya, Application of Gyarmati's variational principle to laminar stagnation flow problem, *Wärme- und Stoffübertragung* **17,** 27 (1983).

6C. A. K. Biń, Mass transfer into a turbulent liquid film, *Int. J. Heat Mass Transfer* **26,** 981 (1983).

7C. M. F. Blair, Influence of free-stream turbulence on turbulent boundary layer heat transfer and mean profile development, Part I—Experimental data, *J. Heat Transfer* **105**(1), 33 (1983).

8C. M. F. Blair, Influence of free-stream turbulence on

turbulent boundary layer heat transfer and mean profile development, Part II—Analysis of results, *J. Heat Transfer* **105**(1), 41 (1983).
9C. L. E. Bobisud, Heat analysis through thin films: approximate analysis and determination of effective heat transfer coefficient, *J. Heat Transfer* **105**(3), 673 (1983).
10C. C.-J. Chen and Y.-H. Yoon, Finite analytic numerical solution axisymmetric Navier–Stokes and Energy Equations, *J. Heat Transfer* **105**(3), 639 (1983).
11C. J. L. S. Chen and J. W. Kearns Jr., Forced convection in non-Newtonian flow past a nonisothermal needle, *J. Heat Transfer* **105**(4), 919 (1983).
12C. A. I. Derzhavina, Thermal waves in a supersonic boundary layer with self-induced pressure, *J. Appl. Math. Mech.* **46**, 503 (1982).
13C. D. G. Drake, Unsteady conjugated heat transfer near a stagnation point, *Appl. Sci. Res.* **40**, 199 (1983).
14C. H. A. Dwyer, R. J. Kee, P. K. Barr and B. R. Sanders, Transient droplet heating at high Peclet number, *J. Fluids Engng* **105**, 83 (1983).
15C. Ye. P. Dyban, A. I. Mazur and A. A. Lesnikovskiy, Comparative efficiency of certain methods of heat transfer augmentation in longitudinal flow over a cylinder, *Heat Transfer—Soviet Res.* **14**(6), 115 (1982).
16C. S. E. Elghobashi and B. E. Launder, Turbulent time scales and the dissipation rate of temperature variance in the thermal mixing layer, *Physics Fluids* **26**(9), 2415 (1983).
17C. L. Fulachier and R. A. Antonia, Spectral relationships between velocity and temperature fluctuations in turbulent shear flows, *Physics Fluids* **26**(8), 2105 (1983).
18C. P. J. Giarratano and W. G. Steward, Transient forced convection heat transfer to helium during a step in heat flux, *J. Heat Transfer* **105**(2), 350 (1983).
19C. R. S. R. Goria, R. Pender and J. Eppich, Heat transfer in micropolar boundary layer flow over a flat plate, *Int. J. Engng Sci.* **21**, 791 (1983).
20C. M. Hiwada, T. Kawamura, I. Mabuchi and M. Kumada, Some characteristics of flow pattern and heat transfer past a circular cylindrical cavity, *Bull. JSME* **26**, 1744 (1983).
21C. B. R. Hollworth, G. Lehmann and J. Rosiczkowski, Arrays of impinging jets with spent fluid removal through vent holes on the target surface, Part 2: local heat transfer, *J. Engng Power* **105**, 393 (1983).
22C. P. Kaiping, Unsteady forced convective heat transfer from a hot film in non-reversing and reversing shear flow, *Int. J. Heat Mass Transfer* **26**, 545 (1983).
23C. H. W. Kim, D. R. Jeng and K. J. DeWitt, Momentum and heat transfer in power-law fluid flow over two-dimensional or axisymmetrical bodies, *Int. J. Heat Mass Transfer* **26**, 245 (1983).
24C. S. Kimura and A. Bejan, The 'heatline' visualization of convective heat transfer, *J. Heat Transfer* **105**(4), 916 (1983).
25C. P. M. Ligrani, R. J. Moffat and W. M. Kays, Artificially thickened turbulent boundary layers for studying heat transfer and skin friction on rough surfaces, *J. Fluids Engng* **105**, 146 (1983).
26C. W. S. Liu, X. X. Du and I. I. Glass, Laminar boundary layers behind detonation waves, *Proc. Royal Soc.* (*A*) **387**(1793), 331 (1983).
27C. R. A. Meric, Boundary integral equation and conjugate gradient methods for optimal boundary heating of solids, *Int. J. Heat Mass Transfer* **26**, 261 (1983).
28C. R. Morel, C. Rey and J. M. Wallace, Evolution of the thermal field in a turbulent wake downstream from an asymmetrically heated plate, *Physics Fluids* **26**(2), 416 (1983).
29C. T. Ota, S. Aiba, T. Tsuruta and M. Kaga, Forced convection heat transfer from an elliptic cylinder of axis ratio 1:2, *Bull. JSME* **26**, 262 (1983).
30C. N. Ramachandran, B. F. Armaly and T. S. Chen, Mixed convection over a horizontal plate, *J. Heat Transfer* **105**(2), 420 (1983).
31C. T. W. Simon and R. J. Moffat, Turbulent boundary layer heat transfer experiments: a separate effects study on a convexly curved wall, *J. Heat Transfer* **105**(4), 835 (1983).
32C. A. L. Snijders, A. M. Koppius and C. Nieuwvelt, An experimental determination of the turbulent Prandtl number is the inner boundary layer for air flow over a flat plate, *Int. J. Heat Mass Transfer* **26**, 425 (1983).
33C. M. Takeuchi, Y. Ishibashi, A. Nagasaka and N. Isshiki, Heat transfer on cylinder covered with close-fitting fabrics, Part II—heat transfer on the stagnation point, *Bull. JSME* **26**, 1373 (1983).
34C. T. Ueda, M. Mizomoto and S. Ikai, Thermal structure of a flat plate turbulent boundary layer diffusion flame, *Bull. JSME* **26**, 399 (1983).
35C. R. L. Zelenka and R. I. Loehrke, Heat transfer from interrupted plates, *J. Heat Transfer* **105**(1), 172 (1983).
36C. V. M. Zhdanov and V. A. Zaznoba, Nonisothermic flow of gas mixture in a channel at intermediate Knudsen numbers, *Appl. Math. Mech.* **45**(6), 801 (1981).

Flow with separated regions
1D. W. Aung, An interferometric investigation of separated forced convection in laminar flow past cavities, *J. Heat Transfer* **105**(3), 505 (1983).
2D. W. Aung, An experimental study of laminar heat transfer downstream of backsteps, *J. Heat Transfer* **105**(4), 823 (1983).
3D. L. W. B. Browne, and R. A. Antonia, Measurements of turbulent Prandtl number in a plane jet, *J. Heat Transfer* **105**(3), 663 (1983).
4D. R. J. Firth and L. Meyer, A comparison of the heat transfer and friction factor performance of four different types of artificially roughened surface, *Int. J. Heat Mass Transfer* **26**, 175 (1983).
5D. T. Igarashi, Fluid flow and heat transfer in the separated region of a circular cylinder with wake control, *Heat Transfer—Jap. Res.* **11**(3), 1 (1982).
6D. S. I. Isatayev, V. V. Polzik, A. M. Tatenow and M. K. Isayev, Heat transfer from a cylinder in crossflow of a bubbly gas–liquid mixture *Heat Transfer—Soviet Res.* **14**(4), 24 (1982).
7D. Y. Kang and K. Suzuki, Numerical study of wall heat transfer in the recirculating flow region of a confined jet, *Heat Transfer—Jap. Res.* **11**(1), 44 (1982).
8D. W. Kühn, Investigation of heat transfer behind an oscillating grid, *Int. J. Heat Mass Transfer* **26**, 1347 (1983).

9D. I. A. Nieva and U. Böhn, Local mass transfer for cross flow through tube banks of square in-line layout at intermediate Reynolds numbers, *Int. J. Heat Mass Transfer* **26,** 1283 (1983).
10D. D. C. T. Pei and G. Hayward, Local heat transfer rates from two adjacent spheres in turbulent flow, *Int. J. Heat Mass Transfer* **26,** 1547 (1983).
11D. E. M. Sparrow, W. Q. Tao and D. D. Radtke, Heat transfer at an array of co-planar slat-like surfaces oriented normal to a forced convection flow, *Int. J. Heat Mass Transfer* **26,** 33 (1983).
12D. E. M. Sparrow and R. F. Prieto, Heat transfer coefficients and patterns of fluid flow for contacting spheres at various angles of attack, *J. Heat Transfer* **105**(1), 48 (1983).
13D. E. M. Sparrow and A. A. Yanezmoreno, Heat transfer in a tube bank in the presence of abrupt upstream cross-sectional enlargement, *Int. J. Heat Mass Transfer* **26,** 1791 (1983).
14D. K. Stephan, S. Lin, M. Durst, F. Huang and D. Seher, An investigation of energy separation in a vortex tube, *Int. J. Heat Mass Transfer* **26,** 341 (1983).
15D. W. F. Tseng and S. P. Lin, Transient heat transfer from a wire in a violently fluctuating environment, *Int. J. Heat Mass Transfer* **26,** 1695 (1983).

Heat transfer in porous media

1DP. A. Bejan, Natural convection heat transfer in a porous layer with internal flow obstructions, *Int. J. Heat Mass Transfer* **26,** 815 (1983).
2DP. A. Bejan, The boundary layer regime in a porous layer with uniform heat flux from the side, *Int. J. Heat Mass Transfer* **26,** 1339 (1983).
3DP. J. C. Benet et P. Jouanna, Non equilibre thermodynami dans les milieuz poreux non satures avec changement de phase, *Int. J. Heat Mass Transfer* **26,** 1585 (1983).
4DP. K. J. Beukema, S. Bruin and J. Schenk, Three-dimensional natural convection in a confined porous medium with internal heat generation, *Int. J. Heat Mass Transfer* **26,** 451 (1983).
5DP. P. A. Blythe, P. G. Simpkins and P. G. Daniels, Thermal convection in a cavity filled with a porous medium: a classification of limiting behaviours, *Int. J. Heat Mass Transfer* **26,** 701 (1983).
6DP. H.-J. Bock, J. Schweinzer and O. Molerus, Fluid dynamics and heat transfer in a fluidized bed with coarse particles, *Chemie-Ingr-Tech.* **55,** 224 (1983).
7DP. V. A. Borodulya, V. L. Ganzha and A. I. Podberezkiy, Heat transfer between a fluidized bed operating under pressure and a vertical wall immersed in it, *Heat Transfer—Soviet Res.* **14**(3), 94 (1982).
8DP. V. A. Borodulya, V. L. Ganzha, A. I. Podberezsky, S. N. Upadhyay and S. C. Saxens, High pressure heat transfer investigations for fluidized beds of large particles and immersed vertical tube bundles, *Int. J. Heat Mass Transfer* **26,** 1577 (1983).
9DP. Yu. A. Buyevich, Flow over immersed bodies and heat and mass transfer in an air-fluidized bed, *Heat Transfer—Soviet Res.* **14**(5), 38 (1982).
10DP. I. D. Chang and P. Cheng, Matched asymptotic expansions for free convection about an impermeable horizontal surface in a porous medium, *Int. J. Heat Mass Transfer* **26,** 163 (1983).
11DP. T.-M. Chiu and E. N. Ziegler, Heat transfer in three-phase fluidized beds, *AIChE J.* **29,** 677 (1983).
12DP. N. Decker and L. R. Glicksman, Heat transfer in large particle fluidized beds, *Int. J. Heat Mass Transfer* **26,** 1307 (1983).
13DP. D. B. Ingham, J. H. Merkin and I. Pop, The collision of free-convection boundary layers on a horizontal cylinder embedded in a porous medium, *Q. J. Mech. Appl. Math.* **36,** 313 (1983).
14DP. K. W. Jackson and W. Z. Black, A unit cell model for predicting the thermal conductivity of a granular medium containing an adhesive binder, *Int. J. Heat Mass Transfer* **26,** 87 (1983).
15DP. A. R. Khan, A. K. A. Juma and J. F. Richardson, Heat transfer from a plane surface to liquids and to liquid–solid fluidised beds, *Chem. Engng Sci.* **38,** 2053 (1983).
16DP. W. B. Krause and A. R. Peters, Heat transfer from single horizontal serrated finned tubes in an air-fluidized, bed of uniformly sized particles, *J. Heat Transfer* **105**(2), 319 (1983).
17DP. V. A. Mayorov and L. L. Vasil'yev, Vaporization in flow through porous metals, *Heat Transfer—Soviet Res.* **14**(2), 92 (1982).
18DP. V. K. Migay, V. N. Shemyakin, V. V. Matsnev and I. V. Zhitomirksaya, Heat transfer on staggered smooth- and finned-tube bundles in fluidized-bed furnaces and in model beds, *Heat Trans.—Soviet Res.* **14**(3), 113 (1982).
19DP. D. Poulikakos and A. Bejan, Natural convection in vertically and horizontally layered porous media heated from the side, *Int. J. Heat Mass Transfer* **26,** 1805 (1983).
20DP. D. Poulikakos and A. Bejan, Unsteady natural convection in a porous layer, *Phys. Fluids* **26**(5), 1183 (1983).
21DP. B. Rubinsky and G. L. Starnes, Experimental comparison of heat transfer data with flow visualization on a flat surface in an air fluidized bed, *J. Heat Transfer* **105**(4), 809 (1983).
22DP. A. E. Saez and B. J. McCoy, Transient analysis of packed-bed thermal storage systems, *Int. J. Heat Mass Transfer* **26,** 49 (1983).
23DP. R. Siegel and A. Snyder, Analysis of coolant entrance boundary shape of porous region to control cooling along exit boundary, *J. Heat Transfer* **105**(3), 513 (1983).
24DP. K. S. Udell, Heat transfer in porous media heated from above with evaporation, condensation, and capillary effects, *J. Heat Transfer* **105**(3), 485 (1983).

Natural convection—internal flow

1F. W. Douglas Baines, A technique for the direct measurement of volume flux of a plume, *J. Fluid Mechanics* **132,** 247 (1983).
2F. W. D. Baines, W. W. Martin and D. M. Smith, Development of stratification in a rectangular tank by horizontal inflow, *J. Fluids Engng.* **105,** 59 (1983).
3F. H. H. Bau and K. E. Torrance, Effects of viscous dissipation and pressure work in free convection loops, *Int. J. Heat Mass Transfer* **26,** 727 (1983).
4F. A. Bejan and R. Anderson, Natural convection at the

interface between a vertical porous layer and an open space, *J. Heat Transfer* **105**(1), 124 (1983).
5F. P. A. Blythe, P. G. Daniels and P. G. Simpkins, Thermal convection in a cavity: the core structure near the horizontal boundaries, *Proc. Royal. Soc.* (*A*) **387**, 91 (1983).
6F. R. D. Boyd, A correlation theory for steady natural convection heat transport in horizontal annuli, *J. Heat Transfer* **105**(1), 139 (1983).
7F. Iu. K. Bratukhin and L. N. Maurin, Stability of thermocapillary convection in a fluid filling a half-space, *J. Appl. Math. Mech.* **46**, 129 (1982).
8F. K. S. Chang, Y. H. Won and C. H. Cho, Patterns of natural convection around a square cylinder placed concentrically in a horizontal circular cylinder, *J. Heat Transfer* **105**(2), 273 (1983).
9F. P. K. B. Chao, H. Ozoe, S. W. Churchill and N. Lior, Laminar natural convection in an inclined rectangular box with the lower surface half-heated and half-insulated, *J. Heat Transfer* **105**(3), 425 (1983).
10F. H. Cordoba, Effects of variable fluid properties and boundary conditions on thermal convection, *Int. Comm. Heat Mass Transfer* **10**(2), 147 (1983).
11F. Stephen J. Cowley and Stephen H. Davis, Viscous thermocapillary convection at high Marangoni number. *J. Fluid Mech.* **135**, 175 (1983).
12F. S. M. ElSheriby, K. G. T. Hollands and G. D. Raithby, Nusselt number distribution in vertical and inclined air layers, *J. Heat Transfer* **105**(2), 406 (1983).
13F. H. W. Emmons, The calculation of a fire in a large building, *J. Heat Transfer* **105**(1), 151 (1983).
14F. M. K. Exeter, N. Hay, J. J. Webster and T. A. Dullforce, Natural convection heat transfer in a confined-flow geometry, *Nucl. Sci. Engng* **83**, 253 (1983).
15F. G. N. Facas and B. Farouk, Transient and steady-state natural convection in a porous medium between two concentric cylinders, *J. Heat Transfer* **105**(3), 660 (1983).
16F. H. Frick, The effect of thermal boundary conditions on the heat transport in vertical channels heated from below, *Int. J. Heat Mass Transfer* **26**, 681 (1983).
17F. H. Frick, F. H. Busse and R. M. Clever, Steady three-dimensional convection at high Prandtl numbers, *J. Fluid Mech.* **127**, 141 (1983).
18F. J. Galan, R. Bruin, M. Fortunato et J. Papin, Fusion et coulee de gaine conduite cylindrique annulaire verticale en presence d'un contre-courant gazeux, *Int. J. Heat Mass Transfer* **26**, 1453 (1983).
19F. H. N. Gari and R. I. Loehrke, A controlled buoyant jet for enhancing stratification in a liquid storage tank, *J. Fluids Engng* **104**, 475 (1982).
20F. R. R. Gay, Gravity-driven flow and heat transfer in a spent nuclear fuel storage pool, *Nucl. Sci. Engng* **83**, 1 (1983).
21F. D. D. Gray, An approximate solution for two dimensional laminar plumes including magnetic effects, *Int. Comm. Heat Mass Transfer* **10**(3), 231 (1983).
22F. M. Haajizadeh and C. L. Tien, Natural convection in a rectangular porous cavity with one permeable endwall, *J. Heat Transfer* **105**(4), 803 (1983).
23F. J. Heestand, C. W. Shipman and J. W. Meader, A predictive model for rollover in stratified LNG tanks, *Amer. Inst. Chem. Engng J.* **29**, 199 (1983).
24F. Judith Y. Holyer, Double-diffusive interleaving due to horizontal gradients, *J. Fluid Mech.* **137**, 347 (1983).
25F. H. Imura, K. Sasaguchi, H. Kozai and S. Mumata, Critical heat flux in a closed two-phase thermo-syphon, *Int. J. Heat Mass Transfer* **26**, 1181 (1983).
26F. M. Kaviany and R. Seban, Analysis of turbulent thermal convection between horizontal plates, *J. Heat Transfer* **105**(4), 789 (1983).
27F. M. Keyhani and F. A. Kulacki, Experiments on transient thermal convection with internal heating—large time results, *J. Heat Transfer* **105**(2), 261 (1983).
28F. M. Keyhani, F. A. Kulacki and R. N. Christensen, Free convection in a vertical annulus with constant heat flux on the inner wall, *J. Heat Transfer* **105**(3), 454 (1983).
29F. G. W. T. Lee, P. Lucas and A. Tyler, Onset of Rayleigh–Benard convection in binary liquid mixtures of ^{3}He in ^{4}He, *J. Fluid Mech.* **135**, 235 (1983).
30F. Yee Lee and Seppo A. Korpela, Multicellular natural convection in a vertical slot, *J. Fluid Mech.* **126**, 91 (1983).
31F. N. N. Lin and A. Bejan, Natural convection in a partially divided enclosure, *Int. J. Heat Mass Transfer* **26**, 1867 (1983).
32F. T. Maxworthy, The dynamics of double-diffusive gravity currents, *J. Fluid Mech.* **128**, 259 (1983).
33F. Trevor J. McDougall, Double-diffusive convection caused by coupled molecular diffusion, *J. Fluid Mech.* **126**, 379 (1983).
34F. Carol L. McTaggart, Convection driven by concentration- and temperature-dependent surface tension, *J. Fluid Mech.* **134**, 301 (1983).
35F. J. C. Mollendorf and K. H. Jahn, Onset of convection in a horizontal layer of cold water, *J. Heat Transfer* **105**(3), 460 (1983).
36F. M. W. Nansteel and R. Greif, Natural convection heat transfer in complex enclosures at large Prandtl numbers, *J. Heat Transfer* **105**(4), 912 (1983).
37F. K. Negishi and T. Sawada, Heat transfer performance of an inclined two-phase closed thermosyphon, *Int. J. Heat Mass Transfer* **26**, 1207 (1983).
38F. T. Hung Nguyen, P. Vasseur, L. Robillard and B. Chandra Shekar, Combined free and forced convection of water between horizontal concentric cylinders, *J. Heat Transfer* **105**(3), 498 (1983).
39F. H. Ozoe, K. Fujii, N. Lior and S. W. Churchill, Long rolls generated by natural convection in an inclined, rectangular enclosure, *Int. J. Heat Mass Transfer* **26**, 1427 (1983).
40F. H. Ozoe, M. Ohmuro, A. Mouri, S. Mishima, H. Sayama and S. W. Churchill, Laser-doppler measurements of the velocity along a heated vertical wall of a rectangular enclosure, *J. Heat Transfer* **105**(4), 782 (1983).
41F. F. Penot et A.-M. Dalbert, Convection naturelle mixte et forcée dans un thermosiphon vertical chauffe à flux constant, *Int. J. Heat Mass Transfer* **26**, 1639 (1983).
42F. C. Perdikis, Free convection flow through a porous medium bounded by two horizontal walls, *Int. Comm. Heat Mass Transfer* **10**(4), 357 (1983).
43F. Dimos Poulikakos and Adrian Bejan, The fluid dynamics of an attic space, *J. Fluid Mech.* **131**, 251 (1983).
44F. D. Poulikakos and A. Bejan, Natural convection

experiments in a triangular enclosure, *J. Heat Transfer* **105**(3), 652 (1983).
45F. D. Poulikakos and A. Bejan, Numerical study of transient high Rayleigh number convection in an attic-shaped porous layer, *J. Heat Transfer* **105**(3), 476 (1983).
46F. R. E. Powe and R. O. Warrington, Jr, Natural convection heat transfer between bodies and their spherical enclosure, *J. Heat Transfer* **105**(3), 440 (1983).
47F. J. Prusa and L. S. Yao, Natural convection heat transfer between eccentric horizontal cylinders, *J. Heat Transfer* **105**(1), 108 (1983).
48F. C. Quon, Convection induced by insulated boundaries in a square, *Phys. Fluids* **26**(3), 632 (1983).
49F. D. C. Reda, Natural convection experiments in a liquid saturated porous medium bounded by vertical coaxial cylinders, *J. Heat Transfer* **105**(4), 795 (1983)
50F. N. Riahi, Nonlinear convection in a porous layer with finite conducting boundaries, *J. Fluid Mech.* **129,** 153 (1983).
51F. Frank M. Richter, Henri-Claude Nataf and Stephen F. Daly, Heat transfer and horizontally averaged temperature of convection with large viscosity variations, *J. Fluid Mech.* **129,** 173 (1983).
52F. N. Rudraiah, T. Masuoka and M. S. Malashetty, Heat transfer by natural convection in a vertical porous layer, *Int. Comm. Heat Mass Transfer* **10**(1), 59 (1983).
53F. K. Sagara, Calculation of turbulent gravity flow, *Heat Transfer—Jap. Res.* **11**(3), 59 (1982).
54F. J. Salan and E. Guyon, Homeotropic nematics heated from above under magnetic fields; convective thresholds and geometry, *J. Fluid Mech.* **126,** 13 (1983).
55F. W. M. M. Schinkel, S. J. M. Linhorst and C. J. Hoogendoorn, The stratification in natural convection in vertical enclosures, *J. Heat Transfer* **105**(2), 267 (1983).
56F. N. Seki, S. Fukusako and A. Yamaguchi, An experimental study of free convective heat transfer in a parallelogrammic enclosure, *J. Heat Transfer* **105**(3), 443 (1983).
57F. A. H. Shaaban and M. N. Ozisik, The effect of nonlinear density stratification on the stability of a vertical water layer in the conduction regime, *J. Heat Transfer* **105**(1), 130 (1983).
58F. E. E. Smith and S. Satija, Release rate model for developing fires, *J. Heat Transfer* **105**(2), 281 (1983).
59F. Marc K. Smith and Stephen H. Davis, Instabilities of dynamic thermocapillary liquid layers. Part 1. Convective instabilities, *J. Fluid Mech.* **132,** 119 (1983).
60F. Marc K. Smith and Stephen H. Davis, Instabilities of dynamic thermocapillary liquid layers. Part 2. Surface-wave instabilities, *J. Fluid Mech.* **132,** 145 (1983).
61F. E. M. Sparrow and M. Charmchi, Natural convection experiments in an enclosure between eccentric or concentric vertical cylinders of different height and diameter, *Int. J. Heat Mass Transfer* **26,** 133 (1983).
62F. E. M. Sparrow, P. Souza Mendes, M. A. Ansari and A. T. Prata, Duct-flow versus external-flow natural convection at a short, wall-attached horizontal cylinder, *Int. J. Heat Mass Transfer* **26,** 881 (1983).
63F. M. Strani, R. Piva and G. Granziani, Thermocapillary convection in a rectangular cavity: asymptotic theory and numerical simulation, *J. Fluid Mech.* **130,** 347 (1983)
64F. G. G. Treshchev, V. A. Sukhov and G. L. Urusov, Experimental study of the onset of self-excited oscillations of flow in a loop with natural circulation of a boiling fluid, *Heat Transfer—Soviet Res.* **14**(4), 35 (1982).
65F. A. B. Tsinober, Y. Yahalom and D. J. Shlien, A point source of heat in a stable salinity gradient, *J. Fluid Mech.* **135,** 199 (1983).
66F. P. Vasseur, L. Robillard and B. Chandra Shekar, Natural convection heat transfer of water within a horizontal cylindrical annulus with density inversion effects, *J. Heat Transfer* **105**(1), 117 (1983).
67F. J. S. Vrentas, C. M. Vrentas, Domain dependence off critical rayleigh numbers, *Int. J. Heat Mass Transfer* **26**(9) (1983).
68F. I. C. Walton, The onset of cellular convection in a shallow two-dimensional container of fluid heated non-uniformly from below, *J. Fluid Mech.* **131,** 455 (1983).
69F. G. S. Wang, F. P. Incropera and R. Viskanta, Mixed convection heat transfer in a horizontal open-channel flow with uniform bottom heat flux, *J. Heat Transfer* **105**(4), 817 (1983).
70F. L. S. Yao, Free and forced convection in the entry region of a heated vertical channel, *Int. J. Heat Mass Transfer* **26,** 65 (1983).

Natural convection—external flows

1FF. N. Afzal, Mixed convection in an axisymmetric buoytant plume, *Int. J. Heat Mass Transfer* **26,** 381 (1983)
2FF. S. Akagi and K. Yoshitani, A study on heat transfer during natural convective heating of a coal oil mixture [COM], *Heat Transfer—Jap. Res.* **11**(4), 1 (1982).
3FF. H. M. Badr, A theoretical study of laminar mixed convection from a horizontal cylinder in a cross stream, *Int. J. Heat Mass Transfer* **26,** 639 (1983).
4FF. J. C. Buell and I. Catton, The effect of wall conduction on the stability of a fluid in a right circular cylinder heated from below, *J. Heat Transfer* **105,** 255 (1983).
5FF. V. P. Carey, A heat transfer correlation for natural convection from a vertical surface in cold water, *J. Heat Transfer* **105,** 658 (1983).
6FF. V. P. Carey, Analysis of transient natural convection flow at high Prandtl number using a matched asymptotic expansion technique, *Int. J. Heat Mass Transfer* **26,** 911 (1983).
7FF. J. L. S. Chen, Natural convection from needles with variable wall heat flux, *J. Heat Transfer* **105,** 403 (1983).
8FF. K. Chen, M. M. Chen and C. W. Sohn, Thermal instability of two-dimensional stagnation-point boundary layers, *J. Fluid Mech.* **132,** 49 (1983).
9FF. T. S. Chen, K. L. Tzuoo and A. Moutsoglou, Vortex instability of horizontal and inclined natural convection flows from simultaneous thermal and mass diffusion, *J. Heat Transfer* **105,** 774 (1983).
10FF. A. M. Clausing, Natural convection correlations for vertical surfaces including influences of variable properties, *J. Heat Transfer* **105**(1), 138 (1983).
11FF. A. O. Demuren, Prediction of the steady surface thermal plane, *Wärme- und Stoffübertragung* **17,** 193 (1983).

12FF. L. M. de Socio, Laminar free convection around horizontal circular cylinders, *Int. J. Heat Mass Transfer* **26,** 1669 (1983).
13FF. R. M. Fand and J. Brucker, A correlation for heat transfer by natural convection from horizontal cylinders that accounts for viscous dissipation, *Int. J. Heat Mass Transfer* **26,** 709 (1983).
14FF. B. Farouk and S. I. Guceri, Natural convection from horizontal cylinders in interacting flow fields, *Int. J. Heat Mass Transfer* **26,** 231 (1983).
15FF. G. Z. Gershuni, E. M. Zhukhovitskii and L. E. Sorokin, On stability of convective flow of a binary mixture with thermal diffusion, *J. Appl. Math. Mech.* **46,** 54 (1982).
16FF. R. J. Goldstein and Kei-Shun Lau, Laminar natural convection from a horizontal plate and the influence of plate-edge extensions, *J. Fluid Mech.* **129,** 55 (1983).
17FF. S. E. Haaland and E. M. Sparrow, Mixed convection plume above a horizontal line source situated in a forced convection approach flow, *Int. J. Heat Mass Transfer* **26,** 433 (1983).
18FF. E. Hasegawa and J. Kojima, Breakup of a thin liquid layer heated from below, *Bull. JSME* **26,** 380 (1983).
19FF. J. M. Higgins and B. Gebhart, The stability of vertical buoyancy-induced flow in cold water, *J. Heat Transfer* **105,** 767 (1983).
20FF. Y. Joshi and B. Gebhart, An experimental study of plane plumes in cold water, *J. Heat Transfer* **105,** 248 (1983).
21FF. A. S. Kabanov and S. N. Netreba, Free convection from a point heat source in a stably stratified medium, *J. Appl. Math. Mech.* **46,** 49 (1982).
22FF. H. Kimoto, A. Kadosuji and T. Hirose, Effect of vibration to the natural convection heat transfer of a horizontal cylinder, *Bull. JSME* **26,** 1154 (1983).
23FF. S. Kimura and A. Bejan, Mechanism for transition in buoyant plume flow, *Int. J. Heat Mass Transfer* **26,** 1515 (1983).
24FF. K. Kishinami and N. Seki, Natural convective heat transfer on an unheated vertical plate attached to an upstream isothermal plate, *J. Heat Transfer* **105,** 759 (1983).
25FF. J. M. Kuhlman and L.-C. Chu, Reynolds number effects on the nearfield mean characteristics of a laboratory model thermal plume, *J. Fluids Engng* **105,** 34 (1983).
26FF. W. M. Lewandowski and P. Kubski, Methodical investigation of free convection from vertical and horizontal plates, *Wärme- und Stoffübertragung* **17,** 147 (1983).
27FF. R. Lochet, D. Lemonnier et Doan-Kim-Son, Correlations enconvection naturelle turbulente. Influence de la pression et de la nature du gaz, *Int. J. Heat Mass Transfer* **26,** 1221 (1983).
28FF. T. J. McDougall, Double-diffusive plumes in unconfined and confined environments, *J. Fluid Mech.* **133,** 321 (1983).
29FF. J. H. Merkin, Free convection boundary layers over humps and indentations, *Q. J. Mech. Appl. Math.* **36,** 71 (1983).
30FF. T. Y. Na and I. Pop, Free convection flow past a vertical flat plate embedded in a saturated porous medium, *Int. J. Engng Sci.* **21,** 517 (1983).
31FF. R. D. Padlog and J. C. Mollendorf, Variable fluid property effects on transport in pure water around the density extremum, *J. Heat Transfer* **105,** 655 (1983).
32FF. I. Pop and T. Y. Na, Free convection flow on a nonisothermal flat plate under nonuniform gravity, *Int. J. Heat Mass Transfer* **26,** 531 (1983).
33FF. A. A. Raptis, Unsteady free convective flow through a porous medium, *Int. J. Engng Sci.* **21,** 345 (1983).
34FF. D. S. Riley and D. G. Drake, Mixed convection in an axisymmetric buoyant plume, *Q. J. Mech. Appl. Math.* **36** (1983).
35FF. C. Rodighiero and L. M. de Socio, Some aspects of natural convection in a corner, *J. Heat Transfer* **105,** 212 (1983).
36FF. M. Sako, T. Chiba, J. M. S. Garza and A. Yanagida, Numerical solution of transient natural convection heat transfer from a horizontal cylinder, *Heat Transfer—Jap. Res.* **11**(4), 27 (1982).
37FF. E. M. Sparrow and M. A. Ansari, A refutation of King's rule for multidimensional external natural convection, *Int. J. Heat Mass Transfer* **26,** 1357 (1983).
38FF. E. M. Sparrow and D. S. Boessneck, Effect of transverse misalignment on natural convection from a pair of parallel, vertically stacked, horizontal cylinders, *J. Heat Transfer* **105,** 241 (1983).
39FF. B. Sunden, Influence of buoyancy forces and thermal conductivity on flow field and heat transfer of circular cylinders at small Reynolds number, *Int. J. Heat Mass Transfer* **26,** 1329 (1983).
40FF. I. Tokura, H. Saito, K. Kishinami and K. Muramoto, An experimental study of free convection heat transfer from a horizontal cylinder in a vertical array set in free space between parallel walls, *J. Heat Transfer* **105**(1), 102 (1983).
41FF. A. B. Tsinober, Y. Yahalom and D. J. Shlien, a point source of heat in a stable salinity gradient, *J. Fluid Mech.* **135,** 199 (1983)
42FF. A. R. Verma, An asymptotic analysis of free convection boundary layer on a horizontal flat plate due to small fluctuations in surface temperature, *Int. J. Engng Sci.* **31,** 35 (1983).
43FF. L. S. Yao, Buoyancy effects on a boundary layer along an infinite cylinder with a step change of surface temperature, *J. Heat Transfer* **105,** 96 (1983).
44FF. L. S. Yao, Natural convection along a vertical wavy surface, *J. Heat Transfer* **105,** 465 (1983).

Convection from rotating surfaces

1G. J. C. Buell and I. Catton, Effect of rotation on the stability of a bounded cylindrical layer of fluid heated from below, *Physics of Fluids* **26**(4), 892 (1983).
2G. C. R. Carrigan and F. H. Busse, An experimental and theoretical investigation of the onset of convection in rotating spherical shells, *J. Fluids Mech.* **126,** 287 (1983).
3G. S. S. Chawla and A. R. Verma, Free convection from a disk rotating in a vertical plane, *J. Fluids Mech.* **126,** 307 (1983).
4G. D. H. Hathaway and R. C. J. Somerville, Three-dimensional simulations of convection in layers with tilted rotation vectors, *J. Fluid Mech.* **126,** 75 (1983).
5G. P. H. Muir, G. Fairweather and M. Vedha-Nayagam, The effect of Prandtl number on heat transfer from an

isothermal rotating disk with blowing at the wall, *Int. Comm. Heat Mass Transfer* **10**(4), 278 (1983).
6G. H. Hamabata and M. Takashima, The effect of rotation on convective instability in a horizontal fluid layer with internal heat generation, *J. Phys. Soc. Japan* **52**(12), 4145 (1983).
7G. G. J. Lucas, J. M. Pfotenhauer and R. J. Donnelly, Stability and heat transfer of rotating cryogens, Pt. 1. Influence of rotation on the onset of convection in liquid ^{4}He, *J. Fluid Mech.* **129,** 251 (1983).
8G. V. U. Sastry and V. Ramohan Rao, Numerical solution of thermal instability of a rotating micropolar fluid layer, *Int. J. Engng Sci.* **21,** 449 (1983).
9G. N. V. Suryanarayana, T. Scofield and R. E. Kleiss, Heat transfer to a fluid in radial, outward flow between two coaxial stationary or corotating disks, *J. Heat Transfer* **105**(3), 519 (1983).
10G. D. Tang and J. L. Hudson, Experiments on a rotating fluid heated from below, *Int. J. Heat Mass Transfer* **26,** 943 (1983).
11G. S. Y. Tsay and C. H. Chou, Laminar convection to rotating disks in non-Newtonian power-law fluids, *Int. Comm. Heat Mass Transfer*, **10,** 377 (1983).

Combined heat and mass transfer
1H. B. T. F. Chung, T. Y. Chang, J. S. Hsiao and C. I. Chang, Heat transfer with ablation in a half-space subjected to time-variant heat fluxes, *J. Heat Transfer* **105**(1), 200 (1983).
2H. R. J. Goldstein and A. Behbahani, Local heat transfer to staggered arrays of impinging circular air jets, *J. Engng for Power* **105,** 354 (1983).
3H. R. J. Goldstein, Y. Kornblum and E. R. G. Eckert, Film cooling effectiveness on a turbine blade, *Israel J. Technol.* **20,** 193 (1982). Proc. XXIV ISR Annu. Conf. Aviation and Astronautics, 1982.
4H. G. Grossman, Simultaneous heat and mass transfer in film absorption under laminar flow, *Int. J. Heat Mass Transfer* **26,** 357 (1983).
5H. G. K. Khoe and J. Van Brakel, Drying characteristics of a draft tube spouted bed, *Can. J. Chem. Engng* **61,** 411 (1983).
6H. S. Kikkawa, T. Onkura and Y. Takeuchi, Three-dimensional film cooling on a flat plate with tangential multi-jets, *Heat Transfer—Jap. Res.* **11**(4), 45 (1982).
7H. J. Kucharski and A. Kmiec, Hydrodynamics, heat and mass transfer during coating of tablets in a spouted bed, *Can. J. Chem. Engng* **61,** 435 (1983).
8H. H. Lang, Heat and mass exchange of a droplet in a polyatomic gas, *Phys. Fluids* **26**(8), 2109 (1983).
9H. N. Laws and J. L. Parry, Mathematical modeling of heat and mass transfer in agricultural grain drying, *Proc. R. Soc.* (*A*) **385**(1788), 169 (1983).
10H. P. J. Loftus and T. V. Jones, The effect of temperature ratios on the film cooling process (82-GT-305), *J. Engng Power* **105,** 615 (1983).
11H. A. I. Mazur, L. A. Rumyantseva, I. G. Golovney and O. N. Kastelin, Local heat transfer on a surface in longitudinal air flow, with air injection from a row of circular impinging jets, *Heat Transfer—Soviet Res.* **14**(6), 125 (1982).
12H. V. Y. Nakoryakov, A. P. Burdukov, N. S. Bufetov, N. I. Grigor'yeva and A. R. Dorokhov, Coefficients of heat and mass transfer in falling wavy liquid films, *Heat Transfer—Soviet Res.* **14**(3), 6 (1982).
13H. G. Reinheimer and H. Beer, Injection of nonisothermal plane jets tangential and normal to a turbulent crossflow for the purpose of film cooling (In German), *Wärme- und Stoffübertragung* **17,** 133 (1983).

Boiling
1J. R. Bartolini, G. Guglieimi and E. Nannei, Experimental study of nucleate boiling of water in vertical upflow and downflow, *Int. J. Multiphase Flow* **9,** 161 (1983).
2J. M. K. Bezrodnyy, Boiling crisis in internally-baffled thermosyphons, *Heat Transfer—Soviet Res.* **14**(3), 24 (1982).
3J. A. M. Bhat, R. Prakash and J. S. Saini, On the mechanism of macrolayer formation in nucleate pool boiling at high heat flux, *Int. J. Heat Mass Transfer* **26,** 735 (1983).
4J. A. M. Bhat, R. Prakash and J. S. Saini, Heat transfer in nucleate pool boiling at high heat flux, *Int. J. Heat Mass Transfer* **26,** 833 (1983).
5J. Z. Bilicki, Latent heat transport in forced boiling flow, *Int. J. Heat Mass Transfer* **26,** 559 (1983).
6J. M. K. Bologa, I. A. Kozhukhar', A. B. Didkovskiy, S. M. Klimov and O. I. Mardarskiy, Heat transfer in vaporization and condensation in an electric field, *Heat Transfer—Soviet Res.* **14**(4), 131 (1982).
7J. M. K. Bologa, G. F. Smirnov and S. M. Klimov, Heat transfer to dielectrics boiling in horizontal slots placed in electric fields, *Heat Transfer—Soviet Res.* **14**(2), 104 (1982).
8J. Y. S. Cha, On bubble cavitation and dissolution, *Int. J. Heat Mass Transfer* **26,** 1173 (1983).
9J. H. R. Chang, R. C. Reid and J. A. Fay, Boiling and spreading of liquid nitrogen and liquid methane on water, *Int. Comm. Heat Mass Transfer* **10**(3), 253 (1983).
10J. Y. Chang and S.-C. Yao, Critical heat flux of narrow vertical annuli with closed bottoms, *J. Heat Transfer* **105,** 192 (1983).
11J. L. C. Chow and J. N. Chung, Evaporation of water into a laminar stream of air and superheated steam, *Int. J. Heat Mass Transfer* **26,** 373 (1983).
12J. M. G. Cooper, The 'mirage' in boiling, *Int. J. Heat Mass Transfer* **26,** 1088 (1983).
13J. M. G. Cooper, K. Mori and C. R. Stone, Behaviour of vapour bubbles growing at a wall with forced flow, *Int. J. Heat Mass Transfer* **26,** 1489 (1983).
14J. M. L. Corradini, Modeling film boiling destabilization due to a pressure shock arrival, *Nucl. Sci. Engng* **84,** 196 (1983).
15J. A. B. Datta and A. K. Majumdar, A calculation procedure for two phase flow distribution in manifolds with and without heat transfer, *Int. J. Heat Mass Transfer* **26,** 1321 (1983).
16J. H. Eickhoff, Prediction of the film vaporization of hydrocarbons in a stream of hot air, *Wärme- und Stoffübertragung* **17,** 233 (1983).
17J. N. E. Fagerholm, A. R. Ghazanfari and K. Kivioja, Boiling heat transfer in a vertical tube with Freon 114, *Wärme- und Stoffübertragung* **17,** 221 (1983).
18J. I. M. Fedotkin, A. V. Gukalov and S. V. Romanovskiy, Effect of gas content and temperature of fluid on the

acoustics of boiling and bubbling, *Heat Transfer—Soviet Res.* **14**(6), 98 (1982).
19J. J. P. Gallagher and R. H. S. Winterton, Effect of pressure on boiling nucleation, *J. Phys. D: Appl. Phys.* **16**, L57 (1983).
20J. S. K. Griffiths and F. A. Morrison, Jr., The transport from a drop in an alternating electric field, *Int. J. Heat Mass Transfer* **26**, 717 (1983).
21J. H. Gürgenci, T. N. Veziróglu and S. Kakaç, Simplified nonlinear descriptions of two-phase flow instabilities in vertical boiling channel, *Int. J. Heat Mass Transfer* **26**, 671 (1983).
22J. E. Hahne, Pool boiling—and the effect of pool geometry, *Wärme- und Stoffübertragung* **17**, 155 (1983).
23J. E. Hahne and J. Müller, Boiling on a finned tube and a finned tube bundle, *Int. J. Heat Mass Transfer* **26**, 849 (1983).
24J. Y. Haramura and Y. Katto, A new hydrodynamic model of critical heat flux, applicable widely to both pool and forced convection boiling on submerged bodies in saturated liquids, *Int. J. Heat Mass Transfer* **26**, 389 (1983).
25J. D. Havazelet and A. Birnboim, Evaporation of metals by a high-energy-density source, *J. Phys. D: Appl. Phys.* **16**(10), 1917 (1983).
26J. G. F. Hewitt, Two-phase flow and its applications: past, present and future, *Heat Transfer Engng* **4**(1), 67 (1983).
27J. M. N. Hutcherson, R. E. Henry and D. E. Wollersheim, Two-phase vessel blowdown of an initially saturated liquid—Part 1: Experimental, *J. Heat Transfer* **105**, 687 (1983).
28J. M. N. Hutcherson, R. E. Henry and D. E. Wollersheim, Two-phase vessel blowdown of an initially saturated liquid—Part 2: Analytical, *J. Heat Transfer* **105**, 694 (1983).
29J. M. Ishii and H. K. Fauske, Boiling and dryout behavior in a liquid–metal fast breeder subassembly bundle under low heat flux and low flow conditions, *Nucl. Sci. Engng* **84**, 131 (1983).
30J. P. K. Jain and R. P. Roy, Stochastic characteristics of vapor fraction and wall pressure fluctuations in boiling flows, *Int. J. Multiphase Flow* **9**, 463 (1983).
31J. B. S. Johnston, A. Sharon, Y. Kozawa and S. G. Bankoff, Boiling heat transfer in a narrow eccentric annulus, Part I: dryout, *J. Engng Power* **105**, 742 (1983).
32J. B. S. Johnston, A. Sharon and S. G. Bankoff, Boiling heat transfer in a narrow eccentric annulus, Part II: heat transfer, *J. Engng Power* **105**, 748 (1983).
33J. G. Joly, J. P. Kernevez and M. Llory, Thermal instability in pool boiling on wires at constant pressure, *SIAM J. Appl. Math.* **43**, 1294 (1983).
34J. M. K. Kasyanenko, V. S. Fokin, V. I. Konvisar, Yu. B. Danilov and V. I. Kibitkin, Heat transfer in concentrators with external boiling zone and natural circulation, *Int. Chem. Engng* **23**, 766 (1983).
35J. I. Kataoka, A. Serizawa and A. Sakurai, Transient boiling heat transfer under forced convection, *Int. J. Heat Mass Transfer* **26**, 583 (1983).
36J. Y. Katto and Y. Haramura, Critical heat flux on a uniformly heated horizontal cylinder in an upward cross flow of saturated liquid, *Int. J. Heat Mass Transfer* **26**, 1199 (1983).
37J. H. Klein and B. Feuerbacher, Marangoni convection induced bubble motion, *Wärme- und Stoffübertragung* **17**, 115 (1983).
38J. G. Kocamustafaogullari and M. Ishii, Interfacial area and nucleation site density in boiling systems, *Int. J. Heat Mass Transfer* **26**, 1377 (1983).
39J. A. A. Kochurkov, L. S. Aksel'rod and N. I. Goncharuk, Certain aspects of desorption in the boiling of liquids in a centrifugal field, *Heat Transfer—Soviet Res.* **14**(4), 28 (1982).
40J. L. D. Koffman and M. S. Plesset, Experimental observations of the microlayer in a vapor bubble growth on a heated solid, *J. Heat Transfer* **105**, 625 (1983).
41J. C. K. B. Lee, Turbulent heat transfer on the surface of a steam-chugging bubble, *Int. Comm. Heat Mass Transfer* **10**(1), 25 (1983).
42J. S. C. Lee and S. G. Bankoff, Stability of steam–water countercurrent flow in an inclined channel: flooding, *J. Heat Transfer* **105**, 713 (1983).
43J. A. Mentes, O. T. Yildirum, H. Gurgenci, S. Kakaç and T. N. Veziroḡlu, Effect of heat transfer augmentation on two-phase flow instabilities in a vertical boiling channel, *Wärme- und Stoffübertragung* **17**, 161 (1983).
44J. J. Mitrović, Das Abreissen von Dampfblasen an festen Heizflächen, *Int. J. Heat Mass Transfer* **26**, 955 (1983).
45J. H. M. Müller, C. Hechler, D. Steiner and E. U. Schlünder, Fouling-phenomena during the evaporation of argon within a horizontal tube, *Wärme- und Stoffübertragung* **17**, 47 (1983).
46J. Y. Onishi and Y. Sone, Kinetic theory of evaporation and condensation for a cylindrical condensed phase, *Phys. Fluids* **26**(3), 659 (1983).
47J. A. G. Ostrogorsky, R. R. Gay and R. T. Lahey, Jr., The prediction of pressure drop and CCFL breakdown in countercurrent two-phase flow, *J. Heat Transfer* **105**, 700 (1983).
48J. D. D. Paul and S. I. Abdel-Khalik, A statistical analysis of saturated nucleate boiling along a heated wire, *Int. J. Heat Mass Transfer* **26**, 509 (1983).
49J. W. C. Peterson and M. G. Zaalouk, Boiling dynamics theory and experiments, *Int. Comm. Heat Mass Transfer* **10**(5), 429 (1983).
50J. V. V. Raj, Experimental investigations on the rewetting of hot horizontal annular channels, *Int. Comm. Heat Mass Transfer* **10**(4), 299 (1983).
51J. M. Renksizbulut and M. C. Yuen, Experimental study of droplet evaporation in a high-temperature air stream, *J. Heat Transfer* **105**, 384 (1983).
52J. M. Renksizbulut and M. C. Yuen, Numerical study of droplet evaporation in a high-temperature stream, *J. Heat Transfer* **105**, 389 (1983).
53J. N. M. Rouai and S. I. Abdel-Khalik, Pool boiling of drag-reducing polymer solutions, *Appl. Sci. Res.* **40**, 209 (1983).
54J. E. U. Schlünder, Heat transfer in nucleate boiling of mixtures, *Int. Chem. Engng* **23**, 589 (1983).
55J. A. Serizawa, Theoretical prediction of maximum heat flux in power transients, *Int. J. Heat Mass Transfer* **26**, 921 (1983).
56J. M. M. Shah, Generalized prediction of heat transfer during subcooled boiling in annuli, *Heat Transfer Engng* **4**(1), 24 (1983).
57J. V. K. Shcherbakov and V. Yu. Kravets, Temperature

and heat transfer in pool boiling of water on microsurfaces, *Heat Transfer—Soviet Res.* **14**(3), 43 (1982).

58J. T. Shigechi, T. Ito and K. Nishikawa, Forced convection film boiling heat transfer from a horizontal cylinder to liquid cross-flowing upward (2nd report, subcooled liquid), *Bull. JSME* **26,** 554 (1983).

59J. L. C. Singal, C. P. Sharma and H. K. Varma, Pressure drop during forced convection boiling of binary refrigerant mixtures, *Int. J. Multiphase Flow* **9,** 309 (1983).

60J. V. N. Slesarenko, A. Ye. Rudakova and G. A. Zakharov, Effect of operating conditions and tube bundle geometry on boiling heat transfer, *Heat Transfer—Soviet Res.* **14**(2), 119 (1982).

61J. E. M. Sparrow, G. K. Kratz and M. J. Schuerger, Evaporation of water from a horizontal surface by natural convection, *J. Heat Transfer* **105,** 469 (1983).

62J. M. A. Styrikovich, A. I. Leont'yev, I. L. Mostinskiy, V. S. Polonskiy and I. M. Chernina, Experimental study of flow and boiling crisis in horizontal steam generating pipes with wire gauze-covered inner walls, *Heat Transfer—Soviet Res.* **14**(2), 86 (1982).

63J. M. Sultan and R. L. Judd, Interaction of the nucleation phenomena at adjacent sites in nucleate boiling, *J. Heat Transfer* **105,** 3 (1983).

64J. K. Suzuki, Y. Hagiwara and T. Sato, Heat transfer and flow characteristics of two-phase two-component annular flow, *Int. J. Heat Mass Transfer* **26,** 597 (1983).

65J. J. R. Thome, Prediction of binary mixture boiling heat transfer coefficients using only phase equilibrium data, *Int. J. Heat Mass Transfer* **26,** 965 (1983).

66J. V. I. Tolubinskiy, V. A. Antonenko and Yu. N. Ostrovskiy, Heat transfer during vaporization in thin films, *Heat Transfer—Soviet Res.* **14**(6), 109 (1982).

67J. T. Ueda, M. Inoue, Y. Iwata and Y. Sogawa, Rewetting of a hot surface by a falling liquid film, *Int. J. Heat Mass Transfer* **26,** 401 (1983).

68J. T. Ueda, S. Tsunenari and M. Koyanagi, An investigation of critical heat flux and surface rewet in flow boiling systems, *Int. J. Heat Mass Transfer* **26,** 1189 (1983).

69J. H. C. Ünal and M. L. G. van Gasselt, Post-dryout heat transfer in steam generator tubes at high pressures, *Int. J. Heat Mass Transfer* **26,** 459 (1983).

70J. J. Wiesman and B. S. Pei, Prediction of critical heat flux in flow boiling at low qualities, *Int. J. Heat Mass Transfer* **26,** 1463 (1983).

71J. D. Woschitz and F. Widmer, Local mass transfer (heat transfer) between walls and fluid in a thin-film evaporator, *Chem.-Ing.-Tech.* **55,** 230 (1983).

72J. Y. M. Yang and J. R. Maa, Pool boiling of dilute surfactant solutions, *J. Heat Transfer* **105,** 190 (1983).

73J. S.-C. Yao and Y. Chang, Pool boiling heat transfer in a confined space, *Int. J. Heat Mass Transfer* **26,** 841 (1983).

74J. S.-C. Yao and A. Salehpour, An investigation of transient boiling heat transfer with conjugate nature, *Int. J. Heat Mass Transfer* **26,** 901 (1983).

75J. T. Ytrehus, Asymmetries in evaporation and condensation Knudsen layer problems, *Phys. Fluids* **26**(4), 939 (1983).

76J. N. Zhang and W.-J. Yang, Evaporative convection in minute drops on a plate with temperature gradient, *Int. J. Heat Mass Transfer* **26,** 1479 (1983).

77J. N. Zhang and W.-J. Yang, Microstructure of flow inside minute drops evaporating on a surface, *J. Heat Transfer* **105,** 908 (1983).

78J. S. A. Zhukov and V. V. Barelko, Nonuniform steady states of the boiling process in the transition region between the nucleate and film regimes, *Int. J. Heat Mass Transfer* **26,** 1121 (1983).

79J. S. A. Zhukov, L. F. Bokova and V. V. Barelko, Certain aspects of autowave transitions from nucleate to film boiling regimes with a cylindrical heat generating element inclined from a horizontal position, *Int. J. Heat Mass Transfer* **26,** 269 (1983).

80J. L. V. Zysin, L. A. Fel'dberg, A. L. Dobkes and A. N. Svetlakov, Experimental and analytic study of the shapes of vapor bubbles in subcooled boiling, *Heat Transfer—Soviet Res.* **14**(3), 64 (1982).

Condensation

1JJ. Th. Adamek, Film condensation on tightly ribbed tubes (In German), *Chem.-Ing. Tech.* **55,** 736 (1983).

2JJ. S. Banerjee, J. S. Chang, R. Girard, and V. S. Krishman, Reflux condensation and transition to natural circulation in a vertical U-Tube, *J. Heat Transfer* **105**(4), 719 (1983).

3JJ. P. V. Deo and D. R. Webb, An experimental investigation of the effect of waves of vapour side heat and mass transfer in filmwise condensation inside a vertical tube, *Int. J. Multiphase Flow* **9,** 37 (1983).

4JJ. E. I. Guyko, L. S. Malkov and I. V. Pershin, Investigation of condensation of water vapor and cyclical defrosting of panels in freeze dryers, *Heat Transfer—Soviet Res.* **14**(2), 75 (1982).

5JJ. H. J. Kim and S. G. Bankoff, Local heat transfer coefficients for condensation in stratified countercurrent steam–water flows, *J. Heat Transfer* **105**(4), 706 (1983).

6JJ. H. Merte, Jr. and C. Yamali, Profile and departure size of condensation drops on vertical surfaces, *Wärme- und Stoffübertragung* **17,** 171 (1983).

7JJ. Z. L. Miropol'skiy, R. I. Shneyerova and V. V. Treputnev, Heat transfer coefficients and pressure drop in the condensation of steam along finned surfaces, *Heat Transfer—Soviet Res.* **14**(3), 12 (1982).

8JJ. H. Shida, M. Kuragasaki and T. Adachi, On the numerical analysis method of flow and heat transfer in condensers, *Bull. JSME* **26,** 785 (1983).

9JJ. H. Sofrata, Experimental study of filmwise condensation considering wave initiation, *Wärme- und Stoffübertragung* **17,** 251 (1983).

10JJ. W. Studzinski, R. A. Zahoransky and S. L. K. Wittig, Analysis of homogeneous unary and binary condensation processes (In German), *Wärme- und Stoffübertragung* **17,** 241 (1983).

11JJ. S. A. Stylianou and J. W. Rose, Drop-to-filmwise condensation transition: heat transfer measurements for ethanediol, *Int. J. Heat Mass Transfer* **26,** 747 (1983).

12JJ. I. Tanasawa and Y. Utaka, Measurement of condensation curves for dropwise condensation of steam at atmospheric pressure, *J. Heat Transfer* **105**(3),

633 (1983).
13JJ. M. Toner, A. Kilic and K. Onat, Comparison and rectangular and triangular fins when condensation occurs, *Wärme- und Stoffübertragung* **17,** 65 (1983).
14JJ. D. R. Webb and R. G. Sardesai, Reply to comments on verification of multi-component mass transfer models for condensation inside a vertical tube, *Int. J. Multiphase Flow* **9,** 455 (1983).

Freezing and wetting

1JM. A. Angelie and J. L. Pelissier, Sodium melting curve calculation. I. Variational approach limitations, *Physica A*, **121A** (1 & 2), 207 (1983).
2JM. K. Aoki, K. Katayama and Y. Hayashi, A study of frost formation (the process of frost formation involving the phenomena of water permeation and freezing), *Bull. JSME* **26,** 87 (1983).
3JM. M. A. Boles and M. N. Özisik, Exact solution for freezing in cylindrically symmetric, porous, moist media, *J. Heat Transfer* **105**(2), 410 (1983).
4JM. A. Bryczkowski, A continuous method of zone melting, *Chem. Engng Sci.* **38,** 1343 (1983).
5JM. I. Catton, W. A. Brinsfield and S. M. Ghiaasiaan, Heat transfer from a heated pool to a melting miscible substrate, *J. Heat Transfer* **105**(3), 447 (1983).
6JM. S. H. Chan, D. H. Cho and G. Kocamustafaogullari, Melting and solidification with internal radiative transfer—a generalized phase change model, *Int. J. Heat Mass Transfer* **26,** 621 (1983).
7JM. M. Chipot and J.-F. Rodriques, On the steady-state continuous casting Stefan problem with nonlinear cooling, *Q. J. Appl. Math.* **40,** 476 (1983).
8JM. S. H. Emerman and D. L. Turcotte, Stokes's problem with melting, *Int. J. Heat Mass Transfer* **26,** 1625 (1983).
9JM. R. Farhadieh, Effect of the fluid media viscosities on the downward heat transfer in a miscible, melting system, *J. Heat Transfer* **105**(2), 398 (1983).
10JM. G. G. Garguichevich, A note on the heat balance integral method applied to the resolution of a one-phase Stefan problem with an increasing prescribed flux on the fixed face, *Int. Comm. Heat Mass Transfer* **10**(4), 349 (1983).
11JM. C. Gau and R. Viskanta, Flow visualization during solid-liquid phase change heat transfer. I. Freezing in a rectangular cavity, *Int. Comm. Heat Mass Transfer* **10**(3), 173 (1983).
12JM. C. Gau, R. Viskanta and C. J. Ho, Flow visualization during solid–liquid phase change heat transfer. II. Melting in a rectangular cavity, *Int. Comm. Heat Mass Transfer* **10**(3), 183 (1983).
13JM. J. M. Hill and A. Kucera, Freezing a saturated liquid inside a sphere, *Int. J. Heat Mass Transfer* **26,** 1631 (1983).
14JM. R. N. Hills, D. E. Loper and P. H. Roberts, A thermodynamically consistent model of a mushy zone, *Q. J. Mech. Appl. Math.* **36,** 505 (1983).
15JM. C. J. Ho and R. Viskanta, Experimental study of solidification heat transfer in an open rectangular cavity, *J. Heat Transfer* **105,** (3), 671 (1983).
16JM. P. J. Huffman and M. A. Dietenberger, Prediction of ice/frost growth on insulated cryogenic tanks, *J. Spacecraft Rockets* **20,** 401 (1983).
17JM. D. A. Huse, A. M. Szpilka and M. E. Fisher, Melting and wetting transitions in the three-state chiral clock model, *Physica A* **121A**(3), 363 (1983).
18JM. Ch. Körber, M. W. Scheiwe and K. Wollhöver, Solute polarization during planar freezing of aqueous salt solutions, *Int. J. Heat Mass Transfer* **26,** 1241 (1983).
19JM. V. J. Lunardini, Approximate phase change solutions for insulated buried cylinders, *J. Heat Transfer* **105**(1), 25 (1983).
20JM. A. S. Menon, M. E. Weber and A. S. Mujumdar, The dynamics of energy storage for paraffin wax in cylindrical containers, *Can. J. Chem. Engng* **61,** 647 (1983).
21JM. J. L. Pelissier, Sodium melting curve calculation. II. A successful perturbation approach, *Physica A* **121A**(1 & 2), 217 (1983).
22JM. P. R. Rao and V. M. K. Sastri, Freezing over a cylinder with finite heat transfer at the wall, *Wärme- und Stoffübertragung* **17,** 181 (1983).
23JM. J. F. Raymond and B. Rubinsky, A numerical study of the thawing process of a frozen coal particle, *J. Heat Transfer* **105**(1), 197 (1983).
24JM. H. Rieger, U. Projahn, M. Bareiss and H. Beer, Heat transfer during melting inside a horizontal tube, *J. Heat Transfer* **105**(2), 226 (1983).
25JM. A. Saito, F. Kuriyama and R. Shimomura, Experimental research on the freezing process with supercooling, *Heat Transfer—Jap. Res.* **11**(2), 70 (1982).
26JM. B. Sammakia and B. Gebhart, Transport near a vertical ice surface melting in water of various salinity levels, *Int. J. Heat Mass Transfer* **26,** 1439 (1983).
27JM. R. Siegel, Cauchy method for solidification interface shape during continuous casting. *J. Heat Transfer* **105**(3), 667 (1983).
28JM. E. M. Sparrow and J. A. Broadbent, Freezing in a vertical tube, *J. Heat Transfer* **105**(2), 217 (1983).
29JM. H. P. Steinhagen and G. E. Myers, Numerical solutions to axisymmetric thawing and heating of frozen logs, *J. Heat Transfer* **105**(1), 195 (1983).
30JM. M. Sugawara, N. Seki and K. Kimoto, Freezing limit of water in a closed circular tube, *Wärme- und Stoffübertragung* **17,** 187 (1983).
31JM. M. Sugawara, N. Seki, S. Kukusako and T. Ota, Snow melting with radiative heating, *Wärme- und Stoffübertragung* **17,** 31 (1983).
32JM. K. Taghavi-Tafreshi and V. K. Dhir, Shape change of an initially vertical wall undergoing condensation-driven melting, *J. Heat Transfer* **105**(2), 235 (1983).
33JM. L. N. Tao, The Stefan problem with an imperfect thermal contact at the interface, *J. Appl. Mech.* **49,** 715 (1982).
34JM. L. N. Tao, The analyticity and general solution of the Cauchy–Stefan problem, *Q. J. Mech. Appl. Math.* **36,** 487 (1983).
35JM. I. Tokura, H. Saito and K. Kishimani, Study on properties and growth rate of frost layers on cold surfaces, *J. Heat Transfer* 105(4), 895 (1983).
36JM. R. Viskanta and C. Gau, Inward solidification of a superheated liquid in a cooled horizontal tube, *Wärme- und Stoffübertragung* **17,** 39 (1983).

37JM. V. R. Voller, Interpretation of the enthalpy in a discretised multidimensional region undergoing a melting/freezing phase change, *Int. Comm. Heat Mass Transfer* **10**(4), 323 (1983).

Radiation

1K. A. Al Abed and J. F. Sacadura, A Monte Carlo-finite difference method for coupled radiation–conduction heat transfer in semitransparent media, *J. Heat Transfer* **105**(4), 931 (1983).
2K. G. C. Alessandretti and P. Violino, Thermometry by CARS in an automobile engine, *J. Phys. D: Appl. Phys.* **16**(9), 1583 (1983).
3K. Yu. N. Antipov, Absolute measurement of spatially inhomogeneous radiation, *Heat Transfer—Soviet Res.* **14**(6), 76 (1982).
4K. P. W. Baker, Atmospheric water vapor differential absorption measurements on vertical paths with a CO_2 lidar, *Appl. Optics* **22**(15), 2257 (1983).
5K. P. Berdahl, M. Martin and F. Sakkal, Thermal performance of radiative cooling panels, *Int. J. Heat Mass Transfer* **26,** 871 (1983).
6K. T. L. Bergman, W. G. Houf and F. P. Incropera, Effect of single scatter phase function distribution on radiative transfer in absorbing–scattering liquids, *Int. J. Heat Mass Transfer* **26**, 101 (1983).
7K. V. A. Borodulya and V. I. Kovensky, Radiative heat transfer between a fluidized bed and a surface, *Int. J. Heat Mass Transfer* **26**, 277 (1983).
8K. R. Braun, S. Fischer and A. Schaber, Elimination of the radiant component of measured liquid thermal conductivities, *Wärme- und Stoffübertragung* **17,** 121 (1983).
9K. C. W. Bruce and N. M. Richardson, Propagation at 10 μm through smoke produced by atmospheric combustion of diesel fuel, *Appl. Optics* **22**(7), 1051 (1983).
10K. E. Burattini, G. Cappuccio, M. Grandolfo, P. Vecchia and Sh. M. Efendiev, Near-infrared refractive index of bismuth germanium oxide ($Bi_{12}GeO_{20}$), *J. Opt. Soc. Am.* **73**(4), 495 (1983).
11K. S. L. Chang and K. T. Rhee, A useful integral function and its application in thermal radiation calculations, *Int. Comm. Heat Mass Transfer* **10**(4), 329 (1983).
12K. L. C. Chang, K. T. Yang and J. R. Lloyd, Radiation–natural convection interactions in two-dimensional enclosures, *J. Heat Transfer* **105**(1), 89 (1983).
13K. V. Ya. Cherepanov, Modulation method of relative measurement of total emissivity, *Heat Transfer—Soviet Res.* **14**(6), 23 (1982).
14K. T. B. Clements and M. N. Özisik, Effects of stepwise variation of albedo on reflectivity and transmissivity of an isotropically scattering slab, *Int. J. Heat Mass Transfer* **26,** 1419 (1983).
15K. V. Emmerich, Experimental studies on the radiant heat of soot-containing flames and flame gases, *Chem.-Ing.-Tech.* **55,** 490 (1983).
16K. J. George and K. S. Joseph, Effect of heating on the electrical and optical properties of tin disulphide thin films, *J. Phys. D: Appl. Phys.* **16**(1), 33 (1983).
17K. R. P. Gupta, T. F. Wall and J. S. Truelove, Radiative scatter by fly ash in pulverized-coal-fired furnaces: application of the Monte Carlo method to anisotropic scatter, *Int. J. Heat Mass Transfer* **26,** 1649 (1983).
18K. M. Hasantani, N. Arai, Y. Band and H. Nakamura, Collection of thermal radiation by a semitransparent fluid layer flowing in an open channel, *Heat Transfer—Jap. Res.* **11**(3), 17 (1982).
19K. M. A. Hassad, A comparison on the stability of horizontal fluid layers with radiative heat generation for two and three porous plates, *Int. Comm. Heat Mass Transfer* **10**(2), 163 (1983).
20K. M. L. Hoke and J. H. Shaw, Parameters of CO_2 bands near 3.6 m, *Appl. Optics* **22**(1), 328 (1983).
21K. Yu. N. Kiselev, I. V. Nemchinov, K. L. Samonin, B. D. Khristoforov and V. V. Shuvalov, Radiation emitted on impact of a high-velocity jet generated by an explosive-driven compressor, *Heat Transfer—Soviet Res.* **14**(5), 88 (1982).
22K. T. Kunitomo, E. Ozaki and K. Kawahara, The effect of dimerization on infrared absorption in the 6.3 μm band and the total emissivity of water vapor, *Heat Transfer—Jap. Res.* **11**(3), 90 (1982).
23K. M. L. Lang and W. L. Wolfe, Optical constants of fused silica and sapphire from 0.3 μm to 25 μm, *Appl. Optics* **22**(9), 1267 (1983).
24K. H. Lee and R. O. Buckius, Reducing scattering to nonscattering problems in radiation heat transfer, *Int. J. Heat Mass Transfer* **26,** 1055 (1983).
25K. I. Lubezky, E. Ceren, Z. Taubenfeld and H. Zipin, Efficient and durable AR coatings for GE in the 8–11.5-μm band using synthesized refractive indices by evaporation of homogeneous mixtures, *Appl. Optics* **22**(12), 1828 (1983).
26K. R. Mihail and Gh. Maria, The emitter–receptor geometrical configuration influence on radiative heat transfer, *Int. J. Heat Mass Transfer* **26,** 1783 (1983).
27K. I. R. Mikk and I. A. Klevtsov, A modified gardon heat flux probe for separate measurement of the components of radiative plus-convection flux, *Heat Transfer—Soviet Res.* **14**(6), 29 (1982).
28K. M. F. Modest, Evaluation of spectrally-integrated radiative fluxes of molecular gases in multi-dimensional media, *Int. J. Heat Mass Transfer* **26,** 1533 (1983).
29K. C. L. Nagendra and G. K. M. Thutupalli, Determination of optical properties of absorbing materials: a generalized scheme, *Appl. Optics* **22**(4), 587 (1983).
30K. Y. S. H. Najjar and E. M. Goodger, Radiation and smoke from the gas turbine combustor using heavy fuels, *J. Heat Transfer* **105**(1), 82 (1983).
31K. H. F. Nelson, Nonequilibrium radiative heating during outer planet atmospheric entry, *J. Spacecraft Rockets* **20,** 407 (1983).
32K. M. A. Ordal, L. L. Long, R. J. Bell, S. E. Bell, R. W. Alexander, Jr. and C. A. Ward, Optical properties of the metals Al, Co, Cu, Au, Fe, Pb, Ni, Pd, Pt, Ag, Ti, and W in the infrared and far infrared, *Appl. Optics* **22**(7), 1099 (1983).
33K. A. C. Ratzell III and J. R. Howell, Two-dimensional radiation in absorbing–emitting media using the *P–N* approximation, *J. Heat Transfer* **105**(2), 333 (1983).
34K. M. M. Razzaque, D. E. Klein and J. R. Howell, Finite element solution of radiative heat transfer in a two-dimensional rectangular enclosure with gray partici-

pating media, *J. Heat Transfer* **105**(4), 933 (1983).
35K. N. B. Rekant, S. A. Demidov, B. A. Khrustalev and A.J. Knyazeva, Instruments for measuring the emissivities of materials and coatings at room temperature, *Heat Transfer—Soviet Res.* **14**(6), 82 (1982).
36K. H. Rosen and T. Novakov, Optical transmission through aerosol deposits on diffusely reflective filters: a method for measuring the absorbing component of aerosol particles, *Appl. Optics* **22**(9), 1264 (1983).
37K. J. A. Roux and B. E. Wood, Infrared optical properties of solid monomethyl hydrazine, N_2O_4, and N_2H_4 at cryogenic temperatures, *J. Opt. Soc. Am.* **73**(9), 1181 (1983).
38K. N. G. Semiletov, An instrument for measuring spherical irradiance, *Heat Transfer—Soviet Res.* **14**(6), 55 (1982).
39K. T. P. Sheahen, Effect of atmospheric attenuation on temperature measurements made using IR scanning systems, *Appl. Optics* **22**(7), 1070 (1983).
40K. G. J. Simonis and R. D. Felock, Index of refraction determination in the near-mm wavelength range using a mesh Fabry–Perot resonant cavity, *Appl. Optics* **22**(1), 194 (1983).
41K. E. B. Singleton and C. T. Shirkey, Optical constants in the IR from thin-film interference and reflectance: the reststrahlen region of muscovite mica, *Appl. Optics* **22**(1), 185 (1983).
42K. I. Ye. Smirnova, G. A. Zhorov and Yu. V. Zenkov, Calorimetric method for measuring the emissivities and absorptivities of materials and films at temperatures between 170 and 370 K, *Heat Transfer—Soviet Res.* **14**(6), 90 (1982).
43K. A. T. Stair, Jr., J. Pritchard, I. Coleman, C. Bohne, W. Williamson, J. Rogers and W. T. Rawlins, Rocketborne cryogenic (10 K) high-resolution interferometer spectrometer flight HIRIS: auroral and atmospheric IR emission spectra, *Appl. Optics* **22**(7), 1056 (1983).
44K. L. P. Stearns and R. F. Peuschel, Infrared optical properties of a coal-fired power plant plume, *Appl. Optics* **22**(12), 1856 (1983).
45K. K. Takarabe, K. Wakamura and T. Ogawa, Infrared reflectivity and crystal structure of InS, *J. Phys. Soc. Japan* **52**(2), 686 (1983).
46K. T. W. Tong and C. L. Tien, Radiative heat transfer in fibrous insulations—part I: analytical study, *J. Heat Transfer* **105**(1), 70 (1983).
47K. T. W. Tong, Q. S. Yang and C. L. Tien, Radiative heat transfer in fibrous insulations—part II: experimental study, *J. Heat Transfer* **105**(1), 76 (1983).
48K. T. Toyoda and M. Yabe, The temperature dependence of the refractive indices of fused silica and crystal quartz, *J. Phys. D: Appl. Phys.* **16**(5), L97 (1983).
49K. F. E. Volz, Infrared specular reflectance of pressed crystal powders and mixtures, *Appl. Optics* **22**(12), 1842 (1983).
50K. Y. S. Yang, J. R. Howell and D. E. Klein, Radiative heat transfer through a randomly packed bed of spheres by the Monte Carlo method, *J. Heat Transfer* **105**(2), 325 (1983).

MHD

1M. A. Bharali and A. K. Borkakati, Magnetohydrodynamic free convection flow with Hall effects, past a porous vertical plate, *J. Phys. Soc. Japan* **52**(1), 6 (1983).
2M. A. K. Borkakati and A. Bharali, Hydromagnetic flow and heat transfer between two horizontal plates, the lower plate being a stretching sheet, *Q. Appl. Math.* **40**, 461 (1983).
3M. A. Raptis and A. K. Singh, MHD free convection flow past an accelerated vertical plate, *Int. Comm. Heat Mass Transfer* **10**(4), 313 (1983).
4M. J. Zachariah and K. R. Singh, MHD flow past a circular cylinder in a channel, *J. Phys. Soc. Japan* **52**(8), 2742 (1983).
5M. J. Zachariah and R. Singh, A two-dimensional MHD flow of an inviscid fluid past an elliptic cylinder, *J. Phys. Soc. Japan* **52**(8), 2751 (1983).

Heat exchangers and heat pipes

1Q. T. Aihara and R. Saga, Performance of a compact cooling unit for semiconductors utilizing air–water mist flow, *J. Heat Transfer* **105**(1), 18 (1983).
2Q. V. M. Borishanskiy, D. I. Volkov, N. I. Ivashchenko, O. P. Krektunov, N. M. Fishman and I. N. Chirkov, Shell-side coefficient of heat transfer from steam contaminated with noncondensible gases, *Heat Transfer—Soviet Res.* **14**(3), 1 (1982).
3Q. T. R. Bott and C. R. Bemrose, Particulate fouling on the gas-side of finned tube heat exchangers, *J. Heat Transfer* **105**(1), 178 (1983).
4Q. P. I. Bystrov, V. F. Goncharov, V. N. Kharchenko and A. N. Shul'ts, Transient heat and mass transfer in liquid–metal heat pipes, *Heat Transfer—Soviet Res.* **14**(3), 18 (1982).
5Q. Chung-Hsung Li, A numerical finite difference method for performance evaluation of a periodic-flow heat exchanger, *J. Heat Transfer* **105**(3), 611 (1983).
6Q. N. Epstein, Thinking about heat transfer fouling—a 5 × 5 matrix, *Heat Transfer Engng* **4**(1), 43 (1983).
7Q. U. Fisher, W. Leidenfrost and J. Li, Hybrid evaporative–condenser is built into the cooling tower, *Heat Transfer Engng* **4**(2), 28 (1983).
8Q. P. J. Heggs and C. L. Render, Transient response of heat exchangers with one infinite capacitance fluid, *Heat Transfer Engng* **4**(2), 19 (1983).
9Q. H. S. Chu, C.-I. Weng and C.-K. Chen, Transient response of a composite straight fin, *J. Heat Transfer* **105**(2), 307 (1983).
10Q. T. Kajikawa, H. Takazawa and M. Mizuki, Heat transfer performance of stainless steel metal fiber sintered surfaces, *Heat Transfer Engng* **4**(1), 57 (1983).
11Q. C. F. Kettleborough and C. S. Hsieh, The thermal performance of the wet surface plastic plate heat exchanger used as an indirect evaporative cooler, *J. Heat Transfer* **105**(2), 366 (1983).
12Q. J. Lach, The exact solution of the Nusselt's model of the cross-flow recuperator, *Int. J. Heat Mass Transfer* **26**, 1597 (1983).
13Q. O. Levenspiel, Design of long heat regenerators by use of the dispersion model, *Chem. Engng Sci.* **38**, 2035 (1983).
14Q. T. F. Lin, K. H. Hawks and W. Leidenfrost, Transient conjugated heat transfer between a cooling coil and its surrounding enclosure, *Int. J. Heat Mass Transfer* **26**,

1661 (1983).
15Q. B. Linnhoff and E. Hindmarsh, The pinch design method for heat exchanger networks, *Chem. Engng Sci.* **38,** 745 (1983).
16Q. A. L. London and R. K. Shah, Costs of irreversibilities in heat exchanger design, *Heat Transfer Engng* **4**(2), 59 (1983).
17Q. M. Manzoor, D. B. Ingham and P. J. Heggs, The one-dimensional analysis of fin assembly heat transfer, *J. Heat Transfer* **105**(3), 646 (1983).
18Q. W. J. Marner, A. E. Bergles and J. M. Chenoweth, On the presentation of performance data for enhanced tubes used in shell and tube heat exchanger, *J. Heat Transfer* **105**(2), 358 (1983).
19Q. A. K. Majumdar, A. K. Singhal and D. B. Spalding, Numerical modeling of wet cooling towers—Part 1: Mathematical and physical models, *J. Heat Transfer* **105**(4), 728 (1983).
20Q. A. K. Majumdar, A. K. Singhal, H. E. Reilly and J. A. Bartz, Numerical modeling of wet cooling towers—Part II: Application to natural and mechanical draft towers, *J. Heat Transfer* **105**(4), 736 (1983).
21Q. Yu. V. Matveyev, V. K. Shchukin, I. I. Mosin and V. N. Voronin, Maximum heat flux through heat pipes with varying cross sections, *Heat Transfer—Soviet Res.* **14**(3), 30 (1982).
22Q. S. Mochizuki, W.-J. Yang, Y. Yagi and M. Ueno, Heat transfer mechanisms and performance in multiple parallel disk assemblies, *J. Heat Transfer* **105**(3), 598 (1983).
23Q. N. A. Nosach, Yu. K. Gontarev, V. F. Prisnyakov, A. G. Yakovenko and A. G. Kostornov, Heat transfer and flow in high-temperature alkali-metal heat pipes, *Heat Transfer—Soviet Res.* **14**(3), 36 (1982).
24Q. A. Pignotti and G. O. Cordero, Mean temperature difference in multipass crossflow, *J. Heat Transfer* **105**(3), 584 (1983).
25Q. A. Pignotti and G. O. Cordero, Mean temperature difference charts for air coolers, *J. Heat Transfer* **105**(3), 592 (1983).
26Q. C. W. Rapley and A. I. C. Webb, Heat transfer performance of ceramic regenerator matrices with sine-duct shaped passage, *Int. J. Heat Mass Transfer* **26,** 805 (1983).
27Q. R. B. Ritter, Crystalline Fouling Studies, *J. Heat Transfer* **105**(2), 374 (1983).
28Q. J. M. Robertson and P. C. Lovegrove, Boiling heat transfer with Freon-11 (R11) in brazed aluminum plate-fin heat exchangers, *J. Heat Transfer* **105**(3), 605 (1983).
29Q. F. E. Romie, Transient response of gas-to-gas crossflow heat exchangers with neither gas mixed, *J. Heat Transfer* **105**(3), 563 (1983).
30Q. M. N. Roppo and E. N. Ganic, Time-dependent heat exchanger modeling, *Heat Transfer Engng* **4**(2), 42 (1983).
31Q. E. K. Ruth, Experiments on a crossflow heat exchanger with tubes of lenticular shape, *J. Heat Transfer* **105**(3), 571 (1983).
32Q. A. D. Snider and A. D. Krause, Recent developments in the analysis and design of extended surface, *J. Heat Transfer* **105**(2), 302 (1983).
33Q. E. M. Sparrow and J. W. Comb, Effect of interwall spacing and fluid flow inlet conditions on a corrugated-wall heat exchanger, *Int. J. Heat Mass Transfer* **26,** 993 (1983).
34Q. J. W. Sutherland, Analysis of mechanical-draught counterflow air/water cooling towers, *J. Heat Transfer* **105**(3), 576 (1983).
35Q. Z. V. Tishchenko and V. N. Bondarenko, Comparison of the efficiency of smooth-finned plate heat exchangers, *Int. Chem. Engng* **23,** 550 (1983).
36Q. N. V. Zozulya, A. A. Khavin and V. I. Leonova, The influence of the sparseness of finned tube bundles on their thermal and aerodynamic characteristics, *Int. Chemie. Engng* **23,** 558 (1983).
37Q. W. Zul and H. De Bruijn, Continuum equations for the prediction of shell-side flow and temperature patterns in heat exchangers, *Int. J. Heat Mass Transfer* **26,** 411 (1983).

General heat transfer applications

1S. C. Casarosa, E. Latrofa and A. Shelginski, The geyser in a two-phase thermosyphon, *Int. J. Heat Mass Transfer* **26,** 933 (1983).
2S. A. E. Delsante, A. N. Stokes and P. J. Walsh, Application of Fourier transforms to periodic heat flow into the ground under a building, *Int. J. Heat Mass Transfer* **26,** 121 (1983).
3S. Ye. P. Dyban, Heat transfer in high-temperature gas turbines, *Heat Transfer—Soviet Res.* **14**(4), 90 (1982).
4S. L. P. Kabanov and O. M. Khachatryan, Rewetting of tubes in a nuclear reactor core by means of a falling film and by a pressurized downflow, *Heat Transfer—Soviet Res.* **14**(6), 102 (1982).
5S. A. Moutsoglou, On the cooling of fibers, *J. Heat Transfer* **105**(4), 830 (1983).
6S. S. M. Pandit and K. P. Rajurkar, A stochastic approach to thermal modeling applied to electro-discharge maching, *J. Heat Transfer* **105**(3), 555 (1983).
7S. B. S. Petukhov, Turbulence in the theory of heat transfer, *Heat Transfer—Soviet Res.* **14**(4), 61 (1982).
8S. N. A. Pokryvaylo, A. K. Nesterov and Yu. Ye. Zberkhovskiy, Effect of polymeric additives on the damping of wall turbulence and on convective heat transfer, *Heat Transfer—Soviet Res.* **14**(2), 52 (1982).
9S. R. D. Rempt, Test/flight thermal data comparisons for Viking and SCATHA, *J. Spacecraft Rockets* **20,** 286 (1983).
10S. I. G. Rice, Steam-cooled gas turbine casings, struts, and disks in a reheat gas turbine combined cycle: Pts I & II—gas generator turbine and power turbine (83-GT-86), *J. Engng Power* **105,** 851 (1983).
11S. H. Saito and N. Seki, Simultaneous transfer of heat and mass occurring in casting process, *Wärme- und Stoffübertragung* **17,** 107 (1983).
12S. Z. P. Shul'man, S. M. Aleynikov and B. M. Khusid, Rheodynamics and heat transfer in the flow of liquids with memory, *Heat Transfer—Soviet Res.* **14**(2), 33 (1982).
13S. E. M. Sparrow, S. B. Vemuri and D. S. Kale, Enhanced and local heat transfer, pressure drop and flow arrays of block-like electronic components, *Int. J. Heat Mass Transfer* **26,** 689 (1983).
14S. Ye. M. Staviskiy, G. I. Savvatimskiy, I. V. Yelkin, E. I.

Loverant and A. P. Proshutinskiy, Investigation of the thermohydraulic situation in a steam-generating tube following a loss of coolant accident (LOCA), *Heat Transfer—Soviet Res.* **14**(2), 111 (1982).
15S. D. A. Throckmorton, Benchmark determination of shuttle orbiter entry aerodynamic heat-transfer data, *J. Spacecraft Rockets* **20,** 219 (1983).
16S. D. A. Throckmorton and E. V. Zoby, Orbiter entry leeside heat-transfer data analysis, *J. Spacecraft Rockets* **20,** 524 (1983).
17S. J. Wolf and S. Moskowitz, Development of the transpiration air-cooled turbine for high-temperature dirty gas systems (83-GT-84), *J. Engng Power* **105,** 821 (1983).
18S. E. V. Zoby, Analysis of STS-2 experimental heating rates and transition data, *J. Spacecraft Rockets* **20,** 232 (1983).

Solar energy

1T. A. Abhat, Low temperature latent heat thermal energy storage: heat storage materials, *Sol. Energy* **30,** 313 (1983).
2T. A. Abhat and T. Q. Huy, Heat and mass transfer considerations in a thermochemical energy storage system based on solid–gas reactions, *Sol. Energy* **30,** 93 (1983).
3T. G. Athanassouli and G. Massouros, The influence of solar and long-wave radiation on the thermal gain of a room with opaque external wall, *Sol. Energy* **30,** 161 (1983).
4T. J. F. Atkinson and D. R. F. Harleman, A wind-mixed layer model for solar ponds, *Sol. Energy* **31,** 243 (1983).
5T. R. E. Bird, R. L. Hulstrom and L. J. Lewis, Terrestrial solar spectral data sets, *Sol. Energy* **30,** 563 (1983).
6T. D. T. Brine and M. Iqbal, Diffuse and global solar spectral irradiance under cloudless skies, *Sol. Energy* **30,** 447 (1983).
7T J. W. Byrd, C. G. Adler, The maximum efficiency of solar energy conversion devices, *J. Appl. Phys.* **54,** 7166 (1983).
8T K. Dhingra and H. C. Bryant, Thermal model of a cylindrically symmetric solar pond, *Sol. Energy* **30,** 589 (1983).
9T. S. M. Eldighidy and I. S. Taha, Optimum mass flow rate of water in a flat plate solar collector coupled with a storage tank and an organic Rankine cycle power loop, *Sol. Energy* **31,** 455 (1983).
10T. R. A. Fisher, Jr. and A. H. Fanney, Thermal performance comparisons for a solar hot water system, *ASHRAE Jl* **25,** no. 8, 27 (1983).
11T. G. Flamant and G. Olalde, High temperature solar gas heating comparison between packed and fluidized bed receivers—1, *Sol. Energy* **31,** 463 (1983).
12T. J. M. Gordon and Y. Zarmi, The utilizability function—II. Validation of theory against data-based correlations, *Sol. Energy* **31,** 537 (1983).
13T. R. Guzzi, G. Lo Vecchio, R. Rizzi and G. Scalabrin, Experimental validation of a spectral direct solar radiation model, *Sol. Energy* **31,** 359 (1983).
14T. N. E. Hager, Jr., Passive solar floors, *ASHRAE Jl* **25**(12), 29 (1983).
15T. C. S. Herrick, An air-cooled solar collector using all-cylindrical elements in a low-loss body, *Sol. Energy* **30,** 217 (1983).
16T. C. S. Herrick, A series connected solar air heating system using the FMTC collector and the rolling cylinder heat store, *Sol. Energy* **31,** 191 (1983).
17T. K. G. T. Hollands, R. G. Huget, A probability density function for the clearness index, with applications, *Sol. Energy* **30,** 195 (1983).
18T. K. G. T. Hollands and J. L. Wright, Heat loss coefficients and effective $\tau\alpha$ products for flat-plate collectors with diathermanous covers, *Sol. Energy* **30,** 211 (1983).
19T. L. D. Jaffe, Dish concentrators for solar thermal energy, *J. Energy* **7,** 304 (1983).
20T. A. Johannsen and G. Grossman, Performance simulation of regenerating type solar collector, *Sol. Energy* **30,** 87 (1983).
21T. J. P. Kenna, The multiple layer solar collector, *Sol. Energy* **30,** 225 (1983).
22T. T. R. Knowles, Proportioning composites for efficient thermal storage walls, *Sol. Energy* **31,** 319 (1983).
23T. E. Kritchman, Solar thermal engine systems using two stage concentrators, *Sol. Energy* **31,** 449 (1983).
24T. B. LeBaron and I. Dirmhirn, Strengths and limitations of the Liu and Jordan model to determine diffuse from global irradiance, *Sol. Energy* **31,** 167 (1983).
25T. C. C. Y. Ma and M. Iqbal, Statistical comparison of models for estimating solar radiation on inclined surfaces, *Sol. Energy* **31,** 313 (1983).
26T. S. L. Marcus, An approximate method for calculating the heat flux through a solar collector honeycomb, *Sol. Energy* **30,** 127 (1983).
27T. M. O. McLinden and S. A. Klein, Simulation of an absorption heat pump solar heat and cooling system, *Sol. Energy* **31,** 473 (1983).
28T. A. T. Mecherikinnel, J. A. Gatlin and J. C. Richmond, Data on total and spectral solar irradiance, *Appl. Optics* **22,** 1354 (1983).
29T. G. L. Morrison and C. M. Sapsford, Long term performance of thermosyphon solar water heaters, *Sol. Energy* **30,** 341 (1983).
30T. J. K. Nayak, N. K. Bansal and M. S. Sodha, Analysis of passive heating concepts, *Sol. Energy* **30,** 51 (1983).
31T. B. J. Novel, Passive cooling strategies, *ASHRAE Jl* **25**(12), 23 (1983).
32T. A. Saito, Y. Utaka, T. Tsuchio and K. Katayama, Transient characteristics of flat-plate solar collector, *Bull. JSME* **26,** 79 (1983).
33T. T. D. M. A. Samuel and N. E. Wijeysundera, Heat withdrawal from multi-layer thermal trap collectors, *Sol. Energy* **30,** 261 (1983).
34T. J. E. Sherry and C. G. Justus, A simple hourly clear-sky solar radiation model based on meteorological parameters, *Sol. Energy* **30,** 425 (1983).
35T. M. S. Sodha, S. N. Shukla and G. N. Tiwari, Transient analysis of a natural circulation solar water heater with a heat exchanger, *J. Energy* **7,** 107 (1983).
36T. M. Sokolov and M. Vaxman, Analysis of an integral compact solar water heater, *Sol. Energy* **30,** 237 (1983).
37T. M. E. Sweat and J. J. Carroll, On the use of spectral radiance models to obtain irradiances on surfaces of arbitrary orientation, *Sol. Energy* **30,** 373 (1983).
38T. P.-H. Theunissen and J.-M. Buchlin, Numerical optimization of a solar air heating system based on encapsulated P.C.M. storage, *Sol. Energy* **31,** 271 (1983).

39T. G. J. van den Brink and C. J. Hoogendoorn, Ground water flow heat losses for seasonal heat storage in the soil, *Sol. Energy* **30,** 367 (1983).
40T. Y. F. Wang and A. Akbarzadeh, A parametric study on solar ponds, *Sol. Energy* **30,** 555 (1983).
41T. B. Window, Heat extraction from single ended glass absorber tubes, *Sol. Energy* **31,** 159 (1983).

Plasma heat transfer

1U. C. J. Allum, Power dissipation in the column of a TIG welding arc, *J. Phys. D: Appl. Phys.* **16**(11), 2149 (1983).
2U. E. Bourdin, P. Fauchais and M. Boulos, Transient heat conduction under plasma conditions, *Int. J. Heat Mass Transfer* **26,** 567 (1983).
3U. X. Chen, Expressions for the specific heat flux to a cylinder in a thermal plasma crossflow, *J. Heat Transfer* **105**(2), 418 (1983).
4U. X. Chen and E. Pfender, Behavior of small particles in a thermal plasma flow, *Plasma Chem. Plasma Proc.* **3**(3), 351 (1983).
5U. X. Chen and E. Pfender, Effect of the Knudsen number on heat transfer to a particle immersed into a thermal plasma, *Plasma Chem. Plasma Proc.* **3**(1), 97 (1983).
6U. H. K. Choi and W. H. Gauvin, Operating characteristics and energy distribution in transferred plasma arc systems, *Plasma Chem. Plasma Proc.* **2**(4), 361 (1982).
7U. L. E. Cram, A model of the cathode of a thermionic arc, *J. Phys. D: Appl. Phys.* **16**(9), 1643 (1983).
8U. P. L. Denbigh, B. F. Jones and D. A. J. Mottram, Relationship between efficacy, arc tube temperature and power dissipation in a high-pressure sodium lamp, *J. Phys. D: Appl. Phys.* **16**(11), 2167 (1983).
9U. G. A. Dyuzhev, G. A. Lyubimov and E. M. Shkol'nik, Conditions of the anode spot formation in a vacuum arc, *IEEE Trans. Plasma Sci.* **PS-11**(1), 36 (1983).
10U. M. T. C. Fang and D. B. Newland, DC nozzle arcs with mild wall ablation, *J. Phys. D: Appl. Phys.* **16**(5), 793 (1983).
11U. S. Goldsmith, S. Shalev and R. L. Boxman, Anode melting in a multicathode-spot vacuum arc, *IEEE Trans. Plasma Sci.* **PS-11**(3), 127 (1983).
12U. K. C. Hsu and E. Pfender, Two-temperature modeling of the free-burning, high-intensity arc, *J. Appl. Phys.* **54**(8), 4359 (1983).
13U. K. C. Hsu, K. Etemadi and E. Pfender, Study of the free-burning high-intensity argon arc, *J. Appl. Phys.* **54**(3), 1293 (1983).
14U. K. C. Hsu and E. Pfender, Analysis of the cathode region of a free-burning high-intensity argon arc, *J. Appl. Phys.* **54**(7), 3818 (1983).
15U. J. C. Ingraham and G. Miller, Infrared calorimeter for time-resolved plasma energy flux measurement, *Rev. Sci. Instr.* **54**(6), 673 (1983).
16U. D. C. Johnson and E. Pfender, The effects of low-ionization-potential contaminants on thermal plasmas, *Plasma Chem. Plasma Proc.* **3**(2), 259 (1983).
17U. Y. Kishimoto and K. Mima, An extension of Spitzer–Härm Theory on thermal transport to steep temperature gradient, case I: general information, *J. Phys. Soc. Japan* **52**(10), 3389 (1983).
18U. H. Konno, T. Kohya, M. Toda, T. Saito, H. Kuriyama and E. Harada, Heat transfer from a thin horizontal wire and large cylinders in a corona wind, *Int. Chem. Engng* **23,** 91 (1983).
19U. C. S. Landram, Measurement of fusion boundary energy transport during arc welding, *J. Heat Transfer* **105**(3), 550 (1983).
20U. J. W. McKelliget and J. Szekely, A mathematical model of the cathode region of a high intensity carbon arc, *J. Phys. D: Appl. Phys.* **16**(6), 1007 (1983).
21U. M. T. Mehmetoglu and W. H. Gauvin, Characteristics of a transferred-arc plasma, *AIChE Jl* **29,** 207 (1983).
22U. P. Meuvus and A. Charette, A heat transfer study for a partially immersed refractory rod in a plasma, *Chem. Engng Sci.* **38,** 1741 (1983).
23U. Y. Pang Tsui and H. Y. Chen, Quenching of air plasma effluents, *Plasma Chem. Plasma Proc.* **2**(4), 387 (1982).
24U. P. R. Smy, R. M. Clements, J. D. Dale, D. Simeoni and D. R. Topham, Efficiency and erosion characteristics of plasma jet igniters, *J. Phys. D: Appl. Phys.* **16**(5), 783 (1983).
25U. M. Vardelle, A. Vardelle, P. Fauchais and M. I. Boulos, Plasma–particle momentum and heat transfer: modelling and measurements, *AIChE Jl* **29,** 236 (1983).
26U. H. Yoshiyasu, S. Murata, S. Yamagata, Y. Wada and Y. Shibuya, Experimental study of electrode vapour flow in a free-burning arc, *J. Phys. D: Appl. Phys.* **16**(11), 2213 (1983).
27U. F. J. Zanner and L. A. Bertram, Behavior of sustained high-current arcs on molten alloy electrodes during vacuum consumable arc remelting, *IEEE Trans. Plasma Sci.* **PS-11**(3), 223 (1983).

HEAT TRANSFER REVIEW - 1984

E.R.G. ECKERT, R. J. GOLDSTEIN, S.V. PATANKAR, E. PFENDER, J.W. RAMSEY, T.W. SIMON, N. DECKER, T.H. KUEHN and H. LEE

Heat Transfer Laboratory, Department of Mechanical Engineering, University of Minnesota, Minneapolis, Minnesota, U.S.A.

INTRODUCTION

THIS REVIEW surveys papers that have been published in the open refereed literature covering various fields of heat transfer during 1984. It is the intention to present a comprehensive survey; however, some sources may have been missed. For this we apologize.

A number of conferences dealt with, or included sessions on heat transfer during 1984.

Two sessions on film cooling, one session on analytical methods, and one on boundary-layer and air foil heat transfer were included in the program of the 29th International Gas Turbine Conference and Exhibit held at Amsterdam, The Netherlands, 3–7 June. Technical papers are available through the ASME order department.

A Heat Transfer Conference at the University of Leeds, U.K., 3–5 July was the first one of a series of U.K. national conferences on heat transfer planned to take place every four years alternating with the existing international heat transfer conference. The proceedings are contained in two volumes published by Pergamon Press. Volume 1 contains, among others, post dry-out and drop heat transfer, thermal hydraulic aspects of accidents and transients, reflood, reactor operational heat transfer, fouling mechanisms, regenerators and fixed beds, heat transfer networks. Volume 2 includes among others enhanced heat transfer, two-phase flow and boiling, high temperature systems and heat transfer in combustion systems.

The highlights of the 22nd National Heat Transfer Conference and Exhibition at Niagara Falls, U.S.A., 5–8 August were the Donald Q. Kern Award lecture on "two-phase flow in heat exchangers and pipe lines" by the recipient Duncan Chisholm and the Max Jakob Award lecture on "measurements of solids and gas motions in air fluidized beds" by the recipient Bei Tse Chao. A birthday celebration was arranged in honor of the 80th birthday of Ernst Eckert. Forty-eight technical sessions dealt with fundamental aspects of heat transfer and with applications like waste heat recovery, heat exchangers, solar energy heat transfer, industrial energy systems, reboilers, heat transfer in glass, and thermal storage heat transfer. Professional development and continuing education programs are already a well-established feature of these conferences. ASME papers are available in reprint and volume form and AIChE papers in the symposium series, Vol. 80, No. 236.

The XVI Symposium of the International Centre for Heat and Mass Transfer at Dubrovnik, Yugoslavia, 3–7 September 1984 was devoted to heat and mass transfer in fixed and fluidized beds and included among others sessions on measurement and application, modeling of fluidized bed reactors, fluidized bed combustion and gasification, and liquid and three-phase fluidization. It was preceded by an advanced course on fluidized bed combustion. The proceedings are contained in a volume distributed by Hemisphere Publishing Corporation, Washington, D.C.

The ETH, Zurich, Switzerland offered an International Seminar on Organic Rankine Cycle and Heat Pump Technology organized by the Verein Deutscher Ingenieure with co-sponsorship from France, Hungary, Japan, Switzerland and the U.S.A. The sessions were devoted to working fluid problems. Session IV was organized around heat exchangers, boilers and augmentation techniques. The proceedings are available as VDI Report No. 539.

The Winter Annual Meeting of the ASME at New Orleans, 9–14 December, included in its program 19 sessions on various topics of heat transfer together with a panel session on high pressure thermophysical data, two poster sessions and a special session on education in heat transfer. Heat transfer in biological media, ice and frost formation, coal slurry mixtures, in heat rejection systems, gas turbine hot sections, microelectronics fabrication and packaging were some special topics. The Heat Transfer Memorial Award was presented to Wen-Jei Yank at the Heat Transfer

Luncheon, which featured as speaker R. G. Watts, talking on "heat transfer and the evolution and variations of the earth's climate". Preprints of the papers are available at ASME Publications Department and many of the papers will also be published in the *Journal of Heat Transfer*.

A number of books have become available during the past year. They are listed under references.

The following highlights illuminate development in heat transfer research during 1984.

Thermal simulation of machining processes comprised a significant portion of the research on conduction with less emphasis on contact resistance.

Channel flow research deals with increasingly complex problems involving fins and other augmentation devices, variable fluid properties, geometrical irregularities, conjugate heat transfer, and non-Newtonian behavior. Several topics in boundary-layer and external flow responded to the need to better understand complex flow pattern and heat transfer in gas turbines, including the effects of larger disturbances on cylinders in crossflow (simulating air foil leading edges), and the effect of the wall on heat transfer from adjacent and perpendicular cylinders (simulating air foil root and tip regions).

A large number of computational papers on natural convection in enclosures filled with a saturated porous medium provide test cases for numerical analysis. They included onset of convection and channelling effects near walls. Unsaturated porous media, phase change systems, or fluidized beds found comparatively little attention.

Papers on experimental techniques and instrumentation were added to the Heat Transfer Review this year. Those chosen contain heat or mass transfer analysis or deal with the measurement of thermal transfer rates or thermal transport properties. Several of the techniques included electro-optics, e.g. holography, lasers.

Natural convection papers included multiphase flow in thermosyphons. Considerable interest was demonstrated in internal energy sources and in the effect of density maxima in fluids such as water.

The number of papers on boiling in nuclear reactors and on fundamental aspects of nucleate boiling decreased substantially. Interest remains strong, however, in boiling of mixtures and in enhanced surface boiling. A greater number of papers deal with film boiling and with the flashing of liquid to vapor upon depressurization. Research remains strong on phenomena in shell side condensation of shell-and-tube condensors. Topics include droplet slipping on the surface, film stability, boundary mixtures and the effects of oils. Several survey papers were offered on condensor performance. Freezing in the thermal entrance region of channel flow and simulations of continuous casting processes were popular topics in freezing and melting. The classical Stefan problem and natural convection-dominated melting continue to receive attention.

Radiative heat transfer calculations in multidimensional geometry were the topic of several papers. The coupling of radiation with conduction/convection and combustion is considered analytically and experimentally. There has also been an emphasis on obtaining relevant property data for various materials and radiating systems.

Developments in numerical methods are directed towards the treatment of arbitrary geometry, reduction of false diffusion errors and improvement of computational efficiency.

Research was focused on some novel methods and devices for measurement of thermal properties of materials. Of special interest were porous materials used in various heat transfer situations.

Over 70% of this year's papers dealing with heat exchangers range from measurements of local heat transfer coefficients to overall performance of exchangers or predicting their performance. Some papers deal with methods of enhancing exchanger performance or of offsetting some of the problems encountered in service, e.g. scaling or fouling of surfaces.

An increasing number of papers demonstrate a strong interest in plasma engineering dealing especially with plasma diagnostics, plasma generation and modeling, transport properties and plasma processing of materials.

To facilitate the use of this review, a listing of the subject headings is made below in the order in which they appear in the text. The letter which appears adjacent to each subject heading is also attached to the references that are cited in the category:

Conduction, A
Channel flow, B
Boundary-layer and external flows, C
Flow with separated regions, D
Heat transfer in porous media, DP
Experimental techniques and instrumentation, E
Natural convection—internal flows, F
Natural convection—external flows, FF

Convection from rotating surfaces, G
Combined heat and mass transfer, H
Change of phase—boiling, J
Change of phase—condensation, JJ
Change of phase—freezing and melting, JM
Radiation in participating media and surface radiation, K
Numerical methods, N
Transport properties, P
Heat transfer applications
Heat pipes and heat exchangers, Q
General, S
Solar Energy, T
Plasma heat transfer, U

CONDUCTION

A variety of theoretical studies of heat conduction have been reported. Some improved analytical solution techniques have been proposed with results given for some basic problems. Solutions are presented for transient and steady multidimensional problems ranging from metal cutting applications to combined conduction, convection and radiation.

The work on fundamentals included a new nonlinear theory of the constitutive equations for electrically and thermally conducting solids [14A]. Analytical solutions were developed for the hyperbolic heat conduction equation describing the wave nature of thermal energy transport in a finite slab with insulated boundaries and a localized volumetric energy sources [31A]. The solution is valid at short time intervals following the beginning of a transient and at temperatures near absolute zero when classical heat diffusion theory breaks down. Macroscopic transport properties were developed for a microscopically disordered system [24A]. The case of a periodic lattice with disorder in the transport properties at the level of individual bonds was considered. Measurements of the effective thermal diffusivity of a copper–niobium alloy were measured from 0 to 400°C [29A]. The rule of mixtures describes the data well. Extensions of available solution methods included the application of Green's function to linear, transient heat conduction problems [4A]. Two thermal boundary value conditions were incorporated in addition to the usual three. Applications in two and three dimensions were also outlined. A sign-count method for determining the roots of some of the transcendental equations associated with the analytic solution of transient heat conduction in a portion of a sphere was demonstrated [30A]. A simple boundary-element formulation employing a numerical solution method was proposed and applied to several examples of steady two-dimensional heat conduction in axisymmetric bodies [36A].

Several studies of transient conduction in homogeneous, isotropic media appeared. A closed-form solution for transient conduction in a slab with periodic surface temperature variation was presented with application to a rotary kiln [17A]. The temperature field in a thermoelastic slab with periodic surface temperatures was also given [12A]. The stability of a reactive slab with periodic surface heat flux was investigated [18A]. An approximate closed-form solution was described for a semi-infinite solid which experiences a surface heat flux pulse [33A]. Heat conduction in a semi-infinite solid with random surface heat flux variations was studied [10A]. The correlation between the interior and the surface conditions was found to decrease as the fourth power of the depth into the solid. Certain similarity was found to exist in the surface heat flux of wedges and semi-infinite solids imposed with a step change in surface temperature [35A]. Several papers considered the interaction of thermal and mechanical stresses which arise in machining processes or with internal heat generation. The transient one-dimensional temperature and stress distributions in a heat generating cylinder with variable thermal conductivity were presented [32A]. Green's function was used to obtain results for surface displacement and stresses due to a line heat source moving at constant speed across the surface of a thermoelastic half plane [1A]. Closed-form solutions were presented for the three-dimensional temperature distributions in a half plane with a moving two-dimensional heat source on the surface [34A]. Superposition could be used to obtain the solution when several discrete surface heat sources exist. Finite-element numerical techniques were used to study an inclined plane heat source moving along the surface of a half plane to estimate the shear-plane temperatures developed in orthogonal machining and grinding operations [11A].

A method of improving contact conductance by using a high conductivity metallic alloy that melts when in service but freezes for disassembly has been developed [9A]. The effect of contact resistance between longitudinal fins and a plane wall was studied [27A]. The contact resistance was found to be negligible when the area of perfect contact is greater than 10% of the total contact area.

Approximate solutions for one-dimensional transient heat conduction within a laminated material subjected to a harmonic surface heat flux were presented [7A]. A separation of variables, Fourier-series method was used to obtain solutions for transient heat conduction in a multilayered slab [15A]. Solutions for steady two-dimensional conduction in anisotropic media were given, using Green's function and a coordinate transformation [23A]. The boundary conditions can be mixed with temperatures prescribed on one surface and heat flux given on the remaining boundaries. Additions to a paper previously published by Atkinson were also reported [6A].

Transient one-dimensional problems include a numerical solution of inverse heat conduction with a radiative boundary condition [28A]. A study of the circular cylinder geometry was made to simulate the quenching of a rod [16A]. A procedure to estimate the spatial decay was presented [20A]. Simplified design charts were developed to approximate the cooling of various shaped solids in an isothermal, uniform convection environment [8A]. Various objects can be placed on the same chart if the characteristic length is chosen as the object's volume to surface area ratio.

Applications include the steady heat loss from a buried pipe network [13A]. Solutions using variable soil thermal conductivity approximate the effect of moisture on the soil thermal properties. The critical thickness of insulation for various shaped objects was determined for both convective and prescribed exterior surface temperature boundary conditions [21A]. A critical perimeter was found to be useful for irregularly shaped objects. The second law of thermodynamics was employed to optimize the number and location of a series of cooled shields located within porous insulation separating a hot and a cold body [5A].

Coupled heat transfer mechanisms or conjugate problems include coupled wall conduction and internal convection in a circular tube [2, 3A]. Conjugate conduction and convection on a circular pin fin has been studied [22A]. An experimental procedure for measuring the effective thermal conductivity of glasses at elevated temperatures was outlined [26A]. The coupling of conduction and radiation heat transfer through cells was studied experimentally with numerical and analytical solutions given for comparison [19A]. A study of a similar problem but including convection was also reported [25A].

CHANNEL FLOW

Confined flows have been studied in a number of different configurations. They include the classical geometries of circular tube and parallel plates; but attention is increasingly given to complex geometries that may enhance or impede heat transfer. The types of flow considered include laminar and turbulent flows, single- and multiphase situations and Newtonian and non-Newtonian behavior. Although the bulk of the literature deals with steady flows, some papers do focus on the time-dependent behavior.

Laminar heat transfer in the developing region of isothermal parallel plates is considered in [30B]. Reference [11B] deals with the parallel-plate geometry with a porous lining. The time scales of the turbulent fluid motion in a channel are investigated in [3B]. For a binary gas mixture, the kinetic phenomena in a channel are considered in [12B]. The unsteady conjugate heat transfer problem for a parallel plate channel is analyzed in [50B].

Among the many studies pertaining to the circular tube geometry, ref. [21B] deals with heat transfer in a laminar flow. The Graetz problem is discussed in [32B]. Reference [28B] considers the time-dependent behavior of the entrance region in a laminar pipe flow in response to a step change in the pumping pressure. The effect of suction or injection on the heat transfer in a turbulent pipe flow is investigated in [56B], while the influence of circumferential variation of heat flux is considered in [7B]. Measurements are reported in a circular tube, in which the turbulent flow is influenced by swirl [44B]. The axial conduction in the fluid and the pipe wall is considered for laminar flow in [13B, 42B] and for turbulent flow in [29B].

Numerical and experimental results are presented for the effect of roughness on heat transfer in tubes [14B]. Reference [24B] deals with the effect of direction of flow and tube position on heat transfer. The forced convection of supercritical *n*-heptane is investigated [25B]. Measurements are given for the heat transfer in a turbulent pipe flow downstream of an orifice [51B]. The turbulence statistics in a heated pipe flow are examined in the presence of suction [16B]. Reference [40B] deals with the turbulent transport in the transition flow region.

Among the studies directed towards the non-Newtonian behavior of flows, ref. [41B] reports the heat transfer enhancement in laminar slurry pipe flow. Heat transfer to drag-reducing turbulent pipe flows is analyzed in [57B] and to viscoelastic fluids in [22B].

Reference [15B] deals with heat transfer to laminar blood flow in a circular pipe. Stratified two-phase flow in circular tubes has been investigated [33B]. The heat and mass transfer in the presence of a mist in laminar duct flow is studied [20B].

Many channel flow papers deal with complex geometries that are encountered in engineering practice. The stability of divergent channel flows is numerically examined in [13B]. An experimental study of pin fins in a turning flow is reported in [47B]. The peripheral variation of wall temperature in a noncircular duct is experimentally found in [6B]. The effect of inlet conditions on turbulent heat transfer coefficients is investigated in [46B]. The introduction of internal circumferential fins in a *laminar* tube flow is shown to reduce the heat transfer instead of augmenting it [38B].

A turbulence model is used to predict the flow and heat transfer in a narrow isosceles triangular duct [37B]. Combined convection and radiation in rectangular ducts is considered [23B]. Fully-developed laminar flow and heat transfer in a triangular duct is analyzed in [34B]. Experiments have been conducted for laminar flow and heat transfer in an elliptical duct [2B]. The fully-developed heat transfer in an elliptic duct is also considered in [9B]. A comparison is made between symmetric and asymmetric periodic disturbances at the walls of a heated flow passage [48B]. Experiments are reported to determine the effect of the edge geometry in a duct flow [45B]. Turbulent flow heat transfer is measured for periodically converging–diverging tubes via a mass transfer analogy [45B]. Reference [18B] considers the effect of rib roughness on heat transfer and friction in channels. The local heat transfer downstream of an abrupt expansion in a circular channel is studied in [8B]. Fully-developed laminar heat transfer in circular sector ducts is analyzed in [53B].

A numerical calculation is presented for the heat transfer in a channel with two right-angled bends [5B]. The flow between two parallel disks is considered in [34B]. Laminar heat transfer in horizontal rectangular ducts is studied in [31B]. A numerical solution is given for the heat transfer in an annulus of periodically-varying cross section [35B]. An investigation deals with the heat transfer behavior downstream of a mixing junction [26B]. Flow and heat transfer in an in-line tube bank are solved numerically in [17B]. Reference [36B] presents an analytical solution for the secondary flow in a square duct. Laminar heat transfer in curved rectangular ducts is analyzed in [27B]. A general mathematical formulation for energy separation in a vortex tube is shown to compare well with experiments [49B]. Heat transfer from a heated concave wall is studied in [39B].

The influence of swirl-flow inducers on heat transfer in a circular duct is investigated in [10B]. Reference [55B] deals with heat transfer augmentation by twisted tapes. The combined use of twisted tapes and internally grooved surfaces is considered in [54B]. The absorption of thermal radiation in an open channel is investigated [19B]. The thermal–hydraulic characteristics in a liquid-metal fast breeder reactor plenum are studied in [52B].

BOUNDARY-LAYER AND EXTERNAL FLOWS

Laminar and transitional flows

Experimental results were presented for free molecular convection on fine metallic filaments [43C]. The dependence of accommodation coefficient on the surface state was found. The effect of buoyancy on Blasius flow over a heated horizontal plate was analyzed from the thermal stability point of view [49C]. Experiments were made to confirm the predicted locations of the onset of longitudinal vortices and the effect of the vortices on heat transfer. Measured and predicted heat flux distributions were compared for the Shuttle orbiter lower fuselage and lower wing centerspans [53C]. Experimental data for heat transfer to a film falling over horizontal cylinders was correlated by assuming complete mixing of the film [11C]. The correlation depends on Reynolds number and film thickness.

An approximate solution was presented for self-similar laminar boundary layers over plane and axisymmetric bodies [36C]. Nusselt numbers were predicted for various pressure gradients over a wide range of Prandtl number. Heat and momentum transport has been numerically evaluated for a pair of spheres in tandem [50C]. The drag coefficient and Nusselt number are always less than those of a single, isolated sphere—the effect is much stronger on the downstream sphere. The effect of heating or cooling on Görtler instability of a boundary layer on a concave wall was analyzed for short-wavelength disturbances [15C]. Critical disturbances are concentrated in thin, viscous layers near the middle of the boundary layer—heating destabilizes the short-wavelength disturbances while cooling does the opposite. Calculated critical Rayleigh numbers and wave numbers for thermally-destabilized boundary layers were found to be independent, or weakly dependent on free-stream acceleration [5C]. Laminar skin friction and surface heat flux on the rotationally oscillating cylinder in crossflow were analytically evaluated using a series method with perturbations [6C]. The unsteadiness effect on skin friction was large and increased with frequency whereas the effect on heat transfer was small near the separation point, and decreased with frequency. An analysis was presented for the heating of a particle which is suddenly introduced into a flowing fluid of different temperature [1C]. Asymptotic ($t \rightarrow \infty$) Nusselt numbers are considerably less than steady-state, isothermal-sphere values. The local Nusselt number is shown to become negative near the rear stagnation point of the sphere. The transient thermal response of a flat plate heated from below and cooled by convection above was analyzed [52C]. A parameter was identified which shows the relative conductive/convective effects. Numerical predictions of Space Shuttle heating were compared with flight data [19C]. Focus was on the effect of angle of attack and catalycity of the surface. An asymptotic solution of the Navier–Stokes equations for hypersonic flow behind bodies was presented [30C]. The solution allowed account of flow resistance and heat and mass transfer from the body. Laminar two-dimensional flow along an isothermal flat plate was analyzed to determine the dependence on fluid properties [16C]. Heat and momentum transfer analyses were presented for flow over surface-mounted parabolic cylinders and paraboloids of revolution—simulating dendrites [10C]. Diffusion extends a greater distance into the flow for parabolic cylinders than for paraboloids of revolution. Also, the boundary-layer assumptions were deemed inappropriate for dendritic surfaces. An analysis of submerged, swirled jets using the method of asymptotic expansion was given [44C]. Results were compared with experiments.

Boundary-layer transition on a heated, underwater body was experimentally investigated [26C]. The concentration of free particles was shown to affect transition by decreasing the effective Reynolds number. Experimental and numerical results were compared for the effect of cooling on transition with supersonic flow [28C]. Cooling stabilizes the first-mode disturbances, destabilizes the second-mode disturbances, and does not affect the acoustic wave interaction.

Turbulent and separated flows

The analogy between spectral distributions of temperature fluctuations and turbulent kinetic energy was experimentally examined [14C]. The analogy applies well except in stable atmospheric surface layers. A discussion of the history of the mixing length model for turbulent transport was presented [25C]. The relationship between heat transfer and skin friction was experimentally evaluated for turbulent wall jets [39C]. A new friction formulation was presented. Turbulence measurements were made in a heated wall jet on a convex wall to investigate the effect of curvature on turbulence structure [9C]. Locations of zero shear stress and streamwise heat flux were displaced, by curvature, further from the point of maximum velocity. The Stanton number is reduced more, by curvature, than the skin friction coefficient and temperature profiles depart from the flat-plate log law. The effect of concave curvature on turbulent boundary-layer heat transfer was experimentally evaluated using liquid crystal techniques [46C]. Large-scale streaks would appear and disappear at random but no stationary, Görtler-type vortices were observed. Experimental results were presented which show the unsteady heat transfer from a uniformly-heated flat plate in an oscillating air flow [12C]. The heat transfer coefficient depends upon the frequency of oscillation. A maximum value was found at a particular frequency. Results of experiments were given for leading edge heat transfer on cylinders with periodic variation of the angle of attack [31C]. The largest effect was a 10% augmentation at the nominal stagnation point. This effect was dependent on the Strouhal number, turbulence level and turbulence length scale. The thickness of the thermal boundary layer on the cylinder wall of a spark-ignition engine was measured using schlieren photographs [27C]. A maximum thickness of 2 mm is observed at the end of expansion. Reduced engine speed resulted in increased thickness. Measurements were presented for unsteady heat transfer from cylinders in an oscillating wake flow caused by an upstream cylinder [24C]. Maximum heat transfer was at the reattachment point of the shear layer which separated from the upstream cylinder.

An analysis was presented for cooling a low-heat-resistance stretching sheet moving through a fluid (a rejoinder) [3C]. Experimental results were given for heat transfer from a cylinder with close-fitting fabric in crossflow [48C]. A dependence on fabric thickness and state of contact was shown. Several fabric types were tested. The effect of finite heated plate width on heat transfer to turbulent flow was investigated [37C].

The effect of inserting heat transfer strips in a boundary layer was experimentally investigated [13C].

The distance between the strip and the wall and the angle of inclination of the strip were varied. Maximum augmentation was found at a particular separation distance and a 45° inclination. The augmentation of turbulent boundary-layer heat transfer on a flat plate due to three-dimensional protuberances was experimentally investigated [21C]. An optimum size/geometry was presented. An analysis was made to determine the optimum configuration of extended surfaces taking radiative interactions into consideration [51C]. Experiments made to evaluate the effects of length-to-diameter ratio and number of rows on the performance of short pin fins showed that the array-averaged performance was enhanced with increasing number of rows [4C]. Effects of pin fin spacing, height and channel height-to-length ratio were experimentally assessed [41C]. Increasing spacing was shown to be a means of reducing pressure drop while keeping heat transfer effective. Two potential geometries for pin fin arrays were experimentally evaluated [33C]. One array was found to augment heat transfer while decreasing pressure drop. Pin surface heat transfer coefficients were approximately twice end-wall values.

Experimental results were presented for evaluating the effects of freestream turbulence intensity on local heat transfer from a circular cylinder in crossflow [54C]. A correlation is proposed for this effect. Heat transfer from a long, thin hot-wire in high-speed subsonic flow was presented using a dimensionless approximation [20C]. Flow slip and thermal accommodation were included. Experiments were made to see the effects of variable properties, transverse wall curvature and freestream turbulence on cylinder external surfaces when the flow is axial [2C]. A correlation which includes these effects was presented. Heat transfer and flow behavior was experimentally investigated for flow over an elliptic cylinder with various angles of attack [40C]. Highest mean heat transfer values were found with angles of attack of 60–90°.

The effect of the wall on heat transfer from a cylinder perpendicular and adjacent to the wall was experimentally studied [17C]. Increases of 90–700% were observed very near the wall. The three-dimensional vortex structure effecting this increase was discussed. Local mass transfer and pressure coefficient data was taken on the side and the exposed end of a short cylinder perpendicular and adjacent to a flat plate [29C]. The effects of slenderness and the boundary-layer thickness were documented. Side-wall Nusselt numbers were reduced from those of an infinite cylinder due to separation. The heat transfer coefficient on the free end was higher than that of the two-dimensional cylinder. Measurements of quasi-local (axially) mass transfer coefficients were made in the wall–cylinder interaction region [47C]. Wall effects were observed within one cylinder diameter of the wall resulting in a decrease of the mass transfer rates of up to 15%.

Measurements of heat transfer from a heated cylinder in crossflow within an array of identical cylinders were performed [45C]. The addition of upstream rows of cylinders increased heat transfer by 50%. Fewer in-line rows were needed to effect this increase than staggered rows. An investigation of the promotion of heat transfer from a heated pipe by corona discharge was made [23C]. The pipe was put into an air jet and into a corona wind—the results were compared and it was found that the two were similar. Experiments and analysis of boundary-layer flow around a heated cylinder in crossflow showed that with increasing wall temperature, local Nusselt numbers increased more if the fluid were propane than if it were nitrogen [35C]. This is due to pyrolysis of propane and the stronger dependence of properties on temperature. The effect of freestream turbulence level on heat transfer from rectangular prisms in crossflow was experimentally investigated [32C]. Turbulence levels of 5% enhanced heat transfer by as much as 55% over that of laminar flow.

Local heat transfer coefficients were measured on a gas turbine nozzle guide vane surface [38C]. This data was presented as verification data for numerical code development. Short-duration heat transfer measurements were made on a flat plate and a gas turbine nozzle air foil at high freestream temperatures [22C]. Concave-surface heat transfer coefficients were high and convex-surface values were much lower than those for a flat plate.

Weak shock waves due to rapid heat addition at the boundary of a confined gas were investigated [7C]. The shock wave has the effect of a piston on the gas. The shock strength is directly related to the duration of heating. A linear acoustic wave is generated in a thick layer adjacent to the surface. The birth and evolution of a strong shock wave due to rapid boundary heating was next analyzed [8C]. The heat flux rises to full value with a modest multiple of the intermolecular collision time. The solution shows a steady shock wave propagating away from the boundary—supporting the moving piston analogy.

Flight data and design prediction values were compared for the shuttle orbiter base heating load

[18C]. Contributions to the heat transfer came from the orbiter main engine and booster plume radiation, freestream air convective cooling, and reversed plume convective heating. A numerical study was made to determine the excessive heating in the tile-to-tile gap due to stepped tiles of the Shuttle Orbiter thermal protective system [42C]. Results show that the tolerance on step height must be tight, or filler material between tiles should be applied. Experimental results on a scale model of the Shuttle Orbiter were presented to show the effects of Reynolds number, Mach number and density ratio on heating distributions upon re-entry [34C]. Results show that the three are essentially negligible for the hypersonic portion of each orbiter flight.

FLOW WITH SEPARATED REGIONS

An analysis considered heat transfer around a heated circular cylinder in a rarified gas at low Mach number covering the range from slip flow to near free molecular flow with results which agree well with previous publications [10D]. Experiments on heat transfer in the separated regions of circular cylinders demonstrate [5D] that the r.m.s. fluctuating pressure at the downstream stagnation line is the dominant parameter determining heat transfer in the separated region. The wake region could be controlled by two small cylinders or a splitter plate. The frequency of vortex shedding in tube bundles was measured and found to be 2–5 times the Strouhal number [11D].

Heat transfer to the separated flow regions was also measured at different angles of attack for a rectangular prism with a side ratio of 4:1 [7D].

The relation

$$Nu_s = 0.0345\, Re_s^{0.46}\, (w/s)^{-0.142}$$

with s denoting the cavity depth and w its length describes the results obtained by a finite-difference analysis of laminar separated forced convection in rectangular cavities [4D]. A heat transfer study [9D] for the heated bottom of a cavity revealed three different regions for the dependence of Nusselt number on Reynolds number when the Reynolds number was based in the channel height downstream of the cavity.

Interesting photos visualizing the character of radial outward flow between two parallel disks [8D] demonstrate that the flow separates alternately from the two disk surfaces creating a vortex street. The flow was found to be rotationally symmetric. Heat transfer coefficients were much larger than values calculated for steady uniform flow.

The heat–mass transfer analogy is discussed for separated flow [6D]. A numerical analysis for flow through a pipe with abrupt expansion demonstrates [1D] that a three-layer model of turbulence gives best agreement with experimental results. A similar study of separating and reattaching flow [2D] concludes that a Reynolds-stress turbulence-closure gives better results than the k–ε model.

Experiments [3D] on temperature separation in vortex tubes (155 and 800 mm diam.) had the aim of determining the shortest length which still results in good separation. It was found that the efficiency can be as high for a length to diameter ratio of 6 as for longer tubes. It is pointed out that compressibility is essential for the energy separation.

HEAT TRANSFER IN POROUS MEDIA

Eighty per cent of the large number of papers in this area deal with convection in a saturated porous medium. Most of them are based on analysis or numerical simulation.

Thermal convection in a medium heated from below is treated extensively: analysis and experiments determine the time required for onset of thermal convection as a function of Rayleigh number [20DP]. The onset of thermal convection is also analyzed for a porous medium with continuous horizontal periodic stratification of the permeability [15DP]. Horizontal temperature gradients are found to inhibit natural convection [36DP]. A time-dependent temperature stratification influences the critical Rayleigh number depending on the ratio of porosity to permeability [21DP]. This Rayleigh number increases linearly with the ratio for values larger than 1000. Counter-rotating cells are formed in a porous medium bounded by a horizontal wall with hot and cold spots [28DP]. The height of the cells above the wall is equal to $\lambda Ra^{1/2}$ with λ denoting the distance between hot and cold spots and Ra denoting the Darcy Rayleigh number. The critical Rayleigh number and the heat transfer are analyzed for a porous medium consisting of alternate thin layers of high permeability and thick layers of small permeability [22DP].

Natural convection was also studied for rectangular enclosures [29DP]. Two cells form in such a cell when it is differentially heated in the horizontal direction. The temperature gradients are concentrated in thermal boundary layers at larger Rayleigh numbers. A numerical simulation [6DP] studied natural convection near 4°C in water filling the voids of a porous medium heated from below. The free surface on a liquid filling a porous slab heated from its sides is determined by numerical calculation for Rayleigh numbers up to 1000 [23DP]. Multicellular convection flow is created in a rectangular porous cavity with vertical walls at different temperatures for certain aspect ratios [31DP]. The result of a numerical calculation of natural convection in a rectangular porous cavity with a constant heat flux on one vertical wall and the other wall isothermally cooled could be expressed by the equation

$$\overline{Nu} = \text{const}\ Ra^m A^n$$

with A denoting the height-to-width ratio [30DP]. The flow was analyzed [10DP] for the corner region between a vertical heated wall and a lower adiabatic wall and for different angles between the two walls. Boundary layers are formed on both of these walls. Transient free convection about a vertical flat plate imbedded in a porous medium was studied [8DP] for a step increase in temperature or in surface heat flux. An analysis of forced convection in a packed bed near an impermeable boundary demonstrates a channelling effect with a velocity maximum near the wall [35DP]. The results reported for low Reynolds number describe thermal convection in a porous medium filling the space between two horizontal eccentric cylinders with the small cylinder either inside or outside of the larger one or the outside cylinder degenerating to a horizontal plane wall above the other cylinder [4DP]. Heat transfer in a horizontal eccentric annulous can be optimized by proper choice of the eccentricity [3DP]. Convective heat transfer was also analyzed for a pipe buried in a semi-infinite porous medium with and without fluid motion through the pipe wall [2DP]. The influence of injection or withdrawal of fluid on free convection around a vertical cylinder has been calculated for a prescribed constant wall temperature or heat flux [40DP]. Transient heat transfer between a fluid and a cylindrical porous bed is increased and the time of approach to steady state is decreased as the flow rate and particle diameter increase [18DP]. The dynamic response of heat transfer to an input pulse was measured and the results were compared with an analysis for a vertical test column with forced flow through it [11DP]. Axial dispersion was found to be important at low Reynolds numbers.

The effect of the ratio of the tube diameter d_t to the particle diameter d_p in the range 3–12 on the radial mass transport was studied experimentally and with a continuum model analysis. The results could be described by the following relation

$$Sh = \left[1.0\text{–}1.5\left(\frac{d_p}{d_t}\right)^{1/5}\right] Sc^{1/3}\, Re^{0.59}$$

in which Sh denotes the Sherwood number, Sc the Schmidt number and Re the Reynolds number [12DP]. The local distribution of the pressure coefficient and the heat transfer coefficient over the surface of a sphere in a bed of packed spheres has been measured [41DP]. An analysis of fluid flow and heat transfer at the edge of a packed bed is based on a model simulating the particles by a row of rectangular bars [9DP]. Results are obtained for Reynolds numbers from 0 to 100 and for Prandtl numbers from 0.72 to 7. The thermal conductance for a granular bed was measured over a wide temperature range for opaque particles at high temperature [42DP]. Axial effective thermal conductivities for flow at low rates were expressed as the sum of a conductivity without flow plus a term depending on the flow rate [37DP]. New correlations for the conductivity were presented for various gases.

A method was presented to calculate the influence of chemical reactions on radial heat transfer in fixed bed reactors [24DP]. It was found that in this case the Soret parameter plays a major role in determining the stability of the system [32DP]. Laser monitoring of the transfer processes in capillary porous bodies is studied analytically [25DP]. The transient response of a packed bed to thermal energy storage was analyzed for various Prandtl numbers [5DP]. The results which compare well with experiments indicate a pronounced effect of the void distribution. A new concept for thermal insulation in flow systems uses a porous segment in the duct causing back radiation [38DP]. It is proposed for the recovery of energy in furnaces or in chimneys.

The ratio of thermal to moisture diffusion coefficients for two unsaturated soils was obtained from the measured temperature and moisture fields [33DP]. Experiments on heat and mass transfer in unsaturated wood establish that total pressure differences in the air space contribute to water movement [1DP].

Semi-empirical equations describe the height of dry-out regions in packed beds heated from below [34DP]; they are based on a counter-current flow model in deep beds and on the boiling crisis in shallow beds and compare them with experimental results [19DP]. An analysis of heat convection on the drying of a porous semi-infinite space heated from a vertical wall assumes a sharp interface between the vapor layer at the wall and the liquid in the bulk of the fluid [17DP]. It thus disregards the capillary effects which make this boundary diffuse.

A small number of papers only treat heat transfer in a fluidized bed. An analysis studying heat transfer between the wall and a fluidized bed at high temperature concludes that radiation adds 15% to heat transfer for opaque particles and 40% for transparent particles at a bed temperature of 900° [13DP]. Wall-to-bed heat transfer coefficients in gas–liquid–solid fluidized beds were measured with flow of water and air through the bed [26DP]. Particles ranged from 0.61 to 6.9 mm in diameter and between 1330 and 3550 kg m^{-3} in density. Experiments measured the thermal resistance between the bed surface and adjacent particles in a fluidized bed for a very rapid transient process (10 ms) [16DP]. No difference was observed between the thermal resistance of a lightly fluidized and a packed bed. Heat transfer from a horizontal tube in a fluidized bed was found [27DP] to be different depending on whether the ratio of tube to particle diameter was smaller or larger than 10–20. Local heat transfer coefficients for a single horizontal tube in a larger particle fluidized bed at a temperature from 800–2000 K exhibit a maximum near 90° [14DP]. The undesirable effect of electric charges on heat transfer between a bed and an immersed wall can be overcome by adding a small amount of fine material [39DP]. An increase of the heat transfer coefficient up to 300% was obtained in experiments. Addition of coarse particles to a fine particle system improves fluidization and increases heat transfer [7DP]. The optimum is 25% coarse particles with seven times the diameter.

EXPERIMENTAL TECHNIQUES AND INSTRUMENTATION

Articles on experimental techniques and instrumentation which specifically include heat transfer or include the measurement of thermal diffusion properties or diffusion rates have been added this year to the heat transfer review.

A method for measuring local thermal conductivities of solids and porous media was presented. Measurements were made such that thermal contact resistance errors cancelled [4E]. A method for measuring thermal conductivity of transparent liquids was presented which employs photon correlation spectroscopy [6E]. Measurement of production rate, thermal diffusivity and conductivity of solids using the principle of Laplace transform was presented [7E]. It has the advantage of being usable under arbitrary heating and cooling conditions. A method for designing experiments for measuring fixed-bed thermal transport properties was given which extends steady-state statistical design concepts to dynamic tests [11E]. A method which uses dynamic light scattering to measure thermal diffusivity of pure and binary fluids was discussed [13E]. Its use over an extended range of states was demonstrated. A simple high-temperature thermal diffusivity measuring apparatus was described which employs a light pipe, a long furnace and a differential thermocouple [16E]. A technique for measuring humidity above 100°C with a double-wick psychrometer was presented [8E].

Temperature gradients in liquids were measured using the phenomenon of laser speckle [14E]. Data was taken where a liquid was heated from below—heat transfer was by conduction. The use of liquid crystal thermography was demonstrated for heat transfer measurements in water tunnel flows [12E]. Error due to wall shear was addressed.

The effect of flow conditions and gage geometry were experimentally examined for a free-standing circular disk heat flux gage [18E]. A correlation of Nusselt number with angle of attack and Reynolds number was presented. A swollen polymer technique was presented for *in situ* measurements of local mass transfer rates [3E]. Holography and electronic speckle pattern were used to measure the degree of swelling. Precise natural convection heat transfer rates were obtained with balance and gradient methods [10E]. Construction and use of the apparatus to do so were discussed.

Theoretical and experimental evaluations of the frequency response of fine-wire thermocouples was presented for heated air flow [5E]. The influence of property variation in the surrounding gas was addressed and the effect of thermocouple geometry was discussed. The transfer functions for cold-wire temperature sensing probes were determined from a dynamic calibration system [9E]. They were then used to improve the dynamic response in turbulent boundary-layer temperature measurements. Experiments were made to demonstrate a new technique for measuring temperatures in flowing gas–solid suspensions with a thermocouple [2E]. Self-heating of ceramic-encapsulated rhodium–iron alloy resistance thermometers was investigated for low temperature operation [1E]. The thermometers were shown to be very stable with repeated thermal cycling.

Experimental results were presented which investigate the measurement of turbulent structure with hot-wires having wire lengths of about one-third the viscous length [15E]. A guide to the evaluation of uncertainties for hot-wire measurements was presented [17E]. Real-time data reduction techniques are shown to have lower uncertainties than time-averaging techniques.

NATURAL CONVECTION—INTERNAL FLOWS

Buoyancy-driven flows continue to be of great interest to the research community, serving as an excellent means for studying fluid-mechanics problems including the basic nature of turbulent flow and transition, as well as being a necessary component of a number of practical engineering problems. Studies on buoyancy-driven flow in differentially-heated plane horizontal, vertical and inclined layers remains of interest, including the effects of a density extremum such as occurs in water near 4°C. Heat transfer to fluids in nonplane cavities including annuli remains of interest. Double-diffusive convection, heat transfer in porous media, the effect of fires on flow, thermosyphons and mixed convection all have gained considerable attention recently.

Plane fluid layers with temperature differences across them approximate a number of physical systems, from windows to solar energy collectors, to planetary atmospheres; even flow in the interior layers of stars can be approximated by such a model. For some years, attention has been and continues to be directed towards horizontal layers heated from below. One key interest relates to getting a better numerical and physical understanding of turbulent flow. Interest has been accentuated recently with descriptions of physical phenomena involving transformation to chaos.

The energy method has been used to predict Nusselt number, temperature distribution and maximum velocity in a horizontal layer heated from below [34F]. Analysis indicates that for a Prandtl number greater than one, a square planform can result when a layer is heated from below if the wall conductivity is of the order of that of the fluid or smaller [35F]. A three-dimensional simulation of Bénard convection indicates the transition from laminar to chaotic flow [16F]. Linear theory leads to a prediction of the turbulence profile in thermal convection which agrees with measurements [28F]. Numerical and analytical predictions of convection in a rectangular layer agree with each other [41F]. The convection in a layer with stress-free boundaries can offer potential for detailed studies of post-stability flow (9F).

The interaction between two horizontal layers, one over the other, and separated by a partially-conducting horizontal plane was examined [11F]. The influence of vertical vorticity on Bénard convection has been described [50F]. The ratio of solid to liquid thickness in a horizontal layer freezing from above determines whether the flow pattern is hexagonal cells or rolls [17F]. Numerical study of convection in water near its maximum density indicates the influence of aspect ratio on transition [7F]. The effect of a density extremum stabilizes a layer of water heated from below [29F].

A linear analysis agrees with experimental results on the initial onset of flow for sudden heating from below [38F]. A nonlinear stability analysis predicts a planform of stable hexagons within a thin fluid layer having a sinusoidal variation of bottom wall temperature [59F]. For a horizontal layer having shear-free upper and lower surfaces and internal energy sources, the time for onset of flow following a decrease in upper surface temperature increases as the Prandtl number of the fluid decreases [39F].

Studies on differentially-heated rectangular layers where the temperature gradient is primarily in the horizontal direction, include stability studies and the influence of turbulence effects. The conditions under which steady state is approached by oscillatory conditions in a low-aspect-ratio layer have been explored [53F]. In a square cavity with differentially heated walls, the k–ε model has been used to predict turbulent transport [45F]. In an open vertical channel with side wall heating, recirculating flow can be drawn into the channel from above [65F]. Relations show the optimal spacing in a vertical channel for heat removal in electronic system [3F]. Flow has been visualized in a cubical cavity, with a transparent top wall heated from any of its vertical walls [8F]; a relatively inactive core was observed surrounded by boundary layers on the walls.

A two-equation model for turbulent flow was extended to natural convection in internal flows [60F]. Visualization using liquid crystals has been useful in studying convection in cavities [57F]. Measurement and analysis indicate three-dimensional natural convection patterns in flow that occurs in glass-making furnaces [15F].

Numerical analysis shows the influence of a semicircular protuberance on the two-dimensional flow in a square cavity [37F]. In a low-aspect-ratio differentially-heated cavity, a vertical partition projecting part way across the channel can strongly influence the flow field [51F].

Measurements with inclined layers of fluid at a Rayleigh number of 2×10^6 show different flow regimes at different angles of inclination [33F]. A series of papers [30F–32F] on convection in an inclined layer with water near its density extremum show the flow pattern and temperature distribution, and provide heat transfer correlations from both experimental and numerical results. In a differentially-heated cavity, wall conduction can stabilize or destabilize the flow depending on the orientation of the heading and cooled surfaces [40F].

A number of studies have been directed toward buoyancy-driven flow inside circular cylinders or pipes. LDV measurements show a region of three-dimensional flow in a horizontal cylinder of low aspect-ratio with different end temperatures [61F]. A numerical solution for buoyancy-driven flows in a vertical tube with a temperature-dependent viscosity shows large effects from variable properties when properties of real fluids are used [72F]. The potential of reduced heat transfer coefficient in the supercritical region due to large variations in properties has been studied in vertical tube flow [47F]. Vortex rolls induced by corona discharge increase free convection inside a vertical cylinder [18F].

Buoyancy-driven flows are often important in the space between an object and an enclosure that surrounds it, such as the annulus between two cylinders, space between two spherical shells, etc. Visualization and numerical calculations indicate the flow and temperature distribution in a vertical annulus between two concentric cylinders in which the inner cylinder is partially heated [22F, 23F]. The effect of variable properties on the heat transfer across the annulus between concentric horizontal cylinders is small for moderate temperature differences [26F]. The transient time for natural convection to start in an annulus when one of the cylinders is suddenly heated is small [68F]. Streamlines indicate three-dimensional flow in the form of a coaxial double helix in an inclined cylindrical annulus [67F]. The Nusselt number–Rayleigh number relationship has been determined for finned tubes in a circular container [5F]; there is little effect of the enclosure for diameter ratios of 25 or more. Convection in a spherical shell has been studied by numerical analysis [19F]; for a radius ratio >41 the heat transfer is essentially equivalent to that from a single sphere in an infinite medium. A perturbation technique [58F] predicts three-dimensional steady convection in a spherical shell. Experiments with one cylinder inside another cylinder show that the Nusselt number is insensitive to the position and orientation of the inner cylinder [66F].

Studies on open and closed large-scale cavities have been used to simulate the progress of fires. A numerical model of buoyancy-driven combusting fluids is suitable for examining the progress of a fire in an enclosure [44F]. Fluid dynamics of large scale fires have been analyzed with a model that simulates the combustion process by a volume heat source [64F]. The velocity profile in a cavity opened on its side has been measured [25F]. The interaction of sprays with buoyant flows is strongly dependent on the relative momentum of the spray droplets and the buoyant flow [1F].

Studies on thermosyphons in which fluid passes through a continuous loop driven by buoyancy from localized heating, include single- and two-phase systems. The variation of friction and heat transfer in a toroidal natural circulation loop has been analyzed for laminar flow [48F]. An analysis of the stability of flow in a closed toroidal loop indicates transition to steady or chaotic flow as opposed to the periodic motion [24F]. A survey of the literature on two-phase thermosyphons has been reported [54F]. The maximum heat transfer in a change-of-phase thermosyphon with internal baffles has been measured to find the effect of area reduction on the flow [6F]. Relations have been developed for determining the maximum heat flux in an evaporating–condensing thermosyphon [70F]. A scale model of a pressurized water reactor was used to study the natural circulation heat removal following loss of pumped flow [63F].

Buoyancy-driven flows in porous media have been studied for a variety of geometries. Within a vertical annulus filled with porous material, the heat transfer in the boundary-layer region is only a weak function of aspect ratio [55F]. An analysis of higher-order corrections to the boundary-layer solutions were presented for laminar vertical plumes and flow along a vertical heated plate [36F]. Numerical analysis of heat transfer in saturated porous media with internal energy sources includes the effect of a liquid layer on top of the porous media [62F]. Flow in porous material with internal energy sources has been analyzed with an expansion technique for low Rayleigh number flow and a linearization method at high Rayleigh number [20F]. With an internal energy source in a horizontal annulus filled with porous media, the maximum temperature can be higher than that which would occur with pure conduction [71F]. A study of three-dimensional convection in a cubical box of porous media includes effects with fluids that have a maximum density at an intermediate temperature in the material [2F].

Double-diffusive convection occurs in fluids in which buoyancy forces are provided by two-diffusion phenomena, usually diffusion of heat due to temperature difference and diffusion of mass due to concentration difference. A conference summarized the state of knowledge and research activities on double-diffusive convection including a description of important applications of such flows [13F]. A theoretical analysis provides insight into thermohaline convection in containers of arbitrary shape [52F]. The development and stability of salt fingers which can occur when a layer of warm salty water is present above a cooler, fresher layer of water was examined [27F]. Experiments on a stratified layer of salt water under a layer of fresh water with heating from the side walls simulates roll-over which can occur in LNG containers [49F]. Numerical calculations and experiments show the transient convection when an initially stably-stratified horizontal layer is heated from below [69F].

Instabilities can develop in an initially thermally-stabilized plane layer when a DC voltage is applied across it [46F]. Pumping produced by traveling electrical waves superimposed on a dielectric fluid with a transverse temperature field has been studied using finite-element methods [42F]. Significant heat transfer enhancement is predicted when ions are created at an electrode by an electrochemical reaction in a fluid layer heated from below [10F].

Combined or mixed convection occurs when the driving force for flow is induced both by buoyancy and by some overall pressure difference which may have been provided by a pump or blower. An analysis of mixed convection in developing flow between parallel walls includes the influence of viscous dissipation [4F]. A numerical analysis of convection in a vertical channel predicts the heat transfer for both aiding and opposing flows [14F]. An analysis of mixed convection in a vertical channel with wavy walls indicates the potential of four different modes of flows [43F]. The relative importance of buoyancy and centrifugal forces in developing flow in a curved horizontal channel depends on the ratio of Dean number to Grashof number [73F]. Significant buoyancy effects can occur with forced flow through a heated spherical annulus at low Reynolds number [56F]. Flow through a reservoir of fluid used for energy storage is strongly influenced by buoyancy forces [12F]. For flow through a low-aspect-ratio channel filled with porous media, there is a considerable range of conditions under which the Nusselt number can be obtained from simple addition of forced and natural convection correlations [21F].

NATURAL CONVECTION—EXTERNAL FLOWS

An asymptotic solution of the Navier–Stokes and energy equations for the problem of free convection from a vertical finite plate is considered [20FF]. The structure of natural convection in flat, variously oriented layers of liquid and on a vertical wall is described [17FF]. Highest-order effects of Darcian free convection boundary-layer flow adjacent to a semi-infinite vertical flat plate with a power law variation of wall temperature were examined theoretically [5FF]. The fire plume along a vertical wall is analyzed using a laminar boundary-layer model, including finite-rate, gas-phase chemical kinetics [4FF]. Axisymmetric free convection flow of a viscous incompressible fluid above a horizontal plane surface is analyzed when the temperature of the plane surface oscillates about a nonzero mean [3FF]. Turbulent convective heat transfer, with appreciable buoyancy effects, over a heated or cooled horizontal flat plate is numerically analyzed by solving four equations for mean square temperature variance, its rate of destruction, turbulent kinetic energy and the rate of kinetic energy dissipation [6FF]. Approximate analytical solutions using a momentum integral method were used to study the effect of longitudinal oscillations on free convection from a horizontal plate [12FF]. Two-dimensional natural convection below an infinite strip and axisymmetric convection below a circular plate are calculated for a uniform surface temperature and a uniform surface heat flux for the special case of liquid metals [25FF]. The effectiveness of a stabilizing solute concentration gradient in delaying the onset of convection in a horizontal fluid layer subjected to transient cooling from the top was investigated [15FF].

A finite-difference numerical scheme was used to study combined free- and forced-convection boundary-layer flow in a porous medium near a vertical wall and the model results show that the inertial and viscous effects have a significant influence on the velocity profiles and heat transfer rate at the surface [23FF]. A description is presented of the heat transfer and fluid flow through a confined porous layer heated and cooled along one of the vertical side walls [22FF]. The boundary-layer solution is described for the problem of heat transfer along a vertical wall facing a porous medium, in the general case where the pore Reynolds number is not necessarily less than order one [2FF].

A comprehensive experimental study was performed to determine the natural convection heat transfer characteristics of a heated horizontal cylinder situated in a vertical channel in air [27FF]. Steady three-dimensional conjugate natural convection heat transfer from a horizontal isothermal cylinder with infinitely large transverse isothermal fins was studied numerically [29FF]. A study was conducted to determine the influence of fin length on steady, conjugate, natural convection heat transfer from an isothermal horizontal cylinder with one vertical longitudinal conducting plate fin [18FF]. The transition from conductive to convective heat transfer has been investigated for horizontal cylinders (platinum wires in various liquids), subjected to internal heat generation which increases linearly with time [10FF].

A formal, unified method is used to derive two sets of approximate equations governing the thermally-driven motion of strongly heated fluids [9FF]. A two-phase mathematical model is proposed for calculating the induced turbulent vertical flow and the temperature field in buoyant plumes driven by rising gas bubbles through a moderately stratified liquid environment [13FF]. An analysis is presented for mixed convection effects in a wall plume flow [16FF].

A solution is presented for self-similar mixed convection flow over a horizontal plate covering a wide range of conditions, beginning with purely natural convection and extending to forced convection separating flows [1FF]. A solution to the problem of mixed convection from an inclined vertical plate to a non-Newtonian fluid was presented [26FF]. Laminar mixed convection from a line thermal source imbedded at the leading edge of an adiabatic vertical surface is analytically investigated for the cases of buoyancy assisting and buoyancy opposing flow conditions [24FF]. An analysis is developed from mixed, forced and free convective combustion on a flat fuel surface of arbitrary inclination that makes use of the laminar boundary-layer approximation to describe the gas flow and of the flame-sheet approximation to describe the gas-phase reactions [19FF]. The desire to predict the convective heat loss from the solar receiver of a solar/electric power plant was the motivation for an experimental investigation of combined heat transfer from an isothermal vertical cylinder to a surrounding gas [7FF]. A finite-difference scheme is used to investigate mixed convection in flow through a finite rod array whose geometrical dimensions simulate those of a CANDU-type pressurized water reactor [8FF].

A mathematical formulation is presented for the transient development of the fluid-flow field and the temperature field in a liquid pool, generated by a spatially variable heat flux falling on an initially solid metal block [21FF]. The convective rates of heat transfer through inclined longitudinal slots are studied for the case where heat is transferred from a lower heated isothermal surface, through the slots, to an upper, cooled isothermal surface [28FF]. Thermal instability and heat transfer by natural convection in an inclined rectangular water layer at low Rayleigh number have been studied using an interferometer [11FF]. The natural convection motion and heat transfer rate in an inclined rectangular air layer in which two opposing isothermal rigid walls were kept at different temperatures were investigated experimentally for various angles of inclination, aspect ratios and Rayleigh numbers [14FF]. Convective thermo-capillary instabilities in liquid bridges have been investigated [30FF].

CONVECTION FROM ROTATING SURFACES

The temperature distribution in a rotating cylinder is considered [16G]. Experiments are reported for a heated rotating cylinder in crossflow [10G]. Numerical solutions are presented for laminar convection between concentric and eccentric heated rotating cylinders [8G].

Heat transfer augmentation on a rotating cylinder by means of surface projections has been considered [12G, 14G]. A theoretical and experimental investigation of turbulent separated flows behind a rotating body is reported [9G]. Flow in circular tubes rotating about a parallel axis has been studied [7G]. Finite-difference solutions are presented in ref. [15G] for heat transfer in a metal-forming roll rotating at high speed. An integral analysis is given for the mixed convection from a rotating cone [6G].

Reference [13G] presents a numerical study of heat transfer in laminar flow through co-rotating parallel disks. Marangoni convection in a rotating fluid layer is considered in [5G]. Measurements are reported for the heat transfer coefficients in a rotating triangular duct [3G]. Finite-difference solutions are given for the flow between two rotating spheres [4G]. The rotating flow over an infinite porous disk is analyzed in [2G]. Heat transfer to rotating cryogens is considered in ref. [11G]. Reference [1G] deals with special wave effects in rotating media.

COMBINED HEAT AND MASS TRANSFER

For the purpose of this review the category of "combined heat and mass transfer" includes studies related to jet–mainstream interactions, such as film cooling and implingement heat transfer, as well as transpiration cooling and simultaneous heat and mass transfer in a number of applications including combustion.

Film cooling has been applied in many important systems, such as gas turbines, to protect surfaces exposed to high temperature gas flows. Experiments on the flow through an array of holes (shower head) near the leading edge of an air foil show the effect of injection (hole) angle and external flow pressure on the coefficient of discharge for the cooling holes [23H]. Injected jets separate from a convex wall at relatively high blowing rates, therefore, providing better cooling of the convex surface at moderate low and blowing rates than occurs on a flat surface [12H]. The characteristics of jets penetrating into a mainstream have been examined experimentally [25H]; these characteristics are important when trying to understand the different film cooling performance for different types of flow. The pressure drop across the large number of holes present in full surface film cooling, and the influence of interaction of heat transfer in the holes on film cooling performance have been studied [1H]. Experimental measurements using mass transfer to determine the effectiveness downstream of an injection slot show the influence of a downstream jet as could occur on a combustor lining [4H]. A model was set up to predict the heat transfer on a permeable surface with injection [18H]. Ablation of Teflon models in a Mach 5.74 tunnel with stagnation temperatures up to 1070° show the importance of the angle of attack on the shape changes in the model [3H].

Experiments on slot jet impingement cooling on the inside of turbine blades indicate a correlation of local Nusselt number with Reynolds number raised to the 0.6 power [16H]. Heat transfer from impinging jets to a flat plate on which a spike and external ring are present gave a much higher local heat transfer rate than on a surface without the protuberance, though the average heat transfer is not significantly changed [9H].

Another study of impingement heat transfer related to turbine blade cooling, indicates the effect crossflow air temperature has on impingement heat transfer as would occur at the mid-chord of a blade [6H]. The influence of external fluid temperature, different from that of the jet temperature on heat transfer has been studied from measurements with single circular jets [8H], and analysis [21H] and measurements [22H] on plane jets. Temperature profiles in jets mixing with a hot crossflow as occurs in a combustor, were studied experimentally and analytically [24H]. The impingement flow heat transfer results for different slot nozzle geometries can be correlated with an effective slot width proportional to the discharge coefficient of the nozzle [7H]. Heat transfer correlations were obtained from mass transfer measurements on a jet impinging on a cylinder, both with the jet in line with the axis of the cylinder [19H] and with it off-set from this axis [20H]. An analysis of the cooling effect of a jet impinging on a liquid surface first predicts the shape of the interface where the jet strikes the liquid [26H].

Experiments indicate the time to reach steady-state heat and mass transfer for natural convection surrounding a cooled horizontal tube in moist air [5H]. An analysis of a falling liquid film shows the influence of heat transfer on absorption; both the Sherwood and the Nusselt numbers reach asymptotic values a short distance from the start of the film [11H].

A special thermocouple probe has been used in a system similar to a thermosyphon to determine the natural convection during crystal growth [2H]. The effect of thermophoresis on particle flow deposition includes the influence of a transpiration cooled wall [10H]. The combined heat and mass transfer to particles in small Péclet number flow has been examined analytically to help understand chemical reaction on the surface of a particle [17H]. Experiments on the turbulent shear layer between adjacent flows of hydrogen and fluorine indicate a very large structure for the region of flow interaction that would influence combustion [15H].

The influence of porosity on the heat transfer from a permeable sphere to a low Reynolds number flow is relatively small if the Péclet number is large but is significant at small Péclet number [13H]. Analysis of the interaction of heat and mass transfer at an interface on which a soundwave is reflected includes a calculation of the time-averaged energy dissipation [14H].

BOILING

Pool boiling and nucleate boiling

A brief and humble personal recollection of the late Professor Nakayama's pioneering work in boiling heat transfer was published [44J], along with an English translation of his original paper of 1934 [45J] which established the fundamental characteristics of the boiling curve.

A kinetic theory approach was used to examine the theory of evaporation and condensation in the limit of continuum behavior [30J]. Mass fluxes predicted were in excess of the Hertz–Knudsen prediction, and the temperature profile appeared inverted from that expected, casting some doubt on the fundamental theory. Simple thermodynamic analysis of two hypothetical cycles was used to determine the relationships governing equilibrium radii of small vapor bubbles and liquid droplets [19J]. Stability and incipience theories for vapor nuclei were confirmed in experiments conducted with a specially prepared glass cavity and uniformly superheated liquid [15J]. New models of bubble growth rate and departure volume were developed for isolated nucleation sites and were shown to compare favorably with new and existing data [58J] while growth rates of groups of bubbles in a superheated liquid were analytically shown to be substantially less than those of isolated bubbles using an asymptotic matching of an inner region solution at the scale of bubble radius and an outer region solution at the scale of the bubble separation distance [8J].

An experimental and analytical investigation of nucleate pool boiling of water at 1 atm on a flat surface with varying angle relative to the horizontal demonstrated and postulated a model to explain that at low heat fluxes, greater angles (up to 175°, or very nearly facing directly down) produced higher heat transfer coefficients, while at higher heat fluxes the angle of inclination had little influence [42J]. Pulsed electrical heating was used to explore transient onset and development of boiling of liquid nitrogen on a horizontal wire and on a vertical plane film [18J]. The transient heating did not appear to induce premature transition to film boiling.

Enhancement of nucleate boiling was the topic of many of the papers reviewed. Enhancement of nucleate pool boiling by electrolytic production of hydrogen at the heat exchange surface was demonstrated [40J]. With simultaneous production of the gas bubbles, low wall superheats can produce large rates of heat transfer. The most prominent feature observed by the accompanying photographic studies was the reduction of the waiting period between bubble departure and the beginning of growth of the next bubble when hydrogen was also being produced at the surface. In a related article [41J] bubble frequency and bubble departure diameter were measured with this electrolytically-augmented boiling. The use of conical nucleation sites with reduced wettability was analytically explored [34J] and experiments were performed with the application of poorly wetting nucleation sites [36J]. Also reported were experiments designed to determine the optimum particle size for porous surfaces for nucleate boiling [16J]. An optimum particle size of 0.126 mm was suggested for use with refrigerant 113. Experiments with boiling of two binary mixtures on a commercially available enhanced boiling surface demonstrated that improvements in thermal performance, comparable to improvements in single component boiling, were achievable with mixture boiling but that activation superheats could be significantly higher than those predicted by a simple mixture interpolation [1J].

There were also several reports of mixture boiling on ordinary surfaces. Extensive experiments were conducted with pool boiling of binary mixtures of sulfur hexafluoride with three refrigerants over a wide range of heat flux densities, pressures and mixture concentrations [6J]. The reductions in heat transfer coefficients relative to those of pure substances were more pronounced at higher heat fluxes and at higher pressures. Concern with oil contamination within refrigeration systems led to experiments with nucleate pool boiling of mixtures of refrigerant 113 and oil [25J]. These experiments confirmed that substantial reductions in heat transfer coefficients result from increased concentrations of oil in such mixtures. A model incorporating the locally increased concentration of oil at the boundary of a growing bubble of refrigerant vapor was appended to the Forster–Zuber correlation, producing reasonable agreement with the experimental data. Oil–refrigerant mixtures were shown experimentally to exhibit dramatic increases in heat flux and wall superheat at critical heat flux as compared with the critical heat flux of pure refrigerant [24J]. Oily contaminants, especially those placed directly on the heating surface, were found to greatly reduce the effectiveness of nucleate pool boiling of water at atmospheric pressure [2J].

In other pool boiling studies, experimentation with varying arrangements of columns of heated and unheated horizontal finned tubes showed considerable enhancement of heat transfer coefficients by neighboring heated tubes at low heat fluxes, but performance approaching that of a single tube at high heat flux [67J]. Heat transfer characteristics of water boiling in a small thermosyphon were found to be comparable to those at 'small' surfaces in pool boiling, where 'small' surfaces have dimensions of the same order as the bubble break-off diameter [26J]. A simple model was developed to explain experimentally-determined variations in heat transfer coefficients and wall temperature during low heat flux heating of a liquid in a narrow rectangular channel with bubbles periodically injected into the channel [31J].

Flow boiling

Recognizing the utility of explicit functions for computer design and performance studies, the F and S functions of the Chen convective boiling correlation were presented in an analytical form [12J]. Local heat transfer was investigated for flow boiling of argon and of nitrogen in a horizontal tube [38J]. Heat transfer in co-current vertical two-phase flows was examined [48J]. Data of several investigations of two-phase heat transfer in two component vertical flow were correlated with a mixed flow Reynolds number determined as the sum of the liquid and gas Reynolds number determined as the sum of the liquid and gas Reynolds numbers [13J]. Another correlation based on momentum considerations was presented for bubbly and bubbly-slug flow in vertical tubes [11J].

An experimental study of flow boiling of refrigerant 22 inside tubes with a smooth wall, a smooth wall with a diametral fin, a microporous wall, and a smooth wall with a microporous diametral fin demonstrated that the microporous wall offered a marked improvement (up to a factor of 7.5) over the heat transfer coefficient of the smooth tube [23J]. In another augmentation study, a simple model was presented to estimate evaporation heat transfer in a channel filled with a highly conductive porous metal, and it was suggested that such a system would undergo no boiling crisis [37J].

Boiling in flow over rod bundles with blockages exhibited significant enhancement of heat transfer coefficients at certain locations downstream from the blockage [11J]. Correlations were developed from the experimental data for magnitudes and locations of the local enhancements. An experimental investigation of two-phase crossflow boiling heat transfer on a single horizontal cylinder showed large increases in heat transfer coefficients as compared with pool boiling [66J]. A correlation similar to the Chen correlation for convective boiling inside tubes was developed for this crossflow case and shown to fit reasonably well the data of the authors and that of two independent investigations.

Critical heat flux

An improved version of the critical heat flux correlation Katto proposed in 1978 has been developed to cover CHF in forced convective boiling in uniformly heated tubes, eliminating discontinuities between successive regimes [28J]. Additional experimentation expanded the data base for CHF in liquid helium to greater length/diameter ratios and demonstrated, contrary to earlier evidence at small length/diameter ratios, that helium also exhibits the 'high pressure character' in which the critical heat flux increases substantially with mass flux at large mass fluxes when vapor density is a large fraction of liquid density [29J]. two articles [60J, 61J] appear to be slightly different translations of the same Soviet source article, in which it is reported that no significant increase in wall temperature was evident at the transition from nucleate to film boiling of water in an 8-mm-diam. tube with pressures above 19 MPa (quality of 0.2), 17 MPa (saturated liquid) and 15 MPa (liquid subcooled by 25°C), suggesting transcritical heat fluxes with no tube damage. An experimental study of critical heat flux and post dryout heat transfer in helical coils was described [59J]. Results of local blockage experiments with water and sodium in model fuel subassemblies were summarized to estimate conditions of boiling and dryout applicable to liquid-metal fast breeder reactor (LMFBR) local fault assessment [69J].

An inclined plane surface used to investigate critical heat flux in flowing liquid films of water and refrigerant 113 produced nucleate boiling curves compatible with those of pool boiling rather than those of flow boiling, but demonstrated critical heat fluxes increasing with increased film velocity, both observations in agreement with other recent studies of liquid films. A model based on the competition between evaporation of a thin subfilm and replenishment of the film by droplets shed by the main film was shown to predict CHF for this system to within a constant [3J]. A correlation was presented for CHF in turbulent falling liquid films of water, ethanol and aqueous glycerol [20J]. Noting the difficulties of detecting dryout in a volumetrically-heated particle bed, an alternative experimental technique is suggested based on the additional time required for such a bed to reflood by gravity after the power input has been halted [56J].

Film boiling

Experiments with film boiling of water covering a pressure range of 20 kPa–2 MPa and heater diameters from 0.7 to 3.0 mm yielded: saturated film boiling heat transfer coefficients about 10% lower than predictions based on Bromley's expression; evidence that liquid/vapor velocity substantially contributes to subcooled film boiling; and evidence that the minimum heat flux and minimum temperature for film boiling are determined by different mechanisms above and below 1.1 MPa [53J]. Taylor instability appeared to govern at low pressure, while the heterogeneous spontaneous nucleation governed at higher pressures. Empirical equations were also given for interfacial wavelength, departing bubble diameter, and frequency. A theoretical model of minimum film boiling temperature and heat flux was proposed [54J] which included the effect of vapor film thickness. Predictions gave lower minimum heat fluxes and temperatures than the model of Berenson, particularly at high pressure. Experiments performed using a pulsed reactor allowed determination of transient subcooled film boiling heat transfer coefficients by way of inverse conduction calculations, showing good agreement with empirical correlations of earlier studies [46J]. A mathematical model of laminar film boiling for conditions with a large density change upon vaporization employed the effect of a pressure gradient caused by the increasing vapor thickness [65J]. The model predicted greater heat transfer coefficients at low flow velocities than the earlier model of Sparrow and Cess. A model based upon steady stagnation flow theory was developed for the case of film boiling on an oxidizing surface and shown to be consistent with available data of oxidation of zirconium particles [14J]. Analysis of flow film boiling over submerged bodies (spheres and cylinders) allowing a quadratic rather than a linear vapor velocity profile predicted thinner vapor layers and higher heat transfer coefficients, closer to those observed in available data [68J].

Droplet evaporation

Kinetic theory was applied to determine the direction of energy flow and mass flow to or from a spherical droplet surrounded by its own vapor, showing that under some circumstances (small Knudsen number and finite Reynolds number) the energy flow and mass flow may be in opposite directions [47J]. Analysis of vaporization of liquid droplets was used to show that the evaporation time of fuel sprays is proportional to the square of the initial mass mean diameter, and that as evaporation proceeds both the mean particle diameter and the Rosin–Rammler exponent characterizing the distribution of droplet sizes increase [9J]. A simple technique for evaluating evaporation time was presented, and it was shown that an optimal fuel droplet size distribution for combustion efficiency may not be optimal for ignition of the spray. A model of fuel droplet vaporization during the intake and compression strokes of a spark-ignition engine [35J] has demonstrated that the reduction in evaporation rate due to increasing pressure approximately offsets the increase in evaporation rate expected due to rising cylinder temperature. The model can also predict the maximum droplet size upon intake which will be completely vaporized at the end of the compression stroke. An analysis was performed on the behavior and evaporation of liquid droplets injected into the gas stream of a centrifugal compressor in the interest of approaching isothermal compression [50J]. Droplet size was determined to be the most crucial parameter. Evaporation from freely falling droplets was also investigated [57J].

Using a laser shadowgraph technique, the interface stability of slowly evaporating liquid droplets on a vertical surface was explored [70J]. Dependent upon liquid properties, the interface behaviors fell into three categories: stable, substable and unstable, with the latter behavior exhibited by liquids with high dielectric constants.

A numerical model was developed to predict two-phase dispersed droplet flow heat transfer in a heated channel, extending previous analyses to include the effects of thermal radiation in systems at high pressure where the vapor was considered to be optically thick [10J]. An experimental investigation of heat transfer from a heated surface to uniformly sized impinging subcooled and nonwetting droplets [55J] illustrated little dependence of heat flux on wall temperature and droplet velocity. A simple model provided a plausible explanation for the lack of influence of wall temperature and good agreement with the data. A model was proposed for evaporation of a liquid droplet heated by a surrounding immiscible liquid [4J]. The bubble growth rate, Nusselt number, and time and height required for complete evaporation were shown to be dependent on Péclet and Jakob numbers, and good agreement was demonstrated with data of other investigators.

Reflooding, quenching and counterflow evaporation

Experiments performed with partially-blocked electrically-heated vertical rod bundles suggested that swelling (ballooning) of cladding does not severely impair the cooling of rod clusters during transient reflooding [22J]. Rewetting of a hot stainless-steel surface by refrigerant 113 was studied with varying liquid subcooling [62J]. A simple expression developed earlier from a one-dimensional wall conduction model reflected well the increase in rewet front velocity with increased subcooling, and heat fluxes in advance of the rewet front were shown to increase with subcooling. It was postulated in ref. [32J] that droplets in two size ranges emanate from the froth at a quench front, with the smaller droplets evaporating so rapidly that they vanish before reaching downstream measuring stations. The large number and rapid evaporation of these small droplets was proposed as the mechanism for the large rate of energy transfer immediately downstream of the quench front and, analogously, downstream of channel restrictions. The height of a two-phase region ahead of the quench front in a reflooding reactor core was measured in a thin 'slab' core [17J] and a predictive model was proposed, sensitive to transverse heat flux (power) profile and pressure-drop boundary conditions but insensitive to crossflow and to axial friction. An experimental investigation demonstrated the effective use of air–water mixture jets in quenching of machine parts [21J]. The technique allows the optimization of quenching rates as well as localized control of finished product properties.

The envelope theory, considering flooding to be the limiting condition for countercurrent flow, was compared with data for flows of condensing steam and water in inclined channels and shown to provide good agreement [33J]. While condensation had negligible effect on bottom flooding, an effective steam flow rate, recognizing the reduction of inlet steam flow by condensation, needs to be considered for top flooding. Liquid entry conditions were observed to have an important role. Experiments were described involving evaporation of water trickling down corrugated plates with an upward flowing air stream [7J].

Flashing

Linearized stability analysis was performed on the case of a rapidly evaporating (highly superheated) liquid surface [51J], suggesting strong instabilities with rapid growth rates. Experiments with sudden depressurization of refrigerant 11 demonstrated that the mass transfer rates during flashing were 10–12 times the rates due to evaporation and that liquid temperatures can fall well below the saturation temperature of the new low pressure [49J]. A combined experimental and analytical investigation of flashing of saturated water within a vessel showed strong dependence of the rate of vapor formation upon the initial level of water, the initial pressure and the diameter of the nozzle through which the vapor was allowed to escape the vessel [39J]. The time constant of the pressure response due to flashing was further related to the vessel's cross-sectional area. A model was suggested for non-equilibrium vapor generation for flashing two-phase flows, requiring *a priori* knowledge of the liquid superheat at the flashing inception point and using the bubble number density at inception as an experimentally-evaluated parameter [52J]. An experimental study of flash evaporation from turbulent water jets indicated that evaporation rates were essentially independent of the thickness of planar jets and that evaporation could be augmented through the use of screens or upward-directed vertical spouts [5J].

Other evaporation-related studies

A mechanistic model of liquid entrainment from a boiling or bubbling pool of liquid was developed [27J]. Good agreement was demonstrated with data from several independent investigations with air and water and with steam and water at various pressures. Experiments were performed to determine the heat transfer coefficient upon a vertical cylinder immersed in a liquid–liquid spray column in which discrete liquid droplets flow through another immiscible liquid [43J]. A simple physical model provided good agreement with the experimental results.

Advantages were demonstrated for the use of a two-phase cooling film for protecting a surface in a high-temperature air stream [64J], and a procedure was provided for calculating the efficiency of cooling of an adiabatic wall by such a film. An abstract of a special report [63J] outlined the use of natural convection cooling in storage of radioactive materials.

CONDENSATION

Dropwise condensation

Local heat transfer coefficients were experimentally determined on a condenser surface which was covered solely by active droplets (too small to be conduction-limited) [31JJ]. High-speed film records of droplet coalescence were used. Experiments were made on the effect of electroplating with silver to enhance dropwise condensation of steam—a significant enhancement was recorded [21JJ]. Silver provides a low-oxygen surface that absorbs trace organics from the surroundings promoting dropwise condensation. Measurements of heat transfer coefficient with dropwise condensation of steam on a vertical surface were made [11JJ]. Characteristics of droplet condensation were clarified. Heat transfer coefficients measured on the external, upper surface of a horizontal tube showed an influence of the frequency of the sweeping action of falling drops [10JJ]. This was dependent upon the tube diameter. Direct contact condensation was experimentally found to be conduction-limited [8JJ]. High heat transfer rates were attributed to deforming oscillations of the droplets due to surface tension—an analysis was presented to support this hypothesis.

A model was developed for droplet formation and growth in condensing binary mixtures—it was an extension of an unary condensation model [29JJ]. Predicted droplet size and density compared well with new experimental results, also presented. An analysis of binary nucleation and heterogeneous droplet growth was presented where a limit of unary condensation was shown [28JJ]. Droplets were shown to attain an equilibrium composition within 10^{-5} s after their metastable state. Comparisons were made with experiments.

Approximations used in the theory of dropwise condensation were assessed for use with mercury [20JJ]. Significant errors can arise when they are used. A numerical analysis was recommended. An analysis of liquid spray injection to a condensable environment showed that the observed reduction in spray angle (compared to injection into a noncondensable environment) is due to a reduced angle of the sheet region [14JJ]. This was confirmed experimentally for a laminar sheet flow. A model for condensation on a liquid droplet in its own vapor was presented [5JJ]. The vapor-phase boundary-layer equations were solved showing that the total rate of condensation varied inversely to the droplet radius. Laminar condensation on a spherical drop in forced flow was analyzed [3JJ]. Because of high radial condensation velocities, the drop was found to not be in the Stokes-flow regime, even though the drop translation velocity was low. Transient laminar condensation was next numerically investigated [4JJ].

Film condensation

Experiments and analysis were performed on the reflux condensation flow phenomena in a closed thermosyphon [6JJ]. The liquid film thickness is increased by vapor shear resulting in reduced heat transfer—Nusselt's solution cannot interpret the observed trends. An analysis of the effects of noncondensable gases on condensation in thermosyphons showed that the one-dimensional diffuse-front model is inappropriate [7JJ]. The importance of radial diffusion depends on the gas load and the ratio of the radial diffusion rate to the condensation rate. An important instability, condensation-induced water hammer, in inclined counter-current flow of steam and cold water was analyzed and a stability map was presented [13JJ]. Heat and mass transfer in turbulent tube flow with film condensation was studied [22JJ]. Condensation in tilted oval tubes was discussed [37JJ].

Condensation of binary mixtures with downward flow in tubes was experimentally investigated [19JJ]. Empirical equations were developed.

Condensation of steam on a subcooled water layer in co-current horizontal channel flow was experimentally investigated [17JJ]. Dependence of surface heat transfer rate on surface waviness was shown and data for a wavy surface was correlated. Experiments were made for downward-flowing vapor and vapor–gas mixtures over horizontal tubes [16JJ]. An approximate equation in boundary-layer form was shown to agree with the measurements. Similar experiments were made to investigate the effect of vapor velocity [15JJ]. Measured heat transfer rates were higher than predicted. It was felt that this is due to the onset of turbulence in the condensate film.

Measurements were made of vapor-phase temperature profiles and rates of condensation of binary vapor mixtures on a small, vertical flat-plate condenser [1JJ]. Data were correlated with a modified Nusselt model.

An analytical study of saturated turbulent filmwise condensation heat transfer was made of the vertical plate geometry [32JJ]. The velocity profile consisted of a laminar sublayer and a turbulent core using the 1/7 power law. Comparisons were made with previous data. Wavy filmwise condensation was analytically and experimentally investigated by assuming a sinewave film profile [27JJ]. Agreement with experiments was good when boundary conditions were properly given.

The effect of pressure gradient due to flow over a curved surface on film condensation was determined analytically [25JJ]. Solutions were not attainable for some locations in the downstream half of the tube. This is thought to indicate instability of the laminar condensate film which leads to enhanced heat transfer. Condensation on a laminar, falling film of immiscible liquid was numerically evaluated [23JJ]. Shorter film lengths were found to yield considerably higher heat transfer rates. Condensation of flowing vapor in a horizontal tube was numerically analyzed—the tube conduction resistance was included [9JJ]. Heat transfer characteristics resemble the uniform wall temperature problem for steam flow and the uniform wall temperature problem for refrigerant vapor flow. Turbulent film condensation in forced convection boundary-layer flow was investigated [33JJ].

Tests were performed to measure the critical conditions of water hammer initiation [2JJ]. The phenomenon is due to the collapse of a steam bubble in subcooled water flow. An analytical prediction model was presented. Arithmetical expressions were developed to describe the effect of excess vapor and the size of dead zones on single-pass gas-cooled condensers [26JJ]. These expressions allow the calculation of two-dimensional temperature and vapor-flow distributions within the condenser. Equations and guidelines were presented for the design of vertical condenser tubes—focus is on the effect of noncondensables [12JJ]. Results of experiments designed to determine the effect of oils on condensation in refrigeration systems were presented [34JJ]. An analysis was given for shell-and-tube condensers with shell-side enhancement using sharp fins [30JJ]. The fins allow a 20% reduction in heat transfer area under the conditions studied.

A survey of recent works studying the separate effects of condensate inundation and vapor shear in shell-side refrigeration condensers was presented [36JJ]. Appropriate correction factors for horizontal tube banks were sought. A survey giving the state-of-the-art of shell-side condensation in refrigeration condensers was made [35JJ]. Issues included vapor shear and inundation, enhanced tubes, noncondensables, oil and bundle geometry. A review of options available to improve heat transfer in film condensation of pure vapors was presented [24JJ]. The evolution of power condenser design and a discussion of present designs were given [18JJ]. Modeling of condenser shell-side heat transfer was discussed.

CHANGE OF PHASE

Freezing and melting

Theoretical and experimental studies of the phase-change process between solid and liquid or vapor phases have been reported for freezing within cavities and in duct flows, melting governed by natural convection in the melt and separation of components by selective freezing. Applications include continuous casting, frost formation in air flows and solid formation in pipe flows.

Basic work on the theory of the melting temperature and volume change vs pressure was developed for several pure metals [26JM–28JM]. The entropy minimum principle was used to develop governing equations for transient one-dimensional Stefan problems [22JM]. Results were obtained for a slab, cylinder and sphere.

Work on freezing included studies of crystal growth which can simulate the crystallization of metals [17JM, 1JM]. An implicit treatment of the temperature field with explicit boundary conditions was used to model the freezing process where more than one solid phase is possible [39JM, 40JM]. A finite-element method was applied to freezing in a corner by mapping the region onto a rectangular domain [19JM]. Another study used a similar method for solving a corner problem and included experimental data for comparison [23JM]. The effect of the boundary conditions on the convergence of finite-element solutions was addressed [41JM]. An experimental study was performed to measure the freezing time of a liquid contained within an inclined cylindrical cavity [18JM]. The inclination was found to have very little effect on the freezing time or heat extraction rate. Approximate solutions for the freezing time within a cube or a long cylinder of square cross section were provided [20JM].

Freezing problems involving a moving solid can be applied to continuous casting. The Cauchy boundary-value method and the product solution method were both used to obtain solutions for the interface of a melt being extruded horizontally between two insulated offset walls [37JM]. A similar study of a vertically extruded melt was made using the Cauchy boundary-value method [38JM]. A numerical solution was presented for the solidification of a steel casting in a rotating cylindrical mold in which the gravity force is negligible [21JM]. A closed-form solution was developed for the Stefan problem that exists when a hot wall moves over a semi-infinite solid [15JM]. Results for large Prandtl number reveal the role viscous heating plays in the melting process.

Solidification in channels has been studied both experimentally and theoretically. Experimental studies of ice formation between two horizontal plates have been made [34JM, 35JM]. The ice surface profile was found to depend on the flow Reynolds number and the instability caused by the ice layer. Numerical solutions were obtained for freezing inside a circular tube for laminar [5JM] and turbulent flow [32JM]. A finite-difference method was used to determine the surface contour and temperature distribution within the solid attached to a cooled plate immersed into a heated fluid [3JM].

A freezing process can be used to separate constituents of a fluid mixture. A one-dimensional Crank–Nicolson procedure was used to model the freezing out of a component from a bicomponent eutectic alloy [10JM]. The effects of various process parameters were examined. The freezing of a sodium chloride salt solution above a horizontal ice layer was studied experimentally with a simple analytical model used to obtain corresponding theoretical results [7JM].

Theoretical solutions were obtained for the freezing of water from a stream of humid air onto a flat plate [6JM]. Both laminar forced and natural convection situations were considered. Observations of frosting during early stages showed that the process begins with condensation of liquid water droplets which freeze and act as nuclei for the subsequent frost formation [33JM]. Freezing of supercooled water droplets suspended at the interface between two immiscible liquids was observed [36JM]. A similar study was performed to observe the freezing of water dispersed within an emulsion [29JM].

One-dimensional melting problems have been solved using a finite-element/Green's function method [25JM]. The inverse heat conduction problem with phase change has also been solved using a finite-element method [16JM]. An analytical procedure using the method of collocation has been used to study the phase change in a heat generating solid confined between two semi-infinite conducting walls [4JM]. The sublimation of a solid has been studied using a numerical approach with variable properties and an analytical solution using Green's function [11JM]. Good agreement between the two methods was obtained when using carbon.

Two-dimensional melting problems are often governed by the natural convection that occurs within the melt region. Melting adjacent to a vertical heated wall has been studied numerically [8JM, 24JM, 12JM]. An experimental study of melting and solidification in a two-dimensional rectangular cavity indicated that short fins were advantageous [13JM]. Temperature fluctuations were used to determine the melt structure and the natural convection flow regimes that existed in a melt of Lipowitz metal [9JM]. Melting of a solid contained within a horizontal cylinder was studied experimentally and theoretically when the solid is fixed [14JM] and when the solid is more dense than the melt and is allowed to sink to the bottom of the cavity [2JM]. Melting around a horizontal cylinder was studied numerically when the heated cylinder had a constant heat flux surface [30JM] and when the surface was maintained at a constant temperature [31JM].

RADIATION IN PARTICIPATING MEDIA AND SURFACE RADIATION

This review of radiation heat transfer literature is organized to first discuss works which treat general radiative heat transfer studies in various geometries. Works dealing with radiation coupled with other phenomenon such as conduction, convection or combustion, are discussed next. Literature on passive solar devices and solar radiation interactions with atmosphere is included. A number of papers on properties is also surveyed. They are roughly grouped into those dealing with radiative properties of thin films, small particles and how they scatter and participate in radiation transfer, gaseous absorption, and data concerning such materials as metals and insulations. A brief mention of instrumentations of interest is also included.

Analytical/numerical works outnumber the experimental work in radiation heat transfer in one-dimensional planar layers. The Galerkin method is used to determine radiation transfer in isotropically scattering [17K] and anisotropically scattering [16K] layers. Vertically inhomogeneous atmospheres with azimuthal dependences are treated by a multilayer discrete ordinate method [93K] and by the matrix operator theory which includes Mie scattering effects [33K]. Analyses of the dissipation of temperature perturbation in finite homogeneous [36K] and inhomogeneous [37K] atmospheres are presented. The question of uniqueness when considering reflection and transmission matrices of homogeneous, plane-parallel layer is considered [85K]. Analytical expressions for some of the integrals involving Legendre polynomials [15K] and Gaussian quadrature formulas for integrals with weight functions $\exp(-|x-t|)$ and $E_n(|x-t|)$ [86K] may be useful in future work. Experimental measurements of vertical and directional distributions of radiance in absorbing–scattering liquids [20K] are compared against discrete ordinate solutions which utilize experimentally-determined properties [46K].

Multidimensional radiative transfer in participating and nonparticipating media is the subject of many analytical and numerical studies. Two-dimensional rectangular enclosure geometry is considered in the following five references. [54K] studies the effect of surface radiant exchange and conduction on natural convection. Heat transfer and temperature profiles are presented for gray participating media at radiative equilibrium [110K]. Simple closed-form approximate expressions for a general multidimensional geometry is also presented and shown to be accurate for rectangular enclosures. Finite-element solutions for heat transfer and temperature distributions are presented for gray absorbing, emitting media where radiation is coupled with conduction [84K]. The S_n discrete ordinate method is presented for gray absorbing, emitting, isotropic scattering media and the results are shown to be very good compared to Hottle's zone and other methods [35K]. Numerical solution for two-dimensional, isotropically scattering, rectangular medium with diffuse incidence is presented [21K]. Finite-element solution for temperature profiles is presented for combined radiation, conduction and convection in a diverging/converging two-dimensional channel [19K]. A stochastic model is developed to handle general configurations with a finite number of isothermal, diffuse or specular, walls enclosing a nonparticipating medium [74K]. This model provides exact solutions if the appropriate view factors are known. [65K] presents view factors for a sphere coaxially placed within a cylinder. An approximate solution technique is developed for a general diffuse walled enclosure with nongray isothermal gases [75K]. In absorbing, emitting and isotropically scattering media, upper and lower limits, as well as explicit analytical expressions for geometric mean transmittance are given [109K].

Along with the references already cited [54K, 84K, 19K], the following consider radiation coupled with other modes of heat transfer. Combined conductive and radiative heat transfer is modeled for optically-thick, porous cordierite slabs and the results are compared with experiments [52K]. Transient coupled conduction and radiation heat transfer in an annular, absorbing–emitting medium is analyzed with a finite-difference technique [40K]. Experimental results are reported for heat transfer in a parallelogram enclosure, where surface radiation and conduction couple with natural convection [6K]. Radiation and convection are coupled in a thin shock layer: ref. [39K] presents an analytical solution for the flow around a body with a surface of zero total curvature, and ref. [94K] considers axisymmetric flow around a 35° sphere cone with chemical nonequilibrium effects included. For situations where the computation of complex flow fields, simultaneous with radiation, is needed, an adaptive-mesh radiation hydrodynamics is proposed [72K, 104K].

In a slightly different vein, thermal transport in spherical targets irradiated by laser beams [105K] and the temperature distribution on flat workpieces irradiated by Gaussian laser beams [87K] are investigated. The effect of radiation heat transfer is considered coupled with internal vibrational heating of polymers [42K], with internal heating in drying coated films [79K], and with endothermic gasification of a solid at depths [11K].

Interest in radiation coupled with combustion is shown in the works involving furnaces and other fundamental studies. Reference [60K] presents a review of models and experimental data for luminous and nonluminous flame radiation in turbine combustion chambers. Spectral normal directional emissivity data of a 16.5-kW test furnace shows up to 30% increase when the flame is loaded with carbon particles [53K]. Measurements and predictions of radiant heat fluxes and temperatures as functions of relative position to a turbulent methane diffusion flame are reported [51K]. On a more fundamental level, ref. [91K] considers the interaction between isotropically scattering particles and combustion gases, CO_2 or H_2O, in one-dimensional planar geometry. Reference [32K] reports on measurements of emission spectrum for carbon particles of 0.5 μm mean diameter which have been added to a 800–900 K nitrogen stream. Attention is also focused on the fly ash particulates: ref. [9K] reports scattering data, and ref. [96K] deals with measured total absorptivity. Study of laser vaporization of small soot particles, less than 100 nm in radius, reveals a decrease in size while the number density remains invariant [22K]. Soot diagnostics based on laser heating is reported sensitive in the 0.02–0.2 μm diameter range [71K].

Other combustion diagnostic techniques are considered. Spatially resolved measurements of OH concentrations using cross-beam saturated absorption spectroscopy are proposed [58K]. OH concentrations and temperature profiles in a premixed propane flame are obtained by laser deflection techniques [55K]. A number of papers report results from CARS technique:

ref. [29K] reports data in augmented jet engine exhausts, [43K] on thermometry in full-rich combustion zones and [56K] from inside a firing internal combustion engine. Uncertainty propagation analysis in combustion diagnostics is discussed in connection with rocket and jet engine exhausts [62K]. Studies of the influence of particulate scattering on such plumes indicate a need for accurate particle size distribution information [76K, 77K].

Passive solar cooling for spacecraft motivates a theoretical study of a number of diffuse radiation shields [88K], as well as a study of those with effectiveness demonstrated by vacuum chamber experiments [7K]. Use of selectively infra-red emitting gases like ammonia, ethylene or ethylene oxide [63K], or coatings of MgO or LiF which have high solar reflectivity [8K] is considered.

A number of articles deal with the actual solar flux on earth and aerosols or participating gases in the atmosphere with which the radiation interacts. References [83K, 108K] report measured sky radiance in Kuwait and the tropics. Atmospheric aerosols have been studied: ref. [78K] reports on spectral extinction coefficient, [82K] on the scattering phase function of cirrus particles, and [34K] on the effect of humidity on aerosol scattering. Spectral absorption of atmospheric water vapor [41K] and NO_2 [69K] are reported. Transmittance band models of nitrous oxide [49K, 80K], SO_2 [81K] and methane [50K] in atmosphere are studied.

Partly due to continuing interest in radiation cells and also because of the importance of coatings in general, many works deal with radiative and optical properties of thin films. Studies that determine optical constants of thin films are reported in [4K, 10K, 103K, 70K]. Reference [10K] further notes that the film optical characteristics differ from the bulk, depending on the conditions during deposition. Reference [13K] discusses the determination of a thin-film absorption coefficient from CO_2 calorimeter data. Reference [26K] discusses the enhancement of solar absorption by having a film with rough top surface and a diffuse reflective bottom. Multilayer coatings for a high-power laser mirror optimized for 10.6 μm are described in [23K]. Specular reflectance of optical black coatings in the far infra-red (12–500 μm) is presented and modeled [92K]. Deposition of infra-red reflection coatings on glass is proposed to improve the efficiency of incandescent lights [106K], while antireflective coatings for glass will produce laser damage resistant coatings [107K].

Small particles and their interaction with radiation are the subject of the following papers. Work is presented to validate Mie emission theory for particles that are small compared to radiation wavelength [31K]. Effects of deep holes in the particle's scattering is studied by solving the Rayleigh–Gans–Debye and anomalous diffraction equations [59K]. It was found that the effect of relatively fine structure on total scattering by randomly oriented particles is small. Light scattering of soil particles is calculated by modeling them as randomly oriented spheroids [45K]. Representations for the extinction and the Mueller matrices are given for a random distribution of nonspherical particles [47K]. Multiple scattering calculations are performed on an aerosol cloud [99K]. Effect of an intense pulsed light beam through absorbing water droplets is modeled [3K]. Scattering of sodium droplets in the cover gas of a sodium-cooled fast reactor is found to reduce radiant heat transfer by less than 10% [95K]. Reference [95K] also compares one-dimensional discrete ordinate method against other approximations. The effect of aerosol coagulation and deposition with time is added to the one-dimensional modeling for the radiation heat flux [100K, 102K]. Importance of anisotropic scattering in such considerations is emphasized [101K].

Information on the properties of nongray gases can be found. Total gas absorption models are proposed based on Elsasser band model and the effect of band contour and low-intensity lines [97K]. Total band absorptance estimates, including the nonoverlapping lines of vibration–rotation bands, are presented [89K, 90K]. Hottel's and Penner's rules for estimating the total absorptivity of homogeneous gases from total emissivity charts or expressions are generalized [30K]. Functional relationships between total and spectral molecular gas properties are considered in the derivation, and pressure scaling is also presented. Absorption band intensities are presented for propane at low temperatures [38K] and for CO_2 [25K]. Three overlapping CO_2 bands near 2.2 μm are analyzed by a nonlinear least-squares fitting procedure [1K]. Correction coefficients are given for the effect of dimerization on the infrared absorption of 6.3 μm water vapor band which is shown to be negligible [57K]. For computational ease, emissivity charts of CO_2 and H_2O have been converted to simple formulas [48K].

Radiation properties of liquids and solids are also studied. Total normal emissivity of liquid sodium is measured in an inert environment to be 0.045–0.057 for surface temperatures of 175–520°C [44K]. A method for determining the real part of the refractive index absorbing condensed media from reflection and transmission measurements is presented [61K]. Approximate formulas are given for the amplitude and the phase of specular reflectance at normal infra-red incidence for a conductor [73K]. Absorptivity of metals and their temperature dependence are considered [5K, 64K]. Optical constants for coal, char and limestone are obtained from combining reflection measurements from the particles dispersed in KBr tablets and the scattering calculations [12K]. Thermal radiative properties of ceramic materials are obtained from measurements made at 0.4–33.3 μm and 290–700 K [66K]. These properties are correlated with the wavelength, temperature, as well as other physical parameters. Results for the extinction coefficient and the first-order term in the scattering phase function are presented for high temperature insulation materials [14K]. Infra-red optical constants are presented for rare-earth doped crystals [111K], NH_3 ice [67K], magnesium fluoride [27K] and ice [98K].

Lastly, a brief review of articles on instrumentation follows. Construction and preliminary application of a thin path length cell, shorter than 10 μm, is reported [2K]. A near infrared remote sensing system for real time absorption measurements is described [18K]. PtSi Schottky infra-red CCD is considered for star detection from 1.2 to 2.5 μm [24K]. Optimization and theoretical ultimate NEP of bolometers are discussed [68K] and possible improvements suggested [28K].

NUMERICAL METHODS

Many problems in heat transfer are analyzed by the use of numerical methods. For the most part, the papers describing the application of a numerical technique to a particular problem are included in the appropriate application category. The papers mentioned in this section focus primarily on the numerical method rather than on the application used to illustrate it.

A combination of the boundary-element method and the analytic function method has been described [6N]. A solution method for the unsteady one-dimensional gas flow has been given in [16N]. An alternative practice to the use of wall functions in a turbulent flow is proposed [8N]; it involves the numerical solution of a simplified form of the turbulence model equations in the near-wall region. Reference [10N] describes multigrid methods for the diffusion equation; multigrid methods for flow are given in [7N]. A numerical procedure is presented in [2N] for compressible flow over a solid body. Some schemes for the flow solutions with nonstaggered grids are given in [12N]. Superposition technique is used to construct a three-dimensional numerical method [14N]. In ref. [5] suggestions are made for the calculation of the entrainment rate, initial values, and transverse velocities in the Patankar–Spalding method for two-dimensional boundary layers. A number of enhancements are described for the SIMPLE method for the calculation of recirculating fluid flow [15N].

The use of nonorthogonal coordinates for convection–diffusion problems is presented in [4N]. Reference [9N] applies the locally analytic differencing scheme to some convection–diffusion problems. The use of an adaptive collocation method for simultaneous heat and mass transfer with phase change is described in [1N]. The two-fluid model equations for boiling are solved by a simple and efficient scheme [3N]. A numerical method is proposed for two-dimensional phase change problems [11N]. A finite-difference method is employed for calculating laminar film boiling [13N].

TRANSPORT PROPERTIES

Research in this area is noteworthy for its development of methods and devices for measuring the thermal properties of materials. Classical transport properties, i.e. coefficients of diffusion, thermal conductivity and viscosity, are the focus of some efforts; in other instances the particular state of a material or its use in special circumstances attracts study.

The apparent density of fine, highly porous particles in a powder is measured using a novel mercury microvolumeter which gives rapid, accurate measure of small volumes of such materials [22P]. The cooling capacity of aqueous aquaplast, an East German produced, polymer-based fluid, is investigated by means of a prismatic heat-flux probe and found to be intermediate between water (high) and oil (low); results have application to the quenching of hot metal parts initially at a temperature $\sim 800°C$ [21P]. For a vertical plate submerged in a high Prandtl number fluid, the transient natural convection resulting from the sudden generation of a uniform heat flux inside the plate is analyzed. Time-dependent expansion solutions are constructed for flows near surfaces with zero and finite thermal capacity [6P].

For binary mixtures of sulphur hexafluoride with noble gases, diffusion and thermal diffusion factors are measured and compared with theoretical values calculated by means of the Chapman–Enskog theory.

Excellent agreement between experimental and theoretical thermal diffusion factors obtains for He–SF_6 and Ne–SF_6 systems, less so for Ar–SF_6 and Kr–SF_6 systems [24P]. Thermal conductivity of gas mixtures continues to be studied by Kestin and associates using a transient, hot-wire instrument [10P, 11P]. The first of these examines binary mixtures of He, Ne, and Ar with CO (at 27.5°C, pressure range 0.7–12 MPa) and uses viscosity data from earlier measurement on these mixtures for calculating contributions of internal degrees of freedom. Despite the essential sameness of the low-density viscosity of these mixtures with those of nitrogen, the thermal conductivity is different, pointing to a significant contribution from rotational degrees of freedom. The second work examines the CH_4–CO system in the same state region; despite this increase in the body of accurate data a predictive algorithm is not yet forthcoming.

Thermal properties for selected porous materials have been studied. An unsteady-state method was developed for determining the thermal diffusivity of a porous sphere. Using theoretical considerations to relate thermal diffusivity to porosity, experimental measurements on four metallic oxides (Cr_2O_3, CuO, Fe_2O_3, V_2O_5) gave results in good agreement with literature values [15P]. For solids with complex pores a model, developed from an analogous diffusion model, is described for predicting pore gas thermal conductivity to be used in conductivity correlations for such materials. The model accounts explicitly for both molecular and Knudsen transport mechanisms and complex pore systems [26P]. The thermal conductivity of the microporous particulate medium, moist silica gel, is studied experimentally to determine the influence of porosity, water content, total gas pressure and temperature by using the transient hot-strip method. The predictions of the effective thermal conductivity of beds of such materials by four simple models agree reasonably well with measurements [4P]. In order to model a packed bed with stagnant fluid ref. [12P] considers energy transport in a fluid-saturated porous medium formed by a cubic array of cylinders, with stagnant fluid. The generalized method allows the overall effective conductivity, volumetric specific energy capacity and effective thermal diffusivity to be derived with good results having only the properties of fluid and solid and void fraction as parameters.

A number of papers deal with techniques for measuring the thermal conductivity of systems. A compact apparatus operates on the linear heat-flow principle using bar-shaped samples held between a heat source and water-cooled sink. It is designed to cover the range 10–380 W m^{-1} K^{-1}, 50–500°C with a $\pm 3.1\%$ agreement of results with certified values. The transient hot-wire method is used for measuring the thermal conductivity of a system in which the wire separates the half spaces of homogeneous materials having different thermal conductivities and diffusion constants. The configuration permits the construction of a measuring probe which simplifies measurements and allows the consideration of anisotropic materials [20P]. The thermal conductivity of pyrotechnic compositions is successfully determined by applying a published method for estimating the thermal conductivity of powders using an electrically-heated copper probe embedded in the sample. The pyrotechnic systems are compositions of antimony and potassium permanganate at various fuel/oxidant ratios [1P]. Large effective thermal conductivities, between two fluid reservoirs maintained at different temperatures and connected to each other by a capillary bundle, are obtained hydrodynamically by oscillating the fluid in the capillaries at high frequency. The hydrodynamic theory developed attributes the enhanced heat conduction to radial heat transfer across very thin Stokes' secondary layers. Results with water show effective thermal diffusivities up to 17,900 times those in the absence of oscillations, exceeding the conductance rates obtained with heat pipes [13P].

The rotating disc viscometer, consisting of a thin copper disc enclosed in a copper cylindrical cavity and activated by two alternating magnetic fields of different strength, requires only a small fluid sample and yields accurate but not absolute measurements in the vicinity of the critical state of the fluid [5P].

In the area of radiation heat transfer properties two papers report results of interest. High temperature fibrous materials are considered in [16P] where an inverse methodology is developed for determining the thermophysical and optical properties for such absorbing, emitting and highly scattering substances. The model accounts for the optical effects: single scattering albedo extinction coefficient, back scattering fraction, index of refraction and the thermal properties: conductivity and specific heat. In instances where day-to-day experiments do not require a standard laboratory-calibrated tungsten strip lamp, ref. [19P] describes a less expensive, uncalibrated lamp–optical pyrometer method.

The influence of temperature-dependent physical properties on the heat transfer in the turbulent boundary layer for parallel flow over a flat plate is considered using a numerical method in order to account discretely for the influence of various properties [23P]. Results indicate that some published methods lead to considerable error. In a related study, ref. [25P] derives the dimensionless groups describing the influence of temperature dependent properties in heat transfer for flow over surfaces generally. For a viscoelastic fluid, ref. [14P] presents estimated eddy diffusivities of momentum and heat.

Efforts continue to obtain from kinetic gas theories improved correlations for predicting dilute gas viscosity and thermal conductivity. The acentric (shape) factor, dipole moment and association parameter are used in [7P] to obtain consistent correlations for 17 pure nonpolar gases and 13 pure polar gases: average accuracy for viscosity, 1.6%, for thermal conductivity, 1.38%. In the instance of gaseous nitrogen, and sulphur hexafluoride and mixtures of these gases [17P] reports calculated transport properties which are in good agreement with published values, experimental and calculated, except for a diffusion coefficient ratio. A separate study on the *P–v–t* behavior of sulphur hexafluoride in the gas–liquid critical region [3P] uses a combination of the Burnett and Isochoric methods to yield results accurate to better than 0.01% in pressure and 0.02% in density and 1 mK in temperature.

For solid systems ref. [18P] describes an accurate versatile system for measuring the linear thermal expansion coefficient. A laser interferometer and parallel spring strip supporting system provides accuracy and general purpose, yielding coefficients accurate to better than 3%.

A useful review article [2P] considers the three types of temperature sensors (resistance thermometers, thermistors and thermocouples), their characteristics and the design of circuitry to convert their outputs to a voltage signal, pointing out the use of a microcomputer as signal processing element to improve accuracy. An interesting examination of Newton's original work in defining a temperature scale, his use of fixed points and his law of cooling in the high temperature range, reconciles his results and conclusions with our current knowledge [9P].

HEAT TRANSFER APPLICATIONS

Heat pipes and heat exchangers

Research on various aspects of heat exchangers is impressive for the variety of circumstances investigated and the detailed examination of the various influences which can affect the transfer of heat. The most novel work [45Q] analyzes the simultaneous exchange of heat and water vapor in a crossflow-type total heat exchanger made of Japanese paper impregnated with a hygroscopic agent. Predictions and experiments for temperature and humidity efficiency show good agreement; the heat transfer rate is dominated by the air rather than the paper, for mass transfer the reverse is true. Disk- and doughnut-type heat exchangers for oil cooling are studied experimentally for various combinations of flow rates and clearances between baffle and shell [18Q]. Another experimental study [43Q] investigates the critical heat flux in helical coils with up and down flow of steam–water mixtures and finds striking differences from the vertical straight-tube arrangement.

Design considerations attract the attention of a number of workers. Two new compact heat exchangers specifically designed using the technique of 'local assumption' are studied and their predicted and measured duties found to agree [14Q]. Multiple parallel passages are used to cool a hot stream of a viscous polymer [24Q]; the analytical treatment of heat transfer and pressure drop leading to a unique exchanger design. For general heat exchanger design, ref. [50Q] introduces a new form of design chart to replace the five which currently exist in the literature; two exchanger configurations are presented (others to follow) with chart displaying both design and performance parameters. Reference [47Q] presents a numerical method for designing the entire tube-bundle heat exchanger using local heat, mass and momentum exchange information. The thermal design of a high-pressure compressor rotor for a turbojet engine uses a small amount of air from the secondary air system to vent the rotor cavities; an improvement in compression is achieved and the rapid thermal response of the rotor is better matched to the casing resulting in reduced clearances, higher compressor efficiencies and larger surge margins [19Q].

A series of analytical investigations using various techniques address a number of the chief concerns in general heat exchanger design. Using the decomposition strategy, ref. [26Q] examines the general problem of n_h hot streams and n_c cold streams with specified mass flows, fluid heat capacities and inlet–outlet target temperatures for the minimum number of heat exchanger units. Matrix formalism is applied to calculate the effectiveness of assemblies of heat exchangers for cases where the complete mixing condition for each fluid in inlet and outlet streams is relaxed [29Q]. The finite-difference method is used to predict exit-fluid temperature responses for a unit step increase in entrance temperature of either fluid [34Q]. A technique is presented for calculating thermo-physical transients in fuel-element assemblies of a reactor cooled by dissociating N_2O_4 [16Q]. Test data and theoretical predictions show good agreement, confirming the validity of a modified selected point matching technique for testing compact heat exchanger surfaces [12Q].

Heat exchange in the presence of phase change engages the interest of [8Q] where the influence of heat and mass transfer resistances on the separation efficiency in molecular distillations is considered; ref. [17Q] has as its goal the development of specific equations for desuperheated design based on experimental measurements and [32Q] considers direct contact heat transfer in the course of motion by a vaporizing two-phase droplet.

Rotary heat exchange is the focus of a group of studies: ref. [9Q] considers numerically and analytically the influence of transient or nonuniform inlet temperature upon the temperature of the outlet stream for such exchangers; [3Q] considers the representation of regeneration fluid carry over from one stream to another due to the motion of the exchanger matrix. Reference [21Q] uses a finite-difference model of a counterflow rotary dehumidifier to determine the influence of the properties of dehumidifier performance. For fluidized bulk solids, ref. [52Q] uses experimental observations to aid the practitioner in laying out a direct-contact heater of rotary-tubes, tumble dryer or plate dryer type.

In the area of heat pipes, ref. [13Q] describes interesting test results for integral heat-pipe sandwich panels which combine the thermal efficiency of the former with the structural efficiency of honeycomb sandwich construction. Within the wick of a heat pipe ref. [38Q] measures the working-fluid distribution. For horizontal operation, the saturation temperature was found to be uniform in the adiabatic and cooling sections but dropped off sharply in the heating section. Dryout of the wick occurred at lower heat flux in the top-heated mode than in horizontal operation. Experiments and analysis assess the performance of a commercial heat pipe under operational conditions [37Q]. Fouling and scaling effects upon heat exchanger performance are a major concern in many applications. Reference [36Q] reviews generally the fouling process. A description of that process, deposition and removal would require an analytical fouling model. Design practice attempts to compensate for this lack of knowledge by overdimensioning of exchangers using estimated fouling factors (with fair success) and operation stresses continuous cleaning systems. An effective scheme for detecting fouling by means of an on-line monitor is described by [49Q] as a potentially useful diagnostic approach for discriminating between different types of deposits *in situ*. Using ring-type turbulence promoters, [15Q] shows that the maximum thickness of scale inside tubes is smaller than in smooth shell-and-tube heat exchanger tubes.

Heat transfer to banks and bundles of tubes is studied under various circumstances. As a first step to obtain basic data for heat transfer and flow around tubes in a staggered tube bank, measurements are obtained for a single tube in three or four staggered tubes in cross-flow of air with constant heat flux [2Q]. Another experimental study shows a slight effect of angle of attack on the average heat transfer coefficient of staggered and in-line tube bundles through which water flows [55Q]; two staggered bundles of twisted oval tubes are studied experimentally for crossflow of air with average heat transfer coefficients 10% higher than a corresponding bundle of smooth circular tubes [44Q]. A related work studies average heat transfer coefficients and hydraulic drag of in-line and staggered tube bundles at various relative pitches [56Q]. To establish the suitability of the commonly used forced convection correlations in the instance of ceramic heat exchangers using SiC tubes, ref. [40Q] measures friction factor and heat transfer data, concluding that either the tube surfaces require smoothing or the heat transfer models need modifications.

Nonsteady effects on heat exchanger performance are reported by [39Q] where pulsatile flows enhance performance beyond the expected for curved tube flows; [48Q] considers analytically and experimentally the steady-state temperature responses of a one- or two-pass exchanger subject to sinusoidal flow rate changes of large magnitude; [31Q] associates the variation of local heat transfer coefficient over the tube perimeter in staggered and in-line configurations with the fluctuation of local air velocity during crossflow. Additional effects considered are: the influence on the performance of a tubular heat exchanger located downstream of a right-angle bend [41Q] and the dynamics of an independently heated heat exchanger with nonuniform axial distribution of the heat flux density [6Q].

The flow pattern in shell-and-tube heat exchangers is revealed in interesting photographs obtained by flow visualization using the oil–lampblack method [28Q]. Heat transfer in shell-and-tube heat exchangers with condensation on the shell side is studied [11Q] and reviewed [25Q]. A model is proposed that predicts the axial temperature distribution in a radiation recuperator based on radiation exchange between a combustion gas and the annular recuperator walls [51Q]. Conditions are determined under which heat exchanger assemblies are invarient to the interchange of the two fluids [30Q]. A simulation of the dynamic character of systems of heat exchangers is compared with experimental results [33Q].

Sponge- and filament-packed heat exchangers received consideration for the special purpose of cooling ^3He [5Q] and in longitudinal counter-flow of air under conditions of evaporative cooling of water in a film-type contact heat exchanger [27Q].

Economic and optimization considerations are integral features of operating heat exchangers. Various levels of optimization are considered in order to facilitate selection and promote economic design of heat exchanger equipment [22Q]. Heat transfer measurements were performed for in-line and staggered tube banks at various spacings, the optimium spacing being that which maximizes the heat transfer per unit of pumping power [1Q]. Using simplifying assumptions, ref. [20Q] gives concise expressions for calculating the dimensions of spiral heat exchangers with minimum annual cost of heating surface and pumping energy. The utility of static mixing devices is examined for use in tubular heat exchangers transferring heat to or from highly viscous liquids [10Q]. A thermodynamic efficiency based on the Second Law is defined for heat exchangers in terms of the mean temperature difference between two fluids exchanging heat and an environmental temperature and used to predict the effect of heat exchanger parameter changes upon efficiency of energy use [53Q].

The use of fins to improve the performance of heat exchangers attracts the interest of a number of investigators. For a two-row tube and plate fin heat exchanger the overall heat transfer coefficient measurements, together with published results for one-row measurements, permits local mass transfer coefficients to be transferred to local heat transfer coefficients [35Q]. To assess the optimal condition for finned heat exchanger ducts, ref. [23Q] considers the case for laminar flow and proposes the optimal configuration as that which provides maximum heat transfer per unit cost of finned duct. Heat transfer and pressure drop measurements are measured for finned tube banks arranged parallel and at specified angles to the flow direction [4Q]. To improve the performance of compact heat exchangers inclined, louvered fins are proposed [46Q]; their likely effect is predicted by means of a simple theoretical model and experimentally studied to determine the specific fin configuration giving optimum heat transfer and pumping loss characteristics. The heat transfer in a crossflow shell-and-tube heat exchanger is augmented by the use of wall radiation [54Q] in both shell and tube sides, with maximum overall coefficients of heat transfer as much as 80% larger than those for the same circumstances but with radiation absent. A simple, accurate (although approximate) calculation procedure adapts the findings to practical use. For designer–practitioners, ref. [42Q] reviews a number of the more noteworthy topical developments in the improvement of heat transfer by convection, in evaporation and condensation.

Using beds of glass beads and sand to represent a fluidized bed, ref. [7Q] measures the heat transfer between such media and horizontal, square, in-line tube bundles of various pitch. Results are compared with those predicted by five models in the literature.

General

Prediction methods for the performance of film-cooled gas turbine blades were compared [8S]. Cascade experiments [11S] determined aerodynamic losses connected with film cooling and optimized [29S] turbine blades with respect to heat transfer. The effect of vibration on the film cooling effectiveness and losses was measured in vane cascades [16S]. End wall flow and heat transfer were determined from flow visualization [9S] and measurements [36S]. An experimental investigation considered gas turbine disk cooling [17S] and the heat flux distribution in a turbine stage [6S, 7S].

A numerical study determined the transient temperature field in a furnace and compared the results with measured values [14S]. Cold model tests of the flow and heat transfer in furnaces of powdered coal boilers were compared with full-scale furnace tests [1S].

A parametric study analyzes heat transfer during a loss of coolant accident to check the validity and limits of the computer model [24S].

Heat transfer was measured in stirred multiphase systems [18S] including suspensions [27S] of micromolecular substances. Heat transfer data were also obtained experimentally for heating helical coils

[13S], for vertical tubes with film evaporation [33S], and for tubes in bubble column reactors [34S]. Dissolved gas was found [15S] to have a negative effect on the critical heat flux in forced flow heat exchangers. The analysis and design of viscous flow coolers was discussed in [23S].

The state of the art in heat and mass transfer was reported for open cycle OTEC systems [4S].

Experiments and analysis using microcomputers consider insulation of attics [28S] and basements [19S, 35S].

A number of papers consider heat transfer in tribology, either for bearings [10S, 20S, 21S] or for sliding contacts [3S].

Numerical calculations consider heat transfer in strip rolling [32S] and in forging [5S].

A number of papers deal with drying of grains [25S, 26S, 31S], of coal [22S], and of wood [2S], as well as of optically semitransparent materials [12S], and in vibro-fluidized beds [30S].

Solar energy

The transient collector performance test method defined in the British Standards Institute Draft for Development DD77: 1982 is shown to be effective in determining the steady-state characteristics of flat-plate collectors from measurements taken under variable conditions [8T]. Flat-plate collectors having copper-tube/copper-sheet and copper-tube/aluminium-sheet absorber plates have been analyzed, fabricated and field tested [24T]. An analytical model was developed to extend the current techniques used in the prediction of the performance of nonmetallic flat-plate solar collectors [18T]. The transient periodic characteristics of flat-plate solar collectors, subjected to solar flux that fluctuates were evaluated experimentally and analytically [21T, 22T]. A comparison of single-equation methods to approximate the top loss coefficient for flat-plate solar collectors is made with the value obtained using the iterative solution of the energy balance equations [9T]. A simple analysis has been developed to predict the transient response of a tubular solar energy collector [1T]. The efficiencies of arrays of evacuated tubular collectors with diffuse reflectors have been determined experimentally using a calorimetric technique, and theoretically using a Monte Carlo ray tracing technique [6T].

The performance of a solar pond as a source of thermal energy has been modeled [20T]. The growth of a nonconvective zone by natural diffusion as a function of time from an initial two-layer configuration in a salt gradient solar pond was analyzed for two limiting cases: (i) complete mixing in the top layer and (ii) no mixing in the top layer [12T]. Transmission spectra and extinction coefficients for different salt solutions have been measured and the effect of pond transparency on the performance of a district solar pond is discussed [15T]. A numerical simulation of ground heat loss from a solar pond shows good agreement with the results obtained from pond measurements [11T].

A mathematical model gives the temperature distribution and the efficiency as functions of the geometrical parameters of a honeycomb structure collector which can be used for space heating in a passive solar system [10T]. The question of designing a nonhomogenous wall to optimize the 'thermal flywheel effect' is considered [7T]. Free convection turbulent heat transfer within the parallel plate channel of a Trombe wall has been studied [4T].

Measurements indicate that natural convection is not a major contributor to the destratification process in a rock bed and that designers of solar systems can safely arrange their ducting for either normal- or reverse-charging, selecting whichever is more convenient [23T]. A method is described of simulating the performance of a solar heating system which can be incorporated into a microcomputer-based system design and sizing procedure [17T]. Important parameters influencing the performance of devices for solar water heating and distillation were analyzed and discussed [5T]. Drying operations under humidistat and temperature control with and without a solar collector were simulated [19T].

The Gaussian Thermal Flux model is a fast thermal radiation submodel which operates with minimal input and predicts flux on sloped surfaces for both clear and cloudy days [13T]. The flux model was tested by sensitivity analysis and compared with observed data and other existing emperical models [14T]. The radiation to and from clear skies is an important solar energy topic. A discussion is presented of the equivalent radiative sky temperature [3T]. A new algorithm has been developed for calculating the thermal radiant temperature of the sky [16T]. In a related paper, the emissivity of clear skies is characterized [2T].

Plasma heat transfer

Plasma heat transfer and related applications are of continuing interest.

Parametric studies of thermal r.f. plasmas show (for

argon and nitrogen at $p = 1$ atm) that the flow and temperature fields in the core region, as well as the heat flux to the wall of the plasma confinement tube, are considerably altered by changes in the torch operating conditions [24U]. Modeling of a free-burning, high-intensity arc at elevated pressures (> 1 atm) shows that the arc contracts as the pressure increases and, as a consequence, the current density, the enthalpy flux and the voltage drop increase also [10U]. Comparisons between theoretical and experimental voltage–current characteristics of ablation-stabilized arcs confined in cylindrical tubes show good agreement [17U]. Nonequilibrium modeling of wall-stabilized AC arcs at low currents indicates that nonequilibrium effects are important near zero current [18U].

Studies of temperature and flow fields of a free-burning arc deflected by a transverse magnetic field indicate that the induced mass flow through the arc cools the arc and flattens its temperature field [7U]. Pressure and heat flux measurements of a thermal argon plasma flow confined in a water-cooled tube are reported. Axial distributions are in fair agreement with analytical predictions [15U]. Numerical results of simulating a thermal plasma flow in a cooled tube are reported [14U]. Experimental studies of the thermal boundary layer in dual-flow plasmas (dual flow orifice nozzle interrupter arrangement) are in reasonable agreement with theory [22U]. Firing of a high-energy plasma jet igniter reveals turbulent structures similar to turbulent puffs. Measurements of the growth rate of these features together with those of their impulse and thermal energy confirm this similarity [32U]. Analytical results based on an optically-thick model for radiative and collisional effects in nonequilibrium argon plasma flows in a circular tube are in reasonable agreement with experiments, indicating the importance of self-absorption effects (29U].

Investigations of the vapor flow patterns in free-burning, high-current arcs indicate two flow patterns depending on the electrode arrangement. Radiation losses exceed 84% of the total energy dissipation in the vapor dominant arc [38U]. Studies of the vapor diffusion from a copper anode in a nitrogen arc indicate diffusion speeds between 10 and 20 m s^{-1} [27U]. High radiation efficiency in high-pressure sodium discharges under controlled vapor conditions is favored by high input power and a mercury-rich amalgam [33U].

Experiments for synthetic gas production from peat using a steam plasma reveal a carbon conversion of 89% and a hydrogen-to-carbon monoxide ratio of 1.8 in the product gas [30U]. Results of thermal decomposition measurements of CO_2 in an argon plasma jet allow an analytical description of this process [11U]. Attempts to reduce TiO_2 by thermal plasma treatment indicate that, at best, 10% (by weight) of TiO_2 can be reduced to TiO by a carbo-thermic reaction [5U].

Coalesced multiple-arc discharges can be used for producing larger and more uniform volumes of ionized gas which is important for heating of gases and for plasma processing of materials [9U]. A 100 kWe magnetoplasmadynamic (MPD) thruster system may be applicable to a wide variety of future space emissions. Roughly 10% of the total electric power delivered to the thruster is converted into heat in the electrodes [28U]. Five optical fibers are heated simultaneously by a 50-Hz AC discharge for fusion mass-splicing of fibers [31U].

Radiation emitted by an SF_6 arc in the range $4000 < \lambda < 6000$ Å is due to radiative recombination of S^+. The values obtained for the Biberman factor are in good agreement with theoretical predictions [13U]. Non-LTE diagnostics of an argon arc shows that LTE is obtained in a 4-mm-diam. arc at atmospheric pressure for electron densities $n_e > 6 \times 10^6$ cm^{-3}. For $n_e < 6 \times 10^{16}$ there is an increasing underpopulation of excited levels as the arc current decreases [25U]. A new two-channel technique for determining electron densities in plasmas from Stark broadening is described and its utility is demonstrated with a laser-produced sodium plasma [3U]. Spectrometric measurements in free-burning argon arcs at $p = 1$ atm seem to indicate that there are departures from Local Thermodynamic Equilibrium (LTE) and deficiencies in the theory describing continuum emission [8U].

Total heat transfer calculations are presented for spherical and cylindrical probes in collisionless plasma flows [4U]. Results of an analysis for the plasma flow around blunt-body electrostatic probes exposed to a flowing argon plasma (arc jet) indicate that the electron temperature at the probe surface may be as much as 30% lower than the freestream electron temperature [2U].

A special arc arrangement has been developed for measuring the thermal conductivity of Cl_2 and O_2 in the dissociation region ($T \leqslant 6000$ K) using optical interferometry (20U, 21U]. Discrepancies between analytical predictions of the thermal conductivity of hydrogen for $T > 4500$ K and experimental results derived from arc experiments are attributed to strong deviations from LTE [26U]. Calculated thermodynamic and transport properties of plasmas produced by the ablation of Teflon, alumina, perspex, and PVC in high current arc switches are presented for temperatures from 5000 to 30,000 K [16U]. Even traces of metallic vapors may exert a strong influence on the properties and the behavior of thermal plasmas and on the associated heat transfer to injected particulate matter during thermal plasma processing [23U]. Calculations of thermodynamic and transport properties of a typical arc furnace plasma show that dissociation of molecular species has a strong effect on the specific heat and on the thermal conductivity of the mixture and the electrical conductivity is strongly affected by metallic vapors from the molten pool and slag cover [34U].

A simplified analytical model for an anchored cathode root in a DC mercury-vapor arc is based on the assumption that both electric current and heat pass through the solid metal anchor rather than through the thin mercury film which wets the solid anchor [19U]. Surface roughness and surface temperature have little effect on the movement of cathode spots in vacuum arcs which is important for establishing cathode heating models [12U]. Studies of the electrode erosion phenomena in high-energy pulsed discharges reveal that a large amount of the erosion is in the form of solid and molten material removed parallel to the electrode surface and from the edge of the macroscopic craters found on the cathode [6U].

The SF_6 spark develops through two different glow phases where the first stage is a low-current (~ 10 mA) phase followed by a second phase in which gas heating leads to dissociation of SF_6 in times of the order of 100 μs with a transition to a higher current (~ 1 A) glow phase [37U]. Studies of microwave absorption and plasma heating due to microwave breakdown in the atmosphere indicate that the microwave field strength in the ionization regions is attenuated to the breakdown value at steady state, and the resulting electron temperature is approx. 2 eV, independent of the incident microwave flux [36U].

MHD free convection flow about a semi-infinite plate has been investigated for the case of uniform heat flux to the plate [35U]. Investigations of magnetoconvection in a rotating fluid between walls of very low thermal conductivity show that in contrast to fluids under the same conditions but with walls of different thermal conductivity, the fluid in the present case is always thermally unstable for any Taylor and Chandrasekhar numbers [1U].

REFERENCES

BOOKS

O. C. Jones and N. M. Farukhi (Editors), *New Experimental Techniques in Heat Transfer—1984*. ASME Order Dept, New York (1984).

L. C. Witte and L. S. Saxens (Editors), *Fundamentals of Natural Convection/Electronic Equipment Cooling*. ASME Order Dept, New York (1984).

J. P. Hartnett and T. F. Irvine, Jr (Editors), *Advances in Heat Transfer*, Vol. 16. Academic Press, New York (1984).

A. Bejan, *Convection Heat Transfer*. John Wiley, New York (1984).

A. J. Chapman, *Heat Transfer*, 4th edn. Macmillan, New York (1984).

T. Cebeci and P. Bradshaw, *Physical and Computational Aspects of Convective Heat Transfer*. Springer Verlag, New York (1984).

F. M. White, *Heat Transfer*. Addison-Wesley, Reading, MA (1984).

U. Grigull and H. Sandner, *Heat Conduction*. Hemisphere, New York (1984).

CONDUCTION

1A. J. R. Barber, Thermoelastic displacements and stresses due to a heat source moving over the surface of a half plane, *J. appl. Mech.* **51**, 636 (1984).

2A. G. S. Barozzi and G. Pagliarini, Conjugated heat transfer in a circular duct with uniform and non-uniform wall thickness, *Heat Technol.* **2**(1), 72 (1984).

3A. G. S. Barozzi and G. Pagliarini, Experimental investigation of coupled conduction and laminar convection in a circular tube, *Int. J. Heat Mass Transfer* **27**, 2321 (1984).

4A. J. V. Beck, Green's function solution for transient heat conduction problems, *Int. J. Heat Mass Transfer* **27**, 1235 (1984).

5A. J. C. Chato and J. M. Khodadadi, Optimization of cooled shields in insulations (83-WA/HT-57), *J. Heat Transfer* **106**(4), 871 (1984).

6A. F. Civan and C. M. Sliepcevich, Some elementary solutions of the heat equation and related equations, *Int. J. Engng Sci.* **22**, 601 (1984).

7A. N. S. Clarke, Heat diffusion in random laminates, *Q. Jl Mech. appl. Math.* **37**(2), 195 (1984).

8A. M. Colakyan, R. Turton and O. Levenspiel, Unsteady state heat transfer to variously shaped objects, *High Temp. Engng* **5**(3–4), 82 (1984).

9A. R. S. Cook, K. H. Token and R. L. Calkins, A novel concept for reducing thermal contact resistance, *J. Spacecraft Rockets* **21**, 122 (1984).

10A. A. Z. Dolginov and N. A. Silantev, Heat transfer and diffusion from the sources with random components, *Int. J. Heat Mass Transfer* **27**, 2111 (1984).

11A. P. R. Dawson and S. Malkin, Inclined moving heat source model for calculating metal cutting temperatures, *J. Engng Ind.* **106**(3), 179 (1984).

12A. W. A. Day, Steady state oscillatory temperatures in coupled quasi-static thermoelasticity, *Q. Jl Mech. appl.*

Math. **37**(4), 581 (1984).
13A. A. Dayan, A. H. Merbaum and I. Segal, Temperature distributions around buried pipe networks in soil with a temperature dependent thermal conductivity, *Int. J. Heat Mass Transfer* **27,** 409 (1984).
14A. Y. Ersoy, A new nonlinear constitutive theory for conducting magnetothermoelastic solids, *Int. J. Engng Sci.* **22,** 683 (1984).
15A. J. Gosse, Sur une mèthode analytique de resolution de la conduction thermique dans les murs composites, *Int. J. Heat Mass Transfer* **27,** 1533 (1984).
16A. T. Hachisu, T. Sasaki and K. Taguchi, Transient temperature distribution in circular cylinders during quenching, *Heat Transfer—Jap. Res.* **10**(1), 52 (1981).
17A. J. C. Hein, Heat conduction in a plane wall subject to periodic temperature changes, *Wärme -u. Stoffübertr.* **18,** 61 (1984).
18A. D. M. Herbert and J. Adler, Thermal stability of a reactive slab with time-periodic heating of one surface, *Z. angew. Math. Phys.* **35**(1), 50 (1984).
19A. K. G. T. Holland, G. D. Raithby, F. B. Russell and R. G. Wilkinson, Coupled radiative and conductive heat transfer across honeycomb panels and through single cells, *Int. J. Heat Mass Transfer* **27,** 2119 (1984).
20A. C. O. Horgan, L. E. Payne and L. T. Wheeler, Spatial decay estimates in transient heat conduction, *Q. appl. Math.* **XLII**(1), 119 (1984).
21A. C. K. Hseih and S. L. Yang, A new theory on the critical thickness of insulation, *J. Heat Transfer* **106**(3), 648 (1984).
22A. Ming Jer Huang and Cha'o-Kuang Chen, Vertical circular pin with conjugated forced convection–conduction flow, *J. Heat Transfer* **106**(3), 658 (1984).
23A. S. C. Huang and Y. P. Chang, Anisotropic heat conduction with mixed boundary conditions, *J. Heat Transfer* **106**(3), 646 (1984).
24A. B. D. Hughes, M. Sahimi, L. E. Scriven and H. T. Davis, Transport and conduction in random systems, *Int. J. Engng Sci.* **22,** 1083 (1984).
25A. D. M. Kim and R. Viskanta, Heat transfer by conduction, natural convection and radiation across a rectangular cellular structure, *Int. J. Heat Fluid Flow* **5**(4), 205 (1984).
26A. T. Kunc, M. Lallemand and J. B. Saulnier, Some new developments on coupled radiative–conductive heat transfer in glasses—experiments and modelling, *Int. J. Heat Mass Transfer* **27**, 2307 (1984).
27A. M. Manzoor, D. B. Ingham and P. J. Heggs, The accuracy of perfect contact fin analyses, *J. Heat Transfer* **106**(1), 234 (1984).
28A. R. C. Mehta, Numerical solution of nonlinear inverse heat conduction problem with a radiation boundary condition, *Int. J. Numer. meth. Engng* **20,** 1057 (1984).
29A. J. E. Ostenson, D. K. Finnemore, J. D. Verhoeven, E. D. Gibson and H. R. Shanks, Thermal diffusivity of *in situ* composites, *J. appl. Phys.* **55**(1), 278 (1984).
30A. M. N. Özişik, M. A. Boles and R. M. Cotta, Eigenvalues basic to diffusion in the part of a sphere cut out by a cone, *J. Heat Transfer* **106**(2), 472 (1984).
31A. M. N. Özişik and B. Vick, Propagation and reflection of thermal waves in a finite medium, *Int. J. Heat Mass Transfer* **27,** 1845 (1984).
32A. K. L. Peddicord, T. L. George and T. C. Kennedy, A model for the thermal–mechanical behavior of mixed-carbide sphere-pac fuel pins, *Nucl. Sci. Engng* **86,** 253 (1984).
33A. M. Slonim and A. Tslaf, An approximating temperature profile of an instantaneous heat source, *J. Heat Transfer* **106**(2), 474 (1984).
34A. Y. Terauchi, H. Nadano and M. Kohno, On the temperature rise caused by moving heat sources (1st report, calculation of the temperature under three-dimensional heat flow), *Bull. JSME* **27,** 831 (1984).
35A. C. Wei and J. T. Berry, Some characteristics of the conduction heat flux at the surface of a wedge enclosure, *J. Heat Transfer* **106**(4), 902 (1984).
36A. F. Yoshikawa and M. Tanaka, A boundary element analysis of steady-state heat-conduction problems in axisymmetric bodies, *Heat Transfer—Jap. Res.* **13**(4), 51 (1984).

CHANNEL FLOW

1B. R. M. Abdel-Wahed and A. E. Attia, Fully developed laminar flow and heat transfer in an arbitrarily shaped triangular duct, *Wärme- u. Stoffübertr.* **18,** 85 (1984).
2B. R. M. Abdel-Wahed, A. E. Attia and M. A. Fifni, Experiments on laminar flow and heat transfer in an elliptical duct, *Int. J. Heat Mass Transfer* **27,** 2397 (1984).
3B. P. H. Alfredsson and A. Johansson, Time scales in turbulent channel flow, *Phys. Fluids* **27**(7), 1974 (1984).
4B. M. J. Allmen and P. M. Eagles, Stability of divergent channel flows: a numerical approach, *Proc. R. Soc.* **A392,** 359 (1984).
5B. R. S. Amano, Laminar heat transfer in a channel with two right-angled bends, *J. Heat Transfer* **106**(3), 591 (1984).
6B. H. Barrow, A. K. A. Hassan and C. Avgerinos, Peripheral temperature variation in a wall of a noncircular duct—an experimental investigation, *Int. J. Heat Mass Transfer* **27,** 1031 (1984).
7B. J. W. Baughn, M. A. Hoffman, B. E. Launder and R. K. Takashashi, Turbulent heat transport in circular ducts with circumferentially varying heat flux, *J. Heat Transfer* **106**(1), 64 (1984).
8B. J. W. Baughn, M. A. Hoffman, R. K. Takahashi and B. E. Launder, Local heat transfer downstream of an abrupt expansion in a circular channel with constant wall heat flux, *J. Heat Transfer* **106**(4), 789 (1984).
9B. M. S. Bhatti, Heat transfer in the fully developed region of elliptical ducts with uniform wall heat flux, *J. Heat Transfer* **106**(4), 895 (1984).
10B. D. Burfoot and P. Rice, Heat transfer and pressure drop characteristics of alternate rotation swirl-flow inducers in a circular duct, *Chem. Engng Res. Des.* **62**(2), 128 (1984).
11B. M. N. Channabasappa, K. C. Umapathy and I. V. Nayak, Convective heat transfer in a parallel plate channel with porous lining, *Wärme- u. Stoffübertr.* **18,** 211 (1983).
12B. V. G. Chernyak, V. V. Kalinin and P. E. Suetin, The kinetic phenonomena in nonisothermal motion of a

binary gas mixture through a plane channel, *Int. J. Heat Mass Transfer* **27,** 1189 (1984).

13B. J. dos Santos Travelho and L. A. Vieria Dias, Temperature calculation in an incompressible permanent laminar fluid flow through a circular pipe with axial conduction and viscosity, *Int. J. Heat Mass Transfer* **27,** 1183 (1984).

14B. M.-R. M. Drižius, V. N. Janušas and R. K. Škema, Coefficients of heat transfer from a rough tube to a flow of water, *Heat Transfer—Sov. Res.* **15**(2), 49 (1983).

15B. A. Dumas and G. S. Barozzi, Laminar heat transfer to blood flowing in a circular duct, *Int. J. Heat Mass Transfer* **27,** 391 (1984).

16B. M. Elena, Suction effects on turbulence statistics in a heated pipe flow, *Phys. Fluids* **27**(4), 851 (1984).

17B. M. Fujii, T. Fujii and T. Nagato, A numerical analysis of laminar flow and heat transfer of air in an in-line tube bank, *Numer. Heat Transfer* **7,** 89 (1984).

18B. J. C. Han, Heat transfer and friction in channels with two opposite rib-roughed walls (83-HT-26), *J. Heat Transfer* **106**(4), 744 (1984).

19B. M. Hasantani, N. Arai, Y. Bando and H. Nakamura, Collection of thermal radiation by a semitransparent fluid layer flowing in an open channel, *Heat Transfer—Jap. Res.* **11**(3), 17 (1982).

20B. Y. Hayashi, A. Takimoto and Y. Yamamoto, Heat and mass transfer with a mist formation in a laminar duct flow, *Heat Transfer—Jap. Res.* **10**(1), 37 (1981).

21B. C.-R. Huang, M. Matlosz, W.-D. Pan and W. Snyder, Jr, Heat transfer to a laminar flow fluid in a circular tube, *A.I.Ch.E. Jl* **30,** 833 (1984).

22B. G. A. Hughmark, Heat transfer with viscoelastic fluids in turbulent flow, *A.I.Ch.E. Jl* **30,** 1033 (1984).

23B. K. H. Im and R. K. Ahluwalia, Combined convection and radiation in rectangular ducts, *Int. J. Heat Mass Transfer* **27,** 221 (1984).

24B. G. I. Isayev, Effect of the direction of flow and tube position on heat transfer in the case of variable physical properties of the flow, *Heat Transfer—Sov. Res.* **15**(2), 1 (1983).

25B. G. I. Isayev, Investigation of forced-convection heat transfer to supercritical *n* heptane, *Heat Transfer—Sov. Res.* **15**(1), 119 (1983).

26B. Y. Kawashima, M. Nakagawa and S. Iuchi, Heat transfer characteristics in a rectangular flow section downstream from a mixing junction, *Heat Transfer—Jap. Res.* **12**(4), 70 (1983).

27B. Y. Komiyana, F. Mikami, K. Okui and T. Hori, Laminar forced convection heat transfer in curved channels of rectangular cross section, *Heat Transfer—Jap. Res.* **13**(3), 68 (1984).

28B. T. F. Lin, K. H. Hawks and W. Leidenfrost, Transient thermal entrance heat transfer in laminar pipe flows with step change in pumping pressure, *Wärme- u. Stoffübertr.* **18,** 201 (1983).

29B. Y. K. Lin and L. C. Chow, Effects of wall conduction on heat transfer for turbulent flow in a circular tube (81-HT-30), *J. Heat Transfer* **106**(3), 597, 1984.

30B. E. Naito, Laminar heat transfer in the entrance region between parallel plates with uniform wall temperature, *Heat Transfer—Jap. Res.* **13**(3), 92 (1984).

31B. H. Nakamura, A. Matsuura, J. Kiwaki, I. Yamada and M. Hasantani, Laminar heat transfer of a high-viscosity Newtonian fluid in a horizontal rectangular ducts, *Heat Transfer—Jap. Res.* **10**(1), 1 (1981).

32B. R. Narayanan, A note on the extended Graetz problem, *Appl. Sci. Res.* **41,** 31 (1984).

33B. L. R. Persen, Stratified two-phase flow in circular pipes, *Int. J. Heat Mass Transfer* **27,** 1227 (1984).

34B. N. Phan-Thien and M. B. Bush, On the steady flow of a Newtonian fluid between two parallel disks, *Z. angew. Math. Phys.* **35,** 912 (1984).

35B. A. T. Prata and E. M. Sparrow, Heat transfer and fluid flow characteristics for an annulus of periodically varying cross section, *Numer. Heat Transfer* **7,** 285 (1984).

36B. M. Rahman and P. Colbourne, Analytic solution for secondary flow with heat transfer in a square duct, *Int. J. Heat Fluid Flow* **5**(3), 167 (1984).

37B. C. W. Rapley and A. D. Gosman, The prediction of turbulent flow and heat transfer in a narrow isosceles triangular duct, *Int. J. Heat Mass Transfer* **27,** 253 (1984).

38B. G. J. Rowley and S. V. Patankar, Analysis of laminar flow and heat transfer in tubes with internal circumferential fins, *Int. J. Heat Mass Transfer* **27,** 553 (1983).

39B. N. Seki, S. Fakusako and M. Yoneta, Heat transfer from the heated concave wall of a return bend with rectangular cross section, *Wärme- u. Stoffübertr.* **17,** 17 (1983).

40B. J. Simonek, Turbulent transport in the transition flow region, *Int. J. Heat Mass Transfer* **27,** 2415 (1984).

41B. C. W. Sohn and M. M. Chen, Heat transfer enhancement in laminar slurry pipe flows with power law thermal conductivities, *J. Heat Transfer* **106**(3), 539 (1984).

42B. H. M. Soliman, Analysis of low-Péclet heat transfer during slug flow in tubes with axial wall conduction, *J. Heat Transfer* **106**(4), 782 (1984).

43B. P. Souza Mendes and E. M. Sparrow, Periodically converging–diverging tubes and their turbulent heat transfer, pressure drop, fluid flow, and enhancement characteristics, *J. Heat Transfer* **106**(1), 55 (1984).

44B. E. M. Sparrow and A. Chaboki, Swirl-affected turbulent fluid flow and heat transfer in a circular tube, *J. Heat Transfer* **106**(4), 766 (1984).

45B. E. M. Sparrow and L. M. Hossfeld, Effect of rounding or protruding edges on heat transfer and pressure drop in a duct, *Int. J. Heat Mass Transfer* **27,** 1715 (1984).

46B. E. M. Sparrow and M. Molki, Turbulent heat transfer coefficients in an isothermal-walled tube for either a built-in or free inlet, *Int. J. Heat Mass Transfer* **27,** 669 (1984).

47B. E. M. Sparrow, A. P. Suopys and M. A. Ansari, Effect of inlet, exit, and fin geometry on pin fins situated in a turning flow, *Int. J. Heat Mass Transfer* **27,** 1039 (1984).

48B. E. M. Sparrow and W. Q. Tao, Symmetric vs asymmetric periodic disturbances at the walls of a heated flow passage, *Int. J. Heat Mass Transfer* **27,** 2133 (1984).

49B. K. Stephan, S. Lin, M. Durst, F. Huang and D. Seher, A similarity relation for energy separation in a vortex tube, *Int. J. Heat Mass Transfer* **27,** 911 (1984).

50B. J. Sucec and A. M. Sawant, Unsteady, conjugated, forced convection heat transfer in a parallel plate duct,

Int. J. Heat Mass Transfer **27,** 95 (1984).
51B. K. Suzuki, Y. Kang,. Sugimoto and T. Sato, Circular tube turbulent heat transfer in the downstream of an orifice, *Heat Transfer—Jap. Res.* **11**(1), 70 (1982).
52B. N. Tanaka, S. Moriya, A. Wada, Thermal–hydraulic characteristics in a liquid-metal fast breeder reactor hot plenum, *Nucl. Sci. Engng* **88,** 445 (1984).
53B. A. C. Trupp and A. C. Y. Lau, Fully developed laminar heat transfer in circular sector ducts with isothermal walls, *J. Heat Transfer* **106**(2), 467 (1984).
54B. U. Usui, Y. Sano and K. Iwashita, Heat transfer enhancement effects by combined use of internally grooved rough surfaces and twisted tapes, *Heat Transfer—Jap. Res.* **13**(4), 19 (1984).
55B. K. Watanabe, T. Taira and Y. Mori, Heat transfer augmentation for tubular flow by twisted tapes at high temperatures and its optimum performance, *Heat Transfer—Jap. Res.* **12**(3), 1 (1983).
56B. V. M. Yeroshenko, A. V. Yershov and L. I. Zaichik, Heat transfer for turbulent flow in a circular tube with uniform suction or injection, *Int. J. Heat Mass Transfer* **27,** 1197 (1984).
57B. H. K. Yoon and A. J. Ghajar, An analysis of the heat transfer to drag reducing turbulent pipe flows, *J. Heat Transfer* **106**(4), 898 (1984).

BOUNDARY LAYER AND EXTERNAL FLOWS

1C. B. Abramzon and C. Elata, Unsteady heat transfer from a single sphere in Stokes flow, *Int. J. Heat Mass Transfer* **27,** 687 (1984).
2C. J.-E. J. Adomaitis, B. A. Cesna and J. V. Vilemas, Experimental study of heat transfer and friction on a cylinder in longitudinal turbulent flow of air with variable physical properties, *Heat Transfer—Sov. Res.* **15**(1), 52 (1983).
3C. N. Afzal, The cooling of a low heat resistance stretching sheet moving through a fluid: a rejoinder, *Wärme- u. Stoffübertr.* **17,** 217 (1983).
4C. B. A. Brigham and G. J. Van Fossen, Length to diameter ratio and row number effects in short pin fin heat transfer (83-GT-54), *J. Engng Pwr* **106,** 241 (1984).
5C. K. Chen and M. M. Chen, Thermal instability of forced convection boundary layers, *J. Heat Transfer* **106**(2), 284 (1984).
6C. E. P. Childs and R. E. Mayle, Heat transfer on a rotationally oscillating cylinder in crossflow, *In J. Heat Mass Transfer* **27,** 85 (1984).
7C. J. F. Clarke, D. R. Kassoy and N. Riley, Shocks generated in a confined gas due to rapid heat addition at the boundary, I. Weak shock waves, *Proc. R. Soc.* **A393,** 309 (1984).
8C. J. F. Clarke, D. R. Kassoy and N. Riley, Shocks generated in a confined gas due to rapid heat addition at the boundary, II. Strong shock waves, *Proc. R. Soc.* **A393**, 331 (1984).
9C. T. Dakos, C. A. Verriopoulos and M. M. Gibson, Turbulent flow with heat transfer in plane and curved wall jets, *J. Fluid Mech.* **145,** 339 (1984).
10C. S. Kumar Dash and W. N. Gill, Forced convection heat and momentum transfer to dendritic structures (parabolic cylinders and paraboloids of revolution), *Int. J. Heat Mass Transfer* **27,** 1345 (1984).
11C. A. R. Dorokhov and V. N. Bochagov, Heat transfer to a film falling over horizontal cylinders, *Heat Transfer—Sov. Res.* **15**(2), 96 (1983).
12C. N. Fujita and T. Tsubouchi, An experimental study of unsteady heat transfer from a flat plate to an oscillating air flow, *Heat Transfer—Jap. Res.* **11**(3), 31 (1982).
13C. H. Fujita, H. Takahama and C. Hattori, Effects of a strip inserted in the boundary layer on heat transfer from a plane surface (distribution of heat transfer coefficients), *Bull. JSME* **27,** 212 (1984).
14C. L. Fulachier and R. A. Antonia, Spectral analogy between temperature and velocity fluctuations in several turbulent flows, *Int. J. Heat Mass Transfer* **27,** 987 (1984).
15C. V. K. Garg and R. C. DiPrima, The effect of heating on the stability of flow over concave walls, *Phys. Fluids* **27**(4), 812 (1984).
16C. K. Gersten and H. Herwig, Momentum and heat transfer for the laminar flat-plate flow with variable fluid properties (in German), *Wärme- u. Stoffübertr.* **18,** 25 (1984).
17C. R. J. Goldstein and J. Karni, The effect of wall boundary layer on local mass transfer from a cylinder in crossflow, *J. Heat Transfer* **106**(2), 260 (1984).
18C. T. F. Greenwood, Y. C. Lee, R. L. Bender and R. E. Carter, Space shuttle base heating, *J. Spacecraft Rockets* **21,** 339 (1984).
19C. R. N. Gupta, J. N. Moss, A. L. Simmonds, J. L. Shinn and E. V. Zoby, Space shuttle heating analysis with variation in angle of attack and catalycity, *J. Spacecraft Rockets* **21,** 217 (1984).
20C. P. Jonas, Transfer of heat from a heated wire at high subsonic gas velocities, *Heat Transfer—Sov. Res.* **15**(1), 112 (1983).
21C. T. Kawamura, M. Hiwada, I. Mabuchi and M. Kumada, Augmentation of turbulent heat transfer on a flat plate with a three-dimensional protuberance (1st report, local heat transfer characteristics and flow pattern around a finite length circular perturbance), *Bull. JSME* **27,** 2787 (1984).
22C. D. W. Kercher, R. E. Sheer, Jr. and R. M. C. So, Short duration heat transfer studies at high freestream temperatures (82-GT-129), *J. Engng Pwr* **105,** 156 (1983).
23C. T. Kohya, I. Goto, E. Harada, M. Toda and H. Konno, Heat transfer from a circular pipe in an air jet and in a corona wind, *Heat Transfer—Jap. Res.* **12**(4), 1 (1983).
24C. S. Kotake and I. Aoki, Heat transfer of a cylinder in large-amplitude oscillating wake flows, *Int. J. Heat Mass Transfer* **27,** 1903 (1984).
25C. S. S. Kutateladze, The mixing length hypothesis in the turbulent theory, *Int. J. Heat Mass Transfer* **27,** 1947 (1984).
26C. G. C. Lauchle and G. B. Gurney, Laminar boundary-layer transition on a heated underwater body, *J. Fluid Mech.* **144,** 79 (1984).
27C. E. J. Lyford-Pike and J. B. Heywood, Thermal boundary layer thickness in the cylinder of a spark-ignition engine, *Int. J. Heat Mass Transfer* **27,** 1873 (1984).
28C. V. I. Lysenko and A. A. Maslov, The effect of cooling on

supersonic boundary-layer stability, *J. Fluid Mech.* **147**, 39 (1984).
29C. I. Mabuchi and M. Kumada, Heat transfer from a finite circular cylinder on a flat plate, *Bull. JSME* **27**, 2430 (1984).
30C. S. V. Manuilovich and E. D. Terentev, On the influence of heat transfer and blowing on the structure of laminar hypersonic flow behind a body, *Appl. Math. Mech.* **47**, 517 (1983).
31C. M. L. Marziale and R. E. Mayle, Effect of an oscillating flow direction on leading edge heat transfer (83-GTJ-3), *J. Engng Pwr* **106**, 222 (1984).
32C. D. C. McCormick, F. L. Test and R. C. Lessmann, The effect of freestream turbulence on heat transfer from a rectangular prism (83-HT-29), *J. Heat Transfer* **106**(2), 268 (1984).
33C. D. E. Metzger, C. S. Fan and S. W. Haley, Effect of pin shape and array orientation on heat transfer and pressure loss in pin fin arrays (83-GTJ-1), *J. Engng Pwr* **106**, 252 (1984).
34C. C. G. Miller, Heating distributions for a 0.006-scale shuttle orbiter over a range of hypersonic flow conditions. *J. Spacecraft Rockets* **21**, 120 (1984).
35C. T. Mizushina, F. Ogino, T. Matsumoto, H. Ueda and K. Baba, Effect of a large temperature difference and pyrolysis on the heat transfer between a cylinder and transverse gas flows of nitrogen and propane, *Int. J. Heat Mass Transfer* **27**, 829 (1984).
36C. A. Nakayana, H. Koyama and S. Ohsawa, Self-similar thermal boundary layers on plane and axisymmetric bodies, *Wärme- u. Stoffübertr.* **18**, 69 (1984).
37C. A. Narumi, S. Kato, K. Terada and R. Izumi, Forced heat convection in turbulent flow over a flat plate of a finite width, *Heat Transfer—Jap. Res.* **13**(3), 17 (1984).
38C. D. A. Nealy, M. S. Mihelc, L. D. Hylton and H. J. Gladden, Measurements of heat transfer distribution over the surfaces of highly loaded turbine nozzle guide vane (83-GT-53), *J. Engng Pwr* **106**, 149 (1984).
39C. P. Y. Nizou, Analogie entre transferts de chaleur et de quantité de mouvement dans un jet pariétal plan turbulent, *Int. J. Heat Mass Transfer* **27**, 1737 (1984).
40C. T. Ota, H. Nishiyama and Y. Taoka, Heat transfer and flow around an elliptic cylinder, *Int. J. Heat Mass Transfer* **27**, 1771 (1984).
41C. Yao Peng, Heat transfer and friction loss characteristics of pin fin cooling configurations (83-GT-123), *J. Engng Pwr* **106**, 246 (1984).
42C. D. H. Petley, D. M. Smith, C. L. W. Edwards, A. B. Carlson and H. H. Hamilton II, Surface step induced gap heating in the shuttle thermal protection system, *J. Spacecraft Rockets* **21**, 156 (1984).
43C. Yu. G. Semyonov, S. F. Borisov and P. E. Suetin, Investigation of heat transfer in rarefied gases over a wide range of Knudsen numbers, *Int. J. Heat Mass Transfer* **27**, 1789 (1984).
44C. Z. P. Shulman, V. I. Korobko and V. K. Shashmin, Heat and mass transfer in viscous swirled jets, *Int. J. Heat Mass Transfer* **27**, 857 (1984).
45C. R. J. Simoneau and G. J. VanFossen, Jr, Effect of location in an array on heat transfer to a short cylinder in crossflow, *J. Heat Transfer* **106**(1), 42 (1984).
46C. J. C. Simonich and R. J. Moffat, Liquid crystal visualization of surface heat transfer on a concavely curved turbulent boundary layer (83-GTJ-7), *J. Engng Pwr* **106**, 619 (1984).
47C. E. M. Sparrow, T. J. Stahl and P. Traub, Heat transfer adjacent to the attached end of a cylinder in crossflow, *Int. J. Heat Mass Transfer* **27**, 233 (1984).
48C. M. Takeuchi, M. Suzuki, Y. Kurosaki and N. Isshiki, Heat transfer on cylinder covered with close-fitting fabrics (Part III, Distribution of local heat transfer coefficients), *Bull. JSME* **27**, 483 (1984).
49C. A. Takimoto, Y. Hayashi and O. Matsuda, Thermal stability of Blasius flow over a horizontal flat plate, *Heat Transfer—Jap. Res.* **12**(1), 19 (1983).
50C. R. Tal (Thau), D. N. Lee and W. A. Sirignano, Heat and momentum transfer around a pair of spheres in viscous flow, *Int. J. Heat Mass Transfer* **27**, 1953 (1984).
51C. S. Tanaka and T. Kunitomo, Optimum fin geometries for simultaneous convective and radiative heat transfer, *Heat Transfer—Jap. Res.* **12**(3), 60 (1984).
52C. C. Treviño and A. Liñan, External heating of a flat plate in a convective flow, *Int. J. Heat Mass Transfer* **27**, 1067 (1984).
53C. S. D. Williams and D. M. Curry, Assessing the orbiter thermal environment using flight data, *J. Spacecraft Rockets* **21**, 534 (1984).
54C. A. A. Zukauskas, P. P. Vaitiekunas and J. J. Ziugzda, Analytic determination of the effect of free-stream turbulence of a viscous fluid on the coefficient of local heat transfer from a circular cylinder, *Heat Transfer—Sov. Res.* **15**(2), 11 (1983).

FLOW WITH SEPARATED REGIONS

1D. R. S. Amano, Development of a turbulence near-wall model and its application to separated and reattached flows, *Numer. Heat Transfer* **7**, 59 (1984).
2D. R. S. Amano and P. Goel, A numerical study of a separating and reattaching flow by using Reynolds-stress turbulence closure, *Numer. Heat Transfer* **7**, 343 (1984).
3D. T. Amitani, T. Adachi and T. Kato, A study on temperature separation in a large vortex tube, *Heat Transfer—Jap. Res.* **12**(2), 41 (1983).
4D. A. Bhatti and W. Aung, Finite difference analysis of laminar separated forced convection in cavities, *J. Heat Transfer* **106**(1), 49 (1984).
5D. T. Igarashi, Correlation between heat transfer and fluctuating pressure in separated region of a circular cylinder, *Int. J. Heat Mass Transfer* **27**, 927 (1984).
6D. V. Kottke, Local detachment fields with temperature or concentration boundary layer on detachment: transfer of heat, mass, and momentum, *Chemie-Ingr-Tech.* **56**(5), 424 (1984).
7D. D. C. McCormick, R. C. Lessmann and F. L. Test, Heat transfer to separated flow regions from a rectangular prism in a cross stream (83-HT-28), *J. Heat Transfer* **106**(2), 276 (1984).
8D. S. Mochizuki and M. Yao, Flow separation and heat transfer in radial flows between two parallel disks, *Heat Transfer—Jap. Res.* **12**(1), 48 (1983).
9D. H. Yamanoto, N. Seki and S. Fukusako, Convection heat transfer from a heated cavity bottom with different

wall heights, *Heat Transfer—Jap. Res.* **13**(3), 30 (1984).
10D. K. Yamamoto, Flow and heat transfer around a rarefied gas, *J. Phys. Soc. Jap.* **53**(1), 167 (1984).
11D. A. A. Žukausakas, R. V. Ulinskas and E. S. Bubelis, Heat transfer in separated flows at *Re* between 1 and $2 \cdot 10^6$ and *Pr* between 0.7 and 10,000 (separation of vortices from tubes), *Heat Transfer—Sov. Res.* **15**(2), 73 (1983).

HEAT TRANSFER IN POROUS MEDIA

1DP. C. Basilico et M. Martin, Approche experimentale des mécanismes de transfert au cours du séchage convectif à haute température d'un bois resineux, *Int. J. Heat Mass Transfer* **27**, 657 (1984).
2DP. H. H. Bau, Convective heat losses from a pipe buried in a semi-infinite saturated porous medium, *Int. J. Heat Mass Transfer* **27**, 2047 (1984).
3DP. H. H. Bau, Thermal convection in a horizontal, eccentric annulus containing a saturated porous medium—an extended perturbation expansion, *Int. J. Heat Mass Transfer* **27**, 2277 (1984).
4DP. H. H. Bau, Low Rayleigh number thermal convection in a saturated porous medium bounded by two horizontal eccentric cylinders, *J. Heat Transfer* **106**(1), 166 (1984).
5DP. D. E. Beasley and J. A. Clark, Transient response of a packed bed for thermal energy storage, *Int. J. Heat Mass Transfer* **27**, 1659 (1984).
6DP. K. R. Blake, A. Bejan and D. Poulikakos, Natural convection near 4°C in a water saturated porous layer heated from below, *Int. J. Heat Mass Transfer* **27**, 2355 (1984).
7DP. P. Chen and D. C. T. Pei, Fluid to particle heat transfer in fluidized beds of mixed-sized particles, *Can. J. chem. Engng* **62**, 469 (1984).
8DP. P. Cheng and I. Pop, Transient free convection about a vertical flat plate embedded in a saturated porous medium, *Int. J. Engng Sci.* **22**, 253 (1984).
9DP. M. T. Dalman, J. H. Merkin and C. McGreavy, Modeling of the fluid flow and heat transfer at the edge of a fixed bed, *Numer. Heat Transfer* **7**, 429 (1984).
10DP. P. G. Daniels and P. G. Simpkins, The flow induced by a uniformly heated vertical wall in a porous medium, *Q. J. Mech. app. Math.* **37**(2), 339 (1984).
11DP. S. C. Dhingra, D. J. Dunn and P. V. Narayanan, The analysis of heat transfer in fixed beds of particles at low and intermediate Reynolds numbers, *Int. J. Heat and Mass Transfer* **27**, 2377 (1984).
12DP. A. G. Dixon, M. A. DiCostanzo and B. A. Soucy, Fluid-phase radial transport in packed beds of low tube-to-particle diameter ratio, *Int. J. Heat Mass Transfer* **27**, 1701 (1984).
13DP. G. Flamant et G. Arnaud, Analyse et modélisation du transfert de chaleur entre une paroi et un lit fluidisé à hàute température, *Int. J. Heat Mass Transfer* **27**, 1725 (1984).
14DP. A. H. George and J. R. Welty, Local heat transfer coefficients for a single horizontal tube in a large-particle fluidized bed at elevated temperature, *A.I.Ch.E. Jl* **30**, 482 (1984).
15DP. K. M. Gjerde and P. A. Tyvand, Thermal convection in a porous medium with continuous periodic stratification, *Int. J. Heat Mass Transfer* **27**, 2289 (1984).
16DP. D. Gloski, L. Glicksman and N. Decker, Thermal resistance at a surface in contact with fluidized bed particles, *Int. J. Heat Mass Transfer* **27**, 599 (1984).
17DP. S. Haber, A. Shavit and J. Dayan, The effect of heat convection on drying of porous semi-infinite space, *Int. J. Heat Mass Transfer* **27**, 2347 (1984).
18DP. H. Inaba and T. Kududa, Transient heat transfer behaviors in cylindrical porous beds at relatively large Reynolds number, *Wärme- u. Stoffübertr.* **18**, 109 (1984).
19DP. S. W. Jones, M. Epstein, S. G. Bankoff and D. R. Pedersen, Dryout heat fluxes in particulate beds heated through the base, *J. Heat Transfer* **106**(1), 176 (1984).
20DP. M. Kaviany, Onset of thermal convection in a saturated porous medium: experiment and analysis, *Int. J. Heat Mass Transfer* **27**, 2101 (1984).
21DP. M. Kaviany, Thermal convective instabilities in a porous medium, *J. Heat Transfer* **106**(1), 137 (1984).
22DP. R. McKibbin and P. A. Tyvand, Thermal convection in a porous medium with horizontal cracks, *Int. J. Heat Mass Transfer* **27**, 1007 (1984).
23DP. C. Y. Loh and H. Rasmussen, The free surface on a liquid filling a porous slab heated from its sides, *Int. J. Heat Mass Transfer* **27**, 921 (1984).
24DP. A. Löwe, Influence of chemical reaction on heat transfer parameters in fixed bed reactors, *Chemie-Ingr-Tech.* **56**(9), 699 (1984).
25DP. O. G. Martynenko and V. V. Levdansky, On laser monitoring of transport processes in capillary porous bodies, *Int. J. Heat Mass Transfer* **27**, 1455 (1984).
26DP. K. Muroyama, M. Fukuma and A. Yasunishi, Wall-to-bed heat transfer coefficient in gas–liquid–solid fluidized beds, *Can. J. chem. Engng* **62**, 199 (1984).
27DP. Y. Nagahashi and N. Hirayama, Heat transfer from a horizontal tube in a fluidized bed, *Heat Transfer—Jap. Res.* **13**(2), 36 (1984).
28DP. D. Poulikakos and A. Bejan, Penetrative convection in porous medium bounded by a horizontal wall with hot and cold spots, *Int. J. Heat Mass Transfer* **27**, 1749 (1984).
29DP. D. Poulikakos, Maximum density effects on natural convection in a porous layer differentially heated in the horizontal direction, *Int. J. Heat Mass Transfer* **27**, 2067 (1984).
30DP. V. Prasad and F. A. Kulacki, Natural convection in a rectangular porous cavity with constant heat flux on one vertical wall, *J. Heat Transfer* **106**(1), 152 (1984).
31DP. V. Prasad and F. A. Kulacki, Convective heat transfer in a rectangular porous cavity—effect of aspect ratio on flow structure and heat transfer, *J. Heat Transfer* **106**(1), 158 (1984).
32DP. P. R. Patil, Thermal diffusion and convective stability of a chemically reacting two component fluid in a porous medium, *Reg. J. Energy Heat Mass Transfer* **6**, 189 (1984).
33DP. D. J. Shah, J. W. Ramsey and M. Wang, An experimental determination of the heat and mass transfer coefficients in moist, unsaturated soils, *Int. J.*

Heat Mass Transfer **27**, 1075 (1984).
34DP. T. Tachiwaki and H. Uyeha, Evaporation in a packed bed heated from the bottom at reduced pressure, *A.I.Ch.E. Jl* **30**, 994 (1984).
35DP. K. Vafai, Convective flow and heat transfer in variable-porosity media, *J. Fluid Mech.* **147**, 233 (1984).
36DP. G. Vaidyanathan and P. R. Patel, Effect of horizontal temperature gradient on thermo-convective stability in a porous medium, *Reg. J. Energy Heat Mass Transfer* **6**, 129 (1984).
37DP. D. Vortmeyer and W. Adam, Steady-state measurements and analytical correlations of axial effective thermal conductivities in packed beds at low gas flow rates, *Int. J. Heat Mass Transfer* **27**, 1465 (1984).
38DP. K. Y. Wang and C. L. Tien, Thermal insulation in flow systems; combined radiation and convection through a porous segment (83-WA/HT-81), *J. Heat Transfer* **106**(2), 453 (1984).
39DP. A. Wolny and I. Opaliński, Heat transfer in an electrostatically charged fluidized bed, *Int. J. Heat Mass Transfer* **27** (2037) (1984).
40DP. A. Ÿucel, The influence of injection or withdrawal of fluid on free convection about a vertical cylinder in a porous medium, *Numer. Heat Transfer* **7**, 483 (1984).
41DP. G. V. Zdanavičius and I. A. Margis, Local aerodynamics and heat transfer in beds of packed spheres, simulating spherical fuel elements, *Heat Transfer—Sov. Res.* **15**(2), 81 (1983).
42DP. D. A. Zumbrunnen, R. Viskanta, F. P. Incropera and D. K. Hennecke, Heat transfer through granular beds at high temperature, *Wärme- u. Stoffübertr.* **18**, 221 (1984).

EXPERIMENTAL TECHNIQUES AND INSTRUMENTATION

1E. L. M. Besley, Use of ceramic-encapsulated rhodium–iron alloy resistance thermometers below 80 K: thermometric properties and stability, *J. Phys. E, Scient. Instrum.* **17**, 778 (1984).
2E. B. S. Brewster and J. D. Seader, Measuring temperature in a flowing gas–solid suspension with a thermocouple, *A.I.Ch.E. Jl* **30**, 676 (1984).
3E. B. L. Button and B. N. Dobbins, Interferometry for *in situ* mass transfer measurements in a wind tunnel, *Exp. Fluids* **2**(4), 203 (1984).
4E. J. Y. Desmons, E. Madieta, M. Martin, R. Torguet et M. Le Ray, Nouvelle méthode de mesure de la conductivité thermique, *Int. J. Heat Mass Transfer* **27**, 511 (1984).
5E. A. Dupont, P. Paranthoen, J. C. Lecordier and P. Gajan, Influence of temperature on the frequency response of fine-wire thermocouples over the range 300 K–800 K in airflows, *J. Phys. E, Scient. Instrum.* **17**, 808 (1984).
6E. M. Hendrix, A. Leipertz, G. Simonsohn, Measurement of thermal conductivity of transparent liquids by photon correlation spectroscopy, *Chemie-Ingr-Tech.* **56**, 334 (1984).
7E. Y. Iida, S. Ohtani and K. Stephan, Experimental method to determine the heat production rate, thermal diffusivity, and conductivity of solids, *Rev. Scient. Instrum.* **55**, 1648 (1984).
8E. A. C. Kent and H. N. Rosen, Measurement of humidity above temperature of 100°C with a double-wick wet bulb psychrometer, *Drying Technol.* **1**(1), 141 (1983/84).
9E. J. C. Lecordier, A. Dupont, P. Gajan and P. Paranthoen, Correction of temperature fluctuation measurements using cold wires, *J. Phys. E, Scient. Instrum.* **17**, 307 (1984).
10E. W. M. Lewandowski and P. Kubski, Effect of the use of the balance and gradient methods as a result of experimental investigations of natural convection action with regard to the conception and construction of measuring apparatus, *Wärme- u. Stoffübertr.* **18**, 247 (1984).
11E. L. E. Murray and E. K. Reiff, Jr, Design of transient experiments for identification of fixed bed thermal transport properties, *Can. J. chem. Engng* **62**, 55 (1984).
12E. T. R. Ogden and E. W. Hendricks, Liquid crystal thermography in water tunnels, *Exp. Fluids* 2(2), 65 (1984).
13E. E. Reile, P. Jany and J. Straub, Measurement of the thermal diffusivity of pure fluids and binary mixtures by dynamic light scattering (in German), *Wärme- u. Stoffübertr.* **18**, 99 (1984).
14E. M. S. Sivasubramanian, R. Cole and P. C. Sukanek, Optical temperature gradient measurements using speckle photography, *Int. J. Heat Mass Transfer* **27**, 773 (1984).
15E. W. W. Willmarth and Lalit K. Sharma, Study of turbulent structure with hot wires smaller than the viscous length, *J. Fluid Mech.* **142**, 121 (1984).
16E. C. Wood and A. Zoitan, Simple high-temperature thermal diffusivity apparatus, *Rev. Scient. Instrum.* **55**, 235 (1984).
17E. S. Yavuzkurt, A guide to uncertainty analysis of hot-wire data, *J. Fluids Engng* **106**(2), 181 (1984).
18E. M. F. Young, Effect of free-stream velocity vector and aspect ratio on the output of a free-standing circular disk heat flux gage, *J. Heat Transfer* **106**(1), 229 (1984).

NATURAL CONVECTION—INTERNAL FLOW

1F. R. L. Alpert, Calculated interaction of sprays with large-scale buoyant flows, *J. Heat Transfer* **106**(2), 310 (1984).
2F. I. Altimir, Convection naturelle tridimensionnelle en milieu poreux sature par un fluide presentant an maximum de densite, *Int. J. Heat Mass Transfer* **27**, 1813 (1984).
3F. A. Bar-Cohen and W. M. Rohsenow, Thermally optimum spacing of vertical, natural convection cooled, parallel plates, *J. Heat Transfer* **106**(1), 116 (1984).
4F. P. M. Beckett and I. E. Friend, Combined natural and forced convection between parallel walls: developing flow at higher Rayleigh numbers, *Int. J. Heat Mass Transfer* **27**, 611 (1984).
5F. H. Berkel, N. K. Mitra and M. Fiebig, Experimental investigation of the free convection heat transfer from horizontal finned tubes in a cylindrical container (in German), *Wärme- u. Stoffübertr.* **18**, 167 (1984).
6F. M. K. Bezrodnyy, S. S. Volkov and D. V. Alekseyenko, Maximum transfer in thermosiphons with separated uptake and downtake flows, *Heat Transfer—Sov. Res.*

15(2), 108 (1983).
7F. K. Blake, D. Poulikakos and A. Bejan, Natural convection near 4°C in a horizontal water layer heated from below, *Phys. Fluids* **27**(11), 2608 (1984).
8F. M. S. Bolhn, A. T. Kirkpatrick and D. A. Olson, Experimental study of three-dimensional natural convection of high-Rayleigh numbers, *J. Heat Transfer* **106**(2), 339 (1984).
9F. F. H. Busse and E. W. Bolton, Instabilities of convection rolls with stress-free boundaries near threshold, *J. Fluid Mech.* **146,** 115 (1984).
10F. A. Castellanos, P. Atten and M. G. Velarde, Oscillatory and steady convection in dielectric liquid layers subjected to unipolar injection and temperature gradient, *Phys. Fluids* **27**(7), 1607 (1984).
11F. I. Catton and J. H. Lienhard, Thermal stability of two fluid layers separated by a solid interlayer of finite thickness and thermal instability, *J. Heat Transfer* **106**(3), 605 (1984).
12F. C. K. Cha and Y. Jaluria, Recirculating mixed convection flow for energy extraction, *Int. J. Heat Mass Transfer* **27,** 1801 (1984).
13F. C. F. Chen and D. H. Johnson, Double-diffusive convection: a report on an engineering foundation conference, *J. Fluid Mech.* **138,** 405 (1984).
14F. L. C. Chow, S. R. Husain and A. Campo, Effects of free convection and axial conduction on forced-convection heat transfer inside a vertical channel at low Péclet numbers (81-WA/HT-66), *J. Heat Transfer* **106**(2), 297 (1984).
15F. N. W. F. Curlet, K. J. Won, L. A. Clomburg, Jr. and A. F. Sarofim, Experimental and mathematical modeling of three-dimensional natural convection in an enclosure, *A.I.Ch.E. Jl* **30,** 249 (1984).
16F. J. H. Curry, J. R. Herring, J. Loncaric and S. A. Orszag, Order and disorder in two- and three-dimensional Bénard convection, *J. Fluid Mech.* **147,** 1 (1984).
17F. S. H. Davis, U. Muller and C. Dietsche, Pattern selection in single-component systems coupling Bénard convection and solidification, *J. Fluid Mech.* **144,** 133 (1984).
18F. M. E. Franke and K. E. Hutson, Effects of corona discharge on free-convection heat transfer inside a vertical hollow cylinder (82-WA/HT-20), *J. Heat Transfer* **106**(2), 346 (1984).
19F. T. Fujii, T. Honda and M. Fujii, A numerical analysis of laminar free convection around an isothermal sphere: finite-difference solution of the full Navier–Stokes and energy equations between concentric spheres, *Numer. Heat Transfer* **7,** 103 (1984).
20F. M. Haajizadeh, A. F. Ozguc and C. L. Tien, Natural convection in a vertical porous enclosure with internal heat generation, *Int. J. Heat Mass Transfer* **27,** 1893 (1984).
21F. M. Haajizadeh and C. L. Tien, Combined natural and forced convection in a horizontal porous channel, *Int. J. Heat Mass Transfer* **27,** 799 (1984).
22F. T. Hanzawa and K. Kato, Heat transfer by natural convection between vertical, enclosed, concentric cylinders with the inner cylinder heated locally, *Kagaku Kogaku Ronbunshu* **9,** 3 (1983), Japan; *Int. chem. Engng* **24,** 695 (1984).
23F. T. Hanzawa and K. Kato, Heat transfer by natural convection between enclosed concentric vertical cylinders—a part of inner cylinder is heated, *Heat Transfer—Jap. Res.* **12**(3), 48 (1983).
24F. J. E. Hart, A new analysis of the closed loop thermosyphon, *Int. J. Heat Mass Transfer* **27,** 125 (1984).
25F. C. F. Hess and R. H. Henze, Experimental investigation of natural convection losses from open cavities, *J. Heat Transfer* **106**(2), 333 (1984).
26F. M. Akbar Hessami, A. Pollard and R. D. Rowe, Numerical calculation of natural convective heat transfer between horizontal concentric isothermal cylinders—effects of the variation of the fluid properties, *J. Heat Transfer* **106**(3), 668 (1984).
27F. J. Y. Holyer, The stability of long, steady two-dimensional salt fingers, *J. Fluid Mech.* **147,** 169 (1984).
28F. J. C. R. Hunt, Turbulence structure in thermal convection and shear-free boundary layers, *J. Fluid Mech.* **138,** 161 (1984).
29F. L.-T. Hwang, W.-F. Lu and J. C. Mollendorf, The effects of the density extremum and boundary conditions on the stability of a horizontally confined water layer, *Int. J. Heat Mass Transfer* **27,** 497 (1984).
30F. H. Inaba and T. Fukuda, An experimental study of natural convection in an inclined rectangular cavity filled with water at its density extremum, *J. Heat Transfer* **106**(1), 109 (1984).
31F. H. Inaba and T. Fukada, Natural convection in an inclined square cavity in regions of density inversion of water, *J. Fluid Mech.* **142,** 363 (1984).
32F. H. Inaba, T. Fukuda, N. Seki and S. Fukusako, Natural convective heat transfer in an inclined rectangular cavity with density inversion of water, *Bull. JSME* **27,** 2440 (1984).
33F. H. Inaba and K. Kanayama, Natural convective heat transfer in an inclined rectangular cavity, *Bull. JSME* **27,** 1702 (1984).
34F. Y. Inoue and R. Ito, Analysis of Bénard convection by the energy integral method, *Heat Transfer—Jap. Res.* **12**(2), 1 (1983).
35F. D. R. Jenkins and M. R. E. Proctor, The transition from roll to square-cell solutions in Rayleigh–Bénard convection, *J. Fluid Mech.* **139,** 46 (1984).
36F. Y. Joshi and B. Gebhart, Vertical natural convection flows in porous median calculations of improved accuracy, *Int. J. Heat Mass Transfer* **27,** 6 (1984).
37F. M. Kaviany, Effect of a protuberance in a square cavity, *J. Heat Transfer* **106**(4), 830 (1984).
38F. M. Kaviany, Onset of thermal convection in a fluid layer subjected to transient heating from below, *J. Heat Transfer* **106**(4), 817 (1984).
39F. M. Kaviany, The onset of thermal convection in a fluid layer with time-dependent temperature distribution, *Int. J. Heat Mass Transfer* **27,** 375 (1984).
40F. D. M. Kim and R. Viskanta, Study of the effects of wall conductance on natural convection in differently oriented square cavities, *J. Fluid Mech.* **144,** 153 (1984).
41F. S. Kimura and A. Bejan, The boundary layer natural convection regime in a rectangular cavity with uniform heat flux from the side, *J. Heat Transfer* **106**(1), 98 (1984).
42F. B. S. Kuo, J. C. Chato and J. M. Crowley, A finite element model for temperature induced electrohydrodynamic

pumping in horizontal pipes (82-WA/HT-23), *J. Heat Transfer* **106**(1), 78 (1984).
43F. G. V. Levina and A. A. Nepomniashchii, On mixed convection modes in a vertical layer with unsteadily deformable boundaries, *Appl. Math. Mech.* **47,** 341 (1983).
44F. C. P. Mao, A. C. Fernandez-Pello and J. A. C. Humphrey, An investigation of steady wall–ceiling and partial enclosure fires, *J. Heat Transfer* **106**(1), 221 (1984).
45F. N. C. Markatos and K. A. Pericleous, Laminar and turbulent natural convection in an enclosed cavity, *Int. J. Heat Mass Transfer* **27,** 755 (1984).
46F. P. J. Martin and A. T. Richardson, Conductivity models of electrothermal convection in a plane layer of dielectric liquid (82-WA/HT-21), *J. Heat Transfer* **106**(1), 131 (1984).
47F. F. Mayinger and M. Scheidt, Heat transfer in the supercritical region with vertical upflow, *Wärme- u. Stoffübertr.* **18,** 207 (1984).
48F. A. Mertol, R. Grief and Y. Zvirin, Two-dimensional analysis of transient flow and heat transfer in a natural circulation loop, *Wärme- u. Stoffübertr.* **18,** 89 (1984).
49F. M. Morioka and S. Enya, Natural convection of density-stratified layers in a vessel. I: Observation of flow patterns, *Heat Transfer—Jap. Res.* **12**(4), 48 (1983).
50F. J. Murphy, J. Lopez, The influence of vertical vorticity on thermal convection, *Aust. J. Phys.* **37**(2), 179 (1984).
51F. M. W. Nansgeel and R. Grief, An investigation of natural convection in enclosures with two- and three-dimensional partitions, *Int. J. Heat Mass Transfer* **27,** 561 (1984).
52F. R. Narayanan, Some differential inequalities for thermohaline convection in bounded containers of arbitrary shape, *Int. J. Engng Sci.* **22,** 927 (1984).
53F. J. C. Patterson, On the existence of an oscillatory approach to steady natural convection in cavities, *J. Heat Transfer* **106**(1), 104 (1984).
54F. I. L. Pioro, Maximum heat transfer capability of two-phase thermosiphons, *Heat Transfer—Sov. Res.* **15**(1), 24 (1983).
55F. V. Prasad and F. A. Kulacki, Natural convection in a vertical porous annulus, *Int. J. Heat Mass Transfer* **27,** 207 (1984).
56F. S. Ramadhyani, M. Zenouzi and K. N. Astill, Combined natural and forced convective heat transfer in spherical annuli, *J. Heat Transfer* **106**(4), 811 (1984).
57F. H. S. Rhee, J. R. Koseff and R. L. Street, Flow visualization of a recirculating flow by rheoscopic liquid and liquid crystal techniques, *Exp. Fluids* **2**(2), 57 (1984).
58F. N. Riahi, Nonlinear convection in a spherical shell, *J. phys. Soc. Japan* **53**(8), 2506 (1984).
59F. M. Roppo, S. Davis, S. Rosenblat, Bénard convection with time-periodic heating, *Phys. Fluids* **27**(4), 796 (1984).
60F. K. Sagara, Calculation of turbulent cavity flow, *Heat Transfer—Jap. Res.* **11**(3), 59 (1982).
61F. G. H. Schiroky and F. Rosenberger, Free convection of gases in a horizontal cylinder with differentially heated end walls, *Int. J. Heat Mass Transfer* **27,** 587 (1984).
62F. T. Schulenberg and U. Muller, Natural convection in saturated porous layers with internal heat sources, *Int. J. Heat Mass Transfer* **27,** 677 (1984).
63F. D. J. Shimeck and G. W. Johnsen, Natural circulating cooling in a pressurized water reactor geometry under accident-induced conditions, *Nucl. Sci. Engng* **88,** 311 (1984).
64F. R. D. Small, D. A. Larson and H. L. Brode, Asymptotically large area fires, *J. Heat Transfer* **106**(2), 318 (1984).
65F. E. M. Sparrow, G. M. Chrysler and L. F. Azevedo, Observed flow reversals and measured-predicted Nusselt numbers for natural convection in a one-sided heated vertical channel, *J. Heat Transfer* **106**(2), 325 (1984).
66F. E. M. Sparrow, P. C. Stryker and M. A. Ansari, Natural convection in enclosures with off-center innerbodies, *Int. J. Heat Mass Transfer* **27,** 49 (1984).
67F. Y. Takata, K. Iwashige, K. Fukuda and S. Hasegawa, Three-dimensional natural convection in an inclined cylindrical annulus, *Int. J. Heat Mass Transfer* **27,** 747 (1984).
68F. Y. T. Tsui and B. Tremblay, On transient natural convection heat transfer in the annulus between concentric, horizontal cylinders with isothermal surfaces, *Int. J. Heat Mass Transfer* **27,** 103 (1984).
69F. H. Ueda, S. Komori, T. Miyazaki, H. Ozoe, Time-dependent thermal convection in a stably stratified fluid layer heated from below, *Phys. Fluid* **27**(11), 2617 (1984).
70F. V. B. Utkin, Heat transfer in thermosiphons, *Heat Transfer—Sov. Res.* **15**(1), 19 (1983).
71F. P. Vasseur, T. Hung Nguyen, L. Robillard and V. K. Tong Thi, Natural convection between horizontal concentric cylinders filled with a porous layer with internal heat generation, *Int. J. Heat Mass Transfer* **27,** 337 (1984).
72F. T. Yamasaki and T. F. Irvine, Jr, Laminar free convection in a vertical tube with temperature-dependent viscosity, *Int. J. Heat Mass Transfer* **27,** 1613 (1984).
73F. L. S. Yao, Heat convection in a horizontal curved pipe, *J. Heat Transfer* **106**(1), 71 (1984).

NATURAL CONVECTION—EXTERNAL FLOWS

1FF. N. Afzal and T. Hussain, Mixed convection over a horizontal plate, *J. Heat Transfer* **106**(1), 240 (1984).
2FF. A. Bejan and D. Poulikakos, The non-Darcy regime for vertical boundary layer natural convection in a porous medium, *Int. J. Heat Mass Transfer* **27,** 717 (1984).
3FF. S. S. Chawla and A. R. Verma, Axisymmetric free convection induced by periodic variation in the temperature of a horizontal plane surface, *Int. J. Engng Sci.* **22,** 741 (1984).
4FF. C. H. Chen and J. S. T'ien, Fire plume along vertical surfaces: effect of finite-rate chemical reactions, *J. Heat Transfer* **106**(4), 713 (1984).
5FF. P. Cheng and C. T. Hsu, Higher order approximations for darcian free convective flow about a semi-infinite vertical flat plate, *J. Heat Transfer* **106**(1), 143 (1984).
6FF. M. K. Chung and H. J. Sung, Four-equation

turbulence model for prediction of the turbulent boundary layer affected by buoyancy force over a flat plate, *Int. J. Heat Mass Transfer* **27,** 2387 (1984).
7FF. A. M. Clausing, K. C. Wagner and R. K. Skarda, An experimental investigation of combined convection from a short vertical cylinder in a crossflow, *J. Heat Transfer* **106**(3), 558 (1984).
8FF. R. Das and A. K. Mohanty, Laminar combined convection in finite circular rod bundles, *J. Heat Transfer* **106**(3), 563 (1984).
9FF. P. C. T. de Boer, Thermally driven motion of strongly heated fluid, *Int. J. Heat Mass Transfer* **27,** 2239 (1984).
10FF. R. E. Faw, R. P. H. Ismuntoyo and T. W. Lester, Transition from conduction to convection around a horizontal cylinder experiencing a ramp excursion in internal heat generation, *Int. J. Heat Mass Transfer* **27,** 1087 (1984).
11FF. R. J. Goldstein and Q.-J. Wang, An interferometric study of the natural convection in an inclined water layer, *Int. J. Heat Mass Transfer* **27,** 1445 (1984).
12FF. T. Govidarajulu, The effect of longitudinal oscillations on free convection from a horizontal plate, *Reg. J. Energy Heat Mass Transfer* **6,** 147 (1984).
13FF. N. A. Hussain and B. S. Narang, Simplified analysis of air-bubble plumes in moderately stratified environments, *J. Heat Transfer* **106**(3), 543 (1984).
14FF. H. Inaba, Experimental study of natural convection in an inclined air layer, *Int. J. Heat Mass Transfer* **27,** 1127 (1984).
15FF. M. Kaviany, Effect of a stabilizing solute gradient on the onset of thermal convection, *Phys. Fluids* **27**(5), 1108 (1984).
16FF. R. Krishnamurthy and B. Gebhart, Mixed convection in wall plumes, *Int. J. Heat Mass Transfer* **27,** 1679 (1984).
17FF. S. S. Kutateladze and V. S. Berdnikov, Structure of thermogravitational convection in flat variously oriented layers of liquid and on a vertical wall, *Int. J. Heat Mass Transfer* **27,** 1595 (1984).
18FF. S. S. Kwon, T. H. Kuehn and A. K. Tolpadi, On natural convection from a short conducting plate fin below a heated horizontal cylinder, *J. Heat Transfer* **106**(3), 661 (1984).
19FF. C.-P. Mao, A. C. Fernandez-Pello and P. J. Pangi, Mixed convective burning of a fuel surface with arbitrary inclination, *J. Heat Transfer* **106**(2), 304 (1984).
20FF. O. G. Martynenko, A. A. Berezovsky and Yu. A. Sokovishin, Laminar free convection from a vertical plate, *Int. J. Heat Mass Transfer* **27,** 869 (1984).
21FF. G. M. Oreper and J. Szekely, Heat- and fluid-flow phenomena in weld pools, *J. Fluid Mech.* **147,** 53 (1984).
22FF. D. Poulikakos and A. Bejan, Natural convection in a porous layer heated and cooled along one vertical side, *Int. J. Heat Mass Transfer* **27,** 1879 (1984.
23FF. P. Ranganathan and R. Viskanta, Mixed convection boundary-layer flow along a vertical surface in a porous medium, *Numer. Heat Transfer* **7,** 305 (1984).
24FF. K. V. Rao, B. F. Armaly and T. S. Chen, Analysis of laminar mixed convection plumes along vertical adiabatic surfaces. *J. Heat Transfer* **106**(3), 552 (1984).
25FF. T. Schulenberg, Natural convection heat transfer to liquid metals below downward facing horizontal surfaces, *Int. J. Heat Mass Transfer* **27,** 433 (1984).
26FF. A. V. Shenoy, Laminar-mixed convection heat transfer from an isothermal inclined flat plate to power law fluid, *A.I.Ch.E. Jl* **30,** 824 (1984).
27FF. E. M. Sparrow and D. R. Pfeil, Enhancement of natural convection heat transfer from a horizontal cylinder due to vertical shrouding surfaces, *J. Heat Transfer* **106**(1), 124 (1984).
28FF. J. G. Symons and M. K. Peck, Natural convection heat transfer through inclined longitudinal slots, *J. Heat Transfer* **106**(4), 824 (1984).
29FF. A. K. Tolpadi and T. H. Kuehn, Conjugate three-dimensional natural convection heat transfer from a horizontal cylinder with long transverse plate fins, *Num. Heat Transfer* **7,** 319 (1984).
30FF. J.-J. Xu, S. H. Davis, Convective thermocapillary instabilities in liquid bridges, *Phys. Fluids* **27**(5), 1102 (1984).

CONVECTION FROM ROTATING SURFACES

1G. S. K. R. Choudhuri, Electro-magneto-thermo-elastic plane waves in rotating media with thermal relaxation, *Int. J. Engng Sci.* **22,** 519 (1984).
2G. P. Bar-Yoseph and S. Olek, Asymptotic and finite element approximations for heat transfer in rotating compressible flow over an infinite porous disk, *Comput. Fluids* **12,** 177 (1984).
3G. R. J. Clifford, W. D. Morris and S. P. Harasgama, An experimental study of local and mean heat transfer in a triangular-sectioned duct rotating in the orthogonal mode (84-GT-142), *J. Engng Pwr* **106,** 661 (1984).
4G. S. C. R. Dennis and L. Quartapelle, Finite difference solution to the flow between two rotating spheres, *Comput. Fluids* **12,** 77 (1984).
5G. R. Friedrich and N. Rudraiah, Marangoni convection in a rotating fluid layer with non-uniform temperature gradient, *Int. J. Heat Mass Transfer* **27,** 443 (1984).
6G. K. Himasekhar and P. K. Sarma, Integral analysis of mixed convective heat transfer from a rotating cone, *Reg. J. Energy Heat Mass Transfer* **6,** 155 (1984).
7G. A. R. Johnson and W. D. Morris, Experimental investigation of the effect of entry conditions and rotation on flow resistance in circular tubes rotating about a parallel axis, *Int. J. Heat Fluid Flow* **5**(2), 121 (1984).
8G. T. S. Lee, Numerical experiments with laminar fluid convection between concentric and eccentric heated rotating cylinders, *Numer. Heat Transfer* **7,** 77 (1984).
9G. A. Nakayama, H. Koyama and S. Ohsawa, Theoretical and experimental study of turbulent separated flows behind a rotating axisymmetric body, *Numer. Heat Transfer* **7,** 259 (1984).
10G. H. Peller, V. Lippig, D. Straub and R. Waibel, Thermofluid-dynamic experiments with a heated and rotating circular cylinder in crossflow. Part 1, *Exp. Fluids* **2**(3), 113 (1984).
11G. J. M. Pfotenhauer, P. G. J. Lucas and R. J. Donnelly,

Stability and heat transfer of rotating cryogens, Part 2, Effects of rotation on heat-transfer properties of convection in liquid ^{4}He, *J. Fluid Mech.* **145,** 239 (1984).
12G. R. Shimada, S. Kimagai and T. Takeyama, Enhancement of heat transfer from a rotating cylinder by means of surface projections, *Heat Transfer—Jap. Res.* **13**(13), 1 (1984).
13G. Y. S. Sim and W.-J. Yang, Numerical study on heat transfer in laminar flow through corotating parallel disks, *Int. J. Heat Mass Transfer* **27,** 1963 (1984).
14G. E. M. Sparrow and D. S. Kadle, Heat transfer from rods or fins which extend radically outward from a rotating shaft, *J. Heat Transfer* **106**(2), 290 (1984).
15G. A. Tseng, Finite-difference solutions for heat transfer in a roll rotating at high speed, *Numer. Heat Transfer* **7,** 113 (1984).
16G. W. Y. D. Yuen, On the steady-state temperature distribution in a rotating cylinder subject to heating and cooling over its surface, *J. Heat Transfer* **106**(3), 578 (1984).

COMBINED HEAT AND MASS TRANSFER

1H. G. E. Andrews and M. C. Mkpadi, Full-coverage discrete hole wall cooling: discharge coefficients (83-GT-79), *J. Engng Pwr* **106,** 183 (1984).
2H. N. Annamalai and T. Gopichand, Natural convection heat transfer during hydrothermal crystal growth, *Chemie-Ingr-Tech.* **56,** 950 (1984).
3H. N. Arai, T. Tani and K. Sato, Transient ablation of blunt bodies at angles of attack, *J. Spacecraft Rockets* **21,** 241 (1984).
4H. B. L. Button, Effectiveness measurements for a cooling film disrupted by a single jet (83-GT-250), *J. Engng Pwr* **106,** 606 (1984).
5H. M. K. El-Riedy, Transient natural convection heat and mass transfer during refrigeration of air by horizontal tubes, *Int. J. Heat Mass Transfer* **27,** 863 (1984).
6H. L. W. Florschuetz, D. E. Metzger and C. C. Su, Heat transfer characteristics for jet array impingement with initial crossflow (83-GT-28), *J. Heat Transfer* **106**(1), 34 (1984).
7H. H. Hardisty and M. Can, An experimental investigation into the effect of changes in the geometry of a slot nozzle on the heat transfer characteristics of an impinging jet, *Mech. Engng Sci.* **197,** 7 (1983).
8H. B. R. Hollworth and S. I. Wilson, Entrainment effects on impingement heat transfer: Part I—measurements of heated jet velocity and temperature distributions and recovery temperatures on target surface, *J. Heat Transfer* **106**(4), 797 (1984).
9H. P. Hrycak, Heat transfer from impinging jets to a flat plate with conical and ring protuberances, *Int. J. Heat Mass Transfer* **27,** 2145 (1984).
10H. S. A. Gokoglu and D. E. Rosner, Correlation of thermophoretically-modified small particle diffusional deposition rates in forced convection systems with variable properties, transpiration cooling and/or viscous dissipation, *Int. J. Heat Mass Transfer* **27,** 639 (1984).
11H. G. Grossman and M. T. Heath, Simultaneous heat and mass transfer in absorption of gases in turbulent liquid films, *Int. J. Heat Mass Transfer* **27,** 2365 (1984).
12H. S.-Y. Ko, J.-Z. Xu, Y.-Q. Yao and F.-K. Tsou, Film cooling effectiveness and turbulence distribution of discrete holes on a convex surface, *Int. J. Heat Mass Transfer* **27,** 1551 (1984).
13H. R. E. Johnson and R. P. Smet, On the heat transfer from a permeable sphere in Stokes flow, *Int. J. Engng Sci.* **22,** 947 (1984).
14H. W. Maier and L. Meinhold-Heerlein, Reflection of sound at an interface, taking into account mass and heat transport across it, *Int. J. Heat Mass Transfer* **27,** 979 (1984).
15H. M. G. Mungal and P. E. Dimotakis, Mixing and combustion with low heat release in a turbulent shear layer, *J. Fluid Mech.* **148,** 349 (1984).
16H. Yu. S. Osherov, Experimental study of heat exchange on leading and trailing edges of cooled nozzle blades of low-flow gas turbines, *Heat Transfer—Sov. Res.* **15**(2), 102 (1983).
17H. A. D. Polyanin, An asymptotic analysis of some nonlinear boundary-value problems of convective mass and heat transfer of reacting particles with the flow, *Int. J. Heat Mass Transfer* **27,** 163 (1984).
18H. P. S. Roganov, V. P. Zabolotsky, E. V. Shishov and A. I. Leontiev, Some aspects of turbulent heat transfer in accelerated flows on permeable surfaces, *Int. J. Heat Mass Transfer* **27,** 1251 (1984).
19H. E. M. Sparrow and A. Alhomoud, Impingement heat transfer at a cicular cylinder due to an offset or non-offset slot jet, *Int. J. Heat Mass Transfer* **27,** 2297 (1984).
20H. E. M. Sparrow, C. A. C. Altemani and A. Chaboki, Jet-impingement heat transfer for circular jet impinging in crossflow on a cylinder, *J. Heat Transfer* **106**(3), 570 (1984).
21H. S. A. Striegel and T. E. Diller, An analysis of the effect of entrainment temperature on jet impingement heat transfer (82-WA/HT-54), *J. Heat Transfer* **106**(4), 804 (1984).
22H. S. A. Striegel and T. E. Diller, The effect of entrainment temperature on jet impingement heat transfer, *J. Heat Transfer* **106**(1), 27 (1984).
23H. E. S. Tillman and H. F. Jen, Cooling airflow studies at the leading edge of a film-cooled airfoil (83-GT-82), *J. Engng Pwr* **106,** 214 (1984).
24H. S. L. K. Wittig, O. W. F. Elbahar and B. E. Noll, Temperature profile development in turbulent mixing of coolant jets with a confined hot crossflow (83-GT-220), *J. Engng Pwr* **106,** 193 (1984).
25H. T. Yoshida and R. J. Goldstein, On the nature of jets issuing from a row of holes into a low Reynolds number mainstream flow (83-GTJ-8), *J. Engng Pwr* **106,** 612 (1984).
26H. S. V. Zhubrin, V. P. Motulevich and E. D. Sergiyevskiy, Heat transfer from a gas jet impinging onto a liquid surface, *Heat Transfer—Sov. Res.* **15**(1), 32 (1983).

CHANGE OF PHASE—BOILING

1J. S. M. Ali and J. R. Thome, Boiling of ethanol–water and ethanol–benzene mixtures on an enhanced boiling surface, *Heat Transfer Engng* **5**(3/4), 70 (1984).

2J. R. W. Alperi and R. D. Mitchell, Effects of oily fluid and surface contaminants on heat transfer to boiling water, *A.I.Ch.E. Jl* **30,** 1028 (1984).

3J. R. P. Baines, M. A. El Masri and W. M. Rohsenow, Critical heat flux in flowing liquid films, *Int. J. Heat Mass Transfer* **27,** 1623 (1984).

4J. P. Battya, V. R. Raghavan and K. N. Seetharamu, Parametric studies on direct contact evaporation of a drop in an immiscible liquid, *Int. J. Heat Mass Transfer* **27,** 263 (1984).

5J. D. Bharathan and T. Penney, Flash evaporation from turbulent water jets, *J. Heat Transfer* **106**(2), 407 (1984).

6J. K. Bier, J. Schmadl and D. Gorenflo, Effect of heat flux density and boiling pressure on heat transfer in pool boiling of binary mixtures, *Verfahrenstechnik*, Vol. 16, No. 9 (1982), F.R.G., *Int. chem. Engng* **24,** 227 (1984).

7J. H. Binder and E. Hahne, Heat and mass transfer at corrugated trickled surface plates, *Chemie-Ingr-Tech.* **56,** 866 (1984).

8J. G. L. Chahine and H. L. Liu, Collective effects on the growth of vapor bubbles in a superheated liquid, *J. Fluids Engng* **106**(4), 486 (1984).

9J. J. S. Chin, R. Durrett and A. H. Lefebvre, The interdependence of spray characteristics and evaporation history of fuel sprays (83-GT-7), *J. Engng Pwr* **106,** 639 (1984).

10J. J. N. Chung and S. I. Olafsson, Two-phase droplet flow convective and radiative heat transfer, *Int. J. Heat Mass Transfer* **27,** 901 (1984).

11J. M. I. Drucker, V. K. Dhir and R. B. Duffey, Two-phase heat transfer for flow in tubes and over rod bundles with blockages, *J. Heat Transfer* **106**(4), 856 (1984).

12J. S. Edelstein, A. J. Perez and J. C. Chen, Analytic representation of convective boiling functions, *A.I.Ch.E. Jl* **30,** 840 (1984).

13J. G. Elamvaluthi and N. S. Srinivas, Two-phase heat transfer in two component vertical flows, *Int. J. Multiphase Flow* **10,** 237 (1984).

14J. M. Epstein, J. C. Leung, G. M. Hauser, R. E. Henry and L. Baker, Jr, Film boiling on a reactive surface, *Int. J. Heat Mass Transfer* **27,** 1365 (1984).

15J. T. W. Forest, A photographic investigation of the stability of a vapor nucleus in a glass cavity, *J. Heat Transfer* **106**(2), 402 (1984).

16J. M. Fujii, Nucleate pool boiling heat transfer from a porous heating surface (optimum particle diameter), *Heat Transfer—Jap. Res.* **13**(1), 76 (1984).

17J. S. M. Ghiaasiaan, I. Catton and R. B. Duffey, Thermal hydraulic and two-phase phenomena in reflooding of nuclear reactor cores. *J. Fluids Engng* **106**(4), 477 (1984).

18J. P. J. Giarratano, Transient boiling heat transfer from two different heat sources: small diameter wire and thin film flat surface on a quartz substrate, *Int. J. Heat Mass Transfer* **27,** 1311 (1984).

19J. H. Glaser, Equilibrium radii of small vapour bubbles and liquid droplets, *Int. J. Heat Mass Transfer* **27,** 1439 (1984).

20J. I. I. Gogonin and A. R. Dorokhov, Correlation of experimental data on critical heat fluxes in falling liquid films, *Heat Transfer—Sov. Res.* **15**(2), 91 (1983).

21J. T. Hachisu and K. Taguchi, Quenching by a water–air mixture jet, *Heat Transfer—Jap. Res.* **12**(4), 25 (1983).

22J. P. Ihle, Heat transfer in rod bundles with severe clad deformations, *Nucl. Sci. Engng* **88,** 206 (1984).

23J. M. Ikeuchi, T. Yumikura, M. Fujii and G. Yamanaka, Heat transfer characteristics of an internal microporous tube with refrigerant 22 under evaporating conditions, *ASHRAE Trans.* **90**(1A), 196 (1984).

24J. M. K. Jensen, Enhancement of the critical heat flux in pool boiling of refrigerant–oil mixtures, *J. Heat Transfer* **106**(2), 477 (1984).

25J. M. K. Jensen and D. L. Jackman, Prediction of nucleate pool boiling heat transfer coefficients of refrigerant–oil mixtures, *J. Heat Transfer* **106**(1), 184 (1984).

26J. V. G. Kashirskiy, Yu. Ya. Pechenogov and Yu. I. Serov, Experimental study of boiling heat transfer to water in a microthermosiphon heated from the end, *Heat Transfer—Sov. Res.* **15**(2), 115 (1983).

27J. I. Kataoka and M. Ishii, Mechanistic modeling of pool entrainment phenomenon, *Int. J. Heat Mass Transfer* **27,** 1999 (1984).

28J. Y. Katto and H. Ohno, An improved version of the generalized correlation of critical heat flux for the forced convective boiling in uniformly heated vertical tubes, *Int. J. Heat Mass Transfer* **27,** 1641 (1984).

29J. Y. Katto and S. Yokoya, Critical heat flux of liquid helium (I) in forced convective boiling, *Int. J. Multiphase Flow* **10,** 401 (1984).

30J. L. D. Koffman, M. S. Plesset and L. Lees, Theory of evaporation and condensation, *Phys. Fluids* **27**(4), 876 (1984).

31J. H. Kusuda, M. Mone, H. Uehara and K. Otubo, Bubble influence on boiling heat transfer in a narrow space (in a low heat-flux region), *Heat Transfer—Jap. Res.* **10**(1), 12 (1981).

32J. R. Lee, J. N. Reyes and K. Almenas, Size and number density change of droplet populations above a quench front during reflood, *Int. J. Heat Mass Transfer* **27,** 573 (1984).

33J. S. C. Lee and S. G. Bankoff, Parametric effects on the onset of flooding in flat-plate geometries, *Int. J. Heat Mass Transfer* **27,** 1691 (1984).

34J. W. Leiner and S. Böhnisch, Study of bubble growth at conical nucleation sites with reduced wettability, *Wärme- u. Stoffübertr.* **18,** 231 (1984).

35J. X. Q. Liu, C. H. Wang and C. K. Law, Simulation of fuel droplet gasification in SI engines, *J. Engng Pwr* **106,** 849 (1984).

36J. K.H. Lösing and W. Leiner, Improvement of heat transfer on boiling by laying poorly wettable nucleation sites on the transfer surfaces, *Chemie-Ingr-Tech.* **56,** 942 (1984).

37J. V. A. Mayorov, Augmentation of evaporative heat transfer in a channel filled with a porous highly conductive metal, *Heat Transfer—Sov. Res.* **15**(1), 99 (1983).

38J. H. Müller-Steinhagen and E.-U. Schlünder, Local and medium circumferential heat transfer on flow boiling of argon and nitrogen in a horizontal tube, *Chemie-Ingr-Tech.* **56,** 220 (1984).

39J. S. Nakamura, T. Matushima and S. Anzai, Transient characteristics of flashing phenomena of saturated water in vessels and their method of analysis, *Heat Transfer—Jap. Res.* **13**(4), 1 (1984).

40J. A. Nakayama and M. Kano, Photographic study of the mechanism of heat transfer enhancement by electrolytic hydrogen gas, *Heat Transfer—Jap. Res.* **12**(1), 64 (1983).
41J. A. Nakayama and M. Kano, Effect of bubble frequency on departure diameter in nucleate pool boiling, *Heat Transfer—Jap. Res.* **12**(3), 32 (1983).
42J. K. Nishikawa, Y. Fujita, S. Uchida and H. Ohta, Effect of surface configuration on nucleate boiling heat transfer, *Int. J. Heat Mass Transfer* **27**, 1559 (1984).
43J. H. Noth and A. Mersmann, Heat transfer phenomena in liquid–liquid spray columns, *Int. J. Heat Mass Transfer* **27**, 2015 (1984).
44J. S. Nukiyama, Memories of my research on boiling, *Int. J. Heat Mass Transfer* **27**, 955 (1984).
45J. S. Nukiyama, The maximum and minimum values of the heat Q transmitted from metal to boiling water under atmospheric pressure, *Int. J. Heat Mass Transfer* **27**, 959 (1984).
46J. N. Ohnishi, K. Ishijima and S. Tanzawa, A study of subcooled film-boiling heat transfer under reactivity-initiated accident conditions in light water reactors, *Nucl. Sci. Engng* **88**, 331 (1984).
47J. Y. Onishi and N. Dohara, The behavior of a vapor gas around its spherical droplet, *J. phys. Soc. Japan* **53**(1), 178 (1984).
48J. T. Oshinowo, R. C. Betts and M. E. Charles, Heat transfer in co-current vertical two-phase flow, *Can. J. chem. Engng* **62**, 194 (1984).
49J. R. J. Peterson, S. S. Grewal and M. M. El-Wakil, Investigations of liquid flashing and evaporation due to sudden depressurization, *Int. J. Heat Mass Transfer* **27**, 301 (1984).
50J. O. Pinkus, Liquid particle dynamics and rate of evaporation in the rotating field of centrifugal compressors (82-GT-86), *J. Engng Pwr* **105**(1), 80 (1983).
51J. A. Prosperetti and M. S. Plesset, The stability of an evaporating liquid surface, *Phys. Fluids* **27**(7), 1590 (1984).
52J. P. Saha, N. Abuaf and B. J. C. Wu, A nonequilibrium vapor generation model for flashing flows, *J. Heat Transfer* **106**(1), 198 (1984).
53J. A. Sakurai, M. Shiotsu, K. Hata, Effect of system pressure on film-boiling heat transfer, minimum heat flux, and minimum temperature, *Nucl. Sci. Engng* **88**, 321 (1984).
54J. M. Shoji and N. Takagi, A theoretical study of the pressure dependence of minimum film boiling for saturated liquids on a horizontal heated surface, *Heat Transfer—Jap. Res.* **12**(1), 34 (1983).
55J. M. Shoji, T. Wakunaga and K. Kodama, Heat transfer from a heated surface to an impinging subcooled droplet (heat transfer characteristics in the non-wetting regime), *Heat Transfer—Jap. Res.* **13**(3), 50 (1984).
56J. C. W. Somerton and I. Catton, The reflood time as a determination for dryout, *J. Heat Transfer* **106**(3), 656 (1984).
57J. J. J. Spillman, Evaporation from freely falling droplets, *Aeronaut. J.* **88**(875), 181 (1984).
58J. N. S. Srinivas and R. Kumar, Prediction of bubble growth rates and departure volumes in nucleate boiling at isolated sites, *Int. J. Heat Mass Transfer* **27**, 1403 (1984).
59J. M. A. Styrikovich, V. S. Polonsky and V. V. Reshetov, Experimental investigation of the critical heat flux and post-dryout temperature regime of helical coils, *Int. J. Heat Mass Transfer* **27**, 1245 (1984).
60J. V. I. Tolubinskiy, A. A. Vasil'yev and A. A. Mitin, Temperature of the pipe wall in transition from nucleate to film boiling, *Heat Transfer—Sov. Res.* **15**(1), 1 (1983).
61J. V. I. Tolubinskiy, A. A. Vasil'yev and A. A. Mitin, Thermal behavior of tube wall in transition from nucleate to film boiling, *Heat Transfer—Sov. Res.* **15**(2), 6 (1983).
62J. T. Ueda and M. Inoue, Rewetting of a hot surface by a falling liquid film—effects of liquid subcooling, *Int. J. Heat Mass Transfer* **27**, 999 (1984).
63J. J. Unger, Freie konvektionsströmungen für industrielle an wendungen, *Brennst.-Wärme-Kraft* **36**, 105 (1984).
64J. A. A. Vasil'yev and V. M. Repukhov, Theory of a two-phase cooling film on an adiabatic wall, *Heat Transfer—Sov. Res.* **15**(1), 42 (1983).
65J. B.-X. Wang and D.-H. Shi, Film boiling in laminar boundary-layer flow along a horizontal plate surface, *Int. J. Heat Mass Transfer* **27**, 1025 (1984).
66J. M. E. Wege and M. K. Jensen, Boiling heat transfer from a horizontal tube in an upward flowing two-phase crossflow, *J. Heat Transfer* **106**(4), 849 (1984).
67J. R. Windisch, V. Kiss and E. Hahne, Heat transfer on boiling at two different arrangements of finned tubes, *Chemie-Ingr-Tech.* **56**, 933 (1984).
68J. L. C. Witte and J. Orozco, The effect of vapor velocity profile shape on flow film boiling from submerged bodies, *J. Heat Transfer* **106**(1), 191 (1984).
69J. K. Yamaguchi, H. Nakamura and K. Haga, Boiling and dryout conditions in disturbed cluster geometry and their application to the liquid–metal fast breeder reactor local fault assessment, *Nucl. Sci. Engng* **88**, 464 (1984).
70J. N. Zhang and W.-J. Yang, Interfacial flow and evaporation of sessile drops on a vertical surface, *J. Heat Transfer* **106**(3), 652 (1984).

CHANGE OF PHASE—CONDENSATION

1JJ. K. Asano and S. Matsuda, Condensation of binary vapor mixtures on a vertical flat plate: an experimental study of the surface temperature of the condensate, *Heat Transfer—Jap. Res.* **11**(1), 1 (1982).
2JJ. R. W. Bjorge and P. Griffith, Initiation of waterhammer in horizontal and nearly horizontal pipes containing steam and subcooled water, *J. Heat Transfer* **106**(4), 835 (1984).
3JJ. J. N. Chung, P. S. Ayyaswamy and S. S. Sadhal, Laminar condensation on a moving drop. Part 1. Singular perturbation technique, *J. Fluid Mech.* **139**, 105 (1984).
4JJ. J. N. Chung, P. S. Ayyaswamy and S. S. Sadhal, Laminar condensation on a moving drop. Part 2. Numerical solutions, *J. Fluid Mech.* **139**, 131 (1984).
5JJ. J. N. Chung and T.-H. Chang, A mathematics model of condensation heat and mass transfer to a moving droplet in its own vapor, *J. Heat Transfer* **106**(2), 417 (1984).
6JJ. S. J. Chen, J. G. Reed and C. L. Tien, Reflux condensation in a two-phase closed thermosyphon,

Int. J. Heat Mass Transfer **27,** 1587 (1984).
7JJ. K. Hijikata, S. J. Chen and C. L. Tien, Non-condensable gas effect on condensation in a two-phase closed thermosyphon, *Int. J. Heat Mass Transfer* **27,** 1319 (1984).
8JJ. K. Hijikata, Y. Mori and S. Kawaguchi, Direct contact condensation of vapor to falling cooled droplets, *Int. J. Heat Mass Transfer* **27,** 1631 (1984).
9JJ. H. Honda and T. Fujii, Condensation of flowing vapor on a horizontal tube—numerical analysis as a conjugate heat transfer problem, *J. Heat Transfer* **106**(4), 841 (1984).
10JJ. T. Hosokawa, T. Kawai, G. Kamatsu and D. Kosaka, Heat transfer in dropwise condensation on the upper surface of a horizontal tube, *Heat Transfer—Jap. Res.* **12**(4), 76 (1983).
11JJ. K. Kaino, M. Izumi, S. Shimizu and T. Takeyama, Heat transfer near the initiation of dropwise condensation, *Heat Transfer—Jap. Res.* **11**(3), 44 (1982).
12JJ. R. Krebs and E.-U. Schlünder, Condensation in the presence of non-condensible gases—equations and guidelines for design of vertical condenser tubes, *Chemie-Ingr-Tech.* **56,** 536 (1984).
13JJ. S. C. Lee and S. G. Bankoff, Stability of steam–water countercurrent flow in an inclined channel: Part II—Condensation-induced waterhammer, *J. Heat Transfer* **106**(4), 900 (1984).
14JJ. S. Y. Lee and R. S. Tankin, Study of liquid spray (water) in a condensable environment (steam), *Int. J. Heat Mass Transfer* **27,** 363 (1984).
15JJ. W. C. Lee, S. Rahbar and J. W. Rose, Film condensation of refrigerant-113 and ethanediol on a horizontal tube—effect of vapor velocity, *J. Heat Transfer* **106**(3), 524 (1984).
16JJ. W. C. Lee and J. W. Rose, Forced convection film condensation on a horizontal tube with and without non-condensing gases, *Int. J. Heat Mass Transfer* **27,** 519 (1984).
17JJ. I. S. Lim, R. S. Tankin and M. C. Yuen, Condensation measurement of horizontal cocurrent steam/water flow, *J. Heat Transfer* **106**(2), 425 (1984).
18JJ. P. J. Marto, Heat transfer and two-phase flow during shell-side condensation, *Heat Transfer Engng* **5**(1/2), 31 (1984).
19JJ. S. Mochizuki, Y. Yagi, R. Tadano and Wen-Jei Yang, Convective filmwise condensation of nonazeotropic binary mixtures in a vertical tube, *J. Heat Transfer* **106**(3), 531 (1984).
20JJ. J. Niknejad and J. W. Rose, Comparisons between experiment and theory for dropwise condensation of mercury, *Int. J. Heat Mass Transfer* **27,** 2253 (1984).
21JJ. G. A. O'Neill and J. W. Westwater, Dropwise condensation of steam on electroplated silver surfaces, *Int. J. Heat Mass Transfer* **27,** 1539 (1984).
22JJ. F. A. Ramadan and U. Renz, Heat and mass transfer on film condensation of pure vapour on turbulent flow through a tube, *Chemie-Ingr-Tech.* **56,** 482 (1984).
23JJ. V. D. Rao and P. K. Sarma, Condensation heat transfer on laminar, falling film, *J. Heat Transfer* **106**(3), 518 (1984).
24JJ. U. Renz, Measures to improve heat transfer for condensation, *Chemie-Ingr-Tech.* **56,** 807 (1984).
25JJ. J. W. Rose, Effect of pressure gradient in forced convection film condensation on a horizontal tube, *Int. J. Heat Mass Transfer* **27,** 39 (1984).
26JJ. H. G. Schrey, Performance of single-pass aircooled or gascooled condensers with inerts (in German), *Wärme- u. Stoffübertr.* **18,** 185 (1984).
27JJ. H. Sofrata, Experimental study of filmwise condensation considering wave initiation, *Wärme- u. Stoffübertr.* **17,** 251 (1983).
28JJ. W. Studziński, R. A. Zahoransky and S. L. K. Wittig, Analysis of homogeneous unary and binary condensation process (in German), *Wärme u. Stoffübertr.* **17,** 241 (1983).
29JJ. W. Studziński, R. A. Zahoransky, S. L. K. Wittig and D. Barschdorff, Droplet formation and growth in condensing binary vapours, *Int. J. Heat Mass Transfer* **27,** 451 (1984).
30JJ. K. Takashashi, W. Nakayama, T. Senshu and H. Yoshida, Heat transfer analysis of shell-and-tube condensers with shell-side enhancement, *ASHRAE Trans.* **90** (1B), 60 (1984).
31JJ. H. Tanaka and T. Tsuriuta, A microscopic study of dropwise condensation, *Int. J. Heat Mass Transfer* **27,** 327 (1984).
32JJ. H. Uehara, H. Kusuda, T. Nakaoka and M. Yamada, Filmwise condensation for turbulent flow on a vertical plate, *Heat Transfer—Jap. Res.* **12**(2), 85 (1983).
33JJ. H. Uehara, T. Nakaoka and M. Kurosaki, Turbulent film condensation in a forced-convective boundary layer flow, *Heat Transfer—Jap. Res.* **13**(4), 89 (1984).
34JJ. J. C. Y. Wang, M. Turaga, S. Lin and P. Fazio, Effects of refrigerant–oil mixtures on condensation heat transfer on the external surface of horizontal tubes in water-cooled shell-and-tube condensers (RP-378), *ASHRAE Trans.* **90**(1B), 26 (1984).
35JJ. R. L. Webb, Shell-side condensation in refrigerant condensers, *ASHRAE Trans.* **90**(1B), 5 (1984).
36JJ. R. L. Webb, The effects of vapor velocity and tube bundle geometry on condensation in shell-side refrigeration condensers, *ASHRAE Trans.* **90**(1B), 39 (1984).
37JJ. J. E. Roth, Heat transfer by condensation in tilted oval tubes, *Forsch. Geb. IngWes.* **50**(3), 102 (1984).

CHANGE OF PHASE—FREEZING AND MELTING

1JM. T. Asano and H. Abe, Melt-crystallization of hexatricontane (n-$C_{36}H_{74}$) in a temperature gradient, *J. phys. Soc. Japan* **53**(3), 1071 (1984).
2JM. M. Bareiss and H. Beer, An analytical solution of the heat transfer process during melting of an unfixed solid phase change material inside a horizontal tube, *Int. J. Heat Mass Transfer* **27,** 739 (1984).
3JM. G. E. Bell, An inverse solution for the steady temperature within a solidifed layer, *Int. J. Heat Mass Transfer* **27,** 2331 (1984).
4JM. F. B. Cheung, T. C. Chawla and D. R. Pedersen, The effects of heat generation and wall interaction on freezing and melting in a finite slab, *Int. J. Heat Mass*

Transfer **27,** 29 (1984).
5JM. K. Chida, Heat transfer in steady laminar pipe flow with liquid solidification, *Heat Transfer—Jap. Res.* **12**(1), 81 (1983).
6JM. P. M. Chung and R. Bywater, Role of the liquid layer in ice accumulation on flat surfaces, *J. Heat Transfer* **106**(1), 5 (1984).
7JM. L. J. Fang, F. B. Cheung, J. H. Linehan and D. R. Pedersen, Selective freezing of a dilute salt solution on a cold ice surface, *J. Heat Transfer* **106**(2), 385 (1984).
8JM. A. Gadgil and D. Gobin, Analysis of two-dimensional melting in rectangular enclosures in presence of convection, *J. Heat Transfer* **106**(1), 20 (1984).
9JM. G. Gau and R. Viskanta, Melting and solidification of a metal system in a rectangular cavity, *Int. J. Heat Mass Transfer* **27,** 113 (1984).
10JM. L. Gradón and D. Orlicki, Liquid mixtures separation in a freezing-out process, *Int. J. Heat Mass Transfer* **27,** 1141 (1984).
11JM. A. M. Hassanein and G. L. Kulcinski, Simulation of rapid heating in fusion reactor first walls using the Green's function approach, *J. Heat Transfer* **106**(3), 486 (1984).
12JM. C.-J. Ho and R. Viskanta, Heat transfer during melting from an isothermal vertical wall, *J. Heat Transfer* **106**(1), 12 (1984).
13JM. C.-J. Ho and R. Viskanta, Inward solid–liquid phase-change heat transfer in a rectangular cavity with conducting vertical walls, *Int. J. Heat Mass Transfer* **27,** 1055 (1984).
14JM. C.-H. Ho and R. Viskanta, Heat transfer during inward melting in a horizontal tube, *Int. J. Heat Mass Transfer* **27,** 705 (1984).
15JM. S. C. Huang, Melting of semi-infinite region with viscous heating, *Int. J. Heat Mass Transfer* **27,** 1337 (1984).
16JM. M. A. Katz and B. Rubinsky, An inverse finite-element technique to determine the change of phase interface location in one-dimensional melting problems, *Numer. Heat Transfer* **7,** 269 (1984).
17JM. W. F. Kaukler, Hot stage and sample cell design for the solidification of transparent materials with and without forced convection, *Rev. scient. Instrum.* **55,** 1643 (1984).
18JM. E. D. Larson and E. M. Sparrow, Effect of inclination on freezing in a sealed cylindrical capsule, *J. Heat Transfer* **106**(2), 394 (1984).
19JM. R. W. Lewis, K. Morgan and P. M. Roberts, Application of an alternating-direction finite-element method to heat transfer problems involving a change of phase, *Numer. Heat Transfer* **7,** 471 (1984).
20JM. S. Lin C. K. Kwok, S. M. Ni and W. L. Dai, A one-dimensional approach for solving freezing or melting processes in cube and long square boxes, *ASHRAE Trans.* **90**(1A), 3 (1984).
21JM. Ya. P. Minosyan, A. A. Ryadno, Yu. G. Solov'yev and V. G. Gerasimov, Mathematical simulation of centrifugal casting of pipes, *Heat Transfer—Sov. Res.* **15**(1), 134 (1983).
22JM. K. Morikawa and K. Kawashimo, Analysis of heat conduction fields with moving boundary, *Bull. JSME* **27,** 1938 (1984).
23JM. T. Nishimura, Y. Kawamura, S. Ito and H. Tsuboi, Analysis of two-dimensional freezing problems by the finite element method, *Heat Transfer*–Jap. Res. **12**(4), 12 (1983).
24JM. M. Okada, Analysis of heat transfer during melting from a vertical wall, *Int. J. Heat Mass Transfer* **27,** 2057 (1984).
25JM. A. T. Patera, A finite-element/Green's function embedding technique applied to one-dimensional change-of-phase heat transfer, *Numer. Heat Transfer* **7,** 241 (1984).
26JM. J. L. Pelissier, Calculation of the melting curve of lead, *Physica A* **126A**(1 & 2), 271 (1984).
27JM. J. L. Pelissier, High pressure calculation of the melting curve: Na, K, Rb and Cs, *Physica A* **126A**(3), 474 (1984).
28JM. J. L. Pelissier, Melting of aluminum: a model potential approach up to 2 megabars, *Physica A* **128A**(1 & 2), 363 (1984).
29JM. P. Pissis, A thermally stimulated depolarisation technique for studying the freezing of water dispersed within emulsion, *J. Phys. D, Appl. Phys.* **17**(4), 787 (1984).
30JM. J. Prusa and L. S. Yao, Melting around a horizontal heated cylinder: Part 1—Perturbation and numerical solutions for constant heat flux boundary condition, *J. Heat Transfer* **106**(2), 376 (1984).
31JM. J. Prusa and L. S. Yao, Melting around a horizontal heated cylinder: Part II—Numerical solution for isothermal boundary condition, *J. Heat Transfer* **106**(2), 469 (1984).
32JM. M. S. Sadeghipour, M. N. Özişik and J. C. Mulligan, The effect of external convection on freezing in steady turbulent flow in a tube, *Int. J. Engng Sci.* **22,** 135 (1984).
33JM. H. Saito, I. Tokura, K. Kishinami and S. Uemura, A study on frost formation (dimensionless parameters correlating density and thickness of frost layer), *Heat Transfer—Jap. Res.* **13**(4), 76 (1984).
34JM. N. Seki, S. Fukusako and G. W. Younan, A transition phenomenon of ice formation in water flow between two horizontal parallel plates, *Wärme- u. Stoffübertr.* **18,** 117 (1984).
35JM. N. Seki, S. Fukusako and G. W. Younan, Ice-formation phenomena for water flow between two cooled parallel plates, *J. Heat Transfer* **106**(3), 498 (1984).
36JM. R. Shimomura, K. Ichimiya, T. Miyazawa, Y. Hanada and Y. Shintani, The supercooling of water droplets, *Heat Transfer—Jap. Res.* **12**(2), 72 (1983).
37JM. R. Siegel, Solidification interface shape for continuous casting in an offset mold—two analytical models, *J. Heat Transfer* **106**(1), 237 (1984).
38JM. R. Siegel, Two-region analysis of interface shape in continuous casting with superheated liquid, *J. Heat Transfer* **106**(3), 506 (1984).
39JM. E. M. Sparrow and W. Chuck, An implicit/explicit numerical solution scheme for phase-change problems, *Numer. Heat Transfer* **7**(1), 1 (1984).
40JM. E. M. Sparrow and W. Chuck, Planar or radial freezing with solid–solid transitions and convective heating at the solid–liquid interface, *Numer. Heat*

Transfer **7**(1), 17 (1984).
41JM. D. Tiba and M. Tiba, Regularity of the boundary data and the convergence of the finite element discretization in two-phase Stefan problems, *Int. J. Engng Sci.* **22,** 1225 (1984).

RADIATION IN PARTICIPATING MEDIA AND SURFACE RADIATION

1K. M. S. Abukakar and J. H. Shaw, Analysis of carbon dioxide bands near 2.2 μm, *Appl. Optics* **23**(19), 3310 (1984).
2K. R. Aoshima, T. Chiba and H. Asai, Thin-layer optical cell with light path length shorter than 10 μm for absorption and fluorescence measurement of highly absorbent solutions, *Appl. Optics* **23**(4), 597 (1984).
3K. R. L. Armstrong, Aerosol heating and vaporization by pulsed light beams, *Appl. Optics* **23**(1), 148 (1984).
4K. D. P. Arndt *et al.*, Multiple determination of the optical constants of thin-film coating materials, *Appl. Optics* **23**(20), 3571 (1984).
5K. G. S. Arnold, Absorptivity of several metals at 10.6 μm: empirical expressions for the temperature dependence computed from Drude theory, *Appl. Optics* **23**(9), 1434 (1984).
6K. Y. Asako and H. Nakamura, Heat transfer in a parallelogram shaped enclosure (4th report, combined free convection, radiation and conduction heat transfer), *Bull. JSME* **27,** 1144 (1984).
7K. S. Bard, Advanced passive radiator for spaceborne cryogenic cooling, *J. Spacecraft Rockets* **21,** 150 (1984).
8K. P. Berdahl, Radiative cooling with MgO and/or LiF layers, *Appl. Optics* **23**(3), 370 (1984).
9K. S. A. Boothroyd and A. R. Jones, Radiative transfer scattering data relevant, *J. Phys. D, Appl. Phys.* **17**(6), 1107 (1984).
10K. J. P. Borgogno, F. Flory, P. Roche, B. Schmitt, G. Albrand, E. Pelletier and H. A. Macleod, Refractive index and inhomogeneity of thin films, *Appl. Optics* **23**(20), 3567 (1964).
11K. W. Börsch-Supan, L. W. Hunter and J. R. Kuttler, Endothermic gasification of a solid by thermal radiation absorbed in depth, *Int. J. Heat Mass Transfer* **27,** 1171 (1984).
12K. M. Q. Brewster and T. Kunitomo, The optical constants of coal, char, and limestone, *J. Heat Transfer* **106**(4), 678 (1984).
13K. A. Bubenzer and P. Koidl, Exact expressions for calculating thin-film absorption coefficients from laser calorimetric data, *Appl. Optics* **23**(17), 2886 (1984).
14K. R. Caps, A. Trunzer, D. Büttner, J. Fricke and H. Reiss, Spectral transmission and reflection properties of high temperature insulation materials, *Int. J. Heat Mass Transfer* **27,** 1865 (1984).
15K. Y. A. Cengel, M. N. Özişik, Integrals involving Legendre polynomials that arise in the solution of radiation transfer, *J. quant. Spectrosc. radiat. Transfer* **31**(3), 215 (1984).
16K. Y. A. Cengel and M. N. Özişik, The use of the Galerkin method for radiation transfer in an anisotropically scattering slab with reflecting boundaries, *J. quant. Spectrosc. radiat. Transfer* **32**(3), 225 (1984).
17K. Y. A. Cengel, M. N. Özişik and Y. Yener, Determination of angular distribution of radiation in an isotropically scattering slab, *J. Heat Transfer* **106**(1), 248 (1984).
18K. K. Chan, H. Ito and H. Inaba, Remote sensing system for near-infrared differential absorption of CH_4 gas using low-loss optical fiber link, *Appl. Optics* **23**(19), 3415 (1984).
19K. T. J. Chung and J. Y. Kim, Two-dimensional, combined-mode heat transfer by conduction, convection, and radiation in emitting, absorbing, and scattering media-solution by finite elements, *J. Heat Transfer* **106**(2), 448 (1984).
20K. T. D. Craig and F. P. Incropera, Radiation transfer in absorbing–scattering liquids—I. Radiance and flux measurements, *J. quant. Spectrosc. radiat. Transfer* **31**(2), 127 (1984).
21K. A. L. Crosbie and R. G. Schrenker, Radiative transfer in a two-dimensional rectangular medium exposed to diffuse radiation, *J. quant. Spectrosc. radiat. Transfer* **31**(4), 339 (1984).
22K. C. J. Dasch, Continuous-wave probe laser investigation of laser vaporization of small soot particles in a flame, *Appl. Optics* **23**(13), 2209 (1984).
23K. F. Demichelis, E. Mezzetti-Minetti, L. Tallone and E. Tresso, Optimization of optical parameters and electric field distribution in multilayers, *Appl. Optics* **23**(1), 165 (1984).
24K. E. L. Dereniak, J. P. Britt, A. M. Fowler, R. R. Joyce, G. Boreman and W. S. Ewing, Evaluation of PtSi Schottky infrared CCD for astronomy, *Appl. Optics* **23**(6), 889 (1984).
25K. V. M. Devi, C. P. Rinsland and D. C. Benner, Absolute intensity measurements of CO_2 bands in the 2395–2680 cm^{-1} region, *Appl. Optics* **23**(22), 4067 (1984).
26K. V. Dixit, J. Lodenquai and J. McTavish, Radiation flux enhancement and absorption in thin films, *J. opt. Soc. Am.* **A1**(3), 278 (1984).
27K. M. J. Dodge, Refractive properties of magnesium fluoride, *Appl. Optics* **23**(12), 1980 (1984).
28K. M. Dragovan and S. H. Moseley, Gold absorbing film for a composite bolometer, *Appl. Optics* **23**(5), 651 (1984).
29K. A. C. Eckbreth, G. M. Dobbs, J. H. Stufflebeam and P. A. Tellex, CARS temperature and species measurements in augmented jet engine exhausts, *Appl. Optics* **23**(9), 1328 (1984).
30K. D. K. Edwards and R. Matavosian, Scaling rules for total absorptivity and emissivity of gases (83-HT-36), *J. Heat Transfer* **106**(4), 684 (1984).
31K. W. G. Egan and T. W. Hilgeman, Radiation emission from small particles, *Appl. Optics* **23**(7), 1021 (1984).
32K. J. D. Eversole, Thermal radiation from carbon particles in a heated nitrogen, *Appl. Optics* **23**(19), 3429 (1984).
33K. J. Fischer and H. Grassl, Radiative transfer in an atmosphere-ocean system: an azimuthally dependent matrix-operator approach, *Appl. Optics* **23**(7), 1032 (1984).
34K. J. W. Fitzgerald, Effect of relative humidity on the aerosol backscattering coefficient at 0.694- and 10.6-μm wavelengths, *Appl. Optics* **23**(3), 411 (1984).

35K. W. A. Fiveland, Discrete-ordinates solutions of the radiative transport equation for rectangular enclosures (83-HT-51), *J. Heat Transfer* **106**(4), 699 (1984).

36K. C. Gay, Boundary conditions in the radiative relaxation problem, *J. quant. Spectrosc. radiat. Transfer* **31**(3), 423 (1984).

37K. C. Gay, Radiative temperature dissipation in a finite atmosphere—II. The inhomogeneous case, *J. quant. Spectrosc. radiat. Transfer* **31**(5), 439 (1984).

38K. L. P. Giver, P. Varanasi and F. P. J. Valero, Propane absorption band intensities and band model parameters from 680 to 1580 cm^{-1} at 296 and 200 K, *J. quant. Spectros. radiat. Transfer* **31**(3), 203 (1984).

39K. V. N. Golubkin, On the asymptotic theory of the three-dimensional flow of a hypersonic stream of radiating gas around a body, *Appl. Math. Mech.* **47**, 493 (1983).

40K. F. Gordoninejad and J. Francis, A finite difference solution to transient combined conductive and radiative heat transfer in an annular medium, *J. Heat Transfer* **106**(4), 888 (1984).

41K. R. Guzzi and R. Rizzi, Water vapor absorption in the visible and near infrared results of field measurements, *Appl. Optics* **23**(11), 1583 (1984).

42K. I. S. Habib, Radiation effects on vibrational heating of polymers, *J. Spacecraft Rockets* **21**, 496 (1984).

43K. R. J. Hall and L. R. Boedeker, CARS thermometry in fuel-rich combustion zones, *Appl. Optics* **23**(9), 1340 (1984).

44K. N. Hattori, H. Takasu and T. Iguchi, Emissivity of liquid sodium, *Heat Transfer—Jap. Res.* **13**(1), 30 (1984).

45K. S. C. Hill, A. C. Hill and P. W. Barber, Light scattering by size shape distributions of soil particles and spheroids, *Appl. Optics* **23**(7), 1025 (1984).

46K. F. P. Incropera, T. D. Craig and W. G. Houf, Radiation transfer in absorbing–scattering liquids—II. Comparisons of measurements with predictions, *J. quant. Spectrosc. radiat. Transfer* **31**(2), 139 (1984).

47K. A. Ishimanu and C. W. Yen, Matrix representations of the vector radiative-transfer theory for randomly distributed nonspherical particles, *J. opt. Soc. Am.* **A1**(4), 359 (1984).

48K. N. K. Jain, Conversion of emissivity graphs into formulas for stream and carbondioxide, *Reg. J. Energy Heat Mass Transfer* **6**, 195 (1984).

49K. J. M. Jarem, J. H. Pierluissi and M. Maragoudakis, Numerical methods of band modeling and their application to atmospheric nitrous oxide, *Appl. Optics* **23**(3), 406 (1984).

50K. J. M. Jarem, J. H. Pierlussi and W.-L. Ng, Validated infrared transmittance band model for methane in the atmosphere, *Appl. Optics* **23**(19), 3331 (1984).

51K. S.-M. Jeng and G. M. Faeth, Radiative heat fluxes near turbulent buoyant methane diffusion flames, *J. Heat Transfer* **106**(4), 886 (1984).

52K. K. Kamiuto, Y. Miyoshi, I. Kinoshita and S. Hasegawa, Combined conductive and radiative heat transfer in an optically thick porous body (case of coorierite porous bodies), *Bull. JSME* **27**, 1136 (1984).

53K. D. Karman and F. R. Steward, The radiation spectrum from a test furnace, *Int. J. Heat Mass Transfer* **27**, 1357 (1984).

54K. D. M. Kim and R. Viskanta, Effect of wall conduction and rádiation on natural convection in a rectangular cavity, *Numer. Heat Transfer* **7**, 449 (1984).

55K. S. W. Kizirnis, R. J. Brecha, B. N. Ganguly, L. P. Goss and R. Gupta, Hydroxyl (OH) distributions and temperature profiles in a premixed propane flame obtained by laser deflection techniques, *Appl. Optics* **23**(21), 3873 (1984).

56K. D. Klick, K. A. Marko and L. Rimai, Optical multichannel analysis with rapid mass storage of spectra: application to CARS measurements of temperature fluctuations, *Appl. Optics* **23**(9), 1347 (1984).

57K. T. Kunitomo, E. Ozaki and K. Kawahara, The effect of dimerization on infrared absorption in the 6.3 μm band and the total emissivity of water vapor, *Heat Transfer—Jap. Res.* **11**(3), 90 (1984).

58K. G. Kychakoff, R. D. Howe and R. K. Hanson, Spatially resolved combustion measurements using cross-beam saturated absorption spectroscopy, *Appl. Optics* **23**(9), 1303 (1984).

59K. P. Latimer, Light scattering by a structure particle: the homogeneous sphere with holes, *Appl. Optics* **23**(11), 1844 (1984).

60K. A. H. Lefebvre, Flame radiation in gas turbine combustion chambers, *Int. J. Heat Mass Transfer* **27**, 1493 (1984).

61K. W. Leupacher and A. Penzkofer, Refractive-index measurement of absorbing condensed media, *Appl. Optics* **23**(10), 1554 (1984).

62K. C. C. Limbaugh and K. R. Kneile, Uncertainties propagation for combustion diagnostics using infrared band models, *J. quant. Spectrosc. radiat. Transfer* **31**(2), 161 (1984).

63K. E. M. Lushiku and C.-G. Granqvist, Radiative cooling with selectively infrared-emitting gases, *Appl. Optics* **23**(11), 1835 (1984).

64K. K. Machida, M. A. Lind and J. L. Stanford, Temperature dependence of infrared reflectivity of chromium-realization of soliton lattice structure of spin density wave, *J. phys. Soc. Japan* **53**(11), 4020 (1984).

65K. B. Mahbod and R. L. Adams, Radiation view factors between axisymmetric subsurfaces within a cylinder with spherical centerbody, *J. Heat Transfer* **106**(1), 244 (1984).

66K. T. Makino, T. Kunitomo, I. Sakai and H. Kinoshita, Thermal radiation properties of ceramic materials, *Heat Transfer—Jap. Res.* **13**(4), 33 (1984).

67K. J. V. Martonchik, G. S. Orton and J. F. Appleby, Optical properties of NH_3 ice from the far infrared to the near ultraviolet, *Appl. Optics* **23**(4), 541 (1984).

68K. J. C. Mather, Bolometers: ultimate sensitivity, optimization, and amplifier coupling, *Appl. Optics* **23**(4), 584 (1984).

69K. B. B. McMahon, Interpretation of NO_2 absorption in twilight sky spectra, *J. quant. Spectrosc. radiat. Transfer* **32**(1), 9 (1984).

70K. R. C. McPhedran, L. C. Botten, D. R. McKenzie and R. P. Netterfield, Unambiguous determination of optical constants of absorbing films by reflectance and

transmittance measurements, *Appl. Optics* **23**(8), 1197 (1984).
71K. L. A. Melton, Soot diagnostics based on laser heating, *Appl. Optics* **23**(13), 2201 (1984).
72K. D. Mihalas, K.-H. A. Winkler and M. L. Norman, Adaptive-mesh radiation hydrodynamic—II. The radiation and fluid equations in relativistic flows, *J. quant. Spectrosc. radiat. Transfer* **31**(6), 479 (1984).
73K. F. A. Modine, D. Y. Smith, Approximate formulas for the amplitude and the phase of the infrared reflectance of a conductor, *J. opt. Soc. Am.* **A1**(12), 1171 (1984).
74K. M. H. N. Naraghi and B. T. F. Chung, A stochastic approach for radiative exchange in enclosures with nonparticipating medium, *J. Heat Transfer* **106**(4), 690 (1984).
75K. D. A. Nelson, Band radiation of isothermal gases within diffuse-walled enclosures, *Int. J. Heat Mass Transfer* **27**, 1759 (1984).
76K. H. F. Nelson, Influence of particulates on infrared emission from tactical rocket exhausts, *J. Spacecraft Rockets* **21**, 425 (1984).
77K. H. F. Nelson, Influence of scattering on infrared signatures of rocket plumes, *J. Spacecraft Rockets* **21**, 508 (1984).
78K. R. W. Nicholls, Wavelength-dependent spectral extinction of atmospheric aerosols, *Appl. Optics* **23**(8), 1142 (1984).
79K. M. Nishimura, M. Kiraishi, Y. Bando and H. Jojima, Influence of the spectral distribution of incident radiation on the effect of internal heating in infrared drying of coated films, *Heat Transfer—Jap. Res.* **13**(2), 30 (1984).
80K. J. H. Pierluissi, J. M. Jarem and W.-L. W. Ng, Validated transmittance band model for atmospheric nitrous oxide in the infrared, *Appl. Optics* **23**(3), 402 (1984).
81K. J. H. Pierluissi, J. M. Jarem and C. Maragoudakis, Validated transmittance band model for SO_2 in the infrared, *Appl. Optics* **23**(19), 3325 (1984).
82K. C. M. R. Platt and A. C. Dilley, Determination of the cirrus particle single-scattering phase function from lidar and solar radiometric data, *Appl. Optics* **23**(3), 380 (1984).
83K. M. F. Quinn, Q. Alameddine and K. Al-Jamal, Experimental measurements of the downward infrared sky radiation in Kuwait, *Appl. Optics* **23**(3), 419 (1984).
84K. M. M. Razzaque, J. R. Howell and D. E. Klein, Coupled radiative and conductive heat transfer in a two-dimensional rectangular enclosure with gray participating media using finite elements, *J. Heat Transfer* **106**(3), 613 (1984).
85K. W. A. de Rooij and H. Domke, On the nonuniqueness of solutions for nonlinear integral equations in radiative transfer theory, *J. quant. Spectrosc. radiat. Transfer* **31**(4), 285 (1984).
86K. R. D. Rosenwald, H. A. Hill and J. D. Logan, Radiative transfer: Gaussian quadrature formulas for integrals with the weight functions $\exp(-|x-t|)$ and $E_n(|x-t|)$, *J. quant. Spectrosc. radiat. Transfer* **31**(3), 221 (1984).
87K. D. J. Sanders, Temperature distributions produced by scanning Gaussian laser beams, *Appl. Optics* **23**(1), 30 (1984).
88K. D. L. Shealy, Theoretical analysis of infrared radiation shields of spacecraft, *J. Spacecraft Rockets* **21**, 293 (1984).
89K. G. M. Shved, A. G. Ishov and A. A. Kutepov, Universal functions for estimating total vibration–rotation band absorptance—I. Linear molecules and spherical tops, *J. quant. Spectrosc. radiat. Transfer* **31**(1), 35 (1984).
90K. G. M. Shved and A. G. Ishov, Universal functions for estimating total vibration–rotation band absorptance—II. Symmetric tops, *J. quant. Spectrosc. radiat. Transfer* **31**(1), 47 (1984).
91K. R. D. Skocypec and R. O. Buckus, Total hemispherical emittance for CO_2 or H_2O including particulate scattering, *Int. J. Heat Mass Transfer* **27**, 1 (1984).
92K. S. M. Smith, Specular reflectance of optical-black coatings in the far infrared, *Appl. Optics* **23**(14), 2311 (1984).
93K. K. Stamnes and P. Conklin, A new multi-layer discrete ordinate approach to radiative transfer in vertically inhomogeneous atmospheres, *J. quant. Spectrosc. radiat. Transfer* **31**(3), 273 (1984).
94K. S. N. Tiwari, K. Y. Szema, J. N. Moss and S. V. Subramanian, Convective and radiative heating of a Saturn entry probe, *Int. J. Heat Mass Transfer* **27**, 191 (1984).
95K. J. S. Truelove, Radiant heat transfer through the cover gas of a sodium-cooled fast reactor, *Int. J. Heat Mass Transfer* **27**, 2085 (1984).
96K. T. F. Wall and H. B. Becker, Total absorptivities and emissivities of particulate coal ash from spectral band emissivity measurements, *J. Engng Gas Turbines Pwr* **106**, 771 (1984).
97K. Wang Ying-Jian, An analytical model for total absorption of a band, *J. quant. Spectrosc. radiat. Transfer* **31**(5), 457 (1984).
98K. S. G. Warren, Optical constants of ice from the ultra-violet to the microwave, *Appl. Optics* **23**(8), 1206 (1984).
99K. S. A. Weisrose and G. Shadmon, Radiative transfer calculations through an aerosol cloud, *J. quant. Spectrosc. radiat. Transfer* **31**(1), 63 (1984).
100K. M. M. R. Williams, The transmission of radiation through a coagulating aerosol, *J. Phys. D, Appl. Phys.* **17**(3), 509 (1984).
101K. M. M. R. Williams, The effect of anisotropic scattering on the radian heat flux through an aerosol, *J. Phys. D, Appl. Phys.* **17**(8), 1617 (1984).
102K. M. M. R. Williams, The transmission of radiation through a coagulating aerosol, *J. Phys. D, Appl. Phys.* **17**(9), 1913 (1984).
103K. M. M. Wind, J. Vlieger, Optical properties of thin films on rough surfaces, *Physica A* **125A**(1), 75 (1984).
104K. K.-H. A. Winkler, M. L. Norman, D. Mihalas, Adaptive-mesh radiation hydrodynamics—I. The radiation transport equation in a completely adaptive coordinate system, *J. quant. Spectrosc. radiat. Transfer* **31**(6), 473 (1984).
105K. B. Yaakobi, J. Delettrez, L. M. Goldman, R. L. McCrory, R. Marjoribanks, M. C. Richardson, D. Shvarts, S. Skupsky, J. M. Soures, C. Verdon, D. M. Villeneuve, T. Boehly, R. Hutchinson and S. Letering, Thermal transport measurements in 1.05 μm laser irradiation of

spherical targets, *Phys. Fluids* **27**(2), 516 (1984).
106K. B. E. Yoldas and T. O'Keefe, Deposition of optically transparent IR reflective coatings on glass, *Appl. Optics* **23**(20), 3638 (1984).
107K. B. E. Yoldas and D. P. Partlow, Wide spectrum antireflective coating for fused silica and other glasses, *Appl. Optics* **23**(9), 1418 (1984).
108K. L. A. Young, Atmospheric transmittance determination from sky radiance in teal time in the tropics, *Appl. Optics* **23**(3), 396 (1984).
109K. W. W. Yuen, A limiting approach for the evaluation of geometric mean transmittance in a multidimensional absorbing and isotropically scattering medium, *J. Heat Transfer* **106**(2), 441 (1984).
110K. W. W. Yuen and L. W. Wong, Analysis of radiative equilibrium in a rectangular enclosure with gray medium, *J. Heat Transfer* **106**(2), 433 (1984).
111K. Y. Zhan and P. D. Coleman, Far-IR optical properties of $Y_3Al_5O_{12}$, $LiYF_4$, $Cs_2NaDyCl_6$, and Rb_2NaYF_6, *Appl. Optics* **23**(4), 548 (1984).

NUMERICAL METHODS

1N. T. C. Chawla, D. R. Pedersen, G. Leaf, W. J. Minkowycz and A. R. Shouman, Adaptive collocation method for simultaneous heat and mass diffusion with phase change, *J. Heat Transfer* **106**(3), 491 (1984).
2N. D. Degani, Numerical algorithm conjugating steady and transient, separated, compressible flow and a solid body having arbitrarily distributed heat sources, *Numer. Heat Transfer* **7**, 396 (1984).
3N. R. C. Dykhuizen, R. P. Roy and S. P. Kalra, Numerical method for the solution of simultaneous nonlinear equations and application to two-fluid model equations of boiling flow, *Numer. Heat Transfer* **7**, 225 (1984).
4N. M. Faghri, E. M. Sparrow and A. T. Prata, Finite-difference solutions of convection–diffusion problems in irregular domains, using a nonorthogonal coordinate, *Numer. Heat Transfer* **7**, 183 (1984).
5N. S. E. Halland, Calculation of entrainment rate, initial values, and transverse velocity for the Patankar–Spalding method, *Numer. Heat Transfer* **7**(1), 39 (1984).
6N. T. V. Hromadka II, Linking the complex variable boundary-element method to the analytic function method, *Numer. Heat Transfer* **7**, 235 (1984).
7N. M. Lacroix, R. Camarero and A. Tapucu, Multigrid scheme for thermohydraulic flow, *Numer. Heat Transfer* **7**, 375 (1984).
8N. B. E. Launder, Numerical computation of convective heat transfer in complex turbulent flows: time to abandon wall functions?, *Int. J. Heat Mass Transfer* **27**, 1485 (1984).
9N. C. Prakash, Application of the locally analytic differencing scheme to some test problems for the convection–diffusion equation, *Numer. Heat Transfer* **7**, 165 (1984).
10N. R. E. Phillips and F. W. Schmidt, Multigrid techniques for the numerical solution of the diffusion equation, *Numer. Heat Transfer* **7**, 251 (1984).
11N. P. Ramakrishna Rao and V. M. K. Sastri, Efficient numerical method for two-dimensional phase change problems, *Int. J. Heat Mass Transfer* **27**, 2077 (1984).
12N. T. M. Shih and A. L. Ren, Primitive-variable formulations using nonstaggered grids, *Numer. Heat Transfer* **7**, 413 (1984).
13N. J. Srinivasan and N. S. Rao, Numerical study of heat transfer in laminar film boiling by the finite-difference method, *Int. J. Heat Mass Transfer* **27**, 77 (1984).
14N. J. Thibault, A three-dimensional numerical method based on the superposition principle, *Numer. Heat Transfer* **7**, 127 (1984).
15N. J. P. Van Doormaal and G. D. Raithby, Enhancement of the SIMPLE method for predicting incompressible fluid flows, *Numer. Heat Transfer* **7**, 147 (1984).
16N. M. D. Warren, Open and outflow boundary approximations for the solution of the equations of unsteady one-dimensional gas flow by the Lax–Wendroff method, *Int. J. Heat Fluid Flow* **5**(1), 43 (1984).

TRANSPORT PROPERTIES

1P. M. W. Beck, M. E. Brown and N. J. H. Heideman, Application of a probe technique for measuring the thermal conductivity of pyrotechnic compositions, *J. Phys. E, Scient. Instrum.* **17**, 793 (1984).
2P. J. P. Bentley, Temperature sensor characteristics and measurement system design, *J. Phys. E, Scient. Instrum.* **17**, 430 (1984).
3P. S. N. Biswas, N. J. Trappeniers and J. H. B. Hoogland, *PVT* properties of sulfurhexafluoride in the gas–liquid critical region, *Physica* **126A**(3), 384 (1984).
4P. H. Bjurstrom, E. Karawacki and B. Carlson, Thermal conductivity of a microporous particulate medium: moist silica gel, *Int. J. Heat Mass Transfer* **27**, 2025 (1984).
5P. L. Bruschi, M. Santini and G. Torzo, Rotating disc viscometer, *J. Phys. E, Scient. Instrum.* **17**, 312 (1984).
6P. V. P. Carey, Surface thermal capacity effects in transient natural convection flows at high Prandtl number, *Int. J. Heat Mass Transfer* **27**, 419 (1984).
7P. T.-H. Chung, L. L. Lee and K. E. Starling, Application of kinetic gas theories and multiparameter correlation for prediction of dilute gas viscosity and thermal conductivity, *Ind. Engng Chem. Fundam.* **23**, 8 (1984).
8P. J. M. Corsan, A compact thermal conductivity apparatus for good conductors, *J. Phys. E, Scient. Instrum.* **17**, 800 (1984).
9P. U. Grigull, Newton's temperature scale and the law of cooling, *Wärme- u. Stoffübertr.* **18**, 195 (1984).
10P. N. Imaishi and J. Kestin, Thermal conductivity of three noble gases with carbon monoxide, *Physica* **126A** (1/2), 98 (1984).
11P. N. Imaishi and J. Kestin, Thermal conductivity of methane with carbon monoxide, *Physica* **126A**(3), 301 (1984).
12P. E. F. Jaguaribe and D. E. Beasley, Modeling of effective thermal conductivity and diffusivity of a packed bed with stagnant fluid, *Int. J. Heat Mass Transfer* **27**, 399 (1984).
13P. U. Kurzweg and L. De Zhao, Heat transfer by high-frequency oscillations; a new hydrodynamic technique for achieving large effective thermal conductivities, *Phys. Fluids* **27**(11), 2624 (1984).

14P. E. Y. Kwack and J. P. Hartnett, Estimated eddy diffusivities of momentum and heat of viscoelastic fluids, *Int. J. Heat Mass Transfer* **27,** 1525 (1984).
15P. F. Lin and J. D. Ford, Thermal diffusivities of some porous oxides, *Can. J. chem. Engng* **62,** 125 (1984).
16P. L. K. Matthews, R. Viskanta and F. P. Incropera, Development of inverse methods for determining thermophysical and radiative properties of high-temperature fibrous materials, *Int. J. Heat Mass Transfer* **27,** 487 (1984).
17P. J. P. Novak and M. F. Frechette, Transport coefficients of SF_6 and SF_6–N_2 mixtures from revised data, *J. appl. Phys.* **55**(1), 107 (1984).
18P. M. Okaji and H. Imai, A practical measurement system for the accurate determination of linear thermal expansion coefficients, *J. Phys. E, Scient. Instrum.* **17,** 669 (1984).
19P. R. Pon and J. Hessler, Spectral emissivity, of tungsten: analytic expressions for 340 nm to 2.6 μm spectral region, *Appl. Optics* **23**(7), 963 (1984).
20P. P. Prelovšek and B. Uran, Generalized hot wire method for thermal conductivity measurements, *J. Phys. E, Scient. Instrum.* **17,** 674 (1984).
21P. A. V. Shustov and Yu. G. Eysmond, Investigation of the cooling capacity of aqueous solutions of polymers, *Heat Transfer—Sov. Res.* **15**(1), 124 (1983).
22P. R. Snel, A simple, inexpensive microvolumeter for the determination of the apparent density of fine, highly porous particles in a powder, *J. Phys. E, Scient. Instrum.* **17,** 342 (1984).
23P. E. Specht and R. Jeschar, Dimensionless groups for the description of the influence of temperature-dependent properties with heat transfer on affluxed bodies (in German), *Wärme- u. Stoffübertr.* **18,** 75 (1984).
24P. R. D. Trengove, H. L. Robjohns and P. J. Dunlop, Diffusion and thermal diffusion in binary mixtures of sulphur hexafluoride with noble gases, *Physica* **128A**(3), 486 (1984).
25P. F. Wehle and F. Brandt, Influence of temperature dependent physical properties on the heat transfer in the turbulent boundary layer of a parallel affluxed flat plate (in German), *Wärme- u. Stoffübertr.* **18,** 141 (1984).
26P. E. P. Wonchala and J. R. Wynncyckyj, The effective thermal conductivity of solids with complex pores in the transition of Knudsen regions, *Can. J. chem. Engng* **62,** 719 (1984).

HEAT TRANSFER APPLICATIONS

Heat pipes and heat exchangers

1Q. S. Aiba and H. Tsuchida, A criterion for tube arrangement and optimum tube spacing in tube banks, *Heat Transfer—Jap. Res.* **12**(3), 86 (1983).
2Q. S. Aiba, H. Tsuchida and T. Ota, Heat transfer around a tube in a staggered tube bank, *Heat Transfer—Jap. Res.* **11**(1), 1 (1983).
3Q. P. J. Banks, The representation of regeneration fluid carryover by bypass flows, *J. Heat Transfer* **106**(1), 216 (1984).
4Q. M. Baran and M. Pronobis, Convection heat transfer and pressure drop in cross flow over finned tubes, *Wärme- u. Stoffübertr.* **18,** 149 (1984).
5Q. R. F. Berg and G. G. Ihas, Direct comparison of copper, silver and palladium heat exchangers for cooling ^{3}He, *Rev. scient. Instrum.* **55,** 1174 (1984).
6Q. Yu. S. Borchevkin and B. P. Korol'kov, Dynamics of an independently heated heat exchanger with nonuniform axial distribution of the heat flux density, *Heat Transfer—Sov. Res.* **15**(2), 58 (1983).
7Q. V. A. Borodulya, V. L. Ganzha, A. I. Podberezsky, S. N. Upadhyay and S. C. Saxena, Heat transfer between fluidized beds of large particles and horizontal tube bundles at high pressures, *Int. J. Mass Transfer* **27,** 1219 (1984).
8Q. A. Bose and H. J. Palmer, Influence of heat and mass transfer resistances on the separation efficiency in molecular distillations, *Ind. Engng Chem. Fundam.* **23,** 459 (1984).
9Q. M. J. Brandemuehl and P. J. Bansk, Rotary heat exchangers with time varying or nonuniform inlet temperatures, *J. Heat Transfer* **106**(4), 750 (1984).
10Q. M. Bueb and P. Walzel, Tubular heat exchangers with static mixing devices for highly viscous liquids, *Chemie-Ingr-Tech.* **56**(7), 539 (1984).
11Q. R. S. Burlingame, Mean temperature difference for shell-and-tube heat exchangers with condensing on the shell side, *High Temp. Engng* **5**(3/4), 34 (1984).
12Q. Z. H. Cal, M. L. Li, Y. W. Wu and H. S. Ren, A modified selected point matching technique for testing compact heat exchanger surface, *Int. J. Heat Mass Transfer* **27,** 971 (1984).
13Q. C. J. Camarda and A. Basiulis, Radiant heating tests of several liquid-metal heat-pipe sandwich panels, *J. Spacecraft Rockets* **21,** 4 (1984).
14Q. C. C. Chen and J. W. Westwater, Application of the local assumption for the design of compact heat exchangers for boiling heat transfer, *J. Heat Transfer* **106**(1), 204 (1984).
15Q. G. A. Dreytser, V. I. Gomon and I. Z. Aronov, Comparison of fouling in tubes with annular turbulence promoters and in smooth tubes of shell-and-tube heat exchangers, *Heat Transfer—Sov. Res.* **15**(1), 87 (1983).
16Q. V. I. Fedorov, Z. A. Martsenyuk and M. D. Kas'yanova, A technique for calculating thermophysical transients in fuel-element assemblies cooled with dissociating N_2O_4, *Heat Transfer—Sov. Res.* **15**(1), 7 (1983).
17Q. V. M. Forostov, Yu. A. Korolenko and M. A. Yurtayev, Operation of injection-type desuperheaters, *Heat Transfer—Sov. Res.* **15**(2), 122 (1983).
18Q. K. P. Goyal and B. K. Gupta, An experimental performance evaluation of a disk and doughnut type heat exchanger, *J. Heat Transfer* **106**(4), 759 (1984).
19Q. D. K. Hennecke, Thermal analysis of a high-pressure compressor rotor of an aeroengine—venting as a means for life improvement, *Wärme- u. Stoffübertr.* **18,** 227 (1984).
20Q. A. B. Jarzebski, Dimensioning of spiral heat exchangers to give minimum costs, *J. Heat Transfer* **106**(3), 633 (1984).
21Q. J. J. Jurianak and J. W. Mitchell, Effect of matrix properties on the performance of a counterflow rotary dehumidifier, *J. Heat Transfer* **106**(3), 638 (1984).
22Q. J. Kern, Economic aspects of heat exchanger design,

Chemie-Ingr-Tech. **56**(6), 432 (1984).
23Q. M. Kovarik, Optimization of finned ducts in laminar flow, *J. Heat Transfer* **106**(3), 586 (1984).
24Q. S. Lynn and C. F. Oldershaw, Analysis and design of a viscous-flow cooler, *Heat Transfer Engng* **5**(1/2), 85 (1984).
25Q. P. J. Marto, Heat transfer and two-phase flow during shell-side condensation, *Heat Transfer Engng* **5**(1/2), 31 (1984).
26Q. D. Mocsny and R. Govind, Decomposition strategy for the synthesis of minimum-unit heat exchanger networks, *A.I.Ch.E. Jl* **30,** 853 (1984).
27Q. V. A. Myshenko and A. A. Khavin, Heat and mass transfer on filament-type packing in contact-type exchangers, *Heat Transfer—Sov. Res.* **15**(1), 94 (1983).
28Q. J. A. Perez and E. M. Sparrow, Patterns of fluid flow in a shell-and-tube heat exchanger, *Heat Transfer Engng* **5**(3/4), 56 (1984).
29Q. A. Pignotti, Matrix formalism for complex heat exchangers, *J. Heat Transfer* **106**(2), 352 (1984).
30Q. A. Pignotti, Stream-symmetry of heat exchangers and heat exchanger assemblies, *Heat Transfer Engng* **5**(3/4), 89 (1984).
31Q. P. S. Poškas and V. J. Survila, Fluctuations of velocity of cross flow of air in the space between tubes in bundles, *Heat Transfer—Sov. Res.* **15**(1), 75 (1983).
32Q. G. K. Raina, R. K. Wanchoo and P. D. Grover, Direct contact heat transfer with phase change motion of evaporating droplets, *A.I.Ch.E. Jl* **30,** 835 (1984).
33Q. A. Rolf, Simulation des nichtlinearen, dynamischen verhaltens von Wärme-tauschern sowie ihrer komplexan Schaltungen im Kraftwerksbau mit einem semianalytischen Berechnungsverfahren, *Brennst.-Wärme-Kraft* **36,** 228 (1984).
34Q. F. E. Romie, Transient response of the counterflow heat exchanger, *J. Heat Transfer* **106**(3), 620 (1984).
35Q. E. C. Rosman, P. Carajilescov and F. E. M. Saboya, Performance of one- and two-row tube and plate fin heat exchangers, *J. Heat Transfer* **106**(3), 627 (1984).
36Q. H. Schnell and B. Slipcevic, Causes and effects of foulding and scaling of heat transfer surfaces, *Chemie-Ingr-Tech.* **56**(6), 441 (1984).
37Q. Z. S. Shang, Z. Tanaka and T. Takahashi, Performance of a commercial heat pipe under operational conditions, *Heat Transfer—Jap. Res.* **12**(2), 17 (1983).
38Q. I. Shishido, T. Matsuda, T. Iida and S. Ohtani, Measurement of the working fluid distribution within a heat pipe wick, *Heat Transfer—Jap. Res.* **13**(2), 65 (1984).
39Q. H. A. Simon, S. Manafzadeh and J. C. F. Chow, Aspects of coiled tube heat transfer, *Wärme- u. Stoffübertr.* **18,** 215 (1984).
40Q. K. O. Smith, Internal heat transfer characteristics of silicon carbide heat exchanger tubes, *J. Heat Transfer* **106**(3), 672 (1984).
41Q. E. M. Sparrow and Y. S. Berman, Heat exchanger situated downstream of a right-angle bend, *Int. J. Heat Mass Transfer* **27,** 1649 (1984).
42Q. K. Stephan and J. Mitrović, Measures for intensifying heat transfer, *Chemie-Ingr-Tech.* **56**(6), 427 (1984).
43Q. M. A. Styrikovich, V. S. Polonsky and V. V. Reshetov, Experimental investigation of the critical heat flux and post-dryout temperature regime of helical coils, *Int. J. Heat Mass Transfer* **27,** 1245 (1984).
44Q. V. J. Survila, R. I. Jankauskas and L. A. Ašmantas, Hydraulic drag and heat transfer in bundles of tightly-packed twisted oval tubes, *Heat Transfer—Sov. Res.* **15**(2), 31 (1983).
45Q. O. Tanaka, An analysis of simultaneous heat and water vapor exchange through a flat plate crossflow total heat exchanger, *Int. J. Heat Mass Transfer* **27,** 2259 (1984).
46Q. T. Tanaka, M. Itoh, M. Kudoh and A. Tomita, Improvement of compact heat exchangers with inclined louvered fins, *Bull. JSME* **26.** 219 (1984).
47Q. P. Theissing, Improved design of tube-bundle heat exchangers considering local heat-, mass- and momentum exchange, *Chemie-Ingr-Tech.* **56**(8), 650 (1984).
48Q. I. Todo, Response of multipass heat exchangers to sinusoidal flow rate changes of large amplitude, *Bull. JSME* **27,** 87 (1984).
49Q. M. Turakhia, W. G. Characklis and N. Zelver, Fouling of heat exchanger surface: measurement and diagnosis, *Heat Transfer Engng* **5**(1/2), 93 (1984).
50Q. R. Turton, C. D. Ferguson and O. Levenspiel, Performance and design charts for heat exchangers, *J. Heat Transfer* **106**(4), 893 (1984).
51Q. D. D. Voelkel and F. P. Incropera, Performance model for a radiation recuperator, *Heat Transfer Engng* **5**(1/2), 74 (1984).
52Q. B. Vosteen, Dry heat transfer in flowable bulk solids in an indirectly heated rotary tube, a tumble dryer, and a plate dryer, *Chemie-Ingr.-Tech.* **56**(11), 858 (1984).
53Q. L. C. Witte and N. Shamsundar, A thermodynamic efficiency concept for heat exchange devices, *J. Engng Pwr* **105,** 199 (1983).
54Q. Y. Yanada, M. Akai and Y. Mori, Shell-and-tube side heat transfer augmentation by the use of wall radiation in a crossflow shell-and-tube heat exchanger, *J. Heat Transfer* **106**(4), 735 (1984).
55Q. A. A. Žukauskas, V. I. Katinas and E. E. Perednis, Average heat transfer coefficients of tube bundles in flows of viscous fluids at different angles of attack, *Heat Transfer—Sov. Res.* **15**(2), 22 (1983).
56Q. A. A. Žukauskas, R. V. Ulinskas and K. F. Marcinauskas, Average heat transfer coefficient and hydraulic drag of tube bundles in the flow of water at *Re* to $2 \cdot 10^6$, *Heat Transfer—Sov. Res.* **15**(2), 39 (1983).

General

1S. F. Adrian, Gedanken zu fragen der strömung, verbrennung und wärme-übertragund in feuerräumen großer kohlegefeuerter dampferzeuger, *Brennst.-Wärme-Kraft* **36,** 82 (1984).
2S. J. N. Beard, Jr, H. N. Rosen and B. A. Adesanya, Temperature distributions and heat transfer during the drying of lumber, *Drying Technol.* **1**(1), 117 (1983/84).
3S. G. A. Berry and J. R. Barber, The division of frictional heat—a guide to the nature of sliding contact, *J. Tribology* **106,** 405 (1984).
4S. D. Bharathan, F. Kreith, D. Schlepp and W. L. Owens, Heat and mass transfer in open-cycle OTEC systems, *Heat Transfer Engng* **5**(1-2), 17 (1984).
5S. P. Dadras and W. R. Wells, Heat transfer aspects of

nonisothermal axisymmetric upset forging (84-WA/Prod-11), *J. Engng Ind.* **106**(3), 187 (1984).
6S. M. G. Dunn, W. J. Rae and J. L. Holt, Measurement and analysis of heat flux data in a turbine stage: Part I—Description of experimental apparatus and data analysis (83-GT-121), *J. Engng Pwr* **106,** 229 (1984).
7S. M. G. Dunn, W. J. Rae and J. L. Holt, Measurement and analyses of heat flux data in a turbine stage: Part II—Discussion of results and comparison with predictions (83-GT-122), *J. Engng Pwr* **106,** 234 (1984).
8S. E. R. G. Eckert, Analysis of film cooled and full-coverage film cooling of gas turbine blades (83-GTJ-2), *J. Engng Pwr* **106,** 206 (1984).
9S. R. E. Gaugler and L. M. Russell, Comparison of visualized turbine endwall secondary flows and measured heat transfer patterns (83-GT-83), *J. Engng Pwr* **106,** 168 (1984).
10S. B. Gecim and W. O. Winder, Steady temperature in a rotating cylinder subject to surface heating and convective cooling, *J. Tribology* **106,** 120 (1984).
11S. B. R. Haller and J.-J. Camus, Aerodynamic loss penalty produced by film cooling transonic turbine blades (83-GT-77), *J. Engng Pwr* **106,** 198 (1984).
12S. M. Hssantani, N. Arai, Y. Itaya and N. Onoda, Drying of optically semitransparent materials by combined radiative–convective heating, *Drying Technol.* **1**(2), 193 (1983/84).
13S. Y.-C. Hsu and R.-F. Shih, Heat transfer from a heating helical coil immersed in a jet-stirred vessel, *Can. J. chem. Engng* **62,** 474 (1984).
14S. Y. Jaluria, Numerical study of the thermal processes in a furnace, *Numer. Heat Transfer* **7,** 211 (1984).
15S. Yu. A. Kalayda, I. M. Pavlenkov and V. I. Maksimov, Effect of mass velocity on the critical heat fluxes in forced flow of high-pressure, high-temperature, gas-saturated coolant, *Heat Transfer—Sov. Res.* **15**(2), 128 (1983).
16S. S. Kikkawa and M. Nonogaki, The effect of vibration on film cooling over a gas turbine blade, *Heat Transfer—Jap. Res.* **13**(2), 1 (1984).
17S. M. Kobayashi, M. Matsumato and M. Shizuya, An experimental investigation of a gas turbine disk cooling system (83-GT-78), *J. Engng Pwr* **106,** 136 (1983).
18S. P. Kurpiers, A. Steiff, P.-M. Weinspach, Heat transfer in stirred multiphase systems, *Chemie-Ingr-Tech.* **56**(3), 232 (1984).
19S. T. Kusuda and J. W. Bean, Simplified methods for determining seasonal heat loss from uninsulated slab-on-grade floors, *ASHRAE Trans.* **90**(1B), 611 (1984).
20S. J. W. Lund and P. K. Hansen, An approximate analysis of the temperature conditions in a journal bearing, Part I: Theory, *J. Tribology* **106,** 228 (1984).
21S. J. W. Lund and J. Tonnesen, An approximate analysis of the temperature conditions in a journal bearing. Part II: Application, *J. Tribology* **106,** 237 (1984).
22S. R. W. Lyczkowski and Y.-T. Chao, Comparison of Stefan model with two-phase model of coal drying, *Int. J. Heat Mass Transfer* **27,** 1157 (1984).
23S. S. Lynn and C. F. Oldershaw, Analysis and design of a viscous-flow cooler, *Heat Transfer Engng* **5**(1/2), 85 (1984).
24S. N. C. Markatos, S. M. Rawnsley and D. B. Spalding, Heat transfer during a small-break loss-of-coolant accident in a pressuarized water reactor—a parametric study for a 4 in. lower-planeum break, *Int. J. Heat Mass Transfer* **27,** 1379 (1984).
25S. S. Matsumoto and D. C. T. Pei, A mathematical analysis of pneumatic drying of grains—I. Constant drying rate, *Int. J. Heat Mass Transfer* **27,** 843 (1984).
26S. S. Matsumoto and D. C. T. Pei, A mathematical analysis of pneumatic drying of grains—II. Falling rate drying, *Int. J. Heat Mass Transfer* **27,** 851 (1984).
27S. A. Matějiček, J. Seidl, J. Eichler and F. Krška, Heat transfer in stirring of suspensions of macro-molecular substances, *Chemie-Ingr-Tech.* **56**(2), 157 (1984).
28S. F. C. McQuiston, S. L. Der, S. B. Sandoval, Thermal simulation of attic and ceiling spaces (RP-250), *ASHRAE Trans.* **90**(1A), 139 (1984).
29S. J. H. Nicholson, A. E. Forest, M. L. G. Oldfield and D. L. Schultz, Heat transfer optimized turbine rotor blades—an experimental study using transient techniques (82-GT-304), *J. Engng Pwr* **106,** 173 (1984).
30S. D. U. Ringer and A. S. Mujumdar, Analysis of aerodynamics and heat transfer in vibro-fluidized beds, *Drying Technol.* **2**(4), 449 (1983–84).
31S. G. C. Shove, Energy consumed in low temperature corn drying, *Drying Technol.* **2**(4), 503 (1983–84).
32S. A. A. Tseng, A numerical heat transfer analysis of strip rolling, *J. Heat Transfer* **106**(3), 512 (1984).
33S. L. Wassner, Wärmeübergangsmessungen für Kondensation und Verdampfung am lotrechten Fallfilmverdampferapparat, *Brennst.-Wärme-Kraft* **36,** 258 (1984).
34S. R. Wendt, F. Deuerler, A. Steiff, P.-M. Weinspach, Study of heat transfer on single tubes and crossflow bundled tubes in bubble column reactors, *Chemie-Ingr-Tech.* **56,** 946 (1984).
35S. D. C. Yard, M. Morton-Gibson and J. W. Mitchell, Simplified dimensionless relations for heat loss from basements, *ASHRAE Trans.* 90(1B), 633 (1984).
36S. R. E. York, L. D. Hylton and M. S. Milhelc, An experimental investigation of endwall heat transfer and aerodynamics in a linear vane cascade (83-GT-52), *J. Engng Pwr* **106,** 159 (1984).

Solar energy

1T. N. K. Bansal and A. K. Sharma, Transit theory of a tubular solar energy collector, *Sol. Energy* **32,** 67 (1984).
2T. P. Berdahl and M. Martin, Emissivity of clear skies, *Sol. Energy* **32,** 663 (1984).
3T. X. Berger, D. Buriot and F. Garnier, About the equivalent radiative temperature for clear skies, *Sol. Energy* **32,** 725 (1984).
4T. T. R. Borgers and H. Akbari, Free convective turbulent flow within the trombe wall channel, *Sol. Energy* **33,** 253 (1984).
5T. L. S. Cheema, Solar water heating and distillation, *Reg. J. Energy Heat Mass Transfer* **6,** 31 (1984).
6T. S. P. Chow, G. L. Harding, B. Window and K. J. Cathro, Effect of collector components on the collection efficiency of turbular evacuated collectors with diffuse reflectors, *Sol. Energy* **32,** 251 (1984).
7T. R. J. Duffin and G. Knowles, A passive wall design to minimize building temperature swings, *Sol. Energy* **33,** 337 (1984).
8T. M. Enery and B. A. Roges, On a solar collector thermal

performance test method for use in variable conditions, *Sol. Energy* **33,** 117 (1984).
9T. H. P. Garg and G. Datta, The top loss calculation for flat plate solar collectors, *Sol. Energy* **32,** 141 (1984).
10T. L. J. Ho, G. Le Palec and M. Daguenet, Theoretical study of honeycomb structure collector for space heating, *Sol. Energy* **32,** 349 (1984).
11T. J. R. Hull, K. V. Liu, W. T. Sha, Jyoti Kamal and C. E. Nielsen, Dependence of ground heat loss upon solar pond size and perimeter insulation. Calculated and experimental results, *Sol. Energy* **33,** 25 (1984).
12T. V. Joshi, V. V. N. Kishore, A. S. Metha and S. D. Gomkale, Establishment of natural concentration gradient in a solar pond, *Reg. J. Energy Heat Mass Transfer* **6,** 161 (1984).
13T. R. F. Kamada and R. G. Flocchini, Gaussian thermal flux model—I. Theory, *Sol. Energy* **32,** 505 (1984).
14T. R. F. Kamada and R. G. Flocchini, Gaussian thermal flux model—II. Validation, *Sol. Energy* **32,** 515 (1984).
15T. P. D. Lund and R. S. Keinonen, Radiation transmission measurements for solar ponds, *Sol. Energy* **33,** 237 (1984).
16T. M. Mario and P. Berdahl, Characteristics of infrared sky radiation in the United States, *Sol. Energy* **33,** 321 (1984).
17T. A. E. McGarity, C. S. ReVelle and J. L. Cohon, Analytical simulation models for solar heating system design, *Sol. Energy* **32,** 85 (1984).
18T. F. C. O'Brien-Bernini and J. G. McGowan, Performance modeling of nonmetallic flat plate solar collectors, *Sol. Energy* **33,** 305 (1984).
19T. B. Pálancz, Analysis of solar-dehumidification drying, *Int. J. Heat Mass Transfer* **27,** 647 (1984).
20T. H. Rubin, B. A. Benedict, S. Bachu, Modeling the performance of a solar pond as a source of thermal energy, *Sol. Energy* **32,** 771 (1984).
21T. A. Saito, Y. Utaka, M. Ishida, Y. Saito and K. Katayama, Transient characteristics of flat-plate solar collector (2nd report, estimation method of outlet water temperature amplitude and efficiency under periodic intensity variation), *Bull. JSME* **27,** 1422 (1984).
22T. A. Saito, Y. Utaka, T. Tsuchio and K. Katayama, Transient response of flat plate solar collector for periodic solar intensity variation, *Sol. Energy* **32,** 17 (1984).
23T. H. F. Sullivan, K. G. T. Hollands and E. C. Shewen, Thermal destratification in rock beds, *Sol. Energy* **33,** 227 (1984).
24T. T. C. Tripathi and J. S. Vaishya, Flat plate collector: design and performance studies, *Reg. J. Energy Heat Mass Transfer* **6,** 103 (1984).

Plasma heat transfer

1U. L. Antonio and O. Davalos, Magnetoconvection in a rotating fluid between walls of very low thermal conductivity, *J. phys. Soc. Jap.* **53**(7), 2173 (1984).
2U. T. K. Bose and R. V. Seeniraj, Electrostatic probes for dense flowing plasma, *Phys. Fluids* **27**(6), 1561 (1984).
3U. M. Cappelli and R. Measures, Two-channel technique for Stark measurements of electron density within a laser-produced sodium plasma, *Appl. Optics* **23**(13), 2107 (1984).
4U. Jen-Shih Chang, R. Godard and R. Scott, Total heat transfer to cylindrical collectors in collisionless plasma flow, *Phys. Fluids* **27**(3), 579 (1984).
5U. D. Degout, F. Kassabji and P. Fauchais, Titanium dioxide plasma treatment, *Plasma Chem. Plasma Proc.* **4**(3), 179 (1984).
6U. A. L. Donaldson, M. O. Hagler, M. Kristiansen, G. Jackson and L. Hatfield, Electrode erosion phenomena in a high-energy pulsed discharge, *IEEE Trans. Plasma Sci.* **PS-12**(1), 28 (1984).
7U. Z.-Y. Guo, The temperature and flow field of a free burning arc deflected by a transverse magnetic field, *Int. J. Heat Mass Transfer* **27,** 383 (1984).
8U. G. N. Haddad and A. J. D. Farmer, Temperature determinations in a free-burning arc: I. Experimental techniques and results in argon, *J. Phys. D, Appl. Phys.* **17**(6), 1189 (1984).
9U. J. E. Harry and R. Knight, Investigation of the intensity distribution of large volume multiple-arc discharges, *J. Phys. D, Appl. Phys.* **17**(2), 343 (1984).
10U. K. C. Hsu and E. Pfender, Modeling of a free-burning, high-intensity arc at elevated pressures, *Plasma Chem. Plasma Proc.* **4**(3), 219 (1984).
11U. A. Huczko and A. Szymanski, Thermal decomposition of carbon dioxide in an argon plasma jet, *Plasma Chem. Plasma Proc.* **4**(1), 59 (1984).
12U. B. Juttner, H. Pursch and V. A. Shilov, The influence of surface roughness and surface temperature on arc spot movement in vacuum, *J. Phys. D, Appl. Phys.* **17**(2), L31 (1984).
13U. H. Kafrouni, S. Vacquie, Study of the recombination continuum in an SF_6 arc discharge, *J. quant. Spectrosc. radiat. Transfer* **32**(3), 219 (1984).
14U. M. Kobayashi, T. Hondo and A. Kanzawa, Numerical analysis of a thermal plasma flow in a tube, *Heat Transfer—Jap. Res.* **13**(1), 64 (1984).
15U. M. Kobayashi, T. Honda and A. Kanzawa, Static pressure and heat flux measurements of a thermal plasma flow in a water-cooled tube, *Plasma Chem. Plasma Proc.* **4**(1), 43 (1984).
16U. P. Kovitya, Thermodynamic and transport properties of ablated vapors of PTFE, alumina, perspex, and PVC in the temperature range 5000–30,000 K, *IEEE Trans. Plasma Sci.* **PS-12**(1), 38 (1984).
17U. P. Kovitya and J. J. Lowke, Theoretical predictions of ablation-stabilised arcs confined in cylindrical tubes, *J. Phys. D, Appl. Phys.* **17**(6), 1197 (1984).
18U. P. Kovitya and S. Simpson, Non-equilibrium model of wall-established AC arcs at low currents, *J. Phys. D, Appl. Phys.* **17**(9), 1829 (1984).
19U. T. Kubono, A simplified analytical model for an anchored cathode root in a DC mercury vapour arc, *J. Phys. D, Appl. Phys.* **17**(10), 1991 (1984).
20U. A. Kurtz and J. Mentel, The thermal conductivity of Cl_2 and O_2 in the dissociation region measured by optical interferometry I: Experiment, *J. Phys. D, Appl. Phys.* **17**(7), 1335 (1984).
21U. A. Kurtz and J. Mentel, The thermal conductivity of Cl_2 and O_2 in the dissociation region measured by optical interferometry II: Evaluation of the thermal conductivity from interferometric data, *J. Phys. D, Appl. Phys.* **17**(7), 1343 (1984).

22U. S. M. Mahajan, J. Ravi and D. M. Benenson, The thermal boundary layer in dual flow arc plasmas, *IEEE Trans. Plasma Sci.* **PS-12**(4), 246 (1984).
23U. J. Mostaghimi and E. Pfender, Effects of metallic vapor on the properties of an argon arc plasma, *Plasma Chem. Plasma Proc.* **4**(2), 129 (1984).
24U. J. Mostaghimi, P. Proulx and M. Boulos, Parametric study of the flow and temperature fields in an inductively coupled r.f. plasma torch, *Plasma Chem. Plasma Proc.* **4**(3), 199 (1984).
25U. K. P. Nick, J. Richter and V. Helbig, Non-LTE diagnostic of an argon arc plasma, *J. quant. Spectrosc. radiat. Transfer* **32**(1), 1 (1984).
26U. Y. Ohguchi, Estimation of thermal conductivity of hydrogen gas through arc studies, *J. phys. Soc. Japan* **53**(12), 4179 (1984).
27U. A. Rahal, B. Rahhaoui and S. Vacquie, Copper vapour diffusion in a nitrogen arc chamber, *J. Phys. D, Appl. Phys.* **17**(9), 1807 (1984).
28U. L. K. Rudolph and D. Q. King, 100 kWe magnetoplasmadynamic thruster system design, *J. Spacecraft Rockets* **21**, 563 (1984).
29U. H. Shiral, K. Tabei and H. Koaizawa, Optically thick model for radiative and collisional effects in non-equilibrium argon plasma flows in a circular tube, *Bull. JSME* **27**, 1414 (1984).
30U. P. R. Stuart, R. J. Munz and W. H. Gauvin, Synthesis gas production from peat using a steam plasma, *Plasma Chem. Plasma Proc.* **4**(4), 297 (1984).
31U. M. Tachikura, Fusion mass-splicing for optical fibers using electric discharges between two pairs of electrodes, *Appl. Optics* **23**(3), 492 (1984).
32U. D. R. Topham, R. M. Clements and P. R. Smy, Turbulent mixing in a pulsed plasma-jet exhaust, *J. Fluid Mech.* **148**, 207 (1984).
33U. D. Wharmby, Energy balance of high-pressure sodium discharges under controlled vapour conditions, *J. Phys. D, Appl. Phys.* **17**(2), 367 (1984).
34U. H. Wilhelmi, W. Lyhs and E. Pfender, Calculation of thermodynamic and transport properties of a typical arc furnace plasma, *Plasma Chem. Plasma Proc.* **4**(4), 315 (1984).
35U. G. Wilks and R. Hunt, MHD free convection flow about a semi-infinite plate at whose surface the heat flux is uniform, *Z. angew. Math. Mech.* **35**(1), 34 (1984).
36U. Wee Woo and J. S. DeGroot, Microwave absorption and plasma heating due to microwave breakdown in the atmosphere, *Phys. Fluids* **27**(2), 475 (1984).
37U. G. Woolsey, P. Stimson and H. Swan, The transition between two glow discharge phases in the SF_6 spark, *J. Phys. D, Appl. Phys.* **17**(11), 2255 (1984).
38U. H. Yoshiyasu, S. Murata, Y. Wada, Y. Shibuye and Y. Morai, Vapour flow patterns observed in free-burning arcs, *J. Phys. D, Appl. Phys.* **17**(7), 1381 (1984).

HEAT TRANSFER REVIEW - 1985

E.R.G. ECKERT, R. J. GOLDSTEIN, E. PFENDER, W.E. IBELE, J.W. RAMSEY, T.W. SIMON, N. A . DECKER, T.H. KUEHN, H.O. LEE and S.L. GIRSHICK

Heat Transfer Laboratory, Department of Mechanical Engineering, University of Minnesota, Minneapolis, Minnesota, U.S.A.

INTRODUCTION

THIS REVIEW surveys papers that have been published in the open literature covering various fields of heat transfer during 1985. The literature search was made as inclusive as possible, however, the great number of publications made selections in some of the review sections necessary.

Several conferences were devoted to heat transfer or included heat transfer topics in their sessions during 1985. A symposium on *Off Shore Mechanics and Arctic Engineering* was held 17–21 February at Dallas, TX. It included a session on arctic thermal design analysis and icing problems were discussed in other papers. The proceedings are available as a bound volume. Information on those may be obtained at the Energy-sources Technology Conference and Exhibition, P.O. Box 59489, Dallas, TX, or through the ASME Order Department. The *30th International Gas Turbine Conference and Exhibit* 17–21 at Houston, TX, was sponsored by the Gas Turbine Division of the American Society of Mechanical Engineers. Four of the sessions were devoted to external and internal heat transfer as well as to film cooling. The 1985 Gas Turbine Award was presented to Chunill Han for the paper "A Navier–Stokes analysis of three-dimensional turbulent flow inside turbine blade rows at design and of design conditions" and the 1985 Tom Sawyer Award was presented to Anselm Franz in recognition of his pioneering contributions to the Gas Turbine Industry. Papers may be obtained at the ASME Order Department. A *Symposium on Transport Phenomena in Rotating Machinery*, held in Honolulu, Hawaii, 28 April–3 May and sponsored by American and Japanese Engineering Societies as well as the Universities of Michigan and Hawaii, included in its program papers on turbine blade cooling, channel flow heat transfer, rotational effects, liquid and evaporation cooling, heat transfer in seals, and superconducting generators. The *29th Heat Transfer and Fluid Mechanics Institute*, held 20–21 June at Sacramento, CA, included in its proceedings papers on melting, vortex flow, combustion of solid propellants, vapor and droplet transport, transient and oscillatory heat transfer, and two-phase flow. Proceedings are published by the California State University, Sacramento, CA. The 1985 *International Solar Energy Congress and Exhibition* sponsored by the American Section of the International Solar Energy Society was held 28 July–1 August at Los Angeles, CA. Sessions dealing with solar radiation and solar furnaces, solar collectors and energy storage, solar heating and cooling of buildings, drying and distillation, focusing collectors, and solar thermal power were included in the program. The *Twenty-third National Heat Transfer Conference*, sponsored by the American Institute of Chemical Engineers and the American Society of Mechanical Engineers held 4–7 August at Denver, CO, was organized in 61 sessions, many of them dealing with heat transfer in specific technology processes like combustion, hazardous waste incinerators, drying and dehumidification, glass, alternative energy technologies, steam generators, nuclear reactors, aerodynamic systems, fibers, buildings and structures. Courses on numerical solutions and augmentation of heat transfer, thermal analysis, control of aerodynamic equipment, and compact heat exchangers as parts of the professional development program rounded off the proceedings. The Max Jakob Memorial Award was presented to L. London and the Donald Q. Kern Award to D. Gidaspow. The papers at this conference sponsored by the ASME are available in paper or volume form through the ASME Order Department. AIChE papers are contained in the volume "Heat Transfer/Denver 1985" of the AIChE Symposium Series. The *20th Intersociety Energy Conversion Engineering Conference*, held 18–23 August at Miami Beach, FL, did not include in its program specific sessions on heat transfer but this topic is touched upon in various of their lectures.

Conference Proceedings are available from SAE, 400 Commonwealth Drive, Warrendale, PA. The *17th International Symposium of the International Centre for Heat and Mass Transfer* held 26–30 August at Dubrovnik, Yugoslavia, was devoted to the topic of high temperature heat exchangers. An introduction lecture by Y. Mori, Japan, discussed future development in high temperature heat exchangers, eight invited lectures were given by experts from the Federal Republic of Germany, Great Britain, Japan, U.S.A., U.S.S.R., and Yugoslavia and 41 lectures discussed various topics of heat exchangers and their applications for gas turbines, Stirling cycle engines, metallurgical processes and future power plants. The Proceedings of the conference are collected in a volume, published by Hemisphere Publishing Corporation. The *1985 Beijing International Gas Turbine Symposium and Exposition*, 1–7 September at Beijing, People's Republic of China, sponsored by the Gas Turbine Division of ASME and by several Chinese professional societies, brought together researchers and engineers from the Pacific Basin. Two sessions on heat transfer dealt with recouperators, turbine blade cooling, liquid and vapor cooling, transpiration and film cooling. Inquiries about papers should be directed to the ASME Order Department. An *International Symposium on Refined Flow Modelling and Turbulence Measurements*, held 16–18 September at Iowa City, IA, was sponsored by the Iowa Institute of Hydraulic Research in cooperation with a number of engineering societies. Two sessions were devoted to modeling and turbulence effects on heat transfer. The proceedings will be published in book form by Hemisphere Publishing Corporation. The *ASME Winter Annual Meeting*, held 17–22 November at Miami Beach, FL, included in its program 17 sessions with discussions on a variety of heat transfer processes. Six symposia dealt with stability in convective flows, mixed convection, fouling and corrosion, augmentation of heat transfer, in energy systems, nuclear steam generators, and heat exchangers. A panel discussion, moderated by R. K. Shah, inquired into opportunities and challenges in heat transfer. In the Robert Henry Thurston lecture, Y. C. Fang discussed the interplay of mechanical engineering and health science and at the heat transfer luncheon, Raymond Colladay talked about thermal management in space stations. At the same luncheon, the heat transfer memorial awards were presented to Ralph Grief and Virgil E. Schrock. Preprints of the papers are available at the ASME Publications Department and many of the papers will also be published in the *Journal of Heat Transfer*.

A number of books have become available during the past year. They are listed under references.

HIGHLIGHTS

The following highlights describe areas of research which have found special attention during last year.

In heat conduction, the emphasis was on nonlinear behavior, presence of internal heat sources, and geometrical irregularities.

Many of the channel flow papers dealt with increasingly complex flow fields created by modern heat transfer augmentation devices, i.e. roughening ribs and fins. Also noted were flows with non-Newtonian fluid behavior, as with polymers and micropolar fluids.

Heat transfer in supersonic and hypersonic flows has emerged as a significant portion of the boundary-layer section. Some papers on this topic include the effects of non-equilibrium chemical reactions. Numerous papers on heat transfer variations due to unsteady flow were published. An example would be the effect of passing wakes on gas turbine airfoil heat transfer. Boundary-layer transition research continues to increase in popularity; common applications which were noted are gas turbine airfoils and heat transfer in high velocity flight.

Heat transfer studies in porous media are beginning to consider seriously departures from Darcy flow. Many papers dealt with fluidization, including several involving three-phase fluidized beds. Simultaneous heat and moisture transfer through porous material was the subject of a number of articles as was the modeling of reactor core debris bed heat transfer.

The development of experimental techniques included the application of thin films for heat flux and temperature measurements, impressive advances in the use of laser-based and other optical diagnostics for making temperature measurements in high-temperature gases, continued work in the area of liquid crystals, and numerous refinements in the use and interpretation of hot-wire anemometry for heat transfer research.

Many studies on natural convection in internal flow include nonlinear phenomena in horizontal fluid layers at relatively low Rayleigh number as well as numerical studies for differentially heated layers. There was an increase in studies on Marangoni or surface tension driven flows and the effect of large property variations on natural convection. The need for turbulent transport parameters in buoyancy driven flows was emphasized.

The vertical flat plate geometry dominates the natural convection boundary-layer literature. Variations include coupled conductive or radiative heat transfer, double-diffusive convection, aiding or opposing mixed convection and the effects of plate motion.

Rotational effects were studied especially in connection with gas turbines. Highlights for combined heat and mass transfer include a very large number of studies related to film cooling particularly for gas turbine systems; the influence of large density difference, pressure gradient, and curvature were studied using an array of mass transfer analogies.

Papers on boiling reviewed this year continued to display our interest in the fundamentals of nucleate boiling. Experiments with boiling of mixtures and with boiling at enhanced surfaces reflected continuing interest in new applications of boiling heat transfer. Still well represented in the literature are studies relating boiling heat transfer to nuclear reactor safety and the service life of steam generators.

Noteworthy developments in condensation heat transfer included a systematic series of experiments on film condensation on horizontal cylinders and tube banks, and several analytical studies of free condensation heat transfer to nucleated droplets.

New analytical methods continue to evolve describing the solidification process. A significant amount of work had been done on freezing within molds, however discussion of continuous casting processes is not emphasized as much as in previous years. Some interesting work on the movement of a heat source as it melts its way through a solid has also been presented.

Radiative transfer combined with other modes of heat transfer and/or phase change received increased attention. There was also a renewed interest in participating one-dimensional media, while the research in multi-dimensional media continued. Surface radiation was still considered in a number of studies, but the papers on participating media greatly outnumber those considering surface exchange.

Developments in numerical methods continue to be directed towards the treatment of arbitrary shapes, removal of false diffusion, and improvement of computational speed.

Significant interest was demonstrated in the thermal conductivity of composite and nonisotropic materials: idealized systems used as analytical models and those used in industrial/commercial application. A number of nonsteady-state techniques are proposed for obtaining thermal conductivities and overall coefficients of heat transfer of sufficient accuracy for everyday commercial practice.

Papers dealt with economical ways of enhancing heat transfer by extended surfaces, flow modification, and heat pipes (thermosyphon). Heat transfer solutions formerly restricted to scientific and advanced technology areas are increasingly found at various levels of application. General applications of heat transfer are scattered in a large number of journals. A cross-section of those can be recognized in Section S.

Plasma heat transfer research was primarily associated with plasma processing of materials.

To facilitate the use of this review, a listing of the subject headings is made below in the order in which they appear in the text. The letter which appears adjacent to each subject heading is also attached to the references that are cited in the category:

Conduction, A
Channel flow, B
Boundary layer and external flows, C
Flow with separated regions, D
Heat transfer porous media, DP
Experimental techniques and instrumentation, E
Natural convection—internal flows, F
Natural convection—external flows, FF
Convection from rotating surfaces, G
Combined heat and mass transfer, H
Change of phase—boiling, J
Change of phase—condensation, JJ
Change of phase—freezing and melting, JM
Radiation in participating media and surface radiation, K
Numerical methods, N
Transport properties, P
Heat transfer application—heat pipes and heat exchangers, Q
Heat transfer applications—general, S
Solar energy, T
Plasma heat transfer and MHD, U

CONDUCTION

Most of the literature on heat conduction deals with theoretical studies, although some experimental investigations have been reported. Solutions have been presented for steady and unsteady problems in multi-dimensional geometries with nonlinearity, anisotropy, and complex boundary conditions.

A variational method has been proposed for solving two-dimensional heat conduction with finite internal heat source [12A]. For axisymmetric bodies, a boundary element method is used to solve steady heat conduction [47A]. A method is described for the shape optimization of a fin to arrive at minimum volume for a given heat loss [43A]. References [14A] describes a flux formulation for hyperbolic heat conduction. A method has been developed for the analysis of heat conduction by a dual difference equation [19A]. For heat conduction with chemical reaction, a method has been described for treating discontinuous boundary conditions [28A]. A moving-mesh finite-element analysis has been applied to calculate the temperature field with a moving heat source [22A].

A number of complex multi-dimensional geometries have been treated. A new class of functions are used to solve for the temperature distribution in a rectangular parallelpiped with longitudinal cylindrical cavities [31A]. The same type of method is used for a rectangular plate with inner circular boundaries [32A]. Heat conduction is analyzed for a plate with a square cut [21A]. An investigation is made of the effect of circumferential variation of heat transfer coefficient and gas radiation on the temperature distribution in a multilayered cylinder [9A]. A numerical method is described for the solution of the heat conduction equation with internal heat generation and thermal anisotropy [18A]. A variational principle is used to obtain a new solution to the heat equation in composite media [8A].

Numerical solutions have been used to determine the critical radius of insulation when circumferential variations of the heat transfer coefficient are present [40A]. Heat conduction analysis is applied to the rolling mill process where a three-layer composite strip is compressed by two rotating cylinders [48A]. The finite-element method is used to determine the temperature distribution in a helix as influenced by the curvature and torsion [38A]. Nonlinear conduction analysis is used to study the self-similar expansion of dense matter [34A].

Nonlinear boundary conditions for heat conduction problems have been considered in refs. [36A] and [41A]. Some papers focus on the solution of the inverse heat conduction problems. A direct analytical approach for these problems is given in ref. [2A], while the solution of three-dimensional inverse problems is discussed in ref. [3A]. Reference [1A] deals with optimal control in the solution of certain inverse heat conduction problems. A numerical method is used to determine the optimum locations of temperature sensors for the solution of inverse conduction problems [7A].

Research continues on the problem of thermal contact resistance. An analytical model is developed and experimentally verified for the resistance of the line contact between a circular cylinder and a smooth flat surface [6A]. Theory and experiment are presented for the contact resistance between spherical and plane surfaces with micro-irregularities [42A]. Reference [4A] shows that the contact resistance can be reduced by the use of metallic coatings. Use of fine powders for producing high thermal resistance is evaluated in ref. [46A].

A number of papers have dealt with unsteady heat conduction. References [20A, 33A] present an analysis of unsteady nonlinear problems. Unified solutions for heat conduction with a surface film of finite heat capacity are given in ref. [27A]. A study is made of the evolution of temperature in the core of a high voltage electric cable [37]. Transient conduction in a multi-layer medium is considered in ref. [10A]. Reference [30A] deals with periodic heat transfer in triangular fins. The unsteady conduction in a hollow cylinder is analytically calculated in ref. [45A]. A numerical technique is presented for the unsteady conduction in nuclear fuel rods [16A]. Transient conduction with convective and radiative boundary conditions is solved by a semi-analytical method [11A]. Some two-dimensional unsteady conduction processes are optimized in ref. [35A]. A coupled conduction and convection problem in the unsteady mode is considered in ref. [15A]. The thermal effects caused by placing a fiber in an ideal gas through which an acoustic wave propagates are described in ref. [24A]. Solutions are obtained for the enhanced heat conduction in fluids within tubes subjected to sinusoidal oscillations [25A]. Reference [17A] deals with unsteady heat conduction with mixed boundary conditions. A variational formulation is used to solve for the unsteady temperature field in heat generating bars of doubly connected cross section [13A].

Experimental work related to heat conduction is mainly directed towards the measurement of thermal conductivity and diffusivity. Reference [39A] proposes a method for the simultaneous measurement of these properties. Measurement of diffusivity is described in ref. [29A]. For multi-layer insulations, the equivalent conductivity is measured [23A]. References [5A, 6A] deal with the measurement of diffusivity of liquids. Heat conduction through a layered pressure vessel wall is investigated in ref. [44A].

CHANNEL FLOW

Heat transfer rates in a laminar duct flow of a non-Newtonian fluid were numerically evaluated [23B]. Viscous dissipation was considered. Experimental results were presented showing the turbulence structure in the vicinity of the wall [54B]. The fluids were water and weak polymer solutions. The addition of polymers leads to an increase of the characteristic time scale determined from the frequency of temperature excursions. Measured Nusselt numbers for laminar flow of a viscoelastic fluid in a rectangular channel were found to be much larger than if the fluid were purely viscous or Newtonian [31B]. The effect was traced to secondary flows which are induced by normal force differences at the boundaries which are unique to elastic fluids. The influence of a thermal field imposed on a laminar pipe flow of pseudoplastic fluid was numerically evaluated [62B]. Correlations were presented. Numerical solutions were found for the Graetz problem for a Bingham plastic in laminar tube flow [6B]. This work was motivated by the use of aqueous foams for drilling wells in high-temperature formations. Numerical results were presented for turbulent flow of an incompressible, initially low-viscosity liquid which undergoes a thermal or induced polymerization in a long tubular reactor [29B]. Various flow regions along the tube, due to the continuing reaction, were discussed. A new approximate substitution principle was presented for flows in which both heat transfer and momentum interchange by viscous stresses are significant [27B]. Calculated results were compared with experiments. Heat transfer was analyzed for compressible flow of a fluid of varying thermophysical properties in a plane-parallel channel [64B]. An analytic solution was proposed for the effect of a resonant acoustic field on heat transfer to laminar channel flow [59B]. An analysis was presented for heat transfer in a porous wall channel with a micropolar fluid [5B]. Temperature distributions were found for several values of suction rate, convection rate and micropolar parameter. An analysis was made of thermally induced flow in a shallow two-dimensional slot [11B]. The extent of surface stagnation due to nondiffusing surfactant was discussed. When the absorbed species are insufficient to fully cover the surface, the interface is either clean and subject to constant stress or contaminated and no-slip is observed. The region between these two types was discussed. Heat transfer in a rarefied polyatomic gas was analyzed [53B]. The problem was transformed to a system of integral equations which were solved numerically. Dependence of heat transfer on Knudsen number, internal energy, Eucken factor and collision relaxation number was discussed. The F//N method for computation of heat transfer in a rarefied gas was applied to flow between parallel plates to demonstrate its efficiency [74B]. Total heat flux, temperature and density profiles, and molecular distribution functions were computed. The mechanisms of heat and mass transfer in a two-component medium were presented using a system of alternating adjacent plates [9B]. The stability of pressure-driven flow in a vertical U-shaped loop with one heated leg was evaluated experimentally and numerically [79B]. When heating is sufficient, buoyancy opposes the driving pressure, oscillation begins and, eventually, the flow reverses. The stability limits are identified. Tests were run with a single-phase fluid and with a packed bed. A numerical model was presented for axisymmetric film flow from a slot taking into account the evaporation and deposition of the substance [36B]. Thermal equilibrium of air/water flow through an annulus was investigated [67B]. Equilibrium depends upon axial length, gas and liquid mass fluxes and void distribution. Heat transfer to an air–solid suspension flow was discussed [42B]. Heat transfer rates from horizontal pipe walls to forced convection of water and suspended glass beads (or ion exchange resin beads) were measured [30B]. When the beads were larger than 0.35 mm, heat transfer was enhanced; when less than 0.15 mm, heat transfer was reduced. Convective heat transfer for steady laminar flow of a radioactive slurry between two concentric pipes with walls heated independently and subject to parabolic heat generation was analyzed [3B]. The parabolic heat generation was shown to have both positive and negative effects on Nusselt numbers. A numerical solution was presented for the heat transfer in turbulent pipe flow of supercritical carbon dioxide under cooling conditions [55B]. The effect of buoyancy on turbulent flow of mercury was numerically evaluated [17B]. One heating effect, also measured, was the reversal of the turbulent axial flux. Measurements of heat transfer to laminar tube flow of mercury showed close agreement with theory; and, in the turbulent regime, measurements agreed well with nonwetted tube literature [28B]. The separation theory of horizontal thermal diffusion columns with external refluxes was derived using an orthogonal expansion technique [73B]. The method could be applied to continuous-contact countercurrent separation processes. A numerical study was made of heat transfer to turbulent flow of supercritical helium [16B]. The mixing length turbulence model was shown to give more accurate results than the k or $k-\varepsilon$ models. The Yaskin correlation was superior to the Nusselt correlation for near-pseudocritical temperatures. Measurements of heat transfer, critical heat flux and pressure drop in forced flow of cryogens were made [70B]. Effects of operating conditions on heat transfer in superconducting devices were discussed and correlations were presented. The correction factor for property variations for laminar convective transport in tubes was discussed [25B]. The exponent on the bulk-to-wall Prandtl number ratio is shown to be a function of 11 dimensionless parameters. An asymptotic theory was presented which accounts for property variations [32B]. Numerical results showed the influence of variable properties on laminar fully-developed pipe flow [33B]. Dependence on the properties was expressed as a Taylor series. Analytical results were given for laminar convection in open and closed channels with large wall-to-wall temperature differences [13B]. The solutions are valid where the hot and cold wall boundary layers have merged. It was shown that variations in static pressure are more sensitive to property variations than are variations in heat flux. Results of experiments were presented which show that channels of reduced density created in the atmosphere can diffuse by turbulence [26B]. Uniform (axisymmetrically or axially) channels diffuse by conduction, whereas nonuniform channels diffuse by turbulent mixing. An analysis was made showing the convective heat transfer in a side-heated planar, two-layer system [65B].

Numerical results were presented for heat transfer in curved channels of various aspect ratio [43B]. At moderate values of Dean number, a double-vortex secondary flow pattern was established. The change of the vortex pattern with increasing Dean number was documented. The influence of buoyancy on fully-developed laminar flow in curved tubes was studied numerically [50B]. Buoyancy acts to increase the average Nusselt number and rotate the orientation of the secondary flow. Density effects following a change in inlet temperature to a U-shaped pipe were computed [76B]. Results computed using the $k-\varepsilon$ model of turbulence were shown to be accurate. The three-dimensional flow and heat transfer situation in a rectangular, 90° miter-bend was studied to investigate the flow curvature effect in a corrugated-wall heat exchanger channel. A detailed examination of the heat transfer mechanism was presented [46B]. Reynolds number and aspect ratio effects were also investigated [47B]. Local heat transfer coefficients were measured for air flow around a 180°, square-section bend [37B]. Consistencies between velocity and temperature fields were noted. Measurements and calculations were made for heat exchange in short cylindrical chambers with swirl [61B]. Velocity and pressure distributions were measured. Turbulence data for swirling flows along an annulus were presented [15B]. Measured eddy diffusivity values for various axial and radial positions were given. The hot-wire anemometer techniques were discussed. An analysis of a screw-type swirl promoter was presented leading to an equation for optimum geometry [41B]. Unsteady, swirling, boundary-layer heat transfer in a compressible fluid flowing through a nozzle or diffuser was numerically evaluated [44B]. The heat transfer rate was found to be sensitive to free-stream velocity, variation in density-viscosity product across the boundary layer, mass transfer (suction) and swirl.

Temperature fluctuations and heat flux values were measured in grid-generated isotropic turbulence in which streamwise and transverse mean temperature gradients were imposed [8B]. The skewness of the temperature fluctuation derivatives was found to be nonzero even though the mean shear was zero. Experimental measurements were made on the enhancement of heat transfer due to laminar flow oscillation within a capillary bundle [45B]. Axial conduction heat flow was shown to be at least four orders of magnitude larger than that without oscillations. Analysis and generalizations of experimental data were presented for unsteady heat transfer and flow in tubes under conditions of heating/cooling and variations in flow rate, heat release in the walls and entrance flow temperature [38B]. Calculation methods for real, unsteady thermal processes were suggested. An analysis was presented for density-wave instabilities in heated, ventilated channels [72B]. Applications include boiling flow instabilities in boiling water reactor cores. Nonisothermal, quasistationary gas flow in an inclined rectangular tube was theoretically investigated [7B]. Dependencies were found for pressure, temperature and the discharge fluctuations. The effect of natural convection on heat transfer in a Graetz-type entry flow was computed [34B]. An oscillatory flow pattern of one-pair and two-pairs of counter-rotating vortices appeared for $Ra \geqq 5 \times 10^5$. Eventually, with increasing Ra, a pair of counter-rotating vortices prevails. Measurements were made of heat transfer in a tube with large-amplitude velocity pulsations and shock waves radiating from the open end [21B].

Results of an analysis of the dynamic aspects of spontaneous ignition were presented [24B]. Effects of three factors—ambient temperature, reactant temperature, and the order of the reaction were discussed. A gaseous mass transfer model was proposed for evaluation of chemical chainback reaction with volatile species in a boiling channel [35B]. The results show that gas release and absorption rates were not consistent with Henry's law. Numerical solutions were presented for isothermal laminar convective transport of reaction products in a rectangular duct with irreversible heterogeneous and homogeneous reactions [12B].

Approximate solutions of the Graetz problem with axial conduction were presented [48B]. The results of the approximate method were more easily computable than those of the exact solution. For Péclet numbers between 1 and 10, the approximate method gives satisfactory results. Results of an analysis for laminar, developing-flow heat transfer between parallel plates were presented [52B]. A fourth-order polynomial was used for the profiles' shapes to improve on the accuracy of a previous study. The effect of substantial amounts of heat addition on the convergence characteristics of implicit algorithms was investigated for convergent–divergent nozzle flow [51B]. The Beam and Warming algorithm was shown to converge more readily than MacCormack's algorithm. Real gas effects were shown to be minor. An experimental study of boundary layer turbulence assisted by flow visualization in a plane diffuser showed that two peaks in turbulence energy generation could be found; one near the wall and the other in the shear layer [40B]. Convective heat transfer rates were measured under forced convection, developing laminar flow conditions in short elliptical channels [14B]. The effects of geometry, properties and free convection were discussed. An analysis was made of steady turbulent flow in an annular channel with an internal wall moving independent of the fluid [20B]. Results were compared with experiments. A numerical analysis including radiative heat transfer was made of turbulent flow in a planar duct [63B]. The selectivity of the gas was taken into account by using a set of four gray gases.

Measurements of heat transfer were made on one wall of a rectangular channel whose other wall was covered by an array of block-like elements [68B]. Nusselt numbers compared well with correlations when the hydraulic diameter was based on the rectangular section between the tops of the elements and the opposite wall. A conversion technique for extending heat transfer data in air-heated annuli with rough inner tubes to fully-rough channels was presented [66B]. The converted data was in agreement with data on rough channels and bundles of rough tubes. An experimental investigation was made of heat transfer to gas flow in annuli with an inner rough tube [75B]. In contrast to smooth tubes, the effect of variable physical properties depends on the roughness size and the Reynolds number, diminishing with increases in these parameters.

Analytical results were presented for two-dimensional heat conduction in a rectangular solid with a cylindrical hole filled with a gas which is transparent to thermal radiation [58B]. Results of a parametric study were presented.

Laminar, fully developed flow in an internally radial-finned, vertical annulus was numerically evaluated [57B]. Accounting for buoyancy effects improves the finned passage as an effective heat exchanger device. Laminar, forced convection heat transfer coefficients for the entrance region of an internally-finned duct were numerically evaluated [56B]. Results for several fin geometries were presented. The convective cooling problem of laminar pipe flow with radiation at the wall and conduction within the wall was analyzed [19B]. Streamwise wall conduction was shown to have an effect on the results. A general procedure which combines the superposition principle with a finite-element method was applied to the conjugate heat transfer problem of laminar pipe flow with external heating [4B]. The importance of axial conduction was discussed. An energy balance was presented for single-phase flowing fluids [60B]. Heat transfer characteristics were numerically analyzed for a constant property fluid flowing laminarly through a parallel plate channel with staggered, transverse ribs [77B]. Significant augmentation by the ribs was found for high Prandtl number fluids. Also, conduction in the wall plays a large beneficial role in enhancing heat transfer. Heat transfer coefficients were measured for laminar and turbulent flow in rectangular passages open along the lateral edges but closed at the downstream end [69B]. Streamwise uniformity of heat transfer coefficient was promoted by shrouding and by increasing the interwall spacing and Reynolds number. Numerical results were presented for turbulent flow in channels of complex geometry as found in heat exchangers [10B]. Wall-bounded flows were predicted well, but jets and external flows required a modification to the modeling of the turbulent length scale. Local heat transfer rates in periodically corrugated wall channels were computed for both laminar and turbulent flow [2B]. The effects of separation; deflection, recirculation and reattachment were discussed. Patterns were shown to drastically change between laminar and turbulent flow. A numerical method for three-dimensional transport processes was presented for laminar flow to power-law fluids flowing in ducts of arbitrary, but uniform, cross section [49B]. Demonstrations were given for square, pentagonal and trapezoidal ducts.

Forced convection heat transfer measurements were made in a narrow vertical channel with upflow and downflow—both laminar and turbulent [71B]. Existing correlations were tested. The channel simulates a subchannel of a nuclear reactor. A simulation of mass transfer between cells in reactor rod bundles for longitudinal flow was presented [78B]. It was shown how results from experiments of this type would be used for calculation of enthalpies of the coolant.

A methodology was presented for the inverse design of interior coolant flow passages with application to turbine cascade blade design [39B]. The user may specify surface temperature and the wall thickness variation is found by the technique.

The transient momentum and energy equations were solved to find the transient temperatures in a thrust bearing [18B]. Analytical results were presented for the problem of two-dimensional transient heating of a circular source in a semi-infinite surface [22B]. The heating transient and a convective cool-down transient were studied. These transients simulate the thermal behavior of a single asparity in an apparent area of contact in a bearing. A numerical study was presented of the steady-state bubble-thermohydrodynamic behavior of rigid circular pad thrust bearings [1B]. The influence of air/gas bubbles included in the lubricant on important design criteria was discussed.

BOUNDARY-LAYER AND EXTERNAL FLOWS

A survey on aeroassisted orbit transfer was presented [88C]. Analytical, computational and experimental methods used in aerothermodynamic design were discussed [56C]. Comparison between flight tests and prediction show that these methods are satisfactory for design. Aerothermodynamic problems associated with hypervelocity flight were described, including their severity [40C]. Problems that require experimental and theoretical research were highlighted. Chemically reacting viscous flow in complex geometries were analyzed using a two temperature model [81C]. Results were almost identical to one-temperature results—an almost negligible difference in electron concentration across the shock layer was observed. Laminar heating was measured at hypersonic conditions on straight and bent-nose (simulating the proposed interplanetary aeroassisted vehicle) biconics [58C]. Equilibrium density ratios were as large as 19. Measurements were made of the effect of Reynolds number, angle of attack and nose shape on the aerothermal heating of a cone at Mach 6.7 [62C]. Surface pressure data were independent of Reynolds number and laminar heating on the windward surface scaled on stagnation point heat transfer. Enhancement by free-stream turbulence of convective heat transfer to the stagnation region of a hemisphere was studied [37C]. Predictions agreed well with experimental results. The effect of free-stream turbulence on the stagnation point velocity gradient was evaluated. An inviscid flow field method was extended to include heat transfer and deal with angle-of-attack conditions [96C]. It can be applied to flow over bodies of revolution of several geometries and was shown to agree well at even high incidence angles for flow over sphere-cones. The results of a study of the shock wave on the front surface of a sphere in supersonic flow were given [32C]. Increases in Reynolds number and cooling of the surface lead to the formation of a vortex ahead of the body. Real-gas, inviscid supersonic flow fields over three-dimensional configurations were computed for three Mach numbers and two altitudes [11C]. The results compared well with previous results. Hypersonic heat transfer and aerodynamic loads to a Mars aerocapture vehicle in a CO_2 atmosphere were evaluated [26C]. Contemporary low-density ablative thermal protection materials were selected. Heat transfer to simulated shuttle tiles was measured at Mach 7 and 3300°R total temperature [9C, 10C]. Sensitivity to flow angle and gap width on local and average heat flux was documented. Computational results were presented for three-dimensional nonequilibrium flow over the windward surface of the shuttle [48C]. The model accounted for the finite rate chemical reactions of multicomponent ionizing air. Three-dimensional effects were shown. Phase-change paint was used to measure heat transfer on space shuttle orbiter models with differentially deflected elevons [92C]. These deflections affected the wing heat transfer about 40% (or less) but the windward elevon heat transfer coefficient could be many times that without deflections. Flight-derived aerodynamic heat transfer data for the shuttle orbiter wing lower surface was compared with ground-based experimental data and computational results [84C]. An interference heat transfer region correlated with the region of minimum static enthalpy predicted by computation. A review of nonequilibrium flow field computations and methods was presented where the results were compared to temperature and heat flux measurements on the space shuttle [76C]. Three thermal protection systems are evaluated for protection of surfaces of future space transportation systems [77C]. They were shown to be viable up to 2300°F. An analysis of radiative heating during aeroassisted orbital transfer was made to determine the impact of the computational chemistry [23C]. The analysis showed that predictions of radiative intensity, exitation rates, reaction rate constants, particle ionization and spectral line widths could be made. The effect of chemical nonequilibrium on the flow field around an aeroassisted orbital transfer vehicle was numerically evaluated [78C]. A vehicle experiences a lower ratio of nonequilibrium/equilibrium heating than computed for the shuttle. This is due to higher velocities of the AOTV. The three-dimensional flow of a hypersonic stream around bodies was analyzed, with the inclusion of radiation, using the method of an optically transparent shock layer [33C]. As an example, radiation heat transfer was determined for a symmetric conical body. The effect of an exothermic reaction within the boundary layer on convective heat transfer was measured and computed [30C]. The effect is negligible until the wall temperature becomes sufficiently high that the exothermic reaction takes place in the boundary layer. Particle to gas heat transfer rates were calculated for an endothermic cyclohexane dehydrogenation reaction [65C]. Heat transfer rates were lower than without reaction, as one might expect for an endothermic reaction. A study was made of the flow of a multicomponent gas in a three-dimensional boundary layer near a surface where there is a heterogeneous chemical reaction [31C]. A formula for heat flow to the surface was proposed. Joint probability density functions of temperature and its dissipation are measured in the reacting zone of a plane jet [5C]. Correlation between the two is weak and their independence increases with distance from the jet centerline. LDA measurements in the flow over a wall heated by combustion were made to investigate the contributions due to a burst in low-momentum fluid from the wall [22C]. Changes in Reynolds stress due to combustion were seen in the sublayer of the combustion products and were not found where the combustion takes place. The nonequilibrium thermodynamics model given by the governing principle of dissipative processes was applied to solve for the characteristics of two-dimensional laminar, constant-property flow with nonuniform wall temperature and main-stream velocity [79C]. The theory of viscous dissipation of high Prandtl number fluid in laminar flow was reviewed [50C].

Measurements were made to show the effect of large-scale obstructions on the heat transfer and flow field of a turbulent boundary layer [18C]. An example is at the leading-edge of a gas turbine airfoil. Measurements and visualization studies were made to investigate mechanisms which cause free-stream turbulence and surface roughness to increase heat transfer in the stagnation region of turbine blades [86C]. A correlation was found between vortices caused by mean velocity gradients and spanwise variations in heat transfer. Heat transfer measurements were made on cooled, flat plates with various degrees of free-stream turbulence intensity, pressure gradient and cooling intensity [74C]. Special emphasis was placed on boundary-layer transition. It is shown that free-stream turbulence and pressure gradients are of primary importance. Measurements were made of channel heat transfer with repeating ribs as used in gas turbine cooling [36C]. Ribs oriented at oblique angles to the flow gave higher heat transfer rates and reduced pumping power. Measurements were made using the laser two-point technique of the effect of impingement of stator wakes on rotor blade heat transfer [17C]. The effect of the wake of a nozzle guide vane on rotor heat transfer under transonic conditions was measured [7C]. The interaction of the rotor and the wakes results in variations in the incidence angle. The shock waves from the moving wake have a major effect on instantaneous heat transfer rates on the rotor surface. Measurements of local heat transfer coefficients were obtained on a modern gas turbine blade with and without film cooling injection using a shock tube [25C]. The presence of an upstream rotor was shown to have a large effect on downstream heat transfer but little effect in the leading edge and midchord locations. Experimental results for transition on gas turbine blades were reviewed [68C]. The effect of free-stream disturbance level was discussed. Data sets, which include transition, were reviewed to extract information about the transition process [29C]. This information may be useful in turbine applications. It shows that free-stream turbulence and pressure gradient have strong and opposite effects on both the location of the start of transition and on the transition length. Measurements were made of heat transfer in boundary layers undergoing transition [91C]. A late-laminar state was documented. Turbulent Prandtl number in late-transitional flow was shown to be higher than that measured in fully turbulent flows. The flow on the leading edge of a turbine blade was experimentally investigated and the effects of Reynolds number, compressibility, incidence and free-stream turbulence were discussed [38C]. The flow is extremely complex even at design conditions, showing separation and reattachment and transition and relaminarization. The effect of heating the surface on transition and the growth of instabilities was studied—the rise in critical Reynolds number with heating was given [47C]. The effect of sinusoidal flow pulsations on heat transfer from a cylinder in cross-flow was measured for pulsation amplitudes as large as 25% and frequencies above and below the shedding frequency [3C]. Calculations of gas turbine blade heat transfer coefficients were made using a low Reynolds number k–ε turbulence model [71C]. Good agreement was found, even in the complex flow on the pressure surface which is transitional over most of the length. An exact solution was presented for heat transfer from an impulsively started plate cooled by transpiration cooling [70C]. Injection was shown to amplify the temperature and momentum boundary-layer growth. Computational predictions were made of the external convective heat transfer on gas turbine blades using a boundary layer program and the k–ε turbulence model [90C]. Free-stream turbulence and transition were included in the model. They agreed with measurements on the pressure side but discrepancies were noted between computation and experiments on the prediction of transition.

A method of multiple scales was proposed to determine the partial differential equations which describe the modulation in amplitude of three-dimensional waves propagating into two-dimensional heated liquid layers [8C]. A new criterion for the most likely frequency was presented. Measurements were made of velocity and temperature fluctuations in the wall region of fully-developed, unstably-stratified flow between horizontal parallel plates [28C]. The stratification effect on turbulent intensity, skewness and flatness factors and bursting speed were presented. Experimental results were presented for the effect of destabilizing heating on Goertler vortex instability in a laminar boundary layer on a concave surface [44C]. When nonlinear effects become important, the vortices become distorted and meander in the lateral direction. The convective character of inviscid instabilities in parallel shear flows was determined by examining branch-point singularities of the dispersion relation for complex frequencies and wave numbers [41C]. A flow is convectively unstable when the branch-point singularities are in the lower half complex-frequency plane. Spacially growing waves can only be observed if the mixing layer is convectively unstable, otherwise, the instability develops temporally. Businger's universal functions were used to compute the friction velocity and the Monin–Obukhov length in unstable boundary layers [75C]. It was shown that the functions behave well outside the normal range. Flow visualization was used to study the flow resulting when a potential vortex rotates normal to a stationary horizontal disc [66C]. Vorticity removes the singularity on the vortex axis leading to the development of a three-dimensional boundary layer. The boundary layer transcends to turbulence then relaminarizes with increasing Reynolds number. Heat transfer from an isothermal cylinder to time-independent cross-flow with fluctuations larger than the mean flow was analyzed and an example with sinusoidal flow was presented [13C]. Results were presented of an experimental, theoretical investigation of the upstream-side heat transfer from a sphere or transversely oriented cylinder in a flow which experiences a step change in temperature [80C]. Mass transfer in converging or diverging channels of equal or different wall concentrations was analyzed [12C]. The analysis involved an infinite series of eigenfunctions. A two-region model was used to analyze the heat transfer problem of laminar flow of a low Prandtl number fluid passing a wedge [21C]. An asymptotic low *Pr* expression for local Nusselt number was given. An approximate analytical solution was presented for the transient problem of a constant temperature plate moving at constant velocity initially at rest and in thermal equilibrium with the surrounding flow [14C]. The convergence of the method was discussed. Results of an experimental investigation of very low Reynolds number forced convective heat transfer to a sphere were presented [93C]. The new experimental technique used for these measurements was discussed. The effect of interface waviness, due to instability, on mass transfer rates was investigated for the subcritical and critical flow regimes [51C]. A convection analysis of a negative pressure gradient laminar boundary layer with suction or blowing was made [49C]. An analysis of laminar flows over surfaces of roughness length less than the boundary-layer thickness was presented [19C].

Heat transfer of a micropolar fluid past a moving semi-infinite porous plate was computed [83C]. The effects of suction and microrotation were discussed. The convective stability of a horizontal layer of micropolar fluid heated from below and rotating about a vertical axis were analyzed [16C]. Rotation has a destabilizing effect, contrary to earlier assertions, and, in the absence of rotation, microinertia has a destabilizing effect. A set of equations are found and solved for the two-dimensional flow of an incompressible micropolar fluid near a stagnation point [15C]. Viscous dissipation effects were included. Heat transfer from a linearly stretching, continuous surface with a power law temperature distribution was solved to give closed-form solutions for several specified conditions [34C]. A boundary-layer approach was applied to evaluate the heat transfer coefficient when a viscoelastic fluid flows past a cylinder [73C]. It is shown that the heat transfer coefficient becomes independent of Reynolds number for large values of Reynolds number. Analytical solutions were provided for Stokes flow fields induced by instantaneous thermal sources in a liquid field and at the center of a solid spherical surface [2C]. The evolution of the flow field was presented. Experimental and theoretical results were given for slow translation of a hot sphere through a fluid at rest [4C]. Separation of the thermal layer ceases when the drag is controlled by the most viscous fluid. The problem of dissipative heating of a fluid under a solid, isothermal plate was considered [54C]. A new analytic technique applicable to the outer region was presented for calculating the momentum and heat transfer in flat plate and circular cylinder turbulent boundary layers [59C]. A new version of the k–W model was proposed for computation of a wide range of fluid dynamic and heat transfer problems [42C]. The main advantage over previous models is the removal of modifications to near-wall constants. Measurements were made of fluctuating velocity and temperature in a turbulent plane jet mixing layer [20C]. Where Reynolds stresses and turbulent heat flux were high, turbulent Prandtl number was 0.4, indicating strong coherence in the flow. The effect of isolated surface projections on heat transfer from a plate to a turbulent flow was measured [95C]. Zones of steep decrease in heat transfer and increase in skin friction were identified. The surface rejuvenation model was applied to the field of mass transfer [53C]. Theoretical expressions for Sherwood and Stanton numbers compared well with experiments. Heat transfer enhancement in turbulent flow on a plate due to edge effect secondary flow was measured [61C]. The effect was always present and depended mostly on the plate thickness. Computations were made to study the effect of outer flow turbulence on turbulent transport using the two-parameter turbulence model [94C]. The results were compared with experimental data. A similarity solution was given for the local Nusselt number on a flate plate with uniform heat flux and a small Prandtl number fluid [27C]. A heton, two discrete baroclinic geostrophic vortices in a two-layer system, was defined [39C]. Hetons transport heat, repel similar hetons (i.e. hot repels hot) and attract dissimilar hetons. A two-heton engine exhibits vortex splitting, loss of potential energy and meridional heat transport. Measurements were taken of the heat transfer from a circular cylinder which is normal to the flow and parallel and near a planar surface [1C]. The effect of separation distance was discussed. Average heat transfer rates were measured for forced convection air flow over inclined and yawed rectangular plates [60C]. It was found that the average heat transfer coefficient is insensitive to aspect ratio but does vary with the angle of attack and Reynolds number. Correlations were given. Experiments showing heat transfer on a disk which is perpendicular to the flow showed a minimum coefficient at the stagnation point increasing to twice the stagnation value at the edge [82C]. Data collapsed when scaled on the root of the Reynolds number. Optimum nozzle-to-plate spacing for heat transfer to a plate with a normal impinging jet was experimentally evaluated [45C]. Maximum stagnation point heat transfer was found when the plate was at the end of the potential core of the jet. A generalized model for predicting optimal spacing was presented. A computational model was presented for predicting the flow and heat transfer of a single semi-confined turbulent jet on an impermeable isothermal surface [67C]. Effects of jet Reynolds number, nozzle-to-surface spacing, and jet turbulence level were discussed. Comparisons with data were favorable; but, further refinement of the turbulence closure model is needed. The problem of a lighter fluid impinging upon a heavier, otherwise quiescent, fluid was analyzed [89C]. The convective heat transfer coefficient was found. The effect of viscous dissipation on mixed convection flow about a rotating sphere was analyzed [69C]. A method was presented for calculating the hydrodynamics and scalar transport at the end-wall of a vortex chamber [87C]. The parameters varied were boundary-layer thickness, local friction, heat transfer rate and gas flow rate.

Laminar flow heat transfer in tube assemblies was computed where the assemblies were inclined slightly to the flow [6C]. The effect of the transverse component is strong and nearly-axial flow shows much higher heat transfer than purely-axial flow. Measurements were made to document the effect of surface roughness on local and average heat transfer coefficients for tube banks in cross flow [97C]. General relations for prediction of local heat transfer, local skin friction and efficiency were presented. Direct measurements of turbulent Reynolds analogy factors were made for 15 nonplanar surface microgeometries including riblets, wavy walls, etc. [52C]. Increases in this term are of the value in optimizing heat exchanger design. Measurements were made of flat plate heat transfer to turbulent flow when there is a cylinder parallel to the plate and perpendicular to the flow but imbedded in the boundary layer [55C]. An optimum separation distance was found. Heat transfer augmentation was due to the turbulence in the lower shear layer of the cylinder wake. Heat transfer from a flat plate, influenced by an array of cylinders normal to the flow and parallel to the plate, was measured [46C]. Optimum values of position and spacings were made. Comparisons were made with the case of a single cylinder inserted in the boundary layer. Convection and radiation heat transfer rates from fixed, rotary metallic and brick type regenerators were computed [43C]. The nature of the variation of heat transfer coefficient was important in evaluating average parameters.

Results were given of research on the behavior of liquid films on heat transfer surfaces as in condensers and film evaporators [63C]. Incentive is provided by a need to strip the water film from the wall of a steam main prior to sampling for steam quality. Experiments were made to measure the thermal effect of burst-type fission gas release [35C]. The surface upon which the gas impinges showed extreme temperature rises. With gas injection, the liquid sodium heat transfer coefficient reduced to as low as 5% of that with single-phase flow. Experimental results and supporting analyses were used to study the rewetting of a hot surface with impinging water jet [24C]. The history of the process was documented with photographs and measurements of the wall temperature history. It was shown that a thin, electrically insulating coating could be used in ionic liquids [57C]. Key to its use was the measurement of the heat transfer resistance. An analytical study of the behavior of a gas in a pressurized container during the early stages of expansion was made [64C]. The outflow was choked. A key quantity, heat transfer from the cylinder wall, was included in the analysis. Combined radiative–convective heat transfer in a laminar boundary layer on a permeable plate was analyzed [72C]. The effect of scattering was presented. The convection Stefan problem with Newton's radiation condition was solved with a new technique which formulates the moving boundary problem as one with a fixed boundary [85C]. The convergence was shown to be improved over the classical series solution.

FLOW WITH SEPARATED REGIONS

Experiments of the effects of sinusoidal flow pulsations on the heat transfer from a cylinder in cross flow at $Re = 50{,}000$ showed no significant increase in the heat transfer due to the pulsations in either the wake or attached boundary-layer region [2D]. In order to examine the effect of the roughness characteristics on icing, a boundary-layer model of heat transfer from the front half of a rough circular cylinder is proposed [11D]. An experimental study has been conducted on the effect of the clearance between a circular cylinder and a plane surface on the heat transfer from the cylinder to a cross flow [1D]. The mass transfer characteristics of a subliming flat naphthalene surface on which a protruding cylinder is mounted were studied for two cylinder height to diameter ratios, 1 and 12. The cylinder height influences the local mass transfer distribution downstream of the cylinder, but has little effect on the upstream region [6D]. Local and mean heat transfer characteristics were measured on an elliptic cylinder of an axis ratio 1:3, on which particles of silica scale are uniformly distributed in order to clarify fouling effects of geothermal water scale on heat transfer [14D]. Experimental investigations on the local and average heat transfer from a square prism to an air stream were carried out in the range of subcritical Reynolds numbers and the average heat transfer coefficients at zero and 45° angles of attack were found to be as high as 40% of the well-known values of Hilpert [9D].

Correlations for heat transfer and pressure drop for in-line and staggered banks of tubes are suggested for Re from 1 to 2×10^6, Pr from 0.7 to 10^4 and a wide range of relative transverse and longitudinal pitches [21D]. The Nusselt number and pressure loss coefficient were found to be highly sensitive to the transverse pitch in a two-row array of finned tubes [19D].

Measurements of the local heat transfer coefficient for a separating water flow past a step in a rectangular channel are presented [13D]. A viscous–inviscid interaction procedure is described for predicting the heat transfer in separating flow in a rearward-facing step/asymmetric channel expansion [8D]. Results of numerical calculations for heat transfer in turbulent recirculating flow over two-dimensional, rearward-facing steps and sudden pipe expansions are presented [7D]. Theoretical results concerning the flow and heat transfer in laminar flow past a backstep show that the size of the initial boundary layer can have opposing effects on the reattachment distance, depending on whether the Reynolds number is held constant [4D]. Combined heat transfer and fluid dynamic measurements in a separated and reattaching boundary layer demonstrate that fluctuating skin friction controls the heat transfer rate near reattachment, while the conventional Reynolds analogy applies in the redeveloping boundary layer [20D].

A theoretical investigation of unsteady marginal separations in a classical, unsteady boundary layer suggest that the nonlinear properties of such separations may be largely responsible for the central features often observed experimentally in the abrupt turbulent reattachment of eddies that follow laminar separation near the leading edges of aerofoils [17D]. Arguments are presented to prove identical results in determining the most thermohydraulically efficient surface in three comparison studies: the heat flux at equal pumping power and area; the pumping power at equal heat flux and surface area; and the surface area at equal heat flux and pumping power [10D].

The average and local heat transfer at the base surface of a cylindrical cavity whose opening is perpendicular to an oncoming free stream were found experimentally to decrease sharply with increasing cavity depth [18D]. In an attempt to investigate numerically buoyancy-induced enclosure flows in which the fluid is radiatively participating, energy conservation equations are derived over a finite, not infinitesimal, control volume [16D]. Heat transfer to and from slender cylinders (filaments, fibers and wires) travelling along the axis of a vertical duct with heated walls was investigated experimentally and it was found that there is a difference between the heat transfer coefficients of monolithic filaments and of multistrand fibers [5D].

Heat transfer studies were conducted experimentally on axisymmetric rocket nose cones with separationless and separated flows using sounding rockets and the results were compared to predictions [15D]. Distributions of the temperature–pressure gradient correlations obtained from measured budgets of the longitudinal and lateral heat fluxes in the self-preserving region of a slightly heated turbulent plane jet into still air are consistent with a model which includes only the turbulence interaction contribution to the pressure fluctuations [3D]. A numerical investigation of the unsteady heat transfer to a droplet suspended in an electric field predicts there is a maximum steady-state Nusselt number equal to about 30 [12D].

HEAT TRANSFER IN POROUS MEDIA

The largest single category of papers in this section deals with saturated porous media. An extensive overview of steady solutions was published in three parts [60DP, 61DP, 62DP]. A very large number of works in this area deal with natural convection heat transfer, the widely used Boussinesq approximation for which was evaluated for its suitability [26DP]. A less restrictive variant of the Boussinesq approximation was compared with the more common assumption, and while the flow field calculations were sensitive to their differences, overall heat transfer quantities were insensitive. Similarity solutions were obtained for free convection boundary-layer flow adjacent to long vertical plate fins in porous media under conditions of power law variations of the fin conductivity–thickness product [63DP]. A numerical and experimental investigation of natural convection in saturated porous media in the annulus between vertical concentric cylinders with small aspect ratio, L/D, was reported [69DP]. As radius ratio (outside/inside) was increased, smaller aspect ratios were required to introduce multicellular flow structures. Solutions were developed for free convection flow and heat transfer from a nearly horizontal, semi-infinite, isothermal, heated surface in a saturated porous medium [73DP]. An asymptotic solution was found for positively inclined surfaces, and solutions were found near the leading edge for both positively and negatively inclined surfaces. Natural convection about a line source at the vertical interface of two saturated porous media of differing permeabilities and diffusivities was analyzed [80DP]. Natural convection from a two-dimensional horizontal line source embedded in saturated porous media has been analyzed using matched asymptotic expansions suitable for use at lower Rayleigh numbers than a previous boundary-layer solution [1DP]. Stably heated corners of saturated porous media were explored by numerical experiments and scale analysis [42DP]. Experiments performed in shallow gas–solid packed beds of a regular packing of spheres between horizontal isothermal surfaces confirmed a value of $4\pi^2$ as the critical Rayleigh number [17DP]. Linear stability analysis was used to determine the critical onset time at which convection starts in a porous layer saturated with water initially at a uniform temperature near, but above, the temperature of maximum density, 3.98°C, when the upper surface is impulsively lowered to a temperature below 3.98°C [66DP]. A numerical model was described for natural convection in saturated porous medium with simultaneous motion of the pore fluid and the matrix, which was treated as a viscous medium [19DP].

Linear stability analysis procedures employed in double diffusion were applied to explore the effect of a third diffusing component in a saturated porous medium providing general expressions for various stability boundaries [67DP]. The effects of simultaneous mass transfer upon buoyancy-driven heat and fluid flow from a concentrated source in a saturated porous medium were analyzed for low Rayleigh numbers for both transient and steady solutions [65DP]. For a porous horizontal cylinder immersed in a gas-filled porous medium, a boundary-layer analysis was carried out for the coupled free convective heat and mass transfer [29DP]. For a vertical surface, natural convection heat and mass transfer was studied through scaling arguments and similarity solutions of the boundary layer approximation to the governing equations [3DP]. Both numerical and scale analyses were presented [89DP] for combined heat and mass transfer by natural convection in finite layers of porous media.

Mixed convection was investigated for special cases of plane surfaces in media with variable permeability [8DP,9DP] yielding similarity solutions for horizontal and inclined surfaces. A nonlocal similarity method was employed to determine effects of buoyancy and wall temperature variations about nonisothermal spheres and cylinders in mixed convection circumstances in porous media [51DP]. Analysis was also performed [72DP] upon a saturated porous media flow with natural convection at a permeable, absorbing boundary and simultaneously a forced oscillatory flow parallel to but distant from that boundary.

An analysis was presented [40DP] of the temperature fields of a single-phase coolant and a porous channel through which it flows. A method was proposed for computer-aided mathematical modeling of a steam-thawing process for treating permafrost in advance of building construction [50DP]. Transient conduction and radiation in nonconvective planar porous materials was analyzed for the case of sudden increases in heat generation at one boundary [86DP], applicable to transient methods of measuring properties of porous insulation, and an analytic method was proposed for calculating unsteady heat exchange in soil applicable to energy analysis of pipeline transportation systems [56DP].

Developing flows through porous media in channels bounded by isothermal parallel plates were analyzed using a modified Darcy model [38DP]. Boundary, convective, and inertia effects were shown to become more important in natural convection in saturated media adjacent to a vertical plate as porosity increases [33DP]. These effects all reduce the heat transfer relative to the conventional solutions based on Darcy flow. Using matched asymptotic solutions for a three-layer model for natural convection at a semi-infinite vertical plate in saturated porous media the Brinkman model was incorporated [34DP]. The boundary effect became more important at larger Rayleigh numbers and higher Darcy numbers. Inertial (non-Darcy) effects at high Rayleigh number in a vertical porous layer subjected to uniform heat fluxes at the sides were numerically evaluated using the Forchheimer model [64DP]. Numerical experiments in a porous layer heated from the side were used to confirm the non-Darcy flow prediction of an analytical solution [68DP] and to illustrate the regime between Darcy flow and flow affected by inertia. Effects of inertia were demonstrated to be significant in natural convection at realistically large values of the Rayleigh number in thin layers of saturated porous media with small Prandtl number [55DP]. A Brinkman-extended Darcy model, incorporating the no-slip condition at an impermeable surface, was shown to give superior predictions of natural convection heat transfer in vertical porous enclosures for large modified Reynolds number and large aspect ratios [87DP]. Steady non-Darcy free convection was analyzed for wedges of saturated porous media bounded by a flat heated surface at an arbitrary angle from the vertical and by another surface which is either at ambient temperature or insulated [74DP].

Boundary and variable porosity effects were demonstrated to be significant in forced convection through packed beds in an experimental and numerical study [91DP]. Simple thermal resistance models were developed for use in predicting steady-state, packed-bed heat transfer characteristics [21DP]. The Blake–Kozeny model for pressure drop in a granular bed was used to develop a model for heat and mass transfer from a fluid to a particle in a granular bed [39DP]. Analytic expressions were presented for asymptotic radial heat transfer coefficients for packed bed reactors [23DP]. Comparisons were made of numerical solutions for the transient gas–solid heat transfer in packed beds using a two-phase model and a single-phase model [32DP], displaying the advantages and limiting parameters of the latter. A new model was developed [45DP] and experimentally supported [46DP] for transient thermal response of a packed bed, employing the method of spatial averaging. An analytical method was proposed for predicting transient temperature distributions in a porous bed subject to a step change in fluid inlet temperature [88DP]. It is suggested that the method could be extended to other high-order partial differential equations resulting from coupled phenomena. Measured transient temperature distributions in a high temperature pebble bed regenerative heat exchanger were shown to be well predicted by a numerical model [94DP]. Heat and mass transfer processes between gas and particles in a fixed bed dryer were numerically investigated, showing that moisture diffusion taking place inside the particles can control the drying process even at moderate Biot numbers [58DP]. Heat and mass transfer from an object immersed in a packed bed was studied, incorporating the effect of the high porosity zone near the surface of the body [6DP]. Experiments were performed and correlations presented for heat transfer and pressure drop of particulate laden gas flow through a tube packed with spheres [27DP].

Orthogonal collocation was demonstrated [59DP] to be a simple and useful method for predicting the combined heat and mass transfer process of drying of porous media. Evaporation in porous media was analyzed with a kinetic model, replacing the collision integral with an external force, the collective interaction of gas molecules with the stationary media [48DP]. A theoretical model was proposed [44DP] for drying and sorption processes in unsaturated porous media. Theoretical and experimentally determined rates of property changes were presented for the heating of a cylindrical sample of water-saturated limestone in hot air [93DP]. A two-dimensional, finite-element formulation was presented for coupled heat and mass transfer in partly saturated soils [85DP]. Modeling and experiments identified void fraction and its distribution, fiber radius, directionality, and tortuosity as properties controlling liquid diffusion in fibrous insulation material [53DP]. Another combined theoretical and experimental investigation of heat and moisture transfer in a fibrous insulant displayed consistent distributions of moisture but overestimations of heat transfer [18DP]. Experiments demonstrated that heavy porous material (brick walls) undergo strong variations of apparent thermal transmission related to slowly stabilizing moisture profiles [5DP]. Analytical evaluation of unsteady filtration in hydrofractured low-permeability rock produced a parameter which may be used as a feasibility test for extracting geothermal energy from specific geological conditions [49DP]. A mathematical model was proposed for describing heat and mass transport within a porous granule of a catalyst which, itself, assists in forming a two-phase mixture within its pores [22DP]. A descriptive review of the processes and consequences of the drying of wood, a distinctly nonisotropic porous medium, was published [82DP].

A dryout model for a particle debris bed was developed with focus on the inter-facial resistance between countercurrently flowing vapor and liquid [10DP] and shown to fit available data well. Safety analysis of liquid metal fast breeder reactors motivated the development of equations governing the transient vapor formation of a hypothetical debris bed [11DP] and those governing particle melting and penetration of the guard vessel lining of an overheated, dried out debris bed [12DP]. A study of countercurrent flooding limitations in porous media led to the conclusion that differences in dryout heat flux data from shallow and deep volume heated beds of large particle diameter and differences in bottom and volume heated beds can be attributed to the role played by coupling between the particulate bed and the overlying liquid layer [47DP]. Physical experiments simulating core debris beds with internal heat generation [41DP] likewise indicated that significant coupling exists between convective flow within the bed and an overlying liquid layer. Boiling of water in an unconstricted bed of granular material was studied visually and with temperature profile measurements [24DP]. Analysis of heat transfer and two-phase flow in porous media with porosity varying with height showed strong influence of the porosity profile upon the height of the two-phase zone and the temperature difference [15DP]. Analysis and experiments were addressed to the length of and the dryout heat flux of the two-phase zone in porous media [90DP].

The suitability of gas fluidized beds for use in heat recovery systems was explored and experiments were performed with inclined distributor plates intended to help the particles to flow through the system [31DP]. Mass transfer coefficients were measured from both freely moving and fixed spheres to a dense phase fluidized bed with results showing no dependence upon the superficial velocity, dependence instead upon the minimum fluidizing velocity, and a 20–50% increase in mass transfer rates for fixed spheres relative to freely moving spheres [70DP]. A correlation was developed for longitudinal dispersion coefficients determined by the tracer impulse response method in fluidized beds which were similar in properties to biofilm reactors [57DP]. The effects of inert particle size and density upon circulation rate and heat transfer were explored for drying equipment of various scales incorporating fluidized beds [81DP]. A mathematical model for drying of particles in an inclined, vibrated, fluidized bed was shown to predict quite well the temperature and moisture variations observed in an extensive experiment with a three-stage unit [30DP]. Vibration considerably enhanced the drying ability of the device. Design considerations for fluidized bed drying equipment were reviewed [25DP]. A computer model was developed to simulate the drying process in a rotary dryer [36DP].

A composite dense phase and lean phase model for heat transfer to tubes immersed in gas fluidized beds has been demonstrated to fit extensive experimental data [7DP]. A relationship was demonstrated between contact resistance models and a property boundary-layer model of transient dense-phase heat transfer to surfaces in fluidized beds [43DP]. A boundary-layer model of heat transfer between fluidized beds and immersed surfaces was presented [13DP], and estimates of maximum Nusselt numbers thereby developed agreed reasonably with data from the literature. Mass transfer from an immersed surface to a fluidized bed, previously considered to bear little similarity to heat transfer, was shown to have strong similarity to heat transfer when particles have large specific mass capacity [16DP]. A modified packet model was shown to agree with data from beds of large and small specific mass capacity. A mechanistic model was proposed for heat transfer between a slugging fluidized bed and a vertical immersed surface, more suitable for long vertical surfaces than previous models [2DP]. It gave good agreement with time-averaged data. Experiments on heat transfer from V-thread finned tubes to gas fluidized beds and analysis of heat conduction in the fins led to correlations applicable to single finned tubes and finned tube bundles [28DP]. An experimental technique involving batch-wise drying of wet porous particles in fluidized beds of identical dry particles provided measures of gas–particle and particle–particle heat transfer rates in large particle systems [20DP]. The gas–particle heat transfer rates measured were rather insensitive to fluidizing velocity. Heat transfer coefficients and local voidage on immersed cylindrical surfaces in a tube bundle were measured in a large particle pressurized fluidized bed [4DP], displaying a large effect of the pressure upon the heat transfer coefficients. An extensive comparison of data with models and correlations for heat transfer coefficients on immersed surfaces in pressurized fluidized beds produced recommendations of preferred predictions for three distinct regimes of fluidization segregated by bounding Reynolds and Archimedes numbers [78DP]. Heat transfer coefficients on a vertical tube in fluidized beds of small particles at elevated temperatures were found to increase with increased bed temperature but were quite insensitive to a change in bed diameter from 25 to 15 cm [79DP]. Experiments of heat transfer from horizontal cylinders at various heights in liquid fluidized beds of glass beads revealed the concentration profiles of segregated regions of different sized particles [35DP].

Using a continuum model to simulate fluidized bed hydrodynamics, predicted heat transfer coefficients compared favorably with a previously performed experiment thereby lending credence to the hydrodynamic simulation [84DP]. Experiments with a fluidized bed furnace revealed considerable axial and radial temperature gradients at low fluidizing velocities [92DP]. Heat transfer coefficients from a fluidized bed to a porous distributor plate were experimentally measured, correlated, and found to vary in a similar way as bed-wall coefficients of heat transfer [95DP].

A comprehensive review article was published covering the status of understanding of fundamentals of gas–liquid–solid fluidization with recommendations of areas for further research [54DP]. Experiments upon wall-to-bed heat transfer in three-phase fluidized beds led to a generalized correlation equation for heat transfer coefficients in liquid–solid and in three-phase fluidized beds, based upon the equality of the product of the heat transfer coefficient and liquid holdup in three-phase beds with the corresponding product for liquid–solid beds [14DP]. A semi-theoretical correlation was proposed for heat transfer coefficients in three-phase fluidized beds based on a correlation for bubble column wall heat transfer and constants were evaluated to fit data from existing literature [83DP]. Heat transfer coefficients were measured at the wall of a three-phase fluidized bed with upward co-current flow of liquid and gas with the liquid at low velocities [77DP]. Previous correlations inadequately predicted these coefficients and a new correlation was proposed.

Other fluid–particle systems involving the transfer of heat are here loosely combined. A variational method was used to predict the time development of coating thickness in fluidized bed coating of a thin metallic substrate [71DP]. A theoretical model was developed and a simple design chart presented for analysis of heat transfer and pressure drop in vibro-fluidized beds [75DP]. Small glass and aluminum particles suspended in gas streams were reported to reduce heat transfer coefficients inside vertical and horizontal tubes except at small loading ratios and moderately low (11,000 to 13,000) Reynolds numbers [37DP]. Experiments and analysis were reported regarding a raining packed bed heat exchanger [76DP]. A numerical solution confirmed that laser-induced thermophoresis improves the efficiency of deposition of aerosol particles on a surface [52DP].

EXPERIMENTAL TECHNIQUES AND INSTRUMENTATION

A review of several new experimental methods and diagnostic techniques for heat transfer research was presented by the faculty of the A. V. Luikov Heat and Mass Transfer Institute of the Byelorussian Academy of Sciences [5E]. Pitfalls in the correlation of engineering data by the use of dimensionless groups were discussed [54E].

One avenue of improvement in heat transfer instrumentation lies in the application of thin-film coatings. In experiments where a uniform wall heat flux is desired, but where the heat transfer coefficient is low, the use of an electrically heated, thin, gold-coated plastic sheet appears to provide a method for providing the flux without having wall conduction overwhelm the results [3E]. Another advantage of thin films is in their improved dynamic response; a gold thin-film resistance thermometer plated onto a non-conducting substrate was developed for heat flux measurements at a moving surface subjected to rapidly varying heat flux; measurements were reported with a resolution of about 0.002 K at sampling rates up to 5 kHz [50E]. A heat flux sensor/thermocouple was constructed by sputtering an electrically insulating, thin-film coating over most of the instrument surface, leaving bare metal exposed only at locations where temperature measurement was desired; minimal disturbance of boundary-layer temperature and wall heat flux was claimed for measurements over a flat plate tested to temperatures up to 911 K [31E]. Thin films of fired metal paints were used as resistance thermometers for measuring surface temperatures to a reported accuracy of $\pm$ 0.023 K [44E]. A coaxial thin-film thermocouple was modeled for measuring the time-varying temperature on the surface of the combustion chamber wall of an internal combustion engine [11E]. Results were presented of heating rates in hypersonic wind tunnels in which it was claimed that the principal uncertainty associated with the use of thin-film gages on MACOR substrates had been resolved (namely, the thermal properties of MACOR) [36E].

Important advances are occurring in the application of optical techniques for temperature measurement, especially for high-temperature environments. A planar laser-induced fluorescence method was developed for obtaining measurements of the two-dimensional temperature field in a combustion flow [46E]. A technique utilizing two optical multichannel analyzers detecting light scattered by two separate mechanisms (e.g. Rayleigh and Raman) makes possible the simultaneous two-dimensional mapping of both temperature and species concentrations in turbulent flames [33E]. An optoacoustic laser beam-deflection technique for measuring flame temperatures at a sampling rate >1 kHz was reported [16E]. A two-color optical pyrometer was developed for measuring the temperature of individual particles (typically 50 μm in diameter) over the temperature range 1200–2500°C, with an accuracy of ± 50 K [23E]. The simultaneous measurement of three quantities—the temperature, size and velocity of individual carbon particles of a burning pulverized fuel—was achieved by an optical method [39E]. The calibration of an optical pyrometer against a tungsten strip lamp at a maximum temperature of 2300°C was extended to 3100°C using a linear extrapolation accurate to 0.65% [21E]. The theory and application of the laser shadow technique was discussed for the case of axisymmetric turbulent flow [14E]; it was shown that three-dimensional spatial resolution could be obtained by use of the Abel transform method.

The application of liquid crystals for temperature measurements was reviewed [15E]. A measurement system consisting of a composite sheet containing a heater element and liquid crystals was reported to provide high-resolution measurements of local heat transfer on a large-scale model turbine blade airfoil [20E]. An intriguing method was reported whereby the velocity and temperature of a thermochromatic fluid are measured simultaneously; the velocity is measured by laser Doppler anemometry while the temperature is measured using the otherwise-discarded intensity of the scattered light [45E]. The temperature can be observed directly by the fluid color change or can be quantitatively measured by a suitable optical detection method. An improved flow visualization technique utilizing photochromic dye and an inexpensive pulse laser was described [22E].

Several improvements in schlieren techniques were reported; a review of color-coding schlieren visualization was presented [47E]. A study of schlieren photography in a water tunnel showed this method to be a viable means of visualization in water flows [12E]. A method was developed to obtain from schlieren fringe patterns information about the propagation of a gravity wave in a density-stratified liquid [41E].

Development of hot-wire anemometry as a tool for heat transfer research continues to receive considerable attention. The static and dynamic response of a hot-wire were examined analytically [10E], and the transient response of the Nusselt number to variations in the Reynolds number over a range from 1 to 40 was investigated in a numerical simulation of a hot-wire [8E]. Hot-wire dynamic behavior was studied in experiments, in which it was found that an alteration in dynamic response can occur at frequencies around 1 Hz [27E]. The influence on hot-wire measurements of the wire diameter and overheat ratio of a wire placed near a conducting wall in a turbulent boundary layer was studied [26E]. Expressions for the response of an inclined hot-wire to an imposed flow were extended to three-dimensional flows of moderate turbulence intensity [42E]. Calibration procedures for inclined hot-wire probes were investigated [7E]; the results show significantly different yaw correction factors depending on the procedure chosen. Yaw, pitch and velocity parameters were derived as functions of the effective cooling velocity of triple hot wires in unsteady flows with large direction changes [35E]. Directional sensitivities of quartz-coated and alumina cylindrical hot-film anemometers were investigated in air and liquid Refrigerant-113 flows [43E]. A simplified hot-wire calibration procedure was described based on the assumption of linear dependence of the fluid temperature on the square of the anemometer voltage [6E]. An experimental investigation of oscillatory phenomena produced by a hot wire located near and below a free surface was reported [53E]. Finally, an alternative whirling hot-wire system was described, including optical data transmission by an LED [52E].

Pyroelectric anemometers received some attention, including one study of geometric considerations in their operation [13E] and another which presented a heat transfer model of their response [19E].

Several new methods for determining thermal properties were reported, including a photoacoustic-phase method for simultaneously determining thermal conductivity and specific heat of a solid [28E]; a digitally controlled heater power supply for measurement of specific heat at cryogenic temperatures [29E]; a calorimeter with mixing-flow cells of aneroid type for determining heat capacity at high pressures and temperatures [48E]; and utilization of the thermal strain of solids to measure the thermal conductivity and thermal diffusivity of solid-phase polymers [2E]. An overview was presented of a workshop at the 9th European Thermophysical Properties Conference, which considered the "effect of radiation on thermal transport measurements" [37E].

Correlations were presented for the response time of chromel–alumel thermocouples with different bead sizes in flowing hot air; the response time was found to be unaffected by the air temperature [40E]. A theoretical study of the dynamic response of thin-wire resistance thermometers found significant nonlinear effects in the measurement of temperature fluctuations under certain velocity conditions [30E]. Reported developments of new heat flux sensors for specific applications include calorimetric loops for determination of thermal loads of steam-generating waterwall tubes [24E] and a temperature-controlled, multiple-chamber indirect calorimeter for measuring heat transfer between animals and the environment [38E]. A monotone-wall calorimeter was developed in which the heat generated by a body is determined by measuring the temperature difference between that body and a reference body which is in the same temperature environment; the detector is shielded from surrounding temperature effects [34E]. A precision *ac* resistance bridge suitable for thermometry was described [9E]. Systems were described for measuring temperatures in optical-fiber preform fabrication [4E] and in liquid metal flows, the latter using a permanent magnet probe [51E]. A rod-like temperature probe for studying heat transfer in flowing polymer melts was examined theoretically [49E]. The error in measuring the heat flux from a furnace wall due to perturbation of the flux by a heat flux transducer was calculated [1E]. The temperature-control system designed for the calorimetric experiment planned for the Spacelab D1 mission was described; the purpose of this experiment is to measure the singularity in specific heat near the gas–liquid critical point under zero-gravity conditions [17E].

Finally, new instrumentation for cryogenic heat transfer studies was reported, including a semiconductor resistance thermometer [32E], rotating cryostats [18E], and a device for investigating heat transfer of a cryogenic liquid under centrifugal forces [25E].

NATURAL CONVECTION—INTERNAL FLOWS

Buoyancy-driven flows in enclosures continue to gain the attention of a number of researchers spanning the disciplines from pure and applied mathematics, to physics, to engineering science, including those interested in nonlinear phenomena and turbulence. Applications include manufacturing, cooling of electronic components, and geophysical and astrophysical flows. As has been true for some time, there is considerable activity in nonlinear phenomena for convection within horizontal layers heated from below (Rayleigh–Bénard or Bénard convection), differential heating of vertical and inclined layers, the influence of internal energy sources on flow, convection in annuli, the influence of surface tension on flows, convection in porous media and mixed convection. A number of studies considered the instability and low Rayleigh number laminar flow in a horizontal layer heated from below. The onset of instability is advanced with temperature fluctuations on the boundary [106F]. Another study [54F] considers time-varying as well as random fluctuations in the bottom surface temperature and their influence on the onset of flow. A study predicts the influence of spatially varying temperature on the lower wall of a horizontal layer of fluid containing two components [11F]. Conditions under which the principle of exchange of stability holds for a viscoelastic fluid layer heated from below were described [96F]. Linear stability theory has been used to predict the instability in a horizontal layer with suction through the boundaries of the container [92F]. Hexagonal cells are found to be preferred for the onset of flow in a horizontal layer of fluid with temperature-dependent properties [57F]. With stress-free boundaries, the maximum Rayleigh number for steady rolls is found to be relatively low [16F]. Three-dimensional numerical solutions for a horizontal layer with stress-free boundaries and a square planform show the heat transport is strongly dependent on Prandtl number [7F]. Square cells are the preferred form of convection in a horizontal layer, with finite conducting boundaries for a very limited range of conditions [80F]. With large variations in viscosity across the layer the transition from square cells to rolls has been predicted [22F]. Another analytical study [39F] of convection in a layer of fluid whose viscosity is strongly dependent on temperature has been reported. The onset of oscillations in a low Prandtl number fluid have been examined analytically to predict instability in superfluid mixtures of helium [35F].

A prediction of the transition to chaos in a fluid layer heated from below has been described [38F]. Another study [64F] on the transition to chaos uses Bénard convection as the example in the analysis.

Large-eddy-scale simulation has been used to study turbulent Rayleigh–Bénard convection up to modest Rayleigh number [36F]. Long-term changes in the temperature of the boundaries with Bénard convection have been analyzed using a nonlinear diffusion equation [40F]. Penetrative convection in water near 4°C is studied assuming the presence of long convection rolls with a constant heat flux boundary condition [81F]. A fluid in which thermal waves travel with finite speed (second sound) has been analyzed to predict Bénard instability [62F].

Experiments on Rayleigh–Bénard flow include the use of a specially designed heat flux meter [32F]. Oscillations in low temperature helium have been studied using a very sensitive thermometer [60F]. Apparatus for precision measurement of Rayleigh–Bénard convection in a cylindrical container include the use of a computer-enhanced shadowgraph imaging system [97F]. Turbulence has been observed at low Rayleigh number in a narrow vertical cylinder heated from below [1F].

Oscillatory instability is predicted when heating a horizontal layer of fluid on which is superimposed another layer of immiscible fluid [77F]. In a related study, a perturbation analysis is used to examine the oscillatory onset of flow [78F]. Experiments performed by heating a layer of liquid water overlaid with ice show different patterns of interactions at the solid–liquid interface [34F].

Several papers consider fluids with internal energy sources in horizontal, vertical or inclined layers. Modeling [5F] of turbulent natural convection in such a horizontal layer indicates different regimes (layers) in terms of temperature profiles. The critical Rayleigh number has been found for a horizontal layer with a linear variation in internal energy source [79F]. In a vertical layer with internal energy sources, instability appears in the form of traveling waves [98F]. An analysis for nonuniform internal energy source predicts the stability that might occur with radiation absorption in a liquid in a vertical slot [88F]. Another paper considering radiation absorption in a vertical slot presents stability and flow calculations [43F]. For an inclined square box containing internal energy sources and also differentially heated, the temperature and velocity fields have been calculated [3F]. A later paper considers the case of internal energy sources and flow if the enclosure is cooled from below [2F].

The breakdown of the conduction region to convection in a differentially heated, vertical slot appears to be quite different at different Prandtl numbers [31F]. A related analysis indicates elongated horizontal rolls in a nonuniformly heated, vertical layer [30F]. The analysis of the flow in a differentially heated, vertical layer for non-Newtonian fluids has been presented [76F]. The laminar convection between vertical plates with fixed outer temperature indicates that the inner surface temperature and the flow are dependent on the relative conductivity of the plates to that of the fluid [21F]. A numerical study for large Rayleigh numbers with a low Prandtl number fluid in a differentially heated, vertical layer has been described [89F]. In experiments on differential heating of a variable-viscosity fluid in a vertical slot, a shadowgraph is used to detect the onset of instability [27F]. The flow of supercritical helium in a heated vertical slot has been analyzed [33F]. The transient temperature distribution in a vertical layer with suddenly heated sidewalls has been described [47F].

Experiments using differential heating across a vertical annulus have been performed over a large range of Rayleigh and Prandtl numbers; turbulence is initiated at a fixed value of the local Grashof number on the inner wall [73F]. Experiments show progressive waves of instability for convection in a differentially heated, vertical annulus [104F].

A number of studies considered square or rectangular enclosures. Numerical finite-difference solution indicates the effect of variable properties on the temperature, velocity distributions and heat transfer in a differentially heated, square cavity [111F]. A numerical and experimental study shows the effect of finite conductivity in walls of a differentially heated, square enclosure [53F]. Convection in a square enclosure when one wall is suddenly cooled, has been studied experimentally and numerically [67F].

Transient heating causes oscillatory flow in a low Prandtl number fluid contained in a cylindrical enclosure [46F]. Stratification develops quite differently with a turbulent boundary layer as compared to a laminar boundary layer in an enclosed heated fluid layer [105F].

Convection in a corner containing a hot vertical wall and a cold horizontal wall yields a single cell over a range of Rayleigh number [55F]. The analysis of flow in a rectangular enclosure when one of the vertical walls is held at one temperature in its upper half and a different temperature in the lower half has been described [71F]. Finite-difference methods for enclosures have been extended to high Rayleigh number; a more refined turbulence model is required at high Rayleigh number [70F]. The velocity distribution in a differentially heated rectangular enclosure has been determined with a laser-Doppler velocimeter [19F].

Several studies consider flow across a differentially heated, inclined layer. One numerical and experimental paper presents results for low Rayleigh number flows when air is the working fluid [85F]. The flow in an inclined layer from a cell immersed in a cooling liquid has been presented [61F]. Experimental and theoretical results are presented for flow when one end-wall of an inclined layer is heated [56F]. Convection in an inclined layer is much weaker with the presence of a partition [4F]. Analysis of the flow in a differentially heated, inclined layer of porous material shows three potential flow structures: two-dimensional unicellular, polyhedral convection cells, and longitudinal coils [23F].

Applications of natural convection occur in the flow near a crystal–melt interface [37F]. Reduced flow occurs in a vertical circular duct when internal fins are inserted, but there is a net increase in heat transfer [72F]. Experiments on convection in a vertical rod bundle indicate decreased Nusselt number for a higher Prandlt number fluid at a fixed Rayleigh number [52F]. The influence of various geometric parameters on convective heat transfer in enclosures with localized heat sources have been experimentally studied to simulate cooling of electronic circuit boards [42F]. Numerical modeling of three-dimensional laminar flow in tanks indicates a circular flow pattern when there is a localized heat source [94F]. A lumped-parameter model analysis is used to predict thermal response of a liquified petroleum gas container exposed to a fire [95F]. Spacing and offset within an array of columns of horizontal cylinders have a significant influence on the overall heat transfer [82F].

The convection in the annulus between two horizontal cylinders has been the subject of a number of investigations. One analysis [6F] includes the effect of internal energy sources and their influence on Nusselt number. Another [44F] includes the influence of variable properties on the differentially heated annulus with large radius ratio; for a radius ratio greater than 10, the Nusselt number appears to be independent of the outer cylinder. A transformation maps the annulus for both concentric and eccentric cylinders into a rectangular region; numerical determination of the flow in this region provides local heat transfer and temperature fields over a range of Prandtl numbers [75F]. An experimental study [87F] indicates that even with a radius ratio of 11.4, the heat transfer from the inner cylinder is not the same as when it is immersed in an infinite medium. Analysis for laminar convection in the region between confocal elliptical cylinders has been performed with different orientations of the line between the two foci [86F]. Three-dimensional flow in a horizontal annulus has been studied for the case of finite cylinders differentially heated on the plane (end) walls [90F].

In a differentially heated spherical annulus, the streamlines are in the form of pairs of double helixes for low Rayleigh number conditions [69F]. Another study on flow in a spherical shell has been performed with stress-free boundaries and a radial gravitational vector [109F]. Experiments for convection between different bodies and external cubical enclosure find a general Nusselt–Rayleigh number correlation [102F]. A design equation has been provided to predict the heat transfer across enclosures of various geometries [10F]. Experiments on natural convection in a horizontal cylindrical cavity included effects of a transient in the wall temperature [28F].

Partially opened enclosures—really open cavities—are important in a number of areas. Laminar convection in open horizontal cavities has been analyzed for small aspect ratios and aspect ratios of unity [25F]. A related study considers different thermal boundary conditions [26F]. Correlation of experimental results in open-ended inclined channels has been provided [9F] for differential heating across the walls. Electrochemical systems have been used to study an open cylindrical cavity at high Rayleigh number [91F].

Flow in porous media is often driven by buoyancy forces. Sidewall heat loss has a strong stabilizing effect on the initiation of convection in a vertical porous slab [51F]. In a saturated porous layer heated from below, three-dimensional flow appears in the form of a square cell [17F]. Experiments on a vertical annulus of saturated porous media indicate the need for a simple model to estimate the effective thermal conductivity of the fluid filled media in a correlation of the results [74F].

Double-diffusive convection occurs when there are two diffusive flows, often one of heat driven by a temperature difference, the other of matter driven by a concentration difference, one example being thermohaline convection. The effect of variable stratification on double-diffusive flow in a horizontal layer has been examined with reference to stability of solar ponds for collecting solar energy [108F]. Thermohaline convection has been studied [8F] to determine the flow in large bodies of water, such as lakes or lagoons. The instability of three-dimensional salt fingers is found to be similar to that encountered with two-dimensional salt fingers [45F]. A differentially heated vertical plane layer has been studied with a stable temperature and unstable salinity gradient to determine the breakdown of steady flow [107F]. Models developed to predict the transient response of salt-stratified, double-diffusive systems demonstrate good agreement with existing data in terms of the mixed layer thickness and temperature distribution [13F]. The propagation of undamped transverse thermal waves in binary mixture have been examined [14F]. Experiments were performed [50F] in a low aspect ratio layer having double-diffusive convection produced by differential heating and an electrochemical reaction.

Surface tension or thermocapillary or Marangoni flow occurs when there is a variation of surface tension or surface free-energy at the interface between two fluids. Marangoni flow near the interface of two liquids has been studied [20F]. The effect of local heating from above has been studied to examine thermocapillary flow in a heavy liquid [100F]. The influence of a deformable upper surface on Marangoni flow has been analyzed [24F]. A steady thermocapillary flow in a square cavity with a free upper surface has been examined in the absence of gravitational forces [110F]. The importance of the convective contribution to heat transport in a liquid column in the absence of gravitational forces has been predicted [12F]. A low gravitation experiment in a rocket shows the importance of thermocapillary convection [59F]. In another study [62F] in a low-gravity environment unsteady Marangoni flow was measured. Analysis of surface tension driven flow has been used to predict performance with electron beam welding [103F]. Rotation can be used to inhibit the instability in a thermocapillary driven flow [84F]. Critical Marangoni numbers have been calculated when surface tension gradients on the top surface of a liquid layer drive the flow [41F].

Mixed convection occurs when buoyancy-driven flows and forced flows are superimposed upon one another. In many geometries this leads to complex three-dimensional patterns even when the flow is laminar. Studies for such flows in the entrance region between horizontal parallel plates include a numerical analysis [65F]. In the transitional and turbulent flow regime, quite different effects are found on the top and bottom plates [68F]. Numerical simulation for the entrance region of a horizontal rectangular duct indicates that the heat transfer can be enhanced by up to 400% when the bottom wall is heated [48F]. The possibility of bifurcation into two different solutions with different vortex flow patterns has been indicated for mixed convection in a horizontal rectangular duct [66F]. A three-dimensional numerical model confirms experimental results for laminar mixed convection in a horizontal heated tube [29F]. In horizontal annuli, constant temperature and constant heat flux boundary conditions have been used in a finite-difference solution [58F]; there is good agreement with experimental results. The stability of mixed convection in a vertical slot with porous walls and cross flow has been analyzed [93F]. Experimental data provide information on up flow and down flow of supercritical helium in vertical tubes where mixed convection is of importance [15F]. Buoyancy can have a significant effect on fully developed laminar flow and heat transfer for flow through shrouded arrays of blocks that simulate electronic devices on a circuit board [18F]. The principle of exchange of stabilities is found valid for some boundary conditions even with a nonlinear temperature profile in the presence of a shear flow [49F]. The onset of instability for a nonuniformly heated horizontal layer is more sudden when there is a weak shear flow imposed on it [101F]. Thermocapillary effects in a thin moving liquid layer with local heating from above can be quite large [83F].

NATURAL CONVECTION—EXTERNAL FLOWS

The vertical flat plate geometry continues to receive considerable attention. A similarity boundary-layer solution for plates of prescribed variable surface temperature or heat flux was presented [32FF]. Approximate solutions to the Navier–Stokes and energy equations were obtained for small Grashof number but large Prandtl number [29FF,52FF]. Experiments in water near 4°C and the use of asymptotic expansion theory were used to show the care which must be used when employing the Boussinesq approximation [16FF]. Experiments simulating the melting of ice were conducted using a salt-stratified fluid [6FF]. Similarity problems were reviewed for a plate of constant wall temperature or uniform heat flux immersed into a stratified fluid [50FF]. Experimental interferograms and finite-difference numerical solutions of natural convection from a heated plate partially in air, partially in water were presented [42FF,43FF]. Reference [17FF] gives numerical results which show the effect of variable fluid properties on the heat transfer. Solutions to the boundary-layer equations were obtained using variable fluid properties near the critical point of CO_2 [58FF]. A review of existing data and heat transfer correlations was made for isothermal flat plates near the critical point of CO_2 or H_2O [12FF]. Two extensive studies on turbulent flow were published. In ref. [51FF], variable property effects were studied in turbulent flow using air up to a Grashof number of 2×10^{12}. Reference [27FF] presented results for turbulent flow in water adjacent to a uniform flux plate. Hydrogen bubbles were used for flow visualization and liquid crystals indicated the local plate temperature up to a modified Rayleigh number of 6×10^{16}. A combination of matched asymptotic expansion and finite-difference technique was used to solve the boundary-layer equations for a step change in plate temperature at low Prandtl number [41FF]. Similarity solutions for a similar problem were obtained as were asymptotic solutions valid at long times after the initial transient [22FF].

Inclined plates and vertical plates with localized heating or insulated boundaries were also investigated. Neutral stability curves were given for a heated, inclined plate facing up [64FF]. Perturbation solutions were given for inclined plates with [18FF] and without [24FF] an unheated surface upstream. The effects of an adiabatic wall parallel to [54FF] and adjacent to the sides of [55FF] a vertical flat plate were studied. An extension of previous work on the effects of localized heaters placed on a vertical adiabatic wall was presented [23FF].

Heat transfer from horizontal surfaces has been studied for heated strips or circular plates facing down [49FF] and for a cooled strip facing upwards in a porous medium [26FF]. Heat loss above a heated plate covered with liquid potassium was studied experimentally [60FF] as a function of potassium layer thickness and evaporative cooling from the upper free surface. A matched asymptotic expansion technique was used to study heat transfer above a heated horizontal plate [1FF]. A similar study was presented for a heated circular disk [33FF] where the curvature effects were found to be restricted near the center of the disk.

Numerical solutions to heat transfer from an insulated horizontal cylinder were obtained to determine the critical radius of insulation under more realistic conditions [56FF]. An approximate integral method was developed for predicting laminar and turbulent natural convection from curved surfaces with a horizontal cylinder used as an example [35FF]. Experimental studies of natural convection from horizontal cylinder arrays include an interferometric study of two vertically aligned cylinders [46FF] and a flow visualization and mean heat transfer coefficient study of two vertical arrays of cylinders bounded by two vertical, adiabatic walls [47FF].

Several papers considered laminar natural convection from a variety of objects of small aspect ratio including a numerical study of short vertical and horizontal plates and prisms [34FF]. A spherical segment [59FF] and a cone frustum [30FF] were also considered. Mean heat transfer coefficients from cubes and spheres were measured in air with varying pressure to change the Grashof number [7FF]. A similar study in air and water was performed [57FF] with the results correlated using a length parameter defined as the surface area divided by the square root of the horizontal projection.

A numerical study was presented for the stability of a natural convection boundary layer adjacent to a vertical flat plate with uniform blowing [11FF]. Theoretical and experimental studies were made on a heated vertical flat plate with blowing or suction through a permeable section at the bottom of the plate [28FF]. Numerical solutions were obtained for blowing or suction through a vertical permeable cylinder surrounded by a standard porous medium [19FF].

A comprehensive study of combined heat and mass transfer from a vertical plate was given in ref. [38FF] where regions of aiding and opposing flows were delineated as a function of Lewis number and buoyancy ratio. A numerical study of particle deposition onto the surface of a vertical cooled plate was given for both laminar and turbulent flow regimes [10FF].

Transient combined heat and mass transfer in air was studied numerically in ref. [53FF]. A semi-analytical approach was used to evaluate the rate of water evaporation from a vertical plate of varying heat flux [36FF]. Three papers by the same authors describe analytical solutions to the combined heat and mass transfer problem for mixed convection from a vertical cone [13FF], and pure natural convection form a horizontal cylinder [14FF] and a sphere [15FF].

Coupled conduction or radiation and natural convection is receiving increased attention. Coupled conduction on a vertical flat plate was solved using transformed boundary-layer equations and the Blasius method [61FF]. Conjugate conduction–natural convection [21FF] and conduction-aiding mixed convection [20FF] were solved numerically for a heated downward facing circular pin fin. An experimental study of coupled conduction–natural convection was performed for a horizontal cylinder with one axial plate fin at various angles [63FF]. A local nonsimilarity technique was used to obtain a solution for natural convection in an optically thick, cold liquid adjacent to a nonemitting, nonreflecting, radiation-transmitting vertical flat plate [9FF].

Mixed convection studies include heat transfer from a suddenly accelerated vertical flat plate [62FF]. Reference [8FF] considers an oscillating vertical heated plate where the natural convective flow is superimposed on the primary Rosenblat flow. Measurements and analyses were made for steady aiding or opposing mixed convection [44FF] and transient aiding flow [48FF] from a vertical flat plate. Laminar aiding flow was studied theoretically for a slender vertical cylinder [5FF] and experimental measurements were made for cross flow [68FF]. The Navier–Stokes and energy equations were solved for cross flow over a heated horizontal cylinder [3FF,31FF]. A similar approach was used for aiding or opposing flow from the cylinder [4FF]. Two papers presented numerical studies of aiding mixed convection from a point heat source [45FF,2FF]. An experimental study produced three-dimensional flow visualization and local surface heat transfer results for mixed convection in a horizontal channel bounded by two asymmetrically heated plates [40FF]. Linear perturbation theory was used to determine the nature of the flow above the downstream of a heated patch [66FF]. Transverse or longitudinal roll cells were formed depending on whether the flow Reynolds number was above or below a critical value. Secondary mixed convective flows were obtained numerically in fully developed flow in a horizontal annulus [37FF].

A numerical study of a plume with positive or negative buoyancy was given [67FF] which included several tables and figures of velocity and temperature profile information. The stability of a buoyant plume above a line heat source was described with analytical wave numbers compared with available experimental data [65FF]. The morphology and ascent rate of a buoyant plume in a homogeneous viscous fluid was measured to predict the hot-spot volcanism activity in the earth's mantle [39FF]. Simple jet trajectory models were used to determine the behavior of a buoyant jet injected into fully developed pipe flow [25FF].

CONVECTION FROM ROTATING SURFACES

Heat transfer from rotating surfaces has been attracting considerable interest, especially in connection with rotary engines.

Studies of laminar heat transfer in a tube rotating about an axis perpendicular to the tube axis show that the secondary flow induced by the Coriolis terms always tends to increase the heat transfer coefficient [18G]. Studies of heat transfer enhancement from a rotating cylinder show that with increasing rotating Reynolds numbers, the heat flux increases also and becomes more uniform [17G]. Numerical solutions for the full Navier–Stokes equations in the nonlinear range for axisymmetric compressible flow in a rotating cylinder with axial convection are presented, and the validity of the present approach is confirmed [23G]. Finite-difference solutions for laminar flow of air in a rotating rectangular duct with aspect ratio 2/1, where the duct wall is subjected to a uniform heat flux, show the effect of rotation on the development of the flow patterns, velocity and temperature variations [13G]. Numerical studies of the effect of sidewall heating in the pressure-driven laminar flow of an incompressible viscous fluid through a rectangular channel that is subjected to a spanwise rotation, show that an increase in the rotation speed leads to an increase in the net heat transfer [22G]. Numerical predictions of the influence of cylinder rotation on turbulence and heat transfer in an annulus are in excellent agreement with available experimental data [1G].

Numerical predictions for the centrifugally driven free convection in a sealed rotating cavity and for the buoyancy-affected flow through a cavity with an inner cylindrical source and an outer cylindrical sink are in good agreement with experimental data [3G]. Finite-difference solutions for forced laminar convection in a rotating cylindrical cavity with radial outflow show that the local Nusselt number changes from that of a free disk near the center of the cavity to that for Ekman layer flow at the larger radii, provided that the fluid enters the cavity from a uniform radial source [4G]. Studies of the source-sink flow inside of a rotating cylindrical cavity using a 'free disk' model for the outflow and a free vortex for the inflow case, are in good agreement with available experimental data [14G].

Studies of turbulent heat transfer in co-rotating annular disks indicate that the disk rotation produces a surge in the local Nusselt number of the near-exit region and a substantial enhancement in the average heat transfer performance [19G]. The thermal effect on the transverse vibration of a high-speed rotating disk is considered under steady-state heat conduction. A new critical speed of disk rotation has been obtained and this critical speed is found to depend on central temperature, thermomechanical anisotropy, etc. [7G]. Studies of laminar cooling of two nonmixing liquids are reported contained between two coaxial disks rotating in opposite directions [15G].

The nature of small amplitude Rayleigh–Bénard convection for a horizontal plane layer of fluid rotating about a vertical axis and heated from below is considered. Steady-state and time-periodic solutions of this system are derived and their stability is discussed [21G]. The stability of a horizontal, viscoelastic fluid layer heated from below is studied for the case of a container rotating with uniform speed and new threshold values for overstability are established [9G]. Studies of the convective stability of a horizontal layer of incompressible micropolar fluid heated from below and rotating about a vertical axis show that the rotation has a destabilizing effect which contradicts an earlier assertion [2G]. The effects of compressible work and frictional heating in axisymmetric swirling flows have been clarified and formulae have been provided for use in engineering calculations [5G]. Experimental studies of the behavior of transient isolated convection in a rigidly rotating fluid revealed an increase of the maximum temperature in the thermal caps with increasing rotation and a more rapid cooling of the buoyancy source [12G]. A unique data base has been established for the assessment of the effects of spin and mass addition on high-angle-of-attack re-entry [11G].

Analytical studies of mixed convection flows generated by a heated rotating sphere reveal interesting flow patterns [6G]. Studies of viscous dissipation on mixed convection about a rotating sphere show that an increase of the viscous dissipation causes a decrease of the heat transfer rate [16G]. A rotating condenser with a scraper is considered as a new device for promoting condensing heat transfer [24G]. The effect of rotation on heat transfer in the channels of moving turbine blades has been determined experimentally with a loop system of cooling [8G].

Overall heat transfer coefficients to water and silicone fluids in cans rotating in a steam retort have been correlated with an analytical expression [20G]. A rotating, packed-drum reactor has been proposed as an immobilized whole cell reactor and its performance for ethanol production has been studied with yeast cells immobilized in a calcium alginate gel [10G].

COMBINED HEAT AND MASS TRANSFER

There has been considerable interest recently in film cooling processes, primarily as they relate to high-temperature gas turbine systems. In addition, transpiration cooling and ablation cooling continue to be studied to some extent as do a variety of processes related to combined heat and mass transfer.

Many studies consider film cooling on a flat surface with or without a pressure gradient along the surface. The flow produced with injection into a mainstream through a vertical slot has been analyzed for its potential influence on film cooling [8H]. Mass transfer from a swollen polymer along with a holographic interferometer is used to measure film cooling on a surface with zero pressure gradient [15H]; a continuation of that work shows little influence of an adverse pressure gradient on film cooling, but a significant effect when a favorable pressure gradient occurs along the surface [16H]. The importance of free-stream turbulence on film cooling downstream of a porous section has been demonstrated experimentally [25H]. The results for film cooling performance with a large temperature difference between the injected coolant and the hot mainstream has been correlated; the measurements were taken in a transient high-pressure tunnel [23H]. Measurements of film cooling following injection through two rows of cooling holes [18H] shows good agreement with earlier studies. On a highly roughened (but flat) surface to simulate actual engine conditions, the film cooling effectiveness can be higher or lower than that on an equivalent flat plate, primarily depending on the magnitude of the blowing parameter [14H]. A two-phase film cooling heat transfer study has been performed [33H] over a range of parameters.

Measurements conducted near the leading edge of a turbine blade demonstrate the importance of different parameters on heat transfer following film cooling [4H]. In a related study [5H], a significant decrease in heat transfer is found with film cooling on the suction side of a simulated turbine blade. In a considerable region along the suction side of a blade near its end wall, the film coolant is swept off the blade by the passage vortex [13H]. Film cooling effectiveness decreases with frequency and amplitude of vibration of a turbine blade on its pressure side; on the suction side, there is an increase in effectiveness upstream of the separation point [21H]. An analysis of the flow distribution through inserts used in gas turbine blades provides information needed to predict film cooling performance [31H]. The film cooling on turbine blades using several sections of porous wall has been calculated to provide a universal correlation [28H]. An analysis [3H] includes the influence of internal convective cooling on the film cooling performance on a turbine blade.

With transpiration cooling, the surface or wall to be protected is permeable with the coolant fluid passing directly through this surface. A study of a low Mach number mainstream with high blowing rate transpiration cooling indicates a reduction in heat transfer with transpiration up to the blow-off condition [11H]. An interferometric study with a porous cylinder demonstrates the major influence of blowing on the heat and mass transfer particularly near the separation point [26H]. Analysis indicates that the instability of transpiration cooling from nonuniform flow can lead to over-heating of a surface [19H]. Analysis shows the importance of thermophoresis on the heat transfer to a transpiration cooled wall from a particle-laden gas flowing over it [12H].

Experiments on transpiration cooling of a surface with a supersonic turbulent boundary layer flowing over it show the interaction of the coolant with an incident shock and the influence of pressure gradient [7H]. Analysis of the heat transfer for a re-entry vehicle show that with large blowing rates in the stagnation region the heat flux can be made very small [29H].

An analytical technique has been developed to predict the influence of a time varying heat flux on an ablating surface [6H]. Perturbation that alters the shape of an ablating body has been studied to see the conditions under which the perturbation would be amplified; generally it is damped [24H].

A turbulence model has been developed to predict the heat transfer of an axisymmetric jet impinging on a flat plate [1H]. The influence of ambient temperature on the heat transfer from a plane surface to an impinging jet has been demonstrated [17H]. An impinging jet has been used to cool the bottom of a cylindrical enclosure [22H]. A flow distribution model for an array of impinging jets helps predict the heat transfer characteristics with nonuniform geometries [9H].

Other studies on combined heat and mass transfer include the development of a similarity solution for mass transfer to a thermally driven boundary layer [20H]. Dehumidification of a flow of moist air over a tube bank has been studied to determine the Nusselt and Sherwood numbers [10H]. Approximations are introduced to predict the influence of gas addition on the transient heat transfer in a stirred liquid [27H]. Studies are reported on the heat transfer to micropolar fluids in a channel with porous walls [2H] and passing over a continuously moving porous plate [30H].

CHANGE OF PHASE—BOILING

Nucleate boiling

An exact analytical solution was presented [1J] for the potential flow due to growth of a spherical bubble at or near a solid wall, applicable to idealized nucleate bubble departure size. A mechanism was proposed [20J] for bubble growth in nucleate boiling within thin falling superheated water films, capable of explaining the much faster observed growth rate of these bubbles than their counterparts in pool boiling. Experimental evaluation of bubble dynamics in subcooled nucleate pool boiling [48J] confirmed bubble growth predictions but illustrated that the models for bubble waiting period diverge from observed behavior at greater subcooling. Transient and permanent nucleation sites were studied by means of high-speed infrared cinephotography [95J], displaying fluctuations in bubble nucleation periods. It was proposed [78J] that short-lived nucleation sites might be the consequence of circulating microbubbles, small enough to be dragged by superheated liquid approaching the surface, but large enough to be thermodynamically stable. Offered as a possible explanation of certain hysteresis effects, this proposal suggests that surface and pool geometry may play a larger role than previously expected in nucleate boiling. Nucleation sites were experimentally shown to interact with one another [16J]; the nature of the interaction appeared to be determined by the dimensionless ratio of the separation distance to the bubble departure diameter. The effect of surface orientation on nucleate boiling, evident only at lower nucleate boiling heat fluxes in some recent experiments, was discussed as exemplary of the Moissis–Berenson transition from an isolated bubble regime to one of slugs and columns: both are considered to be nucleate boiling, but hydrodynamically they are quite distinct [63J]. Two separate heat transfer correlations were presented [11J] respectively for the 'interference' and the 'macrolayer' regions of the nucleate boiling regime. A one-dimensional transient model employing the liquid macrolayer thickness was employed to investigate the effect of heater wall thickness upon nucleate pool boiling heat transfer [89J]. Boiling curves and ebullition cycle periods were measured for subatmospheric pressure pool boiling of potassium on a horizontal surface, varying the height of the liquid level above the heater surface [66J]. The period was found to increase as the level decreased, and liquid levels less than 5 mm seemed to suppress nucleate boiling. Boiling heat transfer experiments with water and organic liquids on microsurfaces, those with dimensions less than a few bubble break-off diameters, led to a correlation for correcting for differences from large surface boiling behavior [58J]. Acoustic phenomena in boiling; the acoustics of single bubbles, acoustic interactions among bubbles, and the sound associated with the approach to critical heat flux were discussed [34J]. A numerical investigation of the one-dimensional transient conduction during hardening of a plate of steel [57J] predicted reasonably well the duration of the nucleate boiling regime; for plate thicknesses greater than a limiting value these durations were invariant with thickness.

Pool boiling

Metal-fiber porous coatings and vibration and impact-machined surfaces were explored as enhancements to boiling of refrigerant 113 at pressures of 1 and 2 atm [28J]; the machined surfaces were found to be superior, while high porosity coated surfaces proved to be less effective than smooth surfaces for subcooled boiling. Pool boiling heat transfer rates of ammonia and refrigerant 22 were shown to be enhanced to considerably different extents by differing enhancement techniques: small patches of thin Teflon layers enhanced boiling of ammonia much more than refrigerant 22, while a porous metallic layer showed the reverse behavior [35J]. Quenching experiments with samples of differing materials, surface roughnesses and surface energies (as indicated by contact angle) illustrated that smaller contact angle had no effect on nucleate boiling but raised both the critical flux and minimum film boiling flux, improving heat transfer throughout the transition regime [26J]. Surface roughness enhanced nucleate boiling but had much smaller effect upon transition boiling. The influence of surface conditions on the pool boiling of liquid metals was analyzed, and the thermal conductivity of the heating surface was shown to have a strong effect on surface temperature distribution [38J]. Pool boiling at atmospheric pressure at porous surfaces of horizontal tubes in water, ethanol and refrigerant 113 exhibited small superheats at bubble initiation, substantial increases in the critical heat flux, and a transition, subject to hysteresis in thick porous layers, from bubble generation near the external surface to the penetration of a vapor film layer through the porous layer [4J]. Preliminary design curves were developed analytically for boiling of water, isopropyl alcohol, and refrigerant 113 on surfaces with longitudinal fins [12J]. Experiments with boiling of refrigerant 113 on finned surfaces produced boiling curves consistent, at near atmospheric pressures, with predictions based upon the use of pool boiling curves of smooth surfaces to prescribe the relationship between heat flux and temperature along the surface of a one dimensional fin [2J]. Poorer agreement at lower pressures shed doubt on the general utility of this predictive method.

A combined experimental and analytical investigation of thermosyphon boiling of water in vertical parallel-walled channels with uniform heat flux demonstrated and explained smaller wall superheats with closer spacing of the channel walls [8J]. Experiments were described for natural convection on the heated internal surface of a horizontal annulus [113J]. Heat transfer coefficients and critical heat flux increased markedly with the annular gap size up to a size at which normal pool boiling correlations were applicable. In another experimental investigation of boiling in annuli, effects of fluid properties, pool subcooling, crevice length and gap size were examined for both concentric and eccentric annular crevices [47J].

The effects of subcooling and binary mixture composition were explored experimentally with focus upon the density of boiling sites [46J]. Experiments and analysis suggested that pool boiling heat transfer coefficients for binary mixtures and pure liquids in strong gravitational fields are reduced by reductions in active nucleation sites [83J]. A monotonic decrease in pool boiling heat transfer coefficients was reported [6J] with increase in centrifugal field strength for water/propanol mixtures at all propanol concentrations. Horizontal wire pool boiling experiments were performed on mixtures of lubricating oils to examine the effects of molecular weight distribution [94J]. Blended oils displayed higher maximum heat transfer coefficients than those of the base oils and typically exhibited two peaks in their boiling curves. Wire superheat at the onset of nucleate boiling increased linearly with increase in the molecular weight of the base oils. Boiling curves were experimentally determined for immiscible water/*n*-nonane mixtures on horizontal tube bundles in pool boiling [82J]. A matrix method was proposed [50J] for calculating interphase flow rates of heat and mass in a multicomponent vapor/liquid system.

Experiments of pool boiling of water and salt-water solutions on smooth and rough horizontal copper plates were performed at atmospheric and subatmospheric pressures [108J]. The smooth surface provided largest heat fluxes for a given superheat in most cases. Results were presented from an experimental study of the effect of a stationary electric field on the probability of boiling of superheated *n*-pentane [120J]. Heat transfer to the horizontal bottom of a volumetrically heated boiling pool was analyzed [24J]. The propagation rate and the width of the front of changes in modes of boiling were investigated and distinctions were made among the dynamic behaviors expected dependent upon the shape of the boiling curve [110J].

Experiments with short intense pulse heating of a fine wire in a pool of water appeared to produce spontaneous nucleation at temperatures consistently 20°C below predicted homogeneous nucleation temperatures [30J]. Heat fluxes up to 120 MW per square meter were induced. Transient experiments were also reported [85J] in closed narrow channels to validate a simulation of behavior of waterlogged reactor fuel elements.

Flow boiling

Subcooled flow boiling experiments using refrigerant 113 showed little influence of mass flux and subcooling upon the incipient boiling wall superheats and suggested a mechanism of departure from nucleate boiling triggered by coalescence of bubbles into larger voids, producing temporary liquid film dryout [44J, 45J]. The effect of turbulent forced flow upon bubble behavior characteristics was developed and compared with experimental findings [114J]. An experimental technique was developed for measuring volumetric vapor contents of forced two-phase refrigerant flows through vertical and horizontal tubes [64J]. Boiling of water in vertical and horizontal pipes at low pressures was studied experimentally [118J]. Improvements in heat transfer coefficients and in stability of the flow were found to result from lining the pipe walls with metallic cloth. A new correlation was proposed [9J] for predicting the onset of 'intensive evaporation' marked by substantial presence of vapor in subcooled flow boiling. Scaling laws for coupled transient heat transfer and two-phase flow were used to derive several scaling numbers applied to investigation of a water cooled nuclear reactor [10J]. Subcooled flow boiling at high heat flux was studied experimentally and modeled [29J]: among the conclusions were that very thin metallic heater surfaces gave lower rates of heat transfer and that microlayer evaporation contributed negligibly to the heat flux. Friction and heat transfer relationships in turbulent gas/liquid flow were developed and employed to permit prediction of conditions which would cause salt to be precipitated on the wall of a steam generating channel [62J]. An improvement was proposed to earlier predictions of pressure and temperature fluctuations in a particular sodium boiling test apparatus [41J], based upon a hypothesized bubble breaking mechanism inferred from experimental data. A simple prediction method was proposed for the critical mass flow rate of initially subcooled liquid in long channels, built upon many empirical relations but successfully demonstrated against experimental data [18J].

A two-layer liquid film model combined with an accounting of liquid droplet entrainment was proposed for predicting heat transfer in annular two-phase flow [33J]. A simplified model was developed and demonstrated for prediction of void fraction in this significant flow regime [106J]. The developing region of two-component annular two-phase flow was studied experimentally and theoretically with emphasis upon the interfacial wave behavior [42J].

A few reports dealt with flow boiling in other geometric configurations. Experiments with bubbly water/air mixtures in cross-flow over a horizontal cylinder were described [107J]. The effects of tube diameter and forced flow velocity were investigated for cross-flow boiling on the outside of horizontal tubes [13J]. While nucleate boiling during flow was only mildly decreased with increasing diameter, the peak heat flux decreased approximately as the 0.28 power of diameter. Falling film evaporation of water over horizontal tubes was shown to be enhanced by both axial grooves and circumferential grooves in both boiling and nonboiling modes [43J]. Effects of surface tension were examined in flow visualization of two-phase flow in a channel with offset strip fins and incorporated in determining the minimum value of the boiling heat transfer coefficient in such a geometry [17J].

A correlation was presented for heat transfer coefficients for flow boiling of gasoline/ethanol mixtures in a vertical tube [59J]. Analysis of the effect of oil in mixture with refrigerants upon their boiling heat transfer coefficients was described [7J]. A semi-empirical formula was developed anticipating a concentration gradient as an additional resistance to vapor generation. Coefficients of heat transfer for boiling mixtures of freon and nitrogen were determined to be lower than those of the pure components while the variations of the coefficients with heat flux were of the same form [98J]. Equations were presented to describe the dynamics of evaporation of a multicomponent mixture flowing through extremely narrow channels [65J].

Film and transient boiling

Natural convection film boiling on a 10.3-cm-tall vertical surface showed no dependence of heat transfer coefficient on distance from the leading edge, and a physical model was proposed consistent with the expectation of laminar flow and the observed vapor/liquid interface behavior [15J]. Asymptotic results for film pool boiling on a vertical surface in a saturated liquid were found using an integral boundary-layer method under conditions of negligible inertia and negligible convection within the vapor film [80J]. An integral method was applied to a two-phase boundary layer in laminar film boiling on a hot plate or cone in a saturated pool, displaying excellent agreement with exact solutions [81J]. Experiments with several cryogenic fluids undergoing free convection film boiling in transient and steady conditions showed substantially smaller heat transfer coefficients for spheres compared with vertical cylinders at large superheats [56J].

Natural convection transition boiling on a vertical surface during slow transient experiments produced two distinct transition boiling curves, one for heating and one for cooling [14J]. A micro thermocouple surface probe was used to measure frequency and duration of liquid–solid contact in film and transition boiling of atmospheric pressure water, with results indicating that these contacts may be the dominant mechanism of heat transfer in transition boiling [61J]. In another experiment [32J] the extent of liquid–solid contact during transition pool boiling was measured by measuring the impedance across a thin dielectric layer deposited on a test surface in water and methanol. A model was proposed [53J] to explain the early (high minimum temperature) transition from film to transition boiling on surfaces insulated by thin layers of low conductivity materials. Experiments with Teflon coatings upon copper supported model predictions to a greater extent than tests with more conductive coatings.

A semi-empirical model was developed for turbulent forced convection film boiling of a subcooled liquid on a flat plate, hypothesizing a vapor–liquid mixing layer between the vapor film and the bulk liquid [116J]. Vapor thickness was analyzed [39J] at the stagnation point during forced convection film boiling on a solid sphere, including the effect of the liquid viscosity which was shown to be important.

An alternately advancing and retreating quench front was assumed in a transition flow boiling model which predicts the proper magnitude of oscillations of wall temperatures [51J]. Using a slow reflood quenching process, nonequilibrium vapor temperature and wall temperature were measured downstream of the quench front in upflow of water in a vertical tube [37J]. Comparison with existing models demonstrated that the models predict wall temperatures reasonably well but fail at predicting the vapor temperatures. Experiments with a large rod bundle provided data on dispersed flow film boiling with which empirical correlations were compared [74J]. Spacer grids caused large rod surface temperature reductions.

Critical heat flux

Theory and experiments were presented [115J] to suggest that the boiling crisis occurs when a maximum superheat is attained in a system, that temperature for which one viable nucleus is generated per second in each cubic centimeter of the liquid. A correlation was developed for pool boiling peak heat flux based upon the principle of corresponding states and employing a function of geometry which is thus applicable to all except extremely low boiling temperature liquids, and which can be extended to flow boiling [97J]. Extensive tables of data supporting the hydrodynamic analogy between nucleate boiling and the bubbling of gas into a liquid at a porous boundary were published [60J], having been presented in implicit form over several years. Although various heater materials and thicknesses of given heater materials altered the critical heat flux of helium on a flat horizontal surface, the film boiling heat transfer coefficient was experimentally shown to be independent of these variations [88J]. Critical heat fluxes for helium in natural circulation in vertical channels of circular and rectangular cross-section were measured and contrasted with the magnitudes of forced convection critical heat fluxes [87J]. In another geometry, the pool boiling crisis of nitrogen in an annular channel was examined with various orientations and gap sizes [54J]. Small, externally induced flows were part of a steady state model presented for predicting tube wall temperatures and potential dryout on the heated inner tube of an eccentric annulus with line contact [112J]. This approximated the conditions of pressurized water reactor steam generator tubes at their support locations. Critical heat flux was also part of the concerns of previously mentioned reports on boiling in annular spaces [47J, 113J].

Critical heat flux and flow instabilities were explored in both upflow and downflow of water in a vertical tube at atmospheric pressures at low flow rates [69J]. Departure from nucleate boiling heat flux [105J] and critical heat flux [68J] were measured in upflow and downflow in vertical rectangular channels. The latter experiments, performed at atmospheric pressure, demonstrated that the critical heat flux at low flow rates can be much lower than that for pool boiling and that the direction of flow is significant. Thermally induced flow instabilities in boiling channels were considered with the inclusion of fluid inertia effects, leading to the conclusion that linear stability boundaries may be inadequate for design of related systems [3J]. Forced convection boiling heat transfer coefficients and critical heat fluxes were experimentally investigated for refrigerant 12 on a locally heated 16-mm-diameter surface mounted flush with a channel wall [119J]. Variations of critical heat fluxes in annular channels with changes in surface structure were reported [111J]. Greater heat fluxes were found possible with most turbulence-promoting roughness geometries, with greatest increases, relative to smooth walls, attained in emulsion and dispersed annular flows. A new correlation of existing data for critical heat flux in the impinging-jet-on-disk configuration was presented [96J]. It is expected to be valid when viscosity and gravity have little influence. Another such correlation was presented [71J] with emphasis on accuracy at moderately low pressures achieved by better accounting for the effect of the liquid/vapor density ratio. The critical heat flux for liquid films in rotating channels was measured experimentally [75J] and modeled as being limited by droplet departure. Increased pressure and rotational speed were both found to increase the critical flux.

Several investigations dealt with hydrodynamics and thermal aspects of reflood, most with application to nuclear reactor safety. A predictive method for determining countercurrent gas/liquid flow in parallel channels was developed and compared with experimental observations [77J]. Experimental wall temperatures and vapor film thicknesses were reported for the inverted annular (liquid core, vapor film) boiling for liquid upflow in a vertical tube [36J]. A true quench temperature was defined as the quench temperature at the onset of rewetting in a flow channel, wherever the rewetting begins, and in the absence of axial conduction [25J]. Experiments were performed so as to measure this temperature rather than apparent quench temperatures, which are always higher. In a one-dimensional analysis, the effects of temperature dependent properties of the solid upon the quench front velocity were shown to be small relative to the uncertainties of rewetting temperature and heat transfer coefficients [86J]. Experimental findings on critical heat flux in closed-bottomed vertical channels were reported for several channel geometries [99J]. Results were reported for experiments with bottom reflooding of a vertical channel including void fraction, quench time, and quench front quality [31J]. Quench promotion and quench delay were both observed, related to high and low reflood velocities, respectively, as a consequence of large scale blockage in reflood tests of the Slab Core Test Facility [100J]. Flow regimes, void fractions, and heat transfer characteristics were reported for upward reflooding with both steady and controlled oscillatory injection of coolant [52J]. Other experiments were performed [5J, 76J, 79J], to validate safety-evaluation codes relating to reflood conditions.

Droplet evaporation

Evaporation rate and heat flux were evaluated for a droplet in a rarefied vapor using kinetic theory analysis [102J, 103J]. Drag coefficients for evaporating and condensing droplets were evaluated numerically [27J], yielding results which compared favorably with experimental data of another investigation. The use of surfactant additives or the presence of traces of surfactant impurities appeared not to significantly alter the electroconvective enhancements predicted for heat and mass transfer from drops and bubbles in electric fields [22J]. Thermocapillary forces acting on a droplet due to its chemical reaction with its surroundings were analyzed [93J], suggesting that conditions could result in either drag or thrust acting on the droplet. Rates of droplet evaporation were modeled in the extremely high temperature medium of a thermal plasma [55J]. Comparison of experimental measurements of evaporating spray characteristics with several models emphasized the importance of finite rates of interphase transport and drop dispersion [104J]. Transient temperatures were measured and modeled for the response of hot metal cylinders to the impingement of a water droplet [73J]. Boiling curves were experimentally obtained for single drops of water 4 mm in diameter impinging on a solid surface at temperatures corresponding to the transient regime of pool boiling [49J]. The maximum heat flux of the boiling curves thus determined was found to lie on an extension of the nucleate boiling segment of the pool boiling curve. Equilibrium and phase-changing configurations of two-phase vapor/liquid bubbles in an immiscible liquid were described and categorized [72J]. Single drops of refrigerant 113 were allowed to fall upon a hotter, less dense, and less volatile liquid and the consequent motions and evaporation times were observed [84J]. Five vaporization modes were observed, including submerged film boiling. A model was developed for evaporation of droplets in direct contact heat exchangers incorporating the effects of sloshing, which was deemed significant for droplets larger than 2 mm diameter [92J].

Flashing and other evaporative processes

The production of bubble nuclei by electrolysis upstream of a nozzle was shown to strongly enhance the evaporation rate and approach to liquid/vapor equilibrium of spray flash evaporation [70J]. Flashing nozzle flows were experimentally found to produce substantially fewer bubbles which are larger and grow at faster rates than predictive models currently suggest [67J]. Heterogeneous nucleation was not observed in the bulk flow; bubble generation appeared exclusively at the wall. Bubble growth, with interaction of surrounding bubbles, was analyzed by a method of matched asymptotic expansions [21J].

Kinetic theory analyses of evaporation were presented [101J, 109J]. The paradoxical inverted temperature profile between opposing evaporating and condensing surfaces was shown to be stable [19J]. Evaporation in the contact line region of thin films of hexane and hexane with small amounts of octane were studied with an optical interference method [117J]. Small changes in the bulk composition had large influence upon the transport processes near the contact line. The net heat flux and evaporative heat flux crossing the air/water interface of a heated body of water were measured and compared unfavorably with existing correlations [23J]. A two-dimensional analysis was presented for heat and mass transfer in the region of the vapor–gas front of a gas-regulable heat pipe [40J]. Steady evaporation of volatile liquid from a partially filled, open-topped adiabatic container was shown analytically to proceed at substantially lower rates than those predicted by the Stefan (isothermal) model since interface temperatures are depressed [90J]. In a related effort, experiments and numerical modeling with a composite rectangular and circular grid addressed the evaporation of liquid from a thin slot cavity by a fully developed laminar flow over the top of the cavity [91J]. The distribution of local evaporation rate at the liquid surface resembled that in a conventional boundary-layer flow.

CHANGE OF PHASE—CONDENSATION

A general integral solution procedure was presented for laminar and turbulent film condensation on an isothermal surface, for bodies of arbitrary geometry [21JJ], and a uniform method of analysis for film flow with superposition of various effects (laminar or turbulent transport, gravity and frictional forces, phase transition and variation of film temperature) was developed [10JJ].

Heat transfer with film condensation on horizontal cylinders received considerable attention. Experimental results for a vapor flowing over a single cylinder indicated that the heat transfer by condensation depends substantially on the magnitude of the cross flow [18JJ]. An analytical/numerical study of film condensation over a horizontal cylinder with combined gravity and laminar forced flow was presented, including solutions in the merging flow region controlled by both driving forces [7JJ]. Film condensation heat transfer on horizontal tube banks was investigated experimentally for both the case of a flowing vapor [17JJ] and that of a stationary vapor with variable condensate flow rate [16JJ]; the ranges of dimensionless parameters were determined within which heat transfer was essentially governed by either the condensate flow rate, the vapor velocity, or the combined action of both factors. Experimental data were obtained for heat transfer with condensation of steam on horizontal wire-wrapped tubes; these results were compared to the performance of smooth tubes [24JJ]. The effects of surface tension were considered in an analysis of heat transfer in laminar film condensation on a horizontal cylinder [15JJ] and in film condensation on tubes with finely fluted fins [1JJ].

Flow pattern data during forced convection condensation of refrigerants R-12 and R-22 inside a horizontal tube were presented [26JJ], as were data on the heat transfer coefficients for these substances flowing through an annulus passage [28JJ]. Experimental results were presented for the flow patterns of condensing steam inside horizontal tubes [22JJ]. All of the horizontal flow pattern correlations tested failed in predicting the wavy-slug transition; the results of adiabatic (gas–liquid) flow in upwardly inclined tubes appear to provide some explanation for this behavior.

The dynamic behavior of a boundary layer (either laminar or turbulent) with condensation along a flat plate was studied both experimentally and numerically [19JJ]; the results were compared to those for the somewhat analogous case of a boundary layer with wall suction. Experiments on laminar film condensation heat transfer on several kinds of downward-facing horizontal surfaces were reported [29JJ]; these results have application to electric devices containing a volatile liquid in a casing.

An approximate relation for the heat transfer coefficient for a fully turbulent falling film at high Prandtl number was derived [4JJ], as was a nonlinear two-wave equation for long waves on the surface of vertical falling films [3JJ]; these waves can exert a significant influence on the heat transfer. Modeling and experimental results were presented for film condensation heat transfer on a general cylindrical surface in the presence of a nonuniform electrostatic field [11JJ]. The problem of simultaneous transient condensation and melting of a vertical surface was investigated using a Nusselt-type analysis [9JJ].

Condensation from multicomponent mixtures to surfaces was analyzed for the case where the mixture included a small amount of noncondensable gas [19JJ] and for the specific case of isopropanol–water mixtures [8JJ].

Dropwise condensation curves were measured for the condensation of propylene glycol on a PTFE-coated surface [27JJ]. The values of the heat transfer coefficient were found to be much smaller than for steam condensing on an oleic acid-promoted copper surface, and the transition mode from dropwise to film-type condensation was quite different.

When a superheated stream flows over a subcooled surface, a sufficiently high supersaturation ratio may initiate homogeneous nucleation in the boundary layer. An analysis for this case was presented [2JJ]; it was found that boundary-layer nucleation always reduces vapor mass transfer to the wall. In laminar flow this deficit is more than compensated by deposition on the wall of nucleated particles, whereas in turbulent flow the net mass transfer decreases, in part because of the diffusion of once-nucleated particles across the boundary layer. Numerical solutions for heat and mass transfer to a free droplet growing through condensation were obtained for droplet Reynolds numbers of about 100 [25JJ]. The effect on transport mechanisms of a condensing droplet subject to contamination with insoluble monolayer surfactants was examined theoretically [5JJ]. The 'monophase boundary-layer approximation' (an alternative approach to classical nucleation theory) was used to analyze the irreversible condensation of a vapor in a binary mixture (the dominant component being inert), free-convective, laminar boundary layer [23JJ]; the findings were applied to an air–water vapor mixture in contact with a vertical isothermal surface at cryogenic temperature.

Other studies of condensation heat transfer included a review of recent results pertaining to nuclear reactor transients [12JJ]; a discussion of condensation in vacuum systems [20JJ]; experimental results on frost formation in finned air coolers [6JJ]; and an experimental investigation of heat transfer from steam/air mixtures to a retort pouch laminate, with particular application to packaging food products [13JJ].

CHANGE OF PHASE—FREEZING AND MELTING

New solution methods continue to be developed for freezing phase change problems. Conformal mapping methods were used to construct explicit solutions for a general Stefan moving-boundary problem [22JM]. An integral approach was found to agree with exact solutions to one-dimensional phase-change problems when $Ste = 5$ [29JM]. Reference [47JM] included crystal growth kinetics. A solution method was developed in which the solution for one phase was obtained using an inverse analysis providing the solution for the other phase was known [12JM]. The effects of isotropic and anisotropic radiation scattering were determined with results given for sapphire [38JM]. Two approximate analytic solutions were obtained for one-dimensional solidification with Newton cooling [19JM]. An approximate solution valid for small Stefan numbers combines the heat balance integral method and time dependent perturbation theory [5JM]. A numerical procedure was developed to model freezing of a liquid phase-change material around a suddenly immersed solid such as a cylinder, sphere or slab [39JM]. Reference [21JM] discussed the conservation laws obeyed by systems containing point defects, dislocations and domain boundaries. Reference [53JM] remarked on the formulation and qualitative behavior or hyperbolic Stefan problems.

Freezing in binary solutions was considered by refs. [55JM, 56JM] in which linear stability analysis was applied to determine the time of first marginal instability as a function of cooling rate for a sodium chloride solution. Reference [26JM] analyzed a similar problem. Analysis and experiments were presented in ref. [23JM] in which the fastest solid growth rate was found to occur when the sodium chloride concentration was 0.9%. Numerical solutions and experiments were performed to evaluate the effectiveness of freezing out one component of a mixture on a rotating, scraped drum [16JM]. A macroscopic numerical model was demonstrated to be especially convenient as it does not require tracking of the interface [1JM]. The theory of binary solid interface motion was used to show that vapor phase transport governs the problem [34JM].

Solidification in forced convective flows included an analytical and experimental study of frost growth on a flat plate with various droplet wetting angles [49JM]. Reference [46JM] described the appropriate dimensionless parameters to use in similar problems. A perturbation solution was used to analyze the freezing and remelting of a liquid flowing parallel to a cool flat plate [8JM]. Experimental results for water freezing inside a pipe showed that the ice layer forms a periodic pattern [20JM]. A numerical study illustrated the shape of the solid layer formed on a cold cylinder in crossflow [7JM]. A simple model and some experimental measurements showed that growth around a single tube is nearly identical to growth around the tubes in an array until 80% of the flow area is blocked by solid [27JM].

Three experimental papers considered the effect of natural convection on the freezing process. The effects on microstructural defects were examined for cadmium–copper alloys [42JM]. Freezing paraffin in a rotating cylinder redistributed the voids and accelerated the freezing process [33JM]. The effect of axisymmetric rotation on the freezing of aluminum–copper alloys was also investigated [41JM].

Solidification in Cartesian, cylindrical or spherical castings was analyzed subject to convective–radiative boundaries [6JM]. A new embedding technique was used to study the solidification within a long cylindrical mold [17JM, 18JM]. A finite element method was applied to a two-dimensional freezing problem that does not include natural convection found in corresponding experiments [35JM]. Another finite-element method was applied to the solidification of magnesium alloy castings [52JM]. Numerical models were employed to simulate the solidification in continuous cast round or square cross-section billets [50JM].

Crystal growth was characterized by reviewing the relevant heat and mass transfer processes governing the transport [4JM]. An analytical model was used to describe the growth of silicon crystals [28JM] and a two-dimensional finite-element model was developed to simulate the growth of gallium arsenide crystals [9JM].

Rapid solidification by melt spinning was investigated experimentally by several investigators to determine the thermal history during ribbon formation [24JM], to measure surface heat flux using a photocalorimetric technique [15JM] and to determine the effect of air jets [3JM]. An experimental study of the freezing of aerosolized aluminum–lead alloy melts was conducted in a nitrogen gas environment [51JM].

Melting adjacent to a heated surface was investigated experimentally in refs. [13JM, 14JM] where gallium was melted and solidified on a horizontal plate. An experimental and numerical study was reported in which natural convection in the melt governed the interface around a heated vertical cylinder [37JM]. Reference [25JM] measured heat transfer enhancements of a factor of five when longitudinal fins were attached to a heated vertical cylinder. A finite-difference numerical study investigated the melting process around a horizontal cylinder where solid subcooling was found to have a significant effect of the melting rate [40JM]. A quasi-three-dimensional solution was obtained for a variable temperature pipe buried in a semi-infinite solid phase-change material [57JM].

Melting inside a vertical rectangular cavity was investigated experimentally and numerically for *n*-octadecane [2JM]. Melting within a horizontal cylinder was measured [43JM] and the effect of the liquid film between the solid and the wall was analyzed using a two-dimensional numerical model [44JM]. An extension of these studies evaluated the effect of internal geometry on the heat removal of an encapsulated phase-change material [45JM]. An experimental study of melting in an inclined cylinder with an aspect ratio near four showed that the horizontal position was best and that the results did not change appreciably until the cylinder axis was within 15° of vertical [54JM].

A hot body melting its way through a solid has been observed using a horizontal heated cylinder and *n*-octadecane [31JM, 32JM]. The fluid motion in the cavity was driven more by the movement of the cylinder than by natural convection currents. An immiscible liquid melting its way through a solid was considered theoretically as an approximation to molten iron melting through the earth's mantle [36JM].

Melting caused by microwave heating of a dielectric material was solved as a boundary value problem [30JM]. Argon-ion laser cutting through silicon was investigated where the propagation of the melt front was controlled by the scan speed, not the laser power [48JM]. A numerical solution was obtained for laser welding which described the size, geometry and motion of the liquid pool [10JM]. Carbon dioxide and YAG lasers were used to cut and weld carbon and stainless-steels and the differences between penetration welding and conduction welding were described [11JM].

RADIATION IN PARTICIPATION MEDIA AND SURFACE RADIATION

In this review, the research in the area of radiation in participating media is described first, followed by the papers on surface radiation. There are a number of papers which are of common interest to both areas. These and other related studies are included at the end. Some attempt is made to include the works dealing with the radiative properties, e.g. nongray gas absorption, anisotropic scattering, and surface properties, but only a small sample of the information that is available is included. There are also many related research fields in which radiative transfer is studied, e.g. atmospheric sciences, aerosol research, remote sensing, combustion diagnostics. Since the intent here is not to review all of radiative transfer research, the focus is toward those works that deal with radiation heat transfer, although exceptions are made where appropriate.

Radiation in participating media

The research papers have been grouped into the following topics for the purposes of the review: one-dimensional media studies, studies in multidimensions, radiation combined with other modes of heat transfer, flame radiation, gas radiation, and scattering media.

(*a*) *One-dimensional media studies.* A wide variety of solution methods is chosen to study the radiative transfer in one-dimensional media. Unless otherwise noted, all the papers deal with one-dimensional, plane parallel media. The Galerkin method is used to obtain the heat flux, the intensity distribution, and the divergence of radiative heat flux in an absorbing, emitting, isotropically scattering sphere with diffuse reflecting, and emitting boundaries [117K]. The Galerkin method is used again to consider a sphere with space-dependent albedo [118K]. Modifications to the matrix operator method for solving the equation of transfer in planar media are proposed. An algorithm for calculating the reflection operator in a semi-infinite medium is developed and used for other optical depths also, reducing the computation time [68K]. Matrix-vector relationships for computing the internal intensity field in emitting and multiply scattering medium are formulated [125K]. Two scaling relations (one for scattering with highly forward peak, and the other for those cases when the scattering is not so peaked) are presented for the Monte Carlo method, to reduce the computation time [9K]. The F_N method is discussed [56K] and modified to solve some benchwork problems [39K]. Results for isotropically scattering spheres and cylinders are given, using the integral transformation techniques and the F_N method [104K]. A n-flux method, which uses a power series approach to deal with the directional dependence of radiation, is presented and recommended for its possibility of choosing n independent of the overall dimensions [60K]. Comparisons of the biangular reflectance of an absorbing and isotropically scattering medium, obtained with a discrete ordinate method and the actual data, are presented [94K].

A number of other solution methods is also utilized. A modified diffusion approximation is used to obtain the radiation characteristics of a plane layer of strongly scattering medium [111K]. Improvement on the accuracy of the diffusion approximation for the optically thin problems, is proposed by including the exact formulation of the integrated intensity at the interfaces [5K]. The finite analytical method, used to reduce the nonlinearity in fluid flow and convection calculations, is applied to radiation in absorbing, emitting, and gray medium with good accuracy [105K]. An anisotropically scattering layer with position-dependent albedo is analyzed by expanding the albedo in a Legendre polynomial series [21K]. Two sets of experimental data on attenuation in water, are compared with the solutions obtained using a finite-difference scheme similar to the S_n method [49K]. Numerical results for the radiation transfer by a finite, inhomogeneously emitting layer [97K], and reflection from scattered radiation in dispersive media [127K], are obtained by using the Pade approximation. A spherical harmonics formulation, which can be used for extreme forward-peaked scattering, is developed for a slab [129K]. In studying the effect of boundary reflection on the resulting temperature distribution and the heat flux, ref. [109K] uses the projection method, for which convergence is guaranteed. A linear systems method is used to solve some radiative transfer problems in a plane parallel layer [43K]. For the integrals arising in the solution of radiative transfer, involving an exponential integral function and exponentials, ref. [116K] presents numerical techniques and expressions for improved accuracy. To study the unsteady radiation field in an optically thick medium, ref. [38K] uses the Laplace time transform of the intensity in a conservative-unsteady problem and the intensity in the corresponding non-conservative-steady problem. Unsteady heat propagation by radiation is considered, and the solution is shown to be self-similar [89K].

The following papers seek to solve the integral equation resulting from radiative transfer in one-dimensional layers. The integral equation for a gray, emitting, absorbing and scattering layer with reflecting boundaries, is solved by a collocation method, using Chebyshev polynomials as basis functions [110K]. A mathematical consideration of an infinite, purely scattering layer (isotropic), looks at the Neumann solution with the theory of singular integrals of the Cauchy type [96K]. The searchlight problem in radiative transfer is considered by removing the singular components before numerically taking the inverse Fourier transforms [103K]. Results in a finite layer are also obtained by using the integral transformation technique and the F_N method [34K]. Singularity subtraction and the Gauss quadrature are used to obtain a system of linear equations to solve the singular Fredholm integral equation, describing the transfer in isotropically scattering layer [28K]. A two-step method to solve the Fredholm integral equation of the second kind, requires taking the first-order Taylor series expansion of the Kernel, and then using the collocation method [22K].

The next four papers consider the radiative transfer in one-dimensional media to obtain radiation property data, by using the solutions in conjuction with experimental data. The extinction coefficient and the solid thermal conductivities of glass fiber insulations are obtained from calorimetric measurements [92K]. The four-flux model solution of radiative transfer and the classical theory of lattice vibration, are combined with reflectance measurements of ceramic materials to yield the thermal radiation properties [67K]. Study of the sensitivity on the measurement errors, when using the multiple-scattering inversion methods to obtain the properties, shows that the method can be ill conditioned [82K]. Reference [51K] numerically tests another inverse method, this one using pulsed multiple-scattering experiments with the azimuth- and time-dependent one-dimensional solution, and shows that the scattering albedo, and the asymmetry factor, and perhaps a third parameter, could be obtained from as little as three azimuth angle measurements.

(*b*) *Studies in multidimensions.* Some of the papers on multidimensional radiative transfer also consider other modes of heat transfer. Such works are then included in the following section. Investigations in two-dimensional radiative transfer are reviewed first, followed by three-dimensional transfer in the next paragraph. Reference [40K] presents numerical solutions for two-dimensional, azimuthally symmetric, steady-state equation of transfer. Collimated laser incidence into a finite thickness, linear anisotropically scattering, cylindrical medium, is considered [27K]; numerical results are presented for the source function, heat flux, and the intensity at the boundaries. Reference [29K] models a uniform collimated incidence into a two-dimensional rectangular medium. A numerical solution is obtained by solving a set of algebraic equations resulting from singularity subtraction and quadrature formula. The results show that a one-dimensional solution is a good approximation except near the wall. Another study in two-dimensional rectangular geometry considers an enclosure of gray medium with internal heat generation [131K]. The solution obtained from a point allocation technique, shows that both the optical thickness and the enclosure geometry affect the resulting temperature distribution, while the average heat transfer to the boundaries is primarily affected by the enclosure geometry. Limiting expressions for the scattering contribution to the geometric-mean transmittance in a multi-dimensional absorbing and anisotropically scattering medium are presented [130K]; by using the kernel substitution technique, closed form expressions for selected two-dimensional geometries are also presented.

Three-dimensional radiative transfer is considered in the following four papers. Exact expressions for the radiative flux density and the energy source term are developed for absorbing, emitting medium in a box-shaped enclosure [101K]. The resulting numerical results are compared with data from a large scale experimental furnace. Alternative, but equivalent, exchange factors for the zonal analysis are developed in ref. [61K]. The advantage with the new formulation is that these exchange factors can be experimentally measured for different geometries. Scattering medium in a planar geometry, with inhomogeneities in the x, y and z directions is considered [69K]. For the general problem of intensity field incident from above and a surface below, the technique proposed in ref. [69K] uses the spatial Fourier transforms and extends the matrix-operator theory developed for one-dimensional planar layer analysis. Reference [77K] considers inhomogeneous, anisotropically scattering media in a three-dimensional rectangular enclosure, by the first and the third spherical harmonics approximation. The purpose is to be able to model furnaces, and the numerical results are compared with those based on the Hottel's zone method.

(*c*) *Radiation combined with other modes of heat transfer*. In many heat transfer applications, one often finds more than one mode of heat transfer; phase changes may also be taking place. Since such situations arise in a variety of applications, it is likely that the review of research papers in this section is not complete. This section is organized to deal with radiation/conduction with possible phase change first, followed by radiation coupled with forced convection, and lastly the coupling with natural convection.

Transient response by combined conduction and radiation heat transfer in porous, highly scattering, semitransparent, high temperature material to intense solar irradiation is studied [72K] using the two-flux method. The numerical results show that the radiation is an important component even for optically thick layers (optical depth > 100). It is in fact, the dominant mode of heat transfer in the front third of a layer, and the location of the maximum temperature is in the interior of the layer. Analysis of the combined conduction/radiation in a gas within a cylindrical enclosure with reflecting boundaries, uses the zone method and the weighted sum of gray gases model, to include the radiation [57K]. The axial and the radial gas temperature distribution, and the wall heat flux and temperatures, are then presented for either the temperature specified or the heat flux specified boundary condition. Thermal conductive/radiative modeling of one- and two-dimensional structural elements is the subject of ref. [4K]. Apparent thermal conductivity of evacuated SiO_2–aerogel tiles is determined by guarded hot plate measurements [99K]. Combined radiation/conduction through multilayer insulation, with spacers that scatter or absorb, is studied both numerically and experimentally [78K]. The next two papers deal with the unsteady response of liquid layers exposed to incident radiation. Reference [53K] shows that the absorption of the liquid is a more important factor than whether the incident radiation is collimated or diffuse. Reference [52K] considers collimated incidence onto a liquid layer seeded with fine particles and shows that for the conditions under study, scattering is negligible. References [86K, 88K] consider phase changes associated with radiative transfer. The effect of anisotropic scattering on the melting of semi-infinite, semi-transparent media is such that a forward scattering fosters melting while backward scattering retards it [86K]. Theory of evaporation of water droplets suspended in air when there is radiation incident, including the infrared participation of the evaporated water, is discussed in ref. [88K].

Radiation coupled with forced convection also

received much attention. A very thorough experimental and numerical study of the thermal conditions in an irradiated, slowly moving liquid layer, is presented in ref. [50K]. A one-dimensional planar radiation model is coupled with laminar three-dimensional convection. The experimental and the numerical models both show a stably stratified upper layer, a core region of roughly uniform temperature, and a thin bottom layer of large, unstable temperature gradient. Similar geometry of a liquid layer in motion—molten salt in a gravity-driven flow—is analyzed [128K]; an optical depth of two is recommended for optimum solar collector efficiency. Process fluid in mixed channel flow (forced and free convection) is irradiated through an upper enclosure which holds a transparent fluid [18K]; the velocity and the temperature fields of both the fluids are numerically obtained. The results show that the lower flow section is strongly affected by the optical thickness of the absorbing fluid. Reference [106K] considers fully developed laminar and turbulent flow of real gases in a cylindrical enclosure where the radiation is included by the zone method. Experimental and theoretical studies of the thermal radiation in two-phase flow show that the condensed phase radiation is dominant, compared to the gaseous component [59K]. The calculation of radiative–convective heat exchange in hypersonic flow about a body is examined [93K]. In particular, the effect of using several models of optical properties of the gaseous medium in a compressed layer, is estimated. Reference [65K] obtains a modified local heat transfer coefficient of a plate fin which transfers heat to the surrounding fluid by radiation and convection.

Radiation coupled with natural convection is the topic of the next papers. References [31K, 7K] are concerned with stability analysis which includes scattering and natural convection. Reference [31K] seeks a realistic base flow solution in a vertically stratified layer for stability analysis; the first order spherical harmonics ($P-1$) approximation is used to include the radiation. Reference [7K] considers the effect of anisotropic scattering on the stabilizing influence of thermal radiation. The scattering is included by using an extended Eddington approximation, and the results show that the critical Rayleigh number and wave number are reduced. Reference [108K] presents experimental heat transfer results of convective–radiative transfer from a horizontal finned tube with sidewall/floor interactions. The effect on the heat transfer is the greatest with the sidewall/floor interaction, and the reduction can be up to 40%.

(*d*) *Flame radiation*. Since it would be difficult to review all the papers in the area of combustion research, only those which explicitly deal with radiation heat transfer in or from flames, are mentioned. Since flames often include scattering particles as well as participating nongray gases, some related papers may also be found in the following two sections. References [123K, 124K] utilize a three-dimensional formulation to determine the incident radiant heat flux on a cylinder that is engulfed in flames. Reference [126K] inverts the infrared emission measurements from a flame to determine the spatial temperature distribution. Reference [83K] presents results from instantaneous, monochromatic infrared radiation measurements from combustion flames of a 4-stroked, direct injection, no swirl diesel engine. The measurements show a continuous thermal emission by the soot and gas bands at 1.9, 2.9, and 4.2 μm. References [48K, 76K] both consider obtaining radiation properties of soot radiation in flames. Noting that the thermal radiation intensity from flares is more accurately modeled when the flame source is considered as a surface, ref. [30K] considers the jet diffusion flame from flare stacks. Reference [23K] reports on total and spectral flame radiation measurements made in a tubular-can combustor. Reference [87K] reports on using yttrium–beryllium oxide thermocouple coatings for flame temperature measurements.

(*e*) *Gas radiation*. Radiative exchange in gaseous media is characterized by the highly spectral nature of gaseous absorption. Of the many papers that deal with the properties of nongray gases, only the ones that deal with common combustion gases are considered here. Reference [115K] presents a semi-empirical, total line shape model for water vapor. References [17K, 46K] present measurements and calculations for the bandstrength and the linewidths of CO. For the CO_2 4.3 μm band, ref. [24K] presents measured absorptivity in a O_2–methane flame at 2900 K; refs. [63K, 25K] report on the temperature dependence of the absorption in the far wing of the band; ref. [19K] considers the spectral line shapes for the band from experimental measurements. CO_2 line positions and the strengths in the 1200–1430 cm^{-1} region are also reported [119K]. References [16K, 15K] report narrow band parameters based on the Elsasser model as well as the correlations for Edwards wide band model for acetylene and methane.

With the nongray gas properties, heat transfer analyses can be performed. Reference [107K] tests the validity of the narrow band models and the random statistical models for CO_2 and H_2O when calculating the temperature and the heat flux distribution for coupled radiation and conduction. Reference [47K] applies the total transmittance nonhomogeneous radiation model to methane combustion, and shows that this method is about 500 times faster than using a narrow band model and still is within 10% accuracy. Reference [54K] reports on modification of the expressions relating the surface temperature and the heat fluxes in an enclosure with isothermal gas, to obtain direct expressions for the heat flux which can be numerically calculated. In different applications ref. [90K] recommends a solar energy receiver–thruster with a rocket propellent of H_2 gas with alkali metal vapors for volumetric gas absorption.

(*f*) *Scattering media.* Many of the papers already reviewed, consider radiative transfer analysis in scattering media. This section includes the studies on scatterings separate from a heat transfer analysis. Since scattering is such a widely studied subject, only a small selection of the works dealing with scattering is included here. Notably, some of the detailed scattering analyses and the size distribution retrieval work are not included. Reference [26K] examines the Dirac delta approximation to the Mie calculations of large dielectric spheres. Reference [75K] recommends an experimental procedure to determine the scattering similarity parameter from *in situ* measurements within a cloud. It is not possible to distinguish experimentally between the forward scattering component and the transmitted beam, but for spherical particles Mie theory can be used to calculate a correction. Reference [13K] considers the correction for nonspherical particles, and finds that although oblate spheroids are similar to equal-area spheres, the correction factor for prolate spheroids can be as much as 20% less then the equal-area spheres. Reference [1K] reports on the calculations and experimental data on scattering from large dielectric cylinders. Reference [6K] finds that for large conducting spheres with rough surface, the scattering cross section calculated can be significantly different than if the surface is smooth. Reference [44K] presents a simple analytical method to approximating Mie scattering, which is an extension of the Rayleigh–Debye approximation. A theoretical method for predicting scattering from random aggregates is tested on latex spheres [62K]. Optical constants of soots in the 0.5–0.75 μm range are obtained from polarized angular reflectance measurements [8K]. Reference [55K] presents the single scattering properties of sodium mist from Mie calculations. Anisotropic scattering in insulations made from randomly oriented fibers is investigated theoretically, and used to optimize the fiber for minimum heat flux [20K]. Reference [2K] considers the regularity of the radiation from two-phase media, e.g. gas–particle systems, bounded by reflecting surfaces.

Surface radiation

The research papers dealing with view factor calculations are presented first, followed by the papers on surface exchange. Reference [35K] describes an algorithm for calculating the view factor, by using Hottel's crossed string method, in an enclosure bounded by arbitrary polygons in the plane. Reference [71K] considers the same kind of boundaries, but looks at the computational savings from transforming the surface integrals into line integrals. Reference [102K] compares three computer algorithms used to compute view factors. Reference [36K] considers calculating the view factors for large, sparse structures when curved surfaces may be present. With the appropriate view factors, energy exchange between surfaces can be calculated, but some ambiguities may need to be cleared up [112K]. A transparent enclosure with diffuse and collimated incidence is analyzed to study the absorption of solar energy in a building space [74K]. Reference [80K] considers convection shields for radiative cooling which have high transmittance in the 8–13 μm wavelength range. Reference [66K] evaluates the effect of radiant transfer to the walls and ceilings, from hot stove and chimney pipes, in simulated overfire conditions.

A number of papers concern the surface exchange in cavities and screens. Reference [58K] presents an exact, closed-form reconstruction algorithm for the temperature distribution of a blackbody from spectral measurements. Reference [84K] numerically computes the bandlimited effective emissivity of nonisothermal diffuse radiators, and shows that the nonisothermality causes only a small deviation of integrated emissivity for a detector a little distance away from the cavity. Reference [10K] considers the emissivities of isothermal and nonisothermal cylindro-inner-cones. Reference [122K] obtains a single equation for the spherical cavity radiation conductivity with a general specular-diffuse surface. Reference [91K] considers the transient temperature field of a cavity with cooled walls and an aperture that is permeable to radiation, to model a receiver of thermal solar tower plant. Reference [95K] utilizes optico-geometric functions to compute heat losses through a system of finite screens. Reference [121K] proposes the construction of a photoelectric apparatus for light model ing, to determine the geometric and the radiation properties of semitransparent screens.

Other related studies

There are a number of papers that discuss the instrumentation needed for experimentation. Reference [85K] shows that the resolution, in photodiode array Fourier transfrom spectroscopes, can be improved by adding dispersing optical elements. The integrating sphere is the topic of the next three papers [113K, 37K, 41K]. Reference [113K] discusses the possibility of increasing the detector signal by adding elliptic flux concentrators, when the field of view is limited. Reference [37K] discusses possible misinterpretations when using an integrating sphere. Reference [41K] uses an integrating sphere with a Fourier transform infrared spectroscope to make reflection and emission measurements. The radiation detector is the topic of the next three papers [100K, 32K, 12K]. Reference [100K] discusses using sodium salicylate as a combined fluorescent and reflective coating on a photomultiplier tube to extend the range of the detector into the near infrared. Reference [32K] discusses the signal to noise advantage of a resonant bolometer. Reference [12K] considers the ideal coating for radiometric detectors—low reflectance, independent of wavelength and direction—and shows that the coating thickness is a critical parameter. Reference [45K] is a general paper that discusses the theory and implications of an integrating nephelometer which is used to make scattering measurements. Reference [70K] analyzes a transient calorimeter used to measure the total hemispherical emittance, and shows the importance of rigorously evaluating the heat loss through the thermocouple leads.

This last group of papers includes radiation in combination with various other effects, mostly as a boundary condition. References [14K, 3K, 79K] deal with radiation effects encountered while measuring thermal properties. Reference [14K] accounts for the radiant energy penetration when finding the thermal diffusivity of layered and porous polymers by the flash technique. Reference [3K] discusses a similar experimental setup for determining the thermal diffusivity of liquids, by heating the front surface of a cell with radiation. Reference [79K] is a summary article of the discussions in a workshop during the 9th Thermophysical Properties Conference, on the topic of the effect of radiation on thermal transport measurements. Reference [11K] discusses an application where one can achieve refrigeration by using reverse-biased semiconductor diodes. Reference [73K] predicts an improved performance when liquid droplets, rather than solid surfaces are used as radiators for cooling in space; an experimental evaluation of the properties of Dow 705 fluid shows that the weight would be about 10 times less when the temperature range is 275–335 K. Reference [64K] studies the transient response of water subject to intense, focused laser pulses of nanosecond duration. Reference [33K] considers the transient response of the temperature distribution of a sphere, with directed heat flux and radiative cooling boundary conditions. Reference [98K] considers the same type of boundary conditions on a rotating hollow cylinder, and shows that quasi-steady temperatures are achieved within one revolution. Reference [42K] numerically solves the heat conduction in a semi-infinite layer with boundary heating and radiative cooling. Reference [81K] considers the application of infrared radiation to drying films, and shows that the internal heating due to the penetration of radiation has significant effects on the drying rate and the temperatures. Reference [120K] considers the response of transluscent materials to high energy radiation fluxes, and concludes that the heat exchange process depends on whether the external radiation is diffuse or collimated. Reference [114K] considers the temperature rise of a surface subject to moving heat sources, radiative cooling to the environment, and convective cooling to the lubricating oil, and concludes that the effect of radiation is negligible.

NUMERICAL METHODS

The use of numerical methods for heat transfer is very common. In general, the papers dealing with the application of a numerical technique to a particular problem are reviewed in the appropriate application category. The papers mentioned in this section are primarily concerned with the numerical method rather than with the application used to illustrate it.

A literature survey of numerical heat transfer for 1982–83 has been published [50N]. The progress of numerical heat transfer has been reviewed in ref.

[51N], where some suggestions for future work are offered.

Heat conduction still continues to be the main testing ground for new formulations. Nine different numerical methods for the solution of three-dimensional heat conduction have been compared in the respect of their performance on a common problem [53N]. A finite-difference scheme has been proposed for the heat conduction equation [33N]. Reference [4N] describes a new explicit scheme for heat conduction. Some numerical algorithms for one-dimensional heat conduction are compared in ref. [28N]. A control-volume finite-element method using a nine-node quadratic element is developed [46N]. A collocation technique is described for the solution of the Laplace equation [7N]; it is suitable for implementation on a personal computer.

Unsteady heat conduction problems have been solved by the use of boundary-fitted coordinates for multiconnected arbitrary-shaped domains [54N]. The question of numerically solving hyperbolic heat conduction is discussed in ref. [21N]. The use of direct and indirect boundary integral methods is evaluated in refs. [3N, 2N]. The boundary element method is discussed in ref. [57N]. The effect of stringent boundary conditions on the solution of heat conduction is investigated in ref. [10N]. Numerical modeling of heat conduction in sand castings has been studied in refs. [11N, 12N]. Reference [22N] describes a new self-adaptive grid method.

A two-dimensional inverse conduction problem is solved by the method of dynamic programming [8N]. A finite-element formulation for the inverse problem is given in ref. [56N]. The Stefan problem with multiple free boundaries is solved in ref. [14N], while ref. [45N] presents a new implicit solution procedure for the problem. An efficient algorithm has been developed for problems with melting and solidification [27N]. An unconditionally stable finite-difference scheme has been proposed for conduction with phase change [39N]. Multidimensional moving boundary problems have been treated by coordinate transformation [23N]. Reference [24N] describes a stochastic modeling of transient heat flow through walls.

Considerable research effort has been directed towards the formulation of convection and diffusion so as to produce an accurate and stable numerical scheme. A streamline-upwind finite-element method for convection-dominated flows has been developed and applied [35N, 36N]. A monotone streamline-upwind method has been proposed [47N]. Reference [15N] proposes a generalized Galerkin method for convection–diffusion problems. Finite-difference and Galerkin techniques have been compared by applying them to the combined convection–diffusion problem resulting from the propagation of pollutants from a line source [49N]. A potential and stream function analysis has been presented for the convection–diffusion problem [55N]. Reference [52N] describes a solution scheme for the convection–diffusion equation, while ref. [37N] presents exact and approximate solutions for the equation. The problem of convection and diffusion has been solved by the boundary element technique [26N]. A new explicit scheme has been proposed for the problem [18N]. Reference [43N] introduces a control-volume finite-element method using quadrilateral elements, while ref. [40N] employs the multigrid technique for the convection–diffusion equation. The flux-corrected transport method has been compared with other schemes in ref. [34N]. Reference [38N] presents a critical evaluation of seven discretization schemes for the convection–diffusion problem.

Prediction of convective heat transfer requires the calculation of the underlying fluid flow. A significant part of the literature is devoted to the numerical methods for the calculation of fluid flow. An adaptive grid technique has been developed for parabolic flow problems [1N]. For compressible gas flow with chemical reaction, an optimized implicit scheme for the gas flow has been proposed [13N]. Reference [41N] describes a multilevel multigrid technique for recirculating flow. A coupled strongly implicit procedure has been formulated for the calculation of velocity and pressure [58N]. Further performance characteristics of marching methods for elliptic equations have been described [48N]. For laminar forced convection, a boundary integral equation method has been presented [44N].

A new cubic spline alternating direction implicit method has been employed to predict natural convection in a cavity [31N]. A formulation for the Navier–Stokes equations using a coupled solution of velocity and vorticity has been described [19N]. Reference [32N] uses velocity and vorticity as variables for solving mass transfer equations. A residual method of finite differencing has been presented for the solution of elliptic fluid flow [5N]. A control-volume finite-element method has been developed for the solution of the Navier–Stokes equations [42N]; the method uses equal-order interpolation for velocity and pressure. Reference [9N] describes continuation techniques for a penalty approximation for fluid flow. A family of stable algorithms has been developed for the solution of the generalized Boussinesq equation by a finite-element approximation [20N].

Matters concerning boundary conditions are dealt with in some papers. Reference [25N] describes a finite-element procedure for buoyancy driven flows with cyclic boundary conditions. A new type of boundary condition in convective heat transfer problems has been proposed in ref. [17N]. Reference [16N] discusses errors arising in convective heat transfer problems in connection with boundary conditions of the third kind.

Comparisons of different methods provide useful guidance in selection of an appropriate method. A comparative analysis of numerical methods for hypersonic re-entry flows has been presented [6N]. Reference [30N] gives the results of a comparison of pressure–velocity coupling solution algorithms. A comparative study has been made of various grid generation techniques in connection with a parabolized Navier–Stokes code [29N].

TRANSPORT PROPERTIES

Research in heat transfer properties includes precise scientific measurements and the expeditious determination of engineering values for industrial application or special materials.

For thermal conductivity and diffusivity a number of new experimental techniques are described. A new design [12P] simultaneously measures thermal conductivity and diffusivity of unconsolidated materials by the transient hot-wire method. For liquids, a novel precision resistance-measuring scheme yields absolute conductivities [36P]; for thin films, ref. [15P] presents results obtained by an AC calorimetric method and [19P] a new technique for simultaneously measuring thermal conductivity and heat capacity. Fibers are the focus of two studies. The first [32P] examines carbon fibers in the range 3–300 K; the second [16P] the effect of misoriented short fibers upon the thermal conductivity of a composite. A dispersion of fully penetrable spheres is examined for effective thermal conductivity [50P], and ref. [43P] reports thermal diffusivity information of partially crystalline polymers. Continuing the study of specialized materials, ref. [41P] reports apparent thermal conductivity of evacuated SiO_2–aerogel tiles under variation of radiative boundary conditions; ref. [25P] gives results for iron–titanium powders; ref. [22P] indicates the variation of thermal conductivity with density for polyurethane foam using a guarded hot-plate; and ref. [28P] records the thermal conductivity of multicomponent saturated refrigerants.

For gases at high pressure, ref. [37P] employs a Pitzer-type generalized correlation for predicting thermal conductivities. For liquid and vapor hydrocarbon systems (pentanes and heavier) at low pressures the normal boiling point and specific gravity correlate the thermal conductivity [38P]. For multicomponent dense gas mixtures, ref. [47P] obtains an approximate formula for computing the heat conduction coefficient. Additional studies examine the thermal conductivity of specific systems: a semi-theoretical equation relates conductivity to volumetric foam density [24P]; an extended correlation technique allows thermal conductivity predictions to be made for coal-derived liquids and petroleum fractions [4P]; ref. [13P] examines the effective thermal conductivity of a suspension; and ref. [31P] measures the thermal conductivity of alumina and silicon carbide in the range 4–300 K.

Diffusion coefficients are calculated in two works: ref. [51P] considers certain dilute binary gaseous systems and employing several asymmetric potentials predicts, for the temperature range 195–400 K, diffusion and thermal diffusion factors; ref. [3P] calculates the coefficient of self diffusion for steam over the pressure range 0.07–27 bar using a theoretical model. A technique for simultaneously determining solute diffusion coefficient and partial pressure is described in ref. [54P].

For electrical conductors, ref. [53P] describes a pulse-heating method for measuring heat capacities from room temperature to 1300 K and gives results for zirconium. For amorphous metal in the range 1.5–10 K, ref. [23P] also uses a pulse technique requiring a small sample and yielding results for copper within ± percent. The specific heat of ultrafine particles of silver, prepared by the gas evaporation technique, appears in ref. [34P].

A concern with systems undergoing pliable change during heat transfer leads to the consideration of the molar latent entropy change as a more fundamental rationale for the thermodynamic generalizations employed in two-phase heat transfer [44P]. The need to predict the thermal regime in arctic offshore structures led to an investigation of the latent heat of frozen saline coarse-grained soil [52P].

Precise viscosity measurements continue to flow from well-known sources. Results are reported for sulphur hexafluoride up to 100 bar using a capillary-flow viscometer [18P] and for light and heavy water and their mixtures by the oscillating viscometer [20P]. An empirical correlation using available experimental saturated liquid viscosity data is presented for 10 refrigerants [45P]. For measuring liquid viscosities a vibrating sphere yields density as well [14P], while ref. [7P] utilizes an automated falling-cylinder, high-pressure laser-Doppler viscometer.

Velocity of sound measurements are reported for argon gas up to 10^3 bar [21P] and in reactive liquids at high pressure and temperature [42P].

A number of papers are focused sharply upon the needs of certain technical endeavors. Thus the calculation of transport properties of mixtures encountered in synthetic fuels process design are discussed in ref. [49P] and mass/heat transfer aspects of anaerobic reactor [33P]. Nuclear engineering activities provide the impetus to consider the thermal conductivity of simulated radioactive waste glass [40P], the measurement of the thermal conductivity of liquid urania [11P] and lithium-containing ceramic breeder material [48P], the thermal properties of stabilized zirconia [9, 29P], and the role of organic coolants for nuclear reactors [39P]. In the textile and fiber areas a comparison is made of the heat and mass (water) transport through cotton and polypropylene [10P]. The thermal insulation properties of 12 fabrics in high-intensity heat exposure were tested in ref. [5P] and the effect of free water in fibrous insulating materials on thermal and physical properties examined in ref. [30P]. Rocket propulsion systems cause the thermal transport properties of ammonium perchlorate to be examined [46P]. The efficacy of metallized plastic sheeting in providing thermal protection is studied in ref. [8P]. The influence of soil tillage on its thermal properties is measured as having a 20% effect on the thermal conductivity using the line source heat-probe method [35P].

In the area of building materials *in situ* measurements of *u*-values for a built structure are obtained by recording heat flows and temperature differences over a period of days and calculating the *u*-value as the quotient of means [27P]. For insulated sandwich panels penetrated by structural members a simple method is presented of calculating the theoretical *u*-value to account for the losses caused [26P].

Measurements on several special systems involve a radiant heat flux apparatus for measuring the thermal response of polymeric materials to high temperatures [17P]. The thermal conductivity of several evacuated load-bearing porous insulations at temperatures between 300 and 650 K [6P], and the analysis of heat transfer during freezing of slab-shaped food commodities [2P]. An inverse method of thermophysical properties and heat fluxes is given in ref. [1P].

HEAT TRANSFER APPLICATIONS—HEAT PIPES AND HEAT EXCHANGERS

Heat transfer applications range widely across a variety of technical fields embracing analysis and testing, modeling and design, heat exchange enhancement and optimization, including individual influences as well as overall performance. Considerable interest continues to be shown toward extended surfaces for improving heat transfer and the heat pipe.

Computer techniques are used to analyze shell-and-tube exchangers [72Q]; multipass heat exchanger networks [64Q]; transient, three-dimensional single-phase flow in nuclear reactor rod bundles [18Q]; the dynamics of transient states of the counterflow heat regenerator [110Q]; and the optimization of heat transfer equipment [66Q].

The electrical enhancement of heat transfer is considered [17Q] for a gas–solid suspension heat exchanger utilizing such a field and in ref. [51Q] where resonant oscillation of drops is induced. For an extended surface, ref. [102Q] determines its temperature distribution for a nonuniform heat transfer coefficient. With optimization as the goal, a number of studies consider specific systems: annular fins with variable thermal parameters [68Q]; arrays of longitudinal, rectangular fins in convective heat transfer [10Q]; choking and optimization of extended surface arrays [56Q]; natural convection and radiation heat transfer from highly populated pin fin arrays [92Q]; and longitudinally-finned cross-flow tube banks and their hydrothermal characteristics [90Q]. A combined analysis and optimization of extended heat transfer surfaces is given in ref. [48Q]. For rectangular fins protruding from a vertical rectangular base, a three-part study [60Q, 61Q, 62Q] reports the optimal fin separation and thermal performance for this configuration. For compact heat exchangers, ref. [86Q] considers the effect of contact resistance in the composite construction used in semiconductor cooling systems; ref. [52Q] treats the efficacy of turbulence promoters; and ref. [95Q] models an offset-strip-fin type exchanger for use at low Reynolds number. In a two-part study [32Q, 33Q] the hydro-thermal performance of finned tube bundles and typical air cooler finned tubes are examined.

The use of finned surfaces for systems undergoing phase change receives consideration by a number of investigators. An analytical model for predicting condensate retention on horizontal integral-fin tubes is presented in ref. [82Q]; the effect of fin spacing on the performance of such a configuration in condenser service is covered in ref. [111Q] and the prediction of the condensation coefficient in ref. [108Q]. Other investigations explore condensation enhancement of vertical spiral double fin tubes with drainage gutters [97Q] and heat transfer coefficients for a vertical fluted-plate refrigerant condenser [101Q]. Two boiling papers consider the optimum use of longitudinal fins of rectangular profiles in boiling liquids [14Q], and surface tension effects on convective boiling in compact heat exchangers with offset strip fins [23Q].

The fouling of heat exchangers is the focus of some investigators. Estimating friction factors in a fouled annulus [67Q]; the causes and effects of sedimentation in condensers and heat exchangers [100Q]; a simulation procedure for chemical reaction fouling in heat exchangers [35Q], a high-intensity ultrasonic method of scale and deposit removal [75Q]; and a general discussion of the problem [16Q] all address this aspect of heat transfer equipment.

For heat exchangers, ref. [80Q] considers the transient response of the parallel flow type, ref. [113Q] calculates the temperature fields in cross-flow including the effects of inlet non-uniformities and ref. [57Q] seeks to characterize the 'thermo-hydraulically efficient' exchanger surface. Specific influences on exchanger performance are considered by ref. [76Q]—viscosity induced non-uniform flow in laminar flow; ref. [50Q]—turbulators in fire tube boilers; ref. [3Q]—unsteady-state transfer and mixing due to flow twisting; ref. [40Q]—the effect of manufacturing inaccuracies and deformation of tube bundles; and ref. [27Q]—two-phase flow in exchangers and pipelines. For compact heat exchangers, single-blow transient testing of surfaces is discussed [63Q], construction of and measurements with extremely compact, cross-flow types is described [96Q] and the general nature and purpose of such devices [52Q].

For plate exchangers, ref. [11Q] discusses the temperature distribution in parallel channels; ref. [12Q] the temperature distribution in a four channel exchanger; and ref. [34Q] the effect of corrugation inclination angle on exchanger performance. To evaluate and coordinate experimental studies of such exchangers, ref. [46Q] proposes a mathematical model. The design of a plate-type evaporator for heat pumps is described by ref. [45Q].

The use of porous materials joins three studies. In a two-part investigation, refs. [37Q, 38Q] examine the improvement in efficiency by the use of such material in a shell and tube design and a single tube-shell and tube exchanger; ref. [22Q] considers the effect of a porous layer on the hydrodynamic behavior of the coolant.

The design of exchangers receives consideration [55Q] through a mathematical analysis governed by a single similitude criterion. Other papers stressing design are directed toward specific applications: refrigerant exchanger [93Q] and steam generator [13Q].

Solid–gas exchangers are investigated in the form of 'raining bed' exchangers for high temperatures [19Q] and an experimental study of the high temperature performance of a pebble bed exchanger [112Q].

Studies stressing the chemical aspects of exchangers include the modeling of an exchanger in which phase changes are possible in both multicomponent streams [106Q]; a novel processing technique combining separation of a mixture into its components with heat transfer [59Q]; the thermal character of a heat exchanger–chemical reactor with coolant in cross-flow [47Q]; and a single fluid, multi-pass exchanger and combustor for low-heating-value gas [28Q].

Specialized exchangers are studied for nuclear reactors: very high-temperature gas-cooled reactor in cross-flow [98Q]; and heat removal in fast reactor cores [114Q]. Oil is the medium of interest in ref. [107Q]—optimum exchanger design for crude oil stabilization; and ref. [83Q]—experimental performance of an oil–air exchanger. Evaporative cooling exchangers are considered by simulation [109Q] analyzed for the packed bed, recirculated-liquid gas cooler–humidifier [58Q]. Air–molten salt direct-contact exchange is examined in ref. [15Q]; a scraper to augment condensation heat transfer is reported [65Q] and the role of the exchanger in power systems is generally discussed in ref. [39Q].

Research on the heat-pipe or thermosyphon continues to attract a high level of interest. The modeling of such a system is considered by ref. [42Q] for a closed-loop type including the Soret effect and by ref. [24Q] for the transient operating characteristics of a low-temperature device. A method of analysis is given for exchangers involving the principle by ref. [49Q]; a variable area one- and two-phase steady-state loop in ref. [77Q]; and a heat-pipe thermodynamic cycle in ref. [53Q]. Mixed boundary conditions for the loop syphon is considered by ref. [43Q].

The stability of these systems is treated in a number of studies. Steady-state behavior and stability thresholds for a closed two-phase type is investigated in ref. [29Q]; oscillatory instability in closed-loops by ref. [25Q]; complex oscillations in ref. [44Q]; and throughflow effects and the effect of motion onset in refs. [118Q, 117Q], respectively. Measurements of working fluid distribution within the wick are reported [87Q] and natural convection in the concentric tube system is treated by ref. [94Q]. The toroidal configuration is the focus of ref. [81Q] for high Graetz and Grashof numbers; ref. [84Q] for the inclined device and the steady-state velocity; and ref. [85Q] for known heat flux. Discrete influences or special circumstances mark the following works: improving maximum heat transfer for axially grooved pipes [9Q]; two-phase operation for fluid near the critical point [41Q]; double-wall artery, high capacity pipe [74Q]; effect of liquid plug in vapor channel [78Q]; design of air–water, co-axial exchanger [5Q]; and a water–air, gravity assisted exchanger [4Q]; low-temperature systems [105Q]; and use in a metallurgical furnace [26Q]. The device is reviewed from its inception in 1973 to the present with its many applied forms [79Q].

Regenerator performance is predicted using a non-linear analogy method [7Q, 8Q] and the Galerkin method applied to solve the symmetric and balanced counter-flow problem [6Q]. Experiments are performed [69Q] in order to determine the thermal stage performance for the device in semi-continuous operation. Quick design and evaluation [31Q], approximate methods for evaluation efficiency in periodic operation [30Q], and the effect of flow-through capacity on the hydro-thermal performance of a Stirling engine type [71Q] conclude the work on this device.

For heat transfer in tubes, attention is directed toward the augmentation of laminar flow exchange by wire coil inserts [104Q], correlations for oscillations in long, sodium-heated, steam-generating tubes [103Q], impedance heating [54Q], and the measurement of heat and mass transfer from humid air to a horizontal tube in a false bank [36Q]. Tube assemblies are examined for the influence of surface roughness on heat flow and drag for crossflow [116Q]; for conditions of laminar axial, transverse and inclined flow [2Q]; and under conditions of turbulent inclined flow [1Q]. Correlations for heat flow and pressure drop for in-line and staggered banks are suggested [115Q]. Proper selection, inspection, maintenance, and failure analysis for HVAC heat transfer coils are discussed in ref. [88Q]. A review of North American usage for estimating refrigerant-side heat transfer and pressure drop for direct expansion coils is given [99Q] and a cost-estimating method provided for double-pipe and multi-tube exchangers with bare tubes and external-longitudinally-finned tubes [73Q].

Shell and tube exchangers are treated in a review on current practice for shell-side two-phase heat transfer [89Q], a prediction method for pressure drop in the instance of roughened tubes and smooth shell surface [20Q], and the determination of shell-side heat transfer coefficients by the naphthalene sublimation technique [70Q, 91Q]. The characteristics of dry cooling towers for condensate cooling are considered by ref. [21Q].

HEAT TRANSFER APPLICATIONS—GENERAL

A large number of papers has been published on heat transfer occurring in specific areas of engineering activity. They are scattered in many journals. Approximately one-half of those known to us have been included in this section of the review and are discussed with the aim of providing a feeling for the various fields of engineering activity in which heat transfer is encountered.

A considerable number of papers deal with *metal and polymer processing*. A good fraction of them are analytical. The temperature fields are calculated for the heating [54S] and cooling [62S] of steel casting because of the pronounced effect which it has on the microstructure of the steel. Friction welding is modeled [27S] and unipolar induction heating is considered [58S] as well as electromagnetic turbulent stirring in furnaces [61S] and impedance pipe heating [38S]. Models for the analysis [45S] and experimental results [18S] are reported for polymer molding. Differences up to 30°C in the cavity surface temperature have been observed in cyclic injection moldings [2S]. Several models describing the effect of heat transfer on the spinning of glass fibers were examined [44S] and experiments established heat transfer in the heating of molten glass banks agitated by jets [9S].

Heat transfer in *chemical processing* is discussed in a number of papers dealing with topics like mathematical modeling used in the scaling of fixed-bed [24S] and agitated [41S] reactors. The penetration theory is used to study heat transfer in stirred reactors [40S, 42S] with immersed heating elements. Heat transfer in boiling of contaminated sulphuric acid [65S] is discussed and a number of papers [49S, 56S] deals with heating and devolatilation and with freezing of coal.

Ground heat transfer studies included methods to calculate the daily soil heat flux [31S] and the subsurface temperature distribution [46S] and provide a model to calculate shallow soil heat extraction [43S] as well as the use of geologic formations for seasonal thermal energy storage [64S]. Greenhouse heating by underground pipes has been analyzed [4S]. Surface heat flow measurements at hot spots of the Kilauea volcano in Hawaii [22S] were used to locate magma deposits. The growth and decay of ice on the Lawrence river [30S] and the undulation of water on an icy Siberean surface [29S] were modeled and described as well as flow and temperature distribution in aquifers [53S] and in wells [5S].

Modeling results of *heat transfer in buildings* were compared with experimental results for two Dutch dwellings [1S] and included the effects of wind direction on the energy consumption [12S]. Building envelopes using new energy efficient materials were optimized [34S] for heating and cooling load. The effect of phase-change component materials in thermal storage walls on their thermal performance is investigated [13S] by a time-dependent analysis. The use of transparent films to transmit solar and reflect infrared radiation in greenhouses is discussed [6S].

Several papers deal with heat transfer in *nuclear power plants* and some for *coal-fired power plants*. A survey compares results of computer codes with experiments and plant operation to study natural circulation cooling in pressurized water reactors [7aS]. Heat transfer from a partially uncovered reactor core was modeled [63S]. The temperature distribution in fuel rods [59S] and scaling criteria in nuclear reactors [35S] were also considered. An analysis studied the performance of coal-fired power plant spray-cooling systems [52S].

The spreading of *fire* in buildings and in nature was studied by analysis and experiments for a fire plume impinging on a ceiling [68S] or moving through several rooms [33S]. The spread of forest fires [37, 20S] was analyzed. The onset of a firestorm [11S] is postulated to require the pre-existence of a swirl.

Experiments and analysis in *bioengineering* considered the temperature distribution in the heart [3S] and in the head [51S] caused by blood flow.

Thermal insulation of *garments* was measured during wear trials [32S] and during burning [50S].

Survey papers described the pioneering contributions of Luikov [48S] and of Sherwood [15S, 17S] to heat and mass transfer research in *drying*. Other contributions deal with wood [55S, 21S], fibrous sheet material [37S], paper [19S] and humid sandstone [66S].

Three papers are examples of studies on heat transfer in *combustion engines*. Two [67S, 25S] deal with diesel engines and one [57S] with gas turbines.

Fusion power plants will experience extreme conditions for heat transfer. These are discussed in papers like [8S, 10S, 26S].

A mathematical model for simulation of industrial *refrigeration* plants is tested against experimental data [57S].

Papers concerned with *aerospace* applications include computer results for the Knudsen layer of a highly cooled blunt body in rarified flow at Mach number 26 [14S] which differ markedly from previous calculations and a detailed heat transfer analysis [59S] to determine the temperature limitations for the thermal control system of a space shuttle payload.

Examples of papers in the field of *electrical engineering* describe experiments on a rotating test rotor simulating superconducting generator cooling [23S], on heat loading of components in multi-megawatt ion sources [47S] and on turbulent convective heat transfer in forced cooled underground electric transmissions using oil [28S].

SOLAR ENERGY

The energy loss mechanisms are analyzed for a parabolic dish/cavity receiver configuration and it was found that varying the concentrator rim angle and cavity geometry can greatly affect the cavity power profile without a large effect on system efficiency [18T]. A semi-empirical equation was developed for the heat loss factor for a tubular absorber with a concentric glass cover, employed as the target of a linear solar concentrator [6T]. A study of the interaction between concentrated solar radiation and a honeycomb matrix or a bed of particles cooled by a gas shows evidence of overheating of the solid near the irradiated surface [31T]. A thermal analysis of a V-trough solar concentrator shows that this design is more efficient and provides higher temperatures than a flat-plate collector [1T]. A 1.5 × CPC type collector described and tested in detail exhibits an optical and thermal behavior which make it an excellent candidate for thermal applications up to 100°C [10T].

Testing of a set of Sanyo STC-CU 250 evacuated tubular collectors and subsequent data reduction demonstrated that large variations in thermal performance characteristics can result among a collector type taken from a production batch [32T]. An investigation is reported of heat transfer between the glass absorber tubes of an all-glass evacuated collector and fluid-in-metal manifolds designed for heat extraction from the tubes [39T].

The performance of a flat-plate solar collector combining a selective surface on the absorber plate and a plastic honeycomb in the air gap can be improved by leaving an air gap of about 10 mm between the plate and the honeycomb [19T]. A theoretical analysis of the regimes of convective heat transfer in a horizontal solar converter shows that beehive-type designs can minimize convective losses by maintaining a conductive regime at Rayleigh numbers as high as 10^6 [14T].

Analytical approximations of the temperatures within a flat-plate collector at a moment $t = 0$, are derived. They are expressed as functions of the time-varying behavior of the solar flux, the ambient temperature and the fluid outlet and inlet temperatures [24T]. A discrete-gradient algorithm is used to identify the parameters in a one-node and two-node capacitance model of a flat-plate collector [20T]. An analytical solution is presented of a matrix differential equation for heat removal from an absorber plate with a serpentine tube [38T]. The effects of noncontinuous bonding of the energy removal tube to the absorber plate of a flat-plate collector are analyzed [11T]. An analytical study of the laminar heat transfer in the ducts of a solar collector indicates that the use of an average Nusselt number for the constant wall temperature boundary condition in a radially lumped solution predicts efficiency factors that are within 6% of those obtained using a more detailed calculation procedure [37T]. It is shown that the multiple-pass air heater performs well for low flow rates and large plate lengths but for small plate lengths and high flow rates, the addition of cover plates does not affect the performance of two- or three-pass air heaters [15T].

An interferometric technique was used to study coupled convective heat transfer for an inclined solar collector [25T]. An experimental study provides data for convective heat transfer coefficients on the front surface of a free-standing array of solar collectors mounted on a flat roof and exposed to the wind [28T]. Studies of collectors operating in a wind tunnel indicate that recent empirical relations for calculating top losses provide good estimates at low wind speeds but underestimate the top loss for wind speeds greater than about $1\ \mathrm{m\,s^{-1}}$ [13T].

A simplified simulation method for solar thermosyphon collectors was developed using the Hottel–Bliss–Whillier theory for the absorber, Close's model of a thermally stratified tank, and a newly defined loop resistance relation [21T]. A numerical simulation model for thermosyphon solar water heaters has been developed and used to study the characteristics of vertical and horizontal tank thermosyphon systems [30T]. In a side-by-side comparison of a pressurized and a nonpressurized solar water heating thermosyphon system, it was found that the nonpressurized system had the higher daily average efficiency [35T]. An integral compact solar water heater proved to have high collection efficiency. However, it also proved to have high losses during the period when there was no incident solar flux [36T].

Steady-state calculations prove that only little sophistication in design and technology of solar collectors is required with favorable weather conditions but that unfavorable weather conditions require collectors with technological refinements [17T]. A study implies that the sophistication and expense involved in a searching for and implementing optimal variable collector flow rate strategies for single-pass, open-loop solar thermal energy systems may not be worth the effort [16T]. The benefits of using phase-change slurries as enhanced heat transfer/storage working fluids in solar energy and waste heat utilization systems are investigated [26T]. Numerical simulation and experimental measurement are used to examine the performance of a greenhouse-type solar timber dryer that dries about $5\ \mathrm{m}^3$ of timber from green to equilibrium in about three weeks [34T]. A thermodynamic analysis of a solar zeolite refrigeration system showed that the system performance depends primarily on the solar collector type and the zeolite absorption properties [8T].

An analytical and numerical study of the thermal and fluid flow effects of heat rejection to the surface layer of a salt-gradient solar pond, by means of a recirculating thermal discharge, is carried out [23T]. The use of an empirical transmittance function in the modeling of a solar pond was found to improve the results of the modeling [2T]. An analytic solution is presented that calculates the heat loss from the bottom of a solar pond to a soil that contains a moving water table [22T].

Measurements performed for the 1981–1982 heating season showed that the solar collection efficiency was 64% and the solar fraction was 0.40 for a passive solar modular house located in Knoxville, Tennessee [33T]. The thermal performance of three walls, a 40-cm-thick concrete wall, an 8-cm-thick hard paraffin wall and an 8-cm-thick soft paraffin wall, was monitored for a three-year period [5T]. A quantitative assessment is presented of the extent to which massive building envelopes can increase solar heating fractions by moderating the storage and the release of solar gain [12T]. A computer model of a general integral collection–storage passive solar component was created to be compatible with the transient system program TRNSYS and was used to develop and verify the simpler monthly performance prediction method [40T]. An efficient method is presented for the computer analysis in the frequency domain of multizone passive solar buildings [3T].

The transient thermal response of a rock bed with no net fluid flow is examined following all-day charging and the results indicate that natural convection may occur in the upper regions of the bed [4T]. A three-dimensional numerical model for seasonal heat storage in the ground using vertical heat exchanger pipes is presented [29T]. An analytical and numerical study demonstrates that the harvesting of solar energy by a solar collector installation with temporary storage can be maximized by the proper selection of the daily regime of operation [9T].

Analytical and experimental studies were performed for night sky radiators as a function of power dissipation, surface properties and water vapor pressure [27T]. Three models of spectral direct irradiance of the sun at groundlevel are found to be in satisfactory agreement with experimental measurements, although their behavior is quite different in the absorption bands [7T].

PLASMA HEAT TRANSFER AND MHD

Papers in plasma heat transfer are primarily concerned with heat transfer in arcs and applications such as thermal plasma processing of materials.

The results of a two-dimensional analysis of free burning arcs in argon (100–400 A) show good agreement with experimental data. The importance of energy transfer due to electron convection currents is demonstrated [28U]. Radial temperature profiles in radiation-dominated, wall-stabilized arcs are calculated by using equations derived from the first two moments of the radiative transfer equation for determining radiation losses [31U]. A transient two-dimensional analysis of forced convection-stabilized arcs (gas-blast circuit breakers) shows good agreement with experiments for the upstream region of steady-state arcs in nitrogen and hexafluoride [34U]. A previously described method for computing radial temperature distributions in wall-stabilized arcs is applied to the treatment of D-line emission from a sodium arc lamp [32U]. Different methods for measuring electron densities in a SF_6 arc are in good agreement indicating departures from LTE due to a demixing effect [48U]. Experimental data indicate that the gas dynamics of vortex-stabilized plasma guns control their energy transfer, electric performance, and service life [40U]. The presence of metal vapor in an argon arc plasma at the $p = 1$ atom leads in the anode region to a temperature decrease, a significant reduction of the electric field, and a large increase of the electrical conductivity [9U].

An improved relaxation method for measuring gas temperatures in arc plasmas is described, based on different relaxation times of density and electron temperature after arc interruption [41U]. Based on a numerical solution of the radiation transfer equation, temperature determination in high-pressure plasmas is possible in cases where no optically thin lines are available [47U]. A new procedure has been developed for determining the line broadening at the axis of an optically thick arc discharge, and the influence of the radial variation of the line broadening on the plasma temperature is determined for the case of symmetric self-reversal [22U]. Temperature measurements in nitrogen seeded, nonequilibrium argon plasmas, using the intensity distributions of the N_2^+ first negative bands, show that at lower pressures, the rotational temperatures are much lower than the vibrational and electron temperatures. At higher pressures they assume the same values since LTE is attained [42U]. By combining the technique of statistical spectroscopy with a probabilistic theory of radiative transport, a new method has been developed for evaluating the radiative power emitted from a thermal plasma by spectral lines [12U]. A new, simple elliptical Abel inversion is described for calculating spatial distributions of electron densities in elliptically shaped plasmas [20U].

Thermodynamic and transport properties of hydrogen and copper plasmas are reported for temperatures from 5000 to 60,000 K and pressures from 1 to 10^4 atm [27U]. Measurements of the thermal conductivity of hydrogen above 4500 K with electric arcs must take deviations from LTE into account [38U]. Simple and accurate expressions for the electrical conductivity in steady-state weakly and highly ionized plasmas are in good agreement with exact numerical calculations [30U]. Computer models are introduced for calculating the ambipolar diffusion coefficient and the reactive thermal conductivity of a two-temperature argon plasma containing higher ionized species [7U]. The nonlinear heat conductivities of a dilute plasma in a magnetic field deviate markedly from the linear theory prediction as the Debye length increases [17U]. A theoretical investigation of plasmas contaminated by various metal vapors shows the importance of the first ionization of these contaminants on the plasma properties [2U]. The theory of thermal diffusion in ionized gases is reconsidered taking into account quantal effects and effects caused by dynamic shielding [35U].

In connection with heat and mass transfer studies of particles injected into thermal plasmas, a new integral mean diffusivity is introduced for calculating mass transfer rates from such particles [43U]. The ambipolar diffusion coefficients in a nitrogen thermal plasma for $4000 \leqslant T \leqslant 15{,}000$ K depend on actual population number densities encountered and not only on temperature [4U]. The electrical conductivity of partially ionized monatomic gases has been calculated for a wide range of electron temperatures and degrees of ionization. Knowledge of the electrical conductivity allows for easy determination of electron mobility and diffusion coefficients in plasmas [16U].

The average cooling energy of electrons emitted from an arc cathode root and the effective work function are strongly dependent on the temperature and the electric field at the location of electron emission [29U]. Studies of the effect of anode vapor (Cu) on the anode region of a high-intensity argon arc show a flattening of the current density distribution which will also affect the heat flux distribution at the anode [15U]. Modeling of the transport phenomena in the anode region of high-intensity arcs indicates that nitrogen arcs tend to form highly constricted anode arc roots in contrast to argon arcs [52U]. An improved heat conduction model is proposed which provides better predictions of the anode voltage drops for electric discharges in dielectric liquids, as for example, in EDM machining [6U]. Cooling of the expanding metal plasma emitted from the cathode spot of a vacuum arc leads to a strong variation of the electrical conductivity across the thermal boundary layer [13U].

At the boundary between a hot and cold plasma, a double layer structure is formed on the hot-plasma side with a potential difference given approximately by $\phi_{DL} \sim T_{eh}/2e$ where T_{eh} is the hot electron temperature and e the elementary charge [21U]. Experimental studies of heat and mass transfer between a plasma jet and a permeable wall cooled by injection of a two-phase flow show that the temperature of the wall undergoes self-induced oscillation due to a shift of the vaporization zone [1U]. A numerical analysis of thermal plasma flow in a pipe indicates that the velocity and energy fields are different from ordinary pipe flow and vary characteristically depending on the state in the pipe and the initial conditions [23U, 24U]. The heat flux to the wall of a water-cooled tube carrying a thermal argon plasma flow changes little with cold gas injection upstream of the location where heat flux measurement were taken [25U]. A correlation for heat transfer between a hydrogen plasma jet and the walls of a cylindrical channel represents the data within 22% [54U].

Computer experiments show the relative importance of various effects on the motion and on heat and mass transfer of a particle injected into a thermal plasma [44U]. The motion of a small spherical particle injected into a thermal plasma may be affected by a number of effects which are not present in an ordinary gas [45U]. The results of heat and mass transfer studies of particles in thermal plasma flows indicate that convective heat transfer coefficients have to be modified due to strongly varying transport properties. In addition, particle heat transfer under noncontinuum conditions is governed by individual contributions from the species in the plasma and by particle charging effects [33U]. Plasma–particle interactions in r.f. induction plasmas are considered using argon at $p = 1$ atm as the plasma gas interacting with copper and alumina particles. Substantial local cooling rates are found for high particle injection rates [46U]. Heat transfer to a copper particle immersed into an argon plasma is substantially affected by the change of the transport coefficients due to copper vapor, except for cases of very high plasma temperatures [10U]. Heat transfer to particles injected into plasmas at reduced pressures ($p < 1$ atm) is severely reduced due to the Knudsen effect [11U].

Theoretical studies of the plasma-activated chemical vapor deposition (PCVD) process, used for the preparation of optical fibers, indicate that the static deposition profile of silica can be computed by considering convection, radiation and diffusion processes inside the glass tube [51U]. Investigations of the pyrolytic decomposition of methane in a plasma indicate that an excess of hydrogen causes a decrease in the overall rate constants for the methane decomposition [14U]. Natural phosphate may be dissociated in a thermal plasma and in this way concentrated phosphorus fertilizer may be produced [36U]. Plasma heat transfer and chemical reactivity are important for an understanding of the surface effects on solids and liquids for refining, nitriding, etc. [3U]. Modeling of plasmas, calculation of plasma transport properties, and mixing of plasmas with cold gases are discussed in the context of plasma–particle momentum and heat transfer. In addition, experimental data are presented and compared with analytical results [19U]. The successful generation and processing of fine powders in thermal plasmas is directly linked to the understanding of heat and mass transfer processes [55U].

Two different numerical approaches are used for modeling the flow and temperature fields in an inductively coupled plasma. The results show that the method may have a substantial influence on the stability and the convergence of the solution [37U]. Investigations of the phenomenon of exploding wires indicate that for energies $E > E_e$ (E_e is the energy density required for complete vaporization), explosive vaporization occurs, resulting in a high-temperature plasma [50U]. During the explosion phase, plasma temperatures may exceed 10 eV. A super high density ($n_e = 5 \times 10^{22}\,cm^{-3}$), high temperature ($T_e \approx 400$ eV) plasma can be produced by an aluminum wire explosion [18U]. The gas heating of the neutral molecules in a glow to spark transition in air and nitrogen gives rise to a neutral depopulation in the core of the discharge [5U]. The heat transfer coefficient from a flowing hydrogen microwave plasma to the plasma discharge tube wall increases with increasing flow rate and pressure and is independent of the absorbed power of the mircowave plasma [8U]. A heated three-chamber pneumatic shock tube is a convenient instrument for investigating a dense plasma composed of various chemical elements over a wide range of parameters [26U].

Low Prandtl number MHD convection studies about a heated semi-infinite horizontal flat plate reveal downstream of the leading edge an unusual structure [53U]. Studies of the spectral emission from MHD combustion gas mixed with pulverized coal ($0.38 < \lambda < 2.7\,\mu m$) revealed many lines of K, Li, Na, Rb, and Ca, but no emission from coal-ash particles was observed [39U]. The combined effects of free forced convection on the flow of an incompressible conducting visco-elastic fluid between two horizontal insulated parallel walls are considered with linear axial temperature variation under the influence of a uniform transverse magnetic field [49U].

BOOKS

R. I. Soloukhin and N. H. Afgan, *Measurement Techniques in Heat and Mass Transfer*. Hemisphere, Washington, DC (1985).

S. Kakac and R. Viskanta (Editors), *Natural Convection*. Hemisphere, Washington, DC (1985).

Xue-jun Chen and T. Nejat Vezirogulu (Editors), *Two-phase Flow and Heat Transfer: China–U.S. Progress*. Hemisphere, Washington, DC (1985).

A. Zukauskas and J. Ziugzda (Edited by G. F. Hewitt), *Heat Transfer of a Cylinder in a Crossflow* (translated from Russian). Hemisphere, Washington, DC (1985).

B. F. Armaly and A. F. Emery (Editors), *HTD-Vol. 49, Radiation Heat Transfer*. ASME Order Dept., New York (1985).

S. Oktay and R. J. Moffat (Editors), *HTD-Vol. 48, Heat Transfer in Electronic Equipment—1985*. ASME Order Dept., New York (1985).

P. J. Bishop (Editor), *HTD-Vol. 41, Heat Transfer in Buildings and Structures*. ASME Order Dept., New York (1985).

W. M. Rohsenow, J. P. Hartnett and E. N. Ganic (Editors), *Handbook of Heat Transfer*, 1st Vol., *Fundamentals*, 2nd Vol., *Applications*. McGraw-Hill, New York (1985).

J. Sucec, *Heat Transfer*. Brown, Dubuque, IA (1985).

Jasbir Singh and B. V. Badger, *Heat Transfer Fluids and Systems for Processs and Energy Applications*. Marcel Dekker, New York (1985).

S. M. Shenkman, J. E. O'Brien, I. S. Habib and J. A. Kohler, *HTD-Vol. 43, Advances in Enhanced Heat Transfer—1985*. ASME Order Dept., New York (1985).

J. T. Pearson and J. B. Kitto, Jr. (Editors), *HTD-Vol. 44, Two-phase Heat Exchanger Symposium*. ASME Order Dept., New York (1985).

G. K. Law, Y. Jaluria, W. W. Yuen and K. Miyasaks, *HTD-Vol. 45, Heat Transfer in Fire and Combustion Systems*. ASME Order Dept., New York (1985).

V. Dembovský, *Plasma Metallurgy, The Principles*. Elsevier,

Amsterdam (1985).

CONDUCTION

1A. A. A. Aleksashenko, Optimal control in solution of certain inverse heat-conduction problems, *Pwr Engng* **23**(1), 98 (1985).

2A. N. M. Alnajem and M. N. Özişik, Direct analytical approach for solving linear inverse heat conduction problems, *J. Heat Transfer* **107**(3), 700 (1985).

3A. N. M. Alnajem and M. N. Özişik, On the solution of three-dimensional inverse heat conduction in finite media, *Int. J. Heat Mass Transfer* **28**(11), 2121 (1985).

4A. V. W. Antonetti and M. M. Yovanovich, Enhancement of thermal contact conductance by metallic coatings: theory and experiment, *J. Heat Transfer* **107**(3), 513 (1985).

5A. N. Araki, K. Kobayasi and M.-A. Ochi, Measuring thermal diffusivity of liquids by stepwise heating (the second report, heat loss characteristics and measurements on molten salts), *Heat Transfer—Japan. Res.* **14**(1), 56 (1985).

6A. N. Araki and K. Natsui, Measuring thermal diffusivity of liquids by stepwise heating (the first report, principle of measuring method and theoretical analysis of errors), *Heat Transfer—Japan. Res.* **14**(1), 36 (1985).

7A. E. A. Artyukhin, Experimental design of measurements for the solution of coefficient-type inverse heat-conduction problems, *J. Engng Phys.* **48**(3), 372 (1985).

8A. J. Baker-Jarvis and R. Inguva, Heat conduction in heterogeneous materials, *J. Heat Transfer* **107**(1), 39 (1985).

9A. B. T. F. Chung, M. M. Kermani, M. J. Braun, J. Padovan and R. C. Hendricks, Heat transfer in thermal barrier coated rods with circumferential and radial temperature gradients, *J. Engng Gas Turbine Pwr* **107**(1), 135 (1985).

10A. R. M. Clever and A. T. Wassel, Three-dimensional transient heat conduction in a multilayer medium, *J. Spacecraft Rockets* **22**(2), 211 (1985).

11A. T. W. Davies, Transient conduction in a plate with counteracting convection and thermal radiation at the boundaries, *Appl. Math. Modell.* **9**(5), 337 (1985).

12A. O. V. Dikhtiyevskiy, A variational method for solving the two-dimensional problem of heat conduction with a finite-size internal source, *Heat Transfer—Sov. Res.* **16**(4), 81 (1985).

13A. L. Ercoli, P. A. A. Laura, R. O. Grossi and G. Sanchez Sarmiento, Unsteady thermal field in bars of doubly connected cross section with heat generation, *Nucl. Engng Des.* **88**(3), 377 (1985).

14A. J. I. Frankel, B. Vick and M. N. Özişik, Flux formulation of hyperbolic heat conduction, *J. appl. Phys.* **58**(9), 3340 (1985).

15A. J. C. Friedly, Transient response of a coupled conduction and convection heat transfer problem, *J. Heat Transfer* **107**(1), 57 (1985).

16A. S. M. Ghiaasiaan, A. T. Wassel, J. L. Farr, Jr. and S. M. Divakaruni, Heat conduction in nuclear fuel rods, *Nucl. Engng Des.* **85**(1), 89 (1985).

17A. S. C. Huang, Unsteady-state heat conduction in semi-infinite regions with mixed-type boundary conditions, *J. Heat Transfer* **107**(2), 489 (1985).

18A. F. W. Jones and M. E. Ertman, Heat flow through thermally anisotropic media: a numerical method and its application to an area of the southwest Scottish Highlands, *Geologie en Mijnbouw* **64**(3), 251 (1985).

19A. Y. Kawashima and K. Nishimoto, A method for analyzing the thermal conduction problems by a dual difference equation, *Bull. J.S.M.E.* **28**(237), 444 (1985).

20A. K. Kawashimo, K. Morikawa and E. Nemoto, Mathematical analysis of nonlinear unsteady heat conduction (1st report, general fundamental idea), *Bull. J.S.M.E.* **28**(244), 2341 (1985).

21A. Yu. M. Kolyano, V. I. Gromovyk and Ya. T. Karpa, Heat-conduction problem for a plate with a square cut, *J. Engng Phys.* **48**(2), 233 (1985).

22A. Z.-B. Kuang and S. N. Atluri, Temperature field due to the moving heat source: a moving mesh finite element analysis, *J. appl. Mech.* **52**(2), 274 (1985).

23A. A. A. Kulkarni and K. G. Narayankhedkar, Measurement of equivalent thermal conductivity of multi-layer insulations, *Indian J. Technol.* **23**(1), 13 (1985).

24A. H. L. Kuntz and N. D. Perriera, Oscillatory conductive heat transfer for a fiber in an ideal gas, *J. Heat Transfer* **107**(1), 52 (1985).

25A. U. H. Kurzweg, Enhanced heat conduction in fluids subjected to sinusoidal oscillations, *J. Heat Transfer* **107**(2), 459 (1985).

26A. G. R. McGee, M. H. Schankula and M. M. Yovanovich, Thermal resistance of cylinder–flat contacts: theoretical analysis and experimental verification of a line-contact model, *Nucl. Engng Des.* **86**(3), 369 (1985).

27A. M. D. Mikhailov and M. N. Özişik, Unified solutions of heat diffusion in a finite region involving a surface film of finite heat capacity, *Int. J. Heat Mass Transfer* **28**(5), 1039 (1985).

28A. P. L. Mills, S. S. Lai and M. P. Dudukovic, Solution methods for problems with discontinuous boundary conditions in heat conduction and diffusion with reaction, *Ind. Engng Chem. Fund.* **24**(1), 64 (1985).

29A. T. Miura, J. Fukai and S. Ohtani, Measurement of thermal diffusivities at a continuous heating rate, *Heat Transfer—Japan. Res.* **13**(1), 20 (1984).

30A. A. M. Mujahid and M. Abdulaziz, Periodic heat transfer in triangular fins, *Heat Technol.* **3**(1), 61 (1985).

31A. A. K. Naghdi, Heat conduction in a rectangular parallelopiped with multiple cylindrical cavities, *J. Heat Transfer* **107**(2), 491 (1985).

32A. A. K. Naghdi, Heat transfer in a rectangular plate fin with inner circular boundaries, *Trans. Can. Soc. mech. Engrs* **9**(1), 11 (1985).

33A. E. Nemoto and K. Kawashimo, Analysis of unsteady state problems in nonlinear heat conduction materials, (1st report, one-dimensional exponential models of x = constant), *Bull. J.S.M.E.* **28**(241), 1459 (1985).

34A. R. Pakula and R. Sigel, Self-similar expansion of dense matter due to heat transfer by nonlinear conduction, *Phys. Fluids* **28**(1), 232 (1985).

35A. E. Ya. Rapoport, Optimization of two-dimensional processes of nonstationary heat conduction, *Pwr Engng* **23**(1), 76 (1985).

36A. C. Rogers and R. Sawatsky, Heat conduction in an inhomogeneous half-space subject to a nonlinear boundary condition. Application of Bergman-type series, *Int. J. Engng Sci.* **23**(4), 415 (1985).
37A. J. Saint Jean Paulin, Study of the evolution of temperature in the core of high voltage electric cable, *Int. J. Engng Sci.* **23**(8), 839 (1985).
38A. D. D. Sayers and M. C. Potter, Effect of curvature and torsion on the temperature distribution in a helix, *J. appl. Mech.* **52**(3), 529 (1985).
39A. R. Singh, N. S. Saxena and D. R. Chaudhary, Simultaneous measurement of thermal conductivity and thermal diffusivity of some building materials using the transient hot strip method, *J. Phys. D: appl. Phys.* **18**(1), 1 (1985).
40A. E. M. Sparrow and A. T. Prata, Effect of circumferential conduction on the critical radius of insulation for forced convection crossflow, *J. Heat Transfer* **107**(4), 960 (1985).
41A. L. N. Tao, Analyticity of solutions on the heat equation with nonlinear boundary conditions, *J. Mech. appl. Math.* **38**(3), 447 (1985).
42A. T. Tsukada, S. Noguchi and K. Yanagi, Contact heat transfer between spherical and plane surfaces with microirregularities, *J. Japan Soc. Lubrication Engrs* **30**(1), 60 (1985).
43A. Y. Tsukamoto and Y. Seguchi, Shape optimization problem for minimum volume fin, *Heat Transfer—Japan. Res.* **1**(1), 1 (1984).
44A. M. Ueda, M. Kinugawa and Y. Hara, Heat conduction through a layered pressure vessel wall, *Hitachi Zosen Tech. Rev.* **46**(2), 68 (1985).
45A. Yu. V. Vidin, On eigenvalue calculation in some hollow cylinder heat conduction problems, *Sov. Aeronaut.* **28**(2), 138 (1985).
46A. D. W. Yarbrough, T. W. Tong and D. L. McElroy, Use of fine powders for high thermal resistance, *High Temp. Sci.* **19**(2), 213 (1985).
47A. F. Yoshikawa and M. Tanaka, A boundary element analysis of steady-state heat-conduction problems in axisymmetric bodies, *Heat Transfer—Japan. Res.* **13**(4), 51 (1985).
48A. W. Y. D. Yuen, On the heat transfer of a moving composite strip compressed by two rotating cylinders, *J. Heat Transfer* **107**(3), 541 (1985).

CHANNEL FLOW

1B. L. A. Abdel-Latif, H. Peeken and J. Benner, Thermohydrodynamic analysis of thrust-bearing with circular pads running on bubbly oil (BTHD-theory), *J. Tribol. Technol.* **107**(4), 527 (1985).
2B. R. S. Amano, A numerical study of laminar and turbulent heat transfer in periodically corrugated wall channel, *J. Heat Transfer* **107**(3), 564 (1985).
3B. O. A. Arnas, M. A. Ebadian and H. C. Topakoglu, Convective heat transfer in annular tubes carrying radioactive slurry, *Heat Technol.* **3**(2), 47 (1985).
4B. G. S. Barozzi and G. Pagliarini, A method to solve conjugate heat transfer problems—the case of fully developed laminar flow in a pipe, *J. Heat Transfer* **107**(1), 77 (1985).
5B. R. Bhargava and M. Rani, Numerical solution of heat transfer in micropolar fluid flow in a channel with porous walls, *Int. J. Engng Sci.* **23**(4), 409 (1985).
6B. B. F. Blackwell, Numerical solution of the Graetz problem for a Bingham plastic in laminar tube flow with constant wall temperature, *J. Heat Transfer* **107**(2), 466 (1985).
7B. V. A. Bruk, Nonisothermal quasistationary gas flow in tubes, *High Temp.* (English translation of *Teplofiz. Vysok. Temp.* **23**(2), 262 (1985).
8B. R. Budwig, S. Tavoularis and S. Corrsin, Temperature fluctuations and heat flux in grid-generated isotropic turbulence with streamwise and transverse mean-temperature gradients, *J. Fluid Mech.* **153,** 441 (1985).
9B. A. S. Buevich and A. I. Filippov, Transfer phenomena accompanying oscillations in a two-component medium, *J. Engng Phys.* (English translation of *Inzh.-Fiz. Zh.*) **48**(2), 158 (1985).
10B. N. I. Buleev and G. A. Zinina, Kinetic model for turbulent transfer in fluid flows, *J. appl. Mech. Tech. Phys.* **26**(1), 42 (1985).
11B. B. Carpenter and G. M. Homsy, The effect of surface contamination on thermocapillary flow in a two-dimensional slot, part 2—partially contaminated interfaces, *J. Fluid Mech.* **155**, 429 (1985).
12B. C.-Y. Chang and C.-F. Lin, Convective transfer in a rectangular duct with heterogenous and homogeneous reactions of arbitrary order, *Chem. Engng Jl, Lausanne* **30**(3), 147 (1985)
13B. D. R. Chenoweth and S. Paolucci, Gas flow in vertical slots with large horizontal temperature differences, *Phys. Fluids* **28**(8), 2365 (1985).
14B. C. Chiranjivi and C. V. Raghavaiah, Developing laminar flow forced convective heat transfer in short elliptical channels, *Indian J. Technol.* **23**(6), 223 (1985).
15B. B. R. Clayton and Y. S. M. Morsi, Determination of principal characteristics of turbulent swirling flow along annuli, part 2—measurement of turbulence components, *Int. J. Heat Fluid Flow* **6**(1), 31 (1985).
16B. M. C. M. Cornelissen and C. J. Hoogendoorn, Forced convection heat transfer to supercritical helium, *Appl. scient. Res.* **42**(2), 161 (1985).
17B. M. J. S. de Lemos and A. Sesonske, Turbulence modeling in combined convection in mercury pipe flow, *Int. J. Heat Mass Transfer* **28**(6), 1067 (1985).
18B. R. Doo, C. M. Rodkiewicz and R. N. Gupta, Development of transient temperatures in a thrust bearing, *J. Tribol. Technol.* **107**(2), 280 (1985).
19B. I. L. Dunin, V. V. Ivanov and V. V. Savenko, Radiative-convective cooling of fluids in pipes with allowance for axial heat flow, *Pwr Engng* (English translation of *Izv. Akad. Nauk SSR, Energetika i Transport*) **23**(1), 93 (1985).
20B. N. M. Galin, F. I. Mashlenko and V. V. Raz'nyak, Heat transfer in turbulent liquid flow in annular channels with a moving internal wall, *High Temp.* (English translation of *Teplofiz. Vysok. Temp.*) **23**(2), 297 (1985).
21B. R. G. Galiullin and G. G. Khalimov, Experimental study of heat transfer in a tube with large-amplitude

resonance oscillations, *Sov. Aeronaut.* (English translation of *Izv. VUZ, Aviat. Tekh.*) **28**(2), 15 (1985).
22B. B. Gecim and W. O. Winer, Transient temperatures in the vicinity of an asperity contact (84-Trib-15), *J. Tribol. Technol.* **107**(3), 333 (1985).
23B. J. C. Gottifredi and A. F. Flores, Extended Leveque solution for heat transfer to non-Newtonian fluids in pipes and flat ducts, *Int. J. Heat Mass Transfer* **28**(5), 903 (1985).
24B. P. Gray and W. Kordylewski, Dynamic aspects of spontaneous ignition—thermal runaway induced by increasing ambient temperature, *Proc. R. Soc.* **A**(1811), 257 (1984).
25B. R. Gregorig, Above for the correction factor $k = (Pr_b/Pr_w)^p$ which considers influence of temperature dependent physical characteristics of fluids on convective heat transfer (in German), *Wärme- u. Stoffübertr.* **19**(3), 235 (1985).
26B. J. R. Greig, R. E. Pechacek and M. Raleigh, Channel cooling by turbulent convective mixing, *Phys. Fluids* **28**(8), 2357 (1985).
27B. E. M. Greitzer, R. W. Paterson and C. S. Tan, An approximate substitution principle for viscous heat conducting flows, *Proc. R. Soc.* **A**(1820), 163 (1985).
28B. B. Sturm, U. Grigull and H. Schmucker, Fluid flow and heat transfer of liquid mercury in wetted and nonwetted tubes (in German), *Wärme- u. Stoffübertr.* **19**(3), 217 (1985).
29B. A. M. Grishin and V. B. Nemirovskii, Nonsteady flow and heat transfer of polymerizing liquids in a long tubular reactor, *J. appl. Mech. Tech. Phys.* **26**(1), 64 (1985).
30B. E. Harada, M. Toda, M. Kuriyama and H. Konno, Heat transfer between wall and solid–water suspension flow in horizontal pipes, *J. Chem. Engng—Japan* **18**(1), 33 (1985).
31B. J. P. Hartnett and M. Kostic, Heat transfer to a viscoelastic fluid in laminar flow through a rectangular channel, *Int. J. Heat Mass Transfer* **28**(6), 1147 (1985).
32B. H. Herwig, An asymptotic theory that accounts for the effects of variable properties in momentum and heat transfer problems, *Forsch. Ing.* **51**(6), 186 (1985).
33B. H. Herwig, The effect of variable properties on momentum and heat transfer in a tube with constant heat flux across the wall, *Int. J. Heat Mass Transfer* **28**(2), 423 (1985).
34B. G.-J. Hwang and M.-J. Lin, Natural convection effects on laminar heat transfer in the thermal entrance region of horizontal isothermal tubes, *Chung-kup King C'eng Hsueh K'an* (*J. Chin. Inst. Engrs*) **8**(4), 343 (1985).
35B. E. Ibe and S. Uchida, Water radiolysis model in a circulating flow system with a boiling region and its application to hydrogen alternate water chemistry of boiling water reactors, *Nucl. Sci. Engng* **90**(2), 140 (1985).
36B. K. P. Ivanov, A. A. Karpov, A. K. Orlov and V. Ya. Rivkind, Liquid flow in a slot, *Heat Transfer—Sov. Res.* **16**(3), 63 (1984).
37B. R. W. Johnson and B. E. Launder, Local Nusselt number and temperature field in turbulent flow through a heated, square-sectioned U-bend, *Int. J. Heat Fluid Flow* **6**(3), 171 (1985).
38B. E. K. Kalinin and G. A. Dreitser, Unsteady convective heat transfer for turbulent flows of gases and liquids in tubes, *Int. J. Heat Mass Transfer* **28**(2), 361 (1985).
39B. S. R. Kennon and G. S. Dulikravich, The inverse design of internally cooled turbine blades (84-GT-7), *J. Engng Gas Turbine Pwr* **107**(1), 123 (1985).
40B. Ye. M. Khabakhpasheva and G. I. Yefimenko, Structure of a turbulent flow in a plane diffuser, *Heat Transfer—Sov. Res.* **16**(2), 1 (1984).
41B. A. A. Khalatov and A. N. Borroto, Optimal geometries of screw-type swirl promoters with axial ducts, *Heat Transfer—Sov. Res.* **16**(5), 107 (1985).
42B. Y.-K. Kim, Heat transfer from the wall of a packed tube to an air–solid suspension flow and pressure drop, *Forsch. Ing.* **51**(4), 122 (1985).
43B. Y. Komiyama, F. Mikami, K. Okui and T. Hori, Laminar forced convection heat transfer in curved channels of rectangular cross section, *Heat Transfer—Japan. Res.* **13**(3), 68 (1984).
44B. M. Kumari and G. Nath, Unsteady compressible boundary layer swirling flow in a nozzle and a diffuser, *Int. J. Engng Sci.* **23**(7), 759 (1985).
45B. U. H. Kurzweg and L. Zhao, Experimental observations on enhanced heat transfer in fluids subjected to high frequency oscillations, *Ching Hua Ta Hsueh Hsueh Pao* (*J. Ching Hua Univ.*) **25**(1), 52 (1985).
46B. G. Kushida, H. Yamashita and R. Izumi, Study on three-dimensional flow and heat transfer in miter-bend (2nd report, analysis of heat transfer in laminar region), *Bull. J.S.M.E.* **28**(243), 1992 (1985).
47B. G. Kushida, H. Yamashita and R. Izumi, Study on three-dimensional flow and heat transfer in miter-bend (3rd report, effects of Reynolds number and aspect ratio), *Bull. J.S.M.E.* **28**(243), 2000 (1985).
48B. C. Laohakul, C. Y. Chan, K. Y. Look and C. W. Tan, On approximate solutions of the Graetz problem with axial conduction, *Int. J. Heat Mass Transfer* **28**(3), 541 (1985).
49B. A Lawal and A. S. Mujumdar, Laminar flow and heat transfer in power-law fluids flowing in arbitrary cross-sectional ducts, *Numer. Heat Transfer* **8**(2), 217 (1985).
50B. J.-B. Lee, H. A. Simon and J. C. F. Chow, Buoyancy in developed laminar curved tube flows, *Int. J. Heat Mass Transfer* **28**(3), 631 (1985).
51B. C. L. Merkle, A. Gulati and Y.-H. Choi, Effect of strong heat addition on the convergence of implicit schemes, *A.I.A.A. Jl* **23**(6), 847 (1985).
52B. E. Naito, Laminar heat transfer in the entrance region between parallel plates with uniform wall temperature, *Heat Transfer—Japan. Res.* **13**(3), 92 (1984).
53B. N. Pazooki and S. K. Loyalka, Heat transfer in a rarefied polyatomic gas—I. Plane parallel plates, *Int. J. Heat Mass Transfer* **28**(11), 2019 (1985).
54B. B. V. Perepelitsa, Properties of large-scale structures in the turbulent flow of water and weak polymer solution, *Heat Transfer—Sov. Res.* **16**(2), 27 (1984).
55B. V. N. Popov and N. E. Petrov, Calculation of heat transfer and flow resistance in the turbulent pipe flow of cooled carbon dioxide in the super critical region,

High Temp. (English translation of *Teplofiz. Vysok. Temp.*) **23**(2), 255 (1985).
56B. C. Prakash and Y.-D. Liu, Analysis of laminar flow and heat transfer in the entrance region of an internally finned circular duct, *J. Heat Transfer* **107**(1), 84 (1985).
57B. C. Prakash and P. Renzoni, Effect of buoyancy on laminar fully developed flow in a vertical annular passage with radial internal fins, *Int. J. Heat Mass Transfer* **28**(5), 995 (1985).
58B. P. Ranganathan and R. Viskanta, Heat transfer through a rectangular solid with a gas-filled cylindrical hole, *Numer. Heat Transfer* **8**(4), 469 (1985).
59B. V. B. Repin, Heat transfer associated with viscous two-dimensional channel flow in a transverse acoustic field, *Fluid Dynam.* **20**(3), 369 (1985).
60B. W. Roetzel and J. Ahrendts, Derivation of the energy equation for single phase flowing fluids (in German), *Wärme- u. Stoffübertr.* **19**(3), 203 (1985).
61B. E. N. Saburov and Yu. L. Leukhin, Aerodynamics and heat exchange of a swirled stream in a cylindrical channel, *J. Engng Phys.* (English translation of *Inzh.-Fiz. Zh.*) **48**(3), 259 (1985).
62B. V. Scirocco, R. Devienne and M. Lebouche, Ecoulement laminaire et transfert de chaleur pour un fluide pseudo-plastique dans la zone d'entrée d'un tube, *Int. J. Heat Mass Transfer* **28**(1), 91 (1985).
63B. V. I. Shcherbinin and F. R. Shklyar, Radiative–convective heat transfer in turbulent flow of gases in a flat duct, *Heat Transfer—Sov. Res.* **16**(5), 113 (1985).
64B. Y. I. Shvets, I. F. Virozub, V. F. Vishnyak, N. M. Kondak, V. N. Panchenko and G. A. Frolov, Friction and heat transfer in the turbulent flow of a compressible gas in a plane channel, *J. Engng Phys.* (English translation of *Inzh.-Fiz. Zh.*) **48**(1), 12 (1985).
65B. I. B. Simanovskii, Steady-state and oscillatory convection motions in a side heated two layer system, *High Temp.* (English translation of *Teplofiz. Vysok. Temp.*) **23**(2), 251 (1985).
66B. V. M. Simonis, Ye. V. Uspuras and P. S. Poskas, Conversion of data and heat transfer and drag of annuli with rough inner tube for purposes of prediction of hydraulic and thermal behavior of completely rough annuli and tube bundles, *Heat Transfer—Sov. Res.* **16**(6), 60 (1984).
67B. M. M. Sorour, M. B. Madi and M. S. El-Beshbeeshy, Thermal equilibrium of two-phase gas–liquid bubbly flow in a vertical heated annulus, *Appl. scient. Res.* **42**(4), 319 (1985).
68B. E. M. Sparrow and D. R. Otis, Jr., Ductflow heat transfer at a smooth wall which faces a wall covered by protuberances, *Int. J. Heat Mass Transfer* **28**(7), 1317 (1985).
69B. E. M. Sparrow, P. C. Stryker and C. A. C. Altemani, Heat transfer and pressure drop in flow passages that are open along their lateral edges, *Int. J. Heat Mass Transfer* **28**(4), 731 (1985).
70B. V. I. Subbotin, V. I. Deev, A. Pridantsev, V. K. Andreev, V. V. Arkhipov, V. N. Novikov, A. N. Savin and V. V. Solodovnikov, Heat transfer and hydrodynamics in cooling channels of super-conducting devices, *Cryogenics* **25**(5), 261 (1985).
71B. Y. Sudo, K. Miyata, H. Ikawa, M. Ohkawara and M. Kaminaga, Experimental study of differences in single-phase forced-convection heat transfer characteristics between upflow and downflow for narrow rectangular channel, *J. nucl. Sci. Technol.* **22**(3), 202 (1985).
72B. R. Taleyarkhan, M. Podowski and R. T. Lahey, Jr., An instability analysis of ventilated channels, *J. Heat Transfer* **107**(1), 175 (1985).
73B. S.-W. Tsai and H.-M. Yeh, Study of the separation efficiency in horizontal thermal diffusion columns with external refluxes, *Can. J. chem. Engng* **63**(3), 406 (1985).
74B. D. Valougeorgis and J. R. Thomas, Jr., *F//N*-method in kinetic theory: II. Heat transfer between parallel plates, *Transport Theory statist. Phys.* **14**(4), 497 (1985).
75B. J. V. Vilemas and V. M. Simonis, Heat transfer and friction of rough ducts carrying gas flow with variable physical properties, *Int. J. Heat Mass Transfer* **28**(1), 59 (1985).
76B. P.-L. Viollet, Experimental study and numerical modeling of density currents resulting from thermal transients in a non-rectilinear pipe flow, *Electricité de France, Bulletin de la Direction des Etudes et Recherches*, Ser. B: Reseaux Electriques, *Materials Electriques* **2**, 33 (1985).
77B. B. W. Webb and S. Ramadhyani, Conjugate heat transfer in a channel with staggered ribs, *Int. J. Heat Mass Transfer* **28**(9), 1679 (1985).
78B. V. A. Yefimov, Simulation of mass transfer between cells in rod bundles in the case of longitudinal flow through such bundles, *Heat Transfer—Sov. Res.* **16**(1), 75 (1984).
79B. K. Y. Zhou and H. H. Bau, On the stability and flow reversal of pressure-driven flow in an asymmetrically heated U-shaped tube, *J. Heat Transfer* **107**(1), 112 (1985).

BOUNDARY-LAYER AND EXTERNAL FLOWS

1C. S. Aiba, Heat transfer around a circular cylinder near a plane surface, *J. Heat Transfer* **107**(4), 916 (1985).
2C. O. O. Ajayi and A. Falade, Time-dependent Stokes flow induced by instantaneous sources, *J. appl. Mech.* **52**(3), 718 (1985).
3C. C. E. Andraka and T. E. Diller, Heat-transfer distribution around a cylinder in pulsating crossflow, *J. Engng Gas Turbine Pwr* **107**(4), 976 (1985).
4C. A. Ansari and S. Morris, The effects of a strongly temperature-dependent viscosity on Stokes' drag law—experiments and theory, *J. Fluid Mech.* **159**, 459 (1985).
5C. F. Anselmet and R. A. Antonia, Joint statistics between temperature and its dissipation in a turbulent jet, *Phys. Fluids* **28**(4), 1048 (1985).
6C. K. A. Antonopoulos, Heat transfer in tube assemblies under conditions of laminar axial, transverse and inclined flow, *Int. J. Heat Fluid Flow* **6**(3), 193 (1985).
7C. D. A. Ashworth, J. E. LaGraff, D. L. Schultz and K. J. Grindrod, Unsteady aerodynamic and heat transfer processes in a transonic turbine stage (85-GT-128), *J. Engng Gas Turbine Pwr* **107**(4), 1022 (1985).

8C. W. Asrar and A. H. Nayfeh, Nonparallel stability of heated two-dimensional boundary layers, *Phys. Fluids* **28**(5), 1263 (1985).
9C. D. E. Avery, Experimental aerodynamic heating to simulated space shuttle tiles in laminar and turbulent boundary layers with variable flow angles at a nominal Mach number of 7, *NASA Tech. Paper* 2307, 48 (1985).
10C. D. E. Avery, P. A. Kerr and A. R. Wieting, Experimental aerodynamic heating to simulated shuttle tiles, *J. Spacecraft Rockets* **22**(4), 417 (1985).
11C. A. Balakrishnan, W. C. Davy and C. K. Lombard, Real-gas flowfields about three-dimensional configurations, *J. Spacecraft Rockets* **22**(1), 46 (1985).
12C. H. F. Bauer, Mass transport in a two-dimensional diffusor or confusor, *Wärme- u. Stoffübertr.* **19**(2), 167 (1985).
13C. B. J. Bayly, Heat transfer from a cylinder in a time-dependent cross flow at low Peclet number, *Phys. Fluids* **28**(12), 3451 (1985).
14C. P. G. Berardi and A. Pozzi, Unsteady momentum and energy boundary layers in laminar flat plate flow, *Int. J. Heat Mass Transfer* **28**(1), 101 (1985).
15C. R. Bhargava and M. Rani, Heat transfer in micropolar boundary layer flow near a stagnation point, *Int. J. Engng Sci.* **23**(12), 1331 (1985).
16C. S. P. Bhattacharyya and M. Abbas, On the stability of a hot rotating layer of micropolar fluid, *Int. J. Engng Sci.* **23**(3), 371 (1985).
17C. A. Binder, W. Foerster, H. Kruse and H. Rogge, Experimental investigation into the effect of wakes on the unsteady turbine rotor flow, *J. Engng Gas Turbine Pwr* **107**(2), 458 (1985).
18C. M. F. Blair, Heat transfer in the vicinity of a large-scale obstruction in a turbulent boundary layer, *J. Propul. Pwr* **1**(2), 158 (1985).
19C. V. V. Bogolepov and I. I. Lipatov, Locally three-dimensional laminar flows, *J. appl. Mech. Tech. Phys.* **26**(1), 24 (1985).
20C. A. J. Chambers, R. A. Antonia and L. Fulachier, Turbulent Prandtl number and spectral characteristics of a turbulent mixing layer, *Int. J. Heat Mass Transfer* **28**(8), 1461 (1985).
21C. Y.-M. Chen, Heat transfer of a laminar flow passing a wedge at small Prandtl number—a new approach, *Int. J. Heat Mass Transfer* **28**(8), 1517 (1985).
22C. R. K. Cheng and T. T. Ng, Conditional Reynolds stress in a strongly heated turbulent boundary layer with premixed combustion, *Phys. Fluids* **28**(2), 473 (1985).
23C. D. M. Cooper, R. L. Jaffe and J. O. Arnold, Computational chemistry and aeroassisted orbital transfer vehicles, *J. Spacecraft Rockets* **22**(1), 60 (1985).
24C. M. De Salve and B. Panella, Interaction between a liquid jet and a wall at high temperatures, *Termotecnica, Milan* **39**(2), 61 (1985).
25C. M. G. Dunn, Turbine heat flux measurements—influence of slot injection on vane trailing edge heat transfer and influence of rotor on vane heat transfer (84-GT-175), *J. Engng Gas Turbine Pwr* **107**(1), 76 (1985).
26C. D. E. Florence, Aerothermodynamic design feasibility of a Mars aerocapture vehicle, *J. Spacecraft Rockets* **22**(1), 74 (1985).
27C. R. L. Frederick, Heat transfer correlation for the laminar boundary layer with uniform wall heat flux, *Int. J. Mech. Engng Ed.* **13**(4), 289 (1985).
28C. K. Fukui and M. Nakajima, Unstable stratification effects on turbulent shear flow in the wall region, *Int. J. Heat Mass Transfer* **28**(12), 2343 (1985).
29C. R. Gaugler, Review and analysis of boundary layer transition data for turbine application, *NASA Tech. Memo* 86880 (1985).
30C. R. Germerdonk and N. N. Nguyen, Increase of the local heat transfer coefficient by 'convection vive' phenomenon, *German chem. Engng* **8**(2), 81 (1985).
31C. E. A. Gershbein and S. V. Peigin, Three-dimensional multicomponent boundary layer on a catalytic surface near the critical point, *High Temp.* **23**(3), 415 (1985).
32C. Yu. P. Golovachev and N. V. Leont'eva, Circulation flow at the front surface of a sphere in a supersonic wake, *Fluid Dynam.* **20**(3), 459 (1985).
33C. V. N. Golubkin, On the asymptotic theory of the three-dimensional flow of a hypersonic stream of radiating gas around a body, *J. appl. Math. Mech.* **47**(4), 493 (1983).
34C. L. J. Grubka and K. M. Bobba, Heat transfer characteristics of a continuous, stretching surface with variable temperature, *J. Heat Transfer* **107**(1), 248 (1985).
35C. K. Haga and Y. Kikuchi, Experimental investigation of coolability degradation by fission gas release into flowing sodium in a fuel pin bundle, *Nucl. Technol.* **70**(2), 220 (1985).
36C. J. C. Han, J. S. Park and C. K. Lei, Heat transfer enhancement in channels with turbulence promoters, *J. Engng Gas Turbine Pwr* **107**(3), 628 (1985).
37C. G. K. Hargrave, M. Fairweather and J. K. Kilham, Turbulence enhancement of stagnation point heat transfer on a body of revolution, *Int. J. Heat Fluid Flow* **6**(2), 91 (1985).
38C. H. P. Hodson, Boundary-layer transition and separation near the leading edge of a high-speed turbine blade (84-GT-179), *J. Engng Gas Turbine Pwr* **107**(1), 127 (1985).
39C. N. G. Hogg and H. M. Stommel, The heton, an elementary interaction between discrete baroclinic geostropic vortices, and its implications concerning eddy heat-flow, *Proc. R. Soc.* **A**(397), 1 (1985).
40C. J. T. Howe, Introductory aerothermodynamics of advanced space transportation systems, *J. Spacecraft Rockets* **22**(1), 19 (1985).
41C. P. Huerre and P. A. Monkewitz, Absolute and convective instabilities in free shear layers, *J. Fluid Mech.* **159**, 151 (1985).
42C. J. O. Ilegbusi and D. B. Spalding, An improved version of the k–W model of turbulence, *J. Heat Transfer* **107**(1), 63 (1985).
43C. N. K. Jain, Variation of heat transfer coefficient by radiative fluid injection in brick-lined regenerators, *Reg. J. Energy Heat Mass Transfer* **7**(2), 111 (1985).
44C. Y. Kamotani, J. K. Lin and S. Ostrach, Effect of destabilizing heating on Goertler vortices, *J. Heat Transfer* **107**(4), 877 (1985).

45C. K. Kataoka, Optimal nozzle-to-plate spacing for convective heat transfer in nonisothermal, variable-density impinging jets, *Drying Technol.* **3**(2), 235 (1985).

46C. Y. Kawaguchi, K. Suzuki and T. Sato, Heat transfer promotion with a cylinder array located near the wall, *Int. J. Heat Fluid Flow* **6**(4), 249 (1985).

47C. A. V. Kazakov, M. N. Kogan and V. A. Kuparev, Stability of the subsonic boundary layer during heating of the surface of a flat plate near the leading edge, *Fluid Dynam.* **20**(3), 394 (1985).

48C. M. D. Kim and C. H. Lewis, Three-dimensional nonequilibrium viscous flow over the shuttle orbiter with catalytic surface effects, *J. Spacecraft Rockets* **22**(2), 97 (1985).

49C. M. N. Kiselev, V. P. Motulevich and E. D. Sergievskii, Effect of the directed blowing and suction on characteristics of the laminar boundary layers, taking account of the negative pressure gradient, *Tekhnicheskikh Nauk* **4**, 94 (1985).

50C. S. A. Korpela, On the viscous dissipation in the boundary layer of a high Prandtl number fluid in laminar flow over a flat plate, *Z. angew. Math. Phys.* **36**(4) 624 (1985).

51C. A. M. Lapin and O. Yu. Tsvelodub, Mass transfer in the flow of a thin wavy film of liquid, *J. appl. Mech. Tech. Phys.* **26**(2), 230 (1985).

52C. A. M. Lindemann, Turbulent Reynolds analogy factors for nonplanar surface microgeometries, *J. Spacecraft Rockets* **22**(5), 581 (1985).

53C. K. F. Loughlin, M. A. Abul-Hamayel and L. C. Thomas, The surface rejuvenation theory of wall turbulence for momentum, heat and mass transfer–application to moderate and high Schmidt (Prandtl) fluids, *A.I.Ch.E. Jl* **31**(10), 1614 (1985).

54C. V. I. Maron and V. N. Nikolaevskii, Dissipative heating in shear flow of the asthenosphere, *Fluid Dynam.* (English translation of *Izv. Akad. Nauk SSR, Mekh. Zhidkosti Gaza*) **20**(2), 212 (1985).

55C. E. Marumo, K. Suzuki and T. Sato, Turbulent heat transfer in a flat plate boundary layer distributed by a cylinder, *Int. J. Heat Fluid Flow* **6**(4), 241 (1985)

56C. S. McAlees, Jr. and R. C. Maydew, Aerothermodynamic design of high speed rockets, *J. Spacecraft Rockets* **22**(3), 309 (1985).

57C. S. McHale and J. F. Richardson, Heat transfer from immersed surfaces to aqueous liquids, *Chem. Engng Sci.* **40**(11), 2154 (1985).

58C. C. G. Miller III, J. R. Micol and P. A. Gnoffo, Laminar heat-transfer distributions on biconics at incidence in hypersonic-hypervelocity flows, *NASA Tech. Paper* 2213, 144 (1985).

59C. V. P. Motulevich, E. D. Sergiyevskiy, S. V. Zhubrin and L. K.-J. Lukosiavicius, Analysis of skin friction and heat transfer in the boundary layer of a turbulent flow, *Heat Transfer—Sov. Res.* **16**(4), 99 (1985).

60C. D. G. Motwani, U. N. Baitonde and S. P. Sukhatme, Heat transfer from rectangular plates inclined at different angles of attack and yaw to an air stream, *J. Heat Transfer* **107**(2), 307 (1985).

61C. A. Narumi, S. Kato, K. Terada and R. Izumi, Forced heat convection in turbulent flow over a flat plate of a finite width, *Heat Transfer—Japan. Res.* **13**(3), 17 (1984).

62C. R. J. Nowak, C. W. Albertson and L. R. Hunt, Aerothermal tests of 12.5 degree cone at Mach 6.7 for various Reynolds numbers, angles of attack, and nose shapes, *NASA Tech. Paper* 2345, 88 (1985).

63C. I. Owen and D. J. Ryley, Flow of thin liquid films around corners, *Int. J. Multiphase Flow* **11**(1), 51 (1985).

64C. S. Paolucci, Heat transfer during the early expansion of gas in pressurized vessels, *Int. J. Heat Mass Transfer* **28**(8), 1525 (1985).

65C. B. Pathangey and S. Kovenklioglu, Effect of chemical reaction on particle to gas heat transfer, *Chem. Engng Sci.* **40**(7), 1205 (1985).

66C. W. R. C. Phillips, On vortex boundary layers, *Proc. R. Soc.* **A**(1819), 253 (1985).

67C. S. Polat, A. S. Mujumdar and W. J. M. Douglas, Heat transfer distribution under a turbulent impinging jet—a numerical study, *Drying Technol.* **3**(1), 15 (1985).

68C. W. J. Priddy and F. J. Bayley, Effects of free stream turbulence on the distribution of heat transfer around turbine blade sections, *Int. J. Heat Fluid Flow* **6**(3), 181 (1985).

69C. R. Rajasekaran and M. G. Palekar, Viscous dissipation effects of mixed convection about a rotating sphere, *Int. J. Engng Sci.* **23**(8), 789 (1985).

70C. S. T. Revankar and P. C. Chakravarti, Effects of transpiration on flow past impulsively started plate, *A.I.Ch.E. Jl* **31**(11), 1923 (1985).

71C. W. Rodi and G. Scheuerer, Calculation of heat transfer to convection-cooled gas turbine blades, *J. Engng Gas Turbine Pwr* **107**(3), 620 (1985).

72C. N. A. Rubtsov and N. N. Ponomarev, Heat transfer in a laminar boundary layer of an absorbing emitting and scattering gas on a permeable plate, *Heat Transfer Sov. Res.* **16**(3), 104 (1984).

73C. E. Ruckenstein and A. Ramagopal, Anomalous heat transfer and drag in laminar flow of viscoelastic fluids, *J. Non-Newtonian Fluid Mech.* **17**(2), 145 (1985).

74C. K. Rued and S. Wittig, Free-stream turbulence and pressure gradient effects on heat transfer and boundary layer development on highly cooled surfaces, *J. Engng Gas Turbine Pwr* **107**(1), 54 (1985).

75C. R. San Jose, J. L. Casanova, R. E. Viloria and J. Casanova, Evaluation of the turbulent parameters of the unstable surface boundary layer outside Businger's range, *Atmos. Envir.* **19**(10), 1555 (1985).

76C. C. D. Scott, Effects of nonequilibrium and wall catalysis on shuttle heat transfer, *J. Spacecraft Rockets* **22**(5), 489 (1985).

77C. J. L. Shideler, G. L. Webb and C. M. Pittman, Verification tests of durable thermal protection system concepts, *J. Spacecraft Rockets* **22**(6), 598 (1985).

78C. J. L. Shinn and J. J. Jones, Chemical nonequilibrium effects on flowfields for aeroassisted orbital transfer vehicles, *J. Spacecraft Rockets* **22**(1), 54 (1985).

79C. P. Singh and S. A. Raj, Analytical study of laminar boundary layers with non-uniform main stream velocity and wall temperature, *J. Non-Equilibrium Thermodynam.* **10**(4), 287 (1985).

80C. B. M. Smolsky, L. A. Sergeyeva and V. L. Sergeyev,

Investigation of the unsteady-state heat transfer in the region of a stagnation point in plane and axisymmetric flows, *Int. J. Heat Mass Transfer* **28**(1), 7 (1985).

81C. D. J. Song, S. Swaminathan and C. H. Lewis, High-altitude effects on three-dimensional nonequilibrium viscous shock-layer flows, *J. Spacecraft Rockets* **22**(6), 614 (1985).

82C. E. M. Sparrow and G. T. Geiger, Local and average heat transfer characteristics for a disk situated perpendicular to a uniform flow, *J. Heat Transfer* **107**(2), 321 (1985).

83C. H. S. Takhar and V. M. Soundalgekar, Flow and heat transfer of a micropolar fluid past a continuously moving plate, *Int. J. Engng Sci.* **23**(2), 201 (1985).

84C. D. A. Throckmorton and L. C. Hartung, Shock-interaction-induced heat transfer to the orbiter wing lower surface, *J. Spacecraft Rockets* **22**(5), 516 (1985).

85C. N. Tokuda, A convection Stefan problem by Lagrange–Burmann expansions (I) small time solution, *J. Phys. Soc. Japan* **54**(12), 4513 (1985).

86C. G. J. VanFossen, Jr. and R. J. Simoneau, Preliminary results of a study of the relationship between free stream turbulence and stagnation region heat transfer, *NASA Tech. Memo* 86884, 15 (1985).

87C. E. P. Volchkov, S. V. Semyonov and V. I. Terekhov, The end wall boundary layer in a vortex chamber, *Heat Transfer—Sov. Res.* **16**(5), 1 (1985).

88C. G. D. Walberg, A survey on aeroassisted orbit transfer, *J. Spacecraft Rockets* **22**(1), 3 (1985).

89C. C. Y. Wang, Stagnation flow on the surface of a quiescent fluid—an exact solution of the Navier–Stokes equations, *Q. appl. Math.* **43**(2), 215 (1985).

90C. J. H. Wang, H. F. Jen and E. O. Hartel, Airfoil heat transfer calculation using a low Reynolds number version of a two-equation turbulence model (84-GT-261), *J. Engng Gas Turbine Pwr* **107**(1), 60 (1985).

91C. T. Wang, T. W. Simon and J. Buddhavarapu, Heat transfer and fluid mechanics measurements in transitional boundary layer flows, *J. Engng Gas Turbine Pwr* **107**(4), 1007 (1985).

92C. W. L. Wells, Heating measurements on space shuttle orbiter models with differentially deflected elevons, *J. Spacecraft Rockets* **22**(2), 126 (1985).

93C. S. L. Yelistratov, Experimental investigation of forced-convection heat transfer to a sphere at low Reynolds numbers, *Heat Transfer—Sov. Res.* **16**(6), 71 (1984).

94C. V. P. Zabolotskii, P. S. Roganov and E. V. Shishov, Influence of outer flow turbulence level on heat transfer and friction in the turbulent boundary layer, *Sov. Aeronaut.* (English translation of *Izv. VUZ, Aviat. Tekh.*) **28**(1), 23 (1985).

95C. G. P. Zigmantas, Transfer of heat from a plate to a flow exhibiting perturbations, *Heat Transfer—Sov. Res.* **16**(3), 25 (1984).

96C. E. V. Zoby and A. L. Simmonds, Engineering flowfield method with angle-of-attack applications, *J. Spacecraft Rockets* **22**(4), 398 (1985).

97C. A. Zukauskas, R. Ulinskas and P. Daunoras, Influence of surface roughness on the heat transfer and drag of tube banks in crossflow, *Heat Technol.* **3**(2), 1 (1985).

FLOW WITH SEPARATED REGIONS

1D. S. Aiba, Heat transfer around a circular cylinder near a plane surface, *J. Heat Transfer* **107**(4), 916 (1985).

2D. C. E. Andraka and T. E. Diller, Heat transfer distribution around a cylinder in pulsating crossflow (85-GT-67), *J. Engng Gas Turbine Pwr* **107**(4), 976 (1985).

3D. R. A. Antonia, On a heat transport mode for a turbulent plane jet, *Int. J. Heat Mass Transfer* **28**(10), 1805 (1985).

4D. W. Aung, A. Baron and F.-K. Tsou, Wall independency and effect of initial shear-layer thickness in separated flow and heat transfer, *Int. J. Heat Mass Transfer* **28**(9), 1757 (1985).

5D. V. R. Borovskiy, L. M. Mishnayevskiy and V. A. Shelimanov, Heat transfer to and from slender moving cylinders, *Heat Transfer—Sov. Res.* **16**(6), 121 (1984).

6D. R. J. Goldstein, M. K. Chyu and R. C. Hain, Measurement of local mass transfer on a surface in the region of the base of a protruding cylinder with a computer-controlled data acquisition system, *Int. J. Heat Mass Transfer* **28**(5), 977 (1985).

7D. A. M. Gorray, C. B. Watkins and S. Aung, Turbulent heat transfer computations for rearward-facing steps and sudden pipe expansions, *J. Heat Transfer* **107**(1), 70–76 (1985).

8D. E. J. Hall and R. H. Pletcher, Application of a viscous–inviscid interaction procedure to predict separated flows with heat transfer, *J. Heat Transfer* **107**(3), 557 (1985).

9D. T. Igarashi, Heat transfer from a square prism to an air stream, *Int. J. Heat Mass Transfer* **28**(1), 175 (1985).

10D. Yu. N. Kuznetsov and V. E. Lipatov, A study of high-performance convective surfaces for viscous fluids, *High Temp.* **6**(1), 31 (1985).

11D. L. Makkonen, Heat transfer and icing of a rough cylinder, *Cold Reg. Sci. Technol.* **10**(2), 105-116 (1985).

12D. D. L. R. Oliver, T. E. Carleson and J. N. Chung, Transient heat transfer to a fluid sphere suspended in an electric field, *Int. J. Heat Mass Transfer* **28**(5), 1005 (1985).

13D. V. V. Orlov, V. V. Ovchinnikov, N. N. Mukhina and V. M. Karsten, Investigation of turbulence and heat transfer in a separating flow behind a step, *Heat Transfer—Sov. Res.* **16**(2), 58 (1984).

14D. T. Ota and H. Nishiyama, Fouling effects of geothermal water scale upon heat transfer around an elliptic cylinder, *Wärme- u. Stoffübertr.* **19**(1), 93 (1985).

15D. A. M. Pavlyuchenko, Heat transfer on axisymmetric rocket nose cones with separationless and separated flows, *Heat Transfer—Sov. Res.* **16**(4), 115 (1985).

16D. T. M. Shih and A. L. Ren, Combined convection and radiative recirculating flows in enclosures, *Numer. Heat Transfer* **8**(2), 149 (1985).

17D. F. T. Smith, F. R. S. and J. W. Elliott, On the abrupt turbulent reattachment downstream of leading-edge laminar separation, *Proc. R. Soc.* **A**(1820), 1 (1985).

18D. E. M. Sparrow and G. T. Geiger, Heat transfer at the base of a cylindrical cavity oriented perpendicular to a freestream flow, *Int. J. Heat Mass Transfer* **28**(12), 2353 (1985).

19D. E. M. Sparrow and F. Samie, Heat transfer and

pressure drop results for one- and two-row arrays of finned tubes, *Int. J. Heat Mass Transfer* **28**(12), 2247 (1985).

20D. J. C. Vogel and J. K. Eaton, Combined heat transfer and fluid dynamic measurements downstream of a backward-facing step, *J. Heat Transfer* **107**(4), 922 (1985).

21D. A. Zukauskas and R. Ulinskas, The efficiency parameters for heat transfer in tube banks, *High Temp.* **(6)**1, 19 (1985).

HEAT TRANSFER IN POROUS MEDIA

1DP. N. Afzal, Two-dimensional buoyant plume in porous media: higher-order effects, *Int. J. Heat Mass Transfer* **28**, 2029 (1985).

2DP. G. F. Barreto, G. Donsi, A. Lancia and G. Volpicelli, Modelling of heat transfer between a slugging fluidized bed and immersed vertical probes, *Powder Technol.* **41**, 41 (1985).

3DP. A. Bejan and K. R. Khair, Heat and mass transfer by natural convection in a porous medium, *Int. J. Heat Mass Transfer* **28**, 909 (1985).

4DP. H.-J. Bock and J. Schweinzer, Heat transfer on horizontal internals in gas/solid pressure fluidized beds, *Chem.-Ing.-Tech.* **57**, 486 (1985).

5DP. P. Bondi, Thermo-hygrometric transients in heat transfer measurements with a hot-box apparatus, *Termotecnica* **39**(2), 57 (1985).

6DP. Yu. A. Buevich and E. B. Perminov, Stationary exchange between an infiltrated granular bed and a body immersed in it, *J. Engng Phys.* **48**, 26 (1985).

7DP. R. Chandran and J. C. Chen, A heat transfer model for tubes immersed in gas fluidized beds, *A.I.Ch.E. Jl* **31**, 244 (1985).

8DP. B. C. Chandrasekhara, Mixed convection in the presence of horizontal impermeable surfaces in saturated porous media with variable permeability, *Wärme- u. Stoffübertr.* **19**, 195 (1985).

9DP. B. C. Chandrasekhara and P. M. S. Namboodiri, Influence of variable permeability in combined free and forced convection about inclined surfaces in porous media, *Int. J. Heat Mass Transfer* **28**, 199 (1985).

10DP. S. H. Chang and S. H. Kim, Derivation of a dryout model in a particle debris bed with the drift-flux approach, *Nucl. Sci. Engng* **91**, 404 (1985).

11DP. T. C. Chawla, D. R. Pedersen and W. J. Minkowycz, Governing equations for heat and mass transfer in heat-generating porous beds—I. Coolant boiling and transient void propagation, *Int. J. Heat Mass Transfer* **28**, 2129 (1985).

12DP. T. C. Chawla, D. R. Pedersen and W. J. Minkowycz, Governing equations for heat and mass transfer in heat-generating porous beds—II. Particulate melting and substrate penetration by dissolution, *Int. J. Heat Mass Transfer* **28**, 2137 (1985).

13DP. P. Chen and D. C. T. Pei, A model of heat transfer between fluidized beds, and immersed surfaces, *Int. J. Heat Mass Transfer* **28**, 675 (1985).

14DP. T.-M. Chiu and E. N. Ziegler, Liquid holdup and heat transfer coefficient in liquid-solid and three-phase fluidized beds, *A.I.Ch.E. Jl* **31**, 1504 (1985).

15DP. Y. K. Chuah and V. P. Carey, Analysis of boiling heat transfer and two-phase flow in porous media with non-uniform porosity, *Int. J. Heat Mass Transfer* **28**, 147 (1985).

16DP. J. Ciborowski and J. Kopec, Mechanism of mass transfer from an immersed surface to a high mass capacity fluidized bed, *Chem. Engng J.* **31**(2), 63 (1985).

17DP. D. J. Close, J. G. Symons and R. F. White, Convective heat transfer in shallow, gas-filled porous media: experimental investigation, *Int. J. Heat Mass Transfer* **28**, 2371 (1985).

18DP. P. Crausse, G. Bacon and C. Langlais, Experimental and theoretical study of simultaneous heat and moisture transfer in a fibrous insulant, *J. Therm. Insul.* **9**, 46 (1985).

19DP. P. R. Dawson and D. F. McTigue, A numerical model for natural convection in fluid-saturated creeping porous media, *Numer. Heat Transfer* **8**, 45 (1985).

20DP. C. Delvosalle and J. Vanderschuren, Gas-to-particle and particle-to-particle heat transfer in fluidized beds of large particles, *Chem. Engng Sci.* **40**, 769 (1985).

21DP. A. G. Dixon, Thermal resistance models of packed-bed effective heat transfer parameters, *A.I.Ch.E. Jl* **31**, 826 (1985).

22DP. V. I. Drobyshevich and V. A. Kirillov, Mathematical model for the description of heat and mass transfer inside a porous granule in the presence of phase transitions, *J. appl. Mech. Tech. Phys.* **26**, 58 (1985).

23DP. S. I. Pereira Duarte, O. M. Martinez, O. A. Ferretti and N. O. Lemcoff, Theoretical prediction of heterogeneous one-dimensional heat transfer coefficients for fixed-bed reactors, *A.I.Ch.E. Jl* **31**, 868 (1985).

24DP. E. R. G. Eckert, R. J. Goldstein, A. I. Behbahani and R. Hain, Boiling in an unconstricted granular medium, *Int. J. Heat Mass Transfer* **28**, 1187 (1985).

25DP. J. Escott, Design considerations for fluid-bed dryers, *Proc. Engng London* **66**(10), 77 (1985).

26DP. D. K. Gartling and C. E. Hickox, Numerical study of the applicability of the Boussinesq approximation for a fluid-saturated porous medium, *Int. J. Numer. Meth. Fluids* **5**, 995 (1985).

27DP. H. Glaser and Y.-K. Kim, Heat transfer from the wall of a tube packed with spheres to a stream of air containing particulate matter and the associated pressure drop, *Chem.-Ing.-Tech.* **57**, 524 (1985).

28DP. N. S. Grewal, T. K. Cheung and S. C. Saxena, Heat transfer between horizontal finned tubes and a gas-solid fluidized bed, *Ind. Engng Chem. Proc. Des. Dev.* **24**, 458 (1985).

29DP. M. Hasan and A. S. Mujumdar, Transpiration-induced buoyancy effect around a horizontal cylinder embedded in a porous medium, *Int. J. Energy Res.* **9**(2), 151 (1985).

30DP. M. Hasatani, N. Arai and K. Hori, Drying of granular particles in a multistage inclined fluidized bed with mechanical vibration, *Drying Technol.* **3**, 39 (1985).

31DP. H. Hattori and J. R. Howard, Gas-to-gas heat recovery using fluidized bed technology, *J. Heat*

Recovery Syst. **5,** 535 (1985).
32DP. C. R. B. Hoerger and W. F. Phillips, Limitations of the single-phase model for solid–gas heat transfer in packed beds, *J. Heat Transfer* **107,** 260 (1985).
33DP. J. T. Hong, C. L. Tien and M. Kaviany, Non-Darcian effects on vertical-plate natural convection in porous media with high porosities, *Int. J. Heat Mass Transfer* **28,** 2149 (1985).
34DP. C. T. Hsu and P. Cheng, The Brinkman model for natural convection about a semi-infinite vertical plate in a porous medium, *Int. J. Heat Mass Transfer* **28,** 683 (1985).
35DP. A. K. A. Juma and J. F. Richardson, Heat transfer to cylinders from segregating liquid–solid fluidised beds, *Chem. Engng Sci.* **40,** 687 (1985).
36DP. F. A. Kamke and J. B. Wilson, Analysis of a rotary dryer: drying of wood particles, *For. Prod. J.* **35**(9), 32 (1985).
37DP. R. S. Kane and R. Pfeffer, Heat transfer in gas–solids drag-reducing flow, *J. Heat Transfer* **107,** 570 (1985).
38DP. M. Kaviany, Laminar flow through a porous channel bounded by isothermal parallel plates, *Int. J. Heat Mass Transfer* **28,** 851 (1985).
39DP. Y. Kawase and J. J. Ulbrecht, New approach for heat and mass transfer in granular beds based on the capillary model, *Ind. Engng Chem. Fund.* **24,** 115 (1985).
40DP. V. L. Khitrik, Distribution of mean temperatures of a porous material and coolant in a channel with a porous insert, *J. Engng Phys.* **48,** 301 (1985).
41DP. H. S. Kim and S. I. Abdel-Khalik, Convective cooling of simulated core debris beds, *Nucl. Technol.* **69,** 268 (1985).
42DP. S. Kimura and A. Bejan, Natural convection in a stably heated corner filled with porous medium, *J. Heat Transfer* **107,** 293 (1985).
43DP. J. Kubie, Heat transfer between gas fluidized beds and immersed surfaces, *Int. J. Heat Mass Transfer* **28,** 1345 (1985).
44DP. P. S. Kuts and N. N. Grinchik, Phenomenological approach to the problems on non-isothermal drying, sorption and filtration under incomplete saturation conditions, *Heat Technol.* **3**(2), 102 (1985).
45DP. J. Levec and R. G. Carbonell, Longitudinal and lateral dispersion in packed beds, part I: Theory, *A.I.Ch.E. Jl* **31,** 581 (1985).
46DP. J. Levec and R. G. Carbonell, Longitudinal and lateral thermal dispersion in packed beds, part II: Comparison between theory and experiment, *A.I.Ch.E. Jl* **31,** 591 (1985).
47DP. J. S. Marshall and V. K. Dhir, Effect of overlying liquid layer on dryout heat flux measurements, *Nucl. Sci. Engng.* **91,** 109 (1985).
48DP. O. G. Martynenko, V. G. Leitsina, V. V. Levdansky and N. V. Pavlyukevich, The kinetic model of one-component gas flow during evaporation in a porous layer, *Drying Technol.* **3,** 483 (1985).
49DP. E. I. Mezlyakov and P. G. Kuntsevskiy, Transient filtration and heat transfer in a geothermal rock fracture produced by hydraulic fracturing, *Heat Transfer—Sov. Res.* **16**(3), 138 (1984).
50DP. M. A. Minkin and S. P. Dmitrieva, Computer-aided mathematical modeling of the steam-thaw process of permafrost, *Soil Mech. Foundation Engng* **22**(2), 73 (1985).
51DP. W. J. Minkowycz, P. Cheng and C. H. Chang, Mixed convection about a nonisothermal cylinder and sphere in a porous medium, *Numer. Heat Transfer* **8,** 349 (1985).
52DP. T. F. Morse, C. Y. Wang and J. W. Cipolla, Jr., Laser-induced thermophoresis and particulate deposition efficiency, *J. Heat Transfer* **107,** 155 (1985).
53DP. S. Motakef and M. A. El-Masri, Liquid diffusion in fibrous insulation, *J. Heat Transfer* **107,** 299 (1985).
54DP. K. Muroyama and L.-S. Fan, Fundamentals of gas–liquid–solid fluidization, *A.I.Ch.E. Jl* **31,** 1 (1985).
55DP. D. A. Nield and D. D. Joseph, Effects of quadratic drag on convection in a saturated porous medium, *Phys. Fluids* **28,** 995 (1985).
56DP. V. N. Novakovskii, Analytical method for calculating nonstationary heat exchange in soil with a pipeline, *Pwr Engng* **23**(1), 139 (1985).
57DP. A. Ohhashi, K. Momonoi and H. Harada, Characteristics of longitudinal dispersion of liquid in fluidized bed biofilm reactor, *Doboku Gakkai Rombun Hokokushu* **357**(II-3), 247 (1985).
58DP. B. Palancz, Modeling and simulation of heat and mass transfer in a packed bed of solid particles having high diffusion resistance, *Comput. chem. Engng* **9,** 567 (1985).
59DP. B. Palancz, Simulation of heat and mass transfer in a finite porous medium using orthogonal collocation method, *Wärme- u. Stoffübertr.* **19,** 229 (1985).
60DP. D. Polisevski, Steady convection in porous media—I. The solutions and their regularity, *Int. J. Engng Sci.* **23,** 733 (1985).
61DP. D. Polisevski, Steady convection in porous media—II. The case of low Rayleigh numbers and asymptotic expansions, *Int. J. Engng Sci.* **23,** 741 (1985).
62DP. D. Polisevski, Steady convection in porous media—III. The structure of the solutions, *Int. J. Engng Sci.* **23,** 751 (1985).
63DP. I. Pop, J. K. Sunada, P. Cheng and W. J. Minkowycz, Conjugate free convection from long vertical plate fins embedded in a porous medium at high Rayleigh numbers, *Int. J. Heat Mass Transfer* **28,** 1629 (1985).
64DP. D. Poulikakos, Departure from the Darcy model in boundary layer natural convection in a vertical porous layer with uniform heat flux from the side, *J. Heat Transfer* **107,** 716 (1985).
65DP. D. Poulikakos, On buoyancy induced heat and mass transfer from a concentrated source in an infinite porous medium, *Int. J. Heat Mass Transfer* **28,** 621 (1985).
66DP. D. Poulikakos, Onset of convection in a horizontal porous layer saturated with cold water, *Int. J. Heat Mass Transfer* **28,** 1899 (1985).
67DP. D. Poulikakos, The effect of a third diffusing component on the onset of convection in a horizontal porous layer, *Phys. Fluids* **28,** 3172 (1985).
68DP. D. Poulikakos and A. Bejan, The departure from Darcy flow in natural convection in a vertical porous

layer, *Phys. Fluids* **28,** 3477 (1985).
69DP. V. Prasad and F. A. Kulacki, Natural convection in porous media bounded by short concentric vertical cylinders, *J. Heat Transfer* **107,** 147 (1985).
70DP. W. Prins, T. P. Casteleijn, W. Draijer and W. P. M. van Swaaij, Mass transfer from a freely moving single sphere to the dense phase of a gas fluidized bed of inert particles, *Chem. Engng Sci.* **40,** 481 (1985).
71DP. V. D. Rao, P. K. Sarma and G. J. V. J. Ragu, Biot's variational method to fluidized-bed coating on thin plates, *J. Heat Transfer* **107,** 258 (1985).
72DP. A. A. Raptis and C. P. Perdikis, Oscillatory flow through a porous medium by the presence of free convective flow, *Int. J. Engng Sci.* **23,** 51 (1985).
73DP. D. A. S. Rees and D. S. Riley, Free convection above a near horizontal semi-infinite heated surface embedded in a saturated porous medium, *Int. J. Heat Mass Transfer* **28,** 183 (1985).
74DP. D. S. Riley and D. A. S. Rees, Non-Darcy convection from arbitrarily inclined heated surfaces in saturated porous media, *Q. Jl Mech. appl. Math.* **38,** 277 (1985).
75DP. D. U. Ringer and A. S. Mujumdar, Analysis of aerodynamics and heat transfer in vibro-fluidized beds, *Drying Technol.* **2,** 449 (1983–84).
76DP. E. Saatdjian and J. F. Large, Heat transfer simulation in a raining packed bed exchanger, *Chem. Engng Sci.* **40,** 693 (1985).
77DP. M. Saberian-Broudjenni, G. Wild, N. Midoux and J. C. Charpentier, Contribution à l'étude du transfert de chaleur à la paroi dans les réacteurs à lit fluidisé gaz–solide à faible vitesse de liquide, *Can. J. chem. Engng* **63,** 553 (1985).
78DP. S. C. Saxena and V. L. Ganzha, Dependence of heat transfer coefficient for immersed surfaces in a gas fluidized bed on pressure, *Powder Technol.* **44,** 115 (1985).
79DP. S. C. Saxena and A. Mathur, Hydrodynamic and heat-transfer studies in fine particle gas-fluidized beds, *Energy* **10,** 57 (1985).
80DP. D. C. Shaw and R. A. Dawe, Effects of a vertical discontinuity in a porous medium on a plane convection plume at high Rayleigh numbers, *Int. J. Heat Mass Transfer* **28,** 789 (1985).
81DP. V. A. Shul'chishin, N. B. Rashkovskaya and L. G. Domracheva, Calculating circulation and heat-transfer rates for a drying apparatus containing a fluidized bed of inert material, *J. appl. Chem. U.S.S.R.* **58**(3), 505 (1985).
82DP. W. T. Simpson, Drying wood: a review. I, *Drying Technol.* **2,** 235 (1983–84).
83DP. I. S. Suh, G. T. Jin and S. D. Kim, Heat transfer coefficients in three phase fluidized beds, *Int. J. Multiphase Flow* **11,** 255 (1985).
84DP. M. Syamlal and D. Gidaspow, Hydrodynamics of fluidization: prediction of wall to bed heat transfer coefficients, *A.I.Ch.E. Jl* **31,** 127 (1985).
85DP. H. R. Thomas, Modeling two-dimensional heat and moisture transfer in unsaturated soils, including gravity effects, *Int. J. numer. analyt. Meth. Geomech.* **9,** 573 (1985).
86DP. T. W. Tong, D. L. McElroy and D. W. Yarbrough, Transient conduction and radiation heat transfer in porous thermal insulations, *J. therm. Insul.* **9,** 13 (1985).
87DP. T. W. Tong and E. Subramanian, A boundary-layer analysis for natural convection in vertical porous enclosures—use of the Brinkman-extended Darcy model, *Int. J. Heat Mass Transfer* **28,** 563 (1985).
88DP. I. Toovey and J. Dayan, Analytical solution for the heat transfer problem of fluid flowing through a packed bed of porous solids, *J. Heat Transfer* **107,** 713 (1985).
89DP. O. V. Trevisan and A. Bejan, Natural convection with combined heat and mass transfer buoyancy effects in a porous medium, *Int. J. Heat Mass Transfer* **28,** 1597 (1985).
90DP. K. S. Udell, Heat transfer in porous media considering phase change and capillarity—the heat pipe effect, *Int. J. Heat Mass Transfer* **28,** 485 (1985).
91DP. K. Vafai, R. L. Alkire and C. L. Tien, Experimental investigation of heat transfer in variable porosity media, *J. Heat Transfer* **107,** 642 (1985).
92DP. V. Vesely, M. Soldat, J. Cermak, A. Mrnkova and M. Kralova, Measurement of heat transfer coefficients in a fluidized bed furnace operating under steady-state conditions, *Chemicky Prum.* **35(60)**(8), 399 (1985).
93DP. C. K. Wei, H. T. Davis, E. A. Davis and J. Gordon, Heat and mass transfer in water-laden sandstone: convective heating, *A.I.Ch.E. Jl* **31,** 1338 (1985).
94DP. K. Yoshikawa, H. Kajiyama, T. Okamura, S. Kabashima, H. Yamasaki, S. Shioda, M. Yamada, T. Yokota, M. Ishimura, K. Nakayama, K. Nakamoto and T. Sasaki, Experimental studies on thermal performance of a pebble bed high-temperature heat exchanger, *J. Heat Transfer* **107,** 722 (1985).
95DP. G.-T. Zhang and F. Ouyang, Heat transfer between the fluidized bed and the distributor plate, *Ind. Engng Chem. Proc. Des. Dev.* **24,** 430 (1985).

EXPERIMENTAL TECHNIQUES AND INSTRUMENTATION

1E. T. Baba, A. Ono and S. Hattori, Analysis of the operational error of heat flux transducers placed on wall surfaces, *Rev. scient. Instrum.* **56,** 1399 (1985).
2E. V. S. Batalov, Utilization of the thermal strain of solids for determining the constants of internal heat transfer of solid-phase polymers, *J. Engng Phys.* **48,** 64 (1985).
3E. J. W. Baughn, R. K. Takahashi, M. A. Hoffman and A. A. McKillop, Local heat transfer measurements using an electrically heated thin gold-coated plastic sheet, *J. Heat Transfer* **107,** 953 (1985).
4E. G. E. Bodeep, Temperature measurement systems for optical-fiber preform fabrication, *Rev. scient. Instrum.* **55,** 1957 (1984).
5E. V. S. Bogachev, E. T. Brook-Levinson, A. A. Bulavko, V. V. Gil, Yu. M. Dmitrenko, V. L. Dragun, V. I. Kordonsky, V. I. Lemesh, O. G. Lysenko, O. G. Martynenko, L. N. Novichyonok, S. E. Perez Guerra, N. A. Pokryvailo, A. D. Solodukhin, R. I. Soloukhin, N. N. Stolovich, N. A. Fomin, I. V. Khodan and V. K. Shchitnikov, Methods of experimental investigation of heat and mass transfer processes, *Heat Transfer—Sov. Res.* **16**(5), 40 (1985).

6E. K. Bremhorst, Effect of fluid temperature on hot-wire anemometers and an improved method of temperature compensation and linearization without use of small signal sensitivities, *J. Phys. E: scient. Instrum.* **18,** 44 (1985).
7E. H. H. Bruun and C. Tropea, The calibration of inclined hot-wire probes, *J. Phys. E: scient. Instrum.* **18,** 405 (1985).
8E. K. J. Bullock, M. A. Ledwich and J. C. S. Lai, Numerical simulation of transient response of heat transfer from a hot-wire anemometer transducer, *Int. J. Heat Fluid Flow* **6,** 57 (1985).
9E. M. R. Corson, Computer-compatible, self-contained, precision AC resistance bridge suitable for thermometry, *Rev. scient. Instrum.* **56,** 2310 (1985).
10E. T. W. Davies, Static and dynamic response of a hot-wire anemometer, *Int. J. Heat Fluid Flow* **6,** 226 (1985).
11E. Y. Enomoto and S. Furuhama, Study on thin film thermocouple for measuring instantaneous temperature on surface of combustion chamber wall in internal combustion engine, *Bull. J.S.M.E.* **28**(235), 108 (1985).
12E. H. Fiedler, K. Nottmeyer, P. P. Wegener and S. Raghu, Schlieren photography of water flow, *Exp. Fluids* **3,** 145 (1985).
13E. J. R. Frederick, J. N. Zemel and N. Goldfine, Pyroelectric anemometers: experimental geometric considerations, *J. appl. Phys.* **57,** 4936 (1985).
14E. T. Grandke, Theory and application of the laser shadow technique, Part 2, *Exp. Fluids* **3,** 277 (1985).
15E. G. W. Gray, The Clifford Paterson lecture, 1985. Liquid crystals: an arena for research and industrial collaboration among chemists, physicists and engineers, *Proc. R. Soc.* **A402**(1822), 1 (1985).
16E. D. F. Grosjean, B. Sarka and L. P. Goss, Time-resolved temperature measurements in a laboratory flame using the optoacoustic beam-deflection method, *Opt. Lett.* **10,** 324 (1985).
17E. M. Grubic and K. Kemmerle, A precision lock-in amplifier for temperature control in a spacelab calorimetric experiment, *J. Phys. E: scient. Instrum.* **18,** 572 (1985).
18E. L. E. Haseler, H. O. Lorch, M. G. Rao, R. G. Scurloc, F. A. Stevens and D. B. Utton, Rotating cryostats for heat transfer and fluid flow studies on the helium cooling of superconducting generator rotors, *Cryogenics* **25,** 355 (1985).
19E. P. Hesketh, B. Gebhart and J. N. Zemel, Heat transfer model for the pyroelectric anemometer, *J. appl. Phys.* **57,** 4944 (1985).
20E. S. A. Hippensteele, L. M. Russell and F. J. Torres, Local heat transfer measurements on a large scale-model turbine blade airfoil using a composite of a heater element and liquid crystals, *J. Engng Gas Turbine Pwr* **107,** 953 (1985).
21E. H. A. Hoff, S. J. Rothman and J. N. Mundy, Calibration of an optical pyrometer above 2300°C, *J. Phys. E: scient. Instrum.* **18,** 197 (1985).
22E. G. R. Johnson, E. Marschall and J. H. Esdorn, Improved flow visualization technique for quantitative velocity measurement in small test volumes, *Rev. scient. Instrum.* **56,** 264 (1985).
23E. F. R. A. Jorgensen and M. Zuiderwyk, Two-color pyrometer measurement of the temperature of individual combusting particles, *J. Phys. E: scient. Instrum.* **18,** 486 (1985).
24E. V. V. Kholshchev, Determination of thermal loads of steam generating waterwall tubes (in Russian), *Teploenergetika* **4,** 72 (1985).
25E. Y. A. Kirichenko, R. I. Fedorov, V. V. Tsybulsky and S. I. Boxhidai, Device for investigation of hydrodynamics and heat transfer during cryogenic liquid motion under centrifugal forces, *Cryogenics* **25,** 279 (1985).
26E. L. V. Krishnamoorthy, D. H. Wood, R. A. Antonia and A. J. Chambers, Effect of wire diameter and overheat ratio near a conducting wall, *Exp. Fluids* **3,** 121 (1985).
27E. W. Kuhn and B. DreBler, Experimental investigations on the dynamic behavior of hot-wire probes, *J. Phys. E. scient. Instrum.* **18,** 614 (1985).
28E. A. Lachaine, Thermal analysis by photoacoustic phase measurements: effect of sample thickness, *J. appl. Phys.* **57,** 5075 (1985).
29E. P. C. Lanchester and K. A. Mohammed, A low-power digitally controlled heater power supply for calorimetry, *J. Phys. E: scient. Instrum.* **18,** 499 (1985).
30E. J. C. Lecordier, D. Moreau and P. Paranthoen, Prong-wire thermal interaction effect induced harmonics for the thin-wire resistance thermometer, *J. Phys. E: scient. Instrum.* **18,** 571 (1985).
31E. C. H. Liebert, R. Holanda, S. A. Hippensteele and C. A. Andrachio, High-temperature thermocouple and heat flux gauge using a unique thin film-hardware hot junction, *J. Engng Gas Turbine Pwr* **107,** 938 (1985).
32E. S. P. Logvinenko and O. A. Rossoshanskii, Low-temperature semiconductor resistance thermometer with heat conduetor, *Cryogenics* **25,** 249 (1985).
33E. M. B. Long, P. S. Levin and D. C. Fourguette, Simultaneous two-dimensional mapping of species concentration and temperature in turbulent flames, *Opt. Lett.* **10,** 267 (1985).
34E. D. M. Mathews and D. Morris, A monotone-wall heat flow calorimeter, *J. Phys. E: scient. Instrum.* **18,** 1007 (1985).
35E. K. Mathioudakis and F. A. E. Breugelmans, Use of triple hot wires to measure unsteady flows with large direction changes, *J. Phys. E: scient. Instrum.* **18,** 414 (1985).
36E. C. G. Miller, Refinement of an 'alternate' method for measuring heating rates in hypersonic wind tunnels, *A.I.A.A. Jl* **23,** 810 (1985).
37E. K. C. Mills and W. A. Wakeham, Effect of radiation on thermal transport measurements, *High Temp.—High Press.* **17,** 343 (1985).
38E. J. A. Nienaber and A. L. Maddy, Temperature controlled multiple chamber indirect calorimeter-design and operation, *Trans. Am. Soc. agric. Engrs (Gen. Ed.)* **28**(2), 55 (1985).
39E. S. Niksa, R. E. Mitchell, K. R. Hencken and D. A. Tichenor, Optically determined temperatures, sizes, and velocities of individual carbon particles under typical combustion conditions, *Combust. Flame* **60,** 183 (1985).
40E. D. K. Pandey, Response time correlations for chromel-

constantan, *J. Phys. E: scient. Instrum.* **18,** 712 (1985).

41E. F. Peters, Schlieren interferometry applied to a gravity wave in a density-stratified liquid, *Exp. Fluids* **3,** 261 (1985).

42E. W. R. C. Phillips, Higher-order response equations for an inclined hot-wire, *J. Phys. E: scient. Instrum.* **18,** 314 (1985).

43E. A. Raman, V. S. Krishnan and R. P. Roy, Directional sensitivity of cylindrical hot-film velocity sensors, *J. appl. Mech.* **52,** 199 (1985).

44E. J. E. Roth, Thin films of fired metal paints as resistance thermometers for heat transfer measurement (in German), *Wärme- u. Stoffübertr.* **19**(4), 305 (1985).

45E. O. Sano and R. Takaki, Simultaneous measurement of velocity and temperature by the use of laser doppler velocimeter, *Phys. Fluids* **28,** 818 (1985).

46E. J. M. Seitzman, G. Kychakoff and R. K. Hanson, Instantaneous temperature field measurements using planar laser-induced fluorescence, *Opt. Lett.* **10,** 439 (1985).

47E. G. S. Settles, Color-coding schlieren techniques for the optical study of heat and fluid flow, *Int. J. Heat Fluid Flow* **6,** 3 (1985).

48E. I. Tomaszkiewicz and S. L. Randzio, A high-pressure and elevated-temperature flow calorimeter, *J. Phys. E: scient. Instrum.* **18,** 92 (1985).

49E. J. Van Dam and H. Janeschitz-Kriegl, Temperature measurement and heat transfer in flowing polymer melts, *Int. J. Heat Mass Transfer* **28,** 395 (1985).

50E. A. R. P. Van Heininghen, W. J. M. Douglas and A. S. Mujmdar, A high sensitivity, fast response heat flux sensor, *Int. J. Heat Mass Transfer* **28,** 1657 (1985).

51E. T. von Weissenfluh, Probes for local velocity and temperature measurements in liquid metal flow, *Int. J. Heat Mass Transfer* **28,** 1563 (1985).

52E. M. D. Walker and M. R. Maxey, A whirling hot-wire anemometer with optical data transmission, *J. Phys. E: scient. Instrum.* **18,** 516 (1985).

53E. M. E. Weill, M. Rhazi and G. Gouesbet, Experimental investigation of oscillatory phenomena produced by a hot wire located near and below a free surface, *J. Phys.* **46,** 1501 (1985).

54E. D. Wilkie, The correlation of engineering data considered, *Int. J. Heat Fluid Flow* **6,** 99 (1985).

NATURAL CONVECTION—INTERNAL FLOWS

1F. J. R. Abernathey and F. Rosenberger, Time-dependent convective instabilities in a closed vertical cylinder heated from below, *J. Fluid Mech.* **160,** 137 (1985).

2F. S. Acharya, Natural convection in an inclined enclosure containing internal energy sources and cooled from below, *Int. J. Heat Fluid Flow* **6**(2), 113 (1985).

3F. S. Acharya and R. J. Goldstein, Natural convection in an externally heated vertical or inclined square box containing internal energy sources, *J. Heat Transfer* **107**(4), 855 (1985).

4F. S. Acharya and C. H. Tsang, Natural convection in a fully partitioned, inclined enclosure, *Numer. Heat Transfer* **8**(4), 407 (1985).

5F. N. Afzal, Turbulent natural convection in a horizontal fluid layer with volumetric energy sources: an intermediate layer, *Int. J. Heat Mass Transfer* **25**(1), 291 (1985).

6F. O. A. Arnas and M. A. Ebadian, Convective heat transfer in a circular annulus with various wall heat flux distributions and heat generation, *J. Heat Transfer* **107**(2), 334 (1985).

7F. W. Arter, Nonlinear Rayleigh–Bénard convection with square planform, *J. Fluid Mech.* **152,** 391 (1985).

8F. T. Asaeda, N. Tamai and Y. Takahasi, Investigation of convection induced by thermal instability, *Trans. Japan Soc. Civil Engrs* **15,** 541 (1985).

9F. L. F. A. Azevedo and E. M. Sparrow, Natural convection in open-ended inclined channels, *J. Heat Transfer* **107**(4), 893 (1985).

10F. R. F. Babus'Haq, S. D. Probert and M. J. Shilston, Natural convection across cavities: design advice, *Appl. Energy* **20**(3), 161 (1985).

11F. V. A. Batischev, V. V. Kolesov, S. K. Slitinskaya and V. I. Yudovich, Effect of spatial modulation of the temperature distribution on the stability of two-dimensional steady flow in a horizontal layer of a two-component liquid, *J. appl. Mech. Tech. Phys.* **26**(2), 226 (1985).

12F. H. F. Bauer, Heat transport in a liquid column in weightless conditions due to thermal Marangoni, *Numer. Heat Transfer* **8**(1), 81 (1985).

13F. T. L. Bergman, F. P. Incropera and R. Viskanta, A differential model for salt-stratified, double-diffusive systems heated from below, *Int. J. Heat Mass Transfer* **28**(4), 779 (1985).

14F. S. N. Bhattacharyya and A. S. Gupta, Thermoconvective waves in a binary mixture, *Phys. Fluids* **28**(11), 3215 (1985).

15F. V. A. Bogachev, V. M. Yeroshenko, O. F. Synitina and L. A. Yaskin, Measurements of heat transfer to super-critical helium in vertical tubes under forced and mixed convection conditions, *Cryogenics* **25**(4), 198 (1985).

16F. E. W. Bolton and F. H. Busse, Stability of convection rolls in a layer with stress-free boundaries, *J. Fluid Mech.* **150,** 487 (1985).

17F. B. Borkowska-Pawlak and W. Kordylewski, Cell-pattern sensitivity to box configuration in a saturated porous medium, *J. Fluid Mech.* **150,** 169 (1985).

18F. M. E. Braaten and S. V. Patankar, Analysis of laminar mixed convection in shrouded arrays of heated rectangular blocks, *Int. J. Heat Mass Transfer* **28**(9), 1699 (1985).

19F. D. G. Briggs and D. N. Jones, Two-dimensional periodic natural convection in a rectangular enclosure of aspect ratio one, *J. Heat Transfer* **107**(4), 850 (1985).

20F. Yu. A. Buevich, Theory of interfacial convection, *J. Engng Phys.* **48**(2), 163 (1985).

21F. T. Burch, T. Rhodes and S. Acharya, Laminar natural convection between finitely conducting vertical plates, *Int. J. Heat Mass Transfer* **28**(6), 1173 (1985).

22F. F. H. Busse and H. Frick, Square-pattern convection

in fluids with strongly temperature-dependent viscosity, *J. Fluid Mech.* **150,** 451 (1985).
23F. J. P. Caltagirone and S. Bories, Solutions and stability criteria of natural convective flow in an inclined porous layer, *J. Fluid Mech.* **155,** 267 (1985).
24F. J. L. Castillo and M. G. Velarde, Marangoni convection in liquid films with a deformable open surface, *J. Colloid Interface Sci.* **108**(1), 264 (1985).
25F. Y. L. Chan and C. L. Tien, A numerical study of two-dimensional laminar natural convection in shallow open cavities, *Int. J. Heat Mass Transfer* **28**(3), 603 (1985).
26F. Y. L. Chan and C. L. Tien, A numerical study of two-dimensional natural convection, *Numer. Heat Transfer* **8**(1), 65 (1985).
27F. C. F. Chen and S. Thangam, Convective stability of variable-viscosity fluid in a vertical slot, *J. Fluid Mech.* **161**(12), 161 (1985).
28F. C. Cianfrini, D. M. Fontana and F. Mattiucci, Natural convection in a cavity with isothermal wall in transient conditions, *Energie Alternative Habitat Territorio Energia* **7**(33), 63 (1985).
29F. J. Pascal Coutier and R. Greif, An investigation of laminar mixed convection inside a horizontal tube with isothermal wall conditions, *Int. J. Heat Mass Transfer* **28**(7), 1293 (1985).
30F. P. G. Daniels, Stationary instability of the buoyancy-layer flow between heated vertical planes, *R. Soc. Proc.* **A**(1820), 145 (1985).
31F. P. G. Daniels, Transition to the convective regime in a vertical slot, *Int. J. Heat Mass Transfer* **28**(11), 2071 (1985).
32F. J. A. Dantzig, Experimental method of measuring energy in the Bénard–Rayleigh problem, *Rev. scient. Instrum.* **56**(5), 740 (1985).
33F. Yu. M. Dashevsky and V. I. Malkovsky, Heat exchange with laminar upflow of supercritical helium in a gravitational field, *Cryogenics* **25**(11), 658 (1985).
34F. C. Dietsche and U. Muller, Influence of Bénard convection on solid–liquid interfaces, *J. Fluid Mech.* **161**(12), 249 (1985).
35F. B. F. Edwards and A. L. Fetter, Onset of oscillations in Rayleigh–Bénard convection: horizontally unbounded slab, *Phys. Fluids* **27**(12), 2795 (1984).
36F. E. M. Eidson, Numerical simulation of the turbulent Rayleigh–Bénard problem using subgrid modelling, *J. Fluid Mech.* **158,** 245 (1985).
37F. O. T. Fang, M. E. Glicksman, S. R. Coriell, G. B. McFadden and R. F. Boisvert, Convective influence on the stability of a cylindrical solid–liquid interface, *J. Fluid Mech.* **151,** 121 (1985).
38F. A. P. Fein, M. S. Heutmaker and J. P. Gollub, Scaling at the transition from quasiperiodicity to chaos in a hydrodynamic system, *Physica Scripta* **T9** (1985).
39F. A. C. Fowler, Fast thermoviscous convection, *Stud. appl. Math.* **72**(3), 189 (1985).
40F. A. C. Fowler, Secular cooling in convection, *Stud. appl. Math.* **72**(2), 161 (1985).
41F. D. Getachew and S. Rosenbalt, Thermocapillary instability of a viscoelastic liquid layer, *A Mech.* **55**(1–2), 137 (1985).
42F. G. Guglielmini, G. Milano and M. Misale, Electronic cooling by natural convection in partially confined enclosures, *Heat Technol.* **3**(3–4), 43 (1985).
43F. M. A. Hassab, The stability of steady convective motion in a vertical slender slot with non-uniform volumetric energy sources and unequal surface temperature, *Int. J. Heat Mass Transfer* **28**(2), 351 (1985).
44F. M. A. Hessami, A. Pollard, R. D. Rowe and D. W. Ruth, A study of free convective heat transfer in a horizontal annulus with a large radii ratio, *J. Heat Transfer* **107**(3), 603 (1985).
45F. J. Y. Holyer, The stability of long steady three-dimensional salt fingers to long-wavelength perturbations, *J. Fluid Mech.* **156,** 495 (1985).
46F. J. M. Hyun, Effect of the Prandtl number on heatup of a stratified fluid in an enclosure, *J. Heat Transfer* **107**(4), 982 (1985).
47F. J. M. Hyun, Thermally-forced stratification build-up in an initially isothermal, continued fluid, *J. Phys. Soc. Japan* **54**(3), 942 (1985).
48F. F. P. Incropera and J. A. Schutt, Numerical simulation of laminar mixed convection in the entrance region of horizontal rectangular ducts, *Numer. Heat Transfer* **8**(6), 707 (1985).
49F. A. D. W. Jones, The stability of buoyancy-driven rolls aligned with a shear flow when the temperature gradient is nonlinear, *Phys. Fluids* **28**(1), 31 (1985).
50F. Y. Kamotani, L. W. Wang, S. Ostrach and H. D. Jian, Experimental study of natural convection in shallow enclosures with horizontal temperature and concentration gradients, *Int. J. Heat Mass Transfer* **28**(1), 165 (1985).
51F. D. R. Kassoy and B. Cotte, The effects of sidewall heat loss on convection in a saturated porous vertical slab, *J. Fluid Mech.* **152,** 361 (1985).
52F. M. Keyhani, F. A. Kulacki and R. N. Christensen, Experimental investigation of free convection in a vertical rod bundle—a general correlation for Nusselt numbers, *J. Heat Transfer* **107**(3), 611 (1985).
53F. D. M. Kim and R. Viskanta, Effect of wall heat conduction on natural convection heat transfer in a square enclosure, *J. Heat Transfer* **107**(1), 139 (1985).
54F. K.-H. Kim and M.-U. Kim, Effect on time-dependent heating on the onset of Rayleigh–Bénard convection, *J. Phys. Soc. Japan* **54**(3), 933 (1985).
55F. S. Kimura and A. Bejan, Natural convection in a differentially heated corner region, *Phys. Fluids* **28**(10), 2980 (1985).
56F. A. G. Kirdyashkin, Gravitational heat convection in a planar inclined layer with adiabatic surfaces and a lateral heat supply, *High Temp.* **23**(1), 104 (1985).
57F. G. H. Knightly and D. Sather, Selection principle for Bénard-type convection, *Arch. Ration. Mech. Anal.* **88**(2), 163 (1985).
58F. S. Kotake and N. Hattori, Combined forced and free convection heat transfer for fully-developed laminar flow in horizontal annuli, *Int. J. Heat Mass Transfer* **28**(11), 2113 (1985).
59F. M. C. Limbourg-Fontaine, G. Petre and J. C. Legros, Texus 8 experiment: effects of a surface tension

minimum on thermocapillary convection, *Physico-Chem. Hydrodynam.* **6**(3), 301 (1985).

60F. Y. Maeno, H. Haucke, R. E. Ecke and J. C. Wheatley, Oscillatory convection in a dilute **3H superfluid **4He solution, *J. Low Temp. Phys.* **59**(3–4), 305 (1985).

61F. V. I. Malkovski, Hydrodynamics and heat transfer within an inclined cell immersed in a cooling liquid, *Cryogenics* **25**(1), 37 (1985).

62F. C. L. McTaggart and K. A. Lindsay, Nonclassical effects in the Bénard problem, *SIAM J. appl. Math.* **45**(1), 70 (1985).

63F. R. Monti and G. Mannara, Texus experiment on the convective heat transfer induced by marangoni flows, *Acta astronaut.* **12**(7–8), 511 (1985).

64F. H. Mori, K. Shobu, B. C. So and H. Okamoto, Spectral structure and universality of intermittent chaos, *Physica Scripta* **T9** (1985).

65F. E. Naito, The effect of buoyancy on laminar flow and heat transfer in the entrance region between horizontal parallel plates, *Heat Transfer—Japan. Res.* **13,** 80 (1984).

66F. K. Nandakumar, J. H. Masliyah and H.-S. Law, Bifurcation in steady laminar mixed convection flow in horizontal ducts, *J. Fluid Mech.* **152,** 145 (1985).

67F. V. F. Nicolette, K. T. Yang and J. R. Lloyd, Transient cooling by natural convection in a two-dimensional square enclosure, *Int. J. Heat Mass Transfer* **28**(9), 1721 (1985).

68F. D. G. Osborne and F. P. Incropera, Experimental study of mixed convection heat transfer for transitional turbulent flow between horizontal parallel plates, *Int. J. Heat Mass Transfer* **28**(7), 1337 (1985).

69F. H. Ozoe, K. Fujii, T. Shibata, H. Kuriyama and S. W. Churchill, Three-dimensional numerical analysis of natural convection in a spherical annulus, *Numer. Heat Transfer* **8**(4), 383 (1985).

70F. H. Ozoe, A. Mouri, M. Ohmuro, S. W. Churchill and N. Lior, Numerical calculations of laminar and turbulent natural convection in water in rectangular channels heated and cooled isothermally on the opposing vertical walls, *Int. J. Heat Mass Transfer* **28**(1), 125 (1985).

71F. D. Poulikakos, Natural convection in a confined fluid-filled space driven by a single vertical wall with warm and cold regions, *J. Heat Transfer* **107**(4), 867 (1985).

72F. C. Prakash and Ye-Di Liu, Buoyancy induced flow in a vertical internally finned circular duct, *J. Heat Transfer* **107**(1), 118 (1985).

73F. V. Prasad and F. A. Kulacki, Free convective heat transfer in a liquid-filled vertical annulus, *J. Heat Transfer* **107**(2), 596 (1985).

74F. V. Prasad, F. A. Kulacki and M. Keyhani, Natural convection in porous media, *J. Fluid Mech.* **150,** 89 (1985).

75F. U. Projan and H. Beer, Prandtl number effects on natural convection heat transfer in concentric and eccentric horizontal cylindrical annuli, *Wärme- u. Stoffübertr.* **19**(4), 249 (1985).

76F. K. R. Rajagopal and T. Y. Na, Natural convection flow of a non-Newtonian fluid between two vertical flat plates, *A Mech.* **54**(3–4), 239 (1985).

77F. Y. Renardy and D. D. Joseph, Oscillatory instability in a Bénard problem of two fluids, *Phys. Fluids* **28**(3), 788 (1985).

78F. Y. Renardy and M. Renardy, Perturbation analysis of steady oscillatory onset in a Bénard problem with similar liquids, *Phys. Fluids* **28**(9), 2699 (1985).

79F. N. Riahi, Nonlinear convection in a horizontal layer with an internal heat source, *J. Phys. Soc. Japan* **53**(12), 4169 (1984).

80F. N. Riahi, Nonlinear thermal convection with finite conducting boundaries, *J. Fluid Mech.* **152,** 113 (1985).

81F. A. J. Roberts, An analysis of near-marginal, mildly penetrative convection with heat flux prescribed on the boundaries, *J. Fluid Mech.* **158,** 71 (1985).

82F. H. Saito, I. Tokura and K. Okada, Experimental study on free convection heat transfer from vertically lined horizontal cylinders set between parallel walls, *Bull. J.S.M.E.* **28**(244), 2347 (1985).

83F. Yu. V. Sanochkin, Thermocapillary cell in a horizontal moving liquid layer with heating from above, *J. appl. Mech. Tech. Phys.* **26**(2), 222 (1985).

84F. G. S. R. Sarma, Effects of inter-facial curvature and gravity waves on the onset of thermocapillary convective instability in a rotating liquid layer subjected to a transverse magnetic field, *PhysicoChem. Hydrodynam.* **6**(3), 283 (1985).

85F. W. M. M. Schinkel and C. J. Hoogendoorn, Core stratification effects on natural convection in inclined cavities, *Appl. scient. Res.* **42**(2), 109 (1985).

86F. W. C. Schreiber and S. N. Singh, Natural convection between confocal horizontal elliptical cylinders, *Int. J. Heat Mass Transfer* **28**(4), 807 (1985).

87F. M. Schulte, N. K. Mitra and M. Fiebig, Heat transfer by free convection from a horizontal tube in concentric and eccentric cylindrical annulus (in German), *Wärme- u. Stoffübertr.* **19**(4), 241 (1985).

88F. A. H. Shaaban and M. N. Özisik, Thermal stability of a vertical fluid layer with volumetric energy source, *J. Heat Transfer* **107**(2), 589 (1985).

89F. W. A. Shay and D. H. Schultz, Second-order approximation to natural convection for large Rayleigh numbers and small Prandtl numbers, *Int. J. numer. Meth. Fluids* **5**(5), 427 (1985).

90F. C. Smutek, P. Bontoux, B. Roux, G. H. Schiroky, A. C. Hurford, F. Rosenberger and G. de Vahl Davis, Three-dimensional convection in horizontal cylinders: numerical solutions and comparison with experimental and analytical results, *Numer. Heat Transfer* **8**(5), 613 (1985).

91F. E. F. C. Somerscales and M. Kassemi, Electrochemical mass transfer studies in open cavities, *J. appl. Electrochem.* **15**(3), 405 (1985).

92F. M. M. Sorour, M. A. Hassab and F. A. Elewa, Effect of suction on the stability of fluid between horizontal plates at differing temperatures, *Proc. Inst. mech. Engrs, Part C, Mech. Engng Sci.* **199**(2), 145 (1985).

93F. M. M. Sorour, M. A. Hassab and F. A. Elewa, Stability of combined natural and forced cross flow in a vertical slot, *Int. J. Heat Mass Transfer* **28**(5), 987 (1985).

94F. A. C. M. Sousa, G. V. Hadjisophocleous and J. E. S. Venart, Numerical simulation of 3-D laminar free convection in tanks with localized heat sources, *Heat Technol.* **3**(1), 49 (1985).
95F. A. C. M. Sousa, P. R. Wentzell and J. E. S. Venart, Thermal modelling of LPG tanks engulfed in flames, *Heat Technol.* **3**(3–4), 1 (1985).
96F. J. Stastna, Convection and overstability in a viscoelastic fluid layer, *J. Non-Newtonian Fluid Mech.* **18**(1), 61 (1985).
97F. V. Steinberg, G. Ahlers and D. S. Cannell, Pattern formation and wave-number selection by Rayleigh–Bénard convection in a cylindrical container, *Physica Scripta* **T9**, 97 (1985).
98F. M. Takashima and H. Hamabata, The stability of natural convection in a vertical fluid layer having side walls of different temperatures and internal heat generation, *J. Phys. Soc. Japan* **54**(5), 1782 (1985).
99F. Yu. V. Val'tsiferov and V. P. Dronov, Numerical investigation of the convective heat transfer in a thin-walled cylindrical vessel with hemispherical ends with different heat-flux distributions over the contours, *High Temp.* **23**(1), 99 (1985).
100F. S. I. Vybornov and Yu. V. Sanochkin, Thermocapillary cell in a layer of a heavy fluid heated from above, *Fluid Dynam.* **20**(1), 153 (1985).
101F. I. C. Walton, The effect of a shear flow on convection in a layer heated non-uniformly from below, *J. Fluid Mech.* **154,** 303 (1985).
102F. R. O. Warrington, Jr. and R. E. Powe, The transfer of heat by natural convection between bodies and their enclosures, *Int. J. Heat Mass Transfer* **28**(2), 319 (1985).
103F. P. S. Wei and W. H. Giedt, Surface tension gradient-driven flow around an electron beam welding cavity, *Welding J.* **64**(9), 251 (1985).
104F. P. D. Weidman and G. Mehrdadtehranfar, Instability of natural convection in a tall vertical annulus, *Phys. Fluids* **28**(3), 776 (1985).
105F. M. Grae Worster and Alison M. Leitch, Laminar free convection in confined regions, *J. Fluid Mech.* **156,** 301 (1985).
106F. W. Wulbrand and H. Kagermann, The influence of random temperature modulation on the convective instability, *Physica* **131A**(1), 182 (1985).
107F. S. Yanase and K. Kohno, The effect of a salinity gradient on the instability of natural convection in a vertical fluid layer, *J. Phys. Soc. Japan* **54**(10), 3747 (1985).
108F. F. Zangrando and L. A. Bertram, The effect of variable stratification on linear doubly diffusive stability, *J. Fluid Mech.* **151,** 55 (1985).
109F. A. Zebib, A. K. Goyal and G. Schubert, Convective motions in a spherical shell, *J. Fluid Mech.* **152,** 39 (1985).
110F. A. Zebib, G. M. Homsy and E. Meiburg, High Marangoni number convection in a square cavity, *Phys. Fluids* **28**(12), 3467 (1985).
111F. Z. Y. Zhong, K. T. Yang and J. R. Lloyd, Variable property effects in laminar natural convection in a square enclosure, *J. Heat Transfer* **107**(1), 133 (1985).

NATURAL CONVECTION—EXTERNAL FLOWS

1FF. N. Afzal, Higher order effects in natural convection flow over a uniform flux horizontal surface, *Wärme- u. Stoffübertr.* **19**(3), 177 (1985).
2FF. N. Afzal, Mixed convection plume above a point heat source in a vertical free stream, *Int. J. Heat Mass Transfer* **28**(11), 2043 (1985).
3FF. M. Amaouche and J.-L. Peube, Convection mixte stationnaire autour d'un cylindre horizontal, *Int. J. Heat Mass Transfer* **28**(7), 1269 (1985).
4FF. H. M. Badr, On the effect of flow direction on mixed convection from a horizontal cylinder, *Int. J. numer. Meth. Fluids* **5**(1), 1 (1985).
5FF. M. N. Bui and T. Cebeci, Combined free and forced convection on vertical slender cylinder, *J. Heat Transfer* **107**(2), 476 (1985).
6FF. J. Calman, Convection at a model ice edge, *Johns Hopkins APL Tech. Digest* (*appl. Phys. Lab.*) **6**(3), 211 (1985).
7FF. M. J. Chamberlain, K. G. T. Hollands and G. D. Raithby, Experiments and theory on natural convection heat transfer from bodies of complex shape, *J. Heat Transfer* **107**(3), 624 (1985).
8FF. S. S. Chawla, Free convection from a torsionally oscillating vertical plane, *Int. J. Engng Sci.* **23**(6), 611 (1985).
9FF. M. M. Elsayed and K. A. Fathalah, Natural convection over a vertical flat plate due to absorption of thermal radiation, *Wärme- u. Stoffübertr.* **19**(2), 121 (1985).
10FF. M. Epstein, G. M. Hauser and R. E. Henry, Thermophoretic deposition of particles in natural convection flow from a vertical plate, *J. Heat Transfer* **107**(2), 272 (1985).
11FF. V. M. Eroshenko, L. I. Zaichik and V. A. Pershukov, Stability of the free-convection boundary layer on a vertical permeable plate, *J. Engng Phys.* **48**(3), 271 (1985).
12FF. S. Ghorbani-Tari and A. J. Ghajar, Improved free convective heat-transfer correlations in the near-critical region, *A.I.A.A. Jl* **23**(10), 1647 (1985).
13FF. M. Hasan and A. S. Mujumdar, Simultaneous mass and heat transfer under mixed convection along a vertical cone, *Int. J. Energy Res.* **9**(2), 129 (1985).
14FF. M. Hasan and A. S. Mujumdar, Simultaneous heat and mass transfer in free convection from a horizontal cylinder, *A.I.A.A. Jl* **23**(10), 1602 (1985).
15FF. M. Hasan and A. S. Mujumdar, Combined heat and mass transfer in free convection about a sphere, *Can. J. chem. Engng* **63**(3), 412 (1985).
16FF. H. Herwig, Asymptotic approach to free-convection flow at maximum density, *Chem. Engng Sci.* **40**(9), 1709 (1985).
17FF. H. Herwig, G. Wickern and K. Gersten, T he influence of variable fluid properties to free convection laminar flows (in German), *Wärme- u. Stoffübertr.* **19**(1), 19 (1985).
18FF. C. T. Hsu and P. Cheng, Effects of upstream geometry on natural convection of a Darcian fluid about a semi-infinite inclined heated surface, *J. Heat Transfer*

107(2), 283 (1985).
19FF. M.-J. Huang and C.-K. Chen, Effects of surface mass transfer on free convection flow over vertical cylinder embedded in a saturated porous medium, *J. Energy Res. Technol. Trans. Am. Soc. mech. Engrs* **107**(3), 394 (1985).
20FF. M.-J. Huang and C.-K. Chen, Conjugate mixed convection and conduction heat transfer along a vertical circular pin, *Int. J. Heat Mass Transfer* **28**(3), 523 (1985).
21FF. M.-J. Huang, C.-K. Chen and J. W. Cleaver, Vertical circular pin with conjugated natural convection-conduction flow, *J. Heat Transfer* **107**(1), 242 (1985).
22FF. D. B. Ingham, Flow past a suddenly heated vertical plate, *Proc. R. Soc.* **A402**(1822), 109 (1985).
23FF. Y. Jaluria, Interaction of natural convection wakes arising from thermal sources on a vertical surface, *J. Heat Transfer* **107**(4), 883 (1985).
24FF. X. K. Kakatsios, Heat released by a flat plate at various angles of inclination, *Chem.-Ing.-Tech.* **57**(10), 865 (1985).
25FF. J. H. Kim, An analytical mixing model for buoyant jet injected into pipe flow, *J. Heat Transfer* **107**(3), 630 (1985).
26FF. S. Kimura, A. Bejan and I. Pop, Natural convection near a cold plate facing upward in a porous medium, *J. Heat Transfer* **107**(4), 819 (1985).
27FF. K. Kitamura, M. Koike, I. Fukuoka and T. Saito, Large eddy structure and heat transfer of turbulent natural convection along a vertical flat plate, *Int. J. Heat Mass Transfer* **28**(4), 837 (1985).
28FF. A. K. Kolar and V. M. K. Sastri, Numerical and experimental investigations into the upstream transpiration problem in free convection, *Wärme- u. Stoffübertr.* **19**(4), 297 (1985).
29FF. A. A. Kovkova, Singular perturbations in the problem of free convection of high-viscosity fluids on a vertical plate, *Heat Transfer—Sov. Res.* **16**(4), 106 (1985).
30FF. H. Koyama, A. Nakayama, S. Ohsawa and H. Yamada, Theoretical and experimental study of free convection from a vertical frustum of a cone of a finite length, *Int. J. Heat Mass Transfer* **28**(5), 969 (1985).
31FF. S.-C. Lee, K.-L. Wong and C.-K. Chen, The finite element solution of laminar combined convection from a horizontal cylinder, *Comp. Meth. appl. Mech. Engng* **50**(2), 147 (1985).
32FF. J. H. Merkin, Note on the similarity solutions for free convection on a vertical plate, *J. Engng Math.* **19**(3), 189 (1985).
33FF. J. H. Merkin, Free convection above a uniformly heated horizontal circular disk, *Int. J. Heat Mass Transfer* **28**(6), 1157 (1985).
34FF. M. Miyamoto, Y. Katoh, K. Kurima, S. Kurihara and K. Yamashita, Free convection heat transfer from vertical and horizontal short plates, *Int. J. Heat Mass Transfer* **28**(9), 1733 (1985).
35FF. A. Nakayama and H. Koyama, An analysis of turbulent free convection about bodies of arbitrary geometrical configurations, *Wärme- u. Stoffübertr.* **19**(4), 263 (1985).
36FF. S. B. Nasrallah, G. Arnaud et A. Trad, Evaporation en convection naturelle sur une plaque verticale chauffée a flux variable. *Z. angew. Math. Phys.* **36**(1), 105 (1985).
37FF. A. O. Nieckele and S. V. Patankar, Laminar mixed convection in a concentric annulus with horizontal axis, *J. Heat Transfer* **107**(4), 902 (1985).
38FF. R. H. Nilson, Countercurrent convection in a double-diffusive boundary layer, *J. Fluid Mech.* **160,** 181 (1985).
39FF. P. Olson and H. Singer, Creeping plumes, *J. Fluid Mech.* **158,** 511 (1985).
40FF. D. G. Osborne and F. P. Incropera, Laminar, mixed convection heat transfer for flow between horizontal parallel plates with asymmetric heating, *Int. J. Heat Mass Transfer* **25**(1), 207 (1985).
41FF. K. H. Park and V. P. Carey, Analysis of transient natural convection flow near a vertical surface at low Prandtl number, *Numer. Heat Transfer* **8**(3), 317 (1985).
42FF. U. Projahn and H. Beer, Theoretical and experimental study of transient and steady-state natural convection heat transfer from a vertical flat plate partially immersed in water, *Int. J. Heat Mass Transfer* **28**(8), 1487 (1985).
43FF. U. Projahn and H. Beer, Theoretical and experimental studies on steady-state and transient heat transfer behavior as a result of natural convection on a vertical heating surface partially immersed in water, *Chem.-Ing.-Tech.* **57**(9), 808 (1985).
44FF. M. Ramachandran, B. F. Armaly and T. S. Chen, Measurements and predictions of a laminar mixed convection flow adjacent to a vertical surface, *J. Heat Transfer* **107**(3), 636 (1985).
45FF. K. V. Rao, B. F. Armaly and T. S. Chen, Mixed convection plumes arising from a thermal point source, *J. Heat Transfer* **107**(3), 720 (1985).
46FF. P. Razelos, An interferometric investigation of the effect of separation distance and temperature imbalance on natural convection for two horizontal cylinders at moderate Rayleigh numbers, *Wärme- u. Stoffübertr.* **19**(4), 255 (1985).
47FF. H. Saito, I. Tokura and K. Okada, An experimental study on free convection heat transfer from vertically lined horizontal cylinders set between parallel walls, *Bull. J.S.M.E.* **28**(244), 2347 (1985).
48FF. B. Sammakia, V. P. Carey and B. Gebhart, Measurements and calculations of transient mixed convection in air, *Int. J. Heat Mass Transfer* **28**(10), 1837 (1985).
49FF. T. Schulenberg, Natural convection heat transfer below downward facing horizontal surfaces, *Int. J. Heat Mass Transfer* **28**(2), 467 (1985).
50FF. V. I. Semenov, Similar problems of steady-state laminar free convection on a vertical plate, *Heat Transfer—Sov. Res.* **16**(2), 69 (1984).
51FF. D. L. Siebers, R. F. Moffat and R. G. Schwind, Experimental, variable properties natural convection from a large, vertical, flat surface, *J. Heat Transfer* **107**(1), 124 (1985).
52FF. Yu. A. Sokovishin and A. A. Kovkova, Free convec-

tion of highly viscous liquid along a vertical finite plate with constant heat flux, *J. Engng Phys.* **48**(1), 46 (1985).

53FF. V. M. Soundalgekar and P. Ganesan, Transient free convection with mass transfer on a vertical plate with constant heat flux, *Int. J. Energy Res.* **9**(1), 1 (1985).

54FF. E. M. Sparrow and L. F. A. Azevedo, Vertical-channel natural convection spanning between the fully-developed limit and the single-plate boundary-layer limit, *Int. J. Heat Mass Transfer* **28**(10), 1847 (1985).

55FF. E. M. Sparrow and L. F. A. Azevedo, Lateral-edge effects on natural convection heat transfer from an isothermal vertical plate, *J. Heat Transfer* **107**(4), 977 (1985).

56FF. E. M. Sparrow and S. S. Kang, Two-dimensional heat transfer and critical radius results for natural convection about an insulated horizontal cylinder, *Int. J. Heat Mass Transfer* **28**(11), 2049 (1985).

57FF. E. M. Sparrow and A. J. Stretton, Natural convection from variously oriented cubes and from other bodies of unity aspect ratio, *Int. J. Heat Mass Transfer* **28**(4), 741 (1985).

58FF. K. Stephan, M. Durst and R. Windisch, Natural convection heat transfer in the near critical region of pure fluids (in German), *Wärme- u. Stoffübertr.* **19**(3), 187 (1985).

59FF. W. E. Stewart, Jr. and J. C. Johnson, Experimental natural convection heat transfer from isothermal spherical zones, *J. Heat Transfer* **107**(2), 463 (1985).

60FF. N. Takenaka, O. Takahashi and I. Michiyoshi, Natural convection heat transfer from a horizontal plate to potassium, *Mem. Fac. Engng, Kobe Univ.* **32,** 145 (1985).

61FF. J. Timma et J.-P. Padet, Etude théorique du couplage convection-conduction en convection libre laminaire sur une plaque plane verticale, *Int. J. Heat Mass Transfer* **28**(6), 1097 (1985).

62FF. M. G. Timol and N. L. Kalthia, Similarity solution of flow past an accelerated vertical infinite plate in the presence of free convection currents, *Reg. J. Energy Heat Mass Transfer* **7**(1), 67 (1985).

63FF. A. K. Tolpadi and T. H. Kuehn, Experimental investigation of conjugate natural convection heat transfer from a horizontal isothermal cylinder with a non-isothermal longitudinal plate fin at various angles, *Int. J. Heat Mass Transfer* **28**(1), 155 (1985).

64FF. K. L. Tzuoo, T. S. Chen and B. F. Armaly, Wave instability of natural convection flow on inclined surface, *J. Heat Transfer* **107**(1), 107 (1985).

65FF. S. Wakitani, Non-parallel-flow stability of a two-dimensional buoyant plume, *J. Fluid Mech.* **159,** 241 (1985).

66FF. I. C. Walton, The effect of a shear flow on convection near a two-dimensional hot-patch, *Q. Jl Mech. appl. Math.* **38**(4), 561 (1985).

67FF. G. Wilks, R. Hunt and D. S. Riley, The two-dimensional laminar vertical jet with positive or adverse buoyancy, *Numer. Heat Transfer* **8**(4), 449 (1985).

68FF. M. F. Young and T. Ozel, Mixed free and forced convection from a short vertical cylinder placed in a laminar horizontal flow, *J. Heat Transfer* **107**(1), 250 (1985).

CONVECTION FROM ROTATING SURFACES

1G. M. M. M. Abou-Ellail and S. M. Morcos, Influence of cylinder rotation on turbulence modeling of flow and heat transfer in an annulus, *Heat Technol.* **3**(3–4), 90 (1985).

2G. S. P. Bhattacharyya and M. Abbas, On the stability of a hot rotating layer of micropolar fluid, *Int. J. Engng Sci.* **23**(3), 371 (1985).

3G. J. W. Chew, Computation of convection laminar flow in rotating cavities, *J. Fluid Mech.* **153,** 339 (1985).

4G. J. W. Chew, Computation of forced laminar convection in rotating cavities, *J. Heat Transfer* **107**(2), 277 (1985).

5G. J. W. Chew, Effect of frictional heating and compressive work in rotating axisymmetric flow, *J. Heat Transfer* **107**(4), 984 (1985).

6G. B. Farouk, Mixed convective flows around a slowly rotating isothermal sphere, *J. Heat Transfer* **107**(2), 431 (1985).

7G. N. C. Ghosh, Thermal effect on the transverse vibration of high-speed rotating anisotropic disk, *J. appl. Mech.* **52**(3), 543 (1985).

8G. K. M. Iskaov and V. A. Trushin, Effect of rotation on heat transfer in the radial cooling channels or turbine blades, *Thermal Engng* **32**(2), 93 (1985).

9G. J. Jimenez and M. G. Velarde, Thermoconvective instability of a rotating viscoelastic fluid layer heated from below, *J. Non-Equil. Thermodynam.* **10**(4), (1985).

10G. C. W. Lee and H. N. Chang, Examination of immobilized cells in a rotating packed drum reactor for the production of ethanol from D-glucose, *Enzyme Microbial Tech.* **7**(11), 561 (1985).

11G. A. M. Morrison and C. Fiscina, Effects of spin and mass addition on high-angle-of-attack re-entry, *J. Spacecraft Rockets* **22**(1), 68 (1985).

12G. M. R. Muller and J. N. Burch, An experimental study of the behavior of transient isolated convection in a rigidly rotating fluid, *Exp. Fluids* **3**(1), 17 (1985).

13G. S. Neti, A. S. Warnock, E. K. Levy and K. S. Kannan, Computation of laminar heat transfer in rotating rectangular ducts, *J. Heat Transfer* **107**(3), 575 (1985).

14G. J. M. Owen, J. R. Pincombe and R. H. Rogers, Source-sink flow inside a rotating cylindrical cavity, *J. Fluid Mech.* **155,** 233 (1985).

15G. J. Pecheux et C. Boutin, Etude d l'ecoulement laminaire de deux liquides non miscibles entre deux disques coaxiaux tournanent en sens contraires, *Z. angew. Math. Phys.* **36**(2), 238 (1985).

16G. R. Rajasekaran and M. G. Palekar, Viscous dissipation effects on mixed convection about a rotating sphere, *Int. J. Engng Sci.* **23**(8), 789 (1985).

17G. R. Shimada, S. Kumagai and T. Takeyama, Enhancement of heat transfer from a rotating cylinder by means of surface projections, *Heat Transfer—Japan. Res.* **13**(3), 1 (1984).

18G. R. Siegel, Analysis of buoyancy effect of fully developed laminar heat transfer in a rotating tube, *J. Heat Transfer* **107**(2), 338 (1985).

19G. Y. S. Sim and W. J. Yang, Turbulent heat transfer in corotating annular disks, *Numer. Heat Transfer* **8**(3), 299 (1985).

20G. C. L. Soule and R. L. Merson, Heat transfer coefficients to Newtonian liquids in axially rotated cans, *J. Food Process Engng* **8**(1), 33 (1985).
21G. A. M. Soward, Bifurcation and stability of finite amplitude convection in a rotating layer, *Physica D: Non-linear Phenomena* **14D**(2), 227 (1985).
22G. S. Thangam and C. G. Speziale, Numerical study of thermal convection in rotating channel flow, *Int. J. numer. Meth. Fluids* **5**(2), 133 (1985).
23G. M. Ungarish and M. Israeli, Axisymmetric compressible flow in a rotating cylinder with axial convection, *J. Fluid Mech.* **154,** 121 (1985).
24G. H. Yamashita, R. Izumi, H. Maeda and Y. Sugiura, Study on a rotating condenser with a scraper (1st report, analysis of condensation on outer surface of horizontal cylinder), *Bull. J.S.M.E.* **28**(239), 868 (1985).

COMBINED HEAT AND MASS TRANSFER

1H. R. Amono and S. Sugiyama, Investigation on turbulent heat transfer of an axisymmetric jet impinging on a flat plate, *Bull. J.S.M.E.* **28**(235), 74 (1985).
2H. R. Bhargava and M. Rani, Numerical solution of heat transfer in micropolar fluid flow in a channel with porous walls, *Int. J. Engng Sci.* **23**(4), 409 (1985).
3H. Ye. N. Bogomolov and V. I. Orlova, Analysis of the overall efficiency of cooling of a perforated wall of a stationary gas-turbine blade, *Heat Transfer—Sov. Res.* **16**(2), 111 (1984).
4H. C. Camci and T. Arts, Experimental heat transfer investigation around the film-cooled leading edge of a high-pressure gas turbine rotor blade (85-GT-114), *J. Engng Gas Turbine Pwr* **107**(4), 1016 (1985).
5H. C. Camci and T. Arts, Short duration measurements and numerical simulation of heat transfer along the suction side of a film-cooled gas turbine blade (85-GT-111), *J. Engng Gas Turbine Pwr* **107**(4), 991 (1985).
6H. B. T. F. Chung and J. S. Hsiao, Heat transfer with ablation in a finite slab subjected to time-variant heat flux, *A.I.A.A. Jl* **23**(1), 145 (1985).
7H. V. M. Epifanov, A. I. Leont'ev, A. M. Pavlyuchenko and A. V. Pashutov, Critical permeability parameter in supersonic gradient flow, *High Temp.* **23**(2), 279 (1985).
8H. A. D. Fitt, J. R. Ockendon and T. V. Jones, Aerodynamics of slot-film cooling: theory and experiment, *J. Fluid Mech.* **160,** 15 (1985).
9H. L. W. Florschuetz and H. H. Tseng, Effect of nonuniform geometries on flow distributions and heat transfer characteristics for arrays of impinging jets, *J. Engng Gas Turbine Pwr* **107**(1), 68 (1985).
10H. T. Fujii, T. Nagata and K.-Ei Shinzato, Experimental study of heat and mass transfer from humid air to a horizontal tube in false tube banks, *Bull. J.S.M.E.* **28**(237), 436 (1985).
11H. D. P. Georgiou and J. F. Louis, Transpired turbulent boundary layer in various pressure gradients and the blow-off condition, *J. Engng Gas Turbine Pwr* **107**(3), 636 (1985).
12H. S. A. Gokuglu and D. E. Rosner, Thermophoretically enhanced mass transport rates to solid and transpiration-cooled walls across turbulent (law-of-the-wall) boundary layers, *Ind. Engng Chem. Fund.* **24**(2), 208 (1985).
13H. R. J. Goldstein, E. R. G. Eckert, H. D. Chiang and E. Elovic, Effect of surface roughness on film cooling performance (84-GT-41), *J. Engng Gas Turbine Pwr* **107**(1), 111 (1985).
14H. R. J. Goldstein and P. H. Chen, Film cooling on a gas turbine blade near the end wall (84-GT-42), *J. Engng Gas Turbine Pwr* **107**(1), 117 (1985).
15H. N. Hay, D. Lampard and C. L. Saluja, Effects of cooling films on the heat transfer coefficient on a flat plate with zero mainstream pressure gradient (84-GT-40), *J. Engng Gas Turbine Pwr* **107**(1), 105 (1985).
16H. N. Hay, D. Lampard and C. L. Saluja, Effects of the condition of the approach boundary layer and of mainstream pressure gradients on the heat transfer coefficient on film-cooled surfaces (84-GT-47), *J. Engng Gas Turbine Pwr* **107**(1), 99 (1985).
17H. B. R. Hollworth and L. R. Gero, Entrainment effects on impingement heat transfer: Part II—local heat transfer measurements, *J. Heat Transfer* **107**(4), 910 (1985).
18H. B. Jubran and A. Brown, Film cooling from two rows of holes inclined in the streamwise and spanwise directions (84-GT-286), *J. Engng Gas Turbine Pwr* **107**(1), 84 (1985).
19H. I. I. Kantorovich, L. I. Sevkovaskay, A. A. Vecher and A. S. Skoropanov, Theory of stability of gas-permeable cooling, *High Temp.* **23**(2), 268 (1985).
20H. K. R. Khair and A. Began, Mass transfer to natural convection boundary layer flow driven by heat transfer, *J. Heat Transfer* **107**(4), 979 (1985).
21H. S. Kikkawa and M. Nonogaki, The effect of vibration on film cooling over a gas turbine blade, *Heat Transfer—Japan. Res.* **13**(2), 1 (1984).
22H. S. B. Koshelev, V. N. Subbotin and V. V. Kharitonov, Components of heat transfer in the jet flow of water in a limited space, *High Temp.* **23**(3), 441–446 (1985).
23H. P. M. Ligrani and C. Camci, Adiabatic film cooling effectiveness from heat transfer measurements in compressible, variable-property flow, *J. Heat Transfer* **107**(2), 313 (1985).
24H. V. V. Lunev and I. N. Polyakov, Stability of steady solution of ablation equation, *Fluid Dynam.* **20**(2), 248 (1985).
25H. B. P. Mironov, V. N. Vasechkin and N. I. Yarygina, Heat transfer behind the injection section at different levels of freestream turbulence, *Heat Technol.* **3**(1), 1 (1985).
26H. K. Nomura and P. V. Farrell, Heat and mass transfer coeffficients for porous horizontal cylinder, *A.I.Ch.E. Jl* **31**(7), 1217 (1985).
27H. M. Salcudean, K. Y. M. Lai and R. I. L. Guthrie, Multidimensional heat, mass and flow phenomena in gas stirred reactors, *Can. J. chem. Engng* **63**(1), 51 (1985).
28H. V. D. Sovershennyy and S. V. Leonov, Cooling of a turbine blade by injection of air through porous zones,

Heat Transfer—Sov. Res. **16**(3), 126 (1984).
29H. C. D. Surma Devi, M. Nagaraj, T. Ramesan and G. Nath, Unsteady compressible three-dimensional stagnation point boundary layers with large injection rates, *Z. angew. Math. Mech.* **65**(8), 385 (1985).
30H. H. S. Takhar and V. M. Soundalgekar, Flow and heat transfer of a micropolar fluid past a continuously moving porous plate, *Int. J. Engng Sci.* **23**(2), 201 (1985).
31H. E. S. Tillman, E. O. Hartel and H. F. Jen, The prediction of flow through leading edge holes in a film cooled airfoil with and without inserts (84-GT-4), *J. Engng Gas Turbine Pwr* **107**(1), 92 (1985).
32H. A. A. Vasil'yev, Efficiency of nonequilibrium two-phase vapor–liquid cooling films on adiabatic walls, *Heat Transfer—Sov. Res.* **16**(5), 119 (1985).

CHANGE OF PHASE—BOILING

1J. M. A. Abdelghaffar and K. Cornwell, On the flow field generated by a growing sphere near a solid plane with application to nucleate boiling, *Int. J. Heat Mass Transfer* **28**, 1131 (1985).
2J. N. Abuaf, S. H. Black and F. W. Staub, Pool boiling performance of finned surfaces in R-113, *Int. J. Heat Fluid Flow* **6**(1), 23 (1985).
3J. J.-L. Achard, D. A. Drew and R. T. Lahey, Analysis of nonlinear density-wave oscillations in boiling channels, *J. Fluid Mech.* **155**, 213 (1985).
4J. N. H. Afgan, L. A. Jović, S. A. Kovalev and V. A. Lenykov, Boiling heat transfer from surfaces with porous layers, *Int. J. Heat Mass Transfer* **28**, 415 (1985).
5J. H. Akimoto, Analysis of TRAC-PF1 calculated core heat transfer for a CCTF test, *Nucl. Engng Des.* **88**, 215 (1985).
6J. K. A. Akiyev, Heat transfer in boiling of azeotropic mixture in inertial fields, *Heat Transfer—Sov. Res.* **16**(6), 89 (1984).
7J. A. I. Anikin, G. N. Danilova and N. I. Mirmov, Generalized correlation for coefficients of boiling heat transfer oil–refrigerant mixtures in tubes, *Heat Transfer—Sov. Res.* **16**(1), 42, (1984).
8J. A. Bar-Cohen and H. Schweitzer, Thermosyphon boiling in vertical channels, *J. Heat Transfer* **107**, 772 (1985).
9J. G. G. Bartolomey, Le Tin Chun and N'Guyen Chih Huo, The dependence of the point of onset of intensive vaporization in subcooled boiling on the operating parameters, *Heat Transfer—Sov. Res.* **16**(4), 60 (1984).
10J. S. Benedek, Scaling laws of homogeneous and two-fluid flow coupled with heat transmission from a heated rod, *Atomkernenergie/Kerntech.* **46**, 257 (1985).
11J. A. M. Bhat, J. S. Saini and R. Prakash, Heat transfer correlations for high heat flux region in nucleate pool boiling, *J. Inst. Engrs (India)*, Part ME, *Mech. Engng Div.* **66**(July), 20 (1985).
12J. S. Biyikli, Optimum use of longitudinal fins of rectangular profiles in boiling liquids, *J. Heat Transfer* **107**, 968 (1985).
13J. R. A. Broussard and J. W. Westwater, Diameter and velocity effects for cross-flow boilling, *A.I.A.A. Jl* **23**, 1615 (1985).
14J. T. D. Bui and V. K. Dhir, Transition boiling heat transfer on a vertical surface, *J. Heat Transfer* **107**, 756 (1985).
15J. T. D. Bui and V. K. Dhir, Film boiling heat transfer on an isothermal vertical surface, *J. Heat Transfer* **107**, 764 (1985).
16J. A. Calka and R. L. Judd, Some aspects of the interaction among nucleation sites during saturated nucleate boiling, *Int. J. Heat Mass Transfer* **28**, 2331 (1985).
17J. V. P. Carey, Surface tension effects on convective boiling heat transfer in compact exchangers with offset strip fins, *J. Heat Transfer* **107**, 970 (1985).
18J. G. P. Celata, M. Cumo, F. D'Annibale and G. E. Farello, A computation method for mass flowrate prediction in critical flows of initially subcooled liquid in long channels, *Heat Technol.* **3**(3–4), 24 (1985).
19J. C. Cercignani, W. Fiszdon and A. Frezzotti, The paradox of the inverted temperature profiles between an evaporating and a condensing surface, *Phys. Fluids* **28**, 3237 (1985).
20J. M. Cerza and V. Sernas, A bubble growth model for nucleate boiling in thin, falling, superheated, laminar water films, *Int. J. Heat Mass Transfer* **28**, 1307 (1985).
21J. G. L. Chahine and H. L. Liu, Singular-perturbation theory of the growth of a bubble cluster in a superheated liquid, *J. Fluid Mech.* **156**, 257 (1985).
22J. L. S. Chang and J. C. Berg, Electroconvective enhancement of mass or heat exchange between a drop or bubble and surroundings in the presence of an interfacial tension gradient, *A.I.Ch.E. Jl* **31**, 149 (1985).
23J. M. Chattree and S. Sengupta, Heat transfer and evaporation from heated water bodies, *J. Heat Transfer* **107**, 779 (1985).
24J. T. C. Chawla and J. D. Bingle, Downward heat transfer in heat-generating boiling pools, *Int. J. Heat Mass Transfer* **28**, 81 (1985).
25J. S. C. Cheng, P. W. K. Lau and K. T. Poon, Measurements of true quench temperature of subcooled water under forced convective conditions, *Int. J. Heat Mass Transfer* **28**, 235 (1985).
26J. S. K. Roy Chowdhury and R. H. S. Winterton, Surface effects in pool boiling, *Int. J. Heat Mass Transfer* **28**, 1881 (1985).
27J. P. Chuchottaworn and K. Asano, Calculation of drag coefficients of an evaporating or a condensing droplet, *J. chem. Engng Japan* **18**, 91 (1985).
28J. G. N. Danilova and A. V. Tikhonov, Enhancement of heat transfer in the boiling of the R-113 refrigerant on various kinds of surfaces, *Heat Transfer—Sov. Res.* **16**(1), 59 (1984).
29J. V. H. Del Valle, M. and D. B. R. Kenning, subcooled flow boiling at high heat flux, *Int. J. Heat Mass Transfer* **28**, 1907 (1985).
30J. K. P. Derewnicki, Experimental studies of heat transfer and vapour formation in fast transient boiling, *Int. J. Heat Mass Transfer* **28**, 2085 (1985).
31J. M. De Salve and B. Panella, Void fraction and quench

time in bottom reflooding experiments, *Heat Technol.* **3**(2), 64 (1985).

32J. D. S. Dhuga and R. H. S. Winterton, Measurement of surface contact in transition boiling, *Int. J. Heat Mass Transfer* **28,** 1869 (1985).

33J. F. Dobran, Heat transfer in an annular two-phase flow, *J. Heat Transfer* **107,** 472 (1985).

34J. B. M. Dorofeev, Acoustic phenomena in boiling, *High Temp.* **23,** 479 (1985).

35J. V. A. Dyundin, A. G. Solov'yev, A. V. Borishanskaya and Yu. A. Vol'nykh, Effect of the kind of surface roughness on the heat transfer coefficient in boiling of refrigerants, *Heat Transfer—Sov. Res.* **16**(1), 53 (1984).

36J. Z. Edelman, E. Elias and D. Naot, Inverted annular boiling in a stainless-steel tube with steady heat sources, *Int. J. Heat Mass Transfer* **28,** 1281 (1985).

37J. D. Evans, S. W. Webb and J. C. Chen, Axially varying vapor superheats in convective film boiling, *J. Heat Transfer* **107,** 663 (1985).

38J. H. Fik, Z. Kolenda, R. Ramirez-Zetina, J. Donizak and E. Kochanska, Influence of surface conditions on boiling heat transfer in New Jersey Zn–Pb–Cd distillation columns, *Trans. Inst. Min. Metall.* Section C **94**(June), 55 (1985).

39J. T. R. Fodemski, The influence of liquid viscosity and system pressure on stagnation point vapour thickness during forced-convection film boiling, *Int. J. Heat Mass Transfer* **28,** 69 (1985).

40J. V. V. Galaktionov and L. P. Trukhanova, Study of the process of heat and mass transfer in the region of the vapor–gas front in a gas-regulable heat pipe, *J. Engng Phys.* **48,** 296 (1985).

41J. A. Galati and S. Migliori, Bubble-breaking effect in a sodium boiling test, *Nucl. Sci. Engng* **90,** 209 (1985).

42J. Y. Hagiwara, T. Miwada and K. Suzuki, Heat transfer and wave structural in the developing region of two-component two-phase annular flow, *PhysicoChem. Hydrodynam.* **6**(1–2), 141 (1985).

43J. JeChin Han and L. S. Fletcher, Falling film evaporating and boiling in circumferential and axial grooves on horizontal tubes, *Ind. Engng Chem. Proc. Des. Dev.* **24,** 570 (1985).

44J. R. Hino and T. Ueda, Studies on heat transfer and flow characteristics in subcooled flow boiling—Part 1. Boiling characteristics, *Int. J. Multiphase Flow* **11,** 269 (1985).

45J. R. Hino and T. Ueda, Studies on heat transfer and flow characteristics in subcooled flow boiling—Part 2. Flow characteristics, *Int. J. Multiphase Flow* **11,** 283 (1985).

46J. T. O. Hui and J. R. Thome, A study of binary mixture boiling: boiling site density and subcooled heat transfer, *Int. J. Heat Mass Transfer* **28,** 919 (1985).

47J. Y.-H. Hung and S.-C. Yao, Pool boiling heat transfer in narrow horizontal annular crevices, *J. Heat Transfer* **107,** 656 (1985).

48J. E. A. Ibrahim and R. L. Judd, An experimental investigation of the effect of subcooling on bubble growth and waiting time in nucleate boiling, *J. Heat Transfer* **107,** 168 (1985).

49J. S. Inada, Y. Miyasaka and K. Nishida, Transient heat transfer for a water drop impinging on a heated surface (1st report, effects of drop subcooling on the liquid–solid contact state), *Bull. J.S.M.E.* **28**(245), 2675 (1985).

50J. Kh. E. Ioarand, I. A. Kamenev and Yu. I Kallas, Conjoint heat and mass transfer in a multicomponent vapor–liquid system, *J. appl. Chem. U.S.S.R.* **58,** 401 (1985).

51J. Y. K. Kao and J. Weisman, A moving-front transition boiling model, *A.I.Ch.E. Jl* **31,** 529 (1985).

52J. M. Kawaji, S. Y. Ng, S. Banerjee and G. Yadigaroglu, Reflooding with steady and oscillatory injection: Part I—Flow regimes, void fraction, and heat transfer, *J. Heat Transfer* **107,** 670 (1985).

53J. Y. Kikuchi, T. Hori and I. Michiyoshi, Minimum film boiling temperature for cooldown of insulated metals in saturated liquid, *Int. J. Heat Mass Transfer* **28,** 1105 (1985).

54J. Yu. A. Kirichenko, K. V. Rusanov and E. G. Tyurina, Effect of orientation of an annular channel on heat exchange in the boiling of nitrogen under conditions of free motion, *J. Engng Phys.* **48,** 121 (1985).

55J. H. Kisley and A. Nir, Droplet evaporation by combined mechanisms in an ICF reactor, *Int. J. Heat Mass Transfer* **28,** 1441 (1985).

56J. V. V. Klimenko, A. A. Kurilenko, V. I. Panevin, V. P. Firsov and A. G. Shelepin, Experimental study of film boiling of cryogenic fluids under free convection, *Heat Transfer—Sov. Res.* **16**(4), 76 (1984).

57J. N. I. Kobasko, N. M. Fialko and N. O. Meranova, Numerical determination of the duration of the nucleate-boiling phase in the course of steel plate hardening, *Heat Transfer—Sov. Res.* **16**(2), 130 (1984).

58J. G. R. Kudritskiy, Experimental data and correlation for boiling heat transfer on microsurfaces, *Heat Transfer—Sov. Res.* **16**(5), 90 (1984).

59J. M. A. Kusumowardhoyo and T. Hardianto, Two phase heat transfer correlation for multi-component mixtures in the saturated flow-boiling regime, *Int. J. Heat Fluid Flow* **6**(4), 265 (1985).

60J. S. S. Kutateladze and I. G. Malenkov, Hydrodynamic analogy between heat transfer and nucleate boiling crisis in boiling and bubbling. Experimental data, *Heat Transfer—Sov. Res.* **16**(4), 1 (1984).

61J. L. Y. W. Lee, J. C. Chen and R. A. Nelson, Liquid–solid contact measurements using a surface thermocouple temperature probe in atmospheric pool boiling water, *Int. J. Heat Mass Transfer* **28,** 1415 (1985).

62J. A. I. Leont'yev, Extension of limiting friction and heat transfer relationships to turbulent gas–liquid flow, *Heat Transfer—Sov. Res.* **16**(3), 1 (1984).

63J. J. H. Lienhard, On the two regimes of nucleate boiling, *J. Heat Transfer* **107,** 262 (1985).

64J. A. A. Malyshev, V. M. Azarskov, G. N. Danilova, B. B. Zemskov and V. K. Kirin, True volumetric vapor content of two-phase freon flows, *J. Engng Phys.* **48,** 128 (1985).

65J. V. L. Malyshev, L. A. Uvarova and A. S. Fel'dblym, Theoretical calculation of gas flows during isothermal vaporization of multicomponent liquids in narrow channels of different cross sections, *High Temp.* **23,** 274 (1985).

66J. I. Michiyoshi, N. Takenaka, T. Murata, T. Shiokawa and O. Takahashi, Effects of liquid level on boiling heat transfer in potassium layers on a horizontal plane heater, *J. Heat Transfer* **107,** 468 (1985).
67J. R. S. Miller, Photographic observations of bubble formation in flashing nozzle flow, *J. Heat Transfer* **107,** 750 (1985).
68J. K. Mishima and H. Nishihara, Effect of flow direction and magnitude on CHF for low pressure water in thin rectangular channels, *Nucl. Engng Des.* **86,** 165 (1985).
69J. K. Mishima, H. Nishihara and I. Michiyoshi, Boiling burnout and flow instabilities for water flowing in a round tube under atmospheric pressure, *Int. J. Heat Mass Transfer* **28,** 1115 (1985).
70J. O. Miyatake, T. Tomimura and Y. Ide, Enhancement of spray flash evaporation by means of the injection of bubble nuclei, *J. Sol. Energy Engng* **107,** 176 (1985).
71J. M. Monde, Critical heat flux in saturated forced convective boiling on a heated disk with an impinging jet, *Wärme- u. Stoffübertr.* **19,** 205 (1985).
72J. Y. H. Mori, Classification of configurations of two-phase vapor/liquid bubbles in an immiscible liquid in relation to direct-contact evaporation and condensation processes, *Int. J. Multiphase Flow* **11,** 571 (1985).
73J. A. Moriyama, K. Araki and T. Okouchi, Heat transfer from hot metal surface to an impinging water droplet in transitional boiling regime, *Trans. Iron Steel Inst. Japan* **25**(3), 198 (1985).
74J. D. G. Morris, C. B. Mullins and G. L. Yoder, Experimental study of rod bundle dispersed-flow film boiling with high-pressure water, *Nucl. Technol.* **69,** 82 (1985).
75J. I. A. Mudawwar, M. A. El-Masri, C. S. Wu and J. R. Ausman-Mudawwar, Boiling heat transfer and critical heat flux in high-speed rotating liquid films, *Int. J. Heat Mass Transfer* **28,** 795 (1985).
76J. Y. Murao, K. Fujiki and H. Akimoto, Experimental assessment of evaluation model for safety analysis on reflood phase of PWR-LOCA, *J. nucl. Sci. Technol.* **22,** 890 (1985).
77J. M. Murase and H. Suzuki, Evaluation of countercurrent gas/liquid flow in parallel channels with restricted ends, *Nucl. Technol.* **68,** 408 (1985).
78J. J. E. Myers, Short-lived sites in nucleate boiling, *A.I.Ch.E. Jl* **31,** 1441 (1985).
79J. H. Nagasaka, M. Katoh, T. Kagawa and H. Aoki, Thermal hydraulic behavior in upper plenum during refill-reflood phase of BWR LOCA, (I): CCFL and CCFL breakdown phenomena in upper plenum, *J. nucl. Sci. Technol.* **22,** 903 (1985).
80J. A. Nakayama and H. Koyama, On asymptotic results for film pool boiling on vertical surfaces, *Int. J. Multiphase Flow* **11,** 565 (1985).
81J. A. Nakayama and H. Koyama, Two-phase boundary layers in laminar film boiling, *Wärme- u. Stoffübertr.* **19,** 211 (1985).
82J. P. J. Nelson and B. M. Burnside, Boiling the immiscible water/*n*-nonane system from a tube bundle, *Int. J. Heat Mass Transfer* **28,** 1257 (1985).
83J. E. I. Nesis and K. A. Akiev, Heat exchange in the boiling of mixtures in strong gravitational fields, *J. Engng Phys.* **48,** 6 (1985).
84J. T. Nosoko and Y. H. Mori, Vaporization of drops of a denser, volatile liquid dropped onto a surface of another liquid, *J. Heat Transfer* **107,** 384 (1985).
85J. M. Ochiai, Experiment on transient heat transfer in closed narrow channel, *J. nucl. Sci. Technol.* **22,** 47 (1985).
86J. S. Olek and Y. Zvirin, Effect of temperature dependent properties on the rewetting velocity, *Int. J. Multiphase Flow* **11,** 577 (1985).
87J. Yu. M. Pavlov, V. I. Antipov and V. A. Leonov, The boiling crisis in natural circulation of helium in vertical channels, *Heat Transfer—Sov. Res.* **16**(4), 64 (1984).
88J. S. A. Potekhin, V. A. Shugayev and V. I. Babich, Heat transfer and film-boiling crisis with helium on a flat horizontal surface, *Heat Transfer—Sov. Res.* **16**(4), 70 (1984).
89J. N. Rajendra Prasad, J. S. Saini and R. Prakash, The effect of heater wall thickness on heat transfer in nucleate pool-boiling at high heat flux, *Int. J. Heat Mass Transfer* **28,** 1367 (1985).
90J. A. T. Prata and E. M. Sparrow, Diffusion-driven nonisothermal evaporation, *J. Heat Transfer* **107,** 239 (1985).
91J. A. T. Prata and E. M. Sparrow, Forced convection evaporation from a cavity containing a liquid whose surface is curved by capillarity: computations in interlocking rectangular and cylindrical domains, *Numer. Heat Transfer* **8,** 667 (1985).
92J. G. K. Raina and P. D. Grover, Direct contact heat transfer with change of phase: theoretical model incorporating sloshing effects, *A.I.Ch.E. Jl* **31,** 507 (1985).
93J. Yu. S. Ryazantsev, Thermocapillary motion of a reacting droplet in a chemically active medium, *Fluid Dynam.* **20,** 491 (1985).
94J. T. Sakai and F. Hirano, Effect of molecular weight distribution of mineral oils on their boiling heat transfer behavior, *Wear* **104**(3), 259 (1985).
95J. J. E. Sgheiza and J. E. Myers, Behavior of nucleation sites in pool boiling, *A.I.Ch.E. Jl* **31,** 1605 (1985).
96J. A. Sharan and J. H. Lienhard, On predicting burnout in the jet-disk configuration, *J. Heat Transfer* **107,** 398 (1985).
97J. A. Sharan, J. H. Lienhard and R. Kaul, Corresponding states correlations for pool and flow boiling burnout, *J. Heat Transfer* **107,** 392 (1985).
98J. L. A. Shneider and Yu. I. Landa, Examination of heat transfer in boiling of cryogenic agents of expansion microcryogenic systems, *Chem. Pet. Engng* **21**(1–2), 19 (1985).
99J. Ye. L. Smirnov, Critical heat flux in flooding in vertical channels, *Heat Transfer—Sov. Res.* **16**(3), 19 (1984).
100J. M. Sobajima and H. Adachi, Coolability study on two-bundle scale flow blockage in the reflood process, *Nucl. Engng Des.* **86,** 345 (1985).
101J. T. Soga, A kinetic theory analysis of evaporation and condensation of a diatomic gas, *Phys. Fluids* **28,** 1280 (1985).
102J. T. Soga, A kinetic theory analysis of evaporation from a droplet, *J. phys. Soc. Japan* **54,** 2494 (1985).

103J. T. Soga, Kinetic analysis of evaporation from a droplet in a rarefied vapor, *Trans. Japan Soc. aeronaut. Space Sci.* **27**(78), 228 (1985).

104J. A. S. P. Solomon, J.-S. Shuen, Q.-F. Zhang and G. M. Faeth, Measurements and predictions of the structure of evaporating sprays, *J. Heat Transfer* **107,** 679 (1985).

105J. Y. Sudo, K. Miyata, H. Ikawa, M. Kaminga and M. Ohkawara, Experimental study of differences in DNB heat flux between upflow and downflow in vertical rectangular channels, *J. nucl. Sci. Technol.* **22,** 604 (1985).

106J. T. N. Tandon, H. K. Varma and C. P. Gupta, A void fraction model for annular two-phase flow, *Int. J. Heat Mass Transfer* **28,** 191 (1985).

107J. A. M. Tatenov, Local heat transfer from a cylinder to a flow of a bubbly water–air mixture, *Heat Transfer—Sov. Res.* **16**(3), 49 (1984).

108J. P. K. Tewari, R. K. Varma, M. P. S. Ramani and S. P. Mahajan, Studies on nucleate boiling of sodium chloride solutions at atmospheric and sub-atmospheric pressures, *Desalination* **52,** 335 (1985).

109J. J. R. Thomas, Jr., F//N-method in kinetic theory: I. Half-space problems, *Transport Theory statist. Phys.* **14,** 485 (1985).

110J. N. P. Timchenko, S. A. Zhukov, K. G. Shkadinskiy and V. I. Fedorov, Effect of the shape of the boiling curve on the dynamics of self-generated waves, *Heat Transfer—Sov. Res.* **16**(1), 67 (1984).

111J. V. I. Tolubinskiy, Ye. D. Domashev and V. F. Godunov, Boiling crisis in annuli with artificially-induced roughness, *Heat Transfer—Sov. Res.* **16**(5), 96 (1984).

112J. C. R. Tong, M. J. Tan and S. G. Bankoff, Model for convective boiling in a narrow eccentric annular gap, *J. Engng Gas Turbine Pwr* **107,** 613 (1985).

113J. V. B. Utkin, Experimental study of boiling in horizontal annuli, *Heat Transfer—Sov. Res.* **16**(3), 35 (1984).

114J. S. J. D. Van Strahlen and W. M. Sluyter, Mass diffusion controlled bubble behavior in boiling and electrolysis and effect of bubbles on ohmic resistance, *J. appl. Electrochem.* **15,** 527 (1985).

115J. A. N. Vasil'yev and N. I. Khramov, Experimental investigation of the thermodynamic boiling crisis, *Heat Transfer—Sov. Res.* **16**(3), 42 (1984).

116J. B.-X. Wang and D.-H. Shi, A semi-empirical theory for forced-flow turbulent film boiling of subcooled liquid along a horizontal plate, *Int. J. Heat Mass Transfer* **28,** 1499 (1985).

117J. P. C. Wayner, Jr., C. Y. Tung, M. Tirumala and J. H. Yang, Experimental study of evaporation in the contact line region of a thin film of hexane, *J. Heat Transfer* **107,** 182 (1985).

118J. V. V. Yagov and O. N. Kaban'kov, Experimental study of boiling heat transfer in pipes at below-atmospheric pressure, *Heat Transfer—Sov. Res.* **16**(4), 53 (1984).

119J. V. V. Yagov and V. A. Puzin, Critical heat fluxes in forced-convection boiling of refrigerant-12 under conditions of local heat sources, *Heat Transfer—Sov. Res.* **16**(4), 47 (1984).

120J. A. I. Zanin and E. N. Sinitsyn, Boiling of superheated n-pentane in an electric field, *J. Engng Phys.* **48,** 125 (1985).

CHANGE OF PHASE—CONDENSATION

1JJ. Th. Adamek, Berechnung des bezuglich dem Warmeubergang durch filmkondensation optimalen wellenprofils mit hilfe der variationsrechnung, *Z. angew. Math. Phys.* **36,** 669 (1985).

2JJ. R. K. Ahluwalia and K. H. Im, Mass transfer to trace species from a superheated stream to a subcooled surface, *Int. J. Heat Mass Transfer* **28,** 2061 (1985).

3JJ. S. V. Alekseenko, V. Ye. Nakoryakov and B. G. Polusaev, Wave formation on a vertical liquid flim, *A.I.Ch.E. Jl* **31,** 1446 (1985).

4JJ. V. P. Carey, Note on heat transfer to turbulent liquid falling films at high Prandtl number, *A.I.Ch.E. Jl* **31,** 1575 (1985).

5JJ. T.-H. Chang and J. N. Chung, Effects of surfactants on the motion and transport mechanisms of a condensing droplet in a high Reynolds number flow, *A.I.Ch.E. Jl* **31,** 1149 (1985).

6JJ. M. N. Chepurnoi, V. N. Lomakin, V. E. Shnaider and V. M. Chepurnoi, Study of the process of frost formation in finned air coolers, *J. Engng Phys.* **48,** 34 (1985).

7JJ. M. di Marzo and M. H. Casarella, Film condensation over a horizontal cylinder for combined gravity and forced flow, *J. Heat Transfer* **107,** 687 (1985).

8JJ. D. Fullarton, E. U. Schluender and L. Yueksel, Condensation of isopropanol–water mixtures (in German), *Chem. Engng Proc.* **19**(2), 103 (1985).

9JJ. D. Galamba and V. K. Dhir, Transient simultaneous condensation and melting of a vertical surface, *J. Heat Transfer* **107,** 812 (1985).

10JJ. J. Huhn, Film flow with superposition of various effects, *Wärme- u. Stoffübertr.* **19,** 131 (1985).

11JJ. F. M. Joos and R. W. L. Snaddon, Electrostatically enhanced film condensation, *J. Fluid Mech.* **156,** 23 (1985).

12JJ. W. Kirchner and S. G. Bankoff, Condensation effects in reactor transient, *Nucl. Sci. Engng* **89,** 310 (1985).

13JJ. W. S. Kisaalita, K. V. Lo, L. M. Staley and M. A. Tung, Condensation heat and mass transfer from steam/air mixtures to a retort pouch laminate, *Can. agric. Engng* **27,** 137 (1985).

14JJ. S. Kotake, Effects of a small amount of noncondensable gas on film condensation of multicomponent mixtures, *Int. J. Heat Mass Transfer* **28,** 407 (1985).

15JJ. R. Krupiczka, Effect of surface tension on laminar film condensation on a horizontal cylinder, *Chem. Engng Proc.* **19**(4), 199 (1985).

16JJ. S. S. Kutateladze, I. I. Gogonin and V. I. Sosunov, The influence of condensate flow rate on heat transfer in film condensation of stationary vapor on horizontal tube banks, *Int. J. Heat Mass Transfer* **28,** 1011 (1985).

17JJ. S. S. Kutateladze and I. I. Gogonin, Investigation of heat transfer in film condensation of flowing vapor on horizontal tube banks, *Int. J. Heat Mass Transfer* **28,** 1831 (1985).

18JJ. S. S. Kutateladze and I. I. Gogonin, Heat transfer in condensation of flowing vapor on a single horizontal cylinder, *Int. J. Heat Mass Transfer* **28,** 1019 (1985).

19JJ. F. Legay-Desesquelles and B. Prunet-Foch, Dynamic

behavior of a boundary layer with condensation along a flat plate: comparison with suction, *Int. J. Heat Mass Transfer* **28**, 2363 (1985).
20JJ. K. Mangnall and D. Webb, Vacuum condensation, *Chem. Engng Lond.* **420**, 37 (1985).
21JJ. A. Nakayama and H. Koyama, An integral treatment of laminar and turbulent film condensation on bodies of arbitrary geometrical configuration, *J. Heat Transfer* **107**, 417 (1985).
22JJ. M. M. Rahman, A. M. Fathi and H. M. Soliman, Flow pattern boundaries during condensation: new experimental data, *Can. J. chem. Engng* **63**, 547 (1985).
23JJ. D. P. Sekulic, Irreversible condensation conditions near the cryosurface, *Int. J. Heat Mass Transfer* **28**, 1205 (1985).
24JJ. R. Sethumadhavan and M. Raja Rao, Condensation of steam on single start and multistart spiral wire-wound horizontal tubes, *Ind. Engng Chem. Proc. Des. Dev.* **24**, 783 (1985).
25JJ. T. Sundararajan and P. S. Ayyaswamy, Heat and mass transfer associated with condensation on a moving drop; solution for intermediate Reynolds numbers by a boundary layer formulation, *J. Heat Transfer* **107**, 409 (1985).
26JJ. T. N. Tandon, H. K. Varma and C. P. Gupta, Prediction of flow patterns during condensation of binary mixtures in a horizontal tube, *J. Heat Transfer* **107**(2), 424 (1985).
27JJ. Y. Utaka, A. Saito and T. Tani, Study on dropwise condensation curves, *Bull. J.S.M.E.* **28**(240), 1150 (1985).
28JJ. H. Yamaguchi, M. Satomura and K. Matsuki, Refrigerant flow condensation in annulus passage (in Japanese), *Shapu Giho* **31**, 49 (1985).
29JJ. M. Yanadori, K. Huikata, Y. Mori and M. Uchida, Fundamental study of laminar film condensation heat transfer on a downward horizontal surface, *Int. J. Heat Mass Transfer* **28**, 1937 (1985).

CHANGE OF PHASE—FREEZING AND MELTING

1JM. V. Alexiades, D. G. Wilson and A. D. Solomon, Macroscopic global modeling of binary alloy solidification processes, *Q. appl. Math.* **43**(2), 143 (1985).
2JM. C. Bénard, D. Gobin and F. Martinez, Melting in rectangular enclosures: experiments and numerical simulations, *J. Heat Transfer* **107**(4), 794 (1985).
3JM. S. Bulik, Air blow devices in melt spinning machines for chemical fibers: Part I. Phenomena in the sub-die zone dependent on spinning conditions, *Polimery* **30**(7), 257 (1985).
4JM. S. Carra, Problems of fluid dynamics in crystal growth, *Prog. Crystal Growth Character.* **11**(1), 1 (1985).
5JM. Ch. Charach and P. Zoglin, Solidification in a finite, initially overheated slab, *Int. J. Heat Mass Transfer* **28**(12), 2261 (1985).
6JM. C.-K. Chen, Y.-M. Chang and J. W. Cleaver, Variational embedding technique to inward solidification subject to convection and radiation, *J. Energy Res. Technol., Trans. Am. Soc. mech. Engrs* **107**(1), 135 (1985).
7JM. K. C. Cheng and P. Sabhapathy, Determination of local heat transfer coefficients at the solid–liquid interface by heat conduction analysis of the solidified region, *J. Heat Transfer* **107**(3), 703 (1985).
8JM. F. B. Cheung, Analysis of freeze coating on a nonisothermal moving plate by a perturbation method, *J. Heat Transfer* **107**(3), 549 (1985).
9JM. J. J. Derby, R. A. Brown, F. T. Geyling, A. S. Jordan and G. A. Nikolakopoulou, Finite element analysis of a thermal-capillary model for liquid encapsulated Czochralski growth, *J. electrochem. Soc.* **132**(2), 470 (1985).
10JM. J. Dowden, M. Davis and P. Kapadia, The flow of heat and the motion of the weld pool in penetration welding with a laser, *J. appl. Phys.* **57**(9), 4474 (1985).
11JM. C. Esposito, G. Daurelio and A. Cingolani, On cutting and penetration welding processes with high power lasers, *Opt. Lasers Engng* **6**(1), 1 (1985).
12JM. D. Frederick and R. Greif, A method for the solution of heat transfer problems with a change of phase, *J. Heat Transfer* **107**(3), 520 (1985).
13JM. C. Gau and R. Viskanta, Effect of natural convection on solidification from above and melting from below a pure metal, *Int. J. Heat Mass Transfer* **28**(3), 573 (1985).
14JM. C. Gau and R. Viskanta, Effect of crystal anisotropy on heat transfer during melting and solidification of a metal, *J. Heat Transfer* **107**(3), 706 (1985).
15JM. A. G. Gillen and B. Cantor, Photocalorimetric cooling rate measurements on a Ni–5wt%Al alloy rapidly solidified by melt spinning, *Acta Metall.* **33**(10), 1813 (1985).
16JM. L. Gradón and D. Orlicki, Separation of liquid mixtures in the freezing-out process—mathematical description and experimental verification, *Int. J. Heat Mass Transfer* **28**(11), 1983 (1985).
17JM. S. C. Gupta, Two-dimensional solidification in a cylindrical mold with imperfect mold contact, *Int. J. Engng Sci.* **23**(9), 901 (1985).
18JM. S. C. Gupta, Axisymmetric solidification in a long cylindrical mold, *Appl. scient. Res.* **42**(3), 229 (1985).
19JM. J. M. Hill and A. Kucera, An approximate solution for planar solidification with Newton cooling of the surface, *Z. angew. Math. Phys.* **36**(5), 637 (1985).
20JM. T. Hirata and M. Ishihara, Freeze-off conditions of a pipe containing a flow of water, *Int. J. Heat Mass Transfer* **28**(2), 331 (1985).
21JM. A. Holz, On the significance of conservation laws in two-dimensional melting, *Physica* **131A**(1), 212 (1985).
22JM. S. D. Howison, J. R. Ockendon and A. A. Lacey, Singularity development in moving-boundary problems, *Q. Jl Mech. appl. Math.* **38**(3), 343 (1985).
23JM. J. S. Huang and A. J. Barduhn, The effect of natural convection on ice crystal growth rates in salt solutions, *A.I.Ch.E. Jl* **31**(5), 747 (1985).
24JM. S. C. Huang, R. P. Laforce, A. M. Ritter and R. P. Goehner, Rapid solidification characteristics in melt

spinning a Ni-base superalloy, *Metall. Trans. A* (*Phys. Metall. Mater. Sci.*) **16A**(10), 1773 (1985).

25JM. B. Kalhori and S. Ramadhyani, Studies on heat transfer from a vertical cylinder, with or without fins, embedded in a solid phase change medium, *J. Heat Transfer* **107**(1), 44 (1985).

26JM. H. Kehtarnavaz and Y. Bayazitoglu, Solidification of binary mixture in a finite planar medium: saline water, *J. Heat Transfer* **107**(4), 964 (1985).

27JM. G. S. H. Lock and T. M. V. Kaiser, Icing on submerged tubes: a study of occlusion, *Int. J. Heat Mass Transfer* **28**(9), 1689 (1985).

28JM. Kh. I. Makeev, V. A. Vaisberg and F. I. Selitskii, Growing cylindrical silicon crystals cooled by immersion in fluid, *Tsvetnye Metally* **2**, 54 (1985).

29JM. J. Mennig and M. N. Özişik, Coupled integral equation approach for solving melting or solidification, *Int. J. Heat Mass Transfer* **28**(8), 1481 (1985).

30JM. A. G. Merzhanov, V. A. Raduchev and E. N. Rumanov, Interaction between a nonuniformly heated dielectric and a microwave field, *J. appl. Mech. Tech. Phys.* **26**(1), 5 (1985).

31JM. M. K. Moallemi and R. Viskanta, Experiments on fluid flow induced by melting around a migrating heat source, *J. Fluid Mech.* **157**, 35 (1985).

32JM. M. K. Moallemi and R. Viskanta, Melting around a migrating heat source, *J. Heat Transfer* **107**(2), 451 (1985).

33JM. J. S. Nelson and E. M. Sparrow, Freezing in the presence of rotation, *J. Heat Transfer* **107**(4), 804 (1985).

34JM. F. A. Nichols, Kinetics of interfacial motion and bubble movement by vapor transport for two-component solid solution, *J. appl. Phys.* **58**(1), 287 (1985).

35JM. T. Nishimura, S. Ito, H. Tsuboi and Y. Kawamura, Analysis of two-dimensional freezing by the finite-element method, *Int. chem. Engng* **25**(1), 105 (1985).

36JM. J. R. Ockendon, A. B. Taylor, S. H. Emerman and D. L. Turcotte, Geodynamic thermal runaway with melting, *J. Fluid Mech.* **152**, 301 (1985).

37JM. M. Okada, Heat transfer during melting around a vertical cylinder (1st report, analysis and experiments of the melting without subcooling), *Bull. J.S.M.E.* **28**(243), 2007 (1985).

38JM. F. O. Oruma, M. N. Özişik and M. A. Boles, Effects of non-grayness on melting and solidification of a semi-infinite, semi-transparent medium, *J. Franklin Inst.* **319**(5), 459 (1985).

39JM. P. V. Padmanabhan and M. V. Krishnamurthy, Freezing around a finite heat sink immersed in an infinite phase change medium, *Int. J. Heat Fluid Flow* **6**(2), 104 (1985).

40JM. J. Prusa and L. S. Yao, Effects of density change and subcooling on the melting of a solid around a horizontal heated cylinder, *J. Fluid Mech.* **155**, 193 (1985).

41JM. G. S. Reddy and J. A. Sekhar, Microstructure refinement with forced convection in aluminium and superalloys, *J. Mater. Sci.* **20**(10), 3535 (1985).

42JM. M. A. Ruggiero and J. W. Rutter, Microstructural defects and their origins in CD-CUCD//3 eutectic, *Mater. Sci. Technol.* **1**(7), 550 (1985).

43JM. A. Saito, Y. Utaka, M. Akiyoshi and K. Katayama, On the contact heat transfer with melting (1st report, experimental study), *Bull. J.S.M.E.* **28**(240), 1142 (1985).

44JM. A. Saito, Y. Utaka, M. Akiyoshi and K. Katayama, On the contact heat transfer with melting (2nd report, analytical study), *Bull. J.S.M.E.* **28**(242), 1703 (1985).

45JM. A. Saito, Y. Utaka, S. Nagakubo and K. Katayama, A heat transfer study on latent heat thermal energy storage systems—discussions of the inner structure to improve the heat removal process, *Heat Transfer—Japan. Res.* **13**(1), 92 (1985).

46JM. H. Saito, I. Tokura, K. Kishinami and S. Uemura, A study on frost formation (dimensionless parameters correlating density and thickness of frost layer), *Heat Transfer—Japan. Res.* **13**(4), 76 (1984).

47JM. J. D. Sauzade and J. Guion, Mono-dimensional model analysis of solidification phenomena with subcooling using the enthalpy method, *Int. J. Heat Mass Transfer* **28**(10), 1859 (1985).

48JM. P. Schvan and R. E. Thomas, Time-dependent heat flow calculation of cw laser-induced melting of silicon, *J. appl. Phys.* **57**(10), 4738 (1985).

49JM. N. Seki, S. Fukusako, K. Matsuo and S. Uemura, An analysis of incipient frost formation, *Wärme- u. Stoffübertr.* **19**(1), 9 (1985).

50JM. D. Shaozhang and Y. Xiaolin, Mathematical model of heat-transfer in continuously cast billets and its numerical calculation method adaptable to micro-computer use, *Kang T'ieh/Iron Steel, Peking* **20**(7), 8 (1985).

51JM. M. Shimizu and H. Takeuchi, Solidification structure of Al-Pb alloy powder by gas atomization process, *J. Japan Soc. Powd. Metall.* **32**(8), 289 (1985).

52JM. T. Soh, T. Nakagawa and Y. Takebayashi, Solidification analysis of castings, *Res. Dev. Kobe Steel, Ltd* **35**(2), 75 (1985).

53JM. A. D. Solomon, V. Alexiades, D. G. Wilson and J. Drake, On the formulation of hyperbolic Stefan problems, *Q. appl. Math.* **43**(3), 295 (1985).

54JM. E. M. Sparrow and T. A. Myrum, Inclination-induced direct-contact melting in a circular tube, *J. Heat Transfer* **107**(3), 533 (1985).

55JM. K. Wollhöver, Ch. Körber, M. W. Scheiwe and U. Hartmann, Unidirectional freezing of binary aqueous solutions: an analysis of transient diffusion of heat and mass, *Int. J. Heat Mass Transfer* **28**(4), 761 (1985).

56JM. K. Wollhöver, M. W. Scheiwe, U. Hartmann and Ch. Körber, On morphological stability of planar phase boundaries during unidirectional transient solidification of binary aqueous solutions, *Int. J. Heat Mass Transfer* **28**(5), 897 (1985).

57JM. Guo-Ping Zhang, L. M. Jiji and S. Weinbaum, Quasi-three-dimensional steady state analytic solution for melting or freezing around a buried pipe in a semi-infinite medium, *J. Heat Transfer* **107**(1), 245 (1985).

RADIATION IN PARTICIPATING MEDIA AND SURFACE RADIATION

1K. M. A. G. Abushagur and N. George, Polarization and wavelength effects on the scattering from dielectric cylinders, *Appl. Opt.* **24**(23), 4141 (1985).

2K. K. S. Adzerikho and S. B. Kostyukevich, Radiation of two-phase media bounded by reflecting and radiating surfaces, *J. Engng Phys.* **48**(1), 82 (1985).

3K. N. Araki, K. Kobayasi and M.-A. Ochi, Measuring thermal diffusivity of liquids by stepwise heating (second report, heat loss characteristics and measurements on molten salts), *Heat Transfer—Japan. Res.* **14**(1), 56 (1985).

4K. C. Arduini, Thermal analytical model of bidimensional and unidimensional structural elements in a space environment, *Acta astronaut.* **12**(7–8), 609 (1985).

5K. F. H. Azad, Differential approximation to radiative transfer in semitransparent media, *J. Heat Transfer* **107**(2), 478 (1985).

6K. E. Bahar and S. Chakrabarti, Scattering and depolarization by large conducting spheres with rough surfaces, *Appl. Opt.* **24**(12), 1820 (1985).

7K. S. Bakan, Thermal stability of radiating fluids: the scattering problem, *Phys. Fluids* **27**(12), 2969 (1984).

8K. C. E. Batten, Spectral optical constants of soots from polarized angular reflectance measurements, *Appl. Opt.* **24**(8), 1193 (1985).

9K. E. Battistelli and P. Bruscaglioni, A. Ismaelli and G. Zaccanti, Use of two scaling relations in the study of multiple-scattering effects on the transmittance of light beams through a turbid atmosphere, *J. opt. Soc. Am. A* **2**(6), 903 (1985).

10K. R. E. Bedford, C. K. Ma, Z. Chu, Y. Sun and S. Chen, Emissivities of diffuse cavities: 4: isothermal and nonisothermal cylindro-inner-cones, *Appl. Opt.* **24**(18), 2971 (1985).

11K. P. Berdahl, Radiant refrigeration by semiconductor diodes, *J. appl. Phys.* **58**(3), 1369 (1985).

12K. D. B. Betts, F. J. J. Clarke, L. J. Cox and J. A. Larkin, Infrared reflection properties of five types of black coating for radiometric detectors, *J. Phys. E: scient. Instrum.* **18**(8), 689 (1985).

13K. C. F. Bohren and G. Koh, Forward-scattering corrected extinction by nonspherical particles, *Appl. Opt.* **24**(7), 1023 (1985).

14K. R. S. Bretzlaff, Radiant-energy penetration effect in the thermal-diffusivity flash technique for layered and porous polymers, *J. appl. Phys.* **58**(8), 2816 (1985).

15K. M. A. Brosmer and C. L. Tien, Infrared radiation properties of methane at elevated temperatures, *J. quant. Spectrosc. radiat. Transfer* **33**(5), 521 (1985).

16K. M. A. Brosmer and C. L. Tien, Thermal radiation properties of acetylene, *J. Heat Transfer* **107**(4), 943 (1985).

17K. G. E. Caledonia, R. H. Krech and T. Wilkerson, Measurements of the CO first overtone bandstrength at elevated temperatures, *J. quant. Spectrosc. radiat. Transfer* **34**(2), 183 (1985).

18K. G. Camera-Roda, M. Bertelà and F. Santarelli, Mixed laminar convection in a participating irradiated fluid—1, *Numer. Heat Transfer* **8**(4), 429 (1985).

19K. M. W. P. Cann, R. W. Nicholls, P. L. Roney, A. Blanchard and F. D. Findlay, Spectral line shapes for carbon dioxide in the 4.3-μm band, *Appl. Opt.* **24**(9), 1374 (1985).

20K. R. Caps, J. Fricke and H. Reiss, Radiative heat transfer in anisotropically scattering fiber insulations, *High Temp.—High Press.* **17**(3), 303 (1985).

21K. Y. A. Cengel and M. N. Özişik, Radiation transfer in an anisotropically scattering plane-parallel medium with space-dependent albedo $\omega(x)$, *J. quant. Spectrosc. radiat. Transfer* **34**(3), 263 (1985).

22K. B. C. Choi and S. W. Churchill, A technique for obtaining approximate solutions for a class of integral equations arising in radiative heat transfer, *Int. J. Heat Fluid Flow* **6**(1), 42 (1985).

23K. R. W. Claus, G. M Neely and F. M. Humenik, Flame radiation and liner heat transfer in a tubular-can combustor, *J. Propul. Pwr* **1**(4), 270 (1985).

24K. A. Coppalle and P. Vervisch, Spectral emissivity of the 4.3-μm CO_2 band at high temperature, *J. quant. Spectrosc. radiat. Transfer* **33**(5), 465 (1985).

25K. C. Cousin, R. LeDoucen, C. Boulet and A. Henry, Temperature dependence of the absorption in the region beyond the 4.3-μm band head of CO_2.2: N_2 and O_2 broadening, *Appl. Opt.* **24**(22), 3899 (1985).

26K. A. L. Crosbie and G. W. Davidson, Dirac-delta function approximations to the scattering phase function, *J. quant. Spectrosc. radiat. Transfer* **33**(4), 391 (1985).

27K. A. L. Crosbie and R. L. Dougherty, Two-dimensional linearly anisotropic scattering in a finite-thick cylindrical medium exposed to a laser beam, *J. quant. Spectrosc. radiat. Transfer* **33**(5), 487 (1985).

28K. A. L. Crosbie and M. Pattabongse, Application of the singularity-subtraction technique to isotropic scattering in a planar layer, *J. quant. Spectrosc. radiat. Transfer* **34**(6), 473 (1985).

29K. A. L. Crosbie and R. G. Schrenker, Multiple scattering in a two-dimensional rectangular medium exposed to collimated radiation, *J. quant. Spectrosc. radiat. Transfer* **33**(2), 101 (1985).

30K. D. M. De Faveri, G. Fumarola, C. Zonato and G. Ferraiolo, Estimate flare radiation intensity, *Hydrocarbon Proc.* **64**(5), 89 (1985).

31K. G. Desrayaud and G. Lauriat, Natural convection of a radiating fluid in a vertical layer, *J. Heat Transfer* **107**(3), 710 (1985).

32K. M. Dragovan and S. H. Moseley, Resonant bolometer—results and an application, *Appl. Opt.* **24**(12), 1770 (1985).

33K. D. Duffy, The temperature distribution within a sphere placed in a directed uniform heat flux and allowed to radiatively cool, *J. Heat Transfer* **107**(1), 28 (1985).

34K. W. L. Dunn and C. E. Siewart, The searchlight problem in radiation transport: some analytical and computational results, *Z. angew. Math. Phys.* **36**(4), 581 (1985).

35K. J.-I. Eichberger, Calculation of geometric configuration factors in an enclosure whose boundary is given by an arbitrary polygon in the plane, *Wärme- u. Stoffübertr.* **19**(4), 269 (1985).

36K. A. F. Emery, H. R. Mortazavi and M. N. Nguyen, Radiation exchange in large space structures and frames, *A.I.A.A. Jl* **23**(6), 947 (1985).

37K. J. Fendley, An analysis of the measuring procedure for the integrating sphere spectrophotometer, *Sol. Energy* **35**(3), 281 (1985).

38K. N. N. Fomin, Unsteady radiation field in an optically thick atmosphere, *Izv. Akad. Nauk SSSR* **21**(1), 42 (1985).

39K. R. D. M. Garcia and C. E. Siewert, Benchwork results in radiative transfer, *Transport Theory statist. Phys.* **14**(4), 437 (1985).

40K. E. Garelis, T. E. Rudy and R. B. Hickman, A numerical solution to the steady-state equation of transfer in azimuthally symmetric geometries, *J. quant. Spectrosc. radiat. Transfer* **34**(5), 417 (1985).

41K. K. Gindele, M. Köhl and M. Mast, Spectral reflectance measurements using an integration sphere in the infrared, *Appl. Opt.* **24**(12), 1757 (1985).

42K. D. E. Glass, M. N. Özişik and B. Vick, Hyperbolic heat conduction with surface radiation, *Int. J. Heat Mass Transfer* **28**(10), 1823 (1985).

43K. D. S. Goodman, Linear systems methods in some problems of radiative transfer between parallel planes, *Appl. Opt.* **24**(19), 3240 (1985).

44K. J. E. Gordon, Simple method for approximating Mie scattering, *J. opt. Soc. Am. A* **2**(2), 156 (1985).

45K. J. I. Gordon and R. W. Johnson, Integrating nephelometer: theory and implications, *Appl. Opt.* **24**(16), 2721 (1985).

46K. S. Green, Calculation of pressure-broadened linewidths for CO in Ar, *J. quant. Spectrosc. radiat. Transfer* **33**(4), 299 (1985).

47K. W. L. Grosshandler and H. D. Nguyen, Application of the total transmittance nonhomogeneous radiation model to methane combustion, *J. Heat Transfer* **107**(2), 445 (1985).

48K. W. L. Grosshandler and J.-P. Vantelon, Predicting soot radiation in laminar diffusion flames, *Combust. Sci. Technol.* **44**(3–4), 125 (1985).

49K. W. S. Helliwell, Finite-difference solution to the radiative-transfer equation for in-water radiance, *J. opt. Soc. Am. A* **2**(8), 1325 (1985).

50K. W. G. Houf, F. P. Incropera and R. Viskanta, Thermal conditions in irradiated, slowly moving liquid layers, *J. Heat Transfer* **107**(1), 92 (1985).

51K. K. K. Hunt and N. J. McCormick, Numerical test of an inverse method for estimating single-scattering parameters from pulsed multiple-scattering experiments, *J. opt. Soc. Am. A* **2**(11), 1965 (1985).

52K. Y. Itaya and M. Hasatani, Effect of isotropic scattering on unsteady heat transfer in a fine-particle semitransparent liquid suspension layer heated by collimated radiation, *J. chem. Engng Japan* **18**(2), 119 (1985).

53K. Y. Itaya, N. Arai and M. Hasatani, Unsteady heat transfer characteristics in a semitransparent medium heated by diffuse and collimated radiation, *J. chem. Engng Japan* **18**(1), 87 (1985).

54K. J. Janata, Radiative heat transfer in isothermal-gas-filled enclosure, *Acta techn. CSAV* **30**(5), 539 (1985).

55K. K. Kamiuto and I. Kinoshita, Radiative properties of sodium mist, *J. nucl. Sci. Technol.* **22**(6), 442 (1985).

56K. C. T. Kelley and T. W. Mullikin, Why does the F_N-method work? *Transport Theory statist. Phys.* **14**(4), 513 (1985).

57K. T. K. Kim and T. F. Smith, Radiative and conductive transfer for a real gas in a cylindrical enclosure with gray walls, *Int. J. Heat Mass Transfer* **28**(12), 2269 (1985).

58K. Y. Kim and D. L. Jaggard, Inverse black body radiation: an exact closed-form solution, *IEEE Trans. Antennas Propagation* **AP-33**(7), 797 (1985).

59K. V. Ya. Klabukov, L. T. Grebenshchikov and E. A. Kosolapov, Experimental-theoretical study of influence of governing parameters on thermal radiation of high-temperature two-phase flow, *Sov. Aeronaut.* **28**(1), 36 (1985).

60K. W. Klose and M. Lent, Prediction of the thermal radiation in atherm media with an extendable *n*-flux-method, *Wärme- u. Stoffübertr.* **19**(1), 73 (1985).

61K. M. E. Larsen and J. R. Howell, The exchange factor method: an alternative basis for zonal analysis of radiating enclosures, *J. Heat Transfer* **107**(4), 936 (1985).

62K. P. Latimer, Experimental tests of a theoretical method for predicting light scattering by aggregates, *Appl. Opt.* **24**(19), 3231 (1985).

63K. R. LeDoucen, C. Cousin, C. Boulet and A. Henry, Temperature dependence of the absorption in the region beyond the 4.3 μm band head of CO_2 1: pure CO_2 case, *Appl. Opt.* **24**(6), 897 (1985).

64K. S. A. Levchenko, G. I. Rudin, N. N. Stolovich and S. I. Shabunya, Nonresonance radiation absorption in a transparent fluid under the influence of a nanosecond laser pulse, *Heat Transfer—Sov. Res.* **16**(3), 117 (1984).

65K. F.-S. Lien, C.-K. Chen and J. W. Cleaver, Radiative effect on the conjugated forced convection-conduction analysis of heat transfer in a plate fin, *A.I.A.A. Jl* **23**(10), 1645 (1985).

66K. J. J. Loftus and R. D. Peacock, Wall and ceiling protection for heating appliances, *Fire Technol.* **21**(3), 213 (1985).

67K. T. Makino, I. Sakai, H. Kinoshita and T. Kumitomo, Thermal radiation properties of ceramic materials, *Heat Transfer—Japan. Res.* **13**(4), 33 (1984).

68K. K. Malachowski, Modification of the matrix-operator method for solving the equation of radiative transfer, *J. quant. Spectrosc. radiat. Transfer* **34**(2), 169 (1985).

69K. J. V. Martonchik and D. J. Diner, Three-dimensional radiative transfer using a Fourier-transform matrix-operator method, *J. quant. Spectrosc. radiat. Transfer* **34**(2), 133 (1985).

70K. H. Masuda and M. Higano, Transient calorimetric technique for measuring total hemispherical emissivities of metals with rigorous evaluation of heat loss through thermocouple leads, *J. opt. Soc. Am. A* **2**(11),

1877 (1985).
71K. F. U. Mathiak, Calculation of form-factors for plane areas with polygonal boundaries, *Wärme- u. Stoffübertr.* **19**(4), 273 (1985).
72K. L. K. Matthews, R. Viskanta and F. P. Incropera, Combined conduction and radiation heat transfer in porous materials heated by intense solar radiation, *J. Sol. Energy Engng* **107**(1), 29 (1985).
73K. A. T. Mattick and A. Hertzberg, Liquid droplet radiator performance studies, *Acta astronaut.* **12**(7–8), 591 (1985).
74K. M. E. McCabe and M. Van Migom, Application of transparent enclosure theory to solar energy absorption by cylindrical tubes in sunspaces, *J. Sol. Energy Engng* **107**(1), 5 (1985).
75K. N. J. McCormick, Methods for estimating the similarity parameter of clouds from internal measurements of the scattered radiation field, *J. quant. Spectrosc. radiat. Transfer* **33**(1), 63 (1985).
76K. M. P. Mengüc and R. Viskanta, On the radiative properties of polydispersions: a simplified approach, *Combust. Sci. Technol.* **44**(3–4), 143 (1985).
77K. M. P. Mengüc and R. Viskanta, Radiative transfer in three-dimensional rectangular enclosures containing inhomogeneous, anisotropically scattering media, *J. quant. Spectrosc. radiat. Transfer* **33**(6), 533 (1985).
78K. R. S. Mikhalchenko, V. F. Getmanets, N. P. Pershin and Yu. V. Batozskii, Theoretical and experimental investigation of radiative–convective heat transfer in multilayer insulation, *Cryogenics* **25**(5), 275 (1985).
79K. K. C. Mills and W. A. Wakeham, Effect of radiation on thermal transport measurements, *High Temp.—High Press.* **17**(3), 343 (1985).
80K. N. A. Nilsson, T. S. Eriksson and C. G. Granqvist, Infrared-transparent convection shields for radiative cooling: initial results on currugated polyethylene foils, *Solar Energy Mater.* **12**(5), 327 (1985).
81K. M. Nishimura, M. Kuraishi, Y. Bando and H. Jojima, Influence of the spectral distribution of incident radiation on the effect of internal heating in infrared drying of coated films, *Heat Transfer—Japan. Res.* **13**(2), 30 (1984).
82K. J. C. Oelund and N. J. McCormick, Sensitivity of multiple-scattering inverse transport methods to measurement errors, *J. opt. Soc. Am. A* **2**(11), 1972 (1985).
83K. T. Oguri, A. Takeda, K. Torii and S. Inaba, Radiation heat transfer of combustion flames in a diesel engine, *Bull. J.S.M.E.* **28**(238), 638 (1985).
84K. Y. Ohwada, Numerical calculation of bandlimited effective emissivity, *Appl. Opt.* **24**(2), 280 (1985).
85K. T. Okamoto, S. Kawata and S. Minami, Optical method for resolution enhancement in photodiode array Fourier transform spectroscopy, *Appl. Opt.* **24**(23), 4221 (1985).
86K. F. O. Oruma, M. N. Özişik and M. A. Boles, Effects of anistropic scattering on melting and solidification of a semi-infinite, semi-transparent medium, *Int. J. Heat Mass Transfer* **28**(2), 441 (1985).
87K. R. C. Peterson and N. M. Laurendeau, Emittance of yttrium–beryllium oxide thermocouple coating, *Combust. Flame* **60**(3), 279 (1985).
88K. V. K. Pustovalov and G. S. Romanov, The theory of heating and evaporation of a spherical particle exposed to optical radiation, *Int. J. Heat Mass Transfer* **28**(1), 277 (1985).
89K. B. V. Putyatin, Exact solution of the heat propagation equations in a radiative medium, *J. appl. Math. Mech.* **48**(3), 369 (1984).
90K. D. F. G. Rault, Radiation energy receiver for solar propulsion systems, *J. Spacecraft Rockets* **22**(6), 642 (1985).
91K. J. Reinkenhof, Transient heat transfer in cooled cavities with apertures permeable to radiation, *Wärme- u. Stoffübertr.* **19**(1), 85 (1985).
92K. H. Reiss and B. Ziegenbein, An experimental method to determine temperature-dependent extinction coefficients and solid thermal conductivities of glass fibre insulations in calorimetric measurements, *Int. J. Heat Mass Transfer* **28**(2), 459 (1985).
93K. M. N. Rolin and F. B. Yurevich, Evaluating certain ambiguities in results of calculations of radiative-convective heat exchange in hypersonic flow about a body, *High Temp.* **23**(2), 282 (1985).
94K. J. A. Roux and A. M. Smith, Biangular reflectance for an absorbing and isotropically scattering medium, *A.I.A.A. Jl* **23**(4), 624 (1985).
95K. S. P. Rusin, Application of optico-geometric functions to the computation of radiative heat transfer in cavity systems, *Izv. Sibirskogo Otdeleniya Akad. Nauk SSSR* **10**, 3 (1985).
96K. B. Rutily and J. Bergeat, The Neumann solution of the multiple scattering problem in a plane-parallel medium—I. The infinite medium, *J. quant. Spectrosc. radiat. Transfer* **33**(4), 373 (1985).
97K. E. A. Saad, A. A. Hendi and S. A. El Wakil, Radiation transfer by a finite-emitting inhomogeneous medium, *J. quant. Spectrosc. radiat. Transfer* **34**(1), 65 (1985).
98K. Y. Sakurai and D. K. Edwards, Transient temperature distributions in a radiantly heated hollow cylinder, *J. Spacecraft Rockets* **22**(6), 665 (1985).
99K. P. Scheuerpflug, R. Caps, D. Buettner and J. Fricke, Apparent thermal conductivity of evacuated, SiO_2–aerogel tiles under variation of radiative boundary conditions, *Int. J. Heat Mass Transfer* **28**(12), 2299 (1985).
100K. S. Seedorf, H. J. Eichler and H. Koch, Detector for the VUV to the IR using sodium salicylate as a combined fluorescent and reflective coating, *Appl. Opt.* **24**(9), 1335 (1985).
101K. N. Selçuk, Exact solutions for radiative heat transfer in box-shaped furnaces, *J. Heat Transfer* **107**(3), 648 (1985).
102K. A. B. Shapiro, Computer implementation, accuracy and timing of radiation view factor algorithms, *J. Heat Transfer* **107**(3), 730 (1985).
103K. C. E. Siewert, On the singular components of the solution to the searchlight problem in radiative transfer, *J. quant. Spectrosc. radiat. Transfer* **33**(6), 551 (1985).

104K. C. E. Siewert and J. R. Thomas, Jr., Radiative transfer calculations in spheres and cylinders, *J. quant. Spectrosc. radiat. Transfer* **34**(1), 59 (1985).
105K. T. F. Smith and S. S. Severin, Development of the finite analytic with radiation method, *J. Heat Transfer* **107**(3), 735 (1985).
106K. T. F. Smith, Z. F. Shen and A. M. Alturki, Radiative and convective transfer in a cylindrical enclosure for a real gas, *J. Heat Transfer* **107**(2), 482 (1985).
107K. A. Soufiani, J. M. Hartmann and J. Taine, Validity of band-model calculations for CO_2 and H_2O applied to radiative properties and conductive–radiative transfer, *J. quant. Spectrosc. radiat. Transfer* **33**(3), 243 (1985).
108K. E. M. Sparrow and M. A. Ansari, Convective–radiative heat transfer from a horizontal finned tube with sidewall or sidewall/floor interactions, *Int. J. Heat Mass Transfer* **28**(3), 641 (1985).
109K. G. Spiga and M. Spiga, Effects of boundary reflection on radiative heat transfer in participating media, *Int. J. Heat Fluid Flow* **6**(4), 235 (1985).
110K. G. Spiga and M. Spiga, Radiant heat transfer in participating media bounded by reflecting surfaces, *Ann. nucl. Energy* **12**(6), 307 (1985).
111K. S. V. Stepanov and M. A. Berkovskii, Radiation characteristics of a plane layer of a strongly scattering, weakly absorbing medium with an arbitrary character of reflection at the boundaries, *High Temp.* **23**(2), 288 (1985).
112K. W. Q. Tao and E. M. Sparrow, Ambiguities related to the calculation of radiant heat exchange between a pair of surfaces, *Int. J. Heat Mass Transfer* **28**(9), 1786 (1985).
113K. H. L. Tardy, Flux concentrators in integrating sphere experiments: potential for increased detector signal, *Appl. Opt.* **24**(22), 3914 (1985).
114K Y Terauchi, H. Nadano and M. Kohno, On the temperature rise caused by moving heat sources (2nd report, calculation of temperature considering heat radiation from surface), *Bull. J.S.M.E.* **28**(245), 2789 (1985).
115K. M. E. Thomas and R. J. Nordstrom, Line shape model for describing infrared absorption by water vapor, *Appl. Opt.* **24**(21), 3526 (1985).
116K. S. T. Thynell and M. N. Özişik, Integrals involving an exponential integral function and exponentials arising in the solution of radiation transfer, *J. quant. Spectrosc. radiat. Transfer* **33**(3), 259 (1985).
117K. S. T. Thynell and M. N. Özişik, Radiation transfer in an isotropically scattering homogeneous solid sphere, *J. quant. Spectrosc. radiat. Transfer* **33**(4), 319 (1985).
118K. S. T. Thynell and M. N. Özişik, Radiation transfer in an isotropically scattering solid sphere with space dependent albedo, $\omega(r)$, *J. Heat Transfer* **107**(3), 732 (1985).
119K. R. A. Toth, Line positions and strengths of CO_2 in the 1200–1430 cm^{-1} region, *Appl. Opt.* **24**(2), 261 (1985).
120K. V. A. Tovstonog, Analysis of heat exchange in light-diffusing materials heated by radiation, *Phys. Chem. Mater. Treat.* **19**(3), 189 (1985).
121K. A. V. Trofimenko, Radiation heat transfer in a system of semitransparent screens, *J. Engng Phys.* **48**(2), 220 (1985).
122K. D. S. Tsai and W. Strieder, Radiation across a spherical cavity having both specular and diffuse reflectance components, *Chem. Engng Sci.* **40**(1), 170 (1985).
123K. M. M. Tunç and A. Karakaş, Three-dimensional formulation of the radiant heat flux variation on a cylinder engulfed in flames, *J. Heat Transfer* **107**(4), 949 (1985).
124K. M. Tunç and J. E. S. Venart, Incident radiation from an engulfing pool fire to a horizontal cylinder—part I, *Fire Safety J.* **8**(2), 81 (1985).
125K. S. Twomey, Green's function formulae for the internal intensity in radiative transfer computations by matrix-vector methods, *J. quant. Spectrosc. radiat. Transfer* **33**(6), 575 (1985).
126K. H. Uchiyama, M. Nakajima and S. Yuta, Measurement of flame temperature distribution by IR emission computed tomography, *Appl. Opt.* **24**(23), 4111 (1985).
127K. S. A. El Wakil, M. H. Haggag, H. M. Machali and E. A. Saad, Radiation transfer in dispersive media, *J. Phys. D: appl. Phys.* **18**(8), 1469 (1985).
128K. B. W. Webb and R. Viskanta, Analysis of heat transfer and solar radiation absorption in an irradiated thin, falling molten salt film, *J. Sol. Energy Engng* **107**(2), 113 (1985).
129K. W. H. Wells and J. J. Sidorowich, Computational techniqes for radiative transfer by spherical harmonics, *J. quant. Spectrosc. radiat. Transfer* **33**(4), 347 (1985).
130K. W. W. Yuen and W. Dunaway, Calculation of the geometric-mean transmittance in a multi-dimensional absorbing and anisotropically-scattering medium, *Int. J. Heat Mass Transfer* **28**(8), 1507 (1985).
131K. W. W. Yuen and C. F. Ho, Analysis of two-dimensional radiative heat transfer in a gray medium with internal heat generation, *Int. J. Heat Mass Transfer* **28**(1), 17 (1985).

NUMERICAL METHODS

1N. S. Acharya and S. V. Patankar, Use of an adaptive grid procedure for parabolic flow problems, *Int. J. Heat Mass Transfer* **28**(6), 1057 (1985).
2N. G. Athanasiadis, Direct and indirect boundary element methods for solving the heat conduction problem, *Comput. Meth. appl. Mech. Engng* **49**(1), 37 (1985).
3N. G. Athanasiadis, Numerical investigations of direct and indirect integral equations for solving the heat conduction problem, *Comput. Meth. appl. Mech. Engng* **49**(2), 203 (1985).
4N. M. C. Bhattacharya, Explicit conditionally stable finite difference equation for heat conduction problems, *Int. J. numer. Meth. Engng* **21**(2), 239 (1985).
5N. T. K. Bhattacharyya and A. B. Datta, Residual method

of finite differencing for the elliptic transport problem and its application to cavity flow, *Int. J. numer. Meth. Fluids* **5**(1), 71 (1985).

6N. B. A. Bhutta, C. H. Lewis and F. A. Kautz II, Comparative analysis of numerical schemes for hypersonic re-entry flows, *J. Spacecraft Rockets* **22**(5), 541 (1985).

7N. G. Burgess and E. Mahajerin, On the numerical solution of Laplace's equation using personal computers, *Int. J. mech. Engng Educ.* **13**(1), 45 (1985).

8N. H. R. Busby and D. M. Trujillo, Numerical solution to a two-dimensional inverse heat conduction problem, *Int. J. numer. Meth. Engng* **21**(2), 349 (1985).

9N. G. F. Carey and R. Krishan, Continuation techniques for a penalty approximation of the Navier–Stokes equations, *Comput. Meth. appl. Mech. Engng* **49**(3), 265 (1985).

10N. A. Charette, R. T. Bui and G. Simard, Effect of stringent boundary conditions on the solution of a transient heat conduction problem, *Math. Comput. Simulation* **27**(1), 47 (1985).

11N. J. A. Dantzig and S. C. Lu, Modeling of heat flow in sand castings: Part I. The boundary curvature method, *Metall. Trans.* **16B**(2), 195 (1985).

12N. J. A. Dantzig and J. W. Wiese, Modeling of heat flow in sand castings: Part II. Applications of the boundary curvature method, *Metall. Trans.* **16B**(2), 203 (1985).

13N. D. E. Dietrich and J. J. Wormeck, An optimized implicit scheme for compressible reactive gas flow, *Numer. Heat Transfer* **8**(3), 335 (1985).

14N. V. I. Dmitriev and E. N. Solov'eva, Numerical solution method for the multiple-free-boundary Stefan problem with a variable critical temperature, *Moscow Univ. Comput. Math. Cybernet.* **1**, 30 (1985).

15N. J. Donea, T. Belytschko and P. Smolinski, A generalized Galerkin method for steady convection–diffusion problems with application to quadratic shape function elements, *Comput. Meth. appl. Mech. Engng* **48**(1), 25 (1985).

16N. A. Sh. Dorfman, Errors arising in convective-heat-transfer problems when using boundary conditions of the third kind, *High Temp.* **23**(3), 430 (1985).

17N. A. Sh. Dorfman, A new type of boundary condition in convective heat transfer problems, *Int. J. Heat Mass Transfer* **28**(6), 1197 (1985).

18N. D. J. Evans and A. R. Abdullah, New explicit method for the diffusion–convection equation, *Comput. Math. Applic.* **11**(1–3), 145 (1985).

19N. B. Farouk and T. Fusegi, Coupled solution of the vorticity-velocity formulation of the incompressible Navier–Stokes equations, *Int. J. numer. Meth. Fluids* **5**(12), 1017 (1985).

20N. M. Gellert, Discrete numerical solution of the generalized Boussinesq problem, *Int. J. numer. Meth. Engng* **21**(12), 2131 (1985).

21N. D. E. Glass, M. N. Ozişik, D. S. McRea and B. Vick, On the numerical solution of hyperbolic heat conduction, *Numer. Heat Transfer* **8**(4), 497 (1985).

22N. J. B. Greenberg, New self-adaptive grid method, *A.I.A.A. Jl* **23**(2), 317 (1985).

23N. R. S. Gupta and A. Kumar, Treatment of multi-dimensional moving boundary problems by coordinate transformation, *Int. J. Heat Mass Transfer* **28**(7), 1355 (1985).

24N. F. Haghighat, T. E. Unny and M. Chandrashekar, Stochastic modeling of transient heat flow through walls, *J. Sol. Energy Engng* **107**(3), 202 (1985).

25N. J. C. Heinrich, Finite element approximation to buoyancy-driven flows with cyclic boundary conditions, *Comput. Meth. appl. Mech. Engng* **48**(1), 91 (1985).

26N. T. Honma, Y. Tanaka and I. Kaji, Regular boundary element solutions to steady-state convective diffusion equations, *Engng Analysis* **2**(2), 95 (1985).

27N. J. S. Hsiao, An efficient algorithm for finite-difference analyses of heat transfer with melting and solidification, *Numer. Heat Transfer* **8**(6), 653 (1985).

28N. T. J. R. Hughes and T. E. Tezduyar, Analysis of some fully-discrete algorithms for the one-dimensional heat equation, *Int. J. numer. Meth. Engng* **21**(1), 163 (1985).

29N. U. K. Kaul and D. S. Chaussee, Comparative study of the parabolized Navier–Stokes code using various grid-generation techniques, *Comput. Fluids* **13**(4), 421 (1985).

30N. B. R. Latimer and A. Pollard, Comparison of pressure-velocity coupling solution algorithms, *Numer. Heat Transfer* **8**(6), 635 (1985).

31N. G. Lauriat and I. Altimir, New formulation of the SADI method for the prediction of natural convection flows in cavities, *Comput. Fluids* **13**(2), 141 (1985).

32N. M. P. Leonchuk, Z. V. Sivak and Yu. E. Shevtsov, Implicit method of solving mass-transfer equations in the variables velocity–vorticity, *Sov. Atomic Energy* **58**(3), 192 (1985).

33N. E. Livne and A. Glasner, A finite difference scheme for the heat conduction equation, *J. comput. Phys.* **58**(1), 59 (1985).

34N. H. P. Miller, Comparison of the flux-corrected transport method with other low-dispersive schemes for the two-dimensional linear advection equation, *RCA Rev.* **46**(2), 214 (1985).

35N. A. Mizukami, An implementation of the streamline-upwind/Petrov–Galerkin method for linear triangular elements, *Comput. Meth. appl. Mech. Engng* **49**(3), 357 (1985).

36N. A. Mizukami and T. J. R. Hughes, A Petrov–Galerkin finite element method for convection-dominated flows and accurate upwinding technique for satisfying the maximum principle, *Comput. Meth. appl. Mech. Engng* **50**(2), 181 (1985).

37N. R. A. Pasmanter, Exact and approximate solutions of the convection–diffusion equation, *Q. Jl Mech. appl. Math.* **38**(1), 1 (1985).

38N. M. K. Patel, N. C. Markatos and M. Cross, Critical evaluation of seven discretization schemes for convection–diffusion equations, *Int. J. numer. Meth. Fluids* **5**(3), 225 (1985).

39N. Q. T. Pham, A fast, unconditionally stable finite-difference scheme for heat conduction with phase change, *Int. J. Heat Mass Transfer* **28**(11), 2079 (1985).

40N. R. E. Phillips and F. W. Schmidt, Multigrid techniques for the solution of the passive scalar advection–diffusion equation, *Numer. Heat Transfer* **8**(1), 25 (1985).
41N. R. E. Phillips and F. W. Schmidt, A multilevel–multigrid technique for recirculating flows, *Numer. Heat Transfer* **8**(5), 573 (1985).
42N. C. Prakash and S. V. Patankar, Control volume-based finite-element method for solving the Navier–Stokes equations using equal-order velocity–pressure interpolation, *Numer. Heat Transfer* **8**(3), 259 (1985).
43N. S. Ramadhyani and S. V. Patankar, Solution of the convection–diffusion equation by a finite-element method using quadrilateral elements, *Numer. Heat Transfer* **8**(5), 595 (1985).
44N. K. R. Rapagopal and A. M. Sadegh, A boundary integral equation method for the study of some laminar forced convection problems, *Numer. Heat Transfer* **8**(4), 485 (1985).
45N. M. J. Raw and G. E. Schneider, New implicit solution procedure for multidimensional finite-difference modeling of the Stefan problem, *Numer. Heat Transfer* **8**(5), 559 (1985).
46N. M. J. Raw, G. E. Schneider and V. Hassani, Nine-noded quadratic control-volume-based finite element for heat conduction, *J. Spacecraft Rockets* **22**(5), 523 (1985).
47N. J. G. Rice and R. J. Schnipke, A monotone streamline upwind finite method for convection-dominated flows, *Comput. Meth. appl. Mech. Engng* **48**(3), 313 (1985).
48N. P. J. Roache, Additional performance aspects of marching methods for elliptic equations, *Numer. Heat Transfer* **8**(5), 519 (1985).
49N. E. Runca, P. Melli and F. Sardel, An analysis of a finite-difference and a Galerkin technique applied to the simulation of advection and diffusion of air pollutants from a line source, *J. comput. Phys.* **59**(1), 152 (1985).
50N. T. M. Shih, Literature survey on numerical heat transfer (1982–1983), *Numer. Heat Transfer* **8**(1), 1 (1985).
51N. A. K. Singhal, Critical look at the progress in numerical heat transfer and some suggestions for improvement, *Numer. Heat Transfer* **8**(5), 505 (1985).
52N. P. J. Sullivan and H. Yip, A solution-scheme for the convective–diffusion equation, *Z. angew. Math. Phys.* **36**(4), 596 (1985).
53N. J. Thibault, Comparison of nine three-dimensional numerical methods for the solution of the heat diffusion equation, *Numer. Heat Transfer* **8**(3), 281 (1985).
54N. S. Uchikawa and R. Takeda, Use of a boundary-fitted coordinate transformation for unsteady heat conduction problems, in multi-connected regions with arbitrarily shaped boundaries, *J. Heat Transfer* **107**(2), 494 (1985).
55N. P. K. Wang, A potential and stream function analysis of two-dimensional steady-state convective diffusion equations involving Laplace fields, *Int. J. Heat Mass Transfer* **28**(6), 1089 (1985).
56N. T. Yoshimura and K. Ikuta, Inverse heat-conduction problem by finite-element formulation, *Int. J. Syst. Sci.* **16**(11), 1365 (1985).
57N. N. Zamani, Some remarks on the use of the boundary element method in transient heat conduction problems, *Math. Comput. Simulation* **27**(1), 61 (1985).
58N. M. Zedan and G. E. Schneider, A coupled strongly implicit procedure for velocity and pressure computation in fluid flow problems, *Numer. Heat Transfer* **8**(5), 537 (1985).

TRANSPORT PROPERTIES

1P. R. T. Abramov, P. M. Kolesnikor, L. Krejci and I. Vogel, An inverse method of determining linear variations of thermophysical properties and heat fluxes, *Heat Transfer—Sov. Res.* **16**(3), 96 (1984).
2P. F. A. Ansari, V. Charan and H. K. Varma, Measurement of thermophysical properties and analysis of heat transfer during freezing of slab-shaped food commodities, *Int. J. Refrig.* **8**(2), 85 (1985).
3P. F. Bakhtar and K. Zidi, On the self diffusion of water vapour, *Proc. Inst. mech. Engrs, Part C, mech. Engng Sci.* **199**(2), 159 (1985).
4P. M. E. Baltatu, J. F. Ely, H. J. M. Hanley, M. S. Graboski, R. A. Perkins and E. D. Sloan, Thermal conductivity of coal-derived liquids and petroleum fractions, *Ind. Engng Chem. Proc. Des. Dev.* **24**(2), 325 (1985).
5P. R. L. Barker and M. L. Young, Thermal insulation properties of fabrics in high intensity heat exposures, *Tappi J.* **68**(12), 62 (1985).
6P. D. Buettner, J. Fricke and H. Reiss, Thermal conductivity of evacuated load-bearing powder and fiber insulations under variable external load, *High Temp.—High Press.* **17**(3), 333 (1985).
7P. R. K. Y. Chan and D. A. Jackson, An automated falling-cylinder high pressure laser-Doppler viscometer, *J. Phys. E: scient. Instrum.* **18**, 510 (1985).
8P. K. S. Chang, R. T. Farrell and J. W. Snellen, Thermal properties of metallized plastics and polyethylene sheeting: a direct calorimetric assessment, *Ergonomics* **28**(11), 1557 (1985).
9P. J. G. Collins, S. J. Collocott and G. K. White, Thermal properties of stabilized zirconi at low temperatures, *Aust. J. Phys.* **38**(4), 617 (1985).
10P. B. Farnworth and P. A. Dolhan, Heat and water transport through cotton and polypropylene underwear, *Text Res. J.* **55**(10), 627 (1985).
11P. J. K. Fink and L. Leibowitz, Analysis of measurements of the thermal conductivity of liquid urania, *High Temp.—High Press.* **17**(1), 17 (1985).
12P. G. C. Glatzmaier and W. F. Ramirez, Simultaneous measurement of the thermal conductivity and thermal diffusivity of unconsolidated materials by the transient hot wire method, *Rev. scient. Instrum.* **56**(7), 1394 (1985).
13P. A. M. Golovin and V. E. Chizhov, The effective thermal conductivity of a suspension, *J. appl. Math. Mech.* **48**(2), 191 (1984).
14P. Z. Guozhen and X. Laoli, Measurements of viscosity and density by a vibrational sphere, *Rev. scient. Instrum.*

56(8), 1639 (1985).
15P. I. Hatta, Y. Sasuga, R. Kato and A. Maesono, Thermal diffusivity measurement of thin films by means of an AC calorimetric method, *Rev. scient. Instrum.* **56**(8), 1643 (1985).
16P. H. Hatta and M. Taya, Effective thermal conductivity of a misoriented short fiber composite, *J. appl. Phys.* **58**(7), 2478 (1985).
17P. J. B. Henderson and S. C. Hagen, Radiant heat flux apparatus for measuring the thermal response of polymeric materials to high temperatures, *Polymer Composites* **6**(2), 110 (April 1985).
18P. J. H. B. Hoogland, H. R. Van Den Berg and N. J. Trappeniers, Measurements of the viscosity of sulfur hexafluoride up to 100 bar by a capillary-flow viscometer, *Physica* **134A**(1), 169 (1985).
19P. C. C. Huang, J. M. Viner and J. C. Novack, New experimental technique for simultaneously measuring thermal conductivity and heat capacity, *Rev. scient. Instrum.* **56**(7), 1390 (1985).
20P. J. Kestin, N. Imaishi, S. H. Nott, J. C. Nieuwoudt and J. V. Sengers, Viscosity of light and heavy water and their mixtures, *Physica* **134A**(1), 38 (1985).
21P. P. J. Kortbeek, M. J. P. Muringer, N. J. Trappeniers and S. N. Biswas, Apparatus for sound velocity measurements in gases up to 10 kbar: experimental data for argon, *Rev. scient. Instrum.* **56**(6), 1269 (1985).
22P. M. H. Kuok, H. K. Sy and K. L. Tan, Determination of thermal conductivity and its variation with density for polyurethane foam, *Reg. J. Energy Heat Mass Transfer* **7**(1), 17 (1985).
23P. P. C. Lanchester and K. A. Mohammed, Apparatus for heat capacity measurements of amorphous metal, *J. Phys. E: scient. Instrum.* **18,** 581 (1985).
24P. R. Lemlich, Semitheoretical equation to relate conductivity to volumetric foam density, *Ind. Engng Chem. Proc. Des. Dev.* **34**(3), 686 (1985).
25P. T.-H. Lin, J. S. Watson and P. W. Fischer, Thermal conductivity of iron–titanium powders, *J. Chem. Engng Data* **30**(4), 369 (1985).
26P. W. C. McCall, Thermal properties of sandwich panels, *Concrete Int.: Des. Construct.* **7**(1), 35 (1985).
27P. D. A. McIntyre, *In situ* measurements of *U*-values, *Bldg. Services Engng Res. Technol.* **6**(1), 1 (1985).
28P. Y. A. Mikhno and V. Z. Geller, Thermal conductivity of multicomponent saturated refrigerants, *Heat Transfer—Sov. Res.* **16**(1), 12 (1984).
29P. V. V. Mirkovich and T. A. Wheat, Thermal transport properties of CaO-stabilized zirconia with varying amounts of stabilization, *High Temp.—High Press.* **17**(1), 67(1), 67 (1985).
30P. D. K. Modi and S. M. Benner, Moisture gain of spray-applied insulations and its effect on effective thermal conductivity—Pt. I, *J. thermal Insulation* **8,** 259 (1985).
31P. T. Nemoto, S. Sasaki and Y. Hakuraku, Thermal conductivity of alumina and silicon carbide ceramics at low temperatures, *Cryogenics* **25**(9), 531 (1985).
32P. B. Nysten, L. Praus and J.-P. Issi, Thermal conductivity of pitch-derived fibres, *Physica D: appl. Phys.* **18**(7), 1311 (1985).
33P. D. J. O'Neil, Rheology and mass/heat transfer aspects of anaerobic reactor design, *Biomass* **8**(3), 205 (1985).
34P. H. Oya and N. Wada, Specific heat of ultrafine particles of Ag prepared by gas evaporation technique, *J. phys. Soc. Japan* **54**(9), 3228 (1985).
35P. K. N. Potter, R. M. Cruse and R. Horton, Tillage effects of soil thermal properties, *Soil Sci. Soc. Am. J.* **49**(4), 968 (July–August 1985).
36P. M. Rehman, V. G. K. Murti and M. R. Gupta, A novel precision type liquid conductivity measuring system, *J. Phys. E: scient. Instrum.* **18,** 902 (1985).
37P. M. R. Riazi and A. Faghri, Prediction of the thermal conductivity of gases at high pressures, *A.I.Ch.E. Jl* **31,** 164 (1985).
38P. M. R. Riazi and A. Faghri, Thermal conductivity of liquid and vapor hydrocarbon systems: pentanes and heavier at low pressures, *Ind. Engng Chem. Process Des. Dev.* **24**(2), 398 (1985).
39P. I. Rowlands, Organic coolant: key to simpler and cheaper thermal reactors? *Nucl. Engng Int.* **30**(376), 52 (1985).
40P. S. Sato, H. Furuya, Y. Nishino and M. Sugisaki, Thermal conductivity of simulated radioactive waste glass, *Nucl. Technol.* **70**(2), 235 (1985).
41P. P. Scheurpflug, R. Caps, D. Büttner and J. Fricke, Apparent thermal conductivity of evacuated SiO_2–aerogel tiles under variation of radiative boundary conditions, *Int. J. Heat Mass Transfer* **28**(12), 2299 (1985).
42P. G. H. Shaw and D. A. "Buzz" Caldwell, Device for measuring sound speeds in reactive liquids at high pressure and temperature, *Rev. scient. Instrum.* **56**(6), 1402 (1985).
43P. T. Ya. Shimchuk, Thermal diffusivity of partially crystalline polymers, *Heat Transfer—Sov. Res.* **16**(1), 136 (1984).
44P. H. Soumerai, Latent heat of vaporization of a widely diverse class of fluids, *J. Heat Transfer* **107**(2), 449 (1985).
45P. K. Srinivasan and M. V. Krishna Murthy, Corresponding states treatment of saturated liquid viscosity of some halogenated hydrocarbon refrigerants, *Int. J. Refrig.* **8**(1), 13 (1985).
46P. J. A. Stark and R. E. Taylor, Determination of thermal transport properties in ammonium perchlorate, *J. Propul. Pwr* **1**(5), 409 (1985).
47P. V. Z. Svoiskii, Heat-conduction coefficient of gas mixtures at high densities, *J. Engng Phys.* **48**(3), 305 (1985).
48P. S. W. Tam, L. Leibowitz, Y. Y. Liu, R. Blomquist and C. E. Johnson, Saturation phenomena and percolative transitions in the high temperature thermal conductivity of -LiA10//2, *J. Nucl. Mater.* 133–134 August 1985 Fusion Reaction Material, *Proc. 1st Int. Conference*, ICFRM-1, Tokyo (1984).
49P. A. S. Teja, P. A. Thurner and B. Pasumarti, Calculation of transport properties of mixtures for synfuels process design, *Ind. Engng Chem. Proc. Des. Dev.* **24**(2), 144 (1985).
50P. S. Torquato and G. Stell, Bounds of the effective thermal conductivity of a dispersion of fully penetrable spheres, *Int. J. Engng Sci.* **23**(3), 375 (1985).

51P. R. D. Trengove, K. R. Harris, H. L. Robjohns and P. J. Dunlop, Diffusion and thermal diffusion in some dilute binary gaseous systems between 195 and 400 K: tests of several asymmetric potentials using the infinite order Sudeden approximation, *Physica* **131A**(3), 506 (1985).
52P. T. S. Vinson and S. L. Jahn, Latent heat of frozen saline coarse-grained soil, *J. Geotech. Engng* **11**(5), 607 (1985).
53P. M. S. Wire, Z. Fisk and G. W. Webb, High-temperature specific heat by a pulse-heating method, *Rev. scient. Instrum.* **56**(6), 1223 (1985).
54P. R. H. Zee, J. F. Watters and R. D. Davidson, Technique to determine solute diffusion coefficient and partial pressure simultaneously, *Rev. scient. Instrum.* **56**(8), 1635 (1985).

HEAT TRANSFER APPLICATIONS—HEAT PIPES AND HEAT EXCHANGERS

1Q. K. A. Antonopoulos, Heat transfer in tube banks under conditions of turbulent inclined flow, *Int. J. Heat Mass Transfer* **28**(9), 1645 (1985).
2Q. K. A. Antonopoulos, Heat transfer in tube assemblies under conditions of laminar axial, transverse and inclined flow, *Int. J. Heat Fluid Flow* **6**(3), 193 (1985).
3Q. L. A. Ashmantas, B. V. Dzyubenko, G. A. Dreitser and M. D. Segal, Unsteady-state heat transfer and mixing of a heat carrier in a heat exchanger with flow twisting, *Int. J. Heat Mass Transfer* **28**(4), 867 (1985).
4Q. E. Azad, F. Bahar and F. Moztarzadeh, Design of water-to-air gravity-assisted heat pipe heat exchanger, *J. Heat Recovery Syst.* **5**(2), 89 (1985).
5Q. E. Azad and F. Moztarzadeh, Design of air-to-water co-axial heat pipe heat exchanger, *J. Heat Recovery Syst.* **5**(3), 217 (1985).
6Q. B. S. Baclic, The application of the Galerkin method to the solution of the symmetric and balanced counterflow regenerator problem, *J. Heat Transfer* **107**(1), 214 (1985).
7Q. P. J. Banks, Prediction of heat and mass regenerator performance using nonlinear analogy method: Part 1—Basis, *J. Heat Transfer* **107**(1), 222 (1985).
8Q. P. J. Banks, Prediction of heat and mass regenerator performance using nonlinear analogy method: Part 2—Comparison of methods, *J. Heat Transfer* **107**(1), 230 (1985).
9Q. V. L. Barantsevich, S. I. Opryshko and V. Ya. Sasin, Methods for improving the maximum heat transfer capacity of axially grooved heat pipes, *Heat Transfer—Sov. Res.* **16**(4), 92 (1985).
10Q. A. Bar-Cohen and M. Jelinek, Optimum arrays of longitudinal, rectangular fins in convective heat transfer, *Heat Transfer Engng* **6**(3), 68 (1985).
11Q. M. K. Bassiouny and H. Martin, Temperature distribution in parallel channels of plate heat exchangers, *Chem.-Ing.-Tech.* **57**(4), 330 (1985).
12Q. M. K. Bassiouny and H. Martin, Temperature distribution in a four channel plate heat exchanger, *Heat Transfer Engng* **6**(2), 58 (1985).
13Q. I. I. Belyakov, M. A. Kvetnyi and D. A. Loginov, Effect of a ballast zone on the hydraulic stability of a direct-flow steam generator, *Sov. atom. Energy* **58**(3), 177 (1985).
14Q. S. Biykili, Optimum use of longitudinal fins of rectangular profiles in boiling liquids, *J. Heat Transfer* **107**(4), 968 (1985).
15Q. M. S. Bohn, Air molten salt direct-contact heat exchange, *J. Sol. Energy Engng* **107**(3), 208 (1985).
16Q. M. Bohnet, Fouling of heat transfer surfaces, *Chem.-Ing.-Tech.* **57**(1), 24 (1985).
17Q. M. K. Bologa, V. V. Pushkov and A. B. Berkov, Electric field induced heat transfer enhancement in a gas–solid suspension heat exchanger, *Int. J. Heat Mass Transfer* **28**(7), 1245 (1985).
18Q. M. Bottoni, B. Dorr, C. Homann and D. Struwe, BACCHUS-3D/SP, a computer program to describe transient three-dimensional single phase flow in LMFBR rod bundles, *Nucl. Technol.* **71**(1), 43 (1985).
19Q. L. Boumehdi, P. Guigon and J. F. Large, Heat recovery from solids in a raining bed exchanger, *J. Heat Recovery Syst.* **5**(5), 407 (1985).
20Q. O. M. Bragina, V. L. Leichuk, A. G. Sorokin and A. V. Turkin, A method for predicting pressure drop in a shell-and-tube heat exchanger with roughened tubes and a smooth shell surface, *Heat Transfer Engng* **6**(1), 26 (1985).
21Q. J. Buxmann and T. Hoffmann, Trockenkühlturm (Änderung des Wasserdurchsatzes bei geschlossenen Kühlkreisläufen mit Trockenkühlturm), *Brennstoff Wärme Kraft* **37**(6), 259 (1985).
22Q. P. I. Bystrov, V. S. Mikhailov, N. N. Garbuz and V. F. Goncharov, Hydraulic characteristics of heat exchangers on a base of porous structures, *High Temp.* **23**(2), 323 (1985).
23Q. V. P. Carey, Surface tension effects on convective boiling heat transfer in compact heat exchangers with offset strip fins, *J. Heat Transfer* **107**(4), 970 (1985).
24Q. W. S. Chang and G. T. Colwell, Mathematical modeling of the transient operating characteristics of a low-temperature heat pipe, *Numer. Heat Transfer* **8**(2), 169 (1985).
25Q. K. Chen, On the oscillatory instability of closed-loop thermosyphons, *J. Heat Transfer* **107**(4), 826 (1985).
26Q. F. M. Chernomurov, N. G. Parfent'yev and V. N. Krivenko, Industrial test of operation of a heat pipe in a metallurgical furnace, *Heat Transfer—Sov. Res.* **16**(2), 122 (1984).
27Q. D. Chisholm, Two-phase flow in heat exchangers and pipelines, *Heat Transfer Engng* **6**(2), 48 (1985).
28Q. S. W. Churchill and P. Tepper, Characteristics of a single-fluid, double- or triple-pass heat exchanger and combustor for low-heating-value gas, *Ind. Engng Chem. Proc. Des. Dev.* **24**(3), 542 (1985).
29Q. F. Dobran, Steady-state characteristics and stability thresholds of a closed two-phase thermosyphon, *Int. J. Heat Mass Transfer* **28**(5), 949 (1985).
30Q. M. P. Dudukovic and P. A. Ramachandran, Evaluation of thermal efficiency for heat regenerators in periodic operation by approximate methods, *Chem. Engng Sci.* **40**(9), 1629 (1985).
31Q. M. P. Dudukovic and P. A. Ramachandran, Quick

design and evaluation of heat regenerators, *Chem. Engng* **92**(12), 63 (1985).
32Q. P. W. Eckels and T. J. Rabas, Heat transfer and pressure drop of typical air cooler finned tubes, *J. Heat Transfer* **107**(1), 198 (1985).
33Q. P. W. Eckels and T. J. Rabas, Heat transfer and pressure drop performance of finned tube bundles, *J. Heat Transfer* **107**(1), 205 (1985).
34Q. W. W. Focke, J. Zachariades and I. Olivier, The effect of the corrugation inclination angle on the thermohydraulic performance of plate heat exchangers, *Int. J. Heat Mass Transfer* **28**(8), 1469 (1985).
35Q. P. J. Fryer and N. K. H. Slater, Direct simulation procedure for chemical reaction fouling in heat exchangers, *Chem. Engng Jl* **31**(2), 97 (1985).
36Q. T. Fujii, T. Nagata and K. Shinzato, Experimental study of heat and mass transfer from humid air to a horizontal tube in false tube banks, *Bull. J.S.M.E.* **28**(237), 436 (1985).
37Q. K. Fukuda, T. Fukuda, S. Hasegawa, R. Echigo and K. Sanokawa, Improvement of high temperature heat exchanger efficiency by inserted porous material, (I): shell and tube type heat exchanger, *Nippon Genshiryoku Gakkaishi* **27**(2), 159 (1985).
38Q. K. Fukuda, T. Fukuda, T. Yamamoto, S. Hasegawa, R. Echigo and K. Sanokawa, Improvement of high temperature heat exchanger efficiency by inserted porous material, (II): preheating, single tube type shell and tube heat exchanger, *Nippon Genshiryoku Gakkaishi* **27**(3), 243 (1985).
39Q. J. Fürst and J. E. Roth, Wärmeaustauscher, *Brennstroff Wärme Kraft* **37**(4), 178 (1985).
40Q. M. A. Gotovsky, E. D. Fedorovich, E. V. Firsova, A. J. Blagoveshchensky, N. V. Mizonov, V. N. Piskunov and V. F. Vinokurov, Thermal effectiveness of heat exchangers: effect of manufacturing inaccuracies and deformation of tube bundles, *Heat Transfer Engng* **6**(1), 52 (1985).
41Q. U. Gross and E. Hahne, Heat transfer in a two-phase thermosyphon operating with a fluid in the near critical state, *Int. J. Heat Mass Transfer* **28**(3), 589 (1985).
42Q. J. E. Hart, A model of flow in a closed-loop thermosyphon including the Soret effect, *J. Heat Transfer* **107**(4), 840 (1985).
43Q. J. E. Hart, A note on the loop thermosyphon with mixed boundary conditions, *Int. J. Heat Mass Transfer* **28**(5), 939 (1985).
44Q. J. E. Hart, Observations of complex oscillations in a closed thermosyphon, *J. Heat Transfer* **107**(4), 833 (1985).
45Q. H. T. Haukas, Design of a plate-type evaporator for heat pumps, *Aust. Refrig. Air Condit. Heat.* **39**(4), 22 (1985).
46Q. R. Hoffmann, W. Krug and M. Arras, Evaluation of experimental investigations into the heat transfer process in plate-type heat exchangers, *Luft Kaeltetech* **3**(21), 160 (1985).
47Q. I. Hossain and J. Zaman, Thermal characteristics of heat exchanger type chemical reactors, *J. Chem. Engng Japan* **18**(3), 193 (1985).
48Q. A. N. Hrymak, G. J. McRae and A. W. Westerberg, Combined analysis and optimization of extended heat transfer surfaces, *J. Heat Transfer* **107**(3), 527 (1985).
49Q. B. J. Huang and J. T. Tsuei, A method of analysis for heat pipe heat exchangers, *Int. J. Heat Mass Transfer* **28**(3), 553 (1985).
50Q. G. H. Junkhan, A. E. Bergles, V. Nirmalan and T. Ravigururajan, Investigation of turbulators for fire tube boilers, *J. Heat Transfer* **107**(2), 354 (1985).
51Q. N. Kaji, Y. H. Mori and Y. Tochitani, Heat transfer enhancement due to electrically induced resonant oscillation of drops, *J. Heat Transfer* **107**(3), 788 (1985).
52Q. E. K. Kalinin, G. A. Dreitser and E. V. Dubrovsky, Compact tube and plate-finned heat exchangers, *Heat Transfer Engng* **6**(1), 44 (1985).
53Q. Y. Kobayashi, Heat pipe thermodynamic cycle and its applications, *J. Sol. Energy Engng* **107**(2), 153 (1985).
54Q. G. L. Koester III and E. O. Classon, Impedance pipe heating, *Heat Transfer Engng* **6**(3), 53 (1985).
55Q. V. M. Korotayev and N. A. Kolesnikova, Solution of equations of the method of finite fluid characteristics with a single governing similitude criterion and its application to heat exchanger design, *Heat Transfer—Sov. Res.* **16**(2), 105 (1984).
56Q. A. D. Kraus and A. D. Snider, Choking and optimization of extended surface arrays, *J. Heat Transfer* **107**(4), 746 (1985).
57Q. Yu. N. Kuznetsov and V. E. Lipatov, Study of high-performance convective surfaces for viscous fluids, *Heat Transfer Engng* **6**(1), 31 (1985).
58Q. R. S. Larson, On the design of recirculated-liquid gas cooler/humidifier, *Ind. Engng Chem. Proc. Des. Dev.* **24**(4), 1023 (1985).
59Q. R. Lavie, Optimal interaction of heat and mass exchanger for the improvement of separation efficiency, *Chem. Engng Sci.* **40**(11), 2019 (1985).
60Q. C. W. Leung, S. D. Probert and M. J. Shilston, Heat exchanger: optimal separation for vertical rectangular fins protruding from a vertical rectangular base, *Appl. Energy* **19**(2), 77 (1985).
61Q. C. W. Leung, S. D. Probert and M. J. Shilston, Heat exchanger design: optimal uniform separation between rectangular fins protruding from a vertical rectangular base, *Appl. Energy* **19**(4), 287 (1985).
62Q. C. W. Leung, S. D. Probert and M. J. Shilston, Heat exchanger design: thermal performances of rectangular fins protruding from vertical or horizontal rectangular bases, *Appl. Energy* **20**(2), 123 (1985).
63Q. C.-H. Li and R. K. Shah, The maximum slope of single-blow transient testing for compact heat exchanger surfaces, *J. Heat Transfer* **107**(1), 264 (1985).
64Q. Y. A. Liu, F. A. Pehler and D. R. Cahela, Studies in chemical process design and synthesis, Pt. VII: systematic synthesis of multipass heat exchanger networks, *A.I.Ch.E. Jl* **31**(5), 487 (1985).
65Q. H. Maeda, R. Izumi, H. Yamashita and S. Kaga, Study on a rotating condenser with a scraper (3rd report, experiment of condensation on outer surface

of vertical and horizontal cylinders), *Bull. J.S.M.E.* **28**(235), 94 (1985).
66Q. A. A. Mikhalevich, Computer calculation and parameter optimization of heat transfer equipment, *Heat Transfer Engng* **6**(1), 39 (1985).
67Q. L. F. Melo and J. D. Pinheiro, Estimating friction factors in a fouled annulus, *Can. J. chem. Engng* **63**(4), 693 (1985).
68Q. M. N. Netrakanti and C. L. D. Huang, Optimization of annular fins with variable thermal parameters by invariant imbedding, *J. Heat Transfer* **107**(4), 966 (1985).
69Q. C. J. L. Otero, J. L. Otero de la Gandara, Multiple-stage heat regenerator—2: experimental determination of thermal stage efficiency, *Rev. Metal., Madrid* **21**(1), 3 (1985).
70Q. J. A. Perez and E. M. Sparrow, Determination of shell-side heat transfer coefficients by the naphthalene sublimation technique, *Heat Transfer Engng* **6**(2), 19 (1985).
71Q. A. I. Petrov and S. I. Efimov, Effect of flow-through capacity of a Stirling engine regenerator on its thermal and hydraulic characteristics, *Mashinostroenie* **3,** 69 (1985).
72Q. M. J. Pettigrew, L. N. Carlucci, P. L. Ko and G. L. Holloway, Computer techniques to analyze shell-and-tube heat exchangers, *Heat Transfer Engng* **6**(3), 40 (1985).
73Q. G. P. Purohit, Costs of double-pipe and multitube heat exchangers: Pt. 1, *Chem. Engng* **92**(5), 93 (1985).
74Q. R. Ponnappan, J. E. Beam and T. Mahefkey, Improved double-wall artery high-capacity heat pipe, *J. Spacecraft Rockets* **22**(6), 592 (1985).
75Q. K. N. Proskuryakov and V. A. Uruchev, Thermo-hydraulic method of generation of high-intensity ultra-sound for removal of scale and deposits from heat-exchanger tubes, *Heat Transfer—Sov. Res.* **16**(4), 139 (1985).
76Q. G. R. Putman and W. M. Rohsenow, Viscosity induced non-uniform flow in laminar flow heat exchangers, *Int. J. Heat Mass Transfer* **28**(5), 1031 (1985).
77Q. E. Ramos, M. Sen and C. Treviño, A steady-state analysis for variable area one- and two-phase thermo-syphon loops, *Int. J. Heat Mass Transfer* **28**(9), 1711 (1985).
78Q. B. M. Rassamakin and E. V. Gal'perin, Laws of formation of a plug of liquid heat-transfer agent in the vapor channel of a thermal diode heat pipe, *J. Engng Phys.* **48**(2), 134 (1985).
79Q. D. A. Reay, Heat pipes, *Phys. Technol.* **16**(2), 69 (1985).
80Q. F. E. Romie, Transient response of the parallel-flow heat exchanger, *J. Heat Transfer* **107**(3), 727 (1985).
81Q. A. Ronen and Y. Zvirin, The behavior of a toroidal thermosyphon at high Graetz (and Grashof) numbers, *J. Heat Transfer* **107**(1), 254 (1985).
82Q. T. M. Rudy and R. L. Webb, An analytical model to predict condensate retention on horizontal integral-fin tubes, *J. Heat Transfer* **107**(2), 361 (1985).
83Q. R. Sarunac, The experimental determination of thermal characteristics of oil-air type heat exchanger, *Strojarstvo* **27**(4), 221 (1985).
84Q. M. Sen, E. Ramos and C. Treviño, On the steady-state velocity of the inclined toroidal thermosyphon, *J. Heat Transfer* **107**(4), 974 (1985).
85Q. M. Sen, E. Ramos and C. Treviño, The toroidal thermosyphon with known heat flux, *Int. J. Heat Mass Transfer* **28**(1), 219 (1985).
86Q. G. L. Serebryanyi, Influence of contacts on the thermal resistance of a bundle of rectangular fins, *J. Engng Phys.* **48**(2), 236 (1985).
87Q. I. Shishido, T. Matsuda, T. Iida and S. Ohtani, Measurement of the working fluid distribution within a heat pipe wick, *Heat Transfer—Japan. Res.* **13**(2), 65 (1984).
88Q. J. Siderowicz, Heat-transfer coils, *Plant Engng* **39**(12), 84 (1985).
89Q. B. Smith, Up-date on shellside two phase heat transfer, *Chem. Engr* **414,** 16 (1985).
90Q. E. M. Sparrow and S. S. Kang, Longitudinally-finned cross-flow tube banks and their heat transfer and pressure drop characteristics, *Int. J. Heat Mass Transfer* **28**(2), 339 (1985).
91Q. E. M. Sparrow and J. A. Perez, Internal, shellside heat transfer and pressure drop characteristics for a shell and tube heat exchanger, *J. Heat Transfer* **107**(2), 345 (1985).
92Q. E. M. Sparrow and S. B. Vemuri, Natural convection/radiation heat transfer from highly populated pin fin arrays, *J. Heat Transfer* **107**(1), 190 (1985).
93Q. J. Starczewski, Better refrigerant exchanger design, *Hydrocarbon Process.* **64**(4), 93 (1985).
94Q. J. K. Steimke, Natural convection heat transfer for a concentric tube thermosyphon, *J. Heat Transfer* **107**(3), 583 (1985).
95Q. K. Suzuki, E. Hirai, T. Miyake and T. Sato, Numerical and experimental studies on a two-dimensional model of an offset-strip-fin type compact heat exchanger used at low Reynolds number, *Int. J. Heat Mass Transfer* **28**(4), 823 (1985).
96Q. G. Swift, A. Migliori and J. Wheatley, Construction of and measurements with an extremely compact cross-flow heat exchanger, *Heat Transfer Engng* **6**(2), 39 (1985).
97Q. H. Takazawa and T. Kajikawa, Condensing heat transfer enhancement on vertical spiral double fin tubes with drainage gutters, *J. Sol. Energy Engng* **107**(3), 222 (1985).
98Q. T. Takizuka and H. Kaburaki, Research on crossflow in VHTR core, (I) two-block crossflow experiments, *Nippon Genshiryoku Gakkaishi* **27**(4), 347 (1985).
99Q. M. Turaga and R. W. Guy, Refrigerant side heat transfer and pressure drop estimates for direct expansion coils: a review of works in North American use, *Int. J. Refrig.* **8**(3), 134 (1985).
100Q. A. H. Tuthill, Sedimentation in condensers and heat exchangers: causes and effects, *Pwr Engng* **89**(6), 46 (1985).
101Q. H. Uehara, T. Nakaoka and S. Nakashima, Heat transfer coefficients for a vertical fluted-plate condenser with R113 and 114, *Int. J. Refrig.* **8**(1), 22 (1985).

102Q. H. C. Ünal, Determination of the temperature distribution in an extended surface with a non-uniform heat transfer coefficient, *Int. J. Heat Mass Transfer* **28**(12), 2279 (1985).
103Q. H. C. Ünal, Two simple correlations for the inception of density-wave oscillations in long sodium-heated stream generator tubes, *Int. J. Heat Mass Transfer* **28**(7), 1385 (1985).
104Q. S. B. Uttarwar and M. R. Rao, Augmentation of laminar flow heat transfer in tubes by means of wire coil inserts, *J. Heat Transfer* **107**(4), 930 (1985).
105Q. L. L. Vasiliev, Low-temperature heat pipes, *J. Heat Recovery Syst.* **5**(3), 203 (1985).
106Q. D. L. Vinograd, I. B. Zhvanetskii, V. M. Platonov and V. V. Zolotorev, Modeling of a heat exchanger for the synthesis of chemical engineering systems, *Theoret. Found. chem. Engng* **9**(1), 38 (1985).
107Q. R. L. Wagner and S. Sjogren, Optimizing heat exchanger design for crude oil stabilization, *Chem. Engng Prog.* **81**(2), 46 (1985).
108Q. R. L. Webb, T. M. Rudy and M. A. Kedzierski, Prediction of the condensation coefficient on horizontal integral-fin tubes, *J. Heat Transfer* **107**(2), 369 (1985).
109Q. R. L. Webb and A. Villacres, Performance simulation of evaporation heat exchanger (cooling towers, fluid coolers, and condensers), *Heat Transfer Engng* **6**(2), 31 (1985).
110Q. Z. Wozniak, Dynamics of transient states of the counter-flow heat regenerator, *Numer. Heat Transfer* **8**(6), 751 (1985).
111Q. K. K. Yau, J. R. Cooper and J. W. Rose, Effect of fin spacing on the performance of horizontal integral-fin condenser tubes, *J. Heat Transfer* **107**(2), 377 (1985).
112Q. K. Yoshikawa, H. Kajiyama, T. Okamure, S. Kabashima, H. Yamasaki, S. Shioda, M. Yamada, T. Yokota, M. Ishimura, K. Nakayama, K. Nakamoto and T. Sasaki, Experimental studies on thermal performance of a pebble bed high-temperature heat exchanger, *J. Heat Transfer* **107**(3), 722 (1985).
113Q. N. I. Zhidov, V. N. Kozlov and B. A. Makarov, Calculation of temperature fields of a cross-flow heat exchanger, accounting for hydraulic and temperature nonuniformities at the inlet, *Teploenergetika* **4,** 68 (1985).
114Q. A. V. Zhukov, A. P. Sorokin and N. M. Matyukhin, Heat removal in fast reactor cores, *Sov. atom. Energy* **58**(4), 250 (1985).
115Q. A. A. Zukauskas and R. Ulinskas, Efficiency parameters for heat transfer in tube banks, *Heat Transfer Engng* **6**(1), 19 (1985).
116Q. A. A. Zukauskas, R. Ulinskas and P. Daunoras, The influence of surface roughness on the heat transfer and drag of tube banks in crossflow, *Heat Technol.* **3**(2), 1 (1985).
117Q. Y. Zvirin, The instability associated with the onset of motion in a thermosyphon, *Int. J. Heat Mass Transfer* **28**(11), 2105 (1985).
118Q. Y. Zvirin, Throughflow effects on the transient and stability characteristics of a thermosyphon, *Wärme- u. Stoffübertr.* **19**(2), 113 (1985).

HEAT TRANSFER APPLICATIONS—GENERAL

1S. G. G. Achterbosch, P. P. G. de Jong, C. E. Krist-Spit, S. F. van der Meulen and J. Verberne, Development of a convenient thermal dynamic building model, *Energy Bldg* **8**(3), 183 (1985).
2S. M. R. Barone and D. A. Caulk, Nonuniform cavity surface temperature in injection molding, *Polymer Process Engng* **3**(1–2), 149 (1985).
3S. E. Barta, R. Beyar and S. Sideman, Temperature distribution within the left ventricular wall of the heart, *Int. J. Heat Mass Transfer* **28**(3), 663 (1985).
4S. G. Beccali, G. Panno and A. Giaccone, Analysis of greenhouse ground heating by underground pipes, *Energie Alternative Habitat Territorio Energia* **7**(35), 194 (1985).
5S. A. E. Beck and P. V. Shen, Temperature distribution in flowing liquid wells, *Geophys.* **50**(7), 1113 (1985).
6S. K. H. Betts, R. R. Parsons and M. J. Brett, Heat mirrors for greenhouses, *Appl. Opt.* **24**(16), 2651 (1985).
7S. B. E. Boyack, Survey of natural-circulation cooling in U.S. pressurized water reactors, *Nucl. Sci. Engng* **91**(3), 248 (1985).
7aS. B. E. Boyack, Survey of natural-circulation cooling in U.S. pressurized water reactors, *Nucl. Sci. Engng* **91**(3), 248 (1985).
8S. D. R. Boyd, C. P. Wong and Y. S. Cha, Thermal-hydraulic assessment of high heat flux removal techniques for fusion applications, *Nucl. Engng Des.* **88**(1), 103 (1985).
9S. L. A. Brovkin, O. I. Gorinov, V. M. Olabin, V. V. Primaychek and L. P. Olevskaya, Experimental investigation of heat transfer in the heating of cylindrical blanks in a glass-melt bath agitated by jets of combustion products, *Heat Transfer—Sov. Res.* **16**(5), 103 (1985).
10S. A. Cardella, Heat removal from the plasma of fusion reactors, *Heat Technol.* **3**(3–4), 58 (1985).
11S. G. F. Carrier, F. E. Fendell and P. S. Feldman, Firestorms, *J. Heat Transfer* **107**(1), 19 (1985).
12S. P. Cernik, Wind direction effects on energy consumption in buildings, *Energy Bldg* **8**(2), 157 (1985).
13S. S. Chandra, R. Kumar, S. Kaushik and S. Kaul, Thermal performance of a non-air-conditioned building with PCM thermal storage wall, *Energy Conv.* **25**(1), (1985).
14S. G. T. Chrusciel and L. A. Pool, Knudsen layer characteristics for a highly cooled blunt body in hypersonic rarefied flow, *A.I.A.A. Jl* **23**(6), 836 (1985).
15S. A. C. Cleland, Experimental verification of a mathematical model for simulation of industrial refrigeration plants, *Int. J. Refrig.* **8**(5), 275 (1985).
16S. E. W. Comings, Contributions of T. K. Sherwood and associates to the field of drying. II, *Drying Technol.* **2**(4), 117 (1983–84).

17S. E. W. Comings, Contributions of T. K. Sherwood and associates to the field of drying. III, *Drying Technol.* **2**(2), 265 (1983–84).

18S. R. J. Crawford and J. A. Scott, Experimental study of heat transfer during rotational moulding of plastics, *Plastics Rubber Process. Applic.* **5**(3), 239 (1985).

19S. W. K. Cui, W. J. M. Douglas and A. S. Mujumdar, Impingement steam drying of paper, *Drying Technol.* **3**(2), 307 (1985).

20S. G. A. Dorrer, A model of the spreading of a forest fire, *Heat Transfer—Sov. Res.* **16**(6), 39 (1984).

21S. B. Dorri, A. F. Emery and P. C. Malte, Drying rate of wood particles with longitudinal mass transfer, *J. Heat Transfer* **107**(1), 12 (1985).

22S. J. C. Dunn and H. C. Hardee, Surface heat flow measurements at the Phimau hot spot, *Geophys.* **50**(7), 1108 (1985).

23S. P. W. Eckels, J. Buttyan, J. H. Parker, Jr. and A. Patterson, Superconducting generator cooling system simulation, *Cryogenics* **25**(8), 471 (1985).

24S. G. Eigenberger and W. Ruppel, Problems of mathematical modelling for industrial fixed-bed reactors, *Chem.-Ing.-Tech.* **57**(3), 181 (1985).

25S. Y. Enomomoto, S. Furuhama and K. Minakami, Heat loss to combustion chamber wall of 4-stroke gasoline engine (1st report, heat loss to piston and cylinder), *Bull. J.S.M.E.* **28**(238), 647 (1985).

26S. V. Erckmann, G. Janzen, W. Kasparek, G. Mueller, P. G. Schueller, K. Schwoerer, M. Thumm and R. Wilheim, Electron cyclotron resonance heating experiments in the Wendelstein VII-A stellarator, *Fusion Technol.* **7**(2), (1985).

27S. A. Francis and R. E. Craine, On a model for frictioning stage in friction welding of thin tubes, *Int. J. Heat Mass Transfer* **28**(9), 1747 (1985).

28S. R. Ghetzler, J. C. Chato and J. M. Crowley, Turbulent convective heat transfer in forced cooled underground electric transmission systems, *J. Heat Transfer* **107**(2), 327 (1985).

29S. E. S. Gogolev and A. N. Krasavin, Investigation of heat transfer during an undulating motion of water at the icy surface, *Izv. Sibirsk. Otdel. Akad. Nauk SSSR, Seriya Tekkjicheskikh Nauk* **10**(14), (1985).

30S. G. M. Greene and S. I. Outcalt, Simulation model of river ice cover thermodynamics, *Cold Reg. Sci. Technol.* **10**(3), 251 (1985).

31S. M. A. Hares, J. Ben-Asher, A. D. Matthias and A. W. Warrick, Simple method to evaluate daily positive soil heat flux, *Soil. Sci. Soc. J.* **49**(1), 45 (1985).

32S. I. Holmer, Heat exchange and thermal insulation compared in woolen and nylon garments during wear trials, *Textile Res. J.* **55**(9), 511 (1985).

33S. W. W. Jones, Multicompartment model for the spread of fire, smoke and toxic gases, *Fire Safety J.* **9**(1), 55 (1985).

34S. S. A. Jurovics, C.-H. Ho and F. Y. Sorrell, New energy-efficient materials for the building envelope, *Bldg Envir.* **20**(2), 55 (1985).

35S. R. L. Kiang, Scaling criteria for nuclear reactor thermal hydraulics, *Nucl. Sci. Engng* **89**(3), 207 (1985).

36S. Ya. S. Kiselev and A. A. Toporishcev, Kinetics of thermal oxidation of wood, *Heat Transfer—Sov. Res.* **16**(6), 53 (1984).

37S. M. Kobari, Y. Shimizum and M. Endo, Contact drying of fibrous sheet material, *Ind. Engng Chem. Proc. Des. Dev.* **24**(1), 188 (1985).

38S. G. L. Koester III and E. O. Classon, Impedence pipe heating, *Heat Transfer Engng* **6**(3), 53 (1985).

39S. E. V. Konev, Analysis of the spreading of natural and intentionally set forest fires, *Heat Transfer—Sov. Res.* **16**(6), 1 (1984).

40S. P. Kurpiers, A. Steiff and P.-M. Weinspach, Heat transfer and scale-up in stirred single- and multiphase reactors with immersed heating elements, *German Chem. Engng* **8**(5), 267 (1985).

41S. P. Kurpiers, A. Steiff and P.-M. Weinspach, Heat transfer and scale-up of agitated single- and multiphase reactors with immersed heating elements, *Chem.-Ing.-Tech.* **57**(6), 632 (1985).

42S. P. Kurpiers, A. Steiff and P.-M. Weinspach, Heat transfer between reactor wall and fluid in a stirred single- and multiphase reactor on use of one- and two-stage disc stirrers, *Chem.-Ing.-Tech.* **57**(6), 700 (1985).

43S. L.-C. Lundin, Simulated physical effects of shallow soil heat extraction, *Cold Reg. Sci. Technol.* **11**(1), 45 (1985).

44S. R. E. Maebius, Effects of heat transfer in melt spinning, *J. appl. Polymer Sci.* **30**(4), 1639 (1985).

45S. C. L. Macosko and J. L. Lee, Heat transfer and property development in liquid silicone rubber molding, *Rubber Chem. Technol.* **58**(2), 436 (1985).

46S. J.-C. Mareschal, J. P. Cunningham and R. P. Lowell, Downward continuation of heat flow data; method and examples from the western United States, *Geophys.* **50**(5), 846 (1985).

47S. M. M. Menon, C. C. Tsai, J. H. Whealton and D. E. Schechter, Heat loading on the components of multi-megawatt ion sources, *J. appl. Phys.* **58**(9), 3356 (1985).

48S. M. D. Mikhailov, Luikov's contribution to drying, *Drying Technol.* **2**(4), 517 (1983–84).

49S. D. B. Moog and B. Rubinsky, An analytical model of thermal and vapor diffusion in freezing of wet coal, *J. Heat Transfer* **107**(1), 5 (1985).

50S. M. J. T. Norton, S. J. Kadolph, R. F. Jordan and A. Kenneth, Design, construction and use of Minnesota woman, a thermally instrumented mannequin, *Textile Res. J.* **55**(1), 5 (1985).

51S. R. W. Olsen, L. J. Hayes, E. H. Wissler, H. Nikaidoh and R. C. Eberhart, Influence of hypothermia and circulatory arrest on cerebral temperature distributions, *J. biochem. Engng* **107**(4), 354 (1985).

52S. S. J. Palaszewski, L. M. Jiji and S. Weinbaum, Performance of power plant spray cooling modules and systems, *Heat Transfer Engng* **6**(3), 60 (1985).

53S. G. Pantelis, Saturated-unsaturated flow in unconfined acquifers. *Z. angew, Math. Phys.* **36**(5), 648 (1985).

54S. H. Pfeifer, F. N. Fett, H. Schaefer and K.-H. Heinen, Heating-up model for steel casting ladles, *Stahl Eisen* **105**(14–15), 759 (1985).

55S. O. A. Plumb, G. A. Spolek and B. A. Olmstead, Heat and mass transfer in wood during drying, *Int. J. Heat

Mass Transfer **28**(9), 1669 (1985).
56S. S. S. Sandhu and H. R. Hashemi, A model for simultaneous heating and devolatilation of a single coal particle in a hot gas stream in the presence of a thermal radiative field, *A.I.Ch.E. Jl* **31**(10), 1714 (1985).
57S. M. T. Schobeiri, The influence of trailing-edge ejection on the mixing losses behind a cooled blade, *Forsch. Ing.* **51**(1), 25 (1985).
58S. D. Schieber, Heating effects through unipolar induction, *J. Franklin Inst.* **320**(1), 15 (1985).
59S. R. Schönfeld, Temperaturverteilung in Brennstabbündeln, *Brennstoff Wärme Kraft* **37**(7–8), 291 (1985).
60S. J. A. Stark, Thermal control system of the Purdue University space shuttle payload, *J. Spacecraft Rockets* **22**(3), 382 (1985).
61S. E. Taberlet and Y. Fautrelle, Turbulent stirring in an experimental induction furnace, *J. Fluid Mech.* **159**(409), (1985).
62S. M. Tajima, T. Mai and K. Katayama, Study of heat transfer phenomena in quenching of steel (effect of latent heat of phase transformation), *Bull. J.S.M.E.* **28**(242), 1710 (1985).
63S. R. Viskanta and D. M. Kin, Heat transfer from a partially uncovered reactor core, *Nucl. Engng Des.* **85**(1), 71 (1985).
64S. M. Walton, Seasonal thermal energy storage (STES), *Underground Space* **9**(1), 5 (1985).
65S. K.-H. Wehde and J. Stichlmair, Heat transfer in the concentration of *contained* sulphuric acid, *Chem.-Ing.-Tech.* **57**(2), 172 (1985).
66S. C. K. Wei, H. T. Davis, E. A. Davis and J. Gordon, Heat and mass transfer in water-laden sandstone: microwave heating, *A.I.Ch.E. Jl* **31**(5), 842 (1985).
67S. Z. Yan, F. Zhen, L. J. Ni and J. Liu, Heat transfer and thermal loading of air-cooled diesel engines, *Neiranji Xuebao* (*Trans. Chin. Soc. intern. Combust. Engrs*) **3**(1), 77 (1985).
68S. H.-Z. You, Investigation of the fire-plume impingement on a horizontal ceiling. 2 — Impingement and ceiling-jet regions, *Fire Math.* **9**(1), 46 (1985).

SOLAR ENERGY

1T. M. Abdul-Hadi, Thermal analysis of a thermosyphon V-trough solar concentrator, *Heat Technol.* **3**(2), 132 (1985).
2T. R. Almanza and J. Lara, Simulation of solar ponds with an experimental transmittance function, *J. Sol. Energy Engng* **107**(4), 357 (1985).
3T. A. K. Athienitis, M. Chandrashekar and H. F. Sullivan, Modeling and analysis of thermal networks through subnetworks for multizone passive solar buildings, *Appl. Math. Modell.* **9**(2), 109 (1985).
4T. D. E. Beasley, J. A. Clark and M. J. Holstege, Observations on the decay of a thermocline in a rock bed with no net fluid flow, *J. Sol. Energy Engng* **107**(1), 50 (1985).
5T. C. Bénard, Y. Body and A. Zanoli, Experimental comparison of latent and sensible heat thermal walls, *Sol. Energy* **34**(6), 475 (1985).
6T. N. C. Bhowmik and S. C. Mullick, Calculation of turbular absorber heat loss factor, *Sol. Energy* **35**(3), 219 (1985).
7T. V. E. Cachorro, A. M. de Frutos and J. L. Casanova, Comparison between various models of solar spectral irradiance and experimental data, *Appl. Opt.* **24**(19), 3249 (1985).
8T. S. C. Chang and J. A. Roux, Thermodynamic analysis of a solar zeolite refrigeration system, *J. Sol. Energy Engng* **107**(3), 189 (1985).
9T. D. E. Chelghoum and A. Bejan, Second-law analysis of solar collectors with energy storage capability, *J. Sol. Energy Engng* **107**(3), 244 (1985).
10T. M. Collares-Pereira, Description and testing of a non-evacuated 1.5 × CPC collector thermal performance comparison with other-collector types, *J. Sol. Energy Engng* **107**(4), 277 (1985).
11T. S. Crha and G. E. Schneider, Effect of contact spacing on efficiency of flat-plate solar collectors, *J. Spacecraft Rockets* **22**(6), 585 (1985).
12T. D. Feuermann, J. M. Gordon and Y. Zarmi, On massive envelopes in passively-heated solar buildings: an analytic sensitivity study, *Sol. Energy* **35**(3), 271 (1985).
13T. J. L. A. Francey and J. Papaioannou, Wind-related heat losses of a flat-plate collector, *Sol. Energy* **35**(1), 15 (1985).
14T. G. Galassini, A. Arconada and G. Labrosse, A contribution to the study of convective regimes in a horizontal solar converter with modulation of the temperature of the absorbing surface, *Int. Chem. Engng* **25**(3), 447 (1985).
15T. H. P. Garg, V. K. Sharma and A. K. Bhargava, Theory of multiple-pass solar air heaters, *Energy* **10**(5), 589 (1985).
16T. J. M. Gordon and Y. Zarmi, Single-pass open-loop solar thermal energy systems: on maximum energy delivery by variable flow rate, *J. Sol. Energy Engng* **107**(4), 273 (1985).
17T. E. Hahne, Parameter effects on design and performance of flat plate solar collectors, *Sol. Energy* **34**(6), 497 (1985).
18T. J. A. Harris and T. G. Lenz, Thermal performance of solar concentrator/cavity receiver systems, *Sol. Energy* **34**(2), 135 (1985).
19T. K. G. T. Hollands and K. Iynkaran, Proposal for a compound-honeycomb collector, *Sol. Energy* **34**(4/5), 309 (1985).
20T. G. B. Hotchkiss, L. C. Burmeister and K. A. Bishop, Solar collector parameter identification from unsteady data by a discrete-gradient algorithm, *J. Sol. Energy Engng* **107**(1), 70 (1985).
21T. B. J. Huang and C. T. Hsieh, A simulation method for solar thermosyphon collector, *Sol. Energy* **35**(1), 31 (1985).
22T. J. R. Hull, Solar pond ground heat loss to a moving water table, *Sol. Energy* **35**(3), 211 (1985).
23T. Y. Jaluria and C. K. Cha, Heat rejection to the surface layer of a solar pond, *J. Heat Transfer* **107**(1), 99 (1985).
24T. W. Kamminga, The approximate temperatures with a

flat-plate solar collector under transient conditions, *Int. J. Heat Mass Transfer* **28**(2), 433 (1985).
25T. T. G. Karyiannis and J. D. Tarasuk, Influence of temperature characteristics of the upper surface on heat transfer in a solar collector, *Trans. Can. Soc. mech. Engrs* **9**(1), 32 (1985).
26T. K. E. Kasza and M. M. Chen, Improvement of the performance of solar energy or waste heat utilization systems by using phase-change slurry as an enhanced heat-transfer storage fluid, *J. Sol. Energy Engng* **107**(3), 229 (1985).
27T. B. A. Kimball, Cooling performance and efficiency of night sky radiators, *Sol. Energy* **34**(1), 19 (1985).
28T. R. J. Kind and D. Kitaljevich, Wind-induced heat losses from solar collector arrays on flat-roofed buildings, *J. Sol. Energy Engng* **107**(4), 335 (1985).
29T. P. D. Lund and M. B. Östman, A numerical model for seasonal storage of solar heat in the ground by vertical pipes, *Sol. Energy* **34**(4/5), 351 (1985).
30T. G. L. Morrison and J. E. Braun, System modeling and operation characteristics of thermosyphon solar water heaters, *Sol. Energy* **34**(4/5), 389 (1985).
31T. G. Olalde, G. Flamant, D. Schwander and C. Combescure, Advanced high temperature semi-transparent solar absorbers, *Sol. Energy Mater.* **12**(6), 461 (1985).
32T. D. Proctor and S. R. James, Analysis of the Sanyo evacuated tube solar collector test results from the IEA Task III Programme, *Sol. Energy* **35**(5), 387 (1985).
33T. R. L. Reid, B. A. McGraw and A. F. G. Bedinger, Passive solar performance of tech house V, *J. Sol. Energy Engng* **107**(1), 35 (1985).
34T. K. H. Taylor and A. D. Weir, Simulation of a solar timber drier, *Sol. Energy* **34**(3), 249 (1985).
35T. R. Uhlemann and N. K. Bansal, Side-by-side comparison of a pressurized and a nonpressurized solar water heating thermosyphon system, *Sol. Energy* **34**(4/5), 317 (1985).
36T. B. Vaxman and M. Sokolov, Experiments with an integral compact solar water heater, *Sol. Energy* **34**(6), 447 (1985).
37T. N. E. Wijeysundera, Effect of variable Nusselt number and nonuniform wall irradiation on collector performance, *J. Sol. Energy Engng* **107**(2), 148 (1985).
38T. H. F. Zhang and Z. Lavan, Thermal performance of a serpentine absorber plate, *Sol. Energy* **34**(2), 175 (1985).
39T. Y. Zhiqiang, G. L. Harding, S. Craig, R. E. Collins and B. Window, Comparative study of fluid-in-metal manifolds for heat extraction from single ended evacuated glass tubular collectors, *Sol. Energy* **35**(1), 81 (1985).
40T. A. Zollner, S. A. Klein and W. A. Beckman, A performance prediction methodology for integral collection-storage solar domestic hot water systems, *J. Sol. Energy Engng* **107**(4), 265 (1985).

PLASMA HEAT TRANSFER AND MHD

1U. V. Ye. Abaltusov and A. N. Golovanov, Experimental study of heat and mass exchange between a plasma jet and a permeable wall cooled by injection of a two-phase flow, *Heat Transfer—Sov. Res.* **16**(5), 141 (1985).
2U. M. Abbaoui, B. Cheminat et P. Andason, Influence de la nature du metal sur la conductivité d'un plasma argon–metal, *J. Phys. D: appl. Phys.* **18**(10), L159 (1985).
3U. J. Amouroux, A. Gicquel, S. Cavvadias, D. Morvan and F. Arefi, Progress in the applications of plasma surface modifications and correlations with the chemical properties of the plasma phase, *Pure appl. Chem.* **57**(9), 1207 (1985).
4U. J. Bacri, A.-M. Gomes and M. Eymard, Ambipolar diffusion in a nitrogen thermal plasma, *J. Phys. D: appl. Phys.* **18**(12), 2339 (1985).
5U. P. Bayle, M. Bayle and G. Forn, Neutral heating in glow to spark transition in air and nitrogen, *J. Phys. D: appl. Phys.* **18**(12), 2395 (1985).
6U. L. Benzerga, C. Lhiaubet and R. M. Meyer, Improvement of a proposed model for heat conduction in an electrode submitted to an electric discharge: application to the indirect determination of the anodic and cathodic voltage drops, *J. appl. Phys.* **58**(1), 604 (1985).
7U. T. K. Bose, D. Kannappan and R. V. Seeniraj, On reactive heat conduction for multiple ionized two-temperature argon plasma, *Wärme- u. Stoffübertr.* **19**(1), 3 (1985).
8U. R. Chapman, T. Morin, M. Finzel and M. C. Hawley, Measurement of energy distribution in flowing hydrogen microwave plasmas, *J. Spacecraft Rockets* **22**(6), 626 (1985).
9U. B. Chemicat et P. Andanson, La conduction dans la colonne d'un arc electrique contamine par des vapeurs de curivre, *J. Phys. D: appl. Phys.* **18**(11), 2183 (1985).
10U. Xi Chen, Y. P. Chyou, Y. C. Lee and E. Pfender, Heat transfer to a particle under plasma conditions with vapor contamination from the particle, *Plasma Chem. Plasma Process.* **5**(2), 119 (1985).
11U. Xi Chen and E. Pfender, Effect of pressure on heat transfer to a particle exposed to a thermal plasma (84-GT-287), *J. Engng Gas Turbine Pwr* **107**(1), 147 (1985).
12U. L. E. Cram, Statistical evaluation of radiative power losses from thermal plasmas due to spectral lines, *J. Phys. D: appl. Phys.* **18**(3), 401 (1985).
13U. U. Daybelge, Theory of the self-induced electric field at a cathodic arc spot, *Phys. Fluids* **28**(2), 312 (1985).
14U. H. Drost, H.-D. Klotz, G. Schulz and H.-J. Spangenberg, The influence of hydrogen on the kinetics of plasmapyrolytic methane conversion, *Plasma Chem. Plasma Process.* **5**(1), 55 (1985).
15U. K. Etemadi and E. Pfender, Impact of anode evaporation on the anode region of a high-intensity argon arc, *Plasma Chem. Plasma Process.* **5**(2), 175 (1985).
16U. D. A. Erwin and J. A. Kunc, Electron temperature and ionization degree dependence of electron transport coefficients in monatomic gases, *Phys. Fluids* **28**(11), 3349 (1985).
17U. B. C. Eu, Temperature dependence of thermal conductivity of a dilute plasma in a magnetic field, *Physica* **133A**(1, 2), 120 (1985).
18U. A. Ya Faenov, S. Ya Khakhalin, A. A. Kolomensky, S. A. Pikuz, A. I. Samokhin and I. Yu. Skobelev, Superdense high-temperature exploded-wire plasma

observation in a high-current vacuum diode, *J. Phys. D: appl. Phys.* **18**(7), 1347 (1985).
19U. P. Fauchais, A. Vardelle, M. Vardelle, J. F. Coudert and B. Pateyron, Plasma spraying and extractive metallurgy: Comparisons between mathematical modeling and measurements and between applications and development, *Pure appl. Chem.* **57**(9), 1171 (1985).
20U. S.-I. Himeno, M. Seki and H. Mochizuki, Computer method for new elliptical Abel inversion applied to holographic plasma diagnostics, *J. phys. Soc. Japan* **54**(5), 1737 (1985).
21U. S. Ishiguro, T. Kamimura and T. Sato, Double layer formation caused by contact between different temperature plasmas, *Phys. Fluids* **28**(7), 2100 (1985).
22U. D. Karabourniotis and C. Karras, Line broadening in optically thick gas discharges: influence on plasma temperature determination, *J. appl. Phys.* **57**(11), 4861 (1985).
23U. M. Kobayashi, T. Honda and A. Kanzawa, Numerical analysis of thermal plasma flow in a pipe, *Int. Chem. Engng* **25**(2), 308 (1985).
24U. M. Kobayashi, T. Honda and A. Kanzawa, Numerical analysis of a thermal plasma flow in a tube, *Heat Transfer—Japan. Res.* **13**(1), 64 (1984).
25U. M. Kobayashi, T. Honda and A. Kanzawa, The effect of gas injection on a thermal argon plasma flow in a water-cooled tube, *Plasma Chem. Plasma Process.* **5**(1), 39 (1985).
26U. I. S. Kovalenko, V. I. Kucharenko and V. A. Sechenov, Three-chamber heated shock tube for investigating a high-enthalpy dense plasma, *J. appl. Mech. Tech. Phys.* **26**(2), 278 (1985).
27U. P. Kovitya, Physical properties of high-pressure plasmas of hydrogen and copper in the temperature range 5000–60,000 K, *IEEE Trans. Plasma Sci.* **13**(6), 587 (1985).
28U. P. Kovitya and J. J. Lowke, Two-dimensional analysis of free burning arcs in argon, *J. Phys. D: appl. Phys.* **18**(1), 53 (1985).
29U. T. Kubono, Calculation of effective work-function and average cooling energy of electrons emitted from an arc cathode root, *J. Phys. D: appl. Phys.* **18**(6), 1131 (1985).
30U. J. A. Kunc, Conductivity of weakly and highly ionized gases in quasistatic electric fields, *Phys. Fluids* **27**(12), 2859 (1984).
31U. H. Lee, A method for computing the radial temperature profiles in high-pressure, high-current arcs, *J. Phys. D: appl. Phys.* **18**(3), 425 (1985).
32U. H. E. Lee and L. E. Cram, On the theory of radiation-dominated wall-stabilized arcs, *J. Phys. D: appl. Phys.* **18**(8), 1561 (1985).
33U. Y. C. Lee, Y. P. Chyou and E. Pfender, Particle dynamics and particle heat and mass transfer in thermal plasmas — II. Particle heat and mass transfer in thermal plasmas, *Plasma Chem. Plasma Process.* **5**(4), 391 (1985).
34U. R. R. Mitchell, D. T. Tuma and J. F. Osterle, Transient two-dimensional calculations of properties of forced convection-stabilized electric arcs, *IEEE Trans. Plasma Sci.* **13**(4), 207 (1985).
35U. L. Monchick and E. A. Mason, A reconsideration of thermal diffusion in ionized gases: quantal and dynamic shielding effects, *Phys. Fluids* **28**(11), 3341 (1985).
36U. A. L. Mosse, V. V. Pechkovsky, E. D. Dzyuba and L. I. Krasovskaya, Production of condensed phosphate by plasma chemical treatment of phosphate raw material, *Plasma Chem. Plasma Process* **5**(3), 275 (1985).
37U. J. Mostaghimi, P. Proulx and M. I. Boulos, An analysis of the computer modeling of the flow and temperature fields in an inductively coupled plasma, *Numer. Heat Transfer* **8**(2), 187 (1985).
38U. Y. Ohguchi, Estimation of thermal conductivity of hydrogen gas through arc studies, *J. phys. Soc. Japan* **53**(12), 4183 (1985).
39U. K. Onda, K. Takano and K. Kato, Spectral emission from MHD combustion gas mixed with pulverized coal, *J. quant. Spectrosc. radiat. Transfer* **33**(3), 281 (1985).
40U. D. I. Sharakhovskiy and N. A. Kostin, Vortex flows in electric-arc gas heaters, *Heat Transfer—Sov. Res.* **16**(5), 126 (1985).
41U. H. Shindo and T. Tamaru, An improved relaxation method for measuring gas temperature in arc plasmas, *J. quant. Spectrosc. radiat. Transfer* **33**(4), 331 (1985).
42U. H. Shirai, K. Tabei and S. Kato, Temperature measurements of non-equilibrium high-temperature plasmas by means of N_2^+ first negative bands, *Bull. J.S.M.E.* **28**(242), 1718 (1985).
43U. S. V. Joshi, Jin Y. Park, P. R. Taylor and L. S. Richardson, The integral mean diffusivity for particle–gas mass transfer calculations under thermal plasma conditions, *Plasma Chem. Plasma Process.* **5**(2), 143 (1985).
44U. E. Pfender, Heat and momentum transfer to particles in thermal plasma flows, *Pure appl. Chem.* **57**(9), 1179 (1985).
45U. E. Pfender and Y. C. Lee, Particle dynamics and particle heat transfer in thermal plasmas — I. The motion of a single particle without thermal effects, *Plasma Chem. Plasma Process.* **5**(3), 211 (1985).
46U. P. Proulx, J. Mostaghimi and M. I. Boulos, Plasma–particle interaction effects in induction plasma modeling under dense loading conditions, *Int. J. Heat Mass Transfer* **28**(7), 1327 (1985).
47U. H. P. Stormberg and R. Schafer, Determination of temperature profiles from optically thick lines, *J. quant. Spectrosc. radiat. Transfer* **33**(1), 27 (1985).
48U. S. Vacquie, A. Gleizes and H. Kafrouni, Measurements of electron density in a SF_6 arc plasma, *J. Phys. D: appl. Phys.* **18**(11), 2193 (1985).
49U. S. Venkataramana and D. Bathaiah, MHD combined free and forced convection flow of an incompressible conducting viscoelastic fluid through two parallel porous plates, *Reg. J. Energy Heat Mass Transfer* **7**(1), 47 (1985).
50U. T. Vijayan and V. K. Rohatgi, Characteristics of exploding-wire plasmas, *IEEE Trans. Plasma Sci.* **13**(4), 197 (1985).
51U. F. Weling, A model for the plasma-activated chemical vapor deposition process, *J. appl. Phys.* **57**(9), 4441 (1985).
52U. H. Wilhelmi, W. Wimmer and E. Pfender, Modeling

of the transport phenomena in the anode region of high current arcs, *Numer. Heat Transfer* **8**(6), 731 (1985).

53U. G. Wilks and R. Hunt, Low Prandtl number MHD convection about a semi-infinite horizontal plate, *Numer. Heat Transfer* **8**(2), 203 (1985).

54U. Ye. M. Yermolayeva, D. Hebecker, V. K. Zabrodin and A. L. Mosse, Investigation of heat transfer between a hydrogen-plasma jet and the wall of a cylindrical channel, *Heat Transfer—Sov. Res.* **16**(4), 130 (1985).

55U. R. M. Young and E. Pfender, Generation and behavior of fine particles in thermal plasmas—a review, *Plasma Chem. Plasma Process.* **5**(1), 1 (1985).

HEAT TRANSFER REVIEW - 1986

E.R.G. ECKERT, R. J. GOLDSTEIN, E. PFENDER, W.E. IBELE, S.V. PATANKAR, J.W. RAMSEY, T.W. SIMON, N. A. DECKER, T.H. KUEHN, H.O. LEE and S.L. GIRSHICK

Heat Transfer Laboratory, Department of Mechanical Engineering, University of Minnesota, Minneapolis, Minnesota, U.S.A.

INTRODUCTION

THIS REVIEW surveys papers that have been published in the literature during 1986 covering various fields of heat transfer. The literature search was made very inclusive, however, the great number of publications made selections in some of the review sections necessary.

Several conferences were devoted to heat transfer or included heat transfer topics in their sessions during 1986. The *AIAA/ASME 4th Thermophysics and Heat Transfer Conference*, sponsored by the American Institute for Aeronautics and Astronautics and the American Society of Mechanical Engineers, was held in Boston, Massachusetts, 2–4 June 1986. Several sessions dealt with heat transfer like conduction, two-phase systems, cryogenics, convection, radiation, porous media, internal flows, others were directed towards space applications like aerodynamic heating and ablation, high temperature heat pipes, immersion cooling of microaerodynamics, melting and freezing in energy systems and processes, and thermal design of spacecraft. The papers presented at the meeting can be obtained at the AIAA Library, AIAA Technical Information Service, New York. Some are also published in the archival journals of the Institute. The *31st ASME International Gas Turbine Conference and Exhibition* was held in Düsseldorf, Federal Republic of Germany, 8–12 June 1986. Sessions on film cooling and heat transfer with rotation dealt with heat transfer and many of the others included heat transfer considerations. The 1984 Gas Turbine Award was presented to H. P. Hodson for his paper "Measurements of wake generated unsteadiness in the rotor passages of axial flow turbines". Copies of the ASME papers may be obtained through the ASME Order Department. The *Solar Thermal Technology Conference*, sponsored by the U.S. Department of Energy and Sandia National Laboratories was held in Albuquerque, New Mexico, 17–19 June 1986. Among the papers presented may be listed those given at sessions on the Sterling engine, on solar concentrators, and on energy storage. The *Eighth International Heat Transfer Conference and Exhibition* was held in San Francisco, California, 17–22 August 1986. It was organized in keynote sessions, 28 papers were presented on various topics of heat transfer. Individual papers were presented in poster sessions. The opening session included a general lecture by A. W. Trivelpiece on energy research in the U.S. Department of Energy and two plenary lectures—one by E. R. G. Eckert on "The early history of international heat transfer conferences" and one by U. Grigull on "Fahrenheit, a pioneer of exact thermometry". The 1985 Max Jakob Memorial Award and Medal was presented to Frank Kreith and the 1985 Donald Q. Kern Award to Stanley J. Green. A recognition dinner honored Warren M. Rohsenow on his 65th birthday. A professional development short course program provided the possibility to update knowledge in numerical heat transfer and fluid flow in compact heat exchangers, in augmentation of heat transfer, in advances in thermal analysis and control of aerodynamic equipment, in heat transfer in packed, agitated and fluidized beds, in recent developments in heat pipes, and in experimental methods in heat transfer. All of the papers are contained in seven hardcover copies of the proceedings available from Hemisphere Publishing Corporation, Washington, D.C. The *21st Intersociety Energy Conversion Engineering Conference*, sponsored by seven engineering societies, was held in San Diego, California, 25–29 August 1986. Among the various sessions may be mentioned those on Sterling engines, hydrogen energy, seasonal thermal storage, and nuclear energy. Copies of the proceedings may be

†W. E. Ibele and S. V. Patankar also contributed to the preceding reviews. By an oversight the name W. E. Ibele was omitted from the list of authors of the 1984 review and the name S. V. Patankar from the list of authors of the 1985 review.

purchased from the American Chemical Society, Department of Meetings and Divisional Activities, Washington, D.C. The *XIII International Symposium on Heat and Mass Transfer in Cryoengineering and Refrigeration* was organized by the International Centre for Heat and Mass Transfer and held in Dubrovnik, Yugoslavia, 1–5 September 1986. Each of the eight sessions started with one or two invited lectures and then presented several individual papers. The following subjects were covered: thermal insulation, heat transfer in refrigerants, heat and mass transfer in mixtures, thermodynamic and thermophysical properties, freezing and melting, heat and mass transfer at very low temperatures, and cooling of superconducting devices. The majority of the papers will be published in a volume available from Hemisphere Publishing Corporation. The *ASME Winter Annual Meeting*, held in Anaheim, California, 7–12 December 1986, featured eight sessions in the field of heat transfer on topics like heat transfer aspects of fusion and fission reactors, combustion in multiphase systems and modeling of combustion systems, numerical methods in heat transfer, two-phase heat and mass transfer in the environment, heat transfer in waste heat recovery and heat rejection systems. The Karl A. Gardner Memorial Session was devoted to thermal/mechanical heat exchanger design. Preprints of the papers are available from the ASME Publications Department and many of the papers will also be published in the *Journal of Heat Transfer*. The *4th Miami International Symposium on Multi-phase Transport & Particulate Phenomena* was held in Miami Beach, Florida, 15–17 December 1986. Papers of interest to heat transfer researchers covered subjects like: modeling of multiphase transport and heat transfer formulation. Interphase transfer in vapor–liquid systems, drying, boiling and condensation, pool boiling, critical heat flux, freezing and melting, and various measurement techniques. Inquiries about reprints should be directed to the Clean Energy Research Institute, University of Miami, Corral Gables, Florida.

A number of books became available during the past year and they are listed in the references. A new journal also started publication.

HIGHLIGHTS

The following highlights illuminate areas of research which received special attention during the last year.

A considerable number of papers presented computer solutions of various heat transfer problems. Some of them appear to be mainly motivated by the capability of the equipment used.

Conduction in composite media received considerable attention with the majority considering the effect of contact resistance. Several new approaches were presented for the handling of the Stefan problem and the inverse conduction problem. Attention was also focused on conduction in materials subject to transient laser irradiation.

Channel flow studies concentrated on complex geometries, non-Newtonian flows, and the effects of buoyancy forces. *Boundary layer flows* with stagnation regions, particularly on cylinders in crossflow and over axisymmetric bodies, found attention. The reduction in the number of papers on viscous heating of high-speed vehicles and the increase in the number of papers on turbulent heat and flow modeling are noteworthy. Papers on *flows with separated regions* considered cylinders in tandem and banks of cylinders. A surprising number of papers were concerned with acoustic effects on cylinders and plates with separated regions.

Heat transfer in *fluidized beds* is one area for which our knowledge is mostly empirical. All papers except one in this area report the results of experimental studies. The opposite is true for heat transfer in *packed beds and porous media* where the majority of papers are analytical.

Noteworthy *instrumentation* developments include two papers on temperature measurement at ultra high pressure, ten papers on cryogenic temperature measurement and some on Raman scattering techniques developed for supersonic flows and flames. The majority of papers on external *natural convection* considered boundary layer flows over vertical flat plates and consist of numerical solutions to steady laminar flow problems. They include effects of temperature dependent fluid properties, viscoelastic fluids and coupling between wall conduction and adjacent convective flow.

Boiling heat transfer continues to be an active field of experimental research. Boiling of specially prepared surfaces and direct contact boiling processes were studied. Transition boiling and boiling of mixtures are topics of increasing interest. *Condensation* rates of horizontal tubes were studied extensively as well as direct contact condensation to flowing and static liquids. Promotion of dropwise condensation continues to be of interest. A number of analytical and experimental studies were concerned with *properties of materials* for specified applications. The majority of papers on *solar energy* considered the thermal performance of components such as solar collectors, solar ponds, and thermal storage systems. A few papers considered the performance of complete active solar heating and cooling systems, passively heated buildings, and solar radiation data.

Heat transfer in *thermal plasma* has been of particular interest in connection with electric arcs and arc applications.

To facilitate the use of this review, a listing of the subject headings is made below in the order in which they appear in the text. The letter which appears adjacent to each subject heading is also attached to the references that are cited in the category:

Conduction, A
Channel flow, B
Boundary-layer and external flows, C
Flow with separated regions, D
Heat transfer in porous media, DP
Experimental techniques and instrumentation, E
Natural convection—internal flows, F
Natural convection—external flows, FF
Convection from rotating surfaces, G
Combined heat and mass transfer, H
Change of phase—boiling, J
Change of phase—condensation, JJ
Change of phase—freezing and melting, JM
Radiation in participating media and surface radiation, K
Numerical methods, N
Transport properties, P
Heat transfer application—heat pipes and heat exchangers, Q
Heat transfer applications—general, S
Solar energy, T
Plasma heat transfer and MHD, U.

CONDUCTION

A number of studies were published in 1986 on heat transfer in composite media and inhomogeneous materials. The applicability limits of the homogeneous property model were considered in relation to the transient thermal behavior of directional reinforced composites [6A]. It was found that in many situations the homogeneous property model may lead to significant errors. The cooling by convection of an infinite parallel-sided composite slab was solved analytically by the Wiener–Hopf technique [20A]. An equivalent inclusion method, analogous to Eshelby's method in elasticity, was proposed for solving steady-state conduction in composites [31A]. A new method based on the theory of systems was advanced for the calculation of heat flow as a function of surface temperatures in a multi-layer wall with time-dependent boundary conditions [50A]. A method was presented for obtaining the effective thermal conductivity and specific heat in composite materials with an oscillating temperature field [49A]. It was claimed that the apparent thermal conductivity for steady-state conduction is not sufficient to describe the transient behavior of an inhomogeneous medium [10A]. A numerial Laplace inversion technique was employed to calculate the one-dimensional transient conduction within laminated binary materials [56A]. The steady-state thermal behavior in a strip-heated composite slab was considered [37A]. A method was proposed for extrapolating the temperature at a point in a complex structure with a specified heating regime on the basis of experimental data obtained in other regimes [7A]. A technique for analyzing multilayer heat conduction, based on construction of appropriate coordinate systems, was formulated [39A].

The thermal contact resistance across an interface between solid layers is a difficult problem which received some attention. The last decade of work in steady-state contact heat transfer was reviewed, with 146 references [43A]. A review with 143 references considered the problem of heat transfer across contacts under mechanical load [57A]. The Fenech–Rohsenow model was used to analyze the thermal conductance of 240 pairs of uranium dioxide–Zircaloy-4 contacts as a function of contact pressure [28A]; conductance values obtained by this method were found to be lower than with the Ross and Stoute correlation, the difference being attributed to the use of average fluid thickness in the latter method. The fundamental concepts in the theory of composite media were reconsidered in the context of a temperature discontinuity between the constituents [17A]. The effect of geometry on contact conductance of contiguous, rough interfaces was analyzed [53A]. Theoretical and experimental work on rough sphere contact, with microirregularities, between air and mineral oil was reported [64A]. The heat transfer rate from a heated sphere to a matrix containing passive spheres of a different conductivity was calculated using a multipole expansion scheme [21A].

Transient conduction in a cylinder imbedded in an infinite conductive medium with contact resistance was expressed in integral form by a Laplace transform and then solved numerically [61A]. The problem of a cylindrical electrical conductor with temperature-dependent properties, and carrying an alternating current, was formulated and solved numerically [46A]. The question of how long a cylinder ought to be before being considered infinite for thermal conduction was examined [22A]. Heat transfer from a set of parallel tubes of arbitrary diameters and positions, imbedded in a semi-infinite medium, was analyzed [4A]. The steady-state temperature distribution in a circular tube with internal convection and external asymmetrical heat flux was obtained by finite Fourier transform, with Biot number as a parameter, and numerical calculation [58A]. Heat conduction in a thin circular pipe with a circumferentially moving heat source is often approximated as a flat plate problem; significant linear corrections to this approximation were obtained via an asymptotic expansion in terms of the ratio of pipe thickness to radius [48A]. The first ten eigenvalues needed for the computation of transient heat or mass diffusion in a hollow cylinder without azimuthal symmetry were calculated [25A].

Conduction in other circular bodies was analyzed in several works, including a discussion of the existence of a classical solution of a mixed third boundary value problem on a sphere [36A] and a study of steady conduction in arbitrary conical systems [11A]. For conical shells, a method was described which allows one to calculate the average temperature (through the shell thickness at a particular location) and temperature gradient directly from the theory of a Cosserat surface [54A]. The steady temperature distribution for arbitrary paraboloidal systems was investigated [12A].

An exact analytical solution for a non-linear fin problem—a rectangular fin subject to a power law-type temperature dependence—was presented [55A]. A series solution was presented for the one-dimensional temperature distribution along a straight rectangular fin with variable surface heat transfer coefficient [9A]. The optimum dimensions of annular fin assemblies were studied [18A].

The Stefan problem, or the problem of phase change with a moving boundary, was treated by several authors. The melting process in a rectangular enclosure was analyzed, with a coupling between solid-phase heat conduction and natural convection in the melt. The kinetics of the melting process were found to be significantly modified by this coupling [16A]. An integral method was presented which utilizes Galerkin functions for phase-change problems, which can accommodate problems with time-dependent temperatures at the boundary [44A]. The Backlund transformation was applied to reduce a class of moving boundary problems to a form which admits a class of exact solutions [51A]. A new method for solving the Stefan problem was proposed in which two associated conduction problems, each having fixed domains and the same boundary and initial conditions as the original problem, are introduced; each includes a moving source as a supplementary condition [60A]. It was shown that the natural sign conditions in a general Stefan-like problem are sufficient to ensure that a solution exists globally [38A].

Analytical approaches to non-linear heat equations were considered in a work which utilized Bergman-type series solutions [52A] and in another which used group theory [23A]. Two studies considered the hyperbolic heat equation which results from a finite speed of heat propagation [24A,30A].

A tabulation and numbering system for Green's functions was proposed, for rectangular coordinates with 220 different sets of boundary conditions [14A]. Approximate integral transformation techniques for difficult conduction problems were discussed [63A]. Stochastic process theory was applied to heat conduction in solids under various random conditions [35A]. A new decomposition method was proposed for approximate solutions to the heat equation [2A]. A simple method was proposed for accounting for temperature-dependent properties and boundary conditions [15A].

Steady conduction in an orthotropic plate with a foreign inclusion of arbitrary but small thickness was examined [47A]. Periodic on–off heating was contrasted with continuous heating in a numerical study [19A].

The application of laser heating in industrial processes is posing an interesting class of conduction problems, involving both optical and thermal properties of the irradiated medium as well as the intensity profile of the laser beam. An exact solution was presented for the heating of a homogeneous slab induced by time-dependent laser irradiance [26A]. A general solution was obtained for a semi-infinite two-layer system heated with a cw laser beam of either Gaussian or uniform circular intensity profile [1A]. The three-dimensional conduction equation was solved numerically for scanning cw laser-annealed multilayer structures [66A]. In the case of irradiation at high power intensities the usual assumptions made in the Fourier theory may be inadequate, as the heat flux through a given plane depends on the electron energy distribution. A model was introduced for analyzing the conduction problem by means of electron kinetic theory [70A]. An integral solution was described to deal with complicated geometries undergoing both motion and pulsed laser irradiation [45A]. While not referring specifically to lasers, a related work used Fourier transform methods to solve the problem of a heat source moving steadily over a surface [27A].

The problem of finding the temperature distribution on a surface given a set of time-varying measurements at interior points, the 'inverse' conduction problem, received considerable attention. The one-dimensional inverse problem was treated by an initial value technique which does not require iteration and which yields estimates of the surface temperature and heat flux histories [32A]. A procedure was proposed which combines two different methods, the sequential function specification method of Beck and the regularization method of Tikhonov [13A]. A generalized function-specification approach was presented for solving inverse problems which allows unified treatment of multidimensional domains in rectilinear, cylindrical and spherical coordinates [29A]. The inverse problem can be approached through Duhamel's integral, which is transformed into an equation which contains an unknown surface temperature and its derivatives [40A, 41A]. A method was proposed for two-dimensional inverse problems, involving a combination of a splitting procedure and the least squares technique [31A]. It was shown that the accuracy of an analytical solution of an inverse problem depends not only on the accuracy of the data, but also on the dimensions of the space–time domain [62A]. A transformation matrix was obtained for estimating the interface temperature distribution for a cutting tool, based on thermocouple measurements of interior points [69A]. A method was devised by determining the heat transfer coefficients during water cooling of metals, in which the temperature close to the surface was measured and used in an implicit finite difference calculation [8A]. The solution of inverse heat conduction problems with partially unknown system geometries was considered [34A].

Problems in thermoelasticity are often of interest from the heat conduction viewpoint. A sudden change in temperature produces a thermal shock in a material. The transient response of an infinitely long annular cylinder composed of two dissimilar materials, subject to a thermal shock, was investigated analytically [68A]. A general solution was obtained for the transient two-dimensional temperature and stress distributions in a thermally and elastically orthotropic slab with a rectangular boundary [67A]. An insulating half-plane sliding on the surface of a conductor generates heat due to friction; the resulting thermoelastic problem was solved numerically [5A]. A study was reported of the transient thermal and mechanical behavior of an infinite medium with a spherical cavity and a non-uniform heat supply [59A]. Calculations were presented of the temperature field in a bar of finite length, subjected to cyclic bending and unsteady thermal boundary conditions [65A].

The thermal stability of a partially insulated reactive slab was analyzed; this has an interesting application to the situation of exothermically active waste material which is in thermal equilibrium with its surroundings but is suddenly subject to a small change in surface conditions, e.g. redevelopment of land formerly used for domestic or industrial waste [33A]. The thermal stability of a Landau slab was examined under various commonly encountered boundary conditions [42A].

CHANNEL FLOW

Research in channel flow is directed towards heat transfer augmentation in complex geometries, turbulent flow, multiphase effects, and non-Newtonian behavior.

Among experimental studies pertaining to turbulent flow ref. [78B] investigates the convection of helium at supercritical pressure. Turbulent heat transfer in spirally enhanced tubes has been considered in ref. [71B]. Experiments on heat transfer and pressure drop for flow in bends of circular cross-section have been reported in refs. [16B, 74B]. Reference [73B] contains a study of a shrouded fin array. An empirical correlation for heat transfer in circular tubes has been proposed [57B]. A near-wall temperature model has been applied to channel flow [1B]. Heat transfer correlation for liquids near the critical point are given [29B]. Experimental and analytical results for an asymmetrically heated duct are reported [72B].

Turbulent pulsed flow in a duct is numerically analyzed [2B]. The influence of buoyancy effects on turbulent flow are evaluated in ref. [8B]. An indirect method for studying heat transfer in a tube has been proposed [45B]. The mass transfer in the transitional regime of a wavy channel is considered [62B]. Turbulent heat transfer in ducts of cross-shaped cross-section is numerically predicted [60B]. Simultaneous convection and radiation in a turbulent channel flow is investigated [84B]. Heat transfer augmentation due to turbulence promoters is considered [65B]. The role of the structure of turbulence in heat and mass transfer is reviewed [10B]. Predictions have been reported for turbulent inclined flow in rod bundles [3B]. Reference [63B] considers the effect of entrance configurations on tube flow heat transfer. Property variations in turbulent flow are considered in ref. [42B]. Transient heat transfer to supercritical helium is investigated in ref. [7B].

Some papers deal with new aspects of simple channel flows. Transient forced convection with a stepwise variation of wall temperature is treated in ref. [20B]. The effect of axial conduction in channel and tube flows is considered in ref. [81B]. The conjugate problem for the parallel plate channel is solved in ref. [47B]. The effect of viscous dissipation in channel and tube flows is considered in refs. [50B, 51B]. Unsteady convection in a channel is experimentally investigated [22B].

Thermal instability in a tube is considered [59B]. The effect of viscosity variation on the heat transfer in a cooled tube is studied [80B]. Mixed convection in the upward flow in a vertical pipe is investigated [76B]. An analytical solution is presented for the laminar entry region of a pipe [75B]. Mixed convection is considered for a vertical annulus in refs. [32B, 33B] and for a horizontal annulus in ref. [39B].

Steady laminar flow through an elliptic twisted pipe is investigated [44B]. Predictions have been reported for laminar inclined flow through tube banks [4B]. Forced convection with periodically varying inlet temperature is considered [19B]. Reference [38B] deals with the entrance region heat transfer for toluene. The influence of an axially varying heat transfer coefficient is described [79B]. Different mechanisms for convective enhancement in disk systems are investigated [55B]. Enhancement of mass transfer due to a wavy wall is studied in ref. [61B]. Heat transfer due to radial flow between two disks is considered [56B].

The influence of combined convection and radiation in a duct flow is described [70B]. An investigation is reported for heat transfer in corrugated tubes [30B]. Experiments are reported for heat transfer on a finned wall in a rectangular duct [52B]. For a two-dimensional bifurcation, laminar forced convection is numerically analyzed [40B]. Numerical solutions are also reported for forced convection in twisted circular-sector ducts [83B]. Temperature and velocity distributions have been obtained in a square duct [17B]. Computer simulation of a melt flow is presented [82B]. Laminar convection with chemical reactions has been considered in a rectangular duct and circular tube [12B, 13B].

Convective heat transfer under the influence of discrete heat sources is considered [35B]. Reference [58B] focuses on the effect of interwall spacing on heat transfer in a corrugated duct. Convection in a multi-passage circular pipe is studied in ref. [23B]. Heat transfer in smooth and buckled tubes is investigated [69B]. Convection studies have been reported for helically coiled rectangular ducts [36B]. Laminar heat transfer in a rotating duct is presented [49B]. Fully developed flow in plate-fin passages has been numerically analyzed [68B]. Heat transfer enhancement due to oscillatory flow in grooved channels is numerically studied [28B]. A combined experimental and analytical investigation of internally finned triangular ducts has been reported [14B]. References [53B, 54B] consider heat transfer around sharp bends in rectangular channels. Numerical results are presented for convection in triangular plate-fin ducts [5B]. Convection in an elliptic duct is discussed [24B].

The combined radiation and convection in particle-laden flows is described [46B]. Measurements have been reported for mass transfer in the model of the respiratory tract [31B]. Heat transfer to cryogenic liquid under centrifugal forces is considered [41B]. Convection in a flow past two spheres in a cylindrical tube is described in ref. [21B]. Studies have been directed towards internal cooling of turbine blades [9B, 15B].

Among non-Newtonian channel flows, power-law fluids in ducts have been studied in refs. [18B, 48B]. The viscoelastic effects have been taken into account in refs. [34B, 43B]. Heat transfer to non-isothermal non-Newtonian flows is considered in refs. [6B, 25B, 85B]. Heat transfer to highly viscous rheological flows is presented in ref. [25B]. Laminar flow in a stenosis is studied in ref. [77B].

Multiphase flow and heat transfer in channels are the subject of a number of papers [27B, 37B, 63B, 66B]. Three-dimensional modeling has been applied to two-fluid flows in elbows and bends [11B].

BOUNDARY-LAYER AND EXTERNAL FLOWS

Several studies in this category dealt with the fundamentals of convection and turbulent transport. It was shown that care is necessary when comparing data for average heat transfer coefficients due to variations with position, particularly when some of the sources for those data are experimental and some are theoretical [5C]. A numerical solution of the Falkner–Skan equation was made for a laminar boundary layer using optimization techniques [3C]. Recent developments in thermodynamics of irreversible processes were applied to obtain rapid analytical solutions in Blasius flow [59C].

Studies of fundamentals in turbulent flows include a numerical study in a turbulent free shear flow [11C] and a study of turbulent motion modeled by the inviscid two-dimensional motion of point vortices [22C]. The spectrum of the latter was close to a Kolmogorov spectrum. The evolution of the temperature field in a thermal mixing layer downstream of a temperature step in grid turbulence was studied [46C]. The data was shown not to be consistent with self-similarity although second-order quantities such as heat flux coefficient and centerline temperature variance appeared to have reached their equilibrium values. An asymptotic solution was obtained for the Euler equations in the shock layer near the stagnation line for separationless flow past a blunt body in the far wake behind another body [18C]. Measurements were made of Reynolds shear stress profiles and lateral heat flux profiles in a nearly self-preserving region of the turbulent wake of a heated circular cylinder—they were compared with calculations [8C]. The comparisons do not support the need for determining a new virtual origin, as previously suggested. Results of experiments conducted in the thermal wake behind an axisymmetric body were used to evaluate the radial profiles of different terms of the temperature fluctuation balance equation [12C]. It was shown that the relationship between dissipation and turbulent diffusion is roughly the same for the flows compared. The idea of semi-preservation in non-isothermal cascade-wake-flows was presented [17C]. It was described in terms of characteristic velocity and temperature scales. An experimental study of the temperature field in round turbulent jets showed that the rate of fluid entrainment is a strong function of nozzle design [52C].

Axisymmetric and stagnation point studies include an experimental investigation of the unsteady heat transfer on the heated cylinder and blunt body stagnation regions [56C] and supersonic flow about axisymmetric bodies [75C]. The second study showed that substantial errors may result from the use of standard formulas without allowance for streamwise wall temperature gradient effects. A numerical investigation was performed to determine the heat/mass transfer from upstream-facing blunt faces of bodies situated in a uniform flow [35C]. Unsteady laminar compressible flow with mass transfer at the stagnation point was computed for different values of parameters which characterize the unsteadiness in the free-stream velocity, the wall temperature, the mass transfer rate and the variation in gas properties [70C]. A theoretical analysis for heat transfer from axisymmetric bodies in non-Newtonian fluids was presented [58C]. Asymptotic expressions give accurate predictions with high or low Prandtl numbers. Scaling relations were derived to show the influence of ballistic coefficient and lift-to-drag ratio on heating during gliding entry into the atmosphere [66C]. The effect of massive blowing on a laminar boundary layer along the stagnation line of an infinite swept cylinder was studied, where the free-stream velocity and blowing rate vary with time [71C]. Hypersonic finite-rate chemically reacting flows over an ablating carbon surface were analyzed to study the effects of ablating carbon [62C] and the interaction of a premixed flame with the flow near the stagnation point was considered under conditions where the flame stands clear of the boundary layer and pushes the incident flow away from the body [19C]. The influence of the shape of a body in motion in a hydrogen–helium atmosphere on radiant heat transfer was numerically evaluated [38C]. Heat transfer in transitional and turbulent flows along a circular rod was computed [42C].

Several studies dealt with the fundamentals of turbulence and turbulence modeling including one which presented a cascade model of redistribution of energy and temperature fluctuations over the spectrum in two-dimensional flows [20C] and one which discussed the transformation of energy within the boundary layer [65C]. Another presented a three-parameter model of turbulence for use in heat transfer calculations [45C]. A first integral of the equation relating mean temperature and heat flux was obtained which permits relating moments of the heat flux distribution to derivatives of coefficients in the mean temperature representation within a thermal mixing layer [44C]. The structure and development of the temperature field behind a line source in grid turbulence was evaluated [64C]. Molecular diffusion and viscosity had an important influence on temperature fluctuation and, it was shown, must be specifically modeled. A solution algorithm was developed for calculation of turbulent reacting flows [69C]. Calculations were made of a confined turbulent diffusion flame.

Two papers described flows with coherent structures. In one, a three-dimensional and time-dependent model was used to study the non-linear interactions between thermal convective motions, rotation and imposed flows with vertical shear [26C]. The effects of shear on convection produce longitudinal rolls aligned with the shear flow. In a second, measurements in a buoyant plume and vortex pair formed above a heated wire were made [61C].

Management of the flow was the theme of several papers. In the first, measurements of Reynolds analogy factors were made for boundary layer flows altered by stacked arrays of large-eddy breakup devices or 'turbulence manipulators' [43C]. The data show that heat transfer is sensitive to flow history. In the second, experimental results and methods for calculating heat exchange with cavitating turbulators of various geometries were presented [7C]. Heated turbulent air jets were used to study the enhancement of convection heat transfer to the stagnation region of a circular cylinder [25C]. Results show that increases in stagnation point heat transfer are more susceptible to free-stream turbulence than indicated in earlier investigations. Temperature distributions and heat transfer rates were measured for helical ribbon impellers in vessels [57C]. A critical impeller speed for eliminating temperature gradients was established.

A series of papers included mass transfer. In one, the results of laboratory studies of the heat and mass transfer from a heated evaporating liquid into a turbulent unstably stratified impinging cold air flow were presented [74C]. A formula is presented for Stanton numbers with air over smooth and wavy surfaces. Experimental results for mass transfer at the base surface of an open cylindrical cavity and a cavity with a constricted opening (Helmholtz resonator) were presented [63C]. There were two local maxima with increasing cavity depth; the first is due to reattachment and the second may be due to fluid-dynamic oscillations. Deposition rates of submicrometer particles on ribbed surfaces were compared with empirical formulation for heat transfer under similar conditions [53C]. Caution was advised when using this technique.

Gas turbine heat transfer was the topic of the following group of papers. Time-averaged heat-flux distributions obtained for the blade of a full-stage rotating turbine were presented [13C]. In a continuation paper, a detailed description of a technique using thin-film heat flux gauges for obtaining time-resolved heat flux was given [15C]. The magnitude of the heat-flux fluctuation resulting from vane–blade interaction was large by comparison with the time-averaged heat flux. Detailed measurement of heat-flux distributions for the nozzle guide vane airfoil, the hub and tip endwalls and the blade for a low-aspect-ratio turbine stage were presented [14C]. Results were compared with results of local flat-plate prediction techniques. Heat transfer measurements were performed along a cooled flat plate with various free-stream turbulence levels, pressure gradients, and cooling intensities [54C]. Results of an experimental study of the local heat transfer on the end surface of a gas turbine stator were presented [37C]. A single similarity equation accounting for the influence of streamline curvature, three-dimensional nature of the flow, flow acceleration and flow laminarization was obtained.

Several papers dealt with fins, finned surfaces and heat exchangers. A method for a combined convection and fin conduction solution was described [21C] and heat transfer from an array of parallel longitudinal fins to a film passing through the interfin spaces was analytically and experimentally studied [32C]. Results of an experimental investigation of convective transport at the surfaces of a plane transverse fin array exposed to steady turbulent flow were presented [68C]. The effect of yaw on heat transfer from a finned circular cylinder was measured [55C] and laminar flow heat transfer was computed for a situation where fluid moves along a parallel plate channel with one wall insulated [10C].

Papers on manufacturing processes included one in which similarity solutions were developed for heat transfer on a continuously accelerated sheet extruded in an ambient fluid of lower temperature [1C]. Melt-spinning and the cooling of extruded metal plates are applications. Also, heat transfer rates occurring in the laminar boundary layer on a continuous moving surface which has arbitrary surface velocity and temperature were computed [31C] and the conjugate problem was addressed [36C]. Results of a laminar flow, micropolar boundary layer study in axial flow along a continuous moving fiber were presented [24C].

Several papers were presented which deal with heat transfer on the space shuttle and space planes [2C, 6C, 48C, 67C]. One related paper discussed the shuttle's hot-gas jets for ice suppression [60C] and another the carbon ablation on a re-entry body [30C].

Studies on fluid property variation effects included one on the influence of temperature-dependent fluid properties on laminar boundary layer wedge flows [27C] and another where a method was suggested for calculating subsonic gas flows with great temperature drops [47C]. A theoretical analysis was proposed for heat transfer from external surfaces immersed in non-Newtonian fluids where integral equations were transformed into characteristic equations which can readily be solved [50C]. An asymptotic expression was derived for a speedy estimation of heat transfer to non-Newtonian fluids [51C]. The motion of a sphere in a rarefied gas was considered focusing on the role of temperature variation in the Knudsen layer [40C]. Results show that velocity and pressure in the layer and the drag on the sphere are not sensitive to this variation. Molecular–kinetic methods were used to investigate the heat transfer on the surface of a spherical particle for arbitrary Knudsen numbers and temperature differences [23C]. An investigation of a reaction wave igniting initially cold matter, with allowance for the physical processes occurring in a completely ionized medium, was presented; electron heat conduction, radiative losses and energy transfer between electrons and ions were included [41C]. An analysis was presented for heat transfer characteristics of a thermo-micropolar fluid flowing through a channel followed by a convergent or divergent section [4C]. An increase in wall mass transfer was reported for liquid–gas flow of drag-reducing fluids [34C].

Studies of flows influenced by external effects include an experimental study on heat transfer to a single drop translating in an immiscible liquid on which an electric field was imposed [33C]. A remarkable enhancement of heat transfer was obtained due to induced circulation. Heat transfer within counter-oscillating slugs of fluid was examined where the axial heat flux pulsates at twice the base oscillation frequency [39C]. Under tuned conditions, heat transfer was orders of magnitude larger than that present in the absence of oscillations. Unsteady laminar boundary layers contiguous to self-similar flows were studied [16C]. When there is no heat transfer through the wall, the boundary layers are self-similar. The applicability and implications of the finite velocity of transport hypothesis are examined utilizing the large eddy interaction model [29C]. The model explains the manner in which turbulence relaxes following sudden curvature removal. The effect of surface roughness on adiabatic wall temperature was investigated [28C].

Among papers concerned with flow influenced by gravity was one where wave numbers were calculated for axisymmetric Rayleigh–Benard convection [9C] and one in which a continuous transformation algorithm was shown to be appropriate for the solution of a turbulent, buoyant jet [72C]. The heat transfer characteristics of laminar combined convection from an isothermal sphere were numerically evaluated [73C]. The average Nusselt number, the location of separation, the drag and the flow and temperature fields were presented. Transport processes in turbulent liquid films undergoing heating or surface evaporation were analyzed [49C]. Correlations similar to those used in conventional internal and external flows were recommended.

FLOW WITH SEPARATED REGIONS

Within this category, there were many studies dealing with cylinders in crossflow. In one, the Navier–Stokes and energy equations for laminar flow were solved by expressing temperature as well as the stream function in truncated Fourier series [8D] and, in another, the unsteady flow was numerically evaluated [21D]. Experimental results were presented for heat transfer around rectangular cylinders [7D]. The heat transfer on the rear face increases when the side face is lengthened to the point where the separated boundary layer reattaches. Heat transfer from a cylinder in crossflow was measured—a dummy cylinder and an adjacent wall were unheated and the flow near the wall was laminar, then turbulent [24D]. The clearance between the cylinder and the wall was varied. Results of an experiment of heat transfer characteristics of two elliptic cylinders placed in a tandem arrangement were presented [19D]. Their angles of attack were the same in magnitude, but opposite in sign. In a similar test, heat transfer characteristics of two elliptic cylinders in tandem were evaluated [20D]. It was found that angle of attack and separation distance were important. Mass transfer coefficients were evaluated for a cylinder in crossflow where the cylinder had a step in diameter in the center of the test section [14D]. Heat transfer rates around a circular cylinder above a plane boundary were measured [1D]. The mean Nusselt number reached a maximum when the separation distance was equal to about one cylinder radius. Experimental results were presented for heat transfer from a cylinder under cavitation condition [13D]. The heat transfer on the down-flow part of the cylinder surface almost doubled for the cavitation case over one without cavitation. The effect of a relatively high-frequency audible sound field on cylinder heat transfer augmentation was measured [1D]. Acoustic streaming was shown to augment heat transfer; a critical level of 148 db was found. In another study, it was shown that acoustic streaming increases heat transfer mainly on the speaker side with the effect depending more on the streaming velocity than on sound pressure level [11D]. The effect of sound fields on heat transfer from a flat plate having separated and reattaching flow was measured [4D]. Increasing the sound level decreased the length of the separation bubble. The effect of vibrations on local and average heat transfer was evaluated [9D, 10D]. Vibration enhances heat transfer and leads to transition at lower Reynolds number, as does free-stream turbulence. An experimental technique for obtaining local heat flux was presented and data were discussed for heat transfer from a cylinder, cylinders in tandem and cylinders in a tube bank [2D]. For tube banks, the average heat transfer coefficient became invariant by the third row. In a similar study, heat transfer over smooth bundled tubes in transverse flow was discussed [25D]. Numerical results were presented for laminar flow and heat transfer in staggered tube banks and a new expression was proposed for predicting the average heat transfer [5D]. The effects of transversal and longitudinal pitches and the effect of tube roughness were experimentally evaluated [17D, 32D]. Tube roughness has a significant effect. Measurements of heat transfer around a two-dimensional blunt body showed an order of magnitude increase across a small separation bubble, when one exists, where the leading edge joins the body [3D]. Applications include turbomachinery airfoils. Heat transfer rates in the stagnation region of the junction of a circular cylinder perpendicular to a flat plate were measured [6D]. The influence of the cylinder on the flat plate extended beyond three-fourths of a cylinder diameter. Experiments were performed to document the effects of vibration and rotation on heat transfer from spheres [15D]. The authors concluded that air velocity fluctuations and rotational movement of particles on heat transfer cannot be neglected in gas–solid two-phase flows.

Several studies dealt with an abrupt change in surface geometry which leads to separation. In one, heat transfer characteristics of forced convection over an inclined flat plate of finite width were investigated [18D]. They depend considerably on the shapes of the leading edge and the angle of attack. The influence of an unheated frontal end of a cylinder on its heat transfer in longitudinal flow was studied [27D]. Differences were found in the zone of flow separation. The effects of free-stream turbulence and of separation angle on heat transfer characteristics in the reattachment region behind a backward-facing step were investigated [16D]. The effect of turbulence varied with angle. Experimental results were presented for turbulent channel flow with an abrupt area expansion [26D]. The heat transfer characteristics were influenced by the large temperature gradient between the bubble and the main flow in the upstream separated region. A numerical study of this problem used a full Reynolds stress model for the Navier–Stokes equations and an algebraic model for the thermal fluxes [22D]. In a similar study, Mach 2.6 flow passed over a back-step—the compressible effects were documented [23D]. Heat transfer with compressible flow and separation in the presence of a step-cavity profile of the divergent part of the Laval nozzle was presented [28D]. The influence of a protrusion of the surface was measured [31D]. The recovery of equilibrium flow was not reached during the relaxation process after reattachment.

An equation presented for flow in a bubble column was used to estimate both the blending in the liquid phase and the heat transfer on heated or cooled surfaces of bubble columns [29D, 30D].

HEAT TRANSFER IN POROUS MEDIA

A very large number of papers dealt with computer solutions of buoyancy-induced flow in saturated porous media considering various geometries, boundary conditions, for Darcy flow and for conditions where inertia effects cannot be neglected. This section of heat transfer in porous media will be organized in the way that papers studying detailed transfer processes are collected in the first section whereas the following sections enumerate briefly specific geometries and boundary conditions.

Multiple solutions for buoyancy induced flow in saturated fluids are discussed in refs. [1DP, 2DP] and the path through bifurcation to chaos occurring with increasing Rayleigh number is considered in ref. [3DP]. Thermodynamic modeling [4DP] using irreversible thermodynamics establishes that Brinkman's law is justified. Effects of flow inertia are considered in ref. [5DP]. Effective heat transfer parameters developed for transient flow [6DP] are claimed to unify all previous solutions. A boundary layer approximation for natural convection in stratified porous media heated from the side [7DP] is compared with numerical calculations. Radiative and conductive heat transfer [8DP] describe heat transport in unsteady high temperature gas streams. Impulsively heated fluid streams percolated through fixed beds [9DP] spread in the form of temperature waves for air flowing through a bed of glass beads but as steepening waves in the flow of CO_2 through a bed of lead beads.

Conjugate heat transfer from vertical fins to a saturated porous medium are discussed in refs. [10DP, 11DP]. Analyses are also presented for steady and unsteady heat transfer from vertical plates [12DP, 13DP] and vertical cylinders [14DP, 14aDP]. Vertical and horizontal annuli are treated in refs. [15DP–17DP]. Analyses of vertical and horizontal channels are presented in refs. [18DP–21DP]. Free convection in an undulating porous cylinder [22DP] and the stability of natural convection flow are analyzed in refs. [22DP, 23DP] as in the free convection in a two-dimensional porous loop [24DP]. Benard convection is studied at various Prandtl numbers in refs. [25DP, 25aDP]. Geothermal applications suggested the analysis of thermally-driven shallow cavity flows [26DP]. Free boundary flow through a porous body and flow along a heated horizontal cylinder [27DP, 28DP] yielded to numerical calculations.

Much smaller is the number of papers dealing with *unsaturated porous medium*. Unsteady heat and mass transfer creates convective flow in a porous slab [29DP]. Accumulation and migration of moisture in insulations occurs by simultaneous heat and mass transfer [30DP]. An analytical paper is concerned with Soret and Dufour effects and double diffusion [31DP]. The validity of experimental models is investigated by an analysis [32DP]. Unsteady simultaneous heat and mass transfer with phase change occurs in three successive regimes [33DP]. At first the condensate is immobile and accumulates with time, then it becomes mobile generating unsteady concentration and temperature profiles, and finally the situation approaches asymptotically a quasisteady condition. The role of capillarity is demonstrated [34DP] to enhance heat transfer in partially-saturated systems by an order of magnitude. An analysis considers natural convection of gas/vapor mixtures in a horizontal rectangular channel [35DP]. Heat transfer between a granular structure and liquid 3He is studied by a microscopic calculation [36DP]. The hydrodynamic equations describing thermal convection and non-linear effects of superfluid 3He–4He mixture are derived [37DP] and two-fluid effects are calculated.

Considerably smaller is the number of *experimental investigations* which are looking again at different geometries and boundary conditions of porous media. Papers in this field deal with transient and steady conditions in coaxial vertical cylinders [38DP, 39DP], in horizontal cylinders [40DP], in stratified porous media [41DP], in complex internal geometries [42DP], adjacent to a horizontal surface submerged in a porous medium saturated with water at the density extremum [43DP]. Calculations and experiments on a horizontal annulus filled with two concentric layers of glass balls of different size [44DP] resulted in the temperature and moisture field and a relation of Nusselt number as a function of Rayleigh number. Heat transfer in a mixed particle bed formed by two sizes of glass beads was measured in ref. [45DP]. An explanation of the formation of channels in debris beads during boiling of a liquid is offered in ref. [46DP]. Dryout in a bed of inductively heated steel balls of various sizes was measured [47DP] at various pressures. The temperature distribution and the freezing front movement of liquids filling the voids between glass and aluminum spheres was measured [48DP]. A simultaneous analysis gave good agreement for the glass bed but not for the aluminum bed. The heat transfer performance of a trickle flow of gas through regularly stacked packing was measured [49DP]. The results served to develop a model based on single particle flow.

Experimental investigations dominate in the field of heat transfer in *fluidized beds* indicating the situation that no general analytic approach has yet been developed. The papers in this field deal with heat transfer to emersed horizontal tubes [50DP–52DP], with horizontal tube banks [53DP, 54DP], with vertical tubes [55DP, 56DP], and with the characteristics of vertical thermal and capacitance probes [57DP]. New experimental data of the residence time of solid particles close to submerged heating surfaces have been presented [58DP]. The following correlation for heat transfer between immersed and fluidized beds of large particles was obtained from experiments [59DP]

$$Nu = 5.95(1-\varepsilon)^{2/3} + 0.55 Ar^{0.3}\, Re^{0.2}\, Pr^{1/3}$$

in which ε indicates the porosity and Ar the Archimedes number. Wall-to-bed heat transfer measurements [60DP] resulted in data for the effective thermal conductivity and a Colburn type equation describing the heat transfer coefficient [61DP]. The results of heat transfer measurements in two- and three-phase slurry fluidized beds [62DP] demonstrated that the surface renewal model satisfactorily predicts heat transfer. Two analytic papers developed a continuum model

[63DP] for heat transfer in particulate flows and a model for fluidized-bed drying of granular solids [65DP]. A correlation for the local Nusselt number as a function of Grashof number based on previously published data describes heat transfer for flowing packed particle beds in circular tubes [64DP].

EXPERIMENTAL TECHNIQUES AND INSTRUMENTATION

A method was described for linearizing the outputs of standard thermocouples, involving an analog fourth-order polynominal curve-fitting circuit [44E]. An improved design was presented for a thermocouple rake for use in combustion studies [14E]. A comparison was reported of the frequency response of intrinsic thermocouples and encapsulated, beaded thermocouples in liquid sodium flows [27E]. The distortion in the steady two-dimensional temperature field which exists around a surface thermocouple was investigated [3E]. Thermoelectric instabilities in nickel-based thermocouples in the mineral-insulated metal-sheathed format were investigated up to 1100°C [6E].

Thin-film gauges enable the measurement of surface heat transfer rates [17E]. A thin-film gauge was developed which allows measurements of heat flux fluctuations up to 100 kHz [18E]. A thin-film electrical resistance heater embedded in an isolated copper cone was used with a differential thermocouple to measure heat flux from a surface with uniform temperature [5E].

A two-color pyrometer was developed for measuring the surface temperature of metal chips formed during high-speed machining [26E], and a radiation pyrometer with optical fiber was described for measuring the grinding temperature in the surface layer of a workpiece [50E]. The temperature of microscopic areas within a head/tape interface can be measured using an infrared radiometric technique [22E]. Radiometers for measuring radiative heat fluxes in large, sooty pool fires were designed to withstand severe environments and to keep the sensing area free of soot [32E]. The accuracy of a new optical fiber thermometer was demonstrated by measuring the freezing points of gold and silver [16E]. A fiber-optic thermometer based on the change in absorption characteristics of a doped glass plate was demonstrated [21E].

Raman scattering was used to measure the translational temperature as well as the molecular velocity and static pressure of nitrogen in a supersonic wind tunnel [19E]. Pulsed Raman scattering from a turbulent hydrogen jet diffusion flame was used to determine zonal averages, intermittency, and conditional probability density functions for temperature and other quantities [37E]. A method was reported using coherent anti-Stokes Raman spectroscopy to measure temperatures in the near-wake recirculating flow of a bluff-body-stabilized diffusion flame [47E]. An interferometric optical fiber sensor was constructed using a highly birefringent monomode optical fiber sensing element, and was demonstrated for measuring furnace temperatures [25E]. Passive uhf emission was used to measure electron temperature in solid-fuel MHD generators [20E]. A review with 131 references was presented on modern developments in flow visualization [43E].

There is considerable interest in instrumentation for cryogenic heat transfer, particularly in connection with superconducting magnet applications. The stability of 25 Chinese germanium resistance thermometers on thermal cycling between 20 and 239 K was reported, along with their thermometric properties in the range 1–28 K [7E]. Another report investigated the performance of Russian germanium resistance thermometers [8E]. The properties of some RuO_2 thick-film resistors were studied for their use as thermometers at temperatures in the range 0.015–4 K and in magnetic fields up to 5 T [9E]. The resistance and magnetoresistance of thick-film chip resistors was measured in the range 0.015–80 K in magnetic fields up to 20 T [30E]. A thin-film platinum resistance thermometer was investigated for temperature measurements in the range 20–300 K in a 5 T magnetic field [24E]. A glass thermometer and a nuclear orientation thermometer were compared for operation at temperatures below 40 mK in magnetic fields up to 6 T [40E]. It was shown that for carbon resistance thermometers at very low temperatures, the temperature dependence is often well represented by a 1/3-power law [46E]. In the temperature range 77–150 K, Eu^{3+}-doped fluoride glass can be used as a thermo-optic transducer because its absorption at 2.2 μm is temperature-dependent; accuracy of 0.5 K was claimed [33E]. An apparatus was reported for measuring the thermal conductance of pressed brass contacts at liquid helium temperatures [41E]. A four-terminal a.c. bridge was designed for use with low-temperature resistance thermometers [10E].

Ultrahigh pressures present a particular challenge for temperature measurement. A thermometer based on ruby R-line luminescence was shown to be useful in diamond-anvil cells, for temperature measurements in the range 10–100 K and pressures up to 120 kbar [52E]. A rhodium +0.5% iron–chromel thermocouple pair was demonstrated for measuring temperatures in a high pressure cell: it was used in a beryllium–copper cell with pressures up to 1.5 GPa, over the temperature range 4–300 K [23E].

A simple analog thermistor thermometer, using switched bridge components chosen to keep bridge sensitivity constant, was described for temperature measurements in the range 0–100°C [38E]. A diode bridge sensor for temperatures in the range 25–100°C was fabricated and tested [1E]. Response time correlations for industrial temperature sensors were reported [12E]. A method was devised for compensating for the disturbance made by a heat flow meter by placing a separate surface heater on the meter surface [15E]. The errors in surface-mounted heat flux sensors for buildings were analyzed [48E].

In miscellaneous instrumentation developments, a platinum wire probe for measuring temperature turbulence in the atmospheric boundary layer was described [31E]; a survey of the circular and square edge effects provides the information necessary to choose the appropriate apparatus size for a measurement of thermal resistance of a specimen of insulation material of arbitrary thickness [35E]; an encapsulated thermochromic liquid crystal, in a packaging arrangement suitable for gas flows, was described for mapping local heat transfer coefficients [4E] (there was surprisingly little work on liquid crystals reported in 1986); a calorimeter was developed for the purpose of evaluating thermodynamic and kinetic properties in chemical reactions associated with manufacturing processes [45E]; and the analogy between heat and mass transfer allows for the measurement of convective heat and mass transfer coefficients based on a wet-bulb thermometer exposed to an imposed heat flux [28E]. Contact and gap resistances in extruded bi-metallic finned tubes were measured by a new technique [13E].

A number of new experimental techniques were developed for measuring properties associated with heat transfer. An apparatus was designed for measuring the thermal conductivity of high-temperature materials such as stainless steel and molybdenum, over the range 400–1100 K [36E]. The thermal conductivity of liquids was measured by means of a new hot-wire instrument [11E]. A technique was developed for measuring the thermal diffusivity in one of the layers of a two-layer sample by imposing a triangular heat-pulse, as might be obtained from an Nd-glass laser [53E]. A data-acquisition system was discussed for measuring thermal diffusivity and propagation properties of thermal waves, based on the phase shift and amplitude ratio of the temperature oscillation at two fixed points [42E]. A device for *in situ* use in solar collectors was described, which measures the mass flow–specific heat product [39E]. A laser pulse method combined with a transient calorimetric technique was developed for measuring the total hemispherical emissivity of metals; the system was demonstrated for platinum in the range 1200–1600 K [34E]. Laser-pulse calorimetry applied to determining the thermal conductivity of molten oxides at high temperature was reviewed [51E].

Finally, new experimental techniques were described for measuring the thermal conductivity of oyster shells [49E] and the conductivity of the heavy oil sands of Trinidad [29E]; and a method was described for measuring the thermal diffusivity of spherical produce [2E]. It was applied to apples, oranges and potatoes.

NATURAL CONVECTION—INTERNAL FLOWS

Buoyancy driven flows are of interest to researchers from a number of disciplines including Applied Mathematics, Physics, and Engineering (both Science and Applied). Non-linear phenomena, transformation to chaos, application to geological and astrophysical systems, as well as more mundane problems continue to draw attention as indicated by the large number of studies published each year. In the present section, attention is directed to buoyancy driven flows in enclosures. These include horizontal layers heated from below and effects in layers of finite horizontal extent; flows in two-fluid systems where one layer of fluid is superimposed on another; double diffusion; vertical slots and other geometries where differential lateral heating is present; heat transfer from one solid body to its surroundings including concentric and eccentric cylinders; convection in porous media; thermocapillary flows; combined forced and natural (mixed) convection; and other systems as well.

In a horizontal layer heated from below, the low Rayleigh number flow has been predicted using a finite element method [85F]. Another finite element study for low Rayleigh numbers indicates the influence of Prandtl number on the flow [16F]. The stability of a layer with linear and non-linear temperature distributions was determined using a technique developed for optimal control problems [71F]. Non-linear three-dimensional instability has been analyzed for constant heat flux boundary conditions [35F]. In non-linear convection with a variable coefficient of thermal expansion square cells transport more heat than two-dimensional rolls, and rolls are unstable [98F]. A number of flows were studied for fluids with high Prandtl numbers [36F]. The onset of convection in a viscoelastic fluid depends strongly on the particular constitutive relationship that characterizes the material [100F]. New oscillatory instabilities are found which indicate the presence of blobs of hot and cold fluid circulating in convection [20F]. Asymptotic properties of a bifurcation sequence that can generate random flow modes were analyzed and indicate periodic solutions which become unstable [122F].

A numerical simulation model for turbulent flow was applied to thermal convection [121F]. Another model developed for high Rayleigh number flow predicts horizontal motion of plumes [50F]. A scalar model for turbulent energy transfer can be used to predict Nusselt number at high Rayleigh number [74F].

Laboratory experiments to simulate convection in the atmosphere indicate the dynamics of the region underneath a stable or linear stratified layer [70F]. Experimental results for high Rayleigh number unsteady turbulent convection yield correlations for the mean temperature field in terms of wall-layer scales and convection scales [2F]. A new measurement technique has been utilized [30F] to study the evolution of a time-dependent convective flow.

Using a Monte Carlo simulation the dependency of the onset time for convection to start in a suddenly heated layer has been developed [62F]. Two-dimensional rolls and hexagonal cells are found to be stable in a fluid layer with internal energy sources [99F]. An electrochemical cell, in which a porous horizontal membrane separates the fluid into two layers, has been used to simulate heating from below [55F]; at high Rayleigh number, the membrane reduces the Sherwood number [54F].

A conservative lower bound for buoyancy driven convection in bounded geometries show inconsistencies in some earlier published numerical results [87F]. Flow patterns in a rectangular container heated from below, have been studied over a range of Rayleigh numbers and Prandtl numbers [68F]. An analysis uses a realistic side wall boundary condition to show differences in the flow patterns with the horizontal extent of a container [81F]. A three-dimensional numerical solution was obtained using a k–ε model for flow in a cube heated from below and cooled on one vertical wall [93F]. The influence of lateral walls on convection with a micropolar fluid [56F] is different from that found earlier with a Newtonian fluid. An experimental study on the random oscillations in a fluid-filled cubical cavity is related to periods of solar activity [18F]. Flow visualization and heat transfer measurements have been performed for various combinations of heated and cooled vertical and horizontal surfaces in a cubical enclosure [66F].

Visualization of fluid motion in a circular cylinder heated from below shows both axisymmetric and three-dimensional motions at modest Rayleigh number [40F]. Flow in a cylindrical cavity with high frequency vibration in a zero gravity field appears in the form of two vortices [103F]. The temperature and flow distributions have been calculated for a fluid contained in a hemisphere heated from below [5F]. Another study on heat transfer in a hemisphere simulates problems following an accident in a gas cooled reactor [104F].

Measurements of velocity and temperature fluctuations in a horizontal layer with uniform volumetric energy sources compare well with a simple mixing-length analysis [61F].

Coupling of electrodynamic and buoyancy forces has a large impact on heat transfer in an enclosure [88F]. The linear stability of a dielectric fluid under the simultaneous influence of a vertical a.c. field and a vertical temperature gradient was determined using a finite difference method [3F]. The influence of a vertical magnetic field on instability in a circular cylindrical container is such that the critical Rayleigh number increases as the intensity of the magnetic field increases [78F]. Another study on the effects of a magnetic field predicts the velocity and temperature fields in the resulting flow [17F].

The interaction between two fluids, one on top of the other, when heated from below has been considered in several studies. A stability analysis for a two-layer system has been extended [96F] to different Rayleigh numbers. A perturbation method [89F] predicts the onset of flow including the influence of slight curvature around the fluid–fluid interface. Calculations [75F] of the heat transfer across two layers yield the values of the Nusselt number as a function of the Rayleigh number and other parameters. A numerical analysis [63F] includes the effects of Marangoni convection at the interface of two layers. An experimental correlation has been obtained for the heat transfer across two layers heated from below [108F].

Considerable attention has been directed toward convection with double diffusion where the density differences which create and maintain a flow are caused by two different properties of the fluid, generally temperature and composition. With non-uniform temperature and composition, diffusion of heat and mass occur simultaneously. Usually the diffusion of mass is at a much lower rate than the thermal diffusion leading to many interesting and complex phenomena. Analysis and measurement of the onset of flow in a stably stratified fluid heated from below show good agreement [59F]. Two-scale methods are employed to predict the planform in the salt-finger regime of thermohaline convection [95F]. Different mathematical approaches have been compared [84F] for predicting potentially oscillatory flows in thermohaline convection. A correlation has been developed to predict the flow in a salt-stratified fluid heated from below [15F]. The motion in a binary mixture held in a vertical layer heated from the side has been analyzed, including effects of thermodiffusion [107F]. Local heat and mass sources in double-diffusive convection influence the size of the flow cells [97F]. The successive bifurcations for convection in a double-diffusive system are extended to the development of chaos [67F]. A model predicts the transient response of a double-diffusive system, destabilized by radiation adsorption at its bottom boundary [14F]. Salinity stratification in a horizontal fluid layer has a destabilizing effect through coupling with an oscillating temperature field [120F].

Studies of convection in an inclined channel consider the influence of geometric parameters on the flow. The inclination angle has an important role in determining the flow direction in a slanted open rectangular cavity [51F]. At small angles of inclination from the horizontal, the sidewalls of a cavity give rise to secondary flows in the form of stationary transverse rolls with horizontal axes parallel to the shorter side, as opposed to the longitudinal rolls predicted for an infinite layer [80F]. A finite difference analysis for flow in an inclined layer with a partition shows the importance of the location of the partition on the flow and heat transfer characteristics [114F]. An experiment in an inclined layer with partitions shows the effect of partitions on the nature of the flow though they appear to have little influence on the heat transfer across the layer [111F]. The tilt angle and aspect ratio of an inclined layer strongly affect the Rayleigh number for the transition threshold from steady to time varying flow [94F].

Convection in the space between one solid object and a surrounding enclosure have been studied for different geometries; one is for a cylinder, often circular, inside another cylinder; another is the annulus between two spheres; and still others are for a range of different shaped objects in surrounding enclosures. The use of modeling procedures for convection in a horizontal annulus provides good agreement with earlier predictions and measurements [34F]. A special curvilinear coordinate system was developed to study the convection in annuli between concentric and eccentric horizontal cylinders [25F]. Numerical solutions for the flow in annuli between concentric and eccentric horizontal cylinders provide stream lines and isotherms, as well as, average Nusselt numbers for constant heat flux boundary conditions [43F]. An analysis shows that the Boussinesq approximation is useful even at large temperature ratios when predicting overall heat transfer with laminar convection in the annulus between horizontal concentric cylinders [79F]. The use of baffles to decrease the convective heat transfer across horizontal eccentric annuli has been studied over a range of conditions to obtain the optimum baffle orientation [11F].

Linear stability analysis has been applied to the flow in thin spherical annuli with axisymmetric disturbances; different modes of flow are found for different Prandtl number fluids [39F]. Another analysis for convection within thin spherical annuli uses a perturbation method in powers of the relative gap width [118F]. Convective flow in the region between rotating eccentric spheres has been modeled [6F].

Computations and experiments on convection in a circular horizontal cylinder with uniform energy dissipation have been perfomed over a large range of Prandtl numbers and Rayleigh numbers [82F]. Confinement of a heated horizontal cylinder by an adiabatic wall enhanced the heat transfer from the cylinder [58F]. A holographic interferometer was used [106F] to show the influence of plate inclination on heat transfer loss from a plate to a surrounding cylindrical enclosure [106F]. Measurements of the heat transfer from two heated isothermal horizontal cylinders to a surrounding enclosure provide correlations for the overall heat transfer [109F]. Unsteady natural convection from a vertical surface to a surrounding cavity was analyzed numerically [60F]. The Nusselt number for heat transfer between a cavity wall and internal fluid has been measured [123F]. A numerical study indicates the heat transfer between two concentric vertical cylinders of different heights including a case when two immiscible fluids are in the enclosure [86F].

Steady-state two-dimensional results were obtained from numerical calculations for heat transfer in a differentially heated enclosed vertical slot with a sufficiently large temperature difference to provide large variations in thermal properties [26F]. Another study considers variable properties for laminar convection in vertical channels with uniform heat flux boundary conditions [4F]. Natural circulation in a number of parallel channels with heat inputs has been studied to simulate heat transfer in a gas-cooled reactor following loss of coolant [112F]. A semi-empirical model of turbulence was employed to predict convective flow in a differentially-heated vertical layer [116F]. Numerical solutions provide heat transfer results for convection in a vertical channel with corrugated walls [38F]. Flow in a vertical wavy channel was studied using a perturbation technique [115F]. Two-dimensional convection in a rectangular cell with horizontal plate fins was determined numerically over a range of aspect ratios, Prandtl numbers, and fin parameters [91F]. A combined experimental and numerical study describes the heat transfer in a channel with a vent in the form of a slit [10F].

A numerical solution for the quasi-steady natural convection heat transfer in a vertical channel the boundary temperature of which increases linearly with time has been described [76F]. A description of the properties of quasi-steady free convection in a vertical cylinder shows the influence of a number of dimensionless parameters [27F]. Experiments in a vertical trapezoidal enclosure were carried out using water as the working fluid [65F]. The influence of the thermal boundary condition on flow and heat transfer in a vertical cylindrical annulus has been analyzed [77F].

The interaction of vaporization from a thin film running down the walls of a vertical open tube and the natural convection and heat transfer within the tube have been analyzed [24F]. In a vertical differentially heated rectangular channel filled with water near 4°C there is an upward flow on both sides of the enclosure and a downward flow in the middle [72F].

Locally measured velocity profiles were compared with numerical predictions for convection in a differentially heated rectangular enclosure with a free liquid surface [110F]. Experimental results for laminar convection in a differentially heated water filled enclosure were compared with numerical predictions [101F]. The stability of small perturbations superimposed on natural convection motions in a differentially heated fluid layer has been examined using two different numerical methods [22F]. The influence of three-dimensional boundary conditions on the temperature distribution and heat flow in a differentially heated enclosure have been measured [19F]. Experiments using a rectangular container with one vertical wall partially heated and partially cooled show two fluid cells, one above the other [41F]. A numerical algorithm has been developed to obtain three-dimensional solutions for convection in a differentially heated rectangular cavity [117F]. Velocity distributions were measured for convection in rectangular cavities with an applied horizontal temperature gradient [105F].

Numerical solutions for three-dimensional buoyancy-driven flows in differentially-heated horizontal cylinders are relevant to crystal growth by vapor transport [21F]. Surface roughness to enhance heat transfer across cubical cavities has been studied experimentally [7F]. Laser-Doppler velocimetry was used to determine the velocity distribution in a shallow differentially-heated cavity [23F]. Experiments on convection in a rectangular enclosure with a partition extending from a vertically heated wall included flow visualization using dye injection [64F]. An experiment on convection in an enclosure partially filled with glass beads provides information on heat transfer, as well as, flow visualization and temperature distribution [90F].

Thermosyphons have been of interest for many years for their practical application in modeling important flows and for their theoretical interest. Flow reversal has been predicted [73F] in a tilted toroidal thermosyphon. The heat transfer in an enclosed thermosyphon has been determined over a range of inclinations and pressures [44F]. Computation of the flow in a cavity with an internal object includes applications to two-dimensional thermosyphons [1F].

Capillary effects must be considered when there are free surfaces or fluid interfaces and a temperature difference causes a spacial variation in surface tension. These phenomena are often important in thin fluid layers, where the effective surface tension forces can be larger than those due to buoyancy producing what is called Marangoni convection. Marangoni instability in a thin horizontal fluid layer was studied for a non-linear dependence of the surface tension on temperature [31F]. The critical temperature gradient for the formation of hexagonal cells in surface tension driven flow in thin layers heated from below has been measured [69F]. An analysis of the deformation of the interface with thermocapillary convection in a rectangular cavity heated from below has been examined [102F]. Prediction of free boundaries for thermocapillary flow during the growth of a single crystal was determined using a finite element technique [33F]. The characteristics of the flow in a square cavity with simultaneous buoyancy and thermocapillary effects indicate that surface tension can either enhance or reduce the heat transfer rate depending upon the specific input parameters [13F].

Correlation of Nusselt number as a function of Rayleigh number, the radii ratio and the eccentricity has been obtained for high Rayleigh number convection in a horizontal eccentric annulus containing a saturated porous medium [49F]. Numerical calculation of linear stability limits has been described for porous media through-flow [57F]. The influence of non-Darcy effects was demonstrated in a numerical study of convection in a vertical enclosure containing a porous medium [12F].

An area of continuing interest is the study of flows with combined natural and forced convection—often called mixed convection. In these, buoyancy and externally-imposed forces interact to provide, an often rather complex, flow. A numerical study indicates that the hydrodynamic entry length in a vertical parallel plate channel is greatly increased by buoyancy effects [8F]. A related study characterizes the fully developed flow in such a channel, including the conditions under which bi-directional flow occurs [9F]. Another study for a vertical channel provides solutions for temperature and velocity with symmetric heating of the sidewalls, as well as when one plate is heated and the other is adiabatic [46F]. With large temperature differences variable fluid properties can influence the heat transfer and flow reversal in a heated vertical annulus [47F]. A shift in the location of the position of maximum temperature and a smaller Nusselt number are observed for a partially blocked heated vertical channel as compared to a smooth channel [45F].

A numerical solution for mixed convection in the entrance region of a horizontal rectangular channel shows that the Prandtl number influence is dependent on the Rayleigh number range [29F]. Calculations for mixed convection in a horizontal duct containing a power-law fluid indicate dual solutions possible with either two-cell or four-cell patterns [113F]. Experiments on mixed convection in the entrance region of horizontal and inclined rectangular channels provide a correlation of Nusselt number in terms of Rayleigh number, Reynolds number and inclination angle [83F]. A study on mixed convection in a horizontal tube surrounded by a liquid medium includes the influence of natural convection on the outer surface of the tube [32F]. Combined convection in a horizontal concentric annulus shows the effect of a rotating inner cylinder [42F]. Numerical procedures have been developed for studying convection in an inclined open cavity heated from below with a forced flow over it [52F, 53F].

Many studies on buoyancy-driven convection have a direct or indirect reference to applications; a number of these have been cited in the paragraphs above. Here we will mention a few that were aimed very closely to specific needs. The convection characteristics of the gas in a vertical narrow annual gap with its bottom open to a high temperature fluid has been evaluated using a special computer code to predict the flow in an LMFBR [119F]. An analysis shows the importance of oscillations on natural convection in a ship tank [37F]. Convection through doorways in a building has been studied to predict the influence of door height [48F]. A numerical analysis shows the convective flow around a Trombe wall, used in a solar heating system [92F]. Natural convection in liquid gas explosives has been studied to indicate its influence on ignition [28F].

NATURAL CONVECTION—EXTERNAL FLOWS

Natural convection from vertical plates continues to receive considerable attention in both experimental and theoretical studies. A non-similar transformation method was utilized to obtain solutions to flow adjacent to vertical, inclined or horizontal plates with either prescribed temperature or heat flux thermal boundary conditions [8FF]. An analysis is presented in which strong magnetic effects are presented [32FF]. Linear stability theory was used to predict the effects of temperature-dependent viscosity and buoyancy on the stability of the flow adjacent to an isothermal, vertical plate [33FF]. Experiments have been performed to investigate the flow near the maximum density of water [18FF, 29FF]. Matched asymptotic expansions and numerical methods are used to obtain solutions for a vertical heated plate with a room temperature plate attached at an arbitrary angle [25FF]. An experimental mass transfer technique to measure the thickness of a natural convection boundary layer is presented [28FF]. Two-dimensional turbulent natural convection in a vertical slot is simulated using a spectral process [12FF].

Experimental and numerical results are presented for laminar flow on a vertical plate with discrete heated elements [19FF] which includes the effect of conduction in the plate. Pure forced convection and aiding mixed convection are simulated numerically for a vertical, conducting plate fin in a micropolar fluid [23FF]. The coupling of laminar forced convection and external natural convection separated by a conducting vertical wall is considered using elliptic numerical solutions and experiments using air [26FF].

Uniform blowing or suction was modeled numerically on a vertical plate which had two sections of different temperature [20FF]. An analytical solution was obtained for the flow adjacent to a horizontal heated flat plate in which the suction velocity varies in a power law manner [2FF].

A comprehensive set of overall heat transfer correlations was presented for aiding and opposing mixed convection flow adjacent to a vertical flat plate [9FF]. An integral analysis provides results for both aiding and opposing mixed convection in which forced convection dominant and natural convection dominant regimes are identified [35FF]. Numerical solutions of transient mixed convection on a vertical plate exposed to a parallel horizontal flow were obtained which indicate that the forced flow dominates near the vertical leading edge and the natural convection dominates at large values of the horizontal coordinate [31FF]. Aiding mixed convection on a vertical plate with discrete heat sources was simulated using numerical analysis with boundary layer approximations [17FF].

Vortex instability [41FF] and wave instability [42FF] on inclined uniform heat flux plates were analyzed using linear stability theory for a fluid with a Prandtl number of 0.7 or 7. Experiments in air were performed to determine the influence of a parallel adiabatic plate located at various spacing above a heated horizontal plate [36FF] in which the maximum heat transfer occurred when the upper plate was removed. One-dimensional governing equations were solved for flow above a horizontal heated plate in water near its maximum density [11FF]. Numerical solutions were obtained for buoyant flow over an adiabatic horizontal plate with a heated strip in the center [38FF]. Investigations of combined heat and mass transfer in a salt stratified layer above a heated strip indicated that the resulting flow was a combination of Rayleigh–Benard type cells and longitudinal roll cells [5FF].

Experiments in helium showed that natural convection from a horizontal cylinder was affected by the absolute temperature involved [1FF]. A wide range of Prandtl and Rayleigh numbers were covered in a series of experiments of natural convection from a horizontal wire in a viscoelastic fluid [27FF]. The aspect ratio above which the flow from a finite length horizontal cylinder becomes two-dimensional was investigated by comparing three-dimensional numerical solutions at the midplane with corresponding two-dimensional solutions [6FF]. Heat transfer from a vertical array of cylinders in free space or enclosed between vertical parallel walls was investigated experimentally in the boundary layer regime where the Nusselt number was found to remain proportional to the Rayleigh number to the 1/4 power [34FF]. Substantially higher heat transfer rates were observed for natural convection from a horizontal finned cylinder situated in a vertical channel than when the cylinder was suspended in free space [37FF]. The effects of temperature-dependent properties were evaluated for transverse mixed convection about a horizontal cylinder using a finite difference numerical approach [3FF]. Finite element solutions were obtained of transverse mixed convection from a pair of horizontal cylinders located in the same horizontal plane [43FF]. Finite difference solutions were obtained for natural convection from a sphere with blowing or suction with either isothermal or uniform heat flux boundary conditions [24FF].

Measurements including mean velocity, local temperature, turbulence intensity and flow visualization were obtained in a buoyant jet constrained in a shallow layer [4FF]. A self-similar model and corresponding experiments were used to determine the nature of plume flows in viscous fluids that may simulate convection in the earth's mantle [15FF]. A plume adjacent to a vertical wall is studied and the results compared to a free plume [7FF]. An adiabatic wall plume adjacent to a vertical or inclined wall was modeled with a k–e–g turbulence model and the results compared to experimental data [22FF]. New correlations for the temperature distribution of a horizontal adiabatic wall situated above a plume rising from a point heat source were presented [10FF]. Wind tunnel experiments were performed in which argon gas simulated the flow of a liquefied natural gas plume flowing past a circular storage tank [21FF].

Two-dimensional finite difference solutions were

obtained for natural convection about a horizontal rectangular body resting on an insulated base [39FF]. Transverse curvature effects on axisymmetric boundary layer flow of water at 4°C past a vertical cylinder were investigated [13FF]. Two theoretical studies have been made on the flow of a thermally stratified fluid through a heated, vertical open-ended cylinder [16FF, 30FF]. The flow in an infinite vertical slot is less stable when the viscosity is an exponential function of temperature than when it is constant provided $Pr > 100$ [40FF]. The optical heartbeat phenomenon is discussed [14FF] in which a laser beam propagating below a liquid free surface oscillates in intensity due to convective Benard and Marangoni instabilities.

CONVECTION FROM ROTATING SURFACES

An analysis describes the effects of Coriolis force and non-uniform temperature gradient on the Rayleigh–Benard convection in a horizontal rotating fluid [13G]. The Galerkin method was found easy to carry out and to give good agreement with numerical computations.

Steady and transient heat transfer, supported by flow visualization, was measured for natural convection in rapidly spinning containers simulating the conditions in superconducting generators with helium cooling [15G]. A correlation describing the Nusselt number for a wide range of Rayleigh and Prandtl numbers correlated the results. The convection structure, studied for Rayleigh numbers between 10^6 and 10^{11} and Taylor numbers from 10^6 to 10^{12} revealed various flow configurations in the shape of rings and grids. This grid pattern still formed around the vertical rotation axis when the vessel was inclined [2G]. Numerical experiments [8G] of steady axisymmetric flow in a heated rotating shell were performed for a non uniform gravitational field [8G] to simulate an atmospheric circulation experiment planned for the Space Shuttle. A dielectric force is considered to simulate gravity. The flow and temperature fields in the boundary layer of heated and rotating spheres, spheroids and paraboloids were obtained [3G] for fluids with various Prandtl numbers.

Six flow regimes were obtained by a boundary layer analysis for axisymmetric flow in an internally heated fluid through a rotating channel with rectangular cross-section [12G] and the results were compared with experiments. The flow field was measured with a laser-Doppler anemometer in a rotating cylindrical cavity with radial inflow of the fluid [4G]. Heat transfer was also measured and good agreement with the Ekman layer theory was obtained. A similarity analysis of the transient temperature distribution in a partially filled rotating horizontal cylinder shows [1G] that the temperature field is a function of the Fourier number and the bubble eccentricity. A numerical calculation studied convection in the annulus between asymmetric cylinders [5G]. The inner cylinder was rotating and heated. The radius ratio was 2.6 and the eccentricity varied between 0 and 2/3. The Rayleigh number extended to 10^6, the rotating Reynolds number varied between 0 and 1120, and the Prandtl number was 0.7. Thermocapillary and centrifugal-buoyancy-driven motion was analyzed in a rapidly rotating liquid cylinder [14G] as it occurs in liquid crystal formatting. The thermocapillary effect was found to be confined to a thin layer adjacent to the surface for small Ekman number. The buoyancy effect was small but influenced the temperature field in the interior of the cylinder.

Heat transfer from rotating annular fins was studied experimentally in several papers [7G, 9G, 16G].

Local heat transfer in a round jet impinging on a rotating disk was measured [11G] for a range of rotational and jet Reynolds numbers. A finite difference calculation of laminar mixed and free convection over a rotating sphere with blowing and suction [6G] was found to be in good agreement with previous calculations. Experiments with a heated and rotating circular cylinder in crossflow, at Reynolds numbers around 4×10^4 and rotational to approach velocity ratios smaller than 2, established the fact that the location of the separation changes strongly with the cylinder wall temperature [10G].

COMBINED HEAT AND MASS TRANSFER

This section covers a number of processes which play important roles in many heat transfer applications and are often not fully understood in terms of fundamental fluid mechanics and turbulent transport. These include heat transfer from or to impinging jets, film cooling as used to protect solid surfaces over which high temperature gas flows occur, and processes similar to film cooling where mass injection from the surfaces diminishes the heat transport to the solid surface. A few papers related to drying and general combined heat and mass transfer systems are also reviewed.

Numerical analysis of the flow and heat transfer with a circular laminar jet impinging on the bottom of a cavity shows the significant influence of the spent fluid flowing from the cavity counter to the impinging jet [1H]. An impinging laminar slot jet can be used to smooth out the temperature distribution of a high temperature slab by transporting heat from one part of the slab to another [9H]. A study using a liquid jet impinging on the surface of a rotating disk, provides a correlation for the average Nusselt number as a function of three independent parameters [4H]. Measurements of the heat and mass transfer to an air jet impinging on a water surface indicate that the dimensionless transport coefficients are lower than those for impingement on a solid surface [19H].

With film cooling a fluid, usually a gas, is introduced at discrete locations on a surface over which a hot gas stream is flowing. The injected fluid (coolant) decreases the temperature in the boundary layer and thus the heat transfer to the solid surface. Numerical analysis procedures have been used to calculate the influence of film cooling using a row of injection holes [5H]; the results generally agree with available measurements. With injection of steam through one or two rows of inclined holes, a high film cooling effectiveness is found, apparently due to the relatively high specific heat of water vapor [8H]. With full coverage film cooling (in which there are a large number of injection holes distributed on the surface) there is a considerably higher heat (mass) transfer on a concave wall than on a convex surface [15H].

Several studies examine film cooling as it applies in gas turbine combustors. Measurement of the heat transfer on the approaching flow for film cooling injection through a small hole into a gas turbine combustion chamber shows the importance of the heat transfer on the approaching side of the flow [3H]. A simple model has been developed [20H] to predict the influence of slot film cooling as required to protect gas turbine combustors. In a companion study [21H] flow visualization and measurement show the importance of the flow within the slot in obtaining high film cooling effectiveness.

Several studies examine the influence of flow through a porous wall on the heat transfer to a fluid flowing over it. The influence of acceleration and mainstream turbulence on the heat transfer in a short porous tube with injection has been analyzed [11H]. The influence of variable physical properties on heat transfer in a tube with porous walls is much greater in the case of injection through the walls, than in the case of a simple mainflow [6H]. Similarity solutions have been obtained to predict the heat transfer to flow passing over a continuously moving porous plate [18H]. Combined sublimation and heat transfer to a moving permeable wall have been analyzed [7H]. Finite difference methods have been used over a range of Reynolds numbers to predict heat transfer for hypersonic viscous flow on a rotating axisymmetric porous body with fluid transpiring through the walls [13H]. Heat transfer characteristics for forced convection flow of a micropolar fluid over a porous sphere through which fluid transpires has been studied [17H].

An analytical model has been used to predict the high-intensity drying of paper adjacent to a hot surface [2H]. Another analysis predicts the mass transfer to jets used in the drying of paper [14H].

Two-phase combined heat and mass transfer has been studied for counter flowing films of liquid and gas [10H]. Heat and mass transfer to falling laminar films are analyzed to predict the absorption of water vapor in a lithium bromide solution [16H]. Computer experiments on the combined heat and mass transfer in laminar flow of a tubular polymerizer shows a strong influence on non-linear phenomena [12H].

CHANGE OF PHASE—BOILING

Nucleate boiling

The motion of liquid around a vapor bubble in saturated nucleate pool boiling was observed to agree with cited predictions and the consequent stirring effects on heat transfer agreed well with data [85J]. Bubble frequency increased but bubble departure size decreased as pressure was increased in the range of 20–129 kPa for pool boiling of liquid potassium [103J]. Numerical predictions and experimental observation addressed the behavior of a moving vapor bubble subjected to a step change in pressure [111J]. The numerical analysis was able to describe the fragmentation that occurs under some conditions as the bubble implodes. Measurements were made of the superheat at the onset of nucleate boiling in a narrow rectangular channel simulating a subchannel of a fuel element [115J]. Theory and experiment were shown to agree that small gas bubbles may persist within a liquid for extended periods of time in a stable thermodynamic state [131J]. A mathematical model of the influence of an electric field upon the departure size of nucleate boiling bubbles agreed with the experimentally observed trend of reduced bubble sizes with stronger fields and a greater liquid dielectric constant [17J]. Increasing the a.c. electric field intensity in two dielectric liquids led to increasing nucleation site density and smaller bubble departure diameter, while the electric breakdown voltage of these liquids was found to decrease when bubbles were produced at the heating surface [30J]. Liquid side mass transfer resistances were shown to be sufficient to explain reductions in heat transfer and selectivity (separation effect) of boiling binary mixtures [37J]. The augmentation of boiling heat transfer by capillary structures on the heating surface was argued to be due to the reduction of the required superheat for bubbles in equilibrium when they are in cavities [57J]. The characteristics of saturated and subcooled nucleate boiling at and near impingement of submerged jets were explored with refrigerant 113 [67J]. As heat flux was increased, boiling curves produced with different jet velocities tended to merge into a fully developed boiling asymptote which was nearly matched by the extrapolation of pool boiling curves. High-speed cinema studies of bubble departure diameter, growth rate and site density were described in the cylinder–liner cooling space of a diesel engine [107J].

Pool boiling

Measurements were made of macrolayer frequency and formation thickness with various large heat fluxes in pool boiling which compared well with analytically predicted values [11J]. The findings suggested that a large portion of the wall heat flux is transferred by conduction through the macrolayer. Pool boiling from planar surfaces in refrigerant 11 was shown to improve dramatically with surface enhancements which, unfortunately, also produced large temperature overshoot at boiling incipience [52J]. Coefficients of heat transfer generally increased slightly with the angle of inclination relative to the horizontal, up to roughly 165° for moderately low heat fluxes, while no noticeable changes occurred with inclination at higher heat fluxes. Considerable enhancement of heat transfer was also found with roughened surfaces in the transition regime [125J].

A model proposed to correlate heat fluxes to liquids boiling in porous (wick) materials was favorably compared with experimental findings for water, acetone, ethanol and refrigerants 11 and 13, and exhibited proportionality between heat flux and the square of the temperature difference [88J]. The onset of boiling was examined for wick-covered heating surfaces in pool boiling of water at pressures from 0.02 to 0.5 MPa [123J]. Flooded wicks began boiling at lower superheats than plain surfaces, greater wick thicknesses requiring smaller superheats. Pool boiling experiments with water, acetone, and a water–acetone mixture were conducted with varying depths of loose 0.2 mm steel particles on a horizontal heating surface [98J]. Particle layer enhancements of heat transfer coefficients were greatest for the pure liquids when the layer was 6–10 mm thick, while enhancements of mixture boiling, inhibited by mass transfer limitations, peaked at a depth of only 1.5 mm. Experiments with plasma-deposited particle coatings on horizontal cylinders were conducted to explore the effects of particle diameter [97J]. For all fluids tested and for all particle sizes, the enhancement of heat transfer declined with increasing heat flux. Heat transfer coefficient improvements relative to uncoated surfaces for water–alcohol and water–acetone mixtures were small or negative, especially at an intermediate particle size around 0.05 mm. Improvements tended to be larger with alcohol, acetone and their mixtures, monotonically increasing with decreasing particle size to the smallest studied, 0.035 mm.

Pulsating pressure and temperature observed in an earlier experiment with stepwise transient heating of a narrow channel open at one end were analyzed, matching the pulsation period well by modeling the bubble growth and collapse process in the fluctuating pressure field [94J]. Experiments were performed on boiling in a narrow vertical rectangular channel heated from one side with artificially injected bubbles [72J]. Heat transfer coefficients were found to vary with the inverse square root of either the period of bubble injection or, at higher heat fluxes, the period of bubble cluster motion. A model was developed to predict the extent of a dry patch produced in a narrow eccentric annulus and the corresponding temperature distribution, applicable to the crevices between tubes and tube support plates in PWR steam generators [51J]. Atmospheric and subatmospheric pressure potassium pool boiling experiments were described along with a model for the minimum heat flux required to produce an alternating dry patch boiling regime [69J]. The effects of oil in a mixture with refrigerants upon boiling heat transfer on a fine horizontal wire were explored experimentally [74J].

Flow boiling

Bounding values of liquid and vapor superficial velocities separating flow regimes in horizontal boiler tubes according to their ability to preclude the anisothermal conditions which could lead to early tube failure were analyzed and compared with data from several sources [8J]. A new correlation was proposed for forced convection boiling, based on data of water, refrigerants and ethylene glycol, for saturated and subcooled boiling in tubes and annuli [38J]. A summary of current knowledge of heat transfer in saturated flow boiling of cryogenic fluids in vertical and horizontal tubes was presented [114J]. Influences of quality, mass velocity and heat flux on flow boiling heat transfer coefficients for argon in a horizontal tube were reported [76J]. Analytical and experimental studies were reported for forced flow of He II along with the establishment of a condition for two-phase flow to occur in a transfer line [22J]. Flow boiling and CHF data were generated for mixtures of refrigerants 11 and 113 and it was determined that a linear mixing rule for composition was adequate for estimating critical quality but not heat transfer coefficients [80J]. Based on the similarity with gas–liquid adsorption processes, a model was proposed for forced convective annular evaporation (with nucleate boiling suppressed) [21J]. A Reynolds analogy approach to subcooled convective boiling was presented, permitting estimation of bubble size and hydraulic drag [7J].

Experiments with upward flowing refrigerant 113 in vertical, uniformly heated test sections with and without twisted tape inserts showed increasing augmentation of heat transfer coefficients with decreasing tape–twist ratio (twist pitch divided by twice the tube diameter) and increasing quality, pressure and heat flux [50J]. Swirled forced convection boiling heat transfer coefficients were also measured for refrigerant 12 in horizontal tubes with twisted tape inserts [2J]. Heat transfer coefficients with the inserts were found generally to exceed those for plain tubes; enhancement was greatest for small twist ratio and varied considerably with quality, heat flux and mass flow rate.

Cold experiments with air–water mixtures and heated experiments with steam–water mixtures gave evidence of a region of unstable stratified flow with moderate and low mass fluxes upward in a circular tube helically coiled around a vertical axis [27J]. Flow and heat transfer characteristics of an air–water two-phase flow were measured in helical coils with horizontal axes [128J]. Heat transfer coefficients and critical heat fluxes were found to vary considerably around the perimeter of channels carrying boiling cryogens under forced convection while simultaneously rotating around a parallel axis [101J].

Subcooled flow boiling of heptane in an annulus and past a coiled wire was investigated experimentally [77J] and compared with correlations from the literature [78J]. Crossflow boiling heat transfer on horizontal tube bundles was systematically studied by contrasting experimental results of a single heated tube alone with those for a single heated tube in an unheated tube bundle and those for a heated tube within a heated tube bundle [42J]. Heat transfer characteristics of boiling He I in aluminum plate heat exchangers with and without fins were reported [48J]. The finned surfaces displayed an advantage in film boiling while finned and unfinned surfaces had the same characteristics in nucleate boiling.

Inverted annular flow was studied with the help of adiabatic gas–water experiments and a transparent heated test section [47J]. Emphasis was placed on flow regime transition observation and modeling, producing preliminary transition criteria and droplet diameter correlations. Effects of liquid droplets on post-dryout heat transfer were found to be similar to those of solid particles in flowing gas–solid mixtures [127J], and a simplified wall temperature calculation method for post-dryout conditions was shown to reasonably approximate a lengthier numerical process presented earlier. Dispersed flow heat transfer in post-dryout conditions was reviewed in the light of recently obtained data which indicated clearly that nonequilibrium between droplets and vapor is characteristic of post-dryout flows [16J]. Several other studies of post-dryout heat transfer were reported [24J, 39J, 66J, 119J].

Transition and film boiling

Transition boiling analysis, discussed in two ways: temporal fluctuations at given locations and spatial variations at any given time, was modeled in the latter way with the goal of obtaining a universal boiling curve through recognition of both hydrodynamic effects and the surface effects of roughness and wettability [41J]. The impedance across a thin dielectric film deposited on a surface was used as an indicator of liquid–solid contact in film, transition, and nucleate boiling [20J]. Extremely small electric voidage probes were used to explore the extent, duration and frequency of liquid–solid contact at a surface with subcooled pool transition boiling of water [44J]. Theory and experiment were applied in order to understand the mechanism of transient boiling with liquid and vapor contact durations chosen as the quantitative characteristics [58J].

Experimental studies of the effects of transients and of surface thermal conductivity on the minimum heat flux condition on a small, circular, horizontal heating surface showed that, below a minimum thickness, thinner samples and quicker transients resulted in lower fluxes but no change in the superheat at the minimum heat flux point [90J]. Higher superheats and fluxes at the minimum point were induced by thicker layers of insulating PTFE on the surface. Transient film boiling and minimum heat flux were measured on a sphere in water at atmospheric pressure for various subcoolings and immersion depths [91J], suggesting a modification to the equation proposed by Hamill and Baumeister. Minimum heat fluxes for saturated pool boiling of refrigerant 113 on a horizontal surface were measured at pressures above and below atmospheric [110J], with results suggesting inclusion of the liquid–vapor density ratio in the correlation of Berenson.

An integral treatment of the two-phase boundary layer of film boiling was applied to subcooled forced convection on a flat plate [83J], forced convection on plane and axisymmetric bodies [84J], laminar pool boiling on curved surfaces [82J], and combined free and forced convection film boiling [86J, 87J]. Steady film boiling experiments were performed with a sphere immersed in a flowing stream of subcooled refrigerant 11 [96J], displaying two maxima and two minima in the boiling curve, the second pair of extrema being explained through a changed wake pattern. A model was presented for coolant vapor film growth on a hot molten sphere in a subcooled liquid coolant [56J]. The model qualitatively predicted the characteristic oscillation amplitude and time scale of experimental observations recorded elsewhere. Annular film flow boiling experiments were conducted with three fluids in a partially heated vertical channel with offset strip fins, transparent on one side to permit visual study of flow regimes [13J]. Measured coefficients of heat transfer lent credence to a postulated model. Transition and film boiling in a bottom-heated liquid-saturated porous body were studied experimentally to determine effects of the packing bead diameter for water and two refrigerants [32J]. With small particles, no maximum or minimum heat flux was observed; heat flux instead increased monotonically with heater temperature. The 'equilibrium load', a heat flux for simultaneous nucleate and film boiling on a surface, was analyzed and related to the speed of propagation of a wave by which nucleate boiling gives way to film boiling or vice versa [121J].

Critical heat flux

A 'lookup table' has been compiled for estimating critical heat flux, CHF, based upon a data base of more than 15 000 data points [36J]. The table can be employed either as a predictive tool or as a means of checking predictions of present or new correlations. An earlier theoretically based DNB prediction method was extended, by allowing for non-uniform void profiles, to void fractions up to 0.8 in both round tubes and rod bundles [132J]. A theoretically based critical heat flux prediction for low void fraction was presented which, unlike another recent predictive model, appears to be applicable to both high and low vapor densities (high and low system pressures) [130J]. The critical heat flux of subcooled boiling of water in a narrow tube was found to increase with shorter tube length and smaller inside diameter [45J]. A new dryout description, the onset of a dry sheath condition, applicable at high mass fluxes where sharp temperature excursions are not likely to be observed, was proposed as a more appropriate limit for PWR operation [35J].

Limiting heat loads were explored for water, ammonia, and three fluorocarbon refrigerants inside a vertical heated tube bundle of a natural circulation system [65J]. Flow visualizations of regimes of natural convection near critical heat flux in horizontal annular channels were described [59J]. Dryout at low heat flux, induced by flow excursions in natural convection boiling was studied, leading to an approximate, flow regime dependent, limit of safe operating heat flux [55J].

A method was proposed for determining, for any fluid, the saturated liquid pressure which will permit the largest peak nucleate boiling heat flux [113J]. It was further stated that the same pressure condition should permit the largest minimum heat flux for film boiling. In an experimental study of two-phase gas–water flow through vertical U and inverted U bends, the inverted U was shown to accommodate higher gas fluxes and lower water fluxes without large temperature rises, consistent with flow visualizations performed in transparent, adiabatic bends [118J]. Three different characteristic regimes of critical heat flux were observed for an impinging jet on a heated disk [73J], and correlations for each were developed. Critical heat flux on a horizontal cylinder with upward subcooled and low quality two-phase crossflow was measured for refrigerant 113 at 2 and 4 atm [49J]. Critical heat flux decreased linearly as quality increased up to roughly 10%, beyond which a flow regime change rendered critical heat fluxes invariant with quality. Uncovered bundle high pressure boil-off experiments yielded two modes of results; low power boil-off, for which existing steam cooling heat transfer correlations gave good predictions, and high power boil-off, for which modification had to be made to correct correlations for larger droplet cooling effects [62J].

The pressure dependence of dryout heat flux in an inductively (volumetrically) heated bed of particles was explored, displaying a monotonic increase of dryout heat flux with pressure and with particle size [70J]. Free beds of 1 mm particles allowed greater heat fluxes than constrained beds, especially at near-atmospheric pressure. Experiments and a semi-theoretical model dealt with the effects upon dryout at a horizontal surface in pool boiling of particles light enough to be suspended by turbulence [14J]. Under some heat flux and particle load conditions, a decrease in heat flux induced dryout. The dryout of downward facing surfaces in debris beds was also explored [93J].

Reactor applications

Thermo-hydraulic instability as a consequence of vapor blockage was studied with refrigerant 113 in a vertical N-shaped boiling channel with an adiabatic bypass [4J]. Two models, one of them linear, were proposed for density wave instability in parallel boiling channels [5J]. Experimental observations of such instabilities were compared [6J] for two heating methods; electrically and with hot water to stimulate conditions of LMFBR steam generators. Density wave instability thresholds were explored in a liquid sodium heated system [28J] and results contrasted with correlations. Instability thresholds predicted using a linear analysis based on a two-fluid model compared favorably with some existing data from water–steam and refrigerant 113 experiments [23J]. New approaches were presented for the analysis of instability in boiling systems [64J]. Elaboration upon an earlier method of stability analysis led to a simple method for predicting oscillation frequency and stability threshold for density wave oscillations [81J]. Coolant and fuel dynamics of a BWR were coupled in a closed form model for stability analysis of density wave oscillations [100J].

Measurements of disturbance waves in heated steam–water annular two-phase flow and in comparable adiabatic air–water flows demonstrated less pronounced and longer-tailed waves in the boiling flow as well as greater scatter of propagation velocities [79J]. The mean period of dryout point fluctuations was shown to be related to the characteristics of the dryout point location and the relationship was demonstrated with data for refrigerant 12 in a horizontal serpentine evaporator [10J]. A frequency response model of boiling flow systems was illustrated, employing a more convenient computation technique which compared well with experimental findings [106J].

In an experimental study of upward steam flow effects upon PWR reflood quenching [1J] it was determined that top-down quench velocity was delayed more by steam upflow when rod surface temperature was low and water flow rate was small. A parametric numerical study performed with a two-dimensional heat conduction code [12J] examined the effects of cladding properties and initial cladding and coolant conditions upon rewetting velocities with given channel thermo-hydraulics. Rewetting experiments with refrigerant 12 at low flow rates were reported [34J]. Experiments were conducted to evaluate the effects of oscillatory coolant injection on quench front motion and liquid carryover in reflooding [92J]. Oscillation appeared to induce initially larger but eventually smaller quench velocities; overall the quench location was related to accumulated liquid inventory in the same way in both steady and oscillatory injections. It was shown [129J] that the heat flux to vertical surfaces during bottom reflooding transients can be approximated by the wetted area heat fluxes of steady transition boiling. Trial application of the two-fluid model simulation code, SABENA, to a low heat flux sodium boiling experiment was described [89J]. A simple model was described for the transient heating and steady boiling of a solution of fissile material, applicable to accident analysis of nuclear fuel reprocessing systems [108J].

Droplet evaporation

A numerical analysis which agreed well with experimental findings suggested that increases in aspect ratio of spheroids increased drag coefficients at low Reynolds numbers but reduced drag coefficients at Reynolds numbers greater than 50, while average heat and mass fluxes showed a monotonic decline with increasing aspect ratio [19J]. Numerical estimates were made of conjugate unsteady heat transfer from spherical droplets and particles at low Reynolds number with equal thermal diffusivities of the two phases [95J]. A numerically efficient formulation of transient multicomponent droplet vaporization suitable for use in combustion spray analysis was developed and confirmed [124J]. Theoretical variations of drop size with time during evaporation in a vapor–gas mixture were presented for drops of constant temperature and drops warmed by the surrounding gas [53J]. In the latter case, the droplet may initially grow in size due to condensation. Theoretical treatments of instantaneous heat transfer coefficients for condensing two-phase bubbles and total evaporating time of a two-phase drop compared well with experimental data [104J]. Data were obtained for instantaneous growth rate, rise velocity, and heat transfer coefficient of butane droplets evaporating in water [71J]. Drops falling through a gas into a liquid were observed to entrain gas bubbles [25J], having implications for a mechanism of secondary nucleation. Flow and heat transfer characteristics were explored as refrigerant 113 boiled upon injection into flowing hot water [31J]. Photographic studies of high temperature liquid drops falling into volatile liquids displayed very uneven vapor bubble surfaces and four types of pressure–time behaviors [43J].

Liquid drop suspension on an air cushion was examined as an analog of Leidenfrost boiling, leading to a mathematical model which very closely predicted the shape of air-suspended drops and which suggested that the Leidenfrost temperature is a decreasing function of drop volume [33J]. Experiments were reported with subatmospheric pressure droplet evaporation on a surface at and above the Leidenfrost temperature [120J]. Analysis was presented of the main trends in heat and mass transfer during suspension of liquid spheroids above heated or gas permeated surfaces [63J]. Experiments were performed to compare the bubble initiation characteristics of superheated water droplets in contact with four different solid materials [116J]. The frequency of production of critical nuclei was shown to increase with a decrease in the wettability of the surface. Droplet velocity was determined to directly influence non-steady heat transfer in drop cooling of a surface, being proportional to the evaporated mass fraction of the drop [68J]. Cool liquid drops striking the surface of molten tin displayed a splash phenomenon which did not occur upon striking molten salts [117J].

Flashing and other evaporative processes

Experiments were described and correlations presented for evaporative air and water flow over a horizontal tube [105J]. Analysis and experiment suggested that heat transfer coefficients were not substantially enhanced by evaporation from a liquid surface unless other sources of heat input to the liquid were overlooked [61J]. Condensation of some of the diffusing vapor was found to stimulate the evaporation of a heated, but not boiling, liquid film into a saturated stream [3J]. Rates of evaporation of water and of toluene from a partially filled cylindrical cavity in the lower wall of a rectangular air flow channel were found to exhibit a maximum when the liquid level was roughly one half cavity diameter below the plane of the duct wall [102J]. Evaporation in film flows was examined [15J, 18J, 99J]. Conjugate evaporation from a falling film, diffusion through a non-condensable gas and film condensation on a cooler surface was analyzed using the local similar technique [26J].

Models of dynamic and thermal behavior of hot gas bubbles discharged into water were discussed [75J]. Theory and experiment were applied to the growth of bubbles in rapidly decompressed bodies of uniform temperature water [122J]. Approximate methods of determining heat flux to a growing vapor bubble from a surrounding superheated liquid were presented and compared with a numerical computation method [60J]. Heterogeneous nucleation at surfaces under decompression was modelled, based on an analysis of subcooled nucleate boiling [109J]. Experiments demonstrated that a surface instability near the limit of superheat, responsible for large heat and mass fluxes of vapor explosions, was suppressed by increased ambient pressure [29J].

The binary mutual diffusion coefficient was shown theoretically to vanish in a critical state of a binary mixture [126J]. Kinetic theory analysis was applied to unsteady evaporation [112J]. The 'paradox' that slow evaporation/condensation between two parallel liquid surfaces at unequal temperatures can result in a vapor temperature profile opposed to the applied temperature difference was shown not to violate principles of non-equilibrium thermodynamics [40J]. A review of the effects of monolayers on the evaporation of liquids showed that attempts to predict associated resistance values have not been generally successful [9J]. Rates of evaporation and composition of vapors emanating from stirred subatmospheric evaporating binary mixtures were measured and modeled incorporating convection and diffusion in the near-surface liquid [46J]. A previously proposed method of calculating mass transfer in multicomponent two-phase systems has been extended to include convective diffusion and heat transfer [54J].

CHANGE OF PHASE—CONDENSATION

An analytical expression was presented for laminar liquid film flow in the thermal entry region down inner or outer walls of a vertical cylindrical tube [40JJ]. Simple relations were developed for heat transfer from a horizontal smooth tube to a laminar falling film [31JJ], corroborating a much earlier recommendation. A theoretical study was reported of conjugate heat transfer from a condensing fluid on one side of a vertical wall to another fluid in natural convection on the other side [36JJ]. Conjugate film condensation and natural convection at the interface between a porous substance and an open space were analyzed for both condensation in the free space and condensation in the porous material [37JJ]. Two simpler alternatives were presented and contrasted with the more elaborate solution of the conjugate heat transfer problem of condensation on a finned surface, showing satisfactory agreement [1JJ].

Experimental results were presented for condensation of nearly stationary vapor on finned tube bundles and mechanisms of condensate irrigation were identified [17JJ]. An equation was presented correlating data for condensation on horizontal tube bundles for both steam and refrigerant 21 [12JJ]. Travelling waves appearing on the underside of horizontal condenser tubes were investigated [4JJ]. Experiments with condensation of nearly motionless vapor upon single horizontal cylinders of various diameters showed that condensate flow waves, which substantially enhance heat transfer coefficients relative to the Nusselt theory, occur on cylinders of all diameters [16JJ]. The effect of spacing of 1 mm thick, 1 mm high fins upon film condensation on horizontal 19 mm diameter tubes was measured for pure steam condensing at atmospheric and subatmospheric pressures [50JJ]. Maximum enhancement of heat transfer occurred with a spacing of 1.5 mm between fins.

An analytical treatment of film condensation of saturated and superheated binary vapor mixtures on a vertical plate showed that buoyant forces have appreciable influence and that vapor temperatures are strongly affected by the diffusion flux toward the liquid vapor interface [27JJ]. Experiments showed that drainage discs considerably enhanced condensation heat transfer with steam condensing at 1 atm on a vertical fluted tube [18JJ]. Optimal spacing of these discs was 50–100 mm for heat fluxes of 60–100 kW m^{-2}. Analysis and experiments were conducted with condensation on downward facing horizontal surfaces with and without porous drainage strips [23JJ]. Finned surfaces had heat transfer coefficients improved by factors up to 9 and 12, respectively, for refrigerant 113 and methanol with the drainage strips. In a brief review of laminar film condensation, it was argued that this basic phenomenon represents an important benchmark for numerical heat transfer and fluid mechanics modeling [11JJ].

Analysis suggested that forced downflow of vapor condensing over a vertical array of horizontal tubes suffers a greater reduction in heat transfer coefficient due to condensate inundation than gravity-controlled vapor flow over the same geometric configuration [13JJ]. The effects of downward vapor velocity were studied for film condensation on a horizontal tube in crossflow [22JJ]. Condensate film was observed to undergo transitions from a smooth surface to two- and then three-dimensional waves as vapor velocity increased.

An integral method, analyzing the two-phase boundary layer flow in laminar film condensation was presented [32JJ] which allowed for finite vapor density and viscosity and accounted for inertia and convection terms. The method may be extended to axisymmetric bodies. Condensing heat and mass transfer were studied for laminar and turbulent saturated vapor–gas flows on a flat plate with negligible accumulation of condensed liquid [29JJ]. The effects of composition of mixtures of refrigerants 12 and 22 during forced convective condensation inside a horizontal tube were measured [45JJ]. The mist-annular flow transition in high mass flux condensation in tubes was examined with the conclusions that a modified Weber number could discriminate between the two flow regimes, with mist flow present for Weber numbers greater than 30 [41JJ]. A mist flow heat transfer correlation was also proposed from this study. Condensation of vapor on a co-flowing liquid in an adiabatic converging duct was analyzed and compared with experimental results [28JJ]. Motivation was the heating of scale-producing fluids at free surfaces.

Heat transfer coefficients were measured for dropwise condensation of steam at pressures ranging from atmospheric down to 1 kPa [19JJ]. Nucleation site density appeared to be affected by surface physico-chemical condition, and not directly by surface roughness. Although heat transfer coefficients of low pressure steam condensing dropwise on gold and chromium surfaces were nearly the same, measured droplet size distributions were markedly different, with population of the droplets on the chromium surface falling off sharply below 7 μm [20JJ]. A model of dropwise condensation was presented based upon the assumption of existence of a smallest droplet nucleation radius [52JJ]. Microscopic observation and light scattering experiments [26JJ] on 'breath figures', patterns formed when a vapor condensed to liquid on a cold surface, displayed remarkable droplet size uniformity for periods less than 300 s, radius growth between coalescences proportional to time to the power 0.23, and overall radius growth (with coalescence) proportional to time to the power 0.75. The endurance and effectiveness of many organic coatings for promoting dropwise condensation were determined for specimens observed in excess of 12 000 h [30JJ]. The sweeping effect of falling drops in dropwise condensation was explored [7JJ], providing a mechanistic explanation of observed trends of heat transfer coefficient variations with heat flux and with surface length. Dropwise condensation on a horizontal tube was modeled and measured and the predictions compared favorably with experimental results [24JJ], showing heat transfer coefficients increasing with tube diameter at low heat fluxes, independent of diameter at higher fluxes. Dropwise condensation curves and drop to filmwise transition were measured for several fluids on a copper surface treated with an agent to promote dropwise condensation [47JJ]. Experiments performed with a small cooled surface with various patterns of gold coatings to promote dropwise condensation demonstrated that mixed surfaces with dropwise and filmwise condensation taking place side by side could provide greater heat fluxes than totally dropwise condensing surfaces [51JJ].

Experiments with condensing jets of low void fraction mixtures of carbon dioxide in water entering still bodies of water were modeled in three ways; of these, only a stochastic, separated-flow analysis including finite transport rates and bubble/turbulence interactions provided reasonable agreement with measurements [43JJ]. Experiments with steam condensing on turbulent subcooled water free of bulk flow and nearly free of surface waves yielded a correlation of the condensation coefficient proportional to the r.m.s. value of the turbulent velocity [42JJ]. In the same apparatus, unstable short high intensity bursts of condensation were observed when a threshold value of turbulence intensity of the liquid was exceeded [8JJ]. Direct contact condensation of stagnant saturated steam on a horizontal flat surface of slowly moving water was studied [6JJ] while varying liquid flow rates and steam and liquid temperatures. The liquid flow rate had greatest influence on the measured heat transfer coefficients. Forced convection within a BWR containment was modeled for prediction of environmental temperature response during a postulated loss of coolant accident [5JJ]. Diffusion limited heat transfer was assumed and results agreed with experimental findings. Experimentally and analytically, thresholds of pool water subcooling were evaluated that permit induction of pressure oscillations as a steam jet discharges into the pool [2JJ]. Greater subcooling is required to allow higher frequency, chugging oscillations while lower subcooling permits low frequency bubbling. The condensation of vapor bubbles detaching from a surface producing transition boiling in subcooled water was observed and modeled [25JJ]. The 'disappearance velocity' defined as the ratio of initial bubble diameter to collapse time, was proportional to the initial growth velocity of the bubble, the square of the wall heat flux, and the cube of the wall superheat.

A kinetic theory treatment of strong condensation in the presence of a non-condensable gas has been developed, extending an intensive condensation treatment for a pure vapor [35JJ]. Despite large variations in number density and pressure of a non-condensable component gas, a kinetic theory analysis suggested that a small amount of non-condensable gas has no influence on the weakly non-linear condensation or evaporation of a vapor [34JJ]. Results reported of film condensation of mixtures of refrigerants 113 and 114 and refrigerants 113 and 11 were well predicted by a solution considering the two independent resistances of the liquid film and the vapor diffusion boundary layer [21JJ]. An approximate method was described for calculation of required heat exchange area for condensation of gas–vapor mixtures, demonstrating satisfactory agreement with experiments [14JJ].

The unsteady flux of condensation nuclei in an expanding supercooled vapor was analytically determined [39JJ]. One-dimensional nozzle flows with non-equilibrium condensation were modeled [10JJ] and extended to supercritical shocks, for which the latent heat addition exceeds the heat input for thermal choking [9JJ].

Options available for improvement of film condensation of pure vapors were reviewed [38JJ]. Heat transfer coefficients by condensation upon a vertical tube were augmented by as much as a factor of 2.8 through the use of non-uniform electric fields produced with helical wire electrodes [44JJ]. Development of a floating OTEC system prompted experimentation with the effects of low frequency, large amplitude periodic motion of a vertical tube upon external filmwise condensation [33JJ]. Heaving (axial motion) had little effect, swaying (lateral motion) produced enhancements increasing with acceleration, and rolling (simulated by pendulum motion) produced enhancements increasing with angular displacement and decreasing with frequency at low frequency, increasing with higher frequency.

Hydrodynamic behavior of the condensing section of a thermosiphon was experimentally observed to be 'highly turbulized' [3JJ]. Experiments with, and modeling of plate-fin devices for fractionating of mixtures were reported [46JJ]. Moisture accumulation and thermal conductivity increase of fibrous insulation material due to moisture migration were analyzed in a transient manner yielding the wet–dry interface location as one of the predicted variables [48JJ]. A stable and convergent numerical method was described, capable of modeling regenerators in which condensation and evaporation occur [49JJ]. Experiments and analysis were presented for simultaneous melting of a vertical planar substrate and condensation upon it in the presence of a non-condensable gas [15JJ]. The reduction of heat transfer due to the non-condensable component was stronger for melting substrates with lower liquid Prandtl number.

CHANGE OF PHASE—FREEZING AND MELTING

The classical Stefan problem, describing the melting or freezing of spheres, cylinders and slabs, continued to receive attention. One paper addressed the multiphase case, with more than one moving boundary separating distinct phases [11JM]. The problem of freezing a saturated liquid inside an infinite circular cylindrical container was solved by fixing the boundary with a logarithmic transformation and obtaining an iterative analytic series solution [21JM]. Approximate analytical methods [19JM] and integral formulations [10JM] were presented for analyzing Stefan problems.

Melting result correlations, based on experiments, were presented for both pure and impure substances [49JM]. A new and efficient algorithm was proposed to be incorporated with the equivalent heat capacity model for the finite element analysis of melting and freezing problems [24JM]. A comparison was made between an approximate analytical solution and a numerical finite difference solution for the one-dimensional solidification of a phase change material of finite size [2JM]. A finite element numerical method based on enthalpy was proposed for the solution of two-dimensional problems dealing with phase change at fixed temperature [4JM]. An algorithm was described that combines a temperature formulation with a finite element treatment of the differential equation and discontinuous integration within two-phase elements to avoid the necessity of regularization [9JM]. Based on the effective heat capacity characteristics of biological materials during freezing, a short-cut equation for predicting their temperature vs time behavior during freezing and thawing has been derived [38JM].

The problem of the flowfield, heat transfer, and melting rate of a solid body immersed in an otherwise quiescent, hot fluid is an interesting example of free convection, differing from the usual free convection problems in that the dominant buoyancy force is due to the fluid–melt density difference rather than the familiar thermal expansion [6JM].

Simple upper and lower limits for the full solidification (melting) time of isothermally heated slabs, cylinders and spheres were derived [5JM]. The method of cubic spline collocation was employed to solve the governing equations of the processes of inward solidification occurring in slabs, cylinders and spheres bounded by a finite wall subject to a convective boundary condition at the outer surface [42JM]. A boundary fixing series technique, previously used on melting and freezing problems where the material was initially at the fusion temperature, was generalized to initially subcooled melting cylinders and spheres [30JM]. Experiments and supplementary numerical solutions have been performed to study the melting of ice [36JM] and other materials [45JM] encapsulated in a horizontal tube. A methodology was set forth for the numerical solution of transient two-dimensional freezing of a material in a vertical tube and results presented for 99% pure n-eicosane, a material for which experimental results are available [47JM, 48JM]. Timewise measurements of in-tube melting in the presence of circumferentially non-uniform heating enabled identification of the pattern of melting for different types of heating and various tube inclinations [46JM]. Melting of ice in porous media has been investigated experimentally and analytically for horizontal and vertical cylindrical capsules [55JM]. The roles of natural and forced convection during solidification of pure tin in an annular crucible were described [54JM].

A theoretical analysis was presented for the phase change process occurring in a cylindrical annulus in which rectangular, uniformly spaced axial fins, spanning the annulus, are attached to the inner isothermal tube, while the outer tube is kept adiabatic [35JM]. A two-dimensional numerical simulation of outward melting of a material contained in a horizontal cylindrical annulus was performed [23JM]. The solidification of a subcooled metallic sphere was analyzed by an enthalpy method [25JM]. Melting heat transfer in an inclined rectangular enclosure was investigated experimentally [57JM]. A study defined the limitations of using a one-dimensional approximation for analyzing buoyancy and surface tension driven natural convection in rectangular cavities during solidification [34JM].

Heat transfer during the melting of ice around a horizontal cylinder was investigated experimentally [58JM] and numerically [22JM]. An approximate three-dimensional solution for melting or freezing around a buried pipe beneath a free surface was presented [61JM]. Melting around a horizontal finned be was studied [3JM] and a comparison of the phase change about finned and bare tubes was performed [40JM, 41JM].

Results were presented of calculations of the melting rate of a sheet of finite width under the effect of variable power heat flows [52JM]. The two-dimensional steady-state shape of a solidified region, such as a frost layer, was determined analytically for formation on a plate that is convectively cooled [44JM]. The solidification of an infinite liquid slab by linear convection cooling from the adjacent air was considered [20JM]. The phenomenon of melting from a flat plate in a porous medium in the presence of boundary-layer natural convection was analyzed by assuming that the melting takes place at a steady rate [27JM]. The role of natural convection on solid–liquid interface motion and heat transfer was described for melting and solidification of a pure metal on a vertical wall [16JM, 57JM]. The melting process of ice–air composite materials was investigated for the case where heat was supplied from the bottom and lost from the top of the body [1JM]. Experiments were performed to gain an understanding of the convection process occurring in a warm liquid pool as it penetrates into an underlying meltable solid of less dense material [15JM]. The linear convection instabilities of a fluid layer of binary alloy, cooled from above and consequently frozen at the bottom, were described [26JM].

An analysis of one-sided freezing of moist soil included the allowance for the phase transition of moisture in a certain temperature range and moisture migration in the thawed and freezing zones [60JM].

A dependence was proposed for determining the heat transfer coefficient at the boundary of an ice mass and a water film running down the ice in the presence of thawing [17JM]. An experimental investigation of snow melting by showering the snow layer with a calcium chloride aqueous solution clarifies the influences of initial concentration of the solution, amount of showering solution and density of snow sample on the melting rate [29JM]. A paper described the melting of a horizontal ice layer from above by an aqua-solvent with a low solidification point (calcium chloride and urea solutes) [51JM].

The melting of a semi-transparent material by radiation was studied experimentally and analytically [12JM]. A numerical analysis was performed of thermocapillary flow in a rectangular cavity during laser melting [50JM]. In a study of the temperature fields associated with laser melting and subsequent recrystallization of a thin silicon film on a glass substrate, particular attention was paid to the change in the material properties with temperature and to the change in reflectivity which occurs when the silicon changes phase [18JM]. Companion papers described exact solutions to problems associated with laser melting of solids, one for time intervals less or equal to the transit time [13JM] and the other for larger time intervals [14JM].

Studies of close-contact melting [32JM, 33JM], contact melting [39JM], and contact ablative melting [31JM] were performed.

The solidification process during the continuous casting of an ingot by withdrawal from a mold was analyzed assuming a constant convective heat transfer coefficient at the outside surface of the ingot [43JM]. The continuous casting problem was also analyzed via the variational inequalities approach [37JM]. Three papers describe gas melting. A three-dimensional numerical method was presented to simulate the effect of electric boosting on glass melt circulation and heat transfer in a glass melting furnace [53JM]. Mathematical models were constructed to predict the temperature distribution and heat transfer in a glass batch blanket and to simulate the effects of individual factors on the conversion process [59JM] and to calculate Joule heat release, glass flow and heat transfer in electric glass furnaces [8JM].

A theoretical study was conducted to investigate the effect of freezing on the heat transfer characteristics for the turbulent flow of a heat generating fluid in a cooled circular tube [28JM]. The critical condition for solidification blockage of laminar straight pipe flow in an arbitrary piping system was derived by comparing two pressure differences in the cooling section, one of which depends on the system (driving pressure difference), and the other is given by a numerical calculation of steady laminar pipe flow with internal fluid solidification [7JM].

RADIATION IN PARTICIPATING MEDIA AND SURFACE RADIATION

Radiation in participating media

Reference [94K] presents and compares two techniques for evaluating the emission integrals used for discrete ordinates solution. Vector algorithms for Chandrasekhar *H*-functions for inhomogeneous media are reported [55K]. By combining a weighted set of distributed internal sources, the photon path length analysis is extended to planar layers with arbitrary temperature distributions [102K]. A generalized spherical harmonics solution for radiative transfer models, including polarization effects, emphasizes the numerical aspects for accuracy [45K]. The spherical harmonics approximation solution to the equation of transfer in plane-parallel, homogeneous, anisotropic scattering (Rayleigh and Henyey–Greenstein) media are obtained by a Chebyshev collocation method [62K]. For isotropic scattering media of the same geometry, ref. [112K] reports on a new, efficient method of analysis, which uses the natural eigenfunctions of the problem. Four-flux model solutions of special types of Lorenz–Mie scatter centers embedded in a slab are presented [76K]. A generalized Eddington approximation for the slab geometry is presented [124K]. Matrix formulations are presented for planar scattering media using the discrete-ordinates and the matrix-operator methods [83K]. A modified Monte Carlo method is reported [64K], which reduces the computing time and improves the convergence stability.

A modified F_N method is used to study the radiative transport in a planar medium with azimuthally unsymmetrical incidence [67K]. A zonal method is proposed [2K], which takes into account the anisotropic character of volumetric and surface scattering. Solution of the radiative transfer in a two-phase medium for a Markov process is reported [120K]. The Sobolev probabilistic method [121K] is used to obtain an exact solution of the equation of transfer in a gas containing particles of its condensed phase. A linearized differential approximation is invoked to consider the transient flow of optically thick, dusty gas in a vertical channel [12K]. Benchmark solutions are presented, four or five digit accuracy, for the time-dependent reflected photon intensity from an anisotropic scattering semi-infinite medium [44K].

Spherical harmonics approximation is used to model radiative transfer in a finite length cylindrical vessel, containing high temperature aerosols that absorb, emit and scatter [77K]. The integral form of the solution of the equation of transfer is shown to reduce the number of independent variables from three to one for isotropically scattering, inhomogeneous solid cylinders [111K]. Reference [80K] considers numerical solutions for axisymmetric, finite cylindrical enclosures containing radiatively participating gases and particles. A solution method for solving the equation of transfer in an isotropically scattering, inhomogeneous solid sphere is also presented [113K]. The far field solution and the energy flux from a moving source in an anisotropic medium is presented [68K]. A rigorous formulation of the problem of radiative transfer in an anisotropic, fibrous medium shows a strong dependence of the transport properties on the different orientations of fibers [71K].

Theoretical and experimental results of backscattering from an optically thick, scattering medium exposed to a laser beam are presented [87K]. A method based on the principles of invariant imbedding, is presented for solutions in two-dimensional anisotropically scattering medium exposed to arbitrary boundary conditions [106K]. A three-dimensional code, based on the diffusion approximation, is used to model the spatial distribution of radiant energy from volumetric isotropic sources [127K].

A hybrid Galerkin-iterative scheme is used to provide a short time solution for coupled conduction and radiation in a participating slab [107K]. Radiation scaling laws are applied to model combined mode heat transfer in planar geometry [70K]. The transient temperature distribution of an absorbing, emitting multilayer composite wall, suddenly exposed to radiative flux, is influenced by both radiation and conduction in each layer [116K]. Effects of the non-linear heat generation, and combined conduction and radiation treated at the optically thick limit, on the vibrational heating of semitransparent polymers are considered [50K]. The effect of scattering in one layer, when the non-stationary radiation–conduction problem in a system of two layers is considered, is presented [97K].

The combined natural convection and radiation heat transfer from pin-fin arrays in air is measured, and the contribution from the radiation is determined analytically to be in the 25–40% range [104K]. The Milne–Eddington approximation is used to express the two-dimensional radiative transfer inside a horizontal cylindrical annulus filled with absorbing, emitting non-gray Boussinesq fluid [88K]. A numerical study of non-gray gases in a flow system, compares the results with those obtained by using a gray gas assumption [54K]. The effect of radiation on the heat transfer in the boundary layer flow along a flat plate are described as a prescribed function of distance from the slit, where the constant velocity flow is introduced into a fluid at rest [21K]. The presence of both the radiation and the magnetic fields is found to resist the formation of shock in electrically conducting and thermally radiating gases, which are treated as optically thin [100K]. An asymptotic representation of a point-source thermal explosion, dominated by radiative heat transfer, makes it possible to analyze the generation of an isothermal shock wave [95K]. Radiative energy transfer and equilibrium chemical reactions in the compressed shock layer around three-dimensional and axisymmetric bodies are considered [5K].

The Surinov generalized zonal method is applied to solve for the radiative transfer in a conical chamber with absorbing and anisotropically scattering medium [48K]. Locally variable radiative heat transfer between long cylinders that contain emitting and absorbing gases is calculated [60K]. A loss of visibility study considers laser reflection from a scattering medium with a reflecting substrate [74K]. Least squares smoothing of direct-exchange areas in zonal analysis, using Lagrange multipliers, is presented [69K]. A local entropy production expression includes the effect of radiation, expressed in terms of radiative stress [6K]. A study of the characteristics of fine-particle semi-transparent suspensions considers the optical depth and boundary condition effects [59K]. Liquid-droplet radiators for spacecraft thermal control are shown to be 3–5 times lighter than conventional radiators [110K]. Calculations of emitted radiation from boron slurry-fueled jet engine exhaust considers the effect of B_2O_3 particles [86K].

Radiation in combustion systems

A two-dimensional model is considered when the heat ray method is applied to radiative transfer problems in furnaces [53K]. The total effective absorptivity and emissivity of fluidized bed combustors are studied, and the important dimensionless parameters that are needed in the radiative transfer correlations are pointed out [16K]. An axisymmetric cylindrical enclosure, containing inhomogeneous and non-isothermal gases and particles, is studied to model a typical gas-turbine combustor [79K]. Direct exchange areas for the zone method are provided to evaluate the radiative transfer in rectangular, gas-filled furnaces [118K]. An algorithm for solving the conjugate problem, which arises in furnaces with a protective atmosphere, is presented [73K]. The effect of isotropic and anisotropic scattering, on the heat transfer in semi-infinite furnace slabs, are studied by using the zone and the Monte Carlo methods [39K]. The enhancement due to the addition of a solid surface, between parallel plates containing non-gray combustion gases, is shown to be as high as 30% [54K]. The influence of the wall emissivity on the performance of high temperature, radiation dominated furnaces is considered [35K]. A review of the models useful for calculating radiative transfer in furnaces is presented [123K].

The influence, of the burner geometry and the fluid parameters, on the local and the mean emissivity of gas diffusion flames, is studied [115K]. Radiation from a methane–air flame which is electrically augmented to enhance the flame temperature is examined [61K]. Reference [31K] presents analytical solutions for the structure and the burning speeds in rich mixtures of combustible solid particles and gaseous oxidizer, viewed as absorbing–emitting media. Light transmission and emission measurements, of the deflagation of ammonium perchlorate with carbon black and copper chromite catalyst additives, is used to quantify the amount of particle radiative feedback [17K]. The dynamic light scattering technique is used to measure the soot particle size distributions, number density, and the total volume, in premixed methane–oxygen flames [99K]. Reference [40K] establishes a framework for computing the scattering characteristics of agglomerated soot spheres, and considers the effects of radial inhomogeneity and agglomeration on light scattering measurements in flames. Emission measurements of high temperature water vapor, produced as exhaust gas from burning natural gas and oxygen, are compared with predicted values [27K].

Surface radiation

A numerical method for calculating the configuration factor, based on the unit-sphere method, uses the computer graphics technique of ray casting. The method is shown to give accurate results for complex geometries and does not suffer from statistical errors associated with the Monte Carlo methods [78K]. Reference [105K] discusses the reliability of the Monte Carlo method for calculating the view factors. A plating algorithm, where the emissivities of surfaces in an enclosure are made to vary from one to its actual value, is used to obtain the script-F transfer factors [37K]. The concept of multiple Markov chains is applied to develop a stochastic approach for radiative exchange in enclosures with directional–bidirectional properties [84K]. Radiative heat transfer across and down a long cylindrical capillary pore is presented [117K]. A study compares simple, surface radiation exchange models that can be used to simulate the thermal behavior of buildings [3K]. Conductive and radiative heat transfer in circular and longitudinal finned tube systems are measured experimentally and compared with numerical predictions [108K]. An approximate computation scheme is proposed to predict the heating of massive, optically dense bodies in a radiation dominated furnace [32K]. An asymptotic, large time solution for the convection Stefan problem with surface radiation is presented [114K]. A model for the interaction between intense radiation and a liquid, takes the absorption coefficient of the liquid into account [46K]. Heat transfer from a source in a transparent cavity with a conducting shell is considered [91K]. A radiative transfer model for severe fuel damage analysis accounts for anisotropic effects of reflected radiation, with simplified view factor calculations [103K].

Radiative properties

A method, for determining the optical properties of semitransparent materials at high temperatures from two independent measurements, is used to obtain the properties of fused quartz and sapphire at 1000°C and 0.6–13 μm [82K]. Reference [30K] describes an experimental rig which measures the directional–hemispherical transmittance and reflectance of porous materials in the 9–11 μm range. Optical constants of strongly absorbing media are obtained from considering the minimum, parallel polarized reflectance measurements [75K]. Two new techniques for determining the thermal radiative properties of thin fabrics are reported [19K]. A technique, developed for evaluating surface coating effectiveness, leads to an analysis of the characteristics of a simple energy absorption transducer [11K]. The method of moments is used to model conduction and radiation in a translucent scattering medium, and the predicted temperatures are compared with the experimental measurements to obtain the radiative properties [92K]. Analytical error estimates for the time-dependent radiative-transfer inverse problem, used to infer the albedo and the Legendre moments of the phase function, are presented [36K].

Radiative characteristics for fly ash and coal are obtained numerically by modeling them as clouds of spherical particles illuminated by blackbody radiation [14K]. The only reported results of the optical constants of propellent-grade ammonium perchlorate are obtained by a combination of dispersion equation curve-fitting and normal spectral reflectance measurements [89K]. Simple expressions for the temperature dependence of the index of refraction of sapphire are given [109K]. Calculations of the mean hemispherical emissivity of hot glass in various configurations are presented [57K]. Reflective characteristics are used to study the radiation properties of industrial refractories of zirconium dioxide [126K]. A stratified media theory is applied to the results of an experimental study of the emissivities of oxide films on metal surfaces [4K]. Infra-red radiative heat transfer in highly transparent silica aerogel is obtained by measuring the extinction [20K].

A theoretical investigation of the dependent scattering properties are presented, assuming the Rayleigh–Debye scattering approximation, and the scattering efficiencies are shown to be smaller than the independent efficiencies [22K]. An experimental investigation of the dependent scattering compares the measured bidirectional transmittance and reflectance to the theoretical predictions [125K]. Computer calculations, which apply the theory of diffraction, show a range of concentrations where the radiative properties of the dispersion is dependent on the distance between the particles [98K]. A localized approximation to the generalized Lorenz–Mie theory is introduced and compared with the Rayleigh–Gans theory [49K]. A simple numerical scheme, based on the Fraunhofer approximation, is used to calculate the forward scattering by a sphere located anywhere in a Guassian beam [24K]. An experimental study designed to distinguish between the transmitted and the forward scattered intensities, compares the measured results with calculations [10K]. The asymptotic form of the Mie-scattering amplitude, useful for large size parameter calculations, is presented [7K]. Kramers–Kronig relations, which can be written independent of any material constants, are used to infer refractive indices from spectral extinction data, without any knowledge of the index of refraction at any frequency [66K]. Correction charts for particle size distribution measurements, needed due to the multiple scattering effects, are presented [47K].

A perturbative approach to scattering by a single-layered sphere is applied to the case of absorption of light by water droplets contaminated with soot [13K]. Using a full wave approach, the particle surface roughness is shown to have a significant effect on the diffuse specific intensities [9K]. Scattering and depolarization by conducting cylinders with rough surfaces are also considered [8K]. The multiple Laplace transform is applied to study the electromagnetic reflection from arbitrary smooth convex cylinders [56K]. A technique is given for computing the scattering by targets which can be subdivided into circular disks [51K]. The iterative extended boundary condition method is utilized to calculate the scattering and the absorption of aerosols that are modeled by spheroids of high aspect ratio [58K]. A study of the scattering from nonspherical particles, which have radii described as functions of Chebyshev polynomials, shows that concavity enhances the spherical–nonspherical differences [81K]. Irregularities, on the surfaces of particles that are large compared with the incident wavelength, are modeled by a function of fractal type [15K].

The penetration of pulse of laser radiation through the atmosphere is shown to allow for the cooling of the water vapor in the channel around the beam, when the pulse duration is controlled [72K]. A wide-band absorption coefficient integration kernel is introduced, which transforms the wave number integration to a more convenient integration over the absorption coefficient [38K]. The infra-red absorption properties of carbon dioxide are the topic of a number of studies [1K, 28K, 29K, 33K, 41K, 93K, 96K]. Absorption properties of carbon monoxide [43K, 90K], water vapor [33K], and gaseous methane [25K, 26K, 42K] are also considered.

Experimental systems

A unique apparatus is developed, which can be used to measure the emittance of semitransparent, porous and particulate media [119K]. Correction factors, which are necessary for integrating spheres with one port, when the sample is used in place of an open port, is calculated with a finite difference method [63K]. Infra-red lasers in the 11–13 or 14 μm wavelengths operate by pumping NH_3 [65K, 101K]. Characteristics of silicon avalanche photodiodes are examined to assess their suitability for photon correlation measurements [18K]. An optimized geometry of far-IR photoconductive detectors is presented [122K]. A cryogenic charge amplifier system is developed for an InSb photodiode array for use in high-resolution infra-red spectrometers [52K]. Darkening of silver halide optical fibers, transparent in the middle infra-red and useful for transmitting a CO_2 laser wavelength of 10.6 μm, is considered [23K]. Mid-infra-red reflectance spectra of sulfur, gold, KBr, halon, and recommendations to their usefulness as reflectance standards are presented [85K].

NUMERICAL METHODS

Numerical methods are described and used in many papers. In this review, the papers that focus on the application of a numerical method are included in the appropriate application category. The papers that emphasize the details of the numerical method are reviewed in this section.

Among topics of general interest, ref. [76N] surveys the application of parallel and vector computations in heat transfer and fluid flow. The use of various spreadsheet programs for solving heat transfer problems on microcomputers is described in ref. [19N]. Various strategies for solving linear and non-linear simultaneous equations have been described in refs. [13N, 41N, 65N]. Reference [49N] deals with a technique for grid generation.

In the area of heat conduction, a number of methods have been developed for non-linear inverse heat conduction [32N, 33N, 72N, 81N]. Finite element analysis has been applied to non-linear heat conduction problems [74N]. Reference [34N] proposes a unified model for axisymmetric heat conduction. Integral transform techniques have been combined with the finite element method for heat conduction [15N]. New types of finite elements have been proposed for heat conduction [78N, 79N]. Applications of boundary element methods to heat conduction are presented in refs. [1N, 55N, 58N]. Various novel techniques for transient heat conduction problems have been presented in refs. [8N, 12N, 21N, 47N, 77N].

Problems involving phase change have been analyzed numerically. Reference [3N] pertains to the growth of crystals from the melt, while numerical techniques for the Stefan problem are developed in refs. [91N, 93N]. The concept of lumped capacitance has been used in finite element methods for phase change [67N].

The question of numerical accuracy for the combined convection–diffusion problem continues to be a topic of intensive research. Reference [66N] evaluates eight different schemes for the problem. The influence of upstream approximations in curved grids is discussed in ref. [27N]. Tetrahedral elements for convection–diffusion problems are employed in ref. [54N]. A new skew, upwind procedure is presented in ref. [75N]. Weighted finite difference methods are proposed in refs. [17N, 64N]. A new finite element formulation for fluid flow and convection–diffusion problems is developed in refs. [35N–39N]. Additional studies of the advection equation are contained in refs. [5N, 20N, 40N, 59N, 62N, 80N].

Adaptive grid methods have been investigated [51N, 61N]. Multi-grid procedures for the solution of recirculating fluid flow have been developed [52N, 86N–88N] and shown to be significantly more efficient than other available methods. The multi-grid technique is also used for the solution of the Euler equation [45N]. Techniques used for the solution of the Euler or Navier–Stokes equations are compared in refs. [2N, 16N, 89N]. Other studies related to the Euler equation have been reported in refs. [12N, 44N, 63N, 94N]. Reference [70N] presents an implementation of the Osher upward scheme.

A large number of new methods have been proposed for the solution of compressible and incompressible fluid flows. Methods that employ the boundary layer characteristics of the flow are presented in refs. [22N, 31N, 56N, 92N]. The vorticity-based methods are described in refs. [25N, 95N]. Calculations have been presented for compressible flow in refs. [4N, 90N]. Among the new methods proposed for fluid flow refs. [42N, 43N] use operator splitting. The use of general orthogonal coordinates is described in ref. [71N]. The problem of temperature–velocity coupling is addressed in ref. [28N], while an equal-order interpolation method is worked out in ref. [69N]. A number of other techniques for solving the flow equation have been described in refs. [6N, 29N, 30N, 57N, 68N, 73N, 83N, 84N].

Comparisons of a class of numerical techniques for certain test problems are presented in refs. [11N, 46N, 60N]. Various special aspects of numerical methods such as grid skewness, nonlinearity, convergence rate, and mesh refinement are discussed in refs. [7N, 10N, 23N, 26N, 53N]. Different schemes have been compared for solving the flow past a sphere [14N] and around a cylinder [9N]. Analysis of problems with free surfaces has been described in refs. [18N, 50N].

TRANSPORT—PROPERTIES

Thermodynamic

A number of papers address the determination of thermodynamic properties. Acoustic gas analyzers for determining binary-gas composition are described [58P, 76P] and a thermodynamic method for measuring local steam quality given [26P]. An image processing system and diamond-anvil cells are used to obtain pressure–volume measurements [74P], gas expansion to determine volumes [33P], and a densitometer for gaining absolute measures of temperature dependence of density, partial volumes, and thermal expansivity of solids and liquids [17P]. Also reported is an improved method for precisely determining the compressibility factor from measurements of refractive index [11P].

Phase transition effects are considered: first the melting lines of simple substances, their thermodynamic similarity and behavior of thermal properties [69P] and second, the relation between thermophysical properties of new ceramic materials and this phenomenon [86P]. Other techniques for measuring properties include a new procedure for obtaining the heat of solid deformation [49P], a new instrument for acoustically measuring complex adiabatic compressibility in liquids [73P] and a simple procedure for determining the Seebeck coefficient of thermoelectric materials [32P].

For specific substances there is a new equation of state for aluminum [40P], measured high pressure densities for argon–helium and argon–neon mixtures [83P], and a transformed psychrometric chart for use in drying technology [70P]. A dual-slope method is presented for measuring specific heats [62P]; other specific heat data are presented for n-pentane, n-hexane and n-heptane at high pressure [20P]. Ni_3Al at low temperature [39P], and some stainless steels and Fe–Ni alloys [16P]. For black phosphorus thermal and elastic properties are reported [89P].

Transport

Interest in special systems and ordinary ones at extremes of temperature continues. A model and calculations of transport coefficients for polyatomic gases are given [81P, 82P] as well as a discussion of the structure of liquids and their thermal properties [21P]. For two-phase materials, a model is presented for estimating transport quantities [1P]. Another study [30P] shows how thermal properties may be estimated by a non-linear least squares method. For electrically conducting materials thermal transport is studied using the transient hot-strip technique [34P]. For chemical-equilibrium flows of partially dissociated and ionized gas mixtures, the effective transport coefficients are defined and computed [80P]. Transport properties of high purity, polycrystalline titanium diboride are reported [85P] and in the area of drying technology some physico-thermal properties of rice bran will be of interest [24P].

Diffusion rates in solids due to shock compression are reported [46P] as is the development of a new fast response probe for studying diffusion in atmospheric air [10P].

Thermal diffusivity measurements are described at low temperature using the laser-flash method [48P], for solids by the two-beam photoacoustic phase measurement [54P], and the quantity defined for shifting composition systems [23P]. The increasing occurrence of special systems undergoing heat transfer leads to the study of equivalent thermal diffusivities of non-homogeneous materials by numerical means [9P], the report of thermal diffusivity in plasma sprayed multi-coatings (apparent) [31P], powders and porous media [61P], ceramics [65P], and uni-directional fiber-reinforced composites [75P].

A number of experiments for measuring thermal conductivity are presented. For systems under hydrostatic pressure techniques are given for determining thermal conductivity (and heat capacity) [3P]; also under pressure, a d.c.–a.c. hot-wire procedure for determining thermal conductivity (and other thermophysical properties) is described [52P]. A transient technique for measuring thermal conductivity is first theoretically developed [56P] and the apparatus described [57P]. For non-homogeneous samples the apparent thermal conductance is measured [51P]. Using a model of a semi-infinite body with a pulsed annular heat source, thermophysical characteristics of materials may be determined without loss of their integrity [66P, 67P].

Improvement in accuracy is the objective of two studies: an experimental and theoretical investigation of the influence of radiative heat transfer on the effective thermal conductivity of liquids [29P] and a proposal, with special respect to connection and radiative effects, for thermal conductivity standards to be used in establishing liquid thermal conductivities [28P]. The instability of thermal conductivity values due to stratification is considered as well [87P]. When using the transient line-source technique it is demonstrated that the end-effect error can be kept so small as not to require correction [47P]. For guarded hot-plates thermal imbalance errors and effective area are discussed [55P]. The general solution for non-linear steady-state heat conduction in a metal conducting electrical current is solved for any temperature variation in thermal and electrical conductivities, allowing interesting comparisons to be made between the variable and constant conductivity cases [90P].

The variety of systems for which thermal conductivity data is required is apparent in the number of papers having this purpose: the measurement of effective thermal conductivity of porous spheres at high temperature [5P], thermal properties of moist porous media below 0°C [78P], and measured thermal conductivity of packed metal powders at different densities and an associated theoretical model [35P, 53P]. The effect of temperature on viscoplastic pore collapse is also described [14P]. In a related vein effective thermal conductivities of loose granular materials are proposed [64P], a heat pulse, line-source method for determining thermal conductivity of consolidated rocks is reported [27P], and the effect of moisture on the thermal conductivity of lightweight aggregate concrete described [71P].

Composite and multilayer systems also attract keen interest as evident by a monograph for calculating the transverse thermal conductivity of uniaxial composite lamina [43P], coated filler composites thermal conductivity [37P], effective thermal conductivity of a composite material with spherical inclusions [22P], and for fiber-reinforced materials [77P].

In the agricultural area experimental thermal conductivities are reported for grains using a line-source method [41P] and for haylage at various bulk densities and moisture content [45P]. The thermal conductivity of tin between 15 and 500°C is given [38P]. Also at temperatures above atmospheric thermal conductivity values are reported for evacuated, transparent silica aerogel tiles [13P], spinel ferrite (400–1000 K) [25P], and plasma sprayed stabilized zirconia and nickel-based coatings [12P]. Factors having an effect on the mean effective thermal conductivity of a melt of glass–plastics are reported [42P].

In the lower temperature regime thermal conductivity results are given for ethane (110–325 K) [63P], pure ice and hydrates [4P], and the effect of plastic deformation on bismuth alloys between 1.5 and 300 K presented [72P]. A simple model of the thermal impedance present in the heat transfer between dielectric solid and superconducting metal with deformation caused defects is found consistent with new and published data [88P].

In the optical area, thermal conductivities of optical coatings are reported [59P]; in the space-vehicle propulsion area thermophysical properties of propellents are given [68P].

Thermal insulations attract the attention of a number of investigators. The nonsteady-state behavior of such materials is discussed [50P], the local thermal conductivity in inhomogeneous glass fiber insulations analyzed [60P] and the temperature variation of the thermal conductivity of self-pumping multilayer insulation reported [36P]. For spray-applied insulations the moisture gain and influence on effective thermal conductivity assessed [6P]. The efficacy of rigid polyurethane foam is analyzed by examining the heat transfer mechanisms through the foam components (gas voids and solid sections) with the prospect of optimizing the system for maximum insulating efficiency [18P]. The influence of lining properties on the control characteristics of industrial furnaces is modeled to obtain wall temperature oscillations [44P].

Where surface tension is a consideration, a cubic equation for its prediction is described [8P].

For contributions to the knowledge of viscosity there are reports on argon at high densities [79P] and the density influence on benzene and methane [84P]. A description of a capillary viscometer for evaluating low-viscosity solutions at elevated temperatures [19P], one for low frequency, low shear rate measurements [7P], and a dual chamber capillary viscometer to study concentrated polymer solutions at elevated temperatures [2P] comprise the experimental techniques report.

HEAT TRANSFER APPLICATION—HEAT PIPES AND HEAT EXCHANGERS

Interest continues to be shown in various schemes for improving heat exchange under a variety of special applications.

For packed beds a two-part study [101Q, 102Q] examines particle to particle heat exchange first by predicting performance characteristics through modeling and simulations and then by examining the design characteristics and performance of a prototype. Other papers report the results of laboratory-scale tests on gas–solid fluidized bed behavior and the influence on bed properties [1Q]; the heat transfer characteristics of a low-pressure-loss fluidized-bed exchanger with single row tubes [4Q]; and a model for enhanced cooling near the edge of a packed bed [17Q]. A review of heat recovery systems from solid particles or gas with very high temperatures [74Q] should be useful. Related work concerns the liquid content and heat transfer in trickle-bed reactors during intensive gas–liquid interaction [68Q] and the development of a high efficiency packed bed heat exchanger for residential application [93Q].

Devices concerned with condensation heat transfer includes horizontal plain and low-finned condenser tubes, examined for the effect of fin spacing and drainage strips on heat transfer and condensate retention [113Q]; a model of heat and mass exchange on a rectification plate with cold reflux [51Q]; and heat recovery from a spray dryer using a glass tube exchanger [19Q].

The role of contact resistance is taken up by a study of the influence of interface thermal contact resistance on the heat transfer performance of prestressed duplex tubes [8Q]. A related paper examines the thermal constriction resistance at the interface of double tubes used increasingly in solar energy applications [109Q]. The importance of crevices formed between tubes and tube plate on heat exchanger operation is considered [2Q].

A general article describes the features of direct contact heat exchangers noting their increased performance, decreased first cost and increased range of application [13Q]. For liquid–liquid direct-contact heat exchangers the temperature jump is considered [41Q].

The augmentation of heat transfer through the use of extended surfaces is explored in a number of papers. The general efficacy of finned tubes in heat exchangers is considered and shown to nearly double the heat transfer accomplished with plain tubes [71Q]. For convective pins with uniform internal heat generation optimal dimensions are sought [85Q]. In a vertical-tube heat and mass exchanger involving an external water film and internal air flow spaced transverse wires are used to promote the process [110Q]. For the case of the bayonet tube heat exchanger the influence of radiation heat transfer is analyzed [58Q] and for a finned wall located in cross-flowing air in a rectangular duct the heat transfer and pressure drop are studied experimentally [64Q]. For industrial finned tubes performance characteristics are given using dimensional parameters [54Q].

The fouling of heat exchangers during use evades solution since relatively little is known about the underlying mechanisms. The precipitation and particulate fouling on the heat transfer surface is considered [53Q]; the scaling of plain and externally finned heat exchanger tubes reported [99Q]; and the results of new investigations of surface fouling by sedimentation and crystallization described [52Q]. The control of heat exchangers with tube-side chemical reaction fouling is simulated [24Q]. To combat fouling ref. [47Q] describes the fluid bed heat exchanger developed in the Netherlands which depends on fluidized particles in the exchanger tubes to break up boundary layers. Specific applications include the effect of ash deposits on heat transfer coefficients in convective boiler tube banks [81Q] and other heat exchangers [63Q]. To study depositions occurring during the processing of milk experiments were conducted on a platinum wire to gauge the effect of acidity and preholding [5Q].

Plate type heat exchangers receive consideration from a number of investigators. Flow distribution, pressure drop and heat transfer are studied experimentally and theoretically [10Q]; also an array of plates aligned at angles to the flow in a duct [56Q]. The flow mechanisms involved in the performance of interrupted-plate exchangers is studied [70Q] and a calculation method for evaporators with vertical channels [30Q]. In a two-part study the fluid flow [112Q] and heat transfer [55Q] in a plate-fin and square-tube exchanger is analyzed. Matrix heat exchangers are studied in ref. [7Q] where a procedure is proposed for calculating the performance characteristics of an exchanger made of perforated plates and in ref. [16Q] where the performance of laminar flow heat transfer in fine matrices is assessed theoretically and confirmed experimentally. A mathematical model of heat and mass transfer on the rotating evaporator surface in a centrifugal molecular still is developed [37Q] and the results given for a wall-type, air-to-air, counter flow exchanger used in ventilation practice [108Q]. Two studies deal with the design aspects of plate heat exchangers, one describing exchanger dimensions and performance in terms of basic dimensionless groups [22Q], the other outlining a procedure for obtaining simple models from the generalized flow diagrams of the exchanger [84Q].

Tube heat exchangers are joined by a new design which is said to minimize the two most serious equipment problems: surface fouling and gasket leakage through a redesign of the transfer surface geometry and fluid flow pattern [62Q]. Other papers describe a variety of approaches to improve exchanger performance: the use of dented tubes [89Q]; the vortex flow in twisted tubes [18Q]; effect of interbaffle spacing on heat transfer and pressure drop [103Q]; and the effectiveness of series assemblies of divided-flow heat exchangers [77Q]. Thermal expansion of a duplex exchanger tube alters the gap-dependent contact resistance at the interface giving rise to more than one steady-state solution [9Q]. Further special considerations include: exact transient solutions of parallel-current transfer processing [57Q]; the dynamic analysis of a multi-concentric cylindrical heat exchanger [105Q]; the determination of transfer coefficients in a short-tube economizer operating in the transition region [75Q]; and a transient exchanger evaluation test for arbitrary fluid inlet temperature variation and longitudinal core conductance [69Q].

Crossflow heat exchange studies include heat transfer in tube bundles at low Reynolds numbers [36Q], the dynamic behavior of such heat exchangers [23Q], and an analysis of crossflow heat exchangers [107Q]. Related to these studies is one examining two-pass counter crossflow heat exchangers with both fluids unmixed throughout [28Q].

An experimental study [66Q] reports shell-side local heat transfer coefficients in single phase flow.

The following papers are distinguished by the special circumstances attending the investigation: feasibility and performance curves for intermittent earth tube heat exchangers [82Q]; the design of multitubular reactors with heat carrier flowing parallel to tubes [104Q]; heat exchange in oil boilers [98Q], and pressure drop and heat transfer in gas-cooled rod bundles [34Q]. In a similar vein, there are reports on heat transfer from parallel-horizontal cylinder rows under corona discharge [49Q]; the theoretical calculation of the thermal efficiency of an evacuated tubular collector [94Q], and an assessment of the air-side performance of air-cooled heat exchangers [12Q].

For design purposes there are some general discussions [25Q, 26Q] as well as specific topics: computer controlled dimensioning of bundled-tube heat exchangers [91Q], direct calculation of exchanger exit temperatures in concurrent flow [83Q]; the design of an apparatus for heat exchange using saturated and underheated superfluid helium (He II) [11Q] and design criteria for nuclear heat exchangers in advanced high temperature reactors [73Q]. Shell-and-tube exchangers having improved design features [87Q] and charts for estimating the performance and design of heat exchangers [106Q] concludes the design aspects.

Unusual application of heat transfer occurs during the use of an inverse finite element to analyze stationary arc welding processes [33Q], a report on platinum black powder as a heat exchanger material at low temperatures [88Q], and the design of aeroassisted orbital transfer vehicle heat shields for a drag brake [79Q].

Heat pipes are a major area of activity. A review paper [97Q] presents the basic concepts, recent innovations in design and applications; others discuss temperature drops in heat pipes [92Q] and the effect of different configurations on performance of heat pipe recovery systems [6Q]. Fluid mechanics effects concern a number of investigators: analysis of vapor flow in a double-walled concentric heat pipe [20Q], the onset of flows and stabilities in parallel loop devices [117Q], flow pattern observations in a circular open thermosyphon [111Q], vapor flow features in slender, cylindrical units (numerical) [72Q], and the initiation of fluid motion in a toroidal heat pipe [116Q]. In space application, the formation of excess liquid in orbital tests of axially grooved heat pipes is reported [67Q].

For closed thermosyphons, the effects of pressure and inclination over a wide range are noted [27Q]; another work [78Q] stresses the effect of inclination on the limiting heat flux, and a third [3Q] considers the heat transfer efficacy of heat pipes with continuous corrugated wicks. For low temperature devices ref. [50Q] considers the influence of heat flow characteristics upon the vapor flow and ref. [49Q] reports on experiments revealing the operating regimes under nearly isothermal conditions. The calculation of heat pipe parameters at high temperature is considered [59Q] and an outdoor method for testing an evacuated tubular collector with heat pipe for solar application described [44Q]. Also for solar conditions, an analysis of a heat pipe absorber array is given [35Q].

Further specific applications consider a heat pipe economy cycle [95Q], a gravity assisted device used with concrete shell steam condensers [45Q] performance of gravity-assisted potassium units [80Q], and the maximum heat flow for a system using a sulfur–iodine mixture (600–900 K) [100Q].

Regenerators receive the attention of a number of workers. A procedure of selecting the optimum matrix for stacked wire gauges is described [29Q] and the thermal performance of diabatic cyclic devices analyzed [31Q]. To improve the cold-side heat transfer coefficient consideration is given to injecting steam because of its radiation properties [39Q]. Two-dimensional heat transfer analysis is used to study high temperature exchange in a closed power cycle MHD regenerator [114Q]. For a regenerator of finite mass the performance of a Stirling engine is reported [42Q].

The thermodynamics of heat exchangers enters from a number of viewpoints. Such considerations appear in the design of long-life, high temperature devices [115Q], in assessing losses due to exchanger irreversibility [43Q] and the entropy generation in such processes [96Q]. Natural circulating water systems are subject to thermodynamic analysis [40Q]. Using load curves, heat exchanger networks are optimized from a second law viewpoint [15Q]. In a two part study of performance of a scraped-surface heat exchanger the principal models for heat transfer and power consumption are reviewed [60Q] and the effect of the axial diffusion of heat [61Q]. Thermal energy recovery systems generally are subjected to thermodynamic optimization [38Q]. Further work on heat exchanger networks include a combined energy-economics approach [32Q], the synthesis of flexible nets for multiperiod operation [21Q], and the influence of pinch phenomena in providing improved network synthesis [86Q].

Several aspects of vapor generators and evaporators receive study. Correlations for rifled tubing in variable pressure boilers are given [65Q], the heat transfer in critical and supercritical zones of sodium–water vapor generators reported [46Q], and a boiler described [14Q] which uses radiator elements for steam generation from solid fuel. For the stationary heating-plane of indirectly heated dryers heat transfer coefficients are provided [76Q] and a monograph devised for predicting the hydrostatic head effect in process plant evaporators [90Q].

HEAT TRANSFER APPLICATIONS—GENERAL

Several papers dealt with heat transfer in electronic and electrical devices. Empirically derived natural convection correlations extracted from the scientific literature were compared with non-dimensionalized data obtained from physical models of uniformly heated electronic circuit cards vertically mounted in a frame [25S]. A thermal model for a hybrid IC in which the conductive heat transfer takes place from one or two large surfaces of the substrate in any combination with the ends was presented [2S]. A simulation model was used to predict the temperature rise for multichip electronic packages [24S]. Methods of effectively cooling computers and high performance electronic modules were presented along with several examples [49S]. The finite element program ANSYS was used to evaluate the temperature distributions, both steady state and transient, for a squirrel cage motor winding [14S]. A generic three-dimensional thermal model was developed for analyzing the thermal behavior of electric vehicle batteries [37S]. The transient three-dimensional conduction equation with a source term that accounts for the propagating resistive wave front has been solved for a superconducting magnet coil [10S]. Temperatures and current decay have been evaluated during superconducting magnet coil quench [28S].

A review was made of numerical computations of air movement and convective heat transfer within buildings [65S]. The effect of a paint coating on the radiative and convective heat transfer from a surface was investigated by treating the coating as a non-gray absorbing, emitting and anisotropically scattering medium [35S]. Simple models of the coupled heat transfer in a typical light fitting were developed [67S]. The dynamic behavior of a liquid convective diode for space heating of passive solar buildings was explained by theory and experiments [26S]. The general conduction equations were presented and solved for large space structures consisting of bar-like members [56S].

Two papers reported numerical and experimental studies of the heat transfer characteristics of a latent heat storage unit containing a circular-finned tube [22S, 23S]. A one-dimensional numerical model of a single tank for stratified thermal storage included the effects of through-flow conditions, and conduction and turbulent mixing within the water [50S]. The assumption of lumped heat capacity in analyzing a thermal storage system was shown to lead to overestimates of the magnitude of the thermal flux in or out of the storage [59S].

A coupled heat and mass transport model was used to simulate water and heat movement in soil under freezing conditions [34S]. The complex variable boundary element method (CVBEM) has also been applied to the problem of predicting freezing fronts in soils [21S]. A study of the energy balance of drying bare soils indicated that the major errors of using null-point calorimetry to estimate soil heat flux and thermal conductivity are likely to arise from net evaporation in upper soil layers [9S]. A thermal analysis provided information on the effectiveness of open field burning on destroying soil bound microorganisms [6S].

The dynamic response of a single-pass crossflow heat exchanger, with both fluids unmixed, to arbitrary time varying inlet temperatures of fluids was investigated analytically [16S]. Modeling studies compare favorably with experimental results for the performance of a horizontal coil ground-coupled heat pump [7S]. An analytical model of transient, simultaneous heat, water and air transfer was used to study drying front movement near low-intensity, impermeable underground heat sources [8S]. The steady-state rates of heat transfer to a cold horizontal pipeline within an atmospheric pressure, air-filled, horizontal rectangular cavity with relatively hot isothermal walls, have been determined experimentally [3S]. Investigations were carried out to study the boiling heat transfer and flow distributions in an evaporator using an oil–fluorocarbon binary mixture [54S]. In a proposed model of the heat transfer in a liquid–liquid spray column, heat is transferred directly from the dispersed phase to the continuous phase as well as indirectly through the wake [60S].

The high temperature insulating properties of unevacuated horizontal multilayer insulating systems were reported [18S], and two design criteria for optimizing performance were presented [29S]. The insulating ability of a multi-layer insulation system, consisting of a few layers on an aluminum taped 77 K surface, was studied experimentally to understand quantitatively how thermal performance changes with the number of multilayers and vacuum level [58S].

Heat loss and angular momentum decay in a constant volume cylindrical vessel is of interest because the flow has features similar to those in internal combustion engines [63S]. The considerations presented in ref. [4S] lead to a proposed method of analyzing the instantaneous heat transfer between working fluid and combustion chamber and included the influences of the turbulent flow field and radiation. Individual states of heat flux distributions and heat losses of the cylinder head, intake and exhaust valves of each stroke have been determined experimentally for a four-stroke gasoline engine [11S]. The effect of gas temperature on the heat transfer coefficient between the gas mixture and cylinder wall of a four-stroke gasoline engine was studied [69S]. The effects of gas flow (swirl and squish) and fuel spray jet on local instantaneous heat flux on the piston crown of a small size direct injection diesel engine were investigated experimentally [18S]. The piston temperature distribution as a function of engine speed and load was analyzed for a diesel engine [27S]. It was shown that the analysis of heat transfer in tube bundle heat exchangers can be applied with high accuracy in predicting cooling of an inline combustion engine [66S]. Experimental measurements were presented of the fundamental frequency, pulsation pressure amplitude and heat transfer coefficient in a natural-gas-fired pulsating combustor [17S].

Individual papers dealt with a variety of devices. An analytical method was presented for computing the cooling performance and pressure loss in internal convective steam-cooling gas turbine blades of return flow type [48S]. In a study of the heat transfer in an axial compressor casing, a non-uniform distribution of the heat transfer coefficients over the casing segments were observed [39S]. Comparisons were made of the heat transfer coefficients of various oil-cooling systems for commercial-vehicle light machines [12S]. Flow visualization studies were used to help explain the effect of working fluid type on the heat transfer in circular open thermosyphons [68S]. Thin film heat flux gauges and flow visualization techniques were used to investigate heat transfer to the walls of a two-dimensional scramjet combusion chamber [44S]. The temperature rise of a spur gear tooth due to a repeated moving heat source was modeled [47S]. Temperature measurements in full circular bearings show that the maximum temperature increases considerably with increasing speed or lubricant viscosity and with decreasing clearance ratio [42S].

The heat transfer aspects of a wide variety of processes were considered. A numerical simulation was used to analyze the effect of air bubblers on glass-melt circulation and heat transfer in a glass-melting tank [62S]. A numerical simulation of a rotary dryer was described [31S]. The intensity of cooling was measured on three commercial coolers of rubber and the results were interpreted with the aid of a simplified, one-dimensional model [19S]. Temperature distributions in the vortex flow target system of an accelerator breeder reactor were predicted [41S]. In gas-driven hydraulic fractures, as occur in rock blasting and underground nuclear testing, the high temperature gases are radically cooled by heat transfer to the host material [15S].

A mathematical model of heat and mass transfer in the cooking of a meat loaf has been derived [20S]. A dimensionless correlation of the heat transfer coefficient between the barrel wall of a food extruder and the extrudate was presented [38S].

The progress in the development of methods, models, and software for analyzing or simulating the flow of heat in welds was summarized [13S]. The contributions and the importance of the different heat transfer modes during arc welding were assessed [55S]. A mathematical model was developed to predict the velocity, temperature, and current density distributions in inert gas welding arcs [40S]. A method to apply forced convection heat transfer by gas jet impingement to weld metals deposited by the GTA weld process was described [64S]. The problem of steady-state and transient heat transport associated with thin-plate welding was formulated and solved using a finite element method [33S].

Two papers described numerical techniques used to analyze the temperature field in the cutting zone during metal cutting [46S, 51S]. The authors of ref. [5S] addressed the contact thermal problem for a friction pair consisting of tool and structural materials that are in continuous plastic sliding contact. Evaporative cutting of a semi-infinite body with a moving CW laser was considered [43S]. Results were presented of an experimental examination of contact heat exchange during metal forming processes [36S]. An analysis was conducted of the temperature distribution in a tubular specimen subjected to high frequency induction heating [45S]. The convective heat transfer coefficients for cylinders under nozzle fields were determined systematically in a preheating chamber model as a function of the nozzle field parameters for the application of the reheating round billets.

An initial solidification analysis in the vicinity of the meniscus in continuous casting was performed [61S]. Research examined the influence of withdrawal speed and the frequency of mold oscillation on heat transfer through the mold walls [53S].

Transient temperature distributions in human skin and subdermal part exposed to cool environment with negligible perspiration and moderate environment were investigated [57S].

The concept of resonant heat transfer enhancement based on excitation of shear-layer instabilities present in internal separated flows was introduced [52S].

The small-scale structure of forced, turbulent flows developed after Taylor and Kolmogorov was extended to that of buoyancy-driven flows [1S].

SOLAR ENERGY

The U.S. National Bureau of Standards solar collector test program was reviewed and recommendations made regarding the use and limitations of the thermal performance measurements [51T]. An analytical solution is presented for heat transfer in flat plate collectors that accounts for duct design and the thermal developing region [33T]. An analysis is developed for a flat plate air heating collector in which a partial second glazing improves the thermal efficiency for long absorber plates [12T]. Collector efficiency vs Graetz number is given for air flow in the triangular ducts of a V-groove collector [19T]. Simulated performance of a compound parabolic concentrator is compared with corresponding flat plate collectors in which low concentration CPC designs are shown to have superior thermal performance [20T]. Experiments and finite difference numerical analysis are performed to study the effects of eleven design and operating parameters for an evacuated tube collector [40T]. The thermal efficiency of a co-axial heat pipe collector was measured for half-hour periods to determine the effect of mean fluid to ambient temperature difference [4T]. The effect of optical depth on a solar collector in which the solar energy is collected directly by a semi-transparent working fluid was investigated theoretically and experimentally [7T]. A two-dimensional analytical solution is developed which is used to simulate the direct absorption of solar energy within a liquid film flowing downward over a solid wall within a cavity receiver [31T]. Single and two-phase flow heat transfer is studied in the helical absorber tubes at the focal point of a paraboloidal concentrator [49T]. A method is established which can be used to measure complex radiative–convective heat transfer in high temperature solar receivers [36T]. A network heat transfer model is presented which predicts the temperature distribution in terrestrial photovoltaic arrays [24T].

A steady-state method of predicting the monthly thermal performance of solar ponds is compared with detailed hourly numerical simulations [25T]. Good agreement is found for both monthly and annual thermal performance. Heat transfer models for steady and unsteady operation of a shallow solar pond are presented and the results are compared with corresponding experimental data [30T]. The results of a nine month test of a small salt-gradient solar pond are presented [16T]. Experimental temperature and salt concentration measurements are given for various lower convective layer temperatures in 5°C increments [10T]. A comprehensive literature review of double-diffusive effects on solar ponds is made [34T]. One-dimensional numerical simulations are used to determine the optimum thickness of the non-convective zone in a solar pond [8T]. The results indicate that the optimum thickness changes from month to month. The effect on heat loss through various materials below a solar pond is analyzed [9T]. An analytical stability analysis for the non-convective zone of a solar pond is given in which the logarithms of the concentration and thermal gradients are found to be related by a simple equation [11T]. Experiments indicated that instabilities occur at the location of minimum salinity gradient and these instabilities are confined to a narrow region [29T]. Alternate heat removal mechanisms are analyzed which shows that heat extraction can be increased as much as 62% over convectional solar pond heat extraction techniques [32T].

A model is presented for stratified liquid sensible heat storage systems when the fluid enters at the location of zero temperature difference [15T]. Various mathematical models are compared to predict the charging time required for sensible liquid storage tanks [38T]. A storage system is modeled using a random walk procedure to determine the distribution of energy stored [35T]. A study is given in which the geometry and heat transfer rate to a packed bed thermal storage system is analyzed [14T]. Accumulator heat exchangers are studied with applications to solar absorption cooling systems [42T]. The thermal performance of a 30 kW storage unit using polyethylene rods and ethylene glycol was evaluated experimentally [27T]. An explicit one-dimensional finite difference model was employed to simulate the heat transfer characteristics measured in the previous reference [28T]. An experimental study was performed on the heat transfer of calcium chloride hexahydrate surrounding a coaxial water filled heat exchanger pipe [54T]. Two different stacking and baffle arrangements were tested with phase change material rods located under the floor of a solar heated greenhouse [23T].

A theoretical and experimental study of a thermosyphon solar water heater system showed that overall thermal efficiency was higher for a flat plate absorber than a tube-in-sheet absorber [55T]. A theoretical model has been developed to predict the performance of thermosyphon solar water heaters with a heat exchanger in the storage tank under various load conditions [39T]. The lack of thermal insulation on the upper pipe of a thermosyphonic hot water system was found to be very detrimental to system performance as it may result in backflow from the storage tank to the collector [50T]. The thermal performance of a collector array on a residence was determined including the effective array absorptivity and loss coefficient [26T]. Simple analysis of integral solar water heaters were found to be limited in their ability to provide an accurate guide to the thermal performance under actual service conditions [48T]. A lumped parameter model is given for an integral water heating collector which shows good agreement with experimentally measured mean water temperature [56T]. Storage of solar energy in a mixture of sand and iron filings that surrounds a buried water pipe is analyzed to determine the effect of service conditions on the overall efficiency [12T]. A theoretical study of solar thermal storage in the ground is presented with applications to drying [1T].

Activated charcoal and methanol were used to construct a diurnal cycle solar powered ice maker that produced 35 kg of ice on a sunny day with 6 m^2 of collector area [37T]. An analysis is presented of a closed type solar powered adsorption cooling system which showed good agreement with existing experimental data [41T]. A method is devised to model a solar powered absorber–generator to determine the temperature of the absorbent [44T].

A simple algebraic model for evaluation of the average performance of passively heated residences is given which is suitable for both direct gain and storage wall systems [22T]. The development and application of a thermal network model is presented to simulate the thermal performance of passively heated direct gain rooms [3T]. The operating characteristics of a Trombe wall for preheating ventilation air in a farrowing house were measured [45T]. Experimental measurements on a lattice solar wall show that it performs better than a convectional solar wall [53T]. Solutions to thermal networks are given representing simulation of structures with a flat brick roof or a lightweight adobe domed roof with and without moist interior surfaces [5T]. The effect of ventilation air flow rate is found to be significant in passively heated solar structures [47T].

Seasonally dependent correlations are developed to predict the daily ratio of diffuse to total radiation incident on a horizontal surface [46T]. An analytic method is described for calculating daily averages of solar angles and air mass from long-term average data [21T]. A linear correlation between monthly average daily global solar radiation and sunshine duration is presented which agrees with annual values to within 4% [6T]. The distribution of atmospheric aerosol is shown to cause an error in the Langley method of determining the solar constant by up to 3% [43T]. The solar transmissivity as a function of incidence angle is measured with a solar simulator for a variety of materials and coating [17T]. Two models are presented to predict the solar transmission through layers of water and glass representing a stack of glass bottles in a liquid [52T]. Solar radiation propagating through turbid media is modeled analytically to simulate the photon flux through plant tissue [18T]. Relationships between chemical treatments of stainless steel and the optical properties of the coatings are discussed relevant to using the coatings as a solar absorber surface [2T].

PLASMA HEAT TRANSFER

The number of papers related to plasma heat transfer showed a substantial increase during the past year. Many papers dealt with electric arcs and their applications.

The stable configuration of an arc in crossflow is due to heating at the leading edge and enhanced cooling of the trailing section. This is achieved by gradients of ohmic heating and heat conduction across the arc caused by its curvature [48U]. In arcs with superimposed flows, turbulent heat and mass transfer frequently exerts a decisive influence on the shape of the arc and on its energy characteristics [82U]. A semiempirical model of turbulence is proposed for describing the interaction of an electric arc and a turbulent gas flow [4U]. Calculated characteristics of an electric arc burning in a turbulent air flow in a long cylindrical channel are compared with experimental data [3U]. Theoretical studies of the thermal contraction of an arc channel at the anode passing through a turbulent plasma boundary layer, indicate that an instability in the heat balance develops along the falling part of the current–voltage characteristic [36U].

Studies of the physical processes in gas–tungsten arcs (welding arcs) indicate that the addition of 0.1% cerium to an argon arc can lead to marked changes of the arc properties [27U]. Spectrometric measurements of the temperature fields in gas–tungsten arcs using water cooled and molten anodes indicate little temperature variation in spite of anode vapor concentrations of up to 2500 p.p.m. in the vicinity of the anode in the case of a molten anode [21U]. The composition as well as the electric and thermal conductivity of a Cu–air plasma have been calculated at atmospheric pressure for temperatures from 5000 to 14 000 K and for different Cu content. The properties of the plasma change substantially in the presence of metal vapor [61U]. A spectroscopic analysis of the plasma generated by a double-flux tungsten inert gas (TIG) arc torch shows that the plasma is dominated by metallic vapor species in the vicinity of the molten anode, while a nearly pure argon plasma is observed in the cathode region of the arc [14U]. There is a strong correlation between arc voltage and anode erosion in an N_2 arc at atmospheric pressure. Metal vapor in the arc seems to decrease the electrical conductivity resulting in a corresponding voltage increase [68U]. Contamination of an arc plasma by electrode vapor depends strongly on the existence of electrode jets which are responsible for carrying and distributing such contaminants [41U].

A computerized spectroscopic system for analyzing LTE and non-LTE plasmas has been developed with automatic correction for self-absorption [73U]. The magnetic field used to rotate an arc also enhances the spectral lines indicating a non-LTE condition [22U]. An approach is presented for handling radiative transfer in high pressure plasmas where collision broadening dominates self-absorption [83U]. Studies of long free-burning vertical arcs in atmospheric air reveal axis temperatures around 7000 K [47U].

Optical measurements, including spectrometric temperature measurements in SF_6 high current arcs (up to 2.6 kA) are in good agreement with calculations based on LTE [29U]. Using differential interferometry, the temperature distribution in the upstream and downstream regions of a dual flow orifice nozzle arc interrupter have been measured with emphasis on the thermal boundary layer (arc mantle) surrounding the arc [69U]. Results of investigations of the nozzle ablation process in high-power arc circuit interrupters are reported which include radiative emission and absorption measurements [32U]. Local overheating of the contacts of a circuit breaker may be avoided by designing the contacts so that their magnetic field induces rotation of the arc [30U]. An improved model for circuit-breaker arcs makes it possible to predict the breaker thermal short-circuit interruption performance within the entire range of practical interest [50U]. The observed nozzle ablation in SF_6 circuit breakers is mainly determined by the geometry of the arcing device [53U]. An ablation arc model applied to ablation controlled arcs allows to quantify the phenomena related to nozzle clogging in gas-blast circuit breakers, namely flow blocking and reverse flow heating [71U]. The dynamic behavior of nozzle arcs is governed by the close coupling between the arc and its external flow [20U]. The post-arc heat diffusion from an arc developed due to flash over on open insulation may play a decisive role in the recovering of the dielectric strength of the ruptured gap [78U].

A method has been developed for measuring the net heat flux to a hot evaporating cathode in a vacuum arc with diffuse cathode emission [62U]. A dynamic model of the cool-down phase of an inactive cathode crater of a vacuum arc is used for determining the solidification time for Cu and W taking heat conduction, phase changes, the motion of the melt and surface cooling due to evaporation and radiation into account [66U]. Studies of the spot behavior and of local cathode heating in high-current (900–3000 A) and low-current (20–60 A) arcs show that three different conditions can occur leading to local heat accumulation at the cathode: spot grouping, prolonged spot attachment, and repeated return to previous attachment sites of individual spots [9U]. Investigations of the cathode arc plasma flow in a Knudsen layer show that the evaporation rate in the cathode spot depends on the processes providing the ion flux in the cathode region which is required to maintain the thermal balance of the cathode [6U]. Experimental data of cathode crater formation in vacuum arcs are in good agreement with an analytical model taking phase changes, Joule and ionic heating, electron emission, and mass loss due to evaporation and ejection of molten metal into account [67U]. The dominant erosion mechanism of cathodes in a vacuum-arc centrifuge seems to be via ions which return to the cathode surface across a collisionless sheath [65U].

Experimental studies of arc-generated flow phenomena in repetitively pulsed gas flow spark gaps are in qualitative agreement with analytical studies confirming that heated gas convects at the undisturbed gas velocity [45U]. Based on experimental data, a simple model has been developed which describes the recovery of a spark gap with gas flow [46U]. During the arc phase of an automotive spark in air, the maximum temperature of the arc plasma was found to be 4600 K derived from the slope of the continuum radiation [63U]. Studies of the radial variation of plasma parameters in a pulsed low-pressure (100–150 Pa) arc in He indicate a maximum electron density of 3.3×10^{21} m^{-3} and a maximum electron temperature of 3.85 eV [1U]. An electric arc heat source may move with speeds comparable to the rate of heat propagation. Results based on a generalized model of this situation show that the largest value of the temperature is reached considerably sooner than by neglecting the inertia of the heat source [39U].

A transferred arc plasma reactor has been used for arc melting, reprocessing, and upgrading of critical materials and alloys [70U]. Besides other heat transfer mechanisms, conditions due to evaporation processes have been included in the analysis of heat transfer by water cooling of wall segments of arc furnaces with emphasis on design criteria not to exceed the critical heat flux [72U]. Studies of the molten metal pool, in a vacuum arc melting furnace indicate that the pool exerts a strong influence on the ingot, especially when melting of titanium alloys from the compacted electrode is considered [31U]. A model has been developed for describing thermoconvective emissions from an open vault arc furnace for determining the proper procedure and dimensioning of a system for capturing and control of fumes, dust, and energy [58U]. A single-roller rapid quenching device with a plasma-arc torch and a water cooled Cu hearth has been developed for rapid quenching of high melting temperature alloys. Amorphous ribbons of $Nb_{80}Si_{20}$, Nb_3Ge, and Ta-(Si, B) alloys have been obtained and amorphous Nb_3Ge showed a superconducting transition temperature of 18.3 K after heat treatment [44U].

Measurements of heat and momentum transfer of powder particles injected into an atmospheric pressure plasma jet arc compared with predicted values [80U]. A simple analytical method, the Z-potential method, is proposed for particle–gas mass transfer calculations under plasma conditions [33U]. Measurements of heat fluxes to spherical models and plane surfaces exposed to plasma jets are compared with experimental results of other authors [16U]. A numerical model describing melting of small particles injected into a thermal plasma jet is used to predict optimal parameters for plasma spraying [26U]. Heat fluxes to a sphere exposed to a two-temperature plasma in the molecular flow regime are independent of the sphere size and approximately proportional to the gas pressure [12U]. Results are presented of a comparison of the effectiveness of heating disperse polymers in gas flames and in plasma streams [37U]. LDA measurements of the velocities of alumina particles injected into a plasma jet reveal values from 20 to 60 m s^{-1} [49U].

Transport coefficients of a chemically reacting plasma in the presence of a magnetic field may be calculated by using the Chapman–Enskog approach. Sample calculations for a potassium-seeded argon plasma are presented [56U]. Calculations of thermodynamic and transport properties of SF_6–N_2 mixtures in the temperature range from 1000 to 30 000 K and pressures from 1 to 10 atm indicate that chemical reactions exert a strong influence on thermal conduction and heat capacities [24U]. The total rate coefficients used in a simplified collisional-radiative (SCR) model have been calculated for thermal nitrogen plasmas at atmospheric pressure [5U]. Significant departures from standard transport coefficients have been found for the electron current and heat flux in a fully ionized plasma with errors up to 65% [18U]. Results based on a model describing the structure of a plane and stationary ionizing current sheath taking radiation transport into account, indicate that radiative losses can be very high, reaching about 30% of the available power [8U]. Effective recombination and ionization coefficients are calculated for optically thin and optically thick non-equilibrium nitrogen plasmas, using a time-dependent approach based on a collisional-radiative model [77U].

A strong vortex producing a 'gas tunnel' for an electric arc proves to be a useful tool for producing high energy density plasma jets [2U]. Temperature measurements in the jet of an inductively-coupled nitrogen plasma at pressures of 0.1 and 0.3 atm confirm previous conclusions that the rotational temperatures are identical with the gas temperature [23U]. High-pressure argon plasmas with temperatures up to 11 000 K have been produced by gas compression. Comparisons with analytical results show the importance of radiation losses and boundary layer effects [34U]. Numerical results of swirl flow cooling in high-heat-flux particle beam targets and swirl-flow-based plasma limiters indicate that local heat fluxes in excess of 7 kW cm^{-2} occur at the water-cooled surface on the side exposed to the beam [55U].

For an Nd-glass laser ($\lambda = 1.06$ μm, $p \sim 10^{16}$ W cm^{-2}) interacting with a plasma ($n \approx 10^{17}$ cm^{-3}), the induced magnetic field is found to be of the order of 100 T, affecting electron thermal conduction and inhibiting heat transport [11U]. Two types of discharge heated longitudinal Sr^+ recombination lasers have been investigated: a low heat loss configuration using input powers of 350 W, and a high heat loss configuration with 1 kW power input to the discharge tube [51U]. Studies are reported on how the effects of deviations from Maxwell–Boltzman electron energy distribution, that are characteristic of heat transport in laser-produced plasmas, could modify the interpretation of such experiments [19U]. Calculated plasma size, peak temperatures, global absorption characteristics, and thermal efficiencies of laser-sustained plasmas in axisymmetric flows are in good agreement with experiments [25U]. Results of a study of laser-sustained plasmas in flowing argon using a pressurized absorption chamber and a 10 kW CO_2 laser indicate a total absorption of up to 80% and thermal conversion efficiencies in the range from 6 to 25%, depending on pressure, flow rate, and laser power [43U]. Quantitative results of another study of laser-sustained plasmas in argon flow clearly indicate that a perceptive analysis of such laser-sustained plasmas must take into consideration the two-dimensionality of both the flow field and the laser energy distribution in the focused beam [35U].

Synthesis of ultrafine SiC powders in thermal arc plasmas reveals a bimodal distribution of the particle sizes with the majority of the particles falling in a size range from 2 to 40 nm [40U]. The nitriding of niobium and tantalum, with argon–nitrogen and argon–nitrogen–hydrogen plasma jets at pressures of 190 and 240 Torr results in higher nitriding rates than those of thermal nitriding at the same temperature [42U]. Chromic oxide decomposition in r.f. argon plasmas shows that the thermal decomposition conversion of Cr_2O_3 into Cr is approximately eight times higher in the homogeneous gas phase than in the solid phase [54U]. Studies of an r.f. silane plasma show that discharge power as well as electrode heating increase the rotational temperature of silane [28U].

For a better understanding of the cathode behavior in an MPD arc jet, a transient heat transfer analysis has been performed based on solutions of the nonlinear transient energy equation [52U]. Ion thruster performance can be affected by cooling with liquid nitrogen. Good agreement exists between the temperature-induced effects predicted by a simple discharge model and measured data [81U].

The heat transfer coefficients of parallel-horizontal rows of cylinders exposed to a corona discharge show an increase as the corona current is increased [38U]. An analysis of the conduction-phase characteristics of a hydrogen thyratron plasma is based on the simultaneous solutions of the Boltzmann equation, the rate equations, and the radiative transfer equation [64U]. An explicit solution is obtained for the temperature distribution inside a cylindrical rod with an insulated inner core when the rod is allowed to enter into a fluid of large dimensions with uniform speed, and a simple integral expression is derived for the value of the sputtering temperature of the rod at the point of entry [10U]. Experimental Schlieren images of the non-cylindrical plasma generated by a plasma focus device can be simulated by computers, allowing a quantitative analysis of these Schlieren images, resulting in electron densities and electron density gradients [60U]. A plasma-electrolytic heating mechanism is described in which an active electrode is heated by charged particle bombardment and is cooled by heat transfer with the surrounding liquid via a vapor–gas shell [75U].

Studies of MHD heat transfer in cylindrical geometry with a discontinuity in the wall temperature demonstrate that the temperature falls as the Hartmann number is increased, and convection dominates for large values of the Peclet number [74U]. An exact analysis of a generalized MHD Couette flow is presented, taking Hall currents into account [76U]. Considering a steady two-dimensional MHD boundary-layer flow on a flat plate with a uniform transpiration through the constant temperature plate, it is found that heat transfer at the plate increases with Joule heating, the transpiration number, the magnetic interaction number, and the Prandtl number [13U]. Exact solutions for hydromagnetic boundary layer flow and heat transfer over a continuous, moving, flat surface with uniform suction and internal heat generation/absorption are presented [79U]. Studies of unsteady, free convection flow of an incompressible electrically conducting viscous liquid through a porous plate in the presence of a transverse magnetic field show the effect of various dimensionless parameters on velocity and temperature distribution, the skin friction, and the heat flux [17U].

Studies of the simultaneous effects of the magnetic field, mass transfer and heat transfer on the steady incompressible laminar boundary layer flow of an electrically conducting fluid over a non-isothermal cone show that the magnetic field reduces both the skin friction and heat transfer [15U]. The effects of the magnetic interaction parameters, the electromagnetic loading parameter and the Hall currents have been studied on the laminar compressible boundary layer physical parameters such as the wall shear stress, the heat flux at the wall and the thickness of the boundary layer [59U].

Multiplex CARS temperature measurements in a coal-fired MHD environment indicate maximum temperatures of 2500 K [7U]. The MHD effects on liquid metal cooling of fusion reactors has been experimentally studied with emphasis on temperature fluctuations [57U].

BOOKS

Heat Transfer & Fluid Flow Data Books with Updating Service. Genium, 1145 Catalyn St., Schenectady, NY 12303-1836 (1986).

D. P. Dewitt and L. F. Albright (Editors), *Measurements of Temperatures in High-temperature Furnaces and Processes* (AIChE Symposium Series, Vol. 82, No. 249). AIChE, 345 East 47 St., New York, NY 10017 (1986).

W. B. Krantz, D. T. Wasan and R. K. Jain (Editors), *Thin Liquid Film Phenomena* (AIChE Symposium Series, Vol. 82, No. 252). AIChE Publ. Sales Dept. C, 345 East 47 St., New York, NY 10017 (1986), ISBN 0-8169-0397-2, LC80-25573, Pub. S-252.

R. J. Gordon (Series Editor), *Transport Phenomena—Special Topics*(AIChEMI Series C: Transport, Vol. 6). AIChE Publ. Sales Dept. C, 345 East 47 St., New York, NY 10017 (1986), ISBN 0-8169-0397-2, LC80-25573, Pub. J-31.

A. E. Sheidlin, *High Temperature Studies*. Hemisphere, Washington, D.C. (1986).

Y. Mori, A. E. Sheindlin and N. Afgam, *High Temperature Heat Exchangers*. Hemisphere, Washington, D.C. (1986).

M. Tamois, *Radiation and Combined Heat Transfer in Channels*. Hemisphere, Washington, D.C. (1986).

W.-J. Yang (Editor), *Heat Transfer and Fluid Flow in Rotating Machinery*. Hemisphere, Washington, D.C. (1986).

K. J. McNaughton (Editor), *The Chemical Engineering Guide to Heat Transfer*. Hemisphere, Washington, D.C. (1986).

J. W. Palen (Editor), *Heat Exchanger Sourcebook*. Hemisphere, Washington, D.C. (1986).

J. P. Gupta, *Fundamentals of Heat Exchanger and Pressure Vessel Technology*. Hemisphere, Washington, D.C. (1986).

Heat and Technology, Vol. 4. CNR, University of Bologna, ENEA, UIT, Italy (1986).

Y. Jaluria and K. E. Torrance, *Computational Heat Transfer*. Hemisphere, Washington, D.C. (1986).

Encyclopedia of Fluid Mechanics. Gulf, Houston, Texas (1986): Vol. 1, *Flow Phenomena and Measurement*; Vol. 2, *Dynamics of Single-fluid Flows and Mixing*; Vol. 3, *Gas–Liquid Flows*; Vol. 4, *Solids and Gas–Solids Flows*; Vol. 5, *Slurry Flow Technology*.

Handbook of Heat and Mass Transfer. Gulf, Houston, Texas (1986): Vol. 1, *Heat Transfer Operations*; Vol. 2, *Mass Transfer and Reactor Design*.

CONDUCTION

1A. A. Abtahi, P. F. Braunlich and P. Kelly, Theory of transient temperature response of a two-layer system heated with a localized laser beam, *J. Appl. Phys.* **60**(10), 3417 (1986).

2A. G. Adomian, New approach to the heat equation—an application of the decomposition method, *J. Math. Analysis Applic.* **113**(1), 202 (1986).

3A. N. M. Alnajem and M. N. Ozisik, A direct analytical approach for solving two-dimensional linear inverse heat conduction problems, *Wärme- und Stoffübertr.* **20**(2), 89 (1986).

4A. R. V. Arimilli, M. Parang and P. R. Surapaneni, Heat transfer from circular tubes in a semi-infinite medium, *J. Heat Transfer* **108**(3), 703 (1986).

5A. A. Azarkhin and J. R. Barber, Thermoelastic instability for the transient contact problem of two sliding half-planes, *J. Appl. Mech.* **53**(3), 565 (1986).

6A. D. L. Balageas and A. M. Luc, Transient thermal behavior of directional reinforced composites: applicability limits of homogeneous property model, *AIAA J.* **24**(1), 109 (1986).

7A. I. E. Balashova, Identification of a specified thermal regime in a structure on the basis of experimental data obtained in other regimes, *J. Engng Phys.* **50**(3), 346 (1986).

8A. M. Bamberger and B. Prinz, Determination of heat transfer coefficients during water cooling of metals, *Mater. Sci. Technol.* **2**(4), 410 (1986).

9A. H. Barrow, Theoretical solution for the temperature in a straight fin with variable surface heat transfer coefficient, *J. Heat Recovery Sys.* **6**(6), 465 (1986).

10A. G. C. J. Bart, C. J. Hoogendoorn and P. B. J. Schaareman, Stationary and transient heat conduction in a nonhomogeneous material, *Wärme- und Stoffübertr.* **20**(4), 269 (1986).

11A. H. F. Bauer, Steady conduction of heat in paraboloidal systems, *Numer. Heat Transfer* **10**(4), 395 (1986).

12A. H. F. Bauer, Steady conduction of heat in conical systems, *Numer. Heat Transfer* **10**(6), 619 (1986).

13A. J. V. Beck and D. A. Murio, Combined function specification–regularization procedure for solution of inverse heat conduction problem, *AIAA J.* **24**(1), 180 (1986).

14A. J. V. Beck, Green's functions and numbering system for transient heat conduction, *AIAA J.* **24**(2), 327 (1986).

15A. M. V. Belostotskii, Modeling of cooling–heating processes, *Theor. Found. Chem. Engng* **20**(1), 32 (1986).

16A. C. Bernard, D. Gobin and A. Zanoli, Moving boundary problem: heat conduction in the solid phase of a phase-change material during melting driven by natu ral convection in the liquid, *Int. J. Heat Mass Transfer* **29**(11), 1669 (1986).

17A. Y. Benveniste and T. Miloh, The effective conductivity of composites with imperfect thermal contact at constituent interfaces, *Int. J. Engng Sci.* **24**(9), 1587 (1986).

18A. A. Buccini and H. M. Soliman, Optimum dimensions of annular fin assemblies, *J. Heat Transfer* **108**(2), 459 (1986).

19A. M. B. Carroll and S. W. Churchill, A numerical study of periodic on–off versus continuous heating by conduction, *Numer. Heat Transfer* **10**(3), 297 (1986).

20A. A. Chakrabarti, Cooling of a composite slab, *Appl. Scient Res.* **43**(3), 213 (1986).

21A. E. Chang and A. Acrivos, Rate of heat conduction

from a heated sphere to a matrix containing passive spheres of a different conductivity, *J. Appl. Phys.* **59**(10), 3375 (1986).

22A. T. K. Chattopadhyay and A. K. De, Infinity in conduction of heat, *J. Inst. Engrs* (*India*) Part ME **66**(5), 114 (1986).

23A. K. L. Chowdhury, On the isovectors of a class of nonlinear diffusion equations, *Int. J. Engng Sci.* **24**(10), 1597 (1986).

24A. B. H. Coleman, W. J. Hrusa and D. R. Owen, Stability of equilibrium for a nonlinear hyperbolic system describing heat propagation by second sound in solids, *Archs Ration. Mech. Analysis* **94**(3), 267 (1986).

25A. R. M. Cotta and M. N. Ozisik, Eigenvalues basic to diffusion in a hollow cylinder without azimuthal symmetry, *Ann. Nucl. Energy* **13**(2), 93 (1986).

26A. M. K. El-Adawi and E. F. Elshehawey, Heating a slab induced by a time-dependent laser irradiance—an exact solution, *J. Appl. Phys.* **60**(7), 2250 (1986).

27A. D. V. Evans and D. S. Riley, The solution of some diffusion problems by Fourier transform methods, *Int. J. Heat Mass Transfer* **29**(9), 1339 (1986).

28A. H. Fenech, Thermal conductance at solid interfaces: an application of the Fenech–Rohsenow model, *Int. J. Heat Mass Transfer* **29**(8), 1103 (1986).

29A. G. P. Flach and M. N. Ozisik, Whole domain function-specification method for linear inverse heat conduction, *Ann. Nucl. Energy* **13**(6), 325 (1986).

30A. D. E. Glass, M. N. Ozisik, D. S. McRae and B. Vick, Hyperbolic heat conduction with temperature-dependent thermal conductivity, *J. Appl. Phys.* **59**(6), 1861 (1986).

31A. H. Hatta and M. Taya, Equivalent inclusion method for steady state heat conduction in composites, *Int. J. Engng Sci.* **24**(7), 1159 (1986).

32A. E. Hensel and R. G. Hills, An initial value approach to the inverse heat conduction problem, *J. Heat Transfer* **108**(2), 248 (1986).

33A. D. M. Herbert, The thermal stability of a partially insulated reactive slab, *Q. J. Mech. Appl. Math.* **39**(2), 197 (1986).

34A. C. K. Hsieh and A. J. Kassab, A general method for the solution of inverse heat conduction problems with partially unknown system geometries, *Int. J. Heat Mass Transfer* **29**(1), 47 (1986).

35A. D. E. Hutchison and M. P. Norton, Applicability of stochastic process theory to heat conduction in solids with random temperature fields, *Appl. Energy* **22**(4), 241 (1986).

36A. H. L. Johnson, A third mixed boundary value problem on a sphere, *Q. Appl. Math.* **44**(2), 201 (1986).

37A. G. F. Jones, Temperature and heat-flux distributions in a strip-heated composite slab, *J. Heat Transfer* **108**(1), 226 (1986).

38A. P. Knabner, Global existence in a general Stefan-like problem, *J. Math. Analysis Applic.* **115**(2), 543 (1986).

39A. V. A. Kudinov and A. D. Roslyakov, Coordinate system construction in solving heat conduction problems for multilayer structures, *Soviet Aeronaut.* **29**(3), 85 (1986).

40A. T. Kuroyanagi, Transformation of Dahamel's integral in the inverse heat conduction problem, *Bull. J.S.M.E.* **29**(255), 2953 (1986).

41A. T. Kuroyanagi, Surface temperature and surface heat flux determination of the inverse heat conduction problem, *Bull. J.S.M.E.* **29**(255), 2961 (1986).

42A. C. K. Liu, The role of boundary conditions on the thermal stability of a Landau slab, *J. Franklin Inst.* **321**(2), 61 (1986).

43A. C. V. Madhusudana and L. S. Fletcher, Contact heat transfer—the last decade, *AIAA J.* **24**(3), 510 (1986).

44A. M. Mashena and A. Haji-Sheikh, An integral solution of moving boundary problems, *Int. J. Heat Mass Transfer* **29**(2), 317 (1986).

45A. M. F. Modest and H. Abakians, Heat conduction in a moving semi-infinite solid subjected to pulsed laser irradiation, *J. Heat Transfer* **108**(3), 597 (1986).

46A. V. T. Morgan and N. G. Barton, The time-dependent temperature distribution in a cylindrical conductor with Joule heating and temperature-dependent properties, *J. Phys. D: Appl. Phys.* **19**(6), 975 (1986).

47A. V. T. Morgan and N. G. Barton, The time-dependent temperature distribution in a cylindrical conductor with Joule heating and temperature-dependent properties, *J. Phys. D: Appl. Phys.* **19**(6), 975 (1986).

48A. S. Nair, Heat conduction in a thin circular pipe due to a circumferentially moving heat source, *Q. Appl. Math.* **43**(4), 421 (1986).

49A. S. Nomura and T. W. Chou, Heat conduction in composite materials due to oscillating temperature field, *Int. J. Engng Sci.* **24**(5), 643 (1986).

50A. L. Oter Duthoit and P. Thery, Application of the theory of systems to the calculation methods of flow in a multilayer wall in variable conditions, *Revue Gen. Therm.* **25**(290), 89 (1986).

51A. C. Rogers, On a class of moving boundary problems in non-linear heat conduction: application of a Backlund transformation, *Int. J. Non-linear Mech.* **21**(4), 249 (1986).

52A. C. Rogers, D. W. Barclay and D. L. Clements, Pade extension of Bergman-series solutions to nonlinear boundary-value problems in heat conduction, *J. Engng Math.* **20**(2), 145 (1986).

53A. B. J. Rozon, P. F. Galpin, G. E. Schneider and M. M. Yovanovich, Effect of geometry on contact conductance of contiguous interfaces, *J. Spacecraft Rockets* **23**(3), 225 (1986).

54A. M. B. Rubin, Heat conduction in plates and shells with emphasis on a conical shell, *Int. J. Solids Struct.* **22**(5), 527 (1986).

55A. A. K. Sen and S. Trinh, An exact solution for the rate of heat transfer from a rectangular fin governed by a power law-type temperature dependence, *J. Heat Transfer* **108**(2), 457 (1986).

56A. T. Shitara, S. Matsumoto and M. Suzuki, New numerical method for one-dimensional unsteady heat conduction problem within laminated binary materials, *Heat Transfer—Jap. Res.* **15**(4), 1 (1986).

57A. B. Snaith, S. D. Probert and P. W. O'Callaghan, Thermal resistances of pressed contacts, *Appl. Energy* **22**(1), 31 (1986).

58A. M. Spiga, Temperature in circular tubes with azimuthal disuniform heating, *Wärme- und Stoffübertr.*

20(3), 207 (1986).

59A. Y. Tanigawa and M. Kosako, Transient thermal stress analysis of an infinite medium with a spherical cavity due to a nonuniform heat supply, *Int. J. Engng Sci.* **24**(3), 309 (1986).

60A. L. N. Tao, Method for solving moving boundary problems, *SIAM J. Appl. Math.* **46**(2), 254 (1986).

61A. J. R. Thomas, Jr. and M. Gundappa, Transient conduction in a cylinder in an infinite conductive medium with contact resistance, *AIAA J.* **24**(2), 349 (1986).

62A. N. M. Tsirel'man, Correctness of an analytical solution of an inverse problem of heat conduction, *J. Engng Phys.* **49**(6), 1492 (1986).

63A. P. V. Tsoi, Development of methods of calculating nonsteady heat transfer, *High Temp.* **24**(3), 388 (1986).

64A. T. Tsukada, S. Noguchi and K. Yanagi, Contact heat transfer between spherical and plane surfaces with microirregularities, *J. JSLE Int. Edn* No. 7, 5 (1986).

65A. V. B. Veselovskii and N. I. Voitsekhovskii, Calculation of temperature field of cyclically loaded bar, *Soviet Aeronaut.* **29**(3), 62 (1986).

66A. D. Waechter, P. Schvan, R. E. Thomas and N. G. Tarr, Modeling of heat flow in multilayer cw laser-annealed structures, *J. Appl. Phys.* **59**(10), 3371 (1986).

67A. H. Wang and T.-W. Chou, Transient thermal behavior of a thermally and elastically orthotropic medium, *AIAA J.* **24**(4), 664 (1986).

68A. Y. C. Yang and C. K. Chen, Thermoelastic transient response of an infinitely long annular cylinder composed of two different materials, *Int. J. Engng Sci.* **24**(4), 569 (1986).

69A. D. W. Yen and P. K. Wright, A remote temperature sensing technique for estimating the cutting interface temperature distribution, *J. Engng Ind.* **108**(4), 252 (1986).

70A. B. S. Yilbas, Heating of metals at a free surface by laser irradiation—an electron kinetic theory approach, *Int. J. Engng Sci.* **24**(8), 1325 (1986).

CHANNEL FLOW

1B. H. Akbari, A. Mertol, A. Gadgil, R. Kammerud and F. Baumann, Development of a turbulent near-wall temperature model and its application to channel flow, *Wärme- und Stoffübertr.* **20**(3), 189 (1986).

2B. P. Andre, R. Creff and J. Batina, Numerical study in heat transfer for a turbulent pulsed ducted flow, *Numer. Heat Transfer* **9**(2), 201 (1986).

3B. K. A. Antonopoulos and A. D. Gosman, The prediction of laminar inclined flow through tube banks, *Comput. Fluids* **14**(2), 171 (1986).

4B. K. A. Antonopoulos, The prediction of turbulent inclined flow in rod bundles, *Comput. Fluids* **14**(4), 361 (1986).

5B. B. R. Baliga and R. R. Azrak, Laminar fully developed flow and heat transfer in triangular plate-fin ducts, *J. Heat Transfer* **108**(1), 24 (1986).

6B. H. F. Bauer, Nonisothermal mass transport in a flat conduit for flow of non-Newtonian liquid, *Numer. Heat Transfer* **9**(4), 473 (1986).

7B. W. Bloem, Transient heat transfer to a forced flow of supercritical helium at 4.2 K, *Cryogenics* **26**(5), 300 (1986).

8B. C. G. Bowers, D. H. Willits, Jr. and H. D. Bowen, Criteria for determining buoyancy effects on turbulent forced convection in an asymmetrically heated horizontal channel, *Trans. Am. Soc. Agric. Engrs (Gen. Edn)* **29**(2), 550 (1986).

9B. A. N. Boyko, M. N. Galkin, S. Z. Kopelev and I. V. Shevchenko, Effect of type of deflector on the internal heat transfer in blades with crossflow of cooling gas, *Heat Transfer—Sov. Res.* **18**(1), 118 (1986).

10B. D. M. Bushnell, Review of structure of turbulence in heat and mass transfer, *AIAA J.* **24**(3), 544 (1986).

11B. M. B. Carver and M. Salcudean, Three-dimensional numerical modeling of phase distribution of two-fluid flow in elbows and return bends, *Numer. Heat Transfer* **10**(3), 229 (1986).

12B. C.-Y. Chang and Y.-T. Chern, Laminar heat and mass transfer with arbitrary order heterogeneous and homogeneous reactions—II. In a circular tube, *Chem. Engng J.* **32**(2), 117 (1986).

13B. C.-Y. Chang, C.-F. Lin and Y.-J. Yang, Laminar heat and mass transfer with arbitrary order heterogeneous and homogeneous reactions—I. In a rectangular duct, *Chem. Engng J.* **32**(2), 111 (1986).

14B. H. Chegini and S. K. Chaturvedi, An experimental and analytical investigation of friction factors for fully developed flow in internally finned triangular ducts, *J. Heat Transfer* **108**(3), 507 (1986).

15B. T.-L. Chiang and G. S. Dulikravich, Inverse design of composite turbine blade circular coolant flow passages, *J. Turbomachinery* **108**(2), 275 (1986).

16B. G. M. Chrysler and E. M. Sparrow, Turbulent flow and heat transfer in bends of circular cross section: II—pressure distribution experiments, *J. Heat Transfer* **108**(1), 212 (1986).

17B. A. E. Cook and M. Rahman, Temperature and velocity distributions in a square duct, *Appl. Math. Modelling* **10**(3), 221 (1986).

18B. R. M. Cotta and M. N. Ozisik, Laminar forced convection of power-law non-Newtonian fluids inside ducts, *Wärme- und Stoffübertr.* **20**(3), 211 (1986).

19B. R. M. Cotta and M. N. Ozisik, Laminar forced convection inside ducts with periodic variation of inlet temperature, *Int. J. Heat Mass Transfer* **29**(10), 1495 (1986).

20B. R. M. Cotta and M. N. Ozisik, Transient forced convection in laminar channel flow with stepwise variations of wall temperature, *C.J.Ch.E.* **64**(5), 734 (1986).

21B. M. T. Dalman, J. H. Merkin and C. McGreavy, Fluid flow and heat transfer past two spheres in a cylindrical tube, *Comput. Fluids* **14**(3), 267 (1986).

22B. G. A. Dreitser and V. V. Balashov, Experimental study of nonsteady heat convection in a plane channel, *J. Engng Phys.* **50**(2), 145 (1986).

23B. M. A. Ebadian, H. C. Topakoglu and O. A. Arnas, Convective heat transfer for laminar flows in a multipassage circular pipe subjected to an external uniform heat flux, *Int. J. Heat Mass Transfer* **29**(1), 107 (1986).

24B. M. A. Ebadian, H. C. Topakoglu and O. A. Arnas, On the convective heat transfer in a tube of elliptic cross

section maintained under constant wall temperature, *J. Heat Transfer* **108**(1), 33 (1986).
25B. G. B. Froishteter, Hydrodynamics and heat transfer in highly viscous rheological media flow, *Heat Transfer—Sov. Res.* **18**(1), 74 (1986).
26B. G. B. Froishteter and P. B. Begoulev, Heat exchange in circular tubes containing highly viscous non-Newtonian liquids with mixed properties, *Theor. Found. Chem. Engng* **20**(2), 128 (1986).
27B. T. Fukuda and Y. Shoji, Pressure drop and heat transfer for three phase flow (1st report, flow in horizontal pipes), *Bull. J.S.M.E.* **29**(256), 3421 (1986).
28B. N. K. Ghaddar, M. Magen, B. B. Mikic and A. T. Patera, Numerical investigation of incompressible flow in grooved channels. Pt 2. Resonance and oscillatory heat-transfer enhancement, *J. Fluid Mech.* **168**, 541 (1986).
29B. A. J. Ghajar and A. Asadi, Improved forced convective heat-transfer correlations for liquids in the near-critical region, *AIAA J.* **24**(12), 2030 (1986).
30B. Y. Goth, M. Feidt, F. Lauro and A. Bailly, Investigation of heat transfer and pressure drop in water/water single-phase flow in corrugated tubes, *Revue Gen. Therm.* **25**(294), 341 (1986).
31B. L. M. Hanna and P. W. Scherer, Measurement of local mass transfer coefficients in a cast model of the human upper respiratory tract, *J. Biomed. Engng* **108**(1), 12 (1986).
32B. T. Hanzawa, A. Sako and K. Kato, Velocity and temperature distributions in vertical concentric annulus with combined free and forced laminar convection, *J. Chem. Engng Jap.* **19**(2), 96 (1986).
33B. T. Hanzawa, A. Sako, H. Endo, M. Kagawa, T. Sunaga and K. Kato, Combined free and forced laminar convective heat transfer from isothermally heated inner tube in vertical concentric annulus, *J. Chem. Engng Jap.* **19**(1), 78 (1986).
34B. J. P. Hartnett and E. Y. Kwack, Prediction of friction and heat transfer for viscoelastic fluids in turbulent pipe flow, *Int. J. Thermophys.* **7**(1), 53 (1986).
35B. F. P. Incropera, J. S. Kerby, D. F. Moffatt and S. Ramadhyani, Convection heat transfer from discrete heat sources in a rectangular channel, *Int. J. Heat Mass Transfer* **29**(7), 1051 (1986).
36B. V. Kadambi, E. K. Levy and S. Neti, Heat transfer and pressure drop in a helically coiled rectangular duct, *J. Heat Transfer* **108**(2), 343 (1986).
37B. T. Kago, T. Saruwatari, M. Kashima, S. Morooka and Y. Kato, Heat transfer in horizontal plug and slug flow for gas–liquid and gas–slurry systems, *J. Chem. Engng Jap.* **19**(2), 125 (1986).
38B. F. I. Kalbaliev and F. K. Babayev, Heat transfer in the thermal inlet region in laminar flow of toluene at supercritical velocities, *Heat Transfer—Sov. Res.* **18**(1), 138 (1986).
39B. M. Kaviany, Laminar combined convection in a horizontal annulus subject to constant heat flux inner wall and adiabatic outer wall, *J. Heat Transfer* **108**(2), 392 (1986).
40B. J. M. Khodadadi, T. M. Nguyen and N. S. Vlachos, Laminar forced convective heat transfer in two-dimensional 90 deg. bifurcation, *Numer. Heat Transfer* **9**(6), 677 (1986).
41B. Yu. A. Kirichenko, R. I. Fedorov and V. V. Tsybulsky, Heat transfer and hydrodynamics of channelled cryogenic liquids in fields of centrifugal forces, *Cryogenics* **26**(10), 539 (1986).
42B. P. L. Kirillov, Effect of the thermophysical properties of the surface on heat transfer in turbulent flow, *J. Engng Phys.* **50**(3), 359 (1986).
43B. M. Kostic and J. P. Hartnett, Effects of fluid elasticity on laminar flow in rectangular duct, *Z. Angew. Math. Mech.* **66**(4), 239 (1986).
44B. W. P. Kotorynski, Steady laminar flow through a twisted pipe of elliptical cross-section, *Comput. Fluids* **14**(4), 433 (1986).
45B. B. S. Larkin and P. Savic, Indirect method of investigating heat transfer in turbulent flow in a tube, *Trans. Can. Soc. Mech. Engrs* **10**(1), 1 (1986).
46B. J. S. Lee and J. A. C. Humphrey, Radiative–convective heat transfer in dilute particle-laden channel flows, *PCH PhysioChem. Hydrodyn.* **7**(5), 325 (1986).
47B. W. C. Lee and Y.-H. Ju, Conjugate Leveque solution for Newtonian fluid in a parallel plate channel, *Int. J. Heat Mass Transfer* **29**(6), 941 (1986).
48B. L.-P. Leu and T. C. Ju Robert, Solution of the Graetz problem of power law fluids with viscous dissipation by orthogonal collocation-constant wall temperature, *J. Chin. Inst. Chem. Engrs* **17**(1), 19 (1986).
49B. E. Levy, S. Neti, G. Brown, F. Bayat and V. Kadambi, Laminar heat transfer and pressure drop in a rectangular duct rotating about a parallel axis, *J. Heat Transfer* **108**(2), 350 (1986).
50B. C.-T. Liou and F.-S. Wang, Solution of low Peclet number heat transfer with viscous dissipation in the infinite axial region between parallel plates via functional analytic methods, *J. Chin. Inst. Engrs* **9**(5), 521 (1986).
51B. C.-T. Liou and F.-S. Wang, Solution of low Peclet number heat transfer with viscous dissipation in the semi-infinite axial region of a tube via functional analytic methods, *J. Chin. Inst. Engrs* **9**(5), 497 (1986).
52B. M.-G. Martin, A. Amokrane and M. Humbert, Experimental study of heat transfer and pressure drop characteristics of a finned wall situated in a cross flow of air in a rectangular duct: application to a smoke recuperator, *Rev. Fr. Mech.* No. 3, 123 (1986).
53B. D. E. Metzger and M. K. Sahm, Heat transfer around sharp 180-deg turns in smooth rectangular channels, *J. Heat Transfer* **108**(3), 500 (1986).
54B. D. E. Metzger and D. E. Larson, Use of melting point surface coatings for local convection heat transfer measurements in rectangular channel flows with 90 deg turns, *J. Heat Transfer* **108**(1), 48 (1986).
55B. S. Mochizuki and W.-J. Yeng, Three mechanisms of convective enhancement in stationary disk systems, *J. Heat Transfer* **108**(4), 978 (1986).
56B. S. Mochizuki and J. Hiraoka, Heat transfer characteristics in radially outward flow between two parallel disks, *Nippon Kikai Gakkai Ronbunshu, B. Hen* **52**(478), 2425 (1986).
57B. M. Molki and E. M. Sparrow, An empirical correlation for the average heat transfer coefficient in circular tubes, *J. Heat Transfer* **108**(2), 482 (1986).

58B. M. Molki and C. M. Yeun, Effect of interwall spacing on heat transfer and pressure drop in a corrugated-wall duct, *Int. J. Heat Mass Transfer* **29**(7), 987 (1986).
59B. V. I. Naidenov, Convective-thermal motion instability of a viscous fluid in a tube, *High Temp.* **24**(1), 74 (1986).
60B. A. Nakayama and H. Koyama, Numerical prediction of turbulent flow and heat transfer within ducts of cross-shaped cross section, *J. Heat Transfer* **108**(4), 841 (1986).
61B. T. Nishimura, Y. Kajimoto and Y. Kawamura, Mass transfer enhancement in channels with a wavy wall, *J. Chem. Engng Jap.* **19**(2), 142 (1986).
62B. T. Nishimura, Y. Kajimoto, A. Tarumoto and Y. Kawamura, Flow structure and mass transfer for a wavy channel in transitional flow regime, *J. Chem. Engng Jap.* **19**(5), 449 (1986).
63B. M. Ogawa and H. Kawamura, Effects of entrance configuration on pressure loss and heat transfer of transitional gas flow in a circular tube, *Heat Transfer—Jap. Res.* **15**(5), 77 (1986).
64B. M. N. Oguztoreli, M. P. duPlessis and B. Ozeum, Heat transfer in axially dispersed two-phase flow systems, *Numer. Heat Transfer* **9**(3), 349 (1986).
65B. K. Oyakawa, T. Shinzato and I. Mabuchi, Effect of heat transfer augmentation of some geometric shapes of a turbulence promoter in a rectangular duct, *Bull. J.S.M.E.* **29**(256), 3415 (1986).
66B. C. Pfender, Flow behaviour of liquid/liquid two phase annular flow and heat transfer to the flowing liquids, *Chemie-Ingr-Tech.* **58**(8), 692 (1986).
67B. C. Pfender, Fluid mechanics of liquid/liquid annular two-phase and the heat transfer to the fluids, *VDI ForschHft.* 35 (1986).
68B. C. Prakash and R. Lounsbury, Analysis of laminar fully developed flow in plate-fin passages: effect of fin shape, *J. Heat Transfer* **108**(3), 693 (1986).
69B. W. Roetzel and J.-E. Roth, Heat transfer and pressure drop in smooth and buckled tubes (in German), *Wärme- und Stoffübertr.* **20**(2), 97 (1986).
70B. I. Satoh and Y. Kurosaki, Heat transfer combined with convection and radiation in an asymmetrically heated duct, *Nippon Kikai Gakkai Ronbunshu, B. Hen* **52**(476), 1741 (1986).
71B. R. Sethumadhaven and M. Rajo Rao, Turbulent flow friction and heat transfer characteristics of single- and multistart spirally enhanced tubes, *J. Heat Transfer* **108**(1), 55 (1986).
72B. E. M. Sparrow, A. Garcia and W. Chuck, Numerical and experimental turbulent heat transfer results for a one-sided heated rectangular duct, *Numer. Heat Transfer* **9**(3), 301 (1986).
73B. E. M. Sparrow and D. S. Kadle, Effect of tip-to-shroud clearance on turbulent heat transfer from a shrouded longitudinal fin array, *J. Heat Transfer* **108**(3), 519 (1986).
74B. E. M. Sparrow and G. M. Chrysler, Turbulent flow and heat transfer in bends of circular cross section: I—heat transfer experiments, *J. Heat Transfer* **108**(1), 40 (1986).
75B. J. Sucec, Transient heat transfer in the laminar thermal entry region of a pipe: an analytical solution, *Appl. Scient. Res.* **43**(2), 115 (1986).
76B. H. Tanaka, S. Hatano and S. Maruyama, Combined forced and natural convection heat transfer for upward flow in a uniformly heated vertical pipe, *Nippon Kikai Gakkai Ronbunshu* **52**(484), 4014 (1986).
77B. G. Theodorou and D. Bellet, Laminar flows of a non-Newtonian fluid in mild stenosis, *Comput. Meth. Appl. Mech. Engng* **54**(1), 111 (1986).
78B. M. A. Valyuzhinich, V. M. Eroshenko and E. V. Kuznetsov, Experimental study of heat transfer in the turbulent convection of helium at supercritical pressure under strongly nonisothermal conditions, *High Temp.* **24**(1), 80 (1986).
79B. B. Vick and R. G. Wells, Laminar flow with an axially varying heat transfer coefficient, *Int. J. Heat Mass Transfer* **29**(12), 1881 (1986).
80B. A Wada, M. Toda, M. Kuriyama, E. Harada, R. Sato and H. Konno, Convective heat transfer for highly viscous fluids with variable viscosity in cooling pipes, *J. Chem. Engng Jap.* **19**(1), 20 (1986).
81B. N. E. Wijeysundera, Laminar forced convection in circular and flat ducts with wall axial conduction and external convection, *Int. J. Heat Mass Transfer* **29**(5), 797 (1986).
82B. C. L. Woodworth, Computer simulation of steady state nonisothermal melt flow in tubes, *Adv. Polym. Technol.* **6**(3), 251 (1986).
83B. W.-M. Yang, Analysis of fluid flow and heat transfer in twisted circular-sector ducts, *Numer. Heat Transfer* **10**(1), 45 (1986).
84B. Y. Yener and M. N. Ozisik, Simultaneous radiation and forced convection in thermally developing turbulent flow through a parallel-plate channel, *J. Heat Transfer* **108**(4), 985 (1986).
85B. G. V. Zhizhin, Nonisothermal couette flow of a non-Newtonian fluid under a pressure gradient, *J. Appl. Mech. Tech. Phys.* **27**(2), 218 (1986).

BOUNDARY-LAYER AND EXTERNAL FLOWS

1C. T. A. Abdelhafez, Laminar thermal boundary layer on a continuous accelerated sheet extruded in an ambient fluid, *Acta Mech.* **64**(3), 207 (1986).
2C. A. M. Agnone, Supersonic flow over a cruciform configuration at angle of attack, *J. Spacecraft Rockets* **23**(5), 451 (1986).
3C. G. M. R. Assassa, Numerical solution of Falkner and Skan equation by optimization techniques, *Adv. Modell. Simul.* **5**(4), 33 (1986).
4C. A. K. Banarjee and S. Ambal, Nonsimilar solutions in thermomicropolar fluid boundary layer, *Int. J. Engng Sci.* **24**(3), 403 (1986).
5C. H. Barrow, On average heat transfer coefficient, *Int. J. Heat Fluid Flow* **7**(3), 162 (1986).
6C. R. C. Blanchard and G. M. Buck, Rarefied-flow aerodynamics and thermosphere structure from shuttle flight measurements, *J. Spacecraft Rockets* **23**(1), 18 (1986).
7C. M. K. Bologa, S. E. Berzoy and V. P. Goncharuk, Effect of cavitating turbulators on the heat transfer of the plate, *Modell. Simul. Control B* **6**(3), 55 (1986).

8C. L. W. B. Browne and R. A. Antonia, Reynolds shear stress and heat flux measurements in a cylinder wake, *Physics Fluids* **29**(3), 709 (1986).
9C. J. C. Buell and I. Catton, Wavenumber selection in large-amplitude axisymmetric convection, *Physics Fluids* **29**(1), 23 (1986).
10C. A. Campo and A. Salazar, Forced convection—axial conduction between parallel walls with unequal heat fluxes, *Wärme- und Stoffübertr.* **20**(3), 177 (1986).
11C. B. R. Circelli and J. B. McLaughlin, Numerical study of heat transfer in turbulent shear flow, *Numer. Heat Transfer* **9**(3), 335 (1986).
12C. Yu. M. Dmitrenko, V. L. Zhdanov, B. A. Kolovandin and I. A. Labuda, Temperature fluctuation balance in an axisymmetric turbulent wake, *J. Engng Phys.* **50**(2), 123 (1986).
13C. M. G. Dunn, Heat-flux measurements for the rotor of a full stage turbine: Pt. 1—time averaged results, *J. Turbomachinery* **108**(1), 90 (1986).
14C. M. G. Dunn, H. L. Martin and M. J. Stanek, Heat-flux and pressure measurements and comparison with prediction for a low-aspect-ratio turbine stage, *J. Turbomachinery* **108**(1), 108 (1986).
15C. M. G. Dunn, W. K. George, W. J. Rae, S. H. Woodward, J. C. Moller and P. J. Seymour, Heat-flux measurements for the rotor of a full-stage turbine: Pt II—description of analysis technique and typical time-resolved measurements, *J. Turbomachinery* **108**(1), 98 (1986).
16C. A. Dyment, Self similar unsteady boundary layers, *Acta Mech.* **59**(1), 91 (1986).
17C. J. W. Elsner and J. Zielinski, Semi-preservation of momentum and heat transfer in cascade-wake-flows, *Int. J. Heat Mass Transfer* **29**(2), 293 (1986).
18C. I. G. Eremeitsev and N. N. Pilyugin, Heat transfer and drag of a body in a far supersonic wake, *Fluid Dyn.* **21**(2), 218 (1986).
19C. E. Eteng, G. S. S. Ludford and M. Matalon, Displacement effect of a flame in a stagnation-point flow, *Physics Fluids* **29**(7), 2172 (1986).
20C. P. G. Frik, Cascade model of two-dimensional turbulent convection, *Fluid Mech., Sov. Res.* **15**(2), 80 (1986).
21C. V. K. Garg and K. Velusamy, Heat transfer characteristics for a plate fin, *J. Heat Transfer* **108**(1), 224 (1986).
22C. P. I. Geshev and B. S. Ezdin, Modelling turbulent transfer in a channel by means of point vortices, *J. Appl. Mech. Tech. Phys.* **27**(2), 228 (1986).
23C. A. G. Gnedovets, E. M. Ivanov, A. A. Uglov and A. M. Shmakov, Heat transfer between a spherical particle and a gas for arbitrary Knudsen numbers and temperature differences in the particle boundary layer, *High Temp.* **24**(3), 416 (1986).
24C. R. S. R. Gorla, Heat transfer from continuous moving fibers in a micropolar fluid, *Acta Mech.* **59**(1), 123 (1986).
25C. G. K. Hargrove, M. Fairweather and J. K. Kilham, Turbulence enhancement of stagnation point heat transfer on a circular cylinder, *Int. J. Heat Fluid Flow* **7**(2), 89 (1986).
26C. D. H. Hathaway and R. C. J. Somerville, Nonlinear interactions between convection, rotation and flows with vertical shear, *J. Fluid Mech.* **164**, 91 (1986).
27C. H. Herwig and G. Wickern, The effect of variable properties on laminar boundary layer flow, *Wärme- und Stoffübertr.* **20**(1), 47 (1986).
28C. B. K. Hodge, R. P. Taylor and H. W. Coleman, Investigation of surface roughness effects on adiabatic wall temperature, *AIAA J.* **24**(9), 1560 (1986).
29C. S. K. Hong and S. N. B. Murthy, Effective velocity of transport in curved wall boundary layers, *AIAA J.* **24**(3), 361 (1986).
30C. L. W. Hunter, L. L. Perini, D. W. Conn and P. T. Brenza, Calculation of carbon ablation on a reentry body during supersonic/subsonic flight, *J. Spacecraft Rockets* **23**(5), 487 (1986).
31C. D. R. Jeng, T. C. A. Chang and K. J. DeWitt, Momentum and heat transfer on a continuous moving surface, *J. Heat Transfer* **108**(3), 532 (1986).
32C. D. S. Kadle and E. M. Sparrow, Numerical and experimental study of turbulent heat transfer and fluid flow in longitudinal fin arrays, *J. Heat Transfer* **108**(1), 16 (1986).
33C. N. Kaji and Y. H. Mori, Direct contact heat transfer to a drop translating in a transverse electric field, *Wärme- und Stoffübertr.* **20**(1), 11 (1986).
34C. D. D. Kale and D. M. Rao, Heat transfer to two-phase gas–liquid flows of drag reducing fluids, *Chem. Engng Sci.* **41**(9), 2434 (1986).
35C. S. S. Kang and E. M. Sparrow, Heat and mass transfer at an upstream-facing blunt face situated in a uniform oncoming flow, *Numer. Heat Transfer* **9**(4), 419 (1986).
36C. M. V. Karwe and Y. Jaluria, Thermal transport from a heated moving surface, *J. Heat Transfer* **108**(4), 728 (1986).
37C. A. A. Khalatov, K. I. Kapitanchuk, A. S. Kovalenko and A. N. Trufanov, Study of local heat transfer on end surface of stator model, *Soviet Aeronaut.* **29**(2), 78 (1986).
38C. V. G. Kochanov, Influence of the shape of a body in motion in a hydrogen–helium atmosphere on radiant heat transfer, *Fluid Dyn.* **21**(1), 97 (1986).
39C. U. H. Kurzweg, Temporal and spatial distribution of heat flux in oscillating flow subjected to an axial temperature gradient, *Int. J. Heat Mass Transfer* **29**(12), 1969 (1986).
40C. W. S. Law and S. K. Loyalka, Motion of a sphere in a rarefied gas. II. Role of temperature variation in the Knudsen layer, *Physics Fluids* **29**(11), 3886 (1986).
41C. A. S. Leibenzon, Some self-similar solutions of two-temperature hydrodynamic equations with allowance for nonlinear heat conduction and exothermic reactions, *Fluid Dyn.* **21**(2), 292 (1986).
42C. F. S. Lien, Heat transfer due to axial turbulent flow along a circular rod, *AIAA J.* **24**(1), 186 (1986).
43C. A. M. Lindemann, Turbulent Reynolds analogy factors of stacked large-eddy breakup devices, *J. Spacecraft Rockets* **23**(3), 348 (1986).
44C. J. L. Lumley, Evolution of a non-self-preserving thermal mixing layer, *Physics Fluids* **29**(12), 3976 (1986).
45C. V. G. Lushchik, A. A. Pavel'ev and A. E. Yakubenko, Three-parameter model of turbulence: heat transfer calculations, *Fluid Dyn.* **21**(2), 200 (1986).

46C. B.-K. Ma and Z. Warhaft, Some aspects of the thermal mixing layer in grid turbulence, *Physics Fluids* **29**(10), 3114 (1986).
47C. Yu. N. Makarov, Calculation of slow viscous heat conducting gas flows at moderate and large Reynolds numbers, *Heat Transfer—Sov. Res.* **18**(2), 97 (1986).
48C. C. G. Miller, P. A. Gnoffo and S. E. Wilder, Measured and predicted heating distributions for Biconics at Mach 10, *J. Spacecraft Rockets* **23**(3), 251 (1986).
49C. I. A. Mudawwar and M. A. El-Masri, Momentum and heat transfer across freely-falling turbulent liquid films, *Int. J. Multiphase Flow* **12**(5), 771 (1986).
50C. A. Nakayama, A. V. Shenoy and H. Koyama, Analysis for forced convection heat transfer from external surfaces to non-Newtonian fluids, *Wärme- und Stoffübertr.* **20**(3), 219 (1986).
51C. A. Nakayama and H. Koyama, Asymptotic expression for forced convection in non-Newtonian power law fluids, *Int. J. Heat Fluid Flow* **7**(2), 99 (1986).
52C. N. T. Obot, T. A. Trabold, M. L. Graska and F. Gandhi, Velocity and temperature fields in turbulent jets issuing from sharp-edged inlet round nozzles, *Ind. Engng Chem. Fund.* **25**(3), 425 (1986).
53C. I. Owen, A. A. El-Kady and J. W. Cleaver, Note comparing the rates of deposition of submicron particles with the rates of heat transfer for ribbed surfaces exposed to turbulent gas streams, *J. Aerosol Sci.* **17**(2), 145 (1986).
54C. K. Rued and S. Wittig, Laminar and transitional boundary layer structures in accelerating flow with heat transfer, *J. Turbomachinery* **108**(1), 116 (1986).
55C. F. Samie and E. M. Sparrow, Heat transfer from a yawed finned tube, *J. Heat Transfer* **108**(2), 479 (1986).
56C. V. L. Sergeyev and L. A. Sergeyeva, Investigation of unsteady-state heat transfer for plane and axisymmetric flows, *Heat Transfer—Sov. Res.* **18**(1), 1 (1986).
57C. P. Ayazi Shamlou and M. F. Edwards, Heat transfer to viscous Newtonian and non-Newtonian fluids for helical ribbon mixers, *Chem. Engng Sci.* **41**(8), 1957 (1986).
58C. A. V. Shenoy and A. Nakayama, Forced convection heat transfer axisymmetric bodies to non-Newtonian fluids, *C.J.Ch.E.* **64**(4), 680 (1986).
59C. P. Singh and S. Antony Raj, Variational solution of Blasius flow for skin friction and heat transfer, *AIAA J.* **24**(9), 1445 (1986).
60C. A. K. Singhal, L. T. Tam, F. Bachtel and J. Vaniman, Thermal environment around the space shuttle with hot-gas jets for ice suppression, *J. Spacecraft Rockets* **23**(6), 547 (1986).
61C. M. C. Smith, D. A. Haines and W. A. Main, Some characteristics of longitudinal vortices produced by line-source heating in a low-speed wind tunnel, *Int. J. Heat Mass Transfer* **29**(1), 59 (1986).
62C. D. J. Song and C. H. Lewis, Hypersonic finite-rate chemically reacting viscous flows over ablating carbon surface, *J. Spacecraft Rockets* **23**(1), 47 (1986).
63C. E. M. Sparrow and D. L. Misterek, Mass transfer at the base of a cylindrical cavity recessed in the floor of a flat duct, *J. Heat Transfer* **108**(4), 853 (1986).
64C. H. Stapountzis, B. L. Sawford, J. C. R. Hunt and R. E. Britter, Structure of the temperature field downwind of a line source in grid turbulence, *J. Fluid Mech.* **165**, 401 (1986).
65C. M. Strscheletzky, Transformation of energy in the viscous boundary sublayers, *Forsch. Ing. Wes.* **52**(5), 167 (1986).
66C. M. E. Tauber, Scaling relations for heating during gliding entry at parabolic speed, *AIAA J.* **24**(12), 2047 (1986).
67C. J. R. Tewell, E. S. Ewing and D. E. Florence, Re-entry module for a shuttle-derived vehicle launch system, *J. Spacecraft Rockets* **23**(1), 83 (1986).
68C. A. T. R. Tindall and E. A. Vallis, Convective transfer from a transverse fin array to two-dimensional turbulent flow, *Int. J. Heat Fluid Flow* **7**(2), 82 (1986).
69C. S. P. Vanka, Calculation of axisymmetric, turbulent, confined diffusion flames, *AIAA J.* **24**(3), 462 (1986).
70C. R. Vasantha and G. Nath, Unsteady compressible second-order boundary layers at the stagnation point of two-dimensional and axisymmetric bodies, *Wärme- und Stoffübertr.* **20**(4), 273 (1986).
71C. R. Vasantha and G. Nath, Boundary-layer flow past a cylinder with massive blowing, *AIAA J.* **24**(11), 1874 (1986).
72C. G. Wilks and R. Hunt, The axisymmetric, turbulent, buoyant jet—numerical solution for a semiempirical theory, *Numer. Heat Transfer* **9**(4), 495 (1986).
73C. K.-L. Wong, S.-C. Lee and C.-K. Chen, Finite element solution of laminar combined convection from a sphere, *J. Heat Transfer* **108**(4), 860 (1986).
74C. A. G. Yurov, Calculation of the heat and mass transfer coefficient in an unstably stratified turbulent air boundary layer over a water surface, *Gidromekhanika* No. 53, 47 (1986).
75C. V. I. Zinchenko and E. N. Putyatina, Solution of coupled heat-transfer problems in flow about bodies of different shapes, *J. Appl. Mech. Tech. Phys.* **27**(2), 234 (1986).

FLOW WITH SEPARATED REGIONS

1D. S. Aiba and H. Tsuchida, Heat transfer around a circular cylinder near a plane boundary, *Heat Transfer—Jap. Res.* **15**(2), 1 (1986).
2D. J. W. Baughn, M. J. Elderkin and A. A. McKillop, Heat transfer from a single cylinder, cylinders in tandem, and cylinders in the entrance region of a tube bank with a uniform heat flux, *J. Heat Transfer* **108**(2), 386 (1986).
3D. W. J. Bellows and R. E. Mayle, Heat transfer downstream of a leading edge separation bubble, *J. Turbomachinery* **108**(1), 131 (1986).
4D. P. I. Cooper, J. C. Sheridan and G. J. Flood, The effects of sound on forced convection over a flat plate, *Int. J. Heat Fluid Flow* **7**(1), 61 (1986).
5D. M. Fujii, T. Fujii and H. Aramaki, Numerical analyses on laminar flow and heat transfer in staggered tube banks, *Nippon Kikai Gakkai Ronbunshu, B. Hen* **52**(476), 1694 (1986).

6D. J. N. Hinckel and H. T. Nagamatsu, Heat transfer in the stagnation region of the junction of a circular cylinder perpendicular to a flat plate, *Int. J. Heat Mass Transfer* **29**(7), 999 (1986).
7D. T. Igarashi, Fluid flow and heat transfer around rectangular cylinders (the case of a width/height ratio of a section of 2.0 to approximately 4.0), *Nippon Kikai Gakkai Ronbunshu, B. Hen* **52**(480), 3011 (1986).
8D. H. Jafroudi and H. T. Yang, Steady laminar forced convection from a circular cylinder, *J. Comput. Phys.* **65**(1), 46 (1986).
9D. V. I. Katinas, A. A. Markevicius and A. A. Zukauskas, Heat transfer behavior of vibrating tubes operating in crossflow 2. Local and average heat transfer coefficients, *Heat Transfer—Sov. Res.* **18**(2), 10 (1986).
10D. V. I. Katinas, A. A. Markevicius and A. A. Zukauskas, Heat transfer behavior of vibrating tubes operating in crossflow 1. Temperature and velocity fluctuations, *Heat Transfer—Sov. Res.* **18**(2), 1 (1986).
11D. H. Kimoto, Study on heat transfer from a heated circular cylinder, horizontally supported in a progressive sound field, *Nippon Kikai Gakkai Ronbunshu, B. Hen* **52**(480), 3003 (1986).
12D. H. Kimoto and Y. Yoshii, Heat transfer characteristics of a heated circular cylinder in a near sound field of 5 kHz, *Nippon Kikai Gakkai Ronbunshu, B. Hen* **52**(475), 1434 (1986).
13D. H. Kimoto and Y. Sumita, Heat transfer characteristics of a circular cylinder in a conduit under cavitation, *Nippon Kikai Gakkai Ronbunshu, B. Hen* **52**(477), 2312 (1986).
14D. N. W. M. Ko and P. F. Y. Wong, Mass transfer from coaxial cylinders with stepwise changes in diameter, *J. Heat Transfer* **108**(4), 981 (1986).
15D. M. Kuriyama, M. Toda, E. Harada and H. Konno, Heat transfer from vibrating and rotating spheres in an air flow, *Heat Transfer—Jap. Res.* **15**(3), 27 (1986).
16D. I. Mabuchi, T. Murata and M. Kumada, Effect of freestream turbulence on heat transfer characteristics in the reattachment region on the bottom surface of a backward-facing step (for different angles of separation), *Nippon Kikai Gakkai Ronbunshu, B. Hen* **52**(479), 2619 (1986).
17D. G. P. Merker and H. Hanke, Heat transfer and pressure drop on the shell-side of tube-banks having oval-shaped tubes, *Int. J. Heat Mass Transfer* **29**(12), 1903 (1986).
18D. A. Narumi, S. Kato and Y. Simomura, Flow and convective heat transfer in the leading edge region of an inclined flat plate with finite width, *Nippon Kikai Gakkai Ronbunshu, B. Hen* **52**(476), 1702 (1986).
19D. H. Nishiyama, T. Ota and T. Matsuno, Forced convective heat transfer from two elliptic cylinders (tandem arrangement in opposite directions), *Nippon Kikai Gakkai Ronbunshu, B. Hen* **52**(479), 2677 (1986).
20D. T. Ota, N. Nishiyama, J. Kominami and K. Sato, Heat transfer from two elliptic cylinders in tandem arrangement, *J. Heat Transfer* **108**(3), 525 (1986).
21D. M. A. Paolino, R. B. Kinney and E. A. Cerutti, Numerical analysis of the unsteady flow and heat transfer to a cylinder in crossflow, *J. Heat Transfer* **108**(4), 742 (1986).
22D. M. Prud'homme and S. Elghobashi, Turbulent heat transfer near the reattachment of flow downstream of a sudden pipe expansion, *Numer. Heat Transfer* **10**(4), 349 (1986).
23D. M. Samimy, H. L. Petrie and A. L. Addy, A study of compressible turbulent reattaching free shear layers, *AIAA J.* **24**(2), 261 (1986).
24D. R. Shimada, T. Takeyama and K. Takayama, Wall effects on convective heat transfer from cylinders in a uniform flow, *Heat Transfer—Jap. Res.* **15**(6), 44 (1986).
25D. D. Traub and K. Stephan, Heat transfer and pressure drop over smooth bundled tubes in transverse flow, *Chemie-Ingr-Tech.* **58**(2), 124 (1986).
26D. N. Wakisaka, Heat transfer in the reattachment region of a sudden expansion flow in ducts, *Heat Transfer—Jap. Res.* **15**(5), 65 (1986).
27D. L. E. Yushina, Influence of unheated end shape on heat transfer from a longitudinally streamlined cylinder, *Heat Transfer—Sov. Res.* **18**(1), 56 (1986).
28D. E. G. Zaulichnyi and V. M. Trofimov, Investigation of heat transfer in separated regions in a supersonic laval nozzle, *J. Appl. Mech. Tech. Phys.* **27**(1), 88 (1986).
29D. P. Zehner, Momentum mass and heat transfer in bubble columns. Pt. 1. Flow model of the bubble column and liquid velocities, *Int. Chem. Engng* **26**(1), 22 (1986).
30D. P. Zehner, Momentum, mass and heat transfer in bubble columns. Pt. 2. Axial blending and heat transfer, *Int. Chem. Engng* **26**(1), 29 (1986).
31D. G. P. Zigmantas and R. B. Shlyazhas, Flow and heat transfer in flow past barriers in a turbulent boundary layer, *Heat Transfer—Sov. Res.* **18**(2), 91 (1986).
32D. A. A. Zukauskas, R. V. Ulinskas and P. I. Daunoras, Local coefficients of heat transfer from staggered bundles of rough tubes in crossflow, *Heat Transfer—Sov. Res.* **18**(1), 108 (1986).

HEAT TRANSFER IN POROUS MEDIA

1DP. G. F. Barreto, A. Lancia and G. Volpicelli, Heat transfer and fluid dynamic characteristics of gas-fluidized beds under pressure, *Powder Technol.* **46**(2), 155 (1986).
2DP. N. E. Bixler and C. B. Carrigan, Enhanced heat transfer in partially-saturated hydrothermal systems, *Geophys. Res. Lett.* **13**(1), 42 (1986).
3DP. H. J. Bock and J. Schweinzer, Heat transfer to horizontal tube banks in a pressurized gas/solid fluidized bed, *Germ. Chem. Engng* **9**(1), 16 (1986).
4DP. A. L. Burka and E. P. Chirkashenko, Radiation-conductive heat exchange in a porous solid body, *Izv. Sib. Otdeleniya Akad. Nauk SSR Ser. Tek. Nauk* No. 4, 3 (1986).
5DP. D. Burton and C. J. Lambert, Microscopic calculation of the heat transfer between a granular structure and liquid **3He**, *J. Low Temp. Phys.* **64**(1), 21 (1986).
6DP. K. S. Chen and J. R. Ho, Effects of flow inertia on vertical, natural convection in saturated, porous

media, *Int. J. Heat Mass Transfer* **29**(5), 753 (1986).
7DP. P. Cheng and C. T. Hsu, Fully-developed, forced convective flow through an annular packed-sphere bed with wall effects, *Int. J. Heat Mass Transfer* **29**(12), 1843 (1986).
8DP. L.-C. L. Chien and A. L. Fetter, Thermal convection and nonlinear effects of a superfluid **3He–**4He mixture in a porous medium, *J. Low Temp. Phys.* **64**(5), 64 (1986).
9DP. P. G. Daniels, P. A. Blythe and P. G. Simpkins, Thermally driven shallow cavity flows in porous media: the intermediate regime, *Proc. R. Soc. London* **A406**(1831), 263 (1986).
10DP. M. R. Davidson, Natural convection of gas/vapour mixtures in a porous medium, *Int. J. Heat Mass Transfer* **29**(9), 1371 (1986).
11DP. E. DiBenedetto, C. M. Elliot and A. Friedman, Free boundary of a flow in a porous body heat from its boundary, *Nonlinear Anal. Theor. Meth. Appl.* **10**(9), 879 (1986).
12DP. A. G. Dixon and D. L. Cresswell, Effective heat transfer parameters for transient packed-bed models, *A.I.Ch.E. Jl* **32**(5), 809 (1986).
13DP. J. A. Doherty, R. S. Verma, S. Shrivastava and S. C. Saxena, Heat transfer from immersed horizontal tubes of different diameter in a gas-fluidized bed, *Energy* **11**(8), 773 (1986).
14DP. E. R. G. Eckert and M. Faghri, A parametric analysis of moisture migration in an unsaturated porous slab caused by convective heat and mass transfer, *Wärme- und Stoffübertr.* **20**(1), 77 (1986).
15DP. R. M. Fand, T. E. Steinberger and P. Cheng, Natural convection heat transfer from a horizontal cylinder embedded in a porous medium, *Int. J. Heat Mass Transfer* **29**(1), 119 (1986).
16DP. R. S. Figliola, E. G. Suarez and D. R. Pitts, Mixed particle size distribution effects on heat transfer, *J. Heat Transfer* **108**(4), 913 (1986).
17DP. J. G. Georgiadis and I. Catton, Prandtl number effect on Benard convection in porous media, *J. Heat Transfer* **108**(2), 284 (1986).
18DP. A. Goshayeshi, J. R. Welty, R. L. Adams and N. Alavizadeh, Local heat transfer coefficients for horizontal tube arrays in high-temperature large-particle fluidized beds: an experimental study, *J. Heat Transfer* **108**(4), 907 (1986).
19DP. D. B. Ingham and I. Pop, A horizontal flow past a partially heated infinite vertical cylinder embedded in a porous medium, *Int. J. Engng Sci.* **24**(8), 1351 (1986).
20DP. D. B. Ingham and S. N. Brown, Flow past a suddenly heated vertical plate in a porous medium, *Proc. R. Soc. London* **A403**(1824), 51 (1986).
21DP. R. M. Islam and K. Nandakumar, Multiple solutions for buoyancy-induced flow in saturated porous media for large Peclet numbers, *J. Heat Transfer* **108**(4), 866 (1986).
22DP. Y. Kato, M. Koyama, T. Kago and S. Morooka, Heat transfer coefficient between bed and inserted horizontal tube in a three-phase fluidized bed, *Heat Transfer—Jap. Res.* **15**(1), 1 (1986).
23DP. M. Kaviany, Non-Darcian effects on natural convection in porous media confined between horizontal cylinders, *Int. J. Heat Mass Transfer* **29**(10), 1513 (1986).
24DP. S. D. Kim, Y. Kang and H. K. Kwon, Heat transfer characteristics in two- and three-phase slurry fluidized beds, *A.I.Ch.E. Jl* **32**(8), 1397 (1986).
25DP. S. Kimura, G. Schubert and J. M. Straus, Route to chaos in porous-medium thermal convection, *J. Fluid Mech.* **166,** 305 (1986).
26DP. M. Kumari, I. Pop and G. Nath, Non-Darcy natural convection on a vertical cylinder in a saturated porous medium, *Wärme- und Stoffübertr.* **20**(1), 33 (1986).
27DP. G. Lebon and A. Cloot, A thermodynamical modeling of fluid flows through porous media; application to natural convection, *Int. J. Heat Mass Transfer* **29**(3), 381 (1986).
28DP. D.-S. Lin and B. Gebhart, Buoyancy-induced flow adjacent to a horizontal surface submerged in porous medium saturated with cold water, *Int. J. Heat Mass Transfer* **29**(4), 611 (1986).
29DP. J.-Y. Liu, W. J. Minkowycz and P. Cheng, Conjugate mixed convection heat transfer analysis of a plate fin embedded in a porous medium, *Numer. Heat Transfer* **9**(5), 575 (1986).
30DP. J.-Y. Liu and W. J. Minkowycz, The influence of lateral mass flux on conjugate natural convection from a vertical plate fin in a saturated porous medium, *Numer. Heat Transfer* **10**(5), 507 (1986).
31DP. J.-Y. Liu, W. J. Minkowycz and P. Cheng, Conjugated, mixed convection–conduction heat transfer along a cylindrical fin in a porous medium, *Int. J. Heat Mass Transfer* **29**(5), 769 (1986).
32DP. T. Masuoka, K. Narazaki, T. Tsuruta and Y. Tohda, Natural convection in stratified porous media heated from the side, *Nippon Kikai Gakkai Ronbunshu, B. Hen* **52**(474), 866 (1986).
33DP. A. Mathur and S. C. Saxena, A correlation for heat transfer between immersed surfaces and gas-fluidized beds of large particles, *Energy* **11**(9), 843 (1986).
34DP. A. Mathur, S. Saxena and A. Chao, Heat transfer from an immersed vertical tube in a gas-fluidized bed, *Ind. Engng Chem. Process Des. Dev.* **25**(1), 156 (1986).
35DP. J. H. Merkin, On dual solutions occurring in mixed convection in a porous medium, *J. Engng Math.* **20**(2), 171 (1986).
36DP. J. H. Merkin, Free convection from a vertical cylinder embedded in a saturated porous medium, *Acta Mech.* **62**(1), 19 (1986).
37DP. E. E. Michaelides, Heat transfer in particulate flows, *Int. J. Heat Mass Transfer* **29**(2), 265 (1986).
38DP. S. Mingheng, Heat transfer to horizontal tube in freeboard region of fluidized bed, *Huagong Xuebao* **2**(2), 212 (1986).
39DP. K. Miyazaki, K. Murai, T. Ohama and N. Yamaoka, Dryout heat flux for core debris bed (I): effects of system pressure and particle size, *J. Nucl. Sci. Technol.* **23**(8), 702 (1986).
40DP. S. Motakff and M. A. El-Masri, Simultaneous heat and mass transfer with phase change in a porous slab, *Int. J. Heat Mass Transfer* **29**(10), 1503 (1986).

41DP. K. Muralidhar, R. A. Baunchalk and F. A. Kulacki, Natural convection in a horizontal porous annulus with a step distribution in permeability, *J. Heat Transfer* **108**(4), 889 (1986).

42DP. K. Muroyama, M. Fukuma and A. Yasunishi, Wall-to-bed heat transfer in liquid–solids and gas–liquid–solid fluidized beds. Pt. I: liquid–solid fluidized beds, *C.J.Ch.E.* **64**(3), 399 (1986).

43DP. K. Muroyama, M. Fukuma and A. Yasunishi, Wall-to-bed heat transfer in liquid–solid and gas–liquid–solid fluidized beds. Pt. II: gas–liquid–solid fluidized beds, *C.J.Ch.E.* **64**(3), 409 (1986).

44DP. S. A. Nagornov and G. G. Serebrennikov, Fluidized bed heat exchange with a submerged heated surface, *Heat Transfer—Sov. Res.* **18**(2), 81 (1986).

45DP. V. Prasad, F. A. Kulacki and A. V. Kulkarni, Free convection in a vertical, porous annulus with constant heat flux on the inner wall—experimental results, *Int. J. Heat Mass Transfer* **29**(5), 713 (1986).

46DP. V. Prasad, Numerical study of natural convection in a vertical, porous annulus with constant heat flux on the inner wall, *Int. J. Heat Mass Transfer* **29**(6), 841 (1986).

47DP. M. Prat, Analysis of experiments of moisture migration caused by temperature differences in unsaturated porous medium by means of two-dimensional numerical simulation, *Int. J. Heat Mass Transfer* **29**(7), 1033 (1986).

48DP. N. Raghavacharyulu, Combined free and forced convection in vertical rectangular porous channels, *Revue Roum. Phys.* **31**(2), 161 (1986).

49DP. D. C. Reda, Natural convection experiments in a stratified liquid-saturated porous medium, *J. Heat Transfer* **108**(3), 660 (1986).

50DP. D. C. Reda, Natural convection experiments with a finite-length, vertical cylindrical heat source in a water-saturated porous medium, *Nucl. Chem. Waste Mgmt* **6**(1), 3 (1986).

51DP. A. W. Reed, A mechanistic explanation of channels in debris beds, *J. Heat Transfer* **108**(1), 1265 (1986).

52DP. D. A. S. Rees and D. S. Riley, Free convection in an undulating saturated porous layer: resonant wavelength excitation, *J. Fluid Mech.* **166,** 503 (1986).

53DP. L. Robillard, T. H. Nguyen and P. Vasseur, Free convection in a two-dimensional porous loop, *J. Heat Transfer* **108**(2), 277 (1986).

54DP. C. Roizard and D. Tondeur, Heat transfer in percolated fixed beds with temperature-dependent physical properties—compressive and dispersive effects, *Int. J. Heat Mass Transfer* **29**(7), 1021 (1986).

55DP. V. B. Sazhin, A. A. Oygenblik, I. N. Dorokhov and V. V. Kafarov, A mathematical model of fluidized-bed drying of granular solids, *Heat Transfer—Sov. Res.* **18**(2), 71 (1986).

56DP. K. S. Schnoebelen and S. I. Abdel-Khalik, Local Nusselt numbers for flowing packed particle beds in circular tubes with constant wall heat flux, *J. Heat Transfer* **108**(2), 466 (1986).

57DP. P. Singh, J. K. Misra and K. A. Narayan, A mathematical analysis of unsteady flow and heat transfer in a porous medium, *Int. J. Engng Sci.* **24**(3), 277 (1986).

58DP. P. H. Steen, Container geometry and the transition to unsteady Benard convection in porous media, *Physics Fluids* **29**(4), 925 (1986).

59DP. M. E. Taslim and U. Narusawa, Binary fluid convection and double-diffusive convection in a porous medium, *J. Heat Transfer* **108**(1), 221 (1986).

60DP. T. W. Tong and E. Subramanian, Natural convection in rectangular enclosures partially filled with a porous medium, *Int. J. Heat Fluid Flow* **7**(1), 3 (1986).

61DP. O. V. Trevisan and A. Bejan, Mass and heat transfer by natural convection in a vertical slot filled with porous medium, *Int. J. Heat Mass Transfer* **29**(3), 403 (1986).

62DP. A. B. Verver and W. P. M. van Swaaij, Heat-transfer performance of gas–solid trickle flow over a regularly stacked packing, *Powder Technol.* **45**(2), 133 (1986).

63DP. J. A. Weaver and R. Viskanta, Freezing of liquid-saturated porous media, *J. Heat Transfer* **108**(3), 654 (1986).

64DP. P. D. Weidman and D. R. Kassoy, The influence of side wall heat transfer on convection in a confined saturated porous medium, *Physics Fluids* **29**(2), 349 (1986).

65DP. T. R. White, A. Mathur and S. C. Saxena, Effect of vertical boiler tube diameter on heat transfer coefficient in gas-fluidized beds, *Chem. Engng J.* **32**(1), 1 (1986).

66DP. D. A. Zumbrunnen, R. Viskanta and F. P. Incropera, Heat transfer through porous solids with complex internal geometries, *Int. J. Heat Mass Transfer* **29**(2), 275 (1986).

EXPERIMENTAL TECHNIQUES AND INSTRUMENTATION

1E. M. M. Abu-Zeid and A. G. Sasburg, Diode-bridge temperature sensor, *Rev. Scient. Instrum.* **57**(10), 2609 (1986).

2E. F. A. Ansari and A. Afaq, New method of measuring thermal diffusivity of spherical produce, *Int. J. Refrig.* **9**(3), 158 (1986).

3E. M. H. Attia and L. Kops, Distortion in thermal field around inserted thermocouples in experimental interfacial studies, *J. Engng Ind.* **108**(4), 241 (1986).

4E. J. W. Baughn, M. A. Hoffman and D. B. Makel, Improvements in a new technique for measuring and mapping heat transfer coefficients, *Rev. Scient. Instrum.* **57**(4), 650 (1986).

5E. J. W. Baughn, D. Cooper, H. Lacovides and D. Jackson, Instrument for the measurement of heat flux from a surface with uniform temperature, *Rev. Scient. Instrum.* **57**(5), 921 (1986).

6E. R. E. Bentley and T. L. Morgan, Ni-based thermocouples in the mineral-insulated metal-sheathed format: thermoelectric instabilities to 1100 deg C, *J. Phys. E: Scient. Instrums* **19**(4), 262 (1986).

7E. L. M. Besley, Properties of some germanium resistance thermometers of Chinese manufacture, *Cryogenics* **26**(4), 207 (1986).

8E. L. M. Besley, G. Q. Zhang, E. C. Horrigan and A. Szmyrka-Grzebyk, Russian germanium resistance thermometers at low temperatures in magnetic fields, *Cryogenics* **26**(7), 413 (1986).

9E. W. A. Bosch, F. Mathu, H. C. Meijer and R. W. Willekers, Behaviour of thick film resistors (Philips type RC-01) as low temperature thermometers in magnetic fields up to 5 T, *Cryogenics* **26**(1), 3 (1986).
10E. L. Bruschi, R. Storti and G. Torzo, New high-precision bridge for low-resistance thermometry, *Rev. Scient. Instrum.* **57**(9), 2361 (1986).
11E. E. F. Buyukbicer, J. E. S. Venart and R. C. Prasad, An apparatus for measuring the thermal conductivity of fluids, *High Temp.–High Press.* **18**(1), 55 (1986).
12E. R. K. Chohan, Response time correlation for industrial temperature sensors, *J. Phys. E: Scient. Instrums* **19**(10), 786 (1986).
13E. D. Coetzee and D. G. Kroger, Analysis and measurement of contact and gap resistances in extruded bimetallic finned tubes, *J. Heat Recovery Sys.* **6**(6), 503 (1986).
14E. V. A. Cundy, J. S. Morse and D. W. Senser, Constant-tension thermocouple rake suitable for use in flame mode combustion studies, *Rev. Scient. Instrum.* **57**(6), 1209 (1986).
15E. M. Cvetkovic, Thermal flux measurements with a compensated heat flow meter at elevated radiation influence, *High Temp.–High Press.* **18**(1), 49 (1986).
16E. R. R. Dils, J. Geist and M. L. Reilly, Measurement of the silver freezing point with an optical fiber thermometer: proof of concept, *J. Appl. Phys.* **59**(4), 1005 (1986).
17E. J. E. Doorly and M. L. G. Oldfield, New heat transfer gages for use on multilayered substrates, *J. Turbomachinery* **108**(1), 153 (1986).
18E. A. H. Epstein, G. R. Guenette, R. J. G. Norton and V. C. Yuzhang, High-frequency response heat-flux gauge, *Rev. Scient. Instrum.* **57**(4), 639 (1986).
19E. R. J. Exton and M. E. Hillard, Raman doppler velocimetry: a unified approach for measuring molecular flow velocity, temperature, and pressure, *Appl. Optics* **25**(1), 14 (1986).
20E. V. N. Garbuzov, G. L. Iosel'son, V. A. Leikin, L. A. Nazarenko and A. S. Rivtsis, Plasma temperature measurement from thermal uhf radiation in solid-fuel MHD generators, *High Temp.* **24**(2), 257 (1986).
21E. K. T. V. Grattan, R. K. Seilli and A. W. Palmer, Fiber-optic absorption temperature sensor using fluorescence reference channel, *Rev. Scient. Instrum.* **57**(6), 1175 (1986).
22E. R. Gulino, S. Bair, W. O. Winer and B. Bhushan, Temperature measurement of microscopic areas within a simulated head/tape interface using infrared radiometric techniques, *J. Tribol.* **108**(1), 29 (1986).
23E. D. R. P. Guy and R. H. Friend, Temperature measurement in high pressure cells using a rhodium+0.5% iron–chromel thermocouple pair, *J. Phys. E: Scient. Instrums* **19**(6), 430 (1986).
24E. T. Haruyama and R. Yoshizaki, Thin-film platinum resistance thermometer for use at low temperatures and in high magnetic fields, *Cryogenics* **26**(10), 536 (1986).
25E. D. A. Jackson, P. Akhavan Leilabady, J. D. C. Jones and M. Corke, A dual interferometer implemented in parallel on a single birefringent monomode fibre, *J. Phys. E: Scient. Instrums* **19**(2), 143 (1986).
26E. J. P. Kottenstette, Measuring tool–chip interface temperature, *J. Engng Ind.* **108**(2), 101 (1986).
27E. L. Krebs and K. Bremhorst, Comparison of computed frequency responses of intrinsic and encapsulated bead type thermocouples in liquid sodium, *Int. J. Heat Mass Transfer* **29**(9), 1417 (1986).
28E. T. Kumada, A method of measuring local heat and mass transfer coefficients based on the wet-bulb temperature, *Heat Transfer—Jap. Res.* **15**(2), 46 (1986).
29E. L. N. S. Kunar and F. I. Campayne, Method for the determination of the thermal conductivity of the heavy oil sands of Trinidad, *West Indian J. Engng* **11**(2), 30 (1986).
30E. Q. Li, C. H. Watson, R. G. Goodrich, D. G. Haase and H. Lukefahr, Thick film chip resistors for use as low temperature thermometers, *Cryogenics* **26**(8), 467 (1986).
31E. P. R. Madhava Panicker, P. K. Kunhikrishnan, K. Sengupta, S. Muralidharan Nair and K. Narayanan Nair, Temperature turbulence probe for atmospheric boundary layer studies, *J. Phys. E: Scient. Instrums* **19**(1), 67 (1986).
32E. L. Matthews, R. Longenbaugh and R. Mata, Transpiration radiometer in fire characterization, *InTech* **33**(8), 55 (1986).
33E. Y. Ohishi and S. Takahashi, Optical temperature sensor using Eu^{3+}-doped fluoride glass, *Appl. Optics* **25**(5), 720 (1986).
34E. H. Ohta, New transient calorimetric technique for determining the total hemispherical emissivity of metals—an application of laser flash method, *High Temp. Mater. Processes* **7**(2), 185 (1986).
35E. B. Peavy and B. Rennex, Circular and square edge effect study for guarded-hot-plate and heat-flow-meter apparatuses, *J. Therm. Insul.* **9**, 254 (1986).
36E. V. E. Peletskii and L. G. Patrusheva, An apparatus for examining the thermal conductivities of constructional materials at high temperatures, *High Temp.* **24**(1), 122 (1986).
37E. R. W. Pitz and M. C. Drake, Intermittency and conditional averaging in a turbulent nonpremixed flame by Raman scattering, *AIAA J.* **24**(5), 815 (1986).
38E. M. A. Player, A simple wide-range thermistor thermometer, *J. Phys. E: Scient. Instrums* **19**(10), 787 (1986).
39E. D. Proctor, The design of a meter to measure the mass flow-specific heat product of a heat transfer fluid, *J. Sol. Energy Engng* **108**(2), 117 (1986).
40E. P. J. Reijntjes, W. van Rijswijk, G. A. Vermeulen and G. Frossati, Comparison of a glass thermometer against a nuclear orientation thermometer in high magnetic fields, *Rev. Scient. Instrum.* **57**(7), 1413 (1986).
41E. L. J. Salerno, P. Kittel, W. F. Brooks, A. L. Spivak and W. G. Marks, Jr., Thermal conductance of pressed brass contacts at liquid helium temperatures, *Cryogenics* **26**(4), 217 (1986).
42E. A. Santucci, L. Verdini and P. G. Verdini, Data-acquisition system for measurement of thermal diffusivity and propagation properties of thermal waves by a non-steady-state method, *Rev. Scient. Instrum.* **57**(8), 1627 (1986).
43E. G. S. Settles, Modern developments in flow visualization, *AIAA J.* **24**(8), 1313 (1986).

44E. V. Shubha and T. G. Ramesh, Precision analogue lineariser for thermocouples, *J. Phys. E: Scient. Instrums* **19**(11), 916 (1986).
45E. G. W. Stockton, S. J. Ehrlich-Moser, D. H. Chidester and R. S. Wayne, Versatile macroscale heat flow calorimeter for the study of chemical processes, *Rev. Scient. Instrum.* **57**(12), 3034 (1986).
46E. N. S. Sullivan and C. M. Edwards, Carbon resistance thermometers and variable-range hopping, *Cryogenics* **26**(4), 211 (1986).
47E. G. L. Switzer, L. P. Goss, D. D. Trump, C. M. Reeves, J. S. Stutrud, R. P. Bradley and W. M. Roquemore, CARS measurements in the near-wake region of an axisymmetric bluff-body combustor, *AIAA J.* **24**(7), 1155 (1986).
48E. H. Trethowen, Measurement errors with surface-mounted heat flux sensors, *Bldg Environ.* **21**(1), 41 (1986).
49E. N. Tulshian and F. Wheaton, Oyster (*Crassostrea virginica*) shell thermal conductivity: technique and determination, *Trans. Am. Soc. Agric. Engrs (Gen. Edn)* **29**(2), 626 (1986).
50E. T. Ueda, A Hosokawa and A. Yamamoto, Measurement of grinding temperature using infrared radiation pyrometer with optical fiber, *J. Engng Ind.* **108**(4), 247 (1986).
51E. Y. Waseda, Application of laser-pulse technique to determining thermal conductivity of molten oxides at high temperature, *High Temp. Mater. Processes* **7**(1), 55 (1986).
52E. B. A. Weinstein, Ruby thermometer for cryobaric diamond-anvil cell, *Rev. Scient. Instrum.* **57**(5), 910 (1986).
53E. X. Z. Zhang, G. H. He, Z. Wei and B. L. Zhou, Measurement of thermal diffusivity by the flash method for a two-layer composite sample in the case of triangular pulse, *Int. J. Thermophys.* **7**(4), 803 (1986).

NATURAL CONVECTION—INTERNAL FLOWS

1F. J. H. Adlam, Computation of two-dimensional time-dependent natural convection in a cavity where there are internal bodies, *Comput. Fluids* **14**(2), 141 (1986).
2F. R. J. Adrian, R. T. D. S. Ferreira and T. Boberg, Turbulent thermal convection in wide horizontal fluid layers. A review, *Expl Fluids* **4**(3), 121 (1986).
3F. N. Agrait and A. Castellanos, Oscillatory and steady convection in a dielectric viscoelastic layer subjected to a temperature gradient in the presence of an electric field, *J. Non-Newtonian Fluid Mech.* **21**(1), 1 (1986).
4F. T. Aihara and S. Maruyama, Laminar free convective heat transfer in vertical uniform-heat-flux ducts (numerical solutions with constant/temperature-dependent fluid properties), *Heat Transfer—Jap. Res.* **15**(3), 69 (1986).
5F. O. O. Ajayl, Natural convection in a hemispherical bowl with a free surface, *J. Heat Transfer* **108**(4), 970 (1986).
6F. S. Aleyas, L. V. K. V. Sarma and S. N. Majhi, Thermal effects on slow viscous flow between rotating eccentric spheres with axial gravitational field, *Int. J. Engng Sci.* **24**(3), 323 (1986).
7F. R. Anderson and M. Bohn, Heat transfer enhancement in natural convection enclosure flow, *J. Heat Transfer* **108**(2), 330 (1986).
8F. W. Aung and G. Worku, Developing flow and flow reversal in a vertical channel with asymmetric wall temperatures, *J. Heat Transfer* **108**(2), 299 (1986).
9F. W. Aung and G. Worku, Theory of fully developed, combined convection including flow reversal, *J. Heat Transfer* **108**(2), 485 (1986).
10F. L. F. A. Azevedo and E. M. Sparrow, Natural convection in a vertical channel vented to the ambient through an aperture in the channel wall, *Int. J. Heat Mass Transfer* **29**(6), 819 (1986).
11F. R. F. Babus'Haq, S. D. Probert, M. J. Shilston and S. Chakrabarti, Influence of baffles upon natural-convective steady-state heat transfers inwards across horizontal eccentric annuli, *Appl. Energy* **24**(3), 165 (1986).
12F. C. Beckermann, R. Viskanta and S. Ramadhyani, A numerical study of non-Darcian natural convection in a vertical enclosure filled with a porous medium, *Numer. Heat Transfer* **10**(6), 557 (1986).
13F. T. L. Bergman and S. Ramadhyani, Combined buoyancy- and thermocapillary-driven convection in open square cavities, *Numer. Heat Transfer* **9**(4), 441 (1986).
14F. T. L. Bergman, F. P. Incropera and R. Viskanta, Transient behavior of a radiatively heated double-diffusive system, *J. Heat Transfer* **108**(2), 317 (1986).
15F. T. L. Bergman, F. P. Incropera and R. Viskanta, Correlation of mixed layer growth in a double diffusive, salt-stratified system heated from below, *J. Heat Transfer* **108**(1), 206 (1986).
16F. H. Bertin and H. Ozoe, Numerical study of two-dimensional natural convection in a horizontal fluid layer heated from below, by finite-element method; influence of Prandtl number, *Int. J. Heat Mass Transfer* **29**(3), 439 (1986).
17F. N. A. Bezdenezhnykh, V. A. Briskman, L. Ye. Sorokin and Ye. L. Tarunin, Convection in a conducting fluid heated from below and existing in a magnetic field at reduced gravity, *Fluid Mech., Sov. Res.* **15**(2), 1 (1986).
18F. G. P. Bogatyrev and V. D. Zimin, Space–time spectra of random fluctuations in hydrodynamic systems, *Fluid Mech., Sov. Res.* **15**(2), 93 (1986).
19F. M. S. Bohn and R. Anderson, Temperature and heat flux distribution in a natural convection enclosure flow, *J. Heat Transfer* **108**(2), 471 (1986).
20F. E. W. Bolton, F. H. Busse and R. M. Clever, Oscillatory instabilities of convection rolls at intermediate Prandtl numbers, *J. Fluid Mech.* **164**, 469 (1986).
21F. P. Bontoux, C. Smutek, B. Roux and J. M. LaCroix, Three-dimensional buoyancy-driven flows in cylindrical cavities with differentially heated endwalls. Part I. Horizontal cylinders, *J. Fluid Mech.* **169**, 211 (1986).
22F. B. Brenier, B. Roux and P. Bontoux, Comparison of the Tau–Chebyshev and Galerkin methods in the study of stability of natural convection motions problem of spurious eigenvalues, *J. Mec. Theor. Applic.* **5**(1), 95 (1986).

23F. Y. L. Chan and C. L. Tien, Laminar natural convection in shallow open cavities, *J. Heat Transfer* **108**(2), 305 (1986).
24F. C. J. Chang, T. F. Lin and W. M. Yan, Natural convection flows in a vertical, open tube resulting from combined buoyancy effects of thermal and mass diffusion, *Int. J. Heat Mass Transfer* **29**(10), 1543 (1986).
25F. C.-K. Chen and S.-W. Hsiao, Natural convection heat transfer in the annuli between concentric and eccentric horizontal cylinders, *Chung-Kuo Chittsueh Kung Ch'eng Hsueh Pao* **7**(5), 297 (1986).
26F. D. R. Chenoweth and S. Paolucci, Natural convection in an enclosed vertical air layer with large horizontal temperature differences, *J. Fluid Mech.* **169,** 173 (1986).
27F. S. G. Cherkasov, Quasisteady free convection regime in a vertical cylindrical vessel, *Fluid Dyn.* **21**(1), 125 (1986).
28F. F. B. Cheung, The effects of turbulent natural convection on thermal explosion critical conditions, *J. Heat Transfer* **108**(3), 713 (1986).
29F. F. C. Chou and G. J. Hwang, Prandtl number effect on combined free and forced laminar convection in the thermal entrance region of a horizontal rectangular channel, *C.J.Ch.E.* **64**(5), 858 (1986).
30F. S. Ciliberto and F. Simonelli, Spatial structures of temporal chaos in Rayleigh–Benard convection, *Europhys. Lett.* **2**(4), 285 (1986).
31F. A. Cloot and G. Legon, Marangoni convection induced by a nonlinear temperature-dependent surface tension, *J. Phys. (Paris)* **47**(1), 23 (1986).
32F. J. P. Coutier and R. Greif, Mixed laminar convection in a horizontal tube with natural convection around its boundaries, *Int. J. Heat Mass Transfer* **29**(3), 391 (1986).
33F. C. Cuvelier and J. M. Driessn, Thermocapillary free boundaries in crystal growth, *J. Fluid Mech.* **169,** 1 (1986).
34F. A. W. Date, Numerical prediction of natural convection heat transfer in horizontal annulus, *Int. J. Heat Mass Transfer* **29**(10), 1457 (1986).
35F. L. A. O. Davalos and O. Manero, Thermoconvective instability of a second-order fluid, *J. Phys. Soc. Japan* **55**(2), 442 (1986).
36F. P. C. T. De Boer, Thermally driven motion of highly viscous fluids, *Int. J. Heat Mass Transfer* **29**(5), 681 (1986).
37F. S. Doerffer and J. Mikielewicz, The influence of oscillations on natural convection in ship tanks, *Int. J. Heat Fluid Flow* **7**(1), 49 (1986).
38F. M. Faghri and Y. Asako, Periodic, fully developed, natural convection in a channel with corrugated confining walls, *Int. J. Heat Mass Transfer* **29**(12), 1931 (1986).
39F. M. T. Farmer, R. W. Douglass and S. A. Trogdon, The stability of natural convection in narrow-gap spherical annuli to axisymmetric disturbances, *Int. J. Heat Mass Transfer* **29**(10), 1575 (1986).
40F. R. S. Figliola, Convection transitions within a vertical cylinder heated from below, *Physics Fluids* **29**(7), 2028 (1986).
41F. P. Filis and D. Poulikakos, An experimental study on the effect of wall temperature non-uniformity on natural convection in an enclosure heated from the side, *Int. J. Heat Fluid Flow* **7**(4), 258 (1986).
42F. T. Fusegi, B. Farouk and K. S. Ball, Mixed-convection flows within a horizontal concentric annulus with a heated rotating inner cylinder, *Numer. Heat Transfer* **9**(5), 591 (1986).
43F. E. K. Glakpe, C. B. Watkins, Jr. and J. N. Cannon, Constant heat flux solutions for natural convection between concentric and eccentric horizontal cylinders, *Numer. Heat Transfer* **10**(3), 279 (1986).
44F. U. Gross and E. Hahne, Heat transfer in a closed thermosyphon over a wide range of pressures and inclinations, *Chemie-Ingr-Tech.* **58**(2), 136 (1986).
45F. S. Habchi and S. Acharya, Laminar mixed convection in a partially blocked, vertical channel, *Int. J. Heat Mass Transfer* **29**(11), 1711 (1986).
46F. S. Habchi and S. Archarya, Laminar mixed convection in a symmetrically or asymmetrically heated vertical channel, *Numer. Heat Transfer* **9**(5), 605 (1986).
47F. K. Hashimoto, N. Akino and H. Kawamura, Combined forced–free laminar heat transfer to a highly heated gas in a vertical annulus, *Int. J. Heat Mass Transfer* **29**(1), 145 (1986).
48F. D. Hill, A. Kirkpatrick and P. Burns, Analysis and measurements of interzonal natural convection heat transfer in buildings, *J. Sol. Energy Engng* **108**(3), 178 (1986).
49F. K. Himasekhar and H. H. Bau, Large Rayleigh number convection in a horizontal, eccentric annulus containing saturated porous media, *Int. J. Heat Mass Transfer* **29**(5), 703 (1986).
50F. L. N. Howard and R. Krishnamurti, Large-scale flow in turbulent convection: a mathematical model, *J. Fluid Mech.* **170**, 385 (1986).
51F. H. Inaba, Natural convection in an inclined rectangular channel heated from the bottom surface, *J. Heat Transfer* **108**(4), 764 (1986).
52F. H. Inaba, T. Fukada and M. Sugawara, Natural and forced convective heat transfer in a channel flow over a rectangular cavity with a heated bottom surface, *Heat Transfer—Jap. Res.* **15**(5), 1 (1986).
53F. H. Inaba and Y. Wada, Convection heat transfer in an inclined cavity heated from the bottom surface in forced laminar flow, *Heat Transfer—Jap. Res.* **15**(5), 92 (1986).
54F. Y. Inoue, S. Saeki and R. Ito, Experimental study of Benard convection in an electrolytic solution divided by a membrane, *Kagaku Kogaku Ronbunshu* **10**(6) (1984), Japan, *Int. Chem. Engng* **26**(3), 527 (1986).
55F. Y. Inoue and R. Ito, Analysis of Benard convection in an electrolytic solution divided by a membrane, *Kagaku Kogaku Ronbunshu* **10**(6) (1984), Japan, *Int. Chem. Engng* **26**(3), 513 (1986).
56F. S. K. Jena and S. P. Bhattacharyya, The effect of microstructure on the thermal convection in a rectangular box of fluid heated from below, *Int. J. Engng Sci.* **24**(1), 69 (1986).
57F. M. C. Jones and J. M. Persichetti, Convective instability in packed bed with through flow, *A.I.Ch.E. Jl*

32(9), 1555 (1986).
58F. F. Karim, B. Farouk and I. Namer, Natural convection heat transfer from a horizontal cylinder between vertical adiabatic walls, *J. Heat Transfer* **108**(2), 291 (1986).
59F. M. Kaviany and M. Vogel, Effect of solute concentration gradients on the onset of convection: uniform and nonuniform initial gradients, *J. Heat Transfer* **108**(4), 776 (1986).
60F. A. Khalilollahi and B. Sammakia, Unsteady natural convection generated by a heated surface within an enclosure, *Numer. Heat Transfer* **9**(6), 715 (1986).
61F. Y. Kikuchi, T. Shioyama and Z. Kawara, Turbulent heat transport in a horizontal fluid layer heated internally and from below, *Int. J. Heat Mass Transfer* **29**(3), 451 (1986).
62F. K.-H. Kim and M.-U. Kim, The onset of natural convection in a fluid layer suddenly heated from below, *Int. J. Heat Mass Transfer* **29**(2), 193 (1986).
63F. T. Kimura, N. Heya, M. Takeuchi and H. Isomi, Natural convection of fluids with two stratified layers in a rectangular enclosure, *Heat Transfer—Jap. Res.* **15**(4), 31 (1986).
64F. T. Kimura, N. Heya, M. Takeuchi and T. Usui, Natural convection heat transfer in partitioned enclosures, *Heat Transfer—Jap. Res.* **15**(4), 89 (1986).
65F. T. Kimura, N. Heya, M. Takeuchi and T. Usui, Natural convection heat transfer in a trapezoidal enclosure (experimental investigation), *Nippon Kikai Gakkai Ronbunshu, B. Hen* **52**(474), 870 (1986).
66F. A. T. Kirkpatrick and M. Bohn, An experimental investigation of mixed cavity natural convection in the high Rayleigh number regime, *Int. J. Heat Mass Transfer* **29**(1), 69 (1986).
67F. E. Knobloch, D. R. Moore, J. Toomre and N. O. Weiss, Transitions to chaos in two-dimensional double-diffusive convection, *J. Fluid Mech.* **166,** 409 (1986).
68F. P. Kolodner, R. W. Walden, A. Pasner and C. M. Surko, Rayleigh–Benard convection in an intermediate-aspect-ratio rectangular container, *J. Fluid Mech.* **163,** 195 (1986).
69F. E. L. Koschmider and M. I. Biggerstaff, Onset of surface-tension-driven Benard convection, *J. Fluid Mech.* **167,** 49 (1986).
70F. R. Kumar and R. J. Adrian, Higher order moments in the entrainment zone of turbulent penetrative thermal convection, *J. Heat Transfer* **108**(2), 323 (1986).
71F. T. T. Lam and Y. Bayazitoglu, Application of the sequential gradient-restoration algorithm to thermal convective instability problems, *J. Optim. Theory Applic.* **49**(1), 47 (1986).
72F. K. E. Lankford and A. Bejan, Natural convection in a vertical enclosure filled with water near 4 deg C, *J. Heat Transfer* **108**(4), 755 (1986).
73F. A. S. Lavine, R. Greif and J. A. C. Humphrey, Three-dimensional analysis of natural convection in a toroidal loop: effect of tilt angle, *J. Heat Transfer* **108**(4), 796 (1986).
74F. A. Legait, A scalar model of turbulent convection, *Physics Fluids* **29**(11), 3883 (1986).
75F. J. H. Lienhard, V. Catton and I. Catton, Heat transfer across a two-fluid-layer region, *J. Heat Transfer* **108**(1), 198 (1986).
76F. Y. S. Lin and R. C. Akins, Thermal description of pseudosteady-state natural convection inside a vertical cylinder, *Int. J. Heat Mass Transfer* **29**(2), 301 (1986).
77F. D. Littlefield and P. Desai, Buoyant laminar convection in a vertical cylindrical annulus, *J. Heat Transfer* **108**(4), 814 (1986).
78F. T. Maekawa and I. Tanasawa, Convective instability problem of a horizontal layer of an electrically conducting fluid under a vertical magnetic field, *Nippon Kikai Gakkai Ronbunshu, B. Hen* **52**(474), 881 (1986).
79F. D. N. Mahoney, R. Kumar and E. H. Bishop, Numerical investigation of variable property effects on laminar natural convection of gases between two horizontal isothermal concentric cylinders, *J. Heat Transfer* **108**(4), 783 (1986).
80F. T. Masuoka and G. Shimizu, Effects of lateral walls on the stability of natural convection in an inclined fluid layer, *Nippon Kikai Gakkai Ronbunshu, B. Hen* **52**(475), 1350 (1986).
81F. P. Metzener, The effect of rigid sidewalls on nonlinear two-dimensional Benard convection, *Physics Fluids* **29**(1), 1373 (1986).
82F. K. Mitachi, K. Aoki, K. Kitamura and N. Komatsu, Natural convection of heat generating fluid within horizontal cylinder, *Bull. J.S.M.E.* **29**(254), 2547 (1986).
83F. S. M. Morcos, M. M. Hilal, M. M. Kamel and M. S. Soliman, Experimental investigation of mixed laminar convection in the entrance region of inclined rectangular channels, *J. Heat Transfer* **108**(3), 574 (1986).
84F. I. M. Moroz, Amplitude expansions and normal forms in a model for thermohaline convection, *Stud. Appl. Math.* **74**(2), 155 (1986).
85F. V. D. Murty, A numerical investigation of Benard convection using finite elements, *Comput. Fluids* **14**(4), 379 (1986).
86F. T. A. Myrum, E. M. Sparrow and A. T. Prata, Numerical solutions for natural convection in a complex enclosed space containing either air–liquid or liquid–liquid layers, *Numer. Heat Transfer* **10**(1), 19 (1986).
87F. A. Nadarajah and R. Narayanan, A comparison result for convection in bounded geometries, *ZAMP* **37**(2), 280 (1986).
88F. D. A. Nelson and E. J. Shaughnessy, Electric field effects on natural convection in enclosures, *J. Heat Transfer* **108**(4), 749 (1986).
89F. A. A. Nepomnyashchiy and I. B. Simanovskiy, Onset of convection in a two-layer system, *Fluid Mech., Sov. Res.* **15**(2), 9 (1986).
90F. T. Nishimura, Y. Kawamura, T. Takumi and H. Ozoe, Experiments of natural convection heat transfer in rectangular enclosure partially filled with particles, *Heat Transfer—Jap. Res.* **15**(4), 62 (1986).
91F. M. Okada and H. Okano, Natural convection heat transfer in a rectangular cell enclosed with vertical walls and plate-fins, *Nippon Kikai Gakkai Ronbunshu, B. Hen* **52**(477), 2158 (1986).

92F. S. J. Ormiston, G. D. Raithby and K. G. T. Hollands, Numerical predictions of natural convection in a Trombe wall system, *Int. J. Heat Mass Transfer* **29**(6), 869 (1986).

93F. H. Ozoe, A. Mouri, M. Hiramitsu, S. W. Churchill and N. Lior, Numerical calculation of three-dimensional turbulent natural convection in a cubical enclosure using a two-equation model for turbulence, *J. Heat Transfer* **108**(4), 806 (1986).

94F. J. F. Pignatel and J. F. Marcillat, Transition to time-dependent free convection in an inclined air layer, *Int. J. Heat Fluid Flow* **7**(3), 169 (1986).

95F. M. R. E. Proctor and J. Y. Holyer, Planform selection in salt fingers, *J. Fluid Mech.* **168,** 241 (1986).

96F. Y. Renardy, Interfacial stability in a two-layer Benard problem, *Physics Fluids* **29**(1), 356 (1986).

97F. N. Riahi and A. T. Hsui, Nonlinear double-diffusive convection with local heat and solute sources, *Int. J. Engng Sci.* **24**(4), 529 (1986).

98F. N. Riahi, Nonlinear convection with variable coefficient of thermal expansion, *Acta Mech.* **60**(3), 143 (1986).

99F. N. Riahi, Convection in a low Prandtl number fluid with internal heating, *Int. J. Non-linear Mech.* **21**(1), 97 (1986).

100F. S. Rosenblat, Thermal convection in a viscoelastic liquid, *J. Non-Newtonian Fluid Mech.* **21**(2), 201 (1986).

101F. F. W. Schmidt, P. W. Giel, R. E. Phillips and D. F. Wang, A comparison of experimental and predicted results for laminar natural convection in an enclosure, *Int. J. Heat Fluid Flow* **7**(3), 183 (1986).

102F. A. K. Sen, Thermocapillary convection in a rectangular cavity with a deformable interface, *Physics Fluids* **29**(11), 3881 (1986).

103F. A. N. Sharifulin, Supercritical vibration-induced thermal convection in a cylindrical cavity, *Fluid Mech., Sov. Res.* **15**(2), 28 (1986).

104F. Y. Shina and K. Fujimura, Heat transfer by natural convection in a half-sphere heated from below, *J. Nucl. Sci. Technol.* **23**(6), 568 (1986).

105F. P. G. Simpkins and K. S. Chen, Convection in horizontal cavities, *J. Fluid Mech.* **166,** 21 (1986).

106F. P. Singh and J. A. Liburdy, Effect of plate inclination on natural convection from a plate to its cylindrical enclosure, *J. Heat Transfer* **108**(4), 770 (1986).

107F. L. Ye. Sorokin, Nonlinear convective motion of a binary mixture with thermodiffusion, *Fluid Mech., Sov. Res.* **15**(2), 19 (1986).

108F. E. M. Sparrow, L. F. A. Azevedo and A. T. Prata, Two-fluid and single-fluid natural convection heat transfer in an enclosure, *J. Heat Transfer* **108**(4), 848 (1986).

109F. W. E. Stewart, Jr. and J. L. Verhulst, Experimental free convection from piping in district heating utilidors, *J. Energy Resour. Technol.* **108**(2), 173 (1986).

110F. R. W. Stickles, A. Ungan and R. Viskanta, Natural convection in a rectangular enclosure with differentially heated end walls and a free liquid surface, *PCH PhysioChem. Hydrodyn.* **7**(2), 161 (1986).

111F. J. G. Symmons, Experimental investigation of natural convection in partially divided enclosures, *J. Heat Transfer* **108**(3), 554 (1986).

112F. T. Takeda, H. Kawamura and M. Seki, Natural circulation in parallel vertical channels with different heat inputs, *Nippon Kikai Gakkai Ronbunshu, B. Hen* **52**(480), 3035 (1986).

113F. R. Tomcej and N. Nandakumar, Mixed convection flow of a power-law fluid in horizontal ducts, *C.J.Ch.E.* **64**(5), 743 (1986).

114F. C. H. Tsang and S. Acharya, Natural convection in an inclined enclosure with an off-center complete partition, *Numer. Heat Transfer* **9**(2), 217 (1986).

115F. K. Vajravelu and L. Debnath, Nonlinear study of convective heat transfer and fluid flows induced by travelling thermal waves, *Acta Mech.* **59**(3), 233 (1986).

116F. I. I. Vertgeym, Evaluation of turbulent free-convection flow in a planar vertical layer, *Fluid Mech., Sov. Res.* **15**(2), 72 (1986).

117F. R. Viskanta, D. M. Kim and C. Gau, Three dimensional natural convection heat transfer of a liquid metal in a cavity, *Int. J. Heat Mass Transfer* **29**(3), 475 (1986).

118F. J. L. Wright and R. W. Douglass, Natural convection in narrow-gap, spherical annuli, *Int. J. Heat Mass Transfer* **29**(5), 725 (1986).

119F. M. Yamakawa, T. Sakai, H. Kawashima, A. Sakurai and S. Hattori, Analysis of natural convection in narrow annular gaps of LMFN4BR, *J. Nucl. Sci. Technol.* **23**(5), 451 (1986).

120F. S. Yanase, The effect of time-periodic heating on the stability of a horizontal double-diffusive fluid layer, *J. Phys. Soc. Japan* **55**(4), 1130 (1986).

121F. A. Yoshizawa, Statistical modeling of turbulent thermally buoyant flows, *J. Phys. Soc. Japan* **55**(9), 3066 (1986).

122F. M. A. Zaks and D. V. Lyubimov, Alteration of stable modes in a convection problem, *Fluid Mech., Sov. Res.* **15**(2), 57 (1986).

123F. M. Zamora and A. Rey de Luna, Energy of a system formed by a convective fluid and its container, *J. Fluid Mech.* **167,** 427 (1986).

NATURAL CONVECTION—EXTERNAL FLOWS

1FF. G. I. Abramov, S. I. Zakharchenko, L. V. Krupnova and L. M. Fisher, Free-convective heat transfer from a horizontal cylinder in gaseous helium with large temperature heads, *High Temp.* **24**(3), 396 (1986).

2FF. N. Afzal and T. Hussain, Effects of large suction in natural convection boundary layer on a horizontal plate, *Wärme- und Stoffübertr.* **20**(3), 203 (1986).

3FF. M. Amouche et J.-L. Peube, Convection mixte autour d'un cylindre horizontal. Influence des variations des proprietes physiques avec la temperature, *Int. J. Heat Mass Transfer* **29**(9), 1293 (1986).

4FF. J. Andreopoulos, A. Praturi and W. Rodi, Experiments on vertical plane buoyant jets in shallow water, *J. Fluid Mech.* **168,** 305 (1986).

5FF. T. L. Bergman and A. Ungan, Experimental and numerical investigation of double-diffusive convection induced by a discrete heat source, *Int. J. Heat Mass Transfer* **29**(11), 1695 (1986).

6FF. P. Bontoux, B. Roux, G. H. Schiroky, B. L. Markham and F. Rosenberger, Convection in the vertical mid-plane of a horizontal cylinder. Comparison of two-dimensional approximations with three-dimensional results, *Int. J. Heat Mass Transfer* **29**(2), 227 (1986).

7FF. G. M. Carlomagno and P. Lombardi, Convective wall plume induced by a linear mass source, *Heat Technol.* **4**(2), 80 (1986).

8FF. T. S. Chen, H. C. Tien and F. Armaly, Natural convection of horizontal, inclined, and vertical plates with variable surface temperature or heat flux, *Int. J. Heat Mass Transfer* **29**(10), 1465 (1986).

9FF. T. S. Chen, B. F. Armaly and N. Ramachandran, Correlations for laminar mixed convection flows on vertical, inclined, and horizontal flat plates, *J. Heat Transfer* **108**(4), 835 (1986).

10FF. L. Y. Cooper and A. Woodhouse, The buoyant plume-driven adiabatic ceiling temperature revisited, *J. Heat Transfer* **108**(4), 822 (1986).

11FF. I. El-Henawy, B. Gebhart and N. D. Kazarinoff, Multiple steady-state solutions for horizontal buoyant flows in cold water, *Int. J. Heat Mass Transfer* **29**(11), 1655 (1986).

12FF. P. G. Frik, Modeling cascade processes in two-dimensional turbulent convection, *J. Appl. Mech. Tech. Phys.* **27**(2), 221 (1986).

13FF. R. Subba Reddy Gorla and R. A. Stratman, Axisymmetric free convection boundary flow of water at 4 deg C past slender bodies, *Int. J. Heat Fluid Flow* **7**(3), 179 (1986).

14FF. G. Gouesbet, M. E. Weill and E. Lefort, Convective and free surface instabilities provoked by heating below an interface, *AIAA J.* **24**(8), 1324 (1986).

15FF. R. W. Griffiths, Thermals in extremely viscous fluids, including the effects of temperature-dependent viscosity, *J. Fluid Mech.* **166,** 115 (1986).

16FF. J. M. Hyun and J. C. Hyun, Adjustment of a thermally-stratified fluid in a container with vertical through-flow, *Int. J. Heat Mass Transfer* **29**(10), 1487 (1986).

17FF. Y. Jalurai, Mixed convection flow over localized multiple thermal sources on a vertical surface, *Physics Fluids* **29**(4), 934 (1986).

18FF. Y. Joshi and B. Gebhart, Measurements and visualizations of transient and steady-state vertical natural convection flow in cold water, *Int. J. Heat Mass Transfer* **29**(11), 1723 (1986).

19FF. K. Kishinami and H. Saito, Natural convection heat transfer on a vertical plate with discontinuous surface temperature (effect of heat conduction in the plate), *Bull. J.S.M.E.* **29**(253), 2145 (1986).

20FF. A. K. Kolar and V. M. K. Sastri, Influence of step discontinuity in porous plate temperature in free convection with transpiration, *Wärme- und Stoffübertr.* **20**(2), 111 (1986).

21FF. K. M. Kothari and R. N. Meroney, Liquefied natural gas (LNG) plume interaction with storage tanks, *J. Heat Transfer* **108**(2), 475 (1986).

22FF. M.-C. Lai, S.-M. Jeng and G. M. Faeth, Structure of turbulent adiabatic wall plumes, *J. Heat Transfer* **108**(4), 827 (1986).

23FF. F.-S. Lien and C.-K. Chen, Effects of microstructure on the conjugated mixed forced and free convection–conduction analysis of heat transfer in a vertical plate fin, *J. Heat Transfer* **108**(3), 580 (1986).

24FF. F.-S. Lien, C.-K. Chen and J.-W. Cleaver, Analysis of natural convection flow of micropolar fluid about a sphere with blowing and suction, *J. Heat Transfer* **108**(4), 967 (1986).

25FF. P. Luchini, Analytical and numerical solutions for natural convection in a corner, *AIAA J.* **24**(5), 841 (1986).

26FF. S. Mori, S. Gotoh, A. Tanimoto and M. Sakakibara, Conjugate heat transfer between internal forced and external natural convections with wall conduction, *C.J.Ch.E.* **64**(2), 216 (1986).

27FF. M. L. Ng, J. P. Hartnett and E. Y. Kwak, Natural convection from horizontal wires to viscoelastic fluids, *J. Heat Transfer* **108**(4), 790 (1986).

28FF. P. N. Pintauro, An electrochemical method for determining natural convection mass transfer boundary-layer thicknesses, *Int. J. Heat Mass Transfer* **29**(5), 741 (1986).

29FF. Z. H. Qureshi and B. Gebhart, The stability of vertical thermal buoyancy induced flows in cold pure and saline water, *Int. J. Heat Mass Transfer* **29**(9), 1383 (1986).

30FF. L. Raham, On the thermal adjustment of an almost-enclosed fluid region with through-flow, *Int. J. Heat Mass Transfer* **29**(10), 1479 (1986).

31FF. M. M. Rahman and V. P. Carey, Steady and transient mixed convection near a vertical uniformly heated surface exposed to horizontal fluid flow, *Numer. Heat Transfer* **10**(4), 327 (1986).

32FF. A. Raptis and C. Perdikis, Magnetic effects on the flow with a great magnetic Reynolds number by the presence free convection and mass transfer, *Wärme- und Stoffübertr.* **20**(2), 163 (1986).

33FF. P. Sabhapathy and K. C. Cheng, The effects of temperature-dependent viscosity and coefficient of thermal expansion on the stability of laminar, natural convective flow along an isothermal, vertical surface, *Int. J. Heat Mass Transfer* **29**(10), 1521 (1986).

34FF. H. Saito, K. Kishinami, I. Tokura and T. Nagasawa, Experimental study on free convection heat transfer from horizontal cylinders in a vertical duct, *Nippon Kikai Gakkai Ronbunshu* **52**(484), 4036 (1986).

35FF. I. Shai and Y. Barnea, Simple analysis of mixed convection with uniform heat flux, *Int. J. Heat Mass Transfer* **29**(8), 1139 (1986).

36FF. E. M. Sparrow and C. K. Carlson, Local and average natural convection Nusselt numbers for a uniformly heated, shrouded or unshrouded horizontal plate, *Int. J. Heat Mass Transfer* **29**(3), 369 (1986).

37FF. E. M. Sparrow, M. A. Ansari, P. C. Stryker and R. Ruiz, Enhanced heat transfer from a horizontal finned tube situated in a vertical channel, *J. Heat Transfer* **108**(1), 62 (1986).

38FF. M. Takeuchi, N. Heya and T. Kimura, Numerical analysis of natural convection from an upward-facing horizontal surface (partial heating with a uniform heat flux), *Heat Transfer—Jap. Res.* **15**(4), 17 (1986).

39FF. M. Takeuchi, N. Heya, T. Kimura and Y. Tanaka, Natural convection from a vertical rectangular body,

Heat Transfer—Jap. Res. **15**(6), 57 (1986).
40FF. S. Thangam and C. F. Chen, Stability analysis on the convection of a variable viscosity fluid in an infinite vertical slot, *Physics Fluids* **29**(5), 1367 (1986).
41FF. H. C. Tien, T. S. Chen and B. F. Armaly, Vortex instability of natural convection flow over horizontal and inclined plates with uniform surface heat flux, *Numer. Heat Transfer* **9**(6), 697 (1986).
42FF. H. C. Tien, T. S. Chen and B. F. Armaly, Wave instability of natural convection flow on inclined flat plates with uniform surface heat flux, *Numer. Heat Transfer* **10**(2), 179 (1986).
43FF. K.-L. Wong and C.-K. Chen, Finite-element solution of laminar combined convection from two horizontal cylinders in tandem arrangement, *A.I.Ch.E. Jl* **32**(4), 557 (1986).

CONVECTION FROM ROTATING SURFACE

1G. J. J. Blech, I. Green and J. Kopelman, Transient temperature distribution in partially filled rotating horizontal cylinders, *J. Appl. Mech.* **53**(2), 436 (1986).
2G. B. M. Boubnov and G. S. Golitsyn, Experimental study of convection structure in rotating fluids, *J. Fluid Mech.* **167,** 503 (1986).
3G. P. Chandran and P. Kumar, Flow and heat transfer in the boundary layer due to rotating spheres, spheroids and paraboloids, *Int. J. Engng Sci.* **24**(5), 685 (1986).
4G. M. Firouzian, J. M. Owen, J. R. Pincombe and R. H. Rogers, Flow and heat transfer in a rotating cylindrical cavity with a radial inflow of fluid. 2. Velocity, pressure and heat transfer measurements, *Int. J. Heat Fluid Flow* **7**(1), 21 (1986).
5G. T. S. Lee, N. E. Wijeysundera and K. S. Yeo, Convection in eccentric annuli with inner cylinder rotation, *AIAA J.* **24**(1), 170 (1986).
6G. F.-S. Lien, C.-K. Chen and J. W. Cleaver, Mixed and free convection over a rotating sphere with blowing and suction, *J. Heat Transfer* **108**(2), 398 (1986).
7G. C. A. Long and J. M. Owen, The effect of inlet conditions on heat transfer in a rotating cavity with a radial outflow of fluid, *J. Turbomachinery* **108**(1), 145 (1986).
8G. M. G. Macaraeg, Numerical experiments of axisymmetric flow in a nonuniform gravitational field, *AIAA J.* **24**(9), 1483 (1986).
9G. S. Mochizuki and W.-J. Yang, Performance evaluation on rotating disk assemblies by automated transient testing method, *Heat Technol.* **4**(2), 1 (1986).
10G. H. Peller, Thermofluiddynamic experiments with a heated and rotating circular cylinder in crossflow. Part 2.1, *Expl Fluids* **4**(4), 223 (1986).
11G. C. O. Popiel and L. Boguslawski, Local heat transfer from a rotating disk in an impinging round jet, *J. Heat Transfer* **108**(2), 357 (1986).
12G. P. L. Read, Regimes of axisymmetric flow in an internally heated rotating fluid, *J. Fluid Mech.* **168,** 255 (1986).
13G. N. Rudraiah and O. P. Chandna, Effects of Coriolis force and nonuniform temperature gradient on the Rayleigh–Benard convection, *Can. J. Phys.* **64**(1), 90 (1986).
14G. M. K. Smith, Thermocapillary and centrifugal-buoyancy-driven motion in a rapidly rotating liquid cylinder, *J. Fluid Mech.* **166,** 245 (1986).
15G. L. Sobel, M. El-Masri and J. L. Smith, Heat transfer and flow visualization in natural convection in rapidly spinning systems, *J. Heat Transfer* **108**(3), 547 (1986).
16G. E. M. Sparrow and C. S. Preston, Heat transfer from rotating annular fins, *Int. J. Heat Mass Transfer* **29**(6), 831 (1986).

COMBINED HEAT AND MASS TRANSFER

1H. A. I. Abrosimov and A. V. Voronkevich, Heat exchange at the bottom of a dead end bathed by a laminar impinging jet, *J. Engng Phys.* **51**(1), 764 (1986).
2H. F. Ahrens and A. Astrom, High-intensity drying of paper, *Drying Technol.* **4**(2), 245 (1986).
3H. G. E. Andrews, M. Alikhanizadeh, A. A. Asere, C. I. Hussain, M. S. Khoshkbar Azari and M. C. Mkpadi, Small diameter film cooling holds: wall convective heat transfer, *J. Turbomachinery* **108**(2), 283 (1986).
4H. H. J. Carper, Jr., J. J. Saaverdra and T. Suwanprateep, Liquid jet impingement cooling of a rotating disk, *J. Heat Transfer* **108**(3), 540 (1986).
5H. A. O. Demuren, W. Rodi and B. Schonung, Systematic study of film cooling with a three-dimensional calculation procedure, *J. Turbomachinery* **108**(1), 124 (1986).
6H. V. M. Eroshenko, A. V. Ershov and L. I. Zaichik, Effect of variability of physical properties of a gas on turbulent flow and heat transfer in a pipe with permeable walls, *J. Engng Phys.* **50**(2), 135 (1986).
7H. V. V. Faleev, Sublimation in a flat channel with a movable wall and a permeable wall, *J. Engng Phys.* **51**(1), 854 (1986).
8H. J. C. Han and A. B. Mehendale, Flat-plate film cooling with steam injection through one row and two rows of inclined holds, *J. Turbomachinery* **108**(1), 137 (1986).
9H. K. Ichimiya, K. Kobashi and R. Echigo, Fundamental study on the flattening of temperature distribution of a high temperature steel slab (1st report, impingement heat transfer and flow situation around a heated slab caused by a laminar slot jet), *Nippon Kikai Gakkai Ronbunshu, B. Hen* **52**(474), 904 (1986).
10H. L. P. Kholpanov, E. Ya. Kenig and V. A. Malyusov, Combined heat and mass transfer for opposing film flows of liquid and gas, *J. Engng Phys.* **51**(1), 768 (1986).
11H. V. T. Kiril'tsev, V. P. Motulevich and E. D. Sergievskii, Study of a turbulent boundary layer in a short channel under the combined influence of injection through a porous wall and acceleration and turbulence of the main flow, *J. Engng Phys* **50**(3), 285 (1986).
12H. C. Kleinstreuer and S. Agarwal, Coupled heat and mass transfer in laminar flow, tubular polymerizers, *Int. J. Heat Mass Transfer* **29**(7), 979 (1986).
13H. N. A. Krasilov, V. A. Levin and S. A. Yunitskii, Investigation of a hypersonic viscous shock layer on rotating axisymmetric bodies in the presence of blowing, *Fluid Dyn.* **21**(1), 90 (1986).

14H. F. Krizek, Impinging jet systems to support and dry paper, *Drying Technol.* **4**(2), 271 (1986).
15H. M. Kumada, N. Ito, T. Mitsuya, M. Hirata and N. Kasagi, Full-coverage film cooling on curved walls (2nd report, mass transfer coefficients on a concavely curved wall), *Nippon Kikai Gakkai Ronbunshu, B. Hen* **52**(479), 2663 (1986).
16H. H. Le Goff, A. Ramadane et P. Le Goff, Un modele simple de la penetration couplee de chaleur et de matiere dans l'absorption gaz–liquide en film ruisselant laminaire, *Int. J. Heat Mass Transfer* **29**(4), 625 (1986).
17H. F. F. Lien and C. K. Chen, Analysis of forced convection micropolar boundary layer about a permeable sphere, *Int. J. Engng Sci.* **24**(6), 991 (1986).
18H. T. V. R. Murty and Y. V. B. Sarma, Heat transfer in flow past a continuously moving porous flat plate with heat flux, *Wärme- und Stoffübertr.* **20**(1), 39 (1986).
19H. E. M. Sparrow, S. W. Celere and L. F. A. Azevedo, Evaporation at a liquid surface due to jet impingement, *J. Heat Transfer* **108**(2), 411 (1986).
20H. G. J. Sturgess, Design of combustor cooling slots for high film effectiveness: Pt. I—Film general development, *J. Engng Gas Turbine Pwr* **108**(2), 354 (1986).
21H. G. J. Sturgess and G. D. Pfeifer, Design of combustor cooling slots for high film effectiveness: Pt. II—Film initial region, *J. Engng Gas Turbine Pwr* **108**(2), 361 (1986).

CHANGE OF PHASE—BOILING

1J. Y. Abe, M. Sobajima and Y. Murao, Experimental study of effects of upward steam flow rate on quench propagation by falling water film, *J. Nucl. Sci. Technol.* **23**(5), 415 (1986).
2J. K. N. Agrawal, H. K. Varma and S. Lal, Heat transfer during forced convection boiling of R-12 under swirl flow, *J. Heat Transfer* **108**(3), 567 (1986).
3J. R. K. Ahluwalia and K. H. Im, Vaporization of a heated liquid into a saturated stream, *Int. J. Heat Mass Transfer* **29**(2), 203 (1986).
4J. M. Aritomi, A. Inoue and H. Ishida, Thermo-hydraulic instability due to a blockage of vapor to flow in a boiling two-phase flow system (1st report, experimental results concerning occurring mechanism), *Nippon Kikai Gakkai Ronbunshu, B. Hen* **52**(480), 3026 (1986).
5J. M. Aritomi, S. Aoki and A. Inoue, Thermo-hydraulic instabilities in parallel boiling channel systems. Pt. 1. A non-linear and linear analytical model, *Nucl. Engng Des.* **95,** 105 (1986).
6J. M. Aritomi, S. Aoki and A. Inoue, Thermo-hydraulic instabilities in parallel boiling channel systems. Pt. 2, *Nucl. Engng Des.* **95,** 117 (1986).
7J. A. A. Avdeev, Application of the Reynolds analogy to the investigation of surface boiling in conditions of forced motion, *High Temp.* **24**(1), 100 (1986).
8J. A. Bar-Cohen, Z. Ruder and P. Griffith, Development and validation of boundaries for circumferential isothermality in horizontal boiler tubes, *Int. J. Multiphase Flow* **12**(1), 63 (1986).
9J. G. T. Barnes, Effects of monolayers on the evaporation of liquids, *Adv. Colloid Interface Sci.* **25**(2), 89 (1986).
10J. B. T. Beck and G. L. Wedekind, On the mean period of dryout point fluctuations, *J. Heat Transfer* **108**(4), 988 (1986).
11J. A. M. Bhat, J. S. Saini and R. Prakash, Role of macro-layer evaporation in pool boiling at high heat flux, *Int. J. Heat Mass Transfer* **29**(12), 1953 (1986).
12J. J. J. Carbajo, Parametric study on rewetting velocities obtained with a two-dimensional heat conduction code, *Nucl. Engng Des.* **92**(1), 69 (1986).
13J. V. P. Carey and G. D. Mandrusiak, Annular film-flow boiling of liquids in a partially heated, vertical channel with offset strip fins, *Int. J. Heat Mass Transfer* **29**(6), 927 (1986).
14J. V. P. Carey, E. Markovitz and Y. K. Chuah, The effect of turbulence-suspended light particles on dryout for pool boiling from a horizontal surface, *J. Heat Transfer* **108**(1), 109 (1986).
15J. C. Casper, Heat transfer in two-phase film flow at high liquid viscosity in a spiral flow tube, *Chemie-Ingr-Tech.* **58**(1), 58 (1986).
16J. J. C. Chen, A short review of dispersed flow heat transfer in post-dryout boiling, *Nucl. Engng Des.* **95,** 375 (1986).
17J. K. J. Cheng and J. B. Chaddock, Maximum size of bubbles during nucleate boiling in an electric field, *Int. J. Heat Fluid Flow* **7**(4), 278 (1986).
18J. M. N. Chepurnoi, V. E. Schnaider and N. I. Sinyuk, Heat exchange in descending annular gas–liquid flows, *J. Engng Phys.* **50**(2), 155 (1986).
19J. P. Chuchottaworn and K. Asano, Numerical analysis of drag coefficients and heat and mass transfer of spheroidal drops, *J. Chem. Engng Japan* **19**(3), 208 (1986).
20J. D. S. Dhuga and R. H. S. Winterton, Measurement of liquid–solid contact in boiling, *J. Phys. E: Scient. Instrums* **19**(1), 69 (1986).
21J. M. DiMarzo and H. D. Ross, A modelling approach to forced convective annular flow evaporation, *Heat Technol.* **4**(1), 1 (1986).
22J. L. Dresner, A. Kashani and S. W. Van Sciver, Studies of heat transport to forced flow He II, *Cryogenics* **26**(2), 111 (1986).
23J. R. C. Dykhuizen, R. P. Roy and S. P. Kalra, A linear time-domain two-fluid model analysis of dynamic instability in boiling flow systems, *J. Heat Transfer* **108**(1), 100 (1986).
24J. M. M. El-Kassaby and E. N. Ganic, Droplet deposition in two-phase, turbulent flow, *Int. J. Heat Mass Transfer* **29**(8), 1149 (1986).
25J. L. Esmailizadeh and R. Mesler, Bubble entrainment with drops, *J. Colloid Interface Sci.* **110**(2), 561 (1986).
26J. K. A. Fathalah, M. M. Elsayed, I. S. Taha and J. Sabbagh, Numerical study of evaporation–condensation in a vertical diffusion gap, *Wärme- und Stoffübertr.* **20**(4), 301 (1986).
27J. B. S. Fokin, M. Ya. Belen'kii, M. A. Gotovskii and N. L. Mikhailov, Distinctive features of the flow structure and heat transfer in helically coiled steam-generator ducts, *High Temp.* **24**(3), 411 (1986).
28J. D. M. France, R. D. Carlson and R. P. Roy, Measure-

ment and analysis of dynamic instabilities in fluid-heated two-phase flow, *Int. J. Heat Mass Transfer* **29**(12), 1919 (1986).

29J. D. Frost and B. Sturtevant, Effects of ambient pressure on the instability of a liquid boiling explosively at the superheat limit, *J. Heat Transfer* **108**(2), 418 (1986).

30J. M. Fujii, M. Hanasaki, Y. Yasojima and G. Yamanaka, Boiling heat transfer and the electric breakdown of fluorocarbon $C_2Cl_3F_3$ and $C_8F_{16}O$ under a uniform electric field, *Nippon Kikai Gakkai Ronbunshu, B. Hen* **52**(476), 1805 (1986).

31J. Y. Fujita, K. Hirahaya, S. Matsuo and K. Nishikawa, Heat transfer process in direct contact evaporator (2nd report, boiling regime and heat transfer characteristics), *Nippon Kikai Gakkai Ronbunshu, B. Hen* **52**(475), 1379 (1986).

32J. S. Fukusako, T. Komoriya and N. Seki, An experimental study of transition and film boiling heat transfer in liquid-saturated porous bed, *J. Heat Transfer* **108**(1), 117 (1986).

33J. M. A. Goldshtik, V. M. Khanin and V. G. Ligai, Liquid drop on an air cushion as an analogue of Leidenfrost boiling, *J. Fluid Mech.* **166,** 1 (1986).

34J. W. J. Green, An experimental investigation of rewetting phenomena at low flow rates using round tubes and pressurized Freon-12, *Nucl. Engng Des.* **97**(2), 187 (1986).

35J. D. C. Groeneveld, S. C. Chèng and T. Doan, 1986 AECL-UO critical heat flux lookup table, *Heat Transfer Engng* **7**(1), 46 (1986).

36J. D. C. Groeneveld, Onset of dry sheath condition—a new definition of dryout, *Nucl. Engng Des.* **92**(2), 135 (1986).

37J. U. Gropp and E. U. Schlunder, Influence of liquid-side mass transfer on heat transfer and selectivity during surface and nucleate boiling of liquid mixtures in a falling film, *Chem. Engng Proc.* **20**(2), 103 (1986).

38J. K. E. Gungor and R. H. S. Winterton, A general correlation for flow boiling in tubes and annuli, *Int. J. Heat Mass Transfer* **29**(3), 351 (1986).

39J. Y. Hayashi, A. Takimoto, O. Matsuda and M. Isozaki, Convective heat transfer with a vaporization of suspended spray droplets (2nd report, experiments of vertical flat plate), *Nippon Kikai Gakkai Ronbunshu, B. Hen* **52**(475), 1342 (1986).

40J. L. J. F. Hermans and J. J. M. Beenakker, The temperature paradox in the kinetic theory of evaporation, *Physics Fluids* **29**(12), 4231 (1986).

41J. Y. Y. Hsu and M. Massoud, A new treatment of transition boiling, *Heat Technol.* **4**(3), 7 (1986).

42J. T. H. Hwang and S. C. Yao, Forced convective boiling in horizontal tube bundles, *Int. J. Heat Mass Transfer* **29**(5), 785 (1986).

43J. Y. Iida, T. Takashima and R. Akiyoshi, Study of the vapor explosion mechanism with single drops of high temperature liquids and volatile liquids (1st report, experiments with a single drop), *Nippon Kikai Gakkai Ronbunshu, B. Hen* **52**(476), 1777 (1986).

44J. S. Inada, Y. Miyasaka, S. Sakumoto and G. R. Chandratilleke, Liquid–solid contact state in subcooled pool transition boiling system, *J. Heat Transfer* **108**(1), 219 (1986).

45J. F. Inasaka, H. Nariai and T. Shimura, Critical heat flux of subcooled flow boiling in a narrow tube (examination of experimental data on critical heat flux), *Nippon Kikai Gakkai Ronbunshu, B. Hen* **52**(476), 1817 (1986).

46J. M. Inuzuka, I. Saito, T. Moriya, I. Yamada, S. Hiraoka, H. Ishikawa and H. Mori, Vaporization from a binary mixture surface under vacuum, *Heat Transfer—Jap. Res.* **15**(5), 25 (1986).

47J. M. Ishii and G. De Jarlais, Flow regime transition and interfacial characteristics of inverted annular flow, *Nucl. Engng Des.* **95,** 171 (1986).

48J. B. Jager, G. Bon Mardion, G. Claudet and M. Desmaris, Heat transfer in He I for industrially manufactured aluminium plate heat exchangers, *Cryogenics* **26**(4), 222 (1986).

49J. M. K. Jensen and M. Pourdashti, Critical heat flux on a horizontal cylinder in an upward subcooled and low-quality two-phase crossflow, *J. Heat Transfer* **108**(2), 441 (1986).

50J. M. K. Jensen and H. P. Bensler, Saturated forced-convective boiling heat transfer with twisted-tape inserts, *J. Heat Transfer* **108**(1), 93 (1986).

51J. B. S. Johnston and S. G. Bankoff, Boiling heat transfer in a narrow eccentric annulus: part III—a model of dry patch extent and temperature distribution, *J. Heat Transfer* **108**(2), 433 (1986).

52J. D. S. Jung and J. E. S. Venart, Effects of enhanced surfaces and surface orientation on pool boiling heat transfer, *Heat Technol.* **4**(3), 62 (1986).

53J. A. L. Kalyazin and D. I. Lamden, Taking account of the variability of gas properties in calculating drop evaporation, *High Temp.* **24**(2), 236 (1986).

54J. E. Ya. Kenig, L. P. Kholpanv, V. A. Malyusov and N. M. Zhavoronkov, Calculation of heat and mass transfer in two-phase multicomponent systems, *J. Appl. Chem., U.S.S.R.* **59**(5), 964 (1986).

55J. M. Khatib-Rahbar and E. G. Cazzoli, Flow excursion-induced dryout at low heat flux natural convection boiling, *J. Heat Transfer* **108**(2), 425 (1986).

56J. B. J. Kim and M. L. Corradini, Recent film boiling calculations: implication on fuel–coolant interactions, *Int. J. Heat Mass Transfer* **29**(8), 1159 (1986).

57J. S. V. Konev and J. Mitrovic, An explanation for the augmentation of heat transfer during boiling in capillary structures, *Int. J. Heat Mass Transfer* **29**(1), 91 (1986).

58J. V. V. Kostyuk, I. I. Berlin and A. V. Karpyshev, Experimental and theoretical study of the transient boiling mechanism, *J. Engng Phys.* **50**(1), 38 (1986).

59J. Yu. D. Kozhelupenko and A. P. Mirgorodskaya, Experimental study of regimes of low-velocity motion of a subheated liquid in horizontal annular channels with a heat-transfer crisis, *J. Engng Phys.* **50**(2), 151 (1986).

60J. A. E. Kroshilin, V. E. Kroshilin and B. I. Nigmatulin, Growth of vapor bubble in a superheated liquid volume for various laws governing the pressure variation in the liquid, *High Temp.* **24**(3), 406 (1986).

61J. T. Kumada, T. Hirota, N. Tamura and R. Ishiguro, Heat and mass transfer with liquid evaporation into a turbulent air stream, *J. Heat Transfer* **108**(1), 4 (1986).

62J. H. Kumamaru, Y. Koizumi, T. Yonomoto, M. Osakabe and K. Tasaka, Investigation of uncovered-bundle heat transfer under high-pressure boil-off conditions, *Nucl. Engng Des.* **96**(1), 81 (1986).
63J. S. S. Kutateladze, Thermohydrodynamics of quasi-stable floating of a free liquid volume over a solid surface, *Int. J. Heat Mass Transfer* **29**(4), 517 (1986).
64J. R. T. Lahey, Jr., Advances in the analytical modeling of linear and nonlinear density-wave instability modes, *Nucl. Engng Des.* **95,** 5 (1986).
65J. A. I. Leont'ev, O. O. Mil'man and V. A. Fedorov, Limiting thermal loads in the boiling of heat carriers in vertical pipes of low-head natural circulation systems, *High Temp.* **24**(2), 231 (1986).
66J. C. H. Liu, N. S. Liao and T. F. Lin, Theoretical and experimental investigations of heat transfer to steam–water dispersed flow in post-dryout regime, *Chung-Kuo Chi Hsueh Kung Ch'eng Hsueh Pao* **7**(2), 75 (1986).
67J. C.-F. Ma and A. E. Bergles, Jet impingement nucleate boiling, *Int. J. Heat Mass Transfer* **29**(8), 1095 (1986).
68J. M. M. Martynyuk, Role of drop superheating in the nonsteady drop cooling of high-temperature wall, *J. Engng Phys.* **51**(1), 857 (1986).
69J. I. Michiyoshi, N. Takenaka and O. Takahashi, Dry patch formed boiling and burnout in potassium pool boiling, *Int. J. Heat Mass Transfer* **29**(5), 689 (1986).
70J. K. Miyazaki, K. Murai, T. Ohama, N. Yamaoka and S. Inoue, Pressure dependence of the particle bed dryout heat flux, *Nucl. Engng Des.* **95,** 271 (1986).
71J. M. R. Mokhtarzadeh-Dehghan and A. A. El-Shirbini, Dynamics of and heat transfer to a butane droplet evaporating in water, *Wärme- und Stoffübertr.* **21**(1), 69 (1986).
72J. M. Monde, H. Kusuda, H. Uehara and K. Fujishiro, Boiling heat transfer in a narrow vertical rectangular channel, *Nippon Kikai Gakkai Ronbunshu, B. Hen* **52**(474), 835 (1986).
73J. M. Monde, O. Nagae and Y. Ishibashi, Critical heat flux in saturated forced convective boiling on a heated disk with an impinging jet (in a high pressure region), *Nippon Kikai Gakkai Ronbunshu, B. Hen* **52**(476), 1799 (1986).
74J. M. Monde and E. Hahne, Boiling heat transfer on a fine horizontal wire in refrigerant–oil mixtures, *Nippon Kikai Gakkai Ronbunshu, B. Hen* **52**(476), 1793 (1986).
75J. F. J. Moody, Dynamic and thermal behavior of hot gas bubbles discharged into water, *Nucl. Engng Des.* **95,** 47 (1986).
76J. H. Muller-Steinhagen and D. Steiner, Heat transfer at flow boiling of argon within a horizontal tube, *Chem. Engng Proc.* **20**(4), 201 (1986).
77J. H. Muller-Steinhagen, A. P. Watkinson and N. Epstein, Subcooled-boiling and convective heat transfer to heptane flowing inside an annulus and past a coiled wire: part I—experimental results, *J. Heat Transfer* **108**(4), 922 (1986).
78J. H. Muller-Steinhagen, N. Epstein and A. P. Watkinson, Subcooled-boiling and convective heat transfer for heptane flowing inside an annulus and past a coiled wire: part II—correlation of data, *J. Heat Transfer* **108**(4), 928 (1986).
79J. S. Nakanishi, M. Kaji, S. Yamauchi and T. Sawai, An experimental study in disturbance waves in boiling two-phase flow, *Heat Transfer—Jap. Res.* **15**(3), 87 (1986).
80J. S. Nakanishi, M. Kaji, H. Matoba and N. Kaji, Flow boiling in tube of mixtures of refrigerants R-11 and R-113, *Nippon Kikai Gakkai Ronbunshu, B. Hen* **52**(479), 2626 (1986).
81J. S. Nakanishi, M. Kaji and S. Yamauchi, Approximation method for construction of a stability map of density-wave oscillations, *Nucl. Engng Des* **95,** 55 (1986).
82J. A. Nakayama and H. Koyama, An integral method in laminar film pool boiling from curved surfaces, *J. Heat Transfer* **108**(2), 490 (1986).
83J. A. Nakayama and H. Koyama, Integral treatment of subcooled forced convection film boiling on a flat plate, *Wärme- und Stoffübertr.* **20**(2), 121 (1986).
84J. A. Nakayama, Subcooled forced-convection film boiling in the presence of a pressure gradient, *AIAA J.* **24**(2), 230 (1986).
85J. A. Nakayama and M. Kano, Motion of a liquid around a vapor bubble in saturated pool nucleate boiling in the region of isolated bubbles and prediction of heat transfer coefficient, *Nippon Kikai Gakkai Ronbunshu, B. Hen* **52**(477), 2177 (1986).
86J. A. Nakayama and H. Koyama, Analysis of combined free and forced convection film boiling part I: forced and free convection regions, *A.I.Ch.E. Jl* **32**(1), 142 (1986).
87J. A. Nakayama and H. Koyama, Analysis of combined free and forced convection film boiling part II: combined free and forced convection region, *A.I.Ch.E. Jl* **32**(1), 146 (1986).
88J. Yu. V. Navruzov, Heat transfer to liquids boiling in porous structures, *Heat Transfer—Sov. Res.* **18**(2), 123 (1986).
89J. H. Ninokata, Analysis of low-heat-flux sodium boiling test in a 37-pin bundle by the two-fluid model computer code SABENA, *Nucl. Engng Des.* **97**(2), 233 (1986).
90J. S. Nishio, Study on the minimum heat flux point for boiling heat transfer on a horizontal flat plate (effects of transients and thermal conductance of surface), *Heat Transfer—Jap. Res.* **15**(1), 15 (1986).
91J. S. Nishio and M. Uemura, Study on film boiling heat transfer and minimum heat flux condition for subcooled boiling (1st report, pool boiling of water at atmospheric pressure from platinum sphere), *Nippon Kikai Gakkai Ronbunshu, B. Hen* **52**(476), 1811 (1986).
92J. S. Oh, S. Banerjee and G. Yadigaroglu, Reflooding with steady and oscillatory coolant injection: part 2—quench front and liquid carryover behavior, *J. Heat Transfer* **108**(2), 448 (1986).
93J. T. Ohama and K. Miyazaki, Influence of inclination on downfacing burnout heat flux in water pool, *J. Nucl. Sci. Technol.* **23**(12), 1104 (1986).
94J. K. Okuyama, S. Aoki, Y. Kozawa and A. Inoue, Transient boiling heat transfer in a narrow channel (fluctuation phenomena in the channel due to bubble growth and collapse), *Bull. J.S.M.E.* **29**(253), 2122

(1986).
95J. D. L. R. Oliver and J. N. Chung, Conjugate unsteady heat transfer from a spherical droplet at low Reynolds numbers, *Int. J. Heat Mass Transfer* **29**(6), 879 (1986).
96J. J. A. Orozco and L. C. Witte, Flow film boiling from a sphere to subcooled freon-11, *J. Heat Transfer* **108**(4), 934 (1986).
97J. N. Yu. Ostrovskiy, Effect of particle size of plasma-deposited coating on the coefficient of boiling heat transfer, *Heat Transfer—Sov. Res.* **18**(2), 53 (1986).
98J. N. Yu. Ostrovskiy, Effect of thickness of the capillary porous coating on boiling heat transfer to mixtures, *Heat Transfer—Sov. Res.* **18**(2), 118 (1986).
99J. H. Papendieck, Flow and heat-transfer in evaporation from un undulating liquid film with superposed vapour flow, *Chemie-Ingr-Tech.* **58**(3), 248 (1986).
100J. G.-C. Park, M. Z. Podowski, M. Becker, R. T. Lahey, Jr. and S. J. Peng, Development of a closed-form analytical model for the stability analysis of nuclear-coupled density-wave oscillations in boiling water nuclear reactors, *Nucl. Engng Des.* **92**(2), 253 (1986).
101J. B. S. Petukhov, V. M. Zhukov and S. B. Anisimov, Cryogenic liquids forced boiling heat transfer in a rotating channel, *Cryogenics* **26**(4), 226 (1986).
102J. A. T. Prata and E. M. Sparrow, Evaporation of water from a partially filled, cylindrical container to a forced convection air flow, *Int. J. Heat Mass Transfer* **29**(4), 539 (1986).
103J. V. F. Prisnayakov and S. V. Tkachenko, Effect of pressure on the departing bubble and other parameters of boiling of liquid potassium, *Heat Transfer—Sov. Res.* **18**(1), 134 (1986).
104J. G. K. Raina and R. K. Wanchoo, Direct contact heat transfer with phase change: bubble growth and collapse, *Can. J. Chem. Engng* **64**(3), 393 (1986).
105J. R. S. Rana, V. Charan and H. K. Varma, Heat and mass transfer from a horizontal tube of an evaporative heat dissipator, *Int. J. Heat Mass Transfer* **29**(4), 555 (1986).
106J. R. P. Roy, M.-G. Su, R. C. Dykhuizen and S. P. Kalra, Frequency response of boiling flow systems based on a two-fluid model, *Int. J. Heat Mass Transfer* **29**(9), 1349 (1986).
107J. M. A. R. Sarker and A. J. Sherbakov, Surface-boiling heat-transfer in cylinder-liner cooling space of diesel engine, *Mech. Engng Res. Bull.* **9,** 42 (1986).
108J. T. Schulenberg and J. Dohler, Heating-up, nucleating and boiling of a critical solution of fissile material, *Int. J. Multiphase Flow* **12**(5), 759 (1986).
109J. T. S. Shin and O. C. Jones, Jr., Active cavity model for flashing, *Nucl. Engng Des.* **95,** 185 (1986).
110J. M. Shoji and H. Nagano, Minimum heat flux of saturated pool boiling on a horizontal heated surface (an experiment for R 133 at reduced and elevated pressure), *Nippon Kikai Gakkai Ronbunshu, B. Hen* **52**(478), 2431 (1986).
111J. W. M. Sluitjer, D. G. Roddeman and S. J. D. van Stralen, On the behaviour of a translating vapour bubble under the influence of a pressure step—numerical solutions of implosion and fragmentation, *Int. J. Heat Mass Transfer* **29**(12), 1891 (1986).
112J. T. Soga, Fluids, plasmas, and electric discharges—a kinetic theory analysis of unsteady evaporation from a liquid surface with temperature exchange, *J. Phys. Soc. Japan* **55**(5), 1556 (1986).
113J. K. Srinivasan and M. V. Krishna Murthy, Determination of the bulk-saturated liquid condition for maximum heat fluxes at boiling crises, *Int. J. Heat Mass Transfer* **29**(12), 1963 (1986).
114J. D. Steiner, Heat transfer during flow boiling of cryogenic fluids in vertical and horizontal tubes, *Cryogenics* **26**(5), 309 (1986).
115J. Y. Sudo, K. Miyata, H. Ikawa and M. Kaminaga, Experimental study of incipient nucleate boiling in narrow vertical rectangular channel simulating subchannel of upgraded JRR-3, *J. Nucl. Sci. Technol.* **23**(1), 73 (1986).
116J. M. Suzuki and I. Nishiwaki, Ebullition behaviour of superheated water droplets on solid surface (effect of wettability of solid surface), *Heat Transfer—Jap. Res.* **15**(5), 37 (1986).
117J. T. Takashima and Y. Iida, Evaporation of a liquid drop dropped onto a hot liquid surface (3rd report, evaporation behavior on molten salts), *Nippon Kikai Gakkai Ronbunshu, B. Hen* **52**(475), 1372 (1986).
118J. T. Takemura, K. Roko, M. Shiraha and S. Midoriyama, Dryout characteristics and flow behavior of gas–water two-phase flow through U-shaped and inverted U-shaped bends, *Nucl. Engng Des.* **95,** 365 (1986).
119J. A. Takimoto, O. Matsuda and Y. Hayashi, Convective heat transfer with a vaporization of suspended spray droplets (1st report, theoretical analysis), *Nippon Kikai Gakkai Ronbunshu, B. Hen* **52**(475), 1333 (1986).
120J. P. Testa and L. Nicotra, Influence of pressure on the Leidenfrost temperature and on extracted heat fluxes in the transient mode and low pressure, *J. Heat Transfer* **108**(4), 916 (1986).
121J. N. P. Timchenko, Calculation of the equilibrium load for simultaneous nucleate and film boiling, *Heat Transfer—Sov. Res.* **18**(2), 104 (1986).
122J. S. Toda and M. Kitamura, Bubble growth in decomposition fields, *Nippon Kikai Gakkai Ronbunshu, B. Hen* **52**(477), 2228 (1986).
123J. V. I. Tolubinskiy, V. A. Antonenko and G. V. Ivanenko, Onset of boiling on a wick-covered wall, *Heat Transfer—Sov. Res.* 18(2), 110 (1986).
124J. A. Y. Tong and W. A. Sirignano, Multicomponent transient droplet vaporization with internal circulation: integral equation formulation and approximate solution, *Numer. Heat Transfer* **10**(3), 253 (1986).
125J. K. Torikai and K. Suzuki, Pool boiling heat transfer in an upward horizontal rough heating surface, *Nippon Kikai Gakkai Ronbunshu, B. Hen* **52**(480), 2998 (1986).
126J. A. Umemura, Fundamental theory of supercritical evaporation, *Nippon Kikai Gakkai Ronbunshu, B. Hen* **52**(478), 2437 (1986).
127J. A. F. Varone, Jr. and W. M. Rohsenow, Post dryout heat transfer prediction, *Nucl. Engng Des.* **95,** 315 (1986).
128J. O. Watanabe, O. Tajima, T. Nakano and M. Shimoya, Flow and heat transfer of a gas and liquid

two-phase flow in helical coils (for the case when the coil axis is horizontally placed), *Nippon Kikai Gakkai Ronbunshu, B. Hen* **52**(476), 1857 (1986).

129J. J. Weisman and S. H. Ying, Effect of high vapor density on critical heat flux predictions at low void fractions, *Int. J. Multiphase Flow* **12**(3), 493 (1986).

130J. J. Weisman and Y. K. Kao, Transition boiling heat transfer during reflooding transients, *A.I.Ch.E. Jl* **32**(2), 344 (1986).

131J. T. Yamasaki, Study on the relation between liquid pressure and bubble size in water at quasi-static thermal equilibrium conditions, *Heat Transfer—Jap. Res.* **15**(1), 34 (1986).

132J. S. H. Ying and J. Weisman, Prediction of the critical heat flux in flow boiling at intermediate qualities, *Int. J. Heat Mass Transfer* **29**(11), 1639 (1986).

CHANGE OF PHASE—CONDENSATION

1JJ. S. Acharya, K. G. Braud and A. Attar, Calculation of fin efficiency for condensing fins, *Int. J. Heat Fluid Flow* **7**(2), 96 (1986).

2JJ. I. Aya and H. Nariai, Occurrence threshold of pressure oscillations induced by steam condensation in pool water, *Bull. J.S.M.E.* **29**(253), 2131 (1986).

3JJ. M. K. Bezrodnyi and V. F. Moklyak, Heat transfer during condensation in vertical closed thermosiphons, *J. Engng Phys.* **51**(1), 753 (1986).

4JJ. V. M. Budov, V. A. Kir'yanov and I. A. Shemagin, On the nonuniform of condensate runoff from a horizontal tube, *Power Engng* **24**(3), 160 (1986).

5JJ. R. S. Bunker and V. P. Carey, Modeling of turbulent condensation heat transfer in the boiling water reactor primary containment, *Nucl. Engng Des.* **91**(3), 297 (1986).

6JJ. G. P. Celata, M. Cumo, G. E. Farello and G. Focardi, Direct contact condensation of steam on slowly moving water, *Nucl. Engng Des.* **96**(1), 21 (1986).

7JJ. Y. Chiba, M. Suzuki and S. Ohtani, Heat transfer in dropwise condensation of steam—effect of falling drops, *Heat Transfer—Jap. Res.* **15**(3), 39 (1986).

8JJ. J.-H. Chun, M. A. Shimko and A. A. Sonin, Vapor condensation onto a turbulent liquid—II. Condensation burst instability at high turbulence intensities, *Int. J. Heat Mass Transfer* **29**(9), 1333 (1986).

9JJ. J. H. Clarke and C. F. Delale, Supercritical shocks in nozzle flows with nonequilibrium condensation, *Physics Fluids* **29**(5), 1414 (1986).

10JJ. J. H. Clarke and C. F. Delale, Nozzle flows with nonequilibrium condensation, *Physics Fluids* **29**(5), 1398 (1986).

11JJ. M. di Marzo, Laminar film condensation—a mathematical test case, *Heat Technol.* **4**(3), 37 (1986).

12JJ. T. Fujii and K. Oda, Correlation equations of heat transfer for condensate inundation on horizontal tube bundles, *Nippon Kikai Gakkai Ronbunshu, B. Hen* **52**(474), 822 (1986).

13JJ. T. Fujii and K. Oda, Theoretical analysis of condensate inundation for forced convection condensation, *Nippon Kikai Gakkai Ronbunshu, B. Hen* **52**(474), 818 (1986).

14JJ. D. Fullarton and E. U. Schluender, Approximate calculation of heat exchanger area for condensation of gas–vapour mixtures, *Int. Chem. Engng* **26**(3), 408 (1986).

15JJ. D. Galamba, V. K. Dhir and K. Taghavi, Analytical and experimental investigation of simultaneous melting–condensation on a vertical wall in the presence of a noncondensable gas, *Int. J. Heat Mass Transfer* **29**(12), 1869 (1986).

16JJ. I. I. Gogonin, Hydrodynamics and heat transfer during condensation of motionless vapor on a horizontal cylinder, *Izv. Sib. Otdeleniya Akad. Nauk SSR Ser. Tek. Nauk* No. 10, 24 (1986).

17JJ. I. I. Gogonin and O. A. Kabov, Influence of irrigation density on heat transfer during condensation in finned tube bundles, *J. Engng Phys.* **51**(1), 759 (1986).

18JJ. H. Gokce, C. Ozgen and T. G. Somer, Effect of drainage discs on the condensing heat transfer performance of vertical fluted tubes, *Int. J. Heat Fluid Flow* **7**(3), 191 (1986).

19JJ. S. Hatamiya and H. Tanaka, Study on the mechanism of dropwise condensation (1st report, measurement of heat-transfer coefficient of steam at low pressures), *Nippon Kikai Gakkai Ronbunshu, B. Hen* **52**(476), 1828 (1986).

20JJ. S. Hatamiya and H. Tanaka, Study on the mechanism of dropwise condensation (2nd report, condensation coefficient of water at low pressures), *Nippon Kikai Gakkai Ronbunshu, B. Hen* **52**(477), 2214 (1986).

21JJ. K. Hijikata, Y. Mori, N. Himeno, M. Inagawa and K. Takahasi, Film condensation of a binary mixture of vapors, *Nippon Kikai Gakkai Ronbunshu, B. Hen* **52**(477), 2195 (1986).

22JJ. H. Honda, S. Nozu, B. Uchima and T. Fujii, Effect of vapour velocity on film condensation of R-113 on horizontal tubes in a crossflow, *Int. J. Heat Mass Transfer* **29**(3), 429 (1986).

23JJ. H. Honda, S. Nozu and Y. Furukawa, Effect of porous drainage strips on film condensation on downward-facing horizontal surfaces, *Nippon Kikai Gakkai Ronbunshu, B. Hen* **52**(475), 1355 (1986).

24JJ. T. Hosokawa, T. Kawai, A. Fujio and G. Komatsu, Dropwise condensation heat transfer on a horizontal tube, *Bull. J.S.M.E.* **29**(254), 2570 (1986).

25JJ. S. Inada, Y. Miyasaka and R. Izumi, Disappearance of a detached vapor mass in subcooled water, *Nippon Kikai Gakkai Ronbunshu, B. Hen* **52**(477), 2222 (1986).

26JJ. C. M. Knobler, Experimental studies of the dynamics of fluid phase transitions in two dimensions, *Physica A* **140**(1), 198 (1986).

27JJ. S. Koyama, M. Watabe and T. Fujii, Gravity controlled film condensation of saturated and superheated binary vapour mixtures on a vertical plate, *Nippon Kikai Gakkai Ronbunshu, B. Hen* **52**(474), 827 (1986).

28JJ. A. M. Lapin, L. I. Sen' and A. M. Te, Heat transfer in a film flowing over the surface of a convergent duct, *J. Appl. Mech. Tech. Phys.* **27**(3), 366 (1986).

29JJ. F. Legay-Desesquelles and B. Prunet-Foch, Heat and mass transfer with condensation in laminar and turbulent boundary layers along a flat plate, *Int. J. Heat Mass Transfer* **29**(1), 95 (1986).

30JJ. P. J. Marto, D. J. Looney, J. W. Rose and A. S.

Wanniarachchi, Evaluation of organic coatings for the promotion of dropwise condensation of steam, *Int. J. Heat Mass Transfer* **29**(8), 1109 (1986).
31JJ. J. Mitrovic, Heat transfer from a horizontal smooth tube to a laminar falling film, *Can. J. Chem. Engng* **64**(6), 1023 (1986).
32JJ. A. Nakayama and H. Koyama, An integral procedure for a two-phase boundary layer in laminar film condensation, *Int. J. Multiphase Flow* **12**(1), 135 (1986).
33JJ. K. Nishiyama, A. Murata and T. Kajikawa, The study of condensation heat transfer performance due to periodic motion of a vertical tube, *Heat Transfer—Jap. Res.* **15**(6), 32 (1986).
34JJ. Y. Onishi, On the behavior of a noncondensable gas of a small amount in weakly nonlinear evaporation and condensation of a vapor, *J. Phys. Soc. Japan* **55**(9), 3080 (1986).
35JJ. L. Long and G. A. Moses, Vapor condensation in the presence of a noncondensable gas, *Physics Fluids* **29**(6), 1796 (1986).
36JJ. D. Poulikakos, Interaction between film condensation on one side of a vertical wall and natural convection on the other side, *J. Heat Transfer* **108**(3), 560 (1986).
37JJ. D. Poulikakos and P. Sura, Conjugate film condensation and natural convection along the interface between a porous and an open space, *Int. J. Heat Mass Transfer* **29**(11), 1747 (1986).
38JJ. U. Renz, Measures to improve heat transfer in condensation, *Germ. Chem. Engng* **9**(1), 30 (1986).
39JJ. V. A. Shneidman and A. L. Shubenko, Unsteady nucleation in supercooled vapor: analytical description and numerical calculations, *Fluid Dyn.* **21**(2), 314 (1986).
40JJ. V. M. Sobin, Heat transfer in the thermal entry region in laminar downflow of a liquid film over a cylindrical surface, *Heat Transfer—Sov. Res.* **18**(2), 18 (1986).
41JJ. H. M. Soliman, The mist-annular transition during condensation and its influence on the heat transfer mechanism, *Int. J. Multiphase Flow* **12**(2), 277 (1986).
42JJ. A. A. Sonin, M. A. Shimko and J.-H. Chun, Vapor condensation onto a turbulent liquid—I. The steady condensation rate as a function of liquid-side turbulence, *Int. J. Heat Mass Transfer* **29**(9), 1319 (1986).
43JJ. T.-Y. Sun, R. N. Parthansarathy and G. M. Faeth, Structure of bubbly round condensing jets, *J. Heat Transfer* **108**(4), 951 (1986).
44JJ. T. Taketani, A. Yabe, K. Kikuchi, Y. Mori and K. Hijikata, Augmentation of condensation heat transfer by applying non-uniform electric fields (2nd report, augmentation of condensation outside a vertical tube by use of helical wire electrodes), *Nippon Kikai Gakkai Ronbunshu, B. Hen* **52**(477), 2207 (1986).
45JJ. T. N. Tandon, H. K. Varma and C. P. Gupta, Generalized correlation for condensation of binary mixtures inside a horizontal tube, *Int. J. Refrig.* **9**(3), 134 (1986).
46JJ. H.-H. Tung, J. F. Davis and R. S. H. Mah, Fractionating condensation and evaporation in plate-fin devices, *A.I.Ch.E. Jl* **32**(7), 1116 (1986).
47JJ. Y. Utaka, A. Saito, H. Ishikawa and H. Yanagida, Study on dropwise condensation curves (dropwise to filmwise transition of propylene glycol, ethylene glycol and glycerol vapors on a copper surface using a monolayer type promoter) (part 1), *Nippon Kikai Gakkai Ronbunshu, B. Hen* **52**(476), 1835 (1986).
48JJ. K. Vafai and S. Sarkar, Condensation effects in a fibrous insulation slab, *J. Heat Transfer* **108**(3), 667 (1986).
49JJ. J. van Leersum, A numerical description of condensation in nonhygroscopic regenerator matrices, *Numer. Heat Transfer* **9**(3), 365 (1986).
50JJ. A. S. Wanniarachchi, P. J. Marto and J. W. Rose, Film condensation of steam on horizontal finned tubes: effect of fin spacing, *J. Heat Transfer* **108**(4), 960 (1986).
51JJ. A. Yamauchi, S. Kumagai and T. Takeyama, Condensation heat transfer on various dropwise–filmwise coexisting surfaces, *Heat Transfer—Jap. Res.* **15**(5), 50 (1986).
52JJ. O. Zabloudil, Heat transfer of stagnant steam during droplet condensation (in German), *Wärme- und Stoffübertr.* **20**(2), 127 (1986).

CHANGE OF PHASE—FREEZING AND MELTING

1JM. K. Aoki, M. Hattori and S. Chiba, A study of the melting process in ice–air composite materials (in the case where a temperature gradient exists in porous materials), *Bull. J.S.M.E.* **29**(253), 2138 (1986).
2JM. G. C. J. Bart, C. J. Hoogendoorn and P. B. J. Schaareman, A characteristic dimensionless time in phase change problems, *J. Sol. Energy Engng* **108**(4), 310 (1986).
3JM. Th. Betzel and H. Beer, Heat transfer on fusion around a horizontal finned tube, *Chemie-Ingr-Tech.* **58**(12), 982 (1986).
4JM. A. M. Cames-Pintaux and M. Nguyen-Lamba, Finite-element enthalpy method for discrete phase change, *Numer. Heat Transfer* **9**(4), 403 (1986).
5JM. C. Charach and P. B. Kahn, Transient phase-change heat transfer in finite bodies: universal bounds for the solidification (melting) time, *J. Appl. Phys.* **59**(10), 3579 (1986).
6JM. M. M. Chen, R. Farhadieh and L. Baker, Jr., On free convection melting of a solid immersed in a hot dissimilar fluid, *Int. J. Heat Mass Transfer* **29**(8), 1087 (1986).
7JM. K. Chida, Blockage of laminar pipe flow in a piping system by internal fluid solidification, *Heat Transfer—Jap. Res.* **15**(6), 75 (1986).
8JM. M. K. Choudhary, Three-dimensional mathematical model for flow and heat transfer in electric glass furnaces, *IEEE Trans. Ind. Applic.* **22**(5), 912 (1986).
9JM. L. A. Crivelli and S. R. Idelsohn, A temperature-based finite element solution for phase-change problems, *Int. J. Numer. Meth. Engng* **23**(1), 99 (1986).
10JM. J. N. Dewynne and J. M. Hill, Integral formulations and bounds for two phase Stefan problems initially not at their fusion temperature, *Acta Mech.* **58**(3), 201 (1986).
11JM. J. N. Dewynne and J. M. Hill, On a multi-phase Stefan problem, *Q. J. Mech. Appl. Math.* **39**(1), 67 (1986).

12JM. L. A. Diaz and R. Viskanta, Experiments and analysis on the melting of a semitransparent material by radiation, *Wärme- und Stoffübertr.* **20**(4), 311 (1986).

13JM. M. K. El-Adawi, Laser melting of solids—an exact solution for time intervals less or equal to the transit time, *J. Appl. Phys.* 60(7), 2256 (1986).

14JM. M. K. El-Adawi and S. A. Shalaby, Laser melting of solids—an exact solution for time intervals greater than the transit time, *J. Appl. Phys.* **60**(7), 2260 (1986).

15JM. M. Epstein and M. A. Grolmes, Natural convection characteristics of pool penetration into a melting miscible substrate, *J. Heat Transfer* **108**(1), 190 (1986).

16JM. C. Gau and R. Viskanta, Melting and solidification of a pure metal on a vertical wall, *J. Heat Transfer* **108**(1), 174 (1986).

17JM. E. S. Golgolev, Heat emission accompanying thawing of a vertical ice surface, *J. Engng Phys.* **49**(6), 1508 (1986).

18JM. C. P. Grigoropoulos, R. H. Buckholz and G. A. Domoto, A heat transfer algorithm for the laser-induced melting and recrystallization of thin silicon layers, *J. Appl. Phys.* **60**(7), 2304 (1986).

19JM. R. S. Gupta and A. Kumar, Approximate analytical methods for multi-dimensional Stefan problems, *Comput. Meth. Appl. Mech. Engng* **56**(2), 127 (1986).

20JM. L. N. Gutman, On the problem of heat transfer in phase-change materials for small Stefan numbers, *Int. J. Heat Mass Transfer* **29**(6), 921 (1986).

21JM. J. M. Hill and J. N. Dewynne, On the inward solidification of cylinders, *Q. Appl. Math.* **44**(1), 59 (1986).

22JM. C. J. Ho and S. Chen, Numerical simulation of melting of ice around a horizontal cylinder, *Int. J. Heat Mass Transfer* **29**(9), 1359 (1986).

23JM. C. J. Ho and K. C. Lin, Outward melting in a cylindrical annulus, *J. Energy Resour. Technol.* **12**(3), 240 (1986).

24JM. J. S. Hsiao and B. T. F. Chung, An efficient algorithm for finite element solution to two-dimensional heat transfer with melting and freezing, *J. Heat Transfer* **108**(2), 462 (1986).

25JM. J. Y. Jang and A. S. Wu, Inward solidification of a subcooled metallic sphere, *Chung-Kuo Chi Hsueh Kung Ch'eng Hsueh Pao* **7**(2), 67 (1986).

26JM. J. J. Jou, Linear stability of a binary fluid layer with freezing, *Chung-juo Kung Ch'eng Hsueh K'an/J. Chin. Inst. Engng* **9**(2), 141 (1986).

27JM. M. Kazmierczak, D. Poulikakos and I. Pop, Melting from a flat plate embedded in a porous medium in the presence of steady natural convection, *Numer. Heat Transfer* **10**(6), 571 (1986).

28JM. Y. Kikuchi and Y. Shigemasa, Effect of liquid solidification on turbulent flow heat transfer of a heat generating fluid in a cold tube, *Mem. Fac. Engng Kyoto Univ.* **48**(2), 209 (1986).

29JM. K. Kimoto, M. Sugawara, H. Inaba and M. Ishino, Snow melting with low temperature solidification aqueous solution, *Wärme- und Stoffübertr.* **20**(2), 133 (1986).

30JM. A. Kucera and J. M. Hill, On inward solidifying cylinders and spheres initially not at their fusion temperature, *Int. J. Non-linear Mech.* **21**(1), 73 (1986).

31JM. H. Martin, J. Lede, H. Z. Li, J. Villermaux, C. Moyne and A. Degiovanni, Ablative melting of a solid cylinder perpendicularly pressed against a heated wall, *Int. J. Heat Mass Transfer* **29**(9), 1407 (1986).

32JM. M. K. Moallemi and R. Viskanta, Analysis of close-contact melting heat transfer, *Int. J. Heat Mass Transfer* **29**(6), 855 (1986).

33JM. M. K. Moallemi, B. W. Webb and R. Viskanta, An experimental and analytical study of close-contact melting, *J. Heat Transfer* **108**(4), 894 (1986).

34JM. T. Munakata and I. Tanasawa, Buoyancy and surface tension for natural convection in rectangular cavities during solidification, *Nippon Kikai Gakkai Ronbunshu, B. Hen* **52**(474), 876 (1986).

35JM. P. V. Padmanabhan and M. V. Krishna Murthy, Outward phase change in a cylindrical annulus with axial fins on the inner tube, *Int. J. Heat Mass Transfer* **29**(12), 1855 (1986).

36JM. H. Rieger and H. Beer, The melting process of ice inside a horizontal cylinder: effects of density anomaly, *J. Heat Transfer* **108**(1), 166 (1986).

37JM. J.-F. Rodrigues, An evolutionary continuous casting problem of Stefan type, *Q. Appl. Math.* **44**(1), 109 (1986).

38JM. A. Rubiolo de Reinick and H. Schwartzberg, Predicting temperature vs time behavior during the freezing and thawing of rectangular foods, *Biotechnol. Prog.* **2**(3), 164 (1986).

39JM. A. Y. Saito, Y. Utaka, K. Shinoda and K. Katayama, Basic research on the latent heat thermal energy storage utilizing the contact melting phenomena, *Bull. J.S.M.E.* **29**(255), 2946 (1986).

40JM. K. Sasaguchi, H. Imura and H. Furusho, Heat transfer characteristics of a latent heat storage unit with a finned tube (1st report, experimental studies on solidification and melting processes), *Bull. J.S.M.E.* **29**(255), 2978 (1986).

41JM. K. Sasaguchi, H. Imura and H. Furusho, Heat transfer characteristics of a latent heat storage unit with a finned tube (2nd report, numerical analysis of solidification process), *Bull. J.S.M.E.* **29**(255), 2986 (1986).

42JM. H.-J. Shaw and C.-K. Chen, Cubic spline numerical solution for inward solidification with finite wall, *Chung-Kuo Chittsueh Kung Ch'eng Hsueh Pao* **7**(5), 329 (1986).

43JM. R. Siegel, Boundary perturbation method for free boundary problem in convectively cooled continuous casting, *J. Heat Transfer* **108**(1), 230 (1986).

44JM. R. Siegel, Free boundary shape of a convectively cooled solidified region, *Int. J. Heat Mass Transfer* **29**(2), 309 (1986).

45JM. E. M. Sparrow and G. T. Geiger, Melting in a horizontal tube with the solid either constrained or free to fall under gravity, *Int. J. Heat Mass Transfer* **29**(1), 1007 (1986).

46JM. E. M. Sparrow and M. L. Zumbrunnen, In-tube melting in the presence of circumferentially nonuniform heating, *Int. J. Heat Mass Transfer* **29**(11), 1629 (1986).

47JM. E. M. Sparrow and Y. Ohkubo, Numerical analysis

of two-dimensional transient freezing including solid-phase and tube-wall conduction and liquid-phase natural convection, *Numer. Heat Transfer* **9**(1), 59 (1986).

48JM. E. M. Sparrow and Y. Ohkubo, Numerical predictions of freezing in a vertical tube, *Numer. Heat Transfer* **9**(1), 79 (1986).

49JM. E. M. Sparrow, G. A. Gurtcheff and T. A. Myrum, Correlation of melting results for both pure substance and impure substances, *J. Heat Transfer* **108**(3), 649 (1986).

50JM. J. Srinivasan and B. Basu, A numerical study of thermocapillary flow in a rectangular cavity during laser melting, *Int. J. Heat Mass Transfer* **29**(4), 563 (1986).

51JM. M. N. Sugawara, H. Inaba and H. Nishimura, Temperature–concentration combined melting of a horizontal ice layer by an aqua-solvent with a low solidification temperature, *Nippon Kikai Gakkai Ronbunshu* **52**(484), 4041 (1986).

52JM. A. A. Uglov, I. Yu. Smurov and L. V. Karaseva, Melting rate of solids under the effect of variable power heat flows, *Phys. Chem. Math. Treatment* **20**(4), 297 (1986).

53JM. A. Ungan and R. Viskanta, Three-dimensional numerical simulation of circulation and heat transfer in an electrically boosted glass melting tank, *IEEE Trans. Ind. Applic.* **22**(5), 922 (1986).

54JM. C. Vives and C. Perry, Effects of electromagnetic stirring during the controlled solidification of tin, *Int. J. Heat Mass Transfer* **29**(1), 21 (1986).

55JM. J. A. Weaver and R. Viskanta, Melting of frozen, porous media contained in a horizontal or a vertical, cylindrical capsule, *Int. J. Heat Mass Transfer* **29**(12), 1943 (1986).

56JM. B. W. Webb and R. Viskanta, Natural-convection-dominated melting heat transfer in an inclined rectangular enclosure, *Int. J. Heat Mass Transfer* **29**(2), 183 (1986).

57JM. B. W. Webb and R. Viskanta, Analysis of heat transfer during melting of a pure metal from an isothermal vertical wall, *Numer. Heat Transfer* **9**(5), 539 (1986).

58JM. D. White, R. Viskanta and W. Leidenfrost, Heat transfer during the melting of ice around a horizontal, isothermal cylinder, *Expl Fluids* **4**(3), 171 (1986).

59JM. X. Wu and R. Viskanta, Modeling of heat transfer in the melting of a glass batch, *J. Non-cryst. Solids* **80**(1), 613 (1986).

60JM. P. A. Yanitskii, Self-similar solution of a problem of freezing of finely dispersed solids with allowance for moisture migration in thawed and frozen zones, *Fluid Dyn.* **21**(2), 264 (1986).

61JM. G.-P. Zhang, S. Weinbaum and L. M. Jiji, An approximate three-dimensional solution for melting of freezing around a buried pipe beneath a free surface, *J. Heat Transfer* **108**(4), 900 (1986).

RADIATION IN PARTICIPATING MEDIA AND SURFACE RADIATION

1K. M. S. Abubakar and J. H. Shaw, Carbon dioxide band intensities and linewidths in the 8–12 μm region, *Appl. Optics* **25**(7), 1196 (1986).

2K. V. N. Adrianov, Calculation of radiation heat transfer with consideration of the anisotropy of optical characteristics, *Izv. Sib. Otdeleniya Akad. Nauk SSSR, Ser. Tek. Nauk* No. 10, 17 (1986).

3K. F. Allard, C. Inard and A. Roldan, Comparative study of different methods of modelling short and long wave radiative exchanges in a dwelling (in French), *Annls Inst. Tech. Bâtim.* **39**(442), 29 (1986).

4K. J. M. Ane and M. Huetz-Aubert, Stratified media theory interpretation of measurements of the spectral polarized directional emissivity of some oxidized metals, *Int. J. Thermophys.* **7**(6), 1191 (1986).

5K. E. Z. Apshtein, N. V. Vartanyan and V. I. Sakharov, Distribution of radiant heat fluxes over the surface of three-dimensional and axisymmetric bodies in supersonic ideal-gas flow, *Fluid Dyn.* **21**(1), 78 (1986).

6K. V. S. Arpaci, Radiative entropy production, *AIAA J.* **24**(11), 1859 (1986).

7K. P. Attard, M. A. Box, G. Bryant and B. H. J. McKellar, Asymptotic behavior of the Mie-scattering amplitude, *J. Opt. Soc. Am. A* **3**(2), 256 (1986).

8K. E. Bahar and M. A. Fitzwater, Scattering and depolarization by conducting cylinders with rough surfaces, *Appl. Optics* **25**(11), 1826 (1986).

9K. E. Bahar, S. Chakrabarti and M. A. Fitzwater, Extinction cross sections and albedos for particles with very very rough surfaces, *Appl. Optics* **25**(15), 2530 (1986).

10K. E. Battistelli, P. Bruscaglioni, A. Ismaelli, L. Lo Porto and G. Zaccanti, Separation and analysis of forward scattered power in laboratory measurements of light beam transmittance through a turbid medium, *Appl. Optics* **25**(3), 420 (1986).

11K. B. T. Beck and G. L. Wedekind, Characteristics of a simple energy absorption transducer, *J. Heat Transfer* **108**(3), 676 (1986).

12K. A. R. Bestman, Radiative transfer to a dusty gas in a vertical channel, *Modell. Simul. Control B: Mech. Ther. Engng* **7**(2), 47 (1986).

13K. R. Bhandari, Tiny core or thin layer as a perturbation in scattering by a single-layered sphere, *J. Opt. Soc. Am. A* **3**(3), 319 (1986).

14K. S. A. Boothroyd and A. R. Jones, A comparison of radiative characteristics for fly ash and coal, *Int. J. Heat Mass Transfer* **29**(11), 1649 (1986).

15K. C. Bourrely, P. Chiapetta and B. Torresani, Light scattering by particles of arbitrary shape: a fractal approach, *J. Opt. Soc. Am. A* **3**(2), 250 (1986).

16K. M. Q. Brewster, Effective absorptivity and emissivity of particulate media with application to a fluidized bed, *J. Heat Transfer* **108**(3), 710 (1986).

17K. M. Q. Brewster, Particle radiative feedback in ammonium perchlorate deflagration, *AIAA J.* **24**(7), 1141 (1986).

18K. R. G. W. Brown, K. D. Ridley and J. G. Rarity, Characterization of silicon avalanche photodiodes for photon correlation measurements. 1: passive quenching, *Appl. Optics* **25**(22), 4122 (1986).

19K. B. Cain and B. Farnworth, Two new techniques for determining the thermal radiative properties of thin

fabrics, *J. Therm. Insul.* **9,** 301 (1986).
20K. R. Caps and J. Fricke, Infrared radiative heat transfer in highly transparent silica aerogel, *Sol. Energy* **36**(4), 361 (1986).
21K. P. Carragher and L. J. Crane, Radiative heat transfer on a continuous boundary layer, *Z. Angew. Math. Mech.* **66**(3), 185 (1986).
22K. J. D. Cartigny, Y. Yamada and C. L. Tien, Radiative transfer with dependent scattering by particles: part 1—theoretical investigation, *J. Heat Transfer* **108**(3), 608 (1986).
23K. R. Chen, A. Katzir, A. Levite, F. Moser and D. Weiss, Absorption and luminescence of silver halide optical fibers, *J. Opt. Soc. Am. B* **3**(5), 696 (1986).
24K. J.-P. Chevaillier, J. Fabre and P. Hamelin, Forward scattered light intensities by a sphere located anywhere in a Gaussian beam, *Appl. Optics* **25**(7), 1222 (1986).
25K. P. Codastefano, P. Dore and L. Nencini, Far-infrared absorption spectrum of the CH_4–H_2 gaseous mixture, *J. Quant. Spectrosc. Radiat. Transfer* **36**(3), 239 (1986).
26K. P. Codastefano, P. Dore and L. Nencini, Temperature dependence of the far-infrared absorption spectrum of gaseous methane, *J. Quant. Spectrosc. Radiat. Transfer* **35**(4), 255 (1986).
27K. A. Coppalle and P. Vervisch, Spectral emissivities of H_2O vapor at 2900 K in the 1–9 μm region, *J. Quant. Spectrosc. Radiat. Transfer* **35**(2), 121 (1986).
28K. C. Cousin, R. Le Doucen, J. P. Houdeau, C. Boulet and A. Henry, Air broadened linewidths, intensities, and spectral line shapes for CO_2 at 4.3 μm in the region of the AMTS instrument, *Appl. Optics* **25**(14), 2434 (1986).
29K. C. Cousin, R. Le Doucen, C. Boulet, A. Henry and D. Robert, Line coupling in the temperature and frequency dependencies of absorption in the micro windows of the 4.3 μm CO_2 band, *J. Quant. Spectrosc. Radiat. Transfer* **36**(6), 521 (1986).
30K. R. Dadou, B. Billerey, C. Terrier, J. Mangin and G. Morlot, Determination of radiative properties for semitransparent, absorbent and diffusing media, *Revue Gen. Therm.* **25**(294), 349 (1986).
31K. B. Deshaies and G. Joulin, Radiative transfer as a propagation mechanism for rich flames of reactive suspensions, *SIAM J. Appl. Math.* **46**(4), 561 (1986).
32K. S. P. Detkov and O. A. Bryukhovskikh, Short-time heating of a body by a radiant flux, *J. Engng Phys.* **50**(1), 77 (1986).
33K. R. L. Di Leon and J. Taine, Infrared absorption by gas mixtures in the 300–850 K temperature range I—4.3 μm and 2.7 μm CO_2 spectra, *J. Quant. Spectrosc. Radiat. Transfer* **35**(5), 337 (1986).
34K. R. L. Di Leon, J. Taine and S. Marbout, Infrared absorption by gas mixtures in the 500–700 K temperature range II—2.7 μm H_2O spectra, *J. Quant. Spectrosc. Radiat. Transfer* **35**(5), 345 (1986).
35K. P. Docherty and R. J. Tucker, Influence of wall emissivity on furnace performance, *J. Inst. Energy* **59**(438), 35 (1986).
36K. T. Duracz and N. J. McCormick, Analytical error estimates for the time-dependent radiative-transfer inverse problem, *J. Opt. Soc. Am. A* **3**(11), 1871 (1986).
37K. D. K. Edwards, The plating algorithm for radiation script-F transfer factor, *J. Heat Transfer* **108**(1), 237 (1986).
38K. R. C. Estes, R. G. Bill, Jr. and G. A. Domoto, A wide-band absorption coefficient integration kernel, *Int. J. Heat Mass Transfer* **29**(12), 1937 (1986).
39K. I. H. Farag, M. K. Hemming and E. L. Reicker, Effects of radiative scattering on a Monte-Carlo semi-infinite furnace slab, *Heat Technol.* **4**(1), 33 (1986).
40K. J. D. Felske, P.-F. Hsu and J. C. Ku, The effect of soot particle optical inhomogeneity and agglomeration on the analysis of light scattering measurements in flames, *J. Quant. Spectrosc. Radiat. Transfer* **35**(6), 447 (1986).
41K. M. Fukabori, T. Nakazawa and M. Tanaka, Absorption properties of infrared active gases at high pressures—I. CO_2, *J. Quant. Spectrosc. Radiat. Transfer* **36**(3), 265 (1986).
42K. M. Fukabori, T. Nakazawa and M. Tanaka, Intensities of CH_4 bands at the 4900–3700 cm^{-1} region, *J. Quant. Spectrosc. Radiat. Transfer* **36**(3), 271 (1986).
43K. M. Fukabori, T. Nakazawa and M. Tanaka, Absorption properties of infrared active gases at high pressures—II. N_2O and CO, *J. Quant. Spectrosc. Radiat. Transfer* **36**(4), 283 (1986).
44K. B. D. Ganapol and M. Matsumoto, Numerical evaluation of time-dependent reflected intensity from an anisotropically scattering semi-infinite atmosphere, *J. Quant. Spectrosc. Radiat. Transfer* **35**(1), 71 (1986).
45K. R. D. M. Garcia and C. E. Siewert, A generalized spherical harmonics solution for radiative transfer models that include polarization effects, *J. Quant. Spectrosc. Radiat. Transfer* **36**(5), 401 (1986).
46K. S. M. Gol'berg, V. M. Podgaetskii, I. E. Poyurovskaya and V. N. Semin, Numerical modeling of the radiant heating of a liquid, *J. Engng Phys.* **50**(1), 87 (1986).
47K. H. Gomi, Multiple scattering correction in the measurement of particle size and number density by the diffraction method, *Appl. Optics* **25**(19), 3552 (1986).
48K. S. N. Gorshkova, Radiative heat transfer in a truncated-cone chamber in which the heat source lies within the gas space, *Heat Transfer—Sov. Res.* **18**(2), 39 (1986).
49K. G. Grehan, B. Maheu and G. Gouesbet, Scattering of laser beams by Mie scatter centers: numerical results using a localized approximation, *Appl. Optics* **25**(19), 3539 (1986).
50K. I. S. Habib, Radiative and convective effects on the vibrational heating of semitransparent polymers, *J. Spacecraft Rockets* **23**(1), 124 (1986).
51K. R. D. Haracz, L. D. Cohen, A. Cohen and C. Acquista, Light scattering from dielectric targets composed of a continuous assembly and circular disks, *Appl. Optics* **25**(23), 4386 (1986).
52K. D. Hartill, F. Lisi and A. Bettarini, An InSb charge amplifier for use in a spectrometer array, *Appl. Optics* **25**(10), 1701 (1986).
53K. H. Hayasaka, K. Kudo, H. Taniguchi, I. Nakamachi,

T. Omori and T. Katayama, Radiative heat transfer analysis by the radiative heat ray method (analysis in a two-dimensional model) (in Japanese), *Nippon Kikai Gakkai Ronbunshu, B. Hen* **52**(476), 1734 (1986).
54K. M. Hirano, T. Miyauchi and Y. Mori, Enhancement of radiation heat transfer based on non-gray feature of radiative gases (2nd report, combustion gas in the non-flow system) (in Japanese), *Nippon Kikai Gakkai Ronbunshu, B. Hen* **52**(473), 456 (1986).
55K. S. L. Hollis and C. T. Kelley, Vector algorithms for H-equations arising in radiative transfer through inhomogeneous media, *Trans. Theor. Stat. Phys.* **15**(1), 33 (1986).
56K. E. Hyman, R. E. Harrison and D. R. Anderson, Electromagnetic reflection from an arbitrary smooth convex cylinder, *J. Opt. Soc. Am. A* **3**(9), 1395 (1986).
57K. J. O. Isard, Emission of thermal radiation from hot glass. Pt. 1. Emissivity of sheets, spheres, cylinders, and tubes, *Glass Technol.* **27**(1), 24 (1986).
58K. M. F. Iskander, S. C. Olson, R. E. Benner and D. Yoshida, Optical scattering by metallic and carbon aerosols of high aspect ratio, *Appl. Optics* **25**(15), 2514 (1986).
59K. Y. Itaya, N. Arai and M. Hasantani, Characteristics of radiative energy collection by a fine-particle semi-transparent suspension (FPSS) used as a vehicle and storage medium of a "volume heat-trap" type solar collector, *Heat Transfer—Jap. Res.* **15**(4), 77 (1986).
60K. J. Janata, Locally variable radiative heat transfer between long cylinders, *Acta Technica* **31**(6), 727 (1986).
61K. P. D. Johnston, Some theoretical calculations of the radiative output from an electrically augmented flame, *J. Inst. Energy* **59**(440), 123 (1986).
62K. K. Kamiuto, Chebyshev collocation method for solving the radiative transfer equation, *J. Quant. Spectrosc. Radiat. Transfer* **35**(4), 329 (1986).
63K. J. Kessel, Transmittance measurements in the integrating sphere, *Appl. Optics* **25**(16), 2752 (1986).
64K. M. Kobiyama, Study on the reduction of computing time of the Monte Carlo method applied to the radiative heat transfer, *Bull. J.S.M.E.* **29**(255), 3000 (1986).
65K. D. F. Kroeker and J. Reid, Line-tunable cw ortho- and para-NH_3 lasers operating at wavelength of 11–14 μm, *Appl. Optics* **25**(17), 2929 (1986).
66K. J. C. Ku and J. D. Felske, Determination of refractive indices of Mie scatterers from Kramers–Kronig analysis of spectral extinction data, *J. Opt. Soc. Am. A* **3**(5), 617 (1986).
67K. S. Kumar and J. D. Felske, Radiative transport in a planar medium exposed to azimuthally unsymmetric incident radiation, *J. Quant. Spectrosc. Radiat. Transfer* **35**(5), 187 (1986).
68K. H. M. Lai and P. K. Chan, Far field and energy flux caused by a radiating source in an anisotropic medium, *Physics Fluids* **29**(6), 1881 (1986).
69K. M. E. Larsen and J. R. Howell, Least-squares smoothing of direct-exchange areas in zonal analysis, *J. Heat Transfer* **108**(1), 239 (1986).
70K. H. Lee and R. O. Buckius, Combined mode heat transfer analysis utilizing radiation scaling, *J. Heat Transfer* **108**(3), 626 (1986).
71K. S. C. Leel, Radiative transfer through a fibrous medium: allowance for fiber orientation, *J. Quant. Spectrosc. Radiat. Transfer* **36**(3), 253 (1986).
72K. V. A. Levin, A. A. Sorokin and A. M. Starik, Water vapor cooling when radiation of wavelength $\lambda = 2.8$ μm is absorbed, *Fluid Dyn.* **21**(3), 456 (1986).
73K. V. G. Lisienko, F. R. Sklyar, Yu. V. Kryuchenkov, L. N. Toritsyn, V. V. Volkov and A. I. Tikhotskii, Combined solution of the problem of external heat transfer and heat transfer inside a gas radiation tube for thermal furnaces with a protective atmosphere, *J. Engng Phys.* **50**(1), 70 (1986).
74K. D. C. Look, Jr. and M. Vijayan, Laser reflection in a scattering medium with a white diffusively reflecting substrate, *J. Heat Transfer* **108**(1), 242 (1986).
75K. Y. Lu and A. Penzkofer, Optical constants measurements of strongly absorbing media, *Appl. Optics* **25**(2), 221 (1986).
76K. B. Maheu and G. Gouesbet, Four-flux models to solve the scattering transfer equation: special cases, *Appl. Optics* **25**(7), 1122 (1986).
77K. M. P. Manguc and R. Viskanta, Radiation transfer in a cylindrical vessel containing high-temperature corium aerosols, *Nucl. Sci. Engng* **92**(4), 570 (1986).
78K. G. M. Maxwell, M. J. Bailey and V. W. Goldschmidt, Calculations of the radiation configuration factor using ray casting, *Comput. Aided Des.* **18**(7), 371 (1986).
79K. M. P. Menguc, W. G. Cummings III and R. Viskanta, Radiative transfer in a gas-turbine combustor, *J. Prop. Pwr* **2**(3), 241 (1986).
80K. M. P. Menguc and R. Viskanta, Radiative transfer in axisymmetric, finite cylindrical enclosures, *J. Heat Transfer* **108**(2), 271 (1986).
81K. A. Mugnai and W. J. Wiscombe, Scattering from nonspherical Chebyshev particles 1: cross sections, single-scattering albedo, asymmetry factor, and back-scattered fraction, *Appl. Optics* **25**(7), 1235 (1986).
82K. V. H. Myers, A. Ono and D. P. DeWitt, A method for measuring optical properties of semitransparent materials at high temperatures, *AIAA J.* **24**(2), 321 (1986).
83K. T. Nakajima and M. Tanaka, Matrix formulations for the transfer of solar radiation in a plane-parallel scattering atmosphere, *J. Quant. Spectrosc. Radiat. Transfer* **35**(1), 13 (1986).
84K. M. H. N. Naraghi and B. T. F. Chung, A stochastic approach for radiative exchange in enclosures with directional–bidirectional properties, *J. Heat Transfer* **108**(2), 264 (1986).
85K. D. B. Nash, Mid-infrared reflectance spectra (2.3–22 μm) of sulfur, gold, KBr, MgO, and halon, *Appl. Optics* **25**(14), 2427 (1986).
86K. H. F. Nelson and E. O. Tucker, Boron slurry-fueled jet engine exhaust plume infrared signatures, *J. Spacecraft Rockets* **23**(5), 527 (1986).
87K. H. F. Nelson, D. C. Look, Jr. and A. L. Crosbie, Two-dimensional radiative back-scattering from optically thick media, *J. Heat Transfer* **108**(3), 619 (1986).
88K. S. O. Onyegegbu, Heat transfer inside a horizontal

cylindrical annulus in the presence of thermal radiation and buoyancy, *Int. J. Heat Mass Transfer* **29**(5), 659 (1986).
89K. R. S. Patel and M. Q. Brewster, Optical constants of propellant-grade ammonium perchlorate *AIAA J.* **24**(11), 1878 (1986).
90K. S. J. Petuchowski, Calculation of H_2O-broadened CO linewidths, *J. Quant. Spectrosc. Radiat. Transfer* **36**(4), 319 (1986).
91K. Yu. A. Popov, Radiation heat transfer in a cavity, accounting for heat losses through the shell (in Russian), *Izv. Sib. Otdeleniya Akad. Nauk SSSR Ser. Tek. Nauk* No. 4, 20 (1986).
92K. P. V. Prosuntsov and S. V. Reznik, Mathematical model of coefficient reverse problem of radiation-conductive heat transfer in a translucent scattering medium (in Russian), *Izv. Sib. Otdeleniya Akad. Nauk SSSR Ser. Tek. Nauk* No. 10, 3 (1986).
93K. C. P. Rinsland, D. C. Benner and V. M. Devi, Absolute line intensities in CO_2 bands near 4.8 μm, *Appl. Optics* **25**(7), 1204 (1986).
94K. J. W. Rish III and J. A. Roux, Evaluation of emission integrals for the radiative transport equation, *AIAA J.* **24**(12), 2049 (1986).
95K. A. S. Romanov and A. A. Stytsyna, Selection of an initial approximation in an asymptotic representation of the solution of the problem of a point-source thermal explosion in a nonlinearly heat-conducting gas, *J. Appl. Mech. Tech. Phys.* **27**(3), 396 (1986).
96K. L. S. Rothman, Infrared energy levels and intensities of carbon dioxide Pt. 3, *Appl. Optics* **25**(11), 1795 (1986).
97K. N. A. Rubtsov and E. P. Golova, Effect of scattering on nonstationary radiation-conductive heat transfer in a two-layer system (in Russian), *Izv. Sib. Otdeleniya Akad. Nauk SSSR Ser. Tek. Nauk* No. 10, 10 (1986).
98K. T. P. Ryzhkova and L. I. Ryzhkov, Mathematical aspects of the theory of diffraction as applied to thermal radiation transfer, *Heat Transfer—Sov. Res.* **18**(1), 125 (1986).
99K. S. M. Scrivner, T. W. Taylor, C. M. Sorensen and J. F. Merklin, Soot particle size distribution measurements in a premixed flame using photon correlation spectroscopy, *Appl. Optics* **25**(2), 291 (1986).
100K. V. D. Sharma, The development of jump discontinuities in radiative magnetogasdynamics, *Int. J. Engng Sci.* **24**(5), 813 (1986).
101K. K. J. Siemsen, J. Reid and D. J. Danagher, Improved cw lasers in the 11–13 μm wavelength region produced by optically pumping NH_3, *Appl. Optics* **25**(1), 86 (1986).
102K. R. D. Skocypec and R. O. Buckius, Photon path length analysis of radiative heat transfer in planar layers with arbitrary temperature distributions, *J. Quant. Spectrosc. Radiat. Transfer* **35**(2), 109 (1986).
103K. M. S. Sohal, Radiation heat transfer model for the SCDAP code, *Nucl. Technol.* **75**(2), 196 (1986).
104K. E. M. Sparrow and S. B. Vemuri, Orientation effects on natural convection/radiation heat transfer from pin-fin arrays, *Int. J. Heat Mass Transfer* **29**(3), 359 (1986).
105K. P. Stefanizzi, Reliability of the Monte Carlo method in black body view factor determination, *Termotecnica* **40**(6), 29 (1986).
106K. G. L. Stephens, Radiative transfer in spatially heterogeneous, two-dimensional, anisotropically scattering media, *J. Quant. Spectrosc. Radiat. Transfer* **36**(1), 51 (1986).
107K. W. H. Sutton, A short time solution for coupled conduction and radiation in a participating slab geometry, *J. Heat Transfer* **108**(2), 465 (1986).
108K. S. Tanaka, T. Yoshida and T. Kunitomo, Experimental study on simultaneous conductive and radiative heat transfer in circular and longitudinal finned tube systems, *Heat Transfer—Jap. Res.* **15**(1), 72 (1986).
109K. J. Tapping and J. L. Reilly, Index of refraction of sapphire between 24 and 1060°C for wavelengths of 633 and 799 nm, *J. Opt. Soc. Am. A* **3**(5), 610 (1986).
110K. R. T. Taussig and A. T. Mattick, Droplet radiator systems for spacecraft thermal control, *J. Spacecraft Rockets* **23**(1), 10 (1986).
111K. S. T. Thynell and M. N. Ozisik, Integral form of the equation of transfer for an isotropically scattering, inhomogeneous solid cylinder, *J. Quant. Spectrosc. Radiat. Transfer* **36**(6), 497 (1986).
112K. S. T. Thynell and M. N. Ozisik, A new efficient method of solution to radiation transfer in absorbing, emitting, isotropically scattering, homogeneous, finite or semi-finite, plane-parallel media, *J. Quant. Spectrosc. Radiat. Transfer* **36**(1), 39 (1986).
113K. S. T. Tynell and M. N. Ozisik, Radiation transfer due to a point source in an isotropically scattering, inhomogeneous solid sphere, *J. Quant. Spectrosc. Radiat. Transfer* **35**(5), 349 (1986).
114K. N. Tokuda, An asymptotic, large time solution of the convection Stefan problem with surface radiation, *Int. J. Heat Mass Transfer* **29**(1), 135 (1986).
115K. J. Tomeczek and Z. Gebhardt, Radiation properties of natural gas diffusion flames, *Gas Waerme Int.* **35**(1), 39 (1986).
116K. C.-F. Tsai and G. Nixon, Transient temperature distribution of a multilayer composite wall with effects of internal thermal radiation and conduction, *Numer. Heat Transfer* **10**(1), 95 (1986).
117K. D. S. Tsai and W. Strieder, Radiation across and down a cylindrical pore having both specular and diffuse reflectance components, *Ind. Engng Chem. Fund.* **25**(2), 244 (1986).
118K. R. J. Tucker, Direct exchange areas for calculating radiation transfer in rectangular furnaces, *J. Heat Transfer* **108**(3), 707 (1986).
119K. V. T. Vader, R. Viskanta and F. P. Incropera, Design and testing of a high-temperature emissometer for porous and particulate dielectrics, *Rev. Scient. Instrum.* **57**(1), 87 (1986).
120K. D. Vanderhaegen, Radiative transfer in statistically heterogeneous mixtures, *J. Quant. Spectrosc. Radiat. Transfer* **36**(6), 557 (1986).
121K. I. A. Vasilieva, Shape of spectral lines emitted from a gas in the presence of a condensed phase, *J. Quant. Spectrosc. Radiat. Transfer* **35**(3), 171 (1986).
122K. J.-Q. Wang, P. L. Richards, J. W. Beeman, N. M. Haegel and E. E. Haller, Optical efficiency of far-infrared photoconductors, *Appl. Optics* **25**(22), 4127

(1986).
123K. L. Wen and Y. Zheng, Mathematical models and calculating method of radiation heat transfer in furnaces (in Chinese), *Shiyou Xuebao Shiyou Huagong* **2**(1), 57 (1986).
124K. S. J. Wilson and K. K. Sen, Generalized Eddington approximation method for radiative transfer problems in slab medium, *J. Quant. Spectrosc. Radiat. Transfer* **35**(6), 467 (1986).
125K. Y. Yamada, J. D. Cartigny and C. L. Tien, Radiative transfer with dependent scattering by particles: part 2—experimental investigation, *J. Heat Transfer* **108**(3), 614 (1986).
126K. V. N. Zapechnikov, V. T. Pushkin and A. G. Zen' kovskii, Directional radiation properties of industrial refractories of zirconium dioxide, *Refractories* **27**(1), 62 (1986).
127K. A. Zardecki, S. A. W. Gerstl and R. E. DeKinder, Jr., Two- and three-dimensional radiative transfer in the diffusion approximation, *Appl. Optics* **25**(19), 3508 (1986).

NUMERICAL METHODS

1N. T. Aizawa, J. Kihara and T. Umeda, Computer diagnosis system for thermal analysis by BEM aided with simple CAD, *Engng Comput.* **1**(4), 175 (1986).
2N. W. K. Anderson, J. L. Thomas and B. Van Leer, Comparison of finite volume flux vector splittings for the Euler equations, *AIAA J.* **24**(9), 1453 (1986).
3N. N. A. Avdonin and V. A. Smirov, Numerical analysis of heat and mass transfer in the growth of large single crystals from the melt, *Growth Cryst.* **13,** 221 (1986).
4N. M. Barnett and R. T. Davis, Calculation of supersonic flows with strong viscous–inviscid interaction, *AIAA J.* **24**(12), 1949 (1986).
5N. C. Basdevant, M. Deville, P. Haldenwang, J. M. Lacroix, J. Quazzani, R. Peyret, P. Orlandi and A. T. Patera, Spectral and finite difference solutions of the Burgers equation, *Comput. Fluids* **14**(1), 23 (1986).
6N. A. C. Benim and W. Zinser, A segregated formulation of Navier–Stokes equations with finite elements, *Comput. Meth. Appl. Mech. Engng* **57**(2), 223 (1986).
7N. H. Bertin and H. Ozoe, Technique for rapid convergence of the penalty finite-element method with a modified Galerkin scheme and its application to natural convection, *Numer. Heat Transfer* **10**(4), 311 (1986).
8N. M. C. Bhattacharya, New improved finite difference equation for heat transfer during transient change, *Appl. Math. Modelling* **10**(1), 68 (1986).
9N. A. Borthwick, Comparison between two finite-difference schemes for computing the flow around a cylinder, *Int. J. Numer. Meth. Fluids* **6**(5), 275 (1986).
10N. M. Braaten and W. Shyy, A study of recirculating flow computation using body-fitted coordinates: consistency aspects and mesh skewness, *Numer. Heat Transfer* **9**(5), 559 (1986).
11N. M. E. Braaten and W. Shyy, Comparison of iterative and direct solution methods for viscous flow calculations in body-fitted co-ordinates, *Int. J. Numer. Meth. Fluids* **6**(6), 325 (1986).
12N. E. K. Buratynski and D. A. Caughey, An implicit LU scheme for the Euler equations applied to arbitrary cascades, *AIAA J.* **24**(1), 39 (1986).
13N. S.-C. Chang, Solution of elliptic PDEs by fast Poisson solvers using a local relaxation factor, *J. Comput. Phys.* **67**(1), 91 (1986).
14N. K. A. Cliffe and D. A. Lever, A comparison of finite-element methods for solving flow past a sphere, *J. Comput. Phys.* **62**(2), 321 (1986).
15N. F. Colin and A. Floquet, Combination of finite element and integral transform techniques in a heat conduction quasi-static problem, *Int. J. Numer. Meth. Engng* **23**(1), 13 (1986).
16N. S. D. Connell and P. Stow, A discussion and comparison of numerical techniques used to solve the Navier–Stokes and Euler equations, *Int. J. Numer. Meth. Fluids* **6**(3), 155 (1986).
17N. L. Demkowicz and J. T. Oden, An adaptive characteristic Petrov–Galerkin finite element method for convection-dominated linear and nonlinear parabolic problems in two space variables, *Comput. Meth. Appl. Mech. Engng* **55**(1), 63 (1986).
18N. G. Dhatt, A. Soulaimani, Y. Ouellet and M. Fortin, Development of new triangular elements for free surface flows, *Int. J. Numer. Meth. Fluids* **6**(12), 895 (1986).
19N. J. C. Eid, Spreadsheets for thermal analysis, *Mach. Des.* **58**(25), 121 (1986).
20N. D. J. Evans and A. R. Abdullah, Boundary value techniques for finite element solutions of the diffusion–convection equation, *Int. J. Numer. Meth. Engng* **23**(6), 1145 (1986).
21N. R. E. Faw and J . K. Shultis, Transient conduction from a finite cylinder with time-varying internal heat generation, *Numer. Heat Transfer* **9**(6), 731 (1986).
22N. R. G. Friel, Some new methods for numerically solving the compressible laminar boundary layer equations, *Int. J. Numer. Meth. Fluids* **6**(9), 623 (1986).
23N. L. Fuchs, A local mesh-refinement technique for incompressible flows, *Comput. Fluids* **14**(1), 69 (1986).
24N. K. Fujinawa, A 'characteristic' finite element scheme for convective–dispersive transport with non-equilibrium reaction, *Int. J. Numer. Meth. Engng* **23**(6), 1161 (1986).
25N. T. Fusegi and B. Farouk, Predictions of fluid flow and heat transfer problems by the vorticity–velocity formulation of the Navier–Stokes equations, *J. Comput. Phys.* **65**(1), 227 (1986).
26N. P. F. Galpin and G. D. Raithby, Treatment of non-linearities in the numerical solution of the incompressible Navier–Stokes equations, *Int. J. Numer. Meth. Fluids* **6**(7), 409 (1986).
27N. P. F. Galpin, G. D. Raithby and J. P. Van Doormaal, Discussion of upstream-weighted advection approximations for curved grids, *Numer. Heat Transfer* **9**(2), 241 (1986).
28N. P. F. Galpin and G. D. Raithby, Numerical solution of problems in incompressible fluid flow: treatment of the temperature–velocity coupling, *Numer. Heat Transfer* **10**(2), 105 (1986).
29N. N. K. Ghaddar, G. E. Karniadakis and A. T. Peters, A conservative isoparametric spectral element method

for forced convection; application to fully developed flow in periodic geometries, *Numer. Heat Transfer* **9**(3), 277 (1986).

30N. J. Guerra and B. Gustafsson, A numerical method for incompressible and compressible flow problems with smooth solutions, *J. Comput. Phys.* **63**(2), 377 (1986).

31N. A. A. M. Halim, Development of an iterative boundary-layer-type solver for axisymmetric separated flows, *AIAA J.* **24**(8), 1298 (1986).

32N. R. Hills, M. Raynaud and E. Hensel, Surface variance estimates using an adjoint formulation for a one-dimensional nonlinear inverse heat conduction technique, *Numer. Heat Transfer* **10**(5), 441 (1986).

33N. R. G. Hills and E. C. Hensel, Jr., One-dimensional nonlinear inverse heat conduction, *Numer. Heat Transfer* **10**(4), 369 (1986).

34N. T. V. Hromadka II, A unified model of radially symmetric heat conduction, *Numer. Heat Transfer* **9**(1), 43 (1986).

35N. T. J. R. Hughes, L. P. Franca and M. Mallet, A new finite element formulation for computational fluid dynamnics: I. Symmetric forms of the compressible Euler and Navier–Stokes equations and the second law of thermodynamics, *Comput. Meth. Appl. Mech. Engng* **54**(2), 223 (1986).

36N. T. J. R. Hughes, M. Mallet and A. Mizukami, A new finite element formulation for computational fluid dynamics: II. Beyond SUPG, *Comput. Meth. Appl. Mech. Engng* **54**(3), 341 (1986).

37N. T. J. R. Hughes and M. Mallet, A new finite element formulation for computational fluid dynamics: III. The generalized streamline operator for multidimensional advective–diffusive systems, *Comput. Meth. Appl. Mech. Engng* **58**(3), 305 (1986).

38N. T. J. R. Hughes and M. Mallet, A new finite element formulation for computational fluid dynamics; IV, A discontinuity-capturing operator for multidimensional advective-diffusive systems, *Comput. Meth. Appl. Mech. Engng* **58**(3), 329 (1986).

39N. T. J. R. Hughes, L. P. Franca and M. Balestra, A new finite element formulation for computational fluid dynamics: V. Circumventing the Babuska–Brezzi condition: a stable Petrov–Galerkin formulation of the Stokes problem accommodating equal-order interpolations, *Comput. Meth. Appl. Mech. Engng* **59**(1), 85 (1986).

40N. K. Y. Huh, M. W. Golay and V. P. Manno, A method of reduction of numerical diffusion in the donor cell treatment of convection, *J. Comput. Phys.* **63**(1), 201 (1986).

41N. B. R. Hutchinson and G. D. Raithby, A multigrid method based on the additive correction strategy, *Numer. Heat Transfer* **9**(5), 511 (1986).

42N. R. I. Issa, Solution of the implicitly discretised fluid flow equations by operator-splitting, *J. Comput. Phys.* **62**(1), 40 (1986).

43N. R. I. Issa, A. D. Gosman and A. P. Watkins, The computation of compressible and incompressible recirculating flows by a non-iterative implicit scheme, *J. Comput. Phys.* **62**(1), 66 (1986).

44N. A. Jameson and D. Mavriplis, Finite volume solution of the two-dimensional Euler equations on a regular triangular mesh, *AIAA J.* **24**(4), 611 (1986).

45N. A. Jameson and S. Yoon, Multigrid solution of the Euler equations using implicit schemes, *AIAA J.* **24**(11), 1737 (1986).

46N. D. S. Jang, R. Jetli and S. Acharya, Comparison of the PISO, SIMPLER and SIMPLEC algorithms for the treatment of the pressure–velocity coupling in steady flow problems, *Numer. Heat Transfer* **10**(3), 209 (1986).

47N. Y.-H. Ju and W.-C. Lee, A method for solving transient diffusion problems, *Numer. Heat Transfer* **9**(6), 663 (1986).

48N. M. Kanch and T. Ueda, Application of weighted finite difference method to one- or two-dimensional convective diffusion problem, *J. Hydrosci. Hydraul. Engng* **4**(1), 17 (1986).

49N. S. R. Kennon and G. S. Dulikravich, Generation of computational grids using optimization, *AIAA J.* **24**(7), 1069 (1986).

50N. C. Kim, P. V. Desai and J. G. Hartley, Moving free surface heat transfer analysis by continuously deforming finite elements, *Numer. Heat Transfer* **10**(2), 147 (1986).

51N. R. I. Kreis, F. C. Thames and H. A. Hassan, Application of a variational method for generating adaptive grids, *AIAA J.* **24**(3), 404 (1986).

52N. D. W. Landry and B. Kaplan, Addendum to Eigenvalue method for solving transient heat conduction problems, *Numer. Heat Transfer* **9**(2), 247 (1986).

53N. P. Le Tallec and V. Ruas, On the convergence of the bilinear-velocity constant-pressure finite element method in viscous flow, *Comput. Meth. Appl. Mech. Engng* **54**(2), 235 (1986).

54N. B. LeDain Muir and B. R. Baliga, Solution of three-dimensional convection–diffusion problems using tetrahedral elements and flow-oriented upwind interpolation functions, *Numer. Heat Transfer* **9**(2), 143 (1986).

55N. B. Litkouhi and J. V. Beck, Multinode unsteady surface element method with application to contact conductance problem, *J. Heat Transfer* **108**(2), 257 (1986).

56N. X. Liu and R. H. Pletcher, A coupled marching procedure for the partially parabolized Navier–Stokes equations, *Numer. Heat Transfer* **10**(6), 539 (1986).

57N. R. Mei and A. Plotkin, A finite-difference scheme for the solution of the steady Navier–Stokes equations, *Comput. Fluids* **14**(3), 239 (1986).

58N. R. A. Meric, Optimal thermal insulation by the boundary element method, *Numer. Heat Transfer* **9**(2), 163 (1986).

59N. H. P. Miller, A comparison of high resolution schemes for the two-dimensional linear advection equation, *Comput. Fluids* **14**(4), 411 (1986).

60N. K. Muralidhar and S. I. Guceri, Comparative study of two numerical procedures for free-convection problems, *Numer. Heat Transfer* **9**(5), 631 (1986).

61N. K. Nakahashi and G. S. Deiwert, Three-dimensional adaptive grid method, *AIAA J.* **24**(6), 948 (1986).

62N. B. Neta and R. T. Williams, Stability and phase speed for various finite element formulations of the advection equation, *Comput. Fluids* **14**(4), 393 (1986).

63N. R. W. Newsome, Euler and Navier–Stokes solutions

for flow over a conical delta wing, *AIAA J.* **24**(4), 552 (1986).

64N. J. T. Oden, T. Strouboulis and P. Devloo, Adaptive finite element methods for the analysis of inviscid compressible flow: Part I. Fast refinement/unrefinement and moving mesh methods for unstructured meshes, *Comput. Meth. Appl. Mech. Engng* **59**(3), 327 (1986).

65N. J. Padovan and L. Krishna, Multiply scaled constrained nonlinear equation solvers, *Numer. Heat Transfer* **10**(5), 463 (1986).

66N. M. K. Patel and N. C. Markatos, An evaluation of eight discretization schemes for two-dimensional convection–diffusion equations, *Int. J. Numer. Meth. Fluids* **6**(3), 129 (1986).

67N. Q. T. Pham, The use of lumped capacitance in the finite-element solution of heat conduction problems with phase change, *Int. J. Heat Mass Transfer* **29**(2), 285 (1986).

68N. M. Pouagare and B. Lakshminarayana, A space-marching method for viscous incompressible internal flows, *J. Comput Phys.* **64**(2), 389 (1986).

69N. C. Prakash, An improved control volume finite-element method for heat and mass transfer, and for fluid flow using equal-order velocity–pressure interpolation, *Numer. Heat Transfer* **9**(3), 253 (1986).

70N. M. M. Rai and S. R. Chakravarthy, An implicit form for the Osher upwind scheme, *AIAA J.* **24**(5), 735 (1986).

71N. G. D. Raithby, P. F. Galpin and J. P. Van Doormaal, Prediction of heat and fluid flow in complex geometries using general orthogonal coordinates, *Numer. Heat Transfer* **9**(2), 125 (1986).

72N. M. Raynaud and J. Bransier, A new finite-difference method for the nonlinear inverse heat conduction problem, *Numer. Heat Transfer* **9**(1), 27 (1986).

73N. M. Reggio and R. Camarero, Numerical solution procedure for viscous incompressible flows, *Numer. Heat Transfer* **10**(2), 131 (1986).

74N. S. R. Robertson, Finite-element analysis of quasi-steady coupled nonlinear heat and electric conduction problems, *Numer. Heat Transfer* **10**(5), 423 (1986).

75N. G. E. Schneider and M. J. Raw, A skewed, positive influence coefficient upwinding procedure for control-volume-based finite-element convection–diffusion computation, *Numer. Heat Transfer* **9**(1), 1 (1986).

76N. T.-M. Shih, L. J. Hayes, W. J. Minkowycz, K.-T. Yang and W. Aung, Parallel computations in heat transfer, *Numer. Heat Transfer* **9**(6), 639 (1986).

77N. K. N. Shukla and D. Gosh, ADI method for unsteady heat conduction in a conical geometry, *Indian J. Technol.* **24**(1), 36 (1986).

78N. K. S. Surana and P. Kalim, Transition finite elements with temperature and temperature gradients as primary variables for axisymmetric heat conduction, *Comput. Struct.* **24**(2), 197 (1986).

79N. K. S. Surana and P. Kalim, Isoparametric axisymmetric shell elements with temperature gradients for heat conduction, *Comput. Struct.* **23**(2), 279 (1986).

80N. Y. Tanaka, T. Honma and I. Kaji, On mixed boundary element solutions of convection–diffusion problems in three dimensions, *Appl. Math. Modelling* **10**(3), 170 (1986).

81N. D. F. Tandy, D. M. Trujillo and H. R. Busby, Solution of inverse heat conduction problems using an Eigenvalue reduction technique, *Numer. Heat Transfer* **10**(6), 597 (1986).

82N. T. E. Tezduyar and Y. J. Park, Discontinuity-capturing finite element formulations for non-linear convection–diffusion-reaction equations, *Comput. Meth. Appl. Mech. Engng* **59**(3), 307 (1986).

83N. V. M. Theodossiou and A. C. M. Sousa, An efficient algorithm for solving the incompressible fluid flow equations, *Int. J. Numer. Meth. Fluids* **6**(8), 557 (1986).

84N. F. N. van de Vosse, A. Segal, A. A. van Steenhoven and J. D. Janssen, A finite element approximation of the unsteady two-dimensional Navier–Stokes equations, *Int. J. Numer. Meth. Fluids* **6**(7), 427 (1986).

85N. S. P. Vanka, Block-implicit multigrid solution of Navier–Stokes equations in primitive variables, *J. Comput. Phys.* **65**(1), 138 (1986).

86N. S. P. Vanka, Performance of a multigrid calculation procedure in three-dimensional sudden expansion flows, *Int. J. Numer. Meth. Fluids* **6**(7), 459 (1986).

87N. S. P. Vanka, A calculation procedure for three-dimensional steady recirculating flows using multigrid methods, *Comput. Meth. Appl. Mech. Engng* **55**(3), 321 (1986).

88N. S. P. Vanka, Block-implicit multigrid calculation of two-dimensional recirculating flows *Comput. Meth. Appl. Mech. Engng* **59**(1), 29 (1986).

89N. M. R. Visbal and J. S. Shang, Comparative study between two Navier–Stokes algorithms for transonic airfoils, *AIAA J.* **24**(4), 599 (1986).

90N. R. W. Walters, D. L. Dwoyer and H. A. Hassan, A strongly implicit procedure for the compressible Navier–Stokes equations, *AIAA J.* **24**(1), 6 (1986).

91N. R. E. White, Nonlinear parallel algorithm with application to the Sefan problem, *SIAM J. Numer. Analysis* **23**(3), 639 (1986).

92N. T. Wilson and R. Rotunno, Numerical simulation of a laminar end-wall vortex and boundary layer, *Physics Fluids* **29**(12), 3993 (1986).

93N. A. S. Wood, An efficient finite-difference scheme for multidimensional Stefan problems, *Int. J. Numer. Meth. Engng* **23**(9), 1757 (1986).

94N. S. F. Wornom and M. M. Hafez, Implicit conservation schemes for the Euler equations, *AIAA J.* **24**(2), 215 (1986).

95N. H. Yang and R. Camarero, An improved vorticity-potential method for three-dimensional duct flow simulations, *Int. J. Numer. Meth. Fluids* **6**(1), 35 (1986).

TRANSPORT PROPERTIES

1P. A. Acrivos and E. Chang, A model for estimating transport quantities in two-phase materials, *Physics Fluids* **29**(1), 3 (1986).

2P. N. S. Akkarachittoor, A. L. Fricke and J. D. Small, Jr., Dual chamber capillary viscometer for viscosity measurements of concentrated polymer solutions at elevated temperatures, *Rev. Scient. Instrum.* **57**(6), 1182 (1986).

3P. S. Andersson and G. Backstrom, Techniques for

determining thermal conductivity and heat capacity under hydrostatic pressure, *Rev. Scient. Instrum.* **57**(8), 1633 (1986).

4P. F. A. Ansari, Empirical method of measuring thermal diffusivity and surface film conductance, *Trans. Am. Soc. Agric. Engrs* **29**(5), 1492 (1986).

5P. S. Aarcone, Conductivity limitations in single-reflection time-domain reflectometry, *J. Phys. E: Scient. Instrums* **19**(12), 1067 (1986).

6P. T. Ashworth, L. R. Johnson and L.-P. Lai, Thermal conductivity of pure ice and tetrahydrofuran clathrate hydrates, *High Temp.–High Press.* **17**(4), 413 (1985).

7P. S. K. Barua and J. R. Wynnyckyj, Measurement of high-temperature effective thermal conductivity of porous spheres, *C.J.Ch.E.* **64**(4), 695 (1986).

8P. R. P. Behringer, Hydrodynamic heat flow in very dilute superfluid **3He–**4He mixtures, *J. Low Temp. Phys.* **62**(1), 15 (1986).

9P. S. M. Benner and D. K. Modi, Moisture gain of spray-applied insulations and its effect on effective thermal conductivity, *J. Therm. Insul.* **9**, 211 (1986).

10P. R. F. Berg and M. R. Moldover, Viscometer for low frequency, low shear rate measurements, *Rev. Scient. Instrum.* **57**(8), 1667 (1986).

11P. P. O. Biney, W. G. Dong and J. H. Lienhard, Use of a cubic equation to predict surface tension and spinodal limits, *J. Heat Transfer* **108**(2), 405 (1986).

12P. L. Binkele, Thermal conductivity, electric resistivity, and Lorenz function for some transition metals measured by a direct electric heating technique, *High Temp.–High Press.* **17**(4), 437 (1985).

13P. P. Bondi, A. Buizza and G. Vacca, Numerical determination of equivalent thermal diffusivity of non-homogeneous materials, *High Temp.–High Press.* **18**(1), 29 (1986).

14P. B. W. Boreham, A new fast response probe for diffusion studies in atmospheric air, *J. Phys. E: Scient. Instrums* **19**(8), 637 (1986).

15P. T. K. Bose, J. M. St.-Arnaud, H. J. Achtermann and R. Scharf, Improved method for the precise determination of the compressibility factor from refractive index measurements, *Rev. Scient. Instrum.* **57**(1), 26 (1986).

16P. R. Brandt, L. Pawlowski, G. Neuer and P. Fauchais, Specific heat and thermal conductivity of plasma sprayed ytrria-stabilized zirconia and NiAl, NiCr, NiCrAl, NiCrAly, NiCoCrAly coatings, *High Temp.–High Press.* **18**(1), 65 (1986).

17P. D. Buttner, R. Caps and J. Fricke, Thermal conductivity of evacuated transparent silica aerogel tiles, *High Temp.–High Press.* **17**(4), 375 (1985).

18P. M. M. Carroll, K. T. Kim and V. F. Nesterenko, The effect of temperature on viscoplastic pore collapse, *J. Appl. Phys.* **59**(6), 1962 (1986).

19P. N. S. Clarke, On the effective diffusivity model for heat diffusion in inclined random laminates, *Q. J. Mech. Appl. Math.* **39**(1), 137 (1986).

20P. S. J. Collocott and G. K. White, Thermal expansion and heat capacity of some stainless steels and FeNi alloys, *Cryogenics* **26**(7), 402 (1986).

21P. H. Craubner, Densitometer for absolute measurements of the temperature dependence of density, partial volumes, and thermal expansivity of solids and liquids, *Rev. Scient. Instrum.* **57**(11), 2817 (1986).

22P. A. Cunningham and D. J. Sparrow, Rigid polyurethane foam: what makes it the most effective insulant? *Cell. Polym.* **5**(5), 327 (1986).

23P. D. B. Cunningham, P. H. Doe, S. D. Joshi and A. Moradi-Araghi, Capillary viscometer for evaluating low-viscosity solutions at elevated temperatures, *Rev. Scient. Instrum.* **57**(9), 2310 (1986).

24P. I. Czarnota, Heat capacity of n-pentane, n-hexane, and n-heptane at high pressure, *High Temp.–High Press.* **17**(5), 543 (1985).

25P. A. J. Dangre, Structure of liquids and their thermal properties, *High Temp.–High Press.* **18**(1), 61 (1986).

26P. R. H. Davis, Effective thermal conductivity of a composite material with spherical inclusions, *Int. J. Thermophys.* **7**(3), 609 (1986).

27P. A. Degiovanni, Indentification de la diffusivite thermique par l'utilisation des moments temporels partiels, *High Temp.–High Press.* **17**(6), 683 (1985).

28P. D. S. K. Devadattam and A. Chakraverty, Some physico-thermal properties of raw rice bran related to drying technology, *Drying Technol.* **4**(1), 145 (1986).

29P. A. A. El-Sharkawy, A. M. Abousehly, El-Sayed M. Higgy and S. R. Atalia, Specific heat capacity, thermal conductivity, and thermal diffusivity of spinel ferrite Nil + 2xFe2-3xSbxO4 in the temperature range 400–1000 K, *High Temp.–High Press.* **18**(3), 241 (1986).

30P. V. I. Fedorov, N. V. Federov and N. Ye. Balenko, A thermodynamic method for measuring the local percentage of liquid water in steam, *Heat Transfer—Sov. Res.* **18**(2), 143 (1986).

31P. M. Fernandez, E. Banda and E. Rojas, Heat pulse line-source method to determine thermal conductivity of consolidated rocks, *Rev. Scient. Instrum.* **57**(11), 2832 (1986).

32P. S. Fischer, A proposal for thermal conductivity standards with special respect to convection and radiation effects, *Warme- und Stoffubertr.* **20**(3), 183 (1986).

33P. S. Fischer and E. Obermeir, Influence of radiative heat transfer on the effective thermal conductivity of liquids: experimental and theoretical investigation, *High Temp.–High Press.* **17**(6), 699 (1985).

34P. I. Fukai, M. Watanabe, S. Ohtani and T. Miura, Simultaneous estimation of thermal properties by a non-linear least squares method, *Heat Transfer—Jap. Res.* **15**(2), 26 (1986).

35P. S. Garg, H. C. Gupta and B. B. Tripathi, Thermal properties of Ta–W alloys, *J. Phys. Soc. Japan* **55**(2), 554 (1986).

36P. K. E. Gilchrist, The thermal conductivity of boron carbide after fast neutron irradiation, *High Temp.–High Press.* **17**(6), 671 (1985).

37P. F. Gitzhofer, L. Pawlowski, D. Lombard, C. Martin, R. Kaczmarek and M. Boulos, The apparent thermal diffusivity and thermal contact resistance in plasma sprayed multi-coatings, *High Temp.–High Press.* **17**(5), 563 (1985).

38P. H. J. Goldsmid, A simple technique for determining the Seebeck coefficient of thermoelectric materials, *J. Phys. E: Scient. Instrums* **19**(11), 921 (1986).

39P. A. Guillot and P. Dantzer, Determination of volumes

by gas expansion, *J. Phys. E: Scient. Instrums* **19**(12), 1008 (1986).

40P. A. Guillot and P. Dantzer, Determination of volumes by gas expansion, *J. Phys. E: Scient. Instrums* **19**(12), 1008 (1986).

41P. S. E. Gustafsson, E. Karawacki and M. A. Chohan, Thermal transport studies of electrically conducting materials using the transient hot-strip technique, *J. Phys. D: Appl. Phys.* **19**(5), 727 (1986).

42P. G. R. Hadley, Thermal conductivity of packed metal powders, *Int. J. Heat Mass Transfer* **29**(6), 909 (1986).

43P. T. L. Halaczek and J. Rafalowicz, Temperature variation of thermal conductivity of self-pumping multilayer insulation, *Cryogenics* **26**(10), 544 (1986).

44P. H. Hatta and M. Taya, Thermal conductivity of coated filler composites, *J. Appl. Phys.* **59**(6), 1851 (1986).

45P. W. Hemminger, Thermal conductivity of tin between 15 and 500°C, *High Temp.–High Press.* **17**(4), 465 (1985).

46P. J. C. Ho, R. C. Liang and D. P. Dandekar, Low-temperature heat capacities of Ni_3Al, *J. Appl. Phys.* **59**(4), 1397 (1986).

47P. K. S. Holian, A new equation of state for aluminum, *J. Appl. Phys.* **59**(1), 149 (1986).

48P. H. Inaba, Measurement of the effective thermal conductivity of agricultural products, *Int. J. Thermophys.* **7**(4), 773 (1986).

49P. K. B. Isaev, Effect of certain factors on the mean effective thermal conductivity of a melt of glass–plastics, *High Temp.* **24**(3), 360 (1986).

50P. Y. Ishihara, Y. Yoshita and I. Nakada, Specific heat of germanium monotelluride, *J. Phys. Soc. Japan* **55**(6), 1948 (1986).

51P. B. W. James and G. S. Keen, A nomogram for the calculation of the transverse thermal conductivity of uniaxial composite lamina, *High Temp.–High Press.* **17**(4), 477 (1985).

52P. R. Jeschar and R. Alt, Influence of lining properties on control characteristics of industrial furnaces, *Gas Waerme Int.* **35**(1), 10 (1986).

53P. S. Jiang, J. C. Jofriet and G. S. Mittal, Thermal properties of haylage, *Trans. Am. Soc. Agric. Engrs (Gen. Edn)* **29**(2), 601 (1986).

54P. A. Kempf and W. R. B. Martin, Measurement of the thermal properties of $TiFe_{0.85}Mn_{0.15}$ and its hydrides, *Int. J. Hydrogen Energy* **11**(2), 107 (1986).

55P. J. K. Kleiman, P. W. C. Chan, N. Salanasky and I. I. Glass, Chemical-reaction and diffusion rates in solids produced by shock compression, *J. Appl. Phys.* **59**(6), 1956 (1986).

56P. P. G. Knibbe, The end-effect error in the determination of thermal conductivity using a hot-wire apparatus, *Int. J. Heat Mass Transfer* **29**(3), 463 (1986).

57P. Y. Kogure, T. Mugishima and Y. Hiki, Low-temperature thermal diffusivity measurement by laser-flash method, *J. Phys. Soc. Japan* **55**(10), 3469 (1986).

58P. V. G. Lisienko, V. S. Kortov, V. B. Kut'in, B. S. Balabanov, V. A. Kudryavtsev and V. V. Tyukov, Relationship between the emissivity and exoemissive properties of the surface of steels at moderate heating temperatures, *High Temp.* **24**(2), 208 (1986).

59P. R. E. Lyon and R. J. Farris, New technique for measuring the heat of solid deformation, *Rev. Scient. Instrum.* **57**(8), 1640 (1986).

60P. D. Marek, A. Mota and J. C. Weber, Magnetic heat conductance between CMN single crystals and liquid ^{3}He, *J. Low Temp. Phys.* **63**(5), 401 (1986).

61P. D. L. McElroy, R. S. Graves, D. W. Yarborough and T. M. Tong, Nonsteady-state behavior of thermal insulations, *High Temp.–High Press.* **17**(4), 395 (1985).

62P. D. E. McMillan and N. G. Utterback, One-line computer to reduce response time in a viscometer system having analog feedback, *Rev. Scient. Instrum.* **57**(5), 978 (1986).

63P. G. Milano and C. Pisoni, Measurements of apparent thermal conductance of nonhomogeneous samples, *High Temp.–High Press.* **17**(4), 421 (1985).

64P. N. Miyata, M. Hagiwara, Y. Ichiyanagi, K. Kuwahara and H. Matsufuji, The effect of heat treatment on ferromagnetic crystalline anisotropy of PdCo alloys, *J. Phys. Soc. Japan* **55**(2), 961 (1986).

65P. A. Z. Moss, Diffusive gas transport capillaries: a solution to the intermediate case based upon the kinetic theory of gases, *J. Appl. Phys.* **59**(8), 2673 (1986).

66P. M. Nagel, Y. Komazaki and S. Suda, Effective thermal conductivity of a metal hydride bed augmented with a copper wire matrix, *J. Less-common Met.* **120**(1), 35 (1986).

67P. O. Nilsson, O. Sandberg and G. Baeckstreom, dc–ac hot-wire procedure for determining thermophysical properties under pressure, *Rev. Scient. Instrum.* **57**(9), 2303 (1986).

68P. A. M. Okaz, S. A. El-Messih and M. El-Osairy, Thermal conductivity of two-phase system ZnO–graphite and MgO–graphite, *Powder Technol.* **47**(1), 35 (1986).

69P. O. Pessoa, Jr., C. L. Cesar, N. A. Patel, H. Vargas, C. C. Ghizoni and L. C. M. Miranda, Two-beam photoacoustic phase measurement of the thermal diffusivity of solids, *J. Appl. Phys.* **59**(4), 1316 (1986).

70P. Q. T. Pham and C. G. Smith, Thermal imbalance errors and effective area in guarded hot plates, *Rev. Scient. Instrum.* **57**(1), 99 (1986).

71P. E. Piorkowska and A. Galeski, Measurements of thermal conductivity of materials using a transient technique, I. Theoretical background, *J. Appl. Phys.* **60**(2), 485 (1986).

72P. E. Piorkowska and A. Galeski, Measurements of thermal conductivity of materials using a transient technique. II. Description of the apparatus, *J. Appl. Phys.* **60**(2), 493 (1986).

73P. E. Polturak, S. L. Garrett and S. G. Lipson, Precision acoustic gas analyzer for binary mixtures, *Rev. Scient. Instrum.* **57**(11), 2837 (1986).

74P. A. Redondo and J. G. Beery, Thermal conductivity of optical coatings, *J. Appl. Phys.* **60**(11), 3882 (1986).

75P. H. Reiss and B. Ziegenbein, Analysis of the local thermal conductivity in inhomogeneous glass fiber insulations, *High Temp.–High Press.* **17**(4), 403 (1985).

76P. A. B. Rejab, B. Britton and R. Stephens, Thermal diffusivity of powders and porous media, *High Temp.–High Press.* **17**(6), 695 (1985).

77P. S. Riegel and G. Weber, A dual-slope method for specific heat measurements, *J. Phys. E: Scient. Instrums* **19**(10), 790 (1986).

78P. H. M. Roder and C. A. Nieto de Castro, Thermal conductivity of ethane at temperatures between 110 and 325 K and pressures to 70 MPa, *High Temp.–High Press.* **17**(4), 453 (1985).

79P. N. S. Saxena, M. A. Chohan and S. E. Gustaffsson, Effective thermal conductivity of loose granular materials, *J. Phys. D: Appl. Phys.* **19**(9), 1625 (1986).

80P. B. Schulz, Thermal diffusivity in ceramics, *Mater. Chem. Phys.* **15**(3), 349 (1986).

81P. A. G. Shashkov, V. P. Kozlov and A. V. Stankevich, Nonsteady heat-flux distribution in semiinfinite bodies in the presence of a finite heat source at the contact boundary of the bodies, *J. Engng Phys.* **50**(2), 228 (1986).

82P. A. G. Shskov, V. P. Kozlov and V. N. Lipovtsev, Determining the thermophysical characteristics of materials on a model of a semiinfinite body with heat supplied by means of a thin annular heater, *J. Engng Phys.* **50**(3), 335 (1986).

83P. R. L. Shoemaker, J. A. Stark and R. E. Taylor, Thermophysical properties of propellants, *High Temp.–High Press.* **17**(4), 429 (1985).

84P. V. P. Skripov and M. Z. Faizullin, Melting lines of simple substances: thermodynamic similarity and behaviour of thermal properties, *High Temp.–High Press.* **18**(1), 1 (1986).

85P. M. Soininen, A perspectively transformed psychrometric chart and its applications, *Drying Technol.* **4**(2), 295 (1986).

86P. A. D. Stuckes, J. A. Tinker and A. Simpson, Effect of moisture on the thermal conductivity of lightweight aggregate concrete, *Bldg Serv. Engng Res. Technol.* **7**(1), 27 (1986).

87P. C. K. Subramaniam and K. D. Chaudhuri, Effect of plastic deformation of the thermal conductivity of bismuth–thallium and bismuth–lead systems between 1.5 and 300 K, *Cryogenics* **26**(11), 628 (1986).

88P. S. Sudo, T. Nishioka, Y. Miyako and T. Miyadai, Specific heat study of metal–insulator transition in antiferromagnetic phase of $NiS_{1.48}Se_{0.52}$, *J. Phys. Soc. Japan* **55**(6), 1806 (1986).

89P. N. Takahashi and K. Otozai, Estimation of boiling point of radon by radiogas chromatography, *Appl. Radiat. Isot.* **37**(3), 191 (1986).

90P. H. Tanaka, T. Nishi and Y. Wada, New instrument for measuring complex adiabatic compressibility and adiabatic pressure-induced temperature variation of liquids in the frequency range from 0.5 Hz to 2 kHz, *Rev. Scient. Instrum.* **57**(10), 2593 (1986).

91P. K. Tanaka and J. Maeda, Pressure–volume measurements by using diamond-anvil cells and an image processing system, *Rev. Scient. Instrum.* **57**(3), 500 (1986).

92P. R. E. Taylor and B. H. Kelsic, Parameters governing thermal diffusivity measurements of unidirectional fiber-reinforced composites, *J. Heat Transfer* **108**(1), 161 (1986).

93P. J. T. Tinge, K. Mencke, L. Bosgra and A. A. H. Drinkenburg, Ultrasonic gas analyzer for high resolution determination of binary-gas composition, *J. Phys. E: Scient. Instrums* **19**(11), 953 (1986).

94P. S. Torquato and J. D. Beasley, Effective properties of fiber-reinforced materials: I—bounds on the effective thermal conductivity of dispersions of fully penetrable cylinders, *Int. J. Engng Sci.* **24**(3), 415 (1986).

95P. I. A. van Haneghem and F. J. Leij, Thermal properties of moist porous media at temperatures below 0°C, *High Temp.–High Press.* **17**(6), 611 (1985).

96P. P. S. Van der Gulik and N. J. Trappeniers, The viscosity of argon at high densities, *Physica A* **135**(1), 1 (1986).

97P. S. A. Vasil'evskii, I. A. Sokolova and G. A. Tirskii, Definition and computation of effective transport coefficients for chemical-equilibrium flows of partially dissociated and ionized gas mixtures, *J. Appl. Mech. Tech. Phys.* **27**(1), 61 (1986).

98P. S. A. Vasil'evskii, I. A. Sokolova and G. A. Tirskii, Definition and computation of effective transport coefficients for chemical-equilibrium flows of partially dissociated and ionized gas mixtures, *J. Appl. Mech. Tech. Phys.* **27**(1), 61 (1986).

99P. R. M. Velasco and F. J. Urib, Transport coefficients for polyatomic gases, *High Temp.–High Press.* **18**(1), 43 (1986).

100P. R. M. Velasco and F. J. Uribe, Transport coefficients for polyatomic gases, *Physica A* **134**(2), 339 (1986).

101P. D. Vidal, L. Guengant, P. Malbrunot and J. Vermesse, Densities of argon–helium and argon–neon dense mixtures up to 8000 bar, *Physica A* **137**(1), 417 (1986).

102P. E. Vogel, E. Bich and R. Nimz, The initial density dependence of the viscosity of organic vapours: benzene and methanol, *Physica A* **139**(1), 188 (1986).

103P. R. K. Williams, R. S. Graves and F. J. Weaver, Transport properties of high purity, polycrystalline titanium diboride, *J. Appl. Phys.* **59**(5), 1552 (1986).

104P. T.-g. Xi, Y.-l. Wang, L. Duan and B.-m. Wang, Thermophysical properties of new ceramic materials in relation to phase transitions, *High Temp. High Press.* **17**(4), 365 (1985).

105P. C.-S. Yih, Instability resulting from stratification in thermal conductivity, *Physics Fluids* **29**(6), 1769 (1986).

106P. K.-H. Yoo and A. C. Anderson, Thermal impedance to normal and superconducting metals, *J. Low Temp. Phys.* **63**(3), 269 (1986).

107P. M. Yoshizawa, I. Shirotani and T. Fujimura, Thermal and elastic properties of black phosphorus, *J. Phys. Soc. Japan* **55**(4), 1196 (1986).

108P. J. H. Young, Steady state Joule heating with temperature dependent conductivities, *Appl. Scient. Res.* **43**(1), 55 (1986).

109P. I. A. van Haneghem and F. J. Leij, Thermal properties of moist porous media at temperatures below 0°C, *High Temp.–High Press.* **17**(6), 611 (1985).

HEAT-TRANSFER APPLICATION—HEAT PIPES AND HEAT EXCHANGERS

1Q. N. J. Abdelmonem and Y. Yakoub, Laboratory-scale tests on gas–solid fluidized bed behaviour and its influence on bed thermal properties, *Modell. Simul. Control B* **6**(3), 17(1986).

2Q. N. Achten, G. Herbsleb and N. Wieling, Importance of crevices formed between tubes and tube plate for the operational behaviour of heat exchangers, *Chemie-Ingr-Tech.* **58**(9), 732 (1986).

3Q. B. A. Afanas'ev, E. P. Vinogradova and G. F. Smirnov, Heat output and maximum heat transfer of heat pipes with continuous corrugated wicks, *J. Engng Phys.* **50**(1), 46 (1986).

4Q. T. Aihara, S. Maruyama, S. Aya and M. Hongoh, Heat transfer characteristics of a low-pressure-loss fluidized-bed heat exchanger with single row tubes, *Nippon Kikai Gakkai Ronbunshu, B. Hen* **52**(476), 1718 (1986).

5Q. V. P. Aneja and D. R. Sathyamurthy, Scale deposition on heat exchanger surfaces, *J. Inst. Engrs India* Pt. AG **66**(1), 48 (1986).

6Q. E. Azad, F. Mohammedieh and F. Moztarzadeh, Effect of different arrangement on performance of heat pipe heat recovery system, *J. Heat Recovery Sys.* **6**(2), 143 (1986).

7Q. V. N. Babak, T. B. Babak, L. P. Kholpanov and V. A. Malyusov, Procedure for calculating matrix heat exchangers formed from perforated plates, *J. Engng Phys.* **50**(3), 330 (1986).

8Q. J. R. Barber, The influence of interface thermal contact resistance on the heat transfer performance of prestressed Duplex tube, *Int. J. Heat Mass Transfer* **29**(5), 761 (1986).

9Q. J. R. Barber, Nonuniqueness and stability for heat conduction through a duplex heat exchanger tube, *J. Therm. Stresses* **9**(1), 69 (1986).

10Q. M. K. Bassiouny, Experimental and theoretical investigations about flow distribution, pressure drop and heat transfer in plate heat exchangers, *Forsch. Ing. Wes.* **52**(3), 75 (1986).

11Q. V. P. Belyakov, V. A. Shaposhnikov, V. V. Budrik and N. M. Volkova, Design of heat-exchange apparatus with saturated and underheated superfluid helium (He-II), *Pwr Engng* **24**(3), 127 (1986).

12Q. B. Berryman and C. Russell, Assessing air-side performance of air-cooled heat exchangers, *Process Engng* **67**(4), 59 (1986).

13Q. R. F. Boehm and F. Kreith, Direct-contact processes promise high performance, low cost, *Mech. Engng* **108**(3), 78 (1986).

14Q. de. V. Caluianu, F. Chirita, G. Mihalascu and D. Simonescu, Boiler made of radiator elements for steam operating with solid fuel, *Bul. Stiintific* **29**(2), 29 (1986).

15Q. J. C. Chato and C. Damianides, Second-law-based optimization of heat exchanger networks using load curves, *Int. J. Heat Mass Transfer* **29**(8), 1097 (1986).

16Q. W. T. Cross and C. Ramshaw, Process intensification: laminar flow heat transfer, *Chem. Engng Res. Des.* **64**(4), 293 (1986).

17Q. M. T. Dalman, J. H. Merkin and C. McGreavy, A model for enhanced cooling near the edge of a packed bed, *Numer. Heat Transfer* **9**(1), 97 (1986).

18Q. B. V. Dzyubenko and A. V. Sakalauskas, Vortex structure of flow in heat exchanger with twisted tubes, *Pwr Engng* **24**(3), 144 (1986).

19Q. Energy Technology Support Unit, Heat recovery from a spray dryer using a glass tube heat exchanger, *J. Heat Recovery Sys.* **6**(1), 25 (1986).

20Q. A. Faghri, Vapor flow analysis in a double-walled concentric heat pipe, *Numer. Heat Transfer* **10**(6), 583 (1986).

21Q. C. A. Floudas and I. E. Grossmann, Synthesis of flexible heat exchanger networks for multiperiod operation, *Comput. Chem. Engng* **10**(2), 153 (1986).

22Q. W. W. Focke, Selecting optimum plate heat exchanger surfaces, *J. Heat Transfer* **108**(1), 153 (1986).

23Q. G. Frank, Dynamic behaviour of crossflow heat exchangers (in German), *Wärme- und Stoffübertr.* **20**(2), 141 (1986).

24Q. P. J. Fryer and N. K. H. Slater, The simulation of heat exchanger control with tube-side chemical reaction fouling, *Chem. Engng Sci.* **41**(9), 2363 (1986).

25Q. J. Furst, Warmeaustauscher; Heat exchanger *Brennst. Wärme Kraft* **4**, 164 (1986).

26Q. J. Furst, Warmedammungcher, *Brennst. Wärme Kraft* **4**, 167 (1986).

27Q. U. Gross and E. Hahne, Heat transfer in a closed thermosyphon over a wide range of pressures and inclinations, *Germ. Chem. Engng* **9**(5), 292 (1986).

28Q. D. D. Gvozdenac, Two-pass countercrossflow heat exchangers with both fluids unmixed throughout, *Wärme- und Stoffübertr.* **20**(2), 151 (1986).

29Q. K. Hamaguchi, H. Kukita, S. Takahashi and H. Miyabe, On the choice procedures of optimum matrix in regenerator design (case of stacked wire gauzes), *Nippon Kikai Gakkai Ronbunshu, B. Hen* **52**(473), 485 (1986).

30Q. H. T. Haukas, Development of a calculation method for plate-type evaporators with vertical channels, *Aust. Refrig. Air Cond. Heat* **40**(2), 18 (1986).

31Q. P. J. Heggs and E. A. Foumeny, Thermal performance of adiabatic cyclic regenerators, *Numer. Heat Transfer* **9**(2), 183 (1986).

32Q. K. Hesselmann, Heat exchanger networks—an exer-economical approach, *Chemie-Ingr-Tech.* **58**(12), 990 (1986).

33Q. Y. F. Hsu, B. Rubinsky and K. Mahin, An inverse finite element method for the analysis of stationary arc welding processes, *J. Heat Transfer* **108**(4), 734 (1986).

34Q. M. Hudina and M. Huggenberger, Pressure drop and heat transfer in gas-cooled rod bundles, *Nucl. Engng Des.* **97**(3), 347 (1986).

35Q. J. H. Hull, Analysis of heat transfer factors for a pipe absorber array corrected to a common manifold, *J Sol. Energy Engng* **108**(1), 11 (1986).

36Q. T. H. Hwang and S. C. Yao, Crossflow heat transfer in tube bundles at low Reynolds numbers, *J. Heat Transfer* **108**(3), 697 (1986).

37Q. M. Inuzuka, R. Sugiyama, I. Saito, I. Yamada, S. Hiroaka, H. Ishikawa and I. Banno, Analysis of heat and mass transfer in a centrifugal molecular still, *J. Chem. Engng Japan* **19**(1), 14 (1986).

38Q. H. A. Irazoqui, Optimal thermodynamic synthesis of thermal energy recovery systems, *Chem. Engng Sci.* **41**(5), 1243 (1986).

39Q. N. K. Jain, Optimum quantity of steam injected as radiant fluid in contraflow regenerators, *J. Heat Recovery Sys.* **6**(5), 355 (1986).

40Q. J. Jekeric, Thermodynamische Berechnungsmethode eines naturlichen Wasserumlaufsystems, *Brennst.–Wärme–Kraft* **7**, 340 (1986).

41Q. G. Johnson and E. Marschall, On the temperature jump in liquid–liquid direct-contact heat exchangers, *Int. J. Multiphase Flow* **12**(1), 127 (1986).

42Q. J. D. Jones, Performance of a Stirling engine regenerator having finite mass, *J. Engng Gas Turbine Pwr* **108**(4), 669 (1986).

43Q. G. St. Kalitzin and N. G. Kalitzin, Losses due to the irreversibility by the heat exchangers, *Heat Technol.* **4**(3), 32 (1986).

44Q. W. Kamminga, The testing of an evacuated tubular collector with a heat pipe using the Fourier frequency domain, *Int. J. Heat Mass Transfer* **29**(1), 83 (1986).

45Q. N. Kayansayan, The gravity assisted heat pipe with application to concrete shell steam condensers, *J. Heat Recovery Sys.* **6**(5), 389 (1986).

46Q. V. V. Khudasko, D. Yu. Kardash and N. S. Grachev, Heat transfer in the critical and supercritical zones of sodium–water steam generators, *Soviet Atomic Energy* **61**(1), 569 (1986).

47Q. D. C. Klaren, Fluid end to fouling? *Process Engng* **67**(1), 45 (1986).

48Q. N. I. Klyuev, V. V. Kostenko, A. L. Luks, V. I. Mikheev and V. Ya. Chernobrivets, Experimental study of operating regimes of low-temperature nearly isothermal heat pipes, *Soviet Aeronaut.* **29**(3), 77 (1986).

49Q. T. Kohya, M. Kuriyama, M. Toda, E. Harada and H. Konno, Heat transfer from parallel-horizontal cylinder rows under a corona discharge, *Heat Transfer—Jap. Res.* **15**(3), 54 (1986).

50Q. M. V. Korshagina and A. Ya. Shelginskii, Vapor flow in low-temperature heat pipes, *J. Engng Phys.* **50**(2), 158 (1986).

51Q. A. N. Krasil'nikov, V. E. Bogoslovskii and T. I. Levina, Model of heat and mass exchange on a rectification plate with cold reflux, *Theor. Found. Chem. Engng* **20**(2), 162 (1986).

52Q. S. Kraus, New investigations on fouling of heat transfer surfaces by sedimentation and crystallization, *Chemie-Ingr-Tech.* **58**(2), 146 (1986).

53Q. S. Krause, Precipitation and particulate fouling on heat transfer surfaces, *Forsch. Ing. Wes.* **52**(6), 168 (1986).

54Q. D. G. Kroger, Performance characteristics of industrial finned tubes presented in dimensional form, *Int. J. Heat Mass Transfer* **29**(8), 1119 (1986).

55Q. G. Kushida, H. Yamashita and R. Izumi, Fluid flow and heat transfer in a plate-fin and tube heat exchanger (analysis of heat transfer around a square cylinder situated between parallel plates), *Nippon Kikai Gakki Ronbunshu, B. Hen* **52**(476), 1726 (1986).

56Q. Y. N. Lee, Heat transfer and pressure drop characteristics of an array of plates aligned at angles to the flow in a rectangular duct, *Int. J. Heat Mass Transfer* **29**(10), 1586 (1986).

57Q. C.-H. Li, Exact transient solutions of parallel-current transfer processes, *J. Heat Transfer* **108**(2), 365 (1986).

58Q. C.-H. Li, Effect of radiation heat transfer on a bayonet tube heat exchanger, *A.I.Ch.E. Jl* **32**(2), 341 (1986).

59Q. V. I. Lutsenko, A conjugate problem of the calculation of high-temperature heat pipe parameters, *Heat Transfer—Sov. Res.* **18**(2), 136 (1986).

60Q. J. F. Maingonnat and G. Corrieu, A study of the thermal performance of a scraped-surface heat exchanger. Pt. I. Review of the principal models describing heat transfer and power consumption, *Entropie* No. 111 (1983), France, *Int. Chem. Engng* **26**(1), 45 (1986).

61Q. J. F. Maingonnat and G. Corrieu, A study of the thermal performance of a scraped-surface heat exchanger. Pt. II. The effect of the axial diffusion of heat, *Entropie* No. 111 (1983), France, *Int Chem. Engng* **26**(1), 55 (1986).

62Q. T. B. Margittai, New heat exchanger geometry, *Chem. Engng Prog.* **82**(7), 39 (1986).

63Q. W. J. Marner, Gas-side fouling, *Mech. Engng* **108**(3), 70 (1986).

64Q. M. G. Martin, A. Amokrane and M. Humbert, Experimental study of heat transfer and pressure drop characteristics of a finned wall situated in a cross flow of air in a rectangular duct. Application to a smoke recuperator, *Revue Gen. Therm.* **25**(294), 325 (1986).

65Q. T. Matsuo, M. Iwabuchi, M. Kanzaka, H. Haneda and K. Yamamoto, Heat transfer correlations of rifled tubing for boilers under sliding pressure operating condition, *Nippon Kikai Gakki Ronbunshu, B. Hen* **52**(476), 1822 (1986).

66Q. H. Matsushima, W. Nakayama, T. Daikoku and Y. Osawa, Shell-side single-phase flows and heat transfer in shell-and-tube heat exchangers (1st report, experimental study on local heat transfer coefficients, *Heat Transfer—Jap. Res.* **15**(2), 60 (1986).

67Q. R. Meyer, U. Laux and R. Schlitt, Excess liquid formation in orbit test results of axially grooved heat pipes, *J. Spacecraft Rockets* **23**(5), 482 (1986).

68Q. N. Midoux, G. Wild, M. Purwasasmita, J. C. Charpentier and H. Martin, Liquid content and heat transfer in trickle-bed reactors under conditions of intensive interaction between gas and liquid, *Chemie-Ingr-Tech.* **58**(2), 142 (1986).

69Q. R. S. Mullisen and R. I. Loerhke, A transient heat exchanger evaluation test for arbitrary fluid inlet temperature variation and longitudinal core conductance, *J. Heat Transfer* **108**(2), 370 (1986).

70Q. R. S. Mullisen and R. I. Loerhke, A study of the flow mechanisms responsible for heat transfer enhancement in interrupted-plate exchangers, *J. Heat Transfer* **108**(3), 377 (1986).

71Q. K. N. Murty, How effective are finned tubes in heat exchangers? *Chem. Engng* **93**(7), 120 (1986).

72Q. K. B. Narayana, Vapor flow characteristics of slender cylindrical heat pipes—a numerical approach, *Numer. Heat Transfer* **10**(1), 79 (1986).

73Q. H. Nickel, F. Schubert and H. Schuster, Very high temperature design criteria for nuclear heat ex-

changers in advanced high temperature reactors, *Nucl. Engng Des.* **94**(3), 337 (1986).
74Q. J. Niogret and A. Le Lan, Recent developments in heat exchangers with solid–gas circulating fluidized bed, *Industrie Céram.* **806,** 391 (1986).
75Q. O. Ta'eed and B. M. Gibbs, Determination of heat transfer coefficient in a short-tube economiser operating in the transition region, *J. Heat Recovery Sys.* **6**(2), 93 (1986).
76Q. T. Ohmori, M. Okazaki and R. Toei, Heat transfer coefficient in stationary heating-plane type of indirect-heat agitated dryer, *J. Chem. Engng Japan* **19**(3), 167 (1986).
77Q. A. Pignotti, Effectiveness of series assemblies of divided-flow heat exchangers, *J. Heat Transfer* **108**(1), 141 (1986).
78Q. I. L. Pioro, Influence of the inclination angle of a two-phase thermosiphon on the limiting heat flux, *Heat Transfer—Sov. Res.* **18**(2), 35 (1986).
79Q. W. C. Pitts and M. S. Murbach, Thermal design of aeroassisted orbital transfer vehicle heat shields for a conical drag brake, *J. Spacecraft Rockets* **23**(4), 442 (1986).
80Q. F. C. Prenger, E. S. Keddy and J. T. Sena, Performance characteristics of gravity-assisted potassium heat pipes, *J. Spacecraft Rockets* **23**(4), 407 (1986).
81Q. M. Pronobis, Effect of fouling on heat transfer in convective boiler tube banks, *VGB Kraftwerstech* **66**(2), 120 (1986).
82Q. V. M. Puri, Feasibility and performance curves for intermittent earth tube heat exchangers, *Trans. Am. Soc. Agric. Engrs (Gen. Edn)* **29**(2), 526 (1986).
83Q. K. Quick, Direct calculation of exchange exit temperatures in concurrent flow, *Chem. Engng* **93**(19), 92 (1986).
84Q. V. S. R. Rao and Ch. Durgaprasada Rao, Dependence of process parameters on design variables in plate heat exchangers, *Indian J. Technol.* **24**(5), 249 (1986).
85Q. P. Razelos, The optimum dimensions of convective pin fins with internal heat generation, *J. Franklin Inst.* **321**(1), 1 (1986).
86Q. E. Rev and Z. Fonyo, Additional pinch phenomena providing improved synthesis of heat exchange networks, *Hung. J. Ind. Chem.* **14**(2), 181 (1986).
87Q. G. H. Roach and R. M. Wood, Shell-and-tube exchangers having improved design features, *Heat Transfer Engng* **7**(1), 19 (1986).
88Q. P. R. Roach, Y. Takano, R. O. Hilleke, M. L. Vrtis, D. Jin and B. K. Sarma, Platinum black powder as a heat exchanger material at low temperatures, *Cryogenics* **26**(5), 319 (1986).
89Q. J. E. Roth and W. Roetzel, Dented tubes with improved heat transfer for use in heat exchangers, *Chemie-Ingr-Tech.* **58**(2), 150 (1986).
90Q. G. K. Roy, Prediction of hydrostatic head effect in process plant evaporators by nomograph, *Chem. Engng World* **21**(2), 43 (1986).
91Q. P. Rummer and K. Strohmeier, Computer-controlled dimensioning of bundled-tube heat exchangers, *Chemie-Ingr-Tech.* **58**(2), 144 (1986).
92Q. A. M. Saatci, A. M. A. Khalifa and M. Akyurt, Temperature drops in heat pipes, *Appl. Energy* **23**(1), 61 (1986).
93Q. K. Saito, T. Saito and M. Sakamoto, Development of a high efficiency heat exchanger with a packed bed for residential appliances, *Heat Transfer—Jap. Res.* **15**(4), 47 (1986).
94Q. M. Sando, A. Fujii, T. Noguchi and S. Toyama, Theoretical calculation of the thermal efficiency of an evacuated tubular collector, *Heat Transfer—Jap. Res.* **15**(3), 12 (1986).
95Q. M. Scofield and J. R. Taylor, A heat pipe economy cycle, *ASHRAE J.* **28**(10), 35 (1986).
96Q. D. P. Sekulic, Entropy generation in a heat exchanger, *Heat Transfer Engng* **7**(1), 83 (1986).
97Q. S. Seshan and D. Vijayalakshmi, Heat pipes—concepts, materials and applications, *Energy Convers.* **26**(1), 1 (1986).
98Q. E. Sheehan, Heat transfer oil boilers, *Plant Oper. Prog.* **5**(4), 193 (1986).
99Q. R. Sheikholeslami and A. P. Watkinson, Scaling of plain and externally finned heat exchanger tubes, *J. Heat Transfer* **108**(1), 147 (1986).
100Q. I. L. Shilovich, I. E. Veneraki, V. S. Mel'nichenko and Ye. N. Panov, Maximum heat flux transferrable by thermosiphons operating with sulfur–iodine mixture at temperatures of 600 to 900°K, *Heat Transfer—Sov. Res.* **18**(2), 29 (1986).
101Q. W. Simonton and M. Stone, Counterflow particle-to-particle heat exchange. Part II. Design performance and performance of a prototype, *Trans. Am. Agric. Engrs (Gen. Edn)* **29**(3), 874 (1986).
102Q. W. Simonton and M. Stone, Counterflow particle-to-particle heat exchange. Part I. Modeling and simulation for prediction of performance characteristics, *Trans. Am. Soc. Agric. Engrs (Gen. Edn)* **29**(3), 868 (1986).
103Q. E. M. Sparrow and L. G. Reifschneider, Effect of interbaffle spacing on heat transfer and pressure drop in a shell-and-tube heat exchanger, *Int. J. Heat Mass Transfer* **29**(11), 1617 (1986).
104Q. A. Stankiewicz, D. Adamska-Rutkowska, A. Cybulski and Z. Leszczynski, Hydraulic design of multitubular reactors with a heat carrier flowing in parallel to the tubes, *Chem. Engng Process* **20**(2), 79 (1986).
105Q. S. Takashima, H. Kanoh and M. Masubuchi, Dynamic analysis of a multi-concentric cylindrical heat exchanger, *Nippon Kikai Gakki Ronbunshu, B. Hen* **52**(479), 1953 (1986).
106Q. R. Turton, D. Ferguson and O. Levenspiel, Charts for the performance and design of heat exchangers, *Chem. Engng* **93**(16), 81 (1986).
107Q. R. Turton and N. Frederick, Analyze cross-flow heat exchangers, *Chem. Engng* **93**(13), 55 (1986).
108Q. P. N. Walker, E. A. Galis, H. G. Manbeck and D. J. Meyer, Wall surface heat exchanger for ventilation air, *Trans. Am. Soc. Agric. Engrs (Gen. Edn)* **29**(2), 565 (1986).
109Q. J. C. Wang and E. S. Nowak, Thermal constriction resistance at the interface of double tubes, *J. Sol. Energy Engng* **108**(1), 31 (1986).
110Q. R. L. Webb and H. Perez-Blanco, Enhancement of

combined heat and mass transfer in a vertical-tube heat and mass exchanger *J. Heat Transfer* **108**(1), 70 (1986).

111Q. H. Yamagishi, Y. Hanaoka, R. Ishiguro and T. Kumada, Heat transfer mechanism of a circular open thermosyphon Pt. I—Flow pattern observations, *Heat Transfer—Jap. Res.* **15**(6), 88 (1986).

112Q. H. Yamashita, G. Kushida and R. Izumi, Fluid flow and heat transfer in a plate-fin and tube heat exchanger (analysis of fluid flow around square cylinder situated between parallel plates), *Bull. J.S.M.E.* **29**(254), 2562 (1986).

113Q. K. K. Yau, J. R. Cooper and J. W. Rose, Horizontal plain and low-finned condenser tubes—effect of fin spacing and drainage strips on heat transfer and condensate retention, *J. Heat Transfer* **108**(4), 946 (1986).

114Q. K. Yoshikawa, S. Kabashima and S. Shioda, Studies on a high temperature regenerative heat exchanger for closed cycle MHD power generation (2nd report, two-dimensional heat transfer analysis), *Nippon Kikai Gakki Ronbunshu, B. Hen* **52**(476), 1750 (1986).

115Q. R. J. Zollweg, Thermodynamics in the design of long-life, high-temperature devices, *Rev. Scient. Instrum.* **57**(7), 1438 (1986).

116Q. Y. Zivirin, Onset of motion in a toroidal thermosyphon, *J. Engng Math.* **20**(1), 3 (1986).

117Q. Y. Zvirin, Onset of flows and instabilities in a thermosyphon with parallel loops, *Nucl. Engng Des.* **92**(2), 217 (1986).

HEAT TRANSFER APPLICATIONS—GENERAL

1S. V. S. Arpaci, Microscales of turbulence and heat transfer correlations, *Int. J. Heat Mass Transfer* **29**(8), 1071 (1986).

2S. R. R. Babayan, P. I. Retinskii, V. I. Gluschenko, A. F. Bikulov, A. P. Zhukov and N. V. Morozova, Computation of temperature fields in hybrid integrated microcircuits, *Sov. Microelectron* **15**(2), 109 (1986).

3S. R. F. Babus'Haq, S. D. Probert, M. J. Shilston and S. Chakrabarti, Optimizing the location of a district-cooling pipeline in a rectangular trench, *Appl. Energy* **23**(2), 109 (1986).

4S. K. Boulouchos and N. Hannoschoeck, Heat transfer between working fluid and combustion chamber surface, *MTZ Motortech. Z.* **47**(9), 337 (1986).

5S. A. V. Chichinadze, K. G. Shchuchev, A. A. Ryzhkin, A. I. Filipchuk, A. G. Ginzburg and M. M. Klimov, Temperature regime in tool material friction with account for three-dimensional heat release, *Soviet J. Friction Wear* **7**(3), 8 (1986).

6S. S. Chinanzvavana, W. L. Grosshandler and D. C. Davis, Analysis of heat transfer in soil during open field burning, *Trans. Am. Soc. Agric. Engrs* **29**(6), 1797 (1986).

7S. F. Conlin, W. S. Johnson and S. Wix, A TRNSYS/GROCS simulation of a horizontal coil ground-coupled heat pump, *J. Sol. Energy Engng* **108**(3), 185 (1986).

8S. R. J. Couvillion and J. G. Hartley, Drying front movement near low-intensity, impermeable underground heat sources, *J. Heat Transfer* **108**(1), 182 (1986).

9S. D. A. DeVries and J. R. Philip, Soil heat flux, thermal conductivity, and the null-aligned method, *Soil Sci. Soc. Am. J.* **50**(1), 12 (1986).

10S. B. Dorri and V. Kadambi, Thermal analysis of quench propagation in superconducting coils, *Numer. Heat Transfer* **10**(2), 165 (1986).

11S. Y. Enomoto and S. Furuhama, Heat loss into combustion chamber wall of 4-stroke gasoline engine (2nd report, heat loss into cylinder head, intake and exhaust valves), *Bull. J.S.M.E.* **29**(253), 2196 (1986).

12S. H. Foerschner, Comparison of heat transfer coefficients of various oil-cooling systems for commercial-vehicle light machines, *Elektrotechnik MaschBau* **103**(11), 535 (1986).

13S. J. Goldak, M. Bibby, J. Moore, R. House and B. Patel, Computer modeling of heat flow in welds, *Metall. Trans. B* **17**(3), 587 (1986).

14S. J. W. Griffith, R. M. McCoy and D. K. Sharma, Induction motor squirrel cage rotor winding thermal analysis, *IEEE Trans. Energy Convers.* **1**(3), 22 (1986).

15S. S. K. Griffiths, R. H. Nilson and F. A. Morrison, Jr., Hybrid analytical/numerical computation of heat transfer in a gas-driven fracture, *J. Heat Transfer* **108**(3), 585 (1986).

16S. D. D. Gvozdenac, Analytical solution of the transient response of gas-to-gas crossflow heat exchanger with both fluids unmixed, *J. Heat Transfer* **108**(4), 722 (1986).

17S. G. K. Hargrave, J. K. Kilham and A. Williams, Operating characteristics and convective heat transfer of a natural-gas-fired pulsating combustor, *J. Inst. Energy* **59**(493), 63 (1986).

18S. Y. Harigaya, S. Ohyagi, Y. Wakatsuki and M. Yoshida, Instantaneous heat flux on the piston crown of a small size direct injection diesel engine (effects of gas flow and fuel spray jet), *Nippon Kikai Gakki Ronbunshu, B. Hen* **52**(1478), 2503 (1986).

19S. M. Hartman, O. Trnka, K. Svoboda and Z. Beran, Heat transfer in full-scale coolers of rubber, *Int. J. Heat Mass Transfer* **29**(2), 257 (1986).

20S. E. Holtz and C. Skjoldebrand, Simulation of the temperature of a meat loaf during the cooking process, *J. Fd Engng* **5**(2), 109 (1986).

21S. T. V. Hromadka II, Predicting two dimensional steady state soil freezing fronts using the CVBEM, *J. Heat Transfer* **108**(1), 235 (1986).

22S. H. Imura and M. Yoshida, Latent heat storage in a unit with a circular-finned tube (2nd report, experimental study), *Nippon Kikai Gakki Ronbunshu, B. Hen* **52**(476), 1850 (1986).

23S. H. Imura and Y. Masamichi, Latent heat storage in a unit with a circular-finned tube (1st report, numerical analysis in the solidification process), *Nippon Kikai Gakki Ronbunshu, B. Hen* **52**(476), 1843 (1986).

24S. M. Ishizuka and Y. Fukuoka, Transient temperature rise for multi-chip packages, *Nippon Kikai Gakki Ronbunshu, B. Hen* **52**(476), 1772 (1986).

25S. C. E. Johnson, Evaluation of correlations for natural convection cooling of electronic equipment, *Heat Transfer Engng* **7**(1), 36 (1986).

26S. G. F. Jones, Heat transfer in a liquid convective diode, *J. Sol. Energy Engng* **108**(3), 163 (1986).

27S. T. C. Ju, H. W. Wu and C. P. Chiu, Temperature distribution on a piston for an engine with computer simulation and experiment, *Chung-Kuo Chi Hsueh Kung Ch'eng Hsueh Pao* **7**(4), 267 (1986).
28S. V. Kadambi and B. Dorri, Current decay and temperatures during superconducting magnet coil quench, *Cryogenics* **26**(3), 157 (1986).
29S. K. Kamiuto, Further design considerations for unevacuated horizontal multi-layer systems, *Appl. Energy* **23**(4), 297 (1986).
30S. K. Kamiuto and I. Kinoshita, Heat transfer performances of unevacuated horizontal multilayer insulating systems at high temperatures, *Appl. Energy* **23**(3), 155 (1986).
31S. F. A. Kamke and J. B. Wilson, Computer simulation of a rotary dryer. Pt II: heat and mass transfer, *A.I.Ch.E. Jl* **32**(2), 269 (1986).
32S. H. Klammer, R. Klima and U. Spiecker, Convective heat transfer under nozzle fields—operational experience with a preheating chamber around billets, *Stahl Eisen* **106**(4), 41 (1986).
33S. H. G. Kraus, Thermal finite element formulation and solution versus experimental results for thin-plate GTA welding, *J. Heat Transfer* **108**(3), 591 (1986).
34S. S. K. Kung and T. S. Steenhuis, Heat and moisture transfer in a partly frozen non-heaving soil, *Soil Sci. Soc. Am. J.* **50**(5), 1114 (1986).
35S. T. Kunimoto, M. Fujitysa and H. M. Shafey, The effect of paint coating on heat transfer, *Heat Transfer—Jap. Res.* **15**(6), 17 (1986).
36S. E. N. Lanskoi, Y. G. Burov and B. M. Pozdneev, Contact heat exchange in warm and hot working, *Phys. Chem. Mater. Treat.* **20**(1), 32 (1986).
37S. J. Lee, K. W. Choi, N. P. Yao and C. C. Christianson, Three-dimensional thermal modeling of electric vehicle batteries, *J. Electrochem. Soc.* **133**(7), 1286 (1986).
38S. L. Levine and J. Rockwood, Correlation for heat transfer coefficients in food extruders, *Biotechnol. Prog.* **2**(2), 105 (1986).
39S. V. I. Lokai, A. G. Karimova and V. I. Prokop'ev, Heat transfer from working fluid to axial compressor casing, *Soviet Aeronaut.* **29**(3), 94 (1986).
40S. J. McKelliget and J. Szekely, Heat transfer and fluid flow in the welding arc, *Metal Trans. A* **174**(4), 1139 (1986).
41S. K. Mitachi, H. Tachibana and K. Furukawa, Temperature distributions in the vortex flow target system for accelerator breeder (1st report, an analysis for the temperature distributions in the target and blanket irradiated with a proton beam), *Bull. J.S.M.E.* **29**(254), 2554 (1986).
42S. J. Mitsui, Y. Hori and M. Tanaka, An experimental investigation on the temperature distribution in circular journal bearings (85-Trib-40), *J. Tribol.* **108**(4), 621 (1986).
43S. M. F. Modest and H. Abakians, Evaporative cutting of a semi-infinite body with a moving CW laser, *J. Heat Transfer* **108**(3), 602 (1986).
44S. R. G. Morgan and R. J. Stalker, Shock tunnel measurements of heat transfer in a model scramjet, *J. Spacecraft Rockets* **23**(5), 470 (1986).
45S. Y. Niitsu, Analysis of the temperature distribution in a tubular specimen under high frequency induction heating, *Heat Transfer—Jap. Res.* **15**(3), 1 (1986).
46S. A. N. Noshchenko, Numerical analysis of three-dimensional nonstationary nonlinear temperature fields in the cutting zone, *Phys. Chem. Mater. Treat.* **20**(1), 29 (1986).
47S. F. Obata, K. Fujita and M. Fujii, Temperature rise of a spur gear tooth due to repeated moving heat source, *Nippon Kikai Gakki Ronbunshu, B. Hen* **52**(476), 1387 (1986).
48S. M. Obata, J. Yamaga and H. Taniguchi, Investigation of steam-cooled gas turbine blade, *Nippon Kikai Gakki Ronbunshu* **52**(480), 3098 (1986).
49S. S. Oktay, R. Hannemann and A. Bar-Cohen, High heat from a small package, *Mech. Engng* **108**(3), 36 (1986).
50S. F. J. Oppel, A. J. Ghajar and P. M. Moretti, Computer simulation of stratified heat storage, *Appl. Energy* **23**(3), 205 (1986).
51S. V. A. Ostaf'ev, Current methods of calculating temperature fields in the cutting zone, *Phys. Chem. Mater. Treat.* **20**(2), 195 (1986).
52S. A. T. Patera and B. B. Mikic, Exploiting hydrodynamic instabilities. Resonant heat transfer enhancement, *Int. J. Heat Mass Transfer* **29**(8), 1127 (1986).
53S. A. M. Pozhivanov, V. I. Dozhdikov and others, Optimizing process of continuous casting of steel by improving heat transfer in mould, *Steel U.S.S.R.* **16**(7), 324 (1986).
54S. S. Sakaguchi, H. Yamazaki and Y. Akatsu, Flow dynamics and heat transfer of an evaporator using an oil–fluorocarbon binary mixture, *Heat Transfer—Jap. Res.* **1**(2), 75 (1986).
55S. M. Salcudean, M. Choi and R. Greif, A study of heat transfer during arc welding, *Int. J. Heat Mass Transfer* **29**(2), 215 (1986).
56S. P. Santi and A. Palozzi, Heat conduction problems in a large space structure, *Acta Astronaut.* **13**(5), 217 (1986).
57S. V. P. Saxena, D. Arya and J. S. Bindra, Transient heat flow problems in dermal and subdermal tissues, *Indian J. Technol.* **24**(2), 71 (1986).
58S. Q. S. Shu, R. W. Fast and H. L. Hart, Heat flux from 277 to 77 K through a few layers of multilayer insulation, *Cryogenics* **26**(12), 671 (1986).
59S. M. S. Sodha, R. L. Sawhney, R. Verma and N. K. Bansal, Effect of finite thermal conductivity on the thermal performance of a storage medium, *Bldg Environ.* **21**(3), 189 (1986).
60S. D. W. Stamps, D. Barr and J. A. Valenzuela, A model of the heat transfer in a liquid–liquid spray column, *J. Heat Transfer* **108**(2), 488 (1986).
61S. T. Takamoto, T. Takawa, H. Tomono and K. I. Tada, Initial solidification analysis in the vicinity of miniscus in continuous casting mold, *Trans. Iron Steel Inst. Japan* **26**(7), 240 (1986).
62S. A. Ungan and R. Viskanta, Effect of air bubbling on circulation and heat transfer in a class-melting tank, *J. Am. Ceram. Soc.* **69**(5), 382 (1986).
63S. Q. C. Wang and C. R. Ferguson, Multidimensional modeling of the decay of angular momentum and internal energy in a constant volume cylindrical vessel, *J. Heat Transfer* **108**(3), 700 (1986).

64S. M. E. Wells and W. E. Lukens, Effect of forced gas cooling on GTA weld pools, *Welding Jl* **65**(12), 314 (1986).
65S. G. E. Whittle, Computation of air movement and convective heat transfer within buildings, *Int. J. Ambient Energy* 7(3), 151 (1986).
66S. H.-P. Willumeit and P. Steinberg, Heat transfer within the combustion engine, *MTZ Motortech. Z.* **47**(1), 9 (1986).
67S. G. A. Wright and K. H. Oke, Coupled heat transfer in a lighting fitting, *Appl. Math. Modelling* **10**(2), 101 (1986).
68S. H. Yamigishi, R. Ishiguro and T. Kumada, Heat transfer mechanism of circular open thermosyphons (2nd report, working fluids: water and ethylene glycol), *Nippon Kikai Gakki Ronbunshu, B. Hen* **52**(476), 1678 (1986).
69S. M. Yosida, S. Asikari and T. Gomi, On the cooling loss of the four stroke gasoline engine (1st report, the effect of gas-temperature on the heat transfer coefficient), *Nippon Kikai Gakki Ronbunshu, B. Hen* **52**(480), 3091 (1986).

SOLAR ENERGY

1T. Ch. A. Amanov, A. Batyrov, V. A. Kravchenko and A. V. Piskunova, Heat regime in the operating chamber of a combined unit, *Appl. Sol. Energy* **22**(3), 64 (1986).
2T. L. Aries, Y. Baziard, D. Fraysee and J. P. Traverse, High performance selective surfaces from stainless steel for the photothermal conversion of solar energy, *Sol. Energy* **36**(6), 521 (1986).
3T. A. K. Athienitis, H. F. Sullivan and K.G.T. Hollands, Analytical model, sensitivity analysis, and algorithm for temperature swings in direct gain rooms, *Sol. Energy* **36**(4), 303 (1986).
4T. E. Azad, F. Moztarzadeh and N. Zargham, An experimental study of a co axial heat pipe solar collector *J. Heat Recovery Sys.* **6**(3), 255 (1986).
5T. M. N. Bahadori and F. Haghighat, Thermal performance of adobe structures with domed roofs and moist internal surfaces, *Sol. Energy* **36**(4), 365 (1986).
6T. V. Bahel, R. Srinivasan and H. Bakhsh, Solar radiation for Dharan, Saudi Arabia, *Energy* **11**(10), 985 (1986).
7T. Y. Bando, M. Nishimura, M. Kuraishi, T. Kasuga and M. Hasantani, Effect of optical depth on outdoor performance of "volume heat trap" type solar collector, *Heat Transfer—Jap. Res.* **15**(1), 57 (1986).
8T. J. C. Batty, J. P. Riley and N. K. Bhise, Optimum thickness of the nonconvective zone in salt gradient solar ponds, *Sol. Energy* **36**(1), 15 (1986).
9T. R. S. Beniwal, N. S. Saxena and R. C. Bhandari, Parametric study of salt gradient solar ponds, *J. Sol. Energy Engng* **108**(1), 75 (1986).
10T. R. S. Beniwal, R. Singh, N. S. Saxena and R. C. Bhandari, Characterisation and heat losses from a laboratory salt gradient solar pond, *J. Heat Recovery Sys.* **6**(2), 105 (1986).
11T. R. S. Beniwal and R. Singh, Stability analysis for a salt gradient solar pond, *J. Heat Recovery Sys.* **6**(5), 381 (1986).
12T. A. K. Bhargava and G. Rizzi, A two-glaze solar air heater with the second glazing of variable length, *Energy* **11**(8), 765 (1986).
13T. A. K. Bhargava, A comprehensive transient analysis of a solar sand water heater, *Energy* **11**(10), 991 (1986).
14T. D. J. Close and M. K. Peck, High temperature energy storage: a method of achieving rapid charge and discharge rates, *Sol. Energy* **37**(4), 269 (1986).
15T. K. DenBraven, An analytic model of stratification for liquid-based solar systems, *J. Sol. Energy Engng* **108**(2), 105 (1986).
16T. M. A. Elhadidy, B. G. Nimmo and S. Zubair, Operation of a small-scale salt-gradient solar pond: experimental results, *J. Sol. Energy Engng* **108**(1), 55 (1986).
17T. B. W. Fintel and G. S. Jakubowski, Obtaining solar collector cover-plate transmissivities from a solar simulator, *J. Sol. Energy Engng* **108**(4), 316 (1986).
18T. N. Fukshansky-Kazarinova, W. Lork, E. Schafer and L. Fukshansky, Photon flux gradients in layered turbid media: application to biological tissues, *Appl. Optics* **25**(5), 780 (1986).
19T. R. M. S. Gama, J. A. O. Pessanha, J. A. R. Parise and F. E. M. Saboya, Analysis of a V.-groove solar collector with a selective glass cover, *Sol. Energy* **36**(6), 509 (1986).
20T. J. M. Gordon, Low-concentration CPC's for low temperature solar energy applications, *J. Sol. Energy Engng* **108**(1), 49 (1986).
21T. C. Gueymard, Monthly averages of the daily effective optical air mass and solar related angles for horizontal or inclined surfaces, *J. Sol. Energy Engng* **108**(4), 320 (1986).
22T. S. Hassid, A linear model for passive solar evaluation, *Sol. Energy* **37**(2), 165 (1986).
23T. B. K. Huang, M. Toksoy and Y. A. Cengel, Transient response of latent heat storage in greenhouse solar system, *Sol. Energy* **37**(4), 279 (1986).
24T. J. G. Ingersoll, Simplified calculation of solar cell temperatures in terrestrial photovoltaic arrays, *J. Sol. Energy Engng* **108**(2), 91 (1986).
25T. V. Joshi and V. V. N. Kishore, Applicability of steady state equations for solar pond thermal performance predictions, *Energy* **11**(8), 821 (1986).
26T. M. S. Kalashyan, O. S. Popel and S. E. Frid, Results of tests of a system of solar collectors on an experimental residence, *Appl. Sol. Energy* **22**(4), 85 (1986).
27T. M. Kamimoto, Y. Abe, S. Sawata, T. Tani and T. Ozawa, Latent thermal storage unit using form-stable high density polyethylene; Pt. I: performance of the storage unit, *J. Sol. Energy Engng* **108**(4), 282 (1986).
28T. M. Kamimoto, Y. Abe, K. Kanari, Y. Takahashi, T. Tani and T. Ozawa, Latent thermal storage unit using form-stable high density polyethylene; Pt. II: numerical analysis of heat transfer, *J. Sol. Energy Engng* **108**(4), 290 (1986).
29T. A. T. Kirkpatrick, R. F. Gordon and D. H. Johnson, Double diffusive natural convection in solar ponds with nonlinear temperature and salinity profiles, *J. Sol. Energy Engng* **108**(3), 214 (1986).
30T. V. V. N. Kishore, M. R. Ghandi and K. S. Rao, Experimental and analytical studies of shallow solar pond systems with continuous heat extraction, *Sol. Energy*

36(3), 245 (1986).
31T. A. Lazeridis, R. J. Copeland and J. Althof, Temperature distribution in a solar irradiated liquid film flowing over a solid wall, *Sol. Energy* **36**(6), 565 (1986).
32T. D. Lowrey and R. R. Johnson, Simulation of a solar pond using upward flow through the storage zone, *J. Sol. Energy Engng* **108**(4), 325 (1986).
33T. K. O'Ferrall Lund, General thermal analysis of parallel-flow flat-plate solar collector absorbers, *Sol. Energy* **36**(5), 443 (1986).
34T. T. A. Newell and P. M. Von Driska, Double diffusive effects on solar pond gradient zones, *J. Sol. Energy Engng* **108**(1), 3 (1986).
35T. A. Nozari and J. C. Lalli, A random walk approximation for a solar energy storage system, *Sol. Energy* **37**(2), 127 (1986).
36T. L. Ya. Paderin, Mathematical investigation of radiant-energy receivers based in the auxiliary-wall method, *Appl. Sol. Energy* **22**(3), 34 (1986).
37T. M. Pons and J. J. Guilleminot, Design of an experimental solar-powered, solid-adsorption ice maker, *J. Sol. Energy Engng* **108**(4), 332 (1986).
38T. M. Ya. Poz and D. Ya. Kogan, Determination of charging time for thermal-storage tanks in solar heating systems by various mathematical models, *Appl. Sol. Energy* **22**(3), 70 (1986).
39T. C. D. Rakopolous and E. Vazeos, Theoretical and experimental investigation of thermosyphon solar domestic water heaters, with indirect heating, under hot water withdrawal conditions, *Energy Convers. Mgmt* **26**(1), 95 (1986).
40T. A. Saito, Y. Utaka, Y. Saito and K. Katayama, Transient characteristics of evacuated tube collectors (1st report, the amplitude of the outlet water temperature variation and the efficiency for periodic insolation), *Bull. J.S.M.E.* **29**(255), 2993 (1986).
41T. A. Sakoda and M. Suzuki, Simultaneous transport of heat and adsorbate in closed type adsorption cooling system utilizing solar heat, *J. Sol. Energy Engng* **108**(3), 239 (1986).
42T. E. Sandru and F. Chiriac, Aspects of energy storage with application to solar energy refrigerating systems, *Bul. Stiintific* **29**(2), 11 (1986).
43T. R. M. Schotland and T. K. Lea, Bias in a solar constant determination by the Langley method due to structured atmospheric aerosol, *Appl. Optics* **25**(15), 2486 (1986).
44T. S. Shadyev, Engineering method of calculating the heat regime of the generator–absorber in a solar refrigerating unit, *Appl. Sol. Energy* **22**(3), 75 (1986).
45T. T. J. Siebenmorgen, A. C. Dale and H. W. Jones, Trombe wall solar heating system augmented with planar reflectors, *Trans. Am. Soc. Agric. Engrs* **29**(5), 1332 (1986).
46T. R. Srinivasan, V. Bahel and H. Bakhsh, Corrrelation for estimation of diffuse fraction of daily global radiation, *Energy* **11**(7), 697 (1986).
47T. K. Toiliev, Effect of air exchange on the temperature regime of a building with passive solar-heating systems, *Appl. Sol. Energy* **22**(3), 97 (1986).
48T. N. Tully, Short-term performance of integral solar water heaters, *J. Sol. Energy Engng* **108**(1), 6 (1986).
49T. F. N. Vafaie and J. R. Dunn, Forced-convection heat transfer for single and two-phase flow in a solar receiver, *J. Heat Transfer* **108**(1), 76 (1986).
50T. M. Vaxman and M. Sokolov, Effects of connecting pipes in thermosyphonic solar systems, *Sol. Energy* **37**(5), 323 (1986).
51T. D. Waksman and W. C. Thomas, The NBS solar collector reliability/durability test program: summary of results and recommendations for collector testing, *J. Sol. Energy Engng* **108**(1), 35 (1986).
52T. M. J. Witte and T. A. Newell, Transmission of solar radiation through a random medium of water and glass, *J. Sol. Energy Engng* **108**(3), 199 (1986).
53T. Q.-S. Yan, Y.-Z. Li and X.-D. Fang, Experimental research on the lattice solar wall, *Taiyangneng Xuebao* **7**(2), 163 (1986).
54T. M. Yanadori and T. Masuda, Heat transferential study on a heat storage container with phase change material, *Sol. Energy* **36**(2), 169 (1986).
55T. H.-M. Yeh and L.-C. Chen, A study on thermosyphon solar water heater with parallel flat-plate collector, *Energy* **11**(6), 579 (1986).
56T. H.-M. Yeh and L.-C. Chen, Analysis of a built-in storage type of solar water heater, *Energy* **11**(7), 727 (1986).

PLASMA HEAT TRANSFER AND MHD

1U. H . Adler and A. Piel, Experimental investigation of radially resolved plasma parameters in a low-pressure helium arc, *J. Phys. D: Appl. Phys.* **19**(8), 1477 (1986).
2U. Y. Arata and A. Kobayashi, Application of a gas tunnel to high-energy-density plasma beams, *J. Appl. Phys.* **59**(9), 3038 (1986).
3U. V. I. Artemov and O. A. Sinkevich, Numerical modeling of the interaction of an electric arc with a turbulent gas glow. Arc in a long cylindrical channel, *High Temp.* **24**(2), 220 (1986).
4U. V. I. Artemov and O. A. Sinkevich, Interaction of electrical arc and turbulent gas flow model of turbulence, *High Temp.* **24**(1), 85 (1986).
5U. J. Bacri and A. M. Gomes, Calculation of the total rate coefficients for non-equilibrium nitrogen thermal plasmas at atmospheric pressure, *J. Phys. D: Appl. Phys.* **19**(9), 1665 (1986).
6U. I. I. Beilis, Cathode arc plasma flow in a Knudsen layer, *High Temp.* **24**(3), 319 (1986).
7U. E. J. Beiting, Multiplex CARS temperature measurements in a coal-fired MHD environment, *Appl. Optics* **25**(10), 1684 (1986).
8U. L. Bilbao, H. Kelly and H. Bruzzone, Radiation transport in an ionizing current sheath, *Physics Fluids* **29**(12), 4182 (1986).
9U. A. I. Bushik, V. A. Shilov, B. Juttner and H. Pursch, Spot behaviour and local cathode heating, *High Temp.* **24**(3), 326 (1986).
10U. A. Chakrabarti, The sputtering temperature of a cooling cylindrical rod with insulated core, *Appl. Scient. Res.* **43**(2), 107 (1986).
11U. B. Chakraborty, M. Khan and B. Bhattacharyya, Intense magnetization generated by standing waves in a plasma, *J. Appl. Phys.* **59**(5), 1473 (1986).

12U. X. Chen and P. He, Heat transfer from a rarefied plasma flow to a metallic or non-metallic particle, *Plasma Chem. Plasma Proc.* **6**(4), 313 (1986).
13U. H. Chuang, Effects of Joule heating and transpiration on heat transfer in magnetohydrodynamic boundary layers, *J. Appl. Phys.* **60**(9), 3366 (1986).
14U. D. Degout and A. Catherinot, Spectroscopic analysis of the plasma created by a double-flux tungsten inert gas (TIG) arc plasma torch, *J. Phys. D: Appl. Phys.* **19**(5), 811 (1986).
15U. C. D. Devi, H. S. Takhar and G. Nath, MHD flow past a cone, *Acta Technica* **31**(4), 400 (1986).
16U. S. V. Dresvin, S. A. Mel'nik and S. M. Mikhal'kov, Investigation of heat exchange of a plasma jet with spherical bodies and a plane, *Izv. Sib. Otdeleniya Akad. Nauk SSSR Ser. Tek. Nauk* No. 4, 69 (1986).
17U. N. T. El Dabe, Magnetohydrodynamic unsteady free convective flow through a porous medium bounded by an infinite vertical porous plate, *Can. J. Phys.* **64**(1), 84 (1986).
18U. E. M. Epperlein and M. G. Haines, Plasma transport coefficients in a magnetic field by direct numerical solution of the Fokker–Planck equation, *Physics Fluids* **29**(4), 1029 (1986).
19U. R. Epstein, S. Skupsky and J. Delettrez, Effects of non-Maxwellian electron populations in non-LTE simulations of laser-plasma thermal transport and implosion experiments, *J. Quant. Spectrosc. Radiat. Transfer* **35**(2), 131 (1986).
20U. M. T. C. Fang and J. F. Zhang, The dynamic behavior of nozzle arcs, *IEEE Trans. Plasma Sci.* **14**(4), 350 (1986).
21U. A. J. D. Farmer, G. N. Haddad and L. E. Cram, Temperature determinations in a free-burning arc: III. Measurements with molten anodes, *J. Phys. D: Appl. Phys.* **19**(9), 1723 (1986).
22U. R. V. Frierson, T. L. Eddy and V. F. Put'ko, Facility for non-LTE studies of a magnetically rotated arc, *Rev. Scient. Instrum.* **57**(8), 2096 (1986).
23U. E. B. George and M. I. Yakushin, Temperature in a jet of nitrogen plasma at reduced pressures, *Fluid Dyn.* **21**(2), 335 (1986).
24U. A. Gleizes, M. Razafinimanana and S. Vacquie, Calculation of thermodynamic properties and transport coefficients for SF_6–N_2 mixtures in the temperature range 1,000-30,000 K, *Plasma Chem. Plasma Proc.* **6**(1), 65 (1986).
25U. R. J. Glumb and H. Krier, Two-dimensional model of laser-sustained plasmas in axisymmetric flow fields, *AIAA J.* **24**(8), 1331 (1986).
26U. I. Groma and B. Veto, Melting of powder grains in plasma spraying, *Int. J. Heat Mass Transfer* **29**(4), 549 (1986).
27U. G. N. Haddad, A. J. D. Farmer, P. Kovitya and L. E. Cram, Physical processes in gas–tungsten arcs, *IEEE Trans. Plasma Sci.* **14**(4), 333 (1986).
28U. N. Hata, A. Matsuda and K. Tanaka, Silane thermometry in radio-frequency discharge plasma by coherent anti-Stokes Raman spectroscopy, *J. Appl. Phys.* **59**(6), 1872 (1986).
29U. K.-P. Hinz, Optical diagnostics of SF_6 high-current discharges, *J. Phys. D: Appl. Phys.* **19**(12), 2381 (1986).
30U. P. Huhse and H. J. Reinhardt, Arc motion in vacuum circuit breakers, *IEEE Trans. Plasma Sci.* **14**(4), 460 (1986).
31U. H. Ichihashi, A. Yamanaka, R. Baba and T. Ikeda, Heat transfer during melting of titanium alloy in vacuum arc furnace by consumable electrode, *Tetsu To Hagane/J. Iron Steel Inst. Japan* **72**(6), 579 (1986).
32U. G. R. Jones, N. Y. Shammas and A. N. Prasad, Radiatively induced nozzle ablation in high-power circuit interrupters, *IEEE Trans. Plasma Sci.* **14**(4), 413 (1986).
33U. S. V. Joshi, J. Y. Park, P. R. Taylor and L. S. Richardson, Particle–gas mass transfer under plasma conditions, *Int. J. Heat Mass Transfer* **29**(10), 1565 (1986).
34U. H. Kasimir and H. J. Kusch, Plasma generation by mechanical compression of argon by piston compression of argon by piston compressor: determination of kinematic data, pressures and temperatures, *J. Phys. D: Appl. Phys.* **19**(4), 513 (1986).
35U. D. Keefer, R. Welle and C. Peters, Power absorption in laser-sustained argon plasmas, *AIAA J.* **24**(10), 1663 (1986).
36U. V. D. Khait, Theory of thermal contraction of the current at the anode in a turbulent flow of thermal plasma, *High Temp.* **24**(2), 149 (1986).
37U. V. V. Kharitonov and O. R. Yurkevich, Heating of disperse polymer materials in gas-thermal spraying, *J. Engng Phys.* **50**(3), 271 (1986).
38U. T. Kohya, M. Kuriyama, M. Toda, E. Harada and H. Konno, Heat transfer from parallel-horizontal cylinder rows under a Corona discharge, *Heat Transfer—Jap. Res.* **15**(3), 54 (1986).
39U. Yu. M. Kolyano and I. O. Goi, Generalized problem of heat conduction for a half space heated by a moving point heat source, *J. Engng Phys.* **50**(2), 233 (1986).
40U. P. Kong, T. T. Huang and E. Pfender, Synthesis of ultrafine silicon carbide powders in thermal arc plasmas, *IEEE Trans. Plasma Sci.* **14**(4), 357 (1986).
41U. Z. Kotacinski, Plasma contamination with electrode metal vapor jets, *Plasma Chem. Plasma Proc.* **6**(3), 299 (1986).
42U. T. Kotaki and O. Matsumoto, Nitriding of niobium and tantalum with a plasma jet under reduced pressure, *Plasma Chem. Plasma Proc.* **6**(1), 53 (1986).
43U. H. Krier, J. Mazumder, T. J. Rockstroh, T. D. Bender and R. J. Glumb, Continuous wave laser gas heating by sustained plasmas in flowing argon, *AIAA J.* **24**(10), 1656 (1986).
44U. Y. Kubo and H. Igarashi, Single-roller rapid quenching of high melting temperature alloys using plasma-arc melting, *J. Appl. Phys.* **60**(1), 396 (1986).
45U. J. M. Kuhlman, G. M. Molen, S. Srinivasan, S. H. Nam and S. N. Tiwari, Arc-generated flow phenomena in repetitively pulsed gas-flow spark gaps, *IEEE Trans. Plasma Sci.* **14**(3), 228 (1986).
46U. J. M. Kuhlman and G. M. Molen, Performance of high-power gas-flow spark gaps, *AIAA J.* **24**(7), 1112 (1986).
47U. D. J. Latham, Anode column behavior of long vertical air arcs at atmospheric pressure, *IEEE Trans. Plasma Sci.* **14**(3), 220 (1986).

48U. H. H. Maecker and H. G. Stablein, What keeps an arc standing in a cross flow? *IEEE Trans. Plasma Sci.* **14**(4), 291 (1986).

49U. L. Mannik and S. K. Brown, Laser Doppler velocimetry of particles in a plasma torch, *Appl. Optics* **25**(5), 649 (1986).

50U. T. Matsumura, W. T. Oppenlander, C. A. Schmidt-Harms and A. D. Stokes, Prediction of circuit-breaker performance based on a refined cybernetic model, *IEEE Trans. Plasma Sci.* **14**(4), 435 (1986).

51U. C. W. McLucas and A. I. McIntosh, Discharge heated longitudinal Sr^+ recombination laser, *J. Phys. D: Appl. Phys.* **19**(7), 1187 (1986).

52U. R. C. Mehta, Transient heat-transfer analysis of a conical cathode of an MPD arcjet, *AIAA J.* **24**(2), 346 (1986).

53U. R. Meier, F. K. Kneubuhl, R. Coccioni, H. Wyss, E. Fischer and H. J. Schotzau, Investigations of nozzle materials in SF_6 circuit breakers, *IEEE Trans. Plasma Sci.* **14**(4), 390 (1986).

54U. P. Meubus and A. Huczko, A study of chromic oxide decomposition in an RF argon plasma, *Plasma Chem. Plasma Proc.* **6**(2), 159 (1986).

55U. S. L. Milora, S. J. Combs and C. A. Foster, Numerical model for swirl flow cooling in high-heat-flux particle beam targets and the design of a swirl-flow-based plasma limiter, *Nucl. Engng Des. Fusion* **3**(3), 301 (1986).

56U. M. L. Mittal and G. P. Gowda, Transport coefficients of a chemically reacting plasma, *J. Appl. Phys.* **59**(4), 1042 (1986).

57U. K. Miyazaki, H. Inoue and Y. Toru, Heat transfer and temperature fluctuation of lithium flowing under transverse magnetic field, *J. Nucl. Sci. Technol.* **23**(7), 582 (1986).

58U. G. Nardin, D. Pozzetto and S. Tommasi, Theoretical research into thermoconvective emissions from electric arc furnaces during open vault stages, *Termotecnica* **40**(5), 31 (1986).

59U. H. R. Nataraja, M. L. Mittal and B. N. Rao, Laminar MHD compressible boundary layer at a wedge, *Int. J. Engng Sci.* **24**(8), 1303 (1986).

60U. R. Noll, C. R. Haas, B. Weikl and G. Herziger, Computer simulation of schlieren images of rotationally symmetric plasma systems: a simple method, *Appl. Optics* **25**(5), 769 (1986).

61U. H. Ouajji, B. Cheminat and P. Andanson, Composition et conductivité d'un plasma air-cuivre, *J. Phys. D: Appl. Phys.* **19**(10), 1903 (1986).

62U. S. N. Paranin, V. P. Polishchuk, P. E. Sychev, V. I. Shabashov and I. M. Yartsev, Thermal conditions in a hot evaporating cathode in a stationary vacuum arc with diffuse cathode emission, *High Temp.* **24**(3), 307 (1986).

63U. M. Pavlov, S. Djurovic and R. Kobilarov, Temperature measurement of the arc phase of a spark in air, *J. Phys. D: Appl. Phys.* **19**(6), 1041 (1986).

64U. B. M. Penetrante and E. E. Kunhardt, Kinetics of hydrogen thyratron plasmas during the conduction phase, *J. Appl. Phys.* **59**(10), 3383 (1986).

65U. R. R. Prasad, R. Consiglio and M. Krishnan, Power flow and cathode erosion in a vacuum-arc centrifuge, *IEEE Trans. Plasma Sci.* **14**(4), 498 (1986).

66U. J. Prock, Solidification of hot craters on the cathode of vacuum arcs, *J. Phys. D: Appl. Phys.* **19**(10), 1917 (1986).

67U. J. Prock, Time-dependent description of cathode crater formation in vacuum arcs, *IEEE Trans. Plasma Sci.* **14**(4), 482 (1986).

68U. A. M. Rahal, The relation between arc voltage and anode erosion in a N_2 arc at atmospheric pressure, *J. Phys. D: Appl. Phys.* **19**(3), 33 (1986).

69U. J. Ravi and D. M. Benenson, Arc mantle studies in dual flow plasmas, *IEEE Trans. Plasma Sci.* **14**(4), 477 (1986).

70U. W. C. Roman, Application of thermal plasma arc melting for critical alloy ingot preparation, *IEEE Trans. Plasma Sci.* **14**(4), 370 (1986).

71U. C. B. Ruchti and L. Niemeyer, Ablation controlled arcs, *IEEE Trans. Plasma Sci.* **14**(4), 423 (1986).

72U. R. Scholz and U. Reiners, Problems of heat transfer in the water cooling of wall elements in arc furnaces, *Tech. Univ. Clausthal* **106,** 1017 (1986).

73U. A. Sedgjomasab, T. L. Eddy, R. T. Murray and A. V. Larson, Arc facility for investigating non-local thermal equilibrium plasmas, *Rev. Scient. Instrum.* **57**(8), 2093 (1986).

74U. B. Singh and P. K. Agarwal, An alternating direction implicit method for magneto-hydrodynamic heat transfer in cylindrical geometry with discontinuity in wall temperature, *Int. J. Heat Fluid Flow* **7**(4), 307 (1986).

75U. D. I. Slovetskii, S. D. Plekhanov and V. G. Plekhanov, Plasma-electrolyte metal heating mechanism, *High Temp.* **24**(2), 277 (1986).

76U. V. M. Soundalgekar and A. G. Uplekar, Hall effects in MHD couette flow with heat transfer, *IEEE Trans. Plasma Sci.* **14**(5), 579 (1986).

77U. R. D. Taylor and A. W. Ali, Saha decrements and collisional-radiative recombination and ionization coefficients for a nonequilibrium nitrogen plasma, *J. Quant. Spectrosc. Radiat. Transfer* **35**(3), 213 (1986).

78U. A. L. Tslaf, Thermal recovery of an arc track, *IEEE Trans. Plasma Sci.* **14**(1), 24 (1986).

79U. K. Vajravelu, Hydromagnetic flow and heat transfer over a continuous, moving, porous, flat surface, *Acta Mech.* **64**(3), 179 (1986).

80U. A. Vardelle, M. Vardelle and P. Fauchais, Heat and momentum transfer between powder particulates and atmospheric plasma jet, *Rev. Int. Haute Temp. Refract.* **23**(2), 69 (1986).

81U. P. J. Wilbur and J. R. Brophy, The effect of discharge chamber wall temperature on ion thruster performance, *AIAA J.* **24**(2), 278 (1986).

82U. M. F. Zhukov, V. P. Lukashov, B. A. Pozdnyakov and N. M. Shcherbik, Electric arc in flow with well-developed turbulence, *J. Engng Phys.* **50**(3), 253 (1986).

83U. R. J. Zollweg, Radiation emission coefficients for modeling finite high-pressure thermal plasmas, *IEEE Trans. Plasma Sci.* **14,** 300 (1986).

AUTHOR INDEX

Aalders, B.G.M. 79-19Q
Aarcone, S. 86-5P
Abakians, H. 86-45A,43S
Abaltusov, V.Ye. 85-1U
Abassi, M.H. 82-1K
Abauf, N. 76-42D
Abbaoui, M. 85-2U
Abbas, M. 85-16C,2G
Abbate, G. 76-1P; 78-1K
Abdel-Khalik, S.I. 76-1T; 78-1F,1S,14T; 83-48J, 53J; 85-41DP; 86-56DP
Abdel-Latif, L.A. 85-1B
Abdel-Wahed, R.M. 76-1F; 77-27C; 78-94F,1P, 2P; 79-1A; 84-1B,2B
Abdelghaffar, M.A. 85-1J
Abdelhafez, T.A. 86-1C
Abdelmeguid, A.M. 79-1B
Abdelmessih, A.H. 78-50J; 81-1J
Abdelmonem, N.J. 86 1Q
Abdelsalam, M. 80-129J
Abdierahman, M. 79-1T
Abdul Majeed, P.M. 80-1S; 81-1S
Abdul-Hadi, M. 85-1T
Abdul-Hamayel, M.A. 85-53C
Abdulaziz, M. 85-30A
Abdulhadi, R.S. 78-2S,10S
Abdulkadir, A. 76-2L; 77-1L
Abdullah, A.R. 85-18N; 86-20N
Abe, H. 84-1JM
Abe, T. 80-38P,39P; 82-1J
Abe, Y. 86-1J,27T,28T
Abed, A.A. 83-1K
Abell, C.J. 77-21E
Abell, S. 78-46F
Aberg, S. 80-17D
Abernathey, J.R. 85-1F
Abernathy, F.H. 78-8E
Abhat, A. 79-1Q; 83-1T,2T
Abou-Ellail, M.M.M. 83-2B; 85-1G
Abousehly, A.M. 86-29P
Abramenko, A.N. 79-86J
Abramov, G.I. 81-1FF; 86-1FF
Abramov, R.T. 85-1P
Abramzon, B. 80-1C; 84-1C
Abrashin, V.N. 79-34A
Abrosimov, A.I. 86-1H
Abrosimov, Yu.G. 76-41B
Abryutin, A.A. 79-11S
Abtahi, A. 86-1A
Abu-Abdou, K. 81-2S
Abu-Zeid, M.M. 86-1E
Abuaf, N. 78-3P; 79-1P; 83-1B; 84-52J; 85-2J
Abubakar, M.S. 86-1K
Abukakar, M.S. 84-1K
Abushagur, M.A.G. 85-1K
Achard, J.-L. 85-3J
Acharya, M. 77-1C; 79-2P
Acharya, S. 81-1F,29FF; 85-2F,3F,4F,21F,1N; 86-114F,1JJ,46N
Achenbach, E. 77-2C
Achia, B.U. 77-1E
Achten, N. 86-2Q
Achterbosch, G.G. 85-1S
Achtermann, H.J. 86-15P
Acker, A. 79-2A
Ackerberg, R.C. 78-1C
Ackerman, M.C. 78-19J
Ackley, D.E. 77-1T
Ackroyd, J.A.D. 79-146F
Acquista, C. 86-51K
Acrivos, A. 76-44B,6G; 77-16D; 80-1B,2C,1E; 82-41C,40F; 86-21A,1P
Adachi, H. 85-100J
Adachi, S. 77-15J
Adachi, T. 83-8JJ; 84-3D
Adam, W. 84-37DP
Adamek, T. 81-2J
Adamek, Th. 83-1JJ; 85-1JJ
Adams, G. 78-12T
Adams, J.C., Jr. 78-1E
Adams, L. 76-1S
Adams, M.J. 81-1B,3S
Adams, R.L. 82-1D; 84-65K; 86-18DP
Adams, R.W. 82-2K
Adams, W.J. 82-24A
Adamska-Rutkowska, D. 86-104Q
Addy, A.L. 86-23D
Adebiyi, G.A. 76-1B
Adelt, M. 81-2B
Ades, M.J. 82-1A
Adesanya, B.A. 84-2S
Adlam, J.H. 86-1F
Adler, C.G. 83-7T
Adler, H. 86-1U
Adler, J. 84-18A
Adomaitis, E.I. 79-14C
Adomaitis, J.-E. 84-2C
Adomian, G. 86-2A
Adrian, F. 84-1S
Adrian, R.J. 77-1P; 86-2F,70F
Adrianov, V.N. 77-1K; 86-2K
Adzerikho, K.S. 79-1K; 80-3K; 81-1K; 85-2K

Aerov, G.Ye. 82-1B
Afanas'ev, B.A. 86-3Q
Afanas'yev, B.A. 80-1J
Afans'yev, V.N. 78-30C; 82-31C
Afaq, A. 86-2E
Afejuka, W.O. 80-1H
Affimito, D. 79-1U
Afgan, N. 77-10
Afgan, N.H. 78-6,1J; 85-4J
Afsar, M.N. 77-2K
Afzal, N. 76-1C,2C,3C; 78-2F; 80-3C,1FF; 81-2FF; 82-3C; 83-1C,1FF; 84-3C,1FF; 85-1DP,5F,1FF, 2FF; 86-2FF
Agarwal, J.P. 81-17T
Agarwal, P.K. 86-74U
Agarwal, S. 86-12H
Agarwal, V.K. 81-2T
Agdelhakim, H. 80-1U
Aggarwal, J.K. 79-1D
Aggarwalla, B.D. 77-38F
Agnone, A.M. 86-2C
Agrait, N. 86-3F
Agrawal, A.K. 81-4S
Agrawal, D.P. 79-40P
Agrawal, H.L. 80-1M
Agrawal, K.N. 82-2J; 86-2J
Agrawal, M.L. 77-1D
Agrawal, S.S. 81-72F; 82-34FF
Ah, L.L. 82-40T
Ahamdi, G. 79-3A
Ahlers, G. 80-1F; 81-73F; 82-7F; 85-97F
Ahluwalia, R.K. 80-2M; 81-1M,2M,3M,4M; 82-14M; 84-23B; 85-2JJ; 86-3J
Ahmad, T. 78-3F
Ahmadi, G. 76-2F; 77-50A; 78-1A; 80-1T
Ahmed, K. 82-12A
Ahn, B.J. 82-31B
Ahrendts, J. 85-60B
Ahrens, F. 86-2H
Aiba, S. 76-3F,1Q; 79-2D; 80-4C,1D; 81-1C; 82-1C,2C; 83-29C; 84-1Q,2Q; 85-1C,1D; 86-1D
Aihara, T. 79-1H; 83-1Q; 86-4F,4Q
Airey, D.R. 79-2U
Ait-Ali, M.A. 80-1Q
Aizawa, T. 86-1N
Ajayi, O.O. 85-2C
Ajayl, O.O. 86-5F
Akagawa, K. 82-1F
Akagi, S. 83-2FF
Akai, M. 84-54Q
Akamatsu, T. 80-38J
Akashi, K. 77-22U; 81-21U; 82-24U
Akatsu, Y. 86-54S
Akbari, H. 79-1F; 84-4T; 86-1B
Akbarzadeh, A. 82-1T; 83-40T
Akbarzedeh, A. 80-1T
Ake, T. 76-60J
Akhunov, N.Kh. 81-17K
Akidas, A.C. 80-2S
Akiev, K.A. 85-83J
Akimoto, H. 85-5J,76J
Akino, N. 86-47F
Akinrelere, E.A. 81-2C
Akins, R.C. 86-76F
Akinsete, V.A. 82-2F
Akiyev, K.A. 85-6J
Akiyoshi, M. 85-43JM,44JM
Akiyoshi, S. 86-43J
Akiyoshi, T. 80-28FF; 82-21FF
Akkarachittoor, N.S. 86-2P
Akoum, F. 79-1E
Aksel'rod, L.S. 76-4Q; 83-39J
Akyurt, M. 86-92Q
Akzamov, Sh.K. 82-8Q
Al Ali, B.M. 77-2D
Al-Arabi, M. 76-5F; 78-4F; 80-2FF; 82-1FF
Al-Astrabadi, F.R. 77-1A,2A; 79-4A
Al-Hayes, R.A.M. 81-4J
Al-Jamal, K. 84-83K
Al-Sharif, M.M. 81-32C
Alad'yev, I.T. 76-1M; 77-1B,1S,2S; 78-3S; 82-4C
Alameddine, Q. 84-83K
Alamgir, M. 79-2F,10P
Alamgir, Md. 81-3J
Alani, K. 76-35C
Alavizadeh, N. 86-18DP
Alberston, C.W. 85-62C
Albrand, G. 84-10K
Alder, C.G. 81-1T
Aleksandrov, V.V. 77-3K
Aleksashenko, A.A. 85-1A
Alekseenko, D.V. 78-6J,7J
Alekseenko, S.V. 85-3JJ
Alekseyenko, D.V. 84-6F
Alekseyev, O.N. 79-19S
Alemasov, V.Ye. 79-3P
Alessandretti, G.C. 83-2K
Alexander, R.W., Jr. 83-32K
Alexiades, V. 82-52A; 85-1JM,53JM
Alexopoulos, N.G. 78-7L
Aleyas, S. 86-6F
Aleynikov, S.M. 83-12S
Alfredsson, P.H. 84-3B
Alhomoud, A. 84-19H
Ali, A. 81-5J
Ali, A.W. 86-77U
Ali, C.L. 81-13F
Ali, M. 78-1U
Ali, S.M. 84-1J
Ali, S.M.J. 81-10FF
Alifanov, O.M. 79-5A,4P
Alikhanizadeh, M. 86-3H
Alipour-Haghighi, F. 81-8M
Alkidas, A.C. 81-6J; 82-1S,2S
Alkire, R.L. 85-91DP
Allam, T.A. 81-6K
Allan, J.C. 82-1H
Allard, F. 86-3K
Allen, R.W. 76-6D
Allmen, M.J. 84-4B
Allum, C.J. 83-1U
Almanza, R. 85-2T
Almenas, K. 84-32J
Alnajem, N.M. 85-2A,3A; 86-3A
Alohmoud, A.A. 81-7F

Alperi, R.W. 78-2A; 81-1A; 84-2J
Alpert, R.L. 77-39F; 84-1F
Alshamani, K.M.M. 78-2E
Alsmeyer, H. 82-2A
Alt, R. 86-52P
Altan, T. 78-23S
Altemani, C.A.C. 79-71A; 80-2B,21Q; 84-20H; 85-69B
Altenkirch, R.A. 82-5C,2FF
Althof, J. 86-31T
Altimir, I. 84-2F; 85-31N
Alturki, A.M. 85-106K
Amaden, A.A. 76-52J
Amano, R.S. 83-3B; 84-5B,1D,2D; 85-2B
Amano, S. 79-144F; 82-1F
Amanov, Ch.A. 86-1T
Amaouche, M. 85-3FF
Amara, N.Y. 82-26U
Amato, W.S. 76-4F
Ambal, S. 86-4C
Ambrazevicius, A.B. 78-1B
Ambrazyavichyus, A.B. 79-19B
Ambrose, C.W. 81-18Q
Amelin, A.D. 80-51J
Ametistov, E.V. 79-46J; 80-76J
Ametistov, Ye.V. 79-26J
Amitani, T. 84-3D
Amlin, D.W. 79-2K
Ammar, A.S.A. 76-39D
Amokrane, A. 86-52B,64Q
Amono, R. 85-1H
Amouche, M. 86-3FF
Amouroux, J. 81-13U; 82-16U; 85-3U
Amr, A. 76-1D
Amstrong, R.C. 82-5B
Amundson, N.R. 79-55A,56A; 80-60B; 82-35A
Anaka, H. 76-29D
Anand, N.K. 83-13B
Andanson, P. 82-1U; 86-61U
Andason, P. 85-2U
Anderson, A.C. 86-10CP
Anderson, D.R. 86-56K
Anderson, E.C. 80-13H
Anderson, N. 77-3A
Anderson, P. 76-2P; 85-9U
Anderson, R. 80-3FF; 81-4D; 83-4F; 86-7F, 19F
Anderson, R.L. 77-29P
Anderson, W.J. 77-4K
Anderson, W.K. 86-2N
Andersson, S. 86-3P
Andrachio, C.A. 85-31E
Andraka, C.E. 85-3C,2D
Andre, P. 86-2B
Andreas, E.L. 79-5P
Andreev, M.M. 76-3Q
Andreev, P.A. 78-2J
Andreev, V.K. 79-4J; 85-70B
Andreopoulos, J. 80-5C; 83-2C,3C
Andreopoulous, M. 86-4FF
Andrews, D.G. 76-40J
Andrews, G.E. 84-1H; 86-3H
Andreyev, A.P. 82-3J
Andrizhiyevskiy, A.A. 79-3J
Ane, J.M. 86-4K
Aneja, V.P. 86-5Q
Ang, T.K. 77-4A
Angelie, A. 83-1JM
Anikin, A.I. 85-7J
Anisimov, S.B. 86-101J
Anisin, A.K. 77-8Q
Ankudinov, G.A. 82-23JJ
Annamalai, N. 84-2H
Ansari, A. 85-4C
Ansari, F.A. 85-2P; 86-2E,4P
Ansari, M.A. 83-62F,37FF; 84-47B,66F; 85-108K; 86-37FF
Anselmet, F. 85-5C
Antar, B.N. 80-1A
Antipov, V.I. 85-87J
Antipov, Yu.N. 83-3K
Antonenko, V.A. 79-79J,80J,21Q; 83-66J; 86-123J
Antonetti, V.W. 85-4A
Antonia, R.A. 76-1E,2E; 77-3C,2E,3E,26E; 78-2C, 3E,4E,5E,21E,22E; 79-9C,24E,28E; 80-6C,1P; 81-45B,3C,24C,29C; 82-40C,45C; 83-17C,3D; 84-14C; 85-5C,20C,3D,26E; 86-8C
Antonini, G. 78-5F; 83-4B
Antonio, L. 84-1U
Antonishin, N.V. 76-46D; 78-3J
Antonopoulos, K.A. 85-6C,1Q,2Q; 86-3B,4B
Antonovskii, V.I. 77-5K
Antonovskii, V.K. 79-1S
Antoranz, J.C. 79-3F
Antsulevich, V.I. 79-1K
Anufriev, A.A. 76-3P
Anzai, S. 84-39J
Aoki, A. 77-15J
Aoki, H. 85-79J
Aoki, I. 84-24C
Aoki, K. 83-2JM; 86-82F,1JM
Aoki, S. 80-3B; 82-4J; 86-5J,6J,94J
Aoshima, R. 84-2K
Appalaswamy, A.V. 80-4FF
Appel, P.W. 76-2D
Appleby, J.F. 84-67K
Apshtein, E.Z. 86-5K
Apter, D. 82-3K
Arai, N. 77-1R; 78-1R; 79-6A; 82-23K; 83-18K; 84-19B,3H,12S; 85-30DP,53K; 86-59K
Arakawa, E.T. 78-19K
Arakeri, V.H. 80-7C
Araki, K. 76-42A; 85-73J
Araki, N. 76-4P; 85-5A,6A,3K
Aralov, A.D. 82-6C
Aramaki, H. 86-5D
Aranake, R.N. 78-92F
Arata, Y. 86-2U
Archambault, P. 77-5A
Archarya, S. 86-45F,46F
Arconada, A. 85-14T
Arduini, C. 85-4K
Are, A.L. 77-3S,4S
Aref'yev, K.M. 79-1M
Arefi, F. 85-3U
Aries, L. 86-2T

Ariga, I. 79-27C
Arimilli, R.V. 86-4A
Aristov, V.V. 80-2P
Aritomi, M. 80-3B; 81-30C; 82-4J; 86-4J,5J, 6J
Arkhipov, L.I. 78-5J
Arkhipov, V.V. 79-4J; 80-2J; 85-70B
Arledge, R.G. 79-7A
Armaly, B.F. 77-6K,7K; 83-30C; 84-24FF; 85-44FF, 45FF,64FF; 86-9FF,41FF,42FF
Armaly, F. 86-8FF
Armstong, R.L. 84-3K
Arnas, D.A. 86-23B
Arnas, O.A. 85-3B,6F; 86-24B
Arnaud, G. 82-14A; 84-13DP; 85-36FF
Arndt, D.P. 84-4K
Arnold, G.S. 84-5K
Arnold, J.N. 76-6F,8T; 77-1F; 79-36F
Arnold, J.O. 85-23C
Arnon, O. 78-4P
Aroesty, J. 79-22C
Aronov, I.Z. 84-15Q
Aronova, N.A. 78-12D
Arora, C.P. 76-18A
Arpaci, V.S. 76-49F; 86-6K,1S
Arras, M. 85-46Q
Arsen'yev, L.V. 77-1H
Artemov, V.I. 86-3U,4U
Arter, W. 85-7F
Arthurs, A.M. 77-3A
Arts, T. 85-4H,5H
Artyukhin, E.A. 85-7A
Artyukhin, Ye.A. 79-5A
Arunachalam, M. 79-64A
Arunasalam, V. 79-19P
Arya, D. 86-57S
Arya, S.P.S. 77-4E
Asadi, A. 86-29B
Asaeda, T. 85-8F
Asai, H. 84-2K
Asako, Y. 78-70F; 82-3F,4F,51F; 84-6K; 86-38F
Asakura, Y. 78-4J
Asano, K. 82-1JJ; 84-1JJ; 85-27J; 86-19J
Asano, T. 84-1JM
Asere, A.A. 86-3H
Asgarpour, S. 80-3T; 82-7C
Asghar, S.M. 76-34B
Ash, R.L. 76-4R; 80-46P
Ashmantas, L.A. 85-3Q
Ashton, G.D. 80-25B
Ashworth, D.A. 85-7C
Ashworth, T. 86-6P
Asikari, S. 86-69S
Aslanov, P.V. 79-34B
Asmalovskiy, V.A. 79-20S
Asmantas, L.A. 84-44Q
Asrar, W. 85-8C
Assassa, G.M. 79-1C
Assassa, G.M.R. 86-3C
Astashenkova, G.G. 78-3C
Astill, K.N. 76-2B; 77-16A; 83-35B; 84-56F
Astrom, A. 86-2H
Astronov, D.N. 76-5P
Atalia, S.R. 86-29P
Athanasiadis, G. 85-2N,3N
Athanassouli, G. 83-3T
Athey, R.E. 76-2T
Athienitis, A.K. 85-3T; 86-3T
Atimtay, A. 81-1H
Atkinson, J.D. 76-1E
Atkinson, J.F. 83-4T
Atluri, S.N. 85-22A
Attanasio, A. 76-1P
Attar, A. 86-1JJ
Attard, P. 86-7K
Atten, P. 79-56F; 84-10F
Attia, A.E. 84-1B,2B
Attia, M.H. 86-3E
Attwater, C.R. 80-32T
Audunson, T. 76-7F; 80-14P
Auerbach, I. 77-1U
Auguste, J.-C. 79-4B
Aung, S. 85-7D
Aung, W. 78-11G; 79-50P; 83-1D,2D; 84-4D; 85-4D; 86-8F,9F
Auracher, H. 81-81J
Auriault, J.L. 83-1A
Ausman-Mudawwar, J.R. 85-75J
Authier, B. 80-2T
Autsulevich, V.I. 81-1K
Auxilien, J.M. 81-2F
Avdeev, A.A. 86-7J
Avdonin, N.A. 86-3N
Avduyevskiy, V.S. 79-4F
Avedisian, C.T. 81-7J,8J; 82-5J
Avery, D.E. 85-9C,10C
Avery, R.K. 82-4K
Avgerinos, C. 83-6B; 84-6B
Awad, M. 79-6D
Aya, I. 86-2JJ
Aya, S. 86-4Q
Ayeni, R.O. 82-2B
Ayer, R.M. 77-67A
Ayers, D.L. 81-1D
Aynshteyn, V.G. 77-13B
Aytimur, T. 82-8C
Ayyaswamy, P.S. 78-5H; 80-9U; 81-20J,21J; 84-3JJ,4JJ; 85-25JJ
Azad, E. 85-4Q,5Q; 86-6Q,4T
Azad, F.H. 81-2K,3K; 85-5K
Azad, R.S. 77-10E,11E,38P
Azarkhin, A. 86-5A
Azarskov, V.M. 85-64J
Azer, N.Z. 78-23J; 80-3J
Azevedo, L.F. 84-65F
Azevedo, L.F.A. 85-9F,54FF,55FF; 86-10F,108F, 19H
Azia, N. 81-1L
Aziz, A. 76-1A,9B,75F; 77-6A; 79-8A; 80-2A, 3A; 81-2A,3A; 82-3FF
Azrak, R.R. 86-5B
Azzam, R.M.A. 76-6P

Baba, K. 84-35C

Baba, R. 86-31U
Baba, T. 85-1E
Babak, T.B. 86-7Q
Babak, V.N. 86-7Q
Babayan, R.R. 86-2S
Babayev, F.K. 86-38B
Babenko, V.Ye. 77-2P
Babezba, A.V. 81-3FF
Babich, V.I. 85-88J
Babu, D.K. 79-9A
Babus'Haq, R.F. 85-10F; 86-11F,3S
Babuska, I. 82-1N
Bache, T.C. 76-38A
Bachner, F.J. 76-9T
Bachtel, T. 86-60C
Bachu, S. 84-20T
Back, L.H. 82-3S
Backharone, E. 77-10Q
Backstrom, G. 76-2P; 86-3P
Backx, E. 76-4C
Baclic, B. 76-64A
Baclic, B.S. 85-6Q
Bacon, G. 81-4H; 85-18DP
Bacri, J. 85-4U; 86-5U
Badie, J.M. 82-14A
Badr, H.M. 83-3FF; 85-4FF
Badri Narayanan, M.A. 77-5E
Baeckstreom, G. 86-67P
Baer, A.D. 77-2R
Bahadori, M.N. 86-5T
Bahar, E. 85-6K; 86-8K,9K
Bahar, F. 85-4Q
Bahel, V. 86-6T,46T
Bahrami, P.A. 80-44FF,45FF
Baierlein, R. 81-62F
Baikov, V.I. 76-82F
Bailey, D.A. 77-11H
Bailey, M.J. 86-78K
Bailly, A. 86-30B
Baines, R.P. 84-3J
Baines, W.D. 76-5C,7P; 77-74F; 83-1F,2F
Bair, S. 86-22E
Bairamov, R. 76-1J,3T
Baird, J.P. 80-14C
Baitonde, U.N. 85-60C
Bajorek, S.M. 82-5F
Bajura, R.A. 80-38FF
Bakan, S. 85-7K
Baker, C.G.J. 77-11D
Baker, L. 76-6B,2S; 78-24J
Baker, L., Jr. 76-7S; 80-16FF,39J; 84-14J; 86-6JM
Baker, P.W. 83-4K
Baker, R.R. 80-9P
Baker-Jarvis, J. 85-8A
Bakhsh, H. 86-6T,46T
Bakhtar, F. 85-3P
Bakhvalov, B.Yu. 81-13H
Baklastov, A.M. 78-2B,5J,1Q
Bakshi, A.S. 76-50J
Bal'tsevich, Ya.A. 79-15K
Balabanov, B.S. 86-58P
Balachandra Rao, S. 82-24D,26D
Baladi, J.Y. 81-1D
Balageas, D.L. 86-6A
Balakeyevskiy, Yu.I. 80-4J; 82-2JJ
Balakrishna, A. 79-3K
Balakrishna, M. 80-4B
Balakrishnan, A. 85-11C
Balankin, S.A. 79-6P
Balashov, V.V. 86-22B
Balashova, I.E. 86-7A
Balcomb, J.D. 79-35T
Balenko, N.Ye. 86-30P
Balestra, M. 86-39N
Baliga, B.R. 77-50B; 78-54B,26Q; 80-2D; 86-5B, 54N
Balik, E.A. 77-3P
Ball, K.S. 86-42F
Ballal, D.R. 80-2E
Ballik, E.A. 79-13P
Ballintijn, J.F. 78-43B
Balmer, R.T. 78-3A
Balobayev, V.T. 77-5S
Baltatu, M.E. 85-4P
Bamberger, M. 86-8A
Banarjee, A.K. 86-4C
Band, Y. 83-18K
Banda, E. 86-31P
Bando, Y. 82-23K; 84-19B,79K; 85-81K; 86-7T
Bandrowski, J. 76-8F; 77-2F,3F; 78-1D; 80-8E; 81-9J
Banerjee, B. 82-1G
Banerjee, S. 77-10E,11E,14J; 79-15F,16F,7P; 81-11F,14J,15J,16J; 83-2JJ; 85-52J; 86-92J
Bank, N. 77 4P
Bankoff, S.G. 76-57J,2M; 77-1,4; 78-63B; 80-5J, 116J; 81-31J,32J,46J; 83-31J,32J,42J,5JJ; 84-19DP,33J,13JJ; 85-112J,12JJ; 86-51J
Banks, P.J. 77-22Q; 84-3Q; 85-7Q,8Q
Bankston, C.P. 82-3S
Bankston, J.D. 77-8K
Bannerot, R.B. 78-1T; 79-2T
Banno, I. 86-37Q
Bansal, J.S. 80-8C
Bansal, N.K. 83-30T; 84-1T; 85-35T; 86-59S
Bansk, P.J. 84-9Q
Banthiya, N.K. 78-2F
Bapu Rao, M.N. 80-4A
Bar-Avraham, E. 79-1U
Bar-Cohen, A. 77-4F; 81-55A; 84-3F; 85-8J, 10Q; 86-8J,49S
Bar-Yoseph, P. 82-2G; 84-2G
Barabash, P.A. 78-60J
Baran, M. 84-4Q
Baranenko, V.I. 80-6J,7J
Baranov, A.V. 83-14B
Baranovskii, B.V. 79-46B
Barantsevich, V.L. 85-9Q
Baratt, P.J. 81-3F
Barbaro, S. 79-3T
Barber, A.R. 79-8P
Barber, J.R. 76-2A; 81-11A,12A; 83-2A,12A, 13A,26A; 84-1A,3S; 86-5A,8Q,9Q
Barber, P.W. 84-45K
Barclay, D.W. 86-52A

Barclay, W.H. 78-13C
Bard, S. 84-7K
Barduhn, A.J. 85-23JM
Bardylo, V.M.I. 78-5P
Bareiss, M. 80-5A; 83-24JM; 84-2JM
Barelko, V.V. 81-98J; 83-78J,79J
Barker, R.L. 85-5P
Barker, S.J. 81-5C
Barkov, Uu.I. 79-37F
Barkovskaya, Ye.N. 83-37A
Barley, C.D. 79-4T
Barnea, D. 83-5B
Barnea, Y. 86-35FF
Barnes, E.R. 76-3A
Barnes, G.T. 86-9J
Barnes, H.B. 80-3P
Barnes, R.M. 80-5U
Barnett, M. 86-4N
Barnette, D.W. 76-2R
Baron, A. 85-4D
Baron, J.R. 76-7K
Barone, M.R. 85-2S
Baronnet, J.M. 80-34U
Barozzi, G.S. 84-2A,3A,15B; 85-4B
Barr, D. 86-60S
Barr, P.K. 83-14C
Barra, O.A. 79-5T
Barrault, M.R. 76-19U,21U; 79-31U
Barreto, G.F. 85-2DP; 86-1DP
Barrow, H. 83-6B; 84-6B; 86-9A,5C
Barschdorff, W. 84-29JJ
Bart, G.C.J. 86-10A,2JM
Barta, E. 85-3S
Bartkus, S.I. 79-16C,17C
Bartkus, S.L. 76-16C
Bartlett, K.G. 76-8P
Bartolini, R. 83-1J
Bartolomey, G.G. 85-9J
Barton, N.G. 86-46A,47A
Bartz, J.A. 83-20Q
Barua, S.K. 86-7P
Baryshev, Yu.V. 79-2H
Basdevant, C. 86-5N
Base, T.E. 81-2D
Bashkatov, N.V. 76-45P
Bashtovoi, V.G. 77-9A
Bashtovoy, V.G. 79-2M
Basilico, C. 81-3D; 84-1DP
Basiulis, A. 84-13Q
Baskir, D.I. 78-4S
Baskov, V.S. 82-23JJ
Bass, B.R. 76-35A
Bassiouni, M.K. 80-5L
Bassiouny, M.K. 85-11Q,12Q; 86-10Q
Basu, B. 86-50JM
Basu, D.K. 76-2J
Batalov, V.S. 85-2E
Bathaiah, D. 85-49U
Bathe, K.-J. 82-14N
Bathelt, A.G. 79-10A,5F,5J; 80-6A; 81-4A
Bathke, E.A. 79-82A
Batina, J. 86-2B
Batischev, V.A. 85-11F
Batozskii, Yu.V. 85-78K
Batten, C.E. 85-8K
Battistelli, E. 85-9K; 86-10K
Batty, J.C. 86-8T
Battya, P. 84-4J
Batyrov, A. 86-1T
Bau, H.H. 81-4F,5F; 82-3A,4A,6F,70F,6J; 83-11A, 3F; 84-2DP,3DP,4DP; 85-79B; 86-49F
Bauder, N.H. 78-5U
Bauer, H.F. 76-3B; 80-5B; 82-3B; 85-12C,12F; 86-11A,12A,6B
Bauer, K. 80-9C
Bauer, R. 76-3D; 78-2D,3D
Baughn, J.W. 77-2T; 78-3B; 80-19P; 84-7B, 8B; 85-3E; 86-2D,4E,5E
Baumann, F. 86-1B
Baumeister, K.J. 77-1J
Baumeister, P. 78-4P
Baunchalk, R.A. 86-41DP
Bayat, F. 86-49B
Bayazitoglu, Y. 79-145F,4K; 80-24F,25F,71F, 13K,3T; 82-29A,7C; 85-26JM; 86-71F
Bayev, A.R. 79-37F
Baykov, V.I. 77-92F,93F
Bayle, M. 85-5U
Bayle, P. 85-5U
Bayley, F.J. 85-68C
Bayly, B.J. 85-13C
Bazanov, V.A. 79-77F
Baziard, Y. 86-2T
Bazinet, J. 82-13N
Beam, J.E. 85-74Q
Bean, B.L. 79-24K
Bean, J.W. 84-19S
Beard, J.N., Jr. 84-2S
Beard, J.T. 78-2T
Beardsley, R.C. 79-1G
Beasley, D.E. 84-5DP,12P; 85-4T
Beasley, J.D. 86-94P
Beatenbough, P.K. 78-7Q
Beccali, G. 85-4S
Beck, A.E. 85-5S
Beck, B.T. 81-10J; 86-10J,11K
Beck, G. 79-18J
Beck, J.V. 77-26A; 79-11A,63A; 81-5A,6A,26A; 82-2N; 84-4A; 86-13A,14A,55N
Beck, M.W. 84-1P
Becker, H.B. 77-10K; 84-96K
Becker, K.M. 81-11J
Becker, M. 86-100J
Becker, R. 80-1K
Becker, R.I. 82-34A
Beckermann, C. 86-12F
Beckett, P.M. 80-7A,6B; 84-4F
Beckman, W.A. 76-13T; 81-18T; 82-29T; 85-40T
Beckwith, P.J. 82-5K
Becky, T.J. 81-42B
Bedeaux, D. 78-11D
Bedenek, S. 76-3J
Bedford, R.E. 76-1L; 80-24P; 85-10K
Bedinger, A.F.G. 85-33T
Beeman, J.W. 86-122K
Beenakker, J.J.M. 82-18A; 86-40J

Beer, H. 78-2H; 80-5A,18J; 81-44A,14D,51F, 56J; 82-37A; 83-13H,24JM; 84-2JM; 85-75F, 42FF,43FF; 86-3JM,36JM
Beery, J.G. 86-74P
Beese, E. 79-2C
Beevers, C.E. 77-7A
Began, A. 85-20H
Beghin, P. 81-4FF
Begoulev, P.B. 86-26B
Beguier, C. 78-6E
Behan, A. 79-2S
Behbahani, A. 83-2H
Behbahani, A.I. 82-4H; 85-24DP
Behee, R.D. 79-31C
Behnia, M. 79-6F; 82-73F,54K
Behnia, M.M. 76-78F
Behringer, R.P. 81-62F; 82-7F; 86-8P
Beilis, I.I. 86-6U
Beisswenger, H. 81-5S
Beiting, E.J. 86-7U
Bejan, A. 77-8A,1Q; 78-6F,7F,8F,9F,10F,2Q, 5S; 79-2E,7F,8F,9F,10F; 80-30F,3FF; 81-4C, 4D,6F,7F,8F,9F; 82-38C,8F,79F,4FF; 83-30A, 24C,1DP,2DP,19DP,20DP,4F,31F,43F,44F,45F, 23FF; 84-6DP,28DP,7F,41F,2FF,22FF; 85-3DP, 42DP,68DP,89DP,55F,26FF,9T; 86-61DP,72F
Belen'kii, M.Ya. 86-27J
Beljaars, A.C.M. 78-7E,25E
Bell, G.E. 76-4A; 78-4A; 79-12A; 84-3JM
Bell, K.J. 80-101J; 82-21B
Bell, R.J. 83-32K
Bell, S.E. 83-32K
Bellamy-Knights, P.G. 80-1G
Bellet, D. 86-77B
Bellmore, C.P. 83-7B
Bellow, W.J. 86-3D
Beloborod'ko, V.T. 77-21S
Belostotskii, M.V. 86-15A
Belov, I.A. 79-4U; 80-27D; 82-9C
Belov, V.M. 78-30C
Belozerov, V.I. 79-104F
Belskaya, E.P. 82-50A
Belward, J.A. 83-4C
Belyakov, I.I. 85-13Q
Belyakov, V.P. 86-11Q
Belyanskii, L.B. 76-5P; 77-5P
Belyayev, V.M. 80-59B
Belytschko, T. 85-15N
Belyy, A.M. 82-15S
Bemrose, C.R. 83-3Q
Ben-Amoz, M. 76-5A
Ben-Asher, J. 85-31S
Benard, C. 82-2T; 85-2JM,5T
Bendell, M.S. 76-9F; 79-42F
Bender, R.L. 84-18C
Bender, T.D. 86-43U
Benderskaya, S.L. 80-35B
Benedek, S. 85-10J
Benedict, B.A. 84-20T
Benedict, R.P. 78-6P
Benenson, D.M. 76-1U; 77-21U; 80-8U; 84-22U; 86-69U
Benet, J.C. 82-2D; 83-3DP
Benim, A.C. 86-6N
Benisty, K. 79-27U
Beniwal, R.S. 86-9T,10T,11T
Benjamin, M.K. 78-22Q
Benner, D.C. 84-25K; 86-93K
Benner, J. 85-1B
Benner, R.E. 86-58K
Benner, S.M. 85-30P; 86-9P
Bennett, H.E. 80-2K
Bennett, H.S. 76-1K,2K; 78-6S
Bennett, J.C. 79-9P
Bennon, W.D. 81-5FF
Benodekar, R.W. 78-4B
Bensler, H.P. 86-50J
Benson, S.V. 80-19F
Bentley, J.M. 78-7S
Bently, R.E. 86-6E
Bentwich, M. 83-3A
Benvensite, Y. 86-17A
Benzerga, L. 85-6U
Benzies, J.Y. 76-1A
Beran, Z. 86-19S
Berardi, P.G. 85-14C
Berbano, M.C. 77-10D
Berdahl, P. 83-5K; 84-8K,2T,16T; 85-11K
Berdnikov, V.S. 79-76F; 84-17FF
Berezhinskii, R.A. 78-5B
Berezovsky, A.A. 84-20FF
Berg, J.C. 78-41F; 82-14J; 85-22J
Berg, R.F. 84-5Q; 86-10P
Berge, P. 78-27F,28F
Bergeat, J. 85-96K
Bergeles, G. 76-1H,3H; 77-3H; 78-1H
Berger, A.E. 79-60A
Berger, F.P. 77-2B
Berger, S.A. 76-66A; 78-105F; 79-22C
Berger, X. 84-3T
Berghams, J. 76-4J
Berghmans, J. 76-5J
Bergholz, R.F. 78-11F; 79-11F; 80-2F
Bergles, A.E. 76-20B,21B,40F; 80-30B; 81-20B, 33J; 82-7J,31J; 83-18Q; 85-50Q; 86-67J
Bergles, G. 81-2H
Bergman, T.L. 82-5FF; 83-6K; 85-13F; 86-13F, 14F,15F,5FF
Bergmann, H.W. 80-24S
Bergougnou, M.A. 77-11D
Bergquam, J.B. 77-2T
Bergstrom, R.W. 76-22S
Berkel, H. 84-5F
Berkov, A.B. 85-17Q
Berkovskii, M.A. 85-111K
Berkovskiy, B.M. 79-3M
Berkovsky, B. 78-1
Berkovsky, B.M. 76-10F; 77-9A
Berkowitz, D.A. 78-25S
Berkuta, A.D. 77-5C
Berlin, I.I. 86-58J
Berman, A.S. 76-14F
Berman, L.D. 79-6J; 82-8J
Berman, Y.S. 84-41Q
Bernard, C. 86-16A
Bernard, J.E. 79-13A

Bernini, U. 76-1P; 78-1K
Bernstein, M. 77-6C
Berry, G.A. 84-3S
Berry, G.F. 81-5M
Berry, J.T. 84-35A
Berry, R.A. 79-31C; 80-16C; 82-22B
Berry, R.D. 78-9H
Berryman, B. 86-12Q
Bershader, D. 82-40F
Bertela, M. 85-18K
Bertin, H. 86-16F,7N
Bertin, J.J. 76-1R,2R; 78-2R; 82-10C
Bertram, L.A. 83-27U; 85-108F
Bertrand, L. 82-28A
Berzodnyy, M.K. 84-6F
Berzoy, S.E. 86-7C
Beschastnov, S.P. 79-104F
Beschkov, V. 78-4C
Besley, L.M. 77-37P; 84-1E; 86-7E,8E
Besse, J.C. 81-3H
Bessem, J.M. 76-35P
Bessler, W.F. 80-4T
Best, C.H. 79-8H
Best, R. 78-2H; 81-6S
Bestman, A.R. 78-7P; 82-3D; 86-12K
Betta, V. 79-7J
Bettarini, A. 86-52K
Betts, D.B. 85-12K
Betts, K.H. 85-6S
Betts, R.C. 84-48J
Betzel, Th. 86-3JM
Beukema, K.J. 82-2H; 83-4DP
Bevilaqua, P.M. 77-8E
Beyar, R. 85-3S
Beyerlein, S.W. 82-4S
Bezdenezhnykh, N.A. 86-17F
Bezrodnyi, M.K. 78-6J,7J; 86-3JJ
Bezrodnyy, M.K. 80-9J; 83-2J
Bhandari, R. 86-13K
Bhandari, R.C. 86-9T,10T
Bharali, A. 82-1M; 83-1M,2M
Bharathan, D. 84-5J,4S
Bhargava, A.K. 85-15T; 86-12T,13T
Bhargava, R. 85-5B,15C,2H
Bhat, A.M. 83-3J,4J; 85-11J; 86-11J
Bhat, A.N. 80-3M
Bhatkar, V.P. 77-10A
Bhatt, B.L. 77-56J; 80-10J,11J
Bhattacharya, A.K. 82-33C
Bhattacharya, D.K. 83-5C
Bhattacharya, M.C. 85-4N; 86-8N
Bhattacharya, S.C. 76-4D
Bhattacharyya, B. 86-11U
Bhattacharyya, S.N. 85-14F
Bhattacharyya, S.P. 85-16C,2G; 86-56F
Bhattacharyya, T.K. 85-5N
Bhatti, A. 84-4D
Bhatti, M.S. 77-3B,4B; 78-6B; 84-9B
Bhawalkar, D.D. 79-20P
Bhise, N.K. 86-8T
Bhowmik, N.C. 85-6T
Bhushan, B. 86-22E
Bhushan, R. 83-8B
Bhutta, B.A. 85-6N
Bianchini, A. 76-6C
Bibby, M. 86-13S
Bich, E. 86-102P
Bier, K. 76-11F; 81-12J; 82-21J; 84-6J
Biery, J.C. 81-1Q
Biggerstaff, M.I. 86-69F
Bikulov, A.F. 86-2S
Bilbao, L. 86-8U
Bilicki, Z. 83-5J
Bill, R.G., Jr. 79-12F; 86-38K
Billerey, B. 86-30K
Bin, A.K. 83-6C
Binder, A. 85-17C
Binder, H. 84-7J
Bindra, J.S. 86-57S
Biney, P.O. 86-11P
Bingle, J.D. 85-24J
Binkele, L. 86-12P
Biot, M.A. 78-4D
Bird, R.E. 82-3T; 83-5T
Biringen, S. 78-5C
Birkebak, R. 77-1L; 79-10P
Birkebak, R.C. 76-2L; 82-12K
Birkhoff, R.D. 77-6P
Birnboim, A. 83-25J
Bishop, A.A. 77-1
Bishop, E.H. 86-79F
Bishop, K.A. 85-20T
Bishop, R. 83-15A
Bismarck-Nasr, M.N. 78-6C
Biswas, D.R. 77-3T
Biswas, S.N. 78-45P; 84-3P; 85-21P
Bitzer, R. 79-19P
Bixler, N.E. 86-2DP
Biyikli, S. 85-12J
Biykili, S. 85-14Q
Bjerle, I. 80-17D
Bjorge, R.W. 82-9J; 84-2JJ
Bjurstrom, H. 84-4P
Blachowikz, R.A. 80-12J
Black, S.H. 82-43J; 85-2J
Black, W.Z. 76-20P; 81-5H; 83-14DP
Blackburn, P.R. 80-2U
Blackwelder, R.F. 76-3E; 79-3E
Blackwell, B.F. 81-1N; 85-6B
Blagoveshchensky, A.J. 85-40Q
Blair, M.F. 77-11H; 79-30C; 80-19C,4H; 83-7C, 8C; 85-18C
Blaise, J.C. 79-3D
Blake, K. 84-7F
Blake, K.R. 84-6DP
Blakebrough, N. 78-8S
Blanchard, A. 85-19K
Blanchard, R.C. 86-6C
Blasinski, H. 77-2Q; 81-3B
Blay, D. 78-76F
Blech, J.J. 82-2G; 86-1G
Blenke, H. 77-16C; 78-27C,32P
Blickensderfer, R. 77-11K
Bligh, T.P. 78-1P
Block, J.A. 80-13J
Bloem, W. 86-7B

Blokh, A.G. 76-3K; 80-3K
Blomquist, R. 85-48P
Bloss, S. 78-53F,9S
Blums, E.Ya. 79-8M
Blythe, P.A. 81-5D; 83-5DP,5F; 86-9DP
Blyumin, Ya.I. 77-4C
Bobba, K.M. 85-34C
Boberg, T. 86-2F
Bobeth, M. 82-5A,6A
Bobisud, L.E. 83-4A,9C
Bochagov, V.N. 84-11C
Bock, H.-J. 83-6DP; 85-4DP
Bock, H.J. 80-3E; 86-3DP
Bode, K.-H. 80-4P
Bodea, M. 77-59A
Bodeep, G.E. 85-4E
Bodenheimer, J.S. 80-4K
Bodonyl, R.J. 79-3C
Body, Y. 82-2T; 85-5T
Boedeker, L.R. 84-43K
Boehly, T. 84-105K
Boehm, R.F. 77-5F; 80-70J; 86-13Q
Boessneck, D.S. 83-38FF
Boffi, V.C. 77-12K; 79-5K
Bogachev, V.A. 85-15F
Bogachev, V.S. 79-72J; 85-5E
Bogachev, V.V. 77-36B
Bogachuk-Kozachuk, K.A. 77-17H; 78-16H
Bogatyrev, G.P. 86-18F
Bogolepov, V.V. 85-19C
Bogolyubov, Yu.Ye. 79-4C
Bogomolov, E.N. 78-3H
Bogomolov, Ye.N. 85 3H
Bogoslovskii, V.E. 86-51Q
Boguslawski, L. 86-11G
Bohm, U. 79-37B; 81-7D
Bohn, M. 86-7F,66F
Bohn, M.S. 85-15Q; 86-19F
Bohn, U. 83-9D
Bohne, C. 83-43K
Bohnet, M. 85-16Q
Bohnisch, S. 84-34J
Bohnstedt, W. 76-56A
Bohon, H.L. 77-3R
Bohren, C.F. 85-13K
Bohrer, T.H. 78-76J
Boisvert, R.F. 85-37F
Boivin, L.P. 78-8P
Bokhari, A. 80-3U
Bokov, A.Ye. 80-42J
Bokova, L.F. 83-79J
Boles, M.A. 83-3JM; 84-30A; 85-38JM,86K
Boley, B.A. 78-5A
Bolhn, M.S. 84-8F
Bolle, L. 76-6A
Bologa, M.K. 78-8J; 80-2Q; 81-25J; 82-9Q; 83-6J,7J; 85-17Q; 86-7C
Bolton, E.W. 84-9F; 85-16F; 86-20F
Bon Mardion, G. 86-48J
Bonacina, C. 79-8J
Bondarenko, V.N. 83-35Q
Bondi, P. 85-5DP; 86-13P
Bonet, C. 81-7U
Bonin, E.V. 80-4U
Bonn, W. 80-7B,14J
Bonniaud, M.L. 79-35F
Bonouvrie, H.J. 77-27P
Bontous, P. 85-90F
Bontoux, P. 78-80F; 79-112F; 81-20F; 86-22F, 6FF
Booker, J.R. 78-12F; 82-64F
Boom, R.W. 79-23S; 80-19FF
Boon-Long, P. 79-13F
Boothroyd, S.A. 84-9K; 86-14K
Boots, H.M.J. 80-25K
Borchevkin, Yu.S. 84-6Q
Borde, I. 80-1C
Boreham, B.W. 86-14P
Boreman, G. 84-24K
Borgers, T.R. 79-1F; 84-4T
Borgogno, J.P. 84-10K
Borhani, M.A. 79-21E
Bories, S. 80-3F; 81-4H; 85-23F
Borishanskaya, A.V. 82-10J,20J,1Q; 85-35J
Borishanskiy, V.M. 79-1M; 82-3JJ,1Q; 83-2Q
Borishansky, V.M. 82-3J
Borishnanskiy, V.M. 78-9J
Borisov, S.F. 84-43C
Borisova, R.D. 78-7B
Borkakati, A.K. 82-1G,1M; 83-1M,2M
Borkowska-Pawlak, B. 85-17F
Borman, V.D. 82-2M
Borodin, A.S. 76-7C,6J
Borodulya, V.A. 83-7DP,8DP,7K; 84-7Q
Borovskiy, V.R. 79-5C; 85-5D
Borroto, A.N. 85-41B
Borsch-Supan, W. 84-11K
Borshanskiy, V.M. 79-9J; 82-10J
Borthwick, A. 86-9N
Bosch, W.A. 86-9E
Bose, A. 84-8Q
Bose, T.J. 80-10C
Bose, T.K. 77-9U; 79-6C; 81-50A; 82-43A; 84-2U; 85-7U; 86-15P
Bosgra, L. 86-93P
Bosmans, L.D. 83-43B
Bott, T.R. 76-2Q; 77-3Q; 83-3Q
Botten, L.C. 84-70K
Botterill, J.S.M. 77-3D; 78-5D
Bottoni, M. 85-18Q
Boubnov, B.M. 86-2G
Boules, M. 80-3U
Boulet, C. 85-25K,63K; 86-28K,29K
Boulos, M. 83-2U; 84-24U; 86-37P
Boulos, M.I. 76-2U; 78-2U; 80-5U; 83-25U; 85-37U,46U
Boulouchos, K. 86-4S
Boumehdi, L. 85-19Q
Boura, A. 76-12F
Bourdin, E. 83-2U
Bourne, D.E. 76-4B
Bourrely, C. 86-15K
Bousgaribies, J.L. 80-58B
Boutin, C. 85-15G
Boutoux, P. 86-21F
Bowen, H.D. 86-8B

Bowers, C.G. 86-8B
Box, M.A. 81-10T; 86-7K
Boxhidai, S.I. 85-25E
Boxman, R.L. 78-13F; 79-122F; 81-28FF; 83-11U
Boyack, B.E. 85-7S
Boyadjiev, C. 78-4C
Boyd, D.A. 76-4T
Boyd, D.R. 85-8S
Boyd, R.D. 81-6FF; 83-6F
Boyer, E. 76-4S
Boyko, A.N. 86-9B
Braaten, M. 86-10N
Braaten, M.E. 85-18F; 86-11N
Bradley, D. 78-1U
Bradley, E.F. 79-44P
Bradley, R.P. 86-47E
Bradshaw, P. 76-37C; 77-7E; 78-52C; 79-41C; 80-5C
Bragin, B.K. 77-7P
Bragina, O.M. 85-20Q
Bragina, O.N. 78-37B; 82-2Q
Branch, S.J. 80-18H
Brand, K.P. 82-2U,18U
Brandeis, J. 82-3H
Brandemuehl, M.J. 84-9Q
Brandon, H.J. 76-3R
Brandt, F. 82-49C; 84-25P
Brandt, H. 82-15Q
Brandt, R. 86-16P
Brannen, D. 79-9U
Branover, H. 79-4M
Bransier, J. 79-14A; 86-72N
Brassington, D.J. 77-5B; 78-8B
Bratukhin, Iu.K. 83-7F
Braud, K.G. 86-1JJ
Brauer, H. 78-4H; 79-7C,4E,22H; 81-4B
Braun, J.E. 81-3T; 85-30T
Braun, M.J. 85-9A
Braun, R. 83-8K
Braunlich, P.F. 86-1A
Braunstein, M. 77-21P
Brause, W.B. 83-16DP
Brebbia, C.A. 81-68A
Breber, G. 80-15J
Brecha, R.J. 84-55K
Breitkoff, G. 80-5P
Breitkopf, G.E. 80-3S
Bremhorst, K. 77-8P; 78-9P; 81-5B; 85-6E; 86-27E
Brenier, B. 86-22F
Brennan, P. 76-16R
Brennen, C.E. 82-36D
Brentley, J.P. 84-2P
Brenza, P.T. 86-30C
Brett, M.J. 85-6S
Bretzlaff, R.S. 85-14K
Breugelmans, F.A.E. 85-35E
Brewster, B.S. 84-2E
Brewster, M.Q. 82-6K; 84-12K; 86-16K,17K, 89K
Briggs, D.G. 76-4L; 85-19F
Brigham, B.A. 84-4C
Brine, D.T. 83-6T
Bringans, R.D. 77-13K
Brinsfield, W.A. 83-5JM
Briskman, V.A. 86-17F
Brison, J.F. 79-11P
Britt, J.P. 84-24K
Britter, R.E. 79-5E; 81-4FF; 86-64C
Britton, B. 86-76P
Britz, D. 79-28E
Broadbent, J.A. 82-53A; 83-28JM
Brocard, D.N. 76-3S
Brode, H.L. 84-64F
Brodkey, R.S. 80-7G
Brodov, Yu.M. 78-10J; 79-67J
Brojahn, U. 81-51F
Bromley, L.A. 76-55J
Bronson, J.P. 82-22B
Brook-Levinson, E.T. 85-5E
Brooks, B.W. 80-12J
Brooks, W.F. 86-41E
Brophy, J.R. 86-81U
Brosh, A. 79-123F; 82-11C
Brosmer, M.A. 85-15K,16K
Brosseau, T.L. 78-31S
Broughton, J. 76-5D; 77-2D
Brovkin, L.A. 85-9S
Browand, F.K. 79-8C
Browand, F.W. 79-17E
Browell, E.V. 79-6K
Brown, A. 78-7C,14F,10P; 79-3H; 82-12C; 85-18H
Brown, G. 86-49B
Brown, L.F. 76-11D
Brown, M.E. 84-1P
Brown, R.A. 77-46J; 85-9JM
Brown, R.G.W. 86-18K
Brown, S.K. 86-49U
Brown, S.N. 78-15F; 86-20DP
Browne, L.W.B. 79-9C; 83-3D; 86-8C
Browne, W.D. 81-8S
Bruce, C.W. 83-9K
Bruce, R.H. 76-9P
Brucker, G.G. 77-2J
Brucker, J. 83-13FF
Bruges, E.A. 76-10P
Bruggemann, G. 80-4U
Bruin, R. 83-18F
Bruin, S. 77-44F; 82-2H; 83-4DP
Bruk, V.A. 85-7B
Brumfield, L.K. 76-8C
Brundrett, E. 77-11P; 78-15P
Brunier, E. 83-4B
Brunn, P.O. 82-13C
Bruscaglioni, P. 85-9K; 86-10K
Bruschi, L. 84-5P; 86-10E
Brusse, J.C. 76-36P
Bruun, H.H. 85-7E
Bruzzone, H. 86-8U
Bryant, G. 86-7K
Bryant, H.C. 83-8T
Bryczkowski, A. 83-4JM
Bryers, R.W. 78-2
Bryukhovskikh, O.A. 86-32K
Bubelis, E.S. 79-27Q; 84-11D
Bubenzer, A. 84-13K

Bubnov, V.A. 76-7A,37A
Buccini, A. 86-18A
Buchberg, H. 76-13F,5T; 77-14T
Buchko, N.A. 79-3S
Buchlin, J.-M. 83-38T
Buck, G.M. 86-6C
Buckholz, R.H. 86-18JM
Buckius, R.O. 77-14K; 78-2K; 79-19T; 80-5K; 82-7K,8K,36K; 83-24K; 86-70K,102K
Buckus, R.O. 84-91K
Buddhavarapu, J. 85-91C
Budny, R.S. 78-30P
Budov, V.M. 86-4JJ
Budrik, V.V. 86-11Q
Budwig, R. 85-8B
Bueb, M. 84-10Q
Buell, J.C. 83-4FF,1G; 86-9C
Buettner, D. 85-99K,6P
Buevich, A.S. 85-9B
Buevich, A.V. 79-70J
Buevich, Yu.A. 85-6DP,20F
Bufetov, N.S. 83-12H
Buglaev, V.T. 76-3Q
Buhler, K. 78-36P; 82-9F,3G
Bui, M.N. 85-5FF
Bui, R.T. 85-10N
Bui, T.D. 85-14J,15J
Buizza, A. 86-13P
Bukovskaya, O.I. 79-5S
Bulanyi, P.F. 81-1U
Bulavko, A.A. 85-5E
Buleev, N.I. 85-10B
Bulik, S. 85-3JM
Bull, M.K. 76-4E
Bullock, K.J. 78-8E; 85-8E
Bundikul, S. 79-3B
Bunditkul, S. 76-52F
Bung, R. 76-8A
Bunker, R.S. 86-5JJ
Burak, L.D. 76-3K
Burattini, E. 83-10K
Buratyuski, E.K. 86-12N
Burch, D.M. 76-6D
Burch, J.N. 85-12G
Burch, T. 85-21F
Burde, S.S. 77-73A
Burdukov, A.P. 79-12P; 82-11J,12J; 83-12H
Buretta, R.J. 76-14F
Burfoot, D. 84-10B
Burgan, J.R. 81-9N
Burge, D.K. 80-2K
Burgemeister, E.A. 77-6S
Burger, M. 82-62J
Burgess, G. 85-7N
Buriot, D. 84-3T
Burka, A.L. 78-3K; 86-4DP
Burkhard, D.G. 78-3T,4T
Burkhard, D.R. 78-3T
Burley, V.D. 77-30B
Burlingame, R.S. 84-11Q
Burmeister, L.C. 77-24J; 78-11J; 79-15A; 82-4JJ; 85-20T
Burmistrov, A.V. 80-17J
Burns, A. 78-3Q; 79-25Q
Burns, P. 86-48F
Burns, P.J. 77-6F
Burnside, B.M. 85-82J
Burov, Y.G. 86-36S
Burow, P. 80-18J
Burton, D. 86-5DP
Burton, R.C. 78-7C
Busby, H.R. 82-23N; 85-8N; 86-81N
Bush, A.W. 79-16A
Bush, M.B. 84-34B
Bushik, A.I. 86-9U
Bushnell, D.J. 82-15FF
Bushnell, D.M. 79-4H; 86-10B
Bushnell, R.H. 79-6T
Businger, J.A. 77-52F
Busse, C.A. 80-3Q
Busse, F.H. 77-24F,25F; 78-22F; 79-14F,56F; 80-4F; 81-10F,15F; 82-10F; 83-17F,2G; 84-9F; 85-16F,22F; 86-20F
Butler, G.A. 78-18P
Butskii, N.D. 77-56B
Butterworth, D. 82-60J
Buttner, D. 84-14K; 85-41P; 86-17P
Button, B.L. 84-3E,4H
Buttyan, J. 85-23S
Buxmann, J. 85-21Q
Buyevich, Yu.A. 76-7D; 82-13J; 83-9DP
Buyukbicer, E.F. 86-11E
Byrd, J.W. 81-1T; 83-7T
Bystrov, P.G. 81-32B
Bystrov, P.I. 83-4Q; 85-22Q
Bywater, R. 84-6JM

Cabelli, A. 77-4T
Cachorro, V.E. 85-7T
Cadman, T.W. 78-1L
Cahela, D.R. 85-64Q
Cailleteau, J. 78-5U
Cain, B. 86-19K
Cairnie, L.R. 82-6FF
Cairns, D.N.H. 77-5B
Cal, Z.H. 84-12Q
Calafell, D.O., II 77-9T
Caldwell, D.A. 85-42P
Caldwell, D.R. 76-15F
Caledonia, G.E. 85-17K
Calfee, R.F. 77-39P
Calia, C. 82-5JJ
Calka, A. 85-16J
Calkins, R.L. 84-9A
Callahan, G.D. 76-16F
Calm, J.M. 82-18S
Calman, J. 85-6FF
Caltagiorone, J.P. 76-17F; 80-23F,56F
Caltagirone, J.P. 79-18F,92F; 80-5F,6F; 85-23F
Caltagirone, M.P. 78-40B
Camarda, C.J. 81-4Q; 84-13Q
Camarero, R. 84-7N; 86-73N,95N
Camci, C. 85-4H,5H,23H
Camera-Roda, G. 85-18K
Cames-Pintaux, A.M. 86-4JM

Campayne, F.I. 86-29E
Campbell, C.W. 80-33P
Campo, A. 76-9A,3L; 77-11A; 79-4B,5B; 83-5A; 84-14F; 86-10C
Camus, J.-J. 84-11S
Can, M. 84-7H
Cane, R.L.D. 77-8F; 80-40A
Cann, M.W.P. 85-19K
Cannell, D.S. 76-9P; 85-97F
Cannon, J.N. 86-43F
Cannon, J.R. 78-6A
Canon, R.M. 76-7J
Cantor, B. 85-15JM
Capelli, M. 84-3U
Capp, S.P. 79-44F
Cappuccio, G. 83-10K
Caps, R. 84-14K; 85-20K,99K,41P; 86-20K,17P
Carajilescov, P. 76-5B; 84-35Q
Carbajo, J.J. 86-12J
Carbonell, R.G. 85-45DP,46DP
Cardella, A. 85-10S
Carey, G.F. 85-9N
Carey, V.P. 77-9F; 78-16F; 80-5FF,6FF; 81-7FF; 82-7FF,26FF; 83-5FF,6FF; 84-6P; 85-15DP, 41FF,48FF,17J,4JJ,23Q; 86-31FF,13J,14J,5JJ
Carleson, T.E. 82-14J; 85-12D
Carli, B. 77-15K
Carlomagno, G.M. 77-73F; 86-7FF
Carlson, A.B. 84-42C
Carlson, B. 84-4P
Carlson, C.K. 86-36FF
Carlson, R.D. 79-19J; 81-26J; 86-28J
Carlucci, L.N. 85-72Q
Caroll, M.M. 86-18P
Carpenter, B. 85-11B
Carpenter, J.R. 76-4L
Carpenter, K.J. 78-11Q
Carpenters, K.J. 76-6Q
Carper, H.J., Jr. 86-4H
Carra, S. 85-4JM
Carragher, P. 86-21K
Carranco Ortiz, M. 82-32D
Carratelli, E.P. 79-5T
Carrier, G.F. 85-11S
Carrigan, C.B. 86-2DP
Carrigan, C.R. 83-2G
Carroll, J.J. 83-37T
Carroll, M.B. 86-19A
Carter, R.E. 84-18C
Cartigny, J.D. 86-22K,125K
Cary, A.M., Jr. 79-4H
Casanova, J. 85-75C
Casanova, J.L. 85-75C,7T
Casarella, M.H. 85-7JJ
Casarosa, C. 83-1S
Casas-Vasquez, J. 83-23A
Casper, C. 86-15J
Caspi, S. 77-10F; 83-9B
Cassulo, J.C. 80-16FF,39J
Casteleijn, T.P. 85-70DP
Castellanos, A. 84-10F; 86-3F
Castiglia, F. 82-33A
Castillo, J.L. 85-24F
Castinel, G. 77-11F
Castle, G.S.P. 83-32A
Castle, J. 80-34T
Castle, J.N. 77-5J
Catherinot, A. 86-14U
Cathro, K.J. 84-6T
Catipovic, N.M. 82-4D
Caton, J.A. 81-7S
Caton, J.S. 83-10B
Catton, I. 76-6F,13F,35F,8T; 77-4D,1F,110F, 5J; 78-79F,106F,107F,108F; 79-142F,75J; 80-51FF; 81-24J; 82-45F,61F; 83-4FF,1G,5JM; 84-11F, 17J,56J; 86-9C,17DP,75F
Catton, V. 86-75F
Caughey, D.A. 86-12N
Caulk, D.A. 83-6A; 85-2S
Caussade, B.-H. 76-10A
Caussade, B.H. 77-12A
Cavany, L.H. 81-13S
Caver, M.B. 86-11B
Cavvadias, S. 85-3U
Cazzoli, E.G. 86-55J
Cebeci, T. 76-9C,10C; 78-8C; 79-10C,11C,12C; 85-5FF
Celata, G.P. 85-18J; 86-6JJ
Celentano, A. 79-31P,32P
Celere, S.W. 86-19H
Cena, V. 77-4Q
Cengel, Y. 84-15K
Cengel, Y.A. 84-16K,17K; 85-21K; 86-23T
Centeno, M. 82-4T
Cercignani, C. 85-19J
Ceren, E. 83-25K
Cermak, J. 85-92DP
Cernik, P. 85-12S
Cerutti, E.A. 86-21D
Cerza, M. 85-20J
Cesar, C.L. 86-69P
Cesna, B.A. 84-2C
Cess, R.D. 76-4K; 78-10
Cetas, T.C. 78-11P
Ceylan, H.T. 80-8A
Cha, C.K. 84-12F; 85-23T
Cha, Y.S. 81-13J; 83-8J; 85-8S
Chaboki, A. 82-7G,16S; 84-44B,20H
Chaddock, J.B. 86-17J
Chahine, G.L. 84-8J; 85-21J
Chakrabarti, A. 79-13C; 86-20A,10U
Chakrabarti, S. 85-6K; 86-11F,9K,3S
Chakraborty, B. 86-11U
Chakravarthy, S.R. 86-70N
Chakravarti, P.C. 85-70C
Chakraverty, A. 86-28P
Chaksh, S.A. 80-4Q
Chalmers, I.D. 79-5U
Chamberlain, M.J. 85-7FF
Chambers, A.J. 78-3E; 80-1P; 81-3C; 82-7H; 85-20C,26E
Chambers, T.L. 77-6E
Chan, A.M.C. 79-15F,16F; 81-14J,15J,16J
Chan, C. 80-6P
Chan, C.K. 82-6JJ,24JJ
Chan, C.Y. 76-11A; 85-48B

Chan, J.H. 79-13P
Chan, K. 84-18K
Chan, K.L. 77-16K
Chan, P.K. 86-68K
Chan, P.W. 76-5U
Chan, P.W.C. 86-55P
Chan, R.K.Y. 85-7P
Chan, S.H. 78-37K; 79-7K; 80-19J,6K,7K,8K, 10K; 82-15J,50K; 83-6JM
Chan, S.K. 76-3U,4U; 79-6U
Chan, W.S. 78-12P
Chan, W.T. 79-26C
Chan, Y.L. 85-25F,26F; 86-23F
Chan, Y.T. 81-11F
Chanda, O.P. 86-13G
Chandra, A. 80-4M
Chandra, R. 78-34A
Chandra, S. 85-13S
Chandran, P. 86-3G
Chandran, R. 80-3D; 85-7DP
Chandrasekhara, B.C. 85-8DP,9DP
Chandrashekar, M. 85-24N,3T
Chandratilleke, G.R. 86-44J
Chandratilleke, T.T. 81-22J
Chandrupatla, A.R. 77-6B; 78-9B
Chang, C. 77-12F
Chang, C.-Y. 85-12B; 86-12B,13B
Chang, C.E. 76-9J
Chang, C.H. 85-51DP
Chang, C.I. 83-1H
Chang, C.J. 86-24F
Chang, E. 86-21A,1P
Chang, H.N. 85-10G
Chang, I.-D. 79-24F
Chang, I.D. 76-19F; 83-10DP
Chang, J.-S. 78-7A; 84-4U
Chang, J.S. 83-2JJ
Chang, K.-K. 80-5T
Chang, K.C. 78-8C
Chang, K.S. 82-12F; 83-8F; 85-8P
Chang, L.C. 83-12K
Chang, L.S. 82-14J; 85-22J
Chang, M.C. 78-76J
Chang, M.H. 77-49B
Chang, M.W. 80-8B
Chang, S.-C. 86-13N
Chang, S.C. 85-8T
Chang, S.H. 85-10DP
Chang, S.L. 83-11K
Chang, S.S.-H. 79-17F
Chang, T.-H. 84-5JJ; 85-5JJ
Chang, T.C.A. 86-31C
Chang, T.Y. 83-1H
Chang, W.S. 85-24Q
Chang, Y. 83-9J,10J,73J
Chang, Y.-M. 85-6JM
Chang, Y.P. 76-42B; 77-13A,14A,15A; 78-33A; 79-17A,50A,51A; 80-18A; 83-16A; 84-23A
Channabasappa, M.N. 83-11B; 84-11B
Chao, A. 86-34DP
Chao, B.T. 76-15P; 78-56F
Chao, P. 81-14F
Chao, P.K.B. 83-9F
Chao, Y.-T. 84-22S
Chaplinskiy, S.I. 78-60J
Chapman, A. 78-3U
Chapman, C.J. 80-7F
Chapman, R. 85-8U
Chapman, T.W. 78-10B
Charach, C. 86-5JM
Charach, Ch. 85-5JM
Characklis, W.G. 81-2Q; 84-49Q
Charalampopoulos, T.T. 82-15K
Charan, V. 85-2P; 86-105J
Charette, A. 83-22U; 85-10N
Charles, M.E. 84-48J
Charmchi, M. 80-67B,68B,73B,25T; 82-11F; 83-61F
Charnay, G. 76-5E; 79-11P
Charpentier, J.-C. 80-5D
Charpentier, J.C. 85-77DP; 86-68Q
Charrier-Mojtabi, M.C. 79-18F
Charters, W.W.S. 82-1T,15T
Chary, S.P. 76-10J
Chashchin, P.I. 76-11J
Chassaing, P. 76-11C; 77-9P
Chastain, S.R. 81-44B
Chato, J. 78-10S
Chato, J.C. 78-2S; 79-6B; 84-5A,42F; 85-28S; 86-15Q
Chattopadhyay, T.K. 86-22A
Chattree, M. 85-23J
Chaturvedi, S.K. 80-17Q; 86-14B
Chau, T.S. 80-66B
Chaudhary, D.R. 85-39A
Chaudhuri, K.D. 86-87P
Chaudran, R. 76-8J
Chaussee, D.S. 85-29N
Chauvin, G. 81-7U
Chavepeyer, G. 76-68F; 77-83F
Chawla, O.P. 81-3Q
Chawla, S.S. 83-3G; 84-3FF; 85-8FF
Chawla, T.C. 79-7K; 80-20J,6K,7K,8K,9K,10K; 81-5K; 82-15J; 84-4JM,1N; 85-11DP,12DP,24J
Chayka, V.D. 80-28J
Chee, K.K. 76-11P
Cheema, L.S. 76-22T; 84-5T
Cheesewright, R. 78-17F
Chegini, H. 86-14B
Chekhovich, V.Yu. 82-2JJ
Chekhovskoi, V.Ya. 80-7P
Chelghoum, D.E. 85-9T
Chemicat, B. 85-9U
Cheminat, B. 82-1U; 85-2U; 86-61U
Chen, B. 82-25T
Chen, B.H. 82-4B,8C
Chen, C.-J. 81-6B,3N; 83-10C
Chen, C.-K. 83-9Q; 84-22A; 85-19FF,20FF,21FF, 31FF,6JM,65K; 86-73C,25F,23FF,24FF,43FF, 6G,42JM
Chen, C.C. 76-18F; 79-19F; 81-17J,5Q; 84-14Q
Chen, C.F. 77-13F; 78-18F; 80-42F,43F; 82-68F; 84-13F; 85-27F; 86-40FF
Chen, C.H. 79-20F; 84-4FF
Chen, C.J. 77-10P; 79-20F,21F
Chen, C.K. 86-68A,17H
Chen, D.M. 80-6U,7U; 81-2U,3U

Chen, E. 76-29T
Chen, F.F. 80-69F; 81-38FF
Chen, H.Y. 83-23U
Chen, J. 80-58F
Chen, J.C. 80-3D,95J; 82-44J; 84-12J; 85-7DP, 37J,61J; 86-16J
Chen, J.H.C. 77-3P
Chen, J.I. 78-13P
Chen, J.J.J. 83-12B
Chen, J.L.S. 78-9C; 83-11C,7FF
Chen, K. 78-109F; 81-50B; 83-8FF; 84-5C; 85-25Q
Chen, K.C. 79-29E
Chen, K.H. 77-5Q,6Q
Chen, K.S. 86-6DP,105F
Chen, L.-C. 86-55T,56T
Chen, L.-T. 78-12J
Chen, L.D. 82-7JJ
Chen, L.W. 78-95J
Chen, M.M. 79-11F; 81-80J; 83-8FF; 84-41B, 5C; 85-26T; 86-6JM
Chen, P. 80-20FF; 84-7DP; 85-13DP
Chen, P.H. 85-14H
Chen, P.J. 76-12A
Chen, R. 86-23K
Chen, S. 85-10K; 86-22JM
Chen, S.-L. 79-34P
Chen, S.C. 83-13B
Chen, S.J. 84-6JJ,7JJ
Chen, T.S. 76-62F,3H; 77-14F,70F,71F; 78-66F, 67F; 79-22F,23F,93F; 80-7FF,8FF,9FF,10FF, 29FF,47FF; 81-24FF; 82-8FF,49K; 83-30C,9FF; 84-24FF; 85-44FF,45FF,64FF; 86-8FF,9FF,41FF, 42FF
Chen, X. 83-3U,4U,5U; 86-12U
Chen, Xi. 82-3U,4U; 85-10U,11U
Chen, Y.-M. 85-21C
Chenchu Raju, V.C. 77-15G
Cheng, K.C. 76-46B,20F,21F,22F,65F,95F,96F; 77-7B,57B,19F,79F; 78-11B,12B,13B,20F,34F, 48F; 79-23C,24C,141F; 81-7A,24FF; 85-7JM; 86-33FF
Cheng, K.J. 86-17J
Cheng, P. 76-19F,60F; 77-15F,16F,17F,18F, 56F; 78-19F,45F,51F; 79-24F,58F,109F; 80-24C; 81-12F,13F,18J,19J; 83-10DP; 84-8DP,5FF; 85-34DP,51DP,63DP,18FF; 86-7DP,15DP,29DP, 31DP
Cheng, R.K. 82-14C; 85-22C
Cheng, S.C. 77-3J; 78-13J,14J; 81-63J; 85-25J; 86-35J
Cheng, S.M. 81-10A
Chenoweth, D.R. 85-13B; 86-26F
Chenoweth, J.M. 83-18Q
Chepurnoi, M.N. 85-6JJ; 86-18J
Chepurnoi, V.M. 85-6JJ
Chepurnoy, M.N. 77-5C
Cherepanov, V.Ya. 83-13K
Cherkasov, S.G. 86-27F
Chermisinoff, N.P. 79-14J
Chern, S.Y. 79-6B,28C; 81-21FF
Chern, Y.-T. 86-12B
Cherney, W. 81-69A,18N
Cherng, J.C. 76-14T
Chernika, I.M. 81-23B,44J
Chernina, I.M. 83-62J
Chernobay, V.A. 76-35J; 77-36J; 78-51J,52J; 79-55J,56J; 80-21J
Chernobrivets, V.Ya. 86-48Q
Chernomurov, F.M. 85-26Q
Chernyak, V.G. 84-12B
Chesna, B.A. 76-57C; 79-14C
Chesters, A.K. 78-15J
Cheung, F.B. 76-6B; 77-12B,20F; 78-21F; 79-11F; 80-8F,9F,11FF,8K,10K; 81-8A,5K; 82-8B; 83-7A; 84-4JM,7JM; 85-8JM; 86-28F
Cheung, T. 77-23C
Cheung, T.K. 82-7H; 85-28DP
Cheung, Y.K. 82-21N
Chevaillier, J.-P. 86-24K
Cheverev, V.G. 83-37A
Chevray, R. 78-9E,26E
Chevrier, J.C. 77-5A
Chew, J.W. 85-3G,4G,5G
Chi, S.W. 76-1
Chiam, M.F. 82-5T
Chiang, H.D. 85-13H
Chiang, T. 79-19J; 80-13A; 81-26J
Chiang, T.-L. 86-15B
Chiapetta, P. 86-15K
Chiba, S. 86-1JM
Chiba, T. 83-36FF; 84-2K
Chiba, Y. 86-7JJ
Chichinadze, A.V. 86-5S
Chichkan, L.A. 80-7J
Chida, K. 84-5JM; 86-7JM
Chida, T. 79-143F
Chidester, D.H. 86-45E
Chien, L.-C.L. 86-8DP
Chien, Y.K. 80-8U
Chieng, C.C. 79-7B; 80-4D
Chigier, N.A. 78-48P
Child, T.N. 80-18T
Childs, E.P. 84-6C
Chilukuri, R. 80-83B
Chin, J.S. 84-9J
Chinanvavana, S. 86-6S
Chiou, J.N. 76-12C
Chiou, J.P. 78-4Q,5Q; 79-25B,95F; 80-30FF; 82-6T
Chiou, J.S. 81-6B
Chipot, M. 83-7JM
Chiranjivi, C. 85-14B
Chiriac, F. 86-42T
Chirita, F. 86-14Q
Chirkashenko, E.P. 86-4DP
Chirkov, I.N. 83-2Q
Chisholm, D. 79-6Q; 85-27Q
Chistyakov, F.M. 79-10J
Chistyakov, Yu.A. 79-14P
Chiu, C.P. 86-27S
Chiu, P.B. 77-4R
Chiu, T.-M. 83-11DP; 85-14DP
Chizhov, V.E. 85-13P
Chmielewski, C.E. 77-1J
Cho, C. 79-11J
Cho, C.H. 82-12F; 83-8F

Cho, D.H. 76-14J; 80-19J,8K,10K; 83-6JM
Cho, S.H. 81-9A
Cho, S.M. 82-14Q,14S
Cho, Y.I. 79-15P; 81-7B; 82-17B
Chohan, M.A. 86-41P,79P
Chohan, R.K. 86-12E
Choi, B.C. 85-22K
Choi, C.K. 77-31F; 82-34F; 83-48B
Choi, H.K. 76-2M; 83-6U
Choi, I.G. 80-10F; 82-15C
Choi, K.W. 86-37S
Choi, M. 86-55S
Choi, S.H. 79-5M
Choi, Y.-H. 85-51B
Chong, M.S. 79-38P
Chongrungreong, S. 80-22J
Choo, Y.K. 76-7B
Chopra, K.L. 79-15T
Chou, C.H. 83-11G
Chou, F.C. 86-29F
Chou, S.C. 80-15H
Chou, T.-W. 86-67A
Chou, T.W. 86-49A
Chou, Y.S. 78-4K
Choudhary, M.K. 86-8JM
Choudhuri, S.K.R. 84-1G
Choudhury, G. 82-7M
Choudry, A. 81-8FF
Chow, J.C.F. 77-49B; 79-36B; 82-30B; 84-39Q; 85-50B
Chow, L.C. 77-6F; 81-17N; 83-11J; 84-29B, 14F
Chow, S.P. 84-6T
Chowdhury, K.L. 86-23A
Chrisholm, D. 78-6Q
Christensen, A.N. 81-17U
Christensen, J.J. 76-12P
Christensen, L.S. 82-13M
Christensen, R.N. 83-28F; 85-52F
Christiansen, E.B. 77-21F
Christianson, C.C. 86-37S
Christie, E.A. 82-39T
Christsensen, R.N. 83-8B
Chrusciel, G.T. 85-14S
Chrysler, G.M. 81-30FF; 82-33FF; 84-65F; 86-16B, 74B
Chu, C.L. 78-16J
Chu, H.-S. 83-9Q
Chu, H.H.-S. 76-23F
Chu, K.H. 78-5K
Chu, L.-C. 83-25FF
Chu, Z. 85-10K
Chuah, Y.K. 85-15DP; 86-14J
Chuang, H. 86-13U
Chuang, Y.-K. 78-10A
Chuchottaworn, P. 85-27J; 86-19J
Chuck, T.L. 78-17J
Chuck, W. 84-39JM,40JM; 86-72B
Chudnovskiy, A.F. 82-21D
Chukaev, A. 80-69B
Chumak, I.G. 77-33F
Chun, J.-H. 86-8JJ,42JJ
Chun, L.T. 85-9J
Chung, B.T.F. 76-13A; 77-72A,32C; 78-14B; 79-24A,84A; 80-9A; 82-10K,40K; 83-1H; 84-74K; 85-9A,6H; 86-24JM,84K
Chung, D.K. 76-22J
Chung, F.J. 76-12C
Chung, J.N. 78-5H; 81-20J,21J; 83-11J; 84-10J, 3JJ,4JJ,5JJ; 85-12D,5JJ; 86-95J
Chung, K.C. 77-16A; 81-2N; 82-13F
Chung, M.K. 84-6FF
Chung, P.M. 84-6JM
Chung, T.-H. 84-7P
Chung, T.J. 80-32B; 82-9K; 84-19K
Chung, Y.C. 82-16C
Chunkuar, S. 81-78J
Churayev, N.V. 78-11H
Churchill, R.J. 76-5U
Churchill, S.W. 76-13C,23F,66F; 77-17A,8B, 6C,22F,80F,81F; 78-15B,57F; 79-101F; 81-10A, 14F,46F; 83-9F,39F,40F; 85-70F,22K,28Q; 86-19A, 93F
Chyou, Y.P. 85-10U,33U
Chyu, M.C. 82-7J
Chyu, M.K. 82-54A; 85-6D
Ciahi, N. 76-41F
Cianfrini, C. 85-28F
Ciborowski, J. 85-16DP
Cichowias, W.M. 77-35P
Cichy, A. 81-4N
Ciliberto, S. 86-30F
Ciment, M. 79-60A
Cingolani, A. 85-11JM
Cipolla, J.W., Jr. 85-52DP
Circelli, B.R. 86-11C
Civan, F. 84-6A
Claiborne, H.C. 82-4S
Clair, C.R., Jr. 82-2N
Claria, A. 76-11C
Clark, C.F. 79-30P
Clark, J.A. 84-5DP; 85-4T
Clark, R.A. 79-6E
Clarke, F.J.J. 85-12K
Clarke, J.F. 76-24F; 84-7C,8C
Clarke, J.H. 86-9JJ,10JJ
Clarke, N.S. 84-7A; 86-19P
Classon, E.O. 85-54Q,38S
Claudet, G. 86-48J
Claus, R.W. 85-23K
Clausen, C.W. 79-8K
Clausing, A.M. 76-15P; 81-9FF; 83-10FF; 84-7FF
Clayton, B.R. 85-15B
Cleaver, J.-W. 86-24FF
Cleaver, J.W. 85-6JM,65K; 86-53C,6G
Cleland, A.C. 77-18A; 85-15S
Clements, D.L. 83-8A; 86-52A
Clements, R.M. 80-12U; 81-4U; 83-24U; 84-32U
Clements, T.B. 83-14K
Clemons, W.E. 82-5M
Clever, J.W. 85-21FF
Clever, R.M. 77-23F,24F,25F; 78-22F; 79-14F; 80-17F; 81-10F,15F; 82-24F; 83-17F; 85-10A; 86-20F
Cliffe, K.A. 86-14N

Clifford, A.A. 80-8P
Clifford, R.J. 84-3G
Clift, R. 76-8D
Cline, V.A. 76-26P
Clomburg, L.A., Jr. 78-23F; 84-15F
Cloot, A. 86-27DP,31F
Close, D.J. 76-6T,11T; 77-5T; 85-17DP; 86-14T
Cloupeau, M. 82-32J
Clumpner, J.A. 77-56A
Clutterbuck, E.K. 76-50J
Clyne, T.W. 80-10A
Coantic, M.F. 76-38C
Coccioni, R. 86-53U
Cochrane, J.W. 81-4T
Cochrane, T. 78-14P
Codastefano, P. 86-25K,26K
Coetzee, D. 86-13E
Coffeen, D.L. 79-9K
Cogley, A.C. 82-46K,16N
Cohen, A. 86-51K
Cohen, I.M. 80-9U
Cohen, L.D. 86-51K
Cohon, J.L. 84-17T
Colakyan, M. 84-8A
Colbourne, P. 84-36B
Coldeway, R.W. 77-46F
Cole, J.B. 79-16P
Cole, K.D. 78-2L
Cole, R. 78-11; 80-23J; 81-76J; 82-30J; 84-14E
Coleman, B.H. 86-24A
Coleman, H.W. 78-10C; 81-6C; 86-28C
Coleman, I. 83-43K
Coleman, P.D. 84-111K
Coleman, T.A. 82-2F
Colin, F. 86-15N
Colladay, R.S. 76-4H
Collares-Pereira, M. 79-7T,8T,9T,10T; 85-10T
Collier, R.P. 80-117J; 81-74J
Collignon, F. 80-1L
Collins, D.J. 76-14C
Collins, F.G. 80-1A
Collins, J.G. 85-9P
Collins, M.A. 78-6D,19D
Collins, M.W. 80-9B,11C
Collins, R.E. 85-39T
Collins, R.L. 79-12J
Collins, W.W. 78-10B
Collocott, S.J. 85-9P; 86-20P
Colwell, G.T. 85-24Q
Comb, J.W. 83-33Q
Combarnous, M. 77-11F; 80-6F
Combescure, C. 85-31T
Combs, S.J. 86-55U
Comings, E.W. 85-16S,17S
Comini, G. 76-14A,15A; 77-45A,8S; 79-41B, 27F; 80-10B; 81-9B; 82-10A
Comninou, M. 80-11A; 81-11A,12A; 83-2A,26A
Comte-Bellot, G. 76-5E; 79-11P
Conder, J.R. 82-16J
Condiff, D.W. 80-19J
Coneu, J.E.R. 79-13G
Coney, M.W.E. 77-19S
Conklin, P. 84-93K
Conlin, F. 86-7S
Conlisk, A.T. 82-4G
Conn, D.W. 86-30C
Connell, S.D. 86-16N
Conrad, R.W. 78-11U
Consiglio, R. 86-65U
Consigny, H. 78-11S; 82-5S
Cook, A.E. 86-17B
Cook, D.S. 82-33FF
Cook, R.S. 84-9A
Cooper, D. 86-5E
Cooper, D.M. 85-23C
Cooper, F. 77-19A
Cooper, J.R. 85-111Q; 86-113Q
Cooper, L.Y. 76-16A; 77-20A; 82-14F,6S; 86-10FF
Cooper, M.G. 81-22J,23J; 82-17J; 83-12J,13J
Cooper, P.I. 81-5T; 86-4D
Cooper, R.E. 78-8E; 79-103F
Copeland, R.J. 86-31T
Coppalle, A. 85-24K; 86-27K
Coppolino, S. 79-3T
Cordaro, J.V. 81-31B
Cordero, G.O. 83-24Q,25Q
Cordier, H. 79-81A
Cordoba, H. 83-10F
Coriell, S.R. 85-37F
Cornelissen, M.C.M. 85-16B
Cornet, I. 79-42B; 80-2G
Cornish, A.R.H. 81-15FF; 82-14FF
Cornwell, K. 76-9D; 77-4J; 82-18J,19J; 85-1J
Corradini, M.L. 80-24J; 83-14J; 86-56J
Corrieu, G. 86-60Q,61Q
Corrsin, S. 76-21E; 80-21E; 85-8B
Corsan, J.M. 84-8P
Corson, M.R. 85-9E
Cosner, A.A. 81-17A
Cotta, R.M. 84-30A; 86-25A,18B,19B,20B
Cotte, B. 85-51F
Couanon, J.C. 80-38F,39F
Couderc, J.P. 77-27D; 81-7C; 82-12D
Coudert, J.F. 85-19U
Coulson, K.L. 77-29T
Cousin, C. 85-25K,63K; 86-28K,29K
Coutanceau, J. 77-41P
Coutier, J.P. 82-7T; 85-29F; 86-32F
Coutler, B.L. 81-1T
Couvillion, R.J. 86-8S
Cowley, M.D. 76-3U,4U
Cowley, S.J. 83-11F
Cox, G. 82-44F
Cox, G.B., Jr. 76-5S
Cox, L.J. 85-12K
Cox, S.K. 82-9T
Craggs, A. 79-141F
Craig, S. 85-39T
Craig, T.D. 84-20K,46K
Craine, R.E. 85-27S
Cram, L.E. 83-7U; 85-12U,32U; 86-21U,27U
Crane, K.C.A. 82-5K
Crane, L.J. 76-25F; 77-26F; 86-21K
Crank, J. 76-4A; 78-8A; 79-18A
Crapper, P.F. 76-5H
Craubner, H. 86-21P

Crausse, P. 81-4H; 85-18DP
Cravalho, E.G. 76-10L; 79-61A; 80-32A; 82-31A, 32A
Crawford, M.E. 80-2H,3H
Crawford, R.J. 85-18S
Creff, R. 86-2B
Cremers, C.J. 81-12H; 82-7A
Crescitelli, S. 82-17C
Crespo, E. 82-15F
Cresswell, D.L. 86-12DP
Cretu, J. 81-8B
Crha, S. 85-11T
Crivelli, L.A. 86-9JM
Cronenberg, A.W. 79-19A,20A,21A; 80-45J
Crosbie, A.L. 76-5L; 78-18K; 79-10K,11K,12K; 81-10K; 84-21K; 85-26K,27K,28K,29K; 86-87K
Cross, M. 81-63A,64A; 85-38N
Cross, W.T. 86-16Q
Crow, L.I. 76-15S
Crowley, A.B. 78-8A,9A; 79-18A,4S
Crowley, J.M. 84-42F; 85-28S
Crozier, R., Jr. 77-7Q
Crump, M.W. 81-5N
Crupper, G., Jr. 81-74F
Cruse, R.M. 85-35P
Cuffe, K.W. 78-7Q
Cui, W.K. 85-19S
Culver, G.G. 78-15S
Cummings, W.G., III 86-79K
Cumo, M. 85-18J; 86-6JJ
Cundy, V.A. 86-14E
Cunningham, A. 86-22P
Cunningham, D.B. 86-23P
Cunningham, J.P. 85-46S
Cunningham, M.E. 81-13A
Cupp, R.E. 77-39P
Cur, N. 76-55A; 78-16B,8Q; 79-8B; 80-70B; 82-36B; 83-44B
Curle, N. 78-11C
Curlet, N.W.F. 84-15F
Currie, I.G. 80-2L
Curry, D.M. 77-69A,70A; 84-53C
Curry, J.H. 84-16F
Curry, R.L. 80-27T
Curzon, A.E. 77-5L
Curzon, F.L. 82-5U
Custer, J. 78-88F
Custer, J.R. 77-27F,28F
Cuti, F. 81-6Q
Cuvelier, C. 86-33F
Cvetkovic, M. 86-15E
Cybulski, A. 86-104Q
Czarnota, I. 86-24P

D'Annibale, F. 85-18J
D'yachenko, Yu.V. 78-35J
D'yachkov, F.N. 79-16J
Daalder, J.E. 78-4U; 79-7U
DaCosta, L.N. 81-16F
Dadou, R. 86-30K
Dadras, P. 84-5S
Dagan, L. 80-9H
Dagis, L.I. 77-31C; 78-12C,56C
Daguenet, M. 84-10T
Dahiya, R.P. 80-4M
Dai, W.L. 84-20JM
Daikoku, T. 82-56J; 86-66Q
Daily, J.W. 80-6P
Dakhin, O.Kh. 83-14B
Dakin, J.T. 78-1G
Dakos, T. 84-9C
Dalbert, A.-M. 83-41F
Dalbert, A.M. 81-17F
Dalcher, A.W. 78-16J
Dale, A.C. 86-45T
Dale, J.D. 83-24U
Dalle Donne, M. 77-9B
Dallman, R.J. 80-3G
Dalman, M.J. 86-17Q
Dalman, M.T. 84-9DP; 86-21B
Daly, S.F. 83-51F
Damelincourt, J.J. 82-9U
Damerell, P.S. 79-25F
Damianides, C. 86-15Q
Damle, A.S. 78-59J
Danagher, D.J. 86-101K
Dancey, C.L. 76-32D
Dandekar, D.P. 86-46P
Daney, D.E. 76-26F; 79-9B
Dang, V. 83-15B
Dang, V.D. 78-17B; 80-11B
Dang, Vi-D. 77-10B
Dang Duc, H. 82-7U
Dangre, A.J. 86-25P
Danh, H.O. 76-3E
Danh, H.Q. 77-3C,2E,26E; 78-2C
Daniel, K.J. 79-13K
Daniels, K.J. 78-6K
Daniels, L.C. 81-8S
Daniels, P.G. 77-29F; 78-24F,25F,2G; 82-16F, 17F; 83-5DP,5F; 84-10DP; 85-30F,31F; 86-9DP
Daniels, T.C. 78-9Q; 79-13J
Danilevich, S.Yu. 76-8B
Danilov, Yu.B. 83-34J
Danilova, G.N. 82-10J,20J,1Q; 85-7J,28J,64J
Danilova, G.P. 82-10J
Dantzer, P. 86-39P,40P
Dantzig, J.A. 85-32F,11N,12N
Daradimos, G. 81-5S
Darkazalli, G. 81-14A
Das, R. 84-8FF
Das Gupta, M. 79-26F
Dasch, C.J. 84-22K
Dashevsky, Yu.M. 85-33F
Daskavitz, M.J. 78-7Q
Date, A.W. 78-4B; 86-34F
Datta, A.B. 76-28F; 83-15J; 85-5N
Datta, G. 84-9T
Datta, N. 76-27F,5G,12U; 77-2G,2U
Daubert, J. 82-28B
Daujotas, P.M. 79-13D
Daunoras, P. 85-97C,116Q
Daunoras, P.I. 86-32D
Daurelio, G. 85-11JM
Davalos, L.A.O. 86-35F

Davalos, O. 84-1U
Dave, R.N. 82-31T
Davey, A. 77-30F
Davey, G. 81-86J
Davidson, G.W. 85-26K
Davidson, M.R. 86-10DP
Davidson, R.D. 85-54P
Davies, H.G. 76-15C
Davies, J.A. 82-8T
Davies, J.M. 82-9T
Davies, P.A. 76-41P
Davies, T.W. 85-11A,10E
Davis, D.C. 86-6S
Davis, E.A. 85-93DP,66S
Davis, E.J. 77-31F; 79-14J
Davis, G.C. 76-5K
Davis, H.T. 78-17D; 79-25A; 84-24A; 85-93DP, 66S
Davis, J.F. 86-46JJ
Davis, L.R. 76-80F; 77-10S,18S; 78-26F; 82-15FF
Davis, M. 85-10JM
Davis, M.R. 79-7E; 82-11K
Davis, P.L. 76-17A
Davis, P.O.A.L. 79-7E
Davis, R.H. 86-26P
Davis, R.T. 86-4N
Davis, S. 84-59F
Davis, S.H. 80-11F; 81-57F; 83-11F,59F,60F; 84-17F,30FF
Davisson, J.W. 77-22K
Davy, W.C. 85-11C
Dawe, R.A. 85-80DP
Dawood, A.S. 81-10FF
Dawson, P.R. 84-11A; 85-19DP
Day, J.P. 79-2Q
Day, W.A. 83-9A; 84-12A
Dayan, A. 82-5D,6D; 84-13A
Dayan, J. 84-17DP; 85-88DP
De, A.K. 86-22A
De Boer, P.C.T. 84-9FF; 86-36F
De Bruijn, H. 83-37Q
De Caluianu, V. 86-14Q
De Faveri, D.M. 85-30K
De Frutos, A.M. 85-7T
De Graag, D.P. 78-43B
De Groot, H.J. 80-32P
De Jarlais, G. 86-47J
De Jong, P.P.G. 85-1S
De Lemos, M.J.S. 85-17B
De Maerteleire, E. 78-7D
De Maeyer, L.C.M. 82-27A
De Roeck, H. 77-68F
De Ron, A.J. 80-6T
De Rooij, W.A. 84-85K
De Salve, M. 85-24C,31J
De Socio, L. 76-6C
De Socio, L.M. 76-1G,6K; 79-30F; 82-18F; 83-12FF,
35FF
De Vahl Davis, G. 77-60F; 79-31F; 85-90F
De Vries, D.A. 77-54J; 78-25E
De Zhao, L. 84-13P
Dean, R.B. 77-7E
Dean, T.A. 79-6S
Deardorf, D.K. 76-13P
Deardorff, D.K. 77-11K
Deardorff, J.W. 82-18C
Debnath, L. 86-115F
DeBroissia, M. 82-13N
Dechene, R.L. 77-35P
Decker, N. 80-12B,12C; 83-12DP; 84-16DP
Deckwer, W.D. 80-4S
Dedale, C.F. 80-25J
Deering, D.W. 82-42K
Deev, V.I. 79-4J,15J; 80-2J; 85-70B
Degach, A.A. 79-25S
Degani, D. 82-11C; 84-2N
Degiovanni, A. 86-31JM,27P
Degnan, T.F. 80-5Q
Degout, D. 84-5U; 86-14U
DeGroot, J.S. 84-36U
DeGroot, R.L. 82-53K
Degtereva, L.N. 79-34A
Deiwert, G.S. 86-61N
Dekeyser, I. 77-3H,19H; 78-6E
DeKinder, R.E., Jr. 86-127K
Del Giudice, S. 76-14A; 77-45A,8S; 79-10B, 41B,27F; 81-9B
Del Guidice, S. 79-8J; 80-10B
Del Valle, V.H. 85-29J
Del Valle M, V.H. 81-40J
Delale, C.F. 86-9JJ,10JJ
Delaunay, G. 80-5D
Delettrez, J. 84-105K; 86-19U
Delhaye, J.M. 77-19J
Delichatsios, M.A. 79-28F
Delmas, H. 81-7C
Delnore, V.E. 80-12FF
Delsante, A.E. 83-2S
Deltour, A. 80-3F
Delvosalle, C. 85-20DP
Demargue, P. 77-7C
Demchenko, R.Ya. 77-4S
Demchuk, S.A. 79-25E
Dement'eva, K.V. 79-11B
Demichelis, F. 84-23K
Demidov, S.A. 83-35K
Demkowicz, L. 86-17N
Dempsey, M.J. 82-19JJ
Demuren, A.O. 83-11FF; 86-5H
Denbigh, P.L. 83-8U
DenBraven, K. 86-15T
Denisov, M.A. 78-12S
Denloye, A.O.O. 77-3D; 78-5D
Dennis, S.C.R. 79-15C; 84-4G
Denny, V.E. 77-33C
Denton, R.A. 79-29F
Deo, P.V. 83-3JJ
Depooter, K. 77-11P; 78-15P
Der, S.L. 84-28S
Derby, J.J. 85-9JM
Derewnichi, K.P. 85-30J
Dergach, A.A. 78-32S,33S; 81-38C
Derzhavina, A.I. 83-12C
Desai, P. 86-77F
Desai, P.V. 86-50N

Deschamps, J.Y. 80-3L
Deshaies, B. 86-31K
Desjardins, R.L. 77-12P
Desmaris, M. 86-48J
Desmond, R.M. 77-5
Desmons, J.Y. 84-4E
Desrayaud, G. 85-31K
Detkov, S.P. 86-32K
Detz, C.M. 76-12J
Deuerler, F. 84-34S
Devadattam, D.S.K. 86-28P
Devi, C.D. 86-15U
Devi, V.M. 84-25K; 86-93K
Devienne, R. 85-62B
Deville, M. 86-5N
Devloo, P. 86-64N
Devoto, R.S. 78-5U
Devoyno, A.N. 82-1B
DeVries, D.A. 86-9S
DeWitt, D.P. 86-82K
DeWitt, K.J. 76-9K; 78-5G; 79-3G; 82-36FF; 83-23C; 86-31C
Dewynne, J.N. 86-10JM,11JM,21JM
Dey, J. 81-10B,8C; 82-9FF
Dey, S.K. 80-13C
Dhanak, A.M. 76-18D
Dhar, P.L. 76-18A
Dharmadurai, G. 83-10A
Dhatt, G. 86-18N
Dhingra, K. 83-8T
Dhingra, S.C. 84-11DP
Dhir, V. 77-4D
Dhir, V.K. 77-31A,5J; 78-70F; 79-75J; 80-36J, 78J,127J; 81-24J; 82-55J,26JJ; 83-32JM; 84-11J; 85-47DP,14J,15J,9JJ; 86-15JJ
Dhuga, D.S. 85-32J; 86-20J
Di Costanzo, M.A. 84-12DP
Di Leon, R.L. 86-33K,34K
Di Marzo, M. 85-7JJ; 86-11JJ
Dias, F.M. 80-49B
Diaz, L.A. 86-12JM
Diaz-Munio, R.F. 81-6N,7N
DiBenedetto, E. 86-11DP
Dickey, T.D. 79-8E
Didkovskiy, A.B. 78-8J; 83-6J
Didkovsky, A.B. 81-25J
Didlake, J.E. 77-53A
Diener, G. 81-4K; 82-5A,6A
Diep, G.B. 77-18D,19D
Dietenberger, M.A. 83-16JM
Dietrich, D.E. 85-13N
Dietsche, C. 82-26T; 84-17F; 85-34F
DiFelice, R.F., Jr. 83-11A
Dikhtiyevskiy, O.V. 85-12A
Dil'man, V.V. 78-7B
Diller, K.R. 76-17S
Diller, T.E. 84-21H,22H; 85-3C,2D
Dilley, A.C. 84-82K
Dillow, C.F. 77-28K
Dils, R.R. 86-16E
Dimarogonas, A.D. 83-27A
DiMarzo, M. 86-21J
Dimic, 77-6J
Dimoff, K. 79-21U
Dimotakis, P.E. 84-15H
Diner, D.J. 85-69K
Dinger, A.K. 77-7J
Dinguirard, J.P. 80-1U
Dinh, S.M. 82-5B
Dinulescu, H.A. 80-4E
DiPrima, R.C. 84-15C
Dirmhirn, I. 83-24T
Divakaruni, S.M. 85-16A
Dixit, V. 84-26K
Dixon, A.G. 84-12DP; 85-21DP; 86-12DP
Diz, L.A. 81-11B
Djordjevic, N.A. 77-14T
Djurovic, S. 86-63U
Dmitrenko, Yu.M. 85-5E; 86-12C
Dmitriev, V.I. 85-14N
Dmitrieva, S.P. 85-50DP
Doan, K.S. 78-17F
Doan, T. 86-35J
Dobbins, B.N. 84-3E
Dobbins, R.A. 77-17P
Dobbs, G.M. 84-29K
Dobkes, A.L. 80-143J; 83-80J
Dobran, F. 80-27J; 85-33J,29Q
Dobrzhanskiy, V.G. 80-28J
Dobud'ko, V.D. 76-4Q
Docherty, P. 86-35K
Dodd, R. 77-32F
Dodge, C.E. 79-27K
Dodge, M.J. 84-27K
Dodson, B. 77-13P
Doe, P.H. 86-23P
Doerffer, S. 86-37F
Doerner, R.C. 81-20C; 82-24A
Dohara, N. 84-47J
Doherty, J.A. 86-13DP
Dohler, J. 86-108J
Doi, M. 76-10D; 79-9E
Doi, Y. 80-25U
Dolginov, A.Z. 84-10A
Dolhan, P.A. 85-10P
Dolinskiy, A.A. 77-22A
Dolzhenkova, G.G. 79-3J
Domashev, Ye.D. 78-77J; 85-111J
Domingos, H. 76-19A
Domke, H. 84-85K
Domoto, G.A. 77-51F; 86-18JM,38K
Domracheva, L.G. 85-81DP
Donald, C. 77-17T
Donaldson, A.L. 84-6U
Donaldson, C. duP. 76-13E
Donea, J. 85-15N
Donelan, M.A. 78-16P
Dong, W.G. 86-11P
Donizak, J. 85-38J
Donnelly, R.J. 83-7G; 84-11G
Donsi, G. 82-17C; 85-2DP
Doo, R. 85-18B
Doorly, J.E. 86-17E
Dopazo, C. 79-10E
Dor, H. 77-3G
Dorakhov, A.R. 82-8JJ

Dore, P. 86-25K,26K
Dorfman, A.Sh. 77-8C; 78-3G; 79-20C; 85-16N, 17N
Dorning, J.J. 77-30A
Dorofeev, B.M. 85-34J
Dorokhov, A.P. 82-17JJ
Dorokhov, A.R. 79-45J; 80-75J; 82-3Q; 83-12H; 84-11C,20J
Dorokhov, I.N. 86-55DP
Dorr, B. 85-18Q
Dorrer, G.A. 85-20S
Dorri, B. 85-21S; 86-10S,28S
Dos Santos Travelho, J. 84-13B
Doss, E.D. 80-2M; 82-3M,4M,6M
Doss, H. 82-8A
Doty, L.F. 78-17A
Dougherty, R.L. 78-18K; 85-27K
Douglas, W.J.M. 76-54C; 80-89J; 85-67C,50E, 19S
Douglass, R.W. 78-88F,10G; 79-32F,33F,34F; 80-3G; 86-39F,118F
Dowden, J. 85-10JM
Downing, H.D. 82-2K
Dozhdikov, V.I. 86-53S
Dragovan, M. 84-28K; 85-32K
Dragun, V.L. 85-5E
Draijer, W. 85-70DP
Drakakis, M. 82-9U
Drake, D.G. 80-9P; 83-13C,34FF
Drake, E.M. 77-7J
Drake, J. 85-53JM
Drake, M.C. 86-37E
Draper, C.W. 80-10P
DreBler, B. 85-27E
Dreister, G.A. 82-7Q; 85-38B
Dreitser, G.A. 85-3Q,52Q; 86-22B
Dreniak, E.L. 84-24K
Dresner, L. 86-22J
Dresvin, S.V. 86-16U
Drew, D.A. 85-3J
Drew, M.S. 79-11T,12T
Dreyster, G.A. 78-34B; 79-65F; 82-8Q
Dreytser, G.A. 84-15Q
Driessn, J.M. 86-33F
Dring, R.P. 80-4H
Drinkenburg, A.A.H. 86-93P
Driver, P.M. 77-6T
Drizhus, M.R.M. 78-54C
Drizhyus, M.-R.M. 76-16C
Drizhyus, M.R.M. 80-13B
Drizius, M.-R. 79-16C,17C
Drizius, M.-R.M. 84-14B
Drobyshevich, V.I. 85-22DP
Dronov, V.P. 85-99F
Drost, H. 85-14U
Droste, D. 82-7D
Drozdov, O.A. 77-14P
Drucker, M.I. 84-11J
Drummond, J.E. 82-36F
Drummond, J.R. 81-14K
Drumond, J.P. 76-4R
Drzewiecki, T.M. 77-15P
Du, X.X. 83-26C
Dua, S.S. 76-13J; 77-8J; 78-18J
Dubois, M. 78-27F,28F
Dubrovsky, E.V. 85-52Q
DuChateau, P.C. 78-6A
Duck, P.W. 77-23A,51A,23D
Dudderar, T.D. 79-124F; 81-66F
Dudukovic, A.P. 77-28B
Dudukovic, M.P. 85-28A,30Q,31Q
Duffey, R.B. 78-19J; 81-76F,24J,83J,84J; 84-11J, 17J
Duffie, J.A. 76-13T; 80-23T; 82-10T
Duffin, R.J. 84-7T
Duffy, D. 85-33K
Duggan, R.C. 80-23Q
Duhamel, P. 81-6D
Dukler, A.E. 78-16E; 82-34B
Dulieu, B. 79-35F,139F
Dulikravich, G.S. 85-39B; 86-15B,49N
Dullforce, T.A. 81-13FF; 82-13FF; 83-14F
Dumargue, P. 81-3H
Dumas, A. 84-15B
Dunavaht, J.G. 76-3R
Dunaway, W. 85-130K
Duncan, C.H. 82-27T
Dunders, J. 80-11A
Dundurs, J. 76-20A; 81-11A,12A
Dunin, I.L. 85-19B
Dunkle, R.V. 77-5T; 81-5T
Dunlop, P.J. 84-24P; 85-51P
Dunn, D.J. 84-11DP
Dunn, J.C. 85-22S
Dunn, J.R. 86-49T
Dunn, M.G. 79-7S; 82-7S; 84-6S,7S; 85-25C; 86-13C,14C,15C
Dunn, P.F. 80-5M
Dunn, S.S. 76-12R
Dunn, W.E. 76-15P
Dunn, W.L. 85-34K
Dunstan, G.F. 79-14Q
Dunwoody, N.T. 79-12B; 80-14B; 81-12B
DuPlessis, M.P. 86-64B
Dupont, A. 84-5E,9E
Durack, D. 77-24A
Duracz, T. 86-36K
Durbin, P.A. 81-13B
Durrett, R. 84-9J
Durst, F. 76-16P; 77-34B; 79-17P
Durst, M. 83-14D; 84-49B; 85-58FF
Dussault, R. 76-15B
Dutta, S. 79-18P
Dwoyer, D.L. 86-90N
Dwyer, H.A. 76-10S; 83-14C
Dwyer, O.E. 76-3M,4M
Dyban, E.P. 77-9C; 79-19C; 80-6D; 81-14B
Dyban, Ye.P. 78-18B,10E; 79-18C; 82-19C; 83-15C, 3S
Dybbs, A. 76-42P; 80-32D
Dyer, J.R. 78-29F
Dykhuizen, R.C. 84-3N; 86-23J,106J
Dyment, A. 86-16C
Dynayev, S.V. 77-23J
Dyundin, V.A. 82-20J,1Q; 85-35J
Dyuzhev, G.A. 83-9U

Dzyuba, E.D. 85-36U
Dzyubenko, B.V. 82-7Q; 85-3Q; 86-18Q

Eagles, P.M. 80-12F; 84-4B
Eardley, G. 79-8U
Earle, R.L. 77-18A
Earnshaw, J.C. 78-14P
Easby, J.P. 78-30F
East, R.A. 80-14C
Easton, C.R. 76-23T
Eastop, T.D. 79-48P
Eaton, J.K. 85-20D
Eatough, D.J. 76-12P
Ebadian, M.A. 85-3B,6F; 86-23B,24B
Eberhard, W.L. 80-11P
Eberhart, R.C. 85-51S
Echigo, R. 83-21B; 85-37Q; 86-9H
Eckbreth, A.C. 84-29K
Ecke, R.E. 85-60F
Eckels, P.W. 85-32Q,33Q,23S
Eckert, E.R.G. 76-2,1G; 77-16H; 78-3,1P,2P; 80-7D,8D,4E,10H; 82-1J; 83-3H; 84-8S; 85-24DP, 13H; 86-14DP
Eckert, R.E. 80-82B
Eddington, R.I. 78-20J; 79-17J
Eddy, T.L. 86-22U,73U
Edelman, Z. 85-36J
Edelstein, S. 84-12J
Edenburn, M.W. 76-7T
Edenfield, E.E. 77-6R
Edwards, B.F. 85-35F
Edwards, C.L.W. 84-42C
Edwards, C.M. 86-46E
Edwards, D.K. 76-6F,13F,23K,24K,5T,8T; 77-1F; 78-4L; 79-36F,3K; 80-4L; 81-37B,18FF,30T; 82-24K; 84-30K; 85-98K; 86-37K
Edwards, M.F. 81-41B; 82-35B; 86-57C
Efendiev, Sh.M. 83-10K
Effekhar, J. 81-14A
Efimenko, A.N. 79-11S
Efimov, S.I. 85-71Q
Efthimion, P.C. 79-19P
Egan, W.G. 78-7K; 84-31K
Egerton, P. 79-8S; 81-9S
Egli, W. 82-2U,18U
Egorov, B.N. 77-16P
Eguchi, W. 78-43P
Ehrlich, L.W. 77-37B
Ehrlich, O. 78-10A
Ehrlich-Moser, S.J. 86-45E
Eichberger, J.-I. 85-35K
Eichhorn, R. 76-18F; 78-13P; 79-19F,53F,137F; 80-13FF; 81-29J; 82-5C,2FF
Eichler, H.J. 85-100K
Eichler, J. 84-27S
Eickhoff, H. 78-6H; 83-16J
Eid, J.C. 86-19N
Eidson, E.M. 85-36F
Eigenberger, G. 85-24S
Einav, S. 79-45P
Eklund, T.I. 77-17P
El Dabe, N.T. 86-17U
El Derini, Z. 82-4M
El Hadidy, M.A. 82-6B
El Wakil, S.A. 85-97K,127K
El-Adawi, M.K. 86-26A,13JM,14JM
El-Ariny, A.S. 76-9B,75F
El-Baz, H.S. 77-6K,7K
El-Beshbeeshy, M.S. 85-67B
El-Gamal, H.A. 78-13C
El-Genk, M.S. 79-19A,20A,21A
El-Gindy, T.M. 79-22A
El-Hawary, M.A. 79-11E; 80-13F
El-Henaway, I. 82-10FF
El-Henawy, I. 86-11FF
El-Kady, A.A. 86-53C
El-Kaissy, M.M. 76-12D
El-Kassaby, M.M. 86-24J
El-Masri, M. 86-15G
El-Masri, M.A. 78-13S; 84-3J; 85-53DP,75J; 86-49C,40DP
El-Messih, S.A. 76-39D; 86-68P
El-Mistikawy, T.M. 78-14C
El-Nashar, A.M. 78-19B
El-Osairy, M. 86-68P
El-Refaee, M.M. 78-4F
El-Reid, M.K. 76-5F
El-Ridi, M.K. 77-33F
El-Riedy, M.K. 81-11FF; 84-5H
El-Sayed, 86-29P
El-Shaarawi, M.A. 81-1G
El-Shaarawi, M.A.I. 80-14F; 82-7B
El-Shaarawl, M.A. 81-18F
El-Shamy, A. 78-18U
El-Shanawany, M. 78-21J
El-Sharkawy, A.A. 86-29P
El-Shirbini, A.A. 78-21J; 79-51J; 80-17S; 86-71J
El-Wakil, M.M. 79-89F,110F; 81-78J; 82-50F, 28T; 84-49J
Elakshar, F.F. 80-10U
Elamvaluthi, G. 84-13J
Elata, C. 84-1C
Elbahar, O.W.T. 84-24H
Eldering, H.G. 82-3T
Elderkin, M.J. 86-2D
Eldighidy, S.M. 81-27T; 83-9T
Eldridge, B.D. 76-11D
Elena, M. 77-9E; 84-16B
Elewa, F.A. 85-92F,93F
Elghobashi, S. 86-22D
Elghobashi, S.E. 81-9C; 83-16C
Elgin, J.B. 76-7K
Elgobashi, S.E. 80-15C
Elhadidy, M.A. 86-16T
Elias, E. 82-74F; 85-36J
Elins, R.E., III 76-11E
Elliot, C.M. 86-11DP
Elliot, J.M. 79-125F; 81-64F
Elliott, J.W. 85-17D
Elnasher, A.M. 80-80J
Elovic, E. 85-13H
Elrod, H.G., Jr. 77-51F
Elsayed, M.M. 85-9FF; 86-26J
Elsdon, R. 77-5D

Elshehawey, E.F. 86-26A
Elsherbiny, S.M. 78-31F; 82-19F,20F
ElSheriby, S.M. 83-12F
Elsner, J.W. 86-17C
Elson, T.P. 79-2G
Eltayer, I.A. 77-1G
Ely, J.F. 85-4P
Emara, A.A. 77-55F; 80-15F
Emerman, S.H. 83-8JM; 85-36JM
Emery, A.F. 76-10B,7E,6S; 79-23A; 80-15B; 82-8S; 85-36K,21S
Emery, E.M. 77-45P
Emmerich, V. 83-15K
Emmerson, G.S. 78-22J
Emmons, H.W. 83-13F
Emori, S. 79-25P
Endo, H. 86-33B
Endo, K. 76-37B
Endo, M. 82-19A; 85-37S
Endoh, K. 77-43B
Endrenyi, S. 76-6H; 82-8D
Ene, H.I. 82-9D
Enery, M. 84-8T
Engelhorn, H.R. 76-11F
Eninger, J.E. 79-3Q
Enoch, I.E. 82-12K
Enomomoto, Y. 85-25S
Enomoto, Y. 85-11E; 86-11S
Enoto, T. 81-20U
Enro, H. 81-30C
Enya, S. 84-49F
Epanov, Y.G. 81-19D
Epherre, J.-F. 77-34F
Epifanov, V.M. 85-7H
Epik, E.Ya. 77-9C; 78-18B; 81-14B; 82-19C
Epperlein, E.M. 86-18U
Eppich, J. 83-19C
Epshtein, M.I. 79-42P
Epstein, A.H. 86-18E
Epstein, M. 76-11B,12B,14J; 77-25A,11B,12B; 78-20B,63B; 79-24H; 80-35A,16B,29J; 81-5K; 82-8B; 84-19DP,14J; 85-10FF; 86-15JM
Epstein, N. 83-6Q; 86-77J,78J
Epstein, R. 86-19U
Era, Y. 76-17C
Erbs, D.G. 82-10T
Erckmann, V. 85-26S
Eremeitsev, I.G. 86-18C
Eremenko, V.I. 77-32P
Erian, F.F. 79-8G; 82-8G
Eriksson, T.S. 82-13K; 85-80K
Erinov, A.E. 78-39P
Ero, M.I.O. 77-35F,36F; 78-15C
Eroli, L. 85-13A
Eroshenko, V.M. 85-11FF; 86-78B,6H
Ershov, A.V. 86-6H
Erskine, D.J. 80-16FF,39J
Ersoy, Y. 84-14A
Ertman, M.E. 85-18A
Erwin, D.A. 85-16U
Esaki, S. 80-30J,31J
Eschamps, P.Y. 79-12L
Escott, J. 85-25DP
Esdorn, J.H. 85-22E
Eslinger, R.G. 79-24A
Esmailizadeh, L. 86-25J
Esposito, C. 85-11JM
Estalote, E.A. 77-2L,8L
Estes, R.C. 86-38K
Estrin, J. 76-43J
Esvignes, F. 78-17P
Etemadi, K. 82-6U; 83-13U; 85-15U
Eteng, E. 86-19C
Eu, B.C. 85-17U
Evans, D. 85-37J
Evans, D.J. 82-3N; 85-18N; 86-20N
Evans, D.L. 82-11T
Evans, D.M. 79-5H; 81-10T
Evans, D.V. 86-27A
Evans, G.H. 82-21F,11FF
Evans, J.W. 82-1K
Evans, R.G. 81-5U
Evenko, V.I. 77-8Q
Everhart, J. 78-59B
Eversole, J.D. 84-32K
Ewing, E.S. 86-67C
Ewing, W.S. 84-24K
Exeter, M.K. 82-4N; 83-14F
Exton, R.J. 86-19E
Eygenson, Ye.B. 82-10U
Eyler, L.O. 80-5E
Eymard, M. 85-4U
Eysmond, Yu.G. 84-21P
Ezdin, B.S. 86-22C
Ezekwe, C.I. 76-37P
Ezekwe, C.L. 78-11E
Ezzat, H.A. 80-14S

Fabre, J. 86-24K
Facas, G.N. 83-15F
Fadeyev, V.A. 79-32U
Faenov, A.Ya. 85-18U
Faeth, G.M. 78-3F,55F; 79-80F; 82-7JJ; 84-51K; 85-104J; 86-22FF,43JJ
Fagerholm, N.E. 83-17J
Faggiani, S. 80-17B
Faghri, A. 76-45J; 78-49C; 85-37P,38P; 86-20Q
Faghri, M. 78-21B; 80-18B,71B,7D,8D,46FF; 84-4N; 86-14DP,38F
Fairbairn, S.A. 78-19J
Fairweather, G. 83-5G
Fairweather, M. 85-37C; 86-25C
Faizullin, M.Z. 86-84P
Fakusako, S. 84-39B
Falade, A. 85-2C
Falco, R.E. 80-12P
Faleev, V.V. 86-7H
Fallen, M. 82-9B
Fan, C.S. 84-33C
Fan, D.N. 80-8G
Fan, J.C.C. 76-9T
Fan, L.-S. 85-54DP
Fan, L.T. 78-23J; 80-3J
Fand, R.M. 76-16J; 77-37F; 83-13FF; 86-15DP
Fang, L.J. 84-7JM

Fang, M.T.C. 76-3U,4U; 78-15U; 79-6U,9U; 80-11U;
83-10U; 86-20U
Fang, O.T. 85-37F
Fang, X.-D. 86-53T
Fanney, A.H. 80-15T; 83-10T
Farag, I.H. 81-6K; 86-39K
Farahat, M.M. 77-9J
Farber, E.A. 82-7T
Fardzinov, V.K. 77-2S
Farello, G.E. 85-18J; 86-6JJ
Farge, M. 81-6D
Farhadich, R. 86-6JM
Farhadieh, R. 78-24J; 81-15A; 82-9A; 83-9JM
Farmer, A.J.D. 84-8U; 86-21U,27U
Farmer, M.T. 86-39F
Farmer, P.R. 77-19S
Farmer, W.M. 80-3P
Farnia, K. 77-26A
Farnworth, B. 85-10P; 86-19K
Farouk, B. 81-12FF; 82-22F,23F,12FF; 83-15F;
85-6G,19N; 86-42F,58F,25N
Farouk, R.M. 83-14FF
Farr, J.L., Jr. 85-16A
Farrall, G.A. 78-6U
Farrar, L.C. 76-15J
Farrel, R. 76-6L
Farrell, P.V. 85-26H
Farrell, R.T. 85-8P
Farris, R.J. 86-59P
Farrow, R.L. 82-14K
Farschi, M. 80-28F
Fast, R.W. 80-58S
Fath, H.S. 78-25J
Fathalah, K.A. 85-9FF; 86-26J
Fathi, A.M. 85-22JJ
Fauchais, P. 80-34U; 82-14U,27U; 83-2U,25U;
84-5U; 85-19U; 86-16P,80U
Fauske, H.K. 76-12B,57J; 79-25J; 83-29J
Fautrelle, Y. 85-61S
Favier, D. 76-17P
Favin, S. 77-32A; 83-17A
Favorskii, O.N. 81-10S
Faw, R.E. 76-7S; 78-71F; 79-13F; 81-13FF;
82-13FF; 84-10FF; 86-21N
Faxvog, F.R. 77-17K
Fay, J.A. 83-9J
Faynzil'berg, S.N. 80-9J
Fazio, P. 84-34JJ
Fearn, D.R. 80-6M
Fedkiew, P. 78-8D
Fedorov, L.F. 77-10J
Fedorov, N.V. 86-30P
Fedorov, R.I. 85-25E; 86-41B
Fedorov, V.A. 86-65J
Fedorov, V.I. 84-16Q; 85-110J; 86-30P
Fedorov, V.N. 78-23Q,24Q
Fedorovich, E.D. 85-40Q
Fedotkin, I.M. 77-11J; 83-18J
Fehiner, F.P. 80-20T
Feidt, M. 86-30B
Feierabend, T.P. 79-1P
Feijoo, L 79-25A
Fein, A.P. 85-38F
Fein, J.S. 76-29F
Feinberg, B. 76-6U
Feingold, A. 76-21A; 78-3L,25P
Feix, M.R. 81-9N
Fel'dberg, L.A. 80-143J; 83-80J
Fel'dblym, A.S. 85-65J
Fel'dshtein, Ya.M. 79-25S
Fel-dshteyn, Ya.M. 78-33S
Felderhof, B.U. 78-48C
Feldmann, P.S. 85-11S
Felland, J.R. 78-4L
Felock, R.D. 83-40K
Felske, J.D. 77-18K; 80-39FF; 81-7K; 82-15K;
86-40K,66K,67K
Fendell, F.E. 85-11S
Fendley, J. 85-37K
Fenech, H. 86-28A
Fenton, G.G. 82-12T
Ferguson, C.D. 84-50Q
Ferguson, C.R. 86-63S
Ferguson, D. 86-106Q
Fernandes, R. 82-16K
Fernandez, M. 86-31P
Fernandez-Pello, A.C. 80-21K; 84-44F,19FF
Ferraiolo, G. 85-30K
Ferreira, R.T.D.S. 86-2F
Ferretti, O.A. 85-23DP
Fertman, V.E. 76-10F; 79-41F
Fertman, V.Ye. 79-37F
Ferziger, J.H. 79-6E
Fett, F.N. 85-54S
Fetter, A.L. 85-35F, 80-8DP
Feuerbacher, B. 83-37J
Feuermann, D. 85-12T
Fialko, N.M. 85-57J
Fichtl, G.H. 80-1A
Fiebig, M. 84-5F; 85-87F
Fiedler, H. 79-32E; 85-12E
Fifni, M.A. 84-2B
Figliola, R.S. 86-16DP,40F
Figueiredo, A.M. 78-47C
Fijalkow, E. 81-9N
Fik, H. 85-38J
Fikioris, J.G. 78-7L
Filin, V.A. 77-46B
Filipchuk, A.I. 86-5S
Filipchuk, V.Ye. 78-18B
Filippov, A.I. 85-9B
Filis, P. 86-41F
Filla, M. 76-8D
Findlay, F.D. 85-19K
Fink, J.K. 85-11P
Finlayson, B.A. 77-13D; 80-8B
Finnemore, D.K. 84-29A
Finnigan, J.J. 79-44P
Fintel, B.W. 86-17T
Finzel, M. 85-8U
Firouzian, M. 86-4G
Firsov, V.P. 85-56J
Firsova, E.V. 85-40Q
Firth, R.J. 83-4D
Fischer, E. 86-53U

Fischer, P.W. 85-25P
Fischer, S. 83-8K; 86-32P,33P
Fiscina, C. 85-11G
Fisenko, V.V. 80-32J
Fish, R.W. 76-13H
Fisher, A. 79-1U
Fisher, H.N. 76-8S
Fisher, J. 84-33K
Fisher, L.M. 86-1FF
Fisher, M.E. 83-17JM
Fisher, R.A., Jr. 83-10T
Fisher, U. 83-7Q
Fishman, N.M. 79-1M; 83-2Q
Fisk, Z. 85-53P
Fiszdon, J.K. 79-10U
Fiszdon, W. 85-19J
Fitt, A.D. 85-8H
Fitzgerald, C.S. 79-8P
Fitzgerald, J.W. 84-34K
Fitzgerald, T.J. 82-4D
Fitzjarrald, D.E. 76-30F; 81-19F
Fitzwater, M.A. 86-8K,9K
Fiveland, W.A. 84-35K
Flach, G.P. 86-29A
Flack, R.D. 79-38F,39F; 80-16F,63F
Flamant, G. 82-17K; 83-11T; 84-13DP; 85-31T
Flament, G. 79-18J
Flanigan, L.J. 81-74J
Fleming, H.E. 76-12K
Fletcher, L.S. 76-22A; 78-5L; 83-25A; 85-43J; 86-43A
Flocchini, R.G. 84-13T,14T
Flood, G.J. 86-4D
Floquet, A. 77-7S; 86-15N
Florence, D.E. 85-26C; 86-67C
Flores, A.F. 83-18B; 85-23B
Florschuetz, L.W. 79-31C; 80-16C; 81-10C; 82-9JJ,10JJ; 84-6H; 85-9H
Flory, F. 84-10K
Floudas, C.A. 86-21Q
Flower, J.R. 80-6Q
Focardi, G. 86-6JJ
Focke, W.W. 85-34Q; 86-22Q
Fodemski, T.R. 85-39J
Foerschner, H. 86-12S
Foerster, W. 85-17C
Fogeleve, V.A. 82-56K
Fokin, B.S. 86-27J
Fokin, V.S. 83-34J
Folger, R.L. 80-47P
Fomin, N.A. 85-5E
Fomin, N.N. 85-38K
Fomin, W.M. 80-20A
Fomina, V.N. 79-11Q
Fong, E. 78-17U
Fontana, D.M. 85-28F
Fonyo, Z. 86-86Q
Foord, R. 78-10P
Foosnaes, J.A. 79-11U
Ford, J.D. 77-4A; 80-77J; 84-15P
Forest, A.E. 84-29S
Forest, T.W. 84-15J
Forestier, B. 76-47C
Forman, R.A. 76-1K,2K
Forn, G. 85-5U
Forostov, V.M. 84-17Q
Fortes, M. 81-11S
Fortin, M. 86-18N
Fortunato, M. 83-18F
Foster, C.A. 86-55U
Foster, N.W. 80-5H
Foumeny, E.A. 86-31Q
Fourguette, D.C. 85-33E
Fowler, A.C. 85-39F,40F
Fowler, A.M. 84-24K
Fowler, B.W. 78-8K
Fowler, L. 77-45P
Fowler, R.J. 78-7Q
Fraley, S.K. 80-12A
Franca, L.P. 86-35N,39N
France, D.M. 76-7M; 77-3M; 79-19J; 80-13A; 81-26J,52J; 86-28J
Francey, J.L.A. 85-13T
Francis, A. 85-27S
Francis, J. 82-16K; 84-40K
Francis, J.E. 77-19T
Francis, M.S. 78-18P
Frank, G. 86-23Q
Franke, M.E. 84-18F
Frankel, J.I. 85-14A
Fraysee, D. 86-2T
Frechette, M.F. 84-17P
Frederick, D. 85-12JM
Frederick, J.R. 85-13E
Frederick, N. 86-107Q
Frederick, R.L. 85-27C
Frederking, T.H.K. 77-10F; 83-9B
Frediani, H.A., Jr. 77-9Q
Freedman, A. 82-23U
Freeman, R.G. 79-73F
Freymuth, P. 76-6E; 80-13P
Frezzotti, A. 85-19J
Frick, H. 80-17F; 82-24F; 83-16F,17F; 85-22F
Fricke, A.L. 86-2P
Fricke, J. 84-14K; 85-20K,99K,6P,41P; 86-20K, 17P
Frid, S.E. 86-26T
Friedly, J.C. 85-15A
Friedman, A. 86-11DP
Friedman, D. 77-30E
Friedman, J.N. 82-48K
Friedrich, R. 82-25D; 84-5G
Friedrich, R.R. 80-7H
Friehe, C.A. 78-3E
Friel, R.G. 86-22N
Friend, I.E. 84-4F
Frierson, R.V. 86-22U
Frieud, R.H. 86-23E
Frik, P.G. 86-20C,12FF
Frind, G. 78-7U; 79-12U
Frisch, U. 78-12E
Frivik, P.E. 77-8S; 82-10A
Froishteter, G.B. 76-8B; 86-25B,26B
Frolov, B.A. 82-2M
Frolov, G.A. 85-64B
Frolova, A.A. 77-3K

Frolova, G.M. 80-102J
Frolova, N.I. 79-10J
Froment, G.F. 80-26K
Frossati, G. 86-40E
Frost, D. 86-29J
Frost, S. 81-13B
Froyshteter, G.B. 83-16B
Fryer, P.J. 85-35Q; 86-24Q
Fuchs, L. 86-23N
Fuchs, R. 79-13T
Fugiwara, Y. 80-18C
Fujii, A. 86-94Q
Fujii, K. 83-39F; 85-69F
Fujii, M. 76-31F; 79-40F; 80-17C,14FF,15FF; 81-14FF; 84-17B,19F,16J,23J; 86-5D,30J,47S
Fujii, S. 77-9S
Fujii, T. 76-31F; 79-40F; 80-14FF,15FF,33J, 34J,35J,36J,37J; 81-14FF; 82-1F; 84-17B, 19F,9JJ; 85-10H,36Q; 86-5D,12JJ,13JJ,22JJ, 27JJ
Fujikawa, S. 80-38J
Fujiki, K. 85-76J
Fujimura, K. 86-104F
Fujimura, T. 86-107P
Fujinawa, K. 86-24N
Fujio, A. 86-24JJ
Fujishiro, K. 86-72J
Fujita, H. 84-13C
Fujita, K. 86-47S
Fujita, N. 82-20C; 84-12C
Fujita, T. 78-16C; 81-16T
Fujita, Y. 77-35J; 84-42J; 86-31J
Fujitysa, M. 86-35S
Fukabori, M. 86-41K,42K,43K
Fukada, T. 86-52F
Fukai, J. 85-29A; 86-34P
Fukshansky, L. 86-18T
Fukshansky-Kazarinova, N. 86-18T
Fukuda, K. 84-67F; 85-37Q,38Q
Fukuda, T. 84-30F,31F,32F; 85-37Q,38Q; 86-27B
Fukui, K. 80-40F; 83-17B; 85-28C
Fukuma, H. 82-40B
Fukuma, M. 84-26DP; 86-42DP,43DP
Fukuoka, I. 85-27FF
Fukuoka, Y. 86-24S
Fukusako, S. 76-39B,37D,79F; 77-91F; 78-9D, 84F,85F,86F,64J,28K; 79-12D,10L,11L; 80-31D; 81-33F,75J; 82-41B,31F; 83-38B,39B,40B,56F; 84-9D,32F,34JM,35JM; 85-49JM; 86-32J
Fulacher, L. 77-19H
Fulachier, L. 77-3H; 83-17C; 84-14C; 85-20C
Fullarton, D. 85-8JJ; 86-14JJ
Fuller, K.A. 82-2K
Fumarola, G. 85-30K
Fumeaux, P. 79-1T
Funaba, H. 81-53J
Fung, W.S. 77-11C
Furayev, V.S. 79-55J,56J; 80-21J
Furst, J. 80-18Q; 85-39Q; 86-25Q,26Q
Furuhama, S. 85-11E,25S; 86-11S
Furukawa, K. 86-41S
Furukawa, M. 79-1R; 81-12S
Furukawa, Y. 86-23JJ
Furusho, H. 86-40JM,41JM
Furuya, H. 85-40P
Furuya, Y. 76-18C
Fusegi, T. 85-19N; 86-42F,25N
Fussey, D.E. 78-32F

Gabitto, J.F. 81-7D
Gabor, J.D. 80-16FF,39J
Gaddis, E.S. 79-20J,5Q
Gaddy, J.L. 77-12Q
Gadgil, A. 81-63F; 82-69F; 84-8JM; 86-1B
Gaffuri, 76-6K
Gaffuri, G. 82-18F
Gagne, J.M. 82-28A,17U
Gagne, R. 80-5U
Gai, S.L. 77-18P
Gainer, J.L. 80-10S
Gajan, P. 84-5E,9E
Gajanana, B.C. 81-16T
Gajewski, R. 76-4T
Gal'perin, A.G. 78-10J
Gal'perin, E.V. 85-78Q
Gal-Or, B. 76-15H; 77-18H; 79-30E
Galaktionov, V.V. 85-40J
Galamba, D. 85-9JJ; 86-15JJ
Galan, J. 83-18F
Galanga, F.L. 82-5M,6M
Galassini, G. 85-14T
Galati, A. 81-27J; 85-41J
Galershtein, D.M. 81-19D
Galeski, A. 86-71P,72P
Galich, N.E. 80-22K
Galin, N.M. 79-13B,14B,1L; 85-20B
Galis, E.A. 86-108Q
Galitseyskiy, B.M. 78-22B; 82-10B
Galiullin, R.G. 85-21B
Galkin, I.M. 80-78B
Galkin, M.N. 86-9B
Gallagher, J.P. 83-19J
Gallivan, K.A. 81-46J
Galloway, D.J. 79-105F
Galpin, D.F. 86-27N
Galpin, P.F. 86-53A,26N,28N,71N
Gama, R.M.S. 86-19T
Gammon, R.B. 78-46P
Ganapol, B.D. 81-8K; 82-18K
Gandhi, F. 86-52C
Gandhidasan, P. 79-14T
Gane, C.R. 79-26A; 82-3N
Ganesan, P. 81-10H; 85-53FF
Ganeshan, S. 82-11B
Gangal, M.K. 77-38F
Ganguli, A. 81-32J
Ganguly, B.N. 84-55K
Gani, R. 80-28T; 81-6T
Ganic, E.N. 77-12J; 79-21J; 80-40J,141J; 81-50J; 82-52J; 83-30Q; 86-24J
Ganopol, B.D. 86-44K
Gany, A. 81-13S
Ganzha, V.L. 82-10D; 83-7DP,8DP; 84-7Q; 85-78DP
Garbuz, N.N. 85-22Q
Garbuzov, V.N. 86-20E

Garcia, A. 80-10A; 86-72B
Garcia, R.D.M. 85-39K; 86-45K
Gard, H.P. 81-15T
Gardiner, S.R.M. 78-29J
Gardner, E. 80-5S
Garelis, E. 85-40K
Garg, H.P. 84-9T; 85-15T
Garg, S. 86-35P
Garg, V.K. 84-15C; 86-21C
Garguichevich, G.G. 83-10JM
Gari, H.N. 83-19F
Garifulin, F.A. 82-25F
Garling, K.K. 81-25F
Garnier, F. 84-3T
Garonne, X. 81-17FF
Garrett, S.L. 86-73P
Garrett, W.D. 82-34J
Garrison, G.W. 82-13M
Garshin, P.A. 76-46P
Gartling, D.K. 78-33F,17S; 79-24T; 85-26DP
Gartner, D. 76-13B
Garza, J.M.S. 83-36FF
Gasser, R.D. 76-32F
Gates, G.A. 76-24S
Gatlin, J.A. 83-28T
Gatski, T.B. 82-5N
Gau, C. 83-11JM,12JM,36JM; 85-13JM,14JM; 86-117F,
16JM
Gau, G. 84-9JM
Gaugler, R. 85-29C
Gaugler, R.E. 84-9S
Gaultieri, G. 82-19K
Gauvin, W.H. 77-4P; 79-37C,24U; 81-15D; 82-15U, 25U; 83-6U,21U; 84-30U
Gavalas, G.R. 82-61A
Gavrilin, A.I. 80-41J
Gavrilova, T.V. 81-6U
Gay, B. 76-5Q
Gay, C. 84-36K,37K
Gay, R.R. 83-20F,47J
Gdalevich, L.B. 79-41F
Gebhardt, Z. 86-115K
Gebhart, B. 76-7F,9F,12F,33F; 77-47F,48F; 78-58F,59F,42P; 79-12F,42F,43F,83F,107F, 115F,119F,120F; 80-29D,35F,5FF,14P; 81-7FF, 23FF,26FF; 82-7FF,10FF,26FF,27FF; 83-19FF, 20FF,26JM; 84-36F,16FF; 85-19E,48FF; 86-28DP, 11FF,18FF,29FF
Gecim, B. 84-10S; 85-22B
Gee, D.L. 80-19B
Gegorig, R. 81-16B
Geiger, D. 78-5F
Geiger, G.T. 85-82C,18D; 86-45JM
Geirnaert, G. 81-7U
Geist, J. 86-16E
Gel'perin, N.I. 77-13B
Gelbin, D. 76-13D,14D; 78-13D
Geller, V.Z. 85-28P
Gellert, M. 85-20N
Genda, H. 78-19P,20P
Genechev, V.G. 80-42J
Genecli, O.F. 80-17FF
Geoola, F. 81-15FF; 82-14FF
George, A.H. 82-4D; 84-14DP
George, E.B. 86-23U
George, J. 83-16K
George, N. 85-1K
George, S.E. 82-21C
George, T.L. 84-32A
George, W.K. 86-15C
George, W.K., Jr. 77-39F; 79-44F
Georgiadis, J.G. 86-17DP
Georgiou, D.P. 85-11H
Gerashchenko, O.A. 78-21P,22P
Gerasimenko, V.A. 83-14B
Gerasimov, V.G. 84-21JM
Gerhard, E.R. 77-72F
Germerdonk, R. 85-30C
Gero, L.R. 85-17H
Gerofi, J.P. 82-12T
Gershbein, E.A. 85-31C
Gershon, P. 79-4M
Gershuni, G.Z. 79-45F; 83-15FF
Gersten, A. 76-28P
Gersten, K. 78-17C; 79-2C; 84-16C; 85-17FF
Gerstl, S.A.W. 86-127K
Gerum, E. 79-22J
Geshele, V.D. 82-58J
Geshev, P.I. 86-22C
Gessner, F.B. 76-10B,7E,8E
Getachew, D. 85-41F
Getling, A.V. 82-26F
Getmanets, V.F. 85-78K
Gettel, L.E. 82-5U
Geyling, F.T. 85-9JM
Ghaddar, N.K. 86-28B,29N
Ghajar, A.J. 77-14B; 81-16FF; 84-57B; 85-12FF; 86-29B,50S
Ghandi, M.R. 86-30T
Ghazanfari, A.R. 83-17J
Ghetzler, R. 85-28S
Ghez, R. 78-7H
Ghiaasiaan, S.M. 83-5JM; 84-17J; 85-16A
Ghizoni, C.C. 86-69P
Ghorbani-Tari, S. 85-12FF
Ghosh, N.C. 85-7G
Giaccone, A. 85-4S
Giamei, A.F. 81-43A
Gianolio, E. 81-6Q
Giarratano, P.J. 83-18C; 84-18J
Gibbs, B.M. 86-75Q
Gibson, C.H. 76-38C
Gibson, E.D. 84-29A
Gibson, J. 77-6D
Gibson, M.M. 76-19C; 79-12E; 84-9C
Gibson, R.D. 79-16A; 81-47A
Gicquel, A. 85-3U
Gicquel, R. 77-7T
Gidaspow, D. 85-84DP
Giddon, D.B. 80-47P
Giedt, W.H. 77-20H,36K,40P; 82-26A; 85-103F
Giel, P.W. 86-101F
Gil, V.V. 85-5E
Gilchrist, K.E. 86-36P
Gile, D. 81-5C

Giles, C.R. 80-12U
Gill, W.N. 81-1H; 84-10C
Gillen, A.G. 85-15JM
Gillespie, J.B. 77-19K,27K
Gilligan, J.G. 81-66A
Gills, I.A. 80-43J
Gilly, B. 78-80F; 79-112F; 81-20F
Gilmore, D.B. 77-8P; 78-9P
Gilpin, R.R. 76-23A; 77-15B,16B,20K; 78-32C, 20F,34F,48F,26J; 79-23C,24C; 81-7A,15B
Gimbutis, G.I. 79-4D; 81-3FF
Gindele, K. 85-41K
Ginoux, J.J. 78-4
Ginzburg, A.G. 86-5S
Ginzburg, A.S. 78-27J
Ginzburg, I.P. 79-4U
Giorgini, A. 76-34F
Girard, R. 83-2JJ
Girgis, S.S. 80-14A
Gitzhofer, F. 86-37P
Giuletti, M. 82-13Q
Giver, L.P. 84-38K
Giver, L.P.M. 82-51K
Gjerde, K.M. 84-15DP
Gladden, H.J. 84-38C
Gladwell, G.M.L. 83-12A,13A
Glakpe, E.K. 86-43F
Glaser, H. 82-9S; 84-19J; 85-27DP
Glaser, J. 79-31P,32P
Glaser, P.E. 76-5R
Glasner, A. 85-33N
Glass, D.E. 85-42K,21N; 86-30A
Glass, I.I. 83-26C; 86-55P
Glasser, D. 78-11A
Glassman, I. 80-21K; 81-7J,8J
Glatzmaier, G.C. 85-12P
Glazko, V.D. 80-68J
Glazkov, V.V. 79-6H
Glazunov, Yu.T. 80-0E
Gleizes, A. 82-7U,26U; 85-48U; 86-24U
Click, R.L. 79 2R
Glickman, L.R. 81-17Q
Glicksman, L. 80-12B,12C; 84-16DP
Glicksman, L.R. 78-26B,7S; 80-9D; 83-12DP
Glicksman, M.E. 85-37F
Gloski, D. 84-16DP
Gluekler, E.L. 82-6D
Glumb, R.J. 86-25U,43U
Gluschenko, V.I. 86-2S
Gnedovets, A.G. 81-18U
Gnielinski, V. 76-14B; 79-4Q,5Q; 80-10D; 81-70J
Gnoffo, P.A. 85-58C; 86-48C
Gobin, D. 82-2T; 84-8JM; 85-2JM; 86-16A
Godard, R. 84-4U
Godel, M. 77-7S
Godfrey, R. 79-30U
Godunov, V.F. 85-111J
Goehner, R.P. 85-24JM
Goel, B.S. 76-25C
Goel, P. 83-3B; 84-2D
Goel, S.K. 79-20P
Goetz, J. 81-12J; 82-21J
Gogna, P.K. 79-15T
Gogolev, E.S. 85-29S
Gogonin, I.I. 79-44J,45J; 80-75J; 82-8JJ, 17JJ; 84-20J; 85-16JJ,17JJ,18JJ; 86-16JJ, 17JJ
Goi, I.O. 86-39U
Gokce, H. 86-18JJ
Gokhale, S.S. 76-8R
Gokoglu, S.A. 84-10H
Gokuglu, S.A. 85-12H
Gol'berg, S.M. 86-46K
Golay, M.W. 86-40N
Gold, B. 78-15A; 79-28A
Gold, D. 78-18U; 81-7U
Goldak, J. 86-13S
Goldbach, C. 77-3U
Goldberg, B. 77-12T
Goldfine, N. 85-13E
Golding, J.A. 76-15B
Goldman, A. 81-16A
Goldman, L.M. 84-105K
Goldner, F.J. 78-23P
Goldratt, E. 80-15P
Goldschmidt, V.W. 76-27C; 81-11C; 86-78K
Goldshtik, M.A. 86-33J
Goldsmid, H.J. 81-4T; 86-38P
Goldsmith, S. 83-11U
Goldstein, L. 76-8G
Goldstein, L., Jr. 76-16B,17B; 77-17B
Goldstein, R.I. 82-1J
Goldstein, R.J. 76-2,50F,51F,21J; 77-16H; 78-54F,10H; 79-9H,10H,11H; 80-18F,33F,34F, 27FF,10H; 82-11D,4H,5H,6H; 83-16FF,2H,3H; 84-17C,11FF,25H; 85-6D,24DP,3F,13H,14H
Golgolev, E.S. 86-17JM
Golitsyn, G.S. 86-2G
Gollub, J.P. 80-19F; 82-27F; 85-38F
Golova, E.P. 86-97K
Golovachev, Yu.P. 85-32C
Golovanov, A.N. 85-1U
Golovanov, V.P. 80-6D
Goloveyko, A.G. 77-27A
Golovin, A.M. 85-13P
Golovney, I.G. 83-11H
Golubev, A.B. 78-60J
Golubkin, V.N. 84-39K; 85-33C
Gomes, A.-M. 85-4U
Gomes, A.M. 86-5U
Gomi, H. 86-47K
Gomi, T. 86-69S
Gomkale, S.D. 84-12T
Gomon, V.I. 84-15Q
Goncharov, V.F. 83-4Q; 85-22Q
Goncharuk, N.I. 83-39J
Goncharuk, V.P. 86-7C
Gontarev, Yu.K. 83-23Q
Gonzalez-Santalo, J.M. 76-20S
Goodarz, A. 76-20C
Goodger, E.M. 83-30K
Goodman, D.S. 85-43K
Goodrich, L.E. 78-12A
Goodrich, R.G. 86-30E
Goodrich, W.D. 76-1R,2R; 77-4R,5R,11R; 78-2R; 82-10C

Goose, J. 78-25B
Gopichand, T. 84-2H
Goplen, S.P. 79-27J
Gorbenko, V.A. 78-2B,1Q
Gorbis, Z.R. 77-53B,38K
Gordon, I.M. 82-13T
Gordon, J. 85-93DP,66S
Gordon, J.E. 85-44K
Gordon, J.I. 85-45K
Gordon, J.M. 81-7T,8T; 82-14T; 83-12T; 85-12T, 16T; 86-20T
Gordon, R.F. 86-29T
Gordoninejad, F. 84-40K
Gore, W.J.Y. 82-51K
Gorelik, G.Ye. 82-20K
Gorenflo, D. 76-11F,84F; 81-12J; 82-21J; 84-6J
Gori, F. 78-23B,24B; 80-17B; 82-6B
Goria, R.S.R. 83-19C
Gorinov, O.I. 85-9S
Gorla, R.S. 77-4H
Gorla, R.S.R. 77-40F; 78-18C; 82-22C,23C; 86-24C
Gorlov, I.G. 76-5M
Gorman, D.N. 76-13R
Gorogin, I.I. 82-3Q
Gorray, A.M. 85-7D
Gorshkov, C.F. 79-4U
Gorshkov, G.F. 82-9C
Gorshkova, S.N. 86-48K
Gorton, C.W. 76-14C
Gortyshev, Yu.F. 82-47C
Goryachko, I.G. 78-10Q
Gosh, D. 86-77N
Goshayeshi, A. 86-18DP
Gosman, A.D. 76-2G,1H,2H; 77-2H; 78-1H; 80-20F; 81-2H; 84-37B; 86-3B
Goss, L.P. 84-55K; 85-16E; 86-47E
Gosse, J. 84-15A
Goswami, D.Y. 77-21K; 81-9T
Goswami, T.K. 81-9T
Goth, Y. 86-30B
Goto, I. 84-23C
Gotoh, K. 76-61F; 78-35F
Gotoh, S. 86-26FF
Gotovskii, M.A. 86-27J
Gotovskiy, M.A. 82-10J
Gotovsky, M.A. 85-40Q
Gottifredi, J.C. 78-19C; 83-18B; 85-23B
Gottlieb, D. 83-14A
Gouda, J.H. 78-33J
Gouesbet, G. 77-4U; 80-13U; 85-53E; 86-49K, 76K
Gough, D.O. 77-102F
Gounot, J. 81-55F
Gousbet, G. 86-14FF
Govaer, D. 81-8T
Govidarajulu, T. 84-12FF
Govind, R. 84-26Q
Gowda, G.P. 86-56U
Goyal, A.K. 85-109F
Goyal, I.C. 79-68A
Goyal, K.P. 79-46F; 84-18Q
Grabebel, W.P. 81-21F
Graboski, M.S. 81-38A; 85-4P
Grace, J.R. 82-21C
Grachev, N.S. 86-46Q
Gradon, L. 77-13J; 84-10JM; 85-16JM
Grady, P.L. 80-5S
Graham, G.W. 80-12L
Graham, R.W. 76-4
Grainer, M. 82-12D
Grandke, T. 85-14E
Grandolfo, M. 83-10K
Grange, B.W. 76-24A; 77-41F
Granovskiy, V.S. 82-14JJ,15JJ
Granqvist, C.-G. 84-63K
Granqvist, C.G. 82-13K; 85-80K
Grant, I.D.R. 79-6Q
Granville, P.S. 76-9E
Granziani, G. 83-63F
Graska, M.L. 86-52C
Grassl, H. 84-33K
Grassman, P.P. 79-15B,16B
Gratten, K.T.V. 86-21E
Graves, R.A., Jr. 77-12R
Graves, R.S. 86-61P,103P
Gray, D.D. 76-34F; 78-36F; 79-47F; 83-21F
Gray, E.W. 78-9U
Gray, G.W. 85-15E
Gray, P. 85-24B
Graziani, R.A. 80-19C
Grebenschchikov, L.T. 82-30K
Grebenshchikov, L.T. 85-59K
Grechannyi, O.A. 79-20C
Gredovets, A.G. 86-23C
Green, I. 86-1G
Green, M.D. 78-14S
Green, S. 85-46K
Green, W.J. 86-34J
Greenberg, J.B. 85-22N
Greene, G.M. 85-30S
Greenfield, A.J. 80-15P
Greenspan, D. 78-13A
Greenwood, T.F. 84-18C
Gregoric, M. 82-15FF
Gregorig, R. 85-25B
Grehan, G. 86-49K
Greif, R. 76-8K; 77-34K; 78-37F,5K,9K; 79-42B, 149F,9S; 80-2G; 81-42F,43F,7G; 82-49F; 83-36F; 85-29F,12JM; 86-32F,73F,55S
Greig, J.R. 85-26B
Greitzer, E.M. 85-27B
Grewal, N.S. 80-11D,12D,19D; 81-9D; 85-28DP
Grewal, S.S. 84-49J
Greywall, M.S. 80-20B
Grief, R. 79-48F; 84-48F,51F
Griffin, O.M. 77-28A
Griffith, B.J. 77-6R
Griffith, J.W. 86-14S
Griffith, P. 76-48J; 82-26J,5JJ; 84-2JJ; 86-8J
Griffiths, R.W. 79-49F; 81-22F; 86-15FF
Griffiths, S.K. 79-7H; 83-20J; 86-15S
Grigor'ev, V.A. 78-28J
Grigor'eva, N.I. 80-75J
Grigor'yev, V.S. 77-18B; 78-45B
Grigor'yeva, N.I. 80-14H; 83-12H

Grigoropoulos, C.P. 86-18JM
Grigull, U. 76-36F,59F; 77-43A; 78-61C,53F, 9S; 79-86F,22J; 80-23A,9C; 84-9P; 85-28B
Grinbaum, M.B. 77-2P
Grinchik, N.N. 85-44DP
Grindrod, K.J. 85-7C
Grishanova, L.V. 79-34A
Grishchencko, T.G. 83-37A
Grishchenko, T.G. 78-22P
Grishin, A.M. 85-29B
Grishin, V.A. 80-52J
Grishutin, K.S. 80-17K
Grisson, W.M. 81-14S
Grivtsov, V.P. 80-51J
Grochowskiy, L.D. 79-7G
Grochowsky, L.D. 77-10G
Groen, G. 78-43B
Groeneveld, D.C. 81-63J; 86-35J,36J
Groeneweg, F. 78-33J
Groenveld, D.C. 78-14J,29J
Groff, E.G. 79-80F
Groll, M. 79-7Q; 82-4Q,5Q
Grolmes, M.A. 76-12B; 86-15JM
Groma, I. 86-26U
Gromovyk, V.I. 85-21A
Grondin, J.C. 78-80F; 79-112F
Gropp, U. 86-37J
Grosch, C.E. 82-5N
Groshev, A.I. 79-14B
Grosjean, D.F. 85-16E
Gross, J.F. 78-47B,17C
Gross, U. 81-2L; 85-41Q; 86-44F,27Q
Gross-Wilde, H. 78-24P
Grosshandler, W.L. 78-10K; 80-11K; 82-21K; 85-47K,48K; 86-6S
Grossi, R.O. 85-13A
Grossman, G. 77-31T; 78-5T; 79-29T; 83-4H, 20T; 84-11H
Grossman, I.E. 86-21Q
Grossman, J.A. 79-21P
Grotzbach, G. 81-8N
Grover, P.D. 82-59J; 84-32Q; 85-92J
Grubic, M. 85-17E
Grubka, L.J. 85-34C
Gruzintsev, G.I. 78-8G
Gryglaszewski, P. 80-21B,55B
Gryzagoridis, J. 78-38F
Guay, J. 80-7T
Guceri, J. 83-14FF
Guceri, S.I. 76-38J; 77-48A; 81-12FF; 82-22F, 23F,12FF; 86-60N
Gudakov, G.A. 82-15S
Gudmundsson, J.S. 77-3Q
Guenette, G.R. 86-18E
Guengant, L. 86-101P
Guerra, J. 86-30N
Guerrier, B. 82-2T
Gueymard, C. 86-21T
Guglieimi, G. 83-1J
Guglielmini, G. 80-44J; 85-42F
Guiffant, G. 78-5F; 83-4B
Guigon, P. 85-19Q
Guile, A.E. 77-7U; 80-14U
Guillaume, R. 81-17FF
Guilleminot, J.J. 86-37T
Guillot, A. 86-39P,40P
Guion, J. 85-47JM
Gukalov, A.V. 83-18J
Gukhman, A.A. 77-14C; 78-7B; 79-50F
Gulati, A. 85-51B
Gulino, R. 86-22E
Gumkowski, S. 79-29J
Gunchenko, E.I. 79-5G
Gundappa, M. 86-61A
Gungor, K.E. 86-38J
Gunn, D.J. 78-10D; 81-8D; 82-16J
Gunnerson, F.S. 80-45J
Gunther, K. 79-13U; 80-26U
Gunther, R. 80-17E
Guntsev, D.F. 78-58B
Guo, Z.-Y. 84-7U
Guozhen, Z. 85-14P
Guppy, J.G. 81-4S
Gupta, A.K. 76-41D
Gupta, A.S. 76-5G,12U; 77-42F; 79-13C,26F, 94F; 85-14F
Gupta, B.K. 84-18Q
Gupta, B.P. 77-46A,18T
Gupta, C.P. 82-27JJ; 85-106J,26JJ; 86-45JJ
Gupta, H.C. 86-35P
Gupta, J.P. 77-17A; 78-15B; 81-45A; 82-36A
Gupta, K.G. 78-25P
Gupta, M.C. 79-14T
Gupta, M.L. 78-1U
Gupta, M.R. 85-36P
Gupta, P.D. 79-20P
Gupta, R. 84-55K
Gupta, R.G. 80-4M
Gupta, R.N. 84-19C; 85-18B
Gupta, R.P. 83-17K
Gupta, R.S. 81-10N; 85-23N; 86-19JM
Gupta, S.C. 78-66J; 79-27A; 85-17JM,18JM
Gupta, S.K. 78-1C,36K; 82-20T
Gura, L.A. 78-4G
Gurdal, U. 81-43B
Gurgenci, H. 83-21J,43J
Gurney, G.B. 84-26C
Gurr, T. 77-21T
Gurtcheff, G.A. 86-49JM
Gurtman, G.A. 78-21A
Gusev, M.D. 79-6H
Gustaffsson, S.E. 86-79P
Gustafson, K.E. 82-11A
Gustafsson, B. 86-30N
Gustafsson, S.E. 79-22P; 82-12A; 86-41P
Gutfinger, C. 78-90J
Gutfinger, G. 76-42D
Guthrie, K.I. 82-15T
Guthrie, R.I.L. 85-27H
Gutierrez, J. 81-9N
Gutman, L.N. 86-20JM
Gutman, R.G. 76-15D
Gutowski, R. 79-57J
Gutsev, D.F. 76-41B
Guy, A.W. 76-6S
Guy, D.R.P. 86-23E

Guy, R.W. 85-99Q
Guyko, E.I. 83-4JJ
Guymon, G.L. 82-15A
Guyon, E. 79-51F; 83-54F
Guzy, C.J. 79-14J
Guzzi, R. 83-13T; 84-41K
Gvozdenac, D.D. 86-28Q,16S

H'chenko, A.L. 80-20K
Haajizadeh, M. 83-22F; 84-20F,21F
Haaland, S.E. 83-17FF
Haas, C.R. 86-60U
Haase, D.G. 86-30E
Habchi, S. 86-45F,46F
Haber, S. 84-17DP
Haberland, C. 79-21C
Haberstroh, R.D. 76-56F
Habib, I.S. 76-18B; 80-6S; 82-22K; 84-42K; 86-50K
Habib, M.A. 82-12B
Hachisu, T. 84-16A,21J
Haddad, G.N. 84-8U; 86-21U,27U
Haddock, R.L. 76-12R
Haddon, E.W. 81-17B
Hadjisophocleous, G.V. 85-94F
Hadley, G.R. 86-42P
Haegel, N.M. 86-122K
Haering, G.W. 76-7H
Hafez, M.M. 86-94N
Haga, K. 84-69J; 85-35C
Hagen, S.C. 85-17P
Hager, N.E., Jr. 83-14T
Haggag, M.H. 85-127K
Haghighat, F. 85-24N; 86-5T
Hagiwara, M. 86-64P
Hagiwara, Y. 83-64J; 85-42J
Hagler, M.O. 84-6U
Hahne, E. 80-5L; 81-2L; 82-54J; 83-22J,23J; 84-7J,67J; 85-41Q,17T; 86-44F,74J,27Q
Hahne, E.W.P. 80-31FF
Hain, R. 85-24DP
Hain, R.C. 85-6D
Haines, D.A. 86-61C
Haines, M.G. 86-18U
Haji, F. 77-20U
Haji-Sheikh, A. 79-7A; 81-14A; 83-19B; 86-44A
Haji-Sheikh, M.J. 83-19B
Hajiloo, A. 80-72B
Hakomori, K. 76-4P
Hakuraku, Y. 85-31P
Halaczek, T.L. 86-43P
Haldenwang, P. 86-5N
Hale, N.W., Jr. 79-23J; 80-15A
Haley, S.W. 84-33C
Halim, A.A.M. 86-31N
Hall, E.J. 85-8D
Hall, G.R. 82-9J
Hall, P. 78-39F
Hall, R.J. 84-43K
Hall, W.B. 76-1B
Halland, S.E. 84-5N
Haller, B.R. 84-11S
Haller, E.E. 86-122K
Halliwell, N.A. 76-18P
Hamabata, H. 83-6G; 85-98F
Hamad, G. 76-75F
Hamaguchi, K. 86-29Q
Hamblin, P.F. 80-22B
Hamblyn, S.M. 77-5U
Hamdani, A.J. 82-12A
Hamed, A. 77-10Q
Hameed, A. 78-50J
Hamelin, P. 86-24K
Hamer, B.J.G. 79-136F
Hamil, T.A. 80-14B
Hamill, T.A. 79-12B
Hamill, T.Z. 81-12B
Hamilton, H.H., II 84-42C
Hammond, G.P. 77-11C
Hampel, D. 76-29D
Han, J.C. 78-26B; 80-21F,18FF,46J; 81-18B; 84-18B; 85-36C,43J; 86-8H
Han, J.T. 79-52F
Han, L.S. 81-17A
Hanada, Y. 84-36JM
Hanaoka, Y. 86-111Q
Hanasaki, M. 86-30J
Hanawa, J. 80-21C; 81-12C,13C
Hancox, W.T. 77-14J
Haneda, H. 86-65Q
Hanjalic, K. 76-10E
Hanke, H. 86-17D
Hankey, W.L., Jr. 76-49C
Hanley, H.J.M. 85-4P
Hanna, L.M. 86-31B
Hanna, M.C. 81-28A
Hanna, O.T. 78-27B
Hannah, B.W. 77-19P
Hannemann, R. 86-49S
Hannemann, R.J. 76-17J,18J; 77-20P; 78-30J
Hannoschoeck, N. 86-4S
Hanratty, T.J. 76-20J; 77-33B,45B,25E,29J; 79-25H
Hansen, J.O. 76-33A
Hansen, L.D. 76-12P
Hansen, P.K. 84-20S
Hansen, W.N. 77-4K
Hanson, R.K. 84-58K; 85-46E
Hanus, G.J. 77-5H
Hanzawa, T. 84-22F,23F; 86-32B,33B
Hara, Y. 85-44A
Haracz, R.D. 86-51K
Harada, E. 82-26C; 84-23C; 85-30B; 86-80B, 15D,49Q,38U
Harada, H. 85-57DP
Harada, I. 80-43P
Harada, T. 81-21U
Harada, Y. 82-71F
Haraguchi, T. 79-1H
Haramura, Y. 83-24J,36J
Harasgama, S.P. 84-3G
Harda, M. 78-43P
Hardee, H.C. 85-22S
Hardianto, T. 85-59J
Harding, G.L. 82-52K; 84-6T; 85-39T

Hardisty, H. 84-7H
Hares, M.A. 85-31S
Hargrave, G.K. 85-37C; 86-17S
Hargrove, G.K. 86-25C
Harigaya, Y. 82-19S; 86-18S
Harkleroad, M. 82-6S
Harleman, D.R.F. 76-3S; 83-4T
Harley, J.G. 76-20P
Harlow, F.H. 76-52J
Harper, D.A. 76-19P
Harpole, G.M. 76-35F; 79-22C; 80-12K; 81-28J
Harrington, J.A. 77-21P
Harris, J.A. 85-18T
Harris, J.S. 81-59A
Harris, K.R. 85-51P
Harrison, A.J. 82-6FF
Harrison, D. 76-4D
Harrison, R.E. 86-56K
Harry, J.E. 79-14U; 84-9U
Hart, F.X. 80-47J
Hart, H.L. 86-58S
Hart, J.E. 84-24F; 85-42Q,43Q,44Q
Hart, R.M. 76-12P
Hartel, E.O. 85-90C,31H
Hartill, D. 86-52K
Hartley, J.G. 81-5H; 86-50N,8S
Hartline, B.K. 77-43F
Hartman, M. 86-19S
Hartmann, H. 79-19H
Hartmann, J.M. 85-107K
Hartmann, U. 85-55JM,56JM
Hartnett, J.P. 76-5; 77-2; 79-15P; 81-7B; 82-17B; 83-28B; 84-14P; 85-31B; 86-34B,43B, 27FF
Hartranft, R.J. 80-6H
Hartung, L.C. 85-84C
Haruyama, T. 86-24E
Hasan, M. 85-13FF,14FF,15FF
Hasan, M.M. 79-53F; 80-13FF
Hasantani, M. 83-18K; 84-19B; 86-59K,7T
Hasatani, M. 82-23K; 85-30DP,52K,53K
Hasegawa, E. 80-22C; 83-18FF
Hasegawa, S. 80-23B; 84-67F,52K; 85-37Q,38Q
Hasegaws, S. 83-21B
Haseler, L.E. 85-18E
Hasen, M. 85-29DP
Hasen, M.M. 81-29J
Hasen, M.Z. 81-29J
Hashemi, A. 76-8K
Hashemi, H.R. 85-56S
Hashimoto, H. 80-7S,70S
Hashimoto, K. 86-47F
Hass, M. 77-22K
Hassab, M.A. 79-54F,100F; 81-23F; 82-28F; 85-43F,92F,93F
Hassad, M.A. 83-19K
Hassan, A.K.A. 83-6B; 84-6B
Hassan, H.A. 86-51N,90N
Hassanein, A.M. 84-11JM
Hassani, V. 85-46N
Hassard, B. 82-10FF
Hassid, S. 86-22T
Hasted, J.B. 77-2K
Hata, K. 84-53J
Hata, N. 86-28U
Hatakeyama, S. 80-13D
Hatami, R. 80-7Q
Hatamiya, S. 82-39B; 86-19JJ,20JJ
Hatano, S. 86-76B
Hatfield, D.W. 81-18FF; 82-24K
Hatfield, L. 84-6U
Hathaway, D.H. 83-4G; 86-26C
Hatta, H. 85-16P; 86-31A,44P
Hatta, I. 79-23P; 85-15P
Hatton, A.P. 76-19B; 78-28B
Hattori, C. 84-13C
Hattori, H. 85-31DP
Hattori, M. 78-14A; 86-1JM
Hattori, N. 78-40F; 84-44K; 85-58F
Hattori, S. 85-1E; 86-119F
Hau, K.F.F.L. 77-2B
Haucke, H. 85-60F
Hauder, G.M. 80-29J
Hauf, W. 76-36F
Haukas, H.T. 85-45Q; 86-30Q
Hauptmann, E.G. 80-24B
Hause, A. 82-7S
Hausen, H. 80-8Q
Hauser, G.M. 77-11B; 78-20B,63B; 79-24H; 81-5K; 84-14J; 85-10FF
Havazelet, D. 83-25J
Havener, A.G. 77-22P
Hawks, K.H. 83-30B,31B,32B,14Q; 84-28B
Hawley, M.C. 85-8U
Hawlitschek, K. 79-29A
Hay, D.R. 77-101F
Hay, N. 80-1H; 82-4N; 83-14F; 85-15H,16H
Hayasaka, H. 86-53K
Hayashi, T. 81-8U
Hayashi, Y. 76-19J; 77-15J; 79-24J; 83-2JM; 84-20B,49C; 86-39J,119J
Hayashiguchi, E. 82-32K
Hayden, T.E. 82-10C
Hayes, L.J. 85-51S; 80-70N
Haynes, F.D. 80-25B
Hayward, G. 83-10D
Hayward, G.L. 78-20C
He, G.H. 86-53E
He, P. 86-12U
Head, M.R. 76-21C
Healzer, J.M. 78-15H; 81-47J
Heart, J.E. 83-20B
Heath, M.T. 84-11H
Hebecker, D. 85-54U
Heberlein, J.V. 77-6U
Hechler, C. 83-45J
Hedderich, C.P. 82-6Q
Hedge, S. 78-43A
Hedges, J.H. 77-2R
Hedrick, J.K. 78-8H
Heestand, J. 83-23F
Hefner, J.N. 76-8H; 79-4H
Hegemier, G.A. 76-38A; 78-21A
Heggs, P.J. 76-6Q; 78-11Q; 80-16A; 81-18A; 82-13A; 83-8Q,17Q; 84-27A; 86-31Q
Heideman, N.J.H. 84-1P

Heidrick, T.R. 77-10E,11E
Heikal, M.R.F. 78-28B
Heim, A. 77-2Q; 80-8S,9S
Hein, D. 82-22J
Hein, J.C. 84-17A
Heinen, K.-H. 85-54S
Heinrich, J.C. 78-60F; 79-85F; 82-65F; 85-25N
Heisler, M.P. 82-29F
Helbig, V. 84-25U
Helland, K.N. 76-21P; 77-12E; 78-3E; 79-13E
Heller, J.M., Jr. 77-6P
Helliwell, J.B. 79-6M
Helliwell, W.S. 85-49K
Helms, V.T., III 79-24P
Hemming, M.K. 86-39K
Hemminger, W. 86-45P
Hempel, H. 80-7H
Hencken, K.R. 85-39E
Henderson, J.B. 79-38T; 85-17P
Hendi, A.A. 85-97K
Hendricks, E.W. 84-12E
Hendricks, R.C. 85-9A
Hendrie, I.J. 76-4B
Hendrix, C.E. 78-110F
Hendrix, M. 84-6E
Heng, K.T. 77-3J; 78-13J,14J
Hennecke, D.K. 84-42DP,19Q
Hennecke, E. 81-30J
Henry, A. 85-25K,63K; 86-29K
Henry, R. 80-21E
Henry, R.E. 76-12B; 79-25J; 81-13J,46J; 83-27J, 28J; 84-14J; 85-10FF
Hensel, E. 86-32A,32N
Hensel, E.C., Jr. 86-33N
Henstock, W.H. 76-20J
Henze, R.H. 81-19A; 84-25F
Herbert, D.M. 84-18A; 86-33A
Herbsleb, G. 86-2Q
Herman, A. 77-4F
Herman, B.M. 81-10T
Herman, M. 79-12L
Hermans, L.J.F. 86-40J
Hernandez, H. 82-18D
Herndon, R.C. 80-10S
Herrera, O. 79-26J
Herrick, C.S. 82-16T; 83-15T,16T
Herring, H.J. 77-19E
Herring, J.R. 84-16F
Herschcovitch, A.I. 82-11U
Hertzberg, A. 85-73K
Herwig, H. 84-16C; 85-32B,33B,16FF,17FF; 86-27C
Herziger, G. 86-60U
Hesketh, P. 85-19E
Hess, C.F. 79-55F; 82-17T,18T
Hess, S. 77-13E
Hessami, M.A. 84-26F; 85-44F
Hesse, C.F. 84-25F
Hesse, G. 76-21J
Hesselmann, K. 86-32Q
Hessler, J. 84-19P
Hetherington, H.J. 76-37F; 77-78F
Heurtault, S. 82-14A
Heutmaker, M.S. 85-38F
Hewitt, G.F. 83-26J
Hewitt, H.C. 77-38J
Hewko, R.A.D. 77-5L
Heya, N. 86-63F,64F,65F,38FF,39FF
Heyes, D.M. 80-11S
Heyt, J.W. 80-79J
Heywood, J.B. 81-7S; 84-27C
Hickman, R.B. 85-40K
Hickman, R.S. 77-75F
Hickox, C.E. 79-24T; 80-22F; 81-24F,25F; 85-26DP
Hickstep, F.L. 78-2T
Hidberg, H.R. 76-22P
Hiddink, J. 77-44F
Hidetoshi, H. 77-13U
Hieber, C.A. 81-26F; 82-13B
Higano, M. 85-70K
Higenyi, J. 79-4K
Higenyl, J. 80-13K
Higgins, J.M. 83-19FF
Higgy, M. 86-29P
High, M.D. 77-35C
Hignett, P. 81-27F
Hijikata, K. 77-33J,1M; 78-31J,13U; 80-16Q; 81-9Q; 82-53J; 84-7JJ,8JJ; 86-21JJ,44JJ
Hiki, Y. 86-57P
Hilal, M.A. 80-19FF
Hilal, M.M. 86-83F
Hildebrand, R.H. 76-19P
Hildebrandt, A.F. 76-27T
Hilgeman, T. 78-7K
Hilgeman, T.W. 84-31K
Hill, A.C. 84-45K
Hill, D. 86-43F
Hill, H.A. 84-86K
Hill, J.E. 76-10T
Hill, J.M. 83-13JM; 85-19JM; 86-10JM,11JM, 21JM,30JM
Hill, L. 80-2T
Hill, R.J. 79-15U; 80-15U
Hill, S.C. 84-45K
Hillard, M.E. 86-19E
Hilleke, R.O. 86-88Q
Hills, R.G. 79-30A; 86-32A,32N,33N
Hills, R.N. 83-14JM
Himasekhar, K. 82-16FF; 84-6G; 86-49F
Himeno, N. 80-15E; 86-21JJ
Himeno, S.-I. 81-20U; 85-20U
Hinch, E.J. 80-2C
Hinchcliffe, C. 76-15Q; 81-7Q
Hinckel, J.N. 86-6D
Hindmarsh, E. 83-15Q
Hing, S.W. 77-45F
Hinkebein, T.E. 78-41F
Hino, R. 85-44J,45J
Hinrichs, T. 81-30J
Hinz, K.-P. 86-29U
Hippensteele, S.A. 85-20E,31E
Hirabayashi, Y. 76-30D
Hirahaya, K. 86-31J
Hirai, E. 85-95Q
Hiraiwa, Y. 76-11S
Hiramitsu, M. 86-93F
Hirano, F. 85-94J

Hirano, K. 79-25P
Hirano, M. 86-54K
Hiraoka, J. 86-56B
Hiraoka, S. 86-46J
Hirasawa, S. 80-48J
Hirasawa, Y. 82-55A
Hirata, M. 80-26B; 86-15H
Hirata, T. 76-39B,37D; 79-23C,24C; 85-20JM
Hirawawa, S. 81-9Q
Hirayama, N. 84-27DP
Hirleman, E.D. 78-38K
Hiroaka, S. 86-37Q
Hirose, K. 80-40FF; 82-58F
Hirose, T. 83-22FF
Hiroto, T. 86-61J
Hirschberg, A. 77-12C; 80-23C
Hirsh, M. 78-5
Hirsh, R.S. 79-11C
Hirshburg, R.I. 82-9JJ,10JJ
Hishida, K. 80-8H; 82-23J
Hishida, M. 78-29B,26P; 79-17B; 82-30F
Hitchcock, A.H. 77-7U
Hittle, D.C. 83-15A
Hiwada, M. 82-24C,25C; 83-20C; 84-21C
Hlavacek, V. 78-30K
Ho, C.-H. 84-14JM; 85-34S
Ho, C.-J. 84-12JM,13JM
Ho, C.-M. 76-23P
Ho, C.F. 85-131K
Ho, C.J. 83-12JM,15JM; 86-22JM,23JM
Ho, C.M. 78-47P
Ho, J.C. 86-46P
Ho, J.R. 86-6DP
Ho, K.-S. 81-3N
Ho, L.J. 84-10T
Ho, R.C.C. 76-16J
Ho, S.P. 81-29F
Hobson, L. 79-14U
Hobson, N. 80-7A
Hoch, J. 81-14C
Hochreiter, L.E. 79-27K; 80-136J; 82-17Q
Hodge, B.K. 78-1E; 86-28C
Hodgson, J.A. 82-22JJ
Hodson, H.P. 85-38C
Hoebink, J.H.B.J. 80-27B,14D
Hoerger, C.R.B. 85-32DP
Hoff, H.A. 85-21E
Hoffer, O. 76-25A
Hoffman, E.G. 77-49J
Hoffman, H. 79-16U
Hoffman, J.D. 82-6N
Hoffman, M.A. 84-7B,8B; 85-3E; 86-4E
Hoffman, R. 85-46Q
Hoffman, T. 80-132J
Hoffman, T.F. 80-12A
Hoffman, T.W. 79-46P
Hoffmann, T. 85-21Q
Hogg, N.G. 85-39C
Hogg, S.W. 77-8T
Hoke, M.L. 83-20K
Holanda, R. 85-31E
Holder, G.D. 79-21E
Holian, K.S. 86-47P
Holland, K.G.T. 84-19A
Hollands, K.G.T. 76-38F,69F,70F,71F; 77-8F; 78-31F,77F,6T; 80-54F,55F,59F; 82-19F,20F; 83-12F,17T,18T; 84-23T; 85-7FF,19T; 86-92F, 3T
Hollis, S.L. 86-55K
Holloway, G.L. 85-72Q
Hollworth, B.R. 78-9H; 80-9H; 83-21C; 84-8H; 85-17H
Holm, F.W. 79-27J
Holman, J.P. 76-3
Holmberg, R.B. 77-11Q; 79-8Q
Holmer, I. 85-32S
Holstege, M.J. 85-4T
Holt, G.H. 76-67F
Holt, J.L. 84-6S,7S
Holtz, E. 86-20S
Holve, D.J. 79-10S; 82-17A,10S
Holyer, J.Y. 81-28F; 83-24F; 84-27F; 85-45F; 86-95F
Holz, A. 85-21JM
Holzknecht, B. 76-26A,6R; 77-29A
Homann, C. 85-18Q
Homsy, G.M. 76-39F; 77-105F,106F; 78-45F, 100F; 80-11F,27F; 81-57F; 82-33F; 85-11B, 110F
Honda, H. 80-33J,34J,35J; 84-9JJ; 86-22JJ, 23JJ
Honda, T. 80-14FF; 81-14FF,8U; 84-19F,14U, 15U; 85-23U,24U,25U
Honeywell, W.K. 78-42F
Hong, J.T. 85-33DP
Hong, N.S. 77-23P
Hong, S.K. 86-29C
Hong, S.W. 76-20B,21B,40F
Hongoh, M. 86-4Q
Hongpan, C. 80-14L
Honji, H. 76-16D
Honma, T. 85-26N; 86-80N
Hoogendoorn, C.J. 77-13C; 78-6,31B,60B; 79-22Q; 80-28D; 81-38F; 83-55F,39T; 85-16B,85F; 86-10A, 2JM
Hoogland, J.H.B. 84-3P; 85-18P
Hooper, F.C. 80-32T
Hopfinger, E.J. 79-56F; 81-21C,4FF
Horak, W.C. 77-30A
Hordvik, A. 77-24P
Hore, P.S. 78-18A; 79-35A
Horgan, C.O. 84-20A
Hori, K. 77-15J; 85-30DP
Hori, T. 84-27B; 85-43B,53J
Hori, Y. 86-42S
Horigome, T. 77-28T
Horigome, Y. 80-11T
Horikoshi, T. 81-34C,35C
Horman, M.H. 76-24P
Horn, H. 78-51C; 79-24B
Hornbeck, R.W. 76-18S
Horne, R.N. 78-43F,44F; 79-109F; 80-23F
Horrigan, E.C. 86-8E
Horstman, C.C. 77-25P
Horton, J.L. 77-29P
Horton, R. 85-35P

Horton, T.E. 80-8T
Horvay, G. 78-15A; 79-28A
Horwitz, C.M. 78-7T
Hosokawa, A. 86-50E
Hosokawa, I. 76-27E
Hosokawa, T. 84-10JJ; 86-24JJ
Hossain, I. 85-47Q
Hossfeld, L.M. 84-45B
Hotchkiss, G.B. 85-20T
Houdeau, J.P. 86-28K
Houf, W.G. 79-14K; 80-28K; 83-6K; 84-46K; 85-50K
House, R. 86-13S
Houston, C.K. 77-4R
Houzelot, J.L. 78-30B
Hove, D.T. 76-22C; 77-7R
Howard, F.C. 77-6H
Howard, J.R. 76-27A; 77-15D,21D; 85-31DP
Howard, L.N. 86-50F
Howard, P.A. 79-81F
Howarth, J.A. 79-8S; 81-9S
Howatson, A.M. 76-7U
Howe, B.M. 82-7H
Howe, J.T. 85-40C
Howe, R.D. 84-58K
Howell, J.R. 77-3; 78-1T; 79-2T; 82-44K; 83-33K, 34K,50K; 84-84K; 85-61K; 86-69K
Howison, S.D. 85-22JM
Howkins, B. 80-21D
Hromadka, T.V., II 82-15A; 84-6N; 86-34N, 21S
Hrusa, W.J. 86-24A
Hrycak, P. 81-10D; 84-9H
Hrymak, A.N. 85-48Q
Hseih, C.K. 81-20A; 84-21A
Hsiao, J.S. 83-1H; 85-6H,27N; 86-24JM
Hsiao, S.-W. 86-25F
Hsieh, C.K. 77-46F,23K; 79-26P,16T; 80-17A; 81-11T; 82-25K; 86-34A
Hsieh, C.S. 83-11Q
Hsieh, C.T. 85-21T
Hsieh, D.M. 80-41B
Hsieh, D.Y. 79-57F; 81-29F
Hsieh, T.C. 76-8K
Hsiung, T.H. 77-7D
Hssantani, M. 84-12S
Hsu, C.-F. 82-14B
Hsu, C.F. 81-21A,22A,56A,57A
Hsu, C.J. 76-4M
Hsu, C.T. 78-45F; 79-58F; 80-24C,20FF; 84-5FF; 85-34DP,18FF; 86-7DP
Hsu, K.-S. 79-28J
Hsu, K.C. 80-6U; 81-2U; 83-12U,13U,14U; 84-10U
Hsu, P.-F. 86-40K
Hsu, T.R. 81-23A
Hsu, W.K. 80-42B
Hsu, Y.-C. 84-13S
Hsu, Y.F. 86-33Q
Hsu, Y.Y. 76-4; 86-41J
Hsui, A.T. 86-97F
Hua, H.M. 76-23C
Huang, B.J. 79-17T,18T; 85-49Q,21T
Huang, B.K. 80-21T; 86-23T
Huang, C.-R. 84-21B
Huang, C.C. 85-19P
Huang, C.L.D. 79-8H; 80-15D; 85-68Q
Huang, F. 83-14D; 84-49B
Huang, H.J. 81-14N
Huang, J.S. 80-16P; 85-23JM
Huang, M.-J. 84-22A; 85-19FF,20FF,21FF
Huang, P.N.S. 79-13L
Huang, S.C. 80-18A; 83-16A; 84-23A,15JM; 85-17A, 24JM
Huang, T.T. 86-40U
Hubbard, G.L. 76-22J; 78-11K
Hubble, P.E. 80-10S
Huber, E.D. 79-27P
Huczko, A. 84-11U; 86-54U
Hudina, M. 79-18B; 86-34Q
Hudson, J.L. 78-46F; 83-10G
Hudson, P.R.W. 76-28A
Hudson, R.P. 80-17P
Huerre, P. 85-41C
Huetz-Aubert, M. 86-4K
Huey, L.J. 78-27P
Huffman, P.J. 83-16JM
Hugel, H. 80-16U
Huget, R.G. 83-17T
Huggenberger, M. 86-34Q
Huggins, C.E. 80-32A; 82-31A,32A
Hughes, B.D. 84-24A
Hughes, P.J. 76-11T
Hughes, R.O. 80-9T
Hughes, T.J.R. 77-31A; 85-28N,36N; 86-35N, 36N,37N,38N,39N
Hughmark, G.A. 76-17D; 77-19B; 80-16D; 82-24J; 84-22B
Huh, K.Y. 86-40N
Huhn, J. 85-10JJ
Huhse, P. 86-30U
Hui, T.O. 85-46J
Huijser, R.H. 80-25K
Huikata, K. 80-48J; 85-29JJ
Hujikata, K. 80-15E
Hull, J.H. 86-35Q
Hull, J.R. 82-19T; 84-11T; 85-22T
Hulstrom, R.L. 82-3T; 83-5T
Humbert, M. 86-52B,64Q
Humenik, F.M. 85-23K
Humphrey, J.A. 80-83B
Humphrey, J.A.C. 81-19A,24B,30F; 84-44F; 86-46B, 73F
Hung, Y. 82-42B
Hung, Y.-H. 85-47J
Hunn, B.D. 77-9T
Hunt, B. 77-20B
Hunt, J.C.R. 79-5E; 84-28F; 86-64C
Hunt, J.L. 76-10R
Hunt, K.K. 85-51K
Hunt, L.R. 85-62C
Hunt, R. 80-21FF; 84-35U; 85-67FF,53U; 86-72C
Hunt, R.F. 81-20C
Hunter, B.P. 76-4B
Hunter, C. 76-41F
Hunter, J.B. 78-4S
Hunter, L.W. 77-32A; 81-24A; 83-17A,18A;

84-11K; 86-30C
Hunter, T.O. 78-1S
Huntley, E. 76-4B
Huo, N.C. 85-9J
Hupper, H.E. 81-31F
Huppert, H.E. 79-59F; 80-22FF
Hurford, A.C. 85-90F
Hurle, D.T.J. 77-99F
Husain, S.R. 84-14F
Huse, D.A. 83-17JM
Hussain, A.K.M.F. 80-6C,1P
Hussain, C.I. 86-3H
Hussain, N.A. 84-13FF
Hussain, T. 84-1FF
Hussaini, M.Y. 76-3G
Hussein, F.D. 81-11D
Hussian, T. 86-2FF
Hussien, T. 81-2S
Hutcherson, M.N. 83-27J,28J
Hutchinson, B.R. 86-41N
Hutchinson, P. 76-9S
Hutchison, D.E. 86-35A
Hutchison, R. 84-105K
Hutson, K.E. 84-18F
Hutton, A.G. 82-19N
Huy, T.Q. 83-2T
Hwang, B.C. 80-4T
Hwang, D.C. 78-2K; 80-5K
Hwang, G.-J. 85-34B
Hwang, G.J 76-22B
Hwang, G.J. 86-29F
Hwang, K.S. 76-23J
Hwang, L.-T. 84-29F
Hwang, T.H. 86-42J,36Q
Hylton, J.O. 76-25P
Hylton, L.D. 84-38C,36S
Hyman, E. 86-56K
Hyun, J.C. 86-16FF
Hyun, J.M. 85-46F,47F; 86-16FF

Iachetta, F.A. 78-2T
Ibbetson, A. 81-27F
Ibe, E. 85-35B
Ibrahim, E.A. 85-48J
Ibrahim, M.B. 77-21B; 81-33C
Ibrayev, Sh. 76-14U
Ichihashi, H. 86-31U
Ichimiya, K. 80-23B; 83-21B; 84-36JM; 86-9H
Ichiyanagi, Y. 86-64P
Iciek, J. 82-8H,9H
Idar, E.S., III 78-2R
Ide, Y. 80-36J; 85-70J
Idelsohn, S.R. 86-9JM
Idiatulin, Z.G. 77-26P
Ievlev, V.M. 82-7Q
Igarashi, H. 86-44U
Igarashi, T. 78-47F; 82-13D; 83-5D; 84-5D; 85-9D; 86-7D
Ignatenko, V.M. 80-36P
Ignatiev, A. 77-17T
Iguchi, I. 79-135F; 80-49FF
Iguchi, T. 84-44K
Ihas, G.G. 84-5Q
Ihle, P. 84-22J
Ihnatowicz, E. 79-29J
Iida, S. 86-43J
Iida, T. 84-38Q; 85-87Q
Iida, Y. 80-49J,50J; 82-25J; 84-7E; 86-117J
Ikai, S. 80-8H; 82-23J; 83-34C
Ikawa, H. 82-23FF; 85-71B,105J; 86-115J
Ikeda, 86-31U
Ikeuchi, M. 84-23J
Ikonomov, N. 76-22U
Ikuta, K. 85-56N
Il'chenko, A.I. 77-25D
Il'in, I.N. 80-51J,52J
Ilegbusi, J.O. 85-42C
Illarionov, Yu.T. 82-3JJ
Iloeje, O.C. 82-26J
Im, K.H. 81-1M,2M,3M,4M; 84-23B; 85-2JJ; 86-3J
Imai, H. 84-18P
Imaishi, N. 84-10P,11P; 85-20P
Imamura, T. 79-9E
Imazu, S. 80-30U
Imber, M. 79-31A,32A; 81-25A
Imberger, J. 79-9F; 80-45F; 81-7F
Imre, K. 80-38A
Imura, H. 78-34F,48F; 80-53J; 83-25F; 86-40JM, 41JM,22S,23S
Imura, S. 82-55A
Inaba, H. 78-9D,84F,85F,86F; 81-7A,32F,33F; 82-31F,11S,12S; 83-19A; 84-18DP,30F,31F, 32F,33F,14FF,18K; 86-51F,52F,53F,29JM,51JM, 48P
Inaba, S. 85-83K; 86-44J
Inaba, T. 76-39C; 80-30U
Inada, S. 81-15C; 82-27J,28J,29J; 85-49J; 86-25JJ
Inagawa, M. 86-21JJ
Inard, C. 86-3K
Inasaka, F. 86-45J
Inazumi, H. 82-19A
Incropera, F.P. 78-6K; 79-60F,13K,14K; 80-66F, 23FF,28K; 81-5FF; 82-42F,5FF,22T; 83-69F, 6K; 84-42DP,20K,46K,16P,51Q; 85-13F,48F, 68F,40FF,50K,72K; 86-35B,66DP,14F,15F,119K
Inger, G.R. 76-24C,9H; 77-8D,9D
Ingersoll, J.G. 86-24T
Ingham, D.B. 78-49F; 79-15C,61F; 81-18A,34F; 82-13A; 83-13DP,17Q; 84-27A; 85-22FF; 86-19DP, 20DP
Ingraham, J.C. 83-15U
Inguva, R. 85-8A
Inoue, A. 80-3B; 81-30C,31J,32J; 82-4J; 86-4J, 5J,6J,94J
Inoue, H. 86-57U
Inoue, M. 77-52J; 81-89J; 83-67J; 84-62J
Inoue, S. 86-70J
Inoue, T. 77-35B
Inoue, Y. 84-34F; 86-54F,55F
Inque, K. 78-21C
Inui, E. 78-49J
Inuzuka, M. 86-46J,37Q
Ioarand, Kh.E. 85-50J
Iorshina, S.M. 79-45F

Iosel'son, G.L. 86-20E
Iosif, N. 77-59A
Iqbal, M. 80-10T; 83-6T,25T
Irazoqui, H.A. 86-38Q
Irvine, R.D. 78-74J
Irvine, T.F., Jr. 76-5; 77-2; 78-3; 84-72F
Irwin, J.C. 77-5L
Isachenko, V.P. 77-17J,18J; 79-30J; 80-54J
Isaev, K.B. 86-49P
Isaev, S.V. 76-10F
Isard, J.O. 86-57K
Isatayev, S.I. 83-6D
Isayama, Y. 81-90J
Isayev, G.I. 84-24B,25B
Isayev, M.K. 83-6D
Isayev, S.V. 79-2M
Isetti, C. 80-6L
Ishi, M. 80-20J
Ishi-I, K. 76-16D
Ishibashi, Y. 83-33C; 86-73J
Ishida, H. 86-4J
Ishida, M. 80-25C; 84-21T
Ishigai, S. 80-26C,24FF
Ishiguro, R. 86-61J,111Q,68S
Ishiguro, S. 85-21U
Ishihara, M. 85-20JM
Ishihara, Y. 86-50P
Ishii, M. 76-44J; 83-29J,38J; 84-27J; 86-47J
Ishijima, K. 84-46J
Ishikawa, H. 81-14U; 86-46J,47JJ,37Q
Ishimanu, A. 84-47K
Ishimura, M. 85-94DP,112Q
Ishino, M. 86-29JM
Ishizuka, M. 86-24S
Ishov, A.G. 84-89K,90K
Iskander, M.F. 86-58K
Iskaov, K.M. 85-8G
Islam, R.M. 86-21DP
Ismaelli, A. 85-9K; 86-10K
Ismuntoyo, R.P.H. 84-10FF
Isobe, M. 81-34C,35C
Isomi, H. 86-63F
Isozaki, M. 86-39J
Israeli, M. 85-23G
Issa, R.I. 86-42N,43N
Issa Abdul-Hadi, M. 79-1J,2J
Isshiki, N. 80-28B,29B; 83-33C; 84-48C
Issi, J.-P. 85-32P
Itaya, Y. 84-12S; 85-52K,53K; 86-59K
Itin, P.G. 79-11D
Ito, H. 84-18K
Ito, M. 81-41J,14U
Ito, N. 86-15H
Ito, R. 84-34F; 86-54F,55F
Ito, S. 79-9G; 80-10H; 84-23JM; 85-35JM
Ito, T. 83-58J
Ito, Y. 79-17S; 82-1F
Itoh, M. 84-46Q
Itoh, T. 81-11K
Iuchi, S. 83-25B; 84-26B
Ivakin, V.P. 79-75F
Ivanenko, G.V. 86-123J
Ivanitskiy, G.K. 77-22A
Ivanov, A.A. 78-5J
Ivanov, D.G. 77-22D
Ivanov, E.M. 86-23C
Ivanov, K.P. 85-36B
Ivanov, O.P. 82-11JJ
Ivanov, V.I. 79-9Q
Ivanov, V.V. 85-19B
Ivanovic, M. 78-18D; 79-32B
Ivashchenko, M.M. 78-33S; 79-25S
Ivashchenko, N.I. 79-9J,1M; 82-3JJ; 83-2Q
Ivlev, A.A. 77-33A; 78-3S; 82-4C
Iwabuchi, M. 86-65Q
Iwasaka, Y. 80-22C
Iwashige, K. 84-67F
Iwashita, K. 84-54B
Iwata, Y. 83-67J
Iwicki, J. 80-14J
Iyengar, S.R.K. 82-16A
Iyer, P.A. 78-50F
Iyican, L. 80-24F,25F
Iynkaran, K. 85-19T
Izatt, R.M. 76-12P
Izumi, M. 82-13JJ; 84-11JJ
Izumi, R. 77-23Q; 78-30Q; 80-26F; 81-19B, 15C; 82-27J,28J,29J; 83-22B,23B; 84-37C; 85-46B,47B,61C,24G,65Q; 86-25JJ,55Q,112Q

Jabbari, M.Y. 78-10H
Jackman, A.P. 77-27P
Jackman, D.L. 84-25J
Jackson, D. 86-5E
Jackson, D.A. 85-7P; 86-25E
Jackson, G. 84-6U
Jackson, K.W. 83-14DP
Jackson, R. 81-11D
Jacobs, E.W. 81-30F
Jacobs, H.R. 80-70J; 82-12JJ
Jaffe, L.D. 83-19T
Jaffe, R.L. 85-23C
Jafroudi, H. 86-8D
Jager, B. 86-48J
Jaggard, D.L. 85-58K
Jagoda, I.J. 80-17U
Jaguaribe, E.F. 84-12P
Jahn, K.H. 83-35F
Jahn, R.G. 78-14U
Jahn, S.L. 85-52P
Jain, M. 80-17Q
Jain, N.K. 84-48K; 85-43C; 86-39Q
Jain, P.C. 76-25C; 79-62F
Jain, P.K. 80-55J; 83-30J
Jain, S.C. 82-32F
Jakeman, E. 77-99F
Jakubowski, G.S. 86-17T
Jaluria, Y. 76-29A,42F; 77-47F,48F; 79-63F; 80-4FF; 81-45A; 82-36A,16FF,17FF,18FF,29FF, 20T; 84-12F,14S; 85-23FF,23T; 86-36C,17FF
James, B.W. 86-51P
James, D.D. 77-20J; 81-34J
James, S.R. 85-32T
Jameson, A. 86-44N,45N
Jana, R.N. 76-27F; 77-2G,2U

Janata, J. 80-7L; 85-54K; 86-60K
Janeschitz-Kriegl, H. 85-49E
Jang, D.S. 86-46N
Jang, J.Y. 86-25JM
Janicka, J. 78-13E
Jankauskas, R.I. 84-44Q
Janssen, J.D. 86-84N
Janssen, K. 81-5S
Janssen, L.A.M. 78-31B
Janusas, V.N. 84-14B
Jany, P. 84-13E
Janzen, G. 85-26S
Jarem, J.M. 84-49K,50K,80K,81K
Jarnot, R.F. 81-14K
Jarvinen, P.O. 77-10T
Jarvis, G.T. 80-25FF
Jarzebski, A.B. 84-20Q
Jaumotte, A. 77-9H
Javeri, V. 76-23B,24B,8U; 77-22B,23B; 78-32B, 33B
Jayakar, K.M. 77-2R
Jean, B. 79-21U
Jeandel, D. 76-36C
Jeffrey, D.J. 80-2C
Jeglic, F.A. 80-19A
Jekeric, J. 86-40Q
Jelinek, M. 85-10Q
Jelti, R. 86-46N
Jemison, T.R. 82-30J
Jen, H.F. 82-13S; 84-23H; 85-90C,31H
Jena, S.K. 81-19FF; 86-56F
Jeng, D.R. 76-9K; 78-5G; 79-3G; 82-36FF; 83-23C; 86-31C
Jeng, S.-M. 84-51K; 86-22FF
Jenkins, B.G. 81-15S
Jenkins, D.R. 84-35F
Jenkins, J.D. 76-5Q
Jenkins, P.E. 76-26C,27C
Jensen, M.K. 81-33J; 82-31J; 83-3B; 84-24J, 25J,66J; 86-49J,50J
Jensen, P.A. 81-24T
Jeschar, R. 84-23P; 86-52P
Jeter, S.M. 82-21T
Jeuck, P.R., III 81-76F
Jhaveri, B. 80-27F
Jhaveri, B.S. 82-33F
Jian, H.D. 85-50F
Jiang, S. 86-53P
Jiji, L.M. 77-66A; 78-16A; 81-14C,61J; 85-57JM, 52S; 86-61JM
Jimenez, J. 85-9G
Jin, D. 86-88Q
Jin, G.T. 85-83DP
Jischa, M. 79-14E
Jischke, M.C. 77-49F
Jishke, C. 80-28F
Jittani, Y. 79-52J
Joffre, G. 82-32J
Jofriet, J.C. 86-53P
Johannsen, A. 83-20T
Johansson, A. 84-3B
Johary, A. 81-31C
Johnsen, G.W. 84-63F
Johnson, A.R. 84-7G
Johnson, B.S. 83-31J,32J
Johnson, B.V. 79-23H
Johnson, C.E. 85-48P; 86-25S
Johnson, C.H. 78-51F
Johnson, D. 79-17U
Johnson, D.C. 83-16U
Johnson, D.H. 84-13F; 86-29T
Johnson, G. 86-41Q
Johnson, G.R. 85-22E
Johnson, H.L. 86-36A
Johnson, J.C. 85-59FF
Johnson, J.W. 81-13S
Johnson, K.E. 76-28T
Johnson, L.R. 86-6P
Johnson, P.D. 77-16U; 79-18U
Johnson, R.E. 84-13H
Johnson, R.O. 76-22S
Johnson, R.R. 76-11E; 86-32T
Johnson, R.S. 81-23FF
Johnson, R.W. 76-18D; 85-37B,45K
Johnson, T.R. 81-5M
Johnson, W.S. 86-7S
Johnston, B.S. 86-51J
Johnston, P.D. 86-61K
Johnston, S.C. 76-10S
Jojima, H. 84-79K; 85-81K
Jolly, D.C. 82-8U
Joly, G. 83-33J
Jonas, P. 84-20C
Jones, A.D.W. 85-49F
Jones, A.F. 78-20H
Jones, A.M. 77-1A,2A,14S; 79-4A
Jones, A.R. 77-23P; 82-4K; 84-9K; 86-14K
Jones, A.T. 79-23A
Jones, B.F. 79-8U; 83-8U
Jones, D.D. 76-31B
Jones, D.N. 85-19F
Jones, F.W. 85-18A
Jones, G.F. 86-37A,26S
Jones, G.R. 76-21U; 77-20U; 78-3U; 79-15U, 31U; 80-15U; 86-32U
Jones, H.W. 86-45T
Jones, I.P. 79-31F,64F
Jones, J.D. 86-42Q
Jones, J.D.C. 86-25E
Jones, J.J. 85-78C
Jones, M.C. 79-9B; 86-57F
Jones, O.C., Jr. 77-1,4,19J; 78-32J,3P; 80-56J; 83-1B; 86-109J
Jones, R.A. 76-4R
Jones, R.R., III 78-16D
Jones, R.W. 77-6T
Jones, S.W. 84-19DP
Jones, T.V. 83-10H; 85-8H
Jones, W.W. 85-33S
Jongenelen, F.C.H. 78-33J
Joos, F.M. 80-9D; 85-11JJ
Joosten, J.G.H. 78-34J
Jordan, A.S. 85-9JM
Jordan, R.F. 85-50S
Jorgensen, F.R.A. 85-23E
Joseph, D.D. 85-55DP,77F

Joseph, K.S. 83-16K
Joshi, S.D. 80-30B; 81-20B; 86-23P
Joshi, S.V. 85-43U; 86-33U
Joshi, V. 84-12T; 86-25T
Joshi, Y. 83-20FF; 84-36F; 86-18FF
Joslyn, H.D. 80-4H
Jotwani, M.M. 76-16J
Jou, D. 80-7E; 83-23A
Jou, J.J. 86-26JM
Jouanna, P. 82-2D; 83-3DP
Joudi, K.A. 77-20J; 81-34J
Joulin, G. 86-31K
Jovic, L.A. 78-1J; 85-4J
Joyce, R.R. 84-24K
Ju, T.C. 86-27S
Ju, Y.-H. 86-47B,47N
Ju Robert, T.C. 86-48B
Jubran, B. 85-18H
Judd, R.L. 76-23J; 78-25J,68J; 80-57J; 81-5J, 35J; 83-63J; 85-16J,48J
Jullien, F. 78-28P
Juma, A.K.A. 83-15DP; 85-35DP
Jung, D.S. 86-52J
Jung, G. 81-3D
Junkhan, G.H. 78-22C; 85-50Q
Jurianak, J.J. 84-21Q
Jurovics, S.A. 85-34S
Jury, S.H. 77-10D
Justus, C.G. 83-34T
Juttner, B. 80-14U; 81-9U; 84-12U; 86-9U
Juul, N.H. 82-26K

Kababaev, A. 76-12T
Kaban'kov, O.N. 80-58J; 85-118J
Kabanov, A.S. 83-21FF
Kabanov, L.P. 78-41J; 83-4S
Kabashima, S. 85-94DP,112Q; 86-114Q
Kabov, O.A. 86-17JJ
Kaburaki, H. 85-98Q
Kaczenski, E.S. 78-15A
Kaczmarek, R. 86-37P
Kaczmarzyk, G. 78-1D; 80-8E
Kadambi, V. 86-36B,49B,10S,28S
Kader, B.A. 77-24B,14C; 78-7B,23C
Kadle, D.S. 84-14G; 86-73B,32C
Kadolph, S.J. 85-50S
Kadosuji, A. 83-22FF
Kadotani, K. 79-9H,10H,11H
Kafarov, V.V. 86-55DP
Kafengaus, N.L. 76-24J
Kafrouni, H. 82-7U,26U; 84-13U; 85-48U
Kaga, K. 83-29C
Kaga, S. 85-65Q
Kagawa, M. 86-33B
Kagawa, T. 85-79J
Kagermann, H. 85-106F
Kago, T. 86-37B,22DP
Kaguei, S. 77-29D,30D
Kahawita, R. 77-50F
Kahn, P.B. 86-5JM
Kaikoku, T. 80-91J,92J
Kaino, K. 82-13JJ; 84-11JJ
Kaiping, B.R. 83-22C
Kaiser, T.M.V. 85-27JM
Kaizerman, S. 82-74F
Kaji, I. 85-26N; 86-80N
Kaji, M. 86-79J,80J,81J
Kaji, N. 80-9Q; 85-51Q; 86-33C,80J
Kajikawa, T. 83-10Q; 85-97Q; 86-33JJ
Kajimoto, Y. 86-61B,62B
Kajiyama, H. 85-94DP,112Q
Kakac, S. 76-63A; 77-21J; 78-12; 83-21J,43J
Kakatsios, X.K. 85-24FF
Kalache, D. 77-7C
Kalafati, D.D. 78-12Q
Kalashnik, V.N. 79-4F
Kalashyan, M.S. 86-26T
Kalayda, Yu.A. 84-15S
Kalb, C.E. 83-24B
Kalb, H.T. 76-26P
Kalbaliev, F.I. 76-28C; 86-38B
Kale, D.D. 77-25B; 86-34C
Kale, D.S. 83-13S
Kalejs, J.P. 77-51B,24D
Kalganova, I.V. 76-9U
Kalhori, B. 85-25JM
Kalim, P. 86-78N,79N
Kalinin, A.N. 83-20A
Kalinin, E.E. 78-34B
Kalinin, E.K. 79-65F; 82-8Q; 85-38B,52Q
Kalinin, L.G. 77-33F
Kalinin, R.I. 79-9Q
Kalinin, V.V. 84-12B
Kalitzin, G.St. 86-43Q
Kalitzin, N.G. 86-43Q
Kallas, Yu.I. 85-50J
Kalmanoviz, D. 79-29T
Kalra, S.P. 84-3N; 86-23J,106J
Kalthia, N.L. 85-62FF
Kalyazin, A.L. 86-53J
Kamada, R.F. 84-13T,14T
Kamakshaiah, S. 77-8U
Kamal, J. 84-11T
Kamath, P.S. 80-18P
Kamatsu, G. 84-10JJ
Kamel, M.M. 86-83F
Kamenetskii, B.Ya. 80-31B
Kamenev, I.A. 85-50J
Kamimoto, M. 80-11T; 86-27T,28T
Kamimura, T. 85-21U
Kaminaga, M. 85-71B; 86-115J
Kaminga, M. 85-105J
Kamino, K. 82-10H
Kamiuto, K. 84-52K; 85-55K; 86-62K,29S,30S
Kamke, F.A. 85-36DP; 86-31S
Kammerud, R. 86-1B
Kamminga, W. 85-24T; 86-44Q
Kamotani, Y. 76-44F; 79-66F; 85-44C,50F
Kamyab, D. 77-5F
Kan, T. 80-59J
Kanai, Y. 80-48P
Kanarchos, A. 82-7N
Kanari, K. 86-28T
Kanayama, K. 81-33F; 84-33F
Kanbe, M. 76-19J

Kanch, M. 86-48N
Kanda, N. 77-9S
Kane, R.S. 85-37DP
Kaneko, S. 80-18U
Kanemaru, K. 83-21B
Kang, S.S. 85-56FF,90Q; 86-35C
Kang, Y. 82-37B,14D; 83-45B,7D; 84-51B; 86-24DP
Kannan, K.S. 85-13G
Kannapan, D. 77-9U
Kannappan, D. 85-7U
Kannberg, L.D. 77-10S
Kannel, K. 78-29P
Kano, M. 84-40J,41J; 86-85J
Kanoh, H. 80-10Q; 81-16Q; 86-105Q
Kanonchik, L.E. 79-86J
Kantorovich, I.I. 85-19H
Kanury, A.M. 79-10S; 82-17A,10S
Kanzaka, M. 86-65Q
Kanzaki, Y. 81-11U; 82-13U
Kanzawa, A. 78-8U; 79-29U; 81-8U; 84-14U, 15U; 85-23U,24U,25U
Kao, T.-T. 76-45F; 77-51F
Kao, T.T. 76-30A; 77-34A; 82-19FF,14Q,14S
Kao, Y.C. 81-16A
Kao, Y.K. 76-59J; 85-51J; 86-130J
Kapadia, P. 85-10JM
Kapatsyna, Yu.G. 82-9Q
Kapinos, V.M. 78-4G; 81-16S
Kapitanchuk, K.I. 86-37C
Kaplan, B. 86-52N
Kaplan, G.M. 77-15T
Kaplan, R.E. 76-3E
Kapral, R. 78-11D
Kapur, S. 76-40D
Karabourniotis, D. 82-9U; 85-22U
Karac, S. 79-80A
Karakas, A. 85-123K
Karam, R.D. 79-3R
Karasev, A.B. 76-21K
Karaseva, L.V. 86-52JM
Karashima, K. 78-1R; 79-6A
Karasina, E.S. 79-11S
Karawacki, E. 79-22P; 84-4P; 86-41P
Kardash, D.Yu. 86-46Q
Karim, F. 86-58F
Karimi, A. 81-48J
Karimova, A.G. 86-39S
Karlekar, B.V. 77-5
Karlsson, B. 82-27K
Karlsson, H.T. 80-17D
Karman, D. 84-53K
Karmi, A. 77-22J
Karni, J. 84-17C
Karniadakis, G.E. 86-29N
Karnis, J. 78-24C
Karpa, Ya.T. 85-21A
Karpenko, A.I. 76-19D
Karpenko, V.S. 78-96J
Karpov, A.A. 85-36B
Karpov, S.V. 78-28S; 79-10G
Karpul, M.E. 76-27P
Karpyshev, A.V. 86-58J
Karr, G.R. 80-32B
Karras, C. 82-9U; 85-22U
Karshima, K. 77-1R
Karsten, V.M. 85-13D
Kartashov, E.M. 77-35A
Karvinen, R. 78-25C,26C; 80-27C
Karwe, M.V. 86-36C
Karyiannis, T.G. 85-25T
Kas'yanova, M.D. 84-16Q
Kasagi, N. 86-15H
Kashani, A. 86-22J
Kashcheyev, V.M. 80-60J
Kashima, M. 86-37B
Kashinskiy, V.I. 82-33J
Kashirskiy, V.G. 76-37J; 84-26J
Kasimir, H. 86-34U
Kasparek, W. 85-26S
Kaspatek, G. 76-7L
Kassab, A.J. 86-34A
Kassabji, F. 84-5U
Kassemi, M. 85-91F
Kassemi, S.A. 80-9A
Kassner, J.L., Jr. 82-49K
Kassoy, D.R. 76-1D,46F; 77-67F,111F; 78-112F; 79-46F; 84-7C,8C; 85-51F; 86-64DP
Kast, W. 80-11H; 81-6H
Kastelin, O.N. 83-11H
Kastenberg, W.E. 77-21A
Kastner, W. 82-22J
Kasuga, T. 86-7T
Kasyanenko, M.K. 83-34J
Kasza, K.E. 85-26T
Katada, N. 82-22FF
Katagiri, M. 76-29C; 79-67F,68F
Katamura, K. 80-26B
Kataoka, I. 84-27J
Kataoka, K. 77-15C; 85-45C
Katayama, K. 81-12T; 82-42A; 83-2JM,32T; 84-21T, 22T; 85-43JM,44JM,45JM,62S; 86-39JM,40T
Katayama, T. 86-53K
Katinas, V.I. 76-30C,31C; 84-55Q; 86-9D,10D
Katkov, Yu.D. 80-32J
Kato, K. 76-25E; 84-22F,23F; 85-39U; 86-32B, 33B
Kato, R. 85-15P
Kato, S. 76-11S; 84-37C; 85-61C,42U; 86-18D
Kato, T. 84-3D
Kato, Y. 80-37J; 86-37B,22DP
Katoh, M. 85-79J
Katoh, Y. 85-34FF
Katsaros, K.B. 77-52F; 82-34J
Katsuhara, T. 81-40F
Katto, Y. 78-48J; 79-32J,33J,34J,35J; 80-61J, 62J,63J,64J,65J,66J,67J; 81-36J,37J,38J, 39J; 82-35J,36J,37J; 83-24J,36J; 84-28J, 29J
Katz, H. 77-36A
Katz, M.A. 84-16JM
Katzir, A. 86-23K
Kaufman, W.B. 79-10Q
Kaufman, Y.J. 76-28P
Kaukler, W.F. 84-17JM
Kaul, R. 85-97J

Kaul, S. 85-13S
Kaul, U.K. 85-29N
Kaups, K. 76-10C
Kaushik, S. 85-13S
Kaushik, S.C. 79-68A
Kautz, F.A., II 85-6N
Kavaljov, O.B. 80-20A
Kaviany, M. 83-26F; 84-20DP,21DP,37F,38F, 39F,15FF; 85-33DP,38DP; 86-39B,23DP,59F
Kawaguchi, S. 84-8JJ
Kawaguchi, Y. 85-46C
Kawahara, K. 82-34K; 83-22K; 84-57K
Kawahashi, M. 81-17S
Kawahata, K. 80-48P
Kawai, T. 84-10JJ; 86-24JJ
Kawaji, M. 85-52J
Kawall, J.G. 78-30P; 79-5D
Kawamura, H. 77-26B; 78-65J; 79-118F; 80-33B; 82-39B; 86-63B,47F,112F
Kawamura, T. 83-20C; 84-21C
Kawamura, Y. 84-23JM; 85-35JM; 86-61B,62B, 90F
Kawara, Y. 77-13U
Kawara, Z. 86-61F
Kawasaki, T. 82-35F
Kawase, Y. 85-39DP
Kawashima, H. 86-119F
Kawashima, Y. 83-25B; 84-26B; 85-19A
Kawashimo, K. 78-13G; 80-21A; 84-22JM; 85-20A, 33A
Kawata, S. 85-85K
Kayansayan, N. 86-45Q
Kaye, L.A. 79-16S
Kayihan, F. 77-7H
Kays, W.M. 78-15H; 79-23E; 80-2H,3H,19H,20H; 81-6C; 83-25C
Kayser, R., Jr. 78-6S
Kazakov, A.V. 85-47C
Kazarinoff, N. 82-10FF
Kazarinoff, N.D. 86-11FF
Kazenin, D.A. 82-13J
Kazimi, M.S. 76-32F; 82-66J,20JJ
Kazlauskas, I.R. 79-35B
Kazmierczak, M. 86-27JM
Kearns, J.W., Jr. 83-11C
Keck, H. 79-15E,16E
Keck, J.C. 81-16C
Keddy, E.S. 86-80Q
Kedzierski, M.A. 85-108Q
Kee, R.J. 76-47F; 83-14C
Keefer, D. 86-35U
Keen, G.S. 86-51P
Keeper, D. 78-11U
Keffer, J.F. 76-5C,7P; 78-30P; 79-5D
Kegawa, K. 82-71F
Kehtarnavaz, H. 85-26JM
Keinonen, R.S. 84-15T
Kelahan, R.C. 77-12Q
Kelleher, M.D. 82-6Q
Kelley, C.S. 78-12K
Kelley, C.T. 85-56K; 86-55K
Kelly, H. 86-8U
Kelly, P. 86-1A
Kelly, R.E. 77-25F; 78-50F,52F
Kelm, H. 82-8A
Kelsic, B.H. 86-92P
Keltner, N.R. 81-26A,36A
Kemink, R.G. 79-72A,39B,40B,79B,130F,74J; 80-70B; 81-27A
Kemme, J.E. 80-3Q
Kemmerle, K. 85-17E
Kemp, W.R.G. 77-37P
Kempf, A. 86-54P
Kempka, S.N. 81-9FF
Kenig, E.Ya. 86-10H,54J
Kenna, J.P. 83-21T
Kennedy, D.A. 78-18P
Kennedy, J.F. 79-28J
Kennedy, L.A. 77-84F
Kennedy, T.C. 84-32A
Kenneth, A. 85-50S
Kenning, D.B.R. 78-20J,73J; 79-17J; 81-40J; 85-29J
Kennon, S.R. 85-39B; 86-49N
Kent, A.C. 84-8E
Kerby, J.S. 86-35B
Kercher, D.W. 84-22C
Kerimov, R.V. 77-54B
Kermani, M.M. 85-9A
Kern, J. 76-31A,32A,33A; 77-37A; 78-11A; 80-11Q; 81-71J; 84-22Q
Kernevez, J.P. 83-33J
Kerr, P.A. 85-10C
Kersner, R. 83-21A
Keselev, O.V. 79-1S
Kessel, J. 86-63K
Kessler, R. 81-6M
Kestin, J. 78-31P; 79-28P; 80-8P; 84-10P, 11P; 85-20P
Keswani, K.K. 76-16J
Kettleborough, C.F. 83-11Q
Key, J.C., Jr. 77-9R
Keyes, J.W. 76-7R
Keyhani, M. 83-8B,27F,28F; 85-52F,74F
Kezelis, P.M. 78-1B
Kezhyalis, R.M. 79-19B
Khabakhpasheva, Ye.M. 80-34B; 82-35C; 85-40B
Khabeev, N.S. 81-57J
Khabenskiy, V.B. 82-14JJ,15JJ
Khabeyev, N.S. 80-94J
Khachatryan, O.M. 83-4S
Khader, M.S. 77-27B; 81-28A
Khair, K.R. 85-3DP,20H
Khait, V.D. 86-36U
Khakhalin, S.Ya. 85-18U
Khalatov, A.A. 79-11G; 85-41B; 86-37C
Khalifa, A.M.A. 86-92Q
Khalifa, H.E. 79-28P
Khalil, E.E. 76-9S
Khalil, H. 82-28K
Khalil, H.M. 79-33A
Khalilollahi, A. 86-60F
Khalimov, G.G. 85-21B
Khamis, M. 82-1FF
Khan, A.A. 81-35F
Khan, A.R. 83-15DP

Khan, M. 86-11U
Khan, M.N. 79-22P
Khan, S.A. 81-10U
Khan, S.U. 78-12P
Khandurdyev, A. 76-12T
Khandwawala, A.I. 81-3Q
Khanin, V.M. 86-33J
Kharchenko, V.N. 83-4Q
Kharitonov, V.V. 85-22H; 86-37U
Khatib-Rahbar, M. 86-55J
Khavin, A.A. 83-36Q; 84-27Q
Khawaja, E.E. 76-29P
Khinkis, P.A. 79-38J
Khitrik, V.L. 85-40DP
Khodadadi, J.M. 84-5A; 86-40B
Khodan, I.V. 85-5E
Khodyko, Yu.V. 78-15K,16K
Khodzhayev, F.Sh. 82-21D
Khodzhiev, M. 76-3T
Khoe, G.K. 83-5H
Kholpanov, L.P. 86-10H,7Q
Kholpanv, L.P. 86-54J
Kholshchev, V.V. 85-24E
Khor'kov, M.G. 80-68J
Khoshbar Azari, M.S. 86-3H
Khosla, P.K. 82-55F
Khoya, T. 82-26C
Khoze, A.N. 77-23J; 78-35J; 79-36J
Khramov, N.I. 85-115J
Khristroforov, B.D. 83-21K
Khrustalev, B.A. 83-35K
Khudasko, V.V. 86-46Q
Khudokormov, A.D. 77-27A
Khudyakov, V.F. 76-5M
Khusid, B.M. 80-35B,64B; 81-40B; 82-50A; 83-12S
Khvostov, V.I. 82-15D
Kiang, R.L. 85-35S
Kibbee, S.E. 77-24J
Kibitkin, V.I. 83-34J
Kichigin, A.M. 77-50J,28P; 79-84J; 80-131J
Kido, H. 76 25E
Kihara, J. 86-1N
Kihm, K.D. 82-34F
Kikkawa, S. 80-17C; 82-10H,11H; 83-6H; 84-16S; 85-21H
Kikuchi, K. 86-44JJ
Kikuchi, M. 78-4J
Kikuchi, S. 80-18D
Kikuchi, Y. 82-35F; 85-35C,53J; 86-61F,28JM
Kikuda, T. 77-9S
Kilburn, K.D. 80-9P
Kilesso, V.S. 77-16P
Kilham, J.K. 85-37C; 86-25C,17S
Kilic, A. 81-29A; 83-13JJ
Killworth, P.D. 80-29F; 81-27F
Kilmenko, V.V. 85-56J
Kim, B.J. 86-56J
Kim, C. 86-50N
Kim, C.B. 76-48F
Kim, D.M. 80-27K; 84-25A,40F,54K; 85-53F; 86-117F
Kim, H.J. 83-5JJ
Kim, H.S. 85-41DP
Kim, H.W. 83-23C
Kim, J.H. 85-25FF
Kim, J.M. 83-26B
Kim, J.Y. 82-9K; 84-19K
Kim, K.-H. 85-54F; 86-62F
Kim, K.T. 86-18P
Kim, M.-U. 85-54F; 86-62F
Kim, M.D. 85-48C
Kim, M.W. 80-16P
Kim, S. 80-5P
Kim, S.D. 77-11D; 85-83DP; 86-24DP
Kim, S.H. 85-10DP
Kim, T.K. 85-57K
Kim, Y. 85-58K
Kim, Y.-K. 85-42B,27DP
Kimagai, S. 84-12G
Kimball, B.A. 85-27T
Kimes, D.S. 82-29K
Kimoto, H. 83-22FF; 86-11D,12D,13D
Kimoto, K. 83-30JM; 86-29JM
Kimura, H. 81-41J
Kimura, K. 85-26FF
Kimura, S. 77-12D; 80-30F; 81-8F; 83-24C, 23FF; 84-41F; 85-42DP,55F; 86-25DP
Kimura, T. 86-63F,64F,65F,38FF,39FF
Kin, D.M. 85-63S
Kin, S. 82-20D
Kind, R.J. 85-28T
Kinevskiy, O.F. 77-25J
King, D.Q. 84-28U
King, H.H. 76-34A
King, J.R. 76-43P
King, R. 79 19T
King, W.I., Jr. 77-19P
Kingsolver, J.B. 79-3L
Kinney, R.B. 86-21D
Kinoshita, H. 84-66K; 85-67K
Kinoshita, I. 84-52K; 85-55K; 86-30S
Kinsinger, R.E. 78-7U
Kinugawa, M. 85-44A
Kipp, R.M. 76-23S
Kir'yanov, V.A. 86-4JJ
Kiraishi, M. 84-79K
Kirbanov, N. 76-12T
Kirchner, W. 85-12JJ
Kirdyashkin, A.G. 79-75F,76F; 85-56F
Kirichenko, Y.A. 85-25E
Kirichenko, Yu.A. 79-69F,70F; 80-69J; 85-54J; 86-41B
Kiril'tsev, V.T. 86-11H
Kirillov, P.L. 86-42B
Kirillov, V.A. 85-22DP
Kirin, V.K. 85-64J
Kirkpatrick, A. 86-48F
Kirkpatrick, A.T. 84-8F; 86-66F,29T
Kirst, T.R. 78-20T
Kiselev, M.N. 85-49C
Kiselev, O.V. 77-5K
Kiselev, Ya.S. 85-36S
Kiselev, Yu.N. 83-21K
Kiseleva, O.A. 78-11H
Kishimani, K. 83-35JM
Kishimoto, Y. 83-17U

Kishinami, K. 80-26FF; 83-24FF,40FF; 84-33JM; 85-46JM; 86-19FF,34FF
Kishore, K. 76-18R
Kishore, V.V.N. 84-12T; 86-25T,30T
Kisley, H. 85-55J
Kiss, V. 84-67J
Kissalita, W.S. 85-13JJ
Kitaljevich, D. 85-28T
Kitamura, K. 85-27FF; 86-82F
Kitamura, M. 86-122J
Kitano, N. 80-7M
Kitazawa, K. 77-4L
Kito, Y. 80-23U,28U
Kitte, F.J. 81-18S
Kittel, P. 86-41E
Kitzinger, F. 82-15U
Kivioja, K. 83-17J
Kiwaki, J. 82-25B; 84-31B
Kizirins, S.W. 84-55K
Klabukov, V.Y. 82-30K
Klabukov, V.Ya. 85-59K
Klammer, H. 86-32S
Klapisz, C. 79-1E
Klaren, D.C. 86-47Q
Kleiman, J.K. 86-55P
Klein, D.E. 83-34K,50K; 84-84K
Klein, H. 83-37J
Klein, S.A. 76-11T,13T; 77-11T; 81-3T,18T, 29T; 82-10T,29T; 83-27T; 85-40T
Klein, W.H. 77-12T
Kleinstreuer, C. 86-12H
Kleiss, R.E. 83-9G
Kleppe, J. 76-57F
Klett, E.E. 81-9T
Klevchishkin, V.A. 78-39P
Klevtsov, I.A. 83-27K
Klick, D. 84-56K
Klima, R. 86-32S
Kliman, A.W. 82-3T
Klimenko, V.V. 81-42J
Klimkiezwicz, R. 77-39J
Klimov, M.M. 86-5S
Klimov, S.M. 83-6J,7J
Klingenberg, B. 83-27B
Klose, W. 85-60K
Klotz, H.-D. 85-14U
Klotz, S.P. 82-7H
Klubnikin, V.S. 76-9U
Klumov, A.A. 78-21H
Klyschaeva, O. 76-12T
Klyuev, N.I. 86-48Q
Kmiec, A. 83-7H
Knaap, H.F.P. 80-9E; 82-18A
Knabner, P. 86-38A
Knauss, D.T. 81-9M
Kneidel, K.E. 79-8P
Kneile, K.R. 84-62K
Kneubuhl, F.K. 86-53U
Knibbe, P.G. 86-56P
Knight, R. 84-9U
Knightly, G.H. 85-57F
Knobler, C.M. 86-26JJ
Knobloch, E. 80-31F; 81-16F; 86-67F
Knott, R. 76-60J
Knowles, G. 84-7T
Knowles, G.R. 77-18T; 79-20B
Knowles, T.R. 83-22T
Knyazev, O.A. 79-29P
Knyazeva, A.I. 83-35K
Ko, N.W.M. 79-26C; 86-14D
Ko, P.L. 85-72Q
Ko, S.-Y. 84-12H
Koaizawa, H. 84-29U
Koba, A.L. 80-1J
Kobari, M. 82-19A; 85-37S
Kobashi, K. 86-9H
Kobasko, N.I. 85-57J
Kobayashi, A. 80-19U; 86-2U
Kobayashi, K. 76-13S
Kobayashi, M. 80-37B; 82-29J; 84-17S,14U, 15U; 85-23U,24U,25U
Kobayashi, N. 80-22A
Kobayashi, Y. 85-53Q
Kobayasi, K. 85-5A,3K
Kobilarov, R. 86-63U
Kobiyama, M. 86-64K
Koblyakov, A.N. 80-36B
Kocamustafaogullari, G. 83-38J,6JM
Koch, H. 85-100K
Kochan, V.A. 76-50P
Kochanska, E. 85-38J
Kocheryzhenkov, G.V. 80-22D
Kochnov, V.G. 86-38C
Kochurkov, A.A. 83-39J
Kodama, K. 84-55J
Kodres, C.A. 80-70J
Koepke, P. 78-13K
Koester, G.L., III 85-54Q,38S
Koewing, J.W. 79-12K
Koffman, L.D. 83-40J; 84-30J
Kogan, D.Ya. 86-38T
Kogan, M.N. 85-47C
Koguchi, K. 81-75J
Kogure, Y. 86-57P
Koh, G. 85-13K
Kohl, M. 85-41K
Kohler, W. 82-22J
Kohno, K. 85-107F
Kohno, M. 84-34A; 85-114K
Kohya, T. 83-18U; 84-23C; 86-49Q,38U
Koidl, P. 84-13K
Koike, M. 85-27FF
Koizumi, H. 80-38B
Koizumi, Y. 79-31J; 80-71J; 86-62J
Kojima, J. 83-18FF
Kokorev, B.I. 78-13Q
Kolacinski, Z. 80-20U
Kolar, A.K. 81-19D; 85-28FF; 86-20FF
Kolb, C.E. 82-23U
Kolchenogova, I.P. 82-31K
Kolenda, Z. 85-38J
Koler, A.D. 80-19D
Kolesnikor, P.M. 85-1P
Kolesnikov, P.M. 79-34A
Kolesnikova, N.A. 85-55Q
Kolesnikova, N.S. 79-34A

Kolesov, V.V. 85-11F
Kollie, T.G. 77-29P
Kolling, M. 80-23A
Kollmann, W. 77-10U; 78-13E
Kolodner, P. 86-68F
Kolomensky, A.A. 85-18U
Kolovandin, B.A. 78-14E; 81-17C; 86-12C
Kolpashchekov, V.L. 78-15E
Kolyano, Yu.M. 85-21A; 86-39U
Kolykhan, L.I. 82-1B
Komai, T. 77-3G
Komarov, A.G. 77-16P
Komarov, I.A. 83-37A
Komatsu, G. 86-24JJ
Komatsu, N. 86-82F
Komazaki, Y. 86-66P
Kominami, J. 86-20D
Komiyama, Y. 85-43B
Komiyana, Y. 84-27B
Komori, S. 77-27E; 79-35P; 82-15B; 84-69F
Komori, T. 79-24J
Komoriya, T. 86-32J
Komorov, V.I. 80-93J
Komorowska-Kulik, J. 82-13H
Komotori, K. 78-49J; 80-88J,9Q; 81-53J
Kon, N. 79-7D; 80-26D
Kon'kov, A.A. 82-10U
Koncar-Djurdjevic, S.K. 77-28B
Kondak, N.M. 85-64B
Kondrat'yev, N.S. 76-1M
Kondrusik, Ye.A. 80-9J
Konev, E.V. 85-39S
Konev, S.V. 86-57J
Kong, P. 86-40U
Konno, H. 82-26C; 83-18U; 84-23C; 85-30B; 86-80B,15D,49Q,38U
Konopel'ko, I.N. 79-14S
Konopnicki, T.T. 79-38F
Konuma, M. 81-11U
Konvisar, V.I. 83-34J
Konyukh, L.A. 78-14K; 82-56K
Konzhukov, F.I. 80-2P
Kooi, C.F. 79-20T
Kool, C.F. 81-13T
Koop, C.G. 79-17E
Koopman, R.N. 76-32C
Kopec, J. 85-16DP
Kopelev, S.Z. 81-10S; 86-9B
Kopelman, J. 86-1G
Kopp, I.Z. 82-38J,1Q
Kopper, F.C. 76-11H; 77-11H; 79-30C
Koppius, A.M. 76-30P; 83-32C
Kopplin, H. 80-21U
Kops, L. 86-3E
Kopyarkevich, R.M. 79-4F
Koram, K.K. 80-73B
Korber, Ch. 83-18JM; 85-55JM,56JM
Korchner, J.A. 82-29K
Kordonsky, V.I. 79-25E; 85-5E
Kordylewski, W. 85-24B,17F
Korenic, B. 82-12Q
Kornblum, Y. 83-3H
Korneichev, A.I. 78-20J
Korneyev, A.D. 80-72J
Korneyev, S.D. 80-72J
Kornfilt, J. 78-14S
Korobko, E.V. 78-52B
Korobko, V.I. 84-44C
Korol'kov, B.P. 80-73J; 84-6Q
Korolenko, Yu.A. 84-17Q
Korotayev, V.M. 85-55Q
Korpela, S.A. 76-49F; 79-2K,37T; 80-10F; 82-36F; 83-30F; 85-50C
Korshagina, M.V. 86-50Q
Kortbeek, P.J. 85-21P
Kortikov, V.S. 76-4Q
Kortov, V.S. 86-58P
Kortschak, E. 80-21D
Koryagina, G.M. 80-10M
Kosaka, D. 84-10JJ
Kosako, M. 86-59A
Koschmider, E.L. 86-69F
Koseff, J.R. 84-57F
Koshelev, S.B. 85-22H
Kosky, P.G. 76-25J
Kosolapov, E.A. 85-59K
Kosovinos, N.E. 77-53F,54F
Kosoy, V.D. 78-13H
Kosson, R.L. 79-58J
Kostenko, V.V. 86-48Q
Koster, J.N. 82-37F
Kostic, M. 85-31B; 86-43B
Kostin, N.A. 85-40U
Kostornov, A.G. 78-34P; 83-23Q
Kostorogov, Ye.P. 79-71F
Kostov, P.D. 78-8G
Kostyuk, V.V. 86-58J
Kostyukevich, S.B. 85-2K
Kotacinski, Z. 86-41U
Kotake, S. 78-40F,36J; 80-74J; 84-24C; 85-58F, 14JJ
Kotaki, T. 86-42U
Kotaoka, I. 83-35J
Kothari, K.M. 86-21FF
Kotorynski, W.P. 86-44B
Kottenstette, J.P. 86-26E
Kottke, V. 77-16C; 78-27C,32P; 84-6D
Kottowski, H.M. 77-26J; 82-7D
Kotyuk, A.F. 76-31P
Kovachev, S.M. 82-28D
Koval'chuk, N.G. 78-5P
Kovalenko, A.S. 86-37C
Kovalenko, I.S. 85-26U
Kovalenko, V.P. 82-42C
Kovalev, S.A. 78-37J; 82-39J,58J; 85-4J
Kovalev, V.N. 79-33U
Kovarik, M. 84-23Q
Kovarik, V.J. 82-11U
Kovasznay, L.S.G. 76-23P
Kovela, I.M. 76-32P
Kovenklioglu, S. 85-65C
Kovensky, V.I. 83-7K
Kovitya, P. 84-16U,17U,18U; 85-27U,28U; 86-27U
Kovkova, A.A. 85-29FF,52FF
Kowalski, C.J. 76-33C
Koyama, H. 77-4G; 79-27C; 81-2G; 84-36C,9G;

85-30FF,35FF,80J,81J,21JJ; 86-60B,50C,51C, 82J,83J,86J,87J,32JJ
Koyama, M. 86-22DP
Koyama, S. 80-15FF; 86-27JJ
Koyanagi, M. 83-68J
Koz'menko, V.K. 79-26H
Kozai, H. 83-25F
Kozawa, Y. 83-31J; 86-94J
Kozdoba, L.A. 79-5S
Kozhelupenko, Yu.D. 86-59J
Kozhevnikov, A.A. 76-33P
Kozhevnikov, A.V. 79-11G
Kozhukar', I.A. 80-2Q
Kozhukhar, I.A. 83-6J
Kozlenko, V.I. 78-33K
Kozlov, V.N. 85-113Q
Kozlov, V.P. 86-81P,82P
Kozlov, Ye.M. 78-12D
Kraebl, J.S. 80-19P
Kraev, O.A. 78-33P
Kralova, M. 85-92DP
Kraning, K.K. 76-6S
Krasavin, A.N. 85-29S
Krasil'nikov, A.N. 86-51Q
Krasilov, N.A. 86-13H
Kraska, J.R. 79-72F
Krasnoshchekov, L.F. 78-14Q
Krasnov, S.N. 78-53B; 80-65B
Krasovskaya, L.I. 85-36U
Kratz, G.K. 83-61J
Kratzer, W. 82-22J
Kraus, A.D. 78-17A; 80-24A; 81-52A; 85-56Q
Kraus, H.G. 86-33S
Krause, A.D. 83-32Q
Krause, S. 86-52Q,53Q
Kravar, S.V. 78-14E
Kravchenko, V.A. 78-38J; 79-37J; 86-1T
Kravchenko, Ye.V. 76-43D
Kravets, V.Yu. 83-57J
Kravetskiy, L.I. 79-38J
Krebs, L. 77-8P; 86-27E
Krebs, R. 80-7B,14J; 84-12JJ
Krech, R.H. 85-17K
Kreeb, H. 79-19Q
Kreider, J.F. 78-7
Kreis, R.I. 86-51N
Kreith, F. 78-7; 80-34T; 84-4S; 86-13Q
Kreitlow, D.B. 78-15S
Krejci, L. 85-1P
Krektunov, O.P. 82-3JJ; 83-2Q
Kremnev, O.A. 78-12D
Krenk, S. 81-18C
Kress, T.S. 78-54J
Kretschmer, D. 77-30P
Krier, H. 76-8R; 81-1B,3S; 86-25U,43U
Krishan, R. 85-9N
Krishman, V.S. 83-2JJ
Krishna, L. 86-65N
Krishna, R. 79-39J,40J; 81-21B
Krishna, V.S. 85-43E
Krishna Murthy, M.V. 78-22S; 85-45P; 86-113J, 35JM
Krishnamoorthy, L.V. 85-26E
Krishnamurthy, M.V. 85-39JM
Krishnamurthy, R. 84-16FF
Krishnamurti, R. 86-50F
Krishnan, K.N. 78-35B; 82-10Q
Krishnan, M. 86-65U
Krishnaswamy, P.R. 78-21Q
Krishnaswamy, R. 79-7M; 82-27C
Krishnayya, A.V.G. 80-19S
Krist-Spit, C.E. 85-1S
Kristiansen, M. 84-6U
Kritchman, E. 83-23T
Krivenko, V.N. 85-26Q
Kriveshko, A.A. 82-69J
Krizek, F. 86-14H
Krocker, D.F. 86-65K
Kroger, D.G. 86-13E,54Q
Krokhin, Yu.I. 79-42J
Kron'kalns, G.Ye. 79-8M
Kroshilin, A.E. 86-60J
Kroshilin, V.E. 86-60J
Krowech, R.J. 77-26T
Krska, F. 84-27S
Krug, W. 85-46Q
Kruglov, V.I. 78-15K,16K
Krupiczka, R. 85-15JJ
Krupnik, L.I. 77-13B
Krupnova, L.V. 86-1FF
Kruse, H. 85-17C
Krutz, G.W. 78-18A; 79-35A
Kruus, R.A. 79-13Q
Krylov, S.Yu. 82-2M
Krymov, G.A. 82-10U
Kryuchenkov, Yu.V. 86-73K
Ktoka, K. 77-3G
Ku, J. 80-23P
Ku, J.C. 86-40K,66K
Kuang, Z.-B. 85-22A
Kubaczka, A. 81-9J
Kubanek, G.R. 79-41J
Kubie, J. 76-5D,20D; 85-43DP
Kublbeck, K. 78-53F
Kubleck, K. 80-32F
Kubo, S. 76-10K; 82-32K
Kubo, Y. 86-44U
Kubono, T. 78-10U; 84-19U; 85-29U
Kubota, H. 77-25K
Kubski, P. 83-26FF; 84-10E
Kucera, A. 83-13JM; 85-19JM; 86-30JM
Kucharenko, V.I. 85-26U
Kucharski, J. 83-7H
Kuchin, G.P. 77-25D; 80-20K
Kudinov, V.A. 86-39A
Kudo, I. 80-25D
Kudo, K. 86-53K
Kudoh, M. 84-46Q
Kudritskiy, G.R. 85-58J
Kudryavtsev, N.N. 82-33K
Kudryavtsev, V.A. 86-58P
Kududa, T. 84-18DP
Kuehn, T.H. 76-50F,51F; 78-19A,54F; 80-33F, 34F,27FF; 82-39F; 84-18FF,29FF; 85-63FF
Kuhlman, J.M. 83-25FF; 86-45U,46U
Kuhn, W. 83-8D; 85-27E

Kuiken, H.K. 77-29B; 81-20FF
Kukita, H. 86-29Q
Kukusako, S. 83-31JM
Kul'gin, A.A. 80-93J
Kul'gina, L.M. 80-93J
Kulacki, F.A. 77-55F; 79-73F; 80-15F; 83-8B, 27F,28F; 84-30DP,31DP,55F; 85-69DP,52F,73F, 74F; 86-41DP,45DP
Kulacki, F.C. 77-1
Kulakowski, B.T. 82-11Q
Kulcinski, G.L. 84-11JM
Kulic, E. 77-27J
Kulikov, A.S. 79-42J
Kulkarni, A.A. 85-23A
Kulkarni, A.V. 86-45DP
Kumada, M. 83-20C; 84-21C,29C; 86-15H
Kumada, T. 86-28E,61J,111Q,68S
Kumagai, S. 85-17G; 86-51JJ
Kumamaru, H. 86-62J
Kumar, A. 85-23N; 86-19JM
Kumar, D. 81-10N
Kumar, K.L. 79-40P
Kumar, P. 86-3G
Kumar, R. 84-58J; 85-13S; 86-70F,79F
Kumar, S. 77-1G; 80-10E; 86-67K
Kumar, S.K. 81-12D
Kumar Dash, S. 84-10C
Kumari, H. 82-20FF
Kumari, M. 78-28C,43C; 79-4G; 80-12H; 85-44B; 86-26DP
Kumata, M. 86-16D
Kumitomo, T. 85-67K
Kunar, L.N.S. 86-29E
Kunc, J.A. 85-30U
Kunc, T. 84-26A
Kuncewicz, C. 77-2Q
Kuncewiecz, C. 81-3B
Kune, J.A. 85-16U
Kung, H.-C. 77-11S
Kung, S.K. 86-34S
Kunhardt, E.E. 86-64U
Kunhikrishnan, P.K. 86-31E
Kunimoto, T. 82-34K; 86-35S
Kunitomo, T. 78-6L; 80-14K; 83-22K; 84-51C, 12K,57K,66K; 86-108K
Kuno, T. 80-12S
Kuntsevskiy, P.G. 85-49DP
Kuntz, H.L. 85-24A
Kuo, B.S. 84-42F
Kuok, M.H. 85-22P
Kuparev, V.A. 85-47C
Kupriyanova, A.V. 82-10J
Kupsova, V.S. 79-74F
Kuptsov, V.K. 78-10J
Kuragasaki, M. 83-8JJ
Kuraishi, M. 85-81K; 86-7T
Kurata, C. 80-67J
Kurihara, S. 85-34FF
Kurilenko, A.A. 85-56J
Kurima, K. 85-34FF
Kuriyama, F. 82-41A,42A; 83-25JM
Kuriyama, H. 83-18U; 85-69F
Kuriyama, I. 77-1M
Kuriyama, M. 82-26C; 85-30B; 86-80B,15D,49Q, 38U
Kurmazenko, E.A. 77-14P
Kuroda, C. 79-9G
Kurosaki, M. 84-33JJ
Kurosaki, Y. 84-48C; 86-70B
Kuroyanagi, T. 86-40A,41A
Kurpiers, P. 84-18S; 85-40S,41S,42S
Kurth, R.E. 81-74J
Kurtz, A. 84-20U,21U
Kurzweg, U. 84-13P
Kurzweg, U.H. 85-25A,45B; 86-39C
Kuscer, I. 80-9E,26P
Kusch, H.J. 86-34U
Kushida, G. 85-46B,47B; 86-55Q,112Q
Kusonoki, J. 80-48P
Kusuda, H. 82-51J; 84-32JJ; 86-72J
Kusuda, H.K. 84-31J
Kusuda, T. 84-19S
Kusumowardhoyo, M.A. 85-59J
Kuszell, A. 76-20U
Kut'in, V.B. 86-58P
Kutateladze, S.S. 79-75F,76F,77F,43J,44J, 45J,15K; 80-75J; 81-19C; 82-16B,40J,16JJ, 17JJ; 84-25C,17FF; 85-60J,16JJ,17JJ,18JJ; 86-63J
Kutepov, A.A. 84-89K
Kuts, P.S. 78-16S; 85-44DP
Kuttler, J.R. 83-18A; 84-11K
Kuvshinov, G.G. 79-12P; 82-11J,12J
Kuwahara, H. 80-91J,92J; 82-41J
Kuwahara, K. 82-38F; 86-64P
Kuyshinov, S.G. 79-10J
Kuz'menko, G.P. 80-99J
Kuz'min, A.V. 80-108J
Kuz'min, V.A. 82-30K
Kuzay, T.M. 77-5G
Kuzmanova, R.B. 81-88J
Kuznetsov, E.V. 86-78B
Kuznetsov, V.S. 79-4P
Kuznetsov, Yu.N. 78-15Q; 85-10D,57Q
Kvasnyuk, S.V. 79-4J
Kvernvold, O. 79-78F,79F; 81-36F
Kvetnyi, M.A. 85-13Q
Kwack, E.Y. 82-17B; 83-28B; 84-14P; 86-34B, 27FF
Kwok, C.K. 84-20JM
Kwon, H.K. 86-24DP
Kwon, S.S. 82-39F; 84-18FF
Kychakoff, G. 84-58K; 85-46E
Kyle, T.G. 80-40P

L'Ecuyer, M.R. 77-5H
La Graff, J.E. 85-7C
Laananen, D.H. 77-22T
Labrosse, G. 85-14T
Labuntsov, D.A. 79-46J; 80-76J
Lacey, A.A. 85-22JM
Lach, J. 82-20A; 83-12Q
Lachaine, A. 85-28E
Lackner, M.F. 79-17K
Lacovides, H. 86-5E

LaCroix, I.V. 76-59J
LaCroix, J.M. 86-21F,5N
Lacroix, M. 84-7N
Laderman, A.J. 76-12E; 78-29C
Laforce, R.P. 85-24JM
LaFramboise, J.G. 78-7A
Lagarde, J.J. 79-81A
Lago, M.E. 82-18B
Laguerie, C. 77-27D
Lahey, R.T. 85-3J
Lahey, R.T., Jr. 76-62J; 80-18P; 83-47J; 86-64J, 100J
Lahiri, A.K. 79-27A
Lahoti, G.D. 78-23S
Lai, H.M. 86-68K
Lai, J.C.S. 85-8E
Lai, K.Y.M. 85-27H
Lai, L.-P. 86-6P
Lai, M.-C. 86-22FF
Lai, S. 82-2J
Lai, S.S. 85-28A
Laird, A.D.K. 76-77F; 78-46C
Lakh, V.I. 77-42P; 78-5P
Lakshminarayana, B. 86-68N
Lakshminarayana, L. 77-17C
Lal, J. 82-12M
Lal, M. 82-21A
Lal, S. 86-2J
Laley, R.T., Jr. 85-72B
Lallemand, M. 84-26A
Lalli, J.C. 86-35T
Lam, T.T. 86-71F
Lamb, J.P. 76-9R; 80-20D
Lambert, C.J. 86-5DP
Lambert, G.A. 78-63B; 79-24H; 80-35A,16B
Lamden, D.I. 86-53J
Lampard, D. 80-1H,5H; 85-15H,16H
Lanchester, P.C. 85-29E,23P
Lancia, A. 85-2DP; 86-1DP
Landa, Yu.I. 85-98J
Landis, F. 81-43A
Landram, C.S. 76-47F; 83-19U
Landro, B. 80-15K
Landry, D.W. 86-52N
Landry, R.J. 77-31P
Lanfranco, M.J. 77-10R
Lang, H. 83-8H
Lang, M.L. 83-23K
Langer, G. 78-13D
Langlais, C. 85-18DP
Langner, H. 82-42J
Lankevich, V.I. 80-39B
Lankford, D.W. 81-71F
Lankford, K.E. 86-72F
Lanskoi, E.N. 86-36S
Lantsman, F.P. 76-26J
Laohakul, C. 85-48B
Laoli, X. 85-14P
Lapidus, L. 77-17Q
Lapin, A.M. 85-51C; 86-28JJ
Lapko, Ya.K. 79-1K
Lara, J. 85-2T
Large, J.F. 85-76DP,19Q
Larin, G. 82-17U
Larkin, B.S. 86-45B
Larkin, J.A. 85-12K
Larkin, N.A. 80-20A
Larsen, M.E. 85-61K; 86-69K
Larson, A.V. 86-73U
Larson, D.A. 84-64F
Larson, D.C. 80-12T,13T; 81-2T
Larson, D.E. 86-54B
Larson, D.W. 77-26K; 78-17S
Larson, E.D. 81-58A; 82-33D; 84-18JM
Larson, R.S. 85-58Q
Larson, T.A. 77-85F
LaRue, J.C. 77-8H; 78-14D; 80-28C
Laskaris, T.E. 76-25B; 78-18S
Lasocki, S. 81-4N
Lassmann, K. 79-12S
Lastochkin, A.P. 80-32J
Latham, D.J. 80-22U; 86-47U
Latigo, B.O. 79-8C
Latimer, B.R. 85-30N
Latimer, P. 84-59K; 85-62K
Latrofa, E. 83-1S
Latyev, L.N. 80-7P
Lau, A.C.Y. 84-53B
Lau, K.H. 77-56F
Lau, K.S. 83-16FF
Lau, P.W.K. 85-25J
Lau, S.C. 80-29C; 81-22B,27C
Laubitz, M.J. 78-25P
Lauchle, G.C. 84-26C
Launder, B.E. 76-19C,10E,1H,2H; 77-2H; 78-6E, 1H,16Q; 79-7B,18E; 80-4D; 81-2H; 83-16C; 84-7B,8B,8N; 85-37B
Laura, P.A.A. 79-36A,4R; 85-13A
Laurence, R.L. 76-34P
Laurendeau, N.M. 78-6K; 79-13K; 82-37K; 85-87K
Laurent, A. 80-5D
Lauriat, G. 82-35K; 85-31K,31N
Lauro, F. 86-30B
Lauvray, T.L. 78-19S
Laux, U. 86-67Q
Lavan, Z. 78-8T; 85-38T
Lavdas, C.H. 80-57J
Lavie, R. 85-59Q
Lavine, A.S. 86-73F
Law, C.G., Jr. 81-3G
Law, C.H. 76-49C
Law, C.K. 76-34C; 82-28C; 84-35J
Law, D.V. 76-2Q
Law, H.-S. 85-66F
Law, H.K. 82-28C
Law, W.S. 86-40C
Lawal, A. 85-49B
Lawing, P.L. 76-10R
Lawless, W.N. 79-30P
Laws, N. 83-9H
Lawson, K.E. 77-1U
Lawson, M.L. 76-52F
Lazarek, G.M. 82-43J
Lazeridis, A. 86-31T
Le Doucen, R. 86-28K,29K
Le Fevere, E.J. 76-53F

Le Goff, H. 86-16H
Le Goff, P. 82-13Q; 86-16H
Le Lan, A. 86-74Q
Le Mong, S.I. 76-14Q
Le Palec, G. 84-10T
Le Quere, P. 81-24B
Le Ray, M. 84-4E
Le Tallec, P. 86-53N
Le-Cong, P. 80-20P,21P
Lea, T.K. 86-43T
Leach, H. 81-37F
Leaf, G. 80-9K; 84-1N
Leal, L.G. 82-16C
LeBaron, B. 83-24T
Lebedev, D.P. 78-39J
Lebedev, P.D. 78-39J
Lebedev, P.I. 81-43J
Lebedev, V.P. 79-26H
Lebon, G. 79-37A; 80-25A; 86-27DP
Lebouche, M. 85-62B
LeClaire, R. 76-7Q
Lecordier, J.C. 80-22P,34P; 81-30A; 84-5E, 9E; 85-30E
LeDain, B.L. 81-38B
Lede, J. 86-31JM
Lederman, S. 79-31P,32P
LeDoucen, R. 85-25K,63K
Ledoux, M. 77-10H
Leduc, B. 77-9H
Ledwich, M.A. 85-8E
Lee, C.K.B. 82-6JJ; 83-41J
Lee, C.W. 85-10G
Lee, D.N. 84-50C
Lee, E.I. 81-61F
Lee, E.J. 76-9K
Lee, G.W.T. 83-29F
Lee, H. 82-36K; 83-24K; 85-31U; 86-70K
Lee, H.E. 76-10U; 85-32U
Lee, H.Y. 77-10F; 79-32T
Lee, J. 79-19E; 86-37S
Lee, J.-B. 85-50B
Lee, J.C. 81-5N
Lee, J.J. 81-37FF
Lee, J.L. 85-45S
Lee, J.S. 86-46B
Lee, K.M. 81-7K
Lee, L. 76-87F,88F; 82-44J
Lee, L.J. 80-13S
Lee, L.L. 84-7P
Lee, L.Y.W. 85-61J
Lee, M.-H. 78-5G
Lee, M.H. 79-3G
Lee, N. 78-16E
Lee, R. 84-32J
Lee, S.-C. 85-31FF; 86-73C
Lee, S.-L. 82-19B; 83-29B
Lee, S.C. 76-22D; 83-42J; 84-33J
Lee, S.L. 80-26A
Lee, S.S. 77-21J; 78-8
Lee, S.Y. 84-13JJ,14JJ
Lee, T.S. 82-39F,40F; 84-8G; 86-5G
Lee, W.-C. 86-47N
Lee, W.C. 84-15JJ,16JJ; 86-47B
Lee, Y. 82-36F; 83-30F
Lee, Y.C. 84-18C; 85-10U,33U,45U
Lee, Y.N. 86-56Q
Leech, W.J. 82-17Q
Leel, S.C. 86-71K
Leenaerts, R. 77-13Q
Lees, L. 84-30J
Lefebvre, A.H. 84-9J,60K
Lefort, E. 86-14FF
LeFur, B. 82-41F
Legait, A. 86-74F
Legay-Desesquelles, F. 77-18D,19D; 85-19JJ; 86-29JJ
Legkii, V.M. 77-30B; 78-36B
Legkiy, V.M. 78-17Q
Legon, G. 86-31F
Legros, J.C. 85-59F
Lehman, G. 83-21C
Lei, C.K. 85-36C
Leibenzon, A.S. 86-41C
Leibowitz, L. 85-11P,48P
Leichuk, V.L. 85-20Q
Leidenfrost, W. 79-10A,5F,5J; 80-23P; 82-12Q; 83-30B,31B,32B,7Q,14Q; 84-28B; 86-58JM
Leighton, S.B. 80-14T
Leij, F.J. 86-95P,109P
Leikin, V.A. 86-20E
Leilababy, P.A. 86-25E
Leiner, W. 84-34J,36J
Leipertz, A. 84-6E
Leitch, A.M. 85-105F
Leitsina, V.G. 78-40J; 85-48DP
Lekic, A. 80-77J
Lel'chuk, V.I. 78-37B
Lel'chuk, V.L. 80-40B; 82-2Q
Lemcoff, N.O. 81-46B,47B; 85-23DP
Lemesh, V.I. 85-5E
Lemlich, R. 85-24P
Lemonnier, D. 83-27FF
Leninger, T. 82-20D
Lenk, D. 81-31A
Lent, M. 85-60K
Lenykov, V.A. 85-4J
Lenz, T.G. 85-18T
Leohrke, R.I. 79-28T
Leonchuk, M.P. 85-32N
Leone, C. 79-3T
Leonov, S.V. 85-28H
Leonov, V.A. 85-87J
Leonova, V.I. 83-36Q
Leont'ev, A.I. 85-7H; 86-65J
Leont'eva, N.V. 85-32C
Leont'yev, A.I. 78-30C,75J; 79-2H,12H; 82-31C; 83-62J; 85-62J
Leont'yev, G.G. 78-60J
Leontiev, A.I. 79-26E; 80-125J; 81-23B,44J; 84-18H
Lepere, Lt.V.J. 82-49J
Lerner, Y. 81-45J
Leslie, D.C. 78-38B
Lesnikovskiy, A.A. 83-15C
Lesslie, D.A. 82-1T
Lessman, R.C. 81-31C

Lessmann, R.C. 84-32C,7D
Lester, T.W. 78-71F; 79-13F; 82-28K; 84-10FF
Leszczynski, Z. 86-104Q
Letan, R. 81-45J,8Q
Letering, S. 84-105K
Letzter, A. 81-11J
Leu, L.-P. 86-48B
Leukhin, Y.L. 79-10G
Leukhin, Yu.L. 85-61B
Leung, C.W. 85-60Q,61Q,62Q
Leung, J.C. 84-14J
Leung, J.C.M. 81-46J
Leupacher, W. 84-61K
Levchenko, N.M. 80-69J
Levchenko, S.A. 85-64K
Levdansky, V.V. 84-25DP; 85-48DP
Levec, J. 85-45DP,46DP
Levenspiel, O. 77-12D; 83-13Q; 84-8A,50Q; 86-106Q
Lever, D.A. 86-14N
Levi, J. 78-5C
Levin, G.A. 81-9K
Levin, P.S. 85-33E
Levin, R.L. 80-27A; 81-32A
Levin, V.A. 86-13H,72K
Levin, Ye.S. 82-8Q
Levina, G.V. 84-43F
Levina, T.I. 86-51Q
Levine, L. 86-38S
Levitan, L.L. 76-26J
Levite, A. 86-23K
Levy, E. 86-49B
Levy, E.K. 85-13G; 86-36B
Levy, S. 81-47J
Lewandowski, W.M. 83-26FF; 84-10E
Lewellen, W.S. 76-13E; 77-14E
Lewis, C.H. 85-48C,81C,6N; 86-62C
Lewis, G.B. 77-57F
Lewis, L.J. 83-5T
Lewis, M.B. 77-15E
Lewis, R.W. 76-15A,35A; 78-27A; 80-16H; 84-19JM
Lewis, W.T. 82-42F,22T
Lewkowicz, A.K. 76-18P
Lezhenin, F.F. 78-34P
Lhiaubet, C. 85-6U
Li, C.-H. 77-13D; 83-22A,5Q; 85-63Q; 86-57Q, 58Q
Li, C.H. 77-6G
Li, C.P. 77-4R
Li, H.Z. 86-31JM
Li, J. 83-7Q
Li, L.J. 85-67S
Li, M.L. 84-12Q
Li, Q. 86-30E
Li, Y.-K. 80-89J
Li, Y.-Z. 86-53T
Lian, M.S. 76-4K
Liang, R.C. 86-46P
Liao, N.S. 86-66J
Libby, P.A. 76-21D; 77-8H; 78-14D,12H; 80-28C
Liburdy, J.A. 79-80F; 86-106F
Lidburdy, J.A. 78-55F
Lie, T.T. 77-38A
Lieberman, M.L. 77-1U
Liebermann, R.W. 76-11U; 77-19U
Liebert, C.H. 85-31E
Lien, F.-S. 85-65K; 86-23FF,24FF,6G
Lien, F.F. 86-17H
Lien, F.S. 86-42C
Lienhard, J.H. 78-13P; 79-137F,47J; 80-19A, 78J; 81-33A,3J,29J,48J; 82-45J,70J; 84-11F; 85-63J,96J,97J; 86-75F,11P
Lierert, C.H. 77-8R
Ligai, V.G. 86-33J
Ligrani, P.M. 83-25C; 85-23H
Likash, V.P. 78-10L
Likherzak, E.E. 79-46B
Lilleieht, L.U. 78-2T
Lim, H.C. 80-56B,57B
Lim, I.S. 84-17JJ
Limbaugh, C.C. 84-62K
Limbourg-Fontaine, M.C. 85-59F
Lin, C. 81-83J
Lin, C.-F. 85-12B; 86-13B
Lin, C.-Y. 82-30C,17D
Lin, C.L. 76-22D
Lin, D.-S. 86-28DP
Lin, F. 84-15P
Lin, F.N. 78-31C,56F; 79-28C; 81-21FF
Lin, H.-T. 81-25B; 82-29C
Lin, J.C. 76-6S
Lin, J.I. 82-8N
Lin, J.K. 85-44C
Lin, K.C. 86-23JM
Lin, M.J. 85-34B
Lin, N.N. 83-31F
Lin, R.-C. 76-65F
Lin, R.C. 77-100F
Lin, S. 82-16D; 83-14D; 84-49B,34JJ,20JM
Lin, S.-J. 78-57F
Lin, S.H. 78-20A; 79-38A,39A,21B; 80-41B, 42B
Lin, S.P. 79-131F; 83-15D
Lin, S.T. 78-23J; 80-3J
Lin, T.-H. 85-25P
Lin, T.F. 83-30B,31B,32B,14Q; 84-28B; 86-24F, 66J
Lin, Y.K. 84-29B
Lin, Y.S. 86-76F
Linan, A. 76-58F; 84-52C
Lincoln, R.L. 77-11K
Lind, M.A. 84-64K
Lindberg, J.D. 77-19K,27K
Lindemann, A.M. 85-52C; 86-43C
Linden, P.F. 79-59F
Lindgren, E.R. 76-11E
Lindsay, K.A. 76-36A; 79-40A; 85-62F
Lineberry, J.T. 82-5M,6M
Linehan, J.H. 79-24H; 80-35A; 84-7JM
Linford, R.M.F. 77-28K
Ling, S.C. 77-16E
Linhorst, S.J.M. 83-55F
Linnhoff, B. 80-6Q; 83-15Q
Linthorst, S.J.M. 81-38F
Lior, N. 83-9F,39F; 85-70F; 86-93F

Liou, C.-T. 86-50B,51B
Lipatov, I.I. 85-19C
Lipatov, V.E. 85-10D,57Q
Lipinski, R.J. 81-60J
Lipovtsev, V.N. 86-82P
Lippig, V. 84-10G
Lipps, F.B. 76-54F
Lipps, F.W. 78-9T
Lipschutz, R.C. 76-30T
Lipsett, A.W. 78-32C
Lipson, S.G. 86-73P
Lisi, F. 86-52K
Lisienko, V.G. 86-73K,58P
List, E.J. 77-54F
Lister, C.R.B. 77-43F
Lita, M. 77-59A
Litkouhi, B. 82-2N; 86-55N
Litoshenko, A.K. 78-77J
Littlefield, D. 86-77F
Litvinov, V.D. 80-142J
Liu, B.L. 82-6M
Liu, C.H. 77-38B; 78-44B; 79-20Q,19U; 80-43B, 6U; 86-66J
Liu, C.K. 86-42A
Liu, C.L. 80-16K
Liu, H.L. 84-8J; 85-21J
Liu, J. 85-67S
Liu, J.-Y. 86-29DP,30DP,31DP
Liu, K.V. 84-11T
Liu, S.T. 80-15T
Liu, S.W. 80-30C
Liu, W.K. 82-8N
Liu, W.S. 83-26C
Liu, W.T. 77-52F
Liu, X. 86-56N
Liu, X.Q. 84-35J
Liu, Y.-D. 85-56B
Liu, Y.A. 77-17Q; 85-64Q
Liu, Y.Y. 85-48P
Liu, Ye-Di 85-72F
Livne, E. 85-33N
Liyepinya, V.E. 79-33U,34U
Ljuboja, M. 81-22FF
Llebot, J.E. 83-23A
Llory, M. 83-33J
Lloyd, J.R. 77-8K,32K; 82-5F; 83-12K; 85-67F, 111F
Lo, K.V. 85-13JJ
Lo Porto, L. 86-10K
Lobele, H.F. 76-14P
Locatelli, M. 83-24A
Locher, F.A. 79-28J
Lochet, R. 83-27FF
Lock, G.S.H. 85-27JM
Lockwood, F.C. 76-2G
Lodenquai, J. 84-26K
Loehrke, R.I. 76-56F,94F; 83-36B,35C,19F
Loerhke, R.I. 86-69Q,70Q
Lof, G.O.G. 80-31T
Loftus, J.J. 85-66K
Loftus, P.J. 83-10H
Logan, J.D. 84-86K
Logan, P.F. 77-11U
Loginov, D.A. 85-13Q
Logvinenko, S.P. 77-32P; 85-32E
Loh, C.Y. 84-23DP
Loh, J.V. 81-17J,5Q
Lohar, B.L. 79-62F
Lokai, V.I. 79-5G; 86-39S
Lokhov, Yu.N. 81-18U
Lokshin, V. 76-85F
Lokshin, V.A. 79-11Q
Lomakin, V.N. 85-6JJ
Lombard, C.K. 85-11C
Lombard, D. 86-37P
Lombardi, P. 86-7FF
Loncaric, J. 84-16F
London, A.L. 80-12Q; 83-16Q
London, E.J. 80-95J
Long, C.A. 86-7G
Long, L. 86-35JJ
Long, L.L. 83-32K
Long, M.B. 85-33E
Long, R.R. 76-55F
Longenbaugh, R. 86-32E
Look, D.C. 77-29K; 78-17K,18K; 79-16K,17K, 2L; 81-10K
Look, D.C., Jr. 86-74K,87K
Look, K.Y. 85-48B
Looney, D.J. 86-30JJ
Lopers, D.E. 83-14JM
Lopez, J. 83-24A; 84-50F
Lorch, H.O. 85-18E
Lorenz, J.J. 79-81F; 80-141J; 82-46J
Lorenzini, E. 82-22A
Lorenzo, E. 82-23T,24T
Lork, W. 86-18T
Losing, K.H. 84-36J
Lou, Y.S. 76-24D
Loudon, R. 80-11E
Loughhead, J.N. 76-2G
Loughlin, K.F. 85-53C
Louis, J.F. 78-13S; 85-11H
Louis, W.C. 78-10T
Lounsbury, R. 86-68B
Lovberg, R.H. 80-20P,21P
Love, T.J., Jr. 77-19T
Lovegrove, P.C. 83-28Q
Lovelace, R.B. 79-12J
Loverant, E.I. 83-14S
Lovinskii, L.S. 76-31P
Low, T. 77-13P
Lowe, A. 84-24DP
Lowell, R.P. 79-82F; 82-18D; 85-46S
Lowke, J.J. 76-11U; 79-20U; 84-17U; 85-28U
Lowrey, D. 86-32T
Loyalka, S.K. 82-23A; 85-53B; 86-40C
Lozano, C.J. 81-34A
Lozhechnik, F.D. 80-3K
Lu, C.C. 80-79J
Lu, P.C. 80-44B
Lu, S.C. 85-11N
Lu, S.S. 79-33P
Lu, W.-F. 84-29F
Lu, Y. 86-75K
Lubezky, I. 83-25K

Luc, A.M. 86-6A
Luca, M. 77-59A
Lucas, G.J. 83-7G
Lucas, K. 76-27J; 79-48J,49J
Lucas, P. 83-29F
Lucas, P.G.J. 84-11G
Luchay, O.V. 78-22P
Luchini, P. 86-25FF
Luchko, N.N. 81-17C
Lucht, R.P. 82-37K
Luckois, M.L. 80-29P
Ludford, G.S.S. 86-19C
Ludtke, P.R. 79-9B
Ludwig, P.W.P.H. 77-21Q
Luebke, H.W. 77-45P
Luhmann, N.C., Jr. 79-21P
Luikov, A.V. 76-37A,23D
Lukashov, V.P. 86-82U
Lukefahr, H. 86-30E
Lukens, W.E. 86-64S
Lukin, K.D. 78-17H
Lukosiavicius, L.K.-J. 85-59C
Luks, A.L. 86-48Q
Lum, M. 77-37F
Lumley, J.L. 76-97F; 77-17E; 78-17E,27E; 86-44C
Lumsdaine, E. 76-14T
Lunardini, V.J. 83-19JM
Lund, J. 84-21S
Lund, J.W. 84-20S
Lund, K.O. 76-26B
Lund, P.D. 84-15T; 85-29T
Lundardini, V.J. 81-35A
Lunde, P.J. 77-13T; 81-14T
Lundin, L.-C. 85-43S
Lunev, V.V. 85-24H
Luque, A. 82-23T,24T
Lushchik, V.G. 86-45C
Lushiku, E.M. 77-3L; 84-63K
Lustman, L. 83-14A
Lutsenko, V.I. 86-59Q
Lutsik, Ya.T. 77-42P
Lyczkowski, R.W. 84-22S
Lyford-Pike, E.J. 84-27C
Lyhs, W. 84-34U
Lykoudis, P.J. 76-28J
Lykoudis, P.S. 76-23C; 77-8E; 81-93J
Lykova, A.V. 78-13H
Lynch, G.F. 77-33P
Lynn, P.P. 76-35C
Lynn, S. 84-24Q,23S
Lynnworth, L.C. 79-8P
Lyon, R.E. 86-59P
Lyons, D.W. 80-80J
Lysenko, O.G. 85-5E
Lysenko, V.I. 84-28C
Lyubimov, D.V. 86-122F
Lyubimov, G.A. 79-35U; 83-9U
Lyubov, B.Ya. 77-35A

Ma, B.-K. 86-46C
Ma, C.-F. 86-67J
Ma, C.C.Y. 83-25T
Ma, C.K. 76-1L; 80-24P; 85-10K
Maa, J.R. 83-72J
Mabuchi, I. 81-34B; 82-24C,25C; 83-34B,20C; 84-21C,29C; 86-65B,16D
Macagno, E.O. 76-16S
Macaraeg, M.G. 86-8G
Maccamy, R.C. 77-39A
Mach, H. 83-27B
Machali, H.M. 85-127K
Machida, K. 84-64K
Mack, W.M., Jr. 78-25Q
Mackley, N.V. 76-5Q
Macleod, H.A. 84-10K
Macosko, C.L. 85-45S
Macosko, C.W. 76-4S; 80-13S
Macpherson, A.K. 77-7G
MacRae, R.A. 78-19K
Madbouly, E.E. 77-9J
Maddy, A.L. 85-38E
Madejski, J. 76-27B; 81-49J
MadhavaPanicker, P.R. 86-31E
Madhusudana, C.V. 83-25A; 86-43A
Madi, M.B. 85-67B
Madieta, E. 84-4E
Madni, I.K. 77-59F; 81-4S
Madni, K.K. 77-58F
Maebius, R.E. 85-44S
Maecker, H.H. 86-48U
Maeda, H. 85-24G,65Q
Maeda, J. 86-91P
Maeda, M. 80-8H; 82-23J
Maeda, S. 77-63A; 78-45A; 79-78J
Maeda, T. 76-10H
Maekawa, H. 81-12T
Maekawa, T. 86-78F
Maeland, E. 82-43F
Maeno, Y. 85-60F
Maesono, A. 85-15P
Maewal, A. 76-38A; 78-21A
Magallon, D. 81-39F
Magatome, S. 81-89J
Magen, M. 86-28B
Magidey, P.L. 78-20S
Mah, R.S.H. 86-46JJ
Mahajan, R.L. 78-58F,59F; 79-83F; 80-35F
Mahajan, S.M. 84-22U
Mahajan, S.P. 85-108J
Mahajerin, E. 85-7N
Mahan, J.R. 79-3L
Mahbod, B. 84-65K
Mahefkey, T. 85-74Q
Maheshri, J.C. 81-62J
Maheu, B. 86-49K,76K
Mahin, K. 86-33Q
Mahoney, D.N. 86-79F
Mai, H.H. 79-21U
Mai, T. 85-62S
Maier, W. 84-14H
Mailen, G. 77-31J
Main, W.A. 86-61C
Maingonnat, J.F. 86-60Q,61Q
Maiorov, V.A. 79-24S
Mairovo, V.A. 78-15D

Maiti, G. 76-76F
Maiti, M.K. 77-41B
Maitra, P.P. 81-11D
Majhi, S.N. 86-6F
Major, B.H. 82-12JJ
Major, S.J. 77-40A
Majumdar, A.K. 83-15J,19Q,20Q
Makarevicius, V.I. 78-33C
Makarov, B.A. 85-113Q
Makarov, Yu.N. 86-47C
Makatos, N.C.G. 78-14H
Makeev, Kh.I. 85-28JM
Makel, D.B. 86-4E
Maki, E.R. 80-14S
Makino, K. 78-43J
Makino, T. 84-66K; 85-67K
Makkonen, L. 85-11D
Maksimov, V.I. 80-32J; 84-15S
Maksimov, V.K. 79-3P
Maksimov, Ye.A. 79-19S,20S
Maksin, P.L. 78-18E
Maksyutenko, S.N. 79-34B
Mal'tsev, A.P. 77-17J
Malachowski, K. 85-68K
Malashetty, M.S. 83-52F
Malashkin, I.I. 78-75J; 80-125J
Malbrunot, P. 86-101P
Malenkov, I.G. 82-47J,48J; 85-60J
Malhotra, A. 80-24B; 81-15T
Malhotra, R.C. 79-40P
Malik, M.A.S. 79-68A
Malik, M.R. 80-45B; 81-26B
Malin, M.R. 82-44F
Malinin, V.G. 79-74F
Malkin, S. 84-11A
Malkov, L.S. 83-4JJ
Malkovski, V.I. 85-61F
Malkovsky, V.I. 85-33F
Mallet, M. 86-35N,36N,37N,38N
Mallinson, G.D. 77-60F
Malmuth, N.D. 76-39A
Maloney, J. 82-25T
Malte, P.C. 85-21S
Malyshev, A.A. 85-64J
Malyshev, V.L. 85-65J
Malyusov, V.A. 86-10H,54J,7Q
Mamayev, V.V. 77-31B
Mamedov, M. 76-1J
Mamonov, V.M. 82-12H
Manafzadeh, S. 84-39Q
Manbeck, H.G. 86-108Q
Mancini, T.R. 76-56F
Mandal, G. 82-7M
Mandal, K.K. 82-7M
Mandarov, A.A. 77-12S
Mandel, S.W. 79-12Q
Mandrusiak, G.D. 86-13J
Maneri, C.C. 76-30J
Manero, O. 86-35F
Mangin, J. 86-30K
Mangnall, K. 85-20JJ
Manins, P.C. 79-84F; 80-29F
Mankevich, G.P. 76-43D
Mankevich, V.N. 82-13J
Mann, R. 82-1H
Mannara, G. 85-63F
Mannik, E. 82-12T
Mannik, L. 86-49U
Manno, V.P. 86-40N
Manocha, B.L. 81-10FF
Manohar, R. 82-9N
Manrique, J.A. 80-8L
Manton, M.J. 77-61F
Mantz, A. 80-45P
Manuilovich, S.V. 84-30C
Manvi, R. 81-16T
Manzoor, M. 81-18A; 82-13A; 83-17Q; 84-27A
Mao, C.-P. 84-19FF
Mao, C.P. 84-44F
Maqsood, A. 82-12A
Maragoudakis, C. 84-81K
Maragoudakis, M. 84-49K
Maratkanova, Ye.I. 82-30K
Marbout, S. 86-34K
Marcak, H. 81-4N
Marchman, J.F. 77-18C
Marcillat, J.F. 86-94F
Marcinauskas, K.F. 84-56Q
Marcus, B.D. 79-3Q
Marcus, C.J. 81-16T
Marcus, S.L. 83-26T
Marcuse, D. 76-11K
Mardarskiy, O.I. 80-2Q; 83-6J
Marder, B.M. 81-36A
Marek, D. 86-60P
Mareschal, J.-C. 85-46S
Margis, I.A. 84-41DP
Margittai, T.B. 86-62Q
Margolin, L.I. 78-17H
Margolis, S.B. 78-22A,11T
Mari, C. 76-36C
Maria, Gh. 83-26K
Marinichenko, S.K. 82-15D
Marinov, M.I. 78-41J
Mario, M. 84-16T
Marjoribanks, R. 84-105K
Markatos, N.C. 82-44F; 84-45F,24S; 85-38N; 86-66N
Markatos, N.C.G. 78-34C; 79-13H
Markevicius, A.A. 86-9D,10D
Markham, B.L. 86-6FF
Marko, K.A. 84-56K
Markovitz, E. 86-14J
Marks, W.G., Jr. 86-41E
Marner, W.J. 76-16F,57F; 83-18Q; 86-63Q
Maron, D.M. 82-18JJ
Maron, V.I. 85-54C
Marschall, E. 76-15J,31J; 78-42J,12T; 80-15Q; 85-22E; 86-41Q
Marschall, J.S. 85-47DP
Marshall, E. 77-28J
Marshall, J. 77-8G
Marshall, K.N. 78-6T
Marshall, R. 82-26T
Marshall, R.S. 78-60F; 79-85F
Marsters, G.F. 80-21D

Martin, B.W. 82-12C
Martin, C. 86-37P
Martin, H. 80-12E; 82-19D; 85-11Q,12Q; 86-31JM, 68Q
Martin, H.L. 86-14C
Martin, M. 79-3D; 81-3D; 83-5K; 84-1DP,4E, 2T
Martin, M.-G. 86-52B
Martin, M.G. 86-64Q
Martin, P.J. 84-46F
Martin, W.R.B. 86-54P
Martin, W.W. 77-41J; 78-23A; 80-2L; 83-2F
Martin-Moran, C.J. 83-26A
Martinet, J. 79-49P
Martinez, F. 85-2JM
Martinez, O.M. 85-23DP
Marto, P.J. 82-49J; 84-18JJ,25Q; 86-30JJ, 50JJ
Martonchik, J.V. 84-67K; 85-69K
Martsenyuk, Z.A. 84-16Q
Martsinauskas, K.F. 77-24Q
Marty, C.C. 77-12U
Martynenko, O.G. 76-4G; 78-12G; 80-22K; 82-20K; 84-25DP,20FF; 85-48DP,5E
Martynyuk, M.M. 86-68J
Marumo, E. 85-55C
Maruyama, S. 86-76B,4F,4Q
Marwah, R.K. 79-62A
Marziale, M.L. 84-31C
Masamichi, Y. 86-23S
Masek, R.V. 76-3R; 81-4Q
Mashelkar, R.A. 78-89F,90F
Mashena, M. 83-19B; 86-44A
Mashlenko, F.I. 85-20B
Masilyah, J.H. 81-27B,28B
Maske, K.U. 82-12U
Masliyah, J.H. 76-28B; 77-32B,19C; 79-22B, 29C; 85-66F
Maslov, A.A. 84-28C
Mason, E.A. 85-35U
Mason, S.G. 82-39C
Mason, W.H. 76-24C
Mass, E.A. 79-42C
Massey, J.S. 80-47J
Massey, T.H. 78-16Q
Massig, J.H. 76-14P
Massimilla, L. 76-8D; 82-17C
Massoud, M. 86-41J
Massouros, G. 83-3T
Mast, M. 85-41K
Mastanaiah, K. 76-40A,11R; 78-35C; 81-50J
Masterson, R.E. 77-14Q
Mastrullo, R. 79-21T
Masubuchi, M. 81-16Q; 86-105Q
Masuda, H. 77-4L; 80-9L; 85-70K
Masuda, S. 79-27C
Masuda, T. 86-54T
Masuoka, T. 81-40F; 83-52F; 86-32DP,80F
Mata, R. 86-32E
Matalon, M. 86-19C
Matavosian, R. 84-30K
Matejicek, A. 84-27S
Mather, J.C. 84-68K
Mathews, C.W. 78-11U
Mathews, D.M. 85-34E
Mathews, M.K. 78-23K
Mathiak, F.U. 85-71K
Mathieu, A.C. 81-7U
Mathieu, J. 76-36C,5E; 78-19E; 79-6D
Mathieu, P. 79-50J
Mathieu, Ph. 79-37A
Mathioudakis, K. 85-35E
Mathis, W.J. 79-7G
Mathu, F. 86-9E
Mathur, A. 85-79DP; 86-33DP,34DP,65DP
Mathur, M.N. 78-36C; 79-34C; 81-19FF
Matlosz, M. 84-21B
Matoba, H. 86-80J
Matorin, A.S. 78-77J
Matsnev, V.V. 83-18DP
Matsuda, A. 86-28U
Matsuda, K. 79-22E; 80-25P
Matsuda, O. 84-49C; 86-39J,119J
Matsuda, S. 82-1JJ; 84-1JJ
Matsuda, T. 84-38Q; 85-87Q
Matsufuji, H. 86-64P
Matsui, H. 81-12T
Matsui, K. 79-52J
Matsuki, K. 85-28JJ
Matsumato, M. 84-17S
Matsumato, O. 82-13U
Matsumoto, M. 86-44K
Matsumoto, O. 77-13U; 81-11U; 86-42U
Matsumoto, S. 77-63A; 84-25S,26S; 86-56A
Matsumoto, T. 80-7M; 84-35C
Matsumura, S. 79-34P
Matsumura, T. 80-23U; 86-50U
Matsunaga, T. 79-40F
Matsuno, T. 86-19D
Matsuo, K. 85-49JM
Matsuo, S. 86-31J
Matsuo, T. 86-65Q
Matsushima, H. 86-66Q
Matsuura, A. 82-25B; 84-31B
Matsuzaki, A. 81-11K
Matten, P.L. 82-14K
Matthews, L. 86-32E
Matthews, L.K. 84-16P; 85-72K
Matthews, R.K. 77-9R
Matthias, A.D. 85-31S
Matthys, E.F. 82-20B
Mattick, A.T. 85-73K; 86-110K
Mattiucci, F. 85-28F
Matushima, T. 84-39J
Matveyev, S.K. 80-22D
Matveyev, Yu.V. 83-21Q
Matyukhin, N.M. 85-114Q
Maurin, L.N. 83-7F
Maury, C. 82-7U
Mavriplis, D. 86-44N
Maxey, M.R. 85-52E
Maxwell, G.M. 86-78K
Maxworthy, T. 83-32F
May, W.B., Jr. 78-2T
Maydew, R.C. 85-56C

Mayhan, K.G. 78-89J
Mayinger, F. 82-42J,61J; 84-47F
Mayle, R.E. 76-11H; 77-11H; 78-37C; 79-30C; 80-19C; 84-6C,31C; 86-3D
Mayorov, M.M. 79-8M
Mayorov, V.A. 83-17DP; 84-37J
Mazumder, B.S. 76-5G,12U; 77-2G
Mazumder, J. 86-43U
Mazur, A.I. 79-18C,19C; 80-6D; 81-29B; 83-15C,11H
Mazza, A. 81-17U
Mazzei, P. 79-7J,21T
McAlees, S., Jr. 85-56C
McAllister, J.E. 78-11E
McCabe, M.E. 85-74K
McCafferty, R.H. 80-26T
McCall, W.C. 85-26P
McCarrier, A.R. 82-27F
McClelland, J.F. 79-13T
McConachie, P.J. 81-30B
McConaghy, G.A. 77-33B
McCormick, D.C. 84-32C,7D
McCormick, N.J. 85-51K,75K,82K; 86-36K
McCormick, P.G. 80-15K
McCoy, B.J. 82-50J,33T; 83-22DP
McCoy, D.D. 77-29J
McCoy, R.M. 86-14S
McCrory, R.L. 84-105K
McCullough, B.R. 77-38J
McDermit, J.H. 80-8T
McDonald, C.F. 77-15Q
McDonald, D.W. 77-34P
McDonough, J.M. 78-108F; 79-142F; 80-51FF; 82-45F
McDougall, T.J. 79-20E; 83-33F,28FF
McElhiney, J.E. 77-9G
McEligot, D.M. 76-14E; 77-34B; 79-33B
McElroy, D.L. 85-46A,86DP; 86-61P
McFadden, G.B. 85-37F
McGarity, A.E. 84-17T
McGaw, D.R. 76-25D,26D; 77-14D
McGee, G.R. 85-26A
McGowan, J.G. 84-18T
McGraw, B.A. 85-33T
McGreavy, C. 84-9DP; 86-21B,17Q
McHale, S. 85-57C
McIlrath, T.J. 79-6K
McIntosh, A.I. 86-51U
McIntosh, M.K. 77-11U
McIntyre, D.A. 85-27P
Mckay, D.C. 82-8T
McKeen, W.J. 82-37FF
McKellar, B.H.J. 86-7K
Mckelliget, J. 82-14U
McKelliget, J. 86-40S
Mckelliget, J.W. 83-20U
McKenzie, D.P. 80-25FF
McKenzie, D.R. 78-13T; 84-70K
McKibbin, R. 81-41F; 84-22DP
McKillop, A.A. 80-19P; 85-3E; 86-2D
McLain, D.K. 81-12U
McLarnon, F.R. 79-23B
McLaughlin, J.B. 82-46F,47F; 86-11C
McLinden, M.O. 83-27T
McLucas, C.W. 86-51U
McMahon, B.B. 84-69K
McManamey, W.J. 78-8S
McMath, T.A. 77-5L
McMillan, A.F. 82-4B,8C
McMillan, D.E. 86-62P
McMillin, L.M. 76-12K
McMurrin, J.C. 77-14T
McPhedran, R.C. 84-70K
McQuiston, F.C. 84-28S
McRae, D.S. 86-30A
McRae, G.J. 85-48Q
McRea, D.S. 85-21N
McTaggart, C.L. 83-34F; 85-62F
McTavish, J. 84-26K
McTigue, D.F. 85-19DP
McWhorter, J.C., III 80-28A
Meader, J.W. 83-23F
Meador, W.E. 79-18K
Mears, D.T. 79-3L
Measures, R. 84-3U
Mecherikinnel, A.T. 83-28T
Mecherikunnel, A. 82-27T
Medvedev, A.M. 78-13H
Medvedeva, M.A. 78-32S
Meek, C.C. 81-20C; 82-24A
Meena, B.K. 77-20C,13H; 78-61F; 79-6G
Megahed, I.E.A. 81-9C
Mehendale, A.B. 86-8H
Mehmetoglu, M.T. 82-15U; 83-21U
Mehra, V.K. 82-7A
Mehrdadtehranfar, G. 85-104F
Mehta, N.D. 82-21B
Mehta, R.C. 77-41A,42A; 79-41A,22U; 84-28A; 86-52U
Mei, R. 86-57N
Meiburg, E. 85-110F
Meier, R. 86-53U
Meijer, H.C. 86-9E
Meinhardt, Y. 78-26A
Meinhold-Heerlein, L. 84-14H
Meinkohn, D. 80-30A; 81-19S
Mekler, V.Sh. 80-83J
Mel'man, M.M. 78-20K
Mel'nichenko, V.S. 86-100Q
Mel'nik, S.A. 86-16U
Mel'nikov, V.K. 79-4L,33U
Melamed, L. 77-15T
Melli, P. 85-49N
Mellichamp, D.A. 77-7H
Melling, A. 77-18E
Mellor, G.L. 77-19E; 79-8E
Melnichenko, V.V. 82-50A
Melo, L.F. 85-67Q
Melton, L.A. 84-71K
Memedlyayev, Z.N. 77-13B
Menashe, J. 82-38K
Mencke, K. 86-93P
Menguc, M.P. 85-76K,77K; 86-77K,79K,80K
Menning, J. 85-29JM
Menon, A.S. 83-20JM
Menon, M.M. 85-47S

Mentel, J. 84-20U,21U
Mentes, A. 83-43J
Meranova, N.O. 85-57J
Merbaum, A.H. 84-13A
Mercer, A.McD. 82-10N
Meric, R.A. 77-62F; 79-42A,43A; 83-27C; 86-58N
Merilo, M. 77-35P
Merker, G.P. 76-59F; 77-43A; 78-62F; 79-86F; 80-32F; 82-48F; 86-17D
Merkin, J.H. 77-63F,64F; 78-38C; 79-87F,88F; 83-13DP,29FF; 84-9DP; 85-32FF,33FF; 86-21B, 35DP,36DP,17Q
Merkle, C.L. 85-51B
Merklin, J.F. 86-99K
Meroney, R. 77-50F
Meroney, R.N. 76-37C,15E; 86-21FF
Mersman, A. 80-15S
Mersmann, A. 81-18D; 84-43J
Merson, R.L. 85-20G
Merte, H. 81-56J
Merte, H., Jr. 81-59J; 83-6JJ
Merte, M., Jr. 76-24S
Mertol, A. 79-48F; 81-42F; 82-49F; 84-48F; 86-1B
Merz, R.A. 76-27D
Merzhanov, A.G. 79-71F; 81-98J; 85-30JM
Merzlyakov, E.I. 77-21S
Mesler, R. 76-32J; 77-31J; 86-25J
Mesler, R.B. 77-30J,57J
Messerle, H.K. 76-10U; 80-15M,11U
Messina, A.D. 81-51J
Messiter, A.F. 76-58F
Mestayer, P.G. 76-38C
Mestvirishvili, Sh.A. 80-120J
Metcalfe, C.I. 77-15D
Metha, A.S. 84-12T
Metzener, P. 86-81F
Metzger, D.E. 77-10G; 78-8H; 79-7G; 80-16C; 81-31B,10C; 82-22B; 84-33C,6H; 86-53B,54B
Metzger, D.W. 79-31C
Meubus, P. 86-54U
Meuvus, P. 83-22U
Meyer, B.A. 79-89F; 82-50F,28T
Meyer, D.J. 86-108Q
Meyer, G.H. 78-24A; 79-44A
Meyer, K. 79-14H
Meyer, L. 77-9B; 80-46B; 82-32C; 83-4D
Meyer, R. 86-67Q
Meyer, R.M. 85-6U
Meyers, J.P. 82-2S
Meynart, R. 80-37F
Mezlyakov, E.I. 85-49DP
Mezzetti-Minettr, E. 84-23K
Miao, H. 79-66F
Michaelides, E.E. 86-37DP
Michel, L. 77-13P
Michelfelder, S. 78-24S
Michels, A.C. 80-25K
Michiyoshi, I. 76-6M; 78-43J; 82-8M; 85-60FF, 53J,66J,69J; 86-69J
Micol, J.R. 85-58C
Midler, L.S. 78-44J
Midoriyama, S. 86-118J
Midoux, N. 85-77DP; 86-68Q
Migai, V.K. 81-32B
Migai, V.M. 78-18Q
Migay, V.K. 78-39C; 83-18DP
Migliori, A. 85-96Q
Migliori, S. 85-41J
Mihail, R. 83-26K
Mihalas, D. 84-72K,104K
Mihalascu, G. 86-14Q
Mihelc, M.S. 84-38C
Mikailov, M.D. 76-29B
Mikami, F. 84-27B; 85-43B
Mikhailov, M.D. 76-41A; 77-44A,45A; 79-27F; 80-29A; 81-37A,88J; 85-27A,48S
Mikhailov, N.L. 86-27J
Mikhailov, V.S. 85-22Q
Mikhal'kov, S.M. 86-16U
Mikhalchenko, R.S. 85-78K
Mikhalevich, A.A. 85-66Q
Mikhaylov, D.M. 79-147F
Mikheev, V.I. 86-48Q
Mikherjee, S.K. 80-41P
Mikhina, G.F. 77-32P
Mikhno, Y.A. 85-28P
Mikic, B.B. 76-17J,18J,47J; 77-48J; 86-28B, 52S
Mikieiewicz, J. 79-29J
Mikielewicz, J. 81-2B,73J; 86-37F
Mikk, I. 80-13Q
Mikk, I.R. 79-13Q,13S; 80-17K; 83-27K
Miklavic, M. 80-26P
Mil'man, O.O. 80-82J,122J; 82-23JJ; 86-65J
Milano, G. 85-42F; 86-63P
Miles, C.R. 78-15S
Miles, J.C. 76-47F
Miletti, D.L. 79-41J
Milhelc, M.S. 84-36S
Miller, C.G. 84-34C; 85-36E; 86-48C
Miller, C.G., III 85-58C
Miller, C.W. 79-55F; 82-17T,18T
Miller, D.C. 78-10T
Miller, G. 80-23D; 83-15U
Miller, H.P. 85-34N; 86-59N
Miller, J.W. 80-26T
Miller, R.D. 78-7U
Miller, R.S. 85-67J
Millet, J. 81-7U
Mills, A.F. 76-22J; 79-43B,32C; 80-81J; 82-20D
Mills, G.S. 77-14U
Mills, K.C. 85-37E,79K
Mills, P.L. 85-28A
Miloh, T. 77-12H; 78-25A; 86-17A
Milojevic, D.Z. 81-12N
Milora, S.L. 86-55U
Milton, F. 78-35P
Mima, K. 83-17U
Min, T.C. 77-3
Minakami, K. 85-25S
Minami, S. 85-85K
Minardi, A. 80-5T
Minayev, A.N. 80-106J
Mineur, J.M. 79-14Q
Mingaleyev, N.Z. 82-25F

Mingheng, S. 86-38DP
Mingle, J.O. 79-45A; 80-13E
Minkin, M.A. 85-50DP
Minkowycz, W.J. 76-60F,7M; 77-3M; 78-40C; 79-90F; 80-9K; 81-26J,52J; 82-17N; 84-1N; 85-11DP,12DP,51DP,63DP; 86-29DP,30DP,31DP, 76N
Minning, C.P. 76-8L; 77-6L; 79-5L,6L
Minosyan, Ya.P. 84-21JM
Minot, M.J. 76-13K
Minot, W.J. 77-30K
Minota, T. 78-21K
Minyaylenko, N.A. 79-1L
Miranda, L.C.M. 86-69P
Mirels, H. 80-30C
Mirgorodskaya, A.P. 86-59J
Mirkovich, V.V. 77-36P; 85-29P
Mirky, V.N. 78-22K
Mirmov, N.I. 85-7J
Mironov, A.I. 77-46B
Mironov, B.P. 78-41C; 82-12H; 85-25H
Miropol'skii, Z.L. 76-30B
Miropol'skiy, Z.L. 83-7JJ
Miropol'sky, Z.L. 80-83J
Misale, M. 85-42F
Mishima, K. 85-68J,69J
Mishima, S. 83-40F
Mishina, L.V. 82-23B
Mishnayevskiy, L.M. 85-5D
Mishra, P. 78-6G,7G; 80-4G
Mishra, S.P. 78-64F
Miskleglu, I. 78-37A
Misra, J.K. 86-57DP
Misra, U.N. 78-53C,91F
Misterek, D.L. 86-63C
Mitachi, K. 86-82F,41S
Mitchel, G.R. 82-39K
Mitchell, J.W. 76-33C; 79-89F,110F; 80-23T; 81-3T,29T; 82-50F,28T; 84-21Q,35S
Mitchell, R.D. 84-2J
Mitchell, R.E. 85-39E
Mitchell, R.R. 85-34U
Mitchell, R.W. 78-11U
Mitenkov, V.B. 77-59J
Mitin, A.A. 84-60J,61J
Mitky, A.M. 76-12S
Mitor, V.V. 79-14S
Mitra, A.K. 80-50B
Mitra, N.K. 84-5F; 85-87F
Mitrayayev, I.B. 77-1H
Mitrovic, J. 80-84J; 83-44J; 84-42Q; 86-57J, 31JJ
Mitsui, J. 86-42S
Mitsuya, T. 86-15H
Mittal, G.S. 86-53P
Mittal, M.L. 80-3M; 86-56U,59U
Mitwally, E.M. 78-39B
Miura, K. 82-25A
Miura, N. 81-34C,35C
Miura, T. 82-25A; 85-29A; 86-34P
Miwa, T. 80-18C
Miwada, T. 85-42J
Miyabe, H. 86-29Q
Miyachi, I. 80-23U,28U
Miyadai, T. 86-88P
Miyake, T. 85-95Q
Miyake, Y. 76-39C
Miyako, Y. 86-88P
Miyamoto, M. 77-65F; 80-28FF; 82-21FF; 85-34FF
Miyamoto, T. 80-25U
Miyasaka, Y. 81-15C; 82-27J,28J,29J; 85-49J; 86-44J,25JJ
Miyata, H. 80-62F
Miyata, K. 81-20U; 85-71B,105J; 86-115J
Miyata, N. 86-64P
Miyatake, O. 80-36J; 85-70J
Miyauchi, T. 80-15E; 86-54K
Miyazaki, H. 77-21C
Miyazaki, K. 86-39DP,70J,93J,57U
Miyazaki, T. 82-26A; 84-69F
Miyazawa, T. 84-36JM
Miyoshi, Y. 84-52K
Mizomoto, M. 83-34C
Mizonov, N.V. 85-40Q
Mizukami, A. 85-35N,36N; 86-36N
Mizukami, K. 77-66F,32J
Mizuki, M. 83-10Q
Mizushima, J. 76-61F; 78-35F; 80-14E
Mizushina, T. 77-27E; 78-41P; 79-35P; 80-25D, 40F,7M; 82-15B,22FF,23FF; 84-35C
Mkpadi, M.C. 84-1H; 86-3H
Moalem, D. 76-28D,33J; 78-51C
Moalem, D.M. 77-12H
Moalem-Maron, D. 78-26A,47J,97J; 79-24B; 80-85J
Moallemi, M.K. 85-31JM,32JM; 86-32JM,33JM
Mochizuki, H. 85-20U
Mochizuki, S. 80-86J; 81-4G; 83-22Q; 84-8D, 19JJ; 86-56B,9G
Mocsny, D. 84-26Q
Modak, A.T. 78-23K
Modest, M.F. 79-19K,7L; 81-2K,3K; 83-28K; 86-45A,43S
Modi, D.K. 85-30P; 86-9P
Modine, F.A. 84-73K
Moffat, R.F. 85-51FF
Moffat, R.J. 78-10C,15H; 79-23E; 80-2H,3H, 19H,20H; 81-6C; 83-25C,31C; 84-46C
Moffatt, D.F. 86-35B
Mogilveskiy, B.M. 82-21D
Mohamed, M. 80-5S
Mohammadi, S.S. 81-38A
Mohammadian, S. 76-16E
Mohammed, K.A. 85-29E,23P
Mohammedieh, F. 86-6Q
Mohanty, A.K. 77-11G; 84-8FF
Mohanty, H.K. 79-91F
Mojibul Hasan, Md. 79-47J
Mojtabi, A. 78-40B; 79-18F,92F; 80-6F
Mokhatarzadeh, M.R. 79-51J
Mokhtarzadeh-Dehghan, M.R. 86-71J
Moklyak, V.F. 86-3JJ
Moldover, M.R. 86-10P
Molen, G.M. 86-45U,46U
Molerus, O. 80-3E; 83-6DP
Moles, F.D. 81-15S
Moli, M. 81-44B

Molki, M. 82-34D; 83-33B; 84-46B; 86-57B, 58B
Mollendorf, J. 82-10FF
Mollendorf, J.C. 77-9F; 78-16F; 79-43F; 80-5FF, 6FF; 81-23FF; 83-35F,31FF; 84-29F
Moller, J.C. 86-15C
Moller, K.O. 82-27A
Moller, R. 77-27E
Momonoi, K. 85-57DP
Monaco, R. 79-15H
Monchalin, J.P. 82-28A
Monchick, L. 85-35U
Monde, M. 78-48J; 80-87J; 82-51J; 85-71J; 86-72J,73J,74J
Mondt, J.R. 77-16Q
Mone, M. 84-31J
Monkewitz, P.A. 85-41C
Mons, R.F. 78-50C
Monsen, W.A. 81-18T; 82-29T
Monsy, G.M. 76-12D
Montakhab, A. 80-24D,14Q
Monteiro, S.L.P. 82-21K
Montesclaros, M.S. 82-30F
Montgomery, D. 77-22E
Monti, R. 85-63F
Montoya, R.C. 80-96J
Montrose, C.J. 80-11S
Moody, F.J. 86-75J
Moody, H.L. 76-12R
Moog, D.B. 85-49S
Moore, D.R. 76-43F; 86-67F
Moore, F.E. 82-29A
Moore, F.K. 76-8Q; 79-15Q
Moore, J. 86-13S
Moore, J.J. 82-12U
Moorthy, C.S. 76-41D
Moose, R.A. 82-52J
Moradi-Araghi, A. 86-23P
Morai, Y. 84-38U
Morcos, S.M. 83-2B; 85-1G; 86-83F
Moreau, D. 85-30E
Moreaux, F. 79-18J
Morel, R. 79-6D; 83-28C
Moresco, L.L. 77-28J; 78-42J; 80-15Q
Moretti, P.M. 86-50S
Morey, T.F. 76-13R
Morfan, D. 82-16U
Morgan, K. 78-27A; 84-19JM
Morgan, R.G. 86-44S
Morgan, T.L. 86-6E
Morgan, V.T. 86-46A,47A
Mori, A. 76-42A
Mori, H. 85-64F; 86-46J
Mori, K. 83-13J
Mori, N. 83-22B
Mori, S. 77-35B; 80-48B; 86-26FF
Mori, Y. 77-33J,1M; 78-31J,13U; 80-47B,15E, 48J,16Q; 81-9Q; 82-41J,53J,55K; 84-55B,8JJ, 54Q; 85-29JJ; 86-21JJ,44JJ,54K
Mori, Y.H. 78-49J; 80-88J,9Q; 81-53J; 85-72J, 84J,51Q; 86-33C
Morikawa, K. 84-22JM; 85-20A
Morikawa, Y. 79-52J
Morin, T. 85-8U
Morioka, I. 76-21S
Morioka, M. 84-49F
Moriya, S. 84-52B
Moriya, T. 86-46J
Moriyama, A. 85-73J
Morkawa, Y. 76-52C
Morland, L.W. 77-67F
Morlot, G. 86-30K
Morooka, S. 80-19Q; 81-10Q; 86-37B,22DP
Moroz, I.M. 86-84F
Morozov, G.N. 80-10M
Morozov, P.A. 82-14JJ,15JJ
Morozov, V.G. 78-19Q; 79-53J
Morozova, N.V. 86-2S
Morris, D. 85-34E
Morris, D.G. 85-74J
Morris, E.W. 77-37F
Morris, M.J. 81-20C
Morris, S. 85-4C
Morris, W.D. 80-49B,9G; 84-3G,7G
Morrison, A.M. 85-11G
Morrison, D.J. 78-14T
Morrison, F.A., Jr. 77-22C; 83-20J; 86-15S
Morrison, F.S. 79-7H
Morrison, G.L. 78-65F; 80-16T; 81-19T; 83-29T; 85-30T
Morse, J.S. 86-14E
Morse, T.F. 85-52DP
Morsi, Y.S.M. 85-15B
Mortazavi, H.R. 85-36K
Morton, J.B. 76-20E
Morton-Gibson, L.D. 84-35S
Morvan, D. 81-13U; 85-3U
Mosa, M.F. 79-6M
Moseev, G.I. 79-22S
Moseley, S.H. 84-28K; 85-32K
Moser, A. 76-10C
Moser, B. 79-49J
Moser, F. 86-23K
Moses, G.A. 86-35JJ
Moses, H.L. 78-16D
Mosin, I.I. 83-21Q
Moskowitz, S. 83-17S
Moss, A.Z. 86-65P
Moss, J.N. 80-13H; 81-37C; 84-19C,94K
Mosse, A.L. 85-36U,54U
Mosso, S.J. 76-2R
Mostaghimi, J. 84-23U,24U; 85-37U,46U
Mostinskiy, I.L. 83-62J
Mostinsky, I.L. 81-23B,44J
Mostofizadeh, Ch. 81-54J
Mosyak, A.A. 78-8G
Mota, A. 86-60P
Motakef, S. 85-53DP
Motakff, S. 86-40DP
Motschmann, H. 80-21U
Mottaghian, R. 80-53B
Mottram, D.A.J. 79-8U; 83-8U
Motulevich, V.P. 78-3C; 84-26H; 85-49C,59C; 86-11H
Motwani, D.G. 85-60C
Motycka, J. 78-16P

Moult, A. 79-13H
Mouri, A. 83-40F; 85-70F; 86-93F
Mouton, H. 77-68F,69F; 80-38F,39F
Moutsoglou, A. 80-10FF,29FF; 81-24FF; 82-33C; 83-9FF,5S
Moyls, A.L. 78-41B
Moyne, C. 86-31JM
Moztarzadeh, F. 85-4Q,5Q; 86-6Q,4T
Mrnkvova, A. 85-92DP
Mruk, G.K. 76-9R
Muchowski, E. 79-46A
Mucoglu, A. 76-62F; 77-14F,70F,71F; 78-66F, 67F; 79-22F,93F
Mudawwar, I.A. 85-75J; 86-49C
Mueller, G. 85-26S
Mueller, W.K. 79-61J
Mugishima, T. 86-57P
Mugnai, A. 86-81K
Muhuri, P.K. 79-94F
Muir, B.L. 86-54N
Muir, P.H. 83-5G
Mujahid, A.M. 85-30A
Mujawar, B.A. 76-13Q
Mujumdar, A.S. 76-54C; 80-89J; 83-20JM; 84-30S; 85-49B,67C,29DP,75DP,50E,13FF,14FF,15FF, 19S
Mukherjee, D. 80-50B
Mukherjee, D.K. 76-12H
Mukhina, N.N. 85-13D
Mukushi, T. 82-24B
Mulhearn, P.J. 79-44P
Mulholland, G.P. 77-46A; 79-30A
Muller, H.M. 83-45J
Muller, J. 82-54J; 83-23J
Muller, M.R. 85-12G
Muller, R.H. 79-23B
Muller, U. 82-2A,37F; 84-17F,62F; 85-34F
Muller-Steinhagen, H. 84-38J; 86-76J,77J, 78J
Mullick, S.C. 85-6T
Mulligan, C. 82-40A
Mulligan, J.C. 76-31B; 78-59B,68F,74F; 80-46A, 32FF; 84-32JM
Mullikin, T.W. 85-56K
Mullin, T.E. 77-72F
Mullins, C.B. 85-74J
Mullisen, R.S. 86-69Q,70Q
Mumata, S. 83-25F
Mumford, J.C. 79-5E
Munakata, T. 86-34JM
Mundy, J.N. 85-21E
Mungal, M.G. 84-15H
Munier, A. 81-9N
Munson, B.R. 79-32F,33F,34F
Muntasser, M.A. 78-68F
Munz, R.J. 84-30U
Munzel, W.D. 79-7Q
Murai, K. 86-39DP,70J
Murai, Y. 80-25U
Muralidhar, K. 86-41DP,60N
Muramoto, K. 83-40FF
Muranov, Yu.M. 80-60J
Murao, Y. 85-76J; 86-1J
Murase, M. 85-77J
Murata, A. 86-33JJ
Murata, H. 78-44P
Murata, S. 76-39C; 83-26U; 84-38U
Murata, T. 85-66J; 86-16D
Murbach, M.S. 86-79Q
Murgatroyd, W. 78-21J; 80-17S
Muringer, M.J.P. 85-21P
Murio, D.A. 86-13A
Muroyama, K. 84-26DP; 85-54DP; 86-42DP,43DP
Murphy, J. 84-50F
Murray, L.E. 84-11E
Murray, R.T. 86-73U
Murthy, M.S. 80-4B
Murthy, S.N.B. 86-29C
Murti, V.G.K. 85-36P
Murty, K.N. 86-71Q
Murty, N.S. 76-40C; 77-34J
Murty, T.V.R. 86-18H
Murty, V.D. 86-85F
Musici, L. 82-18F
Musienko, A.T. 79-34B
Musil, J. 80-35U
Muska, J.F. 76-13H
Mussulman, R.L. 78-21S
Mustacchi, C. 77-4Q
Muthanna, M. 78-9G
Muzzio, A. 76-43A
Myers, G.E. 78-28A; 80-8A; 83-29JM
Myers, J.E. 78-17J; 85-78J,95J
Myers, V.H. 86-82K
Myrum, T.A. 85-54JM; 86-86F,49JM
Myshenko, V.A. 84-27Q

Na, T.Y. 78-69F; 79-25B,95F; 80-2A,3A,30FF; 81-2A,3A,1L; 82-3FF; 83-30FF,32FF; 85-76F
Nachlinger, R.R. 76-12A,44A
Nachman, A. 79-10H
Nadano, H. 84-34A; 85-114K
Nadarajah, A. 86-87F
Naeri-Neshat, H. 81-3N
Nagae, O. 86-73J
Nagahashi, Y. 84-27DP
Nagai, H. 77-29D
Nagai, K. 81-53J
Nagakubo, S. 85-45JM
Nagamatsu, H.T. 86-6D
Nagamoto, K. 85-112Q
Nagano, H. 86-110J
Nagano, Y. 78-29B,26P; 79-17B; 82-30F
Nagaraj, H.S. 77-13S
Nagaraj, M. 85-29H
Nagaraja, S. 78-22S
Nagasaka, A. 83-33C
Nagasaka, H. 85-79J
Nagasawa, T. 86-34FF
Nagashima, A. 78-44P
Nagasue, H. 81-33B
Nagata, T. 85-10H,36Q
Nagatani, T. 78-31J; 80-16E
Nagato, T. 84-17B
Nagel, H. 79-26B

Nagel, M. 86-66P
Nagendra, C.L. 83-29K
Naghdi, A.K. 78-29A; 81-39A; 85-31A,32A
Nagiev, F.B. 81-57J
Nagiyev, F.B. 80-94J
Nagornov, S.A. 86-44DP
Nagornyy, V.S. 82-15S
Nagpal, V. 78-23S
Nahavandi, A.N. 79-21E
Naidenov, V.I. 86-59B
Naidu, G.M. 81-17T
Naik, A.S. 82-55J
Nair, B.G. 76-9D
Nair, K.N. 86-31E
Nair, S. 86-48A
Nair, S.M. 86-31E
Naito, E. 76-32B; 84-30B; 85-52B,65F
Naitou, T. 82-51F
Najat, Z. 81-55J
Najjar, Y.S.H. 83-30K
Nakada, I. 86-50P
Nakagawa, K. 76-4P; 81-21U
Nakagawa, M. 83-25B; 84-26B
Nakagawa, T. 85-52JM
Nakahashi, K. 86-61N
Nakajima, M. 80-40F; 83-17B; 85-28C,126K
Nakajima, T. 80-91J,92J; 82-56J; 86-83K
Nakamachi, I. 86-53K
Nakamoto, K. 85-94DP
Nakamura, H. 78-70F; 82-25B,3F,4F,51F,23K; 83-18K; 84-19B,31B,69J,6K
Nakamura, I. 76-18C
Nakamura, K. 76-58J; 80-16S
Nakamura, S. 84-39J
Nakamura, T. 80-28FF
Nakamura, Y. 78-29B; 81-11K,14U
Nakanishi, S. 86-79J,80J,81J
Nakano, T. 86-128J
Nakaoka, M. 76-79F
Nakaoka, T. 84-32JJ,33JJ; 85-101Q
Nakashima, S. 85-101Q
Nakayama, A. 77-4G; 80-90J; 81-2G; 84-9G, 40J,41J; 85-30FF,35FF,80J,81J,21JJ; 86-50C, 51C,58C,82J,83J,84J,85J,86J,87J,32JJ
Nakayama, K. 85-94DP,112Q
Nakayama, W. 80-48J,91J,92J; 81-9Q; 82-41J, 56J; 84-30JJ; 86-66Q
Nakayana, A. 84-36C
Nakazatomi, M. 80-51B
Nakazawa, T. 86-41K,42K,43K
Nakoryakov, V.Y. 83-12H
Nakoryakov, V.Ye. 79-33C,12P; 80-14H; 82-12J; 85-3JJ
Nakoyama, A. 86-60B
Nakra, N.K. 77-31K
Nam, S.H. 86-45U
Namatame, K. 76-13S
Namba, T. 79-9S
Namboodiri, P.M.S. 85-9DP
Namer, I. 86-58F
Nanda, R.S. 77-11G
Nandakumar, K. 76-28B; 77-32B; 79-22B; 81-27B, 28B; 85-66F; 86-21DP
Nandakumar, N. 86-113F
Nannei, E. 80-44J,6L; 83-1J
Nansteel, M.W. 81-43F; 83-36F; 84-51F
Naot, D. 77-20E; 85-36J
Napolitano, L.G. 77-73F
Naraghi, M.H.N. 82-10K,40K; 84-74K; 86-84K
Narahara, Y. 83-47B
Narain, J.P. 77-12G
Narang, B.S. 84-13FF
Narasimha, R. 76-2C; 77-5E
Narayan, K.A. 86-57DP
Narayan, K.Y. 78-3R
Narayan, M.V. 78-42C
Narayana, K.B. 86-72Q
Narayanan, P.V. 81-8D; 84-11DP
Narayanan, R. 81-72F; 82-34FF; 84-32B,52F; 86-87F
Narayankhedkar, K.G. 85-23A
Narazaki, K. 86-32DP
Nardin, G. 86-58U
Nariai, H. 86-45J,2JJ
Narumi, A. 84-37C; 85-61C; 86-18D
Narusawa, U. 81-44F,34FF; 82-66F,67F; 86-59DP
Nash, D.B. 86-85K
Nash, R.T. 76-15T
Naso, V. 79-7J
Nasrallah, S.B. 85-36FF
Nasser, A.E. 81-15U
Nataf, H.-C. 83-51F
Natali, F. 77-4Q
Natalizio, L.G. 77-74F
Nataraja, H.R. 86-59U
Nath, G. 76-41C; 77-20C,13H; 78-28C,43C,61F, 9G; 79-138F,4G,6G,7M; 80-12H; 81-10B,8C; 82-27C,20FF; 85-44B,29H; 86-70C,71C,26DP, 15U
Natsui, K. 85-6A
Nauman, E.B. 79-27B
Navruzov, Yu.V. 86-88J
Nawoj, H.J. 77-75F
Nayak, I.V. 83-11B; 84-11B
Nayak, J.K. 81-25T; 83-30T
Nayak, P. 79-80A
Nayak, S.K. 77-76F
Nayfeh, A.H. 80-31A; 85-8C
Nazarchuk, M.M. 79-9Q
Nazarenko, L.A. 86-20E
Nazmeev, Yu.G. 80-52B
Ncubizu, C. 76-8Q
Neal, S.B.H.C. 80-27P,28P
Neale, G. 76-29D
Nealy, D.A. 84-38C
Necmi, S. 76-8M; 77-2M
Nee, V.W. 77-25C
Neely, G.M. 85-23K
Negishi, K. 83-37F
Negishi, N. 80-24U
Negrelli, D.E. 77-32K
Negus, C.R. 79-23U
Neiss, J. 76-45A; 82-30A,22D
Neitzel, G.P. 82-52F
Nejat, Z. 80-53B
Nekrasov, V.P. 79-1K

Nelkin, M. 78-12E
Nelson, D.A. 76-14K,15K; 77-33K; 79-20K,21K; 84-75K; 86-88F
Nelson, G. 77-15U
Nelson, H.F. 78-18K; 81-10K; 83-31K; 84-76K, 77K; 86-86K,87K
Nelson, J.S. 82-37T; 85-33JM
Nelson, P.J. 85-82J
Nelson, R.A. 82-57J; 85-61J
Nemchinov, I.V. 83-21K
Nemira, M.A. 78-61B; 79-28B,29B; 80-54B
Nemirovskii, V.B. 85-29B
Nemoto, E. 85-20A,33A
Nemoto, T. 85-31P
Nemtsev, V.A. 79-3J
Nencini, L. 86-25K,26K
Nepomniashchii, A.A. 84-43F
Nepomnyashchiy, A.A. 86-89F
Nesis, E.I. 85-83J
Nesis, S.Ye. 80-93J
Nesis, Ye.I. 80-93J
Nesterenko, V.F. 86-18P
Nesterov, A.K. 79-39P; 83-8S
Neta, B. 86-62N
Neti, S. 85-13G; 86-36B,49B
Netrakanti, M.N. 85-68Q
Netreba, S.N. 83-21FF
Netterfield, R.P. 84-70K
Neubert, J. 79-17Q
Neuer, G. 86-16P
Neumann, A.W. 78-50J; 81-1J
Neumann, R.J. 80-31FF
Neumann, W. 76-13U
Neverson, A.S. 79-65F
Nevskiy, A.S. 78-20K
Nevstruyeva, Ye.I. 82-33J
Newell, T.A. 86-34T,52T
Newland, D.B. 83-10U
Newman, G.R. 77-17E
Newman, J. 76-2D; 78-8D; 81-3G
Newsome, R.W. 86-63N
Newton, A.B. 80-17T
Ng, K.H. 76-42C
Ng, K.M. 78-17D
Ng, M.L. 86-27FF
Ng, S.Y. 85-52J
Ng, T.T. 82-14C; 85-22C
Ng, W. 77-3J
Ng, W.-L. 84-50K
Ng, W.-L.W. 84-80K
Ng, W.W. 78-14J
Ng, W.W.L. 78-13J
Nghiem, L. 81-56J
Nguyen, H.D. 85-47K
Nguyen, M.N. 85-36K
Nguyen, N.N. 85-30C
Nguyen, T.H. 82-24FF; 83-38F; 84-71F; 86-53DP
Nguyen, T.M. 86-40B
Nguyen, T.T. 77-19C; 79-29C
Nguyen-Lamba, M. 86-4JM
Nguyenchi, H. 79-1Q
Ni, S.M. 84-20JM
Nicholis, R.L. 80-18T
Nicholls, R.W. 84-78K; 85-19K
Nichols, F.A. 85-34JM
Nicholson, J.H. 84-29S
Nick, K.P. 84-25U
Nickel, H. 86-73Q
Nicol, A.A. 82-19JJ
Nicolette, V.F. 85-67F
Nicotra, L. 86-120J
Nieckele, A.O. 85-37FF
Nieh, S. 79-17T,18T
Niehammer, J.E. 82-16S
Nield, D.A. 77-77F; 85-55DP
Nielsen, C.E. 84-11T
Nielsen, P.V. 80-20F
Nielson, P.V. 79-96F
Niemeyer, L. 78-12U; 82-2U; 86-71U
Nienaber, J.A. 85-38E
Niethammer, J.E. 79-129F; 81-31FF
Nieto de Castro, C.A. 86-78P
Nieuwoudt, J.C. 85-20P
Nieuwvelt, C. 76-35P; 83-32C
Nieva, I.A. 83-9D
Nigmatulin, B.I. 86-60J
Nigmatulin, R.I. 80-94J; 81-57J
Niinura, M. 76-5U
Niitsu, Y. 86-45S
Nijhawan, S. 80-95J
Nikaidoh, H. 85-51S
Nikandrov, V.I. 78-22P
Nikanjam, M. 77-34K; 79-9S
Nikitin, A.T. 77-36B
Nikitin, V.S. 78-3J
Nikitin, Ye.D. 80-104J
Nikitin, Yu.M. 80-40B
Nikitopoulos, C.P. 79-21F
Niknejad, J. 81-58J; 84-20JJ
Nikolaev, N.A. 80-52B
Nikolaevskii, V.N. 85-54C
Nikolakopoulou, G.A. 85-9JM
Nikolayev, B.I. 82-2M
Nikolova, B.M. 81-88J
Niksa, S. 85-39E
Nilson, R.H. 80-96J,97J; 81-25FF; 85-38FF; 86-15S
Nilsson, B. 79-22K
Nilsson, N.A. 85-80K
Nilsson, O. 86-67P
Nimmo, B.G. 86-16T
Nimmons, M.J. 81-2Q
Nimz, R. 86-102P
Nin, K.S. 78-71F
Ninokata, H. 86-89J
Niogret, J. 86-74Q
Nir, A. 77-16D; 85-55J
Nirmalan, V. 85-50Q
Nishi, T. 86-90P
Nishida, K. 85-49J
Nishida, M. 80-128J
Nishida, N. 77-17Q
Nishihara, H. 85-68J,69J
Nishihara, K. 80-18K
Nishikawa, K. 83-58J; 84-42J; 86-31J
Nishikawa, M. 79-22E

Nishikawa, N. 77-14H; 79-54J
Nishimoto, K. 85-19A
Nishimura, H. 82-24U; 86-51JM
Nishimura, J. 80-30P
Nishimura, M. 76-30D; 84-79K; 85-81K; 86-7T
Nishimura, S. 76-25E
Nishimura, T. 84-23JM; 85-35JM; 86-61B,62B, 90F
Nishino, Y. 85-40P
Nishio, S. 86-90J,91J
Nishioka, T. 86-88P
Nishiwaki, I. 86-116J
Nishiwaki, N. 79-17S
Nishiyama, H. 84-40C; 85-14D; 86-19D
Nishiyama, K. 86-33JJ
Nishiyama, N. 86-20D
Nishkawa, K. 77-35J
Nistor, L.C. 79-36P
Nistor, S.V. 79-36P
Nitsche, W. 79-21C
Nivoit, M. 81-40A
Niwayama, M. 80-39P
Nixon, G. 86-116K
Nizou, P.Y. 84-39C
Nkemdirim, L.C. 76-14S
No, H.C. 82-20JJ
Noble, M.L. 79-5H
Noble, R.D. 77-27P
Nogotov, E.F. 78-9; 79-41F
Noguchi, S. 85-42A; 86-64A
Noguchi, T. 86-94Q
Noll, B.E. 84-24H
Noll, R. 86-60U
Nollez, G. 77-3U
Nomura, K. 85-26H
Nomura, S. 86-49A
Nonogaki, M. 84-16S; 85-21H
Norden, P.A. 82-25F
Nordstrom, R.J. 85-115K
Norman, M.L. 84-72K,104K
Normand, C. 78-28F; 81-45F
Northover, E.W. 80-27P,28P
Norton, M.J.T. 85-50S
Norton, M.P. 86-35A
Norton, R.J.G. 86-18E
Nosach, N.A. 83-23Q
Noshchenko, A.N. 86-46S
Nosoko, T. 85-84J
Nosov, V.S. 77-31B
Noth, H. 80-15S; 84-43J
Nott, S.H. 85-20P
Nottmeyer, K. 85-12E
Novack, J.C. 85-19P
Novak, J.P. 82-18N; 84-17P
Novakov, T. 83-36K
Novakovskii, V.N. 85-56DP
Novas, J.B. 80-9F
Novel, B.J. 83-31T
Novichyonok, L.N. 82-50A; 85-5E
Novikov, S.S. 82-33K
Novikov, V.N. 85-70B
Novotny, J.L. 77-57F,8K,32K
Nowak, E.S. 86-109Q
Nowak, R.J. 85-62C
Nowak, Z. 80-21B,55B
Nowobilski, J.J. 76-1U
Nozaki, K. 80-18K
Nozari, A. 86-35T
Nozdrin, A.A. 82-10B
Nozu, S. 86-22JJ,23JJ
Nukiyama, S. 84-44J,45J
Numata, T. 80-30P
Nunez-Regueiro, M. 83-24A
Nunziato, J.W. 76-12A,44A
Nusupbekova, D.A. 76-14U
Nyamira, M.A. 76-33B; 82-26B
Nylund, O. 81-11J
Nysten, B. 85-32P
O'Brien, J.E. 80-74B; 82-27B
O'Brien, V. 77-37B
O'Brien, W.F., Jr. 78-16D
O'Brien-Bernini, F.K. 84-18T
O'Callaghan, M.G. 80-32A; 82-31A,32A
O'Callaghan, P.W. 77-1A,2A,11C,14S,15S,16T; 79-4A; 82-51A
O'callaghan, P.W. 86-57A
O'Ferrall Lund, K. 86-33T
O'Gallagher, J. 81-21T
O'Keefe, T. 84-106K
O'Neil, D.J. 85-33P
O'Neill, G.A. 84-21JJ
O'Neill, K. 82-34C
O'Neill, P. 77-17T
O'shea, K.R. 77-3L
O'Sullivan, M.J. 78-43F,44F; 81-41F
Oaki, S. 81-30C
Obata, F. 86-47S
Obata, M. 86-48S
Oberkampf, W.L. 76-15S
Obermeir, E. 86-33P
Obot, N.T. 86-52C
Ochi, M.-A. 85-5A,3K
Ochiai, M. 85-85J
Ockendon, H. 79-30B
Ockendon, J.R. 79-4S; 85-8H,22JM,36JM
Oda, K. 80-35J; 86-12JJ,13JJ
Oden, J.T. 86-17N,64N
Odgers, J. 77-30P
Oehlbeck, D.L. 79-8G
Oelund, J.C. 85-82K
Oertel, H. 82-9F,3G
Oertel, H., Jr. 78-36P
Ofer, S. 82-20D
Ofi, O. 77-78F
Ogawa, K. 79-9G
Ogawa, M. 86-63B
Ogawa, T. 83-45K
Ogden, T.R. 84-12E
Ogino, F. 79-35P; 80-25D,7M; 82-15B,22FF, 23FF; 84-35C
Ogniewicz, Y. 81-13D
Oguchi, H. 77-14H
Oguchi, T. 80-10L
Oguri, T. 85-83K
Oguztoreli, M.N. 86-64B
Oh, S. 86-92J

Ohama, T. 86-39DP,70J,93J
Ohguchi, Y. 84-26U; 85-38U
Ohhashi, A. 85-57DP
Ohishi, Y. 86-33E
Ohkawara, M. 85-71B,105J
Ohkubo, Y. 86-47JM,48JM
Ohlidal, I. 80-11L
Ohlmer, E. 76-17E
Ohmori, T. 86-76Q
Ohmuro, M. 83-40F; 85-70F
Ohnishi, N. 84-46J
Ohno, H. 84-28J
Ohsawa, S. 84-36C,9G; 85-30FF
Ohsawa, S.-I. 81-2G
Ohta, H. 84-42J; 86-34E
Ohtake, K. 78-13U
Ohtani, S. 82-25A; 84-7E,38Q; 85-29A,87Q; 86-7JJ,34P
Ohtsuka, M. 80-24P
Ohue, T. 77-13U
Ohwada, Y. 81-3L; 85-84K
Ohyagi, S. 86-18S
Ojha, S.K. 78-36C; 79-34C
Oka, T. 82-1F
Okada, K. 85-82F,47FF
Okada, M. 78-30A; 84-24JM; 85-37JM; 86-91F
Okaji, M. 84-18P
Okamoto, H. 85-64F
Okamoto, T. 85-85K
Okamoto, Y. 79-22E
Okamura, T. 85-94DP
Okamure, T. 85-112Q
Okano, H. 86-91F
Okayama, H. 78-19P,20P
Okaz, A.M. 86-68P
Okazaki, K. 78-13U
Okazaki, M. 86-76Q
Oke, K.H. 86-67S
Oker, O. 81-59J
Okos, M.R. 81-11S
Okouchi, T. 85-73J
Oktay, S. 86-49S
Okuda, S. 80-25U
Okui, K. 84-27B; 85-43B
Okuyama, K. 86-94J
Olabin, V.M. 85-9S
Olafsson, S.I. 84-10J
Olalde, G. 83-11T; 85-31T
Oldershaw, C.F. 84-24Q,23S
Oldfield, M.L.G. 84-29S; 86-17E
Olek, S. 84-2G; 85-86J
Olesiak, Z. 83-12A
Olevskaya, L.P. 85-9S
Olimpiev, V.V. 79-12G
Oliphant, M.V. 80-19T
Oliver, C.J. 80-31P
Oliver, D.L.R. 85-12D; 86-95J
Oliver, D.R. 76-34B; 78-42B; 79-31B
Oliver, D.S. 82-64F
Oliveri, E. 82-33A
Olivetto, M.E. 79-4R
Olivier, I. 85-34Q
Olmedo, L. 77-4R
Olmstead, B.A. 85-55S
Olmstead, W.E. 77-47A
Olovin, B.A. 77-16S
Olsen, R.W. 85-51S
Olson, D.A. 84-8F
Olson, J.M. 79-97F,98F
Olson, M.D. 79-47C
Olson, P. 85-39FF
Olson, S.C. 86-58K
Omori, T. 86-53K
Onat, K. 81-29A; 83-13JJ
Onda, K. 80-8M; 85-39U
Ondris, L. 82-11N
Onishi, Y. 78-72J; 79-144F; 83-46J; 84-47J; 86-34JJ
Onkura, T. 82-11H; 83-6H
Ono, A. 85-1E; 86-82K
Onoda, N. 84-12S
Onyegegbu, S.O. 80-41F; 86-88K
Ooms, G. 78-43B
Oosthuizen, P.H. 76-63F,64F; 77-23C
Opalinski, I. 84-39DP
Oppel, F.J. 86-50S
Oppenlander, W.T. 86-50U
Opryshko, S.I. 85-9Q
Ordal, M.A. 83-32K
Orell, A. 80-63B
Oreper, G.M. 84-21FF
Orivuori, S. 79-47A
Orlandi, P. 86-5N
Orlicki, D. 84-10JM; 85-16JM
Orloff, K.L. 77-1P
Orlov, A.K. 85-36B
Orlov, L.K. 80-99J
Orlov, V.A. 78-5B
Orlov, V.K. 80-98J
Orlov, V.V. 85-13D
Orlova, N.M. 80-68J
Orlova, V.I. 85-3H
Ormiston, S.J. 86-92F
Ornatskiy, A.P. 76-35J; 77-36J; 78-51J,52J, 53J; 79-55J,56J; 80-100J
Orozco, J. 84-68J
Orozco, J.A. 77-24J; 86-96J
Orszag, S.A. 82-46F,47F; 84-16F
Ortabasi, U. 80-20T
Orton, G.S. 84-67K
Oruma, F.O. 85-38JM,86K
Osaka, H. 76-18C
Osakabe, M. 86-62J
Osawa, Y. 86-66Q
Osborne, D.G. 85-68F,40FF
Osherov, Yu.S. 84-16H
Oshima, Y. 82-38F
Oshinowo, T. 84-48J
Oskam, H. 78-5
Ostaf'ev, V.A. 86-51S
Ostensen, R.W. 81-60J
Ostenson, J.E. 84-29A
Oster, D. 79-32E
Osterle, J.F. 85-34U
Osterwegel, G.G. 80-32P
Ostman, M.B. 85-29T

Ostrach, S. 76-44F; 79-66F,99F; 85-44C,50F
Ostrogorsky, A.G. 83-47J
Ostrovskiy, N.Yu. 86-97J,98J
Ostrovskiy, Yu.N. 77-25J,51J; 78-72F,78J, 79J; 79-37J,79J,80J,81J,82J,83J; 82-69J; 83-66J
Ostrovsky, Yu.N. 79-21Q
Oswatitsch, K. 80-74J
Ota, T. 79-2D,7D; 80-4C,1D,26D; 81-1C; 82-1C, 2C; 83-29C,31JM; 84-40C,2Q; 85-14D; 86-19D, 20D
Oter-Duthoit, L. 86-50A
Otero, C.J.L. 85-69Q
Otero de la Gandara, J.L. 85-69Q
Otis, D.R., Jr. 85-68B
Otozai, K. 86-89P
Ott, E.S. 81-11C
Ott, W.R. 76-22A
Otterman, J. 78-24K
Otubo, K. 84-31J
Ou, J.-W. 76-65F
Ou, J.W. 77-57B,79F; 78-11B
Ouajji, H. 86-61U
Ouellet, Y. 86-18N
Outcalt, S.I. 85-30S
Ouyang, F. 85-95DP
Ovchinnikov, V.V. 85-13D
Owen, D.R. 86-24A
Owen, I. 85-63C; 86-53C
Owen, J.M. 85-14G; 86-4G,7G
Owen, R.B. 80-33P
Owens, W.L. 84-4S
Oya, H. 85-34P
Oyakawa, K. 81-19B,34B; 83-22B,23B,34B; 86-65B
Oygenblik, A.A. 86-55DP
Oyvin, V.N. 78-15Q
Ozaki, E. 82-34K; 83-22K; 84-57K
Ozaki, N. 80-43P
Ozaki, Y. 80-19U
Ozawa, T. 86-27T,28T
Ozawa, Y. 81-20U
Ozboya, N. 77-21J
Ozel, T. 85-68FF
Ozeum, B. 86-64B
Ozgen, C. 86-18JJ
Ozguc, A.F. 84-20F
Ozil, E. 82-12K
Ozisik, M.N. 76-49J,12S; 77-6,48A,47B,48B; 78-31A,54J; 79-48A,54F,100F,11J,25K; 80-29A, 19K,21T; 81-37A,62A,48B,49B,23F; 82-40A, 41K,45K; 83-35A,57F,3JM,14K; 84-30A,31A, 32JM,15K,16K,17K; 85-2A,3A,14A,27A,88F,29JM, 38JM,21K,42K,86K,116K,117K,118K,21N; 86-3A, 25A,29A,30A,18B,19B,20B,84B,111K,112K,113K
Ozoe, H. 76-66F; 77-80F,81F; 79-101F; 81-14F, 46F; 83-9F,39F,40F; 84-69F; 85-69F,70F; 86-16F, 90F,93F,7N

Pachter, M. 82-34A
Padalko, G.A. 78-37P
Paderin, L.Ya. 86-36T
Padet, J.-P. 85-61FF
Padlog, R.D. 83-31FF
Padmanabhan, P.V. 85-39JM; 86-35JM
Padovan, J. 78-32A; 80-33A; 81-41A; 82-12N; 85-9A; 86-65N
Page, R.H. 76-27D
Pagliarini, G. 84-2A,3A; 85-4B
Pagni, P.J. 80-30D
Pai, B.R. 78-24S
Pai, D.H. 82-14S
Paily, P.P. 76-16S
Painter, J.H. 76-15U
Painter, L.R. 77-6P
Pakratov, B.M. 79-4P
Pakula, R. 85-34A
Pal, D. 78-52F
Palancz, B. 76-6H; 84-19T; 85-58DP,59DP
Palaszewski, S.J. 81-61J; 85-52S
Palekar, M.G. 85-69C,16G
Palen, J.W. 80-15J
Paliwal, R.C. 80-42F,43F
Palmer, A.W. 86-21E
Palmer, H.J. 81-62J; 84-8Q
Palozzi, A. 86-56S
Pamadi, B.N. 80-27D
Pammer, Z. 80-34A
Pamplin, B.R. 76-67F
Pan, S.S. 76-69A
Pan, W.-D. 84-21B
Pan, Y.C. 82-6M
Panchal, C.B. 80-101J
Panchenko, V.N. 85-64B
Pande, P.K. 77-1D
Pandey, D.K. 85-40E
Pandey, R.N. 76-34D; 77-57A
Pandit, S.M. 83-6S
Panek, C. 76-20A
Panella, B. 85-24C,31J
Panevin, V.I. 85-56J
Pang, Y. 78-14B
Pangi, P.J. 84-19FF
Pankratov, B.M. 79-5A
Panno, G. 85-4S
Panov, Ye.N. 86-100Q
Panteliou, S. 83-27A
Pantelis, G. 85-53S
Panter, G.B. 77-2P
Pao, C.A. 77-49A
Paolino, M.A. 86-21D
Paolucci, S. 85-13B,64C; 86-26F
Papailiou, K.D. 79-1C
Papaioannou, J. 85-13T
Papazian, J.M. 79-57J,58J,90J; 80-23J
Papendieck, H. 86-99J
Papenfuss, H.-D. 78-17C
Papin, J. 83-18F
Papoutsakis, E. 80-56B,57B; 81-35B
Paradis, M.A. 77-15H
Parang, M. 86-4A
Paranin, S.N. 86-62U
Parantheon, P. 79-37P; 80-22P,34P
Paranthoen, P. 78-35P; 81-30A; 84-5E,9E; 85-30E
Parfent'yev, N.G. 85-26Q
Pargeter, D.M. 77-82F

Parikh, P.G. 82-40F
Parise, J.A.R. 86-19T
Park, E.L., Jr. 76-7J; 78-87J,89J; 81-51J
Park, G.-C. 86-100J
Park, J.K. 82-42K
Park, J.S. 85-36C
Park, J.T. 80-35P
Park, J.Y. 85-43U; 86-33U
Park, K.H. 82-12F; 85-41FF
Park, Y.J. 86-82N
Parker, A.G. 76-36P
Parker, J.D. 81-16FF
Parker, J.H., Jr. 85-23S
Parker, R.O. 77-42B
Parmentier, E.M. 78-73F
Parnas, A.L. 76-46D
Parpal, J.L. 82-28A
Parrish, T.F., Jr. 79-12Q
Parry, J.L. 83-9H
Parsapour, H.B. 82-60F
Parsons, J.R., Jr. 78-74F; 80-32FF
Parsons, R.R. 85-6S
Parthansarathy, R.N. 86-43JJ
Partlow, D.P. 84-107K
Parygin, V.V. 80-2J
Pashnin, V.M. 81-22S
Pashutov, A.V. 85-7H
Pasmanter, R.A. 85-37N
Pasner, A. 86-68F
Pasumarti, B. 85-49P
Patankar, S.V. 76-1F; 77-61A,38B,50B,52B, 27C; 78-41A,44B,54B,55B,18D,75F,94F,26Q; 79-32D,59J; 80-37A,2D; 81-22A,30D,1F,49F, 11N; 82-14B,29D; 84-38B; 85-18F,37FF,1N, 42N,43N
Patel, A.S. 76-38C
Patel, B. 86-13S
Patel, J.M. 81-2D
Patel, M.K. 85-38N; 86-66N
Patel, N.A. 86-69P
Patel, P.D. 76-51J; 78-76J
Patel, P.R. 84-36DP
Patel, R.D. 77-17D; 78-1C
Patel, R.S. 86-89K
Patera, A.T. 84-25JM; 86-28B,5N,52S
Patera, R.P. 80-22T
Paterson, R.W. 85-27B
Pathangey, B. 85-65C
Patil, M.R. 78-93F
Patil, P.R. 80-44F; 84-32DP
Patrick, M.A. 77-82F; 79-10D
Patrick, W.P. 79-23H
Patrusheva, L.G. 86-36E
Pattabongse, M. 85-28K
Pattan, T.D. 76-9D
Pattantyus-H, E. 77-37J
Patten, T.D. 76-37F
Patterson, A. 85-23S
Patterson, J. 80-45F
Patterson, J.C. 84-53F
Patterson, W.C. 76-14R
Patula, E.J. 81-42A
Paul, D.D. 83-48J
Paul, P.H. 79-9M; 81-8M
Paulmier, D. 81-40A
Pavel'ev, A.A. 86-45C
Pavlenkov, I.M. 84-15S
Pavlinov, M.P. 79-2M
Pavlishchev, M.I. 79-38J
Pavlov, B.P. 77-7P
Pavlov, M. 86-63U
Pavlov, P.A. 80-104J
Pavlov, V.A. 76-5P
Pavlov, Yu.M. 80-102J; 85-87J
Pavlovic, B. 76-22U
Pavlovic, M.I. 81-12N
Pavlyuchenko, A.M. 85-15D,7H
Pavlyukevich, N.V. 78-40J; 82-20K; 85-48DP
Pawlowski, L. 86-16P,37P
Payne, L.E. 84-20A
Payre, G. 82-13N
Payvar, P. 77-24C; 79-49A
Pazooki, N. 85-53B
Peacock, N.J. 79-23U
Peacock, R.D. 85-66K
Pearce, B.E. 78-4R
Pearlstein, A.J. 81-47F
Pearson, J.K. 77-20S
Pearson, J.R.A. 76-36J,34P
Peavy, B. 86-35E
Peavy, B.A. 76-6D
Pechacek, R.E. 85-26B
Pechenegov, Yu.Ya. 76-37J
Pechenogov, Yu.Ya. 84-26J
Pecheuxet, J. 85-15G
Pechkovsky, V.V. 85-30U
Pechoc, V. 78-24C
Peck, M.K. 82-39T; 84-28FF; 86-14T
Peddicord, K.L. 81-13A; 82-1A; 84-32A
Pedersen, D.R. 77-16H
Pederson, D.R. 84-19DP,4JM,7JM,1N; 85-11DP, 12DP
Pederson, R.J. 77-39B
Peebles, W.A. 79-21P
Peeken, H. 85-1B
Peerless, S.J. 76-16E
Peev, G. 78-4C
Pehler, F.A. 85-64Q
Pei, B.S. 83-70J
Pei, D.C.T. 77-4A; 78-20C; 83-10D; 84-7DP, 25S,26S; 85-13DP
Peier, W. 76-44C; 77-39J; 78-57J
Peigin, S.V. 85-31C
Peinecke, D.R. 76-9M
Peletskii, V.E. 86-36E
Pelissier, J.L. 83-1JM,21JM; 84-26JM,27JM, 28JM
Peller, H. 84-10G; 86-10G
Pellerin, N. 76-17P
Pelletier, E. 84-10K
Pender, R. 83-19C
Penetrante, B.M. 86-64U
Peng, C.M. 81-84J
Peng, S.J. 86-100J
Peng, Y. 84-41C
Penney, T. 84-5J

Penot, F. 81-17F; 82-53F
Penotet, F. 83-41F
Penzkofer, A. 84-61K; 86-75K
Pepper, D.W. 76-22D; 79-103F
Pera, L. 76-33F
Perdikis, C. 83-42F; 86-32FF
Perdikis, C.P. 85-72DP
Perederiy, A.D. 77-1S
Perednis, E.E. 84-55Q
Pereira Duarte, S.I. 85-23DP
Perepelitsa, B.V. 82-35C; 85-54B
Perepelitsa, N.I. 79-60J; 80-39B
Perez, A.J. 84-12J
Perez, J.A. 84-28Q; 85-70Q,91Q
Perez-Blanco, H. 86-110Q
Perez-Garcia, C. 80-7E; 82-25FF
Perez-Guerra, S.E. 85-5E
Pericleous, K.A. 84-45F
Perini, L.L. 86-30C
Perkin, R.M. 80-105J
Perkins, R.A. 85-4P
Perkov, S.V. 77-36J; 78-51J; 80-21J
Perkowitz, S. 79-24K
Perminov, E.B. 85-6DP
Permyakov, V.A. 78-10J
Perrier, A. 78-69J
Perriera, N.D. 85-24A
Perry, A.E. 77-21E; 79-38P
Perry, C. 86-54JM
Persen, L.R. 84-33B
Pershin, I.V. 83-4JJ
Pershin, N.P. 85-78K
Pershukov, V.A. 85-11FF
Persichetti, J.M. 86-57F
Persons, R.W. 80-23T
Pesaran, A. 82-20D
Pessanha, J.A.O. 86-19T
Pessel, W.O. 80-10P
Pessoa, O., Jr. 86-69P
Petashvili, O.M. 77-26P
Peters, A.R. 83-16DP
Peters, A.T. 86-29N
Peters, C. 86-35U
Peters, F. 85-41E
Peterson, R.C. 85-87K
Peterson, R.J. 84-49J
Peterson, R.S. 78-16D
Peterson, R.W. 77-31P
Peterson, V. 81-5S
Peterson, W.C. 76-38J; 83-49J
Petit, C. 79-37P; 80-22P,34P; 81-30A
Petley, D.H. 84-42C
Petre, G. 85-59F
Petrick, M. 81-5M
Petrie, D.J. 80-35A
Petrie, H.L. 86-23D
Petrov, A.I. 85-71Q
Petrov, N.E. 85-55B
Petrov, V.P. 79-104F
Petrovichev, V.I. 79-15J
Pettigrew, M.J. 85-72Q
Pettit, R.B. 78-38P
Petuchowski, S.J. 86-90K
Petukhov, B.S. 78-45B,18E; 82-36C,58J; 83-7S; 86-101J
Peube, J.-L. 85-3FF; 86-3FF
Peube, J.L. 78-76F; 80-58B; 81-17F
Peuschel, R.F. 83-44K
Pevzner, V.I. 78-15Q
Peyayopanakul, W. 78-55J
Peyker, N.K. 79-2H
Peyret, R. 86-5N
Pfeffer, R. 85-37DP
Pfeffer, R.L. 76-29F
Pfeifer, G.D. 86-21H
Pfeifer, H. 85-54S
Pfeil, D.R. 84-27FF
Pfender, C. 82-28B; 86-66B,67B
Pfender, E. 76-44P,16U,17U; 77-6U; 78-2P, 8U; 79-17U,19U,26U; 80-6U,7U; 81-2U,3U; 82-3U, 4U,6U; 83-4U,5U,12U,13U,14U,16U; 84-10U, 23U,34U; 85-10U,11U,15U,33U,44U,45U,52U, 55U; 86-40U
Pfotenhauer, J.M. 83-7G; 84-11G
Pham, Q.T. 85-39N; 86-67N,70P
Phan, C.N. 82-21T
Phan-Thien, N. 84-34B
Philip, J.R. 82-23D; 86-9S
Philips, W.F. 79-22T; 80-24T
Phillippi, R.M. 77-15P
Phillips, K. 80-10U
Phillips, R.E. 84-10N; 85-40N,41N; 86-101F
Phillips, W.F. 81-20T; 82-30T,31T; 85-32DP
Phillips, W.R.C. 85-66C,42E
Phschke, M. 81-82J
Phukan, B.D. 79-5U
Phuoc, H.B. 80-33FF
Pianarosa, P. 82-17U
Piat, J.F. 81-21C
Pickett, P. 76-14E
Pickett, P.E. 79-33B
Picologlou, B.F. 81-2Q
Picot, J.J.C. 76-7C,6J
Pidsukha, N.A. 77-11J
Pieczka, W. 82-20A
Piel, A. 86-1U
Pientka, K. 79-17H
Pieranski, P. 79-51F
Pierce, F.J. 76-37P; 78-11E
Pierini, P. 81-3G
Pierluissi, J.H. 84-49K,50K,80K,81K
Pierre-Eugene, M.J. 80-58B
Pierson, H.O. 77-1U
Pietrzyk, Z.A. 77-15U
Piggott, B.D.G. 78-19J
Pignatel, J.F. 86-94F
Pignotti, A. 83-24Q,25Q; 84-29Q,30Q; 86-77Q
Pikashov, V.S. 78-39P; 80-20K
Pikuz, S.A. 85-18U
Pilipchuck, Yu.L. 80-36P
Pilipenko, V.N. 78-46B; 79-45B
Pilyugin, N.N. 86-18C
Pimenoy, A.K. 80-1J
Pimenta, M.M. 78-10C; 79-23E
Pincombe, J.R. 85-14G; 86-4G
Pinheiro, J.D. 85-67Q

Pinkley, L.W. 76-16K; 77-35K
Pinkus, O. 84-50J
Pinnes, E.L. 79-61J
Pintauro, P.N. 86-28FF
Piorkowska, E. 86-71P,72P
Pioro, I.L. 84-54F; 86-78Q
Piotukh, S.M. 78-3H
Pipes, J.G. 76-49P
Pisarev, V.Ye. 77-51J; 78-72F,78J,79J; 79-81J, 82J,83J
Piskunov, V.N. 85-40Q
Piskunova, A.V. 86-1T
Pismen, L.M. 76-31D
Pisoni, C. 80-44J; 86-63P
Pissis, P. 84-29JM
Pittman, C.M. 85-77C
Pitts, D.R. 77-7,38J; 86-16DP
Pitts, W.C. 86-79Q
Pitz, R.W. 86-37E
Piva, R. 83-63F
Pizey, G. 81-23A
Platonov, V.M. 85-106Q
Platt, C.M.R. 84-82K
Platten, J.K. 76-68F; 77-83F
Play, D. 77-7S
Player, M.A. 86-38E
Plekhanov, S.D. 86-75U
Plekhanov, V.G. 86-75U
Plesset, M.S. 78-56J,58J; 79-66J; 83-40J; 84-30J,51J
Pletcher, R.H. 76-43C; 77-58F,59F; 80-45B; 81-26B; 85-8D; 86-56N
Pletcher, R.P. 76-32D
Plomdeur, P. 77-3U
Plostins, P. 81-22C
Plotkin, A. 86-57N
Plumb, H.H. 77-37P
Plumb, O.A. 77-84F; 82-21F,11FF; 85-55S
Plummer, D.N. 82-26J
Pnueli, D. 81-48F
Po, J.K. 76 8E
Podberezkiy, A.I. 83-7DP
Podberezsky, A.I. 83-8DP; 84-7Q
Podgaetskii, V.M. 86-46K
Podowski, M. 85-72B
Podowski, M.Z. 86-100J
Podsushnyy, A.M. 80-106J
Poe, G.G. 76-6G
Poggemann, R. 79-15S; 80-21S
Pohanerkar, S.G. 78-93F
Pohl, R.O. 80-44P
Pokhoriler, V.L. 81-22S
Pokryvailo, N.A. 85-5E
Pokryvaylo, N.A. 79-39P; 83-8S
Pokusayev, B.G. 79-33C
Polak, A. 78-44C
Polat, S. 85-67C
Polevikov, V.K. 76-10F
Polisevski, D. 85-60DP,61DP,62DP
Polishchuk, V.P. 86-62U
Poll, D.I.A. 81-23C
Pollard, A. 76-71F; 84-26F; 85-44F,30N
Polonskiy, V.S. 78-75J; 83-62J
Polonsky, V.S. 80-125J; 81-23B,44J; 84-59J, 43Q
Polturak, E. 86-73P
Polusaev, B.G. 85-3JJ
Polyak, M.P. 79-25S
Polyakov, A.F. 77-18B; 78-45B,18E,40P; 82-36C
Polyakov, I.N. 85-24H
Polyakov, S.P. 81-1U
Polyanin, A.D. 82-37C; 84-17H
Polzik, V.V. 83-6D
Pomet'ko, R.S. 79-60J
Pon, R. 84-19P
Ponnappan, R. 85-74Q
Ponomarev, N.N. 85-72C
Pons, M. 86-37T
Ponter, A.B. 76-44C; 77-39J; 78-57J
Ponzi, P.R. 79-16S
Pool, L.A. 85-14S
Poole, L.P. 82-43K
Poon, K.C. 78-33A; 79-17A,50A,51A
Poon, K.T. 85-25J
Poots, G. 79-8S; 81-9S
Pop, I. 77-42F; 79-67F,68F; 80-34FF; 83-13DP, 30FF,32FF; 84-8DP; 85-63DP,26FF; 86-19DP, 26DP,27JM
Popalov, V.V. 78-12Q
Popel, A.S. 78-47B; 79-52A
Popel, O.S. 86-26T
Popiel, C.O. 86-11G
Popiel, Cz.O. 80-28D
Popishvili, V.I. 78-49B
Popov, V.N. 80-59B; 82-29B; 85-55B
Popov, Yu.A. 78-25K; 80-91K
Popovic, M.M. 79-27U
Popovska, F. 77-40B
Porter, R.W. 80-17Q
Poshkas, P.S. 78-20Q
Poskas, P.S. 84-31Q; 85-66B
Post, M.J. 77-39P
Post, T.M. 79-13A
Potchkin, S.A. 80-102J
Potekhin, S.A. 80-103J; 85-88J
Potter, J.M. 80-35FF
Potter, K.N. 85-35P
Potter, M.C. 76-7B,40D; 85-38A
Pouagare, M. 86-68N
Poulikakos, D. 82-38C,8F,79F; 83-19DP,20DP, 43F,44F,45F; 84-6DP,28DP,29DP,7F,2FF,22FF; 85-64DP,65DP,66DP,67DP,68DP,71F; 86-41F, 36JJ,37JJ,27JM
Pourdashti, M. 86-49J
Povkh, I.L. 79-34B
Povsten, S.G. 77-50J; 80-131J
Povsten', S.G. 79-84J
Powe, R.E. 77-85F; 80-46F; 83-46F; 85-102F
Powell, R.L. 82-39C
Poyurovskaya, I.E. 86-46K
Poz, M.Ya. 86-38T
Pozdneev, B.M. 86-36S
Pozdnyakov, B.A. 86-82U
Pozhivanov, A.M. 86-53S
Poznak, V.Ye. 80-99J
Poznyak, A.A. 79-53A

Pozowolski, A.E. 80-9M
Pozzetto, D. 86-58U
Pozzi, A. 76-6C; 85-14C
Prabhu, A. 76-2E; 77-3C
Prahl, J.M. 77-28C
Prakash, C. 80-61F,36FF; 81-36B,49F,67F; 84-9N; 85-56B,57B,72F,42N; 86-68B,69N
Prakash, R. 77-1D; 83-3J,4J; 85-11J,89J; 86-11J
Prakash, S. 80-107J
Prasad, A. 76-46A; 78-5R; 79-54A; 80-36A
Prasad, A.N. 86-32U
Prasad, G. 76-18R
Prasad, K.K. 79-26K
Prasad, N.R. 85-89J
Prasad, R.C. 86-11E
Prasad, R.R. 86-65U
Prasad, V. 84-30DP,31DP,55F; 85-69DP,73F, 74F; 86-45DP,46DP
Prat, M. 86-47DP
Prata, A.J. 84-4N
Prata, A.T. 83-62F; 84-35B; 85-40A,90J,91J; 86-86F,108F,102J
Pratap, V.S. 76-35B
Pratt, T.K. 81-43A
Praturi, A. 86-4FF
Praus, L. 85-32P
Prausnitz, J.M. 77-8
Preckshot, G.W. 77-9G
Preiss, G. 76-39J
Prelec, K. 82-11U
Prelovsek, P. 84-20P
Prenger, F.C. 86-80Q
Prenger, F.C., Jr. 76-14R
Prescott, L.E. 79-12U
Preston, C.S. 86-16G
Price, F.A. 76-48J
Price, H.G. 80-20S; 81-20S
Pridantsev, A. 85-70B
Pridantsev, A.I. 79-15J; 80-2J
Priddy, W.J. 85-68C
Priemer, R. 79-19J
Prieto, R.F. 83-12D
Primaychek, V.V. 85-9S
Prins, W. 85-70DP
Prinz, B. 86-8A
Prisnayakov, V.F. 86-103J
Prisnyakov, V.F. 83-23Q
Pritchard, J. 83-43K
Probert, S.D. 77-1A,2A,11C,14S,15S,17S,16T; 79-4A,18S; 82-51A; 85-10F,60Q,61Q,62Q; 86-57A, 11F,3S
Prock, J. 86-66U,67U
Proctor, D. 81-6T; 85-32T; 86-39E
Proctor, M.R.E. 77-86F; 79-105F; 80-7F; 81-50F, 52F; 84-35F; 86-95F
Profizi, J.L. 81-40A
Projahn, U. 81-44A; 82-37A; 83-24JM; 85-42FF, 43FF
Projan, U. 85-75F
Prokop'ev, V.I. 86-39S
Pronobis, M. 84-4Q; 86-81Q
Proshutinskiy, A.P. 83-14S
Proskuryakov, K.N. 85-75Q
Prosperetti, A. 78-56J,58J; 84-51J
Prosuntsov, P.V. 86-92K
Proulx, P. 84-24U; 85-37U,46U
Prouvost, L. 82-54F
Pruckner, E. 81-12K
Prud'homme, M. 86-22D
Prunet-Foch, B. 77-18D,19D; 82-32J; 85-19JJ; 86-29JJ
Prusa, J. 83-47F; 84-30JM,31JM; 85-40JM
Pryor, T.L. 76-6T
Prziremhel, C.E.G. 76-27D
Puchkov, E.I. 80-40B
Puri, V.M. 79-23T; 86-82Q
Purnell, J.G. 81-9M
Purohit, G.P. 85-73Q
Purohit, N.K. 80-50B
Pursch, H. 84-12U; 86-9U
Purwasasmita, M. 86-68Q
Pushkin, V.T. 86-126K
Pushkov, V.V. 85-17Q
Pust, L. 77-20D
Pustovalov, V.K. 85-88K
Pustovalov, V.N. 78-4G
Put'ko, V.F. 86-22U
Putman, G.R. 85-76Q
Putyatin, B.V. 85-89K
Putyatina, E.N. 86-75C
Puzach, V.G. 79-32U
Puzin, V.A. 85-119J
Puzyrev, Ye.M. 80-108J
Puzyrewski, R. 80-109J
Pyadishyus, A.A. 79-35B

Quarmby, A. 76-18E
Quartapelle, L. 84-4G
Quazzani, J. 86-5N
Quenzel, H. 78-13K
Quick, K. 86-83Q
Quinn, M.F. 84-83K
Quintard, M. 82-54F
Quintiere, J. 82-6S
Quiroga, O.D. 78-19C
Quiroga, Q.D. 83-18B
Quon, C. 77-87F; 79-106F; 83-48F
Qureshi, Z.H. 79-107F; 82-27FF; 86-29FF

Rabadi, N.J. 79-36B; 81-52J; 82-30B
Rabas, T.J. 85-32Q,33Q
Rabinovich, Ya.I. 78-11H
Rabl, A. 76-16T,17T,18T; 78-15T; 79-8T,9T, 10T; 81-21T
Radeke, K.-H. 76-13D,14D
Radhakrishnamacharaya, G. 77-41B
Radle, K. 79-8L
Radtke, D.D. 83-11D
Radtke, R. 80-26U
Raduchev, V.A. 85-30JM
Rae, W.J. 84-6S,7S; 86-15C
Rafalowicz, J. 86-43P
Ragaller, K. 81-16U; 82-2U,18U,22U
Raghava, A.K. 79-40P

Raghavachar, T.S. 77-11G
Raghavacharyulu, N. 86-48DP
Raghavaiah, C.V. 85-14B
Raghavan, V.R. 84-4J
Raghaven, C. 79-99F
Ragheb, H.S. 81-63J
Raghu, S. 85-12E
Ragozzino, E. 76-1P; 78-1K
Ragu, G.J.V.J. 85-71DP
Ragunathan, K. 76-33D
Rahal, A. 84-27U
Rahal, A.M. 86-68U
Raham, L. 86-30FF
Rahbar, S. 84-15JJ
Rahhaoui, B. 84-27U
Rahm, L. 80-47F
Rahman, M. 84-36B; 86-17B
Rahman, M.M. 85-22JJ; 86-31FF
Rahn, L.A. 82-14K
Rai, K.N. 76-34D
Rai, M.M. 86-70N
Raina, G.K. 82-59J; 84-32Q; 85-92J; 86-104J
Raithby, G.D. 76-38F,69F,70F,71F; 77-8F; 78-31F, 77F; 80-54F,55F,59F; 81-53F,70F; 82-19F, 20F,20N; 83-12F; 84-19A,15N; 85-7FF; 86-92F, 26N,27N,28N,41N,71N
Raj, S.A. 85-79C; 86-59C
Raj, V.V. 83-50J
Raja Rao, M. 82-11B; 83-41B; 85-24JJ
Rajagopal, K.R. 85-76F
Rajagopalan, S. 77-5E; 79-24E; 81-24C; 82-40C
Rajakumar, A. 78-21Q
Rajasekaran, R. 85-69C,10G
Raju, B.B. 78-34A
Raju, G.V.R. 80-4M
Raju, K.S. 77-26C
Rajurkar, K.P. 83-6S
Rajvanshi, A.K. 77-23K
Rajvanshi, S.C. 80-5G; 82-6G
Rakhmatulina, I.Kh. 81-64J
Rakich, J.V. 77-10R
Rakopolous, C.D. 86-39T
Rakopoulos, C.D. 80-17S
Raleigh, M. 85-26B
Ram, P.C. 80-1M
Ram, R. 76-19K; 78-26K
Rama Rao, K.V. 80-48F
Ramachandran, M. 85-44FF
Ramachandran, N. 81-45A; 82-36A; 83-30C; 86-9FF
Ramachandran, P.A. 85-30Q,31Q
Ramachandran, P.S. 78-36C
Ramadan, F.A. 84-22JJ
Ramadane, A. 86-16H
Ramadhyani, S. 77-61A; 78-41A,75F; 80-37A; 83-35B; 84-56F; 85-77B,25JM,43N; 86-35B, 12F,13F
Ramagopal, A. 85-73C
Ramakers, F.J.M. 79-94J
Ramakrishna, K. 78-78F; 82-55F
Ramakrishna Rao, P. 84-11N
Ramakrishnan, S. 80-11U
Raman, A. 85-43E
Ramanathan, K.G. 77-2L,7L,8L; 78-8L; 79-9L
Ramani, M.P.S. 85-108J
Ramesan, T. 85-29H
Ramesh, T.G. 86-44E
Ramilson, J.M. 80-29D
Ramirez, W.F. 85-12P
Ramirez-Zetina, R. 85-38J
Ramkrishna, D. 79-25A,55A,56A; 80-56B,57B, 60B; 81-35B; 82-35A
Ramma Subba Reddy Gorla, 81-25C
Ramos, E. 85-77Q,84Q,85Q
Ramsey, J.W. 77-18T; 78-35A,42A,56B,27Q; 79-1A, 57A,72A,42C,108F,130F,62J,74J; 80-21Q,25T; 81-58A,59A,22B; 84-33DP
Ramshaw, C. 86-16Q
Rana, R. 79-109F
Rana, R.S. 86-105J
Ranatunga, D.B.J. 80-16T
Randall, J.D. 78-36A
Randall, K.R. 78-1F; 79-110F
Randzio, S.L. 85-48E
Rane, A. 80-80B,110J,137J; 81-97J
Rane, A.G. 81-65J
Ranganathan, P. 84-23FF; 85-58B
Rani, M. 85-5B,15C,2H
Rani, U. 81-15T
Rannenberg, M. 81-14D
Rao, B. 77-10A
Rao, B.N. 86-59U
Rao, B.V. 76-22K
Rao, Ch.D. 86-84Q
Rao, D.M. 86-34C
Rao, D.P. 70-33T
Rao, K.S. 77-25C; 79-33T; 86-30T
Rao, K.V. 84-24FF; 85-45FF
Rao, M.G. 85-18E
Rao, M.R. 76-13Q; 85-104Q; 86-71B
Rao, M.S. 78-34A
Rao, N.S. 84-13N
Rao, P.K.M. 76-40J
Rao, P.P. 77-19T
Rao, P.R. 83-22JM
Rao, S.B. 80-53F; 83-37B
Rao, S.S. 78-42B; 79-31B
Rao, V.D. 84-23JJ; 85-71DP
Rao, V.R. 83-8G
Rao, V.S.R. 86-84Q
Rapagopal, K.R. 85-44N
Rapley, C.W. 83-26Q; 84-37B
Rapoport, D.A. 78-37P
Rapoport, E.Ya. 85-35A
Raptis, A. 83-3M; 86-32FF
Raptis, A.A. 82-9M; 83-33FF; 85-72DP
Rarity, J.G. 86-18K
Rashkovskaya, N.B. 85-81DP
Rasky, D.J. 81-19K
Rasmussen, H. 84-23DP
Rassamakin, B.M. 85-78Q
Rastogi, A.K. 78-45C
Raterman, K.T. 82-46A
Ratiani, V.G. 80-120J
Ratzel, A.C. 79-24T; 82-44K
Ratzel, A.C., III 83-33K

Rau, R.S. 77-8U
Rault, D.F.G. 85-90K
Rautenberg, T.H., Jr. 77-16U; 79-18U
Ravi, J. 84-22U; 86-69U
Ravigururajan, T. 85-50Q
Raw, M.J. 85-45N,46N; 86-75N
Rawlins, W.T. 83-43K
Rawnsley, S.M. 84-24S
Ray, A. 78-25S
Ray, S.R. 80-21K
Ray de Luna, A. 86-123F
Raymond, J.F. 83-23JM
Raynaud, M. 86-32N,72N
Raz'nyak, V.V. 85-20B
Razafinimanana, M. 86-24U
Razani, A. 77-50A
Razavi, M.D. 78-59J
Razelos, P. 78-22Q; 79-58A,16Q; 80-38A; 85-46FF; 86-85Q
Razzaque, M.M. 83-34K; 84-84K
Read, P.L. 86-12G
Reagan, J.A. 81-10T
Reay, D.A. 85-79Q
Reda, D.C. 83-49F; 86-49DP,50DP
Reddy, C.S. 79-140F; 80-49F
Reddy, G.S. 85-41JM
Reddy, J.N. 80-37FF
Reddy Gorla, R.S. 76-45C,46C,72F; 77-10C
Redman, R.S. 79-41P
Redondo, A. 86-74P
Reed, A.W. 86-51DP
Reed, J.G. 84-6JJ
Reemtsen, R. 81-34A
Rees, D.A.S. 85-73DP; 86-52DP
Rees, D.S. 85-74DP
Rees, N.W. 80-113J
Reeves, C.M. 86-47E
Reggio, M. 86-73N
Rehman, M. 85-36P
Reichert, J.K. 77-38P
Reicker, E.L. 86-39K
Reid, J. 86-65K,101K
Reid, R.C. 77-8,7J; 83-9J
Reid, R.L. 76-25P; 83-7B; 85-33T
Reid, W.H. 77-30F
Reiff, E.K., Jr. 84-11E
Reifschneider, L.G. 86-103Q
Reijntjes, P.J. 86-40E
Reile, E. 84-13E
Reilly, H.E. 83-20Q
Reilly, J.L. 86-109K
Reilly, M.L. 86-16E
Reimann, M. 82-2A
Reiners, U. 86-72U
Reinhardt, H.J. 86-30U
Reinhardt, R.T. 80-31U
Reinheimer, G. 83-13H
Reinkenhof, J. 76-15R; 85-91K
Reisman, O.I. 77-42B
Reiss, H. 84-14K; 85-20K,92K,6P; 86-75P
Reistad, G.M. 78-15S
Rejab, A.B. 86-76P
Rekalic, M. 76-22U
Rekant, N.B. 83-35K
Remizov, O.V. 78-86J,26S
Remorino, M.R. 79-37B
Rempt, R.D. 83-9S
Ren, A.L. 84-12N; 85-16D
Ren, H.S. 84-12Q
Renard, G. 77-12A
Renardy, M. 85-78F
Renardy, Y. 85-77F,78F; 86-96F
Render, C.L. 83-8Q
Renk, F.J. 79-63J,64J
Renksizbulut, M. 83-51J,52J
Rennex, B. 86-35E
Renz, U. 84-22JJ,24JJ; 86-38JJ
Renzoni, P. 85-57B
Repin, V.B. 85-59B
Repukhov, V.M. 76-14H; 77-17H; 78-16H; 84-64J
Reshetov, V.V. 84-59J,43Q
Reshotko, E. 77-28C
Restivo, A. 79-96F; 80-20F
Retinskii, P.I. 86-2S
Reutskiy, V.A. 82-28D
Rev, E. 86-86Q
Revankar, S.T. 85-70C
Revelle, C.S. 84-17T
Revyakin, A.V. 77-14P
Rey, C. 78-19E; 79-6D; 83-28C
Reyes, J.N. 84-32J
Reynolds, A.J. 76-19E
Reynolds, W.C. 79-6E
Rezayat, M. 82-5C
Rezhitskiy, V.G. 83-37A
Reznik, S.V. 86-92K
Rhazi, M. 85-53E
Rhee, H.S. 84-57F
Rhee, K.T. 83-11K
Rhee, S.J. 78-79F; 81-37B
Rheinlander, J. 82-32T
Rhodes, E. 77-27J
Rhodes, T. 85-21F
Rhoom, T.N. 80-12L
Riahi, N. 80-4F,50F,51F; 81-54F; 82-10F,5G, 10M; 83-50F; 84-58F; 85-79F,80F; 86-97F, 98F,99F
Riaz, M. 77-20T,21T; 78-16T
Riaz, M.R. 76-19T
Riazi, M.R. 85-37P,38P
Riba, J.P. 82-12D
Ribando, R.J. 76-73F
Ribbin, C.G. 82-27K
Rice, I.G. 83-10S
Rice, J.G. 85-47N
Rice, P. 84-10B
Richard, J.P. 81-55F
Richards, B.E. 76-4C; 78-11S; 79-18H; 82-5S
Richards, H.K. 76-20E
Richards, P.L. 86-122K
Richardson, A. 81-4L
Richardson, A.T. 84-46F
Richardson, J.F. 76-35D; 83-15DP; 85-57C, 35DP
Richardson, L.S. 85-43U; 86-33U
Richardson, M.C. 84-105K

Richardson, N.M. 83-9K
Richardson, S.M. 79-38B
Richley, E. 80-27U; 82-19U,20U
Richmond, J.C. 80-37P; 83-28T
Richter, F.M. 79-111F; 83-51F
Richter, H.J. 78-88J
Richter, J. 84-25U
Riddiford, C.L. 77-6T
Ridley, K.D. 86-18K
Riegel, S. 86-77P
Rieger, H. 81-44A,51F; 82-37A; 83-24JM; 86-36JM
Rieke, H.B. 79-14E
Riester, J.B. 80-38FF
Rietema, K. 80-27B,14D
Rifert, V.G. 78-60J; 79-65J; 80-111J
Righi, J. 82-77F
Riha, P. 76-47A
Riley, D.A.S. 85-74DP
Riley, D.S. 77-23A,51A; 79-126F; 80-9P; 83-34FF; 85-73DP,67FF; 86-27A,52DP
Riley, J.P. 86-8T
Riley, N. 76-24F; 80-35FF; 81-17B; 84-7C, 8C
Rimai, L. 84-56K
Ringer, D. 80-15S
Ringer, D.U. 84-30S; 85-75DP
Rinkinen, W. 82-6S
Rinsland, C.P. 84-25K; 86-93K
Rish, J.W., III 86-94K
Ristorcelli, J.R., Jr. 79-15Q
Ritter, A.M. 85-24JM
Ritter, R.B. 83-27Q
Rivers, R.J. 82-30J
Rivkind, V.Ya. 85-36B
Rivtsis, A.S. 86-20E
Rizhko, V.A. 77-54B
Rizza, J.J. 81-66J
Rizzi, G. 86-12T
Rizzi, R. 83-13T; 84-41K
Rizzo, F.J. 82-5C
Roach, G.H. 86-87Q
Roach, P.R. 86-88Q
Roache, P.J. 79-31F
Roadman, R.E. 83-36B
Roberstson, H.S. 80-22T
Robert, D. 86-29K
Roberts, A.J. 85-81F
Roberts, C.C., Jr. 79-26T
Roberts, G.T. 79-25T
Roberts, J.M. 78-16J
Roberts, M.J. 77-29P
Roberts, P.H. 83-14JM
Roberts, P.M. 84-19JM
Robertson, C.R. 76-44B
Robertson, J.M. 83-28Q
Robertson, R.B. 77-20K
Robertson, S. 79-43P
Robertson, S.R. 79-59A; 86-74N
Robillard, L. 80-64F; 81-56F; 82-56F,72F, 24FF; 83-38F,66F; 84-71F; 86-53DP
Robinson, D.P. 78-48B
Robinson, J.L. 76-74F
Robjohns, H.L. 84-24P; 85-51P
Roche, P. 84-10K
Rockstroh, T.J. 86-43U
Rockwood, J. 86-38S
Roddeman, D.G. 86-111J
Roder, H.M. 86-78P
Rodi, W. 78-45C; 81-22FF; 85-71C; 86-4FF, 5H
Rodighiero, C. 83-35FF
Rodkiewicz, C.M. 85-18B
Rodrigues, A.E. 82-31B
Rodrigues, J.-F. 83-7JM; 86-37JM
Roehler, F. 80-2G
Roethe, A. 78-13D
Roethe, K.P. 78-13D
Roetzel, W. 77-40J,18Q; 79-17Q; 80-18Q; 85-60B; 86-69B,89Q
Roganov, P.S. 79-26E; 84-18H; 85-94C
Roger, C.R. 79-9L
Rogers, C. 85-36A; 86-51A,52A
Rogers, G. 83-8A
Rogers, J. 83-43K
Rogers, J.C.W. 79-60A
Rogers, J.T. 81-26C
Rogers, R.H. 85-14G; 86-4G
Roges, B.A. 84-8T
Rogge, H. 85-17C
Rognlien, T.D. 81-10U
Rohatgi, V.K. 79-62A; 85-50U
Rohatschek, H. 76-38P
Rohm, H.J. 80-112J
Rohsenow, W.M. 76-47J; 77-12J,48J; 78-26B, 25Q,7S; 79-21J; 80-24J; 82-9J,26J,66J; 83-49B; 84-3F,3J; 85-76Q; 80-127J
Roiland, M. 77-69F; 80-38F,39F
Roizard, C. 86-54DP
Rojas, E. 86-31P
Roko, K. 86-118J
Roldan, A. 86-3K
Rolf, A. 84 33Q
Rolff, K.P. 80-21U
Rolin, M.N. 81-13K; 85-93K
Rolph, W.D., III 82-14N
Roman, M.N. 76-35D
Roman, W.C. 86-70U
Romaniuk, M.S. 78-57C
Romanov, A.S. 86-95K
Romanov, G.S. 85-88K
Romanovskiy, I.M. 82-33J
Romanovskiy, S.G. 78-17H
Romanovskiy, S.V. 83-18J
Romberg, T.M. 80-113J
Romero, L.A. 80-97J
Romie, F.E. 79-18Q; 83-29Q; 84-34Q; 85-80Q
Rondeel, W.G.J. 79-11U
Ronen, A. 85-81Q
Roney, P.L. 85-19K
Ronsheim, P. 81-17U
Rooke, J.H. 79-38F
Roost, E.G. 78-32S; 79-26S
Roppo, M. 84-59F
Roppo, M.N. 80-40J; 83-30Q
Roquemore, W.M. 86-47E
Rosahl, B. 76-13D

Roscoe, H.K. 81-14K
Rose, J.W. 76-41J,8M; 77-2M; 78-61J; 79-35C, 36C; 80-114J,124J; 81-58J,67J; 83-11JJ; 84-15JJ, 16JJ,20JJ,25JJ; 85-111Q; 86-30JJ,50JJ,113Q
Rose, M.E. 82-5N
Rose, W.C. 77-25P
Rosen, H. 83-36K
Rosen, H.N. 84-8E,2S
Rosenbalt, S. 85-41F
Rosenberger, F. 79-97F,98F; 85-1F,90F; 86-6FF
Rosenberger, F.R. 84-61F
Rosenblat, S. 81-57F; 84-59F; 86-100F
Rosenblatt, M. 79-13E
Rosenblatt, S. 82-57F
Rosenwald, R.D. 84-86K
Rosiczkowski, J. 83-21C
Roslyakov, A.D. 86-39A
Rosman, E.C. 84-35Q
Rosner, D.E. 79-20H; 84-10H; 85-12H
Rosnovskiy, S.V. 78-45B
Ross, D.C. 76-17S
Ross, M.D. 86-21J
Rosser, J.B. 76-48A; 77-52A
Rossie, A.N. 81-9F
Rossoshanskii, O.A. 85-32E
Rotenberg, Y. 81-1J
Roth, J.-E. 86-69B
Roth, J.E. 84-37JJ; 85-44E,39Q; 86-89Q
Rothman, L.S. 86-96K
Rothman, S.J. 85-21E
Rotta, J.C. 77-28E
Rottigni, G.A. 78-17T
Rotunno, R. 86-92N
Rouai, N.M. 83-53J
Rouanet, A. 82-14A
Rousar, D.C. 80-20S
Roussard, R.A. 85-13J
Roux, B. 76-47C; 78-80F; 79-112F; 81-20F; 85-90F; 86-21F,22F,6FF
Roux, J.A. 76-49P; 78-27K; 83-37K; 85-94K, 8T; 86-94K
Rowe, R.D. 84-26F; 85-44F
Rowlands, I. 85-39P
Rowley, G.J. 84-38B
Roy, E.L. 78-11U
Roy, G.K. 86-90Q
Roy, R.P. 76-42J; 80-55J; 83-30J; 84-3N; 85-43E; 86-23J,28J,106J
Roy Chowdhury, S.K. 85-26J
Roychowdhury, S. 82-21U
Roychowdhury, U.K. 82-21U
Rozen, B.J. 86-53A
Rozhdestvenskiy, V.I. 79-12H
Ruas, V. 86-53N
Rubeye, H.A. 80-18S
Rubi, J.M. 82-25FF
Rubin, F.L. 81-11Q
Rubin, H. 81-58F; 84-20T
Rubin, M.B. 86-54A
Rubin, S.C. 82-55F
Rubin, S.G. 81-22C
Rubinsky, B. 76-49A; 79-61A; 82-38A,39A; 83-21DP, 23JM; 84-16JM; 85-49S; 86-33Q
Rubiolo de Reinick, A. 86-38JM
Rubtsov, N.A. 79-77F,15K; 85-72C; 86-97K
Ruchenstein, E. 78-81F
Ruchti, C.B. 86-71U
Ruckenstein, E. 80-52F,39FF; 85-73C
Rud'ko, A.P. 78-4G
Rudakova, A.Ye. 83-60J
Rudenko, I.M. 76-9Q
Ruder, Z. 86-8J
Rudin, A. 76-11P
Rudin, G.I. 78-40J; 85-64K
Rudisill, J.E. 77-21P
Rudolph, L.K. 84-28U
Rudraiah, N. 80-53F; 82-24D,25D,26D,27D; 83-52F; 84-5G; 86-13G
Rudraian, N. 83-37B
Rudriah, N. 80-44F
Rudy, T.E. 85-40K
Rudy, T.M. 85-82Q,108Q
Rued, K. 85-74C; 86-54C
Ruggiero, M.A. 85-42JM
Ruiz, R. 82-35D; 86-37FF
Rule, T.T. 82-11T
Rumanov, E.N. 85-30JM
Rummer, P. 86-91Q
Rummler, D.R. 77-3R
Rumyantseva, L.A. 83-11H
Runca, E. 85-49N
Runchi, G.P. 77-45A
Ruppel, W. 85-24S
Rusanov, K.V. 85-54J
Rusanova, K.V. 80-69J
Rusin, S.P. 85-95K
Rusishvili, D.G. 80-120J
Russel, C. 86-12Q
Russel, F.B. 84-19A
Russell, L.M. 76-4H; 84-9S; 85-20E
Ruth, D.W. 79-113F; 80-54F,55F; 81-59F; 85-44F
Ruth, E.K. 83-31Q
Rutily, B. 85-96K
Rutter, J.W. 85-42JM
Rvachev, V.L. 78-49B
Ryabov, A.N. 78-62J
Ryadno, A.A. 84-21JM
Ryazantsev, Yu.S. 85-93J
Rybakova, L.E. 76-1J
Rybanin, S.S. 81-43J
Rybchinskaya, G.B. 78-37J; 82-39J
Rybski, J. 77-3F
Rybski, W. 76-8F
Rychkov, V.I. 78-27S
Rykalin, N.N. 81-18U
Rykova, B.G. 78-8G
Ryley, D.J. 85-63C
Ryzhenko, I.A. 77-21S
Ryzhkin, A.A. 86-5S
Ryzhkov, L.I. 86-98K
Ryzhkov, L.N. 78-27S
Ryzhkova, T.P. 86-98K
Ryzhov, Yu.A. 78-22B
Rzayev, A.I. 76-5M

Saad, A. 77-16E
Saad, E.A. 85-97K,127K
Saaidjian, E. 80-56F
Saarimaa, R. 76-39P
Saatci, A.M. 86-92Q
Saatdjian, E. 85-76DP
Saaverdra, J.J. 86-4H
Sabbagh, J. 86-26J
Sabbagh, S.A. 76-75F
Saber, A.J. 78-14U
Saberian, A. 77-12F
Saberian-Broudjenni, M. 85-77DP
Sabersky, R.H. 78-41B; 82-20B,36D
Sabhapathy, P. 85-7JM; 86-33FF
Saboya, F.E.M. 76-36B,10Q; 84-35Q; 86-19T
Saburov, E.N. 78-28S; 79-10G; 85-61B
Sacadura, J.F. 83-1K
Sadatomi, M. 81-68J,69J; 82-33B
Sadd, M.H. 77-53A; 78-37A; 80-28A
Sadegh, A.M. 85-44N
Sadeghipour, M.S. 82-40A; 84-32JM
Sadhal, S.S. 77-41J; 78-23A; 79-66J; 81-46A; 82-4A; 84-3JJ,4JJ
Sadowski, D. 78-1L
Saeki, S. 86-54F
Saez, A.E. 82-33T; 83-22DP
Safar, Z.S. 80-61B
Saga, R. 83-1Q
Sagadeyev, V.I. 79-3P
Sagara, K. 76-26E; 83-53F; 84-60F
Sage, F.E. 76-43J
Saha, P. 76-44J; 80-115J; 84-52J
Saha, S.K. 80-41P
Sahimi, M. 84-24A
Sahm, M.K. 86-53B
Sahoo, D.G. 78-64F
Sahota, M.S. 80-30D
Said, M.N.A. 79-114F
Saido, K. 77-36K
Saifullah, A.Z.A. 81-12T
Saima, A. 76-17C
Saini, J.S. 83-3J,4J; 85-11J,89J; 86-11J
Saint Jean Paulin, J. 85-37A
Saint-Dizier, J.P. 82-17U
Saitiel, C. 81-26T
Saito, A. 81-12T,12T; 82-41A,42A; 83-25JM, 32T; 84-21T,22T; 85-27JJ,43JM,44JM,45JM; 86-47JJ,40T
Saito, A.Y. 86-39JM
Saito, H. 77-54A; 83-40FF,35JM,11S; 84-33JM; 85-82F,47FF,46JM; 86-19FF,34FF
Saito, I. 86-46J,37Q
Saito, K. 86-93Q
Saito, T. 82-26C; 83-18U; 85-27FF; 86-93Q
Saito, Y. 79-17S; 84-21T; 86-40T
Saitoh, T. 77-89F; 78-38A; 80-40FF; 82-58F
Saiy, M. 76-16E
Sakaguchi, S. 86-54S
Sakai, I. 84-66K; 85-67K
Sakai, T. 85-94J; 86-119F
Sakakibara, M. 76-37B; 77-43B; 86-26FF
Sakakibara, T. 80-23U
Sakalauskas, A.V. 86-18Q
Sakamoto, M. 86-93Q
Sakharov, V.I. 86-5K
Sakkal, F. 83-5K
Sako, A. 86-33B
Sako, M. 83-36FF
Sakoda, A. 86-41T
Sakumoto, S. 82-27J; 86-44J
Sakumoto, Y. 82-4J
Sakurai, A. 77-42J,43J; 83-35J; 84-53J; 86-119F
Sakurai, Y. 85-98K
Sakuta, K. 77-28T
Sakuta, T. 80-28U
Salan, J. 79-51F; 83-54F
Salanasky, N. 86-55P
Salaun-Penquer, G. 81-17FF
Salazar, A. 86-10C
Salcudean, M. 85-27H; 86-11B,55S
Salehpour, A. 83-74J
Salerno, L.J. 86-41E
Salman, Y.K. 80-2FF
Salomatov, V.V. 80-108J
Salomzoda, F. 80-54J
Salt, H. 83-28A,29A
Saltiel, C. 82-34T
Salu, Y. 77-22E
Saluja, C.L. 79-3H; 85-15H,16H
Samaraweera, S.S.A. 79-18E
Samie, F. 81-28C,32FF; 82-62F; 85-19D; 86-55C
Samimy, M. 86-23D
Sammakia, B. 79-115F; 81-26FF; 82-26FF,27FF; 83-26JM; 85-48FF; 86-60F
Samoilov, I.N. 76-31P
Samoilov, L.N. 79-42P
Samokhin, A.I. 85-18U
Samonin, K.L. 83-21K
Sampson, P. 81-47A
Samuel, T.D.M.A. 81-22T; 83-33T
Samuels, M. 77-7Q
San Jose, R. 85-75C
Sanborn, D.M. 77-13S
Sanchez Sarmiento, G. 79-36A,4R
Sanchez-Palencia, 82-9D
Sandall, O.C. 77-7H; 78-27B
Sandberg, O. 86-67P
Sandborn, P.A. 80-41FF
Sandborn, V.A. 76-40P; 77-76F; 80-41FF
Sanders, B.R. 83-14C
Sanders, D.J. 84-87K
Sanderson, P.R. 77-21D
Sandford, R.D. 77-13F
Sandhu, S.S. 85-56S
Sando, M. 86-94Q
Sandoval, S.B. 84-28S
Sandru, E. 86-42T
Sangameswar Rao, K. 79-33T
Sangani, A.S. 82-41C
Sani, R.L. 79-72F
Sano, H. 80-88J
Sano, O. 85-45E
Sano, T. 81-7H; 82-28FF
Sano, Y. 82-32B,40B; 84-54B
Sanochkin, Yu.V. 85-83F,100F

Sanokawa, K. 78-65J; 79-118F; 85-37Q
Santarelli, F. 77-12K; 78-34K; 79-5K; 80-23K; 81-16K; 82-45K; 85-18K
Santi, P. 86-56S
Santini, M. 84-5P
Santo, M. 82-47A
Santucci, A. 86-42E
Sapankevich, A.P. 79-60J; 80-39B
Saperov, Ye.V. 77-2S
Sapola, V.J. 79-4D
Sapozhnikov, V.V. 77-36B
Sapritskii, V.I. 76-31P
Sapsford, C.M. 83-29T
Saraf, J.R. 76-17K
Sardel, F. 85-49N
Sardesai, R.G. 81-95J; 82-60J; 83-14JJ
Sarhan, A. 80-14F; 81-1G; 82-7B
Sarhan, A.A. 81-18F
Sari, S.O. 79-27P
Sarka, B. 85-16E
Sarkar, M.K. 81-48A; 82-35T
Sarkar, S. 86-48JJ
Sarker, M.A.R. 86-107J
Sarma, B.K. 86-88Q
Sarma, G.S.R. 85-84F
Sarma, L.V.K.V. 86-6F
Sarma, P.K. 76-10J; 78-42C,78F; 84-6G,23JJ; 85-71DP
Sarma, T.S. 77-90F
Sarma, Y.V.B. 86-18H
Sarmiento, G.S. 85-13A
Sarofim, A.F. 84-15F
Sarunac, R. 85-83Q
Saruwatari, S. 82-33B
Saruwatari, T. 86-37B
Sasaguchi, K. 83-25F; 86-40JM,41JM
Sasaki, O. 80-38P,39P
Sasaki, S. 85-31P
Sasaki, T. 84-16A; 85-94DP,112Q
Sasburg, A.G. 86-1E
Sasin, V.Ya. 78-23Q,24Q; 85-9Q
Sassen, K. 81-15K
Sastri, K.S. 77-104F; 78-97F
Sastri, V.M. 82-10Q
Sastri, V.M.K. 76-40C; 77-6B,34J; 78-9B,35B; 80-76B; 83-22JM; 84-11N; 85-28FF; 86-20FF
Sastry, M.S. 76-3G
Sastry, V.S. 76-38B
Sastry, V.U. 83-8G
Sastry, V.U.K. 76-28F,76F
Sasuga, Y. 85-15P
Sata, K. 77-4G
Satake, A. 80-37FF
Satge, M. 77-41P
Sather, D. 85-57F
Sathyamurthy, D.R. 86-5Q
Satija, S. 83-58F
Sato, A. 86-32B
Sato, K. 77-1R; 78-1R; 80-10L; 82-19S; 84-3H; 86-20D
Sato, R. 86-80B
Sato, S. 85-40P
Sato, T. 82-37B; 83-45B,64J; 84-51B; 85-46C, 55C,95Q,21U
Sato, Y. 78-41P; 81-68J,69J; 82-33B
Satoh, I. 86-70B
Satomura, M. 85-28JJ
Satyamurthy, P. 79-62A
Satyanarayana, S. 82-29FF
Satyaprakash, B.R. 80-6C
Sauer, H.J. 80-22J
Saulnier, J.B. 84-26A
Saunders, K.D. 79-1G
Sauzade, J.D. 85-47JM
Savage, C.J. 79-19Q; 82-5Q
Savel'ev, R.Z. 79-67J
Savel'yev, R.Z. 78-10J
Savel'yev, V.N. 80-98J,99J
Savenko, V.V. 85-19B
Saverchenko, V.M. 82-15S
Savery, C.W. 77-4B; 78-6B
Savic, P. 86-45B
Savin, A.N. 85-70B
Savvatimskiy, G.I. 83-14S
Sawada, T. 83-37F
Sawai, T. 86-79J
Sawant, A.M. 84-50B
Sawata, S. 77-28T; 86-27T
Sawatsky, R. 85-36A
Sawford, B.L. 86-64C
Sawhney, R.L. 86-59S
Sawyer, R.F. 78-10K
Saxena, N.S. 85-39A; 86-79P,9T,10T
Saxena, S. 81-48A; 82-35T
Saxena, S.C. 76-36D; 80-11D,12D,19D; 81-8B, 9D,19D; 82-10D; 84-7Q; 85-28DP,78DP,79DP; 86-13DP,33DP,34DP,65DP
Saxena, U.C. 78-46C
Saxena, V.P. 86-57S
Saxens, S.C. 83-8DP
Sayama, H. 76-66F; 77-80F,81F; 83-40F
Sayegh, N.N. 79-37C,24U; 81-15D
Sayers, D.D. 85-38A
Sazhenin, A.G. 80-143J
Sazhin, B.S. 82-28D
Sazhin, V.B. 86-55DP
Scalabrin, G. 83-13T
Scanlan, J.A. 77-85F; 80-46F
Schaareman, P.B.J. 86-10A,2JM
Schaber, A. 83-8K
Schachinger, E. 80-39A
Schade, E. 82-2U,22U
Schaefer, H. 85-54S
Schafer, E. 86-18T
Schafer, R. 85-47U
Schaflinger, U. 82-21JJ
Schallig, R. 77-43A
Schanberg, B.C. 76-5U
Schankula, M.H. 85-26A
Scharf, R. 86-15P
Scharmann, A. 80-57F
Schatzmann, M. 78-82F
Schbert, G. 79-132F
Schechter, D.E. 85-47S
Scheidt, M. 84-47F
Scheiwe, M.W. 83-18JM; 85-55JM,56JM

Schenk, J. 77-44F; 82-2H; 83-4DP
Schenkel, G. 80-62B
Scherer, P.W. 86-31B
Scheuerer, G. 85-71C
Scheuerpflug, P. 85-99K
Scheurpflug, P. 85-41P
Schieber, D. 85-58S
Schiestel, R. 78-25B
Schinkel, W.M.M. 81-38F; 83-55F; 85-85F
Schiroky, G.H. 84-61F; 85-90F; 86-6FF
Schisler, I.P. 79-63A
Schlader, A.F. 77-30P
Schlepp, D. 84-4S
Schlesinger, C.B. 78-4S
Schlitt, R. 86-67Q
Schlunder, E.U. 76-3D,11Q; 77-26C,19Q; 78-2D, 3D; 80-14J; 81-70J; 83-45J; 84-12JJ; 85-8JJ; 86-37J,14JJ
Schmadl, J. 84-6J
Schmal, M. 78-47C
Schmiddel, W.P., Jr. 78-17S
Schmidt, F.W. 76-20T; 77-22T; 78-29S,30S; 82-11Q; 84-10N; 85-40N,41N; 86-101F
Schmidt, H. 77-44J,45J
Schmidt, K.G. 77-16C; 78-27C
Schmidt, P.S. 78-92J
Schmidt, R.R. 78-42A,83F
Schmidt-Harms, C.A. 86-50U
Schmitendorf, W.E. 77-47A
Schmitt, B. 84-10K
Schmitz, R. 78-48C
Schmucker, H. 85-28B
Schmucker, R. 76-15R
Schnabel, G. 81-70J
Schnaider, V.E. 86-18J
Schneider, G.E. 76-50A; 77-55A; 80-40A; 81-38B; 82-15N; 85-45N,46N,58N,11T; 86-53A,75N
Schneider, H.W. 78-63J
Schneider, J. 79-43P
Schneider, R.E. 76-30J
Schneider, W. 79-16E,116F
Schnell, H. 84-36Q
Schnipke, R.J. 85-47N
Schnipp, A.I. 78-15E
Schnittger, R.B. 82-61J
Schnizer, B. 80-39A
Schnoebelen, K.S. 86-56DP
Schnurmann, R. 76-41P
Schnurr, N.M. 76-51A,38B; 77-44B
Schobeiri, M.T. 85-57S
Schoenhals, R.J. 78-18A; 81-1D
Schoessow, G.J. 77-1J
Scholin, B. 81-11J
Scholz, R. 86-72U
Schon, J.P. 79-6D
Schonert, K. 78-49A
Schonfeld, R. 85-59S
Schonung, B. 86-5H
Schotland, R.M. 80-11P; 86-43T
Schotzau, H.J. 86-53U
Schotzhu, H.J. 80-31U
Schreiber, W.C. 85-86F
Schrenker, R.G. 84-21K; 85-29K
Schrey, H.-G. 81-71J
Schrey, H.G. 84-26JJ
Schreyer, H.L. 81-49A
Schreyer, J.M. 81-23T
Schrock, V.E. 76-16H; 77-46J
Schroppel, J. 81-72J
Schubert, F. 86-73Q
Schubert, G. 78-95F; 80-72F; 85-109F; 86-25DP
Schueller, P.G. 85-26S
Schuepp, P.H. 81-27FF
Schuerger, M.J. 83-61J
Schulenberg, T. 84-62F,25FF; 85-49FF; 86-108J
Schuller, R.B. 82-19J
Schulman, Z.P. 79-25E
Schulte, M. 85-87F
Schultz, D.H. 85-89F
Schultz, D.L. 84-29S; 85-7C
Schultz, W. 83-30A
Schulz, B. 86-80P
Schulz, G. 85-14U
Schulz-Gulde, E. 80-29U
Schumann, U. 77-23E
Schunder, E.-U. 84-38J
Schunder, E.U. 83-54J
Schuocker, D. 79-25U
Schuster, B.G. 80-40P
Schuster, H. 86-73Q
Schutt, J.A. 85-48F
Schutt Mogro, J. 76-34J
Schvan, P. 85-48JM; 86-66A
Schwabe, D. 80-57F
Schwaim, D. 76-17E
Schwalbe, W. 82-62J
Schwander, D. 85-31T
Schwartz, S.H. 80-38FF
Schwartzberg, H. 86-38JM
Schweinzer, J. 83-6DP; 85-4DP; 86-3DP
Schweitzer, H. 85-8J
Schwerdtfeger, K. 78-10A
Schwiesow, R.L. 77-39P
Schwind, R.G. 85-51FF
Schwoerer, K. 85-26S
Scirocco, V. 85-62B
Scofield, M. 86-95Q
Scofield, T. 83-9G
Scott, C.D. 85-76C
Scott, C.J. 77-5G
Scott, G.D. 76-20K
Scott, J.A. 85-18S
Scott, M.H. 82-5M,6M
Scott, M.J. 80-79B; 81-39B
Scott, R. 84-4U
Scriven, L.E. 78-17D; 84-24A
Scriver, S.M. 86-99K
Scudder, D.W. 77-15U
Scurloc, R.G. 85-18E
Scurlock, R.G. 77-13G
Seader, J.D. 77-2R; 83-24B,26B; 84-2E
Seagrave, R.D. 78-62C
Seban, R. 83-26F
Seban, R.A. 76-78F,45J; 78-49C; 80-12Q; 82-22JJ
Sechenov, V.A. 85-26U
Secrist, D.A. 76-18S

Sedahmed, G.H. 81-60F
Sedgjomasab, A. 86-73U
Sedler, B. 81-73J
Seedorf, S. 85-100K
Seeniraj, R.V. 81-50A; 82-43A; 84-2U; 85-7U
Seeniraj, V. 79-64A
Seetharam, T.R. 79-117F
Seetharamu, K.N. 78-78F; 84-4J
Segal, A. 86-84N
Segal, I. 84-13A
Segal, M.D. 85-3Q
Segami, A. 80-14E
Segel, S.L. 77-33P
Segev, A. 80-116J,117J; 81-74J
Seguchi, Y. 85-43A
Seher, D. 83-14D; 84-49B
Seidl, J. 84-27S
Seifert, G. 80-41A
Seifert, H. 80-17E
Seilli, R.K. 86-21E
Seitzman, J.M. 85-46E
Sekhar, J.A. 85-41JM
Seki, M. 78-65J; 79-118F; 85-20U; 86-112F
Seki, N. 76-39B,37D,3F,79F; 77-54A,91F; 78-84F, 85F,86F,64J,28K; 79-12D,10L,11L; 80-31D; 81-32F,33F,75J; 82-41B,31F,11S,12S; 83-38B, 39B,40B,56F,24FF,30JM,31JM,11S; 84-39B,9D, 32F,34JM,35JM; 85-49JM; 86-32J
Sekins, K.M. 82-8S
Sekoguchi, K. 81-68J,69J
Sekulic, D.P. 85-23JJ; 86-96Q
Selcuk, M.K. 79-27T
Selcuk, N. 76-18K; 79-23K; 85-101K
Selecki, A. 77-13J
Selitskii, F.I. 85-28JM
Selvage, R.B.G. 79-11T,12T
Selyavin, G.F. 78-3G
Semenov, K.N. 82-9Q
Semenov, V.I. 85-50FF
Semenov, V.P. 80-122J
Semenova, N.M. 80-139J
Semiletov, N.G. 83-38K
Semin, V.N. 86-46K
Semyonov, S.V. 85-87C
Semyonov, Yu.G. 84-43C
Sen, A.K. 86-55A,102F
Sen, K.K. 86-124K
Sen, M. 85-77Q,84Q,85Q
Sen', L.I. 86-28JJ
Sena, J.T. 86-80Q
Sengers, J.V. 85-20P
Sengupta, D. 80-41P
Sengupta, K. 86-31E
Sengupta, S. 78-8; 85-23J
Sengupta, S.N. 80-41P
Sennikov, V.V. 79-30J
Senser, D.W. 86-14E
Senshu, T. 84-30JJ
Sentis, R. 80-18E
Sepri, P. 77-24E
Serebrennikov, G.G. 86-44DP
Serebryannyy, G.Z. 82-23B
Serebryanyi, G.L. 85-86Q
Sergeev, V.V. 78-86J
Sergeyev, G.I. 82-42C
Sergeyev, V.L. 79-38C; 85-80C; 86-56C
Sergeyeva, L.A. 82-43C; 85-80C; 86-56C
Sergievskii, E.D. 85-49C; 86-11H
Sergievskiy, E.D. 78-3C
Sergiyevskiy, E.D. 84-26H; 85-59C
Serizawa, A. 76-6M; 83-35J,55J
Sernas, V. 76-4L; 79-39C; 81-61F; 85-20J
Serov, Yu.I. 84-26J
Seshadri, K. 78-50B
Seshadri, R. 80-19S
Seshan, S. 86-97Q
Sesonske, A. 80-5E; 85-17B
Seth, A.K. 79-68A
Sethna, P.P. 77-35K
Sethumadhavan, R. 83-41B; 85-24JJ
Sethumadhaven, R. 86-71B
Settles, G.S. 85-47E; 86-43E
Sevast'yanenko, V.G. 78-29K
Severin, S.S. 85-105K
Sevkovaskay, L.I. 85-19H
Sewall, M.H. 76-43P
Seymour, P.J. 86-15C
Sfeir, A.A. 76-52A; 77-56A
Sfier, A.A. 76-48C
Sforza, P.M. 78-50C; 79-40C
Sgheiza, J.E. 85-95J
Sha, W.T. 82-14Q; 84-11T
Shaaban, A.H. 83-57F; 85-88F
Shabashov, V.I. 86-62U
Shabunya, S.I. 85-64K
Shadmon, G. 84-99K
Shadyev, S. 86-44T
Shaeffer, J.F. 76-15U
Shaelwitz, J.A. 82-46A
Shafey, H.M. 86-35S
Shah, D.J. 84-33DP
Shah, M.M. 79-68J,69J; 80-118J; 83-56J
Shah, R.K. 83-16Q; 85-63Q
Shah, S. 80-61F
Shah, V.L. 76-43B
Shahrestani, H. 78-55B
Shai, I. 82-47A; 86-35FF
Shaikh, M.A. 82-16J
Shakiri, K.J. 76-35D
Shalaby, S.A. 86-14JM
Shalakhov, A.A. 80-54J
Shalev, S. 83-11U
Shalina, K.K. 76-14U
Shalvarova, S.V. 76-9L
Shamir, J. 82-3K
Shamlou, P.A. 86-57C
Shammas, N.Y. 86-32U
Shamsundar, N. 76-53A,88F; 79-65A; 81-51A; 82-48A; 84-53Q
Shang, J.S. 76-49C; 86-89N
Shang, Z.S. 84-37Q
Shankar, N. 81-76J
Shanks, H.R. 84-29A
Shaozhang, D. 85-50JM
Shapiro, A.B. 76-38B; 85-102K
Shapiro, A.P. 76-51A

Shaposhnikov, V.A. 86-11Q
Sharaf, A. 81-15U
Sharan, A. 85-96J,97J
Sharayevskiy, I.G. 78-53J; 80-100J
Sharevskaya, D.I. 77-5P
Sharifulin, A.N. 86-103F
Sharkova, N.A. 79-5C
Sharma, A. 82-46K,16N,17N
Sharma, A.K. 84-1T
Sharma, B.I. 77-14G
Sharma, C.P. 83-59J
Sharma, D.K. 86-14S
Sharma, D.V. 82-11M
Sharma, G.K. 79-117F
Sharma, L.K. 84-15E
Sharma, P.R. 78-66J,67J
Sharma, R.N. 82-29D
Sharma, V.D. 86-100K
Sharma, V.K. 85-15T
Sharma, V.P. 78-53C,91F
Sharon, A. 80-63B; 82-62J; 83-31J,32J
Sharp, K. 82-36T
Sharp, M.K. 79-28T
Shashkov, A.G. 86-81P
Shashmin, V.K. 84-44C
Shaughnessy, E.J. 77-27F,28F; 78-88F,10G; 79-32F,33F,34F; 86-88F
Shaukatullah, H. 76-33F; 78-42P; 79-42F,119F, 120F
Shaumeyer, J.N. 81-62F
Shavakhovskiy, D.I. 85-40U
Shavit, A. 84-17DP
Shaw, D.A. 77-45B,25E
Shaw, D.C. 85-80DP
Shaw, G.H. 85-42P
Shaw, H.-J. 86-42JM
Shaw, J.H. 83-20K; 84-1K; 86-1K
Shay, W.A. 85-89F
Shayler, P.J. 78-15U
Shchelkunov, V.N. 79-70F
Shcherbakov, V.K. 83-57J
Shcherbakova, N.S. 79-71J
Shcherbik, N.M. 86-82U
Shcherbinin, V.I. 85-63B
Shchitnikov, V.K. 85-5E
Shchuchev, K.G. 86-5S
Shchukin, V.K. 77-46B; 79-11G,12G; 82-47C; 83-21Q
She, C.Y. 76-8P
Sheahen, T.P. 83-39K
Shealy, D.L. 78-4T; 84-88K
Shearer, C.J. 77-5D
Shee, Y.T. 82-30FF
Sheehan, E. 86-98Q
Sheer, R.E., Jr. 84-22C
Sheih, C.M. 79-44P
Sheikholeslami, R. 86-99Q
Sheindlin, A.E. 80-10M
Shekar, B.C. 83-38F,66F
Shekriladze, I.G. 80-119J,120J; 81-77J
Shelepin, A.G. 85-56J
Shelginski, A. 83-1S
Shelginskii, A.Ya. 86-50Q
Shelimanov, V.A. 79-5C; 85-5D
Shemagin, I.A. 86-4JJ
Shemilt, L.W. 81-60F
Shemmer, L. 79-45P
Shemyakin, V.N. 83-18DP
Shen, P.V. 85-5S
Shen, S.F. 78-59C
Shen, Z.F. 82-48K; 85-106K
Shenoy, A.V. 78-89F,90F; 80-42FF,43FF; 84-26FF; 86-50C,58C
Shepherd, J.E. 82-63J
Sheptun, V.M. 78-51B
Sherbakov, A.J. 86-107J
Shercliff, J.A. 80-11M
Sheridan, J.C. 86-4D
Sherif, I.I. 76-39D
Sherman, A. 76-16R
Sherman, E.O. 76-28T
Sherman, F.S. 81-24B
Sherry, J.E. 83-34T
Sherstyuk, A.N. 76-7G
Sherwood, A.E. 76-39F
Sherwood, T.K. 77-8
Sheu, J.P. 76-22B
Shevchenko, I.V. 86-9B
Shevchuk, M.S. 78-34P
Shevtsov, Yu.E. 85-32N
Shewen, E.C. 84-23T
Sheynkman, A.G. 80-121J
Shi, D.-H. 84-65J; 85-116J
Shibani, A.A. 77-47B,48B
Shibata, T. 81-46F; 85-69F
Shibayama, S. 80-19Q
Shibuya, Y. 83-26U
Shibuye, Y. 84-38U
Shida, H. 83-8JJ
Shideler, J.L. 77-3R; 85-77C
Shigechi, T. 83-58J
Shigemasa, Y. 86 28JM
Shigina, L.F. 76-11J
Shih, M.T. 80-15H
Shih, R.-F. 84-13S
Shih, T.-M. 86-76N
Shih, T.H. 81-13N
Shih, T.M. 80-16K; 81-14N; 84-12N; 85-16D, 50N
Shih, W.C.L. 77-7R
Shih, Y.-P. 81-25B; 82-29C
Shih, Y.D. 80-16M
Shilov, V.A. 84-12U; 86-9U
Shilovich, I.L. 86-100Q
Shilston, M.J. 79-18S; 85-10F,60Q,61Q,62Q; 86-11F,3S
Shima, A. 79-85J
Shimada, R. 84-12G; 85-17G; 86-24D
Shimchuk, T.Ya. 85-43P
Shimeck, D.J. 84-63F
Shimizu, G. 86-80F
Shimizu, J.I. 77-55B
Shimizu, M. 79-35J; 85-51JM
Shimizu, S. 82-13JJ; 84-11JJ
Shimizu, T. 76-18C; 77-4G; 80-38P,39P
Shimizu, Y. 82-19A

Shimizum, Y. 85-37S
Shimko, M.A. 86-8JJ,42JJ
Shimoda, K. 79-25P
Shimoji, S. 77-9L
Shimomura, R. 82-41A,42A; 83-25JM; 84-36JM
Shimonis, V.M. 82-26B
Shimoya, M. 86-128J
Shimura, K. 76-10H
Shimura, T. 86-45J
Shin, T.S. 86-109J
Shina, Y. 86-104F
Shindin, S.A. 78-40P
Shindo, H. 80-30U; 85-41U
Shing-An, W. 79-30T
Shinkawa, T. 82-64J
Shinn, J.L. 84-19C; 85-78C
Shinoda, K. 86-39JM
Shinohara, T. 77-21F
Shintani, Y. 84-36JM
Shinzato, K. 85-36Q
Shinzato, K.-Ei 85-10H
Shinzato, T. 86-65B
Shioda, S. 85-94DP,112Q; 86-114Q
Shioiri, Y. 82-13U
Shiokawa, T. 85-66J
Shiotsu, M. 77-42J,43J; 84-53J
Shioyama, T. 82-35F; 86-61F
Shiozawa, B. 77-30D
Shipman, C.W. 83-23F
Shipov, Ye.V. 82-31C
Shipp, P. 79-26U
Shiraha, M. 86-118J
Shirai, H. 85-42U
Shiral, H. 84-29U
Shiralkar, G. 81-63F
Shiralkar, G.S. 81-65F; 82-59F
Shiratori, T. 80-86J
Shirazi, M.A. 76-80F; 78-26F
Shires, E. 78-5U
Shirkey, C.T. 83-41K
Shirotani, I. 86-107P
Shishedjiev, B.K. 76-29B; 81-88J
Shishido, I. 84-38Q; 85-87Q
Shishkin, A.A. 76-51P
Shishov, E.V. 79-26E; 84-18H; 85-94C
Shishov, Ye.V. 78-30C
Shitara, T. 86-56A
Shitsman, M.Ye. 78-44J
Shitzer, A. 76-49A; 77-31T; 78-5T; 79-29T
Shizuya, M. 84-17S
Shkadinskiy, K.G. 85-110J
Shkema, R.K. 80-13B
Shkilev, V.D. 80-2Q
Shklover, G.G. 79-70J; 80-82J,122J; 82-23JJ
Shklyar, F.R. 85-63B
Shkol'nik, E.M. 83-9U
Shkraba, V.T. 76-5P
Shlanchyauskas, A.A. 76-53C; 80-13B
Shlanchyauskas, A.S. 76-16C
Shlien, D.J. 76-21E,81F; 78-13F; 79-121F, 122F,123F; 81-28FF; 83-65F,41FF
Shlyazhas, R.B. 86-31D
Shmakov, A.M. 86-23C
Shnaider, V.E. 85-6JJ
Shnayder, V.E. 77-5C
Shneider, L.A. 85-98J
Shneidman, V.A. 86-39JJ
Shneyerova, R.I. 80-83J; 83-7JJ
Shobu, K. 85-64F
Shock, R.A.W. 76-46J; 77-47J; 82-60J
Shoemaker, R.L. 86-83P
Shoenhals, R.J. 79-35A,25F
Shoham, O. 82-34B
Shoji, M. 84-54J,55J; 86-110J
Shoji, Y. 86-27B
Sholes, R.R. 80-42P
Shorcri, M.M. 82-18N
Short, R.E. 76-6S
Shoukri, M. 78-68J
Shove, G.C. 84-31S
Showalter, R.E. 82-49A
Shpakovskiy, V.A. 80-21J
Shpil'rain, E.E. 80-10M
Shrayber, A.A. 78-20E
Shrivastava, S. 86-13DP
Shrock, V.E. 76-77F
Shropshire, W., Jr. 77-12T
Shskov, A.G. 86-82P
Shtessel', E.A. 79-71F
Shu, Q.S. 86-58S
Shubenko, A.L. 86-39JJ
Shubha, V. 86-44E
Shubin, G.S. 78-18H; 82-30D
Shuen, J.-S. 85-104J
Shugayev, V.A. 80-103J; 85-88J
Shukla, K.M. 77-57A
Shukla, S.N. 83-35T
Shul'chishin, V.A. 85-81DP
Shul'ga, V.M. 78-33P
Shul'man, Z.P. 77-92F,93F; 80-64B; 81-40B; 83-12S
Shul'ts, A.N. 83-4Q
Shulka, K.N. 86-77N
Shulman, Z.P. 76-82F; 78-52B; 82-50A; 84-44C
Shultis, J.K. 82-28K; 86-21N
Shumkov, S.Kh. 77-22D
Shumyatskii, B.Ya. 80-10M
Shurchkov, A.V. 78-12D
Shurgay, P.I. 78-34P
Shustov, A.V. 84-21P
Shuvalov, V.V. 83-21K
Shuyskaya, K.F. 78-37B; 82-2Q
Shvab, N.S. 76-11J
Shvarts, D. 84-105K
Shved, G.M. 84-89K,90K
Shvegzhda, A.A. 79-26Q
Shvegzhda, S.A. 76-30C,31C
Shvenchyanas, P.P. 81-3FF
Shvets, I.T. 76-14H
Shvets, Y.I. 85-64B
Shyam, R. 82-11M
Shyu, C.-T. 79-82F
Shyy, W. 86-10N,11N
Siang, H.H. 79-8H
Sibulkin, M. 82-31FF
Siddall, R.G. 76-18K; 77-20S; 79-23K

Sideman, S. 78-51C; 79-24B; 80-85J; 82-18JJ; 85-3S
Siderowicz, J. 85-88Q
Sidorowich, J.J. 85-129K
Siebenmorgen, T.J. 86-45T
Siebers, D.L. 77-23T,24T; 79-31T; 85-51FF
Siegel, D.L. 78-26F
Siegel, R. 76-38D; 77-58A; 78-39A,40A; 81-8H; 82-44A,45A,31D; 83-23DP,27JM; 84-37JM,38JM; 85-18G; 86-43JM,44JM
Siemsen, K.J. 86-101K
Sieniutycz, S. 79-66A; 80-19E; 81-9H; 82-13H
Siess, J. 78-17E
Sievers, A.J. 80-12L
Siewert, C.E. 85-34K,39K,103K,104K; 86-45K
Sifaoui, M.S. 78-69J
Sigel, R. 85-34A
Sih, G.C. 80-6H,15H
Silantev, N.A. 84-10A
Silard, A. 77-59A
Silva, T.M. 79-6S
Silver, J.A. 82-23U
Sim, Y.S. 84-13G; 85-19G
Simanavicius, V.S. 79-13D
Simanovskii, I.B. 85-65B
Simanovskiy, I.B. 86-89F
Simard, G. 85-10N
Simeoni, D. 83-24U
Simkiavicius, V.J. 79-4D
Simkin, A.S. 82-71J
Simmers, D.A. 79-13G
Simmonds, A.L. 80-13H; 84-19C; 85-96C
Simmons, L.D. 76-54A
Simomura, Y. 86-18D
Simon, F.F. 76-21T
Simon, H.A. 77-49B; 79-36B; 82-30B; 84-39Q; 85-50B
Simon, T.W. 83-31C; 85-91C
Simoneau, R.J. 84-45C; 85-86C
Simonek, J. 84-40B
Simonelli, F. 86-30F
Simonenko, A.P. 79-34B
Simonescu, D. 86-14Q
Simonich, J.C. 78-52C; 84-46C
Simonis, G.J. 83-40K
Simonis, V.M. 79-28B,29B; 80-54B; 85-66B, 75B
Simonsohn, G. 84-6E
Simonton, W. 86-101Q,102Q
Simpkins, P.G. 79-124F; 81-5D,66F; 83-5DP, 5F; 84-10DP; 86-9DP,105F
Simpson, A. 86-86P
Simpson, C.J. 82-32FF
Simpson, J.M. 77-17D
Simpson, M.E. 82-24JJ
Simpson, O.A. 79-24K
Simpson, R.J. 77-6T
Simpson, R.L. 78-6D,19D
Simpson, S. 84-18U
Simpson, W.T. 85-82DP
Sinagra, E. 79-3T
Singal, L.C. 83-59J
Singer, H. 85-39FF
Singer, R.M. 76-7M; 77-3M,20Q
Singh, A. 76-47J; 77-48J
Singh, A.K. 83-3M
Singh, A.P. 76-3C
Singh, B. 77-20K; 82-12M; 86-74U
Singh, B.S. 76-42P; 80-32D
Singh, K.P. 81-12Q,13Q
Singh, K.R. 83-4M
Singh, M. 80-5G; 81-7M; 82-6G
Singh, O. 77-5L
Singh, P. 76-83F,22T; 78-53C,91F,99F; 85-79C; 86-59C,57DP,106F
Singh, P.C. 78-6G,7G; 80-4G
Singh, R. 83-5M; 85-39A; 86-10T,11T
Singh, S.N. 77-94F; 79-125F; 80-58F; 81-64F; 82-30FF; 85-86F
Singh, S.S. 80-1M
Singh, U. 81-25T
Singhal, A.K. 83-19Q,20Q; 85-51N; 86-60C
Singleton, E.B. 82-47K; 83-41K
Sinha, M.P. 80-50B
Sinha, S.N. 76-46A
Sinitsyn, E.N. 85-120J
Sinitsyn, Ye.N. 80-104J
Sinkevich, O.A. 86-3U,4U
Sinkule, J. 78-30K
Sinkyavichyus, I.E. 80-24K
Sinyuk, N.I. 86-18J
Sipavicius, C. 79-27Q
Sipavicius, C.-S.J. 79-28Q
Sirignano, W.A. 80-107J; 84-50C; 86-124J
Sissom, L.E. 77-7
Sitnitskii, Yu.I. 76-32P
Sivak, Z.V. 85-32N
Sivasubramanian, M.S. 84-14E
Sjogren, S. 85-107Q
Skarda, R.K. 84-7FF
Skema, R.K. 84-14B
Skladaressis, D. 77-95F
Skjoldebrand, C. 86-20S
Sklyar, F.R. 80-73K
Skobelev, I.Yu. 85-18U
Skocypec, R.D. 84-91K; 86-102K
Skoropanov, A.S. 85-19H
Skorov, D.M. 79-6P
Skowronek, Y. 79-27U
Skripov, V.P. 86-84P
Skupsky, S. 84-105K; 86-19U
Skurin, L.I. 79-8D
Skutova, I.V. 76-4G; 78-12G
Slanciauskas, A.A. 79-16C,17C
Slater, N.K.H. 85-35Q; 86-24Q
Slattery, J.C. 82-30C,17D
Slegel, D.L. 77-18S
Slesarenko, A.P. 78-49B; 79-67A
Slesarenko, V.N. 82-15S; 83-60J
Slianciauskas, A.A. 78-54C
Sliepcevich, C.M. 84-6A
Slipcevic, B. 84-36Q
Slitenko, A.F. 81-16S
Slitinskaya, S.K. 85-11F
Sloan, D.M. 81-3F
Sloan, E.D. 81-38A; 85-4P

Slobozhanin, L.A. 79-71J
Slonim, M. 84-33A
Slovetskii, D.I. 86-75U
Sluitjer, W.M. 86-111J
Sluyter, W.M. 82-65J; 85-114J
Small, J.D. 86-2P
Small, J.G. 80-42P
Small, R.D. 77-60A; 84-64F
Smart, D.R. 80-59F
Smeets, G. 83-27B
Smet, R.P. 84-13H
Smirnov, G.F. 78-70J; 83-7J; 86-3Q
Smirnov, M.S. 78-94J
Smirnov, O.K. 78-53B; 80-65B
Smirnov, R.E. 76-46P
Smirnov, T.F. 80-1J
Smirnov, V.A. 86-3N
Smirnov, V.I. 82-30K
Smirnov, V.P. 80-40B
Smirnov, Ye.L. 85-99J
Smirnova, I.Ye. 83-42K
Smirnova, L.P. 79-4C
Smith, A.C. 80-14A
Smith, A.M. 76-49P; 78-27K,5L; 85-94K
Smith, B. 85-89Q
Smith, B.G. 77-8T
Smith, C.G. 86-70P
Smith, D.H. 76-12R
Smith, D.M. 83-2F; 84-42C
Smith, D.W. 77-25T
Smith, D.Y. 84-73K
Smith, E.E. 83-58F
Smith, F.T. 77-23D; 79-126F; 85-17D
Smith, H.M. 76-43P
Smith, J.L. 76-44P,16U,17U; 86-15G
Smith, J.R. 77-40P
Smith, K.O. 84-40Q
Smith, M.C. 76-40D; 86-61C
Smith, M.K. 83-59F,60F; 86-14G
Smith, M.L. 77-10L
Smith, N. 77-9Q
Smith, R. 81-41B; 82-35B
Smith, R.A. 76-48J
Smith, R.C. 82-66J
Smith, R.M. 82-19N
Smith, R.N. 79-42B; 80-22S
Smith, S.H. 79-14G
Smith, S.M. 84-92K
Smith, T. 77-11L
Smith, T.C. 78-8P
Smith, T.F. 77-31K; 79-8K,9M,32T; 81-8M,24T; 82-48K; 85-57K,105K,106K
Smith, T.J. 81-5G
Smith, T.N. 78-9C
Smits, A.J. 79-41C,38P
Smol'skiy, B.M. 79-72J; 82-43C
Smolderen, J.J. 78-13
Smolinski, P. 85-15N
Smolsky, B.M. 85-80C
Smurov, I.Yu. 86-52JM
Smutek, C. 85-90F; 86-21F
Smy, P.R. 80-12U; 81-4U; 83-24U; 84-32U
Smyth, R. 79-9D; 81-14Q
Snaddon, R.W.L. 85-11JJ
Snaith, B. 82-51A; 86-57A
Snel, R. 84-22P
Snellen, J.W. 85-8P
Snider, A.D. 78-17A; 80-24A; 81-52A; 83-32Q; 85-56Q
Snider, D.M. 79-73J
Snijders, A.L. 83-32C
Snoek, C.W. 78-22J
Snyder, A. 81-8H; 82-31D; 83-23DP
Snyder, T.K. 78-7S
Snyder, W., Jr. 84-21B
So, B.C. 85-64F
So, R.M.C. 76-50C; 82-44C; 84-22C
Sobajima, M. 85-100J; 86-1J
Sobanik, J.B. 82-13S
Sobel, D.R. 83-46B
Sobel, L. 86-15G
Sobin, A. 76-23T
Sobin, V.M. 86-40JJ
Sobol, A.D. 76-11J
Sobolev, A.V. 80-77B
Sobolev, V.D. 78-11H
Sobolevskiy, A.S. 79-39P
Sodha, M.S. 79-68A; 81-25T; 83-30T,35T
Sofrata, H. 80-123J; 81-79J; 83-9JJ; 84-27JJ
Sofrata, H.M. 80-20Q; 81-15U
Soga, T. 85-101J,102J,103J; 86-112J
Sogawa, Y. 83-67J
Soh, T. 85-52JM
Sohal, M.S. 86-103K
Sohda, M.S. 86-59S
Sohn, C.W. 81-80J; 83-8FF; 84-41B
Soin, R.S. 79-33T
Soininen, M. 86-85P
Sokolov, M. 77-30E; 80-85J; 81-3C,26T; 82-34T; 83-36T; 85-36T; 86-50T
Sokolova, I.A. 86-97P,98P
Sokovishin, Yu.A. 84-20FF; 85-52FF
Solan, A. 82-2G
Soland, J.G. 78-25Q
Soldat, M. 85-92DP
Soliman, H.M. 80-66B; 82-25JJ; 84-42B; 85-22JJ; 86-18A,41JJ
Soliman, M.S. 86-83F
Solodov, A.P. 76-11F,84F; 77-17J; 79-30J
Solodovnikov, V.V. 85-70B
Solodukhin, A.D. 85-5E
Sologab, I.S. 80-93J
Solomatnikov, Yu.K. 79-147F
Solomon, A. 81-53A,54A
Solomon, A.D. 79-69A,70A; 80-42A,43A; 82-52A; 85-1JM,53JM
Solomon, A.S.P. 85-104J
Solomov, V.M. 76-85F
Soloukhin, R.I. 81-13K; 85-5E
Soloukhin, R.K. 80-22K
Solov'eva, E.N. 85-14N
Solov'yev, A.G. 85-35J
Solov'yev, G.M. 79-50F
Solov'yev, Yu.G. 84-21JM
Soloviev, I.A. 76-37A
Somer, T.G. 86-18JJ

Somers, R.R., II 77-22T; 80-26T
Somerscales, E.F.C. 82-60F; 85-91F
Somerton, C.W. 82-61F; 84-56J
Somerville, R.C.J. 83-4G; 86-26C
Somma, F. 76-1P; 78-1K
Sommer, H. 77-10U
Son, D.-K. 77-41P
Sone, Y. 78-71J,72J; 83-46J
Song, D.J. 85-81C; 86-62C
Sonin, A.A. 86-8JJ,42JJ
Sonju, O.K. 80-13M
Sonn, A. 81-55A,45J
Sorensen, C.M. 86-99K
Soroka, B.S. 78-31K,32K
Sorokin, A.A. 86-72K
Sorokin, A.G. 78-37B; 82-2Q; 85-20Q
Sorokin, A.P. 85-114Q
Sorokin, L.E. 83-15FF
Sorokin, L.Ye. 86-17F,107F
Sorour, M.M. 82-28F; 85-67B,92F,93F
Sorrell, F.Y. 85-34S
Sosinovich, V.A. 78-14E
Soskov, S.A. 77-18J
Soskova, I.N. 77-59J
Sosoka, D.J. 82-45A
Sosunov, V.I. 79-45J; 82-8JJ; 85-16JJ
Soucy, B.A. 84-12DP
Soufiani, A. 85-107K
Soulaimani, A. 86-18N
Soule, C.L. 85-20G
Soule, D.E. 77-25T
Soumerai, H. 85-44P
Soundalgekar, V.M. 76-86F; 77-30C,96F,97F, 98F; 78-92F,93F; 79-127F,10M; 80-34FF,12M, 14M; 81-5G,10H,7M; 85-83C,53FF,30H; 86-76U
Sources, J.M. 84-105K
Souren, C. 80-14L
Sousa, A.C.M. 85-94F,95F; 86-83N
Souza Mendes, P. 83-62F; 84-43B
Sovershennyy, V.D. 85-28H
Soward, A.M. 79-128F; 80-60F; 85-21G
Sowell, E.F. 80-27T
Spalding, D.B. 76-35B,42C; 77-9,10,95F; 78-34C, 24S; 79-1B; 82-6B; 83-19Q; 84-24S; 85-42C
Spandoni, G. 78-34K
Spangenberg, H.-J. 85-14U
Sparin, V.A. 77-23J
Sparks, M. 77-37K
Sparrow, D.J. 86-22P
Sparrow, E.M. 76-53A,55A,16B,17B,36B,45B, 32C,1F,87F,88F,1G,8G,3H,21J,10Q; 77-61A, 17B,38B,39B,50B,51B,52B,21C,27C,24D,14F, 2J,26T,27T; 78-10,35A,41A,42A,16B,44B,54B, 55B,56B,57B,40C,18D,20D,75F,83F,94F,8Q,26Q, 27Q; 79-1A,57A,71A,72A,8B,20B,32B,39B,40B, 79B,42C,90F,108F,129F,130F,59J,62J,74J,20Q, 34T; 80-2B,18B,43B,67B,68B,69B,70B,71B,72B, 73B,74B,29C,61F,36FF,44FF,45FF,46FF,21Q; 81-21A,22A,27A,56A,57A,58A,59A,22B,42B,43B, 44B,27C,28C,67F,29FF,30FF,31FF,32FF; 82-53A, 54A,27B,36B,32D,33D,34D,35D,11F,62F,63F, 33FF,7G,16S,37T; 83-33B,42B,43B,44B,11D, 12D,13D,61F,62F,17FF,37FF,38FF,61J,28JM, 33Q,13S; 84-35B,43B,44B,45B,46B,47B,48B, 47C,65F,66F,27FF,14G,19H,20H,18JM,39JM,40JM, 4N,28Q,41Q; 85-40A,68B,69B,82C,18D,19D,9F, 54FF,55FF,56FF,57FF,90J,91J,33JM,54JM,108K, 112K,70Q,90Q,91Q,92Q; 86-16B,57B,72B,73B, 74B,32C,35C,55C,63C,10F,86F,108F,36FF,37FF, 16G,19H,102J,45JM,46JM,47JM,48JM,49JM,104K, 103Q
Sparrow, V.M. 79-46C
Specht, E. 84-23P
Speitkamp, L. 79-19H
Spelt, J.K. 82-36D
Spence, D.A. 78-73J; 82-70F
Spencer, D.L. 81-24T
Spencer, J.W. 82-38T
Spencer, R.G. 80-20S
Speziale, C.G. 79-27E; 85-22G
Spiecker, U. 86-32S
Spiegel, E.A. 77-102F
Spiga, G. 79-5K; 80-23K; 81-16D,8K,16K; 82-18K, 45K,17S; 85-109K,110K
Spiga, M. 81-16D; 82-22A,17S; 85-109K,110K; 86-58A
Spillman, J.J. 84-57J
Spindler, K. 81-2L
Spinner, I.H. 78-28Q
Spivak, A.L. 86-41E
Spodyryak, N. 76-9Q
Spodyryak, N.T. 76-9L
Spokoynyy, F.Ye. 77-53B,38K
Spolek, G.A. 85-55S
Spotar', S.Yu. 82-15H
Sproston, J.L. 81-60A
Sreenivasan, K.R. 77-3E,26E; 78-21E,22E; 79-28E; 80-20E,21E
Sreenivasan, S. 79-131F
Sridharan, K. 83-31A
Srimani, P.K. 82-25D
Srinivas, N.S. 84-13J,58J
Srinivasan, J. 84-13N; 86-50JM
Srinivasan, K. 85-45P; 86-113J
Srinivasan, R. 79-65A; 80-75B; 86-6T,46T
Srinivasan, S. 76-19K; 86-45U
Sriramulu, V. 79-14T
Srivastava, S. 79-20H
Srokowski, A.J. 77-6H
St-Arnaud, J.M. 86-15P
Stablein, H.G. 86-48U
Stacey, F.D. 78-74J
Stacharska-Targosz, J. 80-21B,55B
Stachyra, L.J. 79-24H; 80-35A
Stadnyk, B.I. 77-42P
Stafnakos, E.K. 81-9T
Stahl, T.J. 84-47C
Staimach, C.J., Jr. 77-5R,11R
Stair, A.T., Jr. 83-43K
Staley, L.M. 85-13JJ
Stalker, R.J. 77-11U; 80-14C; 86-44S
Stamnes, K. 84-93K
Stamps, D.W. 86-60S
Stanek, M.J. 86-14C
Stanford, J.L. 84-64K
Stankevich, A.V. 86-81P

Stankiewicz, A. 86-104Q
Stapountzis, H. 86-64C
Starczewski, J. 85-93Q
Starik, A.M. 86-72K
Stark, J.A. 85-46P,60S; 86-83P
Starkman, E.S. 77-46J
Starling, K.E. 84-7P
Starner, K.E. 76-12Q
Starnes, G.L. 83-21DP
Starov, V.M. 78-11H
Starr, R.F. 82-13M
Stasi, W. 79-40C
Stastna, J. 85-96F
Statsenko, V.N. 80-106J
Staub, F.W. 80-3D,33D; 85-2J
Stearns, L.P. 83-44K
Steen, P.H. 86-58DP
Steenhuis, T.S. 86-34S
Stefanizzi, P. 86-105K
Stegen, G.R. 77-12E
Steiff, A. 79-15S; 80-21S; 82-14H; 84-18S, 34S; 85-40S,41S,42S
Steimke, J.K. 85-94Q
Stein, W. 76-13D,14D
Steinberg, M. 77-10B
Steinberg, P. 86-66S
Steinberg, V. 85-97F
Steinberger, T.E. 86-15DP
Steiner, A. 76-19S
Steiner, D. 80-7B,14J; 83-45J; 86-76J,114J
Steinhagen, H.P. 83-29JM
Steinman, J.F. 82-27F
Stell, G. 85-50P
Stengel, K.C. 78-12F; 82-64F
Stepanov, A.V. 77-25D
Stepanov, S.I. 82-6C
Stepanov, S.V. 85-111K
Stephan, K. 76-26A,56A,6R; 80-84J; 81-54J, 81J,15Q; 82-28B; 83-14D; 84-49B,7E,42Q; 85-58FF; 86-25D
Stephan, L. 77-49J
Stephens, G.L. 86-106K
Stephens, R. 86-76P
Stephenson, J.W. 82-9N
Stephenson, P.L. 76-40B; 79-26A
Stephenson, S.E. 78-22E
Stepka, F.S. 77-8R
Stesik, L.N. 81-43J
Stevens, F.A. 85-18E
Stevens, P.N. 80-12A
Stevenson, W.H. 76-24A,16P; 77-41F; 79-17P
Steward, F.R. 84-53K
Steward, W.G. 83-18C
Stewart, W.E., Jr. 81-33FF; 85-59FF; 86-109F
Stewartson, K. 76-57A; 78-15F,2G; 79-12C; 82-32FF
Stichlmair, J. 85-65S
Stickford, G.H., Jr. 76-18U
Stickles, R.W. 86-110F
Stiening, R. 76-19P
Stimson, P. 84-37U
Stockton, G.W. 86-45E
Stoddard, F.J. 79-7S
Stojanoff, C.G. 77-10U
Stokes, A.D. 76-10U; 79-28U; 86-50U
Stokes, A.N. 83-2S
Stollery, J.L. 79-18H
Stolovich, N.N. 85-5E,64K
Stommel, H.M. 85-39C
Stone, C.R. 81-23J; 83-13J
Stone, M. 86-101Q,102Q
Stones, P.R. 80-16A
Stopka, K.D. 81-15Q
Storck, A. 80-5D
Stormberg, H.P. 85-47U
Storti, R. 86-10E
Stow, P. 86-16N
Strachan, D.C. 76-19U; 77-17U,20U
Strachyra, L.J. 80-16B
Strada, M. 79-41B; 80-10B,16H; 81-9B; 82-65F
Stradomskiy, M.V. 78-33K; 79-19S,20S
Stramigioli, C. 77-12K; 78-34K; 79-5K; 80-23K; 81-16K; 82-45K
Strani, M. 83-63F
Strarostin, A.D. 80-84B
Stratman, R.A. 86-13FF
Straub, D. 84-10G
Straub, J. 76-59F; 78-53F; 79-22J; 80-9C, 32F; 82-48F; 84-13E
Straughan, B. 76-36A; 79-40A
Straus, J.M. 78-95F; 79-132F; 80-72F; 86-25DP
Straviskiy, Ye.M. 83-14S
Strazisar, A.J. 77-28C
Streed, E.R. 76-10T
Street, R.H. 79-21H
Street, R.L. 82-7H; 84-57F
Stretton, A.J. 85-57FF
Strieder, W. 85-122K; 86-117K
Striegel, S.A. 84-21H,22H
Strike, W.T. 77-6R
Strobel, F.A. 80-7FF,8FF,47FF
Strobel, G.L. 78-3T,4T
Stroeve, P. 80-75B
Strohmeier, K. 86-91Q
Strom, B. 79-73A
Strong, A.B. 77-55A,11P; 78-15P; 81-70F; 82-20N
Strouboulis, T. 86-64N
Strscheletzky, M. 86-65C
Struts, G.V. 78-52J
Struwe, D. 85-18Q
Stryker, P.C. 84-66F; 85-69B; 86-37FF
Stryrikovich, M.A. 84-43Q
Stuart, P.R. 84-30U
Stubley, G.D. 82-20N
Stucheli, A. 76-49J
Stuckes, A.D. 86-86P
Studzinski, W. 80-109J; 83-10JJ; 84-28JJ, 29JJ
Stufflebeam, J.H. 84-29K
Stulov, V.P. 78-22K
Stultz, J.W. 76-24T
Stupin, A.B. 79-34B
Sturgess, G.J. 80-17H; 86-20H,21H
Sturm, B. 85-28B
Sturm, T.W. 81-68F

Sturtevant, B. 82-63J; 86-29J
Sturzenegger, Ch. 80-31U
Stutrud, J.S. 86-47E
Stutzman, R.J. 82-78F
Stylianou, S.A. 80-124J; 83-11JJ
Styrikovich, M.A. 78-75J; 80-125J; 81-23B, 44J; 83-62J; 84-59J
Stytsyna, A.A. 86-95K
Styushin, N.G. 82-67J
Su, C.C. 84-6H
Su, K.C. 79-26P,16T; 80-17A; 81-20A
Su, M.-G. 86-106J
Suarez, E.G. 86-16DP
Suarez, R. 80-8L
Subba Raju, K. 76-33D
Subba Reddy Gorla, R. 86-13FF
Subbotin, V.I. 85-70B
Subbotin, V.N. 85-22H
Subhadra Ramachandran, P. 79-34C
Subrahmaniyam, S. 81-48A; 82-35T
Subramaniam, C.K. 86-87P
Subramanian, C.S. 81-45B,29C; 82-45C
Subramanian, E. 85-87DP; 86-60DP
Subramanian, R.S. 79-90J; 81-76J
Subramanian, S.V. 84-94K
Sucec, J. 77-29C; 79-43C; 86-75B
Suck, S.H. 82-49K
Sucker, D. 78-4H; 79-22H
Suckewer, S. 76-20U
Suda, S. 86-66P
Sudhoff, B. 81-82J
Sudo, S. 86-88P
Sudo, Y. 85-71B,105J; 86-115J
Suen, W.M. 79-74A
Suetin, P.E. 76-45P; 84-12B,43C
Sugavanam, A. 78-30E
Sugawara, M. 77-91F; 78-28K; 79-11L; 83-30JM, 31JM; 86-52F,29JM
Sugawara, M.N. 86-51JM
Sugawara, S. 79-10L
Sugimoto, 84-51B
Sugimoto, T. 82-37B; 83-45B
Suginara, K. 79-23A
Sugisaki, M. 85-40P
Sugiura, Y. 85-24G
Sugiyama, R. 86-37Q
Sugiyama, S. 76-30D; 85-1H
Sugrobov, V.M. 77-1S
Suh, I.S. 85-83DP
Sukamel, A.S. 78-58B
Sukanek, P.C. 84-14E
Sukhatme, S.P. 77-90F; 85-60C
Sukhikh, V.D. 77-32P
Sukhov, V.A. 83-64F
Sukhovich, Ye.P. 79-4L
Sukomel, A.S. 76-41B; 77-54B
Sul'zhik, N.I. 77-25D
Sulem, P.L. 78-12E
Sullery, R.K. 76-41D
Sullivan, C.W. 81-76F
Sullivan, H.F. 84-23T; 85-3T; 86-3T
Sullivan, N.S. 86-46E
Sullivan, P.J. 76-22E,89F; 85-52N
Sultan, M. 83-63J
Sultanian, B.K. 80-76B
Sumaria, V.H. 78-25S
Sumikawa, J. 80-28FF
Sumita, Y. 86-13D
Sun, K.H. 76-20S; 81-83J,84J
Sun, T.-Y. 86-43JJ
Sun, Y. 85-10K
Sunada, J.K. 85-63DP
Sunaga, T. 86-33B
Sundarajan, A. 77-4E
Sundaram, R.K. 80-95J
Sundararajan, T. 80-9U; 85-25JJ
Sunden, B. 79-44C; 83-39FF
Sunderland, J.E. 81-9A
Sung, C.S. 78-8K
Sung, H.J. 84-6FF
Suo, M. 76-13H; 79-23H
Suopys, A.P. 84-47B
Supper, W. 79-7Q; 82-4Q,5Q
Suprunova, Zh.A. 79-69F
Sura, P. 86-37JJ
Surana, K.S. 86-78N,79N
Surapaneni, P.R. 86-4A
Surko, C.M. 86-68F
Surma Devi, C.D. 85-29H
Surpun, T.T. 77-9C
Survila, V.J. 84-31Q,44Q
Survila, V.Yu. 78-20Q
Survila, Y.Yu. 79-14D
Survilla, V.Yu. 79-14C
Survilla, Y.V. 77-37C
Suryanarayana, N.V. 76-58A, 77-02A, 81-11D, 83-9G
Suryanarayana, V.V.S. 77-15G
Suryanarayanan, S. 76-13Q
Susec, J. 78-43A; 81-15N; 84-50B
Suslov, A.I. 78-44A
Suss, G. 76-22P
Suter, P. 79-1T
Sutherland, J.W. 83-34Q
Sutherland, P.J. 76-89F
Sutton, K. 76-17R; 81-37C
Sutton, W.H. 79-25K; 80-19K; 86-107K
Suwanprateep, T. 86-4H
Suwono, A. 80-48FF,6G
Suzukawa, Y. 81-44F; 82-66F
Suzuki, H. 85-77J
Suzuki, K. 82-37B,14D; 83-45B,7D,64J; 84-51B; 85-46C,55C,42J,95Q; 86-125J
Suzuki, M. 77-63A; 78-45A; 79-29U; 81-17S; 84-48C; 86-56A,116J,7JJ,41T
Suzuki, T. 78-19H
Svatteri, C. 77-26J
Svetlakov, A.N. 83-80J
Svoboda, K. 86-19S
Svoiskii, V.Z. 85-47P
Svyatetskiy, N.V. 77-53B,38K
Swallom, D.W. 80-13M
Swaminathan, S. 85-81C
Swan, H. 84-37U
Swarup, S. 78-55C
Swean, T.F. 76-9H

Sweat, M.E. 83-37T
Swedish, M.J. 79-24H
Sweeney, D.W. 82-37K
Sweet, D. 77-99F
Swenson, C.A. 77-43P
Swenson, R.J. 80-44A
Swift, G. 85-96Q
Swift, R. 76-4T
Switzer, G.L. 86-47E
Switzer, K.A. 80-19A
Swords, M.D. 79-16P
Sy, H.K. 85-22P
Syamlal, M. 85-84DP
Sychev, P.E. 86-62U
Symmons, J.G. 86-111F
Symons, J.G. 80-28T; 81-6T; 82-39T; 84-28FF; 85-17DP
Synitina, O.F. 85-15F
Syromyatnikov, N.I. 76-19D; 77-31B; 80-126J
Szego, J. 76-20T; 77-22T; 78-29S,30S
Szekely, J. 82-14U; 83-20U; 84-21FF; 86-40S
Szema, K.Y. 84-94K
Szepe, S. 82-18S
Szetela, E.J. 83-46B
Szmyrka-Grzebyk, A. 86-8E
Szpilka, A.M. 83-17JM
Szuszczewicz, E.P. 79-11M
Szymanski, A. 84-11U
Szymczak, W.G. 82-1N

Ta'eed, O. 86-75Q
Tabakoff, W. 77-10Q
Tabarrok, B. 77-100F
Tabei, K. 84-29U; 85-42U
Taberlet, E. 85-61S
Tabor, H. 78-18T
Taborek, J. 80-15J
Tachibana, H. 86-41S
Tachikura, M. 84-31U
Tachiwaki, T. 84-34DP
Tada, K.I. 86-61S
Tada, Y. 78-43P
Tadaki, T. 79-143F
Tadano, R. 84-19JJ
Tadros, S.E. 82-8G
Taga, M. 79-1H
Tagg, D.J. 79-10D
Taghavi, H. 82-48C
Taghavi, K. 86-15JJ
Taghavi-Tafreshi, K. 79-75J; 80-127J; 82-26JJ; 83-32JM
Taguchi, K. 84-16A,21J
Taha, I.S. 81-27T; 83-9T; 86-26J
Taibi, S. 82-33A
Tailby, S.R. 76-50J
Taine, J. 85-107K; 86-33K,34K
Taira, T. 84-55B
Tairov, E.A. 80-73J
Taitel, Y. 82-34B
Tajima, H. 82-25A
Tajima, M. 85-62S
Tajima, O. 86-128J
Takacz, P.Z. 79-11M
Takagi, N. 84-54J
Takahaghi, M. 80-3B
Takahama, H. 81-6G; 84-13C
Takahashi, M. 81-30C
Takahashi, N. 86-89P
Takahashi, O. 76-6M; 80-128J; 82-8M; 85-60FF, 66J; 86-69J
Takahashi, R.K. 84-8B; 85-3E
Takahashi, S. 82-24B; 86-33E,29Q
Takahashi, T. 84-37Q
Takahashi, Y. 86-28T
Takahasi, K. 86-21JJ
Takahasi, Y. 85-8F
Takaki, R. 85-45E
Takami, T. 77-15C
Takamoto, T. 86-61S
Takano, K. 85-39U
Takano, Y. 86-88Q
Takao, S. 81-34FF; 82-67F
Takarabe, K. 83-45K
Takase, H. 79-54J
Takashashi, K. 84-30JJ
Takashashi, R.K. 84-7B
Takashashi, S. 79-143F
Takashima, M. 76-90F; 83-6G; 85-98F
Takashima, S. 86-105Q
Takashima, T. 80-49J,50J; 86-117J
Takashima, Y. 86-43J
Takasu, H. 84-44K
Takata, S. 78-35K
Takata, Y. 84-67F
Takawa, T. 86-61S
Takayama, K. 86-24D
Takayama, N. 77-9S
Takazawa, H. 83-10Q; 85-97Q
Takebayashi, Y. 85-52JM
Takeda, A. 85-83K
Takeda, R. 85-54N
Takeda, S. 80-48P
Takeda, T. 86-112F
Takegoshi, E. 82-55A
Takemura, T. 86-118J
Takenaka, N. 85-60FF,66J; 86-69J
Takenaka, T. 82-55A
Taketani, T. 86-44JJ
Takeuchi, D.I. 78-8H; 79-31C
Takeuchi, H. 80-25D; 82-23FF; 85-51JM
Takeuchi, M. 76-20F; 78-20F; 83-33C; 84-48C; 86-63F,64F,65F,38FF,39FF
Takeuchi, Y. 83-6H
Takeuti, Y. 82-56A
Takeyama, T. 82-13JJ; 84-12G,11JJ; 85-17G; 86-24D,51JJ
Takhar, H.S. 77-30C; 79-127F,10M; 80-12M, 14M; 81-5G; 85-83C,30H; 86-15U
Takimoto, A. 76-19J; 84-20B,49C; 86-39J,119J
Takizawa, S. 78-44P
Takizuka, T. 85-98Q
Takumi, T. 86-90F
Tal, R. 84-50C
Tal Thau, R. 79-30E
Talaat, M.E. 78-19T

Talbot, L. 79-1D
Talenov, A.M. 85-107J
Taleyarkhan, R. 85-72B
Taliaferro, S. 79-16H
Tallone, L. 84-23K
Tam, C.K.W. 79-29E
Tam, S.W. 85-48P
Tam, T. 86-60C
Tamada, K. 79-45C
Tamai, N. 85-8F
Tamanini, F. 77-39F; 78-23E
Tamaru, T. 85-41U
Tamblyn, R.T. 80-29T,30T
Tambour, Y. 76-15H; 77-18H; 80-22E
Tamonis, M.M. 77-31C; 78-12C,56C; 80-24K; 82-38B
Tamura, N. 78-35K; 86-61J
Tan, C.S. 85-27B
Tan, C.W. 85-48B
Tan, K.B. 80-12J
Tan, K.L. 85-22P
Tan, K.S. 78-28Q
Tan, M.J. 85-112J
Tanaka, H. 76-53J; 77-55B; 79-31J,76J; 80-62F; 81-85J; 82-39B; 84-31JJ; 86-76B,19JJ,20JJ, 90P
Tanaka, K. 86-91P,28U
Tanaka, M. 82-8M; 84-36A; 85-47A; 86-41K, 42K,43K,83K,42S
Tanaka, N. 84-52B
Tanaka, O. 84-45Q
Tanaka, S. 78-6L; 84-51C; 86-108K
Tanaka, T. 77-28T; 80-11T; 84-46Q
Tanaka, Y. 85-26N; 86-39FF,80N
Tanaka, Z. 84-37Q
Tanasawa, I. 83-12JJ; 86-78F,34JM
Tandon, T.N. 82-27JJ; 85-106J,26JJ; 86-45JJ
Tandy, D.F. 86-81N
Tang, D. 78-46F; 83-10G
Tani, T. 77-28T; 80-11T; 82-24U; 84-3H; 85-27JJ; 86-27T,28T
Tanigaki, M. 78-43P
Tanigawa, Y. 82-56A; 86-59A
Taniguchi, H. 86-53K,48S
Tanimoto, A. 77-35B; 80-48B; 86-26FF
Tankin, R.S. 84-14JJ,17JJ
Tanner, R.I. 80-33FF
Tanre, D. 79-12L
Tanzawa, S. 84-46J
Tao, L.N. 78-46A; 79-75A,76A,77A,78A; 80-45A; 81-61A; 83-33JM,34JM; 85-41A; 86-60A
Tao, W.Q. 82-63F,37T; 83-42B,11D; 84-48B; 85-112K
Taoka, Y. 84-40C
Tapping, J. 86-109K
Tapucu, A. 84-7N
Tarasuk, J.D. 81-75F; 82-80F,37FF; 83-32A; 85-25T
Tardos, G.I. 76-42D
Tardy, H.L. 85-113K
Tarin, A.A. 76-45P
Tarnawski, W. 76-59A
Tarr, N.G. 86-66A
Tart, K.R. 78-8S
Tarumoto, A. 86-62B
Tarunin, Ye.L. 86-17F
Tarzia, D.A. 82-57A
Tasaka, K. 86-62J
Taske, M.E. 77-14E
Taslim, M.E. 86-59DP
Tatchell, D.G. 78-34C
Tate, A. 78-21C
Tateno, A. 82-39B
Tatenow, A.M. 83-6D
Tatterson, G.B. 80-7G
Taubenfeld, Z. 83-25K
Tauber, M.E. 86-66C
Tauc, J. 77-1T; 78-20T
Tauchert, T.R. 77-64A
Tauchiva, Y. 76-91F
Taussig, R.T. 86-110K
Tavoularis, S. 80-20E,21E; 85-8B
Taya, M. 85-16P; 86-31A,44P
Taylor, A.B. 85-36JM
Taylor, D.S. 78-48P
Taylor, E. 76-22C
Taylor, J.R. 80-19C; 82-5H; 86-95Q
Taylor, K.H. 85-34T
Taylor, M.F. 76-14E; 77-34B; 79-33B
Taylor, P.R. 85-43U; 86-33U
Taylor, R. 81-11H
Taylor, R.D. 86-77U
Taylor, R.E. 85-46P; 86-83P,92P
Taylor, R.P. 86-28C
Taylor, T.W. 86-99K
Taylor-Reed, S. 79-8S
Tazukov, F.Kh. 82-25F
Te, A.M. 86-28JJ
Teipel, I. 81-18S
Teja, A.S. 85-49P
Telegina, I.I. 79-11B
Telin, V.I. 77-44P
Telionis, D.P. 78-57C
Tellex, P.A. 84-29K
Temkin, A.G. 79-79A
Tempkin, A.G. 76-60A
Temps, R.C. 77-29T
Teng, J.T. 79-42B; 80-2G
Tensi, H.M. 76-8A
Teodorescu, V. 79-36P
Tephan, K. 80-129J
Teplyukh, Z.N. 76-32P
Tepper, P. 85-28Q
Terada, K. 84-37C; 85-61C
Teranishi, T. 80-10L
Terasaka, H. 81-16Q
Terauchi, Y. 84-34A; 85-114K
Terekhov, V.I. 82-15H; 85-87C
Terent'yev, Yu.N. 78-21H
Terentev, E.D. 84-30C
Terpigor'yev, V.S. 82-9C
Terrier, C. 86-30K
Test, F.L. 81-31C; 84-32C,7D
Testa, P. 86-120J
Tewari, P.K. 81-69F; 85-108J
Tewell, J.R. 86-67C

Tezduyar, T.E. 85-28N; 86-82N
Thakur, A.K.S. 83-33A
Tham, L.G. 82-21N
Thambynayagam, R.K.M. 80-18H
Thames, F.C. 86-51N
Thangam, S. 82-68F; 85-27F,22G; 86-40FF
Theissing, P. 84-47Q
Thekaekara, M.P. 76-25T,26T
Theodorou, G. 86-77B
Theodossiou, V.M. 86-83N
Theofanous, T.G. 76-8C,51J; 78-76J; 80-130J
Thery, P. 86-50A
Theunissen, P.-H. 83-38T
Thibault, J. 79-46P; 84-14N; 85-53N
Thiel, H.G. 80-4U
Thirst, T.J. 77-17S
Thodos, G. 77-7D
Thomas, A.S.W. 76-4E
Thomas, E.L. 78-10P
Thomas, H.R. 85-85DP
Thomas, J.L. 86-2N
Thomas, J.R., Jr. 82-23A; 85-74B,109J,104K; 86-61A
Thomas, L. 78-58C
Thomas, L.C. 76-51C,23E; 77-32C; 78-14B,24E; 80-23E; 81-32C,33C; 82-46C; 85-53C
Thomas, M.E. 85-115K
Thomas, R.E. 85-48JM; 86-66A
Thomas, W.C. 78-84J; 80-133J; 86-51T
Thomas, W.R.L. 81-28T
Thomason, S.B. 78-59B; 80-46A
Thome, J.R. 81-86J; 82-68J; 83-65J; 84-1J; 85-46J
Thompson, D.E. 77-1E
Thompson, J. 78-8T
Thorbergsen, E. 77-8S
Thornton, E.A. 78-47A; 80-47A
Thornton, G.K. 77-13G
Thorp, J. 82-25T
Thorsen, R.S. 80-27J
Thorsness, C.B. 79-25H
Thring, R.H. 77-26D
Throckmorton, D.A. 83-15S,16S; 85-84C
Thumm, M. 85-26S
Thurner, P.A. 85-49P
Thutupalli, G.K.M. 83-29K
Thyagaraja, A. 82-22N
Thynell, S.T. 85-116K,117K,118K; 86-111K, 112K,113K
Tiba, D. 84-41JM
Tiba, M. 84-41JM
Tichenor, D.A. 85-39E
Tichy, J. 82-69F
Tichy, J.A. 80-22S
Tiederman, W.G. 76-27P; 77-14B
Tiemann, W. 80-32U
Tien, C. 76-4F
Tien, C.-L. 79-10F
Tien, C.L. 76-66A,13J,20S; 77-6F,8J,14K,18K; 78-7F,8F,9F,10F,18J,11K; 80-31K; 81-13D, 17D,63F,65F; 82-59F,6K; 83-22F,46K,47K; 84-38DP, 20F,21F,6JJ,7JJ; 85-33DP,91DP,25F,26F,15K, 16K; 86-23F,22K,125K
Tien, H.C. 86-8FF,41FF,42FF
Tien, J.S. 84-4FF
Tien, K.K. 77-27T; 79-46C,34T
Tien, R.H. 80-48A
Tikhonov, A.V. 85-28J
Tikhotskii, A.I. 86-73K
Tillman, E.S. 84-23H; 85-31H
Tillman, J.E. 77-52F
Timchenko, N.P. 85-110J; 86-121J
Timma, J. 85-61FF
Timmers, J.F. 82-11D
Timol, M.G. 85-62FF
Timon'kin, V.N. 79-70F
Tindall, A.T.R. 86-68C
Ting, C.L. 77-101F
Tinge, J.T. 86-93P
Tinker, J.A. 86-86P
Tirskii, G.A. 86-97P,98P
Tirumala, M. 85-117J
Tishchenko, Z.V. 77-56B; 83-35Q
Tishkoff, J.M. 79-77J
Titov, G.P. 76-45P
Tiwari, G.N. 79-68A; 81-25T; 83-35T
Tiwari, S.N. 78-36K; 84-94K; 86-45U
Tiykma, T.B. 79-13S
Tjioe, L.E. 82-40T
Tkachenko, B.K. 76-21K
Tkachenko, S.V. 86-103J
Tobias, C.W. 79-23B
Tochitani, Y. 80-9Q; 85-51Q
Toda, M. 79-78J; 82-26C; 83-18U; 84-23C; 85-30B; 86-80B,15D,49Q,38U
Toda, S. 86-122J
Todo, I. 78-29Q; 84-48Q
Todorovic, M. 76-22U
Todorovic, P.S. 76-21U
Todreas, N.E. 76-5B; 80-24J
Toei, R. 86-76Q
Tohda, Y. 86-32DP
Toiliev, K. 76-3T; 86-47T
Token, K.H. 84-9A
Tokoi, H. 80-43P
Toksoy, M. 80-21T; 86-23T
Tokuda, N. 85-85C; 86-114K
Tokuda, S. 80-18F
Tokunaga, H. 76-24E
Tokura, I. 83-40FF,35JM; 84-33JM; 85-82F, 47FF,46JM; 86-34FF
Tolpadi, A.K. 84-18FF,29FF; 85-63FF
Tolubinskaya, L.F. 78-38J
Tolubinskiy, V.I. 77-50J; 78-77J,78J,79J; 79-79J,80J,81J,82J,83J,84J; 80-131J; 82-69J; 83-66J; 84-60J,61J; 85-111J; 86-123J
Tolubinskiy, V.N. 77-51J
Tolubinsky, V.I. 79-21Q
Tomaszkiewicz, I. 85-48E
Tomcej, R. 86-113F
Tomeczek, J. 86-115K
Tomimura, T. 85-70J
Tominaga, A. 83-47B
Tomita, A. 84-46Q
Tomita, Y. 79-85J

Tommasi, S. 86-58U
Tomono, H. 86-61S
Tondeur, D. 86-54DP
Toner, M. 83-13JJ
Tong, A.Y. 86-124J
Tong, C.R. 85-112J
Tong, T.M. 86-61P
Tong, T.W. 83-46K,47K; 85-46A,86DP,87DP; 86-60DP
Tong Thi, V.K. 84-71F
Tonini, R.D. 79-37B; 81-46B,47B
Tonnesen, J. 84-21S
Toomre, J. 77-102F; 86-67F
Toor, J.S. 78-39K
Toovey, I. 85-88DP
Topakoglu, H.C. 85-3B; 86-23B,24B
Topham, D.R. 76-7U; 83-24U; 84-32U
Toporishcev, A.A. 85-36S
Torbaty, M.M. 83-35B
Toren, M. 78-50A
Torguet, R. 84-4E
Tori, K. 81-34C
Torii, K. 81-35C; 85-83K
Torikai, K. 86-125J
Torikoshi, K. 78-64J; 80-31D
Toritsyn, L.N. 86-73K
Torquato, S. 85-50P; 86-94P
Torrance, K.E. 76-73F; 79-133F,134F; 81-4F, 5F,69F; 82-6F,6J; 83-3F
Torres, F.J. 85-20E
Torresani, B. 86-15K
Toru, Y. 86-57U
Torvik, P.J. 76-61A
Torzo, B. 84-5P
Torzo, G. 86-10E
Toscano, W.M. 76-10L
Toth, R.A. 85-119K
Tournie, P. 77-27D
Tovichakchaikul, S. 82-12N
Tovma, A.A. 80-36P
Tovstonog, V.A. 85-120K
Tower, L.K. 79-10Q
Townsend, M.A. 76-51A; 79-12Q
Townsend, P. 79-22A
Toyama, M. 76-58J
Toyama, S. 86-94Q
Toyota, T. 83-48K
Trabold, T.A. 86-52C
Trad, A. 85-36FF
Tran, V.Q. 78-65F
Trappeniers, N.J. 78-45P; 80-25K; 84-3P; 85-18P, 21P; 86-96P
Traub, D. 86-25D
Traub, P. 84-47C
Traverse, J.P. 86-2T
Travis, J.R. 76-52J
Tree, D.R. 83-13B
Trefethan, L.M. 82-13F
Trela, M. 81-87J
Tremblay, B. 84-68F
Trengove, R.D. 84-24P; 85-51P
Trepp, Ch. 80-132J
Treptow, R. 78-18U
Treputnev, V.V. 83-7JJ
Treshchev, G.G. 83-64F
Tresso, E. 84-23K
Trethowen, H. 86-48E
Trevino, C. 84-52C; 85-77Q,84Q,85Q
Trevisan, O.V. 85-89DP; 86-61DP
Trezek, G.J. 77-5Q,6Q
Trifonov, A.Z. 81-88J
Trifonova, N.N. 77-1B
Trines, G.R.M. 76-30P,35P
Trinh, S. 86-55A
Trinite, M. 77-4U; 78-35P
Tripathi, B.B. 86-35P
Tripathi, T.C. 84-24T
Tripses, J.T. 77-29K
Trnka, O. 86-19S
Trofimenko, A.V. 85-121K
Trofimov, A.S. 80-77B
Trofimov, V.M. 86-28D
Trofimov, V.P. 80-3K
Trogdon, S.A. 86-39F
Tropea, C. 85-7E
Troyan, V.I. 82-2M
Troyan, Ye.N. 79-33C
Troyer, S.D. 78-10B
Truelove, J.S. 83-17K; 84-95K
Trufanov, A.N. 86-37C
Trujillo, D.M. 82-23N; 85-8N; 86-81N
Trukhanova, L.P. 85-40J
Truman, C.R. 81-10C
Trumble, T.M. 77-28K
Trump, D.D. 86-47E
Trump, W.L. 77-45P
Trung, D.T. 80-15M
Trunzer, A. 84-14K
Truong, H.V. 78-48A
Truong, V.V. 76-20K
Trupp, A.C. 79-114F; 80-66B; 84-53B
Trushin, V.A. 85-8G
Trusov, V.P. 77-1S,2S
Tsahalis, D.T. 77-28D
Tsai, C.-F. 82-50K; 86-116K
Tsai, C.C. 85-47S
Tsai, D.S. 85-122K; 86-117K
Tsai, S.-W. 85-73B
Tsai, S.S. 78-37K
Tsang, C.H. 85-4F; 86-114F
Tsantrizos, P. 82-25U
Tsay, S.Y. 83-11G
Tscheng, S.H. 79-21S
Tseng, A. 84-15G
Tseng, A.A. 84-32S
Tseng, H.H. 85-9H
Tseng, W.F. 83-15D
Tsibinogin, O.G. 77-26P
Tsimermanis, F.Kh. 78-82J
Tsimermanis, L.B. 78-82J
Tsinober, A.B. 83-65F,41FF
Tsirel'man, N.M. 86-62A
Tske, M.E. 76-13E
Tslaf, A. 77-18U; 78-16U; 80-33U; 84-33A
Tslaf, A.L. 86-78U
Tsoi, P.V. 86-63A

Tsou, F.-K. 84-12H; 85-4D
Tsou, R.C.H. 76-42B; 77-14A,15A; 79-51A
Tsuboi, H. 84-23JM; 85-35JM
Tsubouchi, T. 82-20C; 84-12C
Tsuchida, H. 79-2D; 80-4C,1D; 81-1C; 82-1C, 2C; 84-1Q,2Q; 86-1D
Tsuchio, T. 83-32T; 84-22T
Tsuchiya, M. 82-67F
Tsuei, J.T. 85-49Q
Tsuge, M. 76-25E
Tsui, Y.P. 83-23U
Tsui, Y.T. 84-68F
Tsuji, Y. 76-52C; 79-52J
Tsukada, T. 85-42A; 86-64A
Tsukamoto, Y. 85-43A
Tsunenari, S. 83-68J
Tsuriuta, T. 84-31JJ
Tsurko, V.A. 79-34A
Tsuruno, S. 79-135F; 80-49FF
Tsuruta, T. 83-29C; 86-32DP
Tsvelodub, O.Yu. 85-51C
Tsvetkov, F.F. 77-54B
Tsybulsky, V.V. 85-25E; 86-41B
Tsypulev, Yu.V. 82-36C
Tsyrul'nikov, A.S. 77-21S
Tuann, S.-Y. 79-47C
Tucker, E.O. 86-86K
Tucker, R.J. 86-35K,118K
Tufts, D.B. 82-15Q
Tuge, S. 76-26E
Tugov, A.I. 79-22S
Tukhvatullin, S.G. 81-17K
Tully, N. 86-48T
Tulshian, N. 86-49E
Tuma, D.T. 78-17U; 80-27U; 81-16U; 82-19U, 20U; 85-34U
Tuma, M. 79-15B,16B
Tunc, M. 85-124K
Tunc, M.M. 85-123K
Tung, C.Y. 85-117J
Tung, H.-H. 86-46JJ
Tung, M.A. 85-13JJ
Tupichenkow, A.A. 76-46P
Turaga, M. 84-34JJ; 85-99Q
Turakhia, M. 84-49Q
Turcotte, D.L. 82-70F; 83-8JM
Turian, R.M. 78-11G
Turilina, Ye.S. 76-56J; 77-1B; 78-3S; 82-4C
Turkin, A.V. 85-20Q
Turlays, D.P. 80-52J
Turner, B.L. 80-63F
Turner, J.R. 79-48P
Turner, J.S. 80-22FF; 81-31F
Turoki, T. 81-10Q
Turton, R. 84-8A,50Q; 86-106Q,107Q
Tushin, A.M. 80-126J
Tushin, N.N. 82-23B
Tuthill, A.H. 85-100Q
Tutlite, O.L. 80-24K
Tutu, N.K. 78-9E
Tveitreid, M. 77-103F
Tvietereid, M. 78-96F
Twomey, S. 85-125K
Tyabin, N.Y. 83-14B
Tylee, J.L. 83-34A
Tyler, A. 83-29F
Tyukhay, G.G. 76-43D
Tyukov, V.V. 86-58P
Tyurina, E.G. 85-54J
Tyvand, P.A. 79-78F; 81-36F; 84-15DP,22DP
Tzivanidis, G.J. 82-9M
Tzuoo, K.L. 82-8FF; 83-9FF; 85-64FF

Uchida, M. 85-29JJ
Uchida, S. 78-4J; 84-42J; 85-35B
Uchikawa, S. 85-54N
Uchima, B. 86-22JJ
Uchiyama, H. 85-126K
Udal, H. 83-69J
Udell, K.S. 83-24DP; 85-90DP
Udyma, P.G. 80-43J
Ueda, H. 77-27E; 79-35P; 80-40F; 82-15B; 83-17B; 84-35C,69F
Ueda, M. 82-71F; 85-44A
Ueda, T. 76-53J; 77-52J; 78-16C; 79-31J; 81-89J, 90J; 83-34C,67J,68J; 84-62J; 85-44J,45J; 86-50E,48N
Ueda, Y. 80-25U
Uehara, H. 82-51J; 84-31J,32JJ,33JJ; 85-101Q; 86-72J
Uemura, M. 86-91J
Uemura, S. 84-33JM; 85-46JM,49JM
Ueno, M. 83-22Q
Uenosono, C. 80-25U
Uglov, A.A. 81-18U; 86-23C,52JM
Uhlemann, R. 85-35T
Uhlenbusch, J. 76-23U; 78-24P
Ukeje, E. 82-58A
Ulbrecht, J.J. 85-39DP
Ulbricht, R. 80-26U
Ulinskas, R. 85-97C,21D,115Q,116Q
Ulinskas, R.V. 77-24Q; 79-26Q,27Q,28Q,29Q; 84-11D,56Q; 86-32D
Umapathy, K.C. 84-11B
Umapathy, K.G. 83-11B
Umeda, T. 86-1N
Umemura, A. 86-126J
Unal, H.C. 76-54J; 77-53J,21Q; 78-83J; 81-91J, 92J; 85-102Q,103Q
Underwood, V.M. 76-47P
Ungan, A. 86-110F,5FF,53JM,62S
Ungarish, M. 85-23G
Unger, J. 84-63J
Unny, T.E. 76-38F; 77-8F; 85-24N
Unsal, M. 78-84J; 80-133J
Upadhyay, S.N. 80-10E; 81-12D; 82-10D; 83-8DP; 84-7Q
Uplekar, A.G. 86-76U
Urakawa, K. 76-21S
Uran, B. 84-20P
Urano, Y. 82-59A
Urbach, H.B. 81-9M
Uribe, F.J. 86-99P,100P
Uruchev, V.A. 85-75Q
Urusov, G.L. 83-64F

Uskov, V.S. 80-104J
Uspuras, Ye.V. 85-66B
Ustimenko, B.P. 76-51P
Usui, H. 82-32B,40B
Usui, T. 86-64F,65F
Usui, U. 84-54B
Utaka, Y. 81-12T; 83-12JJ,32T; 84-21T,22T; 85-27JJ,43JM,44JM,45JM; 86-47JJ,39JM,40T
Utkin, V.B. 84-70F; 85-113J
Utsunomiya, K. 77-33J
Uttarwar, S.B. 85-104Q
Utterback, N.G. 86-62P
Utton, D.B. 85-18E
Utzinger, D.M. 81-29T
Uvarov, V.V. 78-39J
Uvarova, L.A. 85-65J
Uyeha, H. 84-34DP
Uzunoglu, N.K. 78-7L
Uzzel, J.C., Jr. 79-48A

Vacca, G. 86-13P
Vacek, V. 76-62A
Vachon, R.I. 77-27B
Vacon, R.I. 77-21K
Vacquie, S. 80-1U; 82-26U; 84-13U,27U; 85-48U; 86-24U
Vader, V.T. 86-119K
Vafai, K. 81-17D; 84-35DP; 85-91DP; 86-48JJ
Vafaie, F.N. 86-49T
Vaidyanathan, G. 84-36DP
Vaisberg, V.A. 85-28JM
Vaishya, J.S. 84-24T
Vaitekunas, P.P. 76-53C
Vaitiekunas, P.P. 84-54C
Vajravelu, K. 77-104F; 78-97F; 80-50FF; 86-115F, 79U
Vakhabov, A.A. 82-8Q
Val', L.I. 78-32K
Val'tsiferov, Yu.V. 85-99F
Valaitis, C.C. 81-2D
Valatkevicius, P.J. 78-1B
Valatkyavichyus, P.Yu. 79-19B
Valentin, P. 77-10H; 80-13U
Valenzuela, J.A. 81-17Q; 86-60S
Valero, F.P.J. 82-51K; 84-38K
Vallis, E.A. 86-68C
Valougeorgis, D. 85-74B
Valuyeva, Ye.P. 80-59B; 82-29B
Valyuzhinich, M.A. 86-78B
Van Atta, C.W. 76-21P; 77-12E; 78-3E,4E,5E; 80-35P; 81-3C
Van Brakel, J. 83-5H
Van Buren, P.D. 80-49A
Van Dam, J. 85-49E
Van de Sande, E. 79-136F
Van de Vosse, F.N. 86-84N
Van den Berg, H.R. 85-18P
Van den Brink, G.J. 83-39T
Van den Hoek, W. 81-19U
Van der Gulik, P.S. 86-96P
Van der Mast, V.G. 76-55J
Van der Meer, Th.H. 78-60B; 80-28D
Van der Meulen, S.F. 85-1S
Van Dongen, M.E.H. 80-23C
Van Doormaal, J.P. 81-70F; 84-15N; 86-27N, 71N
Van Dougen, F.G. 78-25E
Van Eldik, R. 82-8A
Van Fossen, G.J. 84-4C
Van Fossen, G.J., Jr. 84-45C; 85-86C
Van Gasselt, M.L.G. 77-21Q; 81-92J; 83-69J
Van Genuchten, M.Th. 79-9A
Van Hagan, T. 77-15Q
Van Haneghem, I.A. 86-95P,109P
Van Heiningen, A.R.P. 76-54C
Van Heininghen, A.R.P. 85-50E
Van Heugten, W.H.H. 80-23C
Van Leer, B. 86-2N
Van Leersum, J. 86-49JJ
Van Leersum, J.G. 77-22Q; 81-18Q
Van Migom, M. 85-74K
Van Muiswinkel, J.C. 77-12C
Van Noy, J.H. 79-12U
Van Ooijen, H. 79-22Q
Van Rijswijk, W. 86-40E
Van Sciver, S.W. 79-23S; 86-22J
Van Steeg, J.G. 78-98F
Van Steenhoven, A.A. 86-84N
Van Strahlen, S.J.D. 85-114J
Van Stralen, S. 78-11
Van Stralen, S.J.D. 77-54J; 78-34J; 79-94J; 82-65J
Van Straler, S.J.D. 86-111J
Van Swaaij, W.P.M. 85-70DP; 86-62DP
Van Thinh, N. 79-47P
Van Woerkom, P.C.M. 82-53K
Van't Klooster, P. 78-45P
Van't Verlatt, P.M. 81-92J
Van-Hull, L.L. 76-27T
Vanderbroek, M.A. 79-7P
Vanderhaegen, D. 86-120K
Vanderplaats, G.N. 82-6Q
Vandersande, J.W. 80-44P
Vanderschuren, J. 85-20DP
Vaniman, J. 86-60C
Vanka, S.P. 82-14M; 86-69C,85N,86N,87N,88N
Vanoli, R. 79-7J
Vantelon, J.-P. 85-48K
Varanasi, P. 84-38K
Vardelle, A. 80-34U; 82-27U; 83-25U; 85-19U; 86-80U
Vardelle, M. 80-34U; 82-14U,27U; 83-25U; 85-19U; 86-80U
Varejao, L.M.C. 79-57A,108F,62J
Varfolomeyev, I.M. 82-47C
Vargas, H. 86-69P
Varma, H.K. 82-2J,27JJ; 83-59J; 85-106J,26JJ, 2P; 86-2J,105J,45JJ
Varma, R.K. 85-108J
Varone, A.F., Jr. 86-127J
Varshney, B.S. 78-66J,67J
Varshney, I.S. 80-3C
Vartanyan, N.V. 86-5K
Vasanova, L.K. 76-19D; 80-126J
Vasantha, R. 86-70C,71C

Vasechikin, V.N. 78-41C
Vasechkin, V.N. 82-12H; 85-25H
Vasil'eva, L.M. 81-17K
Vasil'evskii, S.A. 86-97P,98P
Vasil'yev, A.A. 79-11D; 84-60J,61J,64J; 85-32H
Vasil'yev, A.F. 76-35J; 77-36J; 78-51J,52J; 80-21J
Vasil'yev, A.N. 85-115J
Vasil'yev, L.L. 83-17DP
Vasil'yev, Ye.P. 78-33K
Vasiliev, L.L. 79-86J,24S; 85-105Q
Vasilieva, I.A. 86-121K
Vasseur, P. 80-64F; 81-56F; 82-56F,72F,24FF; 83-38F,66F; 84-71F; 86-53DP
Vatutin, I.A. 76-4G; 78-12G
Vaughan, J.M. 78-10P
Vaxman, B. 85-36T
Vaxman, M. 83-36T; 86-50T
Vazeos, E. 86-39T
Vecchia, P. 83-10K
Vecchio, G.L. 83-13T
Vecher, A.A. 85-19H
Vedhanayaygam, M. 79-137F
Vedra-Nayagam, M. 83-5G
Veerappa, B. 80-53F; 82-26D
Velarde, M.G. 79-3F; 82-15F; 84-10F; 85-24F, 9G
Velasco, R.M. 86-99P,100P
Velichko, V.I. 76-41B; 78-58B
Velkoff, H.R. 79-30U
Vella, G. 82-33A
Velusamy, K. 86-21C
Vemuri, S.B. 83-13S; 86-104K
Venart, J.E.S. 85-94F,95F,124K; 86-11E,52J
Veneraki, I.E. 86-100Q
Venkatachala, B.J. 79-138F
Venkataramana, S. 85-49U
Venkataramani, K.S. 78-26E
Venkateshan, S.P. 79-26K
Venkatramani, N. 79-62A
Venugopalan, M. 82-21U
Vepa, K. 77-15Q
Verberne, J. 85-1S
Vercammen, H.A.J. 80-26K
Verdini, L. 86-42E
Verdini, P.G. 86-42E
Verdon, C. 84-105K
Verhoeven, J.D. 84-29A
Verhulst, J.L. 86-109F
Verma, A. 76-28T
Verma, A.K. 81-13F,19J
Verma, A.R. 83-42FF,3G; 84-3FF
Verma, R. 86-59S
Verma, R.L. 78-99F
Verma, R.S. 86-13DP
Vermesh, R.T. 76-12J
Vermesse, J. 86-101P
Vermeulen, G.A. 86-40E
Vermuri, S.B. 85-92Q
Verneker, V.R.P. 76-18R
Verollet, E. 77-3H,19H
Verret, D.P. 78-8L
Verriopoulos, C.A. 84-9C
Vertgeym, I.I. 86-116F
Verver, A.B. 86-62DP
Vervisch, P. 85-24K; 86-27K
Veselov, V.P. 79-38C
Veselovskii, V.B. 86-65A
Vesely, V. 85-92DP
Veski, A.Y. 79-13Q
Vest, C.M. 81-6FF
Vestrucci, P. 81-16K
Veto, B. 86-26U
Vevai, J.E. 78-42F
Veziroglu, T.N. 76-63A; 77-11,21J; 78-12; 79-80A; 83-21J,43J
Vick, B. 81-62A,48B,49B; 83-35A; 84-31A; 85-14A, 42K,21N; 86-30A,79B
Victor, S.A. 76-43B
Vidal, D. 86-101P
Vidin, Yu.V. 85-45A
Vidyanidhi, V. 77-15G
Vieria Dias, L.A. 84-13B
Vighnesam, N.V. 79-10M
Vigo, P. 77-73F
Vijayalakshmi, D. 86-97Q
Vijayan, M. 86-74K
Vijayan, S. 76-44C
Vijayan, T. 85-50U
Vilemas, J.V. 79-28B,29B; 80-54B; 84-2C; 85-75B
Vilemas, J.W. 78-61B
Vilemas, Yu.V. 82-7Q
Villacres, A. 85-109Q
Ville, J.P. 78-11S
Villeneuve, D.M. 84-105K
Villermaux, J. 78-30B; 86-31JM
Villoutreix, G. 79-49P
Viloria, R.E. 85-75C
Vilyamas, Y.V. 82-26B
Viner, J.M. 85-19P
Vinograd, D.L. 85-106Q
Vinogradov, Yu.A. 79-12H
Vinogradova, E.P. 86-3Q
Vinokurov, V.F. 85-40Q
Vinson, T.S. 85-52P
Violino, P. 83-2K
Viollet, P.-L. 85-76B
Vira, N.R. 80-8G
Virozub, I.E. 85-64B
Visbal, M.R. 86-89N
Vishnyak, V.F. 85-64B
Viskanta, R. 76-24A,22S; 77-41F,26K,23T,24T; 78-38K,39K; 79-10A,5F,6F,5J,23J,31T; 80-6A, 15A,49A,27K; 81-4A,71F; 82-42F,73F,5FF,54K, 22T; 83-69F,11JM,12JM,15JM,36JM; 84-25A, 42DP,40F,23FF,9JM,12JM,13JM,14JM,54K,16P; 85-58B,13F,53F,13JM,14JM,31JM,32JM,50K,72K, 76K,77K,128K,63S; 86-63DP,66DP,12F,14F,15F, 110F,117F,12JM,16JM,32JM,33JM,53JM,55JM, 56JM,57JM,58JM,59JM,77K,79K,80K,119K,62S
Vislovich, A.N. 79-3M
Visser, J.A. 81-19U
Vityamas, Yu.V. 76-33B
Vives, C. 86-54JM
Vlachos, N. 78-34C
Vlachos, N.S. 86-40B

Vlieger, J. 84-103K
Vliet, G.C. 77-12F; 78-92J
Voelkel, D.D. 84-51Q
Voelker, D. 76-19A
Vogel, C. 82-9T
Vogel, E. 86-102P
Vogel, G.J. 76-36D
Vogel, I. 85-1P
Vogel, J.C. 85-20D
Vogel, M. 86-59F
Vogelpohl, A. 80-112J
Voitsekhovskii, N.I. 86-65A
Vol'nykh, Yu.A. 85-35J
Volchenok, V.F. 80-78B
Volchkov, E.P. 79-26H; 82-15H; 85-87C
Voliers, C.T. 80-80J
Volkov, D.I. 79-9J; 82-3JJ; 83-2Q
Volkov, S.S. 84-6F
Volkov, V.A. 76-21K
Volkov, V.V. 86-73K
Volkova, N.M. 86-11Q
Voller, V. 81-63A,64A
Voller, V.R. 83-37JM
Vollmers, H. 77-28E
Voloshko, A.A. 80-134J
Volpicelli, G. 85-2DP; 86-1DP
Volpicelli, G.G. 82-17C
Volz, F.E. 83-49K
Von Driska, P.M. 86-34T
Von Hoffman, T. 78-85J
Von Weissenfluh, T. 85-51E
Vorob'ev, V.A. 78-86J
Voronin, V.N. 83-21Q
Voronkevich, A.V. 86-1H
Vorontsov, A.V. 82-10U
Vorontsova, L.A. 82-3JJ
Voropayeva, E.N. 77-10J
Vortmeyer, D. 76-7L,14Q; 82-27D; 84-37DP
Voshall, R.E. 77-19U
Voskresenskiy, K.D. 76-56J; 77-1B; 78-3S
Vosteen, B. 84-52Q
Vrentas, C.M. 82-34FF; 83-67F
Vrentas, J.S. 81-72F; 82-34FF; 83-67F
Vrtis, M.L. 86-88Q
Vujanovic, B. 76-64A
Vulkanovic, V. 76-22U
Vullierme, J.J. 79-81A
Vyas, B.D. 79-92J
Vybornov, S.I. 85-100F
Vyskocil, J. 80-35U

Waas, P. 76-59F; 79-86F
Wacholder, E. 82-74F
Wachtler, K.P. 78-57B,20D
Wada, A. 84-52B; 86-80B
Wada, N. 85-34P
Wada, Y. 83-26U; 84-38U; 86-53F,90P
Waechter, D. 86-66A
Waechter, R.T. 76-57A
Wagner, K.C. 84-7FF
Wagner, L.Y. 81-93J
Wagner, R.L. 85-107Q
Wagner, T.R. 80-28K
Wagner, W. 76-23T
Wahi, M.K. 77-55J
Waibel, R. 84-10G
Wakamura, K. 83-45K
Wakao, N. 77-29D,30D
Wakatsuki, Y. 86-18S
Wakeham, W.A. 78-31P; 79-28P; 80-8P; 82-38K; 85-37E,79K
Waki, T. 81-10Q
Wakisaka, N. 86-26D
Wakitani, S. 85-65FF
Waksman, D. 86-51T
Wakunaga, T. 84-55J
Walberg, G.D. 85-88C
Walch, J.-P. 79-139F
Walch, J.P. 79-35F
Walden, R.W. 81-73F; 86-68F
Waldram, K.L. 76-57J
Walin, G. 80-47F
Walker, J.D.A. 82-4G
Walker, K. 77-105F
Walker, K.L. 78-100F
Walker, M.D. 85-52E
Walker, P.L. 80-29K
Walker, P.N. 86-108Q
Walker, V. 77-21B; 78-48B
Walklate, P. 76-19B
Wall, D.L. 80-45P
Wall, K.W. 78-87J
Wall, T.F. 83-17K; 84-96K
Wallace, J.M. 83-28C
Wallin, P. 70-39P
Wallis, G.B. 78-88J
Walmsley, H.L. 77-20U; 79-31U
Walsh, P.J. 83-2S
Walsh, S.K. 79-87J
Walters, R.W. 86-90N
Walton, I.C. 78-39F; 80-65F; 82-75F,76F; 83-68F; 85-101F,66FF
Walton, M. 85-64S
Walzel, M.D. 78-9T
Walzel, P. 84-10Q
Wanchoo, R.K. 84-32Q; 86-104J
Wang, B.-X. 84-65J; 85-116J
Wang, C. 76-24D
Wang, C.-Y. 80-50A
Wang, C.H. 84-35J
Wang, C.S. 81-7G
Wang, C.Y. 82-60A,31FF; 85-89C,52DP
Wang, D.F. 86-101F
Wang, F.-S. 86-50B,51B
Wang, G.S. 83-69F
Wang, H. 86-67A
Wang, H.Z. 82-35B
Wang, J.-Q. 86-122K
Wang, J.C. 86-109Q
Wang, J.C.T. 78-59C
Wang, J.C.Y. 84-34JJ
Wang, J.H. 85-90C
Wang, J.Y. 76-48P
Wang, K.Y. 84-38DP
Wang, L.W. 85-50F

Wang, M. 84-33DP
Wang, P.K. 85-55N
Wang, Q.-J. 84-11FF
Wang, Q.C. 86-63S
Wang, R.C.C. 77-32C
Wang, T. 85-91C
Wang, T.-C. 82-25T
Wang, X.A. 82-35FF,25K
Wang, Y.-J. 84-97K
Wang, Y.-L. 86-104P
Wang, Y.F. 83-40T
Wankat, P.C. 77-106F
Wanniarachchi, A.S. 86-30JJ,50JJ
Waray, W.O. 79-35T
Ward, C.A. 83-32K
Ward, J.R. 78-31S
Ware, B.R. 80-47P
Warhaft, Z. 78-27E; 80-24E; 81-36C; 86-46C
Warn-Varnas, A.C. 79-1G
Warneford, I.P. 78-32F
Warner, D.F. 78-89J
Warner, J.G. 80-16FF,39J
Warnock, A.S. 85-13G
Warren, M.D. 84-16N
Warren, S.G. 84-98K
Warrick, A.W. 85-31S
Warrington, R.O. 80-46F
Warrington, R.O., Jr. 81-74F; 83-46F; 85-102F
Waseda, Y. 86-51E
Wassel, A.T. 76-23K,24K; 77-33C; 79-43B; 80-15C; 85-10A,16A
Wassner, L. 80-135J; 84-33S
Watabe, M. 86-27JJ
Watanabe, I. 79-27C
Watanabe, K. 84-55B
Watanabe, M. 86-34P
Watanabe, O. 86-128J
Watanabe, T. 76-58J; 82-59A
Watanabe, Y. 76-11S
Watkins, C.B. 79-50P; 85-7D
Watkins, C.B., Jr. 76-55C; 86-43F
Watkins, R.W. 76-44B
Watkinson, A.P. 79-21S; 80-22Q; 86-77J,78J, 99Q
Watson, A. 77-65A
Watson, C.H. 86-30E
Watson, F.A. 76-8J
Watson, J.S. 85-25P
Watson, R.D. 77-34C,29E
Watters, J.F. 85-54P
Watts, H.A. 80-22F
Waugh, M.D. 76-22K
Wavre, P.D. 77-97F,98F
Wayne, R.S. 86-45E
Wayner, P.C. 79-63J,64J
Wayner, P.C., Jr. 76-39J,59J; 78-28E,29E; 79-88J; 82-28JJ; 85-117J
Wazzan, A.R. 82-48C
Weaver, F.J. 86-103P
Weaver, J.A. 86-63DP,55JM
Weaver, W.L., III 81-4S
Weaver, W.R. 79-18K
Webb, A.I.C. 83-26Q
Webb, B.W. 85-77B,128K; 86-33JM,56JM,57JM
Webb, D. 85-20JJ
Webb, D.R. 81-95J; 83-3JJ,14JJ
Webb, G.L. 85-77C
Webb, G.W. 85-53P
Webb, R.L. 79-23Q; 80-19B,79B; 81-39B,94J, 19Q; 82-16Q; 84-35JJ,36JJ; 85-82Q,108Q,109Q; 86-110Q
Webb, S.W. 85-37J
Weber, C.F. 81-65A
Weber, G. 86-77P
Weber, J.C. 86-60P
Weber, J.E. 78-101F
Weber, M.E. 79-24Q; 83-20JM
Webster, J.J. 82-4N; 83-14F
Wedekind, G.L. 77-56J; 80-10J,11J; 86-10J, 11K
Wedekind, G.T. 81-10J
Wedel, R.K. 78-6T
Wege, M.E. 84-66J
Wegener, P.P. 85-12E
Wehde, K.-H. 85-65S
Wehle, F. 82-49C; 84-25P
Wehrmann, M. 81-18D
Wei, C. 84-35A
Wei, C.K. 85-93DP,66S
Wei, P.S. 85-103F
Wei, Z. 86-53E
Weichert, R. 78-49A
Weidman, P.D. 85-104F; 86-64DP
Weihs, D. 77-60A
Weikl, B. 86-60U
Weill, M.E. 85-53E; 86-14FF
Weinbaum, S. 77-66A; 78-16A; 81-61J; 85-57JM, 52S; 86-61JM
Weinberg, F.J. 80-17U
Weinbert, F.J. 77-23P
Weinspach, P.-M. 79-15S; 84-18S,34S; 85-40S, 41S,42S
Weinspach, P.M. 80-21S; 81-82J; 82-14H
Weinstein, B.A. 86-52E
Weir, A.D. 85-34T
Weisman, J. 76-60J; 85-51J; 86-129J,130J, 132J
Weisrose, S.A. 84-99K
Weiss, D. 86-23K
Weiss, H. 78-90J
Weiss, H.P. 77-107F
Weiss, N.O. 76-43F; 81-16F,35FF,36FF; 86-67F
Weiss, T.A. 80-31T
Weiting, A.R. 80-47A
Welber, B. 77-46P
Welford, W.T. 80-33T
Weling, F. 85-51U
Welle, R. 86-35U
Wellford, L.C., Jr. 77-67A; 81-6N,7N
Wells, M.E. 86-64S
Wells, P.B. 79-82A
Wells, R.G. 86-79B
Wells, W.H. 85-129K
Wells, W.L. 85-92C
Wells, W.R. 84-5S
Welsey, D.A. 79-83A

Welty, J.R. 76-9M; 78-21B; 82-4D; 84-14DP; 86-18DP
Wen, C.Y. 79-18P
Wen, L. 79-36T; 86-123K
Wendraff, B. 77-24A
Wendt, R. 84-34S
Weng, C.-I. 83-9Q
Wentworth, W.E. 76-29T
Wentzell, P.R. 85-95F
Werele, M.J. 78-14C
Werle, M.J. 78-44C
Werley, K.A. 81-66A
Werther, J. 76-44D
Wesely, M.L. 76-30T
Wesley, D.A. 76-45B
Wessel, R.A. 81-70A
Wesseling, P. 78-98F
West, G. 80-23S; 81-67A
Westerberg, A.W. 85-48Q
Westwater, J.W. 78-55J; 79-93J; 80-140J; 81-17J, 96J,5Q; 84-21JJ,14Q; 85-13J
Wetenkamp, H.R. 76-23S
Wetmore, A.E. 82-49K
Whalley, P.B. 78-91J
Whang, G.J. 80-26A
Wharmby, D. 84-33U
Whealton, J.H. 85-47S
Wheat, T.A. 85-29P
Wheatley, J. 85-96Q
Wheatley, J.C. 85-60F
Wheaton, F. 86-49E
Wheeler, L.T. 84-20A
Whitaker, S. 76-6
Whitaker, W. 80-30K
White, D. 86-58JM
White, G.K. 85-9P; 86-20P
White, H.C. 79-37T
White, J.E. 81-12H
White, J.O. 78-20T
White, K.O. 78-40K
White, R.E. 86-91N
White, R.F. 85-17DP
White, T.R. 86-65DP
Whitehead, G.L. 82-13M
Whitelaw, J.H. 76-9S; 77-18E; 79-11C,96F; 80-20F; 82-12B
Whittle, G.E. 86-65S
Whittmann, D. 80-24Q
Wichterle, K. 78-62B
Wickern, G. 85-17FF; 86-27C
Widmer, F. 83-71J
Wiebelt, J.A. 79-38T
Wieling, N. 86-2Q
Wierum, F.A. 81-14S
Wiese, J.W. 85-12N
Wiesman, J. 83-70J
Wieting, A.R. 78-47A; 85-10C
Wiggert, D.C. 77-68A
Wigley, G. 76-9S
Wijeysundera, N.E. 76-31T; 78-9L; 81-22T; 82-40T; 83-33T; 85-37T; 86-81B,5G
Wilbur, P.J. 86-81U
Wilcox, D.C. 77-6E
Wilcox, W.R. 76-9J; 79-57J,90J; 80-23J
Wild, G. 85-77DP; 86-68Q
Wilde, D.J. 80-1Q
Wilder, S.E. 86-48C
Wilheim, R. 85-26S
Wilhelm, H.E. 79-5M
Wilhelmi, H. 84-34U; 85-52U
Wilkerson, T. 85-17K
Wilkerson, T.D. 79-6K
Wilkie, D. 85-54E
Wilkins, C.A. 80-46P
Wilkinson, R.G. 84-19A
Wilkinson, W.L. 77-40B
Wilks, G. 76-92F,93F; 80-21FF; 84-35U; 85-67FF, 53U; 86-72C
Willekers, R.W. 86-9E
Willems, J.F.H. 80-23C
Williams, A. 77-40A; 79-80A; 86-17S
Williams, D. 76-16K; 77-35K
Williams, F.A. 78-50B
Williams, G.T. 80-32T
Williams, J.C. 78-21D
Williams, J.C., III 76-56C
Williams, J.J. 77-20H
Williams, M.M.R. 84-100K,101K,102K
Williams, M.W. 78-19K
Williams, P.G. 79-12C
Williams, R.J. 78-9Q; 79-13J
Williams, R.K. 86-103P
Williams, R.T. 86-62N
Williams, S.D. 77-69A,70A; 84-53C
Williamson, A.S. 76-65A
Williamson, J.W. 76-15T
Williamson, W. 83-43K
Willis, G.E. 82-18C
Willits, D.H., Jr. 86-8B
Willmarth, W.W. 84-15E
Willmott, A.J. 78-3Q; 79-25Q; 80-23Q; 81-7Q
Willrath, H. 78-46P
Willumeit, H.-P. 86-66S
Wilmott, A.J. 76-15Q
Wilson, D.G. 82-52A; 85-1JM,53JM
Wilson, J.B. 85-36DP; 86-31S
Wilson, N.W. 79-92J; 81-37FF
Wilson, R.E. 78-60C
Wilson, S.D. 78-20H
Wilson, S.D.R. 76-61J; 79-87J,91J
Wilson, S.I. 84-8H
Wilson, S.J. 86-124K
Wilson, T. 86-92N
Wimmer, W. 85-52U
Winchester, D.C. 82-2FF
Wind, M.M. 84-103K
Winder, W.O. 84-10S
Windisch, R. 84-67J
Windish, R. 85-58FF
Window, B. 82-52K; 83-41T; 84-6T; 85-39T
Winer, W.O. 77-13S; 85-22B; 86-22E
Wing, L.D. 76-19R
Winkler, K.-H.A. 84-72K,104K
Winkler, W. 78-61C
Winston, R. 76-19P,18T; 80-33T; 81-21T
Winter, E.R.F. 76-45A; 81-56J; 82-30A,22D

Winter, P. 80-18H
Winterton, H.S. 85-32J
Winterton, R.H.S. 81-4J; 83-19J; 85-26J; 86-20J, 38J
Wire, M.S. 85-53P
Wirtz, R.A. 77-108F; 79-140F; 82-77F,78F
Wirz, H.J. 78-13
Wiscombe, W.J. 86-81K
Wissler, E.H. 85-51S
Witfield, D.L. 77-35C
Witt, C.L. 79-39F
Witte, L.C. 80-24F,25F; 82-70J; 84-68J,53Q; 86-96J
Witte, M.J. 86-52T
Wittig, S. 80-7H; 85-74C; 86-54C
Wittig, S.L.K. 83-10JJ; 84-24H,28JJ,29JJ
Wittig, W. 80-5P
Wix, S. 86-7S
Wojcik, L.A. 80-12L
Wojcik, L.H. 77-2R
Wolf, J. 83-17S
Wolf, L. 77-14Q
Wolfe, W.L. 83-23K
Wolff, C.M. 79-41P
Wolff, H.-J. 76-14D
Wollersheim, D.E. 76-48F; 83-27J,28J
Wollhover, K. 83-18JM; 85-55JM,56JM
Wolny, A. 84-39DP
Won, K.J. 84-15F
Won, Y.H. 83-8F
Wonchala, E.P. 84-26P
Wong, B.K. 76-23A
Wong, C.P. 85-8S
Wong, C.P.C. 78-92J
Wong, H.H. 81-53F
Wong, H.Y. 77-12
Wong, K. 77-21A
Wong, K.-L. 85-31FF; 86-73C,43FF
Wong, L.W. 80-51A,32K; 84-110K
Wong, P.F.Y. 86-14D
Wong, S. 80-136J; 81-50B
Wong, S.-L. 77-7B
Wong, S.L. 78-12B
Wong, S.P. 79-74A
Woo, K.C. 81-17N
Woo, W. 84-36U
Wood, A.S. 86-93N
Wood, B.D. 82-11T
Wood, B.E. 76-49P; 78-5L; 83-37K
Wood, C. 84-16E
Wood, D.H. 85-26E
Wood, I.R. 79-29F
Wood, R.M. 86-87Q
Wood, W.L. 77-71A
Woodhouse, A. 86-10FF
Woodman, T.P. 77-30T
Woodruff, D.W. 79-93J; 81-96J
Woods, J.L. 80-9G
Woods, L.C. 80-25E
Woodward, S.H. 86-15C
Woodworth, C.L. 86-82B
Woolner, K.A. 82-20N
Woolsey, G. 84-37U
Worku, G. 86-8F,9F
Wormeck, J.J. 85-13N
Wornom, S.F. 86-94N
Worrlein, K. 76-16Q
Worster, M.G. 85-105F
Woschitz, D. 83-71J
Wozniak, Z. 85-110Q
Wragg, A.A. 77-82F; 79-10D
Wright, G.A. 86-67S
Wright, J.H. 76-94F
Wright, J.L. 83-18T; 86-118F
Wright, P.K. 86-69A
Wright, R.J. 77-19U
Wright, S.J. 77-109F
Wrobel, L.C. 81-68A
Wu, A.S. 86-25JM
Wu, B.J.C. 83-1B; 84-52J
Wu, C.S. 85-75J
Wu, F. 81-39FF
Wu, H.W. 86-27S
Wu, J.C. 78-30E
Wu, J.L. 77-21U
Wu, P.K.S. 79-5R
Wu, P.S. 79-36F; 81-30T
Wu, R.-S. 76-21F,22F,95F,96F; 77-57B,19F; 79-141F
Wu, R.S. 76-46B; 78-13B
Wu, X. 86-59JM
Wu, Y.C.L. 82-3M,5M,6M
Wu, Y.W. 84-12Q
Wulbrand, W. 85-106F
Wunder, R. 80-15S
Wunderlich, R.W. 80-47P
Wung, T.Y. 78-109F; 79-18T
Wygnanski, I. 77-30E; 78-47P; 79-32E
Wyler, J.S. 78-6P
Wyman, C. 80-34T
Wynncyckyj, J.R. 84-26P
Wynnyckyj, J.R. 86-7P
Wyss, H. 86-53U

Xi, T.-G. 86-104P
Xiaolin, Y. 85-50JM
Xideman, S. 76-33J
Xu, J.-J. 84-30FF
Xu, J.-Z. 84-12H

Ya Epik, E. 78-10E
Yaakobi, B. 84-105K
Yabe, A. 86-44JJ
Yabe, M. 83-48K
Yacoub, N. 83-5B
Yadigaroglu, G. 76-42J,62J; 85-52J; 86-92J
Yaghoubi, M.A. 79-60F; 80-66F,23FF
Yagi, Y. 83-22Q; 84-19JJ
Yaglom, A.M. 77-24B; 78-23C
Yagoda, H.P. 83-36A
Yagov, V.V. 80-58J; 85-118J,119J
Yahalom, Y. 83-65F,41FF
Yakoub, Y. 86-1Q
Yakovenko, A.G. 83-23Q

Yakubenko, A.E. 86-45C
Yakuborskiy, Yu.V. 80-28J
Yakubovskiy, Yu.V. 80-106J
Yakusheva, E.V. 77-18J
Yakushin, M.I. 86-23U
Yamada, H. 85-30FF
Yamada, I. 82-25B; 84-31B; 86-46J,37Q
Yamada, K. 82-71F
Yamada, M. 84-32JJ; 85-94DP,112Q
Yamada, T. 78-13G; 80-25C
Yamada, Y. 80-16Q; 82-55K; 86-22K,125K
Yamaga, J. 86-48S
Yamagata, S. 82-28U; 83-26U
Yamagishi, H. 86-111Q
Yamaguchi, A. 83-56F
Yamaguchi, C. 82-71F
Yamaguchi, H. 85-28JJ
Yamaguchi, K. 76-11S; 84-69J
Yamaguchi, S. 77-23Q; 78-30Q
Yamakawa, M. 86-119F
Yamali, C. 83-6JJ
Yamamoto, A. 86-50E
Yamamoto, H. 79-12D; 82-41B
Yamamoto, K. 76-27E,66F; 77-81F; 78-41P; 79-101F; 84-10D; 86-65Q
Yamamoto, T. 85-38Q
Yamamoto, Y. 84-20B
Yamanaka, A. 86-31U
Yamanaka, G. 84-23J; 86-30J
Yamanoto, H. 84-9D
Yamao, A. 80-18U
Yamaoka, N. 86-39DP,70J
Yamasaki, H. 85-94DP,112Q
Yamasaki, T. 84-72F; 86-131J
Yamashita, H. 77-23Q; 78-30Q; 81-19B; 83-22B, 23B; 85-46B,47B,24G,65Q; 86-55Q,112Q
Yamashita, K. 85-34FF
Yamashita, S. 80-19U
Yamauchi, A. 86-51JJ
Yamauchi, S. 86-79J,81J
Yamazaki, H. 86-54S
Yamazaki, Y. 76-1Q
Yamigishi, H. 86-68S
Yan, M.M. 79-13L
Yan, Q.-S. 86-53T
Yan, W.M. 86-24F
Yan, Z. 85-67S
Yanabu, S. 80-19U
Yanada, Y. 84-54Q
Yanadori, M. 85-29JJ; 86-54T
Yanagi, K. 85-42A; 86-64A
Yanagida, A. 83-36FF
Yanagida, H. 86-47JJ
Yanagihara, H. 82-25C
Yanase, S. 78-35F; 80-67F; 85-107F; 86-120F
Yanenko, N.N. 80-20A
Yanezmoreno, A.A. 83-13D
Yang, C.I. 82-14Q
Yang, H. 86-95N
Yang, H.T. 86-8D
Yang, J. 82-36FF
Yang, J.H. 85-117J
Yang, K.-T. 86-76N
Yang, K.T. 77-25C,57F; 83-12K; 85-67F,111F
Yang, L. 81-8B
Yang, Q.S. 83-47K
Yang, S.L. 84-21A
Yang, W.-J. 79-3B; 80-18FF; 82-38FF; 83-76J, 77J,22Q; 84-13G,70J,19JJ; 86-9G
Yang, W.-M. 86-83B
Yang, W.J. 76-52F; 77-36C; 78-93J; 80-21F; 81-18B,4G; 85-19G
Yang, Wen-Jei 80-46J
Yang, Y.-J. 86-13B
Yang, Y.C. 86-68A
Yang, Y.M. 83-72J
Yang, Y.S. 83-50K
Yanitskii, P.A. 86-60JM
Yao, H.-C. 76-16H
Yao, L.-S. 78-102F,103F,104F,105F,106F,107F, 108F; 79-142F; 80-51FF
Yao, L.S. 76-66A; 77-110F; 80-68F,69F; 81-69A, 38FF,18N; 83-47F,70F,43FF,44FF; 84-73F,30JM, 31JM; 85-40JM
Yao, M. 84-8D
Yao, N.P. 86-37S
Yao, S. 82-42B
Yao, S.-C. 79-44B,27K; 80-80B,110J,137J; 81-65J, 97J; 83-10J,73J,74J; 85-47J
Yao, S.C. 82-17Q; 86-42J,36Q
Yao, Y.-Q. 84-12H
Yarborough, D.W. 86-61P
Yarbrough, D.W. 85-46A,86DP
Yard, R.E. 84-35S
Yarden, A.L. 81-21S
Yaron, I. 76-45D
Yartsev, I.M. 86-62U
Yarygina, N.I. 78-41C; 82-12H; 85-25H
Yaskin, L.A. 80-81B,84B; 85-15F
Yasojima, Y. 86-30J
Yasuda, A. 80-48P
Yasuda, H. 81-30J
Yasukawa, Y. 80-138J
Yasunaga, T. 82-53J
Yasunishi, A. 84-26DP; 86-42DP,43DP
Yasutomo, Y. 81-20U
Yatskevich, G.M. 82-56K
Yau, K.K. 85-111Q; 86-113Q
Yaundalders, S.R. 80-51J
Yavlov, Yu.M. 80-103J
Yavuzkurt, S. 80-19H,20H; 84-17E
Yazaki, T. 83-47B
Ybarra, R.M. 80-82B
Yee, G. 80-83B
Yefimenko, G.I. 85-40B
Yefimov, A.L. 78-2B,1Q
Yefimov, V.A. 85-78B
Yefimova, S.V. 77-13B
Yefremova, I.S. 78-94J
Yefremova, Ye.A. 78-33K
Yeh, H.-M. 85-73B; 86-55T,56T
Yeh, L.T. 76-13A; 77-72A; 79-84A
Yeh, S.H. 77-7L,8L
Yelistratov, S.L. 85-93C
Yelkin, I.V. 83-14S

Yellott, J.L. 78-21T
Yen, C.W. 84-47K
Yen, D.W. 86-69A
Yen, S.H. 79-9L
Yen, Y.-C. 80-70F
Yener, N. 82-41K
Yener, Y. 84-17K; 86-84B
Yeo, K.S. 86-5G
Yeo, Y.K. 83-48B
Yeremenko, G.P. 79-15K
Yermakov, G.V. 80-139J
Yermalayev, I.K. 79-32U
Yermolayeva, Ye.M. 85-54U
Yeroshenko, V.M. 78-21H; 80-84B; 81-13H; 84-56B; 85-15F
Yershov, A.V. 84-56B
Yershov, Ye.D. 83-37A
Yeshurun, Y. 80-15P
Yesin, V.M. 79-1L
Yeun, C.M. 86-58B
Yewell, R. 82-79F
Yi-Heng, C. 82-24N
Yih, C.-S. 81-39FF; 86-105P
Yih, S.-M. 78-62C
Yilbas, B.S. 86-70A
Yildirum, O.T. 83-43J
Yilmaz, S. 80-140J
Yim, A. 77-12B; 78-63B
Yin, S.H. 78-109F
Ying, S.H. 86-129J,132J
Yip, F.C. 76-17Q
Yip, H. 85-52N
Yoder, G.L. 85-74J
Yoder, G.L., Jr. 83-49B
Yokoda, T. 85-94DP
Yokosawa, H. 81-6G
Yokota, T. 85-112Q
Yokote, Y. 81-40F
Yokoya, S. 82-37J; 84-29J
Yokoyama, M. 80-7M
Yoldas, B.E. 84-106K,107K
Yonehara, J. 79-78J
Yonemoto, T. 79-143F
Yoneta, M. 83-38B,39B,40B; 84-39B
Yonomoto, T. 86-62J
Yoo, J.K. 76-11E
Yoo, J.Y. 82-34F
Yoo, K.-H. 86-106P
Yoon, H.K. 84-57B
Yoon, S. 86-45N
Yoon, Y.-H. 83-10C
York, R.E. 84-36S
Yortsos, Y.C. 82-61A,37D
Yoselevich, V.A. 79-45B
Yoshida, D. 86-58K
Yoshida, H. 82-64J; 84-30JJ
Yoshida, M. 82-19S; 86-18S,22S
Yoshida, T. 77-22U; 81-21U; 82-6H,24U; 84-25H; 86-108K
Yoshii, Y. 86-12D
Yoshikawa, F. 84-36A; 85-47A
Yoshikawa, K. 85-94DP,112Q; 86-114Q
Yoshimura, T. 79-5B; 85-56N
Yoshita, Y. 86-50P
Yoshitani, K. 83-2FF
Yoshiyasu, H. 82-28U; 83-26U; 84-38U
Yoshizaki, R. 86-24E
Yoshizawa, A. 76-28E; 79-31E; 86-121F
Yoshizawa, M.Y. 86-107P
Yosida, M. 86-69S
Yosinobu, H. 79-144F
You, H.-Z. 85-68S
Younan, G.W. 84-34JM,35JM
Young, J.H. 76-24S; 86-108P
Young, K. 79-74A
Young, L.A. 84-108K
Young, M.F. 81-11C; 84-18E; 85-68FF
Young, M.L. 85-5P
Young, R.M. 85-55U
Young, S.T.B. 79-41C
Yousef, W.W. 81-75F; 82-80F,37FF
Yovanovich, M.M. 76-67A,71F; 77-55A,73A; 80-40A; 85-4A,26A; 86-53A
Ytrehus, T. 83-75J
Yu, C.-L. 77-57J
Yu, C.P. 78-110F
Yu, J.S. 77-58B; 81-51B
Yu, S.K. 77-19S
Yuan, H.-H. 80-7G
Yucel, A. 79-145F; 80-71F; 84-40DP
Yudayev, B.N. 82-6C
Yudovich, V.I. 85-11F
Yueksel, L. 85-8JJ
Yuen, M.C. 78-95J; 83-51J,52J; 84-17JJ
Yuen, P. 79-7P
Yuen, W.W. 80-51A,31K,32K,13L; 81-70A,18K, 19K; 84-109K,110K; 85-130K,131K
Yuen, W.Y.D. 84-16G; 85-48A
Yuh, C.F. 79-23F; 80-9FF,10FF
Yule, A.J. 78-48P
Yumikura, T. 84-23J
Yuncu, H. 76-63A
Yung, D. 80-141J; 82-46J
Yunitskii, S.A. 86-13H
Yurevich, F.B. 81-13K; 82-56K; 85-93K
Yurkevich, O.R. 86-37U
Yurkin, A.V. 79-8D
Yurov, A.G. 86-74C
Yurtayev, M.A. 84-17Q
Yusa, H. 78-4J
Yushina, L.E. 86-27D
Yuta, S. 85-126K
Yuzhang, V.C. 86-18E

Zaalouk, M.G. 83-49J
Zaberzhinskii, L.D. 81-22S
Zabloudil, O. 86-52JJ
Zabolotskii, V.P. 85-94C
Zabolotskiy, V.P. 82-31C
Zabolotsky, V.P. 84-18H
Zabrodin, V.K. 85-54U
Zabrodsky, S.S. 76-46D; 81-19D
Zaccanti, G. 85-9K; 86-10K
Zachariades, J. 85-34Q
Zachariah, J. 83-4M,5M

Zadiraka, V.Yu. 79-65J
Zagaynova, R.V. 77-53B,38K
Zahoransky, R.A. 83-10JJ; 84-28JJ,29JJ
Zaichik, L.I. 81-13H; 84-56B; 85-11FF; 86-6H
Zaixiang, C. 80-14L
Zake, M.V. 79-33U,34U
Zakerullah, Md. 79-146F
Zakharchenko, S.I. 86-1FF
Zakharov, A.S. 78-35J; 79-36J
Zakharov, G.A. 83-60J
Zakharov, V.G. 80-121J
Zakirov, S.G. 82-8Q
Zakkay, V. 80-23D
Zaks, M.A. 86-122F
Zal'tsgendler, E.A. 77-92F,93F
Zaletnev, A.F. 82-71J
Zalmanovich, S. 82-11C
Zalsgendler, E.A. 76-82F
Zaman, J. 85-47Q
Zamani, N. 85-57N
Zamora, M. 86-123F
Zangrando, F. 85-108F
Zanichkovskaya, L.V. 76-50P
Zanin, A.I. 85-120J
Zanner, F.J. 83-27U
Zanoli, A. 85-5T; 86-16A
Zapechikov, V.N. 86-126K
Zapparov, F.I. 82-25F
Zardecki, A. 86-127K
Zargham, N. 86-4T
Zaric, Z.P. 78-14
Zaritskii, D.P. 80-36P
Zarling, J.P. 77-59B
Zarmi, Y. 81-8T; 82-13T,41T; 83-12T; 85-12T, 16T
Zaryankin, A.E. 79-46B
Zaslouk, M.G. 76-38J
Zaulichnyi, E.G. 86-28D
Zaurayev, V.P. 78-41K
Zaytsev, A.A. 79-50F
Zaznoba, V.A. 83-30C
Zberkhovskiy, Yu.Ye. 83-8S
Zdanavichyus, G.B. 76-57C; 77-37C
Zdanavicius, G.V. 84-41DP
Zebib, A. 77-67F,111F; 78-111F,112F; 82-68F; 85-109F,110F
Zebib, Z. 76-46F; 80-72F; 81-35F
Zedan, M. 85-58N
Zee, R.H. 85-54P
Zehner, P. 86-29D,30D
Zeigarnick, Yu.A. 80-142J
Zekster, M.P. 79-35U
Zelenev, Yu.V. 82-31K
Zelenka, R.L. 83-35C
Zelepuga, A.S. 78-96J
Zelver, N. 84-49Q
Zeman, O. 76-97F; 78-17E
Zemel, J.N. 85-13E,19E
Zemskiy, I.F. 80-20K
Zemskov, B.B. 85-64J
Zen'kovskii, A.G. 86-126K
Zenkov, Yu.V. 83-42K
Zenouzi, M. 84-56F
Zhan, Y. 84-111K
Zhang, G.-P. 85-57JM; 86-61JM
Zhang, G.-T. 85-95DP
Zhang, G.Q. 86-8E
Zhang, H.F. 85-38T
Zhang, J.F. 86-20U
Zhang, N. 83-76J,77J; 84-70J
Zhang, Q.-F. 85-104J
Zhang, X.Z. 86-53E
Zhang, Z. 82-38FF
Zhao, L. 85-45B
Zhavoronkov, N.M. 86-54J
Zhdanov, V.L. 86-12C
Zhdanov, V.M. 83-36C
Zheltukhin, V.A. 79-147F
Zhen, F. 85-67S
Zheng, Y. 86-123K
Zhestkov, B.A. 78-10L; 79-6H
Zhidov, N.I. 85-113Q
Zhiqiang, Y. 85-39T
Zhitomirksaya, I.V. 83-18DP
Zhizhin, G.V. 78-10Q; 86-85B
Zhong, Z.Y. 85-111F
Zhorov, G.A. 83-42K
Zhorzholiani, G.I. 80-120J
Zhou, B.L. 86-53E
Zhou, K.Y. 85-79B
Zhubrin, S.V. 84-26H; 85-59C
Zhukauskas, A.A. 76-16C,30C,31C,53C,57C; 77-31C, 37C,24Q; 78-56C,20Q; 79-35B,26Q
Zhukauskas, Z.Z. 78-12C
Zhukhovitskii, E.M. 79-45F; 83-15FF
Zhukov, A.P. 86-2S
Zhukov, A.V. 85-114Q
Zhukov, I.V. 80-17K
Zhukov, M.F. 86-82U
Zhukov, S.A. 81-98J; 83-78J,79J; 85-110J
Zhukov, V.M. 86-101J
Zhurin, V.V. 76-21K
Zhvanetskii, I.B. 85-106Q
Zhyugzhada, I.I. 76 30C,31C,57C
Zick, A.A. 83-38A
Zidi, K. 85-3P
Ziegenbein, B. 85-92K; 86-75P
Ziegler, E.N. 83-11DP; 85-14DP
Zielinski, J. 86-17C
Zien, T.-F. 76-17H
Zien, T.F. 76-68A
Zienkiewicz, O.C. 78-27A,60F; 79-85F
Zigmantas, G.P. 85-95C; 86-31D
Zigrang, D.J. 80-24S
Ziji, W. 77-54J; 78-47J,97J
Zijl, W. 79-94J; 82-65J
Zimin, V.D. 86-18F
Zimmer, G.A. 78-3P; 79-1P
Zinchenko, V.I. 86-75C
Zinina, G.A. 85-10B
Zinser, W. 86-6N
Zinsmeister, G.E. 76-69A; 78-48A
Zipin, H. 83-25K
Zirilli, F. 82-77F
Ziugzda, J.J. 84-54C
Zmeykov, V.N. 76-51P

Zoby, E.V. 77-12R; 81-37C; 83-16S,18S; 84-19C; 85-96C
Zoglin, P. 85-5JM
Zoitan, A. 84-16E
Zollner, A. 85-40T
Zollveg, R.J. 78-19U
Zollweg, R.J. 81-12U; 86-115Q,83U
Zolotar, I.A. 77-22S
Zoloterev, V.V. 85-106Q
Zonato, C. 85-30K
Zoulalian, Z. 82-31B
Zozulya, N.V. 83-36Q
Zrodnikov, V.V. 80-1J
Zscherpel, W. 81-31A
Zubair, S. 86-16T
Zuber, N. 76-44J; 78-32J
Zuckler, K. 80-21U
Zuiderwyk, M. 85-23E
Zukauskas, A. 85-97C,21D
Zukauskas, A.A. 79-13D,14D,27Q,28Q,29Q; 84-54C, 11D,55Q,56Q; 85-115Q,116Q; 86-9D,10D,32D
Zul, W. 78-34J; 83-37Q
Zumbrunnen, D.A. 84-42DP; 86-66DP
Zumbrunnen, M.L. 86-46JM
Zverkhovskiy, Yu.Ye. 80-85B
Zvirin, Y. 77-31T; 78-50A,5T; 79-48F,148F, 149F,29T; 81-42F,76F; 82-49F; 84-48F; 85-86J, 81Q,117Q,118Q; 86-116Q,117Q
Zyl, W. 77-58J
Zysin, L.V. 83-80J
Zysina-Molozhen, L.M. 77-59J; 78-32S,33S; 79-25S,26S; 81-38C
Zyson, L.V. 80-143J
Zyszkowski, W. 76-63J
Zyugzda, I.I. 79-14D

SUBJECT INDEX

APPLICATIONS
Ablation 76-6R,12R; 77-1R,6R; 78-5R
Ablation materials 76-11R; 78-1R
Additive effects 83-8S
Aerospace 85-14S
Air ducts 78-19S
Airfoils and wings 78-3R
Alloys 79-4S
Atmospheric studies 82-4T
Battery vehicles 86-37S
Bearings 80-7S,14S,22S,70S; 84-3S,10S,20S, 21S; 86-42S
Bio-engineering 85-3S,51S
Biological studies 76-6S,17S,24S
Building model 85-1S,6S,12S,13S,34S
Casting 83-11S; 85-54S; 86-39S,53S,61S
Cavities 79-18S
Chemical processing 85-24S,36S,40S,41S, 42S,49S,56S,65S
Circulations 86-62S
Coating 77-8R
Coil heating 84-13S
Condensers 86-113Q
Constant resistance 86-8Q,109Q
Control systems 85-60S
Cooking 86-20S
Cooler design 84-23S
Coolers 86-19S
Cooling coils 78-22S
Cooling of electronic equipments 86-25S, 49S
Cooling of generators 85-23S
Cooling systems 76-13S,20S; 78-7S; 81-1S; 82-9S
Correlations 85-103Q
Critical heat flux 84-15S
Cryogenic thermal control 76-16R
Cutting 86-43S,46S,51S
Desalination 82-15S
Diodes 86-26S
Drying 84-2S,12S,22S,25S,26S,31S; 85-16S, 17S,19S,21S,37S,48S,55S,66S; 86-8S
Drying equipments 78-8S,16S,27S; 86-31S
Electric cables 78-2S,10S
Electric systems 85-28S,47S
Engines 76-18S,23S; 79-9S,19S,20S; 80-2S; 81-7S; 85-25S,67S; 86-4S,11S,17S,18S,27S, 63S,66S,69S
Epoxy 80-24S
Erosion of metals 81-9S,13S
Evaporators 80-16S; 86-37Q,54S
Experimental studies 84-33S
Extruded products 81-11S
Fibers (cooling) 83-5S
Fires 79-10S; 82-6S,10S; 85-11S,20S,33S, 39S,68S
Flow visualization 82-1T
Fluidized bed 81-5S; 84-30S
Food extruders 86-38S
Forging 78-23S; 79-6S; 84-5S
Forming 86-36S
Fractures 86-15S
Furnaces or combustion chambers 76-9S; 78-6S, 12S,24S,28S; 81-6S,15S,20S; 82-1S,2S,3S, 19S; 84-1S,14S; 86-44S
Garments 85-32S,50S
Gears 86-47S
Generators 78-25S,26S
Geothermal problems 77-1S,2S,21S; 78-15S
Grassland 77-12S
Green houses 81-19T
Grinding 79-17S
Guns 78-31S; 81-3S
Hartman-Sprenger tubes 81-17S
Heat exchangers
- Air coolers 81-6Q; 83-25Q
- Boilers 85-50Q; 86-14Q,98Q
- Condensation 79-23Q; 85-65Q
- Condensers 81-11Q
- Cooling coils 83-14Q
- Cooling towers 76-8Q; 80-14Q; 81-17Q; 83-7Q,19Q,20Q,34Q; 85-21Q
- Correlations 77-5Q,7Q; 86-65Q
- Corrugated wall 83-33Q
- Counter flow 84-27Q,34Q; 85-110Q; 86-108Q
- Cracks 86-2Q
- Critical heat flux 84-43Q
- Cross flow 77-23Q; 79-27Q,28Q,29Q; 80-5Q, 10Q,18Q; 83-12Q,24Q,31Q; 84-44Q,45Q; 85-96Q,113Q; 86-23Q,28Q,36Q,107Q,16S

Cylinders 86-49Q
Design 77-1Q,4Q,12Q,15Q,17Q,19Q; 78-14Q; 80-1Q,6Q,11Q,20Q; 82-6Q,13Q; 83-15Q, 16Q,35Q; 84-14Q,19Q,20Q,22Q,24Q,26Q, 42Q,47Q,50Q; 85-13Q,55Q,58Q,93Q,107Q; 86-11Q,25Q,26Q,62Q,73Q,87Q,104Q,106Q, 115Q
Direct contact 85-15Q; 86-13Q,41Q
Disc and doughnut type 84-18Q
Dryers 86-76Q
Dynamic response of 78-15Q,28Q,29Q
Effectiveness 78-5Q; 84-29Q; 86-77Q
Electric field 80-9Q
Entropy generation 86-96Q
Evaporation 82-12Q; 85-109Q
Evaporators 81-15Q; 86-90Q
Finned type 76-16B,36B,4Q,7Q,10Q,16Q; 77-16Q; 78-1Q,13Q,17Q,19Q,25Q,26Q; 79-12Q, 14Q; 80-13Q,21Q; 81-1Q,5Q,9Q; 82-3Q; 83-9Q,17Q,32Q; 84-4Q,23Q,35Q,46Q; 85-10Q, 14Q,23Q,32Q,33Q,48Q,52Q,56Q,60Q,61Q, 62Q,68Q,73Q,82Q,86Q,90Q,92Q,95Q,97Q, 101Q,102Q,108Q,111Q; 86-54Q,64Q,71Q, 85Q
Fluidized bed 86-1Q,4Q,74Q
Fouling effects 76-12Q; 77-3Q; 80-22Q; 81-2Q,13Q,19Q; 83-3Q,6Q,27Q,36Q; 84-36Q, 49Q; 85-16Q,35Q,67Q,75Q,100Q; 86-5Q, 24Q,47Q,52Q,53Q,63Q,81Q,82Q,83Q,99Q
Freon 83-10Q,28Q
General 76-11Q; 78-2Q; 80-7Q,12Q,17Q; 84-5Q,6Q,10Q,12Q,17Q,30Q,40Q,48Q; 85-39Q, 83Q
Heat recovery 85-19Q
Heat transfer enhancement 85-51Q
Helical or spiral coils 77-2Q; 82-7Q; 85-88Q,99Q
Inclined flow 85-1Q,2Q
Irreversibility 86-43Q
Mixtures 85-59Q
Modelling 84-51Q
Multi-concentric cylinders 86-105Q
Multipass 85-28Q,64Q
Multiple disk system 83-22Q
Normal flow 76-2Q
Nuclear reactors 77-14Q,20Q,21Q
Oblique flow 77-8Q
Optimization 85-66Q
Packed bed 86-17Q,68Q,93Q
Parallel flow 81-16Q; 85-80Q; 86-57Q
Performance and efficiency 78-4Q,30Q; 79-17Q,20Q,24Q,26Q; 81-8Q,12Q; 82-16Q; 83-1Q; 84-39Q,41Q,53Q; 85-57Q,76Q,112Q; 86-12Q,60Q,61Q
Periodic flow 83-5Q
Phase change 84-8Q,32Q; 85-106Q
Plastic plates 83-11Q
Plate type 78-8Q; 85-11Q,12Q,34Q,45Q, 46Q; 86-7Q,10Q,16Q,22Q,30Q,55Q,56Q,70Q, 84Q,112Q
Platinum black powder 86-88Q
Pressure drop 78-27Q
Pulsating flow 82-10Q
Radiation effects 80-16Q; 86-58Q
Reactors 84-16Q; 85-47Q,98Q,114Q
Regenerative type 76-6Q,14Q,15Q; 77-11Q, 22Q; 78-3Q,7Q,11Q,22Q; 79-8Q,16Q,18Q, 25Q; 80-8Q,23Q; 81-3Q,7Q,18Q; 82-11Q; 83-13Q,26Q; 85-6Q,7Q,8Q,30Q,31Q,69Q, 71Q; 86-29Q,31Q,39Q,42Q,114Q
Rotary type 77-18Q; 79-2Q; 84-3Q,9Q,21Q, 52Q
Roughness effects 82-1Q,8Q,9Q; 85-116Q
Simulation 84-33Q
Space shuttle 86-79Q
Spherical annulus type 82-15Q
Spray cooling 77-6Q,9Q; 80-15Q
Stability 86-9Q
Steam generators 86-46Q
Suspension 85-17Q
Thermodynamic analysis 86-15Q,21Q,32Q, 38Q,40Q,86Q
Transition region 86-75Q
Tube and shell type 76-5Q; 79-5Q,6Q; 80-4Q; 81-14Q; 82-14Q; 83-2Q,18Q,37Q; 84-11Q, 15Q,25Q,28Q,54Q; 85-20Q,70Q,72Q,89Q, 91Q; 86-18Q,66Q,89Q,103Q
Tube banks 84-1Q,2Q; 85-115Q
Tube bundles 76-1Q,3Q; 77-13Q,24Q; 78-12Q, 16Q,18Q,20Q; 79-4Q,9Q,11Q,13Q; 80-24Q; 82-2Q,17Q; 84-7Q,31Q,55Q,56Q; 85-40Q; 86-34Q,91Q
Tubular collectors 86-94Q
Turbine agitator 76-13Q
Turbine rotor 77-10Q
Two-phase flow 85-27Q
Unsteady state 83-8Q,29Q,30Q; 85-3Q,18Q, 63Q; 86-69Q
V-shaped 79-15Q
Vertical tubes 86-110Q
Welding 86-33Q
Heat pipes 76-17Q; 78-6Q,9Q,10Q,21Q,23Q,

24Q; 79-1Q,3Q,7Q,10Q,19Q,21Q,22Q; 80-2Q, 3Q,19Q; 81-4Q,10Q; 82-4Q,5Q; 83-4Q,21Q, 23Q; 84-13Q,37Q,38Q; 85-4Q,5Q,9Q,24Q,26Q, 49Q,53Q,74Q,78Q,79Q,87Q,105Q; 86-3Q,6Q, 20Q,27Q,35Q,44Q,45Q,48Q,50Q,59Q,67Q,72Q, 78Q,80Q,92Q,95Q,97Q,100Q,111Q,116Q,117Q
Heat pumps 86-7S
Heat transfer enhancement 83-13S; 85-104Q; 86-52S
Heat transmission fluids 82-18S
Heat treatment 80-12S
Housing 77-9S,11S,15S,17S,20S; 84-19S,28S, 35S; 86-65S
Human body 82-8S; 86-57S
Ignition dynamics 76-8R
Impedance heating 85-54Q,38S
Induction 85-58S,61S; 86-45S
Ingot and mold 76-21S
Insulation 86-29S,30S,58S
Kilns 79-21S
Leading edge heat transfer 76-2R
Light fitting 86-67S
Line contacts 80-11S,18S
Liquid metal 81-4S
Liquid spray 80-1S,6S
Magnetic energy storage 79-23S
Mass transfer 84-4S; 85-36Q
Melting 76-8S; 85-9S,41S
Mills 78-4S
Mixers 80-8S,9S
Modules 82-16S
Molding 76-4S; 80-13S; 85-2S,18S,45S
Molten steel 79-8S
Multiphase systems 84-18S
Non-Newtonian fluids 81-18S,19S; 83-12S
Nozzles 76-15R; 86-32S
Nuclear reactors 76-2S,7S; 78-1S,14S,17S; 79-12S,16S; 80-4S,23S; 82-4S; 83-4S; 86-41S
Oil cooling systems 86-12S
Optimization 81-21S
Particle heat transfer 86-101Q,102Q
Pavement 77-8S
Periodic heat flow 83-2S
Permafrost effects 77-4S,5S,22S
Phase change 76-4R,7R
Pipelines 78-21S; 86-3S
Planetary atmospheres, entry of 76-17R
Plating 77-11R
Pollution 76-3S,12S,14S,15S,16S,22S
Power plants 78-20S; 85-7S,8S,10S,26S,35S, 52S,59S,63S
Preheater 76-9Q
Propellant burning rate 76-18R
Pyrolysis phenomena 77-2R
Quenching 77-19S; 85-62S; 86-10S,28S
Radiation transfer 86-35S
Reactors 84-34S
Rectification plates 86-51Q
Refrigeration 85-15S
Rocket
  Propellant 78-4R; 79-2R,4R,5R
Rolling 84-32S
Roughness effects 78-2R
Snow cover effects 77-3S
Soil 77-16S,18S; 79-3S; 85-4S,5S,22S,29S, 30S,31S,43S,46S,53S,64S; 86-6S,9S,21S, 34S
Solar energy collectors 76-1T,3T,4T,5T, 7T,8T,10T,14T,16T,17T,18T,21T,22T,31T; 77-1T,2T,6T,7T,8T,10T,13T,14T,15T,17T, 18T,19T,21T,23T,24T,25T,26T,27T,30T,31T; 78-1T,2T,3T,4T,5T,9T,10T,12T,13T,15T,17T, 18T,20T; 79-1T,2T,7T,9T,10T,15T,16T,17T, 20T,21T,22T,23T,24T,25T,26T,27T,30T,31T, 32T,33T,34T,36T,38T; 80-2T,3T,6T,7T,8T, 9T,12T,13T,16T,19T,20T,22T,24T,25T,26T, 28T,33T,34T; 81-2T,5T,6T,7T,8T,11T,14T, 15T,17T,21T,22T,23T,24T,26T,30T; 82-5T, 6T,11T,15T,16T,23T,24T,28T,30T,34T,37T, 40T; 83-9T,15T,16T,18T,19T,20T,21T,23T, 26T,32T,33T,41T; 84-1T,6T,8T,9T,18T,21T, 22T,24T; 85-1T,6T,10T,11T,13T,14T,17T, 18T,19T,20T,21T,24T,25T,28T,31T,32T,37T, 38T,39T; 86-4T,7T,19T,20T,24T,26T,31T, 33T,36T,40T,49T,51T
Solar energy studies
  General 76-5R,9T,24T,26T; 77-9T; 78-6T, 7T; 79-3T,8T,14T,19T; 80-1T,4T,5T,10T, 14T,21T,31T; 81-10T; 82-2T,3T,12T,25T, 29T,41T; 83-3T,7T,11T,12T,17T; 84-2T, 3T,4T,7T,13T,14T,16T,19T,23T; 85-7T, 27T; 86-2T,6T,17T,18T,21T,43T,46T,52T
  Passive 79-5T,11T,13T,29T,35T
  Power generation 76-2T,19T,23T,27T
Solar heating and cooling systems 76-12T, 13T,15T; 78-21T; 79-4T,6T,12T,18T; 80-15T, 17T; 81-1T,4T,18T,28T; 82-14T,31T; 83-10T, 27T,29T,30T,31T,35T,36T,38T; 84-5T,10T, 17T; 85-3T,4T,8T,12T,15T,29T,30T,34T,35T, 36T; 86-3T,5T,12T,13T,22T,39T,45T,47T, 48T,50T,53T,55T,56T
Solar insolation 76-25T,30T; 77-11T,12T,

29T; 82-13T,36T; 85-33T
Solar irradiation 82-8T,9T,10T,21T,27T, 38T; 83-6T,28T
Solar ponds 81-13T; 82-19T,22T,32T; 83-4T, 8T,40T; 84-11T,12T,15T,20T; 85-2T,22T, 23T; 86-8T,9T,10T,11T,16T,25T,29T,30T, 32T,34T
Solar radiation modelling 81-9T; 83-13T, 24T,25T,34T,37T
Solar spectral analysis 83-5T
Solar still 81-25T
Solar transmittance 82-39T
Solar-powered systems 86-37T,41T,44T
Space radiators 81-12S
Space shuttle 76-1R,9R,14R; 77-3R,4R,5R, 7R,10R; 79-1R,3R; 83-9S,15S,16S,18S
Spray columns 86-60S
Spray cooling 81-14S
Steam boilers 79-1S,11S,13S,14S,22S
Steam generators 82-14S; 83-14S
Structures 86-56S
Submerged port discharges 77-10S
Superconducting coils 78-3S,18S
Suspensions 84-27S
Tables (air) 76-1S
Tanks 76-19R
Temperature measurement 78-9S; 79-5S; 81-2S; 86-2S,14S,24S
Thermal design
of Landers 76-13R
Thermal energy storage systems 76-6T,11T, 20T,28T,29T; 77-3T,4T,5T,16T,20T,22T,28T; 78-5S,29S,30S,8T,11T,14T,16T,19T; 79-28T, 37T; 80-11T,18T,23T,27T,29T,30T,32T; 81-3T, 12T,16T,20T,27T,29T; 82-11S,12S,17S,7T, 17T,18T,20T,26T,33T,35T; 83-1T,2T,14T, 22T,39T; 85-5T,9T,16T,26T,40T; 86-22S, 23S,50S,59S,1T,14T,15T,23T,27T,28T,35T, 38T,42T,54T
Thermal instabilities 76-10S
Thermal insulations 76-19S; 79-2S; 80-5S, 19S
Thermal modeling and simulation 83-6S
Thermal protection system (spacecraft) 76-3R, 10R
Thermal resistance 77-6S,7S,13S,14S
Thermosyphons 83-1S; 85-25Q,29Q,41Q,42Q, 43Q,44Q,77Q,81Q,84Q,85Q,94Q,117Q,118Q; 86-68S
Tools 76-11S; 86-5S
Turbine cooling 82-13S; 83-10S,17S; 84-16S; 86-48S
Turbine systems 76-5S; 78-11S,13S,32S,33S; 79-7S,25S,26S; 80-3S,20S; 81-8S,10S,16S, 22S; 82-5S,7S; 83-3S; 84-6S,7S,8S,9S,11S, 17S,29S,36S; 85-57S
Turbulence 77-12R; 83-7S; 86-1S
Two-phase flow 80-15S
Valves 80-10S
Vapor generators 80-17S
Vessels 79-15S; 80-21S
Welding 85-27S; 86-13S,33S,40S,55S,64S
Windtunnels 77-9R

BOUNDARY LAYER FLOW

Ablation 86-62C
Analytic studies 76-41C; 78-5C,21C; 79-21C; 81-17C; 82-14C,18C,28C,40C; 85-2C,42C, 59C,64C,79C,87C; 86-41C,53C,59C
Anisotropy 76-38C
Approximate methods 79-22C
Axisymmetric bodies 84-36C
Benard convection 86-9C
Blunt bodies 86-35C
Cavities (cylindrical) 83-20C
Chemical reactions 85-31C,65C
Coherent structures 86-26C,61C
Combustion 85-22C
Comparison studies 86-5C
Compressible flow 76-9C,24C,36C,47C,49C; 77-1C,17C,34C; 78-3C,29C,35C; 80-8C,23C; 81-19C; 82-48C; 83-12C; 84-7C,8C,28C,30C; 85-9C,10C,11C,26C,32C,33C,58C,62C,81C, 96C; 86-18C,75C
Cones 76-4C
Conjugate problems 82-11C; 83-9C,27C
Convective Stefan problems 85-85C
Cooling effects 77-4C
Correlations 82-35C; 85-27C
Creeping flow 80-1C
Curvature effects 76-1C,37C,39C,45C,46C, 50C; 79-3C,11C,30C,34C,41C; 82-44C; 83-31C; 84-46C; 86-29C
Cylinders 76-2C,3C,15C,23C,25C,29C,30C, 31C,57C; 77-2C,6C,7C,12C,16C,18C,21C; 78-9C, 18C,24C,36C,46C,47C; 79-2C,5C,10C,14C, 39C,44C,47C; 80-12C; 82-15C,24C,25C,26C, 41C,43C; 83-15C,29C,33C; 84-2C,6C,10C, 11C,35C,40C,45C,48C,54C; 85-1C,3C,13C, 55C; 86-42C,71C

Density interface 81-21C
Design 85-56C
Diffusion 76-14C,44C; 78-37C
Discs 85-82C
Dispersion 82-39C
Divergent flow 78-30C
Effects of inserting strips 84-13C
Energy transformation 86-65C
Experimental studies 78-13C; 81-5C; 85-35C
Extrusion 86-1C
Fins 81-28C; 82-38C; 84-4C,33C,41C,51C; 86-10C,21C,32C,55C,68C
Flat plates 76-13C,51C,53C; 77-24C,27C; 78-25C,26C,38C; 79-36C; 80-14C,29C; 81-12C, 13C; 82-20C; 83-1C,19C,32C,34C,35C; 84-12C, 16C,37C,52C; 85-14C,61C,70C,95C
Flow with obstruction 85-18C
Fluctuating flow 80-25C; 84-14C; 86-39C
Fluidized bed 82-21C
Free stream pressure gradient effects 76-42C, 52C; 78-11C,23C; 81-6C; 82-3C,4C,12C,31C; 85-74C
Free stream temperature effects 79-6C
Free stream turbulence effects 77-9C,33C, 37C; 78-7C,52C; 80-9C; 83-7C,8C; 84-32C; 85-94C
Geometric effects 86-38C
Grid turbulence 81-36C; 86-46C
Heat transfer enhancement 84-21C,23C; 85-30C, 36C,46C
Heat transfer measurement 78-12C,61C; 85-57C
Heated plates 81-29C
Heatline visualization 83-24C
Horizontal tubes 81-26C
Hydromagnetic flow 79-13C
Ice layer 79-23C,24C
Impellers 86-57C
Inclined plates 79-42C; 85-60C
Inclined tubes 85-6C
Instantaneous heat transfer 82-17C
Integral methods 76-21C; 78-58C; 79-1C, 43C; 81-32C; 82-5C; 83-4C
Jets 76-5C,11C,17C,19C,26C,27C,32C,40C, 48C,54C; 77-10C,11C,13C,15C,19C,20C,26C; 78-15C,32C,50C; 79-18C,26C,29C,31C,33C, 38C,40C,45C; 80-6C,16C,18C,26C; 81-9C, 10C,11C,14C,15C,24C; 82-6C,9C; 83-2C,21C; 84-9C,39C,44C; 85-5C,24C,45C,67C; 86-52C, 72C
Knudsen number effects 83-36C
Leading edge 84-31C
Liquid films 76-7C,8C; 77-5C,23C; 78-4C, 16C,42C,49C,51C,62C; 80-17C; 82-30C; 85-63C; 86-49C
Liquid metal 81-20C
Mass transfer 76-34C; 77-14C; 78-1C,27C; 82-42C; 83-6C; 84-17C,29C,47C; 85-12C, 51C,53C; 86-33C,63C
Micropolar flow 85-15C
Mixed free and forced convection 80-11C; 81-8C; 83-30C
Mixing layers 79-8C; 83-16C
Mixing length studies 84-25C
Moving surfaces 79-20C; 80-3C; 82-33C; 84-3C; 86-24C,31C,36C
Non-Newtonian fluids 82-23C; 83-11C,23C; 85-73C; 86-50C,51C,58C
Non-uniformity 76-18C
Optimization techniques 86-3C
Origin of turbulence 81-4C
Particles 80-10C; 86-23C
Phase change 82-7C
Porous plates 85-83C
Prandtl number effects 76-16C,28C; 77-35C; 78-17C,31C; 79-28C,32C,35C; 83-26C; 85-20C
Radiation effects 85-72C
Rarefied gas 84-43C; 86-40C
Rectangular bodies 81-31C
Regenerators 85-43C
Reynolds analogy 78-39C,60C; 85-52C; 86-43C
Roughness effects 78-8C,10C,54C; 79-16C, 17C; 82-10C,32C; 83-25C; 85-19C,97C; 86-28C
Scaling relations 86-66C
Separation 81-2C
Similarity studies 78-2C
Skin friction 76-43C; 81-23C
Slot injection 80-28C
Solar collectors 81-27C
Sources 82-34C; 86-64C
Space shuttles 84-18C,19C,34C,42C; 85-23C, 48C,76C,77C,78C,84C,92C; 86-2C,6C,30C, 48C,60C,67C
Space transportation 85-40C
Spectral analysis 83-17C
Spheres and hemispheres 76-33C; 78-20C, 48C,55C; 79-7C,15C,37C; 80-2C; 81-7C,18C, 34C,35C; 82-13C,16C; 84-1C,50C; 85-4C, 93C; 86-73C
Stability 79-27C; 80-22C,24C; 84-5C,15C, 49C; 85-8C,16C,41C,44C,47C,75C
Stagnation regions 76-22C; 79-4C,12C,19C; 81-22C; 82-22C,27C; 83-5C,13C; 85-37C,

80C,86C,89C; 86-19C,25C,70C
Stratified flow 81-30C; 82-36C; 85-28C; 86-74C
Suction or blowing effects 85-49C
Supersonic and hypersonic flow 80-15C; 81-37C; 82-47C
Survey 85-88C
Thermal mixing layer 86-44C
Thermal wakes 86-12C
Thermo-micropolar fluids 86-4C
Thickness 84-27C
Three dimensional flow 76-10C,56C; 82-8C
Transition 80-7C; 81-3C; 84-26C; 85-91C; 86-54C
Tube banks 80-4C; 81-1C; 82-1C,2C
Turbine blades 80-19C; 81-38C; 84-22C,38C
Turbine systems 85-7C,17C,25C,29C,38C,68C, 71C,90C; 86-13C,14C,15C,37C
Turbulators 86-7C
Turbulence modelling 82-45C,46C; 86-20C, 45C
Turbulent burst 81-33C
Two-phase flow 86-34C
Unsteady state 76-6C,12C,55C; 77-8C,22C, 28C,29C,30C,32C,36C; 78-19C,22C,28C,43C, 53C,57C,59C; 80-21C,27C,30C; 81-16C,25C; 82-29C,37C; 83-14C,18C,22C; 84-24C; 86-16C, 56C
Upstream flow effects 78-41C
Variable boundary conditions 77-3C; 79-9C; 83-3C; 85-34C
Variable flow direction 79-46C
Variable property effects 77-31C; 78-33C, 56C; 82-49C; 86-27C,47C
Velocity profiles 80-5C
Viscosity effects 82-19C; 86-69C
Viscous dissipation effects 85-50C,54C, 69C
Vortex 84-53C; 85-39C,66C; 86-22C
Wakes 83-28C; 86-8C,17C
Wavy surfaces 78-44C
Wedge flow 85-21C
Wires 84-20C

CHANGE OF PHASE

Boiling
Acceleration effects 80-140J
Acoustics 83-18J; 85-34J
Additive effects 76-11J
Alkali metals 77-26J
Analytical studies 79-22J; 84-12J,30J; 86-12J,106J
Annuli 76-35J; 80-30J,60J,131J,142J; 83-31J, 32J; 85-47J,113J; 86-21J,38J,51J
Binary mixtures 77-47J
Boilers 80-59J,113J; 86-8J
Boiling curve 77-3J; 78-13J,55J; 85-66J, 110J
Bubble dynamics 76-28J,51J,54J; 77-6J, 31J,35J,37J,39J,44J,45J,48J,54J,58J; 78-15J,34J,47J,48J,54J,58J,68J,73J,76J, 89J; 80-18J,84J,103J,108J,134J,143J; 81-22J,35J,45J,68J,69J,86J; 82-5J,13J, 17J,19J,47J,65J,69J; 83-8J,12J,13J,40J, 44J,80J; 84-8J,19J,31J,41J,58J; 86-17J, 85J,103J,111J,131J
Bubble-breaking effects 85-41J
Concentrators 83-34J
Conduction effects 80-14J
Contact angle effects 82-18J
Contamination effects 81-30J,34J; 84-2J
Correlations 76-16J; 78-66J,67J,85J; 79-32J; 80-44J,119J; 82-9J,67J
Corrosion effects 82-33J
Critical heat flux 76-26J,48J; 77-36J, 42J,43J; 78-16J,28J,44J,51J,52J,57J, 62J,77J,78J,79J; 79-4J,15J,33J,34J,35J, 47J,53J,55J,56J,60J,68J,84J; 80-32J, 61J,62J,63J,64J,65J,66J,67J,68J,87J, 118J,130J; 81-26J,29J,33J,37J,38J,39J, 44J,46J,55J,73J,84J,89J,90J,92J; 82-31J, 35J,36J,51J; 83-10J,24J,36J,62J,68J, 70J; 84-3J,20J,28J,29J,56J,59J,69J; 86-14J, 35J,36J,45J,49J,55J,59J,62J,65J,70J, 73J,93J,113J,118J,130J,132J
Cryogenics 80-69J,98J,99J; 86-101J,114J
Design 78-38J,87J; 82-60J
Desorption 83-39J
Drops 76-57J; 80-115J
Dryout 78-2J,41J,75J,86J,91J; 79-31J, 73J; 81-11J,36J,49J,50J
Electric field effects 76-4J; 80-12J; 86-30J
Electro-diffusion 82-11J
Engines 86-107J
Entrance region 81-65J
Evaporative heat transfer 80-71J
Evaporators 86-10J
Experimental studies 76-30J; 78-18J; 81-27J
Film boiling 76-2J,5J,15J,21J,31J; 77-1J,

28J,38J; 78-11J,29J,42J,64J; 79-46J, 79J,87J; 80-29J,45J,78J; 81-19J,31J, 32J,42J,78J; 82-26J; 83-14J,58J; 84-14J, 46J,53J,54J,65J,68J; 86-13J,20J,32J, 56J,91J,121J
Finned surfaces 79-26J; 82-54J,58J; 83-23J; 84-67J; 85-2J,12J
Flow boiling 86-2J,22J,27J,50J,76J
Forced convection 77-14J,30J
Free convection 77-9J,50J
Freons 83-17J
Gap 82-4J
Geometric effects 76-47J
Heat exchangers 81-17J; 86-48J
Heat transfer coefficient 84-23J
Heat transfer enhancement 85-28J; 86-57J
Helium 80-2J,76J,102J
Horizontal cylinders 80-126J; 84-38J
Inclined surfaces 82-15J
Initiation 77-20J
Integral methods 86-82J
Interface phenomena 79-88J; 81-47J
Leidenfrost phenomena 78-22J; 79-7J
Liquid metals 83-29J
Liquid nitrogen 83-9J
Low pressure 85-118J
Marangoni effects 83-37J
Mass transfer 84-7J
Microlayer 78-4J,25J,97J
Minimum heat flux 80-25J
Mist 80-137J
Mixed free and forced convection 86-86J, 87J
Mixtures 80-22J; 81-23J,80J; 82-68J; 83-54J, 59J,65J; 84-1J,6J,24J,25J; 85-6J,46J, 83J; 86-37J,74J,80J
Nucleate boiling 76-23J,32J; 77-32J,53J, 57J; 78-12J,17J; 79-80J; 80-41J,51J, 57J,104J; 81-5J,51J,59J,77J,81J,94J; 82-10J,40J,45J; 83-1J,3J,4J; 84-42J; 85-1J,11J,20J,48J,57J,58J,63J,89J; 86-67J, 115J
Nucleate sites 77-4J,11J; 78-20J; 81-40J; 82-30J; 83-19J,38J,48J,63J; 84-34J,36J; 85-16J,78J,95J
Outset of boiling 86-123J
Photographic studies 84-40J
Pool boiling 76-40J; 78-37J,49J; 79-43J; 80-6J,23J,26J,39J,90J,120J,121J,128J, 129J; 81-12J,93J,98J; 82-21J,27J,28J, 29J; 83-7J,22J,33J,53J,72J,73J; 84-26J, 27J; 85-24J,50J,82J,94J,108J,120J; 86-11J, 52J,69J,72J,110J,125J
Post-dryout 82-22J,52J,61J; 83-69J; 86-16J, 24J,39J,66J,119J,127J
Pressure effects 80-28J,58J,100J; 81-8J; 83-5J,30J
Pressure excursions 77-24J
Pseudo-boiling 76-24J
Quenching 81-24J; 84-17J,21J; 86-1J
Rate of vaporization 77-7J
Reactors 80-20J,136J
Reflooding 84-22J,32J,62J; 86-92J
Reviews 84-44J,45J
Rewetting 77-8J; 80-116J; 83-50J,67J; 86-34J
Reynolds analogy 86-7J
Roughness effects 80-125J; 82-38J
Sea water 80-106J
Stability 76-7J; 80-46J,127J; 82-39J, 62J; 84-15J; 86-4J,5J,6J,23J,28J,64J, 81J,100J
Steam generators 78-21J; 81-91J
Subcooled 78-56J; 82-12J,25J,71J; 83-56J, 57J; 84-55J; 85-9J,10J,18J,29J,44J,45J; 86-77J,78J,83J,84J,96J
Surface effects 80-1J,17J,52J,72J,91J, 92J,132J,138J; 85-26J,35J,38J; 86-97J, 98J
Swirling flow 82-2J
Thermosyphons 78-6J,7J; 81-75J; 83-2J; 85-8J
Transition region 77-12J,49J; 78-14J; 79-19J; 81-63J; 82-44J,70J; 83-49J,78J, 79J; 84-60J,61J; 86-41J,44J,47J,58J
Tube bundles 83-60J; 86-42J
Tubes (inside) 82-16J,37J,43J
Tubes (outside) 82-64J
Turbulence effects 85-114J
Turbulence promoters 80-3J,21J
Two-phase flow 86-79J,128J
Unsteady state 76-38J; 77-51J; 79-29J, 81J,82J,83J,91J; 81-56J; 83-35J,55J, 74J; 84-18J; 85-30J,85J; 86-90J,94J, 108J
Vapor density effects 86-129J
Vapor explosion 79-25J
Vapor generation 78-83J; 80-73J
Vapor superheat 80-95J
Variable property effects 82-48J
Vibration effects 80-93J
Void fraction 80-31J,55J

Condensation
Acceleration effects 80-86J
Analytical studies 78-71J,72J; 86-13JJ
Annuli 85-28JJ
Binary mixtures 76-46J; 79-40J,72J; 82-1JJ; 85-13JJ,23JJ,26JJ; 86-14JJ,21JJ,27JJ, 45JJ
Bubble condensation 80-38J,94J; 82-12JJ
Capillary condensation 82-28JJ
Condensers 79-13J; 80-10J,11J; 82-5JJ; 83-8JJ; 84-18JJ,30JJ,35JJ,36JJ
Correlations 79-6J,9J
Cylinders 86-40JJ
Defrosting 83-4JJ
Design 81-2J
Direct contact condensation 80-13J,77J; 82-6JJ; 86-6JJ
Dropwise 76-1J,12J,17J,18J,41J,56J; 77-17J, 27J,41J,59J; 78-8J,30J,50J,61J; 79-1J, 2J,66J,76J,93J; 80-54J,79J,124J,139J; 81-1J,20J,21J,67J,85J,96J; 82-13JJ; 83-6JJ, 10JJ,12JJ; 84-3JJ,4JJ,5JJ,8JJ,10JJ,11JJ, 14JJ,20JJ,21JJ,28JJ,29JJ,31JJ; 85-5JJ, 25JJ,27JJ; 86-7JJ,19JJ,20JJ,24JJ,26JJ, 30JJ,47JJ,51JJ,52JJ
Film condensation 76-14J,27J,43J; 77-22J, 33J; 78-5J,10J,36J,59J,60J,84J; 79-44J, 48J,49J,69J; 80-37J,40J,74J,75J,123J, 133J; 81-18J,25J,54J,58J,71J,72J,74J, 79J; 82-4JJ,9JJ,10JJ,16JJ,21JJ,22JJ; 83-1JJ,3JJ,9JJ; 84-1JJ,2JJ,6JJ,7JJ,9JJ, 15JJ,16JJ,17JJ,19JJ,22JJ,23JJ,25JJ,26JJ, 27JJ,32JJ,33JJ,37JJ; 85-1JJ,3JJ,4JJ, 7JJ,10JJ,11JJ,15JJ,16JJ,17JJ,18JJ,21JJ, 29JJ; 86-11JJ,22JJ,23JJ,28JJ,31JJ,32JJ, 36JJ,37JJ,38JJ
Fins 79-59J; 80-48J; 83-7JJ,13JJ; 85-6JJ; 86-1JJ,17JJ,50JJ
Flat plates 85-19JJ; 86-29JJ
Fog formation 80-19J
Heat transfer coefficient 82-20JJ
Heat transfer enhancement 84-24JJ; 86-44JJ
Helium flow 77-56J
Jets 82-2JJ,7JJ,24JJ; 86-43JJ
Mass transfer 80-81J; 81-95J
Melting 82-26JJ
Mist 76-19J
Mixtures 76-25J; 81-9J; 84-12JJ; 85-8JJ, 14JJ; 86-15JJ,34JJ,35JJ,46JJ
Modelling 86-5JJ
Moisture accumulation 86-48JJ
Non-condensing gas 78-9J,90J
Nozzle flow 86-9JJ,10JJ
Nuclear reactors 85-12JJ
Precursory cooling effects 76-13J
Rate of disposition
Droplet 77-29J
Rotating flow 76-10J
Saturated liquid 77-34J
Stability 84-13JJ
Steam bubbles 77-2J
Steam condensation 77-18J; 79-30J,36J, 65J,67J,70J; 80-4J,5J,82J; 85-2JJ,24JJ; 86-2JJ,8JJ,42JJ
Stratified flow 83-42J,5JJ
Thermosyphons 80-7J,9J,53J; 86-3JJ
Transition region 83-11JJ; 86-41JJ
Tube bundles 78-35J; 80-122J
Tubes (inside) 77-52J; 78-23J; 82-14JJ, 15JJ,18JJ,25JJ; 83-2JJ,14JJ; 85-22JJ
Tubes (outside) 76-8J; 80-15J,27J,33J, 34J,35J,111J,114J; 82-3JJ,17JJ,19JJ, 23JJ; 84-34JJ; 86-4JJ,12JJ,16JJ,18JJ, 33JJ
Two-phase flow 79-20J,50J
Unsteady state 76-42J,61J; 85-9JJ; 86-39JJ
Vacuum systems 85-20JJ
Vapor disappearance 86-25JJ
Vapor mixtures 79-39J,45J; 80-101J; 82-11JJ
Vapor structure 80-109J
Vertical plates 82-8JJ,27JJ
Evaporation
Air streams 86-61J
Analytical studies 79-61J; 81-6J; 85-101J, 109J; 86-9J,40J
Bubbles 81-4J,13J,52J,57J; 86-60J,75J, 122J
Cavities 85-91J; 86-102J
Condensed phase 83-46J
Control 78-33J
Cooling systems 84-63J
Counterflow evaporation 84-33J
Cylinders 82-41J; 84-43J
Diffusive resistance 82-50J
Direct contact evaporation 82-59J,66J
Dissipators 86-105J
Dropwise 77-13J; 78-43J,65J,93J,94J,95J; 79-8J,38J,51J,77J; 80-49J,50J,85J,107J, 110J; 81-28J,43J,53J,61J,66J,76J,82J, 97J; 82-3J,14J,32J; 83-51J,52J,76J,77J; 84-4J,9J,10J,35J,47J,50J,57J,70J; 85-22J, 27J,49J,55J,72J,73J,84J,92J,93J,102J,

103J,104J; 86-19J,25J,31J,33J,43J,53J, 63J,68J,71J,95J,104J,116J,117J,120J, 124J
Drying 80-105J; 81-60J,83J
Electric field effects 83-6J
Emulsions 80-88J
Entrance effects 76-50J
Evaporators 79-10J,86J; 80-141J; 82-46J
Explosion 86-29J
Finned surface 79-16J,41J
Flash evaporation 80-36J,56J,83J; 82-63J; 84-5J,39J,49J,52J; 85-21J,67J,70J; 86-109J
Foams 80-43J
Fouling 83-45J
Geometric effects 76-39J; 79-17J,71J
Heat transfer enhancement 81-62J; 84-37J
Interface phenomena 76-59J; 81-88J; 85-23J, 40J,90J
Ion wind effects 80-47J
Knudsen layer 83-75J
Liquid films 77-40J; 80-70J,117J,135J; 81-41J,70J; 82-34J; 83-16J,66J,71J; 85-117J; 86-15J,18J,26J,99J
Liquid-gas system 83-41J
Meniscus 79-63J,64J
Metal surfaces 83-20J,25J
Mist flow 81-87J; 82-23J,53J
Mixtures 86-46J
Moisture 80-89J
Multicomponent liquids 85-65J
Pressure effects 81-3J,7J,48J
Roughness effects 79-37J
Stability 79-18J; 80-42J; 84-51J
Steam 86-3J
Sublimation 77-5J
Supercritical evaporation 86-126J
Temperature profiles 85-19J
Tubes 79-3J; 81-14J,15J,16J
Two-phase flow 80-24J; 86-54J
Unsteady state 81-10J,64J; 86-112J
Vapor-gas systems 82-8J
Water 83-11J,61J
Two-phase flow
Analytical studies 76-52J; 79-27J; 85-62J, 22JM
Annuli 82-24J,42J; 83-64J; 85-33J,36J, 42J,106J; 86-23JM,35JM
Blowdown 83-27J,28J
Bubbles 78-31J,32J; 79-58J,85J
Channel heat transfer 77-21J,23J; 79-14J
Channel pressure drop and velocity field 79-21J, 78J
Choking phenomena 78-88J
Convective boiling 85-7J,13J,17J,59J
Cooling systems 84-64J
Critical heat flux 85-3J,54J,60J,68J, 69J,71J,75J,77J,87J,88J,96J,97J,99J, 105J,111J,112J,115J,119J
Cryogenics 85-98J
Crystal growth 85-4JM,9JM,28JM
Cylinders 79-54J; 84-66J; 85-107J
Electromagnetic field effects 86-54JM
Entrance effects 76-49J
Evaporators 76-58J
Experimental studies 76-3J,44J; 79-12J, 52J
Falling films 76-22J,45J,53J,55J
Film boiling 85-15J,37J,39J,56J,74J,80J, 81J,116J
Freezing 83-3JM,13JM,14JM,18JM,19JM,22JM, 23JM,25JM,28JM,29JM,30JM; 84-4JM,6JM, 7JM,10JM,18JM,20JM,23JM,29JM,32JM; 85-8JM, 16JM,20JM,33JM,35JM,39JM,41JM,42JM,55JM, 56JM; 86-26JM,38JM,47JM,48JM,60JM
Frost 77-15J; 78-63J; 83-2JM,35JM; 84-33JM; 85-46JM,49JM
Furnaces 86-8JM
Fusion 86-3JM
Gas-liquid system 78-19J
Grooves 85-43J
Ice formation 78-26J; 85-27JM
Interface phenomena 76-20J; 78-1J; 83-27JM; 84-38JM; 85-7JM,34JM
Interline heat flux 79-90J
Liquid metal 78-92J
Liquid nitrogen 77-25J
Mass transfer 80-112J
Measurement techniques 76-6J,63J
Melting 78-24J,74J; 83-1JM,4JM,5JM,6JM, 8JM,9JM,11JM,12JM,16JM,20JM,21JM,24JM, 26JM,31JM,32JM; 84-1JM,2JM,8JM,9JM,12JM, 14JM,15JM,24JM,26JM,27JM,28JM,30JM,31JM; 85-2JM,3JM,21JM,29JM,30JM,31JM,32JM, 36JM,37JM,38JM,40JM,43JM,44JM,45JM,48JM, 54JM,57JM; 86-1JM,6JM,12JM,13JM,14JM, 15JM,16JM,17JM,18JM,22JM,24JM,27JM,29JM, 31JM,32JM,36JM,39JM,45JM,46JM,49JM,50JM, 52JM,53JM,55JM,56JM,57JM,58JM,59JM,61JM
Mixtures 76-34J
Multi-paths 83-15J
Multiphase 86-11JM
Nozzles 77-46J

Particle effects 77-55J
Phase change 79-5J,11J,23J,24J,28J,57J, 62J,74J,75J,92J; 83-7JM,10JM,15JM,17JM, 33JM,34JM,36JM,37JM; 84-3JM,5JM,13JM, 16JM,17JM,19JM,21JM,22JM,25JM,34JM,35JM, 36JM,37JM,39JM,40JM,41JM; 85-12JM,19JM, 25JM,47JM,53JM; 86-2JM,4JM,9JM,10JM, 19JM,20JM,37JM,40JM,41JM
Pressure effects 76-60J; 83-47J
Quenching 82-57J; 85-25J,31J,100J
Reflooding 85-5J,52J,76J,79J
Review 83-26J
Ripple effects 76-33J
Similarity solutions 76-36J
Solidification 85-1JM,5JM,6JM,13JM,14JM, 15JM,17JM,18JM,23JM,24JM,26JM,50JM,51JM, 52JM; 86-5JM,7JM,21JM,25JM,28JM,30JM, 33JM,34JM,42JM,43JM,44JM,51JM
Spectral noise analysis 78-53J
Stability 83-21J,43J
Sublimation 84-11JM
Suspensions 76-37J
Transition region 85-51J
Tube bundles 84-11J
Tubes 77-10J
Unsteady state 77-19J; 85-14J,32J,53J, 61J
Vapor content 85-64J
Variable property effects 85-86J
Vertical flow 84-13J,48J
Void fraction 76-62J
Welding 85-10JM,11JM

CHANNEL FLOW

General
Analytical studies 80-1B,56B,60B,76B; 85-27B,58B,60B
Annuli 81-17B; 82-2B,28B,33B,42B; 84-35B; 85-3B,66B,67B,75B
Binary systems 84-12B
Boiling 81-18B
Branching channel 80-22B
Bubbles 80-51B; 82-4B
Buoyancy effects 80-81B
Capillary flow 81-35B
Cavities 81-24B
Channel with fins 82-22B
Channel with obstacles 76-39B; 78-57B
Chemical reaction 82-3B
Compressible flow 77-59B; 85-21B
Conjugate problems 86-47B
Converging or diverging ducts 81-10B; 84-4B,43B; 85-51B
Cooling 76-18B
Corrugated channels 81-19B; 82-11B,27B; 83-22B,23B; 85-2B; 86-30B,58B
Critical heat flux 80-7B,53B
Cross-flow effects 81-31B
Cryogenics 85-70B; 86-41B
Curvature effects 79-11B,22B,36B; 80-19B
Curved channels 82-21B; 83-25B,33B,38B, 39B,40B,43B; 86-11B,53B,54B,69B
Deposit formation 83-46B
Dispersed flow 83-49B
Edge effects 84-45B
Experimental studies 77-5B; 80-36B
Finned channels 79-32B; 80-39B,66B,79B; 81-42B; 86-14B
Fins 78-54B; 84-47B
Fluctuation 82-10B
Fluidized bed 81-8B
Freezing 76-6B,11B,12B,22B,27B,31B; 77-7B, 11B,12B,15B,16B,29B,47B; 78-12B,20B, 59B,63B; 82-24B
Heat transfer enhancement 76-21B; 77-17B, 28B,39B,50B; 78-16B,25B,26B,51B,60B; 79-8B,27B,39B,40B; 80-68B,73B; 83-4B, 41B,42B; 84-54B,55B
Heat transport 83-47B
Helical ducts 78-31B; 80-47B
Helium 83-9B
Heptanes 84-25B
Ice formation 81-15B
Ignition 85-24B
Immersed cylinders 80-12B,48B
Inclined cylinders 81-5B
Liquid metal 80-3B; 81-6B; 84-52B
Magnetohydrodynamic effects 77-22B,23B
Mass transfer 80-27B; 81-29B; 82-18B; 85-35B,78B; 86-31B,61B,62B
Mass transfer enhancement 79-24B
Micropolar fluids 85-5B
Mist flow 80-80B; 84-20B
Mixed convection and radiation 86-46B, 70B
Mixed free and forced convection 82-7B, 13B; 86-76B
Mixing junctions 84-26B
Multiphase flow 86-27B,37B,64B,66B,67B
Non-circular ducts 84-23B,39B; 85-71B;

86-17B,24B,35B,36B,52B
Non-Newtonian fluids 76-8B,34B,43B,44B; 77-6B,14B,25B,33B; 78-9B,23B,24B,39B, 62B; 79-12B,21B,34B,37B,38B,42B,45B; 80-11B,14B,21B,30B,34B,35B,41B,42B,55B, 64B,78B; 81-3B,7B,12B,20B,25B,40B,41B, 46B,47B; 82-5B,14B,17B,20B,35B; 83-14B, 15B,16B,18B,28B; 85-6B,23B,29B,31B,54B, 62B; 86-6B,18B,25B,26B,34B,43B,48B,77B, 85B
Nozzles 80-28B,29B; 83-1B
Nuclear reactors 80-40B,77B; 82-31B
Nusselt number studies 77-8B,52B
Orifice 82-37B; 83-45B
Parallel discs 77-18B; 84-34B; 86-56B
Parallel plate channels 76-26B; 77-4B; 80-23B,43B,72B; 81-34B; 83-11B,34B,36B; 84-3B,11B,30B,50B
Phase change 82-8B
Prandtl number effects 83-19B,20B
Pressure effects 81-2B
Pulsating flow 77-31B,41B,49B; 79-9B, 15B,16B; 82-9B,30B
Radiation effects 84-19B; 85-19B
Rarefied gas 85-53B,74B
Reactor 77-9B
Rod bundles 80-16B; 81-50B
Secondary flow 76-5B,17B; 84-36B
Separation 83-10B; 85-73B
Similarity 82-16B
Slug flow 82-34B
Spheres 86-21B
Spiral band 80-13B
Stability 76-7B,46B; 85-72B,79B; 86-59B
Sudden change of cross-section 84-8B
Surface contamination effects 85-11B
Suspension flow 83-5B,21B,26B; 85-30B, 42B
Thermal resistance effects 78-30B; 80-70B
Thermal stability 83-48B
Transition flow 84-40B; 86-7B,63B
Tube flow 80-44B
Turbine blades 85-39B; 86-9B,15B
Turbulence or separation promoters 82-32B
Twisted ducts 80-31B,52B; 86-83B
Two component medium 85-9B
Two-phase flow 84-33B
Unsteady state 77-26B; 78-34B,49B; 80-65B; 81-1B; 83-27B; 85-18B,22B; 86-22B
Upstream surface 81-43B
Variable boundary conditions 78-3B,44B; 79-13B; 84-6B
Variable cross section 77-38B; 79-6B; 80-74B
Variable property effects 77-40B; 78-1B, 61B; 84-24B; 86-80B
Laminar
Acceleration effect 77-55B
Analytical studies 76-4B; 85-59B
Annuli 83-35B
Axial conduction effects 77-57B; 81-4B, 33B,48B,49B; 84-13B,42B; 86-81B
Bends 84-5B
Blood flow 84-15B
Buoyancy effects 85-50B
Cavities 82-41B
Chemical reactions 86-12B,13B
Circular sector ducts 84-53B
Circular tubes 84-21B
Compressible flow 80-20B
Conjugated problems 85-4B
Critical heat flux 81-23B
Curvature effects 85-46B,47B; 86-40B
Curved channels 84-27B
Diffusion 81-32B,51B
Entrance effects 76-2B,3B,9B,15B,20B, 23B,24B,25B,29B,32B,42B; 77-3B; 78-6B, 21B,32B,33B,36B,40B,47B; 79-10B,41B, 44B; 80-10B,83B; 81-37B; 83-2B,24B,30B; 84-28B,32B; 85-34B,48B,52B; 86-38B,75B
Film breakdown 80-63B
Finned ducts 76-28B; 77-32B; 85-56B,57B; 86-68B
Fins 84-38B
Gap effects 78-4B
Heat transfer enhancement 85-45B
Interrupted passage 81-36B
Joule-Thomson effects 78-11B
Liquid metal 85-28B
Mass transfer 77-10B,20B,51B,58B; 78-10B, 17B,50B; 80-5B,85B
Micro-structure studies 77-27B
Mixed free and forced convection 86-32B, 33B,39B
Mixed radiation and convection 82-38B
Multipass channels 86-23B
Non-circular ducts
General 77-56B; 78-52B; 79-31B; 80-62B, 67B,69B; 81-38B; 82-25B; 84-1B,2B, 9B,31B; 85-12B,14B,43B,69B,77B; 86-5B, 44B,49B
Mass transfer 79-23B

Non-Newtonian fluids 85-49B
Radiation boundary condition 79-4B; 80-18B
Temperature dependent heat generation 77-35B
Three dimensional flow 81-9B
Tube banks 84-17B
Tube bundles 86-3B
Twisted pipes 81-27B,28B
Unsteady state 83-13B,31B,32B; 86-20B
Variable boundary conditions 77-42B; 78-55B; 85-13B; 86-79B
Variable inlet temperature 86-19B
Variable property effects 77-34B; 78-13B, 35B,42B; 80-9B; 81-16B; 85-25B,32B,33B, 76B
Viscous dissipation effects 86-50B,51B
Turbulent
Analytical studies 76-13B; 81-11B; 82-23B; 86-45B
Annuli 76-33B; 79-18B,28B,29B; 81-26B; 85-20B
Axial conduction effects 77-43B; 78-5B; 79-20B; 82-19B; 84-29B
Buoyancy effects 76-1B; 79-1B; 82-29B; 85-17B; 86-8B
Channels with obstacles 81-44B
Chemical reaction effects 78-43B; 82-1B
Compressible flow 85-64B
Convection velocities 81-30B
Converging or diverging ducts 85-40B
Cooled gas 85-55B
Correlations 78-8B,15B; 80-37B; 81-14B; 82-26B; 83-8B,12B
Curvature effects 85-37B
Diffusers 82-12B
Disturbance 84-48B
Entrance effects 76-38B,40B,41B; 77-30B; 78-18B,58B; 79-3B,5B,7B,19B,79B; 80-15B, 26B,57B,75B; 82-6B
Experimental studies 80-25B; 86-78B
Finned channels 81-39B
Free surface 82-15B
Heat generation 79-25B
Heat transfer enhancement 86-16B,29B, 57B,65B,71B,73B,74B
Helium 85-16B
Inlet conditions 84-46B
Liquid metal 83-29B
Lubrication 80-61B
Mass transfer 77-2B,19B,45B; 78-19B,48B; 80-58B; 81-21B; 84-56B
Mixing length model 76-10B,19B
Mixtures 78-41B
Models 80-33B; 86-1B
Non-circular ducts 76-37B; 78-2B,14B, 28B,29B,56B; 79-26B,35B; 80-2B,6B,17B, 38B,45B,46B,49B,54B; 82-36B,40B; 83-44B; 84-37B; 85-7B,10B,63B,68B; 86-60B,72B
Non-Newtonian fluids 84-22B,41B,57B
Orifices 84-51B
Parallel plates 82-39B
Particle-laden gas 77-13B,36B,46B,53B, 54B
Pipe flow 76-45B
Plenum length effects 81-22B
Prandtl number effects 77-48B; 78-7B
Pressure fluctuations 78-22B
Pulsed flow 86-2B
Radiation effects 86-84B
Reviews 86-10B
Rod bundles 78-37B; 86-4B
Roughness effects 77-24B; 78-38B,46B; 84-14B,18B
Sources 81-13B
Suction effects 84-16B
Swirling flow 84-10B,44B; 85-15B,41B, 44B,61B
Temperature fluctuations 85-8B
Turbine systems 80-4B
Turbulent diffusion 80-8B; 85-26B,65B
Turbulent intensity 79-17B
Two-phase flow 80-50B
Unsteady state 85-38B
Variable boundary conditions 78-45B,53B; 81-45B; 83-6B; 84-7B
Variable cross section 83-3B
Variable property effects 76-14B,30B; 77-21B; 78-27B; 79-30B,33B,43B; 80-59B, 84B; 86-42B
Velocity measurement 80-24B
Viscous dissipation effects 77-1B; 80-82B
Vortex 79-46B; 83-17B; 84-49B

COMBINED HEAT AND MASS TRANSFER

Film cooling
Airfoils 79-23H; 84-23H
Analytical studies 77-4H
Correlation 77-10H
Curvature effects 86-15H
Effectiveness 76-11H,12H; 77-11H,16H, 17H; 78-2H,16H; 81-2H; 84-4H,12H; 85-23H,

32H
Experimental studies 76-1H; 77-8H,9H, 15H; 79-14H; 85-15H
Free stream pressure gradient effects 76-7H; 78-3H; 85-16H
Free stream turbulence effects 85-25H
Injected normal to surface 81-3H
Injection angle effects 76-9H; 77-2H
Jets 77-1H; 78-1H; 84-25H
Liquid films 79-17H; 83-4H,12H; 84-11H
Mass transfer 83-3H
Minimum coolant flow 78-8H
Multiple slots 76-13H; 77-6H; 78-10H; 80-1H; 82-5H,6H; 83-6H
Nozzles 77-20H; 79-18H,26H
Performance 79-9H,10H,11H
Porous wall 77-3H,19H
Pressure drop effects 84-1H
Single slot 79-3H; 85-8H
Slot geometry effects 76-4H,8H; 80-2H, 3H,5H,19H,20H; 85-18H
Slot inclination 86-8H
Stagnation flow 77-13H
Tangential injections 79-4H; 82-11H; 83-13H
Three dimensional effects 86-5H
Tubes 84-5H
Turbine blades 76-10H,14H; 77-5H; 80-4H, 7H,10H,17H; 82-10H; 85-3H,4H,5H,13H, 14H,21H,28H,31H; 86-3H,20H,21H
Wall temperature calculation 83-10H
Mass transfer cooling
Ablation 77-18H; 79-24H; 80-13H; 82-15H; 83-1H; 84-3H; 85-6H,24H
Absorption 82-14H
Agricultural products 82-2H
Analytical studies 78-7H; 79-5H,15H; 80-6H, 18H; 81-6H; 84-17H
Coating 83-7H
Correction factors 79-19H
Crystal growth 84-2H
Cylinders 78-20H
Dehumidification 85-10H
Distillation 77-7H
Droplets 76-16H; 77-12H; 78-5H; 79-7H; 83-8H
Drying 78-18H; 81-5H; 83-5H,9H; 86-2H, 14H
Evaporation 78-6H,19H
Experimental studies 78-15H,21H
Flat plates 79-22H; 80-8H,14H; 81-10H
Free convection effects 85-20H
Frost formation 81-12H
Heat transfer enhancement 79-1H
Jets 78-9H; 80-9H; 82-4H,8H,9H; 83-2H, 11H; 84-6H,7H,8H,9H,16H,19H,20H,21H, 22H,24H,26H; 85-1H,9H,17H,22H; 86-1H, 4H,9H,19H
Liquid films 86-10H,16H
Multi-component systems 76-6H,15H; 81-11H
Non-Newtonian fluids 79-16H
Perforated plates 79-6H
Propagation 81-9H
Radiation effects 79-21H
Reactors 85-27H
Review 78-4H
Roughness effects 82-7H
Shear layers 84-15H
Slabs 79-8H
Sound 84-14H
Stagnation flow 81-7H
Stirring 82-1H
Sublime 77-14H; 78-12H
Surface injection 81-13H
Thermal diffusion 81-1H
Thermodynamic approach 82-13H
Turbulence effects 76-5H
Variable property effects 79-20H
Wavy surfaces 79-25H
Transpiration cooling
Analytical studies 80-11H; 85-12H
Compressible flow 82-3H; 85-7H
Correlations 84-10H
Experimental studies 76-3H
Free stream pressure effects 85-11H
Free stream turbulence effects 85-26H
Integral methods 76-17H
Stability 85-19H
Stagnation regions 80-12H; 85-29H
Turbine blade cooling 81-8H

CONDUCTION
Analog method 81-31A
Analytical studies 76-21A; 77-7A,39A; 78-1A, 6A,22A; 79-22A,36A,40A,49A,55A,56A,63A; 80-4A,21A,25A,28A,29A,30A,41A,44A,50A; 81-1A,14A,40A,42A,67A,70A; 82-22A,34A, 35A; 83-4A,5A,7A,9A,10A,14A,21A,23A,34A, 35A; 84-1A; 85-19A,27A,28A
Anisotropic media 77-4A,13A,14A,15A,46A; 79-17A,50A,51A; 81-37A,44A; 83-8A,16A; 84-23A; 85-18A

Annuli 78-2A; 83-11A
Axial conduction effects 77-65A
Axisymmetric bodies 84-36A; 85-47A
Buried cylinders 84-13A
Cavities 85-31A
Chemical reactivity effects 82-8A,11A,17A, 27A
Composite materials 76-5A,38A; 78-15A,21A, 33A,34A,37A,48A; 79-25A,28A,74A; 80-11A, 18A,31A,39A; 81-17A; 83-1A,15A,28A,29A, 36A; 85-8A,48A; 86-6A,7A,10A,20A,31A,37A, 39A,49A,50A,56A
Conformal mapping 77-33A
Conjugated heat transfer problems 84-2A, 3A,19A,22A,25A,26A
Convective boundary conditions 78-19A; 79-2A; 81-10A; 82-60A
Crack 76-2A; 83-2A,26A,31A
Critical radius of insulation 76-54A; 78-3A; 84-21A; 85-40A
Curvature effects 85-38A
Cut 85-21A
Cylinders 78-29A; 82-26A; 85-9A,45A
Discs 76-65A; 79-83A
Disordered system 84-24A
Electric cables 85-37A
Energy conservation 83-30A
Fiber 85-24A
Fins 76-3A,6A,9A,11A,18A,43A,51A,52A,58A; 77-3A,6A,11A,50A,62A; 78-17A; 79-8A,15A, 24A,58A; 80-2A,3A,16A,24A,38A; 81-2A,3A, 18A,29A,52A,55A,56A; 82-13A,54A; 83-22A; 85-30A,32A,43A
Freezing 76-14A,23A,35A,49A,57A
Gases 82-18A,23A
Heat propagation 83-27A
Hyperbolic conduction equation 76-7A,12A, 17A,30A,36A,37A,44A,60A; 77-9A,27A,68A; 79-66A; 84-31A; 85-14A
Insulation 83-3A
Integral methods 76-68A; 77-22A,23A,28A, 37A,51A,56A; 81-28A; 83-6A
Interface phenomena 82-46A
Inverse problems 79-5A,7A,30A,31A,32A,35A, 41A,79A; 80-13A,17A; 81-20A,65A; 85-1A, 2A,3A,7A
Magnetothermoelastic solids 84-14A
Microcontinuum model 76-47A
Mixed boundary conditions 85-11A,15A
Moving boundary problems 79-34A,53A; 81-41A
Moving cylinders 77-32A
Non-Fourier effects 77-34A,53A
Non-linear boundary condition 85-36A,41A
Nuclear reactor 82-20A
Optimization 85-35A
Oscillating bodies 85-25A
Periodic boundary conditions 84-12A,18A
Phase change 76-8A,13A,24A,25A,26A,40A, 41A,42A,46A,53A,59A,61A; 77-5A,17A,18A, 21A,24A,25A,29A,35A,36A,48A,54A,58A,61A, 66A,67A,72A; 78-4A,5A,9A,10A,11A,12A,13A, 14A,16A,24A,27A,30A,31A,35A,36A,38A,39A, 40A,41A,42A,44A,46A,50A; 79-1A,6A,10A, 12A,14A,18A,19A,20A,21A,27A,44A,45A,48A, 54A,57A,60A,61A,65A,69A,70A,72A,75A,76A, 77A,78A,84A; 80-1A,5A,6A,7A,10A,15A,20A, 22A,23A,26A,27A,32A,35A,36A,42A,43A,45A, 46A,48A,49A; 81-4A,7A,8A,9A,15A,19A,21A, 22A,23A,24A,27A,32A,34A,35A,43A,45A,47A, 48A,50A,51A,53A,54A,57A,58A,59A,60A,61A, 63A,64A,69A; 82-2A,7A,9A,10A,14A,15A,29A, 31A,32A,33A,36A,37A,38A,39A,40A,41A,42A, 43A,44A,45A,48A,49A,52A,53A,57A,61A
Plates, slabs and shells 77-16A; 79-42A
Radiation boundary conditions 84-28A
Rectangles 80-40A
Self-similar expansion 85-34A
Shape factor 76-56A; 81-33A,39A
Shields 84-5A
Sources 77-43A,60A; 78-23A,25A,26A,49A; 79-52A; 80-19A; 82-3A,4A; 84-10A,11A,33A, 34A; 85-12A,22A
Spheres 81-13A; 83-38A
Temperature measurement errors 79-62A
Thermal stresses 82-21A,56A,58A
Thin foil 83-32A
Unsteady state 76-16A,19A,27A,29A,31A,32A, 33A,39A,45A,62A,64A,66A; 77-8A,10A,19A, 20A,26A,38A,41A,42A,44A,45A,47A,49A,57A, 59A,63A,69A; 78-20A,43A,45A; 79-3A,9A, 13A,29A,37A,38A,46A,64A,67A,68A,82A; 80-9A, 14A; 81-5A,6A,26A,30A,36A,38A,66A,68A; 82-16A; 83-33A; 84-4A,7A,8A,15A,16A,17A, 20A,30A; 85-10A,13A,16A,17A,20A,33A
Variable boundary conditions 84-6A
Variable property effects 76-1A,10A; 78-7A; 81-25A; 84-32A
Wedges 84-35A

CONVECTION FROM ROTATING SURFACES

Analytical studies 78-5G
Annuli 77-5G; 78-2G,8G; 79-13G; 81-1G,5G; 82-1G,4G,6G
Axisymmetric flow 78-3G; 84-9G; 85-5G; 86-12G
Benard convection 86-13G
Boxes 82-3G
Cavities 82-7G; 85-3G,4G,14G; 86-4G
Chemical reactions 85-10G
Compressible flow 78-9G; 79-4G
Cones 78-11G; 84-6G
Cooling 77-13G
Cryogenics 84-11G
Cylinders 76-7G; 78-12G,13G; 79-1G,2G,10G; 80-1G,4G,5G,9G; 84-8G,10G,12G,14G,16G; 85-1G,17G,20G,23G,24G; 86-1G,5G,10G,14G
Discs 77-10G,14G; 78-4G,6G; 79-7G,8G,12G, 14G; 80-8G; 81-4G,7G; 82-8G; 83-3G,9G, 11G; 84-13G; 85-7G,15G,19G; 86-11G
Enclosures 76-1G,2G,8G; 77-12G
Entrance effects 77-9G
Fins 86-7G,9G,16G
Flow visualization 86-15G
Heating from below 83-10G
Inclined vessels 86-2G
Marangoni convection 84-5G
Mass transfer 77-3G; 78-7G; 80-2G; 81-2G, 3G; 85-11G
MHD flow 77-1G,2G,7G
Non-circular ducts 84-3G; 85-13G,22G
Non-Newtonian fluids 85-9G
Nozzles 79-6G
Particles 76-6G
Porous discs 84-2G
Porous plates 76-5G
Prandtl number effects 83-5G
Rolls 84-15G
Rotating layers 82-5G
Rotors 82-2G
Separation 77-8G
Shells 83-2G
Spheres 76-3G; 80-6G; 84-4G; 85-6G,16G; 86-3G,6G
Spherical annuli 78-10G; 80-3G
Stability 83-1G,6G,7G,8G; 85-2G,21G
Tubes 76-4G; 77-15G; 78-1G; 79-9G,11G; 81-6G; 84-7G; 85-18G
Turbines 79-5G; 80-7G; 85-8G
Unsteady state 85-12G
Variable property effects 77-6G
Variable surface temperature 79-3G
Wave effects 84-1G
Wedge 77-4G,11G

FLOW WITH SEPARATED REGIONS

Fluidized beds
Analytical studies 76-42D,46D; 82-10D
Bubble flow 76-7D,20D,44D; 80-14D,33D
Combustions 77-6D,15D,25D,26D
Contaminants 77-21D
Effective shear viscosity 78-11D
Electric and magnetic field effects 77-5D, 22D
Electrical resistance 76-43D
Entrainment 76-8D
Experimental studies 82-4D,12D
Flat plates 76-15D
Gas-to-particle heat transfer 78-1D
Granular materials 82-36D
Heat exchangers 76-25D,26D
Heat flux measurement 81-18D
Mass transfer 77-27D; 78-13D; 82-8D,19D
Maximum heat transfer 81-9D
Multi-phase 77-11D
Non-Newtonian fluids 81-12D
Onset of fluidized bed 76-38D
Particle size effects 80-23D
Particle-to-fluid heat transfer 78-10D; 80-31D
Particles 81-8D,11D
Particulate fluidized bed 77-17D
Pulsed flow 76-4D
Radiation effects 80-19D
Reactors 76-36D
Shallow fluidized bed 77-2D,14D
Stability 76-19D
Surface tension effects 78-7D
Thermal conductivity and diffusion 76-33D; 78-2D,3D
Tubes or other bodies in beds 80-3D,11D, 12D; 81-19D
Unsteady state 78-12D; 82-1D
Voidage waves 76-12D
Wall heat transfer 76-35D
Packed beds
Analytical studies 78-5D
Correlations 77-4D
Experimental studies 77-3D
Heat exchangers 77-20D
Length effects 77-13D
Mass transfer 76-2D,17D; 77-7D; 78-4D,

8D
Particle-to-fluid heat transfer 77-29D, 30D
Reactors 77-12D
Recuperators 77-10D
Simulations 76-30D
Surveys 76-14D
Temperature distributions 76-24D
Thermal conductivity and diffusion 76-3D, 13D; 77-16D
Tube banks 83-9D,13D
Single bodies
Analogy 84-6D
Blockage plates 78-20D
Bubble columns 86-29D,30D
Cavities 79-1D,12D; 81-6D; 84-4D,9D; 85-18D
Cones 85-15D
Corner flow 77-23D
Correlations 80-20D
Cylinders 76-16D,21D,22D; 79-2D,4D,5D, 13D,14D; 80-1D,4D,13D,18D,26D; 81-2D, 3D,10D,15D; 82-13D,34D,35D; 83-5D,6D, 15D; 84-5D,10D,11D; 85-1D,2D,5D,6D,9D, 11D,14D,19D,21D; 86-1D,2D,3D,5D,6D,7D, 8D,9D,10D,11D,12D,13D,14D,17D,19D,20D, 21D,24D,25D,27D,32D
Discs 84-8D
Enclosures 78-18D; 85-16D
Fins 82-33D
Flat plates 86-4D,18D
Grids 83-8D
Heat transfer enhancement 77-24D
Jets 80-6D,17D,21D,22D,25D,27D,28D; 82-11D, 14D,20D,29D; 83-3D,7D; 85-3D
Mass transfer 77-8D; 80-16D
Moving walls 77-28D
Normal surfaces 83-11D
Nozzles 86-28D
Perforated plates 82-32D
Protrusion effects 86-31D
Pumping power 85-10D
Reattachment 76-1D,18D; 77-18D,19D; 79-7D, 9D; 84-1D,2D; 85-17D; 86-16D,22D,23D, 26D
Rectangular bodies 84-7D
Roughness effects 79-11D; 83-4D
Separation 76-32D
Spheres 76-40D; 83-10D,12D; 85-12D; 86-15D
Sudden change of cross sections 79-10D
Surface steps 76-37D; 79-3D; 83-1D,2D; 85-4D,7D,8D,13D,20D
Surveys 77-9D
Three-dimensional studies 78-21D
Turbulent shear layers 78-6D,19D
Vortex tubes 83-14D; 84-3D
Wake phenomena 76-27D,41D; 77-1D; 78-14D; 79-6D,8D

LIQUID METALS

Boiling 76-4M,5M
Bubbles 76-2M,3M,7M
Condensation 76-1M,8M; 77-2M
Magnetic field effects 76-6M
Natural convection 76-9M
Reactors 77-1M,3M

MEASUREMENT TECHNIQUES

Accomodation coefficients 80-26P
Balance and gradient methods 84-10E
Baths 80-16P
Boiling 77-28P,33P,35P
Calorimeters 76-12P; 85-17E,34E,38E,48E
Cameras 78-11P
Cold wire 84-9E
Contact resistance 84-4E; 86-13E
Correlations 85-54E
Cryogenic heat transfer 85-18E,25E,32E
Data processing 79-4P
Density 84-22P
Diffusion coefficients
Gases 84-24P
Discharge coefficient measurements 78-6P
Drop-size measurements 78-13P
Enthalpy probes 78-29P
Flow meters 76-32P,41P,46P
Flow visualization 76-36P; 80-12P; 86-43E
Heat capacities 79-23P
Heat flux 76-44P; 80-19P,29P; 84-18E; 85-1E, 3E,24E,31E,36E,44E,50E
Heat transfer coefficients 78-32P; 86-4E
Holographic measurements 77-22P
Hot wire (film) anemometers
Calibration of 79-2P,5P,47P,50P
General 76-7P,17P,21P,30P,37P,40P,51P; 77-4P,8P,9P,17P,18P,25P,41P; 78-9P,18P, 26P,28P,30P,40P,42P,47P; 79-9P,11P,24P, 28P,33P,38P,44P,48P; 80-22P,34P,35P; 84-15E,17E; 85-6E,7E,8E,10E,26E,27E,

30E,35E,42E,43E,52E,53E
Humidity 84-8E
Ice 80-14P
Infrared radiometers 86-22E
Interferometry 76-43P; 77-19P; 78-24P,36P; 80-25P,48P; 86-25E
Laser shadow 85-14E
LDA/LDV 76-8P,16P,26P,27P; 77-1P,3P,23P, 39P; 78-10P,14P,41P; 79-16P,17P,31P,32P, 35P,45P; 80-3P,11P,20P,21P,31P,33P,38P, 39P,43P,47P
Liquid crystal thermography 84-12E
Manometers 80-23P,30P
Mass flow 86-39E
Mass transfer 80-18P,32P; 84-3E; 85-5E; 86-28E
Molecular diffusion 80-9P
Non-Newtonian fluids 84-14P
Optical detectors 76-15P
Optical fiber thermometer 86-16E,21E
Optical systems 76-19P
Pitot tube 76-4P,18P
Plasma diagnostics 79-19P,20P,21P,25P,34P, 43P; 80-41P
Plumes 80-40P
Polarization 76-6P
Pressure 76-22P; 78-27P
Pyroelectric anemometers 85-13E,19E
Pyrometers 78-37P,39P; 85-21E; 86-26E,50E
Radiation properties 76-1P,3P,13P,25P,29P, 31P,49P; 77-6P,21P,24P,46P; 78-4P,12P, 21P,33P,38P,46P; 79-26P,27P,36P,41P,42P; 80-7P,27P,28P,37P,45P; 84-16P,19P; 85-37E; 86-34E
Radiometers and radiation detectors 78-8P; 79-10P; 86-32E
Resistance-thermometer errors 78-5P
Review 80-13P,17P
Schlieren techniques 85-12E,41E,47E
Sensible heat flux 77-12P
Shadowgraph 76-23P
Shear stresses 77-11P,38P; 78-15P
Specific heat 76-9P,10P; 84-6P,21P
Speckle photography 84-14E
Spectrometer 76-14P
Temperature in flowing fluids 76-34P; 84-2E
Temperature measurement
  Devices 77-13P,15P,20P
  Errors 76-39P,42P; 77-10P,29P,31P,34P
  General 76-5P,24P,28P,33P,48P,50P; 77-2P, 30P,40P,42P; 78-17P,22P,45P; 80-1P,2P, 6P; 84-9P; 85-4E,15E,16E,23E,33E,39E, 45E,46E,49E,51E; 86-1E,12E,20E,23E,31E, 38E,47E,52E
Thermal conductivity
  Compositions 84-1P
  Gases 78-31P; 80-8P; 84-7P,10P,11P,17P
  General 77-45P; 79-6P,30P; 80-4P; 84-8P, 20P
  Liquids 78-44P; 79-49P; 84-6E; 86-11E, 51E
  Packed bed 84-12P,13P
  Solids 76-2P,20P,38P; 78-1P,2P,19P,20P, 25P,34P,43P; 79-14P,22P; 80-46P; 84-7E, 26P; 85-2E,28E; 86-29E,36E,41E,49E
  Specific heat 85-29E
Thermal diffusivity
  General 77-16P,36P; 78-23P; 80-44P; 84-13E, 16E
  Solids 86-2E,53E
  Thermal waves 86-42E
Thermal expansion coefficients 84-18P
Thermal flux measurement 86-15E,48E
Thermal properties 84-11E
Thermal resistance 86-35E
Thermocouples
  Characteristics and calibration 77-7P, 26P; 78-48P; 79-29P; 80-5P,24P,36P; 84-5E; 85-11E,40E
  Construction 80-10P
  Frequency response 86-27E
  Linearization 86-44E
  Rake 86-14E
  Stability 86-6E
  Thermal distortion 86-3E
Thermodynamic properties 84-3P; 86-45E
Thermometry 76-35P,45P; 77-5P,14P,27P,32P, 37P,43P,44P; 78-35P; 79-37P; 80-15P,42P; 84-1E,2P; 85-9E; 86-7E,8E,9E,10E,24E,30E, 33E,40E,46E
Thin film gauge 86-5E,17E,18E
Turbine blades 85-20E
Turbulence measurement 86-37E
Two-phase flow 79-1P,3P,7P,8P,12P,18P,46P
Variable property effects 84-23P,25P
Velocimetry 86-19E
Velocity and velocity gradient 78-3P; 79-13P, 39P,40P
Velocity measurement 85-22E
Velocity probes 78-16P
Viscometer 76-11P,47P

Viscosity 78-7P; 79-15P

MHD
Analytical studies 81-1M; 82-11M
Annuli 80-7M; 82-8M
Boundary layers 86-59U
Burst of power 80-9M
Coals 85-39U
Combustion systems 80-4M,8M
Cones 86-15U
Couette flow 79-10M
Cylinders 79-8M,11M; 83-4M,5M; 86-74U
Double diffusion 80-1M
Entrance effects 79-4M; 80-3M
Ferromagnetic fluids 79-2M,3M
Generators 79-1M,9M; 80-13M,15M; 81-5M, 6M,8M,9M; 82-3M,4M,5M,6M,13M
Hall effects 82-1M,7M; 86-76U
Horizontal plates 83-2M; 85-53U
Joule heating 86-13U
Liquid metal 80-11M
Mass transfer 81-3M; 82-9M
Mixed free and forced convection 80-12M; 81-7M; 85-49U
Mixed radiation and convection 81-2M,4M
Onset of separation 79-5M
Polyatomic gases 82-2M
Power plants 80-10M
Prandtl number effects 82-10M
Radiation effects 79-6M
Spheres 80-6M
Stability 80-16M
Stagnation region 79-7M
Temperature measurement 86-7U,57U
Three dimensional flow 80-2M; 82-14M
Two-phase flow 80-5M
Variable wall temperature 82-12M
Vertical plates 83-1M,3M
Viscous dissipation effects 80-14M

NATURAL CONVECTION

Analogy 81-11FF
Analytical studies 76-30F,55F; 81-20FF; 84-21FF; 85-81F,95F
Annuli 76-50F,51F,79F; 78-54F,88F; 79-10F, 18F,71F,92F,125F,136F; 80-10F,28F,33F, 34F,52F,65F,68F; 81-18F,51F,60F; 82-11F, 12F,23F,39F,24FF; 83-6F,28F,47F,61F,66F; 85-6F,44F,73F,75F,86F,87F,90F,104F; 86-11F, 25F,34F,39F,43F,47F,49F,77F,79F,118F
Benard convection 77-29F,83F; 82-17F,26F, 27F,45F,71F; 84-16F,17F,34F,35F,59F; 85-34F, 40F,62F; 86-54F,55F,68F,81F
Binary systems 85-14F; 86-107F
Boussinesq approximation 76-34F
Buildings 86-48F
Buoyancy forces 76-12F
Cavities 78-1F,9F,65F,80F,85F; 79-1F,69F, 70F; 80-2F,30F,45F,46F; 81-2F,7F,33F,62F, 65F,66F; 82-2F,4F,13F,31F,53F,56F,72F; 83-5F; 84-25F,37F,41F,45F,53F,60F; 85-10F, 25F,26F,28F,91F; 86-5F,18F,21F,23F,66F, 103F,104F,105F,117F,123F
Cells 79-49F,72F,111F,124F; 81-50F,73F
Circulation 77-45F
Complex shape 85-7FF,57FF
Compressible flow 80-25FF
Concentric spheres 80-6F,58F; 81-64F
Cones 85-30FF
Convection from immersed bodies 83-14F
Convective rolls 78-22F,27F,28F
Cooling 81-68F; 84-63F
Corrugated plate 78-4F,31F
Coutte flow 77-30F
Critical Rayleigh number 76-59F,83F,94F; 82-75F
Cryogenics 81-1FF
Curvature effects 80-16FF,26FF,30FF
Curved pipes 84-73F
Cylinders 81-10FF,12FF,21FF,30FF,31FF,33FF
Dissipation effects 79-148F; 81-36F
Double diffusive convection 80-31F,42F, 43F,49F,50F,12FF; 81-16F,31F; 82-15F,66F, 67F,68F,76F; 83-24F,29F,32F,33F,28FF; 84-13F; 85-13F,50F,108F; 86-14F,15F,67F,97F,120F, 5FF
Droplets 80-18FF
Earth's mantle 80-72F
Eccentric spheres 86-6F
Electric effects 79-147F
Enclosures 76-23F,26F,47F; 77-44F,60F,85F, 100F,108F; 78-23F,39F,47F,60F; 79-7F,31F, 38F,52F,64F,85F,96F,112F,114F; 80-16F, 20F,24F,25F; 81-14F,43F,45F,61F,63F,67F, 70F,74F; 82-3F,5F,59F,62F,69F; 83-31F, 36F,43F,44F,46F,55F; 84-15F,44F,51F,66F; 85-19F,46F,53F,55F,67F,70F,71F,102F,105F, 111F; 86-7F,12F,13F,19F,26F,27F,41F,56F, 64F,65F,72F,86F,88F,90F,91F,93F,101F,110F

Entrance effects 76-22F; 77-21F,79F; 85-65F
Experimental studies 76-9F,70F,71F; 80-35F; 81-76F,8FF; 83-37FF; 86-2F,30F,70F
Finned tubes 84-5F
Fins 77-33F; 81-29FF
Fires 84-64F
Flames 82-2FF,31FF
Flow reversal 84-65F; 86-9F
Flow visualization 79-137F; 80-23FF; 83-56F; 84-57F
Froude number effects 83-2F
Geophysical phenomena 77-56F
Heat transfer enhancement 77-38F; 84-10F
Heated fluids 84-9FF
Heaters 80-69F
Heating from above 85-83F,100F
Heating from below 76-66F,67F; 77-19F,23F, 52F,80F,81F,99F,102F; 78-101F; 79-11F, 15F,29F,59F,89F,97F,98F; 80-1F,9F; 81-15F, 19F,54F; 82-24F,37F,46F,48F,25FF; 83-16F, 23F,18FF; 84-7F,29F; 85-1F; 86-16F,40F, 53F,61F,85F
Heating from cylinders 80-66F
Heating from sidewalls 77-87F; 78-84F,109F; 80-63F; 81-44F; 82-18F,36F
Helium 85-60F
Hemisphere 77-75F
Horizontal cones 79-2F,95F,142F
Horizontal cylinders 76-20F,36F,69F; 77-22F, 37F,42F,63F,64F,70F,71F,76F,89F,90F,110F; 78-56F,68F,70F,74F; 79-5F,40F,60F,61F, 67F,104F,108F,130F; 80-2FF,27FF; 82-58F, 12FF,30FF,32FF,33FF; 83-62F,2FF,3FF,7FF, 12FF,13FF,14FF,22FF,36FF,38FF,40FF; 84-42F, 61F,10FF,18FF,27FF,29FF; 85-82F,35FF,46FF, 47FF,56FF; 86-58F,109F,1FF,3FF,6FF,27FF, 34FF,43FF
Horizontal discs 79-146F
Horizontal layers 80-18F,27F,37F,62F; 81-57F; 85-5F,79F
Horizontal plates 76-5F,8F,24F; 77-2F,3F, 20F,31F,107F; 78-5F,99F; 79-37F; 80-12F, 8FF; 81-13FF,18FF; 82-8FF,13FF,37FF; 83-26F, 9FF,16FF,42FF; 84-3FF,6FF,12FF; 85-80F, 1FF,60FF; 86-2FF,11FF,36FF,38FF
Horizontal surfaces 79-42F,86F; 85-26FF, 33FF,49FF
Horizontal tubes or channels 76-11F,40F, 44F,84F; 77-95F; 78-8F,10F,75F,83F; 80-45FF
Hot-wire anemometer 79-63F
Ice walls 81-7FF,37FF
Inclination effects 86-106F
Inclined cavities 84-30F,31F,32F,33F,40F
Inclined cylinders 78-76F; 80-48FF; 82-1FF
Inclined enclosures 82-50F,65F,77F; 83-9F, 48F; 86-114F
Inclined layers 76-6F,38F; 80-54F,55F,59F; 83-12F,39F; 84-11FF,14FF; 85-2F,3F,4F, 23F,56F,61F,85F; 86-80F,94F,111F
Inclined slots 81-38F,59F; 84-28FF
Inclined surfaces 78-18F,32F,50F; 79-35F, 36F,45F,54F,100F,101F,110F; 80-29FF; 81-4FF; 82-20F; 85-18FF,23FF,24FF,54FF,55FF,64FF
Inclined tubes or channels 76-65F,75F,85F; 81-23F,46F; 85-9F; 86-51F
Integral methods 82-34FF
Internally finned ducts 85-72F
Jets 76-78F; 77-53F,54F,58F,74F,109F; 78-26F, 53F,82F; 80-38FF; 81-22FF,34FF; 82-15FF, 23FF,29FF; 83-13F,19F,53F,58F; 85-25FF, 67FF; 86-4FF
LDA 79-39F,55F,77F; 83-40F
Liquid metals 77-28F; 82-29F; 84-25FF
Loops 82-49F
Magnetic field effects 76-10F,27F; 78-42F; 79-26F,47F,105F,118F,128F; 81-35FF,36FF; 86-17F,32FF
Magnetohydrodynamic effects 78-110F; 83-54F
Marangoni flow 85-12F,20F,21F,63F,110F
Mass transfer 76-16F; 77-57F,82F; 79-23F, 57F,74F,90F,126F,129F; 80-7FF,9FF,10FF, 47FF; 81-27FF; 85-10FF,13FF,14FF,15FF, 19FF,28FF,36FF,38FF,53FF; 86-28FF
Melting ice 80-70F,5FF,22FF
Mixed conduction and free convection 85-20FF, 21FF,61FF,63FF
Mixed free and forced convection 76-1F, 21F,57F,62F,72F,76F,77F,86F,87F,93F,95F; 78-2F,30F,34F,40F,48F,66F,67F,81F,103F, 104F,105F,106F,108F; 79-9F,22F,27F,33F, 62F,65F,66F,68F,93F,116F,127F,135F,143F; 80-13F,14F,40F,13FF,21FF,41FF,49FF; 81-1F, 26F,30F,49F,75F,2FF,3FF,5FF,24FF; 82-21F, 22F,30F,80F,7FF,9FF,11FF,18FF,20FF,26FF, 35FF; 83-38F,69F,70F,1FF,17FF,34FF; 84-4F, 12F,14F,21F,43F,56F,1FF,7FF,8FF,19FF,23FF, 24FF,26FF; 85-15F,18F,29F,48F,58F,66F, 93F,2FF,3FF,4FF,5FF,8FF,31FF,37FF,40FF, 44FF,45FF,48FF,62FF,66FF,68FF; 86-29F, 32F,42F,45F,46F,52F,82F,83F,113F,9FF,17FF, 31FF,35FF
Molecular diffusion 76-56F

Moving plates 77-96F
Multilayers 86-63F,75F,108F
Natural convection reduction 77-1F,8F
Non-Newtonian fluids 76-4F,48F,82F; 77-40F, 92F,93F; 78-73F,90F; 80-42FF,43FF; 82-25F
Nuclear reactors 76-35F; 83-18F
Onset of convection 76-17F,32F,46F,88F; 78-86F; 79-145F; 80-17FF,32FF; 82-33F; 83-35F,68F,4FF; 84-38F,39F,15FF; 85-35F, 54F,57F,78F; 86-59F,62F,89F,100F
Open channel 82-32F
Oscillatin effects 86-37F
Parallel plate (vertical) 81-71F
Phase change 78-107F
Plumes and thermals 76-3F,33F,80F,81F,89F; 77-5F,26F,35F,36F,39F,59F,61F,101F; 78-13F, 36F,94F; 79-17F,20F,21F,28F,80F,81F,84F, 121F,122F,123F,133F,144F; 80-4FF; 81-6FF, 17FF,23FF,28FF,39FF; 82-14F,38F,44F,4FF, 16FF; 83-1F,21F,65F,11FF,20FF,23FF,25FF, 41FF; 84-4FF,13FF,16FF; 85-39FF,65FF; 86-7FF, 10FF,21FF,22FF
Prandtl number effects 77-27F,86F; 80-8F, 37FF; 81-10F,21F; 83-17F; 85-7F; 86-36F
Radiation effects 82-19F; 85-9FF
Rayleigh-Benard convection 85-32F,36F,97F
Reservoirs 81-8F,9F
Reviews 79-56F
Rod bundles 85-52F
Rotating flow 76-29F,41F,49F; 78-25F,46F; 79-3F,32F,34F,50F,106F,138F; 80-41F,47F, 60F; 81-27F,37F
Roughness effects 77-12F
Saline effects 79-43F
Sidewall effects 78-15F,24F; 83-67F
Similarity 76-60F; 77-14F,16F,18F; 78-102F
Simulation 76-54F; 77-10F; 85-42F,94F; 86-50F, 74F,121F
Solar collectors 76-13F
Sources 78-21F,96F; 79-13F; 80-15F; 82-1F, 35F,70F
Spheres 77-94F; 79-19F; 80-14FF,19FF,35FF; 81-34F,14FF,15FF; 82-14FF,28FF; 84-19F; 85-59FF; 86-24FF
Spherical shells 82-10F; 84-58F; 85-69F, 109F
Spray interactions 84-1F
Stability 76-2F,15F,28F,39F,68F,74F,90F, 96F; 77-13F,24F,25F,106F; 78-62F,71F; 79-14F, 51F; 80-17F,32F,48F,71F,20FF; 81-3F,28F, 29F,47F,48F,52F; 82-9F,16F,28F,34F,52F, 38FF; 83-7F,34F,57F,59F,60F,8FF,15FF,19FF; 84-9F,11F,27F,46F,30FF; 85-8F,11F,16F, 37F,45F,49F,77F,92F,96F,101F,106F,107F, 11FF; 86-3F,20F,22F,35F,57F,71F,78F,96F, 98F,99F,122F,14FF,29FF,41FF,42FF
Stratified fluids 78-11F,35F; 79-140F; 82-42F, 73F,5FF,22FF; 83-21FF; 84-49F,69F; 86-16FF, 30FF
Surface tension effects 77-91F; 78-41F; 79-30F,76F,131F; 80-11F,21F,57F; 83-11F; 85-103F; 86-31F,33F,69F,102F
Survey 86-87F
Thermal boundary condition effects 80-4F, 7F,29F,51F
Thermal explosion 86-28F
Thermocapillary convection 83-63F; 85-41F, 59F,84F
Thermohaline convection 77-41F; 84-52F; 86-84F,95F
Thermosyphons 79-25F,134F; 81-42F; 82-74F; 83-3F,25F,37F,41F,64F; 84-6F,24F,54F,70F; 86-1F,44F
Three dimensional flow 76-43F,63F,64F; 79-16F, 103F; 80-40FF; 84-8F; 86-73F
Transformation method 86-8FF
Transition to turbulence 80-19F; 82-47F, 60F; 85-38F,64F,68F
Transition to unsteady state 77-4F
Turbulence structure 84-28F
Unsteady state 77-55F,68F,97F,98F; 78-20F; 79-6F,48F,91F,149F; 80-64F,11FF,50FF; 81-5F, 32F,56F; 82-79F,27FF; 83-27F,33FF; 84-48F; 86-60F
Variable boundary conditions 78-52F
Variable property effects 78-12F; 80-6FF, 33FF; 81-9FF,16FF; 82-64F; 83-10F,51F, 10FF,31FF; 85-22F,39F; 86-15FF
Vertical annuli 84-22F,23F,26F,67F,68F
Vertical cylinders 76-25F,37F; 80-67F,15FF, 51FF; 81-72F,38FF; 82-7F,43F,57F,3FF; 83-39FF, 43FF; 84-18F; 86-76F,13FF
Vertical layers 81-17F,20F,53F; 85-43F, 47F,88F,98F
Vertical parallel plates 76-61F,91F; 80-44FF; 82-78F; 84-3F
Vertical plates 76-7F,18F,31F,42F,45F,53F, 58F,92F; 77-9F,46F,47F,48F,49F,51F,65F, 66F,69F,73F,78F,84F,104F; 78-3F,14F,16F, 17F,38F,49F,55F,57F,58F,59F,61F,64F,69F, 72F,77F,78F,89F,91F,92F,93F; 79-12F,44F, 53F,75F,83F,94F,107F,113F,115F,117F,119F,

120F; 80-1FF,3FF,24FF,28FF,34FF,36FF,39FF, 46FF; 81-19FF,26FF,32FF; 82-6FF,10FF,17FF, 19FF,21FF,36FF; 83-5FF,6FF,24FF,26FF,27FF, 29FF,30FF,35FF,44FF; 84-5FF,17FF,20FF; 85-21F,30F,76F,89F,6FF,12FF,16FF,17FF, 22FF,27FF,29FF,32FF,34FF,41FF,42FF,43FF, 50FF,51FF,52FF,58FF; 86-12FF,18FF,19FF, 20FF,23FF,25FF,26FF,33FF
Vertical slots 83-30F; 85-27F,31F,33F; 86-40FF
Vertical tubes or channels 77-32F,62F,72F; 78-29F,37F; 79-41F,99F; 80-26F,38F,39F, 61F; 81-39F; 82-40F,51F,55F,63F; 84-47F, 72F; 86-4F,8F,10F,24F,38F,112F,115F,116F, 37FF
Vertical wires 80-31FF
Vortex 77-50F; 84-50F
Wavy wall 78-97F

NUMERICAL METHODS

Applications
General 84-24S
Steam generators 81-5N
Boiling 84-3N,13N
Boundary layer flow
General 76-20C,35C; 77-25C; 78-6C,14C, 34C,40C,45C; 80 13C; 81 14N; 83 10C; 86-11C,22N,31N,56N,92N
Unsteady state 84-16N
Boussinesq problems 85-20N
Change of phase
Boiling 86-89J
Condensation 86-49JJ
Cylinders 81-18N
Evaporation 79-94J
General 82-14N; 84-11N; 86-3N,67N,91N, 93N
Temperature distribution 81-4N
Two-phase flow 76-9J
Variable time step 81-10N
Channel flow
General 77-37B,44B; 79-14B; 80-32B,71B
general 81-15N
General 83-7B; 85-1B,36B; 86-28B,82B
Laminar 76-35B; 81-12N
Combined heat and mass transfer
Film cooling 76-2H
Mass transfer cooling 78-14H; 79-13H; 86-12H
Compressible flow 84-2N,15N; 85-13N; 86-4N, 90N
Condensation
General 85-7N,28N,33N,46N,53N
Integral methods 85-2N,3N,10N,11N,12N, 22N,57N
Inverse problems 85-8N,56N
Phase change 85-14N,23N,27N,39N,45N
Unsteady state 85-21N,24N,54N
Conduction
Axisymmetric 86-34N,78N,79N
Boundary element method 86-1N,55N,58N
Cylindrical coordinates 82-18N
General 76-4A,15A,34A,48A,50A,69A; 77-12A, 30A,31A,52A,64A,70A,71A; 78-8A,18A,28A, 32A,47A; 79-23A,26A,33A,39A,43A,47A, 59A,71A,73A; 80-8A,12A,33A,34A,37A,47A, 51A; 81-16A,49A; 82-12N,15N,21N; 83-17A; 85-4N
Integral technique 86-15N
Inverse problems 81-1N,17N; 82-2N; 86-32N, 33N,72N,81N
Nonlinear 86-74N
Unsteady state 82-23N; 86-8N,12N,21N, 47N,52N,77N
Convection from rotating surfaces 83-4G; 86-8G
Convective diffusion problems 81-6N,7N; 84-4N,9N; 85-15N,18N,26N,34N,35N,36N,37N, 38N,40N,43N,47N,49N,52N,55N; 86-5N,17N, 20N,27N,35N,36N,37N,38N,39N,40N,48N,54N, 59N,62N,64N,66N,75N,80N
Convective dispersive problems 86-24N
Diffusion 82-3N
Euler equation 86-2N,94N
Flow around cylinders 86-9N
Flow around spheres 86-14N
Flow with separated regions
Single bodies 78-16D; 80-2D
Forced convection
Diffusion 82-1N,7N
General 82-20N; 85-16N,17N; 86-29N
Integral methods 85-44N
Free surface flow 86-18N,50N
General 81-2N,3N,9N,11N; 82-4N,5N,6N,8N, 9N,10N,11N,13N,17N,19N,22N,24N; 84-5N, 6N,7N,10N,12N,14N; 85-1N,5N,6N,41N,58N; 86-11N,23N,42N,43N,45N,49N,51N,53N,61N, 71N,73N,82N,83N
Liquid metal
Temperature distributions 81-8N

Mass transfer 84-1N
Natural convection
Cavities 85-31N
General 76-97F; 78-33F,98F; 83-8F,32FF; 85-99F,25N; 86-92F,119F,39FF,7N,60N
Horizontal cylinders 81-13N
Navier-Stokes equation
General 85-9N,29N,42N; 86-6N,16N,26N, 30N,57N,63N,85N,89N
Unsteady state 86-84N
Osher upwind scheme 86-70N
Parallel computations 86-76N
Plasma heat transfer 77-22U; 80-29U; 81-20U; 84-14U; 86-55U
Pressure-velocity coupling 85-30N
Radiation
General 78-8K,29K; 85-40K,49K,97K,129K; 86-64K
Scattering media 82-16N
Recirculating flow 86-10N,44N,87N,88N
Simultaneous equations 86-13N,41N,65N
Space marching methods 86-68N
Spreadsheets 86-19N
Sudden expansion flow 86-86N
Surveys 85-50N,51N
Temperature-velocity coupling 86-28N
Transfer mechanisms 78-30E
Turbulent flow 84-8N
Velocity-pressure coupling 86-46N,69N
Vorticity equation 85-19N,32N; 86-25N,95N

PLASMA HEAT TRANSFER

Ablation 84-16U,17U; 86-20U,32U,71U
Accomodation coefficients 81-8U
Alloys 81-17U
Analytical studies 82-13U,16U,21U; 83-17U
Anode contraction 80-7U; 81-3U
Anode cooling 76-13U
Anode heat transfer 76-17U; 77-8U; 78-14U; 79-17U,25U; 81-1U
Anode region 79-5U,19U; 80-16U; 82-8U,19U; 83-9U,11U; 85-6U,15U,52U; 86-47U
Arc cathode 81-9U
Arc ignition 80-2U
Arc mantle studies 86-69U
Arc welding 77-12U,14U
Arcing damages 78-9U
Argon plasmas 84-25U,29U; 85-7U,28U; 86-35U, 54U
Boundary layers 84-22U
Calorimeters 83-15U
Cathode heat transfer 79-35U
Cathode ignition 82-11U
Cathode region 83-7U,14U,20U; 84-19U; 85-29U
Chemical reactivity 85-3U
Circuit breakers 80-11U; 86-30U,50U
Conduction characteristics 86-64U
Corona discharge 79-30U; 86-38U
Current change 80-23U
Cylinders 83-3U,18U,22U
Decay of plasma 77-10U
Discharge power 86-28U
Double layers 85-21U
Efficiency 80-3U
Electric and magnetic fields 76-6U; 79-11U; 80-18U
Electric arc 76-1U,3U,4U,10U,19U,21U; 77-18U, 20U,21U; 78-6U,10U,12U,16U; 79-20U,27U; 80-4U,19U,21U,25U,33U; 83-12U,13U,21U, 27U; 84-18U; 85-5U,9U,20U,22U,27U,31U, 32U,34U,40U,41U,47U,48U; 86-3U,4U,6U,9U, 36U,45U,46U,48U,62U,63U,65U,66U,67U,68U, 78U,82U
Electric conductivity 78-3U,5U,7U; 85-16U
Electron density 80-26U,35U; 84-3U
Electron temperature 77-9U,15U
Energy balance 77-7U; 78-18U
Energy distribution 83-6U; 85-8U
Energy transport 82-5U,25U
Erosion 79-1U,7U; 80-14U; 84-6U
Experimental studies 79-12U; 82-26U; 86-1U, 39U
Explosion 85-18U,50U
Free recovery 80-15U,27U; 82-2U,18U,20U, 22U
Furnaces 77-13U; 81-7U; 84-34U; 86-58U, 72U
Fusion 84-31U
Gas blast arc 79-28U,31U
Gas injection 80-6U
Gas-tungsten arc 86-27U
Heat flux 76-16U; 81-10U
Heat loss 86-51U
Heat transfer, overall 76-5U
Heating 86-37U,43U,75U
Hybrid plasmas 82-24U
Hydrogen effects 85-14U
Hydrogen-argon plasmas 81-14U
Igniters 83-24U
Intensity distribution 84-9U

Interacting phenomena 86-11U
Internal recirculation 78-2U
Ion thruster 86-81U
Ionization 80-10U; 83-16U
Joule heating effects 78-8U
LTE studies 77-6U; 85-42U; 86-19U,22U,73U
Magnetic pumping 76-2U
Mass transfer 85-1U; 86-33U
Melting 81-13U; 86-31U,44U,70U
Metal processing 82-12U
Metal vapor effects 78-4U; 86-41U,61U
Metal-halide arc 79-8U
MHD 76-8U,12U; 77-19U; 78-13U; 79-22U; 84-1U, 35U
Microwave 84-36U
Mixed free and forced convection 77-2U
Mixtures 81-15U
Modelling 81-2U; 86-25U
Multiple arc 79-14U
Multitemperature plasmas 82-19U
Natural convection effects 80-5U
Nitrogen plasmas 86-77U
Non-equilibrium effects 80-30U
Nozzles 79-6U,9U; 80-32U; 82-23U; 83-10U; 86-53U
Particles 82-3U,4U,27U; 83-4U,5U,25U; 85-10U, 11U,33U,44U,45U,46U,55U; 86-40U
Performance 76-15U
Phosphate production 85-36U
Plasma generation 86-34U
Plasma jets 77-1U; 79-10U,24U,26U,32U,33U, 34U; 80-17U,34U; 81-11U; 84-11U,32U; 85-54U; 86-2U,16U,23U,42U,80U
Plasma spraying 85-19U; 86-26U
Pressure effects 79-15U,23U; 81-12U,19U; 84-10U,15U,33U
Probes 78-1U; 80-12U,24U; 84-2U,4U
Pulsed arc 77-17U
Quenching 83-23U
Radial cooling 78-17U
Radiation transfer 76-11U,18U,20U; 77-3U, 11U; 78-11U,15U,19U; 79-2U,13U,16U; 81-5U, 6U; 82-10U; 84-13U; 85-12U; 86-5U,8U,83U
Refractive index 79-21U
Reignition 80-20U
Roughness effects 84-12U
Self-absorption 82-17U
Shock tubes 85-26U
Similarity solutions 81-16U
Simulations 86-60U
Sodium arc 79-18U
Spheres 76-9U; 86-12U
Stability 76-7U; 80-9U,22U
Stagnation region 79-4U
Static deposition 85-51U
Stationary arc 82-7U
Steam plasmas 84-30U
Submicron particles 81-18U
Surveys 76-23U; 77-5U
Temperature measurement 76-14U; 77-4U,16U; 80-8U,13U,28U; 82-1U,6U,9U,14U,15U; 83-8U; 84-7U,8U; 85-37U; 86-10U,21U,29U
Thermal conductivity 76-22U; 84-20U,21U, 26U; 85-2U,17U,30U,38U
Thermal diffusion 85-4U,35U,43U
Thermal interruption 80-31U
Thermochemical equilibrium 79-29U
Thermodynamic properties 86-24U
Thruster systems 84-28U
Titanium reduction 84-5U
Torches 81-21U; 84-23U,24U; 86-14U,49U
Transition region 81-4U; 84-37U
Transport coefficients 86-18U,56U
Tube flow 85-23U,24U,25U
Unsteady state 83-2U; 86-52U
Vapor addition 80-1U
Vapor flow 82-28U; 83-26U; 84-27U,38U
Welding 83-1U,19U

POROUS MEDIA

Analytical studies 82-25D
Applications
  General 79-24S
  Heat exchangers 85-22Q,37Q,38Q
Boiling 82-7D; 85-15DP,24DP
Boundary condition effects 82-18D,26D,31D
Boundary layers 83-2DP,23DP
Buildings 85-50DP
Cavities 82-23D; 83-5DP
Cellular convection 82-24D
Change of phase
  Boiling 78-3J,69J,70J; 82-1J,6J,7J,20J, 49J,55J,56J; 84-16J; 85-4J; 86-88J
  Condensation 80-96J,97J
  Evaporation 79-42J; 80-80J
  General 82-2D; 83-24DP; 85-90DP
  Two-phase flow 78-27J,39J,40J,82J,96J
Channel flow 83-37B
Chemical reaction 76-31D
Combined heat and mass transfer

Mass transfer cooling 78-11H,13H,17H; 79-2H,12H; 80-15H,16H; 81-4H; 84-13H, 18H; 85-2H,30H
Transpiration cooling 82-12H; 86-6H,7H, 11H,13H,17H,18H
Complex geometry 86-66DP
Contact resistance 76-5D
Cooling 85-41DP
Diffusion constant 76-10D
Double diffusive problems 86-59DP
Drying 82-28D; 84-17DP
Evaporation 76-9D; 83-3DP,17DP
Filtration 85-49DP
Flow with separated regions
Boiling 81-14D
Buoyancy effects 80-29D
Condensation 81-13D
Fluidized beds 78-15D
General 80-32D
Heat flux measurement 81-4D
Mass transfer 80-5D,9D,10D,30D; 81-7D
Moisture migration 80-7D,8D,15D
Packed beds 78-9D,17D
Two-phase flow 81-16D
Unsteady state 80-24D; 81-1D
Variable property effects 81-5D
Velocity fields 81-17D
Fluidized beds
General 83-6DP,7DP,8DP,9DP,11DP,12DP, 15DP,16DP,18DP,21DP; 84-7DP,13DP,14DP, 16DP,26DP,27DP,39DP; 85-2DP,4DP,7DP, 13DP,14DP,16DP,20DP,25DP,28DP,30DP,31DP, 35DP,36DP,43DP,54DP,57DP,70DP,71DP,75DP, 77DP,78DP,79DP,81DP,83DP,84DP,92DP,95DP; 86-1DP,2DP,3DP,13DP,18DP,22DP,33DP,34DP, 37DP,38DP,39DP,42DP,43DP,44DP,51DP,55DP, 65DP
Multi-phase 86-24DP
Variable property effects 86-54DP
Forced convection
Cylinders 86-19DP
Packed beds 84-35DP; 86-7DP
Freezing 86-63DP
Granular beds 84-42DP
Heat sources 83-4DP
Hydraulic conductivity 82-22D
Integral methods 76-34D
Laminar flow 85-38DP
Mass transfer 82-6D; 84-1DP; 85-10DP,11DP, 12DP,18DP,22DP,39DP,44DP,47DP,48DP,53DP, 59DP,82DP,85DP,93DP; 86-40DP
Measurement techniques
Thermal conductivity 84-4P,5P
Thermal diffusivity 84-15P
MHD 86-17U,79U
Microscopic analysis 86-5DP
Mixed free and forced convection

Fins 86-29DP,31DP
General 85-8DP,9DP,51DP; 86-35DP
Vertical channels 86-48DP
Vertical plates 86-30DP
Mixed radiation and convection 84-38DP; 85-86DP; 86-4DP
Mixtures 86-8DP
Moisture migration 82-5D,21D; 86-14DP,47DP
Natural convection
Annuli 84-3DP,4DP; 86-41DP,45DP,46DP
Cavities 84-30DP,31DP
Creeping flow 85-19DP
Cylinders 84-2DP,40DP; 85-29DP; 86-15DP, 23DP,25DP,26DP,27DP,36DP
Enclosures 85-87DP; 86-60DP
General 76-14F,19F,52F,73F; 77-6F,11F, 15F,17F,34F,43F,67F,77F,103F,105F,111F; 78-6F,7F,19F,43F,44F,45F,51F,79F,95F, 100F,111F,112F; 79-4F,8F,24F,46F,58F, 73F,78F,79F,82F,87F,88F,109F,132F,139F, 141F; 80-3F,5F,22F,23F,44F,53F,56F; 81-4F, 6F,11F,12F,13F,22F,24F,25F,35F,40F,41F, 55F,58F,69F,25FF; 82-17D,6F,8F,41F,54F, 61F; 83-1DP,10DP,13DP,19DP,20DP,4F,15F, 20F,22F,42F,45F,49F,50F,52F; 84-6DP, 22DP,23DP,28DP,29DP,2F,20F,36F,55F,62F, 71F,2FF,22FF; 85-17DP,26DP,42DP,60DP, 61DP,62DP,17F,51F,74F; 86-6DP,9DP,10DP, 11DP,21DP,49DP,52DP,53DP
Heating from sides 84-10DP; 86-32DP,64DP
Horizontal plates 85-72DP,73DP,74DP; 86-28DP
Mass transfer 85-3DP,65DP,89DP; 86-61DP
Onset of convection 85-66DP,67DP
Plumes 85-1DP
Prandtl number effects 85-55DP
Stability 84-21DP,36DP
Vertical cylinders 85-69DP; 86-50DP
Vertical plates 84-8DP; 85-33DP,34DP, 63DP,64DP,68DP,80DP
Onset of convection 84-15DP,20DP
Packed beds
General 84-5DP,9DP,12DP,19DP,24DP,34DP, 37DP,41DP; 85-6DP,21DP,23DP,27DP,32DP,

45DP,46DP,58DP,76DP,88DP,91DP,94DP;
86-56DP
Temperature distributions 76-6D
Unsteady state 86-12DP
Particles 85-37DP; 86-16DP
Permeability effects 82-27D
Pore geometry 76-11D
Prandtl number effects 86-17DP
Pulsatile flow 82-3D
Radiation 78-10L; 80-30K; 83-19K; 85-14K
Reactors 76-28D
Reynolds number effects 84-11DP
Soil 84-33DP; 85-56DP
Stability 82-37D; 84-32DP; 85-5DP
Temperature distribution 85-40DP
Thermal conductivity 76-39D; 82-30D; 83-14DP
Thermal dissipation 82-9D
Thermal energy storage systems 83-22DP
Thermophoresis 85-52DP
Transfer processing 84-25DP
Transport phenomena 76-23D,29D
Transport properties 86-95P,109P
Two-phase flow 76-45D; 82-16D; 86-62DP
Unsteady state 84-18DP; 86-57DP,58DP
Vaporization 82-15D
Vertical plates 86-20DP

RADIATION
Absorbing media 78-4K; 83-29K,33K; 85-105K; 86-112K
Absorbing particles 80-5K
Absorptance 77-11K; 81-18K; 84-26K
Absorption 78-39K; 84-2K
Absorption coefficients 76-2K,3K; 77-22K; 78-11K,40K; 84-13K; 86-25K,26K,38K,42K, 43K,90K
Absorptivity
General 84-5K
Spectral
Gas 77-36K; 79-19K; 84-69K
Total
Gas 78-33K; 84-30K,38K
Surface 82-25K; 84-89K,90K
Aerosols 84-3K,34K,78K,82K,100K,101K,102K; 86-77K
Analytical studies 76-1K,6K; 77-10K,16K; 80-22K,25K,28K; 81-4K; 82-19K; 84-1K,18K, 24K,48K,86K; 85-9K,34K,39K,42K,56K,60K, 68K,69K,103K,125K; 86-7K,45K,53K,62K,67K, 69K,76K,86K,102K,124K
Angle factors 76-6L,8L; 77-6L; 78-3L,9L; 79-5L,6L,7L; 80-5L,13L; 81-2L,4L; 82-9K, 10K,26K,40K; 83-26K; 84-65K; 85-35K,36K, 71K,102K,112K; 86-37K,78K,105K
Anisotropic media 77-20K; 78-2L; 80-4L; 85-27K,130K; 86-2K,103K
Annuli 86-88K
Anti-reflection 77-35K
Ashes 86-14K
Atmosphere transfer 79-9K,12L; 80-15K; 82-13K; 83-4K; 84-33K,36K,37K,93K
Attenuation coefficients 82-21K,56K; 83-39K
Axial radiation effects 76-5K
Band absorption 76-8K
Band model 76-15K
Band radiation 79-21K
Black body 80-12K; 85-58K
Bolometers 84-28K,68K
Boundary layer flow 86-21K
Boundary reflection effects 85-109K,110K
Buildings 86-3K
Cavities 84-54K; 85-10K,91K,95K,122K; 86-91K, 117K
Ceramic materials 84-66K
Clouds 84-99K
Combustion 84-43K,58K,62K
Comparison studies 78-23K
Compressible flow 78-21K,22K,26K,41K; 79-26K; 86-5K
Configuration factors see Angle factors
Cryogenics 83-43K; 84-7K; 86-52K
Cylinders 79-1L; 82-22K; 85-74K,98K,104K; 86-8K,56K,60K
Deflagration 80-18K
Dimerization effects 83-22K
Dispersive media 85-127K
Emission 80-29K
Emissivity
General 76-4K; 78-35K; 80-24K; 81-6K, 3L; 84-44K; 85-84K; 86-94K
Spectral
Gas 77-1L; 78-25K; 85-24K; 86-121K
Surface 76-2L,10L; 86-4K
Total
Gas 82-33K
General 83-13K
Surface 76-1L,5L; 77-2L,4L,7L,8L; 78-8L; 79-2L,9L; 80-6L,12L,14L; 83-35K,42K
Emittance 82-52K
Enclosures 79-20K,3L; 80-13K,2L; 83-34K; 84-35K,74K,75K,84K,110K; 85-54K,77K; 86-80K,

84K
Energy diffusion 86-127K
Engines 83-2K; 84-29K,56K
Evaporation 81-15K
Experimental studies 81-12K,17K; 82-11K, 51K; 83-18K; 85-3K,92K
Fibrous media 86-19K,71K
Finned surface 78-6L
Fins 76-3L; 80-9L; 81-1L
Flames 77-14K,17K; 80-1K,21K; 82-14K,37K; 83-15K; 84-22K,51K,55K,60K; 85-23K,30K, 48K,76K,83K,87K,123K,124K,126K; 86-17K, 27K,31K,40K,61K,99K,115K
Fluidized bed 82-17K; 83-7K; 86-16K
Furnaces or combustion chambers 76-18K; 77-18K; 78-10K; 79-23K; 80-16K,17K,26K; 82-30K; 83-17K,30K; 84-9K,32K,53K,91K, 96K; 85-61K,101K; 86-32K,35K,39K,73K,118K, 123K
Gas dynamics 83-21K
Gas heat transfer 76-10K; 77-3K,33K,34K; 78-20K; 79-3K; 82-48K; 83-28K; 84-8K,63K; 86-12K
General radiation effects 85-79K
Glass 86-57K
Gold film 76-20K
Heat exchange processes 85-120K
Heat exchangers 82-55K
Heat transfer enhancement 86-54K
Heating (radiation effects) 84-42K,79K
Heterogenous media 86-120K
Humidity 79-22K
Index of refraction 77-19K,27K,30K; 78-1K, 19K; 79-24K; 80-4K; 82-3K,4K; 83-10K,40K, 48K; 84-61K; 86-66K,109K
Infrared radiative properties 78-5K,4L; 81-14K; 82-31K,34K,39K,47K,49K,53K; 83-37K, 45K,49K; 84-50K,81K,88K; 85-15K,16K,81K; 86-1K,20K,28K,29K,33K,34K,41K,93K,96K
Inhomogeneous radiation 83-3K; 86-55K
Insulation 83-46K,47K
Integral methods 80-9K; 83-11K; 84-15K, 85K; 85-22K,116K; 86-111K
Integrating spheres 85-37K,41K,113K
Intensity 77-12K,28K; 83-20K; 84-25K; 86-24K
Inverse problems 86-36K
Irradiation 83-38K
Laser beams 86-49K
Laser fluorosensor 82-43K
Laser heating 84-71K
Laser irradiation 84-105K
Laser pulses 85-64K
Laser, temperature measurement 84-87K
Lasers 86-65K,101K
Light scattering 86-51K
Losses 76-11K,22K
Magnetic field effects 86-100K
Mean beam length 76-23K
Measurement techniques 85-51K,82K,85K
Melting 78-28K
Mixed radiation and conduction 78-3K,27K, 38K; 79-7K,8L; 80-20K,32K; 82-12K,16K, 24K,38K,44K,50K; 83-1K,8K; 84-6K,40K,52K; 85-4K,57K,72K,78K,99K; 86-70K,97K,107K, 108K,116K
Mixed radiation and convection 76-9K,14K, 24K,4L; 77-8K,25K,26K,31K,32K,38K; 78-9K, 14K,31K,32K; 79-8K,13L; 80-7K,8K,10K,7L; 81-3K,5K,13K; 82-35K; 83-12K,27K; 84-39K, 72K,94K,104K; 85-7K,18K,31K,50K,59K,65K, 93K,106K,108K,128K; 86-50K,104K,114K
Mixtures 82-15K
Modelling 77-1K; 78-30K; 79-18K; 86-46K
Molecular interaction 76-17K
Multilayer systems 80-2K
Multiple scattering 78-17K
Non-gray gases 78-37K; 81-7K,19K; 84-97K; 85-17K,19K,46K,47K,107K,119K
Normal shock 77-10L
Open channel flow 82-23K
Optical constants 77-2K,4K,13K,3L,5L,11L; 82-2K,27K; 83-23K,32K,41K; 84-4K,10K,12K, 27K,67K,70K,98K,103K,111K; 86-75K,89K
Optical depth 78-12K
Optical fibers 86-23K
Optical windows 79-2K
Packed bed 76-7L; 82-1K,6K; 83-50K
Parallel plates 78-1L; 79-4K; 80-31K; 85-43K
Particles 81-9K; 84-31K,45K,47K,59K; 86-9K, 15K,22K,47K,81K,98K,125K
Phase change 84-11K; 85-88K
Photodiodes 86-18K,122K
Planet 83-31K
Plumes and thermals 83-9K,44K; 84-76K,77K
Polarization effects 80-1L,3L
Ponds 79-13K
Power plants 80-3K
Pulse radiation 86-72K
Pyrometers 82-5K
Radiation detectors 85-12K,32K,100K
Radiation entropy 86-6K
Radiation measurement devices 85-45K,70K

Radiators 85-73K; 86-110K
Reactors 84-95K
Receptors 76-9L
Reduction of reflectance 76-13K
Reflectance 80-11L
Reflectivity
General 82-42K; 84-64K; 85-67K; 86-126K
Spectral 76-16K; 78-7K,5L; 80-10L; 82-29K; 84-73K; 86-85K
Total 79-25K; 80-14K
Reflectors 77-15K,37K
Refrigeration 85-11K
Rod bundles 79-27K
Scattering media 77-6K,7K,21K,29K; 78-2K, 6K,18K,34K,7L; 79-5K,10K,11K,12K,16K,17K; 80-6K,19K,23K; 81-2K,10K; 82-7K,8K,28K, 36K,41K,45K,46K; 83-6K,14K,24K; 84-16K, 17K,19K,20K,21K,46K,109K; 85-1K,2K,6K, 8K,13K,20K,21K,26K,28K,29K,44K,55K,62K, 75K,94K,96K,111K,117K,118K; 86-44K,74K, 83K,87K,92K,106K,113K
Semi-transparent media 85-5K,52K,53K,86K, 121K; 86-30K,58K,59K,82K,119K
Shape factors see Angle factors
Shields 85-80K
Shock 76-7K,19K,21K
Slag effects 80-27K
Solar energy collectors 77-23K; 85 90K
Solar flux 84-83K,108K
Solar irradiation 78-24K
Sources 85-114K,131K; 86-48K,68K
Spectrometer 81-11K
Spheres 85-33K; 86-13K
Stability 79-15K
Stoves 85-66K
Stratified flow 82-54K
Surface effects 77-5K; 79-10L,11L; 80-8L; 83-5K,25K; 84-23K,92K,106K,107K; 86-11K
Suspension fluids 79-14K
Temperature measurement 77-9L
Thermal explosion 86-95K
Transmissivity
General 83-36K; 84-14K,49K,80K; 86-10K
Spectral 78-13K,36K
Total 81-16K
Transmittance 76-12K; 80-11K; 86-63K
Turbine systems 86-79K
Two-phase flow 79-1K; 81-1K
Unsteady state 79-4L; 81-8K; 82-18K,32K; 85-38K,89K
Variable optical properties 83-16K
Variable property effects 82-20K; 85-25K, 63K
Vibrational radiation 78-15K,16K
Water vapor 79-6K; 84-41K,57K; 85-115K

THERMODYNAMIC AND TRANSPORT PROPERTIES

Thermodynamic properties

Binary mixtures 86-93P
Ceramics 86-104P
Compressibility 86-15P,90P
General 86-25P,28P,30P,34P,35P,85P
Haylage 86-53P
Heat capacity 86-24P,46P
Insulation 86-22P
Pressure 86-91P
Similarity 86-84P
Specific heat 85-23P,34P,44P,52P,53P; 86-50P,77P,88P
State equation 86-47P
Surface tension 86-11P
Thermal expansion
General 86-39P,40P
Solids 86-21P
Stainless steel 86-20P
Thermal imbalance 86-70P
Transport properties
Analytical studies 86-2A,23A,24A,30A, 35A,52A
Boiling point 86-89P
Building materials 85-27P
Cones 86-11A,12A
Cottons 85-10P
Cylinders 86-4A,22A,25A,48A,58A
Diffusion coefficients
Gases 85-51P; 86-65P
General 85-54P; 86-14P
Liquids 85-3P
Solids 86-55P
Fins 86-9A,18A,55A
Freezing process 85-2P
Furnaces 86-52P
General 85-49P; 86-41P,54P,59P,67P,82P
Heat flux 86-81P
Heat treatment 86-64P
Insulation 85-5P; 86-61P
Inverse problems 85-1P; 86-3A,8A,13A, 29A,32A,34A,40A,41A,62A,69A
Laser heating 86-1A,26A,27A,45A,66A,70A

Mass transfer 85-33P
Metals 86-103P,106P
Mixtures 86-8P,73P,97P,98P,101P
Organic coolants 85-39P
Panels 85-26P
Phase change 86-16A,38A,44A,51A,60A
Phosphorus 86-107P
Plastics 85-8P
Polyatomic gases 86-99P,100P
Pore collapse 86-18P
Pressure vessels 85-44A
Propellants 86-83P
Radiant heat flux 85-17P
Radiation properties 86-58P
Seebeck coefficients 86-38P
Shells 86-54A
Soil 85-35P
Spheres 86-36A
Thermal conductivity
  Agricultural products 86-48P
  Anisotropic materials 82-55A
  Bismuth 86-87P
  Boron 86-36P
  Ceramics 85-31P
  Composite materials 86-26P,44P,51P
  Concrete 86-86P
  Contact resistance 76-20A,22A,28A,63A, 67A; 77-1A,2A,40A,55A,73A; 79-4A,11A, 16A,80A,81A; 81-11A,12A,46A,62A; 82-5A, 6A,24A,28A,47A,51A,59A; 83-12A,13A, 18A,25A; 84-9A,27A; 85-4A,26A,42A, 46A
  Crystals 86-60P
  Ethane 86-78P
  Ferrite 86-29P
  Fibers 85-16P,32P
  Foams 85-22P,24P
  Frozen soils 83-19A
  Gases 85-37P,38P,46P,47P
  General 86-17A,21A,28A,43A,53A,57A, 64A,3P,5P,56P
  Glass 76-55A; 85-40P; 86-49P
  Glass fiber 86-75P
  Granular materials 86-79P
  Graphites 86-68P
  Ice 86-6P
  Insulation 85-6P,30P; 86-9P,12P,43P
  Liquids 85-4P,36P; 86-33P
  Measurement of 82-1A,12A; 85-23A,39A
  Metals 86-66P
  Non-homogeneous materials 86-63P
  Optical coating 86-74P
  Particles 82-50A
  Plasmas 86-16P
  Porous media 82-19A; 83-37A; 86-7P
  Powders 85-25P; 86-42P
  Pulverized coal 82-25A
  Radiation effects 83-24A
  Refrigerants 85-28P
  Reinforced materials 86-94P
  Rocks 86-31P,32P
  Soil 82-30A
  Solids 83-20A; 85-12P,41P,48P,50P; 86-69P, 71P,72P
  Stability 86-105P,108P
  Suspensions 85-13P
  Tiles 86-17P
  Tin 86-45P
  Urania 85-11P
Thermal diffusivity
  Ceramics 86-80P
  Coating 86-37P
  Composite materials 86-92P
  General 85-19P; 86-4P,57P
  Laminates 86-19P,27P
  Liquid 85-5A,6A
  Measurement of 85-29A
  Non-homogeneous materials 86-13P
  Polymers 85-43P
  Porous media 86-76P
  Solids 84-29A
  Thin films 85-15P
Thermal stability 86-33A,42A
Thermo-elasticity 86-5A,67A,68A
Two-phase flow 86-1P
Unsteady state 86-14A,47A,59A,61A,63A, 65A
Variable boundary conditions 86-19A
Variable property effects 86-15A,46A
Velocity 85-21P,42P
Viscometers 86-62P
Viscosity
  Argon 86-96P
  General 86-23P
  Liquids 85-7P,14P,18P
  Low shear rate 86-10P
  Mixtures 85-20P
  Polymer solutions 86-2P
  Refrigerants 85-45P
  Vapors 86-102P
Zirconia 85-9P,29P

TRANSFER MECHANISMS

Analytical studies 78-4E
Buoyancy effects 78-17E
Chemical reactions 78-13E
Composite materials 80-16E
Compressible flow 76-24E,26E; 78-1E; 80-25E
Correlations 78-2E
Eddy diffusivities 77-27E; 78-24E
Entrainment 77-8E
Entropy generation 79-2E
Experimental studies 78-6E,7E,14E; 79-4E
Flames 80-2E,17E
Fluctuating heat flux 80-7E
Fluctuating temperature or velocity 76-5E; 77-2E,26E; 78-3E,8E,27E; 80-11E,21E,24E
Hologram interferometry 77-1E
Hot wire (film) turbulence measurement 76-3E; 77-5E,10E,11E,16E,24E,30E; 80-5E
Interline 78-28E,29E
Jets 76-2E; 78-5E,9E,21E,22E,23E
Magnetic effects 79-25E
Mass transfer 77-25E; 80-1E,3E,6E,12E,19E
Mixing layer 76-16E
Moisture migration 80-4E
Molecular analysis 80-9E,13E,15E
Non-circular duct 77-7E,18E
Non-Newtonian fluids 80-10E
Pneumatic systems 80-8E
Pressure gradient effects 76-4E
Roughness effects 77-21E
Self-diffusion 77-22E
Shear stresses 76-12E
Stability 78-18E
Statistical methods 76-28E; 78-10E
Sublayer 77-9E
Theories of turbulence 77-15E,17E; 79-3E, 8E,9E,10E,13E,19E,24E,31E; 80-18E,20E
Thermal diffusion 80-22E
Thermoconvective waves 79-15E,16E
Thermodynamics of turbulence 78-15E
Turbulence diffusion 79-1E
Turbulence measurement 76-25E; 79-5E,7E, 17E,20E,22E,23E,26E,28E,32E
Turbulence modelling 76-1E,7E,8E,9E,10E, 15E,18E,23E; 77-4E,6E,14E,19E,20E,23E, 28E,29E; 78-12E,20E,25E; 79-6E,11E,12E, 14E,21E,27E,29E,30E
Turbulent density fluctuations 76-13E
Turbulent diffusion 76-21E,22E; 78-16E
Turbulent energy dissipation 76-27E; 77-3E
Turbulent Prandtl number 76-14E,19E; 78-19E
Turbulent transport process 79-18E
Turbulent velocity and temperature measurement 76-6E, 11E,17E,20E; 77-12E; 78-11E,26E; 80-23E
Variable property effects 80-14E
Viscosity 77-13E